Advances in Intelligent Systems and Computing

Volume 1196

AF883738

The series "Advances in Intelligent Systems and Computing" contains publications on theory, applications, and design methods of Intelligent Systems and Intelligent Computing. Virtually all disciplines such as engineering, natural sciences, computer and information science, ICT, economics, business, e-commerce, environment, healthcare, life science are covered. The list of topics spans all the areas of modern intelligent systems and computing such as: computational intelligence, soft computing including neural networks, fuzzy systems, evolutionary computing and the fusion of these paradigms, social intelligence, ambient intelligence, computational neuroscience, artificial life, virtual worlds and society, cognitive science and systems, Perception and Vision, DNA and immune based systems, self-organizing and adaptive systems, e-Learning and teaching, human-centered and human-centric computing, recommender systems, intelligent control, robotics and mechatronics including human-machine teaming, knowledge-based paradigms, learning paradigms, machine ethics, intelligent data analysis, knowledge management, intelligent agents, intelligent decision making and support, intelligent network security, trust management, interactive entertainment, Web intelligence and multimedia.

The publications within "Advances in Intelligent Systems and Computing" are primarily proceedings of important conferences, symposia and congresses. They cover significant recent developments in the field, both of a foundational and applicable character. An important characteristic feature of the series is the short publication time and world-wide distribution. This permits a rapid and broad dissemination of research results.

**** Indexing: The books of this series are submitted to ISI Proceedings, EI-Compendex, DBLP, SCOPUS, Google Scholar and Springerlink ****

More information about this series at http://www.springer.com/series/11156

Andrzej Bartoszewicz · Jacek Kabziński ·
Janusz Kacprzyk
Editors

Advanced, Contemporary Control

Proceedings of KKA 2020—The 20th Polish Control Conference, Łódź, Poland, 2020

Set 1

Springer

Editors
Andrzej Bartoszewicz
Institute of Automatic Control
Lodz University of Technology
Lodz, Poland

Jacek Kabziński
Institute of Automatic Control
Lodz University of Technology
Lodz, Poland

Janusz Kacprzyk
Systems Research Institute
Polish Academy of Sciences
Warsaw, Poland

ISSN 2194-5357 ISSN 2194-5365 (electronic)
Advances in Intelligent Systems and Computing
ISBN 978-3-030-50935-4 ISBN 978-3-030-50936-1 (eBook)
https://doi.org/10.1007/978-3-030-50936-1

This Springer imprint is published by the registered company Springer Nature Switzerland AG
The registered company address is: Gewerbestrasse 11, 6330 Cham, Switzerland

Foreword

This book comprises a collection of contributed papers submitted and, after a thorough peer review process, accepted for presentation at the 20th Polish Control Conference (PCC 2020). This jubilee event is another in a series of triennial congresses organized in Poland since 1958. This year, the conference has been organized jointly by the Committee on Automation and Robotics of the Polish Academy of Sciences and the Institute of Automatic Control of Łódź University of Technology, Poland. The scope of the conference traditionally encompasses all topics related to automatic control and robotics including: design, modelling, identification and analysis of automation systems; calculation methods, data processing and communication in control procedures; mechatronics; robotics and automated manufacturing in industrial systems; motion planning and control for robots; various control problems in transportation and vehicle systems; and also social impact of automation. However, this year the long-established areas of interest of the conference have been enhanced by several, relatively new topics, like: control-theoretic approach to biological, biomedical and medical systems; application of non-integer calculus in control and automation; cyber-physical systems; intelligent vehicles; non-industrial robotics (i.e. application of robots in emerging areas, for example, related to social, assistive or rehabilitation functions of robots); and embedded artificial intelligence and autonomy of both surface and underwater marine vessels. This enhancement was, in no small part, possible thanks to those members of our control engineering community who proposed, promoted and successfully organized a number of invited sessions. The editors of this volume wish to thank all those persons who greatly contributed to the quality, broad coverage and originality of this monograph. In this respect, we particularly acknowledge the contribution of professors: Artur Babiarz, Marek Długosz, Anna Fabijańska, Jarosław Figwer, Zhiqiang Gao, Wojciech Giernacki, Witold Gierusz, Krzysztof Kozłowski, Maciej Ławryńczuk, Rafał Madoński, Maciej Michałek, Krzystof Okarma, Krzysztof Oprzędkiewicz, Ewa Pawłuszewicz, Dariusz Pazderski, Witold Repondek, Paweł Skruch, Piotr Skrzypczyński, Rafał Stanisławski, Ralf Stetter, Piotr Szymak, Roman Śmierzchalski, Andrzej Świerniak, Mirosław Tomera and Marcin Witczak. Indeed, those individuals made an impressive work when identifying new techniques and areas of applications, essentially assisting in the review process and sharing with us their expert knowledge at the editorial stage. Thanks to the joint effort of those people, numerous contributors and hundreds of anonymous reviewers, the conference managed to successfully combine its more than sixty years long tradition with modern approach to broadly understood problems of control engineering. We also thank Dr. Paweł Latosiński and Dr. Piotr Leśniewski for their tireless proofreading of the many drafts of individual chapters. Their exceptional proofreading skills have been of great importance for the final form of this book.

Finally, we wish to thank the entire team of Springer publications for allowing the preparation of this monograph to proceed. We hope that the result of our joint effort will be of true interest to the control engineering community working on various aspects of control theory, automation, robotics, mechatronics, automated manufacturing and related problems of contemporary engineering.

April 2020

Andrzej Bartoszewicz
Jacek Kabziński
Janusz Kacprzyk

Contents

Design of Control Systems - Methods and Applications

Industrial Systems

Modelling, Identification, and Analysis of Automation Systems

Advanced Control Design

Extremal Problems for Infinite Order Parabolic Systems with Time-Varying Lags

Adam Kowalewski(✉)

Institute of Automatics and Robotics, AGH University of Science and Technology, al. Mickiewicza 30, 30-059 Cracow, Poland
ako@agh.edu.pl

Abstract. Extremal problems for infinite order parabolic systems with time-varying lags are presented. An optimal boundary control problem for infinite order parabolic systems in which time-varying lags appear in the state equations and in the boundary conditions simultaneously is solved. The time horizon is fixed. Making use of Dubovicki-Milutin scheme, necessary and sufficient conditions of optimality for the Neumann problem with quadratic performance functionals and constrained control are derived.

Keywords: Boundary control · Infinite order parabolic systems · Time-varying lags

1 Introduction

Extremal problems are now playing an ever-increasing role in applications of mathematical control theory. It has been discovered that notwithstanding the great diversity of these problems, then can be approached by a unified functional-analytic approach, first suggested by Dubovicki and Milutin. The general theory of extremal problems has been developed so intensely recently that its basic concepts may now be considered complete.

Extremal problems were the object of mathematical research at the very earliest stages of the development of mathematics.

The first results were then systematized and brought together under the heading of the calculus of variations with its innumerable applications to physics, automatic control, and mechanics.

In 1962 Dubovicki and Milutin found a necessary condition for an extremum in the form of an equation set down in the language of functional analysis. They were able to derive, as special cases of this condition, almost all previously known necessary extremum conditions and thus to recover the lost theoretical unity of the calculus of variations.

For example, in the paper [7] the Dubovicki-Milutin method was applied for solving optimal control problems for parabolic-hyperbolic systems. Making use of

A. Bartoszewicz et al. (Eds.): PCC 2020, AISC 1196, pp. 3–15, 2020.
https://doi.org/10.1007/978-3-030-50936-1_1

the Dubovicki-Milutin method necessary and sufficient conditions of optimality for the Dirichlet problem with quadratic performance functional and constrained control are derived. The flow chart of the algorithm which can be used in the numerical solving of certain optimization problems for parabolic-hyperbolic systems is also presented.

Moreover, the extremal problems for time lag parabolic systems [6] and for parabolic systems involving time-varying lags [8,12], multiple time-varying lags [13] and integral time lags [15] were examined.

In the papers [6,8,12,13,15] the Dubovicki-Milutin method was applied for solving boundary optimal control problems for the case of time lag parabolic equations [6] and for the case of parabolic equations involving time-varying lags [8,12], multiple time-varying lags [13] and integral time lags [15] in the Neumann boundary conditions.

Sufficient conditions for the existence of a unique solution of such parabolic equations [6,8,12,13,15] are presented.

Consequently, in the papers [6,8,12,13,15] linear quadratic problems of optimal control for the case of parabolic systems with time lags given in various forms (constant time lags [6], time-varying lags [8,12], multiple time-varying lags [13] and integral time lags [15], etc.) were solved.

Extremal problems for time-varying lag infinite order parabolic systems are investigated. The purpose of this paper is to show the use of Dubovicki-Milutin method [6] in solving optimal control problems for infinite order parabolic systems in which time-varying lags appear both in the state equations and in the Neumann boundary conditions.

As an example, an optimal boundary control problem for a system described by a linear infinite order partial differential equation of parabolic type with time-varying lags appearing both in the state equation and in the Neumann boundary condition is considered. Such an infinite order parabolic system can be treated as a generalization of the mathematical model for a plasma control process.

The performance functional has the quadratic form. The time horizon is fixed. Finally, we impose same constraints on the boundary control. Making use of the Dubovicki-Milutin theorem [6], necessary and sufficient conditions of optimality with the quadratic performance functionals and constrained control are derived for the Neumann problem.

2 Preliminaries

Let Ω be a bounded open set of R^n with smooth boundary Γ.

We define the infinite order Sobolev space $H^{\infty}\{a_\alpha, 2\}(\Omega)$ of functions $\Phi(x)$ defined on Ω [1,2] as follows

$$H^{\infty}\{a_\alpha, 2\}(\Omega) = \left\{ \Phi(x) \in C^{\infty}(\Omega) : \sum_{|\alpha|=0}^{\infty} a_\alpha \parallel \mathcal{D}^{\alpha}\Phi \parallel_2^2 < \infty \right\} \tag{1}$$

where: $C^{\infty}(\Omega)$ is a space of infinite differentiable functions, $a_\alpha \geq 0$ is a numerical sequence and $\|\cdot\|_2$ is a norm in the space $L^2(\Omega)$, and

$$\mathcal{D}^{\alpha} = \frac{\partial^{|\alpha|}}{(\partial x_1)^{\alpha_1}\dots(\partial x_n)^{\alpha_n}}, \tag{2}$$

where: $\alpha = (\alpha_1, \dots, \alpha_n)$ is a multi-index for differentiation, $|\alpha| = \sum_{i=1}^{n} \alpha_i$.

The space $H^{-\infty}\{a_\alpha, 2\}(\Omega)$ [1,2] is defined as the formal conjugate space to the space $H^{\infty}\{a_\alpha, 2\}(\Omega)$, namely:

$$H^{-\infty}\{a_\alpha, 2\}(\Omega) = \left\{ \Psi(x) : \Psi(x) = \sum_{|\alpha|=0}^{\infty} (-1)^{|\alpha|} a_\alpha \mathcal{D}^{\alpha} \Psi_\alpha(x) \right\} \tag{3}$$

where: $\Psi_\alpha \in L^2(\Omega)$ and $\sum_{|\alpha|=0}^{\infty} a_\alpha \| \Psi_\alpha \|_2^2 < \infty$.

The duality pairing of the spaces $H^{\infty}\{a_\alpha, 2\}(\Omega)$ and $H^{-\infty}\{a_\alpha, 2\}(\Omega)$ is postulated by the formula

$$\langle \Phi, \Psi \rangle = \sum_{|\alpha|=0}^{\infty} a_\alpha \int_{\Omega} \Psi_\alpha(x) \mathcal{D}^{\alpha} \Phi(x)\, dx, \tag{4}$$

where: $\Phi \in H^{\infty}\{a_\alpha, 2\}(\Omega)$, $\Psi \in H^{-\infty}\{a_\alpha, 2\}(\Omega)$.

From above, $H^{\infty}\{a_\alpha, 2\}(\Omega)$ is everywhere dense in $L^2(\Omega)$ with topological inclusions and $H^{-\infty}\{a_\alpha, 2\}(\Omega)$ denotes the topological dual space with respect to $L^2(\Omega)$ so we have the following chain:

$$H^{\infty}\{a_\alpha, 2\}(\Omega) \subseteq L^2(\Omega) \subseteq H^{-\infty}\{a_\alpha, 2\}(\Omega). \tag{5}$$

3 Existence and Uniqueness of Solutions

Consider now the distributed parameter system described by the following infinite order parabolic equation

$$\frac{\partial y}{\partial t} + Ay + y(x, t - h(t)) = u \qquad x \in \Omega,\ t \in (0, T) \tag{6}$$

$$y(x, t') = \Phi_0(x, t') \qquad x \in \Omega,\ t' \in [-h(0), 0) \tag{7}$$

$$y(x, 0) = y_0(x) \qquad x \in \Omega \tag{8}$$

$$\frac{\partial y}{\partial \eta_A} = y(x, t - h(t)) + v \qquad x \in \Gamma,\ t \in (0, T) \tag{9}$$

$$y(x, t') = \Psi_0(x, t') \qquad x \in \Gamma,\ t' \in [-h(0), 0) \tag{10}$$

where: Ω has the same properties as in the Sect. 2.

$$y \equiv y(x,t;v),\ u \equiv u(x,t),\ v \equiv v(x,t),$$

$$Q = \Omega \times (0,T),\ \bar{Q} = \Omega \times [0,T],\ Q_0 = \Omega \times [-h(0),0)$$

$$\Sigma = \Gamma \times (0,T),\ \Sigma_0 = \Gamma \times [-h(0),0)$$

$h(t)$ is a function representing a time-varying lag, Φ_0 is an initial function defined on Q_0, Ψ_0 is an initial function defined on Σ_0.

The operator $\frac{\partial}{\partial t} + A$ in the state equation (6) is an infinite order parabolic operator and A is given by

$$Ay = \sum_{|\alpha|=0}^{\infty} (-1)^{|\alpha|} a_\alpha \mathcal{D}^{2\alpha} y(x,t) \tag{11}$$

and

$$\sum_{|\alpha|=0}^{\infty} (-1)^{|\alpha|} a_\alpha \mathcal{D}^{2\alpha} \tag{12}$$

is an infinite order elliptic partial differential operator [3].

The Eqs. (6)–(10) constitute a Neumann problem. Then the left-hand side of (8) is written in the form

$$\frac{\partial y}{\partial \eta_A} = \sum_{|w|=0}^{\infty} (\mathcal{D}^{w} y(v)) \cos(n, x_i) = q(x,t) \qquad x \in \Gamma,\ t \in (0,T) \tag{13}$$

where $\frac{\partial}{\partial \eta_A}$ is a normal derivative at Γ, directed towards the exterior of Ω , $\cos(n, x_i)$ is an i-th direction cosine of n, n-being the normal at Γ exterior to Ω and

$$q(x,t) = y(x, t - h(t)) + v(x,t) \tag{14}$$

Let $t \to t - h(t)$ be a strictly increasing function on $[0,T]$, $h(t)$ being non-negative in $[0,T]$ and also being a C^1 function. Then, there exists the inverse function of $t \to t - h(t)$.

Let us denote $r(t) \hat{=} t - h(t)$, then the inverse function of $r(t)$ has the form $t = f(r) = r + s(r)$, where $s(r)$ is a time-varying prediction.

Let $f(t)$ be the inverse function of $t \to t - h(t)$. Thus we define the following iteration $f_j(t) \hat{=} \underbrace{f ... [f[f(t)]]}_{j}$ such that $f_0(t) = t$, where $f_j(t)$ is a j-th iteration of the operation $f(t)$ for $j = 0, 1.... $.

First we shall prove sufficient conditions for the existence of a unique solution of the mixed initial-boundary value problem (6) - (10) for the case where the boundary control $v \in L^2(\Sigma)$.

For this purpose we introduce the Sobolev space $H^{\infty,1}(Q)$ ([17], vol. 2, p. 6) defined by

$$H^{\infty,1}(Q) = H^0(0,T; H^\infty\{a_\alpha,2\}(\Omega)) \cap H^1(0,T; H^0(\Omega)) \tag{15}$$

which is a Hilbert space normed by

$$\left(\int_0^T \| y(t) \|^2_{H^\infty\{a_\alpha,2\}(\Omega)} dt + \| y \|^2_{H^1(0,T;H^0(\Omega))} \right)^{1/2}$$

where: the space $H^1(0,T; H^0(\Omega))$ is defined in Chapter 1 of [17], vol. 1 respectively.

The existence of a unique solution for the mixed initial-boundary value problem (6) - (10) on the cylinder Q can be proved using a constructive method, i.e., first, solving (6) - (10) on the subcylinder Q_1 and in turn on Q_2, etc. until the procedure covers the whole cylinder Q. In this way the solution in the previous step determines the next one.

For simplicity, we introduce the notation

$$E_j \mathrel{\hat{=}} (f_{j-1}(0), f_j(0)), \quad Q_j = \Omega \times E_j, \quad Q_0 = \Omega \times [-h(0), 0)$$

$$\Sigma_j = \Gamma \times E_j, \quad \Sigma_0 = \Gamma \times [-h(0), 0) \text{ for } j = 1, \ldots.$$

Then the following result is fulfilled:

Theorem 1. *Let y_0, Φ_0, Ψ_0, v and u be given with $y_0 \in H^\infty\{a_\alpha, 2\}(\Omega)$, $\Phi_0 \in H^{\infty,1}(Q_0)$, $\Psi_0 \in L^2(\Sigma_0)$, $v \in L^2(\Sigma)$ and $u \in (H^{\infty,1}(Q))'$. Then, there exists a unique solution $y \in H^{\infty,1}(Q)$ for the mixed initial-boundary value problem (6) - (10). Moreover, $y(\cdot, f_j(0)) \in H^\infty\{a_\alpha, 2\}(\Omega)$ for $j = 1, \ldots$.*

We refer to Lions and Magenes ([17], vol.2) for the definition and properties of $H^{r,s}$ and $(H^{r,s})'$ respectively.

In the sequal, we shall fix $u \in (H^{\infty,1}(Q))'$.

4 Problem Formulation. Optimization Theorems

We shall restrict our considerations to the case of the boundary control. Therefore we shall formulate the optimal control problem in the context of Theorem 1.

Let us denote by $Y = H^{\infty,1}(Q)$ the space of states and by $U = L^2(\Sigma)$ the space of controls. The time horizon T is fixed in our problem.

The performance functional is given by

$$I(v) = \lambda_1 \int_Q | y(x,t;v) - z_d |^2 \, dxdt + \lambda_2 \int_0^T \int_\Gamma (Nv)v \, d\Gamma dt \tag{16}$$

where: $\lambda_i \geq 0$ and $\lambda_1 + \lambda_2 > 0$; z_d is a given element in $L^2(Q)$ and N is a strictly positive linear operator on $L^2(\Sigma)$ into $L^2(\Sigma)$. We note from Theorem 1 that for any $v \in U_{ad}$ the cost function (16) is well-defined since $y(v) \in H^{\infty,1}(Q) \subset L^2(Q)$.

We assume the following constraints on controls: $v \in U_{ad}$ is a closed, convex subset of U with non-empty interior,

$$\text{a subset of } U \tag{17}$$

The optimal control problem (6)–(10), (16), (17) will be solved as the optimization one in which the function v is the unknown function.

Making use of the Dubovicki-Milutin theorem [6] we shall derive the necessary and sufficient conditions of optimality for the optimization problem (6)–(10), (16), (17).

The solution of the stated optimal control problem is equivalent to seeking a pair $(y^0, v^0) \in E = H^{\infty,1}(Q) \times L^2(\Sigma)$ that satisfies the Eq. (6)–(10) and minimizing performance functional (16) with the constraints on control (17).

We formulate the necessary and sufficient conditions of the optimality in the form of Theorem 2.

Theorem 2. *The solution of the optimization problem (6)–(10), (16), (17) exists and it is unique with the assumptions mentioned above; the necessary and sufficient conditions of the optimality are characterized by the following system of partial differential equations and inequalities:*

$$\frac{\partial y^0}{\partial t} + Ay^0 + y^0(x, t - h(t)) = u \qquad (x,t) \in \Omega \times (0,T) \tag{18}$$

$$y^0(x,t') = \Phi_0(x,t') \qquad (x,t') \in \Omega \times [-h(0), 0) \tag{19}$$

$$y^0(x,0) = y_0(x) \qquad x \in \Omega \tag{20}$$

$$\frac{\partial y^0}{\partial \eta_A} = y^0(x, t - h(t)) + v^0 \qquad (x,t) \in \Gamma \times (0,T) \tag{21}$$

$$y^0(x,t') = \Psi_0(x,t') \qquad (x,t') \in \Gamma \times [-h(0), 0) \tag{22}$$

Adjoint equations

$$-\frac{\partial p}{\partial t} + A^* p + p(x, t + s(t))(1 + s'(t)) = \lambda_1(y^0 - z_d) \quad (x,t) \in \Omega \times (0, T - h(T)) \tag{23}$$

$$-\frac{\partial p}{\partial t} + A^* p = \lambda_1(y^0 - z_d) \quad (x,t) \in \Omega \times (T - h(T), T) \tag{24}$$

$$\frac{\partial p}{\partial \eta_{A^*}} = p(x, t + s(t))(1 + s'(t)) \quad (x,t) \in \Gamma \times (0, T - h(T)) \tag{25}$$

$$\frac{\partial p}{\partial \eta_{A^*}} = 0 \quad (x,t) \in \Gamma \times (T - h(T), T) \tag{26}$$

$$p(x,T) = 0 \quad x \in \Omega \tag{27}$$

Maximum condition

$$\int_0^T \int_\Gamma (p + \lambda_2 N v^0)(v - v^0)\, d\Gamma dt \geq 0 \qquad \forall v \in U_{ad} \tag{28}$$

We can also notice that

$$\left.\begin{aligned} \frac{\partial p(v)}{\partial \eta_{A^*}}(x,t) &= \sum_{|w|=0}^{\infty} (\mathcal{D}^w p(v)) \cos(n, x_i) \\ A^* p &= \sum_{|\alpha|=0}^{\infty} (-1)^{|\alpha|} a_\alpha \mathcal{D}^{2\alpha} p(x,t) \end{aligned}\right\} \tag{29}$$

Outline of the proof:
According to the Dubovicki-Milutin theorem [6], we approximate the set representing the inequality constraints by the regular admissible cone, the equality constraint by the regular tangent cone and the performance functional by the regular improvement cone.

a) Analysis of the equality constraint

The set Q_1 representing the equality constraint has the form

$$Q_1 = \left\{ \begin{array}{ll} \dfrac{\partial y}{\partial t} + Ay + y(x, t - h(t)) = u & (x,t) \in \Omega \times (0,T) \\ y(x,t') = \Phi_0(x,t') & (x,t') \in \Omega \times [-h(0), 0) \\ y(x,0) = y_0(x) & x \in \Omega \\ \dfrac{\partial y}{\partial \eta_A} = y(x, t - h(t)) + v & (x,t) \in \Gamma \times (0,T) \\ y(x,t') = \Psi_0(x,t') & (x,t') \in \Gamma \times [-h(0), 0) \end{array} \right\} \tag{30}$$

We construct the regular tangent cone of the set Q_1 using the Lusternik theorem (Theorem 9.1 [5]). For this purpose, we define the operator P in the form

$$\begin{aligned} P(y,v) = \Bigg(&\frac{\partial y}{\partial t} + Ay + y(x, t - h(t)) - u,\ y(x,t') - \Phi_0(x,t') \\ &y(x,0) - y_0(x), \frac{\partial y}{\partial \eta_A} - y(x, t - h(t)) - v, \\ &y(x,t') - \Psi_0(x,t') \Bigg) \end{aligned} \tag{31}$$

The operator P is the mapping from the space $H^{\infty,1}(Q) \times L^2(\Sigma)$ into the space $(H^{\infty,1}(Q))' \times H^{\infty,1}(Q_0) \times H^{\infty}\{a_\alpha, 2\}(\Omega) \times L^2(\Sigma) \times L^2(\Sigma_0)$.

The Fréchet differential of the operator P can be written in the following form:

$$P'(y^0,v^0)(\overline{y},\overline{v}) = \Bigg(\frac{\partial \overline{y}}{\partial t} + A\overline{y} + \overline{y}(x,t-h(t)),\ \overline{y}\big|_{Q_0}(x,t'), \overline{y}(x,0), \frac{\partial \overline{y}}{\partial \eta_A} - \overline{y}(x,t-h(t)) - \overline{v},\ \overline{y}\big|_{\Sigma_0}(x,t')\Bigg) \tag{32}$$

Really, $\frac{\partial}{\partial t}$ (Theorem 2.8 [18]), A [3] and $\frac{\partial}{\partial \eta_A}$ (Theorem 2.3 [17]) are linear and bounded mappings.

Using Theorem 1, we can prove that P' is the operator "one to one" from the space $H^{\infty,1}(Q) \times L^2(\Sigma)$ onto the space $(H^{\infty,1}(Q))' \times H^{\infty,1}(Q_0) \times H^{\infty}\{a_\alpha, 2\}(\Omega) \times L^2(\Sigma) \times L^2(\Sigma_0)$.

Considering that the assumptions of the Lusternik's theorem are fulfilled, we can write down the regular tangent cone for the set Q_1 in a point (y^0, v^0) in the form

$$RTC(Q_1, (y^0, v^0)) = \{(\overline{y}, \overline{v}) \in E, P'(y^0,v^0)(\overline{y},\overline{v}) = 0\} \tag{33}$$

It is easy to notice that it is a subspace. Therefore, using Theorem 10.1 [5] we know the form of the functional belonging to the adjoint cone

$$f_1(\overline{y},\overline{v}) = 0 \qquad \forall (\overline{y},\overline{v}) \in RTC(Q_1,(y^0,v^0)) \tag{34}$$

b) Analysis of the constraint on controls

The set $Q_2 = Y \times U_{ad}$ representing the inequality constraints is a closed and convex one with non-empty interior in the space E.

Using Theorem 10.5 [5] we find the functional belonging to the adjoint regular admissible cone, i.e.

$$f_2(\overline{y},\overline{v}) \in [RAC(Q_2,(y^0,v^0))]^*$$

We can note if E_1, E_2 are two linear topological spaces, then the adjoint space to $E = E_1 \times E_2$ has the form

$$E^* = \{f = (f_1, f_2);\ f_1 \in E_1^*, f_2 \in E_2^*\}$$

and

$$f(x) = f_1(x_1) + f_2(x_2)$$

So we note the functional $f_2(\overline{y},\overline{v})$ as follows

$$f_2(\overline{y},\overline{v}) = f_1'(\overline{y}) + f_2'(\overline{v}) \tag{35}$$

where:
$f_1'(\overline{y}) = 0\ \forall y \in Y$ (Theorem 10.1 [5])
$f_2'(\overline{v})$ is a support functional to the set U_{ad} in a point v_0 (Theorem 10.5 [5]).

c) Analysis of the performance functional

Using Theorem 7.5 [5] we find the regular improvement cone of the performance functional (16)

$$RFC(I,(y^0,v^0)) = \{(\overline{y},\overline{v}) \in E, I'(y^0,v^0)(\overline{y},\overline{v}) < 0\} \tag{36}$$

where: $I'(y^0,v^0)(\overline{y},\overline{v})$ is the Fréchet differential of the performance functional (16) and it can be written as

$$I'(y^0,v^0)(\overline{y},\overline{v}) = 2\lambda_0\lambda_1 \int_Q (y^0 - z_d)\overline{y}\, dxdt + 2\lambda_0\lambda_2 \int_0^T \int_\Gamma (Nv^0)\overline{v}\, d\Gamma dt\ .$$

On the basis of Theorem 10.2 [5] we find the functional belonging to the adjoint regular improvement cone, which has the form

$$f_3(\overline{y},\overline{v}) = -\lambda_0\lambda_1 \int_Q (y^0 - z_d)\overline{y}\, dxdt - \lambda_0\lambda_2 \int_0^T \int_\Gamma (Nv^0)\overline{v}\, d\Gamma dt \tag{37}$$

where: $\lambda_0 > 0$.

d) Analysis of Euler-Lagrange's equation

The Euler-Lagrange's equation for our optimization problem has the form

$$\sum_{i=1}^{3} f_i = 0 \tag{38}$$

Let $p(x,t)$ be the solution of (23)–(27) for (v^0,y^0) and denote by $\overline{y}$ the solution of $P'(\overline{y},\overline{v}) = 0$ for any fixed $\overline{v}$. Then taking into account (34), (35) and (37) we can express (38) in the form

$$\begin{aligned} f_2'(\overline{v}) = \lambda_0\lambda_1 \int_Q (y^0 - z_d)\overline{y}\, dxdt + \lambda_0\lambda_2 \int_0^T \int_\Gamma (Nv^0)\overline{v}\, d\Gamma dt \\ \forall(\overline{y},\overline{v}) \in RTC(Q_1,(\overline{y},\overline{v}))\ . \end{aligned} \tag{39}$$

We transform the first component of the right-hand side of (39) introducing the adjoint variable by adjoint equations (23)–(27).

After transformations we get

$$\lambda_0\lambda_1 \int_Q (y^0 - z_d)\overline{y}\, dxdt = \ldots\ \lambda_0 \int_0^T \int_\Gamma p\ \overline{v}\, d\Gamma dt \tag{40}$$

Substituting (40) into (39) gives

$$f_2'(\overline{v}) = \lambda_0 \int_0^T \int_\Gamma (p + \lambda_2 N v^0)\overline{v}\; d\Gamma dt \tag{41}$$

Using the definition of the support functional [5] and dividing both members of the obtained inequality by λ_0, we finally get

$$\int_0^T \int_\Gamma (p + \lambda_2 N v^0)(v - v^0)\; d\Gamma dt \geq 0 \quad \forall v \in U_{ad} \tag{42}$$

The last inequality is equivalent to the maximum condition (28).

In order to prove the sufficiency of the derived conditions of the optimality, we use the fact that constraints and the performance functional are convex and that the Slater's condition is satisfied (Theorem 15.3 [5]). Then, there exists a point $(\widetilde{y}, \widetilde{v}) \in \text{int}\; Q_2$ such that $(\widetilde{y}, \widetilde{v}) \in Q_1$.

This fact follows immediately from the existence of non-empty interior of the set Q_2 and from the existence of the solution of the Eq. (6)–(10) as well.

This last remark finishes the proof of Theorem 2.

One may also consider analogous optimal control problem with the performance functional

$$\widehat{I}(y, v) = \lambda_1 \int_\Sigma |\; y(v)\; |_\Sigma - z_{\Sigma d}\; |^2\; d\Gamma dt + \lambda_2 \int_0^T \int_\Gamma (Nv)v\; d\Gamma dt \tag{43}$$

where: $z_{\Sigma d}$ is a given element in $L^2(\Sigma)$.

From Theorem 1 and the trace theorem ([17], vol. 2, p. 9) for such $v \in L^2(\Sigma)$, there exists a unique solution $H^{\infty,1}(Q)$ with $y\Big|_\Sigma \in L^2(\Sigma)$. Thus $\widehat{I}(y, v)$ is well-defined. Then the solution of the formulated optimal control problem is equivalent to seeking a pair $(y^0, v^0) \in E = H^{\infty,1}(Q) \times L^2(\Sigma)$ that satisfies the Eq. (6)–(10) and minimizing the cost function (43) with the constraints on controls (17).

We can prove the following theorem:

Theorem 3. *The solution of the optimization problems (6)–(10), (43), (17) exists and it is unique with the assumptions mentioned above; the necessary and sufficient conditions of the optimality are characterized by the following system of partial differential equations and inequalities:*
State Eqs. (18)–(22).
Adjoint equations

$$-\frac{\partial p}{\partial t} + A^* p + p\,(x, t + s(t))(1 + s'(t)) = 0 \quad (x, t) \in \Omega \times (0, T - h(T)) \tag{44}$$

$$-\frac{\partial p}{\partial t} + A^* p = 0 \quad (x, t) \in \Omega \times (T - h(T), T) \tag{45}$$

$$\frac{\partial p}{\partial \eta_{A^*}} = p(x, t + s(t))(1 + s'(t)) + \lambda_1(y^0 - z_{\Sigma d}) \quad (x,t) \in \Gamma \times (0, T - h(T)) \tag{46}$$

$$\frac{\partial p}{\partial \eta_{A^*}} = \lambda_1(y^0 - z_{\Sigma d}) \quad (x,t) \in \Gamma \times (T - h(T), T) \tag{47}$$

$$p(x, T) = 0 \quad x \in \Omega \tag{48}$$

Maximum condition

$$\int_0^T \int_\Gamma (p + \lambda_2 N v^0)(v - v^0)\, d\Gamma dt \geq 0 \qquad \forall v \in U_{ad} \tag{49}$$

The idea of the proof of the Theorem 3 is the same as in the case of the Theorem 2.

We must notice that the conditions of optimality derived above (Theorems 2 and 3) allow us to obtain an analytical formula for the optimal control in particular cases only (e.g. there are no constraints on boundary control). It results from the following: the determining of the function $p(x,t)$ in the maximum condition from the adjoint equation is possible if and only if we know that $y^0(x,t)$ will suit the control $v^0(x,t)$. These mutual connections make the practical use of the derived optimization formulas difficult. Thus we resign from the exact determining of the optimal control and we use approximation methods.

In the case of performance functionals (16) and (43) with $\lambda_1 > 0$ and $\lambda_2 = 0$, the optimal control problem reduces to the minimizing of the functional on a closed and convex subset in a Hilbert space. Then, the optimization problem is equivalent to a quadratic programming one [10,11,14] which can be solved by the use of the well-known algorithms, e.g. Gilbert's [4,10,11,14] ones.

The practical application of Gilbert's algorithm to optimal control problem for a parabolic system with the boundary condition involving a time lag is presented in [14]. Using of the Gilbert's algorithm a one dimensional numerical example of the plasma control process is solved.

5 Conclusions and Perspectives

The derived conditions of the optimality (Theorems 2 and 3) are original from the point of view of application of the Dubovicki-Milutin theorem [6] for solving optimal control problems for infinite order parabolic systems in which time-varying lags appear both in the state equations and in the Neumann boundary conditions. Such infinite order parabolic systems constitute a generalization of the mathematical model for a plasma control process [14].

Consequently, a numerical search for an approximate solution to the optimal boundary control problem (6)–(10), (16), (17) is very sophisticated and constitutes the open research problem. This is due to the infinite order parabolic operator (6) which substantially complicates a numerical procedure.

The results presented in the paper can be treated as a generalization of the results obtained in [12] to the case of infinite order parabolic systems with time-varying lags.

The proved optimization results (Theorems 2 and 3) constitute a novelty of the paper with respect to references [9–11,14] concerning application of the Lions scheme [16] for solving linear quadratic parabolic problems for the case of the Neumann problem.

Moreover, the optimization problems presented here constitute a generalization of optimal control problems considered in [6] for parabolic systems with constant time lags appearing in the state equations and in the boundary conditions simultaneously.

One may also derive the necessary and sufficient conditions of optimality for infinite order parabolic system with more complex boundary conditions involving integral time lags.

According to the author similar optimal control problems can be solved for infinite order hyperbolic systems.

Acknowledgements. The research presented here was carried out within the research programme AGH University of Science and Technology, No. 16.16.120.773.

References

1. Dubinskii, J.A.: Sobolev spaces of infinite order and behavior of solution of some boundary value problems with unbounded increase of the order of the equation. Matiematiczeskii Sbornik **98**, 163–184 (1975)
2. Dubinskii, J.A.: Non-trivality of Sobolev spaces of infinite order for a full Euclidean space and a Tour's. Matiematiczeskii Sbornik **100**, 436–446 (1976)
3. Dubinskii, J.A.: About one method for solving partial differential equations. Dokl. Akad. Nauk SSSR **258**, 780–784 (1981)
4. Gilbert, E.S.: An iterative procedure for computing the minimum of a quadratic form on a convex set. SIAM J. Control **4**(1), 61–80 (1966)
5. Girsanov, I. V.: Lectures on the Mathematical Theory of Extremal Problems. Publishing House of the University of Moscow, Moscow (1970). (in Russian)
6. Kowalewski, A., Miśkowicz, M.: Extremal problems for time lag parabolic systems. In: Proceedings of the 21st International Conference of Process Control (PC), 6-9 June pp. 446–451, Strbske Pleso, Slovakia (2017)
7. Kowalewski, A.: On optimal control problem for parabolic-hyperbolic system. Probl. Control Inf. Theory **15**(5), 349–359 (1986)
8. Kowalewski, A.: Extremal Problems for distributed parabolic systems with boundary conditions involving time-varying lags. In: Proceedings of the 22nd International Conference on Methods and Models in Automation and Robotics (MMAR), Międzyzdroje, Poland, 28–31 August, pp. 447–452 (2017)
9. Kowalewski, A.: Optimal control of parabolic systems with time-varying lags. IMA J. Math. Control and Inf. **10**(2), 113–129 (1993)
10. Kowalewski, A.: Boundary control of distributed parabolic system with boundary condition involving a time-varying lag. Int. J. Control **48**(6), 2233–2248 (1988)

11. Kowalewski, A.: Optimal Control of Infinite Dimensional Distributed Parameter Systems with Delays. AGH University of Science and Technology Press, Cracow (2001)
12. Kowalewski, A.: Extremal problems for parabolic systems with time-varying lags. Arch. Control Sci. **28**(1), 89–104 (2018)
13. Kowalewski, A.: Extremal problems distributed parabolic systems with multiple time-varying lags. In: Proceedings of 23rd International Conference on Methods and Models in Automation and Robotics (MMAR), Międzyzdroje, Poland, 27–30 August, pp. 791–796 (2018)
14. Kowalewski, A., Duda, J.: On some optimal control problem for a parabolic system with boundary condition involving a time-varying lag. IMA J. Math. Control Inf. **9**(2), 131–146 (1992)
15. Kowalewski, A., Miśkowicz, M.: Extremal problems for integral time lags parabolic systems. In: Proceedings of the 24th International Conference on Methods and Models in Automation and Robotics (MMAR), Międzyzdroje, Poland, 26–29 August, pp. 7–12 (2019)
16. Lions, J.L.: Optimal Control of Systems Governed by Partial Differential Equations. Springer, Berlin (1971)
17. Lions, J.L., Magenes, E.: Non-Homogeneous Boundary Value Problems and Applications, vols. 1 and 2. Springer, Berlin (1972)
18. Maslov, V.P. : Operators Methods, Moscow (1973). (in Russian)

Stability of Singularly Perturbed Systems with Delay on Homogeneous Time Scales

Ewa Pawluszewicz[1(✉)] and Olga Tsekhan[2]

[1] Bialystok University of Technology, Bialystok, Poland
e.pawluszewicz@pb.edu.pl
[2] Yanka Kupala State University of Grodno, Grodno, Belarus
tsekhan@grsu.by

Abstract. Conditions for the exponential stability of a linear singularly perturbed system with the small parameter defined on homogeneous time scales are presented. To this aim given system is decomposed onto two subsystems of smaller dimensions than the original one, i.e. onto slow and fast subsystems. It is shown that exponential stability conditions for the system do not depend on small parameter.

Keywords: Time scale · Exponential stability · Slow subsystem · Fast subsystem · Small parameter

1 Introduction

It can be observed that a lot of real phenomena lead to dynamical systems described by systems of differential or difference equations. Moreover, some components of these systems can change in time rapidly and other – relatively slowly. Such behavior follows from the fact that their description , as well continuous- as discrete-time cases, a small parameter appears, see for example [9,10,12]. Such class of systems is called *singularly perturbed system*.

As well in the continuous-time as in discrete-time some results concerning, for example, decomposition of singularly perturbed system onto two subsystems with slow and with fast motions, looks similarly, see [12,14,15]. So, a natural question is if some properties of such system can be unify to a general model of time, i.e. on time scales $\mathbb{T}$. In this context a time can be not only continuous (i.e. $\mathbb{T} = \mathbb{R}$) or discrete (i.e. $\mathbb{T} = h\mathbb{Z}$ with $h > 0$), but also can be, for example, an interval model (partially continuous and partially discrete), quantum model, harmonic, etc., for more examples see [4,5]. Originally the ideas of time scales was introduce by Stefan Hilger [8], as a proper language for unification of continuous-time and discrete-time theories.

Classically, in the context of decomposition of perturbed systems into slow and fast subsystems, the phrase *time scale* is used, not as a model of time, but for the stressing some properties of obtained subsystems. Also the system defined on time domain $h\mathbb{Z}$, $h > 0$, differs from the classical discrete-time system, see [2].

A. Bartoszewicz et al. (Eds.): PCC 2020, AISC 1196, pp. 16–26, 2020.
https://doi.org/10.1007/978-3-030-50936-1_2

In this paper the problem of stability of linear singularly perturbed systems with a constant delay on homogeneous time scales is studied. The stability problem of dynamic equations, without delays and perturbations, on time scale was studied in [13]. In [3] necessary and sufficient conditions for the exponential stability of a linear retarded time-delay system defined on a homogeneous time scale were proved. Conditions are formulated in terms of a characteristic equation associated with the given system.

In order to discuss conditions of exponential stability of the linear singular perturbed system with a constant delay on time scales we decomposed it onto slow and fast subsystems. This decomposition is based on non-degenerate change of variables given in [16]. The decoupling transformation is presented in the form of asymptotic series on $\mathbb{T}$. On this way spectrum-type sufficient conditions of exponential stability are obtained.

2 Stability of Systems with Delay on Time Scales

A *time scale* $\mathbb{T}$ is an arbitrary nonempty closed subset of the set $\mathbb{R}$ of real numbers. As a standard cases $\mathbb{R}$ and $\tau\mathbb{Z}$, $\tau > 0$, are time scales. As examples of time scales one can take $P_{a,b} = \bigcup_{k\in\mathbb{Z}}[k(a+b); k(a+b)+a]$ or $Q_{a,b} = \bigcup k \in \mathbb{Z}\{t \in [ka; kb]\} \cup \{0\}$ for any $a < b$. But also $[-2,0] \cup [3,4]$ is a time scale.

For $t \in \mathbb{T}$ the *forward* and *backward jump operators* $\sigma, \rho : \mathbb{T} \to \mathbb{T}$ are defined by $\sigma(t) = \inf\{s \in \mathbb{T}|\ s > t\}$, and $\rho(t) = \sup\{s \in \mathbb{T}|\ s < t\}$, respectively. In addition we set $\sigma(\max\mathbb{T}) = \max\mathbb{T}$ if there exists a finite $\max\mathbb{T}$, and $\rho(\min\mathbb{T}) = \min\mathbb{T}$ if there exists a finite $\min\mathbb{T}$. The *graininess* functions $\kappa : \mathbb{T} \to [0,\infty)$ are defined by[1] $\kappa(t) = \sigma(t) - t$ and $\nu(t) = t - \rho(t)$ for all $t \in \mathbb{T}$. A time scale is called *homogeneous* if both κ and ν are constant functions.

Let $\mathbb{T}^\iota$ denote truncated set consisting of $\mathbb{T}$ except for a possible maximal point t such that $\rho(t) < t$. The *delta (Hilger) derivative* of a function $f : \mathbb{T} \to \mathbb{R}$ at $t \in \mathbb{T}^\iota$ is the real number $f^\Delta(t)$ (provided it exists) such that for each $\varepsilon > 0$ there exists a neighborhood $U(\varepsilon)$ of t, $U(\varepsilon) \subset \mathbb{T}$ such that for all $\tau \in U(\varepsilon)$, $|(f(\sigma(t)) - f(\tau)) - f^\Delta(t)(\sigma(t) - \tau)| \leq \varepsilon|\sigma(t) - \tau|$. Moreover, we say that f is *delta differentiable* on $\mathbb{T}^\iota$ provided $f^\Delta(t)$ exists for all $t \in \mathbb{T}^\iota$.

The *Cauchy integral* is defined as

$$\int_r^s f(t)\Delta t = F(s) - F(r)$$

for all $r, s \in \mathbb{T}^\iota$ and $f : \mathbb{T} \to \mathbb{R}$ such that $F^\Delta(t) = f(t)$, i.e. F is *antiderivative* of function f.

Let $A(t)$ denote an $n \times n$ matrix-valued function on $\mathbb{T}$ and I_n is $n \times n-$identity matrix. Matrix A is called *regressive* provided $I_n + \kappa(t)A(t)$ is invertible for all $t \in \mathbb{T}^\iota$. In fact, A is regressive if and only if its eigenvalues $\lambda_i(t)$ are regressive for all $i = 1, \dots, n$, see [4].

[1] For the purpose of these studies, we use a different notation than the standard in the time scales theory, namely the (forward) graininess function is denoted as κ instead of the standard used μ. μ later on will serve as a parameter.

Let us make the following:
Assumption A. $\mathbb{T}$ is such time scale that if $t, h \in \mathbb{T}$ then also $t - h \in \mathbb{T}$.

Let a, b be real numbers such that $a < b$. By $[a,b]_{\mathbb{T}} = [a,b] \cap \mathbb{T}$ we will denote an interval on time scale $\mathbb{T}$. Let $h > 0$. We will assume that $[-h,0]_{\mathbb{T}}$ contains at least 0. Moreover, by $(C[a,b]_{\mathbb{T}}, \mathbb{R}^n)$ we will denote the Banach space of right hand continuous functions such that $[a,b]_{\mathbb{T}} \to \mathbb{R}^n$ with the topology of uniform convergence. If $\phi \in C([-h,0]_{\mathbb{T}}, \mathbb{R}^n)$, then

$$||\phi||_h := sup_{s\in[-h,0]_{\mathbb{T}}}||\phi(s)||.$$

For $x : \mathbb{T} \to \mathbb{R}^n$ and fixed $h \in (0,+\infty)_{\mathbb{T}}$ define function $x_t : [-h,0]_{\mathbb{T}} \to \mathbb{R}^n$ as $x_t(s) := x(t+s)$ for $s \in [-h,0]_{\mathbb{T}}$. Let $t_0 \in \mathbb{T}$ and $x(t_0) = x_0 \in \mathbb{R}^n$, $\phi \in (C[a,b]_{\mathbb{T}}, \mathbb{R}^n)$. Consider the following system with delay:

$$\begin{aligned} &x^{\Delta} = Ax(t) + \textstyle\int_{-h}^{0} [\Delta A(\xi)]x(t+\xi) + f(t),\ t \in [t_0, t_1]_{\mathbb{T}^\kappa}, \\ &x_{t_0} = \phi,\ t_0 \in \mathbb{T}. \end{aligned}$$

with A – constant matrix, $A(\xi)$ – a matrix-valued function of bounded variation on $[-h,0]_{\mathbb{T}}$, and left continuous at 0, $f : \mathbb{T} \to \mathbb{R}^n$ – bounded right continuous function.

Define

$$\mathcal{A}(z) := A + \int_{-h}^{0} [\Delta A(\xi)]e_{\ominus z}(-\xi, 0), \tag{1}$$

where $e_{\ominus z}(\xi,0)x(t) := x(t-\xi)$, such that

$$\mathcal{A}(z)x_t = Ax(t) + \int_{-h}^{0} [\Delta A(\xi)]x(t+\xi).$$

Then system with delay can be rewritten in the operator form as:

$$\begin{aligned} &x^{\Delta} = \mathcal{A}(z)x_t + f(t),\ t \in [t_0, t_1]_{\mathbb{T}^\kappa}, \\ &x_{t_0} = \phi,\ t_0 \in \mathbb{T}. \end{aligned} \tag{2}$$

The map $x \in (C[t_0 - h, t_1], \mathbb{R}^n)$ is a *solution of (2)* on $[t_0; t_1]$ if $x(t)$ is an antiderivative of $\mathcal{A}(z)x_t + f(t)$ on $[t_0; t_1]_{\mathbb{T}^\kappa}$, and satisfies $x_{t_0}(s) = \phi(s)$ for all $s \in [-h; 0]_{\mathbb{R}}$. If we consider the free solution of (2), then initial conditions should be assumed to be different from zero.

Theorem 1. *Let $t_0 \in \mathbb{T}$ and $A(t)$ is an $n \times n$ matrix-valued function on $\mathbb{T}$ and $f : \mathbb{T} \to \mathbb{R}^n$ is right continuous. Then the initial value problem (2) has the unique solution $x(t)$ on $[t_0, t_1]_{\mathbb{T}}$.*

Proof. The validity of the thesis follows from the fact that the system (2) satisfies all conditions of Theorem 4.6 in [11]. □

For (1) let us introduce the notion

$$\sigma(\mathcal{A}) := \{\lambda \in \mathbb{C} : \det(\lambda I_n - \mathcal{A}(\lambda)) = 0\}.$$

The characteristic equation for system (2) is defined as

$$\chi(\lambda) := \det\left(\lambda I_n - \mathcal{A}(\lambda)\right) = 0. \tag{3}$$

Then the spectrum σ of the system (2) consists of all complex roots of Eq. (3), i.e. $\sigma = \sigma(\mathcal{A})$.

For arbitrary $t_0 \in \mathbb{T}$ let

$$\mathcal{S}_{\mathbb{C}}(\mathbb{T}) := \{\lambda \in \mathbb{C} : \limsup_{T\to\infty} \frac{1}{T - t_0} \int_{t_0}^{T} \lim_{s\to\kappa(t)} \frac{\log|1+s\lambda|}{s} \Delta t < 0\}$$

and

$$\mathcal{S}_{\mathbb{R}}(\mathbb{T}) := \{\lambda \in \mathbb{R} : \forall \tau \in \mathbb{T}\, \exists t \in \mathbb{T}, t > \tau : 1 + \kappa(t)\lambda = 0\}.$$

If $\mathbb{T} = \mathbb{R}$, then, $\mathcal{S}_{\mathbb{R}}(\mathbb{R}) = \emptyset$ and $\mathcal{S}(\mathbb{T}) = \{\lambda \in \mathbb{C} | Re\lambda < 0\}$. If $\mathbb{T} = \tau\mathbb{Z}, \tau > 0$, then $\mathcal{S}_{\mathbb{R}}(\tau\mathbb{Z}) = \{-\frac{1}{\tau}\}$ and $\mathcal{S}(\tau\mathbb{Z}) = \mathcal{B}_{\frac{1}{\tau}}(-\frac{1}{\tau})$, where $\mathcal{B}_{\frac{1}{\tau}}(-\frac{1}{\tau})$ denotes the disc with the centre at $(\frac{-1}{\tau}, 0)$ and with the radius of $\frac{1}{\tau}$, see [1,3].

We say that system (2) with $f(t) \equiv 0$ is *exponentially stable* if there exist $a > 0$ (called the decay rate) and $K \geq 1$ such that $||x_t||_h \leq Ke^{-a(t-t_0)}||x_0||_h$ for any $t \geq t_0$ and any initial function x_{t_0}. The *set of exponential stability* for time scale $\mathbb{T}$ is the set $\mathcal{S}(\mathbb{T}) := \mathcal{S}_{\mathbb{C}}(\mathbb{T}) \cup \mathcal{S}_{\mathbb{R}}(\mathbb{T})$. The decay rate a can be equal the most poorly damped (i.e. lying most closely to stability border) eigenvalue of the state operator.

Theorem 2. *[3] Let $\mathbb{T}$ be a homogeneous time scale. Suppose that system (2) is such that $f(t) \equiv 0$. Then this system is exponentially stable if and only if all solutions of the characteristic equation (3) lie in $\mathcal{S}(\mathbb{T})$.*

3 Singularly Perturbed Linear Systems with Delay on Time Scales

Consider a singularly perturbed linear time-invariant system with delay (shortly noted as SPLTISD) on time scale $\mathbb{T}$, is system of the form

$$x^{\Delta}(t) = \mu A_1 x(t) + \mu \int_{-h}^{0} [\Delta A_1(\xi)]\, x(t+\xi) + \mu A_2 y(t), \tag{4}$$

$$y^{\Delta}(t) = A_3 x(t) + \int_{-h}^{0} [\Delta A_3(\xi)]\, x(t+\xi) + A_4 y(t), \tag{5}$$

where $\mu \in (0, \mu^0]$, $\mu^0 \ll 1$, i.e. μ is a small parameter, $t \in [t_0, +\infty)_{\mathbb{T}}$, $h \in \mathbb{T}$, $x \in \mathbb{R}^{n_1}$ is called a *slow variable*, $y \in \mathbb{R}^{n_2}$ – a *fast variable*, $x(\theta) = \phi(\theta)$, $\theta \in [t_0 - h, t_0]_{\mathbb{T}}$, $y(t_0) = y_0 \in \mathbb{R}^{n_2}$, ϕ is n_1-dimensional vector-function with rd-constant entries. Moreover, $A_i(\xi)$, $\xi \in [-h, 0]_{\mathbb{T}}$, $i = 1, 3$ are matrix-valued functions of bounded variation on $[-h, 0]_{\mathbb{T}}$ and are left continuous at 0. A_2, A_4 are constant matrices of appropriate dimensions. The *characteristic equation* of system (4)–(5) is defined as

$$\chi(\mu, z) := \det\left(zI_{n_1+n_2} - \begin{pmatrix} \mu\mathcal{A}_1(z) & \mu A_2 \\ \mathcal{A}_3(z) & A_4 \end{pmatrix}\right) = 0,$$

and its spectrum is

$$\sigma(\mu) := \{\lambda(\mu) \in \mathbb{C} : \chi(\mu, \lambda) = 0\}.$$

Let $A(\xi) \equiv 0$, $i = 2, 4$. Then, using (1), the SPLTISD (4)–(5) can be rewritten in the equivalent form as:

$$\begin{pmatrix} x^{\Delta}(t) \\ y^{\Delta}(t) \end{pmatrix} = \mathcal{A}(\mu, z) \begin{pmatrix} x_t \\ y_t \end{pmatrix} \tag{6}$$

where

$$\mathcal{A}(\mu, z) = \operatorname{diag}\{\mu I_{n_1}, I_{n_2}\} \begin{pmatrix} \mathcal{A}_1(z) & \mathcal{A}_2(z) \\ \mathcal{A}_3(z) & \mathcal{A}_4(z) \end{pmatrix}.$$

Definition 1. *For a given* $\mu \in (0, \mu^0]$ *the SPLTISD (4)–(5) is* exponentially stable *if there exist* $a > 0$ *and* $K \geq 1$ *such that* $||\{x_t, y_t\}||_h \leq Ke^{-a(t-t_0)}||\{x_0, y_0\}||_h$ *for any* $t \in [t_0; +\infty)_{\mathbb{T}}$ *and any initial conditions* $x_{t_0}, y(t_0)$.

Theorem 3. *For a given* $\mu \in (0, \mu^0]$ *SPLTISD (4)–(5) defined on homogeneous time scale* $\mathbb{T}$ *is exponentially stable if and only if* $\sigma(\mu) \subset \mathcal{S}(\mathbb{T})$.

Proof. Note that equivalent systems (4)–(5) and (6) have the same spectrum, therefore they are stable or unstable at the same time. The thesis follows from application Theorem 2 to system (6). □

Suppose that $\det A_4 \neq 0$. With $(n_1 + n_2)$ – dimensional system (4)–(5) two subsystems can be associated: the slow and the fast ones. The criterion for the decomposition of the system into *fast* and *slow* parts is separation of its own eigenspectrum into two clusters with large and small eigenvalues. Large eigenvalues correspond to fast dynamics while small eigenvalues correspond to the slow dynamics. Thus, the system possesses two-time scale (slow and fast) properties if the smallest eigenvalue of the *fast* part is larger than the largest eigenvalue of the *slow* part. The small parameter μ represents the order of magnitude of the ratio of the large and small eigenvalues.

The slow subsystem is of the form

$$\begin{aligned} \bar{x}^{\Delta}(t) &= \mu \mathcal{A}_0(z)\bar{x}_t, \quad t \in [t_0, +\infty)_{\mathbb{T}}, \\ \bar{x}_{t_0} &= \phi, t_0 \in \mathbb{T}, \end{aligned} \tag{7}$$

where

$$\mathcal{A}_0(z) := \mathcal{A}_1(z) - A_2 A_4^{-1} \mathcal{A}_3(z), \tag{8}$$

and $\bar{x} \in \mathbb{R}^{n_1}$. The slow subsystem (7) is a linear stationary n_1–dimensional system with delay. For this subsystem the characteristic equation is

$$\chi_s(\mu, z) := \det(zI_{n_1} - \mu \mathcal{A}_0(z)) = 0.$$

Then

$$\sigma_s(\mu) := \{\lambda \in \mathbb{C} : \chi_s(\mu, \lambda) = 0\} = \{\mu\lambda : \lambda \in \sigma(\mathcal{A}_0)\} \tag{9}$$

denotes the spectrum of slow subsystem (7).

The fast subsystem is the subsystem of (4)–(5) of the form

$$\tilde{y}^{\Delta}(t) = A_4\tilde{y}(t), \qquad (10)$$
$$\tilde{y}(t_0) = y_0 - \bar{y}(t_0),$$

where $\tilde{y}(t_0) := A_4^{-1}\mathcal{A}_3(z)\phi$, $\tilde{y}(t) \in \mathbb{R}^{n_2}$. The characteristic equation of this subsystem is

$$\chi_f(\lambda) := \det(\lambda I_{n_2} - A_4) = 0.$$

Then

$$\sigma_f := \{\lambda \in \mathbb{C} : \chi_f(\lambda) = 0\} \qquad (11)$$

denotes the spectrum of the fast subsystem (10).

In order to study stability of SPLTISD (4)–(5) let us consider the following system with matrix operators $L(\mu, z)$ and $H(\mu, z)$ such that:

$$A_3(z) - A_4L(\mu, z) + \mu L(\mu, z)(A_1(z) - A_2L(\mu, z)) = 0, \qquad (12)$$
$$\mu(A_1(z) - A_2L(\mu, z))H(\mu, z) - H(\mu, z)(A_4 + \mu L(\mu, z)A_2) + A_2 = 0.$$

where $\mu \in \mathbb{R}, z \in \mathbb{C}$.

Let $O(\mu)$ denote a vector function $f(t, \mu)$, $t \in [t_1, t_2]_{\mathbb{T}}$, with the property: there exist positive constants μ^* and c such that $|f(t, \mu)| \leq c\mu$ for all $\mu \in (0, \mu^*]$ and $t \in [t_1, t_2]_{\mathbb{T}}$.

Lemma 1. *[16] Suppose that the matrix A_4 is nonsingular. Then there exists a real positive constant μ^* such that for all $0 < \mu \leq \mu^*$ there is a solution $L(\mu, z)$ and $H(\mu, z)$, continuously depending on μ, of Eqs. (12) that could be represented in the asymptotic series form*

$$L(\mu, z) = \sum_{i=0}^{k} \mu^i L^i(z) + O(\mu^{k+1}), \qquad (13)$$

$$H(\mu, z) = \sum_{i=0}^{k} \mu^i H^i(z) + O(\mu^{k+1}), \qquad (14)$$

where

$$L^0(z) = A_4^{-1}A_3(z), \quad L^1(z) = A_4^{-2}A_3(z)A_0(z), \qquad (15)$$
$$H^0 = A_2A_4^{-1}, \quad A_0(z) = A_1(z) - A_2A_4^{-1}A_3(z),$$

and other terms of series (13), (14) satisfy the following recursive relations:

$$L^{k+1}(z) = A_4^{-1}L^k(z)A_1(z) - A_4^{-1}\sum_{j=0}^{k} L^{k-j}(z)A_2L^j(z),$$

$$H^{k+1}(z) = A_4^{-1}A_1(z)H^k(z) - A_4^{-1}A_2\sum_{i=0}^{k} L^i(z)H^{k-i}(z) - A_4^{-1}\sum_{i=0}^{k} H^i(z)L^{k-i}(z)A_2.$$

Similarly to [7,9] define

$$K(\mu, z) = \begin{pmatrix} I_{n_1} & \mu H(\mu, z) \\ -L(\mu, z) & I_{n_2} - \mu L(\mu, z) H(\mu, z) \end{pmatrix}.$$

Transformation $\begin{pmatrix} \xi \\ \eta \end{pmatrix} = K^{-1} \begin{pmatrix} x \\ y \end{pmatrix}$, $\xi \in \mathbb{R}^{n_1}, \eta \in \mathbb{R}^{n_2}$, brings SPLTISD (4)–(5), (regarding its representation (6)), to the form

$$\xi^{\Delta}(t) = \mu \mathcal{A}_{\xi}(\mu, z)\xi_t, \tag{16}$$

$$\eta^{\Delta}(t) = \mathcal{A}_{\eta}(\mu, z)\eta_t, \tag{17}$$

where

$$\mathcal{A}_{\xi}(\mu, z) = \mathcal{A}_1(z) - A_2 L(\mu, z), \quad \mathcal{A}_{\eta}(\mu, z) = A_4 + \mu L(\mu, z) A_2.$$

Taking into account (8), (13), (15) we obtain

$$\mathcal{A}_{\xi}(\mu, z) = \mathcal{A}_0(z) + O(\mu), \quad \mathcal{A}_{\eta}(\mu, z) = A_4 + O(\mu). \tag{18}$$

In fact, (16)–(17) presents the exact decomposition of the system (4)–(5) into two separated subsystems with two separated motions: slow given by (16) and fast given by (17). The spectrum of the slow subsystem (16) is

$$\sigma_x(\mu) = \{\lambda(\mu) \in \mathbb{C} : \det(\lambda I_{n_1} - \mu \mathcal{A}_{\xi}(\mu, \lambda)) = 0\}, \tag{19}$$

while he spectrum of the fast subsystem (17) is

$$\sigma_y(\mu) = \{\lambda(\mu) \in \mathbb{C} : \det(\lambda I_{n_2} - \mathcal{A}_{\eta}(\mu, \lambda)) = 0\}. \tag{20}$$

Proposition 1. *Decoupled system (16)–(17) on time scale $\mathbb{T}$ satisfying assumption A is $O(\mu)$-close to the slow subsystem (7) and the fast subsystem (10).*

Proof. The thesis is a consequence of (18). □

Proposition 2. *Suppose that time scale $\mathbb{T}$ satisfies assumption A. For sufficiently small $\mu \in (0, \mu^0]$ the spectrum $\sigma(\mu)$ of the system (4)–(5) is separated into two parts $\sigma_x(\mu)$ and $\sigma_y(\mu)$ such that $\sigma(\mu) = \sigma_x(\mu) \cup \sigma_y(\mu)$ and $\sigma_x(\mu) \cap \sigma_y(\mu) = \emptyset$. The spectrum of the slow subsystem consists of elements $\mu\lambda_i(\mu)$ from set (19) such that*

$$\lim_{\mu \to 0} \lambda_i(\mu) = \lambda_i \in \sigma(\mathcal{A}_0).$$

The spectrum of the fast subsystem consists of elements $\lambda_i(\mu)$ from set (20) such that

$$\lim_{\mu \to 0} \lambda_i(\mu) = \lambda_{fi} \in \sigma_f.$$

If in the spectrum σ_s given by (9) of the slow subsystem (7) there are no multiple values and in the spectrum σ_f given by (11) of fast subsystem (10) there are no multiple values then the eigenvalues of the (4)–(5) are approximated as

$$\lambda_i(\mu) = \mu\lambda_i + O(\mu^2), \; \lambda_i \in \sigma(\mathcal{A}_0), \; \forall \lambda_i(\mu) \in \sigma_x, \tag{21}$$

$$\lambda_i(\mu) = \lambda_{fi} + O(\mu), \; \lambda_{fi} \in \sigma_f, \; \forall \lambda_i(\mu) \in \sigma_y. \tag{22}$$

Proof. For the time scale satisfying assumption A the reasoning is similar to the reasoning in the continuous-time given in [17]. □

Definition 2. *For a given* $\mu > 0$ *the* slow subsystem *(7)* is exponentially stable *if there exist* $a_s(\mu) > 0$ *and* $K_s(\mu) \geq 1$ *such that* $||\bar{x}_t||_h \leq K_s(\mu)e^{-a_s(\mu)(t-t_0)}||x_{t_0}||_h$ *for any* $t \geq 0$ *and any initial function* $\bar{x}_{t_0}$. *The* fast subsystem *(10)* is exponentially stable *if there exists a positive constant* a_f *such that for every* $t_0 \in \mathbb{T}$ *there exists* $k_f = k_f(t_0) > 1$ *with* $||e_{A_4}(t, t_0)\tilde{y}(t_0)|| \leq k_f e^{-a_f(t-t_0)}||\tilde{y}(t_0)||$ *for any* $t \in [t_0, \infty)_{\mathbb{T}}$.

Proposition 3. *Let* $\mathbb{T}$ *be homogeneous time scale. If* $\sigma(\mathcal{A}_0) \subset \mathcal{S}(\mathbb{T})$ *then* $\sigma(\mu\mathcal{A}_0 + O(\mu^2)) \subset \mathcal{S}(\mathbb{T})$ *for all sufficient small* $\mu > 0$.

Proof. For homogeneous time scale it is sufficient to consider two cases: $\mathbb{T} = \mathbb{R}$ and $\mathbb{T} = \tau\mathbb{Z}$.

1. For $\mathbb{T} = \mathbb{R}$ we have $\mathcal{S}(\mathbb{T}) = \{\lambda \in \mathbb{C} : Re\lambda < 0\}$. So for $\lambda \in \sigma(\mathcal{A}_0) \subset \mathcal{S}(\mathbb{T})$ it is true $Re\lambda < 0$ and then $Re(\mu\lambda + O(\mu^2)) < 0$ for all sufficient small $\mu > 0$. Thus for $\mathbb{T} = \mathbb{R}$ the proposition is true.
2. For $\mathbb{T} = \tau\mathbb{Z}$ we have $\mathcal{S}(\mathbb{T}) = \mathcal{B}_{\frac{1}{\tau}}(-\frac{1}{\tau})$. Since the set $\mathcal{S}(\mathbb{T})$ is open and elements $\lambda(\mu) \in \sigma(\mu\mathcal{A}_0)$ continuously depend on μ, then $\sigma(\mu\mathcal{A}_0 + O(\mu^2)) \subset \mathcal{S}(\mathbb{T})$ for all sufficient small $\mu > 0$. □

Theorem 4. *Let* $\mathbb{T}$ *be a homogeneous time scale.*

i. For a given $\mu > 0$ *slow subsystem (7) is exponentially stable if and only if its spectrum is contained in* $\mathcal{S}(\mathbb{T})$.
ii. Fast subsystem (10) is exponentially stable if and only if its spectrum is contained in $\mathcal{S}(\mathbb{T})$.

Proof. Item *i.* is the consequence of applying Theorem 2 for the slow subsystem (7). Item *ii.* follows from the exponential stability of delta-derivative system on time scales for fast subsystem (10), see [13]. □

4 Examples

Example 1. Consider the following system:

$$x^{\Delta}(t) = \mu x(t-\tau) + \mu x(t-2\tau) + \mu y(t), x \in \mathbb{R},\ y \in \mathbb{R},\ t \in [0, \infty)_{\mathbb{T}},$$
$$y^{\Delta}(t) = -x(t-\tau) - 0.5x(t-2\tau) - 0.5y(t),\ \mu \in \left(0, \mu^0\right],\ \mu^0 \ll 1, \quad (23)$$

defined on time scale $\mathbb{T} = \tau\mathbb{Z}$ with $\tau = 3$, $h = 2\tau$, $A_1 = 0$, $A_2 = 1$, $A_3 = 0$, $A_4 = -0.5$ and

$$A_1(\xi) = \begin{cases} 0,\ \xi \leq -h, \\ -1,\ -h < \xi < 0, \\ 0,\ \xi \geq 0, \end{cases} \qquad A_3(\xi) = \begin{cases} 0.1,\ \xi \leq -h, \\ 0.6\ -h < \xi < 0, \\ 0,\ \xi \geq 0. \end{cases}$$

The system (23) is a particular case of (4)–(5) on $\mathbb{T} = \tau\mathbb{Z}$. Parameters of the slow subsystem (7) are:

$$A_0 = 0,\ \mathcal{A}_0(z) = \int_{-h}^{0} [\Delta A_0(\xi)] e_{\ominus z}(-\xi, 0),$$
$$A_0(\xi) = \begin{cases} 0.2,\ \xi \leq -h, \\ 0.2,\ -h < \xi < 0, \\ 0,\ \xi \geq 0. \end{cases}$$

So, finally we have the slow subsystem:

$$\bar{x}^{\Delta}(t) = -0.2\mu\bar{x}(t - \tau),\ t \in [0, \infty)_{\tau\mathbb{Z}}, \tag{24}$$

and the fast subsystem:

$$\tilde{y}^{\Delta}(t) = -0.5\tilde{y}(t),\ t \in [0, \infty)_{\tau\mathbb{Z}}, \tag{25}$$

For the considered case we have $\mathcal{S}(\mathbb{T}) = \mathcal{S}(\tau\mathbb{Z}) = \mathcal{B}_{\frac{1}{3}}(-\frac{1}{3})$. The spectrum of (25) is $\sigma_f = \{-0.5\}$ and $\sigma_f \subset \mathcal{S}(\mathbb{T})$. From [3], (Lemma 1) it follows that for (24) we have $\sigma(\mathcal{A}_0) = \{\frac{1}{30}(-5 + \imath\sqrt{35}), \frac{1}{30}(-5 - \imath\sqrt{35}), -\frac{1}{3}\}$, so $\sigma(\mathcal{A}_0) \subset \mathcal{B}_{\frac{1}{\tau}}(-\frac{1}{\tau})$ (see Fig. 1).

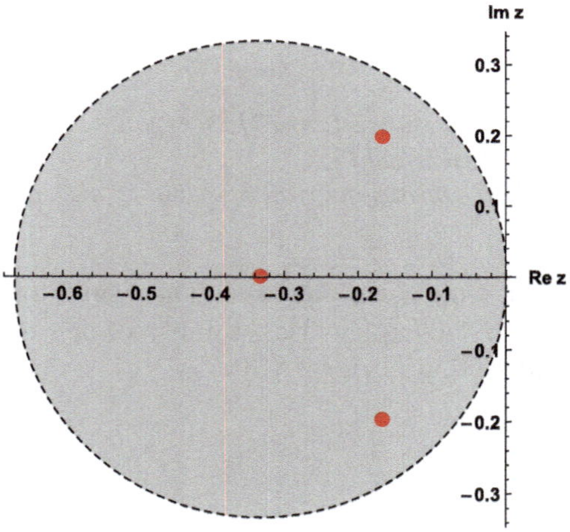

Fig. 1. Example 1: $\sigma(\mathcal{A}_0) \subset \mathcal{B}_{\frac{1}{\tau}}(-\frac{1}{\tau})$

Then, there exists a $\mu^* \in (0, \mu^0]$ such that the SPLTISD (23) is exponentially stable for all $\mu \in (0, \mu^*]$.

Example 2. Consider the following system:

$$x^{\Delta}(t) = \mu x(t) - \mu y(t), x \in \mathbb{R},\ y \in \mathbb{R},\ t \in [0, +\infty)_{\mathbb{T}},$$
$$y^{\Delta}(t) = kx(t - 2) - y(t),\ k \in \mathbb{R},\ \mu \in \left(0, \mu^0\right],\ \mu^0 \ll 1 \tag{26}$$

defined on time scale $\mathbb{T} = \mathbb{R}$ with parameters $h = 2, t_0 = 0$, and scalar functions $A_1 = 0,\ A_2 = -1,\ A_3 = 0,\ A_4 = -1$

$$A_1(\xi) = \begin{cases} -1,\ \xi \leq -2, \\ -1,\ -2 < \xi < 0, \\ 0,\ \xi \geq 0, \end{cases} \quad A_3(\xi) = \begin{cases} -k,\ \xi \leq -2, \\ 0,\ -2 < \xi < 0, \\ 0,\ \xi \geq 0. \end{cases}$$

The system (26) is a particular case of (4)–(5) on $\mathbb{T} = \mathbb{R}$. Parameters of the slow subsystem are:

$$A_0 = 0,\ \mathcal{A}_0(z) = \int_{-2}^{0} [\Delta A_0(\xi)] e_{\ominus z}(-\xi, 0),$$
$$A_0(\xi) = \begin{cases} k-1,\ \xi \leq -2, \\ -1,\ -2 < \xi < 0, \\ 0,\ \xi \geq 0. \end{cases}$$

Note that in this case $e_{\ominus z}(h, 0) = e^{-hz}$ and then $\mathcal{A}_0(z) = 1 - ke^{-2z}$. So, finally we have the slow subsystem:

$$\bar{x}^{\Delta}(t) = \mu\bar{x}(t) - \mu k\bar{x}(t-2),\ t \in [0, +\infty) \tag{27}$$

that is a particular case of system (17) from [6], p. 125 and the fast subsystem:

$$\tilde{y}^{\Delta}(t) = -\tilde{y}(t),\ t \in [0, +\infty). \tag{28}$$

For the considered case we have $\mathcal{S}(\mathbb{T}) = \{\lambda \in \mathbb{C} : Re\lambda < 0\}$. The spectrum of (25) is $\sigma_f = \{-1\} \subset \mathcal{S}(\mathbb{T})$. The spectrum of (27) for $\mu = 1$ lies in $\mathcal{S}(\mathbb{T})$ if and only if $|k| < 1$, see [6]. So, if $|k| < 1$ then there exists a $\mu^* \in (0, \mu^0]$ such that the SPLTISD (26) is exponentially stable for all $\mu \in (0, \mu^*]$.

5 Conclusions

The problem of the exponential stability of linear singularly perturbed system with a small parameter on homogeneous time scales has been studied. It was shown that given system is exponentially stable if and only if its spectrum is contained in the stability set on time scale. From Proposition 3 it follows that conditions of Theorem 4 valid for $\mu = 1$ are sufficient for the stability of the singularly perturbed system under consideration for all sufficiently small $\mu > 0$.

Acknowledgement. The work of Olga Tsekhan was partially supported under the state research program "Convergence-2020" of Republic of Belarus: Task 1.3.02. The work of Ewa Pawluszewicz was supported by grant No. WZ/WM/1/2019 of Bialystok University of Technology, financed by Polish Ministry of Science and Higher Education.

References

1. Bartosiewicz, Z., Piotrowska, E., Wyrwas, M.: Stability, stabilization and observers of linear control systems on time scales. In: Proceedings of the 46th IEEE Conference on Decision and Control New Orleans, LA, USA, pp. 2803–2808 (2007)

2. Bartosiewicz, Z., Kotta, U., Pawłuszewicz, E.: Equivalence of linear control systems on time scales. Proc. Est. Acad. Sci. Phys. Math. **55**(1), 43–52 (2006)
3. Belikov, J., Bartosiewicz, Z.: Stability and stabilizability of linear time-delay systems on homogeneous time scales. Proc. Est. Acad. Sci. **66**(2), 124–136 (2017)
4. Bohner, M., Peterson, A.: Dynamic Equations on Time Scales. Birkhäser, Boston (2001)
5. Davis, J.M., Gravagne, I.A., Marks II, R.J.: Convergence of unilateral Laplace transforms on time scales. Circ. Syst. Signal Process. **29**(5), 917–997 (2010)
6. Elsgolts, L.E., Norkin, S.B.: Introduction to the Theory of Differential Equations with Deviating Argument. Nauka, Moscow (1971)
7. Gajic, Z., Shen, X.: Parallel reduced-order controllers for stochastic linear singularly perturbed discrete systems. IEEE Trans. Autom. Control **36**(1), 87–90 (1991)
8. Hilger S.: Ein Maßkettenkalkumit Anwendung auf Zentrumsmannigfaltigkeiten. Ph.D. thesis, Universitat Wurzburg (1988)
9. Kokotovic, P.V., Khalil, H.K., O'Reilly, J.: Singular Perturbation Methods in Control: Analysis and Design. Academic Press, London (1986)
10. Kurina, G.A., Dmitriev, M.G., Naidu, D.S.: Discrete singularly perturbed control problems (a survey). Dyn. Contin. Discrete Impulsive Syst. Ser. B Appl. Algorithms **24**, 335–370 (2017)
11. Liu, X., Zhang, K.: Existence, uniqueness and stability results for functional differential equations on time scales. Dyn. Syst. Appl. **25**(4), 501–530 (2016)
12. Naidu, D.S., Price, D.B.: Singular perturbations and time scales in the design of digital flight control systems, NASA Technical report 2844 (1988)
13. Pötzsche, C., Siegmund, S., Wirth, F.: A spectral characterization of exponential stability for linear time-invariant systems on time scales. DCSD-A **9**(5), 1223–1241 (2003)
14. Su, W.C., Gajic, Z., Shen, X.M.: The exact slow-fast decomposition of the Algebraic Riccati equation of singularity perturbated systems. IEEE Trans. Autom. Control **37**(9), 1456–1459 (1992)
15. Tsekhan, O.B.: Decoupling transformation for linear stationary singularly perturbed system with delay and its applications to analysis and control of spectrum. Vesnik Yanka Kupala State Univ. Grodno. Ser. 2 Math. Phys. Inform. Comput. Technol. Its Control **7**(1), 50–61 (2017). (in Russian)
16. Tsekhan, O., Pawluszewicz, E.: Slow-fast decomposition of singularly perturbed system with delay on time scales. In: The 20th International Carpathian Control Conference, Kraków-Wieliczka, Poland (2019)
17. Tsekhan, O.: Complete controllability conditions for linear singularly perturbed time-invariant systems with multiple delays via Chang-type transformation. Axioms **8**(71), 1–19 (2019)

Trajectory Following Quasi-Sliding Mode Control for Arbitrary Relative Degree Systems

Katarzyna Adamiak(✉)

Institute of Automatic Control, Łódź University of Technology, 18/22 Stefanowskiego St., 90-924 Łódź, Poland
katarzyna.adamiak@dokt.p.lodz.pl

Abstract. This study presents a new approach to discrete-time sliding mode control, based on a new trajectory following reaching law for arbitrary relative degree output systems. We divide the control design to two stages. First, we generate a desired profile of the sliding variable with a conventional switching type reaching law of Gao *et al.* Next, utilizing arbitrary relative degree sliding variable, we introduce a trajectory following reaching law for the real disturbed system. Our strategy ensures that the representative point of the system approaches the sliding plane monotonically and crosses it in finite time. After the sign of the sliding variable has changed for the first time, it changes again in each consecutive time step and its absolute value does not exceed an a priori known constant. Moreover, the strategy guarantees a reduction of the quasi-sliding mode band width, which results in improved robustness of the system. The paper is concluded with a simulation example.

Keywords: Discrete-Time systems · Arbitrary relative degree · Reaching law · Sliding mode control · Trajectory generator

1 Introduction

The concept of sliding mode control originates from the works of Emelyanov and Utkin from the mid-20th century [1, 2] and remains present in the control theory till these days. Sliding mode control quickly gained numerous followers due to its simplicity and unique system properties. Namely, the idea is to drive the representative point of the system to a predesigned surface in the state space, which ensures stable steady state behavior. For that purpose the structure of the controller must be switched according to the current position of the representative point relative to the sliding plane. As a result, the closed-loop system's dynamics become restricted to one surface in the state space, which reduces the dimensions of the dynamical problem. Moreover, due to continuous switching of the control signal, the system becomes insensitive to external disturbances and model uncertainties [3].

However, with the rapid development of electronic measurements and digital control systems, in the 1980s more attention was paid to adjustment of the sliding mode concept for discrete-time systems. As discretization introduces delays in the switching process,

A. Bartoszewicz et al. (Eds.): PCC 2020, AISC 1196, pp. 27–38, 2020.
https://doi.org/10.1007/978-3-030-50936-1_3

the discrete-time systems may not slide on the sliding surface, but only in its vicinity, in the so-called quasi-sliding mode. The precursors of the method were Milosavljević [4], Sarpturk, Istefanopulos and Kaynak [5], who determined the conditions for the existence and stability of the quasi-sliding mode. The idea was later developed by Utkin and Drakunov [6] and Bartolini *et al.* [7], who established the equivalent control design. However, the real milestone for the discrete-time sliding mode was brought by the study of Gao, Wang and Homaifa in 1995 [8]. The authors for the first time clearly defined the properties of the quasi-sliding mode and proposed a new control design method, which will be utilized further in this paper. According to Gao *et al.* the quasi-sliding mode exists when the representative point of the system approaches the sliding plane monotonically and crosses it in finite time. After the sliding plane was crossed for the first time, the sign of the sliding variable changes in each consecutive step and its absolute value never exceeds an a priori known constant. Moreover, the authors defined a reaching function, which controls the evolution of the sliding variable in the entire course of motion, providing the aforementioned properties. Later on numerous authors have adopted this design method and several new switching type [10–15] and nonswitching type [16–20] reaching laws have been introduced. Nowadays, the main branches of quasi-sliding mode control are optimal sliding modes [21], multirate output feedback sliding mode control [22, 23] and event-triggered sliding mode control [24, 25].

One of the latest achievements is the concept of designing the control signal not to influence the sliding variable in the nearest time step but a few steps ahead. The relation between the control signal and the system's output is called the relative degree. Traditionally, the control applied to the system in step k, gives immediate result in the sliding variable in step $k + 1$, so the relative degree is one. However, recently the research has shown that involving higher relative degree sliding variables [26–29] may have positive influence on the system's robustness. Based on the previous studies in this field, this paper proposes a new trajectory following reaching law for discrete time disturbed systems with arbitrary relative degree (r) sliding variable. Therefore, we focus on a situation when the control signal in the discrete time instant k does not have an impact on the sliding variable in steps $k + 1, k + 2, \ldots, k + r - 1$, but in step $k + r$.

The reminder of this paper is organized as follows. We first introduce the dynamical problem. Next, we propose the generation of the demand evolution of the sliding variable in Sect. 3.1 and introduce new model following reaching laws for relative degree one and arbitrary relative degree (r) sliding variable in Sect. 3.2. Finally, we present a simulation example in Sect. 4 and conclude our paper with final remarks in Sect. 5.

2 System Description

This study takes into account a discrete-time system described by the following state equation:

$$\boldsymbol{\eta}(k+1) = \boldsymbol{\Phi}\boldsymbol{\eta}(k) + \boldsymbol{\Gamma}u(k) + \boldsymbol{\Gamma}f(k), \tag{1}$$

where $\boldsymbol{\eta}$ is the $n \times 1$ state vector, $\boldsymbol{\Phi}$ is the state matrix, $\boldsymbol{\Gamma}$ is the input distribution vector, $u(k)$ is a scalar control signal and $f(k)$ represents a scalar disturbance. The initial conditions are denoted with $\boldsymbol{x}(0) = \boldsymbol{x}_0$. The aim is to drive the system to the desired state $\boldsymbol{x}_d$. The disturbance is lower and upper bounded as follows:

$$-f_{\max} \leq f(k) \leq f_{\max}. \tag{2}$$

In order to design higher relative degree sliding mode control, we first transform the system to the so-called Frobenius form:

$$\boldsymbol{x}(k+1) = \boldsymbol{A}\boldsymbol{x}(k) + \boldsymbol{b}u(k) + \boldsymbol{b}f(k), \tag{3}$$

where

$$\boldsymbol{A} = \begin{bmatrix} 0 & 1 & 0 & \cdots & 0 \\ 0 & 0 & 1 & \cdots & 0 \\ \vdots & \vdots & \vdots & \ddots & \vdots \\ 0 & 0 & 0 & \cdots & 1 \\ -a_0 & -a_1 & -a_2 & \cdots & -a_{n-1} \end{bmatrix} \tag{4}$$

and

$$\boldsymbol{b} = \begin{bmatrix} 0 \; 0 \cdots 0 \; 1 \end{bmatrix}^{\mathrm{T}}. \tag{5}$$

The appropriate matrices are obtained with a transformation operator $\boldsymbol{T} = \boldsymbol{S}_{\boldsymbol{\Phi}}\boldsymbol{S}_A^{-1}$, where $\boldsymbol{S}_{\boldsymbol{\Phi}}$ and $\boldsymbol{S}_A$ are the appropriate controllability matrices: $\boldsymbol{S}_{\boldsymbol{\Phi}} = [\boldsymbol{\Gamma}\ \boldsymbol{\Phi}\boldsymbol{\Gamma} \ \ldots\ \boldsymbol{\Phi}^{n-1}\boldsymbol{\Gamma}]$ and $\boldsymbol{S}_A = [\boldsymbol{b}\ \boldsymbol{A}\boldsymbol{b} \ \ldots\ \boldsymbol{A}^{n-1}\boldsymbol{b}]$. Having calculated the transformation matrix, we get $\boldsymbol{A} = \boldsymbol{T}\boldsymbol{\Phi}\boldsymbol{T}^{-1}$ and $\boldsymbol{b} = \boldsymbol{T}\boldsymbol{\Gamma}$. In the next chapter, we will design the demand evolution of the sliding variable for the system (3). Next, we will present a new reference trajectory following reaching law for relative degree one and for arbitrary relative degree system.

3 Quasi-Sliding Mode Control Strategy

3.1 Trajectory Generator

In the first stage of the control design we need to generate a non-disturbed reference sliding variable profile, which satisfies the conditions of the quasi-sliding mode as defined by Gao *et al.* [8]. Therefore, we denote the desired sliding variable with $s_d(k)$ and choose the sliding surface:

$$s_d(k) = 0. \tag{6}$$

We set the initial condition for the reference sliding variable same as for the plant's sliding variable. For $s_d(k)$ we apply the reaching law of Gao *et al.* [8] in the form:

$$s_d(k+1) = (1-q)s_d(k) - \varepsilon \operatorname{sgn}[s_d(k)], \tag{7}$$

where $0 < q < 1$ and $\varepsilon > 0$. For the sake of clarity, we assume that:

$$\operatorname{sgn}(z) = \begin{cases} 1 & \text{for } z \geq 0 \\ -1 & \text{for } z < 0 \end{cases}. \tag{8}$$

From the reaching law (8) we get:From the reaching law (8) we get:

$$s_d(k) - s_d(k+1) = qs_d(k) + \varepsilon \operatorname{sgn}[s_d(k)]. \tag{9}$$

Therefore, as long as the sign of s_d remains unchanged, the rate of change of the demand sliding variable satisfies:

$$|s_d(k) - s_d(k+1)| \geq \varepsilon. \tag{10}$$

Consequently, there exists a finite moment $k_{0d} \leq \text{floor}[|s_d(0)|/\varepsilon] + 1$, when:

$$\operatorname{sgn}[s_d(k_{0d})] = -\operatorname{sgn}[s_d(k_{0d} - 1)]. \tag{11}$$

Therefore, the sliding plane (6) is crossed for the first time not later than between steps $k_{0d} - 1$ and k_{0d}. Moreover, from (7), we conclude that for any $k \geq k_{0d}$ the absolute value of the sliding variable satisfies:

$$|s_d(k)| \leq \varepsilon. \tag{12}$$

Therefore, for any $k \geq k_{0d}$ the sliding variable belongs to an interval $[-\varepsilon; \varepsilon]$ and its sign changes in each successive step.

3.2 Trajectory Following Reaching Law

Relative Degree One System

In this section we will use the obtained reference profile of the sliding variable s_d to control the real disturbed system (3). We begin with the relative degree one sliding variable, chosen as:

$$s_1(k) = \boldsymbol{c}_1[\boldsymbol{x}_d - \boldsymbol{x}(k)]. \tag{13}$$

Our goal is to drive the sliding variable s_1 to the sliding plane $s_1(k) = 0$. As the system (3) is influenced by external disturbance $f(k)$, we denote its impact on the sliding variable with $d_1(k) = \boldsymbol{c}_1\boldsymbol{b}f(k)$. From (2) we notice that, for any $k \geq 0$, $d_1(k)$ satisfies:

$$|d_1(k)| \leq d_{\text{max}} = \boldsymbol{c}_1\boldsymbol{b}f_{\text{max}}. \tag{14}$$

We propose a trajectory following reaching law. We aim to drive the system's sliding variable $s_1(k)$ in each step to its desired position $s_d(k)$ with accuracy to the disturbance. The reaching function providing this motion is as follows:

$$s_1(k+1) = s_d(k+1) + d_1(k). \tag{15}$$

Conventional reaching laws calculate the next value of the sliding variable $s_1(k+1)$ utilizing $s_1(k)$, which holds the disturbance influence. Therefore, the disturbance impact is accumulated through the whole control process. The trajectory following reaching law, on the contrary, limits the disturbance influence to one control step. With application of (15) $s_1(k+1)$ bears the influence of $d_1(k)$ only.

From (3), (13) and (15) we obtain the following control signal:

$$u(k) = (\boldsymbol{c}_1\boldsymbol{b})^{-1}[\boldsymbol{c}_1\boldsymbol{x}_\mathbf{d} - \boldsymbol{c}_1\boldsymbol{A}\boldsymbol{x}(k) - s_d(k+1)]. \tag{16}$$

However, the impact of external disturbance causes the system to deviate from the reference. Therefore, the control parameters must be chosen in a specific way to ensure the existence of the quasi-sliding mode as defined by Gao *et al.* [8].

Theorem 1

If the control parameters satisfy

$$\varepsilon > \frac{d_{\max}}{q}, \tag{17}$$

then for any $k \geq k_{0d} + 2$ the sign of s_1 satisfies:

$$\operatorname{sgn}[s_1(k)] = -\operatorname{sgn}[s_1(k-1)] \tag{18}$$

and the absolute value of the sliding variable belongs to the interval $[-\varepsilon + d_{\max}; \varepsilon + d_{\max}]$.

Proof

Using the reaching law (15) we begin with expressing $s_1(k+2)$:

$$s_1(k+2) = s_d(k+2) + d_1(k+1). \tag{19}$$

Inserting the reaching law for the reference sliding variable (7) we obtain:

$$s_1(k+2) = (1-q)^2 s_d(k) - (1-q)\varepsilon \operatorname{sgn}[s_d(k)] - \varepsilon \operatorname{sgn}[s_d(k+1)] + d_1(k+1). \tag{20}$$

As it has already been shown by Gao *et al.* the reference sliding variable reaches the quasi-sliding mode in step k_{0d}, that is for any $k \geq k_{0d}$ the sign of $s_d(k)$ changes and the absolute value of $s_d(k)$ does not exceed ε. Considering (11), for any $k \geq k_{0d}$ we obtain:

$$\operatorname{sgn}[s_1(k+2)]|s_1(k+2)| = \operatorname{sgn}[s_d(k)]\Big[(1-q)^2|s_d(k)| + q\varepsilon\Big] + d_1(k+1). \tag{21}$$

From (21) one may notice that the term in brackets is always positive, while the sign of the disturbance is unknown. Taking into account (14) we get:

$$\operatorname{sgn}[s_1(k+2)]|s_1(k+2)| \leq \operatorname{sgn}[s_d(k)]\Big[(1-q)^2|s_d(k)| + q\varepsilon\Big] + d_{\max}. \tag{22}$$

Consequently, if (17) holds, then:

$$(1-q)^2|s_d(k)| + q\varepsilon > d_{\max}. \tag{23}$$

Considering that the sign of $s_d(k)$ changes for any $k \geq k_{0d}$, from (22) and (23) we conclude that:

$$\operatorname{sgn}[s_1(k+2)] = \operatorname{sgn}[s_d(k)] = -\operatorname{sgn}[s_1(k+1)] \tag{24}$$

for any $k \geq k_{0d}$. Therefore, the plant's sliding variable crosses the sliding plane for the first time not later than between steps $k_{0d} + 1$ and $k_{0d} + 2$ and its sign changes in all the following control steps. Furthermore, considering (12), (14) and (15) we get:

$$|s_1(k)| \leq \varepsilon + d_{\max} = \delta_1 \tag{25}$$

for any $k \geq k_{0d} + 2$, which ends the proof. ■

Arbitrary Relative Degree System

In general case the reference trajectory following sliding mode control strategy may be applied to systems with arbitrary relative degree sliding variable s_r, where $r \leq n$ and n is the order of the system. For the general case we select the sliding variable as

$$s_r(k) = \boldsymbol{c}_r[\boldsymbol{x}_d - \boldsymbol{x}(k)], \tag{26}$$

so that $\boldsymbol{c}_r\boldsymbol{A}^l\boldsymbol{b} = 0$, for $l = 1, 2, \ldots, r - 2$, and $\boldsymbol{c}_r\boldsymbol{A}^{r-1}\boldsymbol{b} \neq 0$. Therefore, the control signal $u(k)$ acts directly on the sliding variable $s_r(k + r)$. The sliding variable in steps $k + 1$, $k + 2$, …, $k + r - 1$ is determined by the system's state at time k only. For $l = 1, 2, \ldots, r - 1$, from (3) and (26) we get:

$$s_r(k + l) = \boldsymbol{c}_r\boldsymbol{x}_d - \boldsymbol{c}_r\boldsymbol{A}^l\boldsymbol{x}(k), \tag{27}$$

while for $l = r$ we obtain:

$$s_r(k + r) = \boldsymbol{c}_r[\boldsymbol{x}_d - \boldsymbol{x}(k + r)] = \boldsymbol{c}_r\boldsymbol{x}_d - \boldsymbol{c}_r\boldsymbol{A}^r\boldsymbol{x}(k) - \boldsymbol{c}_r\boldsymbol{A}^{r-1}\boldsymbol{b}u(k) - \boldsymbol{c}_r\boldsymbol{A}^{r-1}\boldsymbol{b}f(k). \tag{28}$$

In this case the disturbance impact on the system is denoted with $d_r(k) = \boldsymbol{c}_r\boldsymbol{A}^{r-1}\boldsymbol{b}f(k)$. For any $k \geq 0$ the disturbance satisfies:

$$|d_r(k)| \leq d_{r\max} = \boldsymbol{c}_r\boldsymbol{A}^{r-1}\boldsymbol{b}f_{\max}. \tag{29}$$

We aim to drive the system to the sliding plane $s_r(k) = 0$ and for that purpose we introduce a reaching law:

$$s_r(k + r) = s_d(k + r) + d_r(k). \tag{30}$$

Taking into account (3), (26) and (30) the control signal for the system becomes:

$$u(k) = \left(\boldsymbol{c}_r\boldsymbol{A}^{r-1}\boldsymbol{b}\right)^{-1}\left[\boldsymbol{c}_r\boldsymbol{x}_d - \boldsymbol{c}_r\boldsymbol{A}^r\boldsymbol{x}(k) - s_d(k + r)\right]. \tag{31}$$

Theorem 2

If the parameter $\varepsilon > d_{r\max}/q$, then for any $k \geq k_0 = k_{0d} + 2$

$$\text{sgn}[s_r(k)] = -\text{sgn}[s_r(k - 1)] \tag{32}$$

and for any $k \geq k_{0d} + 2$ the sliding variable is restricted by the following inequality:

$$|s_r(k)| \leq \varepsilon + d_{r\max} = \delta_r. \tag{33}$$

Moreover, if $|\boldsymbol{c_r A}^{r-1}\boldsymbol{b}| < |\boldsymbol{c}_1\boldsymbol{b}|$, then

$$d_{r\max} < d_{\max} \tag{34}$$

and consequently

$$\delta_r < \delta_1. \tag{35}$$

Proof
We have already shown that for any $k \geq k_{0d}$, $\text{sgn}[s_d(k)] = -\text{sgn}[s_d(k-1)]$. Therefore, if $\text{sgn}[s_r(k+r)] = \text{sgn}[s_d(k)]$, then the change of the sign of s_r will be guaranteed as well. Using (7) and (30) we express $s_r(k+r)$ as:

$$\begin{aligned} s_r(k+r) = (1-q)\big\{(1-q)s_d(k+r-2) - \varepsilon\,\text{sgn}\big[s_d(k+r-2)\big]\big\} \\ -\varepsilon\,\text{sgn}\big[s_d(k+r-1)\big] + d_r(k). \end{aligned} \tag{36}$$

Considering that $\text{sgn}[s_d(k)] = -\text{sgn}[s_d(k-1)]$ for any $k \geq k_{0d}$, from (36) we get:

$$s_r(k+r) = \text{sgn}[s_d(k+r-2)]\Big[(1-q)^2|s_d(k+r-2)| + q\varepsilon\Big] + d_r(k). \tag{37}$$

The term in the brackets in (37) is never smaller than $q\varepsilon$. Using (29) and (30) we conclude that if

$$\varepsilon > \frac{d_{r\max}}{q}, \tag{38}$$

then $\text{sgn}[s_r(k+r)] = \text{sgn}[s_d(k+r-2)]$. Therefore, considering the change of the sign of $s_d(k)$, for any $k \geq k_0 = k_{0d} + 2$ we get:

$$\text{sgn}[s_r(k)] = -\text{sgn}[s_r(k+1)]. \tag{39}$$

Moreover, taking into account (12), (29) and (30), for any $k \geq k_{0d} + 2$, the absolute value of $s_r(k)$ is limited by $\varepsilon + d_{r\max}$. Furthermore, from (14) and (29), we conclude that as long as $|\boldsymbol{c_r A}^{r-1}\boldsymbol{b}| < |\boldsymbol{c}_1\boldsymbol{b}|$ then $\delta_r < \delta_1$, which ends the proof. ■

We have shown that the combination of the reference trajectory following control with higher relative degree sliding variable concept ensures all the properties of the quasi-sliding mode. Moreover, appropriate selection of the control vector $\boldsymbol{c_r}$ allows to scale the impact of disturbances on the system and therefore improve its robustness.

4 Simulation Example

To verify the proposed control strategy we carry out a short simulation example. We consider a third order discrete-time system described by the following state equation:

$$\boldsymbol{x}(k+1) = \begin{bmatrix} -1 & 1.2 & 0 \\ -0.5 & 0.5 & -1 \\ -2 & 0.5 & 1 \end{bmatrix}\boldsymbol{x}(k) + \begin{bmatrix} 0 \\ 0.05 \\ 1 \end{bmatrix}[u(k) + d(k)]. \tag{40}$$

The system's discretization period is equal to 1 s. In the first step we transform the system (40) to the Frobenius form. With the transformation matrix:

$$T = \begin{bmatrix} -0.8657 & 0.9482 & 0.9526 \\ 1.1529 & -0.9070 & 0.0454 \\ -0.7568 & -0.0442 & 0.0022 \end{bmatrix} \tag{41}$$

we obtain the system:

$$\boldsymbol{x}(k+1) = \begin{bmatrix} 0 & 1 & 0 \\ 0 & 0 & -1 \\ 2 & -0.1 & 0.5 \end{bmatrix} \boldsymbol{x}(k) + \begin{bmatrix} 0 \\ 0 \\ 1 \end{bmatrix} [u(k) + d(k)]. \tag{42}$$

We assume that the initial conditions of the system are $\boldsymbol{x_0} = [500\ 10\ 10]^{\mathrm{T}}$ and we aim to drive the system to the desired state $\boldsymbol{x_d} = [0\ 0\ 0]^{\mathrm{T}}$. The disturbance $d(k)$ equals to 1 for $k \in [0, 20]$ and -1 for $k \in [21, 40]$.

Conventionally, the system (42) could be controlled with the most popular discrete-time sliding mode reaching law of Gao *et al.* [8], in the form:

$$s_1(k+1) = (1-q)s_1(k) - \varepsilon \operatorname{sgn}[s_1(k)] + d_1(k). \tag{43}$$

Therefore, to present the benefits of the trajectory following strategy we compare the two methods. For the Gao's reaching law, to provide all the properties of the quasi-sliding mode, the control parameters must satisfy:

$$\varepsilon > \frac{(2-q)d_{\max}}{q}, \tag{44}$$

as studied in [9], and the control results in the quasi-sliding mode band from $-\varepsilon + d_{\max}$ to $\varepsilon + d_{\max}$. From the comparison of (17) and (44) one may see that in the trajectory following strategy smaller values of ε are admissible, which results in a reduction of the quasi-sliding mode band width. Furthermore, as stated in Theorem 2, the width of this band gets further reduced when a higher relative degree sliding variable is chosen. In simulations we compare the Gao's strategy with the trajectory following control with relative degree three sliding variable. For the reaching law (43) we choose the sliding vector $\boldsymbol{c_1} = [0\ 0\ 1]$, which locates all the eigenvalues of the closed-loop system at the origin of the complex plane. The disturbance is bounded by $d_{\max} = 1$. We choose $q = 0.6$ and $\varepsilon = 2.34$, which results in the quasi-sliding mode band $[-3.34, 3.34]$. For the trajectory following method, we choose the sliding variable of order three. Therefore, we set vector $\boldsymbol{c_3} = [0.1\ 0\ 0]$, which results in $d_{3\max} = 0.1$. For a fair comparison, to preserve a similar pace of convergence, we set $q = 0.6$ as well and set $\varepsilon = 0.167$ to obtain the minimum quasi-sliding mode band $[-0.267, 0.267]$. The results of our simulations are presented in Figs. 1, 2, 3, 4 and 5. Plots obtained with the Gao's strategy have been shown in blue, while the results of our control method with RD3 sliding variable were plotted in red.

Figure 1 shows the evolution of the sliding variables. It is clearly visible that in the RD1 system the sliding variables change from the beginning of the control process,

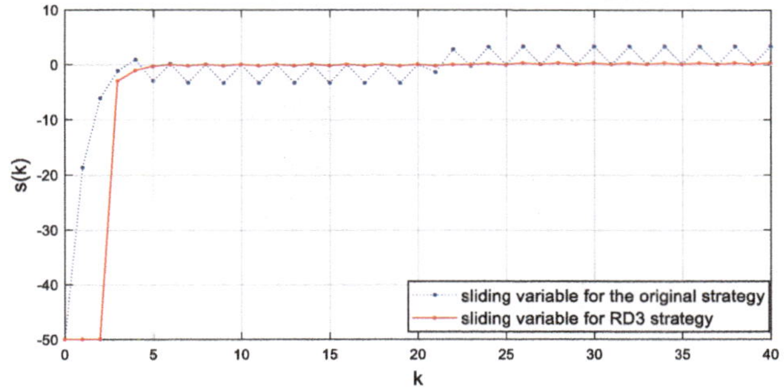

Fig. 1. Evolution of the sliding variable

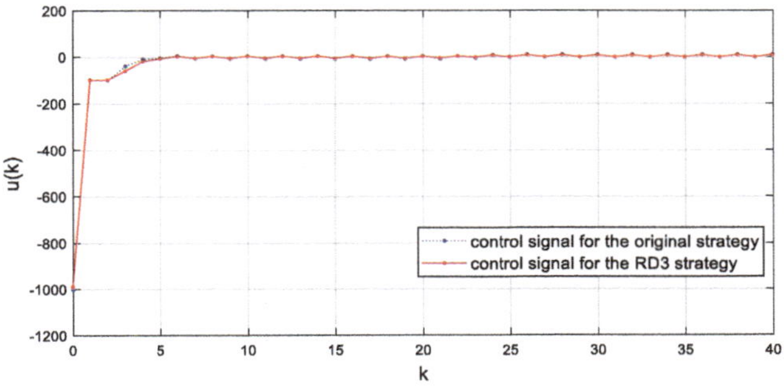

Fig. 2. Evolution of the control signal

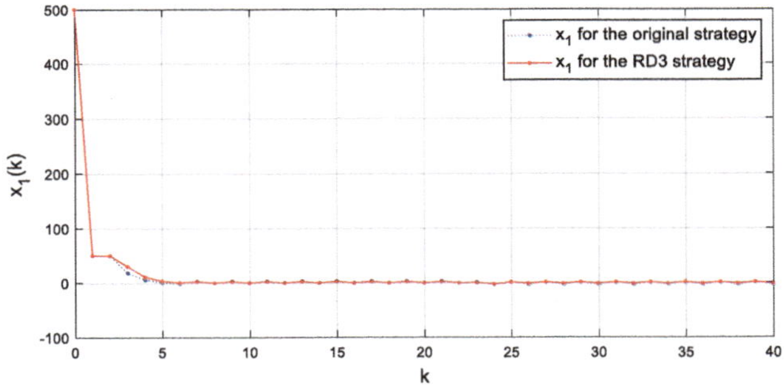

Fig. 3. Evolution of the first state variable x_1

while in the RD3 case the sliding variable begins to change after three control steps. The original control strategy resulted in the quasi-sliding mode band width of 3.34, while in our new method with RD3 sliding variable of 0.267. The price to pay for this improvement is a slight prolongation of the reaching phase. In both control methods the maximum magnitude of the control signal is the same. However, the trajectory following strategy allowed to reduce the control values in the sliding phase. Finally, Figs. 3, 4 and 5 present the evolution of the state variables of the system. It is clear that with the new

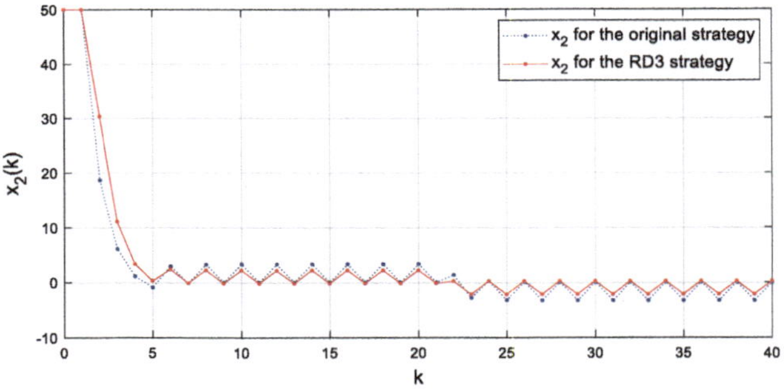

Fig. 4. Evolution of the second state variable x_2

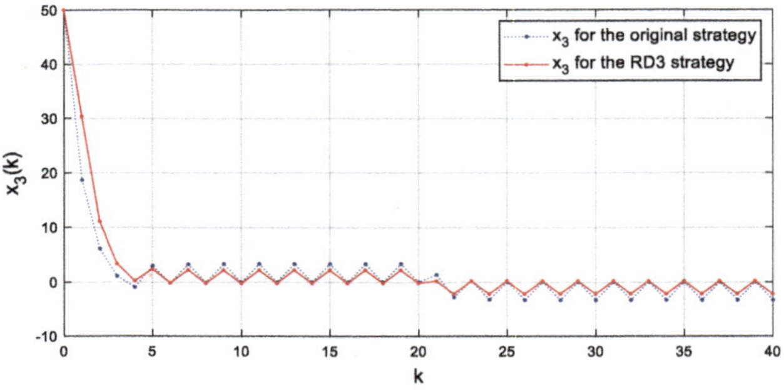

Fig. 5. Evolution of the third state variable x_3

control strategy all the state variable errors have been limited and therefore the robustness of the system has been improved.

5 Conclusions

The paper introduced a trajectory following reaching law for arbitrary relative degree output systems. We considered discrete-time systems described in the so-called Frobenius form. For those systems, we first generated the desired evolution of the sliding variable, using relative degree one sliding variable and a traditional switching type reaching law of Gao. Next, we used the reference profile to control the real disturbed system. We applied a trajectory following reaching law with sliding variable of arbitrary relative degree. We proved that this strategy ensures all the properties of the quasi-sliding mode as defined by Gao. Moreover, the application of the reference trajectory reduces the impact of external disturbances on the system and therefore, improves the system's robustness. The results were verified with a simulation example of relative degree three system.

References

1. Emelyanov, S.V.: Variable-Structure Control Systems. Nauka, Moscow (1967). (in Russian)
2. Utkin, V.I.: Variable structure systems with sliding modes. IEEE Trans. Autom. Control **22**(2), 212–222 (1977)
3. Draženović, B.: The invariance conditions in variable structure systems. Automatica **5**(3), 287–295 (1969)
4. Milosavljević, D.: General conditions for the existence of a quasi-sliding mode on the switching hyperplane in discrete variable structure systems. Autom. Rem. Control **46**(3), 307–314 (1985)
5. Sarpturk, S.Z., Istefanopulos, Y., Kaynak, O.: On the stability of discrete-time sliding mode control systems. IEEE Trans. Autom. Control **22**(10), 930–932 (1987)
6. Utkin, V.I., Drakunov, S.V.: On discrete-time sliding modes. IFAC Proc. Vol. **22**(3), 273–278 (1989)
7. Bartolini, G., Ferrara, A., Utkin, V.I.: Adaptive sliding mode control in discrete-time-systems. Automatica **31**(5), 769–773 (1995)
8. Gao, W.B., Wang, Y., Homaifa, A.: Discrete-time variable structure control systems. IEEE Trans. Ind. Electron. **42**(2), 117–122 (1995)
9. Bartoszewicz, A.: Remarks on 'Discrete-time variable structure control systems'. IEEE Trans. Ind. Electron. **43**(1), 235–238 (1996)
10. Veselić, B., Perunicic-Draženović, B., Milosavljević, Č.: Improved discrete-time sliding-mode position control using Euler velocity estimation. IEEE Trans. Ind. Electron. **57**(11), 3840–3847 (2010)
11. Ma, H., Wu, J., Xiong, Z.: A novel exponential reaching law of discrete-time sliding-mode control. IEEE Trans. Ind. Electron. **64**(5), 3840–3850 (2017)
12. Zhang, F.: Switching reaching law based switched sliding mode control. In: Proceedings of Chinese Control Conference (CCC), Chengdu, China, pp. 4735–4739 (2016)
13. Bartoszewicz, A., Latosiński, P.: Discrete time sliding mode control with reduced switching - a new reaching law approach. Int. J. Robust Nonlinear Control **26**(1), 47–68 (2016)
14. Xu, Q., et al.: A novel exponential reaching law for sliding mode control of discrete-time system with disturbance. In: 2018 ANZCC, Melbourne, VIC, Australia, pp. 275–280 (2018)
15. Ma, H., Li, Y.: Multi-power reaching law based discrete-time sliding-mode control. IEEE Access **7**, 49822–49829 (2019)
16. Bartoszewicz, A.: Discrete-time quasi-sliding-mode control strategies. IEEE Trans. Ind. Electron. **45**(4), 633–637 (1998)
17. Golo, G., Milosavljević, Č.: Robust discrete-time chattering free sliding mode control. Syst. Control Lett. **41**, 19–28 (2000)
18. Bartoszewicz, A., Leśniewski, P.: Reaching law-based sliding mode congestion control for communication networks. IET Proc. Control Theory Appl. **8**(17), 1914–1920 (2014)
19. Chakrabarty, S., Bandyopadhyay, B.: A generalized reaching law for discrete time sliding mode control. Automatica **52**, 83–86 (2015)
20. Chakrabarty, S., Bandyopadhyay, B.: A generalized reaching law with different convergence rates. Automatica **63**, 34–37 (2016)
21. Janardhanan, S., Kariwala, V.: Multirate output-feedback-based lq-optimal discrete-time sliding mode control. IEEE Trans. Autom. Control **53**(1), 367–373 (2010)
22. Bandyopadhyay, B., Janardhanan, S.: Discrete-Time Sliding Mode Control: A Multirate Output Feedback Approach. Springer, Heidelberg (2006)
23. Janardhanan, S., Bandyopadhyay, B.: Multirate output feedback based robust quasi-sliding mode control of discrete-time systems. IEEE Trans. Autom. Control **52**(3), 499–503 (2007)

24. Bandyopadhyay, B., Behera, A.K.: Event-Triggered Sliding Mode Control, Studies in Systems, Decision and Control. Springer, Heidelberg (2018)
25. Behera, A.K., Bandyopadhyay, B., Reger, J.: Discrete event-triggered sliding mode control with fast output sampling feedback. In: 2016 14th International Workshop on Variable Structure Systems (VSS), Nanjing, China, pp. 148–153 (2016)
26. Chakrabarty, S., et al.: Discrete sliding mode control for systems with arbitrary relative degree output. In: 14th International Workshop on Variable Structure Systems (VSS), Nanjing, China, pp. 160–165 (2016)
27. Chakrabarty, S., Bartoszewicz, A.: Improved robustness and performance of discrete time sliding mode control systems. ISA Trans. **65**, 143–149 (2016)
28. Bartoszewicz, A., Latosiński, P.: Reaching law for DSMC systems with relative degree 2 switching variable. Int. J. Control **90**(8), 1626–1638 (2017)
29. Bartoszewicz, A., Latosiński, P.: Generalization of Gao's reaching law for higher relative degree sliding variables. IEEE Trans. Autom. Control **63**(9), 3173–3179 (2018)

Model Based Parameterizations of Different Topologies

Csilla Bányász[1](✉), Laszlo Keviczky[1], and Ruth Bars[2]

[1] Institute for Computer Science and Control, Budapest, Hungary
banyasz@sztaki.hu
[2] Budapest University of Technology and Economics, Budapest, Hungary

Abstract. The optimization of simple two-degree-of-freedom control systems is very easy with the new parameterizations, as YOULA, KEVICZKY-BANYASZ, etc. The comparison of their model-based versions is important at the practical applications.

Keywords: Parameterization · Two-degree-of-freedom control systems · Design of control system

1 Introduction

Many different parameterizations of control structures were investigated in our papers and books in the last period. Mostly the influence of a model used is treated. Here below the topologies based on the *KB* and YOULA parameterization will be discussed [6, 7, 9].

The most important element of a closed-loop control system is the regulator C, which has to be determined during the design procedure [1]. Direct and indirect parameterization methods are reviewed below which can help to make this task easier.

The closed system is called internally stable, if, given a bounded excitation at the arbitrary point of the system, the generated signals in any point remain bounded [2]. Thus stable transfer functions must be obtained between any two input-output points. The mathematical condition of this property can be formalised in the simplest way by introducing the transfer matrix $\boldsymbol{T}_{\mathrm{t}}$ of the closed system, which represents the relationships between any two independent outer and two inner signals. A suitable choice for $\boldsymbol{T}_{\mathrm{t}}$ is

$$\boldsymbol{T}_t(P, C) = \frac{1}{1+CP}\begin{bmatrix} P \\ 1 \end{bmatrix}\begin{bmatrix} C & 1 \end{bmatrix} = \boldsymbol{s}\boldsymbol{D}^{-1}\boldsymbol{r}^T = \boldsymbol{D}^{-1}\boldsymbol{s}\boldsymbol{r}^T =$$
$$= \boldsymbol{s}\boldsymbol{r}^T\boldsymbol{D}^{-1} = \begin{bmatrix} \frac{CP}{1+CP} & \frac{P}{1+CP} \\ \frac{C}{1+CP} & \frac{1}{1+CP} \end{bmatrix} = \frac{1}{1+CP}\begin{bmatrix} CP & P \\ C & 1 \end{bmatrix} \tag{1}$$

where $\boldsymbol{s} = \begin{bmatrix} P & 1 \end{bmatrix}^T$, $\boldsymbol{r} = \begin{bmatrix} C & 1 \end{bmatrix}^T$ and $\boldsymbol{D} = (1 + CP)\boldsymbol{I}$. It can be seen that the control loop is internally stable, if and only if, $\boldsymbol{T}_t(P, C) \in \mathcal{S}$, where $\mathcal{S}$ is the set of the CT stable linear processes. Using simple algebraic rearrangements

$$\boldsymbol{T}_t(P, C) = \begin{bmatrix} -1 & 0 \\ 0 & 1 \end{bmatrix}\boldsymbol{H}_t(P, C) + \begin{bmatrix} 1 & 0 \\ 0 & 0 \end{bmatrix} \tag{2}$$

A. Bartoszewicz et al. (Eds.): PCC 2020, AISC 1196, pp. 39–55, 2020.
https://doi.org/10.1007/978-3-030-50936-1_4

is obtained, where

$$\boldsymbol{H}_t(P, C) = \begin{bmatrix} 1 & P \\ -C & 1 \end{bmatrix}^{-1} = \boldsymbol{D}^{-1} \begin{bmatrix} 1 & -P \\ C & 1 \end{bmatrix} \tag{3}$$

The stability of the transfer matrix $\boldsymbol{T}_t(P, C)$ via can be investigated the stability of matrix $\boldsymbol{H}_t(P, C)$. A similar statement can be made for the dual transfer function $\boldsymbol{T}_t(C, P) = \boldsymbol{r}\boldsymbol{D}^{-1}\boldsymbol{s}^T = \boldsymbol{D}^{-1}\boldsymbol{r}\boldsymbol{s}^T = \boldsymbol{r}\boldsymbol{s}^T\boldsymbol{D}^{-1}$ by introducing the matrix

$$\boldsymbol{H}_t(C, P) = \begin{bmatrix} 1 & C \\ -P & 1 \end{bmatrix}^{-1} = \boldsymbol{D}^{-1} \begin{bmatrix} 1 & -C \\ P & 1 \end{bmatrix} \tag{4}$$

2 YOULA Parameterization

Let us find parameterization for the closed-loop which makes all elements of $\boldsymbol{T}_t(P, C)$ stable. Introduce the following transfer function as parameter

$$Q = \frac{C}{1 + CP} \tag{5}$$

by means of which the new form of the transfer matrix is

$$\boldsymbol{T}_t(P, C) = \boldsymbol{T}_t(P, Q) = \begin{bmatrix} P \\ 1 \end{bmatrix} \begin{bmatrix} Q & 1 - QP \end{bmatrix} = \begin{bmatrix} QP & P(1 - QP) \\ Q & 1 - QP \end{bmatrix} \tag{6}$$

It is clear that for the stable process $P \in \mathcal{S}$ and stable parameter $Q \in \mathcal{S}$, all elements of $\boldsymbol{T}_t$ are stable; thus the internal stability of the closed-loop is ensured (here $Q(\omega = \infty)$ is finite and regular). Otherwise Q represents the transfer function of a one-degree-of-freedom control loop between the reference signal r and the actuating signal u. All elements of $\boldsymbol{T}_t$ are linear (therefore convex) in Q. Similarly, the sensitivity functions are also linear

$$T = \frac{CP}{1 + CP} = QP \quad ; \quad S = \frac{1}{1 + CP} = 1 - QP \tag{7}$$

The above procedure is called YOULA parameterization (*YP*), where Q is the YOULA parameter. From (5) the YOULA-parameterized regulator is

$$C = \frac{Q}{1 - QP} \tag{8}$$

The YOULA-parameterized control loop is shown in Fig. 1, where the above notations are used.

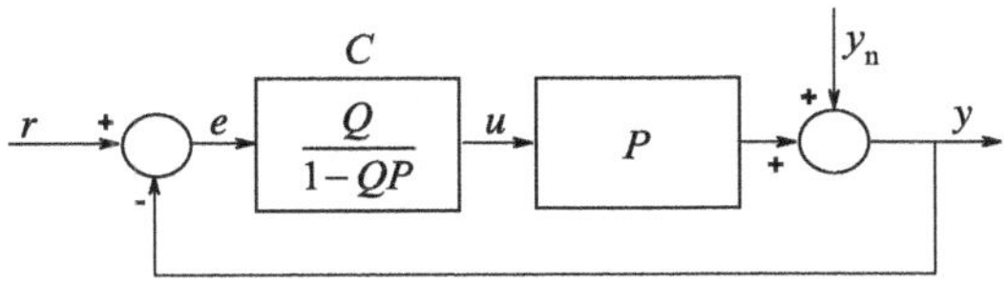

Fig. 1. YOULA-parameterized control loop

The YOULA parameter, as a matter of fact, is a stable (by definition), regular transfer function

$$Q(s) = \frac{C(s)}{1 + C(s)P(s)} \quad \text{or shortly} \quad Q = \frac{C}{1 + CP} \tag{9}$$

where $C(s)$ is a stabilizing regulator, and $P(s)$ is the transfer function of the stable process.

It follows from the definition of the YOULA parameter that the structure of the realizable and stabilizing regulator in the YOULA-parameterized control loop is fixed:

$$C(s) = \frac{Q(s)}{1 - Q(s)P(s)} \quad \text{or shortly} \quad C = \frac{Q}{1 - QP} \tag{10}$$

The YOULA-parameterized control loop is shown in Fig. 1. The sensitivity and complementary sensitivity functions linear in Q of the closed-control systems were defined by (7). It is interesting to observe that the *YP* regulator of (8) can be realized by a simple control loop with positive feedback as shown in Fig. 2.

The relationships between the most important signals of the closed system can be obtained with simple calculations

$$\begin{aligned} u &= Qr - Qy_{\mathrm{n}} \\ e &= (1 - QP)r - (1 - QP)y_{\mathrm{n}} = Sr - Sy_{\mathrm{n}} \\ y &= QPr + (1 - QP)y_n = Tr + Sy_n \end{aligned} \tag{11}$$

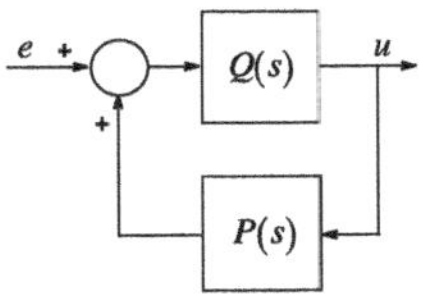

Fig. 2. Realization of *YP* regulator

The effect of r and y_n on u and e is completely symmetrical (not considering the sign). Thus the input of the process depends only on the external signals and $Q(s)$.

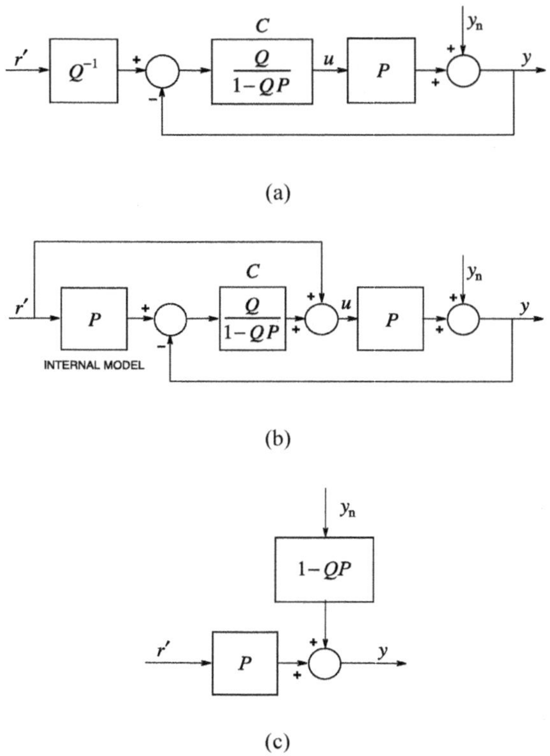

Fig. 3. Block diagrams for opening the closed-loop control systems

Now apply the inverse of Q, connected serial as in Fig. 3a, in the block diagram (Fig. 1) of the YOULA-parameterized control loop. In this way the tracking property becomes independent of Q, i.e., Pr', thus the closed-loop seems to be formally opened. It can be easily checked that this block diagram is equivalent to Fig. 3b. Figure 3c shows the effects of the inputs on the output. The relationships between the most important signals are:

$$
\begin{aligned}
u &= r'Qy_{\mathrm{n}} \\
e &= 0 - (1 - QP)y_{\mathrm{n}} = Sy_{\mathrm{n}} \\
y &= Pr' + (1 - QP)y_{\mathrm{n}} = T'r' + Sy_{\mathrm{n}}
\end{aligned} \tag{12}
$$

The YOULA parameterization extended by Q^{-1}, which formally opens the closed-loop, leaded to the so-called KB parameterization (after the authors KEVICZKY and BÁNYÁSZ [3–5]). As regards the operation of the KB-parameterized loop it should be noted that here the reference signal has a direc effect on the input of the process, and thus it does not go through the regulator and the whole closed-loop. All further control effects (regarding the reference signal) are in operation only when the inner model is not equal to the real process. It can be seen from Fig. 3b that the KB parameterization is independent of the Y

parameterization concerning the reference signal effect, since it opens the closed-loop for any arbitrary regulator C, and thus the overall tracking transfer function is always P. The effect of the disturbance signal, however, can be compensated only by the Y parameterization via a simple linear transfer function $(1 - QP)$ linear in Q. Thus, in the case of two-degree-of-freedom systems (*TDOF*) the two principles have to be jointly applied.

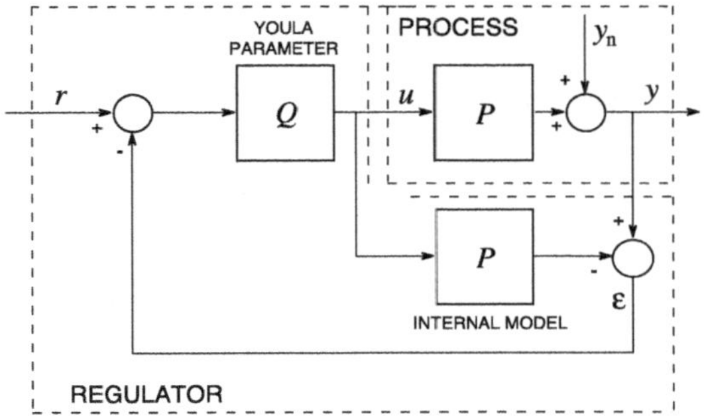

Fig. 4. Equivalent *IMC* loop

Using the scheme in Fig. 2, the YOULA-parameterized closed-loop control shown in Fig. 1 can be redrawn to the equivalent form of Fig. 4. This classical internal model-based scheme is called the *Internal Model Control*: *IMC* [9]. The basic principle of this control is that it has feedback only from the deviation (ε) between the process output and the model output to create the error signal of the control. This error signal is zero in the ideal case when the internal model is completely equal to the process. This case is shown above. But in reality the internal model $\hat{P}(s)$ is only a good approximation of the true process $P(s)$, since the original system is not known exactly (see later). For the sake of simplicity only the ideal case is discussed here first. The relationships between the most important signals of the closed system can be obtained with simple calculations

$$u = Qr - Qy_n$$
$$y = QPr - (1 - QP)y_n = Tr - Sy_n \qquad (13)$$

i.e. the same as in (11).

3 KEVICZKY-BÁNYÁSZ Parameterization

As mentioned previously, the YOULA parameterization extended by Q^{-1}, which formally opens the closed-loop, is called *KB* parameterization after the authors (KEVICZKY and BÁNYÁSZ [3–5]) who suggested the method. The basic scheme of the *KB* parameterization is shown in Fig. 5.

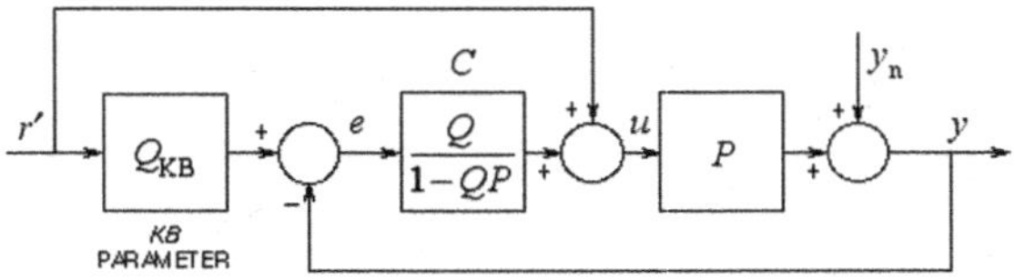

Fig. 5. Scheme of the *KB* parameterization

The basic relationships of the closed-loop are

$$y = (1 - QP + QQ_{\text{KB}})|_{Q_{\text{KB}}=P} Pr' + (1 - QP)y_n = Pr' + (1 - QP)y_n \tag{14}$$

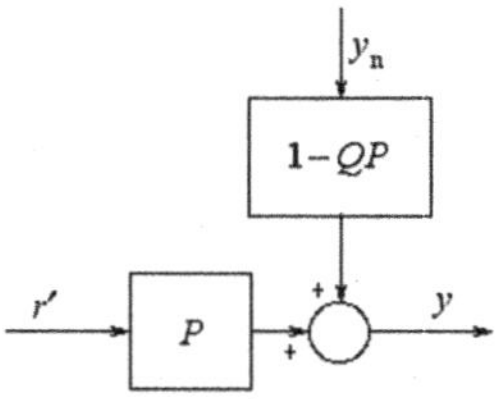

Fig. 6. The ideal choice of the *KB* parameter $Q_{KB} = P$ opens the system

It is worth noting that the reference signal takes effect directly on the input of the process, and thus it does not go through the regulator and the closed-loop. The controlling effect regarding the reference signal operates only if the internal model is not equal to the real process. Considering the reference signal effect the *KB* parameterization shown in Fig. 5 is independent of the *Y* parameterization since it opens the closed-loop even for an arbitrary regulator C, i.e., the overall tracking transfer function is always P if $Q_{\text{KB}} = P$ is chosen. The effect of the disturbance signal is compensated only via the simple transfer function $(1 - QP)$ linear in Q. Thus in the case of *TDOF* systems the two principles have to be applied simultaneously. If $Q_{\text{KB}} = P$ is chosen the equivalent closed-loop corresponds to the open loop in Fig. 6 (which is the same as Fig. 3c).

Although the invention of the *KB* parameterization is inspired by the YOULA parameterization, its validity is more general, since it opens the closed control-loop for any regulator

$$y = (1 + Q_{\text{KB}}C)\frac{P}{1 + CP}y_r + \frac{1}{1 + CP}y_n =$$

$$= \left.\frac{(1 + Q_{\text{KB}}C)P}{1 + CP}\right|_{Q_{\text{KB}}=P} y_r + \frac{1}{1 + CP}y_n = Py_r + \frac{1}{1 + CP}y_n \tag{15}$$

Furthermore

$$u = y_r - \frac{C}{1 + CP}y_n \quad ; \quad e = 0 - \frac{1}{1 + CP}y_n \tag{16}$$

From the last equation in (11) it can be seen that both the YOULA parameterization and the *IMC* have the transfer function QPr concerning the reference signal tracking. If the *KB* parameterization introduced on the figures of Fig. 3 is applied, then the YOULA parameterization can be simply extended for *TDOF* control systems. To do this, let us simply apply a parameter Q_r for the design of the tracking properties, and connect it in serial to the *KB*-parameterized loop of Fig. 3, so the block diagram of Fig. 7 is obtained.

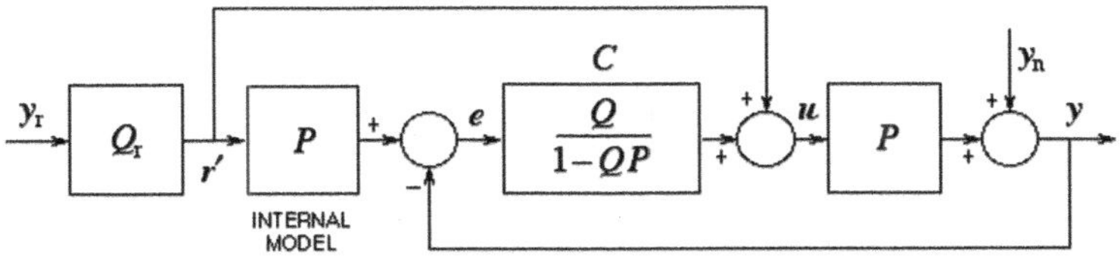

Fig. 7. Two-degree-of-freedom version of the *YP* control loop

The overall transfer characteristics for this system are

$$\begin{aligned} u &= Q_r y_r - Q y_n \\ e &= 0 - (1 - QP) y_n = 0 - S y_n \\ y &= Q_r P y_r + (1 - QP) y_n = T_r y_r + (1 - T) y_n = T_r y_r + S y_n \end{aligned} \tag{17}$$

where the tracking properties can be designed by choosing Q_r in $T_r = Q_r P$, and the noise rejection properties by choosing Q in $T = QP$. These two properties can be handled separately. The reference signal of the whole system is denoted by y_r from here. The conditions for Q_r are the same as for Q. The meaning of T_r is analogous to the meaning of the complementary sensitivity function T of the one-degree-of-freedom control loop for tracking.

The *IMC* of Fig. 4 can be further developed according to Fig. 8. Here the predicted value $\hat{y}_n$ of the output disturbance y_n is constructed from the difference ε between the outputs of the process and the model by the predictor R_n. Similarly the predictor R_r provides the predicted value $\hat{y}_r$ of the reference signal y_r. The noise rejection is performed by giving the predicted value $-\hat{y}_n$ of the disturbance to the process input via the inverse of the process model, thus in the case of accurate estimation the disturbance is eliminated. The reference signal tracking operates in a similar way. Here the operation of R_r can be considered a reference model (desired system dynamics), and therefore the introduced predictors are also called reference models. It is generally required that these predictors (filters) are strictly proper with unit static gain, i.e., $R_n(\omega = 0) = 1$ and $R_r(\omega = 0) = 1$.

By introducing the reference models (or predictors) R_r and R_n the design of the regulator is simplified to the design of these design goals instead of selecting Q_r and Q.

4 Model Based YOULA Regulator

Investigate first the model-based version of the topology on Fig. 3b, i.e.

Here, for the sake of simplicity, the influence of the output disturbance y_n is not investigated (Fig. 9).

$$u = r' - Q\varepsilon \tag{18}$$

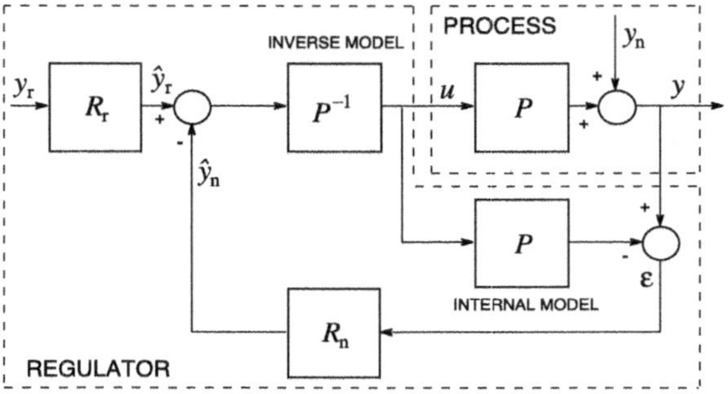

Fig. 8. Extension of the *IMC*-based ideal control loop with reference models

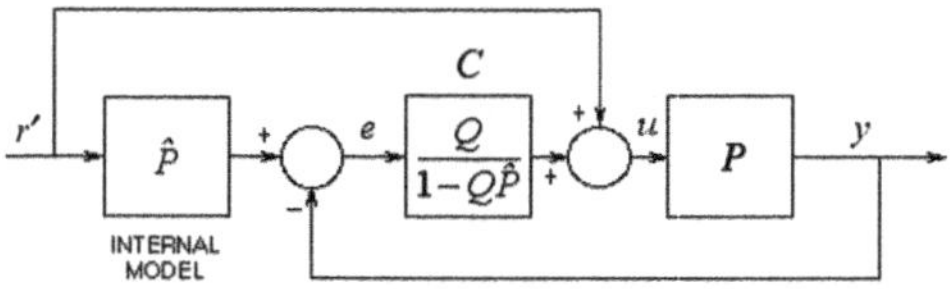

Fig. 9. The model based YOULA regulator with *KB* parameterization

where

$$\varepsilon = y - \widehat{P}u \tag{19}$$

is the model output error (see Figs. 4 and 8). The other important signals in the closed-loop are

$$\begin{aligned} e &= \left(1 - Q\widehat{P}\right)\varepsilon = \widehat{S}\varepsilon \\ y &= \widehat{P}u + \varepsilon = \widehat{P}r' + \left(1 - Q\widehat{P}\right)\varepsilon = \widehat{T}'r' + \widehat{S}\varepsilon \end{aligned} \tag{20}$$

Next investigate the model-based version of the *IMC* topology on Fig. 4 as the Fig. 10 shows.

Here the process input and outputs are

$$\begin{aligned} u &= Qr - Q\varepsilon \\ y &= \widehat{P}u + \varepsilon = Q\widehat{P}r + \left(1 - Q\widehat{P}\right)\varepsilon = \widehat{T}r + \widehat{S}\varepsilon \end{aligned} \tag{21}$$

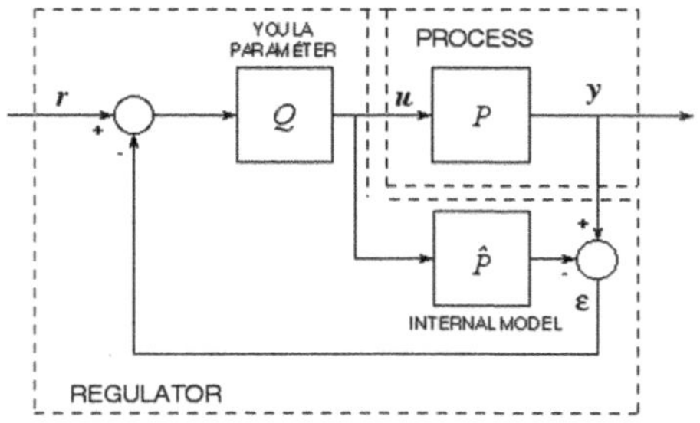

Fig. 10. The model based *IMC* regulator

From the above analysis it can be well seen, that the *KB* parameterization provides zero control error for the ideal case, when $\hat{P} = P$, i.e. the process exactly known. In case of model error the control error is $e = \hat{S}\varepsilon$, i.e. it will depend on the output error ε. The influence of this error can be properly attenuated if $\hat{S}$ is selected well. $\hat{S}$ is generally a high pass filter, so the low frequency domain is always damped.

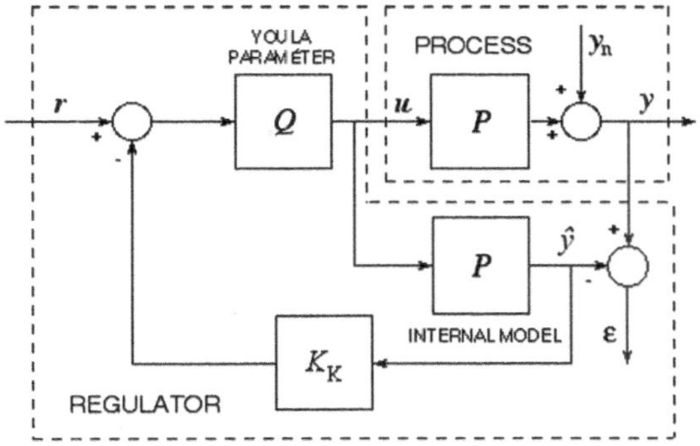

Fig. 11. Combination of the state-feedback principle and the *IMC* regulator

5 Application of the Observer Principle

A possible combination of the state-feedback principle [2] and the *IMC* regulator is shown in Fig. 11. Here the influence of the output disturbance y_n is not investigated. The process input and outputs are

$$u = \frac{Q}{1 + QK_K P} r \quad ; \quad y = \frac{QP}{1 + QK_K P} r \tag{22}$$

The model output error ε is zero now, because $\hat{P} = P$, i.e. the process is exactly known. The resulting block-scheme (Fig. 12) slightly differs from the original *IMC* system

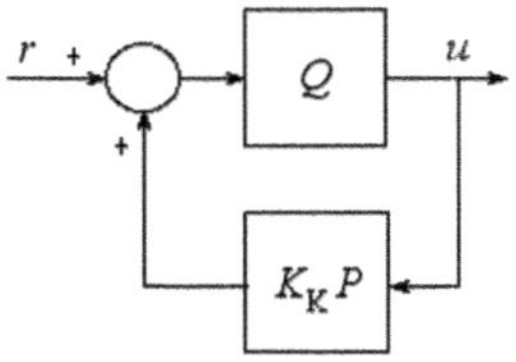

Fig. 12. The equivalent scheme of the state-feedback and the *IMC* regulator

In case of model error the observer principle can be applied and the scheme of Fig. 13 is obtained. The model error now is

$$\varepsilon = y - \hat{y} = \frac{\varepsilon_o}{1 + K_L \widehat{P}} \quad ; \quad \varepsilon_o = y - \hat{y}_o = Pu - \widehat{P}u = \left(P - \widehat{P}\right)u \tag{23}$$

where

$$\hat{y} = \widehat{P} \frac{1 + K_L P}{1 + K_L \widehat{P}}\bigg|_{\widehat{P}=P} u = Pu \tag{24}$$

The process input and outputs are now

$$u = \frac{Q\left(1 + K_L \widehat{P}\right)}{1 + \widehat{P}(K_L + QK_K) + QK_K K_L \widehat{P} P}\Bigg|_{\widehat{P}=P} r = \frac{Q}{1 + QK_K P} r$$

$$y = \frac{QP\left(1 - K_L \widehat{P}\right)}{1 + \widehat{P}(K_L + QK_K) + QK_K K_L \widehat{P} P}\Bigg|_{\widehat{P}=P} r = \frac{QP}{1 + QK_K P} r \tag{25}$$

what prove, if the internal observer-loop is designed well, providing very small ε, then the overall transfer functions will be independent of K_L. For example if very high gain K_L is selected, then

$$u = \frac{Q\left(1 + K_L \widehat{P}\right)}{1 + \widehat{P}(K_L + QK_K) + QK_K K_L \widehat{P} P}\Bigg|_{K_L \to \infty} r = \frac{Q}{1 + QK_K P} r \tag{26}$$

and the effect is the same when the model equals to the process (see (25)), i.e. $\hat{P} = P$.

Theoretically we can obtain the simple *IMC* case (13) back, independently of K_K if

$$K_L = P^{-1} \tag{27}$$

is used in the internal observer-loop [1, 2, 9]. The other system variables are

$$\begin{aligned} \varepsilon &= y - \hat{y} = Pu = y; \ \hat{y} = 0 \\ u &= Qr \\ y &= QPr \end{aligned} \tag{28}$$

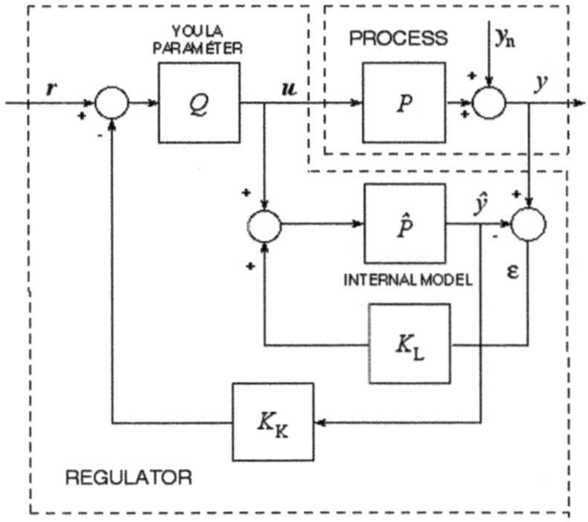

Fig. 13. Combination of the observer principle and the *IMC* regulator

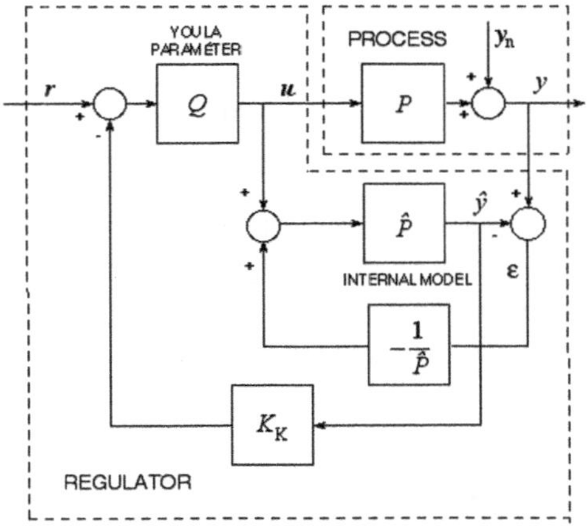

Fig. 14. The optimal combined scheme

The practical realization of the optimal combined scheme is shown in Fig. 14, where $\hat{P}$ is used instead of P in (26) and in Fig. 13. Please note, that $\widehat{P}^{-1}$ is usually a high pass filter.

The model error, the process input and outputs are

$$\varepsilon = y - \hat{y} = \frac{\widehat{P}}{\widehat{P} - 1}\varepsilon_o \tag{29}$$

$$u = \left. \frac{Q}{1 + QK_K \frac{P - \widehat{P}^2}{1 - \widehat{P}}} \right|_{\widehat{P} = P} r = \left. \frac{Q}{1 + QK_K P} \right|_{K_K \to 0} r = Qr$$

$$y = \left. \frac{QP}{1 + QK_K \frac{P - \hat{P}^2}{1 - \hat{P}}} \right|_{\hat{P}=P} r = \left. \frac{QP}{1 + QK_K P} \right|_{K_K \to 0} r = QPr \tag{30}$$

So in the other loop the $K_K \to 0$ condition provides the original transfer functions. As a conclusion the observer loop must be tuned for a high-pass filter, the feedback loop must be a low pass filter. The necessary conditions can be reached only in the high frequency and in the low frequency domain, respectively.

6 Simulation Examples

Example 1

The continuous process to be controlled is given by the transfer function

$$P(s) = \frac{1}{1 + 10s} e^{-30\,s} \tag{31}$$

The process is sampled with sampling period $T_S = 5$ s. At the input zero order hold is applied. Discrete control is realized. The sampled, digitalized output signal is forwarded to the process control computer via A/D converter, the computer calculates the control signal in each sampling point with a discrete algorithm, and forwards it to the process input via a D/A converter.

Let us design a YOULA parameterized controller and a *KB* parameterized controller and compare their behavior for reference signal tracking and output disturbance rejection.

The pulse transfer function is

$$G(z) = \frac{0.3935}{z - 0.6065} z^{-6} = G_+(z) z^{-6} \tag{32}$$

6.1 YOULA Parameterized Controller Design

The reference signal filters according to Fig. 8 are given by transfer functions

$$R_n(s) = \frac{1}{1 + 5\,s} \quad \text{and} \quad R_r(s) = \frac{1}{1 + 8\,s} \tag{33}$$

and their corresponding pulse transfer functions are

$$R_n(z) = \frac{0.6321}{z - 0.3679} \quad \text{and} \quad R_r(z) = \frac{0.4647}{z - 0.5353}. \tag{34}$$

The best control is expected if the inverse model of the plant appears in the forward path of Fig. 8. But as the dead time cannot be inverted, only $G_+^{-1}(z)$ will appear in the YOULA parameter

$$Q = R_n G_+^{-1} = \frac{0.6321}{z - 0.3679} \frac{z - 0.6065}{0.3935} \tag{35}$$

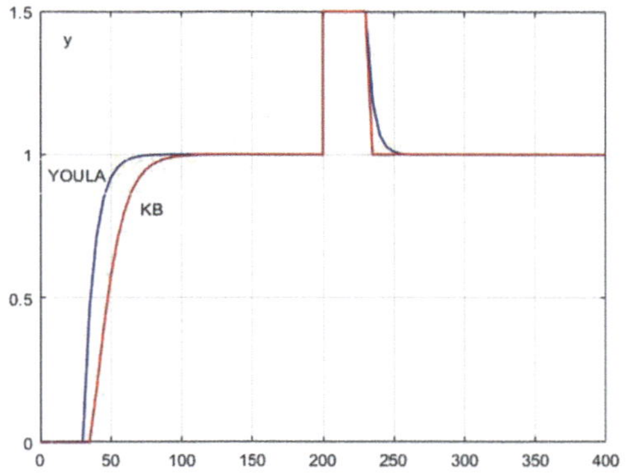

Fig. 15. Output signals of the YOULA and of the *KB* parameterized control systems

6.2 KB Controller Design

According to Fig. 7 in the controller C the parameter $Q = Q_1$ is chosen as the invertible part of the pulse transfer function of the plant.

$$Q_1 = G_+^{-1}/z = \frac{z - 0.6065}{0.3935\,z} \tag{36}$$

and the serial controller is

$$C = \frac{Q_1}{1 - Q_1 \cdot G} = \frac{2.541z^7 - 1.541z^6}{z^7 - 1} \tag{37}$$

Q_1 and so C will influence the disturbance rejection, while Q_r, which is chosen equal to R_r in the YOULA parameterized structure influence the reference signal tracking.

Figure 15 shows the output signals in the two circuits. The reference signal is a unit step starting at $t = 0$ s, and the step disturbance with amplitude 0.5 acts at $t = 200$ s.

Both controllers give a nice control performance. The difference is because in the YOULA parameterized controller two filters influence the control behavior, while in the *KB* structure Q_r influences the tracking performance, while Q_1 separately influences the disturbance rejection.

Figures 16 and 17 show the control signals in the two structures.

Example 2
Let us analyze the performance of a YOULA parameterized continuous control system shown in Fig. 8. The YOULA parameter is (or if the transfer function of the process is not invertible, then only its invertible part is inverted). The input filter is given by. The transfer function of the process is

$$P(s) = \frac{1}{(1 + 5\,s)(1 + 8\,s)(1 + 10\,s)} \tag{38}$$

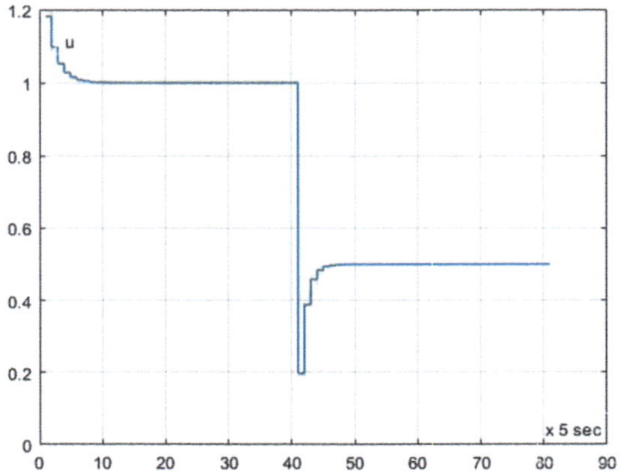

Fig. 16. Control signal in the YOULA parameterized control system

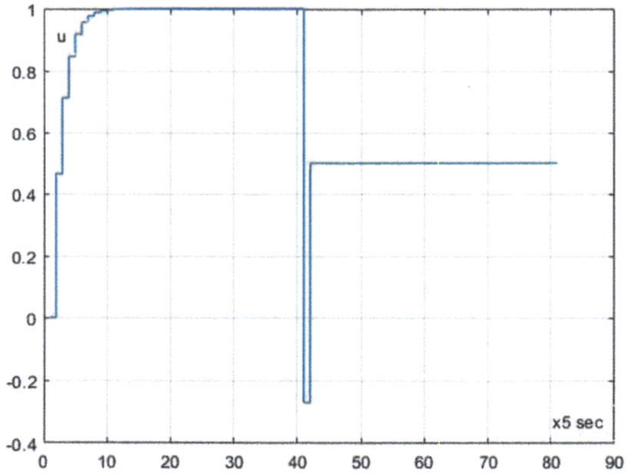

Fig. 17. Control signal in the *KB* parameterized control system

The YOULA parameter is given by

$$Q(s) = \frac{(1 + 5\,s)(1 + 8\,s)(1 + 10\,s)}{(1 + 2\,s)(1 + 3\,s)(1 + 4\,s)} \tag{39}$$

The input filter is chosen as

$$F_1(s) = 1 \quad \text{or} \quad F_2(s) = \frac{1 + 4\,s}{1 + s} \tag{40}$$

Figure 18 shows the step response of the process and the response of the YOULA parameterized control system for unit step reference signal with the two input filters. With the second filter the performance is faster.

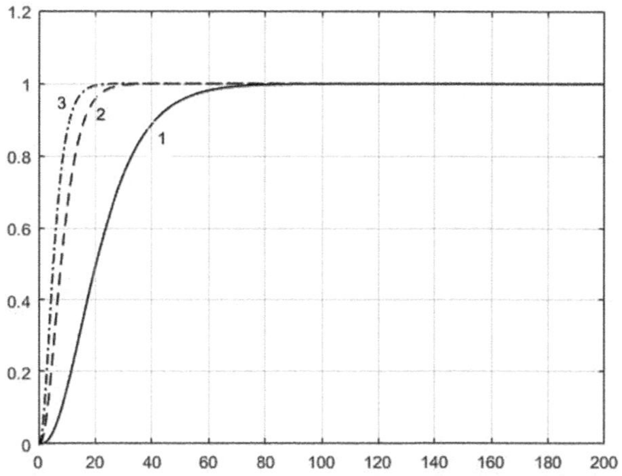

Fig. 18. Step response of the process (curve 1) and reference tracking performance with input filter 1 (curve 2) and input filter 2 (curve 3)

The reference signal is a unit step, while the disturbance signal is a step of amplitude 0.5 acting at time point 100 s. Figure 19 shows the output responses of the control system with the two filters. Disturbance rejection is not influenced by the input filter.

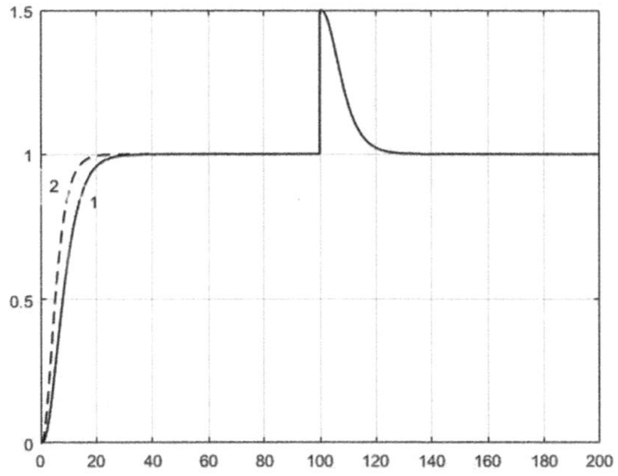

Fig. 19. Reference signal tracking and disturbance rejection with the two filters (curve 1: filter 1, curve 2: filter 2)

Let us analyze now how the control system tolerates plant/model mismatch. The Q YOULA controller is designed for the nominal P process, but the parameters of the real process differ a bit from their nominal values. The input filter now is unity. The considered real processes are

$$\widehat{P}_1(s) = \frac{1}{(1+3\,s)(1+6\,s)(1+8\,s)}; \; \widehat{P}_2(s) = \frac{1}{(1+4\,s)^3};$$

$$\widehat{P}_3(s) = \frac{1}{(1+7\,s)(1+10\,s)(1+12\,s)}; \widehat{P}_4(s) = \frac{1}{(1+11\,s)^3} \tag{41}$$

Figure 20 shows the outputs with the ideal plant and with the not ideal plants $\hat{P}_1(s)$ and $\widehat{P}_2(s)$ with the smaller time constants. With the latter one the behavior is a bit oscillating but still tolerable. Figure 20 gives the outputs with the ideal plant and with real plants $\widehat{P}_3(s)$ and $\widehat{P}_4(s)$ with the bigger time constants. This mismatch is also tolerable.

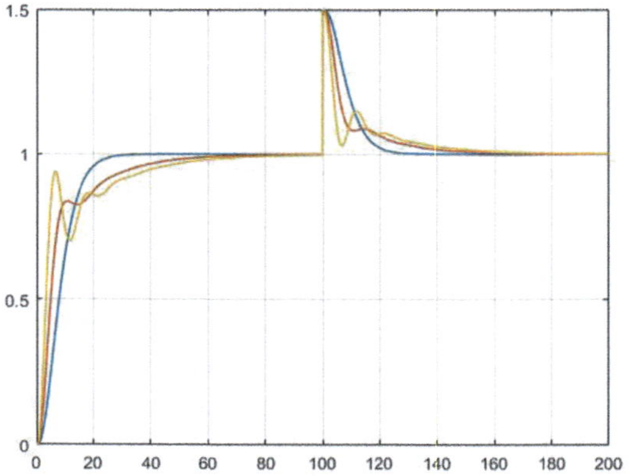

Fig. 20. The output signals considering mismatch with smaller time constants

Figure 20 shows the outputs with the ideal plant and with the not ideal plants $\widehat{P}_1(s)$ and $\widehat{P}_2(s)$ with the smaller time constants. With the latter one the behavior is a bit oscillating but still tolerable. Figure 21 gives the outputs with the ideal plant and with real plants $\widehat{P}_3(s)$ and $\widehat{P}_4(s)$ with the bigger time constants. This mismatch is also tolerable.

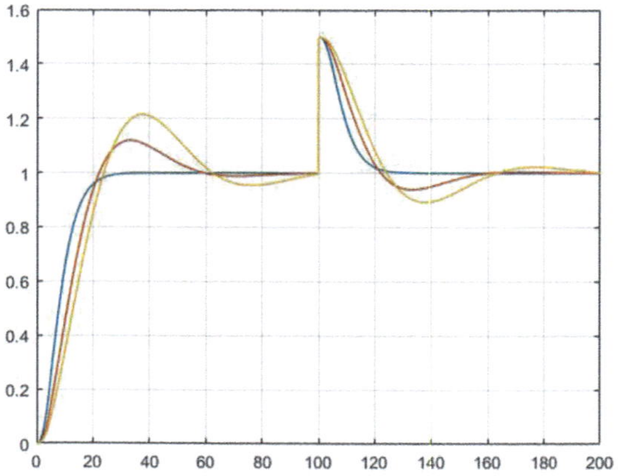

Fig. 21. The output signals considering mismatch with bigger time constants

It is mentioned that the filters can be designed for robust performance.

7 Conclusions

It is shown that the different advanced control-loop parameterization methods have very interesting relationships. If we use model-based forms then these connections are even more sophisticated.

The paper tries to clarify these relationships from the YOULA and *KB* parameterization via the state feedback topologies and the observer principle.

References

1. Åström, K.J., Wittenmark, B.: Computer Controlled Systems, p. 430. Prentice-Hall (1984)
2. Horowitz, I.M.: Synthesis of Feedback Systems. Academic Press, New York (1963)
3. Keviczky, L.: Combined identification and control: another way. (Invited plenary paper.). In: 5th IFAC Symposium on Adaptive Control and Signal Processing, ACASP 1995, Budapest, Hungary, pp. 13–30 (1995)
4. Keviczky, L. and Cs. Bányász: Optimality of two-degree of freedom controllers in $\mathcal{H}_2$- and $\mathcal{H}_\infty$-norm space, their robustness and minimal sensitivity. In: 14th IFAC World Congress, **F**, Beijing, PRC, pp. 331–336 (1999)
5. Keviczky, L., Bányász, C.: Iterative identification and control design using K-B parameterization. In: Åström, K.J., Albertos, P., Blanke, M., Isidori, A., Schaufelberger, W., Sanz, R. (eds.) Control of Complex Systems, pp. 101–121. Springer (2001)
6. Keviczky, L., Bányász, C.: Two-Degree-of-Freedom Control Systems (The Youla Parameterization Approach). Elsevier. Academic Press (2015)
7. Keviczky, L., Bars, R., Hetthéssy, J., Bányász, C.: Control Engineering. Springer (2018)
8. Keviczky, L., Bars, R., Hetthéssy, J., Bányász, C.: Control Engineering: MATLAB Exercises. Springer (2018)
9. Maciejowski, J.M.: Multivariable Feedback Design, p. 424. Addison Wesley (1989)

Rational Transfer Function Approximation Model for 2 × 2 Hyperbolic Systems with Collocated Boundary Inputs

Krzysztof Bartecki(✉)

Institute of Control Engineering, Opole University of Technology,
ul. Prószkowska 76, 45-758 Opole, Poland
K.Bartecki@po.edu.pl

Abstract. Rational transfer function approximation model for distributed parameter systems described by two weakly coupled linear hyperbolic PDEs with the boundary conditions representing two collocated external inputs to the system is considered. Using the method of lines with the backward difference scheme, the original PDEs are transformed into a set of ODEs and expressed in the form of a cascade interconnection of N subsystems, each described by 2×2 rational transfer function matrix. The considerations are illustrated with a parallel-flow double-pipe heat exchanger. The frequency and impulse responses obtained from its original irrational transfer functions are compared with those calculated from its rational approximations of different orders.

Keywords: Distributed parameter system · Hyperbolic equations · Transfer function · Approximation model · Method of lines · Frequency response · Impulse response · Heat exchange

1 Introduction

The complex, infinite-dimensional mathematical models of distributed parameter systems (DPSs), also known as systems with spatio-temporal dynamics, are often replaced, due to different practical applications, by their finite-dimensional approximations. A typical example is the *indirect controller design* procedure, where the first step towards the design of the controller is the finite-dimensional approximation of the original model of the system being controlled [7].

Motivated by this general idea, we consider here finite-dimensional, rational transfer function approximation model developed for a certain class of DPSs, the so-called 2×2 hyperbolic systems of balance laws with a specific configuration of the boundary inputs. As has been shown in the previous works of Author [3,4], their transfer functions are irrational and contain exponential and square root functions of the complex variable s.

A. Bartoszewicz et al. (Eds.): PCC 2020, AISC 1196, pp. 56–67, 2020.
https://doi.org/10.1007/978-3-030-50936-1_5

The outline of the paper is as follows. Section 2 introduces the mathematical model of the considered DPSs in the form of two hyperbolic PDEs with boundary conditions representing the so-called collocated boundary inputs. Next, their input-output models in the form of the irrational transfer functions are recalled based on the results presented in [3,4]. Section 3 introduces the approximation model based on the *method of lines* (MOL), giving as a result the finite-dimensional model in the form of ordinary differential equations (ODEs). The transfer function representation of the approximation model is introduced and its stability analysis is performed. In Sect. 4 the parallel-flow double-pipe heat exchanger is considered as a typical example of 2 × 2 hyperbolic systems with collocated inputs. Assuming typical parameter values of the exchanger system, the frequency- and time-domain responses obtained from its irrational transfer function model are compared with those obtained from rational approximation models of different order. The article concludes with Sect. 5 containing short summary and directions for further research.

2 2 × 2 Hyperbolic Systems

2.1 PDE Representation

We consider dynamical systems which can be mathematically described by the following system of the two weakly coupled linear PDEs of hyperbolic type [4]:

$$\frac{\partial x_1(l,t)}{\partial t} + \lambda_1 \frac{\partial x_1(l,t)}{\partial l} = k_{11} x_1(l,t) + k_{12} x_2(l,t), \tag{1}$$

$$\frac{\partial x_2(l,t)}{\partial t} + \lambda_2 \frac{\partial x_2(l,t)}{\partial l} = k_{21} x_1(l,t) + k_{22} x_2(l,t), \tag{2}$$

where $x_1(l,t), x_2(l,t) : \Omega \times [0,+\infty) \to \mathbb{R}$ are the space- and time-dependent state variables, with $\Omega = [0, L]$ being the domain of the one-dimensional spatial variable l; $[0,+\infty)$ – the domain of the time variable t; $k_{11}, k_{12}, k_{21}, k_{22} \in \mathbb{R}$ – constant entries and $\lambda_1, \lambda_2 \in \mathbb{R} \setminus 0$ – characteristic speeds of the system, usually representing the mass or energy transport rates.

In order to obtain a unique solution of (1)–(2), the appropriate *initial* and *boundary* conditions need to be specified. The initial conditions describe the spatial profiles of both state variables at $t = 0$:

$$x_1(l,0) = x_{10}(l), \quad x_2(l,0) = x_{20}(l), \tag{3}$$

where $x_{10}(l), x_{20}(l) : \Omega \to \mathbb{R}$ describe the initial profiles of the state variables.

The boundary conditions represent the requirements to be met by the solution at the boundary points of Ω, i.e., at $l = 0$ or/and at $l = L$. They can express the boundary reflections and feedbacks (i.e., the interconnections between state variables at $l = 0$ and $l = L$), as well as they can take into account the external *boundary inputs* to the system.

We assume here that the boundary conditions are expressed directly by the two external inputs. We take into account the case of both positive characteristic speeds, $\lambda_1 > 0$ in (1) and $\lambda_2 > 0$ in (2), for which both considered boundary conditions should be imposed at $l = 0$,

$$x_1(0,t) = u_1(t), \quad x_2(0,t) = u_2(t), \tag{4}$$

where $u_1(t), u_2(t) : [0, +\infty) \rightarrow \mathbb{R}$ are the boundary input signals (see [4]).

In addition, we also introduce two *output signals*, given as pointwise "observations" of the state variables performed at the end of the spatial domain, i.e. for $l = L$,

$$y_1(t) = x_1(L,t), \quad y_2(t) = x_2(L,t), \tag{5}$$

which may be considered as *anti-collocated* to the boundary inputs (1).

The above form of the boundary inputs and output signals has its practical motivation which will be presented later in Sect. 4.

2.2 Transfer Function Representation

Contrary to the lumped parameter systems which are described by the rational transfer functions, the transfer functions of distributed parameter systems are irrational (see, e.g., [5]). Transfer function analysis of the considered hyperbolic systems of balance laws in which the mass and/or energy transport phenomena take place was the subject of some of the Author's papers (see, e.g., [2]), as well as of his monograph [4]. Those results relating to the transfer function representation of the considered hyperbolic systems with collocated boundary inputs are recalled below.

Definition 1. *The transfer function matrix $G(s)$ for the considered 2×2 hyperbolic systems given by (1)–(2) with the collocated boundary inputs (4) and the pointwise outputs (5) can be written in the following form:*

$$G(s) = \begin{bmatrix} g_{11}(s) & g_{12}(s) \\ g_{21}(s) & g_{22}(s) \end{bmatrix}, \tag{6}$$

where

$$g_{11}(s) = \frac{y_1(s)}{u_1(s)} = \frac{x_1(L,s)}{x_1(0,s)}, \qquad g_{12}(s) = \frac{y_1(s)}{u_2(s)} = \frac{x_1(L,s)}{x_2(0,s)}, \tag{7}$$

$$g_{21}(s) = \frac{y_2(s)}{u_1(s)} = \frac{x_2(L,s)}{x_1(0,s)}, \qquad g_{22}(s) = \frac{y_2(s)}{u_2(s)} = \frac{x_2(L,s)}{x_2(0,s)}, \tag{8}$$

for zero initial conditions (3), $x_1(l,0) = x_2(l,0) = 0$, with $u_i(s)$ and $y_i(s)$ being the Laplace-transformed input and output signals, respectively[1].

[1] We stick to the notation $f(s)$ assuming that the parameter s alone indicates the Laplace transform of $f(t)$.

Result 1. The analytical expressions for the elements of the transfer function matrix $G(s)$ in (6) take the following form:

$$g_{11}(s) = \frac{\phi_1(s) - p_{22}(s)}{\phi_1(s) - \phi_2(s)} e^{\phi_1(s)L} - \frac{\phi_2(s) - p_{22}(s)}{\phi_1(s) - \phi_2(s)} e^{\phi_2(s)L}, \tag{9}$$

$$g_{12}(s) = \frac{p_{12}}{\phi_1(s) - \phi_2(s)} \left(e^{\phi_1(s)L} - e^{\phi_2(s)L} \right), \tag{10}$$

$$g_{21}(s) = \frac{p_{21}}{\phi_1(s) - \phi_2(s)} \left(e^{\phi_1(s)L} - e^{\phi_2(s)L} \right), \tag{11}$$

$$g_{22}(s) = \frac{\phi_1(s) - p_{11}(s)}{\phi_1(s) - \phi_2(s)} e^{\phi_1(s)L} - \frac{\phi_2(s) - p_{11}(s)}{\phi_1(s) - \phi_2(s)} e^{\phi_2(s)L}, \tag{12}$$

where

$$p_{11}(s) = \frac{k_{11} - s}{\lambda_1}, \quad p_{12} = \frac{k_{12}}{\lambda_1}, \quad p_{21} = \frac{k_{21}}{\lambda_2}, \quad p_{22}(s) = \frac{k_{22} - s}{\lambda_2}. \tag{13}$$

and

$$\phi_1(s) = \alpha(s) + \beta(s), \quad \phi_2(s) = \alpha(s) - \beta(s), \tag{14}$$

with

$$\alpha(s) = \frac{1}{2}\left(p_{11}(s) + p_{22}(s)\right), \quad \beta(s) = \frac{1}{2}\sqrt{\left(p_{11}(s) - p_{22}(s)\right)^2 + 4p_{12}p_{21}}. \tag{15}$$

Proof. Using the Laplace transform method; For details see [3] or [4].

2.3 Frequency- and Time-Domain Responses

Assuming negative semidefiniteness of the sum of the following matrix K and its transpose,

$$K = \begin{bmatrix} k_{11} & k_{12} \\ k_{21} & k_{22} \end{bmatrix}, \qquad \xi^T(K + K^T)\xi \le 0, \quad \forall \xi \in \mathbb{R}^2, \tag{16}$$

where k_{ij} are the constant entries in (1)–(2), we can state that the corresponding hyperbolic system is internally stable (see [4]). Moreover, if we replace the non-strict inequality in (16) by the strict one, which corresponds to the negative-definite matrix $K + K^T$, the system becomes *exponentially stable.* It means that the associated physical plant has no internal energy sources and only dissipates energy, which is satisfied by many chemical engineering systems like tubular reactors and heat exchangers.

It can be shown that the transfer functions of such stable systems are analytic and bounded in the open right half of the complex plane, Re(s) > 0. In order to obtain their frequency responses, one should replace in the expressions for the transfer functions given by Result 1 the operator variable s with $i\omega$, where $\omega \ge 0$ is the angular frequency. Like in the case of LPSs, the frequency responses can be represented in a number of ways, of which two are most commonly used: Nyquist and Bode plots.

On the other hand, the time-domain analysis is also possible using the transfer function representation, bearing in mind that it directly expresses Laplace-transformed impulse responses of the system. The analytical formulas for the impulse responses of the considered hyperbolic systems can be found in the monograph [4].

3 Approximation Model

This section deals with the finite-dimensional approximation of the considered 2×2 hyperbolic systems. Based on the so-called method of lines—which enables reduction of a PDE to a set of ODEs—the approximations of the irrational transfer functions introduced in Sect. 2.2 are derived in the form of rational transfer functions.

3.1 MOL Approximation

The method of lines (MOL) is based on replacing the spatial derivatives in the PDE with their algebraic approximations. The spatial derivatives are no longer stated in terms of the spatial independent variables, and, consequently, only the time variable remains in the resulting equations [6,9].

In order to obtain the approximation model for the considered hyperbolic system, we here use the finite difference (FD) method. For the assumed case of $\lambda_1 > 0$ and $\lambda_2 > 0$ we use the *backward difference* approximation for both (1) and (2). Therefore, the task is to replace the spatial derivatives in these equations with their following algebraic approximations,

$$\frac{\partial x_1(l,t)}{\partial l} \approx \frac{x_{1,n}(t) - x_{1,n-1}(t)}{\Delta l_n}, \quad \frac{\partial x_2(l,t)}{\partial l} \approx \frac{x_{2,n}(t) - x_{2,n-1}(t)}{\Delta l_n}, \tag{17}$$

where

$$x_{1,n}(t) = x_1(l_n, t), \quad x_{2,n}(t) = x_2(l_n, t), \tag{18}$$

represent the values of the state variables at the spatial discretization points l_n for $n = 1, 2, ..., N$, assuming $l_0 = 0$, $l_N = L$, and

$$\Delta l_n = l_n - l_{n-1} \tag{19}$$

being the spatial grid size (nth section size), which, in general, do not have to be of the same size.

As a result, the approximation model takes the form of a system of $2N$ ODEs, with the following two equations representing single nth section, $n = 1, 2, ..., N$:

$$\frac{dx_{1,n}(t)}{dt} = -\frac{\lambda_1}{\Delta l_n} x_{1,n}(t) + \frac{\lambda_1}{\Delta l_n} x_{1,n-1}(t) + k_{11} x_{1,n}(t) + k_{12} x_{2,n}(t), \tag{20}$$

$$\frac{dx_{2,n}(t)}{dt} = -\frac{\lambda_2}{\Delta l_n} x_{2,n}(t) + \frac{\lambda_2}{\Delta l_n} x_{2,n-1}(t) + k_{21} x_{1,n}(t) + k_{22} x_{2,n}(t), \tag{21}$$

where $x_{1,n-1}(t)$ and $x_{2,n-1}(t)$ can be considered as section inputs,

$$x_{1,n-1}(t) = u_{1,n}(t), \quad x_{2,n-1}(t) = u_{2,n}(t), \tag{22}$$

and $x_{1,n}(t)$ with $x_{2,n}(t)$ as its outputs,

$$x_{1,n}(t) = y_{1,n}(t), \quad x_{2,n}(t) = y_{2,n}(t). \tag{23}$$

Therefore, the considered approximation model can be seen as a cascade interconnection of N sections given by (20)–(23) where the ith output of the nth section is connected to the ith input of the section $n+1$, for $i = 1, 2$ and $n = 1, 2, ..., N-1$. In addition, we notice that the inputs of the first section should be identified with the boundary inputs (4) to the system

$$u_{1,1}(t) = u_1(t), \quad u_{2,1}(t) = u_2(t), \tag{24}$$

whereas the outputs of the Nth section will approximate the system output signals (5)

$$y_{1,N}(t) = \hat{y}_1(t), \quad y_{2,N}(t) = \hat{y}_2(t). \tag{25}$$

3.2 Transfer Function Representation

In the following two paragraphs, the transfer function representation of the considered approximation model is presented, for both the single nth section and the resultant N-section model.

Single Section

Definition 2. *The transfer function matrix $G_n(s)$ of the single nth section of the approximation model given by (20)–(23) takes the following form:*

$$G_n(s) = \begin{bmatrix} g_{11,n}(s) & g_{12,n}(s) \\ g_{21,n}(s) & g_{22,n}(s) \end{bmatrix}, \tag{26}$$

where

$$g_{11,n}(s) = \frac{y_{1,n}(s)}{u_{1,n}(s)} = \frac{x_{1,n}(s)}{x_{1,n-1}(s)}, \qquad g_{12,n}(s) = \frac{y_{1,n}(s)}{u_{2,n}(s)} = \frac{x_{1,n}(s)}{x_{2,n-1}(s)}, \tag{27}$$

$$g_{21,n}(s) = \frac{y_{2,n}(s)}{u_{1,n}(s)} = \frac{x_{2,n}(s)}{x_{1,n-1}(s)}, \qquad g_{22,n}(s) = \frac{y_{2,n}(s)}{u_{2,n}(s)} = \frac{x_{2,n}(s)}{x_{2,n-1}(s)}, \tag{28}$$

for zero initial conditions $x_{1,n}(0) = x_{2,n}(0) = 0$, with $u_{i,n}(s)$ and $y_{i,n}(s)$ being the Laplace-transformed input and output signals of the nth section, respectively.

Therefore, the Laplace-transformed output signals of the nth section, can be obtained, assuming zero initial conditions, as

$$y_n(s) = G_n(s)u_n(s) \tag{29}$$

where

$$u_n(s) = \left[u_{1,n}(s) \;\; u_{2,n}(s)\right]^T \quad \text{and} \quad y_n(s) = \left[y_{1,n}(s) \;\; y_{2,n}(s)\right]^T. \tag{30}$$

Result 2. The elements of the transfer function matrix $G_n(s)$ in (26) take the following form:

$$g_{11,n}(s) = \frac{b_{11,1,n}s + b_{11,0,n}}{s^2 + a_{1,n}s + a_{0,n}}, \qquad g_{12,n}(s) = \frac{b_{12,0,n}}{s^2 + a_{1,n}s + a_{0,n}}, \tag{31}$$

$$g_{21,n}(s) = \frac{b_{21,0,n}}{s^2 + a_{1,n}s + a_{0,n}}, \qquad g_{22,n}(s) = \frac{b_{22,1,n}s + b_{22,0,n}}{s^2 + a_{1,n}s + a_{0,n}}, \tag{32}$$

where

$$b_{11,1,n} = \frac{\lambda_1}{\Delta l_n}, \quad b_{11,0,n} = \frac{\lambda_1}{\Delta l_n}\left(\frac{\lambda_2}{\Delta l_n} - k_{22}\right), \quad b_{12,0,n} = k_{12}\frac{\lambda_2}{\Delta l_n}, \tag{33}$$

$$b_{22,1,n} = \frac{\lambda_2}{\Delta l_n}, \quad b_{22,0,n} = \frac{\lambda_2}{\Delta l_n}\left(\frac{\lambda_1}{\Delta l_n} - k_{11}\right), \quad b_{21,0,n} = k_{21}\frac{\lambda_1}{\Delta l_n}, \tag{34}$$

and

$$a_{1,n} = \frac{\lambda_1}{\Delta l_n} + \frac{\lambda_2}{\Delta l_n} - k_{11} - k_{22}, \quad a_{0,n} = \left(k_{11} - \frac{\lambda_1}{\Delta l_n}\right)\left(k_{22} - \frac{\lambda_2}{\Delta l_n}\right) - k_{21}k_{12}. \tag{35}$$

Proof. By applying the Laplace transform to (20)–(21) with zero initial conditions $x_{1,n}(0) = x_{2,n}(0) = 0$, and solving the resulting equations with respect to $x_{1,n}(s)$ and $x_{2,n}(s)$.

Stability Analysis. Each of the transfer functions given by Result 2 has two following poles

$$s_{(1,2),n} = -\frac{a_{1,n}}{2} \pm \frac{\sqrt{a_{1,n}^2 - 4a_{0,n}}}{2}, \tag{36}$$

which are real for $a_{1,n}^2 \geqslant 4a_{0,n}$ and complex for $a_{1,n}^2 < 4a_{0,n}$. In the second case, the complex conjugate pair is asymptotically stable for $a_{1,n} > 0$, which can be written, based on (35), as

$$\frac{\lambda_1 + \lambda_2}{\Delta l_n} > k_{11} + k_{22}, \tag{37}$$

meaning that Δl_n needs to be sufficiently small as compared to the characteristic speeds λ_1, λ_2.

For the case of real poles, the asymptotic stability condition (37) needs to be complemented by the additional requirement $a_{0,n} > 0$, which can be written, based on (35), as

$$\left(k_{11} - \frac{\lambda_1}{\Delta l_n}\right)\left(k_{22} - \frac{\lambda_2}{\Delta l_n}\right) > k_{12} + k_{21}. \tag{38}$$

In addition, we can state that transfer functions $g_{12,n}(s)$ and $g_{21,n}(s)$ do not have zeros, whereas $g_{11,n}(s)$ and $g_{22,n}(s)$ have the zeros at

$$z_{11,n} = -\frac{b_{11,0,n}}{b_{11,1,n}} = k_{22} - \frac{\lambda_2}{\Delta l_n} \quad \text{and} \quad z_{22,n} = -\frac{b_{22,0,n}}{b_{22,1,n}} = k_{11} - \frac{\lambda_1}{\Delta l_n}. \tag{39}$$

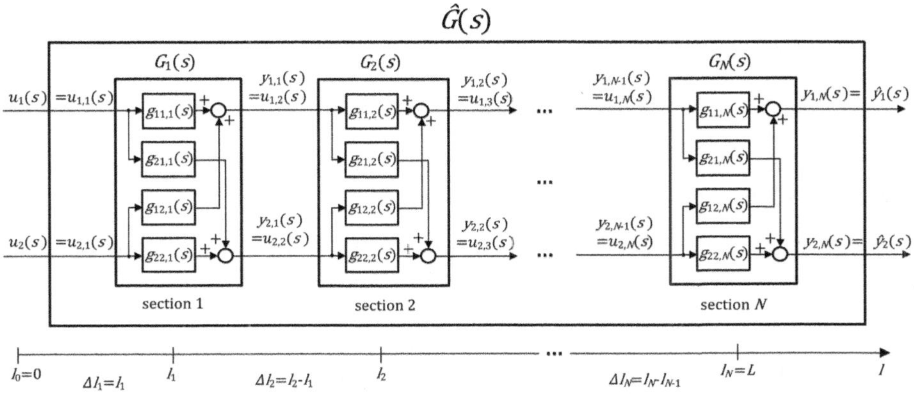

Fig. 1. Block diagram of the rational transfer function approximation model.

***N*-Section Model.** Let consider the approximation model representing the cascade interconnection of individual sections introduced in Sect. 3.1.

Definition 3. *The transfer function matrix $\hat{G}(s)$ of the N-section approximation model is given by*

$$\hat{G}(s) = \begin{bmatrix} \hat{g}_{11}(s) & \hat{g}_{12}(s) \\ \hat{g}_{21}(s) & \hat{g}_{22}(s) \end{bmatrix}, \tag{40}$$

where

$$\hat{g}_{11}(s) = \frac{\hat{y}_1(s)}{u_1(s)} = \frac{x_{1,N}(s)}{x_{1,0}(s)}, \qquad \hat{g}_{21}(s) = \frac{\hat{y}_2(s)}{u_1(s)} = \frac{x_{2,N}(s)}{x_{1,0}(s)}, \tag{41}$$

$$\hat{g}_{12}(s) = \frac{\hat{y}_1(s)}{u_2(s)} = \frac{x_{1,N}(s)}{x_{2,0}(s)}, \qquad \hat{g}_{22}(s) = \frac{\hat{y}_2(s)}{u_2(s)} = \frac{x_{2,N}(s)}{x_{2,0}(s)}, \tag{42}$$

for zero initial conditions, $x_{1,n}(0) = x_{2,n}(0) = 0$, $n = 1, 2, ..., N$, with $u_i(s)$ and $\hat{y}_i(s)$ being the Laplace-transformed input (24) and output (25) signals of the N-section model, respectively.

The Laplace-transformed output signals of the approximation model can be obtained, assuming zero initial conditions, as

$$\hat{y}(s) = \hat{G}(s)u(s) \tag{43}$$

with

$$u(s) = \begin{bmatrix} u_1(s) & u_2(s) \end{bmatrix}^T \quad \text{and} \quad \hat{y}(s) = \begin{bmatrix} \hat{y}_1(s) & \hat{y}_2(s) \end{bmatrix}^T. \tag{44}$$

The graphical structure of the rational transfer function approximation model is presented in Fig. 1.

Result 3. The elements of the transfer function matrix $\hat{G}(s)$ given by Definition 3 can be calculated as

$$\hat{G}(s) = G_N(s)G_{N-1}(s)\dots G_1(s), \tag{45}$$

where $G_n(s)$, $n = 1, 2, ..., N$ are transfer function matrices of individual sections given by Definition 2 and Result 2.

Proof. Based on the transfer function representation of the cascade interconnection of N dynamical subsystems—see, e.g., [1], page 36.

Stability Analysis. We assume here that in the cascaded system of Fig. 1 no pole-zero cancellation occurs, i.e., the number of poles of $\hat{G}(s)$ is the sum of the number of poles in $G_1(s), G_2(s), ..., G_N(s)$, and equals $2N$. In this case, the poles of the approximation model are given by the union of poles of individual sections.

In order to ensure the stability of the approximation model $\hat{G}(s)$, all the transfer functions $G_n(s)$, $n = 1, 2, ..., N$ need to represent stable dynamical subsystems, as considered in Sect. 3.2. This means that the stability condition (37) needs to be fulfilled for each section of the model.

3.3 Frequency- and Time-Domain Responses

The frequency- and time-domain responses of the approximation model can be evaluated in a similar manner to that outlined in Sect. 2.3, based on the transfer function $\hat{G}(s)$. Visual comparison of the responses obtained from the original transfer function model $G(s)$ and its approximations $\hat{G}(s)$ of different orders (i.e., representing different number of sections) can be considered as a simple initial assessment of the approximation quality.

4 Example: Parallel-Flow Double-Pipe Heat Exchanger

As an example we consider a double-pipe heat exchanger shown in Fig. 2. Its mathematical representation can be written, under some simplifying assumptions, in the form of the two following PDEs [4,8,10]:

$$\frac{\partial\vartheta_1(l,t)}{\partial t} + v_1\frac{\partial\vartheta_1(l,t)}{\partial l} = \alpha_1\big(\vartheta_2(l,t) - \vartheta_1(l,t)\big), \tag{46}$$

$$\frac{\partial\vartheta_2(l,t)}{\partial t} + v_2\frac{\partial\vartheta_2(l,t)}{\partial l} = \alpha_2\big(\vartheta_1(l,t) - \vartheta_2(l,t)\big), \tag{47}$$

where $\vartheta_1(l,t)$ and $\vartheta_2(l,t)$ represent the spatio-temporal distribution of the heating and the heated fluids, respectively, and α_1 and α_2 are generalized parameters including heat transfer coefficients, fluid densities, specific heats, and geometric dimensions of the exchanger. As can be seen, the system (46) and (47) is given directly in the weakly coupled hyperbolic form of Eqs. (1) and (2).

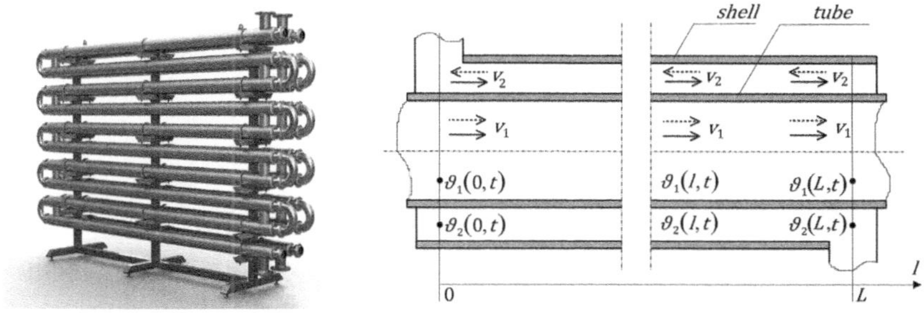

Fig. 2. Schematic of a double-pipe heat exchanger: v_1, v_2—inner and outer pipe fluid velocities; ϑ_1, ϑ_2—inner and outer pipe fluid temperatures; L—heat exchanger length. Solid arrows show flow directions for the parallel-flow mode and dotted ones—for the counter-flow mode.

We consider here the case of the heat exchanger working in the so-called parallel-flow mode, where both fluids flow in the same direction (see solid arrows indicating $v_1 > 0$ and $v_2 > 0$ in Fig. 2). For this configuration we assume the inlet temperatures of both fluids as the input signals,

$$u_1(t) = \vartheta_1(0, t), \quad u_2(t) = \vartheta_2(0, t), \tag{48}$$

which corresponds to the collocated configuration of boundary inputs introduced in (4). The most important from the control point of view are the fluid temperatures measured at the exchanger outflow. Therefore, we assume the following output signals

$$y_1(t) = \vartheta_1(L, t), \quad y_2(t) = \vartheta_2(L, t), \tag{49}$$

which represents the output configuration given by (5). Assuming the spatio-temporal dynamics of the considered plant, its "exact" transfer function representation is given by (9)–(12). In the next section we analyze the rational transfer function approximation model obtained based on the results presented in Sect. 3.2. We assume the following parameter values in (46) and (47): $L = 5$ m, $v_1 = 1$ m/s, $v_2 = 0.2$ m/s, $\alpha_1 = \alpha_2 = 0.05$ 1/s.

4.1 Transfer Function Representation of the Approximation Model

Assuming the division of the considered heat exchanger model into, e.g., $N = 100$ uniform sections, we obtain, for the assumed parameter values, the following section length:

$$\Delta l_n = \frac{L}{N} = 0.05 \text{ m}, \qquad n = 1, 2, \ldots, N, \tag{50}$$

and, consequently, the following transfer functions (31)–(32) of the single section:

$$g_{11,n}(s) = \frac{20s + 81}{s^2 + 24.1s + 82}, \qquad g_{12,n}(s) = \frac{0.2}{s^2 + 24.1s + 82}, \tag{51}$$

$$g_{21,n}(s) = \frac{1}{s^2 + 24.1s + 82}, \qquad g_{22,n}(s) = \frac{4s + 80.2}{s^2 + 24.1s + 82}. \tag{52}$$

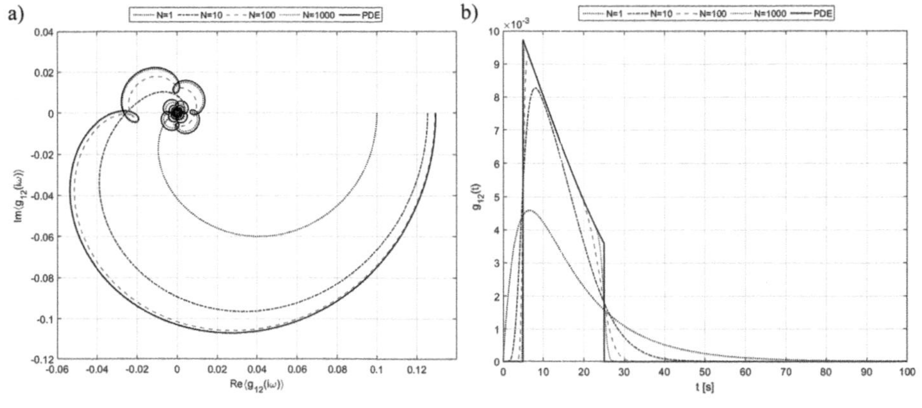

Fig. 3. Nyquist frequency response $g_{12}(i\omega)$ of the PDE model vs. frequency responses $\hat{g}_{12}(i\omega)$ of the approximation models (a) and impulse response $g_{12}(t)$ of the PDE model vs. impulse responses $\hat{g}_{12}(t)$ of the approximation models (b) for the parallel-flow heat exchanger.

The poles of all transfer functions in (51)–(52) are equal and located at $s_{1,n} = -20$ and $s_{2,n} = -4.1$, which means that all considered sections represent stable dynamical subsystems. Therefore, the cascade interconnection of N sections described by the transfer function matrix $G_n(s)$ given by (51)–(52) results in the approximation model $\hat{G}(s)$ with N pairs of stable poles $s_{1,n}$ and $s_{2,n}$ (see analysis in Sect. 3.2).

4.2 Frequency- and Time-Domain Responses

Based on the transfer functions $g(s)$ and $\hat{g}(s)$ of the PDE and high-dimensional ODE models, respectively, it is possible to compare their dynamical properties, expressed both in the frequency and the time domain. Insights obtained from the frequency and time responses are useful for the analysis how the order of the MOL approximation model affects the approximation quality.

Figure 3a shows the Nyquist plot for the original PDE model versus Nyquist plots for the approximation models obtained for the transfer function channel $g_{12}(s)$ for different values of N. One can observe characteristic "loops" on the Nyquist plot which are associated with the resonance-like phenomena taking place inside the heat exchanger. It can be stated that the larger the order of the approximation model, the better the mapping of the original frequency response, together with the above-mentioned oscillations.

Similar conclusions can be drawn based on the analysis of the impulse responses $g_{12}(t)$ for the same input-output channel (Fig. 3b). In order to correctly map the fairly steep slopes of the original impulse response of the PDE model, a relatively high order ODE model is needed which is able to correctly approximate its high-frequency modes.

5 Summary

We have discussed some results concerning finite-dimensional approximation of DPSs described by linear hyperbolic PDEs with the boundary conditions representing collocated boundary inputs. The approximation model is considered here in the form of the cascade interconnection of a number of sections expressed in the transfer function domain, resulting in a high-order rational transfer function model. The quality of such a model can be checked by using frequency-domain norm of the approximation error, such as H_2 or H_∞ norm.

Another task to be performed within the presented framework is to develop the section-based approximation model for the so-called *anti-collocated* boundary input configuration, occurring e.g. in counter-flow heat exchangers. As one can expect, the expressions describing the resulting transfer function model will more complex than in the "parallel" case considered here. This results from the fact that the anti-collocated input configuration will imply the feedback interconnection between the individual sections of the approximation model.

References

1. Albertos, P., Sala, A.: Multivariable Control Systems: An Engineering Approach. Advanced Textbooks in Control and Signal Processing. Springer, London (2004)
2. Bartecki, K.: Computation of transfer function matrices for 2 × 2 strongly coupled hyperbolic systems of balance laws. In: Proceedings of the 2nd Conference on Control and Fault-Tolerant Systems, Nice, France, pp. 578–583, October 2013
3. Bartecki, K.: A general transfer function representation for a class of hyperbolic distributed parameter systems. Int. J. Appl. Math. Comput. Sci. **23**(2), 291–307 (2013)
4. Bartecki, K.: Modeling and Analysis of Linear Hyperbolic Systems of Balance Laws. Studies in Systems, Decision and Control, vol. 48. Springer, Cham (2016)
5. Curtain, R., Morris, K.: Transfer functions of distributed parameters systems: a tutorial. Automatica **45**(5), 1101–1116 (2009)
6. Koto, T.: Method of lines approximations of delay differential equations. Comput. Math. Appl. **48**(1–2), 45–59 (2004)
7. Levine, W.S. (ed.): The Control Systems Handbook: Control System Advanced Methods. Electrical Engineering Handbook. CRC Press, Boca Raton (2011)
8. Maidi, A., Diaf, M., Corriou, J.P.: Boundary control of a parallel-flow heat exchanger by input-output linearization. J. Process Control **20**(10), 1161–1174 (2010)
9. Schiesser, W.E., Griffiths, G.W.: A Compendium of Partial Differential Equation Models: Method of Lines Analysis with Matlab. Cambridge University Press, New York (2009)
10. Zavala-Río, A., Astorga-Zaragoza, C.M., Hernández-González, O.: Bounded positive control for double-pipe heat exchangers. Control Eng. Pract. **17**(1), 136–145 (2009)

Time-Varying Perfect Control Algorithm for LTI Multivariable Systems

Marek Krok(✉), Paweł Majewski, and Wojciech P. Hunek

Opole University of Technology, Proszkowska 76, 45-758 Opole, Poland
m.krok@po.edu.pl
http://po.opole.pl

Abstract. In this paper, a new approach to perfect control design is proposed. The novel application of polynomial degrees of freedom in calculation of generalized matrix inverse has enabled a new branch of perfect control-oriented robust scenarios. Crucially, the obtained closed-loop system has revealed to be internally time-varying even though the general input-output behavior is still time-invariant. Simulation instances made in Matlab/Simulink environment show some interesting peculiarities covering the perfect control speed and energy.

Keywords: Multivariable perfect control · LTV control algorithm · Nonunique matrix inverses · Robustification

1 Introduction

The perfect control algorithm has been under intense scientific investigation for years [1,2]. With distinguished properties, this control law was subjected to studies concerning control speed, energy cost or even robustness [1,3,4]. For nonsquare systems, i.e. plants with different numbers of input and output variables, some properties were almost arbitrarily assigned due to nonuniqueness of received solutions for a discussed problem. In the remarkable part of mentioned study, the different behaviors were obtained by application of various matrix inverses used in the perfect control design techniques [5,6].

Inverse problem for nonsquare matrices is a complex issue that has been undergoing extensive scientific investigation for many years [5,7,8]. Different approaches were proven to entail remarkable development in terms of various inverse applications. With the results covering other than minimum-norm solutions, there is a possibility to show a plethora of useful generalized inverse instances associated with selected examples of so-called degrees of freedom [1,9]. A employment of degrees of freedom, in particular derived from recently introduced nonunique H- and σ-inverse, brings some measurable benefits in control and systems theory [4,10]. As a matter of fact, the lack of stability or robustness for nonminimum-phase perfect control systems has already been overcome. Moreover, it was shown that the properties related to the higher control speed

A. Bartoszewicz et al. (Eds.): PCC 2020, AISC 1196, pp. 68–79, 2020.
https://doi.org/10.1007/978-3-030-50936-1_6

and lower energy consumption of the input variables can mostly be predefined in a simple heuristic way [3,11].

Interestingly, similar properties were obtained for a class of linear filters with the application of some internal time-varying parameters. Moreover, with changing-over-time behaviors, the transient states of low-pass elliptic filter have also been significantly reduced [12]. Additionally, it was shown in Refs. [13,14], that the varying parameters can result in lower energy consumption for some specific test cases.

Taking into account the above issues, a consideration of joint advances coming from the nonunique generalized inverses and time-varying internal part of plant seems to be an interesting proposal. The possible lowering of energy needed to obtain output setpoint is one of the most desired properties, which has already been the subject of numerous studies. However, the phenomena and potential benefits associated with the new approach are still to be explored. It can be assumed, that the crucial target of such time-varying mechanism is to obtain the maximally fast, robust and energy efficient instances of the perfect control algorithm. Although the maximum-speed case connected with the zero-pole placement technique is not the main goal here, the speed higher than in the minimum-norm-oriented approaches is expected to be obtained.

Henceforth, the entire set of the aforementioned peculiarities can be obtained with application of polynomial degrees of freedom of generalized right inverses used in the perfect control design methodology. The application of some time-varying state-feedback gain matrix, strictly derived from nonunique inverses, in control design practice, can efficiently merge both desired energy and speed properties. Thus, the special selected polynomial degrees of freedom provide the internal time-variability gain to the perfect control systems, which together with the accomplished early predefined advantages, constitute the main achievement of this manuscript.

The paper is organized in the following manner. In second Section, the state-space discrete-time system representation is given. Next, in Sect. 3, the perfect control algorithm is shortly described. The crucial nonunique inverses together with the unique encountered benchmarks form the Sect. 4. The main goal of this paper covering the application of polynomial degrees of freedom into perfect control design action is proposed in Sect. 5. Simulation examples presented in the penultimate section reimburse the conducted manuscript's investigation effort. In the end, the last Sect. 6 summarizes the conclusions and open problems.

2 System Description

The mathematical description of a system is the basis for almost every theoretical consideration. Since the perfect control algorithm reveals interesting properties for nonsquare plants, the system representation relations should be reliable for multivariable study. Thus, in the manuscript we are considering the classical LTI MIMO discrete-time state-space objects defined in the following framework

$$\begin{aligned} \mathbf{x}(k+1) &= \mathbf{A}\mathbf{x}(k) + \mathbf{B}\mathbf{u}(k), \quad \mathbf{x}(0) = \mathbf{x}_0, \\ \mathbf{y}(k) &= \mathbf{C}\mathbf{x}(k). \end{aligned} \tag{1}$$

Naturally, the parameter matrices of Eqs. (1) are of dimensions $\mathbf{A} \in \Re^{n \times n}$, $\mathbf{B} \in \Re^{n \times n_u}$, $\mathbf{C} \in \Re^{n_y \times n}$ whilst $\mathbf{x}(k)$, $\mathbf{u}(k)$, $\mathbf{y}(k)$ are the n-state, n_u-input and n_y-output vectors, respectively, in the discrete time k, whereas $\mathbf{x}_0$ denotes an initial state of the considered system. Of course, the stability, observability and controlability are well-known behaviors for the discrete-time linear time-invariant state-space plants, therefore these issues are recalled right now. For such system description covered by the formulas (1), let us switch to the perfect control algorithm in detail, presented in detail in the next section.

3 Multivariable Perfect Control Law

Perfect control algorithm is a very special control structure that ensures the lowest possible control error. Being developed for one-step deterministic output predictor $\mathbf{y}(k+1)$ (under systems with single unit delay, see Eqs. (1)), it provides that the output equals the assumed reference/setpoint $\mathbf{y_{ref}}(k)$ just after time delay $d = 1$. Therefore, the multivariable perfect control law sounds as follows

$$\mathbf{u}(k) = (\mathbf{CB})^{\mathrm{R}}(\mathbf{y_{ref}}(k) - \mathbf{CAx}(k)), \tag{2}$$

where $(\mathbf{CB})^{\mathrm{R}}$ denotes any right inverse of matrix product $\mathbf{CB}$. Naturally, further we are considering the valid right-invertible plants only.

Observe that this general formula allows to obtain an arbitrary output value after time delay deriving from the system description. However, the non-zero reference implies unnecessary non-zero steady states here, so the zero setpoint is considered through conducted study only. Hence, for the zero reference value, the perfect control algorithm can be formulated as

$$\mathbf{u}(k) = -(\mathbf{CB})^{\mathrm{R}}\mathbf{CAx}(k). \tag{3}$$

After a simple transformation, we receive the output formula free from any disturbances or errors. This phenomenon can be described in the following manner

$$\mathbf{y}(k+d) = \mathbf{y_{ref}}(k+d) \ for \ k \geq d. \tag{4}$$

Note, that this statement describes the output behavior, in particular for $d = 1$ corresponding to the typical state-space framework. Naturally, the usage of perfect control formula entails the stable output run even for unstable and nonminimum-phase plants as well (see Fig. 1).

It should be emphasized, that the perfect control law can be rewritten in the state-feedback domain as follows

$$\mathbf{u}(k) = -\mathbf{Kx}(k), \tag{5}$$

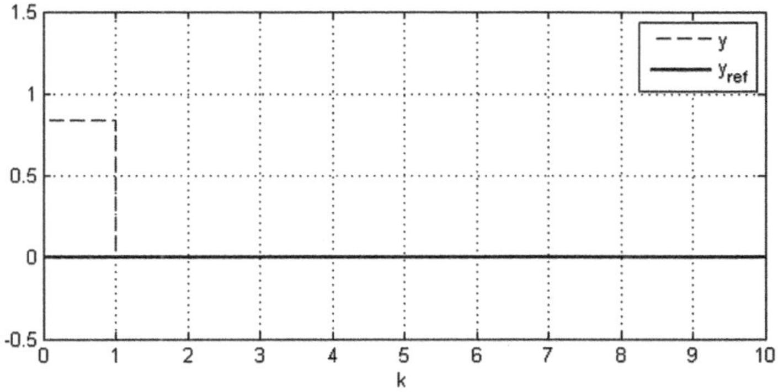

Fig. 1. Perfect control output, exemplary run

knowing that

$$\mathbf{K} = (\mathbf{CB})^{\mathrm{R}}\mathbf{CA}. \tag{6}$$

Naturally, the above formula is in relation with the perfect gain, which is also possible to observe in Eq. (3).

In a consequence, according to the well-known pole-placement method, the closed-loop property, such as stability, can be obtained from the following closed-loop characteristic equation

$$det(z\mathbf{I} - \mathbf{A} + \mathbf{BK}) = 0. \tag{7}$$

Moreover, the state-feedback notion allows to obtain the closed-loop system matrix, which describes some fundamental behaviors of the perfect control plant. Aiming the clarity of further notion, a closed-loop system matrix covering the state-feedback feature sounds in the following manner

$$\mathbf{A}^* = \mathbf{A} - \mathbf{BK}. \tag{8}$$

Again, the closed-loop system stability can similarly be described using the closed-loop system matrix related to the subsequent characteristic equation

$$det(z\mathbf{I} - \mathbf{A}^*) = 0. \tag{9}$$

It is worth emphasizing that the stability does not affect the output behavior in perfect control case. Therefore, the eigenvalues of the closed-loop system influence the state and control signals, allowing to obtain different properties covering, e.g., control robustness, speed or energy optimization [6]. More advanced study concerning stability of perfect control systems, connected with the so-called control zeros, can be found, for instance, in Refs. [1,16].

In some different control strategies, the linear feedback $\mathbf{K}$ is rather related to the unique solution regarding, e.g., the result of discrete-time algebraic Riccati procedure. In fact, it reveals yet another advantage of perfect control resulting from the application of nonunique inverses. It is already shown, that in case of nonsquare plants, there is a possibility to influence the closed-loop system matrix in order to obtain different properties associated with the various perfect control performances. The differentiation has usually been applied in context of different degrees of freedom derived from predefined generalized right inverses. Interestingly, the selection of the mentioned inverses has often been limited to the parameter degrees of freedom only [3,6]. Therefore, an extended approach providing the perfect control-oriented inverses with parameter, and, importantly, polynomial degrees of freedom is presented in the next section.

4 Generalized Right Inverses with Polynomial Degrees of Freedom

As mentioned above, different matrix inverses play a crucial role in the perfect control design. In this section, three of most commonly used inverses are shown in order to clarify their notions and distinctions. The overview starts with the unique T-inverse, defined for full rank matrix $\mathbf{M}$ in the following way

$$\mathbf{M}_0^{\mathrm{R}} = \mathbf{M}^{\mathrm{T}}(\mathbf{M}\mathbf{M}^{\mathrm{T}})^{-1}. \tag{10}$$

Observe, that the approved minimum-norm property of the right T-inverse has granted its wide application throughout different inverse tasks. However, the said uniqueness can sometimes be a drawback, thus another framework for appointing the different instances are given below [15].

Let us switch now to the right σ-inverse be described as follows

$$\mathbf{M}_\sigma^{\mathrm{R}} = \beta^{\mathrm{T}}(\mathbf{M}\beta^T)^{-1}, \tag{11}$$

where β denotes the crucial degree of freedom. Moreover, there is a single limitation imposed on the β matrix. Following the possibility to establish regular inverse of $(\mathbf{M}\beta^T)$, the β selection should presume, near structure $\mathbf{M}$, full rank of matrix $(\mathbf{M}\beta^T)$.

Hence, the usage of polynomial degrees of freedom is possible in this scenario. With assumption of proper form of matrix polynomial $\beta(q^{-1})$, we immediately arrive at

$$\mathbf{M}_\sigma^{\mathrm{R}}(q^{-1}) = \beta^{\mathrm{T}}(q^{-1})(\mathbf{M}\beta^T(q^{-1}))^{-1}, \tag{12}$$

i.e. case which is applicable in robust perfect control design.

On the other hand, the nonunique H-inverse can also be useful in the polynomial degrees of freedom-oriented study. In order to show this potentially useful generalized inverse, the well-known SVD factorization must first be performed as follows

$$\mathbf{M} = \mathbf{U}\mathbf{\Sigma}\mathbf{V}^{\mathrm{T}}, \tag{13}$$

where $\mathbf{\Sigma}$ contains the eigenvalues of $\mathbf{M}$. Additionally, $\mathbf{U}$ and $\mathbf{V}$ are unitary matrices. It is worth mentioning, that in the nonsquare matrix cases, the $\mathbf{\Sigma}$ is also nonsquare and can be concerned as joined zero matrix and diagonal eigenvalue one as follows: $\mathbf{\Sigma} = \begin{bmatrix}\mathbf{\Lambda} \ \mathbf{0}\end{bmatrix}$, with $\mathbf{\Lambda}$ covering generalized eigenvalues of $\mathbf{M}$.

Remark 1. In the considered scenario we are assuming, that the matrix being subject to SVD procedure has no multiple eigenvalues. From now on, $\mathbf{\Lambda}$ is treated as a diagonal matrix [15]. In other cases, there is an opportunity to involve the Jordan blocks in the entire design process, however this instance is omitted here.

Observe, that the crucial degrees of freedom reveal in calculation of right inverse of $\mathbf{\Sigma}$ matrix. As shown in Ref. [4], such inverse sounds in the following form

$$\mathbf{\Sigma}^{\mathrm{R}} = \begin{bmatrix} \mathbf{\Lambda}^{-1} \\ \mathbf{L} \end{bmatrix}, \tag{14}$$

under $\mathbf{L}$ involving parameter crucial modes. Thus, the right H-inverse is presented as

$$\mathbf{M}_H^{\mathrm{R}} = (\mathbf{V}^{\mathrm{T}})^{-1}\mathbf{\Sigma}^{\mathrm{R}}\mathbf{U}^{-1}. \tag{15}$$

However, the degrees of freedom $\mathbf{L}$ can also be in form of matrix polynomial $\mathbf{L}(q^{-1})$. This new proposed phenomenon entails the subsequent structure of innovative right H-inverse

$$\mathbf{M}_H^{\mathrm{R}}(q^{-1}) = (\mathbf{V}^{\mathrm{T}})^{-1}\mathbf{\Sigma}^{\mathrm{R}}(q^{-1})\mathbf{U}^{-1}. \tag{16}$$

Henceforth, the time-dependent full rank H-inverse can be used just like parameter σ-inverse to broadly understand perfect control design tasks.

Remark 2. Note that H-inverse has recently been successfully applied to the process of perfect signal reconstruction existing in almost every telecommunications branch. For more details please see Ref. [9].

Let us now continue with the application of introduced parameter inverses to the new robust time-varying perfect control design methodology.

5 Time Varying Perfect Control

In the previous section, the theoretical basics for potential construction of time-varying perfect control were given. The application of polynomial degrees of freedom derived from our generalized inverses allows to perform the control input variables according to the LTV paradigm, finally to obtain

$$\mathbf{u}(k) = -(\mathbf{CB})^{\mathrm{R}}(q^{-1})\mathbf{CAx}(k). \tag{17}$$

In fact, such control signal entails that the state-feedback formula covering time-varying property as follows

$$\mathbf{u}(k) = -\mathbf{K}(q^{-1})\mathbf{x}(k). \tag{18}$$

Following the considered scenario of multivariable time-varying perfect control, the closed-loop plant can now be described by the polynomial-based framework

$$\mathbf{x}(k+1) = \mathbf{A}^{*}(q^{-1})\mathbf{x}(k), \tag{19}$$

which certainly shows, that the parameters of control matrix $\mathbf{A}^*(q^{-1})$ are now time-dependent.

However, the input-output transfer function of the closed-loop perfect control system for zero reference value is still time-invariant, as the output remains at the setpoint after time delay. Thus, the stability criteria considering the input-output behavior should not be employed here for all subsets of this control law. Of course, similar results are received for perfect control basing on the classical (parameter) degrees of freedom. Knowing that the stability criteria of studied control algorithm constitutes the issue worthy of further consideration even in the typical approach, let us now focus on the stability of time-varying perfect control system in the next section.

6 Stability of Time-Varying Perfect Control

Stability of time-varying systems is a complex problem which has been the subject of intense scientific research for many years. In open literature, at least five different types of stability were identified [17]. In this manuscript, the stability for a time-varying closed-loop perfect control plant with zero reference value is analyzed, so the well-known BIBS technique can be understood in the following manner

$$\mathbf{x}(k), \mathbf{u}(k) \to \mathbf{0} \; for \; k > \gamma. \tag{20}$$

In other words, for a system with non-zero initial states the input response just should converge to zero.

Therefore, in the classical instance covering time-invariant perfect control, the closed-loop stability can be observed in the corresponding sense

$$\mathbf{x}(k+1) = (\mathbf{A}^*)^k \mathbf{x}(0), \tag{21}$$

with eigenvalues of the closed-loop system matrix $\mathbf{A}^*$ being crucial here. However, in the consideration related to the time-varying control systems, the stability criterion is more complex. In our study, the method using an matrix-oriented evolutionary operator $\phi(k, i)$ can be employed in order to determine the stability of the LTV system. Hence, the closed-loop system matrix derived from the state-domain shall be divided into the i time-related pieces as follows

$$\mathbf{x}(k+1) = \mathbf{A}(k)\mathbf{x}(k) + ... + \mathbf{A}(k-i)\mathbf{x}(k-i). \tag{22}$$

For such differential equation, the said evolutionary operator should fulfill the subsequent relationship

$$\phi(k, i) = \mathbf{A}(k)\mathbf{A}(k-1)...\mathbf{A}(k-i), \tag{23}$$

which clearly corresponds to the input response, previously mentioned in Eq. (21). Thus, the transition between the initial $\mathbf{A}(k)$ and final $\mathbf{A}(k-1)$ system matrices is clearly employed by the stability criteria shown, e.g., in Ref. [18] as

$$||\phi(k, i)|| \leq Me^{-\gamma(k-i)}, \tag{24}$$

for some positive M, γ and $k \geq i$. An interpretation of such criteria is immediate. If the evolutionary operator for any parameter i has all eigenvalues inside the unit circle, the time-varying system is exponentially stable. If there are some (k, i), for which the operator provides the eigenvalues grater than unity, the plant is instantly unstable. Naturally, this stability will result in the intuitive feature, that the non-zero initial state and control signals are vanquished in a certain number of steps (for the zero setpoint, naturally).

Having all new issues related to the time-varying perfect control strategy strictly dedicated to the right-invertible LTI MIMO discrete-time state-space plants, let us move onto the simulation instances presented in the next section.

7 Simulation Examples

In this section, a simulation study of newly introduced perfect control scheme is presented. Two, speed- and control-oriented, numerical examples, covering both minimum-norm approach and time-varying σ-inverse-related new solution, confirm the advantages derived from the main goal of this manuscript. At the beginning, we start with the control speed instance.

Example 1: Higher control speed
Let us consider an LTI MIMO discrete-time state-space second-order system described by Eqs. (1) with $\mathbf{A} = \begin{bmatrix} 1.90 & 0.31 \\ 0.67 & -0.49 \end{bmatrix}$, $\mathbf{B} = \begin{bmatrix} -0.38 & 1.21 \\ 0 & -0.40 \end{bmatrix}$ and $\mathbf{C} = \begin{bmatrix} 0.30 & -0.24 \end{bmatrix}$, as well as $\mathbf{x}_0^{\mathrm{T}} = \begin{bmatrix} -2 & 5 \end{bmatrix}$.

After application of the unique minimum-norm right T-inverse, we obtain the perfect control and state signals depicted in Fig. 2.

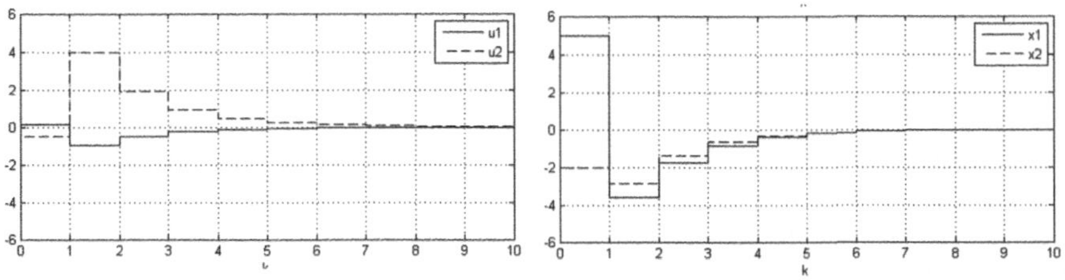

Fig. 2. Runs of perfect control, case: T-inverse

On the other hand, the closed-loop system matrix sounds as follows

$$\mathbf{A}^* = \begin{bmatrix} 0.8047 & -0.2537 \\ 1.0059 & -0.3171 \end{bmatrix},$$

with perfect control poles: $z_1 = 0.4876$ and $z_2 = 0$. With high variability and long-lasting transient states, the total control energy is equal to $E_0 = 22.2121$. Of course, the perfect control output remains at the assumed setpoint (see Fig. 1). On the other side, with the application of right σ-inverse involving polynomial degrees of freedom

$$\beta = \begin{bmatrix} \frac{(22q^{-1})}{5} + \frac{24}{5} & -\frac{(14q^{-1})}{5} - \frac{1}{10} \end{bmatrix},$$

we immediately arrive at the time-varying perfect control system described by the following closed-loop system matrix

$$\mathbf{A}^* = \begin{bmatrix} 0.5581 & -0.3806 \\ 0.6976 & -0.4758 \end{bmatrix} + \begin{bmatrix} 0.7412 & -0.2864 \\ 0.9265 & -0.3580 \end{bmatrix} q^{-1}.$$

It is worth mentioning again, that the input-output relation of time-varying perfect control system is still a time-invariant transfer function in form of z^{-d}. It can immediately be proven by combining the formulas of Eqs. (1) with the control law as in expression (3). The control and state variables supported by the σ-inverse-oriented parameter degrees of freedom are presented in Fig. 3.

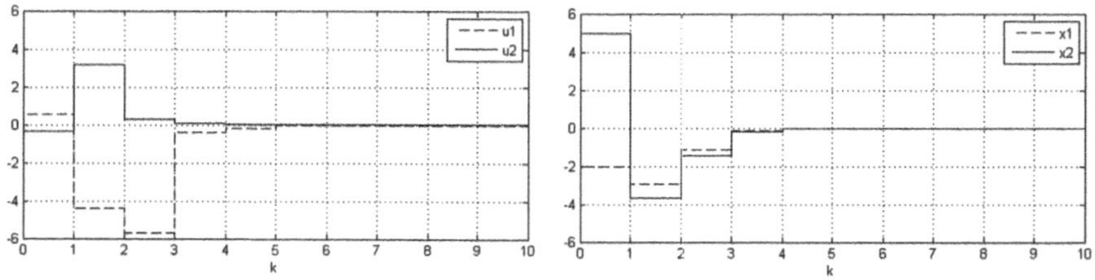

Fig. 3. Runs of perfect control, case: time-varying σ-inverse

The main target of this simulation example was to show that different perfect control instances can be obtained throughout the application of various unique/nonunique inverses. The total energy of the control input variables, received with usage of polynomial degrees of freedom, has reached the value of $E_{\sigma p} = 62.7795$. Hence, both T and σ scenarios can now be compared in terms of speed and control energy. It is clear, that the application of internal time-varying parameters of perfect control plant resulted in substantial improvement in the context of control speed. Although the advantage is not as obvious as in the pole-free/zero-pole scenario [6], the main merit can be observed in terms of control amplitudes. Naturally, the higher speed obtained in this case should rather be connected with higher energy consumption. Concluding, it has already been shown, that for different inverses, the energy- and speed-related properties can be obtained separately. Therefore, in the next simulation example, an attempt to minimize the mentioned control energy for LTV perfect control plant is demonstrated.

Example 2: Lower control energy
As before, we are considering an LTI MIMO state-space system with two state variables. The appropriate matrices are as follows: $\mathbf{A} = \begin{bmatrix} -0.80 & 0.93 \\ 0.42 & 0.72 \end{bmatrix}$, $\mathbf{B} = \begin{bmatrix} 0.71 & -1.43 \\ 0.28 & 1.04 \end{bmatrix}$, $\mathbf{C} = \begin{bmatrix} 0.63 & 0.56 \end{bmatrix}$ and $\mathbf{x}_0^{\mathrm{T}} = \begin{bmatrix} 4 & -3 \end{bmatrix}$. Again, the new approach benchmark is started by the application of minimum-norm right T-inverse into the perfect control design task. In this scenario, the received closed-loop system matrix is equal to

$$\mathbf{A}^* = \begin{bmatrix} -0.2903 & -0.9456 \\ 0.3266 & 1.0638 \end{bmatrix},$$

with a single non-zero eigenvalue: $z = 0.7734$. The state and control variables covering total control energy $E_0 = 63.6263$ (obtained via right T-inverse) are shows in Fig. 4.

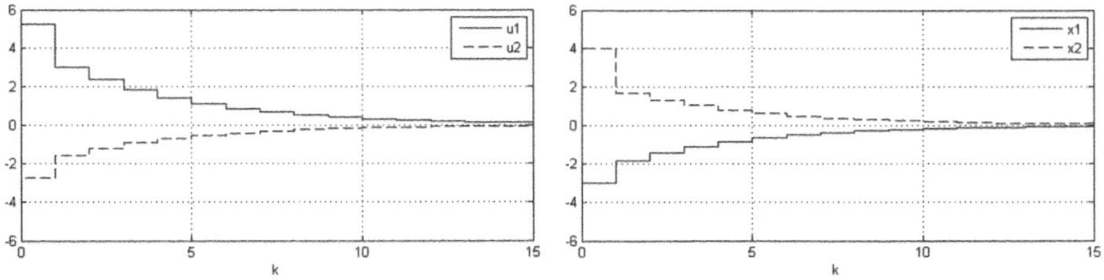

Fig. 4. Runs of perfect control, case: T-inverse

On the other hand, the perfect control design process involvement applying the special selected parameter degrees of freedom in the form of

$$\beta = \left[\frac{(7q^{-1})}{10} + \frac{1}{5}, \frac{1}{10} - \frac{(2q^{-1})}{5} \right],$$

leads to the time-varying instance with the closed-loop system matrix equal to

$$\mathbf{A}^* = \begin{bmatrix} -0.8030 & 0.9411 \\ 0.9034 & -1.0587 \end{bmatrix} + \begin{bmatrix} -0.2778 & -0.9915 \\ 0.3125 & 1.1154 \end{bmatrix} q^{-1}.$$

Corresponding state and input variable runs are depicted in Fig. 5.

In this scenario, the control energy is $E_{\sigma p} = 52.6813$. Interestingly, the cost of obtaining of the perfect output is lower here than in the case of minimum-norm T-inverse. It should now be emphasized, that the minimum-norm inverse has been connected with the minimum-energy property for years, not only under discrete-time state-space framework activities.

Summing up, two different control scenarios allowed to obtain various perfect control properties. With application of polynomial degrees of freedom, a remarkable improvement of considered control strategy can be established. However, every advancement in terms of time-varying perfect control systems is burdened with the higher computational effort.

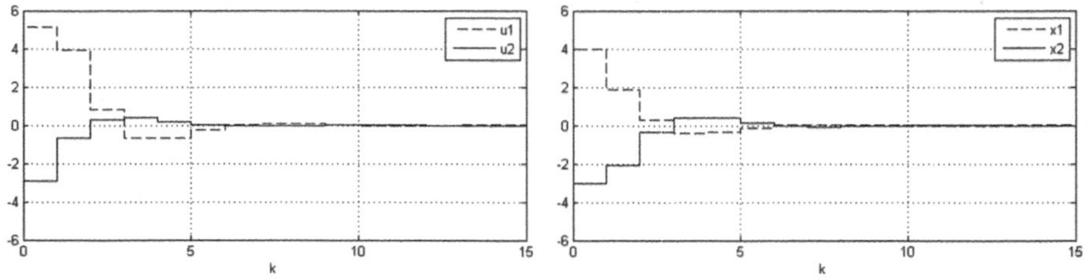

Fig. 5. Runs of perfect control, case: σ-inverse

8 Conclusions and Open Problems

The results presented within this paper clearly show the useful behavior obtained through the application of time-varying generalized inverses into the perfect control design process. Properly developed inverse model control plants, with possible minimum control error, are supported here by the application of various inverse-related polynomial degrees of freedom, which constitutes the novelty of presented work. Measurable gains in the context of control speed and minimum energy of the input variables, assisted by the stability criterion, were obtained in the heuristic scenarios covering new algorithm of time-varying perfect control for LTI MIMO discrete-time state-space systems. Naturally, more studies concerning the input/output-stationary multivariable perfect control plants with internally varying parameters can be conducted in the nearest future. Finally, it would be interesting to give an analytical description of the entire new methodology shown in the manuscript. Also, the classical approach based on the pole placement method should be applicable in case of LTV perfect control. Additionally, a plethora of new simulation instances, in particular those related to the parameter nonunique H-inverse, should clarify and confirm all complex procedures presented in this paper.

References

1. Hunek, W.P.: Towards a General Theory of Control Zeros for LTI MIMO Systems. Opole University of Technology Press, Opole (2011)
2. Latawiec, K.J., Korytowski, A.: A direct solution of the perfect regulation problem for LTI discrete-time systems. In: Proceedings of the 14th National Automatic Control Conference (KKA 2002), Zielona Góra, Poland, pp. 165–168 (2002). (in Polish)
3. Hunek, W.P., Krok, M.: Pole-free perfect control for nonsquare LTI discrete-time systems with two state variables. In: Proceedings of the 13th IEEE International Conference on Control and Automation (ICCA 2017), Ohrid, Macedonia, pp. 329–334 (2017). https://doi.org/10.1109/ICCA.2017.8003082

4. Hunek, W.P.: New SVD-based matrix H-inverse vs. minimum-energy perfect control design for state-space LTI MIMO systems. In: Proceedings of the 20th IEEE International Conference on System Theory, Control and Computing (ICSTCC 2016), Sinaia, Romania, pp. 14–19 (2016). https://doi.org/10.1109/ICSTCC.2016.7790633
5. Karampetakis, N.P., Tzekis, P.: On the computation of the generalized inverse of a polynomial matrix. IMA J. Math. Control Inf. **18**(1), 83–97 (2001). https://doi.org/10.1093/imamci/18.1.83
6. Hunek, W., Krok, M.: Parameter matrix σ-inverse in design of structurally stable pole-free perfect control for second-order state-space systems. In: Proceedings of the 24th International Conference on Automation and Computing (IEEE ICAC 2018) (2018). https://doi.org/10.23919/IConAC.2018.8748977
7. Stanimirović, P.S., Petković, M.D.: Computing generalized inverse of polynomial matrices by interpolation. Appl. Math. Comput. **172**(1), 508–523 (2006). https://doi.org/10.1016/j.amc.2005.02.031
8. Ben-Israel, A., Greville, T.N.E.: Generalized Inverses. Theory and Applications, 2nd edn. Springer, New York (2003)
9. Hunek, W.P., Majewski, P.: Perfect reconstruction of signal - a new polynomial matrix inverse approach. EURASIP J. Wirel. Commun. Networking **2018**(107), 8 (2018). https://doi.org/10.1186/s13638-018-1122-5
10. Noueili, L., Chagra, W., Ksouri, M.: New iterative learning control algorithm using learning gain based on σ inversion for nonsquare multi-input multi-output systems (2018). https://doi.org/10.1504/IJMIC.2018.095829
11. Hunek, W.P., Krok, M.: A study on a new criterion for minimum-energy perfect control in the state-space framework. Proc. Inst. Mech. Eng. Part I J. Syst. Control Eng. **233**(7), 779–787 (2019)
12. Gutierrez de Anda, M.A., Sarmiento Reyes, A., Hernandez Martinez, L., Piskorowski, J., Kaszynski, R.: The reduction of the duration of the transient response in a class of continuous-time LTV filters. IEEE Trans. Circuits Syst. II: Express Briefs **56**(2), 102–106 (2009)
13. Ozgun, M., Tsividis, Y., Burra, G.: Dynamic power optimization of active filters with application to zero-if receivers. IEEE J. Solid-State Circuits **41**(6), 1344–1352 (2006)
14. Tsividis, Y., Krishnapura, N., Palaskas, Y., Toth, L.: Internally varying analog circuits minimize power dissipation. IEEE Circuits Devices Mag. **19**(1), 63–72 (2003)
15. Petersen, K.B., Pedersen, M.S.: The Matrix Cookbook. Technical University of Denmark, November 2012
16. Tokarzewski, J.: Finite Zeros in Discrete Time Control Systems. Lecture Notes in Control and Information Sciences, vol. 338. Springer-Verlag (2006). https://doi.org/10.1007/11587743
17. Niezabitowski, M.: Numerical characteristics of discrete hybrid system. Ph.D. thesis, Faculty of Automatic Control, Electronics and Computer Science, Silesian University of Technology, Gliwice, Poland (2014)
18. Orlowski, P.: Applications of SVD-DFT decomposition. part 2: feedback stability analysis for time-varying systems. Pomiary Automatyka Kontrola **53**(2), 44–47 (2007)

An Algebraic Approach to Solving the Problem of Identification by the Use of Modulating Functions and Convolution Filter. Glass Conditioning Process

Witold Byrski and Michał Drapała(✉)

Department of Automatic Control and Robotics, AGH University of Science and Technology, Al. Mickiewicza 30, 30-059 Kraków, Poland
{wby,mdrapala}@agh.edu.pl

Abstract. The paper presents an application of the modulating functions method (MFM) for the identification of an industrial process of glass conditioning, which involves cooling the glass mass between its melting and further processing. The process is carried out in glass forehearths after leaving a melting unit. Precise temperature control is essential during the process, hence much effort was put into its modelling. Linear model of the process can be obtained on-line based on the input and output signal measurements. The model can be used for predicting molten glass temperature changes in the real forehearth. For the first time, the results obtained by the MFM are compared with the simulation results for the partial differential equation model and with the real process data.

Keywords: System identification · Modulating functions · Exact state observer · Glass forehearth

1 Introduction

The paper concerns the problem of glass conditioning process modeling. Molten glass from a glass furnace flows into a working end part of the installation and then into forehearths. Its temperature has to be decreased between the melting and forming processes. Conditioning takes place mainly in the forehearths. The forehearth is a ceramic channel typically divided into several zones. It is longer than wide. Desired temperature can be adjusted in each zone by changing gas-air mixture pressure for gas burners. Cooling valves and dampers are usually installed in the initial zones of the forehearth. The mentioned heating and cooling devices are usually controlled by PID controllers (separate controller for each zone). Ensuring the proper temperature decrease profile in the forehearth is very important for the glass homogeneity and the quality of final products [10].

The amount of glass flowing through the forehearth within 24 h is called pull rate and can be significantly changed depending on the current product type.

A. Bartoszewicz et al. (Eds.): PCC 2020, AISC 1196, pp. 80–91, 2020.
https://doi.org/10.1007/978-3-030-50936-1_7

The industry experience proves that obtaining the desired temperature set points for new operating conditions is often difficult and time consuming. Commonly used PID controllers do not have a predictive capability and utilizing them often results in temperature overshooting. Therefore, the problem of modelling the forehearth process dynamics has a great practical significance.

The most evident process description with the use of the partial differential equation (PDE) was presented in the papers [5–7]. Authors utilized the obtained model for developing new control strategies that were tested on the real glass forehearth and improved its control quality. PDE models synthesis for control purposes was also discussed in [4]. Special attention was given to the problem of using the implemented models in real-time control systems.

A strictly different approach involves on-line model identification based on the real process data, assuming minimal input information about the modelled system. An interesting example of this kind was described in [3], where an identification method based on Laguerre functions was used for the predictive controller synthesis. Similar problem, but for discrete models, was presented in [11].

The paper is organised as follows. The process dynamics is described in the next section with the use of the PDE. Theoretical basis of the modulating functions method (MFM) are presented in Sect. 3. Application of the method for the glass conditioning problem is described in Sect. 4. Results of the performed simulations are presented in Sect. 5. Section 6 gives the short conclusion of the paper.

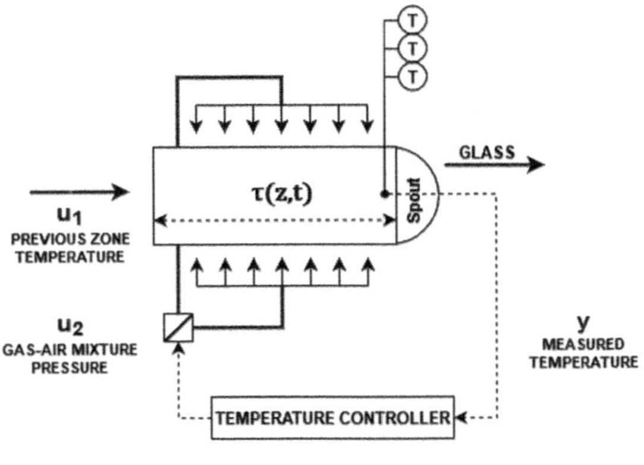

Fig. 1. Diagram of the typical forehearth zone control system.

2 The PDE Model of the Process

The analysed forehearth zone control system was schematically presented in Fig. 1. Measured glass temperature in the previous forehearth zone and gas-air mixture pressure can be treated as the system inputs, while glass temperature in the current zone is the system output. Assuming that the glass flow is laminar [9], the molten glass temperature $\tau(z,t)$, dependent on the position $z \in (0,l)$ and time t, can be described with the use of the PDE (1), like in the works [5–7]:

$$C\rho A\frac{\partial\tau(z,t)}{\partial t} + C\rho Av(t)\frac{\partial\tau(z,t)}{\partial z} + k_1\tau(z,t) = k_2u(z,t), \quad (1)$$

where: C is the heat capacity, ρ is the density, v is the velocity of the glass smelt, u is the control function, A is the cross-section area of the foreheart zone. The first two parameters are constant for the analysed range of temperatures. After simplification, Eq. (1) has the form:

$$\frac{\partial\tau(z,t)}{\partial t} + v(t)\frac{\partial\tau(z,t)}{\partial z} + K_1\tau(z,t) = K_2u(z,t). \quad (2)$$

The system output y corresponds to the simulated glass temperature at the end of the zone:

$$y(t) = \tau(l,t). \quad (3)$$

The parameter l denotes the zone length. The boundary condition refers to the impact between the temperature measured in the previous forehearth zone and the glass temperature in the current zone:

$$\tau(0,t) = u_1(t). \quad (4)$$

The parameters K_1 and K_2 were identified, using the Matlab *fmnisearch* method, based on the historical data. For this reason, the time interval with the constant glass pull rate and significant changes of the measured temperatures was selected. For the analysed forehearth, this parameter can be changed in the range 50–75 t/24h. The minimal velocity of the glass smelt is close to 2.22 mm/s. Assuming that this velocity $v(t)$ depends proportionally on the current pull rate, the parameter value was calculated as 2.4 mm/s for the pull rate equal to 54.08 t/24h. The process historical data, for which the model was obtained, are depicted in Sect. 5. The identified values of the model parameters are presented in Table 1.

Table 1. Identified PDE model parameters.

Parameter value	
K_1	K_2
$2.9024e-05$	0.0041269

3 Description of the MFM for the MISO Systems

The MFM was firstly introduced by Shinbrot [8]. In the form described in the paper, it enables to transform the differential equation for the multiple input single output (MISO) system with K inputs and a single output:

$$\sum_{i=0}^{n} a_i y^{(i)}(t) = \sum_{k=1}^{K} \sum_{j=0}^{m_k} b_{kj} u_k^{(j)}(t), \tag{5}$$

into the algebraic equation with the same parameters:

$$\sum_{i=0}^{n} a_i y_i(t) = \sum_{j=0}^{m_1} b_{1j} u_{1j}(t) + \ldots + \sum_{j=0}^{m_K} b_{Kj} u_{Kj}(t) + \epsilon(t). \tag{6}$$

The input and the output signals: $u_1, \ldots, u_K, y$ are given on the interval $[t_0, T_{ID}]$. Their derivatives: $u_1^{(j)}, \ldots, u_K^{(j)}, y^{(i)}$ are unknown. There are n output derivatives and m_k derivatives for the k-th input, where $m_k \leq n$. The parameters a and b are unknown and should be identified.

The new functions $u_{1j}, \ldots, u_{Kj}, y_i$ are created, in the interval $[t_0 + h, T_{ID}]$, by convoluting the original signals with the modulating function ϕ, meeting the following conditions:

- ϕ has the compact support of width h (closed and bounded),
- $\phi \in C^{n-1}[0, h]$,
- $\phi^{(i)}(0) = \phi^{(i)}(h) = 0$ for $i = 0, \ldots, n-1$,
- $y * \phi = 0 \Rightarrow y = 0$ on the interval $[t_0 + h, T_{ID}]$ - the condition of uniqueness.

These criteria are met by the Loeb and Cahen function used in the described experiments:

$$\phi(t) = t^N (h-t)^M, \tag{7}$$

where $N \neq M$.

The new functions, obtained after the convolution procedure, have the form:

$$y_i = y(t) * \phi^{(i)}(t) = \int_{t-h}^{t} y(\tau) \phi^{(i)}(t-\tau) d\tau. \tag{8}$$

The equation error value $\epsilon(t)$ in the Eq. (6) represents the square difference between its both sides and can be noted as (9):

$$\epsilon(t) = \mathbf{c}^T(t)\boldsymbol{\theta} = [y_0(t), \ldots, y_n(t), -u_{10}(t), \ldots,$$

$$-u_{1m_1}, \ldots, -u_{K0}, \ldots, -u_{Km_K}(t)] \begin{bmatrix} a_0 \\ \vdots \\ a_n \\ b_{10} \\ \vdots \\ b_{1m_1} \\ \vdots \\ b_{K0} \\ \vdots \\ b_{Km_K} \end{bmatrix}. \tag{9}$$

The squared norm of the equation error (10) is minimized during the identification procedure:

$$\min_{\theta} J^2 = \min \|\epsilon(t)\|^2_{L^2[t_0+h,T]} = \min \|\mathbf{c}(t)^T\ \boldsymbol{\theta}\|^2_{L^2}, \tag{10}$$

with the general form of the linear constraint (11) (to avoid a trivial solution):

$$\boldsymbol{\eta}^T\boldsymbol{\theta} = 1. \tag{11}$$

The norm (10) can also be written down as the inner product (12):

$$J^2 = \langle \mathbf{c}^T(t)\boldsymbol{\theta}, \mathbf{c}^T(t)\boldsymbol{\theta}\rangle_{L^2} = \boldsymbol{\theta}^T\langle \mathbf{c}(t), \mathbf{c}^T(t)\rangle\boldsymbol{\theta} = \boldsymbol{\theta}^T\boldsymbol{G}\boldsymbol{\theta}. \tag{12}$$

The square real Gram matrix $\boldsymbol{G}$ consists of the inner products:

$$\langle y_i, u_j\rangle = \int_{t_0+h}^{T_{ID}} y_i(\tau)u_j(\tau)d\tau, \tag{13}$$

and can be expressed as:

$$
\boldsymbol{G} = \left[\begin{array}{ccccccccccc}
\langle y_0, y_0\rangle & \cdots & \langle y_0, y_n\rangle & -\langle y_0, u_{10}\rangle & \cdots & -\langle y_0, u_{1m_1}\rangle & \cdots & -\langle y_0, u_{K0}\rangle & \cdots & -\langle y_0, u_{Km_K}\rangle \\
\vdots & \ddots & \vdots & \vdots & \ddots & \vdots & \cdots & \vdots & \ddots & \vdots \\
\langle y_n, y_0\rangle & \cdots & \langle y_n, y_n\rangle & -\langle y_n, u_{10}\rangle & \cdots & -\langle y_n, u_{1m_1}\rangle & \cdots & -\langle y_n, u_{K0}\rangle & \cdots & -\langle y_n, u_{Km_K}\rangle \\
-\langle u_{10}, y_0\rangle & \cdots & -\langle u_{10}, y_n\rangle & \langle u_{10}, u_{10}\rangle & \cdots & \langle u_{10}, u_{1m_1}\rangle & \cdots & \langle u_{10}, u_{K0}\rangle & \cdots & \langle u_{10}, u_{Km_K}\rangle \\
\vdots & \ddots & \vdots & \vdots & \ddots & \vdots & \cdots & \vdots & \ddots & \vdots \\
-\langle u_{1m_1}, y_0\rangle & \cdots & -\langle u_{1m_1}, y_n\rangle & \langle u_{1m_1}, u_{10}\rangle & \cdots & \langle u_{1m_1}, u_{1m_1}\rangle & \cdots & \langle u_{1m_1}, u_{K0}\rangle & \cdots & \langle u_{1m_1}, u_{Km_K}\rangle \\
\vdots & \ddots & \vdots & \vdots & \ddots & \vdots & \cdots & \vdots & \ddots & \vdots \\
-\langle u_{K0}, y_0\rangle & \cdots & -\langle u_{K0}, y_n\rangle & \langle u_{K0}, u_{10}\rangle & \cdots & \langle u_{K0}, u_{1m_1}\rangle & \cdots & \langle u_{K0}, u_{K0}\rangle & \cdots & \langle u_{K0}, u_{Km_K}\rangle \\
\vdots & \ddots & \vdots & \vdots & \ddots & \vdots & \cdots & \vdots & \ddots & \vdots \\
-\langle u_{Km_K}, y_0\rangle & \cdots & -\langle u_{Km_K}, y_n\rangle & \langle u_{Km_K}, u_{10}\rangle & \cdots & \langle u_{Km_K}, u_{1m_1}\rangle & \cdots & \langle u_{Km_K}, u_{K0}\rangle & \cdots & \langle u_{Km_K}, u_{Km_K}\rangle
\end{array}\right]. \tag{14}
$$

The optimal parameters vector $\boldsymbol{\theta}$ can be obtained with the use of the Lagrange multiplier technique:

$$L = \boldsymbol{\theta}^T \boldsymbol{G} \boldsymbol{\theta} + \lambda(\boldsymbol{\eta}^T \boldsymbol{\theta} - 1), \tag{15}$$

in the form:

$$\boldsymbol{\theta}^o = \frac{\boldsymbol{G}^{-1}\boldsymbol{\eta}}{\boldsymbol{\eta}^T \boldsymbol{G}^{-1}\boldsymbol{\eta}}. \tag{16}$$

Comprehensive description of the presented methodology, with particular emphasis on the optimal constraint vector $\boldsymbol{\eta}$ selection can be found in [1].

4 Application of the MFM for the MISO Glass Conditioning Process Modelling

The main idea of utilizing the MFM for the problem of forehearth process dynamics identification was to develop a method that could be easily implemented for real glass melting installations and used on-line. The measured signals were divided into equal time intervals of width T. The first identification interval should be possibly short, in order to reduce the time without the process model. It was assumed that the model parameters should be identified for the initial interval and then this model can be used in order to predict the system output for a short time period. Integral sate observer was utilized to obtain the initial condition for the system simulation in each interval, except the first one.

In the described problem, the MISO system consists of two single input single output (SISO) subsystems:

- SISO_1 with the input u_1 (temperature measured in the previous forehearth zone) and the simulated output y_1,
- SISO_2 with the input u_2 (adjusted gas-air mixture pressure) and the simulated output y_2.

In order to provide the possibility to simulate the system output with non-zero initial condition, the MISO system space-state representation (17) was introduced based on the identified vector of parameters $\boldsymbol{\theta}$:

$$\boldsymbol{A} = \underset{(K\cdot n\times K\cdot n)}{\begin{bmatrix} \begin{bmatrix} 0 \dots 0 & -\frac{a_{10}}{a_{1n}} \\ 1 \ddots \vdots & \vdots \\ \vdots \ddots 0 & -\frac{a_{1n-2}}{a_{1n}} \\ 0 \dots 1 & -\frac{a_{1n-1}}{a_{1n}} \end{bmatrix} & & 0 \\ & \ddots & \\ 0 & & \begin{bmatrix} 0 \dots 0 & -\frac{a_{k0}}{a_{kn}} \\ 1 \ddots \vdots & \vdots \\ \vdots \ddots 0 & -\frac{a_{kn-2}}{a_{kn}} \\ 0 \dots 1 & -\frac{a_{kn-1}}{a_{kn}} \end{bmatrix} \end{bmatrix}}, \quad \boldsymbol{B} = \underset{(K\cdot n\times K)}{\begin{bmatrix} \begin{bmatrix} \frac{b_{10}}{a_{1n}} \\ \vdots \\ \frac{b_{1n-1}}{a_{1n}} \end{bmatrix} & & 0 \\ & \ddots & \\ 0 & & \begin{bmatrix} \frac{b_{k0}}{a_{kn}} \\ \vdots \\ \frac{b_{kn-1}}{a_{kn}} \end{bmatrix} \end{bmatrix}}, \tag{17}$$

$$\boldsymbol{C} = \underset{(1\times K\cdot n)}{\big[\,[0 \dots 1] \dots [0 \dots 1]\,\big]}, \quad \boldsymbol{D} = \underset{(1\times K)}{[0 \dots 0]}.$$

4.1 Decomposition Idea for the MISO Model

The model parameters obtained after the first step of the identification procedure (MFM applied for MISO systems described in Sect. 3) have the same denominators, hence the obtained system (17) is not observable. For this reason, in the second step of the method, each SISO subsystem is re-identified.

In order to re-identify the k-th SISO subsystem, its virtual output $y_k(t)$ is calculated as the difference between the real system output $y(t)$ and the sum of simulated outputs for the other subsystems in the identification interval:

$$y_k = y(t) - \sum_{j=1,\ldots K: j \neq k} y_j(t). \tag{18}$$

For the MISO system with K inputs, all SISO model re-identification is performed K times, each time starting from the different initial k-th sub-model, to obtain the K MISO models. Another performance index, defined as the square error, is calculated for these models:

$$E = (y(t) - \sum_{k=1,\ldots K} y_k(t))^2. \tag{19}$$

The model with the least performance index (19) value is selected for further simulations. This methodology was inspired by the Gauss-Seidel method. Another example of its application for glass forehearths modeling can be found in the paper [2].

4.2 Application of the Exact State Observer

As previously mentioned, the exact state observer is used in order to obtain the initial condition at the start of the p-th time interval, except the first interval when zero initial condition is assumed (input and output signals should not change significantly near the t_0 point).

This methodology enables to consider the information about the real system output in the described procedure. In the case of significant differences between the simulated and the real system outputs, the simulation error is prevented from increasing in the subsequent intervals. In each of them, the new initial condition is obtained. This initial condition is close to the real system state, provided that the current process model is accurate and its state can be calculated precisely.

The formula for the observed state is given as:

$$x(t_p) = \int_{t_p-T}^{t_p} \boldsymbol{G_1}(T - t_p + t) y(t) dt + \int_{t_p-T}^{t_p} \boldsymbol{G_2}(T - t_p + t) u(t) dt, \tag{20}$$

where:

$$\boldsymbol{M_0} = \int_0^T e^{\boldsymbol{A}'\tau} \boldsymbol{C}' \boldsymbol{C} e^{\boldsymbol{A}\tau} d\tau, \tag{21}$$

$$\boldsymbol{G_1}(t) = e^{\boldsymbol{A}T} \boldsymbol{M_0}^{-1} e^{\boldsymbol{A}'t} \boldsymbol{C}', \tag{22}$$

$$\boldsymbol{G_2}(t) = e^{\boldsymbol{A}T} \boldsymbol{M_0}^{-1} \left[\int_0^t e^{\boldsymbol{A}'\tau} \boldsymbol{C}' \boldsymbol{C} e^{\boldsymbol{A}\tau} d\tau \right] e^{-\boldsymbol{A}t} \boldsymbol{B}, \tag{23}$$

and the successive time moments are:

$$t_p = t_0 + (p-1) \cdot T, \quad p = 2, 3, \ldots. \tag{24}$$

5 Experimental Results

The simulation experiment was performed to test the utility of the developed method. Obtained results were compared with the real system output and the results obtained for the PDE model. Parameters used in the implemented procedure are presented in Table 2.

Table 2. Identification method parameters values.

Parameter	Description	Value
N, M	Loeb and Cahen function parameters	3, 4
h	Filtering function support width	75
n	SISO submodels rank	2
η	Linear constraint vector	[1...1]
T	Interval width	1000

Historical data used in the experiment are shown in Fig. 2.

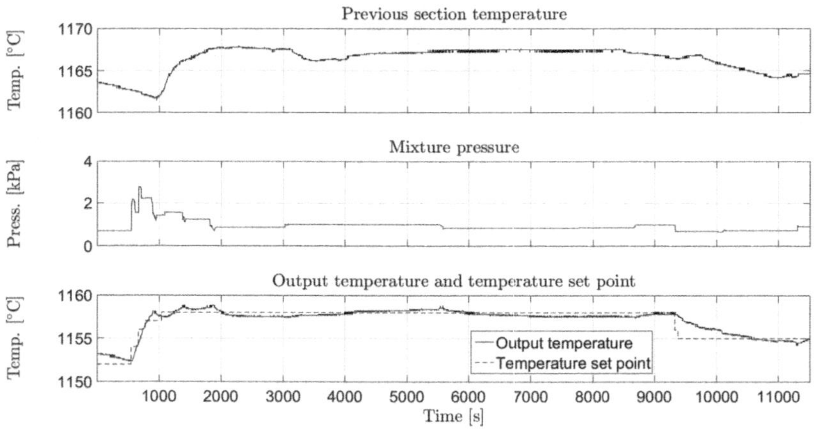

Fig. 2. Historical data used in the experiment.

It must be noted that the parameters of both models (linear and PDE) were identified for the same interval denoted as T_{ID}. In the case of the model obtained with the use of the MFM, the time delay was adopted as 387 s for the $SISO_1$ submodel. This value results from the time in which the molten glass flows through the analysed zone (of length 930 mm) with the assumed velocity $v(t) = 2.4$ mm/s.

The obtained simulation results are shown in Fig. 3 and the identified model parameters are presented in Table 3. Calculated mean square error values for both models are depicted in Table 4. In both cases, their values are less than 1 °C, which is sufficient enough for control applications.

Visible offset between the simulation results for the MFM model and the real system output may be noticed for most of the intervals. This effect results from the selection of the observation window length. In the implemented method, signals from the whole previous interval of length T were used for the state estimation. If shorter interval was selected, the estimated state value could be closer to the real state of the plant, however problems with the numerical stability of the observer algorithm could also occur.

Table 3. Identified linear model parameters values.

Parameter	$SISO_1$	$SISO_2$
a_{k0}	$7.7588e-4$	$1.1238e-4$
a_{k1}	$4.9661e-3$	$2.7289e-2$
a_{k2}	$9.9365e-1$	$9.7184e-1$
b_{k0}	$6.1259e-4$	$6.2307e-4$

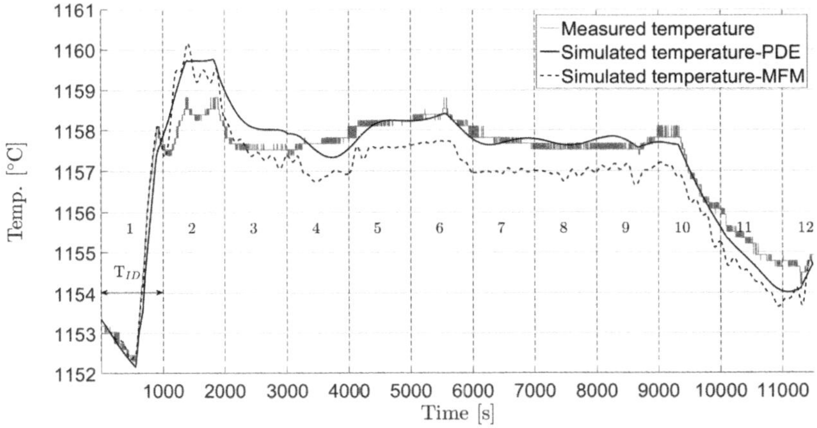

Fig. 3. Obtained simulation results.

It can be claimed that the simulation results of the developed method, for the system linearised near a selected operating point, can be comparable to those obtained with the PDE model, with much less computational effort. The calculation time on a PC (Intel i7-6700 2.6 GHz and 16 GB RAM) was 20.62 s in comparison with 820.82 s for the PDE model.

Table 4. Mean square error values for the identified models.

MSE values	
MFM	PDE
0.8275	0.2168

6 Conclusion

The described methods seems to be useful for the problem of glass conditioning process identification. The linear models can successfully replace the PDE model in the neighbourhood of the defined operating points. In that case, the simulation time necessary for predicting the system output is significantly shorter. This feature, in particular, enables the method to be applied on-line for industrial applications. Among the issues that can be analysed in the future is the tolerance of the identified model for pull-rate changes and application of feed-forward or predictive controllers based on the obtained MFM models.

Acknowledgement. This work was supported by the scientific research funds from the Polish Ministry of Science and Higher Education and AGH UST Agreement no 16.16.120.773 and was also conducted within the research of EC Grant H2020-MSCA-RISE-2018/824046.

References

1. Byrski, W., Byrski, J.: The role of parameter constraints in EE and OE methods for optimal identification of continuous LTI models. Int. J. Appl. Math. Comput. Sci. **22**(2), 379–388 (2012). https://doi.org/10.2478/v10006-012-0028-3
2. Byrski, W., Drapała, M., Byrski, J.: An adaptive method based on the modulating functions technique and exact state observers for modeling and simulation of a nonlinear MISO glass melting process. Int. J. Appl. Math. Comput. Sci. **29**(4), 739–757 (2019). https://doi.org/10.2478/amcs-2019-0055
3. Gough, B.P., Matovich, D.: Predictive-adaptive temperature control of molten glass. In: IEEE Industry Applications Society Dynamic Modeling Control Applications for Industry Workshop, Vancouver, BC, Canada, pp. 51–55 (1997)
4. Grega, W., Piłat, A., Tutaj, A.: Modelling of the glass melting process for real-time implementation. Int. J. Model. Optim. **5**(6), 366–373 (2015). https://doi.org/10.7763/IJMO.2015.V5.490
5. Kharitonov, A., Sawodny, O.: Modeling and control strategies for heating processes in the glass industry. In: Proceedings of the 2004 IEEE International Conference on Control Applications, Taipri, Taiwan, 2–4 September 2004, pp. 1026–1031. https://doi.org/10.1109/CCA.2004.1387506
6. Kharitonov, A., Henkel, S., Sawodny, O.: Two degree of freedom control for a glass feeder. In: Proceedings of the European Control Conference 2007, Kos, Greece, 2–5 July 2007, pp. 4079–4086. https://doi.org/10.23919/ECC.2007.7068450
7. Malchow, F., Sawodny, O.: Model based feedforward control of an industrial glass feeder. Control Eng. Pract. **20**(1), 62–68 (2012). https://doi.org/10.1016/j.conengprac.2011.09.004

8. Shinbrot, M.: On the analysis of linear and nonlinear systems. Trans. Am. Soc. Mech. Eng. J. Basic Eng. **79**, 522–547 (1957)
9. Sorg, H., Sims, R.: Method for conditioning and homogenizing a glass stream. U.S. Patent 5,573.569 (1996)
10. Sorg, H., Sims, R.: Method and apparatus for conditioning and homogenizing a glass stream. U.S. Patent 5,634.958 (1997)
11. Wang, Q., Chalaye, G., Thomas, G., Gilles, G.: Predictive control of a glass process. Control Eng. Pract. **5**(2), 167–173 (1997). https://doi.org/10.1016/S0967-0661(97)00223-2

Boundary Observers for Boundary Control Systems

Zbigniew Emirsajłow(✉)

West Pomeranian University of Technology, Al. Piastów 17, 70-310 Szczecin, Poland
zbigniew.emirsajlow@zut.edu.pl

Abstract. The paper studies the output observation problem for a class of linear distributed parameter systems where both the plant and the observer are modelled as boundary input/output systems. The presented results provide sufficient conditions for the plant and the observer which guarantee the asymptotic output observation to hold. These conditions have the form of three linear operator equations, with one of them being a homogeneous algebraic Sylvester equation, and provide some indication for the general observer design procedure. The results partly extend and generalize the classical approach to linear observer design. The one-dimensional heat equation is considered as an example.

Keywords: Boundary control systems · Boundary observers

1 Introduction

The overall idea of the general output observation problem is clearly expressed in Fig. 1.

Σ_P is the plant, Σ_O is the observer, $u(t)$ is the control input and $y(t)$ is the output of the plant, only $y(t)$ is a *measured signal available to the observer*, $z(t)$ is the *unmeasured output* of the plant (i.e., the *output to be observed*), $z_O(t)$ is the output of the observer. There are no disturbance signals. The *objective* of the interconnection is to track asymptotically the plant output $z(t)$ by the observer output $z_O(t)$, i.e., the *observation error* $e(t) = z_O(t) - z(t)$ has to decay asymptotically

$$\lim_{t \to \infty} e(t) = 0\,, \tag{1}$$

which we call the *output observation condition.*

In the finite-dimensional setup this observation problem and several related problems have been investigated since the sixties of the twentieth century and for the state of the art of the theory we recommend the monograph [10] and references cited therein. In this paper we are interested in an extension of the general output observation problem to linear infinite-dimensional systems understood as in monographs [1] or [11]. There have been several attempts to develop the infinite-dimensional observer theory and a short survey can be found in

A. Bartoszewicz et al. (Eds.): PCC 2020, AISC 1196, pp. 92–104, 2020.
https://doi.org/10.1007/978-3-030-50936-1_8

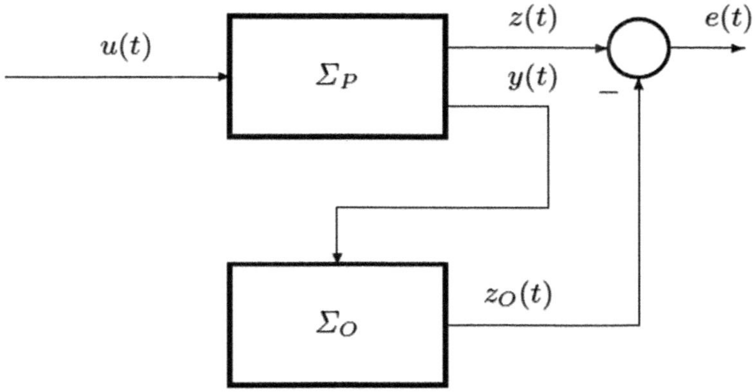

Fig. 1. Interconnection of the plant and the observer

[8]. Since the modelling of infinite-dimensional systems involves very sophisticated mathematical techniques most of researches restrict themselves to special cases or employ special methods. These cover state space models with bounded input and output operators (e.g. [4–6]), second order state space models (e.g. [2]) backstepping (e.g. [9]), sliding mode method (e.g. [3]). In the present paper we use boundary input/output linear systems to model both the plant and the observer. Good sources for mathematical fundamentals of such models can be found in [7,11]. The closest approach to the one developed in this paper can be found in [12].

2 Models of the Plant and the Observer

We model the infinite-dimensional plant and the infinite-dimensional observer using the abstract boundary input/output system description. Our presentation is based on approaches developed in [7,11].

2.1 Plant Model

In order to describe the plant model precisely, we need to introduce the following spaces (Hilbert spaces with appropriate scalar products $\langle\cdot,\cdot\rangle$ and induced norms $\|\cdot\| := \langle\cdot,\cdot\rangle^{1/2}$; identified with their duals):

X is the *plant state space*, $\langle\cdot\,,\,\cdot\rangle_X$, $\|\cdot\|_X$,
U is the *control input space* (input unavailable to the observer), $\langle\cdot\,,\,\cdot\rangle_U$, $\|\cdot\|_U$,
Y is the *measured output space* (output available to the observer), $\langle\cdot\,,\,\cdot\rangle_Y$, $\|\cdot\|_Y$,
Y_z is the *observed output space* (output to be observed), $\langle\cdot\,,\,\cdot\rangle_{Y_z}$, $\|\cdot\|_{Y_z}$.

The *plant dynamics* is described by the abstract boundary input/output system

$$\Sigma_P : \begin{cases} \dot{x}(t) = M_A x(t)\,, \quad x(0) = x_0 \\ K_A x(t) = u(t) \\ z(t) = C_1 x(t) \\ y(t) = C_2 x(t) \end{cases}, \tag{2}$$

where $(x(t))_{t \geq 0} \subset X$ is the *state*, $(u(t))_{t \geq 0} \subset U$ is the *control input* (unavailable to the observer), $(z(t))_{t \geq 0} \subset Y_z$ is the *output to be observed* and $(y(t))_{t \geq 0} \subset Y$ is the *measured output* (available to the observer). The control input signal $u(t)$ enters the system at the boundary and the outputs $y(t)$ and $z(t)$ leave the system at the boundary. Furthermore, we have to impose several assumptions on the operators involved in the model (2).

1. $(M_A, \mathscr{D}(M_A))$ is a linear, unbounded and closed operator on X, $Z_A = \mathscr{D}(M_A)$ is a Banach space equipped with the graph norm, so $M_A \in \mathscr{L}(Z_A, X)$ and is called the *plant formal system operator.*
2. $K_A \in \mathscr{L}(Z_A, U)$ is called the *plant input boundary operator.*
3. $C_1 \in \mathscr{L}(Z_A, Y_z)$ is called the *output to be observed boundary operator.*
4. $C_2 \in \mathscr{L}(Z_A, Y)$ is called the *measured output boundary operator.*
5. A is a linear, unbounded operator on X, called the *plant system operator*, defined as follows: $\mathscr{D}(A) := \ker K_A$, $Ah := M_A h$, $h \in \mathscr{D}(A)$, A generates a strongly continuous semigroup $(T(t))_{t \geq 0} \subset \mathscr{L}(X)$, $\rho(A) \subset \mathbb{C}$ denotes its resolvent set.
6. $X_1 := \mathscr{D}(A)$ is a Hilbert space, equipped with the scalar product $\langle \cdot, \cdot \rangle_{X_1} := \langle (\mu I - A) \cdot, (\mu I - A) \cdot \rangle_X$, where $\mu \in \rho(A)$. Part of A in X_1, denoted by A_1, generates a semigroup $(T_1(t))_{t \geq 0} \subset \mathscr{L}(X_1)$, a restriction of $(T(t))_{t \geq 0}$ to X_1.
7. X_{-1} is a Hilbert space, defined as the completion of X with respect to the scalar product $\langle \cdot, \cdot \rangle_{X_{-1}} := \langle (\mu I - A)^{-1} \cdot, (\mu I - A)^{-1} \cdot \rangle_X$. $(A_{-1}, \mathscr{D}(A_{-1}) = X)$ and $(T_{-1}(t))_{t \geq 0} \subset \mathscr{L}(X_{-1})$ are extensions of A and $(T(t))_{t \geq 0}$ to X_{-1}.

It follows that $X_1 \subset Z_A$ and it is a closed subspace of Z_A and the embeddings $Z_A \subset X \subset X_{-1}$ are dense and continuous.

With the plant model (2) we associate an abstract *plant boundary value problem* (PBVP). In order to make this precise we assume that for every $\mu \in \rho(A)$ and every $u \in U$ the following system of equations

$$(PBVP) : \begin{cases} (M_A - \mu I)x = 0 \\ K_A x = u \end{cases} \tag{3}$$

has a solution $x \in Z_A$, where $(M_A - \mu I) \in \mathscr{L}(Z_A, X)$ and $K_A \in \mathscr{L}(Z_A, U)$. Since $\mu \in \rho(A)$ the solution $x \in Z_A$ is unique and we can define the so-called *abstract Green map* $G_A^\mu : U \to Z_A$ such that

$$x := G_A^\mu u \quad \text{and} \quad G_A^\mu \in \mathscr{L}(U, Z_A)\,. \tag{4}$$

The first result concerning the state $(x(t))_{t \geq 0}$ and the outputs $(z(t))_{t \geq 0}$, $(y(t))_{t \geq 0}$ of the plant Σ_P, described by (2), has the following form.

Theorem 1. *If $u(\cdot) \in C^2([0,\infty);U)$, $u(0)=0$ and $x_0 \in X_1$, then there exists a unique function*

$$x(\cdot) \in C([0,\infty);Z_A) \cap C^1([0,\infty);X)\,, \tag{5}$$

satisfying the first two equations in (2) for every $t \in [0,\infty)$. This function is explicitly given in form

$$x(t) = T(t)x_0 + \int_0^t T_{-1}(t-r)(\mu I - A_{-1})G_A^{\mu}u(r)\mathrm{d}r\,, \quad t \in [0,\infty)\,. \tag{6}$$

Consequently, the output functions, defined by two last equations of (2), satisfy

$$z(\cdot) \in C([0,\infty);Y_z)\,, \quad y(\cdot) \in C([0,\infty);Y)\,. \tag{7}$$

For the proof of this theorem see, e.g., [7].

Remark 1. If we now define the operator

$$B := (\mu I - A_{-1})G_A^{\mu} \in \mathscr{L}(U, X_{-1})\,, \tag{8}$$

then (6) shows that the plant state trajectory $(x(t))_{t\geq 0} \subset Z_A \subset X$ can be interpreted as a mild solution of the differential equation (understood in X_{-1})

$$\dot{x}(t) = A_{-1}x(t) + Bu(t)\,, \quad x(0) = x_0\,, \tag{9}$$

with the outputs

$$\begin{aligned} z(t) &= C_1 x(t)\,,\\ y(t) &= C_2 x(t)\,. \end{aligned} \tag{10}$$

2.2 Observer Model

For the observer description we also need to introduce some Hilbert spaces (with appropriate scalar products and induced norms and assume they are identified with their duals):

V is the *observer state space*, $\langle\cdot\,,\,\cdot\rangle_V$, $\|\cdot\|_V$,
Y_z is the *observer output space* (the same as the plant observed output space).

The *observer dynamics* is described by the abstract boundary input/output system

$$\Sigma_O : \begin{cases} \dot{v}(t) = M_E v(t)\,, \quad v(0) = v_0 \\ K_E v(t) = w(t) \\ z_O(t) = Gv(t) \end{cases}, \tag{11}$$

where $(v(t))_{t\geq 0} \subset V$ is the *observer state*, $(w(t))_{t\geq 0} \subset Y$ is the *observer input* and $(z_O(t))_{t\geq 0} \subset Y_z$ is the *observer output.* The signal $w(t)$ enters the system at the boundary and the output signal $z_O(t)$ leaves the system at the boundary. Recall that, according to the interconnection shown in Fig. 1, $(w(t))_{t\geq 0} \subset Y$ is going to be the plant measured output $(y(t))_{t\geq 0} \subset Y$, but at this moment it is just an independent input signal.

As in the case of the plant, we have to impose several assumptions on the operators involved in the model (11).

1. $(M_E, \mathscr{D}(M_E))$ linear, unbounded and closed operator on V, $Z_E = \mathscr{D}(M_E)$ is a Banach space equipped with the graph norm, so $M_E \in \mathscr{L}(Z_E, V)$ and is called the *observer formal system operator.*
2. $K_E \in \mathscr{L}(Z_E, Y)$ is called the *observer input boundary operator.*
3. $G \in \mathscr{L}(Z_E, Y_z)$ is called the *observer output boundary operator.*
4. E is a linear, unbounded operator on V, called the *observer system operator,* defined as follows: $\mathscr{D}(E) := \ker K_E$, $Eh := M_E h$, $h \in \mathscr{D}(E)$, E generates a strongly continuous *exponentially stable* semigroup $(S(t))_{t\geq 0} \subset \mathscr{L}(V)$, i.e. with the growth bound $\omega_0(S) < 0$, $\rho(E) \subset \mathbb{C}$ denotes its resolvent set.
5. $V_1 := \mathscr{D}(E)$ is a Hilbert space, equipped with the scalar product $\langle \cdot, \cdot \rangle_{V_1} := \langle (\gamma I - E)\cdot, (\gamma I - E)\cdot \rangle_V$, where $\gamma \in \rho(E)$. Part of E in V_1, denoted by E_1, generates a semigroup $(S_1(t))_{t\geq 0} \subset \mathscr{L}(V_1)$, a restriction of $(S(t))_{t\geq 0}$ to V_1.
6. V_{-1} is a Hilbert space, defined as the completion of V with respect to the scalar product $\langle \cdot, \cdot \rangle_{V_{-1}} := \langle (\gamma I - E)^{-1}\cdot, (\gamma I - E)^{-1}\cdot \rangle_V$. $(E_{-1}, \mathscr{D}(E_{-1}) = V)$ and $(S_{-1}(t))_{t\geq 0} \subset \mathscr{L}(V_{-1})$ are extensions of E and $(S(t))_{t\geq 0}$ to V_{-1}.

It follows that $V_1 \subset Z_E$ and it is a closed subspace of Z_E and the embeddings $Z_E \subset V \subset V_{-1}$ are dense and continuous.

With the observer model (11) we associate an abstract *observer boundary value problem* (OBVP). In order to make this precise we assume that for every $\gamma \in \rho(E)$ and every $w \in Y$ the following system of equations

$$(OBVP): \begin{cases} (M_E - \gamma I)v = 0 \\ \qquad\quad K_E v = w \end{cases} \tag{12}$$

has a solution $v \in Z_E$, where $(M_E - \gamma I) \in \mathscr{L}(Z_E, V)$ and $K_E \in \mathscr{L}(Z_E, Y)$. Since $\gamma \in \rho(E)$ the solution $v \in Z_E$ is unique and we can define the *abstract Green map* $G_E^\gamma : Y \to Z_E$ such that

$$v := G_E^\gamma w \quad \text{and} \quad G_E^\gamma \in \mathscr{L}(Y, Z_E). \tag{13}$$

The first result concerning the state $(v(t))_{t\geq 0}$ and the output $(z_O(t))_{t\geq 0}$ of the observer Σ_O, described by (11), has the following form (the same as for the plant).

Theorem 2. *If* $w(\cdot) \in C^2([0,\infty); Y)$, $w(0) = 0$ *and* $v_0 \in V_1$, *then there exists a unique function*

$$v(\cdot) \in C([0,\infty); Z_E) \cap C^1([0,\infty); V), \tag{14}$$

satisfying the first two equations in (11) *for every* $t \in [0,\infty)$. *This function is explicitly given in the form*

$$v(t) = S(t)v_0 + \int_0^t S_{-1}(t-r)(\gamma I - E_{-1})G_E^\gamma w(r)\mathrm{d}r, \quad t \in [0,\infty). \tag{15}$$

Consequently, the output function, defined by the last equation of (11), *satisfies*

$$z_O(\cdot) \in C([0,\infty); Y_z). \tag{16}$$

Remark 2. As for the plant also for the observer we can define the operator

$$F := (\gamma I - E_{-1})G_E^{\gamma} \in \mathscr{L}(Y, V_{-1}), \tag{17}$$

and then it follows from (15) that the observer state trajectory $(v(t))_{t\geq 0} \subset Z_E \subset V$ can be interpreted as a mild solution of the differential equation (understood in V_{-1})

$$\dot{v}(t) = E_{-1}v(t) + Fw(t), \quad v(0) = v_0, \tag{18}$$

with the output

$$z(t) = Gv(t). \tag{19}$$

2.3 Interconnection

After aggregation of Σ_P and Σ_O we obtain the following overall boundary input/output model which can be written in the matrix form

$$\Sigma_{P+O}: \begin{cases} \begin{bmatrix} \dot{v}(t) \\ \dot{x}(t) \end{bmatrix} = \begin{bmatrix} M_E & 0 \\ 0 & M_A \end{bmatrix} \begin{bmatrix} v(t) \\ x(t) \end{bmatrix}, \quad \begin{bmatrix} v(0) \\ x(0) \end{bmatrix} = \begin{bmatrix} v_0 \\ x_0 \end{bmatrix} \\ \begin{bmatrix} K_E & 0 \\ 0 & K_A \end{bmatrix} \begin{bmatrix} v(t) \\ x(t) \end{bmatrix} = \begin{bmatrix} w(t) \\ u(t) \end{bmatrix} \\ \begin{bmatrix} z_O(t) \\ z(t) \\ y(t) \end{bmatrix} = \begin{bmatrix} G & 0 \\ 0 & C_1 \\ 0 & C_2 \end{bmatrix} \begin{bmatrix} v(t) \\ x(t) \end{bmatrix} \end{cases}. \tag{20}$$

Assuming that $\lambda \in \rho(A) \cap \rho(E)$ and combining Theorems 1 and 2 we get immediately the following result.

Corollary 1. *If* $\begin{bmatrix} w(\cdot) \\ u(\cdot) \end{bmatrix} \in C^2([0,\infty); Y \times U)$, $v(0) = 0$, $u(0) = 0$ *and* $\begin{bmatrix} v_0 \\ x_0 \end{bmatrix} \in V_1 \times X_1$, *then there exists a unique function*

$$\begin{bmatrix} v(\cdot) \\ x(\cdot) \end{bmatrix} \in C([0,\infty); Z_E \times Z_A) \cap C^1([0,\infty); V \times X), \tag{21}$$

satisfying the first two equations in (20) *and explicitly given in the form*

$$\begin{bmatrix} v(t) \\ x(t) \end{bmatrix} = \begin{bmatrix} S(t) & 0 \\ 0 & T(t) \end{bmatrix} \begin{bmatrix} v_0 \\ x_0 \end{bmatrix} + \int_0^t \begin{bmatrix} S_{-1}(t-r) & 0 \\ 0 & T_{-1}(t-r) \end{bmatrix} \begin{bmatrix} F & 0 \\ 0 & B \end{bmatrix} \begin{bmatrix} w(r) \\ u(r) \end{bmatrix} \mathrm{d}r. \tag{22}$$

Consequently, the output function, defined by the last equation in (20), *satisfies*

$$\begin{bmatrix} z_O(\cdot) \\ z(\cdot) \\ y(\cdot) \end{bmatrix} \in C([0,\infty); Y_z \times Y_z \times Y). \tag{23}$$

Remark 3. It is also clear that the state trajectory $\left(\begin{bmatrix} v(t) \\ x(t) \end{bmatrix}\right)_{t\geq 0} \subset Z_E \times Z_A \subset V \times X$ (of the overall system) can be interpreted as a mild solution of the differential equation (understood in $V_{-1} \times X_{-1}$)

$$\begin{bmatrix} \dot{v}(t) \\ \dot{x}(t) \end{bmatrix} = \begin{bmatrix} E_{-1} & 0 \\ 0 & A_{-1} \end{bmatrix} \begin{bmatrix} v(t) \\ x(t) \end{bmatrix} + \begin{bmatrix} F & 0 \\ 0 & B \end{bmatrix} \begin{bmatrix} w(t) \\ u(t) \end{bmatrix}, \qquad \begin{bmatrix} v(0) \\ x(0) \end{bmatrix} = \begin{bmatrix} v_0 \\ x_0 \end{bmatrix}, \tag{24}$$

with the output

$$\begin{bmatrix} z_O(t) \\ z(t) \\ y(t) \end{bmatrix} = \begin{bmatrix} G & 0 \\ 0 & C_1 \\ 0 & C_2 \end{bmatrix} \begin{bmatrix} v(t) \\ x(t) \end{bmatrix}. \tag{25}$$

Since Σ_O is an output observer of the plant Σ_P, then according to Fig. 1 we have to assume that

$$w(t) = y(t) = C_2 x(t) \tag{26}$$

and for the observation error we have

$$e(t) = z_O(t) - z(t) = Gv(t) - C_1 x(t)\,. \tag{27}$$

Taking the Eqs. (26) and (27) into account and then using (20) we obtain a description of the interconnection in Fig. 1 which has the form of the following abstract boundary input/output system (which will be denoted by Σ)

$$\Sigma : \begin{cases} \begin{bmatrix} \dot{v}(t) \\ \dot{x}(t) \end{bmatrix} = \begin{bmatrix} M_E & 0 \\ 0 & M_A \end{bmatrix} \begin{bmatrix} v(t) \\ x(t) \end{bmatrix}, \quad \begin{bmatrix} v(0) \\ x(0) \end{bmatrix} = \begin{bmatrix} v_0 \\ x_0 \end{bmatrix} \\ \begin{bmatrix} K_E & -C_2 \\ 0 & K_A \end{bmatrix} \begin{bmatrix} v(t) \\ x(t) \end{bmatrix} = \begin{bmatrix} 0 \\ u(t) \end{bmatrix} \\ e(t) = \begin{bmatrix} G & -C_1 \end{bmatrix} \begin{bmatrix} v(t) \\ x(t) \end{bmatrix} \end{cases}. \tag{28}$$

For Σ we consider the formal system operator $(\mathcal{M}, \mathscr{D}(\mathcal{M}))$ on $V \times X$ defined by

$$\mathcal{M} := \begin{bmatrix} M_E & 0 \\ 0 & M_A \end{bmatrix} \in \mathscr{L}(Z_E \times Z_A, V \times X)\,, \quad \mathscr{D}(\mathcal{M}) := Z_E \times Z_A\,, \tag{29}$$

and the boundary operator

$$\mathcal{K} := \begin{bmatrix} K_E & -C_2 \\ 0 & K_A \end{bmatrix} \in \mathscr{L}(Z_E \times Z_A, Y \times U)\,, \tag{30}$$

with the kernel

$$\ker \mathcal{K} = \{ \begin{bmatrix} v \\ x \end{bmatrix} \in Z_E \times Z_A : K_A x = 0 \text{ and } K_E v - C_2 x = 0 \}\,. \tag{31}$$

Next we define the *system operator* $(\mathcal{A}, \mathscr{D}(\mathcal{A}))$ as follows

$$\mathcal{A} \begin{bmatrix} v \\ x \end{bmatrix} := \mathcal{M} \begin{bmatrix} v \\ x \end{bmatrix}, \quad \begin{bmatrix} v \\ x \end{bmatrix} \in \mathscr{D}(\mathcal{A}) := \ker \mathcal{K}\,, \tag{32}$$

Since $\mathscr{D}(\mathcal{A}) = \ker \mathcal{K}$ is a closed subspace of $Z_E \times Z_A$, then the output operator $\begin{bmatrix} G & -C_1 \end{bmatrix} \in \mathscr{L}(Z_E \times Z_A; Y_z)$ restricts to a bounded operator from $\ker \mathcal{K}$ to Y_z.

On the operators $\mathcal{M}$, $\mathcal{K}$ and $\mathcal{A}$ we *impose* the following assumptions:

1. $\mathcal{A}$ generates a strongly continuous semigroup $(\mathcal{T}(t))_{t \geq 0} \subset \mathscr{L}(V \times X)$.
2. For every $\lambda \in \rho(\mathcal{A})$ and $\begin{bmatrix} w \\ u \end{bmatrix} \in Y \times U$ the abstract boundary value problem

$$(ABVP): \begin{cases} (\mathcal{M} - \lambda I) \begin{bmatrix} v \\ x \end{bmatrix} = 0 \\ \mathcal{K} \begin{bmatrix} v \\ x \end{bmatrix} = \begin{bmatrix} w \\ u \end{bmatrix} \end{cases} \tag{33}$$

has a solution $\begin{bmatrix} v \\ x \end{bmatrix} \in Z_E \times Z_A$.

Since $\lambda \in \rho(\mathcal{A})$ the solution $\begin{bmatrix} v \\ x \end{bmatrix} \in Z_E \times Z_A$ is unique and we can define the *abstract Green map* $\mathcal{G}^\lambda : Y \times U \to Z_E \times Z_A$ such that

$$\begin{bmatrix} v \\ x \end{bmatrix} := \mathcal{G}^\lambda \begin{bmatrix} w \\ u \end{bmatrix} \quad \text{and} \quad \mathcal{G}^\lambda \in \mathscr{L}(Y \times U, Z_E \times Z_A). \tag{34}$$

For the operator $(\mathcal{A}, \mathscr{D}(\mathcal{A}))$, generating the semigroup $(\mathcal{T}(t))_{t \geq 0} \subset \mathscr{L}(V \times X)$, we define (in an obvious way) the spaces $\mathcal{X}_1$, $\mathcal{X}_{-1}$, operators $\mathcal{A}_1$, $\mathcal{A}_{-1}$ and semigroups $(\mathcal{T}_1(t))_{t \geq 0} \subset \mathscr{L}(\mathcal{X}_1)$, $(\mathcal{T}_{-1}(t))_{t \geq 0} \subset \mathscr{L}(\mathcal{X}_{-1})$.

As previously we can state analogous results concerning the state trajectory $\left(\begin{bmatrix} v(t) \\ x(t) \end{bmatrix}\right)_{t \geq 0}$ and the observation error $(e(t))_{t \geq 0}$ of Σ, described by (28).

Theorem 3. *If* $u(\cdot) \in C^2([0, \infty); U)$, $u(0) = 0$ *and* $\begin{bmatrix} v_0 \\ x_0 \end{bmatrix} \in \mathcal{X}_1$, *then there exists a unique function*

$$\begin{bmatrix} v(\cdot) \\ x(\cdot) \end{bmatrix} \in C([0, \infty); Z_E \times Z_A) \cap C^1([0, \infty); V \times X), \tag{35}$$

satisfying the first two equations in (28) *for every* $t \in [0, \infty)$. *This function is explicitly given in the form*

$$\begin{bmatrix} v(t) \\ x(t) \end{bmatrix} = \mathcal{T}(t) \begin{bmatrix} v_0 \\ x_0 \end{bmatrix} + \int_0^t \mathcal{T}_{-1}(t-r)(\lambda I - \mathcal{A}_{-1}) \mathcal{G}^\lambda \begin{bmatrix} 0 \\ u(r) \end{bmatrix} \mathrm{d}r, \quad t \in [0, \infty). \tag{36}$$

Consequently, the observation error, defined by the last equation of (28), *satisfies*

$$e(\cdot) \in C([0, \infty); Y_z). \tag{37}$$

Remark 4. As before we can define the operator

$$\mathcal{B} := (\lambda I - \mathcal{A}_{-1})\mathcal{G}^{\lambda} \in \mathscr{L}(Y \times U, \mathcal{X}_{-1}), \tag{38}$$

and then it follows from (36) that the state trajectory $(\begin{bmatrix} v(t) \\ x(t) \end{bmatrix})_{t \geq 0} \subset Z_E \times Z_A \subset V \times X$ of Σ can be interpreted as a mild solution of the differential equation (understood in $\mathcal{X}_{-1}$)

$$\begin{bmatrix} \dot{v}(t) \\ \dot{x}(t) \end{bmatrix} = \mathcal{A}_{-1} \begin{bmatrix} v(t) \\ x(t) \end{bmatrix} + \mathcal{B} \begin{bmatrix} 0 \\ u(t) \end{bmatrix}, \quad \begin{bmatrix} v(0) \\ x(0) \end{bmatrix} = \begin{bmatrix} v_0 \\ x_0 \end{bmatrix}, \tag{39}$$

with the observation error

$$e(t) = \begin{bmatrix} G & -C_1 \end{bmatrix} \begin{bmatrix} v(t) \\ x(t) \end{bmatrix}. \tag{40}$$

3 Main Result

We are now ready to derive sufficient conditions which guarantee that the output observation condition (1) holds for the system Σ.

Theorem 4. *Given the system Σ, described by* (28)*, and assumptions of Theorem 3 hold. If the operators (M_E, K_E, G) of the observer Σ_O are such that the following* observer equation

$$\begin{cases} M_E \Pi = \Pi M_A \\ K_E \Pi = C_2 \\ G\Pi = C_1 \end{cases} \tag{41}$$

has a solution $\Pi \in \mathscr{L}(X, V) \cap \mathscr{L}(Z_A, Z_E)$, then for $x_0 \in X_1$, $v_0 \in V_1$ and $v_0 - \Pi x_0 \in V_1$ the observation error satisfies the condition

$$\|e(t)\|_{Y_z} \leq C e^{\omega t}, \quad \omega > \omega_0(S). \tag{42}$$

Proof. For a start we transform the state trajectory $(\begin{bmatrix} v(t) \\ x(t) \end{bmatrix})_{t \geq 0} \subset Z_E \times Z_A \subset V \times X$ of Σ as follows

$$\begin{bmatrix} p(t) \\ x(t) \end{bmatrix} = \begin{bmatrix} I & -\Pi \\ 0 & I \end{bmatrix} \begin{bmatrix} v(t) \\ x(t) \end{bmatrix}, \tag{43}$$

for some operator $\Pi \in \mathscr{L}(X, V) \cap \mathscr{L}(Z_A, Z_E)$. We have that $\begin{bmatrix} I & -\Pi \\ 0 & I \end{bmatrix} \in \mathscr{L}(V \times X) \cap \mathscr{L}(Z_E \times Z_A)$ and it is easy to check that

$$\begin{bmatrix} I & -\Pi \\ 0 & I \end{bmatrix}^{-1} = \begin{bmatrix} I & \Pi \\ 0 & I \end{bmatrix} \in \mathscr{L}(V \times X) \cap \mathscr{L}(Z_E \times Z_A). \tag{44}$$

Substituting (43) into (28) we arrive at an equivalent description of Σ

$$\Sigma : \begin{cases} \begin{bmatrix} \dot{p}(t) \\ \dot{x}(t) \end{bmatrix} = \begin{bmatrix} M_E & M_E \Pi - \Pi M_A \\ 0 & M_A \end{bmatrix} \begin{bmatrix} p(t) \\ x(t) \end{bmatrix}, \quad \begin{bmatrix} p(0) \\ x(0) \end{bmatrix} = \begin{bmatrix} v_0 - \Pi x_0 \\ x_0 \end{bmatrix} \\ \begin{bmatrix} K_E & K_E \Pi - C_2 \\ 0 & K_A \end{bmatrix} \begin{bmatrix} p(t) \\ x(t) \end{bmatrix} = \begin{bmatrix} 0 \\ u(t) \end{bmatrix} \\ e(t) = \begin{bmatrix} G & G\Pi - C_1 \end{bmatrix} \begin{bmatrix} p(t) \\ x(t) \end{bmatrix} \end{cases} . \tag{45}$$

If we now assume that $\Pi \in \mathscr{L}(X, V) \cap \mathscr{L}(Z_A, Z_E)$ satisfies the following system of equations

$$\begin{cases} M_E \Pi = \Pi M_A \\ K_E \Pi = C_2 \\ G\Pi = C_1 \end{cases} , \tag{46}$$

then (45) becomes

$$\Sigma : \begin{cases} \begin{bmatrix} \dot{p}(t) \\ \dot{x}(t) \end{bmatrix} = \begin{bmatrix} M_E & 0 \\ 0 & M_A \end{bmatrix} \begin{bmatrix} p(t) \\ x(t) \end{bmatrix}, \quad \begin{bmatrix} p(0) \\ x(0) \end{bmatrix} = \begin{bmatrix} v_0 - \Pi x_0 \\ x_0 \end{bmatrix} \\ \begin{bmatrix} K_E & 0 \\ 0 & K_A \end{bmatrix} \begin{bmatrix} p(t) \\ x(t) \end{bmatrix} = \begin{bmatrix} 0 \\ u(t) \end{bmatrix} \\ e(t) = \begin{bmatrix} G & 0 \end{bmatrix} \begin{bmatrix} p(t) \\ x(t) \end{bmatrix} \end{cases} , \tag{47}$$

and hence we obtain

$$\begin{cases} \dot{p}(t) = M_E p(t), \quad p(0) = v_0 - \Pi x_0 \\ K_E p(t) = 0 \\ e(t) = Gp(t) \end{cases} , \tag{48}$$

which for $p(0) = v_0 - \Pi x_0 \in V_1$ simplifies to the form

$$\begin{cases} \dot{p}(t) = Ep(t), \quad p(0) = v_0 - \Pi x_0 \\ e(t) = Gp(t) \end{cases} . \tag{49}$$

Thus

$$e(t) = Gp(t) = GS_1(t)(v_0 - \Pi x_0), \quad v_0 - \Pi x_0 \in V_1 \subset Z_E , \tag{50}$$

and since the operator E generates an exponentially stable semigroup $(S(t))_{t\geq 0} \subset \mathscr{L}(V)$ with the growth bound $\omega_0(S) < 0$, then also the semigroup $(S_1(t))_{t\geq 0} \subset \mathscr{L}(V_1)$ is exponentially stable with the growth bound $\omega_0(S_1) = \omega_0(S)$ and we have (42). □

4 Example

The observer equation provides a simple design procedure for a stable (full order) observer under the assumption that the pair (M_A, C_2) determines a generator of an exponentially stable semigroup.

1. Check that $(M_A|_{\ker C_2}, \ker C_2)$ generates an asymptotically stable semigroup.
2. Take an arbitrary operator $\Pi \in \mathscr{L}(X, V)$ which is invertible.
3. Compute the observer parameters from (41) as follows

$$\begin{aligned} M_E &= \Pi M_A \Pi^{-1}\,, \quad \mathscr{D}(M_E) := \Pi \mathscr{D}(M_A)\,, \\ K_E &= C_2 \Pi^{-1}\,, \\ G &= C_1 \Pi^{-1}\,. \end{aligned} \tag{51}$$

We assume $\xi \in [0,1]$, $t \geq 0$ and, as an example, consider the plant

$$\Sigma_P : \begin{cases} \dot{x}(\xi,t) = x''(\xi,t)\,, \quad x(\xi,0) = x_0(\xi)\,, \\ x'(0,t) = u_0(t)\,, \quad x'(1,t) = u_1(t)\,, \\ z(t) = x(a,t)\,, \\ y_0(t) = x(0,t)\,, \quad y_1(t) = x(1,t)\,, \end{cases} \tag{52}$$

where $(x(\cdot,t))_{t\geq 0} \subset L_2(0,1)$ is the state, $x_0(\xi)$ is an unknown initial state, $(\begin{bmatrix} u_0(t) \\ u_1(t) \end{bmatrix})_{t\geq 0} \subset \mathbb{R}^2$ is an unknown input, $a \in (0,1)$ is fixed and $(z(t))_{t\geq 0} \subset \mathbb{R}$ is an unmeasured output, $(\begin{bmatrix} y_0(t) \\ y_1(t) \end{bmatrix})_{t\geq 0} \subset \mathbb{R}^2$ is the measured output. This model may describe a temperature distribution $x(\xi,t)$ in a finite rod with an unknown initial distribution $x_0(\xi)$, unknown heat fluxes $(\begin{bmatrix} u_0(t) \\ u_1(t) \end{bmatrix})_{t\geq 0} \subset \mathbb{R}^2$ at both ends and with temperatures $(\begin{bmatrix} y_0(t) \\ y_1(t) \end{bmatrix})_{t\geq 0} \subset \mathbb{R}^2$ measured at these ends. For this plant we construct a stable observer which uses the measured temperatures and produces a signal $z_O(t)$ tracking asymptotically the temperature $z(t)$ at the fixed point $\xi = a$ of the rod.

On defining the spaces $X = L_2(0,1)$, $U = Y = \mathbb{R}^2$, $Y_z = \mathbb{R}$ and the operators

$$\begin{aligned} M_A &= (\;)''\,, \quad \mathscr{D}(M_A) = H_2^2(0,1) \quad (L_2(0,1) = H_2^0(0,1))\,, \\ K_A &= \left[(\;)'_{\xi=0} \; (\;)'_{\xi=1} \right]^T \quad \text{(Neumann b.c.)}, \\ C_1 &= (\;)_{\xi=a} \quad \text{(point evaluation)}\,, \\ C_2 &= \left[(\;)_{\xi=0} \; (\;)_{\xi=1} \right]^T \quad \text{(Dirichlet b.c.)}\,, \\ A &= (\;)''\,, \quad \mathscr{D}(A) = \{h(\xi) : h \in H_2^2(0,1) \text{ and } h'(0) = 0\,,\; h'(1) = 0\}\,, \end{aligned} \tag{53}$$

we obtain that $(A, \mathscr{D}(A))$ generates a strongly continuous semigroup on X with $\omega_0(T) = 0$ and the system (52) fits into our framework. Since the operator $(A^C, \mathscr{D}(A^C)) := (M_A|_{\ker C_2}, \ker C_2)$, given by

$$A^C = (\;)''\,, \quad \mathscr{D}(A^C) = \{h(\xi) : h \in H_2^2(0,1) \text{ and } h(0) = 0\,,\; h(1) = 0\}\,, \tag{54}$$

generates an exponentially stable semigroup on X with $\omega_0(T^C) = -\pi^2$, then we can assume $V = X$, $\Pi = I$ (identity) and, according to (51), obtain an output observer $\Sigma_O = (M_E, K_E, G) = (M_A, C_2, C_1)$ for the plant (52), i.e.

$$\Sigma_O : \begin{cases} \dot{v}(\xi,t) = v''(\xi,t)\,, \quad v(\xi,0) = v_0(\xi)\,, \\ v(0,t) = y_0(t)\,, \quad v(1,t) = y_1(t)\,, \\ z_O(t) = v(a,t)\,. \end{cases} \tag{55}$$

It works for $u_0(\cdot), u_1(\cdot) \in C^2([0,\infty;\mathbb{R})$ with $u_0(0) = u_1(0) = 0$, $x_0(\cdot) \in \mathscr{D}(A)$ and $v_0(\cdot) \in \mathscr{D}(A^C)$.

5 Final Remarks

In the paper we have derived sufficient conditions for the infinite-dimensional plant and the infinite-dimensional observer to satisfy the output observation condition. Both the plant and the observer were modelled as abstract boundary input/output systems. In principle, we have dealt with strong solutions to the differential equations describing the systems. Where appropriate we have also pointed out how to generalize the approach to cover mild solutions. Such development would require to include some notions like admissibility and well-posedness. The observer equation (41) is not yet a ready to use recipe for the observer design however it can be regarded as a good indication.

References

1. Curtain, R., Zwart, H.: An Introduction to Infinite-Dimensional Linear Systems Theory. Springer, New York (1995)
2. Demetriou, M.: Natural second-order observers for second-order distributed parameter systems. Syst. Control Lett. **51**, 225–234 (2004)
3. Drakunov, S.V., Reyhanoglu, M.: Hierarchical sliding mode observers for distributed parameter systems. J. Vib. Control **17**, 1441–1453 (2011)
4. Emirsajłow, Z.: Infinite-dimensional Sylvester equations: basic theory and applications to observer design. Int. J. Appl. Math. Comput. Sci. **22**(2), 245–257 (2012)
5. Emirsajłow, Z.: Remarks on functional observers for distributed parameter systems. In: Proceedings of the 23rd International Conference on Methods and Models in Automation and Robotics, Miedzyzdroje, Poland, 27–30 August 2018
6. Emirsajłow, Z.: Output observers for linear infinite-dimensional control systems. In: Kulczycki, P., Korbicz, J., Kacprzyk, J., (eds.) Control, Robotics and Information Processing, SN Series: Studies in Systems, Decision and Control. Springer (2020, to appear)
7. Emirsajłow, Z., Townley, S.: From PDEs with boundary control to the abstract state equation with an unbounded input operator: tutorial. Eur. J. Control **7**, 1–23 (2000)
8. Hidayat, Z., Babuska, R., De Schutter, B., Nunez, A.: Observers for linear distributed-parameter systems: a survey. In: Proceedings of the 2011 IEEE International Symposium on Robotic and Sensors Environments, Montreal, Canada, 17–18 September 2011, pp. 166–171 (2011)
9. Smyshlyaev, A., Krstic, M.: Backstepping observer for a class of parabolic PDEs. Syst. Control Lett. **54**, 613–625 (2005)

10. Trinh, H., Fernando, T.: Functional Observers for Dynamical Systems. Springer, Berlin (2012)
11. Tucsnak, M., Weiss, G.: Observation and Control for Operator Semigroups. Birkhäuser Advanced Texts, Birkhäuser (2009)
12. Vries, D., Keesman, K.J., Zwart, H.: Luenberger boundary observer synthesis for Sturm-Liouville systems. Int. J. Control **83**, 1503–1514 (2010)

Design of Control Systems - Methods and Applications

Constraining State Variables in Continuous Time Sliding Mode Control

Marek Jaskuła(✉) and Piotr Leśniewski

Institute of Automatic Control, Łodź University of Technology, 18/22 Stefanowskiego Street, 90-924 Łodź, Poland
marek.jaskula@dokt.p.lodz.pl, piotr.lesniewski@p.lodz.pl

Abstract. In this paper the continuous-time system of arbitrary order was taken into consideration. The sliding mode controller was computed using the reaching law approach. The time-varying convergence rate of the sliding variable was designed in such a manner to fulfill the control signal and state variables constraints for the whole regulation time. The sufficient condition guaranteeing the fastest, finite-time, monotonic convergence of the representative point to the predefined switching hyperplane in the presence of given limitations was stated and formally proved.

Keywords: Sliding mode control · Continuous-time systems · State constraint · Control signal constraint · Reaching law

1 Introduction

Nowadays, among the variety of variable structure control techniques, the sliding mode control approach is very popular due to its effectiveness, robustness and computational efficiency. Despite the fact, that the roots of this regulation method were connected only with continuous-time domain [10,18], in time is has been successfully extended also for the discrete-time systems [1,3,16,19]. Over the years sliding mode control approach has found to be useful in many practical applications, for example in power electronics [4,11,21].

In this regulation method we distinguish two different phases. In the first one, called the reaching phase, the representative point is driven onto the predefined switching hyperplane. Further, the sliding motion is obtained along the mentioned manifold up to the desired state. Therefore, the design of the sliding mode controller, starts with the selection of the switching hyperplane parameters, in order to determine the system dynamics during the sliding phase. Furthermore, the control signal can be computed using one of two different methods. In the first of them we propose a controller, and then prove the stability of the sliding motion, whereas in the second one we establish the prior determination of the desirable dynamics of the sliding variable by implementing the so-called reaching law.

A. Bartoszewicz et al. (Eds.): PCC 2020, AISC 1196, pp. 107–114, 2020.
https://doi.org/10.1007/978-3-030-50936-1_9

Origins of the reaching law approach can be found in [12], were it was used for the continuous-time system. Soon after, it has been extended for the discrete-time domain by Gao, Wang and Homaifa in [13]. Since then this method has been studied and developed by many researchers [2,5–9,17,20].

In spite of the benefits, the sliding mode control technique does not guarantee a priori known limits of the state variables and the control signal. Therefore, in this paper we extend our previous approach, based on the discrete-time domain [14,15] onto the continuous-time systems. We demand the fastest, finite-time, monotonic convergence of the representative point to the sliding hyperplane simultaneously satisfying the state and input restrictions.

This paper is organized as follows. Section 2 contains the selection of the switching manifold parameters and the design of the controller. Then, in Sect. 3, the state and control signal constraints are taken into consideration. In Sect. 4 we derive and prove the sufficient condition for the fastest, finite-time, monotonic convergence of the representative point to the predefined sliding hyperplane while satisfying given restrictions. The simulations results are presented in Sect. 5 and finally Sect. 6 comprises the remarks and conclusions.

2 Sliding Mode Controller Design

First of all, let us consider the continuous-time plant of arbitrary order described by the following state equation

$$\dot{\boldsymbol{x}}(t) = \boldsymbol{A}\boldsymbol{x}(t) + \boldsymbol{b}u(t), \tag{1}$$

where $\boldsymbol{x}(t) = [x_1(t), \ldots, x_n(t)]^T$ is the state vector, $\boldsymbol{A}$ is the state matrix, and vector $\boldsymbol{b} = [b_1, \ldots, b_n]^T$ where $b_i \neq 0, i \in \{1, \cdots, n\}$ is connected with the scalar input $u(t)$. Moreover, the switching hyperplane is defined as follows

$$s(t) = \boldsymbol{c}^T\boldsymbol{x}(t) = 0. \tag{2}$$

Vector $\boldsymbol{c}^T = [c_1, \ldots, c_{n-1}, 1]$ is selected so that condition $\boldsymbol{c}^T\boldsymbol{b} \neq 0$ is fulfilled. For the purpose of the designing the continuous-time sliding mode controller we introduce the following reaching law

$$\dot{s}(t) = -K\operatorname{sgn}[s(t)], \tag{3}$$

where convergence rate $K > 0$. From (1), (2) and the above equation we get the following formula for control input

$$u(t) = -\left(\boldsymbol{c}^T\boldsymbol{b}\right)^{-1}\left\{\boldsymbol{c}^T\boldsymbol{A}\boldsymbol{x}(t) + K\operatorname{sgn}\left[\boldsymbol{c}^T\boldsymbol{x}(t)\right]\right\}. \tag{4}$$

Furthermore, from the fact that $\dot{s}s < 0$, the monotonic, finite-time convergence of the state to the sliding manifold is obtained when

$$|\dot{s}(t)| \geq \lambda > 0 \tag{5}$$

is fulfilled up to the sliding phase, where λ is an arbitrarily small, strictly positive value. Using the previously defined reaching law we can write (5) in the form

$$K \geq \lambda > 0. \tag{6}$$

3 Control Signal and State Constraints

In this section, the convergence rate K is selected so that the limits of the control input and of the state variables are guaranteed. We begin with the control signal constraint i.e. demanding that

$$-r_u \leq u(t) \leq r_u \tag{7}$$

is true for $t \geq 0$, where r_u is a positive number. Combining (4) with the above inequality we get

$$-r_u \leq -\left(\boldsymbol{c}^T \boldsymbol{b}\right)^{-1}\left[\boldsymbol{c}^T \boldsymbol{A}\boldsymbol{x}(t) + K\text{sgn}\left(\boldsymbol{c}^T \boldsymbol{x}(t)\right)\right] \leq r_u. \tag{8}$$

Rewriting (8) one can get

$$r_u - \left(\boldsymbol{c}^T \boldsymbol{b}\right)^{-1} \boldsymbol{c}^T \boldsymbol{A}\boldsymbol{x}(t) \geq \left(\boldsymbol{c}^T \boldsymbol{b}\right)^{-1} K\text{sgn}\left(\boldsymbol{c}^T \boldsymbol{x}(t)\right) \geq -r_u - \left(\boldsymbol{c}^T \boldsymbol{b}\right)^{-1} \boldsymbol{c}^T \boldsymbol{A}\boldsymbol{x}(t). \tag{9}$$

Taking into account the fact that in the reaching phase the sign of the sliding variable is constant and considering signs of $\boldsymbol{c}^T \boldsymbol{b}$ and $\boldsymbol{c}^T \boldsymbol{x}(t)$ we obtain that the highest convergence rate (denoted by K_u) for which the absolute value of the control signal is equal to r_u is

$$K_u = \left|\boldsymbol{c}^T \boldsymbol{b}\right| r_u - \text{sgn}\left(\boldsymbol{c}^T \boldsymbol{x}(t)\right) \boldsymbol{c}^T \boldsymbol{A}\boldsymbol{x}(t). \tag{10}$$

Subsequently, we move to the problem of limiting all state variables. Therefore, we demand that conditions

$$-r_i \leq x_i(t) \leq r_i \tag{11}$$

are met for each $i \in \{1, \ldots, n\}$ and $t \geq 0$. From (1) and (4) we have

$$\dot{\boldsymbol{x}}(t) = \boldsymbol{A}\boldsymbol{x}(t) - \boldsymbol{b}\left(\boldsymbol{c}^T \boldsymbol{b}\right)^{-1} \boldsymbol{c}^T \boldsymbol{A}\boldsymbol{x}(t) - \boldsymbol{b}\left(\boldsymbol{c}^T \boldsymbol{b}\right)^{-1} K\text{sgn}\left(\boldsymbol{c}^T \boldsymbol{x}(t)\right). \tag{12}$$

Multiplying both sides the by the i–th versor $\boldsymbol{e}_i$ and equating derivative of i–th state variable to zero we get

$$0 = \boldsymbol{e}_i \boldsymbol{A}\boldsymbol{x}(t) - b_i\left(\boldsymbol{c}^T \boldsymbol{b}\right)^{-1} \boldsymbol{c}^T \boldsymbol{A}\boldsymbol{x}(t) - b_i\left(\boldsymbol{c}^T \boldsymbol{b}\right)^{-1} K\text{sgn}\left(\boldsymbol{c}^T \boldsymbol{x}(t)\right). \tag{13}$$

Calculating the convergence rate and denoting it as K_i we obtain

$$K_i = \text{sgn}\left(\boldsymbol{c}^T \boldsymbol{x}(t)\right) b_i^{-1}\left(\boldsymbol{c}^T \boldsymbol{b}\right) \boldsymbol{e}_i \boldsymbol{A}\boldsymbol{x}(t) - \text{sgn}\left(\boldsymbol{c}^T \boldsymbol{x}(t)\right) \boldsymbol{c}^T \boldsymbol{A}\boldsymbol{x}(t). \tag{14}$$

Applying convergence rate (14) into control signal (4) when i–th state variable is at the level r_i results in maintaining this variable at its constraint. However, the change in the value of the convergence rate may not cause the violation of given restriction. Using a different convergence rate will result in either exceeding the state variable constraint, or in decreasing the absolute value of this variable below the limit. Taking into account the signs of factors that are multiplied with K in equality (13) and the sign of $x_i(t)$ while $|x_i(t)| = r_i$ we obtain that if

$$\operatorname{sgn}\left[b_i \boldsymbol{c}^T \boldsymbol{b} \boldsymbol{c}^T \boldsymbol{x}(t)\, x_i(t)\right] = 1 \tag{15}$$

is satisfied, then increasing the convergence rate will not results in exceeding the constraint. Thus, decreasing K will maintain the representative point in the admissible set. Let us note that left hand side of (15) cannot be equal zero in the reaching phase. Therefore, in the case when (15) is true we can apply the biggest convergence rate that enables us the control signal limitation i.e. K_u.

4 Control Strategy and Sufficient Condition

In this section we will formulate and prove the sufficient condition for monotonic, finite-time convergence of the representative point to the switching hyperplane in the presence of given state and control signal constraints. Subsequently, we will present the control strategy that summarizes our theoretical considerations. First of all, we will require that the input limitation allows us to converge to given sliding hyperplane i.e.

$$K_u \geq \lambda > 0. \tag{16}$$

For that purpose we formulate following theorem:

Theorem 1. *In order to fulfill (16) it is sufficient that*

$$\left|\boldsymbol{c}^T \boldsymbol{b}\right| r_u > sgn\left(\boldsymbol{c}^T \boldsymbol{x}(t)\right) \boldsymbol{c}^T \boldsymbol{A}\boldsymbol{x}(t) + \lambda \tag{17}$$

is satisfied in the set given by state variables constraints.

Proof. Subtracting from both sides term $\operatorname{sgn}\left(\boldsymbol{c}^T \boldsymbol{x}(t)\right) \boldsymbol{c}^T \boldsymbol{A}\boldsymbol{x}(t)$ in (17) and using (10) we get $K_u \geq \lambda > 0$.

Now, we will focus on the main issue of this work. For the sake of clarity, let us denote $K_i = K_i^+$ when (15) is satisfied and $K_i = K_i^-$ otherwise. Let us observe that if $K_i = K_i^+$ then we can increase the convergence rate without violating $i-$th state constraint, on the other hand for $K_i = K_i^-$ we can diminish the convergence rate below K_i. First of all, we need to avoid the situation in which at the intersection of the restrictions inequality $K_i^- < K_j^+$ is true, because in this case one of the constraints must be violated. Furthermore, in order to obtain the monotonic, finite-time convergence the following inequalities must be met

$$K_u \geq K_i^+, \tag{18}$$

$$K_i^- \geq \lambda > 0 \tag{19}$$

for every $i \in \{1, \cdots, n\}$. Inequality (18) shows that if we arrive at the $i-$th state limit, the admissible control signal magnitude is sufficiently large to either make the state "slide" along that constraint or to make it enter the admissible state region again. Inequality (19) implies that on those state limits, on which we cannot exceed the convergence rate K_i, the state will "slide" along that limit towards (and not away from) the switching hyperplane.

Theorem 2. *In order to fulfill conditions (6), (7) and (11) it is sufficient that for every $i, j \in \{1, \ldots, n\}$, $i \neq j$, $|x_i| = r_i, |x_j| = r_j$ the inequalities are true*

$$sgn(x_i)\left(\left|b_i^{-1}\right|\left|\boldsymbol{c}^T\boldsymbol{b}\right|\boldsymbol{e}_i - sgn\left(b_i\boldsymbol{c}^T\boldsymbol{b}\right)\boldsymbol{c}^T\right)\boldsymbol{A}\boldsymbol{x}(t) \leq -\lambda < 0 \tag{20}$$

for $\boldsymbol{x}(t)$ such that K_i does not satisfy (15),

$$\left(\left|b_i^{-1}\right| sgn(x_i)\,\boldsymbol{e}_i + \left|b_j^{-1}\right| sgn(x_j)\,\boldsymbol{e}_j\right)\boldsymbol{A}\boldsymbol{x}(t) \leq 0 \tag{21}$$

for $\boldsymbol{x}(t)$ such that $b_i b_j x_i x_j < 0$ and

$$|b_i|\, r_u > sgn(x_i)\,\boldsymbol{e}_i\boldsymbol{A}\boldsymbol{x}(t) \tag{22}$$

for $\boldsymbol{x}(t)$ such that K_i satisfies (15).

Proof. We begin with obtaining property (18). For that purpose, we multiply both sides of inequality (22) by $\left|b_i^{-1}\right|\left|\boldsymbol{c}^T\boldsymbol{b}\right|$ getting

$$\left|\boldsymbol{c}^T\boldsymbol{b}\right| r_u > \operatorname{sgn}(x_i)\left|b_i^{-1}\right|\left|\boldsymbol{c}^T\boldsymbol{b}\right|\boldsymbol{e}_i\boldsymbol{A}\boldsymbol{x}(t). \tag{23}$$

From the fact that (15) is fulfilled the sign of $b_i\left(\boldsymbol{c}^T\boldsymbol{b}\right)\operatorname{sgn}\left(\boldsymbol{c}^T\boldsymbol{x}(t)\right)$ equals the sign of x_i. Hence,

$$\left|\boldsymbol{c}^T\boldsymbol{b}\right| r_u > b_i^{-1}\left(\boldsymbol{c}^T\boldsymbol{b}\right)\operatorname{sgn}\left(\boldsymbol{c}^T\boldsymbol{x}(t)\right)\boldsymbol{e}_i\boldsymbol{A}\boldsymbol{x}(t). \tag{24}$$

Subtracting from both sides the term $\operatorname{sgn}\left(\boldsymbol{c}^T\boldsymbol{x}(t)\right)\boldsymbol{c}^T\boldsymbol{A}\boldsymbol{x}(t)$ and using definitions of K_u and K_i we obtain property (18). Let us note that without (18) the control signal constraint may not be satisfied simultaneously with the state variable constraint.

Further, we focus on proving (19). Let us consider (20). Taking into account that for this inequality condition (15) is not satisfied we have $\operatorname{sgn}\left(\boldsymbol{c}^T\boldsymbol{x}(t)\right) = -\operatorname{sgn}\left(b_i\left(\boldsymbol{c}^T\boldsymbol{b}\right)x_i\right)$ and $\operatorname{sgn}(x_i) = -\operatorname{sgn}\left(b_i\left(\boldsymbol{c}^T\boldsymbol{b}\right)\left(\boldsymbol{c}^T\boldsymbol{x}(t)\right)\right)$. Therefore, we can rewrite (20) as

$$\left[-\operatorname{sgn}\left(\boldsymbol{c}^T\boldsymbol{x}(t)\right)b_i^{-1}\boldsymbol{c}^T\boldsymbol{b}\boldsymbol{e}_i + \operatorname{sgn}\left(\boldsymbol{c}^T\boldsymbol{x}(t)\right)\boldsymbol{c}^T\right]\boldsymbol{A}\boldsymbol{x}(t) \leq -\lambda < 0. \tag{25}$$

Multiplying all sides by -1 we obtain property (19). This condition guarantees that when the representative point slides along given state variable constraint it also converges to the sliding manifold.

Now let us analyze (21). In this case we have $b_i b_j x_i x_j < 0$ and $|x_i| = r_i, |x_j| = r_j$, $i \neq j$. In order for $b_i b_j x_i x_j < 0$ to be true the equation (15) has to be satisfied for one of i, j and not true for the other one. As we consider all combinations of i, j we can assume that (15) is satisfied for i and false for j, without the loss of generality. Using (14) to calculate K_i, K_j and adding to both sides the expression $\operatorname{sgn}\left(\boldsymbol{c}^T\boldsymbol{x}(t)\right)\boldsymbol{c}^T\boldsymbol{A}\boldsymbol{x}(t)$ we get

$$\operatorname{sgn}\left(\boldsymbol{c}^T\boldsymbol{x}(t)\right)b_i^{-1}\left(\boldsymbol{c}^T\boldsymbol{b}\right)\boldsymbol{e}_i\boldsymbol{A}\boldsymbol{x}(t) > \operatorname{sgn}\left(\boldsymbol{c}^T\boldsymbol{x}(t)\right)b_j^{-1}\left(\boldsymbol{c}^T\boldsymbol{b}\right)\boldsymbol{e}_j\boldsymbol{A}\boldsymbol{x}(t). \tag{26}$$

Rewriting the above inequality we have

$$\left(\left(\boldsymbol{c}^T\boldsymbol{b}\right)\operatorname{sgn}\left(\boldsymbol{c}^T\boldsymbol{x}\left(t\right)\right)b_i^{-1}\boldsymbol{e}_i-\left(\boldsymbol{c}^T\boldsymbol{b}\right)\operatorname{sgn}\left(\boldsymbol{c}^T\boldsymbol{x}\left(t\right)\right)b_j^{-1}\boldsymbol{e}_j\right)\boldsymbol{A}\boldsymbol{x}\left(t\right)>0. \tag{27}$$

Let us observe that in this case $\operatorname{sgn}\left(b_i^{-1}\left(\boldsymbol{c}^T\boldsymbol{b}\right)\left(\boldsymbol{c}^T\boldsymbol{x}\left(t\right)\right)\right)=-\operatorname{sgn}\left(x_i\right)$ and also $\operatorname{sgn}\left(b_j^{-1}\left(\boldsymbol{c}^T\boldsymbol{b}\right)\left(\boldsymbol{c}^T\boldsymbol{x}\left(t\right)\right)\right)=\operatorname{sgn}\left(x_j\right)$. Thus, we can write

$$\left(-\left|\boldsymbol{c}^T\boldsymbol{b}\right|\left|b_i^{-1}\right|\operatorname{sgn}\left(x_i\right)\boldsymbol{e}_i-\left|\boldsymbol{c}^T\boldsymbol{b}\right|\left|b_j^{-1}\right|\operatorname{sgn}\left(x_j\right)\boldsymbol{e}_j\right)\boldsymbol{A}\boldsymbol{x}\left(t\right)>0. \tag{28}$$

Dividing both sides of the above inequality by $-\left|\boldsymbol{c}^T\boldsymbol{b}\right|$ we get condition (22). Therefore, $K_i^- > K_j^+$. That ends the proof.

In the proposed control strategy, if $|x_i| = r_i$ and (15) is false, the convergence rate is selected as K_i, otherwise we select the convergence rate K_u.

5 Simulation Example

For the purpose of verifying the theoretical results we carry out the computer simulations. Let us consider the continuous-time system described by (1) where

$$\boldsymbol{A}=\begin{bmatrix}1 & 0\\ 1 & -1\end{bmatrix}, \qquad \boldsymbol{b}=\begin{bmatrix}1\\ 1\end{bmatrix}, \qquad \boldsymbol{c}^T=\begin{bmatrix}1 \; 1\end{bmatrix}. \tag{29}$$

Our goal is to drive the representative point from the initial state $\boldsymbol{x}_0 = [70, -30]^T$ to the switching line $\boldsymbol{c}^T\boldsymbol{x}\left(t\right) = 0$ monotonically and in finite-time. What is more, we require the fulfillment of the following limitations: $|x_1\left(t\right)| \leq 70$, $|x_2\left(t\right)| \leq 35$ and $|u(t)| \leq 180$ for $t \geq 0$. One can observe that conditions (20), (21), (22), (17) are fulfilled. Therefore, the state and control signal constraints are guaranteed as can been verified in Fig. 1 and Fig. 2. Moreover, the stable sliding motion is obtained and the demand state is reached.

The evolution of the first state variable is shown in Fig. 3. It can be seen that it is a monotonically decreasing function. We can also observe two minor changes of fall rate on the graph, which represents reaching second state variable constraint (approximately t = 0.06) and sliding hyperplane (about t = 0.85). From Fig. 4 we can see that for the very brief moment second state variable drops and then it remains stable on its limitation r_2 = 35. It is caused by the use of the convergence rate K_u, in very beginning of the control process, and subsequently convergence rate K_2 up to the sliding phase.

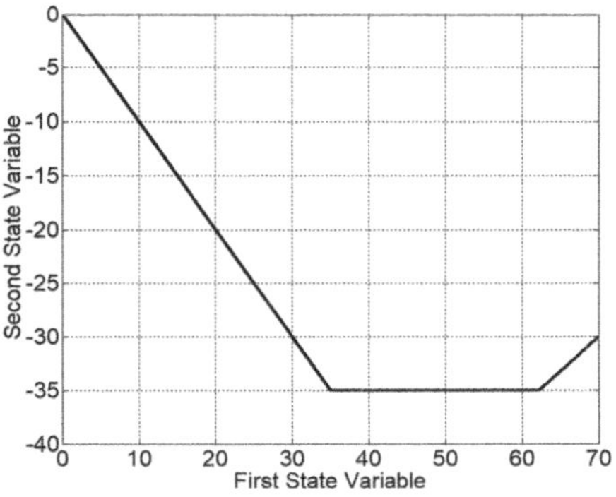

Fig. 1. State Trajectory

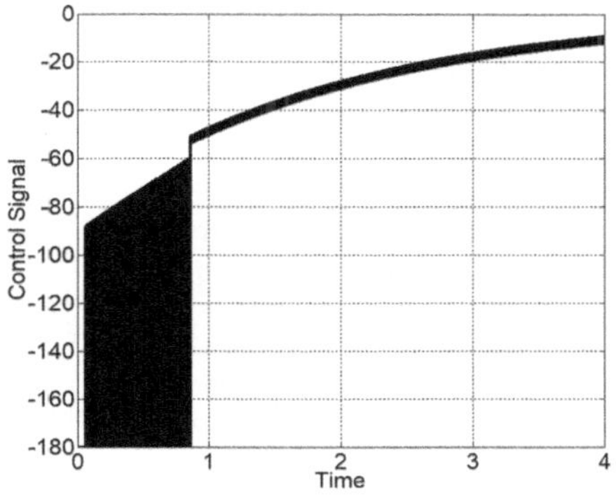

Fig. 2. Control Signal

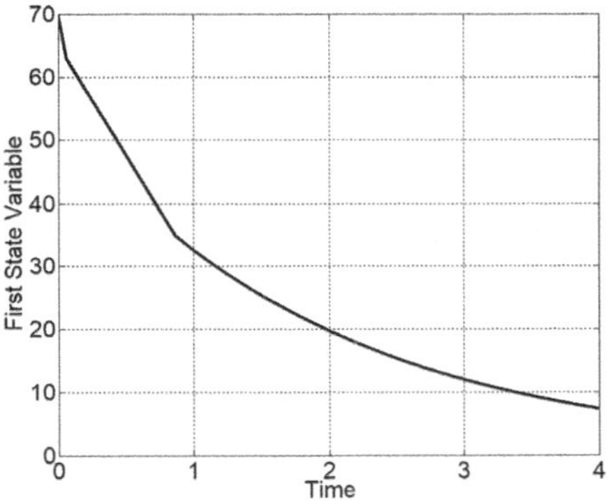

Fig. 3. First State Variable

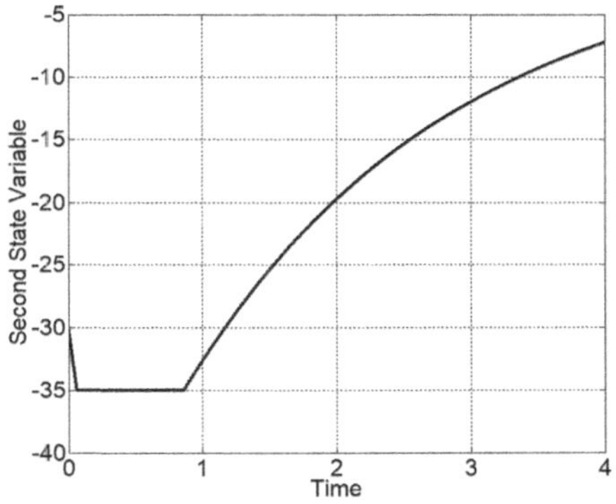

Fig. 4. Second State Variable

6 Conclusions

In this article the continuous-time system of arbitrary order has been considered and the controller was designed using the reaching law technique. The issue of constraining not only the control input but simultaneously also all state variables was analyzed. In order to obtain the fastest, finite-tine, monotonic convergence of the representative point to the predefined switching hyperplane in the presence of given constraints the time-varying convergence rate of sliding variable was designed and the sufficient condition guaranteeing this property was introduced. Simulation results verified theoretical considerations. In future research we are going to include the influence of the external disturbances on the system dynamics.

References

1. Bartolini, G., Ferrara, A., Utkin, V.I.: Adaptive sliding mode control in discrete-time systems. Automatica **31**, 769–773 (1995)

2. Bartoszewicz, A.: A new reaching law for sliding mode control of continuous time systems with constraints. Trans. Inst. Measur. Control **37**, 515–521 (2015)
3. Bartoszewicz, A.: Discrete-time quasi-sliding-mode control strategies. IEEE Trans. Ind. Electron. **45**, 633–637 (1998)
4. Bartoszewicz, A., Kaynak, O., Utkin, V.I. (eds.): Sliding mode control in industrial applications. Special Section IEEE Trans. Ind. Electron. **55**, 3805–4103 (2008)
5. Bartoszewicz, A., Latosiński, P.: Generalization of Gao's reaching law for higher relative degree sliding variables. IEEE Trans. Autom. Control **63**, 3173–3179 (2018)
6. Bartoszewicz, A., Latosiński, P.: Reaching law for DSMC systems with relative degree 2 switching variable. Int. J. Control **90**, 1626–1638 (2017)
7. Bartoszewicz, A., Leśniewski, P.: Reaching law approach to the sliding mode control of periodic review inventory systems. IEEE Trans. Autom. Sci. Eng. **11**, 810–817 (2014)
8. Bartoszewicz, A., Leśniewski, P.: Reaching law-based sliding mode congestion control for communication networks. IET Proc. Control Theory Appl. **8**, 1914–1920 (2014)
9. Chakrabarty, S., Bandyopadhyay, B.: A generalized reaching law with different convergence rates. Automatica **63**, 34–37 (2016)
10. Emelyanov, S.V.: Variable Structure Control Systems. Nauka, Moscow (1967)
11. Fallaha, C.J., Saad, M., Kanaan, H.Y., Al-Haddad, K.: Sliding-mode robot control with exponential reaching law. IEEE Trans. Ind. Electron. **58**, 600–610 (2011)
12. Gao, W., Hung, J.C.: Variable structure control of nonlinear systems: a new approach. IEEE Trans. Ind. Electron. **40**, 45–55 (1993)
13. Gao, W., Wang, Y., Homaifa, A.: Discrete-time variable structure control systems. IEEE Trans. Ind. Electron. **42**, 117–122 (1995)
14. Jaskuła, M., Pietrala, M., Leśniewski, P.: The problem of state constraints in designing the discrete time sliding mode controller. Pomiary Automatyka Robotyka **4**, 15–22 (2017)
15. Jaskuła, M., Leśniewski, P.: Discrete time sliding mode control in the presence of state and control signal constraints. In: 24th International Conference on Methods and Models in Automation and Robotics, Miedzyzdroje, Poland, pp. 201–204 (2019)
16. Milosavljević, Č.: General conditions for the existence of a quasi-sliding mode on the switching hyperplane in discrete variable structure systems. Autom. Remote Control **46**, 307–314 (1985)
17. Niu, Y., Ho, D.W.C., Wang, Z.: Improved sliding mode control for discrete-time systems via reaching law. IET Control Theory Appl. **4**, 2245–2251 (2010)
18. Utkin, V.I.: Variable structure systems with sliding modes. IEEE Trans. Autom. Control **22**, 212–222 (1977)
19. Utkin, V.I., Drakunow, S.V.: On discrete-time sliding mode control. In: IFAC Conference Nonlinear Control, pp. 484–489 (1989)
20. Wang, A., Jia, X., Dong, S.: A new exponential reaching law of sliding mode control to improve performance of permanent magnet synchronous motor. IEEE Trans. Magn. **49**, 2409–2412 (2013)
21. Zhang, X., Sun, L., Zhao, K., Sun, L.: Nonlinear speed control for PMSM system using sliding-mode control and disturbance compensation techniques. IEEE Trans. Power Electron. **28**, 1358–1365 (2013)

Modification of the Firefly Algorithm for Improving Solution Speed

Ryszard Klempka(✉)

Faculty of Electrical Engineering, Automatics, Computer Science and Biomedical Engineering, AGH-University of Science and Technology, 30-059 Krakow, Poland
klempka@agh.edu.pl

Abstract. The demand for fast and intelligent optimization methods is constantly growing. Owing to this, new methods based on the behaviour of living organisms are being developed. This article proposes a modification of the classic firefly algorithm based on the acceptance of each movement of a firefly, that provide more accurate values of an objective function. In addition, the algorithm also involves a reduced value of a coefficient, α, in each of its iterations. The effectiveness of the modification is examined using typical test functions. The modification allows for finding the correct solution faster and more accurately. This improvement is achieved at the expense of the algorithm's sensitivity to a selection parameter, α_{damp}, which affects the speed at which the value of α decreases.

Keywords: Firefly algorithm · Variability of random component · Natural inspired optimization algorithm

1 Introduction

The firefly algorithm (FA) was proposed by Xin-She Yang in 2008 [1] and later in 2010 [2]. Over the years, the FA has been used for several types of optimization. The applications of the FA to mixed continuous/discrete optimization problems are presented in [3]. The method has also been used for optimization with constraints [4, 5] and with changing constraints depending on objective functions [6]. The FA has been tested in nonlinear multicriteria optimization with constraints [7].

The variety of applications of the FA has led to the testing of different variations in the basic FA to increase the efficiency and speed of finding a solution. The application of the elitism variable was presented in [8], in which an additional set of random firefly movements was used and a firefly was moved towards the best solution (if it was more accurate than the existing solution). An upgraded FA was used in [9], where the operator of a single-point crossover of two fireflies (the worst with the best firefly) became active after a certain number of generations. Similar modifications were described in [10] as an "enhanced FA". Another modification of the FA, referred to as a "hybrid FA", was based on the combination of the FA with a genetic algorithm. After completing

A. Bartoszewicz et al. (Eds.): PCC 2020, AISC 1196, pp. 115–124, 2020.
https://doi.org/10.1007/978-3-030-50936-1_10

the phase of determining the new positions of fireflies, the algorithm selected the best fireflies using tournament selection and then crossed them over, creating offspring [11]. An adaptive modified FA was presented in [12]. It was noted that a coefficient, α, had a large impact on convergence to an optimal solution. The value of α was decreased in each iteration depending on the number of generations. Changing α was also proposed in [13], but a different equation was used to define its value. The modification of the random movements of fireflies depending on iterations was proposed in [14]. This was an attempt to balance the emphasis of the algorithm between exploration and exploitation. The same objective achieved in a different manner was implemented in a "crazy FA" [15]. In the standard equation of firefly movement, an element responsible for random part was added.. Several aspects of the FA were comprehensively described in [16]. The effectiveness of the algorithm was tested using several test functions and practical applications [17, 18].

In the present study, a modification of the FA was suggested to increase solution speed, This is the goal of continuous progress as the demand for faster and more accurate algorithms is constantly growing. The modified FA includes a reduced value of α in each iteration. Furthermore, special attention was paid to the selection of the initial value of a parameter, α_{damp}, which influences the speed at which the value of α decreases.

The rest of the paper is organized as follows: First, the standard FA and the proposed modification are described. Then, the tests of effectiveness of the modified algorithm performed using several selected functions is presented. Finally, the conclusions are provided.

2 Firefly

2.1 Classical Algorithm

The FA, like several other intelligent methods, is based on the observation of nature. The effectiveness of these methods stems, among others, from the fact that they are based on swarm algorithms. Each firefly represents a potential solution to a given optimization problem, and its correct position represents a solution. Each firefly glows with light, whose brightness is proportional to the quality of the indicated solution. The firefly with the brightest light becomes more attractive to a larger number of fireflies, thus drawing them in its direction. In each iteration of the FA, fireflies move towards better individuals. Therefore, the area with the largest number of the brightest fireflies is more thoroughly searched to find the global optimum. The brightness of a firefly decreases as the distance between fireflies increases.

The attractiveness of firefly A in relation to firefly B depends on the following three parameters:

- brightness of both fireflies (the brighter firefly is more attractive),
- light absorption rate γ,
- distance between the fireflies r.

$$\beta = \beta_0 \cdot \exp(-\gamma \cdot r^m) \tag{1}$$

where:
β_0 – base value of the attraction coefficient,
γ – light absorption coefficient,
r – distance between fireflies A and B,
m – distance exponent.

Equation 1 shows that the attraction of a firefly decreases exponentially with the mth power of distance r, which is defined by the Euclidean distance given by Eq. 2.

$$r = \sqrt{\sum_{i=1}^{k} (\mathbf{x_A}(i) - \mathbf{x_B}(i))^2} \tag{2}$$

where k is the number of decision variables (dimensions) of the optimized function.

The potential movement of firefly A towards the more attractive firefly B is determined by Eq. 3.

$$\mathbf{x_{A\_new}} = \mathbf{x_A} + \beta \cdot \text{rand}(0, 1) \cdot (\mathbf{x_B} - \mathbf{x_A}) + \alpha \cdot \text{rand}(-1, 1) \cdot (\mathbf{max} - \mathbf{min}) \tag{3}$$

where:
$\mathbf{x_A}$ – position of firefly A (vector),
$\mathbf{x_{A\_new}}$ –potential new position of firefly A (vector),
$\mathbf{x_B}$ – position of firefly B, towards which firefly A moves (vector),
rand(a, b) – random number from the range (a, b),
α – mutation coefficient,
max, **min** – minimum and maximum values of decision variables (vectors).

Potential movements towards every better firefly are determined for each firefly. Out of all the determined potential positions, the position that provides the most significant improvement in the quality of the proposed solution is selected, and the firefly is moved there. If there are no movements that improve the solution, the firefly remains in place.

Equation 3 presents the deterministic part of the movement towards the second firefly with coefficient β and the random element with coefficient α. At the beginning of the algorithm's operation, the random element has a strong influence on the movements of fireflies. However, with each iteration of the algorithm, α is reduced according to Eq. 4. This decreases the area of random penetration.

$$\alpha = \alpha \cdot \alpha_{\text{damp}} \tag{4}$$

Coefficient α_{damp} is one of the parameters of the FA defined at the beginning. The pseudocode of the FA is given below.

```
Firefly algorithm(N,MaxIt,γ,β0,α,αdamp,m,VarMin,VarMax)
1.  Initialization of population N of fireflies with k
    coordinates from the range (VarMin, VarMax)
2.  Assessment of each firefly f(x), x=(x1, x2, ..., xk)
3.  for it = 1 to MaxIt do
4.    for i =1 to N do (for each firefly)
5.      for j = 1 to N do
6.          Determine the movement of firefly xnew(j)
7.          Asses the movement F(j) = f(xnew(j))
8.      end for j
9.      Select the best F(j) and move firefly i to posi-
        tion xnew(j)
10.   end for i
11.   t = t + 1
12.   α = α * αdamp
13. end for it
```

2.2 Modification of the Firefly Algorithm

When using the FA for practical applications [17, 18], a possibility for improving the procedures of optimization has been examined to improve solution speed and accuracy.

In the classic FA, potential firefly movements are generated, one of them is selected as the best, and a firefly is moved to the given position. This article proposes a modification that consists of creating each movement of a firefly while generating potential movements, if such a movement improves the solution represented by the firefly. The pseudocode of the modified FA is given below.

```
Firefly algorithm(N,MaxIt,γ,β0,α,αdamp,m,VarMin,VarMax)
Initialization of population N of fireflies with k
coordinates from the ranges (VarMin, VarMax)
Assessment of each firefly f(x), x=(x1, x2, ..., xk)
for it = 1 to MaxIt do
  for i = 1 to N do (for each firefly)
    for j = 1 to N do
      if f(xj) < f(xi) then (if firefly j is better)
        Determine the movement of firefly xnew
        if f(xnew) < f(xi) then
           move firefly i to position xnew
           f(xi) = f(xnew)
        end if
      end if
    end for j
  end for i
  t = t + 1
  α = α * αdamp
end for it
```

The next section presents the effectiveness of the modified algorithm and the impact of α_{damp} on the effectiveness.

3 Effectiveness of the Modified Firefly Algorithm

The improvement introduced to the FA is designed to accelerate the search for the global optimum and to improve the accuracy of the solution. The quality of the modified algorithm was tested considering the typical test functions given in Table 1 as examples.

Each of the above functions was optimized using the classic and modified algorithms. Different values of αdamp were tested. The test was repeated 50 times for each value of this coefficient. Both FAs started from the same initial population.

The parameters of both FAs were as follows:

- range of variation in decision variables under the terms of the task – Table 1,
- number of fireflies *N* – Table 1,
- number of iterations *MaxIt* – Table 1,
- attraction coefficient base value $\beta_0 = 2$,
- light absorption coefficient $\gamma = 1$,
- exponent distance $m = 1$,
- mutation coefficient $\alpha = 1$.

Table 1. Selected test functions.

F	Formula	Minimum	Parameters
F1	Ackley $y = -20 \cdot \exp\left(-0.2\sqrt{\frac{1}{k}\sum_{i=1}^{k} x_i^2}\right) -$ $\exp\left(\frac{1}{k}\sum_{i=1}^{k}\cos(2 \cdot \pi \cdot x_i)\right) + 20 + \exp(1)$ $x_i \in (-10; 10), \quad k = 10$	$f_{\min} = 0$ $x_i = 0$	$MaxIt = 200$ $N = 50$
F2	Easom $y = \left(\prod_{i=1}^{k}\cos(x_i)\right) \cdot \exp\left(-\sum_{i=1}^{k}(x_i - \pi)^2\right)$ $x_i \in (-20; 20), \quad k = 10$	$f_{\min} = -1$ $x_i = \pi$	$MaxIt = 50$ $N = 20$
F3	Michalewicz $y = -\sum_{i=1}^{k}\left\{\sin(x_i) \cdot \left[\sin\left(i \cdot x_i^{\frac{2}{\pi}}\right)\right]^{20}\right\}$ $x_i \in (0; \pi), \quad k = 2$	$f_{\min} \approx -1.8013034$ $x_1 \approx 2.20283514$ $x_2 \approx 1.570867116$	$MaxIt = 30$ $N = 10$
F4	Rastrigin $y = 10 \cdot n + \sum_{i=1}^{k}\left(x_i^2 - 10 \cdot \cos(4 \cdot \pi \cdot x_i)\right)$ $x_i \in (-5.12; 5.12), \quad k = 50$	$f_{\min} = 0$ $x_i = 0$	$MaxIt = 200$ $N = 50$
F5	Sphare $y = \sum_{i=1}^{k} x_i^2$ $x_i \in (-5.12; 5.12), \quad k = 50$	$f_{\min} = 0$ $x_i = 0$	$MaxIt = 200$ $N = 50$

Figure 1 shows the graphs for the minimization of functions F2 and F3. It is seen that the lines are not continuous. This is because the correct smallest value was found (minimum of the function). The graphs for these functions are shifted upwards to set the minimum at zero. Owing to this, the values in the graphs indicate the distance from the minimum value. In the case of functions F1, F2, and F5, there is a range of variations in α_{damp} for which the modified FA provides significantly more accurate results than the classic algorithm. Both algorithms are comparable for function F3, while the modified algorithm is slightly less effective for function F4. It is worth noting that the classic algorithm is resistant to changes in α_{damp} and finds a similar solution in a wide range

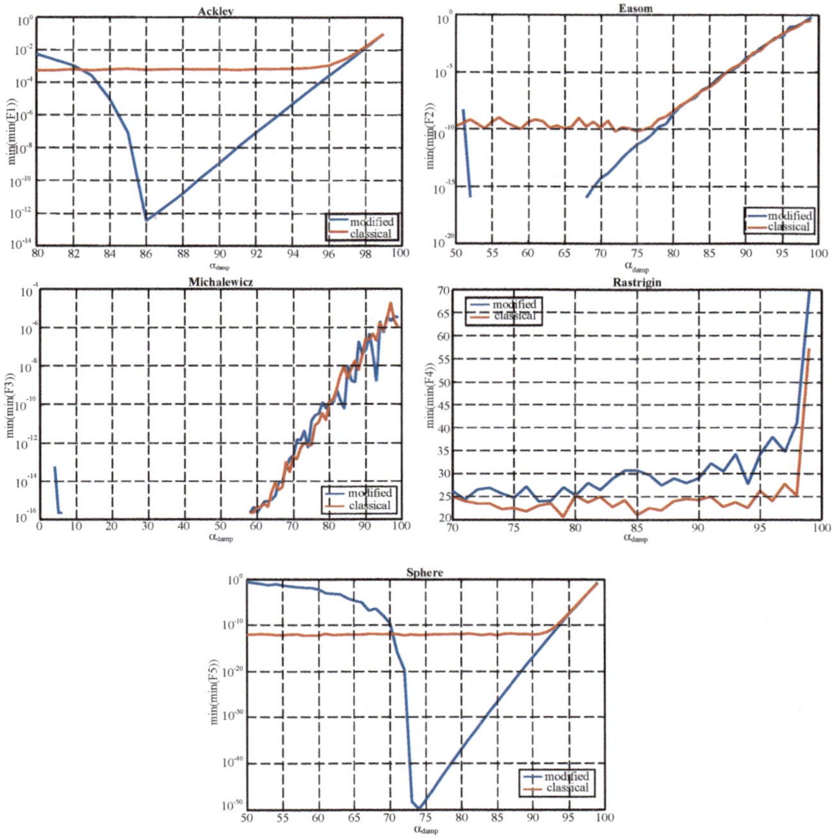

Fig. 1. Dependency of the quality of solutions on α_{damp} for different optimized test functions

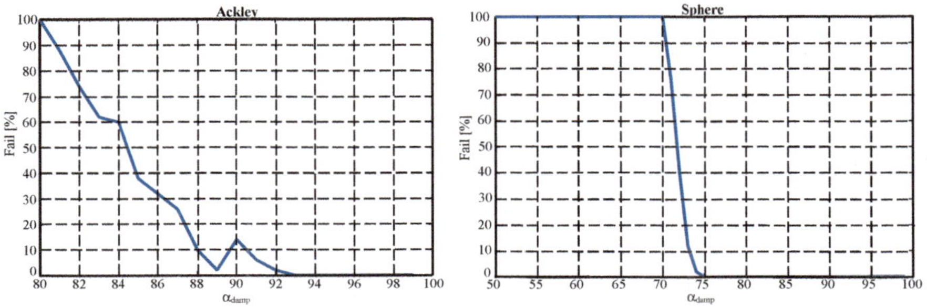

Fig. 2. Stopping of modified firefly algorithm at points that are not the minima of the function, depending on α_{damp}

of values of this parameter. Both algorithms find a solution with less accuracy for large values of α_{damp}.

At the values of α_{damp} for which the modified algorithm is inferior to the classic algorithm, the algorithm stops much before reaching the minimum of the function. This phenomenon can be observed for functions F1 and F5 and for smaller values of α_{damp}.

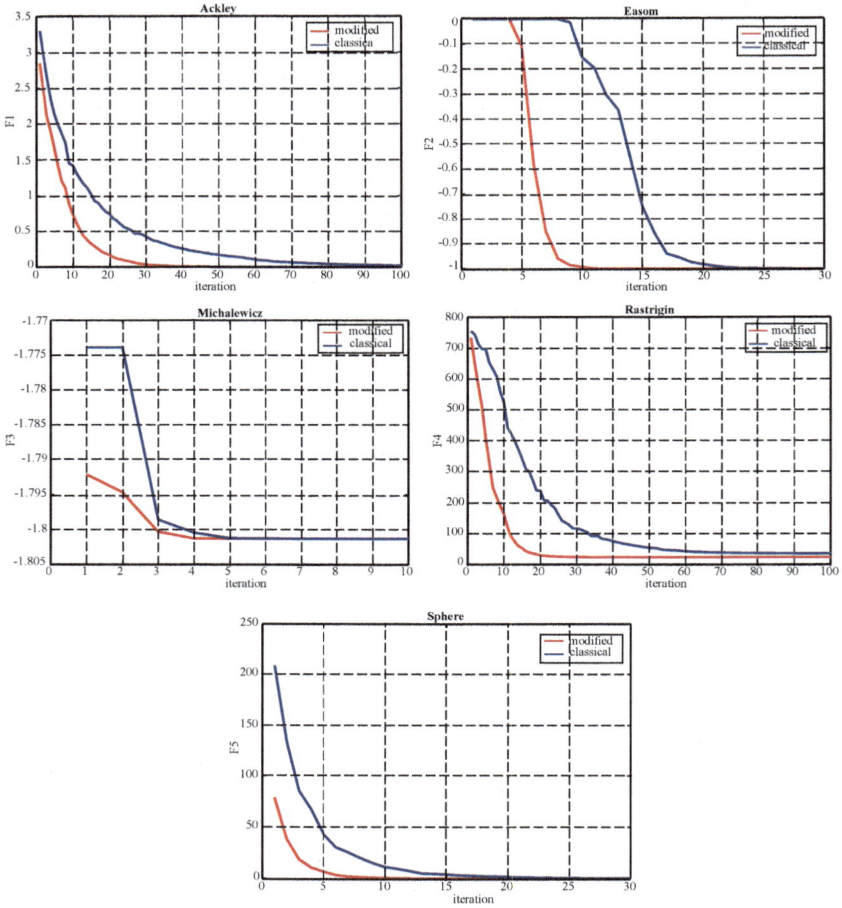

Fig. 3. Comparison of the speed of the optimization process using the classic and modified firefly algorithms

Figure 2 shows the percentage of times at which the modified algorithm stops at points that are not minima of the function, depending on α_{damp}. The phenomenon of stopping is eliminated for large values of α_{damp}. In the remaining functions, the minimum values for both algorithms are similar or those for the modified algorithm are more accurate. The phenomenon of the algorithm stopping is not observed for these functions (F2, F3, and F4).

In addition to finding a more accurate solution (in a certain range of variations in α_{damp}), the modified FA is faster. Figure 3 shows the optimization process of the selected test functions.

As shown in Fig. 3, if the modified algorithm does not stop outside the minimum of the function, the optimization process is faster.

4 Conclusion

This article presents a comparison of efficacy and speed of the classic and modified FAs. In the modified algorithm, a firefly is moved in each proposed direction instead of being moved towards one of the best improvements in each iteration. This ensures an improvement in the quality of the solution.

The modified algorithm finds a more accurate solution in a shorter period of time, provided that appropriate values of α_{damp} are selected. An area with a minimum is found in the early phase of the search.

The condition for obtaining more accurate results is the appropriate selection of αdamp. This constitutes a disadvantage of this method as the algorithm is sensitive to this parameter. Incorrect selection can be the reason for failing to find the right solution. The classic FA is resistant to the value of α_{damp}.

The proposed modified algorithm was also tested using practical optimisation issues [17, 18]. The selection of α_{damp} allowed for the application of the algorithm, which rapidly found the desired solution.

Even though the modified algorithm is faster, it cannot be used automatically. It requires careful selection of the value of α_{damp}. Further work will involve making the algorithm robust against unsuitable values of α_{damp}.

References

1. Yang, X.-S.: Nature-Inspired Metaheuristic Algorithms, 2nd edn. Luniver Press, Bristol (2008)
2. Yang, X.-S.: Firefly algorithm, stochastic test functions and design optimization. Int. J. Bio-Inspired Comput. **2**(2), 78–84 (2010)
3. Gandomi, A.H., Yang, X.-S., Alavi, A.H.: Mixed variable structural optimization using firefly algorithm. Comput. Struct. **89**(23–24), 2325–2336 (2011)
4. Lukasik, S., Zak, S.: Firefly algorithm for continuous constrained optimization tasks. Lect. Notes Comput. Sci. **5796**, 97–106 (2009)
5. Kwiecień, J., Filipowicz, B.: Comparison of firefly and cockroach algorithms in selected discrete and combinatorial problems. Bull. Pol. Acad. Sci. Tech. Sci. **62**, 797–804 (2014)
6. Gomes, H.M.: A firefly metaheuristic structural size and shape optimization with natural frequency constraints. Int. J. Metaheuristics **2**(1), 38–55 (2012)
7. Amjady, N., Naderi, M.: Multi-objective environmental/economic dispatch using firefly technique. In: 11th Environment and Electrical Engineering (EEEIC) Conference, Venice, pp. 461–466 (2012)
8. Tilahun, S.L., Ong, H.C.: Modified firefly algorithm. J. Appl. Math. **2012**, Article ID 467631 (2012)

9. Tuba, M., Bacanin, N.: Upgraded firefly algorithm for portfolio optimization problem. In: 16th International Conference on Computer Modelling and Simulation (UKSim), pp. 112–117. IEEE (2014)
10. Strumberger, I., Bacanin, N., Tuba, M.: Enhanced firefly algorithm for constrained numerical optimization. In: IEEE Congress on Evolutionary Computation (CEC), San Sebastian, pp. 2120–2127 (2017)
11. Zhu, X., Qi, S., Zhang, H.: A hybrid firefly algorithm. In: 2nd Advanced Information Technology, Electronic and Automation Control Conference (IAEAC), Chongqing, pp. 287–291 (2017)
12. Tjahjono, A., Anggriawan, D.O., Faizin, A.K., et al.: Adaptive modified firefly algorithm for optimal coordination of overcurrent relays. IET Gener. Transm. Distrib. **11**(10), 2575–2585 (2017)
13. Trivedi, R., Padhy, P.K.: Design of fractional $\mathrm{PI}\lambda\mathrm{D}\mu$ controller via modified firefly algorithm. In: 11th International Conference on Industrial and Information Systems (ICIIS), Roorkee, pp. 172–177 (2016)
14. Kaur, K., Salgotra, R., Singh, U.: An improved firefly algorithm for numerical optimization. In: International Conference on Innovations in Information, Embedded and Communication Systems (ICIIECS), Coimbatore, pp. 1–5 (2017)
15. Sarangi, S.K., Panda, R., Sarangi, A.: Crazy firefly algorithm for function optimization. In: 2nd International Conference on Man and Machine Interfacing (MAMI), Bhubaneswar, pp. 1–5 (2017)
16. Fister, I., Fister Jr., I., Yang, X.-S., Brest, J.: A comprehensive review of firefly algorithms. Swarm Evol. Comput. **13**, 34–46 (2013)
17. Klempka, R., Waradzyn, Z., Skala, A.: Application of the firefly algorithm for optimizing a single-switch class E ZVS voltage-source inverter's operating point. Adv. Electric. Comput. Eng. **18**(2), 93–100 (2018)
18. Klempka, R., Filipowicz, B.: Optimization of a DC motor drive using a firefly algorithm. In: International Symposium on Electrical Machines (SME), Andrychów, pp. 1–6 (2018)

On the Optimal Topology of Time-Delay Control Systems

Ruth Bars[1], Csilla Bányász[2], and Laszlo Keviczky[2(✉)]

[1] Budapest University of Technology and Economics, Budapest, Hungary
[2] Institute for Computer Science and Control, Budapest, Hungary
keviczky@sztaki.hu

Abstract. It is shown that the SMITH predictor is a subclass of the YOULA parameterization based generic two-degree of freedom controllers. Comparing the algorithms the application of the new approach is suggested.

Keywords: Smith predictor · Youla parameter · Time-delay

1 Introduction

It is clear for control engineers that handling time delay requires special attention from the early days of the control history. The time delay is an uncancelable, invariant property of the process. The early goals tried to find design procedures which allow the selection of the regulator quasi independently from the delay. An early success story was the SMITH predictor or regulator [1].

Consider a continuous time delay process given by its transfer function

$$P(s) = P_+(s)\bar{P}_-(s) = P_+(s)e^{-sT_d} \quad ; \quad P = P_+\bar{P}_- = P_+e^{-sT_d} \tag{1}$$

where T_d is the time delay, P_+ is stable and $\bar{P}_- = e^{-sT_d}$ is the *Inverse-Unstable-Unrealizable* (*IUU*) part of the process, respectively. The original SMITH predictor is shown in Fig. 1, where r is the reference signal and y is the process output.

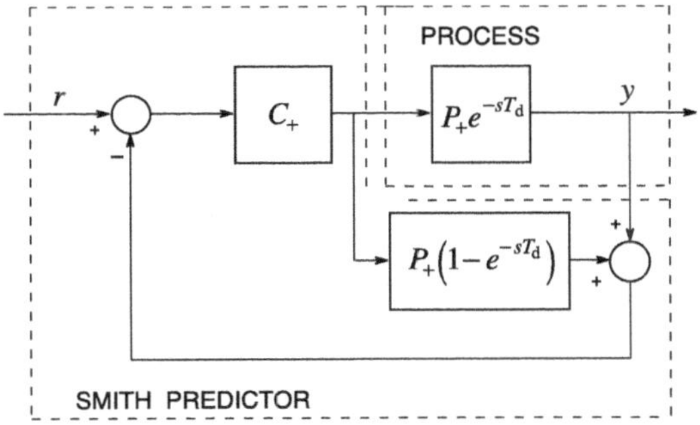

Fig. 1. The block-scheme of the SMITH predictor

A. Bartoszewicz et al. (Eds.): PCC 2020, AISC 1196, pp. 125–136, 2020.
https://doi.org/10.1007/978-3-030-50936-1_11

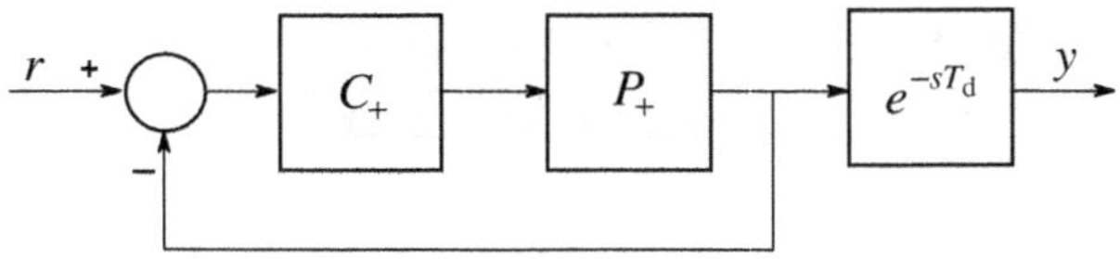

Fig. 2. Equivalent block-scheme of the SMITH predictor

It is easy to check that the SMITH predictor is equivalent to the scheme shown in Fig. 2. This figure clearly shows that the regulator C_+ can be designed to the delay free P_+, independently of the time delay T_d. This scheme explains why the SMITH predictor is also called SMITH regulator [8–10]. The whole procedure is, of course, not independent of T_d, because the predictor scheme contains block depending on the delay.

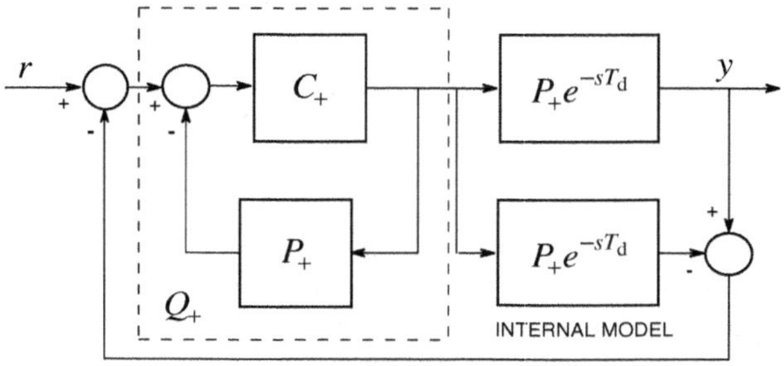

Fig. 3. *IMC* form of the SMITH predictor

It is possible to redraw the SMITH predictor into further schemes, which allow special interpretations. Figure 3 shows another equivalent scheme what corresponds to the well known *Internal Model Control* (*IMC*) scheme and principle. Figure 4 presents the resulting closed-loop with the serial regulator C_s equivalent to the application of the SMITH predictor.

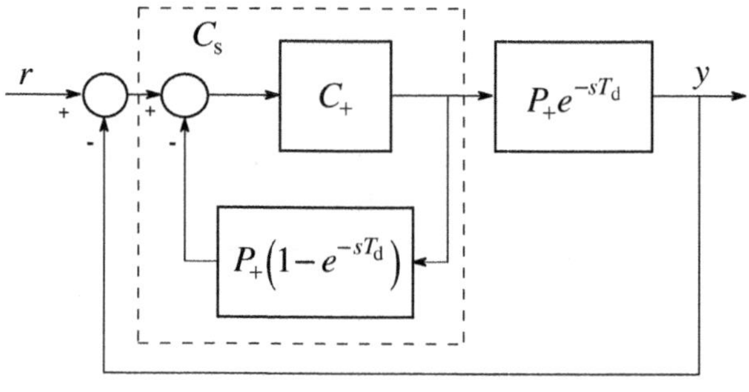

Fig. 4. The resulting closed-loop of the SMITH predictor

2 The YOULA Parameterization

A YOULA-parameterized (*YP*) closed-loop [4, 8] is shown in Fig. 5, where *e* is the error, *u* is the regulator output and *w* is the output disturbance signal, respectively.

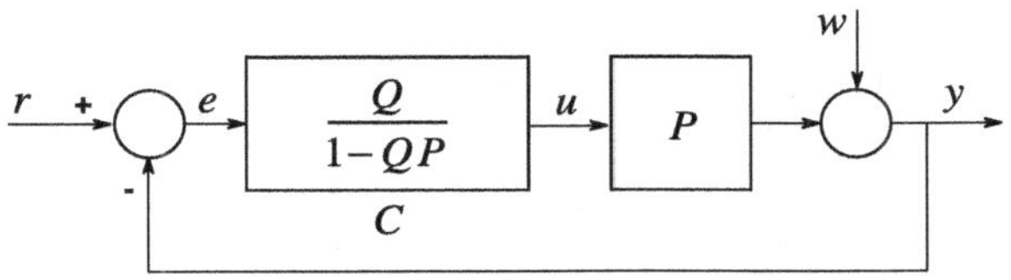

Fig. 5. YOULA-parameterized closed-loop

Here the plant *P* is stable and the *All-Realizable-Stabilizing* (*ARS*) regulator is

$$C = \frac{Q}{1 - QP} \tag{2}$$

The closed-loop transfer function or *Complementary Sensitivity Function* (*CSF*)

$$T = \frac{CP}{1 + CP} = QP \tag{3}$$

which is linear in the stable YOULA parameter *Q*.

It is well known that the *YP* regulator corresponds to the classical *IMC* structure shown in Fig. 6, where *r* is the reference signal, *u* is the regulator output, *y* is the output signal and *w* is the output disturbance signal, respectively. If there is no disturbance and the internal model is equal to the process transfer function, the signal fed back to the reference signal is zero, and the forward path *QP* determines the reference signal tracking. The feedback loop rejects the effect of the disturbance and of the plant/model mismatch.

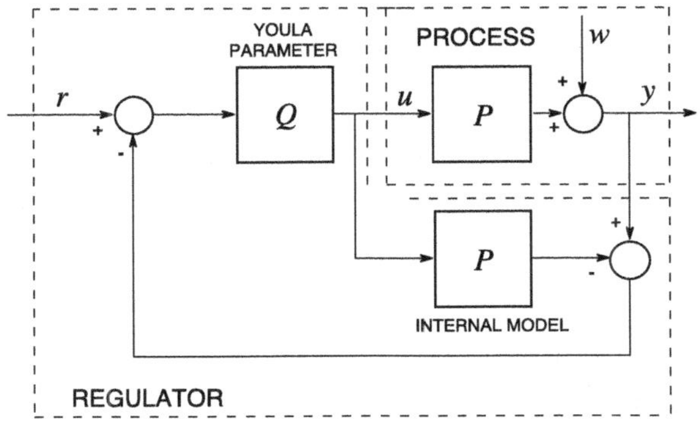

Fig. 6. *IMC* form of the *YP* closed-loop

It can also be well seen that Q_+ in Fig. 3 corresponds to the YOULA parameter. For a more detailed comparison consider the extension of *YP* regulator for more general case next.

3 A *G2DOF* Controller for Stable Linear Plants

The first systematic method introducing the *generic two-degree of freedom* (*G2DOF*) scheme was presented in [5, 8–10] when the process is open-loop stable and it is allowed to cancel the stable process poles, which case occurs at many practical tasks. *2DOF* in this approach means that the dynamics of reference signal tracking and that of disturbance rejecting are different. This framework and topology is based on the *YP* providing *ARS* regulators for open-loop stable plants and capable to handle the plant time-delay, too.

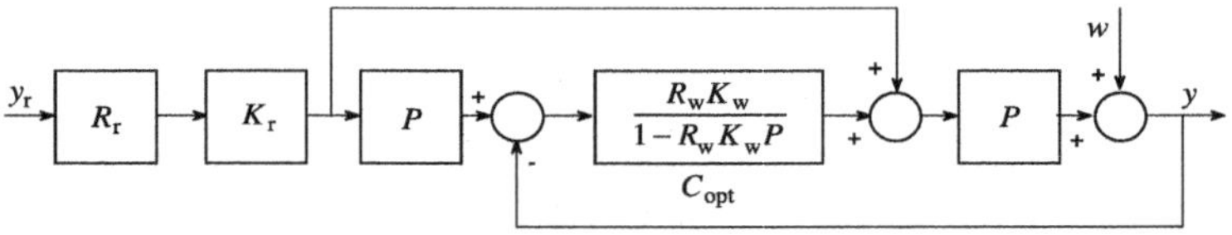

Fig. 7. The *generic 2DOF* (*G2DOF*) control system

A *G2DOF* control system is shown in Fig. 7 for the stable process

$$P = P_+ \bar{P}_- = P_+ P_- e^{-sT_d} \tag{4}$$

which is more general than what was used in (1), because here P_+ is stable and *Inverse-Stable-Realizable* (*ISR*), P_- is *Inverse-Unstable-Unrealizable* (*IUU*).

The optimal *ARS* regulator of the *G2DOF* scheme can be given by an explicit form

$$C_{opt} = \frac{R_w K_w}{1 - R_w K_w P} = \frac{Q_o}{1 - Q_o P} = \frac{R_w G_w P_+^{-1}}{1 - R_w G_w P_- e^{-sT_d}} \tag{5}$$

where

$$Q_o = Q_w = R_w K_w = R_w G_w P_+^{-1} \tag{6}$$

is the associated optimal *Y-parameter.* Furthermore

$$Q_r = R_r K_r = R_r G_r P_+^{-1} \quad ; \quad K_w = G_w P_+^{-1} \quad ; \quad K_r = G_r P_+^{-1} \tag{7}$$

The *YP* regulator (5) can be considered the generalization of the TRUXAL-GUILLEMIN [2, 8–10] method for stable processes.

It is interesting to see how the transfer characteristics of the closed-loop look like:

$$y = R_r K_r P y_r + (1 - R_w K_w P) w = R_r G_r P_- e^{-sT_d} y_r + \left(1 - R_w G_w P_- e^{-sT_d}\right) w = y_t + y_d \tag{8}$$

where y_t is the tracking (servo) and y_d is the regulating (or disturbance rejection) independent behaviors of the closed-loop response, respectively. So the delay e^{-sT_d} and

P_- cannot be eliminated, consequently the ideal design goals R_r and R_w are biased by G_rP_- and G_wP_-. Here R_r and R_w are assumed stable and usually strictly proper transfer functions, that are partly capable to place desired poles in the tracking and the regulatory transfer functions, furthermore they are usually referred as reference signal and output disturbance predictors. They can even be called as reference models, so reasonably $R_r(\omega = 0) = 1$ and $R_w(\omega = 0) = 1$ are selected. The unity gain of R_w ensures integral action in the regulator, which is maintained if the applied optimization provides $G_wP_-(\omega = 0) = 1$.

The role of R_r and R_w (predictors or filters) is threefold. They prescribe the tracking and regulatory properties of the control loop. They influence the magnitude of the actuating signal and also influence the robustness properties of the control system.

An interesting result was found [6] that the optimization of the *G2DOF* scheme can be performed in $\mathcal{H}_2$ and $\mathcal{H}_\infty$ norm spaces by the proper selection of the serial embedded filters G_r and G_w attenuating the influence of the invariant process factor P_-. Using $\mathcal{H}_2$ norm a *Diophantine-equation* (*DE*) should be solved to optimize these filters. If the optimality requires a $\mathcal{H}_\infty$ norm, then the NEVANLINNA-PICK (*NP*) approximation is applied.

After some straightforward block manipulations the *G2DOF* control system can be transformed to another form shown in Fig. 8, which is the generalized version of the classical *IMC* scheme in Fig. 6.

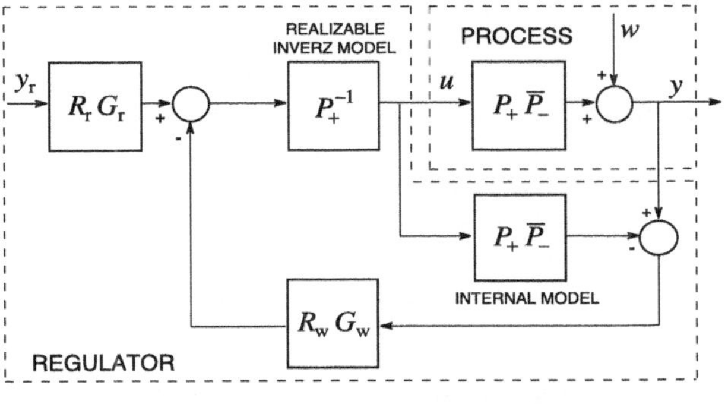

Fig. 8. The generalized *IMC* form of the *G2DOF* control system

4 SMITH Predictor as a Subclass of *G2DOF* Controllers

The previous two sections clearly show that the SMITH predictor is a special subclass of the *G2DOF* controllers with a *YP* parameterized regulator

$$Q_+ = \frac{C_+}{1 + C_+P_+} = \frac{C_+P_+}{1 + C_+P_+}P_+^{-1} = \frac{L_+}{1 + L_+}P_+^{-1} = R_+P_+^{-1} \tag{9}$$

if C_+ is stabilizing P_+, i.e., the delay free part of the process. Here the special *CSF*

$$T_+ = R_+ = \frac{L_+}{1 + L_+} \tag{10}$$

characterizing the closed-loop in Fig. 2 is the reference model R_+ and $L_+ = C_+P_+$ is its loop transfer function.

It is also easy to see that the resulting serial regulator of the SMITH predictor in Fig. 4 is

$$C_{\mathrm{s}} = \frac{Q_+}{1 - Q_+P_+e^{-sT_{\mathrm{d}}}} = \frac{C_+}{1 + C_+P_+\left(1 - e^{-sT_{\mathrm{d}}}\right)} = C_+K_{\mathrm{S}} \tag{11}$$

This formula presents the possible way of realization for a continuous-time (CT) case. Here K_{S} denotes a serial factor modifying the original C_+ regulator of the SMITH predictor

$$K_{\mathrm{S}} = \frac{1}{1 + C_+P_+\left(1 - e^{-sT_{\mathrm{d}}}\right)} = \frac{1}{1 + L_+\left(1 - e^{-sT_{\mathrm{d}}}\right)} \tag{12}$$

At the stability limit cross over frequency ω_{c}, where $L_+ = -1$ the factor K_{S} takes a considerable positive phase advance into the closed-loop

$$K_{\mathrm{S}} = \frac{1}{1 + (-1)\left(1 - e^{-sT_{\mathrm{d}}}\right)} = \frac{1}{1 - 1 + e^{-sT_{\mathrm{d}}}} = e^{sT_{\mathrm{d}}}\Big|_{\omega_{\mathrm{c}}} = e^{j\omega_{\mathrm{c}}T_{\mathrm{d}}} \tag{13}$$

This is the simple physical explanation of the success of the SMITH predictor [3].

Some early evaluations state that unfortunately the SMITH predictor is only good for tracking and not for disturbance rejection. This evaluation is wrong. The SMITH regulator was proposed for a one-degree of freedom (*1DOF*) closed-loop, so it is naturally not for *2DOF* purposes. The real problem of the SMITH regulator is that it allows the design of the closed-loop only via an indirect way by selecting $R_+ = T_+$, while the design procedure of the *G2DOF* scheme gives a direct procedure to design the independent tracking and disturbance rejection properties. This means that the original idea of SMITH was that a classical design of T_+ is necessary for the proper application. One must know that the YOULA parameterization and its application for regulator design was unknown for Otto SMITH when he invented his predictor.

5 The Discrete-Time Version of *G2DOF* Controllers

Although (11) suggests a proper way how to realize the SMITH regulator, it is not realistic to build any regulator containing the $e^{-sT_{\mathrm{d}}}$ delay element for continuous-time case. In the practice only the discrete-time (DT) version can be applied by computer realization. Consider the DT model of the CT process in the form of its pulse transfer function given by

$$\begin{aligned} G\left(z^{-1}\right) &= G_+\left(z^{-1}\right)\bar{G}_-\left(z^{-1}\right) = G_+\left(z^{-1}\right)G_-\left(z^{-1}\right)z^{-d} \\ G &= G_+\bar{G}_- = G_+G_-z^{-d} \end{aligned} \tag{14}$$

where G_+ is stable and *ISR*, G_- is *IUU* and z^{-d} corresponds to the discrete time-delay, where d is the integer multiple of the sampling time. (In a practical case the factor G_-

can incorporate the underdamped zeros and the neglected poles providing realizability, too). The optimal *ARS* regulator of the *G2DOF* scheme can be given now by

$$C_{\mathrm{o}} = \frac{R_{\mathrm{w}} K_{\mathrm{w}}}{1 - R_{\mathrm{w}} K_{\mathrm{w}} S} = \frac{Q_{\mathrm{o}}}{1 - Q_{\mathrm{o}} G} = \frac{R_{\mathrm{w}} G_{\mathrm{w}} G_{+}^{-1}}{1 - R_{\mathrm{w}} G_{\mathrm{w}} G_{-} z^{-d}} \tag{15}$$

which corresponds to the CT case of (5), furthermore (6) and (7) are formally exactly the same for DT case. The transfer characteristics of the closed-loop is now

$$y = R_{\mathrm{r}} K_{\mathrm{r}} G y_{\mathrm{r}} + (1 - R_{\mathrm{w}} K_{\mathrm{w}} G) w = R_{\mathrm{r}} G_{\mathrm{r}} G_{-} z^{-d} y_{\mathrm{r}} + \left(1 - R_{\mathrm{w}} G_{\mathrm{w}} G_{-} z^{-d}\right) w = y_{\mathrm{t}} + y_{\mathrm{d}} \tag{16}$$

Because the optimization of the embedded filters G_{r} and G_{w} requires special knowledge and practice of getting the solution from a *DE* and *NP* approximation, suboptimal design is mostly applied assuming $G_{\mathrm{r}} = G_{\mathrm{w}} = 1$. In such cases the influence of the invariant process factors are not attenuated at all, so they appear in the closed-loop characteristics (15) directly. Such *G2DOF* control scheme is shown in Fig. 9.

It follows from the above discussion that it is not necessary to apply the classical SMITH predictor principle, instead it is more effective to use the regulator design procedure of the *G2DOF* controller scheme.

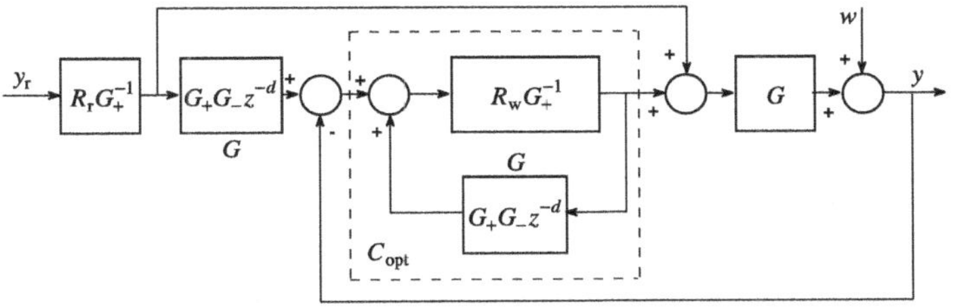

Fig. 9. Discrete-time *G2DOF* control system for the suboptimal $G_{\mathrm{r}} = G_{\mathrm{w}} = 1$ case

6 Simple Examples

Example 1
Consider a very simple first order time-delay process.

$$P = \frac{1}{1+10s} e^{-5s} \quad ; \quad P_{+} = \frac{1}{1+10s} \quad ; \quad \bar{P}_{-} = e^{-5s} \quad ; \quad P_{-} = 10 \tag{17}$$

The tracking and disturbance rejection reference models are

$$R_{\mathrm{r}} = \frac{1}{1+4s} \text{ and } R_{\mathrm{w}} = \frac{1}{1+2s} \tag{18}$$

Here $P_{-} = 1$, therefore $G_{\mathrm{r}} = G_{\mathrm{w}} = 1$ is the optimal selection for the embedded filters.

Design a YOULA-paramererized optimal regulator.

$$C_{\mathrm{opt}} = \frac{R_{\mathrm{w}} G_{\mathrm{w}} P_{+}^{-1}}{1 - R_{\mathrm{w}} G_{\mathrm{w}} P_{-} e^{-sT_{\mathrm{d}}}} = \frac{1}{1 - R_{\mathrm{w}} e^{-sT_{\mathrm{d}}}} R_{\mathrm{w}} P_{+}^{-1} = \frac{1}{1 - \frac{1}{1+2s} e^{-5s}} \frac{1+10s}{1+2s} = \frac{(1+2s)(1+10s)}{1+2s-e^{-5s}} \tag{19}$$

and the optimal serial compensator is

$$R_{\mathrm{r}} K_{\mathrm{r}} = R_{\mathrm{r}} G_{\mathrm{r}} P_{+}^{-1} = R_{\mathrm{r}} P_{+}^{-1} = \frac{1+10s}{1+4s} \tag{20}$$

Both transfer functions are realizable. Because $C_{\mathrm{opt}}(s=0) = \infty$ the regulator is integrating obtained from the condition $R_{\mathrm{w}}(s=0) = 1$. The optimal final closed-loop is shown in Fig. 10. Although all blocks are realizable in this scheme it is very unrealistic that the real CT models of the true process are applied in a practical application. Here the real difficulty is the realization of the time-delay. So this example stands only to represent the *YP* based *G2DOF* design procedure.

It is easy to check that the closed-loop characteristics is

$$y_{\mathrm{opt}} = R_{\mathrm{r}} e^{-sT_{\mathrm{d}}} y_{\mathrm{r}} + \left(1 - R_{\mathrm{w}} e^{-sT_{\mathrm{d}}}\right) w = \frac{1}{1+4s} e^{-5s} y_{\mathrm{r}} + \left(1 - \frac{1}{1+2s} e^{-5s}\right) w \tag{21}$$

according to the general theory.

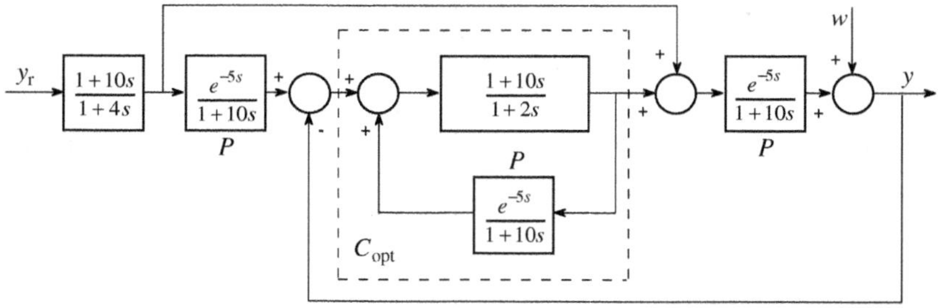

Fig. 10. The designed optimal closed-loop of the example

Example 2

Consider the DT model of a very simple first order time-delay process

$$G = \frac{0.2z^{-1}}{1-0.8z^{-1}} z^{-3} = \frac{0.2z^{-4}}{1-0.8z^{-1}} \quad ; \quad G_{+} = \frac{0.2z^{-1}}{1-0.8z^{-1}} \text{ and } G_{-} = 1 \tag{22}$$

It is required to speed up the process by a closed-loop.

Design a *YP* controller. Select the reference models

$$R_{\mathrm{r}} = \frac{0.8z^{-1}}{1-0.2z^{-1}} \text{ and } R_{\mathrm{w}} = \frac{0.5z^{-1}}{1-0.5z^{-1}} \tag{23}$$

Because $G_- = 1$, there is no optimization task, so the selections $G_r = 1$ and $G_w = 1$ are optimal. The optimal regulator is

$$C_{\text{opt}} = \frac{R_w G_w G_+^{-1}}{1 - R_w G_w G_- z^{-d}} = \frac{1}{1 - R_w z^{-d}} R_w G_+^{-1}$$
$$= \frac{1}{1 - \frac{0.5z^{-1}}{1-0.5z^{-1}} z^{-3}} \frac{0.5z^{-1}}{1 - 0.5z^{-1}} \frac{1 - 0.8z^{-1}}{0.2z^{-1}} = \frac{2.5(1 - 0.8z^{-1})}{1 - 0.5z^{-1} - 0.5z^{-4}} \tag{24}$$

and the serial compensator is

$$R_r G_+^{-1} = \frac{0.8z^{-1}}{1 - 0.2z^{-1}} \frac{1 - 0.8z^{-1}}{0.2z^{-1}} = \frac{4(1 - 0.8z^{-1})}{1 - 0.2z^{-1}} \tag{25}$$

$$R_r G_+^{-1} = \frac{0.8z^{-1}}{1 - 0.2z^{-1}} \frac{1 - 0.8z^{-1}}{0.2z^{-1}} = \frac{4(1 - 0.8z^{-1})}{1 - 0.2z^{-1}} \tag{25}$$

The optimal final closed-loop is shown in Fig. 11. Observe that $R_w(z = 1) = 1$, i.e. the regulator is an integrating one, which follows from the condition $R_w(z = 1) = 1$.

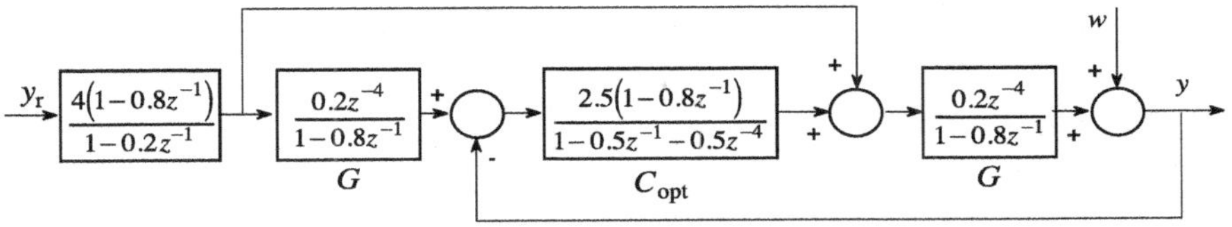

Fig. 11. The designed optimal closed-loop of the example

The closed-loop characteristics is

$$y_{\text{opt}} = R_r z^{-d} y_r + \left(1 - R_w z^{-d}\right) w = \frac{0.8z^{-1}}{1 - 0.2z^{-1}} z^{-3} y_r + \left(1 - \frac{0.5z^{-1}}{1 - 0.5z^{-1}} z^{-3}\right) w$$
$$= \frac{0.8z^{-4}}{1 - 0.2z^{-1}} y_r + \left(1 - \frac{0.5z^{-4}}{1 - 0.5z^{-1}}\right) w \tag{26}$$

which exactly corresponds to our design goals.

This example shows that there is no applicability problem for DT regulator design. These filters are easy to be realized in a computer controlled system.

Example 3

The continuous first order plant with significant time delay is given by the transfer function

$$P(s) = \frac{1}{1 + 10s} e^{-30s} \tag{27}$$

The plant is sampled with sampling time $T_s = 5$ s and a zero order hold is applied at its input. Let us design a *PI* controller ensuring about 60° of phase margin, a Smith predictor and a YOULA-parameterized controller. Compare the reference signal tracking and output disturbance rejection behaviour of the three control systems. Demonstrate the effect of time delay mismatch.

The pulse transfer function of the plant is

$$G(z) = \frac{0.3935}{z - 0.6065} z^{-6} \tag{28}$$

The pulse transfer function of the *PI* controller [7] applying pole cancellation with a gain ensuring the required phase margin is

$$C_{\mathrm{PI}}(z) = 0.204 \frac{z - 0.6065}{z - 1} \tag{29}$$

The SMITH predictor controller C_+ is designed for the delay free process as a *PI* controller and it is obtained as

$$C_+(z) = 2.5 \frac{z - 0.6065}{z - 1} \tag{30}$$

Then it is transformed to the SMITH predictor form according to the discretized version of (11).

$$C_s(z) = \frac{2.5z^7 - 1.516z^6}{z^7 - 0.01636z^6 - 0.9837} \tag{31}$$

In the case of the YOULA parameterized controller let us choose the disturbance filter

$$R_w(s) = \frac{1}{1 + 5s} \tag{32}$$

and rhe reference filter as

$$R_r(s) = \frac{1}{1 + 8s} \tag{33}$$

whose pulse transfer functions are

$$R_w(z) = \frac{0.6321}{z - 0.3679} \text{ and } R_r(z) = \frac{0.4647}{z - 0.5353}, \tag{34}$$

respectively.

The YOULA parameter supposing $G_r = G_w = 1$ is

$$Q(z) = R_w(z) G_+^{-1}(z) = \frac{0.6321}{z - 0.3679} \cdot \frac{z - 0.6065}{0.3935} \tag{35}$$

Figure 12 shows the step response and a shifted step disturbance rejection of the three controllers.

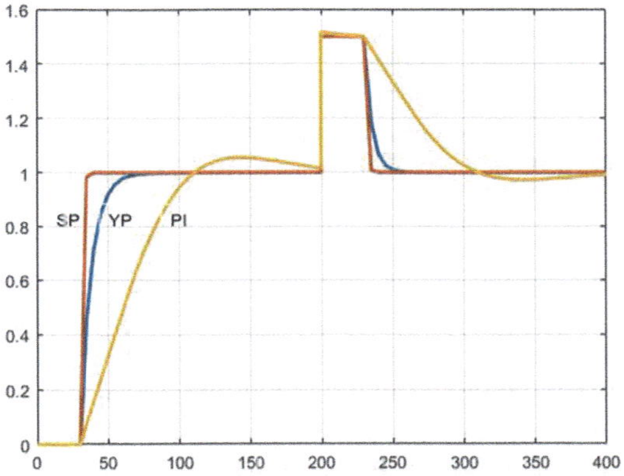

Fig. 12. Step response and disturbance rejection dynamics of the *PI*, SMITH and YOULA controllers

It is seen that in case of significant time delay SMITH predictor and the YOULA parameterized controllers ensure significant acceleration compared to the *PI* controller.

Figure 13 demonstrates the effect of time delay mismatch in the case of the SMITH and the YOULA controllers. The time delay of the model is 30, while the time delay of the process is 33.

It is seen that the YOULA parameterized controller tolerates much better the inaccuracy of the parameter than the SMITH predictor. While the SMITH predictor is very sensitive to the inaccuracies in the parameters (it is not robust), the filters in the YOULA parameterized controller can be designed for robust behaviour [11].

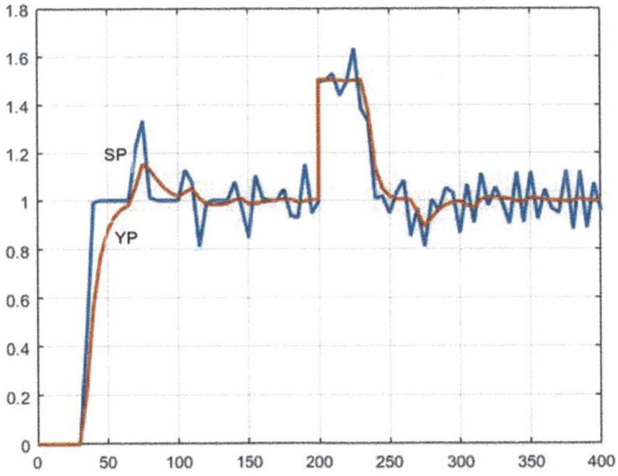

Fig. 13. The effect of time delay mismatch in case of the SMITH and the YOULA controllers

These are, of course, very simple examples standing only to present the simplicity of the *G2DOF* controller scheme, which should replace the classical approach of a SMITH predictor.

7 Conclusions

The SMITH predictor is a classical method of handling time-delay in closed-loop control design. It is shown that this method is a subclass of the *YP* based *G2DOF* control scheme. An obvious drawback of the SMITH predictor is that the closed-loop properties can not be designed directly using simple algebraic methods, which is possible in the *G2DOF* structure. The *G2DOF* scheme allows even the optimal attenuation of the invariant process factors. The appropriate choice and design of the filters allows to influence such important properties as performance and robustness. So the paper suggests to use the newer methodology to design DT controllers for time-delay processes.

The role of the SMITH predictor remains important in the history of control engineering, because it was one of the first, easy to use and widely applied method to simply eliminate the influence of the delay in the design of closed-loop control properties. Nevertheless this method is sensitive to the accurate knowledge of the time delay.

The recent theoretical developments and easily applicable algebraic design methods allow to use more effective and more general controller design procedures.

References

1. Smith, O.J.M.: Closed control of loops with dead time. Chem. Eng. Proc. **53**, 217 (1957)
2. Horowitz, I.M.: Synthesis of Feedback Systems. Academic Press, New York (1963)
3. Åström, K.J., Wittenmark, B.: Computer Controlled Systems, p. 430. Prentice-Hall, Englewood Cliffs (1984)
4. Maciejowski, J.M.: Multivariable Feedback Design. Addison Wesley, Wokingham (1989)
5. Keviczky, L.: Combined identification and control: another way (invited plenary paper). In: 5th IFAC Symposium on Adaptive Control and Signal Processing, ACASP 1995, Budapest, Hungary, pp. 13–30 (1995)
6. Keviczky, L., Bányász, Cs.: Optimality of two-degree of freedom controllers in $\mathcal{H}_2$- and $\mathcal{H}_\infty$-norm space, their robustness and minimal sensitivity. In: 14th IFAC World Congress, F, pp. 331–336. PRC, Beijing (1999)
7. Tan, N.: Computation of stabilizing PI and PID controllers for process with time delay. ISA Trans. **44**, 213–223 (2005)
8. Keviczky, L., Bányász, Cs.: Two-Degree-of-Freedom Control Systems (The Youla Parameterization Approach). Elsevier, Academic Press, Bányász (2015)
9. Keviczky, L., Bars, R., Hetthéssy, J., Bányász, Cs.: Control Engineering. Springer, Singapore (2018)
10. Keviczky, L., Bars, R., Hetthéssy, J., Bányász, Cs.: Control Engineering: MATLAB Exercises. Springer, Singapore (2018)
11. Bányász, Cs., Keviczky, L., Bars, R.: Influence of time delay mismatch for robustness and stability. In: IFAC TDS, Budapest, Hungary, pp. 248–253 (2018)

The Use of a Torque Meter to Improve the Motion Quality at Very Slow Velocity of a Servo Drive with a PMSM Motor

Bogdan Broel-Plater(✉), Krzysztof Jaroszewski, and Daniel Figurowski

West Pomeranian University of Technology Szczecin, Szczecin, Poland
bbp@zut.edu.pl

Abstract. The article shows that by using a torque meter in the servo structure, the quality of motion control can be significantly improved. The classic cascade structure controlling a servo drive with a PMSM motor when non-linear friction with a strong Stribeck friction occurs does not allow to achieve the accuracy of motion required in many applications. The proposed structure overcomes this problem, provides the required motion quality without the need to accurately identify friction characteristics. The second chapter of the article presents the mathematic model of the PMSM motor used during simulation experiments and its implementation in the Matlab/Simulink environment. Section 2.3 of the article presents the modification of the control structure using information from the torque meter and the structure of the correction block itself. The last part shows the waveforms obtained in the simulation for the classical and modified systems. The authors' attention focuses on the problem of precise control with minimum speeds of mechatronic systems, in which there is strong Stribeck friction and the presented paper is the result of work aimed at preliminary testing of a certain concept of modifying the structure of the control system.

Keywords: Servo drive · Braking torque · Precise slow motion control · Torque meter

1 Introduction

Servo drives are the basic element used to build all types of machine tools. In the case of machining, the drives work with variable loads (masses, moments of inertia) and in the presence of strong interference (cutting forces and torques), which makes it difficult to obtain high quality movement. A lot of work is devoted to this subject, which describes the change of design or control to ensure the desired motion quality. The authors of this article deal with various modifications of the servo drive system in order to achieve the most effective implementation of the servo motion trajectory, in particular for its low rotational speeds [2–6]. In this subject there are leading researches using classical control theory [1, 9, 11]. However, in this paper software-hardware solutions is described.

A specific group of machines are measuring machines, which are required to have high precision motion at slow velocity. In the case of such machines, fortunately, there

A. Bartoszewicz et al. (Eds.): PCC 2020, AISC 1196, pp. 137–147, 2020.
https://doi.org/10.1007/978-3-030-50936-1_12

is a lack of distortion and stability of aggravating torques. In this work, the servo drive control in just such a machine has been presented.

Regardless of the machine type its movement is affected by non-linear friction, and especially by Stribeck's friction in the case of low-speed motion.

Minimizing the impact of Stribeck friction by modifying the machine structure is very difficult, so the only chance is to properly design the control system.

2 Servo Drive Structure

The subject of the study was the mechatronic structure, which is a fragment of the measuring machine, shown in Fig. 1, where: 1 - PMSM motor, 2 - torque meter, 3 - measuring table, 4 - lead screw, 5 - gibs, 6 - control system, X – position measuring table, V – velocity measuring table, Iq – current in q axis, Id – current in d axis, Uq – voltage in q axis, Ud – voltage in d axis.

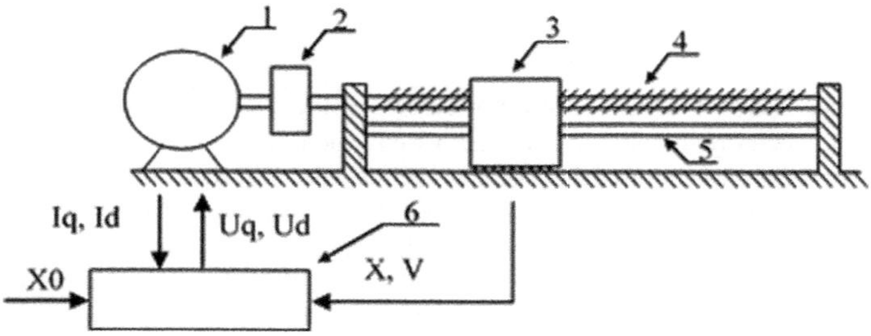

Fig. 1. Studied mechatronic system

2.1 The Motor PMSM

Permanent magnet synchronous motors (PMSMs) are today widely used in various industrial solutions, including servo drives [8–10]. Equations - a two-phase peripheral model written in the d-q system, converted to measurement table coordinate system, developed using a number of simplifying assumptions and using the theory of spatial vectors - describing this type of motor:

$$dX / dt = V \tag{1}$$

$$m\, dV / dt = K_t I_q - F_H(V, X) \tag{2}$$

$$dI_q / dt = 1 / L_s U_q - R_s / L_s I_q - a_d V I_d - K / L_s V \tag{3}$$

$$dI_d / dt = 1 / L_s U_d - R_s / L_s I_d + a_d V I_q \tag{4}$$

where: m – moving mass, R_s – winding resistance, L_s – winding inductance, F_H – the sum of all system resistance forces, K_t, a_q, a_d – constants characterize the motor.

The motor diagram based on this description, implemented in the Matlab/Simulink environment, is shown in Fig. 2. The average values of the parameters (for drive motors used in measuring machines) appearing in the equations, after calculations, are visible in the corresponding blocks in Fig. 2.

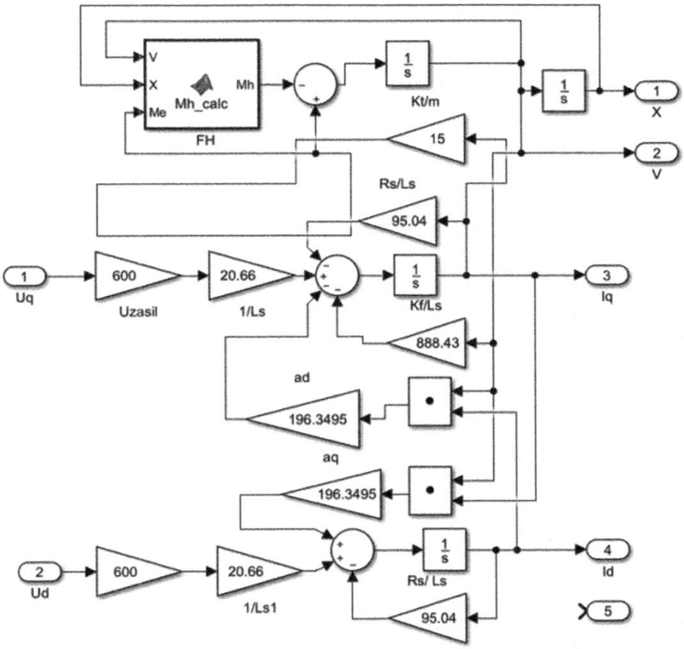

Fig. 2. The PMSM motor diagram based on Eqs. (1)–(4), implemented in the Matlab/Simulink environment

Due to the fact that the work involved controlling the linear movement of the measuring table, it was assumed that performing one rotation of the motor shaft shifts the measuring table by 1 mm; so linear motion equations can be written in the form of 1–4 with the values of the coefficients converted to the axis system related to linear motion.

Occurring in Fig. 2, in the block FH, the function called *Mh_calc* performs non-linear and discontinuous characteristics of the braking torque. The implementation (code) of this function in the Matlab environment is shown below:

```
function FH = Mh_calc(V, X, Me)
if abs(V) <= 0.00000001
   Mn = Me;
  if Mn > (A1*exp(A2*V))
   FH = A1*exp(A2*V) + A4*V + A5*V*V;
elseif Mn <= (-A1*exp(-A2*V))
```

```
    FH = −A1*exp(−A2*V) + A4*V−A5*V*V;
  else
    FH = Mn;
  end;
elseif V > 0.0
    FH = A1*exp(A2*V) + A4*V + A5*V*V;
else
    FH = −A1*exp(−A2*V) + A4*V−A5*V*V;
  end;
```

The code describing the friction function shows the method used to account for Stribeck friction, the effect of which appears only when the velocity of motion *V* is different from zero. This function reflects the simultaneous occurrence of various types of friction, catching torque, as well as the asymmetry of mounting the rotor on the stator. Coefficients *A1*, *A2* affect Stribeck friction, *A4*, *A5* - two forms of viscous friction. During simulation experiments, the coefficients appearing in this function were given the values:

$$A1 = 48.7; A2 = -100000; A4 = 0.915; A5 = 0.915;$$

The friction characteristics used in the simulation experiments with the coefficient values given above are shown in Fig. 3, for (a) the entire speed range and (b) a very narrow speed range that was of interest during testing (showing influence of Stribeck friction on friction characteristics).

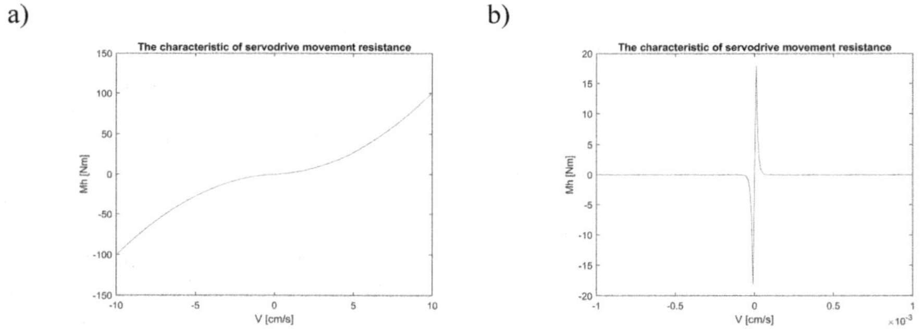

Fig. 3. The friction characteristics used in the simulation experiments for a) the entire velocity range, b) a velocity range that was of interest during testing

2.2 Classic Servo Drive Structure

The structure of a typical servo drive with an synchronous motor contains a cascade of four PID controllers, shown in Fig. 4. In the external loop, the position of the servo motor is controlled by the PID *X* controller, it develops the set point for the PID *V* - servo motor velocity controller, which in turn develops the set point for PID *Iq* controller - current regulator. All PID controllers are described:

$$PTD(s) = P \cdot (1 + I/s + (D \cdot N \cdot s)/(s + N)) \quad (5)$$

Moreover for each controller the output signal is limited to the range of <−1, 1> and back-calculation anti-windup with coefficient *Kb* in PID *X* and PID *V* controllers is used. The parameter values of each controller were selected so that there would be no overshoot in the event of large changes in the set point, and are presented in Tab. 1. The parameters values of the position (PID *X*) and speed controllers (PID *V*) were determined as described in [7], while in the case of current controllers it was assumed that reducing control algorithm to type P with a high gain value is sufficient to achieve the required control quality. All controllers operate on the relative values of control errors determined taking into account the measuring ranges of the sensors (Table 1).

Table 1. PID's coefficients parameters.

	P	*I*	*D*	*N*	*Kb*
PID *X*	27	3.35	0	100	10
PID *V*	0.39797	0.72357	0	100	10
PID *Iq*	100	0	0	100	–
PID *Id*	100	0	0	100	–

In this structure, an additional loop is created by the PID *Id* current controller, and most often it works with a set point equal to zero ($Id0 = 0$). Using such a classic structure, it is difficult to obtain very high motion quality, especially for movements at low velocity or minimal displacements if the resistance characteristics are strongly non-linear, especially with degressive non-linearity (Stribeck friction), as shown in the relevant figures: Figs. 9, 10, 11, 12, 13, 14 and 15.

Blocks, named in Fig. 4.: *Z_x* and *Z_v*, *Z_Id* and *Z_Iq* define the appropriate limit values of corresponding variables. The conversion of measured absolute values (*V*, *I_q*, *I_d*, *X*) into relative values taking into account the measuring ranges of the relevant sensors (*Z_x*, *Z_Iq*, *Z_Id*, *Z_x*) was introduced. Such conversion take into account the influence of the sensor measurement on the operation of the control system.

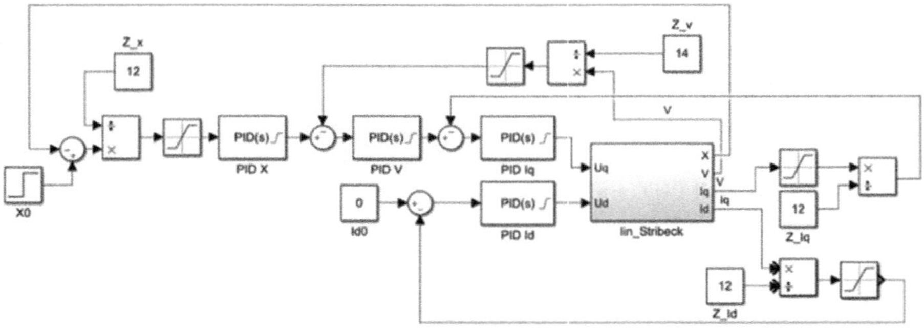

Fig. 4. The structure of a typical servo drive with an executive synchronous motor

2.3 Modified Servo Drive Structure

In order to improve the quality of servo drive control with the PMSM motor, it was decided to check the advisability of using a torque meter mounted between the motor and the load. As a result of testing many applications of signal (*dM*, shown in Fig. 5) from this meter, it was found that high quality control is ensured by using this signal to correct the current set point (*Iq0*) in the q axis. This signal, however, must be pre-processed, which is ensured by the block called *Korektor*. The *Korektor* block is visible in the modified structure of the control system – Fig. 5.

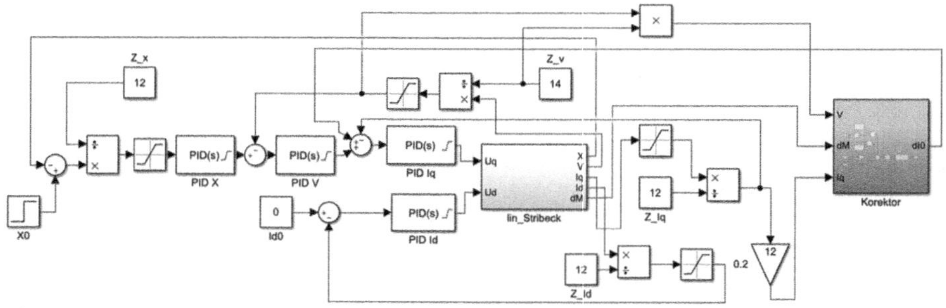

Fig. 5. The modified structure of servo drive with *Korektor* block

This block is a non-linear dynamic block with the structure shown in Fig. 6. The structure of this corrector only includes linear (*K(v)*) and quadratic (*K(v·v)*) viscous friction, bypassing Stribeck friction. The values of all coefficients in corrector's blocks were adjusted experimentally on the simulation way.

However, studies have shown that high quality motion will be ensured by using even linear (very simplified) friction characteristics in the corrector. It is easy to determine based on the measurement of the current consumed by a motor running at a constant speed. The occurrence of Stribeck friction means that such a linear characteristic is easiest to determine for a high (constant) speed of movement, where this friction no

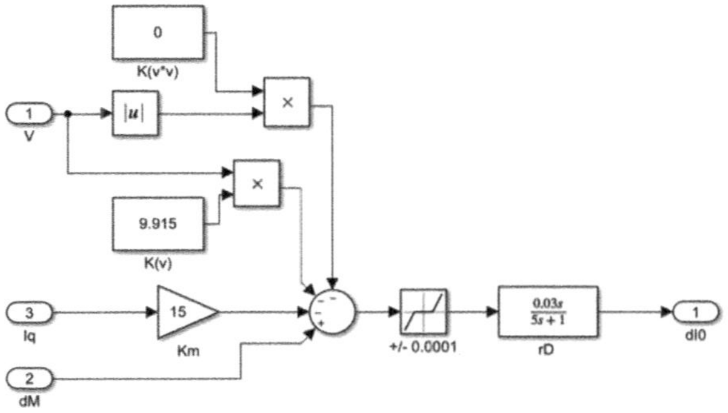

Fig. 6. The *Korektor* block

longer affects the movement resistance. Figure 7 shows the real (including, among others, Stribeck friction) and the corrector (linear) friction characteristics.

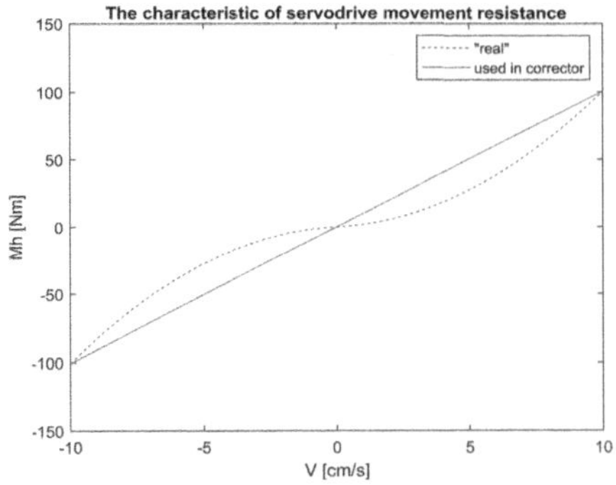

Fig. 7. The real (including, among others, Stribeck friction) and the equalizer (linear) friction characteristics

During the simulation, the signal from the torque meter (*dM*) was obtained by appropriate modification of the engine structure, as shown in Fig. 8.

3 Simulation Results

The research was started with an experiment aimed at choosing the values of the parameters of the controllers ensuring no overshoot after a large step change in the *X0* setpoint. The condition of the lack of overshoot is crucial for the correct operation of measuring machines. The course of the system response for such selected settings is shown in Fig. 9. The response (a) of the output value *X* in the classical system clearly shows the influence of Stribeck friction - when the adjustment error is small. Modifying the structure has definitely improved the quality of control (b).

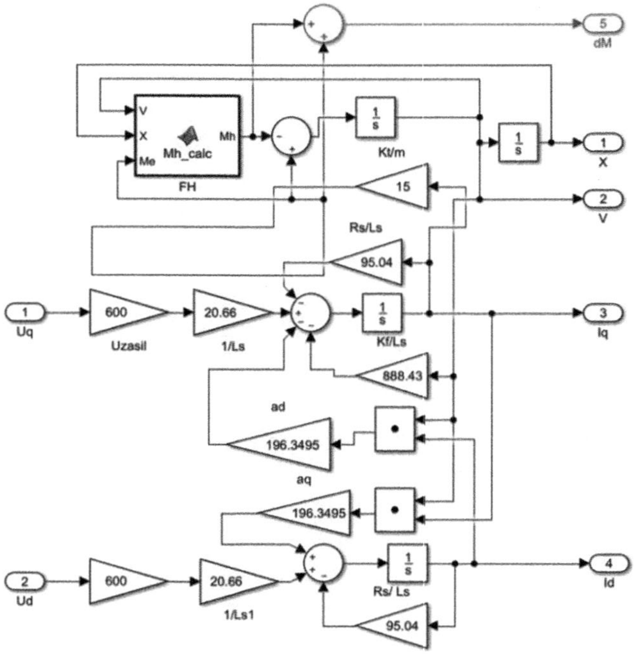

Fig. 8. Structure of the block generating in simulation signal from the torque meter

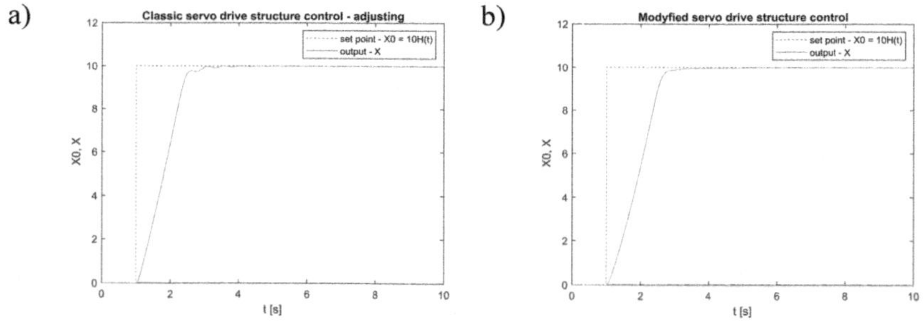

Fig. 9. The high (10) step set point change and response, a) the structure of classical, b) the structure of the modified

Figure 10 and Fig. 11 show the waveforms after a small and very small step change in the set point (typical for measuring machines). For a change of 0.001, the servo drive in the classical structure stay motionless.

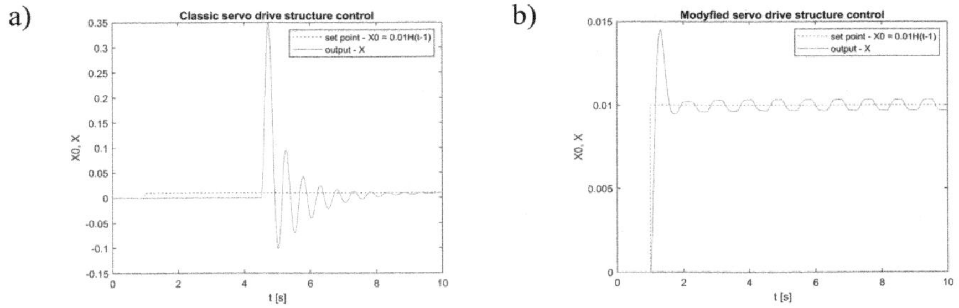

Fig. 10. Small (0.01) a step set point change and response, a) the structure of classical, b) the structure of the modified

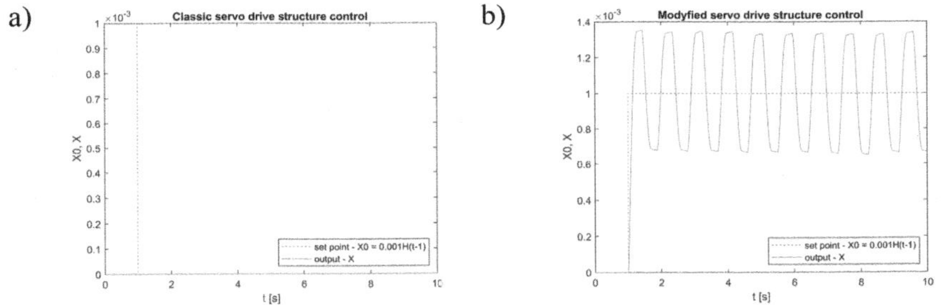

Fig. 11. The smallest (0.001) a step set point change and response, a) the structure of classical, b) the structure of the modified

Figure 12 and Fig. 13 show the controlled variable waveforms after a slow linear change in the set point. In the classical structure, the waveforms were obtained with a long delay and were characterized by very high overshoot and large oscillations. In the modified structure, the system reacted with virtually no delay and with very little overshoot with fast suppression of the transition process.

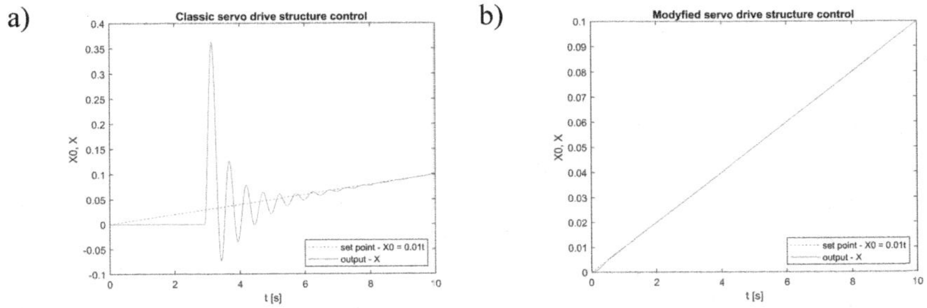

Fig. 12. Small (0.01) a linear change of reference values and response, a) the structure of classical, b) the structure of the modified

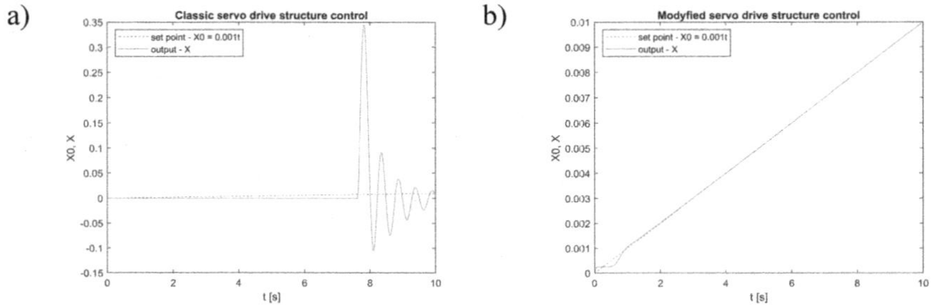

Fig. 13. The smallest (0.001) a linear change of reference values and response, a) the structure of classical, b) the structure of the modified

Figure 14 and Fig. 15 show the results for the sinusoidal reference signal. The classic servo drive for 0.001 amplitude did not move at all, and for 0.01 amplitude it was characterized by a long delay and 3500% overshoot. However, the system with the modified structure kept pace with the changes of the set point with small oscillatory ones.

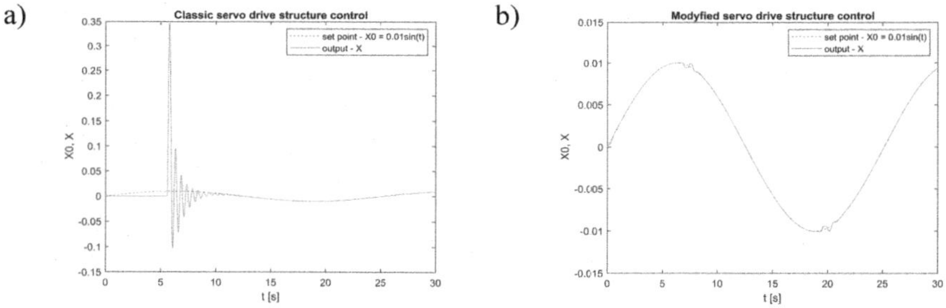

Fig. 14. Small (0.01) the sinusoidal change of reference values and response, a) the structure of classical, b) the structure of the modified

4 Summary

As shown by simulation results, including a torque meter as modification of the classical servo drive structure with PMSM motor allows to significantly improve the motion quality at very slow velocity movements - even in the presence of strong Stribeck friction. The advantage of the proposed method is that the correction signal affects the current controller set point. In many modern servo drives, e.g. B&R ACCOPS, it is possible to include an additional external signal in the current control loop. What is more extensive simulation studies carried out by the authors showed that the modified structure of the control system described in the paper is very little sensitive to the accuracy of the friction model used in the *Korektor* block. For example, the simulation results of a modified control system shown in Fig. 9, 10, 11, 12, 13, 14 and 15 were obtained assuming

a) b)

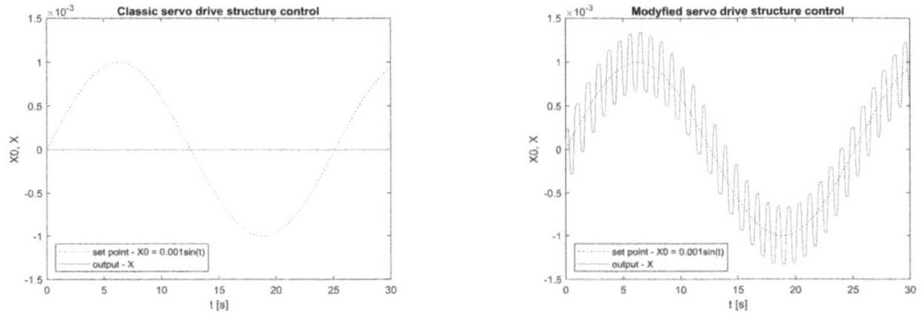

Fig. 15. The smallest (0.001) the sinusoidal change of reference values and response, a) the structure of classical, b) the structure of the modified

that the corrector uses linear friction characteristics (*K(v·v) = 0*) and the "real" friction characteristics (block F_H in Fig. 2) contain the elements linear and quadratic, and a member related to Stribeck friction - a comparison of both characteristics is shown in Fig. 7. In the near future, practical verification of the effectiveness of the modified servo drive structure is planned.

References

1. Brasel., M.: A gain-scheduled multivariable LQR controller for Permanent Magnet Synchronous Motor. In: MMAR 2014, pp. 722–725 (2014). ISBN 978-1-4799-5081-2
2. Broel-Plater, B., Jaroszewski, K.: Comparison of improving servo drive position accuracy methods. In: MMAR 2014, pp. 857–861 (2014). ISBN 978-1-4799-5081-2
3. Broel-Plater, B., Jaroszewski, K.: Fractional order current controller in servo drive control system. In: MMAR 2015, pp. 626–631 (2015). ISBN 978-1-4799-8701-6
4. Broel-Plater, B., Dworak P., Jaroszewski, K.: Fractional order controller in servo drive – case of cogging moment. In: MMAR 2016, pp. 832–837 (2016). ISBN 978-1-5090-1866-6
5. Broel-Plater, B., Jaroszewski, K.: Study of the efficiency of an algorithm that improves the accuracy of slow-motion control of servo drive. In: MMAR 2017, pp. 53–57 (2017). ISBN 978-1-5386-2402-9
6. Broel-Plater, B., Jaroszewski, K.: Improving the quality of the classic servo drive by modifying the motion control algorithm. In: MMAR 2017, pp. 43–46 (2017). ISBN 978-1-5386-2402-9
7. Broel-Plater, B., Jaroszewski, K., Dworak P.: Minimizing the impact of non-linear stribeck friction on positioning of a servo drive. In: MMAR 2018, pp. 870–875 (2018). 978-1-5386-4325-9
8. Hughes, A.: Electric Motors and Drives, 3rd edn. Elsevier, Oxford (2006)
9. Krishnan, R.: Electric Motor Drives. Modeling, Analysis and Control. Prentice Hall, New Jersey (2001)
10. Krishnan, R.: Permanent Magnet Synchronous and Brushless DC Motor Drives. CRC Press, Boca Raton (2010)
11. Younkin, G.W.: Industrial Servo Control Systems. Fundamentals and Applications, 2nd edn. Marcel Dekker Inc., New York (2003)

Design of Terminal Sliding Mode Controllers with Application to Automotive Control Systems with Model Uncertainties

Paweł Skruch(✉)

Department of Automatic Control and Robotics, AGH University of Science and Technology, al. A. Mickiewicza 30/B1, 30-059 Kraków, Poland
pawel.skruch@agh.edu.pl

Abstract. The paper describes the design of terminal sliding model controllers for several types of dynamical systems with model uncertainties. First- and second-order representations of the systems are investigated. These representations are mathematically described by matrix differential equations. Both linear and non-linear systems are considered. Sliding surfaces are defined by equations involving the state of the system and the expected trajectory. Finite-time stability of the corresponding closed-loop systems is proved via Lyapunov stability theory. Applications of the designed controllers are illustrated on automotive control systems.

Keywords: Terminal sliding mode · Sliding surface · Linear system · Non-linear system · Control · Automotive

1 Introduction

1.1 Motivation

Control systems are impacted by many types of uncertainties caused by measurement systems, model parameter identification experiments and working environment. Consequently, the structure of the control system may vary during the control process. Moreover, making right decision in such uncertain environment is extremely difficult, especially in safety critical applications. Thus, one of the main property expected from the control system would be, besides performance, robustness against external disturbances, parameter uncertainties and variations. The concept of the sliding mode control (SMC) can be considered as such robust method to control both linear and non-linear dynamical systems with model uncertainties and external disturbances. The concept requires a definition of so-called sliding surface to which a designed controller shall force the state of the system starting from an initial condition. When a control law is based on linear differential equations, then asymptotic stability of the closed-loop system can be achieved. By describing the controller by non-linear and non-smooth differential

A. Bartoszewicz et al. (Eds.): PCC 2020, AISC 1196, pp. 148–160, 2020.
https://doi.org/10.1007/978-3-030-50936-1_13

equations finite time convergence to the equilibrium is possible – such controller is called as terminal sliding mode controller (TSMC). The paper concentrates on the design of TSMC.

1.2 Related Work

The TSMC-based methods have been investigated until now in many papers. Below short review of the main types of the TSMC-based solutions developed so far is presented. TSMC in [3] concerns to an uncertain linear system described in a first-order form. A linear second-order representation of the non-disturbed dynamical system is investigated in [9] and in [10] where the disturbances are included. TSMC for a single-input nonlinear control system without disturbances is discussed in [7]. In [1] a nonlinear second-order system defined in two-dimensional state space with parameter uncertainties, an external disturbance and constraints imposed on the control signal is considered. Author proposes a sliding control that guarantees system error convergence to zero in finite time, is robust and in addition minimizes the integral of the absolute value of the system error. The TSMC design scheme can be also applied to uncertain linear dynamical systems with the help of some fuzzy logic mechanism as presented in [21]. Another fuzzy TSMC for some class of non-linear systems can be found in [11]. A neural network-based multivariable fixed-time terminal sliding mode control strategy for re-entry vehicles with uncertainties has been proposed in [22]. In [23], a novel fractional-order non-singular TSMC method is described. The paper [24] addresses the TSMC problem for non-linear discrete systems. TSMC techniques can be also applied to the high-order representation of the systems as shown in [28] or to a class of the chaotic systems as shown in [25]. In [4,14] TSMC-based approach is applied to time-delay systems. There is also a lot of works that deal with application of the TSMC strategies to rigid robotic manipulators as the dynamics of such systems is described by matrix differential equations in the second-order form (see e.g., [26,27]). Interested type of TSMC has been designed in [17] for a general class of the disturbed non-linear dynamical systems. These results have been further extended in [18,19]. General conclusion after literature review is that the formal description of the systems has usually different forms depending on a control approach applied. There are not so many paper where TSMC is designed for a general class of the systems.

1.3 Contribution

The main contribution of the paper are control law formulas drawn to force the system trajectory from an initial condition to the defined sliding surface in finite time. The control laws are formulated for four generic representations of the dynamical systems with model uncertainties and under assumption that the control matrix is full size and invertible: (1) first-order uncertain linear systems; (2) second-order uncertain linear systems; (3) first-order uncertain non-linear systems; (4) second-order uncertain non-linear systems. Important part of contribution are the proofs of the finite-time stability done via Lyapunov stability theory.

1.4 Organization of the Paper

The rest of the paper is organized as follows. At the beginning of Sect. 2, preliminaries are provided. After that, first- and second-order representations of both linear and non-linear systems with model uncertainties are described. In Sect. 3, main results are presented which are concerned to the design of the terminal sliding mode controllers for the formulated uncertain systems. Section 4 illustrates potential applications of the designed controllers in the automotive industry domain. Conclusions are given in Sect. 5.

2 Description of the Systems

2.1 Preliminaries

Consider the system

$$\dot{\boldsymbol{x}}(t) = \boldsymbol{f}(t, \boldsymbol{x}), \quad \boldsymbol{x}(0) = \boldsymbol{x}_0, \tag{1}$$

with zero equilibrium point ($\mathbf{0} \in \mathbb{R}^n$), where $\boldsymbol{x}(t) \in \mathbb{R}^n$, $\boldsymbol{f} : \mathbb{R}_+ \times \Omega \to \mathbb{R}^n$ is a non-linear function, $t > 0$ represents time variable, $\Omega \subset \mathbb{R}^n$ is a neighborhood of zero, $\boldsymbol{x}_0 \in \mathbb{R}^n$ is the initial condition, $\mathbb{R}^n$ is the vector space of column vectors with real elements, $n \in \mathbb{N}$.

Definition 1 ([13]). *The equilibrium point $\boldsymbol{x}^* = \mathbf{0}$ of the system (1) is said to be finite-time stable if it is asymptotically stable and any solution $\boldsymbol{x}$ of (1) starting from $\boldsymbol{x}_0 \in \Omega$ reaches the equilibrium at some finite time moment, that is, $\boldsymbol{x}(t) = \mathbf{0}$, $\forall t \geq T(\boldsymbol{x}_0)$, where $T : \mathbb{R}^n \to \mathbb{R}_+ \cup \{0\}$ is the so-called settling-time function. If in addition, the function $T(\boldsymbol{x}_0)$ is bounded by some positive number $T_{\max} > 0$, that is, $T(\boldsymbol{x}_0) \leq T_{\max}, \forall \boldsymbol{x}_0 \in \Omega$, then the equilibrium point $\boldsymbol{x}^* = \mathbf{0}$ is said to be fixed-time stable.*

Lemma 1 ([10,12]). *Consider a continuous positive-definite function $V : \mathbb{R}_+ \to \mathbb{R}$ that satisfies the following inequality*

$$\dot{V}(t) \leq -\alpha V(t) - \beta V(t)^{\kappa}, \quad \forall_{t \geq t_0}, \; V(t_0) \geq 0, \tag{2}$$

where $\alpha > 0$, $\beta > 0$ and κ is a ratio of two odd positive integers such as $0 < \kappa < 1$. Then, for any given time t_0, $V(t)$ converges to zero at least within a finite time calculated as follows

$$t_{\mathrm{r}} = t_0 + \frac{1}{\alpha(1-\kappa)} \ln \frac{\alpha V(t_0)^{1-\kappa} + \beta}{\beta}. \tag{3}$$

Lemma 2 ([13]). *Consider a continuous positive-definite function* $V : \mathbb{R}_+ \to \mathbb{R}$ *that satisfies the following inequality*

$$\dot{V}(t) \leq -\left(\alpha V(t)^p + \beta V(t)^g\right)^k, \tag{4}$$

where $\alpha, \beta, p, g, k \in \mathbb{R}$, $pk < 1$ *and* $gk > 1$. *Then, the equilibrium point of the system* (1) *is fixed-time stable and the settling-time function satisfies*

$$T(\boldsymbol{x}_0) \leq \frac{1}{\alpha^k(1-pk)} + \frac{1}{\beta^k(qk-1)}, \quad \forall \boldsymbol{x}_0 \in \Omega. \tag{5}$$

Lemma 3 ([29]). *Consider a continuous positive-definite function* $V : \mathbb{R}_+ \to \mathbb{R}$ *that satisfies the following inequality*

$$\dot{V}(t) \leq -\alpha V(t)^p - \beta V(t)^k, \tag{6}$$

where $\alpha > 0$, $\beta > 0$, $p > 1$ *and* $0 < k < 1$. *Then, the equilibrium point of the system* (1) *is fixed-time stable and the settling-time function satisfies*

$$T(\boldsymbol{x}_0) \leq \frac{1}{\alpha(p-1)} + \frac{1}{\beta(1-k)}, \quad \forall \boldsymbol{x}_0 \in \Omega. \tag{7}$$

2.2 First-Order Representation of Uncertain Linear Systems

Consider a linear system with model uncertainties whose dynamics is described by a matrix linear differential equation of the first-order

$$\dot{\boldsymbol{x}}(t) + (\boldsymbol{A} + \Delta\boldsymbol{A})\,\boldsymbol{x}(t) = (\boldsymbol{B} + \Delta\boldsymbol{B})\,\boldsymbol{u}(t), \tag{8}$$

with the given initial condition $\boldsymbol{x}(0) = \boldsymbol{x}_0 \in \mathbb{R}^n$, where $\boldsymbol{x}(t) \in \mathbb{R}^n$ denotes the state vector with dimension n, $\boldsymbol{u}(t) \in \mathbb{R}^n$ is the control vector with dimension n, $\boldsymbol{A} \in \mathbb{R}^{n\times n}$ is the state matrix, $\boldsymbol{B} \in \mathbb{R}^{n\times n}$ is the non-singular control matrix, $\Delta\boldsymbol{A} \in \mathbb{R}^{n\times n}$ and $\Delta\boldsymbol{B} \in \mathbb{R}^{n\times n}$ represent model uncertainties.

2.3 Second-Order Representation of Uncertain Linear Systems

A second-order representation of the linear dynamical systems with model uncertainties can have the following form

$$\ddot{\boldsymbol{x}}(t) + (\boldsymbol{F} + \Delta\boldsymbol{F})\,\dot{\boldsymbol{x}}(t) + (\boldsymbol{A} + \Delta\boldsymbol{A})\,\boldsymbol{x}(t) = (\boldsymbol{B} + \Delta\boldsymbol{B})\,\boldsymbol{u}(t), \tag{9}$$

with the initial conditions $\dot{\boldsymbol{x}}(0) = \boldsymbol{x}_{01} \in \mathbb{R}^n$ and $\boldsymbol{x}(0) = \boldsymbol{x}_{02} \in \mathbb{R}^n$. Here, $\boldsymbol{x}(t) \in \mathbb{R}^n$ is the state vector, $\boldsymbol{u}(t) \in \mathbb{R}^n$ is the control vector, $\Delta\boldsymbol{A} \in \mathbb{R}^{n\times n}$, $\Delta\boldsymbol{B} \in \mathbb{R}^{n\times n}$ and $\Delta\boldsymbol{F} \in \mathbb{R}^{n\times n}$ represent model uncertainties for the matrices $\boldsymbol{A} \in \mathbb{R}^{n\times n}$, $\boldsymbol{B} \in \mathbb{R}^{n\times n}$ and $\boldsymbol{F} \in \mathbb{R}^{n\times n}$, respectively. In addition, the important assumption is that the control matrix $\boldsymbol{B}$ shall be full size and invertible.

2.4 First-Order Representation of Uncertain Non-linear Systems

Consider now a non-linear dynamical system with model uncertainties described by the following matrix non-linear differential equation of the first-order

$$\dot{\boldsymbol{x}}(t) + \boldsymbol{G}(\boldsymbol{x}) + \Delta\boldsymbol{G}(\boldsymbol{x}) = (\boldsymbol{H}(\boldsymbol{x}) + \Delta\boldsymbol{H}(\boldsymbol{x}))\,\boldsymbol{u}(t), \tag{10}$$

with the initial condition $\boldsymbol{x}(0) = \boldsymbol{x}_0 \in \mathbb{R}^n$, where $\boldsymbol{x}(t) \in \mathbb{X} \subset \mathbb{R}^n$ is the state vector defined on the state space $\mathbb{X}$, $\boldsymbol{u}(t) \in \mathbb{U} \subset \mathbb{R}^n$ is the control vector defined on the control space $\mathbb{U}$, $\boldsymbol{G} : \mathbb{R}^n \supset \Omega \to \mathbb{R}^n$ and $\boldsymbol{H} : \mathbb{R}^n \supset \Omega \to \mathbb{R}^{n\times n}$ are matrices whose elements are non-linear functions defined on some neighborhood of zero $\Omega \subset \mathbb{R}^n$ ($\boldsymbol{0} \in \mathbb{R}^n$), $\Delta\boldsymbol{G} : \mathbb{R}^n \supset \Omega \to \mathbb{R}^n$ and $\Delta\boldsymbol{H} : \mathbb{R}^n \supset \Omega \to \mathbb{R}^{n\times n}$ represent model uncertainties. All these functions have the following forms: $\boldsymbol{G}(\boldsymbol{\xi}) = [g_i(\boldsymbol{\xi})]_{n\times 1}$, $\Delta\boldsymbol{G}(\boldsymbol{\xi}) = [\Delta g_i(\boldsymbol{\xi})]_{n\times 1}$, $\boldsymbol{H}(\boldsymbol{\xi}) = [h_{ij}(\boldsymbol{\xi})]_{n\times n}$, $\Delta\boldsymbol{H}(\boldsymbol{\xi}) = [\Delta h_{ij}(\boldsymbol{\xi})]_{n\times n}$, $i, j = 1, 2, \ldots, n$. Moreover, the matrix $\boldsymbol{H}(\boldsymbol{x})$ is non-singular for every $\boldsymbol{x} \in \Omega$.

2.5 Second-Order Representation of Uncertain Non-linear Systems

Consider the system described by the matrix second-order non-linear differential equation of the following form

$$\ddot{\boldsymbol{x}}(t) + (\boldsymbol{F}(\dot{\boldsymbol{x}}, \boldsymbol{x}) + \Delta\boldsymbol{F}(\dot{\boldsymbol{x}}, \boldsymbol{x}))\,\dot{\boldsymbol{x}}(t) + \boldsymbol{G}(\boldsymbol{x}) + \Delta\boldsymbol{G}(\boldsymbol{x}) = (\boldsymbol{H}(\boldsymbol{x}) + \Delta\boldsymbol{H}(\boldsymbol{x}))\,\boldsymbol{u}(t), \tag{11}$$

with the initial conditions $\dot{\boldsymbol{x}}(0) = \boldsymbol{x}_{01} \in \mathbb{R}^n$ and $\boldsymbol{x}(0) = \boldsymbol{x}_{02} \in \mathbb{R}^n$, where $\boldsymbol{x}(t) \in \mathbb{X} \subset \mathbb{R}^n$ represents the state of the system, $\boldsymbol{u}(t) \in \mathbb{U} \subset \mathbb{R}^n$ is the input state, $\boldsymbol{F} : \mathbb{R}^n \times \mathbb{R}^n \supset \Omega \times \Omega \to \mathbb{R}^{n\times n}$, $\Delta\boldsymbol{F} : \mathbb{R}^n \times \mathbb{R}^n \supset \Omega \times \Omega \to \mathbb{R}^{n\times n}$, $\boldsymbol{G} : \mathbb{R}^n \supset \Omega \to \mathbb{R}^n$, $\Delta\boldsymbol{G} : \mathbb{R}^n \supset \Omega \to \mathbb{R}^n$, $\boldsymbol{H} : \mathbb{R}^n \supset \Omega \to \mathbb{R}^{n\times n}$, $\Delta\boldsymbol{H} : \mathbb{R}^n \supset \Omega \to \mathbb{R}^{n\times n}$, $\Omega \subset \mathbb{R}^n$ is a neighborhood of zero ($\boldsymbol{0} \in \mathbb{R}^n$), $\mathbb{X}$ is the state space, $\mathbb{U}$ is the control space. The functions $\boldsymbol{F}$, $\boldsymbol{G}$ and $\boldsymbol{H}$ are matrices whose elements are non-linear functions, that is, $\boldsymbol{F}(\boldsymbol{\xi}, \boldsymbol{\eta}) = [f_{ij}(\boldsymbol{\xi}, \boldsymbol{\eta})]_{n\times n}$, $\boldsymbol{G}(\boldsymbol{\xi}) = [g_i(\boldsymbol{\xi})]_{n\times 1}$, $\boldsymbol{H}(\boldsymbol{\xi}) = [h_{ij}(\boldsymbol{\xi})]_{n\times n}$, $i, j = 1, 2, \ldots, n$. In addition, the matrix $\boldsymbol{H}(\boldsymbol{x})$ is non-singular for every $\boldsymbol{x} \in \Omega$. The functions $\Delta\boldsymbol{F}$, $\Delta\boldsymbol{G}$ and $\Delta\boldsymbol{H}$ represent model uncertainties for $\boldsymbol{F}$, $\boldsymbol{G}$ and $\boldsymbol{H}$, respectively.

3 Design of the Controllers

3.1 Sliding Surface

Define a sliding surface

$$\boldsymbol{s}(t) = \boldsymbol{\Lambda}\boldsymbol{e}(t) - \boldsymbol{Q}\boldsymbol{\psi}(t), \tag{12}$$

where $\boldsymbol{\Lambda} = [\boldsymbol{P}\ \boldsymbol{I}_n]$, $\boldsymbol{P} \in \mathbb{R}^{n\times n}$ is a gain matrix, $\boldsymbol{I}_n$ is an identity matrix of the size n, $\boldsymbol{Q} \in \mathbb{R}^{n\times n}$, $\boldsymbol{Q} = \boldsymbol{Q}^{\mathrm{T}} > 0$ is a positive-definite matrix, $\boldsymbol{\psi} : \mathbb{R}^n \to \mathbb{R}^n$ is an $n \times n$ matrix with entries that are nonlinear non-negative functions of $\boldsymbol{e}_{\mathrm{x}}$, $\boldsymbol{e} = [\boldsymbol{e}_{\mathrm{x}}\ \boldsymbol{e}_{\mathrm{dx}}]^{\mathrm{T}}$ is a vector composed of the actual and expected trajectory $\boldsymbol{x}_{\mathrm{d}}$ and $\dot{\boldsymbol{x}}_{\mathrm{d}}$, that is

$$\boldsymbol{e}_{\mathrm{x}}(t) = \boldsymbol{x}(t) - \boldsymbol{x}_{\mathrm{d}}(t), \quad \boldsymbol{e}_{\mathrm{dx}}(t) = \dot{\boldsymbol{x}}(t) - \dot{\boldsymbol{x}}_{\mathrm{d}}(t). \tag{13}$$

Apply the sliding surface (12) to the second-order representations of the systems both in linear (9) and non-linear case (11). For the first-order representations given by the Eqs. (8) and (10), use simplified version of the formula (12), that is,

$$\boldsymbol{s}(t) = \boldsymbol{e}_{\mathrm{x}}(t) = \boldsymbol{x}(t) - \boldsymbol{x}_{\mathrm{d}}(t). \tag{14}$$

In addition, define the matrix $\boldsymbol{S}(t)$ and vector $\boldsymbol{s}_{\eta}(t)$ according to the formulas provided below:

$$\boldsymbol{S}(t) = \mathrm{diag}\left(\mathrm{sgn}\left(s_1(t)\right), \mathrm{sgn}\left(s_2(t)\right), \ldots, \mathrm{sgn}\left(s_N(t)\right)\right), \tag{15}$$

$$\boldsymbol{s}_{\eta}(t) = \left[|s_1(t)|^{\eta}\ |s_2(t)|^{\eta} \ldots |s_N(t)|^{\eta}\right]^{\mathrm{T}}, \tag{16}$$

where $\eta > 0$, $N = 2n$ for the sliding surface (12) and $N = n$ for the sliding surface (14).

3.2 TSMC Design for the First-Order Linear System

Consider the system (8) with the sliding surface (14). Then re-write it to the form

$$\dot{\boldsymbol{x}}(t) + \boldsymbol{A}\boldsymbol{x}(t) = \boldsymbol{B}\boldsymbol{u}(t) + \boldsymbol{z}(t), \tag{17}$$

where the part

$$\boldsymbol{z}(t) = -\Delta\boldsymbol{A}\boldsymbol{x}(t) + \Delta\boldsymbol{B}\boldsymbol{u}(t) \tag{18}$$

can represent a disturbance which is unknown due to model uncertainties. Next, apply to the system the following control signal

$$\boldsymbol{u}(t) = \boldsymbol{B}^{-1}\left(\boldsymbol{A}\boldsymbol{x} + \dot{\boldsymbol{x}}_{\mathrm{d}}(t) - \boldsymbol{S}(t)\boldsymbol{z}^{\max}(t) - \sigma\boldsymbol{S}(t)\boldsymbol{s}_{\eta}(t) - \gamma\boldsymbol{s}(t)\right), \tag{19}$$

where $\sigma > 0$ and $\gamma > 0$ are real positive coefficients, the vector $\boldsymbol{z}^{\max}(t)$ for $t > 0$ contains non-negative elements only and satisfies the inequality $\boldsymbol{z}(t) \leq \boldsymbol{z}^{\max}(t)$. The element $\boldsymbol{z}^{\max}(t)$ in the control law formula (19) can be considered as an active and adaptive compensation of the uncertainties existing in the system under the assumption that their magnitude can be somehow estimated.

Theorem 1. *The trajectory of the system* (17) *when forced by the control* (19) *reaches the sliding surface* (14) *starting from the initial condition* $\boldsymbol{x}_0$ *in finite time and then remains on this surface.*

Proof. Consider the Lyapunov function

$$V(t) = 0.5\boldsymbol{s}(t)^{\mathrm{T}}\boldsymbol{s}(t)\,. \tag{20}$$

Then calculate the time derivative of it along the solution of the closed-loop system

$$\begin{aligned}\dot{V}(t) &= \boldsymbol{s}(t)^{\mathrm{T}}\dot{\boldsymbol{s}}(t) = \boldsymbol{s}(t)^{\mathrm{T}}\left(\dot{\boldsymbol{x}}(t) - \dot{\boldsymbol{x}}_{\mathrm{d}}(t)\right)\\ &= \boldsymbol{s}(t)^{\mathrm{T}}\left(-\boldsymbol{A}\boldsymbol{x} + \boldsymbol{B}\boldsymbol{u}(t) + \boldsymbol{z}(t) - \dot{\boldsymbol{x}}_{\mathrm{d}}(t)\right).\end{aligned} \tag{21}$$

Substitute now (19) into (21)

$$\dot{V}(t) = -\boldsymbol{s}(t)^{\mathrm{T}}\boldsymbol{S}(t)\boldsymbol{z}^{\max}(t) - \sigma\boldsymbol{s}(t)^{\mathrm{T}}\boldsymbol{S}(t)\boldsymbol{s}_{\eta}(t) - \gamma\boldsymbol{s}(t)^{\mathrm{T}}\boldsymbol{s}(t) + \boldsymbol{s}(t)^{\mathrm{T}}\boldsymbol{z}(t). \quad (22)$$

Based on the assumption about the vector $\boldsymbol{z}^{\max}(t)$ conclude that

$$\dot{V}(t) \leq -\gamma\left\|\boldsymbol{s}(t)\right\|^2 - \sigma\left\|\boldsymbol{s}(t)\right\|^{\eta+1} = -2\gamma V(t) - 2^{\frac{\eta+1}{2}}\sigma V(t)^{\frac{\eta+1}{2}}. \quad (23)$$

According to the theses of Lemmas 1–3 the trajectory of the system (17) reaches the sliding surface in finite time and after that remains on this surface.

3.3 TSMC Design for the Second-Order Linear System

Consider the system (9) with the sliding surface (12). The system can be presented as

$$\ddot{\boldsymbol{x}}(t) + \boldsymbol{F}\dot{\boldsymbol{x}}(t) + \boldsymbol{A}\boldsymbol{x}(t) = \boldsymbol{B}\boldsymbol{u}(t) + \boldsymbol{z}(t), \quad (24)$$

where

$$\boldsymbol{z}(t) = -\Delta\boldsymbol{F}\dot{\boldsymbol{x}}(t) - \Delta\boldsymbol{A}\boldsymbol{x}(t) + \Delta\boldsymbol{B}\boldsymbol{u}(t) \quad (25)$$

represents influence of the model uncertainties on the system's dynamics. To reach the sliding surface consider the following control signal

$$\begin{aligned}\boldsymbol{u}(t) = &-\boldsymbol{B}^{-1}\boldsymbol{\Lambda}\boldsymbol{W}\tilde{\boldsymbol{x}}(t) + \boldsymbol{B}^{-1}\boldsymbol{Q}\dot{\boldsymbol{\psi}}(t) + \boldsymbol{B}^{-1}\left(\boldsymbol{P}\dot{\boldsymbol{x}}_{\mathrm{d}}(t) + \ddot{\boldsymbol{x}}_{\mathrm{d}}(t)\right)\\ &-\boldsymbol{B}^{-1}\boldsymbol{S}(t)\boldsymbol{z}^{\max}(t) - \sigma\boldsymbol{B}^{-1}\boldsymbol{S}(t)\boldsymbol{s}_{\eta}(t) - \gamma\boldsymbol{B}^{-1}\boldsymbol{s}(t),\end{aligned} \quad (26)$$

where $\sigma > 0$, $\gamma > 0$, $z_i^{\max}(t) \geq 0$ for $t > 0$ and $i = 1, 2, \ldots, n$, $\boldsymbol{z}(t) \leq \boldsymbol{z}^{\max}(t)$,

$$\boldsymbol{W} = \begin{bmatrix} \boldsymbol{0} & \boldsymbol{I}_n \\ -\boldsymbol{A} & -\boldsymbol{F} \end{bmatrix}, \quad \tilde{\boldsymbol{x}}(t) = \begin{bmatrix} \boldsymbol{x}(t) \\ \dot{\boldsymbol{x}}(t) \end{bmatrix}. \quad (27)$$

Theorem 2. *The trajectory of the system* (24) *when forced by the control* (26) *reaches the sliding surface* (12) *starting from the initial conditions* $\boldsymbol{x}_{01}$ *and* $\boldsymbol{x}_{02}$ *within a finite time and then remains on this surface.*

Proof. Consider the Lyapunov function in the form of (20) and calculate the time derivative of it along the solutions of the closed-loop system

$$\begin{aligned}\dot{V}(t) &= \boldsymbol{s}(t)^{\mathrm{T}}\dot{\boldsymbol{s}}(t) = \boldsymbol{s}(t)^{\mathrm{T}}\left(\boldsymbol{\Lambda}\dot{\boldsymbol{e}}(t) - \boldsymbol{Q}\dot{\boldsymbol{\psi}}(t)\right)\\ &= \boldsymbol{s}(t)^{\mathrm{T}}\left(\boldsymbol{P}\dot{\boldsymbol{x}}(t) - \boldsymbol{P}\dot{\boldsymbol{x}}_{\mathrm{d}}(t) + \ddot{\boldsymbol{x}}(t) - \ddot{\boldsymbol{x}}_{\mathrm{d}}(t) - \boldsymbol{Q}\dot{\boldsymbol{\psi}}(t)\right)\\ &= \boldsymbol{s}(t)^{\mathrm{T}}\left(\boldsymbol{P}\dot{\boldsymbol{x}}(t) - \boldsymbol{P}\dot{\boldsymbol{x}}_{\mathrm{d}}(t) - \boldsymbol{F}\dot{\boldsymbol{x}}(t) - \boldsymbol{A}\boldsymbol{x}(t) + \boldsymbol{B}\boldsymbol{u}(t)\right)\\ &+ \boldsymbol{s}(t)^{\mathrm{T}}\left(\boldsymbol{z}(t) - \ddot{\boldsymbol{x}}_{\mathrm{d}}(t) - \boldsymbol{Q}\dot{\boldsymbol{\psi}}(t)\right).\end{aligned} \quad (28)$$

By inserting the control law (26) into (28) the relationship (23) can be obtained.

3.4 TSMC Design for the First-Order Non-linear System

Consider the system (10) with the sliding surface (14) and present it as

$$\dot{\boldsymbol{x}}(t) + \boldsymbol{G}(\boldsymbol{x}) = \boldsymbol{H}(\boldsymbol{x})\boldsymbol{u}(t) + \boldsymbol{z}(t), \tag{29}$$

where

$$\boldsymbol{z}(t) = -\Delta\boldsymbol{G}(\boldsymbol{x}) + \Delta\boldsymbol{H}(\boldsymbol{x})\boldsymbol{u}(t) \tag{30}$$

is an unknown disturbance. Next, design the control law in the form of

$$\begin{aligned}\boldsymbol{u}(t) = &\boldsymbol{H}(\boldsymbol{x})^{-1}\left(\boldsymbol{G}(\boldsymbol{x}) + \dot{\boldsymbol{x}}_{\mathrm{d}}(t)\right) - \boldsymbol{H}(\boldsymbol{x})^{-1}\boldsymbol{S}(t)\boldsymbol{z}^{\max}(t)\\ &- \sigma\boldsymbol{H}(\boldsymbol{x})^{-1}\boldsymbol{S}(t)\boldsymbol{s}_{\eta}(t) - \gamma\boldsymbol{H}(\boldsymbol{x})^{-1}\boldsymbol{s}(t),\end{aligned} \tag{31}$$

where $\sigma > 0$, $\gamma > 0$, $\boldsymbol{z}(t) \leq \boldsymbol{z}^{\max}(t)$ for $t > 0$ and $z_i^{\max}(t) \geq 0$ for $i = 1, 2, \ldots, n$.

Theorem 3. *The trajectory of the system* (29) *when forced by the control* (31) *reaches the sliding surface* (14) *starting from the initial condition* $\boldsymbol{x}_0$ *in finite time and then remains on this surface.*

Proof. The proof can be obtained by using the Lyapunov function (20). Calculating the time derivative of this function along the solutions of the system (29)

$$\begin{aligned}\dot{V}(t) &= \boldsymbol{s}(t)^{\mathrm{T}}\dot{\boldsymbol{s}}(t) = \boldsymbol{s}(t)^{\mathrm{T}}\left(\dot{\boldsymbol{x}}(t) - \dot{\boldsymbol{x}}_{\mathrm{d}}(t)\right)\\ &= \boldsymbol{s}(t)^{\mathrm{T}}\left(-\boldsymbol{G}(\boldsymbol{x}) + \boldsymbol{H}(\boldsymbol{x})\boldsymbol{u}(t) + \boldsymbol{z}(t) - \dot{\boldsymbol{x}}_{\mathrm{d}}(t)\right),\end{aligned} \tag{32}$$

and next substituting (31) into (32) will lead to the inequality (23). Hence, based on Lemmas 1–3, the thesis of the theorem is proven.

3.5 TSMC Design for the Second-Order Non-linear System

Consider the system (11) with the sliding surface (14). Separate from the system description an unknown part, which is caused by the model uncertainties. Then

$$\ddot{\boldsymbol{x}}(t) + \boldsymbol{F}(\dot{\boldsymbol{x}}, \boldsymbol{x})\dot{\boldsymbol{x}}(t) + \boldsymbol{G}(\boldsymbol{x}) = \boldsymbol{H}(\boldsymbol{x})\boldsymbol{u}(t) + \boldsymbol{z}(t), \tag{33}$$

where

$$\boldsymbol{z}(t) = -\Delta\boldsymbol{F}(\dot{\boldsymbol{x}}, \boldsymbol{x})\dot{\boldsymbol{x}}(t) - \Delta\boldsymbol{G}(\boldsymbol{x}) + \Delta\boldsymbol{H}(\boldsymbol{x})\boldsymbol{u}(t) \tag{34}$$

is an unknown disturbance. As the next step, design a control law to guarantee the existence of the sliding mode in the closed-loop system

$$\begin{aligned}\boldsymbol{u}(t) = &- \boldsymbol{H}(\boldsymbol{x})^{-1}\boldsymbol{\Lambda}\boldsymbol{W}(\dot{\boldsymbol{x}}, \boldsymbol{x}) + \boldsymbol{H}(\boldsymbol{x})^{-1}\boldsymbol{Q}\dot{\boldsymbol{\psi}}(t) + \boldsymbol{H}(\boldsymbol{x})^{-1}\left(\boldsymbol{P}\dot{\boldsymbol{x}}_{\mathrm{d}}(t) + \ddot{\boldsymbol{x}}_{\mathrm{d}}(t)\right)\\ &- \boldsymbol{H}(\boldsymbol{x})^{-1}\boldsymbol{S}(t)\boldsymbol{z}^{\max}(t) - \sigma\boldsymbol{H}(\boldsymbol{x})^{-1}\boldsymbol{S}(t)\boldsymbol{s}_{\eta}(t) - \gamma\boldsymbol{H}(\boldsymbol{x})^{-1}\boldsymbol{s}(t),\end{aligned} \tag{35}$$

where $\sigma > 0$, $\gamma > 0$, $z_i^{\max}(t) \geq 0$ for $t > 0$ and $i = 1, 2, \ldots, n$, $\boldsymbol{z}(t) \leq \boldsymbol{z}^{\max}(t)$,

$$\boldsymbol{W}(\dot{\boldsymbol{x}}, \boldsymbol{x}) = \begin{bmatrix}\dot{\boldsymbol{x}}\\ -\boldsymbol{G}(\boldsymbol{x}) - \boldsymbol{F}(\dot{\boldsymbol{x}}, \boldsymbol{x})\dot{\boldsymbol{x}}\end{bmatrix}. \tag{36}$$

Theorem 4. *The trajectory of the system* (33) *when forced by the control* (35) *reaches the sliding surface* (12) *starting from the initial condition* $\boldsymbol{x}_{01}$ *and* $\boldsymbol{x}_{02}$ *within a finite time and then remains on this surface.*

Proof. Consider the Lyapunov candidate function in the form of (20) and calculate its time derivative along the solution of the system (11)

$$\begin{aligned}\dot{V}(t) &= \boldsymbol{s}(t)^{\mathrm{T}}\dot{\boldsymbol{s}}(t) = \boldsymbol{s}(t)^{\mathrm{T}}\left(\boldsymbol{\Lambda}\dot{\boldsymbol{e}}(t) - \boldsymbol{Q}\dot{\boldsymbol{\psi}}(t)\right) \\ &= \boldsymbol{s}(t)^{\mathrm{T}}\left(\boldsymbol{P}\dot{\boldsymbol{x}}(t) - \boldsymbol{P}\dot{\boldsymbol{x}}_{\mathrm{d}}(t) + \ddot{\boldsymbol{x}}(t) - \ddot{\boldsymbol{x}}_{\mathrm{d}}(t) - \boldsymbol{s}(t)^{\mathrm{T}}\boldsymbol{Q}\dot{\boldsymbol{\psi}}(t)\right) \\ &= \boldsymbol{s}(t)^{\mathrm{T}}\left(\boldsymbol{P}\dot{\boldsymbol{x}}(t) - \boldsymbol{P}\dot{\boldsymbol{x}}_{\mathrm{d}}(t) - \boldsymbol{F}(\dot{\boldsymbol{x}}, \boldsymbol{x})\dot{\boldsymbol{x}}(t) - \boldsymbol{G}(\boldsymbol{x}) + \boldsymbol{H}(\boldsymbol{x})\boldsymbol{u}(t)\right) \\ &+ \boldsymbol{s}(t)^{\mathrm{T}}\left(\boldsymbol{z}(t) - \ddot{\boldsymbol{x}}_{\mathrm{d}}(t) - \boldsymbol{Q}\dot{\boldsymbol{\psi}}(t)\right). \end{aligned} \tag{37}$$

When inserting (35) into (37) it is easy to obtain the inequality (23). Thus, thesis of the theorem is a direct conclusion from Lemmas 1–3.

4 Automotive Applications

Automotive control systems often possess complicated nonlinearities that cannot be neglected in the analysis. These systems are also subject to significant model uncertainties that are introduced during parameter identification experiments as well as to external disturbances. Therefore, sliding mode based techniques might be well suited to these types of the control systems in terms of required control quality, accuracy and robustness. One of the good example is control of the wheel slip that occurs during vehicle acceleration [6] and braking [2]. Sliding mode control finds also applications in the electronic throttle control system [15] to generate a valve opening angle that regulates the air intake of the engine. Vehicle suspension system is another example when the controlled system is subject to model uncertainties and disturbances from the road surface. In order to actively compensate these uncertainties and disturbances, a sliding mode control might be a good solution [16]. The paper [8] investigates sliding mode control scheme to guarantee stability and robustness of the controller for a spark ignition engine in the presence of faults. Good review of automotive applications of sliding mode control can be found in [5].

The sliding mode control can be also applied in adaptive cruise control (ACC) system. The ACC is a control system aims to keep safe distance from the vehicle ahead. Driver sets desired speed and time interval to the car ahead. When system detects slower vehicle the speed is automatically adapted so the vehicle ahead is followed with a setup distance between. Once road is clear again the vehicle returns to the selected speed. The ACC functionality is based mostly on vision and radar sensors. Radar provides range and velocity information and camera provides information about object detection. Fusion algorithms are used to combine both radar and camera data for reliable target detection. Model uncertainties exist in the perception system mainly, what means that with each

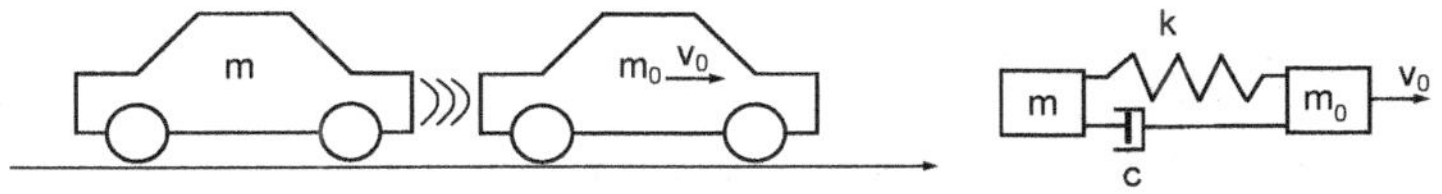

Fig. 1. A vehicle with ACC mode and a mechanical equivalent model

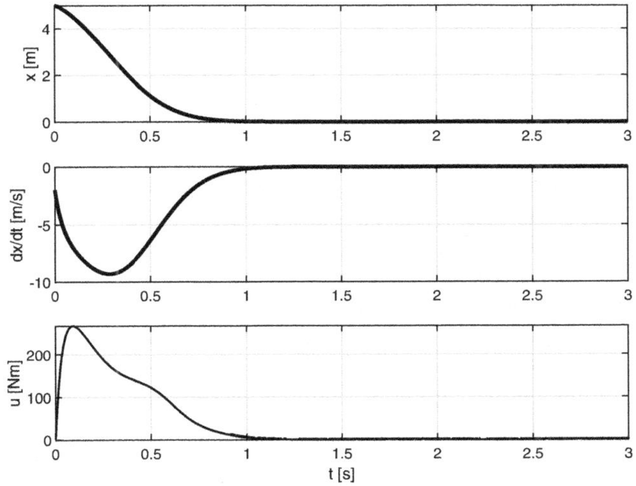

Fig. 2. Trajectory of the closed-loop system and the control signal

object being visible and detectable in the sensor field of view is associated a certain level of probability with which this object is detected. In [20] the dynamic behavior of the series of cars following each other and functioning in ACC mode is mathematically modeled by an equivalent system consisting of a set of masses, springs and dampers (see Fig. 1). Inspired by this concept and including uncertainties introduced by the perception system, the ACC system of a single vehicle following the target vehicle can be modeled by the following equation

$$m\ddot{x}(t) + \left(c(x,\dot{x}) + \Delta c(x,\dot{x})\right)\dot{x}(t) + k(x)x(t) + \Delta k(x) = (b + \Delta b)u(t), \qquad (38)$$

where $x(t)$ represents displacement of the vehicle mass m from the equilibrium state, $c(x,\dot{x})$ is the coefficient representing motion resistance, $k(x)$ is the coefficient related to the driving force, b is the control gain introduced by the ACC controller, $\Delta c(x,\dot{x})$, $\Delta k(x)$, Δb are unknown bounded uncertainties in the system.

The system (38) can be stabilized in finite time using the contoller (35). Indeed, simulation results shown on Fig. 2 confirm that good system performance is achieved using the proposed control scheme. The property of finite time convergence is guaranteed. Simulation experiments have been performed in MATLAB/Simulink environment using the following parameters (units of the parameters have been left out for simplicity): $m = 1500$, initial conditions

$x_{01} = 5.0$, $x_{02} = -2.0$ reflect the situation when the car position and speed are close to the desired speed and time interval to the car ahead, $x_d(t) = 0$, $\dot{x}_d(t) = 0$, $c(x) = 3750(1 + \exp(-0.1\mathrm{x}^2))$, $k(x) = 3.75/(1 + 0.2\exp(-0.1\mathrm{x}^2))$, $b = 1$, $p = 10$, $q = 1.5$, $\gamma = 13$, $\eta = 3/5$, $\sigma = 3$, $z^{\max} = 93.75$, uncertainties have been modeled by bounded disturbance signal in the form of $\sin t$ function. Mathematical representations of the motion resistance and driving force coefficient have been chosen to illustrate effectiveness of the proposed controller only and they might not reflect fully physical aspects of the real phenomena.

5 Conclusions

The following conclusions can be drawn based on the results of this research effort: (a) the controllers (31) and (35) that are used to stabilize the first- and second-order non-linear systems (10) and (11) are more general forms of the controllers (19) and (26) that stabilize the first- and second-order linear systems (8) and (9); (b) the proposed control laws guarantee finite-time stability where the settling-time can be calculated analytically; (c) two representations of uncertain dynamical systems are considered because these classes of the systems are not completely disjunctive and not equivalent, therefore have to be investigated separately; (e) the design process requires for the control matrix to be of full size, that is, $n \times n$ and invertible - this narrows down the possible applications of the proposition, however, if a system can be described equivalently both using first- and second-order representations, then one of these representation can have the control matrix invertible.

References

1. Bartosiewicz, A.: Time-varying sliding modes for second-order systems. IEE Proc. - Control Theory Appl. **143**(5), 455–462 (1996)
2. Drakunov, S., Ozguner, U., Dix, P., Ashrafi, B.: ABS control using optimum search via sliding modes. IEEE Trans. Control Syst. Technol. **3**(1), 79–85 (1995)
3. Feng, Y., Xinghuo. X., Zheng, J.: Nonsingular terminal sliding mode control of uncertain multivariable systems. In: Proceedings of the 2006 International Workshop on Variable Structure Systems, Alghero, Italy, pp. 196–201 (2006)
4. Feng, Y., Yu, X., Zheng, Z.: Second-order terminal sliding mode control of input-delay systems. Asian J. Control **8**(1), 12–20 (2006)
5. Fu, L., Ozguner, U., Haskara, I.: Automotive applications of sliding mode control. IFAC Proc. Vol. **44**(1), 1898–1903 (2011)
6. Haskara, I., Ozguner, U., Winkelman, J.: Wheel slip control for antispin acceleration via dynamic spark advance. Control Eng. Pract. **8**(10), 1135–1148 (2000)
7. Hong, Y., Yang, G., Cheng, D., Spurgeon, S.: A new approach to terminal sliding mode control design. Asian J. Control **7**(2), 177–181 (2005)
8. Kim, Y.-W., Rizzoni, G., Utkin, V.I.: Developing a fault tolerant power-train control system by integrating design of control and diagnostics. Int. J. Robust Nonlinear Control **11**(11), 1095–1114 (2001)
9. Man, Z., Yu, X.: Terminal sliding mode control of MIMO linear systems. IEEE Trans. Circ. Syst. I: Fundam. Theory Appl. **44**(11), 1065–1070 (1997)

10. Mobayen, S., Majd, V., Sojoodi, M.: An LMI-based composite nonlinear feedback terminal sliding-mode controller design for disturbed MIMO systems. Math. Comput. Simul. **85**(11), 1–10 (2012)
11. Mon, Y.J.: Terminal sliding mode fuzzy-PDC control for nonlinear systems. Int. J. Sci. Technol. Res. **2**(4), 218–221 (2013)
12. Moulay, E., Peruquetti, W.: Finite time stability and stabilization of a class of continuous systems. J. Math. Anal. Appl. **323**(2), 1430–1443 (2006)
13. Polyakov, A.: Nonlinear feedback design for fixed-time stabilization of linear control systems. IEEE Trans. Autom. Control **57**(8), 2106–2110 (2012)
14. Rasvan, V.B.: Stability and sliding modes for a class of nonlinear time delay systems. Math. Bohemica **136**(2), 155–164 (2011)
15. Reichhartinger, M., Horn, M.: Application of higher order sliding-mode concepts to a throttle actuator for gasoline engines. IEEE Trans. Industr. Electron. **56**(9), 3322–3329 (2009)
16. Sam, Y., Osman, J.H.S., Ghani, M.R.A.: A class of proportional-integral sliding mode control with application to active suspension system. Syst. Control Lett. **51**(3–4), 217–223 (2004)
17. Skruch, P.: A terminal sliding mode control of disturbed nonlinear second-order dynamical systems. J. Comput. Nonlinear Dyn. **11**(5), 054501 (2016)
18. Skruch, P., Długosz, M.: Design of terminal sliding mode controllers for disturbed non-linear systems described by matrix differential equations of the second and first orders. Appl. Sci. **9**(11), 2325 (2019)
19. Skruch, P., Długosz, M.: A sliding mode controller design for thermal comfort in buildings. Eng. Trans. **64**(4), 475–489 (2019)
20. Skruch, P., Dlugosz, M., Mitkowski, W.: Stability analysis of a series of cars driving in adaptive cruise control mode. In: Mitkowski, W., Kacprzyk, J., Oprzedkiewicz, K., Skruch, P. (eds.) Trends in Advanced Intelligent Control, Optimization and Automation. Advances in Intelligent Systems and Computing, vol. 577, pp. 168-177. Springer, Cham (2017)
21. Tao, C.W., Taur, J.S.: Adaptive fuzzy terminal sliding mode controller for linear systems with mismatched time-varying uncertainties. IEEE Trans. Syst. Man Cybern.—Part B: Cybern. **34**(1), 255–262 (2004)
22. Wang, X., Guo, J., Tang, S.: Neural network-based multivariable fixed-time terminal sliding mode control for re-entry vehicles. IET Control Theory Appl. **12**(12), 1763–1772 (2018)
23. Wang, Y., Luo, G., Gu, L., Li, X.: Fractional-order nonsingular terminal sliding mode control of hydraulic manipulators using time delay estimation. J. Vib. Control **22**(19), 3998–4011 (2016)
24. Xi, Z., Hesketh, T.: On discrete time terminal sliding mode control for nonlinear systems with uncertainty. In: Proceedings of the 2010 Americal Control Conference, Baltimore, USA, pp. 980–984 (2010)
25. Xiang, W., Huangpu, Y.: Second-order terminal sliding mode controller for a class of chaotic systems with unmatched uncertainties. Commun. Nonlinear Sci. Numer. Simul. **15**(6), 3241–3247 (2015)
26. Yang, L., Yang, J.: Nonsingular fast terminal sliding-mode control for nonlinear dynamical systems. Int. J. Robust Nonlinear Control **21**(16), 1865–1879 (2011)

27. Zhihong, M., Paplinski, A.P., Wu, H.R.: A robust MIMO terminal sliding mode control scheme for rigid robotic manipulators. IEEE Trans. Autom. Control **39**(12), 2464–2469 (1994)
28. Zhuang, K., Su, H., Zhang, K., Chu, J.: Adaptive terminal sliding mode control for high-order nonlinear dynamic systems. J. Zhejiang Univ.-Sci. A **4**(1), 58–63 (2003)
29. Zuo, Z.: Nonsingular fixed-time consensus tracking for second-order multi-agent networks. Automatica **54**(4), 305–309 (2015)

Reference Trajectory Based SMC of DCDC Buck Converter

Piotr Leśniewski(✉)

Institute of Automatic Control, Lodz University of Technology, Łódź, Poland
piotr.lesniewski@p.lodz.pl

Abstract. In this paper, discrete time sliding mode control of a DCDC buck converter is considered. It is demonstrated, that using a "traditional" SMC for this task can result in excessive values of the inductor current at the start of the control process. This makes the controller impractical, as one would have to significantly over-engineer the inductor to prevent its damage. On the other hand, using the reaching law approach can minimize this problem, however at the cost of reducing the robustness. Therefore, a reference trajectory following SMC is proposed, which allows to limit the initial value of the inductor current, while maintaining good robustness w.r.t. disturbances, i.e. load changes. These important properties are demonstrated in computer simulations, which take into account all aspects of real application: the PWM modulation, sampling the continuous signals etc.

Keywords: Sliding mode control · Buck converter · Variable structure systems

1 Introduction

DCDC converters are a very common part of modern consumer electronics, photovoltaic systems, etc. Despite of their basic structure being used for quite some time, they remain an interesting and popular research topic. Due to limited space, only a few, recent trends in DCDC converters will be briefly presented. To minimize switching losses, ZVS (zero voltage switching) can be used. This approach was proposed in [26] to obtain a high frequency, high efficiency converter for low power applications. The high frequency used (2 MHz) allows to drastically decrease the dimensions of the output capacitor, thus reducing the dimensions and cost of the whole converter. On the other hand, in [17] a method of online estimation of the losses in boost converters is proposed. This information can then be used for control, as well as fault detection. In [16] a novel way of determining the PWM duty cycle is proposed to reduce the impact of supply voltage changes on the output voltage. In most applications, the PWM frequency is constant. Unfortunately, this makes the switching frequency appear in the output voltage waveform, which is disadvantageous as the load can be sensitive to this particular frequency. To solve this problem, a frequency-hopping [19] approach can be used. In it, the switching frequency rapidly changes between a predefined set of frequencies. Thus, more frequencies appear in the output spectrum, however each of them is significantly lower powered than the original, single one.

A. Bartoszewicz et al. (Eds.): PCC 2020, AISC 1196, pp. 161–170, 2020.
https://doi.org/10.1007/978-3-030-50936-1_14

Sliding mode (SM) control is a well-known variable structure control method [9, 10, 20, 23]. Among its main advantages are the small computational requirements and strong robustness w.r.t. external disturbances. On the other hand, its drawback is, that "basic" SM controllers (meaning not e.g. higher order SM controllers) may require using high frequency control signal switching. Evidently, this drawback is less important for power converter systems, in which high switching frequencies are used irrespective of control type, for e.g. ensuring continuous conduction mode. For the above reasons, SM control is frequently used in power converter systems [11, 13, 14, 18, 21, 22].

The main idea of SM control is to propose a switching hyperplane, to which the state trajectory will be confined. The orientation of this hyperplane will determine the dynamics of the obtained closed loop system. A closely related concept is the so-called sliding variable. This variable is positive on one side of the hyperplane, negative on the other and its absolute value increases as the state gets farther away from the hyperplane. Thus, confining the trajectory to the hyperplane can be simplified to reducing the sliding variable to zero. The next task is postulating a SM controller. A "classical" discrete time SM controller tries to bring the sliding variable to zero in a single control step. Such a task is usually impossible, due to control signal limitations which can lead to undesirable effects such as overshoot. Moreover, at the start of the control process, unacceptably large values of the state variables can occur (e.g. for the DCDC buck converter a large initial value of inductor current). This problem can be addressed by using the so-called reaching laws [1, 5–8, 12], in which the current value of the sliding variable determines its "desired" value in the next time instant. This approach however can reduce the robustness. Therefore, in this work, a reference trajectory [2, 3] SM controller is proposed, in which the sliding variable is made to track a generated *a priori* trajectory. This on one hand prevents overshoots and excessive state variable values at the beginning of control, but on the other hand does not reduce the robustness. This methodology can be viewed as a simplified version of model reference SM control [4, 15], in which the trajectory of the sliding variable is not postulated directly, but obtained from controlling a model of the plant. In this work, control of a DCDC buck converter is considered, however a similar approach could be also applied to other converter types, such as boost or buck-boost.

The remainder of this work is organized as follows. Section 2 presents the mathematical model of a DCDC buck converter. Then, in Sect. 3, a reference trajectory based sliding mode controller is designed. Section 4 comprises of computer simulations, that compare the proposed solution to some known sliding mode controllers. Finally, Sect. 5 presents conclusions and a brief proposal of future research.

2 Model of a DCDC Buck Converter

The scheme of a DCDC buck converter is shown in Fig. 1.

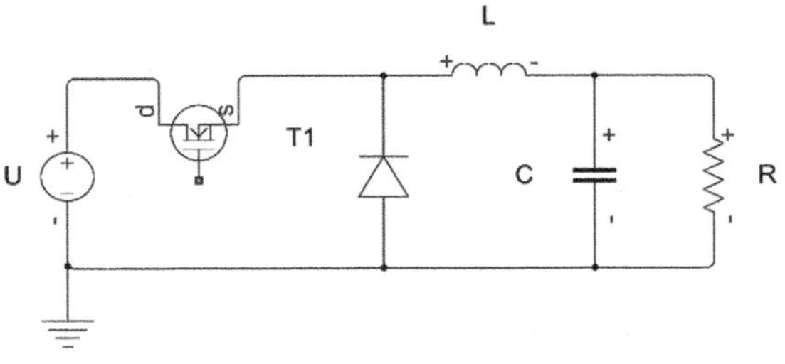

Fig. 1. Scheme of a DCDC buck converter.

The state variables are chosen as the output voltage $x_1(t) = U_C(t)$ and the inductor current $x_2(t) = i_L(t)$. The control input is the duty cycle of transistor T1. The load of the converter is resistive, however the exact value of this resistance is unknown, and it can change during operation. It is assumed, that the converter works in CCM (continuous conduction mode).

The output voltage, irrespective of the transistor state is given by

$$\dot{x}_1(t) = \frac{1}{C}[i_L(t) - i_R(t)] = \frac{1}{C}\left[x_2(t) - \frac{x_1(t)}{R}\right]. \tag{1}$$

For the on-state of transistor T1 the inductor current is described by

$$\dot{x}_2(t) = \frac{1}{L}[U - u_C(t)] = \frac{1}{L}[U - x_1(t)], \tag{2}$$

where U is the supply voltage. When the transistor T1 is switched off, the inductor current flows through the freewheeling diode. Ignoring the forward voltage of this diode, the current is described by

$$\dot{x}_2(t) = -\frac{1}{L}u_C(t) = -\frac{1}{L}x_1(t). \tag{3}$$

Taking into account (2) and (3), one can observe, that if a sufficiently high switching frequency is used, the averaged inductor current can be described by:

$$\dot{x}_2(t) = \frac{1}{L}[u(t)U - x_1(t)], \tag{4}$$

where $u(t) \in [0, 1]$ is the duty cycle. This allows us to formulate the model in the state-space as

$$\boldsymbol{x}(t) = \boldsymbol{A}_c\boldsymbol{x}(t) + \boldsymbol{B}_c\,u(t), \tag{5}$$

where

$$\boldsymbol{A}_c = \begin{bmatrix} -\frac{1}{CR} & \frac{1}{C} \\ -\frac{1}{L} & 0 \end{bmatrix} \boldsymbol{B}_c = \begin{bmatrix} 0 \\ \frac{U}{L} \end{bmatrix}. \tag{6}$$

Next, we calculate the discretization of (5) with the sampling time T and obtain the discrete time state space model as

$$\boldsymbol{x}[(k+1)T] = \boldsymbol{A}\boldsymbol{x}(kT) + \boldsymbol{B}u(kT), \tag{7}$$

where the state vector $\boldsymbol{x}$ has the same elements (output voltage and inductor current) as in the continuous case, and the control signal is the duty cycle in a given discretization period. The elements of the state matrix $\boldsymbol{A}$ and input vector $\boldsymbol{B}$ can be calculated numerically.

3 Controller Design

In this subsection, a sliding mode controller for the system (7) will be designed. The sliding variable is selected as

$$s(kT) = \boldsymbol{c}^T\boldsymbol{e}(kT) = c^T[\boldsymbol{x_d} - \boldsymbol{x}(kT)], \tag{8}$$

where $\boldsymbol{e}(kT)$ is the closed loop system error, $\boldsymbol{c} = [c_1\ c_2]$ is a parameter vector that defines the system dynamics in sliding mode, $\boldsymbol{x_d} = [x_{d1}\ x_{d2}]$ is the demand state. The length of vector $\boldsymbol{c}$ does not alter the system dynamics, only its orientation is important. Thus, it can be assumed, without loss of generality that $\boldsymbol{c} = [1\ c_2]$. This reduces the number of parameters that must be chosen by the designer. The demand state consists of the desired output voltage and the desired inductor current. The output voltage is chosen by the user, appropriately for the needs of the load. The desired inductor current can be calculated from the balance of power. The power drawn from the inverters supply is Ux_2 and the power drawn by the load is x_1i_R, where i_R is the load current. Therefore

$$x_{d2} = \frac{x_1i_R}{U}. \tag{9}$$

The next step is designing the control signal. This signal must ensure, that the state converges to the sliding line or to its close vicinity in finite time. This corresponds to reducing the sliding variable to zero or to some neighborhood of zero. Then the state must move toward the desired state along the sliding line. The simplest approach is deriving a controller that will reduce the sliding variable to zero in the next time instant, namely calculate such a $u(kT)$, that

$$s[(k+1)T] = 0. \tag{10}$$

Unfortunately, such a controller would result in a very large value of the inductor current at the start of the control process. One way of eliminating this problem is to use a reaching law. In this approach, we demand that the sliding variable in the next time instant is a function of its current value, for example

$$s[(k+1)T] = qs(kT) + D(kT), \tag{11}$$

where $q \in (0, 1)$ and $D(kT)$ is the impact of disturbances on the sliding variable. Altering the value of parameter q can smooth-out the reaching phase and reduce the maximum value of the inductor current. However, as the next value of s depends on its current value, this results in unnecessary reduction of robustness w.r.t. external disturbances (this will be demonstrated in more detail in the simulation section). Therefore in this work, a sliding variable reference trajectory, which eliminates this problem is proposed. The controller makes the sliding variable of the plant track its reference value. Since the disturbance is acting on the plant, we cannot require that these values are exactly equal. However, we can require that the only difference is the "last" value of the disturbance, i.e.

$$s[(k+1)T] = s_d[(k+1)T] + D(kT), \tag{12}$$

where s_d is the reference value of the sliding variable, which can be calculated *a priori*.

In order to derive the control signal that enforces (12), the value of the sliding variable in the next time instant is calculated using (7) and (8)

$$s[(k+1)T] = \boldsymbol{c}^T\{\boldsymbol{x_d} - \boldsymbol{x}[(k+1)T]\} = \boldsymbol{c}^T[\boldsymbol{x_d} - \boldsymbol{Ax}(kT) - \boldsymbol{B}u(kT)]. \tag{13}$$

Equating the right hand sides of (12) and (13) the required control signal is equal to

$$u(kT) = \left(\boldsymbol{c}^T\boldsymbol{B}\right)^{-1}\left\{\boldsymbol{c}^T\boldsymbol{x_d} - \boldsymbol{c}^T\boldsymbol{Ax}(kT) - s_d[(k+1)T]\right\}. \tag{14}$$

The last remaining step is to select the reference trajectory $s_d[(k+1)T]$. In this work, we will consider two trajectories, in the first one the sliding variable converges asymptotically to zero

$$s_d(kT) = q^k s_d(0), \tag{15}$$

with $q \in (0, 1)$ and in the second it decreases linearly to zero

$$s_d[(k+1)T] = \begin{cases} s_d(kT) - \delta & \text{for } s_d[(k+1)T] \geq \delta \\ 0 & \text{for } s_d[(k+1)T] < \delta \end{cases}, \tag{16}$$

with $\delta > 0$.

4 Simulation Results

In this section, computer simulations of the DCDC buck converter will be demonstrated. The parameters are $L = 990$ μH, $C = 1$ mF, supply voltage $U = 48$ V, desired output voltage $x_{1d} = 24$ V. It is assumed that the resistance of the load can vary between 2 and 12 Ω. For the controller design, R is assumed equal to its mean value, namely 7 Ω. Therefore, the changes in load resistance affect the system as an unknown disturbance. The actual resistance will be equal to 7 Ω for $t = [0, 0.02$ s$]$, 12 Ω for $t = (0.02$ s, 0.03 s$]$ and 2 Ω for the remainder of the simulation. The simulations depict controlling a continuous model of the converter, using the discrete time controller, according to Fig. 2. It is worth pointing out, that this scheme relates closely to real applications. It takes into account sampling a continuous time signal, implementing the control algorithm in a digital device, and controlling the plant via a PWM signal, not directly the value $u(kT)$.

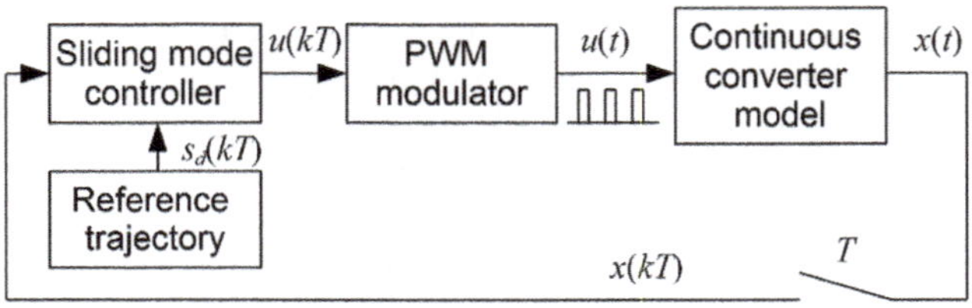

Fig. 2. Scheme of the control loop.

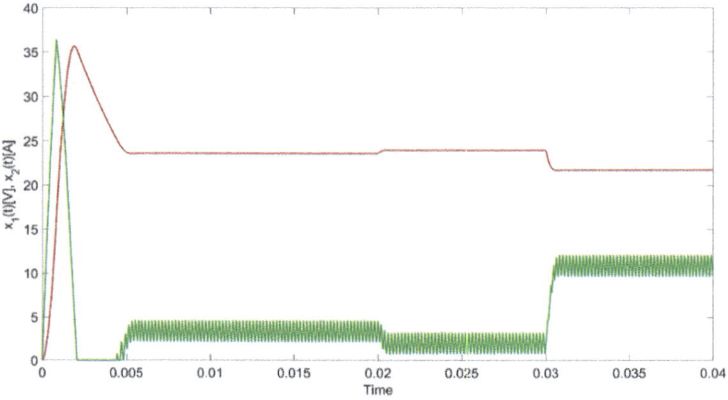

Fig. 3. "Classical" sliding mode controller.

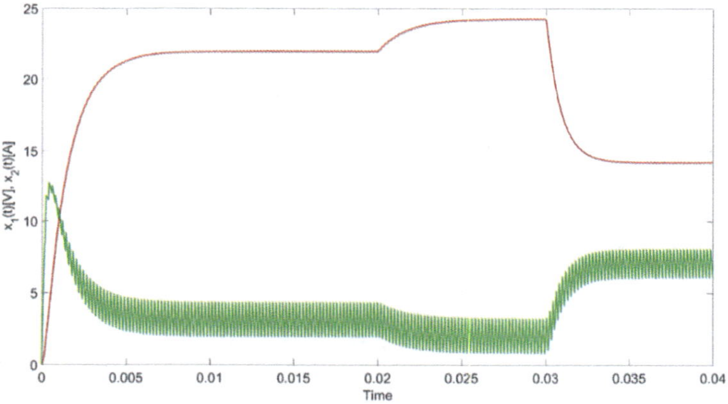

Fig. 4. Reaching law based sliding mode controller.

In practice, the blocks "Sliding mode controller", "Reference trajectory" and "PWM modulator" could all be implemented in a single microcontroller/DSP.

The sampling period is selected as $T = 0.2$ ms (which corresponds to frequency $f = 5$ kHz). The vector $\boldsymbol{c}$ is chosen as $[1 \ \ 0.05]^T$, which corresponds to eigenvalues of the

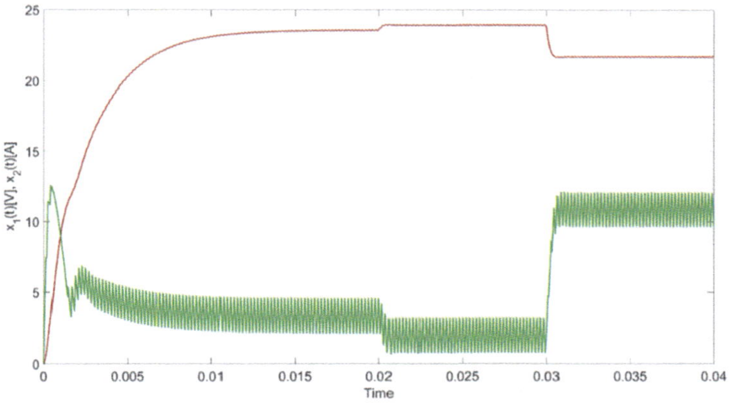

Fig. 5. Sliding mode controller based on reference trajectory (15).

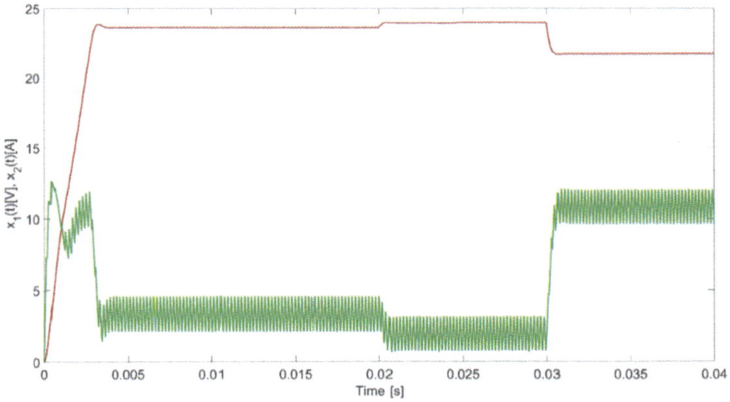

Fig. 6. Sliding mode controller based on reference trajectory (16), $\delta = 1.7$.

closed loop system equal to 0 and -0.33 during sliding mode. A total of four controllers are tested out. For brevity, they will be referred to by the following numbers: the "basic" sliding mode controller, based on Eq. (10) will be called controller 1, the reaching law controller based on (11) controller 2, and two controllers with reference trajectories of the sliding variable (15) and (16) called controllers 3 and 4 respectively. The simulation results are shown in Figs. 3, 4, 5, 6 and 7. The output voltage is shown in red and the inductor current in green. Moreover, Fig. 8 shows a comparison of the sliding variable evolutions for all of the controllers. The controller parameters were chosen so that the initial current does not exceed 13 A, apart from the first one, for which limiting the initial current is not possible. For controller 2, this leads to $q = 0.875$, for controller 3 to $q = 0.925$. For controller 4, limiting the current leads to $\delta = 1.7$, however at such a rapid convergence, an overshoot of the output voltage appears. For some applications, such an overshoot is unacceptable, therefore, a second simulation, with $\delta = 0.7$, was carried out.

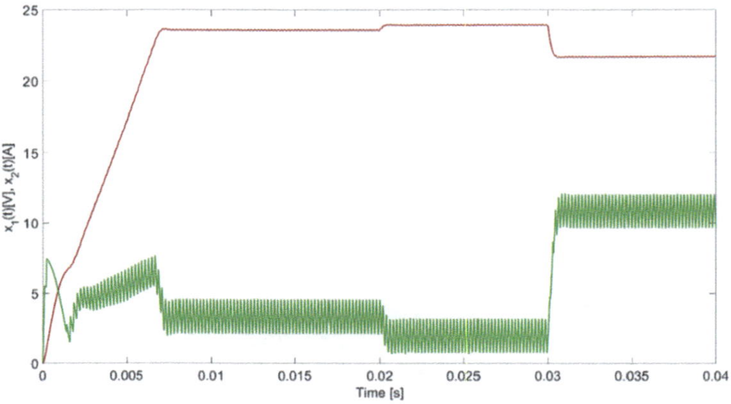

Fig. 7. Sliding mode controller based on reference trajectory (16), $\delta = 0.7$.

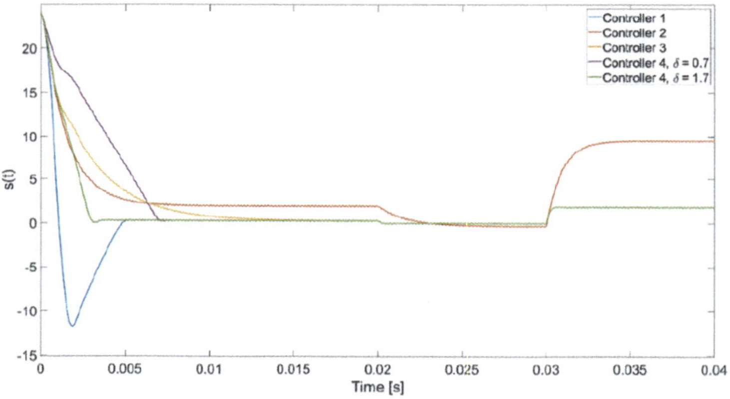

Fig. 8. Comparison of sliding variable values for all of the controllers.

As one can observe, controllers 1, 3 and 4 offer the same robustness w.r.t. changes in the load. This can be easily explained, as for all of them, once the load starts to vary at 0.02 s, the desired value of sliding variable is equal to zero. On the other hand, controller 1 is unable to limit the initial inductor current and leads to a large overshoot of the output voltage. This makes it infeasible in practice. Controller 2 limits the initial value of current, however at the cost of a significant reduction in robustness. Therefore, controllers 3 and 4 outperform both controllers 1 and 2. Comparing controllers 3 and 4, one can observe, that choosing a linear convergence of s_d to zero in place of an asymptotic convergence allows for shorter regulation times. Even for the lower value of $\delta = 0.7$ the voltage settles noticeably faster for controller 4 than for controller 3.

The robustness can also be analyzed using Fig. 8. The smaller the value of s, the closer the representative point is to the sliding line which corresponds to higher robustness. It can be observed, that after about 0.015 s all the values of the sliding variables are virtually equal, apart from controller 2. This corresponds to ensuring the same robustness, as was mentioned in the previous paragraph. Moreover, even though the design of controller 1 was based on bringing the sliding variable to zero in a single step, this does not in fact take place. Since the PWM modulator limits the duty cycle to [0, 1], the control signal magnitude is too small to ensure a single step convergence to zero. It is worth pointing out, that the oscillations visible in all of the figures are not caused by the so-called chattering effect, but arise naturally from the PWM type control. The current obviously must oscillate due to such control, which also causes the voltage to oscillate. Thus the sliding variable, which is a linear combination of the two, also oscillates.

In practice, the robustness of all the controllers would be further improved by introducing an integral of the output voltage error into the sliding variable equation. However, as is well known, adding integral action can lead to oscillatory behavior or even instability. Therefore, it is still beneficial to limit the steady-state error as much as possible "before" the integral action is added, as it allows using larger integration times, thus reducing the risk of these unwanted effects.

5 Conclusions

In this work a reference trajectory based discrete sliding mode controller was proposed for a DCDC buck converter. Two reference trajectories of the sliding variable were tested out: an asymptotical decrease to zero and a linear one. It was demonstrated, that both allow to limit the initial value of the inductor current, without reducing the robustness of the closed loop system. Future work will focus on applying this approach to more complex converter topologies (i.e. ones containing an additional LC input filter [24, 25]) and testing out it in practice.

References

1. Bartoszewicz, A.: A new reaching law for sliding mode control of continuous time systems with constraints. Trans. Inst. Meas. Control **37**(4), 515–521 (2015)
2. Bartoszewicz, A., Adamiak, K.: A reference trajectory based discrete time sliding mode control strategy. Int. J. Appl. Math. Comput. Sci. **29**(3), 517–525 (2019)
3. Bartoszewicz, A., Adamiak, K.: Discrete time sliding mode control with a desired switching variable generator. IEEE Trans. Autom. Control **65**(4), 1807–1814 (2020)
4. Bartoszewicz, A., Adamiak, K.: Model reference discrete-time variable structure control. Int. J. Adapt. Control Signal Process. **32**(10), 1440–1452 (2018)
5. Bartoszewicz, A., Latosiński, P.: Discrete time sliding mode control with reduced switching - a new reaching law approach. Int. J. Robust Nonlinear Control **26**(1), 47–68 (2016)
6. Bartoszewicz, A., Leśniewski, P.: New switching and nonswitching type reaching laws for SMC of discrete time systems. IEEE Trans. Control Syst. Technol. **24**(2), 670–677 (2016)
7. Bartoszewicz, A., Leśniewski, P.: Reaching law approach to the sliding mode control of periodic review inventory systems. IEEE Trans. Autom. Sci. Eng. **11**, 810–817 (2014)
8. Bartoszewicz, A., Leśniewski, P.: Reaching law-based sliding mode congestion control for communication networks. IET Control Theory Appl. **8**(17), 1914–1920 (2014)
9. DeCarlo, R.S., Zak, S., Mathews, G.: Variable structure control of nonlinear multivariable systems: a tutorial. Proc. IEEE **76**, 212–232 (1988)
10. Edwards, C., Spurgeon, S.: Sliding Mode Control: Theory and Applications. Taylor & Francis, London (1998)

11. Erdem, H.: Comparison of fuzzy, PI and fixed frequency sliding mode controller for DC-DC converters. In: ACEMP 2007 and Electromotion 2007 Joint Conference, pp. 684–689 (2007)
12. Gao, W., Wang, Y., Homaifa, A.: Discrete-time variable structure control systems. IEEE Trans. Industr. Electron. **42**(2), 117–122 (1995)
13. Guo, S., Lin-Shi, X., Allard, B., Gao, Y., Ruan, Y.: Digital sliding-mode controller for high-frequency DC/DC SMPS. IEEE Trans. Power Electron. **25**(5), 1120–1123 (2010)
14. Hu, J., Shang, L., He, Y., Zhu, Z.Q.: Direct active and reactive power regulation of grid-connected DC/AC converters using sliding mode control approach. IEEE Trans. Power Electron. **26**(1), 210–222 (2011)
15. Latosiński, P., Bartoszewicz, A.: Reaching law based DSMC with a reference model. IFAC-PapersOnLine **52**(16), 777–782 (2019)
16. Liu, P.J., Chang, C.W.: CCM noninverting buck-boost converter with fast duty-cycle calculation control for line transient improvement. IEEE Trans. Power Electron. **33**(6), 5097–5107 (2018)
17. Renaudineau, H., Martin, J.P., Nahid-Mobarakeh, B., Pierfederici, S.: DC–DC converters dynamic modeling with state observer-based parameter estimation. IEEE Trans. Power Electron. **30**(6), 3356–3363 (2015)
18. Shao, S., Wheeler, P.W., Clare, J.C., Watson, A.J.: Fault detection for modular multilevel converters based on sliding mode observer. IEEE Trans. Power Electron. **28**(11), 4867–4872 (2013)
19. Tao, C., Fayed, A.A.: A buck converter with reduced output spurs using asynchronous frequency hopping. IEEE Trans. Circuits Syst. II Express Briefs **58**(11), 709–713 (2011)
20. Shtessel, Y., Edwards, C., Fridman, L., Levant, A.: Sliding Mode Control and Observation. Springer, New York (2014)
21. Tan, S.C., Lai, Y.M., Tse, C.K., Cheung, M.K.H.: A fixed-frequency pulsewidth modulation based quasi-sliding-mode controller for buck converters. IEEE Trans. Power Electron. **20**(6), 1379–1392 (2005)
22. Utkin, V., Guldner, J., Shi, J.: Sliding Mode Control in Electromechanical Systems. CRC Press, Boca Raton (2009)
23. Utkin, V.: Sliding Modes in Control and Optimization. Springer, Heidelberg (1992)
24. Veerachary, M.: Two-switch semiquadratic buck converter. IEEE Trans. Industr. Electron. **64**(2), 1185–1194 (2017)
25. Veerachary, M.: Analysis of minimum-phase fourth-order buck DC-DC converter. IEEE Trans. Industr. Electron. **63**(1), 144–154 (2016)
26. Xue, J., Lee, H.: A 2-MHz 60-W zero-voltage-switching synchronous noninverting buck-boost converter with reduced component values. IEEE Trans. Circuits Syst. II Express Briefs **62**(7), 716–720 (2015)

A-Weighting for Acoustic Radiation Shaping of a Vibrating Plate

Stanislaw Wrona(✉), Krzysztof Mazur, Jaroslaw Rzepecki, Anna Chraponska, and Marek Pawelczyk

Department of Measurements and Control Systems, Silesian University of Technology, Akademicka 16, 44-100 Gliwice, Poland
stanislaw.wrona@polsl.pl

Abstract. The investigated method enables precise shaping of acoustic radiation of a vibrating plate, i.e. it allows one to relocate or create resonances and anti-resonances for selected frequencies, simultaneously altering their acoustic radiation efficiency in a desired manner. The method can be very beneficial for plates used as noise barriers, both in passive and active applications. The acoustic radiation shaping method involves mounting several additional ribs and masses to the plate surface at locations followed from an optimization process (sensors and actuators can also be included, if active control is considered). The optimization process requires a model of the vibroacoustic system, a cost function corresponding to the considered objective, and an optimization algorithm. In this paper, an introduction of A-weighting to the cost functions, which reflects a human perception of the noise radiated by or transmitted through the plate, is investigated. It follows from the analysis of obtained results that the introduction of A-weighting can provide even 8 dBA better passive noise attenuation.

Keywords: Active structural acoustic control · Active noise control · Passive control · Optimization process · Mathematical model

1 Introduction

Plates are often used in vibroacoustic applications as noise barriers that should block the acoustic noise propagation, both in passive and active control solutions [1–3]. In the latter case, plate vibrations are actively controlled with the use of actuators to enhance passive noise attenuation of the barrier [4,5]. It is also noteworthy that such plates can possess intentional openings with dedicated active noise control systems to allow ventilation and light, without deteriorating the performance of the barrier [6–9]. Nevertheless, in both passive and active applications, it would be very beneficial to have an ability to shape the frequency response of the plate as desired in order to enhance its passive transmission loss.

A method to shape structural frequency response of the plate has been developed by the authors in previous publications [10,11]. The method was later

A. Bartoszewicz et al. (Eds.): PCC 2020, AISC 1196, pp. 171–183, 2020.
https://doi.org/10.1007/978-3-030-50936-1_15

extended to take into account also acoustic radiation estimates. In order to reach desired frequency response, additional masses and ribs arrangement is optimised. General rules are known - additional masses lower the natural frequencies of the plate, whereas stiffeners elevate them. When employed together at appropriate locations, they can precisely alter frequencies of multiple modes as desired. Moreover, even moderate alteration of mode shapes using the attached additional elements may impact to a great extent their acoustic radiation power. This phenomenon can be very useful in application of the considered noise barriers. However, in the previous research, the investigated cost functions considered all natural frequencies with an equal weight, whereas the human ear register lower frequencies with a lesser sensitivity, thus perceiving them as less intense and less disruptive. A-weighting can be introduced into the cost functions in order to reflect the relative loudness perceived by the human ear, thus improving the performance of a barrier in more strongly perceived range of frequency.

In this paper, the impact of A-weighting on performance of designed noise barriers is investigated and discussed. The acoustic radiation shaping approach with A-weighting is employed also for actively controlled barrier. In such a case, structural sensors and actuators are modelled as additional masses, and their arrangement is optimized together with passive masses and ribs, in order to both shape the frequency response and maximize the controllability and observability of the system [12,13].

2 Acoustic Radiation Shaping Method with A-Weighting

This Section describes the proposed shaping method, in which the arrangement of additional elements mounted on a vibrating plate is optimized in order to shape its acoustic radiation, and to enhance controllability and observability measures, if needed. Firstly, an approach used for modelling of the vibroacoustic system is briefly introduced. Then, an optimization problem with appropriate cost functions is defined, and an optimization algorithm employed to find an optimal solution is described.

2.1 Modelling of the Vibroacoustic System

The employed mathematical model is based on a description of the free vibrations of an isotropic rectangular plate with additional elements bonded to its surface, i.e. masses, ribs, vibration actuators and structural sensors. The Kirchhoff-Love theory of thin plates is used for this purpose. In the considered scenarios, the boundary conditions of the plate are assumed to be fully-clamped, although alternative mountings can also be considered in a similar way. The Rayleigh-Ritz method is employed to define an approximate solution, providing natural frequencies ω_i and mode shapes of the vibrating system, where i is the mode number. An appropriate Green's function is used to calculate estimates of the acoustic radiation P_i, according to the following formula

$$P_i = \iint_{S_e} |p_i(x, y, z)|^2 \, \mathrm{d}S_e. \tag{1}$$

where $p_i(x, y, z)$ is the modal sound pressure amplitude, and S_e is a plane parallel to the plate used for integration of the acoustic radiation power. A measure of the acoustic radiation power with applied A-weighting $P_{A,i}$ is also introduced, which is P_i scaled according to the curves defined in the international standard IEC 61672-1:2013 [14]. Then, a state space form of the model is developed, which facilitates the controllability and observability analysis of the system for active control scenarios. Diagonal elements of the controllability and observability Gramian matrices, $\lambda_{c,i}$ and $\lambda_{o,i}$, respectively, are used as measures of controllability and observability of the system.

The derivation of the mathematical model is out of the scope of this paper. It is assumed that the model is accurate, verified, and provides the aforementioned parameters describing the behaviour of the vibroacoustic system. However, a detailed derivation of the model is available in other publications [10,15].

2.2 Optimization Problem

The optimization variables defined for the considered problem are corresponding to the shape and placement of a predefined number of additional elements mounted on a vibrating plate surface. They are optimized according to minimization of an arbitrarily chosen cost function. Four kinds of elements are considered:

1. Additional masses, which are passive elements. The total concentrated mass $m_{m,i}$ and location coordinates $(x_{m,i}, y_{m,i})$ are considered as optimization variables for the i-th additional mass.
2. Ribs, which also are passive elements. The length, location and orientation are defined by coordinates of the first end $(x_{r0,i}, y_{r0,i})$ and the second end $(x_{r1,i}, y_{r1,i})$, which are the optimization variables for the i-th rib. The rib material and cross-section are assumed to be predefined in this paper.
3. Actuators – inertial actuators are considered in this paper. Their shape and mass are assumed to be predefined, hence, only the location coordinates $(x_{a,i}, y_{a,i})$ become the optimization variables for the i-th actuator.
4. Sensors – accelerometers are considered in this research as structural sensors. They are used to provide feedback signals for an active control algorithm. Similarly as in the case of actuators, only the location coordinates $(x_{s,i}, y_{s,i})$ are considered as optimisation variables for the i-th sensor.

Furthermore, the variables specified above are subject to various restraints. The number, weight and dimensions of the elements are limited due to plate shape and other practical constraints dependent on the application. The additional elements should not also overlap. To summarize the formulated problem and assuming that N_m, N_r, N_a and N_s are the total numbers of masses, ribs, actuators and sensors, respectively, the optimization algorithm is required to find a solution in an $(3N_m + 4N_r + 2[N_a + N_s])$–dimensional space.

2.3 Cost Functions

The cost function for the described problem should reflect the discrepancy between the desired and evaluated acoustic radiation of the plate. In a general form, the cost function can be defined as

$$J = f(\omega_i, P_i, P_{A,i}, \lambda_{c,i}, \lambda_{o,i}, \Theta) \ , \tag{2}$$

where Θ is a vector representing optimization variables of the additional elements. Specific definitions of cost functions are presented in Sect. 3, where results for two scenarios are presented and discussed.

2.4 Optimization Algorithm

A Memetic Algorithm (MA) has been chosen to find a solution that satisfies the defined requirements. The MA is a hybrid form of a population-based approach coupled with individual learning [16,17]. MA combines advantages of a global search and local refinement procedures, which enhance converge to the local optima. The MA has been successfully used by the authors in previous applications, e.g. in [18,19]. A flowchart of the employed memetic algorithm is recalled in Fig. 1.

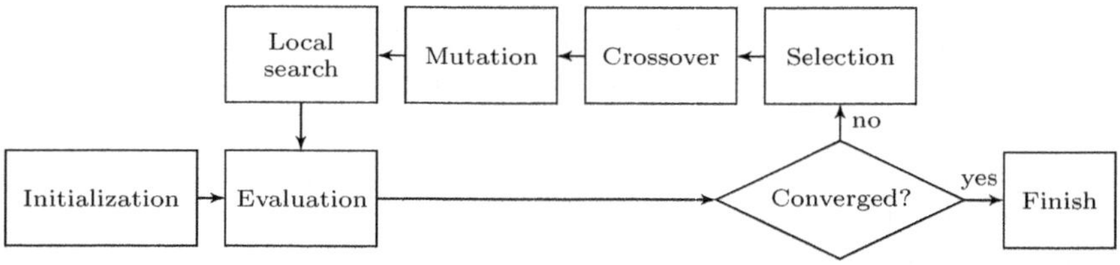

Fig. 1. A memetic algorithm flowchart.

3 Analysis of Shaping Results

The presented shaping results have been obtained for an exemplary steel plate described by the following parameters:

$$a = 0.420 \text{ m}, \quad b = 0.390 \text{ m}, \quad h = 0.001 \text{ m},$$
$$E = 210 \text{ GPa}, \quad \rho_p = 7850 \, \text{kg/m}^3, \quad \nu = 0.3,$$

where a, b and h are the width, height and thickness of the plate, E is the Young's modulus; ν is the Poisson ratio; and ρ_p is the mass density of the plate material. The frequency bandwidth is considered up to 400 Hz, and first 12 eigenmodes of the plate are compared for two selected crucial scenarios: (i) minimization of the acoustic radiation of a passive barrier using additional passive elements and (ii) designing a hybrid passive-active acoustic barrier.

For the unloaded plate, frequency responses obtained from the model are shown in Fig. 2, whereas natural frequencies and acoustic radiation estimates are given in Table 1. Weights of additional masses are limited to maximum value of 0.2 kg. The ribs are assumed to have a constant cross-section of square shape, defined by dimensions: $e_{r,i} = h_{r,i} = 0.006$ m, where $e_{r,i}$ is the width and $h_{r,i}$ is the height of the i-th stiffener. The material of the stiffeners is considered to be the same as of the plate.

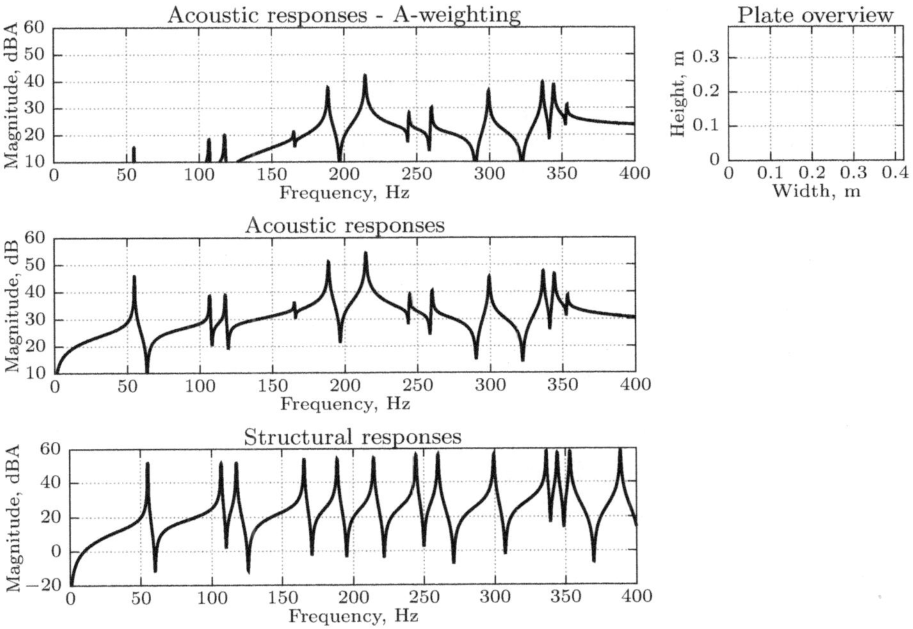

Fig. 2. Acoustic (with and without A-weighting) and structural vibration responses of the unloaded plate (no additional elements attached).

Table 1. Natural frequencies and estimates of modal acoustic power (with and without A-weighting) of the unloaded plate.

Mode	1	2	3	4	5	6	7	8	9	10	11	12
ω_i (Hz)	55.1	106.8	117.6	165.5	188.8	214.4	244.5	260.1	299.4	336.4	344.3	353.2
P_i (dB)	20.1	12.9	13.1	8.4	24.1	27.2	10.8	12.0	17.2	18.4	18.0	8.9
$P_{A,i}$ (dBA)	4.8	2.7	3.6	1.0	17.4	21.1	5.4	6.9	12.7	14.3	14.0	5.1

For each optimization, the population consisted of 300 individuals. Maximum number of generations was set to 15. Probability of crossover, mutation and individual learning was 0.20, 0.30 and 0.06, respectively.

3.1 Minimization of the Acoustic Radiation in a wide Frequency Range

In this Subsection, minimization of the acoustic radiation in a wide frequency range is considered. Such objective is expressed using the following cost function

$$J_1 = \max_i P_i, \qquad \omega_i \leq \omega_{sp}, \tag{3}$$

where ω_{sp} is a set point frequency, limiting the frequency range of interest (only resonances of $\omega_i \leq \omega_{sp}$ are considered in the cost function). The cost function J_1 results in minimization of acoustic radiation of the most radiating mode. The set point frequency was set to $\omega_{sp} = 300\,\text{Hz}$. Only passive elements were considered in this case. The number of additional masses and ribs was set to $N_m = 2$ and $N_r = 2$, respectively. Results of the optimization are given in Fig. 3 and Table 2, where circles and lines represent placements of additional masses and ribs, respectively. In Fig. 3, three frequency responses of the plate are given: a structural response in the bottom, an acoustic response in the middle, and an acoustic response with A-weighting in the top.

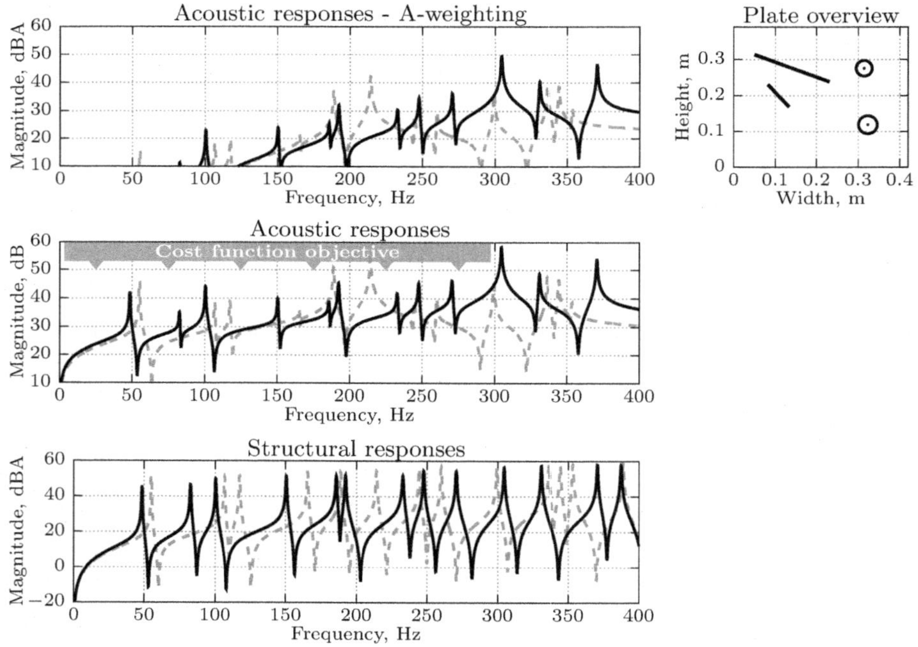

Fig. 3. Acoustic (with and without A-weighting) and structural vibration responses of the plate, obtained for optimization index J_1 (solid line—plate with elements; dashed line—the unloaded plate). The plate overview with the additional elements is also shown (circles—additional masses; lines—ribs).

Table 2. Results of optimization of cost function J_1 for $N_m = 2$ and $N_r = 2$. Natural frequencies and estimates of modal acoustic power of the loaded plate (with and without A-weighting), together with placement of additional elements are presented.

Mode	1	2	3	4	5	6	7	8	9	10	11	12
ω_i (Hz)	48.5	82.7	100.5	150.5	185.8	192.3	233.0	247.7	270.7	304.7	331.0	370.6
P_i (dB)	19.2	11.7	19.2	14.2	12.8	19.4	15.8	18.1	19.0	30.2	19.9	24.7
$P_{A,i}$ (dBA)	2.8	−0.2	8.6	6.2	5.4	12.6	9.9	12.9	14.2	26	15.6	21.3

Masses	$x_{a,i}$ (m)	$y_{a,i}$ (m)	$m_{a,i}$ (kg)
1	0.323	0.119	0.200
2	0.314	0.275	0.146

Ribs	$x_{r0,i}$ (m)	$y_{r0,i}$ (m)	$x_{r1,i}$ (m)	$y_{r1,i}$ (m)
1	0.231	0.238	0.050	0.313
2	0.134	0.169	0.082	0.230

The algorithm reached the solution with cost function value $J_{1,opt} = 19.4$ dB. As a result of the optimization, the magnitudes of the structural response was only slightly altered, whereas the acoustic response magnitudes were strongly reduced, cutting the magnitudes of the highest peaks in the considered frequency range even by 7 dB. It follows from the analysis of the obtained results that the structural vibration resonances can be moved in the frequency domain with additional masses and ribs, whereas it is difficult to reduce the magnitudes. On the other hand, additional masses and ribs strongly affect the mode shapes, which determine to a high extent the acoustic radiation. Thus, the acoustic radiation capability of the plate can be considerably reduced with the use of additional elements, even in a wider frequency range, as shown with the provided results.

Introduction of A-Weighting to the Cost Function

The cost function J_1 takes into account all natural frequencies in the considered frequency range with an equal weight, however, the human ear registers lower frequencies with a lesser sensitivity, thus perceiving them as less intense and less disruptive. To take into account the relative loudness perceived by humans, A-weighting can be introduced to cost function J_1, modifying it into

$$J_2 = \max_i P_{A,i}, \qquad \omega_i \leq \omega_{sp}. \tag{4}$$

The cost function J_2 results in minimization of acoustic radiation of the most radiating mode after the A-weighting, hence the optimization algorithm will adjust the evaluation of the potential solutions according to the human perception, as people are the final recipients of the proposed noise-reducing solutions. Results of the optimization using J_2 are given in Fig. 4 and Table 3.

The algorithm reached the solution with $J_{2,opt} = 6.8$ dB. It follows from the analysis of the results obtained with J_1 and J_2, that the latter cost function makes the algorithm to pay more attention to the higher part of the considered

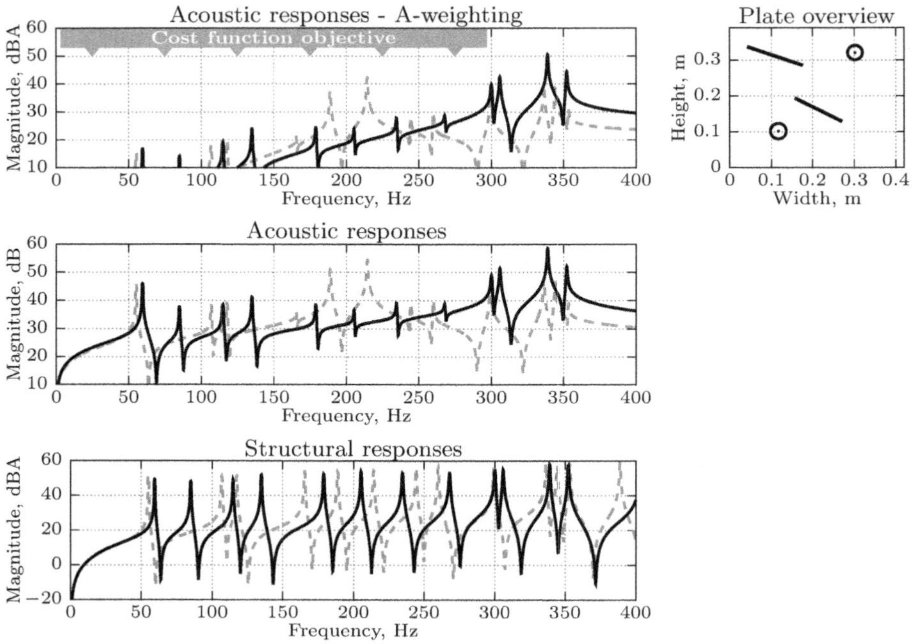

Fig. 4. Acoustic (with and without A-weighting) and structural vibration responses of the plate, obtained for optimization index J_2 (solid line—plate with elements; dashed line—the unloaded plate). The plate overview with the additional elements is also shown (circles—additional masses; lines—ribs).

Table 3. Results of optimization of cost function J_2 for $N_m = 2$ and $N_r = 2$. Natural frequencies and estimates of modal acoustic power of the loaded plate (with and without A-weighting), together with placement of additional elements are presented.

Mode	1	2	3	4	5	6	7	8	9	10	11	12
ω_i (Hz)	59.5	84.9	114.7	134.8	178.9	205.3	234.7	267.9	300.0	305.9	338.6	352.1
P_i (dB)	21.4	13.7	13.9	15.3	12.2	9.1	12.4	11.1	21.3	24.3	29.4	23.5
$P_{A,i}$ (dBA)	6.8	2.0	4.2	6.7	5.2	2.8	6.8	6.0	16.7	19.8	25.4	19.6

Masses	$x_{a,i}$ (m)	$y_{a,i}$ (m)	$m_{a,i}$ (kg)
1	0.117	0.102	0.148
2	0.302	0.321	0.125

Ribs	$x_{r0,i}$ (m)	$y_{r0,i}$ (m)	$x_{r1,i}$ (m)	$y_{r1,i}$ (m)
1	0.157	0.194	0.271	0.129
2	0.177	0.285	0.042	0.337

frequency band, and disregard to some extent the lower part (as the human ear does). Hence, comparing the magnitudes of $P_{A,i}$ for both J_1 and J_2 in Tables 2, 3, the latter cost function provides even 8 dBA better passive noise attenuation. Such solution makes the barrier more fitted in the human-oriented applications.

3.2 Hybrid Passive-Active Acoustic Barrier

A hybrid passive-active acoustic barrier is designed in this Subsection. The objective is expressed using the following cost function

$$J_3 = \prod_i \left\{ \frac{P_i}{\lambda_{c,i} \cdot \lambda_{o,i}} \right\}, \qquad \omega_i \leq \omega_{sp}. \tag{5}$$

The cost function J_3 leads to simultaneous minimization of both acoustic radiation of modes within the considered frequency range, and maximization of their controllability and observability. Vibration actuators and structural sensors are introduced in this scenario, because the active barrier system needs to control vibrations and to sense a feedback signal for its operations. The set point frequency was again $\omega_{sp} = 300\,\text{Hz}$. Actuators, sensors and ribs were used as admissible additional elements, adopting $N_a = 3$, $N_s = 3$ and $N_r = 2$. Results of the optimization are given in Fig. 5 and Table 4.

The obtained solution satisfies all of the stated objectives: (i) the estimates of modal acoustic power of resonances within the frequency range of interest are passively reduced (more than 10 dB for certain frequency bands), while both (ii) controllability and (iii) observability are ensured for all of the considered modes.

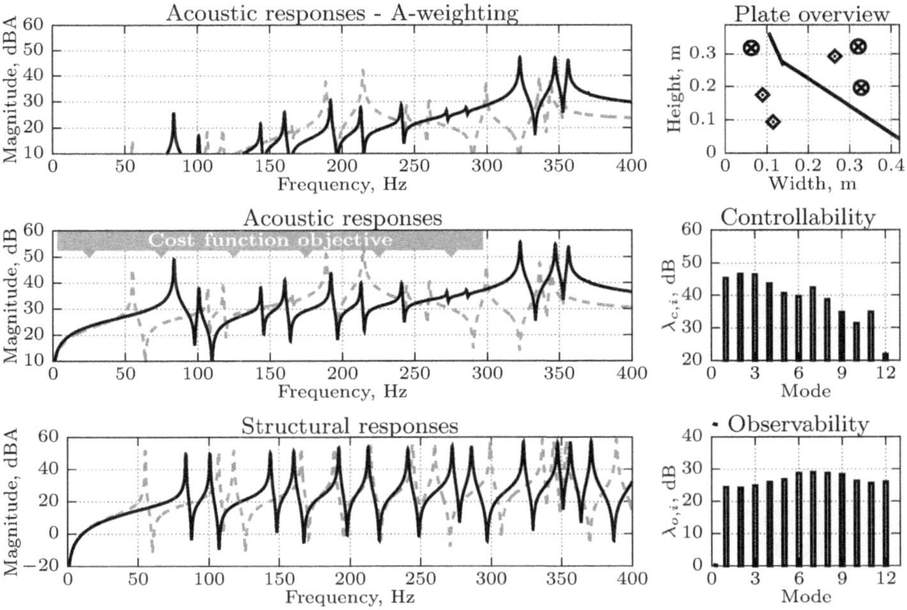

Fig. 5. Acoustic (with and without A-weighting) and structural vibration responses of the plate, obtained for optimization index J_3 (solid line—plate with elements; dashed line—the unloaded plate). The plate overview with the additional elements is also shown (circles with "X" inside—actuators; diamonds—sensors; lines—ribs).

Table 4. Results of optimization of cost function J_3 for $N_a = 3$, $N_s = 3$ and $N_r = 2$. Natural frequencies and estimates of modal acoustic power of the loaded plate (with and without A-weighting), together with placement of additional elements are presented.

Mode	1	2	3	4	5	6	7	8	9	10	11	12
ω_i (Hz)	83.8	100.8	143.8	160.3	192.1	212.8	241.1	272.5	286.1	322.8	347.3	356.3
P_i (dB)	24.4	13.2	12.9	15.5	16.9	13.2	13.4	7.1	7.1	27.1	26.6	25.5
$P_{A,i}$ (dBA)	12.6	2.7	4.7	7.9	10.3	7.1	7.9	2.2	2.4	22.8	22.6	21.7
$\lambda_{c,i}$ (dB)	45.3	46.6	46.4	43.6	40.6	39.6	42.3	38.6	34.7	31.3	34.8	17.8
$\lambda_{o,i}$ (dB)	24.3	24.1	24.8	25.8	26.7	28.7	28.9	28.6	28.3	26.2	25.5	25.9

Actuators	$x_{a,i}$ (m)	$y_{a,i}$ (m)	$m_{a,i}$ (kg)
1	0.321	0.323	0.115
2	0.063	0.319	0.115
3	0.328	0.196	0.115

Ribs	$x_{r0,i}$ (m)	$y_{r0,i}$ (m)	$x_{r1,i}$ (m)	$y_{r1,i}$ (m)
1	0.140	0.264	0.105	0.365
2	0.136	0.277	0.419	0.043

Sensors	$x_{s,i}$ (m)	$y_{s,i}$ (m)	$m_{s,i}$ (kg)
1	0.115	0.094	0.010
2	0.090	0.174	0.010
3	0.265	0.293	0.010

It is noteworthy that only the acoustic power of the first mode was not passively reduced with the obtained solution (cf. Fig. 5). This is due to the fact that the fundamental mode by its nature do not possess any nodal line, what makes it practically impossible to reduce its acoustic radiation by influencing the mode shape (nearly no alteration is possible for such a shape). However, taking into account the A-weighting, the fundamental mode acoustic power could be already weakly perceivable due to its low frequency (and thus neglected) or its frequency can be further decreased to make it perceived as even weaker.

Introduction of A-Weighting to the Cost Function
The cost function J_3, similarly as J_1, can be modified using A-weighting, obtaining

$$J_4 = \prod_i \left\{ \frac{P_{A,i}}{\lambda_{c,i} \cdot \lambda_{o,i}} \right\}, \qquad \omega_i \leq \omega_{sp}. \tag{6}$$

The cost function J_4 pays more attention to modes inherently perceived by human ear as louder, hence tries to reduce more their acoustic radiation and to guarantee more controllability and observability for them. Performance of such hybrid barrier would be better fitted to the human recipient. Results of the optimization using J_4 are given in Fig. 6 and Table 5.

It follows from the analysis of the results obtained with J_3 and J_4, that the latter cost function, by paying more attention to the higher part of the considered frequency band, leads to approximately 3 dBA better passive noise attenuation. Such hybrid passive-active solutions are a novel approach and requires more investigation, however, its potential is clearly visible.

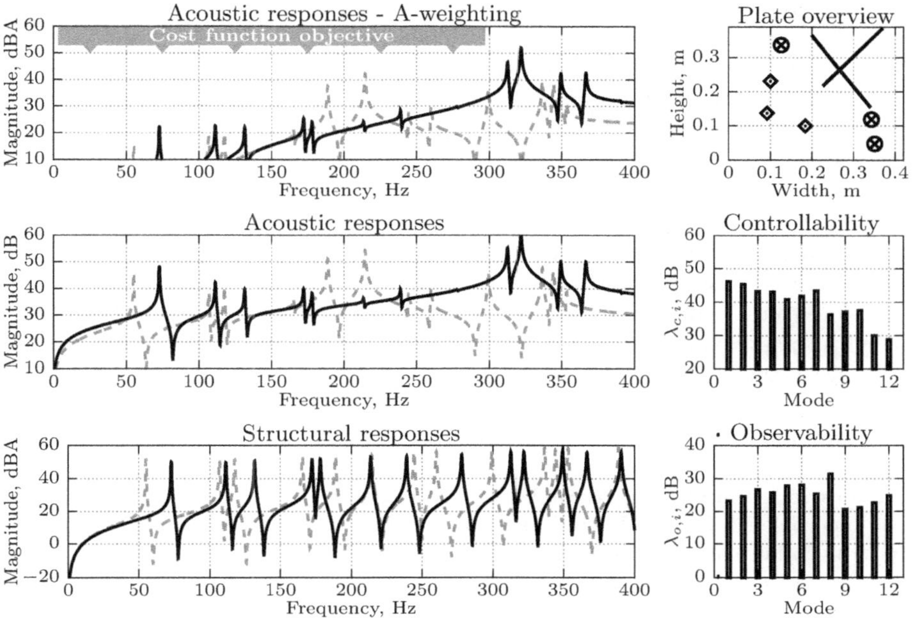

Fig. 6. Acoustic (with and without A-weighting) and structural vibration responses of the plate, obtained for optimization index J_4 (solid line—plate with elements; dashed line—the unloaded plate). The plate overview with the additional elements is also shown (circles with "X" inside—actuators; diamonds—sensors; lines—ribs).

Table 5. Results of optimization of cost function J_4 for $N_a = 3$, $N_s = 3$ and $N_r = 2$. Natural frequencies and estimates of modal acoustic power of the loaded plate (with and without A-weighting), together with placement of additional elements are presented.

Mode	1	2	3	4	5	6	7	8	9	10	11	12
ω_i (Hz)	72.9	111.3	131.6	172.3	178.2	213.7	239.0	278.0	312.6	321.9	349.3	366.5
P_i (dB)	23.0	17.3	14.2	13.8	12.0	7.8	11.8	3.8	27.1	32.4	21.7	22.0
$P_{A,i}$ (dBA)	10.1	7.5	5.4	6.6	5.0	1.7	6.3	−1.1	22.8	28.2	17.7	18.3
$\lambda_{c,i}$ (dB)	46.1	45.4	43.2	43.0	40.8	41.8	43.4	36.3	37.2	37.6	30.0	28.8
$\lambda_{o,i}$ (dB)	23.3	24.7	26.7	25.8	27.9	28.1	25.4	31.5	20.8	21.2	22.7	24.9

Actuators	$x_{a,i}$ (m)	$y_{a,i}$ (m)	$m_{a,i}$ (kg)
1	0.321	0.323	0.115
2	0.063	0.319	0.115
3	0.328	0.196	0.115

Ribs	$x_{r0,i}$ (m)	$y_{r0,i}$ (m)	$x_{r1,i}$ (m)	$y_{r1,i}$ (m)
1	0.140	0.264	0.105	0.365
2	0.136	0.277	0.419	0.043

Sensors	$x_{s,i}$ (m)	$y_{s,i}$ (m)	$m_{s,i}$ (kg)
1	0.115	0.094	0.010
2	0.090	0.174	0.010
3	0.265	0.293	0.010

4 Conclusions

This paper investigates the impact of A-weighting on performance of the designed acoustic barriers. The A-weighting is introduced into cost functions employed for shaping of the acoustic radiation of vibrating plates.

The analysis of obtained results shows that the introduction of A-weighting makes the algorithm to pay more attention to the higher part of the targeted frequency band (as the human ear does), thus providing even 8 dBA better passive noise attenuation for a passive barrier. Such solution makes the barrier more fitted in the human-oriented applications.

Moreover, the optimization of actuators and sensors arrangement, resulting in a hybrid passive-active control systems, can also benefit from the introduction of A-weighting. The optimization process was successfully performed and the passive transmission loss was increased by 3 dBA, while both controllability and observability were ensured for all of the considered modes.

Acknowledgement. The research reported in this paper has been supported by the National Science Centre, Poland, decision number DEC-2017/25/B/ST7/02236, and by State Budget for Science, Poland, in 2019.

References

1. Misol, M., Algermissen, S., Rose, M., Monner, H.P.: Aircraft lining panels with low-cost hardware for active noise reduction. In: 2018 Joint Conference-Acoustics, pp. 1–6. IEEE (2018). https://doi.org/10.1109/ACOUSTICS.2018.8502310
2. Wrona, S., Pawelczyk, M.: Active reduction of device narrowband noise by controlling vibration of its casing based on structural sensors. In: Proceedings of 22nd International Congress on Sound and Vibration. Florence, Italy, 12–16 July (2015)
3. Wrona, S., Mazur, K., Rzepecki, J., Chraponska, A., Pawelczyk, M.: Sound transmission through a thin plate with shaped frequency response. Arch. Acoust. **44**(4), 731–738 (2019). https://doi.org/10.24425/aoa.2019.129728
4. Mao, Q., Pietrzko, S.: Control of Noise and Structural Vibration. Springer, London (2013)
5. Morzyński, L., Szczepański, G.: Double panel structure for active control of noise transmission. Arch. Acoust. **43**(4), 689–696 (2018). https://doi.org/10.24425/aoa.2018.125162
6. Qiu, X.: An Introduction to Virtual Sound Barriers. CRC Press (2019)
7. Lam, B., Elliott, S., Cheer, J., Gan, W.S.: Physical limits on the performance of active noise control through open windows. Appl. Acoust. **137**, 9–17 (2018). https://doi.org/10.1016/j.apacoust.2018.02.024
8. Wang, S., Tao, J., Qiu, X., Pan, J.: A boundary error sensing arrangement for virtual sound barriers to reduce noise radiation through openings. J. Acoust. Soc. Am. **145**(6), 3695–3702 (2019). https://doi.org/10.1121/1.5112502
9. Shi, D., Gan, W.S., Lam, B., Hasegawa, R., Kajikawa, Y.: Feedforward multichannel virtual-sensing active control of noise through an aperture: analysis on causality and sensor-actuator constraints. J. Acous. Soc. Am. **147**(1), 32–48 (2020)

10. Wrona, S., Pawelczyk, M.: Shaping frequency response of a vibrating plate for passive and active control applications by simultaneous optimization of arrangement of additional masses and ribs. Part I: Modeling. Mech. Syst. Sig. Process. **70–71**, 682–698 (2016). https://doi.org/10.1016/j.ymssp.2015.08.018
11. Wrona, S., Pawelczyk, M.: Shaping frequency response of a vibrating plate for passive and active control applications by simultaneous optimization of arrangement of additional masses and ribs. Part II: Optimization. Mech. Syst. Sig. Process. **70–71**, 699–713 (2016). https://doi.org/10.1016/j.ymssp.2015.08.017
12. Wyrwal, J., Zawiski, R., Pawelczyk, M., Klamka, J.: Modelling of coupled vibroacoustic interactions in an active casing for the purpose of control. Appl. Math. Modell. **50**, 219–236 (2017). https://doi.org/10.1016/j.apm.2017.05.002
13. Wyrwal, J.: Simplified conditions of initial observability for infinite-dimensional second-order damped dynamical systems. J. Math. Anal. Appl. **478**(1), 33–57 (2019). https://doi.org/10.1016/j.jmaa.2019.04.066
14. International Electrotechnical Commission: International norm IEC 61672-1:2013: Electroacoustics - Sound level meters - Part 1: Specifications (2013)
15. Rdzanek, W.: Structural vibroacoustics of surface elements [in Polish:Wibroakustyka strukturalna elementów powierzchniowych]. Wydawnictwo Uniwersytetu Rzeszowskiego (2011)
16. Neri, F., Cotta, C., Moscato, P.: Handbook of Memetic Algorithms, vol. 379. Springer, Heidelberg (2012)
17. Nalepa, J., Kawulok, M.: Adaptive memetic algorithm enhanced with data geometry analysis to select training data for SVMs. Neurocomputing **185**, 113–132 (2016). https://doi.org/10.1016/j.neucom.2015.12.046
18. Mazur, K., Wrona, S., Pawelczyk, M.: Active noise control for a washing machine. Appl. Acoust. **146**, 89–95 (2019). https://doi.org/10.1016/j.apacoust.2018.11.010
19. Mazur, K., Wrona, S., Pawelczyk, M.: Design and implementation of multichannel global active structural acoustic control for a device casing. Mech. Syst. Sig. Process. **98C**, 877–889 (2018). https://doi.org/10.1016/j.ymssp.2017.05.025

A New LMI-Based Controller Design Method for Uncertain Differential Repetitive Processes

Robert Maniarski[1(✉)], Wojciech Paszke[2], and Eric Rogers[3]

[1] Institute of Control and Computation Engineering, University of Zielona Góra, Szafrana 2, 65-516 Zielona Góra, Poland
r.maniarski@issi.uz.zgora.pl

[2] Institute of Automation, Electronic and Electrical Engineering, University of Zielona Góra, Szafrana 2, 65-516 Zielona Góra, Poland
w.paszke@iee.uz.zgora.pl

[3] School of Electronics and Computer Science, University of Southampton, Southampton SO17 1BJ, UK
etar@ecs.soton.ac.uk

Abstract. The paper develops new results on stability analysis and control law design for differential linear repetitive processes. These results are based on new dilated LMI characterizations for stability along the pass where auxiliary slack variables with full structure are employed. This provides additional flexibility to the solution. The results are also easily extended to processes with norm-bounded uncertainties. It is also shown that the generalized Kalman-Yakubovich-Popov lemma can be used to obtain stability and controller design procedures in which performance specifications are imposed over finite frequency ranges. Sufficient conditions for the existence of a robust controller in this setting are established. Finally, a simulation example is given to illustrate the merits of the new design.

Keywords: Uncertain repetitive processes · Robust stability and stabilization · Linear matrix inequalities · Finite frequency domain

1 Introduction

Linear repetitive processes repeat the same finite duration operation over and over again, where each repetition is termed a pass and the finite duration the pass length [6]. A physical example is metal rolling operations see, e.g., [7] which, in turn, cites the original work, where in essence deformation of the workpiece takes place between two rolls, multiples times until the desired shape of the workpiece is obtained (or to within an acceptable tolerance). The notation for variables in this paper is of the form $y_k(t)$, $0 \leq t \leq \alpha$ where y is the vector or scalar-valued variable under consideration, $\alpha < \infty$ is the pass length and $k \geq 0$ is the pass number.

A. Bartoszewicz et al. (Eds.): PCC 2020, AISC 1196, pp. 184–196, 2020.
https://doi.org/10.1007/978-3-030-50936-1_16

A stability theory for linear repetitive processes has been developed [6] using a setting that includes all linear time-invariant processes as special cases. Given the unique control problem this theory requires that a bounded initial pass profile y_0 produces a bounded sequence of pass profiles $\{y_k\}_k$, where the boundedness property is defined in terms of the norm on the underlying function space. This stability theory enforces the bounded-input bounded-output property either over the finite and fixed pass length or uniformly with respect to α, where the former property is known as asymptotic stability and the latter stability along the pass.

Stability along the pass, which can be analyzed mathematically by considering $\alpha \to \infty$, is the stronger property. Moreover, asymptotic stability is a necessary condition for stability along the pass. This stability theory has been applied to problems in other areas of control theory, e.g., iterative learning control (ILC) law design, see, e.g. [4] and iterative algorithms for solving nonlinear dynamic optimal control problems based on the maximum principle [5]. Applications include differential processes, where the along the pass dynamics are governed by a linear ordinary differential equation and also discrete dynamics where the along the pass dynamics are governed by an ordinary difference equation. The finite pass length and the structure of the boundary conditions are the main differences with other classes of 2D linear systems, see [6] for a full treatment of this aspect, including repetitive process dynamics to which the existing theory for other classes of 2D linear systems is not applicable even at the stability testing stage.

The scope of this paper is the robust control design for differential repetitive processes with norm-bounded uncertainty. The elimination lemma and the Kalman-Yakubovich-Popov (KYP) Lemma are employed to provide a new conditions for robust stability and controller design. The major outcome is reduced conservatism by the introduction of additional decision variables in the final LMI forms. Also the new conditions are extended to allow control law design and a numerical example is given to demonstrate the effectiveness of the new results.

Throughout this paper the null and identity matrices, respectively, with compatible dimensions are denoted by 0 and I. For a square matrix X, $\mathrm{sym}(X)$ denotes $X + X^T$ and $\rho(\cdot)$ the spectral radius of its matrix argument. Furthermore, $X \succ 0$ ($X \prec 0$) means that the symmetric matrix X is positive definite (negative definite). Also $(*)$ denotes a block entry in a symmetric matrix and $\otimes$ the matrix Kronecker product.

The new results in this paper are developed with the following lemmas that allows that the transformation of non-LMI formulations into LMI form.

Lemma 1. [3] *Let A, B_0 and Θ be given. Then if* $\det(\mathrm{j}\omega I - A) \neq 0$ *for all $\omega \in [0, \infty)$ the following conditions are equivalent:*

i) The frequency domain inequality

$$\begin{bmatrix} (\mathrm{j}\omega I - A)^{-1}B_0 \\ I \end{bmatrix}^* \Theta \begin{bmatrix} (\mathrm{j}\omega I - A)^{-1}B_0 \\ I \end{bmatrix} \prec 0 \quad (1)$$

holds $\forall \omega \in \Omega$ where Ω is the frequency range, i.e. ω belongs to a subset of real numbers denoted by Ω and specified as in Table 1.

ii) There exist symmetric matrices $Q \succ 0$ and P such that

$$\begin{bmatrix} A & B_0 \\ I & 0 \end{bmatrix}^* (\Psi \otimes Q + \Phi \otimes P) \begin{bmatrix} A & B_0 \\ I & 0 \end{bmatrix} + \Theta \prec 0, \tag{2}$$

where

$$\Psi = \begin{bmatrix} \tau & \upsilon \\ \upsilon^* & \varsigma \end{bmatrix}, \ \Phi = \begin{bmatrix} 0 & 1 \\ 1 & 0 \end{bmatrix}. \tag{3}$$

The values of τ, υ and ς for specified choices of $\omega \in \Omega$ are shown in Table 1 and LF, MF and HF denote, respectively, the low, middle and high frequency ranges.

Table 1. Frequency ranges of interest

	LF	MF	HF
Ω	$\|\omega\| < \omega_l$	$\omega_1 \leq \omega \leq \omega_2$	$\|\omega\| > \omega_h$
τ	-1	-1	1
υ	0	$\mathrm{j}\frac{\omega_1+\omega_2}{2}$	0
ς	ω_l^2	$-\omega_1\omega_2$	$-\omega_h^2$

Lemma 2. [1] *Given matrices $\Gamma = \Gamma^T \in \mathbb{R}^{p \times p}$ and two matrices Λ, Σ of column dimension p, there exists an unstructured matrix W that satisfies*

$$\Gamma + \mathrm{sym}\{\Lambda^T W \Sigma\} \prec 0, \tag{4}$$

if, and only if

$${\Lambda_\perp}^T \Gamma \Lambda_\perp \prec 0, \text{ and } {\Sigma_\perp}^T \Gamma \Sigma_\perp \prec 0, \tag{5}$$

where $\Lambda_\perp$ and $\Sigma_\perp$ are arbitrary matrices whose columns form a basis of the nullspaces of Λ and Σ, respectively, Hence $\Lambda\Lambda_\perp = 0$ and $\Sigma\Sigma_\perp = 0$.

Lemma 3. [2] *Given matrices X, Y, $\Phi = \Phi^{\mathrm{T}}$, $\mathcal{F}(t)$ of compatible dimensions, then*

$$\Phi + \mathrm{sym}\{X\mathcal{F}(t)Y\} \prec 0,$$

for all $\mathcal{F}(t)$ satisfying $\mathcal{F}(t)^T\mathcal{F}(t) \preceq I$ if, and only if, there exists $\varepsilon > 0$ such that

$$\Phi + \varepsilon XX^{\mathrm{T}} + \varepsilon^{-1}Y^{\mathrm{T}}Y \prec 0.$$

2 Stability of Differential Linear Repetitive Processes

The differential linear repetitive processes considered in this paper are described by the following state-space model over $0 \leq t \leq \alpha$, $k \geq 0$,

$$\begin{aligned} \dot{x}_{k+1}(t) &= (A + \Delta A(t))x_{k+1}(t) + (B_0 + \Delta B_0(t))y_k(t) + (B + \Delta B(t))u_{k+1}(t), \\ y_{k+1}(t) &= (C + \Delta C(t))x_{k+1}(t) + (D_0 + \Delta D_0(t))y_k(t) + (D + \Delta D(t))u_{k+1}(t), \end{aligned} \tag{6}$$

where $x_k(t) \in \mathbb{R}^n$, $u_k(t) \in \mathbb{R}^m$ and $y_k(t) \in \mathbb{R}^p$, respectively, denote the process state, input and pass profile (output) vectors at time instant t on pass k. Matrices A, B, B_0, C, D and D_0 represent the nominal process dynamics and they are assumed to be time and iteration invariant. $\Delta A(t)$, $\Delta B(t)$, $\Delta B_0(t)$, $\Delta C(t)$, $\Delta D(t)$ and $\Delta D_0(t)$ denote time-varying uncertainties which are assumed to satisfy

$$\begin{bmatrix} \Delta A & \Delta B_0 & \Delta B \\ \Delta C & \Delta D_0 & \Delta D \end{bmatrix} = \begin{bmatrix} H_1 \\ H_2 \end{bmatrix} \mathcal{F}(t) \begin{bmatrix} E_1 & E_2 & E_3 \end{bmatrix}, \tag{7}$$

where E_1, E_2, E_3, E_4, H_1 and H_2 are known real matrices which characterise the structure of the uncertainty and $\mathcal{F}(t)$ is unknown time-varying matrix with Lebesgue measurable elements bounded by

$$\mathcal{F}^T(t)\mathcal{F}(t) \preceq I. \tag{8}$$

The following result is the starting point for the analysis in this paper.

Lemma 4. *[6] A differential linear repetitive process (DLRP) described by (6) is robustly stable along the pass if and only if*

i) $\rho(D_0 + \Delta D_0) < 1$,
ii) all eigenvalues of the matrix $(A + \Delta A)$ *lie in the open left-half of the complex plane, and*
iii) all eigenvalues of $G(s) = (C + \Delta C)(sI - (A + \Delta A))^{-1}(B_0 + \Delta B_0) + (D_0 + \Delta D_0)$, $s = j\omega$, $\forall \omega \geq 0$, *have modulus strictly less than unity.*

The first condition in Lemma 4 is the necessary and sufficient condition for robust asymptotic stability. Also $\rho(D_0 + \Delta D_0) < 1$ describes the direct feedthrough from the previous pass profile to the next. Moreover, the second condition is to be expected since the robust stability property is independent of the finite pass length. The third condition requires frequency attenuation of the previous pass profile dynamics over the complete frequency range. Hence this stability theory has a well defined physical basis.

The key difficulty for robust stability testing and control law design is condition *iii)*, which is equivalent to

$$\rho\left(G(j\omega)\right) < 1, \ \forall \omega \in [0, \infty). \tag{9}$$

Alternatively, this last result requires that, each $\omega \in [0, \infty)$, there exists a $R(j\omega) \succ 0$ such that the following Lyapunov inequality holds

$$G(j\omega)^* R(j\omega) G(j\omega) - R(j\omega) \prec 0, \ \forall \omega \in [0, \infty).$$

but stability along the pass is then characterized by a convex feasibility test over an infinite-dimensional space. Furthermore, the function $R(j\omega)$ depends on ω and hence this inequality cannot be easily solved. However, simple multipliers, e.g., $R(j\omega) = R$ or $R(j\omega) = I$ can be used to avoid computational problems when multipliers with direct dependence on ω are considered.

These simple multipliers allow application of the results of Lemma 1 and this converts the problem to a finite-dimensional convex optimization problem over constraints in terms of LMIs that are comparatively easier to solve and directly lead to the control law design algorithms. To proceed, divide the complete frequency range into N intervals such that

$$[0, \infty) = \bigcup_{i=1}^{N} [\omega_{i-1}, \omega_i), \tag{10}$$

where $\omega_0 = 0$ and $\omega_N = \infty$ and then the result of Lemma 1 is applied to each interval. Furthermore, this allows the use of piecewise constant multipliers over a priori chosen frequency ranges and the following new result can be established.

Theorem 1. *Suppose that the entire frequency range is arbitrarily divided into N different frequency intervals as given in (10). Then, an uncertain differential linear repetitive process described by (6) and (7) is robustly stable along the pass if there exist matrices $S \succ 0$, $P_{2i} \succ 0$, $Q_i \succ 0$ and a symmetric P_{1i} such that the following matrix inequalities*

$$[(A + \Delta A)^T \;\; I](\Phi \otimes S)[(A + \Delta A)^T \;\; I]^T \prec 0, \tag{11}$$

$$\begin{bmatrix} A + \Delta A & B_0 + \Delta B_0 \\ I & 0 \\ 0 & I \end{bmatrix}^T \begin{bmatrix} \Upsilon_{1i} & \begin{bmatrix} 0 \\ (C + \Delta C)^T P_{2i}(D_0 + \Delta D_0) \end{bmatrix} \\ (\star) & (D_0 + \Delta D_0)^T P_{2i}(D_0 + \Delta D_0) - P_{2i} \end{bmatrix} \begin{bmatrix} A + \Delta A & B_0 + \Delta B_0 \\ I & 0 \\ 0 & I \end{bmatrix} \prec 0, \tag{12}$$

where

$$\Upsilon_{1i} = (\Psi_i \otimes Q_i + \Phi \otimes P_{1i}) + \begin{bmatrix} 0 & 0 \\ 0 & (C + \Delta C)^T P_{2i}(C + \Delta C) \end{bmatrix}$$

are feasible for all $i = 1, \ldots, N$ where τ_i, υ_i and ς_i are specified in Table 2.

Table 2. Values of τ_i, υ_i and ς_i

i	1	$1 < i < N$	N
τ_i	-1	-1	1
υ_i	0	$\mathrm{j}\frac{\omega_{i-1}+\omega_i}{2}$	0
ς_i	ω_0^2	$\omega_{i-1}\omega_i$	ω_{N-1}

Remark 1. The entries in Table 2 are those in Table 1 but in this case they also depend on the interval number i, which takes values from 1 to N as in (10).

Proof. Suppose that there exist matrices $S \succ 0$, $P_{2i} \succ 0$, $Q_i \succ 0$ and P_{1i} for all $i = 1, \ldots, N$, such that the inequalities (11)–(12) are feasible. Also, feasibility of (11)–(12) implies

$$(D_0 + \Delta D_0)^T P_{2i}(D_0 + \Delta D_0) - P_{2i} \prec 0, \; (A + \Delta A)S + S(A + \Delta A)^T \prec 0.$$

Equivalently, conditions *i*) and *ii*), respectively of Lemma 4 hold. Next, by routine manipulations each inequality in (12) yields

$$\begin{bmatrix}(A+\Delta A) & (B_0+\Delta B_0)\\ I & 0\end{bmatrix}^T (\Psi\otimes P_{1i}+\Phi_i\otimes Q_i)\begin{bmatrix}(A+\Delta A) & (B_0+\Delta B_0)\\ I & 0\end{bmatrix}$$
$$+\begin{bmatrix}(C+\Delta C) & (D_0+\Delta D_0)\\ 0 & I\end{bmatrix}^T (\Pi\otimes P_{2i})\begin{bmatrix}(C+\Delta C) & (D_0+\Delta D_0)\\ 0 & I\end{bmatrix}\prec 0, \tag{13}$$

where Φ is given in (3), $\Pi=\text{diag}\{1,-1\}$ and Ψ_i is of the form (3), where τ, υ and ς, respectively, are replaced by τ_i, υ_i and ς_i. Direct application of the result of Lemma 1 to each frequency interval with the choice of piecewise constant matrices P_{1i}, $P_{2i}\succ 0$, $Q_i\succ 0$ $\forall i=1,\ldots,N$ yields $G(j\omega)^*P_{2i}G(j\omega)-P_{2i}\prec 0$, where the existence of $P_{2i}\succ 0$ directly implies that $\rho(G(j\omega))<1$, $\forall\omega\geq 0$, i.e., feasibility of (12) guarantees that condition *iii*) of Lemma 4 holds and the proof is complete.

The inequalities in Theorem 1 include terms formed by product of matrix variables and matrices with additive uncertainties and hence the conditions in the above Theorem are not LMIs. The next result reformulates the condition of Theorem 1 to an LMI-based characterization of the robust stability along the pass.

3 Novel LMI-Based Conditions for Robust Stability Along the Pass

The conditions of Theorem 1 can be reformulated to an LMI-based form. Furthermore, some auxiliary slack variables with full structure will be employed to provide additional flexibility to the solution. To proceed, introduce the notation

$$\mathbb{A}=\begin{bmatrix}A & B_0\\ C & D_0\end{bmatrix},\ \Delta\mathbb{A}=\begin{bmatrix}\Delta A & \Delta B_0\\ \Delta C & \Delta D_0\end{bmatrix}=\mathcal{H}\mathcal{F}(t)\mathcal{E}_1,\ \Gamma_{11i}=\begin{bmatrix}\tau_iQ_i & 0\\ 0 & 0\end{bmatrix},\ \Gamma_{22i}=\begin{bmatrix}\varsigma_iQ_i & 0\\ 0 & -P_{2i}\end{bmatrix},$$
$$\mathcal{H}=\begin{bmatrix}H_1H_2\end{bmatrix},\ \ \mathcal{E}_1=\begin{bmatrix}E_1 & E_2\end{bmatrix}.$$

Then the theorem established next reformulates the results of Theorem 1 to give new LMI-based conditions for robust stability along the pass.

Theorem 2. *Suppose that the entire frequency range is arbitrarily divided into N possible different frequency intervals as in (10). Then, an uncertain differential linear repetitive process described by (6) is robustly stable along the pass if there exist matrices $P_{2i}\succ 0$, $Q_i\succ 0$, $S\succ 0$, F_{1i}, F_{2i}, F_{3i}, W_1, W_2, W_3, a symmetric P_{1i} and scalars $\epsilon_1>0$, $\epsilon_2>0$, p and q such that $pq<0$ and the following LMIs*

$$\begin{bmatrix}-\text{sym}\{qW_1\} & (\star) & (\star)\\ S+pW_1+qA^TW_1 & -\text{sym}\{pA^TW_1\}+\epsilon_1E_1^TE_1 & (\star)\\ qH_1^TW_1 & -pH_1^TW_1 & -\epsilon_1I\end{bmatrix}\prec 0, \tag{14}$$

$$\begin{bmatrix} \Gamma_{11i} - \mathrm{sym}\{W_1\} & (\star) & (\star) & (\star) \\ \Gamma_{21i} + \mathbb{A}^T W_1^T - W_2 & \Gamma_{22i} + \mathrm{sym}\{W_2\mathbb{A}\} + \epsilon_2 \mathcal{E}_1^T \mathcal{E}_1 & (\star) & (\star) \\ F_{30i} - W_3 & -F_{12i}^T + W_3\mathbb{A} & P_{2i} - \mathrm{sym}\{F_{3i}\} & (\star) \\ \mathcal{H}^T W_1^T & \mathcal{H}^T W_2^T & \mathcal{H}^T W_3^T & -\epsilon_2 I \end{bmatrix} \prec 0, \tag{15}$$

where

$$\Gamma_{21i} = \begin{bmatrix} P_{1i} + \upsilon_i^* Q_i & F_{1i} \\ 0 & F_{2i} \end{bmatrix}, F_{12i} = \begin{bmatrix} F_{1i} \\ F_{2i} \end{bmatrix}, F_{30i} = \begin{bmatrix} 0 & F_{3i} \end{bmatrix}$$

are feasible for all $i = 1, \ldots, N-1$.

Proof. Assume that the LMIs defined in (11) and (15) are feasible for all $i = 1, \ldots, N$. Then application of Schur's complement formula to (14) gives

$$\begin{bmatrix} -\mathrm{sym}\{qW_1\} & (\star) \\ S + pW_1 + qA^T W_1 & -\mathrm{sym}\{pA^T W_1\} \end{bmatrix} + \epsilon_1 \begin{bmatrix} 0 \\ E_1^T \end{bmatrix} \begin{bmatrix} 0 & E_1 \end{bmatrix} + \epsilon_1^{-1} \begin{bmatrix} qW_1 H_1 \\ -pW_1 H_1 \end{bmatrix} \begin{bmatrix} qH_1^T W_1^T & -pH_1^T W_1^T \end{bmatrix} \prec 0,$$

and by Lemma 3 this last inequality is feasible if and only if

$$\begin{bmatrix} -\mathrm{sym}\{qW_1\} & (\star) \\ S + pW_1 + qA^T W_1 & -\mathrm{sym}\{pA^T W_1\} \end{bmatrix} + \mathrm{sym}\left\{ \begin{bmatrix} qW_1 H_1 \\ -pW_1 H_1 \end{bmatrix} \mathcal{F}(t) \begin{bmatrix} 0 & E_1 \end{bmatrix} \right\} \prec 0.$$

The last inequality can be rewritten as a version of the first inequality of Lemma 2 where

$$\Upsilon = \begin{bmatrix} 0 & S \\ S & 0 \end{bmatrix}, \Lambda = \begin{bmatrix} -I & (\mathbb{A} + \Delta\mathbb{A})^T \end{bmatrix}, \Sigma = \begin{bmatrix} qI & -pI \end{bmatrix}.$$

Since $\Sigma_\perp^T \Upsilon \Sigma_\perp \prec 0$ holds for any p, q satisfying $pq < 0$ then the equivalence between (14) and (11) finally follows from the Lemma 2. Next, application of Schur's complement formula to (15) yields

$$\begin{bmatrix} \Gamma_{11i} & \Gamma_{21i}^T & F_{30i}^T \\ \Gamma_{21i} & \Gamma_{22i} & -F_{12i} \\ F_{30i} & -F_{12i}^T & P_{2i} - \mathrm{sym}\{F_{3i}\} \end{bmatrix} + \mathrm{sym}\left\{ \begin{bmatrix} I & 0 & 0 \\ 0 & I & 0 \\ 0 & 0 & I \end{bmatrix} \begin{bmatrix} W_1 \\ W_2 \\ W_3 \end{bmatrix} \begin{bmatrix} -I & \mathbb{A} + \Delta\mathbb{A} & 0 \end{bmatrix} \right\} \prec 0. \tag{16}$$

Introduce the matrices

$$\Lambda = \begin{bmatrix} I & 0 & 0 \\ 0 & I & 0 \\ 0 & 0 & I \end{bmatrix}, W = \begin{bmatrix} W_1 \\ W_2 \\ W_3 \end{bmatrix}, \Sigma = \begin{bmatrix} -I & \mathbb{A} + \Delta\mathbb{A} & 0 \end{bmatrix}$$

and then (16) can be reformulated by application of Lemma 2 as the second inequality in (5), i.e.

$$\Sigma_\perp{}^T \Gamma \Sigma_\perp \prec 0, \tag{17}$$

where by construction the matrix $\Sigma_\perp$ is

$$\Sigma_\perp = \begin{bmatrix} \mathbb{A} + \Delta\mathbb{A} & 0 \\ I & 0 \\ 0 & I \end{bmatrix}.$$

Since $\Lambda = I$ then $\Lambda_{\perp} = 0$ and hence the first inequality in (5) holds. Furthermore, after some routine matrix manipulations the inequality (17) can be rewritten as

$$\Gamma_{1i} + \text{sym}\left\{ \begin{bmatrix} I & 0 & 0 \\ 0 & I & 0 \\ 0 & 0 & I \end{bmatrix} \begin{bmatrix} F_{1i} \\ F_{2i} \\ F_{3i} \end{bmatrix} \left[(C + \Delta C) \; (D_0 + \Delta D_0) \; -I \right] \right\} \prec 0, \tag{18}$$

where

$$\Gamma_{1i} = \begin{bmatrix} \tau_i(A + \Delta A)^T Q_i (A + \Delta A) + \varsigma_i Q_i + \text{sym}\{P_{1i}(A + \Delta A) + v_i^* Q_i (A + \Delta A)\} \\ \tau_i (B_0)^T Q_i (A + \Delta A) + v_i (B_0 + \Delta B_0)^T Q_i + (B_0 + \Delta B_0)^T P_{1i} \\ 0 \end{bmatrix.}$$
$$\begin{matrix} \tau_i (A + \Delta A)^T Q_i (B_0 + \Delta B_0) + v_i^* Q_i (B_0 + \Delta B_0) + P_{1i}(B_0 + \Delta B_0) & 0 \\ \tau_i (B_0 + \Delta B_0)^T Q_i (B_0 + \Delta B_0) - P_{2i} & 0 \\ 0 & P_{2i} \end{matrix} \Bigg] \prec 0$$

and by Lemma 2, feasibility of (18) implies that the inequality $\Sigma_{1_{\perp}}^T \Gamma_{1i} \Sigma_{1_{\perp}} \prec 0$ must hold where

$$\Sigma_{1_{\perp}} = \begin{bmatrix} I & 0 \\ 0 & I \\ (C + \Delta C) & (D_0 + \Delta D_0) \end{bmatrix}.$$

Finally, this last inequality is equivalent to (12) and by Theorem 1 stability along the pass is ensured.

4 Current Pass State Feedback and Previous Pass Profile Based Control

In this section, the stability results of the previous section are extended to the case of controlled processes. It is assumed that a differential linear repetitive process represented in (6) is subject to a control law described by

$$u_{k+1}(t) = K_1 x_{k+1}(t) + K_2 y_k(t), \tag{19}$$

where K_1 and K_2 are the control law matrices to be found. Application of this control law results in the controlled process state-space model

$$\begin{aligned} \dot{x}_{k+1}(t) &= ((A + \Delta A) + (B + \Delta B)K_1)x_{k+1}(t) + ((B_0 + \Delta B_0) + (B + \Delta B)K_2)y_k(t), \\ y_{k+1}(t) &= ((C + \Delta C) + (D + \Delta D)K_1)x_{k+1}(t) + ((D_0 + \Delta D_0) + (D + \Delta D)K_2)y_k(t). \end{aligned} \tag{20}$$

The aim is to develop LMI-based results that enable the computation of the control law matrices of (19) for a chosen frequency partitioning. To proceed, introduce the following notation

$$\mathbb{B} = \begin{bmatrix} B \\ D \end{bmatrix}, \; \Delta\mathbb{B} = \begin{bmatrix} \Delta B \\ \Delta D \end{bmatrix}, \; \mathbb{K} = [K_1 \; K_2]$$

and then the results of the previous section can be directly used to develop control law design algorithms.

Theorem 3. *Suppose that a control law of the form (19) is applied to an uncertain differential linear repetitive process described by (6). Suppose also that the entire frequency range is arbitrarily divided into N possible different frequency intervals as in (10). Then the resulting controlled process (20) is robustly stable along the pass if there exist matrices $\hat{P}_{2i} \succ 0$, $\hat{Q}_i \succ 0$, $\hat{S} \succ 0$, $\hat{F}_{1i}$, $\hat{F}_{2i}$, $\hat{F}_{3i}$, $\hat{W}_1$, $\hat{W}_2$, N_1, N_2, a symmetric $\hat{P}_{1i}$ and real scalars $\epsilon_1 > 0$, $\epsilon_2 > 0$, p and q such that $pq < 0$ and the following LMIs*

$$\begin{bmatrix} -\operatorname{sym}\{q\hat{W}_1\} + \epsilon_1 q^2 H_1 H_1^T & (\star) & (\star) \\ \hat{S} + p\hat{W}_1 + \Upsilon_1^T q - \epsilon_1 pq H_1 H_1^T & -\operatorname{sym}\{p\Upsilon_1\} + \epsilon_1 p^2 H_1 H_1^T & (\star) \\ 0 & (E_1\hat{W}_1 + E_3 N_1) & -\epsilon_1 I \end{bmatrix} \prec 0, \quad (21)$$

$$\begin{bmatrix} \hat{\Gamma}_{11i} - \operatorname{sym}\{\hat{W}\} + \epsilon_2\mathcal{H} & (\star) & (\star) & (\star) \\ \hat{\Gamma}_{21i} + (\Upsilon - \hat{W})^T + \epsilon_2\mathcal{H} & \hat{\Gamma}_{22i} + \operatorname{sym}\{\Upsilon\} + \epsilon_2\mathcal{H} & (\star) & (\star) \\ \hat{F}_{30i} - \hat{W}_3 & -\hat{F}_{12i}^T + [0\ I]\Upsilon & \hat{P}_{2i} - \hat{F}_{3i} - \hat{F}_{3i}^T + \epsilon_2 H_2 H_2^T & (\star) \\ 0 & \mathcal{E} & 0 & -\epsilon_2 I \end{bmatrix} \prec 0, \quad (22)$$

where

$$\Upsilon_1 = A\hat{W}_1 + BN_1, \Upsilon = \mathbb{A}\hat{W} + \mathbb{B}N, \mathcal{E} = [E_1W_1 + E_3N_1\ E_2W_2 + E_3N_2],$$

$$\mathcal{H} = \begin{bmatrix} H_1 \\ H_2 \end{bmatrix} \begin{bmatrix} H_1 \\ H_2 \end{bmatrix}^T, \hat{W} = \operatorname{diag}\{\hat{W}_1, \hat{W}_2\}, \hat{W}_3 = [0\ \hat{W}_2], N = [N_1\ N_2],$$

$$\Gamma_{21i} = \begin{bmatrix} \hat{P}_{1i} + v_i^*\hat{Q}_i & \hat{F}_{1i} \\ 0 & \hat{F}_{2i} \end{bmatrix}, \hat{\Gamma}_{11i} = \begin{bmatrix} \tau_i\hat{Q}_i & 0 \\ 0 & 0 \end{bmatrix}, \hat{\Gamma}_{22i} = \begin{bmatrix} \varsigma_i\hat{Q}_i & 0 \\ 0 & -\hat{P}_{2i} \end{bmatrix}$$

are feasible for all $i = 1, \ldots, N-1$. Also, if these LMIs are feasible, the required control law matrices K_1 and K_2 of (19) can be calculated as

$$[K_1\ K_2] = N\hat{W}^{-1}. \quad (23)$$

Proof. Suppose that the LMIs in (21) and (22) hold. Then a feasible solution of these inequalities implies that $\hat{P}_{2i} \succ 0$, $\hat{Q}_i \succ 0$, $\hat{S} \succ 0$ and the matrices $\hat{W}_1$ and $\hat{W}_2$ are nonsingular. Next, application of Schur's complement formula to (21) yields

$$\begin{bmatrix} -\operatorname{sym}\{q\hat{W}_1\} & (\star) \\ \hat{S} + p\hat{W}_1 + \Upsilon_1^T q & -\operatorname{sym}\{p\Upsilon_1\} \end{bmatrix} + \epsilon_1 \begin{bmatrix} qH_1 \\ -pH_1 \end{bmatrix} \begin{bmatrix} qH_1^T & -pH_1^T \end{bmatrix}$$
$$+ \epsilon_1^{-1} \begin{bmatrix} 0 \\ (E_1\hat{W}_1 + E_3N_1)^T \end{bmatrix} \begin{bmatrix} 0 & (E_1\hat{W}_1 + E_3N_1) \end{bmatrix} \prec 0,$$

and by Lemma 3 this last inequality holds if and only if

$$\begin{bmatrix} -\operatorname{sym}\{q\hat{W}_1\} & (\star) \\ \hat{S} + p\hat{W}_1 + q((A + \Delta A)\hat{W}_1 + (B + \Delta B)N_1)^T & -\operatorname{sym}\{p(A + \Delta A)\hat{W}_1 + (B + \Delta B)N_1\} \end{bmatrix} \prec 0.$$

Let $\mathcal{W}_1 = \text{diag}\left\{\hat{W}_1^{-1}, \hat{W}_1^{-1}\right\}$; then pre- and post-multiply the above inequality by $\mathcal{W}_1^T$ and $\mathcal{W}_1$, respectively, and set $K_1 = N_1\hat{W}_1$, $S = \hat{W}_1^{-T}\hat{S}\hat{W}_1^{-1}$. Hence, the result of Theorem 2 ensures that the eigenvalues of $(A + \Delta A)+(B + \Delta B)K_1$ lie in the open left-half of the complex plane. As the next step, apply the Schur's complement formula to (22) followed by application of Lemma 3. Furthermore, apply the congruence transformation specified by diag $\left\{\hat{W}^{-1}, \hat{W}^{-1}, \hat{W}_2^{-1}\right\}$ to the last result. Setting $W_1 = W_2 = \hat{W}^{-1}$, $Q_i = \hat{W}_1^{-T}\hat{Q}_i\hat{W}_1^{-1}$, $P_{1i} = \hat{W}_1^{-T}\hat{P}_{1i}\hat{W}_1^{-1}$, $P_{2i} = \hat{W}_2^{-T}\hat{P}_{2i}\hat{W}_2^{-1}$, $F_{1i} = \hat{W}_1^{-T}\hat{F}_{1i}\hat{W}_2^{-1}$, $F_{2i} = \hat{W}_2^{-T}\hat{F}_{2i}\hat{W}_2^{-1}$ and $F_{3i} = \hat{W}_2^{-T}\hat{F}_{3i}\hat{W}_2^{-1}$ and a straightforward application of the same steps as in the proof of (21) allows us to establish equivalence between (22) and and a version of condition *iii*) of Lemma 4 applied to the controlled process of (20). Therefore the controlled process is stable along the pass and the proof is complete.

Remark 2. Comparing Theorems 2 and 3, it is immediate that slack matrix variables W_1 and W_2 must be the same, i.e. $W_1 = W_2$, when using the controller design procedure. Moreover, these matrix variables must have the block diagonal structure and this may introduce a level of conservatism into the design.

Remark 3. The design conditions (22) are LMIs that can be easily and effectively solved via numerical software. In addition, optimal values of the scalar parameters p and q can be sought to reduce the conservatism of the solutions.

5 Case Study

To illustrate the new results in this paper, a metal rolling process from [6] is considered as an example. The nominal state-space model of this process is defined by the matrices

$$A = \begin{bmatrix} 0 & 1 \\ -a_0 & 0 \end{bmatrix}, \ B_0 = \begin{bmatrix} 0 \\ a_0(1-b_2) \end{bmatrix}, \ B = \begin{bmatrix} 0 \\ -c_0 \end{bmatrix}, \ C = \begin{bmatrix} 1 & 0 \end{bmatrix}, \ D_0 = b_2, \ D = 0,$$

and

$$a_0 = \frac{\alpha_1\alpha_2}{M(\alpha_1+\alpha_2)}, \quad b_0 = \frac{\alpha_2}{\alpha_1+\alpha_2}, \quad c_0 = \frac{\alpha_1}{M(\alpha_1+\alpha_2)},$$

where $\alpha_1 = 600$ [N/m] is the stiffness of the adjustment mechanism spring, $\alpha_2 = 2000$ [N/m] is the hardness of the metal strip and $M = 100$ [kg] denotes the lumped mass of the roll-gap adjusting mechanism. The nominal state-space model matrices are therefore

$$\left[\begin{array}{c|c|c} A & B_0 & B \\ \hline C & D_0 & D \end{array}\right] = \left[\begin{array}{cc|c|c} 0 & 1 & 0 & 0 \\ -4.6154 & 0 & 1.0651 & -0.0023 \\ \hline 1 & 0 & 0.7692 & 0 \end{array}\right].$$

Suppose also that the value of the parameter M is uncertain but known to lie in the range from 90 [kg] to 112.5 [kg]. Hence

$$H_1 = \begin{bmatrix} 0 \\ 1 \end{bmatrix}, H_2 = 0, E_1 = [0.5128 \;\; 0], E_2 = [-0.1183], E_3 = [0.0003].$$

The plots of $\rho(G(j\omega))$ against frequency for the nominal model together with those corresponding to the lowest and highest, respectively, extreme values of the M are given in Fig. 1 and confirm that dynamics are unstable along the pass for some frequencies. Suppose therefore the entire frequency range $[0, \infty)$ is divided into 4 frequency intervals, i.e. $[0, \infty) = [0, 1.7) \cup [1.7, 2.29) \cup [2.29, 3) \cup [3, \infty)$, which are marked with dotted lines in Fig. 1. Hence $\omega_0 = 0$ [rad/s], $\omega_1 = 1.7$ [rad/s], $\omega_2 = 2.29$ [rad/s], $\omega_3 = 3$ [rad/s] and $\omega_4 = \infty$. Also the choice of $p = -1$, $q = 1$ gives following control law matrices, obtained by computing the LMIs (21) and (22) and using (23)

$$K_1 = 10^3 \times [0.0429 \;\; 1.1520], K_2 = 492.1181 \; .$$

The controlled process models for the extreme values of M was simulated over 20 passes. This generated the plots shown in Fig. 2, where the initial pass profile ($k = 0$) was taken as $y_0(t) = \sin(2\pi t/10) + 0.5 \sin(4\pi t/10)$, for $0 \leq t \leq 20$.

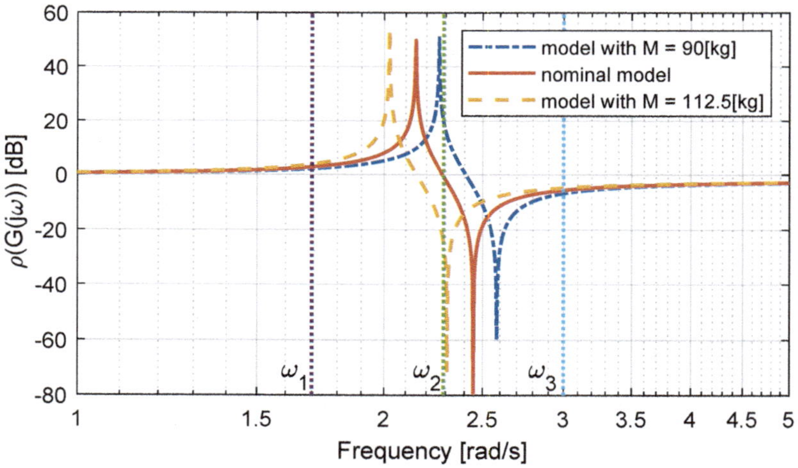

Fig. 1. Plots of $\rho(G(j\omega))$ for the models: nominal and two with the extreme uncertainties

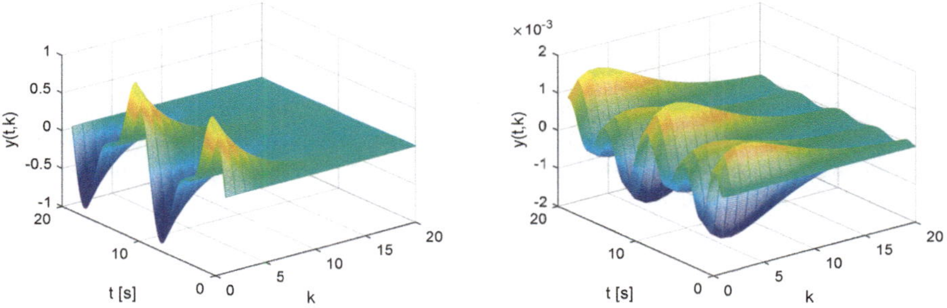

(a) the model with $M = 90[kg]$

(b) the difference between the outputs of the models with the extreme uncertainties

Fig. 2. Response of the controlled processes

6 Conclusions

This paper has developed novel conditions for robust stability along the pass and stabilization of uncertain differential linear repetitive processes. These conditions are given in terms of LMIs and therefore they are numerically trackable. The major benefit of the new results in this paper is the fact that stability tests extend in a direct manner to give control law design algorithms. The advantage over current results in this last aspect is the avoidance of product terms between the state-space model matrices and some Lyapunov/LMI decision matrices. This decoupling has been achieved through the use of slack matrix variables and therefore a reduction in the conservatism of the robust stability tests is possible.

Acknowledgments. This work is partially supported by National Science Centre in Poland, grant No. 2017/27/B/ST7/01874.

References

1. Gahinet, P., Apkarian, P.: A linear matrix inequality approach to $\mathcal{H}_\infty$ control. Int. J. Robust Nonlinear Control **4**, 421–448 (1994)
2. Gu, K., Kharitonov, V.L., Chen, L.: Stability of Time-Delay Systems. Birkhäuser, Boston (2003). https://doi.org/10.1007/978-1-4612-0039-0
3. Iwasaki, T., Hara, S.: Generalized KYP lemma: unified frequency domain inequalities with design applications. IEEE Trans. Autom. Control **50**(1), 41–59 (2005)
4. Paszke, W., Rogers, E., Gałkowski, K.: Experimentally verified generalized KYP lemma based iterative learning control design. Control Eng. Pract. **53**(10), 57–67 (2016)
5. Roberts, P.D.: Two-dimensional analysis of an iterative nonlinear optimal control algorithm. IEEE Trans. Circuits Syst. I Fundam. Theory Appl. **49**(6), 872–878 (2002)

6. Rogers, E., Gałkowski, K., Owens, D.H.: Control Systems Theory and Applications for Linear Repetitive Processes. Lecture Notes in Control and Information Sciences, vol. 349. Springer, Berlin (2007)
7. Rogers, E., Gałkowski, K., Paszke, W., Moore, K.L., Bauer, P.H., Hładowski, L., Dabkowski, P.: Multidimensional control systems: case studies in design and evaluation. Multidimension. Syst. Signal Process. **26**(4), 895–939 (2015). https://doi.org/10.1007/s11045-015-0341-8

Design of a Real Time Path of Motion Using a Sliding Mode Control with a Switching Surface

Jörg Kunkelmoor(✉) and Paolo Mercorelli

Institute of Product and Process Innovation, Leuphana University of Lueneburg, Universitaetsallee 1, 21335 Lueneburg, Germany
joergkunkelmoor1@alice-dsl.net, mercorelli@uni.leuphana.de
https://www.leuphana.de/en/institutes/ppi/staff/paolo-mercorelli.html

Abstract. Due to an increasing variety of tasks in production systems, the programming of robots becomes more complex. The aim of this work is, therefore, to simplify the work involved in programming of different contours as much as possible. Instead of specifying individual points of a contour in code, only one start and one end position are given. The movement between the two points is changed in real time by a robust control scheme, thus simplifying the programming effort for different contours. In this work, the robot is considered as a black box system and the approach to control consists only of considering the error of position and velocity without model. In the presented case, the development of the controller has shown that an Integral Sliding Mode Control (ISMC) strategy does not provide the desired control quality because of the presence of unavoidable saturating actuators in robots. Furthermore, a better result could be achieved with a Sliding Mode Control (SMC) approach that switches between two predefined surfaces. With this approach, good dynamic performances are obtained, in particular, in terms of overshoot which proves to be drastically reduced. Validations of the proposed method are obtained using real measurements realized on an industrial robot.

Keywords: Sliding Mode Control · Robots · Trajectory control · Applications

1 Introduction and Motivation

Optical testing in industry is becoming increasingly important. Production waste should be minimized and faulty products must be detected at an early stage. On the one hand, solutions are in use that statically inspect the object and on the other hand, solutions that dynamically scan the object using an industrial robot or with a linear unit. The industrial robots have several axes and can be programmed in different ways for the respective field of application. An online programming, in which the robot gets the waypoints manually stored by

A. Bartoszewicz et al. (Eds.): PCC 2020, AISC 1196, pp. 197–206, 2020.
https://doi.org/10.1007/978-3-030-50936-1_17

the so-called "Teach In" procedure is used. Another possibility is off line programming, in which the waypoints are taught by a simulation and the 3D CAD model of the object, see [1]. Due to the variety of parts in production and complex contours, it is time-consuming to maintain suitable waypoints with the two methods described. Therefore, in contrast to the two described possibilities of programming the robot, the dynamic adaptation of a path is investigated in real time based on the data of a 3D triangulation sensor. One of the most successful approaches for advanced motion of robots is represented by Sliding Mode Control (SMC), which is proven to be the most useful control strategy. It appeared in the 1950s and is used for incompletely modelled or nonlinear systems since its first appearance. In particular, SMC uses a discontinuous control action switching between two completely different system structures which give the existence of a new system motion, called Sliding Mode, in a specified manifold. Rejection of external disturbances and insensitivity to parameter variations are particular characteristics of the motion in the manifold. A wide range of systems like nonlinear, time-varying, discrete, large-scale, infinite-dimensional, stochastic, and distributed systems use SMC as a new control design method. Robot manipulators, aircraft, underwater vehicles, spacecraft, flexible space structures, electrical motors, power systems, and automotive engines successfully used the SMC in the last years. This paper proposes a switching SMC strategy for motion control of an industrial robot, in which an Integral Sliding Surface (ISS) is proposed to give more robustness to the final reaching phase and to compensate the final error. When the reaching phase is completed, the integral action is switched on. The advantages of the presence of the integral action are known. In fact, it is known that if the initial error is known using an integral action it is possible to reduce the time of the reaching phase to zero. Nevertheless, the integral action can generate oscillations during the sliding motion. Furthermore, the presence of the integral action on the sliding surface can generate windup phenomenon in the presence of saturating actuators. In this context, different contributions are present in the applied literature to avoid saturation in the presence of integral action in a SMC, see [4]. More recent works present advanced theoretical and methodological results to handle this kind of problems also in the presence of integral action, [6,9] and [11], in which Lyapunov stability theory and geometric homogeneity technique are employed to show global finite-time stability in the presence of saturating actuators. More specifically, concerning robots with saturating actuators very recent works, see [10], in which methods to avoid saturation are presented and validated with measured data. Finally in [7] a new sliding surface is first proposed and a robust control is developed for ensuring global approximate fixed-time convergence. Very often the SMC is applied in its Finite-Time variations. The problem here is to ensure that global finite-time stabilization and actuator saturation are not violated [6]. Recently, another integral sliding surface is proposed and a continuous TSM control is presented in [2] for global finite-time tracking of robots, based on the assumption that position, velocity, and acceleration are all available for control design and then are substituted by the output of an exact differentiator. This result is later extended in

[3] to task-space tracking of robot manipulators. As pointed out in [[8], Remark 2], the direct use of acceleration in a feedback control can cause the problem to be ill-posed. The paper is organized in the following way. Section 2 describes the control strategy. In Sect. 3 measured results are shown to validate the method using an industrial robot. The conclusions close the paper.

2 Derivation of a Switching Surface

An ISMC with error of the velocity and the position is considered to define surface $S_1(t)$ as follows:

$$S_1(t) = K_1(x_d(t) - x(t)) + K_2(\dot{x}_d(t) - \dot{x}(t)) + \gamma K_i \int_0^t (x_d(\tau) - x(\tau))d\tau. \tag{1}$$

The parameters K_1, K_2 and K_i are factors to state the dynamics of the motion and γ is a factor to switch between two surfaces, which can either be 1 or 0. This factor is needed to reduce the overshoot caused by the presence of the integral and by the unavoidable saturating effects of the actuators of the robot. The strategy to switch is determined in the following way:

$$\gamma = sat\left(\frac{S_2(t)}{\epsilon_2}\right), \tag{2}$$

with the following surface $S_2(t)$

$$S_2(t) = K_1(x_d(t) - x(t)) + K_2(\dot{x}_d(t) - \dot{x}(t)) \tag{3}$$

and

$$sat\left(\frac{S_2(t)}{\epsilon_2}\right) = \begin{cases} 1 \text{ if } |\frac{S_2(t)}{\epsilon_2}| < 1 \\ 0 \text{ if } \frac{S_2(t)}{\epsilon_2} > 1 \\ 0 \text{ if } \frac{S_2(t)}{\epsilon_2} < -1, \end{cases} \tag{4}$$

where ϵ_2 represents the well-known thickness function to define a region of non-switching. A basic technique for designing a SMC is the definition of a Lyapunov function. This is an important tool to ensure *directly* the stability of the control. More details can be found in [5] in an application context. A quadratic candidate function $V_1(t)$ is therefore, defined for the surface $S_1(t)$.

$$V_1(t) = \frac{1}{2} S_1(t)^2. \tag{5}$$

The stability is obtained when the derivative of the candidate function V(t) is less than zero. This leads to the following condition.

$$\dot{V}_1(t) < 0 \tag{6}$$

and the derivative of the candidate function is

$$\dot{V}_1(t) = S_1(t)\dot{S}_1(t) \tag{7}$$

After inserting the derivative of $S_1(t)$ into the candidate function, the following equation results.

$$\dot{V}_1(t) = S_1(t)[K_1(\dot{x}_d(t) - \dot{x}(t)) + K_2(\ddot{x}_d(t) - \ddot{x}(t)) + \gamma K_i(x_d(t) - x(t))] \tag{8}$$

$\dot{x}(t)$ is the control value of the system. Therefore, the following control law is used to reach the surface.

$$\dot{x}(t) = \frac{K_2\ddot{x}_d(t)}{K_1} - \frac{K_2\ddot{x}(t)}{K_1} + \dot{x}_d(t) + \gamma\frac{K_i\ddot{x}_d(t)}{K_1} - \gamma\frac{K_i\ddot{x}(t)}{K_1} + \beta sat\left(\frac{S_1(t)}{\epsilon_1}\right) + \lambda S_1(t) \tag{9}$$

with the following saturation function:

$$sat\left(\frac{S_1(t)}{\epsilon_1}\right) = \begin{cases} 1 \text{ if } \frac{S_1(t)}{\epsilon_1} > 1 \\ \frac{S_1(t)}{\epsilon_1} \text{ if } |\frac{S_1(t)}{\epsilon_1}| < 1 \\ -1 \text{ if } \frac{S_1(t)}{\epsilon_1} < -1, \end{cases} \tag{10}$$

where ϵ_1 represents the well-known thickness function to define a region of non-switching.

3 The Experimental Results

A control scheme has three conditions to be achieved. Firstly, a stable state should be maintained even if disturbances influence the system, and secondly, these disturbances should be eliminated quickly. In addition, a control difference as small as possible should be left behind. Besides the classical statistical tools, several standards are available for evaluation. In addition, recorded characteristic curves of the nominal and actual values serve as evaluation criteria. The times required by the controller to settle and reach the set point can be read directly from the characteristic curves. In addition, overshoots and visible permanent control differences are often important as a first impression. The standards serve as a criterion for evaluating these data from the characteristics. For example, there are standards for the steady state, i.e. the state without consideration of the transient process, and standards for the transient process, the process in which the transient times and overshoots are also taken into account. In order to characterize a permanent system deviation, there is the ITAE standard, in which the system deviation is weighted with time. In the later course, the control differences are weighted more heavily.

$$A_{ITAE} = \int_0^t |e(t)|t\, dt. \tag{11}$$

One possibility to consider the transient response is $\mathcal{L}_2$ standard. With $\mathcal{L}_2$ standard, the control differences are quadratic in the quality criterion.

$$A_{\mathcal{L}_2} = \int_0^t e(t)^2 \, dt. \tag{12}$$

Large control deviations have a greater impact on the evaluation result. The results should be interpreted in such a way that the lower the values are, the better the control quality is. To test the control, a path was traversed, whereby the 6-axis robot moved in the X direction without control and the Y and Z coordinates were controlled. A 3D image of the respective track can be seen in Fig. 1. The first step is to evaluate the control in the Y-direction and thus also the influence of changing the surface with γ. For a comparison, γ is kept constant in the second series of measurements with

$$\gamma = 1 = const. \tag{13}$$

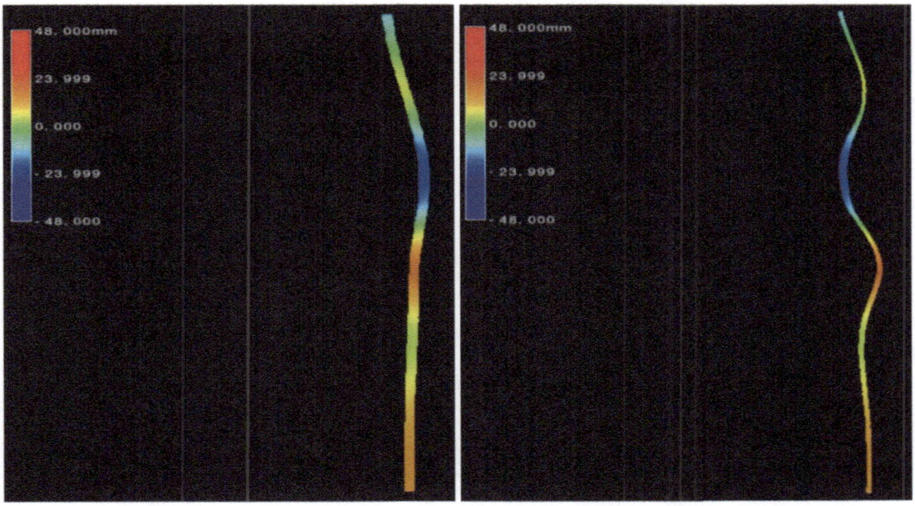

(a) 3D Image Y direction (b) 3D Image Z direction

Fig. 1. The 3D images of the track

It can be seen that the switching function in the sliding surface has resulted in an improvement in the transient response. This is also reflected in the evaluation with the help of $\mathcal{L}_2$ norm, which is worse than the control law obtained using without switching strategy, see Fig. 2. In contrast to the control in the Y-direction, an even clearer improvement can be seen in Fig. 3, in which the evaluation of the $\mathcal{L}_2$ standard for the control in the Z-direction is shown. From the measured error and from the presence of a strong overshoot in case of non-switching strategy, it is possible to see a considerable advantage using the switching strategy. This has to do, among other things, with the different generation

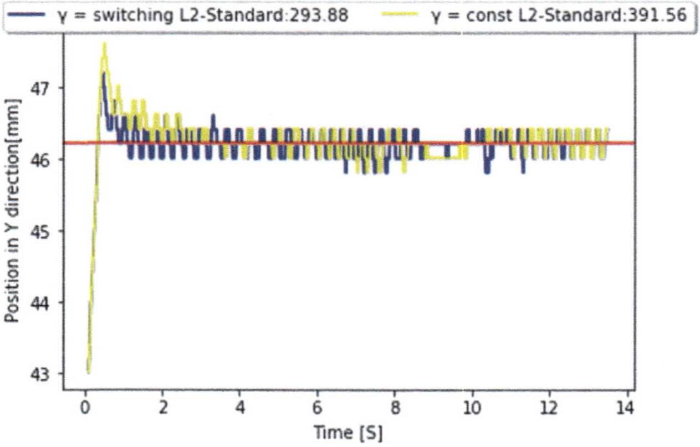

Fig. 2. Switching of γ in Y direction

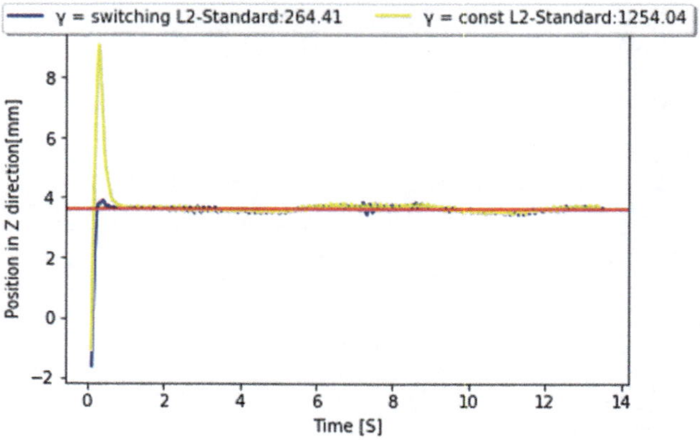

Fig. 3. Switching of γ in Z direction

of data in both directions. In the Y-direction, the sensor data are generated by a fixed data interval of 0.2 mm and the data in the Z-direction are generated continuously with an accuracy of 1 µm. After the transient process, one can see a congruent course of the curves, which is also shown in the evaluation of $\mathcal{L}_2$ standard. In Figs. 4 the values of the errors are visible in the legend of the measured data because they are directly available from the used software. Moreover, in Figs. 4 details of the measured results, which are represented in Fig. 3, are shown with and without switching strategy after the transient. From the measured error it is possible to see a considerable advantage using the switching strategy. In Fig. 5 the basic structure of the system is shown in a diagram. The system is divided into three subsystems that communicate with each other via different interfaces. On the one hand an image processing controller, which is used to process the data of the 3D laser triangulation sensor, sends coordinate

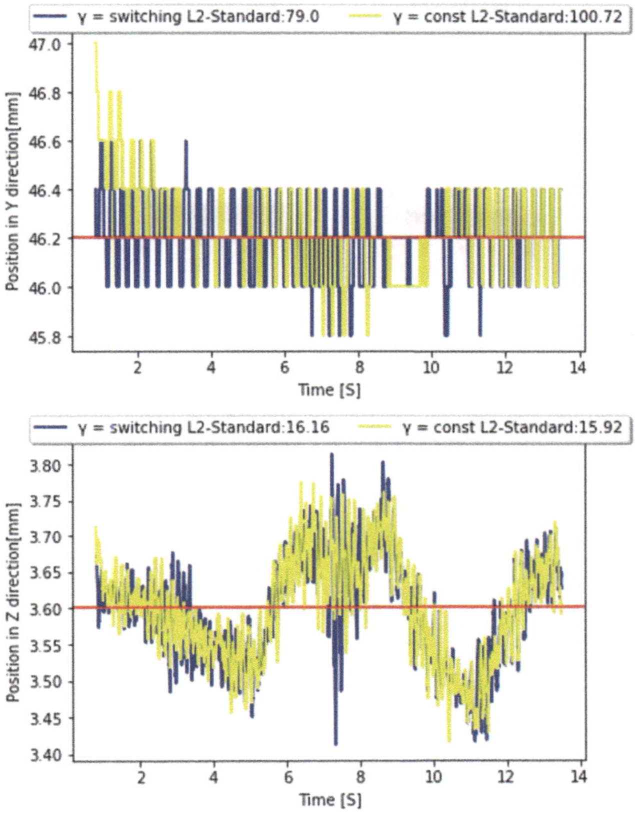

Fig. 4. Comparison of the control after the transient: on the top the Y direction result and on the bottom the Z direction result

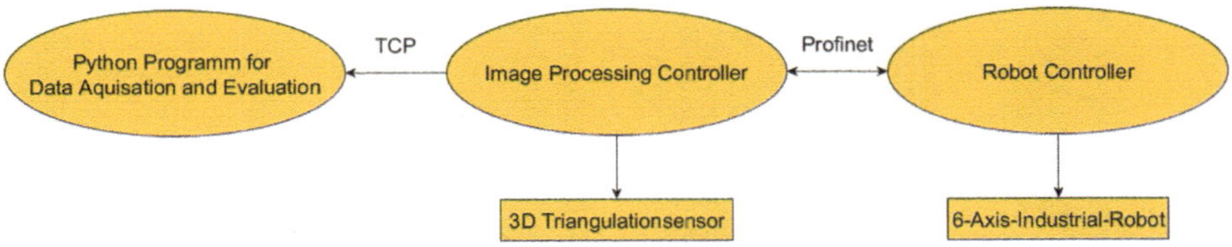

Fig. 5. Basic structure of the system

changes to the robot controller. The 3D laser triangulation sensor is attached to a tool holder of a 6-axis robot. The communication between the vision controller and the robot controller is integrated via a Profinet interface and thus over a fixed interval of 4 ms. Furthermore, the robot controller processes the coordinates of the path in real time and thus adjusts the movement of the 6-axis robot within 4 ms. The Python program is only used to evaluate the data and is

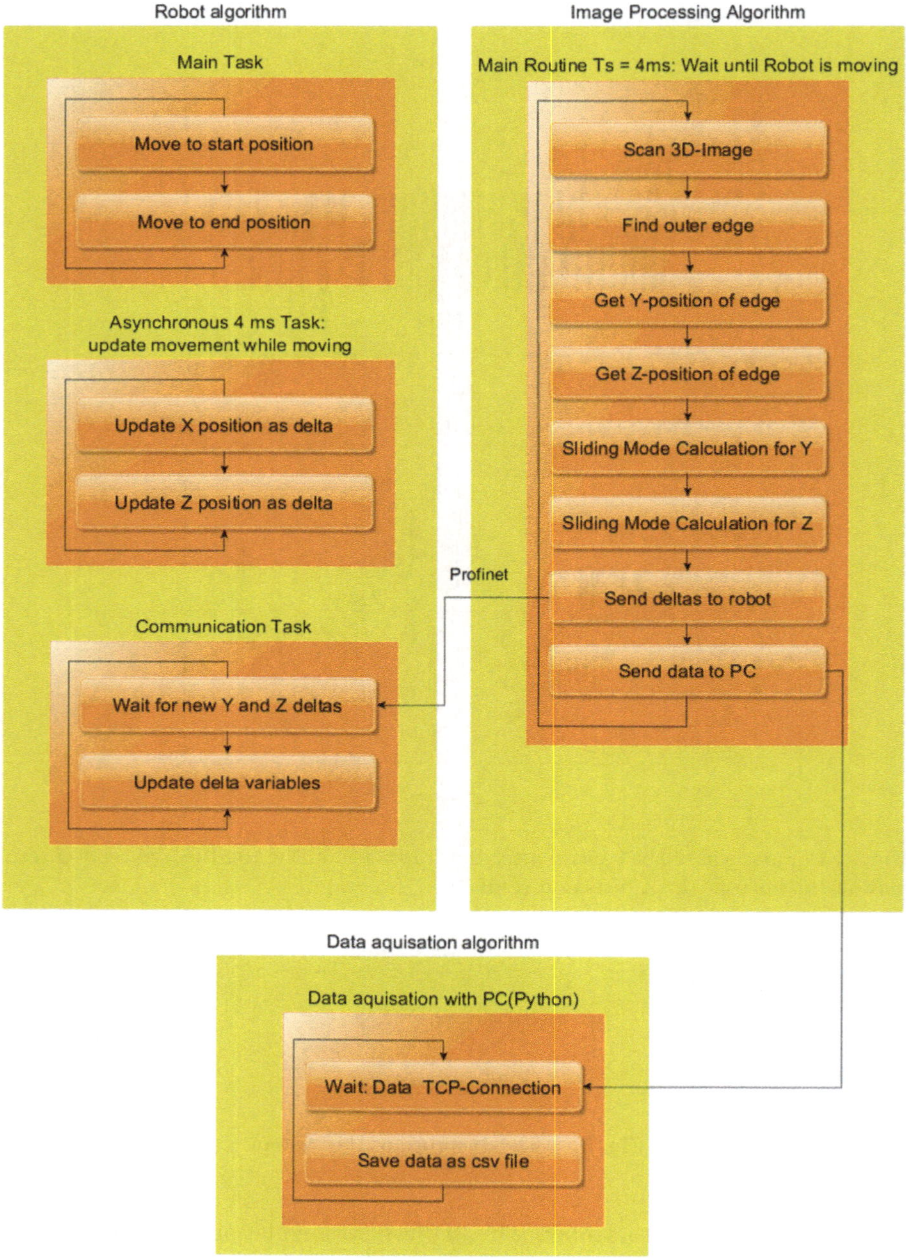

Fig. 6. Flowchart of algorithm

connected to the image processing controller via an Ethernet interface. Figure 6 shows the structure of the software which is implemented. The software structure, like the hardware, consists of several systems. On the one hand, there is the image processing algorithm, which is responsible for recording the contour, processing the data and calculating the SMC. This algorithm is triggered and processed every 4 ms. The captured 3D image consists of eight laser lines. The resulting contour is used to detect the edge and height, which is processed by the controller. After processing, the calculated manipulated variable is sent to the robot via the Profinet interface. The algorithm on the robot has three different tasks, which are processed side by side. The Main Task first moves the robot arm to a start position and then as a linear movement to the end position. Meanwhile, the Communication Task queries the current manipulated variables via a Profinet interface. In an asynchronous task, the movement is updated at 4 ms intervals during the linear movement of the robot arm. The measured values are recorded and evaluated by a Python program on a computer via a TCP connection.

4 Conclusions

This paper proposes a Sliding Mode approach. The development of the controller has shown that an Integral Sliding Mode Controller does not provide the desired control quality. Furthermore, a better result could be achieved with a Sliding Mode Control approach that switches between two surfaces. Good dynamical performances are obtained and validated by experimental results.

References

1. Chang, W.: Hybrid force and vision-based contour following of planar robots. J. Intell. Rob. Syst. **3**(47), 215–237 (2006)
2. Galicki, M.: Finite-time control of robotic manipulators. Automatica **51**, 49–54 (2015)
3. Galicki, M.: Finite-time trajectory tracking control in a task space of robotic manipulators. Automatica **67**, 165–170 (2016)
4. Mercorelli, P.: An anti-saturating adaptive preaction and a slide surface to achieve soft landing control for electromagnetic actuators. IEEE/ASME Trans. Mechatron. **17**(1), 76–85 (2012)
5. Slotine, J.E., Li, W.: Applied Nonlinear Control. Prentice-Hall, Englewood Cliffs (1991)
6. Su, Y., Zheng, C., Mercorelli, P.: Global finite-time stabilization of planar linear systems with actuator saturation. IEEE Trans. Circuits Syst. II Express Briefs **64**(8), 947–951 (2017)
7. Su, Y., Zheng, C., Mercorelli, P.: Robust approximate fixed-time tracking control for uncertain robot manipulators. Mech. Syst. Signal Process. **135**, 106379 (2020)
8. Xian, B., Dawson, D.M., de Queiroz, M.S., Chen, J.: A continuous asymptotic tracking control strategy for uncertain nonlinear systems. IEEE Trans. Autom. Control **49**(7), 1206–1211 (2004)

9. Zheng, C., Su, Y., Mercorelli, P.: Simple relay non-linear PD control for faster and high-precision motion systems with friction. IET Control Theory Appl. **12**(17), 2302–2308 (2018)
10. Zheng, C., Su, Y., Mercorelli, P.: Faster positioning of one degree-of-freedom mechanical systems with friction and actuator saturation. J. Dyn. Syst. Meas. Control Trans. ASME **141**(6), DS-17-1558 (2019). https://doi.org/10.1115/1.4042883
11. Zheng, C., Su, Y., Mercorelli, P.: A simple nonlinear PD control for faster and high-precision positioning of servomechanisms with actuator saturation. Mech. Syst. Signal Process. **121**, 215–226 (2019)

A Review of Sliding Mode Controllers with the Application of Time-Varying Switching Hyperplanes

Mateusz Pietrala(✉)

Institute of Automatic Control, Łodź University of Technology,
18/22 Stefanowskiego St., 90-924 Łodź, Poland
800904@edu.p.lodz.pl

Abstract. Sliding mode control is an effective method of regulation, that provides robustness to external disturbances and modeling uncertainties. Moreover, it is computationally efficient. The application of the time-varying sliding hyperplanes allows one to select parameters of these hyperplanes in order to eliminate the reaching phase. The main advantage of this approach is that we obtain robustness for the whole regulation process, not only after the reaching phase, which is an issue when applying time-invariant switching hyperplanes. This paper presents recent research in the area of time-varying sliding hyperplanes in theory as well as in practice.

Keywords: Sliding mode control · Variable structure systems · Time-varying switching hyperplanes

1 Introduction

Sliding mode control is a popular and effective regulation method, which minimizes the impact of the external disturbances on the dynamics of the system. Furthermore, this strategy is robust with respect to modeling uncertainties and is computationally efficient. Initially, sliding mode control technique was applied to the continuous-time systems [26] and after that the discrete-time systems have been taken into account [12,19,27]. During the controller design, parameters of a so-called sliding hyperplane are selected in order to ensure the desired dynamics of the object. It can be computed using two main methods: pole placement and optimal control. Furthermore, the control signal is introduced. It is important to prove that this control signal ensures the stability of the sliding motion.

The control process can be divided into two main phases: the reaching phase and the sliding phase. In the first one the representative point (state vector) starting from its initial position moves towards a sliding hyperplane or its predefined neighborhood. An important issue is to achieve a finite time of that convergence. If the representative point moves on the switching hyperplane, then such a motion is called a sliding phase. On the other hand, if the state is constrained to the neighborhood of this plane, then we use the term quasi-sliding

A. Bartoszewicz et al. (Eds.): PCC 2020, AISC 1196, pp. 207–218, 2020.
https://doi.org/10.1007/978-3-030-50936-1_18

motion. The representative point is driven to the desired state. However, the system is robust to the external disturbances only in the sliding phase. Therefore, in order to obtain the robustness for the whole regulation process, the time-varying sliding hyperplanes can be adopted. The parameters of these sliding hyperplanes are selected in such a way that this surface crosses the initial state at $t = 0$. After that it starts moving, in order to drive the representative point to the desired state. This results in the elimination of the reaching phase and in consequence our system is robust due to perturbations. In order to achieve a finite-time error convergence to zero, one can introduce a non-linear sliding surface. If a linear sliding hyperplane is applied, we can only obtain an asymptotic convergence. The dynamical performance of the system can be estimated using one of well-known quality indexes such as: regulation time, Integral Absolute Error (IAE), Integral Time Absolute Error (ITAE), Integral Square Error (ISE), Integral Time Square Error (ITSE) and more. When minimizing these indexes it is possible to take into account the limitations imposed on e.g. control signal, position, velocity, acceleration, etc. It is an important issue in order to apply the proposed controller in practice.

The remainder of this paper is organized as follows. Section 2 comprises a review of the selection of the optimal time-varying switching hyperplanes and their application to the second- and the third-order systems. The behavior of the system is evaluated using IAE and ITAE in most cases. The limitations of control signal, velocity and acceleration are taken into account. Furthermore, the theoretical considerations are applied to a hoisting crane model. In Sect. 3 a review of the application of the time-varying sliding hyperplanes in continuous and discrete time systems is introduced. This approach is presented using space exploration, robotic manipulators and data transmission in connection-oriented communication networks as examples. Section 4 comprises the conclusions of the paper.

2 Selection of the Optimal Time-Varying Switching Hyperplanes

This section starts by reviewing the paper in which authors considered second order system. Moreover, they selected parameters of time-varying switching line in an optimal way in order to minimize an optimization index, which evaluates the performance of the system.

Paper [1] comprises the description of three control strategies. In the first one the switching line moves with a constant acceleration. Further, the second strategy, in which the sliding line moves with a constant velocity is adopted. The last approach presents the so-called time-varying terminal slider - a shifting non-linear switching curve, which guarantees that the system's error convergence to zero is obtained in finite time. In every strategy the control signal constraint was included. Moreover, the sliding lines are selected in such a manner to pass through the initial conditions. The optimization index chosen in [1] is IAE. After the selection of the optimal values of the parameters, each strategy was compared to a regular fixed sliding line. It was shown that the system is insensitive to disturbances from the very beginning of the control process. Moreover, the error convergence is faster in proposed strategies, especially when the moving non-linear curve is applied.

In paper [3] authors considered the third-order uncertain, non-linear, time-varying system affected by unknown disturbances. The derivative of the first and the second state variables is equal to the second and the third state variable respectively. The derivative of the third state variable is a function of the control signal, modeling uncertainties and external disturbances. The sliding line was selected in such a manner that it moves and after that it stops and remains fixed. The main goal was to design the sliding mode controller, which guaranteed the insensitivity for the whole regulation process. Again, authors demanded the input constraint. Quality indexes used to evaluate the performance of the proposed controller were IAE and ITAE. The comparison of these strategies shown that when the switching line moves, then the minimization of IAE causes faster convergence than the minimization of ITAE. On the other hand in the second phase, when the sliding line remains fixed, we obtain the opposite result - the minimization of ITAE is more beneficial. The same system was considered in paper [4]. However, the limitation was imposed on the second (velocity) and the third (acceleration) state variables. Authors selected and minimized the IAE quality index.

The work [2] comprises the design of a similar sliding mode controller as in [3]. However, at first authors consider a conventional acceleration constraint and after that the elastic acceleration constraint is taken into account. In this case, they minimize the sum of IAE and penalty function, which imposes the penalty for excessive values of the acceleration.

In the paper [5] all of the previous results were gathered together. Authors considered third-order system and designed the switching plane in the following seven cases: when the control input is constrained, when the acceleration is constrained, when the velocity is constrained, when the acceleration and the velocity are constrained, when the acceleration and control input are constrained, when control input and velocity are constrained and when the acceleration, velocity and control input are constrained. The optimal parameters were calculated in such a manner to minimize the IAE quality index. A similar approach was used in [7]. However, the optimization criteria was ITAE and the considered constraints were imposed on the control signal, velocity and acceleration separately.

The research results described in this section were thoroughly presented in the book [6]. Chapter 2 comprises the sliding mode controller design for the second-order systems, while in Chapter 3 the third-order system was considered. In the second chapter authors derived the control law and the time-varying sliding line. Furthermore, the parameters of this line were selected in such a manner to minimize IAE with the constraints imposed on: input signal, velocity and both control signal and velocity. The same procedure was repeated in order to minimize ITAE quality index. Each case was verified by a simulation example. In the third chapter the minimization on IAE and ITAE was derived in a presence of the limitations of: the control signal, the acceleration, the velocity, the acceleration and the velocity, the acceleration and the control signal, the control signal and the velocity, all of these restrictions combined. Again, simulation example verified each of the mentioned strategy.

The results presented in this section can be implemented in practice. The approach of using the time-varying hyperplane to minimize quality indexes such as IAE or ITAE was tested by using a hoisting crane.

In the paper [21] authors stated the robust control problem with state constraints. They derived the control law with its fractional approximation in which the sign function was replaced by a continuous factor. It allowed them to deal with the chattering problem and obtain the robustness from the very beginning of the control process. Moreover, the monotonic convergence of the error to zero is ensured. Furthermore, two cases were taken into account: a constant deceleration switching line and a constant velocity switching line. In both cases, authors minimized IAE and ITAE quality indexes in order to evaluate the performance of the proposed controller in presence of the velocity and acceleration limitations. Experimental results verified the theoretical considerations contained in the paper.

3 Application of Time-Varying Switching Hyperplanes in Continuous and Discrete Time Systems

In this section we will focus on works in which authors presented the application of time-varying sliding hyperplanes. Firstly, the continuous time systems will be considered and after that we will shortly mention an application in discrete time case. Finally, the practical use of these strategies will be presented.

In the paper [30] the main goal was to select the control input, which drives the state vector to the demanded state, while eliminating the reaching phase. In order to achieve this property authors modified the sliding domain. An exponentially decaying term was added to each state variable error. In order to evaluate the performance of the system a quadratic quality index was chosen. The optimal parameters of the system were selected in such a way to minimize that index. Furthermore, authors estimated the control effort by presenting the minimum and maximum values of the input. The third order non-linear system was selected to verify the theoretical considerations in a simulation example.

The work [11] comprises the application of the time-varying switching hyperplane in order to achieve robustness of a linear uncertain single-input single-output system with saturating actuators i.e. the control input limitations are taken into account. Authors assumed that the absolute value of the external disturbances is bounded from above by a known function. The reaching phase is eliminated by selecting the sliding surface parameters in such a manner that the initial conditions are on this surface at the very beginning of the control process. This results in the robustness of the system for the whole control process. Authors state and prove that parameters can always be selected in such a manner that the feedback controller ensures the system's asymptotic robust stabilization. The simulation example is based on the third-order system. The problem was stated as a missile roll angle regulation. The chattering phenomenon was avoided. It is shown that the initial overshoot is minimized. One drawback of the proposed strategy is that the convergence rate of the state error to zero is marginally decreased.

In the paper [31] authors considered the fourth-order nonlinear system. In this system not only one, but two (second and fourth) state variables are affected by input signal and external disturbances. However, authors propose a decoupled control method, which divides the system to two second-order systems. In this method one of the second-order system (denoted by A) is selected as a priority and the second one (denoted by B) is the secondary goal. The sliding variable in system B depends only on the slope of the switching plane and third and fourth state variables, while the sliding variable in A depends on the slope, first and second state variables and an intermediate signal connected with the system's B sliding hyperplane. Furthermore, the parameters of the sliding variables are selected by using a fuzzy logic method. After that the time-varying sliding surface containing a linear function is presented. It allowed to replace 1D rule bases, which resulted in no obligation to use fuzzy rules, while computing the slope of the sliding hyperplane. Authors analyzed the stability of the proposed decoupling method. In order to prove that this desired property is ensured, it was shown that the product of the first state variable and its derivative is always non-positive. The simulation example was presented using a cart-pole system. These simulations showed that a strategy presented in the paper causes a faster dynamic response and a better dynamic performance by obtaining lower values of quality indexes IAE and ITAE, than in the case, when the time-invariant sliding surfaces are applied. Unfortunately, method presented in this paper causes that the input signal slightly increases.

In paper [22] authors also presented a decoupling method. In this work switching surface moves with a constant speed. Authors applied their approach to an induction motor with a field-oriented control strategy. In the end the comparison between classic VSC (variable structure control) controller and VSC with time-varying switching plane was performed. It was shown that the strategy used in the paper guarantees the robustness with respect to perturbations for the whole regulation process.

Work [24] comprises the design of the sliding mode controller for the classic second order non-autonomous nonlinear uncertain dynamic open-loop system with a single input. The first proposed strategy is the delta neighborhood approach, which involves the selection of the major rotation parameter in a predefined vicinity. The second one is the fuzzy logic tuning approach. The simulation example is presented on a non-linear spring damper system. The compared approaches are: SMC with a constant sliding surface, SMC with a rotating sliding surface using the delta neighborhood of Choi [8], SMC with a rotating sliding surface using the delta neighborhood presented in this paper, SMC with the fuzzy tuning approach and SMC with the rotating sliding surface using a sigmoid function. The control signal chart showed that strategies proposed in the paper decrease the impact of the external disturbances. To evaluate the dynamic performance of the system the following quality indexes was minimized: IAE, ITAE and settling time. The results shown that each of these quality indexes has the smallest value if the fuzzy tuning approach is applied.

In [25] the previous results were extended to higher-order non-autonomous systems. In the simulation example, authors compare four control strategies via minimizing five, different quality indexes. Four of these indexes take the lowest value for the strategy, which removes the reaching time, however, the lowest value of the last quality index- the magnitude of control input is the lowest for the conventional SMC approach.

The paper [20] comprises an application of the time-varying sliding surface, which moves along the trajectory, which movement is connected with a minimum jerk trajectory. First, that so-called smooth trajectory was designed. Further, the controller for the second order system in the presence of disturbances was presented. The stability was proven using a Lyapunov method. The simulation example was conducted for a servo problem for a second order spring-mass-damper system. The results were compared to a conventional linear and ellipsoid approach. It was shown that the proposed method ensures the fastest convergence time. Moreover, the chattering effect was reduced. The system proposed in this paper ensures the lowest sensitivity to perturbations. The energy consumption (integral of the absolute value of the product of the actuation force and velocity of the mass point) is lower than in both conventional methods.

As mentioned before, researchers mainly focus on applying the time-varying sliding hyperplanes to the continuous-time systems. However, in recent years we can find some papers in which the discrete-time systems were considered.

In the work [16] a novel sliding mode control method of single-phase uninterruptible-power-supply (UPS) inverters is introduced. The slope of the time-varying sliding line is dependent on the current error of the output voltage. The coefficient of this sliding hyperplane is computed dynamically. The function of this hyperplane's slope is computed using the approximation from the input/output relationship of the single-input fuzzy logic controller connected with the error variables. Computer simulations and real-life experiment were performed. Comparing the strategy presented in this paper, the total harmonic distortion is much smaller than in the fixed sliding line method.

3.1 Practical Applications

The sliding mode control is known to be computationally efficient and to guarantee robustness to the external, unknown perturbations. That is why it can be easily applied to a wide range of real life systems.

In the paper [9] authors formulated the problem of rigid spacecraft eigenaxis rotation. In order to achieve the robustness, they decided to apply the time-varying sliding mode control strategy. The system was affected by the external perturbations and modeling uncertainties. In order to deal with the chattering phenomenon, a non-continuous sign function was replaced by a saturation function. Moreover, authors presented a disturbance observer-based time-varying sliding mode control algorithm. The simulation example illustrated the theoretical considerations. Authors considered the modulation of the on-off thrusters by applying the pulse-width pulse-frequency technique.

Work [10] comprises the extension of the previous results in order to control a group of space satellites. This group is divided into some independent sets, which allows them to act irrespectively from each other. This causes, that the fast communication between sets is not necessary. However, the communication between satellites in one set has to be guaranteed. The simulation example presented the formation communication topology of four spacecrafts in which every set communicates with two different ones. In order to test the proposed control strategy authors applied three kinds of attitude synchronization situations: formation building, Rest-to-Rest formation maneuver and formation tracking.

Another approach of applying sliding mode control into the space exploration may be found in [23]. Authors stated a problem by presenting a translational motion for reentry flight as six-degree-of-freedom equations connected with the parameters of the system. This flight is divided into three main parts: reentry phase, TAEM (terminal area energy management) and auto landing phase. Authors focused on the reentry phase. The goal was to reach the TAEM phase and obtain the desired motion while the system is affected by perturbations and path constraints. These restrictions are imposed on heating rate, dynamic pressure and load factor. The main parts of the control process were: the design of the reentry trajectory in order to fulfill the listed limitations, design of the control law in order to control the desired trajectory of the object and the design of the attitude controller. Furthermore, authors proposed a high order sliding mode controller, which is proven to be stable using the Lyapunov method. In order to verify theoretical considerations, authors used a model of six-degree-of-freedom reusable launch vehicle. The desired trajectory is followed by an object, however when the switching manifold reaches its desired position, the high value of the control signal occurs. In order to deal with this problem authors used the virtual control method. In a consequence, the control signal became one of the state variables. Therefore, the order of the system was higher. Nevertheless, it allowed authors to control the rate change of the input and made it possible to limit it.

Studying some papers, one can observe that authors found an application of the time-varying switching hyperplanes in control of the robotic manipulators. These manipulators are implemented in practice, therefore it has become an interesting scientific field of research in recent years.

An application of time-varying non-singular terminal sliding mode controller to the rigid robot manipulators can be found in [13]. The presence of perturbations and modeling uncertainties was included. Again the Lyapunov method was used to prove the stability of the system. The reaching phase was eliminated, which guaranteed robustness due to perturbations. Moreover, the finite-time convergence was achieved as a result of specific sliding hyperplane parameters selection. Furthermore, a modified time-varying non-singular terminal sliding mode control law with time-varying gain was introduced. In a consequence the convergence rate was improved and the control signal value was reduced. The computer simulations were carried out on a two-link manipulator. Authors compared a strategy presented in the paper with a time-invariant sliding mode controller. The time-varying controller ensured shorter convergence time of the error to zero. Moreover, the maximum value of the control signal was not exceeded.

In the paper [32] authors proposed a similar approach as in [13]. The time-varying sliding mode control was used to ensure that the tracking error converges to zero at the predefined time during the reaching phase and the nonsingular terminal sliding mode control ensured that the tracking error was equal to zero during the sliding phase. Again, the system was robust due to the perturbations for the whole regulation process. The novel approach includes the fact that the convergence time was no longer computed, but it was considered as a parameter of the system. That means that it can selected in such a way to solve a scientific problem contained in this paper.

Another approach to control the robotic manipulator was presented in [18]. A switching variable was presented as a linear combination of the state errors and their integrals. Furthermore, authors presented a time-varying sliding hyperplane, which ensured the finite time convergence of the representative point to the demanded state. In order to evaluate the performance of the system, quality indexes IAE and ITAE were introduced. The theoretical considerations were verified by computer simulations using the model of a two-link direct-drive manipulator. It was shown that the proposed strategy ensures fast and smooth response and the maximum value of the control signal is lowered.

In the paper [14] authors proposed a sliding mode controller for a two rigid link manipulator system. A so-called gain-scheduled variable hyperplane was presented. It was designed in such a way to react to the environmental stiffness, which is estimated in real time. Again, the Lapunov method was applied in order to prove the stability of the system. In the simulation results the proposed system was compared to the traditional sliding mode controller with a fixed sliding hyperplane. It was shown that the conventional sliding mode controller does not adapt to the environmental stiffness variations. Moreover, the proposed strategy generates much smaller overshoot of the contact force. Furthermore, authors presented a practical comparison between their controller and the impedance controller. They selected two different environments: rubber and sponge and concluded that the controller proposed in the paper has adaptability for the large variety of the environment. On the other hand, the impedance controller does not adapt to the sponge environment. Moreover, it was stated that the proposed controller is robust to the large parameter variation.

In the work [28] authors combined the sliding mode control and the fuzzy logic approach to propose a novel control method. Furthermore, the so-called non-chattering robust sliding mode controller was proposed. The stability was proven using the Lyapunov method. Moreover, the slope of the presented sliding hyperplane is adaptive, which means that it changes dynamically by the fuzzy logic unit. The switching plane follows the system, i.e. it moves in the same direction as the system. However, the sliding line can only move in such a way to remain in the second and the fourth quadrant of the state space, in order to ensure the stable motion in the sliding phase. To illustrate theoretical considerations authors presented a planar robot with two revolute joints. In order to evaluate the performance of the proposed system, a comparison to PID controller and a classical sliding mode controller with fixed sliding line was proposed. The additional mass was attached to the second link of the robot. It was shown that the proposed controller provides better robustness. Moreover, both sliding mode controllers had better resistance to noise than the PID controller.

Paper [29] comprises a similar approach as in [28]. Again, the main goal was to adjust the slope of the switching plane in a dynamic way. A non-chattering robust fuzzy sliding mode controller was designed in such a way that the negative quotient of the derivative of error and error was its input and its output was the slope constant of the sliding line. Authors presented a four-degree-of-freedom nonlinear half-car model. Moreover, an active suspension system with nonlinear spring and piecewise linear damper with dry friction was applied. In order to evaluate the behavior of the proposed controller, authors focused on improving the ride comfort in vehicles. It was shown that the magnitudes of the body displacement and pitch motions dropped and the resonance peak due to the car body was eliminated. Moreover, the energy consumption was reduced and the robustness of the controller was achieved.

A practical approach of applying time-varying sliding hyperplanes into discrete-time systems can be found in [17]. Authors proposed a novel time-varying sliding hyperplane in order to deal with a problem of obtaining large values of the control signal caused by the fast convergence of the representative point to the vicinity of the switching plane. The new virtual state variable was presented as a negative product of a vector of constants connected with sliding variable, a state matrix and a state vector. It was stated that the influence of that state variable on the quasi-sliding motion decreases in time and finally reaches zero. The new switching hyperplane was derived. However, it was shown that the modified sliding hyperplane is of the same form as the original one. Authors proposed a class of functions connected with a control signal, which provides that for any system's state, the control signal is bounded from above by a well-known parameter. A strategy presented in this paper was applied in order to control the data flow in a connection-oriented communication network. It was proven that the amount of requested data is non-negative. Moreover, the bottleneck queue length is always smaller than or equal to a well known parameter, which results in full utilization of an available bandwidth. The control signal limitation was fulfilled.

Another application of the time-varying sliding hyperplane into discrete-time system is described in [15]. Authors described the system with unknown perturbations, which satisfy the matching condition. The main goal of the paper was to eliminate undesirable chattering phenomenon. The stability was proven by using a Lyapunov function. Further, two fixed sliding lines were proposed. Authors stated that the time-varying switching line can only rotate between these two fixed lines. The simulation example was presented for the vertical driving arm. At first, the conventional sliding line was derived and it was shown that it causes a big chattering effect. Further, the time-varying switching line was considered. In this case, the chattering decreased, but the input signal took higher values. In the end, authors presented the approach described in this paper. It was shown that both chattering phenomenon and the control signal decreased.

4 Conclusions

This paper comprises recent results in the area of the application of the time-varying hyperplanes to the sliding mode controllers. The main advantage of the proposed strategy is that the system becomes robust to the external disturbances and modeling uncertainties for the whole regulation process and not only after the reaching phase, which is one of the issues of the conventional sliding mode control with time-invariant sliding hyperplanes. Moreover, the selection of the time-varying sliding hyperplanes ensures the minimization of the initial overshoot. Firstly, papers in which the optimal time-varying sliding hyperplanes were applied to the second- and the third-order systems were described. After that the real life application to the hoisting crane was presented. Furthermore, some papers in which the time-varying switching hyperplanes were applied to continuous and discrete time systems, were introduced. In the end, the application to space exploration, robotic manipulators and data transmission in connection-oriented communication networks was described.

References

1. Bartoszewicz, A.: Time-varying sliding modes for second-order systems. Proc. IEE Part D Control Theory Appl. **143**(5), 455–462 (1996)
2. Bartoszewicz, A., Nowacka-Leverton, A.: SMC without the reaching phase - the switching plane design for the third-order system. Proc. IET Part D Control Theory Appl. **1**(5), 1461–1470 (2007)
3. Bartoszewicz, A., Nowacka, A.: Optimal design of the shifted switching planes for VSC of the third order system. Trans. Inst. Meas. Control **28**(4), 335–352 (2006)
4. Bartoszewicz, A., Nowacka, A.: Reaching phase elimination in variable structure control of the third order system with state constraints. Kybernetika **42**(1), 111–126 (2006)
5. Bartoszewicz, A., Nowacka, A.: Sliding mode control of the third-order system subject to velocity, acceleration and input signal constraints. Int. J. Adapt. Control Signal Process. **21**(8–9), 779–794 (2007)
6. Bartoszewicz, A., Nowacka-Leverton, A.: Time-Varying Sliding Modes for Second and Third Order Systems. Springer Verlag, Heidelberg (2009)
7. Bartoszewicz, A., Nowacka-Leverton, A.: ITAE optimal sliding modes for third order systems with input signal and state constraints. IEEE Trans. Autom. Control **55**(8), 1928–1932 (2010)
8. Choi, S.B., Cheong, C.C., Park, D.W.: Moving switching surfaces for robust control of second order variable structure systems. Int. J. Control **58**(1), 229–245 (1993)
9. Cong, B., Liu, X., Chen, Z.: A precise and robust control strategy for rigid spacecraft eigenaxis rotation. Chin. J. Aeronaut. **24**(4), 484–492 (2011)

10. Cong, B., Liu, X., Chen, Z.: Distributed attitude synchronization of formation flying via consensus-based virtual structure. Acta Astronaut. **68**(11–12), 1973–1986 (2011)
11. Corradini, M.L., Orlando, G.: Linear unstable plants with saturating actuators: robust stabilization by a time varying sliding surface. Automatica **43**(1), 88–94 (2007)
12. Furuta, K.: Sliding mode control of a discrete system. Syst. Control Lett. **14**, 145–152 (1990)
13. Geng, J., Sheng, Y., Liu, X.: Time-varying nonsingular terminal sliding mode control for robot manipulators. Trans. Inst. Meas. Control **36**(5), 604–617 (2014)
14. Iwasaki, M., Tsujiuchi, N., Koizumi, T.: Force control for unknown environment using sliding mode controller with gain-scheduled variable hyperplane. In: IEEE Annual Conference of the Industrial Electronics Society, pp. 1812–1817 (2002)
15. Kanai, Y., Kasahara, M., Mori, Y.: On sliding mode control with time varying switching hyperplane for vertical driving arm. In: 2009 4th IEEE Conference on Industrial Electronics and Applications, pp. 23–28. IEEE (2009)
16. Komurcugil, H.: Rotating-sliding-line-based sliding-mode control for single-phase UPS inverters. IEEE Trans. Ind. Electron. **59**(10), 3719–3726 (2012)
17. Latosiński, P., Bartoszewicz, A.: Sliding mode control with time-varying switching hyperplane for data transmission networks. In: 2018 23rd International Conference on Methods & Models in Automation & Robotics (MMAR). IEEE (2018)
18. Lu, Y.S., Chiu, C.W., Chen, J.S.: Time-varying sliding-mode control for finite-time convergence. Electr. Eng. **92**(7), 257–268 (2010)
19. Milosavljević, Č.: General conditions for the existence of a quasi-sliding mode on the switching hyperplane in discrete variable structure systems. Autom. Remote Control **46**, 307–314 (1985)
20. Mizoshiri, T., Mori, Y.: Sliding mode control with a linear sliding surface that varies along a smooth trajectory. In: 2016 SICE International Symposium on Control Systems (ISCS). IEEE (2016)
21. Nowacka-Leverton, A., Michałek, M., Pazderski, D., Bartoszewicz, A.: Experimental verification of SMC with moving switching lines applied to hoisting crane vertical motion control. ISA Trans. **51**(2), 682–693 (2012)
22. Sivert, A., Betin, F., Faqir, A., Capolino, G.A.: Robust control of an induction machine drive using a time-varying sliding surface. In: Proceedings of the IEEE International Symposium on Industrial Electronics, pp. 1369–1374 (2004)
23. Tian, B., Fan, W., Zong, Q., Wang, J., Wang, F.: Nonlinear robust control for reusable launch vehicles in reentry phase based on time-varying high order sliding mode. J. Franklin Inst. **350**(7), 1787–1807 (2013)
24. Tokat, S., Eksin, I., Guzelkaya, M.: New approaches for on-line tuning of the linear sliding surface slope in sliding mode controllers. Turkish J. Electr. Eng. Comput. Sci. **11**(1), 45–59 (2003)
25. Tokat, S., Eksin, I., Guzelkaya, M.: Linear time-varying sliding surface design based on co-ordinate transformation for high-order systems. Trans. Inst. Meas. Control **31**(1), 51–70 (2009)
26. Utkin, V.I.: Variable structure systems with sliding modes. IEEE Trans. Autom. Control **22**, 212–222 (1977)
27. Utkin, V.I., Drakunow, S.V.: On discrete-time sliding mode control. In: IFAC Conference on Nonlinear Control, pp. 484–489 (1989)
28. Yagiz, N., Hacioglu, Y.: Fuzzy sliding modes with moving surface for the robust control of a planar robot. J. Vib. Control **11**(7), 903–922 (2005)

29. Yagiz, N., Hacioglu, Y., Taskin, Y.: Fuzzy sliding-mode control of active suspensions. IEEE Trans. Ind. Electron. **55**(11), 3883–3890 (2008)
30. Yilmaz, C., Hurmuzlu, Y.: Eliminating the reaching phase from variable structure control. J. Dyn. Syst. Meas. Control **122**(4), 753–757 (2000)
31. Yorgancioglu, F., Komurcugil, H.: Decoupled sliding-mode controller based on time-varying sliding surfaces for fourth-order systems. Expert Syst. Appl. **37**(10), 6764–6774 (2010)
32. Zhao, Y., Sheng, Y., Liu, X.: A novel finite time sliding mode control for robotic manipulators. In: IFAC World Congress, Cape Town, Republic of South Africa, pp. 7336–7341 (2014)

Identification of Linear Models of a Tandem-Wing Quadplane Drone: Preliminary Results

Michał Okulski(✉) and Maciej Ławryńczuk

Institute of Control and Computation Engineering, Warsaw University of Technology, ul. Nowowiejska 15/19, 00-665 Warsaw, Poland
mikkichz2@gmail.com, M.Lawrynczuk@ia.pw.edu.pl

Abstract. A tandem-wing quadplane drone has been built to study control strategies and develop high-performance onboard controllers. In hover flight, the quadplane behaves like a classic quadcopter. Highly non-linear dynamics of the orientation stabilization need a state-of-the-art Model Predictive Controller (MPC). To develop such a controller, an accurate model of the drone needs to be identified – ideally, a linear model. This paper present preliminary results of identifying two linear models: a State-Space Model derived from Newton dynamic principles and a novel Recurrent Neural Network based linear model.

Keywords: Quadplane · Drone · MPC · Model Predictive Controller · State-space model · Recurrent Neural Network · RNN

1 Introduction

A wide variety of Unmanned Aerial Vehicles (UAV) has been already developed and studied. The most typical small size drones are multicopters (e.g. quadcopters). Multicopters usually have great maneuverability, but not necessarily have long range or high cruise speed. Therefore many hybrid drones have been proposed [1] e.g. a quadplane. Some types of quadplanes dynamics and control strategies are described in [2]. Generally, the multicopter dynamics is non-linear and the typical control strategy is based on a Proportional-Integration-Derivative (PID) Controllers [3,4]. This paper focuses on modeling a slightly different type of drone: a tandem-wing quadplane. In this preliminary research however, only the first phase of flight is modeled: a hover flight which is same as for a quadcopter in terms of dynamics and control (Figs. 1 and 2).

2 A Tandem-Wing Quadplane Drone

A custom-designed drone has been built for broad research purposes. It is a kind of a quadplane (a combination of a plane and a quadcopter), but here a

A. Bartoszewicz et al. (Eds.): PCC 2020, AISC 1196, pp. 219–228, 2020.
https://doi.org/10.1007/978-3-030-50936-1_19

Fig. 1. Tied to a chassis – part of the safe experimental environment

Fig. 2. The drone during an experimental free-flight test

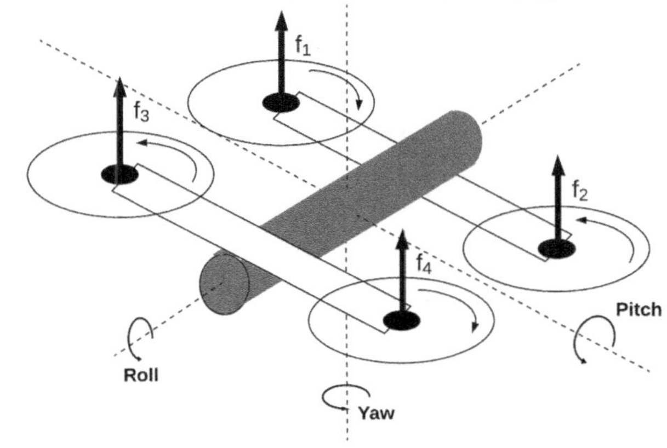

Fig. 3. A quadplane drone model

tandem-wing concept has been used instead of a classic wing and a tail. There is no need to attach additional motor holders which can affect aerodynamic flow in a forward flight because the brushless motors has been installed at the tips of the wings. An additional fifth brushless motor with a pusher propeller is used to provide thrust in a forward flight. After reaching a certain speed, the four quadcopter-like motors can be turned off, because all the lift is provided by the wigs. In this preliminary research however, only the hover flight is considered and modeled. Details of the drone construction:

- Carbon-fiber chassis, full balsa wood wings (reinforced with carbon-fiber bars)
- 6S3P Li-Ion battery
- 4x T-Motor MN3110 KV470 + 12 × 4 prop, 1x T-Motor F60 PRO II KV1750 + 5 × 5 prop
- 1x NVidia Jetson TX2, 2x STM32F7 as onboard computers
- Inertial Measurement Unit (IMU), Laser and Ultrasonic Range-Finders, barometer, differential barometer for a Prandtl tube
- a long-range telemetry radio
- the drone ready to fly weights ca. 3200 g

3 First Principle Drone Orientation Model

Let's consider the quadplane drone dynamics around the hover state. It can be treated then as a classic quadcopter, where the differential thrust produced by its motors affects the drone attitude. The drone contains four identical propulsion units: each consists of a brushless motor and a direct drive propeller. Two propellers rotate clockwise and the remaining two rotate counter-clockwise as in Fig. 3. Each propulsion unit generates thrust force (lift) and torque which are both almost linear around 50% of the full-throttle according to [5]. The drone described in this paper is designed to hover at ca. 50% of throttle and experimental flights confirmed this assumption. Full quadcopter dynamics and kinematics has been described in [6]. Quadcopter reaction torques for *roll*, *pitch* and *yaw* $(\tau_\phi, \tau_\theta, \tau_\psi)$ for the drone can be expressed as:

$$\begin{bmatrix} F_L \\ \tau_\phi \\ \tau_\theta \\ \tau_\psi \end{bmatrix} = \begin{bmatrix} 1 & 1 & 1 & 1 \\ -d & d & -d & d \\ d & d & -d & -d \\ c_y & -c_y & -c_y & c_y \end{bmatrix} \begin{bmatrix} f_1 \\ f_2 \\ f_3 \\ f_4 \end{bmatrix} \tag{1}$$

where $F_L = f_1 + f_2 + f_3 + f_4$ is the total motors thrust force (lift), d is the offset from the center of mass to each rotor and c_y is the coefficient relating the *yaw* reaction moment by air resistance to the rotor. Please note the torques are treated here as scalar values to simplify model considerations. The Eq. 1 shows that each reaction torque of the drone can be controlled individually, at least around the hover state, where additional aerodynamic forces like drag force and lift force produced by wings, can be neglected. For convenience, three control signals u_ϕ, u_θ, u_ψ are introduced – each responsible for corresponding reaction torque. The thrust of each motor can be then calculated as follows:

$$\begin{bmatrix} f_1 \\ f_2 \\ f_3 \\ f_4 \end{bmatrix} = \begin{bmatrix} 1 & -1 & 1 & 1 \\ 1 & 1 & 1 & -1 \\ 1 & -1 & -1 & -1 \\ 1 & 1 & -1 & 1 \end{bmatrix} \begin{bmatrix} u_h \\ u_\phi \\ u_\theta \\ u_\psi \end{bmatrix} \tag{2}$$

where u_h is the throttle control signal needed to hover (typically about 50% of the full motor power). To keep the drone stable when hovering, all the reacting torques should be $\tau_\phi = \tau_\theta = \tau_\psi = 0$ and the orientation of the drone (expressed as Euler angles) should be $\phi = \theta = 0$, $\psi = \text{const}$ (i.e. roll and pitch set to 0 and yaw set to any desired heading direction). The relation between torque value and the drone orientation (Euler angles) can be derived from classic Newton mechanics. To simplify further considerations, each rotation angle is calculated separately with respect to the thrust forces and torque values produced by the drone propulsion systems. The moment of inertia I is given by:

$$I = mr^2 \tag{3}$$

where m is the total mass of the drone, and r is the distance to the pivot point. Let r_ϕ, r_θ, r_ψ be the distances from the drone's center of gravity to the points where combined thrust forces are applied (see Fig. 3). Then the drone dynamics can be decomposed into scalar equations:

$$\begin{cases} \dot{\phi} = \omega_\phi \\ \ddot{\phi} = \dot{\omega}_\phi = \alpha_\phi = \tau_\phi I_\phi = \tau_\phi m r_\phi^2 \\ \dot{\theta} = \omega_\theta \\ \ddot{\theta} = \dot{\omega}_\theta = \alpha_\theta = \tau_\theta I_\theta = \tau_\theta m r_\theta^2 \\ \dot{\psi} = \omega_\psi \\ \ddot{\psi} = \dot{\omega}_\psi = \alpha_\psi = \tau_\psi I_\psi = \tau_\psi m r_\psi^2 \end{cases} \tag{4}$$

where ω is the angular speed, and α is the angular acceleration (around the corresponding axis). Since the mass m of the drone is constant over time and the center of mass is fixed (which means r is constant too), then from Eq. 4 can be deduced that the angular accelerations $\ddot{\phi}$, $\ddot{\theta}$ and $\ddot{\psi}$ are proportional to the corresponding reaction torques and the torques are proportional to the corresponding control signals u_ϕ, u_θ, u_ψ:

$$\begin{aligned} \ddot{\phi} &\propto \tau_\phi \propto u_\phi \\ \ddot{\theta} &\propto \tau_\theta \propto u_\theta \\ \ddot{\psi} &\propto \tau_\psi \propto u_\psi \end{aligned} \tag{5}$$

4 Identification of Models

4.1 Experimental Set-Up

Due to safety reasons, early experiments were conducted with limited flight freedom conditions, e.g. the drone was locked in roll, and yaw axes and only the pitch axis was free running. The limited freedom experiment chassis contains an optical encoder to measure the drone's onboard Inertial Measurement Unit (IMU) accuracy. It has been confirmed that the IMU accuracy is good enough to use the IMU readings as a reference orientation of the drone. Further experiments used onboard high-frequency closed-loop PID controllers (manually tuned) in a free flight (hover) conditions.

As shown in Eq. 4 each rotation angle can be calculated and forced (Eq. 2) separately from the other angles. Therefore further model identifications will focus on the *pitch* angle θ as an example.

4.2 Problem Statement

Experimental flights with the PID controllers proved that the drone is able to hover safely following classic quadcopter control principles. However the quality of that simple control strategy was far from being precise – the drone struggled with control overshooting and oscillations. It has been decided to replace the simple PIDs with a Model Predictive Controller (MPC) to control the drone in hover conditions. To implement an MPC, a model needs to be identified. Ideally, if a good enough linear model is found, then the General Predictive Controller (GPC) variant of the MPC could be easily implemented. GPC is optimal and very efficient, so this preliminary research focuses on identifying a linear model.

In the GPC a linear model is used to predict an output of a plant for a certain (fixed) time horizon. In this research, it has been decided that at least 12 discrete time steps will be predicted by the identified linear model. Drone telemetry data is logged at 50 Hz, which means that a single discrete time step equals to $k = 20$ ms, and the target prediction horizon equals to $12k = 240$ ms. A pitch-related telemetry data that has been recorded during one of the experimental flights is shown in Fig. 4. The dataset has been divided into two subsets: the learn dataset (5749 samples) and the validation dataset (1931 samples).

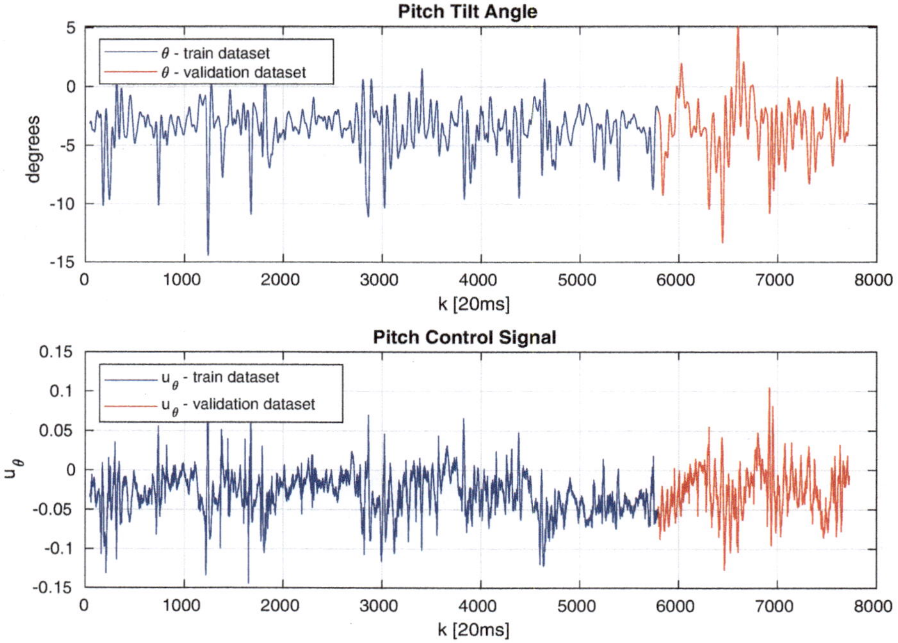

Fig. 4. Experimental flight dataset: hover conditions

To assess identified models, a *Loss Function* needs to be defined. Simplified drone dynamics described above shows that the drone is an inertial plant. Therefore it is more important to predict the direction of future drone rotation correctly, i.e. predict the correct sign of the angular speed rather than the right tilt angle. To stress that assumption, the *Loss Function* has been defined as follows:

$$E = \sum_{k=0}^{N} e(\omega_\theta(k), \hat{\omega}_\theta(k))|\theta(k) - \hat{\theta}(k)| \tag{6}$$

where N is the amount of samples in the dataset, θ and ω_θ are the ground truth pitch tilt angle and the pitch angular speed (according to Eq. 4) and similarly $\hat{\theta}$ and $\hat{\omega}_\theta$ are the pitch angle and the pitch angular speed predicted by the model. Additionally, there is a penalization function $e(\omega_1, \omega_2)$ defined as:

$$e(\omega_1, \omega_2) = \begin{cases} 0.05, & \text{if } \operatorname{sgn}(\omega_1) = \operatorname{sgn}(\omega_2) \\ 1, & \text{if } \operatorname{sgn}(\omega_1) \neq \operatorname{sgn}(\omega_2) \end{cases} \tag{7}$$

4.3 State-Space Linear Model

A discrete state-space model is given by the following equations:

$$\begin{cases} X(k+1) = AX(k) + BU(k) \\ y(k) \qquad = CX(k) + DU(k) \end{cases} \tag{8}$$

Based on Newton dynamic principles and the fact shown in Eq. 5 matrices of a simplified linear model are defined as follows:

$$A = \begin{bmatrix} a_1 & 1 \\ a_2 & a_3 \end{bmatrix}, \quad B = \begin{bmatrix} 0 & 0 \\ b & 0 \end{bmatrix}, \quad C = \begin{bmatrix} 1 & 0 \end{bmatrix}, \quad D = \begin{bmatrix} 0 & d \end{bmatrix},$$
$$X(k) = \begin{bmatrix} \theta(k) \\ \omega_\theta(k) \end{bmatrix}, \quad U(k) = \begin{bmatrix} u_\theta(k) \\ 1 \end{bmatrix} \tag{9}$$

where a_1, a_2, a_3, b, d are model parameters to be identified. Model identification has been conducted using Matlab's *fmincon* function [7]. Results of identified model predictions for validation set are shown in Fig. 5.

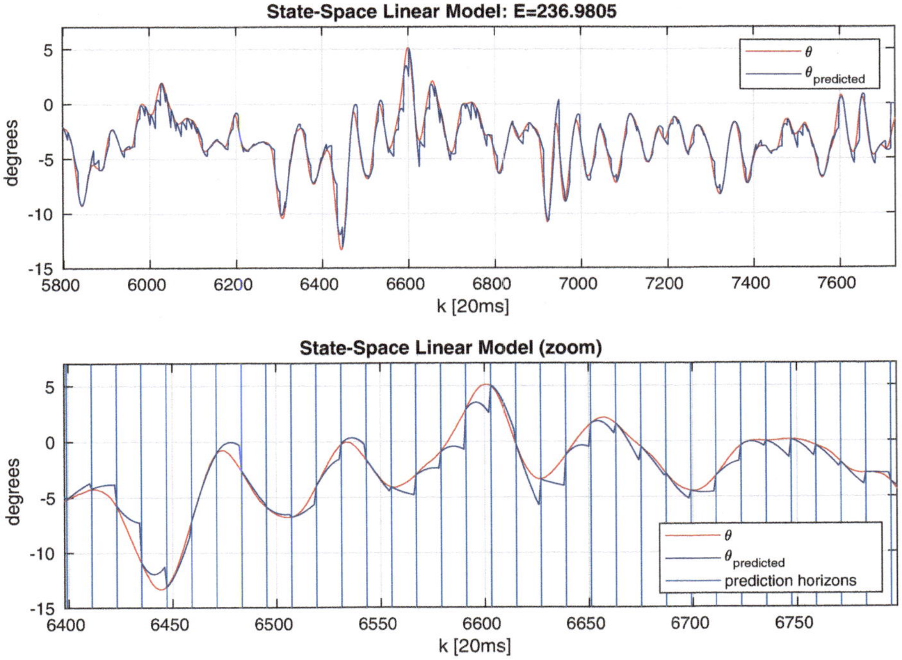

Fig. 5. Linear State-Space Model: Predictions for the validation dataset

4.4 Linear Model Based on a Recurrent Neural Network

A dedicated Neural Network has been designed (see Fig. 6) to act as a Linear Model. There is a significant advantage of using Neural Network: such a network can be easily built, trained and evaluated using modern python frameworks like Keras [8] and TensorFlow r2.1 [9]. The Neural Network consists of:

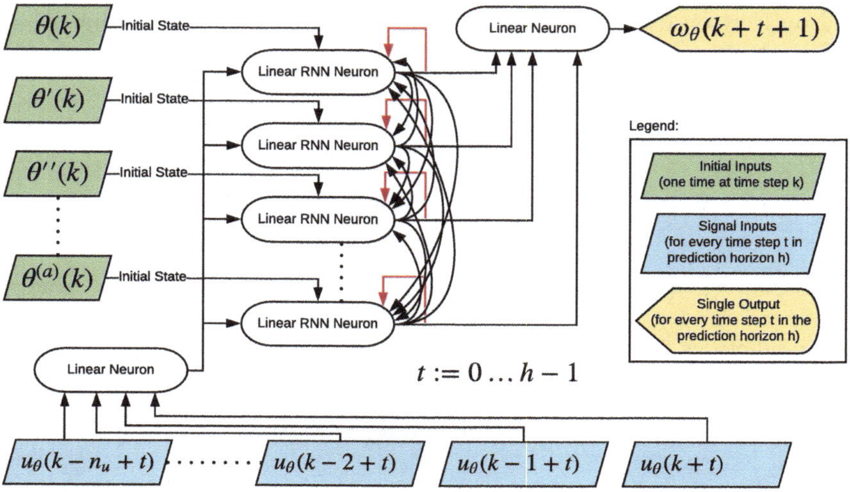

Fig. 6. Structure of the RNN-based Linear Model

1. An input layer with $a + 1$ inputs: a pitch tilt angle $\theta(k)$, a first derivative of the pitch angle $\theta'(k)$ (i.e. the angular speed), a second derivative of the pitch angle $\theta''(k)$ (i.e. the angular acceleration) and so on, up to the a-th derivative of the pitch angle $\theta^{(a)}(k)$. Input from this layer is passed once to initialize the hidden states of the $a + 1$ Recurrent Neurons directly.
2. An Input layer with n_u inputs containing the u_θ signals: from the past $(k + t - n_u)$ up to the current time step in the prediction horizon $(k + t)$, where $t := 0 \dots h - 1$.
3. A single Linear Neuron producing a weighted sum of all the past u_θ signals – it can be considered like a attention layer which is trained to take only the most important part of the past input signals – in fact, it shows what is the input signal delay.
4. A Recurrent Layer of $a + 1$ Linear RNN neurons. Each neuron provides an output signal at every time step t which is then passed to the next layer, but also using recurrent connections to each other neuron (including self).
5. A single Linear Neuron which produces a weighted sum of all the RNN neurons for every time step t. That value is the pitch angular speed at certain time step in the prediction horizon: $\omega_\theta(k + t + 1)$.

An output pitch angular speed at a certain time step $\omega_\theta(k + t + 1)$ can be calculated as follows:

$$
\begin{aligned}
&\omega_\theta(k + t + 1) = \sum_{i=1}^{a+1} b_0^2 + \\
&\sum_{i=1}^{a+1} w_i^2 \left(b_i^1 + w_i^1 \left[b^0 + \sum_{n=1}^{n_u} \left(w_n^0 u_\theta(k + t - n) \right) \right] + \sum_{j=1}^{a+1} \left(w_i^r h_i(t - 1) \right) \right)
\end{aligned}
\tag{10}
$$

where w_n^0, b^0 are the weights and the bias of the linear neuron connected to the signal input layer, b_i^1 is the bias of each RNN neuron, w_i^1 are the RNN input weights, w_i^r are the recurrent weights connected to each RNN hidden state from previous time step and w_i^2, b^2 are the weights and the bias of the output linear neuron.

The value of a and n_u has been found experimentally and set to: $a = 6$, $n_u = 10$. These values give the best trade-off between amount of parameters to tune and the prediction results to the validation dataset. The results are presented in Fig. 7.

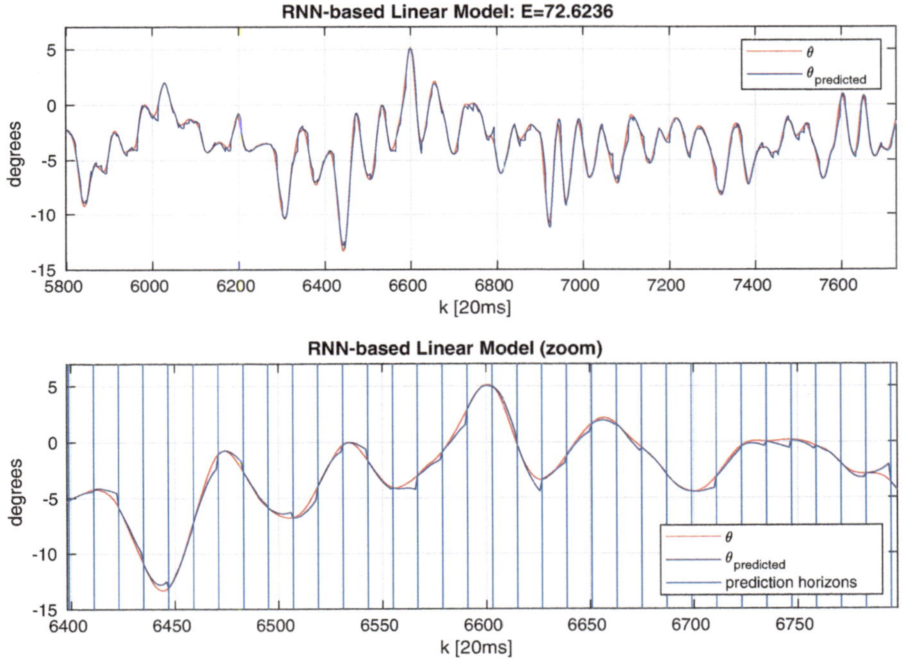

Fig. 7. RNN-based Linear Model: Predictions for the validation dataset

5 Summary

Both presented preliminary models can predict the pitch tilt angle of the drone over the fixed prediction horizon. Loss values are presented in the Table 1. The Recurrent Neural Network based model, however, performs much better in terms of accuracy but has a much higher amount of parameters to tune.

Table 1. Comparison of the two linear models

Model type	Train loss	Validation loss	# Parameters
State-Space Linear Model	171.34	236.98	**5**
RNN-based Linear Model	**75.89**	**72.62**	82

References

1. Saeed, A.S., Younes, A.B., Cai, C., Cai, G.: A survey of hybrid unmanned aerial vehicles. Prog. Aerosp. Sci. **98**, 91–105 (2018)
2. Govdeli, Y., Muzaffar, S.M.B., Raj, R., Elhadidi, B., Kayacan, E.: Unsteady aerodynamic modeling and control of pusher and tilt-rotor quadplane configurations. Aerosp. Sci. Technol. **94**, 105421 (2019)
3. Sámano, A., Castro, R., Lozano, R., Salazar, S.: Modeling and stabilization of a multi-rotor helicopter. Intell. Robot Syst. **69**(1–4), 161–169 (2013)
4. Yang, H., Lee, Y., Jeon, S., Lee, D.: Multi-rotor drone tutorial: systems, mechanics, control and state estimation. Intell. Serv. Robot. **10**(2), 79–93 (2017)
5. Boyang, L., Weifeng, Z., Jingxuan, S., Chih-Yung, W., Chih-Keng, C.: Development of model predictive controller for a tail-sitter VTOL UAV in hover flight. In: Unmanned Aerial Vehicle Networks, Systems and Applications (2018)
6. Mahony, R., Kumar, V., Corke, P.: Multirotor aerial vehicles: modeling, estimation, and control of quadrotor. IEEE Robot. Autom. Mag. **19**(3), 20–32 (2012)
7. Find minimum of constrained nonlinear multivariable function - MATLAB fmincon. https://www.mathworks.com/help/optim/ug/fmincon.html
8. Keras: The Python Deep Learning library. https://keras.io/
9. API Documentation—TensorFlow Core r2.1. https://www.tensorflow.org/api_docs/
10. Haykin, S.: Neural Networks: A Comprehensive Foundation. Prentice Hall, Upper Saddle River (1998)
11. Mandic, D.P., Chambers, J.: Recurrent Neural Networks for Prediction: Learning Algorithms Architectures and Stability. Wiley, New York (2001)

Floating Oil Platform Model with Dynamic Positioning and Reference System

Jakub Wieczorek[1(✉)] and Patryk Chaber[2]

[1] Institute of Electronic Systems, Warsaw University of Technology, ul. Nowowiejska 15/19, 00-665 Warsaw, Poland
jakub.wieczorek2.stud@pw.edu.pl
[2] Institute of Control and Computation Engineering, Warsaw University of Technology, ul. Nowowiejska 15/19, 00-665 Warsaw, Poland
P.Chaber@ia.pw.edu.pl

Abstract. The article describes the process of construction and testing a floating offshore drilling platform model with focus on cooperation of its diverse subsystems in order to stabilise the position and control the movement of the platform. In real-life applications, advanced off-shore oil-rigs are indeed floating without any scaffolding or concrete basis, so the waves and shore current continuously disrupt their position. Since the place of drilling is fixed, those movements are able to damage or even break the drill, which leads to serious environment contamination by the emerging oil. This paper considers PID controller as a dynamic positioning system with a reference system based on image recognition.

Keywords: Oil platform · Dynamic positioning · Reference system · PID

1 Introduction

In the real-life applications there are two types of oil platforms: fixed and floating. First of them are connected to the ground by the scaffolding or a different fixed construction, thus securing the position of the drill. They can operate on the relative shallow waters not crossing a few hundreds meters deep. The second type of the oil rigs are floating vessels like semi-submersible oil platform or spar buoys [2]. They can be anchored to the ground, but the common feature is that they don't have fixed connection with the bottom of the sea. Accordingly it is necessary to provide the drilling rig stabilisation during oil platform drifting on the water surface, to prevent damaging or even breaking the drill. It can be achieved by the construction itself or by control system called dynamic positioning (DP). Control system for the proper position correction requires information about the vessel location and exerted forces, e.g. by waves. That information is given by the reference system, which can be implemented e.g. as vision system [6], GNSS system [8] or as a fusion of various systems [3].

A. Bartoszewicz et al. (Eds.): PCC 2020, AISC 1196, pp. 229–240, 2020.
https://doi.org/10.1007/978-3-030-50936-1_20

PID controllers are commonly used as a dynamic position algorithm [1]. The simplicity of the algorithm, and its robustness make this algorithm great for initial and often even final method of process variables control [12,13]. However with the great technological advancement, increasing number of modern dynamic positioning systems started to utilise Model Predictive Control (MPC) algorithms [10,11]. The advantage of the MPC approach is that it provides higher control quality, but as a downside, it requires model which have to be periodically updated during platform exploitation.

The aim of this paper is to construct the semi-submersible oil platform model with dynamic positioning and reference system, in order to research advanced methods of position control. Because this paper focuses mainly on the problem of the vessel positioning, no actual drilling equipment is provided for the considered oil platform model.

1.1 PID Controller

Proportional-Integral-Derivative controller is a well established algorithm that utilises information about direct control error, its integral and its derivative part to determine new control signal $u(k)$ (manipulated variable). By applying control signal values to the process of control in consecutive discrete time instants k, it is expected to bring the controlled variable $y(k)$ (output of the process of control) to the set point value $y^{\mathrm{sp}}(k)$, thus minimising control error $e(k) = y^{\mathrm{sp}}(k) - y(k)$. The current value of manipulated variable is calculated using following equations:

$$\begin{aligned} u_{\mathrm{P}}(k) &= K_{\mathrm{P}} e(k) \\ u_{\mathrm{I}}(k) &= u_{\mathrm{I}}(k-1) + K_{\mathrm{P}} \frac{T_{\mathrm{s}}}{2T_{\mathrm{I}}}(e(k-1) + e(k)) \\ u_{\mathrm{D}}(k) &= K_{\mathrm{P}} \frac{T_{\mathrm{D}}}{T_{\mathrm{s}}}(e(k) - e(k-1)) \\ u(k) &= u_{\mathrm{P}}(k) + u_{\mathrm{I}}(k) + u_{\mathrm{D}}(k) \end{aligned} \tag{1}$$

where K_{P}, T_{I}, T_{D} denote tuning parameters of PID controller: proportional gain, integral time and derivative time respectively. Because of the discrete form of the presented PID controller, sampling time is denoted as T_{s}. In discrete representation the integral part has to be approximated by a sum – for example of rectangles or trapezoids. For this paper the second approach has been exploited since is characterised by a smaller error rate comparing to the rectangular rule. Finally control signal consists of three components that corresponds to proportional ($u_{\mathrm{P}}(k)$), integral ($u_{\mathrm{I}}(k)$) and derivative ($u_{\mathrm{D}}(k)$) part of the PID controller.

2 Process of Control

Firstly, a physical platform model which would be able to operate in the water environment had to be created. Figure 1 shows the general view of the construction.

Fig. 1. Physical platform model – basic view

The model has cuboid shape (25 × 25 × 15 cm) and is built from acrylic glass. The construction resembles semi-submersible object, although the hull is placed mostly above the water surface, instead of being partly immersed in the water like in the semi-submersible platform. For this reason the construction is more exposed to the water waves and therefore is prone to inaccuracies in position measurements, due to the swinging of the oil platform's top. In the real-life scenario, on the top of the platform, there would be e.g. GPS receiver mounted on a mast. This would cause erroneous measurements of relative position even when the platform is still in the same place [9]. The position measurement based on the ARToolkit [4] marker placed on the top of the construction, tracked by the implemented reference system via the image recognition, is exposed to the same phenomenon. It is possible to compensate for this inaccuracy with vertical reference systems (VRS), e.g. using accelerometers and gyroscopes [5, 7], although it was not considered in this research.

In the real oil platform rigs, especially in the semi-submersible ones, propulsion consists of the set of azimuth thrusters placed on the corners of the vessel and powered by electric or petrol engines, which can freely rotate around their own axis. In the presented construction thrusters are fixed propellers placed perpendicularly to the two opposite walls. It allows to simplify the construction and also divide the position control task into three separate problems with almost no couplings between them. Since in this paper only one degree of freedom is considered, the movement of this model was externally constraint disallowing it to both rotate, and move in the perpendicular direction.

There are two sets of propellers powered by separate 12 V DC motors. Propulsion system is shown on the left side of Fig. 2. Initial tests showed, that one propeller for the side generates imbalanced angular momentum which causes rotation of the whole vessel. In order to balance the momentum two propellers coupled together with a flat belt are used, of which one is inverted, yielding thrust vector that is almost perpendicular to the wall of the vessel. Engines in one pair are interconnected in series. Due to the constraints regarding positioning of the propellers, a redundant set of DC motors and propellers is used on the opposite wall of the platform. This allows for alleviate the problem of lower thrust generated by a single set of propellers when pulling, compared to pushing. When two sets are utilised, only the pushing one is powered in each time instant, eliminating any asymmetry in the thrust, in relation to the direction of movement.

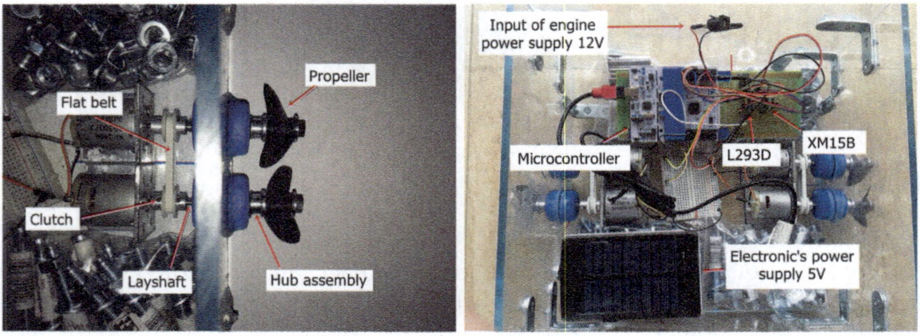

Fig. 2. Oil platform's propulsion system (left) and electronics (right)

On the right side of Fig. 2 the interior of the platform is shown. To control the DC motors, an STM32F103 Nucleo evaluation board is utilised. It contains an STM32F103RBT6 microcontroller based on ARM 32-bit Cortex-M3 RISC core operating at a 64 MHz frequency. Its main task is to receive messages (using Bluetooth transceiver) from the reference system, execute PID control algorithm, and apply this newly calculated control signal to motors.

In order to increase elasticity in terms of priorities assignment, microcontroller executes appropriate actions based on the events. It is worth noting, that interruptions mechanism is still utilised, but action linked with this interrupt is delegated in time, by setting proper event flag. Thanks to this approach interrupt handler functions is as short as possible, so that the main loop (where actions mapped with set events flags are handled) is not interrupted for a long period of time.

Communication between the reference system and the controller is performed using custom protocol over Bluetooth, via XM15B module. The communication is configured as 8N1, with baud rate 9600. Messages received by the microcontroller are stored in a circular buffer with the help of Direct Memory Access mechanism, thus omitting the CPU. The custom protocol assumes that the message contains 7 characters, from which first, and the last one are used to determine beginning and the end of the frame, thus allowing to perform simple correctness check of a single message. If the length of the message is incorrect, it is dropped, and the microcontroller awaits for a new one.

Because the motors require 12 V power supply they cannot be connected directly to the microcontroller's pins, which give 3.3 V and a maximum of 20 mA. Separate power supply and circuit to control the speed of propellers are required. Additionally it is necessary to induce the rotation of only one pair of motors at the same time and to push the construction forward. Thus an H-bridge (L293D) was used for this purpose. To control the rotational speed of propellers a Pulse Width Modulation signal is sent directly to the enable pin of the H-bridge.

In the PID implementation an anti-windup mechanism was utilised, which prevents excessive error accumulation of integral part when the manipulated variable exceeds maximum/minimum value allowed. In that case integral part of the control value is zeroed in the next algorithm iteration. Upper boundary of manipulated variable is set to 1000 which is the maximum value of PWM duty cycle. Lower boundary serves as a deadband implementation. This approach was chosen due to the fact, that there are some values of manipulated variable which are too low to make propellers rotate, mainly because there is not enough torque to overcome the friction. The range of the deadband was determined empirically. For one set of engines it is 300 and for the other 420. Thus manipulated variable values range from 300 to 1000 for one set and from 420 to 1000 for the second one.

3 Reference System

Dynamic positioning system for proper position control of the vessel requires information on the platform's relative placement. The system responsible for that is called reference system. There are many approaches to this task, of which the most popular are based on mechanical phenomena, microwaves, laser and satellite differences, or a combination of above.

Because of the model's size (compared to the real-life oil platform), it is not advised to utilise the same measurements to determine position error as it would be performed in the industrial environment. Commonly used GPS signal is sufficiently precise for the real-life oil platform reference system, while for the small scale model its resolution is far to low. Therefore, even if the reference system based on the image analysis is not applicable to a real-life scenario, it is used in this paper as a substitution of the GPS system. Considered system tracks an ARToolkit marker, visible on Fig. 1, and based on its position relative to the centre of the image, it calculates the position error that will be further compensated by the controller. Image recognition was implemented using `Python` 2.7 with the help of `OpenCV2` library. The reference system is executed on an external PC, which allows to perform necessary calculations in the real-time. Based on the assumption that this system is only a substitution of GPS, it should not introduce any additional delays. The program is highly configurable and customised by `xml` configuration file. In addition to its main task, that is to determine position error, it also exposes a WebAPI with `http` methods `get` and `post` allowing remote access to the controller's both configuration and signals in the real time. This allows to easily supervise the control process, and later perform analysis on collected data.

3.1 Position Error Determination

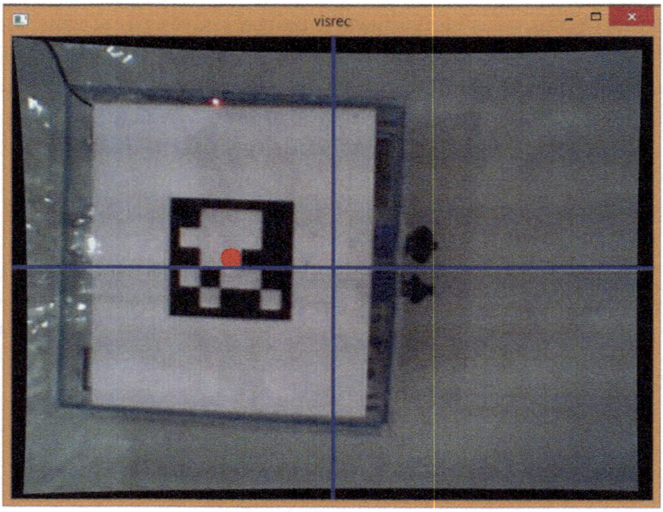

Fig. 3. View of an oil platform model as visible from the vision system – red dot represents the centre of the model (it is considered to be its position), intersection of the blue lines represents drilling point

First step in the image recognition is camera calibration, which allows to remove distortions and yields accurate measurements of position error, based on the prepared ahead of a time chessboard photos. Process of image recognition is performed as a separate thread, as follows: in the while loop the image is captured by the camera, distortions are removed, ARToolkit marker is being found and its centre is determined and marked, by a red dot. Thereafter on the image a cross sign is affixed – this represents the place of theoretical drill (Fig. 3). The difference between a drill's expected position and an actual position of the drill is denoted as a position error. Using the Bluetooth communication, and custom protocol, this data is sent to the platform's dynamic positioning system.

On the microcontroller side the message is received and based on that the control value is calculated and instantly translated into a PWM, which after amplification is used to control the speed of DC motors. This finalises the control loop visible on Fig. 4.

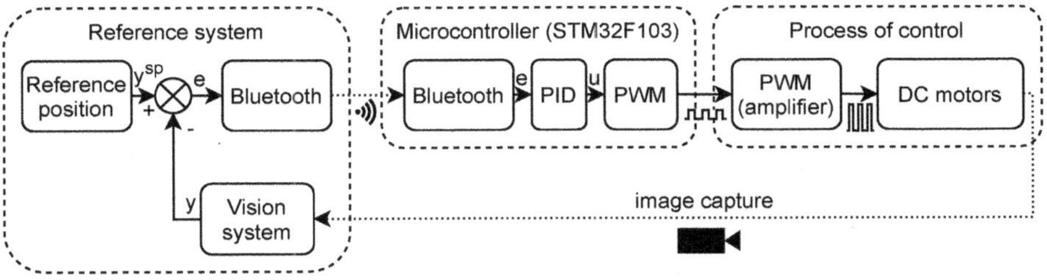

Fig. 4. Scheme of data flow with correspondence to a control loop – dotted lines denotes non physical coupling

3.2 Controller Interface

WebAPI of the vision system acts also as an interface between modelled oil platform and any device that implements aforementioned communication protocol. As the main technology for this purpose, Flask framework and REST API were used. Controller interface allows sending messages to the platform in order to e.g. reset control process or set PID tuning parameters as well as extracting messages from the platform by the clients who will for instance visualise process in the real time.

Vision system consists of multiple threads, thus there is a possibility of a synchronisation problem occurrence. Specifically the messages queue and Bluetooth have to be considered as shared resources in context of this system. In order to solve this problem a `deque` data structure is utilised, which provides atomic operations for performing operations like appending and popping elements. In order to exclude the race conditions in terms of accessing Bluetooth, mutex locking mechanism was implemented using `RLock` class on the send operation.

4 Experiments Results

Experiments were performed in two stages. First, platform was moved by 7 cm from the reference point and control process was recorded for different sets of values of PID tuning parameters. The results have been subjected to a qualitative and quantitative assessment. As a control quality indicator, the sum of squared errors was utilised:

$$E = \sum_{k=1}^{k_{\mathrm{end}}} (y^{\mathrm{sp}}(k) - y(k))^2 \tag{2}$$

For all results presented in this paper sampling time was set to 100 ms and it was assumed that $k_{\text{end}} = 150$. Selected results were presented on Figs. 5, 6, 7, 8 and 9. Set of tuning parameters that yields the best quality of control is further used in the second part of experiments.

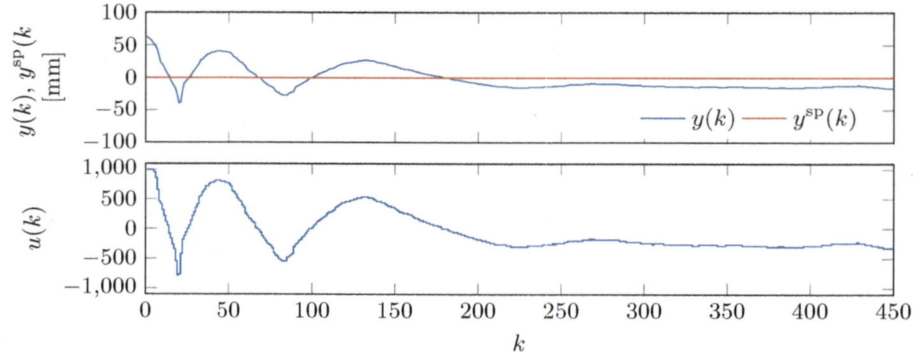

Fig. 5. Control process using P controller ($K_{\text{P}} = 20$) without deadband elimination

As the manipulated variable represents average voltage of DC motors, thus indirectly thrust, it is clear that the process of control contains integrating factor. Therefore simple P controller should be enough to bring control error to a zero. Alas due to the internal friction forces in DC motors and their coupling, only after input power applied exceeds some threshold, those motors start rotating. This is clearly visible on Fig. 5, where the P controller cannot eliminate steady state error.

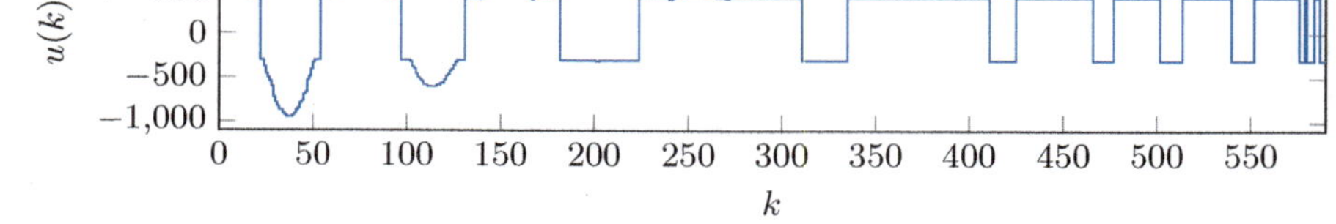

Fig. 6. Control process using P controller ($K_{\text{P}} = 20$) – $E = 92717$

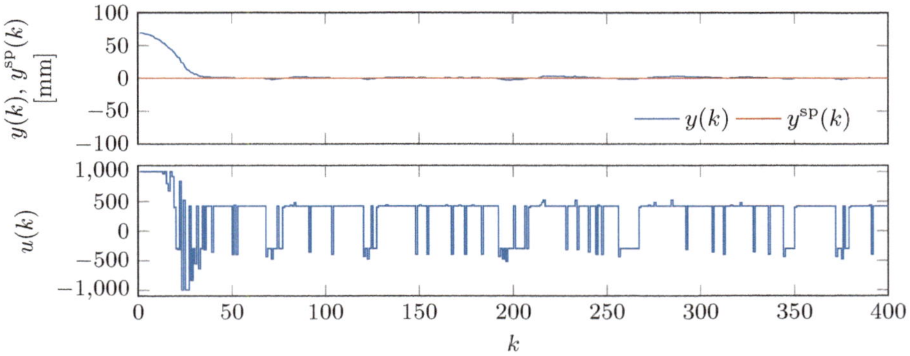

Fig. 7. Control process using PD controller ($K_P = 40$, $T_D = 1$) – $E = 71121$

To mitigate the effect of this phenomenon, manipulated variable, after being determined by the PID algorithm, is further modified. If its value is from range from −300 to 0 or from 0 to 420, its value is set to −300 or 420 respectively. This allow motors to always have an influence on the position of the oil platform model, despite value of manipulated variable as determined by PID. Results of this modification are presented on Fig. 6. Starting from sample 450, absolute value of control error is lower than 2 mm. It has to be noted, that the sudden increase of position at $k = 310$ is a result of image recognition error, and thus erroneous determination of position error.

The aim of further experiment conducted in this part is to find such tuning parameters that result in satisfying overall quality of control. It is worth noting, that all further results are obtained using controllers with implemented dead-band mitigation. On the Fig. 7 a control results for PD controller with tuning parameters $K_P = 40$ and $T_D = 1$ are presented. This configuration allows to achieve set point and stay in its surrounding only after 50 discrete time instants. From this point till the end of the experiment absolute value of position error does not exceed 3 mm. The Deep Water Horizon platform had width of 78 m so it was 312 times bigger than the considered model (25 cm), thus achieved 3 mm control error, would translate in a real off-shore objects to 10 m control error. It stands as a good result for dynamic positioning system.

Finally a full PID controller was tested. Even though process of control is an integrating one, it is worth to see if additional integrating action results in better control quality. Results for tuning parameters $K_P = 30$, $T_I = 10$, $T_D = 1$ are shown on Fig. 8 – no significant change in the quality of control is noticed compared to the previous PD controller.

It has to be added, that the tuning procedure for this model of oil platform is nontrivial due to many physical constraints and nonlinearity of the process of control. Therefore excessive change in one of tuning parameters can cause a major decrease in the control quality e.g. $T_D = 4$ causes overly aggressive

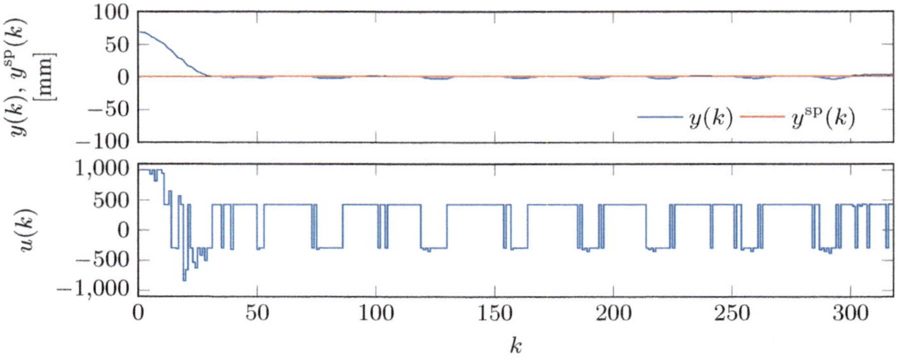

Fig. 8. Control process using PID controller ($K_P = 30$, $T_I = 10$, $T_D = 1$) – $E = 53600$

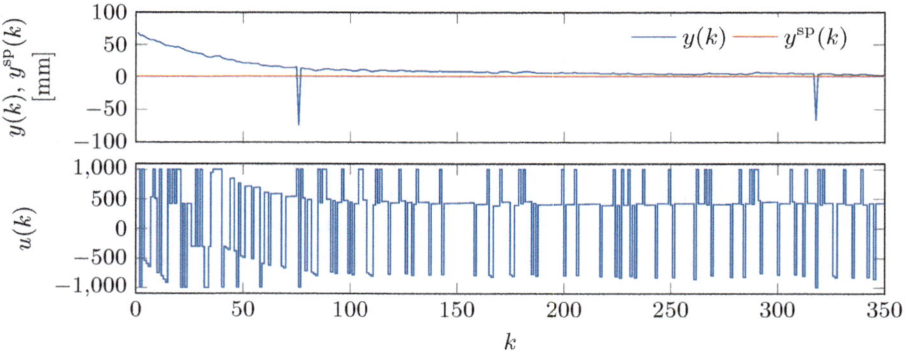

Fig. 9. Control process using PID controller ($K_P = 30$, $T_I = 10$, $T_D = 4$) – $E = 111010$

changes of manipulated variable, and thus lengthens the time required to achieve set point, which is shown on Fig. 9.

Due to the extensive process of PID tuning, only selected results were presented. Based on all considered set of tuning parameters, the lowest sum of squared errors was achieved with the PID controller where $K_P = 30$, $T_I = 10$, $T_D = 1$. This instance of PID controller is also considered the best in terms of the shape of manipulated and controlled variable trajectories, which allows for fast position error elimination, with simultaneous zero overshoot.

Second stage of experiments were focused on disturbance influence on the quality of control. Starting with a position error close to zero, artificially created waves were introduced to the environment, and the reaction was observed and assessed using the same quality indicators. Disturbance in form of waves was generated manually, thus some variation in amplitude and frequency is to be expected. Results of this test are shown on Fig. 10. During the test platform was prevented from maintaining desired position by continues generation of waves

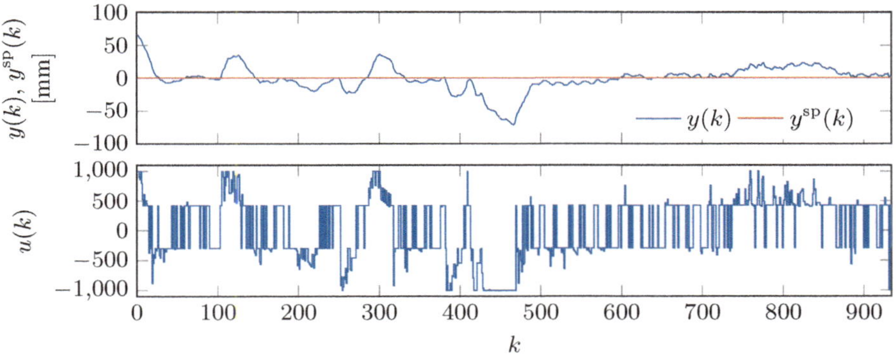

Fig. 10. Control process using PID controller ($K_P = 30, T_I = 10, T_D = 1$) – disturbance influence on control

and sporadically by introducing external force that moves the vessel along the axis in which the dynamic positioning is performed. Unexpected changes in position were introduced in discrete time instants $k = \{105, 185, 250, 382, 413\}$, where the last one introduced the biggest position disruption, leaving the vessel 7 cm away from the reference point. This is reflected by applied control signal, which reached the constraints. During the period from $k = 500$ to $k = 750$ strong waves were generated in the platform's environment. As it was described in the Sect. 2, the construction is highly vulnerable to swinging because of the hull's shape which is placed mostly above the water surface. That swinging causes erroneous readouts, because ARToolkit marker (based on which the position is determined) is placed on top of the platform, and thus moves with each swing, while the bottom of the platform is still in the same position. Results of this phenomenon are clearly visible as oscillations of small amplitude during the aforementioned period.

5 Conclusions

This paper presents physical model of an oil platform, with reference system utilising image recognition to determine current position of the platform and dynamic positioning system based on the PID controller. All tests were performed in natural water environment, which introduced the same disturbances as expected in the real-life application. As it was already mentioned a model presented in this paper can be compared to the Deep Water Horizon platform. In the comparison to the real size off-shore objects control error attained for dynamic positioning system can be treated as satisfactory. Further improvements of this model include consecutive controllers for the other axis and orientation of the platform. Also VRS will be introduced to mitigate the problem of erroneous position determination due to the platform swinging. Lastly advanced control algorithms will be tested and compared to proposed solution.

References

1. Faÿ, H., Delacour, J., Marshall, N.: Dynamic Positioning Systems: Principles, Design, and Applications. Technip, Paris (1990)
2. Gelpke, N.: Oil and gas from the sea. In: Lehmköster, J., Schröder,T. (eds.) World Ocean Review 3 Marine Resources – Opportunities and Risks, chap. 1. Maribus (2014)
3. Jaroś, K., Witkowska, A., Śmierzchalski, R.: Data fusion of GPS sensors using Particle Kalman Filter for ship dynamic positioning system. In: 2017 22nd International Conference on Methods and Models in Automation and Robotics (MMAR), pp. 89–94 (2017)
4. Khan, D., Ullah, S., Rabbi, I.: Factors affecting the design and tracking of ARToolKit markers. Comput. Stand. Interfaces **41**, 56–66 (2015)
5. Küchler, S., Pregizer, C., Eberharter, J.K., Schneider, K., Sawodny, O.: Real-time estimation of a ship's attitude. In: Proceedings of the 2011 American Control Conference, pp. 2411–2416 (2011)
6. Li, C., Gao, J., Wu, P., Feng, L.: Vision-based dynamic positioning system design and experiment for an underwater vehicle. In: 2018 IEEE 8th International Conference on Underwater System Technology: Theory and Applications (USYS)
7. Rogne, R.H., Bryne, T.H., Johansen, T.A., Fossen, T.I.: Fault detection in lever-arm-compensated position reference systems based on nonlinear attitude observers and inertial measurements in dynamic positioning. In: 2016 American Control Conference (ACC), pp. 985–992 (2016)
8. Rogne, R.H., Johansen, T.A., Fossen, T.I.: Observer and IMU-based detection and isolation of faults in position reference systems and gyrocompasses with dual redundancy in dynamic positioning. In: 2014 IEEE Conference on Control Applications (CCA)
9. Rutkowski, G.: Eksploatacja statków dynamicznie pozycjonowanych. Współczesne Technologie Transportu Morskiego, Trademar (2013)
10. Vincenzo, C., Hasnes, G.: Reducing power demand and spikes in dynamic positioning: a model predictive control approach. In: OCEANS 2015 - Genova, pp. 1–5 (2015)
11. Xia, G., Liu, J., Pang, C., Xue, J.: Constrained model predictive control design for dynamic positioning of a supply ship. In: OCEANS 2015 - MTS/IEEE Washington, pp. 1–6 (2015)
12. Xie, D., Han, X., Jia, B., Liu, Y., Zheng, S.: Research on improved PID dynamic positioning system based on nonlinear observer. In: 2018 Chinese Automation Congress (CAC), pp. 698–702 (2018)
13. Xu, L., Liu, Z.: Design of fuzzy PID controller for ship dynamic positioning. In: 2016 Chinese Control and Decision Conference (CCDC), pp. 3130–3135 (2016)

Industrial Systems

Active Power Filter Controller for Harmonics Mitigation of Nonlinear Loads

Krzysztof Kołek(✉)

Department of Automatics and Robotics, AGH University of Science and Technology, Kraków, Poland
kko@agh.edu.pl

Abstract. The paper addresses the problem of the synthesis of a controller which removes harmonics from nonlinear loads connected to the power network. Nonlinear loads, for example containing thyristor circuits, behave in an asymmetrical manner, introduce reactive power and generate harmonic power components, at a frequency that is a multiple of the power grid frequency. While reactive power compensators are designed for the basic grid frequency, the elimination of harmonic components requires tuned circuits designed independently for each harmonic. The article presents the application of an Active Power Filter (APF) for the removal of selected harmonics so that the spectrum of the load current remains within the required limits. Such an approach allows replacing multiple passive filters of a given fixed bandwidth by a single active device, in which compensation bandwidth is tuned in a programming manner. The APF controller is designed as Simulink diagram, which is next applied for the automatic generation of the real-time control application. The experiments show that the amplitude of the selected harmonics is mitigated to a level of 5%–10% of the initial value. This results in a fourfold decrease in the Total Harmonic Distortion index of the load current.

Keywords: Power quality · Active Power Filter · Rapid controller development

1 Introduction

The nonlinear loads are widely applied in high-power industrial applications such as arc furnaces, variable frequency drives, heavy rectifiers, switched-mode power supplies, etc. A nonlinearity is introduced by a switching action and consequently rapid changes of the load current. An example is given in Fig. 1, where a thyristor unit controls power flow to the resistive load. For the power grid the most recommended load current shape should follow the shape of phase voltages, which is not the case for the presented current.

Figure 2 shows the spectrum of the load current. One can notice that beside the basic frequency of 50 Hz there exist other components, of which 250 Hz, 350 Hz and 550 Hz are dominant. The dominant frequencies correspond to the 5th, 7th and 11th harmonic of the grid. As the measure of the distortion of the current, the Total Harmonic Distortion (THD) coefficient can be used. The THD is defined as:

$$\mathrm{THD_N} = \frac{\sqrt{\sum_{i=2}^{N} I_i^2}}{I_1} * 100\% \qquad (1)$$

A. Bartoszewicz et al. (Eds.): PCC 2020, AISC 1196, pp. 243–252, 2020.
https://doi.org/10.1007/978-3-030-50936-1_21

where I_n denotes the RMS of the n-th harmonic of the current. The THD_{11} of the currents presented in Fig. 1 equals to 82.24%. The THD is a kind of global parameter. In addition, some regulations may be specified in the contracts between the grid operator and power consumer, which specify the maximum harmonics level in the relation to the nominal power. Usually the acceptable levels do not exceed a few percent, and are subject to penalty payments when exceeded.

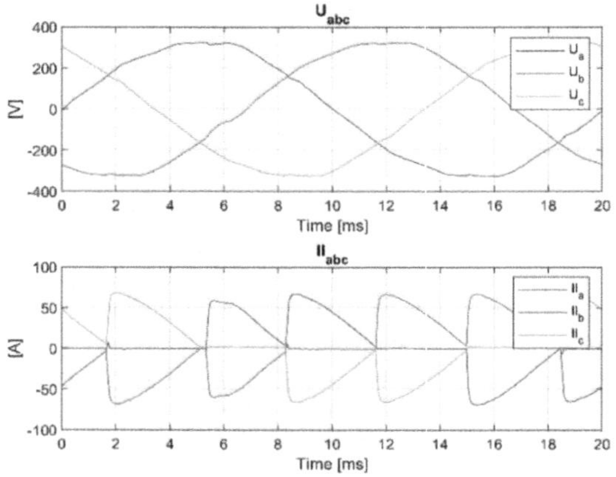

Fig. 1. Nonlinear load: grid voltages (U_a, U_b and U_c) and load currents (Il_a, Il_b and Il_c).

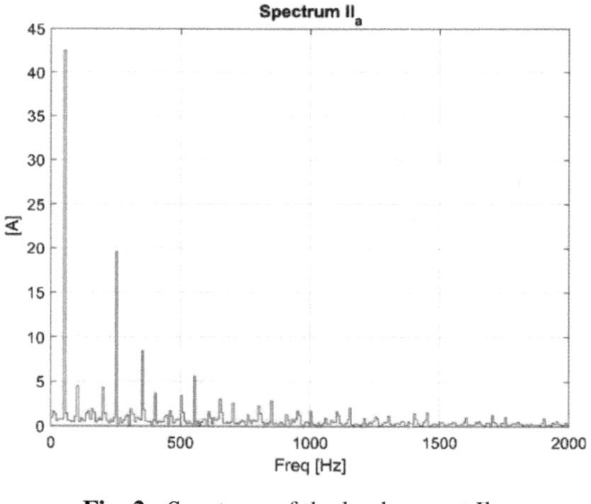

Fig. 2. Spectrum of the load current Il_a.

The load presented in Fig. 1 introduces also reactive power. However, in this paper it is assumed that a passive compensator compensates the reactive power and the controller deals only with the dominant harmonics. Such hybrid configuration is widely applied [1–3] due to the tradeoff between price and flexibility. Fundamental frequency reactive power compensator reduces the power required for active compensation. At the same time, the active compensator can adapt to the load dynamically by changing the compensated frequencies within its nominal power.

The active current compensators operate as Active Power Filter (APF), both in hybrid and shunt configurations [4, 5]. The diagram of 3-wire shunt APF is presented in Fig. 3.

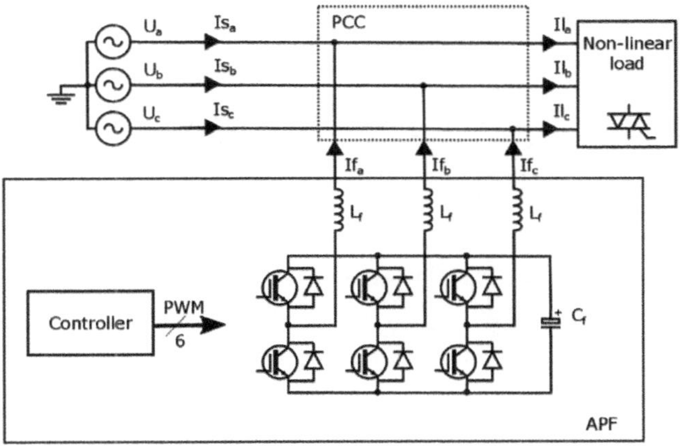

Fig. 3. Shunt active power filter.

The APF contains the Pulse Width Modulated (PWM) Voltage Source Inverter connected to the DC capacitor C_f and reactors L_f at the AC side. The inverter contains six IGBT transistors controller by the PWM waves. The APF and the load are connected at the Point of Common Coupling (PCC). The injection of the APF currents If_a, If_b and If_c to the PCC cancels the selected load current components [6]. The cancellation is complete if the levels of the APF currents and the undesired load components are equal. At the PCC the APF injects the same current as the level of the compensated components, so they disappear from the load of the power grid.

For correct operation of the APF it is also required to keep the constant DC voltage at the C_f capacitor. The DC voltage stabilization is performed by adding an extra APF current, which transfers an active power to the C_f capacitor. The DC voltage stabilization operates as active rectifier and does not introduce adverse current components [7].

The organization of this paper is as follows. Section 2 presents the experimental setup. The APF unit is described, including the description of the APF's control block. Section 3 presents the rapid development path applied for the implementation of the APF's controllers. It focuses on the automatic generation of the real-time APF software. Section 4 contains a description of a real-time experiment. Finally, Sect. 5 gives conclusions.

2 Experimental Setup

The APF setup is presented in Fig. 4. The rated power of the APF is 17 kVA. The filter peak currents are 50 A and the maximum RMS of the currents are 25 A. The filter is connected to the 230 V 50 Hz AC grid.

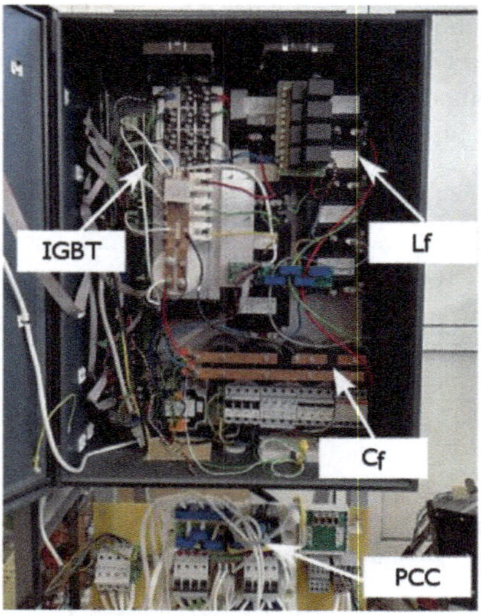

Fig. 4. Active power filter (APF-100/25/3 W – courtesy ELSTA Elektronika, Wieliczka, Poland).

The IGBT transistors are controlled by the PWM signals, which results in ripples of the filter currents. The switching frequency is set to 14.629 kHz. The ripples and the respective levels of the PWM signals are presented in Fig. 5. The filter is designed to keep the ripples below 3.5 A pp.

The APF controller is based on the Xilinx Zynq family of integrated circuits, which include two ARM Cortex A9 processors and reconfigurable FPGA fabric. The Zynq IC is a part of the MicroZed board [8]. The MicroZed is plugged into the carrier board containing A/D converters and an interface to the IGBT drivers. The controller is presented in Fig. 6.

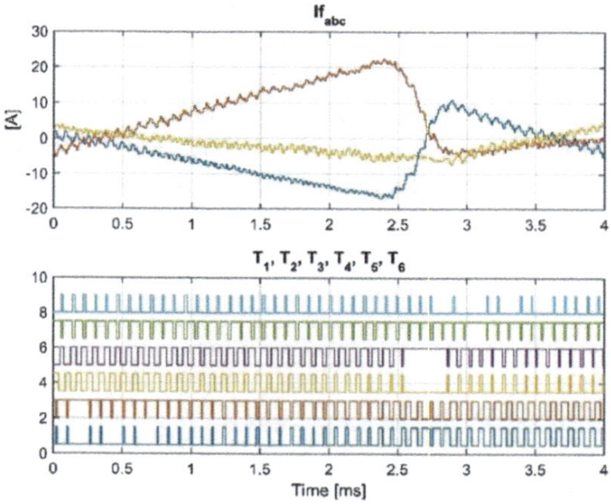

Fig. 5. Ripples of the APF current (sampling 102.4 kHz).

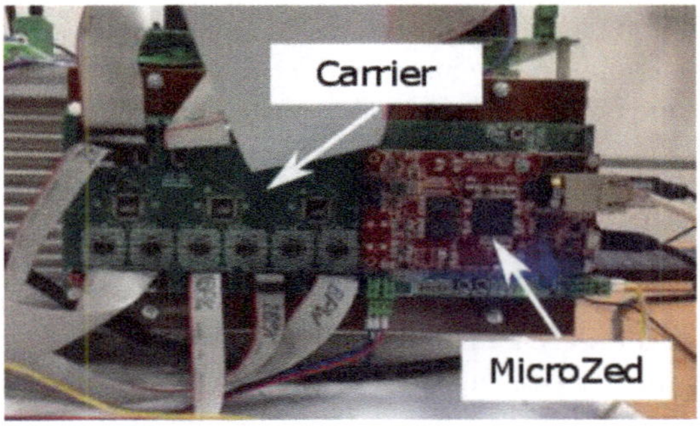

Fig. 6. Controller of the APF.

The most important task is to close the control loop, which consists of measurements, calculation of the reference current and calculation the duty cycle of the PWM signals. This task executes every switching period, which gives approximately 68 ms for calculations. One ARM processor is assigned to run the real-time tasks. It executes a simple Real-Time Operating System (RTOS). FreeRTOS [9] ported to the MicroZed board was applied in this role.

The controller performs tasks with various types of timing requirements. As the Zynq IC contains two processors and an FPGA a natural approach is to split the tasks and perform them on the most suitable platform (see Fig. 7).

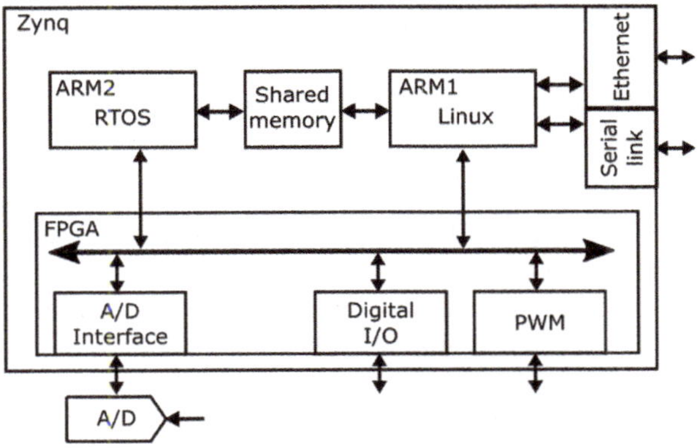

Fig. 7. The architecture of the APF controller.

The second processor runs PetaLinux operating system. It runs all non-real-time tasks like RTOS supervision, parameters tuning, network services (FTP, HTTP, SSH), on-line monitoring and communication with a user and SCADA systems. Asymmetric MultiProcessing (AMP) configuration separates the execution of Linux critical sections from the real-time task, which leads to small jitter level in the execution of real-time tasks. Both processors communicate by a shared memory buffer.

The FPGA unit implements the fastest tasks: interfaces to A/D converters, general-purpose digital signals and PWM generation. The PWM signals are generated with 10-ns precision, which is the unreachable value for processor applications.

3 Rapid Development of the APF Controller

Determination of the harmonic compensation current requires the calculation of the Discrete Fourier Transform (DFT) of the load current for the frequencies corresponding to each selected harmonic (see Fig. 8). For each harmonic, the DFT is calculated for each phase of the load current. New DFT results appear every seven periods of the grid frequency (every 140 ms, which gives 2048 samples) and are used to calculate the amplitudes and phase of each harmonic. The three-phase Phase-Locked Loop (PLL) allows the synchronization of the generated sine signals to the grid. The sum of the sine signals create the three-phase reference filter current $Ifref_{abc}$.

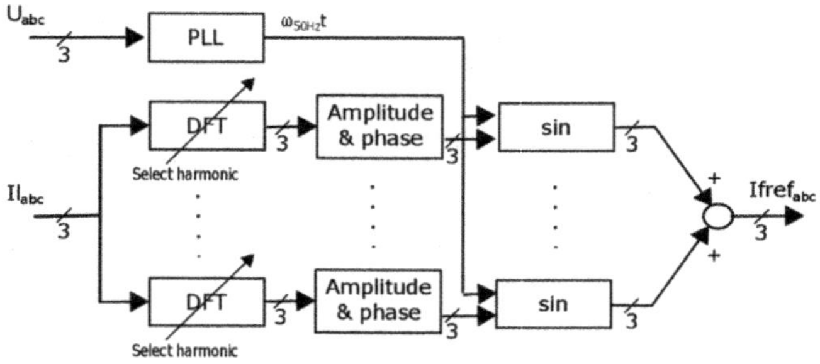

Fig. 8. Calculation of the APF reference current.

The complexity of computing of the DFT for a single spectrum bar is O(n). For the FFT algorithm the complexity is O(n log n), so it is reasonable to apply DFT instead of FFT if only a few harmonics are considered.

The APF controller, including the calculation of the harmonic compensation current, is implemented as a Simulink diagram. The diagram is used for the C-code generation. In order to ensure stable real-time operation, a simple real-time kernel is used – it is FreeRTOS in the presented case (see Fig. 9). After the compilation, a real-time controller is obtained.

The presented approach reflects the model based design paradigm [11]. Model based design focuses on the modelling and validation of the model by a simulation while implementation details are hidden. The design process takes place in a graphic environment offering high abstraction functions. After simulation and validation of the model, a real-time controller, which is functionally equivalent to the model, is automatically generated.

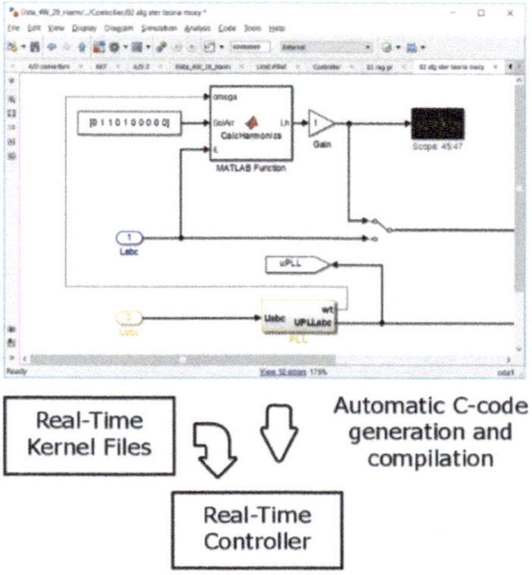

Fig. 9. Rapid controller generation.

4 Real-Time Experiment

To ensure fast APF response to load changes, the controller calculates the duty cycle at the same frequency as the PWM frequency. For the frequency of 14.629 kHz it gives 68.4 ms for calculations. The execution time of the real-time controller generated from the Simulink diagram depends on the number of harmonics selected for compensation. The execution time is 41 ms if harmonics are not calculated and the controller performs only other real-time tasks. The addition of each harmonic increases the execution time by approximately four microseconds. The additional calculation time limits the number of compensated harmonics, as the total time cannot exceed 68.4 ms.

An important parameter of each real-time control system is the punctuality of calculations. The real-time controllers introduce the jitter, which is the deviation from the true periodicity of the presumable constant sampling. A high jitter level can destabilize the system. In the presented AMP architecture, the worst detected deviations are below 400 ns, which is a small level even in comparison with commercial RTOS, where it may exceed a few microseconds. [6, 12].

The compensation algorithm was applied to the load presented in Fig. 1. The algorithm calculated the reference filter currents ($Ifref_a$, $Ifref_b$ and $Ifref_c$) to compensate the 5th, 7th and 11th harmonics. The reference filter currents, filter currents and the grid currents when the compensation algorithm operates are shown in Fig. 10. One can notice that the grid currents look much more similar to sine wave than the original load currents (compare to the bottom diagram in Fig. 1). The selection of the compensated frequencies was made in such a way as to compensate for the strongest harmonics, which in total require the APF current below its nominal value.

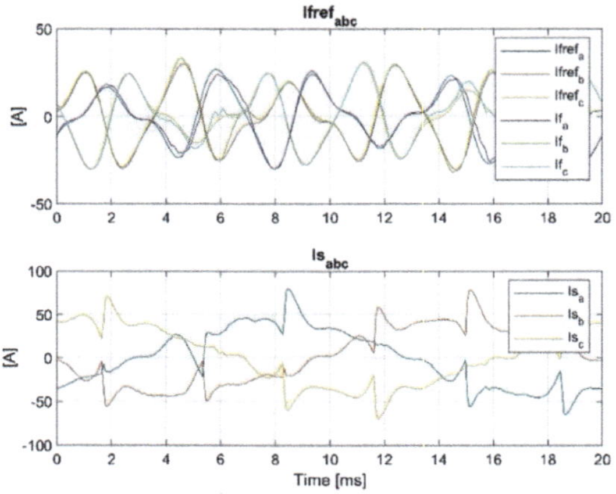

Fig. 10. Compensated current: APF reference current ($Ifref_a$, $Ifref_b$ and $Ifref_c$), APF current (If_a, If_b and If_c) and grid currents (Is_a, Is_b and Is_c).

The spectrum of the compensated current is presented in Fig. 11. Compared to Fig. 2, the frequencies 250 Hz, 350 Hz and 550 Hz are significantly reduced (see Table 1). The mitigation of the dominant harmonics reflects in the THD_{11} level, which for the compensated current lowers from 82.24% to 22.9%.

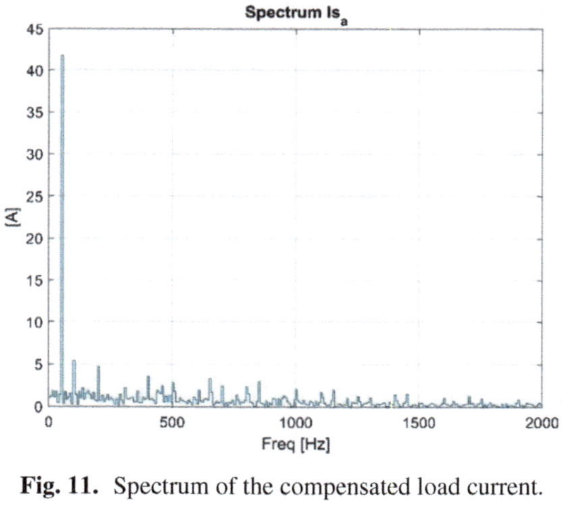

Fig. 11. Spectrum of the compensated load current.

Table 1. Harmonic levels of uncompensated and compensated current.

Harmonic	RMS in load current (Fig. 2)	RMS in compensated load current (Fig. 11)
5th–250 Hz	19.60 A	1.04 A
7th–350 Hz	8.50 A	0.54 A
11th–550 Hz	5.58 A	0.54 A

Beside the frequencies shown in Fig. 11, the APF introduces components related to the switching frequency (see ripples in Fig. 5). The RMS values for frequencies 14.629 kHz and 29.258 kHz are around 1.5 A.

5 Conclusions

The presented application of APF allows effective mitigation of the selected harmonics of the load current. This operation mode focuses only on harmonics, skipping the compensation of reactive power and asymmetry. It is reasonable as long as harmonic compensation can be made within the nominal power of the APF, and compensation of all components would require higher powers.

An important feature is rapid code generation, directly from Simulink diagrams. It allows the quick implementation of new control strategies - if only someone develops a new controller in the form of a diagram, its experimental verification is carried out within a couple of minutes. The presented methodology allows creating real-time controllers that operate in real-time with frequencies of several kHz. The samplings exceeding 20 kHz were tested successfully, still keeping low jitter level.

The extraction of each harmonic lasts approximately four microseconds, which is a significant limitation. Because the total calculation time cannot exceed the sampling period, it limits the number of compensated harmonics to four-five. This limitation can be eliminated by implementing the compensation algorithm (Fig. 8) as a module in the FPGA. This module will fulfill the coprocessor's role, offloading the main processor.

References

1. Motta, L., Faundes, N.: Active/passive harmonic filters: applications, challenges & trends. In: 17th International Conference on Harmonics and Quality of Power (ICHQP) (2016)
2. Kedra, B.: Comparison of an active and hybrid power filter devices. In: 16th International Conference on Harmonics and Quality of Power (ICHQP) (2014)
3. Dekka, A.R., Beig, A.R., Poshtan, M.: Comparison of passive and active power filters in oil drilling rigs. In: 11th International Conference on Electrical Power Quality and Utilisation (2011)
4. Thekkath, P., Prabha, S.: Adaptive hysteresis current controlled Shunt Active Power Filter for power quality enhancement. In: 2013 International Conference on Circuits, Power and Computing Technologies (ICCPCT) (2013). https://doi.org/10.1109/iccpct.2013.6528964

5. Cleary-Balderas, A., Senior, A.M., Cruz-Hernendez, O.: Hybrid active power filter based on the IRP theory for harmonic current mitigation. In: 2016 IEEE International Autumn Meeting on Power, Electronics and Computing (ROPEC) (2016). https://doi.org/10.1109/ropec.2016.7830608
6. Firlit, A., Kolek, K., Piatek, K.: Heterogeneous active power filter controller. In: 2017 International Symposium ELMAR (2017). https://doi.org/10.23919/elmar.2017.8124477
7. Bai, L., Wang, D., Zhou, Y.: Study on the current control strategy of hybrid rectifier based on the PR controller. In: 2013 International Conference on Electrical Machines and Systems (ICEMS) (2013). https://doi.org/10.1109/icems.2013.6713346
8. MicroZed. http://zedboard.org/product/microzed. Accessed 23 Dec 2019
9. The FreeRTOS kernel. https://www.freertos.org/. Accessed 23 Dec 2019
10. PetaLinux Tools. https://www.xilinx.com/products/design-tools/embedded-software/petalinux-sdk.html. Accessed 23 Dec 2019
11. Kołek, K., Piątek, K.: Rapid algorithm prototyping and implementation for power quality measurement. EURASIP J. Adv. Signal Process. **2015**(1), 19 (2015). https://doi.org/10.1186/s13634-015-0192-3
12. A Real Time Operating Systems (RTOS) Comparison. (n.d.). http://citeseerx.ist.psu.edu/viewdoc/summary?doi=10.1.1.584.2009. Accessed 23 Dec 2019

Intelligent Temperature and Vacuum Pressure Control System for a Thermionic Energy Converter

Bartosz Kania(✉), Dariusz Kuś, Piotr Warda, and Jarosław Sikora

Lublin University of Technology, 20-618 Lublin, Poland
b.kania@pollub.pl

Abstract. For a vacuum thermionic energy converter with a dispenser cathode, the cathode temperature affects the vacuum pressure level, especially in the cathode activation process. Adding an adaptive setting system of temporary reference temperature value due to pressure level to a typical digital PID controller ensures the control of the TEC cathode temperature while maintaining the required vacuum pressure level at any time. The results of the cathode temperature tests in the range from 340 K to 1273 K show that at any time the pressure value in a TEC vacuum chamber is kept within the specified range. This work describes dispenser cathode requirements, the adaptive algorithm, its software and hardware implementation and controller test results. The original intelligent controller is highly suitable for powering the dispenser cathode of the thermionic energy converter both during its activation and in standard operation.

Keywords: Intelligent controller · Thermionic energy converter · Dispenser cathode · Temperature · Vacuum pressure

1 Introduction

A thermionic energy converter (TEC) is a static device converting directly heat into electrical energy. A conversion method is based on thermionic emission [1]. This idea was first proposed by W. Schlichter in 1915 [2]. TECs have been studied over a long time, significant development occurred in the 1950s and 1960s, but was limited by the technologies of the time. Dynamic development began in the 2000s due to a new vacuum microsystems technology [3] and new materials with a low work function. TECs can be divided into vacuum, with positive cesium ions and with positive ions of an inert gas such as argon, generated from an auxiliary discharge. A simplified diagram of the vacuum TEC is shown in Fig. 1. Electrodes are placed in a vacuum. The cathode is thermally connected with a heat source, the anode with a heat receiver. The colours of the electrodes reflect the

The original version of this chapter was revised: The information provided in the abstract has been corrected. The correction to this chapter is available at https://doi.org/10.1007/978-3-030-50936-1_130

A. Bartoszewicz et al. (Eds.): PCC 2020, AISC 1196, pp. 253–263, 2020.
https://doi.org/10.1007/978-3-030-50936-1_22

relationship between their temperatures. For TECs, the cathode work function should be higher than the anode. The electrons emitted from the heated cathode are collected on the anode and return to the cathode through a load resistor R. A back emission from the anode is usually relatively low. The output voltage and current can be determined based on an electrostatic potential distribution in an inter-electrode gap [4–6].

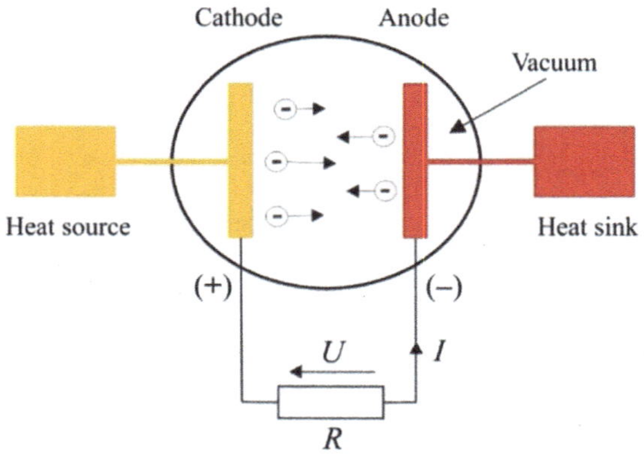

Fig. 1. A simplified diagram of the vacuum thermionic energy converter.

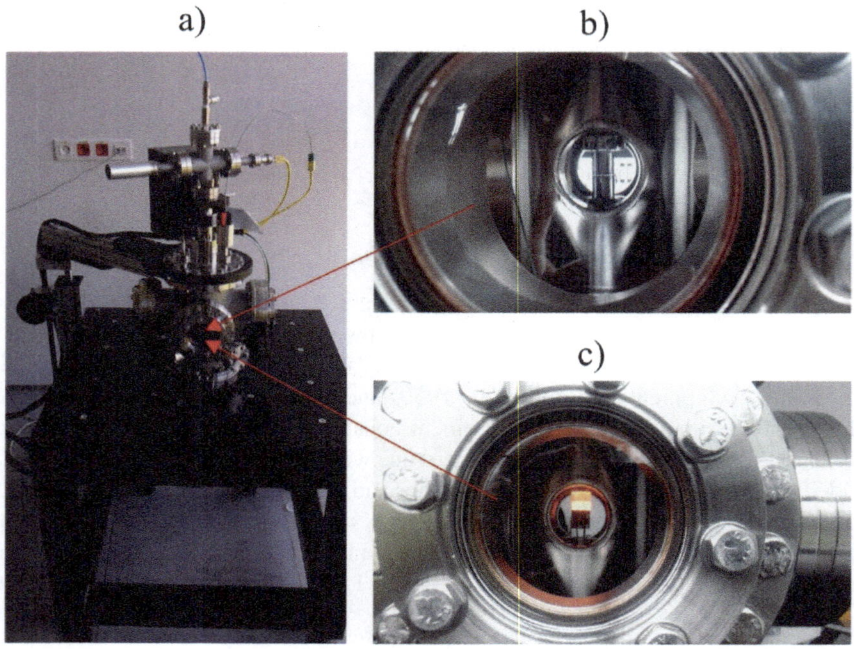

Fig. 2. Experimental thermionic energy converter (a), arrangement of electrodes (b), electrodes during the operation (c).

In our laboratory we have built an experimental thermionic energy converter with a precisely adjustable inter-electrode gap, in order to verify model test results of converter static characteristics. It consists of a table mounted vacuum chamber with viewfinders

(ITL Vacuum Components), set of scroll pump (Edwards nXDS6i) and turbomolecular pump (Leybold TurboVac 350i), electric vacuum-to-ambient couplers, z-axis manipulator and electrodes liquid cooled handlers. The general appearance of the device is shown in Fig. 2a.

The pre-vacuum is produced by a scroll pump, a high vacuum by a turbomolecular one with a constant pumping speed (6 m^3h^{-1}). The arrangement of the electrodes is shown in Fig. 2b and the electrodes during the operation in Fig. 2c. As an emitter we used a dispenser cathode integrated with an electrical tungsten micro-heater (HeatWave Labs). Its work function is equal to 2.01 eV at the temperature of 1450 K.

For the dispenser cathode the vacuum pressure level should be less than $1{,}33 \cdot 10^{-4}$ Pa. During cathode operation, especially in the cathode activation process, an increase in the cathode temperature causes an increase in the pressure level in the vacuum chamber (see text in next section). Hence, the supply of electrical power to the heater and temperature control should be combined with monitoring the vacuum pressure level.

Temperature controllers are widely applied in scientific and industrial apparatus, for example, measurements systems [7], radiant heating atomic layer deposition reactors [8] and energy harvesting systems [9]. Advanced controllers are based on GPC-based closed-loop control [8]. To control nonlinear temperature systems with long-time delay, a time-varying effective particle swarm optimisation algorithm is used [10]. However, these designs cannot be directly applied in control of the cathode temperature and the temperature-dependent pressure level of vacuum chamber, because they don't take into account the pressure value that needs to be controlled in order to prevent cathode damage.

This work describes an original intelligent control system of the cathode temperature and temperature-dependent vacuum pressure level at constant pumping speed. Basing on the analysis of physical phenomena in a TECs vacuum chamber, we designed a system using adaptive setting of a temporary reference value of temperature due to the pressure level. Owing to that, the TEC cathode can be heated to the final temperature, while maintaining the required, user-setable pressure level at any time. The adaptive setting of the temporary reference temperature algorithm in LabVIEW software has been implemented.

2 Design Considerations

Generally, the most common commercial temperature controller is able to control temperature of the TEC cathode, however in our case there barium tungsten dispenser cathode is used [11], so it has special requirements concerning the heating process that cannot be ignored. Before use, the cathode must be baked out and activated, when barium oxide is converted into free barium on the cathode surface. In order to protect the cathode from pollution from the surrounding structure or from the products during bakeout, maintaining a required vacuum level in the vacuum chamber is necessary. Also, the cathode can be permanently polluted by moisture. If a cathode is heated at such a rate that the moisture cannot escape, hydroxides and carbonates can form. They not only reduce emission capabilities, but also cause damage of the tungsten cathode surface. To prevent this, the cathode must be allowed to soak at 200–400 °C long enough to allow complete the water vapour outgassing. The vacuum pressure is probably the best indicator of the gassing rate. Therefore the pressure in the vacuum chamber must be less than $1{,}33 \cdot 10^{-4}$ Pa at all time. When power is applied to the cathode heater after air exposition, the temperature will rise and the pressure will increase due to outgassing. If the pressure goes above $1{,}33 \cdot 10^{-4}$ Pa, there is need to back off the power to the heater until the pressure falls down below this level. Activation after vacuum chamber outgassing and baking out is achieved by converting the barium oxide in the tungsten matrix into free barium on the surface of the cathode. The rate of activation is a function of vacuum chamber cleanliness, cathode pollution, time and temperature. In general, the cathode is activated at or slightly above the operating temperature. The cathode temperature should never exceed 1473 K [12].

Therefore a standard temperature controller must be developed with an algorithm that takes into account heating requirements. We used a commercial temperature controller from HeatWave Labs and an author's software implemented on a PXI unit from National Instruments.

The control system presented here was developed to work with a half-inch diameter cylindrical TEC cathode [11], whose temperature is controlled. The temperature range of the designed control system is up to 1473 K, and the pressure from $1 \cdot 10^{-7}$ to $1 \cdot 10^{-5}$ Pa. The cathode is heated by alternate current in order to avoid constant magnetic field that could affect the TEC efficiency. Figure 3 depicts the concept of the designed temperature and vacuum pressure control system.

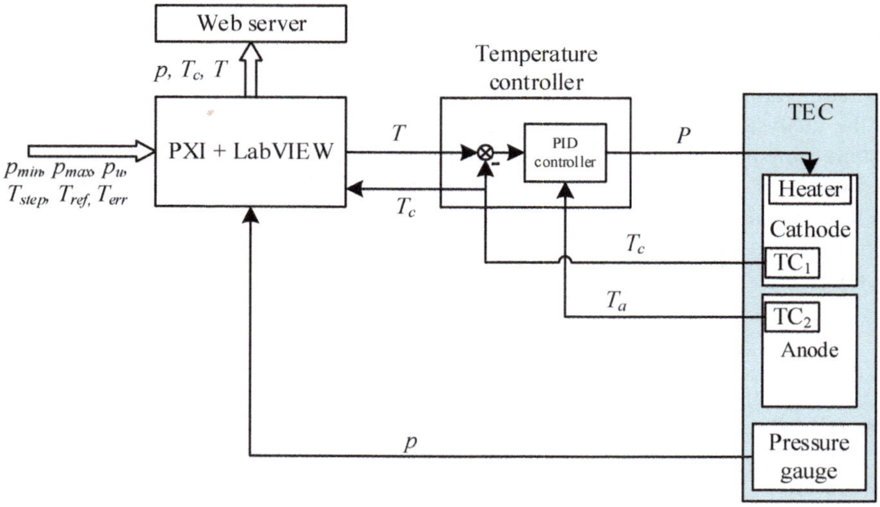

Fig. 3. Design of the temperature and vacuum pressure control system.

The PXI is connected to the temperature controller by the standard serial port to exchange values of important parameters – a temporary reference temperature T and actual temperature of the cathode T_c. It also allows to set up and monitor the controller. The temperature controller includes an industrial digital PID controller from Watlow Corp. with a programmable overheating protection feature, a thermocouple module and a power stage, which make it's a complete temperature controller. The driving output, supplying power P, is connected to the heater embedded in the cathode. The feedback signal T_c is acquired from the thermocouple (TC_1) type K attached to the molybdenum body of the cathode. The signal of the anode temperature T_a from type K thermocouple (TC_2) is also connected to the temperature controller and is used to realise overheating protection. Inside the vacuum chamber a pressure gauge is mounted. The signal of the measured pressure is connected to the PXI unit by the an USB interface. Reference values such as low pressure level p_{min}, high pressure level p_{max}, ultimate pressure level p_u, temperature change step T_{step}, maximum temperature error T_{err} and reference temperature T_{ref}, are set by an user interface. The most important data about baking out, activation, and further work are logged to a web server and it is possible to remotely check the actual parameters of the thermionic energy converter.

3 Hardware

The tungsten dispenser M612 type cathode from HeatWave Labs used in our TEC consists of a porous tungsten matrix impregnated with a mix of barium calcium aluminate. After mounting into the vacuum chamber it has to be activated by heating to achieve high emission properties (about 3 A/cm^2 current density). Such cathodes can be exposed to air and reactivated repeatedly, they are more tolerant to pollution than oxide cathodes and offer more than 40,000 h of life at 3 A/cm^2. A typical operation temperature is 900–1200 °C. Their compact construction results in lower heating power requirements. Moreover, they have aluminium oxide insulation and a non-gassy molybdenum body [13]. The micro-heater is embedded into cathode and does not cause a significant delay in the cathode temperature control process.

The temperature controller is a phase angle fire unit with a power supply. It also offers serial communication (RS-485), a circuit breaker, short-circuit, over-current, open loop and over-temperature protections, as well as a type K thermocouple compensating input. The Watlow PID controller included offers autotuning, alarm mode, internal user-setable high temperature limit and a soft start feature [14].

The PXI is a configurable unit which in our case consists of a PXIe-1078 chassis and a PXIe-8840 Quad Core embedded controller which is a high-performance, compact embedded computer and come with standard features such as an integrated CPU, hard drive, RAM, Ethernet, video, keyboard/mouse, serial, USB, and other peripheral I/Os. There are also installed some modules that will be used to realise future tasks not connected with the temperature control.

The pressure gauge consist of a compact full range Pirani/Bayard-Alpert gauge type PBR 260 and a TPG 361 gauge controller from Pfeiffer Vacuum. It offers a measuring range from $5{\bullet}10^{-8}$ to $1{\bullet}10^5$ Pa, $\pm15\%$ accuracy and 5% repeatability in the range from $1{\bullet}10^{-6}$ Pa to 1 Pa, two configurable relays, connectivity by USB, RS-485 or Ethernet and direct data storage on USB flash memory.

4 Software

The aim of the application is to perform automatic heating of the TEC cathode from the current temperature T_c to the required final temperature T_{ref}. The second monitored parameter is the pressure p in the TEC vacuum chamber. Due to the possibilities of data processing and flexibility of modification of the created code, the LabVIEW programming environment was chosen to develop the program.

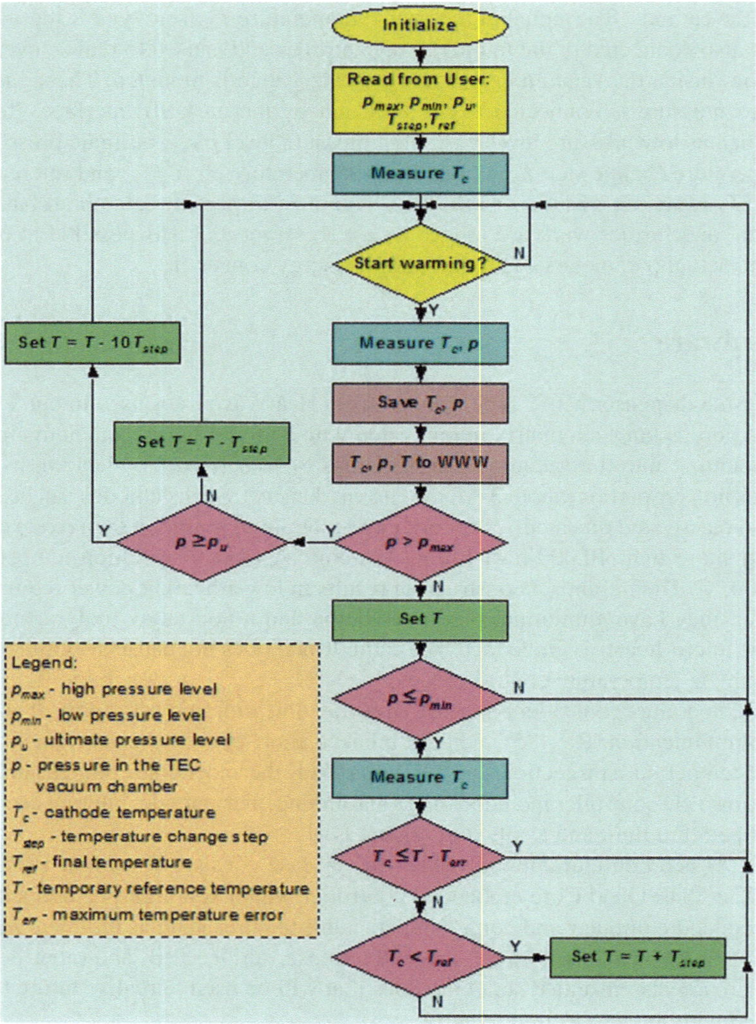

Fig. 4. The control algorithm of the temperature and the vacuum pressure level.

It was assumed that the program realises the control algorithm every second, while the temperature is changed by a step T_{step} of 1 K. The user has possibilities to change both parameters. The operation of the application is shown in Fig. 4.

After initialisation, the values are set by the user: high pressure level p_{max}, low pressure level p_{min}, ultimate pressure level p_u, final temperature T_{ref}, temperature change step T_{step}. Preliminary measurement of the T_c cathode temperature is also made. The system enters the waiting mode for pressing the "Run Warming" button on the front panel of the application, which causes the start of the cathode and the warm–up procedure.

After starting the heating process, the program controls the parameters of the TEC. The pressure p and temperature T_c are read sequentially, after which both values are saved in a text file with the current date and time. Thanks to the information on the system operation, despite the long process of heating the cathode, it is possible to reconstruct the entire heating cycle from the initial moment to the end of work. The name of the file, which will contain the saved data and their location on the hard drive is left to the user's choice. The creation of the data set is shown in Fig. 5.

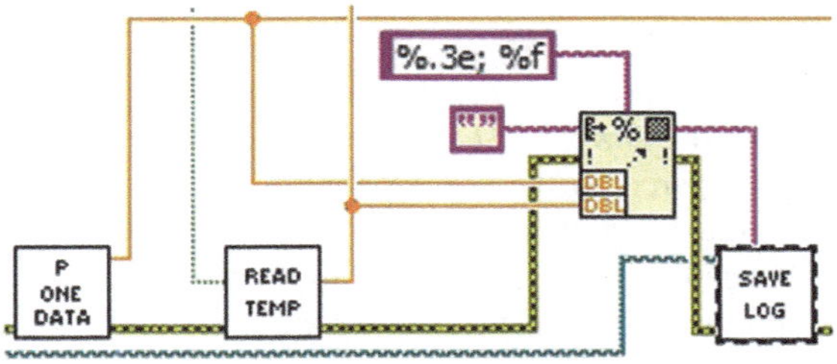

Fig. 5. Part of the diagram, which controls and saves current values of the cathode temperature and vacuum pressure.

In addition, the current pressure p, the temperature T_c and the temporary reference temperature T are transferred to an external application for placement on the website. Thanks to this, it is possible to monitor the work condition of the TEC, the operation of the control system at anytime from anywhere. There is also the possibility of conducting experiments by external entities.

Heating the cathode causes an increase in pressure in the vacuum chamber. For this reason, the initial check has three stages. The first is to check whether there has been a rise in pressure above the high pressure level p_{max}. After the occurrence of such an event, a second step follows, assessing how much the high pressure level p_{max} has been exceeded. If the increase has exceeded the value of ultimate pressure level p_u, the set temperature is reduced by ten times T_{step}. If the pressure does not exceed the ultimate value, then the set temperature is reduced by the value of T_{step}. The implementation of the above fragment of the algorithm is shown in Fig. 6. In addition, the diagram part shows the record of the temporary reference temperature to a local variable and the

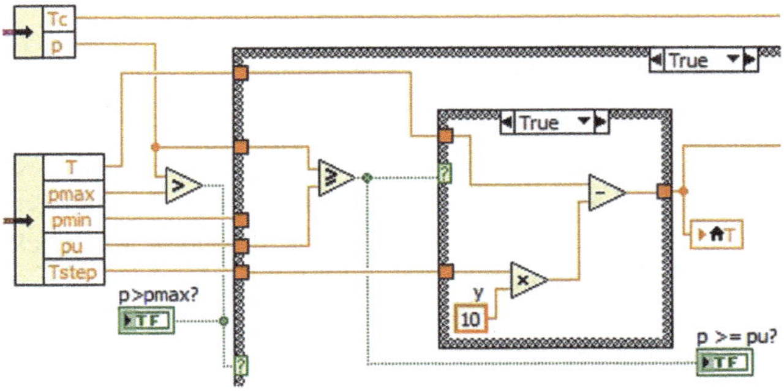

Fig. 6. Part of the diagram implementing pressure control.

logical indicators informing about the relations between the measured pressure p, high pressure level p_{max} and ultimate pressure level p_u.

If the measured pressure does not exceed any of the limit values, the relation between the measured pressure p and low pressure level p_{min} is checked. The application sets the T temperature to the value set in the previous iteration of the program. If the lower pressure is not exceeded, then the transition to the current cathode temperature control T_c takes place, otherwise the program will be waiting for the decreasing of the pressure p below low pressure level p_{min}.

The maximum temperature error T_{err} has an important role in temperature control. The difference between the currently set T temperature and the maximum temperature error T_{err} value is compared to the one currently read from the cathode. If the T_c temperature does not reach a higher value than the calculated difference, the program begins to wait for the cathode to fully heat up monitoring the pressure in the vacuum chamber. The heating process is suspended. In the event that the temperature T_c has reached the calculated difference, it is checked whether the measured temperature T_c is less than the final T_{ref}. If this is not the case, the program stops heating and proceeds to check the pressure and wait for temperature stabilization T_c. If the read temperature of the vacuum chamber is lower than the final one, the program proceeds to setting the new value of temperature T to be reached by the cathode. It is the sum of the T temperature and the T_{step} value. Figure 7 shows a part of the program diagram that increases the temperature T. The SET TEMP subprogram checks the relation between temporary reference value and the cathode temperature. The result of the operation is setting the temperature in the system, additionally passing information to the program about the correctness of the set temperature - green connection to the "$T \gg T_c + T_{step}$" node.

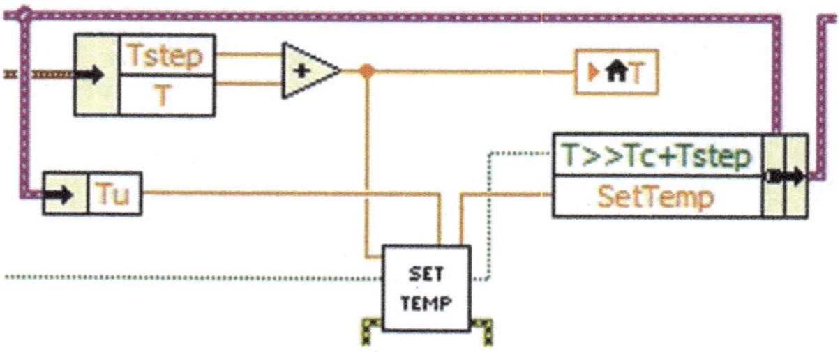

Fig. 7. Part of the diagram, which sets the temporary reference temperature.

The program works in a continuous mode, to reach and maintain temperature T_{ref}.

The application panel (see Fig. 8) allows to set and control all necessary parameters. The "TEST" tab is used to enter the program in the test mode for checking the correct operation. The "Config" tab allows to change the time interval between successive heating cycles and to declare the file name for saving the current TEC parameter's values. In addition, the program informs about the value of the time interval between successive measurements and it signals the moment of starting the heating cycle. A "Save log" switch allows to start logging. At the bottom of the application panel there are indicators of the current values of the pressure p and the temperature T_c.

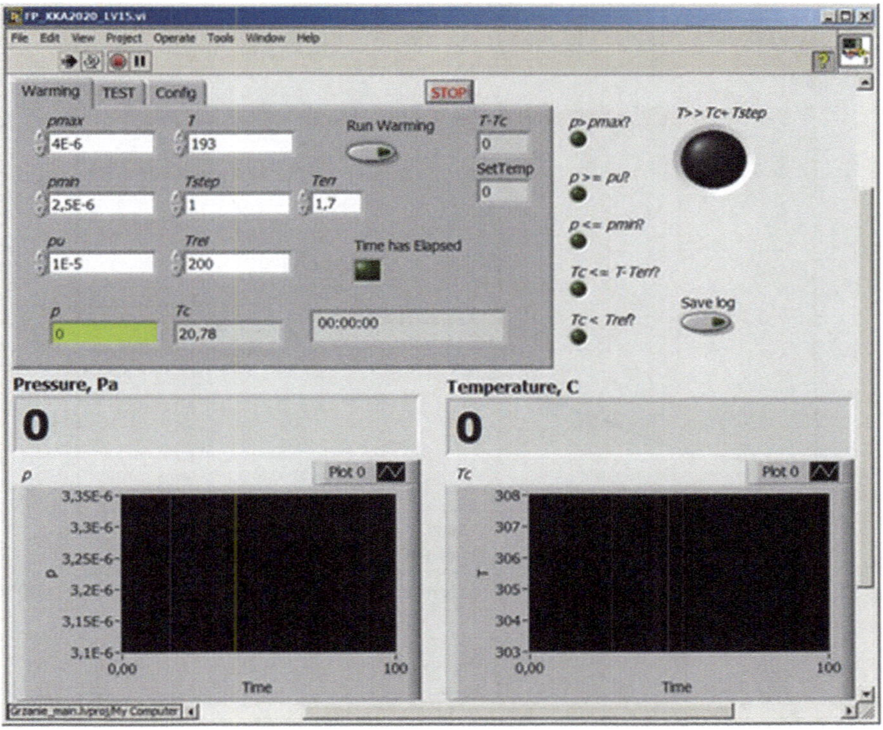

Fig. 8. Panel of the program implementing the heating process of the TEC cathode.

5 Results and Conclusions

The M612 cathode was heated in a programmed temperature range at a sufficiently low vacuum pressure. In the experiment PID parameters were selected manually, the values $p_{min} = 5{\cdot}10^{-6}$ Pa, $p_{max} = 6{\cdot}10^{-6}$ Pa and $p_u = 1{\cdot}10^{-5}$ Pa were adopted for the final temperature at the middle activation temperature. The vacuum pressure p and the cathode temperature T_c during the heating from 67 to 1000 °C is shown in the Fig. 9. Detailed part including the temporary reference temperature T is presented in Fig. 10.

A steady state error in the temperature signal can be observed. It results from principle of operation of Watlow PID controller and it is present for any temperature set. However the error never exceeds 1 K and it doesn't involve the control process. Obtained results show that small change of the cathode temperature cause significant change of the vacuum pressure. This is especially clear during start of heating and in range between 200 and 300 °C when intensive outgassing takes place. Moreover, rate of heating depends on the cathode temperature.

The obtained results confirm the proper operation of the intelligent temperature and vacuum pressure control system for a thermionic energy converter. The system carries out the process of heating and activation of the dispenser cathode in an automatic and maintenance-free manner while maintaining the pressure requirements imposed by

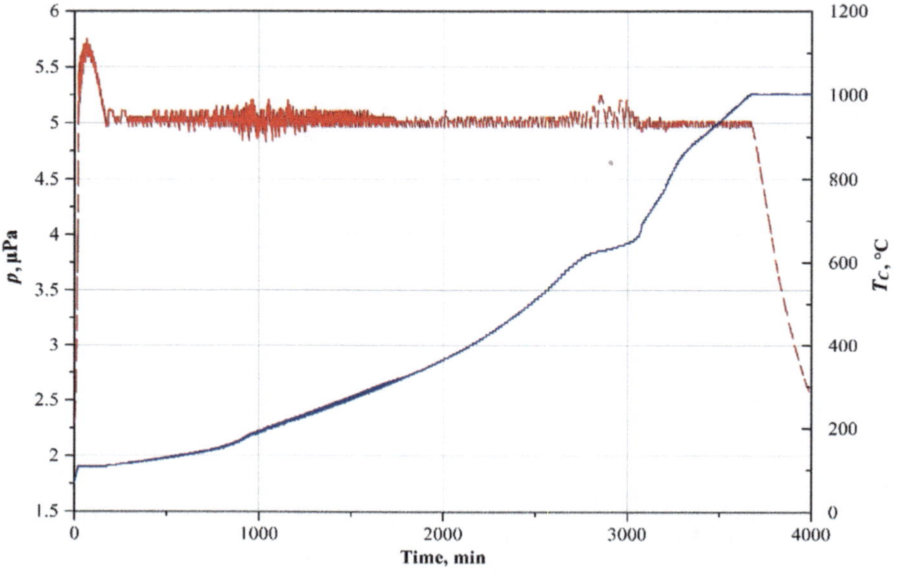

Fig. 9. The vacuum pressure p, and the cathode temperature T_C during heating from 67 to 1000 °C.

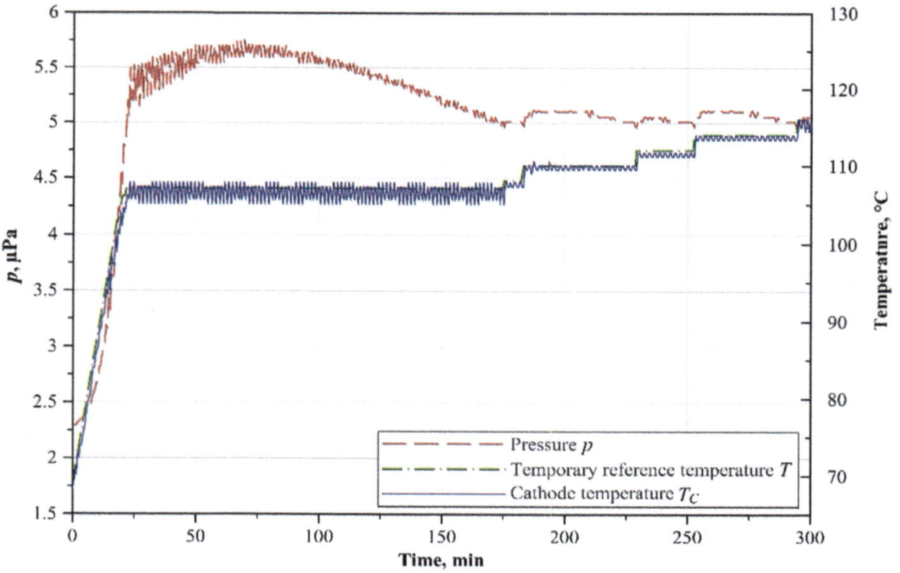

Fig. 10. The vacuum pressure p, the temporary reference temperature T and the cathode temperature T_C during heating, detailed part.

the cathode manufacturer. This promotes a high thermionic emission quality, protects cathode from damage and extends cathode lifetime. Such control system can be also used with other types of the dispenser cathodes as well.

Acknowledgements. Financial support from the Polish Ministry of Science and Higher Education No. 6557/IA/SP/2016 is gratefully acknowledged.

References

1. Richardson, O.W.: (Thermionic) Emission From Hot Bodies. Wexford College Press, Wexford (2003)
2. Schlichter, W.: Die spontane Elektronenemission glühender Metalle und das glühelektrische Element. Ann. Phys. **352**, 573–640 (1915)
3. Howe, R.T.: Vacuum microsystems for energy conversion and other applications. In: 2011 16th International Solid-State Sensors, Actuators and Microsystems Conference, pp. 7–11 (2011)
4. Hatsopoulos, G.N., Gyftopoulos, E.P.: Thermionic Energy Conversion, vol. 1. MIT Press, Cambridge (1973)
5. Lim, I.T., Lambert, S.A., Vay, J.-L., Schwede, J.W.: Electron reflection in thermionic energy converters. Appl. Phys. Lett. **112**, 5 p. (2018). Article no. 073906
6. Sikora, J.: Thermionic Electron Emission Sources: Biasing Conditions [in Polish]. Lublin University of Technology Publishing House, Lublin (2019)
7. Yang, D., Woo, J.-K., Lee, S., Mitchell, J., Challoner, A.D., Najafi, K.: A micro oven-control system for inertial sensors. J. Microelectromechan. Syst. **26**, 507–518 (2017)
8. He, W.-J., Zhang, H.-T., Chen, Z., Chu, B., Cao, K., Shan, B., Chen, R.: Generalized predictive control of temperature on an atomic layer deposition reactor. IEEE Trans. Control Syst. Technol. **23**, 2408–2415 (2015)
9. Kwan, T.H., Wu, X., Yao, Q.: Complete implementation of the combined TEG-TEC temperature control and energy harvesting system. Control Eng. Pract. **95**, 11 p. (2020). Article no. 104224
10. Zhang, Q., Wu, S., Li, Q.: A PSO identification algorithm for temperature adaptive adjustment system. In: 2015 IEEE International Conference on Industrial Engineering and Engineering Management (IEEM), pp. 752–755 (2015)
11. HeatWave Labs, Inc.: Technical Bulletin TB-198 Standard Series Barium Tungsten Dispenser Cathodes (2002). https://www.cathode.com/pdf/tb-198.pdf
12. HeatWave Labs, Inc.: Technical Bulletin TB-147 Processing of Dispenser Cathodes (2004). https://www.cathode.com/pdf/tb-147.pdf
13. HeatWave Labs, Inc.: Technical Bulletin TB-128 Notes on Dispenser and Oxide Cathodes (1999). https://www.cathode.com/pdf/tb-128.pdf
14. HeatWave Labs, Inc.: HeatWave Labs Model 101303-22 Temperature Controller User's Manual. 23

Central Heating Energy Saving Strategies for a Public Building

Krzysztof Kołek(✉)

Department of Automatics and Robotics, AGH University of Science and Technology, Kraków, Poland
kko@agh.edu.pl

Abstract. The paper addresses the problem of saving central heating energy in public buildings. Public buildings operate only on business days during working hours and only at this time it is necessary to guarantee thermal comfort inside the rooms. Outside of business hours, the room temperature can be reduced resulting in a reduction of the heat energy consumed by the building. The article presents two strategies for reducing the energy consumption: globally lowering the supply temperature of central heating system and locally decreasing the room temperatures. The results of real experiments are presented together with estimates of how the described strategies affect energy saving. Depending on the required operating conditions of the building, the results of the experiments show the possibility of daily savings from 1 GJ to 5 GJ of energy. During the 22 days of experiments, the estimated energy savings were about 17% of the total energy supplied to the building.

Keywords: Heat supply of rooms · Thermal energy · Experiment · Energy saving · Building temperature control system · Building automation

1 Introduction

Public buildings, like offices, schools and health clinics, operate only on business days during working hours. In the case of central heating, it is necessary to maintain a temperature that guaranties thermal comfort during working hours. The temperature of thermal comfort, depending on the preferences of the users of the rooms is between 20 °C and 24 °C. One can put the question of whether it is possible to reduce the consumption of thermal energy, resulting reduction in costs, by lowering the temperature of the rooms outside working hours and reproducing the temperature of thermal comfort for the time of beginning the work period.

The implementation of temperature reductions requires mathematical models of the rooms. It is required to determine the conditions for restoring thermal comfort after periods of temperature reduction. In turn, estimating energy savings requires a mathematical model of the entire building. Based on the building model, one can estimate the hypothetical amount of energy consumed in the absence of temperature reductions.

A. Bartoszewicz et al. (Eds.): PCC 2020, AISC 1196, pp. 264–274, 2020.
https://doi.org/10.1007/978-3-030-50936-1_23

The problem of controlling the central heating installation is raised in many papers. In [1] authors describe the architecture design for monitoring and controlling the system of central heating. In [2] central heating control in a small residence is considered. Temperature control implementation is investigated utilizing various controllers: classic, fuzzy [3, 4], iterative learning control [5–7] and model predictive control [8, 9]. In [10] authors present the simulation result considering the guarantee of the thermal comfort in the heated space by controlling the temperature in the heating installation of a non-residential building. Numerous studies were also carried out on temperature modelling in individual rooms: using physical models [11–13], the finite element methods [14], non-linear models [15] and power spectral density method for dynamically modelling of indoor air temperature response [16].

The disadvantage of the available researches is the fact that they either describe simulation tests or experimental tests, but performed for a small number of rooms. The paper presents experimental results performed in the test building. In the building there is a school for about 800 students with a small boarding school for 32 students. The area of the building is about 6000 m^2. The building contains about 70 rooms, heated by about 220 radiators. Electronic radio thermostatic heads have been installed on about 170 radiators to allow individual temperature control in the rooms. The central heating system is supplied from the municipal heating system and the maximum power of the building is 250 kW.

The paper is structured in six sections. Section 3 presents two energy saving strategies. Next section presents the model of single rooms. Section 4 describes how the energy consumed by the building depends on the outdoor temperature. The results of the experiments are presented in Sect. 5. Finally, Sect. 6 gives conclusions.

2 Energy Saving Strategies

Control of heat energy consumption can be done in two ways:

- Globally, by changing the supply temperature of the central heating installation. A PLC controller is installed in the building's heat center to maintain the supply temperature T_{sup} depending on the outside temperature T_o following the formula (1). The T_{ref} value is the source of the setpoint for the T_{sup} controller.

$$T_{ref} = 53\,^{\circ}\mathrm{C} + T_o \tag{1}$$

- Locally, by changing the settings of the thermostatic heads in individual rooms. The valves communicate over Z-Wave network [17]. Both set value and room temperature are available through the Z-Wave network controller. The valves contain local controllers that stabilize the room temperature at a given level.

Energy saving is achieved by decreasing the supply temperature T_{sup} and/or lowering the reference room temperature. In both cases, this leads to lower room temperature. To determine the moment and the mode of restoring thermal comfort, mathematical models of rooms are required. As the beginning of the working hours is known, the models allow determinising the time at which the rooms have to be reheated again or at which the supply temperature has to raise.

An auxiliary algorithm that does not take part in changes in reference temperatures is a mathematical model estimating the amount of heat energy consumed by the building. It allows estimating how much energy would be consumed in the absence of temperature reductions. The energy estimation algorithm determines the validity of the energy saving strategies.

3 Single Room Model

Most of the rooms in the test building are equipped with radiators with electronic valves. Significant differences were observed in room temperature changes depending on their location in the building and depending on the room size. Figure 1 presents temperature in selected rooms in response to a change in the reference temperature level.

For energy saving purposes, we consider only two valve states:

- The reference temperature is set by the user of the room at the appropriate level of thermal comfort. In this case, the model depends on the room temperature T_r, outdoor temperature T_o and supply temperature $T_{sup.}$

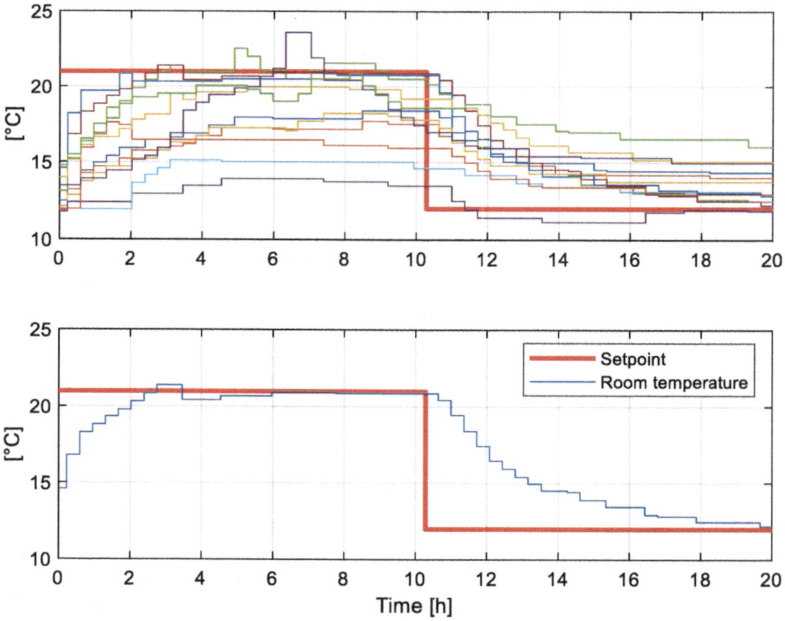

Fig. 1. The temperature in selected rooms as a function of the room's setpoint.

- Outside the operating hours, the temperature setpoint is set to a safe level. During the experiments, the safe temperature level was set to 12 °C. In most cases, this means that the valve will close completely so the model depends only on the room temperature T_r, and outdoor temperature T_o.

Modelling of room temperature requires parts responsible for the heat exchange between the room and internal and external walls, floor, ceiling, windows and radiators [15]. This introduces different time constants to the mathematical model. To recreate thermal comfort, it is sufficient to include only short, several-hour time constants in the models. In this case, modelling the room as a second-order linear system gives satisfactory results.

For a case with a closed thermostatic valve, the model has the form

$$H(z) = \frac{T_r(z)}{T_o(z)} = \frac{C_1 z^{-1}}{1 + C_2 z^{-1} + C_3 z^{-2}} \tag{2}$$

For the case with the active radiator valve, there are two transfer functions affecting the room temperature in total: one for outdoor temperature T_o and the second for the supply temperature T_{sup}.

$$H_{Tout}(z) = \frac{T_{rout}(z)}{T_o(z)} = \frac{C_4 z^{-1}}{1 + C_5 z^{-1} + C_6 z^{-2}} \tag{3}$$

$$H_{Tsup}(z) = \frac{T_{rsup}(z)}{T_{sup}(z)} = \frac{C_7 z^{-1}}{1 + C_8 z^{-1} + C_9 z^{-2}} \tag{4}$$

$$T_r(z) = T_{rout}(z) + T_{rsup}(z) \tag{5}$$

For all transfer functions the sampling period was 60 s.

Figures 2 and 3 present responses of the models for the data sets used for identification as well as for verification data sets. Verification data sets are logged in different operating conditions than that for identification data. Table 1 presents the values of the coefficients for data shown in Fig. 2 and Fig. 3. It is crucial, that the parameters differ for different rooms and the identification must be carried out independently for each room.

Table 1. Coefficients of the H(z), $H_{Tout}(z)$ and $H_{Tsup}(z)$ transfer functions.

H(z)		$H_{Tout}(z)$		$H_{Tsup}(z)$	
C_1	2.398e−7	C_4	4.645e−4	C_7	6.477e−3
C_2	−1.991	C_5	−1.866	C_8	−1.645
C_3	0.9912	C_6	0.8671	C_9	0.6601

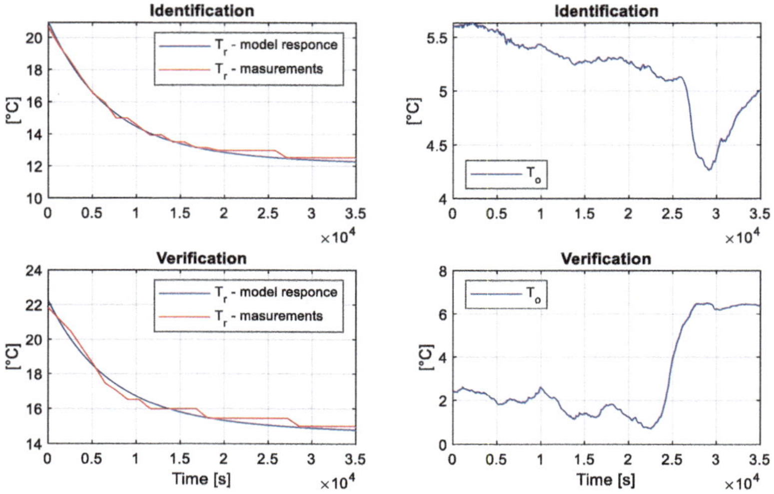

Fig. 2. Identification and verification results for the closed thermovalve (T_r – room temperature, T_o – outdoor temperature).

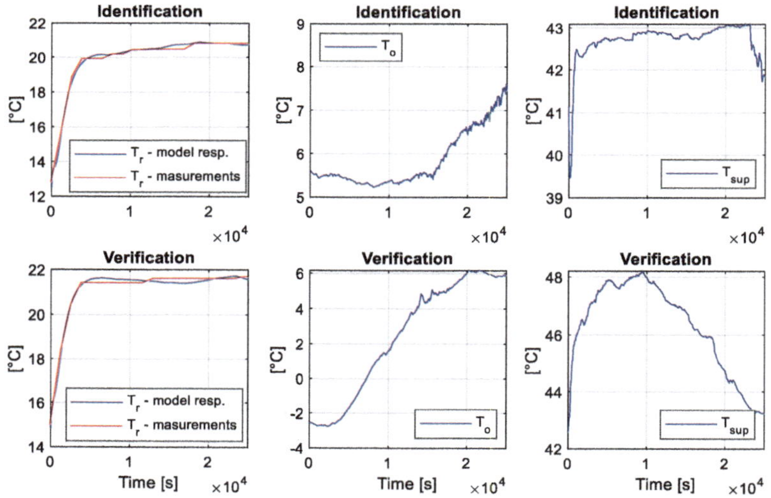

Fig. 3. Identification and verification results for active thermovalve (T_r – room temperature, T_o – outdoor temperature, T_{sup} – supply temperature of central heating system).

4 Building Model

Modelling energy consumption of a building is a complex issue, subject to many disturbances. For example, 800 students provide around 80 kW of energy [18], making a significant contribution to the building's maximum energy of 250 kW. Because the energy estimation algorithm does not take part in planning the moments of temperature changes, a simple linear energy consumption model was made. In Fig. 4 the measurements of energy consumed by the building are presented when the supply temperature T_{sup} does not differ from the reference temperature T_{ref} (1) by more than 2 °C.

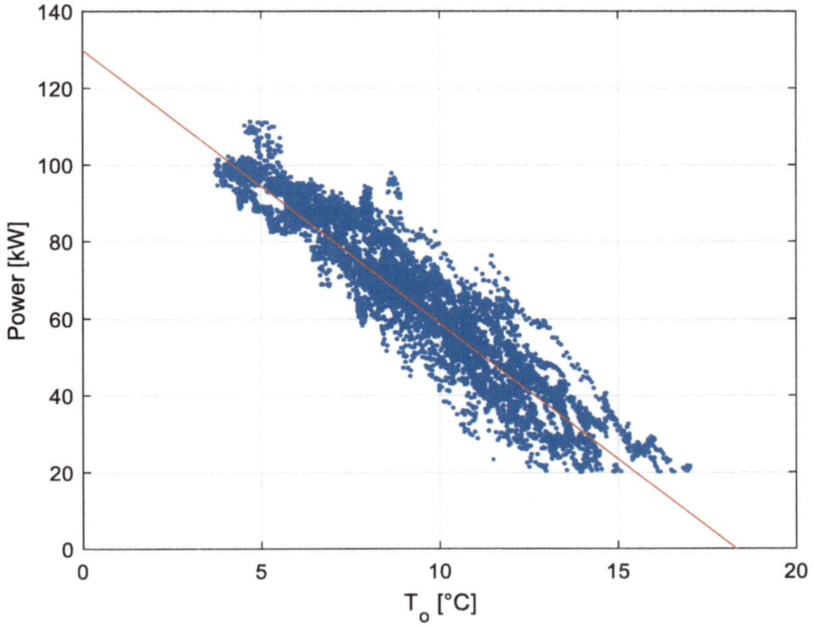

Fig. 4. Central heating power versus outdoor temperature.

The power measurement points have been approximated linearly (6) and this model is used during energy estimations.

$$POWER_{estimated} = 130\,\mathrm{kW} - 7.1T_o \tag{6}$$

Model (6) describes energy consumption in nominal conditions allowing to estimate changes in energy consumption resulting from changes in the supply temperature and/or changes in reference room temperatures.

Figure 5 shows the measured and estimated power in periods when the supply temperature T_{sup} corresponded to the reference temperature T_{ref} and when the supply temperature was lowered (see upper diagram). Model (6) was created only for measurements when the supply temperature corresponded to the temperature determined from formula (1). If temperatures do not match, energy is estimated. The energy estimation points are marked with black points on the bottom graph.

For the periods when supply and reference temperatures match the mean value of the error between the power estimation and the power measurements is 0.52 kW and the standard deviation is 6.16 kW.

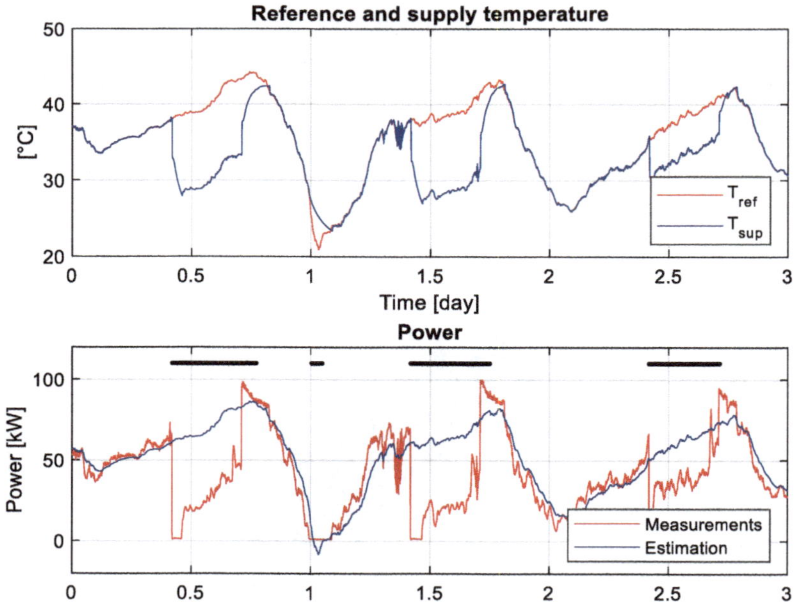

Fig. 5. Central heating power estimation.

5 Experiments

The first experiment aimed to reduce the supply temperature T_{sup} by 10 °C and 5 °C at night. The results are shown in Fig. 6. The bottom left graph shows the cumulated saved energy. The black points on the graph show the periods when the supply temperature significantly differs from the reference temperature. These are the points where energy was estimated according to the model (6). Energy savings associated with reference temperature reductions and energy peaks associated with return to thermal comfort temperature are visible. The energy saved is about 1 GJ/day.

The purpose of the second experiment was to reduce the supply temperature T_{sup} by 15 °C during three days period. The results are shown in Fig. 7. In this case, the save energy is about 15 GJ.

In the third case, the reference temperatures of 170 thermostatic valves installed in the building were lowered. At night-time, these temperatures were set to 12 °C, which results in closing the hot water supply to the radiators. Using room model (2)–(5), the time of turning on the thermostats was calculated. Depending on weather conditions, the turn-on time took place 3–5 h before the building's operation hours. The estimated savings are around 1 GJ/day in this case.

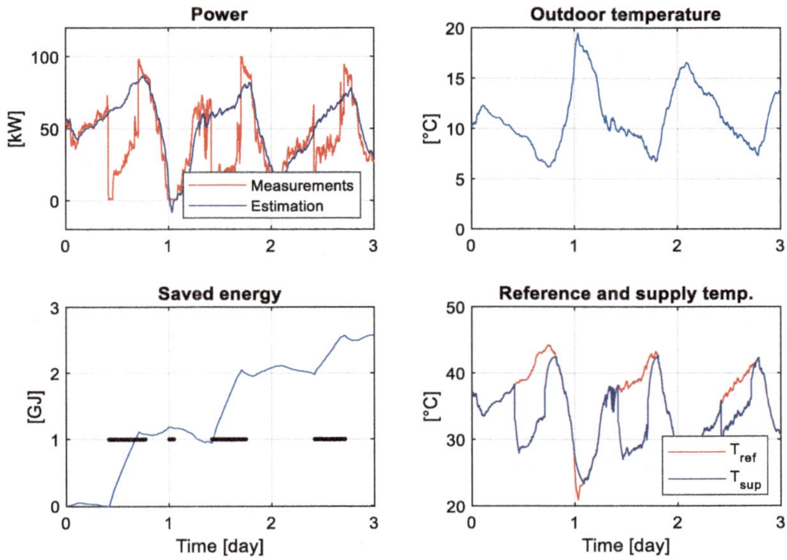

Fig. 6. Saved energy – single night T_{ref} reductions.

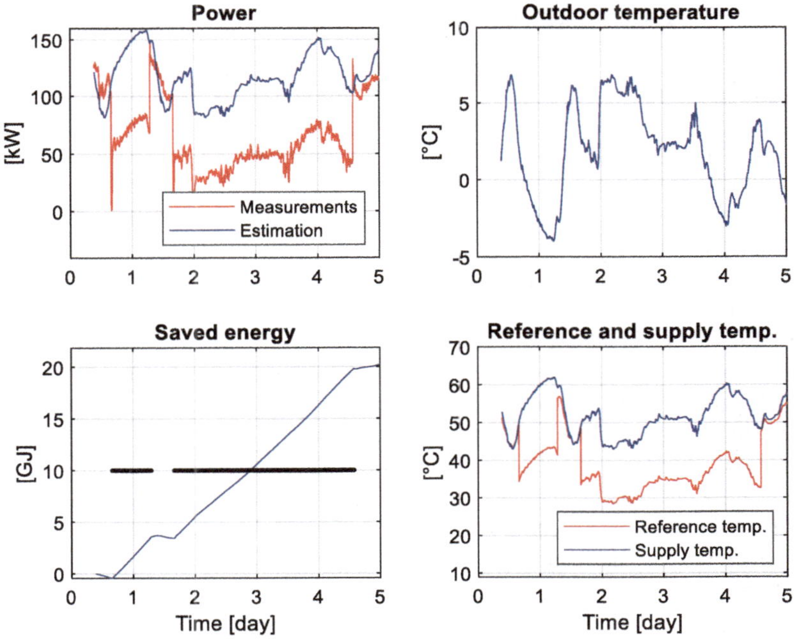

Fig. 7. Saved energy – three days T_{ref} reduction.

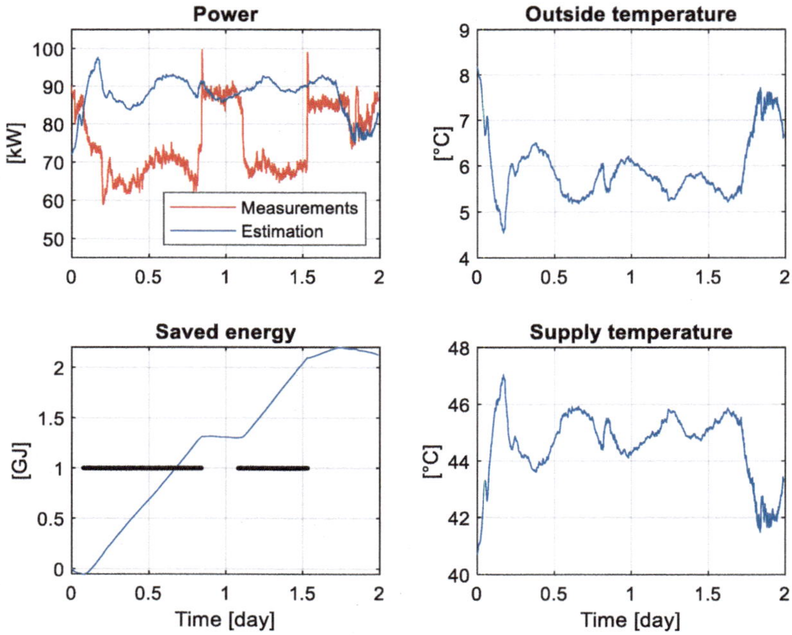

Fig. 8. Saved energy – local rooms temperature reductions.

6 Conclusions

The presented experiments were performed in the period from the 18th of December 2019 to the 10th of January 2020. During the 22 days of the described experiments, the building consumed about 164 GJ of energy. Depending on the selected strategy the energy savings are from 1 GJ/day to 5 GJ/day. It was performed one three-day supply temperature reduction experiment during this period (Fig. 7). On other days, the supply temperature or the reference room temperatures were reduced (Fig. 6 and Fig. 8). According to the estimated level of savings, temperature reductions saved about 34 GJ of energy, so the savings were about 17%. It is worth mentioning that in the described period various experiments were carried out, not necessarily maximizing energy savings while maintaining thermal comfort. It seems that choosing and applying the optimal strategy will increase savings by another few percent.

Unfortunately, during the experimental period, the outdoor temperatures were much higher than the usual minimum winter temperatures. Only in four days, the temperature dropped below zero and its minimum value was −4 °C. The central heating installation in the building is designed for minimum temperatures of −20 °C and until such temperatures occur it is not possible to verify the described algorithms for the entire operating temperature range.

Acknowledgment. The research was supported by the NCBiR project POIR.01.02.00-00-0302/17. The presented data come from the measurement and control system of NG Heat Sp. z O.O., Centrum Energetyki AGH, ul. Czarnowiejska 36 lok. C5/017, 30-054 Krakow, Poland.

References

1. Xiong, X.: The architecture design of monitoring and controlling system of central heating. In: 2011 International Conference on Computer Science and Network Technology (2011)
2. Kaczmarek, P.A.: Central heating temperature control algorithm for systems with condensing boilers. In: 21st International Conference on Methods and Models in Automation and Robotics (2016)
3. Ilhan, İ., Karakose, M., Yavas, M.: Design and simulation of intelligent central heating system for smart buildings in smart city. In: 7th International Istanbul Smart Grids and Cities Congress and Fair (ICSG) (2019)
4. Sun, T., Nie, Z., Liu, R.: Application on fuzzy PID technology for central heating. In: Proceedings of 2011 International Conference on Computer Science and Network Technology (2011)
5. Van Pham, T., Nguyen, D.H., Banjerdpongchai, D.: Design of iterative learning control via alternating direction method of multipliers for building temperature control system. In: 14th International Conference on Electrical Engineering/Electronics, Computer, Telecommunications and Information Technology (ECTI-CON) (2017)
6. Minakais, M., Mishra, S., Wen, J.T.: Database-driven iterative learning for building temperature control. IEEE Trans. Autom. Sci. Eng. **16**(4), 1896–1906 (2019)
7. Van Pham, T., Nguyen, D.H., Banjerdpongchai, D.: Decentralized iterative learning control of building temperature control system. In: SICE International Symposium on Control Systems (2017)
8. Balan, R., Stan, S.-D., Lapusan, C.: A model based predictive control algorithm for building temperature control. In: 3rd IEEE International Conference on Digital Ecosystems and Technologies (2009)
9. Eini, R., Abdelwahed, S.: Distributed model predictive control based on goal coordination for multi-zone building temperature control. In: IEEE Green Technologies Conference (GreenTech) (2019)
10. Popescu, D., Borza, I.: Ensuring the comfort in the heated space by controlling the temperature in the heating installation of a non-residential building. In: 5th International Conference on Smart Cities and Green ICT Systems (SMARTGREENS) (2016)
11. Artem, S., Aliya, A., Ainur, T.: Mathematical model of heat supply of rooms for automated control systems of energy saving. In: 3rd International Conference on Mechatronics, Robotics and Automation (2015)
12. Utama, Y.A.K., Hari, Y.: Design of PID disturbance observer for temperature control on room heating system. In: 4th International Conference on Electrical Engineering, Computer Science and Informatics (EECSI) (2017)
13. Setiawan, N., Mustika, I.W., Cahyadi, A.I., Fikri, M.: State space modeling of thermal in a room for temperature estimation in wireless sensor network. In: 2nd International Conferences on Information Technology, Information Systems and Electrical Engineering (ICITISEE) (2017)
14. Sahyoun, S., Nelson, C., Djouadi, S.M., Kuruganti, T.: Control and room temperature optimization of energy efficient buildings. In: IEEE International Conference on Control Applications (CCA) (2012)
15. Ellis, C., Hazas, M., Scott, J.: Matchstick: a room-to-room thermal model for predicting indoor temperature from wireless sensor data. In: ACM/IEEE International Conference on Information Processing in Sensor Networks (IPSN) (2013)
16. Zhuang, J., Chen, X., Chen, Y.: Dynamic modeling of indoor air temperature based on power spectral density method. In: IEEE 3rd International Conference on Control Science and Systems Engineering (2017)

17. Spirit Z-Wave Plus. https://eurotronic.org/produkte/z-wave-heizkoerperthermostat/spirit-z-wave-plus/. Accessed 10 Jan 2020
18. Zyski ciepła od ludzi. https://www.klimatyzacja.pl/klimatyzacja/artykuly/zyski-ciepla/zyski-ciepla-od-ludzi. Accessed 10 Jan 2020

Approximation Algorithms for Constrained Resource Allocation

Krzysztof Pieńkosz(✉)

Warsaw University of Technology, Institute of Control and Computation Engineering, Nowowiejska 15/19, 00-665 Warsaw, Poland
K.Pienkosz@ia.pw.edu.pl

Abstract. In the paper the problem of effective allocation of a single resource in manufacturing or logistic systems is considered. In order to reduce additional costs, the cardinality constraints are imposed that allow one to allocate the resource only to the limited number of operations. This problem is NP-hard. Two approximation algorithms are proposed and their properties are analyzed. In particular, the worst-case performance of these algorithms is studied, and the results of their experimental comparison are presented.

Keywords: Resource allocation · Cardinality constraints · Approximation algorithms · Worst-case analysis

1 Introduction

Resource allocation is an important strategy which involves a company in deciding where scarce resources could be the most effectively used in the production or services processes. We consider resource allocation problem which concerns the distribution of a single limited resource among manufacturing operations. If the total demand of all operations does not exceed the available amount of the resource, then each operation can consume it as it requires. The essential decision problem arises when there is no enough amount of the resource to satisfy all the demands. Then, we must determine to which operations the resource should be allocated and how large portion of this resource each operation should receive. In order to make this decision, we usually take into account the profits resulting from the allocation of the resource to individual operations.

In many practical situations the selection of operations for the resource allocation may induce some additional cost. Then the solutions in which larger portions of the resource are allocated to a smaller number of operations will be preferred to a solution with a large number of operations with a small amount of the allocated resource. One way of taking into account such preferences is to introduce upper bound k on the number of operations involved in the resource allocation. We will call such a problem the *Resource Allocation Problem with Cardinality Constraints* (RAPCC).

A. Bartoszewicz et al. (Eds.): PCC 2020, AISC 1196, pp. 275–286, 2020.
https://doi.org/10.1007/978-3-030-50936-1_24

RAPCC was studied extensively by de Farias and Nemhauser [3]. They showed that this problem is NP-hard and derived a number of valid inequalities for the convex hull of the set of feasible solutions. These inequalities were used to develop a branch-and-cut algorithm for RAPCC. There were also reported some applications of RAPCC in portfolio optimization, p-median problem, and synthesis of process networks. The structural properties of the optimal solutions of RAPCC were analyzed by Pieńkosz [7]. Some strategies were formulated which can be applied to reduce the size of RAPCC and thus make this problem easier to solve.

In RAPCC it is allowed that the demands for the resource may be satisfied even partially, i.e. an operation may receive smaller amount of the resource than it is ordered. Caprara et al. [1], and Mastrolilli and Hutter [6] analyzed solutions methods for a discrete variant of the resource allocation problem with cardinality constraints. In such a variant it is assumed that each operation selected for the resource allocation must obtain the full required amount of the resource. Caprara et al. [1] also described some applications in solving cutting stock problem and scheduling problems.

The remainder of this paper is organized as follows. In the next section the mathematical formulation of the problem under consideration is presented. There are also described some properties of this problem. Sections 3 and 4 contains the description of two approximation algorithms proposed to solve the problem and the results of their worst-case performance analysis. Some results of the experimental comparison of these two algorithms are presented in Sect. 5.

2 Resource Allocation Problem with Cardinality Constraints

Let $N = \{1, 2, \ldots, n\}$ be the set of operations considered for the allocation of the resource which is available in limited amount c. Each individual operation $j \in N$ requires w_j units of the resource and the profit of such allocation is p_j. We may assign less than w_j units to operation j but then the profit is proportionally smaller than p_j.

In the resource allocation problem with cardinality constraints we try to allocate the resource to at most k operations in such a way that the overall profit is maximized. Formally, RAPCC can be formulated as follows:

$$\text{maximize} \sum_{j \in N} p_j x_j \tag{1}$$

subject to

$$\textstyle\sum_{j \in N} w_j x_j \le c, \tag{2}$$

$$\text{at most } k \text{ variables } x_j \text{ are positive,} \tag{3}$$

$$0 \le x_j \le 1, \quad j \in N. \tag{4}$$

Variables x_j specify how large portion of resource demand w_j is to be satisfied. Note that x_j can be a fractional value. Constraints (2) guarantee that no more than c units of the resource are used and (3) are cardinality constraints limiting the number of operations with the allocated resource. Value k is the parameter of the resource allocation problem with cardinality constraints.

Let us note that RAPCC has the trivial optimal solution if $\sum_{j \in N} w_j \leq c$. In such a case, the best option is to allocate the resource to k operations with the largest profits. Therefore, we assume throughout the paper that

$$\sum_{j \in N} w_j > c. \tag{5}$$

In addition, it is assumed that the operations are ordered according to non-increasing values of their profit to weight ratio p_j/w_j, i.e.

$$\frac{p_1}{w_1} \geq \frac{p_2}{w_2} \geq \ldots \geq \frac{p_n}{w_n}. \tag{6}$$

The profit to weight ratio p_j/w_j will be called the *efficiency* of operation j.

It may happen that RAPCC has more than one optimal solution giving the same profit. In such a case we shall consider the *lexicographically largest optimal solution*. We say that a solution $x^1 = (x_1^1, x_2^1, \ldots, x_n^1)$ is lexicographically larger than $x^2 = (x_1^2, x_2^2, \ldots, x_n^2)$ if $x_j^1 = x_j^2$ for $j = 1, 2, \ldots, i$ and $x_{i+1}^1 > x_{i+1}^2$ for some $i < n$.

Note that due to ordering (6), only at most one decision variable can be fractional in the lexicographically largest optimal solution of RAPCC. The fractional value can have only the last positive variable. The other variables take value 0 or 1. Another property of the lexicographically largest optimal solution of RAPCC is formulated in the following theorem.

Theorem 1. *If $x_i^* = 0$ and $x_j^* > 0$ for some $i < j$ in the lexicographically largest optimal solution x^* of RAPCC, then*

$$w_i < w_j x_j^* \quad \text{and} \quad p_i < p_j x_j^*.$$

Proof. Suppose by contradiction that $x_i^* = 0$, $x_j^* > 0$ and $w_i \geq w_j x_j^*$ for some $i < j$. Then we could assign $w_j x_j^*$ units of the resource to operation i instead to j and thus get better solution than x^* if $p_i/w_i > p_j/w_j$, or lexicographically larger if $p_i/w_i = p_j/w_j$.

Analogously, we obtain a contradictive conclusion if we suppose that $p_i \geq p_j x_j^*$. As $w_i < w_j x_j^*$, so allocating w_i units of the resource to operation i instead of $w_j x_j^*$ units to operation j we would get better solution if $p_i > p_j x_j^*$, or lexicographically larger if $p_i = p_j x_j^*$. □

It was shown by de Farias and Nemhauser [3] that RAPCC is NP-hard. However, if we relax the problem by neglecting constraint (3), we get the *Continuous Knapsack Problem* (CKP) which is very easy to solve. Its optimal solution can be computed based on the following well known theorem (see, e.g. de Farias and Nemhauser [3], Kellerer et al. [4]).

Theorem 2. *The solution*

$$x_j = 1 \quad \textit{for } j = 1, \ldots, s-1, \tag{7}$$

$$x_s = \frac{1}{w_s}(c - \sum_{j=1}^{s-1} w_j) \tag{8}$$

$$x_j = 0 \quad \textit{for } j = s+1, \ldots, n, \tag{9}$$

where

$$s = \min\{i : \sum_{j=1}^{i} w_j > c\} \tag{10}$$

is an optimal solution of CKP.

Note that the solution specified by (7)–(9) is the lexicographically largest optimal solution of CKP. There are at most s variables with positive values in this solution, so if $k \geq s$ then it is also the lexicographically largest optimal solution of RAPCC. The case $k < s$ is considered in the following theorem.

Theorem 3. *If $k < s$ then there are exactly k variables with positive values in the lexicographically largest optimal solution x^* of RAPCC.*

Proof. According to (3), there are no more than k positive variables in the lexicographically largest optimal solution of RAPCC. As $k < s$, we have by (10) that $\sum_{j=1}^{k} w_j \leq c$. Note that if $\sum_{j=1}^{k} w_j = c$ and $k < s$ then solution (7)–(9) is the lexicographically largest optimal solution of RAPCC and it contains exactly k variables with positive values. Suppose that there are less than k such variables in the case when $\sum_{j=1}^{k} w_j < c$. Then it must be $x_i^* = 0$ for some $i \leq k$. Therefore, if $x_j^* = 0$ for each $j > k$ we could allocate some amount of the resource to operation i and thus increase the overall profit which contradicts that x^* is an optimal solution. On the other hand, if there is $x_j^* > 0$ for some $j > k$, we could decrease the value of x_j^* by sufficiently small $\epsilon > 0$, and increase x_i^* by $\epsilon w_j / w_i$, getting better solution if $p_i/w_i > p_j/w_j$, or lexicographically larger if $p_i/w_i = p_j/w_j$. So again we have a contradiction. □

3 Improvement Algorithm

In order to solve RAPCC, we propose approximation algorithm IA. Initially, it starts from the solution in which the resource is allocated to the operations with the largest efficiencies p_j/w_j. Then algorithm IA tries to improve this solution in the consecutive iterations.

Improvement Algorithm (IA)

1. Sort the operations in nonincreasing order of their efficiencies and determine value s from (10). If $k \geq s$, then compute the optimal solution from (7)–(9), otherwise go to Step 2.

2. Allocate the resource to the first k operations
   $Q = \{1, \ldots, k\}$,
   $x_i := 1$ for $i \in Q$ and $x_i := 0$ for $i \in N \setminus Q$,
   $l := k + 1$.
3. For each $i \in Q$ compute $\delta_{il} = \min\{1, (c - \sum_{q \in Q \setminus \{i\}} w_q)/w_l\}$.
   Find operation $j \in Q$ with the largest increment of profit $\Delta z = p_l \delta_{jl} - p_j x_j$.
4. If $\Delta z > 0$, then reallocate the resource between operations j and l.
   $Q := (Q \setminus \{j\}) \cup \{l\}$, $x_j := 0$, $x_l := \delta_{jl}$.
5. If $l < n$ and $\sum_{i \in Q} w_i < c$, then $l := l + 1$ and go to Step 3, otherwise STOP.

In the beginning algorithm IA checks the condition $k \geq s$. If it is satisfied then there are no more than k variables with positive values in the optimal solution of relaxation CKP and thus such a solution is feasible and optimal for RAPCC. If $k < s$, then the resource is allocated to k operations with the largest efficiencies. In the next iterations, the reallocation is done if another operation is found which can provide larger profit. Such improvement is only possible if operations $i \in Q$ do not consume the whole resource, i.e. $\sum_{i \in Q} w_i < c$. When this resource is completely allocated, any exchange with an operation having smaller efficiency than the operations from set Q cannot improve the solution, so algorithm IA can be terminated.

The most time consuming parts of algorithm IA are Steps 1 and 3. In Step 1 sorting of the operations requires $O(n \log n)$ time [2], where n denotes the number of operations. In Step 3, the value of δ_{il} is computed k times in each iteration so altogether at most $k(n-k)$ times. Thus computational complexity of algorithm IA is $O(n^2)$.

The solutions computed by algorithm IA have a similar structure to the lexicographically largest optimal solutions of RAPCC. If $k < s$ then algorithm IA allocates the resource to exactly k operations and only the last operation may have its demand not fully satisfied. These solutions have also analogous properties to those specified in Theorem 1.

Theorem 4. *If $x_i = 0$ and $x_j > 0$ for $i < j$ in solution x computed by algorithm IA, then*

$$(i)\ w_i \leq w_j x_j,$$
$$(ii)\ p_i \leq p_j x_j.$$

Proof. Suppose by contradiction that (ii) is not satisfied. This inequality could be violated only in Step 4 of algorithm IA. Let us suppose that it happened for the first time when $l = b$, as a result of the resource reallocation between operations a and b, where $a < b$. As this reallocation may be done only if $p_a < p_b x_b$, so (ii) could be violated only if $x_j = 1$ and $p_a > p_j$ for some j such that $a < j < b$, or if $x_i = 0$ and $p_i > p_b x_b$ for some i such that $a < i < b$. In the first case, however, the resource would not be taken from operation a in iteration $l = b$ since there are better candidates for reallocation resulting in larger increase of the overall profit, e.g. operation j. In the second case we would have $p_i > p_b x_b > p_a$. Consequently, algorithm IA would not keep the resource assigned to operation a until iteration $l = b$ while this resource is not allocated to more profitable operation i. In both cases we obtain a contradiction.

We have thus proved that if $x_i = 0$ and $x_j > 0$ for $i < j$, then $p_i \leq p_j x_j$. According to (6), there is also $w_i/p_i \leq w_j/p_j$. As a result, we have (i). □

It turns out that quality of solutions produced by algorithm IA depends on the relationship between the value of k and parameter s defined by (10). If $k = 1$ then algorithm IA scans individual operations in the consecutive iterations and selects the one with the largest profit. If $k \geq s$, then the optimal solution of relaxation CKP computed in Step 1 is feasible and thus optimal for RAPCC.

Corollary 1. *If $k = 1$ or $k \geq s$ then algorithm IA computes an optimal solution of RAPCC.*

Analyzing the worst-case performance of algorithm IA for other values of k, we denote by z^{IA} the total profit of the solution computed by algorithm IA and by z^* the optimal solution value of RAPCC.

Theorem 5. *For every instance of RAPCC with $k < s$*

$$\frac{z^{IA}}{z^*} \geq \frac{k}{s} \,. \tag{11}$$

Proof. Let us introduce the new auxiliary notation:

$$\begin{array}{lll} \bar{w}_j = w_j, & \bar{p}_j = p_j & \text{for } j = 1, \ldots, s-1 \\ \bar{w}_s = c - \sum_{j=1}^{s-1} w_j, & \bar{p}_s = p_s \bar{w}_s / w_s. & \end{array}$$

Problem CKP is a relaxation of RAPCC, so by Theorem 2

$$z^* \leq \sum_{j=1}^{s} \bar{p}_j. \tag{12}$$

Let L be the subset consisting of k operations from set $\{1, \ldots, s\}$ with the largest profits $\bar{p}_j$. Then

$$\sum_{j \in L} \bar{p}_j \geq \frac{k}{s} \sum_{j=1}^{s} \bar{p}_j. \tag{13}$$

According to (10), we have $\sum_{j=1}^{s-1} w_j \leq c$, so as long as $l < s$ in algorithm IA, set Q consists of operations $j \leq l$ with the largest profits $\bar{p}_j$. If we could assign only at most $\bar{w}_s$ units of the resource to operation s, then in iteration $l = s$ we would analogously have $Q = L$, so the total profit of the resource allocation would be $\sum_{j \in L} \bar{p}_j$. Since there is no such restriction and $w_s > \bar{w}_s$, then the total profit in this iteration can be larger and it will not decrease in the subsequent iterations. Thus $z^{IA} \geq \sum_{j \in L} \bar{p}_j$. Consequently, taking into account (12) and (13) we have

$$z^{IA} \geq \sum_{j \in L} \bar{p}_j \geq \frac{k}{s} \sum_{j=1}^{s} \bar{p}_j \geq \frac{k}{s} z^*.$$

□

Theorem 6. *For every instance of RAPCC with $k < s$*

$$\frac{z^{IA}}{z^*} \geq \frac{k}{2k-1} . \tag{14}$$

Proof. Let x^* be the lexicographically largest optimal solution of RAPCC, x^l denotes the temporary solution computed by algorithm IA in iteration l, and x^{IA} be its final solution. Taking into account ordering (6), let a be the first operation for which we have $x_i^* = 0$ and $x_i^{IA} > 0$. If such operation does not exist then $z^{IA} = z^*$ by Theorem 3, so (14) is satisfied.

First, we shall prove that

$$z^* - \sum_{i \in A} p_i x_i^* \leq z^{IA}, \tag{15}$$

where $A = \{j : j < a,\ x_j^* > 0,\ x_j^{IA} = 0\}$.

Note that if for each $j > a$ there is $x_j^* > 0$ only if $x_j^{IA} > 0$, then (15) is satisfied. Otherwise, let b be the first iteration of algorithm IA in which $x_d^* > 0$ and $x_d^b = 0$ for some d such that $a < d \leq b$. By Theorem 1 we have $p_d > p_a$. Furthermore, since $x_d^b = 0$ then

(i) either the resource was reallocated from operation d to operation $b > d$ in iteration $l = b$,
(ii) or the resource wasn't allocated to operation d in any iteration of algorithm IA.

In case (i), note that in iteration $l = b$ the resource is allocated to operation a but it not assigned to more profitable operation d. It would be only possible, if assigning the resource to operation d instead of a, we obtain a solution $\bar{x}$ such that $p_d \bar{x}_d \leq p_a x_a^b$. Since $p_d > p_a$, it could only happen if $\bar{x}_d < 1$ and all c units of the resource are consumed in solution $\bar{x}$. If we compare solutions $\bar{x}$ and x^*, we can note that for $a \leq i < b$ if $x_i^* > 0$ then $\bar{x}_i > 0$. Furthermore, according to (6), the profit from operations $j \geq b$ in solution x^* cannot be larger than from operations $i < a$ in solution $\bar{x}$. Therefore, it must be $z^* - \sum_{j \in B} p_j x_j^* \leq \sum_{i=1}^n p_i \bar{x}_i$ where $B = \{j : j < a,\ x_j^* > 0,\ x_j^b = 0\}$. Taking in to account that $\sum_{i=1}^n p_i \bar{x}_i \leq z^{IA}$ and $\sum_{j \in B} p_j x_j^* \leq \sum_{j \in A} p_j x_j^*$ since $B \subseteq A$, we obtain (15).

In case (ii), we can prove (15) in the same way as in case (i) by assuming that $b = d$.

Now, let $S = \{j : j < a,\ x_j^* > 0,\ x_j^{IA} > 0\}$. In both solutions x^* and x^{IA} we have exactly k variables with positive values, so according to definition of a, there must be $k - |S|$ operations $j \geq a$ such that $x_j^{IA} > 0$. By Theorem 4, there is $p_i \leq p_j x_j^{IA}$ for each $i \in A$ and $j \geq a$ such that $x_j^{IA} > 0$. Therefore, if $|A| + |S| = k$, then we have $z^* \leq z^{IA}$, so (14) is satisfied. On the other hand, if $|A| + |S| < k$ then

$$\sum_{i \in A} p_i x_i^* \leq \frac{|A|}{k - |S|}(z^{IA} - \sum_{j \in S} p_j x_j^{IA}) \leq \frac{|A|}{k - |S|} z^{IA} \leq \frac{k - 1 - |S|}{k - |S|} z^{IA} \leq \frac{k-1}{k} z^{IA} .$$

Consequently, it follows from (15) that

$$z^* \leq z^{IA} + \sum_{i \in A} p_i x_i^* \leq z^{IA} + \frac{k-1}{k} z^{IA} = \frac{2k-1}{k} z^{IA},$$

which implies (14). □

The conclusions resulting from Corollary 1, Theorem 5, and Theorem 6 can be summarized as follows.

Corollary 2. *For every instance of RAPCC with $k < s$*

$$\frac{z^{IA}}{z^*} \geq \max\left\{\frac{k}{s}, \frac{k}{2k-1}\right\} \tag{16}$$

and $z^{IA} = z^*$ if $k \geq s$.

It turns out that bound (16) specified in Corollary 2 is tight.

Theorem 7. *The approximation ratio (16) is tight, i.e. there is no real value ρ larger than $\max\{k/s,\ k/(2k-1)\}$ such that $z^{IA}/z^* \geq \rho$ for every instance of RAPCC with $k < s$.*

Proof. It follows from Corollary 1 that (16) is tight for $k = 1$. To proof that it is also tight for $1 < k < s$, we shall show the instances of RAPCC for which approximation ratio z^{IA}/z^* can be arbitrary close to $\max\{k/s,\ k/(2k-1)\}$.

Let $r = \min\{s,\ 2k-1\}$. For a given parameter $M > s$ consider the instance of the problem in which $c = k$ and there are $n = s+1$ operations with the demands and profit coefficients as follows:

$$\begin{array}{lll} w_j = \frac{1}{M}, & p_j = 1 - \frac{1}{M} & \text{for } j = 1, \ldots, r-k \\ w_j = 1, & p_j = 1 & \text{for } j = r-k+1, \ldots, s \\ w_{s+1} = k - \frac{r-k}{M}, & p_{s+1} = k - \frac{r-k+1}{M}. & \end{array}$$

Note that efficiencies p_j/w_j of operations $j \leq r-k$ are larger than 1. They are equal to 1 for operations $j \in \{r-k+1, \ldots, s\}$ and smaller than 1 for $j = s+1$. Therefore, algorithm IA will allocate the resource to operations $j = 1, \ldots, r-k$ first. In the subsequent iterations this resource will be reallocated to operations from set $\{r-k+1, \ldots, r\}$ as they give a larger profit. Consequently, in the final solution of algorithm IA the resource will be allocated only to operations $j \in \{r-k+1, \ldots, r\}$ and thus $z^{IA} = k$.

However, the optimal solution value in this case is not smaller than $(r - k)(1 - \frac{1}{M}) + (k - \frac{r-k+1}{M})$ since such overall profit is obtained by allocating the resource to operations $j \in \{1, \ldots, r-k\} \cup \{s+1\}$. Therefore $z^* \geq (r-k)(1 - \frac{1}{M}) + (k - \frac{r-k+1}{M}) = r - \frac{2r-2k+1}{M}$.

As a result, we have for this instance of the problem

$$\frac{z^{IA}}{z^*} \leq \frac{k}{r - \frac{2r-2k+1}{M}}.$$

On the other hand, according to (16)

$$\frac{z^{IA}}{z^*} \geq \max\left\{\frac{k}{s}, \frac{k}{2k-1}\right\} = \frac{k}{\min\{s, 2k-1\}} = \frac{k}{r}$$

so ratio z^{IA}/z^* may be arbitrary close to $k/r = \max\{k/s,\ k/(2k-1)\}$ if M is sufficiently large. □

4 Relaxation Based Approximation Algorithm

Constraint (3) of RAPCC imposes that the number of variables x_j with positive values is not larger than k. In addition, it follows from (4) that $0 \leq x_j \leq 1$ for each $j \in N$. Thus it must be

$$\sum_{j \in N} x_j \leq k. \tag{17}$$

Replacing constraint (3) by (17) we get another Linear Programming (LP) relaxation of RAPCC.

$$\text{maximize} \sum_{j \in N} p_j x_j \tag{18}$$

subject to

$$\sum_{j \in N} w_j x_j \leq c \tag{19}$$

$$\sum_{j \in N} x_j \leq k \tag{20}$$

$$0 \leq x_j \leq 1 \quad j \in N \tag{21}$$

It is a stronger LP relaxation of RAPCC than CKP since constraint (20) is neglected in CKP. It turns out that in the optimal basic solution of (18)–(21) there can be more than k variables x_j having positive value, but no more than $k+1$.

Theorem 8. *Each basic solution of problem (18)–(21) has at most $k+1$ variables with positive values.*

Proof. If upper bounds on variables x_j are treated implicitly, there will be two basic variables in the basic solution of problem (18)–(21). The other variables will take integer values, so only at most two variables can have fractional values. Consequently, it follows from (20) and (21) that no more than $k+1$ variables can take positive values in the basic solution. □

The second approximation algorithm for RAPCC proposed in the paper, starts form an optimal basic solution of LP relaxation (18)–(21) and tries to modify it as to make it feasible for RAPCC.

Relaxation Based Algorithm (RBA)

1. Compute an optimal basic solution $\bar{x}$ of LP relaxation (18)–(21).

2. If there are no more than k positive values in solution $\bar{x}$, then $\bar{x}$ is the optimal solution of RAPCC. Otherwise go to Step 3.
3. Let $J = \{j \in N : \bar{x}_j > 0\}$. For each $i \in J$ compute profit z_i resulting from optimal allocation of the resource only to the operations from set $J \setminus \{i\}$.
4. Select the solution which gives the largest profit z_i among all $i \in J$.

Algorithm RBA first checks if the optimal basic solution of LP relaxation is feasible for RAPCC since then it also an optimal solution for this problem. Going to Step 3, we know by Theorem 8 that in the optimal basic solution there are exactly $k+1$ variables with positive values. In order to obtain feasible solution for RAPCC one of these variables is set to 0. All $k+1$ options are compared to choose the best solution with the largest profit z_i. Each profit z_i is computed by allocating the resource to the operations from set $J \setminus \{i\}$ in the optimal way. Such a solution can be obtained if the resource is allocated to those operations sequentially in nonincreasing order of their efficiencies.

The most time consuming operations of algorithm RBA are performed in Step 1. Martello and Toth [5] showed that LP relaxation (18)–(21) can be solved in $O(n^2)$ time. Step 3 requires $O(n\log n)$ time for sorting (6) and at most $(k+1)k$ operations to determine the best solution. Thus computational complexity of algorithm RBA is $O(n^2)$.

Let us denote by z^{RBA} the total profit of the solution computed by algorithm RBA.

Theorem 9. *For every instance of RAPCC*

$$\frac{z^{RBA}}{z^*} \geq \frac{k}{k+1}. \tag{22}$$

Proof. Let $\bar{x}$ be an optimal basic solution of problem (18)–(21) and $J = \{j \in N : \bar{x}_j > 0\}$. Since it is a relaxation of RAPCC, so

$$z^* \leq \sum_{j\in J} p_j \bar{x}_j. \tag{23}$$

Obviously (22) is satisfied if $|J| \leq k$. If $|J| = k+1$ then for each $i \in J$

$$z^{RBA} \geq z_i \geq \sum_{j\in J\setminus\{i\}} p_j \bar{x}_j = \sum_{j\in J} p_j \bar{x}_j - p_i \bar{x}_i. \tag{24}$$

Summing up both sides of above inequalities for all $i \in J$ and taking into account (23) we get

$$(k+1)z^{RBA} \geq (k+1)\sum_{j\in J} p_j \bar{x}_j - \sum_{i\in J} p_i \bar{x}_i = k\sum_{j\in J} p_j \bar{x}_j \geq kz^*,$$

which means that $z^{RBA}/z^* \geq k/(k+1)$.

□

5 Computational Results

Although it follows from Corollary 2, Theorem 7, and Theorem 9 that the worst-case performance ratio of algorithm RBA is not worse than for algorithm IA, the experiments show that algorithm RBA not always gives better solution than IA. We examined algorithms IA and RBA on several test instances of RAPCC. In each instance $n = 100$ operations were considered with parameters p_j and w_j generated randomly according to the uniform distribution from the range [1]. The resource limit was set to $c = \frac{1}{2}\sum_{j=1}^{n} w_j$.

Table 1. Experimental comparison of algorithms IA and RBA.

k	$z^{IA} > z^{RBA}$ %	$z^{IA} < z^{RBA}$ %	$z^{IA} = z^{RBA}$ %
$s-19$	0.0	0.0	100.0
$s-18$	0.0	1.0	99.0
$s-17$	4.0	1.0	95.0
$s-16$	2.0	7.0	91.0
$s-15$	1.0	9.0	90.0
$s-14$	3.0	17.0	80.0
$s-13$	2.0	28.0	70.0
$s-12$	10.0	23.0	67.0
$s-11$	8.0	37.0	55.0
$s-10$	12.0	57.0	31.0
$s-9$	6.0	61.0	33.0
$s-8$	7.0	71.0	22.0
$s-7$	7.0	75.0	18.0
$s-6$	9.0	76.0	15.0
$s-5$	10.0	77.0	13.0
$s-4$	8.0	74.0	18.0
$s-3$	9.0	65.0	26.0
$s-2$	12.0	56.0	32.0
$s-1$	14.0	14.0	72.0

Table 1 presents the percentage of cases over 100 examined instances of RAPCC when $z^{IA} > z^{RBA}$, $z^{IA} < z^{RBA}$ and $z^{IA} = z^{RBA}$ for various values of k. When k is small, the resource constraint (2) of RAPCC is usually not active, and both algorithms compute the optimal solution allocating the resource to the most profitable operations. The resource allocation problem with cardinality constraints becomes much harder to solve for larger values of k, and then algorithm RBA much often gave better solution than IA, yet in a fair number of cases it was outperformed by IA. For $k = s - 1$ we got even a similar number of better solutions using algorithm IA. Both algorithms obviously compute an optimal solution for $k \geq s$.

References

1. Caprara, A., Kellerer, H., Pferschy, U., Pisinger, D.: Approximation algorithms for knapsack problems with cardinality constraints. Eur. J. Oper. Res. **123**, 333–345 (2000)
2. Cormen, T.H., Leiserson, C.E., Rivest, R.L., Stein, C.: Introduction to Algorithms. The MIT Press, Cambridge (2009)
3. de Farias Jr., I.R., Nemhauser, G.L.: A polyhedral study of the cardinality constrained knapsack problem. Mathematical Programming Ser. A **96**, 439–467 (2003)
4. Kellerer, H., Pferschy, U., Pisinger, D.: Knapsack Problems. Springer, Heidelberg (2004)
5. Martello, S., Toth, P.: Upper bounds and algorithms for hard 0–1 knapsack problems. Oper. Res. **45**, 768–778 (1997)
6. Mastrolilli, M., Hutter, M.: Hybrid rounding techniques for knapsack problems. Discrete Appl. Math. **154**, 640–649 (2006)
7. Pieńkosz, K.: Reduction strategies for the cardinality constrained knapsack problem. In: 22nd IEEE International Conference on Methods and Models in Automation and Robotics, pp. 945–948 (2017). https://doi.org/10.1109/MMAR.2017.8046956

Analysis of Digital Filtering with the Use of STM32 Family Microcontrollers

Tomasz Marciniak(✉), Kacper Podbucki, Jakub Suder, and Adam Dąbrowski

Faculty of Control, Robotics and Electrical Engineering,
Institute of Automatic Control and Robotics, Division of Electronic Systems and Signal Processing, Poznan University of Technology, Jana Pawła II 24, 60-965 Poznan, Poland
tomasz.marciniak@put.poznan.pl

Abstract. The paper presents an analysis of the realization of digital signal processing algorithms with the use of STM32F4 microcontrollers. Various programming techniques have been demonstrated focusing on the implementation of FIR filters and FFT calculations based on Cortex Microcontroller Software Interface Standard. During the tests we use STM32F407 Discovery module with WM5102 Wolfson Audio Card. The speed of algorithms and the accuracy of calculations were checked. The quality of digital filters were tested using LMS adaptive filter or Analog Discovery 2 device.

Keywords: Digital filters · FFT · CMSIS · STM32 · STM32F407

1 Introduction

Modern microcontrollers have many features previously available only in signal processors. Such features include:

- MAC (Multiply and ACcumulate) calculation unit
- SIMD (Single Instruction, Multiple Data) instructions
- Special addressing modes (circular and with bit reverse).

Manufacturers of microcontrollers offer such features in their systems while maintaining wide communication capabilities or analog-to-digital converters. The combination of all these features allows greater efficiency in typical DSP algorithms, such as FIR and IIR filtration, FFT calculations, PID control or matrix operations (addition, subtraction, multiplication).

Many signal microcontroller manufacturers rely on Cortex-M cores (Cortex-M33, Cortex-M35P, Cortex-M4 and Cortex-M7) ARM Ltd. [1, 2]. The advantages of M4 and M7 cores (in addition to typical DSP features) include the floating point FPU (Floating-Point Unit) arithmetic coprocessor. According to ARM, the Cortex-M7 core is dedicated to autonomous solutions in automation [3].

One of the producers using the above solutions is STMicroelectronics, which, based on ARM cores, offers, among others, families with the designations STM32F4 and

A. Bartoszewicz et al. (Eds.): PCC 2020, AISC 1196, pp. 287–295, 2020.
https://doi.org/10.1007/978-3-030-50936-1_25

STM32F7, using Cortex-M4 and Cortex-M7 [4] cores, respectively. An attractive feature of these families is also the 32-bit data representation that allows for a large dynamic range. The popularity of STM32 solutions results, among others, from the availability of a number of experimental modules Nucleo and Discovery and the possibility of application in automation systems [5–7].

An example of an experimental DSP system is STM32F407 Discovery board with WM5102 Wolfson Audio Card, used by the authors for experimental research and shown in Fig. 1. The STM32F407 Discovery module [8] contains an STM32F407 ARM Cortex-M4 processor with the following features:

- frequency up to 168 MHz, 210 DMIPS/1.25 DMIPS/MHz (Dhrystone 2.1)
- 1 Mbyte of Flash memory
- 192 + 4 Kbytes of SRAM including 64-Kbyte of CCM (core coupled memory) data RAM
- 3 × 12-bit, 2.4 MSPS A/D converters: 24 channels and 7.2 MSPS in triple interleaved mode
- 2 × 12-bit D/A converters.

The module communicates with the host PC via the USB interface, using the ST-LINK tool for programming and debugging. The Keil MDK-ARM environment allows (on the host PC) to compile software written in C/C++, link it and upload to STM32F407 memory. Real-time audio inputs/outputs (I/O) are made available via the Wolfson Pi Audio [9] card (designed primarily for Raspberry Pi modules), which communicates via cable using the I2C and I2S interfaces.

Fig. 1. STM32F407 discovery board with WM5102 Wolfson audio card

2 Cortex Microcontroller Software Interface Standard

Cortex Microcontroller Software Interface Standard - CMSIS, developed by ARM, is a standardized software interface for developers of Cortex-M microcontrollers, independent of their manufacturers [10]. The use of CMSIS libraries creates a hardware independent design environment (HAL - Hardware Abstractions Layer), which facilitates the transfer of programs between microcontrollers with Cortex-M core from different manufacturers. This strategy is based on two elements: hardware and software. On the hardware side, more blocks were integrated in the Cortex-Mx architecture to increase their functional capabilities than in the architecture of earlier ARM7 cores. For example, Cortex-Mx systems from different manufacturers have not only the same processor part (CPU), but also the same vector nested interrupt controller (NVIC - Nested Vectored Interrupt Controller), the same system timer core (SysTick) and the same module logical to detect errors in the program (debugging). This guarantees that all standard software (firmware) will have the same functional properties when launched on all microcontrollers equipped with Cortex-Mx core. Although the system functionality must always be implemented in the program code that is common to all microcontrollers, a reference to CMSIS is required. CMSIS has also been conceived to allow the designer to easily migrate from one software tool provider to another. Therefore, when using the real-time operating system (RTOS) in the application, access to the central unit (CPU) allows an interface independent of the type of microcontroller, including the debug channel.

The CMSIS DSP library contains signal processing functions divided in some categories [10]:

- Basic math functions
- Fast math functions
- Complex math functions
- Filters
- Matrix functions
- Transforms
- Motor control functions
- Statistical functions
- Support functions
- Interpolation functions.

The user can operate on many types of variables such as 32-bit floats and 8-, 16- or 32-bit integers. Signal processing algorithms require full enable of the Cortex-M4 processor capabilities by implementation ARM DSP SIMD (Single Instruction Multiple Data) instruction set and floating-point hardware. The CMSIS DSP library is written in C language and contains source codes for easy implementing own programs for specialized requirements [10].

The X-CUBE-DSPDEMO firmware package allows to demonstrate the use of the DSP library provided with CMSIS [11]. Thanks to this, it is possible to analyze examples of FFT and FIR, as part of full integration with the STM32 family using perfusion devices. It is also possible to run a graphical user interface. During the experiment, the user can adjust settings such as the frequency of the input signal and the adjustment of the data type (fixed or floating point). If the user selects the FIR action, he can change the data type and type of filter (low-pass or high-pass) used.

The firmware is free to download from manufacturer's website. The newest version V1.0.0/17-04-2016 contains examples for STM32F746 and STM32F429. The user have to open it using preferred toolchain, rebuild all files and load to memory of device. Running CMSIS library starts with main program which reset peripherals, initialize required interfaces and SysTick. Then the system clock is configured to run with proper frequency. The next step is calibration of the touch screen after its initialization. The user can choose between examples such as FFT (Q15, Q31, F32) or FIR. It is very important to take care using HAL_Delay() function which provides accurate delay based on variable in SysTick ISR. If this function is called from a peripheral ISR process, then the SysTick interrupt priority have to be higher. In the other case ISR process caller will be blocked. The example directory contains main program, configuration of GUI, FFT example, FIR example, MCU peripherals, RTC, system clock configuration file, interrupt handlers, HAL MSP module, HAL configuration file and required headers files [11].

3 Performance Analysis of FIR and FFT

3.1 Speed of Calculations Using Various Programming Techniques

An interesting analysis of DSP calculation performance can be found in Application note "Digital signal processing for STM32 microcontrolers using CMSIS" [12], which contains analysis for STM32F429 and STM32F746 microcontrollers. Analysis of the performance of FFT calculations shows that thanks to FPU, the calculation time for the floating point representation F32 is comparable with the time for the representation Q15, and even lower than for the Q32 representation. It is even more interesting that in the case of a floating-point representation, the calculation time for FIR filters is lower than in the case of a fixed-point representation [12].

The computational performance of the STM32F407 processor was evaluated for the filter length 81 [13] described in Sect. 3.2. Measurements were made using the Keil uVision environment. Table 1 shows the calculation time for one output sample (F32) using CMSIS (`arm_fir_f32`) and DMA. In addition, the calculation time for direct implementation (C language equations) is illustratively shown. Comparing the calculation time with the processor frequency of 168 MHz and assuming that a filter length of $N = 81$ has been used, it can be estimated that the use of CMSIS and DMA allows to obtain the efficiency of 4 clock cycles per one MAC operation (one tap of filter).

Table 1. Speed performance of FIR filter ($N = 81$).

Calculation type	Execution time (in μs)
C language equations	10.093
C language equations + DMA	11.587
CMSIS	7.208
CMSIS + DMA	1.727

The complex FFT calculation performance for the data block size $N = 128$ was also checked. The time consumption of FFT implementation in C language and the `arm_cfft_f32` function were compared. In addition, the calculation times for direct implementation of DFT and DFTW (twiddle factors previously calculated) are illustratively shown. The result obtained for the CMSIS library makes it possible to assess that real-time calculations can be carried out even for theoretically acquisition rates of over two million samples per second (Table 2).

Table 2. Speed performance of FFT ($N = 128$).

Calculation type	Execution time (in ms)
Direct DFT	2277.8847
Direct DFTW	3.3307
FFT	0.1447
CMSIS FFT	0.0595

3.2 Accuracy of Calculations

The quality of calculations was evaluated by analyzing the frequency characteristics of an example bandpass filter with the following parameters [13]: filter order 80 (float data type), sampling rate 8000 samples / second, cut-off frequency $Fc_1 = 1500$ Hz, $Fc_2 = 2000$ Hz. The filter coefficients can be determined using the Filter Design & Analysis Tool in the Matlab environment.

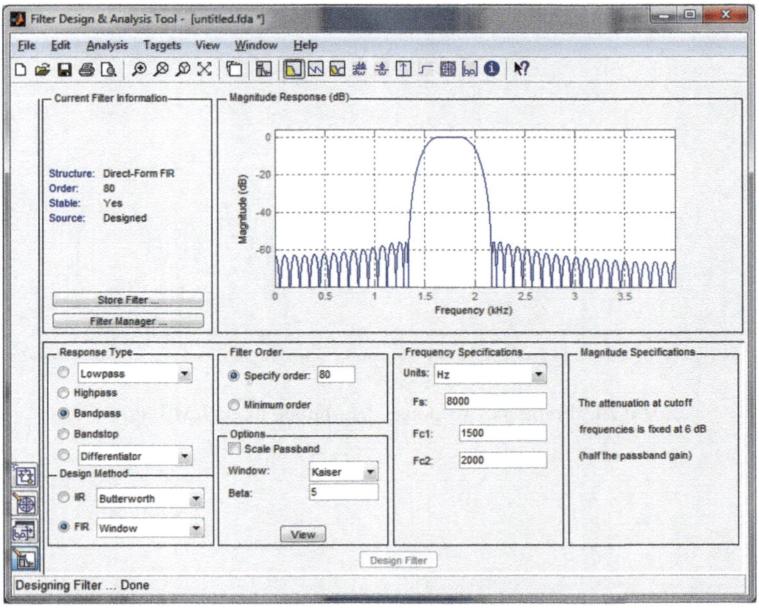

Fig. 2. Project of FIR filter using FDA tool

The filter was tested using two solutions. Figure 3 shows the application of the software noise generator and the LMS adaptive filter (`arm_lms_f32` function with 256 taps number) [13]. In this case, the output from the D/A converter is directly connected to the input of the A/D converter.

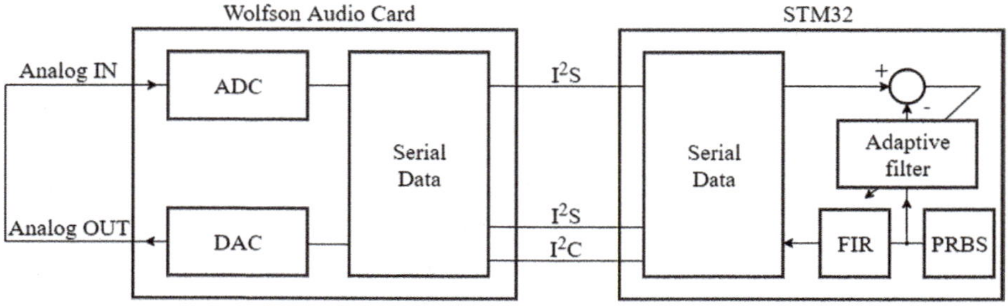

Fig. 3. Application of PRBS generator and LMS filter to determining frequency response

PRBS (pseudo-random bit sequence) function allows to generate noise sequence using bit logic and shift register. It uses short type to return the value (signed 16-bit integer) after XOR operation on XOR of bits 0, 1 and 11, 13. Shift register allows to change value of output and scale noise level which is set up in header file by performing operations byte by byte. It depends on the first byte which is responsible for sign of the variable (positive or negative) (Fig. 4).

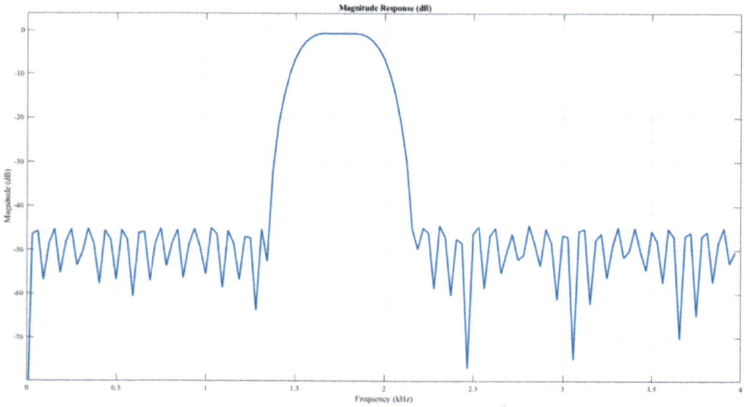

Fig. 4. Frequency response determined using LMS filter

Frequency response was also acquired using the Digilent Analog Discovery 2 device and WaveForms software, which integrates, among others, signal generator and 2-channel oscilloscope, and also allows to determine the frequency characteristics in the range from 1 Hz to 10 MHz [14, 15]. The general connection diagram is shown in Fig. 5, and the resulting frequency characteristics in Fig. 6. Figure 7 shows three frequency responses together. It can be seen that the graph in Fig. 6 is more similar, in terms of stop-bands, compared to the filter design in Fig. 2.

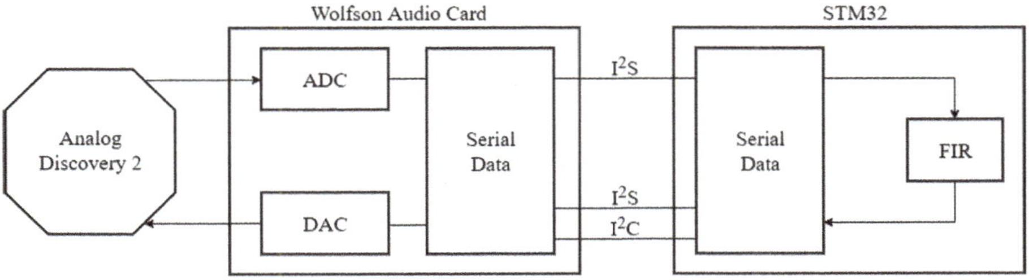

Fig. 5. Connection diagram for identification of magnitude response using Analog Discovery 2

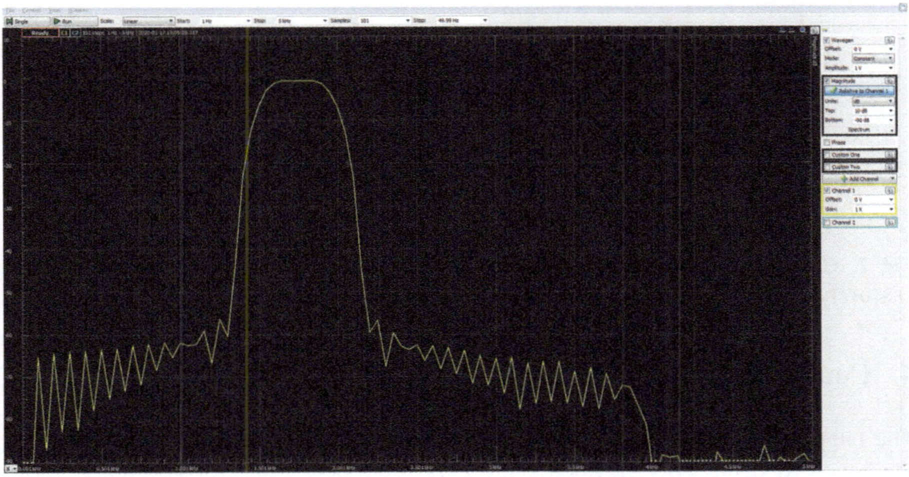

Fig. 6. Frequency response determined using Analog Discovery 2

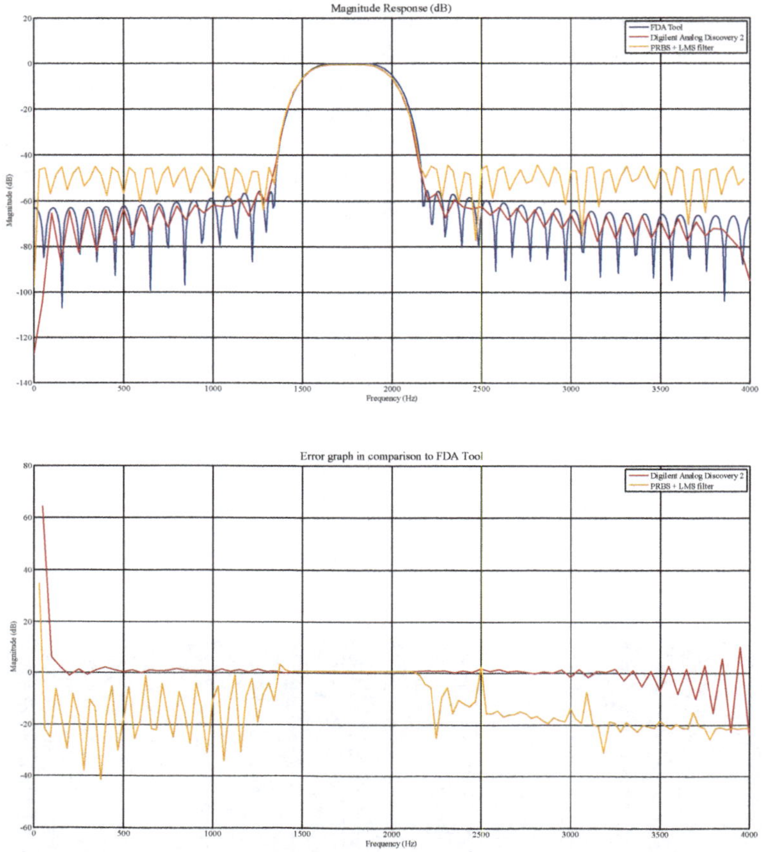

Fig. 7. Comparison of frequency responses of an example bandpass filter measured using Analog Discovery 2 and LMS filter

4 Conclusions

The paper deals with the implementation of algorithms for digital signal processing implemented using STM32 microcontrollers. The Cortex Microcontroller Software Interface Standard library was used to assess the efficiency and accuracy of calculations in FIR and FFT algorithms.

Despite many programming facilities, such as STM32CubeMX, implementation of DSP algorithms requires consideration of microcontroller resources, e.g. DMA channels. As shown when testing the characteristics of FIR filters, the computational capabilities of microcontrollers allow to evaluate the correct operation of such filters without the use of additional measuring equipment such as spectrum analyzer.

Due to the limited size of the paper, only selected aspects of filtration and FFT calculations are shown. The authors plan a comprehensive comparison for different families of STM32 microcontrollers as well as a comparison with the TMS320 families of Texas Instruments.

References

1. Reay, D.S.: Digital Signal Processing Using the ARM Cortex-M4. Wiley, Hoboken (2016)
2. Ünsalan, C., Yücel, M.E., Gürhan, H.D.: Digital Signal Processing using Arm Cortex-M based Microcontrollers: Theory and Practice, Arm Ltd. (2018)
3. Arm Processors for the Widest Range of Devices—from Sensors to Servers. https://www.arm.com/products/silicon-ip-cpu. Accessed 18 Jan 2020
4. STM32 32-bit Arm Cortex MCUs. https://www.st.com/en/microcontrollers-microprocessors/stm32–32-bit-arm-cortex-mcus.html. Accessed 18 Jan 2020
5. Szewczyk, P.: Real-time control of active stereo vision system. In: Mitkowski, W., et al. (eds.) Trends in Advanced Intelligent Control, Optimization and Automation, AISC, vol. 577, pp. 271-280. Springer, Cham (2017). https://doi.org/10.1007/978-3-319-60699-6_26
6. Chaber, P., Ławryńczuk, M.: Automatic code generation of MIMO model predictive control algorithms using transcompiler. In: Mitkowski, W., et al. (eds.) Trends in Advanced Intelligent Control, Optimization and Automation, AISC, vol. 577, pp. 315–324. Springer, Cham (2017). https://doi.org/10.1007/978-3-319-60699-6_30
7. Chaber, P., Ławryńczuk, M.: Implementation of analytical generalized predictive controller for very fast applications using microcontrollers: preliminary results. In: Mitkowski, W., et al. (eds.) Trends in Advanced Intelligent Control, Optimization and Automation. AISC, vol. 577, pp. 378–387. Springer, Cham (2017). https://doi.org/10.1007/978-3-319-60699-6_37
8. STM32F4DISCOVERY Discovery kit with STM32F407VG MCU (2016). https://www.st.com/en/evaluation-tools/stm32f4discovery.html. Accessed 18 Jan 2020
9. Wolfson Microelectronics plc: WM5102 Audio Hub CODEC with Voice Processor DSP, Wolfson Microelectronics plc (2014)
10. CMSIS DSP Software Library. https://arm-software.github.io/CMSIS_5/DSP/html/index.html. Accessed 18 Jan 2020
11. STM Homepage. https://www.st.com/en/embedded-software/x-cube-dspdemo.html. Accessed 02 Jan 2020
12. Digital signal processing for STM32 microcontrollers using CMSIS, Application note, AN4841 (2018)
13. ARM-based Digital Signal Processing Lab-in-a-Box ST Discovery Board and Wolfson Audio Card Edition, ARM University Program (2013)
14. The Analog Discovery 2: A portable USB laboratory for everyone. https://analogdiscovery.com/. Accessed 18 Jan 2020
15. Marciniak, T., Dąbrowski, A., Puchalski, R., Dratwiak, D., Marciniak, W.: Zastosowanie mikrokontrolera STM32F410 do prezentacji zagadnień cyfrowego przetwarzania sygnałów (Application of STM32F410 microcontroller for presentation of digital signal processing). Przegląd Elektrotechniczny R. **95**, 118–120 (2019)

Diagnostics of Processes - Cloud Concept Study

Jan Maciej Kościelny, Michał Bartyś(✉), and Paweł Wnuk

Warsaw University of Technology, A. Boboli 8, 02-525 Warszawa, Poland
{jan.koscielny,michal.bartys,pawel.wnuk}@pw.edu.pl

Abstract. This paper is principally aimed at presenting and discussing the plausibility of implementation of an idea of transferring diagnostics tasks to computational cloud. Particular attention was paid to diagnostics of industrial processes. Some crucial aspects, such as: broadcasting of diagnostic services, revolution in diagnostic management, technical support and cybersecurity issues are discussed, however on academic level.

Keywords: Cloud computing · Diagnostics of processes · Model based diagnostics · Cybersecurity

1 Introduction

The rapidly developing communication and processing technologies have given rise to the concept of cyber-physical networks, in which agents aim at cooperatively solving complex tasks [1]. Doubtless, the diagnostics of industrial processes belongs to such tasks.

The diagnostic systems (*DS*) intended for industrial applications make use of advanced model-based fault prediction, detection and isolation methods [2]. Generally, the fault prediction makes sense in case of incipient faults caused by material wear, sedimentation, obliteration, cocking, fatigue, corrosion, etc. slowly development processes. Such processes involve destruction effects that affect technical characteristics and detoriate the exploitation figures of components, instrumentation and actuators of controlled systems.

The fast detection and accurate fault isolation shall be sought in case of abrupt faults in order to undertake appropriate preventive actions by process operators or by fault tolerant control systems or protective actions provided by safety instrumented systems.

Though, currently the *DS* are not extensively exploited, despite their numerous advantages. The known pilot implementations and industrial trials were mainly conducted by the academic research centers. Up-to-date, the world leading manufacturers of control equipment do not provide any widely recognized and approved solutions of *DS* for industrial processes.

The main reason is the sensitivity of the model based *DS* on non-stationarity of the controlled processes as well as on variation of model parameter caused by

A. Bartoszewicz et al. (Eds.): PCC 2020, AISC 1196, pp. 296–306, 2020.
https://doi.org/10.1007/978-3-030-50936-1_26

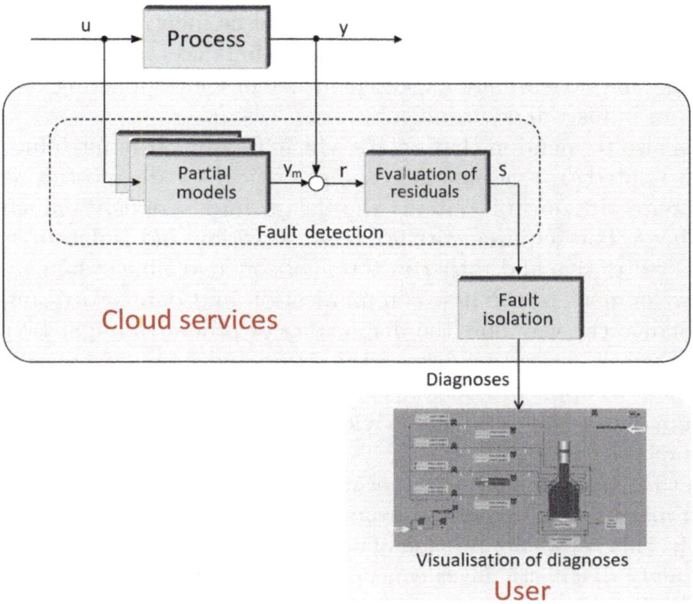

Fig. 1. Conceptual illustration of cloud computing services in a model based diagnostics of processes.

plant renovation actions. Therefore, there is demand on continuous fine-tuning of *DS* models. This, however needs to employ highly-qualified, experienced staff.

It is to state definitely, that there still exists a huge gap between the theory of diagnostics of dynamic systems and the state of the technique. Postulatively, this gap should be filled-in in the near future due to the strong requirements on providing, enforcing and ensuring process safety.

The new opportunities in this regard come with the development of the Industrial Internet of Things (*IIoT*), Internet of Services (*IoS*), as well as the use of cloud computing, data storage and processing that support the idea of Industry 4.0. The development of these technologies provides among others the possibility of convenient storing, processing and making use from big data originated from distributed and heterogenous sources eg. from data acquisition systems, humans, safety instrumented systems etc. This allows for realization of complex data analysis, the use of computing intelligence and machine learning methods. Therefore, it provides tools necessary for the implementation of distributed up-to-date fault detection and isolation algorithms. Figure 1 depicts the example of conceptual diagram of an automatised diagnosing scheme based on partial models of the process [3] with pointed out areas of cloud computing services. There is to mention that model based diagnostics is not the only class of diagnostic approaches which can be considered to be performed in computational cloud. Any other different diagnostic concepts may be also implemented and performed by cloud computing services. However, it should be mentioned, that there is not to expect that performing specific cloud diagnosing tasks will improve the quality of diagnoses. The benefits are expected rather in wide-spreading of diagnostics approaches in industrial environment.

There is also to mention that delays which are obviously introduced by communication related transactions are not so critical for diagnostics as for cloud control systems. In control systems the delays may seriously influence control quality indices. The requirements for fault detection and isolation systems are defined less restrictive and rather in terms of soft real time systems.

The development of the new communication and data processing technologies may change the way how the diagnostics of processes might be performed. Therefore, there is a need to discuss this issue and point out foreseen avenues of development of applied model-based fault detection and isolation approaches at least from the academic point of view. This may be understood as a main motivation of this paper.

Firstly, the new communication technologies allow for making use from wider attempt to information originated from different geographically spread sources. For example, this allows application of advanced high level diagnostic approaches including safety diagnostic layer concept presented in [4].

Secondly, it may change the way, how the diagnostics may be performed. It is supposed that it will strengthen the development of distributed diagnostic approaches. Here, the key-point will be problem of cybersecurity.

Thirdly, it may push forward the number of implementations of diagnostic systems by development of Diagnostics in Cloud (DiC) services served by specialised commercial providers. This will allow to wider attempt to these services by a number of low and middle scale industrial companies.

And, fourthly, the manufacturers of automatic control equipment can participate in DS making use from deep-in knowledge of their own products. Moreover, in a long term, this may help for enhancement of reliability figures of provided products.

This paper discusses the proposition of a general concept and contributes to the discussion on how heavy may be the impact of emerging cloud technologies on the diagnostics of processes. The rapid growth of sensor applications (smart city, smart grid, transport, Industry 4.0, etc.) and the extensively usage of smart sensors in a new generation of machines, devices and technology components makes easier and create new areas for advanced diagnostics. High reliability, functional safety and cybersecurity of automatic control equipment are becoming a common requirements for industrial equipment.

Therefore, it is sought for self-improving, self-adaptation, self-tuning, self-diagnosing, and self-repair generation of the new industrial automatic control devices. Real-time advanced diagnostics are an effective tool for improving the process safety and cybersecurity of all technical systems, especially industrial devices and processes that pose a threat to human beings or environment [4–6]. The significant economic benefits can be gained by means of advanced service strategies providing monitoring of conditional health of technical facilities and predictive maintenance of incipient faults. In this study we adopt an integrated approach to fault diagnosis independently whether the root cause of faults is connected with physical equipment or is result of a cyberattack. This approach to process safety is recommended and referenced in [7,8].

2 The Principles of Diagnostics in Cloud

Advanced real-time diagnostics is one of the highlights of a new approach to the organization and optimization of production in modern technological facilities known as Industry 4.0 [9]. The classic approach to diagnostics relies on the simply checking of limit values of individual physical quantities representing process variables. The advanced fault detection makes use from methods based on quantitative and/or qualitative models of the diagnosed system. Then, the classification or automatic inference methods are used for fault isolation [10,11]. Model-based diagnostics allows early recognition of low size abrupt faults before the negative effects of these faults be revealed. They allows to prognosticate the development of incipient faults (monitoring the degree of degradation degree) as well.

The paper [9] presents the concept of decentralization of diagnostic tasks based on automation control devices and technology apparatuses equipped with embedded or autonomous diagnosing units. The concept, features and examples of applications of the next generation of "smart" diagnosing devices called as diagnozers is presented. By adopting the functional criterion, the diagnozers are classified as: fault sensors, isolators, detectors, analyzers and predictors. These devices can be implemented as the stand-alone or as embedded units.

The Industrial Internet of Things (*IIoT*) and cloud computing technology create quite new opportunities for diagnostic systems, regardless of decentralisation of diagnostic tasks carried out by embedded or autonomous diagnozers.

The general concept of on-line diagnostics of devices, equipment, machinery and processes in computational cloud is illustrated in Fig. 2.

It involves the use of computational cloud in order to store measurement data originating from a wide variety of objects and systems together with diagnostic software configured for particular objects.

It is expected that services of diagnostic software will be provided by a company or companies which are specialized in the design, maintenance and configuration of diagnostic systems. The main task of such company or companies will be to design a knowledge base, development and verification of system models for fault detection and isolation, commissioning of *DS*, operational supervision, modification of *DS* after each repair action, etc. On the other hand, current diagnoses will be transmitted directly to the subscribers e.g. process operators, maintenance services, management, etc.

The diagnostic services may be provided by:

a) manufacturer of the device, machine or technology apparatus. In this case, it supervises its own products located in various production companies around the world;

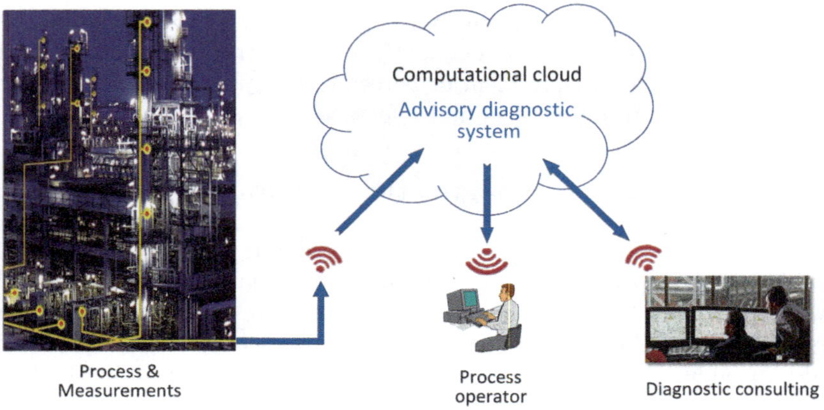

Fig. 2. Illustration of the concept of on-line diagnostics in computational cloud.

b) a specialized company that provides diagnostic outsourcing services for different facilities located along one or more enterprises;
c) specialized department of maintenance services of a large production company.

A company providing diagnostic services should employ high qualified staff possessing specialized competence regarding diagnostics of industrial processes. This is almost impossible to realize in small scale production companies.

Though, such solution breaks down the problem of the lack of trained personnel, which is one of the most important causes of a small number of implementations of advanced diagnostic systems. The staff of such company will be enabled to adequately react and tune the models which are used to fault detection particularly in case of performed repairs, renovations, modifications of hardware as well as by any change of model parameters.

The main purpose of diagnostic tasks performed in the cloud is to make use from software that enables to detect any threats, faults, and cyberattacks. The fundamental difference between fault and cyberattack concerns the source of the threat origin. The cyberattack is deliberately inspired by a humans and intentionally oriented to cause certain losses. On the other hand, the fault arises suddenly or increasingly due to destructive physical processes occurring in the device (aging, wear, etc.) or erroneous (unintended) operation of that device.

Despite different causes, the effects of dangerous faults and cyberattacks can be the same: e.g. shut-off the process, fire, explosion, environmental pollution, destruction of the plant. The same elements of control systems, that are crucial to process safety, can be the target of cyberattacks and regular faults. The alarm systems should not be considered hereas the effective tools for recognizing any of these classes of threats. The symptoms of these threats could be the same or similar. Therefore, it is not possible to carry out fault diagnosis based on symptoms, aside from possible cyberattacks and vice versa. Finally, as shown in

[4], the cyberattacks can be detected by means of the same approaches as faults. All these leads to the conclusion that the diagnosis of faults and cyberattacks should be carried out in the same diagnostic system. Therefore, the separation of detection of cyberattacs and faults is not justified.

The effective recognition of threats in control systems requires implementation of advanced diagnostic approaches of industrial processes. The detection of faults/attacks relies in early detection of discrepancies between actual process outputs and outputs of reference quantitative or qualitative process models. These models are reflecting the behaviour of modelled system in normal (nominal) state. Model-based detection methods allow for early detection of threats. The inference about the causes of observed discrepancies is based on observed diagnostic signals. Diagnostic signals are the outputs of detection algorithms.

The diagnosis is the result of automatic inference based on diagnostic signals. The diagnosis may be understood as the hypothesis concerning faults and/or cyberattacks. Moreover, the diagnostic system in emergency states can generate advises to process operators. This allows them to take effective safeguarding measures. This should bring the process to normal state. As a result, the safety instrumented system (SIS) will not be triggered. This avoids the shut-down of the part or the whole process which, in fact, may cause significant economic losses.

The concept of diagnostics in cloud presented in this paper can be realised in the numerous industrial plants and systems. Below we will address only one chosen example. Particularly interesting are applications concerning monitoring and diagnosing of geographically spread water and sewage networks. Surely, in this case, the main task of diagnostics is fault detection and isolation. Obviously, the serious water loses may result in case of faults which may occur in pipelines, pumps, filters, throttles, valves, instrumentation, data transmission devices, etc. These, may be reason of water outages or shortages for customers. The water and economic losses caused by leaks are enormous high. Wyczółkowski in [12] reported that for urban agglomerations populated with more than 200,000 inhabitants, the average water losses in water supply networks equals 19.9%.

3 Technical Support for Diagnostics in Cloud Concept

According to worldwide recognized research and advisory company Gartner, the global public cloud market is expected to grow by 17.3% in 2019 and reach $206.2 billion. Nowadays in Poland nearly 40% of large scale companies and 27% of middle-scale companies use cloud solutions.

Doubtless, the mile stone will be global implementation of 5G cellular networks. This will promise rapid development of cloud diagnostic services as well as cloud control services of industrial processes.

3.1 Cloud Platforms and Services

Generally, the computational cloud may be seen as an ever-expanding set of cloud computing services. Currently, in industrial practice, the most popular are Software as a Service (*SaaS*) services. However, Infrastructure as a Service (*IaaS*) has also gained a lot over the past few years. Clearly, the Diagnostics in Cloud (*DiC*) concept fits into *SaaS* service model. Also, the *IaaS* model seems to be useful in terms of infrastructure services.

In this regard it is worth mentioning that with the cloud, the computing equipment downtime due to failures and maintenance services is almost non-existent. For a non cloud computing there is also worth to mention the potential usability of the *edge computing*, that is, implementation of diagnostic algorithms on network edge devices such as routers, servers, *PAC*s, etc.

Particularly useful for implementation of *DiC* concept might be MindSphere real-time open cloud platform developed by Siemens [13]. MindSphere provides memory storage services and makes them accessible through digital applications ("MindApps"). Currently, the platform is used in process automation, predictive maintenance and vehicle fleet management. In fact, the predictive maintenance falls directly into (*DiC*) concept. The series of auxiliary MindSphere products (e.g. Data Capture Unit, MindConnect IoT2040 or MindConnect Nano) can be used for real-time data acquisition, communication and big data storage. Moreover, the MindSphere provides a set of open application programming interfaces (*API*) as well as appropriate development tools. This allows to recommend this platform as potentially useful in respect to advanced diagnostic cloud services. Fortunately, up-to-date, there are much more global providers of cloud computing services like Microsoft Azure [14], Google Cloud [15] or AWS from Amazon [16].

Azure cloud computing platform provides: Windows and Linux virtual machines web and mobile application services, massively scalable mass storage, distributed multi-model scalable database, cognitive services of smart *API* capabilities enabling contextual interactions, intelligent *SQL* in the cloud, etc. capabilities. It gives the freedom to build, manage, and deploy applications on a massive, global network using own tools and frameworks.

Secure, high-performance, scalable, and cost-effective cloud platform provides also Google Cloud [15]. As all global cloud computing services it deploys own developed kubernetes services. Kubernetes is a portable, extensible, open-source platform for managing containerized workloads and services. Google Cloud provides: computing, big data stream, batch processing and exploration, *IoT*, artificial intelligence, storage and databases, data transfer, networking, *API* platforms, ecosystems and development tools services.

AWS from Amazon [16] similar to remaining global cloud providers offers: analytical, computing, mobile, *IoT*, networking, machine learning, database, storage, blockchain, robotics, and development tools services. Interesting, in respect to *DiC* concept, seems to be: real-time streaming data service offered by Kinesis, deep-learning interface acceleration offered by Elastic Inference, machine learning inference chip supported by Inferentia services. Particularly valuable for *DiC* seems to be Machine Learning tool. The Amazon provides an example of model based predictive maintenance deploying a machine learning for detection of turbofan degradation.

3.2 Technical Support for Embedded Diagnostics

The idea of smart diagnozers for decentralized diagnostics in Industry 4.0 presented in [9] is based on the concept of dispersing diagnostics among specialised diagnostic units (fault sensors, detectors, isolators, analyzers, predictors) and making use from idea of embedded diagnostics. This however, requires appropriate technological support, particularly when thinking about cloud implementations. Fortunately, nowadays there is available appropriate hardware on the market. Below we introduce an example of a cellular system designed for $IIoT$ applications. It allows direct cellular attempt to computational cloud without necessity of using intermediate access devices.

Low power, nRF9160 cellular IoT System-in-Package (SiP) with integrated $LTE - M$ and $NB - IoT$ wireless modem makes the LTE technology accessible for a wide range of industrial applications. Through the high integration and pre-certification for global operation it solves the complex wireless design. It integrates an application processor, multimode $LTE - M/NB - IoT$ modem, RF front-end, GPS and power management in a relatively tiny package. Built-in GPS module combines location data from the cellular network with GPS satellite trilateration to allow remote monitoring of the device position. A range of analog and digital peripherals supports the powerful application processor. The integrated cryptographic processing enables the $nRF9160$ to meet the demanding SiC security requirements.

4 Security Issues

Nowadays the security within Industrial Control Systems (ICS) is a noticeable problem. Many papers focuses on cyberattack detection [17], resilient control [18], or securing ICS systems [19]. It is usually supposed that one of the basic and effective protective actions against cyberattacks of the OT networks and devices is their separation from other networks and systems. The diagnostic system could be seen as an additional processing layer located outer the standard ICS. Therefore, it is affected by the same threats. In addition, the diagnostic system can't be truly separated from real-time measurements and process variables which are necessary for proper operation of controlled system.

It is noteworthy to mention that true separation of ICS and diagnostic layer is not possible in case if diagnostic and/or maintenance services are moved to the computing cloud. In this case, sensitive process data (measurements, control commands, diagnoses) must be physically transferred outside the plant, and secured against interception by transmission (protect confidentiality), and against distortion of the processing results (protect integrity). From the DiC perspective, the security is the major concern that hampers the adoption of the cloud computing model [20] because:

⋆ enterprises outsource security management to a third parties that hosts their IT assets (loss of control);

⋆ co-existence of assets of different processes and/or enterprises in the same location and using the same instance of the service while being unaware of the strength of security controls used;
⋆ the lack of security guarantees in the current agreements between the cloud consumers and providers;
⋆ the use of publicly available cloud systems, which in itself increase the probability of attacks.

Usually cloud computing model consists of three different layers. Infrastructure as a service allows the tenant to run its own software (including operating system) on a virtualized hardware. For diagnostic system much more interesting are two higher layers: - Platform as a Service (*PaaS*) which serves a platform for implementation and running client applications, and *Saas* providing attempt to complete software applications. It should be taken into account that cloud security architecture can effectively secure the cloud services if and only if the correct defensive actions are carried out at a correct level.

The cloud computing model delivers two key characteristics: multi-tenancy and elasticity. Both of them have serious implications on the cloud model security. Securing cloud systems is a common responsibility of cloud provider and cloud consumer. First of all, - the virtual machines (*VM's*) need to be protected in the same manner as traditional physical servers. The same applies to securing machine images, shared disk space or virtual network. But this is the client responsibility mainly in *PaaS* model. *PaaS* model is based on *SOA* (Service-Oriented Architecture) - what means that inherits all security issues that exists in the *SOA* domain such as *DOS* attacks, replay attacks, man-in-the middle attacks, and many others [20]. In the *SaaS* model enforcing and maintaining security is a shared responsibility among the cloud providers and service providers (software vendors). The *SaaS* model inherits the security issues discussed in the previous two models as it is built on top of both of them including data security management [21] (data locality, integrity, segregation, access, confidentiality, backups) and network security. In this case all security techniques used to securing web applications can be used.

The use of *DiC* sets the new challenges in area of industrial data and systems security. Fortunately, the cloud processing model needed by *DiC* seems to be very similar to *PaaS*/*SaaS*, that allows us to rely on solutions known from *IT* world.

5 Conclusions

The rapidly developing communication and processing technologies has given rise to the concept of diagnostic cyber-physical networks.

The current state of technique makes realistic the implementation of the concept of diagnostics in the cloud. This applies to network services, provision of tools for the development of diagnostic software, as well as to hardware support for the development and application of new generation of diagnostic equipment. This promises to disseminate the idea of *DiC* on a global scale. This gives hope for a wide spread of diagnostics approaches in industry including low and middle scale enterprises and development of diagnostic oriented spin-offs.

Thanks to security assistance and support offered by cloud providers combined with possibilities of assurance of strong cryptographic security of equipment, the degree of vulnerability to cyberattacks is much lower compared to traditional infrastructure.

For the industry, it will be crucial the wide spread of $NB - IoT$ technology together with 5G networks which will substantially support diagnosing as well as control tasks by means of cloud services.

In respect to functional safety, there is important to develop and unify the cloud-to-cloud interfaces allowing for rapid switching between clouds or realizing redundant processing channels.

References

1. Notarstefano, G., Notarnicola, S., Camisa, A.: Distributed optimization for smart cyber-physical networks. Found. Trends® Syst. Control **7**(3), 253–383 (2019). http://dx.doi.org/10.1561/2600000020
2. Korbicz, J., Kościelny, J.M. (eds.): Modeling, Diagnostics and Process Control: Implementation in the DiaSter System. Springer, Heidelberg (2010)
3. Kościelny, J.M.: Diagnostyka on-line procesów przemysłowych dużej skali. In: Kulczycki, P., Korbicz, J., Kacprzyk, J. (eds.) Automatyka, robotyka i przetwarzanie (2020)
4. Kościelny, J.M., Bartyś, M.: The requirements for a new layer in the industrial safety systems. In: 9th IFAC Symposium on Fault Detection, Supervision and Safety of Technical Processes, SafeProcess 2015, Paris, France, 2–4 September, pp. 1333–1338 (2015)
5. Kościelny, J.M., Syfert, M., Wnuk, P.: The idea of on-line diagnostics as a method of cyberattack recognition. In: International Conference on Diagnostics of Processes and Systems. Advanced Solutions in Diagnostics and Fault Tolerant Control, pp. 449–457. Springer (2017)
6. Van Long, D.: Sequential detection and isolation of cyber-physical attacks on SCADA systems. Theèse de doctorat de l"Universite de Technologie de Troyes (2015)
7. Kosmowski, K.T., Śliwiński, M., Barnert, T.: Functional safety and security assessment of the control and protection systems. In: Soares, G., Zio, E. (eds.) Safety and Reliability for Managing Risk. Taylor & Francis Group, London (2006)
8. Śliwiński, M.: Bezpieczeństwo funkcjonalne i ochrona informacji w obiektach i systemach infrastruktury krytycznej, vol. 171. Politechnika Gdańska, seria - monographs, Gdańsk (2018)
9. Kościelny, J.M., Bartyś, M.: The idea of smart diagnozers for decentralized diagnostics in industry 4.0. IEEE Xplore Digital Library (2019). https://doi.org/10.1109/SYSTOL.2019.8864791
10. Kościelny, J.M., Syfert, M.: Application properties of methods for fault detection and isolation in the diagnosis of complex large-scale processes. Bull. Pol. Acad. Sci. Tech. Sci. **62**(3), 571–582 (2014)
11. Isermann, R.: Fault Diagnosis Systems: An Introduction from Fault Detection to Fault Tolerance. Springer-Verlag, New York (2006)

12. Wyczółkowski, R.: Metodyka detekcji i lokalizacji uszkodzeń sieci wodociągowych z wykorzystaniem modeli przybliżonych, vol. 493. Politechnika Śląska, Wydawnictwo Politechniki Śląskiej (2013)
13. https://new.siemens.com/global/en/products/software/mindsphere.html. Accessed 12 Jan 2020
14. https://docs.microsoft.com/en-us/azure/security/azure-security-services-technologies. Accessed 12 Jan 2020
15. https://cloud.google.com/security/products. Accessed 12 Jan 2020
16. https://aws.amazon.com/products/security. Accessed 12 Jan 2020
17. Quevedo, J., Sánchez, H., Rotondo, D., Escobet, T., Puig, V.: A two-tank benchmark for detection and isolation of cyber attacks. IFAC-PapersOnLine **51**(24), 770–775 (2018)
18. Rebaï, S.B., Voos, H., Amin, S., Alamdari, S.: A contribution to cyber-physical systems security: an event-based attack-tolerant control approach. IFAC-PapersOnLine **51**(24), 957–962 (2018)
19. Stouffer, K.A., Falco, J.A., Scarfone, K.A.: SP 800-82. Guide to Industrial Control Systems (ICS) Security: Supervisory Control and Data Acquisition (SCADA) systems, Distributed Control Systems (DCS), and other control system configurations such as Programmable Logic Controllers (PLC). Technical report, National Institute of Standards & Technology, Gaithersburg, MD, United States (2011)
20. Almorsy, M., Grundy, J., Müller, I.: An analysis of the cloud computing security problem. In: Proceedings of the APSEC 2010 Cloud Workshop, Sydney, Australia (2010)
21. Subashini, S., Kavitha, V.: A survey on security issues in service delivery models of cloud computing. J. Netw. Comput. Appl. **34**(1), 1–11 (2011)

Petri Networks for Mechanized Longwall System Simulation

Adam Heyduk(✉)

Silesian University of Technology, Ul. Akademicka 2, 44-100 Gliwice, Poland
Adam.Heyduk@polsl.pl

Abstract. Modern longwall system is a complex electromechanical plant. It consists of many devices that operate together sequentially or in parallel. Petri net – consisting of places and transitions can be used as a convenient modeling language for the description of distributed and concurrent systems. The paper presents subnetwork models of the main longwall system component as shearer movement and powered roof support movement. Petri net offers a graphical notation for these discrete and time-dependent processes. One of the main advantages of the Petri networks is their similarity to the SFC language (GRAFCET) defined in the IEC 61131-3 standard, which facilitates the synthesis of discrete control algorithms implemented with PLCs. Models in the form of the timed Petri network can (with the use of appropriate simulation software) assess the performance of the modelled longwall complex, search for ways to increase this performance, and help with the verification of correct operation of control systems.

Keywords: Petri networks · Longwall system · Performance simulation

1 Introduction

The modern mechanized longwall system is a complex electromechanical (shearer, longwall conveyor) and electrohydraulic (mechanized roof support) system. It includes several dozen or even several hundred devices (roof support sections) cooperating with each other in a manner strictly defined in time and space. Therefore, the assessment of the efficiency of the entire system (which determines the economic efficiency of its application) is a relatively difficult task, usually solved in an approximate manner, with the use of many simplifications, which have a significant impact on the accuracy of this estimation. This problem becomes particularly important in the selection of the mining equipment and form of production organization. Most of mining equipment selection methods are based only on past experiences (ad their statistical generalization) and are not suitable for modern mining equipment – much more powerful and automated. Also modern longwalls are longer than several years ago. It is necessary to take into account the technological downtime related to the interdependence and sequence of individual processes of mining, loading, conveyor movement and supporting the roof by a mechanized roof support sections. There should be also considered a "human factor" e.g. limited speed of operator movement in a confined space alongside the longwall. This is

A. Bartoszewicz et al. (Eds.): PCC 2020, AISC 1196, pp. 307–318, 2020.
https://doi.org/10.1007/978-3-030-50936-1_27

important in the case of modern high-performance shearers (they can move with haulage speed to 30 m/min) and in low profile longwalls – where the space available for operator is even more confined. Petri nets used for modeling many biological, manufacturing, economic and control processes [1–3] can become a convenient tool for this purpose.

2 Basic Principles of Petri-Network- Based Modelling of Discrete Event Systems

The work of the entire mechanized complex is modelled using a set of places, transitions, arcs and markers. Places (represented by circles), transitions (represented by rectangles) and arcs (lines with arrows) are used to model the static structure of individual processes, and markers allow to model their dynamics. Appearance of a marker on all inputs of a particular transition activates that transition. After activation of a given transition all markers appearing on its inputs disappear (are "consumed") and new ones are generated - in the amount depending on the number of outputs of the transition. It is not allowed to directly connect two transitions or two places by arc. The places model resources or states (e.g. the number of mechanized roof support sections alongside the longwall) and the transitions model logical conditions (e.g. the conjunction of all inputs) and the duration of particular activities - e.g. displacement of roof support sections, operator's movement to a new place, time of the shearer's passage through an elementary roof support section in the longwall, etc. The basic elements and frequently encountered configurations (connections) are schematically shown in Fig. 1.

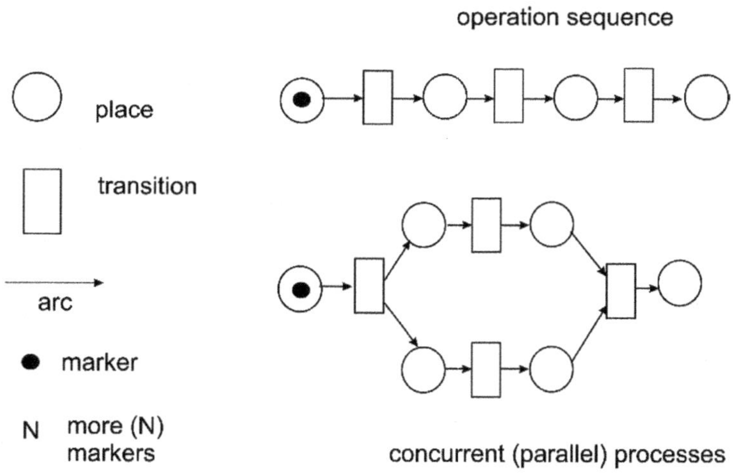

Fig. 1. Basic elements of Petri nets and their example applications.

There are many other useful extensions to Petri nets such as inhibitive arcs - ensuring that a token is not in a place. They can be used for preventing or delaying some action (e.g. in case of any disturbances). For analysis purpose structure of the Petri network cab be described by two matrices:

- The input flow matrix **I** which contains the complete set of input flow weights from places to transitions. Values of this matrix elements are non-negative
- The output flow matrix **O** which contains the complete set of output flow weights from transitions to places. Values of this matrix are also non-negative.

On the basis **I** and **O** matrices there can be calculated an incidence matrix as a difference of these matrices:

$$\mathbf{C} = \mathbf{O} - \mathbf{I}$$

This matrix can contain negative values (unlike the input and output flow matrices), and completely describes the network structure. Detailed analysis of this matrix can provide many information of the network-modelled process (like reachability, coverability). Dynamical behavior of the network can be modelled by a matrix state equation [4, 5].

3 A Petri Net Modelling the Movement of a Longwall Shearer

The continuous movement of a longwall shearer along the face of a longwall (which is one of the most important activities in the process of selecting coal in a longwall system) can be presented in a discrete way as a consecutive crossing of successive elementary intervals of a fixed length. Due to the necessity to model the interdependence between the shearer's movement and the powered roof support movement, it is most convenient if the length of such an elementary interval is equal to the width of one section of the roof support used in the analyzed environment.

In place P0 there is an initial number of markers corresponding to the number of elementary roof support sections to be passed by the shearer alongside the entire longwall.

In the course of the simulation of the longwall operation, this marker number decreases, while newly generated markers begin to appear in the place PK6 mapping the increasing number of elementary length sections passed by the shearer during operation. The transition TK0 models all the initial conditions necessary for the next section movement, the transition TK1 models the shearer passage time of the single roof support section, and the transition TK2 models the time of the operator movement along the section width interval. The transition TK3 models the end of the passage of the next elementary roof support section width interval by the shearer and its operator.

Transition times can be calculated simply as

$$t_{TK1} = \frac{l_s}{v_h}$$

and

$$t_{TK2} = \frac{l_s}{v_o}$$

where l_s denotes width of a single roof support section, v_h – shearer haulage speed, v_o – human operator movement speed.

The place PK3 site is of an auxiliary nature - placing a marker in it enables to initiate the whole process and then its continuation. The whole Petri network modeling the shearer movement is presented in Fig. 2.

Haulage speed $v_h(t)$ can be also modelled as a stochastic variable, and in this case transition times t_{TK1} are also stochastic variables. Also the human operator movement speed can be modelled as a stochastic variable centered about its mean value.

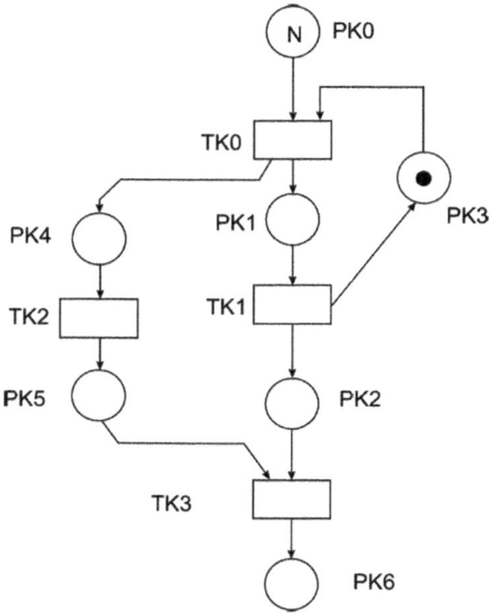

Fig. 2. Petri's network describing the longwall shearer's movement along the entire longwall.

4 Petri Network Modeling the Section Displacement in the Longwall with Mechanized Roof Support

When analyzing the sequence of movements associated with moving the mechanized roof support section, it is necessary to note the fact that the armoured face conveyor section shift can be made only with the legs of roof support expanded, while to move the roof support section it is necessary to lower the position of the girder. Hence, two methods of roof support operation are used in Polish mining:

A) with a step backwards

a. after shearer passage, the legs are lowered (the girder is lowered) and the section is moved (by hydraulic shifters) towards the armoured face conveyor - i.e. towards the longwall face,

b. section leg expansion, which protects the exposed roof;
c. possible correction of section position;
d. displacement of the armoured face conveyor section.

B) without backstepping

a. immediately after shearer's passage, the pivoted canopy is extended and the newly exposed roof is secured at the previous position of the roof support section (legs are all the time extended);
b. displacement of the armoured face conveyor section with hydraulic shifters;
c. lowering the girder, pulling up the powered roof support section towards the conveyor and then re-expanding the whole section legs.

Working without a step back ensures better protection of the ceiling, however, it requires the use of a more complex roof support design with a pivotable or extensible canopy. Hence, it is used primarily in the outermost (initial and final) sections of the longwall, where it is often necessary to additionally strengthen the consistency of the ceiling, while the wall conveyor drives are characterized by greater width.

In the central part of the longwall, simpler sections working with a backward step can be used. A schematic representation of the sequence of individual activities and simple Petri nets corresponding to this sequence for the operation of the roof support with a step backwards is shown in Fig. 3, and for the operation without backstepping in Fig. 4.

Figures 3 and 4. show Petri nets corresponding to the movement sequences of a single powered roof support section. In the whole longwall wall it is necessary to shift N sections. Petri net modelling a sequence of N identical section shifts is shown in Fig. 5.

This paper presents a simplest (no-disturbance) roof support operation sequence. Sometimes there exist external disturbances in particular roof part movements, preventing fast scheduled movement of the whole section. These disturbances are most often connected with the process of canopy extension (need of additional support in the case of too large roof gap) and the process of AFC section movement (in the case of rock particles blocking the movement route) These-appearing stochastically- external disturbances (together with their removal time) can be also modelled by additional inhibitive arcs introduced into the model.

The Petri net describing the sequence of roof support sections shifts, taking into account different operation methods occurring in the extreme sections (including N1 and N3 sections) and in the central part of the wall (including N2 sections) is shown in Fig. 6. As the passage of a given part can follow only after passage of the previous section - lengths of the particular parts of the longwall (expresssed as the number of sections) can be modelled as transition weights. There can be also included additional operations and processes necessary to perform at both ends of the longwall. Generally the roof support rearrangement at both ends of the longwall face is much more time-consuming and cumbersome than in the middle part of the longwall face, as it needs additional space for large electric motors and gearboxes of the armoured face conveyor and there is an additional waiting-time for the reversal of the longwall shearer haulage speed and cutting drums height change.

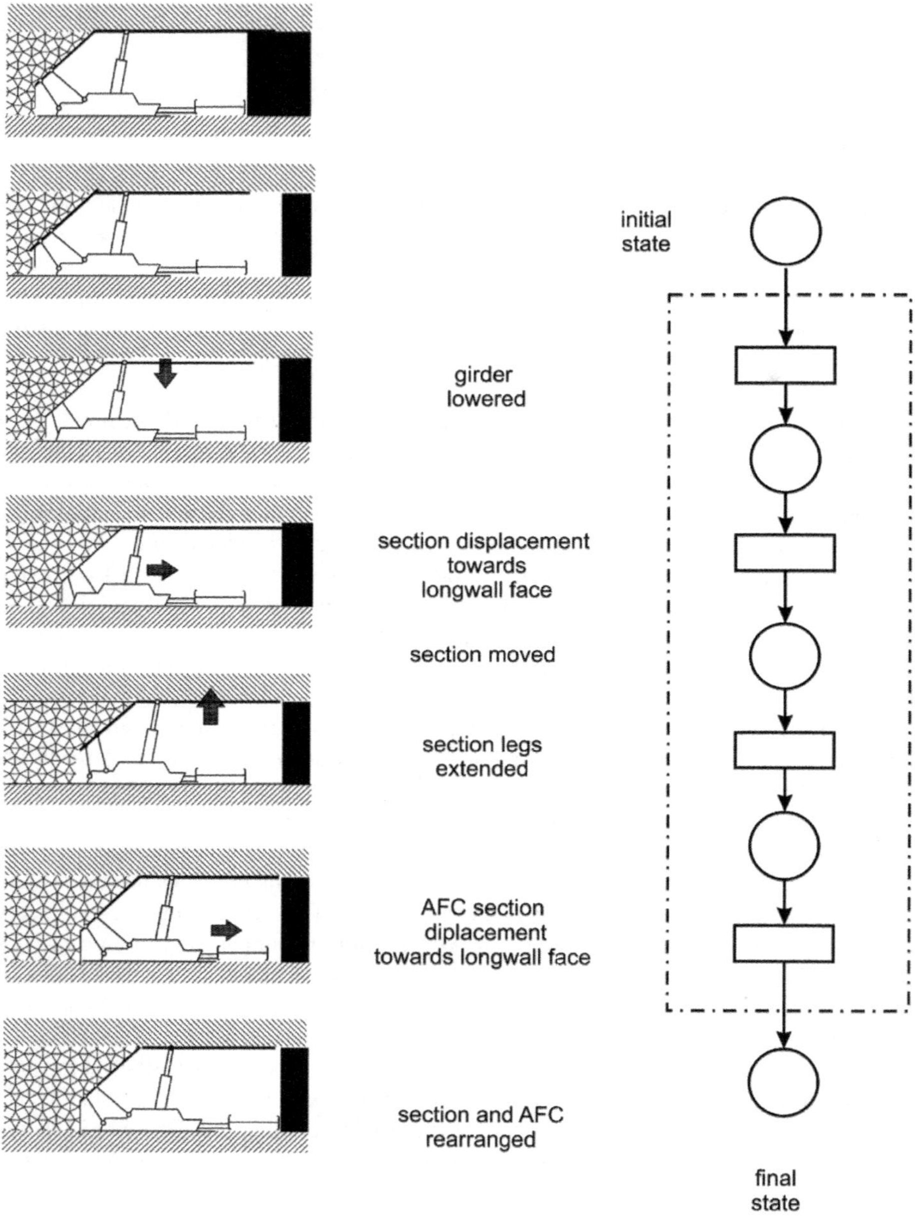

Fig. 3. Individual phases of rearranging single section of powered roof support and the corresponding Petri net for backward step operation mode.

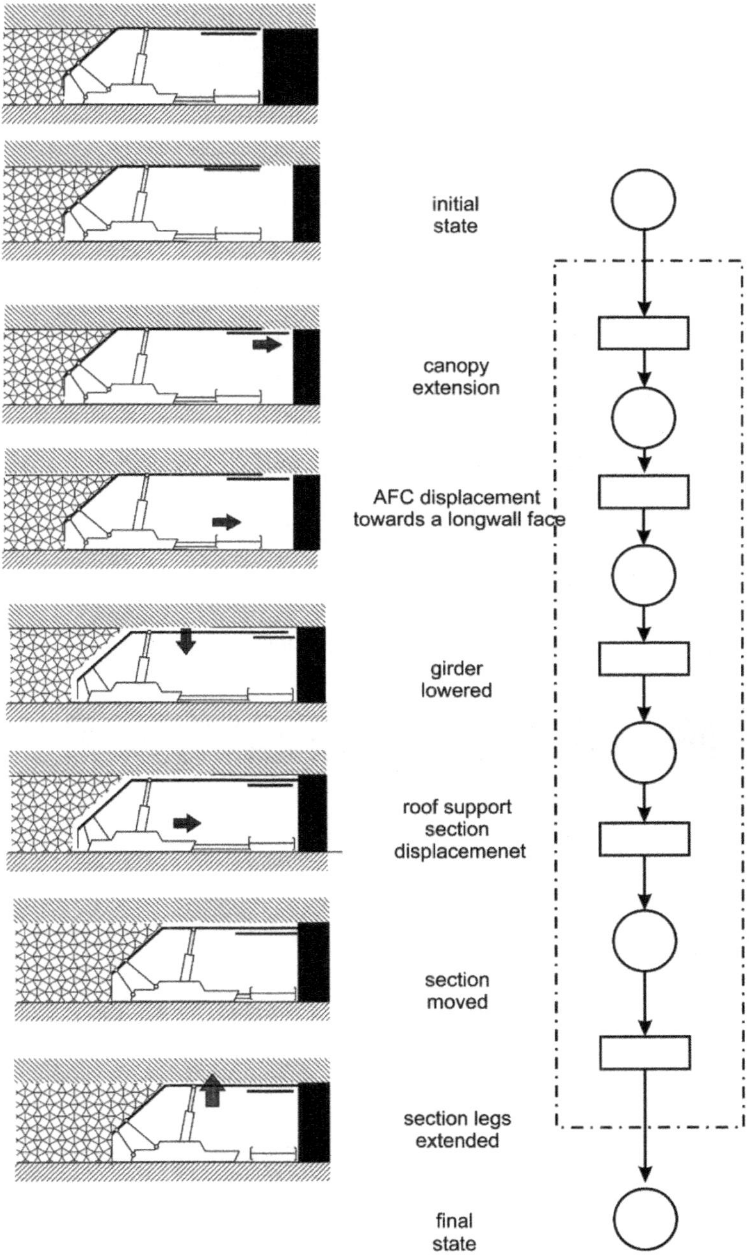

Fig. 4. Individual phases of rearranging single section of powered roof support and the corresponding Petri net for no-backstepping operation mode.

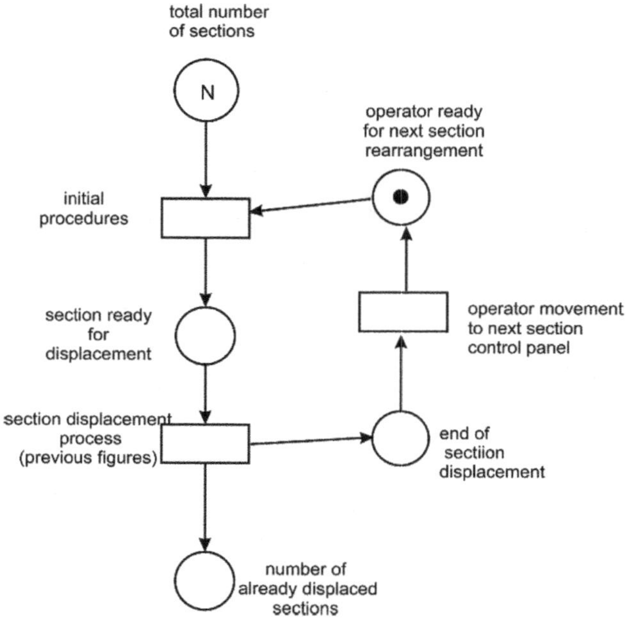

Fig. 5. Petri net describing the process of moving all N sections of the powered support.

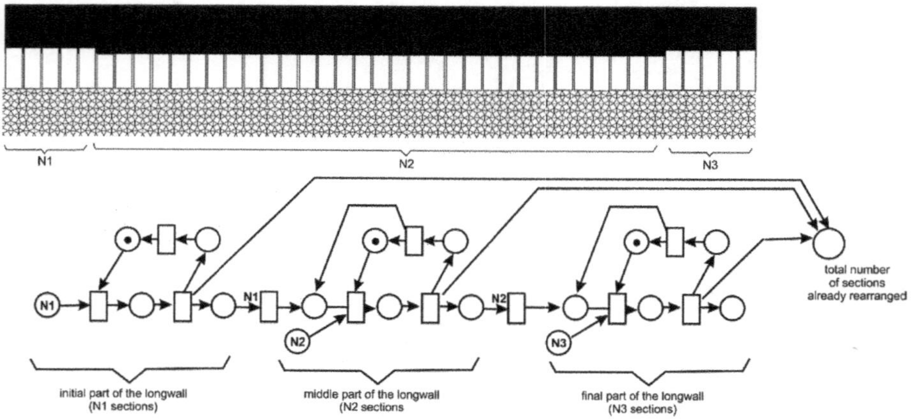

Fig. 6. Petri net describing the process of rearranging all N sections of mechanized roof support, taking into account the different operation modes of the roof support sections in the end and middle parts of the longwall.

5 Interdependence of the Shearer's Movement and the Displacement of the Powered Roof Support Alongside the Longwall Face

During the operation of a mechanized longwall, the processes of coal cutting and supporting the roof are interdependent and must run concurrently to prevent too much exposure of the roof surface. The Petri net describing this correlation is shown in Fig. 7. In place P0 there is the initial number of markers corresponding to the number of elementary sections of the road to be passed by the shearer alongside the entire longwall. During the simulation of the mining process, this number decreases, while newly generated markers begin to appear at the place PK6 reflecting the growing number of elementary sections of the road passed by the shearer during operation.

The transition TK0 models all the initial conditions necessary for the next section displacement, the transition TK1 reflects the time of the elementary section passage of the road by the shearer, and the transition TK2 models the time of operator displacement along the shearer movement route.

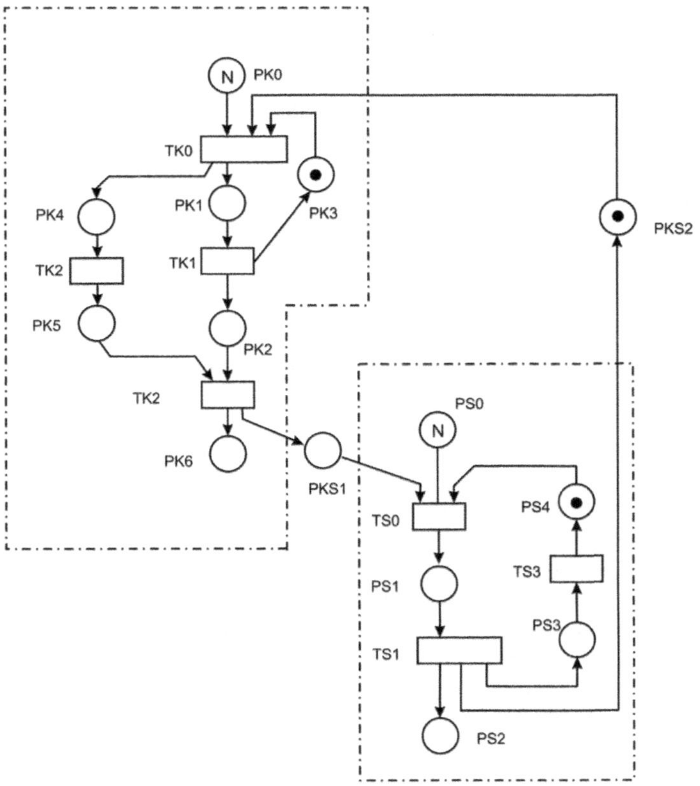

Fig. 7. Petri net describing the interdependence of shearer movement and mechanized roof support section adjustment.

In the PS0 place there is the initial number of markers corresponding to the number of roof support sections to be moved along the whole longwall. During the simulation of work, this number decreases, while newly generated markers begin to appear in the place of PS2 reflecting the growing number of already moved sections. The TS0 transition reflects the fulfillment of all the conditions necessary for moving the next section, the TS1 transition reflects the time of moving one section (which is in itself a sequence of actions according to Fig. 3 or 4), and the TS3 transition reflects the time of operator moving to the next section control panel. The places PKS1 and PKS2 take into account the relationship between the movement of the shearer and the displacement of the powered roof support section - the shearer must travel over the analyzed section to enable section rearrangement and at the same time further shearer movement is possible only when the exposed roof surface is secured.

6 Possible Further Model Extensions

The presented model can be extended with further elements related to the organization of the mining process, and in particular related to activities performed at both ends of the wall necessary for changing the direction of the shearer movement (e.g. rearrangement of cutting drums). It is also necessary to specify the duration of individual operations. The data necessary to create the model can be obtained, for example, from the recording systems of a real operating longwall [8–10]. Accepted durations of individual operations can have a constant value, and can also be treated as random variables with a specific distribution [3, 5–7]. Timed Petri network then becomes a more general stochastic Petri network. Another element of the model intended to extend the network structure as part of further work may be taking into account the length of the shearer (several section widths) - generating a specific delay, and thus affecting the numbering of the currently moved section of the powered roof support.

7 Simulation of a Petri Network Modeling a Mechanized Longwall System

The structure of the Petri net in the simplest form can be modeled in a matrix way using the so-called incidency matrix whose rows correspond to individual places and columns correspond to individual transitions. The values of the matrix elements (at the intersection of the specified row p of the column k correspond to the change in the number of markers stored in the place p after the transition k [7]. However, when specific transitions correspond to non-zero delay times, the analysis of network operation is most conveniently performed using specialized simulation software. A good survey of this available software can be found in [11] For verification of Petri networks presented in this paper, the PIPE simulator (Platform Independent Petri Net Editor [12, 13]) with source code (in Java) available on the *sourceforge* platform was used. An example (corresponding to network presented in Fig. 7) has been presented in Fig. 8.

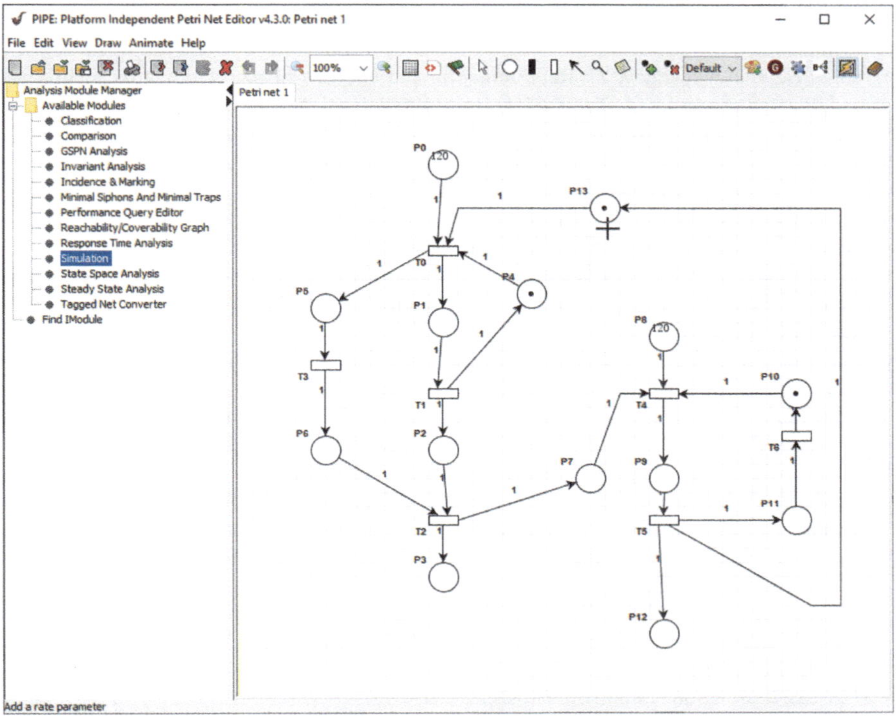

Fig. 8. Diagram of Petri's network modeling the mechanized longwall system (corresponding to the Fig. 7) created in the PIPE simulation program editor

8 Conclusions

Mechanized longwall is a complex system, where Petri nets can be considered as a convenient performance assessment and control algorithm verification tool. Because of the big number of machines operating concurrently and interdependently it is advisable to build the model in a hierarchical way, beginning from simple subnetworks developed separately for each machine type. There is necessary to model their cooperation and time-dependencies. The Petri network model is useful in control system design and verification as it is closely related to GRAFCET (SFC – Sequential Function Chart) language used in PLC programming [14]. Petri-network based simulation, taking into account more technical and organizational factors can provide more reliable results than the simple performance estimates used so far.

References

1. Hruz, B., Zhou, M.C.: Modeling and Control of Discrete-event Dynamic Systems with Petri Nets and Other Tools. Springer Verlag, London (2007)
2. D'Souza, K.A., Khator, S.K.: A survey of petri net applications in modeling controls for automated manufacturing systems. Comput. Ind. **5**(1), 5–16 (1994)

3. Peterson, J.L.: Petri Net Theory and the Modeling of Systems. Prentice-Hall, Englewood Cliffs (1981)
4. Saha, B., Bandyopadhay, S.: Representation and analysis of petrinets via the matrix state equation approach. Int. J. Electr. **65**(1), 1–7 (1988)
5. Staines, A.: Ordinary petri net matrices. In: International Conference AMCSE (2018)
6. Reisig, W.: Understanding Petri Nets Modeling Techniques, Analysis Methods, Case Studies. Springer Verlag, Heidelberg (2013)
7. Zhou, M.C., Venkatesh, K.: Modeling. Simulation and control of flexible manufacturing systems. A Petri Net Approach. World Scientific, Singapore (1998)
8. Brzychczy, E., Trzcionkowska, A.: Process-oriented approach for analysis of sensor data from longwall monitoring system. In: Intelligent Systems in Production Engineering and Maintenance. Springer (2019)
9. Brzychczy, E., Trzcionkowska, A.: Practical aspects of event logs creation for industrial process modelling. Multidisciplinary Aspects Prod. Eng. **1**(1), 77–83 (2018)
10. Brzychczy, E., Trzcionkowska, A.: New possibilities for process analysis in an underground mine. ZN Pol. Śl. nr 1987. Organizacja i zarządzanie z 111, Gliwice (2017)
11. Thong, W.J., Ameedeen, M.A.: A survey of petri net tools. In: Sulaiman, H., Othman, M., Othman,, M., Rahim, Y., Pee, N. (eds) Advanced Computer and Communication Engineering Technology. Lecture Notes in Electrical Engineering, vol. 315. Springer (2015)
12. Bonet, P., LLad, C.M., Puigjaner, R.: PIPE v2.5: a petri net tool for performance modeling (2007)
13. Dingle, N.J., Knottenbelt, W.J., Suto, T.: PIPE2: a tool for the performance evaluation of generalised stochastic petri nets. ACM SIGMETRICS Perform. Eval. Rev. **36**, 34–39 (2000)
14. EN 61131-3, Programmable controllers - Part 3: Programming languages (IEC 61131-3:2013), International Standard, Brussels (2013)

A System for Detection of Pressure Leaks

Andrzej Wojtulewicz(✉) and Maciej Ławryńczuk

Institute of Control and Computation Engineering, Warsaw University of Technology, ul. Nowowiejska 15/19, 00-665 Warsaw, Poland
andrwoj@gmail.com, M.Lawrynczuk@ia.pw.edu.pl

Abstract. This work presents an original software and hardware system whose objective is to detect pressure leaks. Two methods for detection of leaks are considered: the first one is based on an industrial vision system, the second one on a proprietary ultrasonic sensor using Fast Fourier Transformation (FFT). Automation of the measuring process has been done by an industrial six axis robotic arm. For experiments three original laboratory stands have been used.

Keywords: Ultrasound · Pressure leak · Vision system · Industrial robot · PLC · Microprocessor · FFT

1 Introduction

In many industrial applications it is necessary to detect pressure leaks. Industrial plants in the fuel and chemical industries are subject to special restrictions, since leaks in process installations lead to huge safety problems and financial losses. Furthermore, tightness is very important in the food industry since some products require higher pressure during processing and storage [5,7,13,15]. The typical leak detection procedure focuses on controlling critical points of the installation such as welded joints, valves, flanges, fittings and other elements which may be easily damaged. Of course, tightness is not only important in the industry. Other applications include: sea ships, submarines, airplanes, helicopters.

Ultrasonic measurement has a number of advantages over traditional methods of leak detection. The directionality of the measurement should be emphasised. In such a case, in conjunction with an industrial robot, it is possible to precisely determine the place of the leak, spray paint for further repair purposes or directly repair using a welding tool located on the industrial robot. The second important advantage is the lack of invasiveness at the place of measurement using an ultrasonic sensor. It is because the measurements do not have to interfere with the continuous operation of the device. The third important advantage of the measurement is its speed, i.e. the answer is almost immediate [2,4,6,9,16].

Currently existing solutions are based mainly on hand-held portable devices reminiscent of classic electric meters. It is a solution that requires reliable use by its operator and has several disadvantages. The presence of an operator is

A. Bartoszewicz et al. (Eds.): PCC 2020, AISC 1196, pp. 319–331, 2020.
https://doi.org/10.1007/978-3-030-50936-1_28

required. The human factor can lead to negligence and thus a lack of effectiveness of leak detection. Accessibility in hazardous areas is practically impossible without stopping the production process. Periodic inspection of permanent parts of the installation is a task that can introduce a routine of operation and in this case it is very easy to make mistakes.

The described system of automatic leak detection ensures relatively high repeatability of measurements and their reliability. In addition, in areas hazardous for humans, the system is able to work without any problems. The development of an ultrasonic sensor with an algorithm for frequency analysis of the measuring signal allows for precise determination of the level of leakage and its size. The vision system performs detection the "bubble" effect occurring at the gas leak, previously the place is sprayed with the appropriate liquid [1,3,11,14].

2 The System for Detection of Pressure Leaks

2.1 General Assumptions

The most common problem for detection is compressed gas leaks, mainly compressed air installations in industrial plants, where many thousands of kWh are wasted. Sample data for the German market is presented in [10]. Existing 60,000 compressed air systems consume 14,000,000,000 kWh per year. In translating into produced carbon dioxide, very high environmental pollution appears. Even 20% of this production can be lost irretrievably due to leaks in pneumatic networks. Counteracting such large amounts of leaks allows to achieve huge profits. Naturally, large air leaks can be easily diagnosed since they are very characteristically audible through the human ear. However, small leaks with areas smaller than 1 mm^2 are virtually impossible to find. However, the air escaping through such small holes perfectly generates high frequencies in the ultrasound range possible to be easily detected by dedicated devices. It is worth noting that noise in the audible frequency range does not affect the quality of leak detection. Sometimes it may be necessary to perform a preliminary background measurement in an environment of proper leakage to have a reference to the surrounding ultrasound sources. It can find a lot of applications for ultrasonic measurements: leaks in pneumatic and vacuum systems, leaks in pressure containers, leaks in the pneumatic braking system of trucks and trains, leaks in pipe systems, leaks in oxygen installations in hospitals, leaks in steam separators, electrical discharges in damaged insulation, damage of mechanical bearings and damage of mechanical drive transmission systems.

The product described in [8] is one of the representatives of a typical solution in the field of ultrasound detection. The whole device is in the form of a pistol grip with a transparent screen. An ultrasonic microphone allows the detection of various types of ultrasonic radiation. The maximum detection level that the device can obtain is 60 dB. Detection frequency is 40 kHz $\pm$ 2 kHz.

An additional interesting application of ultrasound is the detection of various mechanical damages (bearing damage, mechanical clearances or damage to high voltage installations). A number of examples of spectrum measurements, descriptions and interpretation of results are given in [12]. Frequency analysis of the signal spectrum allows to determine the nature of the damage and draw conclusions before the final damage to the device. For example, the bearing in the engine can be replaced earlier if only a disturbing measurement appears. This is undoubtedly a great financial saving and reasonable management of maintaining the plant in good technical condition. A very interesting work combining measurements from the vision system and the sound detector is presented in [4], a material cutting process is considered.

2.2 System Architecture

Having analysed the available scientific literature and commercial devices, the most important features of the system should be:

- operating frequency of the ultrasonic sensor in the range of 20–100 kHz,
- the dynamics of the signal from the ultrasonic sensor not lower than 70 dB,
- the input signal processing with FFT algorithm with a resolution of 4096 points,
- 200 kHz analog signal sampling frequency,
- use of a vision system for detecting visible leaks,
- integration with an industrial robot for performing measurements,
- three different research objects should be possible.

Figure 1 presents the entire structure of the measuring system. Starting from the top, the MATLAB script is implemented for downloading data from the Programmable Logic Controller (PLC) and displaying it. The configuration of the PLC base board is shown below. For communication with the computer and MATLAB, Ethernet and the Socket Communication protocol are used. The RS485 standard and a dedicated data frame are used to communicate with the ultrasonic sensor. The industrial robot is controlled by a motion module installed on the base board of the controller, communication with the robot drives is carried out by optical fiber. The vision system is connected via Ethernet to trigger the photo and receive the measurement result. The test stand has been designed based on pneumatic technology elements to ensure good quality of compressed air.

2.3 Design of the Ultrasonic Sensor

For the processing of the signal from the ultrasonic sensor, the evaluation board STM32F746G-Discovery has been chosen, which offers a microprocessor with an ARM7 core with a built-in floating point unit. The processor's core is clocked at over 200 MHz, which ultimately allows for very fast signal processing using the DSP library for STM AN4841 processors. A discrete Fourier transform for 1024 points is calculated in 0.4 milliseconds. The entire system configuration is carried out using HAL (Hardware Abstraction Layer) libraries and the generation of the project template by the STM32IDE environment. Then the project

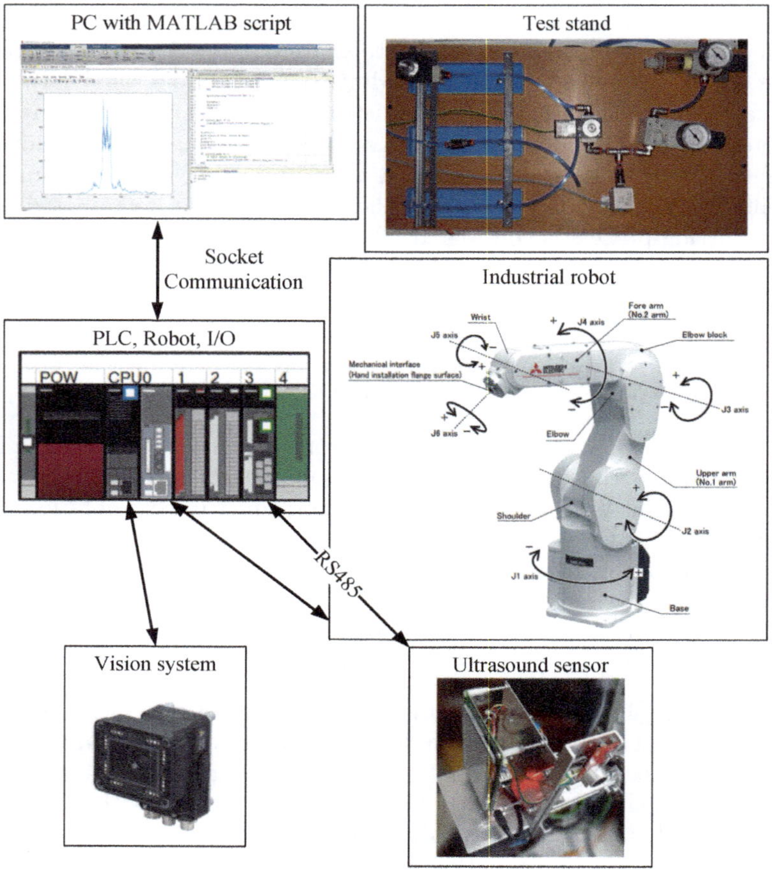

Fig. 1. The entire structure of the system

is supplemented with a code appropriate for implementing signal processing and general microprocessor operation support.

The CMSIS library functions are used to implement the FFT algorithm. The Listing 1.1 shows the basic implementation of the algorithm. A transformation of $N = 4096$ points is applied. The Analog to Digical Converter (ADC) samples with a frequency of 200 kHz, so the maximum spectrum frequency is 100 kHz. The frequency resolution is 48.8 Hz. The ADC has 12 bits which means that one gets a range of digital values from 0 to 4095. The algorithm begins with checking the condition which buffer can be subjected to calculations. Then the measurement data are rewritten into temporary buffers for real and imaginary values. In the presented version the input signal can only be real but in general the algorithm allows to introduce a complex signal as well. The next step is to call the appropriate data processing function. Finally, the amplitude spectrum module is calculated and the results are sent to the PLC.

Listing 1.1. Basic implementation of the algorithm

```
if (Change_buffer != 0)
{
 TimerCount_Start();
 if(Change_buffer == 1)
  k=0;
 if(Change_buffer == 2)
  k=4096;
 for(index_fill_input_buffer=0; index_fill_input_buffer < FFT_Length*2;
        index_fill_input_buffer += 2)
 {
  /* Real part */
  aFFT_Input_f32[(uint16_t)index_fill_input_buffer] =
        (float32_t)aADC1ConvertedValue_s[k] / (float32_t)4096.0f;
  /* Imaginary part */
  aFFT_Input_f32[(uint16_t)(index_fill_input_buffer + 1)] = 0.0f;
  k++;
 }
 /* Process the data through the CFFT/CIFFT module */
 arm_cfft_f32(&arm_cfft_sR_f32_len4096, aFFT_Input_f32,
        FFT_INVERSE_FLAG, FFT_Normal_OUTPUT_FLAG);
 /* Process the data through the Complex Magnitude Module for
        calculating the magnitude at each bin */
 arm_cmplx_mag_f32(aFFT_Input_f32, aFFT_Output_f32, FFT_Length);
}
```

The filter should also have an appropriate slope of transition characteristics between the band pass and the band stop. A standard first-order filter provides a decrease in performance at the rate of 20 dB/dec, a second-order filter gives 40 dB/dec, etc. Unfortunately, practical implementation of high-order analog filters is not simple. Moreover, there are additional changes in the filter characteristics in the band pass and band stop. A compromise should be made between the filter order, the required slope and the cut-off frequency. One should also remember about sufficient attenuation W expressed in dB, which should be met for the frequency $f_p/2$. Assuming an ADC converter with a 12-bit resolution, its dynamic range (signal-to-noise ratio) is up to

$$\mathrm{SNR} = 6.02N + 1.76\ \mathrm{dB} = 6.02 \times 12 + 1.76\ \mathrm{dB} = 74\ \mathrm{dB} \tag{1}$$

where N is the number of bits of ADC.

The ultrasonic sensor is implemented using an ultrasonic microphone and an appropriate level of signal amplification. Further processing is carried out by the ADC. The gain level is used as an anti-aliasing filter. Preliminary calculations also have shown that meeting the condition for dynamics and the frequency band requires a high-order filter of more than 10th order. Hence, it has been decided that these assumptions should be relaxed and the only filter would be the signal amplifier. Non-inverting amplifier configuration with level shift for one 3.3 V power supply is used, it is presented in Fig. 2. An LM324 amplifier with unit gain $f_{\mathrm{T}} = 1$ MHz is used. The gain formula is

$$k_{\mathrm{u}} = \frac{R1}{R6} + 1 \tag{2}$$

It is possible to estimate the cut-off frequency of the whole system (it is the frequency for which the amplitude decrease is 3dB from the maximum value)

$$f_{\mathrm{g}} = \frac{f_{\mathrm{T}}}{k_{\mathrm{u}}} \tag{3}$$

For the 0.124 V_{pp} sensor signal, one gets full dynamics at the ADC input.

2.4 Industrial Robot with Vision System

The Mitsubishi Electric FR-RV2-R industrial robot consists of an arm with six independent axes and a CR800-R drive controller. Motion control is implemented using a dedicated Motion R16RTCPU controller, which can be installed on the base board of the iQ-R controller system. The TeachBox R56TB panel is used to start and diagnose the robot, which allows to set all operating parameters, monitor current positions, track the course of the currently executed program and manually control all axes.

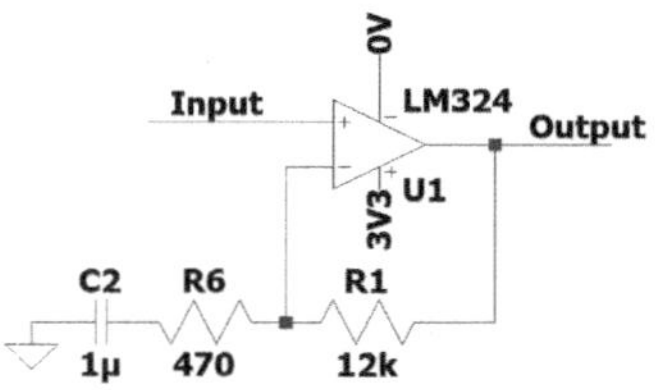

Fig. 2. Amplifier schematic

The vision system is very important. It has been decided to choose a model with a large range of configuration options and functions. The Omron FHV7H-CO16 Smart Camera is a device with an extensive application for configuring all stages of the vision setting. One can control the external illumination, connect Ethernet for communication with the master controller and connect the external configurable output inputs.

The basic procedure of the video processing process is the following. The first stage is, of course, taking the picture. Then, for example, the shape recognition process is carried out and then position compensation is introduced for the found detail. The last vision stage is checking if the object has defects. At the end of the procedure, the object position is sent with confirmation that it is not defective.

2.5 Design of the Complete System

The test plan provides for the construction of a test platform for measurements with both an ultrasonic sensor and a vision system. A pneumatic system has been designed that allows testing three different objects by two methods. The common part is the air preparation, pressure stabilisation, pressure measurement and solenoid valve supplying the right pressure to the selected object.

The object is selected by physically connecting a suitable pneumatic conduit to the output of the 2/2 normally closed distribution valve. The valve is controlled by the PLC to turn on the air only for the time of measurement. The

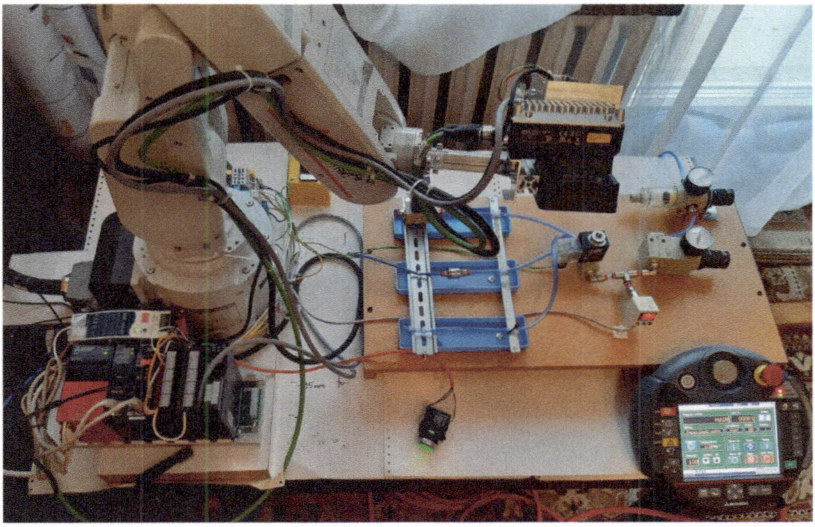

Fig. 3. The entire system: the front view

pressure stabilisation system allows to set the pressure in the range of 0–2 bar with an accuracy of 30 mbar. The Metal Work 9000640 pressure sensor has an LED display for monitoring pressure in the system. A digital output has been configured connected to the PLC. A high logical state is obtained when the pressure in the pneumatic system is at least 0.5 bar. Before tests one has to spill the objects using dedicated liquid to see the bubbling effect. Liquid dripping from the objects is collected in the blue container.

3 Results of Experiments

Figure 3 shows the entire system from the front. The most important elements are: the 24 VDC power supply on the left, the iQ-R PLC controller (with inputs and outputs) on the left, the 6-axis robot arm with cables and tools (camera and ultrasonic sensor) on the left, the middle part vision system being part of robot tool, the middle and right part research objects, the right bottom part is the operator panel for the robot (Teaching Box).

Figure 4 shows a picture of robot tools which include: an aluminum sheet used as insulation and mounting base, the STM32F746G-Discovery board, the mounting plate for the ultrasonic sensor, the ultrasonic sensor, the vision system.

The basic task is the detection of compressed air leaks from three objects which are: a valve, a reduction pipe and a pneumatic pipe. Each object is examined with two measuring devices: a vision system and an ultrasonic sensor.

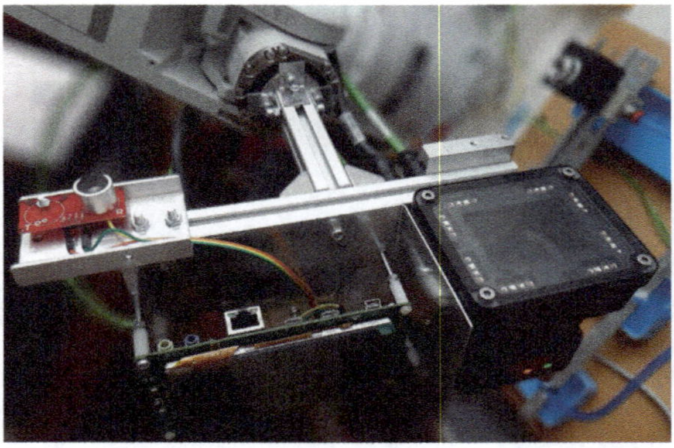

Fig. 4. The tool number 1 (the camera) on the right and the tool number 2 (the ultrasonic sensor) on the left

The vision system examination consists in comparing the current photo with the sample taken in perfect condition. A pattern learning procedure is performed for each object. Before tests, the object is covered with a special liquid that forms bubbles under the influence of air leakage. This study aims to determine the usefulness and usability of the vision system in terms of the ability to detect this type of disorder using pattern comparison mechanisms.

The ultrasonic sensor test is carried out in two versions. The first one is an access to the leakage position with a sensor at a height of Z equal to 10 mm. From this point the movement starts at a constant speed up the distance 200 mm. Position samples and measurement values are saved in the memory every 100 ms. Movement speed is 5 mm/s. The purpose of this test is to determine the nature of the measured value as a function of direct distance from the leak.

The second version is very similar, but the start position of the measurement is the offset in the Y axis by −100 mm from the leakage position, in the Z axis the height above the leakage is 10 mm. The sensor travels in the Y axis 200 mm in the ascending direction. The purpose of this test is to determine the nature of the measured value as a function of the perpendicular position from the leak. It is important to verify the possibility of determining the type of the leak by observing the signal changes on the graph.

For each process, images from the vision system and charts from measurements with an ultrasonic sensor implemented in the MATLAB environment are presented. A stable pressure of 1 bar is applied for ultrasonic measurements.

Any damage to the valve is simulated by gently unscrewing the thread of the pneumatic connection at the valve inlet. An additional rubber seal means that the level of leakage can be set at any level. The left part of Fig. 5 shows an incorrect situation in which a leak is visible in the form of bubbles, the vision system returns NG (Not Good) to the PLC. The right part of Fig. 5 depicts a correct situation, i.e. there is no leakage, the vision system returns OK.

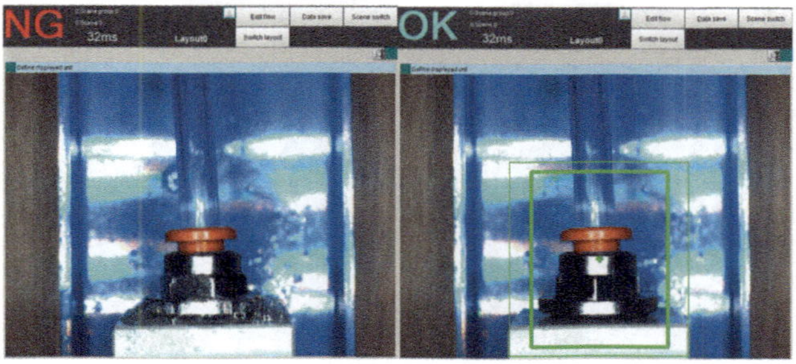

Fig. 5. Vision test with the valve object: NG (not good) on the left, OK (good) on the right

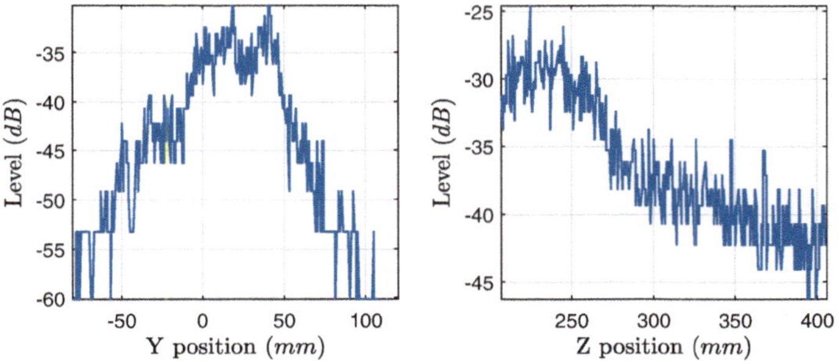

Fig. 6. The result of the signal measurement as a function of displacement in the Y and Z axes for the valve object

Figure 6 presents the result of measuring the signal level (in dB) as a function of the distance of the sensor in the Y and Z axes from the object. The considered position is the actual position of the robot Z axis in relation to its reference zero point. The initial distance in the Z axis from the object is 10 mm. It can be seen that in the initial phase of the sensor is receding, the signal does not change significantly, only at a greater distance begins to weaken significantly. The signal change is not linear. For Y movement one can observe a weak signal level at both extreme measuring ranges, i.e. the sensor is 100 mm from the leak in the other direction. When the leak is zoomed, one see a clear increase in the signal strength from the sensor. The signal increase is uniform, which is due to the nature of the leak, a loose thread causes the air to escape in all directions.

Damage to the reduction is simulated by gently unscrewing the thread of the mechanical connection. An additional rubber seal means that the level of leakage can be set at any level. The left part of Fig. 7 shows the wrong situation (leakage), the right part depicts a correct situation (no leakage).

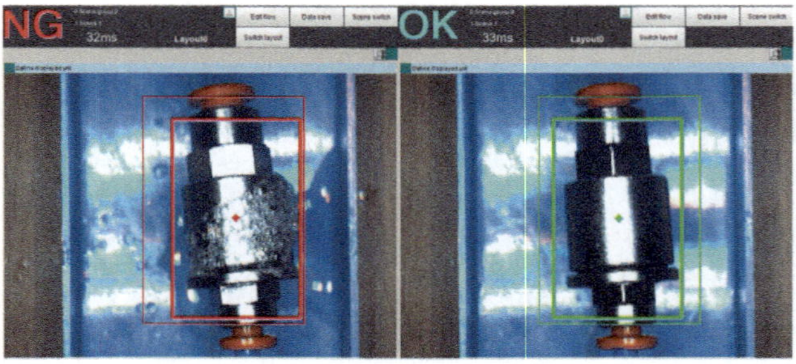

Fig. 7. Vision test with the reduction object: NG (not good) on the left, OK (good) on the right

Figure 8 presents the result of measuring the signal level (in dB) as a function of the distance of the sensor in the Y and Z axes from the object. It can be seen that in the whole range the signal value dropped steadily, but nonlinearly. In addition, the noise level is slightly lower than for the reduction object which may be due to a slightly smaller initial leakage.

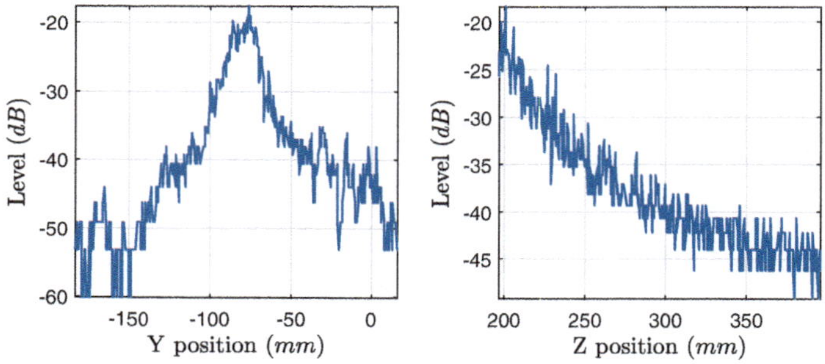

Fig. 8. The result of the signal measurement as a function of displacement in the Y and Z axes for the reduction object

Damage to the pneumatic pipe is simulated by making a hole about 0.2 mm in diameter. In the event of this damage, the air leakage is directional (upwards). The left part of Fig. 9 shows the wrong situation (leakage), the right part depicts a correct situation (no leakage).

Figure 10 presents the result of measuring the signal level (in dB) as a function of the distance of the sensor in the Y and Z axes from the object. It can be seen that in the whole range the signal value dropped steadily but nonlinearly. In addition, the noise level is similar to the reduction object. When the leak is zoomed, one may see a clear increase in the signal strength from the sensor.

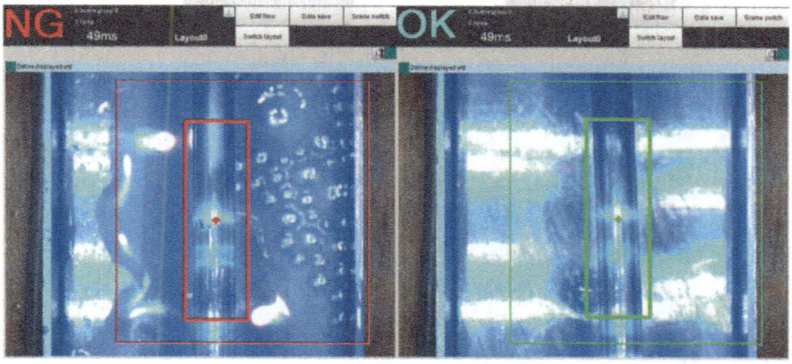

Fig. 9. Vision test with the pneumatic pipe object: NG (not good) on the left, OK (good) on the right

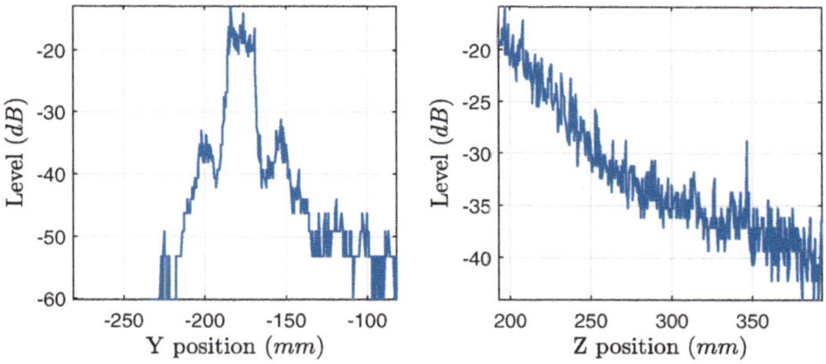

Fig. 10. The result of the signal measurement as a function of displacement in the Y and Z axes for the pneumatic pipe object

4 Summary

The results obtained are very satisfying. The built prototype of the system allows for quick, repeatable diagnostics of high risk systems in which a human may have limited access. Visual measurements show that it is possible to operate with benchmarking methods that effectively recognise leaks. This method has the disadvantage of having to pour additional liquid in the event of a gas leak. It can be used as an option to confirm the initial ultrasonic measurement. Ultrasonic measurements show very high efficiency of leak detection without the need for additional elements. The operation is non-invasive, very sensitive, selective and cannot be disturbed by low-frequency background in the range audible to humans. Spatial selectivity allows to pre-teach the system what kind of leakage it can deal with. It turns out that no frequency change is observed for the used ultrasonic receiver depending on the type of object or the form of ultrasonic measurement. The ultrasonic receiver is tuned to a frequency of about 40 kHz and in a very narrow environment of this frequency remains very sensitive. Techniques call selective frequency band tuning of a given sensor are applied by adding appropriate impedances. However, this method does not provide precise measurements since it is rather intended to slightly extend the frequency range.

There is no confirmation that the loss of sensor dynamics can be a cost. The conclusion for future research is definitely to check the operation of the broadband microphone, especially solutions in the MEMS technology of ultrasonic microphones with PDM (Pulse Density Modulation) coding are very popular. All things considered, the main advantages of the described system are:

- redundancy of operation (vision system and ultrasonic sensor),
- detection using a vision system (spraying a specific point with a liquid solution, followed by a "bubble" effect when there is a gas leak, i.e. taking a picture with a vision system and based on the patterns, the leak is detected)
- a selective ultrasonic sensor has been developed operating in the frequency range from 20 kHz to 100 kHz,
- detection based on the analysis of the FFT spectrogram of the signal obtained from the ultrasonic sensor,
- system integration with the six-axis industrial robot arm for autonomous inspection.

References

1. Alves, T., Oliveir, C., Sanin, C., Szczerbicki, E.: From knowledge based vision systems to cognitive vision. Procedia Comput. Sci. **126**, 1855–1864 (2018)
2. Arifin, B.M.S., Li, Z., Shah, S.L.: Pipeline leak detection using particle filters. IFAC-PapersOnLine **48** (2015)
3. Chen, S., Xiong, J., Guo, W., Bu, R., Zheng, Z., Chen, Y., Yang, Z., Lin, R.: Colored rice quality inspection system using machine vision. J. Cereal Sci. **88**, 87–95 (2019)
4. Chethan, Y., Ravindra, H.V., Krishnegowda, Y.T.: Optimization of machining parameters in turning Nimonic-75 using machine vision and acoustic emission signals by Taguchi technique. Measurement **144**, 144–154 (2019)
5. Giesko, T.: Liquid leak detection using laser triangulation. Institute for Sustainable Technologies, National Research Institute, Radom, vol. 2, pp. 91–97 (2006)
6. Shaoyan, H., Mingyue, D., Ming, Y.: Sparse-view ultrasound diffraction tomography using compressed sensing with nonuniform FFT. Comput. Math. Methods Med. (2014). Article 329350
7. Murvay, P., Silea, I.: A survey on gas leak detection and localization techniques. J. Loss Prev. Process Ind. **25**, 966–973 (2012)
8. Operating manual Leakage detector LD 400. CS Instruments GmbH
9. Pal, B.: Fourier transform ultrasound spectroscopy for the determination of wave propagation parameters. Ultrasonics **73**, 140–143 (2017)
10. Radgen, P.: Efficient Compressed Air Systems. EU-Twinning Project SL04/EN/01 Integrated Pollution Prevention and Control (IPPC) (2006)
11. Rong, D., Xie, L., Ying, Y.: Computer vision detection of foreign objects in walnuts using deep learning. Comput. Electron. Agric. **162**, 1001–1010 (2019)
12. Seeber, S.: Spectral Analysis of Ultrasound. Mid Atlantic Infrared Services
13. Sizeland, E.: Ultrasonic devices improve gas leak detection in challenging environments. Emerson Process Management. World Oil 25 (2014)
14. Sones, R., Novini, A.R.: Machine vision system and method for non-contact container inspection. United States Patent 6172748 (2011)

15. Wang, J., Tchapmi, L., Ravikumara, A., McGuire, B.C., Zimmerle, D., Savarese, S., Brandt, A.: Machine vision for natural gas methane emissions detection using an infrared camera. Manuscript submitted to Applied Energy
16. Xue, H., Wu, D., Wang, Y., Zhao, Z., Chen, T., Teng, Y.: Research on ultrasonic leak detection methods of fuel tank. In: 2015 IEEE International Ultrasonics Symposium (IUS), Taipei, Taiwan, pp. 1–4 (2015)

Modelling, Identification, and Analysis of Automation Systems

Outlier Sensitivity of the Minimum Variance Control Performance Assessment

Kacper Kaczmarek and Paweł D. Domański(✉)

Institute of Control and Computation Engineering, Warsaw University of Technology, ul. Nowowiejska 15/19, 00-665 Warsaw, Poland
01133676@pw.edu.pl, p.domanski@ia.pw.edu.pl

Abstract. Minimum variance (MinVar) control performance assessment (CPA) constitutes one of the most common approaches to the control quality estimation. There are dozens of versions of this method, enriched with practical implementations. However, it should be remembered that the method relies on the same assumptions as the minimum variance control. It is essential that considered disturbance is an independent random sequence. This paper addresses the situations, when loop noise has non-Gaussian properties and is characterized by outliers exhibiting fat-tailed distribution. Sensitivity analysis of minimum variance method against the outliers is conducted using commonly used PID control benchmarks. It is shown that CPA using minimum variance may be significantly biased in non-Gaussian situations, which are very frequent in the industrial reality.

Keywords: CPA · Minimum variance · Robustness · Outliers · PID

1 Introduction

The CPA forms crucial element of the control practice as it supports an engineer with the methodology and indexes that allow to measure control loop performance. Resulting key performance indicators (KPIs) allow to make decisions about loop improvements. There are many different approaches to the CPA, however the minimum variance approach is the most frequent in the research literature. Following the first publication of [14], many researchers have investigated the minimum variance benchmark path. They have proposed several various approaches for different controller algorithms, configurations and performance aspects. The subject is very capacious, so there have been presented extensive descriptions of these method [13,17,22].

The research followed the minimum variance path covering various aspects of the control, as for instance: univariate feedback [5] and feedforward control [6], unstable and nonminimumphase systems [27], multivariate MIMO and MISO cases [11,25], varying setpoint [23], cascaded control [19], aspects of setpoint tracking versus disturbance rejection [26]. The research often followed this path

A. Bartoszewicz et al. (Eds.): PCC 2020, AISC 1196, pp. 335–348, 2020.
https://doi.org/10.1007/978-3-030-50936-1_29

with an introduction of several improvements to the method as for instance Variability Matrix [12], an iterative solution for PID benchmark [20], LQG benchmarks for stationary stochastic disturbances [3] and state space controllers [18], the method based on multi-model mixing time-variant minimum variance [21] and many others.

During 30 years history MinVar approach gained wide acceptance. Software packages have been released [9] supporting plant owners with automatically generated measures. Industrial experience has been collected [2]. Though the method is popular, its limitations were noticed. Contrary to data-driven approaches, MinVar is model-based. It requires not only data, but a model as well. Thereby, *a priori* knowledge of a delay or model order is required. Further method deficiencies, like complexity introduced by a non-linear behavior, assumed representation, and the estimation efficiency have been pointed out [15]. MinVar method uses minimum variance strategy, so it follows its assumptions and inherits its limitations.

This paper focuses on non-Gaussian loop properties and their impact on MinVar. It is assumed that the process is disturbed by an independent random sequence noise in the fundamental method formulation. Industrial practice show that such an assumption is very strong and hardly met. Observation of industrial data [7,8,10] shows that the majority of process industry loops generates time series, that exhibit non-Gaussian, mostly fat-tail properties. Fat tails in data histograms represent data incidents lying far away from the normal performance, which are called outliers or anomalies. They may be generated by real life factors, like varying time-delayed correlations, nonlinearities, non-stationarity, complexity or human interventions. On the other hand regression research [24], shows that the least squares estimation is biased by even a single outlier.

This paper presents results of the simulation research focusing on the subject of minimum variance CPA sensitivity against the outliers. Analysis starts with short introduction to the applied methods in Sect. 2 and 3 and is followed by simulations in Sect. 4. Section 5 concludes the results and identifies areas for further research.

2 Control Performance Assessment

As it has been mentioned above there are many versions of the classical minimum variance method proposed by Harris. The authors have decided to use a detailed algorithm evaluated by method developers [4]. The algorithm is as follows:

1. Perform detrending of the data and remove the mean.
2. Get the state-space model from data.
3. Obtain the time delay d for the process.
4. Determine the infinite response form of the closed-loop model.
5. Determine the variance of the residuals and of the model.
6. Calculate the minimum variance (1)

$$\sigma_{MV}^2 = \left(\sum_{i=0}^{d-1} f_i^2 \right) \sigma_e^2, \tag{1}$$

where σ_e^2 is the variance of the white noise, d is process time delay and f_i are the coefficients of the infinite response model for the closed-loop process.
7. Estimate the actual variance.
8. Calculate the MinVar performance index (2)

$$\eta = \frac{\sigma_{MV}^2}{\sigma_y^2}. \tag{2}$$

The assessment is mostly done with data-driven methods in industry. Two of them are often used: Mean Square Error (MSE) and Integral of Absolute Error (IAE) [9]. These measures are used for comparison with MinVar.

$$\mathrm{MSE} = \frac{1}{n}\sum_{i=1}^{n}\left(\epsilon(i)\right)^2, \tag{3}$$

$$\mathrm{IAE} = \frac{1}{n}\sum_{i=1}^{n}\left|\epsilon(i)\right|, \tag{4}$$

where n is the number of samples and $\epsilon(i)$ the control error.

3 Outliers

The issue of outliers (or anomalies) in data, their identification and impact has been analyzed for years. According to Hawkins [16] an outlier is described as "*an observation which deviates so much from other observations as to arouse suspicions that it was generated by a different mechanism*". The outliers unfortunately significantly impact the process of regression. Mean Square Error (MSE) index is sensitive to any kind of the outliers [24]. Existence of them in data causes changes in the statistical properties of the contaminated data. This effect may be observed by the fat-tails in the time series distributions. Following that probabilistic density functions (PDFs) exhibiting fat tails might be good candidates for a simulation of such process. Research [7,8] has shown that α-stable distribution very frequently well represents industrial control data. Distribution does not have closed PDF and is expressed through the characteristics equation

$$F_{\alpha,\beta,\delta,\gamma}^{\mathrm{stab}}(x) = \exp\left\{i\delta x - |\gamma x|^{\alpha}\left(1 - i\beta l(x)\right)\right\}, \tag{5}$$

where

$$l(x) = \begin{cases} \mathrm{sgn}(x)\tan\left(\frac{\pi\alpha}{2}\right) & \text{for } \alpha \neq 1 \\ \mathrm{sgn}(x)\frac{2}{\pi}\ln|x| & \text{for } \alpha = 1 \end{cases} \tag{6}$$

$0 < \alpha \leq 2$ is called a stability factor or a characteristic exponent, $|\beta| \leq 1$ is a skewness parameter, $\delta \in \mathbb{R}$ is a distribution location (position), $\gamma > 0$ is a scale or a dispersion. Thus, α-stable distribution has been selected as a model for control loop disturbances.

4 Simulations

The analysis is performed using simulations, which are carried out for an univariate template sketched in Fig. 1. Parallel PID form is used (7). There are two disturbances: random measurement noise *d(t)* with relatively low magnitude *0.05* and fat-tailed disturbance modeled with α-stable PDF (Fig. 2) with tail index $\alpha = 1.0$ added before the process. The goal is to find how MinVar index (2) behaves and how it detects poor control in case of non-Gaussian disturbance. Two PID benchmarks proposed by Åström [1] are analyzed (in simulations there are used discrete versions with sample time 0.1 s):

$$G_{PID}(s) = k_p \left(1 + \frac{1}{T_i s} + T_d s \right). \tag{7}$$

– a time-delay and double lag plant

$$G_1(s) = \frac{1}{(0.2s+1)^2} e^{-s}, \tag{8}$$

– an oscillatory transfer function

$$G_2(s) = \frac{1}{(s+1)(0.04s^2 + 0.04s + 1)}, \tag{9}$$

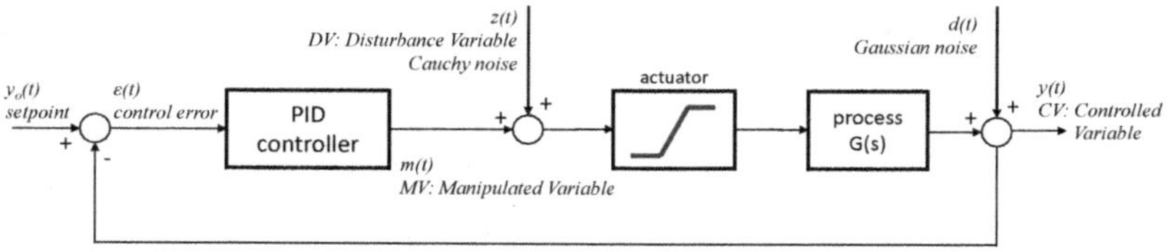

Fig. 1. Simulation environment for noise filter design analysis

The analysis for each considered plant consists of two elements. One set of simulations is run for the undisturbed plant, i.e. without added disturbance before the plant, while in the second set of simulations is disturbed by fat-tailed disturbance before the plant. In both cases Gaussian measurement noise is added. It must be noted that each simulation uses different representation of noise to represent industrial reality and verify method robustness. Each set of simulations includes 2000 runs for different PID parameters. It enables to see the index sensitivity against disturbance. Comparison of the results according to the used indexes is included in Table 1. Optimal tuning is obtained with Matlab script minimizing weighted ITAE criterion ($\mu_1 = 1$), overshoot ($\mu_2 = 10$) and sensitivity ($\mu_3 = 20$). MinVar method shows the values closest to 1.0.

The analysis starts with $G_1(s)$ plant. Control surfaces showing MSE index value for selected PID tuning are sketched in Figs. 3 and 4. The surface reflects

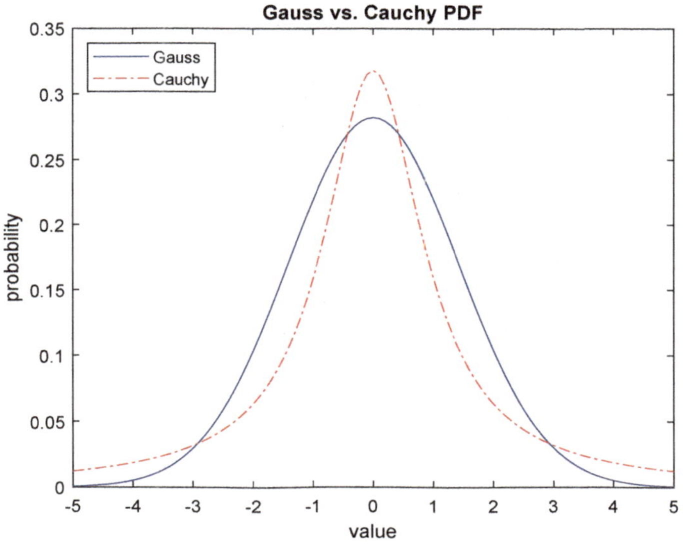

Fig. 2. Comparison of Cauchy ($\alpha = 1.0$) and Gauss PDFs

Table 1. Detection results with different indexes for selected plants

		Optimal	MSE		IAE		MinVar	
			nonDist	Dist	nonDist	Dist	nonDist	Dist
G1(s)	k_p	**0.27**	0.57	0.18	0.12	0.03	1.17	1.17
	T_i	**0.60**	0.80	2.00	0.30	1.00	2.30	2.1
G2(s)	k_p	**0.14**	0.60	0.34	0.22	0.04	0.60	0.86
	T_i	**0.25**	0.35	0.35	0.40	1.95	0.60	0.20

only changes in controller gain k_p and its integration time T_i. In all cases it is assumed that derivative time $T_d = 0.21$ is kept constant. Cyan point ■ always denotes optimal tuning, while the green one ■ the best solution detected by the considered measure. Similar plots for the IAE are shown in Figs. 5 and 6 and minimum variance surfaces are presented in Figs. 7 and 8.

The surfaces for integral indexes are smooth without a disturbance. They are quite similar. Embedded disturbances cause ragged surface with unclear detection ability. Detection with MSE and IAE gives two observations. MSE indicates aggressive tuning as it tends to minimize squares of control error, while IAE tends towards moderate performance. In both cases, disturbance shadows the detection [8].

For minimum variance index both surfaces are slightly ragged. Detected solutions tend to lie close to the stability edge. The other alarming fact is that MinVar index is not within its limits, i.e. in $(0, 1\rangle$. As it is shown in Eq. (2), it consists of

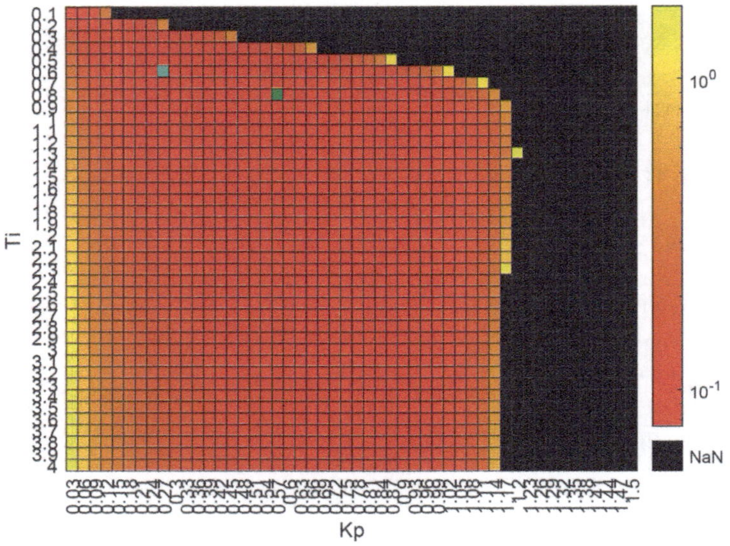

Fig. 3. MSE surface – undisturbed $G_1(s)$ loop

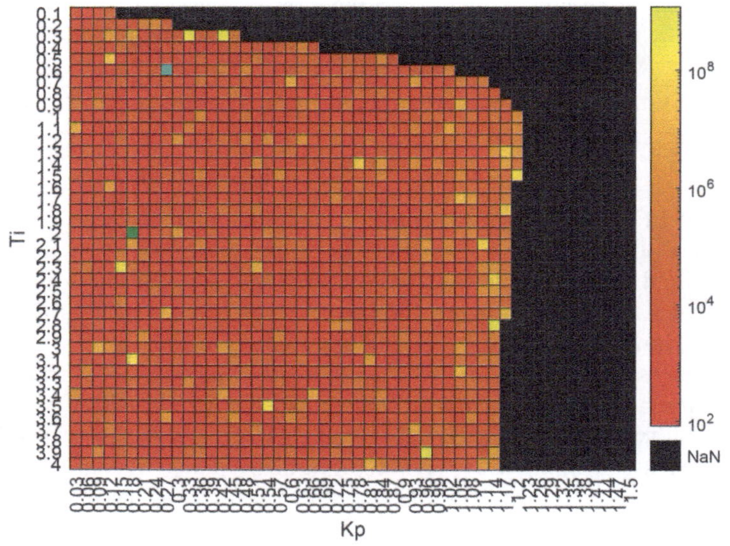

Fig. 4. MSE surface – disturbed $G_1(s)$ loop

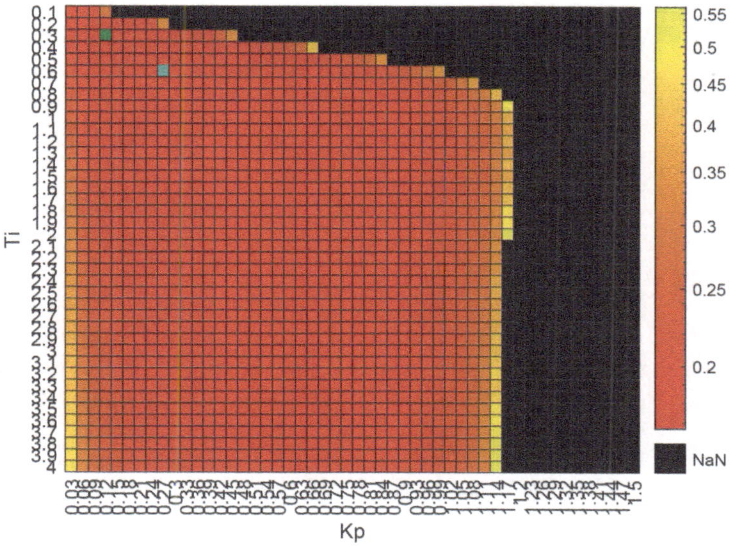

Fig. 5. IAE surface – undisturbed $G_1(s)$ loop

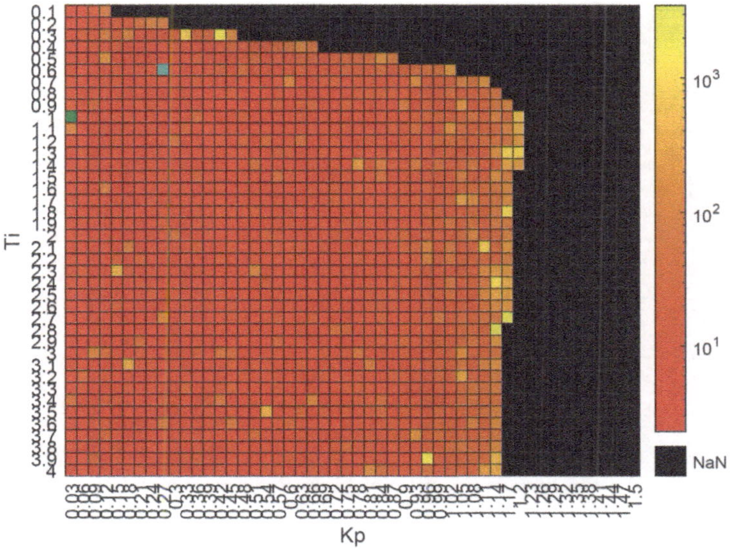

Fig. 6. IAE surface – disturbed $G_1(s)$ loop

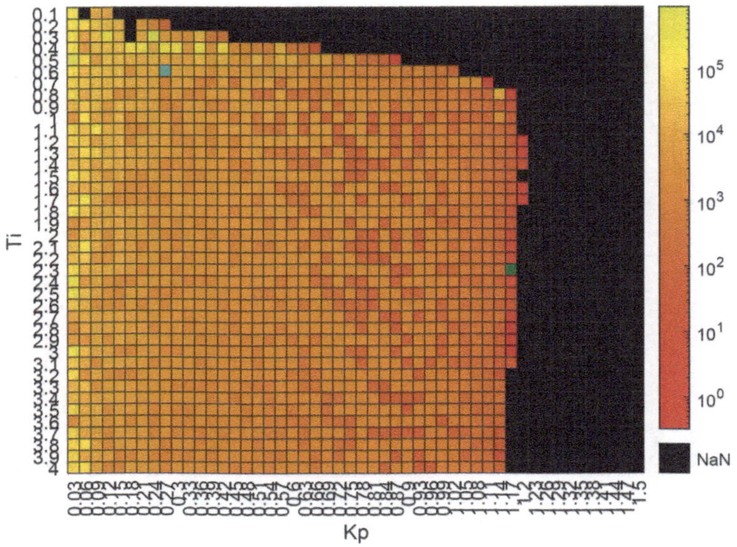

Fig. 7. MinVar – undisturbed $G_1(s)$ loop

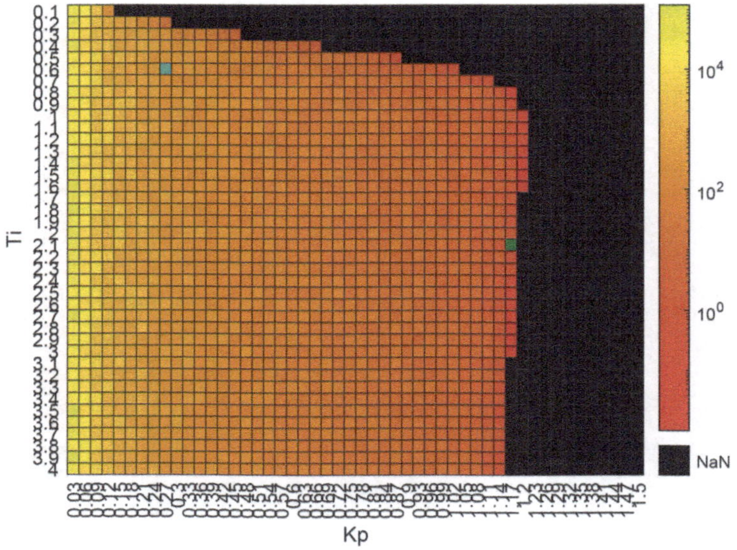

Fig. 8. MinVar – disturbed $G_1(s)$ loop

the ratio between optimal and current process variable variance. Analysis of the algorithm and partial results shows that disturbance significantly deteriorates evaluation of the variance of the model residuals. Additionally, in case of the disturbed case also process variable variance is strongly biased. This observation confirms sensitivity due to the least squares features, and thereby minimum variance assessment.

Next, the second transfer function $G_2(s)$ is analyzed. The surfaces for selected three indexes, i.e. MSE, IAE and MinVar are presented in consecutive Figs. 9, 10, 11, 12, 13 and 14. The surface reflects only changes in k_p and T_i. In all cases it is assumed that $T_d = 0.08$. Identified good tuning for each index and disturbance scenario is shown in Table 1.

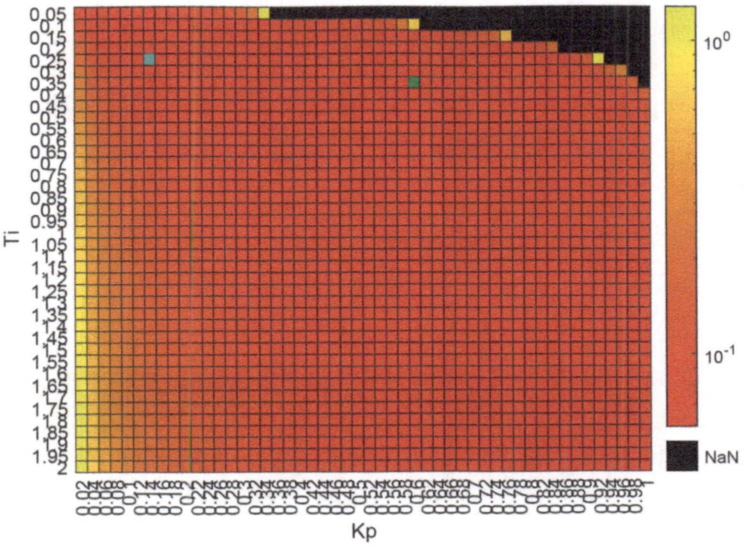

Fig. 9. MSE surface – undisturbed $G_2(s)$ loop

Integral indexes behave in a very similar manner, while the MinVar surfaces are even more ragged. Proper MinVar detection, similarly to the previous plant, is biased and unreliable. The reasons seem to be the same.

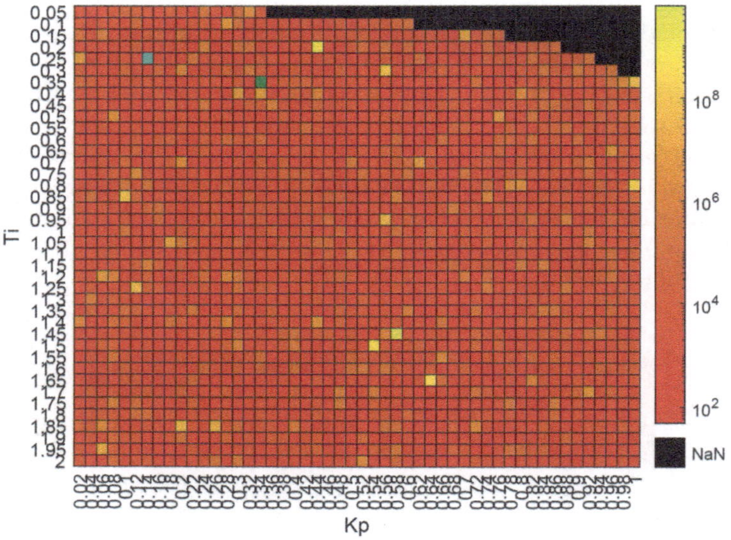

Fig. 10. MSE surface – disturbed $G_2(s)$ loop

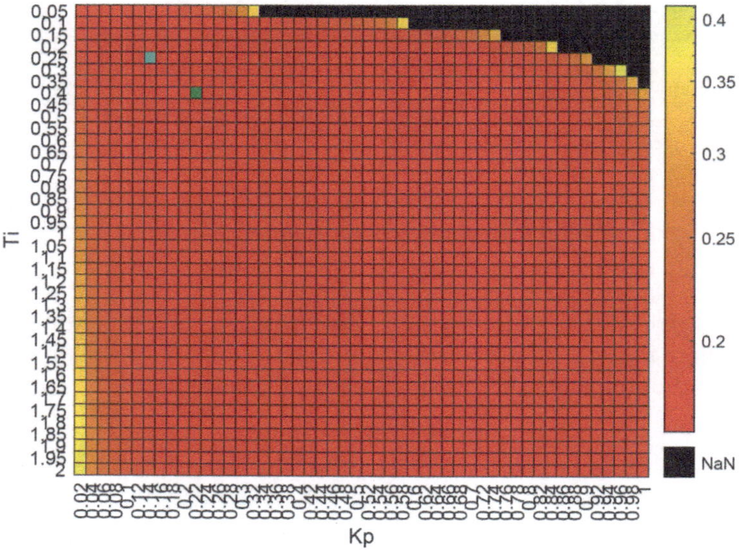

Fig. 11. IAE surface – undisturbed $G_2(s)$ loop

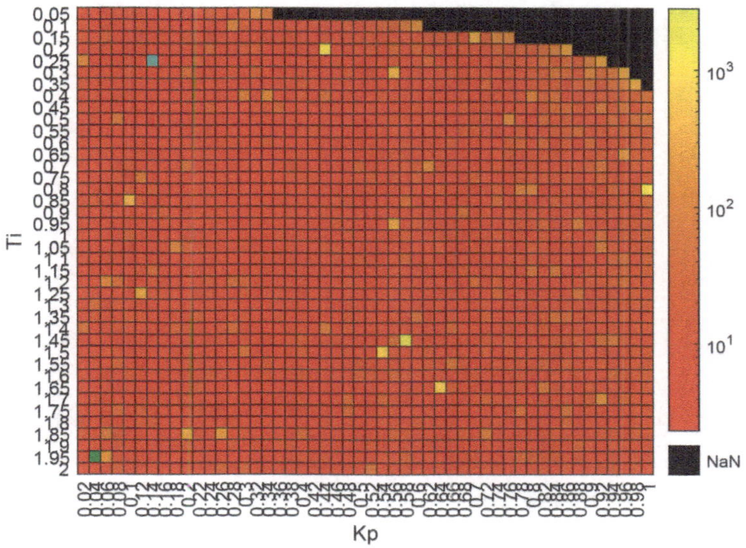

Fig. 12. IAE surface – disturbed $G_2(s)$ loop

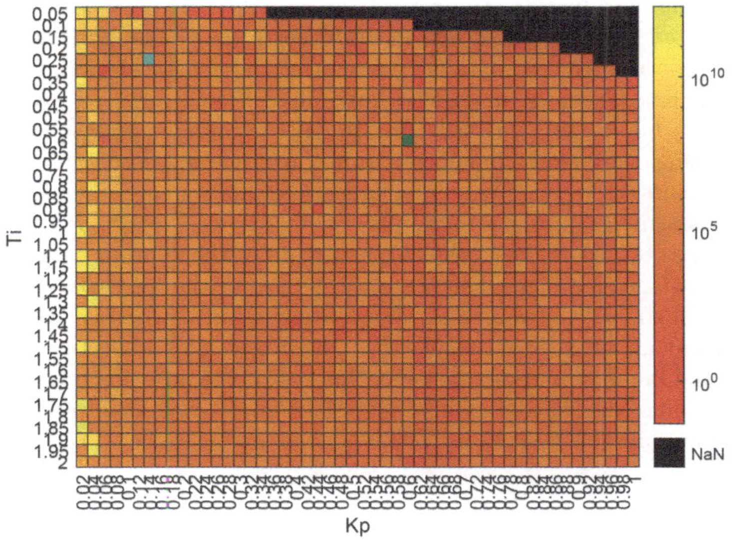

Fig. 13. MinVar – undisturbed $G_2(s)$ loop

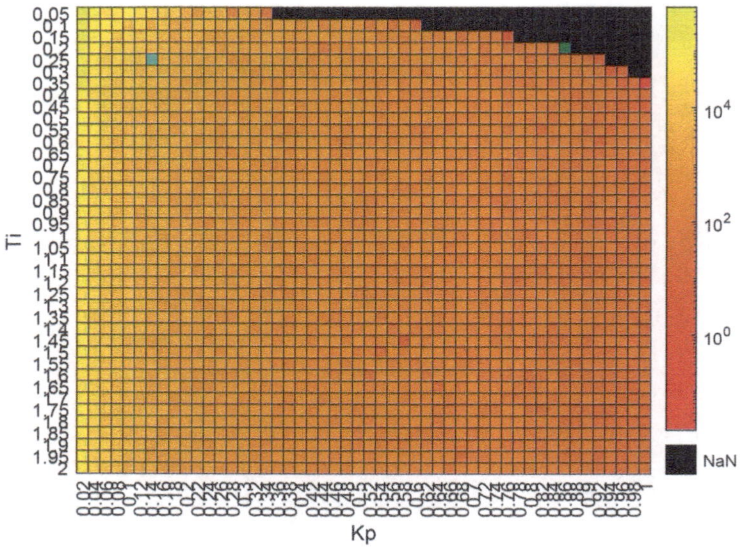

Fig. 14. MinVar – disturbed $G_2(s)$ loop

5 Conclusions and Further Research

Presented analyses have shown that the minimum variance control performance assessment may not be robust in case of the loop disturbances. The analysis has been conducted for two plants. In both cases the outliers introduced into the loop, which are reflected by the fat tails in process variable distribution, significantly deteriorate the assessment making it unreliable. If an effect appears with such a simple raw book-like PID structure, it will be even magnified in industrial structure using dead bands, signal constraints, anti-windup, etc. Literature shows that the outliers strongly impede least squares estimation, and thereby minimum variance assessment approach follows it.

It is worth to observe that the ragged shape of the contour surfaces for the minimum variance assessment is not only in case of the fat-tailed loop, but at the undisturbed one as well. This observation requires further attention and more detailed analysis. It is visible that there is a lot to be done to improve the method. One may propose various solutions to the presented sensitivity, like application of robust statistics and regression or re-evaluation of the method with non-Gaussian PDF scale factors.

References

1. Åström, K.J., Hägglund, T.: Benchmark systems for PID control. In: IFAC Digital Control: Past, Present and Future of PID Control, pp. 165–166 (2000)
2. Bauer, M., Horch, A., Xie, L., Jelali, M., Thornhill, N.: The current state of control loop performance monitoring - a survey of application in industry. J. Process Control **38**, 1–10 (2016)
3. Bialic, G.: Methods of control performance assessment for sampld data systems working under stationary stoachastics disturbances. Ph.D. thesis, Dissertation of Technical University of Opole, Poland (2006)
4. CPC Control Group: Univariate Controller Performance Assessment, Limited Trial Version 2.5. University of Alberta, Computer Process Control Group (2010). https://sites.ualberta.ca/~control/manuals/uvpa.pdf. [downloaded: 04-December-2019]
5. Desborough, L., Harris, T.J.: Performance assessment measures for univariate feedback control. Can. J. Chem. Eng. **70**(6), 1186–1197 (1992)
6. Desborough, L., Harris, T.J.: Performance assessment measures for univariate feedforward/feedback control. Can. J. Chem. Eng. **71**(4), 605–616 (1993)
7. Domański, P.D.: Non-gaussian properties of the real industrial control error in SISO loops. In: Proceedings of the 19th International Conference on System Theory, Control and Computing, pp. 877–882 (2015)
8. Domański, P.D.: Statistical measures for proportional-integral-derivative control quality: simulations and industrial data. Proc. Inst. Mech. Eng. Part I J. Syst. Control Eng. **232**(4), 428–441 (2018)
9. Domański, P.D.: Control Performance Assessment: Theoretical Analyses and Industrial Practice. Springer, Cham (2020)
10. Domański, P.D., Golonka, S., Jankowski, R., Kalbarczyk, P., Moszowski, B.: Control rehabilitation impact on production efficiency of ammonia synthesis installation. Ind. Eng. Chem. Res. **55**(39), 10366–10376 (2016)
11. Ettaleb, L.: Control loop performance assessment and oscillation detection. Ph.D. thesis, University of British Columbia, Canada (1999)
12. Farenzena, M.: Novel methodologies for assessment and diagnostics in control loop management. Ph.D. thesis, Dissertation of Universidade Federal do Rio Grande do Sul, Porto Alegre, Brazil (2008)
13. Gomez, D., Moya, E.J., Baeyens, E.: Control performance assessment: a general survey. In: de Carvalho, A.P.L.F., Rodriguez-Gonzalez, S., De Paz Santana, J.F., Rodriguez, J.M.C. (eds.) Distributed Computing and Artificial Intelligence: 7th International Symposium, pp. 621–628. Springer, Heidelberg (2010)
14. Harris, T.J.: Assessment of closed loop performance. Can. J. Chem. Eng. **67**, 856–861 (1989)
15. Harris, T.J., Yu, W.: Controller assessment for a class of non-linear systems. J. Process Control **17**(7), 607–619 (2007)
16. Hawkins, D.M.: Identification of Outliers. Chapman and Hall, London (1980)
17. Jelali, M.: Control Performance Management in Industrial Automation: Assessment, Diagnosis and Improvement of Control Loop Performance. Springer, London (2013)
18. Kadali, R., Huang, B.: Controller performance analysis with LQG benchmark obtained under closed loop conditions. ISA Trans. **41**(4), 521–537 (2002)
19. Ko, B.S., Edgar, T.F.: Performance assessment of cascade control loops. AIChE J. **46**(2), 281–291 (2000)

20. Ko, B.S., Edgar, T.F.: PID control performance assessment: the single-loop case. AIChE J. **50**(6), 1211–1218 (2004)
21. Liu, M.C.P., Wang, X., Wang, Z.L.: Performance assessment of control loop with multiple time-variant disturbances based on multi-model mixing time-variant minimum variance control. In: Proceeding of the 11th World Congress on Intelligent Control and Automation, pp. 4755–4759 (2014)
22. Ordys, A., Uduehi, D., Johnson, M.A.: Process Control Performance Assessment - From Theory to Implementation. Springer, London (2007)
23. Perrier, M., Roche, A.A.: Towards mill-wide evaluation of control loop performance. In: Proceedings of the Control Systems, pp. 205–209 (1992)
24. Rousseeuw, P.J., Leroy, A.M.: Robust Regression and Outlier Detection. Wiley, New York (1987)
25. Seppala, C.T.: Dynamic analysis of variance methods for monitoring control system performance. Ph.D. thesis, Queen's University Kingston, Ontario, Canada (1999)
26. Thornhill, N.F., Huang, B., Shah, S.L.: Controller performance assessment in set point tracking and regulatory control. Int. J. Adapt. Control Signal Process. **17**(7–9), 709–727 (2003)
27. Tyler, M.L., Morari, M.: Performance assessment for unstable and nonminimum-phase systems. IFAC Proc. Volumes **28**(12), 187–192 (1995)

Model of Aeration System at Biological Wastewater Treatment Plant for Control Design Purposes

Robert Piotrowski and Tomasz Ujazdowski(✉)

Faculty of Electrical and Control Engineering, Gdańsk University of Technology, Gdansk, Poland
robert.piotrowski@pg.gda.pl, tomek-ujazdowski@wp.pl

Abstract. The wastewater treatment plant (WWTP) is a dynamic, very complex system, in which the most important control parameter is the dissolved oxygen (DO) concentration. The air is supplied to biological WWTP by the aeration system. Aeration is an important and expensive activity in WWTP. The aeration of sewage fulfils a twofold role. Firstly, oxygen is provided as the main component for biological processes. Secondly, it supports mixing the sludge with the delivered sewage, which helps to treat the sewage. The paper proposes a model of the aeration system for biological WWTP located in Northeast Poland. This aeration system consists of the blowers, the main collector pipeline, three lines of the aeration with different diameters and lengths and diffusers. This system is a nonlinear dynamic system with faster dynamics compared to the internal dynamics of the DO at the biological WWTP. Control of the aeration system is also difficult in terms of control of the DO. A practical approach to model identification and validation is proposed. Simulation tests for aeration system at Matowskie Pastwiska WWTP are presented.

Keywords: Aeration system · Modelling · Nonlinear system · Wastewater treatment plant

1 Introduction

Wastewater treatment plays a key role for humanity. The waste entering lakes, rivers, and seas deteriorates the daily quality of life. Therefore, it is very important to improve the efficiency of wastewater treatment.

Batch type wastewater treatment plant (WWTP) (named Sequencing Batch Reactor – SBR) is a complex control system due to nonlinear dynamics, large uncertainty, multiple time scales in the internal process dynamics and multivariable structure. In addition, limited measurements are possible during plant operation. In the SBR, all biochemical processes occur in one tank, in the predefined sequence. SBR is a fill-and-draw activated sludge treatment system. This technology is widely used under small wastewater inflow conditions and may be designed using a single tank or a system of multiple tanks working

A. Bartoszewicz et al. (Eds.): PCC 2020, AISC 1196, pp. 349–359, 2020.
https://doi.org/10.1007/978-3-030-50936-1_30

in parallel. A usual work cycle involves following operational phases: filling, biological reactions (aerobic, anaerobic), sedimentation, decantation, and idling [1].

The oxygen is delivered into the SBR by the aeration system composed of blowers, pipes, valves, and diffusers. Aeration is an important and expensive activity in WWTP. The aeration of sewage fulfils a twofold role. Firstly, oxygen is provided as a main component for biological processes (denitrification, nitrification, phosphorus removal). Secondly, it supports mixing the sludge with the delivered sewage, which helps to treat the sewage. Insufficient amounts of oxygen interfere with the proper course of biological processes. At the same time, an excessively high oxygen level does not act profitably either. It does not improve the effectiveness of biological processes while producing higher costs of sewage treatment due to a longer time of aerating.

The variable which is directly related to sewage aeration is the dissolved oxygen (DO) concentration. DO control is important for WWTP energy efficiency. The total energy consumed by aeration processes is the factor deciding about total energy consumption in WWTP (above 50% of total operational cost) [2]. Previous research works presented various algorithms of DO control, e.g. [3–14].

The remainder of this paper is organized as follows. The case study aeration system is described in Sect. 2. Model creation, structure presentation and subsystem division are presented in Sect. 3. The results analysis and verification tests are illustrated in Sect. 4. Section 5 concludes the paper.

2 Description of the Aeration System

WWTP at Matowskie Pastwiska was modernized in 2018. The entire aeration system has changed and a new DO control system has been introduced. WWTP is composed of two identical, independently operating SBRs. Each of them is equipped with a separate aeration system which consists of a blower, diffuser system and pipeline. The required DO level is maintained by the variable speed blower. Blower airflow can be controlled within the range of 150–320 m^3/h. The pumped airflow moves into a pipeline and splits to tube diffuser systems located at the bottom floor of the SBR.

The pipeline consists of the main pipeline with a diameter of 104 mm and a length of the first part (horizontal) equal to 75.6 m and a second (vertical) 6 m that passes into a collector with a diameter of 154 mm and a length of 13 m. 6 branches depart from it (each with a diameter of 69 mm and 5.6 m long) supplying medium to distribution pipes (80 mm $\times$ 80 mm $\times$ 5 m). Figure 1 shows the pipeline layout. The blower station is located on the left side of the pipeline, at the end of a 75 m pipe. The diffuser system is located at the end of six branches. The diffusers are membrane tube type. There are 84 diffusers, separated 14 per branch spread over the entire surface of the SBR. The SBR has dimensions of 14 m length, 5.6 m width and 6 m height. The maximum fill level of the SBR is 5.5 m, while the minimum is 4 m. The level implies a hydrostatic pressure in the tank.

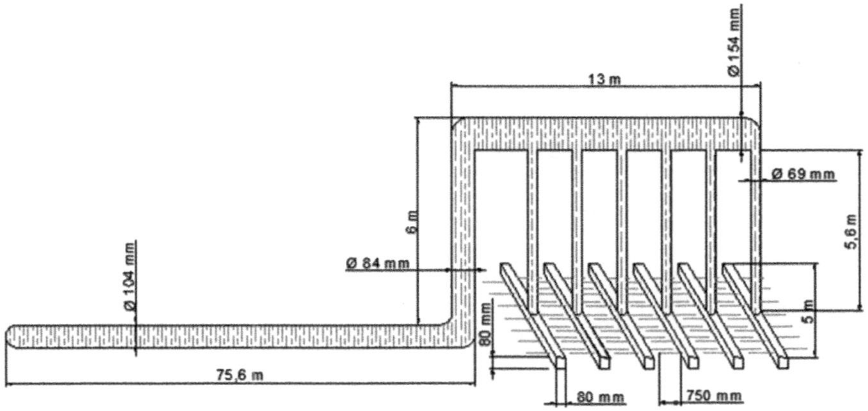

Fig. 1. Schematic diagram of the pipeline.

3 Modelling and Parameters Identification

3.1 Model Structure

The general methodology of aeration system modelling was presented in [14]. The theoretical knowledge, manufacturer data records from case study plant and documentation characteristics of the system elements were applied. The structure of the model was based on a proven approach using an electric analog. This allows processes related to gas flows to be presented in a simpler way. The circuit is shown in Fig. 2. There are three main subsystems in the model: blowers, pipeline and aeration segment units. The blower is represented as a nonlinear current source and is accompanied by the designations - Q_b, Δp_b. Hydrostatic pressure is shown as a voltage source with a pressure drop Δp_h. Resistor R_c corresponds to the total unit pressure losses along the pipeline length, R_z losses resulting from changes in pipe diameter. The pipeline is presented as total fluid-flow capacitance C_c and p_c node. The aeration segment units are described by R_d resistance and C_d capacity. The pressure loss across diffuser is represented as Δp_d.

3.2 Blowers

The blower station compresses air assuring appropriate pressure p_c at the pipeline. The blower operation was modelled based on the characteristics provided by the blower manufacturer. Using the characteristics, three matrices were created: characteristic velocities (Table 1), airflows (Table 2) and pressures (Table 3). The input parameters of the algorithm representing the operation of the blower are pressure p_c and blower rotational speed n. Based on them, the algorithm selects the appropriate mass airflow at the output. The values between the intervals are calculated as a weighted average of two adjacent values.

The blower model is represented as a nonlinear function:

$$Q_b = f(p_c, n) \tag{1}$$

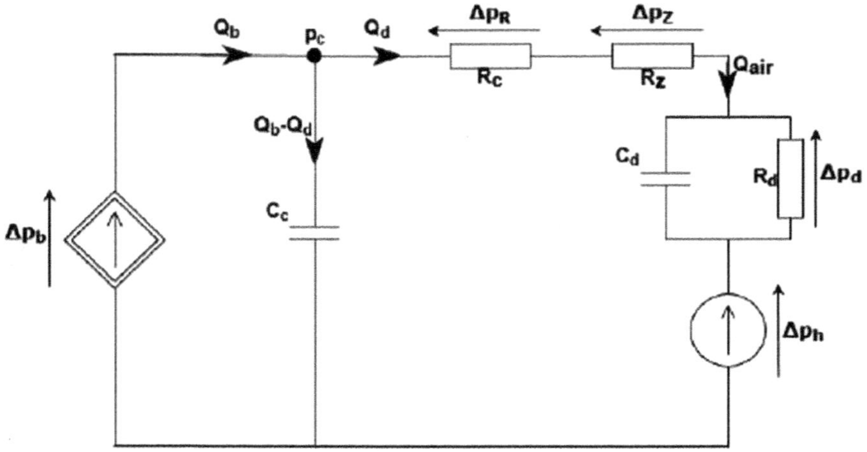

Fig. 2. Electrical analogy of the aeration system model.

where Q_b, p_c, n are the blower output airflow, pressure drop across the blower and motor rotational speed, respectively.

The use of a matrix to build a blower model allows for model adaptation depending on the type of blower in WWTP. The size of the characteristic pressure matrix (Table 3) is related to the use of other types of blowers in previous studies [15]. In order to preserve the universal character of the blower model, no matrix simplifications were applied.

Table 1. Matrix of blower characteristic velocities - N

Column index and values (n [rpm])										
1	2	3	4	5	6	7	8	9	10	11
3000	3200	3400	3600	3800	4000	4200	4400	4600	4800	5000

The current rotational speed and pressure are compared with the table of characteristic values and the nearest elements marked as index are selected. The rotational speed proportionality (k_n) coefficient is obtained according to equation:

$$k_n = \frac{n - \boldsymbol{N}(j_n)}{\boldsymbol{N}(j_n + 1) - \boldsymbol{N}(j_n)} \tag{2}$$

where j_n is the column number of the N matrix with less or equal rotational speed. Intermediate pressure values can be expressed as:

$$p_i = \boldsymbol{P}(j_n) + k_n \cdot (\boldsymbol{P}(j_n + 1) - \boldsymbol{P}(j_n)) \tag{3}$$

Intermediate airflow values are obtained from a similar relationship:

$$f_i = \boldsymbol{B}(j_n) + k_n \cdot (\boldsymbol{B}(j_n + 1) - \boldsymbol{B}(j_n)) \tag{4}$$

Table 2. Matrix of blower characteristic airflows – B

B		Column index and values (Q_b [Nm^3/hm])										
		1	2	3	4	5	6	7	8	9	10	11
Row index and values	1	186.0	202.2	218.4	234.6	250.8	267.0	283.2	299.4	315.6	331.8	348.0
	2	182.4	198.6	214.8	231.0	247.2	263.4	279.6	295.8	312.0	328.2	344.4
	3	178.2	194.4	210.6	226.8	243.0	259.2	275.4	291.6	307.8	324.0	340.2
	4	174.6	190.8	207.0	223.2	239.4	255.6	271.8	288.0	304.2	320.4	336.6
	5	169.8	186.0	202.2	218.4	234.6	250.8	267.0	283.2	299.4	315.6	331.8
	6	166.2	182.4	198.6	214.8	231.0	247.2	263.4	279.6	295.8	312.0	328.2
	7	163.8	180.0	196.2	212.4	228.6	244.8	261.0	277.2	293.4	309.6	325.8
	8	157.2	173.4	189.6	205.8	222.0	238.2	254.4	270.6	286.8	303.0	319.3

Table 3. Matrix of blower characteristic pressures – P

P		Column index and values (p_c [kPa])										
		1	2	3	4	5	6	7	8	9	10	11
Row index and values	1	30	30	30	30	30	30	30	30	30	30	30
	2	35	35	35	35	35	35	35	35	35	35	35
	3	40	40	40	40	40	40	40	40	40	40	40
	4	45	45	45	45	45	45	45	45	45	45	45
	5	50	50	50	50	50	50	50	50	50	50	50
	6	55	55	55	55	55	55	55	55	55	55	55
	7	60	60	60	60	60	60	60	60	60	60	60
	8	70	70	70	70	70	70	70	70	70	70	70

To obtain the final value of the airflow at the blower, the pressure proportionality coefficient is calculated:

$$k_p = \frac{p_c - \boldsymbol{p}_i(j_f)}{\boldsymbol{p}_i(j_f + 1) - \boldsymbol{p}_i(j_f)} \tag{5}$$

where j_f is an element of the flow matrix with less or equal pressure from the **P** matrix.

Finally, the output value of the airflow Q_b is described by the following equation:

$$Q_b = \boldsymbol{f}_i(j_f) + k_p \cdot (\boldsymbol{f}_i(j_f + 1) - \boldsymbol{f}_i(j_f)) \tag{6}$$

Performance test curves based on which matrices were developed were characterized by constant pressure values for which the speed was changed. This allowed to create a simplified model using Eq. (7):

$$Q_b = 0.081 \cdot n - (0.72p_c + 37.2) \tag{7}$$

The equation parameters were calculated using optimization methods. The task of minimizing the objective function was used, where the objective function was defined as the difference between the function sought and the matrix approach. Relative error was examined for characteristic pressure and flow points. The application of the simplified equation to the matrix approach generated an error in the range of 1% to 2%. The blowers are designed to work in the pressure range from 30 to 70 kPa, but the most common operating range is in the range of 45-65 kPa.

3.3 Pipeline

The pipeline has a large fluid flow capacity that significantly changes the dynamics of the model. Air capacity in the pipeline can be represented as:

$$C_c = k_c \cdot V_c \cdot p_c \tag{8}$$

where V_c determines the total volume of the pipeline (9), p_c determines the gas pressure inside the pipeline and kc is the unit conversion coefficient. The total volume is obtained as the sum of five pipeline fragments of different lengths and cross-sections using Eq. (10). Total volume:

$$V_c = V_1 + V_2 + V_3 + V_4 + V_5 \tag{9}$$

$$V_i = \pi \cdot \left(\frac{d_i}{2}\right)^2 \cdot l_i \, ; i \in \{1, 2, 3, 4, 5\} \tag{10}$$

The pressure change in the pipeline is achieved by using the principle of mass conservation in the pipeline node:

$$\frac{dp_c}{dt} = \frac{1}{C_c} \cdot (Q_b - Q_{air}) \tag{11}$$

The pressure losses occurring in the pipeline consist of three elements: unit linear pressure losses over a specified section, local pressure losses due to changes in pipe cross-sections, pressure loss due to height difference.

Unit linear pressure losses depend on the cross-section of the pipeline, as well as gas density and flow value. These changes were modelled according to the Renouard formula [16]:

$$\Delta p_R = 0,776457 \cdot 10^{-8} \cdot \rho \cdot \frac{V^{1,82}}{D^{4,82}} \tag{12}$$

where V is the mass airflow, D is the diameter of the pipeline and ρ is gas density. Individual gas constant for air $r = 287{,}05$ J/kgK and temperature was used to calculate the density:

$$\rho = \mathrm{p_c} / \mathrm{r} \cdot \mathrm{T} \tag{13}$$

Local pressure losses are caused by the influence of the Reynolds number on the value of the local resistance coefficient ξ and gas flow speed w. They are described by Eq. (14) where ξ is presented as (15):

$$\Delta p_Z = \sum \xi_j \cdot \frac{\rho}{2} \cdot w^2 j \in \{1, 2, 3, 4\} \tag{14}$$

$$\xi_j = \left(1 - \frac{A_j}{A_{j+1}}\right)^2 j \in \{1, 2, 3, 4\} \tag{15}$$

where A_j, A_{j+1} are pipe cross section areas before and after narrowing, respectively.

Pressure loss caused by the difference of levels significantly affect the pipelines with large changes in altitude and low gas pressures. The vertical sections influence the pressure change using the difference between the density of the medium, and air density under standard conditions $\rho_p = 1.225$ kg/m^3, according to the equation:

$$\Delta p_H = g \cdot \Delta H \cdot (\rho - \rho_p) \tag{16}$$

where g, ΔH are gravitational acceleration and height difference, respectively.

Due to small changes in height in the pipeline pressure loss caused by the difference of levels have not been used. The total pressure loss in the pipeline used in the model, visible in Fig. 2, can be represented as equation:

$$\Delta p_c = \Delta p_b - \Delta p_R - \Delta p_Z - \Delta p_h \tag{17}$$

3.4 Aeration Segment Units

The diffuser operation depends on the pressure difference (Δp_c) between the pipeline pressure (p_c) and the tank hydrostatic pressure drop (Δp_h) (see Fig. 2). The opening of the diffusers, i.e. the deflection of the membrane allowing air to escape, occurs when a minimum pressure p_{min} difference of 3.9 kPa is reached. In the steady-state the open diffuser airflow – pressure drop link is described by a nonlinear function:

$$Qair = \begin{cases} 0, & \Delta p_c < p_{min} \\ f(\Delta p_c), & \Delta p_c \geq p_{min} \end{cases} \tag{18}$$

The characteristic obtained from the manufacturer's data has been described by a nonlinear function and the diffuser airflow dynamics can be described as:

$$Qair = a_1 \cdot \Delta p_c^4 + a_2 \cdot \Delta p_c^3 + a_3 \cdot \Delta p_c^2 + a_4 \cdot \Delta p_c + a_5 \tag{19}$$

where $a_1 = -0.3484$, $a_2 = 7.1474$, $a_3 = 53.5072$, $a_4 = 178.1669$, $a_5 = -223.0322$. Function parameters were calculated using the least-squares method. The function has been simplified by approximating the first-order equation:

$$Qair = 5.167\, \Delta p_c - 19.6196 \tag{20}$$

Differences between the characteristics of the 1st and 4th order are shown in Fig. 3. The hydrostatic pressure drop Δp_h is described in accordance with the expression:

$$\Delta p_h = \rho_m \cdot g \cdot h \tag{21}$$

where ρ_m, g, h are sewage density in the SBR tank, gravity acceleration and height of diffusers in SBR tank, respectively.

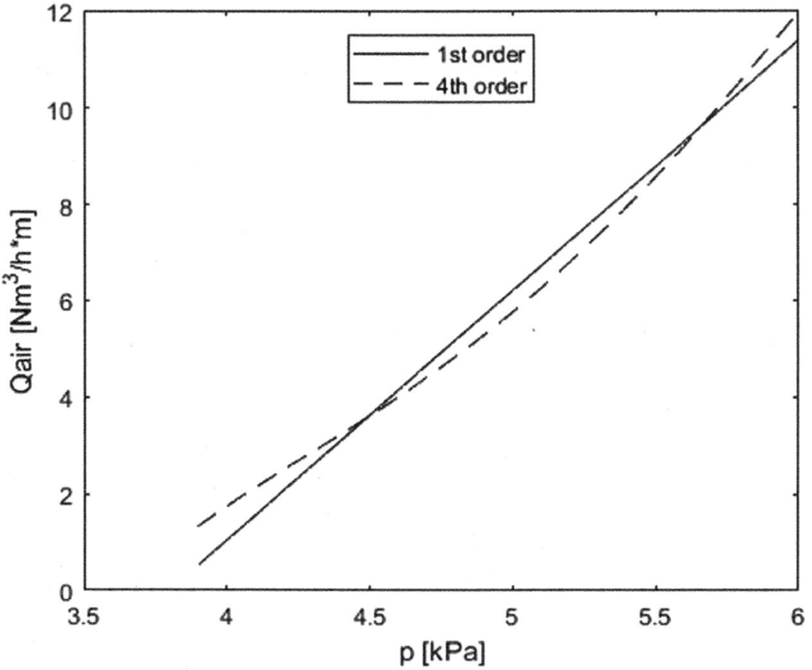

Fig. 3. Approximations of diffuser operation characteristics.

4 Verification Test and Results Analysis

All values of model parameters are: d_1 = 104 mm, d_2 = 84 mm, d_3 = 154 mm, d_4 = 69 mm, d_5 = 80 mm, l_1 = 75.6 m, l_2 = 6 m, l_3 = 13 m, l_4 = 5.6 m, l_5 = 5 m, T = 20°C, g = 9.81 m/s^2, $k_c = 1\ m^2s^4/kg^2$, r = 287,05 J/kgK, ρ_m = 1150 kg/m^3, $p_{min} = 3.9$ *kPa*, ρ_p = 1.225 kg/m^3, V_c = 1.2352 m^3. Model variable units: Q_{air}, Q_b, – Nm^3/hm; p_c, Δp_c, Δp_R, Δp_Z, Δp_H, Δp_h – kPa; n – rpm; h, ΔH – m, ρ – kg/m^3.

The pressure sensors available at WWTP Matowskie Pastwiska were used for the measurements. The blower control system sensors delivered measurements of pressure in pipeline (p_c) and blower frequency, which used to calculate the blower speed. Model

Table 4. Simulation test of the first SBR tank

Frequency [Hz]	Blower [kPa]	Model [kPa]	Error [kPa]	Absolute error
30	54.7	56.22	1.52	2.78%
35	55.1	56.34	1.24	2.25%
40	55.6	56.48	0.88	1.58%
45	56.0	56.61	0.61	1.09%
50	56.6	56.72	0.12	0.21%

Table 5. Simulation test of the second SBR tank

Frequency [Hz]	Blower [kPa]	Model [kPa]	Error [kPa]	Absolute error
30	63.0	64.2	1.52	1.90%
35	63.6	64.7	1.24	1.73%
40	64.0	64.8	0.88	1.25%
45	64.9	65.0	0.61	0.15%
50	66.0	65.1	−0.9	−1.36%

validation tests were carried out on two SBR tanks with different filling levels and medium densities. This resulted in significant differences in hydrostatic pressure, and thus different pressures needed to open the diffusers. In the first SBR tank, the sewage level was 4.1 m, and the wastewater density was 2 g/l. In the second SBR tank, the wastewater level was 4.7 m and the wastewater density was 1.3 g/l. The measurement results were compared to the simulation results for the same conditions.

Comparing the operation of the blower in WWTP and the simulation tests of the model presented in the paper in Table 4 and Table 5, the small error results are about 1–2%.

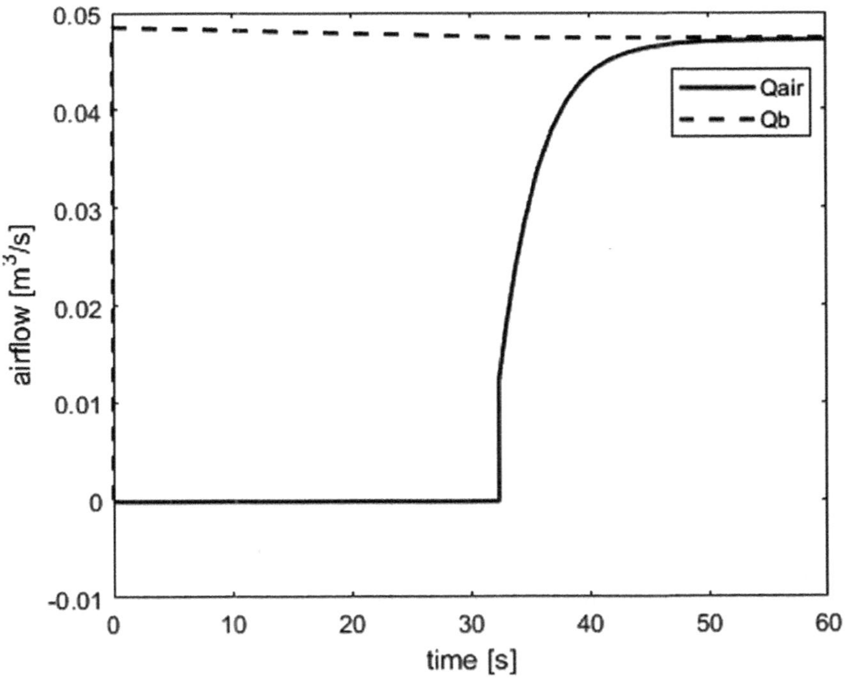

Fig. 4. Model airflows.

The dynamics of blowers and diffusers are illustrated in Fig. 4. The characteristic moment of opening the diffusers after exceeding the pressure threshold is visible here.

The characteristics of the effect of the SBR tank fill level on the Q_{air} at a constant blower speed is shown in Fig. 5. In the simulation test, the level of the medium varies from minimum to maximum while maintaining a constant density.

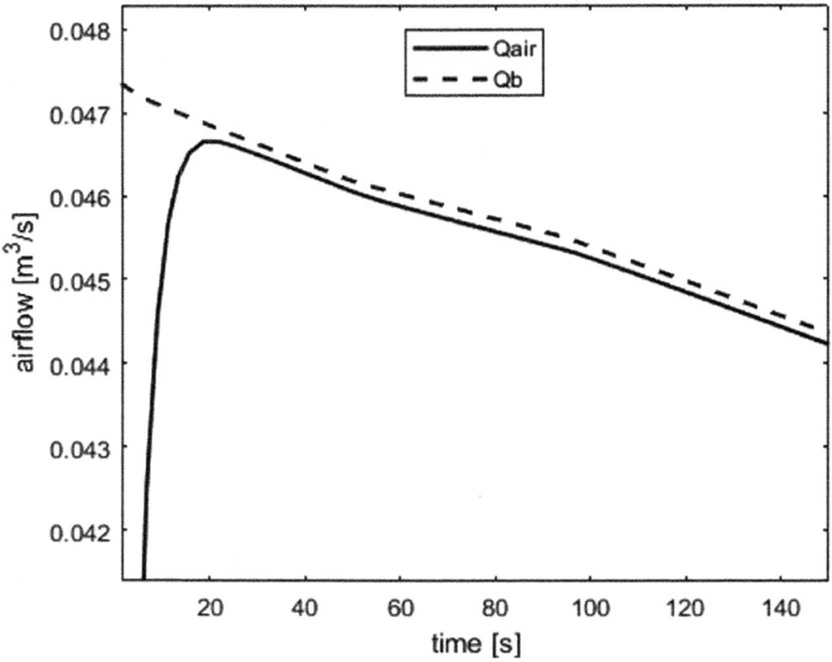

Fig. 5. Hydrostatic pressure test.

In summary, modelling results are verified to be satisfactory and can be used for control purposes.

5 Conclusions

Aeration control systems are evolving towards the latest control solutions. More advanced control algorithms require mathematical models and simulations to work properly. This paper presents the modelling of the aeration system for control purposes. Relative to the previous research work [9], the diffuser approach has changed and equations describing pressure losses in the pipeline were used. The model structure, its parameterisation and approach to the parameter calculation from the manufacturer data have been successfully validated by application to the case study system.

Acknowledgements. The authors would like to thank the staff of the Matowskie Pastwiska WWTP for their help with access to the plant, information and data.

References

1. Wilderer, P.A., Irvine, R.L., Goronszy, M.: Sequencing Batch Reactor Technology. Scientific and Technical Report No. 10, IWA Publishing, London (2001)
2. Jenkins, T.E.: Aeration Control System Design. A Practical Guide to Energy and Process Optimization. Wiley, New Jersey (2013)
3. Belchior, C.A.C., Araújo, R.A.M., Landeck, J.A.C.: Dissolved oxygen control of the activated sludge wastewater treatment process using stable adaptive fuzzy control. Comput. Chem. Eng. **37**(10), 152–162 (2012)
4. Vrečko, D., Hvala, N., Stražar, M.: The application of model predictive control of ammonia nitrogen in an activated sludge process. Water Sci. Technol. **64**(5), 1115–1121 (2011)
5. Åmand, L., Carlsson, B.: Optimal aeration control in a nitrifying activated sludge process. Water Res. **46**(7), 2101–2110 (2012)
6. Błaszkiewicz, K., Piotrowski, R., Duzinkiewicz, K.: A model-based improved control of dissolved oxygen concentration in sequencing wastewater batch reactor. Stud. Inf. Control **23**(4), 323–332 (2014)
7. Yang, T., Qiu, W., Ma, Y., Chadli, M., Zhang, L.: Fuzzy model-based predictive control of dissolved oxygen in activated sludge processes. Neurocomputing **136**, 88–95 (2014)
8. Piotrowski, R.: Two-Level multivariable control system of dissolved oxygen tracking and aeration system for activated sludge processes. Water Environ. Res. **87**(1), 3–13 (2015)
9. Piotrowski, R., Skiba, A.: Nonlinear fuzzy control system for dissolved oxygen with aeration system in sequencing batch reactor. Inf. Technol. Control **44**(2), 182–195 (2015)
10. Santín, I., Pedret, C., Vilanova, R.: Applying variable dissolved oxygen set point in a two level hierarchical control structure to a wastewater treatment process. J. Process Control **28**, 40–55 (2015)
11. Piotrowski, R., Błaszkiewicz, K., Duzinkiewicz, K.: Analysis the parameters of the adaptive controller for quality control of dissolved oxygen concentration. Inf. Technol. Control **45**(1), 42–51 (2016)
12. Ruan, J., Zhang, C., Li, Y., Li, P., Yang, Z., Cheng, X., Huang, M., Zhang, T.: Improving the efficiency of dissolved oxygen control using an on-line control system based on a genetic algorithm evolving FWNN software sensor. J. Environ. Manag. **187**, 550–559 (2017)
13. Du, X., Wang, J., Jegatheesan, V., Shi, G.: Dissolved oxygen control in activated sludge process using a neural network-based adaptive pid algorithm. Appl. Sci. **8**, 261 (2018)
14. Piotrowski, R., Brdyś, M.A., Konarczak, K., Duzinkiewicz, K., Chotkowski, W.: Hierarchical dissolved oxygen control for activated sludge processes. Control Eng. Pract. **16**(1), 114–131 (2008)
15. Krawczyk, W., Piotrowski, R., Brdyś, M.A., Chotkowski, W.: Modelling and identification of aeration systems for model predictive control of dissolved oxygen – Swarzewo wastewater treatment plant case study. In: Conference: Proceedings of the 10th IFAC Symposium on Computer Applications in Biotechnology, Cancun, Mexico, 4–6 June 2007 (2007)
16. Renouard, M.P.: Nouvelles règles à calcul pour la détermination des pertes de charge dans les conduites de gaz. Journal des Usines à Gaz, 337–339 (1952)

Control of a Nonlinear and Linearized Model of Self-balancing Electric Motorcycle

Adam Wonia(✉), Michał Wonia, and Robert Piotrowski

Faculty of Electrical and Control Engineering,
Gdańsk University of Technology, Gdańsk, Poland
adam.wonia@gmail.com, michal.wonia@onet.pl,
robert.piotrowski@pg.edu.pl

Abstract. Self-Balancing Electric Motorcycle (SBEM) is a dynamic and nonlinear electromechanical system. In this paper, the process of mathematical modelling and linearization of SBEM is presented. The model of the control system in Matlab environment is implemented. The control system using the PID controller is designed. The operation of particular structures of the PID controller on the simulation model is compared. Due to simulation research, the most appropriate structure and parameters of the PID controller are chosen.

Keywords: PID controller · Control system · Mathematical modelling · Self-balancing Electric Motorcycle

1 Introduction

Electric vehicles use electric power to work. These vehicles are usually powered by electric motors which can drive each wheel separately or the whole axle. Electric vehicles such as cars, trucks, trains, bikes and bicycles are mainly used to transport people and to travel. An electric vehicle has its own power source like a battery to provide electric power. It can be recharge using solar energy or a charging station. It can also have systems that recover energy from braking to recharge the battery. The advantages of electric vehicles compared to combustion vehicles are: quiet and clean work, environmentally friendly, relative long drive distances, cheaper usage and they can be used indoors.

Self-Balancing Electric Motorcycle (SBEM) can be used to transport people in the desired direction. The advantage of SBEM is that it is possible to stand in a vertical position without using an additional kickstand.

Control and modelling of SBEM can be an interesting subject of research. It has a nonlinear and complex mathematical model shown in [1]. The position stabilization system can be realized by means of various algorithms. The PI controller is applied in [2]. Control algorithm using LQR is introduced in [3]. Self-balancing similar vehicles are common projects. The motorcycle using a flywheel to stabilize its position is shown in [4] and an autonomous bicycle is designed in [5].

The mathematical model is an abstract, simplified, mathematical construct related to a part of reality and created for a particular purpose [6]. The model allows for a better

A. Bartoszewicz et al. (Eds.): PCC 2020, AISC 1196, pp. 360–371, 2020.
https://doi.org/10.1007/978-3-030-50936-1_31

understanding of the process and predicts certain actions under certain conditions and defined input signals. It also can be used to define dangerous states of the process and critical values of signals.

The main purpose of modelling SBEM is to better understand its principle of operation. The mathematical model is used to implement a control system which uses a PID controller in Matlab environment [7]. The final structure of the PID controller and values of its parameters are indicated during the simulation test.

The paper is organized as follows. Section 2 presents the process of mathematical modelling and linearization of SBEM. In this section, the simulation results of a nonlinear and linearized model of SBEM are presented. Section 3 presents the structure of the control system and modelling of the DC motor. In this section, two methods of tuning the PID controller are described. In Sect. 4 the simulation tests and results analysis are conducted. Concluding remarks are listed in the last section.

2 Modelling of SBEM

2.1 Operating Principle

The SBEM principle of operation is derived from the inverted reaction wheel pendulum (Fig. 1). The main element that stabilizes the structure vertically is the reaction wheel, driven by the DC (Direct Current) motor. It uses the change of its angular momentum to bring the structure to a vertical position.

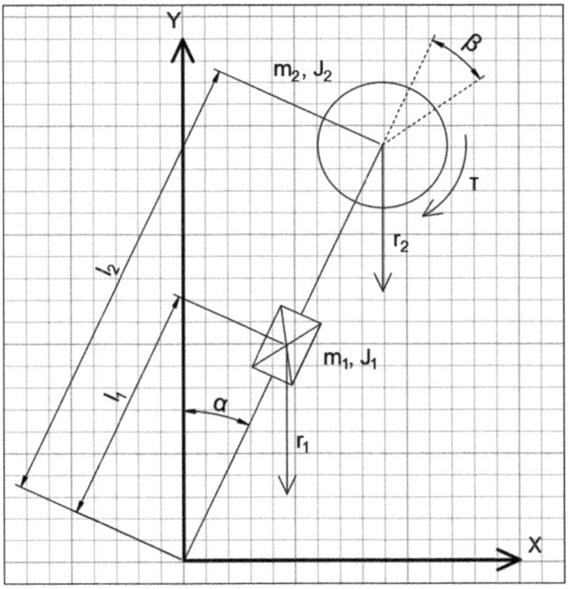

Fig. 1. Inverted reaction wheel pendulum

where m_1 – mass of the construction without mass of the reaction wheel, m_2 – mass of the reaction wheel, l_1 – distance of the centre of the mass m_1 from the coordinate system origin, l_2 – distance of the centre of the mass m_2 from the coordinate system origin, J_1 – moment of inertia of the mass m_1, J_2 – moment of inertia of the mass m_2, r_1 – length of the gravity force arm acting on the mass m_1, r_2 – length of the gravity force arm acting on the mass m_2, α – angle of the pendulum, β – angle of the reaction wheel, τ – torque applied on reaction wheel by DC motor.

The stabilization mechanism is based on Newton's second law for rotational motion. According to this principle, the derivative of the angular momentum of a rigid body equals the torque acting on it. It is given by Eq. (1).

$$\frac{dL(t)}{dt} = J \cdot \frac{d\omega(t)}{dt} = \tau(t) \tag{1}$$

where $L(t)$, J, $\omega(t)$, $\tau(t)$ are the angular momentum, the moment of inertia, the angular velocity and the torque, respectively.

When the equilibrium of the system is disturbed, voltage is applied to the DC motor and the torque of the motor is applied to the reaction wheel causing it to accelerate. According to Eq. (1), the torque acting on the reaction wheel is created. The reaction wheel in turn according to Newton's third law applies the equal amount of the torque to the DC motor, but in the opposite direction. Because it is mounted to motorcycle body, the torque acts on the whole construction, bringing it back to the vertical position [1].

By controlling this reaction torque the motorcycle body can be balanced. The torque of the reaction wheel DC motor should correspond to the moment of gravitational force acting on the vehicles center of mass when deflected from equilibrium point [3].

2.2 Nonlinear Model of SBEM

For the mathematical model of SBEM Euler-Lagrange equations are applied [8]. A description of the total kinetic and potential energy is required. It is given by the following equations:

$$E_{tp}(t) = g \cdot cos\alpha(t) \cdot (m_1 \cdot l_1 + m_2 \cdot l_2) \tag{2}$$

$$\begin{aligned} E_{tk}(t) = {} & \frac{1}{2} \cdot \left(m_1 \cdot l_1^2 + m_2 \cdot l_2^2 + J_1 + J_2\right) \cdot \left(\frac{d\alpha(t)}{dt}\right)^2 \\ & + J_2 \cdot \frac{d\alpha(t)}{dt} \cdot \frac{d\beta(t)}{dt} + \frac{1}{2} \cdot J_2 \cdot \left(\frac{d\beta(t)}{dt}\right)^2 \end{aligned} \tag{3}$$

where $E_{tp}(t)$, g, $E_{tk}(t)$ are the total potential energy, the gravity constant and the total kinetic energy of the system, respectively.

Euler-Lagrange equations use Lagrangian given by Eq. (4). It is a difference between the total kinetic energy and the total potential energy.

$$L\left[\alpha(t), \beta(t), \frac{d\alpha(t)}{dt}, \frac{d\beta(t)}{dt}, t\right] = E_{tk}(t) - E_{tp}(t) \tag{4}$$

where L is Lagrangian and t is time.

Euler-Lagrange equations are given by the following equation:

$$\frac{\partial L}{\partial y_i(x)} - \frac{d}{dt}\left(\frac{\partial L}{\partial \dot{y}_i(x)}\right) = \tau_i(t) \tag{5}$$

where $y_i(x)$ is the function depended on variable x and $\tau_i(t)$ is the generalized torque in the $y_i(x)$ direction.

Lagrangian is given by Eq. (6).

$$L = \frac{1}{2}\cdot\left(m_1\cdot l_1^2 + m_2\cdot l_2^2 + J_1 + J_2\right)\cdot\left(\frac{d\alpha(t)}{dt}\right)^2 + J_2\cdot\frac{d\alpha(t)}{dt}\cdot\frac{d\beta(t)}{dt} + \frac{1}{2}\cdot J_2\cdot\left(\frac{d\beta(t)}{dt}\right)^2 - g\cdot cos\alpha(t)\cdot(m_1\cdot l_1 + m_2\cdot l_2) \tag{6}$$

Euler-Lagrange equations are given by the following equations:

$$\begin{cases} \frac{\partial L}{\partial \alpha(t)} - \frac{d}{dt}\left(\frac{\partial L}{\partial \dot{\alpha}(t)}\right) = 0 \\ \frac{\partial L}{\partial \beta(t)} - \frac{d}{dt}\left(\frac{\partial L}{\partial \dot{\beta}(t)}\right) = \tau(t) \end{cases} \tag{7}$$

where $\tau(t)$ is the torque provided by DC motor.

Using Eqs. (6) and (7) the mathematical model of SBEM is derived:

$$\begin{cases} g\cdot sin\alpha(t)\cdot(m_1\cdot l_1 + m_2\cdot l_2) - \left(m_1\cdot l_1^2 + m_2\cdot l_2^2 + J_1 + J_2\right)\cdot\frac{d^2\alpha(t)}{d^2t} - J_2\cdot\frac{d^2\beta(t)}{d^2t} = 0 \\ J_2\cdot\frac{d^2\alpha(t)}{d^2t} + J_2\cdot\frac{d^2\beta(t)}{d^2t} = \tau(t) \end{cases} \tag{8}$$

The system of Eqs. (8) can be reduced to a single equation of the following form:

$$\frac{d^2\alpha(t)}{dt^2} = -\frac{1}{m_1\cdot l_1^2 + m_2\cdot l_2^2 + J_1}\cdot\tau(t) + \frac{g\cdot(m_1\cdot l_1 + m_2\cdot l_2)}{m_1\cdot l_1^2 + m_2\cdot l_2^2 + J_1}\cdot\sin\alpha(t) \tag{9}$$

2.3 Linearization of SBEM Model

The mathematical model of SBEM is nonlinear due to trigonometric function *sinα(t)*. Because of that, it is difficult for a wide operating range using a PID controller. The process of linearization is needed [9].

The first step is to determine the static duty point that sets out values of system parameters in a steady-state. The static duty point is described as follows:

$$S_0 = \left(\alpha_0, \frac{d^2\alpha_0}{dt^2}, \tau_0\right) = (0, 0, \tau_0) \tag{10}$$

where S_0 is the static duty point, α_0 is the angle of the pendulum in the steady-state and τ_0 is the torque applied by DC motor in the steady-state.

The value of the torque in the steady-state can be derived from Eq. (9) using values of parameters from Eq. (10):

$$f(0,0,\tau_0) = -\frac{1}{m_1 \cdot l_1^2 + m_2 \cdot l_2^2 + J_1} \cdot \tau_0 + \frac{g \cdot (m_1 \cdot l_1 + m_2 \cdot l_2)}{m_1 \cdot l_1^2 + m_2 \cdot l_2^2 + J_1} \cdot \sin 0 = 0 \quad (11)$$

where f is the function which describes the equation of static characteristic.

The value of the torque in a steady-state equals zero. The next step is to expand Eq. (9) into a Taylor series with the operating point and neglect higher-order terms. The linearized equation of SBEM mathematical model is given by Eq. (12).

$$\left(\frac{d^2\alpha(t)}{dt^2} - 0\right) = -\frac{1}{m_1 \cdot l_1^2 + m_2 \cdot l_2^2 + J_1} \cdot (\tau(t) - \tau_0) + \frac{g \cdot (m_1 \cdot l_1 + m_2 \cdot l_2)}{m_1 \cdot l_1^2 + m_2 \cdot l_2^2 + J_1} \cdot (\alpha(t) - \alpha_0) \quad (12)$$

The final step is to use increment variables to describe Eq. (12). It is given by Eq. (13).

$$\Delta\frac{d^2\alpha(t)}{dt^2} = -\frac{1}{m_1 \cdot l_1^2 + m_2 \cdot l_2^2 + J_1} \cdot \Delta\tau(t) + \frac{g \cdot (m_1 \cdot l_1 + m_2 \cdot l_2)}{m_1 \cdot l_1^2 + m_2 \cdot l_2^2 + J_1} \cdot \Delta\alpha(t) \quad (13)$$

The nonlinear (9) and linearized (13) mathematical model of SBEM was implemented in Matlab environment.

2.4 Simulation of SBEM Model

The models implemented in Matlab were tested to determine the accuracy of the linearized model of SBEM. For this purpose, the unit-step responses of derived models were examined and the linearization errors were calculated. The linearization error is the difference between nonlinear and linearized model response.

All values of models parameters are: $m_1 = 0.59$ kg, $m_2 = 0.11$ kg, $l_1 = 0.06$ m, $l_2 = 0.12$ m, $J_1 = 3.26\text{e-}3$ kg $\cdot$ m^2, $J_2 = 403\text{e-}6$ kg $\cdot$ m^2, $g = 9.81$ m/s^2.

The unit-step responses of the models are shown in Fig. 2.

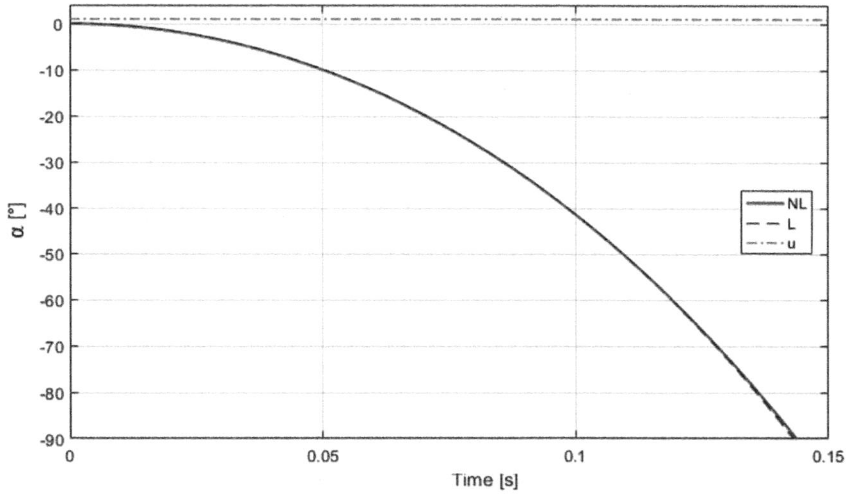

Fig. 2. Unit-step responses of nonlinear and linearized models (*NL* – nonlinear model, *L* – linearized model, *u* – unit-step input)

In Fig. 2 an angle of −90° indicates a situation in which the motorcycle is completely tilted to one side. Linearization error is illustrated in Fig. 3.

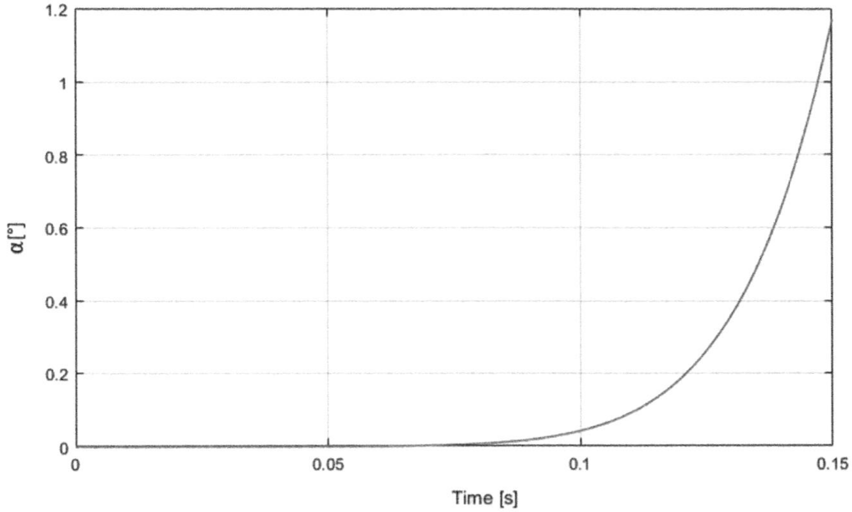

Fig. 3. Linearization error

The linearization errors are due to the fact that the linearized model is only an approximation of a nonlinear model in the specified area of a duty point. This means that it will only behave like a nonlinear model within a certain range of deviations from the duty work point. The further away from the duty point, the greater the linearization error.

3 Design of Control System

3.1 Structure of the Control System

The control structure of the vertical position of the SBEM is illustrated in Fig. 4.

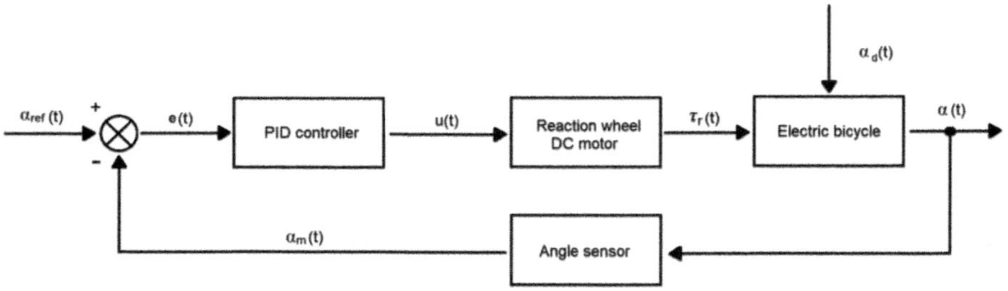

Fig. 4. SBEM control system

where α_{ref} *(t)* – reference angle, *e(t)* – error signal, *u(t)* – voltage signal, $\tau_r(t)$ – reaction torque, α_d *(t)* – equilibrium disturbance, $\alpha(t)$ – angle, $\alpha_m(t)$ – measured angle.

The reference value is the vertical angle of SBEM. In this case, the reference value equals 0°. The error signal is created by subtracting the reference angle and the measured angle. Next, the PID controller generates a control signal fed to the reaction wheel DC motor. When the reaction wheel starts spinning, the reaction torque is generated and the electric bicycle is balanced. The actual angle of SBEM is measured by an angle sensor. The equilibrium disturbance is an external force which causes deviation from the vertical axis of SBEM. Two types of disturbances were considered: impulse and constant. The first corresponds to the application of the force for a short time. It is an equivalent of a short push. The second type corresponds to the application of force for a long time. It can be regarded as placing the mass on one side of SBEM.

3.2 Model of Reaction Wheel DC Motor

To design a control system in Matlab the DC motor mathematical model is needed. To obtain the mathematical model the equivalent scheme of the DC motor is used (Fig. 5).

where *U(t)* – rotor power supply voltage, *i(t)* – motor current, *R* – rotor winding resistance, *L* – rotor winding inductance, *E(t)* – electromotive force of induction, $M_s(t)$ – rotor torque, *J* – rotor shaft moment of inertia, *B* – viscous friction coefficient, $M_L(t)$ – load torque, $\omega_s(t)$ – angular velocity.

Mathematical model adequate to DC motor scheme (Fig. 5) is given by Eqs. (14).

$$\begin{cases} U(t) = R \cdot i(t) + L \cdot \frac{di(t)}{dt} + k_e \cdot \omega_s(t) \\ k_m \cdot i(t) = J \cdot \frac{d\omega_s(t)}{dt} + B \cdot \omega_s(t) + M_L(t) \end{cases} \tag{14}$$

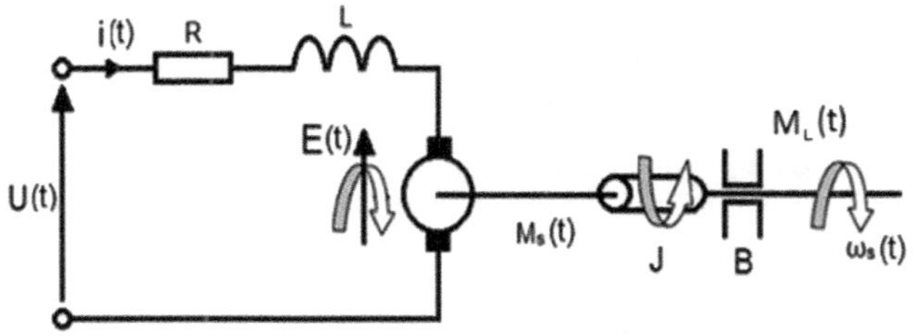

Fig. 5. DC motor equivalent scheme

where k_e is the electromotive constant and k_m is torque constant.

To implement the model of the DC motor in Matlab the Laplace transform is applied. By using the Laplace transform in Eqs. (14) and reorganizing the output equation, the DC motor model is obtained.

$$\Omega_s(s) = \frac{1}{J \cdot s + B} \cdot \left(\frac{U(s) - k_e \cdot \Omega_s(s)}{R + L \cdot s} \cdot k_m - M_L(s) \right) \quad (15)$$

The Eq. (15) was implemented in Matlab environment. All values of DC motor parameters are: $R = 5.71\ \Omega$, $L = 380\text{e-}6$ H, $k_m = 0.80$ N · m/A, $k_e = 0.13$ V · s/rad, $J = 3.6\text{e-}6$ kg · m/s^2, $B = 3.69\text{e-}4$ N · m · s/rad, $M_L = 0.0$ N · m.

3.3 Tuning of PID Controller

To tune the PID controller the second Ziegler-Nichols tuning method is used. The required parameters to calculate the PID controller such as critical gain K_{cr} and period of sustained oscillation T_{cr} are indicated. Using these two parameters, the parameters of the PID controller are computed and presented in Table 1.

Table 1. Parameters for PID controller

Controller structure	K_p	T_i	T_d
P	7.0	–	–
PI	6.3	0.63	–
PID	8.4	0.38	0.09

where K_p, T_i, T_d are proportional gain, integral time and derivative time, respectively.

The second method of tuning the PID controller is based on simulation tests. To determine controller parameters, the following quality control indicators were considered:

- permissible control error < 1% of disturbance amplitude,

- settling time t_s below < 1 s – it is the time required by the response to reach and steady below $|\varepsilon| = 1°$,
- permissible overshoot $< 10\%$ – calculated as a ratio of the absolute value of second peak A_2 to the absolute value of first peak A_1 in percentages:

$$OS_{\%} = \left|\frac{A_2}{A_1}\right| \cdot 100\% \tag{16}$$

All parameters needed to calculate quality control indicators are shown in Fig. 6.

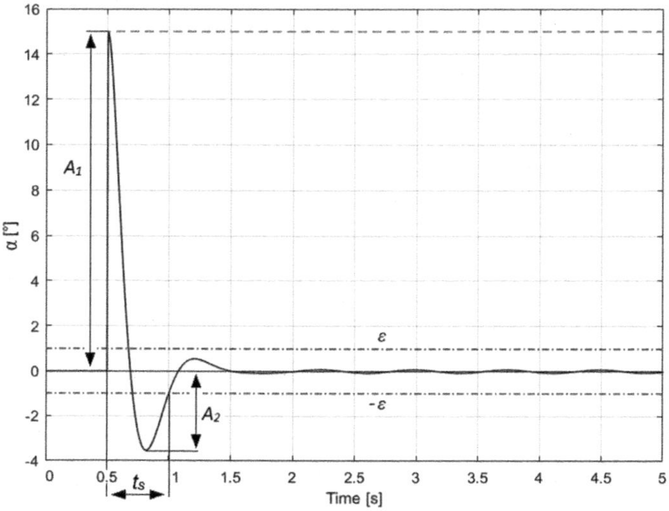

Fig. 6. Parameters of step response needed to calculate quality control indicators

On the basis of the simulation tests, the parameters of the PID controller that meet the mentioned requirements were determined. These values are shown in Table 2.

Table 2. Parameters for PID controller

Controller structure	K_p	T_i	T_d
P	31.8	–	–
PI	32.2	0.04	–
PID	32.0	0.06	0.1

4 Simulation Tests and Results Analysis

In this section, simulation tests were carried out. By analyzing simulation results the most appropriate structure of the PID controller was chosen. The PID structure was determined on the basis of tests.

Simulation tests of the control system of the nonlinear and linearized model of SBEM using two sets of PID controller parameters were conducted. Two types of disturbances: step and impulse were examined. The amplitude of 15° was taken as the maximum deviation. The control results of the second Ziegler-Nichols tuning method are shown in Fig. 7.

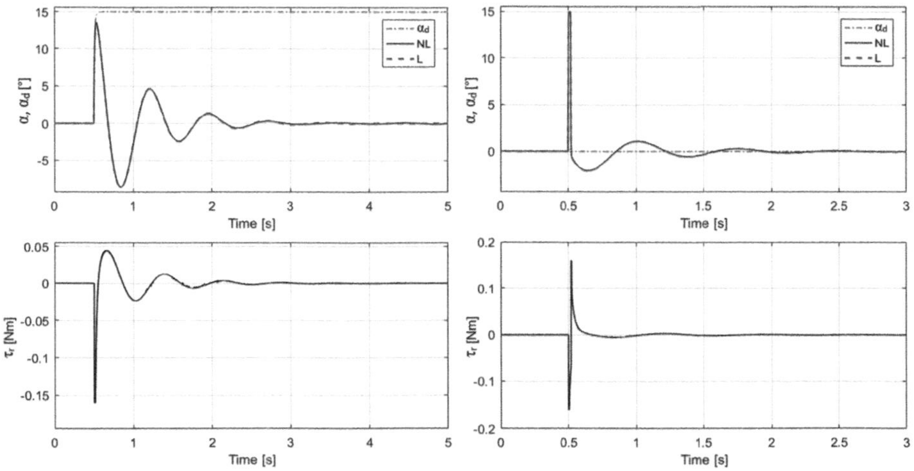

Fig. 7. Control results of the second Ziegler-Nichols tuning method

The system was found to be stable on the basis of the research conducted. However, the required control quality indicators are not met. All values of the mentioned quality indicators are shown in Table 3.

The results of experimental PID controller tuning method are shown in Fig. 8.

The system was found to be stable on the basis of the research conducted. The required control quality indicators are achieved. The shape of the reaction torque τ_r waveforms is caused by upper and lower limits of its value. These are the maximum torque values provided by the reaction wheel DC motor. All values of the mentioned quality indicators are shown in Table 3.

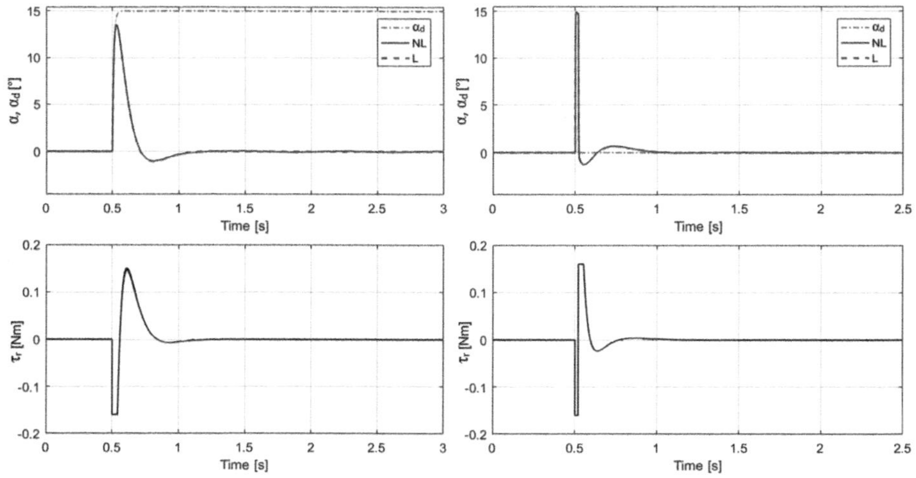

Fig. 8. Control results of the experimental tuning method

Table 3. Values of control quality indicators

Tuning method of PID	II Ziegler-Nichols		Experimental	
Input signal	Step	Impulse	Step	Impulse
Permissible control error [°]	0.08	0.01	0.05	0.02
Settling time [s]	1.53	0.23	0.31	0.27
Permissible overshoot [%]	64.23	13.78	7.46	8.33

5 Conclusions

In this paper, the processes of mathematical modelling and linearization of SBEM were presented. The implementation of the control system and control results analysis were done. Two methods of tuning the PID controller were verified. The operation of particular structures of the PID controller were compared. The most appropriate structure of the PID controller and values of its parameters were obtained. All quality control indicators were achieved using an experimental tuning method of the PID controller.

References

1. Almujahed, A., Deweese, J., Duong, L., Potter, J.: Auto-Balanced Robotic Bicycle (ABRB), ECE-492/3 Senior Design Project, Spring 2009
2. Block, D.J., Astrom, K.J., Spong, M.W.: The Reaction Wheel Pendulum. Synthesis Lectures on Controls and Mechatronics. Morgan & Claypool, San Rafael (2007)

3. Owczarkowski, A.: Application of selected control algorithms for nonlinear systems in unmanned bicycle robot stabilized by an inertial drive. Doctoral Dissertation, Institute of Control and Information Engineering, Faculty of Electrical Engineering, Poznań University of Technology, April 2017
4. Lam, P.Y.: Design and development of a self-balancing bicycle using control moment gyro. Master thesis, Department of Mechanical Engineering, National University of Singapore (2012)
5. An Won, S., Dong, R., Huang, E., Hwang, J., Imsdahl, O., Mi, W., Sharma, A., Wampler, R., Xu, X.: Autonomous Bicycle Project. Mechanical and Aerospace Engineering, Cornell University (2015)
6. Bender, E.: An Introduction to Mathematical Modeling, 1st edn. University of California, San Diego (2000). Kindle Edition
7. Jain, S., Kapshe, S.: Modeling and Simulation Using Matlab – Simulink: For ECE. Wiley, Hoboken (2016)
8. Arfken, G.B., Weber, H.J.: Mathematical Methods for Physicists, 6th edn, pp. 1053–1056. Academic Press, Cambridge (2005)
9. Westphal, L.: Handbook of Control Systems Engineering, pp. 744–758 (2001)

New Delay Product Type Lyapunov-Krasovskii Functional for Stability Analysis of Time-Delay System

Sharat Chandra Mahto[1(✉)], Sandip Ghosh[1], Shyam Krishna Nagar[1], and Pawel Dworak[2]

[1] Department of Electrical Engineering, Indian Institute of Technology (BHU), Varanasi 221005, UP, India
sharatcm.rs.eee15@itbhu.ac.in, {sghosh.eee,sknagar.eee}@iitbhu.ac.in
[2] Department of Control Engineering and Robotics, West Pomeranian University of Technology, Szczecin, Poland
pawel.dworak@zut.edu.pl

Abstract. This paper concerns delay-dependent stability analysis of linear system with time-varying delay. A new delay-product based functional (DPF) is formulated by including the new states introduced in the second order Bessels-Legendre integral inequality. Two delay-dependent stability criteria are derived in terms of linear matrix inequalities by utilizing this DPF in combination with improved reciprocally convex lemma and bounding technique reciprocal lemma. Two numerical examples are considered for demonstrating the improvement provided by the proposed criteria.

Keywords: Stability · Time-delay systems · Lyapunov-Krasovskii functional · LMI

1 Introduction

In many dynamical systems delays are usually time varying in nature. Two approaches available for stability analysis for such systems are Lyapunov-Krasovskii (LK) and Lyapunov-Razumikhin (LR). Both approaches can be useful to handle dynamical systems with time-varying delay. Less conservative results can be obtained using LK method as compared to LR one, since it takes the advantage of using additional information on the derivative of time varying delay [18]. Therefore, in robust stability analysis of system with time-varying delay using LK approach gets lot of attention.

With the aid of Linear matrix inequalities (LMIs), a variety of stability conditions have been proposed to find larger upper bound delay value by ensuring negative definiteness of derivative of the LK functional (LKF). For obtaining less conservative stability criteria of time-delay systems, it is crucial to find a

A. Bartoszewicz et al. (Eds.): PCC 2020, AISC 1196, pp. 372–383, 2020.
https://doi.org/10.1007/978-3-030-50936-1_32

precise bound of the quadratic integral function in the derivative of LKF. For this purpose, there has been a great amount of effort to find an effective inequality and as a result use of some inequalities have been proposed in the literature such as the Jensen inequality (JI) [4], Wirtinger based inequality (WI) [13], auxiliary function related inequality (AFI)[12], Bessel's-Legendre inequality (BLI) [15] and free matrix related inequalities [19,20].

Another way to obtain the allowable maximum time-delay value retaining stability is to formulate a suitable LKF. Mainly, following types of functionals have been proposed to contain more information such as (i) augmented LKF [13,23], where delay term is included in the state vector (ii) delay partitioning approach [1,3], which divides the delay interval into several segments and (iii) Cross-term variables based LKF [5,6], where $x(t), x(s)$ and $\dot{x}(s)$ are used to create quadratic terms, and (iv) multiple integral LKF [2,16]. These LKFs helps to obtain less conservative results to a certain extent. In addition matrix based function is employed to the existing LKF have been developed, which leads to provide fruitful results. Recently, delay product type LKF [17,21] have been introduced such that the information of delays and its derivative has been fully utilized. In this DPF, non-integral quadratic terms have been constructed by augmenting the state vectors used in the WI to exploit newly introduced single integral states. Similar idea has been used in [25] and [22] to form the functional by using the signals present in AFI by transforming into delay and its inverse dependent matrices for passivity analysis. In [7] a new form of functional has been reported in which the integral inequalities such as JI, WI are utilized to form DPF, such that its derivative includes delay variations based integral functions to yield better results.

On the basis of above discussion, this note further investigates delay-dependent stability analysis for linear systems with time-varying delay. The contribution of this paper is that two new states are introduced in the augmented vectors of DPF and in the Lyapunov matrix based quadratic term to formulate the LKF. Using this LKF and second order BLI, two stability criteria are proposed. Finally, Two numerical examples are provided to show the effectiveness of the proposed criteria.

Notations:- In this brief, 0 and I corresponds to $(n \times n)$ zero matrix and identity matrix respectively. For any matrices $P, Q, diag(P, Q)$ stands for the matrix $\begin{bmatrix} P & 0 \\ 0 & Q \end{bmatrix}$. Also, for any square matrix P, we defined $Sym\{P\} = P + P^T$.

2 Problem Formulation

Consider the time-delay system as:

$$\dot{x}(t) = Ax(t) + A_d x(t - d(t)), \tag{1}$$

where $x(t) \in \mathbb{R}^n$ is the state vector; $A, A_d \in \mathbb{R}^{n \times n}$ are the constant system matrices, with continuously differentiable initial condition.The delay function

$d(t)$ and its derivative satisfy

$$0 \le d(t) \le h, \quad \mu_0 \le \dot{d}(t) \le \mu_1 \le 1 \tag{2}$$

where h, μ_0 and μ_1 are constants. To obtain the main results, the following lemmas are needed.

Lemma 1. [23] *For real scalar* $\alpha \in (0,1)$, *symmetric matrices* $R_i(i=0,1) \ge 0$, *and any matrices* S_0 *and* S_1 *such that the following inequality holds*

$$\begin{bmatrix} \frac{1}{\alpha}R_0 & 0 \\ 0 & \frac{1}{1-\alpha}R_1 \end{bmatrix} \ge \begin{bmatrix} R_0 + (1-\alpha)X_0 & (1-\alpha)S_0 + \alpha S_1 \\ * & R_1 + \alpha X_1 \end{bmatrix} \tag{3}$$

where $X_0 = R_0 - S_1 R_1^{-1} S_1^T$ *and* $X_1 = R_1 - S_0^T R_0^{-1} S_0$

Lemma 2. [10] *For any symmetric matrices* $[R_i(i=0,1)]_{n\times n}$ *and any matrices* $[Y_i(i=0,1)]_{2n\times n}$*, the following inequality holds* $\alpha \in (0,1)$

$$\begin{aligned} \begin{bmatrix} \frac{1}{\alpha}R_0 & 0 \\ 0 & \frac{1}{1-\alpha}R_1 \end{bmatrix} &\ge Sym\left(Y_0 \left[I_n \; 0_{n\times n}\right] + Y_1 \left[0_{n\times n} \; I_n\right]\right) \\ &\quad - \alpha Y_0 R_0^{-1} Y_0^T - (1-\alpha) Y_1 R_1^{-1} Y_1^T \end{aligned} \tag{4}$$

Lemma 3. [14,15] *For any constant matrix* $R \ge 0$*, the following inequality holds for all continuously differentiable function* $w \in [a,b] \to \mathbb{R}^n$*;*

$$(b-a)\int_a^b \dot{w}^T(s) R \dot{w}(s) ds \ge \theta_1^T R \theta_1 + 3\theta_2^T R \theta_2 + 5\theta_3^T R \theta_3 \tag{5}$$

$$(b-a)\int_a^b w^T(s) R w(s) ds \ge \vartheta_1^T R \vartheta_1 + 3\vartheta_2^T R \vartheta_2 \tag{6}$$

where
$\theta_1 = w(b) - w(a)$,
$\theta_2 = w(b) + w(a) - \frac{2}{(b-a)}\int_a^b w(s)ds$,
$\theta_3 = w(b) - w(a) - \frac{6}{(b-a)}\int_a^b \delta_{a,b}(s) w(s) ds$,
$\vartheta_1 = \int_a^b w(s)ds, \vartheta_2 = \int_a^b \delta_{a,b}(s) w(s) ds$ *and*
$\delta_{a,b}(s) = 2\left(\frac{s-a}{b-a}\right) - 1$.

Remark 1. The integral inequalities (5) is a particular case of second order Bessel-Legendre inequality of [14]. This inequality overcomes the conservatism provides by Wirtinger-based integral inequality by introducing an extra quadratic term $\theta_3^T R \theta_3$. This additional term gives improvement by considering a new state $\int_a^b \delta_{a,b}(s) w(s) ds$, which contains both single and double integral terms. In this paper this state is used to form the DPF defined later.

3 Main Results

In this section, we construct a new DPF to derive a delay-dependent stability result for Linear system with time varying delay (1) with constraints (2). To simplify the representation, we introduce some notations as follows:

$$\begin{aligned}
&h_d(t) = h - d(t), \qquad x_d(t) = x(t - d(t)),\\
&x_h(t) = x(t-h), \qquad \bar{d} = 1 - \dot{d}(t)\\
&w_1(t) = \frac{1}{d(t)}\int_{-d(t)}^{0} x_t(s)ds, w_2(t) = \frac{1}{h_d(t)}\int_{-h}^{-d(t)} x_t(s)ds\\
&w_3(t) = \frac{1}{d(t)}\int_{-d(t)}^{0} \delta_1(s)x_t(s)ds\\
&w_4(t) = \frac{1}{h_d(t)}\int_{-h}^{-d(t)} \delta_2(s)x_t(s)ds\\
&\xi(t) = col[x(t), x_d(t), x_h(t), \dot{x}_d(t), w_1(t), w_2(t), w_3(t), w_4(t)]\\
&e_i = [0_{n\times(i-1)}, I_n, 0_{n\times(8-i)}], i = 1, 2, ..., 8,\\
&e_s = Ae_1 + A_d e_2, \qquad e_0 = 0_{n\times 8n}
\end{aligned}$$

According to the function $\delta_{a,b}$ given in Lemma 3, the functions $\delta_i, i = 1, 2$ can be expressed as

$$\delta_1(s) = 2\left(\frac{s + d(t)}{d(t)}\right) - 1, \qquad \delta_2(s) = 2\left(\frac{s+h}{h - d(t)}\right) - 1$$

On the basis of above $\delta_1(s)$ and $\delta_2(s)$ as augmented terms, delay product type quadratic terms has been constructed by extending the idea to construct LKF in [21]. Also based on $x(s), \dot{x}(s)$ and $\int_{t-d(t)}^{s} \dot{x}(s)ds$, a cross-term based quadratic term is being constructed similar to [6]. Using the combination of these delay product type and cross-term based terms a new candidate LKF is constructed as

$$V(t) = V_0(t) + V_1(t) + V_2(t) + V_3(t) \tag{7}$$

where

$$\begin{aligned}
V_0(t) &= \varpi_1^T(t)P\varpi_1(t)\\
V_1(t) &= d(t)\varpi_2^T(t)Q_1\varpi_2(t) + h_d(t)\varpi_3^T(t)Q_2\varpi_3(t)\\
V_2(t) &= \int_{t-d(t)}^{t} \varpi_4^T(s)Q_3\varpi_4(s)ds + \int_{t-h}^{t-d(t)} x^T(s)Zx(s)ds\\
V_3(t) &= \int_{-d(t)}^{0}\int_{t+u}^{t} \dot{x}^T(s)R_1\dot{x}(s)dsdu + \int_{-h}^{-d(t)}\int_{t+u}^{t} \dot{x}^T(s)R_2\dot{x}dsdu\\
&+ \int_{-d(t)}^{0}\int_{t+u}^{t} x(s)^T(s)M_1x(s)dsdu + \int_{-h}^{-d(t)}\int_{t+u}^{t} x^T(s)M_2x(s)dsdu
\end{aligned}$$

with

$$\varpi_1(t) = col[x(t), x_d(t), d(t)w_1, h_d(t)w_2, d(t)w_3, h_d(t)w_4]$$
$$\varpi_2(t) = col[x(t), x_d(t), w_1, w_3]$$
$$\varpi_3(t) = col[x(t), x_d(t), w_2, w_4]$$
$$\varpi_4(s) = col[x(s), \dot{x}(s), \int_{t-d(t)}^{s} \dot{x}(s)ds]$$

Remark 2. The terms in $V_3(t)$ are similar to the [11]. The $\dot{x}(s)$ and $x(s)$ dependent double integral quadratic terms are considered separately for the intervals $[t-d(t), t]$ and $[t-h, t-d(t)]$ respectively to exploit the delay range. The time derivative of $V_3(t)$ provides $d(t)$ and $\dot{d}(t)$ related single-integral functions. The estimation of these delay variation based integral terms introduces new and more $d(t)$ and $\dot{d}(t)$ dependent terms, which helps to get improved result.

By employing LKF (7), a delay-dependent stability criterion for system (1) with conditions (2) is as follows:

Theorem 1. *For positive definite matrices* $0 < P \in \mathbb{R}^{6n\times 6n}, 0 < Q_1, Q_2 \in \mathbb{R}^{4n\times 4n}, 0 < Q_3 \in \mathbb{R}^{3n\times 3n}, 0 < Z \in \mathbb{R}^{n\times n}, 0 < R_1, R_2, M_1, M_2 \in \mathbb{R}^{n\times n}$, *and matrices* $S_1, S_2 \in \mathbb{R}^{3n\times 3n}$ *with given scalars* h, μ_0 *and* μ_1, *system (1) is asymptotically stable, if following LMIs satisfy for all* $\dot{d}(t) \in [\mu_0, \mu_1]$ *and* $i = 1, 2$.

$$\begin{bmatrix} \Phi_0(0, \mu_i) - \Phi_1(0, \mu_i) & E_1^T S_2 \\ * & -h\mathcal{R}_2 \end{bmatrix} < 0 \tag{8}$$

$$\begin{bmatrix} \Phi_0(h, \mu_i) - \Phi_1(h, \mu_i) & E_2^T S_1^T \\ * & -h\mathcal{T}_1(\mu_i) \end{bmatrix} < 0 \tag{9}$$

such that the following conditions are satisfied

$$T_1(\mu_i) \geq 0 \tag{10}$$

$$T_2(\mu_i) \geq 0 \tag{11}$$

where

$$\begin{aligned} \Phi_0(d(t), \dot{d}(t)) = {} & Sym\{G_0^T P G_1\} + \dot{d}(t) G_2^T Q_1 G_2 + Sym\{G_2^T Q_1 G_3\} - \dot{d}(t) G_4^T Q_2 G_4 \\ & + Sym\{G_4^T Q_2 G_5\} + G_6^T Q_3 G_6 - \bar{d} G_7^T Q_3 G_7 + Sym\{G_8^T(d(t)) Q_3 G_9\} \\ & + e_s^T[d(t)R_1 + h_d(t)R_2]e_s + e_1^T[d(t)M_1 + h_d(t)M_2]e_1 \\ \Phi_1(d(t), \dot{d}(t)) = {} & E_1^T \frac{(2-\alpha)}{h} \mathcal{T}_1 E_1 + E_2^T \frac{(1+\alpha)}{h} \mathcal{R}_2 E_2 \\ & + \frac{2}{h} E_1^T[(1-\alpha)S_1 + \alpha S_2]E_2 + E_3^T d(t) \mathcal{T}_2 E_3 + E_4^T h_d(t) \mathcal{M}_2 E_4 \end{aligned}$$

$$
\begin{aligned}
G_0 &= [e_1^T, e_2^T, d(t)e_5^T, h_d(t)e_6^T, d(t)e_7^T, h_d(t)e_8^T]^T \\
G_1 &= [e_s^T, \bar{d}e_4^T, e_1^T - \bar{d}e_2^T, \bar{d}e_2^T - e_3^T, e_1^T + \bar{d}e_2^T - (2-\dot{d}(t))e_5^T - \dot{d}(t)e_7^T, \\
&\quad \bar{d}e_2^T + e_3^T - (2-\dot{d}(t))e_6^T + \dot{d}(t)e_8^T]^T \\
G_2 &= [e_1^T, e_2^T, e_5^T, e_7^T]^T, \\
G_3 &= [d(t)e_s^T, d(t)\bar{d}e_4^T, e_1^T - \bar{d}e_2^T - \dot{d}(t)e_5^T, \\
&\quad e_1^T + \bar{d}e_2^T - (2-\dot{d}(t))e_5^T - 2\dot{d}(t)e_7^T]^T \\
G_4 &= [e_1^T, e_2^T, e_6^T, e_8]^T \\
G_5 &= [h_d(t)e_s^T, h_d(t)\bar{d}e_4^T, \bar{d}e_2^T - e_3^T + \dot{d}(t)e_6^T, \\
&\quad \bar{d}e_2^T + e_3^T - (2-\dot{d}(t))e_6 + 2\dot{d}(t)e_8^T]^T \\
G_6 &= [e_1^T, e_s^T, e_1^T - e_2^T]^T, G_7 = [e_2^T, e_4^T, 0_{8n\times n}]^T \\
G_8 &= [d(t)e_5^T, e_1^T - e_2^T, d(t)(e_5^T - e_2)^T]^T, G_9 = [0_{8n\times n}, 0_{8n\times n}, -\bar{d}e_4^T]^T
\end{aligned}
$$

Proof. The derivative of $\int_{-d(t)}^{0} \delta_1(s)x_t(s)ds$ and $\int_{-h}^{-d(t)} \delta_2(s)x_t(s)ds$ can be expressed as

$$
\begin{aligned}
\frac{d}{dt}[\int_{-d(t)}^{0} \delta_1(s)x_t(s)ds] &= x(t) + \bar{d}x_d(t) - (2-\dot{d}(t))w_1(t) - \dot{d}(t)w_3(t) \\
\frac{d}{dt}[\int_{-h}^{-d(t)} \delta_2(s)x_t(s)ds] &= \bar{d}x_d(t) + x_h - (2-\dot{d}(t))w_2(t) + \dot{d}(t)w_4(t)
\end{aligned}
$$

Similarly, the derivative of Q_1, Q_2 and Q_3 dependent quadratic terms in (7) can be obtained as

$$
\begin{aligned}
\frac{d}{dt}[d(t)\varpi_2^T(t)Q_1\varpi_2(t)] &= \dot{d}(t)\varpi_2^T(t)Q_1\varpi_2(t) \\
&\quad + 2d(t)\varpi_2^T(t)Q_1 \begin{bmatrix} \dot{x}(t) \\ \bar{d}\dot{x}_d(t) \\ \frac{x(t)-\bar{d}x_d(t)-\dot{d}(t)w_1}{d(t)} \\ \frac{x(t)+\bar{d}x_d(t)-(2-\dot{d}(t))w_1-2\dot{d}(t)w_3}{d(t)} \end{bmatrix}
\end{aligned} \tag{12}
$$

$$
\begin{aligned}
\frac{d}{dt}[h_d(t)\varpi_3^T(t)Q_2\varpi_3(t)] &= -\dot{d}(t)\varpi_3^T(t)Q_2\varpi_3(t) \\
&\quad + 2h_d(t)\varpi_3^T(t)Q_2 \begin{bmatrix} \dot{x}(t) \\ \bar{d}\dot{x}_d(t) \\ \frac{\bar{d}x_d(t)-x_h+\dot{d}(t)w_2}{h_d(t)} \\ \frac{\bar{d}x_d(t)+x_h-(2-\dot{d}(t))w_2+2\dot{d}(t)w_4}{h_d(t)} \end{bmatrix}
\end{aligned} \tag{13}
$$

$$\frac{d}{dt}\left[\int_{t-d(t)}^{t}\varpi_4^T(s)Q_3\varpi_4(s)\right]=\varpi_4^T(t)Q_3\varpi_4(t)-\bar{d}\varpi_4^T(t-d(t))Q_3\varpi_4(t-d(t))$$
$$+2\begin{bmatrix}d(t)w_1\\x(t)-x_d(t)\\d(t)(w_1-x_d(t))\end{bmatrix}Q_3\begin{bmatrix}0_{n\times 1}\\0_{n\times 1}\\-\bar{d}\dot{x}_d(t)\end{bmatrix}\tag{14}$$

The differentiation of $V_3(t)$ can be expressed as

$$\begin{aligned}\dot{V}_3(t)&=\dot{x}^T(t)[d(t)R_1+h_d(t)R_2]\dot{x}(t)\\&\quad+x^T(t)[d(t)M_1+h_d(t)M_2]x(t)-\wp_1(t)-\wp_2(t)\end{aligned}\tag{15}$$

where

$$\wp_1(t)=\int_{t-d(t)}^{t}\dot{x}^T(s)T_1(\dot{d}(t))\dot{x}(s)ds+\int_{t-h}^{t-d(t)}\dot{x}^T(s)R_2\dot{x}(s)ds$$
$$\wp_2(t)=\int_{t-d(t)}^{t}x^T(s)T_2(\dot{d}(t))x(s)ds+\int_{t-h}^{t-d(t)}x^T(s)M_2x(s)ds$$

$$T_1(\dot{d}(t))=\bar{d}R_1+\dot{d}(t)R_2\tag{16}$$
$$T_2(\dot{d}(t))=\bar{d}M_1+\dot{d}(t)M_2\tag{17}$$

Then, using (14)–(17), one can write derivative of $V(t)$ with respect to time along the trajectory of system (1) as:

$$\begin{aligned}\dot{V}(t)&=\dot{V}_0(t)+\dot{V}_1(t)+\dot{V}_2(t)+\dot{V}_3(t)\\&=\xi^T(t)\Phi_0(d(t),\dot{d}(t))\xi(t)-\wp_1(t)-\wp_2(t)\end{aligned}\tag{18}$$

where $\Phi_0(d(t),\dot{d}(t))$ is defined in (12).

The condition (10) ensures the positive definiteness of $T_1(\dot{d}(t))$, then by applying integral inequality (5) of Lemma 3 to approximate the integral terms in $\wp_1(t)$, one can write

$$\wp_1(t)\geq\xi^T(t)\left[\frac{1}{d(t)}E_1^T\mathcal{T}_1E_1+\frac{1}{h_d(t)}E_2^T\mathcal{R}_2E_2\right]\xi(t)\tag{19}$$

where

$$\begin{aligned}E_1&=col\{e_1-e_2,e_1+e_2-2e_5,e_1-e_2-6e_7\}\\E_2&=col\{e_2-e_3,e_2+e_3-2e_6,e_2-e_3-6e_8\}\\\mathcal{T}_1&=diag\{T_1,3T_1,5T_1\}\\\mathcal{R}_2&=diag\{R_2,3R_2,5R_2\}\end{aligned}$$

By using Lemma 1 with $\alpha=\frac{d(t)}{h}$, $\wp_1(t)$ can be expressed as

$$\begin{aligned}\wp_1(t)\geq\frac{1}{h}\xi^T(t)[&E_1^T(2-\alpha)\mathcal{T}_1E_1+E_2^T(1+\alpha)\mathcal{R}_2E_2+2E_1^T\{(1-\alpha)S_1+\alpha S_2\}E_2\\&-\alpha E_2^TS_1^T\mathcal{T}_1^{-1}S_1E_2-(1-\alpha)E_1^TS_2\mathcal{R}_2^{-1}S_2^TE_1]\xi(t)\end{aligned}$$

Similarly, the condition (11) guaranteed the positive definiteness of $T_2(\dot{d}(t))$, hence utilizing inequality (6) of Lemma 3, $\wp_2(t)$ can be estimated as

$$\wp_2(t) \geq \xi^T(t) \begin{bmatrix} E_3 \\ E_4 \end{bmatrix}^T \begin{bmatrix} d(t)\mathcal{T}_2 & 0 \\ * & h_d(t)\mathcal{M}_2 \end{bmatrix} \begin{bmatrix} E_3 \\ E_4 \end{bmatrix} \xi(t) \tag{20}$$

where

$$E_3 = [e_5^T, e_7^T]^T, E_4 = [e_6^T, e_8^T]^T$$
$$\mathcal{T}_2 = diag\{T_2, 3T_2\}, \mathcal{M}_2 = diag\{M_2, 3M_2\}$$

Replacing and in one can write

$$\begin{aligned} \dot{V}(t) &\leq \xi^T(t)[\Phi_0(d(t),\dot{d}(t)) - \Phi_1(d(t),\dot{d}(t)) + \Pi(d(t),\dot{d}(t))]\xi(t) \\ &= \xi^T(t)\Psi\xi(t), \end{aligned} \tag{21}$$

where $\Phi_1(d(t),\dot{d}(t))$ is defined in (13) and

$$\Pi = \frac{\alpha}{h} E_2^T S_1^T \mathcal{T}_1^{-1} S_1 E_2 + \frac{(1-\alpha)}{h} E_1^T S_2 \mathcal{R}_2^{-1} S_2^T E_1,$$
$$\Psi = \Phi_0(d(t),\dot{d}(t)) - \Phi_1(d(t),\dot{d}(t)) + \Pi$$

Finally, $\dot{V}(t) < 0$, if the matrix Ψ is negative definite for all $d(t) \in [0, h]$ and $\dot{d}(t) \in [\mu_0, \mu_1]$ It is clear that the matrix Ψ in (21) is linear with respect to $d(t)$ and $\dot{d}(t)$. So, if the condition (26) is satisfied at the vertices $[0, \mu_0]$, $[0, \mu_1]$, $[h, \mu_0]$ and $[h, \mu_1]$, then it can also satisfy at all the vertices of $(d(t), \dot{d}(t)) \in [0, h] \times [\mu_0, \mu_1]$. Then using Schur complement the condition (21) can be transformed into linear matrix inequalities of (8) and (9). This completes the proof.

The criterion of Theorem 1 are obtained by utilizing Lemmas 1 and 3. Alternatively, a similar criterion can be developed by using Lemma 2 instead of Lemma 1, which results in next theorem.

Theorem 2. *For positive definite matrices* $0 < P \in \mathbb{R}^{6n\times 6n}, 0 < Q_1, Q_2 \in \mathbb{R}^{4n\times 4n}, 0 < Q_3 \in \mathbb{R}^{3n\times 3n}, 0 < Z \in \mathbb{R}^{n\times n}, 0 < R_1, R_2, M_1, M_2 \in \mathbb{R}^{n\times n}$, *and matrices* $Y_1, Y_2 \in \mathbb{R}^{8n\times 3n}$ *with given scalars* h, μ_0 *and* μ_1, *system* (1) *is asymptotically stable, such that following LMIs satisfy for* $i = 0, 1$.

$$\begin{bmatrix} \Phi_0(0,\mu_i) - \bar{\Phi}_1(0,\mu_i) & Y_2 \\ * & -h\mathcal{R}_2 \end{bmatrix} < 0 \tag{22}$$

$$\begin{bmatrix} \Phi_0(h,\mu_i) - \bar{\Phi}_1(h,\mu_i) & Y_1^T \\ * & -h\mathcal{T}_1(\mu_i) \end{bmatrix} < 0 \tag{23}$$

with conditions

$$T_1(\mu_i) \geq 0, T_2(\mu_i) \geq 0 \tag{24}$$

where $\Phi_0(d(t),\dot{d}(t))$ *is defined in Theorem 1 and*

$$\bar{\Phi}_1(d(t),\dot{d}(t)) = \frac{1}{h} Sym\{Y_1E_1 + Y_2E_2\} + E_3^T d(t)\mathcal{T}_2E_3 + E_4^T h_d(t)\mathcal{M}_2E_4$$

Proof. The proof follows the same lines of reasoning as the proof of Theorem 1. The main modification is the application of Lemma 2 in place of Lemma 1 for estimation of $\wp_1(t)$ term in (18). So, based on Lemma 2, $\wp_1(t)$ can be expressed as

$$\wp_1(t) \geq \frac{1}{h}[Sym\{Y_1E_1 + Y_2E_2\} - \alpha Y_1^T \mathcal{T}_1^{-1} Y_1 - (1-\alpha) Y_2 \mathcal{R}_2^{-1} Y_2^T] \tag{25}$$

Substituting (25) in (21), one can write

$$\dot{V}(t) \leq \xi^T(t)[\Phi_0(d(t),\dot{d}(t)) - \bar{\Phi}_1(d(t),\dot{d}(t)) + \frac{1}{h}\alpha Y_1^T \mathcal{T}_1^{-1} Y_1 + \frac{1}{h}(1-\alpha) Y_2 \mathcal{R}_2^{-1} Y_2^T]\xi(t) \tag{26}$$

Then using Schur complement the condition (26) can be transformed into linear matrix inequalities of (22) and (23). This completes the proof.

4 Numerical Analysis

In this section, two numerical example are considered to show the improvement made in the proposed Theorems.

Example 1. Consider the system (1) with

$$A = \begin{bmatrix} -2 & 0 \\ 0 & -0.9 \end{bmatrix}, \quad A_d = \begin{bmatrix} -1 & 0 \\ -1 & -1 \end{bmatrix} \tag{27}$$

The system (27) has analytical bound $h_{max} = 6.17258$ for constant delay ($\dot{d}(t) = 0, \forall t \geq 0$) case [4]. The largest upper bound h for different μ using Theorem 1 and 2 are listed in Table 1 along with similar existing approaches in the literature. Moreover, the number of LMI variables (NLVs) involved in solving stability

Table 1. Possible upper bound delay h with a variety of μ

Methods	$\mu = -\mu_0 = \mu_1$				NLVs
	0.1	0.2	0.5	0.8	
Theorem 1 [8]	4.8313	4.1428	3.1487	2.7135	$142n^2 + 18n$
Theorem 8 (N = 2) [15]	4.93	4.22	3.09	2.66	$67n^2 + 5n$
Proposition 1 [23]	4.910	4.2166	3.233	2.789	$57.5n^2 + 17n$
Theorem 2 (N = 2) [9]	4.9036	4.1906	3.1652	2.7357	$65n^2 + 8n$
Proposition 1 (N = 1) [24]	4.9192	4.2116	3.1978	2.7656	$72n^2 + 9n$
Proposition 2 (N = 2) [24]	4.9217	4.2157	3.2211	2.7920	$104n^2 + 9n$
Theorem 1	4.9229	4.2205	3.2059	2.7678	$58n^2 + 10n$
Theorem 2	4.9301	4.2361	3.2396	2.8072	$88n^2 + 10n$

criterion are shown in Table 1. It is observed that Theorem 1 yields better results than all the others except Theorem 8 (N = 2) of [15] for slow varying delays ($\mu = 0.2$). However, for fast varying delays Theorem 1 provides better results except Proposition 2 (N = 2) of [24]. But Theorem 8 (N = 2) of [15] and Proposition 2 (N = 2) of [24] utilizes more LMI variables as compared to Theorem 1. Theorem 2 provides better results in comaprison to all the approaches listed in Table 1.

Table 2. Possible upper bound delay h with a variety of μ

Methods	$\mu = -\mu_0 = \mu_1$				NLVs
	0.1	0.2	0.5	0.8	
Theorem 1 [19]	7.148	4.466	2.352	1.768	$65n^2 + 11n$
Theorem 1 [8]	7.167	4.517	2.415	1.838	$142n^2 + 18n$
Theorem 2 (N = 2) [9]	7.2633	4.5914	2.5750	2.0115	$65n^2 + 8n$
Proposition 1 [23]	7.230	4.556	2.5090	1.940	$57.5n^2 + 17n$
Theorem 1	7.2477	4.5571	2.5266	2.0018	$57n^2 + 9n$
Theorem 2	7.2707	4.5979	2.5885	2.0441	$87n^2 + 9n$

Example 2. Consider the system (1) with

$$A = \begin{bmatrix} 0 & 1 \\ -1 & -2 \end{bmatrix}, \quad A_d = \begin{bmatrix} 0 & 0 \\ -1 & 1 \end{bmatrix} \tag{28}$$

For Example 2, the obtained values of h for $\mu \in [0.1, 0.2, 0.5, 0.8]$ are tabulated in Table 2. One can observed that Theorem 1 gives better results as comparison to to all approaches of Table 1 except Theorem 1 [8]. But Theorem 2 provides less conservative results than all approches in Table 1.

5 Conclusion

The problem of stability analysis of linear systems with time-varying delay has been investigated. By constructing a new delay product type Lyapunov-Krasovskii functional using the state vectors of second order Bessels-Legendre integral inequality, two stability criteria have been derived for the system under study in combinations with reciprocal convex approaches. Two examples has been considered to validate the improvement offered by the proposed criteria.

References

1. Briat, C.: Convergence and equivalence results for the jensen's inequality-application to time-delay and sampled-data systems. IEEE Trans. Autom. Control **56**(7), 1660–1665 (2011)

2. Chen, J., Xu, S., Zhang, B.: Single/multiple integral inequalities with applications to stability analysis of time-delay systems. IEEE Trans. Autom. Control **62**(7), 3488–3493 (2016)
3. Gouaisbaut, F., Peaucelle, D.: Delay-dependent stability analysis of linear time delay systems. IFAC Proc. Vol. **39**(10), 54–59 (2006)
4. Gu, K., Chen, J., Kharitonov, V.L.: Stability of Time-Delay Systems. Springer, Heidelberg (2003)
5. Kim, J.H.: Further improvement of jensen inequality and application to stability of time-delayed systems. Automatica **64**, 121–125 (2016)
6. Kwon, O., Park, M., Park, J.H., Lee, S., Cha, E.: Improved results on stability of linear systems with time-varying delays via wirtinger-based integral inequality. J. Franklin Inst. **351**(12), 5386–5398 (2014)
7. Lee, T.H., Park, J.H.: Improved stability conditions of time-varying delay systems based on new lyapunov functionals. J. Franklin Inst. **355**(3), 1176–1191 (2018)
8. Lee, T.H., Park, J.H., Xu, S.: Relaxed conditions for stability of time-varying delay systems. Automatica **75**, 11–15 (2017)
9. Lee, W.I., Lee, S.Y., Park, P.: Affine bessel-legendre inequality: application to stability analysis for systems with time-varying delays. Automatica **93**, 535–539 (2018)
10. Liu, K., Seuret, A., Xia, Y.: Stability analysis of systems with time-varying delays via the second-order bessel-legendre inequality. Automatica **76**, 138–142 (2017)
11. Long, F., Zhang, C.K., Jiang, L., He, Y., Wu, M.: Stability analysis of systems with time-varying delay via improved lyapunov-krasovskii functionals. IEEE Trans. Syst. Man Cybern. Syst. (2019)
12. Park, P., Lee, W.I., Lee, S.Y.: Auxiliary function-based integral inequalities for quadratic functions and their applications to time-delay systems. J. Franklin Inst. **352**(4), 1378–1396 (2015)
13. Seuret, A., Gouaisbaut, F.: Wirtinger-based integral inequality: application to time-delay systems. Automatica **49**(9), 2860–2866 (2013)
14. Seuret, A., Gouaisbaut, F.: Hierarchy of lmi conditions for the stability analysis of time-delay systems. Syst. Control Lett. **81**, 1–7 (2015)
15. Seuret, A., Gouaisbaut, F.: Stability of linear systems with time-varying delays using bessel-legendre inequalities. IEEE Trans. Autom. Control (2017)
16. Sun, J., Liu, G., Chen, J., Rees, D.: Improved delay-range-dependent stability criteria for linear systems with time-varying delays. Automatica **46**(2), 466–470 (2010)
17. Xu, H.T., Zhang, C.K., Jiang, L., Smith, J.: Stability analysis of linear systems with two additive time-varying delays via delay-product-type Lyapunov functional. Appl. Math. Model. **45**, 955–964 (2017)
18. Xu, S., Lam, J.: Improved delay-dependent stability criteria for time-delay systems. IEEE Trans. Autom. Control **50**(3), 384–387 (2005)
19. Zeng, H.B., He, Y., Wu, M., She, J.: Free-matrix-based integral inequality for stability analysis of systems with time-varying delay. IEEE Trans. Autom. Control **60**(10), 2768–2772 (2015)
20. Zeng, H.B., He, Y., Wu, M., She, J.: New results on stability analysis for systems with discrete distributed delay. Automatica **60**, 189–192 (2015)
21. Zhang, C.K., He, Y., Jiang, L., Wu, M.: Notes on stability of time-delay systems: bounding inequalities and augmented lyapunov-krasovskii functionals. IEEE Trans. Autom. Control **62**(10), 5331–5336 (2017)

22. Zhang, X.M., Han, Q.L., Ge, X., Zhang, B.L.: Passivity analysis of delayed neural networks based on Lyapunov-Krasovskii functionals with delay-dependent matrices. IEEE Trans. Cybern., 1–11 (2018)
23. Zhang, X.M., Han, Q.L., Seuret, A., Gouaisbaut, F.: An improved reciprocally convex inequality and an augmented lyapunov-krasovskii functional for stability of linear systems with time-varying delay. Automatica **84**, 221–226 (2017)
24. Zhang, X.M., Han, Q.L., Seuret, A., Gouaisbaut, F., He, Y.: Overview of recent advances in stability of linear systems with time-varying delays. IET Control Theory Appl. **13**(1), 1–16 (2018)
25. Zhang, X.M., Han, Q.L., Wang, J.: Admissible delay upper bounds for global asymptotic stability of neural networks with time-varying delays. IEEE Trans. Neural Netw. Learn. Syst. **29**(11), 5319–5329 (2018)

Adaptive MRAC Controller in the Effector Trajectory Generator for Industry 4.0 Machines

Krzysztof Lalik(✉), Mateusz Kozek, Ireneusz Dominik, and Patryk Łukasiewicz

Faculty of Mechanical Engineering and Robotics, AGH University of Science and Technology, al. Mickiewicza 30, 30-059 Kraków, Poland
klalik@agh.edu.pl

Abstract. Modern industry, referring to the concept of the Fourth Industrial Revolution, sets new standards for automation systems. There are many applications, such as glue and paint application, which can be improved by using a trajectory generator described in g-code language. With such an application, it would be possible to meet one of the requirements of the standard, that is mass unit production. Unfortunately, the g-code is accessible only for cnc devices and some special systems, which can be only semi-automatic. This article shows the project of a PLC based generator for the effector movement trajectory, which interprets the g-code in fully automatic mode. An adaptive regulator was also proposed to reduce the vibration of the effector.

Keywords: Adaptive controller · Industry 4.0 · Industrial manufacturing systems

1 Introduction

The Concept of Industry 4.0 significantly affects manufacturing technology. In parallel with the customization of production and the growing requirements for individual treatment of the product, the requirements for machines on the production line increase. The best example of the objectives of the Fourth Industrial Revolution is 3D printing - a fast-growing technology that provides easy adaptation. Despite the growing market share, additive production is less common than the machining, and modern numerically controlled machines are often used for turning or milling operations [8, 12].

The three-axis manipulator presented in Fig. 1 is based on three servomechanisms. It can be equipped with an executive tip in both the form of a cutting tool, printing tool, and an extruder for incremental production. The approach to date required each time the servo mechanism configuration and PLC configuration was retooled to complete a new trajectory. This approach was rejected by using a program to create a cloud-based g-code generator and the creation of a special g-code interpreter in the PLC.

A. Bartoszewicz et al. (Eds.): PCC 2020, AISC 1196, pp. 384–395, 2020.
https://doi.org/10.1007/978-3-030-50936-1_33

(a) (b)

Fig. 1. (a) and (b): Three-axis servo drive system: (a) general view; (b) close-up of the effector.

With this approach, it is possible to create a customer-oriented application that would allow user to fully personalize the product without having to individually retune the technological line [7,15]. The whole innovation in this approach is the direct adaptation of the g-code interpreter to the industrial PLC. There are many solutions dedicated to CNC machines or 3D printers, but they always require operator intervention. On the other hand, generating a trajectory set for automatic technological sequences requires stopping production each time and uploading a new sequence of motions for the manipulator. A good example is e.g. the process of applying glue to the gasket. Each new series of gasket shapes requires reprogramming of the entire effector trajectory, stopping the technological process and uploading a new trajectory to it. The innovation of the solution described in this article falls under one of the pillars of the Industry 4.0 standard, which is Smart Manufacturing. The device downloads data on the current product from the Database located in the Cloud and performs a given trajectory without maintenance. For an industrial PLC, this means no need to retune and stop mass production. For the entire mass production system, this means the possibility of one-lot production without the need for additional operators. The innovation of this approach therefore means the industrial implementation of Smart Manufacturing, which is one of the pillars of Industry 4.0.

However, with the construction of such an interpreter, difficulties with physical implementation were noticed. The proposed method consists in the most accurate playback of the motion parameters stored in subsequent g-code lines, which means that on the boundary of trajectory fragments, the generator will inflict them with a leap change of speed, assuming that the drive control system will reproduce it quickly enough that there will be no significant distortion of the motion path. Given the parameters of modern servo drives, the assumption is not unfounded, but large, seeking infinity and short-term values of acceleration, have an adverse effect on the structure, as they cause its vibrations.

Abrupt axis speed changes resulting from g-code records are causing instantaneous, large torque on drives that cause vibrations of the structure. They can be observed during the execution of files containing the continuity of the trajectory, as well as in the drawings made, when there is too little pressure force effect or effect to the array (Fig. 2). An open-chain manipulator, hatched for testing, is prone to the occurrence of vibrations, which were not of much importance during laboratory measurements, but in the case of industrial use could cause accelerated failures due to the long operation of the equipment. In addition, they could exceed the permissible tolerances of trajectory execution. The solution to the problem may be to reduce the speed of movement. Then the moments occurring with sudden changes in direction are smaller and vibrations reduced. This would also increase the product's performance time. Another method involves modifying g-code entries so that there are no discontinuity of trajectory, but requires significant interference with the g-code file and changes the motion path. The third solution is to significantly increase the stiffness of the manipulator. However, the optimal solution is to use the adaptive regulator [10,13,14].

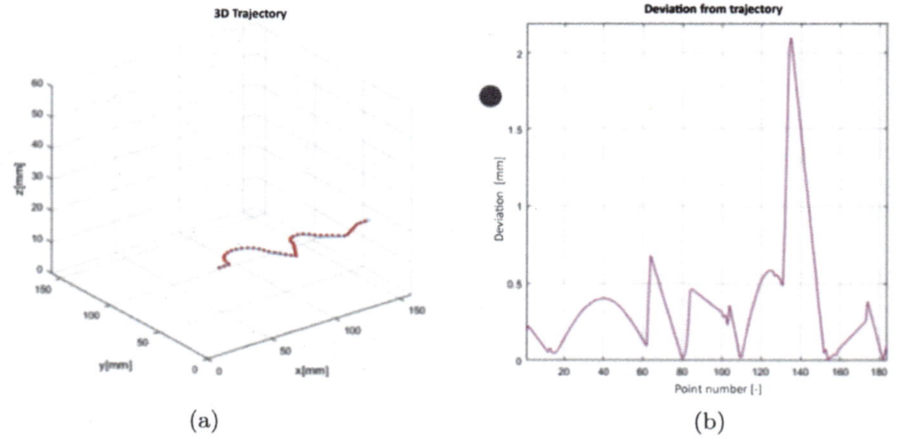

Fig. 2. Results of the trajectory without adaptive regulator, a) Trajectories 3D; b) Deviations.

2 Methodology

Classic adjustment systems with fixed controller parameters can perform their task optimally only if the control object is fully known. Given that there is a problem of incomplete knowledge of the control object or the change in its properties in real systems, the quality of the controller system is usually insufficient. In this case, adaptive adjustment applies, the parameters of which are automatically adjusted to the changing properties of the adjustment object and its surroundings.

The controller parameters are changed by comparing the output signal of the reference model to the output signal of the adjustment object. The controller is designed to have properties similar to the reference model. The block diagram of this type of adjustment and is presented in Fig. 3.

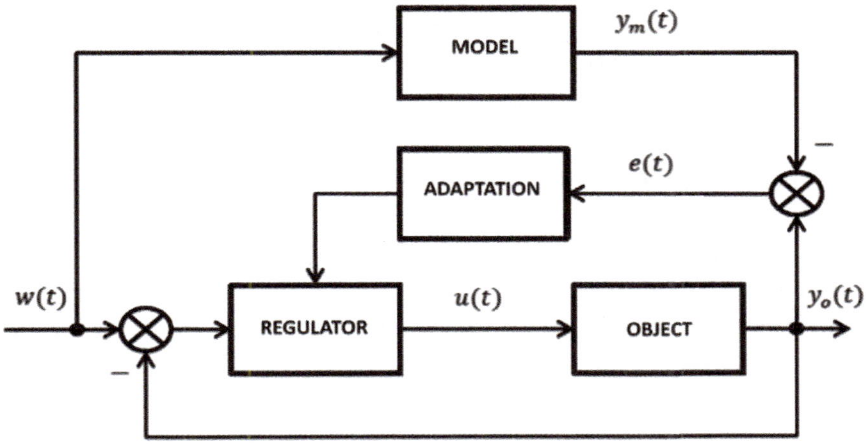

Fig. 3. Control system block diagram with reference model for a classic adjustment system.

2.1 Adapting Mechanism

The proposed control type with feedback is non-dual control. It allows you to optimize the quality of regulation criterion in the system without interference or with stochastic or deterministic interference [2,3,5]. Among the non-dual adaptive regulators, regulators can be distinguished:

- reference model (MRAC – Model Reference Adaptive Control)
- with model identification adaptive control (MIAC – Model Identification Adaptive Control)

The MRAC control system diagram, together with the adapted mechanism output is shown in Fig. 4.

Due to the limited computing power of the industrial PLC controller, which would not be able to manage the current identification of the system, it was decided to use the MRAC regulator. All used parameters are shown in diagram in Fig. 4.

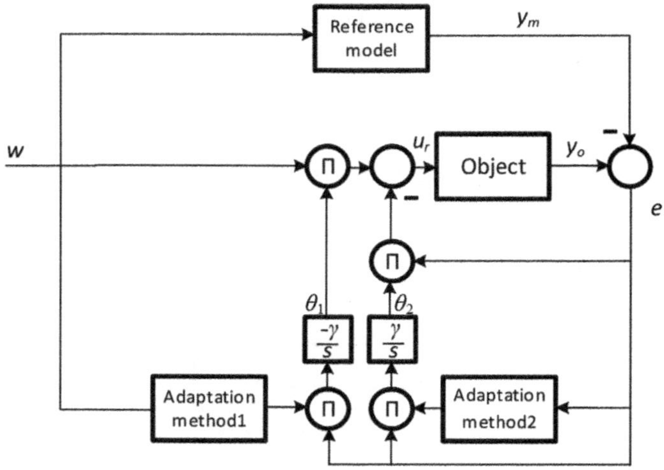

Fig. 4. MRAC control system block diagram

There are many methods for adapting the parameters of the controller [4,6, 11]. This work uses a gradient method.

The cost function described in [1,9,10] is defined as (1).

$$J(\theta) = \frac{1}{2}e^2(\theta) \tag{1}$$

where:
$e = y_0 - y_m$ - adaptation error
θ - adaptation parameter
y_0 - object output signal
y_m - model output signal

It seeks to minimize cost functions by changing the θ parameter appropriately. It follows that the change of θ is proportional to the J gradient, with the opposite character. This is described in the Eq. (2)

$$\frac{d\theta}{dt} = -\gamma\frac{\delta J}{\delta\theta} \tag{2}$$

where:
γ - speed rate of adaptation

The above dependencies show the parameter θ adaptation algorithm that describes the dependency (3).

$$\frac{d\theta}{dt} = -\gamma\frac{\delta J}{\delta e}\frac{\delta e}{\delta\theta} \tag{3}$$

Eventually, the parameter θ adaptation algorithm describes the Eq. (4).

$$\frac{d\theta}{dt} = -\gamma e\frac{\delta e}{\delta\theta} \tag{4}$$

where:
$\frac{\delta e}{\delta \theta}$ - error sensitivity of parameter θ change.

It is assumed that two parameters are necessary – strengthening the feedback loop and the forward loop. Outputs (5, 6) are intended to designate derivatives associated with these parameters.

$$u = \theta_1 w - \theta_2 y_0 \tag{5}$$

where:
u - controller output

$$e = y_0 - y_m = G_0 u - G_m w \tag{6}$$

where:
G_0 - transfer function of the object
G_m - transfer function of the model.

The accepted transfer function of the object is described with a Eq. (7). It was chosen as linear damped second order system due a fact of small amount of computing power in the industrial PLC. The second reason was that the regulator's main task was to damp vibrations when stopping and starting the actuator's movement. Therefore, further complications of the model structure were not necessary.

$$G_0(s) = \frac{b_0}{a_2 s^2 + a_1 s + a_0} \tag{7}$$

The object model transmits the model to describe the Eq. (8). The parameters of the model were determined using parametrical ARMAX method with initial condition zero for both position and velocity of the axis.

$$G_m(s) = \frac{b_0 m}{s^2 + a_1 m s + a_0 m} \tag{8}$$

The output of the object could be described as a dependency of control signal u_r and transfer function G_0 (9)

$$y_0 = G_0 u_r \tag{9}$$

Gathering Eqs. (7), (8) and (9) we obtain Eqs. (10) and (11).

$$y_0 = \frac{b_0}{a_2 s^2 + a_1 s + a_0}(\theta_1 u_r - \theta_2 y_2) \tag{10}$$

$$y_0 = \frac{b_0 \theta_1}{a_2 s^2 + a_1 s + a_0 + b_0 \theta_2} u_r \tag{11}$$

By using outputs (10) and (11), the error can be written as Eq. (12).

$$e = \frac{b_0 \theta_1}{a_2 s^2 + a_1 s + a_0 + b_0 \theta_2} u_r - G_m w \tag{12}$$

Derivatives that determine the sensitivity of the error to change parameters θ_1 and θ_2 can be described by dependencies (13, 14, 15).

$$\frac{\delta e}{\delta\theta_1} = \frac{b_0}{a_2 s^2 + a_1 s + a_0 + b_0\theta_2} u_r \tag{13}$$

$$\frac{\delta e}{\delta\theta_2} = \frac{b_0^2\theta_1}{(a_2 s^2 + a_1 s + a_0 + b_0\theta_2)^2} u_r \tag{14}$$

$$\frac{\delta e}{\delta\theta_2} = \frac{b_0}{a_2 s^2 + a_1 s + a_0 + b_0\theta_2} y_0 \tag{15}$$

The purpose of the MRAC control is to approximate the object's response to the reference model response. You can therefore describe it in the Eq. (16).

$$a_2 s^2 + a_1 s + a_0 + b_0\theta_2 \approx s^2 + a_1 m s + a_0 m \tag{16}$$

Considering also the MIT rule [6] controller parameter are chosen as $\theta_1 = \frac{b_0 m}{b_0}$ and $\theta_2 = \frac{b_0 - b_0 m}{b_0}$. Derived sizing the sensitivity of an error to parameter changes can be described using Eqs. (17) and (18).

$$\frac{\delta e}{\delta\theta_1} = \frac{b_0 m}{s^2 + a_1 m s + a_0 m} u_r \tag{17}$$

$$\frac{\delta e}{\delta\theta_2} = \frac{b_0 m}{s^2 + a_1 m s + a_0 m} y_0 \tag{18}$$

Eventually, you can describe parameter update rules for parameters θ_1 (Eq. 19) and θ_2 (Eq. 20) using the gradient adaptation method.

$$\frac{d\theta_1}{dt} = -\gamma \frac{\delta e}{\delta\theta_1} e = -\gamma \left(\frac{b_0 m}{s^2 + a_1 m s + a_0 m} u_r \right) \tag{19}$$

$$\frac{d\theta_2}{dt} = \gamma \frac{\delta e}{\delta\theta_2} e = \gamma \left(\frac{b_0 m}{s^2 + a_1 m s + a_0 m} y_0 \right) \tag{20}$$

3 Results

The PID controller has been implemented in the generator improvement system in order to obtain a reference control system. Using the recorded load-free object jump response and the PID Tuner/Matlab tool, the PID regulator's presets have been calculated through an iterative optimization algorithm. Zero initial conditions were assumed for both the PID controller and MRAC. This was in line with reality, because each axis during the identification experiment was based to home position. After stopping, the results were recorded when axis started moving to a given position. While tests on a real object, these settings proved to be unsatisfactory. The regulator's final setting was chosen by a heuristic method based on the experience of the object designer, which makes it possible to fine-tune the regulator so that the system's response meets the assumptions.

The code of the PID and MRAC controllers in both cases was executed by a cyclic organization block called by a cyclic interrupt with a frequency of 10 Hz. The adaptation coefficient of MRAC controllers for each of 3 servo–axes were optimized in this cyclic interrupt block with Eqs. (19) and (20). The computational costliness for such a Cartesian plant and the need to reproduce many lines of g-code and support for communication with servos meant that the simplest object model had to be considered here.

The object response for one axis is presented in Fig. 5. A steady state error ob-served despite the presence of Integral part of PID because of the presence of torque dead-zone. The axis was disturbed via different load and different geometry of the plant. So in this case the external load was too heavy for the I-part, to be great enough, to move in preset time. It occurred because the PID controller was tuned with no external load.

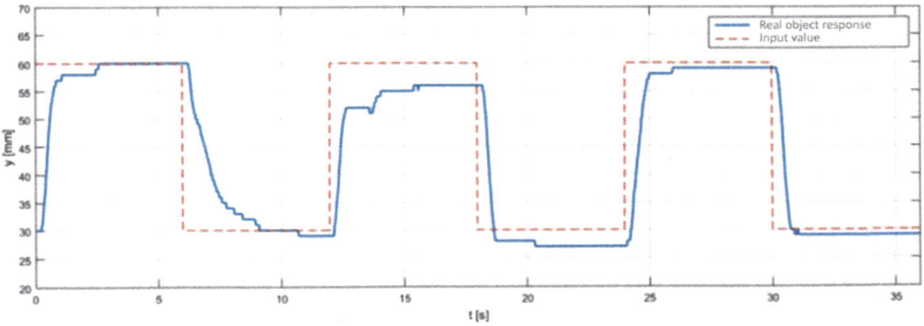

Fig. 5. Actual object response in PID control system

Until 11th second, the adjustable object had no additional load. It can be said that the response of the system is satisfactory – the axis reached the setpoint value in about 3 s. An additional load was added to the object between 10th and 12th second. This is seen in the form of a decrease in the current position of the stroller. Subsequent system responses to the change in settling value are no longer satisfactory in means of quality as in the first cycle. When driving up, the axis does not reach the setpoint value, while when driving the axis down exceeds the set value.

In order to determine the quality of the control, the cost function in the form of an integral square error described in the Eq. (21) is defined.

$$I = \int_0^\infty e^2(t)dt \tag{21}$$

In the shown object, the cost function value is: $I = 127443\,[\text{mm}^2]$. An MRAC algorithm has been used to improve the quality of regulation. As with computer simulation MRAC control, the adaptive speed factor was selected by heuristic method. The best results were achieved for the $\gamma = 0.001$ ratio. The course of such a system response is presented in Fig. 6.

The set value was changed identically to that of the PID regulator. Also, by analogy initially, the control object did not have an additional load. It was added after about 8 s. It is noticeable that there is a slight over-control in each cycle. The position of the trolley almost coincides with the output of the reference model. The stroller achieves its set value in a very short time. The load does not affect the object response. The value of the control indicator in the form of an integral square error for MRAC control with the speed of the $\gamma = 0.001$ is: $I = 104029\,[\mathrm{mm}^2]$.

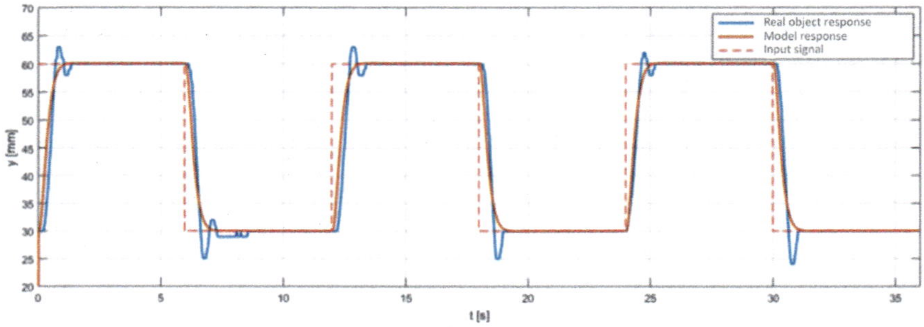

Fig. 6. Actual object response in MRAC control system for $\gamma = 0.001$ adapt speed coefficient

Both the comparison of the actual object response course for PID regulation and MRAC regulation and the combination of the integral indicator value for these systems allow the conclusion that the application of adaptive regulation avoids the phenomenon of mismatching the parameters of the controller to the parameters of the object and thus to avoid deterioration of the quality of the control. With the adaptation mechanisms used, changing the axle load does not change the nature of the output response of the object.

In order to verify the impact of the speed rate of adaptation to the system response, a response was also recorded for $\gamma = 0.01$. It is shown in Fig. 7. In this case, conclusions can be drawn analogous to conclusions from computer simulation – a higher value of the adaptive speed coefficient increases over-regulation and non-extinguishing oscillations.

The value of the control indicator in the form of a integral square error for MRAC control with the speed of the $\gamma = 0.01$ is: $I = 110280\,[\mathrm{mm}^2]$.

As is easy to see, despite the use of the best value of the adapt speed factor, the control indicator value is less than the system with the traditional PID system.

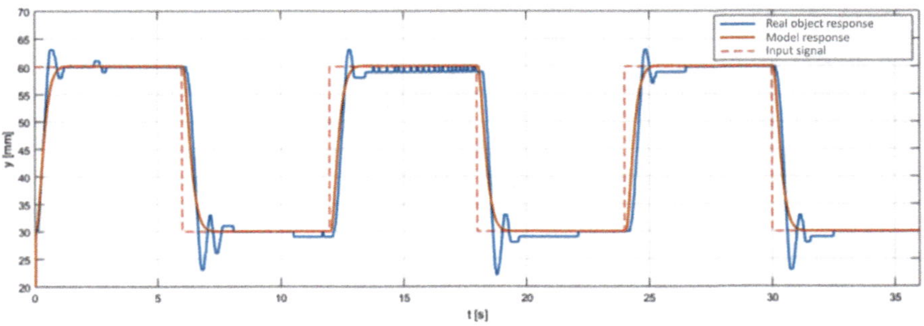

Fig. 7. Actual object response in MRAC control system for $\gamma = 0.01$

4 Conclusions

This article shows the implementation of the device and the control system for a solution similar to the concept of Industry 4.0, as evidenced by the simplicity of the operation of function blocks ("Plug and Play") and the autonomous operation of the controller, which retrieves a g-code file from an FTP server, converts and controls the operation of drives. The approach presented for the industrial production line would greatly increase its adaptability and flexibility, thanks to the possibility of an easy change in the trajectory implemented without interfering with the control systems software. An additional advantage of the generator is its versatility. Since a solution has been adopted in which the speed of all axes is calculated at the interpreter's out-put, it is possible to use any speed control drives and even control the operation of the industrial robot and plot the trajectory with its effector. The restrictions on the use of low-power industrial equipment for its tasks did not disqualify the use of PLC controllers, and the skillful management of processor resources allowed to achieve satisfactory quality of the drawn shapes. All calculations requiring higher computing power have been transferred to the Cloud, which further fits the proposed solution to the Industry 4.0 concept.

This article describes two major innovations. First of all, it describes an interpreter that enables unit production for technological lines, which until now could not be so flexible. The proposed interpreter has no equivalent in any other fully automated process.

Unfortunately, full freedom in generating trajectories caused further inconvenience. First of all, the variable geometry, variable downforce and trajectory unknown during the design of the machine caused vibrations which could not handle the classic PID. This resulted in trajectory disturbances, especially in the case of "sharp" shapes. The authors instead of stiffening the constructions proposed an adaptive MRAC regulation system. Using it for three independent servo axes was an engineering challenge. This algorithm, although it has existed in control theory for a long time, is not present in industrial PLC controllers and hence constitutes a second innovation.

The correct selection of the speed of adaptation was quite time-consuming and required a few iterations on the actual object. Thus, adaptive design requires a lot of experience from the designer, but the benefits of a properly implemented adaptive regulator are huge. The variability of object parameters to a certain extent will not cause the regulator to mismatch the parameters of the object and thus will not degrade the quality of the control.

References

1. Chowdhary, G., Kingravi, H.A., How, J.P., Vela, P.A.: Bayesian nonparametric adaptive control using Gaussian processes. IEEE Trans. Neural Netw. Learn. Syst. **26**(3), 537–550 (2015). https://doi.org/10.1109/TNNLS.2014.2319052
2. Dominik, I.: Advanced controlling of the prototype of SMA linear actuator. Solid State Phenom. **177**(1), 93–101 (2011). https://doi.org/10.4028/www.scientific.net/SSP.177.93
3. Dominik, I.: Type-2 fuzzy logic controller for position control of shape memory alloy wire actuator. J. Intell. Mater. Syst. Struct. **27**(14), 1917–1926 (2016). https://doi.org/10.1177/1045389X15610907
4. Garon, M., Boulet, P.O., Doironz, J.P., Beaulieu, L., Lalonde, J.F.: Real-time high resolution 3D data on the HoloLens. In: Adjunct Proceedings of the 2016 IEEE International Symposium on Mixed and Augmented Reality, ISMAR-Adjunct 2016, pp. 189–191 (2017). https://doi.org/10.1109/ISMAR-Adjunct.2016.0073
5. Huang, K.Y., Chin, H.C., Huang, Y.C.: A model reference adaptive control strategy for interruptible load management. IEEE Trans. Power Syst. **19**(1), 683–689 (2004). https://doi.org/10.1109/TPWRS.2003.821444
6. Jain, P., Nigam, M.J.: Design of a model reference adaptive controller using modified MIT rule for a second order system. Adv. Electron. Electr. Eng. **3**(4), 477–484 (2013)
7. Lee, J., Bagheri, B., Kao, H.A.: A cyber-physical systems architecture for Industry 4.0-based manufacturing systems. Manuf. Lett. **3**(1), 18–23 (2015). https://doi.org/10.1016/j.mfglet.2014.12.001
8. Lele, A.: Industry 4.0. In: Smart Innovation, Systems and Technologies, pp. 205–215. Springer Science and Business Media Deutschland GmbH (2019). https://doi.org/10.1007/978-981-13-3384-2_13
9. Lim, Y., Venugopal, R., Galip Ulsoy, A.: Auto-tuning and adaptive control of sheet metal forming. Control Eng. Pract. **20**(2), 156–164 (2012). https://doi.org/10.1016/j.conengprac.2011.10.006
10. Maiti, S., Chakraborty, C., Hori, Y., Ta, M.C.: Model reference adaptive controller-based rotor resistance and speed estimation techniques for vector controlled induction motor drive utilizing reactive power. IEEE Trans. Industr. Electron. **55**(2), 594–601 (2008). https://doi.org/10.1109/TIE.2007.911952
11. Pankaj, S., Kumar, J.S.: Comparative analysis of MIT rule and Lyapunov rule in model reference adaptive control scheme. Innov. Syst. Des. Eng. **2**(1), 154–163 (2011). https://doi.org/10.1007/0-306-46887-5
12. Reinhard, G., Jesper, V., Stefan, S.: Industry 4.0: building the digital enterprise. 2016 Global Industry 4.0 Survey, vol. 46, issue 1, pp. 35–50 (2016). https://doi.org/10.1080/01969722.2015.1007734
13. Sunwoo, M., Cheok, K.C., Huang, N.J.: Model reference adaptive control for vehicle active suspension systems. IEEE Trans. Industr. Electron. **38**(3), 217–222 (1991). https://doi.org/10.1109/41.87590

14. Zhang, D., Wei, B.: A review on model reference adaptive control of robotic manipulators (2017). https://doi.org/10.1016/j.arcontrol.2017.02.002
15. Zhong, R.Y., Xu, X., Klotz, E., Newman, S.T.: Intelligent manufacturing in the context of Industry 4.0: a review. Engineering **3**(5), 616–630 (2017). https://doi.org/10.1016/J.ENG.2017.05.015

Development and Modelling of a Laboratory Ball on Plate Process

Krzysztof Zarzycki(✉) and Maciej Ławryńczuk(✉)

Institute of Control and Computation Engineering, Warsaw University of Technology, ul. Nowowiejska 15/19, 00-665 Warsaw, Poland
krzysztof.zarzycki93@gmail.com, M.Lawrynczuk@ia.pw.edu.pl

Abstract. This work is concerned with a laboratory ball on plate process developed in the Institute of Control and Computation Engineering, Warsaw University of Technology. Process development, modelling and model simplification are discussed. Although the rudimentary fundamental model of the process is nonlinear, it is shown that very good modelling accuracy is achieved when a simplified linear model is used. Results of laboratory experiments are given to show the effectiveness of the model.

Keywords: Ball on plate process · Modelling · Model simplification · Model validation

1 Introduction

University education in process control must consists of lectures and laboratory tasks [1,7]. Typical examples of laboratory benchmark processes available on the market are: the pendulum on a cart, the inverted pendulum on a cart, the twin-rotor aerodynamical system (a model of a helicopter), the tower crane, the magnetic levitation system, the modular servo [2,5,12]. They make it possible to test various control configurations, e.g. [4,11]. Furthermore, there are numerous examples of custom laboratory processes, developed for particular needs [3,9,10, 15]. For example, the emphasis may be put on Multiple-Input Multiple-Output (MIMO) nature of the process, possibility of fault emulation and accessibility of various communication interfaces [15].

This work is concerned with a laboratory ball on plate process developed in the Institute of Control and Computation Engineering, Warsaw University of Technology. It has been developed as a platform for testing MIMO controllers, e.g. Linear-Quadratic-Gaussian (LQG) or Model Predictive Control (MPC) algorithms. Because good control requires models of the process, formulation, simplification and validation of the state-space model is discussed. Although, typically, the model is derived from Euler-Lagrange equations [13], in this work a simpler and straightforward approach is used in which the model is developed considering all possible forces acting on the ball. In order to use control algorithms based on linear models, a linear approximation of the rudimentary nonlinear model is derived and experimentally validated.

A. Bartoszewicz et al. (Eds.): PCC 2020, AISC 1196, pp. 396–408, 2020.
https://doi.org/10.1007/978-3-030-50936-1_34

2 Process Description

Figure 1 shows the laboratory ball on plate process design in Solidworks. Its main elements are:

1. the touch pad,
2. two digital servomotors,
3. two arms and two servo pushers,
4. the microcontroller development board,
5. additional construction elements:
   (a) the wooden table top – the process base,
   (b) four metal rollers – the process support;
   (c) the wooden table top with walls – the tilted plate,
   (d) the metal rod – the support for the plate,
   (e) the cross joint – to tilt the plate in two axes,
   (f) some elements for attaching servomotors to the process top.

The considered ball on plate process has two degrees of freedom which means that there are two manipulated and two controlled variables. The manipulated variables result in rotation of the plate around the X and Y axes, respectively. The digital Hi-tech HS 5485 HB servomotors are used as the actuators. They do not require the use of an additional control loop for engine rotations. The servomotors are equipped with a proportional controller. The 50 Hz Pulse Width Modulation (PWM) signal is used to control the servomotors.

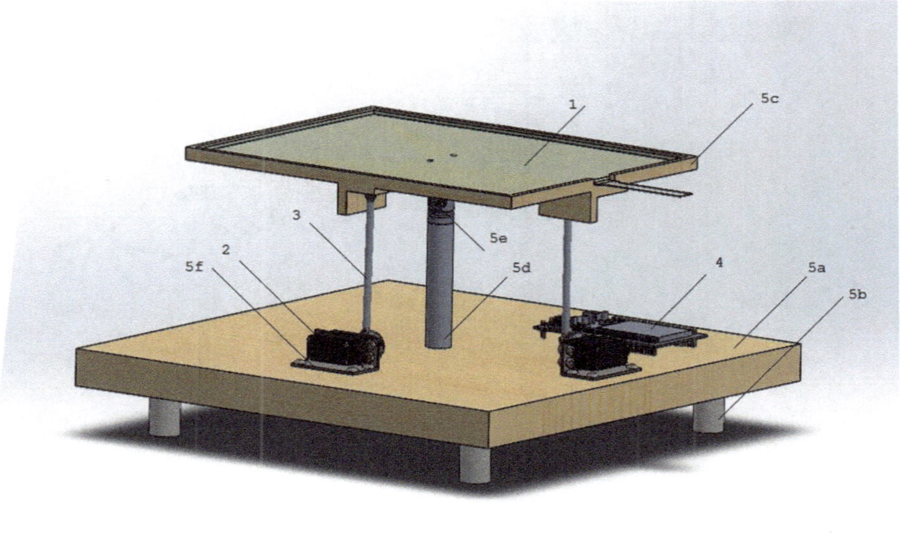

Fig. 1. The laboratory ball on plate process design in Solidworks

The ball's position may be measured using a touch screen [6], an overhead camera [13] or an array of phototransistors [16]. Image processing introduces unavoidable delay whereas the last solution is quite difficult. Hence, the position is measured using a touch screen. In general, there are a few methods used for touch detection: capacitive, infrared, surface acoustic wave and resistive. The resistive touch panel has been chosen. It has the following advantages: the ball may be made of any material, resistance to water and dust pollution, low cost. Additionally, the resistance touch screen is very easily controlled by the microcontroller. The Analog to Digital Converters (ADCs) simply read the current X and Y position of the ball.

The STM32 F401C Disco microcontroller development board is responsible for communication with an external PC computer and will be used in the future as a hardware platform for implementation of control algorithms. Because of the future application, the floating point unit and a significant amount of memory are important features of the chosen microcontroller. The considered development board offers all necessary I/O pins, i.e. two PWM outputs to control the servomotors and two ADCs to read the ball's position in two dimensions.

Width of the panel is 322 mm, its height is 247 mm, which determines the size of the whole process. The developed process is depicted in Fig. 2.

Fig. 2. The developed laboratory ball on plate process

Having completed the process, some problems appeared. They are related to the nature of the resistance touch panel. The ball's mass must be sufficient, not lower than 60–70 g. If the ball's mass is too low, the sensor does not detect the ball's position at all. Additionally, it may also happen that the ball's position is temporarily detected in a different place than it really is. Unfortunately, even for balls that weigh 100 g, such problems also occur. It is because the ball sometimes temporarily detaches from the surface or the gravitational component of the force acting on it becomes lower which results in lower pressure. In order to cope with the mentioned problems, the following filters have been designed:

- The filter that detects the situation when the sensor stops to measure the ball's position. In such a case the touch pad returns position zero or not a number values. When the filter detects such a value, it is ignored and the last reasonable numeric value is taken as the current position value.
- The filter that detects a wrong ball's position. If the position change at a given time is greater than the specified threshold, it is ignored.
- The summation filter for the ball position. The current position of the ball is determined taking into account not only one, but several dozens of averaged measurements. It is possible due to the fact that ADCs operate at a much higher frequency than the servos.
- The Kalman filter, which makes the measurements smoother, even those which are not detected by the aforementioned filters.

When all filters are used, no problems are observed.

3 State-Space Model of the Process

3.1 Derivation of the Fundamental Model

Typically, the model of the process is derived from Euler-Lagrange equations [13]. In this work a simpler and straightforward approach is used. The model is developed considering all possible forces acting on the ball which are depicted in Fig. 3 [8]. At first, let us define the necessary variables and constants:

- m – mass of the ball,
- r – radius of the ball,
- S – area of the ball,
- $J = 2/5\ \mathrm{mr}^2$ – moment of inertia of the ball,
- ω – angular velocity of the ball,
- v – linear velocity of the ball,
- Φ – tilt angle of the plate in relation to the X axis,
- ϕ – rotation angle of the servo arm related to the Φ plate tilt,
- Θ – tilt angle of the plate in relation to the Y axis,
- θ – rotation angle of the servo arm related to the Θ plate tilt,
- k – rolling friction coefficient,
- c – coefficient of aerodynamic resistance of the ball,
- ρ – density of air,
- g – gravitational acceleration,
- d – length of the servo arm,

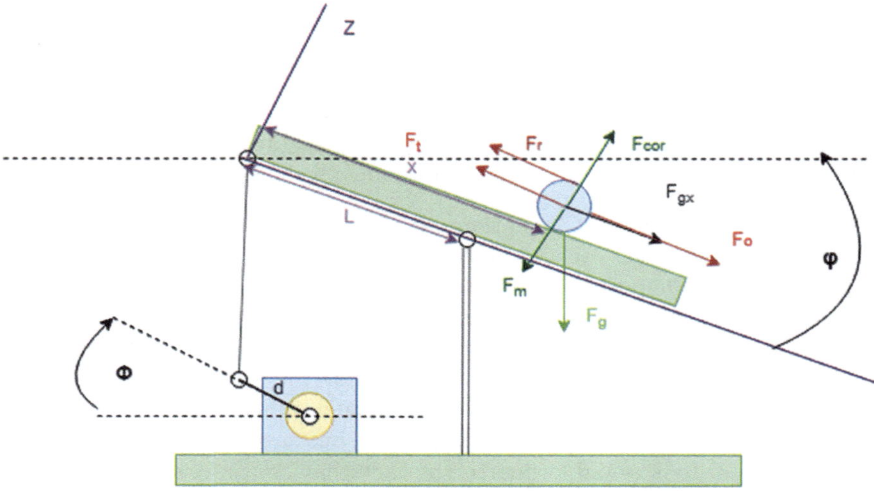

Fig. 3. All forces acting on the ball

- L – distance between the place of attachment of the servo pusher to the plate top and the point on the table top above the cross joint.

Three forces act along the Z-axis. The Coriolis force

$$F_{\mathrm{Cor}} = 2\dot{\Phi} \tag{1}$$

the Euler force

$$F_{\mathrm{Eul}} = m_{\mathrm{b}} x \ddot{\Phi} \tag{2}$$

and the gravity force

$$F_{\mathrm{m}} = mg\cos\Phi \tag{3}$$

Six forces act along the X-axis. The translation force

$$F_{\mathrm{t}} = m\frac{\delta^2 x}{\delta t^2} = m\ddot{x} \tag{4}$$

the rotation force

$$F_{\mathrm{r}} = \frac{J}{r}\frac{\delta\omega}{\delta t} = \frac{J}{r^2}\frac{\delta^2 x}{\delta t^2} = \frac{J}{r^2}\ddot{x} = \frac{2}{5}m\ddot{x} \tag{5}$$

the inertial force

$$F_{\mathrm{o}} = m\omega_{\Phi} = m\left(\frac{\delta\Phi}{\delta t}\right)^2 = m\ddot{\Phi}^2 \tag{6}$$

the component of gravity force

$$F_{\mathrm{gx}} = mg \sin \Phi \tag{7}$$

the friction force

$$T_{\mathrm{t}} = k\omega = \frac{k}{r}\frac{\delta v}{\delta t} = \frac{k}{r}\dot{x} \tag{8}$$

and the air resistance force

$$F_{\mathrm{op}} = c\frac{1}{2}\rho s \frac{\delta x}{\delta t}\left|\frac{\delta x}{\delta t}\right| = c\frac{1}{2}\rho S\dot{x}|\dot{x}| \tag{9}$$

In order to determine the acceleration $\ddot{x}$ of the ball along the X-axis, it is necessary to consider all acting forces

$$F_{\mathrm{t}} + F_{\mathrm{r}} + F_{\mathrm{o}} + T_{\mathrm{t}} + F_{\mathrm{op}} = F_{\mathrm{gx}} \tag{10}$$

which gives

$$m\ddot{x} + \frac{2}{5}m\ddot{x} + m\ddot{\Phi}^2 + my\dot{\Theta} + \frac{k}{r}\dot{x} + c\frac{1}{2}\rho S\dot{x}|\dot{x}| = mg \sin \Phi \tag{11}$$

The result may be expressed in the form

$$m\left(\frac{7}{5}\ddot{x} + \ddot{\Phi} + y\dot{\Theta} + \frac{k}{rm}\dot{x} + \frac{c\rho S\dot{x}|\dot{x}|}{2m}\right) = mg \sin \Phi \tag{12}$$

Similarly, for the Y-axis one obtains

$$m\left(\frac{7}{5}\ddot{y} + \ddot{\Theta}^2 + y\dot{\Phi} + \frac{k}{rm}\dot{y} + \frac{c\rho S\dot{y}|\dot{y}|}{2m}\right) = mg \sin \Theta \tag{13}$$

3.2 Model Simplification

The following simplifying assumptions are made: a) friction is neglected, b) air resistance is neglected, c) angular velocities of the surface plate are very low. In such a case, model Eqs. (12)–(13) are simplified to

$$\frac{7}{5}\ddot{x} = g \sin \Phi, \ \frac{7}{5}\ddot{y} = g \sin \Theta \tag{14}$$

The relation between the angle of rotation of the plate around the X-axis Φ and the angle of rotation of the servo motor arm ϕ is

$$\sin \Phi = \frac{d}{L_{\mathrm{x}}} \sin \phi \tag{15}$$

Similarly, the relation between the angles Θ and θ is given by

$$\sin \Theta = \frac{d}{L_{\mathrm{y}}} \sin \theta \tag{16}$$

Provided that the angles ϕ and θ are low, $\sin\phi \approx \phi$ and $\sin\theta \approx \theta$. Hence, Eqs. (15)–(16) become

$$\sin\Phi = \frac{d}{L_{\mathrm{x}}}\phi, \ \sin\Theta = \frac{d}{L_y}\theta \tag{17}$$

From Eqs. (14) and (17), the model becomes

$$\ddot{x} = \frac{5}{7}g\frac{d}{L_{\mathrm{x}}}\phi = C_{\mathrm{x}}\phi, \ \ddot{y} = \frac{5}{7}g\frac{d}{L_{\mathrm{y}}}\theta = C_{\mathrm{y}}\theta \tag{18}$$

Using the Laplace transform, one gets

$$X(s)s^2 = C_{\mathrm{x}}\phi(s), \ Y(s)s^2 = C_{\mathrm{y}}\theta(s) \tag{19}$$

Hence, the transfer functions are

$$G_{\mathrm{x}}(s) = \frac{X(s)}{\phi(s)} = \frac{C_{\mathrm{x}}}{s^2}, \ G_{\mathrm{y}}(s) = \frac{Y(s)}{\theta(s)} = \frac{C_{\mathrm{y}}}{s^2} \tag{20}$$

For the developed process the value of parameters are: $d = 0.024$ m, $L_{\mathrm{x}} = 0.165$ m, $L_{\mathrm{y}} = 0.135$ m, $C_{\mathrm{x}} = 1.0189$ m/s^2, $C_{\mathrm{y}} = 1.2453$ m/s^2. Hence

$$G_{\mathrm{x}}(s) = \frac{1.0189}{s^2}, \ G_{\mathrm{y}}(s) = \frac{1.2453}{s^2} \tag{21}$$

The simplified model may be expressed in the classical form, typical of linear state-space models

$$\begin{cases} \dot{x}(t) = \boldsymbol{A}x(t) + \boldsymbol{B}u(t) \\ y(t) = \boldsymbol{C}x(t) \end{cases} \tag{22}$$

where the state, input and output vectors are

$$x = \begin{bmatrix} x \\ \dot{x} \\ y \\ \dot{y} \end{bmatrix}, \ u = \begin{bmatrix} \phi \\ \theta \end{bmatrix}, \ y = \begin{bmatrix} x \\ y \end{bmatrix} \tag{23}$$

The model matrices are

$$\boldsymbol{A} = \begin{bmatrix} 0\,1\,0\,0 \\ 0\,0\,0\,0 \\ 0\,0\,0\,1 \\ 0\,0\,0\,0 \end{bmatrix}, \ \boldsymbol{B} = \begin{bmatrix} 0 & 0 \\ C_{\mathrm{x}} & 0 \\ 0 & C_{\mathrm{y}} \\ 0 & 0 \end{bmatrix} = \begin{bmatrix} 0 & 0 \\ 1.0189 & 0 \\ 0 & 1.2453 \\ 0 & 0 \end{bmatrix}, \ \boldsymbol{C} = \begin{bmatrix} 1\,0\,0\,0 \\ 0\,0\,1\,0 \end{bmatrix} \tag{24}$$

4 Model Validation

For model validation a few experiments are performed. In the first experiments one step of the manipulated variable, i.e. the servo arm rotation angle ϕ is used. The comparison of real position (x) and the position estimated from the model (x_{mod}) is depicted in Fig. 4 for different excitations. The model works very well for bigger angles of the servo's arm. For angles lower than 5° plane deflection is

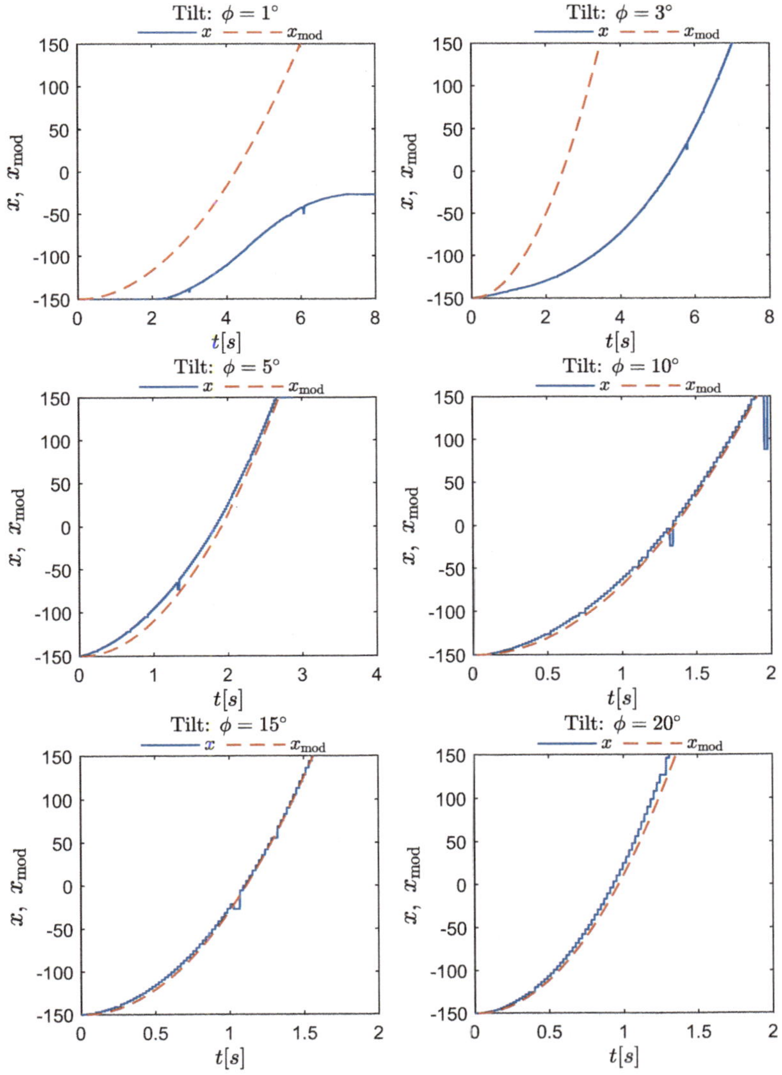

Fig. 4. Validation of the model: one step of the manipulated variable is applied

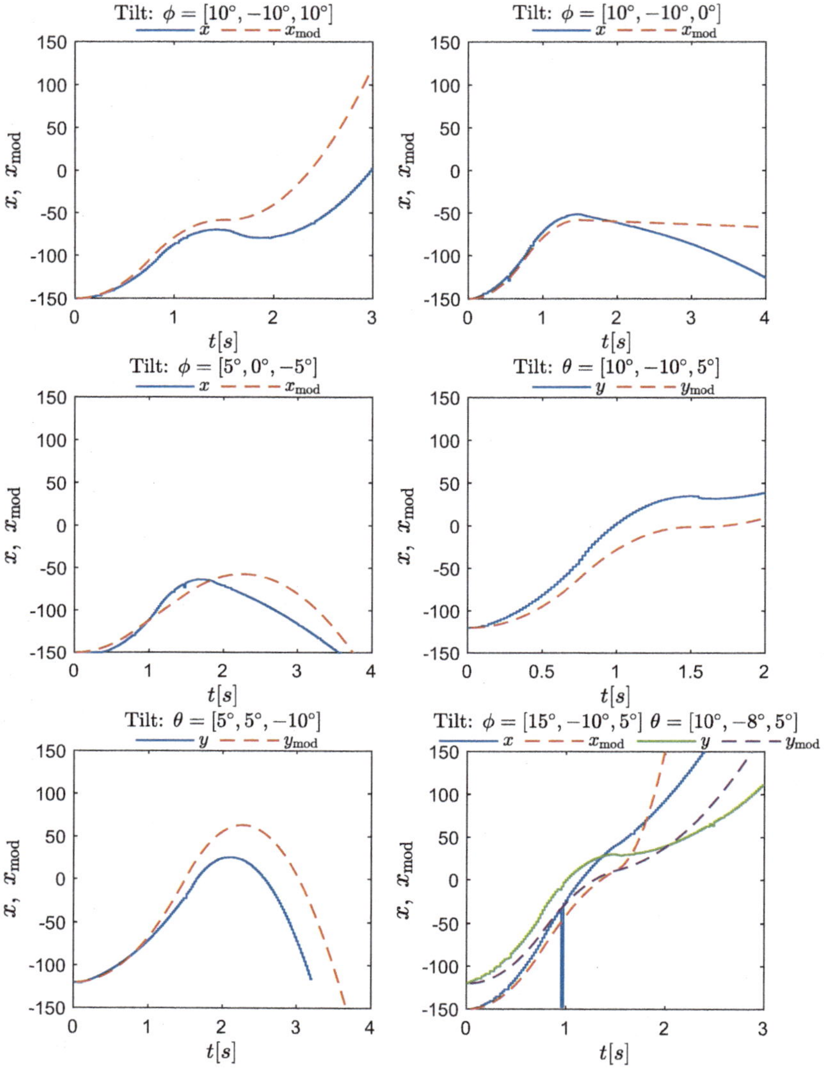

Fig. 5. Validation of the model: a series of three steps of the manipulated variables is applied

also very low. Before the ball begins to roll, it must overcome the static friction force acting on it. This phenomenon is not taken into account by the model. The next experiments are performed with a series of three control signal steps. In some of them only the servo angle ϕ is manipulated while the angle θ is constant and equal to 0, in some other experiments the angle θ is changed while the angle ϕ is constant. During the last series of experiments both angles are manipulated. The results are shown in Fig. 5.

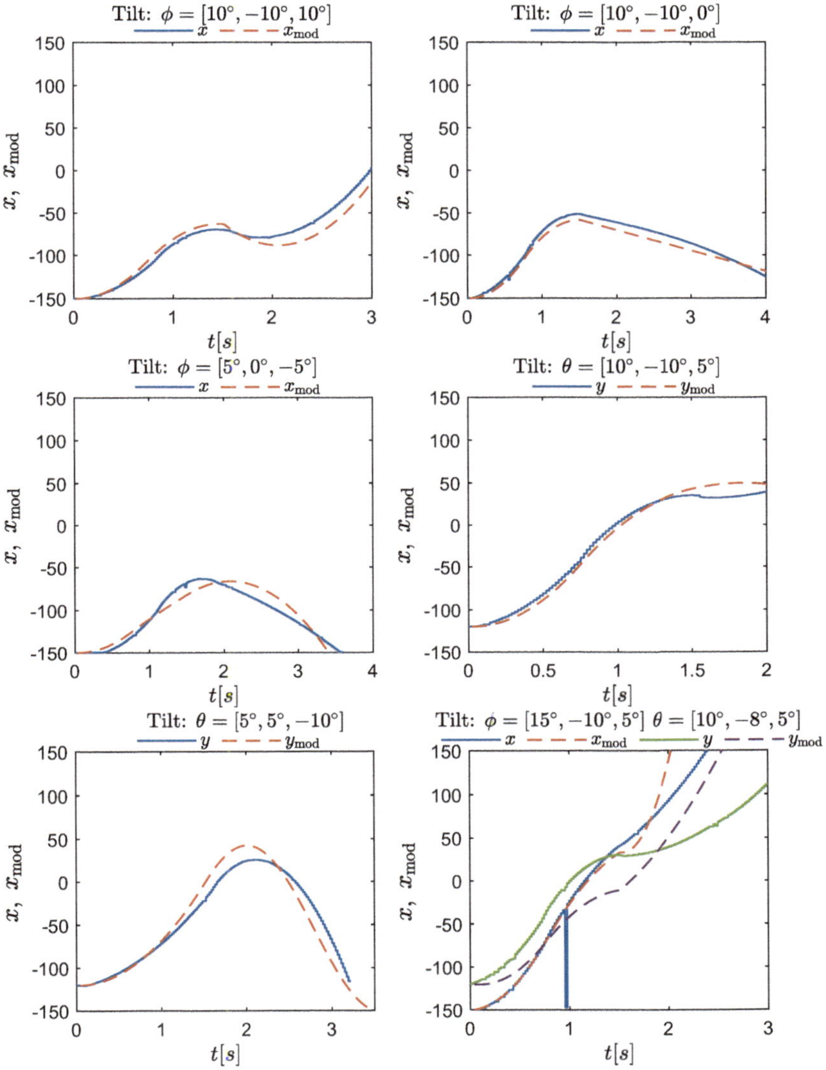

Fig. 6. Validation of the modified model: a series of three steps of the manipulated variables is applied

In Figs. 4, 5 the model and the process trajectories are somehow different. The differences result from neglecting the nonlinear model parts, i.e. $\ddot{\phi}^2$ and $y\dot{\Theta}$. They can be interpreted as the force with which the plane acts on the ball when swinging. The model assumes that angular velocities are very low, so these forces do not occur. During a quick change of the swing angle, especially when its sign changes, the ball accelerates or brakes. An attempt has been made

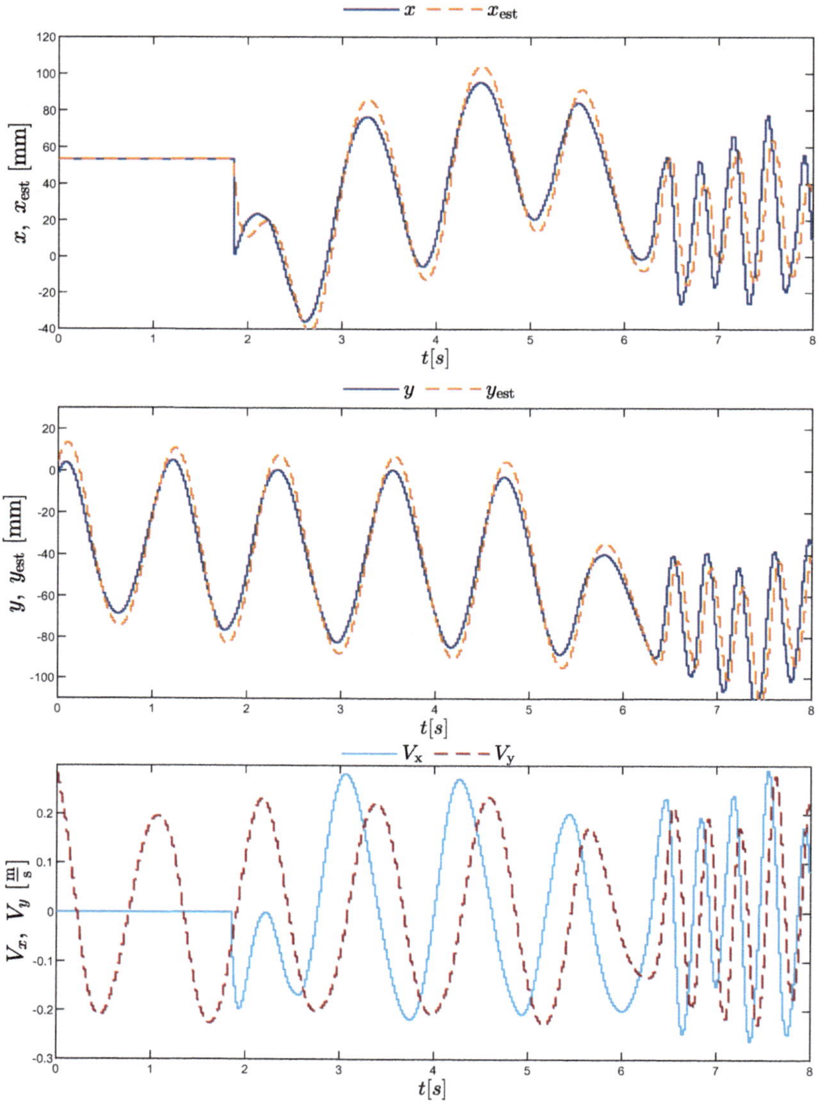

Fig. 7. Validation of the Kalman filter: x position of the ball vs. its estimation (the top panel), y position of the ball vs. its estimation (the middle panel), the estimation of the ball's velocities (the bottom panel)

to take into account information about the speed of change of the plane's tilt angle. It turns out to be troublesome, because it is not measured. In the case of a short series of manipulated variables changes, it is possible to optimise the angular velocity values using optimisation methods. The trajectories of the modified model and the process are compared in Fig. 6. However, during normal process operation, the manipulated variables change constantly. The modified model may give worse results than the simplified one. Hence, the simplified model is chosen for further application in state estimation and control. Figure 7 shows efficiency of the Kalman filter. It estimates the unmeasured states, i.e. the velocities, and makes the measured ones, i.e. the position coordinates, smoother than real measurements.

5 Conclusions

Although the fundamental state-space model of the laboratory ball on plate process is nonlinear, it may be greatly simplified. It is shown in this work that very good modelling accuracy is achieved when a simplified linear model is used. In future research the obtained model will be used in MPC algorithms [14].

References

1. Amadi-Echendu, J.E., Higham, E.H.: Curriculum development and training in process measurements and control engineering. Eng. Sci. Educ. J. **6**, 104–108 (1997)
2. Bytronic. http://www.bytronic.net/
3. Çatalbaç, B., Uyanık, İ.: A low-cost laboratory experiment setup for frequency domain analysis for a feedback control systsms course. IFAC PapersOnLine **50**(1), 15704–15709 (2017)
4. Czerwiński, K., Wojtulewicz, A., Ławryńczuk, M.: Fuzzy controller for laboratory levitation system: real-time experiments using programmable logic controller. Int. J. Control Autom. Syst. **17**, 1507–1514 (2019)
5. Inteco. http://www.inteco.com.pl/
6. Kassem, A., Haddad, H., Albitar, C.: Commparison between different methods of control of ball and plate system with 6DOF Stewart platform. IFAC PapersOnLine **48**, 47–52 (2015)
7. Kheir, N.A., Åström, K.J., Auslander, D., Cheok, K.C., Frankiln, G.F., Masten, M., Rabins, M.: Control systems education. Automatica **32**, 147–166 (1996)
8. Leyko, J.: Mechanics. PWN Scientific Publishing House, Warsaw (2006)
9. Leva, A.: A simple and flexible experimental laboratory for automatic control courses. Control Eng. Pract. **14**, 167–176 (2006)
10. McLoone, S.C., Maloco, J.: A cost-effective hardware-based laboratory solution for demonstrating PID control. In: Proceedings of the UKACC 11th International Conference on Control (CONTROL), Belfast, UK, pp. 1–6 (2016)
11. Piłat, A., Turnau, A.: Neural adapted controller learned on-line in real-time. IFAC Proc. Volumes **42**, 47–52 (2009)
12. Quanser. https://www.quanser.com/
13. Stander, D., Jiménez-Leudo, S., Quijano, N.: Low-cost "ball and plate" design and implementation for learning control systems. In: 2017 IEEE 3rd Colombian Conference on Automatic Control (CCAC), Cartagena, Colombia, pp. 1–6 (2017)
14. Tatjewski, P.: Advanced Control of Industrial Processes. Structures and Algorithms. Springer, London (2007)

15. Wojtulewicz, A., Chaber, P., Ławryńczuk, M.: Multiple-input multiple-output laboratory stand for process control education. In: Proceedings of the 21th IEEE International Conference on Methods and Models in Automation and Robotics MMAR 2016, Międzyzdroje, Poland, pp. 466–471 (2016)
16. Zeeshan, A., Nauman, N., Jawad Khan, M.: Design, control and implementation of a ball on plate balancing system. In: The 9th International Bhurban Conference on Applied Sciences Technology (IBCAST), Islamabad, Pakistan, pp. 22–26 (2012)

Nonlinear Control

Robust Controller Based on Kharitonov Theorem for Bicycle with CMG

Maciej Różewicz(✉) and Adam Piłat

Akademia Górniczo-Hutnicza im. Stanisława Staszica w Krakowie, al. Adama Mickiewicza 30, 30-059 Kraków, Poland
{rozewicz,Adam.Pilat}@agh.edu.pl

Abstract. A new approach to robust control of a nonlinear system with uncertain model parameters is discussed. The problem is devoted to stabilization of unmanned bicycle in a vertical position. A complete research consisting of controller design, simulation, experimental investigations, and comparison with previously realized control algorithms is given.

Keywords: Nonlinear control · Robust control · Bicycle stabilization

1 Introduction

1.1 Bike Stabilization Problem

Bikes and motorbikes are one of the most popular means of transport and recreation. They have a wide popularity around the world. A number of people is using them for over a hundred years. Bicycle is also probably the cheapest means of transportation. It may be noted that, in recent years, thanks to the development of digital technology, more and more attention in practical automotive industry is paid on issues associated with the possibility of autonomous driving cars, and the so-called active safety systems and advanced driver assistance systems (ADAS). Such systems, are in contrast to the passive safety systems. They reduce injuries occurring in moment of an accident. Active safety systems prevents accidents and have a certain activity before and during an accident, if it occurs. Such a system monitors neighborhood and state of host vehicle in real-time. On the basis of collected information's is able to notice potentially dangerous situation. Moreover, such a system is prepared to alert the driver or seize control. For example, brake system is activated automatically when pedestrians are detected ahead of the vehicle (such system is called AEB - Autonomous Emergency Braking).

Also more work is put into autonomous cars and it gives good results, but these solutions are working only for cars. Extension of similar safety systems for other road traffic participants seems to be natural.

The new idea presented by authors in [12] is to create a new active safety algorithms for use with single-track vehicles (shortly STV). Such a system should help to stabilize a bicycle in vertical position. It can be called Active Stabilizing

A. Bartoszewicz et al. (Eds.): PCC 2020, AISC 1196, pp. 411–423, 2020.
https://doi.org/10.1007/978-3-030-50936-1_35

Assistant System (shortly ASAS) and was proposed by authors in [12] and by Chen in [3]. Similar algorithms can be used to enable autonomous vehicle, and this subject is becoming more popular recently. In the case of ADAS algorithms a new factors must be considered, which are not present in four-wheels vehicles. The first thing, which comes to mind, is mainly possibility of loose vertical equilibrium. Because, as it was shown in [2] there exist some limit value and when it is reached, the driver can not do anything to prevent fall. Therefore, some kind of system generating counteractive torque can be very helpful for the driver. This problem is quite similar to the classical problem considered in control theory - stabilization of inverted pendulum. Further application of a such stabilization algorithm can be some kind of city "mini-car". Some similar application was presented by LitMotors [9] company.

In this paper authors present a novel approach to control algorithm robust for model parameters change. This approach takes into account a nonlinear model and is based on Kharitonov theorem for uncertain linear systems.

1.2 Robust Control Problem

The goal of the considered controller is to stabilize the bicycle in vertical position. The fact, that dynamics of a such system depends on the mass of cyclist is obvious. Mainly, it influences mass, moment of inertia, center of gravity location. Therefore, robustness analysis of control algorithm is very important to ensure stability for different people driving a bicycle.

During previous studies a few different types of controllers like PID, LQ, H_∞, Sliding Mode Controller, Feedback Linearization and controller based on nonlinear disturbance observer where tested by authors in simulation scenarios. The best results were obtained by the controller based on Feedback Linearization. But this simulation was made with assumption of ideal model parameters, and as it was shown in [4] and [1] this type of control algorithm cannot be easy combined with H_∞ controller. Some extensions for robustness were presented in [1,4] and [11]. In this paper authors present some Feedback Linearization like controller with application of Kharitonov theorem for robustness analysis.

1.3 Paper Organization

This paper is organized as follows. In Sect. 2 mathematical model of bicycle with Control Moment Gyroscope (shortly CMG) used for controller design and simulation is presented. Next section presents controller design for assumption of exactly known parameters. Section 4 presents some results of the proposed controller and comparison with some previously developed controllers. Experimental results on laboratory model are presented in Sect. 5. Last section contains summary of work.

2 Bicycle with CMG Dynamic Model

In previous publications: [12] and [14] authors derived model of bicycle with CMG dynamic used for control purpose. This model is based on assumption that STV consists of four rigid bodies (see Fig. 1): frame, front wheel, rear wheel, and CMG flywheel. Schematic view with distinguished elements and coordinate systems is presented in Fig. 1.

Mathematical model was derived by application of Euler-Lagrange formalism. This model contains dynamic of CMG, influence of centrifugal force and handlebar movement. Derived model can be presented in nonlinear affine form:

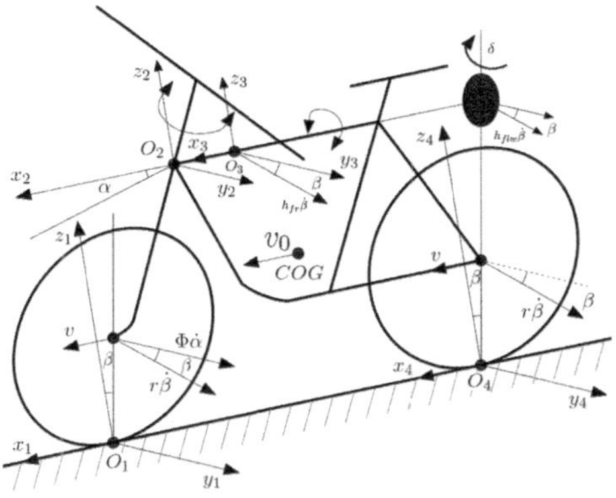

Fig. 1. Bike schematic drawing (based on work [5]).

$$\begin{Bmatrix} \dot{x} = f(x) + g(x)u \\ y = h(x) \end{Bmatrix} \tag{1}$$

where:

- $x = \begin{bmatrix} x_1 \; x_2 \; x_3 \end{bmatrix}^T$ is state, u is control, y is observation,
- x_1 is bicycle frame tilt from vertical position, $x_2 = \dot{x}_1$ is frame tilt rate, x_3 precession angle of CMG,
- u is precession angle rate,
- $f(x) = \begin{bmatrix} f_1(x) \\ f_2(x) \\ 0 \end{bmatrix} = \begin{bmatrix} x_2 \\ G\frac{\sin x_1}{J(x_1,x_3)} \\ 0 \end{bmatrix}$,
- $g(x) = \begin{bmatrix} 0 \\ g_2(x) \\ g_3(x) \end{bmatrix} = \begin{bmatrix} 0 \\ \frac{1}{J(x_1,x_3)}[2(I_p - I_r)x_2 \sin x_3 \cos x_3 + I_p \omega \cos x_3] \\ 1 \end{bmatrix}$,
- $J(x_1, x_3) = 2J_2 + J_4 + h_{fr}^2 m_{fr} + h_{flw}^2 m_{flw} + J_p \cos {x_1}^2 + J_r \sin {x_3}^2$,
- $G = g(h_{fr} m_{fr} + h_{flw} m_{flw} + 2m_w r)$,
- detailed description of all equations parameters listed here can be found in [12] - table with values used by simulations is presented in Appendix A in Table 1,
- $h(x) = x$.

It is important to add, that obtained model is non-minimum phase model with stable, but not asymptotically stable zero dynamic.

3 Control Algorithm

3.1 Proposed Control

Proposed controller design is based on representation of nonlinear model (1) in pseudo-linear form:

$$\dot{x} = A(x)x + g(x)u \tag{2}$$

where $A(x)x = f(x)$.

For this system, the controller can be found in pseudo-linear form $u(x) = K(x)x$, if pair $(A(x), g(x))$ is controllable. In considered system, this conditions are meet for all $g_2(x) \neq 0$. By making some simplification assumptions and by considering some physical limitations of a system, it can be reduced to $g_2(x) > 0$, which means, that $x_3 \in (-\frac{\pi}{2}, \frac{\pi}{2})$. More detailed information about controllability region for nonlinear system can be found by usage of Theorem 13.2 from [7].

Constraints for function $K(x) = \begin{bmatrix} k_1(x) & k_2(x) & k_3(x) \end{bmatrix}$ can be found by stability analysis of matrix $A(x) + g(x)K(x)$. This analysis can lead to provide only constrains for function $K(x)$, which ensures stability or even to define $K(x)$ which sets desired eigenvalues for matrix $A(x) + g(x)K(x)$.

Assumed control law leads to closed loop system in form:

$$\dot{x} = A_3(x)x = \begin{bmatrix} 0 & 1 & 0 \\ \frac{f_2(x)}{x_1} + g_2(x)k_1^3(x) & g_2(x)k_2^3(x) & g_2(x)k_{3^3}(x) \\ k_1^3(x) & k_2^3(x) & k_3^3(x) \end{bmatrix} x. \tag{3}$$

Stability conditions can be found by the analysis of characteristic polynomial of matrix A_3, defined as (4).

$$\alpha_3(\lambda) = det(I\lambda - A_3(x)) = \lambda^3 + a_2^3(x)\lambda^2 + a_1^3(x)\lambda + a_0^3(x) \tag{4}$$

where:

- $a_2^3(x) = -k_3^3(x) - g_2(x)k_2^3(x)$,
- $a_1^3(x) = -g_2(x)k_1^3(x) - \frac{f_2(x)}{x_1}$,
- $a_0^3(x) = \frac{f_2(x)}{x_1}k_3^3(x)$.

Stability of this polynomial can be analysed for example by Hurwitz criteria [10], but this approach gives no information about sensitivity for coefficients variations.

In control system two main types of uncertainty can be specified:

- **structural** - this type of uncertainty is generally some unmodelled dynamic,
- **non-structural** - this type of uncertainty is lack of exact knowledge about model parameters.

In this paper only non-structural uncertainty will be considered.

In the case of uncertain model parameters the coefficients of characteristic polynomials (4) and (10) are uncertain, because:

$$\begin{aligned} \underline{f_2(x)} \leq f_2(x) \leq \overline{f_2(x)} \\ \underline{g_2(x)} \leq g_2(x) \leq \overline{g_2(x)} \end{aligned} \tag{5}$$

For verification stability of such uncertain polynomial Kharitonov theorem [10] and [6] can be used.

Theorem 1. *An interval polynomial:*

$$W(\lambda) = \lambda^n + a_{n-1}\lambda^{n-1} + a_{n-2}\lambda^{n-2} + a_{n-3}\lambda^{n-3} + \cdots$$

where each coefficient $a_i \in R$ *can take any value in the specified intervals:*

$$\underline{a_i} \leq a_i \leq \overline{a_i}$$

is stable if and only if the four so-called Kharitonov polynomials:

$$\begin{aligned} W_1(\lambda) = \lambda^n + \overline{a_{n-1}}\lambda^{n-1} + \underline{a_{n-2}}\lambda^{n-2} + \underline{a_{n-3}}\lambda^{n-3} + \cdots \\ W_2(\lambda) = \lambda^n + \underline{a_{n-1}}\lambda^{n-1} + \overline{a_{n-2}}\lambda^{n-2} + \overline{a_{n-3}}\lambda^{n-3} + \cdots \\ W_3(\lambda) = \lambda^n + \overline{a_{n-1}}\lambda^{n-1} + \overline{a_{n-2}}\lambda^{n-2} + \underline{a_{n-3}}\lambda^{n-3} + \cdots \\ W_4(\lambda) = \lambda^n + \underline{a_{n-1}}\lambda^{n-1} + \underline{a_{n-2}}\lambda^{n-2} + \overline{a_{n-3}}\lambda^{n-3} + \cdots \end{aligned} \tag{6}$$

are stable.

3.2 Controller Design

Analysis of Kharitonov conditions for full state model leads to very complicated inequalities. To simplify controller design process, it was split into two steps:

Step 1 designing the controller for non-zero dynamics only - it is reduced to 2 dimensional system, so $k_1^3(x)$ and $k_2^3(x)$ bounds can be found from this analysis,

Step 2 finding the constraints for k_3^3 in a full state system after applied results of $k_1(x)$ and $k_2(x)$ obtained in Step 1.

3.3 State Without Zero Dynamics

Non-zero dynamics has the form presented by Eq. (7):

$$\begin{bmatrix} \dot{x}_1 \\ \dot{x}_2 \end{bmatrix} = \begin{bmatrix} x_2 \\ f_2(x) \end{bmatrix} + \begin{bmatrix} 0 \\ g_2(x) \end{bmatrix} u. \tag{7}$$

For this system control law can be proposed as (8):

$$u_2(x) = K_2(x)x = k_1^2(x)x_1 + k_2^2(x)x_2 \tag{8}$$

which leads to closed loop system in form (9):

$$\dot{x} = A_2(x)x = \begin{bmatrix} 0 & 1 \\ \frac{f_2(x)}{x_1} + g_2(x)k_1^2(x) & g_2(x)k_2^2(x) \end{bmatrix} x \tag{9}$$

with following (10) characteristic polynomial:

$$\alpha_2(\lambda) = det(I\lambda - A_2(x)) = \lambda^2 + a_1^2(x)\lambda + a_0^2(x) \tag{10}$$

where:

- $a_1^2(x) = g_2(x)k_2^2(x)$,
- $a_0^2(x) = -\frac{f_2(x)}{x_1} - g_2(x)k_1^2(x)$.

It can be found, that coefficients of polynomial (10) can be constrained in following way:

$$\begin{gathered} -\frac{\overline{f_2(x)}}{x_1} - \underline{g_2(x)}k_1^2(x) \le a_0^2(x) \le -\underline{\frac{\underline{f_2(x)}}{x_1}} - \overline{g_2(x)}k_1^2(x) \\ -\underline{g_2(x)}k_2^2(x) \le a_1^2(x) \le -\overline{g_2(x)}k_2^2(x) \end{gathered} \tag{11}$$

Based on this constraints and Theorem 1, conditions for $k_1^2(x)$ and $k_2^2(x)$ are formulated as:

$$\left\{ \begin{array}{l} k_1^2(x) \le -\frac{\overline{f_2(x)}}{\underline{g_2(x)}x_1} \\ k_2^2(x) \le -C_2, \ \ where \ C_2 > 0 \end{array} \right\}. \tag{12}$$

Conditions (12) can be specified to form:

$$\left\{ \begin{array}{l} k_1^2(x) = -\frac{\overline{f_2(x)}}{\underline{g_2(x)}x_1} - C_1, \ \ where \ C_1 > 0 \\ k_2^2(x) = -C_2, \ \ where \ C_2 > 0 \end{array} \right\}. \tag{13}$$

3.4 Added Zero Dynamics

For the full state analysis, an additional simplification assumption can be made, that $k_3^3(x) > 0$. This leads to constraint for polynomial coefficients in the form (14):

$$\begin{gathered} \frac{\underline{f_2(x)}}{x_1}k_3^3(x) \le a_0^3(x) \le \frac{\overline{f_2(x)}}{x_1}k_3^3(x) \\ -\underline{g_2(x)}k_1^3(x) - \frac{\overline{f_2(x)}}{x_1} \le a_1^3(x) \le -\overline{g_2(x)}k_1^3(x) - \frac{\underline{f_2(x)}}{x_1} \\ -k_3^3(x) - \underline{g_2(x)}k_2^3(x) \le a_3^3(x) \le -k_3^3(x) - \overline{g_2(x)}k_2^3(x) \end{gathered} \tag{14}$$

This implies limits for $k_3^3(x)$ function as following:

$$\left\{ \begin{array}{l} k_3^3(x) < \overline{g_2(x)}C_2 \\ k_3^3(x) \ge \frac{\overline{g_2(x)}\underline{g_2(x)}C_1C_2x_1}{\underline{g_2(x)}C_1x_1 + \underline{f_2(x)}} \end{array} \right\}. \tag{15}$$

It can be noted, that those conductions are not contradicted in any case.

3.5 Lyapunov Analysis

If closed loop system matrix A_3 (2) has a negative real parts of eigenvalues, then it can be shown, that Lyapunov equation (17) has solution $W(x) = W(x)^T > 0$, which is continuous function of x. For x which meet condition:

$$X = \left\{x \in R^3 : x_1 \in (-\frac{\pi}{2}, \frac{\pi}{2})\, and\, x_3 \in (-\frac{\pi}{2}, \frac{\pi}{2})\right\}. \tag{16}$$

Equation (17) always has a solution.

$$W(x)A_3(x) + A_3(x)^T W(x) = -G \tag{17}$$

where:

- $G = diag(G_1, G_2, G_3)$,
- $G_i > 0$, for $i = 1, 2, 3$.

Therefore, the Lyapunov functional can be proposed in the form (18):

$$V(x) = x^T W(x) x \tag{18}$$

and it's derivative:

$$\dot{V}(x) = x^T G x < 0,\ \forall x \neq 0. \tag{19}$$

So, by La Salle invariant theorem [8] a set of points which meet condition (16) can be estimated as attraction region.

4 Simulation Results

4.1 Controller Implementation

Designed controller was implemented and verified in MATLAB/Simulink environment. It can be seen, that functions $k_1^2(x)$, $k_1^3(x)$, $k_3^2(x)$, $k_3^3(x)$ proposed as Eqs. (12) and (13) have discontinuity in $x = \begin{bmatrix} 0\ x_2\ x_3 \end{bmatrix}^T$. To avoid this problem a robust controller was implemented in the following form (20):

$$u(x) = -\frac{\overline{f_2}}{\underline{g_2}} - C_1 x_1 - \overline{g_2} C_2 x_2 + k(x) x_3 \tag{20}$$

where:

- $k(x) = \min(\overline{g_2(x)} C_2, c(x))$,
- $c(x) = \begin{cases} \frac{\overline{g_2(x)}\,\underline{g_2(x)} C_1 C_2 x_1}{\underline{g_2(x)} C_1 x_1 + \underline{f_2(x)}} & \text{when: } \lvert x_1 \rvert > \epsilon \\ \frac{\overline{g_2(x)}\,\underline{g_2(x)} C_1 C_2}{\underline{g_2(x)} C_1 + \frac{G}{J(x_1, x_3)}} & \text{when: } \lvert x_1 \rvert \leq \epsilon \end{cases}$,
- $J(x_1, x_3)$ and G are defined in Sect. 2.

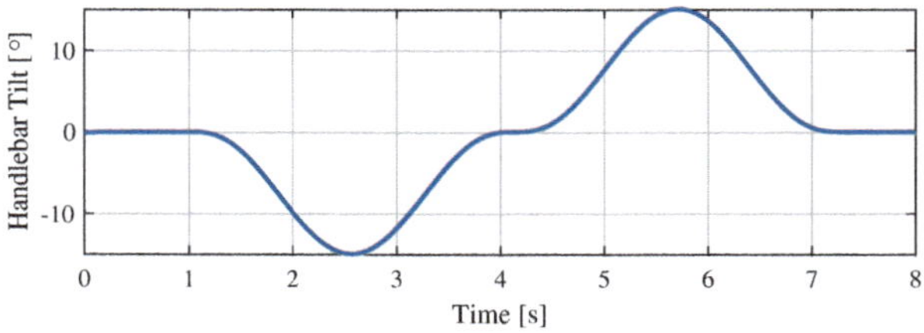

Fig. 2. Disturbance

4.2 Simulation

The controller was designed without any bounds for control signal, but such a situation is not realistic. For this reason during the simulation some limitations for control signal were set. Those limitations are: maximum value of saturation, rate limiter, quantization and some inertia. To test the proposed controller, the following scenario was applied: bicycle starts ride in the upper vertical position with 0 handlebar angle. After a piece of time the handlebar motion is set left, next right and than back to initial position. This handlebar movement is presented in Fig. 2. The proposed controller was compared to results of H_∞ controller, designed for Taylor linearization of the system (1), presented by authors in [13], as well well with the Feedback Linearization controller. To visualize the difference between both algorithms, two kind of plots are given:

- state and control for ideal value of parameters - those results are presented in Fig. 3,
- value of quality measure $Q = \int_0^T x_1^2(t)dt$ for some range of uncertain frame mass (see Table 1) - these results are presented in Fig. 4.

5 Experimental Results

Simulation result were obtained for the static scenario, only. Bicycle velocity $v = 0$, and initial tilt angle x_1^0 were applied. The noisy measurements from the IMU sensors were observed due to high vibrations generated by the spinning flywheel. A few filtering methods were tested without success. Therefore, the measurement of tilt angle was realized by an external laser distance sensor according to Fig. 5 and Eq. (21):

$$\beta = \arctan(\frac{d - d_0}{h}). \tag{21}$$

View of laboratory model in stabilization mode with marked sensor is presented in Fig. 6. The extra attention during further research should be paid on vibrations elimination and accurate signals filtering. First of all, the Fig. 7, presents results of a new controllers. Three different combinations of parameters were tested: $(C_1, C_2) = \{(25, 3), (25, 5), (25, 7)\}$. All combinations have the same value of

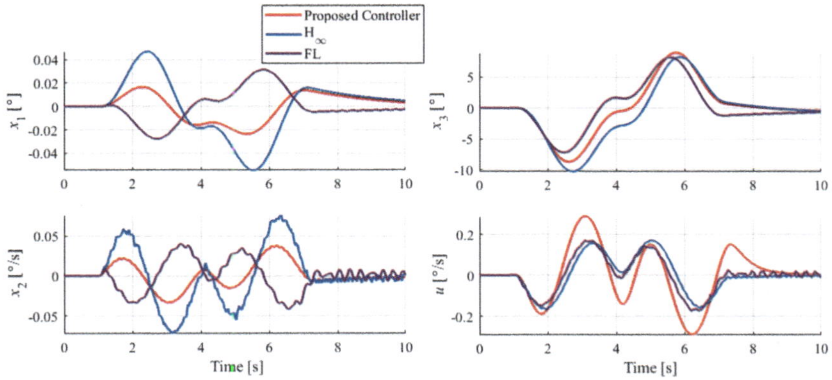

Fig. 3. Results of the proposed controller for exact model parameters.

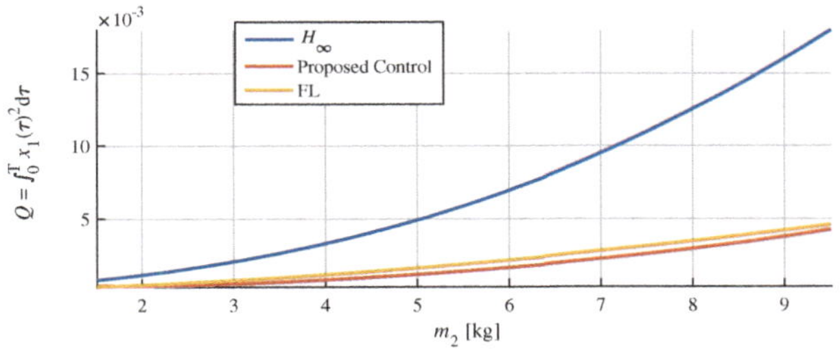

Fig. 4. Comparison of H_∞, Feedback Linearization and proposed controllers for different frame mass m_2, according to selected quality factor Q.

parameter C_1 and only C_2 is different. It can be seen, that for the lowest value, the state has some small oscillations, which are dumped with higher C_2 value, and for $C_2 = 7$ state is aperiodic. The second test was conducted to compare the proposed controller performance with respect to the variable mass m_2. The results for nominal mass and with additional 1 kg load are given in Fig. 8. It can be seen from plots, that performances in both cases have relatively small difference. Finally, a different controllers: (same as tested in simulation study: H_∞, Feedback Linearization and proposed controller) were compared in static scenario. It can be seen, that Linear controller H_∞ has the biggest error in tilt angle. Performance of proposed controller and Feedback Linearization method are very similar, although, the proposed controller has a shorter stabilization time. It should be noted, that there are many ways to tune the presented control algorithms. Many calibrations were tested during experiments, and only the best results according to quality factors were selected (Fig. 9).

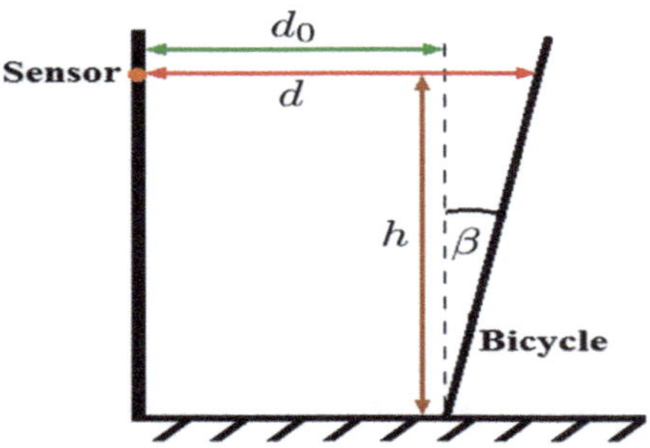

Fig. 5. Schematic view of tilt angle external measurement system.

Fig. 6. Bicycle in stabilization mode with marked external laser sensor.

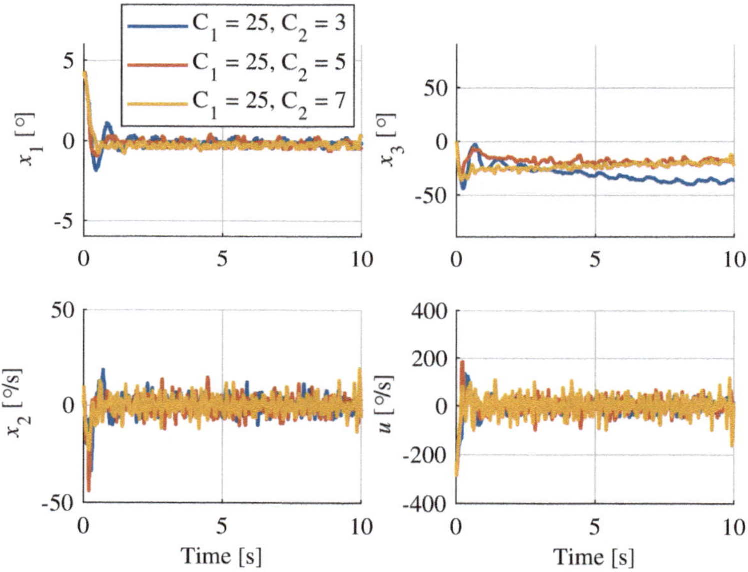

Fig. 7. Experimental results for different parameters.

6 Summary

For the selected controllers, the system was stabilized in assumed interval of parameter m_{fr}. It can be seen, that proposed controller gives better results, according to the selected quality factor, than compared algorithms. Additional advantage of the proposed approach is fact, that it provides a range of possible control functions and easy way of finding Lyapunov Functional, which can be helpful in further stability analysis. The experimental results confirmed simulations. The proposed controller is characterized by the better performance with respect to the discussed ones.

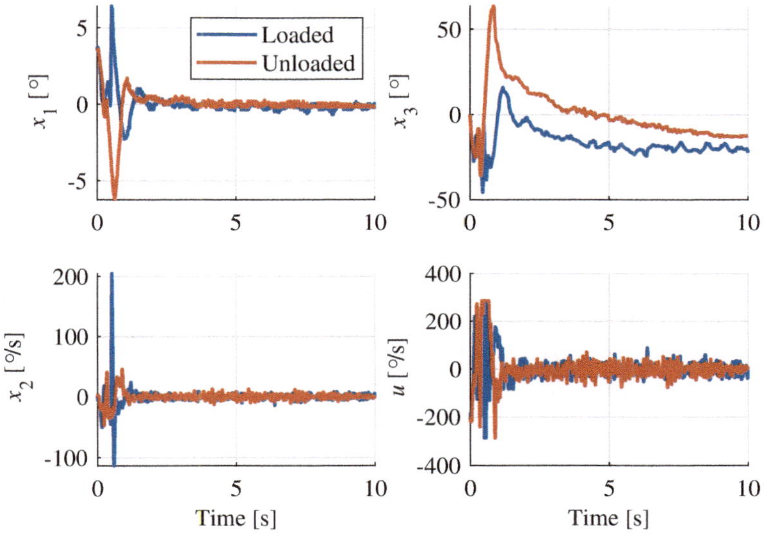

Fig. 8. Comparison of proposed controller for a different masses.

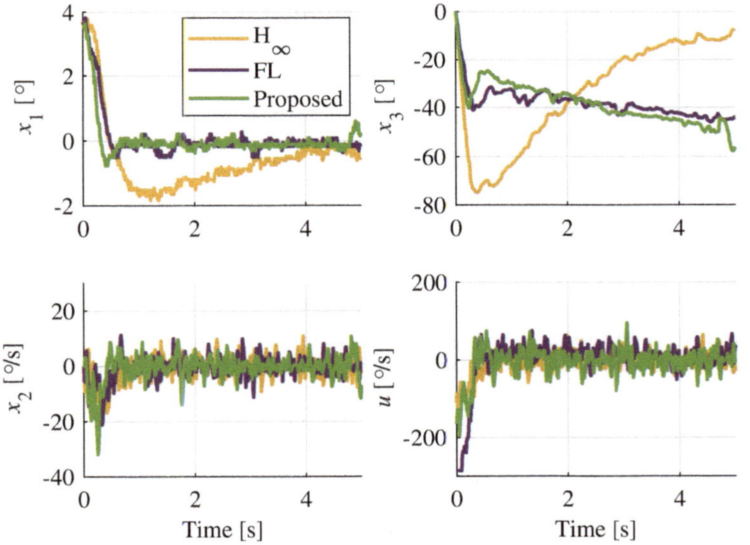

Fig. 9. Experimental comparison of a few controllers: H_∞, Feedback Linearization and proposed one.

A Bicycle Model Parameters

In the simulation, the following set of parameters for real kid-size bike was used:

Table 1. Model parameters.

Description	Symbol	Value
Frame mass	m_{fr}	5.512 [kg]
Wheel mass	m_w	0.909 [kg]
CMG mass	m_{flw}	0.39 [kg]
Frame height	h_{fr}	0.25 [m]
Wheel radius	r	0.15 [m]
Distance between wheels	l	0.54 [m]
CMG radial moment of inertia	J_r	0.001 [kgm^2]
CMG moment of inertia	J_p	0.0005 [kgm^2]
Custer	Φ	0.04 [m]
CMG speed	ω	3000 [rpm]
Velocity	v_{max}	10 [km/h]

References

1. Ataei, M.A., Esmaelizadeh, R., Alizadeh, G.: Robust feedback linearization. In: 4th WSEAS International Conference on Non-Linear Analysis, Non-Linear Systems and Chaos, Sofia, Bulgaria, pp. 114–119 (2005)
2. Bonci, A., De Amicis, R., Longhi, S., Andreucci, A., Scala, G.A.: Motorcycle lateral and longitudinal dynamic modelling in presence of tyre slip and rear traction. In: Conference on Methods and Models in Automation and Robotics, pp. 391–396 (2016)
3. Chen, C.K., Chu, T.D., Zhang, X.D.: Modelling and control of an active stabilizing assistant system for a bicycle. Sensors **19**(2), 248 (2019)
4. Guilliard, H., Bourles, H.: Robust feedback linearization. In: Mathematical Theory of Networks and System Conference, pp. 70–75 (2000)
5. Guo, L., Liao, Q., Wei, S., Huang, Y.: A kind of bicycle robot dynamic modelling and nonlinear control. In: Proceedings of the International Conference on Information and Automation, pp. 1613–1617 (2010)
6. Kaczorek, T.: Teoria sterownia i systemów. Wydawnictwo Naukowe PWN (1993)
7. Khalil, H.K.: Nonlinear Systems, 3rd edn. Prentice Hall, Upper Saddle River (2002)
8. La Salle, J., Lefschetz, S.: Stability by Liapunov's Direct Method with Application. Academic Press, New York (1961)
9. LitMotors (2019). http://litmotors.com/c1
10. Mitkowski, W.: Równania macierzowe i ich zastosowania. Wydawnictwo AGH (2012)
11. Pop, C.I., Dulf, E.H.: Robust feedback linearization control for reference tracking and disturbance rejection in nonlinear systems. In: Recent Advances in Robust Control - Novel Approaches and Design Methods, pp. 273–290 (2011)

12. Różewicz, M., Piłat, A.: Modelling of bike steered by CMG. In: Conference on Methods and Models in Automation and Robotics, pp. 595–600 (2016)
13. Różewicz, M., Piłat, A.: Robust control of bike steered by CMG. In: Conference on Methods and Models in Automation and Robotics, pp. 369–374 (2016)
14. Różewicz, M., Piłat, A.: Study on controller embedding stage using model-based-design for a bike with CMG. In: Conference on Methods and Models in Automation and Robotics, pp. 680–685 (2018)

Dual Kalman Filters Analysis for Interior Permanent Magnet Synchronous Motors

Tanja Zwerger[1(✉)] and Paolo Mercorelli[2]

[1] Rolls Royce, Maybachplatz 1, 88045 Friedrichshafen, Germany
Tanja.Zwerger@ps.rolls-royce.com
[2] Institute of Product and Process Innovation, Leuphana University of Lueneburg, Universitaetsallee 1, 21335 Lueneburg, Germany
mercorelli@uni.leuphana.de
https://www.leuphana.de/en/institutes/ppi/staff/paolo-mercorelli.html

Abstract. This paper deals with an analysis and design of Dual Extended Kalman Filters (DKFs) to estimate parameters and state variables in Permanent Magnet Synchronous Machines (PMSMs) to be utilized in a control structure. A dual estimation problem consists of a simultaneous estimation of states of the dynamical system and its parameters using only noisy output observations. In this paper, the limit of an Augmented and Extended Kalman Filter (AEKF) obtained through standard state augmentation to estimate parameters is shown and, alternatively, a DKF approach which is characterized by the use of the state model descriptions in the output of an AEKF is proposed. The two different approaches are analyzed and compared. These results are supported by simulations.

Keywords: Dual Extended Kalman Filter · Augmented Extended Kalman Filter · Permanent Magnet Synchronous Machine · Parameters estimation

1 Introduction and Motivation

Permanent Magnet Synchronous Machines (PMSMs) are often used in applications because of their advantageous power volume ratio. For this intrinsical optimality they can be applied with their full capability and they are very popular in many applications. The EKF is well known with the sensorless control of PMSMs or the field estimation in Asynchronous Machines (ASMs) as recently presented in [11]. More in general, sensorless control represents one of the most important and applied control techniques in which a minimal number of sensors is intended to be used, see [4–6,8] and [7]. The usage the EKF as an online parameter estimation is another popular area. The main contribution of this paper consists of a comparison of the performance of two approaches which consider Dual Kalman

A. Bartoszewicz et al. (Eds.): PCC 2020, AISC 1196, pp. 424–435, 2020.
https://doi.org/10.1007/978-3-030-50936-1_36

Filters (DKFs). The first approach considers an estimation of constant parameters and state variables using an Augmented Extended Kalman Filter (AEKF). The second approach considers *combined* Dual Extended Kalman Filter (DKF) structures in which each parameter is estimated just considering each parameter as an augmented state but with the dynamics to be estimated *localized* in the output model equation of the EKF. In accordance with the definition of a Dual Extended Kalman Filter (DKF) as reported in Chapter 5 of [1] a dual estimation problem consists of a simultaneous estimation of states of the dynamical system and its parameters using only noisy output observations. In this sense, two possible approaches can be proposed: a Virtual Measured Model Based Dual Extended Kalman Filter (VMBKF) and an *Output Dual Extended Kalman Filter* (ODKF). In this paper, the second approach will be considered. The advantage of this approach is that the calculation load is drastically reduced. In fact, each parameter is estimated using an ODKF with dimension one. The drawback of this approach is that these ODKFs should be interconnected to each other. This generates a phase delay in the estimation. The results between ODKF and the AEKF, as proposed in [14], approaches are discussed and compared. The paper is structured as follows. In Sect. 2 the description of the adopted model is shown. Section 3 describes the proposed two estimation approaches. Simulations with the obtained results and conclusions close the paper.

The main nomenclature

$u_d(t)$: the direct voltage input
$u_q(t)$: the quadrature voltage input
$i_d(t)$: the direct current
$i_q(t)$: the quadrature current
$i_{dd}(t)$: desired direct current
$i_{qd}(t)$: desired quadrature current
$\omega_r(t)$: angular velocity of the rotor
$\omega_{el}(t)$: electrical pulsation on the induced voltage
p: couple of magnetic poles
$\omega_{el}(t) = p\omega_r(t)$
R_s: coil resistance
L_d: the direct axis self-inductance
L_q: the quadrature axis self-inductance
ψ_p: armature (or stator) back Electromotive Force (back EMF) constant in Vs
$T_e(t)$: electromagnetic torque
EE: estimated electrical states
EM: estimated mechanical states
$AEKF$: Augmented and Extended Kalman Filter
$ODKF$: output based Dual Kalman Filter

2 Description of the Physical Systems

Among a variety of models presented in the literature since the introduction of PMSM, the two-axis dq-model, obtained by using Park's dq-transformation, is the most widely used in variable speed PMSM drive control applications [3]. In the electrical machine both fluxes ψ_d and ψ_q share their way through the stator. In (3) and (4) the linear equations for the cross coupling effect are defined without considering the saturation effects. By definition the permanent flux lies in the same direction as ψ_d, like in (4). ψ_p defines the portion of permanent magnetic flux from the rotor over the air gap and the stator. The other portion of magnetic flux which runs in the rotor is not considered in ψ_p. In (9), $i_d(t)$, $i_q(t)$, $u_d(t)$ and $u_q(t)$ are the dq-components of the stator currents and voltages in synchronously rotating rotor reference frame, $\omega_{el}(t)$ is the rotor electrical angular speed, the parameters L_d, L_q, R_s, ψ_p and p are the stator resistance, d-axis and q-axis inductance, the amplitude of the back EMF, and p is the number of couples of permanent magnets, respectively. Considering the following model:

$$u_q(t) = i_q(t)R_s + \frac{d\psi_q(t)}{dt} + \omega_{el}(t)\psi_d(t), \tag{1}$$

$$u_d(t) = i_d(t)R_s + \frac{\psi_d(t)}{dt} - \omega_{el}(t)\psi_q(t), \tag{2}$$

in which, according to [12,13] and [10],

$$\psi_d(t) = L_d i_d(t) + \psi_p, \tag{3}$$

and

$$\psi_q(t) = L_q i_q(t), \tag{4}$$

where $\psi_d(t)$ and $\psi_q(t)$ represent the field and the torque flux respectively.

$$\frac{\psi_q(t)}{dt} = L_q\frac{di_q(t)}{dt} = u_q(t) - i_q(t)R_s - \omega_{el}(t)\big(L_d i_d(t) + \psi_p(t)\big), \tag{5}$$

and thus

$$\frac{di_q(t)}{dt} = \frac{u_q(t)}{L_q} - \frac{i_q(t)R_s}{L_q} - \frac{\omega_{el}(t)L_d i_d(t)}{L_q} - \frac{\omega_{el}(t)\psi_p}{L_q}. \tag{6}$$

In the same way the following relation is obtained:

$$\frac{\psi_d(t)}{dt} = L_d\frac{di_d(t)}{dt}, \tag{7}$$

and thus

$$\frac{di_d(t)}{dt} = \frac{u_d(t)}{L_d} - \frac{i_d(t)R_s}{L_d} + \frac{\omega_{el}L_q i_q(t)}{L_d}. \tag{8}$$

The dynamic electrical model of the synchronous motor in dq-coordinates can be represented as follows:

$$\begin{bmatrix} \frac{di_d(t)}{dt} \\ \frac{di_q(t)}{dt} \end{bmatrix} = \begin{bmatrix} \frac{-R_s}{L_d} & \frac{L_q}{L_d}p\omega_r(t) \\ -\frac{L_d}{L_q}p\omega_r(t) & \frac{-R_s}{L_q} \end{bmatrix} \begin{bmatrix} i_d(t) \\ i_q(t) \end{bmatrix} + \begin{bmatrix} \frac{1}{L_d} & 0 \\ 0 & \frac{1}{L_q} \end{bmatrix} \begin{bmatrix} u_d(t) \\ u_q(t) \end{bmatrix} - \begin{bmatrix} 0 \\ \frac{\psi_p p\omega_r(t)}{L_q} \end{bmatrix} \tag{9}$$

and the electromagnetic torque can be expressed as follows:

$$T_e(t) = \frac{3}{2}p\Big\{i_q(t)\psi_d(t) - i_d(t)\psi_q(t)\Big\}, \tag{10}$$

where the term $(i_q(t)\psi_d(t) - i_d(t)\psi_q(t))$ defines the cross coupling of the PMSM which leads to the effect, that a variation of the current $i_d(t)$ has an impact on the current $i_q(t)$ in the q-axis and reverse. Considering (3) and (4), then the torque can be expressed as follows:

$$T_e(t) = \frac{3}{2}p\Big\{(L_d - L_q)i_d(t)i_q(t) + \psi_p i_q(t)\Big\}. \tag{11}$$

3 Kalman Filter Structures

In this section, background and two EKF structures are shown. In particular, starting from a centralized EKF structure it is possible to see how incrementing the number of parameters to be estimated using a standard augmented EKF, not only the *observability level* of the system decreases, but the system to be estimated becomes unobservable and a standard augmented EKF cannot be utilized to this scope. A different situation is the case in which the model is built in outputs of the EKF. Because very often the parameters to be estimated appear in the same output equation, the structure of the EKF cannot be centralized because of observability problems. Nevertheless, each parameter must be identified in a single ODKF which results of dimension 1×1 and thus all ODKFs must be interconnected to each other.

3.1 EKF Background

The a priori estimation of the state is as follows:

$$\hat{x}^-(k) = f(\hat{x}^-(k-1), u(k-1), w(k-1)) \tag{12}$$

in which function $f(\hat{x}^-(k-1), u(k-1), w(k-1))$ represents a nonlinear field used to model the considered system. The a priori error covariance matrix is as follows:

$$P^-(k) = J(k)P(k-1)J(k)^T + Q, \tag{13}$$

where matrix J represents the well known Jacobian approximation of $f(\hat{x}^-(k-1), u(k-1), w(k-1))$ calculated at the corrected state variable $\hat{x}_(k)$ defined below. Matrix Q is specified by the covariance matrix of the *Process Noise* which is supposed to be a *White Gaussian Noise* which *measures* the reliability of the model. The higher the values of the trace in the matrix are, the higher the reliability of the estimated values is. $P(k-1)$ represents the a posteriori error covariance matrix at step $k-1$ (previous step). Kalman gain $K(k)$ is calculated as:

$$K(k) = P^{-}(k)H^{T}(HP^{-}(k)H^{T} + R_{v})^{-1}, \tag{14}$$

with R_v as the variance and covariance matrix for the measurement noise and H as the output Jacobian indicating which state serves as the measurement in the EKF algorithm. The a posteriori estimation is as follows:

$$\hat{x}(k) = \hat{x}^{-}(k) + K(k)(z(k) - H\hat{x}^{-}(k)), \tag{15}$$

where $z_k(k) = x(k) + v(k)$ represents the measured data in which vector signal $v(k)$ represents the measured *White Gaussian Noise* associated to the sensoring system and which is assumed to be independent of *Process Noise* defined above. Finally, the a posteriori estimation of the error covariance is as follows:

$$P(k) = (I - K_{k}H)P^{-}(k). \tag{16}$$

3.2 Augmented and Extended Kalman Filter

Considering an AEKF in which parameters R_s, L_d and L_q should be estimated characterized by the following Jacobian matrix J_c and map sensor matrix H_c:

$$J_c = \begin{bmatrix} 1 - \frac{T_s R_s}{L_d} & T_s p\omega_r \frac{L_q}{L_d} & \frac{(R_s i_d - \omega_r p i_q L_q - u_d)T_s}{L_d^2} & T_s p\omega_r \frac{i_q}{L_d} & -T_s \frac{i_d}{L_d} \\ -T_s p\omega_r \frac{L_d}{L_q} & 1 - T_s \frac{R_s}{L_q} & -T_s p\omega_r \frac{i_d}{L_q} & \frac{(R_s i_q + p\omega_r i_d L_d + p\omega_r \psi_p - u_q)T_s}{L_q^2} & -T_s \frac{i_q}{L_q} \\ 0 & 0 & 1 & 0 & 0 \\ 0 & 0 & 0 & 1 & 0 \\ 0 & 0 & 0 & 0 & 1 \end{bmatrix} \tag{17}$$

and

$$H_c = \begin{bmatrix} 1\,0\,0\,0\,0 \\ 0\,1\,0\,0\,0 \end{bmatrix}. \tag{18}$$

Calculating the following observability matrix for a linearized system around the a posteriori estimation trajectory in accordance with the analysis reported in [9] for a linear time varying system and considering that in our case

$$\frac{dH_c}{dt} = 0,$$

it is enough to check the observability constructing the observability matrix using J_c and H_c. The following formal expression is obtained:

$$\mathcal{O} = \begin{bmatrix} H_c \\ H_c J_c \\ H_c J_c^2 \\ H_c J_c^3 \end{bmatrix}. \tag{19}$$

Considering the symbolic toolbox of Matlab, the rank of matrix *Observability Test* can be calculated. The result indicates that just 4 states and not 5 are observable. Even though this test does not represent just a sufficient condition and thus the test does not indicate that the linearized system is not observable, the experience indicates that if the *Observability Test* is not satisfied it is difficult to find a combination of values of the tuning matrices $\mathbf{Q}$ and $\mathbf{R}_v$ which guarantee the convergence of the estimation. To conclude, the AEKF obtained using an augmentation of the state variable cannot represent a viable solution to estimate R_s, L_d and L_q which this kind of EKF structure.

3.3 Interconnected Output Dual Kalman Filters

The structure of the ODKF consists of three interconnected ODKFs as shown in Fig. 1. The a priori estimation of $L_q(k)$:

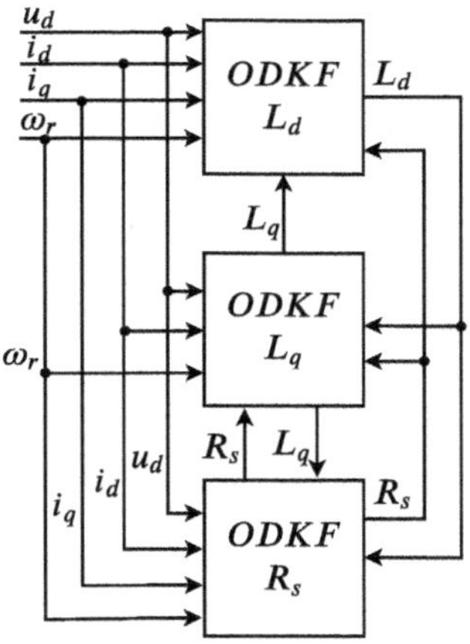

Fig. 1. Structure of ODKF

$$\hat{L}_q^-(k) = \hat{L}_q(k-1) \tag{20}$$

in which the state Jacobian matrix is represented by J = 1 (the unit constant). The following output structure

$$z_{L_q}(k) = h_{L_q}(\hat{x}(k)) \tag{21}$$

is in this case as follows:

$$h_{L_q}(k) = \frac{T_s(p\omega_r(k-1)i_q(k-1)\hat{L}_q^-(k))}{\hat{L}_d(k-1)} - \frac{T_s\hat{R}_s(k-1)i_d(k-1)}{\hat{L}_d(k-1)} + \frac{T_s u_d(k-1)}{\hat{L}_d(k-1)} + i_d(k-1). \tag{22}$$

The a priori estimation of $L_d(k)$:

$$\hat{L}_d^-(k) = \hat{L}_d(k-1) \tag{23}$$

in which the state Jacobian matrix is represented by J = 1 (the unit constant). The following output structure

$$z_{L_d}(k) = h_{L_d}(\hat{x}(k)) \tag{24}$$

which represents the output model current $i_q(k)$ calculated in the a priori estimation of the state and in which $\hat{L}_d(k)$ is estimated using again the three coupled ODKFs. For the $\hat{L}_d(k)$ ODKF is the output structure used as followed:

$$\begin{aligned} h_{L_d}(k) =& \frac{-T_s(p\omega_r(k-1)i_d(k-1)\hat{L}_d^-(k-1))}{\hat{L}_q(k-1)} - \frac{T_s\hat{R}_s(k-1)i_q(k-1)}{\hat{L}_q(k-1)} \\ &+ \frac{T_s u_q(k-1)}{\hat{L}_q(k-1)} + i_q(k-1) - \frac{\psi_p p\omega_r(k-1)}{i_q(k-1)}. \end{aligned} \tag{25}$$

This output structure for the estimation of $\hat{L}_d(k)$ has to be calculated with the output model for $i_q(k)$, otherwise there would be a division almost by zero in the Jacobian. In fact, using Eq. (8) because of the presence of $L_d(k-1)$ in the denominator, a $L_d(k-1)^2$ is present in the corresponding *Jacobian Matrix*. The a priori estimation of $R_s(k)$:

$$\hat{R}_s^-(k) = \hat{R}_s(k-1) \tag{26}$$

in which the state Jacobian matrix is represented by J = 1 (the unit constant). The following output structure

$$z_{R_s}(k) = h_{R_s}(\hat{x}(k)) \tag{27}$$

which represents the output model current $i_q(k)$ of Eq. (6) calculated in the a priori estimation of the state and in which $\hat{R}_s(k)$ is estimated using again the three coupled ODKFs. Its expression is as follows:

$$\begin{aligned} h_{R_s}(\hat{x}(k)) =& i_q(k-1) + T_s\Big[- i_q(k-1)\frac{R_s}{\hat{L}_q} \\ & - p\omega_r(k-1)\frac{\hat{L}_d i_q(k-1)}{\hat{L}_q} - p\omega_r(k-1)\frac{\psi_p}{\hat{L}_q} + \frac{u_q(k-1)}{\hat{L}_q}\Big]. \end{aligned} \tag{28}$$

For the $\hat{R}_s(k)$ ODKF both output structures, whether for output model current $i_q(k)$ or for $i_d(k)$ can be used and the output model current $i_q(k)$ is chosen.

Remark 1. *To sum up, the criteria of the choice of the outputs with respect to the states to be estimated, it is possible to say that Eq. (8) is critical just to estimate the value of* L_d *because this must be done during the transient otherwise* L_d *is not longer estimable because of the presence of the multiplication of* L_d

with the derivative of i_d. *In fact, because of the presence of d/q transformation after the transient the derivative of* i_d *equals zero. In this sense, it is important to associate the output Eq. (6) to the estimation of* L_d. *For the estimation of* L_q *Eq. (8) is considered to avoid a quadratic term in the* Jacobian Matrix. *To estimate* R_s *both equations are possible to be chosen: (8) and (6).*

The Jacobian of this output structures calculated in the a priori estimation of the state is as followed for the estimation of the three different states. For $\hat{L}_q(k)$ the *Jacobian Matrix* is

$$H_{L_q} = \frac{T_s p \omega_r(k-1) i_q(k-1)}{\hat{L}_d(k-1)}; \tag{29}$$

and for $\hat{L}_d(k)$ it is

$$H_{L_d} = -\frac{T_s p \omega_r(k-1) i_d(k-1)}{\hat{L}_q(k-1)}; \tag{30}$$

and the *Jacobian Matrix* in the output for $\hat{R}_s(k)$ is

$$H_{R_s} = -\frac{T_s i_q(k-1)}{\hat{L}_q(k-1)}; \tag{31}$$

The Kalman gain K is calculated in each ODKF for R_s, L_d and L_q in the same way as defined in (14).

4 Simulation Results

For both proposed structures the estimation of the values R_s, L_d and L_q was examined for an initial failure of 10%. In the examination of the ODKF, the interaction of the three EKFs becomes clear. The estimated stator resistance R_s has a bias to the set point which is caused by the time based coupling of the three estimated values and leads to an error like mentioned in Table 2. Table 2 shows the bigger weight of the variance in the system noise by giving more trust to the model. The dynamic grows for a bigger variance in the measurement noise matrix R_v because of the lack of accuracy in the measurement and in the output model. The tuning of both matrices R_v and Q is very sensitive because of strong coupling effects of each single ODKF. The error for all estimated values becomes smaller by introducing a small band-limited white noise with a power of 15×10^{-14} in i_q in the EKF of R_s.

As shown in Table 1, the resulting errors are all in the same range. The tuning of the matrices R_v and Q is not sensitive to changes as in the ODKF.

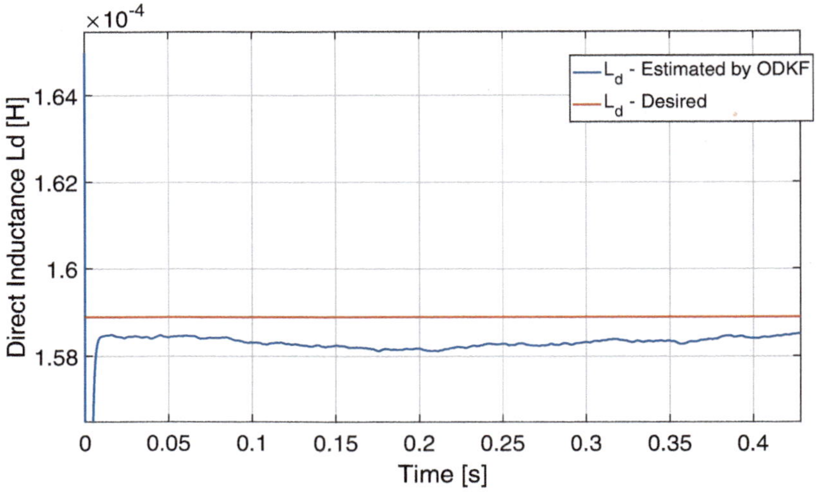

Fig. 2. Estimation of L_d with ODKF

Table 1. Error for AEKF

	L_d	L_q	R_s
Setpoint	0.00016 H	0.00045 H	0.0122 Ω
Actual value	0.00017 H	0.00049 H	0.0154 Ω
Error after 5 s	5.8%	9.8%	26%
Variance R_v	1e0	1e0	1e0
Variance Q	1e-9	1e-9	1e-9

Table 2. Error for ODKF

	L_d	L_q	R_s
Setpoint	0.0001589 H	0.0004528 H	0.0122 Ω
Actual value	0.0001583 H	0.0004524 H	0.0121 Ω
Error after 1 s	0.3%	0.08%	0.8%
Variance R_v	1e4	1e4	1e4
Variance Q	1e-6	1e-8	1e-2

In Table 3 and 4 a comparison to estimate the level of the calculation load as reported in [2] was proposed. Parameter m denotes the number of measured outputs, q is the number of inputs and n indicates the number of estimated states and l is the number of necessary EKFs for estimating the three states in ODKF. It is to notice that in case of ODKF $l = 2$ is considered to compare the

Table 3. Overview of the arithmetic operation requirements for the proposed ADKF structure

	Number of multiplications ($n = 3,\ m = 3,\ q = 5,\ l = 1$)	Number of additions ($n = 3,\ m = 3,\ q = 5,\ l = 1$)
$\hat{\mathbf{x}}_k^-$ Eq. (12)	$(n^2 + nq)l$	$(n^2 - n + nq)l$
$\mathbf{P}_k^-$ Eq. (13)	$(2n^3)l$	$(2n^3 - n^2)l$
$\mathbf{K}_k$ Eq. (14)	$(n^2m + 2nm^2 + m^3)l$	$(n^2m + 2nm^2 + m^3 - 2nm)l$
$\hat{\mathbf{x}}_k^+$ Eq. (15)	$(2nm + qm)l$	$(2nm + qm)l$
$\mathbf{P}_{k-1}^+$ Eq. (16)	$(n^3 + n^2m)l$	$(n^3 + n^2m - n^2)l$
Total	267	234

level of estimation load of the estimation of two parameters as for ADKF. The graphical results of the estimation of L_d, L_q and R_s are shown in Fig. 2, 3 and in 4. The error, based on the deviation from set point to actual value is listed in the table below for the estimation with AEKF, as proposed in [14], and ODKF setting the desired torque which equals 50 Nm and the desired velocity of 2000 rpm. Additionally, the values of the measurement noise matrix Q and the system noise matrix R_v are given in the table below.

Table 4. Overview of the arithmetic operation requirements for the proposed ODKF structure

	Number of multiplications ($n = 1,\ m = 1,\ q = 4,\ l = 3$)	Number of additions ($n = 1,\ m = 1,\ q = 4,\ l = 2$)
$\hat{\mathbf{x}}_k^-$ Eq. (12)	$(n^2 + nq)l$	$(n^2 - n + nq)l$
$\mathbf{P}_k^-)$ Eq. (13)	$(2n^3)l$	$(2n^3 - n^2)l$
$\mathbf{K}_k$ Eq. (14)	$(n^2m + 2nm^2 + m^3)l$	$(n^2m + 2nm^2 + m^3 - 2nm)l$
$\hat{\mathbf{x}}_k^+$ Eq. (15)	$(2nm + qm)l$	$(2nm + qm)l$
$\mathbf{P}_{k-1}^+$ Eq. (16)	$(n^3 + n^2m)l$	$(n^3 + n^2m - n^2)l$
Total	48	28

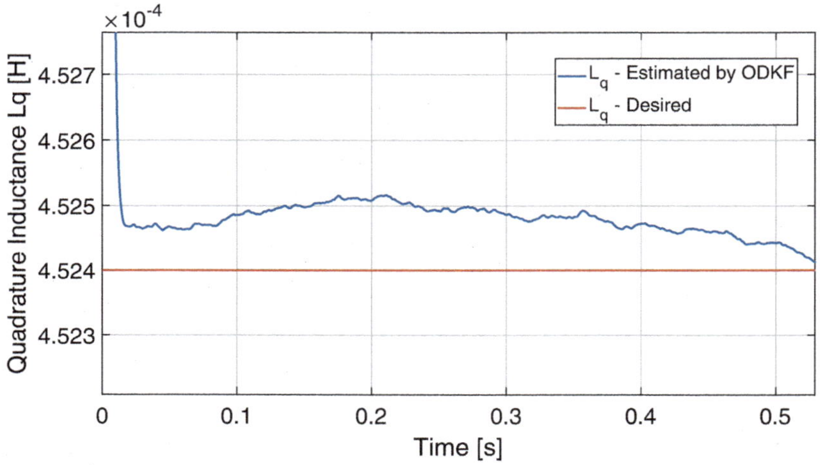

Fig. 3. Estimation of L_q with ODKF

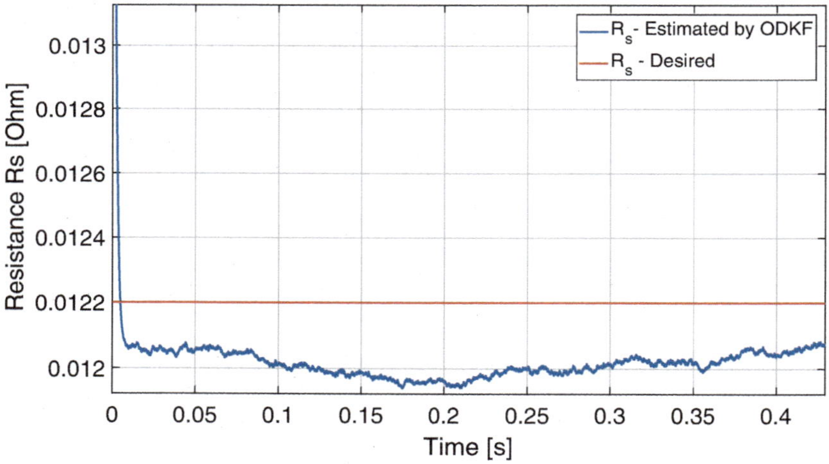

Fig. 4. Estimation of R_s with ODKF

5 Conclusions

This paper proposes a Dual Extended Kalman Filter to be applied for estimation of parameters of Permanent Magnet Synchronous Machines. A particular structure is proposed in which an interconnection between three Dual Extended Kalman Filters is proposed. Performances are compared with an alternative Augmented and Extended Kalman Filter to estimate the same parameters. Computer simulations close the paper.

References

1. Haykin, S.: Kalman Filtering and Neural Networks. Wiley, Boston (2001)
2. Hilairet, M., Auger, F., Berthelot, E.: Speed and rotor flux estimation of induction machines using a two-stage extended Kalman filter. Automatica **45**, 1819–1827 (2009)
3. Rahman, M.A., Vilathgamuwa, D.M., Uddin, M.N., Tseng, K.J.: Nonlinear control of interior permanent magnet synchronous motor. IEEE Trans. Indus. Appl. **39**(2), 408–416 (2003)
4. Mercorelli, P.: Robust feedback linearization using an adaptive PD regulator for a sensorless control of a throttle valve. Mechatronics **19**(8), 1334–1345 (2009)
5. Mercorelli, P.: A hysteresis hybrid extended Kalman filter as an observer for sensorless valve control in camless internal combustion engines. IEEE Trans. Indus. Appl. **48**(6), 1940–1949 (2012)
6. Mercorelli, P.: A two-stage augmented extended Kalman filter as an observer for sensorless valve control in camless internal combustion engines. IEEE Trans. Indus. Electron. **59**(11), 4236–4247 (2012)
7. Mercorelli, P.: A two-stage sliding-mode high-gain observer to reduce uncertainties and disturbances effects for sensorless control in automotive applications. IEEE Trans. Indus. Electron. **62**(9), 5929–5940 (2015)
8. Mercorelli, P.: A switching observer for sensorless control of an electromagnetic valve actuator for camless internal combustion engines. In: Proceedings of the 50th International Conference on Decision and Control (CDC 2011), Orlando, USA, December 2011
9. Sontag, E.D.: Mathematical Control Theory: Deterministic Finite Dimensional Systems. Springer-VerlagBerlin, Heidelberg (1998)
10. Stumberger, B., Stumberger, G., Dolinar, D., Hamler, A., Trlep, M.: Evaluation of saturation and cross-magnetization effects in interior permanent-magnet synchronous motor. IEEE Trans. Indus. Appl. **39**(5), 1264–1271 (2003)
11. Tian, G., Yan, Y., Jun, W., Ru, Z.Y., Peng, Z.X.: Rotor position estimation of sensorless PMSM based on extented Kalman filter. In: Proceedings of the 2018 IEEE International Conference on Mechatronics, Robotics and Automation (ICMRA), pp. 12–16 (2018)
12. Vas, P., Brown, J.E., Hallenius, K.E.: Cross-saturation in smooth-air-gap electrical machines. IEEE Power Eng. Rev. **PER–6**(3), 37 (1986)
13. Vas, P., Hallenius, K.E., Brown, J.E.: Cross-saturation in smooth-air-gap electrical machines. IEEE Trans. Energy Convers. **EC–1**(1), 103–112 (1986)
14. Zwerger, T., Mercorelli, P.: Combining SMC and MTPA using an EKF to estimate parameters and states of an interior PMSM. In: Proceedings of the 20th International Carpathian Control Conference (ICCC), pp. 1–6 (2019)

Neural Network Control by Error-Feedback Learning for Hydrostatic Transmissions with Disturbances and Uncertainties

Ngoc Danh Dang and Harald Aschemann(✉)

Chair of Mechatronics, University of Rostock,
Justus-von-Liebig-Weg 6, 18059 Rostock, Germany
{Ngoc.Dang,Harald.Aschemann}@uni-rostock.de,
https://www.com.uni-rostock.de/

Abstract. This paper presents a decentralized control approach based on a neural network for a hydrostatic transmission. The bent-axis angle of the hydraulic motor is adjusted by a pure feedforward control law based on identified physical parameters, whereas the corresponding motor angular velocity is controlled using a combination of a generalized proportional-derivative (PD) controller and a multilayer perceptron with one hidden layer that is trained by an error-feedback learning approach and uses only measurable input variables. In this observer-free control structure, the neural network learns the inverse dynamics by minimizing the PD controller output and, as a consequence, an accurate tracking of the desired trajectory is achieved. As no physical modelling is required for the motor velocity control design, it can be considered as model-free. The tracking performance shows the robustness of the overall control structure for the hydrostatic transmission despite disturbances and uncertainties. The proposed control scheme is investigated by simulations first. Second, experimental results are presented taken from a dedicated test rig at the Chair of Mechatronics, University of Rostock. Finally, an experimental comparison with results from previous work is provided.

Keywords: Mechatronics · Hydrostatic transmission · Neural network · Learning control · Nonlinear control

1 Introduction

A hydrostatic transmission represents a closed hydraulic circuit that involves a hydraulic pump and a hydraulic motor connected by hydraulic hoses, see Fig. 1. Here, the volumetric displacements of both pump and motor can be altered continuously. The hydraulic pump is typically driven by either a combustion engine or an electric motor.

A. Bartoszewicz et al. (Eds.): PCC 2020, AISC 1196, pp. 436–448, 2020.
https://doi.org/10.1007/978-3-030-50936-1_37

Hydrostatic transmissions are frequently employed in the drive trains of heavy working machines, construction and agriculture machinery and off-road vehicles, cf. [1]. Nowadays, they are present also in applications such as wind turbines, cf. [2], and power-split gearboxes, cf. [3]. Hydrostatic transmissions possess many technical features offering advantages for applications over conventional mechanical gearboxes: they provide a continuously variable transmission ratio, a high power density, a directional reversion without changing gears, and they are able to serve as a wearless braking system, see [4]. The axial piston type is the most popular structure of hydraulic pump and motor. With this structure, the transmission ratio can be adjusted by changing the swashplate angle of the hydraulic pump, by altering the bent-axis angle of the hydraulic motor or by changing both simultaneously by means of the displacement units. As a result, the system becomes a multivariable control system. Typically, either the output torque or the angular velocity of the hydraulic motor represent one of the controlled outputs. Hydrostatic transmissions with flexible connections by reinforced rubber hoses offer also a high flexibility regarding the geometric arrangement, which make hydrostatic transmissions attractive in industrial applications. However, energy efficiency and control issues still need to be addressed properly. From a control point of view, hydrostatic transmissions are highly nonlinear, undergoing disturbances such as load torque and fluid leakage flow and unavoidable uncertainty of physical parameters such as fluid viscosity or elasticity of the hydraulic hoses. These characteristics ask for advanced nonlinear control methods regarding an accurate tracking control.

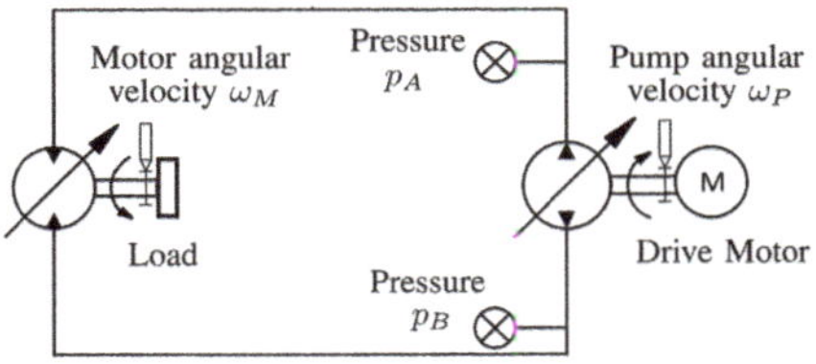

Fig. 1. Hydrostatic transmission system: topology and components.

Fig. 2. Hydrostatic transmission test rig at Chair of Mechatronics, University of Rostock.

For the control of hydrostatic transmissions, gain-scheduled-PID controllers are still predominant in industrial practice, see [1]. If a higher control performance is required, however, conventional PID controllers are not qualified. In the last decade, a lot of alternative nonlinear control approaches have been successfully implemented and validated. A general overview as well as corresponding list of references can be found in [5]. Regarding the control structure, both centralized and decentralized topologies are applicable and enable an accurate tracking control. However, as investigated and shown in [5], decentralized control approaches of the motor bent-axis angle and the motor angular velocity outperform the centralized control approaches: a higher tracking performance can be achieved, and the implementation simplifies in comparison to the centralized approach. The contribution [6] presents a decentralized control approach for the motor angular velocity with three alternative estimator schemes: a disturbance observer, a neural network as suggested in [7] and an online parameter estimation as discussed in [8]. These three approaches perform equivalently and with a comparable tracking performance w.r.t. other classical model-based approaches. The main advantage, however, is given by the fact that less physical model knowledge is necessary: only some characteristics of the system like the relative degree are required, which improves the capability of the controller to deal with disturbances and uncertainties affecting the system.

In this contribution, a new decentralized control scheme is investigated: A simple feedforward control is deployed for the bent-axis angle of the hydraulic motor, whereas a neural network control is designed for the motor angular velocity using error-feedback learning (EFL) as proposed in [9]. The benefits are as follows:

- The neural network using EFL does not require any physical information of the system and is, hence, model-free as opposed to the one in [6].
- A state and disturbance estimator is no longer necessary, which simplifies the implementation.

The tracking control performance is validated by means of both simulation and experiments on a dedicated test rig which is available at Chair of Mechatronics, University of Rostock, see Fig. 2.

The paper is organized as follows: A control-oriented model of a hydrostatic transmission that serves for simulation studies is presented briefly in Sect. 2. The decentralized control structure with the application of the error-feedback learning is discussed in Sect. 3. The results of simulations and experiments are discussed and compared with previous results in Sect. 4 and, finally, the paper finishes with conclusions in Sect. 5.

2 Simulation Model of Hydrostatic Transmission

This section presents briefly the modelling of the HST system. More details can be found in [5].

2.1 Hydraulic Subsystem

The hydraulic subsystem includes models of the hydraulic pump, driven by an electric servo motor, the hydraulic motor and the connecting hydraulic hoses.

Pump Flow Rate. The pump flow rate q_P is determined by a nonlinear function

$$q_P = \frac{V_P(\alpha_P)\omega_P}{2\pi}, \tag{1}$$

with ω_P as the angular velocity of the pump. The nonlinear relationship between the volumetric displacement $V_P(\alpha_P)$ and the tilt angle α_P of the swashplate, see Fig. 3, becomes

$$V_P(\alpha_P) = N_P\, A_P\, D_P\, \tan(\alpha_{P,\max} \cdot \tilde{\alpha}_P), \tag{2}$$

with the normalized swashplate angle $\tilde{\alpha}_P = \alpha_P/\alpha_{P,\max}$. Here, the geometrical parameters are the effective piston area A_P, the diameter D_P of the piston circle, and the number N_P of pistons present in the pump. With the maximum volumetric displacement $\tilde{V}_P = \frac{N_P A_P D_P}{2\pi}$, the pump flow rate can be stated as

$$q_P = \tilde{V}_P \tan\left(\alpha_{P,\max} \cdot \tilde{\alpha}_P\right) \omega_P. \tag{3}$$

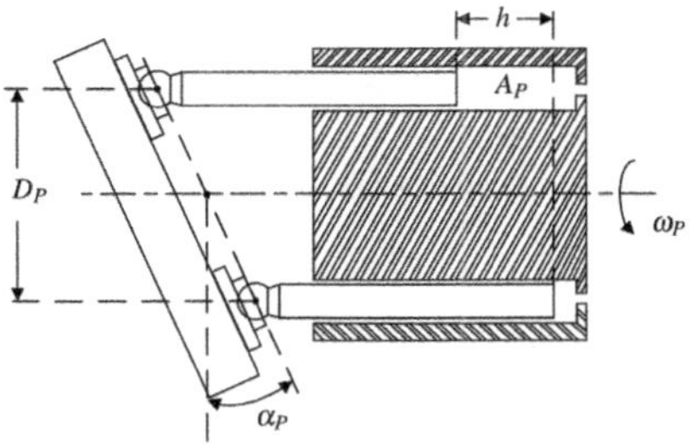

Fig. 3. Swashplate mechanism of the hydraulic pump.

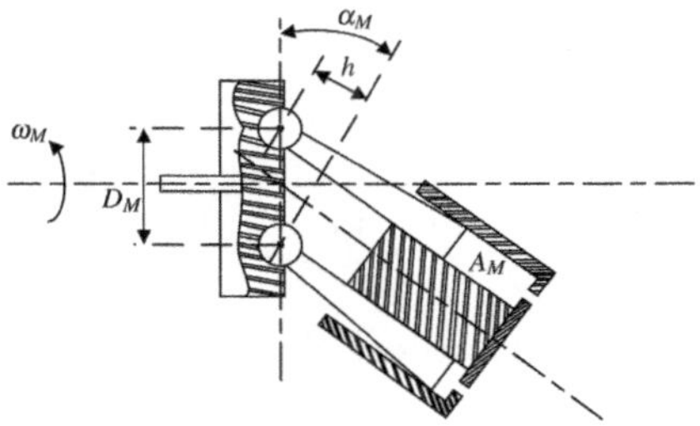

Fig. 4. Bent-axis mechanism of the hydraulic motor.

Motor Flow Rate. The hydraulic motor is of a bent-axis design, see Fig. 4. Therefore, similarly to the pump, the ideal volume flow rate q_M into the hydraulic motor is given by

$$q_M = \frac{V_M(\alpha_M)\omega_M}{2\pi}, \tag{4}$$

where $V_M(\alpha_M)$ represents the nonlinear volumetric displacement of the motor and ω_M the motor angular velocity. Given the geometrical parameters N_M, A_M, and D_M, the volume flow rate can be written as

$$q_M = \tilde{V}_M \sin\left(\alpha_{M,\max} \cdot \tilde{\alpha}_M\right) \omega_M. \tag{5}$$

Here, a normalized bent-axis angle is introduced according to $\tilde{\alpha}_M = \alpha_M/\alpha_{M,\max}$ as well as the maximum volumetric displacement $\tilde{V}_M = \frac{N_M A_M D_M}{2\pi}$, similar to the pump model.

Pressure Dynamics. The pressure dynamics involves the dynamics of the high-pressure and the low-pressure sides of the hydrostatic transmission. For practical reasons, the pressure dynamics is reduced to the dynamics of the difference pressure between the high and low pressure sides. The assumptions made here are symmetric physical conditions and negligible pressure losses in the hydraulic hoses. The differential equation for the difference pressure results in

$$\Delta\dot{p} = \frac{2}{C_H}\left(\tilde{V}_P \tan\left(\alpha_{P,\max}\cdot\tilde{\alpha}_P\right)\omega_P - \tilde{V}_M sin\left(\alpha_{M,\max}\cdot\tilde{\alpha}_M\right)\omega_M\right) - \frac{q_U}{C_H}, \quad (6)$$

where q_U is a lumped disturbance caused by the individual leakage flows, and C_H denotes the hydraulic capacitance.

Actuator Dynamics. The dynamics of the displacement units of the pump and the motor are represented by first-order lag models

$$\begin{aligned} T_{uP}\dot{\tilde{\alpha}}_P + \tilde{\alpha}_P &= k_P u_P, \\ T_{uM}\dot{\tilde{\alpha}}_M + \tilde{\alpha}_M &= k_M u_M. \end{aligned} \quad (7)$$

Here, T_{uP} and T_{uM} stand for the corresponding time constants, k_P and k_M are the proportional gains, and u_P and u_M the input voltages of the servo valves. In the given physical design, the angles are bounded by $\tilde{\alpha}_P \in [-1, 1]$ and $\tilde{\alpha}_M \in [\epsilon_M, 1]$, with $\epsilon_M > 0$.

2.2 Mechanical Subsystem

The mechanical aspects of the HST system are governed by the equation of motion for the motor

$$J_V\dot{\omega}_M + d_V\omega_M = \tilde{V}_M \Delta p \, \sin(\alpha_{M,\max}\cdot\tilde{\alpha}_M) - \tau_U, \quad (8)$$

where d_V is the damping coefficient, J_V the mass moment of inertia and τ_U stands for a lumped disturbance torque covering load disturbances and model uncertainty.

2.3 State-Space Model of the Overall System

Combining all subsystem discussed before, the dynamics of the HST system can be described by four first-order differential equations as follows

$$\begin{bmatrix} \dot{\tilde{\alpha}}_M \\ \dot{\tilde{\alpha}}_P \\ \Delta\dot{p} \\ \dot{\omega}_M \end{bmatrix} = \begin{bmatrix} -\frac{1}{T_{uM}}\tilde{\alpha}_M + \frac{k_M}{T_{uM}}u_M \\ -\frac{1}{T_{uP}}\tilde{\alpha}_P + \frac{k_P}{T_{uP}}u_P \\ \frac{2\tilde{V}_P}{C_H}\tan(\alpha_P)\omega_P - \frac{2\tilde{V}_M}{C_H}\sin(\alpha_M)\omega_M - \frac{q_U}{C_H} \\ -\frac{d_V}{J_V}\omega_M + \frac{\tilde{V}_M}{J_V}\sin(\alpha_M)\Delta p - \frac{\tau_U}{J_V} \end{bmatrix}, \quad (9)$$

where $\alpha_M = \tilde{\alpha}_M \cdot \alpha_{M,\max}$ as well as $\alpha_P = \tilde{\alpha}_P \cdot \alpha_{P,\max}$ have been used and the control inputs are given by u_P and u_M.

3 Decentralized Control Scheme

In the decentralized control scheme, the motor bent-axis angle and the angular velocity of the hydraulic motor are controlled separately.

3.1 Feedforward Control of the Motor Bent-Axis Angle

The dynamics of the hydraulic motor bent angle is characterized by a simple first-order lag system. Hence, in the implementation a feedforward control is sufficient. The inverse dynamics of the motor displacement unit becomes

$$u_M = \frac{1}{k_M}\tilde{\alpha}_M + \frac{T_{uM}}{k_M}\dot{\tilde{\alpha}}_M. \tag{10}$$

The feedforward control signal results in

$$u_{Md} = \frac{1}{k_M}\tilde{\alpha}_{Md} + \frac{T_{uM}}{k_M}\dot{\tilde{\alpha}}_{Md}, \tag{11}$$

where $\tilde{\alpha}_{Md}$ is desired trajectory of motor bent-axis angle, and $\dot{\tilde{\alpha}}_{Md}$ is its first time derivative, which is available from the trajectory generator.

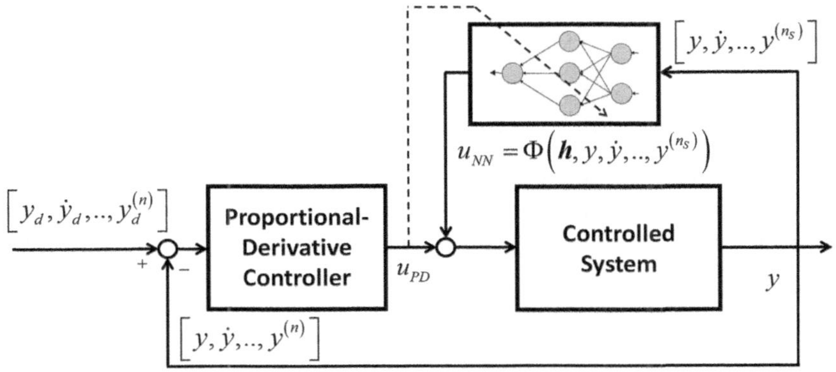

Fig. 5. Control structure using error-feedback learning.

3.2 Tracking Control of the Motor Angular Velocity Using a Neural Network

The angular velocity of the hydraulic motor is controlled using a neural network – a multiple layer perceptron (MLP) with one hidden layer – that is trained online by an error-feedback as proposed in [9]. As shown in Fig. 5, the control structure consists of two main parts - a conventional feedback controller and a neural network controller.

The conventional controller, which is chosen as a generalized proportional-derivative feedback of the tracking error, is designed to provide asymptotic stability and, at the same time, serves as an inverse reference model regarding the response of the physical system. The second controller, which is a neural network, learns the inverse dynamics of the controlled system using the output of the conventional controller as training signal. The neural network also receives the controlled output y and its first n_S time derivatives as the usual synaptic inputs. As the neural network acquires the inverse dynamics model of the physical system through learning, the response of the controlled system follows the desired trajectory and the proportional-derivative controller is relieved. The main benefits are given by the fact that changes in the inverse dynamics can be learned and that the proposed control represents a model-free one.

The output of the neural network, as a part of the overall control input of the controlled system, is a nonlinear function of synaptic inputs – the controlled output y and its first n_S time derivatives $y^{(j)}, j = 1, 2, ..., n_S$ – and the neural network weights $\boldsymbol{h}$. This control part can be expressed by the function

$$u_{NN} = \Phi\left(\boldsymbol{h}, y, \dot{y}, \ddot{y}, .., y^{(n_S)}\right). \tag{12}$$

The update rule for the neural network weights during the learning process is given in general form by the gradient-type law

$$\frac{d\boldsymbol{h}}{dt} = \eta \frac{\partial \Phi}{\partial \boldsymbol{h}} u_{PD}, \tag{13}$$

where the output u_{PD} of the proportional-derivative controller reflects the training error and η represents the learning rate, i.e., the step-size. It is assumed that:

- the learning rate η is small and positive,
- the generalized proportional-derivative controller is designed properly to guarantee the convergence of y towards the desired values y_d during the control process.

The stability of the overall control structure has been proved for a robotic application in a stochastic setting using averaging techniques and Lyapunov's second method, see [9] for more details.

Design of a Generalized Proportional-Derivative Controller. For the hydrostatic transmission, a classical proportional-derivative controller may be designed using the measured output, i.e., the motor angular velocity $y = \omega_M$. As confirmed by simulation and experimental results – in compliance with a relative degree of three of the angular velocity as controlled output –, the generalized proportional-derivative controller allows for a better performance if the first two time derivatives $\dot{\omega}$ and $\ddot{\omega}$ of the output are used as well in the feedback. With this specification, this controller can be stated as

$$u_{PD}(t) = k_0 e(t) + k_1 \dot{e}(t) + k_2 \ddot{e}(t), \tag{14}$$

where the tracking error $e(t)$ is defined by

$$e(t) = \omega_{Md}(t) - \omega_M(t). \tag{15}$$

Here, ω_{Md} is the desired value of the motor angular velocity. This value and its derivatives are provided by a trajectory generator in such a way that they are feasible and no input saturation is attained. The feedback gains k_0, k_1, k_2 are properly chosen by pole placement to obtain a Hurwitz polynomial corresponding to an asymptotically stable third-order error dynamics

$$\left[k_0 + k_1 s + k_2 s^2 + s^3\right] \ e(s) = \Delta(s), \tag{16}$$

which is excited by a lumped disturbance term $\Delta(s)$ accounting for remaining compensation errors.

Design of the Neural Network Controller. In this control structure, a multilayer perceptron (MLP) is employed, which represents a popular universal approximation approach and allows for a straight-forward practical implementation, cf. [10]. The neural network shown in Fig. 6 consists of three layers: an input layer, a hidden layer with a sigmoidal activation function and a linear output layer. These specifications allow to express the output u_{NN} of the neural network as

$$u_{NN} = \Phi\left(\boldsymbol{h}, y, \dot{y}, \ddot{y}, .., y^{(n_S)}\right) = \sum_{j=1}^{L} w_j \sigma\left(\mathbf{v}_j^T \mathbf{x}\right), \tag{17}$$

where L is the number of neurons in the hidden layer, w_j is the j-th weight of the weighting vector $\boldsymbol{w}$ of the output layer, $\mathbf{v}_j$ denotes the j-th column of the input layer weighting matrix $\mathbf{V}$ that corresponds to the j-th neuron. The sigmoidal activation function in the hidden layer is denoted as $\sigma(\cdot)$, and the input vector $\mathbf{x}$ is defined by

$$\mathbf{x} = \left[\omega_M \ \dot{\omega}_M\right]^T, \tag{18}$$

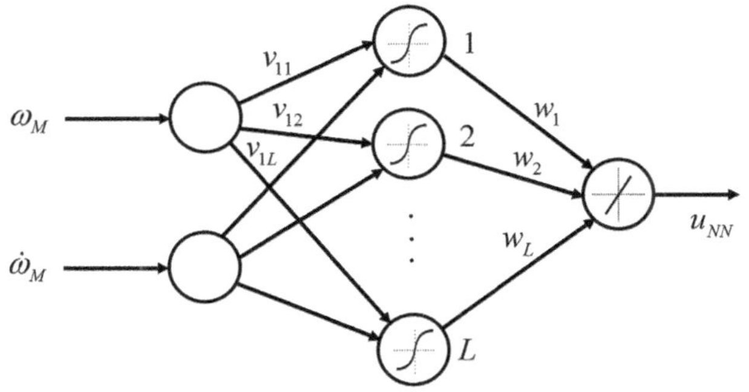

Fig. 6. The implemented multilayer perceptron.

which corresponds to $n_S = 1$. In practice, as indicated in dedicated experiments, second and higher time derivatives may amplify measurement noise and, hence, may deteriorate the learning behaviour of the neural network. To circumvent this problem, in the selected MLP structure only the angular velocity and its first time derivative are used as synaptic inputs as defined by vector $\mathbf{x}$ in (18).

Using the backpropagation technique for the tuning of the weights, the gradients of objective function Φ w.r.t. each individual weight can be derived after some mathematical operations as follows

$$\frac{\partial \Phi}{\partial \mathbf{w}} = \boldsymbol{\delta}_1, \quad \frac{\partial \Phi}{\partial \mathbf{V}} = \mathbf{x} \cdot \mathbf{w}^T \cdot \boldsymbol{\delta}_2. \tag{19}$$

Here, $\boldsymbol{\delta}_1$ and $\boldsymbol{\delta}_2$ are defined by

$$\boldsymbol{\delta}_1 = \sigma\left(\mathbf{V}^T \mathbf{x}\right), \quad \boldsymbol{\delta}_2 = diag\left[\sigma'\left(\mathbf{V}^T \mathbf{x}\right)\right]. \tag{20}$$

In the last expression, $\sigma'(\cdot)$ denotes the derivative of the sigmoidal activation function $\sigma(\cdot)$ w.r.t. its input. The weights tuning rules are evaluated according to (13)

$$\dot{\mathbf{w}} = \eta \cdot \boldsymbol{\delta}_1 \cdot u_{PD}, \quad \dot{\mathbf{V}} = \eta \cdot \mathbf{x} \cdot \mathbf{w}^T \cdot \boldsymbol{\delta}_2 \cdot u_{PD}. \tag{21}$$

The overall control law for the motor angular velocity is given by the sum of the input signals from both the proportional-derivative controller and the neural network $u_P = u_{PD} + u_{NN}$, see Fig. 7, showing the implementation. Please note that no physical knowledge is required anymore in contrast to previous control approaches like those in [5] and [6]. Moreover, also a state and disturbance observer is no longer necessary.

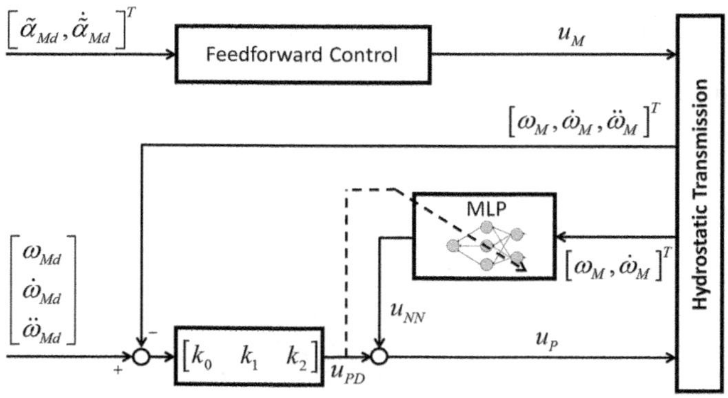

Fig. 7. Block diagram of the overall control scheme: feedforward control of the bent-axis angle and model-free control of the motor angular velocity using the combination of a generalized proportional-derivative output error feedback and a MLP neural network.

4 Simulation Study and Validation

The benefits of the proposed control structure shall be pointed out in the sequel by means of simulations and experiments at a dedicated test rig.

4.1 Simulation Results

The simulations are conducted using the nonlinear system model of the hydrostatic transmission derived in Sect. 2. Here, a simulation step size of 0.02 s is employed. To obtain realistic and reliable results, measurement noise is also added to the output signal. The hidden layer of the MLP neural network consists of 10 neurons.

For the motor bent-axis angle and the motor angular velocity, smooth desired trajectories are designed to avoid any saturation of the displacement units due to the limits of mechanical design. Saturating inputs are not taken into account in this study.

Figure 8 shows a comparison of simulated and desired values for the tilt angle of the hydraulic motor. As can be seen, feedforward control alone is capable to achieve a good tracking of the bent-axis angle of the hydraulic motor. The tracking behaviour regarding the motor angular velocity is shown in Fig. 9, which indicates a high performance of the proposed control structure in comparison to a large increase in the tracking error if the control action of the neural network is deactivated. Using the overall feedback – the sum of the generalized proportional-derivative error feedback and the online-trained neural network–, the absolute value of the maximum tracking error is below 0.5 rad/s and indicates the effectiveness of the model-free EFL approach regarding the compensation of nonlinearities, disturbances and model uncertainty. To assess the control performance of the proposed observer-free control structure using error-feedback learning (EFL), a comparison with the result of previous work, see [6] for details, is provided in the sequel. The alternative approaches in [6] involve simulations of different estimators – a neural network with an adaptive learning law (NN), a disturbance observer (DO) and recursive parameter estimation (PE). Please note that all the solutions presented in [6] estimate an unknown term in the inverse dynamics but still require some physical knowledge of the system and use estimates from a state and disturbance observer. The simulated tracking errors are depicted in Fig. 10 together with the proposed solution of this paper (EFL). Moreover, the root-mean square (RMS) values and the maximum tracking errors are stated in Table 1. The differences between the error signals are small but the implementation effort of the EFL approach is significantly lower.

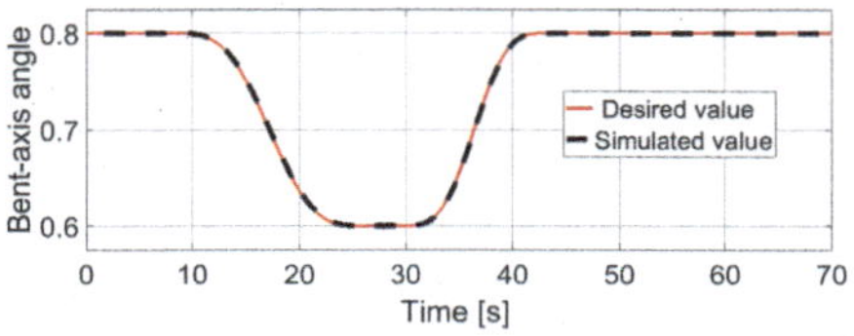

Fig. 8. Comparison of simulated and desired values regarding the motor bent-axis angle $\tilde{\alpha}_M$.

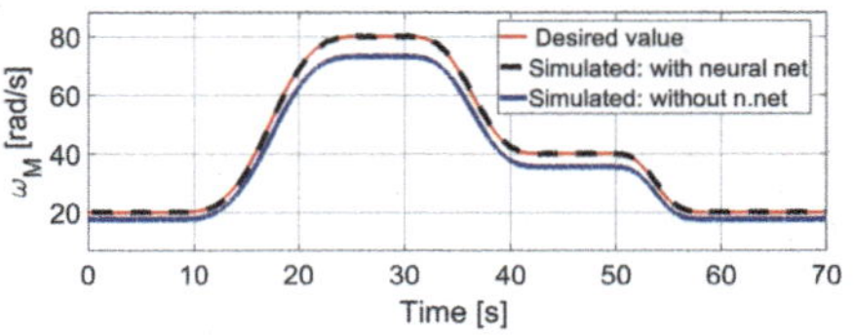

Fig. 9. Comparison of simulated with/without neural network and desired values of the motor angular velocity ω_M.

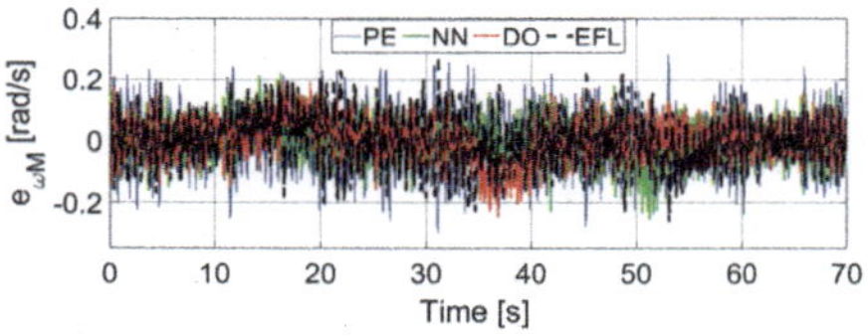

Fig. 10. Simulation results for the achieved tracking errors w.r.t. the motor angular velocity: MLP with error-feedback learning (EFL) in comparison to previous results from [6]: disturbance observer (DO), neural network (NN) using an adaptive learning law, and recursive parameter estimation (PE).

Table 1. Comparison of error measures (simulation results).

	NN	PE	DO	EFL
Max. error in rad/s	0.2534	0.2962	0.2470	0.2809
RMS error in rad/s	0.06378	0.08798	0.0627	0.07702

4.2 Experimental Results and Validation

The complete control structure is implemented on the test rig and runs at a sampling time of 0.02 s using the same desired trajectories. The required time derivatives of the measured angular velocity are calculated by real differentiation in the form of filtered derivatives. The corresponding experimental results are shown in Fig. 11 up to Fig. 13.

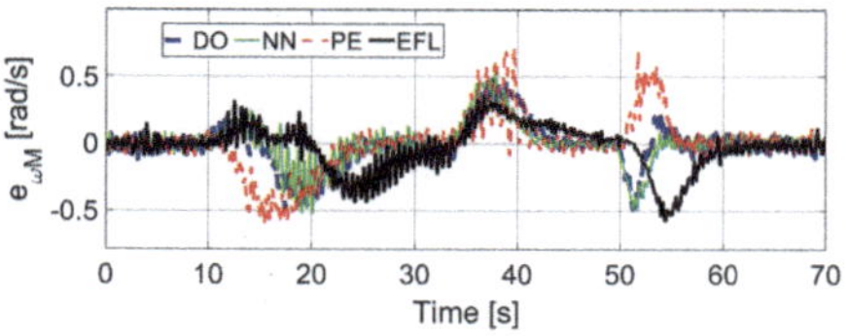

Fig. 11. Experimental results for the achieved tracking errors w.r.t. the motor angular velocity: MLP with error-feedback learning (EFL) in comparison to previous results from [6]: disturbance observer (DO), neural network (NN) using an adaptive learning law, and recursive parameter estimation (PE).

Table 2. Comparison of error measures (experimental results).

	NN	PE	DO	EFL
Max. error in rad/s	0.5469	0.7215	0.5148	0.5852
RMS error in rad/s	0.1492	0.2044	0.1621	0.1685

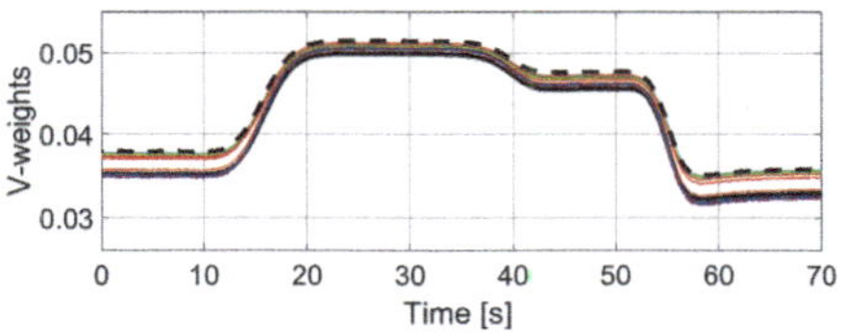

Fig. 12. Variation of **V**-weights during trajectory tracking.

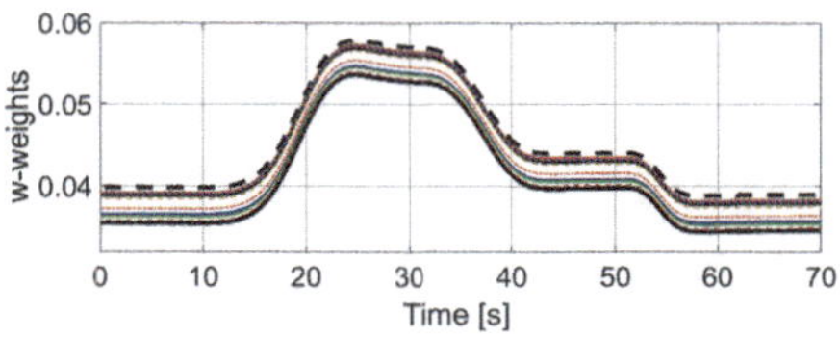

Fig. 13. Variation of **w**-weights during trajectory tracking.

To assess the control performance of the proposed observer-free control structure using error-feedback learning (EFL), an comparison with the result of previous work, cf. [6], is performed based on implementations on the test rig. The tracking errors obtained at the test rig are depicted in Fig. 11 together with the one from the proposed solution of this paper (EFL). Moreover, the root-mean square (RMS) values and the maximum tracking errors are stated in Table 2. Figures 12 and 13 illustrate the corresponding variations of the neural network weights during the tracking process. It can be concluded that the proposed model-free control structure – that does not require any physical knowledge for the control of the angular velocity – attains a similarly high tracking accuracy as the others. Taking into account the reduced implementation effort, the slightly larger maximum tracking error is still acceptable. The observer-free control system offers a good compensation regarding model nonlinearities, disturbances like leakage flows and friction torque as well as model uncertainty.

5 Conclusions

The proposed decentralized control structure consists of a feedforward control for the motor bent-axis angle and a neural network tracking control of the motor angular velocity using error-feedback learning based on backpropagation techniques. The control of the motor angular velocity does not involve any model knowledge and is, hence, model-free. Without any state and disturbance observer or estimator, as required in model-based control approaches investigated in former research, the overall control structure is simplified and, hence, requires less implementation effort. Only the angular velocity needs to be measured, whereas the first time derivative is determined by means of real differentiation. As a result of the online-training of the MLP neural network and the subsequent nonlinearity compensation, this control approach is highly capable of counteracting disturbances, nonlinearities and model uncertainty affecting the system dynamics.

The successful compensation of such effects is shown by a high tracking performance achieved in both simulations and experiments on a dedicated hydrostatic transmission test rig at the Chair of Mechatronics, University of Rostock.

References

1. Schulte, H.: Control-oriented modeling of hydrostatic transmission using Takagi-Sugeno fuzzy systems. In: Proceedings of IEEE International Fuzzy Systems Conference, London, pp. 1–6 (2007)
2. Schulte, H.: Control-oriented description of large scale wind turbines with hydrostatic transmission using Takagi-Sugeno models. In: Proceedings of IEEE Conference on Control Applications (CCA), Antibes, pp. 664–668 (2014)
3. Shamshirband, S., Petković, D., Amini, A., Anuar, N., Nikolić, V., Ćojbašić, Z., Kiah, M., Gani, A.: Support vector regression methodology for wind turbine reaction torque prediction with power-split hydrostatic continuous variable transmission. Energy **67**, 623–630 (2014)
4. Aschemann, H., Sun, H.: Decentralised flatness-based control of a hydrostatic drive train subject to actuator uncertainty and disturbances. In: Proceedings of 18th International Conference on Methods and Models in Automation and Robotics (MMAR), Miedzyzdroje, pp. 759–764 (2013)
5. Sun, H.: Decentralized nonlinear control for a hydrostatic drive train with unknown disturbances. Ph.D, thesis, University of Rostock, Shaker (2015)
6. Dang, N.D., Aschemann, H.: Comparison of estimator-based compensation schemes for hydrostatic transmissions with uncertainties. In: Proceedings of 23rd International Conference on Methods and Models in Automation and Robotics (MMAR), Miedzyzdroje, pp. 692–697 (2018)
7. Lewis, F.L., Jagannathan, S., Yesildirek, A.: Neural Network Control of Robots and Nonlinear Systems. Taylor and Francis, London (1999)
8. Slotine, E., Li, W.: Applied Nonlinear Control. Prentice Hall, Upper Saddle River (1991)
9. Gomi, H., Kawato, M.: Learning control for a closed loop system using feedback-error-learning. In: Proceedings of 29th IEEE Conference on Decision and Control, Honolulu, Hawaii, vol. 6, pp. 3289–3294 (1990)
10. Pinkus, A.: Approximation theory of the MLP model in neural networks. Acta Numerica **8**, 143–195 (1999)

Motion Control with Hard Constraints – Adaptive Controller with Nonlinear Integration

Jacek Kabziński(✉) and Przemysław Mosiołek

Institute of Automatic Control, Lodz University of Technology, Łódź, Poland
jacek.kabzinski@p.lodz.pl

Abstract. The paper presents an adaptive controller designed for a nonlinear servo in the presence of hard state constraints. The proposed approach is based on a nonlinear state-space transformation and adaptive backstepping. It allows achieving UUB tracking of any reference trajectory inside the constraints, in spite of unknown plant parameters. Three control schemes, each using integral action differently, are designed and compared. Several examples demonstrate the main features of the design procedure and prove that it may be applied in practical motion control problems.

Keywords: Motion control · Nonlinear control · Adaptive control · State constraints

1 Introduction

High-accuracy control of position and velocity of various machines, vehicles, servo drives, robotic arms, and other devices remains among the most important problems of control theory and applications. Load and disturbing torques or forces affecting the drive are usually nonlinear (like friction), and the parameters (like mass or inertia) may be unknown or changing. Therefore, nonlinear models of motion are widely used and nonlinear adaptive control is applied to design the controller.

Hard position and velocity constraints are imposed in any practical motion control problem. Safe operation requires preserving the constraints during any possible transient trajectory. Therefore nonlinear adaptive control with hard state constraints is the main tool for motion controller design.

Several control techniques have been developed to control a nonlinear plant with state constraints. Methods based on set invariance [1], admissible set control [2, 3], model predictive control [4, 5] and reference governors [6, 7] lead to complex numerical algorithms and are difficult to apply in practice.

Barrier Lyapunov functions (BLF) approach is used together with the backstepping technique to handle different (state [8, 9] or output [10, 11], constant or varying [12, 13]) constraints. This method is used for motion control in [14, 15]. Unfortunately, if the tracking problem is considered, the BLF approach allows imposing constraints for

A. Bartoszewicz et al. (Eds.): PCC 2020, AISC 1196, pp. 449–461, 2020.
https://doi.org/10.1007/978-3-030-50936-1_38

the error system. Hence, effective constraints for state variables must be derived taking into account the reference trajectory and the controller parameters used in successive control loops. The obtained controller may be useful only if it is possible to find a set of design parameters that satisfies the so-called "feasibility conditions" [16, 17].

The method to handle hard state constraints proposed in this contribution is based on a nonlinear state-transformation, such that if the new state variables are bounded the original state variables are inside the imposed constraints. The desired trajectory tracking problem is solved using the backstepping approach with command control filtering [18, 19]. This approach allows proving the uniform, ultimate boundedness (UUB) [20] of the error system. It is sufficient for the practical applicability of the derived controller, although a bounded quasi-steady-state tracking error is expected. To reduce this error the integrator is used at the initial stage of backstepping. Unfortunately, it is observed that the linear integrator which reduces the quasi-steady-state tracking error results in more aggressive control and higher oscillations, especially at the initial part of the trajectories. Therefore we propose to moderate the integral action by a nonlinear function of the actual tracking error. The influence of such modification is discussed in detail.

2 Problem Statement

The motion dynamics is modelled by differential equations

$$\begin{aligned} \dot{x}_1 &= x_2, \\ J\dot{x}_2 &= A^T \xi(x_1, x_2, t) + gu, \end{aligned} \tag{1}$$

where the angular or linear position is denoted by x_1, the rotational or linear speed by x_2, J corresponds to the system inertia, $g > 0$ represents the transformation of the control input u into the propulsion torque or force. The component $f = A^T \xi(x_1, x_2, t)$ describes all load or disturbance torques/forces like friction, torque/force ripples, etc. The function ξ is known. Unknown, constant parameters in α-dimensional vector A will be approximated by the adaptive parameters. It is assumed that also parameters J and g are unknown, although positive and constant. Such an equation may be used to model rotational or linear motion of various plants with different sources of propulsion.

Hard, inviolable, asymmetric constraints are considered for both state variables:

$$-b_{i,1} < x_i < b_{i,2}, i = 1, 2. \tag{2}$$

The control aim is to follow the desired, smooth trajectory $x_{1d}, \dot{x}_{1d}$ with a sufficient accuracy preserving the constraints (2).

3 Nonlinear State Transformation Preserving the Constraints

New state variables are defined by the following nonlinear transformation:

$$s_i = \ln \frac{b_{i,1} + x_i}{b_{i,2} - x_i}, i = 1, 2. \tag{3}$$

The transformation is invertible:

$$x_i = b_{i,2} - \frac{b_{i,2} + b_{i,1}}{e^{s_i} + 1} = -b_{i,1} + \frac{\left(b_{i,2} + b_{i,1}\right)}{e^{-s_i} + 1} \tag{4}$$

and it is obvious that s_i approaches $-\infty$ if $x_i \to -b_{i,1}$ and ∞ if $x_i \to b_{i,2}$ and vice versa. Therefore, if the new state variables are bounded, the original state variables remain inside the constraints.

The time-derivatives of the transformed state variables are given by

$$\dot{s}_i = \frac{d}{dt} \ln \frac{b_{i,1} + x_i}{b_{i,2} - x_i} = \frac{2 + e^{-s_i} + e^{s_i}}{b_{i,2} + b_{i,1}} \dot{x}_i =: k_i(s_i)\dot{x}_i. \tag{5}$$

It is important that the functions $k_i(s_i)$ are strictly positive for any $-\infty < s_i < \infty$

$$k_i(s_i) = \frac{2 + e^{-s_i} + e^{s_i}}{b_{i,2} + b_{i,1}} \geq \frac{4}{b_{i,2} + b_{i,1}}. \tag{6}$$

The derivatives of $k_i(s_i)$ will be used for the controller design, but fortunately they may be easily obtained as:

$$h_i(s_i) := \frac{d}{ds_i} k_i(s_i) = \frac{-e^{-s_i} + e^{s_i}}{b_{i,2} + b_{i,1}}, \tag{7}$$

$$\frac{d}{dt} k_i(s_i) = h_i(s_i)\dot{s}_i = h_i(s_i)k_i(s_i)\dot{x}_i. \tag{8}$$

A similar transformation is applied to the reference trajectory

$$s_{1d} := \ln\left(\frac{b_{1,1} + x_{1d}}{b_{1,2} - x_{1d}}\right), \quad \dot{s}_{1d} = k_1(s_{1d})\dot{x}_{1d}. \tag{9}$$

4 Controller Derivation

In spite of the original tracking error $e := x_{1d} - x_1$, a similar one for the transformed state variable is defined:

$$\begin{aligned} \epsilon_1 &:= s_{1d} - s_1, \\ \dot{\epsilon}_1 &= k_1(s_{1d})\dot{x}_{1d} - k_1(s_1)\dot{x}_1. \end{aligned} \tag{10}$$

The backstepping scheme [21, 22] is used to derive the controller. The integral action is added at the initial stage, so finally, three stages are necessary to design the control law.

Stage 1

The transformed tracking error modified by the original tracking error is integrated at the initial stage of backstepping. The additional state variable is defined by:

$$p := \int_0^t F\left(e^2(\tau)\right)\epsilon_1(\tau)d\tau, \tag{11}$$

where $F(e^2) \geq 0$ denotes a nonnegative scaling factor. It should be close to zero if the tracking error e is large, and close to 1 if it's argument is small. Hence, $F(e^2)$ may be considered as a kind of nonlinear anti-wind-up factor, excluding high levels of the integrator output during a transient and allowing to minimize quasi-steady-state error. The influence of a particular shape of $F(e^2)$ on the system performance is discussed later.

The motion of p is obviously given by $\dot{p} = F(e^2)\epsilon_1$. Unfortunately, it is impossible to control p by ϵ_1 directly, so the desired trajectory ϵ_{1d} of ϵ_1 is defined and the tracking error is $\epsilon_0 := \epsilon_1 - \epsilon_{1d}$. Under this notation

$$\dot{p} = F\left(e^2\right)\epsilon_1 = F\left(e^2\right)(\epsilon_{1d} + \epsilon_0). \tag{12}$$

Selecting a desired trajectory ϵ_{1d}

$$\epsilon_{1d} = -L_0 p, \;\; L_0 > 0, \tag{13}$$

where $L_0 > 0$ is a design parameter, guaranties the stable motion of p if ϵ_0 is negligible, as it follows from the Lyapunov function:

$$V_0 = \frac{1}{2}p^2 \Rightarrow \dot{V}_0 = -L_0 F\left(e^2\right)p^2 + F\left(e^2\right)p\epsilon_0. \tag{14}$$

Several other choices of ϵ_{1d} are possible, but the one given in (13) is simple and assures the system stability.

Stage 2

As $\dot{\epsilon}_{1d} = -L_0\dot{p} = -L_0 F(e^2)\epsilon_1$, the tracking error from the first loop ϵ_0 moves according to the equation

$$\dot{\epsilon}_0 = \dot{\epsilon}_1 - \dot{\epsilon}_{1d} = k_1(s_{1d})\dot{x}_{1d} - k_1(s_1)x_2 + L_0 F\left(e^2\right)\epsilon_1. \tag{15}$$

A linear filter is used to create a necessary, virtual control in (15). The filter input is the second transformed state variable s_2 and the filter is defined by:

$$\dot{z} = -C(z - s_2), \;\; C > 0, \;\; z(0) = s_2(0), \;\; \rho := s_2 - z. \tag{16}$$

When the filter transition state is over $z \approx s_2$. The gap ρ may be arbitrarily narrowed by a proper choice of the filter parameter C (if $|\dot{s}_2| \leq c < \infty$ than $|\rho| \leq \frac{c}{C}$ [19]). Therefore, it is assumed that $|\rho| \leq \rho_{max} < \infty$.

The signals z and s_2 are added and subtracted in (15). This allows to use s_2 as a virtual control in (15). The desired trajectory for the virtual control s_2 is denoted by s_{2d} and the tracking error by

$$\epsilon_2 := s_{2d} - s_2. \tag{17}$$

So, (15) is transformed into

$$\dot{\epsilon}_0 = k_1(s_{1d})\dot{x}_{1d} - k_1(s_1)x_2 + L_0 F\left(e^2\right)\epsilon_1 - s_{2d} + \epsilon_2 + z + \rho. \tag{18}$$

The stability of the subsystem (12, 18) is obtained by the use of Lyapunov function

$$V_1 = V_0 + \frac{1}{2}\epsilon_0^2. \tag{19}$$

The derivative of V_1 along the subsystem trajectories (p, ϵ_0) is given by

$$\begin{aligned}\dot{V}_1 &= -L_0F\left(e^2\right)p^2 + F\left(e^2\right)p\epsilon_0 + \epsilon_0(k_1(s_{1d})\dot{x}_{1d} - k_1(s_1)x_2)\\ &\quad + \epsilon_0L_0F\left(e^2\right)\epsilon_1 + \epsilon_0(-s_{2d} + \epsilon_2 + z + \rho),\end{aligned} \tag{20}$$

hence, selecting

$$s_{2d} = (k_1(s_{1d})\dot{x}_{1d} - k_1(s_1)x_2) + z + F\left(e^2\right)p + L_0F\left(e^2\right)\epsilon_1 + L_1\epsilon_0, \tag{21}$$

where $L_1 > 0$ is a design parameter, reduces (20) to

$$\dot{V}_1 = -L_0F\left(e^2\right)p^2 - L_1\epsilon_0^2 + \epsilon_0\epsilon_2 + \epsilon_0\rho \tag{22}$$

and transforms (18) into

$$\dot{\epsilon}_0 = -L_1\epsilon_0 - F\left(e^2\right)p + \epsilon_2 + \rho = -L_1\epsilon_0 - F\left(e^2\right)p + s_{2d} - z. \tag{23}$$

Although the derivative of s_{2d} is quite complex:

$$\begin{aligned}\dot{s}_{2d} &= \tfrac{d}{dt}[k_1(s_{1d})\dot{x}_{1d} - k_1(s_1)x_2] + \tfrac{d}{dt}\left[z + F\left(e^2\right)p + L_0F\left(e^2\right)\epsilon_1 + L_1\epsilon_0\right]\\ &= (h_1(s_{1d})k_1(s_{1d})\dot{x}_{1d})\dot{x}_{1d} + k_1(s_{1d})\ddot{x}_{1d} - h_1(s_1)k_1(s_1)\dot{x}_1x_2 - k_1(s_1)\dot{x}_2\\ &\quad + \dot{z} + F\left(e^2\right)\dot{p} + 2epF'\left(e^2\right)\dot{e} + L_0F\left(e^2\right)\dot{\epsilon}_1 + 2L_0e\epsilon_1F'\left(e^2\right)\dot{e} + L_1\dot{\epsilon}_0,\end{aligned} \tag{24}$$

it may be simplified by plugging in expressions for $\dot{x}_1, \dot{x}_2, \dot{\epsilon}_1, \dot{\epsilon}_0$ and represented in a compact form

$$\dot{s}_{2d} = G - \frac{k_1(s_1)}{J}\left(A^T\xi + gu\right). \tag{25}$$

where

$$\begin{aligned}G :=\ & h_1(s_{1d})k_1(s_{1d})\dot{x}_{1d}^2 + k_1(s_{1d})\ddot{x}_{1d} - h_1(s_1)k_1(s_1)x_2^2 + \left[F\left(e^2\right)\right]^2\epsilon_1\\ & - L_1^2\epsilon_0 - L_1p - C(z - s_2) - L_1[z - s_{2d}] + L_0F\left(e^2\right)\left[s_{2d} - z - F\left(e^2\right)p\right]\\ & + L_0F\left(e^2\right)\left[-L_0F\left(e^2\right)\epsilon_1 - L_1\epsilon_0\right] + 2eF'\left(e^2\right)(p + L_0\epsilon_1)(\dot{x}_{1d} - x_2)\end{aligned} \tag{26}$$

is composed of the known signals and parameters.

Stage 3

The error dynamics is given by

$$\begin{aligned}\dot{\epsilon}_2 &= \tfrac{d}{dt}(s_{2d} - s_2) = G - k_2(s_2)\dot{x}_2 - \tfrac{k_1(s_1)}{J}\left(A^T\xi + gu\right)\\ &= G - \tfrac{k_1(s_1)}{J}\left(A^T\xi + gu\right) - \tfrac{k_2(s_2)}{J}\left(A^T\xi + gu\right)\\ &= G - \tfrac{[k_1(s_1) + k_2(s_2)]}{J}\left(A^T\xi + gu\right)\end{aligned} \tag{27}$$

and may be represented by

$$\frac{J}{g}\dot{\epsilon}_2 = P^T\varphi - [k_1(s_1) + k_2(s_2)]u, \tag{28}$$

where

$$\begin{aligned} P^T &:= \left[\tfrac{J}{g}\ \tfrac{1}{g}A^T\right], \\ \varphi &:= \begin{bmatrix} G \\ -[k_1(s_1) + k_2(s_2)]\xi \end{bmatrix} \end{aligned} \tag{29}$$

are unknown constant parameters and known regressor, respectively. The unknown parameters will be substituted by adaptive parameters $\hat{P}$ and the error of adaptation is denoted by $\tilde{P} := P - \hat{P}$.

The final Lyapunov function is constructed using a positive definite matrix Γ

$$V_2 = V_1 + \frac{1}{2}\frac{J}{g}\epsilon_2^2 + \frac{1}{2}\tilde{P}^T\Gamma^{-1}\tilde{P}, \tag{30}$$

and its derivative equals

$$\begin{aligned} \dot{V}_2 = &-L_0F(e^2)p^2 - L_1\epsilon_0^2 + \epsilon_0\epsilon_2 + \epsilon_0\rho \\ &+ \epsilon_2\big(P^T\varphi - [k_1(s_1) + k_2(s_2)]u\big) + \tilde{P}^T\Gamma^{-1}\tfrac{d}{dt}\tilde{P}. \end{aligned} \tag{31}$$

The control

$$u = \frac{1}{k_1(s_1) + k_2(s_2)}\left(\hat{P}^T\varphi + L_2\epsilon_2 + \epsilon_0\right), \tag{32}$$

where $L_2 > 0$ and the identity $\frac{d}{dt}\tilde{P} = -\frac{d}{dt}\hat{P}$, reduce (32) to

$$\dot{V}_2 = -L_0F\left(e^2\right)p^2 - L_1\epsilon_0^2 - L_2\epsilon_2^2 + \epsilon_0\rho + \tilde{P}^T\left(\epsilon_2\varphi - \Gamma^{-1}\frac{d}{dt}\hat{P}\right). \tag{33}$$

If the robust adaptive law

$$\frac{d}{dt}\hat{P} = \Gamma\left(\epsilon_2\varphi - \sigma\|\bar{e}\|\hat{P}\right),\ \sigma > 0,\ \bar{e} = \left[\epsilon_0\ \epsilon_2\right]^T \tag{34}$$

is applied, the well-known equality $\tilde{P}^T\hat{P} = -\left\|\tilde{P}\right\|^2 + \|P\|^2 - \left\|\hat{P}\right\|^2$ and the notation $L_{min} = \min\{L_1, L_2\}$ alows to simplify the Lyapunov function derivative:

$$\begin{aligned} \dot{V}_2 &= -L_0F(e^2)p^2 - L_1\epsilon_0^2 - L_2\epsilon_2^2 + \tilde{P}^T\left(\epsilon_2\varphi - \Gamma^{-1}\Gamma\left(\epsilon_2\varphi - \sigma\|\bar{e}\|\hat{P}\right)\right) \\ &\quad + \epsilon_0\rho = -L_0F(e^2)p^2 - L_1\epsilon_0^2 - L_2\epsilon_2^2 + \epsilon_0\rho - \sigma\|e\|\tilde{P}^T\hat{P} \\ &= -L_0F(e^2)p^2 - L_1\epsilon_0^2 - L_2\epsilon_2^2 + \epsilon_0\rho - \tfrac{\sigma}{2}\|\bar{e}\|\left[-\left\|\tilde{P}\right\|^2 + \|P\|^2 - \left\|\hat{P}\right\|^2\right]. \\ &\le -L_0F(e^2)p^2 - L_{min}\|\bar{e}\|^2 + \rho\max\|\bar{e}\| + \tfrac{\sigma}{2}\|\bar{e}\|\left\|\tilde{P}\right\|^2 - \tfrac{\sigma}{2}\|\bar{e}\|\|P\|^2 \\ &= -L_0F(e^2)p^2 - \|\bar{e}\|\left[L_{min}\|\bar{e}\| - \rho\max + \tfrac{\sigma}{2}\left\|\tilde{P}\right\|^2 - \tfrac{\sigma}{2}\|P\|^2\right] \end{aligned} \tag{35}$$

The component $-L_0F(e^2)p^2$ is negative for $p \neq 0$ if $L_0F(e^2) > 0$ and equals 0 if $L_0F(e^2) = 0$. In the last case, if $L_0 = 0$, the integrator is simply switched off and $p = 0$ may be taken. As it is demonstrated in Fig. 1, for $L_0 = 0$, the variable p is not used by the controller, the $\epsilon_{1d} = -L_0 p = 0$, and $\epsilon_0 = \epsilon_1$. the error system consists of two state variables $\epsilon_1 = \epsilon_0$ and ϵ_2 and the controller derivation starts from stage 2. As the result, the controller structure without the integrator may be obtained from the previously derived by plugging in $L_0 = 0, p = 0$. If $F(e^2) = 0, L_0 > 0$, the integrator output is "frozen" and kept constant until $F(e^2) > 0$.

The component $-\|\bar{e}\|\left[L_{min}\|\bar{e}\| - \rho\max + \frac{\sigma}{2}\left\|\tilde{P}\right\|^2 - \frac{\sigma}{2}\|P\|^2\right]$ is negative for any $\tilde{P}$ if $\|\bar{e}\| > \frac{1}{L_{min}}(\rho\max + \frac{\sigma}{2}\|P\|^2)$. Also it is negative for any $\bar{e}$ if $\left\|\tilde{P}\right\| > \sqrt{\frac{2}{\sigma}\rho\max + \|P\|^2}$

Therefore, the derivation of the controller may be summarized by the corollary which follows from the Lyapunov theorem extensions [20]:

Corollary: Under the proposed control, the adaptive parameter errors $\tilde{P}$ and the tracking errors $\bar{e} = \left[\epsilon_0\ \epsilon_2\right]^T$ are uniformly ultimately bounded (UUB). Increasing the design parameter L_{min} allows reducing the limit set for $\bar{e}$ arbitrarily. The integrator output p remains bounded.

As the state variables $\bar{e} = \left[\epsilon_0\ \epsilon_2\right]^T$ are UUB and p is bounded, also $\epsilon_1 = \epsilon_0 - L_0 p$ is UUB and the design parameter L_{min} may be used to narrow the tracking error.

It follows (from the boundedness of the error variables, under the assumption that the desired trajectory $x_{1d}, \dot{x}_{1d}$ stays inside constraints (2)), that the state variables s_1, s_2 are bounded, and hence x_1 and x_2 stay inside constraints (2).

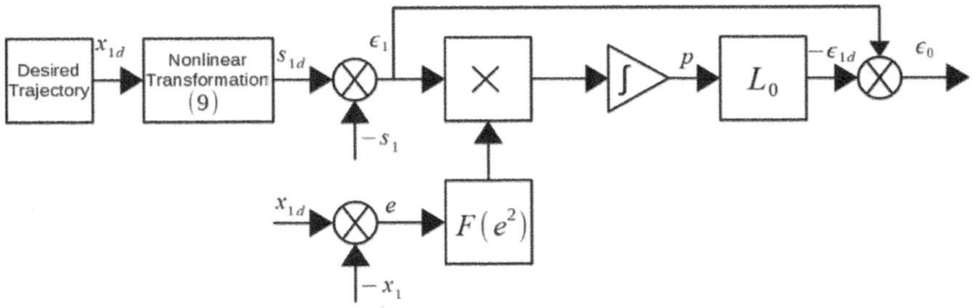

Fig. 1. The scheme of the integrator loop.

5 Examples

The considered plant is an electric motor drive, propelling a rigid arm working against the gravitational force. The model of the plant is described by the equation

$$\begin{aligned} \dot{x}_1 &= x_2, \\ J\dot{x}_2 &= -bx_2 + c\,\sin(x_1) - T_f(x_2) + k_i u, \end{aligned} \tag{36}$$

The friction torque T_f is given by

$$T_f = a_1(\tanh(a_2 x_2) - \tanh(a_3 x_2)) + a_4 \tanh(a_5 x_2). \tag{37}$$

The parameters a_1, a_4, b, c, as well as the motor inertia J and the torque/current constant k_i are unknown. According to the introduced notation the regressor is given by

$$\xi = \begin{bmatrix} -x_2 \\ \sin(x_1) \\ -\tanh(a_2 x_2) + \tanh(a_3 x_2) \\ -\tanh(a_5 x_2) \end{bmatrix}. \tag{38}$$

The nominal values of parameters are $J = 0.1\,[\mathrm{kgm}^2]$, $b = 0.3\,[\mathrm{Nm}\frac{\mathrm{s}}{\mathrm{rad}}]$, $c = 2$, $a_1 = 40\,[\mathrm{Nm}]$, $a_2 = 110$, $a_3 = 100$, $a_4 = 2$, $a_5 = 100$, $k_i = 2\,[\frac{\mathrm{Nm}}{\mathrm{A}}]$.

It is assumed that state variables must fulfill constraints: $b_{11} = b_{12} = 1\,[\mathrm{rad}]$, $b_{21} = b_{22} = 5\,[\frac{\mathrm{rad}}{\mathrm{s}}]$.

The plant was used to compare three controllers:

- (0I) controller without the integrator – $L_0 = 0$,
- (LI) controller with a linear integrator – $L_0 > 0$, $F(e^2) = 1$,
- (NI) controller with a nonlinear integrator gain – $L_0 > 0$, $F(e^2) = 1 - \tanh(Ke^2)$.

For all experiments $\Gamma = \mathrm{diag}(0.1; 0.01; 0.3; 70; 0.07)$, $\sigma = 0.01$ and the initial values of adaptive parameters are 80% of their nominal values.

In the first experiment the stabilization of motion was studied, so $x_{1d} = \dot{x}_{1d} = 0$. The selected gains are: $L_0 = 4$, $L_1 = 10$, $L_2 = 10$. The system trajectories for different initial conditions are shown in Figs. 2, 3 and 4. In spite of the controller used, the state constraints are not violated and the stabilization is achieved. The main differences are visible near the steady state point, as it is shown by the next experiments. Figures 5, 6, 7 and 8 demonstrate the state variables for different controllers and different values of parameters L_i. Controllers 0I and NI provide smaller position and velocity overshoots than the controller with linear integrator.

The aim of the second experiment was to follow the desired trajectory $x_{1d} = 0,9\sin(0,25t)$, $x_{2d} = \dot{x}_{1d}$. The selected gains are: $L_0 = 10$, $L_1 = 10$, $L_2 = 10$. Although the desired trajectory oscillates quite close to the constraints, the actual position and velocity remain inside the constraints. Tracking errors for three controllers are compared in Fig. 9, 10, 11 and 12. The accuracy of tracking is sufficient (error amplitude is smaller than 0.01% of input amplitude), in spite of the used controller. The linear integrator (LI) produces smaller quasi-steady-state error than 0I, but it generates significantly higher overshoot (Fig. 9 and Fig. 11). The proposed nonlinear integrator gain (NI) causes as small overshoot as 0I and the quasi-steady-state error remains as small as for LI.

The norm of error of adaptive parameters is shown in Fig. 13. For all controllers, the adaptive parameters remain bounded. The tracking errors for different values of parameter K (different shapes of $F(e^2)$) are shown in Figs. 14 and 15. The increase of the parameter K allows reducing the overshoot.

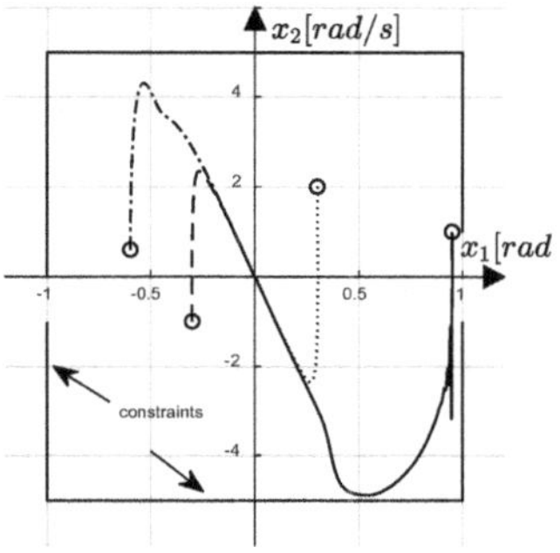

Fig. 2. Trajectories of state variables for several starting points. System without the integrator.

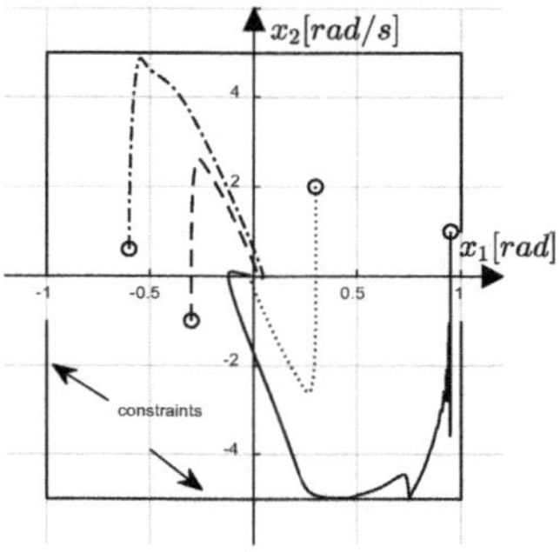

Fig. 3. Trajectories of state variables for several starting points. System with the linear integrator.

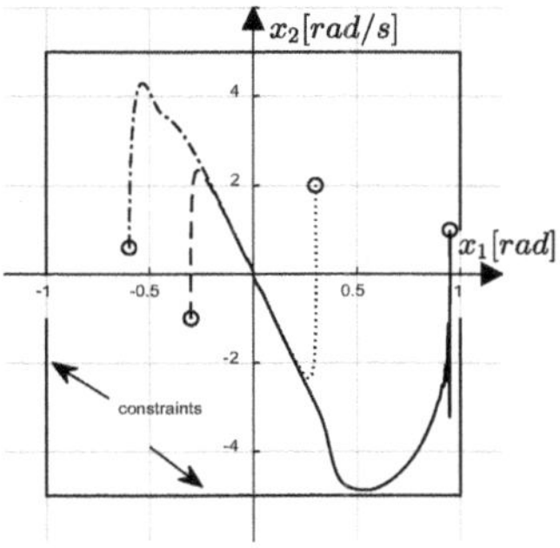

Fig. 4. Trajectories of state variables for several starting points. System with the nonlinear integrator gain.

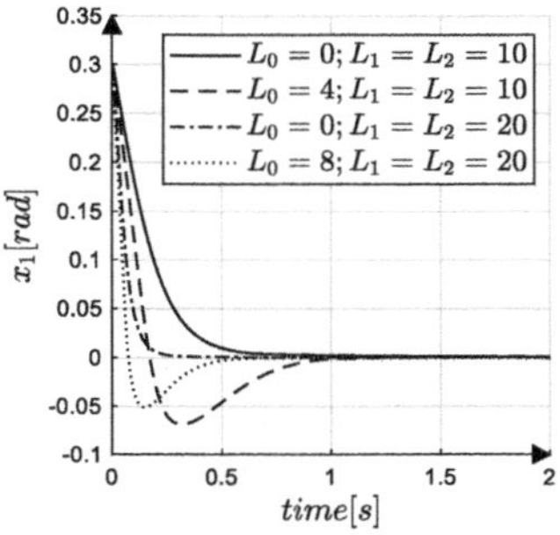

Fig. 5. Stabilization of the position x_1 for different values L_1, L_2. If $L_0 = 0$ the integrator is off, if $L_0, > 0$ – on (LI)

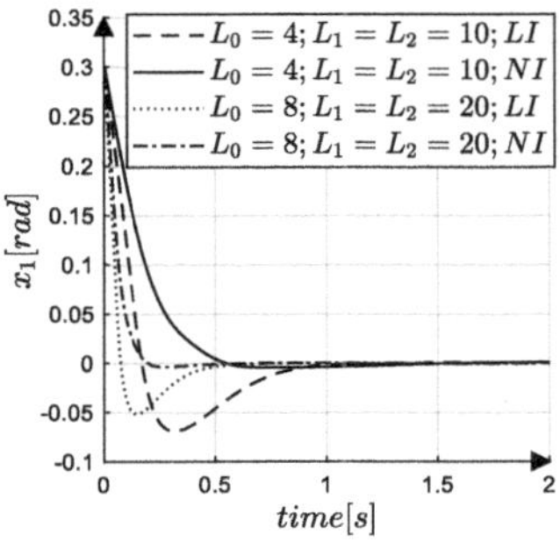

Fig. 6. Stabilization of the position x_1 for different values L_0, L_1, L_2.

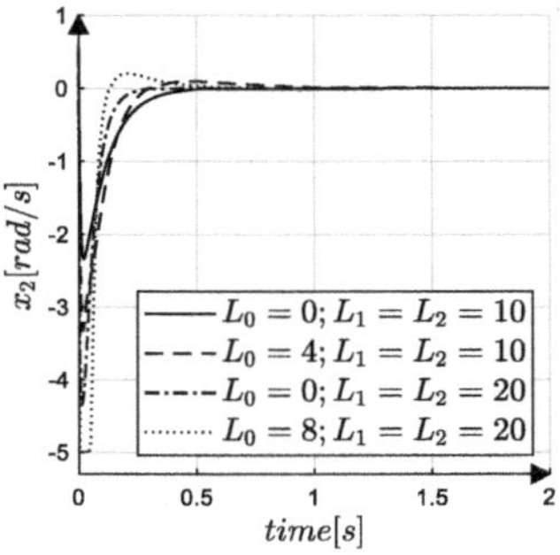

Fig. 7. Stabilization of the velocity x_2 for different values L_1, L_2. If $L_0 = 0$ the integrator is off, if $L_0, > 0$ – on (LI).

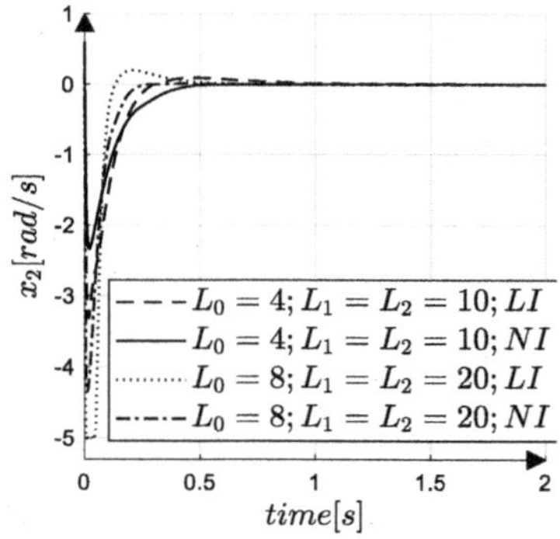

Fig. 8. Stabilization of the velocity x_2 for different values L_0, L_1, L_2.

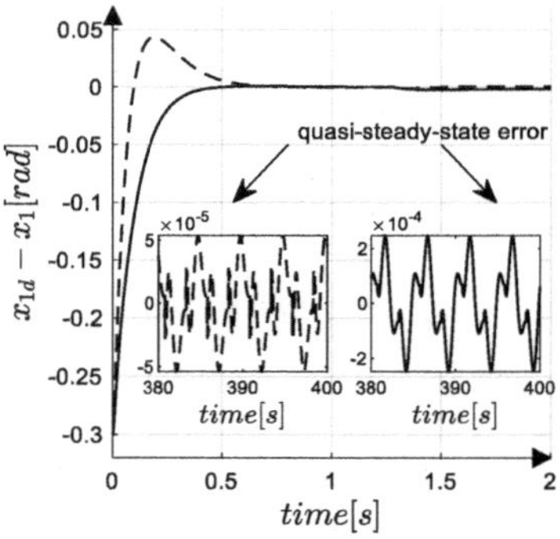

Fig. 9. Tracking error $x_{1d} - x_1$ for: controller with the integrator (LI) – dashed line, controller without the integrator (0I) – solid line.

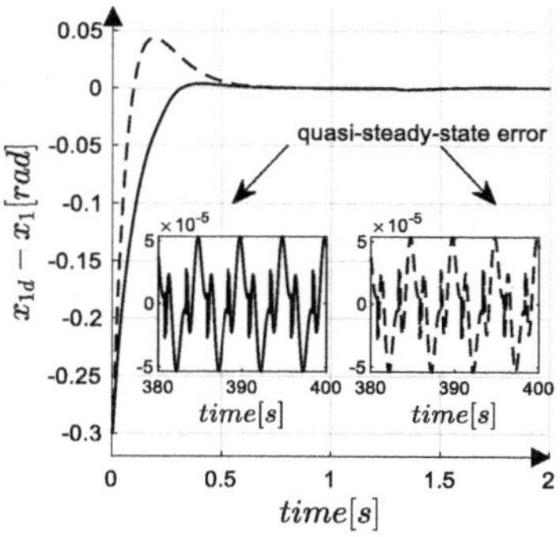

Fig. 10. Tracking error $x_{1d} - x_1$ for: controller with the integrator (LI) – dashed line, controller with the integrator (NI) – solid line.

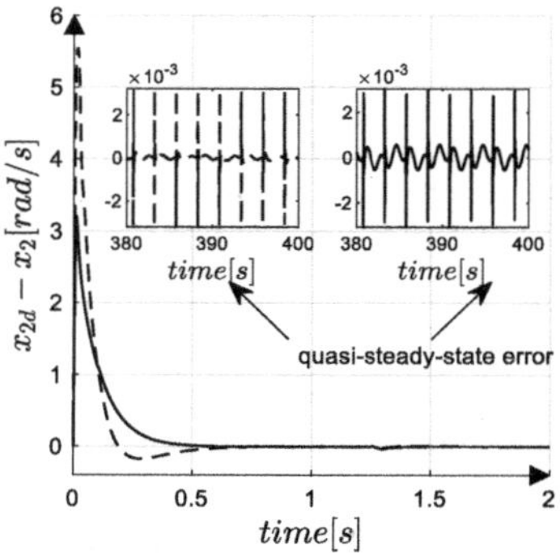

Fig. 11. Tracking error $x_{2d} - x_2$ for: controller with the integrator (LI) – dashed line, controller without the integrator (0I) – solid line.

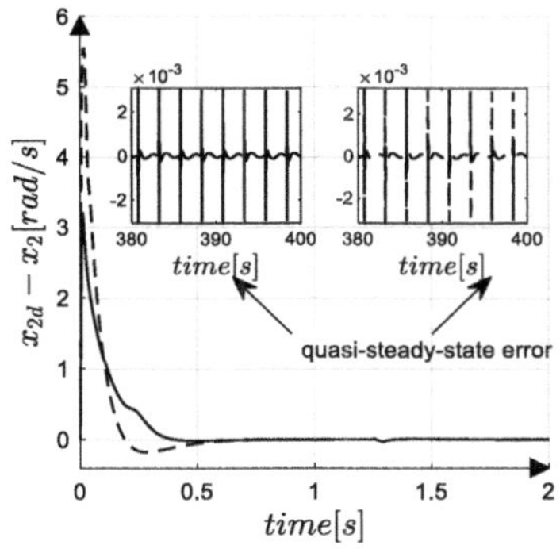

Fig. 12. Tracking error $x_{2d} - x_2$ for: controller with the integrator (LI) – dashed line, controller with the integrator (NI) – solid line.

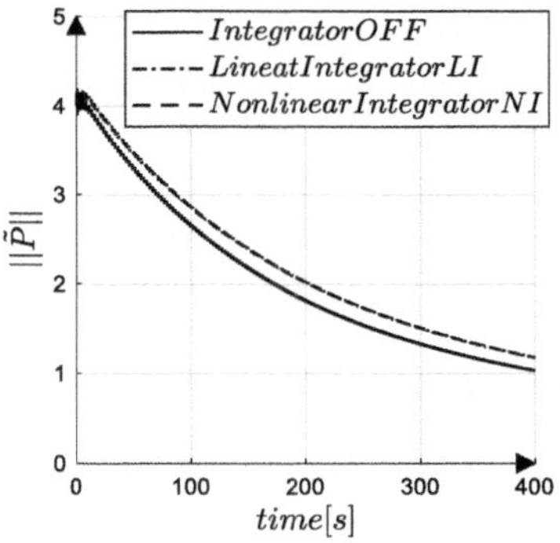

Fig. 13. Norm of adaptive parameters error $||\tilde{P}||$ for different controllers.

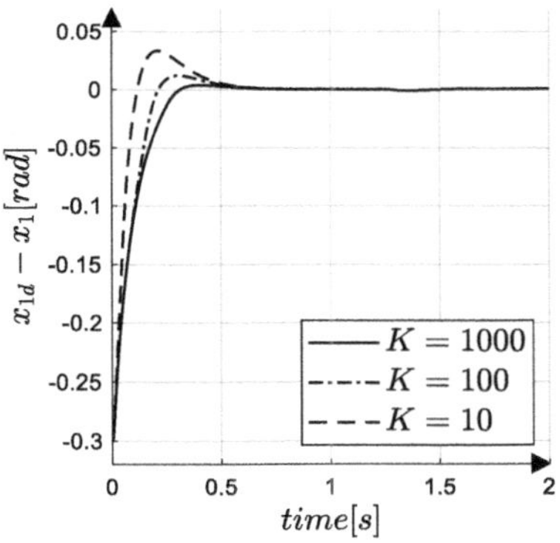

Fig. 14. Tracking error $x_{1d} - x_1$, controller NI, different values of K.

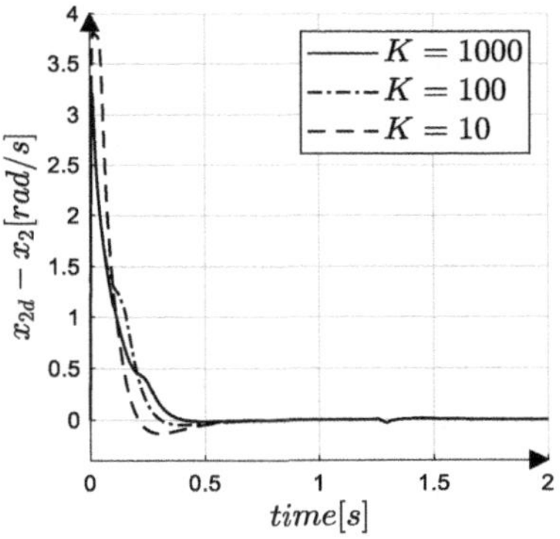

Fig. 15. Tracking error $x_{2d} - x_2$ controller NI, different values of K.

Table 1. Comparison of system performance for different controllers.

Controller	Overshoot	Quasi-steady-state error
Without integrator (0I)	Moderate	Acceptable
With linear integrator (LI)	Significant	Small
Nonlinear integrator (NI)	Small	Small

6 Conclusions

Adaptive backstepping approach with a special nonlinear state transformation proved to be efficient techniques to solve a nonlinear motion control problem in the presence of hard position and velocity constraints. The proposed nonlinear state transformation eliminates difficult feasibility conditions, typical for barrier Lyapunov function approach. Three different controllers are derived and compared. All controllers are easy to tune, for any of them the state variables remain inside the constraints and the tracking errors converge to a small compact set, which may be decreased by increasing gains L_i. The comparison of the controllers is summarized in Table 1. The proposed solution with a nonlinear integrator gain preserves the advantages of both simpler approaches.

References

1. Blanchini, F.: Set invariance in control. Automatica **35**(11), 1747–1767 (1999)
2. Pérez, E., Ariño, C., Blasco, F.X., Martínez, M.A.: Maximal closed loop admissible set for linear systems with non-convex polyhedral constraints. J. Process Control **21**(4), 529–537 (2011)
3. Gilbert, E.G., Tan, K.T.: Linear systems with state and control constraints: the theory and application of maximal output admissible sets. IEEE Trans. Autom. Control **36**(9), 1008–1020 (1991)
4. Wang, R., Bao, J.: A differential Lyapunov-based tube MPC approach for continuous-time nonlinear processes. J. Process Control **83**, 155–163 (2019)
5. Mayne, D.Q., Rawlings, J.B., Rao, C.V., Scokaert, P.O.M.: Constrained model predictive control: stability and optimality. Automatica **36**(6), 789–814 (2000)
6. Kogiso, K., Hirata, K.: Reference governor for constrained systems with time-varying references. Robot. Auton. Syst. **57**(3), 289–295 (2009)
7. Oh-hara, S., Urano, Y., Matsuno, F.: The control of constrained system with time-delay and its experimental evaluations using RC model helicopter. In: 2007 International Conference on Control, Automation and Systems, pp. 2897–2901 (2007)
8. Li, J., Liu, Y.: Control of nonlinear systems with full state constraints using integral Barrier Lyapunov Functionals. In: 2015 International Conference on Informative and Cybernetics for Computational Social Systems (ICCSS), pp. 66–71 (2015)
9. Wang, W., Tong, S.: Adaptive fuzzy containment control of nonlinear strict-feedback systems with full state constraints. IEEE Trans. Fuzzy Syst. **27**(10), 2024–2038 (2019)
10. Sachan, K., Padhi, R.: Output-constrained robust adaptive control for uncertain nonlinear MIMO systems with unknown control directions. IEEE Control Syst. Lett. **3**(4), 823–828 (2019)
11. Luo, S., Song, Y.: Chaos analysis-based adaptive backstepping control of the microelectromechanical resonators with constrained output and uncertain time delay. IEEE Trans. Industr. Electron. **63**(10), 6217–6225 (2016)
12. Wang, C., Wu, Y., Yu, J.: Barrier Lyapunov functions-based adaptive control for nonlinear pure-feedback systems with time-varying full state constraints. Int. J. Control Autom. Syst. **15**(6), 2714–2722 (2017)
13. Wang, C., Wu, Y., Wang, F., Zhao, Y.: TABLF-based adaptive control for uncertain nonlinear systems with time-varying asymmetric full-state constraints. Int. J. Control 1–9 (2019)
14. Yin, Z., Wang, B., Du, C., Zhang, Y.: Barrier-Lyapunov-function-based backstepping control for PMSM servo system with full state constraints. In: 2019 22nd International Conference on Electrical Machines and Systems (ICEMS), pp. 1–5 (2019)
15. Kabziński, J., Mosiołek, P., Jastrzębski, M.: Adaptive position tracking with hard constraints—barrier Lyapunov functions approach. In: Studies in Systems, Decision and Control, pp. 27–52. Springer, Heidelberg (2017)
16. Tang, Z.L., Ge, S.S., Tee, K.P., He, W.: Robust adaptive neural tracking control for a class of perturbed uncertain nonlinear systems with state constraints. IEEE Trans. Syst. Man Cybern. Syst. **46**(12), 1618–1629 (2016)

17. Tee, K.P., Ge, S.S.: Control of nonlinear systems with partial state constraints using a barrier Lyapunov function. Int. J. Control **84**(12), 2008–2023 (2011)
18. Kabziński, J.: Adaptive control of drillstring torsional oscillations. IFAC-PapersOnLine **50**(1), 13360–13365 (2017)
19. Kabziński, J.: Adaptive, compensating control of wheel slip in railway vehicles. Bull. Pol. Acad. Sci. Tech. Sci. **63**, 955–963 (2015)
20. Khalil, H.K.: Nonlinear Systems. Prentice Hall, Upper Saddle River (2002)
21. Krstic, M., Kanellakopoulos, I., Kokotovic, P.V.: Nonlinear and Adaptive Control Design. Wiley, New York (1995)
22. Kabziński, J., Mosiołek, P.: Projektowanie nieliniowych układów sterowania (Nonlinear Control Design). Wydawnictwo Naukowe PWN (2018)

Robotics and Mechatronics

Specification of Agent Based Robotic Systems Using Hierarchical Finite State Automatons

Cezary Zieliński(✉)

Warsaw University of Technology, Institute of Control and Computation Engineering, Nowowiejska 15/19, 00–665 Warsaw, Poland
C.Zielinski@ia.pw.edu.pl

Abstract. The paper assumes the composition of robotic systems out of embodied agents. It presents a utilitarian decomposition of an agent into subsystems. Both subsystem behaviours and their selection can be described in terms of Finite State Automatons (FSA), thus Hierarchic FSAs result. Mathematical formalisation of this description enables the verification of correctness of some aspects of system operation.

Keywords: Robotic system specification methodology · Hierarchical finite state automaton · Hierarchical finite state machine

1 Introduction

The use of the concept of an agent in system design has been gaining popularity over the years [7,18,19,23,28]. This concept has also penetrated both into artificial intelligence [21,22] and robotics [3,30,31]. Although software engineering postulates the separation of concerns, especially the separation of software specification from its implementation (e.g. [9]), many systems are designed without following this rule, especially in robotics [15].

FSAs have been widely employed by general system design methods (e.g. [11,20]), as well as in robot system design (e.g. [1,2,8,12,13,17,25]). The use of FSAs enables the verification of certain properties of the designed system. Software engineering relies on model checking [6] to verify that the designed FSA based system (its requirements or model) possesses certain properties expressed as formulas in certain logic. Automatic model checkers ascertain or falsify a property [27]. To do that they need both the full system model and formally expressed properties, so the checking is performed once the model is complete. However this paper postulates an alternative method. Some properties can be checked at the stage of producing the system model, e.g. FSA disjunction and integrality conditions can be verified mathematically at that stage.

The paper is structured as follows. Section 2 introduces both the structure and operation of an embodied agent. It also defines the Hierarchic FSA (HFSA) describing the operation of an agent's subsystems. Section 3 provides an example of a rudimentary system, presenting the specification of the associated HFSA and the verification of some of its properties. Conclusions are drawn in Sect. 4.

A. Bartoszewicz et al. (Eds.): PCC 2020, AISC 1196, pp. 465–476, 2020.
https://doi.org/10.1007/978-3-030-50936-1_39

2 Embodied Agent Based Robotic Systems

In general, an **embodied agent** is an entity perceiving the physical environment through receptors and acting upon that environment through effectors, having an internal urge to attain a certain goal. This definition is inspired by the one presented in [3]. A more detailed definition requires the presentation of the internal structure of the agent and the way that it operates.

2.1 Specification of Structure

The structure of an agent (Fig. 1) is based on the above definition. There are two paths, one primary and one secondary. The primary loop is formed by the receptors gathering information from the environment. This information is conveyed to the control subsystem, which combines this information with its knowledge about the task that should be executed, to produce the control decisions for the effectors influencing the environment. The reverse path enables the configuration of receptors and acquisition of proprioceptive data from the effectors.

A systematic notation facilitates the description of the structure and operation of an embodied agent [14,32,33,36,37]. Thus, an embodied agent ,a_j where j is its designator, is composed of: **real effectors** $E_{j,m}$, i.e. devices influencing the environment, **virtual effectors** $e_{j,n}$, i.e. effector drivers, **real receptors** $R_{j,l}$, i.e. sensors acquiring the information from the ambience, **virtual receptors** $r_{j,k}$, that aggregate the information obtained from the real receptors, and the **control subsystem** c_j, which supervises the execution of the task; where k, l, m, n are the designators of particular subsystems.

Subsystems process data that arrives in their **input buffers** (designated by a leading subscript x) and is contained in their **internal memory** (no left subscript), in effect producing data that is either dispatched to the other subsystems through **output buffers** (left subscript y) or retained in the internal memory. The central symbol describes the type of the subsystem containing the buffer or memory, i.e. c, e, r, E, R. To reference **any subsystem** in general s is used instead. The left superscript designates the type of the subsystem that the considered subsystem communicates with. T refers to a transfer buffer used by the control subsystem for communication with other agents. The right superscript refers to the discrete time instant at which the contents of the buffer or memory are considered. Generally i is a current instant and $i + 1$ is the next instant. Each subsystem operates at its own sampling rate, however for brevity, i is used as a time stamp of any subsystem. Example pertaining to the agent named robot: ${}^{c}_{x}e^{i}_{\text{robot,rm}}$ denotes the input buffer of the virtual effector named rm (right manipulator), that has obtained at instant i the commands from the control subsystem c_{robot}.

2.2 Specification of Operation

The activities of each subsystem conform to a pattern. Any subsystem s, e.g. named v, of an agent a_j, i.e. $s_{j,v}$, gets data from its input buffers at instant i,

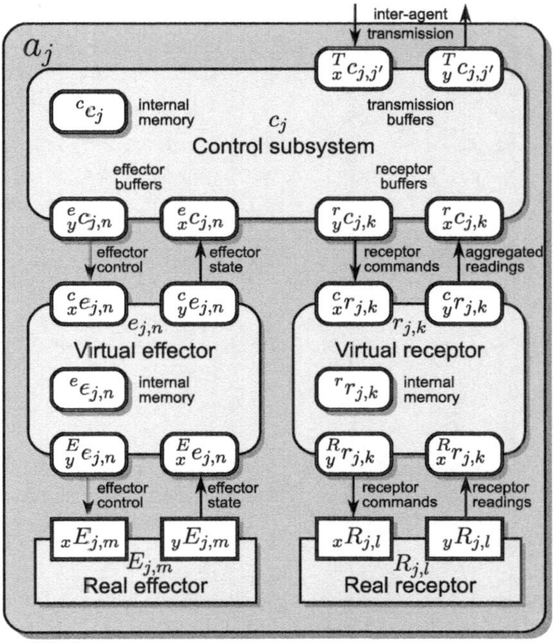

Fig. 1. Internal structure of an embodied agent a_j

computes its transition function, and sends out the results to the other subsystems and its own memory at instant $i+1$. The **transition function** ${}^{s}f_{j,v,\omega}$ is defined as:

$$\left[{}^{s}s_{j,v}^{i+1},\ {}_{y}s_{j,v}^{i+1}\right] := {}^{s}f_{j,v,\omega}({}^{s}s_{j,v}^{i},\ {}_{x}s_{j,v}^{i}), \tag{1}$$

ω – designator of the transition function. The above enumerated operations form an **elementary action** ${}^{s}\mathcal{A}_{j,v,\omega}$ The transition function only processes data. Methods of defining transition functions are presented in [14,30,31,33–35].

Behaviour ${}^{s}\mathcal{B}_{j,v,\omega}$ is an iterative execution of an elementary action ${}^{s}\mathcal{A}_{j,v,\omega}$. Iteration of ${}^{s}\mathcal{A}_{j,v,\omega}$ ceases when either the behaviour ${}^{s}\mathcal{B}_{j,v,\omega}$ **terminal condition** ${}^{s}f^{\tau}_{j,v,\xi}$ or its **error condition** ${}^{s}f^{\varepsilon}_{j,v,\beta}$ is fulfilled. Both of them are predicates taking as arguments the contents of internal memory ${}^{s}s^{i}_{j,v}$ and the input buffers ${}_{x}s^{i}_{j,v}$ at instant i. Sampling time is the period from i to $i+1$. Behaviour ${}^{s}\mathcal{B}_{j,v,\omega}$ of the subsystem $s_{j,v}$ is represented by a **Finite State Automaton** (FSA) ${}^{s}\mathcal{F}^{\mathcal{B}}_{j,v,\omega}$ (Fig. 2):

$${}^{s}\mathcal{B}_{j,v,\omega} \equiv {}^{s}\mathcal{F}^{\mathcal{B}}_{j,v,\omega} = \langle {}^{s}_{-}\hat{\mathcal{S}}_{j,v,\omega}, {}^{s}_{-}\hat{\mathcal{I}}_{j,v,\omega}, {}^{s}_{-}\hat{\mathcal{O}}_{j,v,\omega}, {}^{s}_{-}\Theta_{j,v,\omega}, {}^{s}_{-}\Lambda_{j,v,\omega}\rangle \tag{2}$$

Herein sets are distinguished from other concepts by having a hat over the central symbol. The five **substates** form the **set of substates** ${}^{s}_{-}\hat{\mathcal{S}}_{j,v,\omega}$:

$${}^{s}_{-}\hat{\mathcal{S}}_{j,v,\omega} = \{{}^{s}_{-}\mathcal{S}^{1}_{j,v,\omega}, {}^{s}_{-}\mathcal{S}^{2}_{j,v,\omega}, {}^{s}_{-}\mathcal{S}^{3}_{j,v,\omega}, {}^{s}_{-}\mathcal{S}^{4}_{j,v,\omega}, {}^{s}_{-}\mathcal{S}^{5}_{j,v,\omega}\} \tag{3}$$

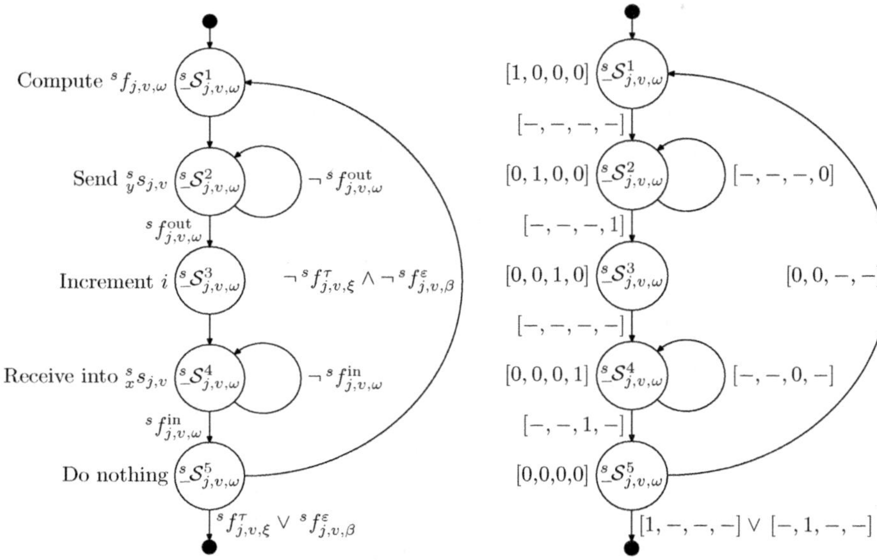

Fig. 2. Graph of the FSA ${}^s\mathcal{F}^{\mathcal{B}}_{j,v,\omega}$ executing a behaviour ${}^s\mathcal{B}_{j,v,\omega}$; Left: explicitly defined subactions and predicates; Right: binary version (dash is a don't care)

The **input vectors** of ${}^s\mathcal{F}^{\mathcal{B}}_{j,v,\omega}$ consist of four predicates. The first two are the values of the terminal and error conditions, while the latter two define the communication mode used by the subsystem when receiving or sending data [32].

$$\underline{{}^s\mathcal{I}}_{j,v,\omega} = [{}^sf^{\mathcal{T}}_{j,v,\xi}, {}^sf^{\varepsilon}_{j,v,\beta}, {}^sf^{\mathrm{in}}_{j,v,\omega}, {}^sf^{\mathrm{out}}_{j,v,\omega}], \qquad \underline{{}^s\mathcal{I}}_{j,v,\omega} \in \underline{{}^s\hat{\mathcal{I}}}_{j,v,\omega} \tag{4}$$

Those vectors form the **input vector set** $\underline{{}^s\hat{\mathcal{I}}}_{j,v,\omega}$. As predicates produce binary values, 4-element binary vectors are the elements of $\underline{{}^s\hat{\mathcal{I}}}_{j,v,\omega}$.

The **output vector set** $\underline{{}^s\hat{\mathcal{O}}}_{j,v,\omega}$ contains **output vectors** $\underline{{}^s\mathcal{O}}_{j,v,\omega}$ composed of four binary values ${}^s\mathfrak{a}^q_{j,v,\omega}$, $q = 1, \ldots, 4$, representing **subactions** of an elementary action:

$$\underline{{}^s\mathcal{O}}_{j,v,\omega} = [{}^s\mathfrak{a}^1_{j,v,\omega}, {}^s\mathfrak{a}^2_{j,v,\omega}, {}^s\mathfrak{a}^3_{j,v,\omega}, {}^s\mathfrak{a}^4_{j,v,\omega}], \qquad \underline{{}^s\mathcal{O}}_{j,v,\omega} \in \underline{{}^s\hat{\mathcal{O}}}_{j,v,\omega} \tag{5}$$

${}^s\mathfrak{a}^1_{j,v,\omega} = 1$, when the transition function is computed, ${}^s\mathfrak{a}^2_{j,v,\omega} = 1$, when the contents of ${}_ys_{j,v}$ are sent to the other subsystems, ${}^s\mathfrak{a}^3_{j,v,\omega} = 1$, when the discrete time i is incremented, and ${}^s\mathfrak{a}^4_{j,v,\omega} = 1$, when new data is received by ${}_xs_{j,v}$. When no subaction is to be executed, $\underline{{}^s\mathcal{O}}_{j,v,\omega} = [0, 0, 0, 0]$. As the subactions are disjoint in all the other cases this vector contains just a single 1, only 5 possibilities appear – as many as the number of ${}^s\mathcal{F}^{\mathcal{B}}_{j,v,\omega}$ states.

FSA state transition function defines transitions between ${}^s\mathcal{F}^{\mathcal{B}}_{j,v,\omega}$ states:

$$\underline{{}^s\Theta}_{j,v,\omega} : \underline{{}^s\hat{\mathcal{S}}}_{j,v,\omega} \times \underline{{}^s\hat{\mathcal{I}}}_{j,v,\omega} \to \underline{{}^s\hat{\mathcal{S}}}_{j,v,\omega}. \tag{6}$$

It should not be mixed up with the transition function (1). Both have different arguments and values. **FSA output function** associates outputs to states:

$$ {}_{-}^{s}\Lambda_{j,v,\omega} : {}_{-}^{s}\hat{\mathcal{S}}_{j,v,\omega} \rightarrow {}_{-}^{s}\hat{\mathcal{O}}_{j,v,\omega}. \tag{7} $$

A Moore type deterministic automaton results [16,26], i.e. a behaviour pattern.

As behaviours ${}^{s}\mathcal{B}_{j,v,\omega}$ are equivalent to ${}^{s}\mathcal{F}^{\mathcal{B}}_{j,v,\omega}$ FSAs, they become the building blocks of the lowest level of a **Hierarchic Finite State Automaton** (HFSA) [29]. A HFSA of a subsystem $s_{j,v}$ is defined recursively as follows:

$$ {}^{s}\mathcal{F}_{j,v} = \langle {}^{s}\hat{\mathcal{S}}_{j,v}, {}_{+}^{s}\hat{\mathcal{S}}_{j,v}, {}^{s}\hat{\mathcal{I}}_{j,v}, {}^{s}\hat{\mathcal{B}}_{j,v}, {}^{s}\Theta_{j,v}, {}^{s}\Lambda_{j,v}, {}_{+}^{s}\Lambda_{j,v} \rangle \tag{8} $$

where ${}^{s}\hat{\mathcal{S}}_{j,v}$ is the set **HFSA states** determining the activities of subsystem $s_{j,v}$, ${}_{+}^{s}\hat{\mathcal{S}}_{j,v}$ is the set of **HFSA superstates**, ${}^{s}\hat{\mathcal{I}}_{j,v}$ is the set of input vectors composed of initial conditions (predicates), and ${}^{s}\hat{\mathcal{B}}_{j,v}$ is the set of behaviours ${}^{s}\mathcal{B}_{j,v,\omega} \in {}^{s}\hat{\mathcal{B}}_{j,v}$. Moreover, 3 functions are defined:

HFSA state transition function:

$$ {}^{s}\Theta_{j,v} : ({}^{s}\hat{\mathcal{S}}_{j,v} \cup {}_{+}^{s}\hat{\mathcal{S}}_{j,v}) \times {}^{s}\hat{\mathcal{I}}_{j,v} \rightarrow ({}^{s}\hat{\mathcal{S}}_{j,v} \cup {}_{+}^{s}\hat{\mathcal{S}}_{j,v}) \tag{9} $$

HFSA output function:

$$ {}^{s}\Lambda_{j,v} : {}^{s}\hat{\mathcal{S}}_{j,v} \rightarrow {}^{s}\hat{\mathcal{B}}_{j,v} \tag{10} $$

and **HFSA superoutput function**:

$$ {}_{+}^{s}\Lambda_{j,v} : {}_{+}^{s}\hat{\mathcal{S}}_{j,v} \rightarrow {}^{s}\mathcal{F}_{j,v} \tag{11} $$

It is assumed that neither direct nor indirect self-recursion is used while composing a HFSA. The HFSA starts its activities in the initial state or superstate. This is a single entry point. It should be noted that after substitution of all HFSAs for superstates of topmost ${}^{s}\mathcal{F}_{j,v}$ and behaviours ${}^{s}\mathcal{F}^{\mathcal{B}}_{j,v,\omega}$ for states a single flat FSA results, thus this is not a composition of separately acting parallel FSAs. The resulting flat FSA is always in a single substate of the currently active behaviour. A single definition (8) would suffice, however a two-level one was assumed, where the lowest level is defined by (2), because a behaviour follows a pattern defined by ${}^{s}\mathcal{F}^{\mathcal{B}}_{j,v,\omega}$, thus its distinction is beneficial from the point of view of its implementation. Moreover, as behaviour is a verified pattern, only the verification of the upper level of the HFSA is necessary.

Figure 3 presents a rudimentary example of a HFSA. To flatten it out to the behaviour level (i.e. without the introduction of substates) the ${}^{s}\mathcal{F}_{j,v,5}$ must be substituted for the superstate ${}_{+}^{s}\mathcal{S}^{5}_{j,v}$. The result is presented in Fig. 4. In general, all directed arcs pointing at a superstate (e.g. ${}_{+}^{s}\mathcal{S}^{5}_{j,v}$) have to point at the state connected to the entrance point of the subautomaton (here ${}^{s}\mathcal{S}^{2}_{j,v}$). All directed arcs going out of the superstate must emerge from the state being the exit point of the subautomaton ${}^{s}\mathcal{F}_{j,v,5}$ (here ${}^{s}\mathcal{S}^{3}_{j,v}$). HFSAs flattened out solely to the state

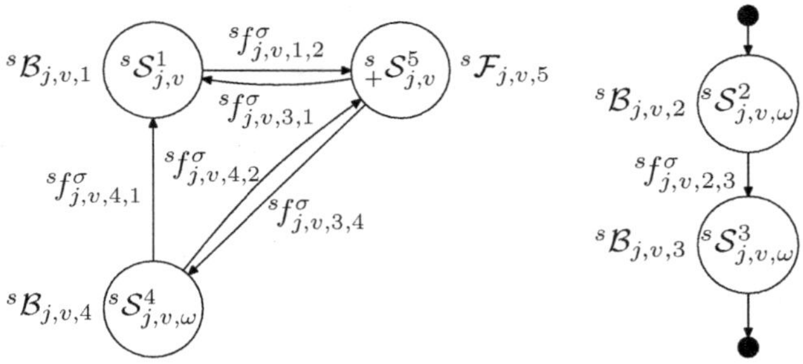

Fig. 3. Exemplary HFSA ${}^s\mathcal{F}_{j,v}$ and a FSA ${}^s\mathcal{F}_{j,v,5}$ substituting superstate ${}^s_+\mathcal{S}^5_{j,v}$

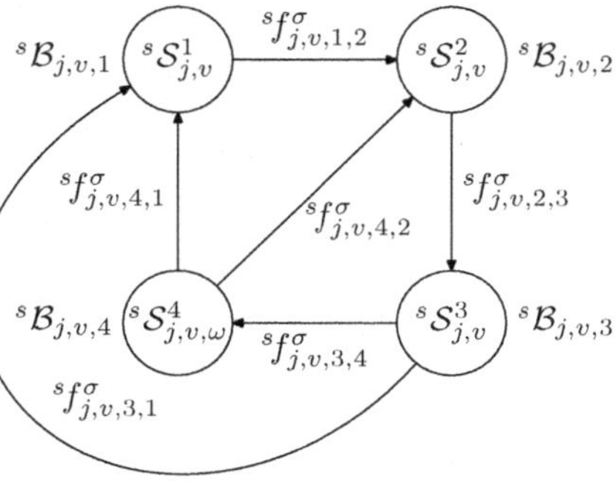

Fig. 4. Flattened out graph of an exemplary HFSA ${}^s\mathcal{F}_{j,v}$ from Fig. 3

level FSA (e.g. Fig. 4), without including the behaviour substates (Fig. 2), will be discussed further on.

When the currently executed behaviour expires, due to either its terminal condition ${}^s f^{\tau}_{j,v,\xi}$ or error condition ${}^s f^{\varepsilon}_{j,v,\beta}$ being satisfied, a next behaviour is selected by the HFSA state transition function ((9)). Assuming that the behaviour ${}^s\mathcal{B}_{j,v,\omega}$, associated with state ${}^s\mathcal{S}^{\omega}_{j,v}$, has just been terminated, the next state is chosen based on the initial conditions ${}^s f^{\sigma}_{j,v,\omega,\gamma}({}^s s^i_{j,v}, {}_x s^i_{j,v})$ labeling the directed arcs emerging from ${}^s\mathcal{S}^{\omega}_{j,v}$ and leading to ${}^s\mathcal{S}^{\gamma}_{j,v}$, where ${}^s\mathcal{S}^{\gamma}_{j,v} \in {}^s\hat{\mathcal{S}}^{\omega}_{j,v}$, i.e. the set of states pointed at by the arcs going out of ${}^s\mathcal{S}^{\omega}_{j,v}$. Considering deterministic automatons, and assuming that ${}^s\mathcal{S}^{\omega}_{j,v}$ is the current state of ${}^s\mathcal{F}_{j,v}$, the values of initial conditions labeling the arcs emerging from ${}^s\mathcal{S}^{\omega}_{j,v}$, at an instant i, in which ${}^s\mathcal{B}_{j,v,\omega}$ has been terminated, must satisfy two conditions:

$$\begin{aligned} &\forall_{\gamma \neq \gamma'}\ {}^s f^{\sigma}_{j,v,\omega,\gamma}({}^s s^i_{j,v}, {}_x s^i_{j,v}) \wedge {}^s f^{\sigma}_{j,v,\omega,\gamma'}({}^s s^i_{j,v}, {}_x s^i_{j,v}) = \text{False and} \\ &\bigvee_{\gamma}\ {}^s f^{\sigma}_{j,v,\omega,\gamma}({}^s s^i_{j,v}, {}_x s^i_{j,v}) = \text{True}, \qquad \text{where } {}^s\mathcal{S}^{\gamma}_{j,v}, {}^s\mathcal{S}^{\gamma'}_{j,v} \in {}^s\hat{\mathcal{S}}^{\omega}_{j,v}. \end{aligned} \tag{12}$$

The first one is called the **disjunction condition** and the second one the **integrality condition**. The first assures that the automaton is deterministic, while the second one that at i there always exists a next state. For example, assuming that the current state of the automaton ${}^{s}\mathcal{F}_{j,v}$ is ${}^{s}\mathcal{S}^{4}_{j,v}$ (i.e. $\omega = 4$) (Fig. 4), and behaviour ${}^{s}\mathcal{B}_{j,v,4}$ has terminated its activities in the instant i, taking into account that ${}^{s}\hat{\mathcal{S}}^{4}_{j,v} = \{{}^{s}\mathcal{S}^{1}_{j,v}, {}^{s}\mathcal{S}^{2}_{j,v}\}$, the initial conditions must fulfil the disjunction condition: ${}^{s}f^{\sigma}_{j,v,4,1}({}^{s}s^{i}_{j,v}, {}_{x}s^{i}_{j,v}) \wedge {}^{s}f^{\sigma}_{j,v,4,2}({}^{s}s^{i}_{j,v}, {}_{x}s^{i}_{j,v}) = \text{False}$, and the integrality condition: ${}^{s}f^{\sigma}_{j,v,4,1}({}^{s}s^{i}_{j,v}, {}_{x}s^{i}_{j,v}) \vee {}^{s}f^{\sigma}_{j,v,4,2}({}^{s}s^{i}_{j,v}, {}_{x}s^{i}_{j,v}) = \text{True}$. It should be noted that conditions (12) do not have to be necessarily fulfilled at an instant other than the one in which ${}^{s}\mathcal{B}_{j,v,\omega}$ terminates its activities. Moreover, even if the integrality condition is fulfilled in each state, that does not necessarily lead to the HFSA being complete. However the resulting incomplete automaton is guaranteed to operate correctly, always having a next state to transit to.

3 Example

To present the specification method employing HFSAs a rudimentary, yet useful, task has been chosen. The task is to track a selected moving object. This is one of the skills that a service robot should exhibit. The task of a manipulator is to approach a selected moving object with an intension of grasping it. The hardware of the system consists of the manipulator having 6 dof and two RGB cameras: a stationary camera located over the workspace (SAC – stand-alone camera) and a moving camera integrated with the gripper (EIH – eye in hand) (Fig. 5 (Left)) [4,5,24].

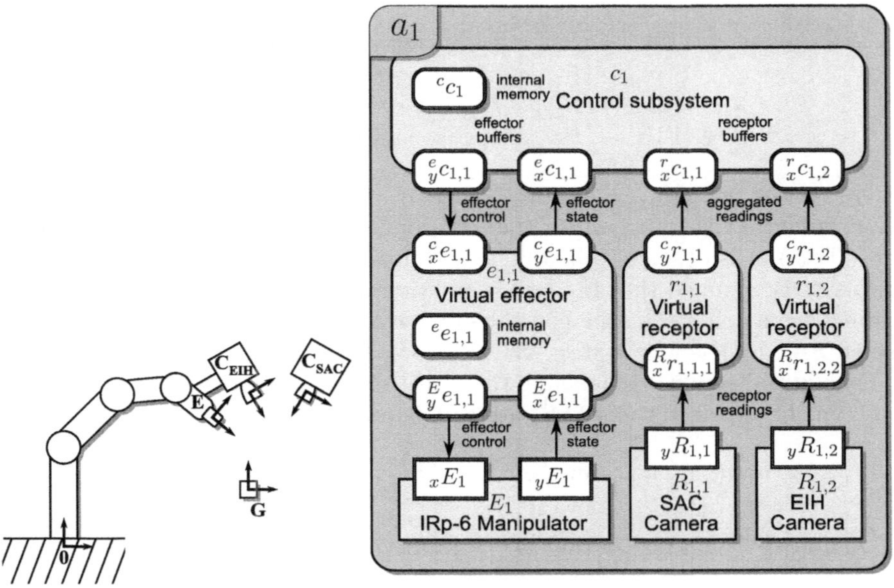

Fig. 5. Left: Robot endowed with SAC and EIH vision; Right: Structure of the embodied agent a_1 representing that robot

The structure of the embodied agent a_1 representing this system is shown in Fig. 5 (Right). The data obtained from each of the cameras is aggregated by an associated virtual receptor: $r_{1,1}$ (from SAC – $R_{1,1}$) and $r_{1,2}$ (from EIH – $R_{1,2}$). As a result the location of the object to be tracked is obtained. This data is used by the control subsystem c_1 to produce the control commands transferred to the virtual effector $e_{1,1}$, which performs the inverse kinematics calculations, the results of which are provided as a control command to the manipulator motor controllers $E_{1,1}$. The present paper focuses only on the definition of the HFSA ${}^{c}\mathcal{F}_1$ of the control subsystem c_1 of the embodied agent a_1. A more comprehensive discussion of the specification and implementation of that task is presented in [14].

To follow a selected object it has to be spotted first. For that purpose SAC is best suited. However to get hold of the object EIH is better suited. As the manipulator gets closer to the object it can obscure the view of SAC and, moreover, a closeup view received from EIH enables a more precise localisation of the object. Yet, just EIH will not suffice, as the view obtained from it depends on the direction that the manipulator is currently pointing its end-effector. Usually initially it is not the direction of the object to be tracked. Hence the system has to cope with two sources of sensor information. In different phases of task execution the object is: invisible, visible by either of the cameras, or by both of them, so the system has to exhibit four behaviours: ${}^{c}\mathcal{B}_{1,0}$ – idle, ${}^{c}\mathcal{B}_{1,1}$ – tracking an object using the stationary camera (SAC), ${}^{c}\mathcal{B}_{1,2}$ – tracking an object using the moving camera (EIH), ${}^{c}\mathcal{B}_{1,3}$ – tracking an object using both cameras. The terminal conditions of those behaviours are:

$$\begin{aligned}
\text{Idle} \rightarrow {}^{c}\mathcal{B}_{1,0} &: {}^{c}f^{\tau}_{1,0} = \mathrm{new}({}^{r}_{x}c^{i}_{1,1}) \vee \mathrm{new}({}^{r}_{x}c^{i}_{1,2})\\
\text{SAC} \rightarrow {}^{c}\mathcal{B}_{1,1} &: {}^{c}f^{\tau}_{1,1} = \neg\mathrm{new}({}^{r}_{x}c^{i}_{1,1}) \vee \mathrm{new}({}^{r}_{x}c^{i}_{1,2})\\
\text{EIH} \rightarrow {}^{c}\mathcal{B}_{1,2} &: {}^{c}f^{\tau}_{1,2} = \mathrm{new}({}^{r}_{x}c^{i}_{1,1}) \vee \neg\mathrm{new}({}^{r}_{x}c^{i}_{1,2})\\
\text{SAC \& EIH} \rightarrow {}^{c}\mathcal{B}_{1,3} &: {}^{c}f^{\tau}_{1,3} = \neg\mathrm{new}({}^{r}_{x}c^{i}_{1,1}) \vee \neg\mathrm{new}({}^{r}_{x}c^{i}_{1,2})
\end{aligned}$$

where new is a predicate deciding whether the data currently present in the input buffer, being its argument (i.e. ${}^{r}_{x}c_{1,1}$ or ${}^{r}_{x}c_{1,2}$), is new (has been not used as yet – True) or is obsolete (already has been used – False). To keep the example simple, it is assumed that the error condition is always False. As there are four distinct behaviours that the control subsystem c_1 has to provide, the graph of the HFSA ${}^{c}\mathcal{F}_1$ has four states: ${}^{c}\mathcal{S}^{0}_{1}, {}^{c}\mathcal{S}^{1}_{1}, {}^{c}\mathcal{S}^{2}_{1}, {}^{c}\mathcal{S}^{3}_{1} \in {}^{c}\hat{\mathcal{S}}_1$ (Fig. 6).

The initial conditions labeling the directed arcs representing the transitions between the states in the set ${}^{c}\hat{\mathcal{S}}_1$ are as follows:

$$\begin{aligned}
{}^{c}f^{\sigma}_{1,0,1} &\triangleq \mathrm{new}({}^{r}_{x}c_{1,1}) \wedge \neg\mathrm{new}({}^{r}_{x}c_{1,2}), & {}^{c}f^{\sigma}_{1,1,0} &\triangleq \neg\mathrm{new}({}^{r}_{x}c_{1,1}) \wedge \neg\mathrm{new}({}^{r}_{x}c_{1,2}),\\
{}^{c}f^{\sigma}_{1,0,2} &\triangleq \neg\mathrm{new}({}^{r}_{x}c_{1,1}) \wedge \mathrm{new}({}^{r}_{x}c_{1,2}), & {}^{c}f^{\sigma}_{1,1,2} &\triangleq \neg\mathrm{new}({}^{r}_{x}c_{1,1}) \wedge \mathrm{new}({}^{r}_{x}c_{1,2}),\\
{}^{c}f^{\sigma}_{1,0,3} &\triangleq \mathrm{new}({}^{r}_{x}c_{1,1}) \wedge \mathrm{new}({}^{r}_{x}c_{1,2}), & {}^{c}f^{\sigma}_{1,1,3} &\triangleq \mathrm{new}({}^{r}_{x}c_{1,1}) \wedge \mathrm{new}({}^{r}_{x}c_{1,2}),\\
{}^{c}f^{\sigma}_{1,2,0} &\triangleq \neg\mathrm{new}({}^{r}_{x}c_{1,1}) \wedge \neg\mathrm{new}({}^{r}_{x}c_{1,2}), & {}^{c}f^{\sigma}_{1,3,0} &\triangleq \neg\mathrm{new}({}^{r}_{x}c_{1,1}) \wedge \neg\mathrm{new}({}^{r}_{x}c_{1,2}),\\
{}^{c}f^{\sigma}_{1,2,1} &\triangleq \mathrm{new}({}^{r}_{x}c_{1,1}) \wedge \neg\mathrm{new}({}^{r}_{x}c_{1,2}), & {}^{c}f^{\sigma}_{1,3,1} &\triangleq \mathrm{new}({}^{r}_{x}c_{1,1}) \wedge \neg\mathrm{new}({}^{r}_{x}c_{1,2}),\\
{}^{c}f^{\sigma}_{1,2,3} &\triangleq \mathrm{new}({}^{r}_{x}c_{1,1}) \wedge \mathrm{new}({}^{r}_{x}c_{1,2}), & {}^{c}f^{\sigma}_{1,3,2} &\triangleq \neg\mathrm{new}({}^{r}_{x}c_{1,1}) \wedge \mathrm{new}({}^{r}_{x}c_{1,2}).
\end{aligned}$$

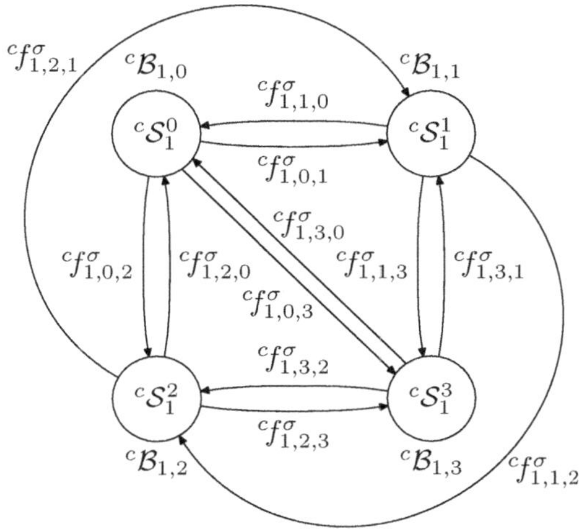

Fig. 6. Graph of the control subsystem HFSA $^{c}\mathcal{F}_1$ of the embodied agent a_1 representing the robot endowed with SAC and EIH vision

The four groups of initial conditions of next states generate the following sets:

1) for $^{c}\mathcal{S}^{0}_{1}$, i.e. the states in the set $^{c}\hat{\mathcal{S}}^{0}_{1} = \{^{c}\mathcal{S}^{1}_{1}, {^{c}\mathcal{S}^{2}_{1}}, {^{c}\mathcal{S}^{3}_{1}}\}$,

2) for $^{c}\mathcal{S}^{1}_{1}$, i.e. the states in the set $^{c}\hat{\mathcal{S}}^{1}_{1} = \{^{c}\mathcal{S}^{0}_{1}, {^{c}\mathcal{S}^{2}_{1}}, {^{c}\mathcal{S}^{3}_{1}}\}$,

3) for $^{c}\mathcal{S}^{2}_{1}$, i.e. the states in the set $^{c}\hat{\mathcal{S}}^{2}_{1} = \{^{c}\mathcal{S}^{0}_{1}, {^{c}\mathcal{S}^{1}_{1}}, {^{c}\mathcal{S}^{3}_{1}}\}$ and

4) for $^{c}\mathcal{S}^{3}_{1}$, i.e. the states in the set $^{c}\hat{\mathcal{S}}^{3}_{1} = \{^{c}\mathcal{S}^{0}_{1}, {^{c}\mathcal{S}^{1}_{1}}, {^{c}\mathcal{S}^{2}_{1}}\}$.

For each state of the FSA $^{c}\mathcal{F}_1$ we need to verify that the initial conditions of its next states satisfy disjunction and integrality conditions (12).

Disjunctions of the initial conditions of the next states of the state $^{c}\mathcal{S}^{0}_{1}$ are:

$$\begin{aligned}
{^{c}f^{\sigma}_{1,0,1}} \wedge {^{c}f^{\sigma}_{1,0,2}} &= \left(\text{new}({^{r}_{x}c_{1,1}}) \wedge \neg\text{new}({^{r}_{x}c_{1,2}})\right) \wedge \left(\neg\text{new}({^{r}_{x}c_{1,1}}) \wedge \text{new}({^{r}_{x}c_{1,2}})\right) \\
&= \text{new}({^{r}_{x}c_{1,1}}) \wedge \neg\text{new}({^{r}_{x}c_{1,1}}) \wedge \text{new}({^{r}_{x}c_{1,2}}) \wedge \neg\text{new}({^{r}_{x}c_{1,2}}) = \text{False} \\
{^{c}f^{\sigma}_{1,0,1}} \wedge {^{c}f^{\sigma}_{1,0,3}} &= \left(\text{new}({^{r}_{x}c_{1,1}}) \wedge \neg\text{new}({^{r}_{x}c_{1,2}})\right) \wedge \left(\text{new}({^{r}_{x}c_{1,1}}) \wedge \text{new}({^{r}_{x}c_{1,2}})\right) \\
&= \text{new}({^{r}_{x}c_{1,1}}) \wedge \text{new}({^{r}_{x}c_{1,1}}) \wedge \neg\text{new}({^{r}_{x}c_{1,2}}) \wedge \text{new}({^{r}_{x}c_{1,2}}) = \text{False} \\
{^{c}f^{\sigma}_{1,0,2}} \wedge {^{c}f^{\sigma}_{1,0,3}} &= \left(\neg\text{new}({^{r}_{x}c_{1,1}}) \wedge \text{new}({^{r}_{x}c_{1,2}})\right) \wedge \left(\text{new}({^{r}_{x}c_{1,1}}) \wedge \text{new}({^{r}_{x}c_{1,2}})\right) \\
&= \neg\text{new}({^{r}_{x}c_{1,1}}) \wedge \text{new}({^{r}_{x}c_{1,1}}) \wedge \text{new}({^{r}_{x}c_{1,2}}) \wedge \text{new}({^{r}_{x}c_{1,2}}) = \text{False}
\end{aligned}$$

hence the disjunction condition is satisfied for the initial conditions of the next states of the state $^{c}\mathcal{S}^{0}_{1}$. The proof of the satisfaction of the disjunction condition for the other states, i.e. $^{c}\mathcal{S}^{1}_{1}$, $^{c}\mathcal{S}^{2}_{1}$ and $^{c}\mathcal{S}^{3}_{1}$, is similar.

The verification of the integrality of initial conditions for each of the states is a bit more involved. The method is presented here on an example of the integrality of initial conditions of the next states of the state ${}^{c}\mathcal{S}_{1}^{0}$. Behaviour ${}^{c}\mathcal{B}_{1,0}$, associated with the state ${}^{c}\mathcal{S}_{1}^{0}$, ends when its terminal condition is satisfied, i.e. ${}^{c}f_{1,0}^{\tau} = \text{new}({}^{r}_{x}c_{1,1}^{i}) \vee \text{new}({}^{r}_{x}c_{1,2}^{i})$, thus this behaviour is realised when the following condition is satisfied: $\neg\,{}^{c}f_{1,0}^{\tau} = \neg\text{new}({}^{r}_{x}c_{1,1}^{i}) \wedge \neg\text{new}({}^{r}_{x}c_{1,2}^{i})$. Hence at the termination of behaviour ${}^{c}\mathcal{B}_{1,0}$ the following is fulfilled: $\neg\,{}^{c}f_{1,0}^{\tau} = \text{False}$. From the following tautology

$$\begin{array}{l}\neg\text{new}({}^{r}_{x}c_{1,1}^{i}) \wedge \neg\text{new}({}^{r}_{x}c_{1,2}^{i}) \vee \text{new}({}^{r}_{x}c_{1,1}^{i}) \wedge \neg\text{new}({}^{r}_{x}c_{1,2}^{i}) \vee \\ \neg\text{new}({}^{r}_{x}c_{1,1}^{i}) \wedge \text{new}({}^{r}_{x}c_{1,2}^{i}) \quad \vee \text{new}({}^{r}_{x}c_{1,1}^{i}) \wedge \text{new}({}^{r}_{x}c_{1,2}^{i}) \quad = \text{True}\end{array}$$

and the fact that $\neg\text{new}({}^{r}_{x}c_{1,1}^{i}) \wedge \neg\text{new}({}^{r}_{x}c_{1,2}^{i}) = \text{False}$ results

$$\text{new}({}^{r}_{x}c_{1,1}^{i}) \wedge \neg\text{new}({}^{r}_{x}c_{1,2}^{i}) \vee \neg\text{new}({}^{r}_{x}c_{1,1}^{i}) \wedge \text{new}({}^{r}_{x}c_{1,2}^{i}) \vee \text{new}({}^{r}_{x}c_{1,1}^{i}) \wedge \text{new}({}^{r}_{x}c_{1,2}^{i}) = \text{True}$$

Thus ${}^{c}f_{1,0,1}^{\sigma} \vee {}^{c}f_{1,0,2}^{\sigma} \vee {}^{c}f_{1,0,3}^{\sigma} = \text{True}$,, hence the integrality condition is satisfied for the initial conditions of the next states of state ${}^{c}\mathcal{S}_{1}^{0}$. The above procedure has to be repeated for each of the other states of the HFSA ${}^{c}\mathcal{F}_{1}$. Once this is done it turns out that for each state of the designed HFSA ${}^{c}\mathcal{F}_{1}$ both the disjunction and the integrality condition are satisfied, thus the assumed HFSA is correct. The proof of the correctness of the virtual effector and virtual receptor HFSAs is much simpler, as the former contains two states and the latter only one state.

4 Conclusions

The presented formalisation of the specification of the control software of a robotic system enables, on the one hand, the design of well structured systems following the principle of separation of concerns [9], and on the other hand, the verification of some properties of the HFSAs of subsystems. This paper has focused on the selection of subsystem behaviours, thus on the overall system operation. It should be noted that the proposed verification method is local, thus can be applied to an incomplete system model, and it avoids combinatorial explosion associated with global property checking. In the presented case the fulfillment of disjunction and integrality conditions is checked for each state of the HFSA, thus the number of tests is equal only to the number of states of that automaton, which is finite *ex definitione*.

Full system verification requires, moreover, a proof of the correctness of each of the behaviours, thus its transition function and the terminal condition, as well as the inter-subsystem communication. Proving the correctness of the transition functions, which depend on the values obtained from the environment, requires establishing whether correct output values are produced for all input values. Moreover it has to be checked, whether the real-time performance of the behaviour is satisfactory, i.e. its sampling time has been chosen correctly.

Although this is not simple in the case of complex systems, the proposed system decomposition facilitates this process, as at every stage it concentrates on relatively small elements of the specification. Those aspects of the verification process are the subject of the currently ongoing investigations.

The system model presented here relies on HFSAs, however alternatively Hierarchical Petri nets [10] can be used.

References

1. Boren, J., Cousins, S.: The SMACH high-level executive. IEEE Rob. Autom. Mag. **17**(4), 18–20 (2010)
2. Brooks, R.A.: New approaches to robotics. Science **253**, 1227–1232 (1991)
3. Brooks, R.A.: Intelligence without reason. Artif. Intell. Crit. Concepts **3**, 107–163 (1991)
4. Chaumette, F., Hutchinson, S.: Visual servo control, part I: basic approaches. IEEE Rob. Autom. Mag. **13**(4), 82–90 (2006)
5. Chaumette, F., Hutchinson, S.: Visual servoing and Visual tracking. In: The Handbook of Robotics, pp. 563–583. Springer, Heidelberg (2008)
6. Clarke, E., Grumberg, O., Kroening, D., Peleg, D., Veith, H.: Model Checking. MIT Press, Cambridge (2018)
7. DeLoach, S., Wood, M., Sparkman, C.: Multiagent systems engineering. Int. J. Softw. Eng. Knowl. Eng. **11**(3), 231–258 (2001)
8. Dhouib, S., Kchir, S., Stinckwich, S., Ziadi, T., Ziane, M.: Robotml, a domain-specific language to design, simulate and deploy robotic applications. In: Noda, I., Ando, N., Brugali, D., Kuffner, J.J. (eds.) Simulation, Modeling, and Programming for Autonomous Robots, pp. 149–160. Springer, Berlin, Heidelberg (2012)
9. Dijkstra, E.: On the role of scientific thought. In: Selected Writings on Computing: A Personal Perspective, pp. 60–66. Springer, Heidelberg (1982)
10. Figat, M., Zieliński, C.: Methodology of designing multi-agent robot control systems utilising Hierarchical Petri Nets. In: 2019 International Conference on Robotics and Automation (ICRA), pp. 3363–3369 (2019)
11. Friedenthal, S., Moore, A., Steiner, R.: A Practical Guide to SysML: The Systems Modeling Language, 3rd edn. Morgan Kaufmann, Burlington (2015)
12. Klotzbücher, M., Smits, R., Bruyninckx, H., De Schutter, J.: Reusable hybrid force-velocity controlled motion specifications with executable domain specific languages. In: IEEE/RSJ International Conference on Intelligent Robots and Systems, San Francisco, USA, 25–30 September 2011, pp. 4684–4689 (2011)
13. Klotzbücher, M., Bruyninckx, H.: Coordinating robotic tasks and systems with rFSM statecharts. J. Softw. Eng. Rob. **3**(1), 28–56 (2012)
14. Kornuta, T., Zieliński, C.: Robot control system design exemplified by multi-camera visual servoing. J. Intell. Rob. Syst. **77**(3–4), 499–524 (2013)
15. Kortenkamp, D., Simmons, R., Brugali, D.: Robotic systems architectures and programming. In: Siciliano, B., Khatib, O. (eds.) Springer Handbook of Robotics, 2nd edn, pp. 283–306. Springer, Heidelberg (2016)
16. Moore, E.F.: Gedanken-experiments on sequential machines. In: Shannon, C.E., McCarthy, J. (eds.) Automata Studies, Annals of Mathematical Studies, no. 34, pp. 129—153. Princeton University Press, Princeton (1956)
17. Nguyen, H., Ciocarlie, M., Hsiao, K., Kemp, C.C.: ROS commander (ROSCo): behavior creation for home robots. In: IEEE International Conference on Robotics and Automation (2013)

18. Nwana, H.S., Ndumu, D.T.: A brief introduction to software agent technology, pp. 29–47. Springer, Heidelberg (1998). https://doi.org/10.1007/978-3-662-03678-5_2
19. Padgham, L., Winikoff, M.: Developing Intelligent Agent Systems: A Practical Guide. Wiley, Hoboken (2004)
20. Pilone, D., Pitman, N.: UML 2.0 in a Nutshell. O'Reilly, Newton (2005)
21. Russell, S., Norvig, P.: Artificial Intelligence: A Modern Approach. Prentice Hall, Upper Saddle River (1995)
22. Rybinski, H., Ryżko, D., Więch, P.: Learning of defaults by agents in a distributed multi-agent system environment, pp. 197–213. Springer, Heidelberg (2013). https://doi.org/10.1007/978-3-642-28699-5_8
23. Shoham, Y.: Agent-oriented programming. Artif. Intell. **60**(1), 51–92 (1993)
24. Staniak, M., Zieliński, C.: Structures of visual servos. Rob. Auton. Syst. **58**(8), 940–954 (2010)
25. Stenmark, M., Malec, J., Stolt, A.: From high-level task descriptions to executable robot code, pp. 189—202. Springer, Heidelberg (2015)
26. Wakerly, J.: Digital Design: Principles and Practices, 3rd edn. Prentice-Hall, Upper Saddle River (2000)
27. Webster, M., Dixon, C., Fisher, M., Salem, M., Saunders, J., Koay, K.L., Dautenhahn, K., Saez-Pons, J.: Toward reliable autonomous robotic assistants through formal verification: a case study. IEEE Trans. Hum.-Mach. Syst. **46**(2), 186–196 (2016)
28. Wooldridge, M.: Intelligent Agents. Multiagent Systems, pp. 27–77. MIT Press, Cambridge (1999)
29. Yannakakis, M.: Hierarchical state machines. In: van Leeuwen, J., Watanabe, O., Hagiya, M., Mosses, P., Ito, T. (eds.) Theoretical Computer Science: Exploring New Frontiers of Theoretical Informatics, pp. 315–330. Springer, Heidelberg (2000)
30. Zieliński, C.: A unified formal description of behavioural and deliberative robotic multi-agent systems. In: 7th International IFAC Symposium on Robot Control (SYROCO), vol. 7, pp. 479–486 (2003)
31. Zieliński, C.: Specification of behavioural embodied agents. In: Kozłowski, K. (ed.) Fourth International Workshop on Robot Motion and Control (RoMoCo 2004), 17–20 June 2004, pp. 79–84 (2004)
32. Zieliński, C.: General robotic system software design methodology. In: Uhl, T. (ed.) 15th IFToMM World Congress Advances in Mechanism and Machine Science, Mechanisms and Machine Science, Kraków, Poland, vol. 73, pp. 2779–2788 (2019)
33. Zieliński, C., Winiarski, T.: General specification of multi-robot control system structures. Bull. Polish Acad. Sci. Tech. Sci. **58**(1), 15–28 (2010)
34. Zieliński, C., Winiarski, T.: Motion generation in the MRROC++ robot programming framework. Int. J. Rob. Res. **29**(4), 386–413 (2010)
35. Zieliński, C.: Transition-function based approach to structuring robot control software. In: Kozłowski, K. (ed.) Robot Motion and Control, Lecture Notes in Control and Information Sciences, vol. 335, pp. 265–286. Springer, Heidelberg (2006)
36. Zieliński, C., Kornuta, T.: Diagnostic requirements in multi-robot systems. In: Korbicz, J., Kowal, M. (eds.) Intelligent Systems in Technical and Medical Diagnostics, Advances in Intelligent Systems and Computing, vol. 230, pp. 345–356. Springer, Heidelberg (2014)
37. Zieliński, C., Kornuta, T., Winiarski, T.: A systematic method of designing control systems for service and field robots. In: 19-th IEEE International Conference on Methods and Models in Automation and Robotics, MMAR, pp. 1–14. IEEE (2014)

Controlling the Posture of a Humanoid Robot

Teresa Zielinska[1(✉)] and Luo Zimin[2]

[1] Faculty of Power and Aerospace Engineering, Warsaw University of Technology, Warsaw, Poland
teresaz@meil.pw.edu.pl
[2] Singapore, Singapore

Abstract. The method for planning the robotic hand trajectory and postural adjustments of a humanoid robot is presented. The planar case is analyzed and the body displacement is represented in the sagittal plane. Such a scenario is typical for the tasks when two hands are equally involved in handling, placing or collecting some objects. The robot is expected to adjust its standing posture according to the designed trajectory of the hands. The geometrical approach to trajectory planning of the robot hands is presented. The damped least-square pseudo-inverse with null space projection is used for calculating the pseudo-inverse of augmented Jacobian. Presented method was tested using the robot model and the real prototype. The paper is ending with conclusions.

Keywords: Humanoid robot · Motion generation · Postural adjustment · Redundant structure

1 Introduction

There are many approached used for designing the postural adjustments of humanoids. Some methods use biological inspirations referring to the neurological fundamentals of human motion control [5]. Special controllers rejecting the disturbances and providing the postural corrections for gaining the postural stability are proposed [4]. Another methods apply the engineering knowledge and mathematical modeling. To that belong the methods using the task space description. Quite common is the approach with task space division when the task is divided into static part which is configuration dependent and to the dynamic part describing the movement [6]. Another example of task decomposition is the work by Grey et al. [3]. The robot body was here divided to the upper and lower part. The inverse kinematic problem is solved first for the lower body, and next it is checked if required position of the upper body is feasible. If not, then the lower body adjusts its posture. Often the classic Jacobian based approaches with matrix pseudo-inverses are used. These approaches are avoiding the problems caused by the kinematics singularities and by the redundancy of the kinematic structures [2]. Arisumi et al. [1] proposed the control law

A. Bartoszewicz et al. (Eds.): PCC 2020, AISC 1196, pp. 477–487, 2020.
https://doi.org/10.1007/978-3-030-50936-1_40

which uses the advantages of manipulated objects dynamics. The saturation of the actuators torque and the singularities of the arms position when lifting the objects are here considered. It is rather simple approach which takes advantage of strong actuators installed in the legs.

This work is presenting the method for planning the humanoid robot hands trajectory that provides human like postural adjustments. The Jacobian pseudo-inverse is used for evaluating the joints position. The discussion of different pseudo-inverses and Jacobian based methods considering the inverse kinematic problem of the humanoid robot was described in [10]. The problem and considered robot are introduced first. Next the hands trajectory planning is described. The feasible end-effector trajectory is obtained using geometric approach assisted by testing the postural stability. The internal coordinates describing the robot posture are obtained numerically using the damped-square pseudo-inverse of the augmented Jacobian. It is the type of pseudo-inverse which was evaluated as the most proper for the given task [10]. Applied approach allows to consider the boundaries and conditions imposed by the added control tasks. The testing examples are presented. The overall strategy of motion generation is outlined. The work ends with conclusion.

2 The Task

The pick and place task is considered. The aim is do design the feasible reference trajectories of the hands (end-effectors) when performing such a task. The humanoid stays in one place (the legs are not displacing). The sequence of planar postural adjustments (in the sagittal plane) following the upper limbs motion must be designed.

The robot kinematic structure creates the open chains. The structure and photographs of considered robot are shown in Fig. 1.

The robot posture is stable when the projection of the overall mass center (CoM) is located within the supporting polygon. The robot must properly adjust the posture and move the upper limbs following the designed trajectory when keeping the postural stability. Carried by the robot object's weight and object's position are considered when checking the postural stability. The orientation of the end-effector (hands) is not pertinent. The robot has 8DOF, 3 in each leg and 1 in each shoulder. The kinematic structure is redundant, the joint space has more degrees of freedom than it is required for performing the given task. It is expected that the achieved posture will resemble a human posture.

Due to the structural redundancy it is not sufficient to use direct solution of the inverse kinematics problem (see Fig. 2), therefore the Jacobian based approach must be applied for evaluating the required joint positions.

The Jacobian matrix denoted as $\mathbf{J}$ relates the task and joint space velocities $\dot{\mathbf{x}}$ and $\dot{\mathbf{q}}$:

$$\dot{\mathbf{x}} = \mathbf{J}\dot{\mathbf{q}} \tag{1}$$

The joint space velocities are obtained using Jacobian inverse:

$$\dot{\mathbf{q}} = \mathbf{J}^{-1}\dot{\mathbf{x}} \tag{2}$$

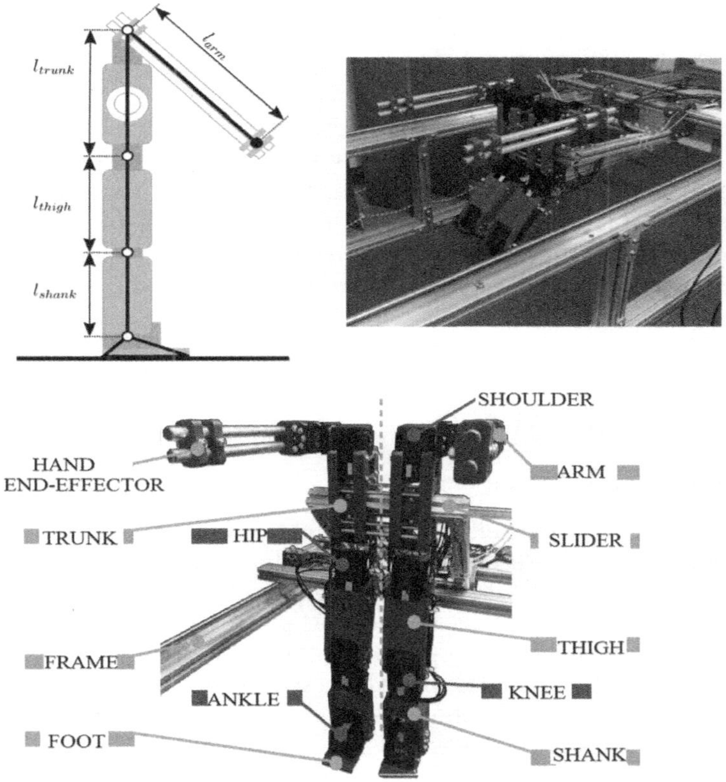

Fig. 1. Simplified view of the robot structure and the robot views.

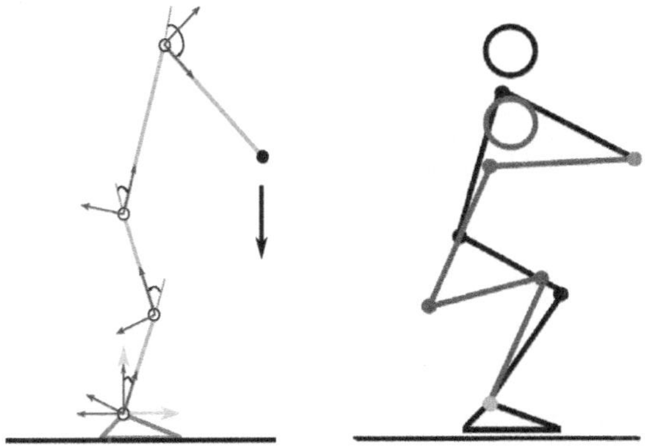

Fig. 2. Robot kinematics and illustration of redundancy.

Instead of regular inverse the pseudo-inverse $\mathbf{J}^{\#}$ is used for managing the redundancy and singularities, and:

$$\hat{\dot{\mathbf{q}}} = \mathbf{J}^{\#}\dot{\mathbf{x}} \tag{3}$$

As it was presented in [10] pseudo-inverse obtained by damped-least square method provides the solution without the numerically caused oscillations as it happens in the other methods. To the cost function $F(\hat{\dot{\mathbf{q}}}) = \|\mathbf{J}\hat{\dot{q}} - \dot{\mathbf{x}}\|^2$ used for minimizing the difference between ($\dot{\mathbf{q}}$) and ($\hat{\dot{\mathbf{q}}}$) is added the term providing the damping of oscillations. In the other words, the damping factor λ is penalizing the "virtual" velocities appearing in the neighborhood of singularities [9]:

$$F(\hat{\dot{\mathbf{q}}}) = \|\mathbf{J}\hat{\dot{q}} - \dot{\mathbf{x}}\|^2 + \|\lambda\hat{\dot{\mathbf{q}}}\|^2 \tag{4}$$

3 Design of the Motion

3.1 End-Effector Trajectory

The end-effector trajectory must be produced and next transformed to the joint trajectories. We start with the given data: – the initial robot configuration $\mathbf{q}_0$, – the hands destination G described in the Cartesian space, – the mass of the manipulated object m (it influences the CoM position). We search for the hands trajectory which, after being converted to generalized trajectory in the joint space Q, will be send to the motors. The designed trajectory of the hands must be such, that the resulting robot poses hold stability resembling the human postures without sharp and big reconfigurations.

Planning the motion in joint space, we determine first the final configuration of the humanoid robot $\boldsymbol{q}_f$ considering the hands designation G expressed in Cartesian space. We need to find such $\boldsymbol{q}_f$ which results in minimal change in the posture. Particularly it concerns the trunk pose between the initial and the final configuration. It is achieved through geometric approach. The advantage of the geometric approach is such, that the requirements for achieving human like posture can be easily manipulated (e.g. it can concern the hip position).

The method of hands trajectory design is following. First, we draw a circle from the target point with the radius equal to the length of the arm l_{arm} to find the admissible shoulder location. There are two possible scenarios: 1) when there is no intersection between the goal and the trunk, including its extension, and 2) the circle and the extension line intersect at b_1 and b_2. b_1 and b_2 are thereby potential shoulder locations that keeps the trunk straight. Subsequently, we compare the distances from the current shoulder location to each point and move the shoulder to the one with shorter distance. In regard to the first case, we have to draw another circle to find the potential hip location. The consecutive procedure is to connect the goal A and the shoulder and we mark the intersection point as a_1, this is also the shoulder position that can reach A and is the closest to the current position. Once a_1 is found, we draw a second circle with the radius l_{trunk} from the center a_1, and mark its intersection with the trunk extension as a_2 and a_3, and move the hip to the intersection point which is closer to the current location.

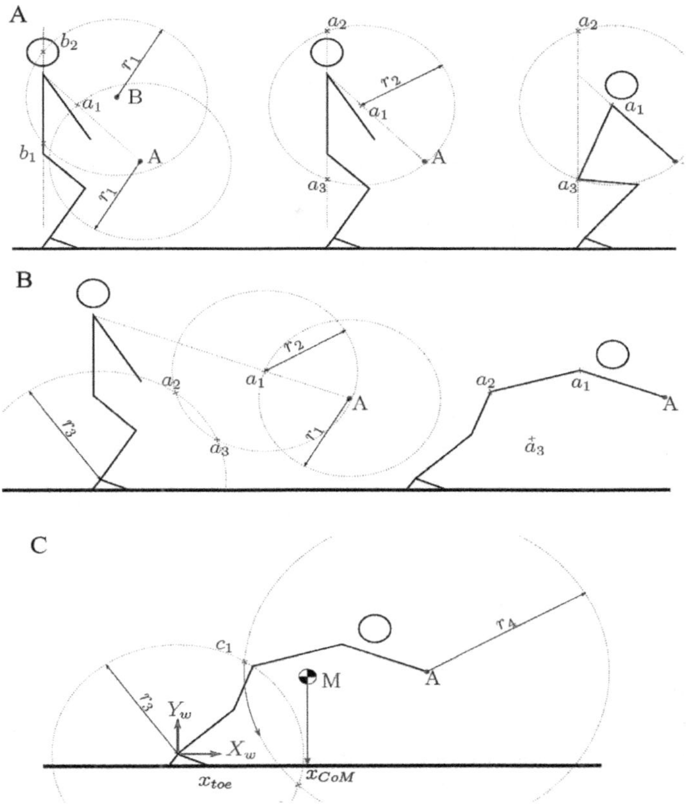

Fig. 3. Planning the hand trajectory: A) the obstacle is close, B) the obstacle is far, C) checking the postural stability.

In the case where the third circle has no intersection with trunk extension, we need to draw a third circle from the ankle with the radius $l_{shank} + l_{thigh}$, which represents the workspace of the lower body mechanism. The overlap area between the second circle and the third one is the collection of admissible hip positions. We choose the leftmost intersection a_2 to be our target hip position, and the final configuration is shown in Fig. 3A. In this figure the target location B is close enough, the robot is able to reach it while keeping its trunk straight; otherwise the trunk will lean forward to reach the goal A. The black dotted line represents the trunk extension, the blue circle is centered at B with radius of $r_1 = l_{arm}$; the red circle in the right figure is centered at a different point A with radius $r_1 = l_{arm}$ and the one in the middle and right figure is centered at a_1 with radius $r_2 = l_{trunk}$. The final pose is shown on the right in a free environment without considering the balance constraint. a_1, b_1 and b_2 are the target shoulder location, a_2 and a_3 are the potential hip location. Points A and B are marking different target locations.

Figure 3B presents the situation When there is no intersection between the second circle and the trunk extension, a third circle is drawn from the ankle with radius $r_3 = l_{shank} + l_{thigh}$. The hip can be positioned in the overlap area between a_2 and a_3 in order to reach the goal A. For simplification purpose, we consider only a_2 and a_3 as potential locations and eventually move the hip joint to a_2 in this exemplary configuration. The final pose in a obstacle-free environment is shown on the right without considering the balance constraint (just in this stage). The notation is following: $r_1 = l_{arm}$, $r_2 = l_{trunk}$, a_1 is the target shoulder joint location, a_2 and a_3 are potential hip joint locations. A denotes the target point.

Checking whether a target point can be achieved while keeping the balance uses also the geometric approach what illustrates Fig. 3C. Here $r_4 = l_{arm} + l_{trunk}$, $r_3 = l_{shank} + l_{thigh}$. The overlap of the two circles represents the possible hip joint locations. We find out the hip joint location by moving the hip from c_1 towards c_2 along the curve and calculate whether the CoM projection is within the support polygon. x_{CoM}, x_{toe} are the CoM and toe coordinates along the horizontal line. x_{CoM} is calculated considering the robot partial masses, the body parts positions and the dimensions. If $x_{CoM} > x_{toe}$ when x_{hip} is at its lowest position, we conclude that the target point A can not be reached by the humanoid robot without losing the balance.

Having determined the initial and final configurations, we subsequently generate the trajectory points. The balance constraint is checked in our procedure, as it is described bellow. Moreover the hands trajectory is planned considering the collision avoidance. It is a simple approach with produces first the obstacle region and defines the intermediate points around the obstacle. The avoidance trajectory consists of the fragments connecting smoothly these points. To assess whether a path point can be reached by the robot while maintaining balance, involves two actions: – testing whether the point can be reached with the current configuration, and if not – modifying the posture and testing whether the point can be reached with modified configuration, as it is presented below.

3.2 Controlling the Configuration

Although the path points' validity was checked and confirmed beforehand, the robot can become imbalanced while following the trajectory due to the redundancy or disturbances, therefore the postural stability must be tested during the motion as well.

The input parameters of the task manipulation algorithm are: – initial configuration $\boldsymbol{q}_0$, – the destination G expressed in the Cartesian space, – the mass of the hold object is m. The final result of our algorithm is the trajectory in the joint space Q which is send to the actuators for realization.

The overall procedure is following:

A. The start position S of the end-effector is computed using Forward Kinematics; if S and G are both valid (are within the work space), described above trajectory generation procedure generates a straight-line with high density of via points. The collision against the obstacles is checked (assuming that the obstacle position and size is known). If the collision occurs between planned trajectory and the obstacle, the avoiding obstacles procedure returns intermediate points (around the obstacle). Considering S and G the parabolic trajectories between these points are generated in the Cartesian space. If the environment is obstacle-free the parabolic trajectory from S to G is directly generated.

B. Once a trajectory is generated, the Jacobian-based inverse kinematics with null-space projection and priority control is applied computing $\Delta \boldsymbol{q}_{i+1}$ basis on the Eq. 5:

$$\Delta \boldsymbol{q}_{i+1}^{i} = \mathbf{J}(q)\boldsymbol{e}_{i+1}^{i} \tag{5}$$

$\Delta \boldsymbol{q}_{i+1}^{i}$ is the difference in the joint space between time instant i and $i+1$; $\boldsymbol{e}_{i+1}^{i}$ is the difference in the Cartesian space between the end-effector position (at i) and the goal position at $i+1$.

C. During the movement the following actions are taken:

1) if the CoM projection x_{CoM} stays in the stability region, then the position is updated: $Q = \boldsymbol{q}_0 + \Delta \boldsymbol{q}_{i+1}^{i}$, where $i = 0, ..., N-1$, and N is the number of the path points including S and G,
2) if the control procedure detects that x_{CoM} will leave the stability region in the next time instant for which the position $\boldsymbol{q}$ was planned, the movement is stopped immediately, the current configuration $\boldsymbol{q}_c$ is stored, as well as the current end-effector position $\boldsymbol{p}_c^i$ (it is the sub-goal position). The pose adjustment procedure modifies the configuration until x_{CoM} reaches the position expected for next time instant. The pose adjustment procedure is as following (note: it is performed by calculations and not in the real robot): the hip position is modified by small backward displacement along horizontal line with holding the end-effector position unchanged. Stability of the new position is tested. If the hip displacement provides unstable position the displacement is disregarded. As the next possible postural adjustment the small hip displacement downwards and along an arc is considered and the whole procedure is repeated till the success. Let us assume that the new successfully reached configuration is denoted by $\boldsymbol{q}_n$. A cubic polynomial trajectory is generated between $\boldsymbol{q}_c$ and $\boldsymbol{q}_n$. In the meantime, a new parabolic trajectory is generated from $\boldsymbol{p}_c^{i-1}$ to $\boldsymbol{p}_c^i$, $\boldsymbol{p}_c^{i-1}$ denotes the previous sub-goal, and for $i = 1$ it is G. The joint angles resulted from both, motion towards the goal and the pose modification generate the path Q.
3) in any case, if one or two joint angles are about to exceed their limits in the next time instant, the Jacobian matrix drops the corresponding ranks and produces zero velocities to these joints.

3.3 Solving the Inverse Kinematics Problem

Solving the inverse kinematics problem we use the augmented Jacobian which incorporates additional boundaries and conditions [7]. It means that the additional tasks with the dimension equal to the degree of redundancy are defined and added to the Jacobian. The additional tasks are prioritized. The highest priority (main) task is located as the first one, next are placed the remaining tasks following their decreasing priorities:

$$\mathbf{J}_A = [\mathbf{J}^{(main)}, \mathbf{J}^{(2)}, \mathbf{J}^{(3)}]^T \tag{6}$$

As the first added task (with highest priority) we defined balance maintenance, which is monitored by checking the CoM projection x_{CoM}. As the second added task we considered a path following requirement in the Cartesian space.

The null space projection is used when evaluating augmented Jacobian pseudo-inverse. The null space describes such motions of the parts of a kinematic chain which are not moving the end-effector. In other words, the joint velocities projected through the range of the Jacobian provides the end-effector motion, but the joint velocities projected through the null space of the Jacobian are not affecting the end-effector motion. It can be describes by:

$$\dot{\mathbf{q}} = \dot{\mathbf{q}}_r + \dot{\mathbf{q}}_{\aleph} \tag{7}$$

where $\dot{\mathbf{q}}_r$ are the velocities that drive the end-effector motion, $\dot{\mathbf{q}}_{\aleph}$ are the null space velocities. The inverse kinematics problem is solved using the augmented Jacobian pseudo-inverse and the null space (see [7,8]):

$$\dot{\mathbf{q}} = \mathbf{J}^{\#}\dot{\mathbf{x}} + (\mathbf{I} - \mathbf{J}^{\#}\mathbf{J})\mathbf{z} \tag{8}$$

Here $(\mathbf{I} - \mathbf{J}^{\#}\mathbf{J})$ is a null space projector, and $\mathbf{z}$ is an arbitrary vector projected to the null space of the Jacobian matrix. This vector not causes any motion in Cartesian space (not affects $\dot{\mathbf{x}}$). Therefore $\mathbf{z}$ is considered as an additional factor in our task control criteria. Following [8] priorities are defined first. The velocities concerning the lower priority tasks are projected into the null space of higher priority tasks. Such projection prevents from interfering the lower priority tasks with the higher priority tasks.

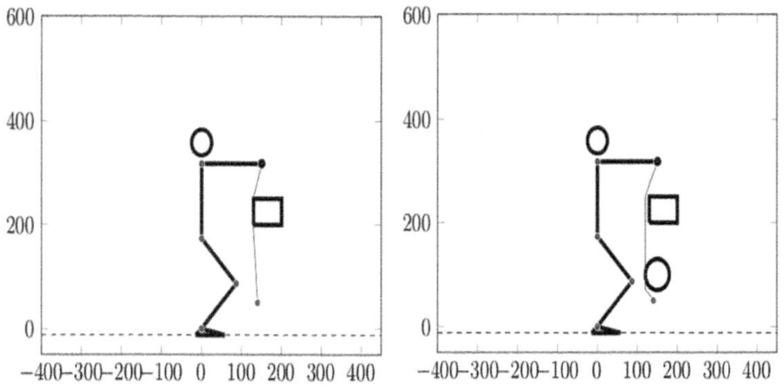

Fig. 4. Hands trajectory with obstacle avoidance: A) rectangular obstacle, B) two obstacles of different shape.

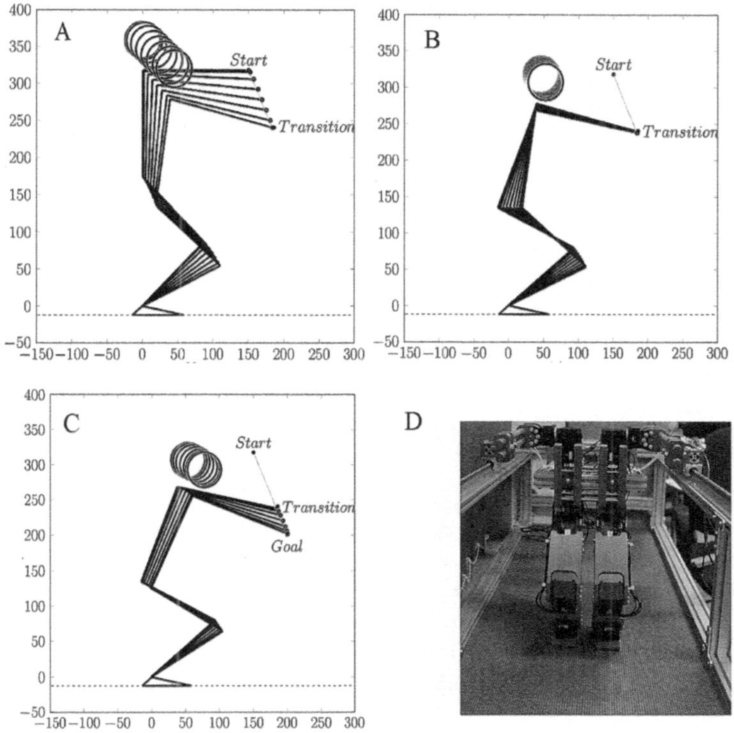

Fig. 5. Postural adjustments when the hands are following the designed trajectory: A) first phase of motion, B) transition, C) last phase of motion, D) real robot starting the designed motion.

4 Example

Doing the tests the parameters of prototype robot including dimensions and body parts masses of each link were used. It was assumed that the left and right side of the body are keeping the identical posture, therefore the 8DOF robot was reduced to 4DOF structure. Therefore the task space Jacobian $\mathbf{J}_{task}$ (2×4) was connecting the hands velocity expressed in the 2D space with the four joint velocities.

The method was successfully tested by simulation and by the experiments. Figure 4 illustrates the designed hands trajectory with obstacles avoidance. Two cases are shown, with one rectangular obstacle and with two obstacles - rectangular and round. Figure 5 is showing the postural adjustments during following the designed trajectory of the end-effector. In Fig. 5A the first phase of motion is presented when the body moves down with trunk leaning to the front, first part of hands trajectory is followed. In the transition point the farther motion by this scheme is impossible because the postural stability will be lost. Figure 5B illustrates the second phase (transition) when the hip position is adjusted for getting the robust stable posture. In Fig. 5C the motion is continued in similar manner as in the first stage (till transition point). Figure 5D is showing the real robot in the beginning of postural adjustments.

5 Conclusions

The problem of postural adjustment of a humanoid robot was studied. The adjustments were allowing the robot hands to follow the given trajectory when lifting the loads. The cases with free space and space with the obstacles were considered. In presented method the robot stops the hands motion at few selected points surrounding the obstacle and adjusts the posture. The final trajectory of the hands should be smoothed what will be the next stage of the research. In presented works it was assumed that the robot's feet are fixed to the ground, however it is reasonable to move one foot forwards or backwards to enhance the postural stability. This could be particularly useful while lifting a heavier objects. Furthermore, the lifting motion can be incorporated with walking motion so that the robot can be close as much as possible to the objects. In presented research the self-collisions are prevented by simple checking of the joint limit constraints, however it is a more complex problem in a practice. The most straightforward way for considering the collisions is using the 3D robot model for defining the collision free joint motion ranges.

References

1. Arisumi, H., Miossec, S., Chardonnet, J.R., Yokoi, K.: Dynamic lifting by whole body motion of humanoid robots. In: IEEE/RSJ International Conference on Intelligent Robots and Systems, pp. 668–675 (2008)
2. Youngjin, C., Kim Doik, O., Yonghwan, Y.B.-J.: Posture/Walking control for humanoid robot based on kinematic resolution of CoM Jacobian with embedded motion. IEEE Trans. Rob. **23**, 1285–1293 (2008)
3. Grey, M., Joo, S., Zucker, M.: Planning heavy lifts for humanoid robots. In: IEEE-RAS International Conference on Humanoid Robot, pp. 640–645 (2015)
4. Lippi, V., Mergner, T.: Human-derived disturbance estimation and compensation (DEC) method lends itself to a modular sensorimotor control in a humanoid robot. Frontiers Neurorob. (2017)
5. Mergner, T., Lippi, V.: Posture control - human-inspired approaches for humanoid robot benchmarking: conceptualizing tests, protocols and analyses. Frontiers Neurorob. **21**(21) (2018)
6. Thomassino, P.: Task-space separation principle: a force field approach to posture and movement planning for redundant manipulators. In: From Human Postural Synergies to Bio-Inspired Motion Planning for Redundant Manipulators, pp. 23–56. Springer Theses (2019)
7. Sciavicco, L., Siciliano, B.: Modelling and Control of Robot Manipulators, 2nd edn. Springer, London (2000)

8. Siciliano, B., Slotine, J.J.E.: A general framework for managing multiple tasks in highly redundant robotic systems. In: 5th International Conference on Advanced Robotics, pp. 1211–1216 (1991)
9. Wampler, C.W.: Manipulator inverse kinematic solutions based on vector formulations and damped least-squares methods. IEEE Trans. Syst. Man Cybern. **16**, 93–101 (1986)
10. Zielinska, T., Zimin, L., Szumowski, M., Ge, W.: Motion planning for a humanoid robot with task dependent constraints. In: Advances in Mechanism and Machine Science. IFToMM World Congress 2019. Mechanisms and Machine Science, vol. 73. Springer (2019)

Tracking Objects Using Stereo Vision System with Vergence and Gaze Control Mechanism

Przemysław Szewczyk(✉)

Department of Robot Control, Institute of Automatic Control,
ul. B. Stefanowskiego 18/22, 90-924 Lodz, Poland
przemyslaw.szewczyk@o2.pl

Abstract. This paper describes Image-Based (IBVS) and Pose-Based (PBVS) Visual Servo control approaches used in stereo vision system with vergence and gaze control mechanism. Using IBVS or PBVS controller in cameras positioning mechanism ensures that cameras are fixed on a common visual target. Keeping the tracking object fixated causes the images of the target to lie near the principal points of both cameras. For this case disparity of target object is very close to zero, so for 3D reconstruction of target neighborhood, the stereo matching algorithm which accepts only limited range of disparities can be used.

Keywords: Stereo vision · Tracking objects · Image-Based Visual Servo · Pose-Based Visual Servo · Vision-based robot control

1 Introduction

The stereo vision has been the subject of research for years and nowadays there are many industries in which it is widely used, from entertainment through construction measurements to autonomous vehicles.

Most of these applications use a canonical system in which the cameras are stationary and their optical axes are parallel. The use of this configuration causes some simplification of calculations [3] but on the other hand, it enforces the use of a stereo matching algorithm, which works in a wide range of disparities. The wide range of discrepancies means that more computer resources are needed to find good correspondences, and this is a major limitation on the use of such solution.

In many cases it is not necessary to calculate the reconstruction of a 3D scene frame by frame, but only to update the position of an object which state has changed or which is in a particular region of interest. By using the stereo vision system with vergence and gaze control mechanisms, it is possible to keep the fixation point close to some visual target. This causes that the tracking object always lies near the zero disparity surface called horopter [4] and it is in the center of the common field of view of both cameras.

A. Bartoszewicz et al. (Eds.): PCC 2020, AISC 1196, pp. 488–499, 2020.
https://doi.org/10.1007/978-3-030-50936-1_41

The stereo matching algorithms are computationally expensive [4]. Reducing the range of disparities by using such fixation mechanism can significantly speed up process of searching stereo correspondences and it became a main motivation to build stereo vision system with vergence and gaze control mechanism. This article describes two different visual servo systems for vergence and gaze control. Detailed information about mechanical construction of system, mechanical linkage parameter optimization and motor-based stereo camera parameter estimation were described in articles [6,7].

2 Vision-Based Control

Two basic configurations can be distinguished considering the location of camera in a robotic system [2,8]:

- *Eye-to-hand(ETH)*, fixed, stand alone camera is observing the target and the motion of the robot effector, Fig. 1a,
- *Eye-in-hand(EIH)*, the camera is attached to the moving robot effector and observing the relative position of the target, Fig. 1b.

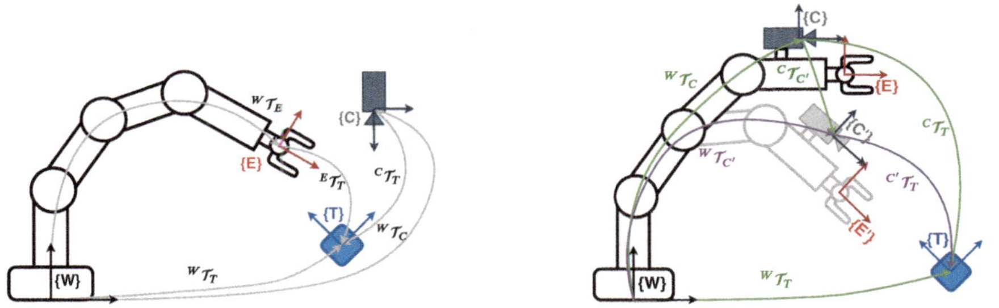

(a) Eye-to-hand (b) Eye-in-hand

Fig. 1. Visual servo main configurations and relevant coordinate frames: base {B}, end-effector {E}, camera {C} and target {T}

Considering type of feedback signal in the control loop, visual servos are classified into the two fundamental types [1,3,5]:

- *Position/Pose-Based Visual Servo (PBVS)*, which uses observed visual features, a calibrated camera and a known geometric model to determine position of the target with respect to the camera,
- *Image-Based Visual Servo (IBVS)*, which omits the pose estimation step and uses the images features directly.

Assuming that both cameras in stereo vision system have similar vision pipelines - the same types of image sensors and optical lenses used, resolutions, exposure settings, etc., and the system also ensures image capture synchronization between cameras, then the pair of cameras can be presented as one image sensor, and the whole system as an eye-in-hand configuration.

2.1 Pose-Base Visual Servo

In a PBVS system the control is performed in task space which is commonly $\mathbb{R}^3$. The error signal is calculated as a difference between current $\{E\}$ and desired $\{E'\}$ end-effector pose with respect to the current object pose in camera coordinate system ${}^{C}\hat{\mathcal{T}}_T$. The relationship between poses is shown in Fig. 1b. The error is therefore a homogeneous transformation matrix defined as follows:

$$\varepsilon_E = {}^{E}\mathcal{T}_{E'} = {}^{E}\mathcal{T}_C \, {}^{C}\hat{\mathcal{T}}_T \, {}^{T}\mathcal{T}_{E'} \tag{1}$$

where:

- ${}^{E}\mathcal{T}_C$ - camera pose in end-effector coordinate system,
- ${}^{C}\hat{\mathcal{T}}_T$ - estimated in vision task target pose in camera coordinate system,
- ${}^{T}\mathcal{T}_{E'}$ - required end-effector pose in target object coordinate system.

Assuming that in EIH configuration ${}^{E}\mathcal{T}_C = {}^{E'}\mathcal{T}_C = const.$, the error can be defined in camera coordinate system:

$$\varepsilon_C = {}^{C}\mathcal{T}_{C'} = {}^{C}\hat{\mathcal{T}}_T \, {}^{T}\mathcal{T}_{C'} \tag{2}$$

where:

- ${}^{T}\mathcal{T}_{C'}$ - required camera pose in object coordinate system.

The change in pose ε_C might be large and it can be realized in small steps by control system:

$${}^{W}\mathcal{T}_C(k+1) = {}^{W}\mathcal{T}_C(k) \, \hat{\varepsilon_C}(k) \tag{3}$$

where:

- $\hat{\varepsilon_C}(k)$ - the subgoal error (increment) between pose $\{C\}$ and $\{C'\}$ with respect to frame $\{T\}$.

The PBVS controller calls (3) repeatedly until $\hat{\varepsilon_C}$ falls below some threshold and the motion is completed. In this way even if robot does not move as requested due to errors or the target object changes position, the controller will take this error into account at the next iteration.

The whole PBVS process can be presented in diagram form as shown in Fig. 2.

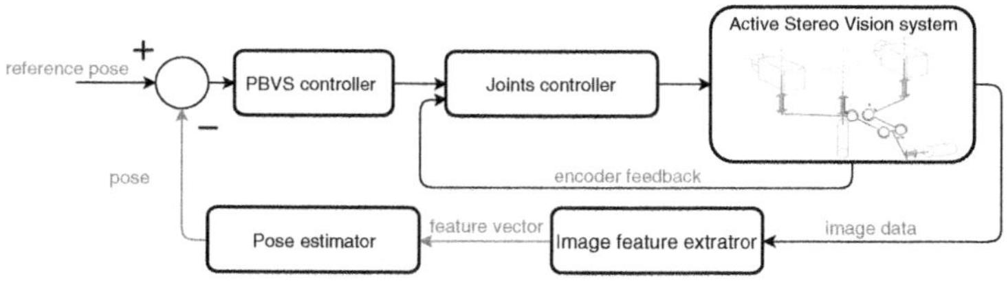

Fig. 2. Pose-Based Visual Servo in stereo vision system

Assuming that stereo vision non-fixed camera system (SnFCS) can independently control gaze and vergence angles by using the mechanism as shown in Fig. 3, the forward kinematics task for this system can be defined as follows:

$$ {}^{W}\mathcal{T}_{Cl} = {}^{W}\mathcal{T}_{B}\,\mathcal{T}(\alpha_G)\,{}^{B}\mathcal{T}_{Cl}\,\mathcal{T}(\alpha_{V+}) \tag{4} $$
$$ {}^{W}\mathcal{T}_{Cr} = {}^{W}\mathcal{T}_{B}\,\mathcal{T}(\alpha_G)\,{}^{B}\mathcal{T}_{Cr}\,\mathcal{T}(\alpha_{V-}) \tag{5} $$

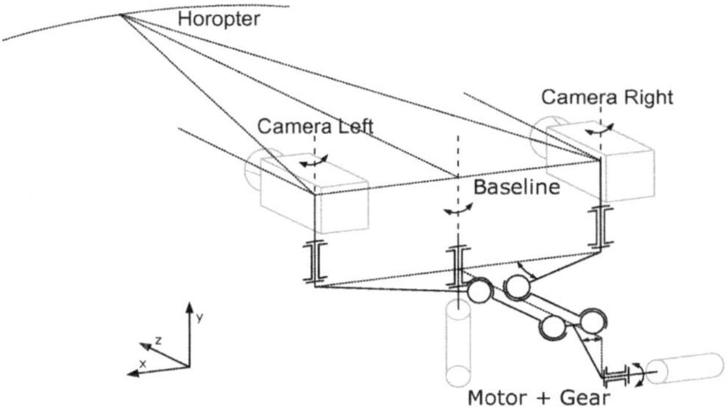

Fig. 3. Vergence control mechanism

where:

- ${}^{W}\mathcal{T}_{Cl}$, ${}^{W}\mathcal{T}_{Cr}$ - pose of the left and right camera in world coordinate system,
- ${}^{W}\mathcal{T}_{B}$ - initial pose of the robot base, as shown in Fig. 4(a, b),
- ${}^{B}\mathcal{T}_{Cl}$, ${}^{B}\mathcal{T}_{Cr}$ - initial pose of the left and right camera in base coordinate system, as shown in Fig. 4(a, b),
- $\mathcal{T}(\alpha_G)$ - rotation of the robot base,
- $\mathcal{T}(\alpha_{V+})$ - rotation of the left camera,
- $\mathcal{T}(\alpha_{V-})$ - rotation of the right camera.

Having the current pose estimation of a tracked object from the stereo processing task, which is typically assigned to one of the cameras (might be also introduced a virtual camera), e.g. left camera ${}^{Cl}\hat{\mathcal{T}}_T$, the object pose can be presented in world coordinate system using formula (4), and next the gaze angle α_G can be calculated:

$$ {}^{W}\hat{\mathcal{T}}_T = {}^{W}\mathcal{T}_{Cl}\,{}^{Cl}\hat{\mathcal{T}}_T = \begin{bmatrix} {}^{W}\mathbf{R}_T & {}^{W}\mathbf{T}_T \\ \mathbf{0} & 1 \end{bmatrix} \tag{6} $$

$$ \alpha_G = \arccos \hat{T}_z \tag{7} $$

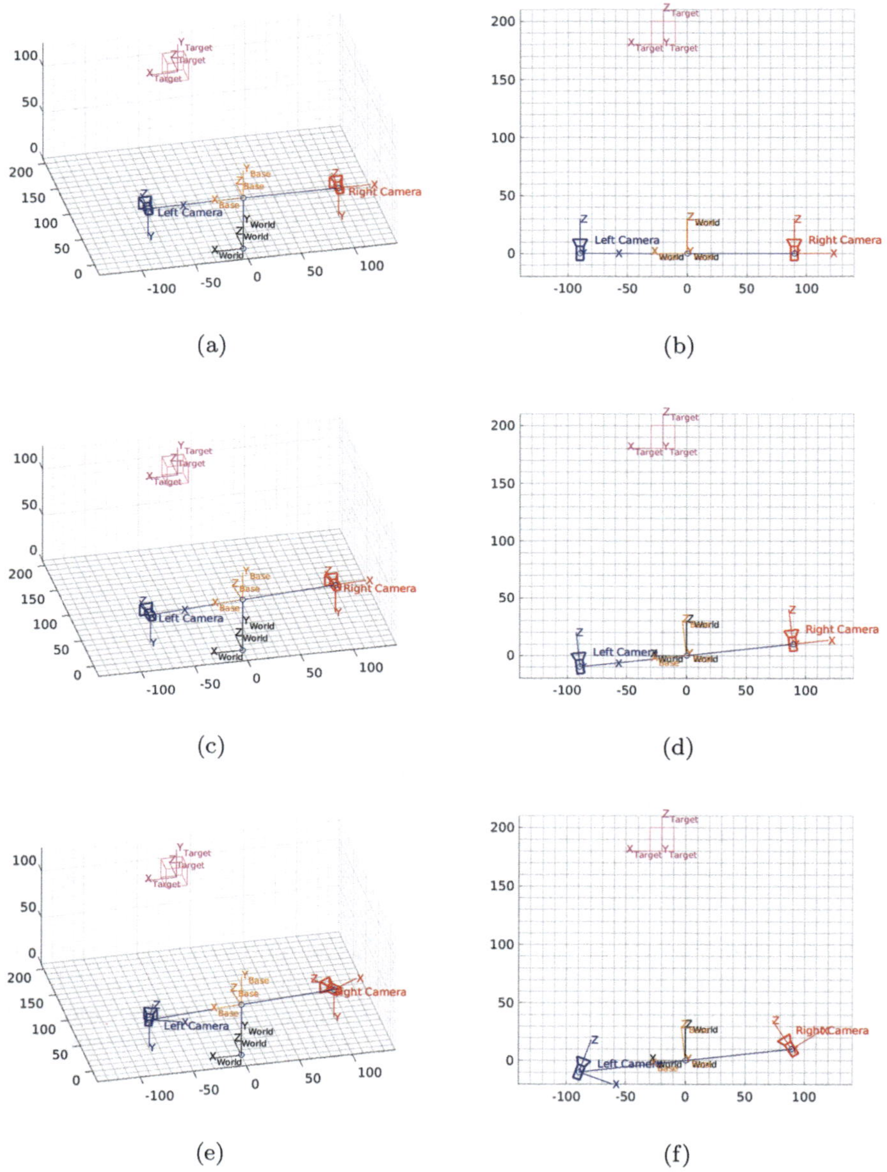

Fig. 4. The use of PBVS in SnFCS - Phase I, Initial (canonical) pose of robot (a, b), Phase II, The robot turned towards target object (c, d), Phase III, the robot and both cameras turned towards target object (e, f)

where:

- ${}^{W}\mathbf{T}_T = [T_x\ T_y\ T_z]^T$ - translation vector component of the matrix ${}^{W}\hat{\mathcal{T}}_T$,
- $\hat{T}_z$ - z component of normalized vector ${}^{W}\hat{\mathbf{T}}_T = \frac{[T_x\ T_z]^T}{\|[T_x\ T_z]\|}$.

For the new α_G estimation (Fig. 4(c, d)), by using formula (4) again, it is possible to determine the new pose of the left camera ${}^{W}\mathcal{T}_{Cl'}$ and right camera ${}^{W}\mathcal{T}_{Cr'}$ and then calculate the desired vergence angle α_V:

$$ {}^{Cl'}\mathcal{T}_T = {}^{W}\mathcal{T}_{Cl'}^{T}\ {}^{W}\hat{\mathcal{T}}_T = \begin{bmatrix} {}^{Cl'}\mathbf{R}_T & {}^{Cl'}\mathbf{T}_T \\ \mathbf{0} & 1 \end{bmatrix} \tag{8} $$

$$ \alpha_{V+} = \arccos \hat{T}_z \tag{9} $$

where:

- ${}^{Cl'}\mathbf{T}_T = [T_x\ T_y\ T_z]^T$ - translation vector component of the homogeneous matrix ${}^{Cl'}\hat{\mathcal{T}}_T$,
- $\hat{T}_z$ - z component of normalized vector ${}^{Cl'}\hat{\mathbf{T}}_T = \frac{[T_x\ T_z]^T}{\|[T_x\ T_z]\|}$.

The vergence angle for the right camera α_{V-} can be determined in the same way, or assuming the perfect system symmetry, use the value $\alpha_{V-} = -\alpha_{V+}$.

The Fig. 4 shows the sequence of steps in determining the desired pose of the camera system relative to the tracked object. In addition to presenting model of the camera system, simulations of right and left camera views were also shown.

The pose estimation of the target object obtained in the vision task, due to detection errors, can be omitted for a single or several images in the sequence. Additionally, it contains some noise. Assuming that the motion of the tracked object can be described in the form of a dynamic linear model and that the measurement noise associated with object detection is a Gaussian noise, the Kalman filter to estimate the position of this object can be used.

The motion model can be described in state-space representation:

$$ \mathbf{x}_k = \mathbf{F}_k \mathbf{x}_{k-1} + \mathbf{B}_k \mathbf{u}_k + \mathbf{w}_k \tag{10} $$

$$ \mathbf{z}_k = \mathbf{H}_k \mathbf{x}_k + \mathbf{v}_k \tag{11} $$

where:

- $\mathbf{x}_k, \mathbf{x}_{k-1}$ - state vector at time k and $k-1$, $\mathbf{z}_k$ - measurement vector, $\mathbf{u}_k$ - control vector,
- $\mathbf{F}_k$ - state-transition model, $\mathbf{B}_k$ - control-input model, $\mathbf{H}_k$ - observation model,
- $\mathbf{w}_k$ - process noise vector, $w_k \sim N(0, \mathbf{Q}_k)$, $\mathbf{Q}_k$ - covariance of the process noise,
- $\mathbf{v}_k$ - measurement noise, $v_k \sim N(0, \mathbf{R}_k)$, $\mathbf{R}_k$ - covariance of the observation noise.

In the general case, it is not known how the tracked object is controlled, so the expression $B_k u_k$ takes the value 0. Assuming the use of *Constant Acceleration Model*, the state vector can be defined as follows:

$$\mathbf{x}_k = [\, x_k \; y_k \; z_k \; \dot{x_k} \; \dot{y_k} \; \dot{z_k} \; \ddot{x_k} \; \ddot{y_k} \; \ddot{z_k} \,]^T \tag{12}$$

where:

- $x_k = p_{xk}$, $y_k = p_{yk}$, $z_k = p_{zk}$ - coordinates of the object at the time k associated with the robot base coordinate system,
- $\dot{x_k} = v_{xk}$, $\dot{y_k} = v_{yk}$, $\dot{z_k} = v_{zk}$ - object velocity components at the moment k,
- $\ddot{x_k} = a_{xk}$, $\ddot{y_k} = a_{yk}$, $\ddot{z_k} = a_{zk}$ object acceleration components at the moment k.

For above state vector, the state-transition model can be defined as follows:

$$\mathbf{F}_k = \begin{bmatrix} 1 & 0 & 0 & \Delta t & 0 & 0 & \frac{(\Delta t)^2}{2} & 0 & 0 \\ 0 & 1 & 0 & 0 & \Delta t & 0 & 0 & \frac{(\Delta t)^2}{2} & 0 \\ \vdots & \ddots & 1 & 0 & 0 & \Delta t & 0 & 0 & \frac{(\Delta t)^2}{2} \\ & & & 1 & 0 & 0 & \Delta t & 0 & 0 \\ & & & & 1 & 0 & 0 & \Delta t & 0 \\ & & & & & 1 & 0 & 0 & \Delta t \\ & & & & & & 1 & 0 & 0 \\ \vdots & & & & & & \ddots & 1 & 0 \\ 0 & \cdots & & & & & \cdots & 0 & 1 \end{bmatrix} \tag{13}$$

The process noise vector from the Eq. 10 takes the form:

$$\mathbf{w}_k = \begin{bmatrix} w_{px} \\ w_{py} \\ w_{pz} \\ w_{vx} \\ w_{vy} \\ w_{vz} \\ w_{ax} \\ w_{ay} \\ w_{az} \end{bmatrix} = \begin{bmatrix} j_x \frac{\Delta t^3}{6} \\ j_y \frac{\Delta t^3}{6} \\ j_z \frac{\Delta t^3}{6} \\ j_x \frac{\Delta t^2}{2} \\ j_y \frac{\Delta t^2}{2} \\ j_z \frac{\Delta t^2}{2} \\ j_x \Delta t \\ j_y \Delta t \\ j_z \Delta t \end{bmatrix} = \begin{bmatrix} \frac{\Delta t^3}{6} & 0 & 0 \\ 0 & \frac{\Delta t^3}{6} & 0 \\ 0 & 0 & \frac{\Delta t^3}{6} \\ \frac{\Delta t^2}{2} & 0 & 0 \\ 0 & \frac{\Delta t^2}{2} & 0 \\ 0 & 0 & \frac{\Delta t^2}{2} \\ \Delta t & 0 & 0 \\ 0 & \Delta t & 0 \\ 0 & 0 & \Delta t \end{bmatrix} \begin{bmatrix} j_x \\ j_y \\ j_z \end{bmatrix} = \mathbf{G}_k \mathbf{j}_k \tag{14}$$

where:

- j_x, j_y, j_z - components of jerk vector $\mathbf{j}_k$ at time k.

The Δt is calculated at each iteration of the algorithm. The jerk is a random vector with an average value of 0 and a standard deviation of σ_{jx}, σ_{jy}, σ_{jz}.

The covariance matrix $\mathbf{Q}_k$ can be determined based on the process noise vector $\mathbf{w}_k$:

$$\mathbf{Q}_k = E[\mathbf{w}_k\mathbf{w}_k^T] = E[\mathbf{G}_k\mathbf{j}_k\mathbf{j}_k^T\mathbf{G}_k^T] = \mathbf{G}_k E[\mathbf{j}_k\mathbf{j}_k^T]\mathbf{G}_k^T \tag{15}$$

$$\mathbf{Q}_k = \mathbf{G}_k \begin{bmatrix} j_x \\ j_y \\ j_z \end{bmatrix} \mathbf{G}_k^T = \mathbf{G}_k \begin{bmatrix} \sigma_{jx}^2 & \sigma_{jxy} & \sigma_{jxz} \\ \sigma_{jxy} & \sigma_{jy}^2 & \sigma_{jyz} \\ \sigma_{jxz} & \sigma_{jyz} & \sigma_{jz}^2 \end{bmatrix} \mathbf{G}_k^T = \mathbf{G}_k\hat{\mathbf{Q}}_k\mathbf{G}_k^T$$

Assuming that the values j_x, j_y, j_z are uncorrelated, covariances σ_{jxy}, σ_{jxz}, σ_{jyz} are equal to zero in the matrix $\hat{\mathbf{Q}}_k$.

In the vision task, only the position of the tracked object is estimated, so the observation model $\mathbf{H}_k$ takes the following form in the Eq. (11):

$$\mathbf{H}_k = \begin{bmatrix} 1 & 0 & \cdots & & \cdots & 0 \\ 0 & 1 & 0 & \cdots & \cdots & 0 \\ 0 & 0 & 1 & 0 & \cdots & 0 \end{bmatrix} \tag{16}$$

and the measurement covariance matrix $\mathbf{R_k}$ in the absence of correlation between the components:

$$\mathbf{R}_k = E[\mathbf{v}_k\mathbf{v}_k^T] = \begin{bmatrix} \sigma_{px}^2 & 0 & 0 \\ 0 & \sigma_{py}^2 & 0 \\ 0 & 0 & \sigma_{pz}^2 \end{bmatrix} \tag{17}$$

The algorithm in each iteration performs the state prediction in the first phase:

$$\mathbf{x}_{k+1} = \mathbf{F}_k\mathbf{x}_k + \mathbf{w}_k \tag{18}$$

$$\mathbf{P}_{k+1} = \mathbf{F}_k\mathbf{P}_k\mathbf{F}_k^T + \mathbf{Q}_k \tag{19}$$

where:

- $\mathbf{P}_k$, $\mathbf{P}_{k+1}$ - covariance matrices for time k and $k+1$.

In the second phase the model is updated:

$$\mathbf{v}_k = \mathbf{z}_k - \mathbf{H}_k\mathbf{x}_{k+1} \tag{20}$$

$$\mathbf{S}_k = \mathbf{H}_k\mathbf{P}_{k+1}\mathbf{H}_k^T + \mathbf{R}_k \tag{21}$$

where:

- $\mathbf{v}_k$ - innovation vector at time k,
- $\mathbf{S}_k$ - innovation covariance at moment k.

In the next step of the model update phase, the gain matrix $\mathbf{K}_k$ is computed and used to update the state vector estimates and the covariance matrix from the first prediction stage according to the scheme:

$$\mathbf{K}_k = \mathbf{P}_{k+1}\mathbf{H}_k^T\mathbf{S}_k^-1 \tag{22}$$

$$\hat{\mathbf{x}}_{k+1} = \mathbf{x}_{k+1} + \mathbf{K}_k\mathbf{v}_k \tag{23}$$

$$\hat{\mathbf{P}}_{k+1} = (\mathbf{1} - \mathbf{K}_k\mathbf{H}_k)\mathbf{P}_{k+1} \tag{24}$$

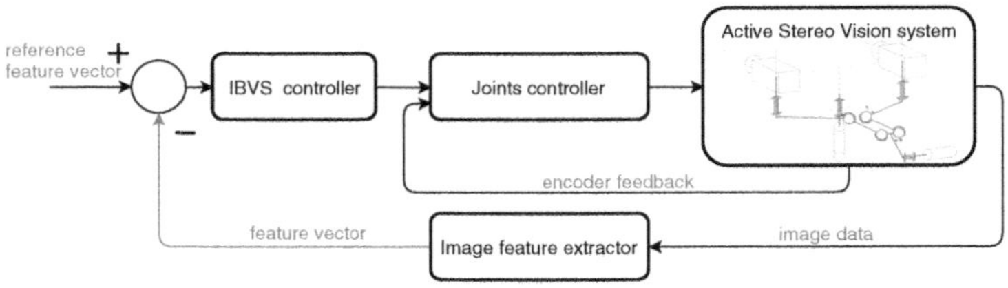

Fig. 5. Image-Based Visual Servo system

2.2 Image-Base Visual Servo

The IBVS control omits the pose estimation step, Fig. 5. The control is performed in image coordinate space $\mathbb{R}^2$. The desired robot pose with respect to the target is defined implicitly by the image feature values at the goal pose. The IBVS control is typically defined in following way [1,3]:

$$\boldsymbol{\nu} = -\lambda \mathbf{J}^{+} \, \dot{\mathbf{p}} \tag{25}$$

where:

- $\boldsymbol{\nu}$ - camera velocity vetor,
- $\lambda \in (0, 1)$ - scaling factor,
- $\mathbf{J}$ - interaction matrix,
- $\dot{\mathbf{p}}$ - feature points velocity.

The purpose of the IBVS control system is keeping the point associated with the object in the image as close as possible to the principal points of both cameras. For this case the vector $\dot{\mathbf{p}}$ can be described in form of point coordinates expressed in pixels relative to the camera principal point:

$$\dot{\mathbf{p}} = \begin{bmatrix} u_{Cl} - u_{Cl0} \\ u_{Cr} - u_{Cr0} \\ v_{Cl} - v_{Cl0} \\ v_{Cr} - v_{Cr0} \end{bmatrix} = \begin{bmatrix} \overline{u}_C l \\ \overline{u}_C r \\ \overline{v}_C l \\ \overline{v}_C r \end{bmatrix} \tag{26}$$

Because the robot cannot rotate cameras around the horizontal axis, the $\overline{v}$ component can be omitted and the $\dot{\mathbf{p}}$ vector can be described as follows:

$$\dot{\mathbf{p}} = [\overline{u}_{CL} \; \overline{u}_{CR}]^T \tag{27}$$

Additionally, in this case, instead of calculating the camera velocity vector, the joints velocities can be estimated directly:

$$\begin{bmatrix} \nu_G \\ \nu_V \end{bmatrix} = \begin{bmatrix} K_G & K_G \\ K_V & -K_V \end{bmatrix} \begin{bmatrix} \overline{u}_{CL} \\ \overline{u}_{CR} \end{bmatrix} \tag{28}$$

where:

- ν_G - gaze angle velocity,
- ν_V - vergence angle velocity,
- K_V, K_G - gains determined empirically.

Similarly to the PBVS controller, a Kalman filter can be used for detection filtering and prediction. Because the error is calculated in the image space, the object model can be limited to one dimension and two independent filter implementations can be used for the right and left cameras, described by matrices:

$$\mathbf{x}_k = \left[\overline{u}_k \; \dot{\overline{u}}_k \; \ddot{\overline{u}}_k\right]^T \tag{29}$$

$$\mathbf{F}_k = \begin{bmatrix} 1 & \Delta t & \frac{\Delta t^2}{2} \\ 0 & 1 & \Delta t \\ 0 & 0 & 1 \end{bmatrix} \tag{30}$$

$$\mathbf{H}_k = \left[1\;0\;0\right] \tag{31}$$

$$\mathbf{Q}_k = \begin{bmatrix} \frac{\Delta t^6}{36} & \frac{\Delta t^5}{12} & \frac{\Delta t^4}{6} \\ \frac{\Delta t^5}{12} & \frac{\Delta t^4}{4} & \frac{\Delta t^3}{2} \\ \frac{\Delta t^4}{6} & \frac{\Delta t^3}{2} & \Delta t^2 \end{bmatrix} \sigma^2_{\dddot{\overline{u}}_k} \tag{32}$$

$$\mathbf{R}_k = \sigma^2_{\overline{u}_k} \tag{33}$$

3 Experiments and Results

The IBVS and PBVS controllers were implemented on host machine as separate ROS nodes as was shown on system component diagram, Fig. 6.

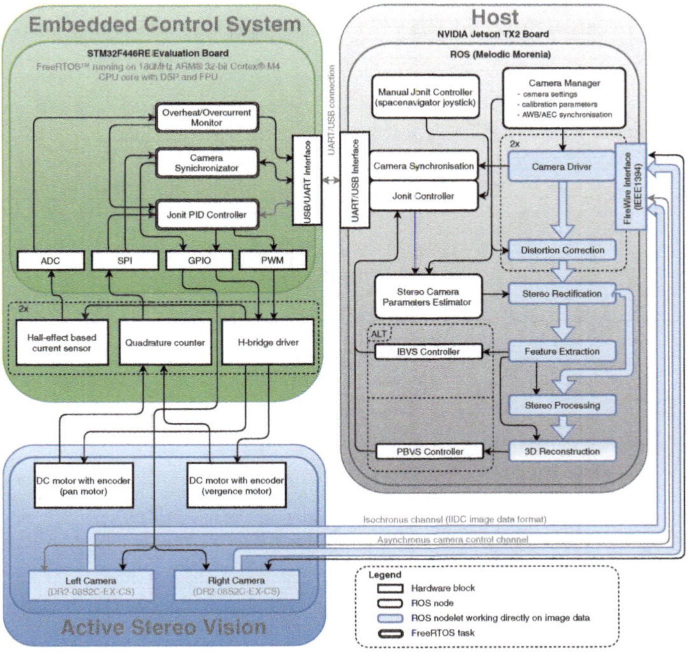

Fig. 6. System component diagram

For testing the system, PlayStation Move motion game controller was used as a target object for tracking. The game controller has an orb at the head which can glow in any of a full range of colors using LEDs. The uniform spherical shape allows the system to simply detect object on image and estimate the center of tracked object, as was shown in Fig. 7.

(a) Rectified images (b) Disparity

Fig. 7. PS Move Controller as a target object

The PlayStation Move controller was then hanged from the ceiling of the room, creating a pendulum. The SnFCS was positioned in relation to pendulum freely, but in such a way that the target motion were visible in all axes, assuming world coordinate system as shown in Fig. 4a. Assuming small oscillation's amplitude, the period of the motion should be around $T_0 = 2\pi\sqrt{l/g}$, so for $l = 2.46$ m gives $T_0 = 3.15$ s. The Fig. 8 shows target object movement estimation done by stereo vision system using the PBVS and IBVS controller respectively (Fig. 8).

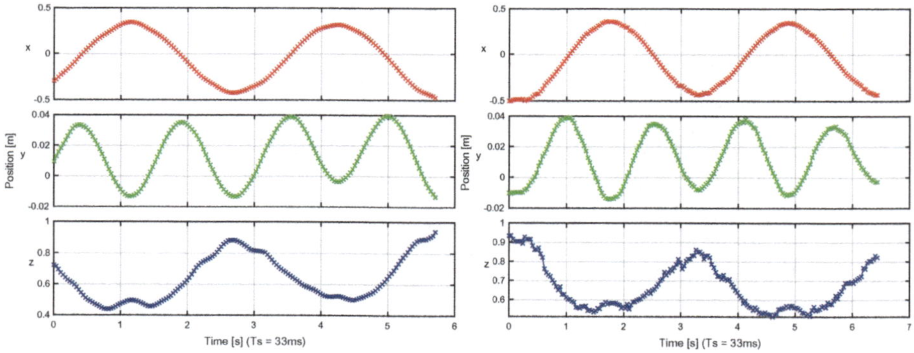

(a) PBVS (b) IBVS

Fig. 8. Target object position

(a) (b)

Fig. 9. Stereo vision system

4 Conclusion

The results described in this paper show that visual servo controller in pose-based and image-based configuration can be used in stereo vision non-fixed camera system to keeping fixation point near some visual target. The IBVS in form presented in this paper are easy to implement, not need information about geometric relation between cameras and object, but it requires tuning of gains from Eq. (28). The PBVS needs more computation in feedback loop, however in this case the control is done in task space, what makes it possible to use a better model in Kalman filter, especially, if the type of target object motion is known.

References

1. Chaumette, F., Hutchinson, S.: Visual servo control. II. Advanced approaches. IEEE Robot. Autom. Mag. **14**(1), 109–118 (2007)
2. Chaumette, F., Hutchinson, S.: Visual servo control. I. Basic approaches. IEEE Robot. Autom. Mag. **13**(4), 82–90 (2006)
3. Corke, P.: Robotics, Vision and Control: Fundamental Algorithms in MATLAB®, vol. 73. Springer, Heidelberg (2011)
4. Hartley, R., Zisserman, A.: Multiple View Geometry in Computer Vision, 2nd edn. Cambridge University Press, New York (2003)
5. Staniak, M., Zieliński, C.: Structures of visual servos. Robot. Auton. Syst. **58**(8), 940–954 (2010)
6. Szewczyk, P.: System stereowizyjny z aktywnym mechanizmem pozycjonowania kamer. Prace Naukowe Politechniki Warszawskiej. Elektronika. Problemy Robotyki z. **175**(t. 1), 231–240 (2010)
7. Szewczyk, P.: Real-time control of active stereo vision system. Adv. Intell. Syst. Comput. **577**, 271–280 (2017)
8. Zieliński, C., Kornuta, T., Boryń, M.: Specification of robotic systems on an example of visual servoing. IFAC Proc. Vol. (IFAC-PapersOnline) **45**(22), 45–50 (2012)

Low-Cost Autonomous UAV-Based Solutions to Package Delivery Logistics

Jacek Grzybowski, Karol Latos, and Roman Czyba(✉)

Silesian University of Technology, Gliwice, Poland
{jacegrz498,karolat245}@student.polsl.pl, roman.czyba@polsl.pl

Abstract. Advancements in technology serve to eliminate inconvenience from our lives since the dawn of mankind. Today, one of the problems in automation is to tackle the logistics behind package delivery using unmanned aerial vehicles (UAVs). To further this area of engineering, we have attempted a solution of our own, using a geographical information system based on QR address codes and an automated dispenser. The drone, equipped with an ADS-B module and a gripping mechanism, lands near the shipment, where it is located by a camera on a package feeder. A 2DoF extension arm directs the payload underneath the vehicle, where the UAV can easily intercept the package and analyze its address code for delivery destination. We found this approach to be economically rewarding, easy to implement and possible to modify for big-scale scenarios.

Keywords: Package delivery · UAV · U-space · Logistics technology

1 Introduction

Unmanned aerial vehicles, commonly known as drones, are becoming increasingly popular all over the world [3,4]. Poland is one of the countries where the drone market is growing at the fastest rate [6]. The popularity of drones is particularly evident in such areas as infrastructure, agriculture, transport, security, media, entertainment, insurance and telecommunications [2,7]. Their popularity is evidenced by the fact that in Poland alone in 2019, flew about 100 000 drones weighing from 250 g to 600 kg [6]. The popularity of drones has become so great that their use began to be regulated by law by ULC (Civil Aviation Authority) and PANSA (Polish Air Navigation Services Agency). Legal regulations, an airspace management system, drone products and services are the main pillars of U-space, making up a new segment of the aviation industry. Launched in 2018 in the Upper Silesian Metropolis, the Central European Drone Demonstrator is a response to the growing drone market and the demand for services performed by unmanned aerial systems (UAS), anticipated in the near future. U-space is a set of new services and specific procedures designed to support safe, efficient and secure access to airspace for large numbers of drones. These services rely on a high level of digitalisation and automation of functions [1], which is part of the

A. Bartoszewicz et al. (Eds.): PCC 2020, AISC 1196, pp. 500–507, 2020.
https://doi.org/10.1007/978-3-030-50936-1_42

development trends of Industry 4.0. In this context, the proposition of an automated package delivery system, based on low-cost and accessible components, perfectly fits the needs of the drone market.

2 Prerequisites to Cargo Interception

The problem of carrying a shipment from one parcel to another is tied with a set of engineering and economic decisions. Particular choices of hardware, software and reproducibility of different components will vary depending on the operation budget, resources available and the scale of the proposed solution. Our goal is to outline a model for package transportation, which is relatively cheap to assemble, yet still reliable and capable of fulfilling its purpose. The solution will be composed of a few subsystems, first of which is the vehicle; it should be capable of carrying the payload for extended amount of time without malfunction. According to the statistical data gathered from 41 countries in 2019, 86% of all e-commerce shipments weighted less than 2 kg [7], therefore we approximate the carrying capacity of the drone to be around 5 kg, including the battery, in order to successfully complete vast majority of ordinary shipments. The UAV used in prototyping the solution was Tarot X6 hexacopter frame with Tarot 5008 340 KV motor and one 22000 mAh Tattu battery. The described combination leaves around 2 kg of loading capability, while still maintaining high power to weight ratio (greater than 2:1). Should the carrying capacity be utilized completely, the vehicle is able to make a 25-min flight, according to the product specification. Another component of the system is a mechanical gripper powered by a servo electric engine.

3 Gripping Mechanism

The presented interception module (Fig. 1, Fig. 2, Fig. 3) has been designed to be easy in production using consumer-grade manufacturing devices, e.g. three-dimensional printers. For the purpose of prototyping we have used a budget fused deposition modeling printer. The mechanism is composed of eight parts, two stationary elements (coloured red in the model render) and six moving parts (coloured blue). While the stationary parts allow for the device to be connected to the UAV, the moving parts create a servomechanism pulling clutches of the contraption together to a minimum distance around 5.5 cm; the device locks, since it is not back-driven. The gripper can be joined with an opensource autopilot, e.g. PX4. To ensure safe grip on the cargo, we propose using either disposable plastic T-shaped adapters mounted on top of the package, or additional packaging with dents or holes on both sides. Inasmuch as the battery is mounted on one side of the drone, the grasping module is attached to the opposite side providing a counterweight to ensure the UAV stability.

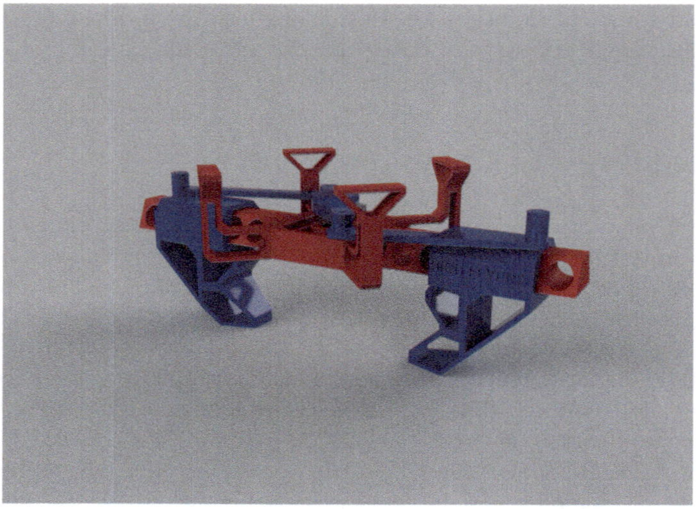

Fig. 1. Gripping mechanism model

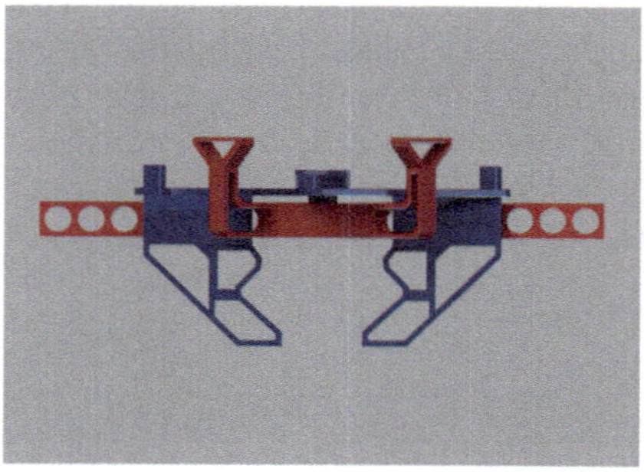

Fig. 2. Gripping mechanism - front

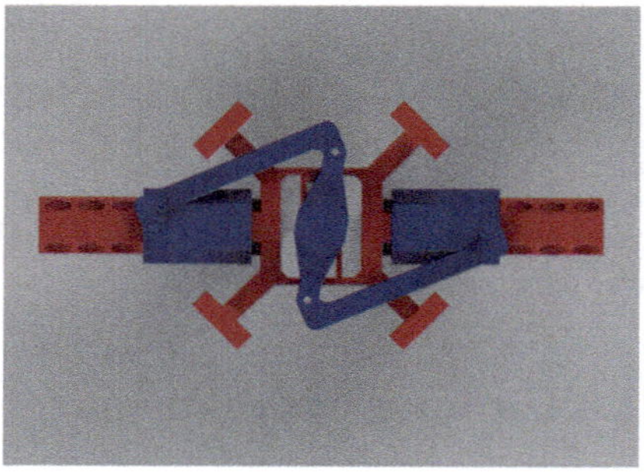

Fig. 3. Gripping mechanism - top

4 Package-Dispensing Unit

Considering the problem of locating the package by the vehicle, the aim is for the UAV to be launched at arbitrary point and then to find the shipment to intercept. Consequently, it becomes necessary to implement positioning modules both in the drone and the package. Since the primary goal of our solution is to minimize costs, we used Pixhawk 2 controller, composed, among other things, of three IMU modules and GPS utilities. Yet, Pixhawk localization system is not accurate enough for the drone to stably hover over the package nor to land exactly above it. High-grade solutions exist to solve this problem—multi-frequency Global Navigation Satellite System or differential GPS; however, not only does their cost rule them out from serious consideration for low-budget operations, external factors like weather can induce too much instability. This calls for an additional accurate and cheap module, which will allow the vehicle to reliably position itself near the shipment in order to collect it. Among other approaches, one is to attach an optical sensor to the drone and a visual marker to the package. This way, when the UAV is approaching the payload, it will locate the shipment using camera-based navigation control. However, this solution yields many inconveniences; by design, the drone sways as it adjusts its position in the air, changing the sensor field of view and rendering new set of instructions. Although similar systems have been implemented [1], they are usually made for large-scale terrain mapping and are not suitable for landing with centimeter-scale precision; fixing this problem would require a stabilizer for the sensor, but equipping every vehicle with a high-resolution camera on stabilizer would needlessly increase the cost and decrease replaceability of the transporting unit. In our approach, the shipment is placed on a feeding mechanism, composed of a ground-based automated extension arm with two degrees of freedom and an optical sensor, which locates the unmanned platform landed nearby. The arm adjusts itself to the position of the vehicle and transports the package underneath it, so that the gripper can intercept the payload. Then the arm retracts, ready for another cargo to be deposited, while the drone is ready to make the shipment.

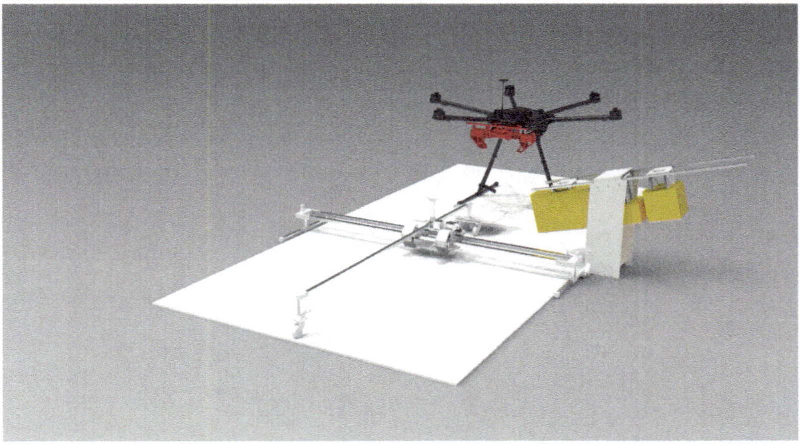

Fig. 4. Feeding mechanism model

The presented model of the feeding mechanism has been designed to use as little materials as possible, in the spirit of our approach, yet still be able to thoroughly fulfill its job (Fig. 4).

5 Destination Decoding

When the vehicle is shipped, its delivery address must be known in order to make the shipment. Since the destination of a flight is individual for every package, it can be stored as a codified version of a destination in a geocode system. We propose using Quick Response geocode, e.g. what3numbers - geographical latitude, altitude and height coded in QR pattern. What emerges from that approach is a necessity for an optical QR scanner, placed on the dispensing module; after loading the shipment onto the cargo feeder, the address pattern is decoded and the package is transported to the gripper working area, where it is intercepted. Were the scanner positioned on the drone, the replaceability of the transporting unit would be reduced.

6 Cargo Interception

In order to control the precision of dispatching, the UAV is equipped with a proximity sensor, verifying the position of the shipment underneath. When the drone lands in the working area of the feeder, that is 1.5 m × 1.5 m, the dispensing unit locates it by scanning the surroundings with a camera. When the vehicle position is known, the 2D cartesian coordinate robot lines up the shipment arm with the drone and slides the cargo until the UAV's proximity sensor signals the presence of a package. The grasping mechanism is then instructed to grip the payload, while the extension arm goes back to the initial state for another cargo to be deposited (Fig. 5).

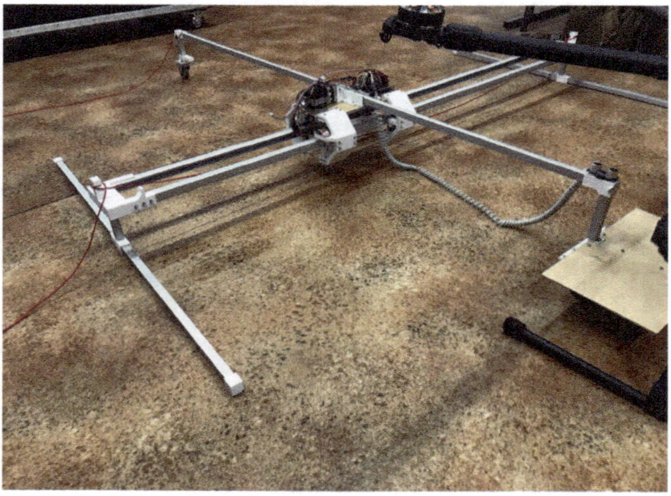

Fig. 5. Feeding mechanism real implementation

7 Delivering the Package

After capturing the payload by the UAV and analyzing the delivery address code by the scanner, the flight is initiated. One way to establish a clear control sequence for all of the modules in our solution is to use Robot Operating System (ROS). For the vehicle to be automatically steered towards given destination, some rules have to be met. The rules differ from one country to another, while some parts of legislature are similar across the nations; e.g. UAVs must not fly within a given radius from airports or obstruct flight of other aircraft. One component ensures both localization and safe travel of the vehicle - automatic dependent surveillance-broadcast (ADS-B), a technology incorporated in some aviation legislature, allowing the UAV to broadcast its position either to a ground control station or other aerial vehicles to avoid collisions. When the ADS-B, QR code scanner and an autopilot module (e.g. PX4 opensource autopilot) work together under ROS, strict cooperation of subsystems is assured.

$$\begin{array}{ccccc} & & \text{Proximity sensor} & & \\ & & \downarrow & & \\ \text{ADS-B} & \longrightarrow & \text{ROS} & \longleftrightarrow & \text{PX4} \\ & & \uparrow & & \\ & & \text{QR scanner} & & \end{array}$$

The location of the vehicle is being broadcast by the ADS-B to the ROS, while the geographical position of the feeder is either manually set-up or given by a GPS module. The data is provided to the PX4, which steers the drone to the package dispenser, utilizing ADS-B to avoid collisions. Afterwards, the UAV lands in the working area of the feeding unit, which locates the vehicle and slides the payload underneath it, until the proximity sensor sends appropriate information to the ROS. The grasping mechanism is then initiated and locked on the cargo. When the package is intercepted, the address code is provided to the system via route *QR scanner—decoder—ROS* and the vehicle starts the delivery.

8 Implementation

Conceptual models reflected the convenience of low cost manufacturing, therefore the production was not excessively expensive. Our implementation of models presented aimed at minimizing the expenses while maintaining efficiency. We printed the grasping mechanism using a costumer-grade three-dimensional printer and PLA material. The resource could be changed to ABS, which is a more durable and lighter substance, to ensure more reliability. The feeding mechanism has been constructed with plywood and simple guide rails with motors. Our proposed system was implemented using the aforementioned hexarotor and the performance of all modules has been tested; the UAV was able to land in the working area of the feeder; the package was successfully dispatched in the range of the gripper; the shipment was stably and safely gripped by the grasping mechanism in a flight test. The complete system still requires further tests to improve and refine it, but the results obtained so far look promising (Fig. 6, Fig. 7). The total estimated cost of the prototype system ranges between $2000 and $6000.

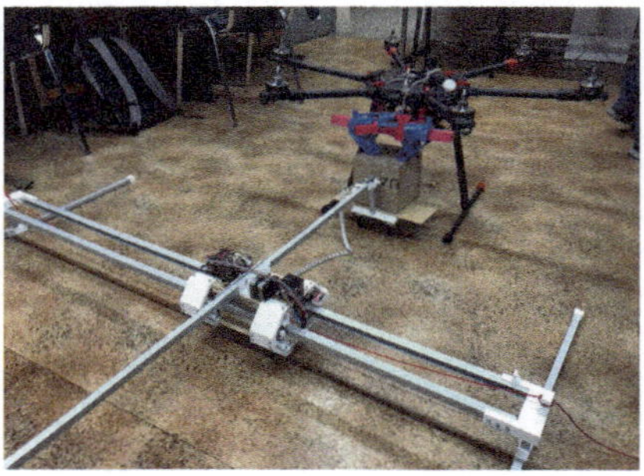

Fig. 6. Implemented mechanisms (1)

Fig. 7. Implemented mechanisms (2)

9 Conclusion

The construction of an autonomous system capable of performing more than simple, mundane tasks sometimes seems like an economical bottomless pit. Restricting the project to accessible, consumer-grade and low-budget modules calls for engineering insight in contriving new solutions. Sometimes, those ideas induce substantial progress in industrial-scale inventions; sometimes, they motivate not only specialists, but also hobbyists, to search for new solutions on the frontier between automation engineering and the problems of everyday. Referring to the idea of Industry 4.0, whose keynote is digital transformation based on the use of modern technologies, such as vehicle autonomy or ubiquitous connectivity, the presented article shows the possibilities of useful CAD modeling, prototyping and constructing unmanned flying systems. The PFR (Polish Development Fund) sectoral program 'Zwirko i Wigura', which has the priority goal of creating an unmanned traffic management system, is part of this thought. The goal of U-Space is to coordinate drone traffic, provide IT support and create infrastructure for UAV users and service providers. In this context, the presented autonomous UAV system for delivering packages fits perfectly into the development trends of Industry 4.0. Finally, we believe that the proposed solution model contradicts the idea that new technologies concerning automation in logistics require an unreasonable budget and are out of reach for mere enthusiasts.

References

1. Engel, J., Sturm, J., Cremers, D.: Camera-based navigation of a low-cost quadrocopter. In: 2012 IEEE/RSJ International Conference on Intelligent Robots and Systems (2012)
2. Austin, I.R.: Unmanned Aircraft Systems - UAVS Design, Development and Deployment. Wiley (2010)
3. Valavanis, K.P.: Advances in Unmanned Aerial Vehicles. Springer, Netherlands (2007)
4. Castillo, P., Lozano, R., Dzul, A.E.: Modelling and Control of Mini-flying Machines. Springer-Verlag (2005)
5. Nonami, K., Kendoul, F., Suzuki, S., Wang, W., Nakazawa, D.: Autonomous Flying Robots: Unmanned Aerial Vehicles and Micro Aerial Vehicles. Spirnger, London (2010)
6. Czyba, R., Szafrański, G., Janusz, W., Niezabitowski, M., Czornik, A., Błachuta, M.: Concept and realization of unmanned aerial system with different modes of operation. In: Proceedings of the 10th International Conference on Mathematical Problems in Engineering, Aerospace And Sciences: ICNPAA 2014, Narvik, Norway, pp. 261–270 (2014)
7. International Post Corporation: Dynata: Cross-Border E-Commerce Shopper Survey 2019, p. 13. IPC 2020 (2020)

Selection of Methods for Intuitive, Haptic Control of the Underwater Vehicle's Manipulator

Tomasz Grzejszczak[1(✉)], Artur Babiarz[1], Robert Bieda[1], Krzysztof Jaskot[1], Andrzej Kozyra[1], and Piotr Ściegienka[2]

[1] Faculty of Automatic Control, Electronics and Computer Science, Silesian University of Technology, Akademicka 16, 44-100 Gliwice, Poland
tomasz.grzejszczak@polsl.pl
[2] SR Robotics sp. z o.o., ul. Karoliny 4, 40-186 Katowice, Poland

Abstract. This paper is the early report of available market and scientific solutions allowing intuitive control. Manipulator control is presented in form of the Human Machine Interaction loop that describes both machine possibilities of sensing the human control and human possibilities of sensing the machine state. The survey is presented in form of the description and discussion of the advantages, disadvantages and usability of the available solutions. The aim of the research is to chose the proper path of development of the new way of intuitive control.

Keywords: Intuitive control · HCI · HMI · Motion capture · Haptics

1 Introduction

Communication is the way of sending information between two individuals. It comes naturally when communicating between humans, however while creating machines, human need to develop the new way of communication. This paper presents and describes the early report of available market and scientific solutions that allows the creation of intuitive control in form of the communication loop between Human and Machine.

The motivation comes from the project focused on research and development of the new way of teleoperation, that is the control of the manipulator attached to the underwater Remotely Operated Vehicle (ROV). The controller should be applicable on the swaying boat, should reduce the fatigue during usage and be easy to use and learn, that means, be intuitive. On the other hand, moving manipulator should present, in form of the feedback, as much information as possible. The most crucial information in case of the manipulator control is the obstacle or resistance encounter.

The following paper is divided into 3 main sections. First, Sect. 2, presents loop of interaction between human and machine, explaining The theoretical background, meaning of HMI, HCI and the differences between interaction and interface. Next chapters are organised by the human senses that are stimulated and

A. Bartoszewicz et al. (Eds.): PCC 2020, AISC 1196, pp. 508–519, 2020.
https://doi.org/10.1007/978-3-030-50936-1_43

machine senses that are artificially developed in the interaction loop. The sensors that are used to implement the artificial machine senses to understand the human control commands are presented in Sect. 3. The feedback systems that stimulates human senses while perception of the machine state are presented in Sect. 4. Final part of the paper is the conclusions used to chose the right system, summary and further research description.

2 HMI and HCI

While looking at the research about ways of communication with machine, the most common abbreviations found are HCI and HMI. The first stage of setting the theoretical background for intuitive control is to explain those abbreviations. In those three letters abbreviations, the least questionable letter is H standing for Human. In the designed system, the human user is the one that tries to establish the communication and needs to research and develop the best ways to do so.

The communication is established with either Computer or Machine, hence there are two possible letters in the abbreviation: C or M. The modern development tends toward Internet of Things concept, thus the machines are becoming less mechanical and are build as computers with actuators. So the question is, do we communicate with the machine or do we communicate with the computer that controls the machine? Due to this blurred border, some research papers use the connected abbreviation "HMI/HCI" [1].

The last letter of the abbreviation I is explained in two ways, as Interaction or Interface. The interface is the way of connecting and sharing information between two individuals or objects. If the communication is mutual and the communication of one is dependent upon the message of the other, then there is an interaction. In other worlds, the interaction is a part of interface. The difference is clear, however in the abbreviation HCI or HMI one does not know whether the subject is about the interface or interaction.

In conclusion, there are two abbreviations and the borders are blurred, however some usual meanings can be noticed. HMI, as an interface, usually describes the operator panel, that is a set of controls and displays for process control. HCI is an interaction topic describing different controllers or human detecting vision systems and their intuitiveness.

2.1 Interface Structure

The interaction structure is presented to properly classify the intuitive control system components. The human computer or machine interface structure is usually presented in form of the interaction loop (Fig. 1) [18].

According to Fig. 1, first, the decision is formulated in **human** brain and expressed with use of neural system and muscles creating **motor responses**. Those responses are recognized by the machine **controls**. Then, the **machine state** changes, according to the control protocol, that is a set of instructions that

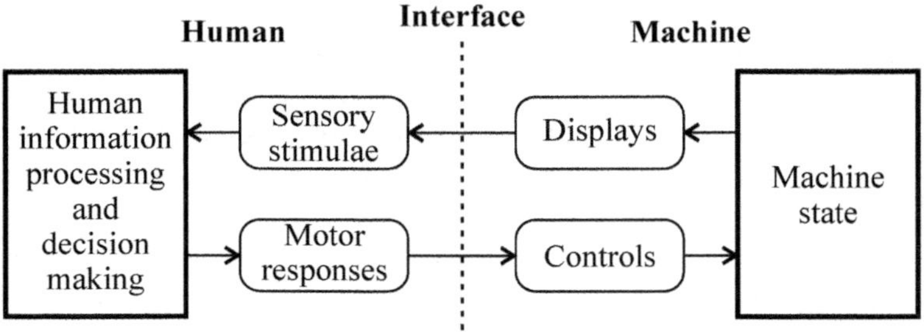

Fig. 1. The human-machine interface. Input devices are the controls humans manipulate to change the machine state. [18]

transfer the machine state according to changes in the controls. The changes in the **machine state** can be visualised with use of the **displays** in form of the values, charts or the camera stream with augmented reality. The data is received by human senses, that is a neural **sensory stimulae**, such as eyes or skin receptors. The ability to observe the machine state encourages **human** to **process the information and make a new decision**, that closes the interaction loop.

Each arrow of the loop is prone to disturbance, leading to miscommunication. For example human neurological illness can disturb the motor responses or sensory stimulae reception. Wrong controls selection can lead to need of unnatural or difficult motor responses. The lack of crucial data display can lead to incomplete information and uncertainty.

In this paper, the two crucial parts of the loop are examined. In the next chapters, some hardware solutions are presented. Section 3 is focused on the sensors that can be put in the **controls** block, with the aim of necessity of most intuitive human motor responses, but also focused on the most reach data that can be used to create a control protocol. Section 4 is focused on the feedback that is represented in form of the **displays** block. The aim of the feedback is to enhance the human sensory stimulae to provide the reach awareness of machine state.

3 Sensors

The controls of the machine are designed to maximize the natural capabilities of human communication. In the human-human communication, we use verbal (speech) and nonverbal (hand gestures, face mimics, micro expressions, body posture) communicates. Those ways of communication are natural, because we have been taught this since we were infants. Thus, this is the way we, as a human race, would like to communicate with machines. The artificial machine senses, in form of sensors, are created to imitate the human senses and are used to control the machine. This leads to the division of this chapter according to the human senses.

3.1 Touch

The oldest ways of machine control is with use of knobs, buttons and joysticks [5]. There are many various types of the devices, but each is working in the same way. Each transfers the human (usually finger) movement, outputting an analog or digital signal. The effort of designing the intuitive controller is to think about the process of translating the controller output signal into machine state changes. Sometimes in the design process the ergonomy is important, meaning that the device is designed according to human body factors, maximizing the performance and minimizing health problems and injuries. Thus, the controller should be both, ergonomic and intuitive, where first focus on optimization of the movement process while the second on optimization of learning and decision making process [3,26]. The examples of controllers imitating machine touch sens are: keyboard, mouse, touch screen, pad, joystick, teach pendant, machine control panel [9]. An example is presented in Fig. 2a.

(a) KUKA SmartPAD teach pendant with keyboard, touch panel and 6D mouse

(b) Kraft force feedback mini-master

(c) CyberGlove III with flex sensors

Fig. 2. Examples of touch-based controllers. Source of images: official manufacturers' web pages.

In some cases, the controller can imitate the shape of the machine and becoming its "model". The operator can change the position of the "model" elements and these movements are translated into the movement of the real machine. Machine control can also consist of controlling the position of the tool that the robot actually moves. In such configurations, the human controls the position of the tool model (e.g. a scalpel model), the model's movements are translated into the movements of the real tool. This type of control requires translation of the movement of the tool into the movement of individual machine elements. In this case, the system is called a master-slave control, where the manipulator is a slave, that follows the movement of the master (controller) [6,10]. The master controller is presented in Fig. 2b.

The machine can directly sense the human full hand movement, when a special data glove [8] or even a full arm exoskeleton [14] is applied. Many solutions rely on transferring hand movement to robot motion commands. In such constructions, it is necessary to measure finger deflection or finger pressure on an object. It can be done e.g. by using simple resistive flex or force sensors. This led for building gloves with mounted sensors (Fig. 2c). Such gloves can be fitted with various types of actuators that provide haptic feedback (see Sect. 4.2).

3.2 Sight

Human movement can also be monitored by cameras, that imitate machine's vision. There is a wide range of vision systems types varying in frame quality, speed of acquisition, spectrum range and number of cameras. With better quality (both optics and sensor size measured in megapixels) comes higher price and more detailed image, but there is more data to process and more computing power necessity. Same with speed of acquisition, that is measured in frames per second. High speed cameras can capture fast reaction gestures [25] or phenomena occurring in tenths of seconds, but usually for the high computational cost or processing after recording (not in real time). Common cameras work in visible light spectrum range, however there are cameras working in infrared, providing ability to sense depth (measuring Time of Flight) [15,23] or temperature (thermovision) [12]. When the vision system is created with multiple cameras [11], the object can be seen from multiple angles, enhancing the accuracy and eliminating veiling of objects. The vision sensor is usually connected to the computer with proper image processing algorithm enabling the recalculation of detected feature points or placed artificial markers into the machine commands. The process of development of proper image processing algorithm is usually the most difficult part of the vision system. Usually, the simplest (in construction, hardest in algorithm development) vision based solutions are using the image from simple, wide accessible camera.

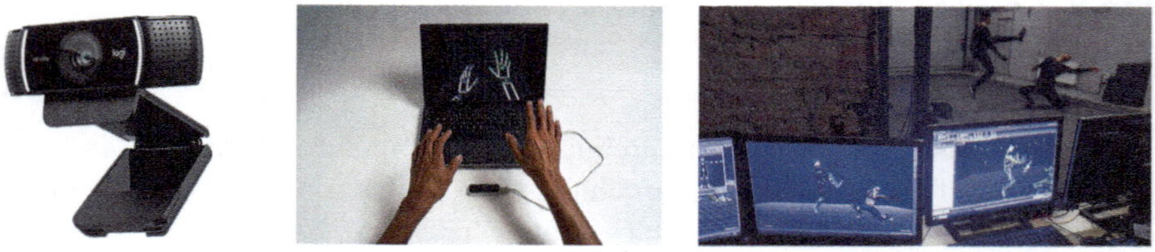

(a) c922 USB camera offered by Logitech

(b) LeapMotion controller for hand landmarks tracking

(c) Motion Capture offered by Platige Image

Fig. 3. Examples of camera-based controllers. Source of images: official manufacturers' web pages.

The simple camera can provide a 2 dimensional image. There is a possibility of detecting depth. The most common approach to obtain a depth map is to use a Time of Flight camera, that recalculates distance from the time that passed from the infrared emission to reception. Depth image is very helpful in image processing for background segmentation and feature extraction. An examples of infrared sensors capable of human movement detection are Kinekt [15,23] for body tracking and LeapMotion [4] for hand tracking [5].

Another, frequently used vision system construction for human movement registration is a Motion Capture laboratory [16]. In a room, there are several cameras located around the workspace, connected to a computer capable of markers position tracking. Knowing the exact place of each camera and knowing the marker detection position on image, the marker position in workspace coordinate system is calculated. The disadvantage of such system is the necessity of precise calibration and that this system is usually stationary (in a special room). Moreover, after recording, the material must be processed in order to eliminate the uncertainty caused by markers that were covered (Fig. 3).

3.3 Proprioception

The proprioception sense is one of the human senses, beyond the basic 5 senses, responsible for reception of body movement. During machine development and implementation of sensors imitating human senses, the one that can bring the machine the values of acceleration and rotation is introduced by Inertial Measurement Unit (IMU). For example, having one IMU on board of the Unmanned Aerial Vehicle can be used to stabilize it, in similar way to human's vestibular sense [2].

(a) Xsens (b) Nansense (c) Perception Neuron

Fig. 4. Examples of IMU based Motion Capture systems. Source of images: official manufacturers' web pages.

On the other hand, multiple IMU senors can be attached in crucial human body places [19] creating the same Motion Capture system, as the one created with use of cameras and markers. The control sequence is usually recalculated depending on the mutual displacements of IMUs. The application of IMU sensors brings two main advantages with comparison to vision based motion capture: it allows the system to work outside the laboratory or studio and the effect of covered markers is not present.

There are some ready to use systems available on the market, that comes with the sensors, costume, communication module and software. Examples of IMU based Motion Capture are presented in Fig. 4. There comparison of specification of those systems are presented in Table 1.

Table 1. Comparison of specification of different IMU based Motion Capture systems

Parameter	Xsense	Nansense	Perception neuron
max IMU count	17	54	32
Accuracy	0.5–1.5°	1°	1–2°
Output rate	30–120 Hz*	24–240 Hz*	60–120 Hz*
Sensor weight	16 g	–	15.8 g
Battery life	6 h	8 h	3.5 h

* For more connected IMU sensors, the maximum output rate drops.

3.4 Other

There are also other types of robot control. In neurological rehabilitation, the EMG signal from muscle is measured, and the rehabilitation robot enhances the patient movement [22] . EMG electrodes can also detect the brain activity [24], however this solution is inaccurate and is still limited to few commands. Voice recognition is also frequently used and can be applied to robot control, however in this approach, the operation is more focused on semi-autonomy and planing than teleoperation and control via direct movement.

4 Feedback

To properly chose the feedback system, one need to think about the human sensory stimulae, so the solutions are divided according to the human senses. Moreover the solutions can be divided into stand alone solution and build into controller.

In case of robot control, the feedback data usually comes from the indirect force measurement that are present in case of obstructed movement (e.g. pushing against a wall). This can be implemented in form of servo current measurement, torque changes, tension postponed in robot elements measured by tensometers [20] or hydraulic pressure changes in case of hydraulic manipulator [20].

4.1 Sight and Hearing

The main aspect of Human Computer Interface, understood as a operational panel is the proper configuration of displays and other audio-visual signals. The full machine state can be modeled in Supervisory Control And Data Acquisition (SCADA) technique. The condition of the machine is displayed on the screen or even can simply be indicated by a light or an audio alarm. An everyday example of the audio-visual feedback is the car parking sensor system, that while reverse drive, depending on the distance to the obstacle, displays the visualisation and modulates the frequency of sound. In order to maximize the perception of machine state and enable the user to immerse into the machine is to use Augmented Reality or as an On Screen Display, for example in flight control.

4.2 Touch

In order to stimulate the human touch sense, the feedback needs to involve a haptic system. Haptic technology transforms the output of the machine into the forces that can be represented in form of vibrations, resistance force [13] of the controller knob, grip tightening [21] or even electric shock [17].

The aim of the haptic feedback in case of manipulator control is the representation of the manipulator's touch. Too much force applied can cause the manipulator to break things. The vibration is the easiest to implement haptic feedback, and can be produced by vibration of a motor with unbalanced shaft. The implementation can be found in everyday devices like mobile phones or console game controllers. When user sense the vibration, this can indicate that more progression of control in this direction can be dangerous. Such feedback also does not require visual observation of the condition of the machine. It allows the operator to focus his sight on other activities.

More advanced haptic systems allows to control the position of operator fingers, so the fingers position corresponds to the position of the gripper. This can be achieved in many ways eg. using a lever with servos attached to the fingers and metacarpus. Such a solution, as can be seen in the Fig. 5a and 5b, can be uncomfortable and heavy. Therefore, solutions that are lighter and more comfortable are sought. Instead of electric servos, lighter pneumatic systems are used (Fig. 5c) or electrostatic adhesive brakes which stiffen fingers in a given position in the glove after applying voltage to the electrostatic element (Fig. 5d) [7,27].

5 Choosing the Right System

In order to create the full Human Machine Interaction system, one need to think about the best way of sending and receiving information and the solution's' advantages and disadvantages. In the case of the manipulator control several problems needs to be discussed.

The first aspect is the definition of what exactly is controlled. The position and orientation of the effector can be directly transferred from the position and

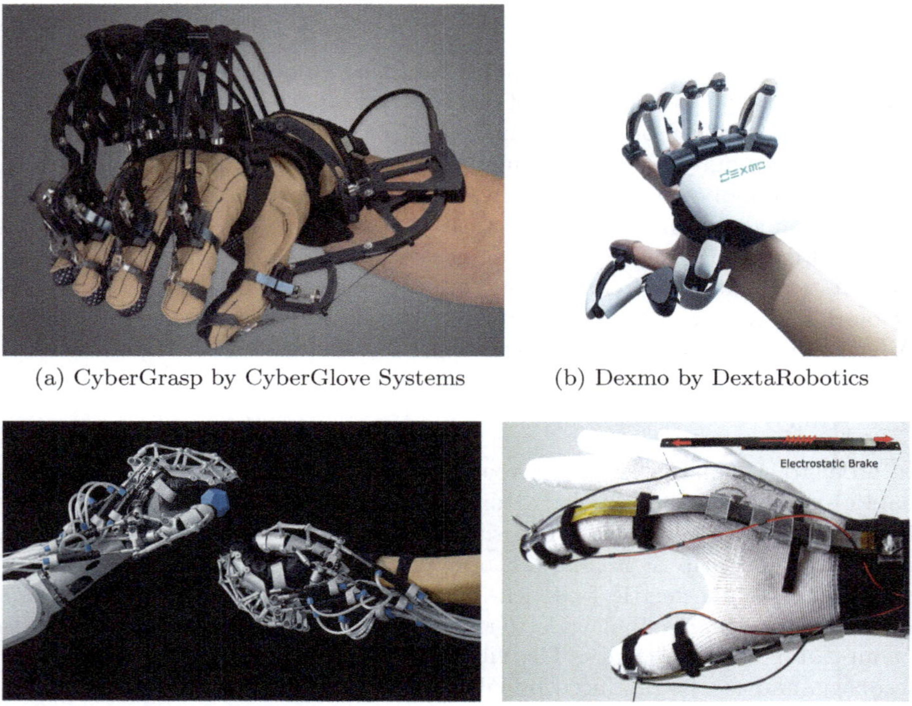

(a) CyberGrasp by CyberGlove Systems (b) Dexmo by DextaRobotics

(c) ExoHand by Festo (d) DextrES

Fig. 5. Examples of IMU based Motion Capture systems. Source of images: official manufacturers' web pages.

orientation of the controller. The control of the effector position as the control sequence of displacement and rotation $(x, y, z, \phi, \theta, \psi)$ is easy to obtain from a controller, but difficult to execute by the robot, due to necessity of inverse kinematics calculations and the related problems such as redundancy or singularities.

On the other hand, control of angles of manipulator's joints can be tiring or unintuitive. This approach can be implemented in all types of controllers, however, even the specially designed master-slave controller (Fig. 2b) is hard to learn and becomes tiring in case of long work.

This brings the next important aspect, that is the comfort of work. Designers of the system need to take into account that the user can be tired after a longer time. How long one can hang the arm in the air, especially, when equipped with additional sensors or gloves? Moreover, the system should be able to pause the tracking to enable user to correct his position.

The final aspect is the possibilities of application. The final product should be applicable on the swaying boat, so the final control and feedback system should be immune to drifting, constant oscillations of the base, electromagnetic field disturbance of the big metal elements of the boat, changes in light conditions, etc.

6 Summary and Further Research

This research paper presents the Human Machine Interface structure and the individual components to create the intuitive manipulator control. The available solutions and products are divided into two categories: sensors that imitate machine's senses and feedback that is used to present the machine state back to user. Each solution is categorised by the sense that is stimulated or intended to imitate.

The aim of the project is to investigate and develop the way of intuitive manipulator control. As the part of the preliminary survey, each of the most common available products and methods are discussed and categorised, so the most promising can be chosen to be tested in further research. In the future research the following tests are planned:

- comparison of the accuracy and intuitiveness of vision Motion Capture and IMU based Motion Capture,
- comparison of the accuracy and intuitiveness of image processing vision system, LeapMotion and data glove,
- tests of the application of the solutions in difficult environments (swaying boat),
- development of the proper feedback for haptic controller.

Acknowledgments. The research is financed by Polish National Centre for Research and Development under project number POIR.01.01.01-00-0266/18: Inteligentny, efektywny system prowadzenia specjalistycznych prac podwodnych (Smart and effective system for performing specialized subsea works) realized by SR Robotics sp. z o.o.

References

1. Adikari, S., McDonald, C.: User and usability modeling for HCI/HMI: a research design. In: 2006 International Conference on Information and Automation, pp. 151–154 (2006). https://doi.org/10.1109/ICINFA.2006.374099
2. Blachuta, M., Grygiel, R., Czyba, R., Szafranski, G.: Attitude and heading reference system based on 3D complementary filter. In: 2014 19th International Conference on Methods and Models in Automation and Robotics (MMAR), pp. 851–856 (2014). https://doi.org/10.1109/MMAR.2014.6957468
3. Blackler, A., Popovic, V., Mahar, D.P.: Intuitive use of products. In: Design Research Society (DSR) International Conference: Common Ground, pp. 1–15. Staffordshire University Press (2002)
4. Bosscher, P.M., Summer, M.D.: Telematic interface with control signal scaling based on force sensor feedback (2014). US Patent 8,918,215
5. Gîrbacia, F., Postelnicu, C., Voinea, G.D.: Towards using natural user interfaces for robotic arm manipulation. In: International Conference on Robotics in Alpe-Adria Danube Region, pp. 188–193. Springer, Heidelberg (2019)
6. Hildebrandt, M., Christensen, L., Kerdels, J., Albiez, J., Kirchner, F.: Realtime motion compensation for ROV-based tele-operated underwater manipulators. In: OCEANS 2009-EUROPE, pp. 1–6. IEEE (2009)

7. Hinchet, R., Vechev, V., Shea, H., Hilliges, O.: Dextres: wearable haptic feedback for grasping in VR via a thin form-factor electrostatic brake. In: The 31st Annual ACM Symposium on User Interface Software and Technology, pp. 901–912. ACM (2018)
8. Jhang, L.H., Santiago, C., Chiu, C.S.: Multi-sensor based glove control of an industrial mobile robot arm. In: 2017 International Automatic Control Conference (CACS), pp. 1–6. IEEE (2017)
9. Katyal, K.D., Brown, C.Y., Hechtman, S.A., Para, M.P., McGee, T.G., Wolfe, K.C., Murphy, R.J., Kutzer, M.D., Tunstel, E.W., McLoughlin, M.P., et al.: Approaches to robotic teleoperation in a disaster scenario: from supervised autonomy to direct control. In: 2014 IEEE/RSJ International Conference on Intelligent Robots and Systems, pp. 1874–1881. IEEE (2014)
10. Kim, T.W., Marani, G., Yuh, J.: Underwater vehicle manipulators, pp. 407–422. Springer, Cham (2016). https://doi.org/10.1007/978-3-319-16649-0_17
11. Kofman, J., Wu, X., Luu, T.J., Verma, S.: Teleoperation of a robot manipulator using a vision-based human-robot interface. IEEE Trans. Ind. Electron. **52**(5), 1206–1219 (2005)
12. Le Ba, N., Oh, S., Sylvester, D., Kim, T.T.H.: A 256 pixel, 21.6 μw infrared gesture recognition processor for smart devices. Microelectron. J. **86**, 49–56 (2019)
13. Li, S., Rameshwar, R., Votta, A.M., Onal, C.D.: Intuitive control of a robotic arm and hand system with pneumatic haptic feedback. IEEE Rob. Autom. Lett. **4**(4), 4424–4430 (2019)
14. Li, Z., Huang, B., Ajoudani, A., Yang, C., Su, C.Y., Bicchi, A.: Asymmetric bimanual control of dual-arm exoskeletons for human-cooperative manipulations. IEEE Trans. Rob. **34**(1), 264–271 (2017)
15. Liang, H., Yuan, J., Thalmann, D., Zhang, Z.: Model-based hand pose estimation via spatial-temporal hand parsing and 3D fingertip localization. Vis. Comput. **29**(6–8), 837–848 (2013)
16. Lu, Z., Zhang, Y., Cheng, D., Wang, S., et al.: Method of dual manipulator human-friendly control based on wireless motion capture technology. In: 2018 11th International Workshop on Human Friendly Robotics (HFR), pp. 31–35. IEEE (2018)
17. Ma, J., Khang, G.: Quantification and adjustment of pressure and vibration elicited by transcutaneous electrical stimulation. Int. J. Precis. Eng. Manuf. **19**(8), 1233–1238 (2018)
18. MacKenzie, I.S.: Input devices and interaction techniques for advanced computing. Virt. Environ. Adv. Interf. Des., 437–470 (1995)
19. Mardiyanto, R., Utomo, M.F.R., Purwanto, D., Suryoatmojo, H.: Development of hand gesture recognition sensor based on accelerometer and gyroscope for controlling arm of underwater remotely operated robot. In: 2017 International Seminar on Intelligent Technology and Its Applications (ISITIA), pp. 329–333 (2017). https://doi.org/10.1109/ISITIA.2017.8124104
20. Nuelle, K., Schulz, M.J., Aden, S., Dick, A., Munske, B., Gaa, J., Kotlarski, J., Ortmaier, T.: Force Sensing, Low-Cost Manipulator in Mobile Robotics. In: 3rd IEEE International Conference on Control, Automation and Robotics (ICCAR), Nagoya, Japan, 22–24 April 2017, pp. 196–201. IEEE (2017)
21. Premarathna, C.P., Ruhunage, I., Chathuranga, D.S., Lalitharatne, T.D.: Haptic feedback system for an artificial prosthetic hand for object grasping and slip detection: a preliminary study. In: 2018 IEEE International Conference on Robotics and Biomimetics (ROBIO), pp. 2304–2309. IEEE (2018)

22. Stańczyk, K., Poświata, A., Roksela, A., Mikulski, M.: Assessment of muscle fatigue, strength and muscle activation during exercises with the usage of robot luna EMG, among patients with multiple sclerosis. In: International Conference on Information Technologies in Biomedicine, pp. 117–128. Springer, Heidelberg (2019)
23. Suau, X., Alcoverro, M., López-Méndez, A., Ruiz-Hidalgo, J., Casas, J.R.: Real-time fingertip localization conditioned on hand gesture classification. Image Vis. Comput. **32**(8), 522–532 (2014)
24. Wang, K.J., Zheng, C.Y., Mao, Z.H.: Human-centered, ergonomic wearable device with computer vision augmented intelligence for VR multimodal human-smart home object interaction. In: 2019 14th ACM/IEEE International Conference on Human-Robot Interaction (HRI), pp. 767–768. IEEE (2019)
25. Yamakawa, Y., Matsui, Y., Ishikawa, M.: Human–robot collaborative manipulation using a high-speed robot hand and a high-speed camera. In: 2018 IEEE International Conference on Cyborg and Bionic Systems (CBS), pp. 426–429. IEEE (2018)
26. Zhang, H., Yan, X., Li, H.: Ergonomic posture recognition using 3D view-invariant features from single ordinary camera. Autom. Constr. **94**, 1–10 (2018)
27. Zhang, K., Follmer, S.: Electrostatic adhesive brakes for high spatial resolution refreshable 2.5 D tactile shape displays. In: 2018 IEEE Haptics Symposium (HAPTICS), pp. 319–326. IEEE (2018)

Applicability of Artificial Robotic Skin for Industrial Manipulators

Piotr Falkowski(✉), Zbigniew Pilat, and Marek Pachuta

ŁUKASIEWICZ Research Network - Industrial Research Institute for Automation and Measurements PIAP, Warsaw, Poland
pfalkowski@piap.pl

Abstract. As the concept of cooperation between robots and humans is becoming more and more popular, companies develop devices transforming industrial manipulators into cobots, or collaborative robots intended to interact with humans. Their interactions with humans can be enabled by equipping the robots with so-called artificial robotic skins, which allow robots to receive and interpret touch. However, it is only in limited conditions that such manipulators comply with international standards. In this paper, the applicability of robotic skins for industrial purposes is assessed based on experimental trials. The procedure involved colliding robotic skin pads with a pendulum. The amount of energy transferred during the impact was calculated from measured deviation. The study showed that industrial robots equipped with artificial robotic skins can be safely used when the velocity of motion is within a very limited range of maximum velocity. It is suggested that this range may be extended to more feasible by decreasing stiffness of the materials used to make the pads or by improving sensitivity of sensors.

Keywords: Artificial robotic skin · Cobots · Industry 4.0 · Soft robotics

1 Introduction

Development of Industry 4.0 brought an idea of cobots. Collaborative robots spread around the factories and are applied for the tasks, which require particular manipulation or validation methods which, until recently, could only be performed by humans. They got especially popular in pharmaceutical and automotive sectors [1].

Nearly all the robots' manufacturers created their own models of collaborative industrial manipulators. According to international standard *ISO/TS 15066*, their maximum impact on a user must not cause energy and momentum transfer higher than stated in it [2]. However, it may be also possible to use typical industrial robots as cobots, which does not require the application of safety barriers and can cooperate with humans.

Artificial robotic skin is a tactile device placed on a manipulator and connected to either its controller or a safety circuit. Such skin should have a structure

A. Bartoszewicz et al. (Eds.): PCC 2020, AISC 1196, pp. 520–528, 2020.
https://doi.org/10.1007/978-3-030-50936-1_44

enabling immediate transmission of a signal resulting from the robot's coming in contact with any object [3]. As such, it can be used as a safety device or additional sensor transforming an industrial manipulator into a cobot.

Robotic skin may consist of either pads or smaller tiles. Therefore, it receives a different number of signals (typically one for every pad or tile). Softer pads partially disperse energy upon collision and therefore have found use in industrial applications where (operator/human) safety is of high concern [4–6]. Tiles on the other hand are typically used in human-computer interaction scenarios, particularly for service robots such as ASIMO or TOMM [7,8].

2 Setup and Experimental Procedure

The main aim of this study was to verify the applicability of artificial robotic skin for industrial purposes. That is, to assess whether and under what conditions could a robot equipped with skin pads meet the requirements stated in *ISO/TS 15066*, and thus be treated as a cobot and cooperate with a human without any safety barriers. In the conducted experiments (iterations) maximum velocity of the robot was varied - from 0.25 m/s to 2.0 m/s. Thus, values of the calculated parameters become discrete functions of maximum velocity set for the motion (from 0.25 m/s up to 2.0 m/s, increased by 0.25 m/s at each trial). Afterwards, they are compared to the corresponding limits for different body parts, presented in the *ISO/TS 15066* standard. Outcomes from the experiment should contribute to the improvement of contemporary solutions without the loss of their compatibility with adequate pieces of legislation. The results may also inform the direction of the ongoing development of such systems.

The tests were conducted on a setup consisting of *KUKA KR/16/KRC4* industrial manipulator with a welding torch assembled as a tool, two *Airskin Module Pads* (200 × 200 mm and 200 × 300 mm) [4] connected directly to the safety circuit of the robot and a pendulum used as a part of a measuring set (see Fig. 1). The had a moment of inertia $J = 0.197\ [\mathrm{kg} \cdot \mathrm{m}^2]$ (as calculated according to the Steiner's theorem), and a distance between its centre of mass and an axis of rotation was $l = 6 \cdot 10^2\ [\mathrm{mm}]$.

The experimental procedure consisted of directing the robot's tool such that it would hit the pendulum with one of its skin pads. The intention was for the *TCP* to move linearly and to achieve the maximum set velocity by the time of the collision. The purely linear motion of *TCP* was to ensure that each time the point of impact was as close to the centre of the pad as possible. This would maximise the reliability of measurements, as the pad's elasticity differs across its surface and that in turn affects the time of the sensor's response to the stimulus. Moreover, the *TCP* should move with the maximum velocity set while colliding with pendulum. This appears within the particular range of velocities and for the long-distance of motion only. Therefore, the results obtained for the velocities higher than 1.25 m/s were significantly different than expected and thus, they were ignored at the further stages of the experiment.

The impact of a pendulum onto the pad's surface would cause the emergency stop circuit to activate and stop the robot's motion immediately. Every collision involved a transfer of the momentum and energy between the robot and the pendulum. Their values were determined from the energy transfer calculation, based on the maximum deviation angle of the pendulum from its vertical equilibrium. On the contrary, the distance travelled by the *TCP* during the braking was computed based on a final deviation angle of the pendulum from its vertical equilibrium. These two angles were the main parameters measured in the experiment. Additionally, an angle between the pendulum's vertical equilibrium and a vertical axis of an observed frame of reference was measured to eliminate any errors caused by the potential displacement of the setup.

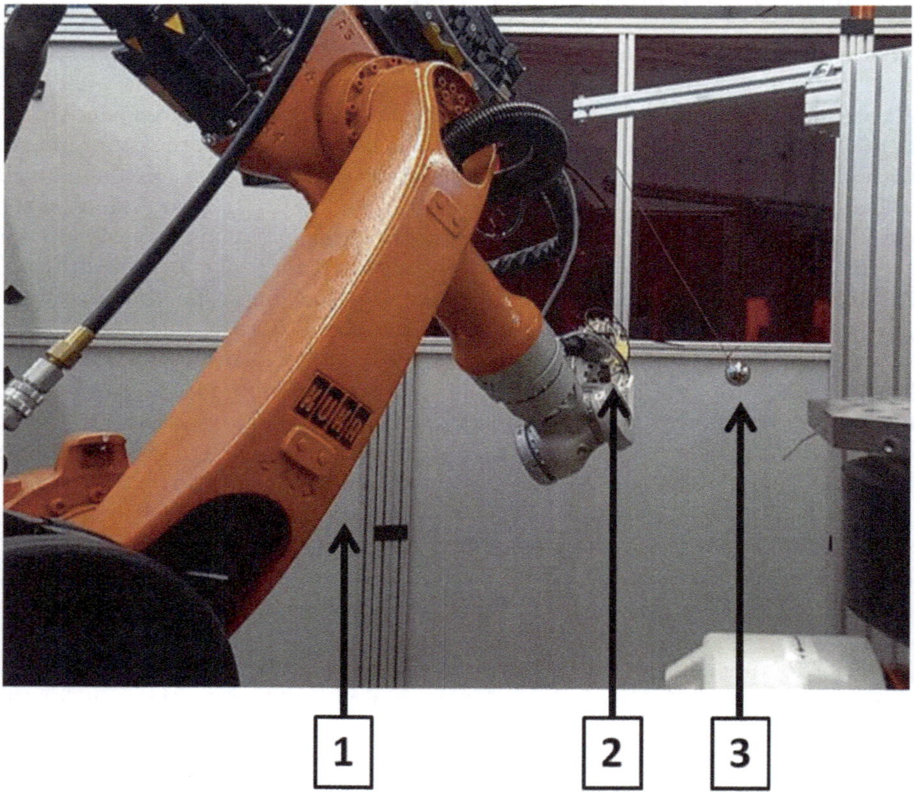

Fig. 1. Setup used for the experiments. 1 - KUKA KR16 robot, 2 - Airskin Module Pads, 3 - pendulum.

Each trial was recorded with a camera (1080p, 120fps) mounted onto a tripod and focussed onto the point of collision between the robot's skin pad and pendulum. The films were studied frame by frame. To do so, pixels were converted into millimetres based on the length of a pendulum and the measured angles.

To reduce measurement errors below the value of 0.5%, all the measured angles were calculated with the trigonometric analysis of difference of vectors.

The objective of these was a further approximation of the parameters dependent on the maximum velocity of the robot during its motion.

3 Experiment

3.1 Estimation of Energy Loss in a Pendulum

A pendulum loses its energy during motion due to aerodynamic damping. Moreover, all the computed parameters are based on the change of pendulum's potential energy during the 1/4 of the period. Therefore, the effect of damping has to be considered in these calculations.

To do calculate a damping coefficient, β, an additional experiment was conducted. The pendulum was displaced from its equilibrium and released. Then, all its maximum deviation angles φ and corresponding periods T were measured. Based on these, energy loss ΔE_T and relative energy loss $\Delta E_{T\%}$ were computed for every period (see Table 1). The average of all values of $\Delta E_{T\%}$ calculated for 9 periods was taken as the damping coefficient, $\beta = 2\%$.

Table 1. Parameters regarding loss of pendulum's energy

Period	$\Delta E_T[mJ]$	$T[s]$	$\Delta E_{T\%}[mJ]$
1	7.65	1.55	2.39%
2	2.93	1.56	0.87%
3	2.87	1.60	0.89%
4	8.14	1.59	2.43%
5	10.04	1.63	3.19%
6	10.90	1.58	3.36%
7	13.53	1.57	4.45%
8	6.57	1.58	2.12%
9	8.23	1.57	2.77%

3.2 Calculation and Estimation of Setup's Parameters After Collision, Based on Trials

Recorded collisions of the artificial skin pad with the pendulum were analysed frame by frame. Freeze-frame shots were used to geometrically calculate maximum deviation angle of the pendulum φ_{max} (see Table 2) and its final deviation angle of the pendulum φ_f (see Table 2), measured after complete stop of the robot. The former parameter was applied to the formula (1) for energy conservation modified to account for the damping coefficient, $\beta = 2\%$, where l is a length of a pendulum's link, m_k is a mass of a pendulum's ball, m_l is a mass of a pendulum's link and g is the gravitational acceleration.

$$E = \frac{gl}{\left(1 - \frac{\beta}{2}\right)} \left(m_k + \frac{m_l}{2}\right) \cdot (1 - \cos\varphi_{max}) \tag{1}$$

Using the moment of inertia, J of the pendulum calculated beforehand, energy E transferred during the collision was computed (see Table 2). Momentum p transferred during the collision was derived based on a formula of momentum conservation (2).

$$p = \sqrt{2JE} \tag{2}$$

Furthermore, a real braking distance d_f of the robot was determined based on Eq. (3) (see Table 2).

$$d_f = l \sin\varphi_f \tag{3}$$

All the parameters discussed above were calculated for the different values of the maximum velocity, v_{max}, of the TCP employed in this study.

Table 2. Computed parameters of setup after collision

$v_{max}\left[\frac{m}{s}\right]$	$\varphi_{max}[rad]$	$\varphi_f[rad]$	$v_t\left[\frac{m}{s}\right]$	$p\left[\frac{kg\cdot m}{s}\right]$	$E[J]$	$d_f[mm]$
0.25	0.165	0.058	2.49	0.13	0.04	34.78
0.5	0.340	0.117	5.10	0.26	0.17	70.04
0.75	0.513	0.174	7.65	0.39	0.38	103.87
1	0.703	0.268	10.39	0.53	0.71	158.88
1.25	0.896	0.33	13.07	0.67	1.12	194.43

Energies transferred in each collision are plotted in Fig. 2. Using the least square error method, a curve was fitted to the data points in order to model the dependence of transferred energy on the robot's velocity at the point of collision.

To estimate the maximum value of energy transferred upon collision, it was assumed to be equal 100% of the kinetic energy of the robot in the moment of collision. Since kinetic energy is proportional to velocity squared, $\hat{E}_{max}$ was expressed as a function of v_{max}^2 to fit the recorded data points according to the formula (4)

$$\hat{E}_{max}\,[J] = 0.7528 v_{max}^2 - 0.0468 v_{max} + 0.0023 \tag{4}$$

This proved to be only applicable to velocities up to 1.25 m/s (see Fig. 2). A relative standard deviation of such approximation is equal to $\sigma_E = 3.5\%$ (calculated according to the formula (5)).

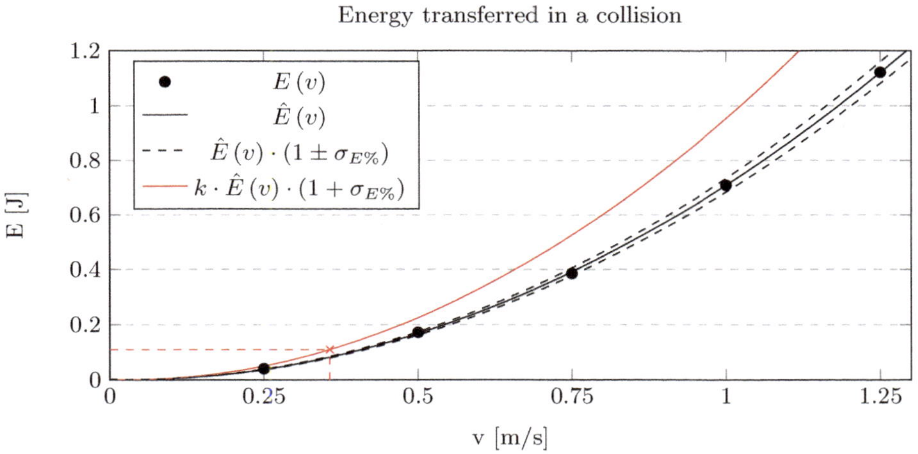

Fig. 2. Correlation between energy transferred in a collision and the maximum velocity of the robot during its motion

$$\sigma_x = 100\% \cdot \sqrt{\frac{1}{n}\sum_{i=1}^{n}\left(\frac{x_i - \hat{x}_i}{x_i}\right)^2} \tag{5}$$

A similar approach was applied towards an approximation of a momentum transferred in a collision $\hat{p}_{max}$ (see Fig. 3). However, for this, a linear function was chosen to comply with theoretical physics' principles. This estimate is expressed with a formula (6) and its relative standard deviation is equal to $\sigma_p = 1.1\%$.

$$\hat{p}_{max}\left[\frac{kg \cdot m}{s}\right] = 0.5328v_{max} - 0.0047 \tag{6}$$

It is worth noticing, that the low values of momentum for the low velocities of the robot does not have a physical interpretation and are a result of an approximation with the least square errors method.

Deceleration of a robot is assumed to remain constant for the short braking time. Thus, braking distance $\hat{d}_{max}$ is approximated with a linear function (7) (see Fig. 4). Relative standard deviation for this approximation is equal $\sigma_d = 3.5\%$.

$$\hat{d}_{max}\,[mm] = 23.66v_{max}^2 + 127.94v_{max} + 0.1486 \tag{7}$$

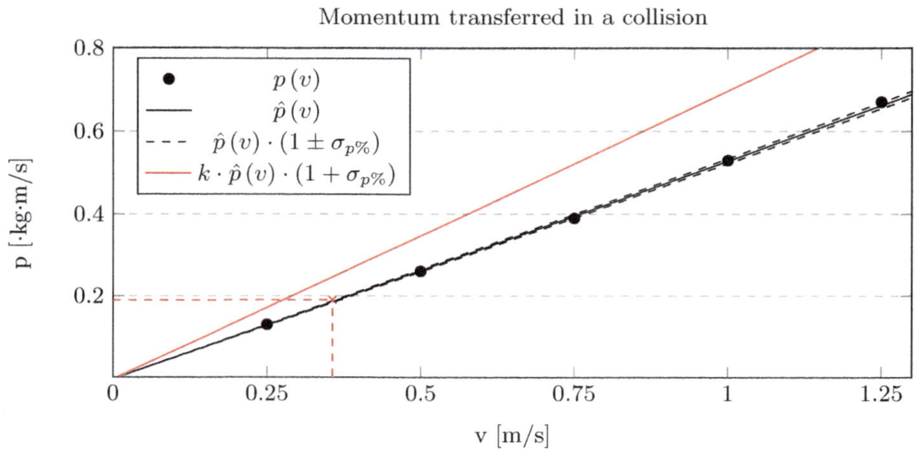

Fig. 3. Correlation between momentum transferred in a collision and the maximum velocity of the robot during its motion

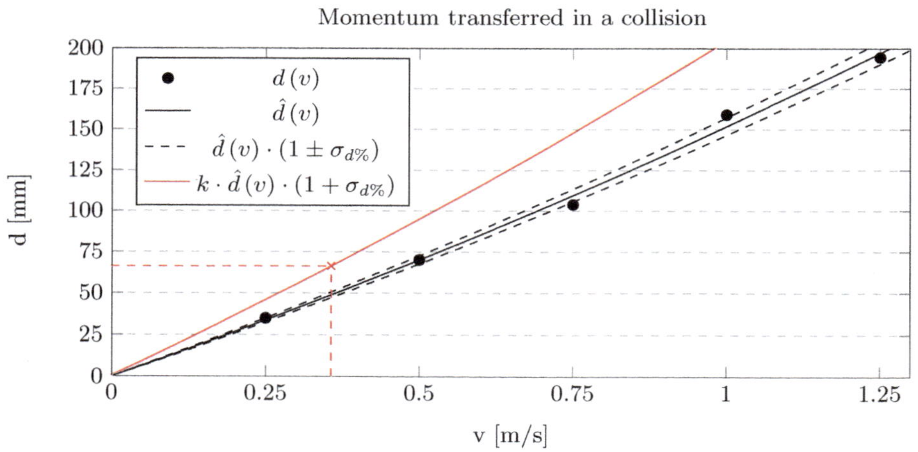

Fig. 4. Correlation between the distance of decellerating after a collision and the maximum velocity of the robot during its motion

4 Necessary Conditions to Use the Setup in Compliance with the ISO/TS 15066 Standard

Maximum velocity of the robot is estimated based on the limits of energy transferred during a collision stated in the *ISO/TS 15066* standard (see Table 3).

The face is the body region for which the maximum allowed energy is the lowest, and thus the value $\hat{E}_{max} = 0.11$ J was taken to calculate the upper limit of permissible velocity of the cobot in question. Relative standard deviation

Table 3. Limits of transferred energy based on the body region of an impact [2]

Body region	Maximum transferred energy [J]
Skull and forehead	0.23
Face	0.11
Neck	0.84
Back and shoulders	2.5
Chest	1.6
Abdomen	2.4
Pelvis	2.6
Arms and elbow joints	1.5
Forearms and wrist joints	1.3
Hands and fingers	0.49
Thighs and knees	1.9
Lower legs	0.52

$\sigma_E = 3.5\%$ and a safety factor $k = 1.3$ are implemented to Eq. (8). As a result the maximum robot's velocity $v_{max} = 0.357\frac{\mathrm{m}}{\mathrm{s}}$ is calculated.

$$\frac{\hat{E}_{max}}{1+\sigma_E} = k\left(0.7528v_{max}^2 - 0.0468v_{max} + 0.0023\right) \tag{8}$$

Values of a maximum momentum transferred $\hat{p}_{max} = 0.188\frac{\mathrm{kg \cdot m}}{\mathrm{s}}$ and a braking distance $\hat{d}_{max} = 65.79\,\mathrm{mm}$ are calculated according to formulas (9) and (10).

$$\frac{\hat{p}_{max}}{1+\sigma_p} = k\left(0.5328v_{max} - 0.00473\right) \tag{9}$$

$$\frac{\hat{d}_{max}}{1+\sigma_d} = k\left(23.66v_{max}^2 + 127.94v_{max} + 0.14863\right) \tag{10}$$

5 Summary

It was shown based on the analysis of conducted experiments, that *KUKA KR16/KRC4* equipped with *Airskin Module Pads* can be used as a cobot and directly collaborate with a human only if its velocity of motion is limited to $v_{max} = 0.357\frac{\mathrm{m}}{\mathrm{s}}$. This limit arises from the safety requirements set in *ISO/TS 15066*. Compared to the maximum speed of 0.25 m/s, this robot can achieve in a manual *T1* mode, this cobot configuration does not offer a significant advantage.

To enable full unrestricted collaboration with a human at higher robot's velocities, a solution needs to be developed that would dissipate impact energy more efficiently and allow faster transmission of signal upon the impact. More easily compressible pads and sensors with shorter reaction times are therefore crucial.

Additional capacitive sensors system could also be implemented to allow the robot to stop when it is about to hit an object or a human operator rather than upon the collision.

A solution, which may enable a full collaboration with human, has to disperse mechanical energy in a more efficient way and react faster on a trigger. This can cause less energy transferred to a hit object while moving with a higher velocity. It seems that using more-compressive pads and sensors with shorter reaction times is necessary. Also, it would be advantageous if the sensors react rather to getting closer to the robot than just to colliding with it.

This idea may be used to design a robotic soft-coat, assembled to the robot while working within its range (e.g. servicing a nearby machine). Its application can significantly decrease the downtime time of a factory and thus, increase the profits of entrepreneurs.

Acknowledgements. The article is based on the results of the IV stage of the multi-annual program "Improving safety and working conditions", financed in 2017–2019, in the scope of scientific research and development works, by the Ministry of Science and Higher Education/National Center for Research and Development.

Program Coordinator: Central Institute for Labour Protection - National Research Institute.

References

1. Vysocky, A., Novak, P.: Human-robot collaboration in industry. MM Sci. J. **2**, 903–906 (2016). https://doi.org/10.17973/MMSJ.2016_06_201611
2. ISO TS 15066-Robots and robotic devices-Collaborative robots (2016)
3. Lamy, X., Colledani, F., Geffard, F., Measson, Y., Morel, G.: Robotic skin structure and performances for industrial robot comanipulation. In: IEEE/ASME International Conference on Advanced Intelligent Mechatronics, pp. 427–432. IEEE (2009). https://doi.org/10.1109/AIM.2009.5229975
4. Airskin Module Pads. https://www.bluedanuberobotics.com/airskin/. Accessed 19 Jan 2020
5. Rexroth's Production Assistants (APAS). https://www.boschrexroth.com/en/xc/products/product-groups/production-assistants-apas/template-overview-9. Accessed 19 Jan 2020
6. Mechavision's Contact Skins. https://www.mecha-vision.com/en/. Accessed 19 Jan 2020
7. Sakagami, Y., Watanabe, R., Aoyama, C., Matsunaga, S., Higaki, N., Fujimura, K.: The intelligent ASIMO: system overview and integration. In: IEEE/RSJ International Conference on Intelligent Robots and Systems, vol. 3, pp. 2478–2483. IEEE (2002). https://doi.org/10.1109/IRDS.2002.1041641
8. TOMM. http://www.ics.ei.tum.de/en/research/platforms/robot-tomm/. Accessed 19 Jan 2020

Hardware in the Loop Control Based on the Open Source Simulation Environment

Damian Wroński(✉) and Grzegorz Granosik

Lodz University of Technology, Institute of Automatic Control, Lodz, Poland
damian.wronski.95.dws@gmail.com, grzegorz.granosik@p.lodz.pl

Abstract. This paper presents an implementation and tests of the hardware in the loop control idea using Scilab-Xcos software – an open source alternative for Matlab-Simulink simulation environment and modular drive system dedicated to brushless motors. Communication infrastructure was based on USB interface and special protocol between hardware and Scilab-Xcos. For the test of the hardware in the loop control, the Field Oriented Control was implemented within Scilab-Xcos. Thanks to this approach, control algorithms do not have to be implemented directly on the device, but in an environment that ensures convenient operation, including quick tests with variable design structure. Software can be tested on both Windows and Linux. Using Linux OS makes this solution pure open source.

Keywords: Robotics · Motor control · Open source software · USB communication · Graphical interface · Hardware in the loop

1 Introduction

Presented article is a continuation of work conducted in area of control algorithms implementation and an enhancement of testing capabilities of general purpose drive module, developed at the Institute of Automatic Control. The following paper demonstrates an attempt to implement and test hardware in the loop control idea with real hardware, in order to speed up elaboration of the control technique process. Versatility of created solutions allows to apply the drive modules in any medium class mobile robot and other propulsions operating in torque, speed or position mode. Demonstrated solution of the drive module is characterized by modular structure of the hardware and open source environment for control algorithm development. Three separate electronic boards (Fig. 1) provide flexible configuration and the same control unit may collaborate with different power boards. Every board supports the same communication standards and sensors feedback signals (phase currents, shaft velocity, shaft position). Properties of created software and graphical environment has been investigated with the FOC (Field Oriented Control) algorithm and USB communication. Of course, the presented solution may also be used to test another kinds of control with real hardware, however this concept is mostly directed to control the objects with rather slow dynamics. Graphical User Interface (GUI) is used to significantly simplify the process of designing control algorithms, and thus speeding up the testing procedures. Additionally, the GUI may be used as a

A. Bartoszewicz et al. (Eds.): PCC 2020, AISC 1196, pp. 529–540, 2020.
https://doi.org/10.1007/978-3-030-50936-1_45

device configurator featuring static and dynamic variables adjustment, serve diagnostics and data plotting. The whole control block is realised by Scilab-Xcos software, which is open source and free alternative for Matlab-Simulink. Open source and open hardware solutions are becoming more and more popular not only in amateur applications, but also educational and commercial products. This proposal is at least partly in line with this trend. In the further part of this article, more details about system, control and communication architecture will be presented, as well as system tests conducted using created control structure.

2 Description of the Hardware

For testing purposes of the implemented FOC algorithm within Scilab-Xcos simulation environment, a Vishan EC4070S-2411 BLDC Motor with 70 W of power was used. It is a single pole-pair and three phase motor with maximum nominal speed of 11000 rpm and a 24 V voltage supply. Additionally, the motor has built-in Hall sensors. Despite the fact that the used motor is nominally DC, the electromotive force (B-EMF) has practically a type of sine waveform, therefore the method of FOC will not have a bad influence on quality of the control and may be applied in this case. The main BLDC motor control module is based on Infineon XMC4800. The microcontroller is ARM cortex-M4 with the maximum of 144 MHz clock frequency [2]. The software for the XMC4800 is written in C programming language, utilizing DAVE IDE software, which is dedicated programming environment for Infineon microcontrollers. The modules are equipped with a USB connector for communication with external devices. Thanks to the implemented communication protocol, the control unit can connect to the Scilab-Xcos graphic layer, that can be launched on both Linux and Windows operating systems. The power module for brushless motor is based on DRV8323 IC and MOSFET transistors. The DRV8323 integrated circuit is designed for applications with voltage supply up to 60 V [3]. The control of the individual transistor gate is accomplished by PWM signals, which come from the main control unit. The encoder circuit is equipped with AS5147 IC. It is a contactless magnetic position sensor with a resolution of 14 bits (0.022° per step) [4]. Digital value of absolute position can be read through the SPI interface or the position can be calculated with encoder quadrature outputs (A, B, Index) and POSIF interface built-in XMC4800.

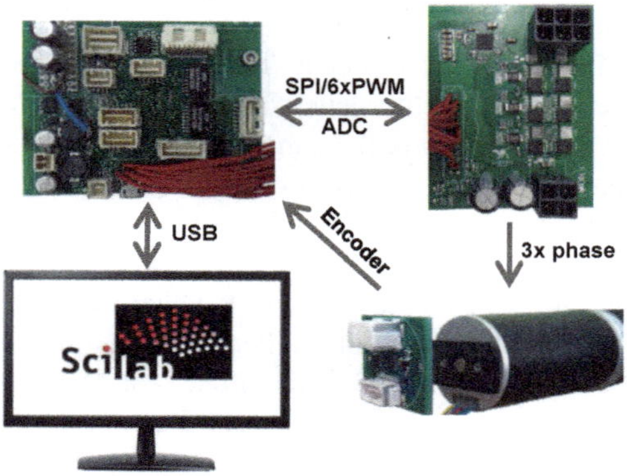

Fig. 1. Structure of the modular drive system.

3 Field Oriented Control Algorithm

Filed Oriented Control is one of the most popular method of controlling three-phase AC motors. The advantage of this approach is accurate control of the internal motor torque, which ensures efficient dynamic properties of the entire drive system [1].

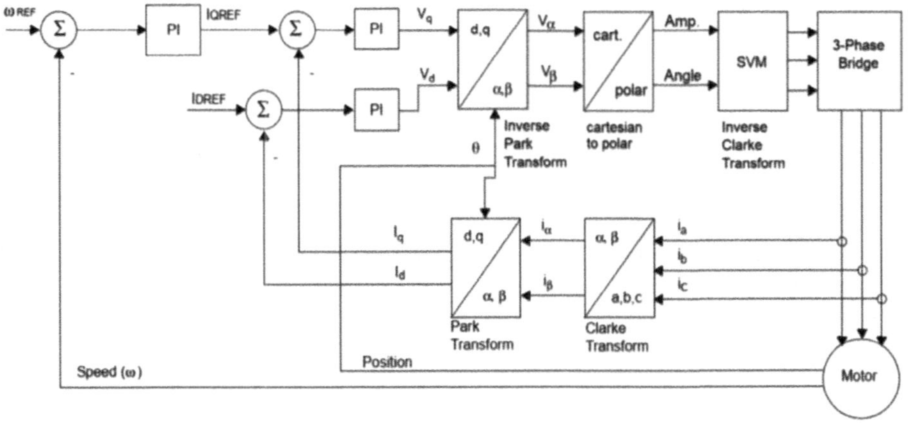

Fig. 2. Block diagram of the control system [6].

This control algorithm involves independent control of two perpendicular to each other components of the stator current space vector: one responsible for generating internal torque in the engine and second responsible for generating an additional magnetic flux energising or de-energising the engine. Thus, it is possible to achieve high dynamics of both: when reproducing changes in the motor shaft velocity set-point and when compensating the sudden changes in load torque occurring on the motor's shaft.

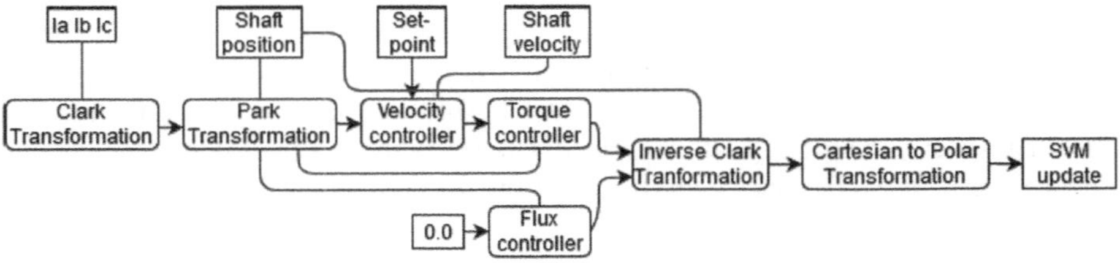

Fig. 3. FOC control diagram.

The diagram (Fig. 3) shows the procedures that are performed during the normally launched process of simulation in Scilab-Xcos. The procedures are performed with accordance to the black arrows. The blue lines represent the information flow between the procedures. Rectangles located outside the outline act as the input/output variables (e.g. set-point, controller gains or output signal) for implemented FOC and controllers.

4 Scilab-Xcos Environment

The whole graphical interface (Fig. 4) for implementation of FOC algorithm was based on block functions, which are available from Palette browser within the Scilab-Xcos software. Main objective of the Scilab-Xcos utilization is to simplify the implementation of algorithms and their testing as much as possible.

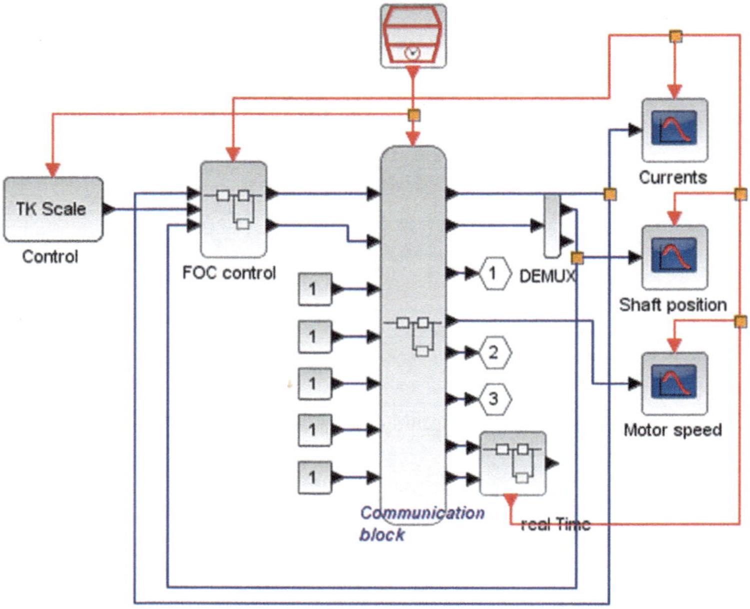

Fig. 4. Implementation of the hardware-in-the-loop control.

Setting the configuration parameters for the control system is done by entering individual coefficients to block with CONST_m designator. Next to each blocks a description label has been places in order to indicate which parameter is configurable. In turn, the display of variables is obtained by using CSCOPE blocks - under each block, there is a label, that specify the type of data displayed. On the right hand side of the communication block there are feedback signals coming from the drive module and on the left hand side, there are control signals.

5 Hardware in the Loop Control Implementation

The algorithm of the FOC (Fig. 5) was entirely created using the basic blocks provided in the Scilab-Xcos software library. The whole structure was based on block diagram presented in the Field Oriented Control algorithm chapter (Fig. 2). The method of time and data synchronization with the drive module was presented in another publications [7, 9].

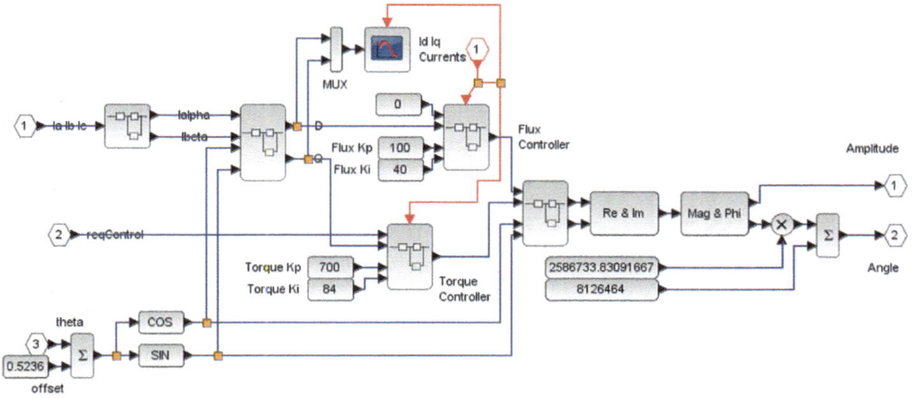

Fig. 5. Implementation of the FOC with Scilab-Xcos.

5.1 Id Iq - Stator Current Space Vector Components Control

In this test of the FOC algorithm, the square wave with an amplitude of 1 A was used as a set-point and a time interval of 2 s. The test was organized according to Table 1. The first test shown, that the coefficients for the controller are improperly matched to the dynamic of the object, because the set point of the Iq component (Fig. 6 - First test) was achieved one second after being forced. Similar situation applies to Id component, but there is approximately half of one second to achieve correct value. In the second test, the coefficients of the Iq controller were modified to ensure faster response to the set-point value. As depicted in the Fig. 6 - Second test, the result is satisfactory sufficiently to proceed to the next step. The last test from the Table 1 also illustrates adjustment of the Id controller to fulfil the condition of a faster reaching the set-point (Fig. 6 - Third test).

Table 1. Table of tested configurations for controllers.

Test section	Flux controller - Id		Torque controller - Iq	
First test	Kp	100	Kp	700
	Ki	40	Ki	80
Second test	Kp	100	Kp	4500
	Ki	40	Ki	90
Third test	Kp	2000	Kp	4500
	Ki	80	Ki	90

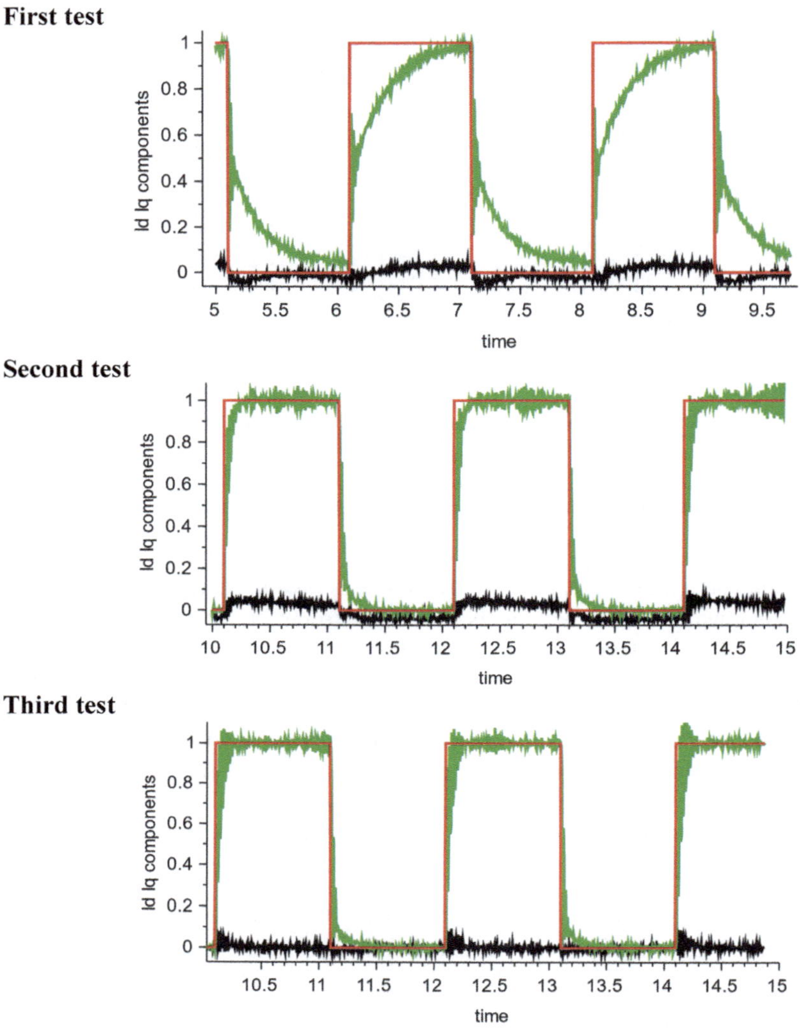

Fig. 6. Green line - Iq component responsible for generation of internal torque in the engine, black line - Id component responsible for generation of an additional magnetic flux, red line - required value of Iq component.

5.2 Motor Shaft Velocity Control Mode

For the control of the motor's shaft velocity the project was modified (Fig. 7) and the square wave with an amplitude of 10 rotation per second was used as a set-point and with a time interval of 2 s. The test as previously was organized according to Table 2. The first test of the velocity control turned out to be unsuccessful (Fig. 8) and due to excessively high gain for the Iq controller, it was quite hard to set coefficients for the velocity controller to work properly. There are also significant current surges when setting the value (Fig. 9), so that the gain for the Iq controller had to be reduced. After this operation the control object started to behave more stable (Fig. 10) and the current spikes have also been mitigated (Fig. 11). At the last step the tuning of the velocity controller was carried out in order to reduce the time of settling to the set-point value. As it is presented in the Fig. 12, the set-point was reached approximately after 250 ms, so it was rather not satisfactory result and the control system needs further parameters tuning to achieve better results.

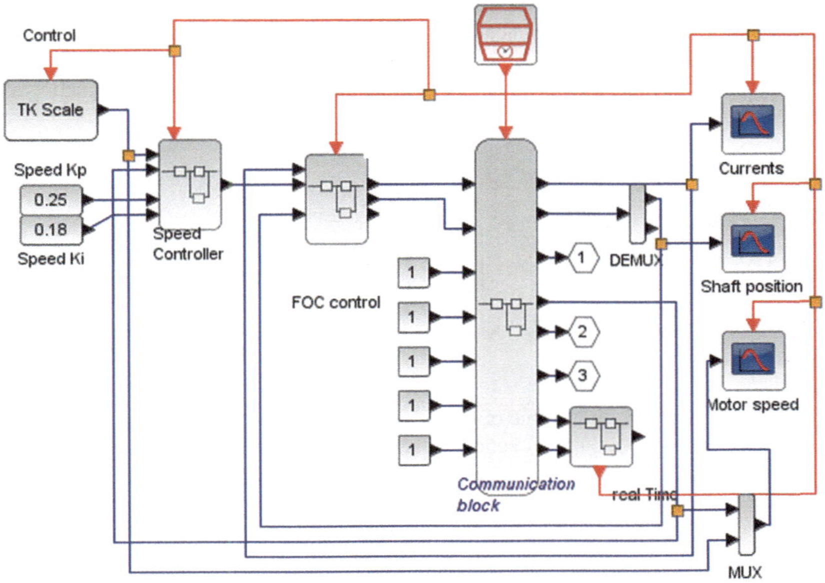

Fig. 7. Modification of the Scilab-Xcos schematic to support velocity mode.

Table 2 Table of tested configurations for controllers.

Test section	Flux controller - Id		Torque controller - Iq		Velocity controller	
First test	Kp	2000	Kp	3000	Kp	0.01
	Ki	80	Ki	90	Ki	0.04
Second test	Kp	2000	Kp	1250	Kp	0.01
	Ki	80	Ki	90	Ki	0.04
Third test	Kp	2000	Kp	1250	Kp	0.25
	Ki	80	Ki	90	Ki	0.18

First test.

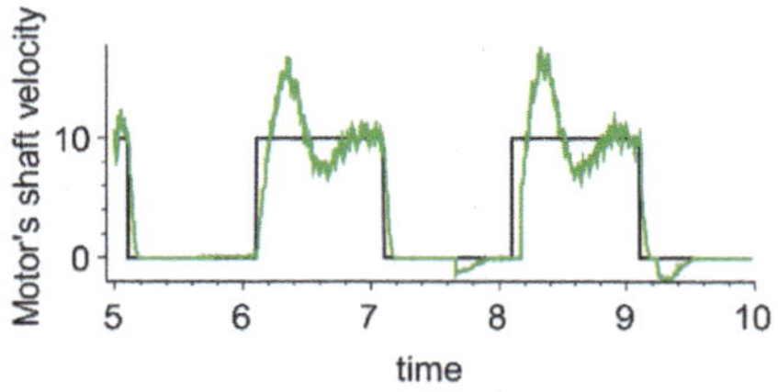

Fig. 8. Black line - required value of the motor's shaft velocity, green line - actual engine shaft velocity.

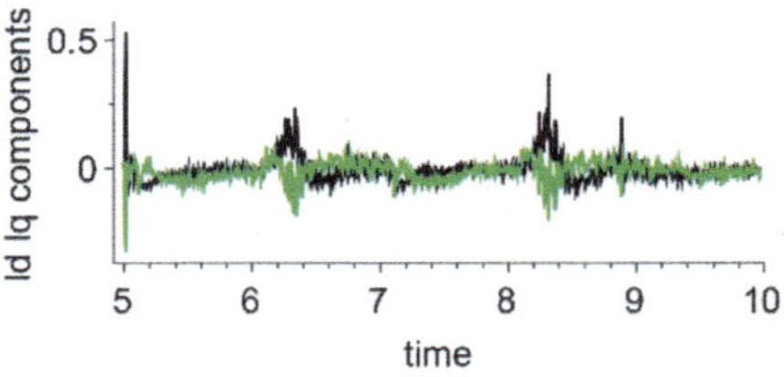

Fig. 9. Green line - Iq component responsible for generation of internal torque in the engine, black line - Id component responsible for generation of an additional magnetic flux

Second test.

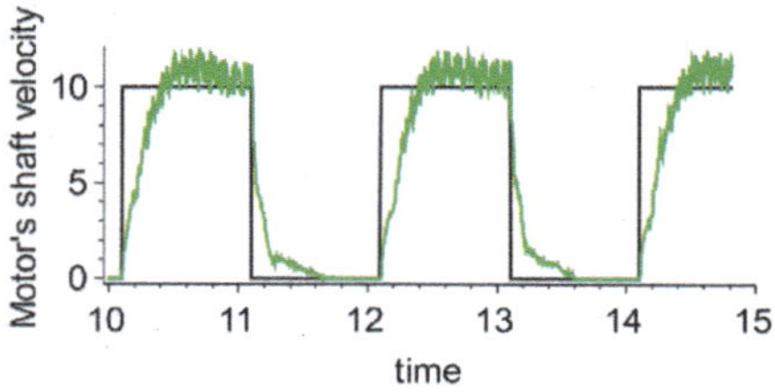

Fig. 10. Black line - required value of the motor's shaft velocity, green line - actual engine shaft velocity.

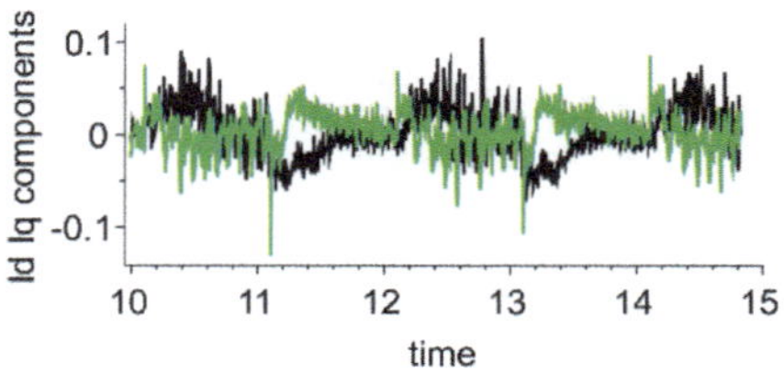

Fig. 11. Green line - Iq component responsible for generation of internal torque in the engine, black line - Id component responsible for generation of an additional magnetic flux

Third test.

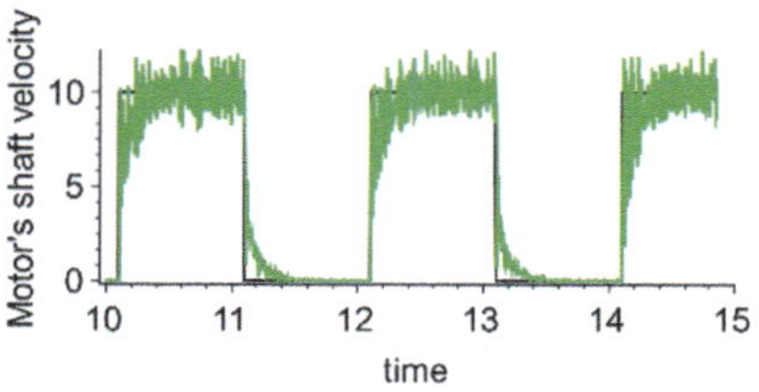

Fig. 12. Black line - required value of the motor's shaft velocity, green line - actual engine shaft velocity.

5.3 Modification of the Current Control Schema to Support Anti-windup Functionality

In this test the functionality of anti-windup for Iq component has been added. For this test, the sine wave was used with an amplitude of 2 A and current cut-off set to value of 1A. In order to determine the saturation values for the controller, the Fig. 13 was necessary for the designation, where the dependency between Iq component and control output indicates. Applied structure of the anti-windup function (Fig. 14) is a fairly typical and well known method of limiting controller output [5, 8]. As it is presented on the Fig. 15, the controller is functioning properly, because with an imposed limit of 1 A, the Iq component is reduced to the limit.

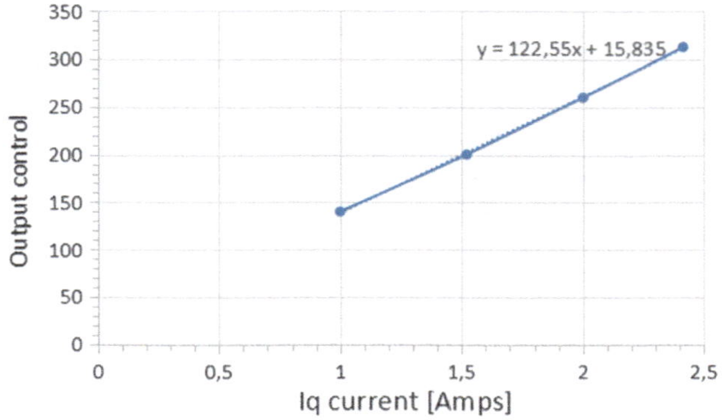

Fig. 13. Dependency graph of the Iq component to the output control.

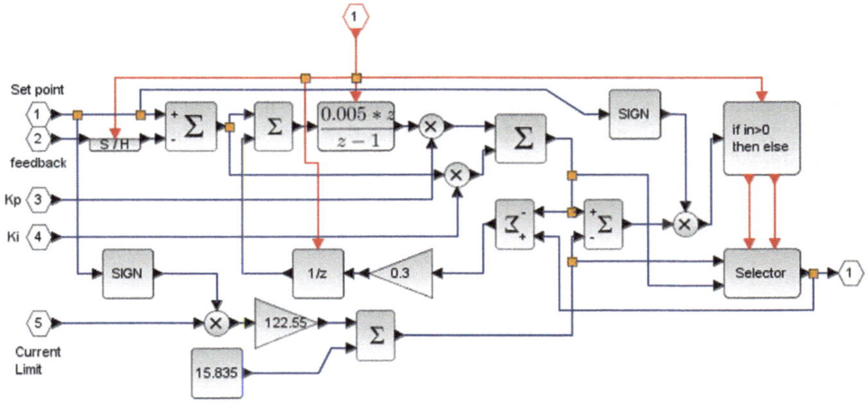

Fig. 14. Modification of the Scilab-Xcos Iq controller to support anti-windup functionality.

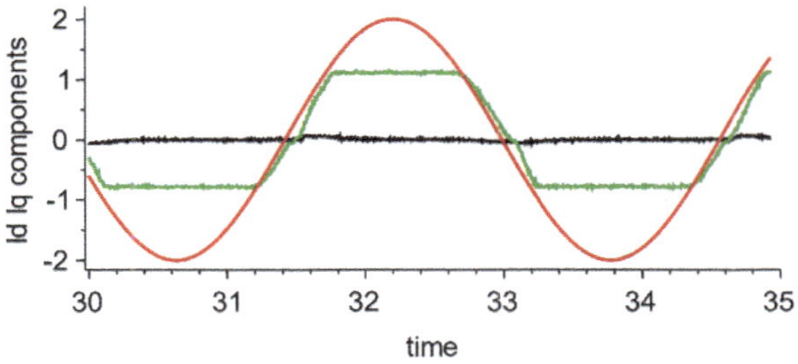

Fig. 15. Green line - Iq component responsible for generation of internal torque in the engine, black line - Id component responsible for generation of an additional magnetic flux, red line - required value of Iq component.

5.4 Modification of the Velocity Control Schema to Support Anti-windup Functionality

After editing the Iq controller and activating the velocity controller, it turns out that after application of additional load to motor's shaft the amplitude of the output signal from the speed controller has increased (Fig. 16, red line) due to the error of the set-point and actual motor's shaft velocity. Therefore, the velocity controller has to be modified, in order to get rid of unwanted issue. The diagrams of the project was modified with simple approach and in accordance with Fig. 17 and Fig. 18. After this operation the integral part of the velocity was disabled after the Iq component limit occurrence (Fig. 19), so the value of integral did not increase any more.

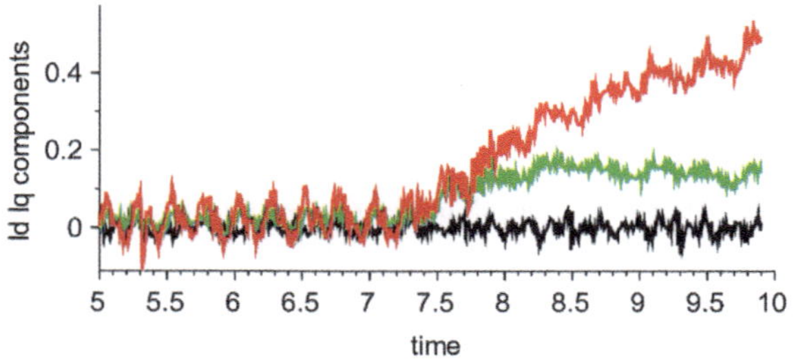

Fig. 16. Green line - Iq component responsible for generation of internal torque in the engine, black line - Id component responsible for generation of an additional magnetic flux, red line - required value of Iq component (velocity controller output).

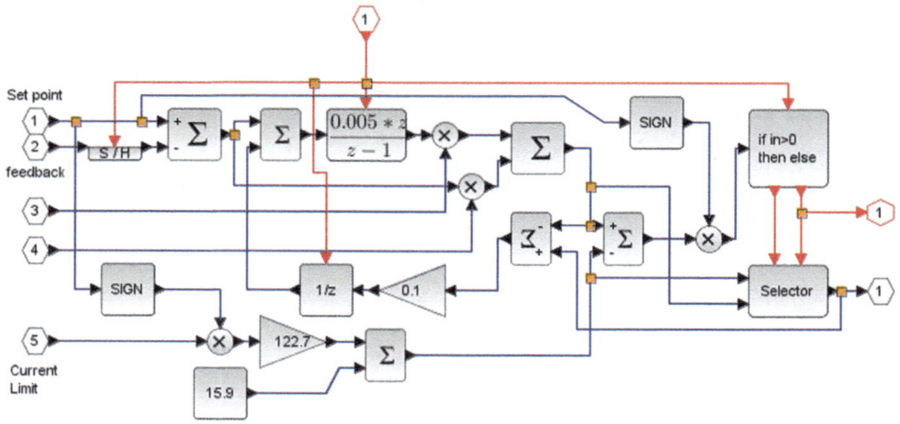

Fig. 17. Modification of the Scilab-Xcos Iq controller to support integral block restriction functionality for velocity controller.

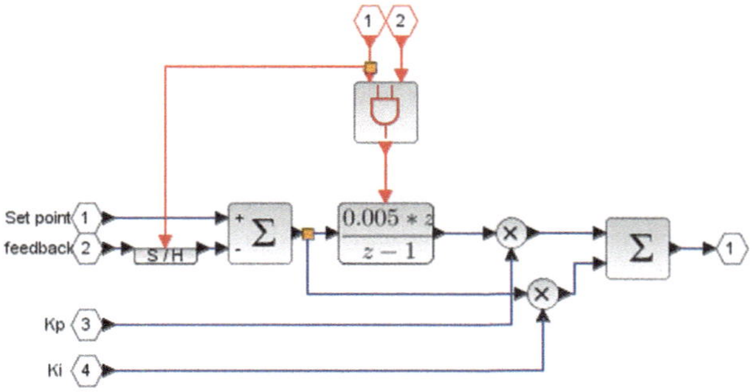

Fig. 18. Modification of the Scilab-Xcos velocity controller to support integral limitation functionality.

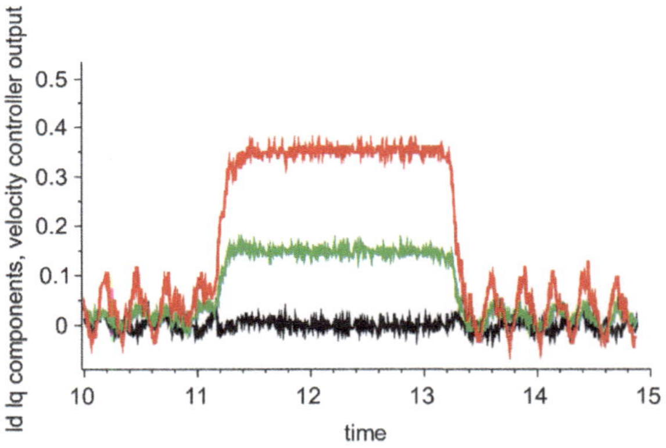

Fig. 19. Green line - Iq component responsible for generation of internal torque in the engine, black line - Id component responsible for generation of an additional magnetic flux, red line - required value of Iq component (velocity controller output).

6 Conclusions

The paper presents a practical implementation with experimental verification of motor control and concerns an important research direction, which is rapid prototyping of electric drives used in devices with increased requirements related to their quality of operation, especially in dynamic states. Authors proposed using an graphical environment based on open-source GUI (graphical user interface), and test its use on a laboratory bench with a real BLDC motor. The results of tests of the experimental electric drive presented in the article have shown the impact on its operation of the proper structure and settings of controllers. This applies to the regulators of the motor stator current vector both components and the shaft speed controller. It also has been shown that in order to improve the quality of this drive in dynamic states, the speed controller must be equipped with an anti-windup system that actively limits the output signal of this controller.

References

1. Lee, S., Lemley, T., Keohane, G.: A comparision study of the commutation methods for the three-phase permanent magnet brushless DC motor. In: Electrical Manufacturing Technical Conference 2009: Electrical Manufacturing and Coil Winding Expo, pp. 49–55 (2009)
2. Infineon, XMC4800 Data Sheet. https://www.infineon.com/dgdl/Infineon-XMC4700-XMC4800-DS-v01_01-EN.pdf?fileId=5546d462518ffd850151908ea8db00b3. Accessed 14 Aug 2019
3. Texas Instruments, DRV832x6 to 60-V Three-Phase Smart Gate Driver, http://www.ti.com/lit/ds/symlink/drv8323r.pdf. Accessed 23 Mar 2019
4. Austria Mikro Systeme, AS514714 Magnetic Rotary Position Sensor. https://ams.com/documents/20143/36005/AS5147_DS000307_2-00.pdf/6921a55b-7cba-bf20-78c0-660d62bd0a5b. Accessed 14 Aug 2019
5. Debowski, A.: Automatyka. Napęd elektryczny, Wydawnictwo WNT (copyright Wydawnictwo Naukowe PWN S.A.), Warszawa (2017)
6. Microchip, Sensorless Field Oriented Control of a PMSM. http://ww1.microchip.com/downloads/en/appnotes/01078b.pdf. Accessed 14 Aug 2019
7. Wroński, D.: Robotyczny system napędowy dla silników bezszczotkowych wraz z interfejsem użytkownika. MSc thesis, Lodz University of Technology (2019)
8. Debowski, A.: Automatyka – podstawy teorii. Wydawnictwo WNT (copyright Wydawnictwo Naukowe PWN S.A.), Warszawa (wyd. II) (2016)
9. Wroński, D., Granosik, G.: Modular drive system for brushless motors with user interface based on open source solutions. In: 2019 SENE Conference (2019)
10. Klaus Weichinger, Scicos Serial-Interface-Block Manual (2012). http://bioe.sourceforge.net/scicosserialinterfaceblock/ScicosSerialInterfaceBlock12.05-Manual.pdf. Accessed 23 Mar 2020
11. Infineon, XMC4800 Reference Manual. https://studio.segger.com/packages/XMC4000/CMSIS/Documents/Infineon-ReferenceManual_XMC4700_XMC4800-UM-v01_03-EN.pdf. Accessed 14 Aug 2019
12. Chattopadhyay, S., Mitra, M., Sengupta, S.: Electric power quality (Chapter 12 Clarke and Park Tranform). Springer (2011). ISBN 978-94-007-0634-7
13. Sundaram, M., Semiconductor, C.: Implementing field oriented control of a brushless DC motor (2012). https://www.eetimes.com/document.asp?doc_id=1279321. Accessed 23 Mar 2020
14. Fisher, P.: High performance brushless DC motor control. School of Engineering & Technology, CQUniversity Australia (2014)
15. http://www.ti.com/lit/an/sprt703/sprt703.pdf. Accessed 23 Mar 2020

Control Approach to Bio-medical Applications

Development of Control Modes Used in Manipulator for Remote USG Examination

Adam Kurnicki[1(✉)] and Bartlomiej Stanczyk[2]

[1] Department of Automation and Metrology, Faculty of Electrical Engineering and Computer Science, Lublin University of Technology, ul. Nadbystrzycka 38a, 20-618 Lublin, Poland
a.kurnicki@pollub.pl

[2] ACCREA Engineering, ul. Hiacyntowa 20, 20-143 Lublin, Poland
b.stanczyk@accrea.com

Abstract. The article focuses on design and implementation of control modes for the control system of a manipulator used as the main part of a robot for remote medical ultrasound examination. This control system has been developed within the ReMeDi (Remote Medical Diagnostician) project. At the beginning of the article, the manipulator kinematics and its control system structure are addressed. The essential components of the control system are discussed in detail. Finally, issues connected with design of control modes and their solutions are presented.

Keywords: Remote medical examination · Medical robot · Manipulator · Control system

1 Introduction

In the modern aging societies there is a growing demand for specialised medical care. Successful medical treatment depends on a timely and correct diagnosis, but the availability of doctors of various specialisations is limited, especially in provincial hospitals or after regular working hours. Medical services performed remotely are emerging, yet current solutions are limited to merely teleconferencing and are insufficient. In order to help in such situations, a number of telerobotic systems were developed: *VGO* [1] for medical teleconsultation, *MEDIROB* [2] for echocardiography examination and *MELODY* [3] for abdominal ultrasonography. However, to the best knowledge of the authors, there are currently no devices allowing a complete remote medical examination (i.e. interview, auscultation, palpation and USG examinations) and diagnosis based on contemporary medical standards, apart from the ReMeDi [4] system. The Remote Medical Dignostician ReMeDi goes beyond classical telepresence concepts. It captures and processes multi-sensory data (integrating visual, haptic, speech) making ReMeDi a diagnostic assistant offering context-dependent and

A. Bartoszewicz et al. (Eds.): PCC 2020, AISC 1196, pp. 543–554, 2020.
https://doi.org/10.1007/978-3-030-50936-1_46

proactive support for the doctor. The first version of this system was designed and made during the ReMeDi project [4] funded by the European Union's Research and Innovation 7th Framework Programme (EU FP7). Nowadays the system is developed within the project: "Research and innovation" activity 1.2 RPO WL 2014–2020 funded by the Lublin Enterprise Support Agency.

The ReMeDi system is a medical grade teleoperation system illustrated in Fig. 1. It consists of two subsystems physically located at different sites. The first subsystem comprises a ReMeDi Robot located on the patient's site e.g. small hospital in a rural area, medical praxis, community without medical specialists, nursing home, offshore platform or remote military area. The second subsystem is called DiagUI (Diagnostician User Interface) and is located at the doctor's site - in another hospital or the doctor's home. The system enables the doctor to carry out remote medical examination of the patient in a way that is essentially identical to the traditional examination. During the examination, the doctor moves and rotates a dummy USG probe held in the palm. The dummy probe is connected to the top of a haptic interface (see left part of Fig. 1). It allows to generate and send position and orientation demands for the real USG probe, which examines the patient. In order to obtain the right ultrasound image, the doctor has to properly position the real probe on the patient's body. This is possible, because the real probe is the end-effector of the manipulator fixed to the mobile ReMeDi robot (see right part of Fig. 1) and follows the required movements. At the same time, the forces acting on the ultrasound probe are transmitted to the dummy probe via the haptic interface so that the doctor feels the stiffness and the geometry of the patient's body.

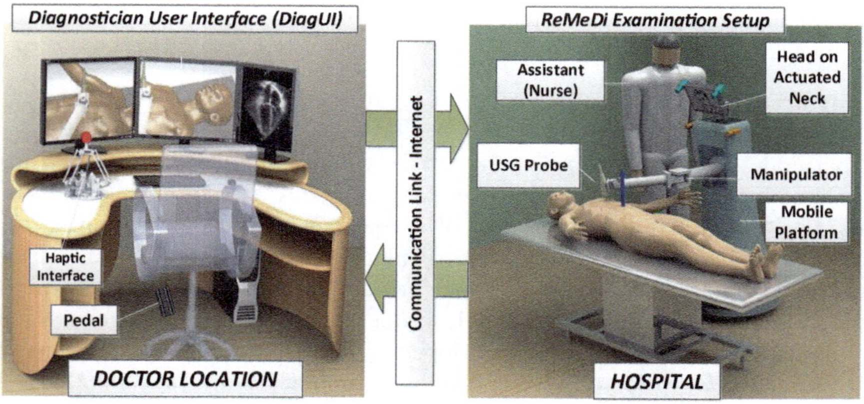

Fig. 1. ReMeDi system overview

An advanced teleconferencing system with quality vision is used for communication between the doctor, the patient and the assistant, as well as to enable the doctor to observe the examination and the moving ultrasound probe.

Bilateral teleoperation with the use of wave variables [5] is used, since all signals are transmitted via the Internet.

2 Manipulator Construction

The ReMeDi manipulator, shown in Fig. 2a, is a 7DoF human-sized arm mounted on the ReMeDi robot. The manipulator is built of two spherical joints with 2 DoFs (shoulder and wrist) and three rotational joints (torso, elbow and probe rotation). Its Denavit-Hartenberg parameters are listed in Fig. 2b and the corresponding set of frames is shown in Fig. 2c. The arm is kinematically redundant.

a)

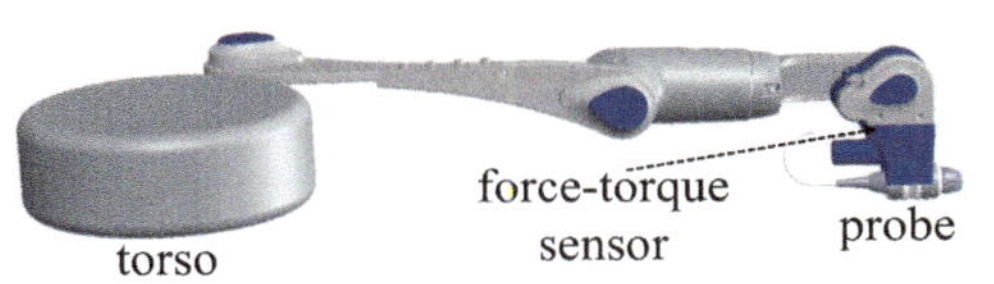

b)

Joint no	Link offset a_i [m]	Link twist α_i [rad]	Link length d_i [m]	Joint angle q_i [rad]
1	0.0	$\pi/2$	0.35	q_1
2	0.0	$-\pi/2$	-0.26	q_2
3	0.4	$\pi/2$	0.0	q_3
4	0.0	$\pi/2$	0.0	$q_4 + \pi/2$
5	0.0	$-\pi/2$	0.32	q_5
6	-0.115	$\pi/2$	0.0	q_6
7	0.0	0.0	0.03	q_7

c)

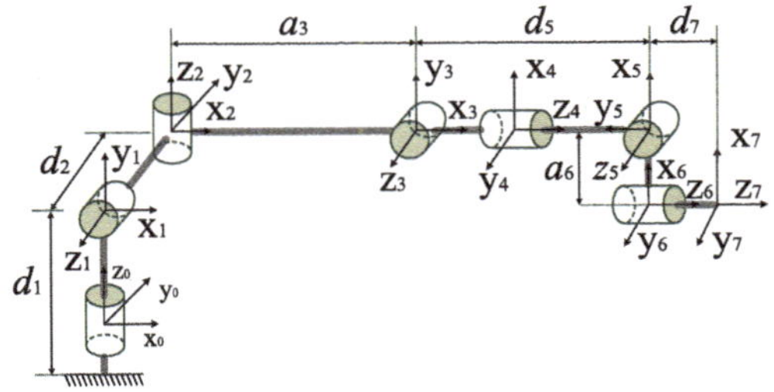

Fig. 2. ReMeDi 7 DoF manipulator a – physical construction, b – Denavit-Hartenberg parameters, c – kinematics structure and link coordinate systems

The manipulator is rigid in relation to the patient's body. The construction is based on aluminium/steel elements and electro-mechanical components like DC motors with harmonic gears and incremental encoders.

In order to provide force feedback to the haptic interface, the arm is equipped with a dedicated custom built six-axis force/torque sensor (FTS) mounted in a series between the last joint of the arm and the probe mounting mechanism.

3 Manipulator Control Architecture

There is a quite wide variety of control system architectures dedicated to manipulators used in tele-echography research. Most of them utilise classical manipulator control architecture with inner position or velocity control loop (usually with a PID controller) at a joint level. Joint references are computed in an outer loop, which performs Cartesian control through velocity-based (less often position-based) inverse kinematics. Mathiassen *et al.* [6] propose to control the manipulator implementing a compliance force control algorithm before velocity based inverse kinematics. The robotic tele-echography system developed by Koizumi *et al.* [7] uses impedance control at the Cartesian position control level, while the orientation is controlled directly with the use of continuous path controller [8]. More advanced and complicated control architecture dedicated to robotic-assisted tele-echography manipulator was developed by Luis Santos and Rui Cortesao. In [9] they applied indirect force control (admittance control) for the Cartesian motion control loop to establish the contact dynamics between the echographic probe and the patient. The orientation is controlled without the admittance control loop. Additionally, the motion controller has a velocity model-reference adaptive control (AOB) in the joint space level, driven by task space posture errors. A two-task hierarchy architecture with posture optimisation is proposed in [10], where Cartesian force control is the primary task, orientation control is secondary, and posture optimisation is performed in the null space of all prioritised tasks.

3.1 Structure of ReMeDi Manipulator Control System

The architecture of the current ReMeDi manipulator control system is an evolution of our previous work [11]. The general block diagram of this architecture is presented in Fig. 3. From the point of view of the position and orientation control strategy, it has a cascading structure, composed of two parts. The internal loop, formed by the four blocks: *ReMeDi Arm*, *Position Controller*, *Joint Demands Filter with Velocity Limiter and Integrator* and *Joint Space Control Modes* (*JSCM*) is used to control the position and orientation of the arm in joint space, i.e. it tracks the reference trajectory $\boldsymbol{q}_{sd} \in \mathbb{R}^7$ by the vector of the actual joint position values $\boldsymbol{q}_s \in \mathbb{R}^7$. The external loop serves the realisation of the

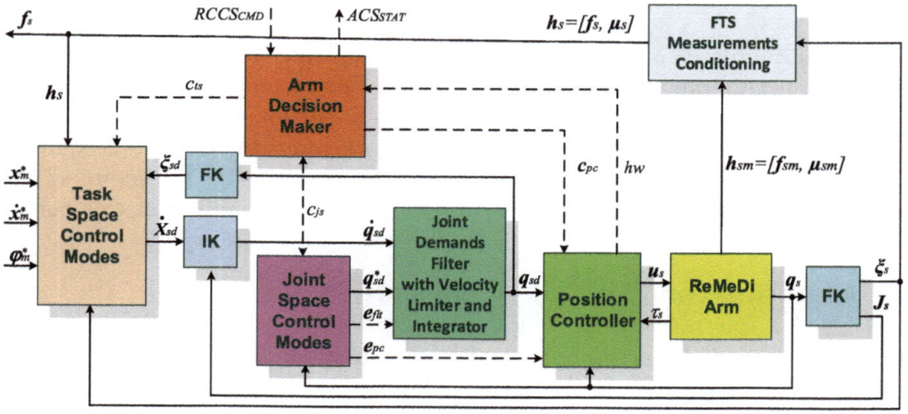

Fig. 3. ReMeDi arm control system architecture

control algorithms in the task space. This part consists of the *IK-Inverse Kinematics*, *Task Space Control Modes* (*TSCM*) and two *FK - Forward Kinematics* blocks.

The *Position Controller* implements seven independent joint position control systems in a specialised unit called JointsController, based on the STM32 microcontroller and manufactured by ACCREA Engineering (for details see [11]). Each of these systems has a form of position-velocity-torque cascade control structure, the details of which can be found in [11]. Every control loop, shown there, is executed with 1 kHz frequency and based on a PID controller with a friction and gravity compensation algorithm.

The algorithms, which realise the functionalities of all other blocks are executed with a frequency of 1 kHz under the real-time Linux PREEMPT system of a PC based controller.

The *TSCM* and the *JSCM* blocks perform actions (modes) required by system users. The first one generates the arm's end-effector desired velocity vector $\dot{\boldsymbol{X}}_{sd} \in \mathbb{R}^6$ on the basis of the following demands: positions $\boldsymbol{x}_m \in \mathbb{R}^3$, velocities $\dot{\boldsymbol{x}}_m \in \mathbb{R}^3$ and orientation angles $\boldsymbol{\varphi}_m \in \mathbb{R}^3$ received from master input device (haptic interface) and forces $\boldsymbol{f}_s \in \mathbb{R}^3$ / torques $\boldsymbol{\mu}_s \in \mathbb{R}^3$ ($\boldsymbol{h}_s = [\boldsymbol{f}_s, \boldsymbol{\mu}_s] \in \mathbb{R}^6$) exerted at the probe's tip. The second one consists of set of switchable applications which generate the demanded joint positions $\boldsymbol{q}^*_{sd} \in \mathbb{R}^7$. Algorithms implemented in both blocks are presented with details in Sect. 4.

The demanded joint positions $\boldsymbol{q}^*_{sd}$ are filtered in the *Joint Demands Filter with a Velocity Limiter and Integrator* block (see Sect. 3.3). The same block limits and integrates demanded joints velocities $\dot{\boldsymbol{q}}_{sd} \in \mathbb{R}^7$ calculated by inverse kinematics (i.e. the *IK* block, presented in Sect. 3.2).

*FK*s are used to calculate the actual $\boldsymbol{\xi}_s$ and demanded $\boldsymbol{\xi}_{sd}$ end-effector pose (i.e. Cartesian position $\boldsymbol{x}_s \in \mathbb{R}^3$ and unit quaternion $\boldsymbol{\mathcal{Q}}_s \in \mathbb{R}^4$ alternatively represented by orientation matrix $\boldsymbol{R}_s \in \mathbb{R}^{3\times3}$) based on the measured $\boldsymbol{q}_s$ and desired $\boldsymbol{q}_{sd}$ position values of the arm joints. The first one additionally calculates the arm's Jacobian $\boldsymbol{J}_s \in \mathbb{R}^{6\times7}$ used by the *IK* algorithm.

The *FTS Measurements Conditioning* block is used to obtain exact values of forces $\boldsymbol{f}_s$ and torques $\boldsymbol{\mu}_s$ from forces $\boldsymbol{f}_{sm} \in \mathbb{R}^3$ and torques $\boldsymbol{\mu}_{sm} \in \mathbb{R}^3$ measured by the FTS, i.e. to eliminate gravity impact on the FTS caused by the weight of the USG probe mechanism. This algorithm is based on the values of the end-effector rotation matrix $\boldsymbol{R}_s$ elements and prameters which specify the centre of mass of the probe mechanism.

There is one additional, very important block, which manages the ReMeDi arm control system - *Arm Decision Maker* (*ADM*). The *ADM* on the basis of hardware status *hw* and control commands $RCCS_{CMD}$ received from the ReMeDi Robot Central Control System (RCCS - see more detailed information in publication [12]), generates control commands: c_{ts} for task space control system part (i.e. for *Task Space Control Modes*), c_{js} for *Joint Space Control Modes* and c_{pc} for *Position Controller*. It was designed in the form of a finite state machine, which operates on four main states: Startup, Active, Shutdown and Failure (see detailed description in [11]). Active state is the most commonly used and very important from the medical examination point of view. It is described with details in Sect. 4. The actual state of the *ADM* is indicated by the ACS_{STAT} signal and is sent to the RCCS.

3.2 Inverse Kinematics

The ReMeDi manipulator control system uses a velocity based (differential) inverse kinematics algorithm to calculate the velocity vector $\dot{\boldsymbol{q}}_{sd}$ from the arm's end-effector desired velocity vector $\dot{\boldsymbol{X}}_{sd} \in \mathbb{R}^6$. The vector $\dot{\boldsymbol{X}}_{sd} = [\dot{\boldsymbol{x}}_{sd}, \boldsymbol{\omega}_{sd}]$ consists of two vectors: the desired linear $\dot{\boldsymbol{x}}_{sd} \in \mathbb{R}^3$ and desired angular $\boldsymbol{\omega}_{sd} \in \mathbb{R}^3$ velocities.

Since the manipulator forms a redundant kinematic chain, to solve the inverse kinematics problem a pseudoinverse control algorithm

$$\dot{\boldsymbol{q}}_{sd} = \boldsymbol{J}_s^{\#} \dot{\boldsymbol{X}}_{sd} + [\boldsymbol{I} - \boldsymbol{J}_s^{\#} \boldsymbol{J}_s] \dot{\boldsymbol{q}}_{sd0}, \tag{1}$$

has been used, where: $\boldsymbol{J}_s^{\#}$ is a Moore–Penrose pseudoinverse of the arm's Jacobian $\boldsymbol{J}_s \in \mathbb{R}^{6\times 7}$ and $\dot{\boldsymbol{q}}_{sd0} \in \mathbb{R}^7$ is an arbitrary value of the joint velocity vector. The first term of (1) is a solution of minimising quadratic cost function of joint velocities (according to the least square method [13]). In order to avoid the least square inverse method's problems with singularities, the weighted dumped least square (WDLS) method has been introduced for the ReMeDi manipulator as a modification of the DLS method [13]. Then minimizing the cost function:

$$\varrho(\dot{\boldsymbol{q}}_{sd}, \dot{\boldsymbol{X}}_{sd}) = \frac{1}{2}(\dot{\boldsymbol{X}}_{sd} - \boldsymbol{J}_s \dot{\boldsymbol{q}}_{sd})^T \boldsymbol{W}_x (\dot{\boldsymbol{X}}_{sd} - \boldsymbol{J}_s \dot{\boldsymbol{q}}_{sd}) + \frac{1}{2} \dot{\boldsymbol{q}}_{sd}^T \boldsymbol{W}_q \dot{\boldsymbol{q}}_{sd}, \tag{2}$$

where: $\boldsymbol{W}_x \in \mathbb{R}^{6\times 6}$ and $\boldsymbol{W}_q \in \mathbb{R}^{7\times 7}$ are symmetric positive-definite weighting matrices associated with the errors in the task space and joint space respectively, gives the following solution:

$$\dot{\boldsymbol{q}}_{sd} = (\boldsymbol{J}_s^T \boldsymbol{W}_x \boldsymbol{J}_s + \boldsymbol{W}_q)^{\#} \boldsymbol{J}_s^T \boldsymbol{W}_x \dot{\boldsymbol{X}}_{sd} + [\boldsymbol{I} - \boldsymbol{J}_s^{\#} \boldsymbol{J}_s] \dot{\boldsymbol{q}}_{sd0}. \tag{3}$$

Matrix $\boldsymbol{W}_q$ has a damping function of the joint velocities with elements calculated as follows:

$$\boldsymbol{W}_q = \boldsymbol{I}_7 \lambda_q, \tag{4}$$

where: λ_q is a variable damping factor, adjusted automatically to have a small value (zero) away from singularities and a large value in the neighbourhood of them. λ_q is chosen as:

$$\lambda_q = \lambda_o \begin{cases} 0 & for\ \sigma \geq \sigma_o \\ (1 - \frac{\sigma}{\sigma_o})^2 & for\ \sigma < \sigma_o \end{cases}, \tag{5}$$

where: $\lambda_o = 0.01$ is a constant damping factor, σ represents virtual distance to singularity and $\sigma_o = 0.1$ specifies an experimentally chosen threshold. The value of σ depends on the arm's Jacobian determinant:

$$\sigma = c_o |det(\boldsymbol{J}_s \boldsymbol{J}_s^T)|, \tag{6}$$

where: $c_o = 40$ is a positive coefficient.

3.3 Joint Demands Filter with Velocity Limiter and Integrator

In order to smooth the rapid changes of demanded joint position $\boldsymbol{q}_{sd}^*$ signals, generated in stepper manner by applications of the *JSCM*, a second order low-pass filter with a cut of frequency $f_{jc} = 2.5$ Hz was implemented. The structure of such a filter used for each joint is shown in Fig. 4. When the task space control is active (e.g. during teleoperation), filtration is switched off by the low state of the logical signal e_{flt}. Then the vector of the demanded positions $\boldsymbol{q}_{sd}$ is obtained by the integration of the limited velocities $\dot{\boldsymbol{q}}_{sd}$ received directly from the *IK* block.

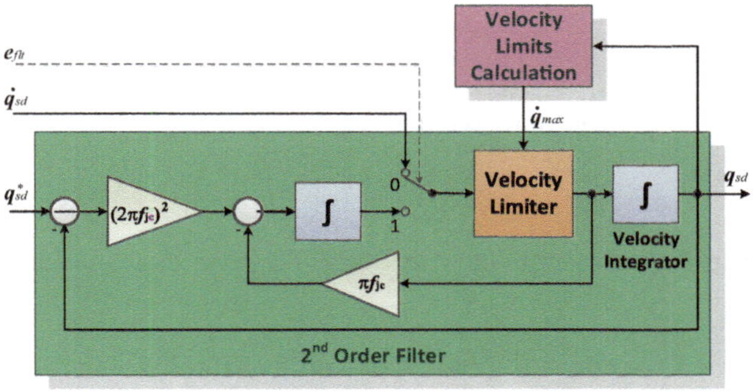

Fig. 4. Joint demands filter with velocity limiter and integrator

The *Velocity Limiter* limits the rate of change of the demanded positions $\boldsymbol{q}_{sd}$ to values $\dot{\boldsymbol{q}}_{max}$ commanded by the *Velocity Limits Calculation* block. The velocity limitation is used for three reasons:

- not to allow the joints faster movement than maximum permitted velocity specified for each joint,
- to reach the physical position limit smoothly,
- not to allow the joints to cross the physical position limits (protection against arm damage).

The velocity limitation algorithm adjusts $\dot{\boldsymbol{q}}_{max}$ to have a small value (zero) close to the physical limit and a large value (max. permitted velocity) away from it.

4 Control Modes

A set of control modes for the arm control system has been developed in order to carry out the operations required by the assistant and physician before, during and just after the remote examination. The algorithms that fulfil these control modes have been implemented in form of switchable functions (applications). Which mode is currently being performed is managed (switched) by the *ADM* (see Sect. 3.1) according to the activity of its states. During normal operation the Active state is the most useful. Its substates are presented in Fig. 5. When the manipulator is switched on, the Active state begins in one of the four states:

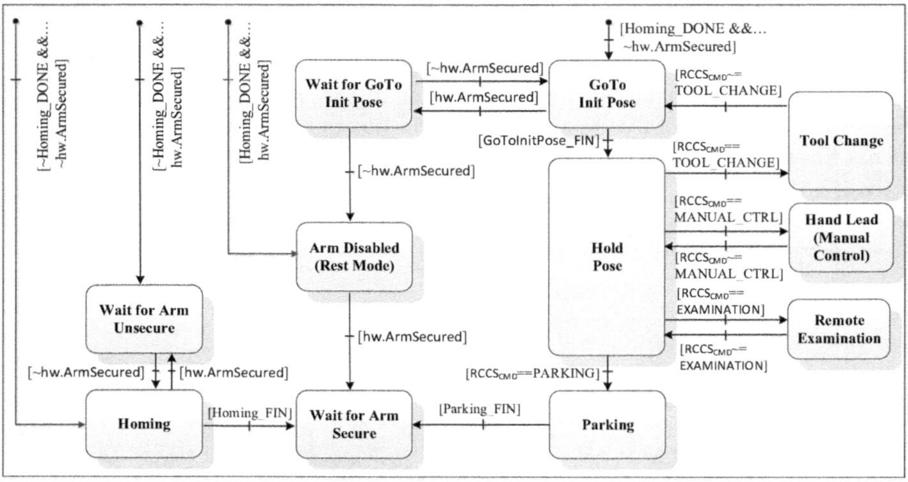

Fig. 5. Active state of Arm Decision Maker state machine

- Wait for Arm Unsecure - the state preceding Homing, activated in case the arm has not been mechanically unlocked, i.e. the sensor signal *ArmSecured* is "1",
- Homing - the procedure of initialisation/reset of the arm's joint encoders, executed when the homing procedure has not yet been done,
- Rest Mode - the joint drives are disabled (relaxed in case there is no need to keep the arm power consumed), activated when homing has already been done and the arm is mechanically locked (i.e. the *ArmSecured* is "0"),
- GoTo Init Pose - the arm goes to the initial position in which it waits for preparation for the examination.

After the GoTo Init Pose procedure is completed, the algorithm goes to the Hold Pose state - the starting point for the other commonly used actions. In this state all the moving parts of the arm are stopped.

The Hand Lead and Tool Change states are used during the arm's preparation for examination. The former helps the assistant to manually position the arm close to the examined body part of the patient. The Tool Change state allows the assistant to change the probe. When the Remote Examination state is active, the doctor remotely operates the arm. After the examination is finished and if the RCCS required parking procedure ($RCCS_{CMD}$== PARKING), the *ADM* goes to the Parking state. Then the arm is led to the parking position. Finally, just after the arm has been secured (*ArmSecured* is set to "1"), the *ADM* activates the Rest Mode.

4.1 Joint Space Control Modes

Almost all modes (except Hand Lead and Remote Examination) are implemented in the *JSCM* block. The **Hold Pose**, **Wait for Arm Secure** and **Wait for Go to Init Pose** are simple modes in which the demanded position equals the value of the arm position latched in the moment a particular mode begins.

The **Go to Init Pose**, **Homing**, **Parking** and **Tool Change** modes have been implemented in the form of multilayer state machines whose highest layer is shown in Fig. 6. Just after the mode has been activated ($c_{js}.ModeActive ==$ "1") the steps grouped in the Mode Execution state are sequentially performed.

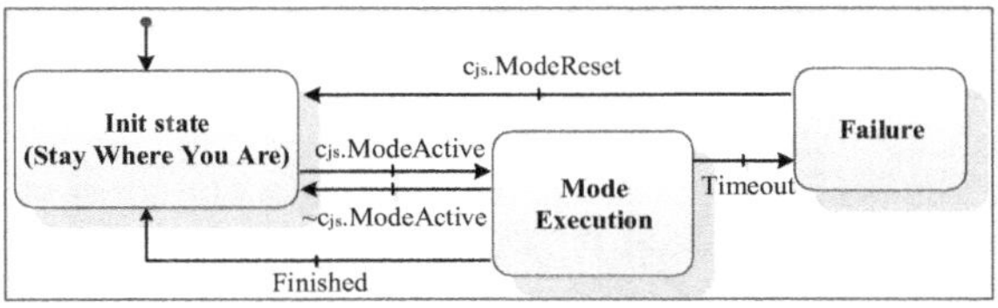

Fig. 6. Main states of Go to Init Pose, Homing, Parking and Tool Change state machins

4.2 Task Space Control Modes

The **Hand Lead** mode allows the assistant to move or rotate (change orientation) the probe by easy (i.e. without the use of too much significant force/torque) hand pushing or rotation of the end-effector. The algorithm which performs this operation, maps gravitally compensated force/torque sensor measurements $\boldsymbol{h}_s = [\boldsymbol{f}_s, \boldsymbol{\mu}_s]$ to the demanded velocity vector $\dot{\boldsymbol{X}}_{sd} = [\dot{x}_{sdx}, \dot{x}_{sdy}, \dot{x}_{sdz}, \omega_{sdx}, \omega_{sdy}, \omega_{sdz}]$ according to the following equations:

$$\dot{x}_{sdi} = \begin{cases} 0 & for\ |f_{si}| \leq f_o \\ c_{fv}(|f_{si}| - f_o)sgn(f_{si}) & for\ |f_{si}| > f_o \end{cases}, \tag{7}$$

$$\omega_{sdi} = \begin{cases} 0 & for\ |\mu_{si}| \leq \mu_o \\ c_{\mu\omega}(|\mu_{si}| - \mu_o)sgn(\mu_{si}) & for\ |\mu_{si}| > \mu_o \end{cases}, \tag{8}$$

where: index i denotes *X*, *Y* and *Z* axis, $c_{fv} = 0.04$ and $c_{\mu\omega} = 4/3$ are the force to Cartesian velocity and the torque to angular velocity mapping coefficients, $f_o = 1\ N$ and $\mu_o = 0.2\ Nm$, are the force and torque sensitivity thresholds.

The **Remote Examination** mode application is much more complex than the Hand Lead algorithm. It consists of the following operations:

- calculation of the USG probe reference pose $\boldsymbol{\xi}_d$ (i.e. the position $\boldsymbol{x}_d \in \mathbb{R}^3$ and unit quaternion $\boldsymbol{\mathcal{Q}}_d = [\eta_d, \boldsymbol{\varepsilon}_d] \in \mathbb{R}^4$ corresponding to the reference orientation matrix $\boldsymbol{R}_d \in \mathbb{R}^{3\times3}$) according to the transformations described with details in [11], on the basis of signals $\boldsymbol{x}_m$ and $\boldsymbol{\varphi}_m$ received from the haptic interface,

- protection arm mechanics against rapid changes of the desired position and orientation commanded by the input master device - use of low-pass filters,
- improving the contact stability and avoiding large contact forces during touching the patient's body by the probe - use of admittance control.

The 2nd order low-pass filters, with the cut-off frequencies f_c and f_{co} for translational and rotational motion respectively, have a structure similar to that presented in Fig. 4. They are implemented together with the admittance control algorithm. For translational motion the filter is expressed as follows:

$$\frac{1}{(2\pi f_c)^2}\ddot{\boldsymbol{x}}_{sdf} + \frac{1}{\pi f_c}\dot{\boldsymbol{x}}_{sdf} + \boldsymbol{x}_{sdf} = \boldsymbol{x}_d. \tag{9}$$

The implemented admittance control algorithm is based on the force-torque $\boldsymbol{h}_s = [\boldsymbol{f}_s, \boldsymbol{\mu}_s] \in \mathbb{R}^6$ measurement recalculated to the end-effector tip. This algorithm establishes a virtual mass-spring-damper system on the end-effector so that the arm becomes compliant. According to this, the filtered desired position $\boldsymbol{x}_{sdf} \in \mathbb{R}^3$ and orientation $\boldsymbol{\mathcal{Q}}_{sdf} \in \mathbb{R}^4$ are corrected by linear $\boldsymbol{x}_{dc} \in \mathbb{R}^3$ and angular (quaternion based) $\boldsymbol{\Delta\mathcal{Q}}_{dc} = [\Delta\eta_{dc}, \boldsymbol{\Delta\varepsilon}_{dc}] \in \mathbb{R}^4$ displacements:

$$\boldsymbol{x}_{sd} = \boldsymbol{x}_{sdf} - \boldsymbol{x}_{dc}, \tag{10}$$

$$\boldsymbol{\mathcal{Q}}_{sd} = \boldsymbol{\mathcal{Q}}_{dc}^{-1} * \boldsymbol{\Delta\mathcal{Q}}_{sdf}, \tag{11}$$

where: "*" denotes the quaternion product [13]. Both displacements are calculated from the impedance equations [14] for translational and rotational motion:

$$M\ddot{\boldsymbol{x}}_{dc} + D\dot{\boldsymbol{x}}_{dc} + K\boldsymbol{x}_{dc} = -\boldsymbol{f}_s, \tag{12}$$

$$M_o\dot{\boldsymbol{\omega}}_{dc} + D_o\boldsymbol{\omega}_{dc} + K_o'\boldsymbol{\Delta\varepsilon}_{dc} = -\boldsymbol{\mu}_s, \tag{13}$$

where: $M = 4\,\text{kg}$, $M_o = 0.04\,\text{kgm}^2$, $D = 63\,\text{Ns/m}$, $D_o = 1.3\,\text{Nms/rad}$, $K = 500\,\text{N/m}$ are inertia, damping and stiffness positive scalar parameters chosen experimentally, K_o' is an equivalent stiffness which is related to an experimentally chosen stiffness $K_o = 10\,\text{Nm/rad}$ according to the relations described in [14].

Determining the $\boldsymbol{x}_{sdf}$ from (9) and $\boldsymbol{x}_{dc}$ from (12) and substituting them in (10) under the assumptions that $1/(2\pi f_c)^2 = M/K$ and $1/\pi f_c = D/K$ gives the final solution for the translational movement:

$$M\ddot{\boldsymbol{x}}_{sd} + D\dot{\boldsymbol{x}}_{sd} = \boldsymbol{f}_s + K(\boldsymbol{x}_d - \boldsymbol{x}_{sd}). \tag{14}$$

Similar transformations, under the assumptions that $1/(2\pi f_{co})^2 = M_o/K_o$ and $1/\pi f_{co} = D_o/K_o$, and based on (11) and (13), lead to the final solution for the rotational movement:

$$M_o\dot{\boldsymbol{\omega}}_{sd} + D_o\boldsymbol{\omega}_{sd} = \boldsymbol{\mu}_s + K_o'\boldsymbol{\Delta\varepsilon}_{d-sd}, \tag{15}$$

where: $\boldsymbol{\Delta\varepsilon}_{d-sd}$ is a vector part of the quaternion product $\mathcal{Q}_{sd}^{-1} * \mathcal{Q}_d$. Systems described by Eqs. (14) and (15) require $\mathbf{x}_{sd}$ and $\mathcal{Q}_{sd}$ as input signals. These signals are parts of the demanded pose $\boldsymbol{\xi}_{sd} = [\boldsymbol{x}_{sd}, \boldsymbol{\varepsilon}_{sd}]$ (see Fig. 3) which is calculated by forward kinematics on the basis of the desired joint positions $\boldsymbol{q}_{sd}$. Solving (14) for $\dot{\mathbf{x}}_{sd}$ and (15) for $\boldsymbol{\omega}_{sd}$ gives the desired velocity vector $\dot{\boldsymbol{X}}_{sd} = [\dot{\mathbf{x}}_{sd}, \boldsymbol{\omega}_{sd}]$, which is used by the inverse kinematics algorithm (3).

To evaluate the performance of the presented control modes, many tele-echography experiments have been performed. The results (i.e. the control algorithm performance in tracking a reference posture during the heart's apical examination) of one of them is shown in Fig. 7. The probe is initially in free space. Contact with the patient's body arises when the physician tries to obtain a good quality USG image, i.e. around the interval (140 ... 200 s).

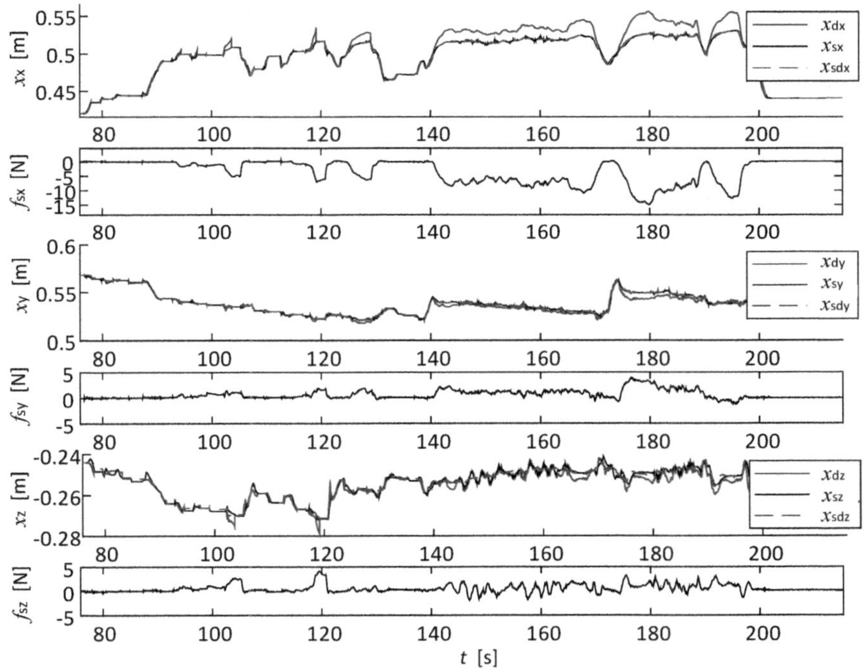

Fig. 7. Position tracking and exerted forces in X, Y and Z direction during the heart's remote USG examination using admittance control

5 Summary

The control system as well as control modes presented in this article, used for control of the position and orientation of the ReMeDi manipulator, has been implemented in a real remote USG examination system. The entire system was subjected to evaluation with the participation of doctors and patient volunteers. The experimental results obtained during evaluations, like those presented in Fig. 7, have confirmed the operational correctness of the designed control modes and the whole control system. The motion controller allows good tracking performance of the compliant reference $\boldsymbol{x}_{sd}$ by the actual position $\boldsymbol{x}_s$. The introduced compliance allows the doctor for stable manipulation during the probe contact with the patient and obtaining of a good quality, stable USG image.

Properly designed control modes increase the comfort of the assistant's and doctor's work and shorten the time of the examination.

References

1. VGo, the manufacturer website. http://www.vgocom.com
2. Medirob, the manufacturer website. http://www.medirob.se
3. AdEchoTech, the manufacturer website. http://www.adechotech.com/products/
4. Remote Medical Diagnostician — ReMeDi Project — FP7 — CORDIS — European Commission. https://cordis.europa.eu/project/rcn/110646/factsheet/en
5. Kurnicki, A., Cholewiński, M., Stańczyk, B., Arent, K.: Implementation and evaluation of a bilateral teleoperation with use of wave variables in the ReMeDi system for remote medical examination. In: Proceedings of the 22nd International Conference on Methods and Models in Automation and Robotics (MMAR 2017), pp. 131–136 (2017). https://doi.org/10.1109/MMAR.2017.8046811
6. Mathiassen, K., Fjellin, J.E., Glette, K., Hol, P.K., Elle, O.J.: An ultrasound robotic system using the commercial robot UR5. Frontiers Rob. AI **3**, 1–16 (2016). https://doi.org/10.3389/frobt.2016.00001
7. Koizumi, N., Warisawa, S.: Construction methodology for a remote ultrasound diagnostic system. IEEE Trans. Rob. **25**(3), 522–538 (2009). https://doi.org/10.1109/TRO.2009.2019785
8. Koizumi, N., Warisawa, S., Hashizume, H., Mitsuishi, M.: Continuous path controller for the remote ultrasound diagnostic system. IEEE/ASME Trans. Mech. **13**(2), 206–218 (2008). https://doi.org/10.1109/TMECH.2008.918530
9. Santos, L., Cortesão, R.: Admittance control for robotic-assisted tele-echography. In: 16th International Conference on Advanced Robotics (ICAR), pp. 1–7 (2013). https://doi.org/10.1109/ICAR.2013.6766502
10. Santos, L., Cortesão, R.: Computed-torque control for robotic-assisted tele-echography based on perceived stiffness estimation. IEEE Trans. Autom. Sci. Eng. **15**(3), 1337–1354 (2018). https://doi.org/10.1109/TASE.2018.2790900
11. Kurnicki, A., Stańczyk, B.: Manipulator control system for remote USG examinantion. J. Autom. Mob. Rob. Intell. Syst. **13**(2), 48–59 (2019). https://doi.org/10.14313/JAMRIS/2-2019/18
12. Stańczyk, B., Kurnicki, A., Arent, K.: Logical architecture of medical telediagnostic robotic system. In: Proceedings of the 21st International Conference on Methods and Models in Automation and Robotics (MMAR 2016), pp. 200–205 (2016). https://doi.org/10.1109/MMAR.2016.7575133
13. Siciliano, B., Sciavicco, L., Villani, L., Oriolo, G.: Robotics. Modelling, Planning and Control. Advanced Textbooks in Control and Signal Processing. Springer, London (2009). https://doi.org/10.1007/978-1-84628-642-1
14. Stańczyk, B., Peer, A., Buss, M.: Development of a high-performance haptic telemanipulation system with dissimilar kinematics. Adv. Rob. **20**(11), 1303–1320 (2006). https://doi.org/10.1163/156855306778792461

Cell Cycle as a Fault Tolerant Control System

Jaroslaw Smieja(✉), Andrzej Swierniak, and Roman Jaksik

Department of Systems Biology and Engineering, Silesian University of Technology, Akademicka 16, 44-100 Gliwice, Poland
Jaroslaw.Smieja@polsl.pl

Abstract. We present models focused on the control mechanisms in cell cycle, allowing to predict the propagation of faults and its consequences for the cell fate. Development of such models is a two-stage process. First a graph representing molecules and interaction among them is built, through an extensive search of bioinformatic databases and publications. Such graph can be subsequently used to find cutting nodes, representing proteins or complexes or cutting edges, representing biochemical processes that are needed by control mechanisms. The second step is modeling and development of a dynamical model, e.g. in the form of ordinary differential equations that describe changes in concentration of the molecules involved in control mechanisms.

Keywords: DNA damage-repair · Fault-tolerant control system · Cell cycle · Mutations · Cancer

1 Introduction

In this paper we discuss a control engineering approach to a problem of DNA damage-repair in human cells. We discuss chosen control systems existing in the cells and demonstrate that they constitute an almost perfect fault tolerant system. Moreover we present some consequences of gaps in this system and explain why it cannot be completely perfect. Each cell consists of cytoplasm and organelles which are used to conduct various metabolic processes required for the cell and organism survival, its ability to develop and reproduce. The central element of the cell is the nucleus which stores the genetic information used to build individual elements of the cell structure and control all of the chemical and biological processes. The information contained in the nucleus is stored using deoxyribonucleic acid (DNA) based on four distinct subunits termed nucleotides. The specific order of nucleotides additionally controls the information availability and rate of transcription, a process in which its specific elements termed genes are copied to a pattern built from ribonucleic acid (RNA). This pattern is used to create multiple copies of a single protein in a process of translation. An essential feature of this mechanism is that its efficiency depends not only on the complexity of the RNA template and availability of the gene region but on the concentration and activity of various other molecules involved in the production process [1]. Additionally each pattern can be used to analyze the activities in a way which enables to maintain the cellular structure [1], catalyze biochemical reactions [2], carry molecules from one place to another [3], control

A. Bartoszewicz et al. (Eds.): PCC 2020, AISC 1196, pp. 555–566, 2020.
https://doi.org/10.1007/978-3-030-50936-1_47

the efficiency of gene expression [4] and control the stability of other molecules [5]. The control system is able to respond to various external signals using a complex mechanism of signal propagation which is based on the balance between production and degradation rate of various molecules (Fig. 1). The basic mechanism of the production rate control utilizes a set of specific proteins termed transcription factors (TF) that initiate the RNA production process by binding to specific regions in the DNA. Their amount, specificity to the DNA and accessibility of binding regions determine the rate of transcription. The negative control is much more complex, occurring on various levels of the gene expression process. Negative regulation is based on control of the mRNA decay, using either RNA binding proteins (RBP) or short non-coding RNAs.

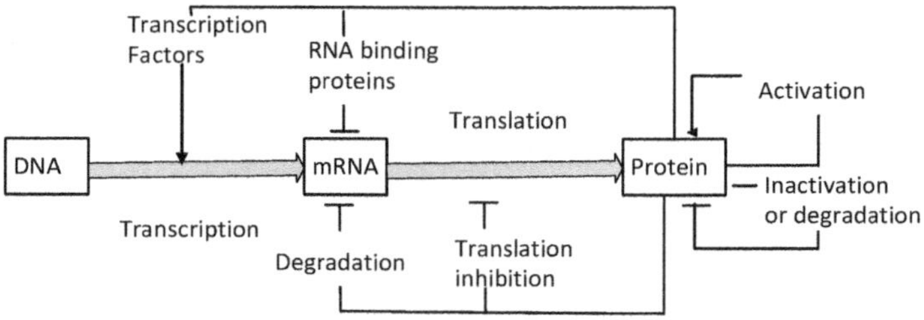

Fig. 1. Control loops in gene expression regulation (arrows and flat endings represent positive and negative feedbacks, respectively)

All of the control mechanisms introduced are based on a target recognition processes that requires a specific nucleotide sequence motif in order for the regulation to occur. The required specificity level varies between protein- or RNA-based regulatory factors and depends on a formation of many weak non-covalent bonds, including ionic bonds, hydrogen bonds, and van der Waals attractions.

Human genome includes over 22 thousand genes which can be used to create over 45 thousand distinct transcripts used as a template for the creation of over 500 thousand distinct proteins, that can be further modified post-transnationally. Post- transnational protein modifications form a next class of the signal propagation system. Chemical modification of the protein can affect its activity, cellular location and stability, which allows to control various intracellular processes. Such processes include cell differentiation, metabolism, or immunological response. Phosphorylation and ubiquitination are two most common protein modifications which can turn on or off specific functions of various proteins by changing their activity or affecting molecule binding capabilities [6].

The control of individual biological processes and levels of particular proteins is possible due to many positive and negative feedback loops built of various RNAs and protein molecules that interact with each other. Negative feedback loops are an essential element of intracellular regulation systems in all complex organisms, allowing to maintain homeostasis by controlling the levels of synthesis and turnover rate of various chemical substances. Positive feedback loops are not as common although they are still an essential element of the regulatory system. Cellular positive feedback loops work usually by double negation where A inhibits an A-inhibiting factor B [6] or, less frequently, either by activating protein A trough protein B which is responsible for protein

A activation. Positive feedbacks are used to invoke a very fast response to the detected DNA damages, for example [7, 8]. Such mechanism is utilized in many signaling pathways and plays an important role in the regulatory processes of the cell cycle, which allows the cell to grow and divide making it one of the most complex regulatory systems found in nature.

2 Control of the Cell-Cycle

One of the most complex process to be regulated is arguably the cell cycle. Not only a particular sequence of synchronized events is to be followed, but also safety mechanisms need to be incorporated in the intracellular machinery, providing detection and repair of DNA damage [9] as well as additional processes, should this repair be ineffective. Therefore, specific control checkpoints evolved, which stop the cycle unless a specific "go" signal is received [10]. This allows to control timing of each stage and their order, additionally making sure that the transition to the next stage will not occur if the previous one was not completed successfully. These stages are called cell cycle phases G1, S, G2 and M (Fig. 2). First, proteins needed for the processes leading to DNA duplication are produced in the G1 phase. This is followed by DNA replication in the S phase. Then, in another growth phase G2 other molecules are produced and activated, leading to the mitosis phase M, resulting in cell division.

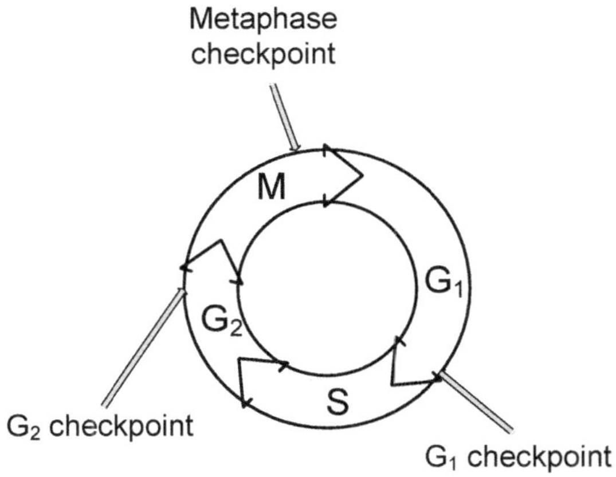

Fig. 2. Cell cycle phases and their checkpoints

Each phase lasts a specific amount of time, which is cell-type dependent, imposing the correct order of processes and preventing one process from starting before the previous one was completed. The transition from one phase to another is controlled by a set of proteins named cyclin-dependent kinases (Cdks) and their regulatory subunits cyclins that control the Cdks activity. Cdks are expresses at constant levels throughout the cell cycle but require cyclins to gain their protein kinase activity. The levels of cyclins, in turn, oscillate in the cell cycle, due to specific control of their production and degradation.

As a result, cyclin-Cdk activity oscillates, determining the time of each cycle step and triggering additional response mechanisms, if abnormal events are detected. Cell cycle events are triggered by the oscillations of the cyclin-Cdk complex activity, for example initiation of S-phase requires high amount of active S-phase cyclin-Cdk complexes. Changes in the activity of cyclin-Cdk complexes affect activity of other proteins that control specific elements of the cell cycle, including mitosis and DNA replication.

Taking all these into account, switching control can be easily observed in cell cycle, together with integral-like and derivative-like control actions (as kinetic rates of some processes are proportional to the accumulated level of specific molecules, while other are initiated by either increasing or decreasing concentration of molecular players), providing robustness of the cell cycle.

Mathematical modeling of the cell cycle and the processes associated with it has a long history. Two main groups of models might be distinguished. The first one comprises statistical models aimed at capturing heterogeneity in a population of cells originating from a single cell predecessor (e.g., [11]). The other one was focused at kinetics of intracellular processes and regulatory mechanisms involved in the control of cell cycle (e.g., [12–14]).

One of the simplest models of the second type mentioned above describes changes of concentrations of a cyclin, fraction of active cdc2 kinase and fraction of active cyclin protease, denoted by C, M and X, respectively. It is given by the following set of ordinary differential equations [15]:

$$\dot{C} = p_1 - p_2 X \frac{C}{K_1 + C} - p_3 C, \tag{1}$$

$$\dot{M} = V_1 \frac{C}{K_2 + C} \frac{1 - M}{K_3 + (1 - M)} - V_2 \frac{M}{K_4 + M}, \tag{2}$$

$$\dot{X} = V_3 \frac{M(1 - X)}{K_5 + (1 - X)} - V_4 \frac{X}{K_6 + X}, \tag{3}$$

where p_i, V_i and K_i are model parameters.

Abnormal cell conditions or environmental stress may result in low level of molecules acting as actuators, pushing cell into the next phase of the cell cycle or DNA damage that should be repaired before being replicated, or to other undesired cell state. Hence, cell cycle checkpoints, mentioned earlier in his section, have evolved (Fig. 2). They ensure that a cell may enter the next phase only if all preceding processes required have been completed successfully and that the cell condition is suitable to proceed to the next phase [10]. Three main checkpoints are distinguished. The first of them, checkpoint R, located at the end of G1 phase, prevents replication of damaged DNA. Such damage, in the form of single- or double-strand breaks, as well as and various chemical modifications like depurinations or cytosine deamination [9] is detected by damage detection proteins, which activate a specific protein kinase – ATM [8] controlling the cell cycle arrest mechanism and initiating the DNA repair processes [16] to prohibit propagation of DNA damages to the daughter cells [17]. These processes involve, among others, Akt, Mdm2 and p53 proteins and are structured into a complex regulatory network, employing both positive and negative feedback loops (a simplified block diagram is presented in the Fig. 3). The simplest model of p53-mdm2 regulatory module has been presented in [18] and consists of three differential equations, describing concentrations of total p53, cytoplasmic and nuclear Mdm2, denoted by P, M_c and M_n, respectively:

$$\dot{P} = k_1 - d_1 P M_c^2 \tag{4}$$

$$\dot{M}_c = k_2 + \frac{k_3 P^3}{k_4 + P^3} - k_5 \frac{M_c}{k_5 + P} \tag{5}$$

$$\dot{M}_n = k_5 \frac{M_c}{k_5 + P} - d_2 M_n, \tag{6}$$

where k_i and d_i are model parameters.

It should be noted, that the models (1)–(3) and (4)–(6) have been developed for completely different purposes and are uncoupled. As a result, any attempt to combine them in order to capture and understand dynamics related to cell cycle, DNA damage and repair and cell fate requires introduction of many additional variables, leading to high-order, nonlinear models, often with stochastic switches. These models are shortly mentioned in the subsequent section.

If the damages are beyond repair or the repair process exceeds the provided time the cell will switch to a non-dividing state termed G0.

The checkpoint at the end of the G2 phase has a similar role of preventing improperly replicated DNA to be passed to daughter cells. The processes involved in this checkpoint involve, among others, inhibition of B1/Cdc2 cyclin, which prevents the cell from entering M-phase until the replication is complete and/or the DNA damages are fixed [17].

The last of three main checkpoints controls the genome integrity after mitosis. It is responsible for ensuring that each daughter cell receives a complete copy of each newly replicated chromosome [19].

Cell cycle checkpoints prevent genomic instability which occurs when the daughter cells receive only a part of the DNA or the DNA is severely damaged. Molecular processes that are involved in these checkpoints have been uncovered only partially and multiple research efforts are focused to gain new knowledge in this area. As experimental investigation with accurate measurements is often not possible in this area, in particular in the case of fast processes, mathematical modeling is often the only tool that might help in testing and verification of hypotheses. However, the models should take into account that mechanisms to be described are highly responsive to checkpoint information at the one hand, and at the same time should be resistant to noise on the other. Cell state is described by concentrations of molecules involved in regulatory actions but at the same time they undergo transition between discrete metastates, determining their ultimate fate. The fault resistance of the system is achieved by the use of multiple negative signals that block the advancement to the next stage rather than positive, that stimulate the cycle progression. Those signals form some kind of a rule-based control system, in which Each of the negative signals is activated independently, if at least one of the conditions is not met, for example the level of required proteins or nutrients is too low for the cell to conduct the process or the growth of the cell and its subcomponents was not completed [1]. From the control theory point of view such approach has a significant advantage since the system is much more fault resistant if it has to detect at least one "stop" signal rather than multiple "go" signals indicating that the previous processes were completed successfully. Those signals form some kind of a rule-based control system, in which the level of molecules involved is compared to some threshold values to decide which control mechanisms should be switched on or off. The effectiveness of this mechanism depends however on the ability of the cell to quickly detect damages to its internal components and most importantly on the ability to control the genome integrity.

3 Damage Recovery

Despite the redundancy in multiple mechanisms described above, the cellular detection and repair system is not failure-free. As a result, mutations, rearrangements and other disruption of genetic information may take place, particularly during DNA replication. They are the most dangerous when they occur in genes that code for proteins controlling either cell cycle or DNA repair processes, since in these cases the error signal propagates through the cell gene regulatory network an later may spread in a population, leading to development of cancer. Mutator phenotype is an extreme example of such process that leads to even 100-fold elevated mutation rate, which in normal cells is estimated at 2.0×10^{-7} mutations/gene/cell division [20]. Mutator phenotypes result from mutations in the genes involved in the maintenance of genomic stability, with a significant role in DNA detection and the mechanisms of damage repair [21]. One of the best-described mutator phenotypes results from mutations in the exonuclease domain of human polymerase ε or δ. Damages in the exonuclease domain impair the proofreading function of DNA polymerases, which is used to check each nucleotide during DNA synthesis and excise those which are different from the template. While this impairment doesn't lead to immediate significant phenotype changes, and in rare cases can even be inherited, the rapid accumulation of new mutations leads to genetic instability and development of cancer or even multiple independent instances of cancer [22].

Two critical elements are necessary in the control mechanisms described in the preceding section. The first one is the subsystem used for fault detection. Though base pair mismatch and single strand breaks might be sensed through several pathways, double strand break recognition relies on ATM molecules. This mechanism is broken in several diseases, including leukemia and lymphoma. The other sensitive component can be found in repair subsystem. Mutations in mismatch repair genes are involved in development of diffuse large B cell lymphomas, colorectal tumors, some forms of breast, ovarian and pancreatic cancer are associated with mutations in recombination modifying genes.

Impaired DNA mismatch repair can lead to microsatellite instability (MSI) which is another example of genetic hypermutability, caused by the appearance of increasing counts of repeat sequences associated with polymerase slippage. Microsatellites, mostly evolve by changing the number of repeat units (usually 1-6 nucleotides long), with mutation rates of 10^{-4}–10^{-2} in humans. Microsatellite instability can lead to gastric, endometrium, ovarian, brain and skin cancers although it is most prevalent in associations with colon cancers [23].

The negative effects of inefficient or improperly working failure detection and repair subsystems can bring even more damage if combined with malfunctioning p53 pathway.

Actions exerted by the tumor suppressor p53 should counteract the accumulation of DNA defects leading to cancer. Therefore, if the p53 protein is dysfunctional, e.g. due to mutation in the gene that codes it, cells carrying wrong genetic information may divide give rise to various forms of human cancer. Damages to the genetic material occur very often, for example Kohn and Bohr [24] report around 55 thousand single strand breaks per cell per day. For this reason cell cycle must be able to withstand damages that can occur during each of the cell division stages and prevent them from being propagated, which could have catastrophic consequence for the organism. DNA damages occur most commonly as a result of exposition to genotoxic substances and factors such as ultraviolet [25] and ionizing radiation [26] or high level of reactive oxygen species [27]. Damage detection is not trivial due to the size of the DNA which in human cells exceeds 3 billion base pairs, requiring a very sensitive detection mechanism, which could not only detect single- or double-strand breaks but also modifications of certain DNA monomers [17]. DNA damages trigger the p53 protein signaling pathway, which controls various processes that protect the cell and if necessary activate its self-degradation [7]. It does that by initiating cell cycle arrest in G1 or G2 phase and triggering the DNA repair machinery if severe damages were detected [7]. One should also remember that environment also contributes to stability of the cell cycle [2]. Extracellular signals from neighboring cells, called mitogens may overcome intracellular mechanisms that block or slow the cell cycle. They act through signaling pathways involving a small GTPase Ras. A mutation in Ras-coding gene may cause it to be permanently active, thus leading to continuous progress of the cell cycle as well as helping to fuel metabolic pathways, supporting growth and division. These mutations are found in about 25% of human cancers and are highly prevalent in hematopoietic malignancies. On the other hand, viral infections may also lead to changes that promote activation of transition to the next phases of the cell cycle under wrong conditions, and, ultimately, result in carcinogenesis. For example, human papilloma virus (HPV) produces oncoproteins E6 and E7, which disrupt, otherwise well-performing, regulatory network. The first of these blocks p-53 mediated activation of the p21 protein, while the other inactivates Rb, thus activating E2F and inducing cell cycle progression independent of the G1-S checkpoint Cdks.

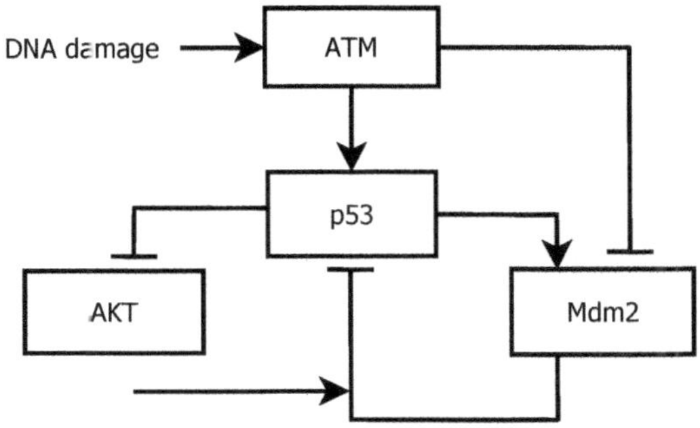

Fig. 3. Simplified model of the p53 protein regulation

The p53 pathway regulation involves two feedback loops one positive and one negative [28], as shown on Fig. 3. The first, negative loop involves Mdm2 protein which is responsible for p53 degradation [29]. Mdm2 is, however, activated by the p53 creating a feedback which maintains a constant level of p53 in the cell. The second, positive loop, involves AKT protein and works thought double negation. AKT is negatively regulated by the p53 and at the same time it mediates the Mdm2 dependent p53 degradation. This mechanism allows p53 to inhibit its own inhibitor, which, if the DNA damages are detected and p53 is triggered by the ATM, significantly increases the p53 concentration [30]. ATM additionally can block the Mdm2-mediated p53 degradation, which allows to significantly increase the p53 level [8]. If activated p53 pathway provides the cell only a certain time to repair the DNA damages. If the repair is ineffective p53 level reaches a very high level triggering the transcription of proapoptotic genes that initiate the programmed cell decay named apoptosis. This mechanism is used to eliminate severely damaged cells, protecting the entire organism. From system engineering point of view this system is a control system with switching parameters (see e.g. [30]). Analysis of this class of hybrid systems and synthesis of switching rules for them belong to the hottest issues in modern control theory see e.g. [31]). The models that describe dynamics of processes behind DNA damage detection and recovery, taking into account specificity of cell cycle phases are, however, large in dimension and their analysis is done mostly through simulation. The smallest of them seem to be the one discussed in [14], which consists of 11 ODEs and captures dynamics of the main molecular players mentioned in the preceding text, i.e. Mdm2, p53, ATM and Wip1. Other models, combining the aforementioned processes with call fate include much more variables - for example, in [13] there are 33 ODEs and 9 algebraic equations and the variables include different forms of p53, responsible for directing the cell into cell cycle arrest or apoptosis with even more complex model introduced in [8, 12].

Replication of damaged DNA is not the only problem that may arise in cell cycle and contribute to the development of cancer. Another one lies in acceleration of the cell cycle, caused by earlier than necessary entry into the S phase (Fig. 4).

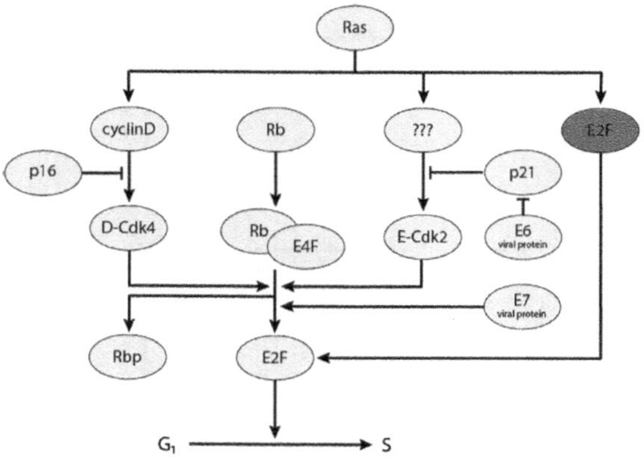

Fig. 4. Interaction network regulating entry into the S phase and sources of its failure in cancer cells: (A) D1 overexpression; (B) p16 inactivation (C) Rb inactivation (D) p21 and (E) p27 failures.

This entry is dependent on increased activity of the E2F family of transcription factors, which when dysregulated can lead to inappropriate S-phase entry or even apoptosis. E2F proteins form a large family of transcription factors containing one or multiple conserved DNA binding domains, allowing them to regulate expression levels of target genes with specific motif in their promoter sequence. E2F proteins are mainly associated with transcriptional activation (E2F1, E2F2 and E2F3A) or repression (E2F3B, E2F4, E2F5, E2F6, E2F7 and E2F8), however genetic or epigenetic alterations can change the function of individual proteins, which is frequently observed in cancer [31]. E2F activators regulate mitosis through multiple regulators, including Sgo1, Nek2, Hec1, BubR1, and Mps1/TTK [32]. Expression of E2F proteins can be promoted either by inducing their transcription or by inactivation of the retinoblastoma protein (Rb) that acts as a brake on the cell-cycle progression. This, in turn, can be achieved through the activation of the G1-Cdk cyclin (cyclin D-Cdk4), preceded by increased cyclin D1 production. The cyclin D1 is frequently overexpressed in a wide range of cancers, sometimes coincident with gene amplification or somatic mutations of the gene coding it. A frequent alternative splicing leads to production of cyclin D1b protein that lacks a specific phosphorylation site required for nuclear export, leading to its accumulation in the nucleus and increased interaction with Rb, and, subsequently, promoting entry into the S phase [1]. While this could be prevented by another control mechanism, based on the p16 protein that blocks the formation of an active D1-cdk4 complex, many cancer cells have either a deleted, inactivated or silenced p16 gene.

Moreover, some mutations in the p16 gene promote cancer metastasis. On top of this, in some cancers p16 is overexpressed and despite that, these cancers may have poor prognoses This suggests that our knowledge of even this, relatively small part of the signaling network, is far from complete and caution is recommended, concerning conclusions drawn from experimental work and mathematical modeling that supports it. Another important regulator of the cell cycle is the p27 protein. Its increased degradation, natural in a normal cell cycle, leads to G1/S-Cdk activation, thus promoting entry into the S phase. It has been found that in some tumors p27 is mutated in a way that reduces its stability. The ultimate result of such mutation is, once again, uncontrolled entry into the S phase and acceleration of the cell cycle.

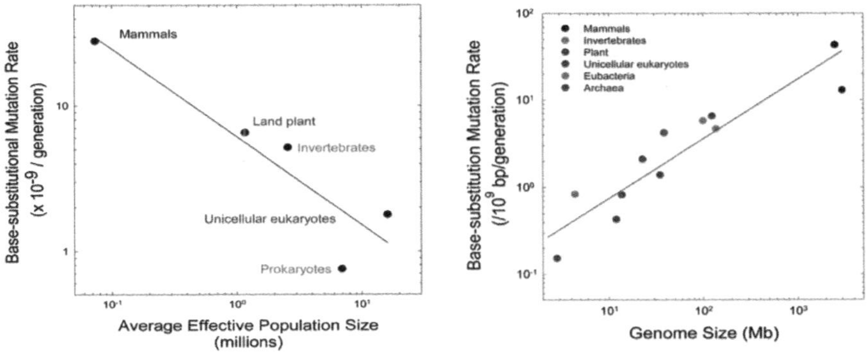

Fig. 5. Mutation rate as a function of population and genome sizes

On the other hand the failure prone system cannot be perfect in order to allow some nucleotide substitutions to occur becoming one of the most important elements of the organism evolution. Permanent changes of the nucleotide sequence occurring as a result of unrepaired DNA damages, named mutations are crucial for the survival of the entire species.

Figure 5 explains why DNA repair mechanism could not and should not be a perfect fault prone system. Nevertheless from a single organism perspective they can be very harmful leading to severe genetic diseases including cancer.

4 Discussion

Analysis of the dynamical properties, and comparison of simulation and experimental results help to find missing elements of the signaling network and identify kinetic parameters of the processes [27]. This, in turn, helps to plan subsequent biological experiments and enhance the analysis of external simulation effects on the signaling pathways [28]. Moreover, the control engineering point of view enables to simplify those models in some stages of their analysis without wasting important control properties of the modeled processes. It opens also new perspectives of controlling these processes by external interventions on the molecular level [33] and references therein). Although such mechanisms have been already discussed in literature, their understanding is far from being complete.

We present the cell cycle as a fault tolerant system which however is not perfect. Nevertheless we try to answer the question: why the gaps in this system should exist and when their existence play negative role.

Acknowledgment. This work was partially supported by Silesian University of Technology internal grant in the year 2020.

References

1. Alberts, B., Johnson, A., Lewis, J., Raff, M., Roberts, K., Walter, P.: Molecular Biology of the Cell, 4th edn. Garland Science, New York (2002)
2. Neurath, H., Walsh, K.A.: Role of proteolytic enzymes in biological regulation (a review). Proc. Natl. Acad. Sci. **73**(11), 3825–3832 (1976)
3. Saier, M.J.: Classification of transmembrane transport systems in living organisms. In: Van Winkle, L. (ed.) Biomembrane transport. Academic Press, San Diego (1999)
4. Todeschini, A.L., Georges, A., Veitia, R.A.: Transcription factors: specific DNA binding and specific gene regulation. Trends Genet. **30**(6), 211–219 (2014)
5. Glisovic, T., Bachorik, J.L., Yong, J., Dreyfuss, G.: RNA-binding proteins and post-transcriptional gene regulation. FEBS Lett. **582**(14), 1977–1986 (2008)
6. Pomerening, J.R.: Positive-feedback loops in cell cycle progression. FEBS Lett. **583**(21), 3388–3396 (2009)
7. Puszynski, K., Hat, B., Lipniacki, T.: Oscillations and bistability in the stochastic model of p53 regulation. J. Theor. Biol. **254**(2), 452–465 (2008)

8. Jonak, K., Kurpas M., Puszynski, K.: Prediction of the behavior of mammalian cells after exposure to ionizing radiation based on the new mathematical model of ATM-Mdm2-p 53 regulatory pathway. In: Piętka, E. Kawa, J., Wieclawek, W. (eds.) Information Technologies in Biomedicine, pp. 349–362. Springer, Cham (2014)
9. De Bont, R., van Larebeke, N.: Endogenous DNA damage in humans: a review of quantitative data. Mutagenesis **19**(3), 169–185 (2004)
10. Houtgraaf, J.H., Versmissen, J., van der Giessen, W.J.: A concise review of DNA damage checkpoints and repair in mammalian cells. Cardiovasc Revasc. Med. **7**(3), 165–172 (2006)
11. Dolbniak, M., Kimmel, M., Smieja, J.: Modeling epigenetic regulation of PRC1 protein accumulation in the cell cycle. Biol. Direct **10**, 62 (2015)
12. Jonak, K., Kurpas, M., Szoltysek, K., Janus, P., Abramowicz, A., Puszynski, K.: A novel mathematical model of ATM/p53/NF- κB pathways points to the importance of the DDR switch-off mechanisms. BMC Syst. Biol. **10**(1), 75 (2016)
13. Hat, B., Kochańczyk, M., Bogdał, M.N., Lipniacki, T.: Feedbacks, bifurcations, and cell fate decision-making in the p53 system. PLoS Comput. Biol. **12**(2), e1004787 (2016)
14. Wang, D.G., Wang, S., Huang, B., Liu, F.: Roles of cellular heterogeneity, intrinsic and extrinsic noise in variability of p53 oscillation. Sci Rep. **9**(1), 5883 (2019)
15. Goldbeter, A.: A minimal cascade model for the mitotic oscillator involving cyclin and cdc2 kinase. Proc. Natl. Acad. Sci. **88**(20), 9107–9111 (1991)
16. Pietenpol, J.A., Stewart, Z.A.: Cell cycle checkpoint signaling: cell cycle arrest versus apoptosis. Toxicology **181–182**, 475–481 (2002)
17. Branzei, D., Foiani, M.: Regulation of DNA repair throughout the cell cycle. Nat. Rev. Mol. Cell Biol. **9**(4), 297–308 (2008)
18. Hat, B., Puszynski, K., Lipniacki, T.: Exploring mechanisms of oscillations in p53 and nuclear factor-β systems. IET Syst. Biol. **3**, 342–355 (2009)
19. Cooper, G., Hausman, R.: The Cell: A Molecular Approach, 6th edn. Sinauer Associates, Sunderland (2013)
20. DeMars, R., Held, K.R.: The spontaneous azaguanine-resistant mutants of diploid human fibroblasts. Humangenetik **16**(1), 87–110 (1972). https://doi.org/10.1007/BF00393992
21. Loeb, L.A.: A mutator phenotype in cancer. Cancer Res. **61**(8), 3230–3239 (2001)
22. Rohlin, A., Zagoras, T., Nilsson. S., Lundstam, U., Wahlström, J., Hultén, L., Martinsson, T., Karlsson, G.B., Nordling, M.: A mutation in POLE predisposing to a multi-tumour phenotype. Int. J. Oncol. **45**(1), 77–81 (2014)
23. Popat, S., Hubner, R., Houlston, R.S.: Systematic review of microsatellite instability and colorectal cancer prognosis. J. Clin. Oncol. **23**(3), 609–618 (2005)
24. Kohn, K., Bohr, V.: Genomic instability and DNA repair. In: Alison, M. (ed.) The Cancer Handbook. Macmillan, Palgrave (2005)
25. Widel, M., Krzywon, A., Gajda, K., Skonieczna, M., Rzeszowska-Wolny, J.: Induction of bystander effects by UVA, UVB, and UVC radiation in human fibroblasts and the implication of reactive oxygen species. Free Radic. Biol. Med. **68**, 278–287 (2014)
26. Puszynski, K., Jaksik, R., Swierniak, A.: Regulation of P53 by Sirna in radiation treated cells: simulation studies. Int. J. Appl. Math. Comput. Sci. **22**(4), 1011–1018 (2012)
27. Cieslar-Pobuda, A., Saenko, Y., Rzeszowska-Wolny, J.: PARP-1 inhibition induces a late increase in the level of reactive oxygen species in cells after ionizing radiation. Mutation Res. Fundam. Mol. Mech. Mutagenesis **732**(1–2), 9–15 (2012)
28. Harris, S.L., Levine, A.J.: The p53 pathway: positive and negative feedback loops. Oncogene **24**(17), 2899–2908 (2005)
29. Haupt, Y., et al.: Mdm2 promotes the rapid degradation of p 53. Nature **387**(6630), 296–299 (1997)

30. Cantley, L.C., Neel, B.G.: New insights into tumor suppression: PTEN suppresses tumor formation by restraining the phosphoinositide 3-kinase AKT pathway. Proc. Natl. Acad. Sci. **96**(8), 4240–4245 (1999)
31. Johnson, D.G., Schneider-Broussard, R.: Role of E2F in cell cycle control and cancer. Front Biosci **3**, 447–448 (1998)
32. Lee, M., Rivera-Rivera, Y., Moreno, C.S., Saavedra, H.I.: The E2F activators control multiple mitotic regulators and maintain genomic integrity through Sgo1 and BubR1. Oncotarget **8**(44), 77649–77672 (2017)
33. Swierniak, A., Kimmel, M., Smieja, J., Puszynski, K., Psiuk-Maksymowicz, K.: System Engineering Approach to Planning Anticancer Therapies. Springer International Publishing, Cham (2016). https://doi.org/10.1007/978-3-319-28095-0

Biological Models' Parameter Estimation Based on Discrete Measurements and Adjoint Sensitivity Analysis

Krzysztof Fujarewicz(✉) and Krzysztof Łakomiec

Department of Systems Biology and Engineering, Silesian University of Technology, Akademicka 16, 44-100 Gliwice, Poland
{krzysztof.fujarewicz,krzysztof.lakomiec}@polsl.pl

Abstract. Mathematical models of biological processes are usually continuous time (CT) and take the form of non-linear ordinary differential equations. On the other hand the estimation of model parameters is done based on discrete time (DT), relatively rare, measurements. Hence, overall problem of parameter estimation has hybrid, continuous-discrete form: it uses CT model and minimise DT performance index depending on DT prediction errors. In our previous works we have published Generalized Back Propagation Through Time (GBPPT) method—a method allowing us to use the adjoint sensitivity analysis for obtained hybrid system, and giving as a result a computationally effective recipe for calculating gradient of the performance index in parameter space. GBPTT specifies rules for construction of the adjoint system, in particular it specifies how to manage elements interfacing between CT and DT parts of the system: ideal sampler (IS) and ideal pulser (IP). Such rules for isolated IS and IP elements has been proposed without strict formal rationale. In this article we deliver a proof of correctness of such rules. Additionally, as an illustration, we present an example of application of GBPTT to parameter estimation of chemical enzymatic reaction which is one of basic biochemical reaction.

Keywords: Parameter estimation · Sensitivity analysis · Ordinary differential equations

1 Introduction

The concept of adjoint systems is exploited in many areas. It is widely used in optimization, identification, automatic differentiation, sensitivity analysis and, recently, in neural network theory. For example in the neural networks theory two basic gradient methods: Backpropagation (BP) for feed-forward neural nets and Backpropagation Through Time (BPTT) for recurrent neural nets, both use the concept of the adjoint system. In the work [1] we presented an extension of the BPTT algorithm called Generalized Backpropagation Through Time

A. Bartoszewicz et al. (Eds.): PCC 2020, AISC 1196, pp. 567–578, 2020.
https://doi.org/10.1007/978-3-030-50936-1_48

(GBPTT) which may be applied for any hybrid, continuous-discrete time system given by a block diagram. The block diagram may be composed of elements typical for continuous-time (CT) and discrete-time (DT) systems and, additionally, elements interfacing between CT and DT part of the system. These interfacing elements are: ideal sampler and ideal (Dirac) pulser. GBPTT method simply gives rules for construction of a sensitivity model and an adjoint system for given block diagram with arbitrary structure. In particular, it specifies equivalent elements that should replace each element from the original system in the sensitivity and adjoint model, respectively. Finally the adjoint system is used for generation of the gradient of the scalar performance index in parameter space. It is done by only one simulation regardless the number of estimated parameters.

The GBPTT method is universal and may be used for different purposes. It has been used for optimization of a fractional order hold [2], identification of a non-linear model of the helicopter based on DT measurements [3] and for identification of non-linear models of cell signaling pathways [4–6]. It has been also further extended on systems with delays [7,8] and on spatial systems [9,10].

2 Parameter Estimation Problem

The classical scheme of the identification experiment is presented in Fig. 1. The real complex system S is modeled by the mathematical model M. Both, the real system and the model, are assumed to be continuous-time. Here we do not specify the topology of the model which may be chosen by the modeler. The system and the model are stimulated by the common and known continuous-time input signal $u_m(t)$. The initial and the final times of the identification experiment are t_0 and t_f respectively. p is a vector of parameters of the model to be identified.

The output of the system $y_s(t)$ is measured at N time moments: $\{t_n\}_{n=1}^{N}, t_n \in [t_0, t_f]$, giving discrete-time measurements: $y_s(t_n), n = 1, 2, \ldots, N$. The identification task involves minimizing the following performance index

$$J = \sum_{n=1}^{N} h(e(t_n), t_n) \tag{1}$$

where

$$e(t_n) = y_m(t_n) - y_s(t_n) \tag{2}$$

is the identification error at time t_n, and $h(e(t_n), t_n)$ is a given nonlinear differentiable function. In most cases $h(\cdot)$ is a quadratic function.

Identification Task 1. *Find optimal p such that the performance index (1) is minimized.*

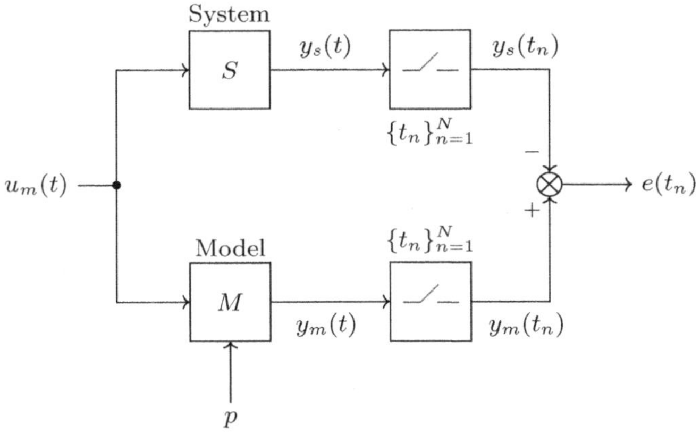

Fig. 1. The block diagram of the identification experiment.

When a gradient-based optimization procedure is used to carry out the task above, then the following problem has to be solved:

Identification Task 2. *Find the gradient of the performance index (1) w.r.t. the vector of the parameters p*

$$\nabla_p J \tag{3}$$

Figure 2 presents the extension of the scheme from Fig. 1. There are additional elements and signals: $p_d(t_n)$ and $y_d(t_n)$.

Signal $p_d(t_n) = p \cdot \delta_K(t_n - t_0)$, where $\delta_K(\cdot)$ is the Kronecker's pulse, is the input signal of the ideal pulser giving the Dirac pulse at time t_0 only. Hence the output signal of the integrator (which transfer function is $\frac{1}{s}$) is constant and equal to p.

On the other hand the sum operator $\sum_n$ (which discrete transfer function is $\frac{z}{z-1}$) is used to calculate the performance index (1).

Based on this one may write:

$$\begin{aligned} p_d(t_0) &= p \\ y_d(t_N) &= J \end{aligned} \tag{4}$$

Hence the gradient (3) reduces to the Jacobi matrix

$$\nabla_p J = \frac{\partial y_d(t_N)}{\partial p_d(t_0)}. \tag{5}$$

The Identification Task 2 now reduces to the problem of finding the sensitivity of the signal y_d at time t_N w.r.t. the signal p_d at time t_0.

The sensitivity function (5) may be obtained by using the sensitivity model presented in Fig. 3. The model is valid for variations of all signals (dashed symbols). $\overline{M}$ is the sensitivity model of the model M and $h_e(t_n)$ is the derivative of the function h w.r.t. e calculated for different discrete-time moments t_n.

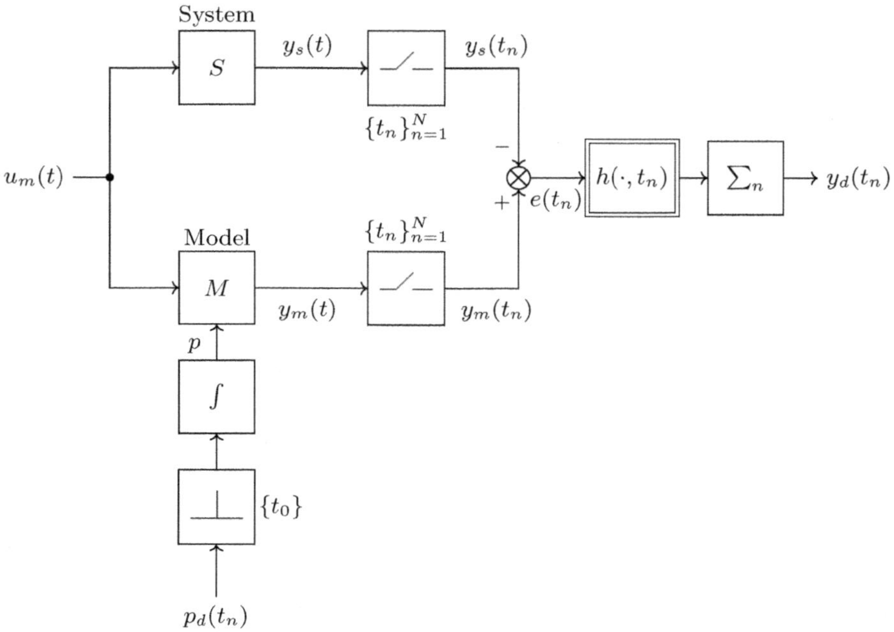

Fig. 2. The extended block diagram of the identification experiment.

Finally, the sensitivity model may be presented as a serial connection of continuous-time (C-T) part and discrete time (D-T) part with the sampler and the pulser, where the input and output signals are discrete-time. See Fig. 4. In this scheme we use the symbol $\overline{y}_c(t)$ instead of $\overline{y}_m(t_n) = \overline{e}(t_n)$ to maintain homogeneity of notation.

The pulser produces one pulse only, at time t_0. The sampler samples the continuous-time signal $\overline{y}_c(t)$ at times $t_1, t_2, \ldots, t_N$ and produces the discrete-time signal $\overline{u}_d(t_n)$.

In [1] we proposed a method for construction of so called modified adjoint system if the original system is given in block-diagram form. The method consists in replacing all elements by their equivalents according to Table 1. Similar structural approaches to construction of adjoint systems have been previously proposed for discrete- or continuous-time systems in the literature. The original contribution of [1] was that rules were proposed for any hybrid (continuous/discrete-time) system containing, in addition to typical CT and/or DT elements, ideal pulsers and samplers—last two rows in the Table 1. In work [1] these two rules have been only proposed, without formal proof of their correctness. In next Sections of present article we will provide strict formal rationale.

Table 1. Rules for construction of the sensitivity model and the modified adjoint system.

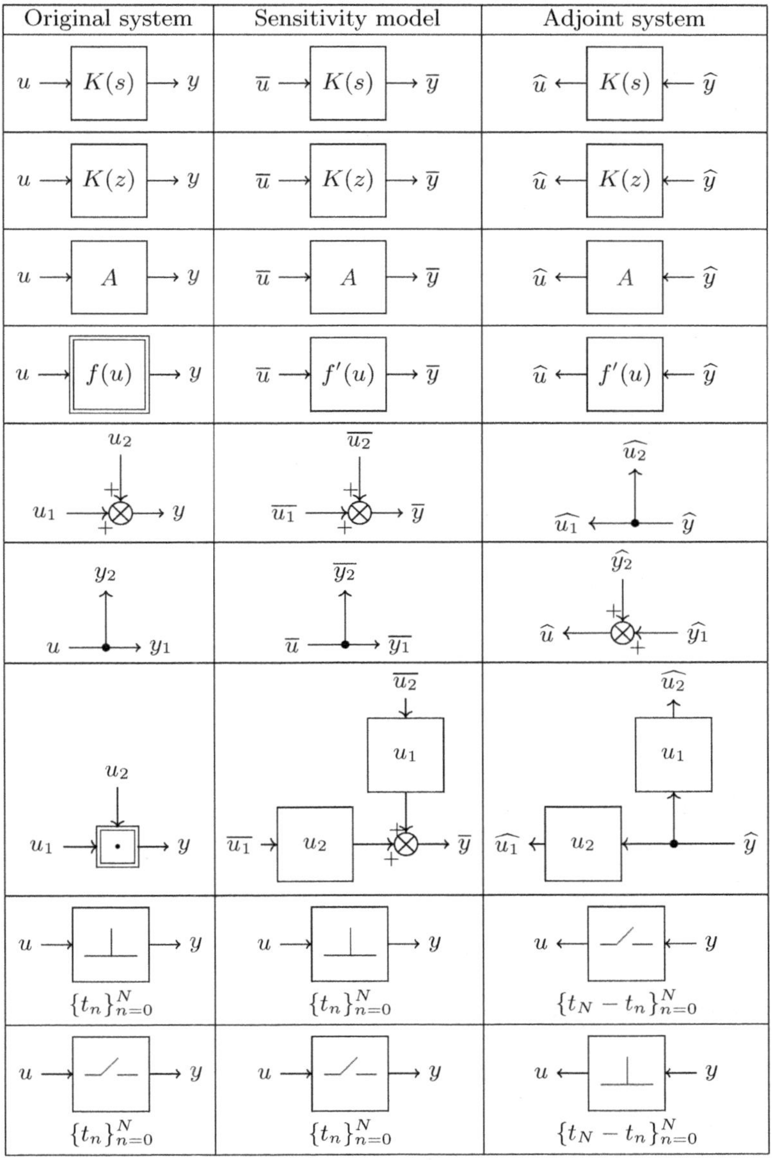

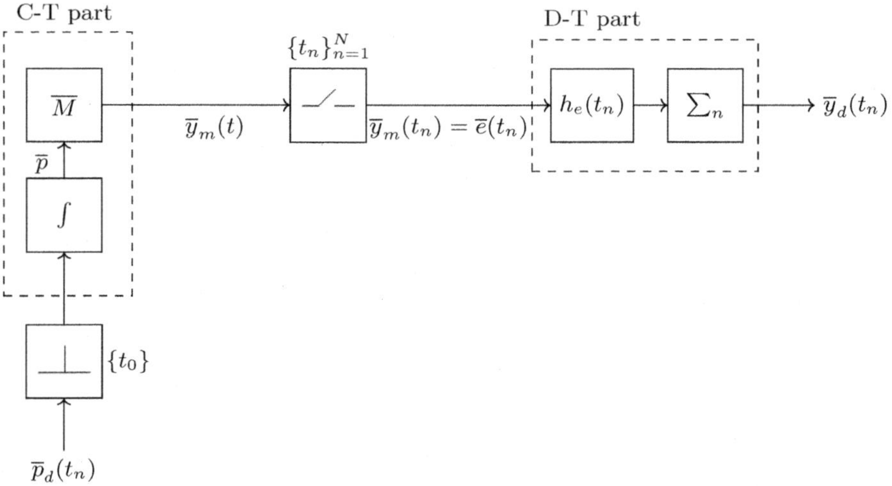

Fig. 3. The sensitivity model of the system from Fig. 2.

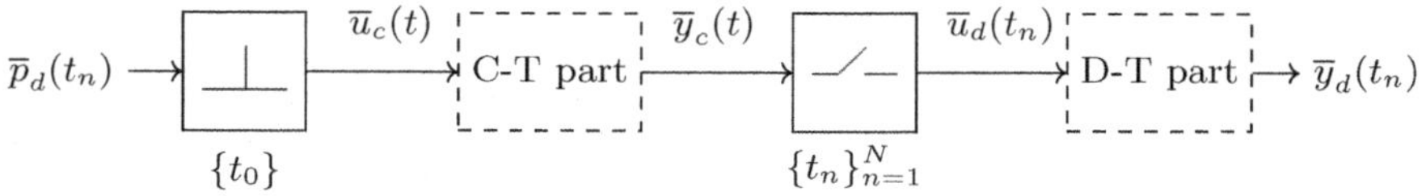

Fig. 4. General scheme of the sensitivity model—the serial connection of continuous- and discrete-time parts.

3 Adjoint Systems

Let us consider two linear systems (operators) $A: \; S_1 \rightarrow S_2$ and $B: \; S_2 \rightarrow S_1$ with input and output signals: $\bar{u}, \hat{y} \in S_1$, $\hat{u}, \bar{y} \in S_1$ as presented in Fig. 5. Let us also assume that there are defined scalar products for pairs of signals $\bar{u}, \hat{y}$ and $\hat{u}, \bar{y}$ respectively.

Definition 1. *System B is said to be adjoint to the system A (and vice versa) when*

$$\langle \bar{u}, \hat{y} \rangle = \langle \hat{u}, \bar{y} \rangle \tag{6}$$

for any $\bar{u}$ and $\hat{u}$.

4 Ideal Sampler and Ideal Pulser

Ideal sampler (IS) and ideal pulser (IP) are idealized elements being interfaces between CT and DT parts of a hybrid system.

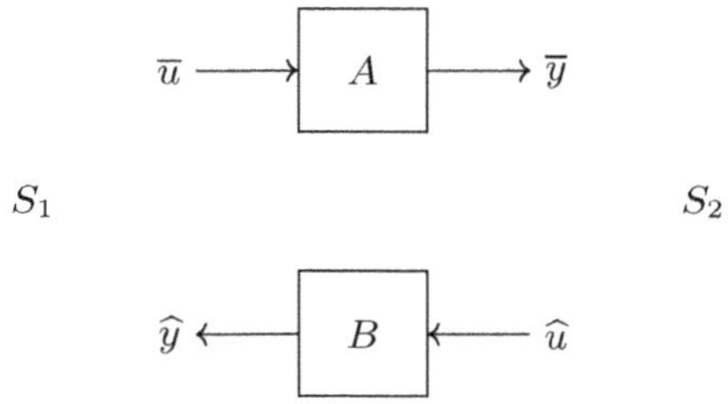

Fig. 5. Two linear systems.

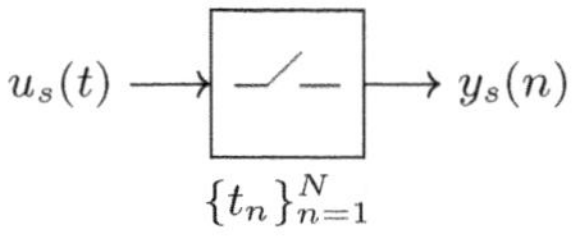

Fig. 6. Ideal sampler (IS).

IS, presented in Fig. 6, samples CT signal $u_s(t)$ and gives DT output signal $y_s(i)$

$$y_s(i) = \sum_{n=1}^{N} u_s(t_n)\delta_K(i-n) \tag{7}$$

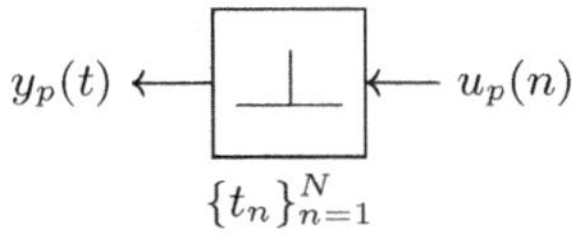

Fig. 7. Ideal pulser (IP).

Similarly, IP, presented in Fig. 7, transforms DT signal $u_p(n)$ into CT signal $y_p(t)$

$$y_p(t) = \sum_{n=1}^{N} u_p(n)\delta_D(t-t_n) \tag{8}$$

Now, we will prove that rules from last two rows of the Table 1 are correct.

Theorem 1. *IP is adjoint to IS and vice versa*

Proof. To prove the theorem one has to show that

$$\langle u_s(t), y_p(t)\rangle = \langle u_p(n), y_s(n)\rangle \tag{9}$$

Right side of (9) equals

$$\langle u_p(n), y_s(n) \rangle = \sum_{n=1}^{N} u_p(n) y_s(n) = \sum_{n=1}^{N} u_p(n) u_s(t_n) \tag{10}$$

Left side of (9) equals

$$\begin{aligned} &\langle u_s(t), y_p(t) \rangle \\ &= \int_0^{t_f} u_s(t) y_p(t) dt \\ &= \int_0^{t_f} u_s(t) \sum_{n=1}^{N} (u_p(n) \delta_D(t - t_n)) dt \\ &= \sum_{n=1}^{N} \left(u_p(n) \int_0^{t_f} u_s(t) \delta_D(t - t_n) dt \right) \\ &= \sum_{n=1}^{N} u_p(n) u_s(t_n) \end{aligned} \tag{11}$$

which is equal to the formula obtained in (10) and which ends the proof.

Theorem 1 for the sake of simplicity of notation are formulated and proved for sensitivity and adjoint models both running forward of time. The only difference in rules presented in last two rows of Table 1 is that the adjoint system in GBPTT method is defined for time running backward.

5 Construction of the Adjoint System for Parameter Estimation Problem According to GBPTT Method

Let us construct the system adjoint to the sensitivity model presented in Fig. 4. according to Table 1. The result is presented in Fig. 8.

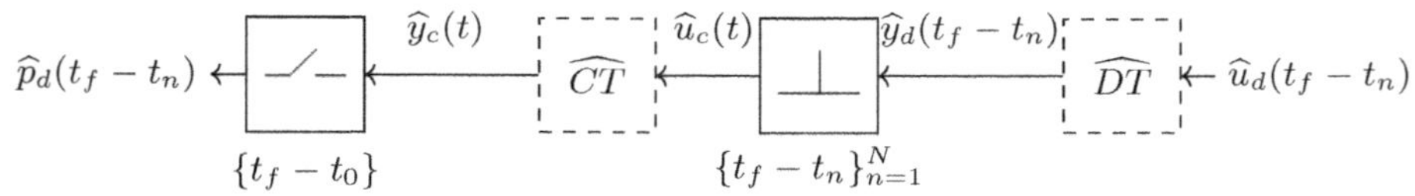

Fig. 8. The system created according to the rules of Table 1 based on the system of Fig. 4. Elements $\widehat{CT}$ and $\widehat{DT}$ are adjoint systems to the continuous- and discrete-time parts of the sensitivity model.

The assignments of instantaneous values of input and output signals in Fig. 4 and Fig. 8 are as follows

$$\begin{aligned} \overline{p}_d(t_n) &\longleftrightarrow \widehat{p}_d(t_f - t_n) \\ \overline{y}_d(t_n) &\longleftrightarrow \widehat{u}_d(t_f - t_n) \end{aligned} \tag{12}$$

The modified adjoint system may be used in order to find the i-th component of the gradient (9) sought, instead of the sensitivity model.

6 Example of Application

As an example we used the adjoint system obtained from GBPTT method to parameter estimation of enzymatic reaction model [11]. The general scheme for enzymatic reaction can be presented as follows:

$$\mathrm{S} + \mathrm{E} \underset{k_2}{\overset{k_1}{\rightleftharpoons}} \mathrm{C} \xrightarrow{k_3} \mathrm{E} + \mathrm{P}$$

where S, E, C, P are the concentrations of substrate, enzyme, substrate-enzyme complex and product, respectively. k_1, k_2, k_3 are kinetic parameters for particular reactions.

In short, enzyme catalyses the conversion of substrate to product. Substrate binds to enzyme and creates a temporary substrate-enzyme complex. This complex can dissolve to enzyme and product or dissolve back to substrate and enzyme. This general scheme leads to following system of ordinary differential equations

$$\frac{dS}{dt} = -k_1 SE + k_2 C \tag{13}$$

$$\frac{dE}{dt} = -k_1 SE + (k_2 + k_3)C \tag{14}$$

$$\frac{dC}{dt} = k_1 SE - (k_2 + k_3)C \tag{15}$$

$$\frac{dP}{dt} = k_3 C \tag{16}$$

The parameters of the model are

$$p = [k_1, k_2, k_3] \tag{17}$$

To create the experimental data we simulated the model (13–16) with known values of parameters, and we added to the simulation results normally distributed noise with standard deviation equal to 0.2. Parameters used to obtain the experimental data was acquired from work [11]: $k_1 = 0.18, k_2 = 0.02, k_3 = 0.23$. We assumed that only the substrate-enzyme complex (variable C in the model) can be measured at discrete time moments. The created experimental data used to estimate the parameters (17) is shown on Fig. 9.

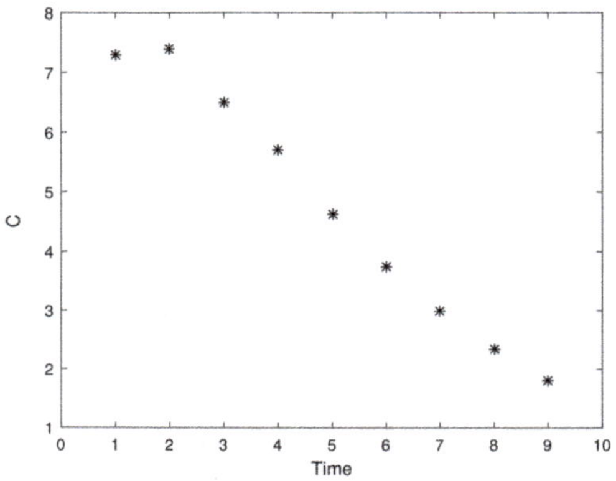

Fig. 9. Experimental data used to estimate parameters of the model (13–16).

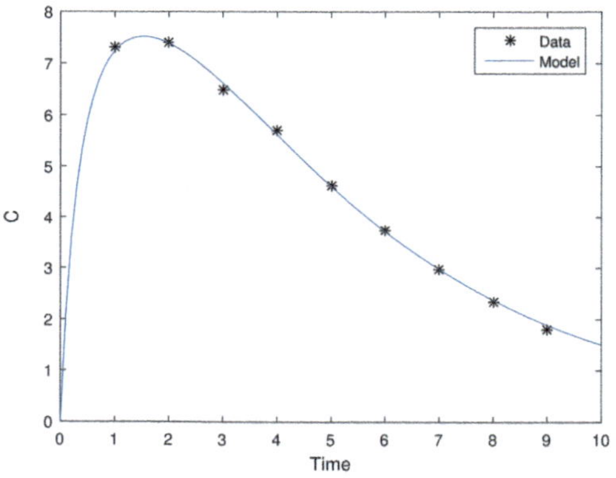

Fig. 10. Graphical fit of the model (13–16) to the experimental data.

The initial conditions for the model were set to $S(0) = E(0) = 12$, and $C(0) = P(0) = 0$. The final time of simulation was set to 10.

The objective function was defined as

$$J = \sum_{i=1}^{N} (y(t_i) - d(t_i))^2 \tag{18}$$

where $d(t_i)$ is the experimental data measured at time t_i, and $y(t_i)$ is the output of the model at time t_i. The gradient of the objective function J with respect to the estimated parameters p was calculated using adjoint system constructed using GBPTT method. The minimization of J was done using the trust-region method implemented in a function *fmincon* from MATLAB environment. The initial values of parameters for the trust-region method were set to zero. Fit of the model to the experimental data after estimation process is shown on Fig. 10.

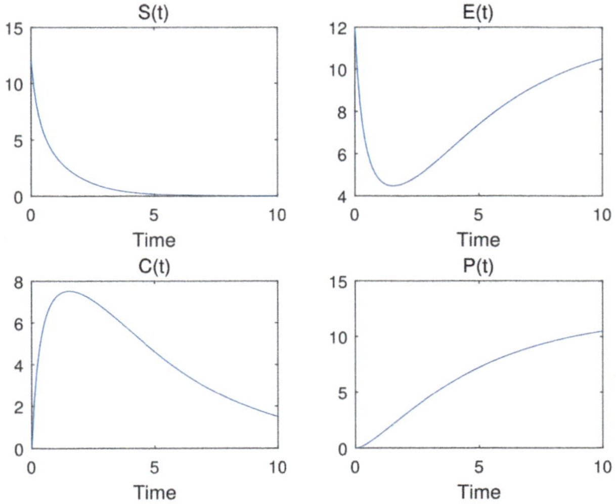

Fig. 11. Simulation of the model (13–16).

Simulation of the model using estimated parameters is shown on Fig. 11. Value of the objective function J after estimation process was equal to 0.03. Values of the parameters were estimated as $k_1 = 0.19, k_2 = 0.03, k_3 = 0.23$.

7 Conclusions

The paper presented the mnemonic method for construction of the adjoint systems for nonlinear continuous-discrete dynamical systems given in a block-diagram form. The correctness of the approach was proven. The provided examples shown its usefulness for continuous-time models identification based on discrete-time measurements. Nevertheless, its possible application is much wider. It can be also used for hybrid systems optimization under integral or sum performance index.

Acknowledgements. This work was supported by the Silesian University of Technology and by the Polish National Science Centre under grants UMO-2018/29/B/ST7/02550 (K.F.), DEC-2016/21/B/ST7/02241 (K.Ł.). Calculations were performed using the infrastructure supported by the computer cluster Ziemowit (https://www.ziemowit.hpc.polsl.pl/en/) funded by the Silesian BIO-FARMA project No. POIG.02.01.00-00-166/08 and expanded in the POIG.02.03.01-00-040/13 in the Computational Biology and Bioinformatics Laboratory of the Biotechnology Centre at the Silesian University of Technology.

References

1. Fujarewicz, K., Gałuszka, A.: Generalized backpropagation through time for continuous time neural networks and discrete time measurements. Lecture Notes in Computer Science, pp. 190–196. Springer, Heidelberg (2004)
2. Fujarewicz, K.: Optimal discrete-time control of continuous-time systems with fractional order hold. In: Proceedings of 4th International Carpathian Control Conference, Slovakia , pp. 857-860 (2003)
3. Fujarewicz, K.: Adjoint sensitivity analysis for identification of continuous-time models based on discrete-time measurements. In: Proceedings of 11th International Conference on Methods and Models in Automation and Robotics, Międzyzdroje, Poland, pp. 519–524 (2005)
4. Fujarewicz, K., Kimmel, M., Lipniacki, T., Świerniak, A.: Adjoint systems for models of cell signalling pathways and their application to parametr fitting. IEEE/ACM Trans. Comput. Biol. Bioinf. **4**(3), 322–335 (2007)
5. Łakomiec, K., Kumala, S., Hancock, R., Rzeszowska-Wolny, J., Fujarewicz, K.: Modeling the repair of DNA strand breaks caused by γ-radiation in a minichromosome. Phys. Biol. **11**(4), 045003 (2014)
6. Łakomiec, K., Fujarewicz, K.: Parameter estimation of non-linear models using adjoint sensitivity analysis. In: Sobecki, J., Boonjing, V., Chittayasothorn, S. (eds.) Advanced Approaches to Intelligent Information and Database System, pp. 59–68. Springer, Heidelberg (2014)
7. Fujarewicz, K., Łakomiec, K.: Parameter estimation of systems with delays via structural sensitivity analysis. Discrete Continuous Dyn. Syst.-Ser. B **19**(8), 2521–2533 (2014)
8. Fujarewicz, K.: Estimation of initial functions for systems with delays from discrete measurements. Math. Biosci. Eng. **14**(1), 165–178 (2017)
9. Fujarewicz, K., Łakomiec, K.: Adjoint sensitivity analysis of a tumor growth model and its application to spatiotemporal radiotherapy optimization. Math. Biosci. Eng. **13**(6), 1131–1142 (2016)
10. Fujarewicz, K., Łakomiec, K.: Spatiotemporal sensitivity of systems modeled by cellular automata. Math. Methods Appl. Sci. **41**(18), 8897–8905 (2018)
11. Kutalik, Z., Cho, K.H., Wolkenhauer, O.: Optimal sampling time selection for parameter estimation in dynamic pathway modeling. Biosystems **75**, 43–55 (2004)

Systems Approach Based on Petri Nets as a Method for Modeling and Analysis of Complex Biological Systems Presented on the Example of Atherosclerosis Development Process

Kaja Gutowska[1,2], Dorota Formanowicz[3], and Piotr Formanowicz[1,2](✉)

[1] Institute of Computing Science, Poznan University of Technology, Piotrowo 2, 60-965 Poznan, Poland
Piotr.Formanowicz@cs.put.poznan.pl
[2] Institute of Bioorganic Chemistry, Polish Academy of Sciences, Noskowskiego 12/14, 61-704 Poznan, Poland
[3] Department of Clinical Biochemistry and Laboratory Medicine, Poznan University of Medical Sciences, Rokietnicka 8, 60-806 Poznan, Poland

Abstract. Living organisms, as well as their functional blocks, are complex systems. From this follows that for their deep understanding it is necessary to use systems methods for their analysis. The basis for such an analysis is a precise mathematical model. It seems that models expressed as Petri nets are especially promising in the area of modeling and analysis of biological systems. In this paper a Petri net based approach for such an analysis is presented on the example of a complex process of atherosclerosis development. Classic methods of analysis based on MCT sets and t-clusters were used. Additionally, extended methods of analysis such as significance analysis and knockout analysis were applied. Significance analysis can be used to determine which elementary subprocesses are key to the functioning of the modeled system, while knockout analysis can be used to confirm the results of the significance analysis.

Keywords: Petri nets · Modeling · t-invariants · Biological systems · Atherosclerosis

1 Introduction

Atherosclerosis is a very complex disease and the use of Petri nets for its modeling and analysis allows to systematize existing knowledge about this phenomenon and for new discoveries. This is especially important because the studied process is very common, the mechanisms that contribute to it are still studied and therefore new data is coming. Based on it, models of this process have been created and then analyzed in detail. Analysis of the models expressed as Petri nets have been based on t-invariants. They correspond to subprocesses occurring in the

A. Bartoszewicz et al. (Eds.): PCC 2020, AISC 1196, pp. 579–586, 2020.
https://doi.org/10.1007/978-3-030-50936-1_49

modeled system which do not change its state. Searching for similarities among t-invariants may allow to identification of subprocesses which interact with each other. This, in turn, may lead to discoveries of some previously unknown properties of the biological system. Such an analysis can be based on clustering of t-invariants. In addition, identification of MCT sets, i.e., sets containing transitions corresponding to exactly the same t-invariants, helps to identify important functional blocks of the system. Moreover, as a complement to the analysis, it was determined which elementary subprocesses are crucial for the functioning of the entire system. This is possible through obtaining the results of the significance analysis and the knockout analysis. In this work the application of this methodology to the analysis of processes related to atherosclerosis is presented and discussed. The presented model is small and general because this paper is focused on the strength of the methodology and approaches to the analysis of Petri net based models of biological systems. More detailed models of selected aspects of atherosclerosis development were presented in [1,2] and the extended methods of analysis were presented in another paper concentrated on the model of abdominal aortic aneurysm development [3].

2 Methods

2.1 Petri Nets Basis

In formal definition, a Petri net is defined as a 5-tuple of the form: $Q = (P, T, F, W, M_0)$, where: $P = \{p_1, ..., p_n\}$ is a finite set of places, $T = \{t_1, ..., t_m\}$ is a finite set of transitions, $F \subseteq (P \times T) \cup (T \times P)$ is a finite set of arcs, $W : F \to \mathbb{Z}^+$ is a weight function, $M_0 : P \to \mathbb{N}$ is some initial marking, $P \cap T = \emptyset \wedge P \cup T \neq \emptyset$ [4].

Petri net structure is a weighted directed bipartite graph which consists of two disjoint subsets of vertices, i.e., places and transitions. Arcs connect vertices which belong to different subsets. It means that arcs connect a place with a transition or a transition with a place [4–6]. Other important elements are tokens, which are located in places and represent quantities of passive components of a system modeled by the net. The distribution of tokens over places corresponds to a state of the modeled system. Tokens can flow from one place to another via transitions what correspond to flow of information, substances etc. in the modeled system [4,5].

Graphical representation of Petri nets is very intuitive, i.e., places are represented by circles, transitions are represented by rectangles, arcs as arrows and tokens as dots (or positive integer number located in places). This representation helps to understand the structure of the model and also its behaviour during simulation. However, for a formal analysis of Petri nets other representation, called an incidence matrix is used. Such matrix A is composed of n rows corresponding to places and m columns corresponding to transitions. Entry a_{ij} of matrix A is an integer number equal to the difference between the numbers of tokens present in place p_i before and after firing transition t_j [4]. On the basis of incidence matrix t-invariants (transition invariants) can be calculated, which play a crucial role in an analysis of models of biological system [7,8].

2.2 Modeling and Analysis of Biological Systems

In the case of biological processes modeling, places correspond to passive elementary components like biological or chemical substrates or products, while transitions correspond to active elementary components like chemical reactions [8,9].

Tokens represent an amount of substrates and products of chemical reaction or other interaction between passive components. The flow of tokens is governed by a transition firing rule: transitions is active, when the number of tokens in each place which directly precedes this transition is equal to or greater than a weight of the arc connecting the place with the transition. An active transition can be fired, what means that tokens flow from each place directly preceding the transition to each place directly succeeding it. The number of flowing tokens is equal to the weight of a given arc [4,9,10].

The analysis of Petri net based models of biological systems can be based on t-invariants. Such an invariants is vector $x \in \mathbb{N}^m$ which is a solution of equation $A \cdot x = 0$. A t-invariant represents some subprocess occurring in the modeled system which does not change its state. With every t-invariant x there is associated set $supp(x) = \{t_j : x_j > 0, j = 1, 2 \ldots, m\}$, called its support. Firing every transition $t_j \in supp(x)$ x_j times does not change the distribution of tokens in the net.

The analysis of t-invariants may lead to discoveries of some previously unknown properties of the modeled system. This analysis is based on similarities between t-invariants, what corresponds to similarities between subprocesses. For a pair of t-invariants $x^{(1)}$ and $x^{(2)}$ there correspond subprocesses $\alpha^{(1)}$ and $\alpha^{(2)}$. Moreover, for every transition in the net there corresponds some elementary process in the biological system. Hence, if set $supp(x^{(1)}) \cap supp(x^{(2)})$ is non-empty, their elements correspond to common elementary processes of subprocesses $\alpha^{(1)}$ and $\alpha^{(2)}$. These common processes may lead to interactions between $\alpha^{(1)}$ and $\alpha^{(2)}$, what, in turn, may be a source of some properties of the analyzed biological system. This is the reason that the analysis based on t-invariants is so important.

In practice, searching for the above mentioned similarities may be difficult, especially in the case, where the number of t-invariants is huge. In such a case these invariants should be grouped into sets called t-clusters. However, decisions have to be made which clustering algorithm should be used and also what similarity measure and what number of t-clusters are the most appropriate. In general, these decisions depend on the particular biological system and its model.

Despite that t-invariants can be grouped into t-clusters also transitions can be grouped into maximal Common Transition sets (MCT sets). Every such a set contains transitions belonging to supports of exactly the same t-invariants. Both, MCT sets and t-clusters, correspond to some functional blocks of the modeled system, whose biological meaning should be determined [8–10].

As an extension to the analysis of MCT sets and t-cluster analysis, the significance analysis and the knockout analysis were performed [3]. The significance analysis allows to distinguished which subprocesses are more crucial for the functioning of the modeled system. It can be possible through determination for each transition, its attendance frequency in all supports of t-invariants. To complement the significance analysis the knockout analysis was also performed. The knockout analysis allows to estimate which subprocesses have been excluded in consequence of knockout of selected transitions. Thus, it allows to estimate how crucial selected transition is, based on the number of excluded subprocesses.

3 The Example Modeled System

3.1 The Model

The proposed model of the biological system has been built using Snoopy software [11]. The model is presented in Fig. 1 and it is divided into many parts ($a-j$), which correspond to important subprocesses of the modeled phenomenon: a) nitric oxide (NO) synthesis, b) positive role of NO (when quantity of NO is sufficient for proper action), c) proliferation caused by growth factor and cytokines, d) endothelial damage caused by various factors, e) negative role of NO, f) inflammation response, g) development of atherosclerosis (plaque formation), h) block the coronary blood vessels, i) respiratory burst, j) low-density lipoprotein (LDL) oxidation.

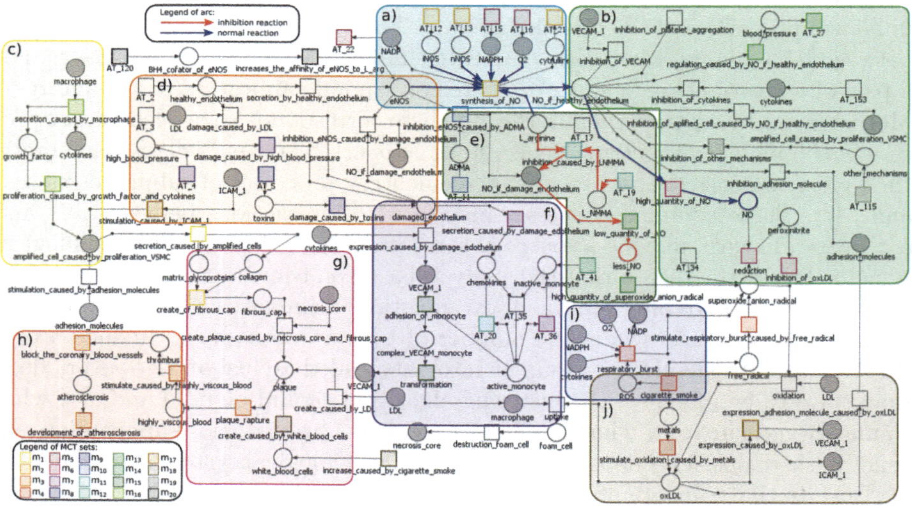

Fig. 1. The proposed Petri net model includes 20 MCT sets (marked with coloured transitions) and it is divided into many subprocesses ($a - j$) marked with coloured frames.

The model presented in Fig. 1 includes also 20 MCT sets, which are indicated as various colours of transitions. Moreover, it includes, e.g., transition called "AT_12", which is the abbreviation for "auxiliary transition number 12". Two types of auxiliary transitions has been distinguished: input and output ones. Input transitions correspond to reactions engaged in production of some components and output transitions correspond to reactions engaged in other processes, which are not important for the modeled diseases. Additionally, in the presented Petri net an inhibition reactions have been modeled. In Fig. 1 one example of inhibition reactions is indicated as red arc and normal reaction is indicated as blue arcs. Precisely, in healthy condition all necessary components are present, which results in proper NO synthesis. What means that transition called "synthesis of NO" can be fired when tokens are in preceding places ("eNOS", "iNOS", "nNOS", "NADPH", "O_2", "L-arginine"). Firing transition called "synthesis of NO" enables flow of tokens to following places and in this case it leads to correct NO synthesis (quantity of NO is sufficient to correct work) - this process is indicated as blue arcs in Fig. 1. In opposite situation, NO synthesis can be inhibited by different factors, e.g., inhibition of L-arginine caused by L-NMMA. When tokens are in places called "L-NMMA" (inhibitor of L-arginine) and "L-arginine" it leads to firing transition called "inhibition caused by L-NMMA", which leads to limited NO synthesis - this process is indicated as red arcs. In this inhibition process, tokens are taken form place "L-arginine", which prevents correct NO synthesis.

3.2 Analysis of Model

The analysis of the presented model is based on t-invariants (whose number is dependent on the nature of the modeled system). The model includes 49 places, 70 transitions, 20 MCT sets and it is covered by 124 t-invariants, which have been grouped into 19 t-clusters. The biological meaning of non-trivial MCT sets (i.e., those ones containing more then one transition) is as follows:

- $m_1 = \{t_{37}, t_{40}, t_{41}, t_{46}, t_{47}\}$ and $m_{18} = \{t_{57}, t_{58}\}$: plaque formation,
- $m_2 = \{t_{11}, t_{12}, t_{13}, t_{20}\}, m_5 = \{t_{14}, t_{15}, t_{21}\}$ and $m_{20} = \{t_{61}, t_{62}\}$: NO synthesis,
- $m_3 = \{t_{42}, t_{43}, t_{44}, t_{49}\}$: block the coronary blood vessels,
- $m_4 = \{t_{48}, t_{50}, t_{53}, t_{56}\}$ and $m_{17} = \{t_{52}, t_{54}\}$: stimulation of LDL oxidation,
- $m_6 = \{t_{22}, t_{29}, t_{30}\}, m_{13} = \{t_{23}, t_{26}\}$ and $m_{19} = \{t_{60}, t_{65}\}$: positive role of NO,
- $m_7 = \{t_1, t_{34}\}$ and $m_{12} = \{t_{19}, t_{39}\}$: inflammation response,
- $m_8 = \{t_4, t_{63}\}$ and $m_9 = \{t_5, t_{64}\}$: endothelial damage,
- $m_{10} = \{t_9, t_{10}\}$ and $m_{11} = \{t_{17}, t_{18}\}$: inhibition of NO,
- $m_{14} = \{t_{27}, t_{28}\}$: negative role of NO,
- $m_{15} = \{t_{35}, t_{36}\}$: transformation to macrophages,
- $m_{16} = \{t_{38}, t_{45}\}$: proliferation of vascular muscle cells,

For finding the best clustering different algorithms and similarity measures have been applied. The used algorithms were: average linkage, centroid method, complete linkage, McQuitty's method, median method, single linkage, Ward's method and the applied similarity measures were: binary, euclidean, manhattan, canberra, maximum, Minkowski, centered and uncentered Pearson. The best clustering has been found on the basis of Mean Split Silhouette (MSS) index (cf. [9]). MSS evaluates a fit of each t-invariant to its cluster and an average quality of a given clustering [12]. To find the best number of clusters Calinski-Harabasz (C-H) coefficient has been used (the optimal number of clusters is indicated by the highest value of C-H coefficient) [13]. This coefficient has been calculated for the number of clusters in the range [2, 20] (cf. [9]).

This analysis indicated that the best clustering has been obtained using Pearson similarity measure and average linkage method and consists of 19 t-clusters, whose biological meaning is as follows:

- c_1, c_4, c_5 and c_6: inflammation response on endothelial damage,
- c_2 and c_3: development of atherosclerosis (plaque formation caused by cigarette smoke and caused by high level of LDL),
- c_7, c_{10}, c_{16} and c_{19}: dual role of NO: positive role when quantity of NO is sufficient to prevention of LDL oxidation and negative role of NO when harmful mechanisms lead to limited NO synthesis,
- c_8 and c_9: correct NO synthesis which results in inhibition of harmful subprocesses,
- c_{11}, c_{12}, c_{13} and c_{14}: endothelial damage, inflammation, LDL oxidation are main subprocesses which result in plaque formation and atherosclerosis development (almost all subprocesses are included in these t-clusters),
- c_{15}, c_{17} and c_{18}: negative role of NO (inhibition of NO synthesis) results in promotion of LDL oxidation and atherosclerosis development.

In order to find the most key reactions for the functioning of the whole system, significance analysis was performed. The results for the most important reactions

Table 1. The results of the significance analysis of the Petri net model (124 t-inv).

Significance analysis			
No.	Name of transition	t-inv	Frequency trans/t-inv
t14	AT_15 (source of NADPH)	98	79,03%
t15	AT_16 (source of O2)	98	79,03%
t16	AT_17 (source of L-arginine)	98	79,03%
t21	AT_22 (NADP - product of NO synthesis)	98	79,03%
t11	AT_12 (source of iNOS)	96	77,42%
t12	AT_13 (source of nNOS)	96	77,42%
t13	Synthesis of NO	96	77,42%
t20	AT_21 (citrulline - product of NO synthesis)	96	77,42%
t3	AT_3 (source of LDL)	84	67,74%
t31	Oxidation (LDL to oxLDL)	78	62,90%

(transitions) are presented in Table 1. Table 1 contains those reactions whose percentage contribution in all support of t-invariants is higher than 60%. Most of the elementary subprocesses (transitions) from Table 1 are associated with synthesis of nitric oxide. The remaining ones are associated with reaction of LDL oxidation. These subprocesses are crucial for the development of atherosclerosis.

The knockout analysis confirms the significance of the subprocesses listed in Table 1. Excluding of transitions associated with NO synthesis (t_{11}, t_{12}, t_{13}, t_{14}, t_{15}, t_{16}, t_{20}, t_{21}) leads to turn off 100 out of 124 subprocesses (t-invariants), which consists about 81% of all modeled subprocesses. Nevertheless, excluding of transitions associated with LDL oxidation (t_3, t_{31}) leads to turn off 84 of 124 subprocesses, which consists about 68% of all subprocesses.

4 Conclusions

In this study systems approach for modeling biological processes has been presented. The process of atherosclerosis development has been used as an example The main purpose of the article was to present a methodology rather than a detailed analysis of the model. Using Petri net based approach and analysis based on t-invariants may lead to discovery of unknown properties of the modeled system and also to better understanding of dependencies occurring between subprocesses. Short conclusions from the analysis of the exemplary model allow to notice that the analysis of t-cluster was a bit too general and for this reason more detailed analysis of particular t-invariants has been used. The results confirm that respiratory burst, endothelial damage and inflammation lead to atherosclerosis development and these processes can influence each other. Important here is NO bioavailability, because NO can play a dual role. When its concentration is not sufficient to correct work, then NO has direct influence on atherosclerosis by promotion of LDL oxidation. On the other hand, it revealed many cytoprotective properties. The systems approach allowed to better understand the modeled phenomenon and distinguished important subprocesses. A more detailed model and analysis of selected aspects of atherosclerosis development can be found in [1].

Acknowledgement. This research has been partially supported by the National Science Centre (Poland) grant No. 2012/07/B/ST6/01537 and by the statutory funds of Poznan University of Technology.

References

1. Chmielewska, K., Formanowicz, D., Formanowicz, P.: The effect of cigarette smoking on endothelial damage and atherosclerosis development - modeled and analyzed using Petri nets. Arch. Control Sci. **27**(2), 211–228 (2017)
2. Formanowicz, D., Gutowska, K., Formanowicz, P.: Theoretical studies on the engagement of interleukin 18 in the immuno-inflammatory processes underlying atherosclerosis. Int. J. Mol. Sci. **19**(11), 3476 (2018)

3. Gutowski, Ł., Gutowska, K., Pioruńska-Stolzmann, M., Formanowicz, P., Formanowicz, D.: Systems approach to study associations between OxLDL and abdominal aortic aneurysms. Int. J. Mol. Sci. **20**(16), 3909 (2019)
4. Murata, T.: Petri nets: properties, analysis and applications. Proc. IEEE **77**(4), 541–580 (1989)
5. Reising, W.: Understanding Petri Nets. Modeling Techniques, Analysis Methods, Case Studies. Springer, Heidelberg (2013)
6. David, R., Alla, H.: Discrete, Continuous, and Hybrid Petri Nets. Springer, Heidelberg (2005)
7. Grafahrend-Belau, E., Schreiber, F., Heiner, M., Sackmann, A., Junker, B.H., et al.: Modularization of biochemical networks based on classification of Petri net t-invariants. BMC Bioinform. **9**(1), 90 (2008)
8. Koch, I., Reisig, W., Schreiber, F. (eds.): Modeling in Systems Biology. The Petri Net Approach. Springer, Heidelberg (2010)
9. Formanowicz, D., Kozak, A., Głowacki, T., Radom, M., Formanowicz, P.: Hemojuvelin - hepcidin axis modeled and analyzed using Petri nets. J. Biomed. Inform. **46**(6), 1030–1043 (2013)
10. Sackmann, A., Heiner, M., Koch, I.: Application of Petri net based analysis techniques to signal transduction pathways. BMC Bioinform. **7**(1), 482 (2006)
11. Heiner, M., Herajy, M., Liu, F., Rohr, C., Schwarick, M.: Snoopy - a unifying Petri net tool. In: International Conference on Application and Theory of Petri Nets and Concurrency, pp. 398–407. Springer, Heidelberg (2012)
12. Kaufman, L., Rousseeuw, P.J.: Finding Groups in Data: An Introduction to Cluster Analysis, vol. 344. Wiley, New York (1990)
13. Calinski, T., Harabasz, J.: A dendrite method for cluster analysis. Commun. Stat. **3**(1), 1–27 (1974)

Influence of the Number of Thresholds on the Dynamics of Models with Switchings of the Biological Systems

Magdalena Ochab and Krzysztof Puszynski(✉)

Department of Systems Biology and Engineering, Silesian University of Technology, Akademicka 16, 44-100 Gliwice, Poland
{magdalena.ochab,krzysztof.puszynski}@polsl.pl

Abstract. When the mathematical model of the biological system is developed, especially the one concerning intracellular processes, usually one of two approaches is applied: deterministic model based on ordinary differential equations or stochastic based on Gillespie algorithm. Usually, in both cases only numerical analysis is possible. This inconvenience may be overcome by the models with switchings, especially the one using linear description of the dynamics inside the domains. However, in such case, the question arises concerning the number and location of thresholds for the particular variables. In the presented work we are dealing with this problem, showing how the response of the simple model of the p53 signalling pathway depends on the thresholds used.

Keywords: Mathematical models · Systems witch switchings · Switching thresholds

1 Introduction

In recent years, with the growing number of biological findings concerning cells functioning, especially dynamics of its intracellular processes, the problem arises with proper handling and interpretation of the more and more detailed data. It becomes clear that the biologist may require help from mathematicians, especially statisticians and specialist of systems dynamics to build and analyse the mathematical models of the considered biological processes. These mathematical models may not only explain the observed phenomena [1] but also propose a new hypothesis of the cell functioning [2] and even the therapeutic protocols [3].

During the years two main approaches to the mathematical modelling of the biological systems become "gold standard" as long as the spatial localization is neglected. In case when the considered system is deterministic, most scientists use Ordinary Differential Equations (ODE) [4]. When the stochastic system is considered usually Gillespie Algorithm is used [5]. Both approaches base on strongly non-linear formulas and give good numerical results but are hard or even impossible to analyse analytically. Even the simplest task which is to find the equilibrium points may be impossible to solve.

A. Bartoszewicz et al. (Eds.): PCC 2020, AISC 1196, pp. 587–598, 2020.
https://doi.org/10.1007/978-3-030-50936-1_50

The solution to this problem may be to apply the system with switchings for modelling the biological systems, especially the ones with linear description of the system dynamics in particular domains [6–8]. As we showed in [9], for such models we can not only determine analytically the location and type of equilibrium points but also the switching times and in some cases the whole trajectory from the given initial point to the equilibrium.

However, with the proposed approach another question arises. How many thresholds one has to use and how to set their location to correctly reflect the behaviour of the modelled system, at least qualitatively. In some cases, when the considered dynamics is enzymatic type i.e. has the shape of sigmoid the threshold location is easy to determine, but when it is linear or quadratic type, which is common in biological systems, the problem become more complex - see Fig. 1.

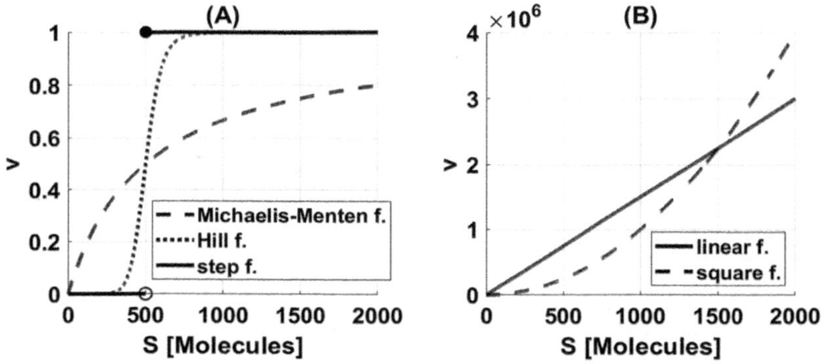

Fig. 1. The reaction speed as a function of molecules number for different function types.

Another question is how many thresholds we should put to the protein production terms in the developed model. Usually, every cell has two copies of each gene, which can be activated by transcription factors, so we have two options to consider. In the first case with only one threshold, both genes are inactive when the transcription factor level is to low i.e below threshold, so only spontaneous, very small production is present. If transcription factor level is high, i.e. above the threshold, it activates both genes simultaneously and we have full production. The second case includes two thresholds. The transcription factor level may be to low (i.e below the first threshold) to activate any gene copy so only spontaneous production is present. Then it may be at the middle level i.e between thresholds - in this state the transcription factor level is high enough to activate genes, but because of spontaneous detachment of transcription factor from the promoter region and thus gene deactivation, only one gene copy is active most of the time, so the production is at the medium level. Finally, when the transcription factor level is very high, above the second threshold, its concentration around the promoter region is so high that any spontaneous detachment is immediately followed by another particle attachment. As a result, both gene copies are active most of the time and production goes with full speed.

To investigate the main problem, which is the number of thresholds in the systems with switchings, we considered a simple mathematical model of the p53 signalling pathway. We worked with this pathway before, so we have the complex model of the p53 signalling pathway [1] and we know the pathway dynamics. Thus we created a simple, 4-equations model of p53, which qualitatively has the same dynamics as a complex model. In short, the p53 signalling pathway dynamic is as follow: without the stress factor, such as irradiation (R), which is the input to the system, level of the p53 protein is kept low by its inhibitor Mdm2, which marks p53 particles for degradation. Stress causes DNA damage and following rapid Mdm2 degradation and p53 stabilization thus its level grows. P53 is a transcription factor responsible for production of many proteins involved in the cell cycle blockade, DNA repair processes and apoptosis. Among them is protein Mdm2, which constitutes a negative feedback loop. As a result, after stress, p53-Mdm2 level oscillations are observed in the biological experiments [10]. This is the time when cells stop its cell cycle and try to repair the damaged DNA. The important part of the p53-Mdm2 interaction is the import of newly synthesized Mdm2 to the nucleus where p53 is located. This requires phosphorylation of Mdm2 in cytoplasm, which may be blocked by the other signalling pathway leading by protein PTEN. Because PTEN is transcriptionally dependent on p53 we have a positive feedback loop which works through double negation: p53 produces PTEN which inhibits the action of p53 inhibitor Mdm2. Because the positive feedback loop is much longer than the negative one, it works as a clock giving the cell time to repair DNA [2]. For the higher irradiation dose, when DNA damages are too extensive and the cell cannot repair them in the given time, the positive feedback is activated, which blocks the negative feedback. Thus the p53-Mdm2 oscillations stop and p53 goes to higher level. This high p53 level is the signal for the cell to initiate apoptosis [11].

To simplify the model we neglected the DNA damage repair and use only four main variables: p53 (P), cytoplasmic Mdm2 (M), nuclear Mdm2 (N) and PTEN (T). As a base model, we used the one with one threshold on the p53 level so the gene activation is all or nothing and two thresholds on the nuclear Mdm2 level so the p53 degradation is slow, medium or high. This model properly reflects the dynamics of the system observed biologically and modelled by us in the full model [1]. Then we checked if and how the system dynamics changes in the meaning of qualitatively change when the Mdm2 and PTEN production will depend on two thresholds of the p53 level. The second experiment involved the change of two existing thresholds on nuclear Mdm2 level into the one located in the position of the original lower threshold, higher threshold and in the middle between them. In all cases three levels of the input signal were considered: $R = 0$ [a.u] which means we have no stress and thus no DNA damage, $R = 5$ [a.u.], which is medium level of stress, where we expect cell cycle blockade but not apoptosis in response and finally $R = 8$ [a.u.], which is high stress dose and thus high DNA damage level, so we expect that cells should commit apoptosis.

2 Considered Models

The base model contains 4 threshold values: one threshold for p53 (θ_P), two threshold for nuclear Mdm2 (θ_{N1} and θ_{N2}) and one for PTEN (θ_T). The model equations took the following form:

$$\frac{dP(t)}{dt} = p_1 - (d_{10} + d_{11}\, Z_{N1} + d_{12}\, Z_{N2})\, P(t), \tag{1}$$

$$\frac{dM(t)}{dt} = p_{20} + p_{21}\, Z_P - d_2\,(1+R)\, M(t) - (k_{10} - k_{11}\, Z_T)\, M(t), \tag{2}$$

$$\frac{dN(t)}{dt} = (k_{10} - k_{11}\, Z_T)\, M(t) - d_2\,(1+R)\, N(t), \tag{3}$$

$$\frac{dT(t)}{dt} = p_{30} + p_{31}\, Z_P - d_3\, T(t), \tag{4}$$

where

$$Z_P = \begin{cases} 1, & \text{if } P \geq \theta_P \\ 0, & \text{otherwise} \end{cases} \qquad Z_{N1} = \begin{cases} 1, & \text{if } N \geq \theta_{N1} \\ 0, & \text{otherwise} \end{cases}$$

$$Z_{N2} = \begin{cases} 1, & \text{if } N \geq \theta_{N2} \\ 0, & \text{otherwise} \end{cases} \qquad Z_T = \begin{cases} 1, & \text{if } T \geq \theta_T \\ 0, & \text{otherwise} \end{cases}$$

with assumption that $k_{10} > k_{11}$. The parameter values for the model are presented in Table 1.

Expanded Model. Then we investigated how considering two-stage gene activation will influence achieved results. We added second threshold value for p53, so the production of proteins Mdm2 and PTEN can have three values: basic, medium (1 gene active) and fast (2 genes active). The equations for M and T are changed by:

$$\frac{dM(t)}{dt} = p_{20} + p_{21}^{exp}(Z_{P1} + Z_{P2}) - d_2\,(1+R)\, M(t) - (k_{10} - k_{11} Z_T)\, M(t), \tag{5}$$

$$\frac{dT(t)}{dt} = p_{30} + p_{31}^{exp}\,(Z_{P1} + Z_{P2}) - d_3\, T(t), \tag{6}$$

where

$$Z_{P1} = \begin{cases} 1, & \text{if } P \geq \theta_{P1}^{exp} \\ 0, & \text{otherwise} \end{cases} \qquad Z_{P2} = \begin{cases} 1, & \text{if } P \geq \theta_{P2}^{exp} \\ 0, & \text{otherwise.} \end{cases}$$

Table 1. Parameters, thresholds and initial values in model with switchings.

Parameter	Value	Parameter	Value
p_1	$8.8\ \text{s}^{-1}$	θ_P	$4.5 \cdot 10^4$ molecules
p_{20}	$2.4\ \text{s}^{-1}$	θ_{N1}	$4 \cdot 10^4$ molecules
p_{21}	$21.6\ \text{s}^{-1}$	θ_{N2}	$8 \cdot 10^4$ molecules
p_{30}	$0.5172\ \text{s}^{-1}$	θ_T	10^5 molecules
p_{31}	$3.6204\ \text{s}^{-1}$	P_0	$2.6858 \cdot 10^4$ molecules
d_{10}	$9.8395 \cdot 10^{-5}\ \text{s}^{-1}$	M_0	$1.1166 \cdot 10^4$ molecules
d_{11}	$6.5435 \cdot 10^{-5}\ \text{s}^{-1}$	N_0	$1.5438 \cdot 10^4$ molecules
d_{12}	$9.8395 \cdot 10^{-5}\ \text{s}^{-1}$	T_0	$1.7240 \cdot 10^5$ molecules
d_2	$1.375 \cdot 10^{-5}\ \text{s}^{-1}$		
d_3	$3 \cdot 10^{-5}\ \text{s}^{-1}$		
k_{10}	$1.5 \cdot 10^{-4}\ \text{s}^{-1}$		
k_{11}	$1.4713 \cdot 10^{-4}\ \text{s}^{-1}$		

The values of the expanded model parameters are presented in the Table 2 and the unchanged parameters are in the Table 1:

Table 2. Parameters and thresholds values in expanded model.

Parameter	Value	Parameter	Value
p_{21}^{exp}	$10.8\ \text{s}^{-1}$	θ_{P1}^{exp}	$3.36 \cdot 10^4$ molecules
p_{31}^{exp}	$1.8102\ \text{s}^{-1}$	θ_{P2}^{exp}	$4.5 \cdot 10^4$ molecules

Reduced Model. In the next case we created the model with 3 threshold values, to check if the division for slow, medium and fast p53 degradation is needed to achieve proper dynamics. In this model the equation describing change of the p53 level took the form:

$$\frac{dP(t)}{dt} = p_1 - (d_{10} + d_{11}^{red}\, Z_N)\, P(t), \tag{7}$$

where

$$Z_N = \begin{cases} 1, & \text{if } N \geq \theta_N^{red} \\ 0, & \text{otherwise} \end{cases}$$

and the remaining equations are the same as in the base model. We investigated 3 cases of Mdm2 threshold locations:

- in the location of lower threshold in the original model: $\theta_N^{red} = \theta_{N1}$,
- in the location of upper threshold in the original model: $\theta_N^{red} = \theta_{N2}$,
- in the middle between two original thresholds: $\theta_N^{red} = (\theta_{N1} + \theta_{N1})/2$,

and 3 cases of parameters values:

- neglect fast stage: $d_{11}^{red} = d_{11}$,
- neglect medium stage: $d_{11}^{red} = d_{11} + d_{12}$,
- create average stage: $d_{11}^{red} = d_{11} + d_{12}/2$.

The numerical approach to the simulation is based on the gold standard of Runge-Kutta 4th–5th order algorithm. To add the switchings, in each step of the deterministic simulation we check the state of the system and determine values of parameters. The obtained time courses are presented in next section.

To compare the created models, we simulated them with many different parameters and analysed the achieved results. We analysed type of response and level of protein in stationary points based on time courses.

3 Results

We assumed that the single threshold on the p53 level (θ_P) separates the regions in which both genes copies of Mdm2 and PTEN are inactive and active. Thus during the extension of the original model, the second threshold on p53 level has to be below the original θ_P. We found that the best results are obtained when the lower threshold will be at level $3.36 \cdot 10^4$ molecules. It is in agreement with the biological knowledge as it is not required that the transcription factor level increases two times to cause the activity of both genes instead of one on average. As one can notice in case of the expanded model, we have qualitatively the same response for all considered doses (Fig. 2). Without stress the p53 level stays constant at the $3.3 \cdot 10^4$ molecules. The medium level of stress ($R = 5$ [a.u.]) causes oscillations of the p53 level between 3.3 and $4.9 \cdot 10^4$ molecules which is enough to stop the cell cycle but not enough for apoptosis. After the high stress ($R = 8$ [a.u.]) we can observe that after few initial oscillations p53 level goes much higher than in the other cases reaching the level of $9 \cdot 10^4$, which is enough to initiate apoptosis. One can notice that the period of the oscillations changes as well as the number of oscillations before apoptotic decision is made in the case of $R = 8$ [a.u.]. Also the time to take the apoptotic decision is much shorter in the case of the expanded model. What is important all these differences depend on the location of the second threshold thus the threshold locations in such cases may be adjusted to the biological findings.

On the Fig. 3 we presented the 3D plots of the sample trajectory in the case of $R = 8$ [a.u.]. As one can notice both trajectories, for the original and the expanded model, start at the same point and ends almost at the same. But the expanded case shows that the trajectory moves much faster through the T axis, which causes fewer oscillations before the apoptotic decision is made.

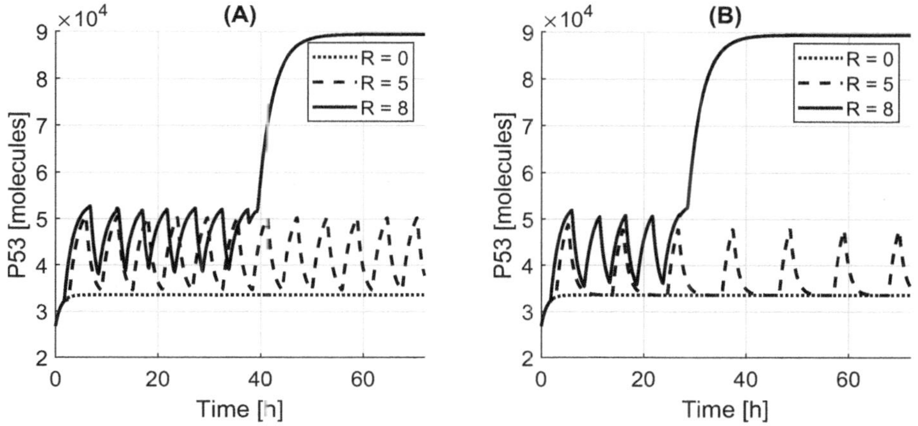

Fig. 2. Time courses of the p53 protein level after various input signals. Panel A. Base model with 4 thresholds. Panel B. Expanded model with 5 thresholds.

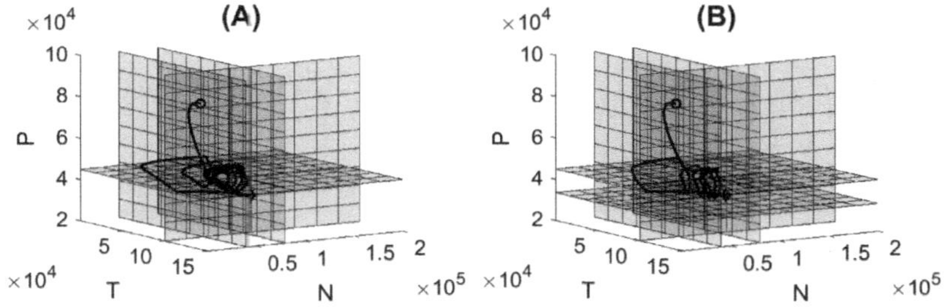

Fig. 3. Sample trajectory of proteins level after the input signal $R = 8$ [a.u.]. Red dot indicates the beginning and green one end of the trajectory. Panel A. Base model with 4 thresholds. Panel B. Expanded model with 5 thresholds.

The second considered variant of the original model modification is more complicated. In the original ODE model, which was inspiration for the model with switchings, the p53 level dependency on Mdm2 has a quadratic form of $-P \cdot N^2$. In the original model with switchings, we used two thresholds on the Mdm2 level. First $\theta_{N1} = 4 \cdot 10^4$ molecules and second $\theta_{N2} = 8 \cdot 10^4$ molecules. If we want to replace them with just one threshold, the question arises: should we leave only lower one (θ_{N1}), upper one (θ_{N2}) or maybe create one in the middle? The next question, closely related to the first one, is the degradation coefficient of d_{11}^{red} value. Should it be as in the original model when p53 crossing the first threshold or the second or it should take the medium value? We considered all these cases and found that the correct results may be obtained only when degradation parameter takes the value which original model has after the crossing the upper threshold (Fig. 4 vs Fig. 6).

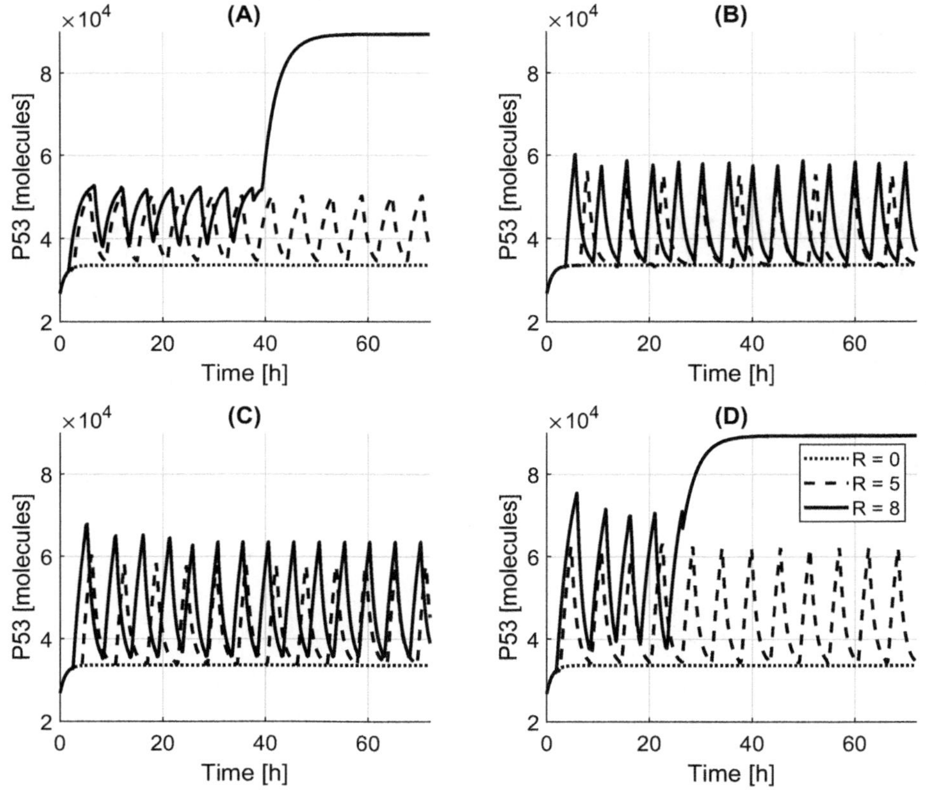

Fig. 4. Time courses of the p53 protein level after various input signals in the case $d_{11}^{red} = d_{11} + d_{12}$. Panel A - Base model. Panel B - Reduced model with $\theta_N^{red} = \theta_{N1}$. Panel C - Reduced model with $\theta_N^{red} = (\theta_{N1} + \theta_{N2})/2$. Panel D - Reduced model with $\theta_N^{red} = \theta_{N2}$

As we showed on the Fig. 4 only in the case when we replace the two thresholds with one in the position of upper original threshold (panel D), the system response for all inputs is qualitatively correct. In all cases without the input signal we have stable low level of p53 and with medium input we have oscillations. However, in the cases when the new threshold is in the location of the original lower one (panel B) or between the original thresholds (panel C) we do not see the apoptotic solution. As one can notice the oscillations in all cases have a slightly higher amplitude than in the original model but not enough to cause apoptosis. This result may indicate that some cells can become cancerous because of the threshold location change. It may be when p53 or Mdm2 is mutated in such way, that p53 becomes more sensitive to Mdm2. Also, the mutation which causes over expression of the Mdm2 may be considered as the specific threshold location shift and it is biologically known that such mutation

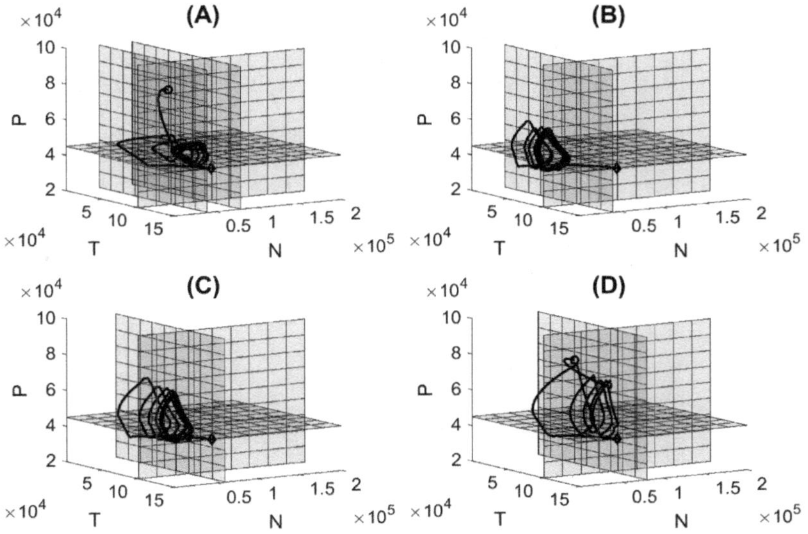

Fig. 5. Sample trajectory of proteins level after the input signal $R = 8$ [a.u.]. Red dot indicates the beginning and green one end of the trajectory. Panel A - Base model. Panel B - Reduced model with $\theta_N^{red} = \theta_{N1}$. Panel C - Reduced model with $\theta_N^{red} = (\theta_{N1} + \theta_{N2})/2$. Panel D - Reduced model with $\theta_N^{red} = \theta_{N2}$.

causes cancer [12]. Also, in this case, one can see that the higher value of the threshold the faster oscillations are. The period of the oscillations is the closest to the original one in the case of the medium threshold location.

On the Fig. 5 one can see that the trajectories for various threshold locations differ by the meaning of the number of oscillations and their size. In the case with a single threshold, they are significantly bigger in the meaning of the amplitude, especially on the P axis. The second noticeable property is the same as in the case of the expanded model - with growing threshold location oscillations are slower in the meaning they have greater period.

When the degradation coefficient takes medium ($d_{11}^{red} = d_{11} + d_{12}/2$) value (Fig. 6) one can notice that the qualitatively the system response seems similar to the case with high value of it (Fig. 4). However, the problem is with the lowest level of p53. Comparing Figs. 4 and 6 we see a significantly higher level of the p53 protein even in the case with no input signal (green line). In the real cells, it may cause the persistent cell cycle blockade, which means that the whole organ will not function as intended.

In the case of low value of degradation coefficient ($d_{11}^{red} = d_{11}$) the p53 level is even higher than in the previous case. Moreover we do not see the apoptotic solution for $R = 8$ [a.u.] even in the case with $\theta_N^{red} = \theta_{N2}$. Because of the problems mentioned above we conclude that the only possible solution is to set $d_{11}^{red} = d_{11} + d_{12}$.

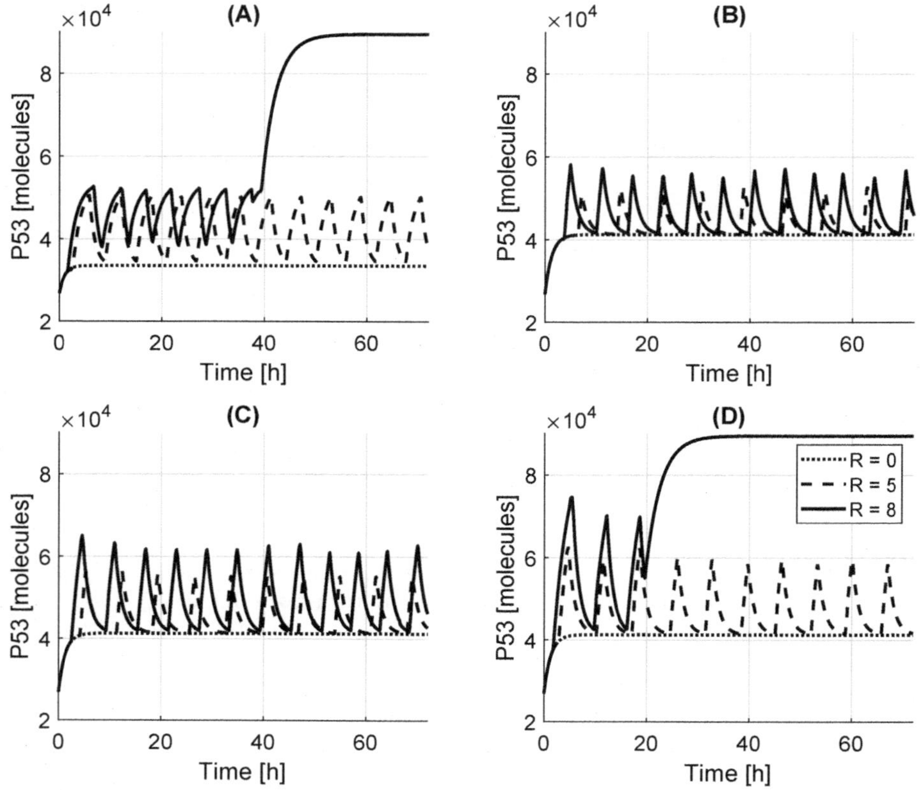

Fig. 6. Time courses of the p53 protein level after various input signals in the case $d_{11}^{red} = d_{11} + d_{12}/2$. Panel A - Base model. Panel B - Reduced model with $\theta_N^{red} = \theta_{N1}$. Panel C - Reduced model with $\theta_N^{red} = (\theta_{N1} + \theta_{N2})/2$. Panel D - Reduced model with $\theta_N^{red} = \theta_{N2}$.

4 Discussion

The mathematical modelling of the biological systems is challenging, especially when one wants to consider the intracellular processes. The number of the various kinds of involved molecules and their variants causes, that even when an only small part of the complex system is considered, the models consist of few or tens of variables and thus strongly non-linear equations. This makes them very hard or even impossible to analyse in a different way than by numerical simulations - deterministic or stochastic. The next step for the better understanding of those systems may be made by using a well-known methodology for their modelling, which is systems with switchings [13,14], especially with the linear description of the dynamics inside the particular domains. But to build the mathematical model with switchings we have to decide on the number and location of thresholds, which is not easy, especially when modelled dynamics do not have the switch-like shape. In this work, we showed that the number and location of the thresholds may significantly change the observed dynamics, especially when the quantitate results are compared. However, the nature of the biological experiments, on which the developed models are based, is usually qualitative not quantitative. We can distinguish the small, medium or large levels of proteins and we can define the change of the protein levels in comparison to the earlier stage but usually, we cannot determine their exact number. In such cases, we can consider the mathematical model as valid as long as the qualitative results produced by it are in agreement with the biological experiments.

Two of the basic processes of the intracellular pathways are protein production and degradation. In our work, we showed that when protein production is considered, the thresholds of the transcription factor level may be associated with the gene activation and deactivation. Usually, each gene has two copies which change their state independently, so it seems natural to use two thresholds. As we showed in this work, it is not required, because the results obtained with just one threshold, which means both genes are on or off simultaneously, are in agreement with these for two. The other situation is when we consider degradation process. Usually, the degradation in the biological systems is linear or quadratic. In such cases, the number and location of the thresholds may play important role.

What is also worth to notice is the fact, that different location of thresholds and parameter values may reflect the various abnormal states in cells, caused by mutations. These mutations lead to the change in the system dynamics which may lead, for example, to skip the apoptotic equilibrium. As a result, the cells may become cancerous. Analysis of the threshold numbers and location on the observed dynamics may reveal that.

Acknowledgments. The presented work was supported by the grant funded by National Science Centre Poland, with number 2016/23/B/ST6/03455

References

1. Kozlowska, E., Puszynski, K.: Application of bifurcation theory and siRNA-based control signal to restore the proper response of cancer cells to DNA damage. J. Theor. Biol. **408**, 213–221 (2016)
2. Jonak, K., Kurpas, M., Szoltysek, K., Janus, P., Abramowicz, A., Puszynski, K.: A novel mathematical model of ATM/p53/NF-kB pathways points to the importance of the DDR switch-off mechanisms. BMC Syst. Biol. **10**(1), 1–12 (2016). https://doi.org/10.1186/s12918-016-0293-0
3. Puszynski, K., Gandolfi, A., d'Onofrio, A.: The Pharmacodynamics of the p53-Mdm2 targeting drug Nutlin: the role of gene-switching noise. PLOS Comput. Biol. **10**(2), e1003991 (2014)
4. Polynikis, A., Hogan, S.J., di Bernardo, M.: Comparing different ODE modelling approaches for gene regulatory networks. J. Theor. Biol. **261**(4), 511–530 (2009)
5. Gillespie, D.T.: Exact stochastic simulation of coupled chemical reactions. J. Phys. Chem. **81**(25), 2340–2361 (1977)

6. Ochab, M., Puszynski, K., Swierniak, A., Klamka, J.: Variety behavior in the piecewise linear model of the p53-regulatory module. In: International Conference on Bioinformatics and Biomedical Engineering, pp. 208-219. Springer, Cham (2017)
7. Omholt, S.W., Plahte, E., Øyehaug, L., Xiang, K.: Gene regulatory networks generating the phenomena of additivity, dominance and epistasis. Genetics **155**(2), 969–980 (2000)
8. Casey, R., De Jong, H., Gouze, J.L.: Piecewise-linear models of genetic regulatory networks: equilibria and their stability. J. Math. Biol. **52**(1), 27–56 (2006)
9. Ochab, M., Puszynski, K., Swierniak, A.: Influence of parameter perturbations on the reachability of therapeutic target in systems with switchings. Biomed. Eng. Online **16**, 1–77 (2017)
10. Geva-Zatorsky, N., Rosenfeld, N., Itzkovitz, S., Milo, R., Sigal, A., Dekel, E., Alon, U.: Oscillations and variability in the p53 system. Mol. Syst. Biol. **2**(1), 0033 (2006)
11. Kracikova, M., Akiri, G., Georg, A., Sachidanandam, R., Aaronson, S.A.: A threshold mechanism mediates p53 cell fate decision between growth arrest and apoptosis. Cell Death Differ. **20**, 576–588 (2013)
12. Oliner, J.D., Kinzler, K.W., Meltzer, P.S., George, D.L., Vogelstein, B.: Amplification of a gene encoding a p53-associated protein in human sarcomas. Nature **358**, 80–83 (1992)
13. Liberzon, D., Hespanha, J.P., Morse, A.S.: Stability of switched systems: a Lie-algebraic condition. Syst. Control Lett. **37**(3), 117–122 (1999)
14. Sun, Z.: Switched Linear Systems: Control and Design. Springer Science & Business Media, London (2006)

Geometric Methods in Nonlinear Control

Normal Forms of a Free-Floating Space Robot

Krzysztof Tchoń(✉)

Wroclaw University of Science and Technology, Wroclaw, Poland
krzysztof.tchon@pwr.edu.pl
https://kcir.pwr.edu.pl/~tchon/

Abstract. This paper is devoted to the dynamics of a free-floating space robot consisting of a mobile base (a spacecraft) and an on-board manipulator. Lagrangian equations of motion are derived. Special attention is paid to the equation resulting from the conservation of the angular momentum. This equation is represented in the form of a control system driven by joint velocities of the on-board manipulator. The main result of this paper consists in showing that this control system can be transformed by feedback to the chained form. Explicit form of the feedback transformations has been found for the case when the on-board manipulator is mounted at an arbitrary point of the base, and compared with the previously studied case of mounting point fixed at the center of mass of the base.

Keywords: Space robot · Dynamics · Feedback · Chained form

1 Introduction

The synthesis of control algorithms for non-linear systems benefits considerably from the existence of normal forms of the systems' equations. A normal form is usually understood as a simple control system that is equivalent to the original system. A natural equivalence of control systems is defined by feedback that includes a change of state coordinates and a state-dependent transformation of the controls [1]. Due to the invertibility of the equivalence it is possible to transfer the control problem from the original system to its normal form, design a control algorithm for the normal form, and finally transfer the algorithm back to the original system. The feedback linearization is perhaps the best known example of using normal forms in control [2]. Another instance of a normal form is the chained form that has dedicated control algorithms [3,4]. In the motion planning of space robots a normal form approach has been employed in [5,6].

In this paper we shall study the dynamics of a specific free-floating space robot, and derive feedback transformations establishing the feedback equivalence of the robot's dynamics to the chained form. This robot has been recently designed in the Space Research Center of the Polish Academy of Science, to be used to collect and remove the space debris [7]. The robot, referred to as the

A. Bartoszewicz et al. (Eds.): PCC 2020, AISC 1196, pp. 601–610, 2020.
https://doi.org/10.1007/978-3-030-50936-1_51

SRC space manipulator, consists of a planar mobile base equipped with a 2DOF planar on-board manipulator. In the paper we derive the Lagrangian equations of motion of the robot, and then concentrate on the equation resulting from the conservation of angular momentum. Depending on whether the conserved angular momentum is non-zero or zero, the dynamics of the SCR space manipulator are described either by a control-affine or a driftless control system.

The SRC space manipulator has been studied in [8] under assumption that the on-board manipulator is mounted at the center of mass of the base. Relying on this assumption, normal forms of the robot dynamics have been derived, both for the non-zero and the zero conserved angular momentum. These normal forms have been exploited in a motion planning algorithm presented in [9].

In this paper we make a more realistic assumption that the on-board manipulator has been attached at an arbitrary point of the base, and confine to the zero angular momentum of the robot. Under these two assumptions we prove that the dynamics of the SRC space manipulator can be converted by feedback to the chained form with three state variables and two controls, and provide explicitly the feedback transformations. These transformations are valid outside a set of posture singularities involving joint positions of the on-board manipulator. For comparison, we refer to the SRC space manipulator with on-board manipulator fixed at the center of mass of the base. We re-derive one of the normal forms obtained in [8] and then specify to this case the normal form invented here. It is shown that the feedback transformations in the case of arbitrary mounting point of the on-board manipulator are relatively complex.

The composition of this paper is the following. In Sect. 2 we derive the Lagrangian dynamics of the SRC space manipulator, introduce its control system representation, and define its posture singularities. Section 3 presents the SRC space manipulator with the on-board manipulator attached at the center of mass of the base. Main results of this paper, stated as Theorem 1 and Corollary 1, are presented in Sect. 4. Section 5 concludes the paper.

2 Space Robot

The SRC space manipulator is shown in Fig. 1 (a view and a schematic). Details concerned with the design of the SRC space manipulator can be found in e.g. [7]. The SRC space manipulator has been studied in [8] under assumption that the on-board manipulator has been mounted at the center of mass of the base ($a, b = 0$), see Fig. 1. Our basic concern in this paper is the SRC space manipulator with arbitrary position of the mounting point. We shall show that shifting the mounting point away of the center of mass complicates considerably the analysis of the SRC space manipulator's dynamics in comparison to the previous case.

2.1 Lagrangian

The motion of the SRC manipulator will be described by coordinates $q = (\bar{x}, \bar{y}, \phi, \theta_1, \theta_2) \in \mathbb{R}^2 \times \mathbb{T}^3$, denoting, respectively, the position of the center of mass

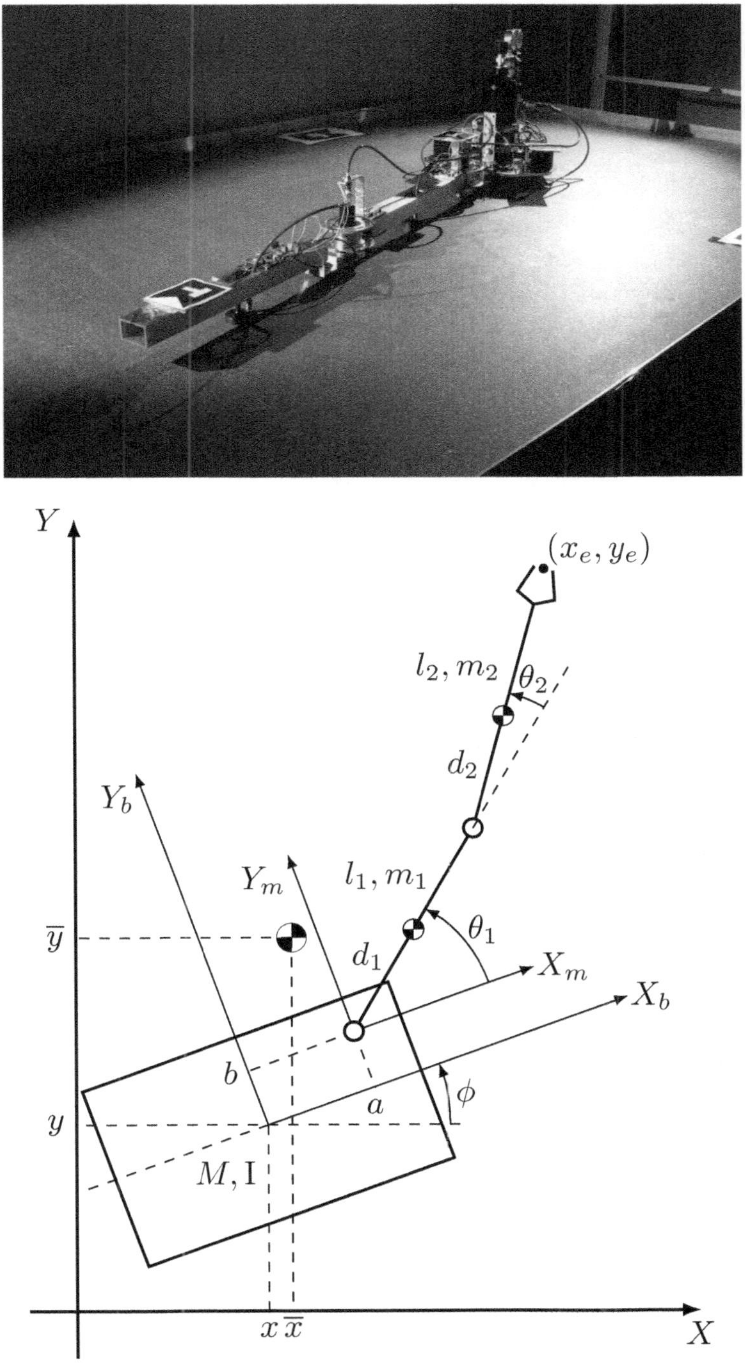

Fig. 1. SRC space manipulator

of the robot and the orientation of its base with respect to the inertial frame, and joint positions of the on-board manipulator. Assuming that the potential energy of the robot is zero due to the lack of gravity forces, the Lagrangian is equal to the robot's kinetic energy [10].

$$L(q, \dot{q}) = \frac{1}{2}m_c(\dot{\bar{x}}^2 + \dot{\bar{y}}^2) + \frac{1}{2}\mathrm{I}_P\dot{\phi}^2 + \frac{1}{2}\mathrm{A}(\dot{\phi} + \dot{\theta}_1)^2 + \frac{1}{2}\mathrm{B}(\dot{\phi} + \dot{\theta}_{12})^2$$
$$+ \mathrm{C}(\dot{\phi} + \dot{\theta}_1)(\dot{\phi} + \dot{\theta}_{12})c_2 + \mathrm{D}\dot{\phi}(\dot{\phi} + \dot{\theta}_1)(ac_1 + bs_1) + \mathrm{E}\dot{\phi}(\dot{\phi} + \dot{\theta}_{12})(ac_{12} + bs_{12}). \quad (1)$$

The meaning of symbols used in (1) is the following: M, I the mass and inertia of the base, m_i, l_i, d_i, $i = 1, 2$, the mass, length and the position of the center of mass of link i, a, b coordinates of the mounting point. We also use abbreviations

$$\theta_{12} = \theta_1 + \theta_2,\ \ s_i = \sin\theta_i,\ c_j = \cos\theta_j,\ s_{12} = \sin\theta_{12},\ c_{12} = \cos\theta_{12}, \quad (2)$$

and define

$$\begin{aligned} m_c &= M + m_1 + m_2,\ \mathrm{A} = \tfrac{m_1m_2(l_1-d_1)^2+M(m_1d_1^2+m_2l_1^2)}{m_c}, \\ \mathrm{B} &= \tfrac{(M+m_1)m_2d_2^2}{m_c},\ \mathrm{C} = \tfrac{m_1m_2(l_1-d_1)d_2+Mm_2l_1d_2}{m_c}, \mathrm{D} = \tfrac{M(m_1d_1+m_2l_1)}{m_c}, \\ \mathrm{E} &= \tfrac{Mm_2d_2}{m_c},\ \mathrm{I}_P = \mathrm{I} + \tfrac{M(m_1+m_2)(a^2+b^2)}{m_c}. \end{aligned} \quad (3)$$

The coordinates of the center of mass are computed as ($s_\phi = \sin\phi$, $c_\phi = \cos\phi$ etc)

$$\begin{aligned} \bar{x} &= \tfrac{Mx+m_1(x+ac_\phi-bs_\phi+d_1c_{\phi+\theta_1})+m_2(x+ac_\phi-bs_\phi+l_1c_{\phi+\theta_1}+d_2c_{\phi+\theta_{12}})}{m_c}, \\ \bar{y} &= \tfrac{My+m_1(y+as_\phi+bc_\phi+d_1s_{\phi+\theta_1})+m_2(y+as_\phi+bc_\phi+l_1s_{\phi+\theta_1}+d_2s_{\phi+\theta_{12}})}{m_c}. \end{aligned} \quad (4)$$

Numeric values of parameters of the SRC space manipulator have been given in [8]. As a consequence of free-floating of the robot, the Lagrangian does not depend on the position of the center of mass of the robot or the orientation of the base, resulting in the conservation of linear and angular momenta. This being so, the Euler-Lagrange equations of motion of the SRC manipulator can be written in the following form

$$m_c\dot{\bar{x}} = \mathrm{p}_1, \quad m_c\dot{\bar{y}} = \mathrm{p}_2 \quad (5)$$

$$F\dot{\phi} + G\dot{\theta}_1 + H\dot{\theta}_2 = \mathrm{p} \quad (6)$$

$$\frac{d}{dt}\frac{\partial L}{\partial \dot{\theta}_i} - \frac{\partial L}{\partial \theta_i} = \tau_i, \quad i = 1, 2. \quad (7)$$

Constants p_1, p_2 and p appearing on the right hand side of Eqs. (5) and (6) represent the conserved momenta. Equation (7) describes the dynamics of the on-board manipulator whose joints are driven by torques τ_1 and τ_2. It follows from (5) that the center of mass of the robot moves uniformly and rectilinearly, in a way completely independent of the remaining motions of the robot. Therefore, a pivotal role in the description of motion of the robot is played by Eq. (6) that yields a robot's trajectory $(\phi(t), \theta_1(t), \theta_2(t))$. Given this trajectory, Eq. (7)

just computes the required torques at the joint. This being so, we shall restrict our attention exclusively to Eq. (6) expressing the angular momentum conservation. Coefficients appearing in this equation depend on the joint positions in the following way

$$\begin{aligned} F(\theta_1,\theta_2) &= \mathrm{I}_P + \mathrm{A} + \mathrm{B} + 2\mathrm{C}c_2 + 2\mathrm{D}(ac_1 + bs_1) + 2\mathrm{E}(ac_{12} + bs_{12}), \\ G(\theta_1,\theta_2) &= \mathrm{A} + \mathrm{B} + 2\mathrm{C}c_2 + \mathrm{D}(ac_1 + bs_1) + \mathrm{E}(ac_{12} + bs_{12}), \\ H(\theta_1,\theta_2) &= \mathrm{B} + \mathrm{C}c_2 + \mathrm{E}(ac_{12} + bs_{12}). \end{aligned} \tag{8}$$

It has been checked that for parameters of the SRC space manipulator presented in [8] function F is positive for any values of a and b.

2.2 Control System

Equation (6) can be equivalently represented as a control-affine system of the following form

$$\dot\phi = \frac{\mathrm{p}}{F} - \frac{G}{F}u_1 - \frac{H}{F}u_2, \quad \dot\theta_1 = u_1, \quad \dot\theta_2 = u_2 \tag{9}$$

or as a control-affine system

$$\dot q = f(q) + g_1(q)u_1 + g_2(q)u_2, \tag{10}$$

where $q = (\phi, \theta_1, \theta_2) \in \mathbb{T}^3$, the drift vector field $f(q) = (\frac{\mathrm{p}}{F}, 0, 0)^\top$, and control vector fields $g_1(q) = (-\frac{G}{F}, 1, 0)^\top$ and $g_2(q) = (-\frac{H}{F}, 0, 1)^\top$. When the conserved angular momentum is zero (i.e. at the initial time instant neither the base nor the joints are in motion), we get $f(q) = 0$, and system (10) converts to the driftless control system

$$\dot q = g_1(q)u_1 + g_2(q)u_2 \tag{11}$$

determined by two control vector fields. Taking into account the form of the control vector fields we find that the Lie bracket $g_{12} = [g_1, g_2] = (\frac{2\sigma}{F^2}, 0, 0)^\top$, where

$$\sigma(q) = (\mathrm{BD}-\mathrm{CE})(-as_1+bc_1)+(\mathrm{CD}-\mathrm{AE})(-as_{12}+bc_{12})+(\mathrm{DE}(a^2+b^2)-\mathrm{CI}_P)s_2. \tag{12}$$

It follows that outside posture singularities

$$S_{ab} = \{(\theta_1,\theta_2)|\sigma(q) = 0\} \tag{13}$$

vector fields g_1, g_1 and g_{12} make system (11) to satisfy the Lie Algebra Rank Condition. Furthermore, it can be demonstrated that system (11) can be transformed by feedback to the chained normal form. A suitable feedback will be computed in Sect. 4.

3 The Case of $a = b = 0$

Normal forms of the SRC space manipulator carrying on board a manipulator mounted at the center of mass of the base are known, see [8]. For the sake of completeness we shall re-state them here confining to the case of zero conserved angular momentum represented by the driftless control system (11). Observe that the substitution of $a = b = 0$ yields

$$F(\theta_2) = \mathrm{I}_P + \mathrm{A} + \mathrm{B} + 2\mathrm{C}c_2,\ G(\theta_2) = \mathrm{A} + \mathrm{B} + 2\mathrm{C}c_2,\ H(\theta_2) = \mathrm{B} + \mathrm{C}c_2 \quad (14)$$

thus simplifying the control vector fields of (11). Simultaneously, the posture singularities restrict to

$$S = \{(\theta_1, \theta_2) | s_2 = 0\}. \quad (15)$$

To proceed we shall first convert system (11) to so-called pre-normal form, and then define an additional feedback transforming the pre-normal form to the chained form. Let us choose new coordinates

$$\alpha = (\mathrm{I}_P + \mathrm{A} + \mathrm{B})\phi + (\mathrm{A} + \mathrm{B})\theta_1 + \mathrm{B}\theta_2 + \mathrm{C}s_2,\ \beta = \phi + \theta_1,\ \theta_2 = \theta_2 \quad (16)$$

followed by feedback

$$w_1 = \frac{\mathrm{I}_P}{F}u_1 - \frac{H}{F}u_2,\ w_2 = u_2. \quad (17)$$

The pre-normal form of system (11) under (16) and (17) is

$$\dot{\alpha} = -2\mathrm{C}c_2 w_1,\ \dot{\beta} = w_1,\ \dot{\theta}_2 = w_2. \quad (18)$$

In the second step we apply to the pre-normal form the coordinate change

$$z_1 = \beta,\ z_2 = -2\mathrm{C}c_2,\ z_3 = \alpha \quad (19)$$

and feedback

$$v_1 = w_1,\ v_2 = 2\mathrm{C}s_2 w_2, \quad (20)$$

that results in the chained form system

$$\dot{z}_1 = v_1,\ \dot{z}_2 = v_2,\ \dot{z}_3 = z_2 v_1. \quad (21)$$

Note that the transformations (19) and (20) are valid outside the set of posture singularities (15). Having composed both these transformations we obtain the coordinate change and feedback conveying the dynamics of the SRC space manipulator to the chained form as

$$\begin{aligned} &z_1 = \phi + \theta_1,\ z_2 = -2\mathrm{C}c_2,\ z_3 = (\mathrm{I}_P + \mathrm{A} + \mathrm{B})\phi + (\mathrm{A} + \mathrm{B})\theta_1 + \mathrm{B}\theta_2 + \mathrm{C}s_2, \\ &v_1 = \frac{\mathrm{I}_P}{F}u_1 - \frac{H}{F}u_2,\ v_2 = 2\mathrm{C}s_2 u_2. \end{aligned} \quad (22)$$

4 The Case of $a \neq 0$, $b \neq 0$

We shall work outside posture singularities of the SRC space manipulator represented by the driftless control system (11) with functions F, G, H defined by (8). The following result establishes a feedback equivalence between system (11) and a chained form system.

Theorem 1. *Set*

$$P = -\mathrm{C}s_2 + \mathrm{E}(-as_{12} + bc_{12}),\ U = \mathrm{I}_P + \mathrm{A} + \mathrm{B} + 2\mathrm{D}(ac_1 + bs_1),\ Z = U - 2\mathrm{B},$$
$$f = \sqrt{\mathrm{C}^2 + 2\mathrm{CE}(ac_1 + bs_1) + \mathrm{E}^2(a^2 + b^2)},\ d = U^2 - 4f^2, \tag{23}$$

and let $\sigma = \sigma(q)$ be defined by (12). Then, in the region of $\sigma \neq 0$ and $P \neq 0$ system (11) is feedback equivalent to chained form (21) by a coordinate change

$$\begin{aligned} z_1 &= h_1(q) = \theta_1, \\ z_3 &= h_2(q) = -\mathrm{sign}(P)\phi - \frac{1}{2}\arcsin\frac{F-U}{2f} + \frac{Z}{2\sqrt{d}}\arcsin\frac{FU-d}{2fF}, \\ z_2 &= h_3(q) = \mathrm{sign}(P)\frac{G}{F} + \frac{\partial h_2(q)}{\partial \theta_1} \end{aligned} \tag{24}$$

and feedback

$$v_1 = u_1,\ v_2 = \frac{\partial h_3(q)}{\partial \theta_1}u_1 + \mathrm{sign}(P)\frac{2\sigma}{F^2}u_2. \tag{25}$$

Hereabout sign *denotes the signum function.*

Proof. We proceed in accordance with the result of Murray and Sastry [3], pp. 369–370, and associate with system (11) three distributions

$$\Delta_0 = span\{g_1, g_2, g_{12}\},\ \Delta_1 = span\{g_2, g_{12}\},\ \Delta_2 = span\{g_2\}. \tag{26}$$

It is easily checked that outside posture singularities these distributions have dimensions 3, 2 and 1, and are involutive. Using this fact we introduce coordinate functions h_1, h_2 and h_3 satisfying the following conditions

$$dh_1\Delta_1 = 0,\ dh_1 g_1 = 1,\ dh_2\Delta_2 = 0,\ h_3 = dh_2 g_1. \tag{27}$$

Taking into account the form of vector fields g_1, g_2 and g_{12} we get

$$h_1 = \theta_1,\ -H\frac{\partial h_2}{\partial \phi} + F\frac{\partial h_2}{\partial \theta_2} = 0. \tag{28}$$

The partial differential equations defining h_2 can be solved by means of the method of characteristics. Following this method we are looking for a first integral of vector field $X = (-H, 0, F)^\top$ that results in the differential equation

$$-d\phi = \frac{H}{F}d\theta_2 = \frac{\mathrm{B} + \mathrm{C}c_2 + \mathrm{E}(ac_{12} + bs_{12})}{\mathrm{I}_P + \mathrm{A} + \mathrm{B} + 2\mathrm{C}c_2 + 2\mathrm{D}(ac_1 + bs_1) + 2\mathrm{E}(ac_{12} + bs_{12})}d\theta_2 \tag{29}$$

that needs to be considered for fixed θ_1. In what follows we shall use notations defined in the statement of Theorem 1. First, observe that $F = Z+2H$, Z being independent of θ_2, so the right hand side of this equation can be written as

$$\frac{F-Z}{4F}\frac{dF}{P}, \tag{30}$$

where $P = -\mathrm{C}s_2 + \mathrm{E}(-as_{12} + bc_{12})$. Now, a crucial observation is that

$$P^2 + \frac{(F-U)^2}{4} = f^2, \tag{31}$$

f depending only on θ_1 or, equivalently,

$$P = \mathrm{sign}(P)\sqrt{f^2 - \frac{(F-U)^2}{4}}. \tag{32}$$

All these observations finally yield the following differential equation

$$-\,\mathrm{sign}(P)d\phi = \frac{1}{2}\frac{(F-Z)dF}{F\sqrt{4f^2-(F-U)^2}} \tag{33}$$

whose solution relies on computing two elementary integrals

$$I_1 = \int \frac{dF}{\sqrt{4f^2-(F-U)^2}}, \quad \text{and} \quad I_2 = \int \frac{dF}{F\sqrt{4f^2-(F-U)^2}} \tag{34}$$

and taking their combination $I_1 - ZI_2$. Finally, in accordance with the method of characteristics, we obtain the coordinate function $h_2(q)$. Having found h_2, we set $h_3 = dh_2g_1$ that establishes the coordinate change. To identify the feedback, we compute

$$\begin{aligned}
\dot z_1 &= \dot\theta_1 = u_1 = v_1,\\
\dot z_2 &= \frac{\partial h_3}{\partial \phi}\dot\phi + \frac{\partial h_3}{\partial \theta_1}u_1 + \frac{\partial h_3}{\partial \theta_2}u_2 = \frac{\partial h_3}{\partial \theta_1}u_1 + \mathrm{sign}(P)\frac{\partial}{\partial \theta_2}\left(\frac{G}{F}\right)u_2 + \frac{\partial^2 h_2}{\partial\theta_2\partial\theta_1}u_2\\
&= \frac{\partial h_3}{\partial \theta_1}u_1 + \mathrm{sign}(P)\left(\frac{\partial}{\partial \theta_2}\left(\frac{G}{F}\right) - \frac{\partial}{\partial \theta_1}\left(\frac{H}{F}\right)\right)u_2 = \frac{\partial h_3}{\partial \theta_1}u_1 + \mathrm{sign}(P)\frac{2\sigma}{F^2}u_2 = v_2,\\
\dot z_3 &= \frac{\partial h_2}{\partial \phi}\dot\phi + \frac{\partial h_2}{\partial \theta_1}u_1 + \frac{\partial h_2}{\partial \theta_2}u_2 = -\mathrm{sign}(P)\left(-\frac{G}{F}u_1 - \frac{H}{F}u_2\right) + \frac{\partial h_2}{\partial \theta_1}u_1\\
&- \mathrm{sign}(P)\frac{H}{F}u_2 = \left(\mathrm{sign}(P)\frac{G}{F} + \frac{\partial h_2}{\partial \theta_1}\right)u_1 = z_2v_1.
\end{aligned} \tag{35}$$

In computing the feedback we have used identities $\frac{\partial h_3}{\partial \phi} = 0$ and $\frac{\partial h_2}{\partial \theta_2} = -\mathrm{sign}(P)\frac{H}{F}$.

The region of well definiteness of the feedback transformation has been visualized in Fig. 2 below (right). Additionally, a plot of function F is shown (left). The inversion of coordinate change (24) can be performed in the following way. Given

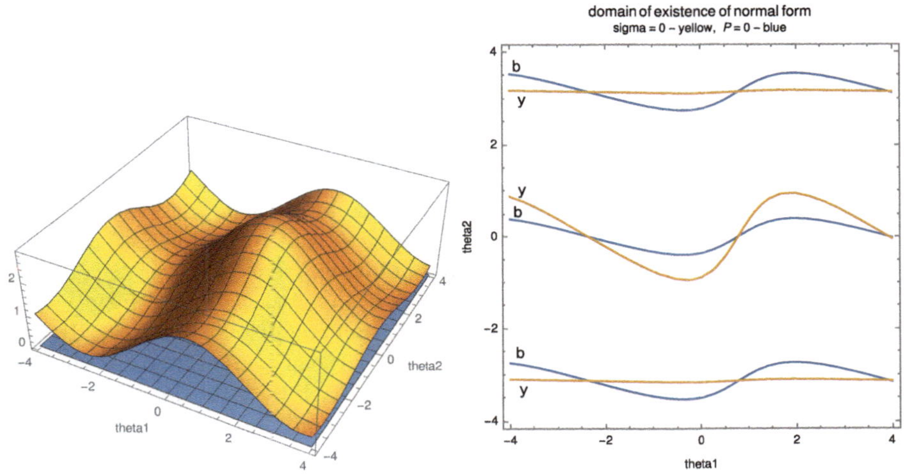

Fig. 2. Plot of function F (left) and of the domain of existence of normal form (right), $a = b = 0.2$

z_1, we directly obtain $\theta_1 = z_1$. Next, from the identity $z_2 = h_3(q)$, we compute θ_2 And finally, given θ_1 and θ_2, we find ϕ from $z_3 = h_2(q)$. By the implicit function theorem, the second identity is solvable for θ_2 on condition that

$$\frac{\partial h_3}{\partial \theta_2} = \mathrm{sign}(P)\left(\frac{\partial}{\partial \theta_2}\left(\frac{G}{F}\right) - \frac{\partial}{\partial \theta_1}\left(\frac{H}{F}\right)\right) = 2\mathrm{sign}(P)\frac{\sigma}{F^2} \neq 0 \qquad (36)$$

that is equivalent to the non-vanishing of P and σ.

In the case of zero displacement of the on-board manipulator, $a = b = 0$, we get $P = -\mathrm{C}s_2$ and $\sigma = -\mathrm{CI}_P s_2$, so the non-vanishing conditions of P and σ coincide. In consequence, the domain of existence of the chained form comprises two horizontal strips $(-\pi, 0)$ and $(0, \pi)$ in θ_2 angle. A specification of the transformations established in Theorem 1 to this case leads to another feedback equivalence of system (11) to the chained normal form. This is described by the following

Corollary 1. *Suppose that in (11) $a = b = 0$. Then, the coordinate change*

$$z_1 = \theta_1,\ z_2 = -\mathrm{sign}(s_2)\frac{G}{F},$$
$$z_3 = \mathrm{sign}(s_2)\phi - \frac{1}{4}\pi + \frac{1}{2}\theta_2 + \frac{1}{2}\frac{U - 2B}{\sqrt{U^2 - 4C^2}} \arcsin\frac{2\mathrm{C} + Uc_2}{U + 2\mathrm{C}c_2} \qquad (37)$$

and feedback

$$v_1 = u_1,\ v_2 = \frac{2\mathrm{CI}_P|s_2|}{(U + 2\mathrm{C}c_2)^2}u_2 \qquad (38)$$

convert the system to chained form (21). Hereabout F and G are defined by (14), and $U = \mathrm{I} + \mathrm{A} + \mathrm{B}$.

5 Conclusion

We have studied the dynamics of the SRC space manipulator and derived feedback transformations of the dynamics to the chained form. These transformations have resulted from an application of well known conditions of equivalence to the chained form of systems with 2 inputs. The usefulness of this normal form for the control of the SRC space manipulator requires a thorough examination, especially in view of a relatively complex feedback transformation (24) and (25), and a quite complicated shape of the domain of its existence (see Fig. 2). Besides, it has been established that in the case of $a = b = 0$ the chained form can be obtained more directly, in two steps, starting from the pre-normal form. It is an open question whether a relevant pre-normal form can be designed for the general case of $a, b \neq 0$, possibly simplifying the feedback transformations provided in this paper. Also, a normal form applicable around the zero locus of σ and P remains to be discovered. Last but not least, the case of non-zero conserved angular momentum needs to be investigated.

References

1. Jakubczyk, B.: Equivalence and invariants of nonlinear control systems. In: Sussmann, H.J. (ed.) Nonlinear Controllability and Optimal Control. M. Dekker, New York (1998)
2. Krener, A.J.: Feedback Linearization of Nonlinear Systems. In: Baillieul, J., Samad, T. (eds.) Encyclopedia of Systems and Control. Springer, London (2013)
3. Murray, R.M., Li, Z., Sastry, S.S.: A Mathematical Introduction to Robotic Manipulation. CRC Press, Boca Raton (1994)
4. Jiang, Z.P., Nijmeijer, H.: A recursive technique for tracking control of nonholonomic systems in chained form. IEEE Trans. Autom. Control **44**, 265–279 (1999)
5. Papadopoulos, E., Tortopidis, I., Nanos, K.: Smooth planning for free-floating space robots using polynomials. In: Proceedings of the 2005 IEEE International Conference on Robotics and Automation, Barcelona, Spain , pp. 4283–4288 (2005)
6. Tortopidis, I., Papadopoulos, E.: On point-to-point motion planning for underactuated space manipulator systems. Rob. Auton. Syst. **55**, 122–131 (2007)
7. Rybus, T., et al.: Application of a planar air-bearing microgravity simulator for demonstration of operations required for an orbital capture with a manipulator. Acta Astronautica **155**, 211–229 (2019)
8. Tchoń, K., Respondek, W., Ratajczak, J.: Normal forms and configuration singularities of a space manipulator. J. Intell. Rob. Syst. **93**, 621–634 (2019)
9. Tchoń, K. Ratajczak, J. Jakubiak, J.: Normal forms of robotic systems with affine Pfaffian constraints: a case study. In: Lenarcic, J., Parenti-Castelli, V. (eds) Advances in Robot Kinematics 2018, pp. 250–257. Springer, Heidelberg (2019)
10. Krzykała, A.: Modeling and Control of Free-Floating Space Robots. Wroclaw University of Science and Technology, Wroclaw (2019). Diploma project, (in Polish)

Construction of a Homogeneous Approximation

Grigorij Sklyar[1(✉)] and Svetlana Ignatovich[2]

[1] Institute of Mathematics, University of Szczecin, Wielkopolska 15, 70-451 Szczecin, Poland
sklar@univ.szczecin.pl

[2] V. N. Karazin Kharkiv National University, Svobody sqr. 4, Kharkiv 61022, Ukraine
ignatovich@ukr.net

Abstract. We discuss the concept of homogeneous approximation and describe a method for constructing a homogeneous approximation for driftless systems, which uses free algebraic approach developed by the authors in their previous papers.

Keywords: Nonlinear control system · Homogeneous approximation · Series of iterated integrals · Core Lie subalgebra

1 Introduction

The homogeneous approximation problem for nonlinear control systems attracts a great attention during several decades. Different approaches were proposed, developing various ideas from analysis, algebra, differential geometry, [1–4], which led to general descriptions of homogeneous approximations [5–11] as well as to concrete constructing algorithms including applications to different control problems.

One of the most important and widely used methods of studying qualitative properties of nonlinear systems is a method of the first approximation. It concerns differential equations arising in mechanics, in particular, when studying their stability. For example, it is well known that if the first approximation of some autonomous system is asymptotically stable, then the system itself is asymptotically stable as well. Moreover, the instability of the first approximation is also informative: if the first approximation has an exponentially growing solution, then the initial system is unstable. So, only an intermediate case cannot be studied by use of the first approximation, namely, when the first approximation is stable but not asymptotically stable and does not have an exponentially growing solution. As an example, we mention two differential equations $\dot{x} = -x^3$

The work was financially supported by Polish National Science Centre grant no. 2017/25/B/ST1/01892.

A. Bartoszewicz et al. (Eds.): PCC 2020, AISC 1196, pp. 611–624, 2020.
https://doi.org/10.1007/978-3-030-50936-1_52

and $\dot{x} = x^3$. The first one is stable while the second one is unstable at the origin. Their common linear approximation at the origin $\dot{x} = 0$ gives very few information about the initial equations.

As for controllability, another situation appears. Namely, if the first approximation of a control system is controllable, then the system itself is controllable. However, if the first approximation is uncontrollable, no conclusion about the controllability of the initial system can be made.

We explain this point more specifically, using the class of systems that are commonly studied in the field, namely, driftless systems of the form

$$\dot{x} = \sum_{i=1}^{m} u_i X_i(x), \quad x \in U(0) \subset \mathbb{R}^n, \; u_1, \ldots, u_m \in \mathbb{R}, \tag{1}$$

where $X_1(x), \ldots, X_m(x)$ are real analytic vector fields in a neighborhood of the origin. The controls $u_i(t)$ are assumed to be bounded by some pre-given constants; without loss of generality we suppose $|u_i(t)| \le 1$, $i = 1, \ldots, m$. Such systems arise in mechanics when studying kinematic models, for example. We observe that the first approximation (at the origin) of the system (1) has the form

$$\dot{x} = \sum_{i=1}^{m} u_i X_i(0).$$

It is controllable only if the number of controls equals n, i.e., $m = n$ (the least interesting case) and, additionally, the vectors $X_1(0), \ldots, X_m(0)$ are linearly independent. Clearly, this is almost never the case. Thus, the controllability property for a system of the form (1) is provided just by its nonlinearity. This distinguishes such systems from systems with drift

$$\dot{x} = X_0(x) + \sum_{i=1}^{m} u_i X_i(x)$$

arising, for example, in mechanics when considering dynamical models, for which the first approximation method can be successfully applied.

Thus, the first approximation method is an inappropriate tool for studying driftless systems (1). As an alternative to the first approximation, the following idea can be developed: to approximate a complicated nonlinear system by another *nonlinear* system which is, however, essentially simpler. Naturally, such a "simplicity" requirement is rather uncertain and should be clarified. In the control theory, one of the most important interpretations is connected with a homogeneity property.

In fact, as *simple* systems of the class (1), it is natural to consider systems with polynomial right-hand sides. However, *the simplest* systems correspond to *homogeneous* polynomials. (Maybe, a more suitable term is "quasi-homogeneous", however, we use the word "homogeneous" keeping in mind the term "homogeneous approximation".) Namely, suppose that certain integers $1 \le w_1 \le \cdots \le w_n$ are given; the polynomial system

$$\dot{x}_k = \sum_{i=1}^{m} u_i P_i^k(x), \ k = 1, \ldots, n, \tag{2}$$

is called homogeneous (at the origin) if

$$P_i^k(x) = \sum p_{i,s_1,\ldots,s_{k-1}}^k x_1^{s_1} \cdots x_{k-1}^{s_{k-1}},$$

where the sum in the right hand side is taken over a set of collections of non-negative integers $\{s_1, \ldots, s_{k-1}\}$ such that

$$w_1 s_1 + \cdots + w_{k-1} s_{k-1} = w_k - 1$$

and $p_{i,s_1,\ldots,s_{k-1}}^k \in \mathbb{R}$. (We assume $x_i^0 = 1$ and $P_i^1(x) \equiv const.$) In other words, the weight w_k is assigned to the coordinate x_k. These weights are added under coordinates' multiplication, and the weight decreases by one when differentiating.

It follows from the definition that a homogeneous system is feedforward, i.e., the right hand side of the k-th equation contains $x_1, \ldots, x_{k-1}$ only. This means that, if we know $u_i(t)$, we can find all components of the trajectory $x(t; u)$ one by one without solving differential equations, using integration only.

We mention one more advantage of such systems. Let $x = x(t; u)$ be a trajectory of the system (2) with $u = u(t) = (u_1(t), \ldots, u_m(t))$, starting at the origin, $x(0; u) = 0$. For any $\theta > 0$, let us consider a reachable set at the time θ

$$R_\theta = \{x(\theta; u) : |u_i(t)| \le 1, t \in [0, \theta], i = 1, \ldots, m\}.$$

If the system is homogeneous, R_θ satisfies the homogeneity condition

$$R_\theta = H_\theta(R_1), \quad \theta > 0,$$

where $H_\theta(x)$ is a dilation the form $H_\theta(x) = (\theta^{w_1} x_1, \ldots, \theta^{w_n} x_n)$. That is, all reachable sets are of the same shape. In particular, this allows considering a homogeneous system in the whole space $\mathbb{R}^n$ rather than in a neighborhood of the origin.

It turns out that for any system (1) there exists a homogeneous system which can be regarded to be an approximation of (1). Roughly speaking, the approximation property means that, after some change of variables in the initial system, the trajectories of the initial system and of the approximating system with the same control are close. More precisely, their k-th components differ by a value of order of smallness higher than w_k for small t. In these terms, the homogeneous approximation problem is to determine weights $w_1, \ldots, w_k$, polynomials $P_i^k(x)$ defining an approximating system, and the corresponding change of variables in the initial system. This can be done by applying differential-geometric technique. The main results were summarized in the paper of A. Bellaïche [7], where a method of finding a homogeneous approximation for control systems (1) was proposed. The approach suggests that a homogeneous approximation results after finding special "privileged" coordinates, which can be constructed step-by-step.

As a practical example, we mention the kinematic model of a car with trailers, which is well known in robotics [12]. This model, being the first nonholonomic system studied in robotics, was a source of many ideas in the control theory [13,14], in particular, for developing homogeneous approximation approach [15]. Namely, a car pulling a chain of n trailers is considered. The car has two wheels, which are controlled independently. Suppose their velocities are v_1 and v_2; it is convenient to consider $u_1 = \frac{1}{2}(v_1 + v_2)$ and $u_2 = \frac{1}{2}(v_1 - v_2)$ as controls (i.e., velocities of driving and of turning). For simplicity, let us consider the case $n = 2$, then the control system reads

$$\begin{aligned}
\dot{x}_1 &= u_1 \cos x_3, \\
\dot{x}_2 &= u_1 \sin x_3, \\
\dot{x}_3 &= u_2, \\
\dot{x}_4 &= u_1 \sin(x_3 - x_4), \\
\dot{x}_5 &= u_1 \cos(x_3 - x_4) \sin(x_4 - x_5),
\end{aligned}$$

where x_1 and x_2 are coordinates of the car, x_3 is the angle between the moving direction of the car and the x_1-axis, and x_4, x_5 are the angles between the moving directions of the two trailers and the x_1-axis respectively (in some of cited works slightly different models are considered). In solving path-planning problems, such as the parking problem, it proved to be useful to substitute this difficult nonlinear system by its homogeneous approximation [16]. It turned out that the homogeneous approximation depends on the point, thus, it could be applied only locally. Namely, in the above-mentioned example, two principally different approximations arise depending whether $x_3 - x_4$ equals $\pm\frac{\pi}{2}$ or not; the condition $x_3 - x_4 = \pm\frac{\pi}{2}$ means that the car's and the first trailer's directions are perpendicular to each other. For $n > 2$, the situation becomes much more complicated. This observation inspired deep theoretical studies in the field during the last 25 years [17–19], mainly based on differential geometric tools.

Another perspective technique was developed by the authors of the present paper. Namely, we apply the approach proposed by M. Fliess [20]; some useful discussions can be found in [21]. This allows attracting definitions and results from free algebras and "combinatorics of words" [22] to describe and classify homogeneous approximations [4,9–11,23–25]. As one of by-products of our technique, finding privileged coordinates and constructing a homogeneous approximation become two independent tasks. In the rest of the present paper we describe a coordinate-free algebraic method of finding homogeneous approximations of systems (1) based on our approach.

2 Series of Iterated Integrals: How Free Algebras Arise

Let us consider a system of the form (1), where the vector fields $X_1(x), \ldots, X_m(x)$ are real analytic in a neighborhood of the origin. We assume that admissible controls belong to the set

$$B^\theta = \{u(t) = (u_1(t), \ldots, u_m(t)) : |u_i(t)| \le 1,\ i = 1, \ldots, m,\ t \in [0, \theta]\}.$$

Then there exists $T > 0$ such that, for any $\theta \in [0, T]$ and an arbitrary $u \in B^\theta$, the trajectory $x(t; u)$ of the system (1) starting at the origin, $x(0; u) = 0$, with the control $u = u(t)$, $t \in [0, \theta]$, is well defined for $t \in [0, \theta]$. Below we consider $\theta \in [0, T]$ as an arbitrary parameter.

Let us denote by $\mathcal{E}_{X_1,\ldots,X_m}$ the operator that takes a pair (θ, u) to the end point of this trajectory, i.e., $\mathcal{E}_{X_1,\ldots,X_m}(\theta, u) = x(\theta; u)$. We call $\mathcal{E}_{X_1,\ldots,X_m}$ *the endpoint map (at the origin) of the system* (1). Below we suppose that the system (1) is controllable, what means that the set $\mathcal{E}_{X_1,\ldots,X_m}(\theta, B^\theta)$ has a nonempty interior if $\theta \in (0, T)$.

The endpoint map admits the following representation in the form of a series of iterated integrals [20],

$$\mathcal{E}_{X_1,\ldots,X_m}(\theta, u) = \sum_{k=1}^{\infty} \sum_{1 \le i_1,\ldots,i_k \le m} c_{i_1\ldots i_k} \eta_{i_1\ldots i_k}(\theta, u), \tag{3}$$

which is absolutely convergent for any $\theta \in (0, T') \subset (0, T)$ and any $u \in B^\theta$, where

$$\eta_{i_1\ldots i_k}(\theta, u) = \int_0^\theta \int_0^{\tau_1} \cdots \int_0^{\tau_{k-1}} u_{i_1}(\tau_1) u_{i_2}(\tau_2) \cdots u_{i_k}(\tau_k) d\tau_k \cdots d\tau_2 d\tau_1 \tag{4}$$

are "iterated integrals" and $c_{i_1\ldots i_k}$ are constant vector coefficients, which can be found by

$$c_{i_1\ldots i_k} = X_{i_k} X_{i_{k-1}} \cdots X_{i_1} E(0). \tag{5}$$

Here $E(x) = x$ is the identity map and the vector fields X_i are considered as differential operators of the first order acting on a vector function f as $X_i f = Df \cdot X_i$, $i = 1, \ldots, n$.

In these terms we introduce the following definition of a homogeneous approximation, which is equivalent to the definition of [7]. Namely, a controllable system

$$\dot{x} = \sum_{i=1}^{m} u_i \widehat{X}_i(x) \tag{6}$$

with the series representation

$$(\mathcal{E}_{\widehat{X}_1,\ldots,\widehat{X}_m}(\theta, u))_j = \sum_{1 \le i_1,\ldots,i_{w_j} \le m} (\widehat{c}_{i_1\ldots i_{w_j}})_j \eta_{i_1\ldots i_{w_j}}(\theta, u), \quad j = 1, \ldots, n, \tag{7}$$

where $1 \le w_1 \le \cdots \le w_n$ are integers, is called *a homogeneous approximation of the system* (1) if there exists a (analytic) change of variables $y = F(x)$ that transforms the system (1) to the system

$$\dot{y} = \sum_{i=1}^{m} u_i Y_i(y)$$

with the series representation

$$(\mathcal{E}_{Y_1,\ldots,Y_m}(\theta, u))_j = (\mathcal{E}_{\widehat{X}_1,\ldots,\widehat{X}_m}(\theta, u))_j + \rho_j(\theta, u), \quad j = 1, \ldots, n,$$

where $\rho_j(\theta,u)$ contains integrals of multiplicity greater than w_j. Since here we deal with transformations of series, let us look into them more thoroughly.

First, let us turn to iterated integrals (4). For any $\theta>0$, they are linearly independent as functionals of u, which runs through the set B^θ [20]. Hence, they form a basis of the *free* associative algebra with the "concatenation" as a product operation. In order to describe the set of iterated integrals, let us introduce m abstract independent elements $\eta_1,\dots,\eta_m$ ("letters") and consider all finite sequences of these elements, $\eta_{i_1\dots i_k}=\eta_{i_1}\cdots\eta_{i_k}$, $k\ge 1$, $1\le i_1,\dots,i_k\le m$ ("words"). All linear combinations of words with real coefficients form a free associative algebra

$$\mathcal{F}=\mathrm{Lin}\{\eta_{i_1\dots i_k}:k\ge 1,\ 1\le i_1,\dots,i_k\le m\}$$

with the concatenation as an algebraic operation, $\eta_{i_1\dots i_k}\eta_{j_1\dots j_s}=\eta_{i_1\dots i_k j_1\dots j_s}$. The linear span of iterated integrals is, therefore, a realization of such a free algebra. The algebra $\mathcal{F}$ possesses a natural grading,

$$\mathcal{F}=\sum_{k=1}^{\infty}\mathcal{F}^k,\quad \text{where } \mathcal{F}^k=\mathrm{Lin}\{\eta_{i_1\dots i_k}:1\le i_1,\dots,i_k\le m\},\quad k\ge 1,$$

which corresponds to control restriction $|u_i(t)|\le 1$, $i=1,\dots,m$, adopted above. Having in mind coefficients (5), let us define the linear operator $c:\mathcal{F}\to\mathbb{R}^n$ by equalities

$$c(\eta_{i_1\dots i_k})=c_{i_1\dots i_k},\ k\ge 1, 1\le i_1,\dots,i_k\le m.$$

In these terms, an abstract analog of the series (3) reads

$$\mathcal{E}_{X_1,\dots,X_m}=\sum_{k=1}^{\infty}\sum_{1\le i_1,\dots,i_k\le m}c(\eta_{i_1\dots i_k})\eta_{i_1\dots i_k},\tag{8}$$

where $\eta_{i_1\dots i_k}\in\mathcal{F}$. The right hand side of (8) is a *formal* series of elements of $\mathcal{F}$ with constant vector coefficients.

Now let us introduce the free graded Lie algebra generated by the same m abstract independent elements $\eta_1,\dots,\eta_m$,

$$\mathcal{L}=\sum_{k=1}^{\infty}\mathcal{L}^k,\quad \text{where } \mathcal{L}^1=\mathrm{Lin}\{\eta_1,\dots,\eta_m\},\ \mathcal{L}^{k+1}=[\mathcal{L}^1,\mathcal{L}^k],\ k\ge 1,$$

where $[\cdot,\cdot]$ denotes the Lie brackets operation, $[\ell_1,\ell_2]=\ell_1\ell_2-\ell_2\ell_1$. Let us denote by φ the anti-homomorphism from $\mathcal{L}$ to the Lie algebra of vector fields generated by $X_1(x),\dots,X_m(x)$ defined by $\varphi(\eta_i)=X_i$, $i=1,\dots,m$; therefore,

$$\varphi([\eta_{i_1},[\eta_{i_1},\cdots[\eta_{i_{k-1}},\eta_{i_k}]\cdots]])=[[\cdots[X_{i_k}(x),X_{i_{k-1}}(x)]\cdots,X_{i_2}(x)],X_{i_1}(x)]$$

for all $k\ge 2$, $1\le i_1,\dots,i_k\le m$. Then obviously $c(\ell)=\varphi(\ell)E(0)$. This connection allows us to formulate the Rashevsky-Chow condition, which means the controllability of the system, in the form

$$c(\mathcal{L})=\mathbb{R}^n,\tag{9}$$

and the realizability conditions for series of iterated integrals [26,27] as

$$\text{if } c(\ell) = 0 \text{ for some } \ell \in \mathcal{L}, \text{ then } c(a\ell) = 0 \text{ for all } a \in \mathcal{F}. \qquad (10)$$

Actually, it is sufficient to require property (10) for all $a = \eta_{i_1} \cdots \eta_{i_k}$, $k \geq 1$, $1 \leq i_1, \ldots, i_k \leq m$; together with (9) this directly leads to the realizability condition expressed via the Lie rank [26,27].

We say that $a \in \mathcal{F}$ is *homogeneous* if $a \in \mathcal{F}^k$ for some k; in this case we write $\mathrm{ord}(a) = k$. We also consider the unitary algebra $\mathcal{F}^e = \mathcal{F} + \mathbb{R}$, assuming 1 is a unit with respect of the multiplication introduced above. Now we take into account this grading in order to specify the property (10).

For the system (1), let us consider the subspaces

$$\mathcal{P}^1 = \{\ell \in \mathcal{L}^1 : c(\ell) = 0\}, \quad \mathcal{P}^k = \{\ell \in \mathcal{L}^k : c(\ell) \in c(\mathcal{L}^1 + \cdots + \mathcal{L}^{k-1})\}, \ k \geq 2.$$

In particular, $\mathcal{P}^1$ contains those elements $\sum_{i=1}^m \alpha_i \eta_i$ for which $\sum_{i=1}^m \alpha_i X_i(0) = 0$. The subspaces $\mathcal{P}^k$ can be described in the following way. It is well known [28] that values of Lie brackets of vector fields $X_1(x), \ldots, X_m(x)$ at the origin provide directions in which a trajectory $x(t;u)$ with an appropriate control can move; the length of the bracket defines the velocity of moving. In order to provide all n directions, one checks elements from $\mathcal{L}^1$ first, then from $\mathcal{L}^2$ and so on. In this sense, the subspaces $\mathcal{P}^k$ contain elements that *do not define new directions.*

Let us denote

$$\mathcal{L}_{X_1,\ldots,X_m} = \sum_{k=1}^{\infty} \mathcal{P}^k.$$

It can be shown that $\mathcal{L}_{X_1,\ldots,X_m}$ is a graded Lie subalgebra of $\mathcal{L}$ [11,24]. We call it *the core Lie subalgebra* of the system (1). The condition (9) implies that $\mathcal{L}_{X_1,\ldots,X_m}$ is of codimension n in $\mathcal{L}$. Moreover, an arbitrary graded Lie subalgebra of $\mathcal{L}$ of codimension n is the core Lie subalgebra of some system [11,24].

Let us introduce the (graded) left ideal generated by $\mathcal{L}_{X_1,\ldots,X_m}$, i.e.,

$$\mathcal{J}_{X_1,\ldots,X_m} = \mathrm{Lin}(\mathcal{F}^e \mathcal{L}_{X_1,\ldots,X_m}) = \mathrm{Lin}\{a\ell : a \in \mathcal{F}^e, \ell \in \mathcal{L}_{X_1,\ldots,X_m}\}.$$

We call $\mathcal{J}_{X_1,\ldots,X_m}$ *the left ideal induced by the system* (1). This ideal, like the core Lie subalgebra, is invariant w.r.t. changes of variables in the system. It can be shown that $\mathcal{J}_{X_1,\ldots,X_m} \cap \mathcal{L} = \mathcal{L}_{X_1,\ldots,X_m}$, hence, the left ideal $\mathcal{J}_{X_1,\ldots,X_m}$ defines the core Lie subalgebra uniquely.

The multiplication of iterated integrals (4) corresponds to the *shuffle product* in $\mathcal{F}^e$ defined as

$$\eta_{i_1\ldots i_k} \sqcup\!\sqcup\, \eta_{j_1\ldots j_r} = \eta_{i_1}(\eta_{i_2\ldots i_k} \sqcup\!\sqcup\, \eta_{j_1\ldots j_r}) + \eta_{j_1}(\eta_{i_1\ldots i_k} \sqcup\!\sqcup\, \eta_{j_2\ldots j_r}), \quad k, r \geq 1,$$

where $1 \sqcup\!\sqcup\, a = a \sqcup\!\sqcup\, 1 = a$ for any $a \in \mathcal{F}^e$. (Here we assume $\eta_{i_s\ldots i_k} = 1$ if $s > k$.) The term "shuffle" is used since two sequences $\{i_1, \ldots, i_k\}$ and $\{j_1, \ldots, j_r\}$ are *shuffled* in all possible ways; for example, $\eta_1 \sqcup\!\sqcup\, \eta_2 = \eta_{12} + \eta_{21}$, $\eta_1 \sqcup\!\sqcup\, \eta_{23} = \eta_{123} + \eta_{213} + \eta_{231}$, $\eta_{12} \sqcup\!\sqcup\, \eta_{34} = \eta_{1234} + \eta_{1324} + \eta_{1342} + \eta_{3124} + \eta_{3142} + \eta_{3412}$. One can show that

$$\eta_{i_1\ldots i_k}(\theta, u)\, \eta_{j_1\ldots j_r}(\theta, u) = (\eta_{i_1\ldots i_k} \sqcup\!\sqcup\, \eta_{j_1\ldots j_r})(\theta, u), \qquad (11)$$

where, in the right hand side, we first find the shuffle product of elements in $\mathcal{F}$ and then substitute the corresponding iterated integrals. For example, the following product

$$\eta_1(\theta,u)\cdot\eta_{23}(\theta,u)=\int_0^\theta u_1(\tau_1)d\tau_1\cdot\int_0^\theta\int_0^{\tau_2}u_2(\tau_2)u_3(\tau_3)d\tau_2 d\tau_3$$

equals the integral of $u_1(\tau_1)u_2(\tau_2)u_3(\tau_3)$ over the union of three sets described as follows: $0\le\tau_3\le\tau_2\le\tau_1\le\theta$, $0\le\tau_3\le\tau_1\le\tau_2\le\theta$, and $0\le\tau_1\le\tau_3\le\tau_2\le\theta$, which in turn correspond to three integrals $\eta_{123}(\theta,u)$, $\eta_{213}(\theta,u)$, and $\eta_{231}(\theta,u)$. Therefore,

$$\eta_1(\theta,u)\cdot\eta_{23}(\theta,u)=\eta_{123}(\theta,u)+\eta_{213}(\theta,u)+\eta_{231}(\theta,u)=(\eta_1 ⧢ \eta_{23})(\theta,u).$$

Finally, introduce the *inner product* $\langle\cdot,\cdot\rangle$ in $\mathcal{F}$ assuming the basis $\{\eta_{i_1\dots i_k}\}$ is orthonormal, i.e., $\langle\eta_{i_1\dots i_k},\eta_{j_1\dots j_r}\rangle=1$ if $k=r$ and $i_s=j_s$ for $s=1,\dots,k$ and $\langle\eta_{i_1\dots i_k},\eta_{j_1\dots j_r}\rangle=0$ otherwise.

Now we recall three facts, which are used below.

(A) Suppose $\{\ell_j\}_{j=1}^\infty$ is an arbitrary basis of $\mathcal{L}$. The Poincaré-Birkhoff-Witt Theorem [22] states that

$$\{\ell_{j_1}^{p_1}\cdots\ell_{j_s}^{p_s}:1\le j_1<\cdots<j_s,\ p_1,\dots,p_s\ge1,\ s\ge1\}\tag{12}$$

is a basis of $\mathcal{F}$, where $\ell^p=\ell\cdots\ell$ (p times). If $\{\ell_j\}_{j=1}^\infty$ is homogeneous, then elements of (12) are homogeneous as well.

(B) Consider the *dual basis* $d_{i_1\dots i_r}^{q_1\dots q_r}$ for (12), i.e., a basis of $\mathcal{F}$ such that

$$\langle\ell_{j_1}^{p_1}\cdots\ell_{j_s}^{p_s},d_{i_1\dots i_r}^{q_1\dots q_r}\rangle=\begin{cases}1 & \text{if } s=r,\ j_t=i_t, \text{ and } p_t=q_t,\ t=1,\dots,s,\\ 0 & \text{otherwise.}\end{cases}$$

It exists since elements of (12) are homogeneous and subspaces $\mathcal{F}^k$ are finite-dimensional. As is known [29], elements of the dual basis can be expressed as

$$d_{i_1\dots i_r}^{q_1\dots q_r}=\frac{1}{q_1!\cdots q_r!}\,d_{i_1}^{⧢ q_1}⧢\cdots⧢ d_{i_r}^{⧢ q_r},\tag{13}$$

where $d_i=d_i^1$, $i\ge1$, and $d^{⧢ q}=d⧢\cdots⧢ d$ (q times). Therefore, in order to find the dual basis, it is sufficient to find elements d_i^1, $i\ge1$. In the general case, d_i^1 may not belong to $\mathcal{L}$; see Sect. 5 for an example.

(C) Now let us choose a (arbitrary) homogeneous basis $\{\ell_j\}_{j=n+1}^\infty$ of the core Lie subalgebra $\mathcal{L}_{X_1,\dots,X_m}$ and let $\ell_1,\dots,\ell_n\in\mathcal{L}$ be (arbitrary) homogeneous elements complementing $\{\ell_j\}_{j=n+1}^\infty$ to the basis of $\mathcal{L}$, i.e., such that

$$\mathcal{L}=\mathrm{Lin}\{\ell_1,\dots,\ell_n\}+\mathcal{L}_{X_1,\dots,X_m}\ \text{ and }\ \mathrm{ord}(\ell_i)\le\mathrm{ord}(\ell_j)\ \text{ if }\ 1\le i<j\le n.$$

Then one can show [11] that the set

$$\{d_1^{⧢ q_1}⧢\cdots⧢ d_n^{⧢ q_n}:q_1,\dots,q_n\ge0,\ q_1+\cdots+q_n\ge1\}\tag{14}$$

forms a basis of $\mathcal{J}_{X_1,\dots,X_m}^\perp$ (the orthogonal complement of $\mathcal{J}_{X_1,\dots,X_m}$). This fact is a generalization of R. Ree's Theorem [30] about the connection of Lie elements and the shuffle product.

3 Homogeneous Approximation Described in Algebraic Terms

As follows from facts (A) and (B), the series in the right hand side of (8) can be re-expanded as

$$\mathcal{E}_{X_1,\dots,X_m} = \sum_{r=1}^{\infty} \sum_{\substack{i_1,\dots,i_r \geq 1 \\ q_1,\dots,q_r \geq 1}} \frac{1}{q_1! \cdots q_r!} c(\ell_{i_1}^{q_1} \cdots \ell_{i_r}^{q_r})\, d_{i_1}^{\sqcup\!\sqcup q_1} \sqcup\!\sqcup \cdots \sqcup\!\sqcup d_{i_r}^{\sqcup\!\sqcup q_r}. \tag{15}$$

In this expansion, if $i_r \geq n+1$, then $\ell_{i_r} \in \mathcal{L}_{X_1,\dots,X_m}$, therefore, $\ell_{i_1}^{q_1} \cdots \ell_{i_r}^{q_r} \in \mathcal{J}_{X_1,\dots,X_m}$. This means that all terms, in which $i_r \geq n+1$, cannot form a "principal part" of the series since their coefficients belong to the linear span of coefficients of elements of less order. On the other hand, if $i_r \leq n$, then $i_1 < \cdots < i_r \leq n$. Hence, the series (15) can be rewritten as

$$\mathcal{E}_{X_1,\dots,X_m} = \mathcal{S} + \mathcal{T},$$

where

$$\mathcal{S} = \sum_{\substack{q_1,\dots,q_n \geq 0 \\ q_1+\cdots+q_n \geq 1}} \frac{1}{q_1! \cdots q_n!} c(\ell_1^{q_1} \cdots \ell_n^{q_n})\, d_1^{\sqcup\!\sqcup q_1} \sqcup\!\sqcup \cdots \sqcup\!\sqcup d_n^{\sqcup\!\sqcup q_n}$$

and $\mathcal{T}$ contains all terms with $i_r \geq n+1$. This means that we may, paying no attention on $\mathcal{T}$, transform only $\mathcal{S}$.

Now recall that the shuffle product corresponds to the usual product of iterated integrals. Hence, there exists a transformation $z = F(x)$, which reduces $\mathcal{S}$ to the triangular form

$$(F(\mathcal{S}))_i = d_i + \rho_i, \quad i = 1,\dots,n,$$

where ρ_i contains terms of order greater than the order of d_i. As a result,

$$(F(\mathcal{E}_{X_1,\dots,X_m}))_i = d_i + \widehat{\rho}_i, \quad i = 1,\dots,n,$$

where $\widehat{\rho}_i$ contains terms of order greater than the order of d_i. This means that the series $\widehat{\mathcal{E}}$ such that

$$\widehat{\mathcal{E}}_i = d_i, \quad i = 1,\dots,n,$$

could be considered as a homogeneous approximation of the series $\mathcal{E}_{X_1,\dots,X_m}$. Such a transformation represents Bellaïche's privileged coordinated, which allows extracting a homogeneous approximation of the system.

However, the considerations mentioned above suggest that a homogeneous approximation can be found *without finding the transformation* F. In fact, as was shown, d_i are elements of the dual basis (13), hence, they can be found directly by a linear algebraic procedure.

The question remains, how to re-construct the system whose series coincides with $\widehat{\mathcal{E}}$. In other words, we regard $\widehat{\mathcal{E}}$ as the series (7) and look for the corresponding system (6). This can be done easily by the following procedure [11], which uses the fact (C) mentioned in the previous section.

4 Reconstructing an Approximating System

Obviously, $d_1 \in \mathcal{F}^1$, i.e., $d_1 = \sum_{i=1}^m \alpha_i \eta_i$, where $\alpha_i \in \mathbb{R}$. Hence, for $\mathcal{E}_1 = d_1$, the appropriate choice of the first components of $\widehat{X}_i$ is

$$(\widehat{X}_i(x))_1 = \alpha_i, \; i = 1, \dots, m.$$

Suppose that k first components of $\widehat{X}_i$ are chosen so that $\mathcal{E}_j = d_j$, $j = 1, \dots, k$, then k first components of the trajectory of the approximating system are

$$x_j(\theta; u) = d_j(\theta, u), \; j = 1, \dots, k. \tag{16}$$

Let us consider d_{k+1} and rewrite it as $d_{k+1} = \sum_{i=1}^m \eta_i a_i$, where $a_i \in \mathcal{F}^e$. If $a_i \in \mathbb{R}$ (i.e., $d_{k+1} \in \mathcal{F}^1$), the choice of $(\widehat{X}_i(x))_{k+1}$ is as above.

Suppose $a_i \in \mathcal{F}$. One easily proves that $a_j \in \mathcal{J}^{\perp}_{X_1,\dots,X_m}$, $j = 1, \dots, m$. In fact, for any $b \in \mathcal{J}_{X_1,\dots,X_m}$ we have

$$\langle a_j, b \rangle = \langle \eta_j a_j, \eta_j b \rangle = \langle \sum_{i=1}^m \eta_i a_i, \eta_j b \rangle = \langle d_{k+1}, b \rangle = 0,$$

since $d_{k+1} \in \mathcal{J}^{\perp}_{X_1,\dots,X_m}$. However, the set (14) is a basis of $\mathcal{J}^{\perp}_{X_1,\dots,X_m}$. Besides, a_i is of order less than d_{k+1}. Hence, a_i is a shuffle polynomial of $d_1, \dots, d_k$, i.e.,

$$d_{k+1} = \sum_{i=1}^m \eta_i P_{i,k+1}(d_1, \dots, d_k),$$

where

$$P_{i,k+1}(d_1, \dots, d_k) = \sum p^i_{q_1 \cdots q_k} d_1^{\sqcup\!\sqcup q_1} \sqcup\!\sqcup \cdots \sqcup\!\sqcup d_k^{\sqcup\!\sqcup q_k},$$

and the sum in the right hand side is taken over a set of collections of nonnegative integers $\{q_1, \dots, q_k\}$ such that $q_1 \mathrm{ord}(d_1) + \cdots + q_k \mathrm{ord}(d_k) = \mathrm{ord}(d_{k+1})$ and $p^i_{q_1 \cdots q_k} \in \mathbb{R}$. Let us show that the $(k+1)$-th component of $\widehat{X}_i(x)$ can be chosen as

$$(\widehat{X}_i(x))_{k+1} = P_{i,k+1}(x_1, \dots, x_k) = \sum p^i_{q_1 \dots q_k} x_1^{q_1} \cdots x_k^{q_k}, \quad i = 1, \dots, m. \tag{17}$$

In fact, taking into account (11) and (16) for $\theta = t$ (recall that $\theta \in [0, T]$ is arbitrary), we get

$$x_1^{q_1}(t; u) \cdots x_k^{q_k}(t; u) = (d_1^{\sqcup\!\sqcup q_1} \sqcup\!\sqcup \cdots \sqcup\!\sqcup d_k^{\sqcup\!\sqcup q_k})(t, u).$$

The definition (4) implies that for any $a \in \mathcal{F}$

$$\int_0^\theta u_i(t) a(t, u) dt = (\eta_i a)(\theta, u), \quad i = 1, \dots, m.$$

Therefore, substituting (17), we get

$$\begin{aligned}
x_{k+1}(\theta; u) &= \int_0^\theta \sum_{i=1}^m u_i(t) P_{i,k+1}(x_1(t;u), \ldots, x_k(t;u))dt \\
&= \sum_{i=1}^m \int_0^\theta u_i(t) P_{i,k+1}(d_1, \ldots, d_k)(t,u)dt \\
&= \left(\sum_{i=1}^m \eta_i P_{i,k+1}(d_1, \ldots, d_k) \right)(\theta, u) = d_{k+1}(\theta, u).
\end{aligned}$$

It remains to notice that all the constructions include only finite number of elements $\ell_1, \ldots, \ell_N$, which is a basis of the subspace $\mathcal{L}^1 + \ldots + \mathcal{L}^p$, where p is the minimal number such that $\dim c(\mathcal{L}^1 + \cdots + \mathcal{L}^p) = n$.

So, the way for constructing a homogeneous approximation for a given system (1) is as follows.

1. Find coefficients of the series (3) up to order p described above.
2. Choose a homogeneous basis $\ell_{n+1}, \ldots, \ell_N$ of the subspace $\mathcal{P}^1 + \ldots + \mathcal{P}^p$. Choose homogeneous elements $\ell_1, \ldots, \ell_n$ that complement this basis to the basis of $\mathcal{L}^1 + \ldots + \mathcal{L}^p$.
3. Find elements $d_i = d_i^1$, $i = 1, \ldots, n$, of the dual basis (13). When finding d_i, only elements from the basis (12) such that $j_1 < \cdots < j_s \le N$ may need.
4. Use the method described above to reconstruct an approximating system. This system is unique in the sense that all other approximating systems can be obtained by applying polynomial changes of variables, which keep the order.

We emphasize that this algorithm does not require any "guessing" of changes of variables. It is quite formal and includes, in fact, only manipulations of linear algebraic kind (finding a basis, finding decomposition with respect to a basis in a linear subspace etc.).

5 Example

As an illustrative example, let us consider the system

$$\dot{x}_1 = u_1, \quad \dot{x}_2 = u_2 + x_1 u_2, \quad \dot{x}_3 = \tfrac{1}{2} x_1^2 u_2 + x_2 u_2. \tag{18}$$

Obviously, this system is not homogeneous. It is clear that the term $x_1 u_2$ in the right hand side of the second equation is small as compared with the term u_2. If one drops this term, he obtains the system

$$\dot{x}_1 = u_1, \quad \dot{x}_2 = u_2, \quad \dot{x}_3 = \tfrac{1}{2} x_1^2 u_2 + x_2 u_2.$$

Now, the term $\frac{1}{2}x_1^2u_2$ in the right hand side of the third equation is small as compared with x_2u_2. Dropping it, we get the system

$$\dot{x}_1 = u_1, \quad \dot{x}_2 = u_2, \quad \dot{x}_3 = x_2u_2,$$

which is *not controllable* since $\frac{d}{dt}(x_3 - \frac{1}{2}x_2^2) = 0$. Thus, even in such a simple example the construction of a homogeneous approximation is not obvious.

Let us find a homogeneous approximation of the system (18) using the way described above. Finding coefficients (5), we get $p = 3$, since the first linearly independent coefficients of Lie brackets are $c(\eta_1) = e_1, c(\eta_2) = e_2, c([\eta_1, [\eta_1, \eta_2]]) = e_3$. This means that one can choose $\ell_1 = \eta_1$, $\ell_2 = \eta_2$, $\ell_3 = [\eta_1, [\eta_1, \eta_2]]$. Since $c([\eta_2, \eta_1]) = e_2$, $c([\eta_2, [\eta_2, \eta_1]]) = e_3$, we choose $\ell_4 = [\eta_2, \eta_1]$, $\ell_5 = [\eta_1, [\eta_1, \eta_2]] - [\eta_2, [\eta_2, \eta_1]]$. Obviously, $d_1 = \ell_1 = \eta_1$, $d_2 = \ell_2 = \eta_2$. Let us find d_3. Recall that $d_3 \in \mathcal{F}^3$ should be orthogonal to $\ell_1^3 = \eta_{111}$, $\ell_1^2\ell_2 = \eta_{112}$, $\ell_1\ell_2^2 = \eta_{122}$, $\ell_2^3 = \eta_{222}$, $\ell_1\ell_4 = \eta_{121} - \eta_{112}$, $\ell_2\ell_4 = \eta_{221} - \eta_{212}$, and $\ell_5 = \eta_{112} - 2\eta_{121} + \eta_{211} - \eta_{221} + 2\eta_{212} - \eta_{122}$, and satisfy the equality $\langle d_3, \ell_3 \rangle = 1$. These conditions can be expressed, in fact, as a system of linear equations with respect to coefficients of the representation $d_3 = \sum_{i,j,k=1}^{2} \alpha_{ijk}\eta_{ijk}$. In this way, we obtain 8 equations for 8 coefficients. Solving this system, we get, $d_3 = \eta_{211} - \eta_{221} - \eta_{212}$; note that $d_3 \notin \mathcal{L}$. The latter equation of the system, $\langle d_3, \ell_3 \rangle = 1$, is not so important to satisfy since it is sufficient to find d_i up to a nonzero constant; only the condition $\langle d_3, \ell_3 \rangle \neq 0$ is essential.

Now let us reconstruct an approximating system. Obviously, the first two equations are $\dot{x}_1 = u_1$, and $\dot{x}_2 = u_2$. Finally, since $d_3 = \eta_{211} - \eta_{221} - \eta_{212} = \eta_2(\eta_{11} - \eta_{21} - \eta_{12}) = \eta_2(\frac{1}{2}\eta_1^{⧢2} - \eta_1 ⧢ \eta_2)$, we get $\dot{x}_3 = \frac{1}{2}x_1^2u_2 - x_1x_2u_2$.

References

1. Crouch, P.E.: Solvable approximations to control systems. SIAM J. Control Optimiz. **22**, 40–54 (1984). https://doi.org/10.1137/0322004
2. Hermes, H.: Nilpotent approximations of control systems and distributions. SIAM J. Control Optimiz. **24**, 731–736 (1986). https://doi.org/10.1137/0324045
3. Agrachev, A.A., Gamkrelidze, R.V., Sarychev, A.V.: Local invariants of smooth control systems. Acta Appl. Math. **14**, 191–237 (1989). https://doi.org/10.1007/BF01307214
4. Sklyar, G.M., Ignatovich, S.Yu.: Approximation of time-optimal control problems via nonlinear power moment min-problems. SIAM J. Control Optim. **42**, 1325–1346 (2003). https://doi.org/10.1137/S0363012901398253
5. Bianchini, R.M., Stefani, G.: Graded approximation and controllability along a trajectory. SIAM J. Control Optim. **28**, 903–924 (1990). https://doi.org/10.1137/0328050
6. Hermes, H.: Nilpotent and high-order approximations of vector field systems. SIAM Rev. **33**, 238–264 (1991). https://doi.org/10.1137/1033050
7. Bellaïche, A.: The tangent space in sub-Riemannian geometry. In: Sub-Riemannian Geometry. Progress in Mathematics, vol. 144, pp. 1–78 (1996). https://doi.org/10.1007/978-3-0348-9210-0_1

8. Agrachev, A.A., Marigo, A.: Nonholonomic tangent spaces: intrinsic constructions and rigid dimensions. Electr. Res. Ann. Amer. Math. Soc. **9**, 111–120 (2003). https://doi.org/10.1090/S1079-6762-03-00118-5
9. Sklyar, G.M., Ignatovich, S.Yu.: Description of all privileged coordinates in the homogeneous approximation problem for nonlinear control systems. C. R. Math. Acad. Sci. Paris **344**, 109–114 (2007). https://doi.org/10.1016/j.crma.2006.11.016
10. Sklyar, G.M., Ignatovich, S.Yu.: Fliess series, a generalization of the Ree's theorem, and an algebraic approach to a homogeneous approximation problem. Int. J. Control **81**, 369–378 (2008). https://doi.org/10.1080/00207170701561427
11. Sklyar, G.M., Ignatovich, S.Yu.: Free algebras and noncommutative power series in the analysis of nonlinear control systems: an application to approximation problems. Dissertationes Math. (Rozprawy Mat.) **504**, 1–88 (2014). https://doi.org/10.4064/dm504-0-1
12. Laumond, J.-P.: Controllability of a multibody mobile robot. IEEE Trans. Robot. Autom. **9**, 755–763 (1993). https://doi.org/10.1109/70.265919
13. Tilbury, D., Laumond, J.-P., Murray, R., Sastry, S., Walsh, G.: Steering car-like systems with trailers using sinusoids. In: Proceedings of the 1992 IEEE International Conference on Robotics and Automation, pp. 1993–1998 (1992). https://doi.org/10.1109/ROBOT.1992.219988
14. Rouchon, P., Fliess, M., Lévine, J., Martin, P.: Flatness, motion planning and trailer systems. In: Proceedings of 32nd IEEE Conference on Decision and Control, vol. 3, pp. 2700–2705 (1993). https://doi.org/10.1109/CDC.1993.325686
15. Bellaïche, A., Jean, F., Risler, J.J.: Geometry of nonholonomic systems. In: Robot Motion Planning and Control. Lecture Notes in Control and Information Science, vol. 229, pp. 55–91 (1998). https://doi.org/10.1007/BFb0036071
16. Vendittelli, M., Laumond, J.-P., Oriolo, G.: Nilpotent approximation of nonholonomic systems with singularities: a case study. In: IFAC Proceedings Volumes, vol. 31, issue 17, pp. 751–756 (1998). https://doi.org/10.1016/S1474-6670(17)40429-0
17. Jean, F.: The car with N trailers: characterization of the singular configurations. ESIAM Control Optimis. Calc. Var. **1**, 241–266 (1996). https://doi.org/10.1051/cocv:1996108
18. Pasillas-Lépine, W., Respondek, W.: On the geometry of Goursat structures. ESIAM Control Optimis. Calc. Var. **6**, 119–181 (2001). https://doi.org/10.1051/cocv:2001106
19. Mormul, P., Pelletier, F.: Symmetries of special 2-flags. J. Singul. **21**, 187–204 (2020). https://doi.org/10.5427/jsing.2020.21k
20. Fliess, M.: Fonctionnelles causales non linéaires et indéterminées non commutatives. Bull. Soc. Math. France **109**, 3–40 (1981)
21. Kawski, M.: Nonlinear control and combinatorics of words. In: Geometry of Feedback and Optimal Control, Monographs and Textbooks in Pure and Applied Mathematics, vol. 207, pp. 305–346. Dekker, New York (1998)
22. Reutenauer, C.: Free Lie Algebras. Clarendon Press, Oxford (1993)
23. Sklyar, G.M., Ignatovich, S.Yu.: Representations of control systems in the Fliess algebra and in the algebra of nonlinear power moments. Syst. Control Lett. **47**, 227–235 (2002). https://doi.org/10.1016/S0167-6911(02)00201-3
24. Ignatovich, S.Yu.: Realizable growth vectors of affine control systems. J. Dyn. Control Syst. **15**, 557–585 (2009). https://doi.org/10.1007/s10883-009-9075-y
25. Ignatovich, S.Yu.: Normalization of homogeneous approximations of symmetric affine control systems with two controls. J. Dyn. Control Syst. **17**, 1–48 (2011). https://doi.org/10.1007/s10883-011-9109-0

26. Fliess, M.: Realizations of nonlinear systems and abstract transitive Lie algebras. Bull. Amer. Math. Soc. **2**, 444–446 (1980). https://doi.org/10.1090/S0273-0979-1980-14760-6
27. Jakubczyk, B.: Local realizations of nonlinear causal operators. SIAM J. Control Optim. **24**, 230–242 (1986). https://doi.org/10.1137/0324013
28. Jakubczyk, B.: Introduction to geometric nonlinear control; controllability and Lie bracket. Lecture Notes, Summer School on Mathematical Control Theory, Trieste, 3–28 September 2001. http://users.ictp.it/~pub_off/lectures/lns008/Jakubczyk/Jakubczyk.pdf
29. Melançon, G., Reutenauer, C.: Lyndon words, free algebras and shuffles. Canad. J. Math. **41**, 577–591 (1989). https://doi.org/10.4153/CJM-1989-025-2
30. Ree, R.: Lie elements and an algebra assotiated with shuffles. Ann. Math. **68**, 210–220 (1958). https://doi.org/10.2307/1970243

On Linearizability Conditions for Non-autonomous Control Systems

Katerina Sklyar[1(✉)] and Svetlana Ignatovich[2]

[1] Institute of Mathematics, University of Szczecin, Wielkopolska 15, 70-451 Szczecin, Poland
jekatierina.sklyar@usz.edu.pl
[2] V. N. Karazin Kharkiv National University, Svobody sqr. 4, Kharkiv 61022, Ukraine
ignatovich@ukr.net

Abstract. The paper deals with the problem of mappability of nonlinear non-autonomous control systems to linear non-autonomous systems with analytic matrices. We study the existence of non-local linearizing map of class C^2 for nonlinear systems of class C^1. In the paper *K. Sklyar, On mappability of control systems to linear systems with analytic matrices. Systems Control Lett. 134 (2019), 104572*, linearizability conditions were obtained under the additional requirement concerning existence of a non-local driftless form of the system. The goal of the present paper is to reduce this requirement.

Keywords: Linearizability problem · Non-autonomous systems · Driftless form · Non-local first integrals

1 Introduction

The linearizability problem attracts a lot of attention in the contemporary control theory. Almost fifty years ago, the first results on linearizability, [1] and [2], defined main ways for development. Namely, V. I. Korobov initialized considering special classes of nonlinear systems that are linearizable; the main achievements are connected with triangular systems [3–6]. At the same time, A. Krener suggested applying Lie brackets technique, which turns out to be an indispensable tool in the field [7–13]. As a rule, such investigations require systems that are infinitely smooth or even analytic. On the other hand, an important advantage of Korobov's approach is the possibility to consider systems under weak smoothness requirements; in fact, the class C^1 is appropriate. These ideas inspired studies of linearizability of general nonlinear systems of class C^1 [14–18].

At the same time, an absolute majority of researches on linearizability deal with autonomous systems. The linearizability problem for non-autonomous systems is more complicated, however, from the theoretical point of view it promises

The work was financially supported by Polish National Science Centre grant no. 2017/25/B/ST1/01892.

A. Bartoszewicz et al. (Eds.): PCC 2020, AISC 1196, pp. 625–637, 2020.
https://doi.org/10.1007/978-3-030-50936-1_53

nice mathematical results. As a practical example, we mention Hill's differential equation with control

$$\ddot{x} + p(t)x = u, \tag{1}$$

where $p(t)$ is a periodic function. Since such linear non-autonomous equations play an important role in different applications, a non-autonomous linearizability problem is of undoubted interest.

The present paper deals with a version of linearizability for non-autonomous systems proposed recently [19]. Our goal is to improve the linearizability conditions obtained there. In Sect. 2 we recall the definition, give some explanations, and formulate the linearizability conditions from [19]. Section 3 contains the main results of the present paper (Theorem 2 and Corollary 1). Also we give three illustrative examples.

2 Background

Let us consider single-input systems of the form

$$\dot{x} = f(t, x, u), \; x \in Q \subset \mathbb{R}^n, \; u \in \mathbb{R}, \; t \in [\alpha, \beta], \tag{2}$$

where Q is a given domain and $[\alpha, \beta]$ is a given time interval. We are interested in mapping of such a system to a linear non-autonomous system

$$\dot{z} = A(t)z + b(t)u, \tag{3}$$

by use of a change of variables assuming that this change of variables also is time-dependent, i.e., $z = F(t, x)$. One possible way is to extend the state space adding the time as an extra coordinate, $x_{n+1} = t$. Then systems (2) and (3) become autonomous but *uncontrollable* (since this new coordinate is uncontrollable). Moreover, the linear system (3) becomes *nonlinear* w.r.t. this new variable. Hence, the well known linearizability conditions for the autonomous case cannot be applied directly. We mention the following important difference between the autonomous and non-autonomous cases. As a canonical controllable autonomous linear system, the following chained form is commonly used,

$$\begin{aligned} \dot{z}_1 &= z_2, \\ &\dots \\ \dot{z}_{n-1} &= z_n, \\ \dot{z}_n &= p_1 z_1 + \cdots + p_n z_n + p_0 + u; \end{aligned} \tag{4}$$

therefore, all the diversity of linear autonomous systems is defined by at most $n + 1$ scalar parameters. On the contrary, non-autonomous linear systems are described by functional parameters; in [19] the corresponding invariants are given for linear system with real analytic matrices. Complete description of the set of invariants as well as conditions of linearizability for non-autonomous systems are open questions.

However, one class of non-autonomous nonlinear systems admits linearizability conditions close to those for the autonomous case. We mean *affine driftless systems*, i.e., systems (2) with $f(t, x, u) = b(t, x)u$. Below we explain their role in the problem under consideration.

In [19], the following simpler problem was studied: to find conditions of mappability of a nonlinear system to a *preassigned* linear non-autonomous system. We emphasize that for autonomous systems the solution of such a problem easily follows from the general result, which is unknown in the non-autonomous case.

Let us formulate the precise meaning of linearizability we adopt in the paper. Here and below sub-indexes x and t denote partial derivatives, i.e., $F_x = \partial F/\partial x$ and $F_t = \partial F/\partial t$.

Definition 1. *We say that a nonlinear control system of the form* (2) *is locally analytically mappable in the domain* Q *on the time interval* $[\alpha, \beta]$ *to a preassigned linear controllable system* (3) *where* $A(t)$, $b(t)$ *are analytic on* $[\alpha, \beta]$ *if there exists a change of variables*

$$z = F(t, x) \in C^2([\alpha, \beta] \times Q) \tag{5}$$

satisfying the condition

$$\det F_x(t, x) \neq 0, \quad (t, x) \in [\alpha, \beta] \times Q, \tag{6}$$

which reduces the system (2) *to the system* (3).

Remark 1. It is almost obvious that, if the system (2) is locally analytically mappable to a linear system, then it has an affine form

$$\dot{x} = a(t, x) + b(t, x)u, \tag{7}$$

where $a(t, x), b(t, x) \in C^1([\alpha, \beta] \times Q)$ [19]. In the rest of the present paper we consider only affine nonlinear systems (7).

Remark 2. In Definition 1, the word "analytically" means "analytically with respect to the time". The word "locally" means that the map $z = F(t, x)$ being defined in the whole domain $[\alpha, \beta] \times Q$ is locally invertible with respect to x at any $t \in [\alpha, \beta]$. We call such a map "change of variables" having in mind that this is a *local* change of variables. Such a definition allows strengthening the results, while conditions of a purely local mappability can be directly obtained as corollaries. This definition follows [14].

Remark 3. We recall that the linear system (3) is called *controllable* if for any points x^0, x^1 there exists a control $u(t)$ steering the system from x^0 to x^1 on the time interval $[\alpha, \beta]$. This is the case iff the components of the vector function $\Phi^{-1}(t)b(t)$ are linearly independent at *some* point $t \in [\alpha, \beta]$ and therefore, due to analyticity, at *any* point $t \in [\alpha, \beta]$, where $\Phi(t)$ is a fundamental matrix of the system $\dot{x} = A(t)x$. This condition is equivalent to the following one, which is used below:

$$\operatorname{rank}\{\Delta^k(t)\}_{k=0}^{\infty} = n, \quad t \in [\alpha, \beta], \tag{8}$$

where

$$\Delta^k(t) = (-A(t) + \tfrac{d}{dt})^k b(t), \;\; k \geq 0. \tag{9}$$

Remark 4. Instead of (4), *driftless* systems with $A(t) \equiv 0$ can be considered as canonical non-autonomous linear systems. For them, the controllability condition (8), (9) is simplified, namely, it includes only derivatives of $b(t)$ in t. Clearly, any analytic linear system (3) can be mapped to a driftless analytic linear system

$$\dot{z} = \widehat{g}(t)u \tag{10}$$

by use of the linear (and analytic in t) change of variables $z = \Phi^{-1}(t)x$, where $\Phi(t)$ is a fundamental matrix of the system $\dot{x} = A(t)x$. Hence, we get $\widehat{g}(t) = \Phi^{-1}(t)b(t)$. Therefore, it is sufficient to study mappability of systems (7) to linear driftless systems (10) only.

The main new contribution of the paper [19] was to transfer this idea to nonlinear systems, i.e., to consider driftless affine systems

$$\dot{y} = g(t, y)u \tag{11}$$

as a kind of a canonical form of nonlinear control systems. Any affine system (7) can be transformed to a driftless form, at least locally; this is essentially used below in Theorem 2.

In order to formulate the conditions of local analytic mappability, let us introduce the following notation. For the system (7), denote by $\mathcal{R}$ the operator acting as

$$\mathcal{R}\varphi(t, x) = \varphi_t(t, x) + \varphi_x(t, x)a(t, x) - a_x(t, x)\varphi(t, x).$$

If $a(t, x)$ and $b(t, x)$ are of class C^1, the vector field $\mathcal{R}b(t, x)$ is well defined, however, generally it is of class C^0. If $\mathcal{R}b(t, x)$ is of class C^1, the vector field $\mathcal{R}^2b(t, x)$ is well defined, and so on. Below we assume that all vector fields $\mathcal{R}^kb(t, x)$, $k = 1, \ldots, n$, exist, therefore, we can introduce the matrix

$$R(t, x) = (b(t, x), \mathcal{R}b(t, x), \ldots, \mathcal{R}^{n-1}b(t, x)).$$

Below, by $[\cdot, \cdot]$ we define Lie brackets of non-autonomous vector fields, namely, we set $[c(t, x), d(t, x)] = d_x(t, x)c(t, x) - c_x(t, x)d(t, x)$.

For the linear driftless system (10), denote

$$\widehat{K}(t) = (\widehat{g}(t), \dot{\widehat{g}}(t), \ldots, \widehat{g}^{(n-1)}(t)). \tag{12}$$

The determinant $\det \widehat{K}(t)$ is an analytic function on $[\alpha, \beta]$; it is nonzero since the system (10) is controllable. Hence, due to analyticity, the matrix $\widehat{K}(t)$ is invertible everywhere on $[\alpha, \beta]$ except maybe a finite number of points, say, $\{t_k\}_{k=1}^N$ such that $\alpha = t_0 \leq t_1 < t_2 < \cdots < t_N \leq t_{N+1} = \beta$. Then the components of the vector function $\widehat{K}^{-1}(t)\widehat{g}^{(n)}(t)$ are meromorphic on $[\alpha, \beta]$ with poles in $\{t_k\}_{k=1}^N$.

Theorem 1 ([19, Theorem 3]). *A nonlinear system* (7) *is mapped to a preassigned linear controllable system* (10) *if and only if*

(i) it is mappable to a driftless form, i.e., there exists a change of variables $y = G(t,x) \in C^2([\alpha,\beta]\times Q)$ *such that* $\det G_x(t,x) \neq 0$ *and the system* (7) *in the new variables has the form* (11);

(ii) all vector functions $\mathcal{R}b(t,x), \mathcal{R}^2 b(t,x), \dots, \mathcal{R}^n b(t,x)$ *exist, belong to the class* $C^1([\alpha,\beta]\times Q)$*, and satisfy the conditions*

$$[\mathcal{R}^k b(t,x), \mathcal{R}^j b(t,x)] = 0,\ \ k,j = 0,\dots,n-1,\quad t\in[\alpha,\beta],\quad x\in Q, \tag{13}$$

$$\operatorname{rank} R(t,x) = n,\quad t\in[\alpha,\beta]\backslash\{t_i\}_{i=1}^N,\ x\in Q, \tag{14}$$

and

$$R^{-1}(t,x)\mathcal{R}^n b(t,x) = \widehat{K}^{-1}(t)\widehat{g}^{(n)}(t),\quad t\in[\alpha,\beta]\backslash\{t_i\}_{i=1}^N,\ x\in Q. \tag{15}$$

Requirements (13) and (14) are non-autonomous versions of the autonomous linearizability conditions [10], [14] while the condition (15) is naturally connected with the preassigned linear system. On the contrary, condition (i) seems to be of technical kind. It was essentially used in the proof of the theorem in [19], however, the question remains if it follows from the other conditions of the theorem. The main problem with condition (i) is that it hardly can be checked by definition.

The goal of the present paper is to show that condition (i) can be omitted under the additional requirement $a(t,x)\in C^2([\alpha,\beta]\times Q)$.

3 Main Results

Theorem 2. *Let a nonlinear system* (7) *be given,* $a(t,x)\in C^2([\alpha,\beta]\times Q)$, $b(t,x)\in C^1([\alpha,\beta]\times Q)$. *It is locally analytically mappable in the domain* Q *on the time interval* $[\alpha,\beta]$ *to a preassigned linear controllable system* (10) *if and only if all vector functions* $\mathcal{R}b(t,x), \mathcal{R}^2 b(t,x), \dots, \mathcal{R}^n b(t,x)$ *exist, belong to the class* $C^1([\alpha,\beta]\times Q)$*, and satisfy the conditions* (13), (14), (15).

Proof. Necessity follows from Theorem 1. Let us prove *sufficiency.*

Suppose that $a(t,x)\in C^2([\alpha,\beta]\times Q)$ and $b(t,x)\in C^1([\alpha,\beta]\times Q)$, the vector functions $\mathcal{R}b(t,x),\dots,\mathcal{R}^n b(t,x)$ exist and belong to the class $C^1([\alpha,\beta]\times Q)$, and conditions (13), (14), (15) hold. We are going to find a change of variables (5) satisfying the condition (6) which reduces the system (2) to the system (10). Our construction includes three steps.

(A) First, we consider an arbitrary point $(\widetilde{t},\widetilde{x})\in[\alpha,\beta]\times Q$. Since $a(t,x)\in C^2([\alpha,\beta]\times Q)$, the system without control

$$\dot{x} = a(t,x) \tag{16}$$

locally, in a neighborhood of the point $(\widetilde{t},\widetilde{x})$, has n independent first integrals of class C^2; this can be proved by use of the straightening theorem [20, Chapter 2, § 7, § 10]. Thus, there exists a neighborhood $\widetilde{U} = U(\widetilde{t},\widetilde{x}) = [\widetilde{\alpha},\widetilde{\beta}]\times\widetilde{Q}$ such that the system (16) has n independent first integrals $\Phi_1(t,x),\dots,\Phi_n(t,x)\in C^2(\widetilde{U})$. Let us consider the change of variables $y = \Phi(t,x) = (\Phi_1(t,x),\dots,\Phi_n(t,x))^{\mathsf{T}}$.

Then $\det \Phi_x(t,x) \neq 0$, therefore, the transformation $y = \Phi(t,x)$ is locally invertible w.r.t. x for any fixed $t \in [\widetilde{\alpha}, \widetilde{\beta}]$. Since $\Phi_j(t,x)$ are first integrals of (16), we get $\Phi_t(t,x) + \Phi_x(t,x)a(t,x) = 0$ for any $(t,x) \in \widetilde{U}$. Therefore, applying the change of variables $y = \Phi(t,x)$ to the system (7), we get

$$\dot{y} = g(t,y)u, \quad g(t,y) = \Phi_x(t,x)b(t,x)|_{x=\Phi^{-1}(t,x)}.$$

This means that the system (7) is mappable to a driftless form in $\widetilde{U}$.

Now, let us apply Theorem 1 in the domain $\widetilde{U} = [\widetilde{\alpha}, \widetilde{\beta}] \times \widetilde{Q}$, Condition (i) is just proved and condition (ii) follows from our assumptions, since equalities (13)–(15) are satisfied in the wider domain $[\alpha, \beta] \times Q \supset \widetilde{U}$. Therefore, due to Theorem 1, there exists a change of variables

$$z = \widetilde{F}(t,x) \in C^2(\widetilde{U}) \tag{17}$$

such that

$$\det \widetilde{F}_x(t,x) \neq 0, \quad (t,x) \in \widetilde{U}, \tag{18}$$

which reduces the system (7) to the system (10). This implies

$$\widetilde{F}_x(t,x)b(t,x) = \widehat{g}(t), \quad (t,x) \in \widetilde{U}, \tag{19}$$

$$\widetilde{F}_t(t,x) + \widetilde{F}_x(t,x)a(t,x) = 0, \quad (t,x) \in \widetilde{U}. \tag{20}$$

Let us prove that the vector functions $\mathcal{R}^k b(t,x)$ exist for all $k \geq 0$ and satisfy the equalities

$$\widetilde{F}_x(t,x)\mathcal{R}^k b(t,x) = \widehat{g}^{(k)}(t), \quad k \geq 0, \quad (t,x) \in \widetilde{U}. \tag{21}$$

We argue by induction on k. For $k = 0$, (21) coincides with (19). Now suppose that (21) holds for some $k \geq 0$ and $\mathcal{R}^k b(t,x) \in C^1(\widetilde{U})$. Since the right hand side of (21) depends only on t, we get from (21)

$$\begin{aligned}&(\widetilde{F}_x(t,x)\mathcal{R}^k b(t,x))_x a(t,x)\\ &= \widetilde{F}_{xx}(t,x)a(t,x)\mathcal{R}^k b(t,x) + \widetilde{F}_x(t,x)(\mathcal{R}^k b(t,x))_x a(t,x) = 0,\end{aligned} \tag{22}$$

$$(\widetilde{F}_x(t,x)\mathcal{R}^k b(t,x))_t = \widetilde{F}_{xt}(t,x)\mathcal{R}^k b(t,x) + \widetilde{F}_x(t,x)(\mathcal{R}^k b(t,x))_t = \widehat{g}^{(k+1)}(t). \tag{23}$$

On the other hand, (20) gives

$$\begin{aligned}&\widetilde{F}_{tx}(t,x)\mathcal{R}^k b(t,x) + (\widetilde{F}_x(t,x)a(t,x))_x \mathcal{R}^k b(t,x)\\ &= \widetilde{F}_{tx}(t,x)\mathcal{R}^k b(t,x) + \widetilde{F}_{xx}(t,x)\mathcal{R}^k b(t,x)a(t,x) + \widetilde{F}_x(t,x)a_x(t,x)\mathcal{R}^k b(t,x) = 0.\end{aligned} \tag{24}$$

Adding (22) and (23), subtracting (24), and taking into account equalities of mixed partials of $\widetilde{F}(t,x) \in C^2(\widetilde{U})$, we get

$$\widetilde{F}_x(t,x)\left((\mathcal{R}^k b(t,x))_t + (\mathcal{R}^k b(t,x))_x a(t,x) - a_x(t,x)\mathcal{R}^k b(t,x)\right) = \widehat{g}^{(k+1)}(t),$$

therefore,

$$\widetilde{F}_x(t,x)\mathcal{R}^{k+1}b(t,x) = \widehat{g}^{(k+1)}(t).$$

Since $\widetilde{F}_x(t,x) \in C^1(\widetilde{U})$ is invertible, we get

$$\mathcal{R}^{k+1}b(t,x) = (\widetilde{F}_x(t,x))^{-1}\widehat{g}^{(k+1)}(t) \in C^1(\widetilde{U}).$$

Applying induction arguments, we get that all vector functions $\mathcal{R}^k b(t,x)$ exist, belong to the class $C^1(\widetilde{U})$, and satisfy equalities (21).

Since all these results are obtained for any $(\widetilde{t},\widetilde{x}) \in [\alpha,\beta] \times Q$, we have proved that for all $k \geq 0$ the vector functions $\mathcal{R}^k b(t,x)$ exist and belong to the class $C^1([\alpha,\beta] \times Q)$. Moreover, (21) implies that for any $m_1,\ldots,m_n \geq 0$ and any $(t,x) \in [\alpha,\beta] \times Q$

$$\operatorname{rank}(\mathcal{R}^{m_1}b(t,x),\ldots,\mathcal{R}^{m_n}b(t,x)) = \operatorname{rank}(\widehat{g}^{(m_1)}(t),\ldots,\widehat{g}^{(m_n)}(t)). \quad (25)$$

Recall that we are looking for a change of variables $F(t,x)$ defined in the whole domain $[\alpha,\beta] \times Q$. In part (B) we find changes of variables existing within a finite number of "strips" and in part (C) we show how to glue them. To this end, we essentially use that all $\mathcal{R}^k b(t,x)$ exist and satisfy equalities (25).

(B) By the assumption of Theorem 2, $\det R(t,x) \neq 0$ for all $t \in [\alpha,\beta] \backslash \{t_i\}_{i=1}^N$, $x \in Q$. Let us consider the points $t_1,\ldots,t_N$. By our supposition, the linear system (10) is controllable. Applying the condition (8), (9) to the system (10) we get

$$\operatorname{rank}\{\widehat{g}^{(k)}(t)\}_{k=0}^{\infty} = n, \quad t \in [\alpha,\beta].$$

Therefore, for any t_i there exists a set of integers $(m_{i1},\ldots,m_{in})$ such that

$$\det(\widehat{g}^{(m_{i1})}(t_i),\ldots,\widehat{g}^{(m_{in})}(t_i)) \neq 0.$$

Then there exists an interval $(t_i - \varepsilon_i, t_i + \varepsilon_i)$ such that

$$\det(\widehat{g}^{(m_{i1})}(t),\ldots,\widehat{g}^{(m_{in})}(t)) \neq 0, \quad t \in (t_i - \varepsilon_i, t_i + \varepsilon_i),$$

therefore, due to (25),

$$\det(\mathcal{R}^{m_{i1}}b(t,x),\ldots,\mathcal{R}^{m_{in}}b(t,x)) \neq 0, \quad (t,x) \in (t_i - \varepsilon_i, t_i + \varepsilon_i) \times Q. \quad (26)$$

For simplicity let us suppose $t_1 > \alpha$ and $t_N < \beta$ (if one of the latter inequalities is not true, the further proof can be easily modified). Without loss of generality we assume

$$t_{i-1} < t_i - \varepsilon_i < t_i + \varepsilon_i < t_{i+1}, \quad i = 1,\ldots,N,$$

and

So, we have a sequence of intervals, (t_0,t_1), $(t_1 - \varepsilon_1, t_1 + \varepsilon_1)$, (t_1,t_2), $\ldots$, $(t_N - \varepsilon_N, t_N + \varepsilon_N)$, (t_N,t_{N+1}), and each two successive intervals intersect. For the sake of convenience, let us denote them by (α_i,β_i), namely,

$$\alpha_{2j+1} = t_j, \quad \beta_{2j+1} = t_{j+1}, \quad j = 0,\ldots,N,$$
$$\alpha_{2j} = t_j - \varepsilon_j, \quad \beta_{2j} = t_j + \varepsilon_j, \quad j = 1,\ldots,N.$$

Let us introduce the domains $U^1 = [\alpha_1, \beta_1) \times Q$, $U^{2N+1} = (\alpha_{2N+1}, \beta_{2N+1}] \times Q$, and $U^{2j+1} = (\alpha_{2j+1}, \beta_{2j+1}) \times Q$, $j = 1, \ldots, N-1$. Now we are going to define a change of variables in each domain U^{2j+1}. To this end, for any $p = 1, \ldots, n$ in each such domain we consider a system of partial differential equations

$$\begin{aligned} &\varphi_x(t,x)\mathcal{R}^r b(t,x) = \widehat{g}_p^{(r)}(t), \ r = 0, \ldots, n-1, \\ &\varphi_t(t,x) + \varphi_x(t,x)a(t,x) = 0 \end{aligned} \tag{27}$$

(where $\widehat{g}_p^{(r)}(t)$ is the p-th component of the vector function $\widehat{g}^{(r)}(t)$).

Analogously, we introduce domains $U^{2j} = (\alpha_{2j}, \beta_{2j}) \times Q$, $j = 1, \ldots, N$. In order to define a change of variables in U^{2j}, for any $p = 1, \ldots, n$ in each domain U^{2j} we consider a system of partial differential equations

$$\begin{aligned} &\varphi_x(t,x)\mathcal{R}^{m_{jr}} b(t,x) = \widehat{g}_p^{(m_{jr})}(t), \ r = 1, \ldots, n, \\ &\varphi_t(t,x) + \varphi_x(t,x)a(t,x) = 0. \end{aligned} \tag{28}$$

As was shown in part (A), each of these systems is locally solvable in the corresponding domain. Actually, $\widetilde{F}(t,x)$ are their solutions at any neighborhood $\widetilde{U}$, what follows from equalities (20) and (21) for $k = 0, \ldots, n-1$ or for $k = m_{j1}, \ldots, m_{jn}$ respectively. Since any domain $U^1, \ldots, U^{2N+1}$ is simply connected, each of these systems has a solution of the class C^2 in the whole domain due to the Frobenius Theorem [21, Chapter VI]. For any $i = 1, \ldots, 2N+1$ and any $p = 1, \ldots, n$, let us denote by $\varphi_{ip}(t,x)$ the solution of the corresponding system in the domain U^i. We consider vector functions $F^i(t,x) = (\varphi_{i1}(t,x), \ldots, \varphi_{in}(t,x))^{\mathsf{T}}$; due to (14) and (26), $F^i_x(t,x)$ is locally invertible. Moreover, analogously to part (A) it can be shown that $F^i(t,x)$ satisfies all the equalities

$$F^i_x(t,x)\mathcal{R}^k b(t,x) = \widehat{g}^{(k)}(t), \quad k \geq 0, \quad (t,x) \in U^i. \tag{29}$$

This means that each map $z = F^i(t,x)$ is an appropriate change of variables defined in U^i. Actually, for any $i = 1, \ldots, 2N+1$ we have

$$\begin{aligned} &F^i_x(t,x)b(t,x) = \widehat{g}(t), \quad (t,x) \in U^i, \\ &F^i_t(t,x) + F^i_x(t,x)a(t,x) = 0, \quad (t,x) \in U^i, \end{aligned}$$

therefore,

$$\dot{z} = F^i_t(t,x) + F^i_x(t,x)a(t,x) + F^i_x(t,x)b(t,x)u = \widehat{g}(t)u.$$

(C) Let us consider two domains, $U^1 = [\alpha_1, \beta_1) \times Q$ and $U^2 = (\alpha_2, \beta_2) \times Q$, with nonempty intersection $U^1 \cap U^2 = (\alpha_2, \beta_1) \times Q$. Then, due to (29), for any $(t,x) \in U^1 \cap U^2$ we have

$$F^1_x(t,x)\mathcal{R}^k b(t,x) = F^2_x(t,x)\mathcal{R}^k b(t,x), \ k = 0, \ldots, n-1,$$

$$F^1_t(t,x) + F^1_x(t,x)a(t,x) = F^2_t(t,x) + F^2_x(t,x)a(t,x).$$

Due to the condition (14), this implies

$$(F^1(t,x) - F^2(t,x))_x = 0, \quad (F^1(t,x) - F^2(t,x))_t = 0, \quad (t,x) \in U^1 \cap U^2.$$

Since $U^1 \cap U^2$ is connected, we get

$$F^1(t,x) = F^2(t,x) + C_1, \quad (t,x) \in U^1 \cap U^2,$$

where C_1 is a constant vector. Hence, the vector function

$$\widehat{F}^2(t,x) = \begin{cases} F^1(t,x), & (t,x) \in U^1, \\ F^2(t,x) + C_1, & (t,x) \in U^2 \setminus U^1, \end{cases}$$

is defined in the union of two domains $U^1 \cup U^2$ and is of class $C^2(U^1 \cup U^2)$.

Further, we consider $F^3(t,x)$, which is defined in $U^3 = (\alpha_3, \beta_3) \times Q$. Analogously,

$$\widehat{F}^2(t,x) = F^3(t,x) + C_2, \quad (t,x) \in (U^1 \cup U^2) \cap U^3,$$

where C_2 is a constant vector. Hence, the change of variables

$$\widehat{F}^3(t,x) = \begin{cases} \widehat{F}^2(t,x), & (t,x) \in U^1 \cup U^2, \\ F^3(t,x) + C_2, & (t,x) \in U^3 \setminus (U^1 \cup U^2), \end{cases}$$

is defined in the domain $U^1 \cup U^2 \cup U^3$ and is of class $C^2(U^1 \cup U^2 \cup U^3)$.

Continuing in such a way, after $2N$ steps we obtain the change of variables $z = F(t,x) = \widehat{F}^{2N+1}(t,x)$ defined in $U^1 \cup \cdots \cup U^{2N+1} = [\alpha, \beta] \times Q$ of class $C^2([\alpha,\beta] \times Q)$, which is locally invertible and satisfies the equalities

$$F_x(t,x)b(t,x) = \widehat{g}(t), \quad F_t(t,x) + F_x(t,x)a(t,x) = 0.$$

Therefore, it maps the system (7) to the system (10).

The theorem is proved.

Taking into account Remark 4, we easily obtain the following

Corollary 1. *Let a nonlinear system* (7) *be given,* $a(t,x) \in C^2([\alpha,\beta] \times Q)$, $b(t,x) \in C^1([\alpha,\beta] \times Q)$. *It is locally analytically mappable in the domain* Q *on the time interval* $[\alpha,\beta]$ *to a preassigned linear controllable system* (3) *if and only if all vector functions* $\mathcal{R}b(t,x), \mathcal{R}^2 b(t,x), \ldots, \mathcal{R}^n b(t,x)$ *exist, belong to the class* $C^1([\alpha,\beta] \times Q)$, *and satisfy the conditions* (13), (14), *and*

$$R^{-1}(t,x)\mathcal{R}^n b(t,x) = K^{-1}(t)\Delta^n(t), \quad t \in [\alpha,\beta] \setminus \{t_i\}_{i=1}^N, \ x \in Q,$$

where $K(t) = (\Delta^0(t), \ldots, \Delta^{n-1}(t))$, *vector functions* $\Delta^k(t)$ *are defined by* (9), *and* $\{t_i\}_{i=1}^N$ *are those points at which the matrix* $K(t)$ *is singular.*

As was shown in [19], the vector function $K^{-1}(t)\Delta^n(t)$ is invariant with respect to linear non-autonomous changes of variables in the system (3). It is an interesting and open problem to describe all such vector functions.

Example 1. Let us consider an oscillator with nonlinear and time-dependent dissipation,

$$\ddot{x} + k(t,\dot{x}) + x = u,$$

which corresponds to the system

$$\begin{aligned}\dot{x}_1 &= x_2,\\ \dot{x}_2 &= -x_1 - k(t, x_2) + u.\end{aligned} \tag{30}$$

Let us consider the following question: could a system (30) be linearizable in the sense of Definition 1 for *some* function $k(t, x_2)$? For the system (30) we have

$$a(t,x) = \begin{pmatrix} x_2 \\ -x_1 - k(t,x_2) \end{pmatrix}, \quad b(t,x) = \begin{pmatrix} 0 \\ 1 \end{pmatrix}, \quad \mathcal{R}b(t,x) = \begin{pmatrix} -1 \\ k_{x_2}(t,x_2) \end{pmatrix}.$$

Therefore, the condition $[b(t,x), \mathcal{R}b(t,x)] = 0$ holds only if $k_{x_2x_2}(t,x_2) = 0$, i.e., if $k(t,x_2)$ is linear in x_2. This means that all systems of the form (30), except of linear ones, are not linearizable in the sense of Definition 1.

Example 2. Let us consider the following system

$$\begin{aligned}\dot{x}_1 &= x_2 + p(t)(\tfrac{1}{4}x_2^7 - x_1x_2^2|x_2|) + x_2^2|x_2|u,\\ \dot{x}_2 &= p(t)(\tfrac{1}{4}x_2^3|x_2| - x_1) + u,\end{aligned} \tag{31}$$

where $p(t)$ is a real analytic function. In this case

$$a(t,x) = \begin{pmatrix} x_2 + p(t)(\frac{1}{4}x_2^7 - x_1x_2^2|x_2|) \\ p(t)(\frac{1}{4}x_2^3|x_2| - x_1) \end{pmatrix}, \quad b(t,x) = \begin{pmatrix} x_2^2|x_2| \\ 1 \end{pmatrix}$$

are of the class $C^2(\mathbb{R}^3)$,

$$\mathcal{R}b(t,x) = \begin{pmatrix} -1 \\ 0 \end{pmatrix}, \quad \mathcal{R}^2b(t,x) = \begin{pmatrix} -p(t)x_2^2|x_2| \\ -p(t) \end{pmatrix}.$$

Therefore, $[b(t,x), \mathcal{R}b(t,x)] = 0$ and

$$R(t,x) = \begin{pmatrix} x_2^2|x_2| & -1 \\ 1 & 0 \end{pmatrix}, \quad R^{-1}(t,x)\mathcal{R}^2b(t,x) = \begin{pmatrix} -p(t) \\ 0 \end{pmatrix}.$$

It is not hard to find a linear system (3) such that $K^{-1}(t)\Delta^2(t) = (-p(t), 0)^{\mathsf{T}}$; for example, we can take

$$\begin{aligned}\dot{z}_1 &= z_2,\\ \dot{z}_2 &= -p(t)z_1 + u,\end{aligned} \tag{32}$$

which corresponds to the Eq. (1). Thus, all the conditions of Theorem 2 are satisfied, and therefore the system (31) is locally analytically mappable (in any domain $Q \subset \mathbb{R}^2$ on any time interval $[\alpha, \beta] \subset \mathbb{R}$) to the system (32). Actually, the linearizing map is $z_1 = x_1 - \frac{1}{4}x_2^3|x_2|$, $z_2 = x_2$.

Example 3. As an illustrative example of greater dimension, let us consider the system

$$\begin{aligned}\dot{x}_1 &= u,\\ \dot{x}_2 &= -\tfrac{1}{3}x_4^2|x_4|\sin t + (t + x_4|x_4|\cos^2 t)u,\\ \dot{x}_3 &= \tfrac{1}{3}x_4^2|x_4|\cos t + (\sin t + x_4|x_4|\sin t\cos t)u,\\ \dot{x}_4 &= \cos t u.\end{aligned} \tag{33}$$

Here

$$a(t,x) = \begin{pmatrix} 0 \\ -\frac{1}{3}x_4^2|x_4|\sin t \\ \frac{1}{3}x_4^2|x_4|\cos t \\ 0 \end{pmatrix}, \quad b(t,x) = \begin{pmatrix} 1 \\ t + x_4|x_4|\cos^2 t \\ \sin t + x_4|x_4|\sin t\cos t \\ \cos t \end{pmatrix},$$

i.e., $a(t,x) \in C^2(\mathbb{R}^5)$, $b(t,x) \in C^1(\mathbb{R}^5)$. Analogously to the previous example, one can show that it satisfies the conditions (13) and (14) and $R^{-1}(t,x)\mathcal{R}^4 b(t,x) = (0,0,-1,0)^\mathsf{T}$. Since the driftless system

$$\dot{x}_1 = u, \quad \dot{x}_2 = t\,u, \quad \dot{x}_3 = \sin t\,u, \quad \dot{x}_4 = \cos t\,u \tag{34}$$

satisfies the condition $\widehat{K}^{-1}(t)\widehat{g}^{(4)}(t) = (0,0,-1,0)^\mathsf{T}$, we conclude that the system (33) is locally analytically mappable (in any domain $Q \subset \mathbb{R}^4$ on any time interval $[\alpha,\beta] \subset \mathbb{R}$) to the system (34). In this case the linearizing map is $z_1 = x_1$, $z_2 = x_2 - \frac{1}{3}x_4^2|x_4|\cos t$, $z_3 = x_3 - \frac{1}{3}x_4^2|x_4|\sin t$, $z_4 = x_4$.

4 Conclusion

The paper is devoted to the linearizability problem in the non-autonomous case, which has been studied much less than the autonomous case, though linear non-autonomous systems are important models for applications. Namely, conditions for mappability of an affine system (7) of class C^1 to a preassigned analytic controllable linear system (10) are studied. A linearizing change of variables $z = F(t,x)$ is supposed to be locally invertible and of class $C^2([\alpha,\beta] \times Q)$, where the domain $[\alpha,\beta] \times Q$ is pre-given. In the paper [19] such conditions were proposed under the requirement that the system (7) was mappable to a driftless form (11) in the whole domain $[\alpha,\beta] \times Q$; a transformation, which maps to a driftless form, also was supposed to be of class $C^2([\alpha,\beta] \times Q)$. In the general case of systems (7) of class C^1, such a transformation exists locally and is of class C^1; its non-local existence and additional smoothness degree are conditions that hardly can be checked without explicit finding the transformation itself. Thus, the question arises if these conditions follow from other suppositions of the theorem, or if there are some other sufficient conditions that can be verified easily. In the present paper we have shown that the mentioned requirement can be omitted if $a(t,x)$ is of class C^2. In the case $a(t,x) \in C^1$ the question is still open.

References

1. Korobov, V.I.: Controllability, stability of some nonlinear systems. Differ. Uravnenija **9**, 614–619 (1973). (in Russian)
2. Krener, A.: On the equivalence of control systems and the linearization of nonlinear systems. SIAM J. Control **11**, 670–676 (1973). https://doi.org/10.1137/0311051

3. Celikovsky, S.: Global linearization of nonlinear systems – a survey. In: Geom. in Nonlin. Control and Diff. Incl., Warszawa, pp. 123–137 (1995). https://doi.org/10.4064/-32-1-123-137
4. Celikovsky, S., Nijmeijer, H.: Equivalence of nonlinear systems to triangular form: the singular case. Syst. Control Lett. **27**, 135–144 (1996). https://doi.org/10.1016/0167-6911(95)00059-3
5. Korobov, V.I., Pavlichkov, S.S.: Global properties of the triangular systems in the singular case. J. Math. Anal. Appl. **342**, 1426–1439 (2008). https://doi.org/10.1016/j.jmaa.2007.12.070
6. Korobov, V.I., Sklyar, K.V., Skoryk, V.O.: Stepwise synthesis of constrained controls for single input nonlinear systems of special form. Nonlinear Differ. Equ. Appl. 23–31 (2016). https://doi.org/10.1007/s00030-016-0385-y
7. Brockett, R.W.: Feedback invariance for nonlinear systems. In: Proceedings of the Seventh World Congress IFAC, Helsinki, pp. 1115–1120 (1978)
8. Jakubczyk, B., Respondek, W.: On linearization of control systems. Bull. Acad. Sci. Polonaise Ser. Sci. Math. **28**, 517–522 (1980)
9. Su, R.: On the linear equivalents of nonlinear systems. Syst. Control Lett. **2**, 48–52 (1982). https://doi.org/10.1016/S0167-6911(82)80042-X
10. Respondek, W.: Geometric methods in linearization of control systems. In: Mathematical Control Theory, vol. 14, pp. 453–467. Banach Center Publication, PWN, Warsaw (1985)
11. Respondek, W.: Linearization, feedback and Lie brackets. In: Scientific Papers of the Institute of Technical Cybernetics of the Technical University of Wroclaw, no. 70, Conf. 29, pp. 131–166 (1985)
12. Nicolau, F., Respondek, W.: Flatness of multi-input control-affine systems linearizable via one-fold prolongation. SIAM J. Control Optim. **55**(5), 3171–3203 (2017). https://doi.org/10.1137/140999463
13. Li, S., Moog, C.H., Respondek, W.: Maximal feedback linearization and its internal dynamics with applications to mechanical systems on $\mathbb{R}^4$. Int. J. Robust Nonlinear Control **29**(9), 2639–2659 (2019). https://doi.org/10.1002/rnc.4507
14. Sklyar, G.M., Sklyar, K.V., Ignatovich, S.Y.: On the extension of the Korobov's class of linearizable triangular systems by nonlinear control systems of the class C^1. Syst. Control Lett. **54**, 1097–1108 (2005). https://doi.org/10.1016/j.sysconle.2005.04.002
15. Sklyar, K.V., Ignatovich, S.Y., Skoryk, V.O.: Conditions of linearizability for multi-control systems of the class C^1. Commun. Math. Anal. **17**, 359–365 (2014)
16. Sklyar, K.V., Ignatovich, S.Y.: Linearizability of systems of the class C^1 with multi-dimensional control. Syst. Control Lett. **94**, 92–96 (2016). https://doi.org/10.1016/j.sysconle.2016.05.016
17. Sklyar, K.V., Ignatovich, S.Y., Sklyar, G.M.: Verification of feedback linearizability conditions for control systems of the class C^1. In: 25th Mediterranean Conference on Control and Automation, MED, pp. 163–168 (2017). 7984112
18. Sklyar, K.V., Sklyar, G.M., Ignatovich, S.Y.: Linearizability of multi-control systems of the class C^1 by additive change of controls. In: André, C., Bastos, M., Karlovich, A., Silbermann, B., Zaballa, I. (eds.) Operator Theory, Operator Algebras, and Matrix Theory. Operator Theory: Advances and Applications, vol. 267, pp. 359–370. Birkhauser/Springer, Cham (2018)
19. Sklyar, K.: On mappability of control systems to linear systems with analytic matrices. Syst. Control Lett. **134**, 104572 (2019). https://doi.org/10.1016/j.sysconle.2019.104572

20. Arnol'd, V.I.: Ordinary Differential Equations. Springer-Verlag, Heidelberg (1992)
21. Hartman, P.: Ordinary Differential Equations, vol. XIV. Wiley, New York, London, Sydney (1964)

A Classification of Feedback Linearizable Mechanical Systems with 2 Degrees of Freedom

Marcin Nowicki[1,2](✉) and Witold Respondek[1]

[1] Laboratoire de Mathématiques, Normandie Université, INSA de Rouen, 76801 Saint-Etienne-du-Rouvray, France
witold.respondek@insa-rouen.fr

[2] Institute of Automatic Control and Robotics, Poznan University of Technology, Piotrowo 3a, 61-138 Poznań, Poland
marcin.nowicki@put.poznan.pl

Abstract. A classification of feedback linearizable mechanical control system with 2 DOF is proposed. We develop 3 types of linearization and for each we establish a normal form. Then, we characterize each class and calculate linearizing outputs. As a consequence, necessary and sufficient linearizability conditions are formulated for all cases. We illustrate our result by mechanical linearization of the TORA system.

Keywords: Mechanical systems · Feedback linearization · Classification · Normal forms

1 Introduction

In this paper, we present a classification of feedback linearizable (F-linearizable) mechanical control systems with two degrees of freedom and a scalar control. We introduce three classes of F-linearizable mechanical systems based on type of transformations used for linearization. This allows to establish normal forms for each class. This article is organised as follows. In Sect. 2, we study mechanical control systems and derive their equations of motion. Section 3 recalls classical results on feedback linearization and introduces its new variants in case of mechanical systems, namely configuration feedback linearization (CF-linearization) and mechanical feedback linearization (MF-linearization). In Sect. 4, we present the main result concerning the classification. In Sect. 5, we provide an application of our results to, first, MF-linearize and, second, to solve the stabilization control problem for the TORA system.

2 Mechanical Control Systems with 2 Degrees of Freedom

Consider a mechanical control system with 2 degrees of freedom (DOF) and a scalar control u. Its Lagrangian, defined as the difference between the kinetic

A. Bartoszewicz et al. (Eds.): PCC 2020, AISC 1196, pp. 638–650, 2020.
https://doi.org/10.1007/978-3-030-50936-1_54

energy T and the potential energy V, reads $\mathcal{L} = T - V = \frac{1}{2}y^T M(x)y - V(x)$, where $x = (x_1, x_2)$ are coordinates on a two dimensional configuration manifold Q and its velocities are denoted $y = (y_1, y_2) := (\dot{x}_1, \dot{x}_2) \in \mathrm{T}_x Q$. The symmetric positive definite matrix $M(x)$ is the inertia matrix (a metric tensor) of the system. We assume that there is no damping (e.g. friction) in the system and that it is subject to two external forces that are positional: an external control force $\tau(x)u$ and an uncontrolled external force $\tau_0(x)$. The corresponding controlled Euler-Lagrange equations are

$$\frac{d}{dt}\frac{\partial \mathcal{L}}{\partial y} - \frac{\partial \mathcal{L}}{\partial x} = \tau_0(x) + \tau(x)u,$$

giving $M(x)\dot{y} + C(x,y)y + P(x) = \tau(x)u$, where $C(x,y)$ is the Coriolis matrix, $P(x) = -(\frac{\partial \mathcal{L}}{\partial x} + \tau_0)$ is an uncontrolled force (possibly non potential because of τ_0) and $\tau(x)$ is an external force controlled by u. Inverting the inertia matrix $M(x)$ and using the coordinates (x, y) result in the first order system on the tangent bundle $\mathrm{T}Q = \{(x,y) : x \in Q, y \in T_x Q\}$:

$$\begin{aligned} \dot{x}_1 &= y_1 \\ \dot{x}_2 &= y_2 \\ \dot{y}_1 &= -\Gamma^1_{11}(x)y_1^2 - 2\Gamma^1_{12}(x)y_1y_2 - \Gamma^1_{22}(x)y_2^2 + \eta_1(x) + g_1(x)u \\ \dot{y}_2 &= -\Gamma^2_{11}(x)y_1^2 - 2\Gamma^2_{12}(x)y_1y_2 - \Gamma^2_{22}(x)y_2^2 + \eta_2(x) + g_2(x)u, \end{aligned} \tag{MS}$$

where $\Gamma^i_{jk}(x)$ are the Christoffel symbols of the second kind, $\eta(x) = (\eta_1, \eta_2)^T$ is an uncontrolled vector field and $g(x) = (g_1, g_2)^T$ is a controlled vector field, both being vector fields on Q.

Equivalently, the mechanical system (MS) can be written in the form of a control system on the manifold $Z = TQ$ of dimension 4, with coordinates $z = (x, y)$,

$$\dot{z} = F(z) + G(z)u, \tag{1}$$

where $F = \sum_{i=1}^{2} y_i \frac{\partial}{\partial x_i} + \left(-\Gamma^i_{11}(x)y_1^2 - 2\Gamma^i_{12}(x)y_1y_2 - \Gamma^i_{22}(x)y_2^2 + \eta_i(x)\right)\frac{\partial}{\partial y_i}$ and $G = \sum_{i=1}^{2} g_i(x)\frac{\partial}{\partial y_i}$. In the subsequent considerations we assume that the controlled vector field g does not vanish at x_0, around which we work, i.e. $g(x_0) \neq 0$.

3 Classification Problem Statement

We start by recalling classical result of feedback linearization of control systems. Since, in this paper, we deal with control systems with a scalar control, we present results limited to that class. The generalizations to the multi-input case can be found in [1–4]. For mathematical preliminaries concerning the Lie derivative, the Lie bracket, distributions, etc. we refer to [1,2,4] and for a geometrical approach to mechanical systems we refer to [5,6].

Consider two n-dimensional control-affine systems with a scalar control

$$\Sigma: \quad \dot{z} = F(z) + G(z)u, \qquad \tilde{\Sigma}: \quad \dot{\tilde{z}} = \tilde{F}(\tilde{z}) + \tilde{G}(\tilde{z})\tilde{u}, \tag{2}$$

where $z \in Z$, $\tilde{z} \in \tilde{Z}$, $u, \tilde{u} \in \mathbb{R}$. We say that Σ and $\tilde{\Sigma}$ are *feedback equivalent*, shortly F-equivalent, if there exist a diffeomorphism $\Phi: Z \to \tilde{Z}$ and an invertible feedback of the form $u = \alpha(z) + \beta(z)\tilde{u}$, such that $\beta(z) \neq 0$ and

$$\frac{\partial \Phi}{\partial z}(z)\left(F + G\alpha\right)(z) = \tilde{F}\left(\Phi(z)\right) \qquad \text{and} \qquad \frac{\partial \Phi}{\partial z}(z)\left(G\beta\right)(z) = \tilde{G}\left(\Phi(z)\right).$$

Throughout all systems are assumed C^∞-smooth and, in particular, the vector fields $F, G, \tilde{F}, \tilde{G}$ and the functions α and β are C^∞-smooth. Recall that feedback transformations (Φ, α, β) preserve trajectories, that is, if $z\left(t, z^0, u(t)\right)$ is a trajectory of Σ (passing through z^0 and corresponding to a control $u(t)$), then $\tilde{z}\left(t, \tilde{z}^0, \tilde{u}(t)\right) = \Phi\left(z\left(t, z^0, u(t)\right)\right)$ is a trajectory of $\tilde{\Sigma}$ (passing through $\tilde{z}^0 = \Phi(z^0)$ and corresponding to $\tilde{u}(t)$, where $u(t) = \alpha\left(z(t)\right) + \beta\left(z(t)\right)\tilde{u}(t)$.

Definition 1. *Σ is feedback linearizable (F-linearizable) if it is F-equivalent to a controllable linear control system of the form $\dot{\tilde{z}} = A\tilde{z} + b\tilde{u}$.*

In other words, there exist a diffeomorphism $\Phi: Z \to \mathbb{R}^n$ and an invertible feedback of the form $u = \alpha(z) + \beta(z)\tilde{u}$ s.t. the control system Σ, in the new coordinates $\tilde{z} = \Phi(z)$ and with the new controls $\tilde{u}$, reads

$$\dot{\tilde{z}} = \frac{\partial \Phi}{\partial z}\left(F + G\alpha\right)\left(\Phi^{-1}(\tilde{z})\right) + \frac{\partial \Phi}{\partial z}\left(G\beta\right)\left(\Phi^{-1}(\tilde{z})\right)\tilde{u} = A\tilde{z} + b\tilde{u}.$$

In order to formulate the result, we associate with Σ the following sequence of nested distributions $\mathcal{D}^0 \subset \mathcal{D}^1 \subset \mathcal{D}^2 \subset \ldots \subset \mathcal{D}^i \subset \ldots \subset TQ$, where

$$\mathcal{D}^0 = \operatorname{span}\{G\}, \qquad \mathcal{D}^i = \operatorname{span}\left\{ad_F^j G, 0 \leq j \leq i\right\}.$$

If the distributions $\mathcal{D}^i$ are involutive, then they are invariant under feedback transformations of the form $u = \alpha(z) + \beta(z)\tilde{u}$, i.e. they remain unchanged if we replace F and G by, respectively, $F + G\alpha$ and $G\beta$. Main results of F-linearization are summarized as follows, see e.g. [1–4].

Theorem 1. *The following conditions are equivalent, locally around $z_0 \in Z$,*

(i) *Σ is F-linearizable;*
(ii) *Σ satisfies*
  (F1) rank $\mathcal{D}^{n-1} = n$,
  (F2) *$\mathcal{D}^i$ are involutive and of constant rank, for $0 \leq i \leq n-2$;*
(iii) *there exists a smooth function h such that*

$$L_G L_F^j h = \begin{cases} 0 & \text{for } 0 \leq j \leq n-2 \\ \lambda & \text{for } j = n-1, \end{cases}$$

where λ is a nonvanishing function.

Corollary 1. *A system Σ on a 4-dimensional state-space Z is locally F-linearizable if and only if*

(F1)' rank $\mathcal{D}^3 = 4$,
(F2)' $\mathcal{D}^0, \mathcal{D}^1, \mathcal{D}^2$ *are involutive and of constant rank.*

Remark 1. If we consider h as an artificial (dummy) output of Σ, then condition (iii) of Theorem 1 states that there exists a dummy output of Σ whose relative degree is n. Any such h will be called a *linearizing output.*

Among all feedback transformations we distinguish two special types that apply to mechanical control systems of the form (MS). Since mechanical systems evolve on the tangent bundle TQ there is a natural distinction between configurations x and velocities y, which we employ in our classification problem. Inspired by [7], we introduce a class of configuration feedback linearization, where linearizing output functions depend on the positions only, i.e. $h = h(x)$.

Definition 2. *A mechanical system* (MS) *is configuration feedback linearizable (shortly, CF-linearizable) if there exists a linearizing output h satisfying condition (iii) of Theorem 1 such that $h = h(x)$, i.e. depends on configurations x only.*

Another class of feedback transformations is that preserving the mechanical structure of the system (MS) due to the requirement that the diffeomorphism Φ maps positions into new positions and velocities into the corresponding velocities:

$$(\tilde{x}, \tilde{y}) = \Phi(x, y) = \left(\phi(x), \frac{\partial \phi}{\partial x}(x) y\right), \tag{3}$$

called an extended point transformation and is induced by $\phi(x)$ being a diffeomorphism of the configuration manifold Q and $\frac{\partial \phi}{\partial x}(x) : \mathrm{T}_x Q \to \mathrm{T}_{\phi(x)} Q$ being the tangent map (or simply, the Jacobi matrix) of ϕ at x. Moreover, the feedback $u = \alpha + \beta \tilde{u}$ is assumed to preserve the mechanical structure of (MS), i.e. $\beta = \beta(x)$ and α is a polynomial of degree 2 with respect to y, i.e.

$$u = \sum_{j,k=1}^{n} \gamma_{jk}(x) y_j y_k + \varepsilon(x) + \beta(x) \tilde{u}, \tag{4}$$

where γ_{jk}, ε and β are functions depending on configurations.

Definition 3. *A mechanical control system* (MS) *is mechanical feedback linearizable (shortly, MF-linearizable) if there exist a diffeomorphism* (3) *and feedback* (4)*, that transform* (MS) *into a linear controllable mechanical system*

$$\dot{\tilde{x}} = \tilde{y}, \qquad \dot{\tilde{y}} = E\tilde{x} + B\tilde{u}.$$

We study the feedback linearization problem for mechanical systems in Lagrangian form. For the state-space linearization of systems in Lagrangian form see [8] and for the state-space linearization of Hamiltonian systems see [9].

Denoting the class of MF-linearizable (respectively CF- and F-linearizable) mechanical systems (MS) by $\mathcal{MF}$-lin (respectively $\mathcal{CF}$-lin and $\mathcal{F}$-lin) we have the following inclusions (the first holds because any MF-linearizable system always admits a linearizing output depending on x-only) explaining relations between the three types of feedback linearizations

$$\mathcal{MF}\text{-lin} \subset \mathcal{CF}\text{-lin} \subset \mathcal{F}\text{-lin}.$$

An interest in CF- and MF-linearization is that, for the first class, we use linearizing outputs $h = h(x)$, depending on configurations x only, to parametrize all states and controls and, for the second, that the natural splitting into configurations and velocities is preserved.

4 Feedback Linearizable Mechanical Systems with 2 DOF

4.1 A Preliminary Normal Form

In this section, we propose a prenormal form needed for further analysis. In order to simplify a mechanical system in the form (MS) we will rectify the controlled vector field g.

Proposition 1. *A system* (MS) *is locally around* (x_0, y_0) *MF-equivalent to*

$$\begin{aligned} \dot{x}_1 &= y_1 \\ \dot{x}_2 &= y_2 \\ \dot{y}_1 &= -a(x_1,x_2)y_1^2 - 2b(x_1,x_2)y_1y_2 - c(x_1,x_2)y_2^2 + e(x_1,x_2) \\ \dot{y}_2 &= u, \end{aligned} \tag{NF0}$$

that is, locally there exist a mechanical diffeomorphism of the form (3) *and a mechanical feedback of the form* (4) *transforming* (MS) *into* (NF0).

Proof. By $g(x_0) \neq 0$ we may assume $g_2(x_0) \neq 0$ (if not, permute x_1 and x_2) and it follows that there exists, around x_0, a function $\varphi(x)$ that is a solution of the following first order PDE

$$\frac{\partial \varphi}{\partial x_1} g_1 + \frac{\partial \varphi}{\partial x_2} g_2 = 0$$

satisfying $\frac{\partial \varphi}{\partial x_1}(x_0) \neq 0$. Apply the mechanical diffeomorphism $\Phi = (\tilde{x}, \tilde{y}) = (\phi, \frac{\partial \phi}{\partial x} y)$, where $\tilde{x} = \phi(x) = \begin{pmatrix} \varphi(x) \\ x_2 \end{pmatrix}$ and $\frac{\partial \phi}{\partial x} = \begin{pmatrix} \frac{\partial \varphi}{\partial x_1} & \frac{\partial \varphi}{\partial x_1} \\ 0 & 1 \end{pmatrix}$ that brings (MS) into

$$\begin{aligned} \dot{\tilde{x}}_1 &= \tilde{y}_1 \qquad \dot{\tilde{y}}_1 = -\tilde{\Gamma}_{11}^1(\tilde{x})\tilde{y}_1^2 - 2\tilde{\Gamma}_{12}^1(\tilde{x})\tilde{y}_1\tilde{y}_2 - \tilde{\Gamma}_{22}^1(\tilde{x})\tilde{y}_2^2 + \tilde{\eta}_1(\tilde{x}) \\ \dot{\tilde{x}}_2 &= \tilde{y}_2 \qquad \dot{\tilde{y}}_2 = -\tilde{\Gamma}_{11}^2(\tilde{x})\tilde{y}_1^2 - 2\tilde{\Gamma}_{12}^2(\tilde{x})\tilde{y}_1\tilde{y}_2 - \tilde{\Gamma}_{22}^2(\tilde{x})\tilde{y}_2^2 + \tilde{\eta}_2(\tilde{x}) + \tilde{g}_2(\tilde{x})u. \end{aligned}$$

Finally, applying the mechanical feedback $\tilde{u} = -\tilde{\Gamma}^2_{11}\tilde{y}_1^2 - 2\tilde{\Gamma}^2_{12}\tilde{y}_1\tilde{y}_2 - \tilde{\Gamma}^2_{22}\tilde{y}_2^2 + \tilde{\eta}_2 + \tilde{g}_2 u$, brings the system into normal form (NF0) (after dropping the "tildas" and replacing $\tilde{\eta}_1$ by e, and $\tilde{\Gamma}^1_{11}, \tilde{\Gamma}^1_{12}, \tilde{\Gamma}^1_{22}$ by a, b, c, respectively). □

4.2 A Classification Result

Now we will study linearizability conditions $(F1)' - (F2)'$ of Corollary 1. The distribution $\mathcal{D}^0 = \operatorname{span}\{G\}$ is always involutive but assuming involutivity of $\mathcal{D}^1$ allows to further simplify (NF0).

Proposition 2. *Assume that for* (MS) *the distribution* $\mathcal{D}^1$ *is involutive. Then, in its normal form* (NF0) *we have* $c \equiv 0$.

Moreover, either $b \equiv 0$ *implying that* (MS) *is MF-equivalent to*

$$\begin{aligned} \dot{x}_1 &= y_1 \\ \dot{x}_2 &= y_2 \\ \dot{y}_1 &= -a(x_1,x_2)y_1^2 + e(x_1,x_2) \\ \dot{y}_2 &= u, \end{aligned} \tag{NF1}$$

for which $\mathcal{D}^2$ *is also involutive, or* $b \neq 0$ *and then* (MS) *is F-equivalent to*

$$\begin{aligned} \dot{x}_1 &= y_1 d(x_1,x_2) \\ \dot{x}_2 &= y_2 \\ \dot{y}_1 &= -a(x_1,x_2)y_1^2 + e(x_1,x_2) \\ \dot{y}_2 &= u, \end{aligned} \tag{NF2}$$

with $\frac{\partial d}{\partial x_2} \not\equiv 0$.

Remark 2. Note that (MS) is F-equivalent to (NF2) but not MF-equivalent. The reason is that $\tilde{y}_1 = \psi(x_1,x_2)y_1$, see (5) below, is a pseudo-velocity but not a velocity (we cannot find a function $\tilde{x}_1 = \xi(x)$ such that $\tilde{y}_1 = \frac{d}{dt}\xi(x)$).

Proof. By Proposition 1 we know that (MS) is MF-equivalent to (NF0) for which we calculate

$$G = \frac{\partial}{\partial y_2}, \qquad ad_F G = -\frac{\partial}{\partial x_2} + 2\left(by_1 + cy_2\right)\frac{\partial}{\partial y_1}, \qquad [G, ad_F G] = 2c\frac{\partial}{\partial y_1},$$

implying that $\mathcal{D}^1 = \operatorname{span}\{G, ad_F G\}$ is involutive, i.e. $[G, ad_F G] \in \mathcal{D}^1$, if and only if $c \equiv 0$. It is immediate to see that if $b \equiv 0$, then the system is in (NF1) form for which $\mathcal{D}^2$ is, indeed, involutive. If $b \neq 0$ we deal with the system

$$\begin{aligned} \dot{x}_1 &= y_1 \qquad & \dot{y}_1 &= -a(x_1,x_2)y_1^2 - 2b(x_1,x_2)y_1y_2 + e(x_1,x_2) \\ \dot{x}_2 &= y_2 \qquad & \dot{y}_2 &= u. \end{aligned}$$

In order to rectify $ad_F G = -\frac{\partial}{\partial x_2} + 2by_1\frac{\partial}{\partial y_1}$, we solve the following PDE

$$-\frac{\partial \varphi}{\partial x_2} + \frac{\partial \varphi}{\partial y_1} 2by_1 = 0,$$

and get $\varphi = \psi(x)y_1$, where $\psi(x) = \exp\left(2\int_0^{x_2} b(x_1, s)\, ds\right)$. Apply the diffeomorphism

$$\tilde{x}_1 = x_1, \qquad \tilde{x}_2 = x_2, \qquad \tilde{y}_1 = \varphi(x, y) = \psi(x)y_1, \qquad \tilde{y}_2 = y_2, \tag{5}$$

which brings (NF0) into

$$\begin{aligned}
\dot{\tilde{x}}_1 &= y_1 = \frac{\tilde{y}_1}{\psi} = \tilde{y}_1 d(\tilde{x}) \\
\dot{\tilde{x}}_2 &= y_2 = \tilde{y}_2 \\
\dot{\tilde{y}}_1 &= \frac{d}{dt}(y_1\psi(x)) = \left(-ay_1^2 - 2by_1y_2 + e\right)\psi + y_1\left(\frac{\partial\psi}{\partial x_1}y_1 + 2\psi by_2\right) \\
&= -\left(\psi a + \frac{\partial \psi}{\partial x_1}\right)y_1^2 + \psi e = -\tilde{a}(\tilde{x})\tilde{y}_1^2 + \tilde{e}(\tilde{x}) \\
\dot{\tilde{y}}_2 &= \dot{y}_2 = u,
\end{aligned}$$

where (since $\tilde{x} = x$), we have $\tilde{a} = \frac{1}{\psi^2}\left(\psi a + \frac{\partial\psi}{\partial x_1}\right)$ and $\tilde{e} = \psi e$. Dropping the tildes we get

$$\begin{aligned}
\dot{x}_1 &= y_1 d(x_1, x_2) & \dot{y}_1 &= -a(x_1, x_2)y_1^2 + e(x_1, x_2) \\
\dot{x}_2 &= y_2 & \dot{y}_2 &= u,
\end{aligned}$$

which is (NF2)-form with $\frac{\partial d}{\partial x_2} \not\equiv 0$. Indeed, note that $d = \exp\left(-2\int b\, dx_2\right)$ and therefore $\frac{\partial d}{\partial x_2} = -2b\exp\left(-2\int_0^{x_2} b\, ds\right)$, which is not identically zero unless $b \equiv 0$, (the assumption $b \equiv 0$ leads to the first case and form (NF1)). □

Remark 3. Notice that (NF0) is, in general, a local form (since it requires a rectification of g). On the other hand, if (NF0) is global, then the transformations to either (NF1) or (NF2) are global and then both forms hold globally.

The main result describes MF-, CF-, and F-linearizability in terms of the normal forms.

Theorem 2. *For a mechanical system* (MS) *with 2 DOF we have:*

(i) (MS) *is MF-linearizable if and only if* (MS) *is MF-equivalent to* (NF1) *satisfying*

$$\frac{\partial a}{\partial x_2} \equiv 0 \quad \text{and} \quad \frac{\partial e}{\partial x_2}(x_0) \neq 0.$$

Moreover, all linearizing outputs providing MF-linearization are of the form $h(x_1) = k\int_0^{x_1}\exp\left(\int_0^{s_2} a(s_1)ds_1\right)ds_2$, *where* $k \in \mathbb{R}$, $k \neq 0$.

(ii) (MS) *is CF-linearizable if and only if* (MS) *is MF-equivalent to* (NF1) *and* $\frac{\partial a}{\partial x_2}(x_0)y_{10}^2 + \frac{\partial e}{\partial x_2}(x_0) \neq 0$. *Moreover, all linearizing outputs are of the form* $h = h(x_1)$, *with* $h'(x_{10}) \neq 0$.

(iii) (MS) *is F-linearizable, but not CF-linearizable, if and only if* (MS) *is F-equivalent to* (NF2) *with*

$$d(x_1, x_2) = d_1(x_1)e(x_1, x_2) + d_0(x_1)$$
$$c(x_1, x_2) = a_1(x_1)e(x_1, x_2) + a_0(x_1),$$

in which case (NF2) *can be further normalized (with new* y_1, a_0, d_0, d_1, e*) to*

$$\begin{aligned}
\dot{x}_1 &= (d_1(x_1)e(x_1, x_2) + d_0(x_1))\, y_1 \\
\dot{x}_2 &= y_2 \\
\dot{y}_1 &= -a_0(x_1)y_1^2 + e(x_1, x_2) \\
\dot{y}_2 &= u.
\end{aligned} \tag{NF2'}$$

Moreover, all linearizing outputs of (NF2') *are, around points such that* $d_1(x_{10}) \neq 0$, *nontrivial functions of* $h(x_1, y_1) = y_1^2 - 2\int_0^{x_1} \frac{ds}{d_1(s)}$.

Proof. (i). Sufficiency. By Proposition 2, mechanical system (MS) is MF-equivalent to (NF1). Then $\frac{\partial a}{\partial x_2} = 0$ implies

$$\begin{aligned}
\dot{x}_1 &= y_1 \qquad & \dot{y}_1 &= -a(x_1)y_1^2 + e(x_1, x_2) \\
\dot{x}_2 &= y_2 \qquad & \dot{y}_2 &= u.
\end{aligned}$$

Apply the linearizing mechanical diffeomorphism $\Phi = (\phi, \frac{\partial \phi}{\partial x}y)$, with $(\tilde{x}_1, \tilde{x}_2) = \phi(x) = (\varphi, \varphi' e)$, where $\varphi(x_1) = k\int_0^{x_1} \exp\left(\int_0^{s_2} a(s_1)ds_1\right) ds_2$, $k \in \mathbb{R}$, $k \neq 0$, is the solution of the following ODE

$$\varphi'' - \varphi' a = 0.$$

Notice that $\phi = (\varphi, \varphi' e)$ is a diffeomorphism on Q because $\varphi'(x_1) \neq 0$ and $\frac{\partial e}{\partial x_2}(x_1, x_2) \neq 0$. The transformed system reads

$$\begin{aligned}
\dot{\tilde{x}}_1 &= \varphi' y_1 = \tilde{y}_1 \\
\dot{\tilde{x}}_2 &= \varphi'' e y_1 + \varphi' \left(\frac{\partial e}{\partial x_1} y_1 + \frac{\partial e}{\partial x_2} y_2 \right) = \tilde{y}_2 \\
\dot{\tilde{y}}_1 &= \varphi'' y_1^2 - \varphi' a y_1^2 + \varphi' e = (\varphi'' - \varphi' a)\, y_1^2 + \varphi' e = \varphi' e = \tilde{x}_2 \\
\dot{\tilde{y}}_2 &= A_2(x, y) + A_0(x) + B(x)u = \tilde{u},
\end{aligned}$$

where $A_2(x, y)$ is a homogenous polynomial of degree 2 with respect to y, so the feedback $\tilde{u} = A_2 + A_0 + Bu$ is mechanical and, moreover, $h = \varphi = k\int_0^{x_1} \exp\left(\int_0^{s_2} a(s_1)ds_1\right) ds_2$ is indeed an MF-linearizing output.

Necessity. Assume that (MS) is MF-linearizable, that is, MF-equivalent to the linear controllable system

$$\dot{x}_1 = y_1 \qquad \dot{x}_2 = y_2 \qquad \dot{y}_1 = x_2 \qquad \dot{y}_2 = u, \tag{6}$$

All MF-transformations that map (6) into (NF1) are of the form

$$\tilde{x}_1 = \varphi(x_1) \qquad \tilde{y}_1 = \varphi' y_1 \qquad \tilde{x}_2 = \psi(x_1, x_2) \qquad \tilde{y}_2 = \frac{\partial \psi}{\partial x} y$$

followed by $\tilde{u} = \dot{\tilde{y}}_2$. The transformed system is

$$\dot{\tilde{x}}_1 = \tilde{y}_1 \qquad \dot{\tilde{x}}_2 = \tilde{y}_2 \qquad \dot{\tilde{y}}_1 = \frac{\varphi''}{\varphi^2}\tilde{y}_1^2 + \varphi' x_2 = \tilde{a}\tilde{y}_1^2 + \tilde{e},$$

where $\tilde{a} = \frac{\varphi''}{\varphi^2}$ and $\tilde{e} = \varphi' x_2$. Clearly, $\tilde{a} = \tilde{a}(\tilde{x}_1)$ and $\frac{\partial \tilde{e}}{\partial \tilde{x}_2}(\tilde{x}_0) = \varphi'(x_{10}) \neq 0$.

(ii). Sufficiency. If the mechanical system (MS) is MF-equivalent to (NF1), then we will show that any $h = h(x_1)$, $h'(x_{10}) \neq 0$, satisfies condition (iii) of Theorem 1, thus proving that (MS) is CF-linearizable. Having $F = y_1 \frac{\partial}{\partial x_1} + y_2 \frac{\partial}{\partial x_2} + \left(-a y_1^2 + e\right) \frac{\partial}{\partial y_1}$ and $G = \frac{\partial}{\partial y_2}$, a direct calculations shows

$$L_G h = L_G L_F h = L_G L_F^2 h = 0 \quad \text{and} \quad L_G L_F^3 h = h' \left(-\frac{\partial a}{\partial x_2} y_1^2 + \frac{\partial e}{\partial x_2} \right),$$

which is nonvanishing at (x_0, y_0) by the assumption.

Necessity. Let $h = h(x_1, x_2)$ be a CF-linearizing output of (NF0) and thus of the relative degree 4. Then $L_F h = \frac{\partial h}{\partial x_1} y_1 + \frac{\partial h}{\partial x_2} y_2$ and $L_G L_F h = \frac{\partial h}{\partial x_2} = 0$ implying $h = h(x_1)$. Hence $L_F h = h' y_1$, $L_F^2 h = h'' y_1^2 + h'(-a y_1^2 - 2b y_1 y_2 - c y_2^2 + e)$, and $L_G L_F^2 h = -h'(b y_1 + 2c y_2) = 0$ implying that $b \equiv c \equiv 0$ (since the relative degree is 4), which yields form (NF1). A direct calculation for $h = x_1$ gives $\left(\frac{\partial a}{\partial x_2} y_{10}^2 + \frac{\partial e}{\partial x_2} \right) \neq 0$.

(iii). Necessity. If the mechanical system (MS) is F-linearizable then, by Proposition 2, is either CF-linearizable, or F-equivalent to (NF2) for which $\mathcal{D}^1$ is involutive. Therefore, we need to check the involutivity of D^2. For (NF2) we calculate the Lie brackets

$$ad_F G = -\frac{\partial}{\partial x_2}, \qquad ad_F^2 G = \left(\frac{\partial d}{\partial x_2} y_1 \right) \frac{\partial}{\partial x_1} + \left(-\frac{\partial a}{\partial x_2} y_1^2 + \frac{\partial e}{\partial x_2} \right) \frac{\partial}{\partial y_1},$$

$$[G, ad_F G] = \left[G, ad_F^2 G\right] = 0$$

$$\left[ad_F G, ad_F^2 G\right] = -\frac{\partial^2 d}{\partial x_2^2} y_1 \frac{\partial}{\partial x_1} + \left(\frac{\partial^2 \tilde{a}}{\partial x_2^2} y_1^2 - \frac{\partial^2 e}{\partial x_2} \right) \frac{\partial}{\partial y_1}.$$

Involutivity of $\mathcal{D}^2 = \text{span}\left\{G, ad_F G, ad_F^2 G\right\}$ requires $\left[ad_F G, ad_F^2 G\right] \in \mathcal{D}^2$ implying the following set of conditions

$$\frac{\partial^2 d}{\partial x_2^2} = \gamma(x) \frac{\partial d}{\partial x_2} \qquad \frac{\partial^2 a}{\partial x_2^2} = \gamma(x) \frac{\partial a}{\partial x_2} \qquad \frac{\partial^2 e}{\partial x_2^2} = \gamma(x) \frac{\partial e}{\partial x_2},$$

for some smooth function $\gamma(x)$. Denoting $E = \frac{\partial e}{\partial x_2}$, $A = \frac{\partial a}{\partial x_2}$, $D = \frac{\partial d}{\partial x_2}$ and taking the last two equations, we get the equality $\frac{E'}{E} = \frac{A'}{A}$. Integrating both sides with respect to x_2 yields $\ln|A| = \ln|E| + \tilde{a}(x_1)$ and taking exponential gives $A = a_1(x_1)E$. Again, by integration both sides with respect to x_2 we have $a = a_1(x_1)e + a_0(x_1)$. By repeating the same procedure with the first and third equations we conclude the following relations

$$\begin{aligned} a(x_1, x_2) &= a_1(x_1)e(x_1, x_2) + a_0(x_1) \\ d(x_1, x_2) &= d_1(x_1)e(x_1, x_2) + d_0(x_1), \end{aligned}$$

i.e. a and d are affine functions of $e(x_1, x_2)$ with coefficients that are functions of x_1 only. In order to get (NF2'), put $\tilde{y}_1 = \varphi(x_1)y_1$, where φ is a solution of $\varphi' d_1 - \varphi a_1 = 0$. Then we have

$$\begin{aligned} \dot{\tilde{y}}_1 &= \varphi'\left(d_1 e + d_0\right) y_1^2 - \varphi\left(a_1 e + a_0\right) y_1^2 + \varphi e = \frac{\varphi' d_0 - \varphi a_0}{\varphi^2} \tilde{y}_1^2 + \varphi e \\ &= \tilde{a}_0(x_1)\tilde{y}_1^2 + \tilde{e}, \end{aligned}$$

which gives (NF2'). Note that the solution $\varphi = \exp\left(\int_0^{x_1} \frac{a_1}{d_1}(s)ds\right)$ and, since $\varphi = \varphi(x_1)$, there exists $\tilde{x}_1 = \psi(x_1)$ such that $\psi' = \varphi$ and thus the transformation $\tilde{x}_1 = \psi(x_1)$, $\tilde{y}_1 = \varphi(x_1)y_1$ is actually a MF-transformation.

Sufficiency. If (MS) if F-equivalent to (NF2) and thus to (NF2'), then the linearizing output of (NF2') can be found using condition (*iii*) of Theorem 1. Take $h = h(x_1, x_2, y_1, y_2)$, calculate $L_G h = \frac{\partial h}{\partial y_2} = 0$, then the Lie derivative $L_G L_F h(x_1, x_2, y_1) = \frac{\partial h}{\partial x_2} = 0$. It follows $h = h(x_1, y_1)$ and we use it to calculate the next one

$$L_G L_F^2 h(x_1, y_1) = \frac{\partial e}{\partial x_2}\left(\frac{\partial h}{\partial x_1} d_1 y_1 + \frac{\partial h}{\partial y_1}\right) = 0,$$

whose solution is $h(x_1, y_1) = y_1^2 - 2\int_0^{x_1} \frac{ds}{d_1(s)}$ (or any nontrivial function of h), which is a linearizing output around all points, where $L_G L_F^3 h = -2\frac{\partial e}{\partial x_2}\left(d_1 a_0' y_1^4 + \left(3a_0 + d_0' - \frac{d_1' d_0}{d_1}\right) y_1^2 + \frac{d_0}{d_1}\right)$ does not vanish. □

5 Application to the Mechanical TORA System

We will study mechanical feedback linearizability of the TORA (Translational Oscillator with Rotational Actuator) system (see Fig. 1), which is a nonlinear benchmark system studied in the literature, e.g. [10] (however we add gravitational effects). It consists of a spring-mass system, with a mass m_1 and a spring constant k_1, and a pendulum of length L_2, mass m_2, and moment of inertia J_2. The displacement of the system is denoted by x and the angle of the pendulum by θ. The control u is a torque applied to the pendulum. The kinetic energy is $T = \frac{1}{2}(m_1 + m_2)\dot{x}^2 + \frac{1}{2}(J_2 + m_2 L_2^2)\dot{\theta}^2 + m_2 L_2 \cos\theta \dot{x}\dot{\theta}$ and the potential energy is $V = -m_2 L_2 a \cos\theta + \frac{1}{2}k_1 x^2$, where a is the gravitational constant. The equations of dynamics in (x, θ)-coordinates read

$$\begin{aligned} (m_1 + m_2)\ddot{x} + m_2 L_2 \cos\theta\ddot{\theta} - m_2 L_2 \sin\theta\dot{\theta}^2 + k_1 x = 0 \\ m_2 L_2 \cos\theta\ddot{x} + (m_2 L_2^2 + J_2)\ddot{\theta} + m_2 L_2 a \sin\theta = u. \end{aligned}$$

The equations of the mechanical system can be re-written on TQ by denoting coordinates $(x_1, x_2) := (x, \theta)$

$$\begin{aligned} \dot{x}_1 &= y_1 \qquad & \dot{y}_1 &= -\Gamma^1_{22} y_2^2 + \eta_1 + g_1 u \\ \dot{x}_2 &= y_2 \qquad & \dot{y}_2 &= -\Gamma^2_{22} y_2^2 + \eta_2 + g_2 u, \qquad \text{where} \end{aligned} \tag{7}$$

$$\begin{aligned} \Gamma^1_{22} &= \frac{p_0 \sin x_2}{p_1 - p_2 \cos^2 x_2}, & \eta_1 &= \frac{p_4 \sin x_2 \cos x_2 - p_3 x_1}{p_1 - p_2 \cos^2 x_2}, & g_1 &= \frac{-m_{12} \cos x_2}{p_1 - p_2 \cos^2 x_2}, \\ \Gamma^2_{22} &= \frac{p_2 \sin x_2 \cos x_2}{p_1 - p_2 \cos^2 x_2}, & \eta_2 &= \frac{-p_5 \sin x_2 + p_6 x_1 \cos x_2}{p_1 - p_2 \cos^2 x_2}, & g_2 &= \frac{m_{11}}{p_1 - p_2 \cos^2 x_2}, \end{aligned}$$

with constant parameters $m_{11} = m_1 + m_2$, $m_{12} = m_2 L_2$, $m_{22} = m_2 L_2^2 + J_2$, $p_0 = -m_{12} m_{22}$, $p_1 = m_{11} m_{22}$, $p_2 = m_{12}^2$, $p_3 = k_1(m_2 L_2^2 + J_2)$, $p_4 = m_2^2 L_2^2 a$, $p_5 = m_2 L_2 (m_1 + m_2) a$, $p_6 = k_1 m_2 L_2$.

In order to find linearizing coordinates, for $x_2 \neq \pm\pi$, we rectify the control vector field $(g_1, g_2)^T$ to obtain (NF0) via the transformation

$$\begin{aligned} \tilde{x}_1 &= m_{11} x_1 + m_{12} \sin x_2 \qquad & \tilde{y}_1 &= m_{11} y_1 + m_{12} y_2 \cos x_2 \\ \tilde{x}_2 &= -k_1 x_1 & \tilde{y}_2 &= -k_1 y_1 \end{aligned} \tag{8}$$

and calculate the system in new coordinates

$$\begin{aligned} \dot{\tilde{x}}_1 &= m_{11} y_1 + m_{12} y_2 \cos x_2 = \tilde{y}_1 \\ \dot{\tilde{y}}_1 &= m_{11} \dot{y}_1 + m_{12} \left(-y_2^2 \sin x_2 + \dot{y}_2 \cos x_2\right) = -k_1 x_1 = \tilde{x}_2 \\ \dot{\tilde{x}}_2 &= -k_1 y_1 = \tilde{y}_2 \\ \dot{\tilde{y}}_2 &= -k_1 \left(-\Gamma^1_{22} y_2^2 + \eta_1 + g_1 u\right) = \tilde{u}, \end{aligned}$$

which is (NF1), with $a \equiv 0$ and $e = -k_1 x_1$, thus yielding a linear mechanical system. The simulation scheme is shown in Fig. 2. The block "TORA" consists of the equation of dynamics (7), the block "MF-feedback" is the linearization controller which reads $u = \frac{1}{g_1}\left(\Gamma^1_{22} y_2^2 - \eta_1 - \frac{\tilde{u}}{k_1}\right)$. The linear stabilization task controller

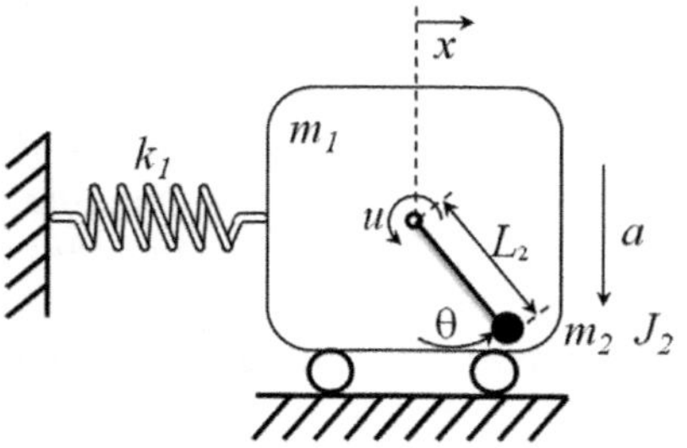

Fig. 1. The TORA system

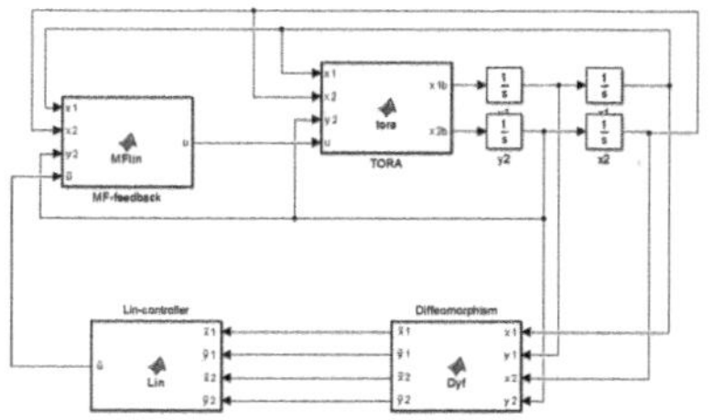

Fig. 2. The simulation scheme for the TORA system

(denoted "Lin-controller" on the scheme) reads $\tilde{u} = \kappa_1 \tilde{x}_1 + \kappa_2 \tilde{x}_2 + \kappa_3 \tilde{y}_1 + \kappa_4 \tilde{y}_2$ and finally the block "Diffeomorphism" consist of change of coordinates (8).

The simulation results are shown in Fig. 3 for the following gain parameters κ_i and initial conditions: $m_1 = 2[kg]$, $m_2 = 0.5[kg]$, $L_2 = 0.5[m]$, $J_2 = 0.1[kg \cdot m^2]$, $a = 9.81[\frac{m}{s^2}]$, $k_1 = 3$, $\kappa_1 = -70.71$, $\kappa_2 = -63.72$, $\kappa_3 = -28.71$, $\kappa_4 = -7.58$, $x_1(0) = y_1(0) = y_2(0) = 0$, $x_2(0) = 0.7$.

6 Summary

We have studied F-linearizable mechanical systems with 2 DOF and single control which is the simplest possible underactuated case. We asserted that there are 3 non-equivalent types distinguished by the class of transformations used in the linearization process. We have studied each class and, using established normal forms, we formulated necessary and sufficient conditions under which a mechanical system is linearizable via, respectively, mechanical, configuration, and general feedback.

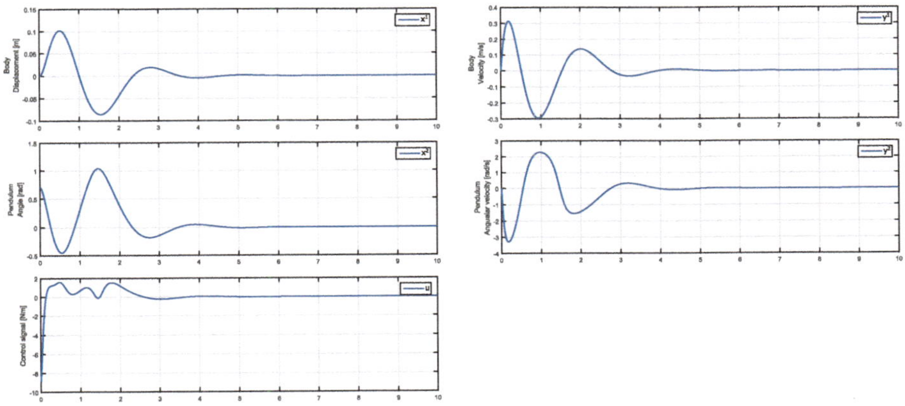

Fig. 3. The simulation results for the TORA system.

References

1. Nijmeijer, H., van der Schaft, A.J.: Nonlinear Dynamical Control Systems. Springer-Verlag, New York (1990). ISBN 978-0-387-97234-3
2. Respondek, W.: Introduction to geometric nonlinear control; linearization, observability and decoupling. In: Mathematical Control Theory No.1, Lecture Notes Series of the Abdus Salam, ICTP, vol. 8, Trieste (2001)
3. Jakubczyk, B., Respondek, W.: On linearization of control systems. Bull. Acad. Polonaise Sci. Ser. Sci. Math. **28**, 517–522 (1980)
4. Isidori, A.: Nonlinear Control Systems, 3rd edn. Springer-Verlag, Berlin (1995)

5. Bullo, F., Lewis, A.D.: Geometric Control of Mechanical Systems. Modeling, Analysis and Design for Simple Mechanical Control Systems. Springer-Verlag, New York (2004)
6. Respondek, W., Ricardo, S.: Equivariants of mechanical control systems. SIAM J. Control Optim. **51**(4), 3027–3055 (2013)
7. Murray, M., Rathinam, M., Sluis, W.: Differential flatness of mechanical control systems: a catalog of prototype systems. In: ASME International Mechanical Engineering Congress and Exposition. Citeseer (1995)
8. Respondek, W., Ricardo, S.: On linearization of mechanical control systems. IFAC Proc. Volumes **45**(19), 102–107 (2012)
9. van der Schaft, A.: Linearization of Hamiltonian and gradient systems. IMA J. Math. Control Inf. **1**, 185–198 (1984)
10. Wan, C., Bernstein, D., Coppola, V.: Global stabilization of the oscillating eccentric rotor. Nonlinear Dyn. **10**, 49–62 (1995)

Stabilization of a 3-Link Pendulum in Vertical Position

Krzysztof Kozłowski(✉), Dariusz Pazderski, Paweł Parulski, and Patryk Bartkowiak

Poznan University of Technology, Poznań, Poland
krzysztof.kozlowski@put.poznan.pl

Abstract. The aim of the paper is to verify the linearizabilty conditions for the triple inverted pendulum driven by 2 inputs, and stabilize it in the upright position. Moreover, the zero dynamics is derived and illustrated graphically.

Keywords: Nonlinear dynamics · Control theory · 3-link pendulum · Linearization · Zero dynamic

1 Introduction

Inverted pendulums are highly unstable nonlinear systems that present challenging control problems. Moreover, they provide an excellent test bench for the evaluation and comparison of different control strategies. The stabilization of single and double inverted pendulums has been the subject of numerous papers. Despite this fact, the analysis and control of n-link inverted pendulum, for $n \geq 3$, is still an interesting topic, investigated by researchers for years. Considered models include both "bare" pendula, as well as the pendulums mounted on a car ([1–3, 7, 11]). None of cited references are based on formal approach using differential geometry along with an attempt to identify a linearizable subsystem of maximal dimension part that is linearized. In this paper authors intent to fill this gap.

This work presents preliminary results on analyzing and controlling the triple inverted pendulum with two actuators. The analyzed system is highly nonlinear, and its largest feedback linearizable subsystem is of dimension 4. As there is not possible to fully linearize the system, one needs to use a partial feedback linearization techniques to control the system. The aim of the paper is to check the linearizability conditions for the triple inverted pendulum driven by 2 inputs, and to stabilize it in the upright position. Moreover, the zero dynamics is going to be derived and verified.

The paper is organized as follows. In Sect. 2 the triple pendulum is analyzed, in a view of its mathematical properties. Section 3 treats the stabilization problem of underactuated nonlinear system which the triple pendulum is, together with a brief analysis of its internal zero dynamics. And finally, in Sect. 4 simulation results of adapted control scheme are provided.

A. Bartoszewicz et al. (Eds.): FCC 2020, AISC 1196, pp. 651–662, 2020.
https://doi.org/10.1007/978-3-030-50936-1_55

2 Mathematical Properties of the Model

In the following section we want to characterize the considered system whether is feedback equivalent to linear one, or what is the largest possible linearizable subsystem of a given system, and how does the output functions looks like, in the context of control.

Consider the system with multiple inputs with dynamics given by:

$$\dot{x} = f(x) + G(x)u, \tag{1}$$

where

$$f(x) = \begin{bmatrix} f_1(x) \\ \vdots \\ f_n(x) \end{bmatrix}, \ G(x) = \begin{bmatrix} g_{11}(x) & \cdots & g_{1m}(x) \\ \vdots & \ddots & \vdots \\ g_{n1}(x) & \cdots & g_{nm}(x) \end{bmatrix} = \begin{bmatrix} g_1(x) \cdots g_m(x) \end{bmatrix}. \tag{2}$$

where $x \in \mathbb{R}^n$, $u \in \mathbb{R}^m$ and $f(x), G(x)$ are of appropriate dimensions. We can define the distributions $\boldsymbol{D}_j = \text{span}\{g_1, ..., g_m, ad_f g_1, ..., ad_f g_m, ..., ad_f^{j-1} g_1, ..., ad_f^{j-1} g_m\}$ (where $ad_f g_i = [f, ad_f^{k-1} g_i](x)$, for any $k \geq 1$, setting $ad_f^0 g_i(x) = g_i(x)$), and let $\bar{\boldsymbol{D}}_j$ denote the involutive closure of $\boldsymbol{D}_j$, which is the smallest involutive distribution containing $\boldsymbol{D}_j$ and $j = 0, 1, ..., n-1$ [9].

In order to define a state feedback and a change of coordinates transforming the system described by (1) into a linear and controllable system, one needs to check several conditions, described in details in [4]. In general, when system is underactuated, full feedback linearisation is not possible. The system should be decomposed into two subsystems, one which is linear, and one which stays still nonlinear. An important issue is the maximal dimension of the linear subsystem that might be obtained. The partial linearization problem is that of when (1) is equivalent to a system of the form

$$\dot{z}_1 = f^1(z^1, z^2) \tag{3}$$

$$\dot{z}_2 = Az^2 + Bv \tag{4}$$

where $z^1 = (z_1^1, ..., z_{n-k}^1)$ and $z^2 = (z_1^2, ..., z_k^2)$ are coordinates on R^n, $v \in R^n$ and the $k \times k$ matrix A and the $(k \times m)$ matrix B are constant. The problem is equivalent to find suitable artificial outputs with relative degrees as highest as possible, sometimes called maximally part-linearizing output [6]. It is interesting to know how to find the largest feedback linearizable subsystem. Taking into account conditions given in [9] it is needed to check the rank of proper distributions, and on the basis of this, evaluate the number of independent maximally part-linearizing outputs and the linearizability defect.

2.1 Robot Model

Considered robot is a connection of $N = 3$ rigid bodies coupled in a tree structure, supported on ground via an actuated frictionless revolute joint. All links

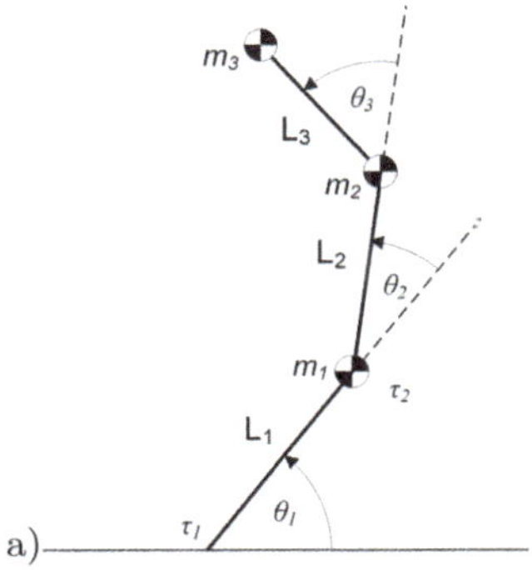

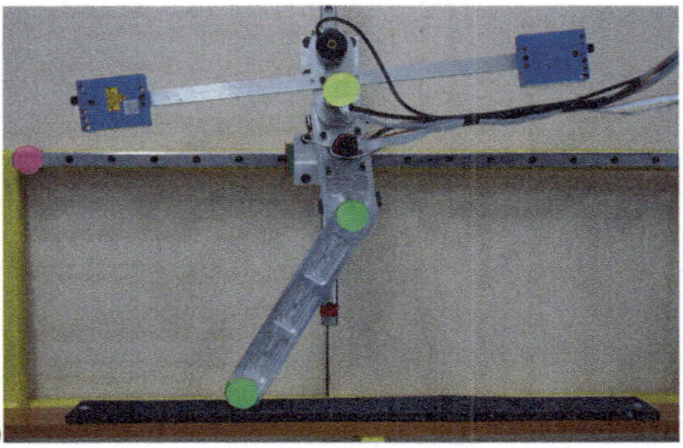

Fig. 1. a) Three link inverted pendulum, with two actuators; b) Experimental test-bed

have non-zero mass and one of the revolute joints connecting them is unactuated. As a result, the system has one degree of underactuation (3 DOF with 2 independent actuators).

In Fig. 1a, the standard 3-link inverted pendulum structure is depicted. The reference frame is attached to the pivot point, and coordinates are indicated as $\theta = [\theta_1\, \theta_2\, \theta_3]^\top \in S^1 \times S^1 \times S^1$. Since we are dealing with revolute joints instead of x_i, an angle θ_i is used. In order to establish the system dynamics one can define Lagrangian $L = K - V$, while $K = \frac{1}{2}\dot{\theta}^T D(\theta)\dot{\theta}$ denotes kinetic energy, with D being a positive definite inertia matrix, and V is the potential energy. Next, taking into account the actuation on the system one obtains

$$\frac{d}{dt}\frac{\partial L}{\partial \dot{\theta}_k} - \frac{\partial L}{\partial \theta_k} = \begin{cases} \tau_k, & k = 1, 2 \\ 0, & k = 3 \end{cases} \tag{5}$$

with $\tau_k \in \mathbb{R}$. The mathematical model of the system dynamics thus takes the following standard form

$$D(\theta)\ddot{\theta} + C(\theta, \dot{\theta})\dot{\theta} + G_r(\theta) = B\tau, \tag{6}$$

where: $M = \begin{bmatrix} m_{11} & m_{12} & m_{13} \\ m_{21} & m_{22} & m_{23} \\ m_{31} & m_{32} & m_{33} \end{bmatrix}, \quad C = \begin{bmatrix} c_{11} & c_{12} & c_{13} \\ c_{21} & c_{22} & c_{23} \\ c_{31} & c_{32} & c_{33} \end{bmatrix}, \quad G_r = \begin{bmatrix} G_1 \\ G_2 \\ G_3 \end{bmatrix},$

$$\begin{aligned} m_{11} &= a_1 + a_2 + a_3 + a_4 + a_5 + 2r_1 + 2r_2 + 2r_3 \\ m_{12} &= a_2 + a_3 + a_4 + r_1 + r_2 + 2r_3 \\ m_{13} &= a_3 + r_1 + r_3 \\ m_{21} &= m_{12} \\ m_{22} &= a_2 + a_3 + a_4 + 2r_3 \\ m_{23} &= a_3 + r_3 \\ m_{31} &= m_{13} \\ m_{32} &= m_{23} \\ m_{33} &= a_3 \end{aligned} \quad , \quad \begin{aligned} c_{11} &= -d_1\dot{\theta}_2 - d_2\dot{\theta}_3 \\ c_{12} &= -d_1(\dot{\theta}_1 + \dot{\theta}_2) - d_2\dot{\theta}_3 \\ c_{13} &= -d_2(\dot{\theta}_1 + \dot{\theta}_2 + \dot{\theta}_3) \\ c_{21} &= d_1\dot{\theta}_1 - d_3\dot{\theta}_3 \\ c_{22} &= -d_3\dot{\theta}_3 \\ c_{23} &= -d_3(\dot{\theta}_1 + \dot{\theta}_2 + \dot{\theta}_3) \\ c_{31} &= d_2\dot{\theta}_1 + d_3\dot{\theta}_2 \\ c_{32} &= d_3(\dot{\theta}_1 + \dot{\theta}_2) \\ c_{33} &= 0 \end{aligned}$$

where: $a_1 = m_1L_1^2, a_2 = m_2L_2^2, a_3 = m_3L_3^2, a_4 = m_3L_2^2, a_5 = (m_2 + m_3)L_1^2, r_1 = L_1L_3m_3\cos(\theta_2 + \theta_3), r_2 = L_1L_2(m_2 + m_3)\cos\theta_2, r_3 = L_2L_3m_3\cos\theta_3$, $b_1 = (m_1 + m_2 + m_3)L_1\cos\theta_1$, $b_2 = (m_2 + m_3)L_2\cos(\theta_1 + \theta_2)$, $b_3 = m_3L_3\cos(\theta_1 + \theta_2 + \theta_3)$, $d_1 = L_1L_3m_3\sin(\theta_2+\theta_3)+L_1L_2(m_2+m_3)\sin\theta_2$, $d_2 = L_1L_3m_3\sin(\theta_2+\theta_3) + L_2L_3m_3\sin\theta_3$, $d_3 = L_2L_3m_3\sin\theta_3$, $B = [[1\ 0\ 0]^\top, [0\ 1\ 0]^\top]^\top$, $\tau \in \mathbb{R}^2$ is the control input, m_i, L_i denotes i-th link mass and length, respectively (see Table 1), $G_1 = g(b_1 + b_2 + b_3)$, $G_2 = g(b_2 + b_3)$, $G_3 = gb_3$, are gravity forces components, g is the gravitational acceleration.

Robot Parameters. In simulations the robot parameters (Table 1) were selected in order to adapt them to the physically existing mechanism, i.e. to the one-legged robot presented in Fig. 1b. This one-legged mechanism can be studied as a triple pendulum. The considered robot was built in the Institute of Automation and Robotics at Poznan University of Technology, as a testbed for a four-legged walking robot investigated in [5,8]. The physical model depicted in

Table 1. Robot parameters

i – Link	m_i – Mass [kg]	Centre of mass [m]	L_i – Length [m]	Inertia [kg m^2]
1	1.118	0.062	0.07	0.0118
2	1.593	0.074	0.15	0.0119
3	0.405	0.134	0.295	0.0117

Fig. 1b is driven by Maxon 200W EC-Powermax 30 brushless motors with planetary gearhead of $N = 53$ reduction. Such drives provide the maximum torque of approximately 6 Nm. This value is considered as a saturation of the control input.

2.2 Linearization Conditions

Rewriting Eq. (6) in a form of first order differential equations one can get

$$\dot{x} = f(x) + G(x)u \tag{7}$$

where

$$f(x) = \begin{bmatrix} w_1 \\ 0 \\ w_2 \\ 0 \\ w_3 \\ -R_3 + J_1(\theta_2,\theta_3)R_1 + J_2(\theta_3)R_2 \end{bmatrix}, G(x) = [g_1, g_2] = \begin{bmatrix} 0 & 0 \\ 1 & 0 \\ 0 & 0 \\ 0 & 1 \\ 0 & 0 \\ J_1(\theta_2,\theta_3) & J_2(\theta_3) \end{bmatrix},$$

for $x = [\theta_1\ w_1\ \theta_2\ w_2\ \theta_3\ w_3]^\top$, where $w_i = \dot{\theta}_i$, $J_1(\theta_2, \theta_3) = -\frac{m_{13}}{a_3}$, $J_2(\theta_3) = -\frac{m_{23}}{a_3}$ with $u = [\ddot{\theta}_1, \ddot{\theta}_2]^\top$ as inputs, and R_1, R_2, R_3 are scalar function depending on variables of state x, and they are not written explicitly due to lack of space.

In order to find the largest linearizable subsystem we propose to analyze the following distributions. First we analyze distribution $D_0 = \text{span}\{g_1, g_2\}$, and obviously D_0 is involutive. Next consider the following distribution $D_1 = \text{span}\{g_1, g_2, ad_f g_1, ad_f g_2\}$, where

$$ad_f g_1 = \begin{bmatrix} -1 \\ 0 \\ 0 \\ 0 \\ -J_1(\theta_2, \theta_3) \\ F_{16}(\theta_2, \theta_3, w_1, w_2, w_3) \end{bmatrix}, \quad ad_f g_2 = \begin{bmatrix} 0 \\ 0 \\ -1 \\ 0 \\ -J_2(\theta_3) \\ F_{26}(\theta_2, \theta_3, w_1, w_2, w_3) \end{bmatrix}, \tag{8}$$

Consequently we check involutivity condition of distribution D_1, with relevant Lie brackets being calculated as follows $[g_1, ad_f g_1] = [0\ 0\ 0\ 0\ 0\ F_{56}(\theta_2, \theta_3)]^\top$, $[g_1, ad_f g_2] = [0\ 0\ 0\ 0\ 0\ F_{66}(\theta_2, \theta_3)]^\top$, $[g_2, ad_f g_1] = [0\ 0\ 0\ 0\ 0\ F_{76}(\theta_2, \theta_3)]^\top$, $[g_2, ad_f g_2] = [0\ 0\ 0\ 0\ 0\ F_{86}(\theta_2, \theta_3)]^\top$, $[ad_f g_1, ad_f g_2] = [0\ 0\ 0\ 0\ F_{95}(\theta_2)\ F_{96}(\theta_2, \theta_3, w_1, w_2, w_3)]^\top$, where F_{ij} are scalar functions depending on state vector components. All scalar functions that appear in considered vector fields were carried out symbolically using MATLAB software and are omitted here.

There are two Lie brackets among $[g_i, ad_f g_i]$ that are independent modulo D_1, thus the D_1 distribution is not involutive. Therefore we calculate its closure $\bar{D}_1$, which is an involutive closure of D_1,

$$\bar{D}_1 = \text{span}\{g_1, g_2, ad_f g_1, ad_f g_2, [g_1, ad_f g_1], [ad_f g_1, ad_f g_2]\} \tag{9}$$

It turns out that the $\bar{D}_1$ distribution is involute (i.e. $\bar{D}_1$ is of rank 6, equal to the dimension of the state vector).

Then one needs to find an output function h that anihilates $\bar{D}_1$, i.e.

$$\left[\frac{\partial h}{\partial x_1}\ \frac{\partial h}{\partial x_2}\ \frac{\partial h}{\partial x_3}\ \frac{\partial h}{\partial x_4}\ \frac{\partial h}{\partial x_5}\ \frac{\partial h}{\partial x_6}\right] \left[g_1\ g_2\ ad_f g_1\ ad_f g_2\ [g_1, ad_f g_1]\ [ad_f g_1, ad_f g_2]\right] = 0$$

As a result we get:

$$\frac{\partial h}{\partial w_1} = 0,\ \frac{\partial h}{\partial w_2} = 0,\ \frac{\partial h}{\partial \theta_1} = 0,\ \frac{\partial h}{\partial \theta_2} = 0,\ \frac{\partial h}{\partial w_3} = 0,\ \frac{\partial h}{\partial \theta_3} = 0. \tag{10}$$

It is trivial that the only solution of Eq. (10) is h = constant because $\bar{D}_1$ is of full rank 6. As a conclusion in the case considered here the largest feedback linearizable subsystem is of dimension 4, (which is obvious and unfortunately this dimension cannot be higher).

3 Stabilization Problem

Due to the fact that the largest feedback linearizable subsystem is of dimension 4 (which seems to be obvious, but we have expected better results to those presented in Sect. 2), we propose here to implement known method introduced by

Spong [10] (namely collocated and noncollocated feedback linearization) instead of making use of quasivelocities in building a state feedback transformation and applying it directly to the system (7). This approach is under current investigation and will be reported shortly.

To our best knowledge Spong's method has not been implemented to 3-link pendulum robot.

The aim of the work is to examine an implementation of a hybrid controller using the formalism presented in [10] to stabilize a pendulum around its top unstable position, taking into account the limitations and constraints resulting from practical conditions (existing robot).

Stabilization will be obtained with the two commonly known approaches which utilizes the collocated and non-collocated methods described in [10]. The additional LQR controller is used to stabilize the system near the equilibrium point.

3.1 Control Algorithm

Let's rewrite the Eq. (6) in the following form

$$\begin{bmatrix} m_{11} & m_{12} & m_{13} \\ m_{12} & m_{22} & m_{23} \\ m_{13} & m_{23} & m_{33} \end{bmatrix} \begin{bmatrix} \ddot{\theta}_1 \\ \ddot{\theta}_2 \\ \ddot{\theta}_3 \end{bmatrix} + \begin{bmatrix} c_{11} & c_{12} & c_{13} \\ c_{21} & c_{22} & c_{23} \\ c_{31} & c_{32} & c_{33} \end{bmatrix} \begin{bmatrix} \dot{\theta}_1 \\ \dot{\theta}_2 \\ \dot{\theta}_3 \end{bmatrix} + \begin{bmatrix} G_1 \\ G_2 \\ G_3 \end{bmatrix} = \begin{bmatrix} \tau_1 \\ \tau_2 \\ 0 \end{bmatrix}. \tag{11}$$

and assume that $C_1 = [c_{11}, c_{12}, c_{13}]\,\dot{\theta}$, $C_2 = [c_{21}, c_{22}, c_{23}]\,\dot{\theta}$, $C_3 = [c_{31}, c_{32}, c_{33}]\,\dot{\theta}$ and other symbols are defined in Subsect. 2.1.

Case I – Non-collocated Linearization w.r.t. $\ddot{\theta}_1$. Let's apply the method of non-collocated feedback linearization to the system (11). Obtaining $\ddot{\theta}_1$ from the last equation of (11) one gets the following relationship

$$\ddot{\theta}_1 = -\frac{m_{23}\ddot{\theta}_2 + m_{33}\ddot{\theta}_3 + C_3 + G_3}{m_{13}}. \tag{12}$$

Substituting $\ddot{\theta}_1$ into the first and second equation of (11), results in

$$\bar{m}_{12}\ddot{\theta}_2 + \bar{m}_{13}\ddot{\theta}_3 + \bar{C}_1 + \bar{G}_1 = \tau_1 \tag{13}$$

$$\bar{m}_{22}\ddot{\theta}_2 + \bar{m}_{23}\ddot{\theta}_3 + \bar{C}_2 + \bar{G}_2 = \tau_2 \tag{14}$$

where: $\bar{m}_{12} = m_{12} - \frac{m_{11}m_{23}}{m_{13}}$, $\bar{m}_{13} = m_{13} - \frac{m_{11}m_{33}}{m_{13}}$, $\bar{C}_1 = C_1 - \frac{m_{11}}{m_{13}}C_3$, $\bar{G}_1 = G_1 - \frac{m_{11}}{m_{13}}G_3$, $\bar{m}_{22} = m_{22} - \frac{m_{12}m_{23}}{m_{13}}$, $\bar{m}_{23} = m_{23} - \frac{m_{12}m_{33}}{m_{13}}$, $\bar{C}_2 = C_2 - \frac{m_{12}}{m_{13}}C_3$, $\bar{G}_2 = G_2 - \frac{m_{12}}{m_{13}}G_3$, (analysis concerning when m_{13} is nonzero is discussed in Appendix) and the linearizing controller $u_h = [\tau_1, \tau_2]^\top$ can be defined by (15) and (16):

$$\tau_1 = \bar{m}_{12}v_2 + \bar{m}_{13}v_3 + \bar{C}_1 + \bar{G}_1 \tag{15}$$

$$\tau_2 = \bar{m}_{22}v_2 + \bar{m}_{23}v_3 + \bar{C}_2 + \bar{G}_2 \tag{16}$$

where v_2 i v_3 are new control inputs defined as follows:

$$v_2 = \ddot{\theta}_2 = \ddot{\theta}_2^d + K_2^D(\dot{\theta}_2^d - \dot{\theta}_2) + K_2^P(\theta_2^d - \theta_2) \quad (17)$$

$$v_3 = \ddot{\theta}_3 = \ddot{\theta}_3^d + K_3^D(\dot{\theta}_3^d - \dot{\theta}_3) + K_3^P(\theta_3^d - \theta_3) \quad (18)$$

where K_2^D, K_2^P, K_3^D i K_3^P are positive gains, and $\theta_2^d, \dot{\theta}_2^d, \ddot{\theta}_2^d, \theta_3^d, \dot{\theta}_3^d, \ddot{\theta}_3^d$ denote desired values at the equilibrium point.

Case II – Non-collocated Linearization w.r.t. $\ddot{\theta}_2$. Similarly to the previous case, calculating $\ddot{\theta}_2$ from the last equation of (11) one gets the following relationship

$$\ddot{\theta}_2 = -\frac{m_{13}\ddot{\theta}_1 + m_{33}\ddot{\theta}_3 + C_3 + G_3}{m_{23}} \quad (19)$$

and applying similar algebra, can be obtained the linearizing controller $u_h = [\tau_1, \tau_2]^\top$ defined by (20) and (21) (analysis concerning when m_{23} is nonzero is discussed in Appendix):

$$\tau_1 = \bar{m}_{11}v_1 + \bar{m}_{13}v_3 + \bar{C}_1 + \bar{G}_1 \quad (20)$$

$$\tau_2 = \bar{m}_{21}v_1 + \bar{m}_{23}v_3 + \bar{C}_2 + \bar{G}_2 \quad (21)$$

where $\bar{m}_{11} = m_{11} - \frac{m_{12}m_{13}}{m_{23}}$, $\bar{m}_{13} = m_{13} - \frac{m_{12}m_{33}}{m_{23}}$, $\bar{C}_1 = C_1 - \frac{m_{12}}{m_{23}}C_3$, $\bar{G}_1 = G_1 - \frac{m_{12}}{m_{23}}G_3$, $\bar{m}_{21} = m_{12} - \frac{m_{13}m_{22}}{m_{23}}$, $\bar{m}_{23} = m_{23} - \frac{m_{22}m_{33}}{m_{23}}$, $\bar{C}_2 = C_2 - \frac{m_{22}}{m_{23}}C_3$, $\bar{G}_2 = G_2 - \frac{m_{22}}{m_{23}}G_3$, and v_1 i v_3 are new control inputs:

$$v_1 = \ddot{\theta}_1 = \ddot{\theta}_1^d + K_1^D(\dot{\theta}_1^d - \dot{\theta}_1) + K_1^P(\theta_1^d - \theta_1) \quad (22)$$

$$v_3 = \ddot{\theta}_3 = \ddot{\theta}_3^d + K_3^D(\dot{\theta}_3^d - \dot{\theta}_3) + K_3^P(\theta_3^d - \theta_3) \quad (23)$$

where K_1^D, K_1^P, K_3^D i K_3^P are controller gains, and $\theta_1^d, \dot{\theta}_1^d, \ddot{\theta}_1^d, \theta_3^d, \dot{\theta}_3^d, \ddot{\theta}_3^d$ denote desired values at the equilibrium point.

Case III – Collocated Linearization. Obtaining $\ddot{\theta}_3$ from the last equation of (11) one gets a following relationship

$$\ddot{\theta}_3 = -\frac{m_{13}\ddot{\theta}_1 + m_{23}\ddot{\theta}_2 + C_3 + G_3}{m_{33}} \quad (24)$$

and making similar calculations (here notice that m_{33} is always nonzero), the following result for linearizing controller $u_h = [\tau_1, \tau_2]^\top$ was obtained

$$\tau_1 = \bar{m}_{11}v_1 + \bar{m}_{12}v_2 + \bar{C}_1 + \bar{G}_1 \quad (25)$$

$$\tau_2 = \bar{m}_{21}v_1 + \bar{m}_{22}v_2 + \bar{C}_2 + \bar{G}_2 \quad (26)$$

where $\bar{m}_{11} = m_{11} - \frac{m_{13}^2}{m_{33}}$, $\bar{m}_{12} = m_{12} - \frac{m_{13}m_{23}}{m_{33}}$, $\bar{C}_1 = C_1 - \frac{m_{13}}{m_{33}}C_3$, $\bar{G}_1 = G_1 - \frac{m_{13}}{m_{33}}G_3$, $\bar{m}_{21} = m_{12} - \frac{m_{13}m_{23}}{m_{33}}$, $\bar{m}_{22} = m_{22} - \frac{m_{23}^2}{m_{33}}$, $\bar{C}_2 = C_2 - \frac{m_{23}}{m_{33}}C_3$, $\bar{G}_2 = G_2 - \frac{m_{23}}{m_{33}}G_3$. and v_1 i v_2 are additional control inputs

$$v_1 = \ddot{\theta}_1 = \ddot{\theta}_1^d + K_1^D(\dot{\theta}_1^d - \dot{\theta}_1) + K_1^P(\theta_1^d - \theta_1) \quad (27)$$

$$v_2 = \ddot{\theta}_2 = \ddot{\theta}_2^d + K_2^D(\dot{\theta}_2^d - \dot{\theta}_2) + K_2^P(\theta_2^d - \theta_2) \quad (28)$$

where K_1^D, K_1^P, K_2^D i K_2^P are positive gains, and $\theta_1^d, \dot{\theta}_1^d, \ddot{\theta}_1^d, \theta_2^d, \dot{\theta}_2^d, \ddot{\theta}_2^d$ denote desired values at the equilibrium point.

Stabilizing Controller. The following linear controller

$$u_{Lin} = -K(x_r - x), \tag{29}$$

which stabilizes the robot at the equilibrium point is proposed. In (29) values $x_r = [\theta_1^d\ \theta_2^d\ \theta_3^d\ \dot{\theta}_1^d\ \dot{\theta}_2^d\ \dot{\theta}_3^d\ ]^\top$ and $K = \begin{bmatrix} k_1\ k_2\ k_3\ k_4\ \ k_5\ \ k_6 \\ k_7\ k_8\ k_9\ k_{10}\ k_{11}\ k_{12} \end{bmatrix}$, stand for the reference state and the controller gains, respectively. The elements of the matrix K were obtained online at each simulation step, with the use of LQR-like controller, for the linear approximation of Eq. (6) at the equilibrium point, making use of the following cost function J, dependent on state and control signal, defined as

$$J = \int_t^\infty \left(x^T Q x + {u_{Lin}}^T R u_{Lin}\right) d\tau, \tag{30}$$

where $Q = \text{diag}[0.5\ 0.5\ 1\ 0.5\ 0.5\ 5]$ and $R = \text{diag}[1\ 1]$ are weight matrices for state and input vectors, respectively, chosen by trial and error method. As both controllers (15–16) and (29) (or (20–21, 29) or (25–26, 29)) have different convergence domains, one can propose to combine them in order to increase the attraction set, obtaining the control law for the pendulum in the following form

$$u = \begin{cases} u_h & \text{for swing,} \\ u_{Lin} & \text{for stabilization.} \end{cases} \tag{31}$$

Consequently, the proposed control law for the pendulum consists of two feedbacks. The first one is a swing-up-like type, whose task is to bring the manipulator near the equilibrium point. When the robot state is in a set containing the equilibrium it switches to the linear controller, Eq. (29), which is designed to keep the system at the equilibrium pose.

3.2 Zero Dynamics

The zero dynamics was obtained for each Case, on the basis of conclusion from Sect. 2.2, i.e. that $h = const.$

For Case I one can assume that $\theta_2^d = 0, \theta_3^d = 0, \dot{\theta}_2^d = 0, \dot{\theta}_3^d = 0, \ddot{\theta}_2^d = 0$ and $\ddot{\theta}_3^d = 0$ and substituting them into Eq. (11) one obtains the relationship for zero dynamics

$$(a_1 + m_3 L_1 L_3 + m_3 L_2 L_3)\ddot{\theta}_1 + g m_3 L_3 \cos\theta_1 = 0 \tag{32}$$

Similarly in Case II one can assume that $\theta_1^d = \frac{\pi}{2}, \theta_3^d = 0, \dot{\theta}_1^d = 0, \dot{\theta}_3^d = 0, \ddot{\theta}_1^d = 0$ and $\ddot{\theta}_3^d = 0$ that leads to zero dynamics in the following form

$$(a_3 + m_3 L_2 L_3)\ddot{\theta}_2 - g m_3 L_3 \sin\theta_2 = 0. \tag{33}$$

Finally, in Case III, assuming that $\theta_1^d = \frac{\pi}{2}, \theta_2^d = 0, \dot{\theta}_1^d = 0, \dot{\theta}_2^d = 0, \ddot{\theta}_1^d = 0$ and $\ddot{\theta}_2^d = 0$, the zero dynamics is calculated as follows

$$a_3\ddot{\theta}_3 - gm_3L_3 \sin\theta_3 = 0. \tag{34}$$

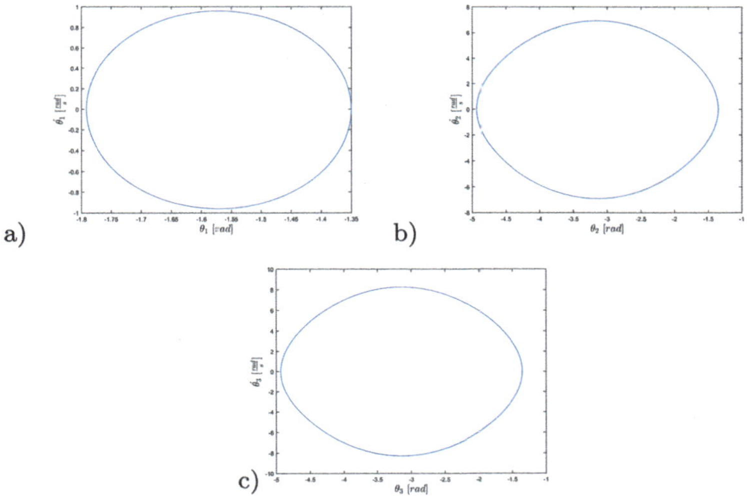

Fig. 2. Zero dynamics, a) noncollocated w.r.t. θ_1, b) noncollocated w.r.t. θ_2, c) collocated.

Each of these zero dynamics and their phase portraits (Fig. 2) are locally stable and formed by closed curves.

4 Simulation Results

This section provides simulation results of implementing control method (31) to the system in a form of (6). The aim of the presented simulations is to check the performance of the controller applied to stabilization of pendulum in the upright position. Another purpose is to verify the convergence domain for the closed-loop system, to define a set of initial conditions that stabilize the robot. The algorithm verification is carried out in two ways, with the use of non-collocated and collocated approach. Taking into account properties of the real robot the torque magnitude is restricted to 6 Nm. Thus, the algorithm is being checked whether is capable of controlling the model of physical robot discussed in Sect. 2.

In simulations it was assumed that the desired stabilization pose is the upright position for which the angles θ_1^d, θ_2^d and θ_3^d were equal $\frac{\pi}{2}, 0$ and 0, respectively. The hybrid controller (31) was considered to verify the effectiveness of the proposed control scheme. To determine which control law should be chosen during the stabilization process, the following switching procedure described by Algorithm 1 was implemented.

Algorithm 1. Switching procedure for Eq. (31)

if $(\sqrt{(\theta_1 - \theta_{1_d})^2 + (\theta_2 - \theta_{2_d})^2 + (\theta_3 - \theta_{3_d})^2} < \epsilon))$ **then**
    $u = u_{Lin}$
else
    $u = u_h$
end if

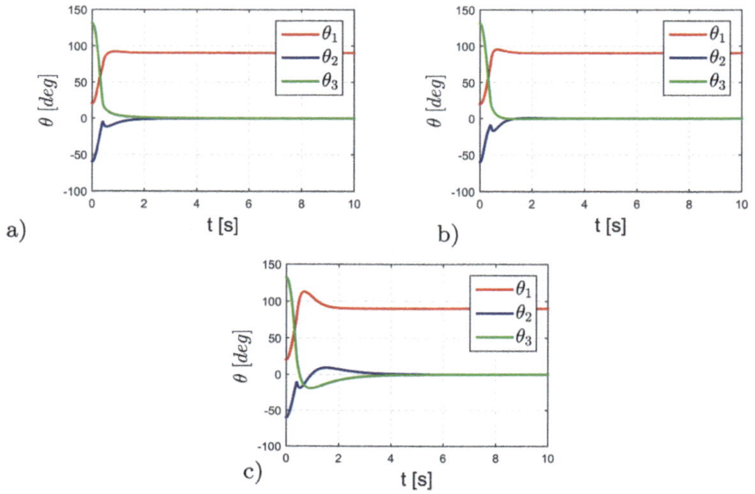

Fig. 3. Angular positions of links, a) non-collocated w.r.t. θ_1, b) non-collocated w.r.t. θ_2, c) collocated.

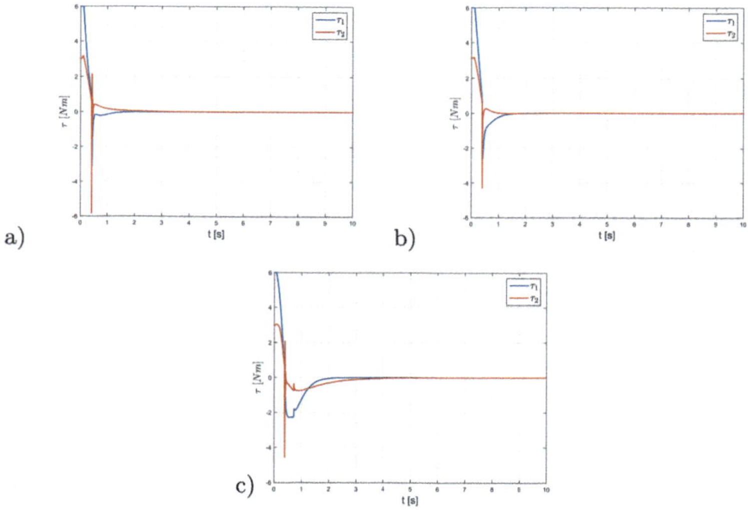

Fig. 4. Motor torque, a) non-collocated w.r.t. θ_1, b) non-collocated w.r.t. θ_2, c) collocated.

The trial and error method was being taken as successful while the error between actual and desired angle was smaller than 1° within simulation time $t = 10\,\mathrm{s}$.

Another restriction imposed on the model were the initial angular velocities equal zero. This situation is not favorable, as there is no any initial kinetic energy in the system at the beginning of the pendulum movement. Hence, it is required to provide a great amount of energy to drive the robot from starting to desired position, and not to fall at the very beginning of the movement.

The convergence area of considered method was analyzed and as a result a set of initial conditions that lead to stable upright position was obtained. It is obvious that not all initial conditions lead to stable position, but due to the limited space a figure corresponding to convergence area was omitted. To see how the algorithm drives the robot to the reference upright position one of the successful trails was chosen. The common exemplary successful initial condition (for Case I, II and III) was chosen as $\theta_{1_0} = 20°$ and $\theta_{2_0} = -60°$ and $\theta_{3_0} = 131°$. The obtained angular trajectories are shown in Fig. 3 for the first (3a), the second (3b) and third case (3c), respectively. Figure 4 depicts the motor torque expended during stabilization process for exemplary initial conditions, for all cases.

5 Conclusions

It was shown that for the triple inverted pendulum with two input signals, the largest feedback linearizable subsystem is of dimension 4. Imposing control signals on the first and second joint results that the output function is trivial namely constant. Further work will consider alternative control schemes, among them using pseudo-velocity scalar variables to linearize four dimensional system along with comparative analysis of different control analysis in studying triple pendulum. Experimental work is in progress and will be a subject of publication in the near future. Placement of two control signals in two alternative places of the three link robot along with zero dynamics analysis will be discussed too.

Acknowledgments. We express our thanks to Prof. W. Respondek for fruitful discussion and useful comments.

Appendix

In order to guarantee that the $\bar{D}_2$ distribution is full rank some scalar functions from Sect. 2.2 need to be nonzero. The F_{95} scalar function is trivial and results in $\theta_2 \neq 2k\pi$. The scalar function F_{56} is complicated and hard to be written analytically. However one can depict it graphically (omitted here) assuming that θ_2 and θ_3 varies form $(-\frac{\pi}{2}, \frac{\pi}{2})$ and thus their zeros can be observed for three link manipulator considered here. Calculations from Sect. 3.1 are valid when the system is not in its singularity, when determinants of matrices: $\det \begin{bmatrix} \bar{m}_{12} & \bar{m}_{13} \\ \bar{m}_{22} & \bar{m}_{23} \end{bmatrix}^{-1} = \frac{m_{13}}{\det M}$, $\det \begin{bmatrix} \bar{m}_{11} & \bar{m}_{13} \\ \bar{m}_{21} & \bar{m}_{23} \end{bmatrix}^{-1} = \frac{m_{23}^2}{\det M}$ and $\det \begin{bmatrix} \bar{m}_{11} & \bar{m}_{12} \\ \bar{m}_{21} & \bar{m}_{22} \end{bmatrix}^{-1} = \frac{m_{33}}{\det M}$, (here $m_{33} > 0$ by definition) must not be equal to zero, i.e. $m_{13} \neq 0$, $m_{23} \neq 0$, and moreover: $J_1 \neq 0$ and $J_2 \neq 0$, respectively, for two cases:

1. $a_3 = m_3L_3(L_1 + L_2)$ for $\theta_2 = 0$, $\theta_3 = \pi + 2k\pi$;
   $a_3 < m_3L_3(L_1 + L_2)$ for solution of the following equation: $a_3 = -r_1 - r_3$.
2. $a_3 = m_3L_2L_3$ for $\theta_3 = \pi + 2k\pi$;
   $a_3 < m_3L_2L_3$ for $\theta_3 = -\arccos(\frac{a_3}{m_3L_2L_3})$.

References

1. Eltohamy, K.G., Kuo, C.Y.: Nonlinear optimal control of a triple link inverted pendulum with single control input. Int. J. Control **69**(2), 239–256 (1998)
2. Furuta, K., Ochiai, T., Ono, N.: Attitude control of a triple inverted pendulum. Int. J. Control **39**(6), 1351–1365 (1984)
3. Glück, T., Eder, A., Kugi, A.: Swing-up control of a triple pendulum on a cart with experimental validation. Automatica **49**(3), 801–808 (2013). https://doi.org/10.1016/j.automatica.2012.12.006
4. Jakubczyk, B., Respondek, W.: On linearization of control systems. Biuletyn Polskiej Akademii Nauk **28**(9–10) (1980)
5. Kozłowski, K., Kowalski, M., Michalski, M., Parulski, P.: Universal multiaxis control system for electric drives. IEEE Trans. Ind. Electron. **60**(2), 691–698 (2013)
6. Li, S., Moog, C., Respondek, W.: Maximal feedback linearization and its internal dynamics with applications to mechanical systems on R^4. Int. J. Robust Nonlinear Control **29**(9), 2639–2659 (2019). https://doi.org/10.1002/rnc.4507
7. Medrano-Cerda, G.A.: Robust stabilization of a triple inverted pendulum-cart. Int. J. Control **68**(4), 849–866 (1997)
8. Michalski, M., Kowalski, M., Pazderski, D.: Quadruped walking robot WR-06 - design, control and sensor subsystems. In: Kozłowski, K. (ed.) Robot Motion and Control 2009, pp. 175–184. Springer, London (2009)
9. Respondek, W.: Partial linearization, decompositions and fibre systems. Theor. Appl. Nonlinear Control Syst. **85**, 137–154 (1986)
10. Spong, M.W.: Partial feedback linearization of underactuated mechanical systems. In: Proceedings of IEEE/RSJ International Conference on Intelligent Robots and Systems (IROS 1994), vol. 1, pp. 314–321 (1994)
11. Xin, X., Zhang, K., Wei, H.: Linear strong structural controllability for an n-link inverted pendulum in a cart. In: 2018 IEEE Conference on Decision and Control (CDC), pp. 1204–1209 (2018)

System Identification and Adaptive Control

Synthesis and Generation of Random Fields in Nonlinear Environment

Jarosław Figwer(✉)

Silesian University of Technology, Gliwice, Poland
Jaroslaw.Figwer@polsl.pl

Abstract. In the paper, a generalisation of the method used for adaptive generation of random fields in linear environment to the case of synthesis and generation of such fields in a nonlinear environment is presented. The random fields to be synthesised and generated are defined by their power spectral density functions. Realisations of the random fields to be generated are obtained using a synthesis and simulation method of power spectral defined random processes based on multisine random time-series. Generation of the corresponding random fields in the nonlinear environment is aided by active noise control systems used to attenuate unwanted random noise present in this environment.

Keywords: Random fields · Nonlinear systems · Multisine random time-series · Active noise control

1 Introduction

Active noise control is a technique allowing not only unwanted noise attenuation (see e.g. [24]) but also generation of local and distributed in the environment random fields with predefined spectral properties [8–11]. This generation is done with an aid of power spectral density defined multisine random time-series [3,15]. Up till now only a case of linear environment, i.e. environment that can be modelled sufficiently precisely by linear dynamic models is described in published research literature. Discussion presented in the sequel is an extension of the results presented in this literature to the case of nonlinear environment, i.e. environment that can be described by nonlinear dynamic models. A key for this extension is an approach to generation of power spectral density time-series in the nonlinear environment that can be approximated by nonlinear block-oriented dynamic models. This generation may be additionally aided by active noise control systems [21] used to attenuate an external unwanted random noise present in the nonlinear environment.

The paper is organised as follows: (1) definition of power spectral density defined multisine random time-series is reminded; (2) an approach to generation of these time-series in a nonlinear environment is proposed; (3) creation of local random fields with predefined spectral properties in a disturbed nonlinear environment with active noise control algorithms is described; (4) issues concerning

A. Bartoszewicz et al. (Eds.): FCC 2020, AISC 1196, pp. 665–677, 2020.
https://doi.org/10.1007/978-3-030-50936-1_56

creation of random fields distributed in nonlinear environment are also included. The presented discussion is illustrated by simulation examples.

2 Power Spectral Density Defined Multisine Random-Times Series

Let $v(i)$ (i denotes consecutive discrete-time instants) be a wide-sense stationary random process with the power spectral density $\Phi_{vv}(\omega T)$, where T is the sampling interval and $\omega T \in [0, 2\pi)$ denotes relative frequency. It is assumed that $\Phi_{vv}(\omega T) < \infty$ for $\omega T \in [0, 2\pi)$ and the corresponding autocorrelation function $R_{vv}(\tau)$ for lags $|\tau| > \tau_0$ satisfies the condition $R_{vv}(\tau) = 0$.

The finite N-sample power spectral density defined multisine random time-series $u(i)$ is defined [3,8,9,15] in the following way:

Definition 1. *The N-sample (N is even) power spectral density defined multisine random time-series $u(i)$ is defined in the time-domain by a sum of $\frac{N}{2}+1$ discrete-time harmonic sines, including a constant component:*

$$u(i) = \sum_{n=0}^{\frac{N}{2}} A_n \sin(\Omega n i + \phi_n), \tag{1}$$

where $\Omega = \frac{2\pi}{N}$ denotes the fundamental relative frequency, $n = 0, 1, \ldots, \frac{N}{2}$ denotes consecutive harmonics of this frequency in the range $[0, \pi]$, $i = 0, 1, \ldots, N-1$ denotes consecutive discrete-time instants, ϕ_n are phase shifts, of which $\phi_0 = \frac{\pi}{2}$ is deterministic and the remaining phase shifts are random, independent and:

- *uniformly distributed on $[0, 2\pi)$ for $n = 1, 2, \ldots, \frac{N}{2} - 1$,*
- *Bernoulli distributed $\mathcal{B}\left(\frac{1}{2}, \left\{\frac{\pi}{2}, \frac{3\pi}{2}\right\}\right)$ for $n = \frac{N}{2}$, i.e.:*

$$P\left\{\phi_{\frac{N}{2}} = \frac{\pi}{2}\right\} = P\left\{\phi_{\frac{N}{2}} = \frac{3\pi}{2}\right\} = \frac{1}{2}, \tag{2}$$

where $P\{X\}$ denotes the probability of an event X.

A_n are deterministic amplitudes of the sine components chosen

- *for $n = 1, 2, \ldots, \frac{N}{2} - 1$ as*

$$A_n = 2\sqrt{\frac{\Phi_{vv}(\Omega n)}{NT}}, \tag{3}$$

- *for $n = 0, \frac{N}{2}$ as*

$$A_n = \sqrt{\frac{\Phi_{vv}(\Omega n)}{NT}}. \tag{4}$$

Realisations of the power spectral density defined multisine random time-series $u(i)$ for the given power spectral density $\Phi_{vv}(\omega T)$ may be obtained by first synthesising its spectrum and then simulating the corresponding realisation by transforming this spectrum to the time-domain with the use of FFT algorithm, see e.g. [3,4,6,15]. These two steps allowing obtaining realisations of multisine time-series $u(i)$ based on the corresponding power spectral density $\Phi_{vv}(\omega T)$ are a part of so called multisine generator.

3 Generation of Power Spectral Density Defined Multisine Random Time-Series in a Nonlinear Environment

Let $v(i)$ be a wide-sense stationary local random field (random process) to be generated around a given point in nonlinear environment. A time-series series being realisation of this random field is generated from wave-sender placed in a given vicinity of this point and the result is measured by a sensor placed at this point. This realisation may be obtained as a realisation of power spectral density defined multisine random time-series but the obtained realisation cannot be used directly to excite the wave-sender. To obtain time-series $u(i)$ at the given point in environment the wave-sender should be excited by an another time-series $u_{in}(i)$ that may be easy and numerically effectively calculated for a linear environment knowing a dynamical model of so called secondary path [8–11]. This path consists of electronic components being the wave-sender channel, environment in which the wave propagates and measurement device. The time-series $u_{in}(i)$ can also be numerically effectively calculated for a nonlinear environment in which the secondary path may be approximated by a nonlinear block-oriented dynamic model. It is obvious that nonlinear block-oriented dynamic model of the secondary path should be identified [1,5,7,13,14,16–18,22,23,25,27,29] before calculating $u_{in}(i)$. Idea how $u_{in}(i)$ is calculated is explained using an example in which the secondary path is modelled as a Wiener system. It is done in this case in the following way: having $u(i)$ in the first step an N-sample input $y_{in}(i)$ of the output static nonlinearity is calculated by solving for each discrete-time instant i $(i = 0, 1, \ldots, N-1)$ a nonlinear equation [28] and then, in the second step, using N-sample $y_{in}(i)$ and the frequency response of dynamical component of Wiener system its N-sample input $u_{in}(i)$ is calculated using FFT algorithm [8–10,16]. This idea is illustrated by the following simulation example.

3.1 Example 1

Let us assume that the secondary path is modelled by a discrete-time Wiener system having the following components:

- an input linear dynamic subsystem described by the following transfer function:

$$K(z^{-1}) = \frac{1.0 - 1.1z^{-1}}{1.0 - 1.5z^{-1} + 0.7z^{-2}}. \tag{5}$$

Its input is $u_{in}(i)$ and output is $y_{in}(i)$;
- the output static nonlinearity:

$$u(i) = y_{in}(i) + y_{in}^3(i). \tag{6}$$

It is assumed that sampling interval T is equal to 1 s. A multisine random time-series at the output of secondary path is defined by the following power spectral density:

$$\Phi_{vv}(\omega T) = e^{2.23\cos(\omega T)}, \tag{7}$$

where $\omega T \in [0, 2\pi)$. This power spectral density is presented in Fig. 1 (blue line). The synthesised and simulated $N = 1024$-sample power spectral density defined multisine random time-series realisation $u(i)$ and the corresponding calculated time-series $u_{in}(i)$ at input of the secondary path are presented in Fig. 2. In Fig. 1 the periodogram [2] of secondary path output (red line) excited by $u_{in}(i)$ and additionally processed by an 8-bit quantizer (time-series $u_{rec}(i)$) is compared with the pattern (7). Variance of the difference between $u(i)$ and $u_{rec}(i)$ was equal to $5.04 \cdot 10^{-4}$. The corresponding variance without this 8-bit quantizer was equal to $7 \cdot 10^{-30}$.

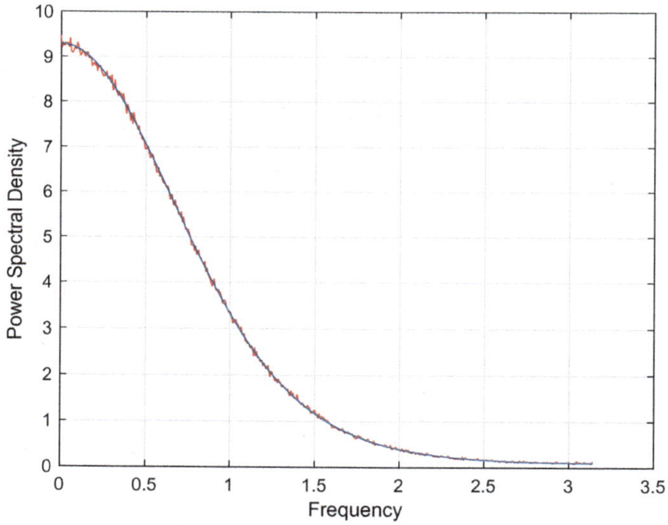

Fig. 1. Pattern of power spectral density defined multisine random time-series at the output of secondary path being a Wiener system (blue line) and periodogram calculated for $u_{rec}(i)$ (red line)

4 Creation of Local Random Fields in a Disturbed Nonlinear Environment

The calculated values of time-series $u_{in}(i)$ processed additionally by a nonlinear secondary path reach, as a wave propagating in a nonlinear environment, the

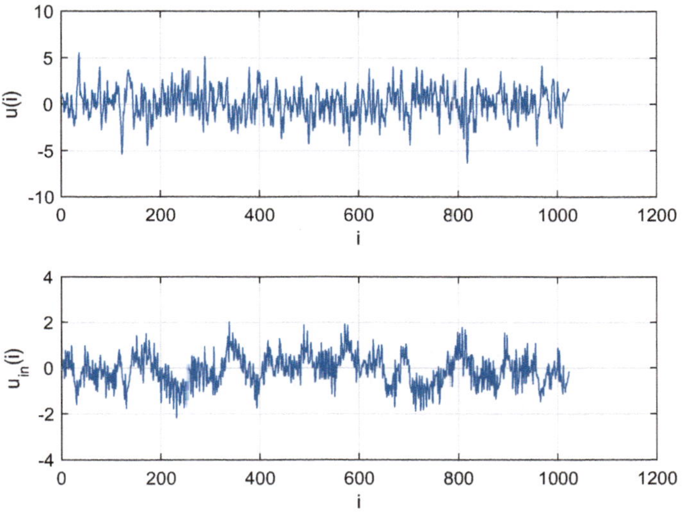

Fig. 2. Input multisine random time-series realisation $u_{in}(i)$ calculated for the output $u(i)$ of the secondary path being a Wiener system

point around which the corresponding local random field is created. Model of the secondary path used to calculate values of $u_{in}(i)$ may change during local random field generation and an update of this model during creation of the local field based on on-line secondary path model identification [5,7,15,23,29] with multisine random excitations may be sometimes necessary. This on-line identification is done using values of time-series $u_{in}(i)$ and time-series $e(i)$ from the sensor placed at point around which the local random field is created. Signal from this sensor may also be used to:

- reduce an influence of random noise present in a nonlinear environment on obtained local random field by using ideas of active noise control techniques. Designing the corresponding active noise control system used to attenuate the mentioned random noise, nonlinearities of primary, secondary and reference signal measurement paths should be taken into account;
- compensate an influence of the residual noise, being the result of using active noise control system, on properties of the obtained local random field. This compensation is done by changing the power spectral density $\Phi_{uu}(\omega T)$ used by the multisine generator to obtaining realisations $u_{in}(i)$ based on identified power spectral density $\Phi_{ee}(\omega T)$ of signal $e(i)$ returned by the sensor around which the local random field is created;
- an additional shaping of the local random field to be created in the case when there is no possibility to obtain the local random field with given pattern, i.e. if for all frequencies $\omega T \in [0, \pi]$ the following condition $\Phi_{vv}(\omega T) \geq \Phi_{ee}(\omega T)$ is not satisfied then there is no possibility to obtain the local random field with given pattern. It implies that the local random field creation system should

return the corresponding information and local random field creation problem may be changed into a local random field shaping problem, i.e. creation of a local random field with properties similar to the given pattern with a scale of similarity.

In Fig. 3 the block diagram of a control system used for creation of local random fields in a noisy nonlinear environment is presented. It has the same structure as the corresponding system designated for adaptive synthesis and generation of local random fields in linear environment [8–10]. It is worth to mention that this control system is a classical active noise control system with an additional piece of software designated for local random field synthesis and generation. Dynamical properties of this control system - rate of convergence and obtained noise attenuation - may be controlled in the same way as it is done for active noise control systems [19,20,26].

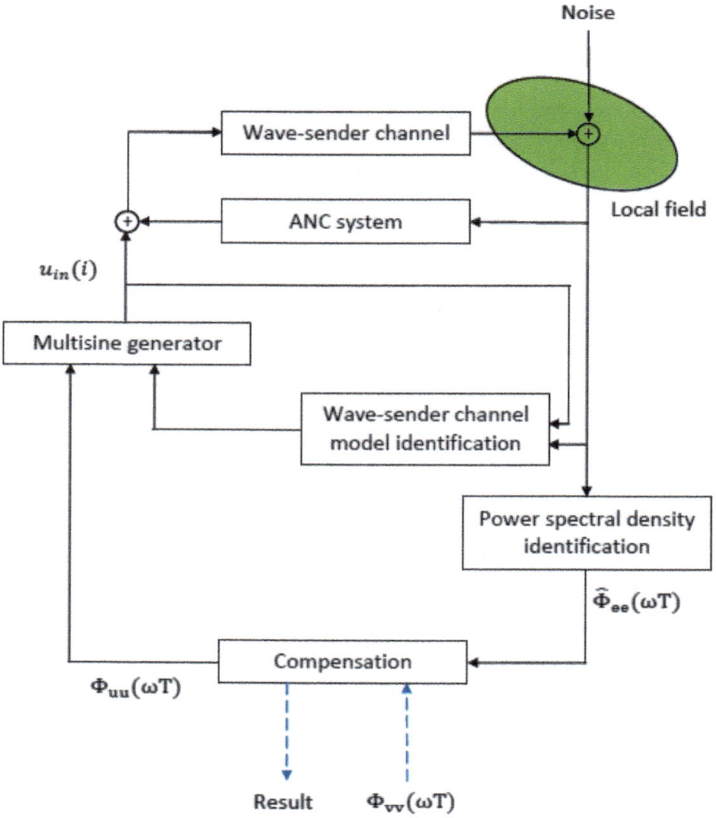

Fig. 3. Block diagram of local random field synthesis and generation system

4.1 Example 2

In this simulation example creation of an acoustic local random field in a slightly nonlinear environment using a hardware working with the constant sampling interval T equal to 0.002 s is discussed. The acoustic local random field with the pattern presented in Fig. 4 (blue line) was created around an error microphone placed in a laboratory enclosure of about 23 m^3 of cubature. This enclosure was disturbed by an external wide-sense zero-mean stationary Gaussian random noise $d(i)$ with the power spectral density presented also in Fig. 4 (red line). This noise on the way from its source to the error microphone was one sample delayed and then processed by a nonlinear transformation of the form: $d(i-1) + 0.1d^3(i-1)$. It was attenuated using a feedforward FxLMS active noise control system adopted to nonlinear environment [7]. The adaptive compensator was a linear FIR filter having 300 tuned parameters and the adaptation algorithm parameter μ was equal to 0.0009. A loudspeaker used to create the acoustic local random field was placed at about 0.6 m far from the error microphone. The corresponding secondary path was simulated and modelled as a Wiener system with the impulse response of dynamic linear component presented in Fig. 5 (it is a FIR filter with 250 elements) and the same nonlinearity as was used to transform the noise to be attenuated. It was assumed during generation of the time-series $u_{in}(i)$ that the secondary path has no nonlinearity - the term $0.1y_{in}^3(i)$ was omitted. Parameter N was assumed as equal to 1024 and after every 1024 sampling intervals new N-sample $u_{in}(i)$ was generated. The resulting power spectral density of obtained acoustic local random field was compared with

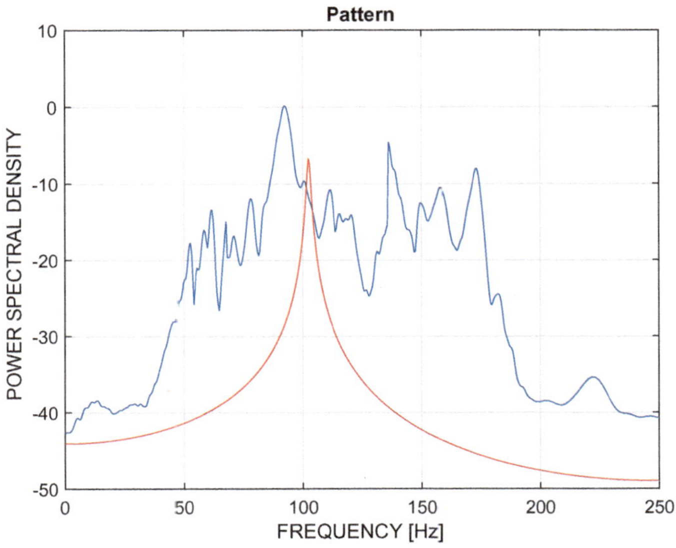

Fig. 4. Power spectral density of random field to be generated (blue line) and of external Gaussian random noise to be attenuated (red line)

the pattern in Fig. 6. This power spectral density was estimated using averaged periodogram. It follows from this example that for slightly nonlinear environment generation of local random fields may be done in the same way as it is done for linear environment. Obtained results are very satisfactory. There is no need to estimate full nonlinear block-oriented dynamic model of the secondary path - only estimate of its linear component [12,15] is necessary.

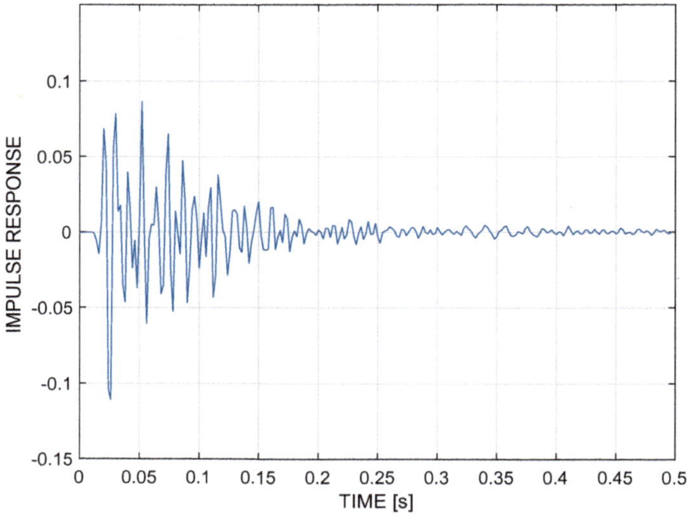

Fig. 5. Impulse response of the linear component of Wiener system modeling secondary path

4.2 Example 3

In this simulation example influence of nonlinearities on functioning of local random field synthesis and generation system was more noticeable than in the previous simulation example. Models of primary and secondary paths were still Wiener systems but with changed static nonlinearities. The changed static nonlinearities were sums of linear component having the coefficient equal to 1 and the third power of this component multiplied by different coefficients. These coefficients were equal to: 1 for the primary path, 0.5 for the simulated secondary path, 0.8 for the simulated model of secondary path and 1 for the model used by algorithm generating time-series $u_{in}(i)$ in a nonlinear environment. The corresponding linear dynamic components of all mentioned Wiener systems were the same as in the previous example. In such defined nonlinear environment local random field synthesis and generation system working correctly in slightly nonlinear environment gave no satisfying results. This system started to work properly when in generation of time-series $u_{in}(i)$ assumed static nonlinearity of the corresponding Wiener system was taken into account. The result of simulation is presented in Fig. 7. In Fig. 8 a result of local random field shaping is

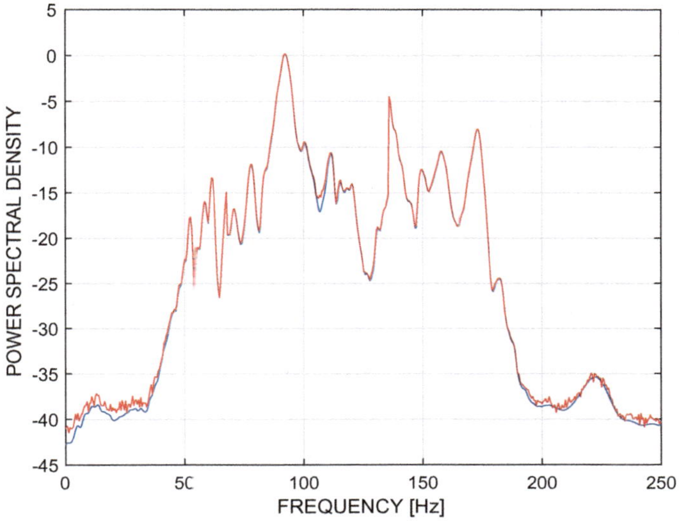

Fig. 6. Comparison of power spectral density of local random field to be generated (blue line) with a result of random field creation for the case of slightly nonlinear environment (red line)

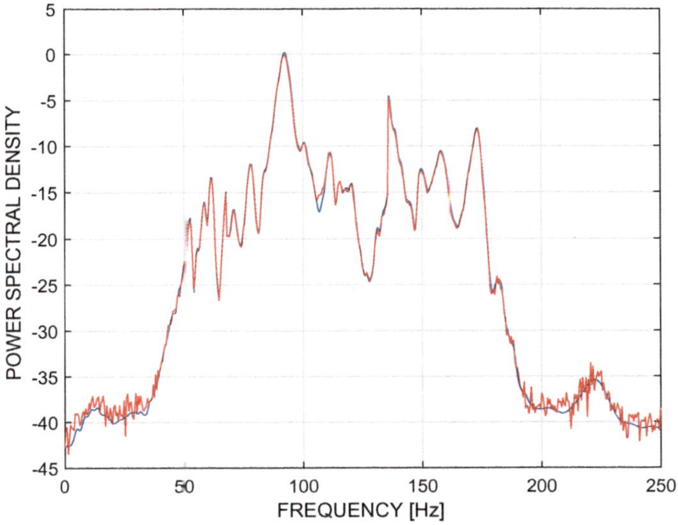

Fig. 7. Comparison of power spectral density of local random field to be generated (blue line) with a result of random field creation for the case of nonlinear environment modelled by a Wiener system (red line)

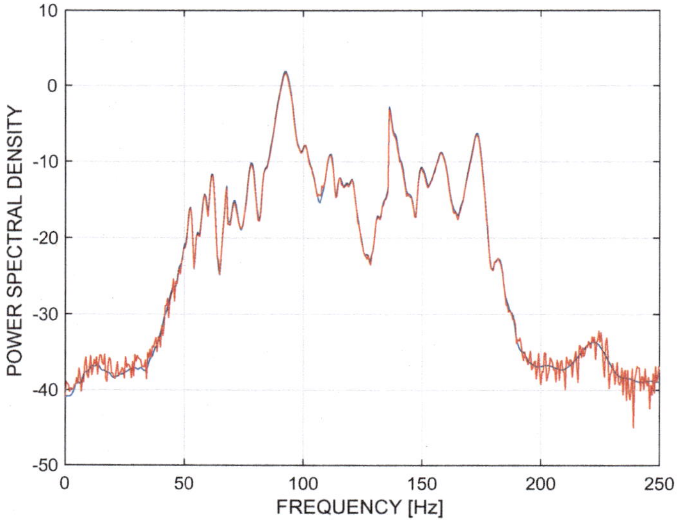

Fig. 8. Comparison of power spectral density of local random field to be generated (blue line) with a result of random field shaping in nonlinear environment modelled by a Wiener system (red line)

presented - the pattern (Fig. 4) was multiplied by the value 1.5. It follows from the presented results of simulation that using the proposed algorithm of power spectral density multisine random time-series generation in nonlinear environment aided by an active noise control system the corresponding local random field may be created very precisely in a noisy nonlinear environment.

5 Creation of Random Fields Distributed in Nonlinear Environment

Described in the previous section local random field synthesis and generation system may be used as a basic element of the corresponding control system used to creation of random fields distributed in nonlinear environment. They are defined by a distributed in the space power spectral densities. Creation of this distributed field is done in the same way as it was done for linear environment [9,11], i.e. problem of synthesis and generation of distributed in nonlinear environment random field is decomposed into a set of the corresponding autonomous local random fields synthesis and generation problems solved using multivariate orthogonal multisine random time-series. The resulting local random field synthesis and generation systems may mutually affect each other. Additionally, these systems are affected by properties of the nonlinear environment. It implies that autonomous local field synthesis and generation systems should be coordinated by a higher level strategy.

6 Conclusions

In the paper, an approach to synthesis and generation of random fields with predefined spectral properties in a nonlinear environment modelled by the corresponding nonlinear block-oriented dynamic systems is presented. The presented approach is based on a synthesis and simulation of power spectral density defined multisine random time-series extended to the case of nonlinear environment. Though the discussion is concentrated on wide-sense stationary local random fields synthesis and generation issues of synthesis and generation of random fields distributed in the nonlinear environment are also mentioned. The presented discussion may be extended to synthesis and generation of piece-wise wide-sense stationary random local and distributed in the nonlinear environment random fields.

Acknowledgments. The partial financial support of this research by The Polish Ministry of Science and Higher Education is gratefully acknowledged.

References

1. Giri, F., Bai, E.W. (eds.): Block-oriented Nonlinear System Identification. Lecture Notes in Control and Information Sciences. Springer, London (2010)
2. Bendat, J.S., Piersol, A.G.: Random Data Analysis and Measurement Procedures. Wiley, New York (1986)
3. Figwer, J.: Synthesis and Simulation of Random Processes, Zeszyty Naukowe Politechniki Ślaskiej, Seria Automatyka, Zeszyt nr 126, Gliwice (1999)
4. Figwer, J.: Multisine Transformation - Properties and Applications, Nonlinear Dynamics, pp. 331–346. Kluwer Academic Publishers, New York (2004)
5. Figwer, J.: Identyfikacja modeli Wienera z wykorzystaniem wielosinusoidalnych losowych sygnałów pobudzajacyc. In: Materiły XV Krajowej Konferencji Automatyki, Warszawa, pp. 343–348 (2005)
6. Figwer, J.: Multisine Random Number Generator. Jacek Skalmierski Computer Studio, Gliwice (2007)
7. Figwer, J.: Nonlinear secondary path model identification. In: Proceedings of the 14th International Congress on Sound and Vibration, Cairns (2007)
8. Figwer, J.: Synthesis and generation of piece-wise stationary random acoustic local fields. In: Proceedings of the 15th International Congress on Sound and Vibration, Daejeon (2008)
9. Figwer, J.: Adaptive Synthesis and Generation of Random Fields. Jacek Skalmierski Computer Studio, Gliwice (2008)
10. Figwer, J.: Adaptive generation of acoustic local fields aided by active noise control systems. In: Proceedings of the Sixteenth International Congress on Sound and Vibration, Kraków (2009)
11. Figwer, J.: A New Approach to Acoustic Distributed Field Shaping, Mechanics, pp. 65–68 (2009)
12. Figwer, J.: Modelowanie systemów nieliniowych. In: Grzech, A., Juszczyszyn, K., Kwaśnicka, H., Nguyen, N.T. (eds.), Inżynieria Wiedzy i Systemy Ekspertowe, pp. 89–100. Akademicka Oficyna Wydawnicza EXIT, Warszawa (2009)

13. Figwer, J.: Secondary path model identification in active noise control. In: Proceedings of the 15th International Conference on Methods and Models in Automation and Robotics, Miedzyzdroje, pp. 110–113 (2010)
14. Figwer, J.: Frequency response identification in the case of periodic disturbances. In: Proceedings of the 16th International Conference on Methods and Models in Automation and Robotics, Miedzyzdroje, pp. 1–4 (2011)
15. Figwer, J.: Wielosinusoidalne procesy losowe. Teoria i zastosowania, Akademicka Oficyna Wydawnicza EXIT, Warszawa (2012)
16. Figwer, J.: Hammerstein system identification with multisine excitations - quantized low-power observations case. In: Proceedings of the 18th International Conference on Methods and Models in Automation and Robotics, Miedzyzdroje, pp. 127–131 (2013)
17. Figwer, J.: Continuous-time nonlinear block-oriented dynamic system identification from sampled step and step-like responses. In: Proceedings of the 23rd International Conference on Methods and Models in Automation and Robotics, Miedzyzdroje, pp. 77–82 (2018)
18. Figwer, J.: Identification of multichannel nonlinear systems excited by realisations of mutivariate orthogonal multisine random time-series. In: Proceedings of the 24th International Conference on Methods and Models in Automation and Robotics, Miedzyzdroje, pp. 496–500 (2019)
19. Figwer J., Michalczyk, M.: On initialization of adaptation in active noise control. In: Proceedings of the 23rd International Conference on Methods and Models in Automation and Robotics, Miedzyzdroje, pp. 533–537 (2018)
20. Figwer J., Michalczyk, M., Główka, T.: Accelerating the rate of convergence for LMS-like on-line identification and adaptation algorithms. Pt. 1, Basic ideas. In: Proceedings of the 22nd International Conference on Methods and Models in Automation and Robotics, Miedzyzdroje, pp. 247–250 (2017)
21. George, N.V., Panda, G.: Advances in active noise control: a survey, with emphasis on recent nonlinear techniques. Signal Process. **92**, 363–377 (2013)
22. Giannakis, G.B., Serpedin, E.: A bibliography on nonlinear system identification. Signal Process. **81**, 533–580 (2001)
23. Główka T., Figwer, J.: Identification of linear subsystem for simple block-oriented nonlinear systems working in closed-loop - higher order spectra approach. In: Proceedings of the 18th International Conference on System Theory, Control and Computing, Sinaia, pp. 909–914 (2014)
24. Hansen, C.H., Snyder, S.D.: Active Control of Noise and Vibration. Cambridge University Press, Cambridge (1997)
25. Maliński, Ł., Figwer, J.: Nonlinear system identification using memetic algorithms. In: The 20th International Conference on Methods and Models in Automation and Robotics, Miedzyzdroje, pp. 1086–1091 (2015)
26. Michalczyk, M.I., Główka, T., Figwer, J.: Adaptation in active noise control - a simulation case study. In: Proceedings of the 21st International Conference on Methods and Models in Automation and Robotics, Miedzyzdroje, pp. 420–425 (2016)
27. Mzyk, G.: Combined Parametric-Nonparametric Identification of Block-Oriented Systems. Lecture Notes in Control and Information Sciences. Springer, Cham (2014)

28. Press, W.H., Teukolsky, S.A., Wetterling, W.T., Flannery, B.P.: Numerical Recipes in C. The Art of Scientific Computing, 2nd edn. Cambridge University Press, Sydney (1992)
29. Schoukens, M., Tiels, K.: Identification of block-oriented nonlinear systems starting from linear approximations: a survey. Automatica **85**, 272–292 (2017)

On Feasibility of Tuning and Testing Control Loops by Nonstandard Inputs

Leszek Trybus and Andrzej Bożek(✉)

Department of Computer and Control Engineering,
Rzeszów University of Technology, ul. Wincentego Pola 2, 35-959 Rzeszów, Poland
{ltrybus,abozek}@kia.prz.edu.pl

Abstract. A simple methodology for tuning and testing PID loops by means of nonstandard inputs such as a slowly varying signal, a sequence of small steps, etc., is presented. A discrete 2nd order transfer function with delay is assumed both as a model of the plant and of the closed-loop system. After smoothing the output data, the models are identified by least-squares enabling reconstruction of corresponding step responses. Given the plant response, PID controller can be tuned in standard ways. Overshoot and settling time of the closed-loop response indicate whether the system satisfies specification.

Keywords: Identification · Least-squares · Step response · PID tuning

1 Introduction

Two classical methods are most common for tuning PID loops, namely step response and relay oscillations. Both use standard inputs to detect plant dynamics, the former a step and the latter an on-off control. Here we consider an extension of the first method to include nonstandard inputs, such as a slowly varying signal, a sequence of small steps or other. By recording the response a simplified model suitable for controller tuning may be identified and the step response reconstructed by simulation. Then the PID settings can be calculated in a number of ways.

Similarly, a step response is needed to test or diagnose a closed-loop system. By looking at the shape of the response a practicing engineer can tell whether the system needs retuning or not. Here also a nonstandard reference may help by reconstructing the step response from an identified model.

In this paper we apply 2nd order models with delay both to identify the plant and to test the closed-loop system. It is assumed that the data acquired from nonstandard inputs are recorded and processed by least-squares (LS) to identify the models and reconstruct the step responses. Possible real-time application is kept in perspective.

A. Bartoszewicz et al. (Eds.): PCC 2020, AISC 1196, pp. 678–688, 2020.
https://doi.org/10.1007/978-3-030-50936-1_57

From vast literature on issues related to the subject the following publications have turned out particularly relevant:

- identification, smoothing, least-squares – [1,4,5],
- step responses of processes with delay – [1,2,12],
- settings of PI and PID controllers – [6,9].

Despite the classical flavour the overdamped processes with delay are still the subject of research, as for instance [8] and [11].

The paper is organized as follows. Simple methodology how to use responses to nonstandard inputs for tuning and testing PID loops is presented in the next section. Section 3 applies the methodology to reconstruct step response of a plant. Simplified models for selection of PI or PID controller and tuning rules are given in Sect. 4. Testing the closed-loop system by means of nonstandard reference is described in Sect. 5. Conclusions and some remarks on real-time application are given in the last section. The same running example is presented throughout the paper.

2 Methodology

It is assumed that step response of a plant, if available, would be an S-shape curve like the one in Fig. 1. Such a curve may be modeled by 1st, 2nd or n-th order transfer function with delay [1,2,12]. Suppose that the plant is excited sufficiently strongly and the input/output data are recorded. The proposed methodology leading to reconstruction of the step response and calculation of controller settings consists of the following steps:

1) smoothing the data to reduce content of the noise component,
2) LS identification of the plant by a discrete 2nd order model with delay,
3) generation of step response from the identified model,
4) creation of two tuning-oriented models, 1st and 2nd order with delay,
5) choosing the one from the two models which approximates the step response better,
6) calculation of PI or PID settings according to the chosen model.

Given the tuned PI or PID controller the feedback loop can be closed.

Assuming that the controller settings are intended to provide responses similar to the ones of 2nd order, the resulting closed-loop loop system may exhibit oscillatory or overdamped behavior shown clearly in the response to a step reference. Such response may also be reconstructed from reaction to a nonstandard reference in the steps 1, 2, 3 above, after some rearrangement of the data for the LS due to integral action of the controller.

3 Plant Identification

Suppose we are given an input sequence u_Δ generated with a cycle Δ for which the plant responds with an output sequence y_Δ. Sample plots of u_Δ, y_Δ, generated by Matlab and used here to identify the plant, are shown in Fig. 2. Since

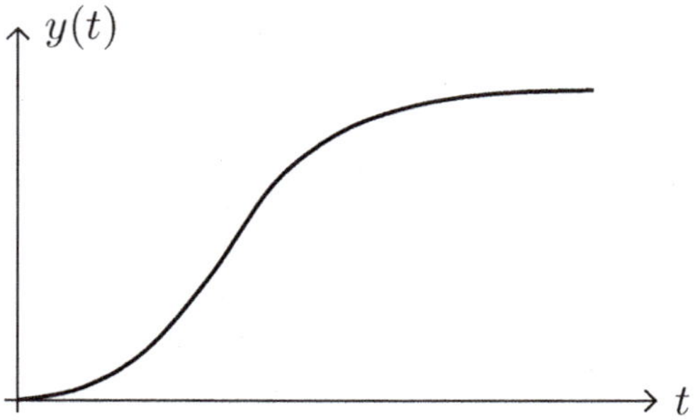

Fig. 1. S-curve as plant step response

Δ is in practice quite small therefore to implement the LS method we need a larger discretization step h. So in the first step of the methodology the original y_Δ is smoothed out by averaging over the intervals h, namely

$$y = \frac{1}{N_h}\left(y_{\Delta 1} + y_{\Delta 2} + \ldots + y_{\Delta N_h}\right) \quad \text{with } N_h = \frac{h}{\Delta} \tag{1}$$

for one of the samples. For consistency, let u be also a smoothed u_Δ. The sequences u and y, each involving N samples, are the data for identification.

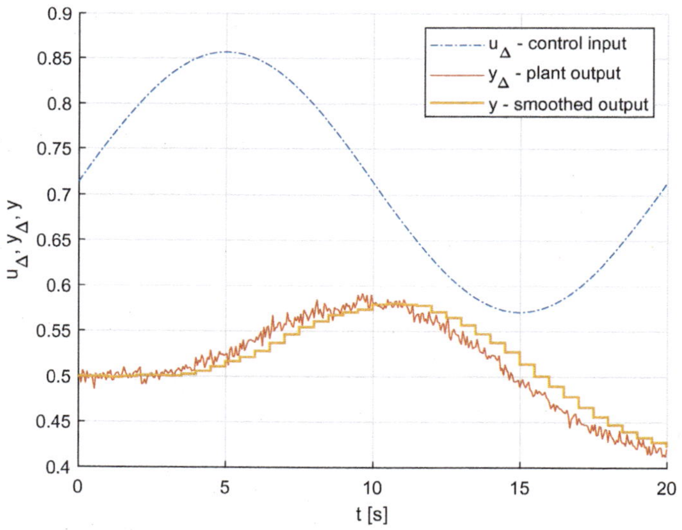

Fig. 2. Input and output of the plant

Assuming that an S-shape curve like the one in Fig. 1 characterizes the plant, the corresponding model may be reasonably represented by a 2nd order transfer function with delay, written below in a discrete form

$$\frac{Y_m(z)}{U(z)} = \frac{\Theta_3 z^{-1} + \Theta_4 z^{-2}}{1 - \Theta_1 z^{-1} - \Theta_2 z^{-2}} z^{-d}. \tag{2}$$

Given the data u, y the problem is to find the estimates $\widehat{\Theta}, \widehat{d}$ that minimize the standard quadratic cost. To do so we assume that the delay d belongs to certain integer interval $D = \langle d_m, d_M \rangle$.

The output y_m of the model (2) calculated for the data u, y at a step k is given by

$$y_{m,k}(\Theta, d) = \Theta_1 y_{k-1} + \Theta_2 y_{k-2} + \Theta_3 u_{k-1-d} + \Theta_4 u_{k-2-d}. \tag{3}$$

The estimates $\widehat{\Theta}, \widehat{d}$ can be found by searching over the interval D for the best LS approximation of y by $y_m(\Theta, d)$. This may be done by the following algorithm:

Data: y, u, N

Result: $\widehat{\Theta}, \widehat{d}$

$J_{\min} := \infty$

foreach $d \in D$ **do**

  `// create the data matrix:`

  $\Phi_d := [y_{k-1}\ \ y_{k-2}\ \ u_{k-1-d}\ \ u_{k-2-d}\ :\ k = 1, 2, \ldots, N]$

  `// calculate the LS estimate for particular` d:

  $\widehat{\Theta}_d := \left(\Phi_d^\top \Phi_d\right)^{-1} \Phi_d^\top y$

  `// compute the cost:`

  $J(\widehat{\Theta}_d, d) := \sum_{k=1}^{N} \left(y_k - y_{m,k}(\widehat{\Theta}_d, d)\right)^2$

  `// find the minimum cost:`

  if $J(\widehat{\Theta}_d, d) < J_{\min}$ **then**

    $J_{\min} := J(\widehat{\Theta}_d, d)$

    $\widehat{\Theta} := \widehat{\Theta}_d$

    $\widehat{d} := d$

  end

end

Using the resulting $\widehat{\Theta}, \widehat{d}$ in the model (2), a unit step response $y_{m,step}$ can be now recursively reconstructed by taking $u = \mathbf{1}$, where

$$\mathbf{1}_k = \begin{cases} 0, & k < 1, \\ 1, & k \geq 1, \end{cases} \qquad k \in \mathbb{Z}.$$

Example. The output plot shown in Fig. 2 is generated by the plant

$$G(s) = \frac{0.7}{(s+1)^5} \tag{4}$$

driven by a “slow” sinusoid

$$u(t) = \frac{0.1}{k_o} \sin\left(\frac{2\pi}{T_u} t\right) = 0.14 \sin\left(\frac{2\pi}{20} t\right) \tag{5}$$

with a step $\Delta = 0.05$ (note the small amplitude). The response is supplemented be a white noise with standard deviation $\sigma = 0.005$, giving the output y_Δ. The noisy y_Δ corresponds more or less to practical cases [3].

The signal y_Δ is subsequently fed to an output filter $1/(0.5\,\mathrm{s}{+}1)$ and smoothed out by averaging over the intervals $h = 0.5$ taken as discretization steps. So every consecutive $N_h = h/\Delta = 10$ values are averaged. Samples of the estimates for a particular run are the following

$$\widehat{\Theta} = [1.0265,\ -0.1649,\ 0.0442,\ 0.0526], \qquad \widehat{d} = 5.$$

Corresponding step responses of the plant (4) and the model (2), i.e. $y_{m,step}$, are shown in Fig. 3a.

Naturally, due to the random noise each simulation run gives slightly different $y_{m,step}$. Range of the reconstructed responses for 100 runs is shown in Fig. 3b and the range of $J(\widehat{\Theta}_d, d)$ plots in Fig. 3c. The estimate $\widehat{d} = 5$ is the most common in Fig. 3c (vertical line) what corresponds to continuous delay $\tau = \widehat{d}\,h = 2.5$ (compare Figs. 3a,b).

4 Tuning PID Controller

The reconstructed $y_{m,step}$ is assumed fairly close to step responses the following tuning-oriented continuous transfer functions

$$\mathrm{I}: \ \frac{k_o}{Ts+1} e^{-\tau s} \qquad \mathrm{II}: \ \frac{k_o}{(Ts+1)^2} e^{-\tau s}. \tag{6}$$

The one that fits $y_{m,step}$ better will be chosen to design PI or PID controller.

The value of k_o is equal to the last value of $y_{m,step}$, whereas T and τ are calculated according to Table 1, where t_{10}, t_{90} denote times for which $y_{m,step}$ reaches 10 or 90% of its final value, respectively. Similar rules, but involving t_{50}, t_{90}, are presented in [1]. To increase accuracy of reading the values t_{10}, t_{90}, a linear interpolation between two adjacent points may be applied.

Having the models (6) with calculated parameters k_o, T, τ, step responses $y_{\mathrm{I},step}$ and $y_{\mathrm{II},step}$ may be generated by corresponding discrete equivalents, i.e.

$$G_{\mathrm{I}}(z) = k_o \frac{\left(1 - e^{-h/T}\right) z^{-1}}{1 - e^{-h/T} z^{-1}} z^{-\lfloor \tau/h \rfloor}$$

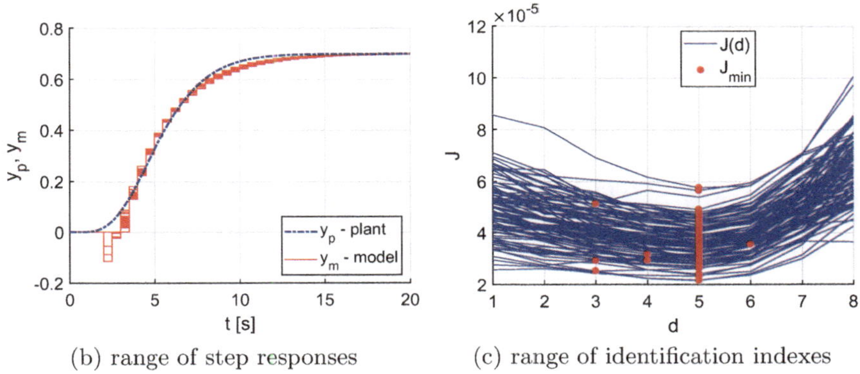

Fig. 3. Results of plant identification for 100 runs

Table 1. Time constant and delay for the models I and II

	T	τ
I : $\dfrac{k_o}{Ts+1}e^{-\tau s}$	$\dfrac{t_{90}-t_{10}}{2.2}$	$t_{10}-0.1T$
II : $\dfrac{k_o}{(Ts+1)^2}e^{-\tau s}$	$\dfrac{t_{90}-t_{10}}{3.3}$	$t_{10}-0.53T$

in case of I. $G_{\mathrm{II}}(z)$ is not shown for brevity. Given $y_{\mathrm{I},step}$ and $y_{\mathrm{II},step}$, quadratic indexes of approximation of the reconstructed $y_{m,step}$ can be computed and the better model from (6) selected. Here it is the model II which for a sample run looks as follows

$$G_{\mathrm{II}}(s) = \frac{0.7}{(2.5s+1)^2}e^{-3.2s}. \tag{7}$$

For the model II a full PID algorithm is required, so the transfer function of the controller holds the form

$$\text{PID}: \; k_p\left(1+\frac{1}{T_i s}+\frac{T_d s}{\frac{T_d}{D}s+1}\right) \cong k_p\left(1+\frac{1}{T_i s}+T_d s\right), \quad D \geq 4,$$

whereas for the model I a PI algorithm would suffice [9] ($D = 5$ is typical for Siemens and $D = 8$ for Honeywell). To determine the settings k_p, T_i, T_d we are given parameters k_o, T, τ and a certain percentage overshoot $p_{\%}$ required for the closed-loop system.

Remarkable number of PID and PI tuning rules is available in [6]. Here however, we use simple analytic rules from [9] based on pole–zero cancellation and Padé approximation. Closed-loop responses resulting from the rules are very similar to those of 2nd order.

For the model II the cancellation requires $T_i = 2T$ and $T_d = T_i/4$ (as in familiar Ziegler–Nichols settings). Then the open-loop transfer function obtains the form

$$G_{open}(s) = k\frac{e^{-\tau s}}{s} \qquad \text{with} \quad k = \frac{k_p k_o}{T_i}.$$

By applying 1st order Padé approximation $e^{-\tau s} \cong (-\tau s + 2)/(\tau s + 2)$, the system becomes of 2nd order, so having a damping coefficient ξ (from $p_{\%}$), the parameter k can be found by algebraic calculations. In particular, for $\xi = 1$ (critical damping, $p_{\%} = 0$) one obtains $k = 2(3 - 2\sqrt{2})/\tau$ [10].

Oscillatory responses for $\xi = 1/\sqrt{2}$ ($p_{\%} = 4.3$) or $\xi = 1/2$ ($p_{\%} = 16.3$) require k to be increased by a factor of 1.56 or 2.23, respectively. Finally $k_p = 2Tk/k_o$. For the sample model (7) and $p_{\%} = 16.3$ we get

$$k_p = 1.71, \quad T_i = 5, \quad T_d = 1.25.$$

Selection of the model I would require a PI controller with $k_p = Tk/k_o$, $T_i = T$.

Further we assume that the PID controller (or PI) has been implemented in the closed-loop system. Step responses for the system with the plant (4) or the model (7) are shown in the following section (Fig. 5).

5 Testing and Diagnostics

After closing the feedback loop we need to check how the response to a step reference looks like. Assuming stability, one may expect an oscillatory or overdamped response, with an overshoot or not. However, generation of sufficiently large response may be sometimes difficult, for instance due to limits imposed on control signal or technological conditions. Hence application of a slowly varying reference and reconstruction of the step response from an identified model may be a solution.

So assume that for some reference r we are given a closed-loop system output y, smoothed out similarly as before. The system is modeled by the discrete transfer function (2), however this time with the condition

$$\Theta_1 + \Theta_2 + \Theta_3 + \Theta_4 = 1,$$

since integral action of the controller provides steady-state accuracy ($y \rightarrow r_{const}$). Hence there are three independent parameters, for instance Θ_1, Θ_2, Θ_3, and the delay d to be identified. Recursive equation of the model obtains the form

$$\begin{aligned} y_{m,k} &= \Theta_1(y_{m,k-1} - r_{k-2-d}) + \Theta_2(y_{m,k-2} - r_{k-2-d}) \\ &\quad + \Theta_3(r_{k-1-d} - r_{k-2-d}) + r_{k-2-d}. \end{aligned} \tag{8}$$

The estimates $\widehat{\Theta}, \widehat{d}$ are identified by a slightly modified for-loop-with-LS algorithm from Sect. 3.

In Fig. 4, a response of the system with the plant (4) for the slow sinusoidal reference $r(t) = k_o u(t)$ is given, where $u(t)$ is defined in (5). Having the estimates $\widehat{\Theta}, \widehat{d}$, a unit step response follows from (8) for $r = \mathbf{1}$.

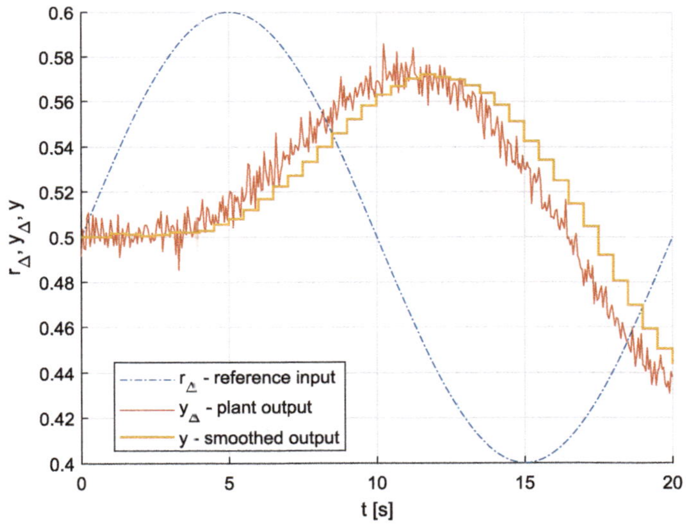

Fig. 4. Reference and response of the closed-loop system

Responses for the system and the model for 100 runs and $p_{\%} = 4.3$, 16.3 are shown in Figs. 5a,b. Consistency seems reasonable although the model exhibits somewhat smaller overshoot.

In some cases, as a consequence of the noise and low order model, the reconstructed step response may have one or two negative values right after the delay $\widehat{d}$. They may be removed while reconstructing by accepting only non-negative values for $y_{m,step}$.

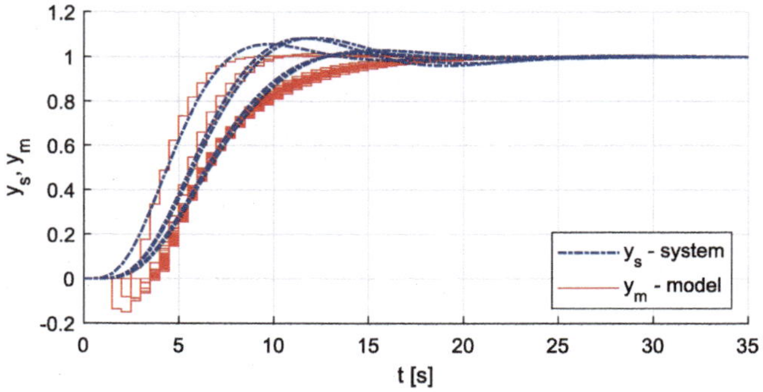

(a) tuned for $p_{\%} = 4.3$

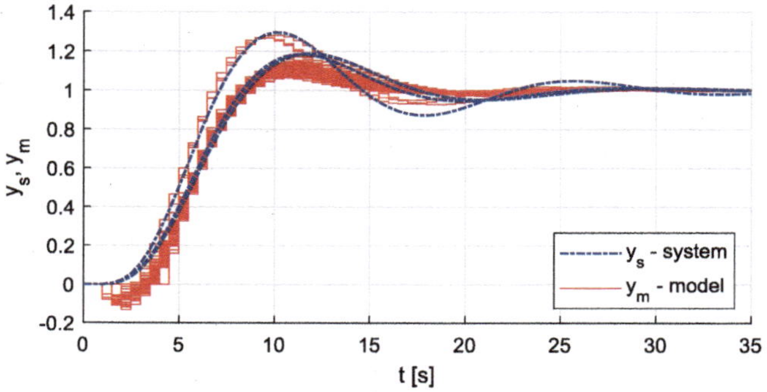

(b) tuned for $p_{\%} = 16.3$

Fig. 5. Step responses of the closed-loop system for 100 runs

Figures 6 show seemingly more practical case where the plant input or the system reference are changed in small steps imitating an activity of the operator. Here also the reconstructed step responses correspond reasonably well to the original ones.

The simulations involving the sinusoidal and small steps tests indicate that the methodology may be possibly applied to diagnose a closed-loop system which has been in use for a long time, so plant dynamics could change. Although reconstruction of reliable step responses from higher order models of such system turns out practically impossible due to the noise (already for 3rd order), particularly in case of undershoots, the 2nd order model provides useful information on actual settling time.

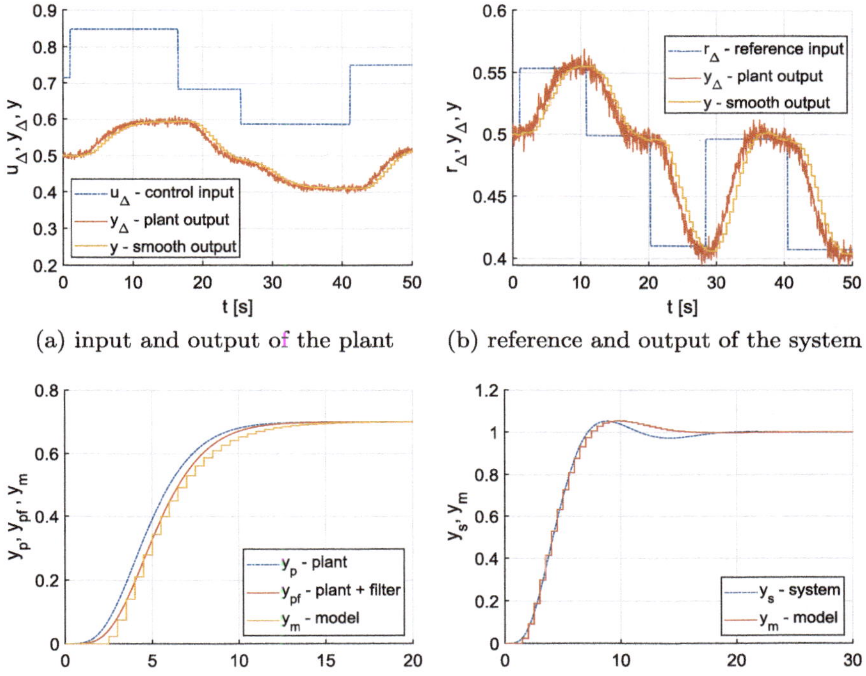

(a) input and output of the plant (b) reference and output of the system

(c) result of plant identification (d) result of system identification

Fig. 6. Identification based on small steps of control/reference input

6 Conclusions

The proposed methodology and the example have shown that tuning and testing PID loops by means of nonstandard inputs is feasible, providing that recorded data are sufficiently smoothed out. By using a discrete 2nd order model with delay identified by least-squares, the plant step response can be reconstructed to calculate PID settings. The closed-loop response unveils an overshoot and settling time, basic indicators of the system behavior.

The authors have also examined issues involved in possible real-time lab implementation of the methodology, assuming that software will be written in a language typical for PLCs, i.e. ST of IEC 61131-3 standard. First of all, to execute in parallel a PID algorithm, plant monitoring, and the steps of the methodology, a multitasking system is needed. A WinController runtime mentioned recently in [7] may be one of possible solutions. The WinController being a Windows service is able to execute complex software composed of projects running with different cycles. The projects are prepared in IEC-compliant CPDev control engineering environment.

Acknowledgments. This project is financed by the Minister of Science and Higher Education of the Republic of Poland within the "Regional Initiative of Excellence" program for years 2019–2022. Project number 027/RID/2018/19, amount granted 11 999 900 PLN.

References

1. Bielińska, E.: Klasyczne metody identyfikacji odpowiedzi skokowej (Classical methods of step response identification). In: Kasprzyk, J. (ed.) Identyfikacja procesów: praca zbiorowa (Process identification: joint publication), pp. 51–62. Wyd. PŚl, Gliwice (1997)
2. Byrski, W.: A new method of multi-inertial systems identification by the Strejc model. In: Mitkowski, W., Kacprzyk, J., Oprzędkiewicz, K., Skruch, P. (eds.) Trends in Advanced Intelligent Control, Optimization and Automation, KKA 2017, Advances in Intelligent Systems and Computing, vol. 577, pp. 536–549. Springer, Cham (2017)
3. Jacheć, M.: Astor. Poradnik automatyka – Teoria sterowania w praktyce – Identyfikacja obiektu (Control engineer handbook – Control theory in practice – Plant identification) (2016). www.astor.com.pl/poradnikautomatyka, Accessed 03 Jan 2020
4. Królikowski, A., Horla, D., Ziętkiewicz, J.: Identyfikacja obiektów sterowania: metody dyskretne parametryczne (Identification of control plants: parametric discrete methods). Wyd. PP, Poznań (2017)
5. Mańczak, K., Nahorski, Z.: Komputerowa identyfikacja obiektów dynamicznych (Computer identification of dynamic plants). PWN, Warszawa (1983)
6. O'Dwyer, A.: Handbook of PI and PID Controller Tuning Rules, 3rd edn. Imperial College Press, London (2009)
7. Rzońca, D., Sadolewski, J., Stec, A., Świder, Z., Trybus, B., Trybus, L.: Developing a multiplatform control environment. J. Autom. Mob. Rob. Intell. Syst. **13**(4) (2019). in print
8. Shin, G.W., Song, Y.J., Lee, T.B., Choi, H.K.: Genetic algorithm for identification of time delay systems from step responses. Int. J. Control Autom. Syst. **5**(1), 79–85 (2007)
9. Trybus, L.: A set of PID tuning rules. Arch. Control Sci. **15**(1), 5–18 (2005)
10. Trybus, L.: Automatyka i sterowanie (Control and regulation)(2017). http://materialy.prz-rzeszow.pl/materialy.php?przedmiot=71, Accessed 03 Jan 2020
11. Yan, R., Liu, T., Chen, F., Dong, S.: Gradient-based step response identification of overdamped processes with time delay. Syst. Sci. Control Eng. **3**(1), 504–513 (2015)
12. Ziętkiewicz, J.: Identyfikacja obiektów sterowania: ćwiczenia laboratoryjne (Identification of control plants: lab experiments). Wyd. PP, Poznań (2018)

Linear High-Gain Correction Observer in Nonlinear Control

Andrzej Latocha(✉)

Faculty of Electrical Engineering, Automatics, Computer Science and Biomedical Engineering, AGH University of Science and Technology, Kraków, Poland
alatocha@agh.edu.pl
https://www.agh.edu.pl/

Abstract. The development of innovation technology requires one to provide control processes with better quality indexes. In modern control theory, the algorithms for calculating linear-quadratic regulator (LQR) for linear dynamic systems are known. The classical LQR algorithms do not take into account transport delays of the signals or uncertainty of measured values and models, which causes the task of stabilizing physical systems by classical LQR algorithm to contain significant errors which decrease quality indexes. Most control systems are non-linear, not isolated from the uncertainty of measured values and models. For this reason, in the paper the possibility of calculating a linear-quadratic regulator for non-linear systems, which will provide stability in a wider environment around the reference point for additive noise based on the high-gain disturbance correction observer, are proposed.

Keywords: High-gain observer · Disturbance observer · Estimation · Correction · Nonlinear control · Tracking

1 Problem Statement and Preliminaries

The present work relates to the latest achievements in nonlinear control theory and high-gain observers (HG) [6,7]. In industrial applications, dual [8] or multi loop controllers [13,16] are often used to control nonlinear systems. Such solutions ensure system stability at low quality indexes. High quality indexes are provided by regulators based on modern control theory. For this reason, the present study focuses on calculation linear-quadratic regulator (LQR) for nonlinear systems on a wider horizon outside the equilibrium point to fulfill the stability condition in the bounded input bounded output (BIBO) sense. In modern control theory, mathematical proofs show a solution to the problem of linear-quadratic regulation for certain classes of nonlinear systems based on modified extended Kalman filters and hybrid algorithms [1,4,17]. The solutions require accurate mathematical models of the nonlinear systems and linear estimators, but they are not sufficient for determining the controllability criterion for some classes of nonlinear systems exposed on additive noise due to the uncertainty of the model. For these reasons, a new approach to non-linear control has

A. Bartoszewicz et al. (Eds.): FCC 2020, AISC 1196, pp. 689–700, 2020.
https://doi.org/10.1007/978-3-030-50936-1_58

been proposed. The exact mathematical model of the system and estimate of the state trend were considered. Consider the wide class of nonlinear single input single output (SISO) systems that are stable in the BIBO sense, described by the following state-space representation:

$$\psi = f(\xi(t), u(t)), \tag{1}$$

$$y(t) = g(\psi), \tag{2}$$

where $\xi(t) \in \mathbb{R}^{\gamma \times l}$ is the system state, $\psi \in \mathbb{R}^{l}$ is the state function, $u(t) \in \mathbb{R}$ is the control signal, and $y(t) \in \mathbb{R}$ is the system output. The problem that should be solved is a linear estimation of state variables that will allow for control of the nonlinear system by using a modified LQR controller for states in wide environments of the linearization interval.

2 Linearization by Projection

The goal of this algorithm is the dynamic linearization of certain classes of nonlinear systems and systems with uncertain parameters at the measurement interval $[t - T, t]$, where t is the current moment, and T is the horizon of data series equally distant in time from the linearization interval of nonlinear system $u(t)$, $y(t)$; $\forall t$. This linearization can be interpreted as a projection of the SISO non-linear system model onto a SISO linear system model with an acceptable margin of error [9]. For the projection, a modified algorithm of the least squares estimation of equation error are used [9–11]. The non-linear model (1, 2) is represented by a digitized series of data on input $u(k)$ and output $y(k)$, which are equally distant in time; this allows one to linearize non-differentiable functions. The series of data $u(k)$, $y(k)$ is used to estimate the coefficients of the AutoRegressive Moving Average with eXogenous input (ARMAX) model in the operator form as follows:

$$y(k) = \frac{B(z^{-1})}{A(z^{-1})}u(k) + \frac{C(z^{-1})}{A(z^{-1})}\varepsilon(k), \tag{3}$$

where $u(k) \in \mathbb{R}$, $y(k) \in \mathbb{R}$, $A \in \mathbb{R}^{n \times n}$, $B \in \mathbb{R}^{n \times 1}$, $C \in \mathbb{R}^{1 \times n}$, and $k \in \mathbb{N}$, $z \in \mathbb{C}$, $\varepsilon(k)$ is the error of measurement. For presented assumptions, a new approach of Least Squares Estimation (LSE) for data that does not meet the assumptions of zero initial conditions leads to data preprocessing as follows. The data preprocessing is composed of an initial value of zero for optimal $u(k)$ first samples to provide convergence of LSE estimation and minimize errors:

$$\widetilde{u}(k) = u[1 \ldots \eta_o] \equiv 0;\ \forall k \leq \eta_o, \tag{4}$$

$$\widetilde{u}(k) = u(k);\ \forall k > \eta_o, \tag{5}$$

where $\eta_0 \in \mathbb{N}$. The errors were added to the LSE equation error as follows:

$$\varepsilon = \frac{\hat{\overline{C}}(z^{-1})}{\hat{\overline{A}}(z^{-1})}\sum_{k=1}^{N}(y_{\xi}(k)-\hat{\overline{y}}(k)) + \frac{\hat{\overline{B}}(z^{-1})}{\hat{\overline{A}}(z^{-1})}\sum_{k=1}^{N}(u(k)-\widetilde{u}(k)) + \varepsilon(k), \qquad (6)$$

where $\hat{\overline{A}}, \hat{\overline{B}}, \hat{\overline{C}}, \hat{\overline{y}}(k)$ estimated discrete linear model, $y_{\xi}(k)$ is the SISO nonlinear system discrete output. This conditioning of the task allows the use of least square estimation of equation error for non-zero initial conditions on any interval as well as for non-differentiable functions. For this reason, equation (3) has been transformed into a difference equation (7), for assumptions (4), (5), (6) and by solving as follow:

$$y_{\xi}(k) + \ldots + \hat{\overline{a}}_{k-n}y_{\xi}(k-n) + \varepsilon = \hat{\overline{b}}_{k-1}\widetilde{u}(k-1) + \ldots + \hat{\overline{b}}_{k-m}\widetilde{u}(k-m), \qquad (7)$$

where $k = 1 \ldots N$, $N = f(T_i)$, $N \in \mathbb{N}$.

$$\hat{\overline{\theta}}_i = [\widetilde{\Phi}_i^T \widetilde{\Phi}_i]^{-1} \widetilde{\Phi}_i^T Y_i, \qquad (8)$$

where $\hat{\overline{\theta}}_i$ is a vector of estimated coefficients (7) from the hyperplane least squares, $\widetilde{\Phi}_i \in \mathbb{R}^{(n+m)\times N_i}$ is a matrix of data samples $\widetilde{u}(k)$, $y(k)$, Y_i is a matrix of the output, $i \in \mathbb{N}$ is the number of the data horizon T_i,

$$\hat{\overline{\theta}}_i = f(\widetilde{u}(k), y_{\xi}(k)), \qquad (9)$$

$$\hat{\overline{\theta}}_o = arg_{(\hat{\overline{\theta}}_i, \eta_{i(o)})} inf(e_i), \qquad (10)$$

$$e_i = \frac{1}{N(i)}\sum_{k=1}^{N(i)}(y_{\xi}(k) - \hat{\overline{y}}(k))^2. \qquad (11)$$

For disturbed Gaussian noise, the pseudo-square data matrix $\widetilde{\Phi}_i$ can be close to row loss, even though it is inverted and its condition generates non acceptable estimation errors (11). The correct conditioning of data is as follows:

$$(\widetilde{\Phi}_i^T \widetilde{\Phi}_i)(\widetilde{\Phi}_i^T \widetilde{\Phi})_i^{-1} = \widetilde{I}_i \approx I, \qquad (12)$$

$$J_{\Phi_i} = \sum_{i=1}^{n+m}\sum_{i=1}^{n+m}(I_{i,i} - \widetilde{I}_{i,i})^2. \qquad (13)$$

Significant perturbations outside the main diagonal (13) reveal the poor conditioning of the solution. To correct this, the horizon T_i of data matrix $\widetilde{\Phi}_i$ should be changed. After projecting the SISO non-linear system model onto the SISO linear model, a sub-optimal linear estimator was obtained at the linearization horizon (14), [10,11]. The solution is an loaded discrete linear estimator in relation to the zero initial conditions as follow:

$$\hat{\overline{G}}_o = \frac{\hat{\overline{b}}_m z^m + \ldots + \hat{\overline{b}}_1 z + \hat{\overline{b}}_0}{z^n + \hat{\overline{a}}_{n-1} z^{n-1} + \ldots + \hat{\overline{a}}_1 z + \hat{\overline{a}}_0}, \tag{14}$$

where $n \geq m > 1$; $m, n \in \mathbb{N}$. The load of the estimator underestimates gains and time constraints. The stability and innacuracy problems of discrete estimators are known and widely described in the literature related to the subject of extended Kalman filters [21]. For this reason, the conversion of a discrete linear estimator into the continuous time domain of the Frobenius control form has been proposed as follows (15), (16), (17), [3]:

$$\hat{\overline{A}}_C = \begin{bmatrix} -\hat{\overline{a}}_{n-1} & -\hat{\overline{a}}_{n-2} & \ldots & -\hat{\overline{a}}_0 \\ 1 & 0 & \ldots & 0 \\ 0 & 1 & \ldots & 0 \\ 0 & 0 & \ldots & 0 \end{bmatrix},$$

$$\hat{\overline{B}}_C = \begin{bmatrix} 1 \\ 0 \\ 0 \\ 0 \end{bmatrix},$$

$$\hat{\overline{C}}_C = [\hat{\overline{c}}_{n-1}\ \hat{\overline{c}}_{n-2} \ldots \hat{\overline{c}}_0], \tag{15}$$

where

$$\gamma = [\gamma_1\ \gamma_2 \ldots \gamma_n] = [\hat{\overline{b}}_0 - \hat{\overline{a}}_0 \hat{\overline{b}}_n\, \hat{\overline{b}}_1 - \hat{\overline{a}}_1 \hat{\overline{b}}_n \ldots \hat{\overline{b}}_{n-1} - \hat{\overline{a}}_{n-1} \hat{\overline{b}}_n], \tag{16}$$

$$\hat{\overline{C}}_C = \left[\gamma_n\ \gamma_{n-1} \ldots \gamma_1\right]. \tag{17}$$

Similar concepts of linearization often appear in the literature [14,19,20]. A significant difference in the proposed algorithm is $u(k)$ signal preprocessing before mapping and the need to take into account model errors and poor numericall conditioning, which is crucial for the stability of the algorithm in nonlinear and uncertain systems.

3 Linear High-Gain Disturbance Correction Observer

Consider the control law for nonlinear systems:

$$u(t) = -\tilde{K}(t)\xi(t) \leq -K\hat{\overline{x}}(t), \tag{18}$$

for

$$\dot{\hat{\overline{x}}}(t) = (\hat{\overline{A}}_C - G\hat{\overline{C}}_C)\hat{\overline{x}}(t) + \hat{\overline{B}}_C u(t) + Gy(t). \tag{19}$$

The overestimation value of state estimates are considered as the form of the observer gain oversizing. The classical high gain is $G_{HG} = \beta G$, $\beta \in \mathbb{R}$, $\beta \gg 1$

where G is an observer gain. The HG observer is stable as shown in the proof [6,7] but is not asymptotically stable. For this reason, a new approach was proposed to apply an oversized gain to the linearization model as follows:

$$\hat{\overline{G}}_{HG}(s) = \kappa_O \hat{\overline{G}}_o(s), \tag{20}$$

where $\hat{\overline{G}}_o(s)$ is a transfer function in a continuous operator form converted from (14), κ_O is the coefficient of the oversizing gain (20). To give the observer properties of the phase correction "*Lead*", it was assumed that the degree of numerator of the source function (14) is $m \in \mathbb{N}[1;n]$, where n is the degree of denominator (14). After transforming to the state space Frobenius control form (15), the observer equations can be obtained:

$$\hat{\overline{C}}_{HG_C} = \kappa_O \hat{\overline{C}}_C;\ \kappa_O \in \mathbb{R}(0;1], \tag{21}$$

$$\dot{\hat{\overline{x}}}(t) = (\hat{\overline{A}}_C - \frac{G}{\kappa_O}\hat{\overline{C}}_{HG_C})\hat{x}(t) + \hat{\overline{B}}_C u(t) + \frac{G}{\kappa_O} y(t). \tag{22}$$

The observer gain G is calculated using the Ackermann formula (23), [3,5] for assumed eigenvalues as follows:

$$\alpha(s) = |sI - \hat{\overline{A}}_C + \frac{G}{\kappa_O}\hat{\overline{C}}_{HG_C}| = (s-\lambda_1)(s-\lambda_2)\ldots(s-\lambda_n), \tag{23}$$

where eigenvalues $\lambda_n \in \mathbb{R}(-\infty;0)$. The auxiliary coefficient κ_O has been calculated using a numerical optimization due to non convex functions (22) for quality function

$$e_{y_\xi} = inf \frac{1}{N} \sum_{k=1}^{N} (y_\xi(k) - y_r(k))^2, \tag{24}$$

on the data horizon T_i. Where y_r is the reference trajectory. For proposed observer the local convergence [15,22], are obtained as follows (Fig. 1):

$$|\hat{\overline{x}}(t)| > |\xi(t)|;\ \forall(u(t) \neq 0,\ y(t) \neq 0,\ \tilde{z} = 0), \tag{25}$$

$$|\hat{\overline{x}}(t)| \rightarrow |\xi(t)| \rightarrow 0;\ \forall(u(t) = 0,\ y(t) = 0,\ \tilde{z} = 0), \tag{26}$$

$$\lim_{t\rightarrow\infty} \hat{\overline{x}}(t) = \hat{\overline{x}}_0;\ \forall(u(t) = 0,\ y(t) = 0,\ \tilde{z} = 0), \tag{27}$$

$$|\hat{\overline{x}}(t) - \xi(t_0)| < \varepsilon_\xi;\ \forall\varepsilon_\xi > 0,\ \exists t_0,\ \forall t > t_0, t \in T_i, \tag{28}$$

where $\tilde{z}$ is additive noise on input. These properties are desired for closed loop control system.

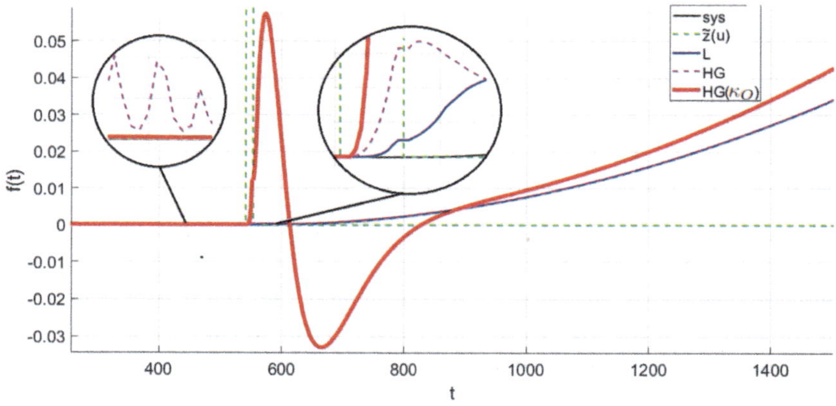

Fig. 1. Observers state estimates for noise disturbance on input: Luenberger, HG, proposed linear high-gain disturbance correction observer.

4 Closed Loop Control

To control a nonlinear system the closed loop system is used as shown in (Fig. 2) for assumptions (18), (19). The stationary linear-square regulator with an infinite horizon was used. For this reason, a Riccati equation for an overestimated linear estimator (20) should be solved:

$$\hat{\overline{A}}_O^T P + P\hat{\overline{A}}_O - P\hat{\overline{B}}_{HG_O} R \hat{\overline{B}}_{HG_O}^T P + Q = 0. \tag{29}$$

In solving this equation, the continuous state space equations in Frobenius observer form can be used (30), [3].

$$\hat{\overline{A}}_O = \begin{bmatrix} 0 & 1 & \dots & \dots \\ 0 & 0 & 1 & \dots \\ \dots & \dots & \dots & 1 \\ -\hat{\overline{a}}_0 & -\hat{\overline{a}}_1 & \dots & -\hat{\overline{a}}_{n-1} \end{bmatrix},$$

$$\hat{\overline{B}}_{HG_O} = \begin{bmatrix} \sigma_1 \\ \sigma_2 \\ \dots \\ \sigma_n \end{bmatrix},$$

$$\hat{\overline{C}}_O = [1\, 0 \dots 0], \tag{30}$$

$$\hat{\overline{B}}_{HG_O} = \begin{bmatrix} \sigma_1 \\ \sigma_2 \\ \dots \\ \sigma_n \end{bmatrix} = \begin{bmatrix} \hat{\overline{b}}_{n-1} - \hat{\overline{a}}_{n-1}\hat{\overline{b}}_n \\ \hat{\overline{b}}_{n-2} - \hat{\overline{a}}_{n-1}\sigma_1 - \hat{\overline{a}}_{n-2}\hat{\overline{b}}_n \\ \dots \\ \hat{\overline{b}}_0 - \hat{\overline{a}}_{n-1}\sigma_{n-1} - \hat{\overline{a}}_{n-2}\sigma_{n-2} - \dots - \hat{\overline{a}}_1\sigma_1 - \hat{\overline{a}}_0\hat{\overline{b}}_n \end{bmatrix} \tag{31}$$

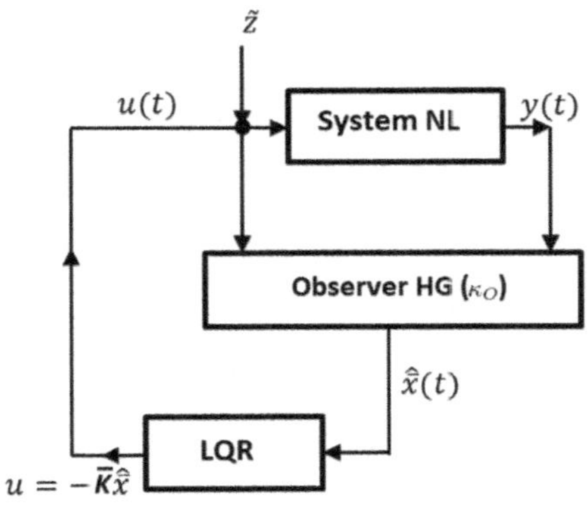

Fig. 2. Nonlinear closed loop control with linear high-gain disturbance correction observer and LQR controller.

$$\overline{K} = R^{-1}\hat{\overline{B}}^{T}_{HG_O}P. \tag{32}$$

where conditions (25), (26), (27), (28), (33), [2] and [15] for the non-linear system controllability should occur

$$Re\lambda(\hat{\overline{A}}_O + \hat{\overline{B}}_{HG_O}\overline{K}) < 0 \,\forall\, Re\lambda(\hat{\overline{A}}_C - G\hat{\overline{C}}_{HG_C}) < 0. \tag{33}$$

The control law occurs as function (34) for assumptions (25), (26), (27), (28),

$$u = -\overline{K}\hat{\overline{x}}(t, \kappa_O, \tilde{z}). \tag{34}$$

The transition of noise disturbances by the proposed disturbance correction observer (Fig. 2) allows estimations of trend profiling (Fig. 1) for the correlated control law (34) and adjustment of the dynamic gain of the observer-controller system. Such solutions can lend to increased quality indexes for closed loop control in relation to the classic correction of the control signal by disturbance observer-based control (DOBC) [12].

5 Application and Discussion

This section presents the results for the proposed algorithms.

5.1 Linear Part of the Nonlinear Model

In the first example, the linear part of the nonlinear system under consideration (35), [18] has been used to calculate the observer and feedback control laws (Fig. 2).

$$\ddot{x}(t) + a_1(1 - x^2)\dot{x}(t) + a_2x(t) = bu(t) \tag{35}$$

where $a_1 = 2.2165$, $a_2 = b = 12.7388$, and $|u(t)| \leq 15$. Such parameterization makes the system (35) globally asymptotically stable at the zero equilibrium point; outside the equilibrium point, the system falls into non-dampening oscillations. Let's consider only the linear part (36) of the system (35) to calculate the linear high-gain disturbance correction observer (22) and correlated control law (34).

$$\ddot{x}(t) + a_1\dot{x}(t) + a_2 x(t) = bu(t) \tag{36}$$

For the linear part (36) of the model (35) and considering the quality function (24), we are looking for a suboptimal coefficient κ_O. For the suboptimal κ_O the stability of the system (35) was tested using the phase plane topological method (Fig. 3) for non zero initial conditions and different work states with additive noise (Fig. 2).

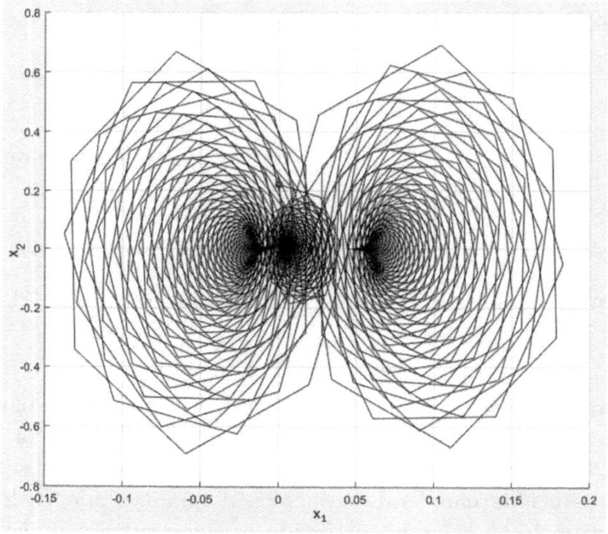

Fig. 3. Nonlinear system (35), (Fig. 2) state trajectories for stabilization task at different work points for calculating linear high-gain correction observer and corelated controller of linear part (36) of the model (35).

5.2 Projecting the Nonlinear System onto Linear Estimator

The second example for the same system for different initial state conditions was tested using the proposed algorithms, projecting the system (35) onto system (14), calculate the linear high-gain disturbance correction observer and correlated control law.

$$J_{HG(\kappa_O)} = 0.01543 \tag{37}$$

$$J_{ARMAX-HG(\kappa_O)} = 0.004112 \tag{38}$$

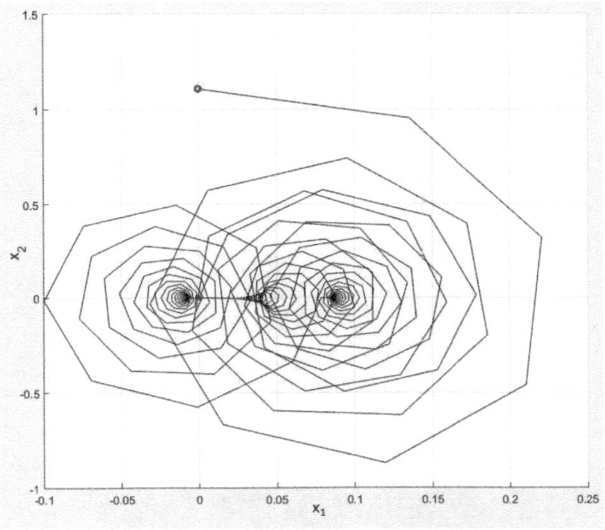

Fig. 4. Nonlinear system (35), (Fig. 2) state trajectories for stabilization task at different work points, for calculating linear high-gain correction observer and corelated controller from projection of the model (35).

The use proposed algorithms obtains significantly better quality indices (38) in comparison to linearization, which only uses the linear part of the model (36) extracted from the non-linear model (35). Stability analysis of the proposed algorithms using the phase plane topological method showed better convergence as confirmed by quality indices (37), (38) for other zero initial state conditions that were outside the attractors area of the first example [15] as seen in (Fig. 3), (Fig. 4).

5.3 Physical System of Magnetic Levitation

The physical system of magnetic levitation was used to test the proposed algorithms as shown in (Fig. 5), (Fig. 6). There is uncertainty in this model and measurement due to their dependence on the control signal. Using the presented algorithms, a projection was perform of the physical model onto a linear estimator (14), and then the high-gain disturbance observer and correlated control law were calculated. As shown in (Fig. 6), the control signal has a lower variance compared to controlling by the PID controller but the output signal has a higher variance. The presented algorithm obtains good quality indices for control of nonlinear systems and systems with model and measurement uncertainty that are subject to disturbed Gaussian noise, such as magnetic levitation.

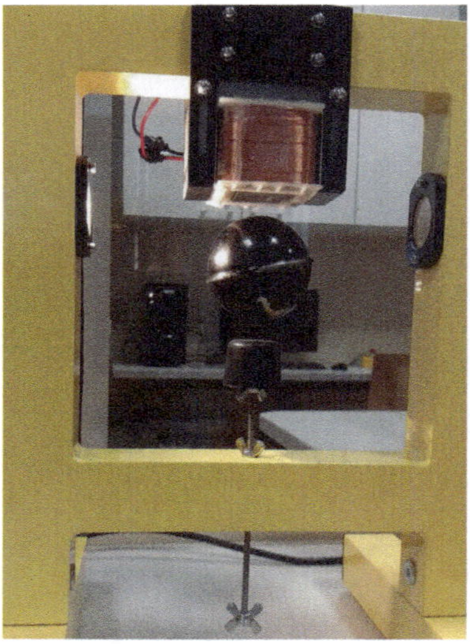

Fig. 5. Physical system of the magnetic levitation control by the presented algorithm.

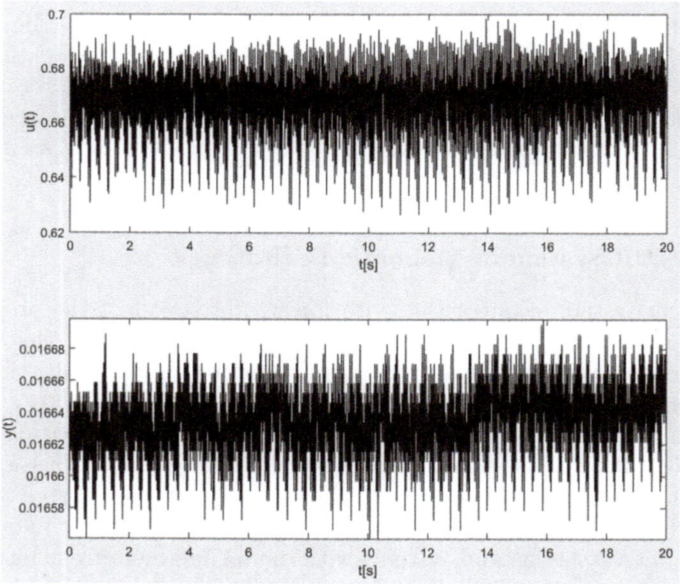

Fig. 6. Control signal and ball stabilization levels for magnetic levitation.

6 Conclusions

In this paper, nonlinear systems controls are realized by linearization through projecting a SISO nonlinear system model onto SISO linear estimator model and calculating the correlated high-gain disturbance observer in a closed loop control. The approach has been investigated for a class of nonlinear systems and systems with uncertain parameters. To achieve locally bounded stability on a wider horizon around a linearization point, the closed loop feedback control used dynamic gains, which were realized through the use of a high-gain disturbance correction observer and corelated LQR controller. The presented approach is characterized by good quality indexes of control, stability, and convergence state trajectories. There are some interesting problems that require further investigation. One of them is the optimal calculation of the high-gain disturbance correction observer and corelated control law. The second issue to be studied is mathematical proof of stability for the proposed solution. The algorithms that have been studied in this paper have shown that it is possible to control and stabilize a wide class of nonlinear systems on a wider horizon around a linearization point.

References

1. Bavdekar, A., Deshpande, A., Patwardhan, C.: Identification of process and measurement noise covariance for state and parameter estimation using extended Kalman filter. J. Process Control **21**(4), 585–601 (2011)
2. Bronsztejn, I., Siemiendiajew, K., Musiol, G., Muhlig, H.: Compendium of Modern Mathematics. PWN, Warszawa (2007)
3. Byrski, W.: Obserwacja i sterowanie w systemach dynamicznych, Uczelniane Wydawnictwo Naukowo-Dydaktyczne AGH, Monografie, vol. 18 (2007)
4. Iyad Hashlamon, I.: A new adaptive extended Kalman filter for a class of nonlinear systems. J. Appl. Comput. Mech. **61**(1), 1–12 (2020)
5. Kailath, T.: Linear Systems. Prentice-Hall Inc., Englewood Cliffs (1980)
6. Khalil, H., Hassan, K.: High-gain observers in feedback control application to permanent magnet synchronous motors. IEEE Control Syst. Mag. **37**(3), 25–41 (2017)
7. Khalil, H.: Nonlinear Control, Global edn. Pearson Education Limited, Edinburgh (2015)
8. Latocha, A.: System sterowania procesami silnie nieliniowymi, Pomiary Automatyka Robotyka, no. 9. Skamer-ACM (1998)
9. Latocha, A.: A robust linear-quadratic moving averaging controller for strongly nonlinear systems. In: NDC 2017 International Conference on Nonlinear Dynamics and Complexity Lodz University of Technology, Post Conference Materials, pp. 1–9 (2017)
10. Latocha, A.: Fast and robust online dynamic system identification. In: Kościelny, J., Syfert, M., Sztyber, A. (eds.) Advanced Solutions in Diagnostics and Fault Tolerant Control, DPS. Advances in Intelligent Systems and Computing, vol. 635. Springer, Cham (2017)
11. Latocha, A.: Robust fault detection, location, and recovery of damaged data using linear regression and mathematical models, IFAC-PapersOnLine **51**(24), 300–306 (2018). ISSN 2405-8963. https://doi.org/10.1016/j.ifacol.2018.09.593. http://www.sciencedirect.com/science/article/pii/S240589631832305X

12. Li, X., Gao, Z., Ai, W., Tian, S.: Differentiator-based disturbance observer. In: IEEE 8th Data Driven Control and Learning Systems Conference (DDCLS), pp. 976–981 (2019). https://doi.org/10.1109/DDCLS.2019.8908853
13. Lurie, B., Enright, P.: Classical Feedback Control. Marcel Dekker, New York (1986)
14. Merola, A., Cosentino, C., Colacino, D., Amato, F.: Optimal control of uncertain nonlinear quadratic systems. Automatica **83**, 345–350 (2017)
15. Mitkowski, W.: Zarys teorii sterowania, Wydawnictwo AGH. Komitet Automatyki i Robotyki PAN (2019)
16. Morari, M., Zafiriou, E.: Robust Process Control, vol. 10. Prince Hall, Englewood Cliffs (1989)
17. Witczak, M., Buciakowski, M., Puig, V., Rotondo, D., Nejjari, F., Jozef Korbicz, J.: A bounded-error approach to simultaneous state and actuator fault estimation for a class of nonlinear systems. J. Process Control **52**, 14–25 (2017). https://doi.org/10.1016/j.jprocont.2017.01.002
18. Zak, S.: Systems and Control. Oxford University Press, New York, Oxford (2003). School of Electrical and Computer Engineering Purdue University
19. Ljung, L., Gunnarsson, S.: Adaptation and tracking in system identification–a survey. Automatica **26**(1), 7–21 (1990)
20. Schön, T.B., Wills, A., Ninness, B.: System identification of nonlinear state-space models. Automatica **47**(1), 39–49 (2011)
21. Ljung, L.: Asymptotic behavior of the extended Kalman filter as a parameter estimator for linear systems. IEEE Trans. Autom. Control **24**(1), 36–50 (1979)
22. Bernard, P., Marconi, L.: Hybrid implementation of observers in plant's coordinates with a finite number of approximate inversions and global convergence. Automatica **111**, 108654 (2020). ISSN 0005-1098. https://doi.org/10.1016/j.automatica.2019.108654

Grey Wolf Optimizer in Design Process of Stable Neural Controller – Theoretical Background and Experiment

Marcin Kaminski(✉)

Wroclaw University of Science and Technology, Wroclaw, Poland
marcin.kaminski@pwr.edu.pl

Abstract. This article deals with an adaptive neural controller applied for a nonlinear plant with time-varying parameters. The structure of the controller is based on Radial Basis Function Neural Network. The output part of the controller (weights) is modified in several iterations of the control structure. In this application, the coefficients of the Gaussian functions are constant (it means the centers and width). The relevance of proper selection of those values is presented in tests performed for a real plant (an electrical drive). Moreover, for optimization of this part of the controller the metaheuristic – Grey Wolf Optimizer – algorithm was applied. The centers were selected in a clustering process. The synthesis of the controller includes stability analysis (using the Lyapunov method). The content of this article can be divided into two basic parts, the first shows theoretical considerations and the second is related to the experimental tests of the analyzed neural controller (executed in a laboratory, for the drive with 0.5 kW nominal power, using dSPACE card).

Keywords: Grey Wolf Optimizer · Radial Basis Function Neural Network · Clustering · Adaptive control · Electric drive

1 Introduction

Control of a group of plants, analyzed in control theory as nonlinear time-varying systems, is a very important task considered in many scientific centers. It is related to the features that represent such plants. It responses to real industrial issues, in which nonlinearities, time delays and changes of parameters are common problems. Assuming mentioned conditions, high quality of control is difficult to obtain [1–4]. A specific example can be here electrical drives. DC motors can be a representation of nonlinear plant. Sometimes a simplified mathematical expression can be considered as a description of those machines. However the friction phenomenon is an important element in the motion equation. Moreover, fluctuation of parameters often can appear. It can be caused by long-time acting of the motor (e.g. an influence of temperature) or changes of work condition (modifications of the whole system). The mentioned plant is also still worth attention because, it is one of the most popular machine used in practice. Substantiation for above fact, is the simplicity of the speed and position control method

A. Bartoszewicz et al. (Eds.): PCC 2020, AISC 1196, pp. 701–712, 2020.
https://doi.org/10.1007/978-3-030-50936-1_59

applied for those machines. It makes possible the implementation of the algorithm in a low-cost processor. Additionally, the construction of power electronics devices (e.g. the H-bridge) is very simple [5, 6].

The basis of the adaptive controller, analyzed in this article, is the radial neural network [7]. However, ultimately the control signal is a combination of a classic (with fixed parameters) and a neural controller. The signal from the neural model and the P-type controller were added, omitting integrals was one of the assumptions, the goal was to reduce implementation problems (in hardware applications) related to signal limitation on these elements. The stability of control systems with neural networks is one of the most important aspects of control systems design [8, 9]. The aforementioned concern, analyzed theoretically, is a key importance for the implementation possibilities of analyzed control structure and ensures safety of the plant with the tested controller. Therefore, at the design stage of the neural controller, dependencies were derived, which are assumptions that guarantee negative values of the proposed Lyapunov function. As mentioned above, the controller analyzed in this project contained a radial neural network. Data processing in such model is one-way - feedforward. However, it is significantly different from the most commonly used MLP (***M****ulti-****L****ayer* ***P****erceptrons*) or RNN (***R****ecurrent* ***N****eural* ***N****etworks*) [10–12]. Here, in the first phase of calculations, the Euclidean distance between the input data set and the centers of neuron activation function is determined. The output values of hidden layer are signals in the calculation of the output neuron (with tunable weighting coefficients). The above methodology of data processing in radial networks has been presented (in a simplified manner) in order to highlight some similarities between the calculation of radial networks and data clustering. The precise determination of the output values in these models is often presented as the separation of the input data space. In radial networks, this is more evident because of specific (the Gaussian) activation functions that perform local calculations. In MLP networks, combinations of sigmoidal functions are used for this purpose. Based on the above observation, it is possible to use data clustering algorithms to designate radial neuron centers in the design stage of the RBFNN (***R****adial* ***B****asis* ***F****unction* ***N****eural* ***N****etwork*) [13, 14].

Metaheuristic algorithms are becoming very popular in the parameters tuning of electric drive control structures. The mentioned methods are becoming more and more popular due to their beneficial features (theoretical assumptions and those related to practical implementation): no need to determine the gradient of the objective function, the possibility of omitting complex elementary operations (performed on the processed data), taking into account the limitations and changing working conditions of the plant. Limiting the considerations to applications in issues related to electric drives, the most popular of them include effective examples of applications: the PSO (***P****article* ***S****warm* ***O****ptimization*) [15], the GWO (***G****rey* ***W****olf* ***O****ptimizer*) [16], the BAT algorithm [17] and the ABC (***A****rtificial* ***B****ee* ***C****olony*) [18]. In addition to the most common tasks, related to the optimization of the parameters of selected elements in the control system, the metaheuristic algorithms are also used for data clustering (e.g. the PSO) [19, 20].

The goal of this paper is the original combination of the Lyapunov method and the clustering algorithm (using the GWO) for design of an adaptive neural controller. An exemplary application was done for a real electrical drive.

2 Synthesis of Neural Controller

In the analyzed implementation, for control, the RBFNN model is used. The considered application assumes on-line data handling (including the weights adaptation, described below), so for the i-th step of calculations, the output y_m is achieved using equation [21]:

$$y_m(i) = y_{RBF}(i) = W^T\varphi = \sum_m w_m \exp\left(-\frac{\|X_{RBF} - C_m\|^2}{2b_m^2}\right), \tag{1}$$

where: y_{RBF} – output of the neural network, W – weight matrix, φ – matrix with values of radial neurons, C_m – centers of activation function, b_m – width of the Gaussian function, X_{RBF} – input vector (n-size). The adaptation of the neural network weights, in several iteration i, can be described as:

$$\Delta w_k(i) = -\eta\frac{\partial E(i)}{\partial w_k(i)} = \eta(y(i) - y_m(i))\varphi_k(i), \tag{2}$$

in the above equation: η is a training constant ($\eta > 0$ and $\eta \in (0, 1)$), E is a cost function, w_k – value of several weight, y_m – output value of RBFNN, φ_k – outputs of following radial neurons and y – a reference output value. Changes of the error in the control structure can be expressed using following expressions:

$$e = x - x_r, \tag{3}$$

$$\dot{e} = \dot{x} - \dot{x}_r, \tag{4}$$

where: x – is an actual system state. Then, for a nonlinear system, change of reference state x_r for input signal u_r assuming the most precise following the reference state of the plant:

$$\dot{x}_r(t) = f(x_r(t), u_r(t)), \tag{5}$$

it can be rewritten as:

$$\dot{e} = f(x(t), u(t)) - f(x_r(t), u_r(t)). \tag{6}$$

For the above defined time-derivative of the error, the Taylor series expansion (at x_r and u_r) was calculated (the t symbol is omitted for easier analysis):

$$\dot{e} = f(x_r, u) + \left.\frac{\partial f(x, u)}{\partial x}\right|_{x=x_r}(x - x_r) + r_{x_r} + f(x, u_r) + \left.\frac{\partial f(x, u)}{\partial u}\right|_{u=u_r}(u - u_r) + r_{u_r} - f(x_r, u) - f(x, u_r), \tag{7}$$

where: r_{xr} and r_{ur} stand for higher order terms. After simple transformations:

$$\dot{e} = \left.\frac{\partial f(x, u)}{\partial x}\right|_{x=x_r}(x - x_r) + r_{x_r} + \left.\frac{\partial f(x, u)}{\partial u}\right|_{u=u_r}(u - u_r) + r_{u_r}. \tag{8}$$

Assuming that r_{xr} and r_{ur} can be removed (for properly adjusted neural controller) and after the linear approximation (using partial derivative of function describing state of the plant at equilibrium points) in (8):

$$\dot{e} = A(x - x_r) + B(u - u_r) = Ae + B(u - u_r). \tag{9}$$

In the proposed structure, the hybrid-parallel model of the controller is used. The proportional part (classical controller u_c) and the neural model u_{RBF} (the RBFNN trained on-line) are combined:

$$u = u_c + u_{RBF}. \tag{10}$$

Such construction of the controller should be taken in (9):

$$\dot{e} = Ae + B(u_c + u_{RBF} - u_r). \tag{11}$$

In formula (11), the neural control signal is calculated using:

$$u_{RBF} = y_{RBF} = \hat{W}^T \varphi, \tag{12}$$

$\hat{W}$ and φ are matrices of calculated weights (real values) and hidden layer outputs, respectively. The first element of (10) is a proportional controller (with gain k_p):

$$u_c = k_p e. \tag{13}$$

The optimal control signal in (11) can be formulated using expression:

$$u_r = y_{RBFo} + \varepsilon = \varepsilon + \sum_m w_{mo} \exp\left(-\frac{\|X_{RBF} - C_{ko}\|^2}{2b_{ko}^2}\right), \tag{14}$$

where: w_{mo} are optimal values of weights, y_{RBFo} is the optimal output of the neural network, C_{ko} are properly selected centers, b_{ko} are optimal widths of the Gaussian functions, ε the error of calculations using the radial network. Precise control means significant reduction of ε, thus bounds ε_b of the training error can be assumed [22]:

$$\varepsilon_b = \sup\|u_r - y_{RBFo}\|. \tag{15}$$

Moreover, the following errors are defined:

$$\tilde{f} = f_o - \hat{f} = W_o^T \varphi + \varepsilon - \hat{W}^T \varphi, \tag{16}$$

$$\tilde{W} = W_o - \hat{W}. \tag{17}$$

where: $\hat{f}$ – is real function realized using RBFNN, f_o – optimal function that should be performed using neural network, W_o – matrix of optimal weights. Now (11) can be updated using (12)–(14):

$$\dot{e} = Ae + B\left(k_p e + \hat{W}^T \varphi - W_o^T \varphi - \varepsilon\right), \tag{18}$$

and also using (16)–(17):

$$\dot{e} = Ae + B\left(k_p e + \hat{W}^T\varphi - W_o^T\varphi - \varepsilon\right) = Ae + Bk_p e - B\varepsilon + B\left(\hat{W}^T\varphi - W_o^T\varphi\right)$$
$$= Ae + Bk_p e - B\varepsilon - B\tilde{W}^T\varphi. \tag{19}$$

For assignation of stable adaptation law, the Lyapunov function should be defined as:

$$L = \frac{1}{2}e^T Pe + \frac{1}{2}tr\left(\tilde{W}^T \Xi \tilde{W}\right), \tag{20}$$

where: P is the positive definite matrix, Ξ is the nonnegative matrix and tr is the trace of matrix. The derivative of the Lyapunov function is presented below:

$$\dot{L} = \frac{1}{2}\dot{e}^T Pe + \frac{1}{2}e^T \dot{P}e + \frac{1}{2}e^T P\dot{e} + tr\left(\tilde{W}^T \Xi \dot{\hat{W}}\right). \tag{21}$$

In following calculations sizes of matrices are neglected for simplification of description. If (19) is applied in (21):

$$\dot{L} = Ae^T Pe + Bk_p e^T Pe - B\varepsilon Pe - B\tilde{W}^T\varphi Pe + tr\left(\tilde{W}^T \Xi \dot{\hat{W}}\right), \tag{22}$$

for:

$$Q = -(AP + Bk_p P), \tag{23}$$

it is possible to obtain:

$$\dot{L} = -e^T Qe - B\varepsilon Pe - B\tilde{W}^T\varphi Pe + tr\left(\tilde{W}^T \Xi \dot{\hat{W}}\right). \tag{24}$$

Since:

$$tr\left(ba^T\right) = a^T b, \tag{25}$$

then:

$$\dot{L} = -e^T Qe - B\varepsilon Pe + tr\left(\tilde{W}^T \Xi \dot{\hat{W}} - B\tilde{W}^T\varphi Pe\right). \tag{26}$$

Let the proposed stable adaptive law be as:

$$\dot{\hat{W}} = -\frac{1}{\Xi}B\varphi Pe. \tag{27}$$

After recalculation of (26), using (27) and (17):

$$\dot{L} = -e^T Qe - B\varepsilon Pe. \tag{28}$$

For stable work of the control structure negative values of $\dot{L}$ are necessary:

$$\dot{L} \le 0. \tag{29}$$

Above condition is met if:

$$\dot{L} \leq -|e|f_{\min}(Q)|e| + |B||\varepsilon_b|f\max(P)|e|, \tag{30}$$

where: | | denotes absolute value, f_{min} is the minimum eigenvalue of matrix, f_{max} maximum eigenvalue of matrix. It should be also noted:

$$|\varepsilon| \leq |\varepsilon_b|. \tag{31}$$

Now, from (29) and (30), following expression can be calculated:

$$-|e|f_{\min}(Q)|e| + |B||\varepsilon_b|f\max(P)|e| \leq 0, \tag{32}$$

$$-|e|f_{\min}(Q)|e| \leq -|B||\varepsilon_b|f\max(P)|e|. \tag{33}$$

The conclusion is following: if (27) is compared with (2), the (33) is information about values of training constant (limitation) which guarantees stability.

3 Data Clustering Using the Grey Wolf Optimizer

In the previous section the parameters of the Gaussian function are constant. In this work, those values were optimized off-line, using clustering of the input data. The mentioned procedure was done using the Grey Wolf Optimizer. It is an algorithm from the metaheuristic group. Mathematically, the data processing is realized in a loop, the whole group is the potential optimal value, but the best one is determined after each iteration and then the solution space is modified using information about the leader. Details of the calculations are presented in [23].

In this application, the Grey Wolf Optimizer is implemented in searching for the middle points of data sets with similar properties. These points are cluster centers. The number of parameters optimized using GWO is the number of clusters. Based on data processing of the RBFNN, it was assumed that the centers obtained can be used as centers of activation functions in the neural controller. It also imposes the number of nodes in the RBFNN structure. For this purpose, the objective function called by the optimization algorithm is defined as follows:

$$f_{GWO} = \sum_{row} \left(\min_{row}\left(D\left(X_{in}, X_{GWO_{row}}\right)\right)\right), \tag{34}$$

where: D is the Euclidean distance, X_{in} is data for analysis, X_{GWOrow} is several *row* of population matrix X_{GWO} (it consists of actually tested centers). Then the distance between X_{GWOrow} and X_{in} is obtained. The smallest values of each row in this matrix are selected. By means of that, the nearest centers for each input value are found. In the last step the values are added together. In general, calculations lead to a minimization of the objective function. In result, based on the above analysis of (34), it should shift centers to the nearest elements in the data space. In this application, the width of clusters are selected as the average value calculated relative to the cluster center.

In this section only tests of data clustering achieved using GWO are presented (Fig. 1). The first test was prepared for randomly generated data. Several group of data was printed using a different color. Each point was collected in a 1-D row vector (Fig. 1a). The centers at starting iteration were chosen as random numbers (cross marker type, using magenta color). The final point of calculations is marked with black circles. Moreover, the dashed lines are borders of the data space regions. Some elements of the set have small values (close to zero), thus for better transparency of results, the part of the chart has been enlarged (in Fig. 1b). In several step of GWO calculations, the distances of the optimized centers to the proper part of data were decreased (Fig. 1c). It forces movement of the centers according to the suitable area (Fig. 1d).

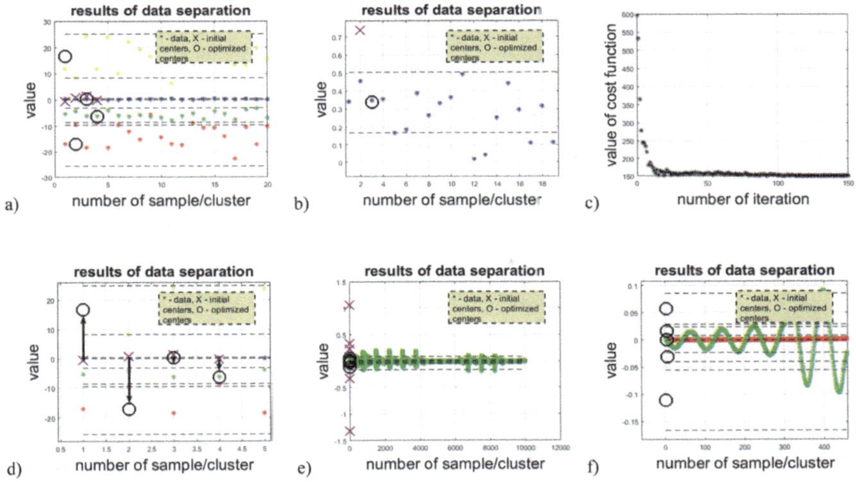

Fig. 1. Clustering of random data set (a)–(d) and results for speed control error signal analysis (e), (f).

Table 1. Results of data clustering.

	Initial values of centers	Centers optimized using GWO
Test 1	−0.641; 0.737; 1.311; −0.204	16.653; −17.050; 0.337; −6.343
Test 2	−1.328; −0.338; 0.225; 0.323; 1.053	−0.110; 0.057; −0.002; 0.016; −0.031

In the second test, an error signal (from present and one previous iteration) from a closed control system (used for electric drive with DC machine) was analyzed (Fig. 1e and Fig. 1f). It can be assumed as an off-line method of centers (of activation functions) selection of the RBFNN controller. The proper determination of these coefficients should lead to an even coverage of the data space. The data for clustering, in this case, can be generated using a model or real data with (even other type) the adaptive controller. In some cases a classical controller can also be applied, because concentration of set points and values are the most important here (special behaviors of the plant can be taken into account). Concluding, in both examples, presented above, an effective data clustering can be observed (precise values are presented in Table 1).

4 Experiment

The present section of this paper contains a short description of the hardware implementation and the experimental results. This article, except the theoretical part, is focused on the experiment. Thus, the mathematical model of the DC motor is omitted in this article. However, details can be found in [24]. It is worth noting the specific attribute of neural controllers, which deals with neglecting of direct identification (mathematical equations, time constants, etc.) of the plant during the design stage. Regarding the implementation of the RBFNN based speed controller, the dSPACE 1103 card was applied. The network consists of two inputs (speed error value and previous sample of this signal), five hidden layer neurons and one output (u_{RBF}). The output weights ware adapted on-line. The centers value are constant (optimized before tests using the GWO algorithm). The gain of the proportional part is equal $k_p = 0.5$. The main motor (with nominal power 0.5 kW) and the load machine were connected using a rigid shaft. This motor was driven by a power converter according to the signal from the processor (reference PWM waveform). In the experiment separately-excited DC motors were used, so the second circuits (of both machines) was connected to a power rectifier. The armature circuit of the load was controlled by a simple switch. The calculating step of the speed control loop is within 1 ms. For the speed measurement the encoders were used (precision of information is defined as 36000 pulses per one rotation). The voltage signal from the current sensor was connected to the ADC converter. Both measured information were further scaled to per-unit system. In Fig. 2 the functional diagram built to achieve the experimental results of the analyzed control structure is shown.

In Fig. 3a the long-time (40 s.) work period of the electrical drive is presented. Due to the limit of article pages, only output value of control structure - speed - was plotted. The reference trajectory forces repeatable speed reversals from 0.25 to −0.25 and then back to a positive level of speed. The values are presented in per-unit system ([p.u.]). The Fig. 3b (upper part of the speed waveform has been posted) and Fig. 3c (final fragment of the measured motor speed) show details of the speed transient. The results prove effective adaptation of the controller and better quality of control after following reversion of the drive (Fig. 3b). However the zoom (Fig. 3c) also shows a lack of overshoots. The first part of the experiment verifies also the stable operation of the speed control system with the adaptive neural model.

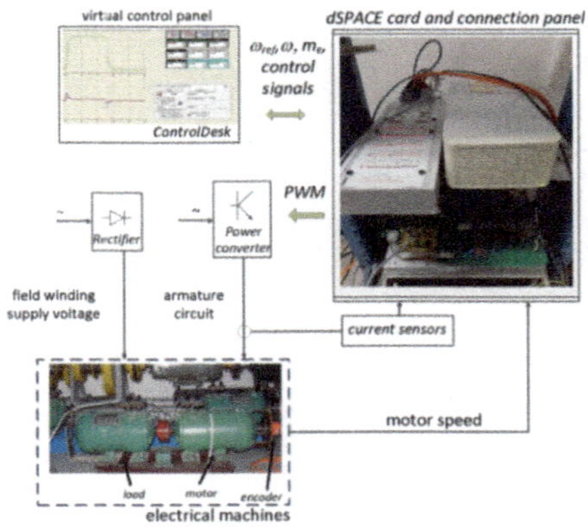

Fig. 2. The scheme of laboratory stand connections.

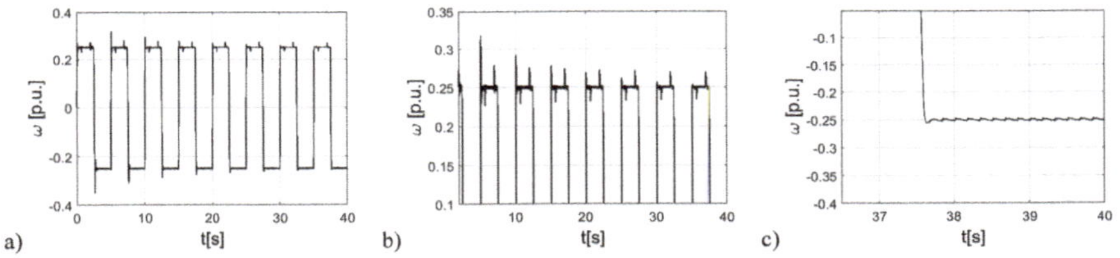

Fig. 3. Response of the control structure for square reference signal.

From the results presented in Fig. 3, the proper work of the electrical drive is observed. For the correct work of the adaptive controller, the parameter used in the training rule is *a priori* selected by user. Now, the significance of this coefficient for the adaptation algorithm has been analyzed (Fig. 4). For appropriate comparisons, identical initial values of weights and centers of activation function were applied for each start of the drive. Comparing, in this context, two tests (for several values of adaptation parameters) - without load torque - the following conclusion can be inferred: in the analyzed range, higher values of the η lead to faster adaptation of the speed controller.

The appropriate designation of the radial function centers of the neural controller is presented in this article as a significant design problem. These parameters of RBFNN were selected using the GWO algorithm. Practical verification of this issue is analyzed throughout this section of the article. But, the importance of this element of the controller is shown in Fig. 5a. Three sample sets of center values were applied for several test. The first of them was optimized (Centers1), the other two were randomly designated (Centers2 and Centers3). There is a clear difference in subsequent tests. The neural controller with centers calculated off-line worked correctly. In the second case (red line), the adaptation algorithm had a problem with elimination of disturbances. In order to protect the system, the drive operation was stopped after about $t = 5$ s.

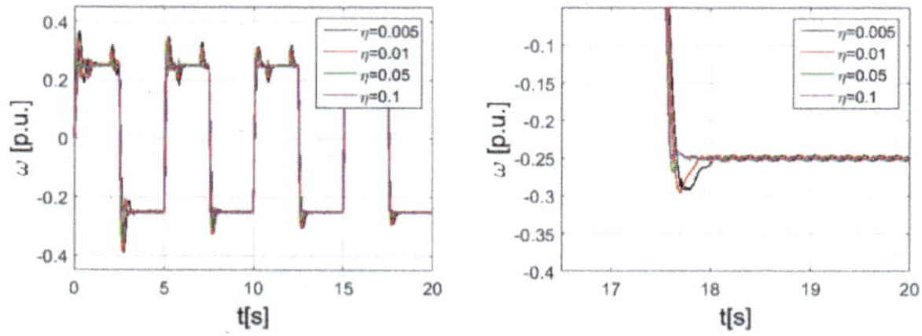

Fig. 4. Influence of the training coefficient η on dynamics of adaptation.

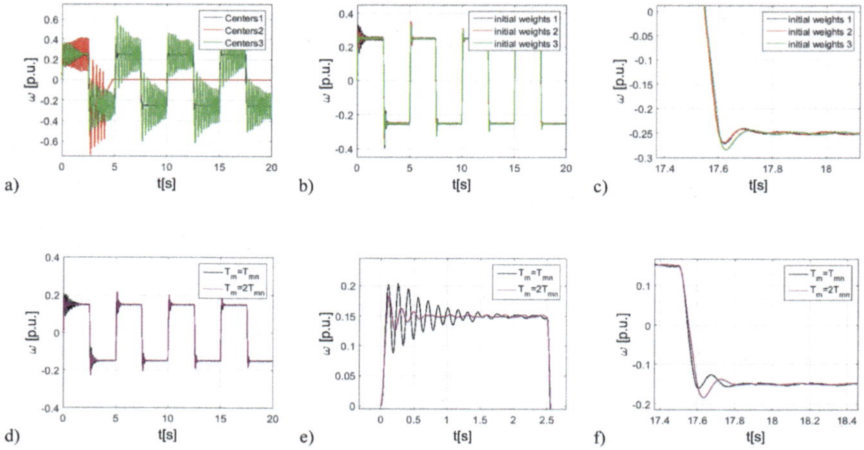

Fig. 5. Transients of state variables in control structure with RBFNN controller – special issues.

Most often, in many applications of neural controllers, the initial values of weights, that are subject of adaptation, are random. The impact of this process on the adaptation process is shown in Fig. 5b and Fig. 5c. The differences are only in the initial part of the speed transients, it is associated with different starting point of the optimization algorithm. However, in a short time, after about two reversions of the drive, all controllers with different initial weights work similarly.

In the last stage of experimental research, the special operation of the neural controller was analyzed (Fig. 5d–Fig. 5f). The parameter of the plant was changed (mechanical time constant). For this purpose, an additional load (steel disc) was attached to the shaft of the machine. After the initial period (before $t = 2.5$ s), after start of the motor, speed oscillations are visible (Fig. 5e). Then, the controller is tuned. Introduced, in construction of the drive, the element leads to more efficient suppression of interferences. However, also in this situation, the adaptation algorithm calculates the parameters of speed controller suitably. As a result, both waveforms are close after about $t = 20$ s.

5 Conclusions

The Radial Basis Function Neural Network and its combination with the proportional controller in the speed control scheme of DC motor was successfully developed and analyzed in this article. The centers of the neural network structure were calculated off-line in the clustering process. The space of input data, from experimental measurements, was divided using the Grey Wolf Optimizer. The introduced method seems to be an alternative to known clustering algorithms. The adaptive controller was implemented and tested in the laboratory. The experimental results show the high-quality tracking performance of the tested controller. During design process, the mechanical parameters and delays in the internal control loop (electromagnetic torque) was not considered. The initial state of weights were selected randomly. However, on-line adaptation of neural coefficients can effectively reduce the speed control error. Future work will include the comparison with selected types of adaptive or classical controllers and optimization of neural network clusters (according the GWO rules) directly in closed control structure of the drive (off-line, based on the model).

References

1. Zhang, C., Yan, H.-S.: Identification and adaptive multi-dimensional Taylor network control of single-input single-output non-linear uncertain time-varying systems with noise disturbances. IET Control Theory Appl. **13**(6), 841–853 (2019)
2. You, F., Chen, N., Zhu, Z., Cheng, S., Yang, H., Jia, M.: Adaptive fuzzy control for nonlinear state constrained systems with input delay and unknown control coefficients. IEEE Access **7**, 53718–53730 (2019)
3. Mazenc, F., Niculescu, S.-I., Bekaik, M.: Stabilization of time-varying nonlinear systems with distributed input delay by feedback of plant's state. IEEE Trans. Autom. Control **58**(1), 264–269 (2013)
4. Louarroudi, E., Pintelon, R., Lataire, J.: Nonparametric tracking of the time-varying dynamics of weakly nonlinear periodically time-varying systems using periodic inputs. IEEE Trans. Instrum. Meas. **61**(5), 1384–1394 (2012)
5. Qi, W., Zong, G., Karimi, H.R.: Sliding mode control for nonlinear stochastic singular semi-Markov jump systems. IEEE Trans. Autom. Control **65**(1), 361–368 (2020)
6. Hekimoğlu, B.: Optimal tuning of fractional order PID controller for dc motor speed control via chaotic atom search optimization algorithm. IEEE Access **7**, 38100–38114 (2019)
7. Yu, H., Xie, T., Paszczynski, S., Wilamowski, B.M.: Advantages of radial basis function networks for dynamic system design. IEEE Trans. Industr. Electron. **58**(12), 5438–5450 (2011)
8. Wang, H., Liu, P.X., Li, S., Wang, D.: Adaptive neural output-feedback control for a class of nonlower triangular nonlinear systems with unmodeled dynamics. IEEE Trans. Neural Netw. Learn. Syst. **29**(8), 3658–3668 (2018)
9. San, P.P., Ren, B., Ge, S.S., Lee, T.H., Liu, J.-K.: Adaptive neural network control of hard disk drives with hysteresis friction nonlinearity. IEEE Trans. Control Syst. Technol. **19**(2), 351–358 (2011)
10. Pajchrowski, T., Zawirski, K., Nowopolski, K.: Neural speed controller trained online by means of modified RPROP algorithm. IEEE Trans. Industr. Inf. **11**(2), 560–568 (2015)
11. Pajchrowski, T.: Porównanie struktur regulacyjnych dla napędu bezpośredniego z silnikiem PMSM ze zmiennym momentem bezwładności i obciążenia. Przegląd Elektrotechniczny **94**(5), 133–138 (2018)
12. Kaminski, M., Orlowska-Kowalska, T.: Adaptive neural speed controllers applied for a drive system with an elastic mechanical coupling - a comparative study. Eng. Appl. Artif. Intell. **45**, 152–167 (2015)

13. Yang, X., Li, Y., Sun, Y., Long, T., Sarkar, T.K.: Fast and robust RBF neural network based on global K-means clustering with adaptive selection radius for sound source angle estimation. IEEE Trans. Antennas Propag. **66**(6), 3097–3107 (2018)
14. Raitoharju, J., Kiranyaz, S., Gabbouj, M.: Training radial basis function neural networks for classification via class-specific clustering. IEEE Trans. Neural Netw. Learn. Syst. **27**(12), 2458–2471 (2016)
15. Calvini, M., Carpita, M., Formentini, A., Marchesoni, M.: PSO-based self-commissioning of electrical motor drives. IEEE Trans. Industr. Electron. **62**(2), 768–776 (2015)
16. Sun, X., Hu, C., Lei, G., Guo, Y., Zhu, J.: State feedback control for a PM hub motor based on gray wolf optimization algorithm. IEEE Trans. Power Electron. **35**(1), 1136–1146 (2019)
17. Premkumar, K., Manikandan, B.V.: Speed control of brushless DC motor using BAT algorithm optimized adaptive neuro-fuzzy inference system. Appl. Soft Comput. **32**, 403–419 (2015)
18. Szczepanski, R., Tarczewski, T., Grzesiak, L.M.: Adaptive state feedback speed controller for PMSM based on artificial bee colony algorithm. Appl. Soft Comput. **83**, 1–12 (2019)
19. Ma, H., Wang, T., Li, Y., Meng, Y.: A time picking method for microseismic data based on LLE and improved PSO clustering algorithm. IEEE Geosci. Remote Sens. Lett. **15**(11), 1677–1681 (2018)
20. Yang, S., Li, C.: A clustering particle swarm optimizer for locating and tracking multiple optima in dynamic environments. IEEE Trans. Evol. Comput. **14**(6), 959–974 (2010)
21. Wang, Z., Hu, C., Zhu, Y., He, S., Yang, K., Zhang, M.: Neural network learning adaptive robust control of an industrial linear motor-driven stage with disturbance rejection ability. IEEE Trans. Industr. Inf. **13**(5), 2172–2183 (2017)
22. Gao, J., Proctor, A., Bradley, C.: Adaptive neural network visual servo control for dynamic positioning of underwater vehicles. Neurocomputing **167**, 604–613 (2015)
23. Mirjalili, S., Mirjalili, S.M., Lewis, A.: Grey wolf optimizer. Adv. Eng. Softw. **69**, 46–61 (2014)
24. Orlowska-Kowalska, T., Szabat, K., Jaszczak, K.: The influence of parameters and structure of PI-type fuzzy-logic controller on DC drive system dynamics. Fuzzy Sets Syst. **131**(2), 251–264 (2002)

Residual Error Shaping in Active Noise Control - A Case Study

Małgorzata I. Michalczyk(✉)

Silesian University of Technology, 44-100 Gliwice, Poland
malgorzata.michalczyk@polsl.pl

Abstract. The aim of the paper is to utilize the possibility of residual error shaping in active noise control system creating spatial zones of quiet in enclosure. The filtered-x least mean squares algorithm that uses additional residual error shaping filter is applied for attenuation of broadband disturbances. Simulations conducted with the use of simplified and real plant models give an insight into the performance and convergence behavior of the FX-LMS ng filter. The conclusions on the behavior of the FX-LMS ng were broadened onto the problem of modeling of anti-aliasing filters application in the error signal measurement path.

Keywords: Active noise control · FX-LMS · Residual noise shaping

1 Introduction

Active noise control (ANC) systems may effectively attenuate the unwanted noise in many applications [1, 2], like active headphones (e.g. [3]) or enclosures [4, 5]. Using microphones, loudspeakers, A/D and D/A converters, some electronic transducers (amplifiers, analogue filters) and a DSP processor with the I/O card, local zones of quiet of even a few meters of diameter can be created in a noisy enclosure [4–6]. The classic approach uses filtered-x least mean squares (FX-LMS) algorithm for adaptation of controller filter coefficients, which allows system initialization and adjusts the system performance to changing-in-time environment conditions. The idea of the FX-LMS algorithm is to minimize the mean square $E\{e^2(i)\}$ of the error signal $e(i)$ [7], thus the overall sound field is attenuated. Unintentional differences in disturbance attenuation are imposed by dynamic properties of the electro-acoustic plant, that is controlled in ANC systems.

In some cases, however, there is a need of residual error shaping, which can be obtained by filtration of the error signal by an additional filter $F_{er}(z^{-1})$, resulting in the FX-LMS ng first proposed in [8]. This algorithm is applied in so called psychoacoustic ANC system to improve the noise attenuation performance in terms of hearing perception [8–10]. In some applications of ANC a signal carrying valuable audible information must be retained; it may be speech or alarm signals. Ramos et al. [11] point also to the need of retaining an engine noise in ANC systems implemented in cars. Further, filtering of the error signal may be also helpful in choosing the frequency range of the system operation,

A. Bartoszewicz et al. (Eds.): PCC 2020, AISC 1196, pp. 713–724, 2020.
https://doi.org/10.1007/978-3-030-50936-1_60

as it may be uneconomical to operate over or under some frequency range [11], which is similar to the application of pass-band anti-aliasing filters.

In the paper the possibility of residual error shaping in the ANC system creating spatial zones of quiet in enclosure is analyzed. The influence of the error signal $e(i)$ filtration through band-pass and stop-band filters on the ANC system behavior, in particular on the adaptive control algorithm convergence, is illustrated with simulations results. In the research the real-world ANC system electro-acoustic plant models are exploited along with the simplified plant models, designed to represent electro-acoustic plant features.

2 ANC System

2.1 ANC Systems

In the feedforward ANC system (Fig. 1) the reference signal $x(i)$, carrying an information about the disturbing noise, is picked up by the reference microphone and then processed by the controller filter. Obtained control signal $u(i)$ must be processed by the electro-acoustic plant, called secondary path, including reconstruction filters, amplifiers and loudspeaker. Then the acoustic signal interferes with the disturbance in an acoustic domain, the resulting sound is picked up by the error microphone and processed by anti-aliasing filters and A/D converters giving the error signal $e(i)$. The error signal is used along with the reference signal by an adaptive control algorithm.

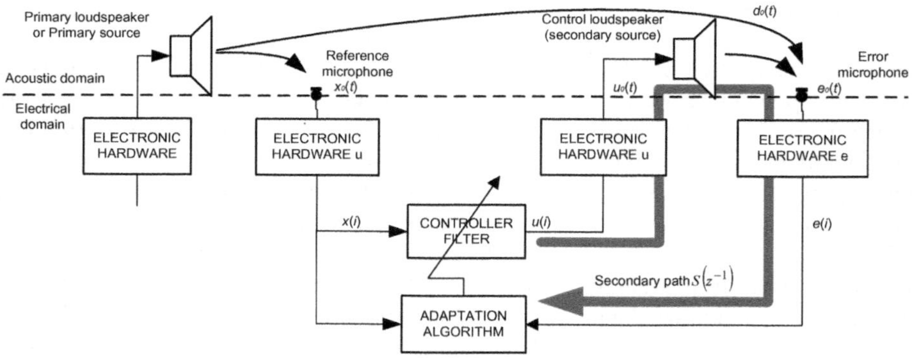

Fig. 1. An adaptive ANC system.

2.2 Adaptive Control Algorithm

Application of an adaptive control algorithm enables automatic system initialization and on-line adjusting of the controller filter $W(z^{-1})$ coefficients to system changes in time of its operation. The most popular adaptive algorithm employed for ANC is the widely described filtered-x least mean squares (FX-LMS) algorithm [1, 2], that is, however, difficult to parameterize, especially in ANC systems creating zones of quiet in enclosures (see e.g., [4]). The filtered-x modification of the LMS algorithm is a filtration of the reference signal $x(i)$ through the model of the secondary path $\hat{S}(z^{-1})$ (Fig. 1), that compensates for the influence of the secondary path dynamics.

Another LMS algorithm modification - the adjoint LMS algorithm [12], called also the FE-LMS algorithm [1, 13] is used in some ANC system applications, e.g. for attenuation of a noise in a duct [14]. Instead of the filtration of the reference signal, as in the FX-LMS algorithm, the error signal $e(i)$ is filtered through the filter, that is the inverse of the secondary path model and the reference signal $x(i)$ is also appropriately delayed. Because the reference signal is delayed [2], convergence speed of the FE-LMS algorithm is lower than that of the FX-LMS algorithm, thus it is rarely applied. However, it is less computationally complex than the FX-LMS algorithm, if applied for a multi-channel system [2, 12, 15].

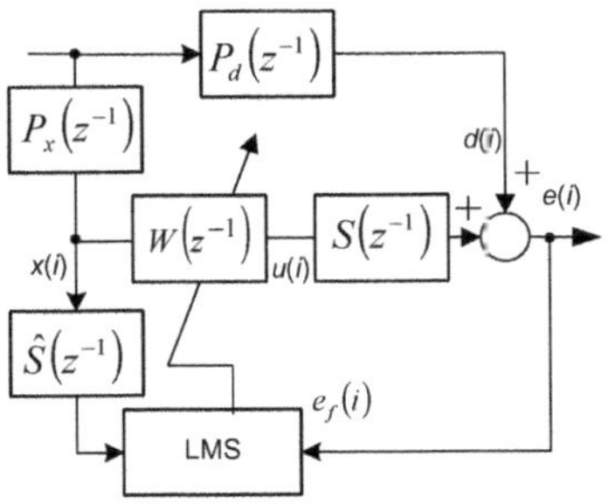

Fig. 2. An adaptive ANC system with the classical FX-LMS algorithm.

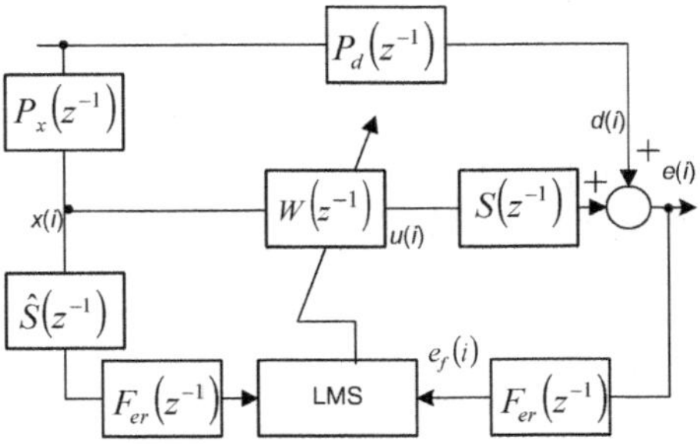

Fig. 3. An adaptive ANC system with the FX-LMS ng.

3 FX-LMS ng

Another version of the FE-LMS algorithm is an error signal filtration applied additionally to the FX-LMS algorithm [8] (Fig. 3). This algorithm, which is the subject of the current study, is further addressed hereby as FX-LMS ng. To assure the control algorithm convergence, the reference signal is also filtered through the $F_{er}(z^{-1})$ filter along with the secondary path model $\hat{S}(z^{-1})$. In such configuration the $F_{er}(z^{-1})$ may be any filter, designed depending on kind of application and according to the ANC system application requirements.

Psychoacoustic ANC. In psychoacoustic ANC systems the shaping filter is designed to improve the error signal properties in terms of hearing perception. It may model an inverse of an A-weighting [8, 9] or ITU-R 468 weighting shape [10]. In [8] it was noticed, that the application of an A-weighting shaping can speed up the convergence speed of some frequency components amplified by the filter, but the overall convergence can be slower than for the classical FX-LMS algorithm. It was contributed to the choice of the shaping filter, however, the research seems to be incomplete (simple shaping filter with high gain, simple plant models and different step-sizes μ applied both algorithms).

Excluding Some Range of Frequencies from the System Operation. If there is a need of passing through the ANC system some audible information (speech, alarm signals, engine noise as a feedback for a driver) a band-stop or a low-pass filter is applied.

Limiting the Frequency Range of the System Operation. It may be uneconomical to operate under or over some frequency range. The effectiveness of active noise control decreases along with the growth of the attenuated frequency (zones of quiet are very small), thus it is justified to exclude higher frequencies from system operation taking into account the specific demands of the particular ANC system. On the other hand, Ramos et al. [11] claimed that some low-frequency components (e.g. 35 Hz) are "scarcely audible" and concentrated the system effort on higher frequencies. In such cases a band-pass or a low-pass filter is applied.

Modeling the Analogue and Digital Filters in the Measurement Path. The last case corresponds to the application of filters in the measurement path in an ANC system. It should be noticed, that there are two error signals in the ANC system with the FX-LMS ng. The algorithm minimizes the mean square the filtered error signal $e_f(i)$. However, the system performance should be evaluated basing on the error signal $e_O(i)$ picked up by the error microphone, corresponding to what the ANC system user can hear. Consequently, observing $e_O(i)$ with application of any low-pass or band-pass filter is like gaining insight into the real residual noise (e_O(t) in Fig. 1), heard by the ANC system user.

4 Simulations

4.1 Simulation Setup

The simulations were conducted to compare the behavior of classical FX-LMS algorithm and the FX-LMS ng. An ANC system placed in a reverberant room of 70 m^3 cubature was built of a disturbing primary loudspeaker, a control loudspeaker, an error and a reference microphones [16]. The models of electro-acoustic plant paths were identified as FIR filters with 150 coefficients with sampling frequency 500 Hz (Fig. 4). Dynamical properties of the electro-acoustic plant paths are very complicated. One of the main problems are high differences in the secondary path $S\left(z^{-1}\right)$ gain for different frequency components, which result in problems in parameterization of FX-LMS-based control algorithms [1, 4].

Therefore, to obtain a clear insight into the system performance with the shaping filter $F_{er}\left(z^{-1}\right)$ the parameterization of the ANC system is to be simplified. The filters modeling ANC system paths are designed as the band-pass filters with the same bandwidth between 30 Hz and 180 Hz and high attenuation in the stop-band, which corresponds to the real plant dynamical properties, however, the flat magnitude in the pass-band is set (Fig. 4); phases are almost linear. In order to reduce an adaptation time, the models' delays are set in the way to assure shorter delay time in the control path $P_x\left(z^{-1}\right)S\left(z^{-1}\right)$ than in the disturbance path $P_d\left(z^{-1}\right)$, that corresponds to picking up the reference signal in maximum possible precedence. Further, the simulation experiments are repeated using the real-world electro-acoustic plant models (Fig. 4).

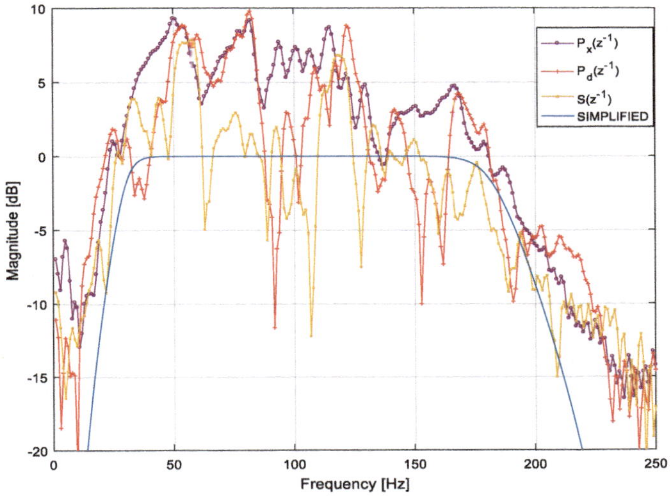

Fig. 4. Magnitudes of the frequency responses of the real-world and a simplified plant models.

In all simulations experiments presented below it is assumed that an acoustic feedback between the control loudspeaker and the reference microphone is perfectly cancelled and the secondary path model is perfect ($\hat{S}(z^{-1}) = S(z^{-1})$). The controller filter $W(z^{-1})$ is an FIR filter with 100, 200 or 300 coefficients. FX-LMS step size μ value is chosen individually for each simulation experiment, to assure fast convergence and system stability, the same for both algorithms. As the performance of the FX-LMS ng is compared with that of the classical FX-LMS algorithm, step-size μ value is chosen once for each pair of simulations. The disturbance is simulated as the white noise signal filtered through the models of the real-world system (see Fig. 2 and 3). Consequently, the reference signal $x(i)$ is obtained as the filtration of the white noise through the model $P_x(z^{-1})$, the disturbance signal $d(i)$ is obtained as the filtration of the white noise through the model $P_d(z^{-1})$.

The signals analyzed are: error signal $e(i)$ for classical FX-LMS algorithm, and for FX-LMS ng: filtered error signal $e_f(i)$ and the observed error signal $e_O(i)$ - error signal picked up by the error microphone, corresponding to what the ANC system user can hear (Fig. 7).

Each simulation experiment was repeated 100 times, each signal power series was exponentially smoothed with the factor 0.99 and then results were averaged over 100 realizations. Each signal power spectral density (PSD) was windowed over the 91000 last samples of one simulation realization. Disturbance attenuation was calculated as the power of the error signal over the power of the disturbance signal ratio averaged over last 10% of the simulation time.

4.2 Application of the Low-Pass and Pass-Band Filters

As the performance of the ANC system with low-pass and band-pass $F_{er}(z^{-1})$ filter is similar, the figures show the behavior of the FX-LMS algorithm with the band-pass shaping filter. The simplified ANC system path models is applied in 1'000'000 iterations simulation, which corresponds to over 33 min of system operation (Fig. 5).

It seems that the FX-LMS ng converges faster than the FX-LMS - filtered error signal $e_f(i)$ reaches the lowest values in the shortest time with the disturbance attenuation of 29 dB. However, if the ANC system is evaluated in terms of observed error signal $e_O(i)$ values (disturbance attenuation of 15 dB), the FX-LMS ng is slower and its performance is worse than that of the FX-LMS algorithm (disturbance attenuation of 23 dB). On the other hand, the PDSs of the analyzed signals in Fig. 6, show that the disturbance attenuation obtained with the FX-LMS ng in the frequency band of interest is higher, power of the error signal $e_O(i)$ is 3–6 dB lower than the power of the error signal $e(i)$ given by the FX-LMS algorithm. To summarize, the amplitude of the observed error signal $e_O(i)$ reaches higher values in the whole frequency range, because the attenuation is not obtained in the frequency ranges under 30 Hz and over 150 Hz, filtered out by the $F_{er}\left(z^{-1}\right)$ filter stop-band.

Additional simulation experiments were conducted to find out, when the classical FX-LMS algorithm outruns the FX-LMS ng in terms of disturbance attenuation for the whole frequency range. The same controller filter with 100 coefficients was tuned with larger step-size value μ (increased to 0.07 from 0.02 to obtain faster algorithm convergence) for 50'000'000 iterations (one realization only), corresponding to almost 28 h of ANC system operation in the unchanged environment. Then, only for a few frequency components around 50 Hz the disturbance attenuation obtained by the FX-LMS is slightly lower than that obtained with the FX-LMS ng (Fig. 6). The overall disturbance attenuation for this simulation experiment is 32.1 dB, 16.7 dB and 38,4 dB at $e(i)$, $e_O(i)$ and $e_f(i)$ signals respectively (Fig. 9).

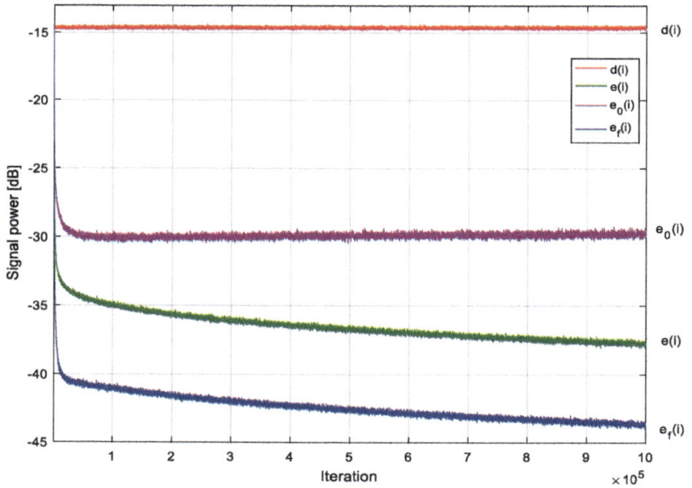

Fig. 5. Learning curves of the mean square error signals - FX-LMS algorithm and FX-LMS algorithm with band-pass residual error shaping filter.

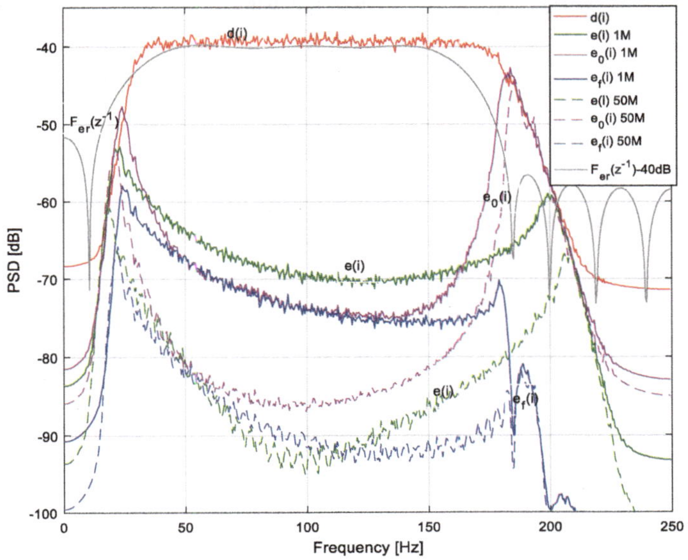

Fig. 6. PSD of the ANC system signals - FX-LMS and FX-LMS algorithm with band-pass residual error shaping filter (its magnitude, shifted −40 dB is shown with a grey line) obtained after 1'000'000 and 50'000'000 iterations.

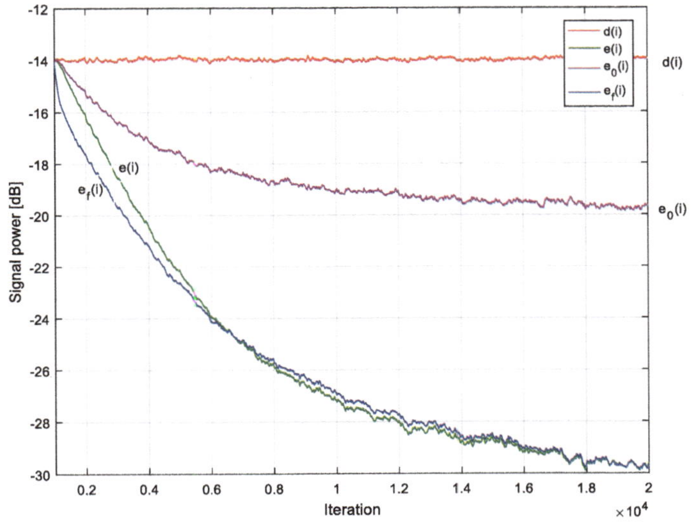

Fig. 7. Learning curves of the mean square error signals - FX-LMS algorithm and FX-LMS algorithm with band-stop residual error shaping filter.

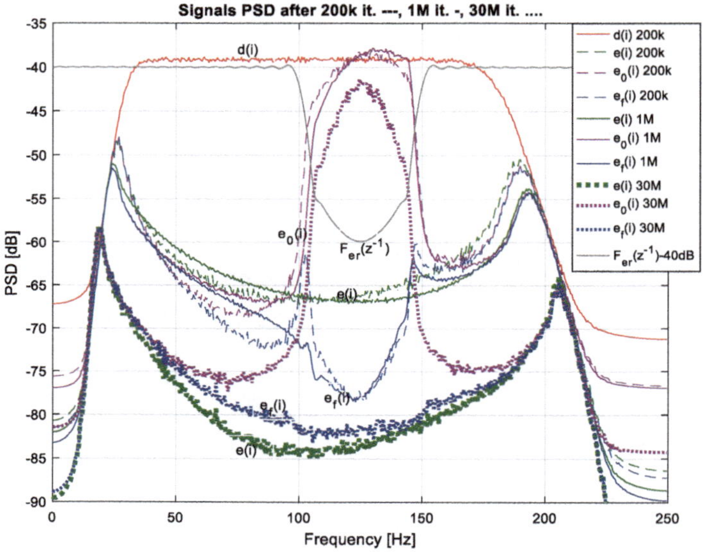

Fig. 8. PSD of the ANC system signals - FX-LMS and FX-LMS algorithm with band-stop residual error shaping filter (its magnitude, shifted −40 dB is shown with a grey line) obtained after 200'000, 1'000'000 and 30'000'000 iterations.

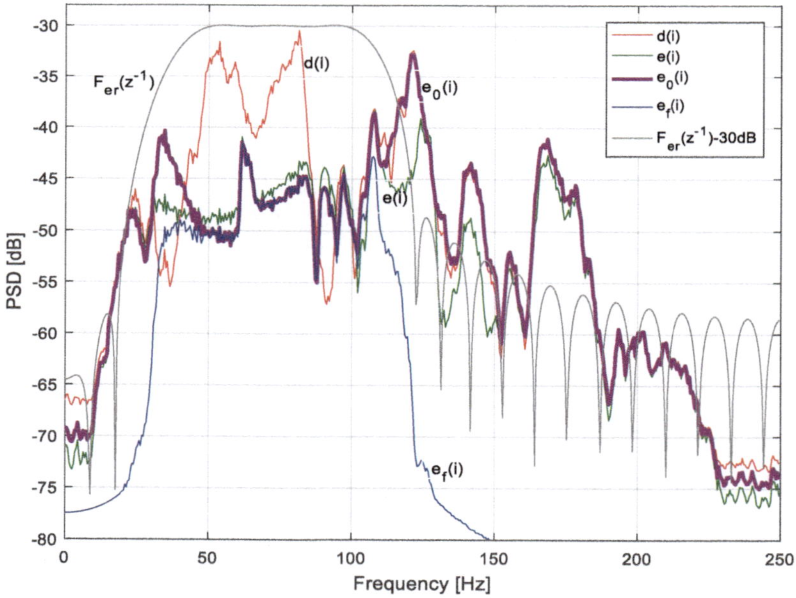

Fig. 9. PSD of the ANC system signals - FX-LMS and FX-LMS algorithm with band-pass residual error shaping filter (its magnitude, shifted −30 dB is shown with a grey line) - simulated with the use of real-world path models

4.3 Application of the Band-Stop Filter

The application of a filter with a narrow stop-band (Figs. 7 and 8) is a more difficult task. After over 33 min (1'000'000 iterations) of ANC system operation disturbance attenuation is 20.4 dB for the FX-LMS algorithm ($e(i)$), and 5.5 dB and 21.6 dB for the FX-LMS algorithm with residual error shaping ($e_O(i)$ and $e_f(i)$).

The filter magnitude shape is difficult to model, therefore the performance of the ANC system with the FX-LMS ng is worse.

There are PSDs of the error signals calculated after 200'000 (almost 7 min.), 1'000'000 (over 33 min) and 30'000'000 (16:40 h) iterations of ANC system operation compared in Fig. 8. Again, similar as in the ANC system with the pass-band residual error shaping filter, the FX-LMS ng converges faster for some frequencies in the frequency range of interest than the classical FX-LMS algorithm. After over 16 h of adaptation the FX-LMS ng allows for passing the signal frequencies in the desired frequency range. However, the observed error signal power $e_O(i)$ is higher than error signal $e(i)$ power obtained using the classical FX-LMS algorithm for all signal frequency components, thus finally the classical FX-LMS outruns the FX-LMS ng in terms on disturbance attenuation. The highest disturbance attenuation (about 45 dB) is obtained for frequencies around 100 Hz using the classical FX-LMS algorithm in the longest simulation experiment. The disturbance attenuation is not the same over the passband of electro-acoustic path models, it gets lower while approaching the bandwidth edges. The overall disturbance attenuation for the longest simulation experiment is 29,9 dB, 10,2 dB and 29,7 dB at $e(i)$, $e_O(i)$ and $e_f(i)$ signals respectively.

The behavior of the FX-LMS algorithm with shaping filter $F_{er}(z^{-1})$ gain different from 1 was studied. The simulations' results show, that increasing of the gain of the shaping filter $F_{er}(z^{-1})$ speeds up the algorithm convergence and has the same effect as the increase of the step-size μ. Consequently, damping of the shaping filter results in slower algorithm convergence.

4.4 Application of the Pass-Band Filter - Real-World Models

The simulations conducted with the use of real-world ANC system plant models confirm the results of the simplified studies (Fig. 9). The band-pass filter with the bandwidth between 50 Hz and 100 Hz is applied for residual error shaping. Controller filter of 300 coefficients is adapted for 1'000'000 iterations and enables disturbance attenuation of 6.8 dB, 4.3 dB and 11 dB at $e(i)$, $e_O(i)$ and $e_f(i)$ signals respectively.

4.5 Modeling of Anti-aliasing Filters Application

The observations of the behavior of the FX-LMS ng may be expanded on modeling of anti-aliasing filters application. What the ANC system user hears is represented by the signal $e_O(t)$ picked up by the error microphone (Fig. 1) and then processed by analogue and digital transducers, that have shaping effect on the resulting error signal $e(i)$. Thus, the existence of analogue and digital filters in the error signal measurement path should have the same effect on the ANC system performance as the application of the residual error shaping filter. It should be expected, that the real broadband disturbance attenuation, heard by the ANC system user, is lower than calculated basing on the error signal $e(i)$ power, however, for frequency components in the bandwidth there can be some improvement (the control algorithm convergence may be faster and for some time the attenuation may be higher).

4.6 Shaping Filter Influence on the Observed Error Signal $e_0(i)$

It is interesting, that the simulations' results show, that the equation for error signal $e(i)$ calculation for FX-LMS ng given by Kuo and Tsai (Eq. 5 in [8]) is wrong for any tested kind of shaping filter $F_{er}(z^{-1})$. In fact, it should be written as:

$$e_0(i) \neq F_{er}^{-1}(z^{-1})e_f(i). \tag{1}$$

It was verified in the frequency domain by averaged the division of the shaping filter magnitude $|F_{er}(j\omega)|$ and the error signals magnitude ratio $|E_f(j\omega)/E_0(j\omega)|$. The result was far from 1, even for the longest simulation experiments, in which high disturbance attenuation was obtained (Fig. 10). The most likely reason is that in the analyzed ANC system exists a nonlinear adaptation feedback, which, consequently, should not be easily neglected in an adaptive control system analysis.

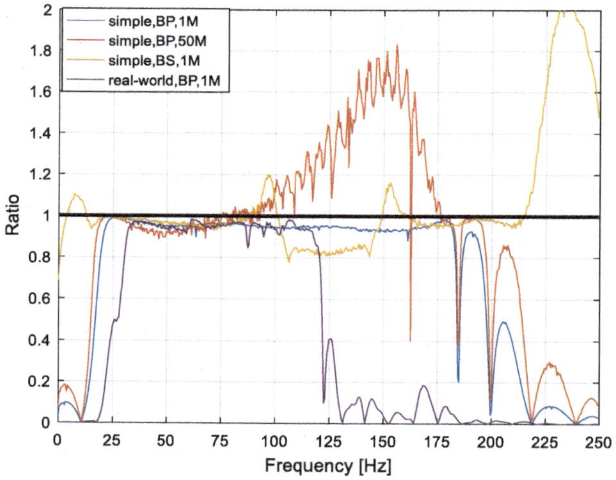

Fig. 10. Averaged $|F_{er}(j\omega)|/|E_f(j\omega)/E_0(j\omega)|$ ratio for simulations experiments using simplified (simple) or real-world plant models, band-pass (BP) and band-stop (BS) residual error shaping filters and lasting 1'000'000 (1 M, averaged over 100 realizations) or 50'000'000 (50M, one realization only) iterations; the black line denotes expected ratio value.

5 Conclusions

The behavior of the FX-LMS algorithm with residual error signal shaping was observed in simulation experiments, conducted with simplified and real-world plant models, modeling the ANC system creating spatial zones of quiet in a reverberant enclosure.

It is shown, that the application of residual error shaping filter, which narrows the bandwidth of ANC system operation, worsens overall disturbance attenuation level, as in the ANC system does not operate in frequency ranges suppressed by the shaping filter. However, in all cases the speed of convergence of error signal frequency components in the shaping filter bandwidth is initially higher for the FX-LMS ng than for the classical FX-LMS algorithm. Finally, after a long adaptation, the classical FX-LMS algorithm reaches even higher disturbance attenuation levels for the whole frequency range. The influence of the shaping filter gain was shown to be the same as the change of the step-size of the FX-LMS algorithm value.

The simulation results showed also that the influence of a nonlinear adaptation feedback should not be neglected when analysing the ANC system performance.

The conclusions on the behavior of the FX-LMS ng were broadened onto the problem of modeling of anti-aliasing filters application in the error signal measurement path.

Acknowledgement. The partial financial support of this research by The Polish Ministry of Science and Higher Education is gratefully acknowledged.

References

1. Kuo, S.M., Morgan, D.R.: Active Noise Control System, Algorithms and DSP Implementations. J. Wiley & Sons Inc., New York (1996)
2. Elliott, S.J.: Signal Processing for Active Control. Academic Press, London (2001)
3. Pawelczyk, M., Latos, M., Michalczyk, M.I., Czyz, K., Mazur, K.: An efficient communication system for noisy environments, mat. In: ICMIC 2011 (2011)
4. Michalczyk, M.I.: Parametrization of LMS-based control algorithms for local zones of quiet. Arch. Control Sci. **15**(1), 5–34 (2005)
5. Michalczyk, M.I.: Active noise control system creating distributed zones of quiet. In: 12th IEEE International Conference on Methods and Models in Automation and Robotics, pp. 309–314 (2006)
6. Michalczyk, M.I.: Multichannel active noise control. In: Proceedings of IFAC Workshop on Programmable Devices and Systems PDS'04, pp. 366–371 (2004)
7. Mitra, S.K., Kaiser, J.F. (eds.) Handbook of digital signal processing. J. Wiley (1993)
8. Kuo, S.M., Tsai, J.: Residual noise shaping technique for active noise control systems. J. Acoust. Soc. Am. **95**(3), 1665–1668 (1994)
9. Bao, H., Panahi, I.M.S.: Using A-weighting for psychoacoustic active noise control. In: 31st Annual International Conference of the IEEE EMBS (2009)
10. Bao, H., Panahi, I.M.S.: Psychoacoustic active noise control with ITU-R 468 noise weighting and its sound quality analysis. In: Annual International Conference of the IEEE EMBS (2010)
11. Ramos, P., Salinas, A., López, A., Masgrau, E.: Practical implementation of a multiple-channel FxLMS Active Noise Control system with shaping of the residual noise inside a van. In: Proceedings of the 2002 International Symposium on Active Control of Sound (2002)
12. Wan, E.A.: Adjoint LMS: an efficient alternative to the filtered-X and multiple error LMS algorithms. In: Proceedings of the 1996 IEEE International Conference on Acoustics, Speech, and Signal Processing, ICASSP-96, pp. 1842–1845 (1996)
13. Elliott, S.J.: Filtered reference and filtered error LMS algorithms for adaptive feedforward control. Mech. Syst. Signal Process. **12**(6), 769–781 (1998)

14. Eriksson, L.J., Allie, M.C., Greiner, R.A.: The selection and application of an IIR adaptive filter for use in active sound attenuation. IEEE Trans. Acoust. Speech Signal Process. **35**(4), 433–437 (1987)
15. Douglas, S.C.: Fast implementations of the filtered-X LMS and LMS algorithms for multichannel active noise control. IEEE Trans. Speech Audio Process. **7**(4), 454–465 (1999)
16. Michalczyk, M.I.: Comparison of feedforward and IMC controllers for active noise control system with moving error microphone: simulation results. In: 9th Conference on Active Noise and Vibration Control Methods, Poland (2009)

Design and Development of Industrial Cyber-Physical System Testbed

Jakub Możaryn[1(✉)], Andrzej Ordys[1], Adam Stec[1], Konrad Bogusz[1], Omar Y. Al-Jarrah[2], and Carsten Maple[2]

[1] Institute of Automatic Control and Robotics, Warsaw University of Technology, Warsaw, Poland
jakub.mozaryn@pw.edu.pl
[2] Warwick Manufacturing Group, The University of Warwick, Coventry CV4 7AL, U.K.

Abstract. Cyber-physical systems (CPS) are integral components of Industry 4.0. However, there is a lack of benchmark systems for the design, development and testing of CPS. To this end, this article presents the design and development of an industrial CPS testbed using a stand of two coupled tanks, a Programmable Logic Controller (PLC), and an Internet-of-Things (IoT) gateway. The testbed can connect to cloud services, combining the industrial and management components together, where a cloud-based service can be used to manage the system. It also can be used to design, develop and evaluate fault diagnosis, fault-tolerant control, and cyber-security algorithms for CPS. Due to its versatility and reconfigurability, the proposed testbed can be used to test various scenarios of possible faults and cyber-attacks in industrial systems.

Keywords: Cyber-physical system · Cloud-based control · Fault diagnosis · Fault-tolerant system · Cyber-security

1 Introduction

Internet of Things (IoT) is a novel paradigm enabled by the integration of several technological and communication solutions, such as wireless sensor and actuator networks and distributed intelligence [1], where physical devices can exchange information via communication networks (e.g., the Internet) [2,5]. The Cyber-Physical Systems (CPS) are, arguably, the next stage of development where the physical components of a system are monitored and controlled by computer-based systems/algorithms, via the Internet and/or some type of local wireless network. Communication between CPS could drastically increase the complexity of tasks that may be performed/achieved.

CPSs have been identified as one of the key research areas by the European Union Research programmes[1] as well as by National Science Foundation in the

[1] https://ec.europa.eu/digital-single-market/en/cyber-physical-systems.

A. Bartoszewicz et al. (Eds.): PCC 2020, AISC 1196, pp. 725–735, 2020.
https://doi.org/10.1007/978-3-030-50936-1_61

USA[2]. This vibrant field of research is likely to dominate the control systems design, due to the rapid progress in communication, computing and networking. Such systems are one of the pillars of Industry 4.0, the concept describing the changes in manufacturing, leading to what is called the 4th industrial revolution. Application examples are numerous including, automotive systems – optimization of power train of hybrid vehicles [15], collaborating robotic systems for smart production [16], robotic systems for medical applications [17], distributed power generation (e.g. wind turbines) [18], smart homes [19], to mention just a few.

Technologies underpinning the CPS include sensors and wireless sensor networks, communication protocols, distributed control systems and cloud computing. In particular:

- sensors are the means by which CPS collects information about its status. By exchanging such information (i.e., the system's status) with other systems, more information about the process and the environment can be obtained, hence improving the quality of control actions;
- communication platforms which involve communication protocols, network architecture and wireless communication devices, are the means to exchange information between different CPS; and
- distributed control systems (e.g., cloud-based control system) issue control commands to individual CPS, using control algorithms, which should be resilient to possible information loss and cyber-attacks.

Currently, more emphasis is placed on the control algorithm itself being located on the cloud which referred to as Control as a Service (CaaS), providing undoubtful benefits in terms of cost, flexibility, and maintenance. Such a solution also introduces new challenges such as resilience of control actions, the security of information flow and information processing.

Major automation companies e.g. Siemens (MindSphere[3]), ABB (Ability[4]), Rockwell Automation (FactoryTalk InnovationSuite[5]), follow this approach they provide different modules, facilties, and services, allowing to create cloud-based control and IoT operating system for industry. Cloud-based control of systems and CaaS have became a focus of interest, recently. However, most existing works, mainly focus on the automation of production lines and tooling machines. Monostori [7] outlined expectations and challenges related to such systems. Goldschmidt et al. [8] and Schlechtendahl et al. [9] explore the term Control as a Service in relation to a programmable logic controller (PLC) used for industrial automation tasks (soft PLC). They focus on communication requirements and on the scalability of the controller. Farokhi et al. [10] and Junsoo et al. [11] propose forms of encryption mechanism, to prevent possible cyber-attacks. Then they discuss the design of the encryption mechanism in such a way that it would not be necessary to de-encrypt the information in order to perform standard

[2] https://www.nsf.gov/funding/pgm_summ.jsp?pims_id=503286.
[3] https://new.siemens.com/global/en/products/software/mindsphere.html.
[4] https://ability.abb.com/.
[5] https://www.rockwellautomation.com/global/products/factorytalk/overview.page.

operations of the controller (addition and multiplication). In relation to process industry, Costa et al. [12] present a cloud-based control of a pilot plant (two tank system). They demonstrate feasibility of applying cloud based self-evolving fuzzy controller acting as supervisory controller for a local Programmable logic controller attached to the plant. Although, this defined a pilot plant to test standard process operations, it didn't consider information security aspects of the problem.

This article presents a laboratory set-up for analyzing the resilience of control algorithms of CPS with continuous dynamics, wireless communication network and cloud service. We follow the example of the two-tank pilot plant [12] but discuss functionalities, communication and connections which would enable tests related to the resilience of control algorithms concerning cyber-attacks and information loss.

The remaining of this paper is organized as follows: The experimental testbed is described in Sect. 2. Next, in Sect. 3, sensors, actuators and communication links are specified, and possible configurations of the system are presented. Proposed test scenarios, of cyber-security and discriminating between the malfunctions of the sensors and malicious changes to values of signals in the communication channels, for the testbed are presented in Sect. 4. Section 5 concludes this work and presents further steps of the project.

2 Description of Experimental Set-Up: Coupled Tanks System

The process part of the laboratory test stand is presented in Fig. 1, whereas the schematic diagram is given in Fig. 2.

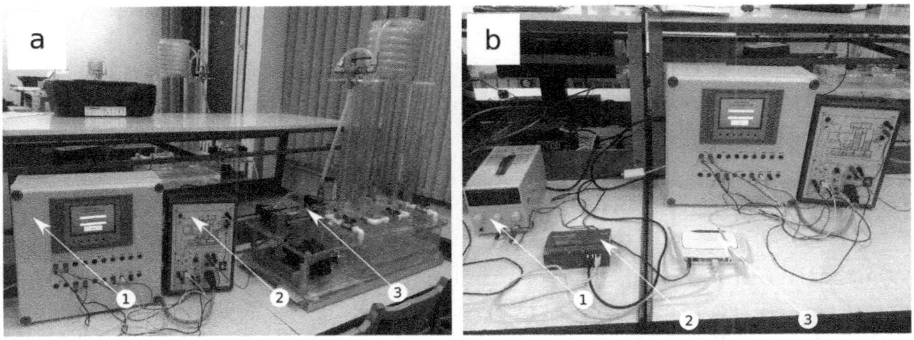

Fig. 1. Laboratory stand: a-1) control cabinet, a-2) control interface, a-3) test stand, b-1) power supply, b-2) SIMAATIC IOT2040 industrial gateway, b-3) Router.

Components of the stand (Fig. 1) are: tanks Z_1 and Z_2, a pump P that is controlled by standard current signal that corresponds to a change of the pump

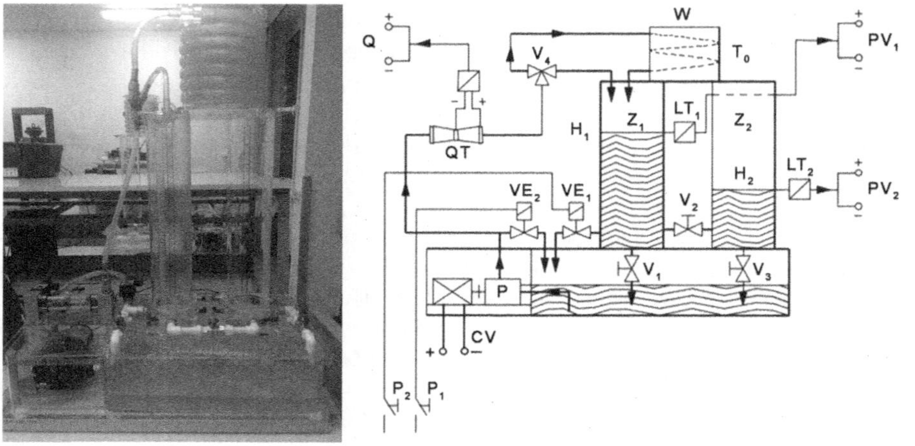

Fig. 2. Schematic diagram of the laboratory test stand.

capacity 0–6.5 [l/min], LT_1, LT_2 are pressure transducers (measuring range 0–500 mm H_20) for measuring a liquid level in each of tanks (H_1, H_2). To introduce the transport delay in the system, an elastic tube W, is used through which the liquid can flow into the tank Z_1. There are manually operated cut-off valves $V1$, $V2$, $V3$, $V4$ which are used to change the way that liquid flows. The tanks are connected with the shut-off valve V_2. The outlet of each tank is connected with the main tank with manually operated, cut-off valves V_1 (tank Z_1), and V_2 (tank Z_2).

There are also two cut-off electro-mechanical valves (VE_1, VE_2), used to introduce disturbances into the process. By ZK_1 we denote leakage from the tank Z_1 (opening the valve VE_1), and by ZK_2 we denote leakage at pump outlet (opening the valve VE_2).

Depending on the configuration of the valve states V_2 and V_4 and the use of a specially constructed elastic tube W at the inlet of tank Z_1, various properties of the control object, as given in Table 1, can be realized.

3 Testbed Configuration

Control Cabinet. The control cabinet contains Siemens SIMATIC S7-1200 controller. This controller has a modular design (up to 8 signal modules, 1 signal board, 3 communication modules, the maximum number of binary and analog inputs is 284 and 51, respectively). The device has Profinet/Industrial Ethernet interfaces that are integrated with support for TCP/IP, ISO-on-TCP and S7 protocols. The controller can diagnose and monitor software through the Ethernet port and can communicate via RS-232, RS-485 and Modbus RTU protocols.

Table 1. Object properties depending on valve configuration (V_2 and V_4)

Object type	Process value	Valve	Valve setup
First-order system	Liquid level H_1 in tank Z_1	V_1	Open
		V_2	Closed
		V_3	Closed
		V_4	Closed - the liquid flows directly to the tank Z_1
First-order system with delay	Liquid level H_1 in tank Z_1	V_1	Open
		V_2	Closed
		V_3	Closed
		V_4	Open - the liquid flows to the tank Z_1 through tube W
Second-order system, overdamped	Liquid level H_2 in tank Z_2	V_1	Closed
		V_2	Open
		V_3	Open
		V_4	Closed - the liquid flows directly to the tank Z_1
Second-order system with delay, overdamped	Liquid level H_2 in tank Z_2	V_1	Closed
		V_2	Open
		V_3	Open
		V_4	Open - the liquid flows to the tank Z_1 through tube W
Second-order system, underdamped	Liquid level difference between tank Z_1 and tank Z_2 $H_d = H_1 - H_2$	V_1	Closed
		V_2	Open
		V_3	Open
		V_4	Closed
Second-order system, underdamped	Liquid level difference between tank Z_1 and tank Z_2, $H_d = H_1 - H_2$	V_1	Closed
		V_2	Open
		V_3	Open
		V_4	Open - the liquid flows to the tank Z_1 through tube W

The laboratory stand is equipped with:

- CPU 1214C unit,
- 6ES7 234 analog input/output expansion module,
- 230VAC/24VDC PS1207 power supply,
- Switch CSM1277.

The hardware connection in the control cabinet is presented in Fig. 3.

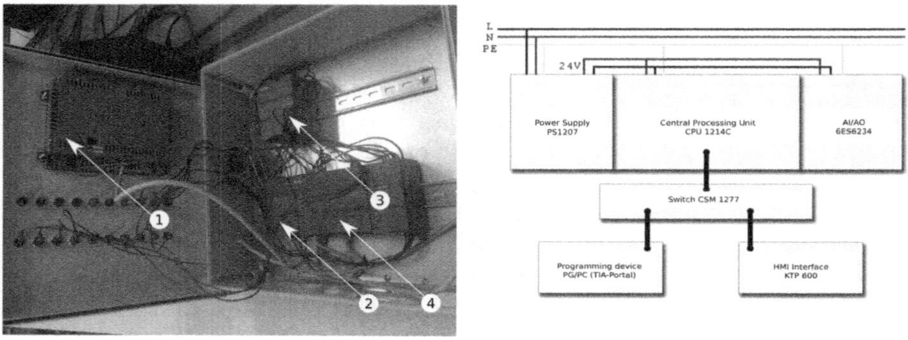

Fig. 3. The hardware configuration in the control cabinet: 1) HMI KTP600, 2) Switch, 3) PLC CPU with I/O modules, 4) Power Supply.

TIA-Portal. Configuration, programming and monitoring of the SIMATIC S7-1200 controller and HMI KTP600 operator panel are performed using TIA-Portal environment [6]. It integrates the STEP 7 Basic software (application for PLC programming) and WinCC Basic (application for HMI programming), which allows data exchange between them. The environment also contains tools for creating new libraries of project objects (process variables, most frequently used functions - e.g. PID) or object graphics (elements of process HMI). A proven concept of OB organizational blocks, FC functions as well as FB function blocks and DB data blocks [6] is used. The TIA Portal environment allows run-time mode, which allows, among others: preview and change of variable values specified in the program, observation of the state of devices configured in the control system, simulation of the operator panels existing in the project, preview and diagnosis of the currently executed program.

3.1 Communication Links

Simatic IOT2040. The laboratory testbed is equipped with IoT framework [5], using Simatic IOT2040 gateway. It is an open platform for collecting, processing and transferring the data between production and IT systems, or clouds, in the production environment. As an intelligent gateway interface, it can be used for transferring data in both directions. This gateway supports programming languages such as Java, Python or C++, and multiple communications protocols such S7 Protocol, OPC UA, Profinet, TCP IP, MQTTL via various interfaces, including RS232/422/485, serial USB, Ethernet or Wi-Fi. It is equipped with Intel Quark x1020 processor, and 1 GB RAM memory. It can be expanded for various applications with Arduino and mini-PCI shields, depending on number and types of inputs and outputs. The high-level application, which is developed in the IoT, enables the gateway to collect the process data and automatically transmits it to the cloud. Such solution is an efficient way to transfer data without sophisticated programming known from conventional embedded systems.

Communication between the PLC and the IoT gateway is performed using Siemens S7 libraries. There is continuous refreshing of PLC inputs and outputs (it will work after possible short-term disconnection from the power supply). The libraries require access through "rack slot 1" and in the programmable interface of TIA Portal there is optimized access for blocks that are already in use. Also, the PUT/GET access in the hardware configuration has to be enabled within the S7-1200 controller.

Moreover, IoT gateway works with SIMATIC IOT2000 IO Arduino Shield, with 5 digital inputs, 2 analog inputs, 2 digital outputs. This configuration allows for an additional read of input/output signals from the laboratory stand.

Node-RED. The IoT gateway gathers all input/output data from the system and sends all information to the cloud service. Communication is established using Node-RED[6] - an open-source flow-based JavaScript-based development tool for the integration of IoT hardware devices, APIs (Application Programming Interfaces) and online services, developed by IBM Emerging Technology. The Node-RED configuration program contains visually connected nodes, that represent two-way operations. Each node provides different functions, such as to monitor the flow as the debug out node, or to read and write with GPIO pins of I/O shield as the node. Created flows are stored using JavaScript Object Notation (JSON). Moreover, Node-RED enables wire-up input, output and processing nodes to create flows for data processing, control signals, and alerts. The programmable interface Node-RED contains modified decision logic when it is possible to change remotely available process signals.

Cloud Technology. Cloud technology provides the computing resources and storage capacity as a service for different groups of end-users [4]. The access to cloud applications is through a web or mobile application, while software and collected data are located on remote servers. There are three types of cloud technology: Software as a Service (SaaS), Platform as a Service (PaaS), Infrastructure as a Service (IaaS).

As mentioned before, Siemens provides MindSphere, the open cloud platform for applications in the context of IoT, operating as a cloud-based PaaS. However, we adopted the Amazon Web Services (AWS)[7] in this research. It is a collection of remote computing services working according to IaaS paradigm, that together make up a cloud computing platform. The most central and well-known of these services are Elastic Compute Cloud (EC2) and Amazon s3 (Simple Storage Service). AWS provides also services for Big Data Analytics (Athena, EMR, Redshift, Kinesis, Elasticsearch, Quicksight) and IoT (IoT Core, IoT Device Defender, IoT Device Management, IoT Analytics, IoT SiteWise, IoT Events, IoT Things Graph). Due to its popularity and availability, AWS platform was adopted in this research.

[6] https://nodered.org/.
[7] https://aws.amazon.com/.

The designed stand uses only the AWS IoT service, which is dedicated to cooperation and collecting data from IoT devices. AWS IoT enables receiving and sending data from and to SIMATIC IOT2040, using the MQ Telemetry Transport (MQTT) protocol. Cloud communication with IOT2040 is implemented as AWS communication with Node-RED, which is configured on the SIMATC IOT2040 gateway.

4 Potential Test Scenarios

The configuration of the system explained in Sect. 3, enables basic tests associated with process control applications. In particular, it is possible to devise mathematical models of the system, to verify the models through simulation and comparison with the real plant, to implement and tune PID controllers for first order and second order systems and to apply more advanced control algorithms. Furthermore, it is possible to devise or/and program the scenarios in which the system structure and the system model changes during its operation (by changing the positions of cut-off valves). Moreover, the system configuration could enable fault detection and tests of cyber-security of all configurations. Model-based fault detection algorithms would compare the system variables obtained from the simulation with those measured on the experimental set-up. The discrepancies would then be mapped on the fault identification algorithm.

Besides, the system could be vulnerable to several cyber-attacks including the following:

- The reading of one or more of the sensors (transducers LT_1 and LT_2) could be falsified;
- The information about the position of electromagnetic cut-off valves (leakage valves) could be falsified;
- The set-point value for the controller could be changed to a non-admissible value;
- The tuning parameters of the controller could be changed (the controller has been de-tuned); and
- The information about the structure of the system (first order, second order-over/underdamped) could be falsified.

In the existing configuration, because the only access point to the system is through the internet connection of the PLC controller, the most likely scenarios are the last three. There are well-developed methods for on-line performance assessment of controllers which could be applied to identify the situation of de-tuning [13], especially during the steady-state operation of a system with a standard controller (e.g. PID). Some extensions to the situation of dynamic responses are also available (e.g. [14]). Such methods may require modification to deal with cyber-attacks where the malfunction is designed in a way that makes the detection difficult.

Cloud technology is still being actively developed, and thus it has many vulnerabilities that can be exploited, such as data vulnerability, cloud API and services vulnerability, shared technology vulnerability, Several cyber-attacks can be launched against cloud computing. This includes malware injection attacks, Denial of Service (DoS) attacks, abuse of cloud services, side-channel attacks, wrapping attacks, man-in-the-cloud attacks and account or service hijacking. Figure 4 presents the information links between TIA Portal/PLC Controller, Node-RED, and AWS IoT Services. It's a two-way connection, allowing sending I/O signals (eg. process values, set-point values) between PLC and cloud services. Therefore, the proposed testbed can be used to create the scenarios, when data (eg. set-point of the PID controller) is maliciously affected in the cloud environment. It can change the behaviour of the industrial control system. Therefore, the diagnosis methods of the improper functioning of the system should differentiate between the system failures due to faults of the sensors or actuators, and failures due to cyber-attacks [20].

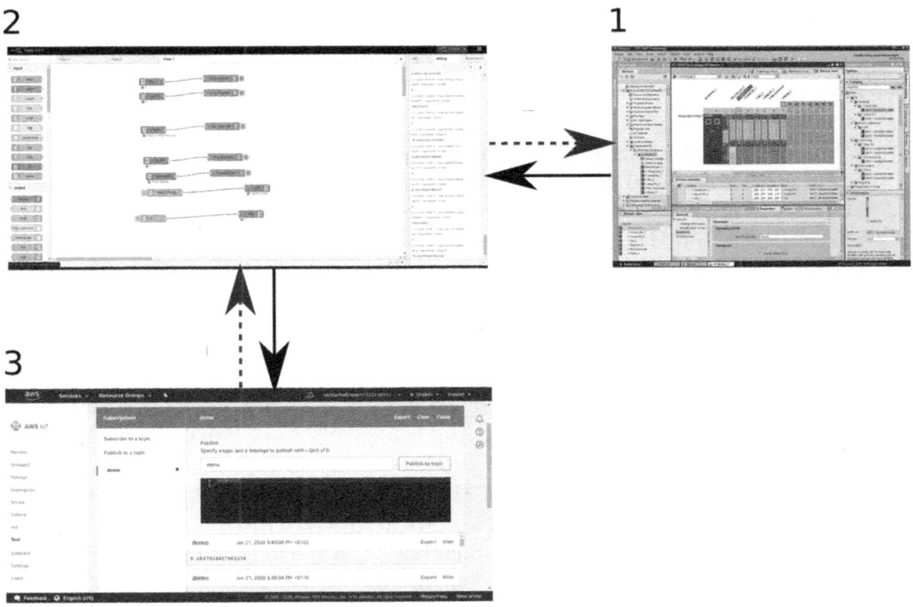

Fig. 4. The communication links between PLC and cloud services: 1) TIA Portal/PLC, 2) Node-RED, 3) AWS IoT cloud services.

5 Conclusions

The paper describes the experimental testbed set-up of industrial CPS. It is based on the two-coupled tanks stand, equipped with PLC controller (Simatic S7-1200), and IoT gateway (SIMATIC IOT2040). The possible configurations of the system are presented from the perspective of tests of cyber-security, discriminating between the malfunctions of the sensors and changes to values of signals in the cloud due to cyber-attacks.

For the future work, it would be interesting to consider that those valves, which currently are operated manually, could be replaced by automated valves operated remotely. This would enable remote reconfiguration of the system and testing various scenarios related to the falsification of the information on the structure of the system. Moreover, the falsification of the information on the model of the system could change the order of the system, which is a potential future research area.

The laboratory facility described in this article is currently operational in the Institute of Automatic Control and Robotics, Faculty of Mechatronics, Warsaw University of Technology. We invite Colleagues to collaborate by proposing and executing remotely tests of cyber-security for this installation.

Acknowledgments. Prof. Andrew Ordys acknowledges support from National Agency of Academic Exchange (NAWA), "Polish Returns" grant No: PPN/PPO/2018/1/00063/U/00001.

References

1. Atzori, L., Iera, A., Morabito, G.: The Internet of Things: a survey. Comput. Netw. **54**(15), 2787–2805 (2010). https://doi.org/10.1016/j.comnet.2010.05.010
2. Maple, C.: Security and privacy in the Internet of Things. J. Cyber Policy **2**(2), 155–184 (2017). https://doi.org/10.1080/23738871.2017.1366536
3. Almada-Lobo, F.: The industry 4.0 revolution and the future of manufacturing execution systems (MES). J. Innov. Manage. **3**(4), 16–21 (2015)
4. Marinescu, D.: Cloud Computing. Elsevier (2018)
5. Buyya, R., Dastjerd, A.V.: Internet of Things, Principles and Paradigms. Morgan Kaufmann Publishers (2016)
6. Stenerson, J., Deeg, D.: Siemens Step 7 (TIA Portal) Programming, a Practical Approach. CreateSpace Independent Publishing Platform (2015)
7. Monostori, L.: Cyber-physical production systems: roots, expectations and R&D challenges. In: Proceedings of the 47th CIRP Conference on Manufacturing Systems (2014). https://doi.org/10.1016/j.procir.2014.03.115
8. Goldschmidt, T., Murugaiah, M.K., Sonntag, C., Schlich, B., Biallas, S., Weber, P.: Cloud-based control: a multi-tenant, horizontally scalable soft-PLC. In: 2015 IEEE 8th International Conference on Cloud Computing, New York, NY, pp. 909–916 (2015). https://doi.org/10.1109/CLOUD.2015.124
9. Schlechtendahl, J., Kretschmer, F., Sang, Z., Lechler, A., Xu, X.: Extended study of network capability for cloud based control systems. Robot. Comput. Integr. Manuf. **43**, 89–95 (2017). https://doi.org/10.1016/j.rcim.2015.10.012
10. Farokhi, F., Shames, I., Batterham, N.: Secure and private cloud-based control using semi-homomorphic encryption. IFAC PapersOnLine **49**(22), 163–168 (2016). https://doi.org/10.1016/j.ifacol.2016.10.390
11. Junsoo, K., Lee, C., Shim, H., Cheon, J.H., Kim, A., Kim, M., Song, Y.: Encrypting controller using fully homomorphic encryption for security of cyber-physical systems. IFAC PapersOnLine **49**(22), 175–180 (2016). https://doi.org/10.1016/j.ifacol.2016.10.392

12. Costa, B., Skrjanc, I., Blazic, S., Angelov, P.: A practical implementation of self-evolving cloud-based control of a pilot plant. In: 2013 IEEE International Conference on Cybernetics (CYBCO), pp. 7–12 (2013). https://doi.org/10.1109/CYBConf.2013.6617464
13. Ordys, A.W., Uduehi, D., Johnson, M.: Process control performance assessment, from theory to implementation. In: Monograph Series: Advances in Industrial Control. Springer Verlag, London (2007). https://doi.org/10.1007/978-1-84628-624-7
14. Ordys, A., Grimble, M.J.: Benchmarking and tuning PID controllers. In: Vilanova, R., Visioli, A. (eds.) PID Control in the New Millennium: Lessons Learned and New Approaches, Springer Verlag (2012). https://doi.org/10.1007/978-1-4471-2425-2_13
15. Krasniqi, X., Hajrizi, E.: Use of IoT technology to drive the automotive industry from connected to full autonomous vehicles. IFAC PapersOnLine **49**(29), 269–274 (2016). https://doi.org/10.1016/j.ifacol.2016.11.078
16. Ruiz Garcia, M.A., Rojas, R., Gualtieri, L., Rauch, E., Matt, D.: A human-in-the-loop cyber-physical system for collaborative assembly in smart manufacturing. Procedia CIRP **81**, 600–605 (2019). https://doi.org/10.1016/j.procir.2019.03.162
17. Patel, A.R., Patel, R.S., Singh, N.M., Kazi, F.S.: Vitality of robotics in healthcare industry: an Internet of Things (IoT) perspective. In: Bhatt, C., Dey, N., Ashour, A. (eds.) Internet of Things and Big Data Technologies for Next Generation Healthcare, Studies in Big Data, vol. 23, pp. 91–109. Springer, Cham (2019). https://doi.org/10.1007/978-3-319-49736-5_5
18. Satuyeva, B., Sauranbayev, C., Ukaegbu I.A., Nunna, H.S.V.S.K.: Energy 4.0: towards IoT applications in Kazakhstan. Procedia Comput. Sci. **151**, 909-915 (2019). https://doi.org/10.1016/j.procs.2019.04.126
19. Mocrii, D., Chen, Y., Musilek, P.: IoT-based smart homes: a review of system architecture software, communications, privacy and security. Internet of Things **1–2**, 81–98 (2018). https://doi.org/10.1016/j.iot.2018.08.009
20. Al-Jarrah, O.Y., Maple, C., Dianati, M., Oxtoby, D., Mouzakitis, A.: Intrusion detection systems for intra-vehicle networks: a review. IEEE Access **7**, 21266–21289 (2019). https://doi.org/10.1109/ACCESS.2019.2894183

Batch Algorithm for Balancing the Air Bearing Platform

Paweł Zagórski(✉), Paweł Król, and Alberto Gallina

AGH University of Science and Technology, al. Mickiewicza 30, 30-059 Krakow, Poland
pzagor@agh.edu

Abstract. This paper discusses the process of balancing the satellite simulator mounted on the spherical air bearing table. In order to accurately simulate the satellite motion with such a test stand it is necessary to bring the Center Of Mass (COM) of the system as close as possible to the Center Of Rotation (COR) of the air bearing by moving the balancing masses. This calibration process reduces the gravity torque influencing the system movements. A new batch method of determining the COM of the balancing platform is proposed, allowing for its later adjustment. The novelty of the method comes from the idea that the freely rotating system model can be divided into rigid part for which the center of mass is constant, and the balancing masses constituting the variable influence on the COM. Advantage of this approach is the fact, that while gathering data for the batch calibration counterweights can be actuated in a known way, which turns out to greatly improve the estimation precision. Potential disadvantage is, that estimate of the masses and paths of movement of the counterweights is required, which in practice constitutes additional sources of error. Sensitivity analysis is performed to asses the viability of this trade-of considering the inaccuracies in the balance masses paths, as well as sensor noises and misalignment.

Keywords: Air bearing table · Small satellites · Attitude determination and control

1 Introduction

Attitude determination and control systems of satellites are fundamental for most satellite missions. In order to reach high-performance and reliable Attitude Determination and Control System (ADCS) designs, experiments have to be conducted on the Earth by means of dedicated test rigs that simulate the (almost) torque-free environment of a spacecraft in orbit. For this purpose platform floating upon a spherical air bearing may be used. However, only a well-balanced system can minimize the gravity torque. The balancing process implies to bring the center of gravity of the floating system very close to its center of rotation. This is achieved by adjusting the position of movable counterweights present

A. Bartoszewicz et al. (Eds.): PCC 2020, AISC 1196, pp. 736–746, 2020.
https://doi.org/10.1007/978-3-030-50936-1_62

on the platform. The operation is usually performed by an automatic systems that guarantees faster and more accurate results than a manual approach. One of the first automatic balancing system based on accelerometer measurements was presented in [1]. Since then, several more advanced algorithms have been proposed, which exploit acceleration and angular rate sensing. There exist feedback control algorithms [2–4] that guarantee on-line operations and off-line batch methods which elaborate sensor measurements recorded for a finite time span. Batch algorithms come in several different implementations. In [5–9] simplifications are made to the dynamic equation of motion to reduce the computational burden required to estimate the center of mass of the floating system. The additional estimation of the inertia tensor is achieved in more elaborated algorithms whose formulations are based on the torque method [10], momentum integral or energy balance [11]. This paper proposes a further variation of batch algorithm. Batch methods need to excite the floating system during measurements for increasing the system dynamics and so regularizing the identification process. It is usually done by activating momentum/reaction wheels mounted on the floating platform. The novelty of the proposed method consists of exploiting the counterweights of the balancing system also for this purpose. In this way balancing system can be made simpler and more compact. This paper presents theoretical aspects of the method and provides results of a sensitivity analysis that shows the robustness of the algorithm to several sources of uncertainties that are expected to appear in the physical model.

2 Balancing System Description

2.1 Mechanical Design

The table top air bearing system under development in the AGH Space Tech Lab is composed of a main platform and three tilted balancing arms. The platform provides a mechanical interface between the air bearing hemisphere, the experimental payload and the three actuated counterweight assemblies. The air bearing is manufactured by Physik Instrumente with a maximum payload of 15 kg. The turbine torque measured experimentally is of the order of 10^{-6} Nm. Each of the three balancing counterweight arms includes a NEMA8 stepper motor that actuates by a lead screw a counterweight that may travel on a straight rail. Each arm is pivoted to the platform and its tilt angle can be changed by substituting the holding slabs. The motor mountings on the arm can be changed from the top to the bottom position so to further lower the vertical position of the center of mass of the system if needed. Most of the mechanical components are made of polyamid 3D printed using Selective Laser Sintering (SLS) technique. An important guideline followed in the design was the minimization of the mass and geometry of the entire system. Currently, two different versions of the platform are being assembled that differ for the quality of the motors and linear rail being used. Figure 1 shows the prototype of the assembled system without electronics.

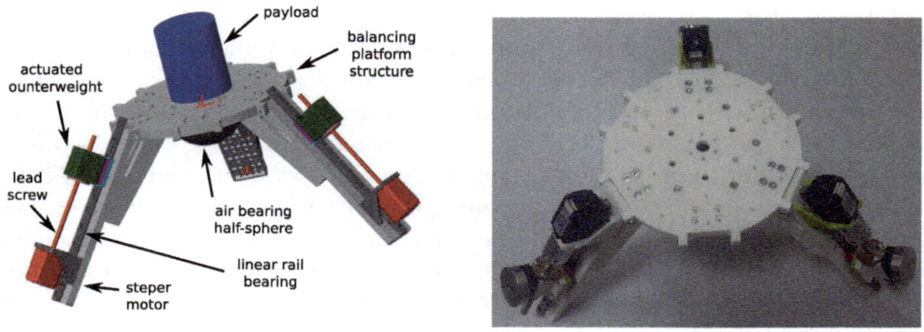

Fig. 1. Balancing platform. Preliminary CAD model (left) and physical model without electronics (right).

2.2 Balancing Process

The balancing process of the system starts with an initial rough manual balancing that brings the Center Of Mass (COM) within a few millimeters distance to the Center Of Rotation (COR). Next, the platform is released from a generic attitude configurations and accelerometer and gyro measurements collected. During the platform motion, the counterweights are also moved on their rails. In this study all counterweights are moved from the bottom to the top of the same amount (7 cm). It should be noted that the aim of shifting the masses in this phase is not to reach a balanced configuration but to record the dynamics behavior of the system at different geometric configurations.

2.3 Mathematical Model

The total mass m_s of the balancing system can be written as

$$m_s = m_p + m_b = m_p + \sum_{i=1}^{n} m_i, \tag{1}$$

where m_p is the mass of the platform, m_b is the accumulated mass of the n balancing masses m_i.

Figure 2 shows the geometry of the system. The center of rotation of the air bearing constitutes the origin of the body reference frame. The z axis is defined as vertical of the platform, the remaining two are selected arbitrarily to complete the right-handed triad. Position $\mathbf{r_i}$ of the i-th balancing mass in that frame is

$$\mathbf{r_i} = \mathbf{p_i} + \mathbf{d_i} u_i, \tag{2}$$

where $\mathbf{p_i}$ is the origin of the balance mass (position of the mass when the displacement u_i is equal to zero), and $\mathbf{d_i}$ is the unit vector describing the mass displacement direction. The center of mass of the balancing masses $\mathbf{r_b}$ is

$$\mathbf{r_b} = \frac{1}{m_b} \sum_{i=1}^{n} \mathbf{r_i} m_i. \tag{3}$$

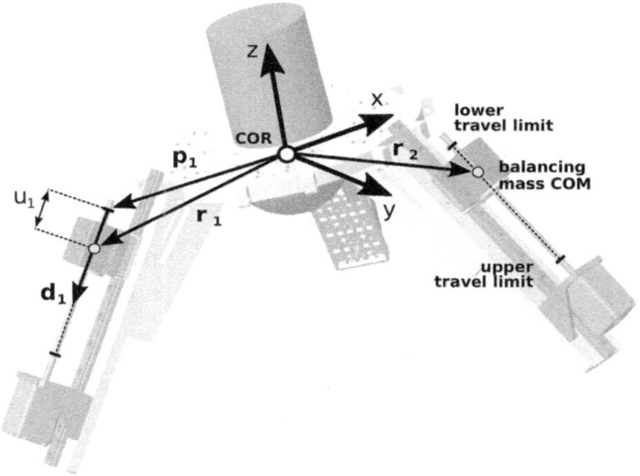

Fig. 2. Reference frames.

In general, inertia of a single balance mass around the system COR changes, depending on its position and weight distribution. It can be described as

$$J_i = R_i J_i' R_i^\top + m_i \left(I \left(\mathbf{r_i}^\top \mathbf{r_i} \right) - \mathbf{r_i} \mathbf{r_i}^\top \right), \tag{4}$$

where I is the identity matrix, J_i' is the inertia of a single counterweight around its own COM, and R_i is the rotation matrix transforming the reference frame in which J_i' is represented to the body reference frame of the system (see Fig. 2). If the counterweights are comparatively small and light with respect to the entire system, we can assume them being point masses. Then, inertia J_i' becomes a zero 3×3 matrix. Then, Eq. (4) simplifies to

$$J_i = m_i \left(I \left(\mathbf{r_i}^\top \mathbf{r_i} \right) - \mathbf{r_i} \mathbf{r_i}^\top \right). \tag{5}$$

Total inertia of all counterweights around the COR is then equal to

$$J_b = \sum_{i=1}^{n} J_i = \sum_{i=1}^{n} m_i \left(I \left(\mathbf{r_i}^\top \mathbf{r_i} \right) - \mathbf{r_i} \mathbf{r_i}^\top \right). \tag{6}$$

2.4 System Parameters

Table 1 lists the physical and geometrical parameters of the system resulting from the CAD model and assumed in this study.

Table 1. Parameters of the balancing system

Parameter	Value	Unit	Comment
$\mathbf{p_1}$	$[48.34 \quad 83.72 \quad 9.34]^\top$	mm	Same for all
$\mathbf{p_2}$	$[-96.67 \quad 0.00 \quad 9.34]^\top$	mm	
$\mathbf{p_3}$	$[48.34 \quad -83.72 \quad 9.34]^\top$	mm	
$\mathbf{d_1}$	$[297.95 \quad 516.06 \quad -803.06]^\top$	mm	
$\mathbf{d_2}$	$[-595.90 \quad -0.00 \quad -803.06]^\top$	mm	
$\mathbf{d_3}$	$[297.95 \quad -516.06 \quad -803.06]^\top$	mm	
$u_{i,max}$	70	mm	
m_s	1593,8	g	Same for all
m_p	1493,0	g	
m_i	33.6	g	
J_p products	$[9054.0 \quad 10284.7 \quad 13133.1]$	kg/mm^2	
J_p moments	$[-238.6 \quad 1045.4 \quad -375.9]$	kg/mm^2	

3 Automatic Balancing Algorithm

The moment of inertia of the system J_s with respect to the body reference frame can be written as

$$J_s = J_p + J_b \tag{7}$$

where J_p is the moment of inertia of the platform including the experimental payload and J_b that of the balancing masses. The inertia of the platform is assumed to be constant, and unknown (due to the unknown payload inertia), while the term J_b in general changes when the balance masses are moved. The rotational motion of the system about COR is governed by the equation

$$(\dot{J_s \omega}) + \omega \times J_s\omega = \mathbf{r_s} \times m_s\mathbf{g}\,, \tag{8}$$

which can be expanded into

$$J_p\dot{\omega} + \omega \times J_p\omega - \mathbf{r_p} \times m_p\mathbf{g} = -(\dot{J_b \omega}) - \omega \times J_b\omega + \mathbf{r_b} \times m_b\mathbf{g}. \tag{9}$$

Balancing masses, total inertia moments and products can be combined into the vector $\bar{J}_b$ where

$$\bar{J}_b = [J^{(b)}_{xx}\ J^{(b)}_{yy}\ J^{(b)}_{zz}\ J^{(b)}_{xy}\ J^{(b)}_{xz}\ J^{(b)}_{yz}]^\top\,. \tag{10}$$

After introducing the analogical vectorized platform inertia estimate $\tilde{J}_p$

$$\tilde{J}_p = [J^{(p)}_{xx}\ J^{(p)}_{yy}\ J^{(p)}_{zz}\ J^{(p)}_{xy}\ J^{(p)}_{xz}\ J^{(p)}_{yz}]^\top, \tag{11}$$

and the corresponding angular rate matrix

$$\Omega = \begin{bmatrix} \omega_x & 0 & 0 & \omega_y & \omega_z & 0 \\ 0 & \omega_y & 0 & \omega_x & 0 & \omega_z \\ 0 & 0 & \omega_z & 0 & \omega_x & \omega_y \end{bmatrix}, \tag{12}$$

we can rewrite (9) as

$$\begin{bmatrix} \dot{\Omega} + [\omega\times]\Omega & m_p[g\times] \end{bmatrix} \begin{bmatrix} \tilde{J}_p \\ \tilde{\mathbf{r}}_\mathbf{p} \end{bmatrix} = -(\dot{\Omega \bar{J}_b}) - [\omega\times]\Omega\bar{J}_b - m_b[g\times]\mathbf{r_b}\,, \tag{13}$$

where $[\bullet\times]$ denotes the matrix

$$[\bullet\times] = \begin{bmatrix} 0 & -\bullet_z & \bullet_y \\ \bullet_z & 0 & -\bullet_x \\ -\bullet_y & \bullet_x & 0 \end{bmatrix}. \tag{14}$$

Note that the values of $\bar{J}_b(t)$ and $\mathbf{r_b}(t)$ can be calculated based on Eqs. (3) and (6). In order to do so, we also need to know the geometry of the system, and mass displacement of each of the n masses $u_i(t)$ at all instants of time to solve (2). Masses m_p and m_b can be measured relatively precisely prior to assembling the system. Angular rate of the platform can be measured directly with rate sensor, and represented as matrix $\Omega(t)$. Similarly, it is possible to use centrally mounted accelerometer to provide values of $[g\times](t)$. Alternatively, the direction of the gravity vector can be determined with IMU sensor via sensor fusion algorithms.

The only two unknowns in (13) are COM components of the platform $\tilde{\mathbf{r}}_\mathbf{p}$ and its inertia $\tilde{J}_p$. As they are both constant in time, and the equation has a form of $\Phi\psi = \gamma$, where ψ is the vector $\begin{bmatrix} \tilde{J}_p & \tilde{\mathbf{r}}_\mathbf{p} \end{bmatrix}^\top$, they can both be estimated. Measured values of the angular rate are noisy, so calculating the derivative $\dot{\Omega}$ is troublesome. Fortunately, after integrating both sides of (13) we get:

$$\phi(t) = \begin{bmatrix} \Omega(t) - \Omega(t_0) + \int_{t_0}^{t} [\omega\times]\Omega dt & m_p \int_{t_0}^{t} [g\times] dt \end{bmatrix}, \tag{15}$$

$$\gamma(t) = -\left(\Omega\bar{J}_b\right)\Big|_{t_0}^{t} - \int_{t_0}^{t} [\omega\times]\Omega\bar{J}_b dt - m_b \int_{t_0}^{t} [g\times]\mathbf{r_b} dt. \tag{16}$$

Although we end up with another derivative, $\dot{\bar{J}}_b$, this time we can calculate it based on the required values of the counterweight position controller, so it is not burdened with measurement noise.

After performing the experiment where platform is set in motion, and the balance masses are moved, we can stack ϕ and γ, so that

$$\Phi = \begin{bmatrix} \phi(t_0) \\ \phi(t_1) \\ \vdots \\ \phi(t_{end}) \end{bmatrix}, \quad \Gamma = \begin{bmatrix} \gamma(t_0) \\ \gamma(t_1) \\ \vdots \\ \gamma(t_{end}) \end{bmatrix}, \quad \psi = \begin{bmatrix} \tilde{J}_p \\ \tilde{\mathbf{r}}_\mathbf{p} \end{bmatrix}. \tag{17}$$

We can arrange the equation $\Phi\psi = \Gamma$ and use it find $\tilde{J}_p$ and $\tilde{\mathbf{r}}_\mathbf{p}$ estimates with the least-square method, where $\psi = (\Phi^\top \Phi)^{-1}\Phi^\top \Gamma$.

4 Numerical Simulations

A Simulink model of the system dynamics has been created. Sensor sensing is also implemented in the model. During the first 10 s of simulations the sensor measurements are recorded and then postprocessed by a Matlab scripts that implements the identification algorithm. As described in Sect. 3, the algorithm is able to estimate the COM position and inertia properties of the system. This work will focus only on the investigation of the COM estimation.

At the beginning, the behavior of the algorithm for different positions of the COM of the system has been analyzed. A 30-sample Monte Carlo simulation was run with COM coordinates defined within a $[-10, -10]$ mm range for x and y components and $[-2, -7]$ mm for the z component. These configurations can be generated by relocating a concentrated mass within the system. The identification error obtained from simulations, expressed in terms of Euclidean distance between the estimated and the true COM, proved to be always below 0.015 mm.

Since a deterministic analysis is not fully descriptive of the reality, then the system behavior under different sources of uncertainties has been examined in the following. Considered uncertainties are:

1. The inherent noise of the sensors.
2. The sensors mounting tolerance in terms of misalignment of the sensors with respect to the body frame and the unbalanced offset of the accelerometer to the COR of the system.
3. The geometric tolerance of mechanical parts producing a drift of the counterweight paths from the nominal ones.

4.1 Sensor Noise

Sensor noise is modeled as a zero-mean Gaussian random variable added to the signal. The assumed standard deviations are in agreement with the specifications given by the sensor manufacturer and presented in Table 2.

Table 2. Gaussian noise properties of the sensors (based on SparkFun MPU-9250 IMU)

Sensor	Standard deviation	Unit
Gyro	0.1	deg/s
Accelerometer	0.08	g

Two 100-sample Monte Carlo analyses have been performed to examine how the sensor noise propagating through the model affects the estimated COM position. The histograms of the Monte Carlo outcomes in Fig. 3 depict two different instances. The plot on the left shows the error obtained by the implementation of the algorithm illustrated in Sect. 3, while that on the right the error obtained when holding the counterweights steady during the sensing phase. The results show the need of actuating the masses during sensing for achieving accurate results. Monte Carlo analysis evidences that also in the presence of sensor noise, the estimate error never exceeds 0.04 mm.

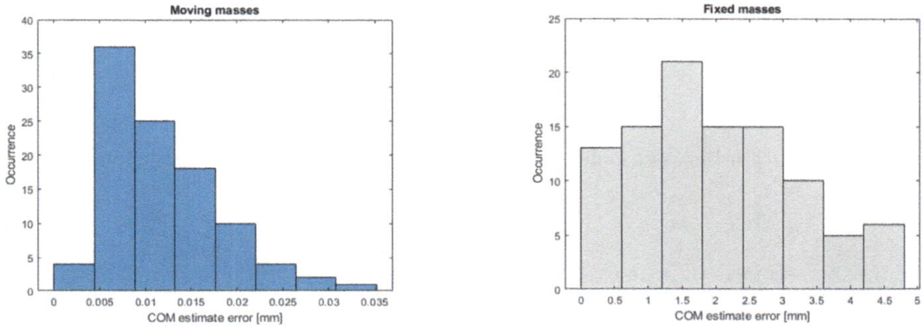

Fig. 3. Effect of sensor noise on COM estimate error when implementing mass motion (left) and holding the masses fixed (right).

4.2 Sensor Mounting Tolerance

Ideally, in the assembled system, the orientation of the sensor frame with respect to the body frame and the position of the accelerometer with respect to COM should be known with exactness. In practice this knowledge is often subjected to uncertainty. Thus, a sensitivity analysis has been conducted to study the influence of geometric mounting tolerance on the estimation error. The undesired sensor and body frame misalignment has been defined by 3 Euler angles (precession, nutation, and spin) with span up to 1°. The influence of the unaccounted accelerometer drift from the COR has been described by assuming a variability range of [0, 50] mm for all three components. The analysis highlights high linear correlation between COR-accelerometer distance and COM estimation error, as shown in Fig. 4. The increase of the estimate error is about 0.04 mm per every unaccounted centimeter of drift. The unexplained variability in the plot is about 0.06 mm and it includes also the effect of frame misalignment. Thus it is expected that after removing from accelerometer measurements the effect of apparent forces the estimation error will not significantly exceeds that caused by the unexplained variability.

4.3 Guide Rail Mounting Tolerance

Due to manufacturing tolerance of 3D printed parts, the position and direction of the counterweight paths can deviate from the nominal ones. Referring to

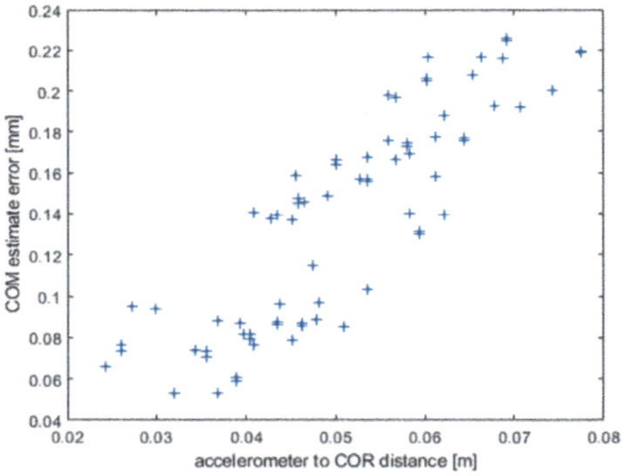

Fig. 4. Correlation between unaccounted accelerometer to COR distance and estimate error.

Fig. 5, this source of uncertainty has been modeled by assuming length and angle tolerances of the vectors $\mathbf{p_1}$ and $\mathbf{d_1}$ previously defined. In particular, length tolerance ΔL is assumed for $\mathbf{p_1}$ and angle tolerances $\Delta\alpha$ and $\Delta\beta$ for $\mathbf{d_1}$. $\Delta\alpha$ is the angle tolerance of the arm tilt α therefore laying on the plane π defined by $\mathbf{p_1}$ and $\mathbf{d_1}$. $\Delta\beta$ is the angle tolerance on the plane perpendicular to π. The ranges of variability assumed for the three parameters are listed in the table of Fig. 5 and they are the same for the three arms.

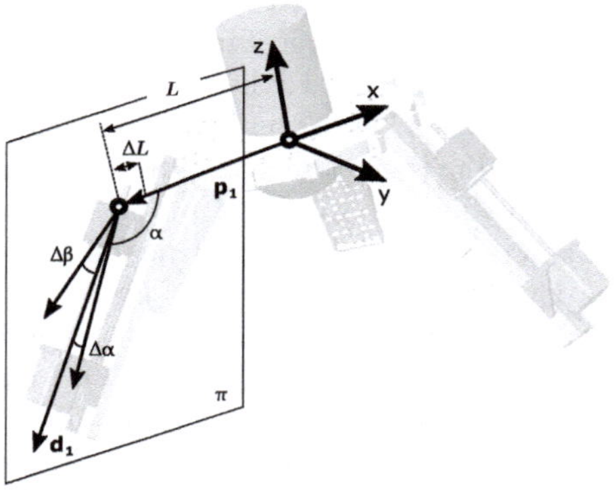

Nr	Parameter	Range	Unit
1-3	ΔL	$[0, 1]$	mm
4-6	$\Delta\beta$	$[0, 1]$	deg
7-9	$\Delta\alpha$	$[0, 1]$	deg

Fig. 5. Parameters defining the geometric tolerance of the counterweight path.

A parameter screening based on the Morris' method [12] has been performed. The Morris' method is a sample-based procedure that calculates, for each analyzed parameter, two indexes, Im and Is. These indexes give qualitative insight into the influence of the corresponding parameter on the response. The index Im says about the global influence, while a high Is value points out non-linearity or interaction of the associated parameter with the other parameters. Results of the analysis are presented in Fig. 6. The plots show that all parameters are highly non-linear and related one-another. However, the two angle tolerances have a significantly larger influence of the estimate error than ΔL.

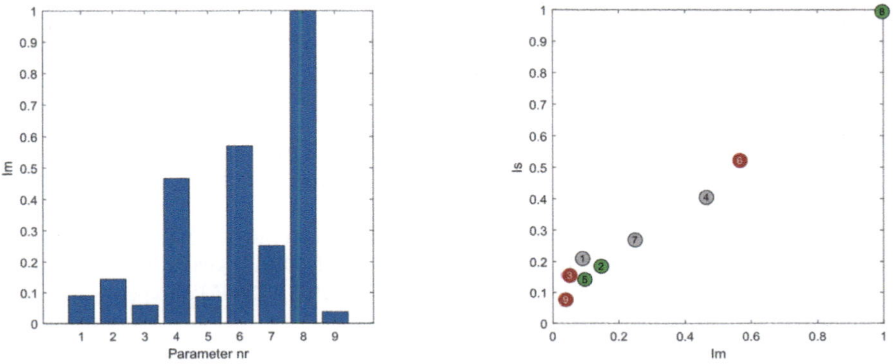

Fig. 6. Parameter screening of parameters analyzing the effect of geometric tolerances of the assembling on the total COM estimate error.

In general, the impact of the aforementioned parameters on the total estimate error is expected to be below 0.1 mm. This assessment is justified by the fact that the maximum estimate error calculated out of 250 Morris samples is less than 0.085 mm.

5 Conclusion

The paper presents a novel automatic balancing algorithm for table top nanosatellite attitude determination and control test stand. The algorithm has been implemented within a numerical simulation framework and test extensively. The robustness of its center of mass estimate against various sources of uncertainty existing in the physical model has been examined. Analyses have evidenced that removing the effect of apparent forces from accelerometer measurements the algorithm should yield estimates within 0.1 mm error, which is the target precision of the system. A prototype of the balancing system is being assembled and tested soon.

References

1. Zajac, F., Small, D.: A linearized analysis and design of an automatic balancing system for the three-axis air bearing table. Technical report, NASA Goddard Space Flight Center (1963)

2. Kim, J.J., Agrawal, B.N.: Automatic mass balancing of air-bearing-based three-axis rotational spacecraft simulator. J. Guid. Control Dyn. **32**(3), 1005–1017 (2009). https://doi.org/10.2514/1.34437
3. Chesi, S., Gong, Q., Pellegrini, V., Cristi, R., Romano, M.: Automatic mass balancing of a spacecraft three-axis simulator: analysis and experimentation. J. Guid. Control Dyn. **37**(1), 197–206 (2014). https://doi.org/10.2514/1.60380
4. Xuan, H.T., Chemori, A., Anh, T.P., Xuam, H.L., Hoai, T.P., Viet, P.V.: From PID to L1 adaptive control for automatic balancing of a spacecraft three-axis simulator. Int. J. Emerg. Technol. Adv. Eng. **6**(1), 77–86 (2016)
5. Young, J.: Balancing of a small satellite attitude control simlator on an air bearing (1998)
6. Keim, J.A., Aqikmese, A.B., Shields, J.F.: Spacecraft inertia estimation via constrained least squares. In: IEEE Aerospace Conference. IEEE (2006). https://doi.org/10.1109/aero.2006.1655995
7. Prado, J., Bisiacchi, G., Reyes, L., Vicente, E., Contreras, F., Mesinas, M., Juarez, A.: Three-axis air-bearing based platform for small satellite attitude determination and control simulation. J. Appl. Res. Technol. **3**(03) (2005). https://doi.org/10.22201/icat.16656423.2005.3.03.563
8. Li, Y., Gao, Y.: Equations of motion for the automatic balancing system of 3-DOF spacecraft attitude control simulator. In: 3rd International Symposium on Systems and Control in Aeronautics and Astronautics. IEEE, June 2010. https://doi.org/10.1109/isscaa.2010.5633650
9. Wolosik, A.T.: Advancements in the design and development of CubeSat attitude determination and control testing at the Virginia tech space systems simulation laboratory. Master's thesis, Virginia Polytechnic Institute (2018)
10. Kato, T., Heidecker, A., Dumke, M., Theil, S.: Three-axis disturbance-free attitude control experiment platform: face. Trans. Jpn. Soc. Aeronaut. Space Sci. **12**, 1–6 (2014). https://doi.org/10.2322/tastj.12.td_1
11. Schwartz, J.L., Hall, C.D.: System identification of a spherical air-bearing spacecraft simulator. In: Proceedings of the AAS/AIAA Space Flight Mechanics Conference, no. AAS 2004-122 (2004)
12. Saltelli, A., Ratto, M., Andres, T., Capolongo, F., Cariboni, J., Gatelli, D., Saisana, M., Tarantola, S.: Global Sensitivity Analysis: The Premier. Wiley, Hoboken (2009)

FxLMS Control of an Off-Road Vehicle Model with Magnetorheological Dampers

Piotr Krauze(✉) and Jerzy Kasprzyk

Silesian University of Technology, 44-100 Gliwice, Poland
{piotr.krauze,jerzy.kasprzyk}@polsl.pl

Abstract. The paper presents a study on adaptive vibration control applied in a semi-active vehicle suspension. The simulation-based analysis is dedicated to laboratory tests conducted for an experimental all-terrain vehicle subjected to mechanical exciters emulating road-induced vibration. The implemented simulation environment consists of a full-car model which exhibits seven degrees of freedom (7 DoFs), equipped with magnetorheological dampers. The MR damper model based on hyperbolic tangent function was included in the vehicle model. The front wheels of the model were subjected to sinusoidal excitation with constant frequency within the range 0.5–25 Hz. The FxLMS (Filtered-x Least Mean Squares) algorithm is adopted and used for controlling the MR dampers where the vertical velocity of excitation is assumed as a reference signal of the control algorithm. The goal of the algorithm is to attenuate vertical velocity of the front middle vehicle body part. The proposed FxLMS was compared to passive suspension controlled with constant control current and Skyhook algorithm. Comparison of different suspension configurations based on transmissibility characteristics and quality indices confirm usefulness of FxLMS with respect to vibration control, suspension deflection and adaptability features.

Keywords: Full-car model · Magnetorheological damper · Semi-active suspension · Vibration control · FxLMS algorithm · Skyhook algorithm

1 Introduction

Ride comfort and driving safety are key factors in vehicles, commonly hard to be reconciled, which need to be taken into account in suspension design process. The active suspension can be applied to vehicle body or vehicle cabin, presented in many studies, e.g. in [1]. It was introduced in road vehicles in order to adjust the suspension parameters to instantaneous road conditions. The active suspension is a type of a hybrid suspension which consists of standard dampers and springs accompanied with force actuators. However, they require additional significant source of power what is especially confusing in mobiles.

Semi-active suspension is a compromise between passive suspension and power-consuming active suspension. In the case of semi-active dampers energy

A. Bartoszewicz et al. (Eds.): PCC 2020, AISC 1196, pp. 747–758, 2020.
https://doi.org/10.1007/978-3-030-50936-1_63

is not added to the system but only dissipation of vibration energy is controlled. Standard control scheme dedicated to semi-active dampers was introduced in [2] as Skyhook algorithm, which can be further generalized to LQ (linear-quadratic) [3] or PI (proportinal-integral) control [4].

Two types of semi-active dampers are generally applied in vehicles, i.e. magnetorheological [5] or servo-valve dampers [2], where the former are favoured for short time response. MR dampers exhibit nonlinear force-velocity characteristics including hysteretic loops [6]. Different MR damper models were proposed in the literature which deal with these nonlinearities and are dedicated to implementation in real-time controllers or to simulation studies, e.g. model based on hyperbolic function [7] or Spencer-Dyke model [5].

The adaptive approach is favoured especially for the possibility to adapt to varying road conditions and vehicle parameters. The presented study is focused on developing direct adaptive control method applied for MR dampers. The FxLMS (Filtered-x Least Mean Squares) algorithm is under consideration, which is an example of adaptive feedforward control algorithm. It was previously applied for control of a half-car (4 DoFs) semi-active suspension accompanied by the Skyhook algorithm [8].

This study is a continuation of the research which involves extension of the study on FxLMS applied to MR dampers in order to implement it in the suspension controller of the experimental off-road vehicle presented in [4]. The control algorithm is implemented in the simulation environment which is fully compatible with the real suspension controller. Majority of features of the laboratory setup are preserved in the simulation environment including dominant dynamics of the vehicle, parameters of the excitation signal as well as available measurement signals used in the control. The paper is organized as follows. Elements of the implemented simulation environment are described in Sect. 2. Tested vibration algorithms are defined in Sect. 3 and results of simulation experiments are discussed in Sect. 4. The study is concluded in Sect. 5.

2 Simulation Environment for Semi-active Control

Vibration control algorithms synthesized for the off-road vehicle are validated in several stages including computer simulation in Matlab and using C programming, real-time simulation, where vehicle model and controller are implemented in separate computational units, laboratory experiments and final tests in terrain. At this stage, computer simulation is performed using C programming in the environment corresponding to the real vehicle.

2.1 Vehicle Suspension Model

Analysis of vehicles with respect to suspension control is commonly based on multi-body models, where only selected vibration modes are mapped. Generally, quarter-car [2], half-car [8] or full-car [1] models are presented in the literature. In the considered experimental vehicle a 7 DoFs model represents vertical motion of each wheel as well as vertical, pitch and roll motion of the body. The off-road vehicle and graphical representation of the model are presented in Fig. 1.

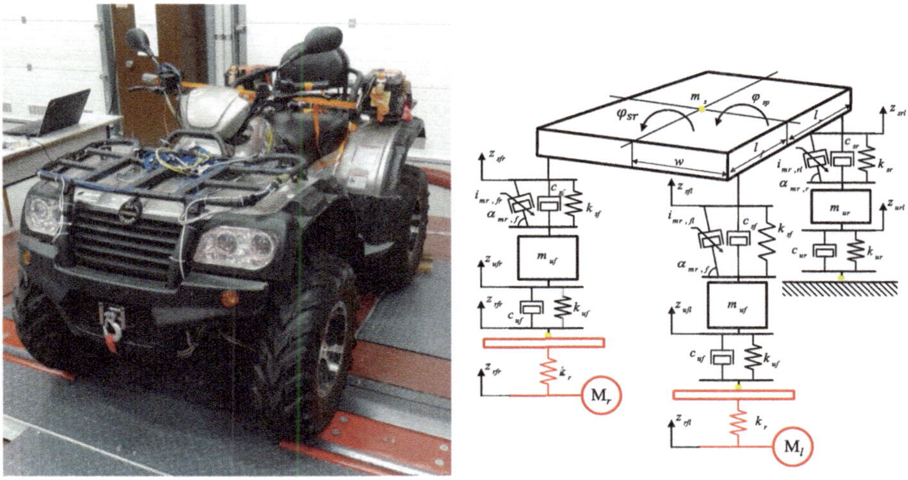

(a) (b)

Fig. 1. The experimental off-road vehicle with MR dampers: a) laboratory setup b) graphical representation of the full-car model.

Assuming small pitch and roll angles, dynamics of the body can be described by the following differential equations :

$$m_s \ddot{z}_s = F_{sfr} + F_{sfl} + F_{srr} + F_{srl}, \tag{1}$$

$$I_{sp}\ddot{\varphi}_{sp} = -l_f(F_{sfr} + F_{sfl}) + l_r(F_{srr} + F_{srl}), \tag{2}$$

$$I_{sr}\ddot{\varphi}_{sp} = -w(F_{sfr} + F_{srr}) + w(F_{sfl} + F_{srl}). \tag{3}$$

Four remaining equations describing vertical dynamics of wheels can be described as follows:

$$m_{u,j}\ddot{z}_{u,j} = F_{u,j} - F_{s,j}, \tag{4}$$

where j denotes a quarter of the vehicle body selected from a set $\{fr, fl, rr, rl\}$. Vertical displacement, velocity and acceleration of the selected part are denoted as z, $\dot{z}$ and $\ddot{z}$, respectively. Similarly, angular quantities are denoted as φ, $\dot{\varphi}$ and $\ddot{\varphi}$. Indices r, u and s correspond to the road-induced excitation, vehicle wheels or body, respectively. Thus, symbols z_s, φ_{sp} and φ_{sr} correspond to vertical displacement of the vehicle body center of gravity as well as its pitch and roll angles, respectively. Dynamics of the body is influenced by its mass m_s, pitch I_{sp} and roll I_{sr} moments of inertia in the center of gravity. Masses of front and rear wheels are denoted as m_{uf} and m_{ur}, respectively. Longitudinal distance of the body center of gravity from its front and rear is defined by symbols l_f, l_r, and its transversal distance from body right and left edges is denoted as w.

Equations (1-4) include forces generated by the suspension $F_{s,j}$ and tires $F_{u,j}$ in each quarter of the vehicle, which are described as follows:

$$F_{s,j} = -k_{sf}(z_{s,j} - z_{u,j}) - c_{sf}(\dot{z}_{s,j} - \dot{z}_{u,j}) + \sin(\alpha_{mr,j})F_{mr,j}, \tag{5}$$

$$F_{u,j} = -k_{uf}(z_{u,j} - z_{r,j}) - c_{uf}(\dot{z}_{u,j} - \dot{z}_{r,j}). \tag{6}$$

Stiffness parameters of the front and rear suspension are denoted as k_{sf} and k_{sr}, respectively. Similarly, corresponding damping parameters are denoted as c_{sf} or c_{sr}. Symbol $F_{mr,j}$ denotes force generated by the MR damper in a selected j quarter of the suspension which is inclined in the actual vehicle at an angle $\alpha_{mr,j}$. Thus, axial motion of the MR damper model is defined as follows:

$$z_{mr,j} = \frac{1}{\sin(\alpha_{mr,j})}(z_{s,j} - z_{u,j}). \tag{7}$$

In order to correctly simulate this full-car model the motion of extreme points of the body need to be evaluated with respect to the assumed degrees of freedom as follows:

$$\begin{aligned} z_{sfr} &= z_s - l_f\varphi_{sp} - w\varphi_{sr}, \quad z_{sfl} = z_s - l_f\varphi_{sp} + w\varphi_{sr} \\ z_{srr} &= z_s + l_r\varphi_{sp} - w\varphi_{sr}, \quad z_{srl} = z_s + l_r\varphi_{sp} + w\varphi_{sr}. \end{aligned} \tag{8}$$

Parameters of the simulated full-car model are listed in Table 1.

Table 1. Parameters of the full-car model and the MR damper Tanh model.

7 DoFs full-car model			
$l_f = 0.681$ m	$l_r = 0.609$ m	$w = 0.475$ m	
$m_s = 343$ kg	$I_{sp} = 47.57$ kgm^2	$I_{sr} = 25.80$ kgm^2	
$k_{sf} = 22.208$ kNm^{-1}	$k_{sr} = 34.857$ kNm^{-1}	$c_{sf} = 300$ Nsm^{-1}	$c_{sr} = 300$ Nsm^{-1}
$m_{uf} = 10$ kg		$m_{ur} = 15$ kg	
$k_{uf} = 70$ kNm^{-1}	$k_{ur} = 70$ kNm^{-1}	$c_{uf} = 130$ Nsm^{-1}	$c_{ur} = 159$ Nsm^{-1}
$\alpha_{mr,f} = 72°$		$\alpha_{mr,r} = 63°$	
Tanh based model of the MR damper			
$i_{mr} \in (0.0; 1.33)$ A	$\alpha_0 = 62.42$ N	$\alpha_1 = 1340$ N$\sqrt{\text{A}^{-1}}$	
$\beta = 39.96$ sm^{-1}	$c_0 = 802.8$ Nsm^{-1}	$c_1 = 488.5$ Nsm$^{-1}\sqrt{\text{A}^{-1}}$	

2.2 MR Damper Model

The MR damper exhibits strongly nonlinear force-velocity characteristics including force saturation and hysteretic loops. These features have decisive influence on suspension control quality especially in the case of feedforward control of MR damper force. The Bouc-Wen or Spencer model [5] is known to accurately describe the MR damper. A Bouc-Wen hysteretic component which is defined by a nonlinear differential equation is a key element of this model. Another group of models take advantage of cyclometric or hyperbolic functions which closely describe the shape of the force-velocity characteristics.

Selection of appropriate MR damper model for application in semi-active vibrating system needs to take into account requirements of the system and complexity of the damper model itself. Taking into account a compromise between limited model complexity and its accuracy, a Tanh model was selected for implementation and simulation tests in the modified form:

$$F_{mr} = -(\alpha_0 + \alpha_1\sqrt{i_{mr}})\tanh(\beta\dot{z}_{mr}) - (c_0 + c_1\sqrt{i_{mr}})\dot{z}_{mr}, \tag{9}$$

which was also studied in [10]. This model includes a viscous damping component described by parameters c_0 and c_1. The hyperbolic tangent function describes shape of the force-velocity characteristics using parameters α_0, α_1 and β. Both model components are dependent on a square root of the control current i_{mr}. Possible values of i_{mr} are limited for the actual MR damper to the range 0–1.33 A. Parameters of the implemented Tanh model are listed in Table 1.

2.3 Excitation of Vehicle Vibration

During laboratory tests the experimental vehicle is subjected to road-induced vibration generated by two mechanical exciters located under wheels (front or rear) and operating in phase. In further studies it is assumed that front wheels are subjected to excitation. These exciters are a part of a modified vehicle diagnostic station. Each of them is powered by an asynchronous motor controlled by a frequency inverter.

Each of exciter plates is being moved by the motor by means of an eccentric shaft. Such construction results in sinusoidal motion of these plates. Frequency value is set on inverters within range of 2–15 Hz. In consequence, the dominant frequency of the vibration excitation allows for analysis of vehicle response in the selected frequency range. Thus, excitation of the front wheels generated by the setup can be described by the following formula:

$$z_{rfr}(t) = z_{rfl}(t) = z_r(t) = \sin(2\pi f t) \cdot 0.05\,\mathrm{m}\frac{2\,\mathrm{Hz}}{f} \tag{10}$$

and excitation of the rear wheels $z_{rrr}(t) = z_{rrl}(t) = 0$. Symbol f denotes frequency of the sinusoidal excitation. Amplitude of excitation of the front wheels is intentionally made inversely proportional to the excitation frequency, where a reference displacement amplitude equal to 0.05 m is obtained for 2 Hz. This assumption results in a constant amplitude of the excitation velocity equal to 0.628 ms^{-1} as well as the amplitude of displacement which decreases with respect to frequency. Consequently, similar range of operation of MR damper models for all analysed frequencies is preserved, since MR damper behaviour depends mainly on its piston relative velocity.

3 Suspension Control Algorithm

The key analysis is conducted on the FxLMS algorithm applied for the semi-active vehicle suspension. Other two configurations, i.e. passive suspension controlled by a constant current within the range 0–0.1 A and the Skyhook algorithm are treated as the references. For the purpose of further study and description of vibration control it is assumed that previously defined time-continuous velocity signals which are sampled in the controller with a finite rate, are further denoted as v.

In order to make comparative analysis FxLMS and Skyhook control algorithms are synthesized based on the same error signal which is to be attenuated. Thus, the error signal is defined as a vertical velocity of the front middle vehicle body as follows:

$$e = v_{sf} = 0.5(v_{sfr} + v_{sfl}). \tag{11}$$

Semi-active control schemes were implemented in a structure where a key part of vibration control generating desired force F_{alg} is distinguished from an inverse MR damper model.

3.1 FxLMS Control

During laboratory tests the measurement and control system of the vehicle is connected to the measurement system of the diagnostic station in order to acquire data of the mechanical exciters for further data preprocessing. For the purpose of FxLMS algorithm the same measurement signal can be used as a reference signal describing disturbance of the plant. Another approach to track the disturbance during experiments is to apply laser sensors, which are available within the test setup, in order to measure vertical displacement of the exciter plate and consequently estimate excitation velocity. Thus, the reference signal, commonly denoted as x in FxLMS, is assumed to be available as $x = v_r$, where v_r is a vertical velocity of excitation.

According to the FxLMS, presented in block diagram in Fig. 2, the control signal F_{alg} is generated by filtering a reference signal v_r using the adaptive filter $H(z^{-1})$ as follows:

$$F_{\text{alg},fr}(n) = F_{\text{alg},fl}(n) = F_{\text{alg}}(n) = H(z^{-1}) \cdot v_r(n). \tag{12}$$

Here, operator z^{-1} denotes a one-sample delay. The implemented full-car model is longitudinally symmetrical. Thus, models of secondary signal paths corresponding to the right and left vehicle side are equivalent. Since the error and reference signals are also the same with respect to both vehicle sides the evaluated control force F_{alg} can be used for both front MR dampers. Parameters of the adaptive FIR (Finite Impulse Response) filter $H(z^{-1})$ of degree $M = 256$, where M was selected experimentally, are updated as follows:

$$\boldsymbol{h}(n+1) = \gamma \boldsymbol{h}(n) - \mu \cdot v_{sf}(n) \cdot \frac{\boldsymbol{r}(n)}{\boldsymbol{r}^T(n) \cdot \boldsymbol{r}(n) + \zeta}, \tag{13}$$

where $\mu = 0.005$ denotes an adaptation step of the FxLMS algorithm and v_{sf} is a vertical velocity of the front middle vehicle body. Symbol $\boldsymbol{r}$ denotes a vector of length M of consecutive samples of signal r. The r signal is generated by filtering the reference signal v_r (vertical velocity of excitation) using the model of the secondary signal path $H_{F_{mr,f} v_{sf}}(z^{-1})$. In order to avoid division by zero in the case of zeroing $\boldsymbol{r}$, a parameter $\zeta = 10^{-15}$ is introduced. Symbol $\gamma = 0.999$ denotes a parameter of leakage of FxLMS algorithm which is used in order to maintain stability of the algorithm. The instability can occur in the case of saturation of control signal which is a common feature in semi-active systems. These parameters were determined experimentally.

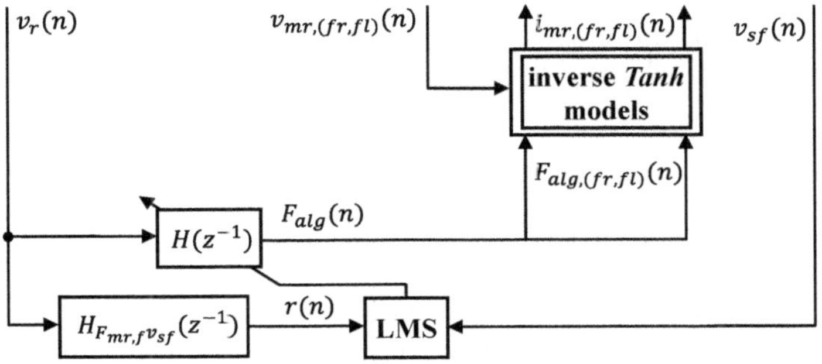

Fig. 2. Block diagram of the FxLMS algorithm applied for semi-active system.

3.2 Skyhook Control

Skyhook algorithm represent a robust method of vibration control in mechanical systems. According to [2] force generated by Skyhook algorithm is equivalent to the force generated by a fictitious viscous damper attached to a reference point in the space with one end and to the sprung mass with the another end. As a result the sprung mass is subjected to force which is directed against its velocity. In the presented study this fictitious damper is attached to the front middle vehicle body and force desired by the algorithm can be evaluated according to the following formula:

$$F_{\text{alg},fr}(n) = F_{\text{alg},fl}(n) = F_{\text{alg}}(n) = -c_{sh} \cdot v_{sf}(n), \tag{14}$$

where c_{sh} denotes the gain of Skyhook control. Similarly to FxLMS, the evaluated control force F_{alg} is used for both front MR dampers.

3.3 Inverse MR Damper Model

The feedforward force control was selected by using an inverse MR damper model. The inverse model which is responsible for transforming the desired control force into the force generated by the damper was assumed to be fully consistent with the MR damper model. Influence of MR damper modelling errors on vibration control quality is not considered in this study. As a result, the desired force is only limited by a dissipative domain of the MR damper model that is an inherent feature of semi-active systems.

The inverse model was evaluated based on Tanh model (9) as follows:

$$i_{mr,j} = \left(\frac{-F_{\text{alg}} - \alpha_0 \tanh(\beta v_{mr,j}) - c_0 v_{mr,j}}{\alpha_1 \tanh(\beta v_{mr,j}) + c_1 v_{mr,j}} \right)^2, \tag{15}$$

where i_{mr} denotes the control current, which should be given by the control algorithm in order to make the MR damper to generate the desired force F_{alg}. It should be highlighted that despite the fact that desired control force was evaluated as equal for both front dampers, the resultant control currents obtained for selected part of the suspension j can differ which is caused by any slight differences between relative damper piston velocities denoted as $v_{mr,j}$.

4 Results

The simulator consists of the vehicle model and the controller. The vehicle dynamics including full-car model and MR damper models were simulated using numerical Runge-Kutta solver of differential equations with variable integration step. The controller operates in discrete-time domain with the sampling interval equal to 2 ms that is consistent to parameters of the actual control of the vehicle.

Furthermore, it was assumed that all velocity signals used in algorithms, especially the reference signal v_r, the error signal v_{sf} and the relative velocities of pistons $v_{mr,j}$ were previously estimated and are available. Velocity estimation is not considered in this study, however, different estimation methods are known, e.g. numerical integration with inertia or numerical differentiation, and can be applied in semi-active systems [4,9].

4.1 Transmissibility Characteristics and Control Quality

Tested algorithms were validated within frequency range from 0.5 to 25 Hz with step equal to $\Delta f = 0.1$ Hz, since generally the analysis of vehicle body vibration and comfort-related indices is performed in this frequency range. Thus, frequency resolution of evaluated transmissibility characteristics is equal to Δf. Velocity transmissibility is defined as follows:

$$T_{v_r, v_{sf}}(f) = \sqrt{\frac{\sum_{n=1}^{N} v_{sf}^2(n)}{\sum_{n=1}^{N} v_r^2(n)}} \Bigg|_f, \tag{16}$$

where the initial samples corresponding to the transient response were excluded from the evaluation and each point of the transmissibility was evaluated for an integer number of cycles (equal to 10) of the sinusoidal excitation.

In order to compare results obtained for a certain control parameter and suspension configuration, a normalized quality index was defined based on integrated square value of velocity transmissiblity $T_{v_r,v_{sf}}$ as follows:

$$J_{v_{sf}} = \sqrt{\frac{\sum_{f=0.5\,\mathrm{Hz}}^{25\,\mathrm{Hz}} \left[T_{v_r,v_{sf}}(f)\right]^2}{\sum_{f=0.5\,\mathrm{Hz}}^{25\,\mathrm{Hz}} \left[T_{0,v_r,v_{sf}}(f)\right]^2}}, \tag{17}$$

where $T_{0,v_r,v_{sf}}$ corresponds to a front vehicle body transmissibility evaluated for the passive suspension at the control current equal to zero.

4.2 Optimization and Validation of Control Algorithms

The optimization process was carried out with respect to the normalized quality index $J_{v_{sf}}$ dedicated to the vertical velocity v_{sf}. The goal of optimization is to find a set of parameters for each configuration of semi-active suspension which represent a minimal value of $J_{v_{sf}}$. The optimization process need to be performed numerically based on simulation since the vehicle model including MR damper models is significantly nonlinear. It can be stated based on preliminary tests that the Skyhook algorithm can represent three cases: undertuned, overtuned or properly tuned with respect to a certain vehicle model. Thus, representative comparison of FxLMS and Skyhook algorithms requires the latter to be initially numerically optimized. A set of transmissiblity characterisics evaluated for Skyhook algorithm for different control parameters c_{sh} are presented in Fig. 3. It can be stated that for low values of c_{sh} the resonance peak located at 2 Hz is slightly mitigated. High values of c_{sh} results in significant mitigation of vibration at 2 Hz and simultaenously in deterioration of vibration damping for higher frequencies. Values of the quality index evaluated for Skyhook algorithm presented in Fig. 4a indicate the optimized value $c_{sh} = 1100$ Nsm^{-1}, since the lower value of quality index $J_{v_{sf}}$ the better vibration damping. For the passive suspension presented in Fig. 4b it can be noticed that each value of control current higher than zero leads to deterioration of velocity vibration control. The same figures confirm that FxLMS offers better control quality than the best case of the Skyhook.

Best cases of all tested suspension configuration are additionally compared in frequency domain using the transmissiblity characteristics in Fig. 5. It can be noticed that despite the lower value of $J_{v_{sf}}$ offered by FxLMS, the results obtained for FxLMS and Skyhook are comparable. However, it should be highlighted that results represented by Skyhook required extensive simulations performed for numerous values of control parameters which is not necessary in the case of FxLMS algorithm.

Validation process of the FxLMS algorithm was extended to the analysis of its performance with respect to suspension deflection. Generally, the lower suspension deflection the lower probability of suspension failure or its wear out.

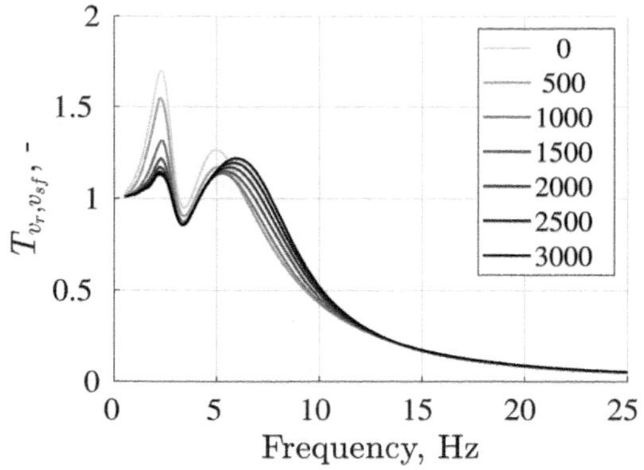

Fig. 3. Skyhook algorithm for different control parameters c_{sh} in Nsm^{-1} based on transmissibilities of vertical velocity of the front middle vehicle body.

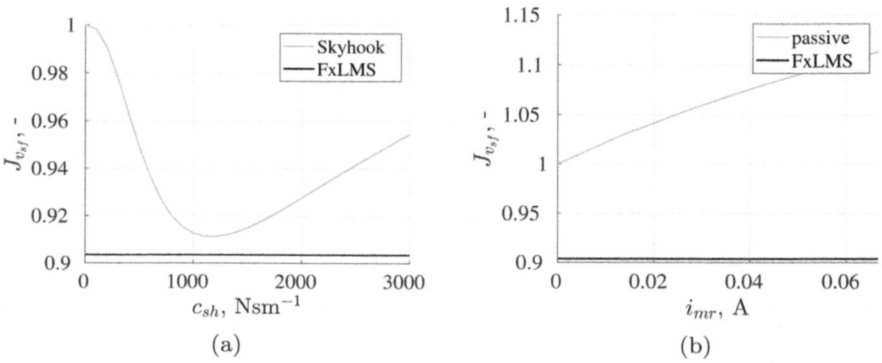

Fig. 4. Comparison of control algorithms based on $J_{v_{sf}}$ quality index: a) FxLMS vs. Skyhook, b) FxLMS vs. passive suspension.

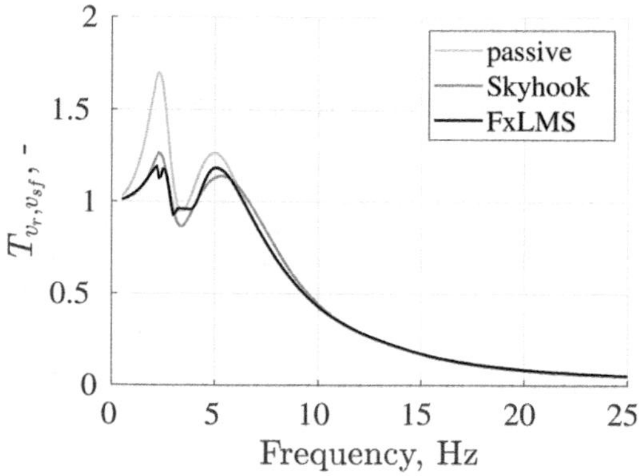

Fig. 5. Optimized cases of FxLMS, Skyhook algorithms and passive suspension based on transmissibilities of vertical velocity of the front middle vehicle body.

Similarly to (17), a quality index dedicated to suspension deflection was evaluated for Skyhook and FxLMS algorithms, and its values are presented in Fig. 6. It can be stated that the proposed FxLMS algorithm improves simultaneously suspension deflection apart from the mitigation of the body vibration.

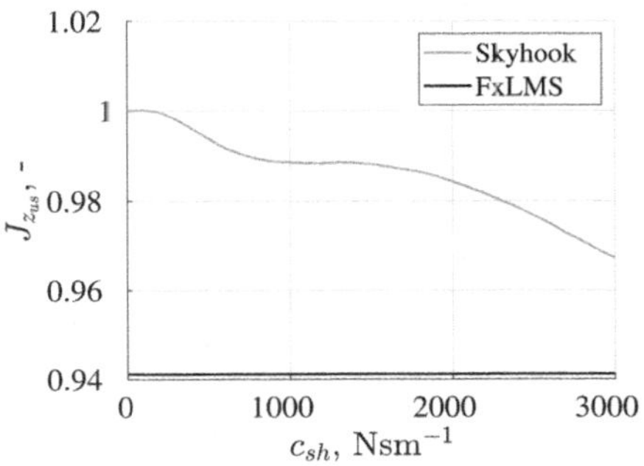

Fig. 6. Values of $J_{z_{us}}$ quality index for FxLMS and Skyhook algorithms.

5 Conclusions

The paper is dedicated to an adaptive vibration control applied for a vehicle suspension model with MR dampers. The analysis corresponds to a laboratory tests conducted for an experimental all-terrain vehicle. During experiments front wheels of the vehicle are subjected to sinusoidal excitation generated by mechanical exciters. The study was conducted for a full-car model with 7 DoFs, which includes MR damper models based on the hyperbolic tangent function. The FxLMS (Filtered-x LMS) algorithm is adopted and used for controlling the MR dampers where the vertical velocity of excitation is used as a reference signal.

The goal of all analyzed configurations of the semi-active suspension is to minimize vertical velocity of the front middle vehicle body part. Results obtained for the FxLMS algorithm were compared to Skyhook algorithm and passive suspension controlled with different values of constant control current. Parameter of the Skyhook algorithm was optimized based on the quality index evaluated for the minimized velocity signal. All algorithms were analysed based on transmissibility characteristics evaluated within frequency range 0.5–25 Hz and based on defined quality index. The usefulness of FxLMS was confirmed with respect to vibration control, suspension deflection and adaptation features. Additionally, the FxLMS do not require extensive tuning in comparison to Skyhook algorithm which need to be tuned to a specific vehicle. This requirement is especially troublesome and time-consuming in real experiments.

Acknowledgments. The partial financial support of this research by the Polish Ministry of Science and Higher Education is gratefully acknowledged.

References

1. Lagraa, N., Boukhetala, D., Bloch, G., Boudjema, F.: Nonlinear control design of active suspension based on full car model. J. Control Sci. **17**(4), 439–457 (2007)
2. Karnopp, D., Crosby, M.J., Harwood, R.A.: Vibration control using semiactive force generators. J. Eng. Ind. **96**, 619–626 (1974)
3. Sibielak, M., Rączka, W., Konieczny, J.: Modified clipped-LQR method for semi-active vibration reduction systems with hysteresis. Solid State Phenom. **177**, 10–22 (2011)
4. Krauze, P., Kasprzyk, J., Kozyra, A., Rzepecki, J.: Experimental analysis of vibration control algorithms applied for an off-road vehicle with magnetorheological dampers. J. Low Freq. Noise Vib. Active Control **37**(3), 619–639 (2018). https://doi.org/10.1177/1461348418782166
5. Spencer, B.F., Dyke, S.J., Sain, M.K., Carlson, J.D.: Phenomenological model of a magnetorheological damper. ASCE J. Eng. Mech. **123**, 230–238 (1997)
6. Sapiński, B.: Magnetorheological Dampers in Vibration Control. AGH University of Science and Technology Press, Cracow (2006)
7. Kasprzyk, J., Wyrwał, J., Krauze, P.: Automotive MR damper modeling for semi-active vibration control. In: IEEE/ASME International Conference on Advanced Intelligent Mechatronics (AIM), pp. 500-505, Besancon, France (2014)
8. Krauze, P., Kasprzyk, J.: Mixed Skyhook and FxLMS control of a half-car model with magnetorhelogical dampers. Adv. Acoust. Vib. **2016**, 1–13 (2016)
9. Krauze, P., Kasprzyk, J., Rzepecki, J.: Experimental attenuation and evaluation of whole body vibration for an of-road vehicle with magnetorheological dampers. J. Low Freq. Noise, Vib. Active Control **38**(2), 852–870 (2019). https://doi.org/10.1177/1461348418756018
10. Krauze, P.: Control of semiactive vehicle suspension system using magnetorheological dampers, Ph.D dissertation. Silesian University of Technology, Gliwice, Poland (2015)

Recent Challenges and Applications of Computer Vision

The Use of a Laser to Measure the Speed of a Production Line

Stanisław K. Musielak[1] and Jerzy Kasprzyk[2(✉)]

[1] CCE Corrugated Consulting Engineering, 92729 Weiherhammer, Germany
[2] Department of Measurements and Control Systems, Silesian University of Technology, Gliwice, Poland
jerzy.kasprzyk@polsl.pl

Abstract. The accuracy and quality of cutting in the cardboard production process have a significant impact on the quality of the final product, and thus the financial effects of the manufacturer. The paper describes the results of research on the impact of using a laser system to measure the speed of a corrugated board production line on cutting accuracy obtained in a rotary cutter. The tests were carried out in various working conditions of the corrugator. In particular, the precision of cutting for constant line speeds, as well as when accelerating or decelerating the work of the corrugator was examined. The cutting accuracy obtained from the laser system as a contactless sensor was compared with the results obtained for a classic transducer based on a measuring roller with an encoder. The influence of sheet formats and cardboard quality on uncertainty of measurements was also investigated. The tests were carried out on corrugators in various factories. Sources of measurement errors that can occur when using the laser system are also discussed.

Keywords: Laser surface velocimeter · Measurement error · Cutting accuracy

1 Introduction

Increasing requirements regarding the accuracy of the cardboard cutting process and high repeatability are the basic criteria for assessing the work of cutters in the corrugator's technological line. The quality of the final product depends, among others, on the achievable accuracy of dimensions and shape of the sheets [1]. Very high demands on the cutting accuracy result from the fact that cardboard sheets are processed into packaging using robots.

The rotary cutoff sheeter is an important element of the corrugated board production line. Its task is to cut cardboard with the required accuracy, which places high demands on the control and drive systems. Proper cutting requires that the cutter shaft cut the cardboard band at a linear speed exactly equal to the speed of the production line. Also, the moment when the knife enters the cardboard is precisely determined by the fixed length of the cardboard being cut, the speed of the line, and the rotation frequency of the knife shaft.

A. Bartoszewicz et al. (Eds.): PCC 2020, AISC 1196, pp. 761–772, 2020.
https://doi.org/10.1007/978-3-030-50936-1_64

It is well known that control quality criteria indices depend significantly on the uncertainty of the process value measurement. In this case, the production process is carried out in such a way that there is no technical possibility to measure online the length of cut sheets. The only thing left is to measure the stack of sheets at the end of the line. Thus, the control system of the knife shaft drive is mainly based on the production line speed information. So, one of the factors that may negatively affect the operation of the cutter is the uncertainty of measuring the speed of the corrugator production line. In a classic solution, this speed is measured using a conventional roller transducer, driven by cardboard moving underneath it. The line speed is determined based on the pulse frequency generated by the encoder. This method of measurement involves a number of different technical problems, which are presented in more detail in Sect. 2.1. Therefore, the authors proposed to measure the line speed using a laser system.

The use of a laser as a contactless sensor is known in the metallurgy [2], textile [3] or paper [4] industries, however, there are few scientific publications on this subject. The first attempt to use the laser in the corrugator line was presented in [5]. Much higher costs of such a solution caused that it was not initially adopted by the manufacturer.

The purpose of this study was to confirm or not, to what extent the replacement of a conventional roller transducer by a laser gauges makes sense in a corrugated board production line. This required the installation of both measuring systems in the corrugator line (see Fig. 1), and checking how the use of each of them affects the cutting accuracy.

Fig. 1. Roller transducer and laser (above) in a corrugator line

As a first step, lasers from different manufacturers had to be tested to select those that meet the requirements. To obtain a fairly certain answer, tests were carried out for several corrugators working in different conditions, i.e. for different line speeds and different types of cardboard. As a result, both methods of measuring line speed were tested on 9 machines of the same manufacturer in different workplaces. The collected data were subjected to statistical analysis in terms of assessment of cutting accuracy and the main causes of errors. In general, the conclusions drawn from these studies are similar. For obvious reasons, only a few selected results can be presented in the paper.

The paper is organized as follows: (1) both measuring systems are described and analysis of measurement error sources is carried out; (2) results of experiments performed for industrial corrugators are presented. Finally, some practical conclusions are drawn.

2 Description of the Measuring Systems

2.1 Roller Transducer

In the conventional system an incremental encoder is used [6]. It generates electrical impulses that result from the rotational movement of the roller. The number of pulses per revolution of the encoder axis is constant and usually equal to 4096 pulses for a typical wheel circumference of 500 mm. The encoder output is a standard TTL signal which is used as a reference signal for the drive control system in the cutter. This sensor is characterized by a low price and easy replacement in the case of failure. However, there are some factors affecting the measurement uncertainty, e.g.:

- impact of the measuring wheel slip phenomenon,
- changes in the measuring wheel diameter - assembly errors, impact of tread wear,
- mechanical damage,
- vibrations of the wheel,
- ambient temperature changes,
- settling of adhesive particles on the measuring wheel,
- material thickness differences,
- various cardboard quality,
- types of cardboard waves.

Speed measurement using a roller transducer is particularly sensitive to changes in the temperature of the cardboard. Changes in the diameter of the measuring roller due to temperature changes can cause cutting errors of several mm for longer cardboard formats, i.e. 3000–6500 mm. This is especially the case during winter and summer periods, when the halls are cold or overheated. To compensate for this, special construction material with a low expansion coefficient is used for the measuring wheel.

2.2 Laser System

As an alternative to the classic system the laser gauge is proposed. The principle of the laser surface velocimeter (LSV) is based on the use of the Doppler effect [7]. The beam from the laser diode is divided into two rays, with the light frequency in one of them shifted by $f_s = 40$ MHz (in the Polytec laser). Both beams cross on the moving surface of the material at an angle φ. As a result of the interference of two rays, light and dark stripes appear on the surface of the material (Fig. 2). They are spaced apart by a value Δs being a function of the angle φ and wavelength λ:

$$\Delta s = \frac{\lambda}{2 \sin \varphi}. \tag{1}$$

The laser rays, passing through the bright fields of interference stripes, bounce off the cardboard and reach the receiver lens. The detector modulates the intensity of light with the Doppler frequency f_D that is proportional to the speed of the material v:

$$f_D = \Delta s \cdot v. \tag{2}$$

In addition, to be able to measure near-zero speed, the so-called base frequency f_B is added to the frequency f_D. Finally, the resulting frequency is processed by the electronic system, giving a pulse signal corresponding to a typical encoder. The number of pulses per unit length is programmable and can be, e.g., 8196 per 1000 mm of material. The output of the electronic laser block consists of 4 channels defined as: A, B and their negations. These signals are coupled to the cutter control system similarly as it is done for the encoder.

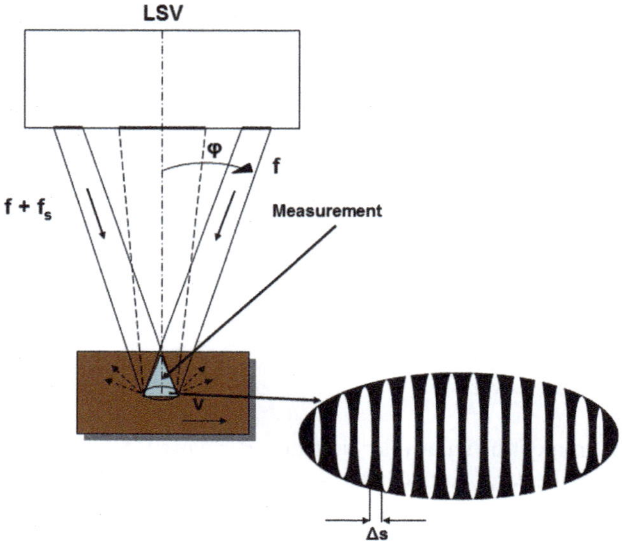

Fig. 2. Measuring principle using Laser Surface Velocimeter (LSV)

In a laser system, the main measurement error may be caused by excessive temperature. Due to the high temperature of the cardboard, depending on the type of cardboard wave and the environment, the temperature of the laser head may rise above the technical specification (45 °C), which allows the laser to work even at temperatures around 200 °C. Also, in a dusty environment, the sensor must be additionally protected, e.g. by enclosing the sensor in a housing and subjecting it to air blows, which improves system reliability.

The laser head is calibrated and with a small change in the distance from the measured surface to the laser head, the uncertainty of the speed measurements does not change. However, in practice, there may be changes in the distance of the laser in relation to the cardboard surface, e.g. when the warp effect occurs, see Fig. 3.

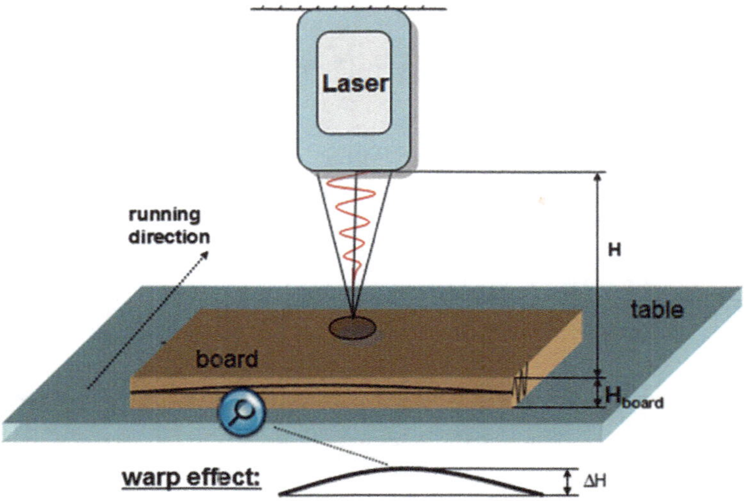

Fig. 3. Cardboard warp

In this case, the laser measuring field may be exceeded, especially if the production process is not carried out correctly. The effect of this phenomenon is the deterioration of the accuracy of cardboard cutting, and even TTL pulses may disappear at the output of the laser electronic block. The solution is to use a stabilizing brush or non-contact roller with the material.

There is also the possibility of additional errors if the laser head is subjected to transverse vibrations in relation to the direction of cardboard movement. Therefore, the sensor should not be mechanically coupled to the machine structure. Besides, due to the fact that the laser has class 3B, an additional tube-shaped cover should be used, which protects the eyes of the operator against the harmful effects of the laser beam.

3 Comparison of Uncertainty of Measurements for Both Methods

An example of a time diagram of the measured line speed using a laser is shown in Fig. 4. The arrow shows the place of the cutting phase in the cutter. This indicates that the laser measurement is very sensitive to any changes in line speed, in this case caused by the interaction of the cutter block with cardboard (cutting strokes). This may suggest that this measurement reflects well the actual line speed.

On the other hand, roller speed measurement does not always match the actual line speed. For example, for a measuring wheel at a speed of 320 m/min, the speed difference may reach ±1.5%, while for a laser it is about ±0.3% (Fig. 5).

Interesting results were obtained using a camera recording up to 1000 images per second. The vibrations of the measuring wheel were observed, which occurred not only at high but also at low speed of the corrugator. The Track & Trace application provided by the camera manufacturer enabled processing of the collected image sequences to detect the position of the measuring roller and track its changes. The algorithm of tracking two

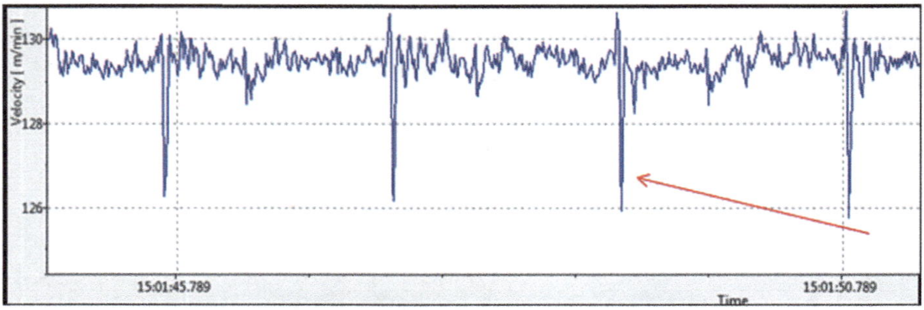

Fig. 4. Time diagram of a line speed measured by the laser

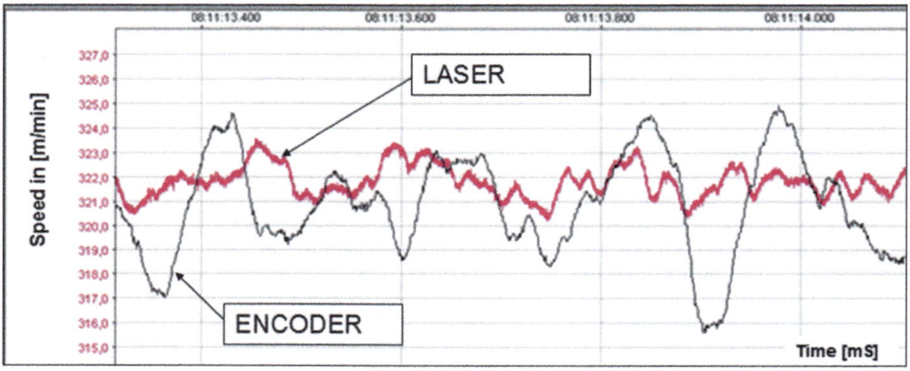

Fig. 5. Comparison of line speed measurements with a laser and encoder

points around the perimeter of the roll was used, determining its horizontal and vertical position. An example photo is shown in Fig. 6, with marked the X and Y axes of the roller center position and measurement points for which deviations from the reference point were calculated. It was stated that changes in the position of the measuring wheel for a horizontal position can reach +0.2 mm.

3.1 Impact of Speed Changes on Cutting Accuracy

One of the main objectives of the research was to check how the change in production line speed affects the accuracy of cutting for both measuring systems. It should be emphasized that in order to obtain production speed the corrugator needs over a dozen seconds. Depending on the type of the corrugator, these values may be different, currently the maximum achieved line speed is 450 m/min.

The behavior of the cutting system in the case of machine acceleration/deceleration was tested. The speed of the machine was changed from 50 to 250 m/min, and then it was reduced back to 50 m/min, Tests were repeated for various types of cardboard and dimensions of sheets. Laser systems from four manufacturers were tested. It turned out that during acceleration and deceleration some of them caused significant cutting errors,

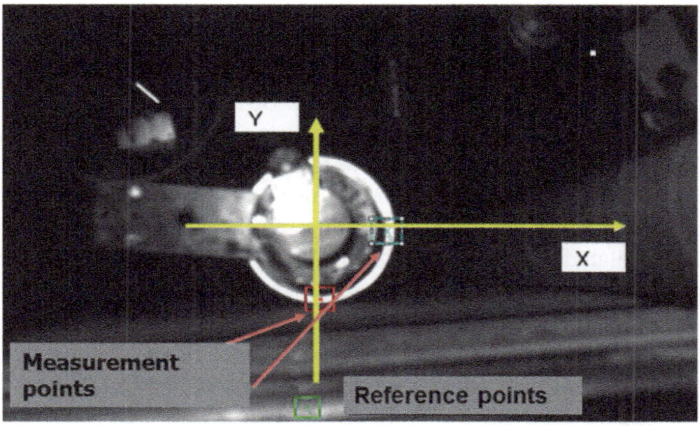

Fig. 6. Displacement of the measuring wheel from the cardboard surface

reaching up to ±10 mm. Hence, finally lasers from two manufacturers were selected as an alternative to the classic system: the Polytec model LSV 6000 [8] and Beta Laser Mike LS 4000 [9]. These gauges ensured that the cutting deviations for large speed changes were similar to the results achieved at a constant speed of the production line. They were within the required tolerance range −1 to +1 mm. Also high cutting repeatability was obtained in various working conditions.

Then, comparative studies of both measurement systems were carried out. Exemplary 50 deviations of the length of cut sheets from the set value are shown in Fig. 7. For the classic measuring roller with encoder, irregularities are visible in a series of cut sheets (upper plot), whereas in the case of the laser gauge sporadic errors of 1 mm occur (lower plot).

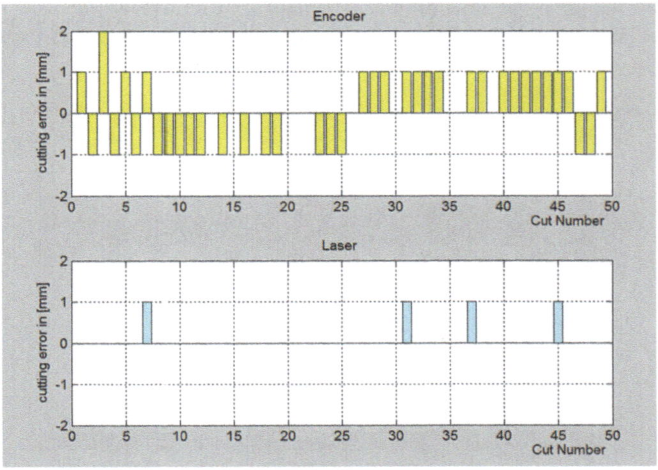

Fig. 7. Cutting errors for encoder and laser speed measurements

3.2 Impact of Various Types of Cardboard and Sheet Formats

Further tests focused on measurements of various lengths of cut cardboard formats, which can take values from 600 up to 6000 mm. Because the length of cartons affects the accuracy of cutting, hence the collected data was divided into two ranges: 600–2000 mm and 2001–6000 mm. The tests were carried out on 9 machines of the same manufacturer, but in different technical condition, as well as different technical solutions, e.g. the drive used. It is also worth noting that the cutting accuracy is influenced by many different factors related to the state of the mechanical components, tuning of the cutter control systems, type of waves in cardboard, its weight, etc. Besides, due to the high costs of such tests, experiments had to be carried out during normal production, so the experimental conditions were not exactly the same. Therefore, it is difficult to compare different machines in terms of production results obtained. However, it was important to qualitatively compare the impact of both tested measuring systems on cutting accuracy, whether using a laser improved the accuracy and to what extent.

The experiments were conducted in such a way that first the signal from the roller transducer was used in the cutter control system, then the experiment with the laser gauge was repeated for similar operating conditions. Several thousand measurements were done on each machine, hence the results are presented in the form of a probability density function (pdf) of cutting errors calculated using the kernel estimator (see e.g. [10]) for both length ranges. The results for two exemplary machines denoted as Corrugator 1 and Corrugator 2 will be discussed.

Figure 8 shows pdf of deviations of cut cardboard determined for sheets of 600–2000 mm, for different line speeds and different cardboard waves. When working with laser measurement, two modes are visible, focused around the central value of 0 mm and the value of +1 mm, that can be considered correct as these values fulfill the requirements. In the case of a measuring wheel with an encoder, pdf is unimodal with higher variance.

Results of cardboard cutting in the range from 2001 mm to 6000 mm are shown in Fig. 9. When working with the roller, a strong flattening of the pdf function is visible. Its maximum reaches a value of approx. 0.13 for cutting deviations of 3 mm. There is also an offset here, which is more common for longer formats when using a roller. In

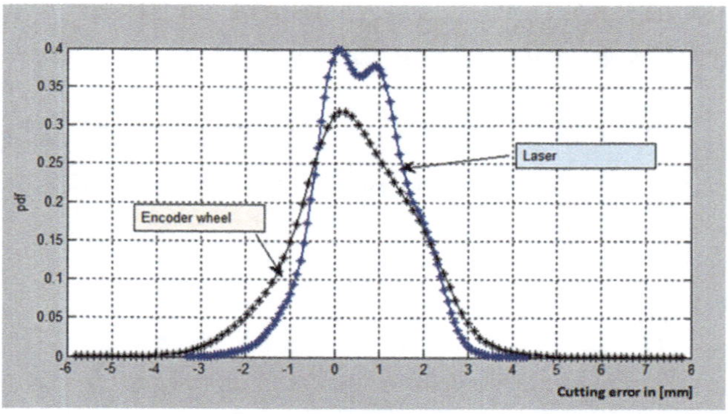

Fig. 8. The probability density function of cutting errors of the sheets within the range of 600–2000 mm (Corrugator 1 with a laser and an encoder)

the case of the laser the pdf function is sharp and its maximum reaches 0.34 for cutting error equal to 1 mm, which is a very good result.

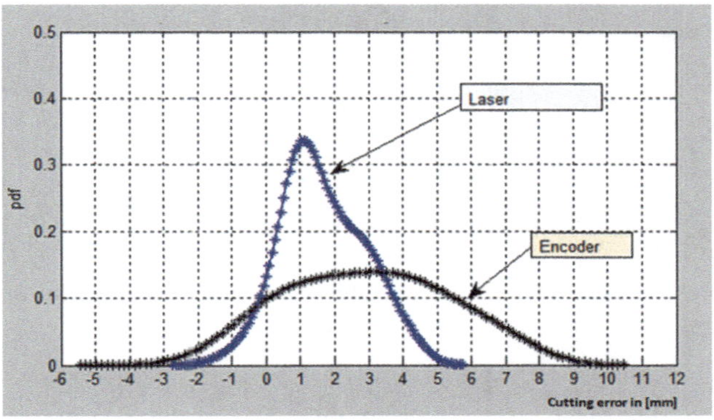

Fig. 9. The probability density function of cutting errors of the sheets within the range of 2001–6000 mm (Corrugator 1 with a laser and an encoder)

Similar results were obtained for Corrugator 2, see Fig. 10 and Fig. 11. Here the longest sheet format is 4000 mm. The advantage of the laser over the measuring wheel is also visible here, especially for a larger range of formats.

Additionally, a statistical analysis of cutting errors was carried out for various types of waves, formats of sheets and line speeds. The total number of tests was over 3000 for the selected measuring system. It has been assumed here that for the entire range of manufactured cardboard the Lower Specification Limit (LSL) is −1 mm for cutting deviations, while the Upper Specification Limit (USL) is +2 mm. Sheets outside of LSL and USL are not acceptable.

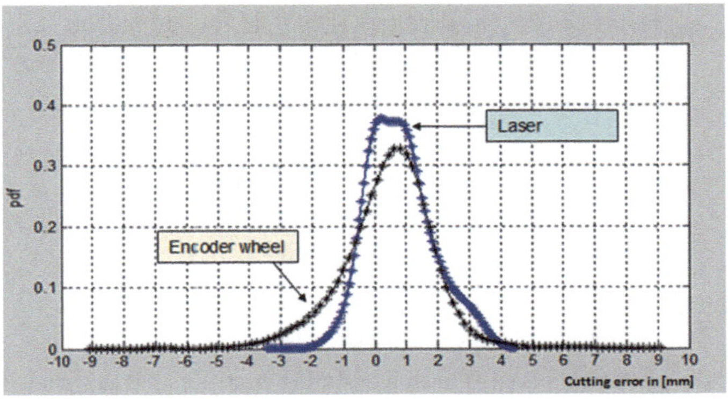

Fig. 10. The probability density function of cutting errors of the sheets within the range of 600–2000 mm (Corrugator 2 with a laser and an encoder)

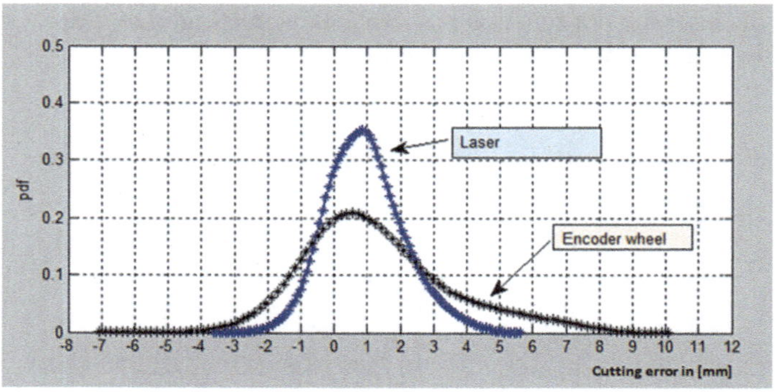

Fig. 11. The probability density function of cutting errors of the sheets within the range of 2001–4000 mm (Corrugator 2 with a laser and an encoder)

It turned out that for the classic system only 71% of the measurement results were within the given range of cutting deviations, whereas for laser gauge this value was 87% (Fig. 12).

However, for the production of shorter sheets the advantage of the laser over the encoder is not so pronounced, although it is still visible. Figure 13 presents results obtained for the corrugator No. 3, which produces a much smaller range of formats. In this case, for the roller transducer 94% of the data met the requirements, whereas for the laser gauge it was as much as 99%.

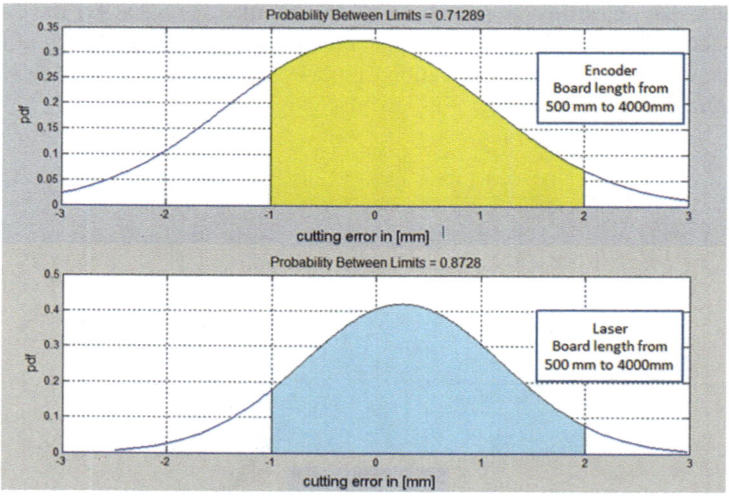

Fig. 12. The probability density function of cutting errors for all data from Corrugator 2

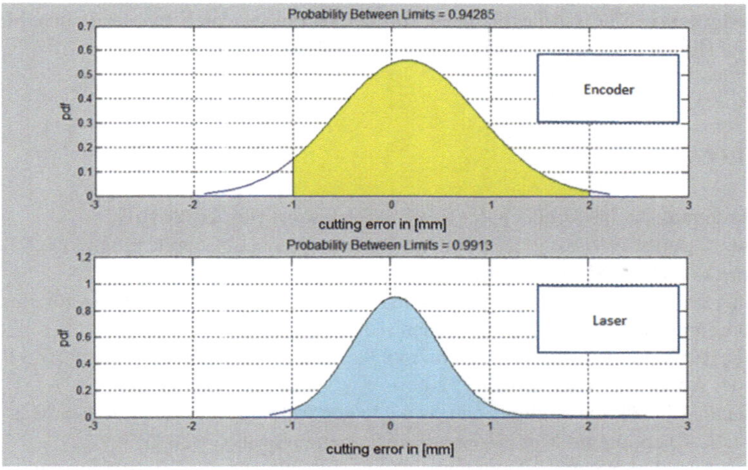

Fig. 13. The probability density function of cutting errors for data from Corrugator 3

4 Conclusions

It is important to emphasize the high costs of such experiments, and the time and effort involved. The presented tests were carried out during normal production, hence the limited scope of research. In addition, many factors influenced the results and it is difficult to precisely determine the impact of each of them on the final effect of cutting cardboard. However, some general conclusions can be drawn from these studies.

Research on the use of a laser gauge to measure the speed of the corrugator production line has demonstrated its undoubted suitability for such application. It turned out that this solution provides less uncertainty of measurements as well as robustness to changing machine operating conditions compared to a classic solution based on a measuring wheel with an encoder. The use of a laser therefore allows more accurate control of the knife drive and better quality of the final product. It seems reasonable to assume that the conclusions drawn from the presented research may apply not only to corrugators, but also to other industries.

Tests show, however, that not every laser gauge is suitable for the mentioned application. Preliminary tests were required to select the appropriate gauges. However, if the production line does not have large speed changes and the line works stable for a long time, then the classic solution is usually sufficient to achieve the required accuracy of the machine. Thus, in this case, the use of a laser, which is associated with high costs, does not necessarily have to be economically justified. Although in this case one can still get an improvement in the quality of production.

We are generally interested in improving the quality of the process. With the decreasing costs of laser gauges, their use can become more and more popular. So it is worth breaking the barriers of technical staff accustomed to certain standard solutions and testing new measurement technologies.

Acknowledgments. The partial financial support of this research by the Polish Ministry of Science and Higher Education is gratefully acknowledged.

References

1. Blechschmidt, J.: Taschenbuch der Papiertechnik. Carl Hanser (2010)
2. Polytec. https://www.polytec.com/eu/velocimetry/areas-of-applications/$steel-aluminum-and-metal/. Accessed 21 Mar 2020
3. Herman, F.J.: Patent. DE 10 2008 004 731 A1 Verfahren und Anordnung zur Bestimmung des Durchmessers eines laufenden Faden (2009)
4. Polytec: Berührungsloser LSV Geschwindigkeits-Sensor eine unterbrechungsfrei Papierproduktion. Applikationsnote LSV 03 (2006)
5. Musielak, S.K.: Geschwindigkeitsmessung in der Wellpappeindustrie. Sensor Mag. **2** (2011)
6. Basler, S.: Encoder und Motor – Feedback Systeme. Springer (2016)
7. Polytec: LSV Laser Surface Velocimeter. Non-contact speed and length measurement. https://www.polytec.com/fileadmin/d/Velocimetrie/OM_PB_LSV_E_42450.pdf. Accessed 21 Dec 2019
8. Polytec: Laser Surface Velocimeter LSV 6000 series
9. Beta LaserMike. http://www.laserspeedgauge.com/phocadownload/olg_brochure_web.pdf. Accessed 21 Mar 2020
10. Silverman, B.W.: Density estimation for Statistics and Data Analysis. Chapman & Hall, New York (1986)

Application of Multi-layered Thresholding Based on Stack of Regions for Unevenly Illuminated Industrial Images

Hubert Michalak(✉) and Krzysztof Okarma

Faculty of Electrical Engineering, West Pomeranian University of Technology in Szczecin, Sikorskiego 37, 70-313 Szczecin, Poland
{michalak.hubert,okarma}@zut.edu.pl

Abstract. Binarization of unevenly illuminated and natural images usually cannot be conducted using typical global thresholding methods. Due to changes of the local contrast and potential presence of some other distortions the application of more computationally demanding adaptive methods is necessary. Nevertheless, to find balance between the global and adaptive methods, some relatively fast region based approaches might be considered providing satisfactory results. Since typical issues for unevenly illuminated industrial images are quite similar to some distortions which may be found in degraded document images, in view of lack of industrial image databases, all numerical experiments have been conducted using recently developed challenging Bickley Diary dataset. Results of experiments, verified for sample unevenly illuminated industrial images containing text information, are promising and confirm usefulness of the proposed multi-layered approach based on the use of stack of regions.

Keywords: Image binarization · Adaptive thresholding · Unevenly illuminated images · Document images

1 Introduction

Industrial applications of machine vision cover wide areas of potential usage of images captured by cameras, particularly in view of the developments of Industry 4.0 solutions. Regardless of the rapid development of advanced colour image analysis methods as well as advances of deep learning based approaches to image recognition and classification, there are still some industrial needs of relatively simple but robust methods, characterised by low computational demands. Such algorithms may be useful especially in embedded systems, robotic applications such as video based control of mobile robots, fast analysis of binary images, etc. Observing the necessity of training of the CNN based methods [25], which requires the availability of large datasets, often leading to data overfitting, there is a danger of their low performance for the images differing from those used previously for training. Hence, a great interest in "explainable" methods based on

A. Bartoszewicz et al. (Eds.): PCC 2020, AISC 1196, pp. 773–784, 2020.
https://doi.org/10.1007/978-3-030-50936-1_65

classical solutions, may be observed mainly in industrial applications, especially when large sets of training data acquired in real situations, e.g. in robotics, are unavailable.

One of the interesting areas of the use of machine vision is the recognition of binary images, utilised e.g. in document text recognition. Considering the necessity of recognition of text data as well as e.g. QR codes from images captured in unknown lighting conditions, an essential step of the image processing is its binarization, where the loss of important information may occur. Since in many systems, the processing time may play a significant role, the multi-stage sophisticated algorithms, proposed e.g. during Document Image Binarization Competitions (DIBCO), may be inappropriate for some tasks.

An interesting example may be the development of robotic book scanners for digital archiving purposes, where fully controlled lightning conditions are ensured to prevent the presence of shadows, especially for thick books. With automatic page turning above 1000 pages per minute and perfect page illumination independent from ambient light, currently available solutions cause no problems with further Optical Character Recognition (OCR). However, the key issue is the control of lighting conditions, impossible to achieve in some other applications, such as e.g. recognition of text from degraded quality books or machine's metal nameplates. Some other interesting examples may be found in the paper [11], being partially the motivation of our research.

Unfortunately, a reliable performance verification of image binarization algorithms intended for industrial purposes would require the development of a dedicated large database containing numerous images together with ground truth (GT) data. A relatively simple possible solution might be the dataset of natural images containing some texts together with respective text data, although in this case the verification should be made by the text recognition accuracy, dependent not only on the thresholding method but also on the applied OCR engine, e.g. Tesseract. To make the assessment fully independent on the OCR engine, the comparisons of each pixel may be applied, although in this case the most suitable image database containing also binary ground truth images should be used.

Although in document image binarization, the most widely used benchmark datasets are DIBCO images [21], they contain some distortions specific to historical documents, which are not common in natural images expected in industrial applications. Therefore, it has been decided to use another database, known as Bickley Diary dataset [4], in this paper. This database contains 92 images being the copies of an about 100 years old diary written by the wife of Bishop George H. Bickley, who was one of the first missionaries in Malaysia. The 1050×1350 pixels images are affected by different ink contrasts, water stains, discolorization as well as photocopying noise, and therefore is it considered as more challenging in comparison to typically used DIBCO datasets. A subset of 7 images (no. 5, 18, 30, 41, 60, 74 and 87) contains additionally the labels used for discrimination of foreground text in difficult (low-contrast) regions. All GT images have been produced using the user-assisted Binarizationshop software [4].

2 Methods of Image Binarization

Although during last several years many image binarization methods have been proposed, there is still a need of finding some new algorithms, especially filling the gap between fast global thresholding and more sophisticated, computationally demanding adaptive solutions, which usually require the analysis of the neighbourhood of all image pixels and – in some cases – some additional post-processing steps.

Probably the most popular global image binarization method was proposed by Nobuyuki Otsu [20], which is based on the minimization of the sum of intra-class variances, being equivalent to the maximization of inter-class variance for the greyscale image subjected to thresholding. Since it works properly for well and evenly illuminated images, some modifications of this method were proposed later, such as AdOtsu [17] or region based approach [13]. A similar method, utilising entropy instead of variance was proposed by Kapur [6], which can also be applied for regions. Another popular global binarization method, known as minimum error thresholding, was presented by Kittler and Illingworth [8], with the assumption of normal distribution of grey level values of image pixels.

Growing availability and popularity of digital cameras have caused increased interest in adaptive thresholding, also considering the document images, due to more problems with uniformity of illumination, e.g. in comparison to flatbed scanners. Some of the most popular methods are based on the original Niblack's method [18], where the local threshold is determined as the local average brightness lowered by $k = 20\%$ of the local standard deviation (however this parameter k as well as the size of the local window are tunable). Further modifications of this method have been proposed by Sauvola and Pietikäinen [23], Feng [5] and Wolf [26]. An overview of such family of algorithms may be found in the paper [7], where the method known as NICK has been proposed as well as in recent survey papers [3,24]. A faster implementation of Niblack's method based on integral image and second order integral image used for the calculation of standard deviation has been recently proposed by Samorodova and Samorodov [22]. An attempt to use the region based approach for Niblack's thresholding has been made by Kulyukin [10]. Nevertheless, due to the application of the Support Vector Machines (SVM) in the post-processing stage, this method cannot be considered as fast.

Another adaptive method has been proposed by Bernsen [1], where the local contrast has been used to distinguish between background and foreground pixels which should exceed the average of the local minimum and maximum values (called local midgrey). A relatively fast adaptive method utilising the idea of integral images has been described by Derek Bradley and Gerhard Roth [2], which is ca. two or three times faster in comparison to popular Sauvola's approach due to the calculation of the local mean only without the direct use of the local standard deviation. The popularity of this method is partially caused by the availability of its implementation in Matlab environment as the *adaptthresh* function included in the Image Processing Toolbox in 2016.

Recently, some alternative methods have been proposed, including the idea of the used of the Monte Carlo method to speed up the computation of the histogram [12] and the application of the Generalized Gaussian Distribution (GGD) for image pre-processing [9]. Another direction of research, investigated also in this paper, is the application of region based approach. However, the use of relatively large blocks may lead to the presence of rapid changes of the obtained binary values near their borders, causing a quite similar effect to well-known JPEG blocking artifacts. To reduce its impact on the overall text recognition accuracy, a multi-layered method has been proposed in one of our recent papers [15], being the starting point for the experiments presented in this paper.

3 Proposed Approach

To find a balance between the simplicity of poor performance global thresholding and more computationally demanding adaptive methods, the reduction of the computational effort can be obtained using the region based methods, where local thresholds are calculated for blocks (regions) instead of the use of sliding windows. Hence, the amount of computations decreases significantly, although due to the presence of some blocking distortions in resulting binary image the overall performance results may not be satisfactory. To overcome this problem the multi-layered method based on the stack of regions may be used based on the calculation of thresholds using the specified formula for the blocks shifted by the defined number of pixels. Therefore, depending on the number of layers, a number of different threshold values may be obtained for each pixel, which belongs to different blocks for different layers [15]. The illustration of the idea of shifting the regions forming the stack for an image initially divided into 16 regions is presented in Fig. 1 with the assumption of 50% and 25% shifting for 2 and 4 layers respectively. The fragments outside the image have not been cropped for better visibility.

Forming the stack of M layers, the shift of each layer by $\frac{1}{M}$ of the size of the block has been assumed with the use of square $N \times N$ pixels blocks. For each of the overlapping blocks the value of the local threshold T is calculated as

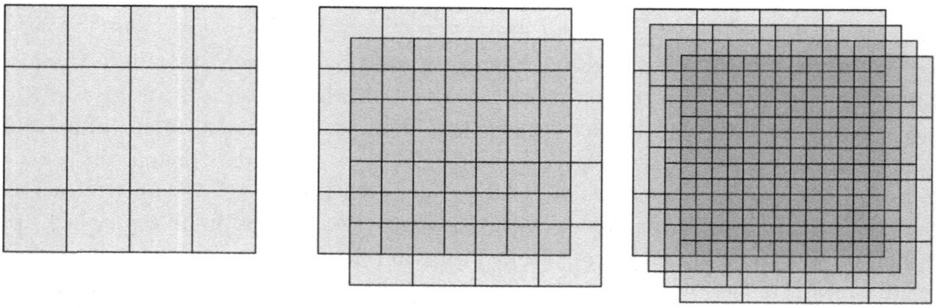

Fig. 1. Illustration of the idea of the stack of shifted regions: by 50% using two layers and by 25% using four layers

$$T = a \cdot mean(X) - b, \tag{1}$$

where $mean(X)$ denotes the average brightness in the block with the assumed optimized values of the parameters a and b. Another parameter subjected to optimization of the size of the block.

The final local threshold value for each pixel belonging to a number of overlapping blocks covering the sub-region is calculated as the average threshold values determined for them. Apart from good binarization results, one of the most relevant advantages of such approach is relatively high robustness to rapid changes of luminance levels as well as relatively high computational speed, considering the potential parallelization of calculations of local thresholds for each layer.

4 Results of Experimental Verification

To find the optimal values of the number of layers and the block size, the binarization performance has been calculated for all images from Bickley Diary dataset, choosing accuracy (described in Sect. 4) as the optimization criterion. The obtained results have been illustrated in Fig. 2. The optimal values of the parameters $a = 0.95$ and $b = 7$ have been the same as in the paper [15], although being optimized there in view of the OCR accuracy. Nevertheless, their changes have not provided better results also for Bickey Diary dataset.

All the calculations have been made setting the block size to 4, 8, 16, 32, 48, 64, 96, 128, 192, 256, 38 and 512 pixels in both dimensions (horizontal and vertical), and the number of layers was set to 1, 2, 4, 6, 8, 10, 12, 14 and 16. In comparison to the paper [15], the most appropriate size of the block has changed from 32×32 into 16×16 pixels. Considering the time necessary for calculations and further results, and potential parallel processing, a reasonable choice is the application of $K = 8$ layers, since the increase of the classification accuracy for the larger stack of regions is marginal. As all images in Bickley Diary dataset have the same size, the dependence of the block size and the number of layers on the image size has not been analysed.

According to typical classification metrics applied for the evaluation of image binarization methods [19], many metrics may be computed based mainly on counting true positives (TP), true negatives (TN), false positives (FP) and false negatives (FN). In this case the true positives are considered as correctly obtained foreground pixels, whereas e.g. false negatives are incorrectly classified background pixels (i.e. pixels which should be classified as foreground).

One of the most popular metrics, known as F-Measure or F1-score can be computed as:

$$FM = 2 \cdot \frac{PR \cdot RC}{PR + RC} = \frac{2 \cdot TP}{2 \cdot TP + FP + FN}, \tag{2}$$

where PR and RC are precision (ratio of true positives to the sum of all positives) and recall (true positives to the sum of true positives and false negatives), respectively, calculated considering the foreground pixels as "ones" and background pixels as "zeros". It is worth to note that in this case foreground pixels representing text are black.

Considering the asymmetry of F-Measure leading to different results for negative images, caused by reverse interpretation of positives and negatives, the additional verification of results can be made by the calculation of accuracy, often considered as the most intuitive performance measure, defined as:

$$ACC = \frac{TP + TN}{TP + TN + FP + FN}, \tag{3}$$

To verify the usefulness of the proposed approach numerically, the accuracy and F-Measure values have been calculated for the whole Bickley Diary database as well as for the subset of 7 annotated images. The results are presented in Table 1 and compared with the methods discussed earlier, as well as with our recently proposed method based on background estimation [16]. The approximate running time has been determined using the same computer with Intel Core-i7 CPU and 16GB of RAM with MATLAB 2018b running on 64-bit Windows 10 operating system and normalized relatively to execution time of Otsu's method as shown in Table 1. To demonstrate the increase of computational demands of the proposed method, the relative running time for the stack of regions has been intentionally calculated using sequential processing.

As can be seen in Table 1, proposed approach outperforms the other metrics both in terms of accuracy and F-Measure, particularly for 7 annotated images. To illustrate the results of the application of some of the methods listed in Table 1, a comparison of the resulting binary images obtained for a sample image from Bickley Diary dataset is presented in Fig. 3.

Analysing the properties of the proposed method, particularly for images from Bickley Diary dataset, a relatively small amount of noise can be noticed in resulting images, although such single isolated pixels are easy to eliminate and should not influence the final OCR accuracy. A relatively large amount of background remaining after the use of some other algorithms may be much more troublesome in some image fragments.

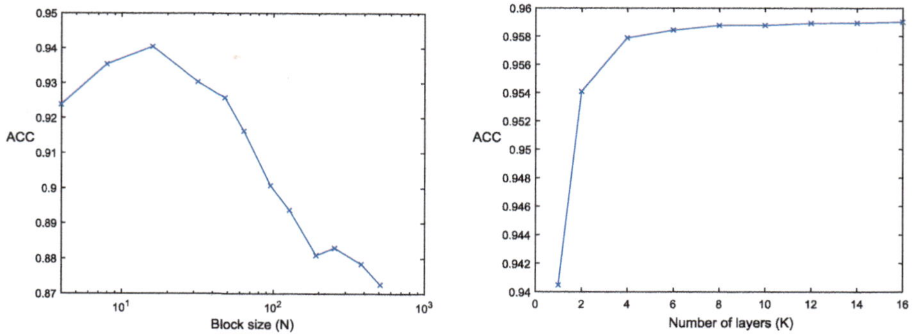

Fig. 2. Illustration of the results of optimization of parameters of the proposed method

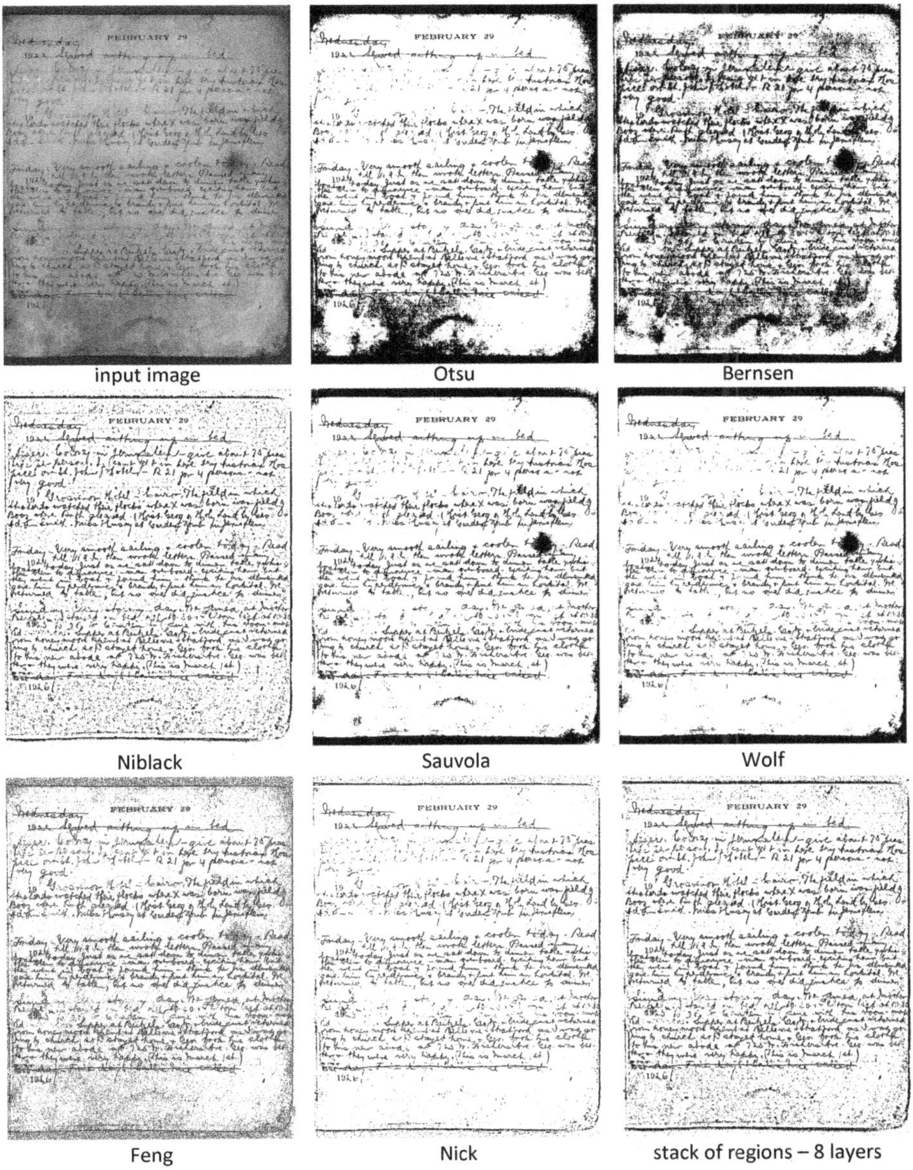

Fig. 3. Experimental results obtained for a sample image from Bickley Diary dataset

Considering the potential usefulness of the proposed approach for binarization of industrial images captured in unknown lighting conditions, the next experiments have been conducted using various images of nameplates, where the GT images are unknown. Nevertheless, a visual comparison of the

Table 1. Comparison of experimental results obtained for Bickley Diary database using various binarization methods (results for the subset of 7 annotated images are shown in parentheses)

Binarization algorithm	Accuracy	F-Measure	Relative running time
Otsu [20]	0.8470 (0.8188)	0.5947 (0.5040)	1.00
Kittler [8]	0.8867 (0.8540)	0.5818 (0.3926)	22.45
Bernsen [1]	0.7907 (0.7826)	0.5552 (0.5288)	150.72
Bradley [2] (mean)	0.8988 (0.8850)	0.7097 (0.6568)	15.40
Bradley [2] (Gaussian)	0.8937 (0.8794)	0.6904 (0.6266)	120.84
Feng [5]	0.8942 (0.8865)	0.6753 (0.6317)	154.13
Niblack [18]	0.9567 (0.9322)	0.8441 (0.7547)	56.80
Sauvola [23]	0.9181 (0.8940)	0.7305 (0.6323)	56.90
Wolf [26]	0.9101 (0.8849)	0.6973 (0.5875)	59.86
NICK [7]	0.9580 (0.9355)	0.8186 (0.7077)	52.51
JUCS [16]	0.9540 (0.9300)	0.8096 (0.6948)	10.88
Single layer [14]	0.9406 (0.9260)	0.7939 (0.7350)	43.88
Stack of regions - 2 layers	0.9541 (0.9324)	0.8361 (0.7532)	97.06
Stack of regions - 4 layers	0.9579 (0.9347)	0.8479 (0.7595)	188.38
Stack of regions - 6 layers	0.9584 (0.9351)	0.8496 (0.7605)	287.64
Stack of regions - 8 layers	0.9587 (0.9353)	0.8506 (0.7611)	390.75
Stack of regions - 12 layers	0.9589 (0.9356)	0.8510 (0.7620)	593.56
Stack of regions - 16 layers	0.9590 (0.9356)	0.8513 (0.7620)	795.78

readability of text data confirms the advantages of the proposed method, as illustrated in Figs. 4 and 5 for two sample representative images.

As can be noticed, results of binarization of the industrial image no. 1, which is strongly non-uniformly illuminated, for the global thresholding as well as dome adaptive methods, e.g. Bernsen or Bradley are unacceptable due to unreadability of text data as presented in Fig. 4. Some other algorithms, such as Sauvola or Wolf, as well as our earlier method based on background estimation, marked as JUCS, cause the loss of text data. Application of Feng thresholding leaves a noticeable amount of noise near the alphanumerical signs, which may cause some troubles in further character recognition procedure. A reasonable readability of alphanumerical data is ensured by Niblack and NICK methods, however the application of the proposed method based on the 8-layers stack of regions leads to visually best results.

Analysing the application of the same algorithms for the image no. 2, containing the light reflecting metal plate, presented in Fig. 5, too high sensitivity on the light reflections may be noticed not only for Otsu global thresholding, but also for Bernsen, Sauvola, Wolf and Bradley methods. Much better results,

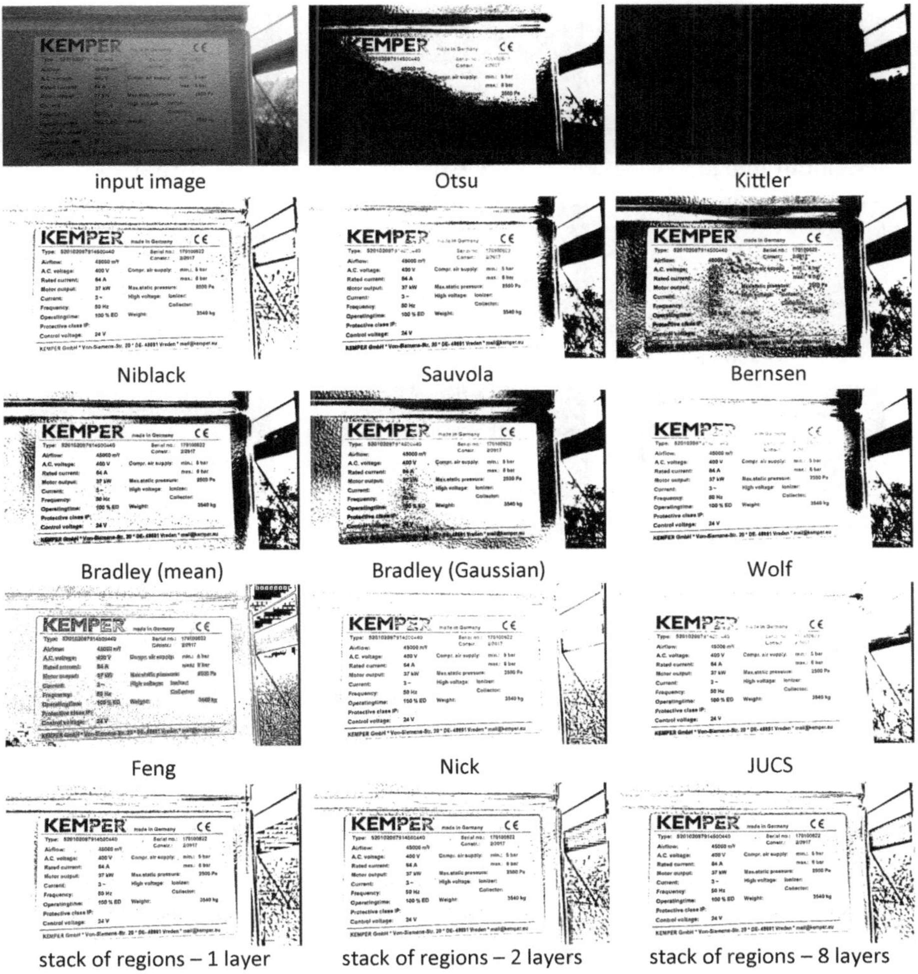

Fig. 4. Experimental results obtained for a sample industrial image no. 1

although noisy, may be observed for adaptive thresholding methods proposed by Niblack and Feng. The best results may be obtained using NICK method as well as faster one based on the 8-layers stack of regions. Visually slightly worse results may be achieved using the method marked as JUCS. Nevertheless, considering the running time of the proposed method and its dependence on the number of layers, reasonable results may also be achieved using two layers.

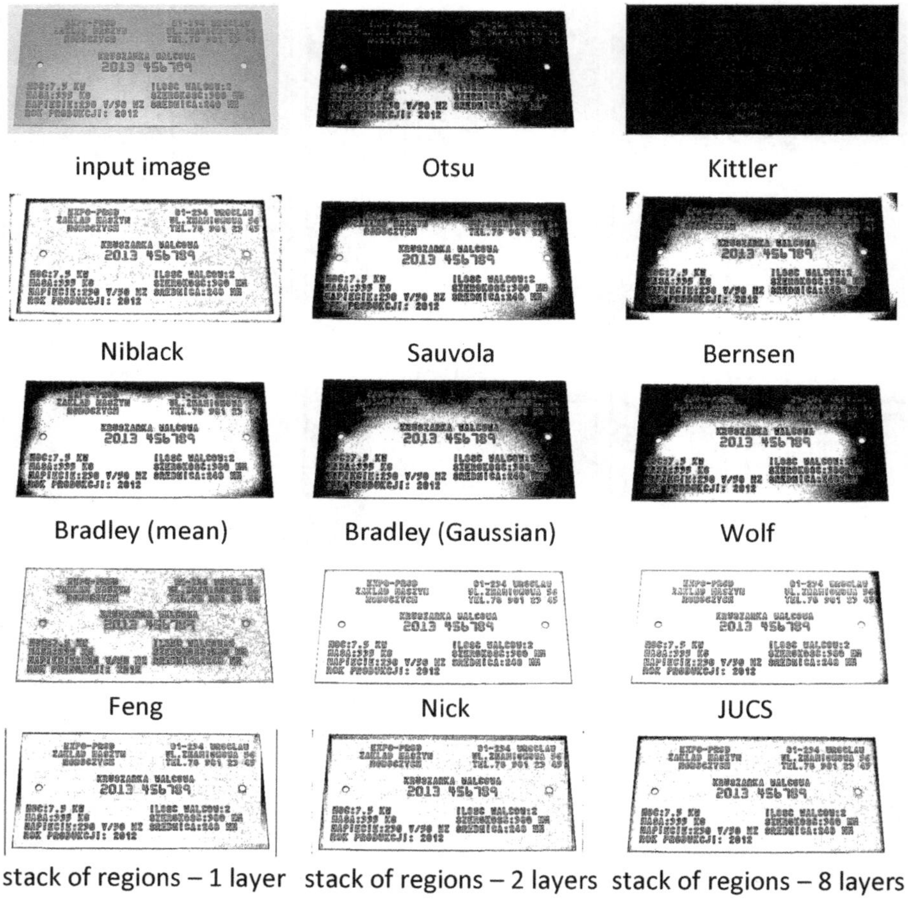

Fig. 5. Experimental results obtained for a sample industrial image no. 2

5 Concluding Remarks

The approach to multi-layered region based image binarization presented in this paper, originally proposed for the preprocessing of unevenly illuminated images for further text recognition [15], may be successfully adopted not only for the binarization of degraded historical document images, but also for some industrial images acquired in unknown lighting conditions. Due to the lack of appropriate datasets containing such real world images, the optimization of parameters and numerical validation of results may be conducted using document image datasets containing also the binary ground truth images.

Since the verification with the use of probably the most challenging and the largest currently available document image database, namely Bickley Diary dataset, confirms the advantages of the proposed scheme, considering its relatively high processing speed in comparison to adaptive algorithms, assuming the use of parallel processing, it may be successfully applied in some industrial systems, e.g. mobile robotics. Depending on the required quality, the number of layers may be increased in some applications.

A visual verification of the results of binarization with the use of natural unevenly illuminated images of the nameplates additionally proves the validity of the proposed approach. Nevertheless, our future experiments should be made towards further optimization of the proposed procedure, e.g. by the automatic elimination of low variance blocks, which do not contain text data or additional increase of the processing speed by the use of parallel processing or calculations based on integral images.

References

1. Bernsen, J.: Dynamic thresholding of grey-level images. In: Proceedings 8th International Conference on Pattern Recognition (ICPR), pp. 1251–1255 (1986)
2. Bradley, D., Roth, G.: Adaptive thresholding using the integral image. J. Graph. Tools **12**(2), 13–21 (2007). https://doi.org/10.1080/2151237X.2007.10129236
3. Chaki, N., Shaikh, S.H., Saeed, K.: Exploring image binarization techniques. In: Studies in Computational Intelligence, vol. 560. Springer, New Delhi (2014). https://doi.org/10.1007/978-81-322-1907-1
4. Deng, F., Wu, Z., Lu, Z., Brown, M.S.: BinarizationShop: a user assisted software suite for converting old documents to black-and-white. In: Proceedings Annual Joint Conference on Digital Libraries, pp. 255–258 (2010)
5. Feng, M.L., Tan, Y.P.: Adaptive binarization method for document image analysis. In: Proceedings 2004 IEEE International Conference on Multimedia and Expo (ICME), vol. 1, pp. 339–342 (2004). https://doi.org/10.1109/ICME.2004.1394198
6. Kapur, J., Sahoo, P., Wong, A.: A new method for gray-level picture thresholding using the entropy of the histogram. Comput. Vis. Graph. Image Process. **29**(3), 273–285 (1985). https://doi.org/10.1016/0734-189X(85)90125-2
7. Khurshid, K., Siddiqi, I., Faure, C., Vincent, N.: Comparison of Niblack inspired binarization methods for ancient documents. In: Document Recognition and Retrieval XVI, vol. 7247, pp. 7247–7247–9 (2009). https://doi.org/10.1117/12.805827
8. Kittler, J., Illingworth, J.: Minimum error thresholding. Pattern Recogn. **19**(1), 41–47 (1986). https://doi.org/10.1016/0031-3203(86)90030-0
9. Krupiński, R., Lech, P., Tecław, M., Okarma, K.: Binarization of degraded document images with Generalized Gaussian Distribution. In: Rodrigues, J.M.F.E. (ed.) Computational Science – ICCS 2019. Lecture Notes in Computer Science, vol. 11540, pp. 177–190. Springer (2019). https://doi.org/10.1007/978-3-030-22750-0_14
10. Kulyukin, V., Kutiyanawala, A., Zaman, T.: Eyes-free barcode detection on smartphones with Niblack's binarization and Support Vector Machines. In: Proceedings 16th International Conference on Image Processing, Computer Vision, and Pattern Recognition (IPCV 2012), vol. 1, pp. 284–290. CSREA Press (2012)
11. Lam, O., Dayoub, F., Schulz, R., Corke, P.: Text recognition approaches for indoor robotics: a comparison. In: Proceedings of Australasian Conference on Robotics and Automation (ACRA), Melbourne, Australia (2014). Paper no. 138

12. Lech, P., Okarma, K.: Optimization of the fast image binarization method based on the Monte Carlo approach. Elektronika Ir Elektrotechnika **20**(4), 63–66 (2014). https://doi.org/10.5755/j01.eee.20.4.6887
13. Lech, P., Okarma, K., Wojnar, D.: Binarization of document images using the modified local-global Otsu and Kapur algorithms. Przegląd Elektrotechniczny **91**(1), 71–74 (2015). https://doi.org/10.15199/48.2015.02.1
14. Michalak, H., Okarma, K.: Region based adaptive binarization for optical character recognition purposes. In: Proceedings of International Interdisciplinary PhD Workshop (IIPhDW), pp. 361–366. Świnoujście, Poland (2018). https://doi.org/10.1109/IIPHDW.2018.8388391
15. Michalak, H., Okarma, K.: Adaptive image binarization based on multi-layered stack of regions. In: Vento, M., Percannella, G. (eds.) Computer Analysis of Images and Patterns. Lecture Notes in Computer Science, vol. 11679, pp. 281–293. Springer, Cham (2019). https://doi.org/10.1007/978-3-030-29891-3_25
16. Michalak, H., Okarma, K.: Fast binarization of unevenly illuminated document images based on background estimation for optical character recognition purposes. J. Univ. Comput. Sci. **25**(6), 627–646 (2019). https://doi.org/10.3217/jucs-025-06-0627
17. Moghaddam, R.F., Cheriet, M.: AdOtsu: an adaptive and parameterless generalization of Otsu's method for document image binarization. Pattern Recogn. **45**(6), 2419–2431 (2012). https://doi.org/10.1016/j.patcog.2011.12.013
18. Niblack, W.: An Introduction to Digital Image Processing. Prentice Hall, Englewood Cliffs (1986)
19. Ntirogiannis, K., Gatos, B., Pratikakis, I.: Performance evaluation methodology for historical document image binarization. IEEE Trans. Image Process. **22**(2), 595–609 (2013). https://doi.org/10.1109/TIP.2012.2219550
20. Otsu, N.: A threshold selection method from gray-level histograms. IEEE Trans. Syst. Man Cybern. **9**(1), 62–66 (1979). https://doi.org/10.1109/TSMC.1979.4310076
21. Pratikakis, I., Zagoris, K., Kaddas, P., Gatos, B.: ICFHR 2018 Competition on Handwritten Document Image Binarization (H-DIBCO 2018). In: 2018 16th International Conference on Frontiers in Handwriting Recognition (ICFHR), pp. 489–493 (2018). https://doi.org/10.1109/ICFHR-2018.2018.00091
22. Samorodova, O.A., Samorodov, A.V.: Fast implementation of the Niblack binarization algorithm for microscope image segmentation. Pattern Recogn. Image Anal. **26**(3), 548–551 (2016). https://doi.org/10.1134/S1054661816030020
23. Sauvola, J., Pietikäinen, M.: Adaptive document image binarization. Pattern Recogn. **33**(2), 225–236 (2000). https://doi.org/10.1016/S0031-3203(99)00055-2
24. Saxena, L.P.: Niblack's binarization method and its modifications to real-time applications: a review. Artif. Intell. Rev. **51**(4), 673–705 (2019). https://doi.org/10.1007/s10462-017-9574-2
25. Tensmeyer, C., Martinez, T.: Document image binarization with fully convolutional neural networks. In: 14th IAPR International Conference on Document Analysis and Recognition, ICDAR 2017, Kyoto, Japan, 9–15 November 2017, pp. 99–104 (2017). https://doi.org/10.1109/ICDAR.2017.25
26. Wolf, C., Jolion, J.M.: Extraction and recognition of artificial text in multimedia documents. Formal Pattern Anal. Appl. **6**(4), 309–326 (2004). https://doi.org/10.1007/s10044-003-0197-7

An Algorithm of Pig Segmentation from Top-View Infrared Video Sequences

Paweł Kielanowski(✉) and Anna Fabijańska

Institute of Applied Computer Science, Lodz University of Technology, 18/22 Stefanowskiego Street, 90-924 Lodz, Poland
{pkielanowski,an_fab}@iis.p.lodz.pl

Abstract. This paper considers the problem of pig automatic segmentation from infrared top view images of a pen. Particularly, an algorithm for accurate delineation of pig's contour is presented. The method consists of two main steps. In the first step, a rough contour is determined using standard image processing methods. Next, the initial contour is gradually deformed so that it reflects the actual contour of the pig as much as possible. This effect is obtained by attracting initial contour points to the nearest local gradient peaks. In the last step, the contour is refined and smoothed by removing loops. This step incorporates analysis of the angles between contour segments passing through the consecutive contour points. Results of the proposed approach for sample infrared images of pigs in a pen are presented and discussed. They reveal that the method performs reasonably well with the average DICE score exceeding the level of 0.97 and the average Jaccard index above 0.95.

Keywords: Computer vision · Agriculture · Pig segmentation · Image processing

1 Introduction

The automation of industrial processes is an essential part of the current world progress. The increasing number of solutions have been invented to optimize selected production cycles. This progress is associated with the so-called another industrial revolution. It is now driven by the Internet of Things (IoT). The main idea of the IoT world is to support humans in every step. There are, however, some areas in which this development is not as dynamic as in others. The example is agriculture, especially pig farming. In 2018 there were about 1 billion pigs in the world. Recently, new technologies supporting breeders have appeared in the market. Examples of these solutions are sorting scales, devices automatically selecting the amount of feeding. These technologies are, however, not very popular among breeders because they mostly require large financial outlays for implementation.

Until now, most of the scientists and researchers have focused on tracking an animal to extract their behavior, including movement, time spent on eating,

A. Bartoszewicz et al. (Eds.): PCC 2020, AISC 1196, pp. 785–796, 2020.
https://doi.org/10.1007/978-3-030-50936-1_66

drinking water, and sleeping [1,4]. This kind of data may be provided by RFID sensors or beacon sensors, which produce the localization heatmap on the output. However, not every type of animal can wear earrings with these sensors. An additional limitation is the cost of these sensors and the way of applying them to animals. For these reasons, the majority of solutions for automatic animal behavior analysis are based on digital cameras. Mainly, a typical way to determine the behavior of the animals is to install the camera with the top view on the pen to monitor the whole herd. For these solutions, the biggest challenge is to distinguish single animals, especially when they have direct contact with each other without spaces between them [7]. On the other hand, to generate a heat map of animal mobility, only a general outline of their silhouette is needed.

There also exist some attempts to estimate pig weight based on the 2D image and digital image processing. For example, it could be done by extracting structure from motion [5]. Image-based approaches to pig weight estimation could be competitive to the new commercial solutions enabling contactless monitoring of the pigs' weights based on the 3D cameras [2], which are still too expensive for most of the breeders.

The estimation of pig weight is essential for several reasons. First and the simplest one is that the breeders would like to know which pigs are bigger than the others to sell the heaviest animals first and thus reduce feed consumption and increase income. The second reason is the logistics process in companies which have slaughterhouse. Recently finisher contracting has become popular on the market. The finisher contract is when pork meat producers give the pigs, feed, and veterinary care to breeders paying them a fixed fee for this service. The meat producers usually have heard of more than tens of thousands of pigs. In the case of such an amount of animals, there is a need for logistics optimization. These companies expect some repeatability in the case of pig weights because the whole production cycle could be most effective. The knowledge about the weight of the animals in the specific pen would be appreciated then because it could tell meat producers where to expect the particular total weight of the pigs.

With the above-mentioned in mind, this paper presents the method of precise segmentation of the pigs from the top-view infrared images. This task is an essential step in a pipeline aiming at pig weight estimation based on shapes of their silhouettes.

The following part of this paper is organized as follows. First, in Sect. 2 data used in this study is characterized. Then follows in Sect. 3 the detailed description of the proposed pig segmentation approach. The results of pig segmentation from sample movie frames are presented in Sect. 4 and discussed in Sect. 5. Finally, Sect. 6 concludes the paper.

2 Input Data

In this study, infrared video sequences of pigs in the pen were used. The sequences were obtained from the OVERMAX OV-CAMSPOT 4.5 infrared digital camera mounted still above the pen with the top view on the pigs. Unlike the previously reported works, the camera field of view covered only a part of the pen due to the excessive pen size (typical for piggeries in Poland). An infrared camera was selected to be a video source to facilitate pig image acquisition constantly, also at night when the light sources are not available.

A total of 70 h of video signal were registered, consisting of about 5 million video frames in total, each of resolution 720×1024 pixels. Frames were monochrome, whit a bit depth of 8bpp. Sample infrared video frames are presented in Fig. 1.

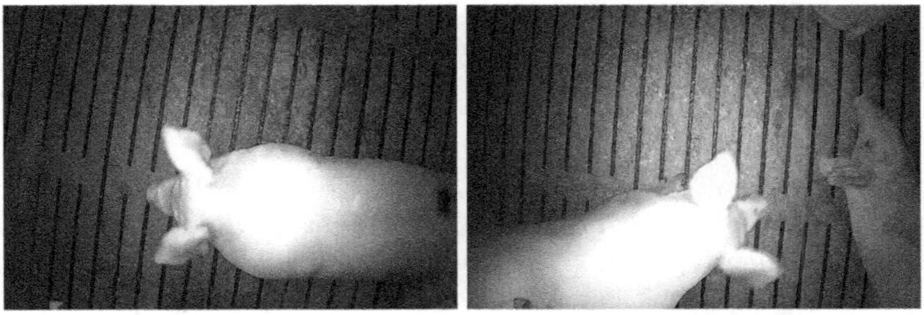

Fig. 1. Sample top-view infrared images of pigs in a pen.

3 Proposed Approach

The segmentation of pigs from infrared top-view images is challenging due to several reasons. The most important ones are not uniform intensity distribution within a pig area and distortions within the floor region. Particularly, due to infrared signal acquisition, the image intensity is the highest in the center of the pig region and decreases towards the edges (see Fig. 1). As a result, pig segmentation cannot be performed directly via image thresholding. Therefore, the approach proposed in this paper combines the latter technique with edge-based segmentation to perform precise pig segmentation via contour delineation. The proposed method of pig segmentation consists of several steps described in the following subsections.

3.1 Background Removal

Although a pig is visible on the contrasting background, its edges cannot be precisely delineated with classical edge detectors, since there are also many edges in the floor region. These edges hinder image processing. Also, nonuniform object intensity causes problems with distinguishing edges distant from pig center where the infrared light captured by a camera has lower intensity. This effect can be seen in Fig. 2 where the pig body was delineated, but the resulting edge is discontinuous, irregular, and distorted by floor elements.

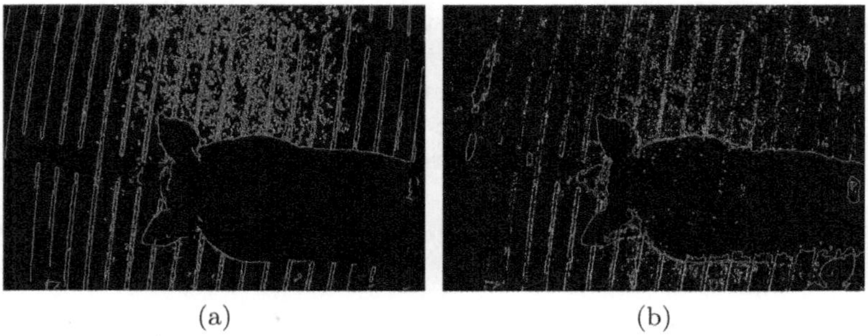

Fig. 2. The results of edge detection applied to a sample pig image; a) Canny edge detector, b) - Satoshi Suzuki's method [6]

To alleviate these problems and unify intensity distribution, a background is removed from the input image in the first step of the algorithm. Since the camera is permanently in one place, the background image is easy to obtain as a frame showing a pen without pigs (see Fig. 3a). Particularly, multiple images of an empty pen are averaged to construct an approximate background frame.

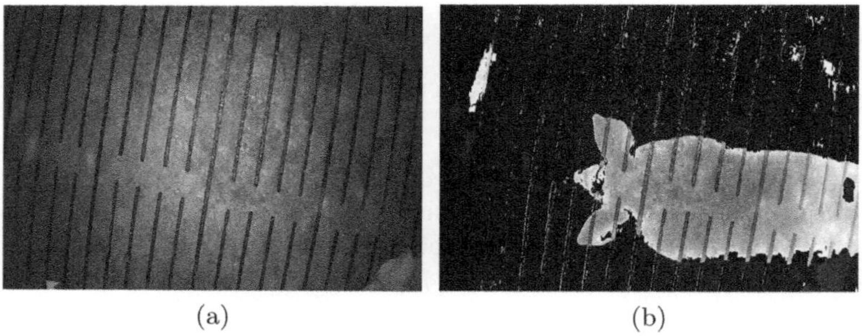

Fig. 3. Background subtraction from a sample pig image a) background image - frame without pigs, b) sample frame with background removed.

Background removal is then performed by image subtraction. One can see (Fig. 3b) that this operation removes most of the floor noise. The background subtraction does not resolve the problem entirely but changes the intensity distribution. This effect is presented in Fig. 4, where subfigures (a) and (b) present histograms of image prior and after background subtraction.

3.2 Binarization and Morphological Processing

After the background subtraction, the image histogram becomes bimodal (see Fig. 4), and the threshold that coarsely separates pig from the background can

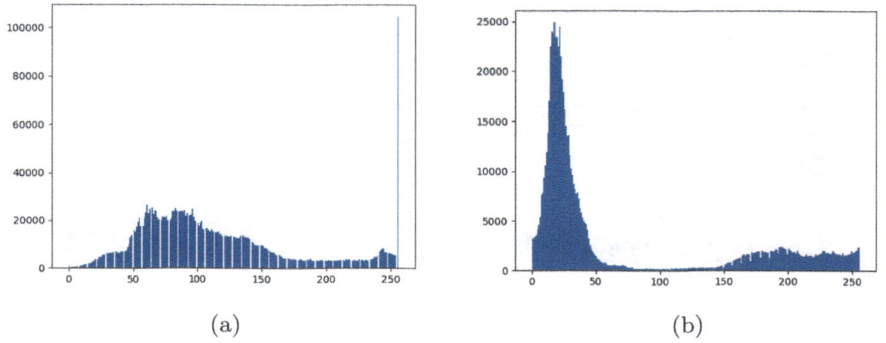

Fig. 4. Pixel intensity distribution in a sample input frame; a) histogram for a raw frame, b) histogram after background subtraction.

easily be found. From Fig. 4b, one can see that the threshold should be set between 50 and 150. The valley between histogram peaks is flat. However, lower thresholds result in fewer holes in the pig body, whereas higher thresholds provide more regular edges than lower thresholds. This effect is presented in Fig. 5, where results of global thresholding with thresholds equal to 50 and 140 respectively are shown. Li and Lee's method based on minimum cross-entropy is applied [3] to segment a pig region coarsely. Using this method allows finding the optimal threshold universally for all frames.

Fig. 5. The results of image thresholding with a threshold that equals 50 (a) and 140 (b).

Once the image is binarised, it is subjected to the morphological opening, which aims at removing outliers but preserving pig dimension. The results of these operations are presented in Fig. 6.

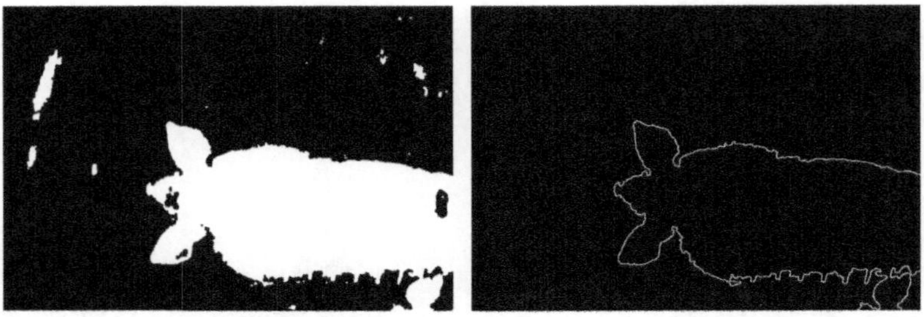

(a) (b)

Fig. 6. The results of morphological processing; a) image opening; b) coarse contour.

3.3 Coarse Contour Extraction and Refinement

The coarse contour of a pig is obtained with the use of the Satoshi Suzuki's approach for border tracking in binary images [6]. The largest connected component in the image after morphological opening is considered at this step. In Fig. 6b one can see that although this contour delineates a pig reasonably accurately, it is still irregular and, in some regions, deviates from the pig body. Therefore, in the next step, the coarse contour is refined. Particularly, the contour is attracted to the nearest sharp local intensity change. The latter is determined by using the Laplacian operator (see Fig. 7a). Intensity values in the resulting image are close to 1 and below 10, as shown in Fig. 8, therefore to obtain the map of local intensity changes, global thresholding is performed with a threshold equal to 10 (see Fig. 7b).

In the refining step, a neighborhood of 20×20 pixels around each pixel of the coarse contour is searched, and the pixel is moved towards the nearest intensity change masked by thresholded Laplacian image. The outcome of this procedure is presented in Fig. 9 where the coarse contour is shown in blue while the refined contour is shown in red.

Contour refined via attracting to gradient maximum is sharpened; however, there are still some local fluctuations, self-intersections, and loops. Therefore in the last step, these distortions are removed, and the contour is smoothed. In the first step, the contour is divided into an even number of sections (see Fig. 10b). If two consecutive segments intersect, the points between them are removed from the contour (see Fig. 10c). The results of this step are presented in Fig. 11a. Afterward, the contour is smoothed based on the angle between segments sharing contour points. The pig shape is cylindrical, that is why the inner angles could not be obtuse above $220°$; otherwise, the middle point with the obtuse angle is missing.

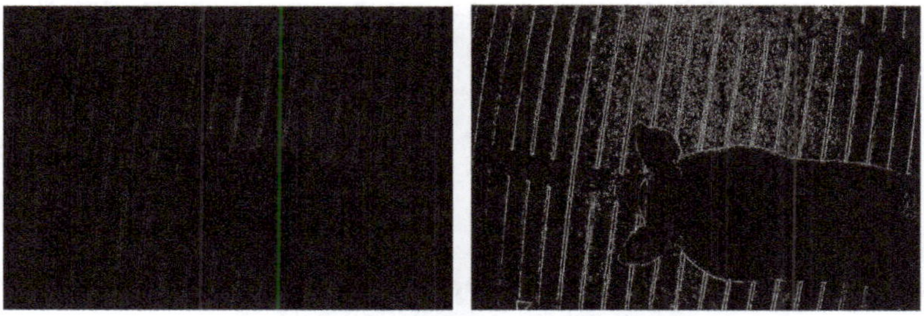

(a) (b)

Fig. 7. Map of the local intensity changes; a) Laplacian image; b) its thresholded equivalent.

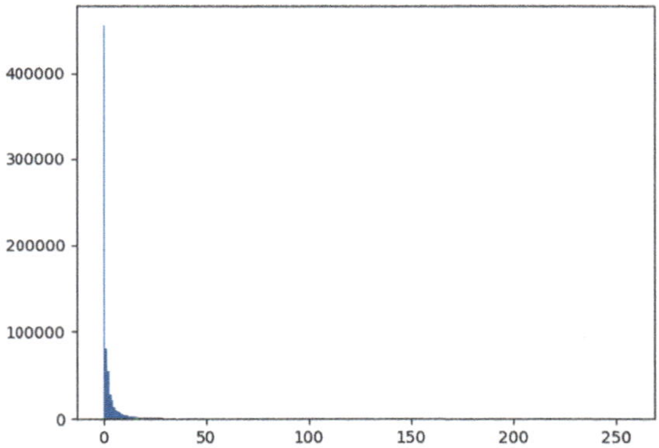

Fig. 8. Image histogram after applying threshold on gradient.

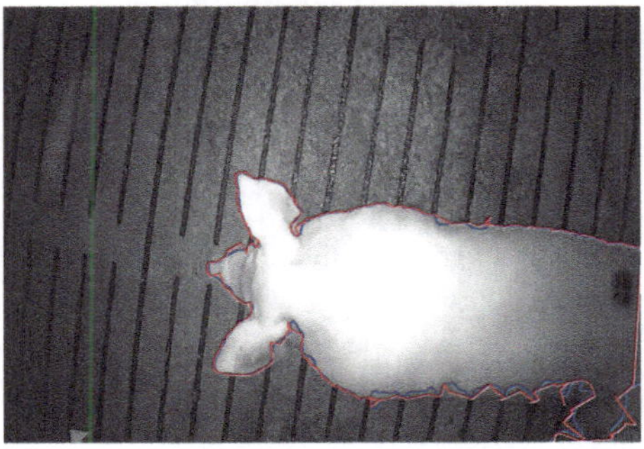

Fig. 9. The results of contour refinement. The blue line corresponds to coarse contour while red line extends the basic outline to the nearest sharp intensity change.

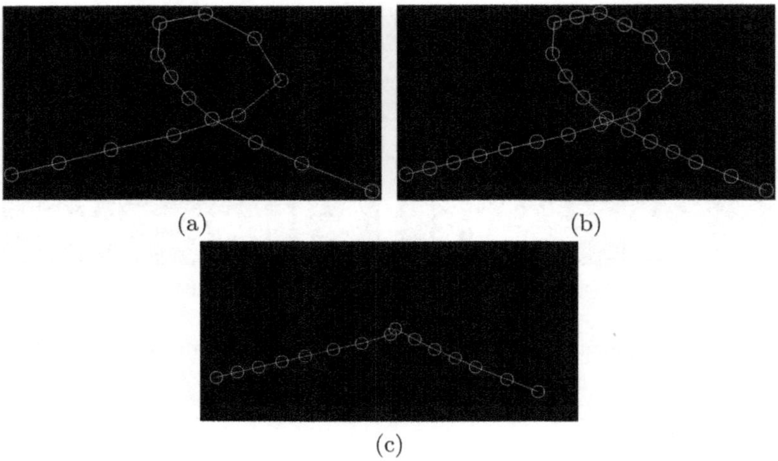

Fig. 10. The consecutive steps of the algorithm that removes loops and self-intersections of contour; a) basic contour with a circle denoting contour nodes; b) contour with additional nodes added based on the average distance between nodes in the basic contour; c) contour without self-intersection. If the distance between any points is smaller than the average distance between any pair of points, then it is checked whether the sections intersect. If there is an intersection, then all nodes in between are removed.

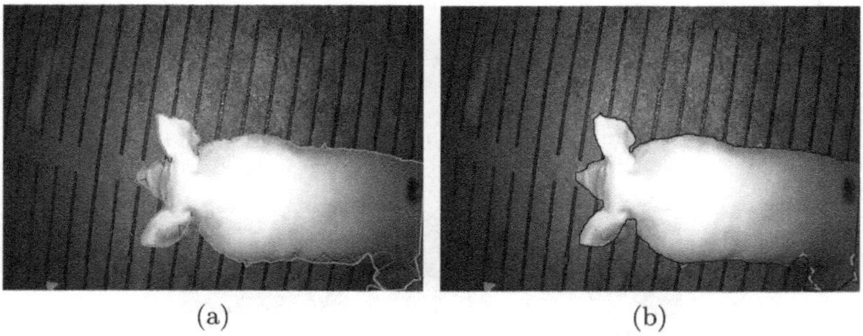

Fig. 11. The results of contour refinement; a) the red line corresponds to contour after being attracted to local sharp intensity changes, and the green line presents the contour after removing loops and smoothing angles; b) black color is the contour with coordinates smoothed with a median filter.

Finally, the coordinates of consecutive contour points are smoothed with a median filter of size 9 (selected empirically). The results of this step are shown in Fig. 11b.

4 Results

For assessment of the accuracy of the proposed approach, the results of pig segmentation were compared with ground truth results. The latter were prepared manually in an image editor.

The sample visual results of the proposed pig segmentation algorithm are presented in Fig. 12. The green outline was obtained using the proposed approach, whereas yellow contour refers to ground truth results.

The numerical assessment was performed employing the Sorensen-DICE coefficient and Jaccard index defined by Eqs. 1 and 2, respectively.

$$DIC(A,B) = \frac{2|A \cap B|}{|A| + |B|} \tag{1}$$

$$JCC(A,B) = \frac{|A \cap B|}{|A \cup B|} \tag{2}$$

where A and B denote the obtained segmentation results and the ground truth respectively.

Values of the above image segmentation performance measures obtained for sample cases from Fig. 12 are presented in Table 1.

Table 1. Numerical assessment of pig image segmentation accuracy.

Case ID	DICE	JCC
Fig. 12a	0.964	0.931
Fig. 12b	0.981	0.965
Fig. 12c	0.978	0.956
Fig. 12d	0.982	0.966
Fig. 12e	0.955	0.914
Fig. 12f	0.980	0.961
Fig. 12g	0.986	0.972
Fig. 12h	0.974	0.951
Fig. 12i	0.975	0.952
Fig. 12j	0.975	0.952
Fig. 12k	0.981	0.963
Fig. 12l	0.964	0.931
Mean	0.974	0.951

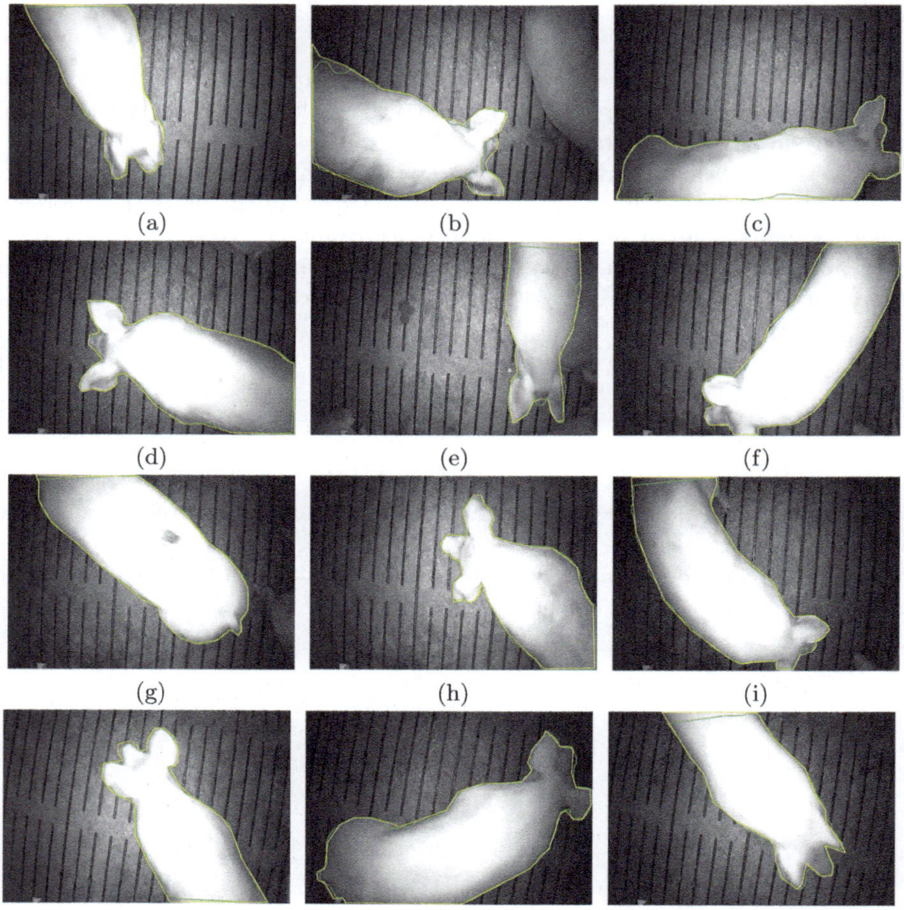

(a) (b) (c) (d) (e) (f) (g) (h) (i) (j) (k) (l)

Fig. 12. The results of the proposed algorithm. The yellow colour of the contour was made manually by the human and green one was made by algorithm.

5 Discussion

Based on visual assessment, it can be seen that the proposed approach accurately delineates pig silhouette. The resulting contour is smooth; however, it can also capture shape details like pig ears and snout.

The worst results among the selected test images were obtained for cases shown in Fig. 12e and Fig. 12l. An unbalanced light caused it. In such a case, the algorithm experienced problems in regions at pig edges, since it is challenging to find gradient there. For those images, the results were respectively: 0.955 and 0.964 for Sorensen-DICE coefficient and 0.914, 0.931 for the Jaccard index, which is still a good result. For other considered cases, the results were significantly better, resulting in the average values of the DICE coefficient and Jaccard index at the level of 0.97 and 0.95, respectively.

Several issues were discovered during this study and need consideration in the future to improve the performance of the proposed segmentation approach. The first issue is the ambient noise that affects image quality and, thus, pig segmentation accuracy. During image acquisition with an infrared camera, occasionally, there were effects similar to a snowstorm caused by infrared rays reflected from dust particles. This observation was compounded by more intensive animal movement within the building. The second issue is the necessity to ensure that the camera is not able to accidentally move since it's every shift hampers pig segmentation by inaccurate background subtraction. The next issue is the camera quality. The uneven illumination of the observed environment with an infrared beam is typical for such cameras; however, depending on the quality of the digital camera, the differences in background illumination may vary. An important issue is also that to standardize measurements, the image even with proper illumination - for example, in the daytime - is still recorded using infrared.

Last but not least is an optimization of the algorithm. Only a few piggeries have a high-speed internet connection, which means that mostly the data needs to be processed locally. The current data processing is time-consuming and could be improved in the future.

6 Conclusions

The proposed algorithm of pig segmentation from top-view infrared pen images precisely outlines the pig body. In our future works, this step is to precede contactless pig weight determination based on selected shape descriptors. However, the proposed segmentation algorithm can potentially contribute to works in precise animal tracking. Accurate segmentation could also enable more accurate determination of pig characteristics in the behavioral studies considering animal movement. If implemented, both of the above development directions allow greater automation during pig farming, which directly leads to production optimization and higher income for the farmers.

References

1. Ju, M., Choi, Y., Seo, J., Sa, J., Lee, S., Chung, Y., Park, D.: A kinect-based segmentation of touching-pigs for real-time monitoring. Sensors (Basel) **18**(6), 891–921 (2018). https://doi.org/10.3390/s18061746
2. Kongsro, J.: Estimation of pig weight using a Microsoft Kinect prototype imaging system. Comput. Electron. Agric. **109**, 32–35 (2014). https://doi.org/10.1016/j.compag.2014.08.008
3. Li, C., Lee, C.: Minimum cross entropy thresholding. Pattern Recogn. **30**, 617–625 (1993). https://doi.org/10.1016/0031-3203(93)90115-D
4. Nilsson, M., Ardo, H., Astrom, K., Herlin, A., Bergsten, C., Guzhva, O.: Learning based image segmentation of pigs in a pen. In: 22nd International Conference on Pattern Recognition (ICPR) (2014)

5. Pezzuolo, A., Milani, V., Zhu, D., Hao, G., Guercini, S., Marinello, F.: On-barn pig weight estimation based on body measurements by structure-from-motion (SfM). Sensors **18**, (2018). https://doi.org/10.3390/s18113603
6. Satoshi, S., Keiichi, A.B.: Topological structural analysis of digitized binary images by border following. Comput. Vis. Graph. Image Process. **30**, 32–46 (1985). https://doi.org/10.1016/0734-189X(85)90016-7
7. Zhang, L., Gray, H., Ye, X., Collins, L., Allinson, N.: Automatic individual pig detection and tracking in surveillance videos (2018)

Contour Classification Method for Industrially Oriented Human-Robot Speech Communication

Piotr Skrobek and Adam Rogowski(✉)

Warsaw University of Technology, ul. Narbutta 85, 02-524 Warsaw, Poland
piotrskrobekwipair@gmail.com, adam.rogowski@pw.edu.pl

Abstract. This paper describes research dealing with application of image recognition as a tool supporting efficient speech communication between humans and collaborative robots in industrial environment. Image-based recognition of objects in robot's workspace may provide a context for voice commands. In this way the commands can be shorter and more concise. As the robots "understand" abstract technical terms used by human operators, a user-friendly speech communication can be provided. In order to recognize objects properly, classification of their contours must usually take into account the fact that some objects described by one abstract term may differ in dimensions and shapes, whereas some other objects described by different terms may be very similar. Since the object classification rules are usually application-specific, it is impossible to develop general algorithm applicable in all situations. This problem can be solved using Flexible Editable Contour Templates (FECT). However, the crucial factor determining applicability of contour classification method to speech communication is rapidity of the algorithm used for comparison of real contours against the FECTs. Currently, a computationally-expensive algorithm of segment matching is used. In this paper, we propose an alternative method, based on artificial neural networks (ANN).

Keywords: Contour classification · Man-machine communication · Collaborative robots

1 Introduction

Speech is the most natural way of communication for humans. Therefore it seems to be reasonable to use it also for man-machine interaction [1], particularly in collaborative robotics or in systems with diversified levels of automation where the humans are involved in manual activities hindering simultaneous control of machines using traditional means like buttons, touch screens, teach pendants etc. An example thereof is the voice control system for a surgical robot [2]. The surgeon manipulates laparoscopic instruments using both hands and controls the robotic endoscope holder with simple voice commands. Although this example does not refer directly to industrial robots, similar solutions could be introduced also in the industry, particularly in situations where the humans work together with robots side-by-side.

A. Bartoszewicz et al. (Eds.): PCC 2020, AISC 1196, pp. 797–808, 2020.
https://doi.org/10.1007/978-3-030-50936-1_67

Various aspects of collaborative robotics are currently researched e.g. new methods for robot programming [3], human-robot collaboration safety [4, 5], or ergonomic aspects [6]. There are also reports on successful implementations in the industry [7]. As far as speech-based man-machine communication is concerned, it may be applied in collaborative robotics [8] but also in areas like conventional industrial robotics [9], medicine [10], assistant robots [11], and others.

An effective speech communication is usually associated with multimodality providing appropriate context to verbal expressions The most popular source of the context are vision systems recognizing objects, humans, as well as human gestures. Various image recognition methods have been developed, aimed at e.g. recognition of faces [12], facial expressions [13], human body motions and hand gestures [14–17]. Much less attention has been paid to recognition of objects manipulated by industrial collaborative robots so far.

As far as industrial environment is concerned, the voice commands for machines should be as concise as possible. This condition can be fulfilled only when the machine "understands" technical terms used by the human operator. In the case of collaborative robotics, those terms will usually pertain to the objects like workpieces or tools to be handled by robot. The context for voice commands can be derived from images of those objects taken by the camera. Generally, object identification may be based on various features [18] like brightness, color, shape factors etc. In industrial environment, however, recognition and classification of object contours will play the crucial role.

2 Contour Recognition and Classification

Various methods are used for recognition and classification of contours: chain codes [19], polygon approximation [20], shape signature [21, 22], Fourier descriptors [23], wavelet descriptors [24], deformable templates (elastic matching) [25], and others. However, all those methods suffer from a crucial disadvantage that hinders their effective use in industrial applications. Namely, they cannot cope with the problem of two contradictory requirements: selective flexibility of the template and selective precision requirements. In objects belonging to one class some contour fragments may be of almost any shape but some other fragments must be strictly determined. Some objects described by one abstract term may differ in dimensions and shapes, whereas some objects described by different terms may be very similar. Figure 1 depicts this issue. Besides, the criteria for classification of workpieces in industrial practice are usually application-specific. Therefore it is impossible to develop general algorithms applicable in all situations. A solution to this problem is application of Flexible Editable Contour Templates (FECT) because they are based on human experience and take into account the specific character of individual applications [26]. Therefore the two contradictory requirements mentioned before can be fulfilled.

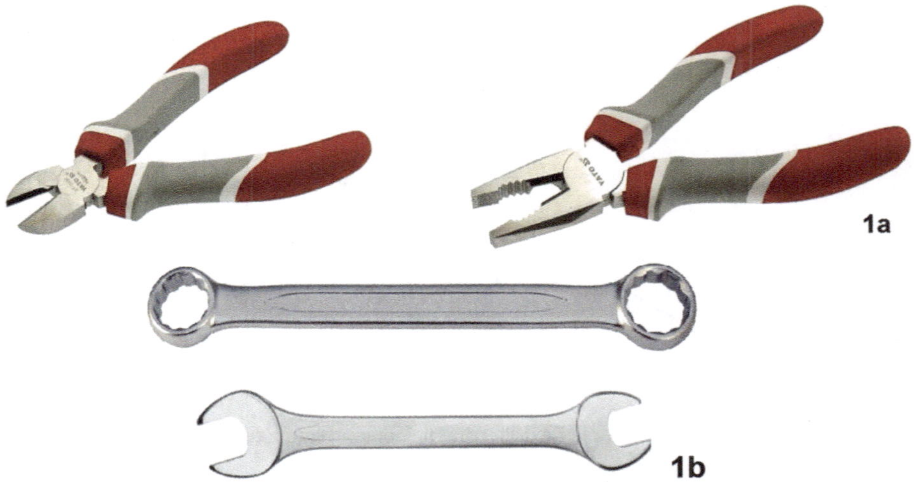

Fig. 1. Similarity of shapes belonging to different classes (a) vs. variability of shapes within one class (b).

In order to provide a suitable tool for creating FECTs, a special format FCD (Flexible Contour Description) was developed. Its detailed description can be found in [26]. FCD lets describe contours of the objects in a manner similar to movement instructions used in CAM system but it has much more flexible character due to the use of variables and ranges of values. A sample FECT depicted in Fig. 2 is presented below in FCD format.

```
# cnt hammer
        line: a
        go: left/angle = 90
        line:b
        go: left/angle = 90
        line: c
        arc: d/bend: right, angle: beta = 1,15
        go: right/angle:90-beta
        line: e
        go: left/angle:90
        line: f
        go: left/angle:90
        line: g /length: 1
        arc:h /bend:left, angle:gamma = 30,90
        go: left/angle:180-gamma,170
        arc: i/bend:right
```

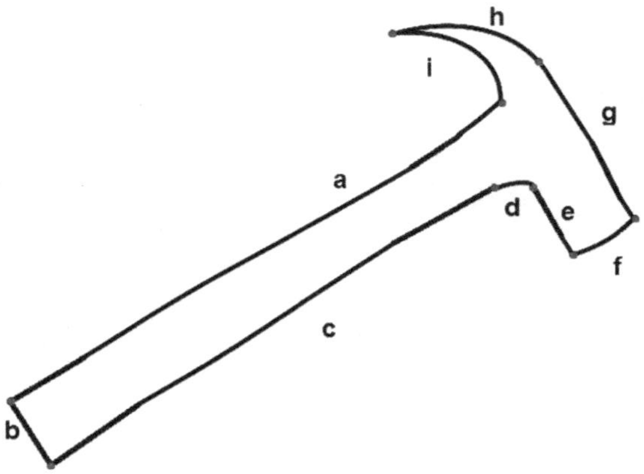

Fig. 2. A sample flexible editable contour template (FECT)

As can be seen, the template consists of flexibly defined elementary contour segments like lines and arcs i.e. geometric primitives constituting contours of workpieces commonly encountered in industrial robotized manufacturing systems (although Bezier curves can be also included in FECTs).

However, the crucial factor determining robustness of contour classification is the algorithm used for comparing the real contours derived from vision system against the FECTs. Currently, an algorithm of segment matching is used. In this paper, we propose an alternative method, based on artificial neural network (ANN).

3 Shortages of Segment Matching Algorithm

The segment matching algorithm used for classification of contours consists in comparing individual contour segments derived from the image against individual FECT segments. Prescinding from details which can be found in [26], this algorithm consists of the following steps:

- Calculate contour signature for all points belonging to the image contour
- Determine the number m of segments constituting the FECT
- Based on signature profile, extract the feature points F_j $(j = 1..n)$
- For each F_j $(j = 1..n)$ as a starting point, consider all m-combinations from the set of n feature points
- For each combination try to match m segments of image contour against corresponding FECT segments

Feature points are the points in which elementary contour segments join. Their extraction is necessary to perform subsequent steps of segment matching. Examples of feature points are the corners which can be extracted using e.g. Harris detector. However, there are also feature points where the segments fluently pass into each other. In order to cope with this problem, a more general method was used by Bielecka in contour classifier (based partially on syntactic approach) where the first and second derivatives of contour were used [27].

The approach represented by segment matching algorithm is somehow similar. The signature has the form of contour curvature and its derivative. Feature points are the points where the curvature and/or its derivative change in substantial manner. Extraction algorithm based on analysis of curvature and its derivative is very simple and very fast one. This is of big importance because delays in feature point extraction may influence effectiveness of speech communication. However, if the recognition of feature points were 100% reliable (i.e. the number of recognized feature points were equal to the number of segments in the FECT), the whole algorithm would not be computationally expensive: the number of passes consisting in comparison of segments would be simply n^2. In reality, due to the noise and image distortions, some points of image contour can be falsely recognized as feature points. Therefore the number of detected feature points may be much greater. The redundant feature points make the algorithm for contour classification computationally very expensive: the number of passes n_p would be:

$$n_p = n^2 \times n! \,/\, m! \,/\, (n-m)! \tag{1}$$

In order to evaluate the reliability of feature point detection, an experiment has been conducted using the image shown in Fig. 3. FECT corresponding to the wrenches shown in the image is following:

```
#cnt wrench
     line:a
     arc:b/bend:right,angle:alpha = 1,45
     arc:c/bend:left,angle:beta,radius:r
     go:left/angle:gamma = 90,150
     line:d/length:l
     go:right/angle:90
     line:e
     go:right/angle:90
     line:f/length:l
     go:left/angle:delta = 90,150
     arc:g/bend:left,angle:epsilon
     arc:h/bend:right,angle:-alpha + beta + gamma + delta +
epsilon-360
     line:i
     arc:j/bend:right,angle:phi = 1,45
     arc:k/bend:left,angle:180 + 2*phi
     arc:l/bend:right,angle:phi
```

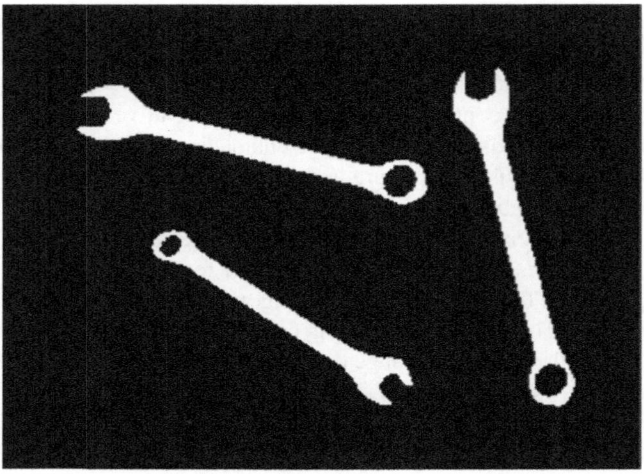

Fig. 3. Binarized image of tools handled by robot

As can be seen, this FECT consists of 12 segments. After feature point extraction algorithm was performed, the number of detected points for three objects in the image was 14, 15, 17 respectively. The biggest number 17 corresponds to 5 redundant points. In this case the number of passes would be:

$$n_p = 17^2 \times 17! / 12! / 5! = 1788332 \quad (2)$$

It means that – in comparison with 100% reliable feature point detection – the time of contour recognition would be more than 1.5 mln times greater. This could substantially influence the time of robot reaction to voice command. In order to cope with this problem, two different solutions can be applied: either to increase the feature point detection reliability or to find a way to perform contour matching without necessity to search for the feature points. Therefore the second solution employs an algorithm based on ANN.

4 Artificial Neural Networks in Image Recognition

Neural networks are characterized by the possibility of inference based on a set of different, unrelated data and the ability to learn on examples and generalize acquired knowledge. Appropriate teaching of the network allows objects to be identified based on incomplete data [28]. One of very popular areas where ANNs are applied, is image recognition. For example, Elsalamony [29] used a neural network to classify the images of red blood cell distortions in order to diagnose anemia. The optimized image was input to an algorithm separating individual red blood cells and the quality of such blood cells was checked based on shape indicators. Each deviation from the circle was analyzed and entered into statistical data. After passing through all the tested samples, the final result of healthy (diseased) blood cells was decided. The neural network consisted of three layers. A simple network structure (six inputs, ten neurons in a hidden layer and one output signal) in combination with a classification algorithm, allowed obtaining results with very high accuracy (up to 100%).

A very popular way of image classification consists in application of so called convolutional neural network (CNN) which is a class of deep neural network. The level of advancement of the neural network structure must be tailored to the task the network faces. As a rule, increasing the number of neurons and layers results in increased accuracy, but this also involves higher requirements and computation time. Cires et al. [30] stated that complexity of neural network structure has an impact on quality of results. The study consisted of classification of shapes depicted on the photos. When using a network consisting of a big number of layers (they applied 10 layers), the error has been significantly reduced. Other studies on the use of neural networks in image recognition confirm that this method is very useful, providing large possibilities and relatively low level of complexity. Researchers use neural networks in various fields: medicine [28], agricultural [31, 32] and others [33]. The experiences gained confirm that there is no one good neural network structure, all tests were preceded by in-depth analyzes of research objects as well as analyzes regarding algorithms for controlling of neural networks.

However, in our research we did not intend to apply artificial neural networks in a standard way used in existing image recognition systems. Our aim was to combine ANNs with flexible editable contour templates in order to take advantage of strong points characterizing both methods. ANNs helped us to eliminate the phase of feature point detection which induced the matching procedure to be very time-expensive when redundant points occurred.

5 Application of ANN for FECTs

Our goal was to invent a method alternative to segment matching that would provide high robustness and would be less computationally expensive than segment matching algorithm. Faster contour recognition provides a shorter reaction time to voice commands. Therefore we decided to combine neural network with FECTs. Based on flexible editable contour templates, it is possible to automatically generate a huge number of sample contours. As a result, a big set of contour signatures is available and can be used to train the neural network. However, our approach faces the challenge consisting in the fact that contours used for network training are of various lengths. An individual FECT may describe contours of quite different dimensions. On the other hand, the number of neural network inputs must remain constant and unchanged for all contours in the image. The most obvious solution is to divide all contours into the same number of short sections. The number of ANN inputs must be, of course, equal to the number of those sections. Contour signature is then calculated for each individual section as a mean value of curvature for this section. This value constitutes input to ANN.

However, the next problem arises: possible big variability of signature value within one sections. This may be of particular importance when the contour contains very short consecutive segments characterized by different angular orientation Taking into account only the mean value of the signature could result in neglecting little but important details. Such details may be sometimes crucial when two objects belonging to different classes can be distinguished using those details only. For example, the objects shown in Fig. 1a are two different tools, but the difference in shape occurs at one location only. Generally, their shapes are very similar, or almost the same, hence the system must be very precise and pay special attention to such type of differences. In order to avoid this problem, three neural network inputs are dedicated to each contour section. Three parameters of the section are delivered to those inputs:

- mean value of curvature,
- maximum value of curvature,
- minimum value of curvature.

The algorithm that generates input data for neural network performs the following steps:

- Divide the contour into pre-defined number of sections n_s
- For each section perform the following steps:
  - Calculate contour curvature for all points belonging to the section
  - Calculate the mean value of the curvature
  - Calculate the maximum value of the curvature
  - Calculate the minimum value of the curvature
  - Deliver the mean, maximum, and minimum values of the curvature as input values to three consecutive network entries.

The number of sections can be calculated as:

$$n_s = max\{\,(l_{ci}/\,l_{si}) \times k : i = 1..m\,\} \tag{3}$$

where:

- k is constant factor (e.g. 4) determining number of sections corresponding to the shortest contour segment
- m is the number of templates
- l_{ci} is the maximum possible contour length for template i
- l_{si} is the minimum possible contour length for template i

During network training this algorithm must be repeated for all possible combinations of segment dimensions provided by FECT. During image contour classification it must be run only once for each contour derived from the image. Comparing with segment matching algorithm, the algorithm based on ANN has almost constant, predictable computational time complexity. Of course, the use of ANN itself is the source of big computational complexity, but only in the phase of network training (which is performed off-line). Once the network has been trained, the algorithm for contour classification can be run in order to classify the image contour effectively.

In order to provide comparison between both algorithms (segment matching algorithm and algorithm based on ANN), an experiment has been conducted. We used 6 templates with various numbers of feature points. Neural network consisted of 4 layers.

The results are summarized in Table 1 which contains the times (in seconds) needed by both algorithms to classify contours. As can be seen, the ANN-based algorithm is generally much faster.

Table 1. Contour classification time for both methods

Template	Number of feature points	Recognition time for matching algorithm	Number of ANN inputs	Recognition time for ANN-based algorithm
1	10	7.95 s	220	0.03 s
2	17	1.92 s	500	0.17 s
3	10	0.42 s	350	0.09 s
4	12	3.18 s	460	0.14 s
5	7	0.15 s	300	0.06 s
6	8	0.15 s	320	0.06 s

For both methods the classification time is determined by different factors. For segment matching algorithm, the redundant feature points have crucial influence. Their number depends mainly on image quality. The influence of FECT structure (e.g. the number of template feature points) is minor. It is confirmed by comparison of results for object 1 and 3 in Table 1. The number of feature points is equal for both templates but the recognition times differ significantly.

In contrary, the time needed to classify contours by the ANN-based algorithm is mainly determined by number of network inputs i.e. by the number of sections the contour is divided into. It depends on FECT and not on image quality. Therefore ANN is not only faster but also more predictable. This is an obvious advantage from the point of view of speech communication.

6 Conclusions and Future Plans

The main contribution of research presented in this paper to the area of human-machine communication is development of a novel method for contour identification based on combination of flexible editable contour templates (FECT) and artificial neural network (ANN). This method allows to combine advantages of both FECTs and ANN. The use of FECTs provides selective flexibility of the template and fulfillment of selective precision requirements. Application of ANN results in substantially shorter and predictable recognition time. Although the main disadvantage of ANN is time-consuming network training, nevertheless the contour recognition itself is very fast. This is crucial when the image recognition system is used in combination with speech recognition system because fast reactions to voice commands are required in order to provide effective human-robot communication. Combination of both methods make it possible to quickly compare the features of the object to be recognized and, based on response from neural network, determine whether it is the object the operator asks for.

Further works include testing in a real robotized system equipped with an industrial robot, milling machine, vision system and speech recognition system. As a result, it will be possible to state with certainty the correctness of system's operation and its reliability. In addition, placing it in an industrial environment will allow to check whether the system is resistant to interference from noise or imperfect lighting. On this basis, a decision will be made on the suitability and further development paths.

References

1. Pires, J.N.: Robot-by-voice: experiments on commanding an industrial robot using the human voice. Ind. Rob. **32**, 505–511 (2005). https://doi.org/10.1108/01439910510629244
2. Zinchenko, K., Wu, C.Y., Song, K.T.: A study on speech recognition control for a surgical robot. IEEE Trans. Ind. Inform. **13**, 607–615 (2017). https://doi.org/10.1109/TII.2016.2625818
3. Rozo, L., Calinon, S., Caldwell, D.G., Jiménez, P., Torras, C.: Learning physical collaborative robot behaviors from human demonstrations. IEEE Trans. Robot. **32**, 513–527 (2016). https://doi.org/10.1109/TRO.2016.2540623
4. Maurtua, I., Ibarguren, A., Kildal, J., Susperregi, L., Sierra, B.: Human-robot collaboration in industrial applications: safety, interaction and trust. Int. J. Adv. Robot. Syst. **14**, 1–10 (2017). https://doi.org/10.1177/1729881417716010
5. Matthias, B., Kock, S., Jerregard, H., Källman, M., Lundberg, I.: Safety of collaborative industrial robots: certification possibilities for a collaborative assembly robot concept. In: Proceedings of the 2011 IEEE International Symposium on Assembly and Manufacturing, ISAM, Tampere, pp. 1–6 (2011). https://doi.org/10.1109/isam.2011.5942307
6. Zanchettin, A.M., Bascetta, L., Rocco, P.: Acceptability of robotic manipulators in shared working environments through human-like redundancy resolution. Appl. Ergon. **44**, 982–989 (2013). https://doi.org/10.1016/j.apergo.2013.03.028
7. El Makrini, I., Elprama, S.A., Van Den Bergh, J., Vanderborght, B., Knevels, A.J., Jewell, C.I.C., Stals, F., De Coppel, G., Ravyse, I., Potargent, J., Berte, J., Diericx, B., Waegeman, T., Jacobs, A.: Working with walt: how a cobot was developed and inserted on an auto assembly line. IEEE Robot. Autom. Mag. **25**, 51–58 (2018). https://doi.org/10.1109/MRA.2018.2815947
8. Rogowski, A., Skrobek, P.: Direct human-robot collaboration in welding. Przegląd Spawalnictwa **90**, 9–14 (2018). https://doi.org/10.26628/ps.v90i1.847
9. Rogowski, A.: Can we talk to a welding robot? Przegląd Spawalnictwa **88**, 5–8 (2016). https://doi.org/10.26628/wtr.v88i1.557
10. Gundogdu, K., Bayrakdar, S., Yucedag, I.: Developing and modeling of voice control system for prosthetic robot arm in medical systems. J. King Saud Univ. – Comput. Inf. Sci. **30** (2017). https://doi.org/10.1016/j.jksuci.2017.04.005
11. Jayawardena, C., Watanabe, K., Izumi, K.: Controlling a robot manipulator with fuzzy voice commands using a probabilistic neural network. Neural Comput. Appl. **16**, 155–166 (2007). https://doi.org/10.1007/s00521-006-0056-8
12. Hamdan, B., Mokhtar, K.: Face recognition using angular radial transform. J. King Saud Univ. – Comput. Inf. Sci. **30**, 141–151 (2018). https://doi.org/10.1016/j.jksuci.2016.10.006
13. Barman, A., Dutta, P.: Facial expression recognition using distance and shape signature features. Pattern Recogn. Lett., 1–8 (2017). https://doi.org/10.1016/j.patrec.2017.06.018
14. Tsarouchi, P., Athanasatos, A., Makris, S., Chatzigeorgiou, X., Chryssolouris, G.: High level robot programming using body and hand gestures. Procedia CIRP **55**, 1–5 (2016). https://doi.org/10.1016/j.procir.2016.09.020

15. Van Delden, S., Umrysh, M., Rosario, C., Hess, G.: Pick-and-place application development using voice and visual commands. Ind. Rob. **39**, 592–600 (2012). https://doi.org/10.1108/01439911211268796
16. Maurtua, I., Fernández, I., Tellaeche, A., Kildal, J., Susperregi, L., Ibarguren, A., Sierra, B.: Natural multimodal communication for human-robot collaboration. Int. J. Adv. Robot. Syst. **14**, 1–12 (2017). https://doi.org/10.1177/1729881417716043
17. Pleshkova, S., Zahariev, Z.: Development of system model for audio visual control of mobile robots with voice and gesture commands. In: Proceedings of the International Spring Seminar on Electronics Technology, pp. 1–4 (2017). https://doi.org/10.1109/isse.2017.8000979
18. Tasevski, J., Nikolić, M., Mišković, D.: Integration of an industrial robot with the systems for image and voice recognition. SERBIAN J. Electr. Eng. **10**, 219–230 (2013). https://doi.org/10.2298/SJEE1301219T
19. Karczmarek, P., Kiersztyn, A., Pedrycz, W., Dolecki, M.: An application of chain code-based local descriptor and its extension to face recognition. Pattern Recogn. **65**, 26–34 (2017). https://doi.org/10.1016/j.patcog.2016.12.008
20. Prematilake, C., Ellingson, L.: Evaluation and prediction of polygon approximations of planar contours for shape analysis. J. Appl. Stat. **45**, 1227–1246 (2018). https://doi.org/10.1080/02664763.2017.1364716
21. Elsalamony, H.A.: Detection and matching of object using proposed signature. In: Emerging Trends in Image Processing, Computer Vision, and Pattern Recognition, pp. 583–596. Elsevier Inc. (2015). https://doi.org/10.1016/b978-0-12-802045-6.00038-7
22. Elsalamony, H.A.: Automatic object detection and matching based on proposed signature. In: Proceedings of International Conference on Audio, Language and Image Processing, Shanghai, pp. 68–73 (2012). https://doi.org/10.1109/icalip.2012.6376588
23. Tomakova, R., Komkov, V., Emelianov, E., Tomakov, M.: The use of Fourier descriptors for the classification and analysis of peripheral blood smears image. Appl. Math. **8**, 1563–1571 (2017). https://doi.org/10.4236/am.2017.811114
24. Cai, Z.X., Gu, M.Q.: Traffic sign recognition algorithm based on shape signature and dual-tree complex wavelet transform. J. Cent. South Univ. **20**, 433–439 (2013). https://doi.org/10.1007/s11771-013-1504-0
25. Attalla, E., Siy, P.: Robust shape similarity retrieval based on contour segmentation polygonal multiresolution and elastic matching. Pattern Recogn. **38**, 2229–2241 (2005). https://doi.org/10.1016/j.patcog.2005.02.009
26. Rogowski, A., Skrobek, P.: Object identification for task-oriented communication with collaborative industrial robots. Sensors **20**, 1773–1792 (2020). https://doi.org/10.3390/s20061773
27. Bielecka, M.: Syntactic-geometric-fuzzy hierarchical classifier of contours with application to analysis of bone contours in X-ray images. Appl. Soft Comput. **69**, 368–380 (2018). https://doi.org/10.1016/j.asoc.2018.04.038
28. Firpi, H.A., Ucar, D., Tan, K.: Discover regulatory DNA elements using chromatin signatures and artificial neural network. Bioinformatics **26**, 1579–1586 (2010). https://doi.org/10.1093/bioinformatics/btq248
29. Elsalamony, H.A.: Healthy and unhealthy red blood cell detection in human blood smears using neural networks. Micron **83**, 32–41 (2016). https://doi.org/10.1016/j.micron.2016.01.008
30. Cires, D.C., Meier, U., Schmidhuber, J.: Multi-column deep neural networks for image classification. https://arxiv org/pdf/1202.2745.pdf. Accessed 10 Mar 2019
31. Singh, K.: SVM-BDT PNN and Fourier moment technique for classification of leaf shape. Int. J. Signal Process. Image Process. Pattern Recogn. **3**, 67–78 (2011)

32. Selvaraj, A., Shebiah, N., Ananthi, S., Varthini, S.: Detection of unhealthy region of plant leaves using neural network. Agric. Eng. Int.: CIGR J. **15**, 211–217 (2013)
33. Thanda, A., Venkatesan, S.M.: Audio visual speech recognition using deep recurrent neural networks. In: Multimodal Pattern Recognition of Social Signals in Human-Computer-Interaction: 4th IAPR TC 9 Workshop, pp. 98–109 (2016). https://doi.org/10.1007/978-3-319-59259-6_9

Foreground Object Segmentation in RGB–D Data Implemented on GPU

Piotr Janus, Tomasz Kryjak(✉), and Marek Gorgon

AGH University of Science and Technology, Kraków, Poland
{piojanus,tomasz.kryjak,mago}@agh.edu.pl

Abstract. This paper presents a GPU implementation of two foreground object segmentation algorithms: Gaussian Mixture Model (GMM) and Pixel Based Adaptive Segmenter (PBAS) modified for RGB–D data support. The simultaneous use of colour (RGB) and depth (D) data allows one to improve segmentation accuracy, especially in case of colour camouflage, illumination changes and shadow occurrence. Three GPUs were used to accelerate computations: embedded NVIDIA Jetson TX2 (Maxwell architecture), mobile NVIDIA GeForce GTX 1050m (Pascal architecture) and efficient NVIDIA RTX 2070 (Turing architecture). Segmentation accuracy comparable to previously published works was obtained. Moreover, the use of a GPU platform allowed us to get real-time image processing. In addition, the system has been adapted to work with two RGB–D sensors: RealSense D415 and D435 from Intel.

Keywords: Foreground object segmentation · Background subtraction · RGB–D · GPU · GMM · PBAS · Intel RealSense

1 Introduction

Foreground object segmentation is one of the most important components of modern Advanced Video Surveillance Systems (AVSS). It can be used in a variety of vision systems such as object detection and tracking, as well as human behaviour analysis. Moreover, it is a key element of applications like abandoned luggage detection or forbidden zone protection [5].

The simplest group of foreground object detection algorithms is based on subtracting subsequent frames from a video sequence. More advanced approaches involve the so-called background modelling. Usually, for each pixel a dedicated model is assigned that describes the background appearance in a given location. Then, depending on the used algorithm, the new pixel value is compared to the background model and classified as foreground, background and sometimes also as shadow. The model is usually updated to incorporate changes in the scene like slow or fast light variations and movement of background objects e.g. a chair. The paper [3] provides a complete survey of the traditional and recent approaches in background modelling. Available resources, datasets and libraries are also presented.

A. Bartoszewicz et al. (Eds.): PCC 2020, AISC 1196, pp. 809–820, 2020.
https://doi.org/10.1007/978-3-030-50936-1_68

However, some situations are difficult to handle by the proposed approach. Examples involve: bootstrapping (model initialization), colour camouflage (object are very similar to the background), illumination changes, intermittent motion (stopped or removed objects), background motion (like flowing water) and shadows – a more comprehensive discussion on this issue can be found in [11].

Some of the mentioned issues can be solved with the use of a depth sensor. Information about the scene geometry can be obtained in several ways. The most straightforward is passive stereovision – the use of two or more cameras and appropriate image processing algorithms allows to obtain a 3D representation of the scene. Recently active sensors are gaining more and more attention: LiDAR, Time-of-Flight (ToF) cameras, structured light 3D scanners or active IR (InfraRed) stereo. The last technology uses an IR emitter and one or two IR cameras. The emitter displays an irregular pattern of dots. Then the IR camera registers the infrared light reflected from the subjects. Finally, the use of advanced image processing allows to estimate the depth map. This approach is used in Microsoft Kinect (mono IR) and Intel RealSense (stereo IR) devices.

On the other hand, the use of depth information causes problems in specific situation like: depth camouflage (objects very close to the background), depth shadows, transparent or semi-transparent materials (like windows), out of sensor range – a more detailed discussion can be found in [11]. Therefore the majority of approaches involve combine colour (RGB) with depth (D) data. This type of image is usually called RGB–D (or RGBD).

In this paper two commonly used foreground segmentation algorithms (FOS) Gaussian Mixture Model (GMM) and Pixel–Based–Adaptive–Segmenter (PBAS) have been modified to include depth information. GMM is one of the most popular FOS algorithm and is available in Matlab and OpenCV library. PBAS represents a good trade-off between computational complexity and performance. We used the *Intel RealSense* D415 and D435 sensors for colour and depth image acquisition. Three computing platforms were considered: NVIDIA Jetson TX2 (embedded), NVIDIA GeForce GTX 1050m (mobile) and NVIDIA RTX 2070 (high-end). GPU acceleration allowed to obtain real-time RGD–D data processing. Moreover, we evaluated our approach on a commonly used and publicly available dataset.

The remainder of this paper is organized as follows. In Sect. 2 previous works related to use of RGB–D sensor for foreground object segmentation are briefly discussed and papers concerning GMM and PBAS acceleration using GPU are presented. Section 3 describes the proposed versions of GMM and PBAS methods. In Sect. 4 the designed heterogeneous system is presented. The evaluation of the proposed algorithms is discussed in Sect. 5. The paper ends with a conclusion and discussion of future research directions.

2 Previous Works

Over the years, several solutions for foreground object segmentation with the use of a RGB–D sensor have been proposed. An excellent and quite recent (2018) review is presented in [11]. Here we limit our discussion to papers that use algorithms comparable with our approaches i.e. GMM and PBAS.

One on the first works on using RGB–D data in foreground segmentation was [4]. The approach was originally applied to stereovision data. It was based on the Gaussian Mixture Models (also known as Mixture of Gaussians – MoG) concept. The authors assumed that colour and depth features are independent. They also divided the depth data into "valid" and "invalid". In the first case, the depth was used to estimate the background, as usually it is behind the foreground (an exception are occlusions). In the second case, the typical colour-based GMM algorithm was used. During segmentation the depth data was used to influence the colour-based matching criterion. For reliable depth data, the criterion was relaxed to avoid camouflage errors. In the other case, the criterion was harder, to avoid segmentation errors due to shadows and illumination changes. No information about processing time was provided.

The GMM approach was also described in [18]. It was originally applied to data obtained by a ToF sensor, which provided depth data and a near infrared image. In contrast to [4], the authors used two separate models for depth and IR image and obtained two foreground masks. The final segmentation was based on fusion of these masks with additional information about depth gradient to separate overlapping foreground objects. No information about processing time was provided.

Another GMM based algorithm adapted to work with a RGB–D sensor was proposed in [16]. The authors used two separate models and combined their output to obtain the final segmentation result. The evaluation was done on sequences recorded by the authors. It showed that the proposed approach works well in case of colour camouflage. No information about processing time was provided.

A different algorithm – ViBE (Visual Background Extractor) – was used in the work [10]. It should be noted that ViBE is the predecessor of PBAS. It was applied to ToF data and two separate models were used. Moreover, motion information was also included. The obtained foreground masks were combined and post-processed with morphological operations. No information about processing time was provided.

In the paper [12] an algorithm, named SCAD, based on ViBE and combination of colour, texture and depth information was proposed. The final segmentation was obtained using graph cuts. The solution was implemented in C++ as a single thread application on a Intel Xeon @ 3.7 GHz with 32 GB RAM. The system processed, on average, 640×480@1.95 frame per second.

A similar approach was presented in [20]. The authors fused segmentations results from two ViBE models: for colour and depth. The solution was implemented in C++/OpenCV. On a Intel Core Duo 2 CPU E7500 platform with 2.00 GB of RAM, a 640×480@30 fps performance was obtained.

There are several published papers on GPU acceleration of foreground object detection algorithms like: GMM, ViBE and PBAS. In [13] GMM was implemented on a low end GPU – GeForce 9600GT. This allowed to process up to 50

HD frames per second. In [7] a NVIDIA Tesla K20 GPU was used to accelerate PBAS. For a 320 × 240 video 646 fps was reported. In the paper [19], a GPU implementation of the ViBE algorithm was presented. The algorithm was tested on a PC equipped with *Intel Core Quad Q8400* and *Nvidia GTX 650Ti*. For a 960 × 540 video stream resolution the achieved performance is 1.8 fps for CPU and 26 fps for the GPU implementation. In [9] another variant of the GMM algorithm with connected component labelling and morphological operations for post–processing is described. The authors presented a PU implementation which achieves significant speed-ups of 15 times for the GMM algorithm comparing to *Intel Xeon processor*. The proposed system is able to process 22.3 frames per second for HD video stream.

3 The Considered Algorithms

In this research, we implemented two different foreground object segmentation algorithms. The standard RGB version were modified to take benefit of depth data. The first algorithm is an extended version of *Gaussian Mixture Models* [17], while the second one is a modification of *Pixel Based Adaptive Segmenter* [6]. Both are similar regarding the background model concept. It is independent for each pixel and dynamically updated after every frame. In the following subsections a description of both methods is presented.

3.1 GMM Algorithm with RGB–D Data

Gaussian Mixture Models (GMM) [17] is one of the most commonly used method for background modelling. In this approach each pixel is modelled by k Gaussian distributions characterized by three parameters (ω, μ, σ^2), where ω is the normalized weight (range 0–1) of the Gaussian distribution, μ is the means vector of each colour component of a particular pixel – (r_{mean}, g_{mean}, b_{mean}) is case of RGB, and σ^2 is the variance of given Gaussian distribution – a single value is used for each colour component. Usually, it is assumed that RGB components are independent, which allows to use three σ^2 values instead of a covariance matrix. In this work, a version partially based on [16] and the open source image processing library OpenCV was applied. As a detailed description of the method is quite long and available in multiple research papers (starting with [17]), we only discuss the adaptation to RGB–D data.

There are two options when applying a RGB background model to RGB–D data. The first is to incorporate the depth data into the model. In the context of GMM this results in extending the μ and σ^2 parameters. The second is to use two different models and then combine the segmentation results. In the presented research we followed the second approach. During our experiments we found out that such a solution provides better protection against noise from depth image and illumination changes.

```
ComputeProbability(pixel_in, model, classification_result) {
  p = sη(pixel_in, model.μ, model.σ);

  if (classification_result == BACKGROUND) {
    if (p < 255)
      return p;
    return 0;
  }
  if (p < 255) {
    return 255;
  }
  return 255-p;
}
```

```
PixelClassification(input_rgb, input_depth, model) {
  classification_result = GmmClassification(input_rgb, model);
  p_rgb = ComputeProbability(input_rgb, model, classification_result);

  if (IsDepthValueValid(input_depth)) {
    p_depth = ComputeProbability(input_depth, model, classification_result);
    if (p_rgb*p_depth > 255)
      return BACKGROUND;
    return FOREGROUND;
  }
  return classification_result;
}
```

Fig. 1. GMM – pseudocode of computing probability and classification

The classification procedure is based on a probability density function, which depends on the pixel value X_t at time t:

$$\eta(X_t, \mu, \sigma) = \frac{1}{2\pi\sigma} e^{-\frac{d(X_t,\mu)^2}{2\sigma}} \quad (1)$$

The pseudo code of this operation is presented in Fig. 1. The s parameter is used for scaling the probability density and its default value is 10000. The process of computing the probability factor is the same for both models. Then the product of two values is computed and final classification is made according to the code. In addition the depth model is considered only if the depth value obtained from sensor is valid (greater than 0). Otherwise only RGB classification is performed. In the final implementation the following algorithm parameters are used: number of Gaussians – 7 for RGB model and 3 for depth, learning rate – 0.001, Gaussian parameters are represented as a 64-bit float number.

3.2 PBAS Algorithm with RGB–D Data

The Pixel Based Adaptive Segmenter (PBAS) algorithm [6] is an extension of the Visual Background Extractor (ViBE) method proposed in [1]. Both algorithms use a similar background model, however PBAS involves a more advanced foreground classification and model update procedure. As a detailed description of the method is quite long and available in [6], we only discuss here the adaptation to RGB–D data.

In the PBAS method the background model is composed of two parts. The first one is a buffer of N samples from the analysed video sequence. In our approach a single sample consists both of a RGB and a depth value. Since in segmentation and model update each component is processed separately, the addition of depth data is straightforward. Moreover, as in case of the GMM algorithm, depth data is utilized only if depth value from the sensor is valid, otherwise classification is based only on RGB data. In the final implementation a model containing $N = 20$ samples is used.

4 Algorithms Implementation

4.1 Hardware Setup

We used two RGB–D video sources. To compare our solution with other approaches, a publicly available dataset was used – [15]. Also to demonstrate a vision system working in real-time we used Intel RealSense Depth Cameras D415 and D435. Both sensors provide a Full HD resolution (1920 × 1080) for the RGB image and HD resolution (1280 × 720) for the depth map. They are able to distinguish objects in range 10 cm to 10 m from the camera lens. A well known alternative to *RealSense* sensors is Microsoft Kinect. It has been used for a long time as an entry level device for RGB–D image analysis, also in the used dataset [15]. Unfortunately it was discontinued by Microsoft in 2017 and it is no longer available on the market. Moreover, its technical specification is significantly inferior than Intel RealSense sensors, as the maximum resolution for RGB camera and depth map is only 640 × 480@30 Hz. Nowadays D415 and D435 devices are affordable, provide a reasonable price to value ratio and are used in a range of robotic applications like drone navigation [2].

For GPU implementation three different platforms were used. The first one was NVIDIA Jetson TX2 – an embedded GPU, equipped with a 64-bit ARM Cortex A57 CPU and a NVIDIA Maxwell GPU with 256 CUDA cores. The second platform was a laptop with a Intel Core i7–7700HQ (4 cores/8 threads @ 2.8 GHz) CPU and a NVIDIA GeForce GTX 1050m GPU based on Pascal architecture. The third was a PC equipped with a Intel Core i7–9700k (8 cores @ 4.5 GHz) CPU and a NVIDIA RTX 2070 GPU (Touring architecture).

To accelerate RGB–D data processing the CUDA (*Compute Unified Device Architecture*) platform was used. It is developed by NVIDIA and can be used only with GPUs based on their architecture. The CUDA itself is a parallel computing platform with an API, which allows to use NVIDIA graphics processing units for general purpose computing. Thanks to this approach, it is straightforward to port our implementation to different computing platforms like laptops, PCs or embedded GPUs. The only requirement is the use of an NVIDIA GPU.

4.2 Application Architecture

The distribution of computing tasks between CPU and GPU is one of the key elements when implementing a heterogeneous system. The host (CPU) is responsible for image acquisition (from the sensor or hard drive) and copying image data to the shared GPU memory (DRAM). Moreover, the memory allocation for the background model is also done by the host. The communication between host and GPU is done over a PCI bus. A general overview of this architecture is shown in Fig. 2.

The GPU architecture allows to process each pixel in parallel, independently, using separate threads. The output of the system is a binary mask containing foreground objects. This mask has to be copied from the GPU DRAM memory to CPU RAM and then forwarded to an external display or stored on a hard

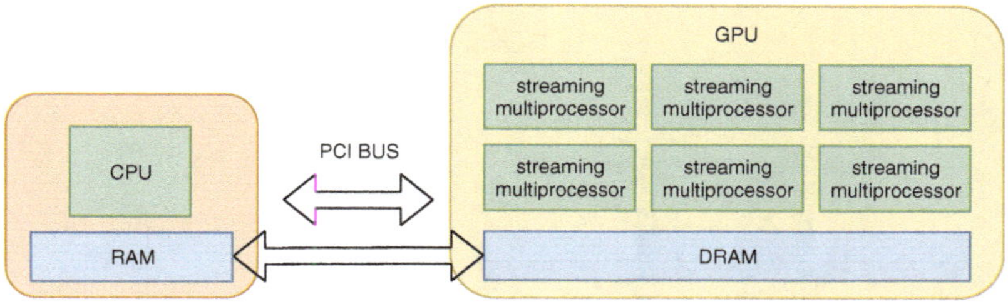

Fig. 2. Communication between host (CPU) and GPU

Table 1. Performance

	480p/480p	720p/480p	720p/720p	1080p/720p
Jetson TX2	28 fps	11 fps	9 fps	6 fps
i7 7700hq + GTX 1050m	30 fps	18 fps	16 fps	10 fps
i7 9700k + RTX 2070	30 fps	30 fps	30 fps	30 fps

drive. The flow of exchanging data and executing operations on the CPU and GPU sides is presented in Fig. 3.

The CPU part was implemented in *C++* language. Image acquisition was done with the use of a dedicated SDK provided by Intel for RealSense RGB–D sensors. It allows to acquire both RGB images and depth maps in real-time with a maximum frequency limited to 30 fps. The depth map is received as a 16-bit unsigned integer (a bigger number means a greater distance from the camera), so it needs to be converted to 8-bit per pixel format for compatibility with the RGB background model based on 8-bit numbers. As it was mentioned before, according to the hardware specification, the minimal range of the depth sensor is 10 cm, while the maximum is 10 m. After rescaling from the range 0–65535 to 0–255, the depth accuracy will be about 4 cm, which is enough for the considered vision system. Finally each pixel is represented by 32 bits, 8 bits for each colour component and depth.

4.3 Performance

Computing performance on particular GPUs has been measured for different resolutions: 480p/480p, 720p/480p, 720p/720p, 1080p/720p, where values represent RGB camera and depth resolution respectively. Results are presented in Table 1 and a sample RGB image, depth map and segmentation output is shown in Fig. 4. It should be noted, that similar results were obtained for the GMM and PBAS methods.

Analysing the obtained results, it can be noticed that only the RTX 2070 GPU is able to provide real-time processing for all considered resolutions. It should be emphasized that for the RealSense sensor, both the RGB image and

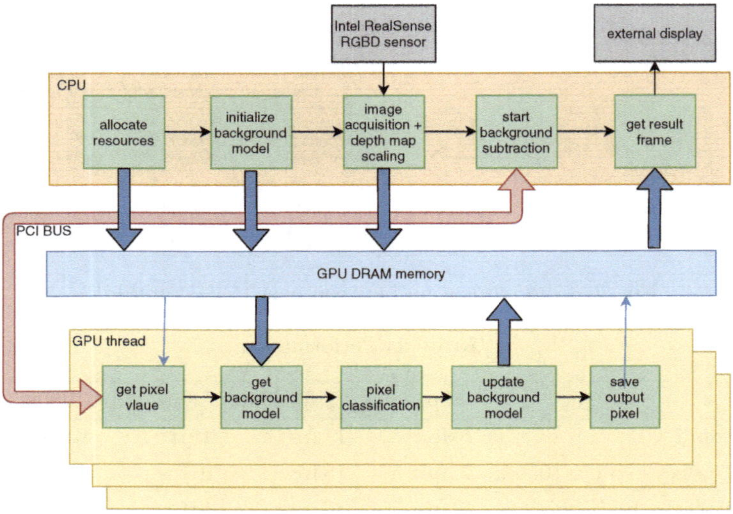

Fig. 3. Exchanging data between host and GPU – flow diagram

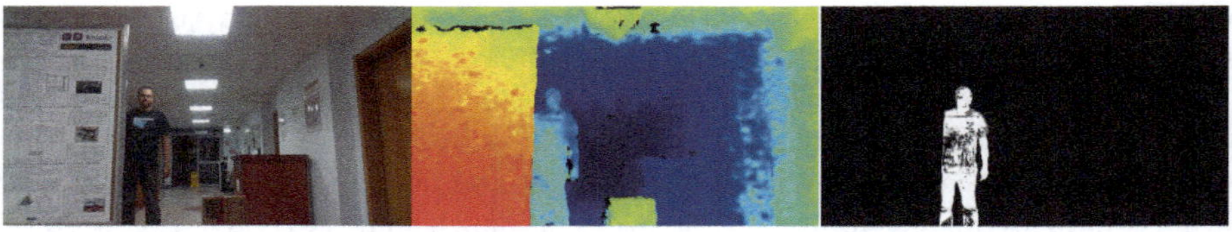

Fig. 4. Exemplary RGB, depth input and segmentation output for the GMM algorithm

the depth map are acquired 30 times per second. Thus, 30 fps is for the considered system the maximum value.

In the case of mobile GTX 1050m and embedded Jetson TX2 GPUs, only for 480p resolution it is possible to obtain real-time image processing. However, it is worth noting that these are definitely more energy efficient platforms than RTX GPUs with the Touring architecture. The maximum power consumption for the considered platforms is 7.5 W, 75 W and 215 W respectively.

A decrease in performance can be seen when different resolutions for the depth map and image are used. This is due to the depth map upscaling to the same resolution as the RGB image.

5 Evaluation

Two experiments were performed to test the implemented algorithms. In the first one, short videos were recorded using the Intel RealSense D435 sensor. They contained situations when the object's colour was similar to the background (colour camouflage) and when the object was close to the background

Table 2. Evaluation results on sequences registered with RealSense

Algorithm	PWC	FNR	FPR	Si
GMM	5.85	0.1287	0.0423	0.51
GMM + RGBD	4.21	0.0812	0.0323	0.61
PBAS	1.23	0.0398	0.0123	0.87
PBAS + RGBD	1.20	0.0289	0.113	0.89

(depth camouflage). For each recording a ground truth has been prepared. In the second experiment, test sequences from the SBM RGBD [15] dataset were used. This allowed to compare the proposed methods with those described in the literature [11].

A typical evaluation methodology was used. Based on the comparison of the segmentation result and the ground-truth mask, the following factors were determined:

TP: the number of pixels correctly classified as the foreground (true positive),
TN: number of pixels correctly classified as background (true negative),
FN: number of pixels incorrectly classified as a background (false negative),
FP: the number of pixels incorrectly classified as the foreground (false positive).

Then four quality indicators were determined:

1. Percentage of Wrong Classifications (PWC): $100(FN + FP)/(TP + FN + FP + TN)$
2. False Negative Rate (FNR): $FN/(TP + FN)$
3. False Positive Rate (FPR): $FP/(FP + TN)$
4. Similarity (Si): $TP/(TP + FP + FN)$

The test results for the sequences registered with the RealSense sensor are presented in Table 2. On their basis, it can be concluded that in the case of the GMM algorithm, the obtained results are definitely better when using the RGB–D sensor. For the PBAS method, adding depth data only slightly improved the segmentation.

In the next experiment, four sequences from the SBM RGBD dataset belonging to different categories were used: illumination changes, colour camouflage, depth camouflage and shadows. We decided to select the four most common sequences from the dataset, because other algorithms were usually tested on this subset. The obtained results are presented in Table 3 and compared with previously proposed solutions *MoG4D* [4], *ViBeRGB+D* [10], *MoGRGB+D* [18] (evaluation results taken from [11]). The selected algorithms are comparable to ours i.e. with similar background model and computational complexity. It should also be noted that the SBM RGBD set contains more sequences.

The results presented in Table 3 indicate that using depth information allows to obtain better performance in each of the considered categories. The biggest

Table 3. Evaluation on the SBM RGBD dataset

Sequence	Method	PWC	FNR	FPR	Si
Illumination changes	GMM	4.49	0.0248	0.4783	0.73
	PBAS	3.75	0.0212	0.0412	0.75
	GMM + RGBD	3.60	0.0183	0.0376	0.78
	PBAS + RGBD	3.32	0.0131	0.0319	0.80
	MoG4D	1.93	0.0063	0.0209	0.79
	ViBeRGB+D	12.39	0.0065	0.1385	0.44
	MoGRGB+D	2.03	0.1701	0.0016	0.79
Color camouflage	GMM	18.89	0.8001	0.0154	0.19
	PBAS	17.87	0.7702	0.0112	0.19
	GMM + RGBD	10.01	0.0160	0.0124	0.72
	PBAS + RGBD	9.02	0.0125	0.0930	0.76
	MoG4D	3.49	0.0038	0.0613	0.91
	ViBeRGB+D	6.94	0.0017	0.1269	0.81
	MoGRGB+D	38.47	0.8287	0.0075	0.22
Depth camouflage	GMM	7.24	0.5108	0.4734	0.26
	PBAS	6.91	0.4912	0.0438	0.31
	GMM + RGBD	7.22	0.5001	0.0465	0.27
	PBAS + RGBD	6.89	0.4832	0.0435	0.32
	MoG4D	2.11	0.1525	0.0131	0.61
	ViBeRGB+D	9.31	0.0548	0.0955	0.30
	MoGRGB+D	3.57	0.6087	0.0009	0.32
Shadows	GMM	14.60	0.6754	0.0603	0.23
	PBAS	10.24	0.33	0.04	0.32
	GMM + RGBD	9.12	0.1409	0.0412	0.43
	PBAS + RGBD	8.99	0.1023	0.0298	0.46
	MoG4D	3.94	0.0059	0.0450	0.77
	ViBeRGB+D	7.15	0.0001	0.0834	0.66
	MoGRGB+D	3.43	0.2351	0.0008	0.75

difference can be seen in the case of colour camouflage. As expected, in this situation depth information gives the greatest benefits. A clear improvement was also seen in the sequence containing a lot of shadows. In the case of lighting changes and depth camouflage, the benefits of using depth maps were not so impressive.

We noticed, that the PBAS algorithm allows to obtain, depending on the test sequence, slightly or clearly better results than the GMM method. Among the analysed methods, the best results are obtained by the *MoG4D* algorithm, most likely due to a more advanced method of analysing incorrect depth values.

The methods implemented in this paper obtain comparable results to *MoGRGB+D* and *ViBERGB+D*.

6 Conclusion

In this paper the implementation of GMM and PBAS algorithms adapted to RGB–D data has been presented. In both cases, hardware acceleration with CUDA parallel computing platform was used. The system was launched on three GPUs with different levels of performance from embedded Jetson TX2, through GTX 1050m, ending with RTX 2070 with Touring architecture. The last of the mentioned platforms allowed to obtain real-time processing for 1080p data (30 fps). The performed evaluation showed that using the RGB–D sensor provides an increase in segmentation accuracy. As expected, the largest improvement was reached for the "colour camouflage" case, when objects have similar colour to the background.

As part of future work, the proposed algorithms could be improved by adding more advanced fusion of RGB and depth data, as well as detection of static objects. For example, the approach proposed in [8] could be applied. In addition, it is worth to consider preparing a set of sequences registered with various sensors: Kinect (like SBM RGBD), RealSense, a stereo camera and a ToF sensor. This would allow to compare different algorithms on different RGB–D data and evaluate which solution is best for foreground object segmentation. Another research direction could be the acceleration of RGB–D algorithms using FPGA devices, as this could potentially allow real-time processing of a stream with a resolution of 1080p with significantly lower power consumption than the RTX 2070 GPU.

Acknowledgements. The work presented in this paper was supported by the AGH University of Science and Technology project no. 16.16.120.773.

References

1. Barnich, O., Van Droogenbroeck, M.: ViBe: a powerful random technique to estimate the background in video sequences. In: 2009 IEEE International Conference on Acoustics, Speech and Signal Processing. IEEE (2009)
2. Campos-Macías, L., Aldana-López, R., de la Guardia, R., Parra-Vilchis, J.I., Gómez-Gutiérrez, D.: Autonomous navigation of MAVs in unknown cluttered environments. J. Field Robot. (2020). https://doi.org/10.1002/rob.21959. ISSN 1556-4967
3. Garcia-Garcia, B., Bouwmans, T., Silva, A.J.R.: Background subtraction in real applications: challenges, current models and future directions. Comput. Sci. Rev. **35** (2020). https://doi.org/10.1016/j.cosrev.2019.100204
4. Gordon, G.G., Darrell, T., Harville, M., Woodfill, J.: Background estimation and removal based on range and color. In: Proceedings of the 1999 Conference on Computer Vision and Pattern Recognition (CVPR 1999), Ft. Collins, CO, USA, pp. 2459–2464 (1999)

5. Guler, P., Emeksiz, D., Temizel, A., Teke, M., Temizel, T.T.: Real-time multi-camera video analytics system on GPU. J. Real-Time Image Proc. **11**(3), 457–472 (2016). ISSN 1861–8219
6. Hofmann, M., Tiefenbacher, P., Rigoll, G.: Background segmentation with feedback: the pixel-based adaptive segmenter. In: 2012 IEEE Computer Society Conference on Computer Vision and Pattern Recognition Workshops. IEEE (2012)
7. Karahan, Ş., Sevilgen, F.E.: CUDA implementation of the pixel based adaptive segmentation algorithm. In: 2015 23rd Signal Processing and Communications Applications Conference (SIU), Malatya, pp. 2505–2508 (2015). https://doi.org/10.1109/SIU.2015.7130393
8. Kryjak, T., Komorkiewicz, M., Gorgon, M.: Real-time foreground object detection combining the PBAS background modelling algorithm and feedback from scene analysis module. In. J. Electron. Telecommun. **60**(1), 61–72 (2014)
9. Kumar, P., Singhal, A., Mehta, S., et al.: Real-time moving object detection algorithm on high-resolution videos using GPUs. J. Real-Time Image Proc. **11**, 93–109 (2016). https://doi.org/10.1007/s11554-012-0309-y
10. Leens, J., Piérard, S., Barnich, O., Van Droogenbroeck, M., Wagner, J.M.: Combining color, depth, and motion for video segmentation. In: Proceedings of the Computer Vision Systems: 7th International Conference on Computer Vision Systems (ICVS 2009), Liège, Belgium (2009)
11. Maddalena, L., Petrosino, A.: Background subtraction for moving object detection in RGBD data: a survey. J. Imaging **4**, 71 (2018)
12. Minematsu, T., Shimada, A., Uchiyama, H., Taniguchi, R.: Simple combination of appearance and depth for foreground segmentation. In: Proceedings of the New Trends in Image Analysis and Processing (ICIAP 2017), Catania, Italy (2017)
13. Pham, V., Vo, P., Vu, H.T., Le, B.: GPU implementation of extended gaussian mixture model for background subtraction. In: IEEE RIVF International Conference on Computing & Communication Technologies, Research, Innovation, and Vision for the Future (RIVF), Hanoi, pp. 1–4 (2010). https://doi.org/10.1109/RIVF.2010.563400
14. Qin, L., Sheng, B., Lin, W., Wu, W., Shen, R.: GPU-accelerated video background subtraction using Gabor detector. J. Vis. Commun. Image Represent. **32**, 1–9 (2015)
15. SBM-RGBD Dataset. http://rgbd2017.na.icar.cnr.it/SBM-RGBDdataset.html. Accessed 20 Jan 2020
16. Song, Y., Noh, S., Yu, J., Park, C., Lee, B.: Background subtraction based on Gaussian mixture models using color and depth information. In: The 2014 International Conference on Control, Automation and Information Sciences (2014)
17. Stauffer, C., Eric, W., Grimson, L.: Adaptive background mixture models for real-time tracking. In: Proceedings of the 1999 IEEE Computer Society Conference on Computer Vision and Pattern Recognition (Cat. No PR00149), vol. 2. IEEE (1999)
18. Stormer, A., Hofmann, M., Rigoll, G.: Depth gradient based segmentation of overlapping foreground objects in range images. In: Proceedings of the 2010 13th International Conference on Information Fusion, Edinburgh, UK, pp. 1–4 (2010)
19. Qin, L., Sheng, B., Lin, W., Wu, W., Shen, R.: GPU-accelerated video background subtraction using Gabor detector. J. Vis. Commun. Image Represent. **32**, 1–9 (2015). https://doi.org/10.1016/j.jvcir.2015.07.010
20. Zhou, X., et al.: Improving video segmentation by fusing depth cues and the visual background extractor (ViBe) algorithm. Sensors **17**(5), 1177 (2017)

Particle Filter for Reliable Estimation of the Ground Plane from Depth Images in a Travel Aid for the Blind

Mateusz Owczarek, Piotr Skulimowski, and Pawel Strumillo(✉)

Institute of Electronics, Lodz University of Technology, Lodz, Poland
pawel.strumillo@p.lodz.pl

Abstract. This paper presents a reliable method for the segmentation of image sequences of 3D scenes used in an electronic travel aid for the blind. We propose an implementation of the particle filtering (PF) algorithm to estimate and track the orientation and position of the ground plane free of obstacles. We explain how the state vector in the PF algorithm was defined, and verify the results on a large set of indoor and outdoor sequences, shot by a moving stereovision camera. The movement of the camera is not restricted, and is capable of six degrees of freedom (DoF). We show that ground plane orientation root mean square (RMS) error estimates do not exceed 2 or 3°, for the roll and pitch angles, respectively. The overlap between the mean values for the Jaccard similarity coefficient of the detected ground-truth ground plane regions and the ground plane regions was 0.94. Although the method was developed for use in an electronic travel aid for the blind, it could also find applications in the automatic navigation of autonomous vehicles and unmanned aerial vehicles (UAVs).

Keywords: Object tracking · Particle filtering · Depth images

1 Introduction

Loss of vision deprives a person of the most important sensory modality for spatial orientation and mobility (O&M). Limited mobility affects almost every aspect of the daily life of a blind person. Primary travel aids (such as a white cane or a guide dog) can help, but only with the micro-navigation tasks such as obstacle detection and avoidance, or with solving local navigation tasks such as recognizing landmarks and shorelines. However, in a new, unfamiliar environment the visually impaired are unable independently to solve macro-navigation challenges, such as identifying a route or finding a safe path from an arbitrary current location A to the destination point B [1].

Building an electronic travel aid (ETA), whether dedicated to micro- or macro-navigation tasks, has proven to be a difficult interdisciplinary challenge [2]. The main contribution of this work is to build a simple model of 3D scenes for the purpose of presenting the environment nonvisually to the visually impaired. In particular, we present a method of particle filtering (PF), which enables reliable detection of the ground plane and recursive tracking of its orientation in a sequence of depth images of 3D scenes.

A. Bartoszewicz et al. (Eds.): PCC 2020, AISC 1196, pp. 821–833, 2020.
https://doi.org/10.1007/978-3-030-50936-1_69

2 Related Work

2.1 Electronic Travel Aids

Numerous solutions for personal assistive aids for the blind have been proposed over recent decades [3, 6]. Those solutions can generally be subdivided into two groups. The first consists of systems for solving specific, usually isolated problems, such as recognizing particular elements of the environment [4], or distinguishing colours and patterns on clothing [5]. There are also more complex solutions, such as "TapTapSee" or the crowdsource-based "Be My Eyes" and "VizWiz" applications, which use human and machine intelligence to generate detailed descriptions of scenes captured using a mobile phone camera. The second group consists of wearable systems which assist the blind and visually impaired to navigate in a known or an unknown environment, either indoors and outdoors [6]. This group can be divided into three main categories: 1) Electronic travel aids (ETAs), 2) Electronic orientation aids (EOAs), and 3) Position locator devices (PLDs). ETAs and EOAs are assistive devices built with the aim of enhancing travel and mobility of the visually impaired and blind people by providing information about the environment by means of non-visual representation, whereas PLDs use GPS receivers and an appropriate interfaces to aid the visually impaired in independent wayfinding task.

Most ETAs of the past and present use the concept of sensory substitution [2]. A sensory substitution system converts one modality of perception, such as visual stimulation, into a different one, such acoustic and/or haptic signals. At the Institute of Electronics in Lodz University of Technology (Poland), we have built various classes of ETA systems (www.naviton.pl). We also contributed, as a technical partner, to the recent European Horizon 2020 project "Sound of Vision," devoted to building assistive solutions for non-visual spatial perception by sound and haptics (soundofvision.net) [7].

2.2 Segmentation and Analysis of Depth Images

Different sensing techniques are employed in ETAs for reconstructing the 3D geometry of the environment. These technologies can be grouped into two major types [6]:

1. Active depth estimation techniques (laser rangefinders, structured light, time of flight devices),
2. Passive depth estimation techniques (stereovision and mono-camera solutions, employing depth of field techniques, structure from motion or shape from shading methods).

An example image of a 3D scene acquired by a Structure Sensor (an active depth camera) and the corresponding depth map are shown in Fig. 1. In an earlier study, we used an off-the-shelf stereo vision camera, the Stereolabs ZED. This stereovision comes with factory calibrated geometry of the two camera and corrected geometric distortions of individual cameras.

In advanced ETA solutions that process depth images of 3D scenes, the main image processing problem is how to ensure reliable segmentation of a ground plane (GP) region

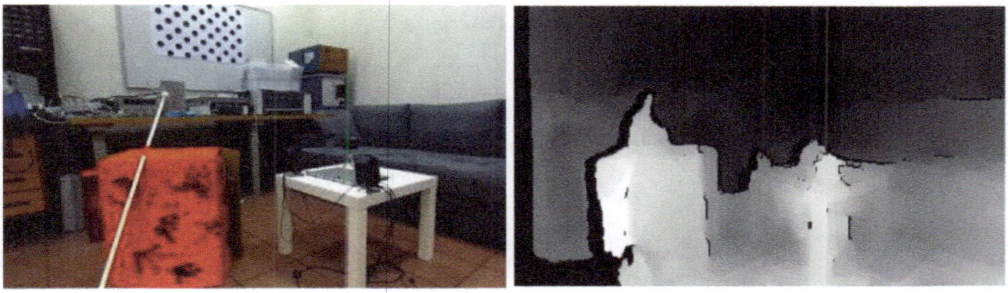

(a) (b)

Fig. 1. Sample images of an indoor 3D scene acquired using a stereovision camera: a) scene image, b) depth image (the brighter the image region, the closer the scene region to the camera).

free of obstacles. By 3D scene segmentation, further scene modelling and analysis tasks can be simplified. Retrieving information about positioning and orientation of the GP in 3D scenes is also a fundamental task in navigating robots, autonomous road vehicles, Unmanned Aerial Vehicles (UAVs), systems for retrieving digital terrain models (DTMs), and robust object tracking techniques.

A general overview of different depth segmentation methods, with an emphasis on obstacle detection and identification of the GP region, is presented in Table 1.

Table 1. Overview of depth segmentation methods for distinct categories aimed at segmentation of ground plane (GP) regions and obstacles.

Ref.	Category	Description
[8]	Edge-based	Two-tone cartoon technique combined with an edge depth map results in a cartoon-like edge depth map (the Cartoon Depth Map), which reduces the quantity of information about the imaged scene
[9]	Region-based	Ensemble Empirical Mode Decomposition (EEMD is used to decompose depth images, by filtering out any function that forms a complete and nearly orthogonal basis of the modelled image regions. The disparity map is then filtered to eliminate higher frequencies (i.e., noise and fine details) by eliminating the selected Intrinsic Mode Functions (IMF)
[10]	Region-based	Erroneously calculated and uncertain values ("black holes") in the depth map are detected, resulting in the detection of "negative obstacles" (i.e. large potholes, ditches, stairwells and other depressions)

(continued)

Table 1. (*continued*)

Ref.	Category	Description
[11]	Clustering (unsupervised)	Super-pixel segmentation and semantic labelling can be performed successfully on colour or depth images only. However, the fusion of colour and depth information (i.e., a multimodal approach) improves the detection of scene objects, including the GP
[12]	Classifiers (supervised)	In this simple method, the depth map is pre-processed by means of filling in occlusions. Under the assumption that the GP is located in the lower part of the depth image, Principal Component Analysis (PCA) is performed in the selected Region of Interest (RoI). The computed components define the normal vector of the GP and the point at which the GP originates
[13]	Neural Network	Semantic segmentation of indoor scenes by means of depth-adaptive Deep Neural Networks (DNN) have produced promising results
[14]	V-depth map processing and camera motion estimation	Analysis of the V-depth histogram, i.e. row-histogram of the depth map in search for ground surface, superpixel segmentation of depth map, post-processing by morphological operations and application of camera motion estimation for tracking the detected ground plane

The main drawback of methods [8–13] for the segmentation of depth images is that they lack inter-frame consistency among consecutive images. As a consequence, in the majority of cases detection is performed for each frame independently, without any a priori data on the position of objects in the preceding scenes being included. The main aim of this study, therefore, was to develop robust method for identifying and tracking a GP that changes its position in consecutive images. The only method that provides inter-frame consistency of the detected ground plane is the method reported in [14], which is, however, computationally complex and requires setting of large number of parameters for reliable performance for a given environment. The method we propose is conceptually simpler and proved reliable both in indoor and outdoor environments.

3 Particle Filter-Based Estimation and Tracking of Ground Plane Orientation

Implementation of object tracking functionality in computer vision systems has presented a challenge to researchers for decades. The general problem of object tracking in sequences of images encompasses three major tasks: 1) object detection, 2) identifying the flow of object location in consecutive images, and 3) analysis of the tracked trajectory data. These tasks are particularly difficult due to the varying illumination of the imaged scene, varying size and shape of the tracked object, occlusions of the tracked objects, and finally the motion of the tracking camera (meaning that both the tracked object and the background move in relation to the camera).

There have been numerous methods proposed for object tracking in image sequences. The main categories are methods based on point tracking and region tracking. Methods based on motion estimation (current state estimation) have gained popularity because they are robust against occlusions and cumulative position estimation errors. The most prominent and widely used technique for estimating the state of a dynamic system from noisy measurements (e.g. tracking the position of a moving object) is referred to as the Kalman filter (KF). However, the drawback of the KF is its poor tracking performance against temporarily occluded objects or object shapes which are not represented by a continuous body object (i.e. that form non-uniform image regions).

A conceptually different approach uses the Particle Filtering (PF) algorithm, often referred to as the Sequential Monte Carlo (SMC) [15] method, a simulation-based technique which stems from the Monte Carlo method. The PF method is a simple yet effective way of searching for an optimal solution to multidimensional problems by randomly generating large numbers of possible system states. This enables observation of overall system behaviour and selection of the best solution. In a PF-based approach, the system state is estimated by a set of so-called particles. Each particle is represented by a vector:

$$\mathbf{c}_i^{(n)} = \begin{pmatrix} \mathbf{s}_t^{(n)} \\ \pi_t^{(n)} \end{pmatrix} \quad (1)$$

where the superscript (n) denotes a particle number ranging from 1 to N, the latter being the size of the particle set, and t denotes a discrete time instant. Vector $\mathbf{s}_t^{(n)}$ represents the system state and $\pi_t^{(n)}$ is particle weight, i.e. a value that reflects how accurately a given particle estimates the system state.

The posterior distribution of the system state $\mathbf{s}_t^{(n)}$ is estimated by the probability mass function:

$$p(\mathbf{s}_t|\mathbf{y}_{0:t}) \approx \sum_{n=1}^{N} \delta\left(\mathbf{s} - \mathbf{s}_t^{(n)}\right)\pi_t^{(n)} \quad (2)$$

where $\delta(\cdot)$ is the Dirac delta function. Each particle is weighted in terms of the observations, and the weighting function is specific to the application. A number of studies have been reported in which PF was applied to road tracking in automotive applications. However, the degrees of freedom (DoF) of the cameras were limited [16].

In the approach presented here, there is no restriction on camera movement. The implementation of the PF algorithm for GP detection and tracking of the state vector representing the GP parameters can be defined as follows:

$$\mathbf{s}_t^{(n)} = \left(x, y, z, a, b, c, \frac{dx}{dt}, \frac{dy}{dt}, \frac{dz}{dt}, \frac{da}{dt}, \frac{db}{dt}, \frac{dc}{dt}, \frac{d^2a}{dt^2}, \frac{d^2b}{dt^2}, \frac{d^2c}{dt^2}\right) \quad (3)$$

where:

- $\left(\frac{dx}{dt}, \frac{dy}{dt}, \frac{dz}{dt}\right)$ represents motion of point $\mathbf{x}_h^{(n)}$,
- $\left(\frac{da}{dt}, \frac{db}{dt}, \frac{dc}{dt}\right)$ *and* $\left(\frac{d^2a}{dt^2}, \frac{d^2b}{dt^2}, \frac{d^2c}{dt^2}\right)$ represent first and second order rates of change of the components of the normal vector $\mathbf{n}_h^{(n)} = (a, b, c)^T$ to a GP (inclusion of the second-order derivatives into the state vector allows better reproduction of swift changes in the orientation of the GP over time).

Further computation steps in the PF algorithm are performed according to a recursive scheme, described below and also shown as a block diagram in Fig. 2.

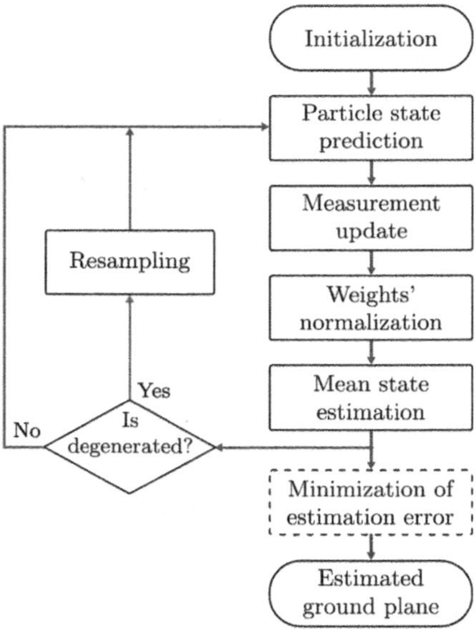

Fig. 2. Block diagram of the PF-based estimation of GP orientation.

Particle State Prediction
In every new frame, the state of each particle $\mathbf{s}_t^{(n)}$ is generated by a linear difference equation based on prior observations $\mathbf{s}_{t-1}^{(n)}$:

$$\mathbf{s}_t^{(n)} = \mathbf{A}s_{t-1}^{(n)} + \mathbf{w}_{t-1}^{(n)} \text{for } n = \{1, \cdots, N\} \quad (4)$$

where:

- **A** denotes a transition matrix and describes the relationship between the particular components of the state vector $\mathbf{s}_t^{(n)}$,
- $\mathbf{w}_{t-1}^{(n)}$ denotes the stochastic component of the equation (each particle is perturbed with this individually generated vector of multivariate normally distributed random variates).

Measurement Update
Each measurement $\mathbf{z}_t^{(n)}$ updates the weights of the particles according to the equation:

$$\pi_t^{(n)} = \pi_{t-1}^{(n)} p\left(\mathbf{z}_t | \mathbf{s}_t^{(n)}\right) \tag{5}$$

where $p\left(\mathbf{z}_t | \mathbf{s}_t^{(n)}\right)$ is a conditional probability density of measuring $\mathbf{z}_t^{(n)}$ given that the particle state is $\mathbf{s}_t^{(n)}$ [14].

Normalization of Weights
All the particle weights $\pi_t^{(n)}$ are normalized so that they sum to unity:

$$\pi_t^{(n)} = \frac{\pi_t^{(n)}}{\sum_{n=1}^{N} \pi_t^{(n)}} \tag{6}$$

Mean State Estimation
The estimated GP state is the weighted average of all particle states (see Fig. 3a):

$$\bar{\mathbf{s}}_t = \mathrm{E}\left[\mathbf{s}_t^{(n)}\right] = \sum_{n=1}^{N} \mathbf{s}_t^{(n)} \pi_t^{(n)} \tag{7}$$

Resampling
After a number of algorithm iterations, all but a few particles have negligible weights and therefore do not participate in the simulation effectively. This situation can be detected by calculating the degeneration indicator, expressed by:

$$d_t = \frac{1}{L \sum_{n=1}^{N} \left(\pi_t^{(n)}\right)^2} \tag{8}$$

If d_t falls below a given threshold, then a process called resampling is evoked and a new set of particles is created. Resampling causes the probability density function of $\mathbf{s}_t^{(n)}$ to be refined in the next iterations of the algorithm. Thus, a better estimate can be found.

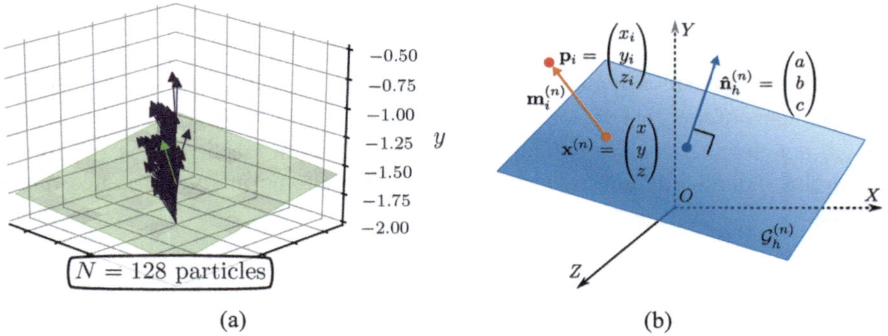

Fig. 3. Visualization of the mean particle state $\mathbf{s}_{30}^{(n)}$ for an arbitrary test scene image with a semi-transparent surface symbolizing the estimated GP (a), visualization of an arbitrary point $\mathbf{p}_i = (x_i, y_i, z_i)^T$ versus a hypothetical GP $g_h^{(n)}$ (b).

Error Minimization
In the proposed solution, the probability of a hypothetical GP is determined by a collection of points that may be supposed to lie on a plane. The coordinates of the points (hereinafter referred to as sampling points) are not known a priori; however, their projection onto the analyzed image forms a grid that should at least partially cover the GP region. The Euclidean distance of an arbitrary sampling point $\mathbf{p}_i = (x_i, y_i, z_i)^T$ to hypothetical GP $g^{(n)}$ is given by (see Fig. 3b):

$$D_t^{(n)}\left(p_i, g_h^{(n)}\right) = \frac{|ax_i + by_i + cz_i + d|}{\sqrt{a^2 + b^2 + c^2}} \tag{9}$$

Points $\mathbf{p}_i = (x_i, y_i, z_i)^T$ are assigned weights $w_t^{(n)}$ that increase with a decreasing distance $D_t^{(n)}$ to a GP. The weight of particle $\pi_t^{(n)}$ is then defined as the mean of all point weights: $\pi_t^{(n)} = \sum_{i=1}^{M} w_t^{(n)}(\mathbf{p}_i)$, where M is the number of considered points. Because the estimated GP normal vector is the result of a weighted average of all states (not the best state), there is a random (non-systematic) error present in the estimation. The error that has to be minimized to accurately and precisely detect the GP. The method we applied for minimization of the estimation error (i.e. estimation of the best plane position) is based on a procedure for fitting a plane to noisy points [17].

4 Results

A set of 38 stereovision test sequences were prepared to evaluate the performance of the PF-based method for estimating and tracking the GP. The image sequences contain a chequerboard (as shown in Fig. 4a), which allows a reference method to be applied for estimating the position of the GP (note that the depth map is computed with subpixel accuracy in the applied stereovision camera). The chequerboard is made of a thick layer of Polyvinyl chloride (PVC) and consists of 35 squares, each 25 × 25 cm in size.

The test image sequences were shot for the following camera movements versus the chequerboard: A) approaching the chequerboard, B) panning, C) simultaneous tilting and panning, D) extreme sideways rotations. The accuracy of GP orientation estimation (in terms of roll (α) and pitch (γ) angles) was verified by comparing PF-estimations (from depth images) and GP estimations obtained from a precise method based on computing the position on the chequerboard. A summary of the test results is presented in Table 2.

Table 2. Results of GP orientation estimation for pitch and roll angles.

Camera movement	Pitch (α) [°]		Roll (γ) [°]	
	ε_{RMS}	mean ± st. dev.	$\varepsilon_{RMS}(\gamma)$	mean ± st. dev.
A) Approaching the chequerboard	2.47	0.51 ± 0.40	0.90	0.26 ± 0.19
B) Panning	1.67	0.53 ± 0.45	1.36	0.47 ± 0.32
C) Simultaneous tilting and panning	2.54	1.17 ± 0.69	1.70	0.76 ± 0.38
D) Extreme sideways rotations	2.19	0.62 ± 0.50	1.81	0.58 ± 0.46

Note that the root mean square (RMS) errors do not exceed 2° or 3° for the roll and pitch angles, respectively. In order to verify how such precision in GP orientation estimation translates into accurate detection of the GP region, further tests were performed using indoor and outdoor scenes (see an example in Fig. 4a).

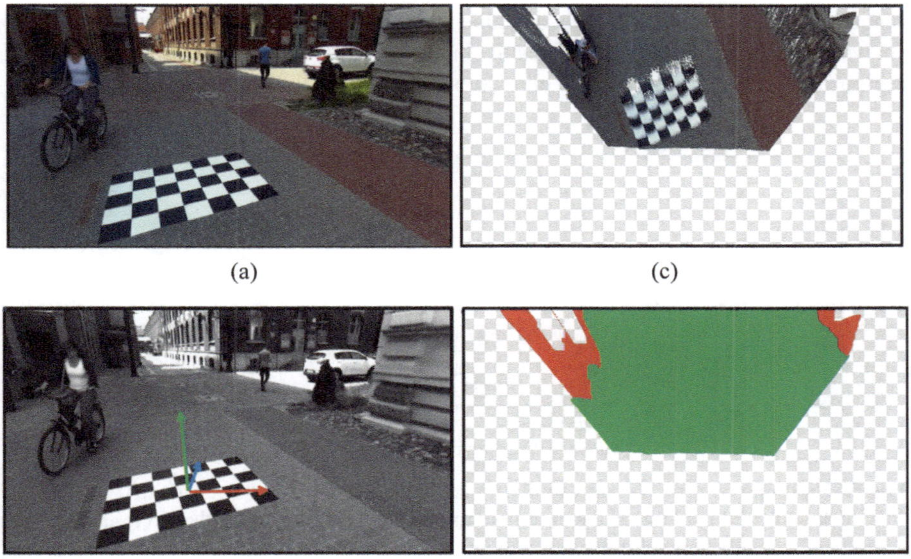

(a) (c) (b) (d)

Fig. 4. Procedure for computation of the ground truth region of the ground plane, used to verify the PF algorithm for tracking the ground :plane in depth map sequences a) photo of the scene, b) detection of the position and orientation of the chequerboard, c) top-down view of the scene, d) detection of the ground plane region marked in green colour.

In the adopted verification procedure, a reference (ground truth) region of the GP is first determined. A top-down view of the reference image is then computed (see Fig. 4c). This computation is based on the position and angle orientation of a large-scale chequerboard located on the ground (Fig. 4b). Knowledge about the GP equation expressed in the camera coordinate system allows the Euclidean distance of points in the 3D scene to be specified relative to the GP. If this distance is below a certain threshold (± 10 cm in this case), the corresponding point in the top-down view is assigned as belonging to the GP. Otherwise, it is marked as belonging to an obstacle. Distant parts of the scene (further than 4 m) are not taken into account. As a result, an occupancy grid map is built in which the GP is symbolized by a green colour and obstacles are shown in red (Fig. 4d).

The computed ground truth regions of GP for each of the camera movement scenarios A÷D were compared to the detections of GP carried out by applying the PF approach.

The Jaccard similarity coefficient was computed for comparing region $\mathcal{M}_{gt}$ of the GP detected by the reference (ground truth) method and region $\mathcal{M}_d$ obtained from the PF-based method:

$$JC(\mathcal{M}_d, \mathcal{M}_{gt}) = \frac{|\mathcal{M}_d \cap \mathcal{M}_{gt}|}{|\mathcal{M}_d \cup \mathcal{M}_{gt}|} = \frac{|\mathcal{M}_d \cap \mathcal{M}_{gt}|}{|\mathcal{M}_d| + |\mathcal{M}_{gt}| - |\mathcal{M}_d \cap \mathcal{M}_{gt}|} \in [0, 1] \qquad (5)$$

If there is a perfect overlap of the regions $JC = 1$, and if there is no overlap $JC = 0$.

Figure 5 illustrates violin plots for A÷D test scenarios. The shaded areas in the plots represent the density distribution of the errors. Note that the distribution of the JC values range from 0.74 to 0.99, with a mean value at 0.94 ± 0.04 and a median at 0.95. Figure 6 shows an example GP detection result, in which the boundaries of the $\mathcal{M}_{gt}$ and $\mathcal{M}_d$ regions are plotted in orange and blue colours, respectively.

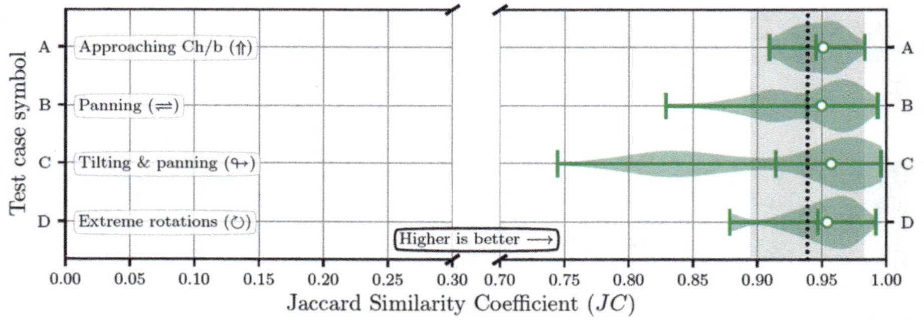

Fig. 5. Violin plots of the Jaccard similarity coefficient (JC) for test scenarios A, B, C, and D. The thick vertical dotted line is the ensemble mean for all test scenarios, the circles are the mean values and vertical short solid lines denote minimum, mean, and maximum values, respectively, computed for each test separately.

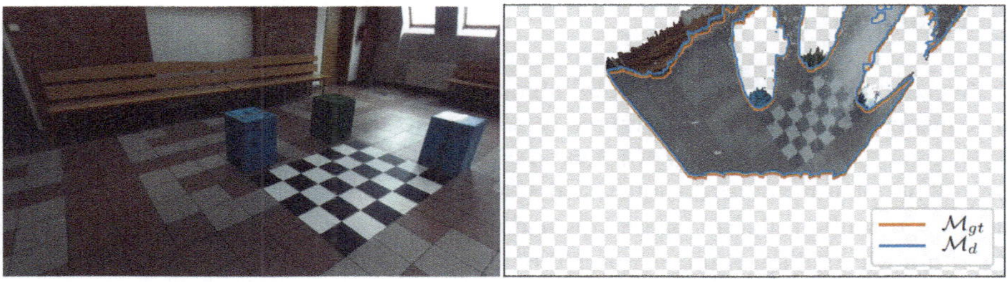

(a) reference image (b) top-down view

Fig. 6. An example scene in which the true positive rate of the GP segmentation reaches the smallest value at 0.8.

In total, 38 indoor and outdoor testing sequences using the PF-based method were recorded, and detections of the GP were computed. The test sequences were recorded in different lighting and weather conditions An online document available at https://doi.org/10.6084/m9.figshare.7694651 contains supplementary video material, showing the proposed method operating in typical situations for a visually impaired pedestrian using an assistive system, such as to avoid obstacles. Two snapshots of the sequences are shown in Fig. 7.

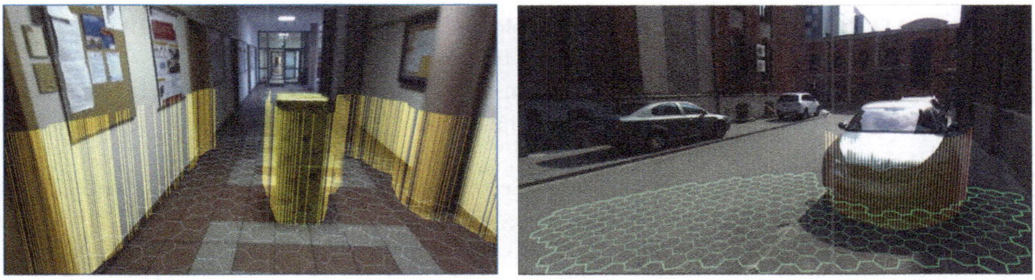

(a) (b)

Fig. 7. Examples of segmentation results of indoor (a) and outdoor (b) scenes. The detected GP region is marked by a hexagonal grid and the detected obstacles are represented by yellow vertical sticks approx. 1 m in height.

The proposed PF algorithm for estimation of plane orientation and position has been successfully implemented on graphics processing units (GPUs) using CUDA technology. The execution time of the processing pipeline is approximately 43 ms (on NVidia Geforce GTX 960 M with 640 CUDA cores), which corresponds to a processing speed of 23 image frames per second. Deployment of the PF on GPU hardware enabled the computations to be sped up by a factor of about 2.5 in comparison to computations on a general purpose multicore processor.

5 Conclusions

In this study, we have presented a PF algorithm for estimating the positions of planes in sequences of depth images of three-dimensional scenes, shot using a freely moving depth camera (with 6 degrees of freedom). Tests showed that the RMS errors for pitch and roll orientation angles do not exceed 3° and 2° respectively. A large overlap between the ground-truth ground plane regions and the ground plane detected by the PF-based algorithm were achieved for indoor and outdoor test image sequences (the mean value of the Jaccard similarity measure $JC = 0.94$). Achieving reliable detection of ground plane regions free of obstacles is an important processing step in many computer vision applications, such as for navigating autonomous vehicles or aerial objects (drones). The method has also been shown to enable the detection of obstacles, as applied in our ETA for the visually impaired (see Fig. 7). After further processing, the simplified model of the 3D scene geometry is converted into a non-visual (auditory or haptic) presentation of the environment. Early trials of our ETA system are presented in [18]. In the future work we will address the problem of segmenting the ground plane to acquire information about the type of surface, e.g. pavement, grass, soil, water etc. Also, we envision that our method can be applied for detecting unevenness of the ground surface, e.g. potholes.

References

1. Hersh, M., Johnson, M.: Assistive Technology for Visually Impaired and Blind People. Springer, London (2008). https://doi.org/10.1007/978-1-84628-867-8
2. Maidenbaum, S., Abboud, S., Amedi, A.: Sensory substitution: closing the gap between basic research and widespread practical visual rehabilitation. Neurosci. Biobehav. Rev. **14**, 3–15 (2014)
3. Dakopoulos, D., Bourbakis, N.G.: Wearable obstacle avoidance electronic travel aids for blind: a survey. IEEE Trans. Syst. Man Cybern. Part C Appl. Rev. **40**(1), 25–35 (2010)
4. Matusiak, K., Skulimowski, P., and Strumillo, P.: Object recognition in a mobile phone application for visually impaired users. In 2013 6th International Conference on Human System Interactions (HSI), pp. 479–484 (2013)
5. Yang, X., Yuan, S., Tian, Y.: Assistive clothing pattern recognition for visually impaired people. IEEE Trans. Hum. Mach. Syst. **44**(2), 234–243 (2014)
6. Tapu, R., Mocanu, B., and Zaharia, T.: Wearable assistive devices for visually impaired: a state of the art survey. Pattern Recogn. Lett. (2018). https://doi.org/10.1016/j.patrec.2018.10.031
7. Strumillo, P., Bujacz, M., Baranski, P., Skulimowski, P., Korbel, P., Owczarek, M., Tomalczyk, K., Moldoveanu, A., Unnthorsson, R.: Different approaches to aiding blind persons in mobility and navigation in the "Naviton" and "Sound of Vision" projects. In: Pissaloux, E.E., Velazquez, R. (eds.) Mobility in Visually Impaired People - Fundamentals and ICT Assistive Technologies. Springer, Cham (2018). https://doi.org/10.1007/978-3-319-54446-5_15
8. Picton, P., Capp, M.: Relaying scene information to the blind via sound using cartoon depth maps. Image Vis. Comput. **24**(4), 570–577 (2007)
9. Costa, P., Fernandes, H., Martins, P., Barroso, J., Hadjileontiadis, L.J.: Obstacle detection using stereo imaging to assist the navigation of visually impaired people. Procedia Comput. Sci. **14**, 83–93 (2012). Proceedings of the 4th International Conference on Software Development for Enhancing Accessibility and Fighting Infoexclusion (2012)

10. Herghelegiu, P., Burlacu, A., Caraiman, S.: Negative obstacle detection for wearable assistive devices for visually impaired. In: 2017 21st International Conference on System Theory, Control and Computing (ICSTCC), pp. 564–570 (2017)
11. Caraiman, S., Morar, A., Owczarek, M., Burlacu, A., Rzeszotarski, D., Botezatu, N., Herghelegiu, P., Moldoveanu, F., Strumillo, P., Moldoveanu, A.: Computer vision for the visually impaired: the sound of vision system. In: The IEEE International Conference on Computer Vision (ICCV) (2017)
12. Kim, J.: Obstacle Detection. Source code available at: https://github.com/joowonkim/Obstacle detection (2016). Accessed 10 Jan 2019
13. Kang, B., Lee, Y., Nguyen, T.Q.: Depth-adaptive deep neural network for semantic segmentation. IEEE Trans. Multimedia **20**(9), 2478–2490 (2018)
14. Caraiman, S., Zvoristeanu, O., Burlacu, A., Herghelegiu, P.: Stereo vision based sensory substitution for the visually impaired. Sensors **19**, 2771 (2019)
15. Liu, J.S., Chen, R.: Sequential monte carlo methods for dynamic systems. J. Am. Stat. Assoc. **93**, 1032–1044 (1998)
16. Kwon, J., Dragon, R., Gool, L.V.: Joint tracking and ground plane estimation. IEEE Signal Process. Lett. **23**(11), 1514–1517 (2016)
17. Ernerfeldt, E.: Fitting a plane to noisy points in 3D (2017). www.ilikebigbits.com/2017_09_25_plane_from_points_2.html. Accessed 10 Jan 2019
18. Skulimowski, P., Owczarek, M., Radecki, A., Bujacz, M., Rzeszotarski, D., Strumillo, P.: Interactive sonification of U-depth images in a navigation aid for the visually impaired. J. Multimod. User Interfaces **13**, 219–23 (2019)

GPSR Compliance

The European Union's (EU) General Product Safety Regulation (GPSR) is a set of rules that requires consumer products to be safe and our obligations to ensure this.

If you have any concerns about our products, you can contact us on ProductSafety@springernature.com

In case Publisher is established outside the EU, the EU authorized representative is:

Springer Nature Customer Service Center GmbH
Europaplatz 3
69115 Heidelberg, Germany

Batch number: 10761593

Printed by Printforce, the Netherlands

Advances in Intelligent Systems and Computing

Volume 1196

The series "Advances in Intelligent Systems and Computing" contains publications on theory, applications, and design methods of Intelligent Systems and Intelligent Computing. Virtually all disciplines such as engineering, natural sciences, computer and information science, ICT, economics, business, e-commerce, environment, healthcare, life science are covered. The list of topics spans all the areas of modern intelligent systems and computing such as: computational intelligence, soft computing including neural networks, fuzzy systems, evolutionary computing and the fusion of these paradigms, social intelligence, ambient intelligence, computational neuroscience, artificial life, virtual worlds and society, cognitive science and systems, Perception and Vision, DNA and immune based systems, self-organizing and adaptive systems, e-Learning and teaching, human-centered and human-centric computing, recommender systems, intelligent control, robotics and mechatronics including human-machine teaming, knowledge-based paradigms, learning paradigms, machine ethics, intelligent data analysis, knowledge management, intelligent agents, intelligent decision making and support, intelligent network security, trust management, interactive entertainment, Web intelligence and multimedia.

The publications within "Advances in Intelligent Systems and Computing" are primarily proceedings of important conferences, symposia and congresses. They cover significant recent developments in the field, both of a foundational and applicable character. An important characteristic feature of the series is the short publication time and world-wide distribution. This permits a rapid and broad dissemination of research results.

**** Indexing: The books of this series are submitted to ISI Proceedings, EI-Compendex, DBLP, SCOPUS, Google Scholar and Springerlink ****

More information about this series at http://www.springer.com/series/11156

Andrzej Bartoszewicz · Jacek Kabziński ·
Janusz Kacprzyk
Editors

Advanced, Contemporary Control

Proceedings of KKA 2020—The 20th Polish Control Conference, Łódź, Poland, 2020

Set 2

Springer

Editors
Andrzej Bartoszewicz
Institute of Automatic Control
Lodz University of Technology
Lodz, Poland

Jacek Kabziński
Institute of Automatic Control
Lodz University of Technology
Lodz, Poland

Janusz Kacprzyk
Systems Research Institute
Polish Academy of Sciences
Warsaw, Poland

ISSN 2194-5357 ISSN 2194-5365 (electronic)
Advances in Intelligent Systems and Computing
ISBN 978-3-030-50935-4 ISBN 978-3-030-50936-1 (eBook)
https://doi.org/10.1007/978-3-030-50936-1

This Springer imprint is published by the registered company Springer Nature Switzerland AG
The registered company address is: Gewerbestrasse 11, 6330 Cham, Switzerland

Foreword

This book comprises a collection of contributed papers submitted and, after a thorough peer review process, accepted for presentation at the 20th Polish Control Conference (PCC 2020). This jubilee event is another in a series of triennial congresses organized in Poland since 1958. This year, the conference has been organized jointly by the Committee on Automation and Robotics of the Polish Academy of Sciences and the Institute of Automatic Control of Łódź University of Technology, Poland. The scope of the conference traditionally encompasses all topics related to automatic control and robotics including: design, modelling, identification and analysis of automation systems; calculation methods, data processing and communication in control procedures; mechatronics; robotics and automated manufacturing in industrial systems; motion planning and control for robots; various control problems in transportation and vehicle systems; and also social impact of automation. However, this year the long-established areas of interest of the conference have been enhanced by several, relatively new topics, like: control-theoretic approach to biological, biomedical and medical systems; application of non-integer calculus in control and automation; cyber-physical systems; intelligent vehicles; non-industrial robotics (i.e. application of robots in emerging areas, for example, related to social, assistive or rehabilitation functions of robots); and embedded artificial intelligence and autonomy of both surface and underwater marine vessels. This enhancement was, in no small part, possible thanks to those members of our control engineering community who proposed, promoted and successfully organized a number of invited sessions. The editors of this volume wish to thank all those persons who greatly contributed to the quality, broad coverage and originality of this monograph. In this respect, we particularly acknowledge the contribution of professors: Artur Babiarz, Marek Długosz, Anna Fabijańska, Jarosław Figwer, Zhiqiang Gao, Wojciech Giernacki, Witold Gierusz, Krzysztof Kozłowski, Maciej Ławryńczuk, Rafał Madoński, Maciej Michałek, Krzystof Okarma, Krzysztof Oprzędkiewicz, Ewa Pawłuszewicz, Dariusz Pazderski, Witold Repondek, Paweł Skruch, Piotr Skrzypczyński, Rafał Stanisławski, Ralf Stetter, Piotr Szymak, Roman Śmierzchalski, Andrzej Świerniak, Mirosław Tomera and Marcin Witczak. Indeed, those individuals made an impressive work when identifying new techniques and areas of applications, essentially assisting in the review process and sharing with us their expert knowledge at the editorial stage. Thanks to the joint effort of those people, numerous contributors and hundreds of anonymous reviewers, the conference managed to successfully combine its more than sixty years long tradition with modern approach to broadly understood problems of control engineering. We also thank Dr. Paweł Latosiński and Dr. Piotr Leśniewski for their tireless proofreading of the many drafts of individual chapters. Their exceptional proofreading skills have been of great importance for the final form of this book.

Finally, we wish to thank the entire team of Springer publications for allowing the preparation of this monograph to proceed. We hope that the result of our joint effort will be of true interest to the control engineering community working on various aspects of control theory, automation, robotics, mechatronics, automated manufacturing and related problems of contemporary engineering.

April 2020

Andrzej Bartoszewicz
Jacek Kabziński
Janusz Kacprzyk

Contents

Design of Control Systems - Methods and Applications

Noninteger Calculus in Automation

On a Solution of an Optimal Control Problem for a Linear Fractional-Order System

Mikhail I. Gomoyunov[1,2](✉)

[1] Krasovskii Institute of Mathematics and Mechanics, The Ural Branch of the Russian Academy of Sciences, Ekaterinburg 620990, Russia
m.i.gomoyunov@gmail.com
[2] Ural Federal University, Ekaterinburg 620002, Russia

Abstract. We consider an optimal control problem for a dynamical system described by a linear differential equation with the Caputo fractional derivative of an order $\alpha \in (0, 1)$. A cost functional to be minimized evaluates a deviation of a system's terminal state from a given target point. In order to construct a solution, we turn from the considered problem to an auxiliary optimal control problem for a first-order linear system with concentrated delays, which approximates the original system and, after that, we reduce this auxiliary problem to an optimal control problem for an ordinary differential system. Moreover, on this basis, we propose a feedback scheme of optimal control of the original system. The efficiency of the approach is illustrated by an example, and the results of numerical simulations are presented.

Keywords: Optimal control problem · Linear system · Caputo fractional derivative · Approximation · Feedback control · Numerical method

1 Introduction

We consider an optimal control problem for a dynamical system whose motion is described by a linear differential equation with the Caputo fractional derivative of an order $\alpha \in (0, 1)$. The time interval of the control process is fixed and finite. Control actions are subject to geometric constraints. The goal of control is to minimize a cost functional that evaluates a deviation of a system's terminal state from a given target point.

The research follows the game-theoretical approach [9–11]. Some elements of this approach were developed for fractional-order systems in [4,17]. In the paper, based on the results from [3,4], we turn from the considered problem to an auxiliary optimal control problem for a first-order linear system with concentrated delays, which approximates the original system. Then, relying on the results from [5,14], we reduce the auxiliary problem to an optimal control problem for an ordinary differential system. Further, on this basis, we propose a

A. Bartoszewicz et al. (Eds.): PCC 2020, AISC 1196, pp. 837–846, 2020.
https://doi.org/10.1007/978-3-030-50936-1_70

feedback scheme of optimal control of the original system that uses the approximating system as a guide [11]. In this scheme, a control in the approximating system is formed with the help of an optimal positional control strategy [10] from the reduced problem. Thus, the presented approach allows us to apply the methods developed in control theory for ordinary differential systems to construct solutions of control problems for fractional-order systems. In particular, the considered problem is effectively solved using the results from [13]. We also note that a feature of the proposed scheme of optimal control is that it naturally extends to control problems under conditions of disturbances or counteractions.

Nowadays, optimal control problems for linear systems with the Caputo fractional derivatives are studied quite intensively. Various statements are considered including linear-quadratic problems (see, e.g., [6]), problems of minimization of an integral cost functional (see, e.g., [7]), problems of transition of a system into a given state in the shortest time or with the minimal norm of a control (see, e.g., [8,12,15]). In most cases, in order to find a solution, suitable variants of the maximum principle, methods of variational calculus and convex analysis, and methods related to the problem of moments are applied. The present paper focuses mainly on reducing of control problems for linear fractional-order systems to control problems for ordinary differential systems.

2 Optimal Control Problem

Let a motion of a dynamical system on a time interval $[t_0, \vartheta]$ be described by the linear fractional differential equation

$$\begin{gathered} ({}^C D^\alpha x)(t) = A(t)x(t) + B(t)u(t), \\ x(t) \in \mathbb{R}^n, \quad u(t) \in \mathbb{U} \subset \mathbb{R}^r, \quad t \in [t_0, \vartheta], \end{gathered} \tag{1a}$$

under the initial condition

$$x(t_0) = x_0. \tag{1b}$$

Here, t is the time, $x(t)$ is the state of the system at the time t, $u(t)$ is the current control action; $x_0 \in \mathbb{R}^n$ is a fixed initial state of the system; $\mathbb{U}$ is a compact set; by $({}^C D^\alpha x)(t)$, we denote the Caputo fractional derivative of the order $\alpha \in (0,1)$ at the time t (see, e.g., [1, Sect. 3.1]):

$$({}^C D^\alpha x)(t) = \frac{1}{\Gamma(1-\alpha)} \frac{\mathrm{d}}{\mathrm{d}t} \int_{t_0}^{t} \frac{x(\tau) - x(t_0)}{(t-\tau)^\alpha} \, \mathrm{d}\tau,$$

where Γ is the gamma-function. The functions $A(t) \in \mathbb{R}^{n \times n}$ and $B(t) \in \mathbb{R}^{n \times r}$, $t \in [t_0, \vartheta]$, are assumed to be continuous.

Let $AC^\alpha([t_0, \vartheta], \mathbb{R}^n)$ denote the set of functions $x(t) \in \mathbb{R}^n$, $t \in [t_0, \vartheta]$, for each of which there exists a measurable and essentially bounded function $\varphi(t) \in \mathbb{R}^n$, $t \in [t_0, \vartheta]$, such that the following equality holds:

$$x(t) = x(t_0) + \frac{1}{\Gamma(\alpha)} \int_{t_0}^{t} \frac{\varphi(\tau)}{(t-\tau)^{1-\alpha}} \, \mathrm{d}\tau, \quad t \in [t_0, \vartheta].$$

In other words, the function $x(\cdot)$ can be represented as the sum of the initial value $x(t_0)$ and the Riemann–Liouville integral of the order α of the function $\varphi(\cdot)$ (see, e.g., [16, Sect. 2.3] and [1, Sect. 2.1]). Let us note that, for such a function $x(\cdot) \in AC^\alpha([t_0, \vartheta], \mathbb{R}^n)$, the Caputo derivative $({}^C D^\alpha x)(t)$ exists for almost every $t \in [t_0, \vartheta]$, and the equality below is valid (see, e.g., [16, Theorem 2.4]):

$$x(t) = x(t_0) + \frac{1}{\Gamma(\alpha)} \int_{t_0}^{t} \frac{({}^C D^\alpha x)(\tau)}{(t - \tau)^{1-\alpha}} \, \mathrm{d}\tau, \quad t \in [t_0, \vartheta].$$

By an admissible (open-loop) control, we mean any measurable function $u(t) \in \mathbb{U}$, $t \in [t_0, \vartheta)$. Let $\mathcal{U}$ be the set of all such controls. A motion of system (1a), (1b) that corresponds to a control $u(\cdot) \in \mathcal{U}$ is defined as a function $x(\cdot) \in AC^\alpha([t_0, \vartheta], \mathbb{R}^n)$ that satisfies the equality in (1b) and, together with $u(\cdot)$, satisfies the differential equation in (1a) for almost every $t \in [t_0, \vartheta]$. Due to the assumptions made, such a motion exists and is unique (see, e.g., [2, Theorem 3.1]), and we denote it by $x(\cdot \mid u(\cdot))$.

Let a target vector $c \in \mathbb{R}^n$, a matrix $K \in \mathbb{R}^{d \times n}$, where $d \in \overline{1, n}$, and a norm $\mu(s) \in \mathbb{R}$, $s \in \mathbb{R}^d$, be given. The goal of control is to minimize the cost functional

$$\gamma(u(\cdot)) = \mu\Big(K\big(x(\vartheta \mid u(\cdot)) - c\big)\Big), \quad u(\cdot) \in \mathcal{U}. \tag{1c}$$

The value of the optimal result in optimal control problem (1) is defined by

$$\rho = \inf_{u(\cdot) \in \mathcal{U}} \gamma(u(\cdot)).$$

For $\zeta > 0$, a control $u_0(\cdot) \in \mathcal{U}$ is called ζ-optimal if

$$\gamma(u_0(\cdot)) \leq \rho + \zeta.$$

In the paper, we develop an approach for finding the value ρ and ζ-optimal controls $u_0(\cdot)$ via approximation of problem (1) by an auxiliary optimal control problem for a dynamical system which motion is described by a first-order linear differential equation with concentrated delays. In Sects. 3 and 4, we consider open-loop ζ-optimal controls. In Sect. 5, we study the question of forming such controls by using feedback control schemes.

3 Approximating Optimal Control Problem

The idea of approximation goes back to the Grünwald–Letnikov approach to fractional differentiation (see, e.g., [16, Sect. 20.4] and [1, Sect. 2.4]). For a direct relation between the Caputo and Grünwald–Letnikov fractional derivatives, on which the results below are based, the reader is referred to [3].

Let us fix an approximation parameter $h > 0$. In what follows, we assume that $\vartheta - t_0 = Nh$ for some $N \in \mathbb{N}$. Denote

$$\vartheta_i = \vartheta - ih, \quad k_i(t) = \begin{cases} (-1)^i \binom{1-\alpha}{i} h^{\alpha-1}, & t \in [t_0 + ih, \vartheta], \\ 0, & \text{otherwise}, \end{cases} \quad i \in \overline{0, N}, \tag{2}$$

where $\binom{1-\alpha}{i}$ are the binomial coefficients.

We consider an auxiliary optimal control problem for the dynamical system

$$\dot{y}(t) = A(t) \sum_{i=0}^{N} k_i(t) y(t - ih) + B(t) p(t) + A(t) x_0, \\ y(t) \in \mathbb{R}^n, \quad p(t) \in \mathbb{U}, \quad t \in [t_0, \vartheta], \tag{3a}$$

under the initial condition

$$y(t_0) = 0 \tag{3b}$$

and the cost functional

$$\gamma_y^{(h)}(p(\cdot)) = \mu\Big(K\Big(x_0 + \sum_{i=0}^{N} k_i(\vartheta) y(\vartheta_i \mid p(\cdot)) - c\Big)\Big), \quad p(\cdot) \in \mathcal{U}. \tag{3c}$$

Here, $y(t)$ is the state of the auxiliary system at the time t; $p(t)$ is the current control action; $\dot{y}(t) = \mathrm{d}y(t)/\mathrm{d}t$. Let us note that, for every $i \in \overline{1, N}$, the values $y(t - ih)$ for $t - ih < t_0$ are used in (3a) just formally, since $k_i(t) = 0$ due to (2). Therefore, in particular, as initial condition (3b), it is sufficient to specify the value $y(t_0)$ only. In (3c), by $y(\cdot \mid p(\cdot))$, we denote a motion of the auxiliary system corresponding to a control $p(\cdot) \in \mathcal{U}$, which is defined as an absolutely continuous function $y(t) \in \mathbb{R}^n$, $t \in [t_0, \vartheta]$, that satisfies the equality in (3b) and, together with $p(\cdot)$, satisfies the differential equation in (3a) for almost every $t \in [t_0, \vartheta]$. Under the considered assumptions, such a motion exists and is unique (see, e.g., [5, 14]). The goal of control is to minimize cost functional (3c).

The value of the optimal result in optimal control problem (3) is

$$\rho_y^{(h)} = \inf_{p(\cdot) \in \mathcal{U}} \gamma_y^{(h)}(p(\cdot)).$$

For $\zeta > 0$, a control $p_0(\cdot) \in \mathcal{U}$ is ζ-optimal in this problem if

$$\gamma_y^{(h)}(p_0(\cdot)) \leq \rho_y^{(h)} + \zeta.$$

The proposition below gives a connection between problems (1) and (3).

Proposition 1. *For any $\zeta > 0$, there exist $h_* > 0$ and $\zeta_* > 0$ such that, for every $h \in (0, h_*]$, the inequality $|\rho - \rho_y^{(h)}| \leq \zeta$ is valid, and each ζ_*-optimal control in auxiliary problem (3) is ζ-optimal in original problem (1).*

This statement follows directly from a uniform closeness [3, Theorem 2] between motions $x(\cdot \mid u(\cdot))$ and $y(\cdot \mid p(\cdot))$ of the original and approximating systems corresponding to the same controls, i.e., when $p(\cdot) = u(\cdot)$, and a uniform boundedness [2, Proposition 5.1] of motions $x(\cdot \mid u(\cdot))$, $u(\cdot) \in \mathcal{U}$. It should be noted that Proposition 1 deals with ζ-optimal controls only, and, in general, it does not allow to find the optimal ones.

Thus, problem (1) is approximated by problem (3). The next section is devoted to a reduction of problem (3) to an optimal control problem for a dynamical system which motion is described by an ordinary differential equation.

4 Reduced Optimal Control Problem

Let $E \in \mathbb{R}^{n\times n}$ stand for the identity matrix. Let the function $Y^{(h)}(\tau,t) \in \mathbb{R}^{n\times n}$, $\tau \in [t_0,\vartheta]$, $t \in [t_0,\vartheta]$, be such that, for every $\tau \in [t_0,\vartheta]$, it is absolutely continuous in t on the interval $[t_0,\tau]$, and almost everywhere in this interval it satisfies the differential equation

$$\frac{\partial}{\partial t}Y^{(h)}(\tau,t) = -\sum_{i=0}^{N} k_i(t+ih)Y^{(h)}(\tau,t+ih)A(t+ih) \tag{4}$$

under the conditions

$$Y^{(h)}(\tau,\tau) = E, \quad Y^{(h)}(\tau,t) = 0, \quad t \in (\tau,\vartheta].$$

By analogy with (3a), for every $i \in \overline{1,N}$, the values $Y^{(h)}(\tau,t+ih)$ and $A(t+ih)$ for $t+ih > \vartheta$ are used in (4) just formally, since $k_i(t+ih) = 0$ due to (2). Set

$$B^{(h)}(t) = K\sum_{i=0}^{N} k_i(\vartheta)Y^{(h)}(\vartheta_i,t)B(t),$$
$$f^{(h)}(t) = K\sum_{i=0}^{N} k_i(\vartheta)Y^{(h)}(\vartheta_i,t)A(t)x_0, \quad t \in [t_0,\vartheta].$$

We consider an optimal control problem for the dynamical system

$$\begin{aligned} \dot{z}(t) &= B^{(h)}(t)p(t) + f^{(h)}(t), \\ z(t) &\in \mathbb{R}^d, \quad p(t) \in \mathbb{U}, \quad t \in [t_0,\vartheta], \end{aligned} \tag{5a}$$

under the initial condition

$$z(t_0) = K(x_0 - c) \tag{5b}$$

and the cost functional

$$\gamma_z^{(h)}(p(\cdot)) = \mu\big(z(\vartheta \mid p(\cdot))\big), \quad p(\cdot) \in \mathcal{U}. \tag{5c}$$

Here, $z(t)$ is the state of the system at the time t. The goal of control is to minimize (5c), where $z(\cdot \mid p(\cdot))$ denotes the motion of system (5a), (5b), generated by a control $p(\cdot) \in \mathcal{U}$.

In optimal control problem (5), we define the value of the optimal result by

$$\rho_z^{(h)} = \inf_{p(\cdot)\in\mathcal{U}} \gamma_z^{(h)}(p(\cdot))$$

and ζ-optimal controls $p_0(\cdot) \in \mathcal{U}$:

$$\gamma_z^{(h)}(p_0(\cdot)) \le \rho_z^{(h)} + \zeta. \tag{6}$$

Due to Proposition 1 and a connection between problems (3) and (5) (see, e.g., [5, Lemma 1] and the equalities in (11) below), we have

Proposition 2. *For any $\zeta > 0$, there exist $h_* > 0$ and $\zeta_* > 0$ such that, for every $h \in (0, h_*]$, the inequality $|\rho - \rho_z^{(h)}| \leq \zeta$ is valid, and each ζ_*-optimal control in reduced problem (5) is ζ-optimal in original problem (1).*

So, problem (1) is reduced to problem (5). In the next section, the proposed constructions are applied in order to form ζ-optimal controls in original problem (1) by using feedback control schemes.

5 Feedback Control Schemes

In accordance with [10], by a positional control strategy in reduced problem (5), we mean any function

$$P(t, z, \varepsilon) \in \mathbb{U}, \quad t \in [t_0, \vartheta], \quad z \in \mathbb{R}^d, \quad \varepsilon > 0,$$

where ε is an accuracy parameter. Let Δ be a partition of the interval $[t_0, \vartheta]$:

$$\Delta = \{\tau_j\}_{j \in \overline{1,k+1}}, \quad \tau_1 = t_0, \quad \tau_j < \tau_{j+1}, \quad j \in \overline{1,k}, \quad \tau_{k+1} = \vartheta. \tag{7}$$

The triple $\{P, \varepsilon, \Delta\}$ is called a control law. This law forms in system (5a), (5b) a piecewise constant control by the following step-by-step feedback rule:

$$p(t) = P(\tau_j, z(\tau_j), \varepsilon), \quad t \in [\tau_j, \tau_{j+1}), \quad j \in \overline{1,k}. \tag{8}$$

A strategy $P_0^{(h)}$ is optimal in problem (5) if, for any $\zeta > 0$, there exist a number $\varepsilon_* > 0$ and a function $\delta_*(\varepsilon) > 0$, $\varepsilon \in (0, \varepsilon_*]$, such that, for every $\varepsilon \in (0, \varepsilon_*]$ and partition Δ of type (7) satisfying the condition

$$\max_{j \in \overline{1,k}} (\tau_{j+1} - \tau_j) \leq \delta_*(\varepsilon), \tag{9}$$

the determined by the law $\{P_0^{(h)}, \varepsilon, \Delta\}$ control $p(\cdot)$ is ζ-optimal, i.e., satisfies (6). Let us note that, under the considered conditions, such optimal strategy $P_0^{(h)}$ exists (see, e.g., [10, Theorem 9.2]).

On the basis of the law $\{P_0^{(h)}, \varepsilon, \Delta\}$, we form a control in approximating system (3a), (3b) in the following way. Let $j \in \overline{1,k}$, and let $y_{\tau_j}(t) = y(t)$, $t \in [t_0, \tau_j]$, be the history of the motion of this system that has been realized by the time τ_j. Then, in accordance with (8), we set

$$p(t) = P_0^{(h)}\big(\tau_j, w^{(h)}(\tau_j, y_{\tau_j}(\cdot)), \varepsilon\big), \quad t \in [\tau_j, \tau_{j+1}). \tag{10}$$

Here,

$$w^{(h)}(\tau_j, y_{\tau_j}(\cdot)) = K\Big(x_0 + \sum_{i=0}^{N} k_i(\vartheta) w_i(\tau_j, y_{\tau_j}(\cdot)) - c\Big),$$

where, for every $i \in \overline{0,N}$, in the case $\tau_j < \vartheta_i$, we define

$$w_i(\tau_j, y_{\tau_j}(\cdot)) = Y^{(h)}(\vartheta_i, \tau_j) y(\tau_j) + \sum_{q=0}^{N} \int_{\tau_j}^{\tau_j + qh} k_q(\tau) Y^{(h)}(\vartheta_i, \tau) A(\tau) y(\tau - qh)\, \mathrm{d}\tau,$$

and, otherwise, we have

$$w_i(\tau_j, y_{\tau_j}(\cdot)) = y(\vartheta_i).$$

Following [5,10,13,14], the vector $w^{(h)}(\tau_j, y_{\tau_j}(\cdot))$ is called an informational image of the pair $(\tau_j, y_{\tau_j}(\cdot))$. Let us note that initial condition (5b) has been chosen in view of the equalities

$$w_i(t_0, y(t_0)) = 0, \quad i \in \overline{0, N}, \quad w^{(h)}(t_0, y(t_0)) = K(x_0 - c), \tag{11}$$

which are valid owing to (2) and (3b).

In view of [5, Theorem 1], we obtain

Proposition 3. *For any $h > 0$ and $\zeta > 0$, there exist a number $\varepsilon_* > 0$ and a function $\delta_*(\varepsilon) > 0$, $\varepsilon \in (0, \varepsilon_*]$, such that, for every $\varepsilon \in (0, \varepsilon_*]$ and partition Δ of type (7), (9), the determined by rule (10) control $p(\cdot)$ is ζ-optimal in approximating problem (3).*

In original system (1a), (1b), we perform a feedback control using a guide (see, e.g., [11, Sect. 8.2]), for which we choose the optimally controlled approximating system (3a), (3b). Namely, taking parameters $h > 0$ and $\varepsilon > 0$ and a partition Δ of type (7), we form controls $u(\cdot) \in \mathcal{U}$ and $p(\cdot) \in \mathcal{U}$ in the original and approximating systems, respectively, according to the following step-by-step rule. Let $j \in \overline{1, k}$, and let the state $x(\tau_j)$ of the original system and the history of the motion $y_{\tau_j}(\cdot)$ of the approximating system have been realized by the time τ_j. Then, in the next step, for $t \in [\tau_j, \tau_{j+1})$, we set

$$u(t) = u_j \in \underset{u \in \mathbb{U}}{\operatorname{argmin}} \left\langle x(\tau_j) - x_0 - \sum_{i=0}^{N} k_i(\tau_j) y(\tau_j - ih), B(\tau_j) u \right\rangle, \tag{12}$$

and determine $p(t)$ by (10). In (12), $\langle \cdot, \cdot \rangle$ stands for the inner product of vectors.

Proposition 4. *For any $\zeta > 0$, one can choose $h_* > 0$ such that, for each $h \in (0, h_*]$, there exist a number $\varepsilon_* > 0$ and a function $\delta_*(\varepsilon) > 0$, $\varepsilon \in (0, \varepsilon_*]$, such that, for every $\varepsilon \in (0, \varepsilon_*]$ and partition Δ of type (7), (9), the control $u(\cdot)$ formed by control procedure with a guide (10), (12) is ζ-optimal in problem (1).*

The validity of this statement follows from Proposition 3 and [4, Theorem 1]. Due to a specific form of cost functional (5c), for the value of the optimal result $\rho_z^{(h)}$ and an optimal positional control strategy $P_0^{(h)}$ in reduced problem (5), we have the following formulas (see, e.g., [13] and the references therein):

$$\begin{gathered}\rho_z^{(h)} = \max_{l \in G} \left(\langle l, K(x_0 - c) \rangle + \psi^{(h)}(t_0, l) \right), \\ P_0^{(h)}(t, z, \varepsilon) \in \underset{u \in \mathbb{U}}{\operatorname{argmin}} \, \langle l_0^{(h)}(t, z, \varepsilon), B^{(h)}(t) u \rangle, \quad t \in [t_0, \vartheta], \quad z \in \mathbb{R}^d, \quad \varepsilon > 0,\end{gathered} \tag{13}$$

where

$$\begin{gathered} l_0^{(h)}(t, z, \varepsilon) \in \underset{l \in G}{\operatorname{argmax}} \left(\langle l, z \rangle + \psi^{(h)}(t, l) - \sqrt{\left(\varepsilon + (t - t_0)\varepsilon\right)(1 + \|l\|^2)} \right), \\ \psi^{(h)}(t, l) = \int_t^{\vartheta} \min_{u \in \mathbb{U}} \langle l, B^{(h)}(\tau) u + f^{(h)}(\tau) \rangle \, \mathrm{d}\tau, \\ G = \left\{ l \in \mathbb{R}^d : \max_{s \in \mathbb{R}^d : \, \mu(s) \le 1} \langle l, s \rangle \le 1 \right\}. \end{gathered}$$

Thus, Propositions 2 and 4 allow us to apply formulas (13) in order to find the value of the optimal result ρ and construct ζ-optimal controls by using feedback control schemes in original problem (1).

6 Example

In order to illustrate the results obtained in the paper, we consider an optimal control problem for the dynamical system

$$\begin{cases} ({}^C D^\alpha x_1)(t) = x_2(t) + (0.5 - t)u_1(t), \\ ({}^C D^\alpha x_2)(t) = -x_1(t) - \sin(\pi t)x_2(t) + u_2(t), \end{cases} \\ x(t) = (x_1(t), x_2(t)) \in \mathbb{R}^2, \quad u_1^2(t) + u_2^2(t) \le 1, \quad t \in [0, 1], \tag{14a}$$

under the initial condition

$$x(0) = x_0 = (-1.5, 1.5) \tag{14b}$$

and the cost functional to be minimized

$$\gamma(u(\cdot)) = \sqrt{x_1^2(1 \mid u(\cdot)) + x_2^2(1 \mid u(\cdot))}, \quad u(\cdot) \in \mathcal{U}. \tag{14c}$$

Let us present the results of numerical simulations. The calculations were carried out for various orders of differentiation $\alpha = 0.5$, $\alpha = 0.7$, and $\alpha = 0.9$, when choosing the approximation parameter $h = 0.005$, the accuracy parameter $\varepsilon = 0.01$, and the uniform partition Δ of type (7) with the step $\delta = 0.001$.

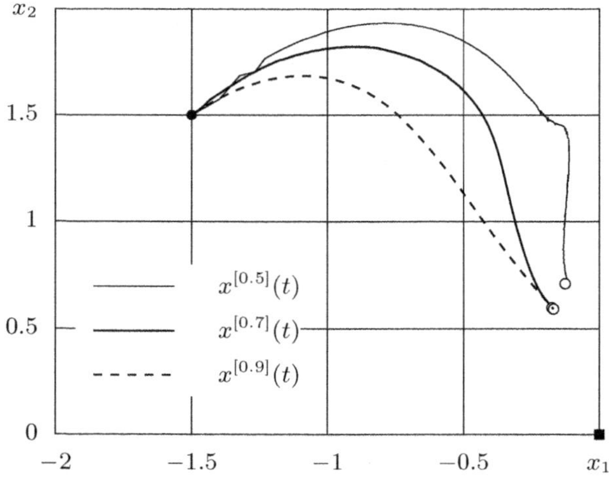

Fig. 1. The motions of system (14a), (14b) generated by control procedure with a guide (10), (12) for various orders of differentiation $\alpha = 0.5$, $\alpha = 0.7$, and $\alpha = 0.9$.

In each case, a control was formed by control procedure with a guide (10), (12). The corresponding motions of system (14a), (14b) are shown in Fig. 1.

The found values of the optimal result in problem (14) and the realized values of cost functional (14c) are the following:

$$\begin{aligned} \rho^{[0.5]} &\approx 0.668, & \gamma^{[0.5]} &\approx \sqrt{(-0.126)^2 + (0.709)^2} \approx 0.72, \\ \rho^{[0.7]} &\approx 0.613, & \gamma^{[0.7]} &\approx \sqrt{(-0.174)^2 + (0.597)^2} \approx 0.622, \\ \rho^{[0.9]} &\approx 0.613, & \gamma^{[0.9]} &\approx \sqrt{(-0.168)^2 + (0.592)^2} \approx 0.615. \end{aligned}$$

7 Conclusion

In the paper, we have considered an optimal control problem for a dynamical system described by a linear differential equation with the Caputo fractional derivative of an order $\alpha \in (0, 1)$. In order to find the value of the optimal result and ζ-optimal controls, we have proposed a reduction of the considered problem to an optimal control problem for an ordinary differential system. The latter can be solved by applying a wide range of the methods developed in control theory for ordinary differential systems. An illustrative example has been considered.

References

1. Diethelm, K.: The Analysis of Fractional Differential Equations. Springer, Heidelberg (2010). https://doi.org/10.1007/978-3-642-14574-2
2. Gomoyunov, M.I.: Fractional derivatives of convex Lyapunov functions and control problems in fractional order systems. Frac. Calc. Appl. Anal. **21**(5), 1238–1261 (2018). https://doi.org/10.1515/fca-2018-0066
3. Gomoyunov, M.I.: Approximation of fractional order conflict-controlled systems. Progr. Fract. Differ. Appl. **5**(2), 143–155 (2019). https://doi.org/10.18576/PFDA/050205
4. Gomoyunov, M.I.: Solution to a zero-sum differential game with fractional dynamics via approximations. Dyn. Games Appl. 1–27 (2019). https://doi.org/10.1007/s13235-019-00320-4
5. Gomoyunov, M.I., Lukoyanov, N.Yu.: Guarantee optimization in functional-differential systems with a control aftereffect. J. Appl. Math. Mech. **76**(4), 369–377 (2012). https://doi.org/10.1016/j.jappmathmech.2012.09.002
6. Idczak, D., Walczak, S.: On a linear-quadratic problem with Caputo derivative. Opuscula Math. **36**(1), 49–68 (2016). https://doi.org/10.7494/OpMath.2016.36.1.49
7. Kamocki, R., Majewski, M.: Fractional linear control systems with Caputo derivative and their optimization. Optim. Control Appl. Meth. **36**(6), 953–967 (2015). https://doi.org/10.1002/oca.2150
8. Kaczorek, T.: Minimum energy control of fractional positive electrical circuits with bounded inputs. Circuits Syst. Signal Process. **35**(6), 1815–1829 (2016). https://doi.org/10.1007/s00034-015-0181-7
9. Krasovskii, N.N., Kotelnikova, A.N.: Stochastic guide for a time-delay object in a positional differential game. Proc. Steklov Inst. Math. **277**(suppl. 1), 145–151 (2012). https://doi.org/10.1134/S0081543812050148

10. Krasovskii, N.N., Krasovskii, A.N.: Control Under Lack of Information. Birkhäuser, Berlin (1995). https://doi.org/10.1007/978-1-4612-2568-3
11. Krasovskii, N.N., Subbotin, A.I.: Game-Theoretical Control Problems. Springer, New York (1988)
12. Kubyshkin, V.A., Postnov, S.S.: Optimal control problem for a linear stationary fractional order system in the form of a problem of moments: problem setting and a study. Autom. Remote Control. **75**(5), 805–817 (2014). https://doi.org/10.1134/S0005117914050014
13. Lukoyanov, N.Yu., Gomoyunov, M.I.: Differential games on minmax of the positional quality index. Dyn. Games Appl. **9**(3), 780–799 (2019). https://doi.org/10.1007/s13235-018-0281-7
14. Lukoyanov, N.Yu., Reshetova, T.N.: Problems of conflict control of high dimensionality functional systems. J. Appl. Math. Mech. **62**(4), 545–554 (1998). https://doi.org/10.1016/S0021-8928(98)00071-9
15. Matychyn, I., Onyshchenko, V.: Optimal control of linear systems with fractional derivatives. Fract. Calc. Appl. Anal. **21**(1), 134–150 (2018). https://doi.org/10.1515/fca-2018-0009
16. Samko, S.G., Kilbas, A.A., Marichev, O.I.: Fractional Integrals and Derivatives. Theory and Applications. Gordon and Breach Science Publishers, Amsterdam (1993)
17. Surkov, P.G.: Dynamic right-hand side reconstruction problem for a system of fractional differential equations. Differ. Equ. **55**(6), 849–858 (2019). https://doi.org/10.1134/S0012266119060120

The Quickly Adjustable Digital FOPID Controller

Klaudia Dziedzic(✉) and Krzysztof Oprzędkiewicz

Department of Automatic Control and Robotics, AGH University,
Al. Mickiewicza 30, 30-059 Krakow, Poland
{kdz,kop}@agh.edu.pl

Abstract. This study investigates an accuracy estimation of CFE approximation describing the switchable Fractional Order PID controller (FOPID). Its idea consists of the use of predefined fractional CFE models stored in an array. The set of models describes the range of fractional orders between 0 and 1 with predefined quantization step. In the paper, the accuracy analysis of the proposed approach is presented. The influence of various factors is examined during the operation of the switching mechanism between fractional orders. Results are verified by simulations and tests on PLC.

Keywords: Digital fractional order PID controller · CFE approximation · Accuracy analysis

1 Introduction

The concept of fractional calculus is being recognized for a better modeling and control of many dynamical systems in the last decade [5,10,12]. By extending the order of derivatives and integrals from integer to non-integer order, the advantages of resistance and better control are visible [6]. These benefits have resulted in the renewed interest in several applications of fractional order (FO) control. The conventional PID controllers are still widely applied in the industry, however, it is worth ensuring that their quality is improved. The FOPID, proposed by Podlubny in [14] is a generalization of classical PID with two additional parameters the order of fractional integration α and the order of fractional derivative β.

The implementation of fractional order operators used in FOPID on digital platforms can be done by approximation methods [1]. Many suggestions are proposed in the literature [15]. Regardless of the platform, these operations generate a heavy computational load, for example implementation on PLC [11] or on the microcontroller [7]. As a result of comparisons made in the article [8], the Continuous Fraction Expansion (CFE) approximation is used in this work.

The change of a fractional order during real-time work of a controller requires recalculating all the coefficients of the approximation describing the fractional order. That is why, the method proposed by the authors uses the switchable

A. Bartoszewicz et al. (Eds.): PCC 2020, AISC 1196, pp. 847–856, 2020.
https://doi.org/10.1007/978-3-030-50936-1_71

FOPID controller. Its idea consists of the use of predefined fractional CFE models stored in an array. Adjusting the selected fractional order from range 0.0–1.0 consists in the selection of suitable coefficients from memory. The influence of various factors should be examined during the operation of the switching mechanism between fractional orders. The crucial problem is the accuracy of the proposed method.

The paper is organized as follows. At the beginning, theoretical background of FO systems is recalled. The next chapter discusses the basic concept of a quickly adjustable FOPID controller. In addition, the experimental system is described. Later, the accuracy analysis of the proposed approach with simulations is presented. Furthermore, time calculations are conducted on a real system. The last section concludes this work with some insights.

2 Preliminaries

2.1 Fractional Order Systems - Definitions

The basic definition for non-integer systems is the differential-integer operator ${}_aD_t^\alpha$ [16], defined as follows:

$$ {}_aD_t^\alpha = \begin{cases} \frac{d^\alpha}{dt^\alpha} & \text{when } \alpha > 0 \\ 1 & \text{when } \alpha = 0 \\ \int_a^t (d\tau)^\alpha & \text{when } \alpha < 0 \end{cases}, \tag{1} $$

where a and t are the limits of integration and α is the order of operation. There is a lot of definitions of fractional derivative-integral operator, e.g. [3,4]. The most commonly used is the Grünwald and Letnikov definition of the FO operator expressed as

$$ {}_aD_t^\alpha f(t) = \lim_{h \to 0} h^{-\alpha} \sum_{j=0}^{[\frac{t-a}{h}]]} (-1)^j \binom{\alpha}{j} f(t - jh), \tag{2} $$

$$ \binom{\alpha}{j} = \begin{cases} \frac{\alpha(\alpha-1)\ldots(\alpha-j+1)}{j!} & j > 0 \\ 1 \quad j = 0 \end{cases}. \tag{3} $$

Another important definition is the one from Caputo, which is given as

$$ {}_aD_t^\alpha(t) = \frac{d^\alpha f(t)}{dt^\alpha} = \frac{1}{\Gamma(p-\alpha)} \int_0^t \frac{f^p(\tau)}{(t-\tau)^{\alpha+1-p}} d\tau, \tag{4} $$

where $\Gamma()$ is the gamma function of Euler Eq. (5), p is a positive integer satisfying limitation $p - 1 < \alpha < p$.

$$ \Gamma(x) = \int_0^\infty t^{x-1} e^{-t} dt \tag{5} $$

2.2 The Laplace Transform

The Laplace transform is a useful tool for analyzing and solving ordinary and partial differential equations. It can also be used for numerical solutions of fractional systems. The Laplace transform for Caputo operator is expressed as

$$\begin{aligned}
&\mathcal{L}({}_0^C D_t^\alpha f(t)) = s^\alpha F(s), \quad \alpha < 0 \\
&\mathcal{L}({}_0^C D_t^\alpha f(t)) = s^\alpha F(s) - \sum_{k=0}^{n-1} s^{\alpha-k-1}{}_0 D_t^k f(0), \\
&\alpha > 0, \quad n-1 < \alpha \leq n \in N
\end{aligned} \tag{6}$$

Thus, the inverse Laplace transform for a fractional order function is expressed as follows

$$L^{-1}(s^\alpha F(s)) ={}_0 D_t^\alpha f(t) + \sum_{k=0}^{n-1} \frac{t^{k-1}}{\Gamma(k-\alpha+1)} f^{(k)}(0^+), \qquad n-1 < \alpha \leq n, \quad n \in N \tag{7}$$

The Laplace transform of used later fractional order expression is defined by:

$$L^{-1}\left\{\frac{1}{(s)^m}\right\} = \frac{t^{m-1}}{\Gamma(m)} \tag{8}$$

2.3 CFE Approximation

In order to implement the operator s^α, we need to approximate it to integer order. For this calculation, we can use the Continuous Fraction Expansion Approximation. The CFE method is based on expanding expression written as:

$$\begin{aligned}
G(s) &\simeq a_0(s) + \cfrac{b_1(s)}{a_1(s) + \cfrac{b_2(s)}{a_2(s) + \cfrac{b_3(s)}{a_3(s) + \ldots}}} \\
&= a_0(s) + \frac{b_1(s)}{a_1(s)} + \frac{b_2(s)}{a_2(s)} + \ldots
\end{aligned} \tag{9}$$

The resulting finite dimensional approximation of fractional operators in the z-domain, can be expressed as:

$$\begin{aligned}
&\left(\omega(z^{-1})\right)^\gamma = G_{CFE}(z^{-1}, \gamma) = \left(\tfrac{1+a}{h}\right)^\gamma CFE\{\left(\tfrac{1-z^{-1}}{1+az^{-1}}\right)^\alpha\}_{M,M} \\
&= \tfrac{P_{\gamma M}(z^{-1})}{Q_{\gamma M}(z^{-1})} = \left(\tfrac{1+a}{h}\right)^\gamma \tfrac{CFE_N(z^{-1},\gamma)}{CFE_D(z^{-1},\gamma)} = \frac{\sum_{m=0}^{M} w_m z^{-m}}{\sum_{m=0}^{M} v_m z^{-m}}.
\end{aligned} \tag{10}$$

In (10) a is the coefficient depending on approximation type (for example: $a = 0$ for Euler approximation, $a = 1$ for Tustin approximation), h denotes the sample time, M is the order of approximation. Numerical values of coefficients w_m and v_m and different values of parameter a can be calculated for example with the use of MATLAB function given by Petras in [13]. This MATLAB function was applied in experiments described in the next section If the Tustin approximation is considered ($a = 1$) then $CFE_D(z^{-1}, \alpha) = CFE_N(z^{-1}, -\alpha)$ and the polynomial $CFE_D(z^{-1}, \gamma)$ can be given in the direct form (see [2]). Examples of polynomial $CFE_D(z^{-1}, \alpha)$ for $M = 1, 3, 5$ are given in the Table 1.

Table 1. Coefficients of CFE polynomials $CFE_{N,D}(z^{-1}, \gamma)$ for Tustin approximation with respect to [2]

Order M	w_m	v_m
$M = 1$	$w_1 = -\gamma$	$v_1 = \gamma$
	$w_0 = 1$	$v_0 = 1$
$M = 3$	$w_3 = -\frac{\gamma}{3}$	$v_3 = \frac{\gamma}{3}$
	$w_2 = \frac{\gamma^2}{3}$	$v_2 = \frac{\gamma^2}{3}$
	$w_1 = -\gamma$	$v_1 = \gamma$
	$w_0 = 1$	$v_0 = 1$
$M = 5$	$w_5 = -\frac{\gamma}{5}$	$v_5 = \frac{\gamma}{5}$
	$w_4 = \frac{\gamma^2}{5}$	$v_4 = \frac{\gamma^2}{5}$
	$w_3 = -\left(\frac{\gamma}{5} + \frac{2\gamma^3}{35}\right)$	$v_3 = -\left(\frac{-\gamma}{5} + \frac{-2\gamma^3}{35}\right)$
	$w_2 = \frac{2\gamma^2}{5}$	$v_2 = \frac{2\gamma^2}{5}$
	$w_1 = -\gamma$	$v_1 = \gamma$
	$w_0 = 1$	$v_0 = 1$

The parameters of discrete transfer function (10) can be also calculated using Matlab function *dfod1* available in [13]. The time response of the fractional operator based on (10) in k-th time moment is expressed as underneath:

$$y^+_{CFE}(k) = \frac{1}{v_0}\left[-\sum_{m=1}^{M} v_m y^+(k-m) + \sum_{m=1}^{M} w_m u^+(k-m)\right], \tag{11}$$

where $y^+_{CFE}(k-m)$ is the output signal and $u^+(k-m)$ is the input signal in $k-m$-th time moments respectively. v_m and v_m are coefficients of CFE polynomials from Table 1. It will be used in simulations and experiments presented in this paper.

3 The Quickly Adjustable FOPID

The FOPID controller is described by the following transfer function:

$$G_c(s) = k_p + k_i s^{-\alpha} + k_d s^{\beta}, \tag{12}$$

where k_p, k_i and k_d are parameters describing the proportional, integral and derivative parts, α, β are fractional orders of integral and derivative actions. The analytical form of the step response of the FOPID is as follows (see [8]):

$$\begin{aligned} y_a(t) = L^{-1}\left\{\frac{1}{(s)}(k_p + \frac{k_I}{s^\alpha} + k_D s^\beta)\right\} = \\ k_p + k_I \frac{t^\alpha}{\Gamma(\alpha+1)} + k_D \frac{t^{-\beta}}{\Gamma(1-\beta)} \end{aligned}, \tag{13}$$

where $\Gamma(..)$ is the complete Gamma function.

The main difficulty during practical use of FOPID (12) is caused by the fact that each change of fractional order α or β requires to re-calculate coefficients of CFE approximation with respect to Table 1. This takes a lot of time and causes problems during real time work of a control system (see [9]).

An idea of the proposed Quickly Adjustable FOPID controller (QAFOPID) is to apply "quantized" fractional orders with precalculated parameters w and v. Each set of parameters is prepared with respect to Table 1 and fixed sample time h. Next it is stored in non retentive memory of a controller.

Adjusting of a particular order α consists of choosing a proper approximation (10) describing the "quantized" fractional order nearest to the adjusted one. Imagine that we need to approximate the order α. Its approximation is defined by γ_k equal:

$$\gamma_k = k\Delta. \tag{14}$$

where:

$$k\Delta \leq \alpha \leq (k+1)\Delta,\ k = 1, ..K-1. \tag{15}$$

In (14) and (15) $\Delta = 1/K$ is the smallest quantization step. The increasing K decreases the quantization error of fractional order and simultaneously increases the use of memory. Next, the change of K or sample time h requires to recalculate the whole array with predefined, quantized approximations.

Consequently, the basic FO element s^{γ_k} can be expressed as the integer order element p^k where $p = s^\Delta$. This allows to express the transfer function of QAFOPID controller in the following form:

$$G_{QF}(p) = k_p + k_i \frac{1}{p^{k_\alpha}} + k_d p^{k_\beta}. \tag{16}$$

where $\alpha = k_\alpha \Delta$, $\beta = k_\beta \Delta$, $k_\alpha, k_\beta \in Z$. The transfer function (16) is integer order transfer function and its form allows to analyse the stability of the considered FO control system using Matignon theorem (see for example [1], pp. 21–22). When switching to the discrete transfer function (10) the pre-calculated transfer functions describing α and β are employed.

4 Accuracy Analysis

The accuracy of the proposed solution can be analyzed for PI and PD parts separately. Firstly PI part was examined. The approximation error as a function of the time $\Delta_{PI}(t)$ is expressed as follows:

$$\Delta_{PI}(t) = k_I \left(\frac{t_{PI}^{\alpha}}{\Gamma(\alpha+1)} - \frac{t_{PI}^{\alpha+\Delta}}{\Gamma(\alpha+1+\Delta)} \right). \tag{17}$$

The maximum error for the PI part Δ_{PImax} can be calculated as follows:

$$\frac{d\Delta_{PI}(t)}{dt} = 0.$$

The time moment t_{0PI} when the maximum is achieved equals to:

$$t_{0PI} = k_I \left(\frac{\alpha\Gamma(\alpha+1+\Delta)}{(\alpha+\Delta)\Gamma(\alpha+1)} \right)^{\frac{1}{\Delta}}. \tag{18}$$

and finally we obtain:

$$\Delta_{PImax}(t_{0PI}) = k_I \left(\frac{\alpha\Gamma(\alpha+1+\Delta)}{(\alpha+\Delta)\Gamma(\alpha+1)} \right)^N \frac{\Delta}{(\alpha+\Delta)\Gamma(\alpha+1)}. \tag{19}$$

The maximum approximation error Δ_{PDmax} for the PD part can be calculated analogically. This is shown underneath.

$$\Delta_{PD}(t) = k_D \left(\frac{t_{PD}^{-\beta}}{\Gamma(1-\beta)} - \frac{t_{PD}^{-\beta-\Delta}}{\Gamma(1-\beta-\Delta)} \right), \tag{20}$$

$$\frac{d\Delta_{PD}(t)}{dt} = 0,$$

$$t_{0PD} = k_D \left(\frac{(\beta+\Delta)\Gamma(1-\beta)}{\beta\Gamma(1-\beta-\Delta)} \right)^{\frac{1}{\Delta}}, \tag{21}$$

$$\Delta_{PDmax}(t_{0PD}) = k_D \left(\frac{(\beta+\Delta)\Gamma(1-\beta)}{\beta\Gamma(1-\beta-\Delta)} \right)^N \frac{\Delta}{\beta\Gamma(1-\beta-\Delta)}. \tag{22}$$

The both errors (17) and (20) as functions of time are shown in Figs. 1 and 2.

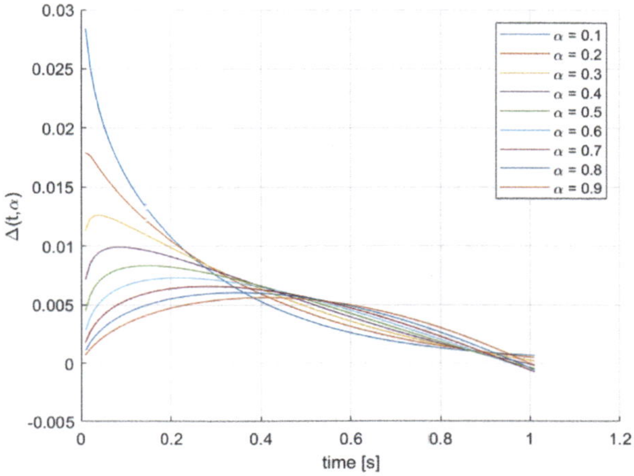

Fig. 1. The approximation error (17) as a function of time.

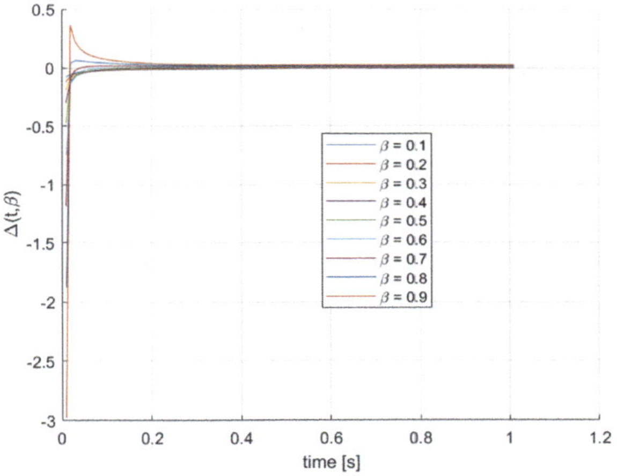

Fig. 2. The approximation error (20) as a function of time.

5 Experiments

Experiments were executed using controller PLC SIEMENS 1500 with CPU 1516. The hardware configuration is shown in Fig. 3. Collecting data and controlling is done by PROFINET in this laboratory stand. Each experiment contains two steps:

1. Firstly the fractional order is required to be set. In the "classic" situation the function calculating CFE coefficients is called. If the QAFOPID is applied, here the suitable set of parameters is read from memory.
2. Next the FB instance with these coefficients is executed.

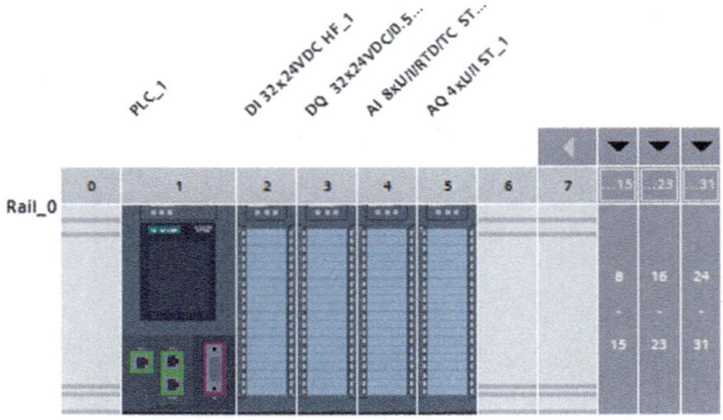

Fig. 3. The hardware configuration

Table 2. Components of the software.

POU	Name of component	Description
Organisation block	Main OB1	Main organisation block. It measures the runtime of specific organization blocks
System blocks	RT_INFO FC804	System function is calculating the time of the process
Organisation block	Cyclic OB30	Cyclic interrupt organization block. It is responsible for calling instance to calculate FOPID
Function block	FO_controller FB	Function block instance is calculating the response of FOPID controller. Depending on the version, it calls the appropriate instance of CFE method
Data block	FO_controller DB	Data block saves variables needed to calculate the response of FOPID with the CFE method
Function	Coeff FC	This function is calculating the coefficients of CFE approximation method according to [13]
Function block	CFE_calc FB2	Function block instance calculates the response of CFE approximation method according to (11). Coefficients are being calculated online
Data block	CFE_DB DB2	Data block correlated with FB2. It saves all variables needed to calculate the CFE approximation
Data block	CFE_wsp_DB DB3	The data block which stores previously calculated CFE approximation coefficients for QAFOPID
Function block	CFE_calc_QAFOPID FB3	Function block instance is calculating the response of CFE approximation method according to (11). It takes coefficients from CFE_wsp_DB
Data block	CFE_calc_QAFOPID DB4	Data block correlated with FB3. It saves all variables needed to calculate the CFE approximation response

The software was prepared using TIA PORTAL v13 with respect to the IEC61131.3 standard. The runtime of the user program was calculated by RT_INFO function.

In TIA Portal, two types of FOPID controllers have been implemented. The first one employes a dedicated function to calculate coefficients of CFE approximation while working. The second one uses a data block containing precalculated coefficients. A detailed description of the software in TIA Portal is provided in Table 2.

Table 3 shows the results of duration of the whole cyclic interruption organization block OB30, where the programs are executed. It is noticeable that when we use QAFOPID the duration is almost twice shorter in each case.

Table 3. Durations of calculation during the use of QAFOPID vs FOPID with directly calculated parameters.

α & β	Method	Duration [μs]
−0.1 & 0.9	CFE_table	**135**
	CFE_calculated	297
−0.5 & 0.5	CFE_table	**125**
	CFE_calculated	211
−0.9 & 0.1	CFE_table	**124**
	CFE_calculated	305
−0.4 & 0.6	CFE_table	**132**
	CFE_calculated	235
−0.9 & 0.9	CFE_table	**128**
	CFE_calculated	224

6 Conclusions

A quick adjustable FOPID controller has been proposed. The simulations illustrate the influence of various factors during the operation of the switching mechanism between fractional. The obvious benefits from experiments are that the QAFOPID is twice faster than the FOPID with coefficients calculated before each use. This approach allows to use the proposed controller in time-critical applications with variable fractional orders.

References

1. Caponetto, R., Dongola, G., Fortuna, L., Petras, I.: Fractional Order Systems: Modeling and Control Applications. World Scientific Series on Nonlinear Science, Series A, vol. 72. World Scientific Publishing, Singapore (2010)

2. Chen, Y.Q., Moore, K.L.: Discretization schemes for fractional order differentiators and integrators. IEEE Trans. Circuits Syst. - I Fundam. Theory Appl. **49**(3), 263–269 (2002)
3. Das, S.: Functional Fractional Calculus for System Identification and Controls. Springer, Heidelberg (2008)
4. Kaczorek, T.: Selected Problems of Fractional System Theory. Springer Verlag, Heidelberg (2011)
5. Majka, Ł., Klimas, M.: Diagnostic approach in assessment of a ferroresonant circuit. Electr. Eng. (2019). https://doi.org/10.1007/s00202-019-00761-5
6. Matignon, D.: Stability results for fractional differential equations with applications to control processing. Arch. Electr. Eng. (2019). https://doi.org/10.24425/aee.2019.129342
7. Matusiak, M., Ostalczyk, P.: Problems in solving fractional differential equations in a microcontroller implementation of an FOPID controller. In: IMACS-SMC Proceedings, Lille, France (1996)
8. Oprzędkiewicz, K.: Accuracy estimation of digital fractional order PID controller. In: Theory and Applications of Non-integer Order Systems (2017). https://doi.org/10.1007/978-3-319-45474-0_24
9. Oprzędkiewicz, K., Mitkowski, W., Gawin, E., Dziedzic, K.: The Caputo vs. Caputo-Fabrizio operators in modeling of heat transfer process. Bull. Pol. Acad. Sci. Tech. Sci. **66**, 501–507 (2018)
10. Oziablo, P., Mozyrska, D., Wyrwas, M.: A digital PID controller based on Grünwald-Letnikov fractional-, variable-order operator. In: Conference: 2019 24th International Conference on Methods and Models in Automation and Robotics (2019). https://doi.org/10.1109/MMAR.2019.8864688
11. Petras, I.: Realization of fractional order controller based on PLC and its utilization to temperature control. Transfer inovaci nr **14**, 34–38 (2009)
12. Petras, I.: Fractional derivatives, fractional integrals and fraction differential equations. Technical University of Kosice (2012)
13. Petras, I.: http://people.tuke.sk/igor.podlubny/USU/matlab/petras/dfod1.m
14. Podlubny, I.: Fractional Differential Equations. Academic Press, Cambridge (1999)
15. Sahoo, P.: Optimizing Current Strategies and Applications in Industrial Engineering, India (2019). https://doi.org/10.4018/978-1-5225-8223-6
16. Tepljakov, A., Alagoz, B.B., Yeroglu, C., Gonzalez, E., HosseinNia, S.H., Petlenko, E.: Fopid controllers and their industrial applications: a survey of recent results. IFAC-PapersOnLine **51**(4), 25–30 (2018)

Control of the Inverted Pendulum Using Quickly Adjustable, Discrete FOPID Controller

Krzysztof Oprzędkiewicz(✉), Klaudia Dziedzic, Maciej Rosół, and Jakub Żegleń

Department of Automatic Control and Robotics, AGH University, Al. Mickiewicza 30, 30-059 Krakow, Poland
{kop,kdz,mr}@agh.edu.pl, zeglenjakub@gmail.com

Abstract. In the paper the control of inverted pendulum by discrete, Quickly Adjustable Fractional Order PID (QAFOPID) controllers is addressed. The fractional order parts of the both controllers are approximated using CFE approximation. The fractional orders can be easily switched using predefined CFE coefficients loaded from memory. The QAFOPIDs were tuned using GWO optimizer and simulations. Results of simulations and experiments show that the use of QAFOPID controllers allows one to obtain good control performance in the sense of the considered cost function.

Keywords: Fractional order systems · Fractional PID control · Inverted pendulum · Real time system · GWO optimizer

1 Introduction

A control of an inverted pendulum has been presented by many Authors, for example in papers: [1,10,21]. This object is an unstable system with nonlinear dynamics. It is an example of the under actuated system – there are less inputs than numbers of degrees of freedom [13].

There are many articles presenting different implementations of controllers for the inverted pendulum. Among the most popular are the PID [13] and the LQR [18] controllers. There are also other articles discussing more advanced methods like the Lyapunov theorem based controller [8] and the fuzzy neural network controller [19].

The solution of FOPID proposed in this paper uses quickly adjustable FOPID controllers (QAFOPID) in control loops for pendulum and cart. The use of the proposed approach makes possible very quick switch the fractional orders of integration and derivative without calculation coefficients of the CFE approximation because these coefficients are loaded from memory. This allows saving time necessary to tuning the controllers during hard real-time work of the control system. The proposed solution is the test of use the QAFOPID in time

A. Bartoszewicz et al. (Eds.): PCC 2020, AISC 1196, pp. 857–869, 2020.
https://doi.org/10.1007/978-3-030-50936-1_72

critical systems. The paper is organized as follows. At the beginning the CFE approximation and the GWO optimizer are recalled. The QAFOPID algorithm is also proposed. Next, the inverted pendulum and its FOPID control system are recalled. Finally, tuning of the control system using simulations and experimental verification of results are given.

2 Preliminaries

2.1 The CFE Approximation

An implementation of the elementary operator s^γ at each digital platform (PLC, microcontroller) requires to apply an integer order, finite dimensional, discrete approximant. The most known are PSE (Power Series Expansion) and CFE (Continuous Fraction Expansion). They allow to estimate a non integer order element using digital FIR or IIR filter. The PSE approximant bases directly on a discrete version of GL definition and it has the form of FIR filter containing only zeros. However, its digital implementation to keep a good quality requires to apply long memory buffer (high order of the filter). The CFE approximant has the form of IIR filter containing both poles and zeros. It is faster convergent and easier to implement due to its relatively low order, typically not higher than 5. It is expressed as discrete transfer function $G_{CFE}(z^{-1},\gamma)$.

The discretization of fractional order element s^γ, $\gamma \in \mathbb{R}$ can be done with the use of the so called generating function $s \approx \omega(z^{-1})$. The new operator raised to power γ has the following form (see for example [3], [14], p. 119):

$$\begin{aligned}\left(\omega(z^{-1})\right)^\gamma &= G_{CFE}(z^{-1},\gamma) = \left(\tfrac{1+a}{h}\right)^\gamma CFE\{\left(\tfrac{1-z^{-1}}{1+az^{-1}}\right)^\alpha\}_{M,M} \\ &= \frac{P_{\gamma M}(z^{-1})}{Q_{\gamma M}(z^{-1})} = \left(\frac{1+a}{h}\right)^\gamma \frac{CFE_N(z^{-1},\gamma)}{CFE_D(z^{-1},\gamma)} = \frac{\sum_{m=0}^{M} w_m z^{-m}}{\sum_{m=0}^{M} v_m z^{-m}}\,. \end{aligned} \tag{1}$$

In (1) a is the coefficient depending on approximation type (for example: $a = 0$ for Euler approximation, $a = 1$ for Tustin approximation), h denotes the sample time, M is the order of approximation. Numerical values of coefficients w_m and v_m and different values of parameter a can be calculated for example with the use of MATLAB function *dfod1* given by Petras in [15]. This function was applied in simulations and experiments in this paper. If the Tustin approximation is considered ($a = 1$) then $CFE_D(z^{-1},\alpha) = CFE_N(z^{-1},-\alpha)$ and the polynomial $CFE_D(z^{-1},\alpha)$ can be given in the direct form (see [3]). Examples for the polynomial $CFE_D(z^{-1},\alpha)$ for $M = 1, 3, 5$ are given in Table 1. The approximator using the Muir recursion is presented for example in [17]. The detailed analysis of various forms of the CFE approximators has been given by [16].

Table 1. Coefficients of CFE polynomials $CFE_{N,D}(z^{-1}, \alpha)$ for Tustin approximation.

Order M	w_m	v_m
$M = 1$	$w_1 = -\alpha$	$v_1 = \alpha$
	$w_0 = 1$	$v_0 = 1$
$M = 3$	$w_3 = -\frac{\alpha}{3}$	$v_3 = \frac{\alpha}{3}$
	$w_2 = \frac{\alpha^2}{3}$	$v_2 = \frac{\alpha^2}{3}$
	$w_1 = -\alpha$	$v_1 = \alpha$
	$w_0 = 1$	$v_0 = 1$
$M = 5$	$w_5 = -\frac{\alpha}{5}$	$v_5 = \frac{\alpha}{5}$
	$w_4 = \frac{\alpha^2}{5}$	$v_4 = \frac{\alpha^2}{5}$
	$w_3 = -\left(\frac{\alpha}{5} + \frac{2\alpha^3}{35}\right)$	$v_3 = -\left(\frac{-\alpha}{5} + \frac{-2\alpha^3}{35}\right)$
	$w_2 = \frac{2\alpha^2}{5}$	$v_2 = \frac{2\alpha^2}{5}$
	$w_1 = -\alpha$	$v_1 = \alpha$
	$w_0 = 1$	$v_0 = 1$

2.2 The Quickly Adjustable Fractional Order PID Controller (QAFOPID Controller)

The Fractional Order PID controller (FOPID) controller is described by the following fractional order transfer function:

$$G_p(s) = k_p + k_i \frac{1}{s^\alpha} + k_d s^\beta. \tag{2}$$

In (2) k_p, k_i and k_d denote coefficients of proportional, integral and derivative actions respectively, α and β denote non integer orders of the both actions.

The main difficulty during practical use of FOPID (2) is caused by the fact that each change of fractional order α or β requires to re-calculate coefficients of CFE approximation with respect to Table 1. This takes a lot of time and causes problems during real time work of a control system (see [11]).

An idea of the proposed Quickly Adjustable FOPID controller (QAFOPID) is to apply "quantized" fractional orders with precalculated parameters w and v. Each set of parameters is prepared with respect to Table 1 and fixed sample time h. Next it is stored to non retentive memory of a controller.

Adjusting of a particular order α consists in choose a proper approximation (1) describing the "quantized" fractional order nearest to the adjusted one. Imagine that we need to approximate the order α. Its approximation is defined by γ_k equal:

$$\gamma_k = k\Delta_\gamma. \tag{3}$$

where:

$$k\Delta_\gamma \leq \alpha \leq (k+1)\Delta_\gamma,\ k = 1, ..K - 1. \tag{4}$$

In (3) and (4) $\Delta_\gamma = 1/K$ is the smallest quantization step. The increasing K decreases the quantization error of fractional order and simultaneously increases the use of memory. Next, the change of K or sample time h requires to recalculate the whole array with predefined, quantized approximations.

Consequently, the basic FO element s^{γ_k} can be expressed as the integer order element p^k where $p = s^{\Delta_\gamma}$. This allows to express the transfer function of QAFOPID controller in the following form:

$$G_{QF}(p) = k_p + k_i \frac{1}{p^{k_\alpha}} + k_d p^{k_\beta}. \tag{5}$$

where $\alpha = k_\alpha \Delta_\gamma$, $\beta = k_\beta \Delta_\gamma$, $k_\alpha, k_\beta \in \mathbb{Z}$. The transfer function (5) is integer order transfer function and its form allows to analyse the stability of the considered FO control system using Matignon theorem (see for example [2], pp. 21–22). During going to the discrete transfer function (1) the pre-calculated transfer functions describing α and β are employed.

2.3 The GWO Optimizer

In this paper, the optimization problem occurs. To minimize a cost function in order to find the optimum, the GWO algorithm is used. There is a wide range of optimization algorithms, the advantage of the GWO algorithm has been shown in [4].

The wolf hunting mechanism and a strictly defined herd hierarchy have become the basis of the Grey Wolf Optimizer algorithm [9]. The GWO algorithm starts with a random selection of initial solutions from the predetermined range. The population of wolves is analyzed every iteration in GWO in order to achieve the individual (best combination of features) that has the least cost function (best fitness).

In the herd, a social dominant hierarchy is held. The wolves are divided into four types of wolf: alpha, beta, delta and omega. The leader is the alpha wolf who is a decision-maker. The second best individual is the beta wolf. The beta helps the alpha. The next one is for the delta wolves, they occupy various positions in the herd like hunters, guards, and scouts. The rest of the individuals are considered as the omega wolves.

The hunting method involves three procedures.

1. Tracking the victim
   While hunting, grey wolves follow the prey. Each agent tries to update its position to the location of the victim. It is shown by equations:

$$S = 2o * r_1 - o \tag{6}$$

$$U = 2 * r_2 \tag{7}$$

$$W = |U * X_p(t) - X(t)| \tag{8}$$

$$X(i+1) = X_p(i) - W * S \tag{9}$$

where X_p indicates the prey location vector in the ith iteration, and X represents the agent position, S, U and W mean vector coefficients, t is current iteration, o is proportional coefficient, r_1 and r_2 are random numbers between [0,1].

2. Harassment, approaching the victim
   Individuals can identify the location of the victim. In the GWO, the hunting is guided by the alpha, the beta and the delta. Mathematically, we assume that the most important wolves have the best solutions. The omega agents are forced to relocate themselves towards the given locations. The grey wolf position can be expressed by

$$X(i+1) = \frac{X_\alpha + X_\beta + X_\delta}{3}, \tag{10}$$

   where $X(i+1)$ is the position of wolf in next iteration, X_α, X_β, X_δ are the locations of the alpha, the beta and the delta wolves.
3. Attack of prey
   When the prey is trapped, grey wolves end the hunt. For the mathematical purpose, the coefficient a linearly decreases from 2 to 0 with the increasing of iteration number.

There exist several stop criteria that can be used in heuristic algorithms. The maximum number of iterations will be used for this work.

3 The Inverted Pendulum and Its Models

The construction of the inverted pendulum was presented for example in [20]. Considered inverted pendulum on a cart can move only in one direction on rails (see Fig. 1). This system has an infinite number of equilibrium points, but in the considered case we are interested only in the one unstable equilibrium point when the pendulum is in the upper vertical position. The goal of control is to stabilize the system in its unstable, upper position. The mathematical model is described by determining the kinetic and potential energy in the system. Define the state vector as follows:

$$x = [\theta \quad \dot{\theta} \quad x \quad \dot{x}]^T = [x_1, x_2, x_3, x_4]^T \tag{11}$$

Then the non linear state equation (see [20]) takes the following form:

$$\begin{cases} \dot{x}_1 = x_2 \\ \dot{x}_2 = \frac{1}{q}(FmL\cos x_1 - bmL\dot{x}_3\cos x_1 - (M+m)mLg\sin x_1 + m^2L^2\dot{x}_1^2\sin x_1\cos x_1 \\ \dot{x}_3 = x_4 \\ \dot{x}_4 = \frac{1}{q}((I+mL^2)(-F+b\dot{x}_3 - mL\dot{x}_1^2\sin x_1) + m^2L^2g\sin x_1\cos x_2) \end{cases} \tag{12}$$

where:

$$q = m^2L^2\cos^2 x_1 - (M+m)(I+mL^2) \tag{13}$$

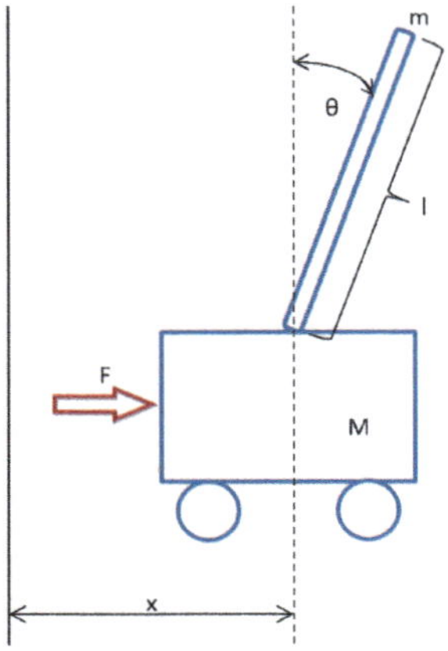

Fig. 1. The inverted pendulum [13].

In (12) and (13) M denotes mass of the cart, m is the mass of the pendulum, θ is the angular position of the pendulum, g denotes the standard gravity, L is the length of the pendulum, I is moment of inertia of the pendulum, F describes the force applied to the cart and x is cart position.

The above model is non-linear, so it can be linearized around unstable equilibrium point $x_0 = [0 \quad 0 \quad 0 \quad 0]^T$. For small values of x_1 it can be assumed that $\sin x_1 \approx x_1$ and finally we obtain the following linear model:

$$\ddot{x}_1 = \frac{(M+m)g}{ML}x_1 + \frac{b}{ML}\dot{x}_3 - \frac{F}{ML} \tag{14}$$

$$\ddot{x}_3 = -\frac{mg}{M}x_1 - \frac{b}{M}\dot{x}_3 + \frac{F}{M} \tag{15}$$

The linear model (14), (15) can be expressed as the following state equation:

$$\begin{cases} \dot{x} = Ax + Bu \\ y = Cx \end{cases}. \tag{16}$$

where A, B and C are defined as follows:

$$A = \begin{bmatrix} 0 & \frac{(M+m)g}{ML} & \frac{b}{ML} & 0 \\ 1 & 0 & 0 & 0 \\ 0 & -\frac{mg}{M} & -\frac{b}{M} & 0 \\ 0 & 0 & 1 & 0 \end{bmatrix}. \tag{17}$$

$$B = \begin{bmatrix} -\frac{1}{ML} \\ 0 \\ \frac{1}{M} \\ 0 \end{bmatrix}. \tag{18}$$

$$C = \begin{bmatrix} 0\,1\,0\,0 \\ 0\,0\,0\,1 \end{bmatrix}. \tag{19}$$

The above model will be implemented at SIMULINK and employed to tune the control system.

4 The Control of the Pendulum

4.1 Control Algorithms

In the considered case the variable structure control system is applied. The moving from the lower to the upper position is done by the swing-up controller. Its idea is to react on changes in the total value accumulated in the mechanical energy system [7,20]. This information is obtained from the feedback loop.

When energy is properly added to the system, it is possible to move the pendulum to the upper, unstable equilibrium point. Around it, the control algorithm is switched to the QAFOPID. Its task is to keep the pendulum around the unstable equilibrium point.

The energy E of the system in the unstable equilibrium point is equal to 0. This value can be calculated by the following formula:

$$E = \frac{\dot{\theta} I}{2} + mgl(\cos\theta - 1) \tag{20}$$

This allows to calculate the control signal of the swing-up controller as follows:

$$u = -\dot{\theta}\cos\theta \tag{21}$$

In order to properly control the pendulum around the equilibrium point, two controllers are needed. The first one uses information about the position of the pendulum while the other uses the position of the cart. Next, both control signals are summarized. The whole control system is shown in the Fig. 2.

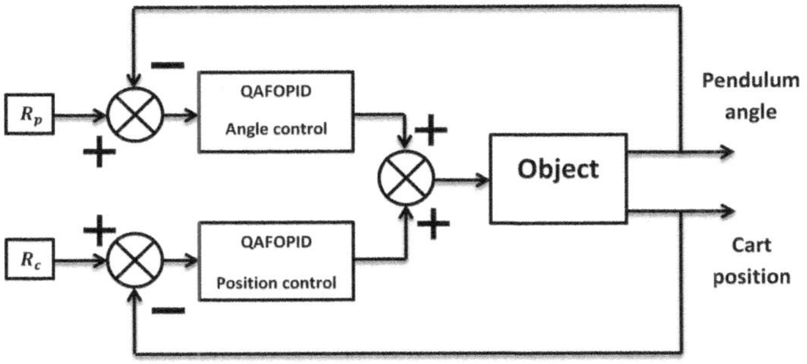

Fig. 2. Control system of the pendulum [1].

In the control system shown in the Fig. 2 the QAFOPID controllers described by (5) are used. The first controller cooperates with the pendulum, the second with the cart:

$$G_{cp}(s) = k_{pp} + k_{pi}\frac{1}{s^{\alpha_p}} + k_{pd}s^{\beta_p}. \tag{22}$$

$$G_{cc}(s) = k_{cp} + k_{ci}\frac{1}{s^{\alpha_c}} + k_{cd}s^{\beta_c}. \tag{23}$$

In (22) and (23) $k_{(p,c)(p,i,d)}$ denote coefficients of proportional, integral and derivative actions for the pendulum and cart controllers respectively, $\alpha_{p,c}$, $\beta_{p,c}$ denote orders of integration and derivative actions for the both controllers respectively. The use of QAFOPID allows to express all the fractional orders as follows:

$$\begin{aligned} \alpha_p &= k_{\alpha p}\Delta\gamma, \\ \beta_p &= k_{\beta p}\Delta\gamma, \\ \alpha_c &= k_{\alpha c}\Delta\gamma, \\ \beta_c &= k_{\beta c}\Delta\gamma. \end{aligned} \tag{24}$$

This notation uniquely describes each fractional order as a number of elementary steps $\Delta\gamma$ and it will be employed during further considerations. All the parameters of both controllers determine the control quality in the system. They can be collected as the following vector q:

$$q = [k_{pp}, k_{pi}, k_{\alpha p}, k_{\beta p}, k_{cp}, k_{ci}, k_{\alpha c}, k_{cd}, k_{\beta c}]^T. \tag{25}$$

4.2 GWO Tuning of QAFOPID Controllers

The quality of the control can be estimated by the following cost function, determined by parameters of both controllers:

$$I_v(q) = \int_0^\infty \left(\dot{\theta}^2(t) + \dot{x}^2(t)\right) dt. \tag{26}$$

The cost function describes the positions of pendulum and cart. The cost functions was optimized using GWO algorithm presented previously and simulations. The results are presented in the next section.

5 Simulations and Experiments

Simulations were done using the simulink model presented in the Fig. 3. Next results were verified experimentally using real model, produced by INTECO company [5] and shown in the Fig. 4. The QAFOPID controllers were used to control both cart and pendulum position. All parameters of the both controllers are collected in the vector q, defined by (25). The vector q_0 optimal in the sense of the cost function (26) was found using GWO algorithm described by (8)–(10).

The number of wolfs was equal to 5, the number of iterations was equal to 30. Results are presented by the Table 2. This table presents cost function (26) for simulations and experiments as well as duration of calculations during tuning the control system using GWO. Trends of all signals during simulations and experiments are shown in Figs. 5 and 6.

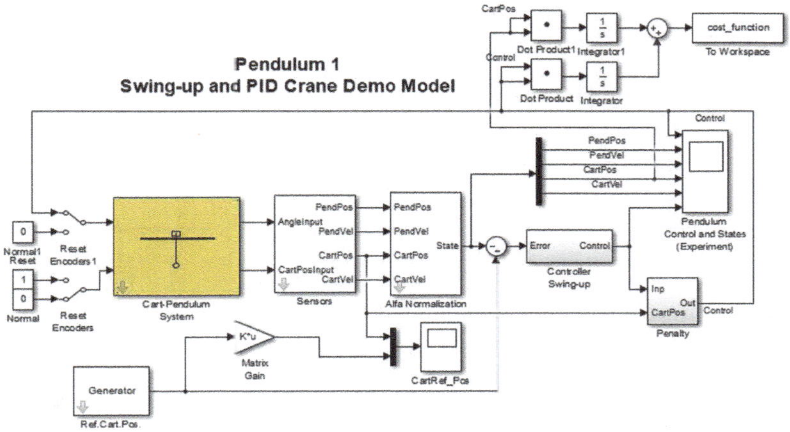

Fig. 3. The simulink based control system

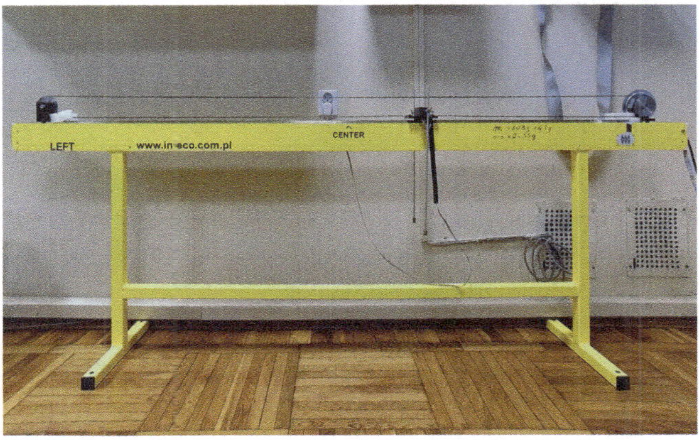

Fig. 4. The experimental system

Table 2. Control system parameters minimizing cost function (26)

Vector q_0	$[4.1327, 55.1710, 94, -1582.17614, 10, 64.4704, 9.7304, 9, 76.7544, 5]$
Cost function-simulations	0.0151
Cost function-experiment	0.002053
Duration of tuning [s]	604.2823

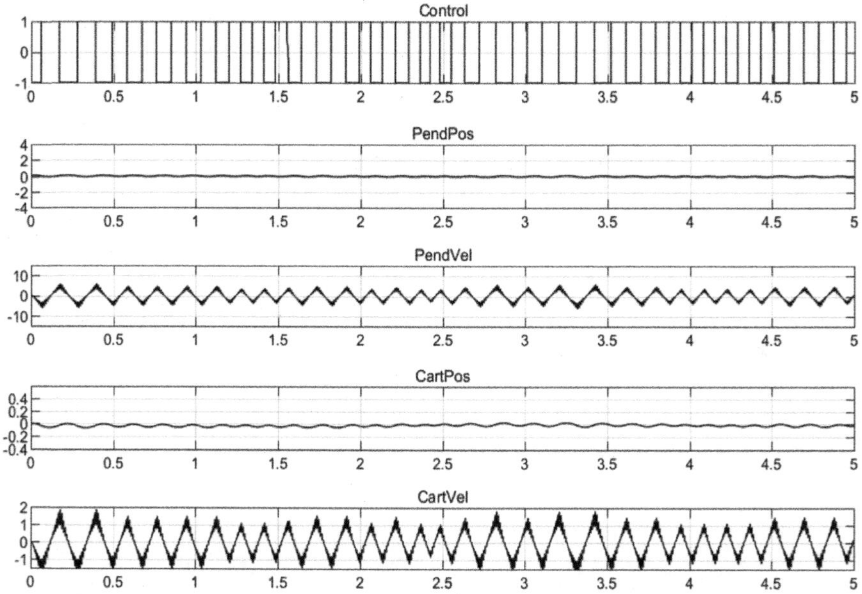

Fig. 5. Trends of all variables for system optimal in the sense of cost function (26), simulations performed in Simulink.

From Table 2 and Figs. 5, 6 the good compliance experiments to simulations can be observed as well as the good control performance in the sense of cost function (26).

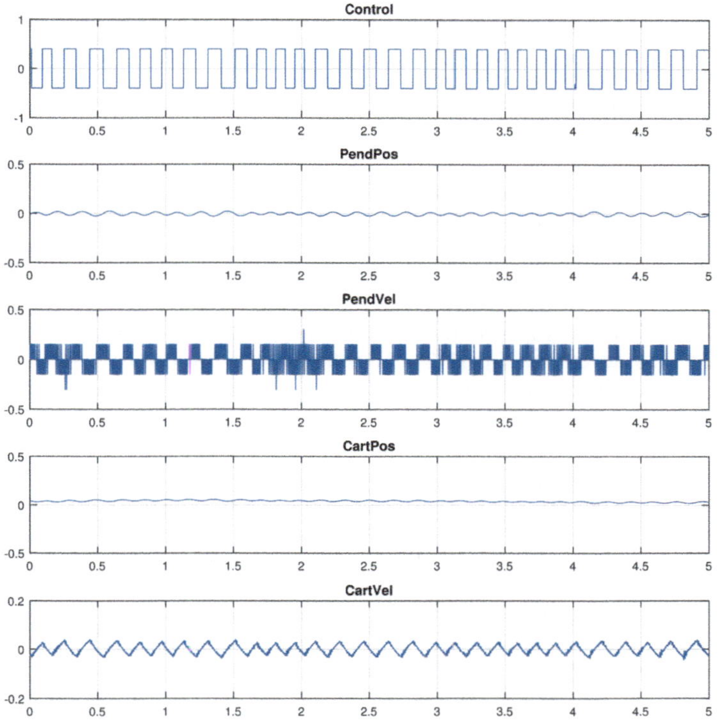

Fig. 6. Trends of all variables for system optimal in the sense of cost function (26), experiment performed on real object.

6 Conclusions

The main conclusion from the paper is that the proposed QAFOPID controller is able to assure good control performance in the sense of considered cost function. The compliance simulations to experiments is also good.

The future work will cover u.a. comparison to classic PID controller, the stability analysis as well as PLC implementation of the proposed control algorithm.

Acknowledgment. This paper was sponsored by AGH project no 16.16.120.773.

References

1. Agarwal, H., Singh, A.P., Srivastava, P.: Fractional order controller design for inverted pendulum on a cart system (POAC). WSEAS Trans. Syst. Control **10**, 172–178 (2015). E-ISSN 2224-2856
2. Caponetto, R., Dongola, G., Fortuna, l., Petras, I.: Fractional Order Systems. Modeling and Control Applications. World Scientific Series on Nonlinear Science, Series A, vol. 72. World Scientific Publishing

3. Chen, Y.Q., Moore, K.L.: Discretization schemes for fractional order differentiators and integrators. IEEE Trans. Circ. Syst. I Fundam. Theory Appl. **49**(3), 263–269 (2002)
4. Dziedzic, K.: Identification of fractional order transfer function model using biologically inspired algorithms. In: Automation 2019 (2019). https://doi.org/10.1007/978-3-030-13273-6_5
5. http://www.inteco.com.pl/products/pendulum-cart-control-system/
6. Jia-Jun, W.: Position and speed tracking control of inverted pendulum based on double PID controllers. In: 2015 34th Chinese Control Conference (CCC), Hangzhou, pp. 4197–4201 (2015). https://doi.org/10.1109/ChiCC.2015.7260286
7. Lam, J.: Control of an Inverted Pendulum (2018). https://pdfs.semanticscholar.org/b6b2/f079bb6011a28ccd5d28d08df00f57ef5697.pdf
8. Maruki, Y., Kawano, K., Suemitsu, H., Matsuo, T.: Adaptive backstepping control of wheeled inverted pendulum with velocity estimator. Int. J. Control Autom. Syst. **12**, 1040–1048 (2014)
9. Mirjalili, S., Mohammad Mirjalili, S., Lewis, A.: Grey wolf optimizer. Adv. Eng. Softw. **69**, 46–61 (2014)
10. Mishra, S.K., Chandra, D.: Stabilization and tracking control of inverted pendulum using fractional order PID controllers. J. Eng. (2014). Article ID 752918, 9 pages
11. Oprzedkiewicz, K., Gawin, E., Gawin, T.: Real-time PLC implementations of fractional order operator. In: Szewczyk, R., Zielinski, C., Kaliczynska, M. (eds.) Automation 2018: Innovations in Automation, Robotics and Measurement Techniques. Advances in Intelligent Systems and Computing, vol. 743, pp. 36–51. Springer, Cham (2018). ISSN 2194-5357
12. Ostalczyk, P.: Discrete Fractional Calculus. Applications in Control and Image Processing. Series in Computer Vision, vol. 4. World Scientific Publishing, Singapore (2016)
13. Paliwal, S., Pathak, V.K.: Analysis & control of inverted pendulum system using PID controller. J. Eng. Res. Appl. **7**(5), 01–04 (2018). ISSN 2248-9622. http://www.ijera.com/papers/Vol7_issue5/Part-4/A0705040104.pdf
14. Petráš, I.: Fractional - order feedback control of a DC motor. J. Electr. Eng. **60**(3), 117–128 (2009)
15. Petráš, I. http://people.tuke.sk/igor.podlubny/USU/matlab/petras/dfod1.m
16. Stanislawski, R., Latawiec, K.J., Lukaniszyn, M.: A comparative analysis of Laguerre-based approximators to the Grunwald-Letnikov fractional-order difference. Math. Probl. Eng. (2015). Article ID 512104, 10 pages. https://doi.org/10.1155/2015/512104
17. Vinagre, B.M., Podlubny, I., Hernandez, A., Feliu, V.: Some approximations of fractional order operators used in control theory and applications. Fract. Cal. Appl. Anal. **3**(3), 231–248 (2000)
18. Wang, H., Dong, H., He, L., Shi, Y., Zhang, Y.: Design and simulation of LQR controller with the linear inverted pendulum. In: 2010 International Conference on Electrical and Control Engineering, Wuhan, pp. 699–702 (2010). https://doi.org/10.1109/iCECE.2010.178
19. Yu, L.H., Jian, F.: An inverted pendulum fuzzy controller design and simulation. In: 2014 International Symposium on Computer, Consumer and Control, Taichung, pp. 557–559 (2014). https://doi.org/10.1109/IS3C.2014.151

20. Żegleń, J.: Advanced control methods of an inverted pendulum on the card implemented in PLC (Zaawansowane metody sterowania modelem laboratoryjnym wahadła na wózku implementowane w sterowniku PLC). Master thesis at AGH University under supervision M. Rosół (2018)
21. Żegleń, J.: The application of an adaptive controller combined with the LQR controller for the inverted pendulum. Pomiary Automatyka Robotyka **4**(2019), 47–54 (2019). https://doi.org/10.14313/PAR_234/47

On Stabilization of Linear Descriptor Control Systems with Multi-order Fractional Difference of the Caputo-Type

Ewa Pawluszewicz(✉)

Bialystok University of Technology, Bialystok, Poland
e.pawluszewicz@pb.edu.pl

Abstract. The descriptor linear control systems described by the Caputo-type h-difference multi-order fractional operator are considered. Problems of stability and stabilizability for these class of systems are discussed. Conditions for stability and stabilizability of given system are discussed.

Keywords: Fractional control system · Caputo-type h-difference fractional operator · Descriptor system · Stability · Stabilization

1 Introduction

It is known that there exists a wide class of control systems that are not stable, but after implementing some feedbacks they become stable. On the other hand, in nature a lot of physical phenomena can be modelled by fractional or integral order systems, see for example [4,15]. Comparing with the classical integer models, fractional models provide a better description of behavior real phenomena and processes, see for example [1,2,17]. A question that appears when considering the practical usage of fractional differential or difference operators is which kind of operator should be taken into consideration in the given case. Usually the answer depends on a given problem and observed data. But anyway, the right choice is not easy. The fractional order Caputo integro-differential operator usually is used in real problems related to zero initial conditions and in problems in which there is no need to take into account the effect of theirs initialization/history.

On the other hand, considering the use of digital tools, approximation or discretization of fractional order operators should be used in practice. To this aim their generalizations to fractional difference forms is used.

Our goal is to study stability and stabilizability problems of descriptor difference systems of fractional multi-order with Caputo-type difference and with a sampling step h. In fact descriptor (singular) systems can be met for example in electronic, economic, dynamic balances of mass and energy, see for example [3,5]. The problem of stability for h-fractional order difference nonsingular

A. Bartoszewicz et al. (Eds.): PCC 2020, AISC 1196, pp. 870–878, 2020.
https://doi.org/10.1007/978-3-030-50936-1_73

system was consider in [11,13,18]. The conditions for delay-independent stability of the continuous-time linear singular systems with the Caputo differential operator was given in [19]. The stabilization problem of linear multi-parameter fractional difference control systems with Grünwald-Letnikov–type fractional h-difference operator was touch in [10]. Conditions for stabilization of positive descriptor systems with Grünwald-Letnikov fractional operator by decentralized controller were given in [7,16]. Taking into account relation between fractional order operators (see [8]), these conditions need not be valid for systems with Caputo fractional difference operator. Now, conditions under which the descriptor system with fractional muli-order Caputo type h-difference is stable and next stabilizable are study.

2 Preliminaries

Let $h > 0$ and $a \in \mathbb{R}$. Then $(h\mathbb{N})_a := \{a, a+h, a+2h, ...\}$. Consider a function $x : (h\mathbb{N})_a \to \mathbb{R}$. The forward h-difference operator is classically defined as $(\Delta_h x)(t) = \frac{x(t+h)-x(t)}{h}$. An extension of q-fold application of the difference operator Δ_h leads to the fractional h-sum of order $\alpha > 0$ for function $x : (h\mathbb{N})_a \to \mathbb{R}$.

Let

$$\varphi_\mu(n) = \binom{n+\mu-1}{n}, \quad \text{for } n \in \mathbb{N}_0;$$

defines a family of binomial sequences on $\mathbb{Z}$ parameterized by $\mu > 0$. Following [12], define the convolution of sequences φ_μ and $\overline{x}(s) := x(a+sh)$, $n \in \mathbb{N}_0$ as follows:

$$(\varphi_\mu * \overline{x})(n) := \sum_{s=0}^{n} \binom{n-s+\mu-1}{n-s} \overline{x}(s)$$

The *Caputo–type h-difference fractional order operator* ${}_a\Delta_{h,*}^{\alpha}$ *(of order α)* for a function $x : (h\mathbb{N})_a \to \mathbb{R}$ is defined by

$$\left({}_a\Delta_{h,*}^{\alpha} x\right)(t) = h^{-\alpha}\left(\tilde{\varphi}_{1-\alpha} * \Delta_{h=1}\overline{x}\right)(n)$$

for any $t = a + (1-\alpha)h - nh$ and $\overline{x}(n) = x(a+nh)$.

Proposition 1. *[12] Let $a \in \mathbb{R}$ and $\alpha \in (0,1)$. Define $y(n) := \left({}_a\Delta_h^{\alpha} x\right)(t)$, where $t \in (h\mathbb{N})_{a+(1-\alpha)h}$ and $t = a + (1-\alpha)h + nh$. Then*

$$\mathcal{Z}\left[\left({}_a\Delta_{h,*}^{\alpha} x\right)(t)\right](z) = z\left(X(z) - \frac{z}{z-1}x(a)\right)\left(\frac{hz}{z-1}\right)^{-\alpha},$$

where $X(z) = \mathcal{Z}[\overline{x}](z)$ is the classical $\mathcal{Z}$-transform of function $\overline{x}(n)$.

It is known that in the solution of a linear fractional order difference equations naturally occurs *Mittag-Leffler function*:

$$\mathcal{E}_{(\alpha,\beta)}(\lambda, n) := \sum_{k=0}^{\infty} \lambda^k \varphi_{k\alpha+\beta}(n-k)$$

Letter on a special attention will be paid on the Mittag–Leffler function that arises from solution of fractional order h-difference equations with Caputo-type operator, i.e. on function $\mathcal{E}_{(\alpha,1)}$.

Proposition 2. *[9] Let $\alpha \in (0,1]$ and $a = (\alpha - 1)h$. Initial value problem $({}_a\Delta^{\alpha}_{h,*}x)(t) = Ax(t+a)$, $x(a) = x_0$, where $A \in \mathbb{R}^{n\times n}$, has the unique solution given by*

$$x(t) = \mathcal{E}_{(\alpha,1)}(Ah^{\alpha}, n)x_0$$

for any $t \in (h\mathbb{N})_a$.

3 Linear Descriptor Systems with *h*-Difference Fractional Caputo-Type Operator

Let us consider the following fractional order discrete-time control system

$$E\left({}_{t_0}\Delta^{\alpha_i}_{h,*}x_i\right)(t+h) = \sum_{j=1}^{n} a_{ij}x_j(t) + \sum_{l=1}^{m} b_{il}u_l(t)\,, \tag{1}$$

where $\alpha_i \in (0,1]$, $i = 1,\dots,n$, $h > 0$, $x : (h\mathbb{N})_{t_0} \to \mathbb{R}^n$ denotes a state vector, $u : (h\mathbb{N})_{t_0} \to \mathbb{R}^m$ - a control vector and $a_{ij} \in \mathbb{R}$, $b_{il} \in \mathbb{R}$, $E \in \mathbb{R}^{n\times n}$ is a real matrix with constant coefficients such that rank$E = r \leq n$. If $\det E = 0$ then system (1) is called the *singular (descriptor)* system. The vector $x_0 = \left[x_1(t_0) \dots x_n(_0)\right]^T \in \mathbb{R}^n$ is said to be a *consistent initial state* if the system (1) posses at least one solution.

Let

$$A := [a_{ij}] \tag{2}$$

The descriptor system is solvable if and only if the matrix pair (E, A) is regular, i.e. $\det(\lambda E - \gamma A) \neq 0$ for some $(\lambda_\iota, \gamma_\iota) \in \mathbb{C}^2 \setminus \{(0,0)\}$. *Generalized eigenvalues* of the regular matrix pair (E, A) are pairs $(\lambda_\iota, \gamma_\iota) \in \mathbb{C}^2$, such that $\det(\lambda_\iota E - \gamma_\iota A) = 0$, $\iota = 1,\dots,n$. If $E \in \mathbb{C}^{n\times n}$ the range of E, denoted by $\mathcal{R}(E)$, and the null space of E, $\{x : Ex = 0\}$, by $\mathcal{N}(E)$ then $\dim\mathcal{R}(E) + \dim\mathcal{N}(E) = n$ The *index* of the system (1), denoted by $Ind(E, A)$, is defined as the degree of nilpotency, i.e as the least nonnegative integer ν such that $\mathcal{N}(E^{\nu}) = \mathcal{N}(E^{\nu+1})$. For systems that are regular and of index at most one there exists a unique solution for all admissible controls with a consistent initial condition.

Recall that for any matrix $E \in \mathbb{R}^{n\times n}$ there exists a unique matrix *Drazin inverse* of E denoted as E^D, i.e. matrix E^D such that $E^DE + EE^D = E^D$, $E^DEE^D = E^D$ and $E^DE^{\nu+1} = E^{\nu}$, $\nu = Ind(E)$. The Driazin inverse is unique. Methods of determination Driazin inverse can be found for example in [3,6,20].

Let matrix A be given by (2) and

$$\hat{E} = (\lambda E - hA)^{-1}E, \quad \hat{A} = (\lambda E - hA)^{-1}A, \quad \hat{B} = (\lambda E - hA)^{-1}B.$$

Theorem 1. *Let $\alpha \in (0,1]$ and $x : (h\mathbb{N})_{t_0} \to \mathbb{R}^n$. If (E, A) is a regular pair and of the index at most one, then system*

$$E\left({}_{t_0}\Delta_{h,*}^{\alpha_i} x_i\right)(t+h) = \sum_{j=1}^{n} a_{ij} x_j(t),\ j = 1, \ldots, n \tag{3}$$

with initial condition $x(t_0) = x_0 = [x_{01}, \ldots, x_{0n}]^T$ has the unique solution $x(t) = [x_1(t), \ldots, x_n(t)]^T$ where

$$x_j(t) = \mathcal{E}_{(\alpha_j,1)}\left(h^{\alpha_j} \hat{E}^D \hat{A}, \frac{t-t_0}{h}\right) \hat{E}\hat{E}^D x_{0_j}, \tag{4}$$

where $j = 1, \ldots, n$. A vector x_0 is the consistent initial state for the equation (3) if and only if $x_0 = \hat{E}\hat{E}^D x_0$.

Proof. Since the pair $(E\ A)$ is regular, then there exists a nonsingular matrix T such that $T^{-1}ET = \begin{bmatrix} C & 0 \\ 0 & N \end{bmatrix}$ and $T^{-1}AT = \begin{bmatrix} \lambda C - I & 0 \\ 0 & \lambda N - I \end{bmatrix}$ where C is nonsingular and N is nilpotent. Putting $x = T\xi = T\begin{bmatrix} \xi_1 \\ \xi_2 \end{bmatrix}$, system (3) can be decomposed as

$$C\left(\Delta_{h,*}^{\alpha} \xi_1\right)(t+h) = (\lambda C - I)\xi_1(t) \tag{5}$$

$$N\left(\Delta_{h.*}^{\alpha} \xi_2\right)(t+h) = (\lambda N - I)\xi_2(t). \tag{6}$$

By similar arguments as in [9] and by Proposition 2 it follows that

$$\xi_{1j}(t) = \mathcal{E}_{(\alpha_j,1)}\left(h^{\alpha_j} C^{-1}(\lambda C - I), \frac{t-t}{h}\right) \xi_{1j}(0).$$

for $j = 1, \ldots, \nu$ and $\xi_1 = [\xi_{11}, \ldots, \xi_{i\nu}]$.

Multiplying Eq. (6) by $N^{\nu-i}$, $i = 1, \ldots, i = \nu - 1$, we find $\xi_2 = 0$. Next, by properties of the Driazin's matrix, and by some arguments as in [3], it follows that for $j = 1, \ldots, n$ one has

$$x_j(t) = \mathcal{E}_{(\alpha_j,1)}\left(h^{\alpha_j} \hat{E}^D \hat{A}, \frac{t-t_0}{h}\right) \hat{E}\hat{E}^D x_0,$$

and $x(t) = [x_{11}, \ldots, x_{1n}]$. Characterization of the consistent initial state follows directly from the form of solution (4). □

4 Stability

Recall that vector $x^{\mathrm{eq}} = [x_1^{\mathrm{eq}}, \ldots, x_n^{\mathrm{eq}}]^T$ is an *equilibrium point* of the system

$$E\left({}_{t_0}\Delta_{h,*}^{\alpha_i} x_i\right)(t+h) = \sum_{j=1}^{n} a_{ij} x_j(t),\ \imath = 1, \ldots, n \tag{7}$$

where $t \in (h\mathbb{N})_{t_0}$ if and only if

$$E\left({}_{t_0}\Delta^{\alpha_i}_{h,*} x_i^{\text{eq}}\right)(t+h) = \sum_{j=1}^{n} a_{ij} x_j^{\text{eq}}$$

for all $n \in \mathbb{N}_0$. The equilibrium $x^{\text{eq}} = 0$ of system (7) is said to be

(a) *stable* if, for each $\epsilon > 0$, there exists $\delta = \delta(\epsilon) > 0$ such that $\|x_0\| < \delta$ implies $\|\overline{x}(t+a)\| < \epsilon$, for all $k \in \mathbb{N}_0$.
(b) *asymptotically stable* if it is stable and there exists $\delta > 0$ such that $\|x_0\| < \delta$ implies $\lim_{t\to\infty} \overline{x}(t) = 0$.

System (7) is called *stable* (respectively *asymptotically stable*) if their equilibrium $x^{\text{eq}} = 0$ is stable (respectively asymptotically) stable.

Theorem 2. *Let $\alpha_i \in (0,1]$, $i = 1,\dots,n$. Suppose that (E,A), where A is given by (2), is the regular pair and $Ind(E,A) \le 1$. Let R be the set of all roots of the equation*

$$\det\left(E - \frac{1}{z}\Lambda_\alpha \hat{E}^D \hat{A}\right) = 0.$$

where $\Lambda_\alpha = \text{diag}\{\left(\frac{hz}{z-1}\right)^{\alpha_i} : i = 1,\dots,n\}$. If all elements from R are strictly inside the unit circle, then the system (7) is asymptotically stable.

Proof. From the Weierstrass-Kronecker canonical form if follows that system (1) can be decomposed onto fractional h-difference subsystem and pure algebraic subsystem. Then only fractional subsystem affects on stability. So, the thesis follows from Theorem 1 and properties of the Driazin's inverse of E by using the same reasoning as in [12]. □

Proposition 3. *Suppose that (E,A), where A is given by (2), is the regular pair of the index at most one and with finite eigenvalues $\chi_\iota = \frac{\lambda_\iota}{\gamma_\iota}$, $\iota = 1,\dots,n$. Then the system (7) is asymptotically stable if and only if for each $\chi_\iota \in Spec(E,A)$ holds the following conditions*

(a) $\arg\chi_\iota \in \left[\pi\alpha(2n+\frac{1}{2}), 2\pi(1+n\alpha-\alpha\frac{1}{4}\right]$ *for all* $\iota = 1,\dots,n$
(b) $|\chi_\iota| < |w_\iota|$ *for all* $\iota = 1,\dots,n$

where $\arg\chi_\iota$ and $|\chi_\iota|$ are respectively the main argument and modulus of $\chi_\iota \in Spec(\hat{E}^D\hat{A})$ and $|w_\iota| = \left(\frac{2}{h}\left|\sin\left(\frac{\arg\chi_\iota}{2-\alpha} - \frac{\pi}{2}\frac{\alpha+4(n+s)}{2-\alpha}\right)\right|\right)^\alpha$ for all natural s.

Proof. Let k_ι be algebraic multiplicities of finite eigenvalues $v_\iota \in Spec(\hat{E}^D\hat{A})$ and $p_i \in \mathbb{N}$ be theirs geometric multiplicities. Then $Spec(E,A)$ can be determinate by $Spec J$ in the Weierstrass-Kronecker canonical form. So, the result is the direct consequence of Proposition 10 in [11]. □

5 Stabilizability

Consider an open loop control system of the form (1), i.e. system of the form

$$E\left({}_{t_0}\Delta^{\alpha_i}_{h,*}x_i\right)(t+h)=\sum_{j=1}^{n}a_{ij}x_j(t)+\sum_{l=1}^{m}b_{il}u_l(t) \tag{8}$$

where $x:(h\mathbb{N})_{t_0}\to\mathbb{R}^n$ is a state vector, $u:(h\mathbb{N})_{t_0}\to\mathbb{R}^m$ - a control sequence and $A=[a_{ij}]\in\mathbb{R}^{n\times n}$, $B=[B_{il}]\in\mathbb{R}^{n\times m}$, $E\in\mathbb{R}^{n\times n}$ are real constant matrices such that $\text{rank}E=r\leq n$.

The system (8) is called *stabilizable* if there exists a linear state–feedback controller with matrix gain $F\in\mathbb{R}^{m\times n}$, i.e. $u(t)=Fx(t)$, such that the closed loop system

$$E\left(\Delta^{\alpha_i}_{h}x_i\right)(t+h)=\sum_{j=1}^{n}k_{ij}x_j(t)\,,\quad t\in(h\mathbb{N})_{t_0} \tag{9}$$

where $K=[k_{ij}]=A+BF$, is asymptotically stable.

Proposition 4. *Let matrix $F\in\mathbb{R}^{m\times n}$ be such that the pair $(E,A+BF)$ is regular and $Ind(E,A+BF)\leq 1$. If R is the set of all roots of the equation*

$$\det\left(E-\frac{1}{z}\Lambda_\alpha\hat{E}^D\hat{M}\right)=0\,, \tag{10}$$

where $\hat{M}=(\lambda_1E-A+BF)(A+BF)^{-1}$ for some $\lambda_1\in\mathbb{C}$, are strictly inside the unit circle, then system (8) is stabilizable by the state-feedback controller $u(t)=Fx(t)$, $t\in(h\mathbb{N})_{t_0}$.

Proof. The result is consequence of Theorem 2. □

The method of solving Eq. (10) is discussed in [14]. As immediate consequence of Proposition 3 one obtains the following.

Corollary 1. *Suppose that $(E,A+BF)$ is the regular pair of the index at most one and with finite eigenvalues $\bar{\chi}_\iota=\frac{\bar{\lambda}_\iota}{\bar{\gamma}_\iota}$, $\iota=1,\ldots,n$. Then the system (8) is stabilizable if and only if for each $\bar{\chi}_\iota\in Spec(E,A+BF)$ the following holds*

(a) $\arg\bar{\chi}_\iota\in\left[\pi\alpha(2n+\frac{1}{2}),2\pi(1+n\alpha-\alpha\frac{1}{4}\right]$ *for all* $\iota=1,\ldots,n$
(b) $|\bar{\chi}_i|<|\omega_\iota|$ *for all* $\iota=1,\ldots,n$

where $\arg\bar{\chi}_\iota$ and $|\chi_\iota|$ are respectively the main argument and modulus of $\bar{\chi}_\iota\in Spec(\hat{E}^D\hat{M})$ and $|\omega_\iota|=\left(\frac{2}{h}\left|\sin\left(\frac{\arg\bar{\chi}_i}{2-\alpha}-\frac{\pi}{2}\frac{\alpha+4(n+s)}{2-\alpha}\right)\right|\right)^\alpha$ for all natural s.

Below we consider fractional system (11) with initial state $x(t_0)=0$. Only for this initial conditions obtained results are also valid for fractional order linear control system with the Riemann-Liouville- or Grünwald-Letnikov-type h difference operators, see [8].

Example 1. Consider the following open loop system on $(h\mathbb{N})_0$ of the form

$$E\left(\Delta_h^{0.9}x\right)(t+h) = Ax(t) + Bu(t)\,, \tag{11}$$

with step $h = 0.5$, matrices $E = \begin{bmatrix} 1 & 0 & 0 \\ 0 & 1 & 0 \\ 0 & 0 & 0 \end{bmatrix}$, $A = \begin{bmatrix} -3 & -3 & 1 \\ 1 & -3 & 1 \\ 1 & 1 & -1 \end{bmatrix}$, $B = \begin{bmatrix} 0.5 \\ 0.5 \\ 0 \end{bmatrix}$.

Then, $\hat{E} = \begin{bmatrix} 0.4 & -0.2 & 0 \\ 0.2 & 0.4 & 0 \\ 0.6 & 0.2 & 0 \end{bmatrix}$, $\hat{A} = \begin{bmatrix} 0.4 & -0.2 & 0.2 \\ 0.2 & 0.4 & 0.6 \\ 0.6 & 0.2 & 2.8 \end{bmatrix}$, $\hat{B} = \begin{bmatrix} 0.1 \\ 0.3 \\ 0.4 \end{bmatrix}$. Observe that as $rank(\hat{E}) = rank(\hat{E})^2 = 2$, then $\hat{E}^D = \begin{bmatrix} 2 & 1 & 0 \\ -1 & 2 & 0 \\ 1 & 3 & 0 \end{bmatrix}$. It can be check that the given open loop system is not stable. If we take gain matrix $F = \begin{bmatrix} -0.3 \\ 0 \\ 0 \end{bmatrix}$, then for $\lambda = 0$ it holds $\hat{M} = (-A + BF)(A + BF)^{-1} = \begin{bmatrix} -0.925 & -0.075 & 0 \\ 0.075 & -1.075 & 0 \\ 0 & 0 & -1 \end{bmatrix}$ and

$$\det\left(I - \frac{1}{z}\left(\frac{hz}{z-1}\right)^{\alpha}\hat{E}^D\hat{M}\right)$$

$$= \left(z^2 + 2.063163915505...z\left(\frac{z}{z-1}\right)0.9 + 1.4358729439...\left(\frac{z}{z-1}\right)^{1.8}\right)z^{-2}$$

with roots $R = \{-0.1672793016 \pm 0.5646735185i\}$ and $|r| = 0.5889300020$. One can check that for example for $\alpha = 0.9$ and $x_1(0) = x_2(0) = 0{,}2$, $x_3(0) = 0{,}4$ one gets stable trajectory, see Fig. 1 for x_1x_2 plot as $x_3 = x_1 + x_2$.

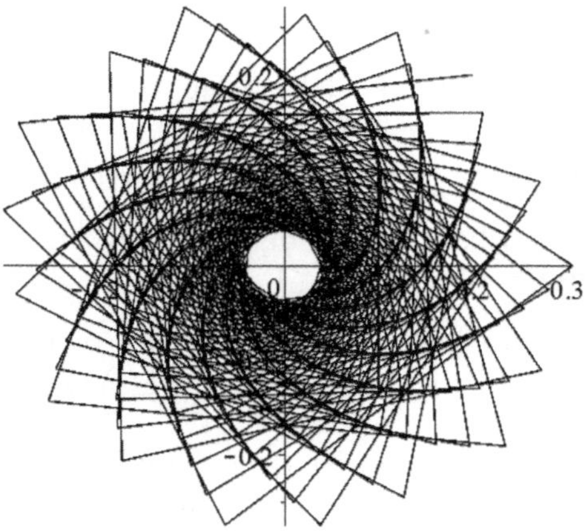

Fig. 1. The graph of x_1x_2 – stable trajectory of system (11) ($T = 300$ steps.)

6 Conclusions

Problems of stability and stabilization of the linear fractional multi-order descriptor control system with h-difference Caputo-type operator have been studied. The formula for trajectories for this class of systems has been given. This formula is valid only for consistent initial state. Basing on it, it is was shown that fractional order descriptor control system with h-difference Caputo-type operator is stabilizable if there exists a linear state-feedback controller with the gain matrix F such that roots of the complex multivalued equation (10) are strictly inside the unit circle. Obtained results extends on descriptor fractional order systems with Caputo-type operator stability conditions presented in [11].

Acknowledgments. The work has been carried out in the framework of Bialystok University Technology grant No WZ/WM/1/2019 and financed from the funds for science by the Polish Ministry of Science and Higher Education.

References

1. Ambroziak, L., Lewon, D., Pawluszewicz, E.: The use of fractional order operators in modeling of RC-electrical systems. Control Cybern. **45**(3), 275–288 (2016)
2. Bandyopadhyay, B., Kamal, S.: Stabilization and control of fractional order systems: a sliding mode approach. In: Lecture Notes in Electrical Engineering, vol. 317, pp. 55–90, Springer (2015)
3. Campbell, S.L.: Singular Systems of Differential Equations. Research Notes in Mathematics. Pitman Publishing, San Francisco (1980)
4. Das, S.: Functional Fractional Calculus for System Identification and Controls. Springer, Heidelberg (2008)
5. Darouach, M., Boutat-Baddas, L.: Observers for a class of nonlinear singular systems. IEEE Trans. Autom. Control **53**(11), 2627–2633 (2008)
6. Kaczorek, T.: Driazin inverse matrix method for fractional descriptor discrete-time linear system. Bull. Pol. Acad. Sci. Tech. Sci. **64**(2), 395–399 (2016)
7. Kaczorek, T.: Decentralized stabilization of fractional positive descriptor continuous-time linear systems. Int. J. Appl. Math. Comput. **28**(1), 135–140 (2018)
8. Mozyrska, D., Girejko, E., Wyrwas, M.: Comparison of h-difference fractional operators. In: Lecture Notes in Electrical Engineering, vol. 257, pp. 191–197. Springer (2013)
9. Mozyrska, D., Pawluszewicz, E., Wyrwas, M.: Local observability and controllability of nonlinear discrete-time fractional order systems based on their linearisatin. Int. J. Syst. Sci. **48**(4), 788–794 (2017)
10. Mozyrska, D., Wyrwas, M., Pawluszewicz, E.: Stabilization of linear multi-parameter fractional difference control systems. In: International Conference on Methods and Models in Automation and Robotics MMAR 2015, Poland, pp. 315–319 (2015)
11. Mozyrska, D., Wyrwas, M.: Stability of discrete fractional linear systems with positive orders. In: Preprints of the 20th World Congress of the International Federation of Automatic Control, Toulouse, France (2017)
12. Mozyrska, D., Wyrwas, M.: The $\mathcal{Z}$-transform method and delta-type fractional difference operators. Discrete Dyn. Nat. Soc. (2015). Article ID 852734

13. Mozyrska, D., Wyrwas, M.: Stability by linear approximation and the relation between the stability of difference and differential fractional systems. Mathe. Methods Appl. Sci. **40**(11), 4080–4091 (2017)
14. Pawluszewicz, E.: Perfect observers for fractional discrete-time linear systems. Kybernetika **52**(6), 914–928 (2016)
15. Podlubny, I.: Fractional Differential Systems. Academic Press, San Diego (1999)
16. Sajewski, L.: Stabilization of positive descriptor discrete-time linear system with two different fractional orders by decentralized controller. Bull. Pol. Acad. Sci. Tech. Sci. **65**(5), 709–714 (2017)
17. Sierociuk, D., Dzieliński, A., Sarwas, G., Petras, I., Podlubny, I., Skovranek, T.: Modelling heat transfer in heterogenous media using fractional calculus. Philos. Trans. Roy. Soc. Math. Phys. Eng. Sci. **371**(1990) (2013). Article Number: 20120146
18. Wyrwas, M., Pawluszewicz, E., Girejko, E.: Stability of nonlinear h-difference systems with n fractional orders. Kybernetika **51**(1), 112–136 (2015)
19. Zhang, H., Wu, D., Cao, J., Zhang, H.: Stability Analysis for fractional-order linear singular delay differential systems. Discrete Dyn. Nat. Soc. (2014). Article ID 850279
20. Zhang, L.: A chacterization of the Drazin inverse. Linear Algebra Appl. **335**, 183–188 (2001)

Fast Evaluation of Grünwald-Letnikov Variable Fractional-Order Differentiation and Integration Based on the FFT Convolution

Mariusz Matusiak(✉)

Institute of Applied Computer Science, Lodz University of Technology, 18/22 Stefanowskiego Street, 90-924 Lodz, Poland
mmatusiak@iis.p.lodz.pl

Abstract. The main topic of this research is the development of a new efficient method for solving fractional differential equations in the time domain using the Grünwald-Letnikov (GL) definition of the differintegral operator. The goal is to reduce or, in some cases, eliminate the necessity of introducing the maximum number of samples of the approximation, which is always a trade-off between accuracy, memory consumption, and computational speed. The algorithm involving Fast Fourier Transform and Fast Convolution operations has been proposed. The implementation in two different environments - the MATLAB/Simulink on a PC and on a hardware platform with the STM32H743 microcontroller - is described in this paper and the results of two iterations of experiments are presented. Fast Convolution algorithm is proven to be highly effective for processing block lengths $N \geq 128$. In the most complex analyzed case ($N = 8192$ samples) on the Intel® Core™ i5-8250U CPU reduction of the computation time reached 85%, compared to the implementation of the classic definition characterized by the $\mathcal{O}(N^2)$ complexity.

Keywords: Fractional-order calculus · Fast Convolution · Fast Fourier Transform · Grünwald-Letnikov VFOD/I · Arm® Cortex-M7 · Microcontroller implementation

1 Introduction

Non-local nature of the fractional-order differintegral operator involves processing all past samples of the input signal in the procedure of evaluating the output of the fractional differentiation or integration [6,14,16,18]. In practical realizations, e.g., control algorithms in the closed-loop systems (CLS) or synthesis of dynamical models, a maximum number of previous samples is established first (requirement of the so-called *short memory* techniques) to reduce the computation time and memory consumption on the target architecture at the cost of the higher approximation error. Well-known and applied with the Grünwald-Letnikov (GL) differintegral operator is the Short Memory Principle introduced

A. Bartoszewicz et al. (Eds.): PCC 2020, AISC 1196, pp. 879–890, 2020.
https://doi.org/10.1007/978-3-030-50936-1_74

by I. Podlubny in [18]. The alternative approach based on processing samples acquired in specific adaptive time steps was proposed and verified by MacDonald *et al.* in [10]. This method, however, requires a function $f(t)$ to be smooth since picking up the samples in varying time steps may lead to loss of accuracy. Another type of method is based on the Laplace transform and approximation of the Laplace operator of non-integer order s^{ν}, for example, the Oustaloup Recursive filter approximation (ORA), introduced by A. Oustaloup in [17]. For a rational function representation of the operator, the continued fraction expansion (CFE) approach is employed. Direct discretization of the system in the form of infinite impulse response filter (IIR) is performed with the CFE of a generating function $s \approx w(z^{-1})$, evaluated, e.g., for Tustin's, Euler's, or Al-Alaoui operators [14,24].

The above approximations are widely applied in systems of constant fractional orders but are not convenient for variable ones. In the following paper, a method for optimization of evaluation of Grünwald-Letnikov variable fractional-order derivatives and integrals based on Fast Fourier Transform (FFT) convolution is presented and examined in two iterations of experiments. The paper is organized as follows: first, the essential mathematical definitions and theorems are recalled. Next, the description of the algorithm and the example implementation in MATLAB is presented. Finally, the results from the experiment conducted on a separate embedded hardware platform are discussed. In the last section pros and cons of the method are summarized, and the conclusions are drawn.

2 Mathematical Preliminaries

Definition 1. *Grünwald-Letnikov fractional-order backward difference/sum (GL-FOBD/S).*

Let us consider the time $t = kh$ where $k \in \mathbb{N}$, $h \in \mathbb{R}_+$ is the finite and constant sampling period, and $\nu \in \mathbb{R}_+$ denotes the fractional order. The GL-FOBD is then defined as [16,18]:

$$ {}^{GL}\Delta_k^{\nu} f(t) = \sum_{j=0}^{\infty} (-1)^j \binom{\nu}{j} f((k-j)h) \tag{1}$$

where $\binom{\nu}{j}$ is a binomial coefficient, represented and evaluated as:

$$\binom{\nu}{j} = \frac{(\nu(\nu-1)\dots(\nu-j+1))}{j!} = \frac{\Gamma(\nu+1)}{j!\Gamma(\nu-j+1)} \tag{2}$$

Symbol Γ denotes the Euler's Gamma function. For order values $\mu \in \mathbb{R}_-$, e.g. $\mu = -\nu$, the definition above turns into GL-FOBS. Since in practical applications some starting point t_0 is considered before which ($t \in (-\infty, t_0)$) the function $f(t)$ is undefined, a simplification described in Definition 2 is employed.

Definition 2. *Truncated Grünwald-Letnikov fractional-order backward difference.*

GL-FOBD evaluated for the previous $N = \lfloor \frac{t-t_0}{h} \rfloor$ *values of the function* $f(t)$ *is defined as [23]:*

$$ {}^{GL}_{k-N}\Delta^{\nu}_{k} f(t) = \sum_{j=0}^{N} (-1)^j \binom{\nu}{j} f((k-j)h) \tag{3}$$

The operation can be also expressed as the partial discrete convolution of signals a^{ν}_j *and* $f(t)$*:*

$$ {}^{GL}_{k-N}\Delta^{\nu}_{k} f(t) = a^{\nu}_j * f(t) \tag{4}$$

where $a^{\nu}_j = (-1)^j \binom{\nu}{j}$*.*

Definition 3. *Recursive formula for oblivion coefficients.*

In order to avoid numerical issues caused by the overflow of numbers in formulas (1)–(3), a recursive function for obtaining the binomial coefficients is recommended [14]:

$$ a^{\nu}_j = (-1)^j \binom{\nu}{j} = \begin{cases} 1 & \text{for } j = 0 \\ a^{\nu}_{j-1}\left(1 - \frac{1+\nu}{j}\right) & \text{for } j = 1, 2, \ldots \end{cases} \tag{5}$$

Definition 4. *Grünwald-Letnikov fractional-order derivative/integral (GL-FOD/I).*

Based on Definition 2 one obtains fractional-order derivative and integral with the formula [5, 23]:

$$ {}^{GL}_{k-N}D^{\nu}_{k} f(t) = \lim_{h \to 0^+} \frac{{}^{GL}_{k-N}\Delta^{\nu}_{k} f(t)}{h^{\nu}} \tag{6}$$

which for small h *is approximated by:*

$$ {}^{GL}_{k-N}D^{\nu}_{k} f(t) \approx \frac{{}^{GL}_{k-N}\Delta^{\nu}_{k} f(t)}{h^{\nu}} \tag{7}$$

Definition 5. *Discrete Fourier Transform (DFT) and Inverse Discrete Fourier Transform (IDFT).*

The N-point Discrete Fourier Transform (DFT) of a finite input signal $x[n]$ *is defined as [9]:*

$$ \begin{aligned} X[m] = DFT\{x[n]\} &= \sum_{n=0}^{N-1} x[n] W_N^{nm} = \sum_{n=0}^{N-1} x[n] e^{\frac{-j2\pi nm}{N}} \\ &= \sum_{n=0}^{N-1} x[n]\left(\cos\frac{2\pi nm}{N} - j\sin\frac{2\pi nm}{N}\right) \end{aligned} \tag{8}$$

where $W_N^{nm} = e^{\frac{-j2\pi nm}{N}}$ *is known as the twiddle factor,* m *denotes the harmonic sample index and* $X[m]$ *are complex spectrum values of* $x[n]$*. Similarly the N-point IDFT is defined as:*

$$\begin{aligned} x[n] = IDFT\{X[m]\} &= \frac{1}{N}\sum_{m=0}^{N-1} x[n]W_N^{-nm} = \frac{1}{N}\sum_{m=0}^{N-1} x[n]e^{\frac{j2\pi nm}{N}} \\ &= \frac{1}{N}\sum_{n=0}^{N-1} x[n](\cos\frac{2\pi nm}{N} + j\sin\frac{2\pi nm}{N}) \end{aligned} \tag{9}$$

Definition 5 is characterized by the $N(N-1)$ complex additions and N^2 complex multiplications [13]. Hence, it is usually replaced with the recursive Fast Fourier Transform algorithm, requiring $\frac{N}{2}log_2N$ multiplications and $Nlog_2N$ additions at the additional cost of zero-padding the input signal and kernel vectors to a common length L [9,21]. Well-known Cooley-Tukey Radix-2 divide and conquer algorithm [3] requires the vectors of the length being the next power of two.

Definition 6. *Cooley-Tukey FFT Radix-2*

$X[o]$ spectrum samples, according to the Cooley-Tukey method, are obtained with the formula [3]:

$$X[o] = \begin{cases} DFT_{\frac{N}{2}}\{x[n] + x[n+\frac{N}{2}]\} & \textit{for even index } o = 2m \\ DFT_{\frac{N}{2}}\{x[n] - x[n+\frac{N}{2}]\}e^{\frac{-j2\pi n}{N}} & \textit{for odd index } o = 2m+1 \end{cases} \tag{10}$$

At the 1st stage of the algorithm the input signal $x[n]$ is split into two signals of length $N/2$, according to formula (10) and the $X[o]$ sample is a combination of two $N/2$-point DFTs. The algorithm is performed recursively until the sequence of 2-point DFTs is reached (refer to Fig. 1). Additionally, the output samples are bit-reversed so the reordering of input (decimation in time, DIT) or output samples (decimation in frequency, DIF) may be necessary.

Theorem 1. *Convolution Theorem.*

Convolution Theorem states that multiplication in the frequency domain corresponds to the circular convolution operation in the time domain [4,9,15].

$$x[n] \circledast h[n] = y[n] \tag{11}$$

$$X(\omega)H(\omega) = Y(\omega) \tag{12}$$

Hence, the inverse Discrete Fourier Transform of the product of Discrete Fourier Transforms of signals $x[n]$ *and* $h[n]$ *equals the circular convolution of those signals, modulo length* N*:*

$$y[n] = IDFT_N\{DFT_N\{x[n]\} \cdot DFT_N\{h[n]\}\} = \sum_{m=0}^{N-1} x[m]h[n-m]_{modN} \tag{13}$$

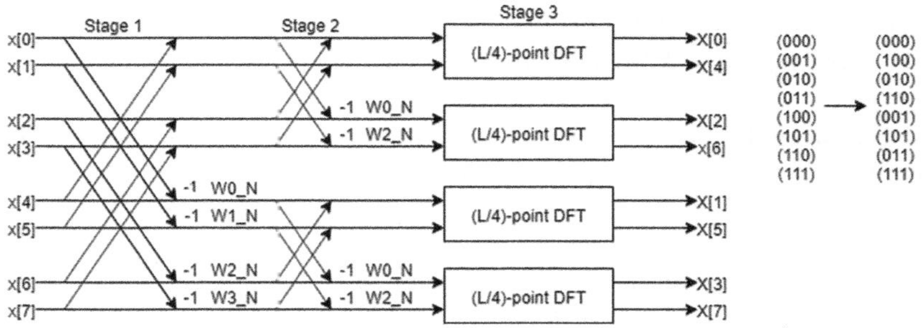

Fig. 1. FFT Cooley-Tukey butterfly algorithm for $L = 8$ samples.

Result signal of the discrete convolution $y[n]$ has length $L = M + N - 1$ where M and N are lengths of vectors $h[n]$ and $x[n]$, respectively. Assuming the same length $N = M$, for particular n the naive implementation of the operation from Definition 2 requires N multiplications and $N - 1$ additions with the total cost $T_c(N) = N(2N - 1)$. Complexity in terms of the Big O asymptotic annotation for N-size problem [7] is of range $\mathcal{O}(N^2)$. Experiment conducted in MATLAB environment on a PC (Intel® Core™ i5-8250U CPU @ 1.60GHz, 4 physical cores, MS Windows 10 Pro OS) revealed that the execution time of the script calculating the result of formula (3) using the *conv()* procedure, grew from $t = 2.96\,\mu s$ at $N_1 = 32$ to $t = 160.64\,\mu s$ at $N_2 = 1024$ samples.

Replacing DFT with FFT in (13) is known as Fast Convolution. It involves: 1) the element-wise multiplication of two Fast Fourier Transforms of signals $x[n]$ and $h[n]$ and 2) the Inverse Fast Fourier Transform of the obtained product. Since the obtained result is a circular convolution, in order to obtain a linear convolution from Definition 2 and to eliminate aliasing of two input signals, one must firstly zero-pad the vectors of lengths N and M to the common length $L_{min} = M + N - 1$, and secondly, extend them to the required length of the selected FFT implementation. For Radix-2, the quadratic complexity of the classic convolution is then reduced to $\mathcal{O}(N \log N)$ with total $T_{fc}(N) = 45Nlog_2 3N + 18N$ operations. This makes the operation more efficient for long signals [4]. According to the research on the accuracy of evaluation of GL-FOBD conducted in [1,2], the calculation tail of at least $N = 600$ samples is required for sufficient approximation of fractional-order backward difference. However, the final accuracy still depends on the characteristics of the input signal and the value of fractional order. In control systems with discrete fractional-order PID controllers reaching the limit of the short buffer during the transient state (especially for high sampling frequencies f_s) may introduce a significant error [12]. With at least 128 samples (or even 60 according to the results presented in [21]), the Fast Convolution algorithm is already more effective than the classic one [20]. In the next section, the proposed algorithm and the MATLAB simulations are described.

3 Grünwald-Letnikov Differintegral Based on Fast Convolution Algorithm

We can now apply formula (13) to (3) and obtain the algorithm of Fast Convolution for GL differintegral operator (see Algorithm 1).

Algorithm 1: GL differentiation and integration based on Fast Convolution.

Step 1. Calculate the length L of the convolution product $L_{min} := M + N - 1$, where M, N denotes the lengths of the vectors a_j^{ν} and $f(t)$, respectively. Since in (3) $M = N$, we can rewrite this as $L_{min} := 2N - 1$.

Step 2. Zero-pad the vector of the oblivion coefficients a_j^{ν} to the length of L_{min}. To use the faster Radix-2 FFT algorithm extend it to the next power of two $q := \lceil \log_2(L_{min}) \rceil$, $L := 2^q$.

Step 3. (Optional) For varying order $\nu(t)$ in the algorithm of a variable fractional-order derivative/integral recalculate the $a_j^{\nu(t)}$ coefficients. For the constant value of order ν, the coefficients shall be precomputed at the beginning of the program or stored in Flash memory in a lookup table (LUT).

Step 4. Calculate L-point Fast Fourier Transforms of $a_j^{\nu(t)}$ and $f(t)$. These operations can be executed simultaneously on the condition that a multicore architecture or a real-time operating system is used.

$$A_j^{\nu(t)}(\omega) = FFT_L\{a_j^{\nu(t)}\} \tag{14}$$

$$F(\omega) = FFT_L\{f(t)\} \tag{15}$$

Step 5. Perform element-wise multiplication and the Inverse L-point Fast Fourier Transform of its product.

$$\hat{y}(kh) = IFFT_L\{F(\omega)A_j^{\nu(t)}(\omega)\} \tag{16}$$

Step 6. Limit the output samples to the initial length N

$$y(kh) = \hat{y}(kh)[1 : N] \tag{17}$$

Complexity of the classic and Fast Convolution algorithms determined for maximum 1024 samples is presented in Fig. 2. The responses of fractional-order backward differences of the Heaviside unit step signal $u(k)$ for the example order values $\nu_1 = 0.1, \nu_2 = 0.5, \nu_3 = 0.9$ and the vector length $N = 32$ are shown in Fig. 3. Simulations were performed in MATLAB using *conv()* routine for the evaluation of formula (4) and the implementation of Algorithm 1. As can be noticed, the output characteristics for the order ν_1 tends to zero slowly (fractional

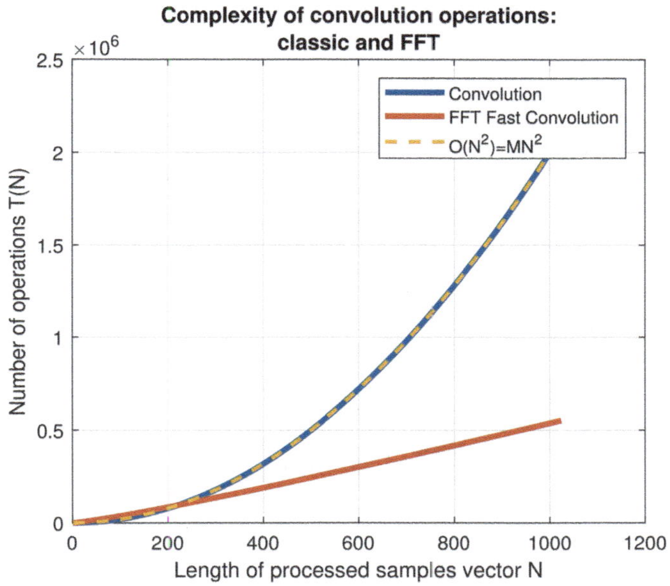

Fig. 2. Comparison of complexity of classic and Fast Convolution algorithms (value of constant $M = 2$).

differintegral property $D^{\nu} f(t) = f(t)$ for $\nu = 0$). For the order ν_3 shape of the response is similar to the classic first-order derivative of the step function, that is the Dirac delta function $\delta(t)$.

The average computation times $t_{CPU,avg}$ for both methods were obtained based on a number of measurements for two sets of lengths N: a) multiplies of 10 ($N_a = \{20, 50, 100, 200, 500, 1000, 2000, 4000, 8000\}$) and b) powers of 2 ($N_b = \{32, 64, 128, 256, 512, 1024, 2048, 4096, 8192\}$) using MATLAB routines *tic, toc.* Due to significant discrepancies in measured times on a PC, the procedure was repeated I = 1000 times. Results are presented in Fig. 4 and the scripts are provided in [11].

Finally, to analyze the accuracy of the proposed approach the approximation of the fractional order derivative (7) for $h = 0.01$ and $N = 512$ samples was compared with the derivative evaluated analytically employing the complex operator s^{ν} [19,23]. Mean squared error (MSE) indicator (18) was used for this purpose.

$$MSE = \frac{1}{N} \sum_{k=0}^{N-1} (y(kh) - y_{ref}(kh))^2 \tag{18}$$

For zero initial conditions the solution is of the form:

$$y_{ref}(t) = \mathcal{L}^{-1}\{s^{\nu-1}\} = \frac{t^{-\nu}}{\Gamma(1-\nu)} = \frac{t^{-0.5}}{\Gamma(0.5)}|_{\nu=0.5} \tag{19}$$

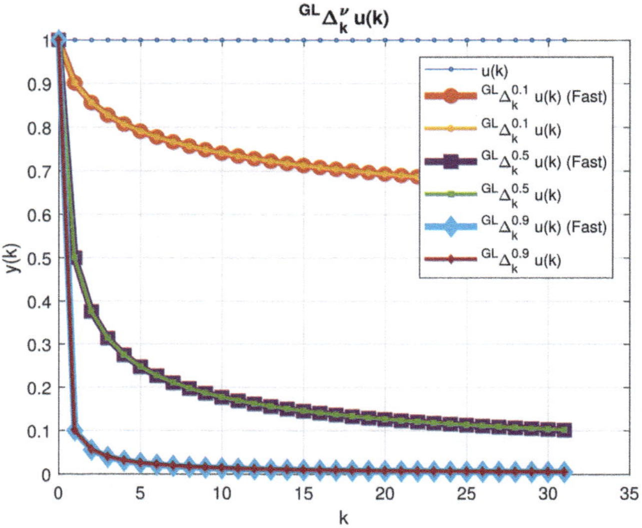

Fig. 3. Fractional-order backward difference of the Heaviside step function, $\nu_1 = 0.1, \nu_2 = 0.5, \nu_3 = 0.9$, $N = 32$.

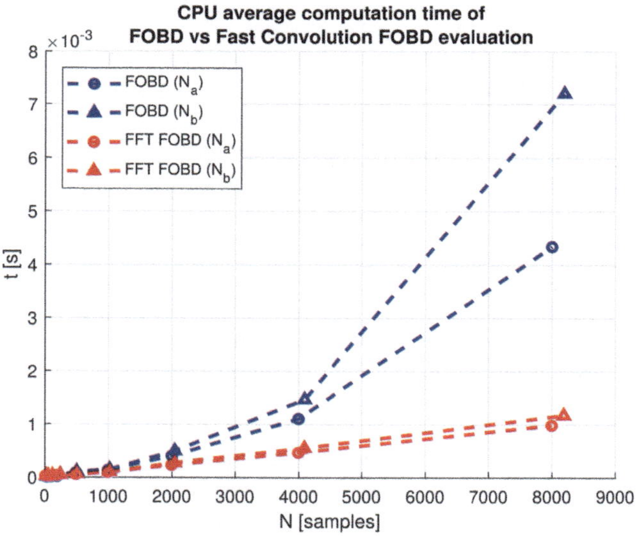

Fig. 4. Comparison of average computation time for various lengths of signals.

Values of the MSE for $t > 0$ and different orders ν_1, ν_2, ν_3 are presented in Table 1.

Table 1. Mean squared error (MSE) for fractional-order derivative approximation.

ν	MSE
0.1	$1.0105e-05$
0.5	$9.8878e-04$
0.9	$2.2180e-04$

4 Hardware Implementation

Experiment with hardware implementation of the algorithm was performed on the NUCLEO-H743 board with the STM32H743 microcontroller. The unit is developed on the basis of the high-end 32-bit Arm® Cortex-M7 RISC architecture. It is characterized by the 1MB of internal SRAM memory, availability of the hardware double-precision floating point unit, and the maximum clock frequency of $f_{CPU} = 480$ MHz [22]. The number of CPU cycles required to complete the operation was measured using the Data Watchpoint and Trace peripheral. Values of the $a_j^{0.5}$ coefficients were precomputed at the initialization stage of the program. Different conditions, including the varying lengths of buffers $N \in [20..8192]$, were examined. Three methods for calculating the FOBD were implemented: 1) classic convolution, based on Eq. (3), 2) partial convolution from the Arm® CMSIS-DSP library [8], and 3) the Fast Convolution algorithm based on the FFT implementation in CMSIS-DSP. The library contains optimized signal processing functions for fixed and floating point arithmetics

Table 2. Number of CPU cycles required to calculate FOBD on STM32H743 using different algorithms.

	Algorithm		
Samples length N	Convolution	CMSIS-DSP partial convolution	Fast convolution
32	67355	17289	46365
64	247845	56353	78697
128	938453	215759	205953
256	3799857	799389	389606
512	15286793	3526993	699141
1024	57390660	12960951	1573763
2048	238666016	49529088	3256317

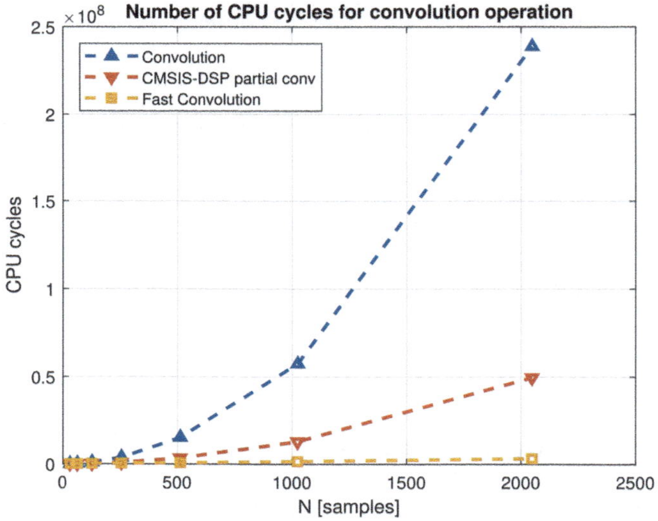

Fig. 5. Number of CPU cycles required to calculate FOBD on STM32H743 using different algorithms.

in format notations Q7, Q15, Q31, and F32. Therefore, the computations were performed using the single-precision floating point type (*float32_t*). The program was compiled with Arm® GCC Toolchain without any additional optimizations performed by the compiler (-O0). Results are presented in Table 2 and Fig. 5.

5 Conclusions

Improving the performance of the numerical evaluation of fractional-order equations on a microcontroller allows increasing the accuracy of the non-local operator approximation due to a higher number of processed samples during the same constant sampling period h. This is especially important for applications based on low-cost microcontrollers with no integrated hardware floating-point units due to the high complexity of the computations. The method proposed in the paper eliminates the quadratic complexity of the discrete convolution, which for samples length $N \geq 128$ resulted in shorter computation time in the described implementations. To fully benefit from the FFT algorithm, the allocation of additional memory space and the extension of the input signal may be required, e.g., to the next power of two in the Radix-2 algorithm. The method can be further optimized by precomputing the set of binomial coefficients at the beginning of the software execution if values of fractional orders are constant. The paper serves as a good starting point for future research on integrating Overlap-Add (OLA) and Overlap-Save (OLS) methods [21].

Acknowledgments. This work was supported by Polish funds of the National Science Center under grant DEC-2016/23/B/ST7/03686.

References

1. Brzeziński, D.W., Ostalczyk, P.: The Grunwald-Letnikov formula and its equivalent Horner's form accuracy comparison and evaluation for application to fractional order PID controllers. In: 2012 17th International Conference on Methods and Models in Automation and Robotics, MMAR 2012 (2012). https://doi.org/10.1109/MMAR.2012.6347821
2. Brzeziński, D.W., Ostalczyk, P.: About accuracy increase of fractional order derivative and integral computations by applying the Grünwald-Letnikov formula. Commun. Nonlinear Sci. Numer. Simul. (2016). https://doi.org/10.1016/j.cnsns.2016.03.020
3. Cooley, J.W., Tukey, J.W.: An algorithm for the machine calculation of complex fourier series. Math. Comput. (1965). https://doi.org/10.2307/2003354
4. Fix, J.: Efficient convolution using the Fast Fourier Transform, Application in C++ (2013). https://github.com/jeremyfix/FFTConvolution
5. Garrappa, R., Kaslik, E., Popolizio, M.: Evaluation of fractional integrals and derivatives of elementary functions: overview and tutorial. Mathematics **2**(1), 1–21 (2019). https://doi.org/10.3390/math7050407
6. Kilbas, A.A., Srivastava, H., Trujillo, J.J.: Theory and Applications of Fractional Differential Equations, vol. 204. Elsevier Science Inc., New York (2006). https://doi.org/10.1016/S0304-0208(06)X8001-5
7. Knuth, D.E.: Chapter 1.2.11. Asymptotic representations. In: The Art of Computer Programming. Volume 1: Fundamenal Algorithms, 3rd edn. pp. 107–123. Addison Wesley Longman Publishing Co., Inc. (1997). https://doi.org/10.5555/260999. https://dl.acm.org/doi/book/10.5555/260999
8. Lorenser, T.: The DSP capabilities of Arm® Cortex-M4 and Cortex-M7 Processors. DSP feature set and benchmarks (2016)
9. Lyons, R.G.: Understanding Digital Signal Processing. Prentice Hall PIR, Upper Saddle River (2004)
10. MacDonald, C.L., Bhattacharya, N., Sprouse, B.P., Silva, G.A.: Efficient computation of the Grünwald-Letnikov fractional diffusion derivative using adaptive time step memory. J. Comput. Phys. **297**, 221–236 (2015). https://doi.org/10.1016/J.JCP.2015.04.048
11. Matusiak, M.: Fast evaluation of Grünwald-Letnikov variable fractional-order differentiation and integration based on the FFT Convolution - Appendix (2020). https://dx.doi.org/10.24433/CO.8416219.v1
12. Matusiak, M., Ostalczyk, P.: Problems in solving fractional differential equations in a microcontroller implementation of an FOPID controller. Arch. Electr. Eng. **68**(3), 565–577 (2019). https://doi.org/10.24425/aee.2019.129342
13. Matusiak, R.: Implementing Fast Fourier Transform algorithms of real-valued sequences with the TMS320 DSP platform (2001). http://scholar.google.com/scholar?hl=en&btnG=Search&q=intitle:Implementing+Fast+Fourier+Transform+Algorithms+of+Real-Valued+Sequences+With+the+TMS320+DSP+Platform#0
14. Monje, C.A., Chen, Y., Vinagre, B.M., Xue, D., Feliu, V.: Fractional-Order Systems and Controls Fundamendals and Applications. Advances in Industrial Control. Springer, London (2010). https://doi.org/10.1007/978-1-84996-335-0

15. Nussbaumer, H.J.: Fast convolution algorithms. In: Fast Fourier Transform and Convolution Algorithms, 2nd edn., chap. 3, pp. 32–79. Springer, Heidelberg (1982). https://doi.org/10.1007/978-3-642-81897-4_3
16. Oldham, K.B., Spanier, J.: The Fractional Calculus: Theory and Applications of Differentiation and Integration to Arbitrary Order. Academic Press, New York (1974)
17. Oustaloup, A.: La commande CRONE: Commande robuste d'ordre non entier. Hermes (1991)
18. Podlubny, I.: Fractional Differential Equations. An Introduction to Fractional Derivatives, Fractional Differential Equations, to Methods of their Solution and some of their Applications. Academic Press, San Diego (1999)
19. Scherer, R., Kalla, S.L., Tang, Y., Huang, J.: The Grünwald-Letnikov method for fractional differential equations. Comput. Math. Appl. **62**(3), 902–917 (2011). https://doi.org/10.1016/J.CAMWA.2011.03.054
20. Smith, J.I.: MUS421 Lecture 2 Review of the Discrete Fourier Transform (DFT) (2019)
21. Smith, S.W.: The Scientist and Engineer's Guide to Digital Signal Processing, 2nd edn. California Technical Publishing, San Diego (1997)
22. STMicroelectronics: STM32H742xI/G STM32H743xI/G 32-bit Arm® Cortex(R)-M7 480MHz MCUs, up to 2MB Flash, up to 1MB RAM, 46 com. and analog interfaces. Datasheet - production data (2019)
23. Valério, D., Trujillo, J.J., Rivero, M., Machado, J.A.T., Baleanu, D.: Fractional calculus: a survey of useful formulas. Eur. Phys. J. Spec. Top. **222**(8), 1827–1846 (2013). https://doi.org/10.1140/epjst/e2013-01967-y
24. Vinagre, B.M., Podlubny, I., Hernandez, A., Feliu, V.: Some approximations of fractional order operators used in control theory and applications. Fract. Calc. Appl. Anal. **3**, 231–248 (2000). https://doi.org/10.1089/vbz.2015.1837

Fractional-Order Linear System Transformation to the System Described by a Classical Equation

Piotr Ostalczyk(✉)

Lodz University of Technology, Lodz, Poland
postalcz@p.lodz.pl

Abstract. In the paper, a method of "intigeration" of a linear time-invariant continuous or discrete-time system described by fractional-order differential/difference equations is proposed. The word *intigeration* means a procedure of connecting in series to the fractional plant a fractional element called further an "intigerator" such that the resulting two block system is described by the classical integer order differential/difference equation. The *intigerator* synthesis method is given. The stability conditions of the integer system are given. The proposed procedure enables to use classical methods of PID control tuning. It may also be used to tune the variable-, fractional – order PID controller.

Keywords: Fractional calculus · Fractional-order transfer function · Linear fractional-order continuous system · Linear fractional-order discrete system · PID control

1 Introduction

Mathematical models based on the fractional-order differential [12,13,16, 17]/difference equations [15] for over five decades have been a popular choice in modeling and designing closed - loop control systems [21]. This is caused by more accurate modeling of the dynamical behaviour of systems. The fractional model adopts overlooked physical phenomena such as friction (in mechanical systems) or electrostatic and electromagnetic couplings (in electrical systems). Fractional models describe more sophisticated transients of dynamic systems. Now, the most important dynamic properties of fractional systems (stability, controllability, observability etc.) for continuous and discrete-time systems are solved.

The PID tuning controllers in the closed-loop systems with the fractional plant are not yet satisfactionily solved. In the 100-year history of PID controllers, many tuning methods have been developed [2]. Although many attempts have been undertaken to synthesise the fractional-order PID controller (FOPID) [1, 3,5] or variable-, FOPID (VFOPID) it seems that the problem is still open. In the paper a method enabling to describe a closed-loop system with fractional

A. Bartoszewicz et al. (Eds.): PCC 2020, AISC 1196, pp. 891–903, 2020.
https://doi.org/10.1007/978-3-030-50936-1_75

plant by a classical integer-order differential or difference equation is proposed. It is possible by introducing a linear fractional block - *intigerator* between the plant and a controller. The stability of the new system is analysed.

The paper is organised as follows. After a short introduction to the fractional calculus and fractional linear time-invariant continuous and discrete system description given in Sects. 2 and 3 the main result - synthesis of the *intigerator* is presented. Discussions relating to the classical and variable-, fractional-order PID controller synthesis are given.

2 Mathematical Preliminaries

In this Section some fundamental notions which will be used in the paper will be recalled. The fundament creates the left-hand fractional-order derivative and its link to the fractional-order backward difference. The former is applied in the fractional-order dynamical systems description whereas the latter in the discrete-time system analysis and synthesis.

2.1 Fractional-Order Integral and Derivative

There are several definitions of the fractional-order integrals and derivatives. For a finite interval $[t_0, t]$ on the real axis $\mathbb{R}$ the Riemann - Liouville integral of order $\nu \in \mathbb{C}$ with $(\Re\{\mathbb{C}\})$ is defined as follows

Definition 1. *For a continuous and integrable real-valued function f the left-sided integral is defined as a sum*

$$\left({}_{t_0}^{RL}I_t^{(\nu)} f\right)(t) := \frac{1}{\Gamma(\nu)} \int_{t_0}^{t} \frac{f(x)}{(t-x)^{1-\nu}} dx \tag{1}$$

where $\Gamma(\nu)$ is the Gamma function.

The Riemann-Liouville left-derivative of an order ν and $n = [\Re\{\nu\}] + 1$ ($[\Re\{\nu\}]$ denotes the integral part of $\Re\{\nu\}$) defined over $[t_0, t]$ is defined as a n integral

Definition 2. *For a continuous and summable real-valued function f the left-sided integral is defined as a sum*

$$\begin{aligned}\left({}_{t_0}^{RL}D_t^{(\nu)} f\right)(t) &:= \left(\frac{d}{dt}\right)^n \left({}_{t_0}^{RL}I_t^{(n-\nu)} f\right)(t) \\ &= \frac{1}{\Gamma(n-\nu)} \left(\frac{d}{dt}\right)^n \int_{t_0}^{t} \frac{f(x)}{(t-x)^{\nu-n+1}} dx \end{aligned} \tag{2}$$

For $t_0 = 0^+$ and transformable function $|f(t)| \leqslant Ce^{ct}$ in a Laplace transform sense the Laplace transform of the fractional-order derivative yields

$$\left(\mathcal{L}{}_0^{RL}D_t^{(\nu)}f\right)(s) = s^\nu\left(\mathcal{L}f\right)(s) - \sum_{k=0}^{n-1} s^{n-k-1}\frac{d^k}{dt^k}\left({}_0^{RL}I_t^{(n-\nu)}f\right)(0^+) \tag{3}$$

where $\Re(s) > c$. Assuming that $\frac{d^k}{dt^k}\left({}_0^{RL}I_t^{(n-\nu)}f\right)(0^+) = 0$ for $k = 0, 1, \cdots, n-1$ the formula given above simplifies to the form

$$\left(\mathcal{L}{}_0^{RL}D_t^{(\nu)}f\right)(s) = s^\nu\left(\mathcal{L}f\right)(s) \tag{4}$$

which is very useful in the description of the fractional-order SISO time-invariant systems. To the Riemann-Liouville left-hand derivative the so called Grünwald - Letnikov fractional-order derivative and integral are related.

Definition 3 (Equivalent definition of the Grünwald - Letnikov fractional derivative). *The Grünwald - Letnikov fractional-order derivative is defined as the following limit*

$$\left({}_{t_0}^{GL}D_t^{(\nu)}f\right)(t) := \lim_{\substack{h \to 0 \\ kh = t - t_0}} = \frac{\left({}_{k_0}^{GL}\Delta_k^{(\nu)}f\right)(t)}{h^\nu} \tag{5}$$

where

$$\left({}_{k_0}^{GL}\Delta_k^{(\nu)}f\right)(t) = \sum_{i=0}^{k} a^\nu(i)f(t-hi) \tag{6}$$

$$a^\nu(i) = \begin{cases} 1 & \text{for } \quad i = 0 \\ (-1)^i\frac{\nu(\nu-1)(\nu-2)\cdots(\nu-i+1)}{i!} & \text{for } i = 1, 2, \cdots \end{cases} \tag{7}$$

2.2 Grünwald - Letnikov Fractional-Order Backward Sum and Difference

For a discrete-time bounded function f one defines the fractional-order backward difference as a following sum

Definition 4.

$$\left({}_{k_0}^{GL}\Delta_k^{(\nu)}f\right)(k) = \sum_{i=0}^{k-k_0} a^\nu(i)f(k-k_0-i) \tag{8}$$

Remark 1. In Definition 4 an order ν is assumed to be constant. As a generalization it can be treated as a function $\nu \in \mathbb{R}_+$ of a discrete variable k. Such a fractional-order backward difference will be further called the Grünwald - Letnikov variable-, fractional order backward difference.

Remark 2. The Grünwald - Letnikov fractional-order backward sum defined by Formula (8) is evaluated for $-\nu > 0$ and it can be also generalized for an order function $-\nu(k) < 0$.

For $k_0 = 0$ and zero initial conditions one evaluates the one-sided $\mathcal{Z}$ transform of the fractional-order backward difference

$$\left(\mathcal{Z}_0^{GL}\Delta_k^{(\nu)} f\right)(z) = \left(1 - z^{-1}\right)^{\nu} \tag{9}$$

3 Linear Time-Invariant Commensurate Fractional-Order SISO System Description

Now a short review of linear continuous and discrete-time system description forms is given. Using the classic convention one starts with the continuous-time systems.

3.1 Continuous Linear Time-Invariant Commensurate Fractional-Order SISO System Description

It is assumed that the plant is described by a linear time-invariant fractional-order differential equation of the form

$$\sum_{i=0}^{n} a_i \left({}_{t_0}^{RL}D_t^{(\nu_i)} y_F\right)(t) = \sum_{j=0}^{m} b_j \left({}_{t_0}^{RL}D_t^{(\mu_j)} u\right)(t) \tag{10}$$

where $a_n = 1$, $a_i, b_j \in \mathbb{R}$ for $i = 0, 2, \cdots, n-1$, $j = 0, 1, \cdots, m$, u is a transformable function and the subscript F emphasises the response fractionality. Here, the fractional orders are expressed in a form

$$\begin{aligned} \nu_i &= \frac{i}{q} \text{ for } q \in \mathbb{N}, i = 0, 1, \cdots, n \\ \mu_j &= \frac{j}{q} \text{ for } j = 0, 1, \cdots, m \end{aligned} \tag{11}$$

Comments

- Every rational number can be represented by a constant fraction $\frac{1}{q}$ $(q \in \mathbb{N})$ multiplied by an integer. This means that only commensurate fractional order equations are considered
- To preserve fitting the mathematical model represented by equation (12) to the real system it is assumed $m \leqslant n$.

Applying now the one-sided Laplace transform to the differential equation presented above one gets

$$\sum_{i=0}^{n} a_i s^{\frac{i}{q}} Y_F(s) = \sum_{i=0}^{m} b_i s^{\frac{i}{q}} U(s) \tag{12}$$

where $Y_F(s), U(s)$ are the Laplace transforms of the system response and input signals, respectively. From Eq. (12) one derives the plant fractional transfer function

$$P_F(s) = \frac{Y_F(s)}{U(s)} = \frac{\sum_{i=0}^{m} b_i s^{\frac{i}{q}}}{\sum_{i=0}^{n} a_i s^{\frac{i}{q}}} = \frac{n_F(s)}{d_F(s)} \tag{13}$$

Denoting

$$w_q = s^{\frac{1}{q}} \text{ for } q \in \mathbb{N} \tag{14}$$

the denominator of (13) can be transformed to the form

$$\begin{aligned} d_F(s) &= w_q^n + a_{n-1} w_q^{n-1} + \cdots + a_1 w_q + a_0 \\ &= \prod_{i=1}^{n_R} (w - p_{Ri})^{n_{Ri}} \prod_{i=1}^{n_C} (w - p_{Ci})^{n_{Ci}} \prod_{i=1}^{n_C} (w - p_{Ci}^*)^{n_{Ci}} \end{aligned} \tag{15}$$

$$\begin{aligned} n_F(s) &= w_q^m + b_{m-1} w_q^{m-1} + \cdots + m_1 w_q + m_0 \\ &= \prod_{i=1}^{m_R} (w - z_{Ri})^{m_{Ri}} \prod_{i=1}^{m_C} (w - z_{Ci})^{m_{Ci}} \prod_{i=1}^{m_C} (w - z_{Ci}^*)^{m_{Ci}} \end{aligned} \tag{16}$$

where $n = n_R + 2n_C$, $p_{Ri} \in \mathbb{R}$, $p_{Ci}, p_{Ci}^* \in \mathbb{C}$, $m = m_R + 2m_C$, $z_{Ri} \in \mathbb{R}$, $z_{Ci}, z_{Ci}^* \in \mathbb{C}$ play a role of "pseudo-poles and zeros" of transfer function (13), respectively.

3.2 Discrete Linear Time-Invariant Commensurate Fractional-Order SISO System Description

Similarly to the continuous system case one defines the difference equation modelling the discrete-time system

$$\sum_{i=0}^{n} a_i^{'} \left(\frac{{}^{RL}_{k_0 h}\Delta^{(\nu_i)}_{kh}}{h^{\nu_i}} y_F \right)(kh) = \sum_{j=0}^{m} b_j^{'} \left(\frac{{}^{RL}_{k_0 h}\Delta^{(\mu_j)}_{kh}}{h^{\mu_j}} u \right)(kh) \tag{17}$$

with the same assumptions concerning orders, coefficients and input signal u. Next, one introduces a simplified notation $a_i = \frac{a_i^{'}}{h^{\nu_i}}$ and $b_j = \frac{b_j^{'}}{h^{\mu_j}}$ one gets

$$\sum_{i=0}^{n} a_i \left({}^{RL}_{k_0 h}\Delta^{(\nu_i)}_{kh} y_F \right)(kh) = \sum_{j=0}^{m} b_j \left(\frac{{}^{RL}_{k_0 h}\Delta^{(\mu_j)}_{kh}}{h^{\mu_j}} u \right)(kh) \tag{18}$$

Now, a direct application of the one-sided $\mathcal{Z}$ transform to (16) yields

$$\sum_{i=0}^{n} a_i \left(1 - z^{-1}\right)^{\frac{i}{q}} Y_F(z) = \sum_{i=0}^{m} b_i \left(1 - z^{-1}\right)^{\frac{i}{q}} U(z) \tag{19}$$

where $Y_F(z), U(z)$ are the one-sided $\mathcal{Z}$ transforms of the system response and input signals, respectively. The plant fractional transfer function is as follows

$$P_F(z) = \frac{Y_F(z)}{U(z)} = \frac{\sum_{i=0}^{m} b_i \left(1 - z^{-1}\right)^{\frac{i}{q}}}{\sum_{i=0}^{n} a_i \left(1 - z^{-1}\right)^{\frac{i}{q}}} = \frac{n_F(z)}{d_F(z)} \tag{20}$$

Following the procedure performed for a continuous case a following notation will be introduced

$$v_q = \left(1 - z^{-1}\right)^{\frac{1}{q}} \text{ for } q \in \mathbb{N} \tag{21}$$

The above substitution causes that the polynomial of fractional exponentials is transformed to the classical one. Hence the denominator of (20) is as follows

$$\begin{aligned} d_F(z) &= v_q^n + a_{n-1} v_q^{n-1} + \cdots + a_1 v_q + a_0 \\ &= \prod_{i=1}^{n_R} (v - p_{Ri})^{n_{Ri}} \prod_{i=1}^{n_C} (v - p_{Ci})^{n_{Ci}} \prod_{i=1}^{n_C} (v - p_{Ci}^*)^{n_{Ci}} \end{aligned} \tag{22}$$

$$\begin{aligned} n_F(z) &= v_q^m + b_{m-1} v_q^{m-1} + \cdots + b_1 v_q + b_0 \\ &= \prod_{i=1}^{m_R} (v - z_{Ri})^{m_{Ri}} \prod_{i=1}^{m_C} (v - z_{Ci})^{m_{Ci}} \prod_{i=1}^{m_C} (v - z_{Ci}^*)^{m_{Ci}} \end{aligned} \tag{23}$$

where $n = n_R + 2n_C$, $p_{Ri} \in \mathbb{R}$, $p_{Ci}, p_{Ci}^* \in \mathbb{C}$, $m = m_R + 2m_C$, $z_{Ri} \in \mathbb{R}$, $z_{Ci}, z_{Ci}^* \in \mathbb{C}$ play a role of "pseudo-poles and zeros" of transfer function (20), respectively. There is another useful form of (23).
Noting that

$$\left(1 - z^{-1}\right)^{\frac{i}{q}} = \sum_{i=0}^{+\infty} a^{\frac{i}{q}}(i) z^{-i} \tag{24}$$

the discrete transfer function (20) takes a form

$$P_F(z) = \frac{Y_F(z)}{U(z)} = \frac{\sum_{i=0}^{m} b_i \sum_{i=0}^{+\infty} a^{\frac{i}{q}}(i) z^{-i}}{\sum_{i=0}^{n} a_i \sum_{i=0}^{+\infty} a^{\frac{i}{q}}(i) z^{-i}} = \frac{b_0^{'} + b_1^{'} z^{-1} + b_2^{'} z^{-2} + \cdots}{1 + a_1^{'} z^{-1} + a_2^{'} z^{-2} + \cdots} \tag{25}$$

4 Intigerator Synthesis

First, crucial property of polynomials (15) and (21) will be proved.

Proposition 1. *Given a polynomial $p_q(x) = x^{\frac{1}{q}} - p_i$ where $x \in \{s, z\}$, $q \in \mathbb{N}$. There exist a polynomial $i_p(x)$ with related fractional exponents*

$$u_p(x) = \sum_{j=1}^{q} p_i^{j-1} x^{1-\frac{j}{q}} \tag{26}$$

such that

$$p_q(x) u_p(x) = x - p_i^q \tag{27}$$

Proof. Direct multiplication of both polynomials show

$$\begin{aligned} & x\left(x^{\frac{1}{q}} - p_i^q\right)\left(x^{-\frac{1}{q}} + p_i x^{-\frac{2}{q}} + p_i^2 x^{-\frac{3}{q}} + \cdots + p_i^{q-2} x^{\frac{1-q}{q}} + p_i^{q-1} x^{-\frac{q}{q}}\right) \\ & = x\left(x^{\frac{1}{q}} - p_i^q\right) \sum_{j=1}^{q-1} p_i^j x^{-j-1} = x - p_i^q \end{aligned} \tag{28}$$

Remark 3. The number of terms of polynomial (26) linearly depends on q.

The noun form "*intigerator*" will be used to describe a dynamic system (represented by SISO linear time-invariant fractional-order system) which connected in series with the fractional plant gives a resulting system described by classical (integer orders) differential/difference equations. It is presented in Fig. 1

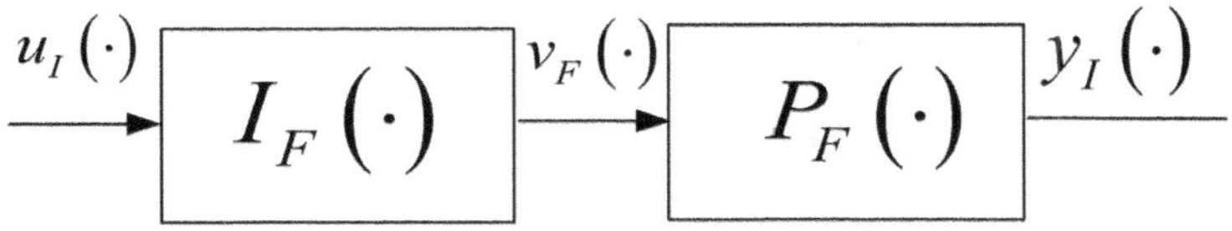

Fig. 1. Block diagram containing the fractional plant P, fractional intigerator I

Now, one defines a fractional-order linear system described by a transfer function (13) or (20). To generalize considerations further one substitute w_q, v_q by x. The result is formulated by the following Proposition.

Proposition 2. *For a fractional-order plant (13) or (20) with x as a variable there exists a fractional-order intigerator described by a fractional-order transfer function*

$$
\begin{aligned}
I_F(x) &= \frac{V_F(x)}{U_I(x)} \\
&= \frac{\prod_{i=1}^{m_R}\left(\sum_{j=1}^{q-1} z_{Ri}^{j} x^{-j-1}\right)^{m_{Ri}} \prod_{i=1}^{m_C}\left(\sum_{j=1}^{q-1} z_{Ci}^{j} x^{-j-1}\right)^{m_{Ci}}}{\prod_{i=1}^{n_R}\left(\sum_{j=1}^{q-1} p_{Ri}^{j} x^{-j-1}\right)^{n_{Ri}} \prod_{i=1}^{n_C}\left(\sum_{j=1}^{q-1} p_{Ci}^{j} x^{-j-1}\right)^{m_{Ci}}} \cdots \qquad (29)\\
&\cdots \frac{\prod_{i=1}^{m_C}\left(\sum_{j=1}^{q-1} z_{Ci}^{*j} x^{-j-1}\right)^{m_{Ci}}}{\prod_{i=1}^{n_C}\left(\sum_{j=1}^{q-1} p_{Ci}^{*j} x^{-j-1}\right)^{m_{Ci}}}
\end{aligned}
$$

A connection in series of the above fractional-order element to the fractional-order plant leads to the integer order equivalent system

$$
\begin{aligned}
P_F(\cdot)I_F(\cdot) &= \frac{V_F(\cdot)}{U_I(\cdot)} \\
&= \frac{\prod_{i=1}^{m_R}(x - z_{Ri}^{q})^{m_{Ri}} \prod_{i=1}^{m_C}(x - z_{Ci}^{q})^{m_{Ci}} \prod_{i=1}^{m_C}(x - z^{*q}_{Ci})^{m_{Ci}}}{\prod_{i=1}^{n_R}(x - p_{Ri}^{q})^{n_{Ri}} \prod_{i=1}^{n_C}(x - p_{Ci}^{q})^{n_{Ci}} \prod_{i=1}^{n_C}(x - p^{*q}_{Ci})^{n_{Ci}}} \qquad (30)
\end{aligned}
$$

Proof. Applying the results of the Proposition 1 to appropriate pairs of factors in numerator and denominator one immediately gets the stated result.

The connection of considered fractional-order systems is presented in Fig. 1.

Example 1 (Intigerator synthesis). Consider the fractional-order system characterised by a transfer function

$$
\begin{aligned}
P_F(s) &= \frac{1}{s^{\frac{4}{3}} - s^{\frac{2}{3}} + 2s^{\frac{1}{3}} + 2} = \frac{1}{\left(s^{\frac{1}{3}} + 1\right)^2 \left(s^{\frac{2}{3}} - 2s^{\frac{1}{3}} + 2\right)} \\
&= \frac{1}{\left(s^{\frac{1}{3}} + 1\right)^2 \left(s^{\frac{1}{3}} - 1 - j\right)\left(s^{\frac{1}{3}} - 1 + j\right)} \qquad (31)
\end{aligned}
$$

Following the result formulated in Proposition 1 one has

$$
\begin{aligned}
\left(s^{\frac{1}{3}} + 1\right)\left(s^{\frac{2}{3}} - s^{\frac{1}{3}} + 1\right) &= s + 1 \\
\left(s^{\frac{1}{3}} - 1 + j\right)\left(s^{\frac{2}{3}} + (1 - j)s^{\frac{1}{3}} - 2j\right) &= s + 2 + 2j \qquad (32)\\
\left(s^{\frac{1}{3}} - 1 - j\right)\left(s^{\frac{2}{3}} + (1 + j)s^{\frac{1}{3}} + 2j\right) &= s + 2 - 2j
\end{aligned}
$$

Hence,

$$I_F(s)$$
$$= \frac{1}{\left(s^{\frac{2}{3}} - s^{\frac{1}{3}} + 1\right)^2 \left(s^{\frac{2}{3}} + (1-j)s^{\frac{1}{3}} - 2j\right)\left(s^{\frac{2}{3}} + (1+j)s^{\frac{1}{3}} + 2j\right)} \tag{33}$$
$$= \frac{1}{\left(s^{\frac{2}{3}} - s^{\frac{1}{3}} + 1\right)^2 \left(s^{\frac{4}{3}} + 2s + 2s^{\frac{2}{3}} + 4s^{\frac{1}{3}} + 4\right)} \tag{34}$$

and finally

$$P_F(s)I_F(s) = \frac{1}{(s+1)^2\,(s^2 - 4s + 4)} \tag{35}$$

4.1 Intigerator Realizability

In the proposed method there is a need to connect in series an intigerator $I_F(x)$ with the plant $P_F(x)$. Below some comments relating to its realizability are presented.

Continuous System. In the considered case there are two approaches to a physical realization of the synthesized element.

In the first one linear elements characterised by a fractional dynamic behaviour are used in the electrical circuit [7,9]. Here, one should mention: the supercapacitor [18], the memrystor and the coil described by the fractional-order differential equation [19]. The analysis of electrical circuits is given in [10].

The other approach to the intigerator realization problem is to build an approximate realization with classical electric elements: resistors, capacitors and coils. The fractional-order transfer function is realised due to many methods. Here, one should mention [4,16,21].

Discrete System. The much easier intigerator realijsation is in the discrete system case. Its action is realized by an appropriate algorithm in micro-controller program. One should have in mind that the fractional-order difference equation describing the mentioned algorithm in every consecutive step performs linearly growing number of operations. This leads to the lack of time to perform calculations in the constant operation cycle and the amount of microcontroller memory.

4.2 Integer-Order System Stability

One can easily realise that the resulting system $P_F(x)I_F(x)$ (30) may be unstable. The stability conditions is formulated by the following Proposition.

Proposition 3. *Consider the asymptotically stable continuous fractional-order system 13 with denominator 16. The integer order system $P_F(s)I_F(s)$ is asymptotically stable if and only if*

$$\frac{\pi}{2} < q \arg\{p_i^q\} < \frac{3\pi}{2} \tag{36}$$

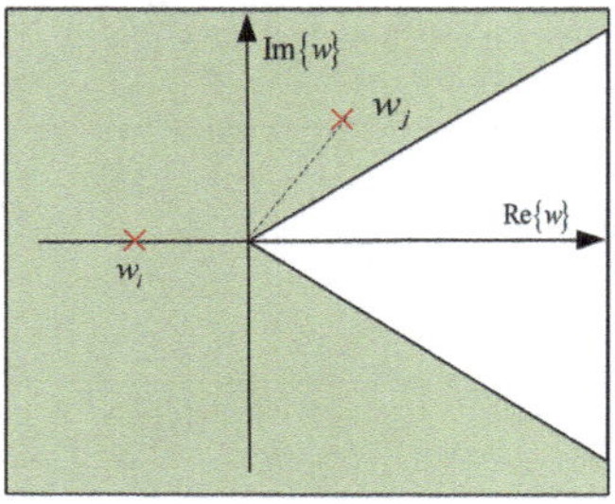

Fig. 2. Stability region (in olive) of the fractional linear system for $\nu < 0.5$

Proof. The stability area for fractional-order system is presented in Fig. 2. It may be expressed in a form of inequalities

$$\frac{\pi}{2}\nu = \frac{\pi}{2q} < \arg(p_i) < 2\pi - \frac{\pi}{2q} \tag{37}$$

Multiplying both sides of (36) by q one gets the thesis.

The unstable integer-order system $P_F(s)I_F(s)$ may be stabilized using different methods. Here, a solution presented in Fig. 3 is proposed. For an unstable integer-order system described by a transfer function $P_F(s)I_F(s)$ it is to find an integer-order linear stabilizing element $S_I(s)$ such that the closed-loop system

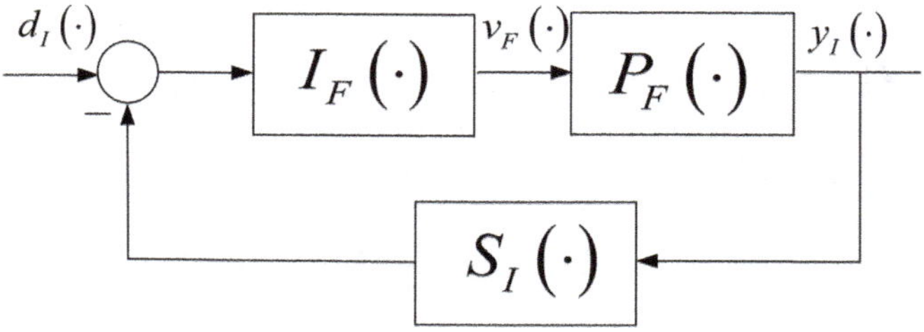

Fig. 3. Block diagram containing the fractional plant P, fractional intigerator I and stabilizing element S

$$G(s) = \frac{Y_I(s)}{D_I(s)} = \frac{P_F(s)I_F(s)}{1 + P_F(s)I_F(s)S_I(s)} \tag{38}$$

is asymptotically stable.

4.3 Closed-Loop Discrete System Variable-, Fractional/Integer-Order Synthesis

To the integer-order system given in Fig. 4 one can apply any continuous or discrete PID controller [2] tuning parameters procedure

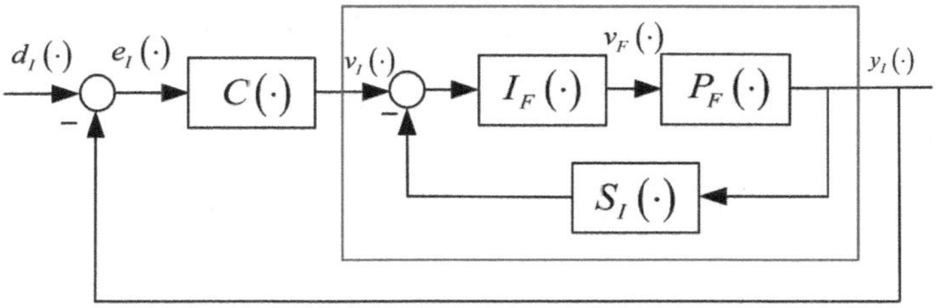

Fig. 4. Block diagram containing the integer plant $P_F I_F$ and a PID controller

In general, one can use fractional-order PID controller [5,6] or its generalization to variable-, fractional-order PID one described by the equation

$$v_I(k) = K_P e_I(k) + K_I {}_0^{GL}\Delta_k^{\mu(k)} e_I(k) + K_D {}_0^{GL}\Delta_k^{-\mu(k)} e_I(k) \tag{39}$$

where K_P, K_I, K_D are non-negative controller constants and $\nu(k), \mu(k) > 0$ for $k = 0, 1, \cdots$. As a special case the order functions can chosen functions

$$\nu(k) \in \{0, 1\},\ \mu(k) \in \{-1, 0\} \text{ for } i = 0, 1, 2, \cdots \tag{40}$$

Hence, the closed-loop system can be treated as described by the integer-order (classical) difference equation.

5 Final Conclusions

In the proposed method there are still open problems which are mentioned below.

- The stability conditions of the *intigerized* discrete system should be established
- The physical realizability of the integer continuous system needs further investigations
- The controllability and observability of the intigerized systems should be analysed in context of the *intigerizer* and plant transfer functions.

Acknowledgments. The work was supported by funds of the Polish National Science Center granted on the basis of decision DEC-2016/23/B/ST7/03686.

References

1. Azarmi, R., Tavakoli-Kakhiki, M., Sedigh, A.K., Fatehi, A.: Robust fractional order PI controller tuning based on Bode's ideal transfer function. IFAC-PaperOnLine **49**(9), 158–163 (2016)
2. Åström, K.J.: PID controllers: theory, design and tuning. In: Instrument Society of America (1995). https://doi.org/1556175167
3. Barbosa, R.S., Silva, M.F., Machado, J.A.: Tuning and application of integer and fractional order PID controllers, pp. 245–255. Springer, Heidelberg (2009)
4. Chen, Y.Q., Vinagre, B.M., Podlubny, I.: Using continued fraction expansion to discretize fractional order derivatives. In: Nonlinear Dynamics, Special Issue on Fractional Derivatives and Their Applications, pp. 1 - 18 (2003)
5. Das, S., Saha, S., Das, S., Gupta, A.: On the selection of tuning methodology of FOPID controllers for the control of higher order processes. ISA Trans. **50**(3), 376–388 (2011). https://doi.org/10.1016/J.ISATRA.2011.02.003
6. Dastranj, M.R., Rouhani, M., Hajipoor, A.: Design of optimal fractional order PID controller using PSO algorithm. Int. J. Comput. Theory Eng. (2012). https://doi.org/10.7763/ijcte.2012.v4.499
7. Elkwakil, A.S.: Fractional-order circuits and systems: an emerging disciplinary research area. IEEE Circ. Syst. Mag. **10**(4), 40–50 (2010)
8. Gligor, A., Dulău, T.M.: Fractional order controllers versus integer order controllers. Procedia Eng. **181**, 538–545 (2017). https://doi.org/10.1016/j.proeng.2017.02.431
9. Jakubowska, A., Walczak, J.: Electrical realizations of fractional-order elements: i synthesis of the arbitrary order elements. Poznan Univ. Technol. Acad. J. **85**, 137–148 (2016)
10. Kaczorek, T., Rogowski K.: Fractional linear systems and electrical circuits. In: Studies in Systems, Decision and Control, vol. 13. Springer, Cham (2015). https://doi.org/10.1007/978-3-319-11361-6
11. Kai-Xin, H., Ke-Qin, Z.: Mechanical analogies of fractional elements. Chin. Phys. Lett. **26**(10), 108301 (2009). https://doi.org/10.1088/0256-307X/26/10/108301
12. Kilbas, A.A., Srivastava, H.M., Trujillo, J.J.: Theory and Applications of Fractional Differential Equations, vol. 204. Elseiver, Amsterdam (2006). https://doi.org/10.1016/S0304-0208(06)X8001-5
13. Miller, K.S., Ross, B.: An Introduction to the Fractional Calculus and Fractional Differential Equations. Wiley, New York (1993)
14. Oldam, K.B., Spanier, J.: The Fractional Calculus: Theory and Applications of Differential and Integration of Arbitrary Order. Dover Publications, New York (1995)
15. Ostalczyk, P.: Discrete Fractional Calculus. Some applications in control and image processing. Series in Computer Vision, vol. 4. World Scientific Publishing Co. Pte. Ltd., Singapore (2016)
16. Oustaloup, A.: La dérivation non entière: théorie, synthèse et applications, Traité des Nouvelles Technologies séroe automatique, Hermes, Paris (1995)
17. Podlubny, I.: Fractional Differential Equations. Academic Press, London (1999)
18. Rafik, F., Gualous, H., Gallay, Y.: Frequency, thermal and voltage supercapacitor characterisation and modelling. J. Power Sour. **165**, 928–934 (2007)
19. Schafer, J., Kruger, K.: Modeling of coils using fractional derivatives. J. Magn. Magn. Mater. **307**, 91–98 (2006)

20. Sheng, H., Chen, Y.Q., Qiu, T.S.: Fractional Processes and Fractional-Order Signal Processing: Techniques and Applications. Springer, Singapore (2012)
21. Valerio, D., da Costa, J.: An Introduction to Fractional Control. The Institution of Engineering and Technology, London (2013)
22. Vinagr, B.M., Podlubny, I., Hernandez, A., Feliu, V.: Some approximations of fractional order operators used in control theory. Fractional Calc. Appl. Anal. **3**(3), 231–248 (2000)

Comparison of Non-integer PID, PD and PI Controllers for DC Motor

Wojciech Mitkowski(✉) and Waldemar Bauer

Department of Automatics and Biomedical Engineering, AGH University of Science and Technology, Al. Mickiewicza 30, 30-059 Kraków, Poland
{wojciech.mitkowski,bauer}@agh.edu.pl
http://sdts.agh.edu.pl/

Abstract. In this work we will present a control method for DC motors based on non-integer PID, PD and PI controllers. The original element in this paper consists of a comparative analysis of various controllers stabilizing the position of the motor shaft To design all controller types we are using global optimization method simulated annealing.

Keywords: DC motor · Noninteger PID · Noninteger PD · Noninteger PI · Time domain Oustaloup

1 Introduction

Non-integer controllers are a broadly researched topic. Questions of great importance are the design of non-integer order controllers and rules of tuning this type of controller (see [16,17]). In this paper, the authors focus on the comparison result of numerical experiments for non-integer PID, PD and PI controler for DC Motor.

Electric drive plays crucial role in science and technology. Electric machine is an interesting object converting electricity to mechanical power or mechanical power to electricity. There is a vast literature on both modeling and control of various types of electric drives, e.g. [2,13,15,18]. The bibliography information in both works provides further references. In this work, we will consider a special problem in control of a DC motor. It will be analyzed how the non-integer and classical PID controller influence the efficiency of the system. Despite different concepts of "smart controllers" (e.g. neural controllers, fuzzy controllers, etc.), our research confirms that properly tuned PID controller is still useful in controlling the physical processes of various types, even control objects with distributed parameters. A formal control law for what we now call PID or three-term control was first developed using theoretical analysis, by Russian American engineer Nicolas Minorsky at 1922. Minorsky was researching and designing automatic ship steering for the US Navy and based his analysis on observations of a helmsman.

A. Bartoszewicz et al. (Eds.): PCC 2020, AISC 1196, pp. 904–913, 2020.
https://doi.org/10.1007/978-3-030-50936-1_76

The current capabilities of computer techniques allow the analysis of very complex dynamics of controlled systems using a classical and non-integer PID, that is, one that takes into account the actual mathematical models of amplifiers, integrating and differentiating elements.

The problem of controlling a DC motor was widely analyzed in AGH. For example, use LQ controller see [3,4], LQR controller [10], controller on discrete time [12], parameter identification [9], state estimation [5,12], adaptive control [10], speed feedback. Some of the results were verified in the laboratory. Implementation and properties of noninteger order PID was analyzed in [6–8,11].

This paper is organized as follows. At the beginning, we presented a simplified model of DC motor in form of state space and transfer function. Then the different type of nonitiger controller and implementation method of this controller. The last part consists of optimization results and planned future works.

2 Simplified Model of a DC Motor

In this paper to analyse author use the control system with separately excited DC motor (see Fig. 1). The motor is controlled with signal $u(t)$. Assuming zero inductance L of the armature, the current in the armature is constant. The motor shaft is connected via a gear mechanism with an adjustable potentiometer with voltage output $y(t)$. The angular position of the motor shaft $x_1(t)$ affects both the position of directional antenna and the position of the brush of the potentiometer giving $y(t)$. Therefore, the voltage $y(t)$ describes explicitly the angular position of directional antenna.

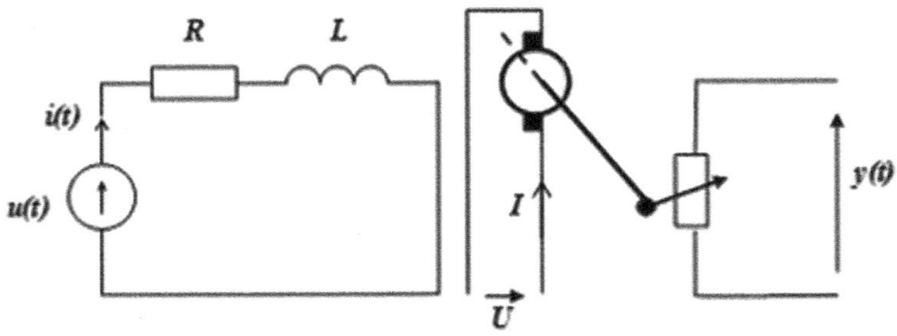

Fig. 1. Control of position $y(t)$ in a DC motor

The angular velocity of the shaft is denoted by $x_2(t) = \dfrac{dx_1(t)}{dt}$ and J is the moment of inertia of the shaft. Using the simplifying procedure (e.g. [2]), the control system can be described with the state equation:

$$\begin{cases} \dot{\mathbf{x}}(t) = \mathbf{A}\mathbf{x}(t) + \mathbf{B}u(t) \\ \mathbf{y}(t) = \mathbf{C}\mathbf{x}(t) \end{cases} \tag{1}$$

where:

$$\mathbf{x}(t) = \begin{bmatrix} x_1(t) \\ x_2(t) \end{bmatrix}, \ \mathbf{A} = \begin{bmatrix} 0 & 1 \\ 0 & -f \end{bmatrix}, \ \mathbf{B} = \begin{bmatrix} 0 \\ b \end{bmatrix}, \ \mathbf{C} = \begin{bmatrix} c \ 0 \end{bmatrix},$$

$$c \neq 0, \ f = \frac{\mu}{J} = \frac{1}{T} \text{ and } b = \frac{k_1 k_2 I}{JR} = KR.$$

Moreover, k_1 and k_2 are constant values describing simplified ($L = 0$ and $I = \text{const}$) dependencies between flux, voltage $u(t)$ and torque, μ denotes friction coefficient of rotation of the motor shaft and R is excitation circuit resistance. If we assume that $L \neq 0$, the state matrix A will be in $\mathrm{R}^{3\times3}$.

Instead of state equations, it is possible to use the transfer function $G_o(s)$ or spectral transfer function $G_o(j\omega)$, $j^2 = -1$:

$$G_o(s) = \mathbf{C}[s\mathbf{I} - \mathbf{A}]^{-1}\mathbf{B} = \frac{cb}{s(s+f)} = \frac{K}{s(Ts+1)} \tag{2}$$

3 Noninteger Controllers

Podlubny proposed a generalisation of the PID, namely the $\mathrm{PI}^\lambda\mathrm{D}^\mu$ controller, involving an integrator of order λ and a differentiator of order μ. In this paper we consider reduced form with standard differentiation order 1. Transfer function for the $\mathrm{PI}^\lambda\mathrm{D}^\mu$ controller's output has the form (see [17]):

$$G_{PI^\lambda D^\mu}(s) = K_p + K_i s^{-\lambda} + K_d s^\mu \tag{3}$$

Where:

- K_p is proportional gain
- K_i is integral gain
- K_d is derivative gain
- $\lambda \in (0, 1)$
- $\mu \in (0, 1)$

This formula described more generalized Non-integer PID controller. How we can easily see when the value of K is equal 0 then regulator have form PD^μ and analogously when $K_d = 0$ then regulator have form PI^λ.

3.1 Implementation of Fractional s^α

To implement controller, the non-integer transfer function s^γ have to be approximated with an integer function. There are several methods to approximate non-integer order transfer function arbitrarily close in the specified range $[\omega_{min}, \omega_{max}]$ (see [1]). The Oustaloup continuous integer approximation is given by Eq. (4) [14]:

$$s^{\gamma} \approx K \prod_{k=1}^{N} \frac{s + \omega_k'}{s + \omega_k}, \quad \gamma > 0, \tag{4}$$

where poles, zeros and gain can be evaluated as:

$$\begin{aligned} \omega_k' &= \omega_{min} \omega_u^{(2k-1-\gamma)/N} \\ \omega_k &= \omega_{min} \omega_u^{(2k-1+\gamma)/N} \\ K &= \omega_{max}^{\gamma} \\ \omega_u &= \sqrt{\frac{\omega_{max}}{\omega_{min}}} \end{aligned} \tag{5}$$

Approximation is designed for frequencies range $\omega \in [\omega_{min}, \omega_{max}]$ and N is the order of the approximation. As it can be seen, its representation takes form of a product of a series of stable first order linear systems. As one can observe choosing a wide band of approximation results in large ω_u and high order N result in spacing of poles spacing from close to $-\omega_h$ to those very close to $-\omega_b$. This spacing is not linear (there is a grouping near $-\omega_b$) and causes problems in discretisation process.

Time Domain Approximation. This approach is to realize every block of the transfer function (4) in form of a state space system. Those first order systems will be then collected in a single triangular matrix resulting in full matrix realization. This continuous system of differential equations will be then discretized (see [7]).

For zero initial condition we can transform Oustaloup approximation to state space system:

$$\frac{s + \omega_k'}{s + \omega_K} \iff \begin{cases} \dot{x}_k = A_k x_k + B_k u_k \\ y_k = x_k + u_k \end{cases}$$

where

$$\begin{aligned} A_k &= -\omega_k \\ B_k &= \omega_k' - \omega_k \\ C_k &= 1 \\ D_k &= 1 \end{aligned}$$

Finally the time domain Oustaloup approximation is given by matrix equation:

$$\dot{\mathbf{x}} = \begin{bmatrix} A_1 & 0 & 0 & \dots & 0 \\ B_2 & A_2 & 0 & \dots & 0 \\ B_3 & B_3 & A_3 & \dots & 0 \\ \vdots & \vdots & \vdots & \ddots & \vdots \\ B_N & B_N & \dots & B_N & A_N \end{bmatrix} \mathbf{x} + \begin{bmatrix} KB_1 \\ KB_2 \\ KB_3 \\ \vdots \\ KB_N \end{bmatrix} u \tag{6}$$
$$y = \begin{bmatrix} 1 \; 1 \dots 1 \; 1 \end{bmatrix} \mathbf{x} + Ku$$

What can be immediately observed is that the state matrix is triangular. This is extremely important in the discretization process. In this case, we can use directly Tustin method to discretization this representation of system in the time domain. In result we dispose numerical stable representation of s^{α}.

4 Comparison of Non-integer PID, PI and PD Controllers

In this section will be presented a description of the controllers designed method based on the optimization process. On the end, the comparison between non-integer PID, PI and PD controller has been done.

4.1 Optimization Process

To the optimization process the Simulated Annealing method has been chosen. Simulated Annealing is a minimization technique for solving an unconstrained and a bound-constrained optimization problems, which gives a good results in finding a local minimum. At each iteration of the Simulated Annealing algorithm, a new point is randomly generated. The distance of the new point from the current point is based on a probability distribution (temperature function) with a scale proportional to the temperature. The algorithm accepts all new points that lower the quality function value, but also, with a certain probability, points that raise the quality function value. By accepting points that raise the quality function, the algorithm avoids being trapped in local minima in early iterations and is able to explore globally for better solutions.

To the experiments has been chosen quality index described as:

$$Q(t) = \int_0^t e^2 \tag{7}$$

end

$$Q(t) = \int_0^t |e| \tag{8}$$

4.2 Experiments Results

The optimisation process minimises an chosen quality index for all considered non integer controller types. In comparison process authors use classical parameters described step response.

All evaluation function for formula (7) tuning process all controllers type are shown in Fig. 2 and for (8) on Fig. 3. The final settings of non-integer controllers are given in Table 1.

Time response on the DC motors for step signal show plots Fig. 4, as can be seen in the chart, the PD controller has the shortest adjustment time with settings from the quality function described by (8). The PI regulator behaves the worst in this respect. The PI controller also has the largest overshoot, while the PID and PD controllers are similar in this respect.

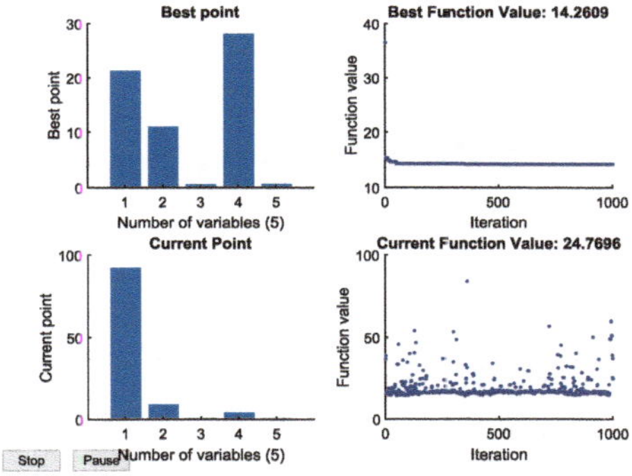

(a) Non-integer PID

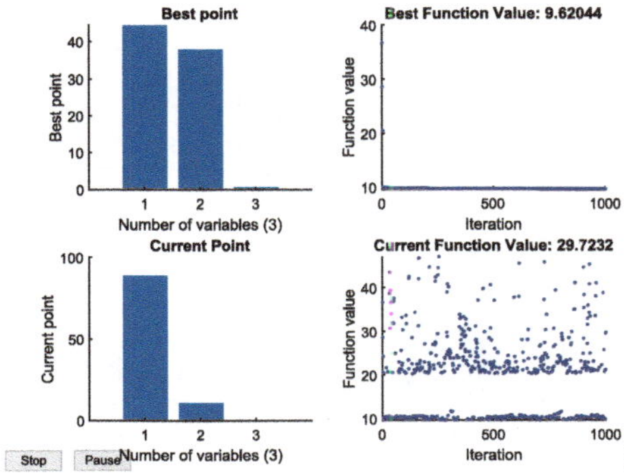

(b) Non-integer PD

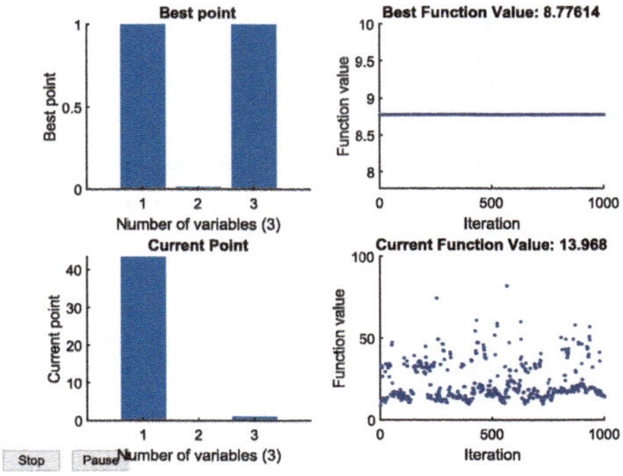

(c) Non-integer PI

Fig. 2. Comparison optimalization process for quality index (7)

Table 1. Comparison optimalization results

	Controller					
	PID		PI		PD	
Quality index	$\int_0^t e^2$	$\int_0^t \|e\|$	$\int_0^t e^2$	$\int_0^t \|e\|$	$\int_0^t e^2$	$\int_0^t \|e\|$
Quality index value	17.84	18.33	12.44	12.50	29.29	10.72
K_p	21.33	12.74	1	0.88	44.41	28.03
K_i	11.02	18.04	0.01	0.001	–	–
λ	0.62	0.95	1	0.99	–	–
K_d	28.16	17.52	–	–	37.73	24.31
μ	0.70	0.87	–	–	0.62	0.73

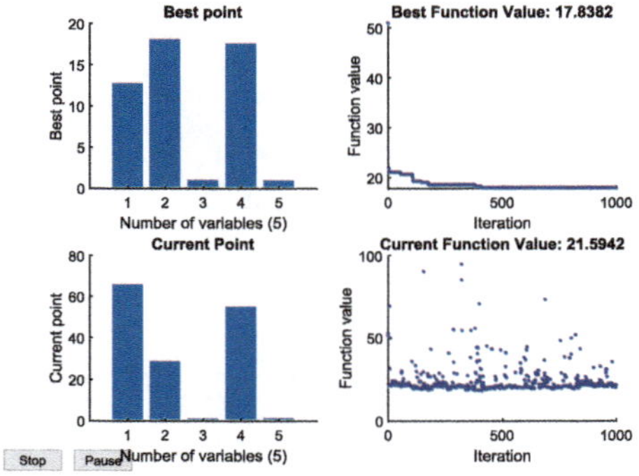

(a) Non-integer PID

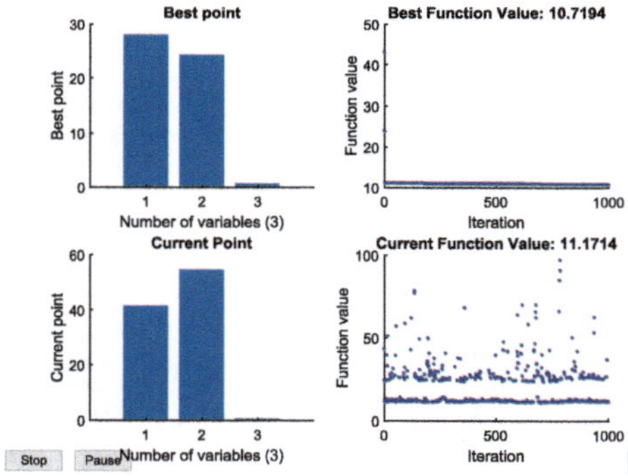

(b) Non-integer PD

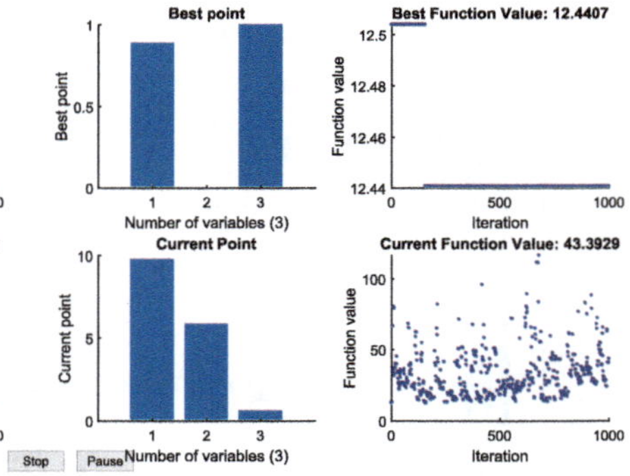

(c) Non-integer PI

Fig. 3. Comparison optimalization process for quality index (8)

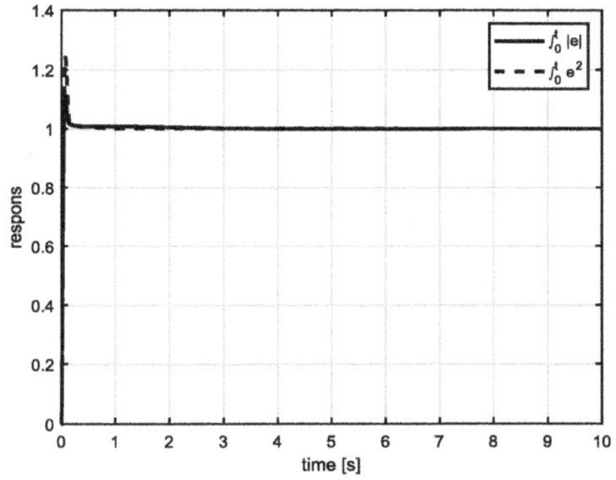

(a) Non-integer PID response

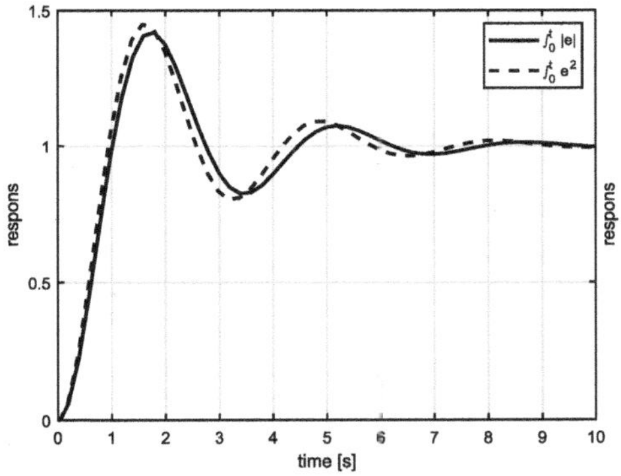

(b) Non-integer PI response

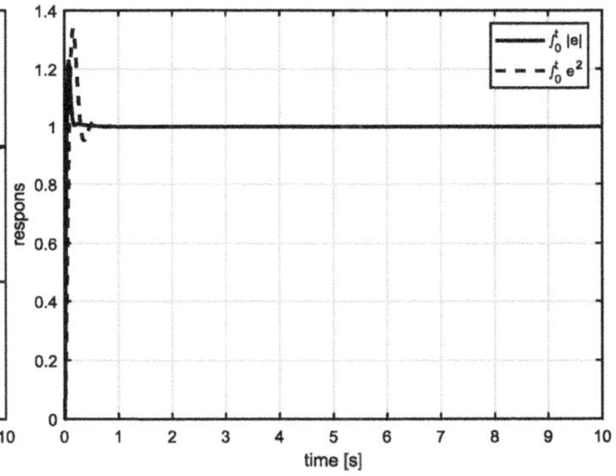

(c) Non-integer PD response

Fig. 4. Comparison step response non-integer controllers for different quality index

5 Conclusion

In this article, the comparison between implementation non-integer controller and tuning methods for two quality function is described.

Experiments show that non-integer PD controller with parameters from function (8) is most promising. The presented methodology can have applicability in control processing in digital control systems. Author show that implementation of non-integer order controller for different structure of controller and tuning methods is possible and give good result.

Further work will include different non-integer order prototypes, methods of tuning, methods of transfer function realisation, methods of discretisation and implementation on hardware platforms.

References

1. Åström, K.J.: Model uncertainty and robust control. In: Lecture Notes on Iterative Identification and Control Design, pp. 63–100 (2000)
2. Baranowski, J.: Projektowanie obserwatora dla serwomechanizmu prądu stałego. In: Materiały VIII Międzynarodowych Warsztatów Doktoranckich OWD 2006, vol. 2, pp. 373–378 (2006)
3. Baranowski, J., Długosz, M., Ganobis, M., Skruch, P., Mitkowski, W.: Applications of mathematics in selected control and decision processes. Matematyka Stosowana: pismo Polskiego Towarzystwa Matematycznego **12/53**(spec. issue), 65–90 (2011)
4. Baranowski, J., Długosz, M., Mitkowski, W.: Remarks about DC motor control. Arch. Control Sci. **18(LIV)**(3), 289–322 (2008)
5. Baranowski, J., Długosz, M., Mitkowski, W.: Nonlinear observer based control of DC servo. In: Materiały XXXII Międzynarodowej konferencji z podstaw elektrotechniki i teorii obwodów IC-SPETO, Ustroń, pp. 115–116 (2009). Extended version on CD
6. Bauer, W.: Implementation of Non-integer $PI^{\lambda}D^{\mu}$ controller for the ATmega328P Micro-controller. In: 2016 21st International Conference on Methods and Models in Automation and Robotics (MMAR), pp. 118–121 (2016)
7. Bauer, W., Baranowski, J., Dziwiński, T., Piątek, P., Zagórowska, M.: Stabilisation of magnetic levitation with a $pi^{\lambda}d^{\mu}$ controller. In: 2015 20th International Conference on Methods and Models in Automation and Robotics (MMAR), Międzyzdroje, Poland, pp. 638–642 (2015)
8. Bauer, W., Rydel, M.: Application of reduced models of non-integer order integrator to the realization $PI^{\lambda}D$ controller. In: 2016 39th International Conference on Telecommunications and Signal Processing, pp. 611–614 (2016)
9. Długosz, M., Lerch, T.: Komputerowa identyfikacja parametrów silnika prądu stałego. Przegląd Elektrotechniczny (2010)
10. Długosz, M., Mitkowski, W.: Adaptive LQR controller for angular velocity stabilisation in series DC motor. In: XXXI IC-SPETO-2008 (2008)
11. Dziwiński, T., Bauer, W., Baranowski, J., Piątek, P., Zagórowska, M.: Robust non-integer order controller for air heater. In: 2014 19th International Conference on Methods and Models in Automation and Robotics (MMAR), Międzyzdroje, Poland, pp. 434–438 (2014)
12. Mitkowski, W., Baranowski, J.: Observer design for series dc motor – multi output approach. In: Materiały XXX Międzynarodowej konferencji z podstaw elektrotechniki i teorii obwodów IC-SPETO, Ustroń, pp. 135–136 (2007). Extended version on CD
13. Mitkowski, W., Zagórowska, M., Bauer, W.: Comparative analysis of DC motor control system. In: Propulsion Systems, Mechatronics and Communication. Applied Mechanics and Materials, vol. 817, pp. 111–121. Trans Tech Publications Ltd. (2016). https://doi.org/10.4028/www.scientific.net/AMM.817.111
14. Oustaloup, A., Levron, F., Mathieu, B., Nanot, F.M.: Frequency-band complex noninteger differentiator: characterization and synthesis. IEEE Trans. Circ. Syst. I: Fundam. Theory Appl. **47**(1), 25–39 (2000)
15. Paszek, S., Berhausen, S., Boboń, A., Majka, Ł., Nocoń, A., Pasko, M., Pruski, P., Kraszewski, T., Szuster, D.: Estymacja parametrów dynamicznych generatorów synchronicznych. Prace Naukowe Politechniki Śląskiej. Elektryka **1**, 105–108 (2015)
16. Petráš, I.: Fractional-Order Nonlinear Systems: Modeling, Analysis and Simulation. Nonlinear Physical Science. Springer, Heidelberg (2011)

17. Podlubny, I.: Fractional Differential Equations: An Introduction to Fractional Derivatives, Fractional Differential Equations, to Methods of Their Solution and Some of Their Applications. Mathematics in Science and Engineering, vol. 198. Elsevier Science, Amsterdam (1998)
18. Sieklucki, G., Bisztyga, B., Zdrojewski, A., Orzechowski, T., Sykulski, R.: Modele i Zasady Sterowania Napędami Elektrycznymi. Wydawnictwo AGH (2014)

Trajectory Planning and Motion Control for Mobile Robots and Intelligent Vehicles

Lining-Up Stabilizers for Pusher and Puller Articulated Vehicles

Maciej Marcin Michałek(✉)

Institute of Automatic Control and Robotics, Poznan University of Technology, Poznań, Poland
maciej.michalek@put.poznan.pl

Abstract. The paper presents a derivation, a local stability analysis, and a numerical validation of control systems with lining-up feedback stabilizers for two kinds of kinematic structures widely used for articulated vehicles in a transportation practice: the tractor-driven vehicle pulling a trailer with a steerable axle, and the wagon-driven vehicle pushing a non-driven prime-mover. Simulation results reveal effectiveness of the stabilizers for various motion strategies of the vehicles (backward, forward, and oscillatory). The stabilizers can be used in the intelligent articulated buses and tractor-trailer vehicles in a fully automated mode or as an advanced driver assistance system.

Keywords: Pusher/puller articulated vehicle · Tractor-trailer · Kinematics · Steerable wheels · Lining-up control · Intelligent vehicle

1 Introduction

A contemporary trend in promoting the large-capacity transportation solutions leads to applications of (multi-body) articulated ground vehicles for translocation of goods and people, [3,4,9]. Due to substantial dimensions and complex dynamics of articulated structures, maneuvering with this kind of vehicles is difficult and burdening. Therefore, various motion tasks are being (semi-)automated to help human drivers, or even replace them, in complex maneuvering – see, e.g., [7,8,10,12,14]. Lining up the bodies of an articulated vehicle is often a maneuver preceding or finishing various motion tasks in the freight and public transportation. The common practice to line up a multi-body vehicle is to drive forward along a line with a prime-mover until a chain of a vehicle's bodies straightens up with a sufficient precision. One can call it a *passive* lining-up strategy, which directly refers to structural stability of joint-angle dynamics observed for most articulated vehicles in forward motion. Under these conditions, the transients decay with a rate inversely proportional to lengths of the vehicle's bodies, see [11], therefore the resultant passive lining-up maneuver can be excessively long both in time and in a distance travelled by a prime mover. Hence, if the workspace is limited and/or the maneuver should be performed fast enough, the

A. Bartoszewicz et al. (Eds.): PCC 2020, AISC 1196, pp. 917–927, 2020.
https://doi.org/10.1007/978-3-030-50936-1_77

passive lining-up strategy is not effective and more advanced control concept is expected in this case.

We propose to solve the lining-up problem in an *active* manner by designing the joint-angle feedback stabilizers for two types of two-body *pusher* and *puller* articulated vehicles widely used in the freight and public transportation today, [2,4]. For these two kinematic structures, we derive two cascade control laws by using Taylor's approximations of the joint-angle kinematics at the proper working points. In contrast to the solution presented in [11] for the N-trailers, the stabilizers proposed in a sequel locally guarantee asymptotic lining-up of vehicles' segments for all possible strategies of motion, that is, for the forward, backward, and the oscillatory (forward-backward) motion. The lining-up stabilizers can be applied in the intelligent (automated) articulated buses and tractor-trailer vehicles.

2 Vehicle Kinematics and Problem Statement

We will consider kinematics of two types of two-body articulated vehicles depicted in Fig. 1 in the form of single-track schemes (i.e., with only effective wheels depicted). The vehicles comprise a car-like prime-mover (the segment number 0), equipped with a front steering wheel and a fixed (non-steerable) rear wheel, and a wagon or a trailer (the segment number 1). The prime-mover is interconnected with the wagon/trailer by a passive rotary joint. Characteristic kinematic parameters of the vehicles are: the prime-mover's length $L_0 > 0$, the

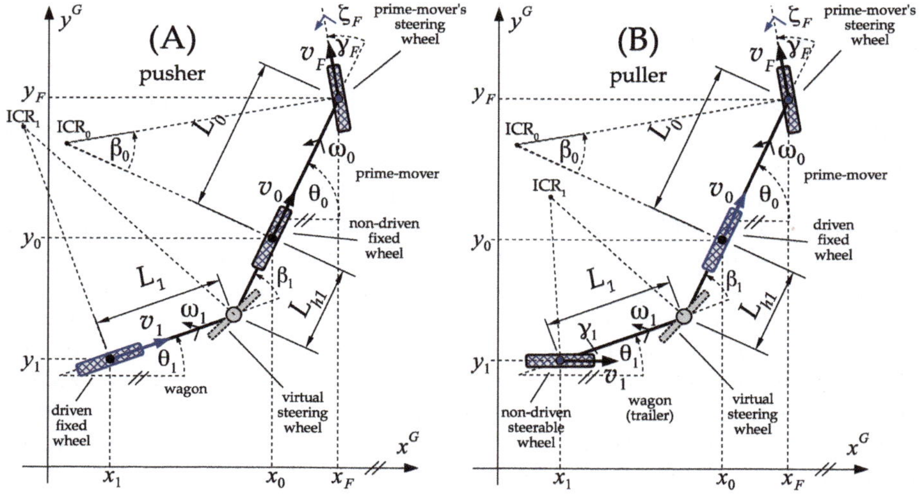

Fig. 1. Single-track schemes of articulated vehicles considered in the paper (independent control inputs are highlighted in blue): (A) the *pusher* articulated vehicle with a non-steerable but driven wagon's axle, and (B) the *puller* articulated vehicle with a driven prime-mover's rear axle and a steerable wagon's axle.

wagon's/trailer's length $L_1 > 0$, and the hitching offset $L_{h1} \neq 0$, $|L_{h1}| < L_1$. In the case of a wagon-driven vehicle of structure (A) – shortly called the *pusher* – the wagon's wheel is driven, while both the prime-mover's wheels are non-driven. This kind of a kinematic structure characterizes contemporary low-deck articulated urban buses. On the other hand, a vehicle of structure (B) – shortly called the *puller* – is characterized by a non-driven but steerable wheel of a wagon/trailer, while the main drive of a vehicle is located in a rear wheel of a prime-mover. Kinematic structure (B) is used in articulated (trolley-)buses and tractor-trailer vehicles.

2.1 Kinematics of the Pusher Articulated Vehicle

A configuration of the pusher vehicle can be represented by

$$\boldsymbol{q}_A \triangleq [\gamma_F \ \beta_1 \ \boldsymbol{q}_j^\top]^\top \in \mathcal{Q}_A = [-\bar{\gamma}_F; \bar{\gamma}_F] \times [-\bar{\beta}_1; \bar{\beta}_1] \times \mathbb{R}^2 \times \mathbb{S}^1, \tag{1}$$

where γ_F is a prime-mover's steering angle, β_1 is a joint angle, whereas $\boldsymbol{q}_j = [x_j \ y_j \ \theta_j]^\top$, $j \in \{0, 1\}$, is a pose of a selected vehicle's segment. The upper bounds $\bar{\gamma}_F, \bar{\beta}_1 \in (0; \frac{\pi}{2})$ represent physical limitations imposed by a mechanical construction of a vehicle. A full kinematic model of the pusher, relating two independent components of the control input $\boldsymbol{u}_A = [\zeta_F \ v_1]^\top$ – including the prime-mover's steering rate ζ_F and the wagon's longitudinal velocity v_1 (see Fig. 1) – with time derivatives of configuration (1) has the form of a driftless system $\dot{\boldsymbol{q}}_A = \boldsymbol{S}_A(\boldsymbol{q}_A)\boldsymbol{u}_A$ – see [13] for a systematic derivation. From this system, let us excerpt the sub-model of the steering and joint kinematics which (using the same general notion as proposed in [13] for a better reference) can be written in the form

$$\begin{bmatrix} \dot{\gamma}_F \\ \dot{\beta}_1 \end{bmatrix} = \begin{bmatrix} 1 & 0 \\ 0 & \boldsymbol{c}^\top \boldsymbol{\Gamma}_1(\beta_1)\boldsymbol{J}_1^{-1}(\beta_1) \left[\begin{smallmatrix} \kappa_1(\gamma_F, \beta_1) \\ 1 \end{smallmatrix}\right] \end{bmatrix} \begin{bmatrix} \zeta_F \\ v_1 \end{bmatrix} \quad \Leftarrow \quad \dot{\boldsymbol{x}}_A = \boldsymbol{H}_A(\boldsymbol{x}_A)\boldsymbol{u}_A, \tag{2}$$

with a new configuration sub-vector $\boldsymbol{x}_A \triangleq [\gamma_F \ \beta_1]^\top$, where

$$\boldsymbol{c}^\top \boldsymbol{\Gamma}_1(\beta_1)\boldsymbol{J}_1^{-1}(\beta_1) \left[\begin{smallmatrix} \kappa_1(\gamma_F, \beta_1) \\ 1 \end{smallmatrix}\right] = -\left(1 + \tfrac{L_1}{L_{h1}} \mathrm{c}\beta_1\right) \kappa_1(\gamma_F, \beta_1) + \tfrac{1}{L_{h1}} \mathrm{s}\beta_1, \tag{3}$$

while

$$\kappa_1(\gamma_F, \beta_1) = \tfrac{1}{L_1} \tan\left(\beta_1 - \arctan\left(\tfrac{L_{h1}}{L_0} \tan \gamma_F\right)\right) \tag{4}$$

is a motion curvature of the wagon. One can observe that $P_A = (\boldsymbol{x}_{Ae}, \boldsymbol{u}_{Ae}) \triangleq (\boldsymbol{0}, [0 \ \mathrm{v}_1]^\top)$ is one of the equilibria of dynamics (2) for any bounded (possibly time-varying) input $v_1 = \mathrm{v}_1 \not\equiv 0$.

2.2 Kinematics of the Puller Articulated Vehicle

A configuration of the puller vehicle can be represented by

$$\boldsymbol{q}_B \triangleq [\gamma_F \ \beta_1 \ \gamma_1 \ \boldsymbol{q}_j^\top]^\top \in \mathcal{Q}_B = [-\bar{\gamma}_F; \bar{\gamma}_F] \times [-\bar{\beta}_1; \bar{\beta}_1] \times [-\bar{\gamma}_1; \bar{\gamma}_1] \times \mathbb{R}^2 \times \mathbb{S}^1, \tag{5}$$

where γ_F is a prime-mover's steering angle, β_1 is a joint angle, γ_1 is a wagon's (trailer's) steering angle, whereas $\boldsymbol{q}_j = [x_j\ y_j\ \theta_j]^\top$, $j \in \{0,1\}$, is a pose of a selected vehicle's segment. The upper bounds $\bar{\gamma}_F, \bar{\beta}_1, \bar{\gamma}_1 \in (0; \frac{\pi}{2})$ represent physical limitations imposed by a mechanical construction of a vehicle. A full kinematic model of the puller in the driftless form $\dot{\boldsymbol{q}}_B = \boldsymbol{S}_B(\boldsymbol{q}_B)\boldsymbol{u}_B$ (see [13]) relates three independent components of the control input $\boldsymbol{u}_B = [\zeta_F\ v_0\ \zeta_1]^\top$ – including the prime-mover's steering rate ζ_F, the prime-mover's longitudinal velocity v_0 (see Fig. 1), and the wagon's steering rate ζ_1 – with time derivatives of configuration (5). From the full kinematic model, let us excerpt the sub-model of the steering and joint kinematics which (using the notion proposed in [13]) can be written in the form

$$\begin{bmatrix}\dot{\gamma}_F\\ \dot{\beta}_1\\ \dot{\gamma}_1\end{bmatrix} = \begin{bmatrix}1 & 0 & 0\\ 0 & \boldsymbol{c}^\top \boldsymbol{\Gamma}_1(\beta_1,\gamma_1)\left[\begin{smallmatrix}\kappa_0(\gamma_F)\\ 1\end{smallmatrix}\right] & 0\\ 0 & 0 & 1\end{bmatrix}\begin{bmatrix}\zeta_F\\ v_0\\ \zeta_1\end{bmatrix} \quad \Leftarrow \quad \dot{\boldsymbol{x}}_B = \boldsymbol{H}_B(\boldsymbol{x}_B)\boldsymbol{u}_B, \tag{6}$$

with a new configuration sub-vector $\boldsymbol{x}_B \triangleq [\gamma_F\ \beta_1\ \gamma_1]^\top$, where

$$\boldsymbol{c}^\top \boldsymbol{\Gamma}_1(\beta_1,\gamma_1)\left[\begin{smallmatrix}\kappa_0(\gamma_F)\\ 1\end{smallmatrix}\right] = \left(1 + \frac{L_{h1}\mathrm{c}(\beta_1-\gamma_1)}{L_1\mathrm{c}\gamma_1}\right)\kappa_0(\gamma_F) - \frac{\mathrm{s}(\beta_1-\gamma_1)}{L_1\mathrm{c}\gamma_1} \tag{7}$$

while

$$\kappa_0(\gamma_F) = \tfrac{1}{L_0}\tan\gamma_F \tag{8}$$

represents a motion curvature of the prime mover. One observes that $P_B = (\boldsymbol{x}_{Be}, \boldsymbol{u}_{Be}) \triangleq (\boldsymbol{0}, [0\ \mathrm{v}_0\ 0]^\top)$ is one of the equilibria of dynamics (6) for any bounded (possibly time-varying) input $v_0 = \mathrm{v}_0 \not\equiv 0$.

2.3 Control Problem Formulation

Problem 1. Let us consider kinematics $\dot{\boldsymbol{x}} = \boldsymbol{H}(\boldsymbol{x})\boldsymbol{u}$ taking the special form (2) or (6) for the pusher or puller vehicle, respectively. The control problem relies on designing, independently for particular types of considered vehicles, a feedback stabilizer $\boldsymbol{u} = \boldsymbol{u}(\boldsymbol{x}, \mathrm{v}_i)$ which guarantees asymptotic stability of the equilibrium point $\boldsymbol{x}_e = \boldsymbol{0}$, in the sense that a response of the closed-loop dynamics $\dot{\boldsymbol{x}} = \boldsymbol{H}(\boldsymbol{x})\boldsymbol{u}(\boldsymbol{x}, \mathrm{v}_i)$ is bounded and $\lim_{t\to\infty} \boldsymbol{x}(t) = \boldsymbol{x}_e$ for a bounded driving velocity $v_i = \mathrm{v}_i \not\equiv 0$, where $i \in \{0,1\}$.

In a physical interpretation, the stabilizers $\boldsymbol{u} = \boldsymbol{u}(\boldsymbol{x}, \mathrm{v}_i)$ should asymptotically line-up the articulated vehicle's segments, from any initial configuration $\boldsymbol{x}(0)$ in some vicinity of equilibrium $\boldsymbol{x}_e$, with a non-zero initial joint angle $\beta_1(0)$.

3 Design of Lining-Up Feedback Stabilizers

3.1 Stabilizer for the Pusher Articulated Vehicle

Taylor's approximation of kinematics (2) at the working point $P_A \triangleq (\boldsymbol{0}, [0\ \mathrm{v}_1]^\top)$ leads to the linear dynamics

$$\begin{bmatrix}\dot{\gamma}_F\\ \dot{\beta}_1\end{bmatrix} \cong \begin{bmatrix}0 & 0\\ \frac{\mathrm{v}_1}{L_1}\eta & -\frac{\mathrm{v}_1}{L_1}\end{bmatrix}\begin{bmatrix}\gamma_F\\ \beta_1\end{bmatrix} + \begin{bmatrix}1\\ 0\end{bmatrix}\zeta_F, \tag{9}$$

where v_1 is now a parametrizing variable (which can be constant or time-varying), and $\eta = (L_1 + L_{h1})/L_0$. By rewriting the second row of (9) as

$$\dot{\beta}_1 = \tfrac{\mathrm{v}_1}{L_1}\left(\eta\gamma_F - \beta_1\right), \tag{10}$$

one can treat γ_F as a virtual control input and formulate a postulate

$$\eta\gamma_F - \beta_1 := -\alpha L_1 \mathrm{v}_1 \beta_1, \qquad \alpha > 0, \tag{11}$$

where α is a design coefficient. Satisfaction of postulate (11) leads to the desirable closed-loop dynamics $\dot{\beta}_1 = -\alpha \mathrm{v}_1^2 \beta_1$, which ensures exponential convergence $\beta_1(t) \to 0$ as $t \to \infty$ for any bounded and persistently exciting (PE) driving velocity satisfying $\lim_{t\to\infty}\int_0^t \mathrm{v}_1^2(\tau)\mathrm{d}\tau = \infty$. Since the postulate (11) can be rewritten in the form $\gamma_F := \frac{L_1}{\eta}(-\alpha \mathrm{v}_1 + \frac{1}{L_1})\beta_1$, we define the desired steering angle for a prime mover as follows

$$\gamma_{Fd} \triangleq \mathrm{sat}\left[\gamma_{Fc}(\beta_1, \mathrm{v}_1), \bar{\gamma}_F\right], \qquad \gamma_{Fc}(\beta_1, \mathrm{v}_1) \triangleq \tfrac{L_1}{\eta}\left(-\alpha \mathrm{v}_1 + \tfrac{1}{L_1}\right)\beta_1, \tag{12}$$

where $\mathrm{sat}[z, \bar{z}] \triangleq \min\{|z|, \bar{z}\}\cdot\mathrm{sign}(z)$ is a conventional saturation function. Using the saturation function in the above definition comes from the configuration domain $\mathcal{Q}_A$ introduced in (1). Now, upon the first row of kinematics (9), let us propose the following steering controller

$$\zeta_F \triangleq k_a(\gamma_{Fd} - \gamma_F) + \dot{\gamma}_{Fd}, \qquad k_a > 0, \tag{13}$$

where k_a is a design coeficient, whereas the feedforward term

$$\dot{\gamma}_{Fd} = \begin{cases} \dot{\gamma}_{Fc} & \text{for } |\gamma_{Fc}| \le \bar{\gamma}_F \\ 0 & \text{for } |\gamma_{Fc}| > \bar{\gamma}_F \end{cases}, \qquad \dot{\gamma}_{Fc} = \tfrac{L_1}{\eta}\left[-\alpha\dot{\mathrm{v}}_1\beta_1 + \left(-\alpha \mathrm{v}_1 + \tfrac{1}{L_1}\right)\dot{\beta}_1\right]. \tag{14}$$

Note that (14) depends on time derivatives of the applied driving velocity $v_1 = \mathrm{v}_1$ and the joint-angle β_1; the latter results from the second row of (2).

Proposition 1. *Locally, in a small vicinity of $\boldsymbol{x}_A = [\gamma_F\ \beta_1]^\top = \mathbf{0}$, the joint-angle stabilizer defined by (13) and (12) solves the Problem 1 for the pusher articulated vehicle if a bounded driving velocity $\mathrm{v}_1 \in \mathcal{C}^1$ is non-zero for almost all $t \ge 0$ and satisfies the PE condition $\lim_{t\to\infty}\int_0^t \mathrm{v}_1^2(\tau)\mathrm{d}\tau = \infty$.*

Let us briefly justify Proposition 1. For small values of angles β_1 and γ_F, holds $\gamma_{Fd} \equiv \gamma_{Fc}$ (no saturation occurs in (12)), and the closed-loop joint kinematics can be approximated by $\dot{\beta}_1 \cong -\alpha \mathrm{v}_1^2\beta_1 - \frac{\mathrm{v}_1}{L_1}\eta e_F$, where $e_F \triangleq (\gamma_{Fd} - \gamma_F)$ is a steering error. Furthermore, after substituting (13) into (9), one obtains the closed-loop steering error dynamics $\dot{e}_F + k_a e_F = 0$ which implies: $\forall t \ge 0\ |e_F(t)| \le |e_F(0)|$ and $e_F(t) \to 0$ exponentially as $t \to \infty$. Next, introducing a positive definite function $V \triangleq \frac{1}{2}\beta_1^2$, one can assess its time derivative along a solution of the closed-loop dynamics as follows

$$\dot{V} = \beta_1\dot{\beta}_1 \le -\alpha \mathrm{v}_1^2(1-\nu)\beta_1^2 + \left[\tfrac{|\mathrm{v}_1|}{L_1}\eta\,|e_F| - \nu\alpha \mathrm{v}_1^2\,|\beta_1|\right]|\beta_1|, \tag{15}$$

where $\nu \in (0;1)$ is a majorization constant. Upon (15) and the boundedness of $e_F(t)$, and by recalling the input-to-state stability (ISS) result formulated, e.g., in [5], one can conclude that $\forall t \geq 0$ $|\beta_1(t)| < \infty$ and $\limsup_{t\to\infty} |\beta_1(t)| \leq \limsup_{t\to\infty} \frac{\eta}{L_1 \nu \alpha |\mathrm{v}_1(t)|} |e_F(t)| = b_1$. Since $e_F(t) \to 0$ with an exponential rate, and since $\mathrm{v}_1(t)$ is non-zero for almost all $t \in [0, \infty)$ and is allowed to (possibly) terminally converge to zero strictly slower than $e_F(t)$ does (by satisfying the PE condition), one concludes that $b_1 = 0$. As a consequence, one claims $\beta_1(t \to \infty) \to 0$. Boundedness of $\beta_1(t)$ and of $\mathrm{v}_1(t)$ implies boundedness of $\gamma_{Fd}(t) \equiv \gamma_{Fc}(t)$ for $t \geq 0$ (see (12)). The latter, together with boundedness of $e_F(t)$ directly imply boundedness of $\gamma_F(t)$. Upon (12), $\gamma_{Fc}(\beta_1 \to 0, \mathrm{v}_1) \to 0$, and since $e_F(t \to \infty) \to 0$, one infers $\gamma_F(t) \to 0$ as $t \to \infty$.

Remark 1. It is worth stressing, that the result formulated in Proposition 1 is valid for any sign of the driving velocity $\mathrm{v}_1(t)$, admitting both forward and backward motion of a vehicle. Conservative local character of the result in Proposition 1 is a direct consequence of linear approximation (9) which is valid in a sufficiently small neighbourhood of the equilibrium P_A. On the other hand, the joint stabilizer (13) has been defined with a saturation function in (12) – suggesting admissibility of large values for γ_F during a stabilization process – in order to show effectiveness of the stabilizer also far away from the equilibrium point $\boldsymbol{x}_A = \mathbf{0}$ – it has been illustrated by simulation results provided in Sect. 4.

3.2 Stabilizer for the Puller Articulated Vehicle

Taylor's approximation of model (6) at the working point $P_B \triangleq (\mathbf{0}, [0\ \mathrm{v}_0\ 0]^\top)$ leads to the linear dynamics

$$\begin{bmatrix} \dot{\gamma}_F \\ \dot{\beta}_1 \\ \dot{\gamma}_1 \end{bmatrix} \cong \begin{bmatrix} 0 & 0 & 0 \\ \frac{\mathrm{v}_0}{L_1}\eta & -\frac{\mathrm{v}_0}{L_1} & \frac{\mathrm{v}_0}{L_1} \\ 0 & 0 & 0 \end{bmatrix} \begin{bmatrix} \gamma_F \\ \beta_1 \\ \gamma_1 \end{bmatrix} + \begin{bmatrix} 1 & 0 \\ 0 & 0 \\ 0 & 1 \end{bmatrix} \begin{bmatrix} \zeta_F \\ \zeta_1 \end{bmatrix}, \tag{16}$$

where v_0 is treated now as a parametrizing variable, and $\eta = (L_1 + L_{h1})/L_0$. By rewriting the second row of (16) in the form

$$\dot{\beta}_1 = \frac{\mathrm{v}_0}{L_1} (\eta\gamma_F - \beta_1 + \gamma_1), \tag{17}$$

one can treat γ_F and γ_1 as two virtual control inputs of the above dynamics. Let us formulate a postulate

$$\eta\gamma_F - \beta_1 + \gamma_1 := -\alpha L_1 \mathrm{v}_0 \beta_1, \qquad \alpha > 0, \tag{18}$$

where α is a design coefficient. Satisfaction of postulate (18) leads to the desirable closed-loop dynamics $\dot{\beta}_1 = -\alpha \mathrm{v}_0^2 \beta_1$, which ensures exponential convergence $\beta_1(t) \to 0$ as $t \to \infty$ for any bounded and persistently exciting (PE) driving velocity satisfying $\lim_{t\to\infty} \int_0^t \mathrm{v}_0^2(\tau)\mathrm{d}\tau = \infty$. By rewriting (18) in the form

$$\eta\gamma_F + \gamma_1 := (-\alpha L_1 \mathrm{v}_0 + 1)\beta_1, \qquad \alpha > 0 \tag{19}$$

one observes that satisfaction of postulate (18) leads to the *control allocation* problem, [6], where we have two virtual control inputs, γ_F and γ_1, and only a single controlled variable β_1. In this context, we propose to introduce an allocation factor $\mu \in [0; 1]$, which will allow splitting the control action between the virtual control inputs. Taking into account the configuration domain $\mathcal{Q}_B$ introduced in (5), let us define the desired steering angles:

$$\gamma_{1d} \triangleq \text{sat}\left[\gamma_{1c}(\beta_1, \text{v}_0), \bar{\gamma}_1\right], \quad \gamma_{1c}(\beta_1, \text{v}_0) \triangleq \mu\left(-\alpha L_1 \text{v}_0 + 1\right)\beta_1, \tag{20}$$

$$\gamma_{Fd} \triangleq \text{sat}\left[\gamma_{Fc}(\beta_1, \text{v}_0), \bar{\gamma}_F\right], \quad \gamma_{Fc}(\beta_1, \text{v}_0) \triangleq \tfrac{1}{\eta}\left[(-\alpha L_1 \text{v}_0 + 1)\beta_1 - \gamma_{1d}\right], \tag{21}$$

where $\mu \in [0; 1]$ determines a *portion of the right-hand side* of (19) which should be affected by the wagon's/trailer's steering angle. Now, we define the steering controllers

$$\zeta_1 \triangleq k_1(\gamma_{1d} - \gamma_1) + \dot{\gamma}_{1d}, \qquad \zeta_F \triangleq k_b(\gamma_{Fd} - \gamma_F) + \dot{\gamma}_{Fd}, \qquad k_b, k_1 > 0, \tag{22}$$

where k_b, k_1 are design parameters, while the feedforward terms

$$\dot{\gamma}_{id} = \begin{cases} \dot{\gamma}_{ic} & \text{for } |\gamma_{ic}| \leq \bar{\gamma}_i \\ 0 & \text{for } |\gamma_{ic}| > \bar{\gamma}_i \end{cases}, \qquad i \in \{1, F\}, \tag{23}$$

with the corresponding time derivatives

$$\dot{\gamma}_{1c} = \mu\left[-\alpha L_1(\dot{\text{v}}_0\beta_1 + \text{v}_0\dot{\beta}_1) + \dot{\beta}_1\right], \tag{24}$$

$$\dot{\gamma}_{Fc} = \tfrac{1}{\eta}\left[-\alpha L_1 \dot{\text{v}}_0 \beta_1 + (-\alpha L_1 \text{v}_0 + 1)\dot{\beta}_1 - \dot{\gamma}_{1d}\right]. \tag{25}$$

A form of the time derivative $\dot{\beta}_1$, present in the above formulas, comes from the second row of (6).

Proposition 2. *Locally, in a small vicinity of* $\boldsymbol{x}_B = [\gamma_F \ \beta_1 \ \gamma_1]^\top = \mathbf{0}$, *the joint-angle stabilizer defined by (22) and (20)–(21) solves the Problem 1 for the puller articulated vehicle if a bounded driving velocity* $\text{v}_0 \in \mathcal{C}^1$ *is non-zero for almost all* $t \geq 0$ *and satisfies the PE condition* $\lim_{t\to\infty} \int_0^t \text{v}_0^2(\tau)\text{d}\tau = \infty$.

Let us, in brief, justify Proposition 2. For small values of angles β_1, γ_1, and γ_F, holds $\gamma_{Fd} \equiv \gamma_{Fc}$ and $\gamma_{1d} \equiv \gamma_{1c}$ (no saturation occurs in (20) and (21)), and the closed-loop joint kinematics can be approximated by $\dot{\beta}_1 \cong -\alpha \text{v}_0^2 \beta_1 - \frac{\text{v}_0}{L_1}(\eta e_F + e_\gamma)$, where $e_F \triangleq (\gamma_{Fd} - \gamma_F)$ and $e_1 \triangleq (\gamma_{1d} - \gamma_1)$ are the steering errors of the prime mover and the wagon/trailer, respectively. After substituting (22) into (16), one obtains the closed-loop steering error dynamics: $\dot{e}_F + k_b e_F = 0$ and $\dot{e}_1 + k_1 e_1 = 0$. They imply: $\forall t \geq 0 \ |e_F(t)| \leq |e_F(0)|$ and $|e_1(t)| \leq |e_1(0)|$ together with the exponential convergence $e_F(t) \to 0$ and $e_1(t) \to 0$ as $t \to \infty$. Introducing a positive definite function $V \triangleq \frac{1}{2}\beta_1^2$, one can assess its time derivative along a solution of the closed-loop dynamics as follows

$$\dot{V} = \beta_1\dot{\beta}_1 \leq -\alpha \text{v}_0^2(1 - \nu)\beta_1^2 + \left[\tfrac{|\text{v}_0|}{L_1}(\eta\,|e_F| + |e_\gamma|) - \nu\alpha \text{v}_0^2\,|\beta_1|\right]|\beta_1|, \tag{26}$$

where $\nu \in (0;1)$ is a majorization constant. Upon (26) and the boundedness of steering errors $e_F(t)$ and $e_1(t)$, and by recalling the input-to-state stability (ISS) result, one can conclude that $\forall t \geq 0$ $|\beta_1(t)| < \infty$ and $\limsup_{t\to\infty} |\beta_1(t)| \leq \limsup_{t\to\infty} \frac{\eta}{L_1\nu\alpha|\mathrm{v}_0(t)|} |e_F(t)| + \limsup_{t\to\infty} \frac{1}{L_1\nu\alpha|\mathrm{v}_0(t)|} |e_1(t)| = b_2$. Since $e_F(t)$ and $e_1(t)$ converge to zero with an exponential rate, and since $\mathrm{v}_0(t)$ is non-zero for almost all $t \in [0,\infty)$ and is allowed to (possibly) terminally converge to zero strictly slower than both steering errors do (by satisfying the PE condition), one concludes that $b_2 = 0$. As a consequence, one claims $\beta_1(t \to \infty) \to 0$. Boundedness of $\beta_1(t)$ and of $\mathrm{v}_0(t)$ imply boundedness of $\gamma_{1d}(t) \equiv \gamma_{1c}(t)$ and of $\gamma_{Fd}(t) \equiv \gamma_{Fc}(t)$ for $t \geq 0$ (see (20)–(21)). The latter, together with boundedness of $e_F(t)$ and $e_1(t)$ directly imply boundedness of $\gamma_F(t)$ and $\gamma_1(t)$. Now, according to (20)–(21), $\gamma_{1d}(\beta_1 \to 0, \mathrm{v}_0) \to 0$ and $\gamma_{Fd}(\beta_1 \to 0, \mathrm{v}_0) \to 0$. As a consequence, and since $e_F(t \to \infty) \to 0$ and $e_1(t \to \infty) \to 0$, one infers $\gamma_F(t) \to 0$ and $\gamma_1(t) \to 0$ as $t \to \infty$.

Remark 2. Also in the case of the puller vehicle, the lining-up stabilizer formulated in Proposition 2 is valid for any sign of the driving velocity $\mathrm{v}_0(t)$, admitting both forward and backward motion of a vehicle. Despite a local formulation of the result in Proposition 2, the joint stabilizer (22) has been defined with saturation functions used in (20)–(21) in order to show effectiveness of the control law also far away from the equilibrium point $\boldsymbol{x}_B = \mathbf{0}$ – it can be observed upon simulation results presented in Sect. 4.

Remark 3. Definition of $\gamma_{1c}(\beta_1, \mathrm{v}_0)$ in (20) reflects the proposed control allocation strategy, according to which we directly impose the allocation factor $\mu \in [0;1]$ for the wagon's/trailer's steering wheel to determine its workload in the process of lining-up control. The prime mover's steering wheel takes over the rest workload of a control process, also in the case when the wagon's steering wheel saturates (in practical constructions usually $\bar{\gamma}_1 < \bar{\gamma}_F$). By taking $\mu > 0.5$, a designer decides to make the wagon's/trailer's steering wheel dominating in the lining-up control process, up to the saturation limit $\bar{\gamma}_1$.

4 Exemplary Simulation Results

Numerical validation of the proposed lining-up strategies has been conducted using kinematic parameters and limiting values of the *Urbino 18* articulated bus (consult [1]), namely: $L_0 = 5.9\,\mathrm{m}$, $L_1 = 4.17\,\mathrm{m}$, $L_{h1} = 1.83\,\mathrm{m}$, $\bar{\gamma}_F = 45° \cdot (\pi/180°)\,\mathrm{rad}$, $\bar{\beta}_1 = 54° \cdot (\pi/180°)\,\mathrm{rad}$, and $\bar{\gamma}_1 = 30° \cdot (\pi/180°)\,\mathrm{rad}$. In all cases, the design coefficients of the stabilizers have been selected as $\alpha = 25$, and $k_a = k_b = k_1 = 10$. Two driving scenarios have been addressed for both, the pusher and the puller vehicle, namely: the monotonic backward driving with a constant driving velocity $\mathrm{v}_j = -0.3\,\mathrm{m/s}$, $j \in \{1, 0\}$, and the non-vanishing oscillatory forward-backward driving with driving velocity $\mathrm{v}_j = 0.2\sin(0.3t)\,\mathrm{m/s}$, $j \in \{1, 0\}$. The zero initial configurations of the pusher and puller vehicles have been chosen, except the joint angle, which in all the cases has been initialized as $\beta_1(0) = 20° \cdot (\pi/180°)\,\mathrm{rad}$. For the puller bus, the allocation factor has been set to $\mu = 0.3$.

The results of simulations obtained for the pusher are provided in Fig. 2, while for the puller in Fig. 3.

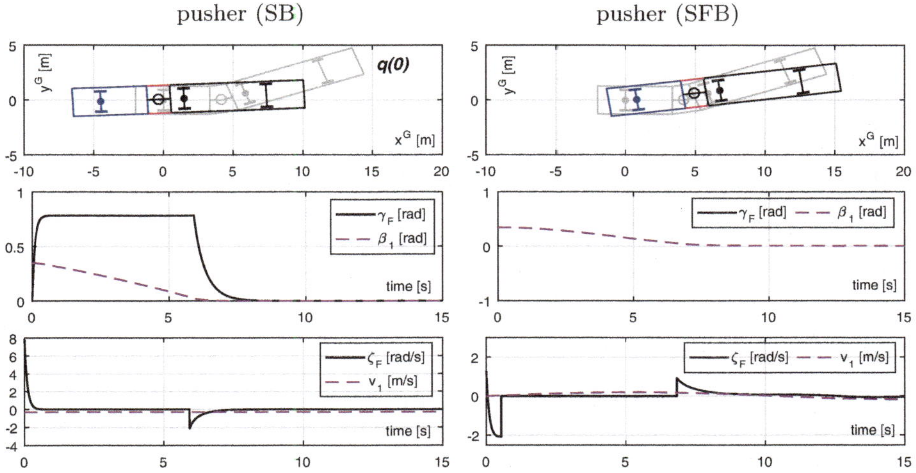

Fig. 2. Simulation results of the lining-up maneuvers obtained for the pusher articulated vehicle (with fixed wagon's wheels) in two driving scenarios: moving monotonically backward (SB), and moving oscillatory forward-backward (SFB); initial configuration $\boldsymbol{q}(0)$ of a vehicle is drawn in the light-grey colour.

Table 1 contains the values of duration times T_d obtained for the lining-up maneuvers in four cases of control strategies: moving forward in the open loop with zero steering angles (SF-passive), moving forward with the proposed lining-up stabilizer (SF-active), moving backward with the proposed lining-up stabilizer (SB-active), moving in the oscillatory manner (forward-backward) with the proposed lining-up stabilizer (SFB-active). In all the cases, except SFB-active, the absolute driving velocity was selected as $|\mathrm{v}_j| = 0.3\,\mathrm{m/s}$, $j \in \{1, 0\}$. The finite duration times T_d, obtained upon simulations, corresponded to the approximated lining-up maneuvers characterized by satisfaction of the following condition: $|\beta_1(t \geq T_d)| \leq 0.5° \cdot (\pi/180°)\,\mathrm{rad}$.

Table 1. Finite duration times T_d (in seconds) obtained for various strategies of the (approximate) lining-up maneuvers.

Strategy →	SF-passive	SF-active	SB-active	SFB-active
T_d [s]: **pusher**	51.00	4.04	6.36	13.01
T_d [s]: **puller**	51.46	3.27	3.88	7.69

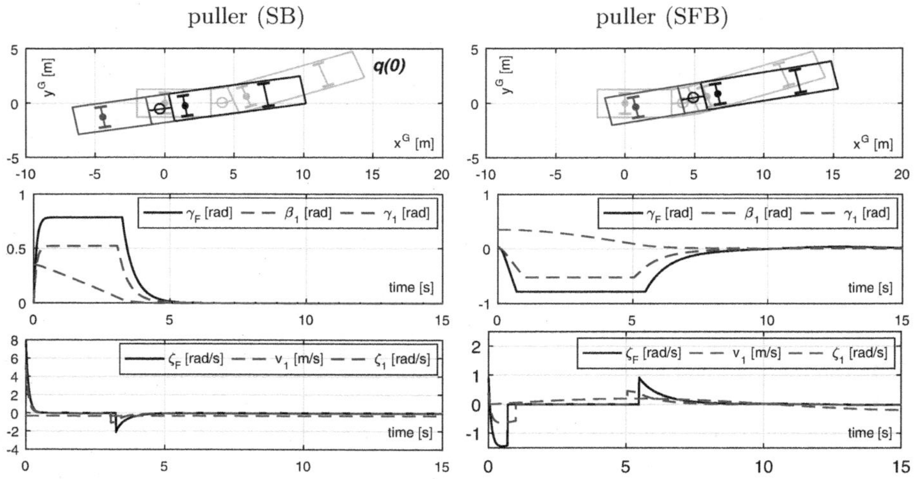

Fig. 3. Simulation results of the lining-up maneuvers obtained for the puller articulated vehicle (with steerable wagon's wheels) in two driving scenarios: moving monotonically backward (SB), and moving oscillatory forward-backward (SFB); initial configuration $\boldsymbol{q}(0)$ of a vehicle is drawn in the light-grey colour.

According to the presented results, one can observe what follows:

- The oscillatory forward-backward driving strategy allows accomplishing the lining-up maneuver occupying smaller subset of a vehicle workspace; it concerns both, the pusher and the puller vehicle (compare the initial configurations $\boldsymbol{q}(0)$ highlighted in light grey with final configurations $\boldsymbol{q}(15)$ denoted in colours in Figs. 2–3). One may expect that a resultant size of a workspace occupied by the vehicle during maneuvering depends on the amplitude and frequency of the driving velocity. As a consequence, selection of a suitable driving strategy can let the vehicle accomplish the lining-up maneuver even in highly cluttered workspaces.
- Despite the sustained saturations of the steering angles within a transient stage, an asymptotic stability of the zero equilibrium has been preserved.
- Forcing the allocation factor $\mu < 0.5$ makes the control effort of the wagon's steering angle γ_1 smaller relative to the effort required for the prime-mover's steering angle γ_F. The less control effort implies the time intervals of a saturation for the wagon's steering angle shorter when compared to the saturation duration observed for the prime mover's steering angle.
- Usage of the wagon's steering wheel in the lining-up control process leads to a faster convergence of the joint angle toward zero when compared to the lining-up maneuver performed by the pusher vehicle.
- The lining-up maneuvering performed in the active manner (that is, with the proposed stabilizers) lead to a much faster convergence of the joint angle $\beta_1(t)$ when compared to the conventional SF-passive strategy.

5 Conclusions

In the paper, two lining-up feedback controllers for the pusher and puller two-body articulated vehicles have been designed. The control laws lead to a local asymptotic stabilization of the joint angle at zero for both forward and backward driving motion strategies in the persistently exciting driving conditions. Simulation results reveal large basins of asymptotic stability for the closed-loop systems with the proposed stabilizers, even in the presence of steering angles saturations. The proposed lining-up controllers may be applied as a part of the advanced driver assistance systems (ADAS) in urban articulated buses and tractor-trailers or as a part of a control system for the highly automated articulated vehicles.

Acknowledgement. The work was supported in part by the National Centre for Research and Development (NCBR), Poland, as a grant No. POIR.04.01.02-00-0081/17, and in part by the research subvention No. 0211/SBAD/911.

References

1. Solaris Bus & Coach: Alternative powertrain. Product catalogue (2018)
2. Dang, H.A., Kovanda, J.: Determination of trajectory of articulated bus turning along curved line. Trans. Transport Sci. **7**(1), 35–44 (2014)
3. El-Geneidy, A.M., Vijayakumar, N.: The effects of articulated buses on dwell and running times. J. Public Transport. **14**(3), 63–86 (2011)
4. Hemily, B., King, R.D.: Uses of higher capacity of buses in transit service. Transportation Research Board, Washington, D.C. (2008)
5. Isidori, A.: Nonlinear Control Systems II. Springer, London (1999)
6. Johansen, T.A., Fossen, T.I.: Control allocation - a survey. Automatica **49**, 1087–1103 (2013)
7. Jujnovich, B.A., Cebon, D.: Path-following steering control for articulated vehicles. ASME J. Dyn. Sys. Meas. Cont. **135**(031006), 1–15 (2013)
8. Kim, Y.C., Yun, K.H., Min, K.D.: Automatic guidance control of an articulated all-wheel-steered vehicle. Vehicle Sys. Dyn. **52**(4), 456–474 (2014)
9. Leduc, G.: Longer and heavier vehicles. An overview of technical aspects. Technical report EUR 23949 EN, European Commission, Joint Research Centre, Institute for Prospective Technological Studies, Luxembourg (2009)
10. Ljungqvist, O., Evestedt, N., Axehill, D., Cirillo, M., Pettersson, H.: A path planning and path-following control framework for a general 2-trailer with a car-like tractor. J. Field Robot. **36**(8), 1345–1377 (2019)
11. Michałek, M.: Lining-up control strategies for N-trailer vehicles. J. Intell. Robot. Syst. **75**(1), 29–52 (2014)
12. Michałek, M.M.: Agile maneuvering with intelligent articulated vehicles: a control perspective. IFAC PapersOnLine **52**(8), 458–473 (2019)
13. Michałek, M.M.: Modular approach to compact low-speed kinematic modelling of multi-articulated urban buses for motion algorithmization purposes. In: 2019 IEEE Intelligent Vehicles Symposium (IV), Paris, France, pp. 1803–1808 (2019)
14. Montes, H., Salinas, C., Fernandez, R., Armada, M.: An experimental platform for autonomous bus development. Appl. Sci. **7**(1131), 1–22 (2017)

Evaluation of Parameterizations in Local Lie-Algebraic Motion Planning

Ignacy Duleba(✉) and Arkadiusz Mielczarek

Department of Cybernetics and Robotics, Wroclaw University of Science and Technology, Janiszewski St. 11/17, 50-372 Wroclaw, Poland
ignacy.duleba@pwr.edu.pl

Abstract. In this paper some parameterizations of controls are examined in a Lie algebraic method of motion planning for driftless nonholonomic systems. The purpose of the examination is to establish how numerous the parameterization should be and which items of a harmonic basis are to be included into the parameterization. An algorithm is presented to evaluate parameterizations without (or reduced) impact of a local, desired direction of motion.

Keywords: Nonholonomic systems · Motion planning · Control parameterization · Evaluation

1 Introduction

In most tasks of motion planning there is a need to generate off-line a desired trajectory (or controls) for a control module responsible for on-line tracking the trajectory [1]. As controls are functions determined on a fixed time horizon thus they belong to an infinite dimensional search space [2]. Unfortunately, computers cannot process infinite dimensional objects effectively therefore there is a practical need to decrease dimension of the search space while avoiding to lose some desired properties (controllability). Usually, in order to express controls, a functional basis is chosen and a finite representation (parameterization) is selected (alternative, infinite dimensional approaches have also been tested [3]). When the basis is selected, another problem is encountered: how many and which items of the basis should enter a parameterization. This problem arises in global motion planning methods (for example in the endogenous configuration space method [2]) as well as in local ones [4]. In this paper we examine an impact of parameterizations on a local performance of a Lie algebraic, local method of motion planning for driftless nonholonomic systems.

The paper is organized as follows. In Sect. 2 some mathematical preliminaries are provided. Drifless nonholonomic systems analyzed in this paper are defined by a set of vector fields called generators. Using a generalized Campbell-Baker-Hausdorff-Dynkin formula (gCBHD) a local, around a current configuration, motion towards a goal point is expressed as a combination of generators and

A. Bartoszewicz et al. (Eds.): PCC 2020, AISC 1196, pp. 928–940, 2020.
https://doi.org/10.1007/978-3-030-50936-1_78

their descendants multiplied by some control dependent functions. Then, controls are expressed in a parametric form. Consequently, the motion planning can be viewed as a redundant inverse kinematics task with known displacement towards a goal and unknown values of parameters of controls. This task will be solved using the Newton algorithm with an optimization in the null space of the Jacobian matrix derived from the kinematics. The optimization is aimed at determining the best possible set of parameters with respect to a selected quality function. In Sect. 3, it is discussed how to evaluate parameterizations locally. An algorithm is provided to evaluate selected parameterizations at a single configuration. In Sect. 4 simulations performed on a unicycle model are provided and a few parameterizations are evaluated at some points within the configuration space. In Sect. 5 concluding remarks are collected.

2 Theoretical Background

Driftless nonholonomic systems result from nonholonomic constraints in the Pfaff form [5] and they are described by the equation

$$\dot{\boldsymbol{q}} = \sum_{i=1}^{m} \boldsymbol{g}_i(\boldsymbol{q})u_i = \boldsymbol{A}(t)(\boldsymbol{q}(t)), \tag{1}$$

where $\boldsymbol{q}$ is a configuration vector that belongs to the configuration space $\mathbb{Q}$, $\boldsymbol{g}_i(\boldsymbol{q})$ are vector fields (later on called generators) and u_i are controls. For illustration purposes and to cover common and the most difficult case, it is assumed that $m = 2$ and vector fields are denoted as $\boldsymbol{g}_1 = \boldsymbol{X}$, $\boldsymbol{g}_1 = \boldsymbol{Y}$. Locally, using the generalized Campbell-Baker-Hausdorff-Dynkin formula, a trajectory initialized at a given configuration $\boldsymbol{q}_0$ can be approximated as follows

$$\boldsymbol{q}(t) \simeq \boldsymbol{z}(t)(\boldsymbol{q}_0) + \boldsymbol{q}_0, \tag{2}$$

where a shift operator $\boldsymbol{z}(t)(\boldsymbol{q}_0)$, for $t \to 0$, is described by a series [6]

$$\boldsymbol{z}(t)(\boldsymbol{q}_0) \simeq \sum_{r=1}^{\infty} \int_{T_r(t)} \left(\sum_{\boldsymbol{\sigma} \in P_r} c(\boldsymbol{\sigma}) \boldsymbol{E}_{\boldsymbol{\sigma}} \right) d\boldsymbol{s}^r, \tag{3}$$

where $\int_{T_r(t)} = \int_{s_r=0}^{t} \int_{s_{r-1}=0}^{s_r} \cdots \int_{s_1=0}^{s_2}$ is an r-dimensional simplex, $d\boldsymbol{s}^r = ds_1 ... ds_r$; P_r is a set of all permutations derived from the set $\{1, \ldots, r\}$ and

$$\boldsymbol{E}_{\boldsymbol{\sigma}} = [[\ldots[\boldsymbol{A}(s_{\sigma(1)}), \boldsymbol{A}(s_{\sigma(2)})], \ldots], \boldsymbol{A}(s_{\sigma(r)})], \quad c(\boldsymbol{\sigma}) = (-1)^{e(\boldsymbol{\sigma})} / \{r^2 \binom{r-1}{e(\boldsymbol{\sigma})}\}. \tag{4}$$

$c(\boldsymbol{\sigma}) \in \mathbb{R}$ and $e(\boldsymbol{\sigma})$ counts the number of errors in consecutive pairs of integers in the permutation $\boldsymbol{\sigma} = \{\sigma(1), \sigma(2), \ldots, \sigma(r)\}$, for example $e((1,2,3)) = 0$, $e((2,1,3)) = 1$, $e((3,2,1)) = 2$.

The operation $[\boldsymbol{V}, \boldsymbol{Z}]$ used many times in Eq. (4) denotes a Lie bracket of vector fields $\boldsymbol{V}, \boldsymbol{Z}$ [7] and, in coordinates, it is described by the formula

$$[\boldsymbol{V}, \boldsymbol{Z}] = \frac{\partial \boldsymbol{Z}}{\partial \boldsymbol{q}}\boldsymbol{V} - \frac{\partial \boldsymbol{V}}{\partial \boldsymbol{q}}\boldsymbol{Z}. \tag{5}$$

Without discovering their nature, vector fields can be considered as more general objects, Lie monomials. To each Lie monomial its degree can be assigned by counting the number of generators within the monomial. Consequently, for generators $\deg(\boldsymbol{X}) = \deg(\boldsymbol{Y}) = 1$, while for composed Lie monomials $\deg([\boldsymbol{X}, \boldsymbol{Y}]) = 2$, $\deg([\boldsymbol{X}, [\boldsymbol{X}, \boldsymbol{Y}]]) = 3$. A layer contains all Lie monomials sharing the same degree. Roughly speaking, each vector field belonging to a given layer is as energy costly to generate as other vector fields from the same layer. Vector fields with smaller degrees are less energy expensive to generate than those with higher degrees.

The Lie bracket operation (5) is bi-linear, clearly anti-symmetric and satisfies also the Jacobi identity [7]. Consequently, it is desirable to exclude dependent Lie monomials and to work with independent ones, collected in a basis of a Lie algebra. The most popular is the Ph. Hall basis (PHB) which can be computed effectively using the algorithm presented in [8]. The first three layers of the PHB initialized with generators $\boldsymbol{X}, \boldsymbol{Y}$ are the following

$$(\overbrace{\boldsymbol{H}_1^1, \boldsymbol{H}_2^1}^{\boldsymbol{H}^1}, \overbrace{\boldsymbol{H}_1^2}^{\boldsymbol{H}^2}, \overbrace{\boldsymbol{H}_1^3, \boldsymbol{H}_2^3}^{\boldsymbol{H}^3}, \ldots) = (\boldsymbol{X}, \boldsymbol{Y}, [\boldsymbol{X}, \boldsymbol{Y}], [\boldsymbol{X}, [\boldsymbol{X}, \boldsymbol{Y}]], [\boldsymbol{Y}, [\boldsymbol{X}, \boldsymbol{Y}]], \ldots) \tag{6}$$

where $\boldsymbol{H}_d^e$ denotes the d-th item within the e-th layer of the PHB. As the system (1) is nonholonomic thus it satisfies the Lie algebra rank condition [9] which can be equivalently expressed in terms of the Ph. Hall basis as follows

$$\forall \boldsymbol{q} \in \mathbb{Q} \quad \operatorname{rank}(\mathrm{PHB}(\boldsymbol{q})) = n. \tag{7}$$

Condition (7) states that the system (1) can evolve at any configuration in any direction. In practice, it is desirable to generate a minimal number of layers $i^\star$ to satisfy (7). Condition (7) implies that $r = \sum_{i=1}^{i^\star} \#\boldsymbol{H}^i \geq n$ and here $\#$ denotes the cardinality of a given set.

As $\boldsymbol{A}(s_{\boldsymbol{\sigma}})$ in (4) depends on controls $\boldsymbol{u}$ (cf. Eq. (1)) thus the shift operator $\boldsymbol{z}(t)(\boldsymbol{q}_0)$ can be expressed as a combination of vector fields multiplied by control-depended coefficients. For the two-input system (1), the shift operator (3) is expressed as a series of control-depended coefficients $\boldsymbol{\alpha}$ multiplying vector fields (PHB elements) evaluated at a current configuration $\boldsymbol{q}_0$

$$\begin{aligned} \boldsymbol{z}(t) =& \boldsymbol{X}\alpha_1^1 + \boldsymbol{Y}\alpha_2^1(t) + [\boldsymbol{X}, \boldsymbol{Y}]\alpha_1^2 + [\boldsymbol{X}, [\boldsymbol{X}, \boldsymbol{Y}]]\alpha_1^3 + [\boldsymbol{Y}, [\boldsymbol{X}, \boldsymbol{Y}]]\alpha_2^3 + \ldots = \\ & [\boldsymbol{H}^1 \ldots \boldsymbol{H}^{i^\star}] \cdot \boldsymbol{\alpha} = \boldsymbol{H} \cdot \boldsymbol{\alpha}. \end{aligned} \tag{8}$$

where the matrix $\boldsymbol{H}$ is composed of r columns and coefficients $\boldsymbol{\alpha}$ are time-dependent, via controls $\boldsymbol{u}(s)$, $s \in [0,t]$, and equal to [8]

$$\begin{aligned}
&\alpha_1^1(t) = \int_{T_1(t)} u_1(s_1)\boldsymbol{s}^1, \quad \alpha_2^1(t) = \int_{T_1(t)} u_2(s_1)\boldsymbol{s}^1,\\
&\alpha_1^2(t) = \tfrac{1}{2}\int_{T_2(t)} (u_1(s_1)u_2(s_2) - u_2(s_1)u_1(s_2))d\boldsymbol{s}^2,\\
&\alpha_1^3(t) = \tfrac{1}{6}\int_{T_3(t)} \{u_1(s_1)u_1(s_2)u_2(s_3) - 2u_1(s_1)u_2(s_2)u_1(s_3) + u_2(s_1)u_1(s_2)u_1(s_3)\}d\boldsymbol{s}^3\\
&\alpha_2^3(t) = \tfrac{1}{6}\int_{T_3(t)} \{-u_1(s_1)u_2(s_2)u_2(s_3) + 2u_2(s_1)u_1(s_2)u_2(s_3) - u_2(s_1)u_2(s_2)u_1(s_3)\}d\boldsymbol{s}^3\\
&\ldots
\end{aligned} \tag{9}$$

The formula (8) is local (valid in a small neighborhood of a current configuration $\boldsymbol{q}_0$) as only in this case the tail of the series (3) composed of vector fields from layers higher than $\boldsymbol{H}^{i^\star}$ is negligible.

In Eq. (9) controls have not been determined yet. Usually, it is assumed that $t = T$ is fixed and T denotes one-step control horizon. At that time, it is checked whether controls generate a desired shift towards the goal configuration $\boldsymbol{q}_f$

$$\boldsymbol{z}(T)(\boldsymbol{q}_0) = \xi \cdot (\boldsymbol{q}_f - \boldsymbol{q}_0). \tag{10}$$

In Eq. (10) a small parameter ξ is selected to preserve a small displacement and reliability of the approximation (8) (all necessary vector fields are evaluated at $\boldsymbol{q}_0$). Usually, controls are selected in a parametric form and each control is composed of N_i items

$$u_i(t) = \sum_{j=1}^{N_i} \phi_{ij}(t) \cdot p_{ij}, \quad i = 1, \ldots, m. \tag{11}$$

The functions $\phi_{ij}(t)$ belong to an orthonormal functional basis defined on the interval $[0,T]$. All parameters p_{ij}, collected in a vector, form $\boldsymbol{p}$ ($\dim \boldsymbol{p} = \sum_{i=1}^m N_i$) which uniquely determines controls $\boldsymbol{u}$ (when the orthonormal basis and a parameterization of each control are fixed). A minimal requirement to satisfy controllability condition (7) is that $\dim \boldsymbol{p} = n$ but a small redundancy ($\dim \boldsymbol{p} - n > 0$) is desirable.

Equation (11) substituted into Eq. (9) and then into Eq. (8) produces the equation

$$\boldsymbol{z}(T)(\boldsymbol{q}_0) = \xi \cdot (\boldsymbol{q}_f - \boldsymbol{q}_0) = \boldsymbol{H}(\boldsymbol{q}_0)\boldsymbol{F}(\boldsymbol{p}) \tag{12}$$

with $\boldsymbol{F}(\boldsymbol{p}) = \boldsymbol{\alpha}(\boldsymbol{u}(\cdot, \boldsymbol{p})) \in \mathbb{R}^{i^\star}$, an unknown variable $\boldsymbol{p}$ and known the left hand side. The mapping resembles forward kinematics for manipulators [10],

$$\boldsymbol{z} = \boldsymbol{k}(\boldsymbol{p}) = \boldsymbol{H}(\boldsymbol{q}_0)\boldsymbol{F}(\boldsymbol{p}), \tag{13}$$

with the input space $\mathbb{P} \ni \boldsymbol{p}$ and the task-space $\mathbb{Z} \ni \boldsymbol{z}$. Thus any technique to solve inverse kinematics can be applied to find $\boldsymbol{p}^\star$ such that Eq. (12) is satisfied. To achieve the goal point $\boldsymbol{z}(T)(\boldsymbol{q}_0)$ in the task-space, a Jacobian matrix is defined

$$\boldsymbol{J} = \boldsymbol{J}_{\boldsymbol{q}_0}(\boldsymbol{p}) = \frac{(\partial \boldsymbol{H}(\boldsymbol{q}_0)\boldsymbol{F}(\boldsymbol{p}))}{\partial \boldsymbol{p}} = \boldsymbol{H}(\boldsymbol{q}_0)\frac{\partial \boldsymbol{F}(\boldsymbol{p})}{\partial \boldsymbol{p}}. \tag{14}$$

Then, the Newton algorithm with an optimization in the null space of the Jacobian matrix is applied [10]

$$\boldsymbol{p}_{i+1} = \boldsymbol{p}_i + \boldsymbol{J}^{\#}(\boldsymbol{p}_i)\cdot \boldsymbol{z}(t)(\boldsymbol{q}_0) + (\boldsymbol{J}^{\#}(\boldsymbol{p}_i)\boldsymbol{J}(\boldsymbol{p}_i) - \boldsymbol{I})\frac{\partial w(\boldsymbol{p})}{\partial \boldsymbol{p}}|_{\boldsymbol{p}=\boldsymbol{p}_i}, \tag{15}$$

where $\boldsymbol{J}^{\#} = \boldsymbol{J}^T(\boldsymbol{J}\cdot \boldsymbol{J}^T)^{-1}$ denotes a pseudo-inverse of the Jacobian matrix, $\boldsymbol{I}$ is a $(\dim \boldsymbol{p} \times \dim \boldsymbol{p})$ identity matrix and $w(\boldsymbol{p})$ is a function to be optimized. The algorithm (15) should be initialized with a selected value of $\boldsymbol{p}_0$.

A quite natural quality function to minimize is an energy of motion given by the formula

$$w(\boldsymbol{u}(\cdot)) = \int_0^T \boldsymbol{u}^T(t)\boldsymbol{u}(t)dt = \sum_{i=1}^{m}\sum_{j=1}^{N_i} p_{ij}^2. \tag{16}$$

However, the function (16) does not reflect the fact that vector fields from different layers are more or less energy expensive to generate. A new quality function should prefer a motion along lower degrees vector fields corresponding to smaller indices i of components in $\boldsymbol{F}(\boldsymbol{p}) = (F_1(\boldsymbol{p}),\ldots,F_r(\boldsymbol{p}))^T$

$$w(\boldsymbol{u}(\cdot)) = \sum_{i=1}^{r} c_i F_i^2(\boldsymbol{p}) \tag{17}$$

with weight coefficients c_i equal to each layer. Later on, it is selected $c_i = T$ for coefficients of the first layer, $c_i = 4\sqrt{T}$ for the second and $c_i = 8\sqrt[3]{T}$ for the third. The values correspond to energies of motion on time horizon $[0,T]$ along vector fields from particular layers. For example, to generate $T[\boldsymbol{X},\boldsymbol{Y}]$ the four segments of the length $\sqrt{T}$ with amplitude of controls ± 1 can be applied $\sqrt{T}\boldsymbol{X} \to \sqrt{T}\boldsymbol{Y} \to -\sqrt{T}\boldsymbol{X} \to -\sqrt{T}\boldsymbol{Y}$ which costs $4\sqrt{T}$ energy units.

Below, some examples of $\boldsymbol{F}(\boldsymbol{p})$ are provided for two-input system (1) with controls (u_1,u_2) given in the interval $t\in[0,T=1]$. Controls

$$(p_{11} + p_{13}\cos(2\pi t), p_{21} + p_{22}\sin(2\pi t))^T, \tag{18}$$

generate the function $\boldsymbol{F}(\boldsymbol{p})$ in the form

$$\boldsymbol{F}(\boldsymbol{p}) = (p_{11}, p_{21}, \frac{1}{4\pi}(-2p_{11}p_{22} + p_{13}p_{22}))^T, \tag{19}$$

while for controls

$$(p_{11} + p_{12}\sin(2\pi t) + p_{15}\cos(4\pi t), p_{21} + p_{23}\cos(2\pi t) + p_{24}\sin(4\pi t))^T, \tag{20}$$

it takes the form

$$\boldsymbol{F}(\boldsymbol{p}) = (p_{11}, p_{21}, \frac{1}{8\pi}(4p_{12}p_{21} - 2p_{12}p_{23} - 2p_{11}p_{24} + p_{15}p_{24}))^T. \tag{21}$$

It is easy to notice, cf. Eqs. (9), (19), (21), that components of $\boldsymbol{F}(\boldsymbol{p})$ corresponding to the i-th layer are just sums of products of control parameters and degree of each product is equal to the layer number.

3 Evaluation of Parameterizations

In the previous section one step motion was planned for a selected parameterization of controls. The aim of this section is to compare and evaluate some parameterizations. A quite natural idea is to evaluate parameterizations by applying them to solve either a given motion planning task or a set of such tasks. Unfortunately, local methods generate different trajectories to the goal point and sometimes a bad initial control decision can result in a good overall performance. To avoid this phenomenon, the evaluation will be performed only locally for one step motion. To make the evaluation also independent on local direction of motion towards the goal, a set of possible directions will be generated and overall performance averaged over all results. Steps of the evaluation algorithm are given in Listing 1. It should be noted that the initial value of vector $\boldsymbol{p}_0$ should contain only constant terms for both controls. As each control u_i has to contain a constant term to perform a motion, cf. Eqs. (19), (21), and parameterizations can differ in selected items of a functional basis therefore such a choice sets the same initial conditions for all tested parameterizations.

Algorithm 1. Evaluation of selected parameterizations

Step 1. Read in a nonholonomic system (1), an initial configuration $\boldsymbol{q}_0$, a small length of one-step motion δ. Determine appropriate number of layers $i^\star$ to satisfy controllability condition (7) and a set of K_1 parameterizations of appropriate size, i.e. satisfying the condition $\dim \boldsymbol{p} \geq r$. Fill out with zeroes a table of total marks for each parameterization.
Step 2. Generate randomly (uniformly distributed) K_2 directions of motion $\Delta \boldsymbol{q}^i$, $i = 1, \ldots, K_2$, each of the length equal to δ
Step 3. for $i = 1, \ldots, K_2$ repeat Steps 4-5
Step 4. Using each parameterization and applying Algorithm (15), initialized with the same $\boldsymbol{p}_0$, determine a cost of motion in the direction $\Delta \boldsymbol{q}^i$, i.e. the motion towards the goal configuration $\boldsymbol{q}_f = \boldsymbol{q}_0 + \Delta \boldsymbol{q}^i$, and collect results in an auxiliary table.
Step 5. Sort the auxiliary table in an increasing order, to establish which parameterization is the best (index 1) and which is the worst one (index K_1) For each parameterization add its index within sorted auxiliary table to the table of total marks.
Step 6. Output all tested parameterization together with the averaged mark (their total marks divided by the number of trials K_2).

4 Simulations

Simulations were performed on the unicycle robot, described by a driftless nonholonomic system

$$\dot{\boldsymbol{q}} = \begin{bmatrix} \dot{x} \\ \dot{y} \\ \dot{\theta} \end{bmatrix} = \begin{bmatrix} \cos(\theta) \\ \sin(\theta) \\ 0 \end{bmatrix} u_1 + \begin{bmatrix} 0 \\ 0 \\ 1 \end{bmatrix} u_2 = \boldsymbol{X}(\boldsymbol{q}) u_1 + \boldsymbol{Y}(\boldsymbol{q}) u_2, \tag{22}$$

where (x, y) denote position of the vehicle on a plane while θ its orientation. Control u_1/u_2 is a linear/angular velocity. Without loosing generality, it was assumed the initial configuration $\boldsymbol{q}_0 = (0, 0, 0)^T$ (other initial configurations can be transformed to $\boldsymbol{q}_0$ by an appropriate change of coordinates).

After calculating $[\boldsymbol{X}, \boldsymbol{Y}] = (\sin(\theta), -\cos(\theta), 0)^T$ it can be checked that two layers ($i^\star = 2$) of vector fields $(\boldsymbol{H}^1, \boldsymbol{H}^2) = (\boldsymbol{X}, \boldsymbol{Y}, [\boldsymbol{X}, \boldsymbol{Y}]) = \boldsymbol{H}(\boldsymbol{q})$ are enough to span the configuration space everywhere as (cf. Eqs. (6), (7))

$$\forall \boldsymbol{q} \; \det(\boldsymbol{H}(\boldsymbol{q})) = 1 \quad \Rightarrow \quad \mathrm{rank}(\boldsymbol{H}(\boldsymbol{q})) = 3 = n, \tag{23}$$

and $r = n$. The number of control parameters should be equal to or bigger than n.

Nine parameterizations of controls ($K_1 = 9$) to compare were selected within an orthonormal harmonic basis given on the interval $[0, T]$. They differ in the number of parameters varying form 4 to 10. First five items (and their codes) of the basis

$$\phi_i(t) = \frac{\beta_i}{\sqrt{T}} \psi_i(t), \tag{24}$$

and the parameterization themselves are presented in Table 1a and Table 1b, respectively.

Table 1. a) The first five items of the harmonic basis and their codes (i), b) Codes of controls for selected parameterizations

a)

code i	β_i	$\psi_i(t)$
1	1	1
2	$\sqrt{2}$	$\sin(\omega t)$
3	$\sqrt{2}$	$\cos(\omega t)$
4	$\sqrt{2}$	$\sin(2\,\omega t)$
5	$\sqrt{2}$	$\cos(2\,\omega t)$
$\omega = 2\pi/T$		

b)

No.	u_1	u_2
#1	(1,2)	(1,3)
#2	(1,3)	(1,2)
#3	(1,4)	(1,5)
#4	(1,5)	(1,4)
#5	(1,2,3)	(1,2,3)
#6	(1,4,5)	(1,4,5)
#7	(1,2,5)	(1,3,4)
#8	(1,3,4)	(1,2,5)
#9	(1,2,3,4,5)	(1,2,3,4,5)

Two values of one-step motion length were selected, relatively short and long with the values of $\delta = 0.1$ and $\delta = 0.5$, respectively. In order to get a statistically valuable data, for each of them, 500 random directions of motion were generated in the configuration space using the formula

$$\Delta \boldsymbol{q}^i = \delta \cdot \frac{\boldsymbol{q}_{rand}}{||\boldsymbol{q}_{rand}||}, \quad \boldsymbol{q}_{rand} = (\mathrm{rand}(-1, 1), \mathrm{rand}(-1, 1), \mathrm{rand}(-1, 1))^T,$$

where rand(a, b) generates uniformly distributed random numbers in the range $[a, b]$. In Fig. 1 projections on the xy-plane of generated directions towards end-points are presented. All tasks were initialized with a vector of parameters $\boldsymbol{p}_0$ with all but constant terms, equal to $p_{11} = p_{21} = 1$, set to the zero value.

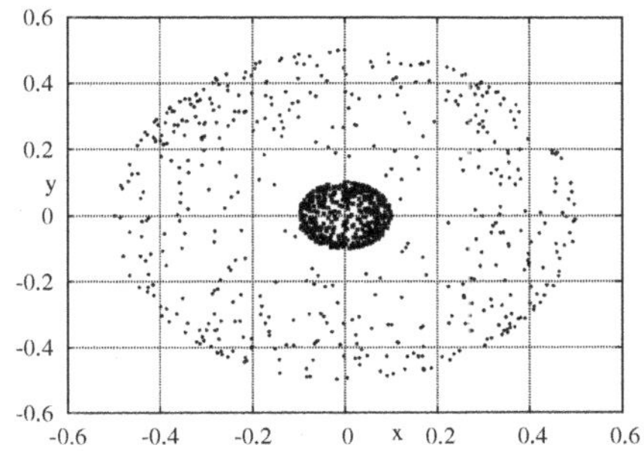

Fig. 1. The xy-plane projection of end-points of directions $\Delta\boldsymbol{q}$ generated in the configuration space.

An infinite series $\boldsymbol{z}(t)$ given in Eq. (8) is approximated with its finite representation, Eq. (12) restricted to the very first $i^\star$ layers thus a final set of parameters derived with Algorithm (15) may cause some inaccuracies to reach the goal point $\boldsymbol{q}_f$. In fact, when the set of parameters is determined, controls $\boldsymbol{u}$, cf. Eq. (11), are applied to the system (1) initialized at $\boldsymbol{q}_0$ a real final configuration $\boldsymbol{q}_{freal}$ is reached. In order to evaluate inaccuracy in reaching a target, the condition

$$\|\boldsymbol{q}_f - \boldsymbol{q}_{freal}\| = \|\boldsymbol{q}_0 + \Delta\boldsymbol{q} - \boldsymbol{q}_{freal}\| < 0.1 \cdot \delta. \tag{25}$$

is checked for any tested parameterization and a direction of motion. It should be mentioned that failing to meet Condition (25) in application of the method in motion planning warns that a length of one-step motion is too large and it should be decreased. In this case the dropped tail of series in $\boldsymbol{z}(t)$ becomes smaller and smaller.

The results of running Algorithm 1 are collected in Table 2 for one-step parameter set to $\delta = 0.1/0.5$, respectively. In the column *fail* of the tables, the percentage of cases failed to meet Condition (25) is provided. The remaining columns contain values of an average mark for each parameterization, i.e. for each task (direction of motion) results for all parameterizations were ordered from 1 (the best) to 9 (the worst) according to the following criteria: the shortest length of motion (len) or the smallest value of a quality function (qual). Extra numeric data added to the criteria show the cases when:

1. there is no restriction resulting from Condition (25),
2. all parameterizations failed to meet Condition (25) are marked as the worst (9),

3. a tested point is neglected if any parameterization fails to meet Condition (25), at the point.

Other aspects of running Algorithm 1 for tested cases are presented in Table 3. Columns contain averaged values of the length of motion (len) and the quality function (qual) with extra number indicating:

1. no restrictions,
2. only results meeting Condition (25) are considered.

The column avg_fail contains the averaged value of $\|\boldsymbol{q}_0 + \Delta\boldsymbol{q} - \boldsymbol{q}_{freal}\|$ for each parameterization.

In Fig. 2 paths on the xy-plane are presented to reach the exemplary goal configuration $q_f = (0.026, 0.097, 0)^T$ using different parameterizations. Columns of the table of figures are constructed as follows:

- the first column corresponds to a parameterization containing a full set of harmonics in both controls,
- the second and the third column correspond to proper subset (some items are excluded) of controls from the first column parameterization.

The first/second/third row in Fig. 2 corresponds to 0-1/0-2/0-1-2 harmonics, respectively. Based on results collected in Tables 2, 3 and depicted in Figs. 2 some remarks can be formulated:

- the best quality function in terms of the length of motion and inaccuracy in reaching the goal is the energy function Eq. (16),
- a value of the quality function Eq. (17) does not dependent on parameterization used,
- using the quality function Eq. (16), the best results both in energy efficiency and length of paths are obtained by using a full two harmonics parameterization (1,2,3,4,5)(1,2,3,4,5). However using parameterizations that contains at least one of first-level harmonic functions nearly the same result can be obtained. The worst results are generated with parameterizations with only high-level harmonic functions.
- In both cases of no quality function and the quality function Eq. (17), results in terms of path length are very similar, slightly better for the quality function Eq. (17). In both cases the worst parameterization are (1,4)(1,5) and best parameterization are (1,3)(1,2), but in the best case inaccuracy in reaching goal points is relatively big, especially for the parameterization (1,3)(1,2). Surprisingly, parameterizations with the biggest average lengths have also the smallest inaccuracy in reaching target points.
- When some low level harmonics are missed from controls, one can expect more cusps in resulting paths (as switching in directions of motion due to passing through zero of higher harmonics appears frequently). Also a volume of a resulting path within a configuration space is likely to be smaller than for the case of low level harmonics present. This observation can be utilized in a motion planning in environments with obstacles.

Table 2. Results of running Algorithm 1) (rank of each parameterization) averaged from 1000 random directions.

fun	par	$\delta = 0.1$							$\delta = 0.5$						
		len1	len2	len3	qual1	qual2	qual3	fail [%]	len1	len2	len3	qual1	qual2	qual3	fail [%]
no	#1	6.7	5.0	6.3	—	—	—	0.0	5.8	5.8	6.2	—	—	—	51.0
	#2	1.4	5.3	1.8	—	—	—	28.0	1.6	8.0	2.7	—	—	—	75.8
	#3	8.9	7.1	8.8	—	—	—	23.0	8.8	6.6	8.3	—	—	—	75.8
	#4	5.1	5.8	5.2	—	—	—	0.0	6.0	7.9	5.8	—	—	—	53.8
	#5	3.8	4.9	3.7	—	—	—	47.6	3.7	7.6	3.4	—	—	—	85.6
	#6	7.8	6.7	7.7	—	—	—	40.4	7.8	7.6	7.0	—	—	—	83.8
	#7	5.4	3.7	4.8	—	—	—	0.0	5.5	6.2	5.1	—	—	—	50.8
	#8	2.5	5.3	3.0	—	—	—	9.4	2.6	8.0	3.6	—	—	—	63.6
	#9	3.4	4.3	3.7	—	—	—	27.0	3.3	7.4	3.0	—	—	—	75.2
Eq. (16)	#1	4.5	4.0	4.7	4.3	3.8	4.5	0.0	5.1	6.1	6.2	5.7	6.3	6.3	51.0
	#2	4.1	5.5	4.1	4.3	5.8	4.5	13.6	4.7	7.9	4.3	5.7	8.0	4.8	71.8
	#3	8.2	7.2	8.2	8.1	7.2	8.1	13.4	8.2	6.8	8.3	6.7	6.5	8.6	69.8
	#4	8.0	7.6	8.0	8.1	7.7	8.1	0.0	8.1	7.7	7.4	6.6	7.8	7.5	52.8
	#5	3.3	3.9	3.1	3.3	3.9	3.0	28.0	2.3	7.0	2.2	2.9	7.1	2.3	80.0
	#6	7.8	7.0	7.7	7.7	7.0	7.7	28.0	6.8	7.5	6.2	6.3	7.4	5.8	78.2
	#7	3.7	3.1	3.8	3.6	3.1	3.7	0.0	3.9	5.8	4.3	4.7	6.0	5.0	50.6
	#8	2.9	4.6	2.9	3.5	5.1	3.6	1.8	3.6	7.6	3.6	4.6	7.7	3.8	66.0
	#9	2.8	3.5	2.6	2.0	2.8	1.7	11.6	2.3	6.8	2.4	1.8	6.6	1.0	68.8
Eq. (17)	#1	6.8	5.3	6.7	5.0	4.0	4.7	0.0	5.6	5.8	6.2	5.6	5.9	5.4	50.8
	#2	1.2	4.5	1.3	5.1	6.7	5.0	24.4	1.9	8.0	2.7	5.5	8.3	4.7	75.8
	#3	9.0	7.4	9.0	7.4	6.1	7.4	24.2	8.7	6.6	8.2	8.0	6.5	7.3	75.2
	#4	5.4	6.0	5.6	7.5	7.6	7.5	0.0	6.2	7.8	5.6	8.0	8.1	7.1	51.2
	#5	3.5	4.6	3.6	2.9	4.4	3.1	41.4	3.7	7.6	3.4	2.5	7.5	3.6	84.8
	#6	7.9	6.8	7.8	6.3	5.6	6.4	38.0	7.8	7.6	7.2	6.3	7.2	6.0	83.0
	#7	5.7	4.2	5.5	4.3	3.5	4.3	0.0	5.4	6.0	5.2	3.8	5.7	4.2	50.6
	#8	2.2	4.8	2.3	4.3	6.0	4.3	6.8	2.4	7.9	3.3	3.8	8.1	4.0	63.0
	#9	3.4	4.4	3.2	2.2	3.9	2.4	23.6	3.4	7.4	3.0	1.5	7.2	2.6	74.4

Table 3. Average path lengths, quality function values and fail to meet Condition (25) of Algorithm 1 obtained from 500 random directions of motion with $\delta = 0.1$.

qual_fun	par	avg_len1	avg_len2	avg_qual1	avg_qual2	avg_fail
no	#1	0.6493	0.4843	—	—	0.0019
	#2	0.3668	0.2934	—	—	0.0101
	#3	0.8896	0.7019	—	—	0.0019
	#4	0.5343	0.4016	—	—	0.0070
	#5	0.5020	0.3709	—	—	0.0067
	#6	0.7202	0.5459	—	—	0.0051
	#7	0.5914	0.4248	—	—	0.0040
	#8	0.4204	0.3414	—	—	0.0092
	#9	0.4965	0.3716	—	—	0.0062
Eq. (16)	#1	0.4871	0.4227	0.6500	0.4933	0.0021
	#2	0.4828	0.4181	0.6433	0.4873	0.0071
	#3	0.6890	0.5980	1.2918	0.9773	0.0021
	#4	0.6844	0.5931	1.2824	0.9688	0.0052
	#5	0.4634	0.3934	0.6006	0.4362	0.0051
	#6	0.6684	0.5723	1.2334	0.9089	0.0039
	#7	0.4824	0.4166	0.6479	0.4906	0.0022
	#8	0.4797	0.4141	0.6412	0.4847	0.0070
	#9	0.4608	0.3903	0.5954	0.4303	0.0051
Eq. (17)	#1	0.6145	0.4897	0.0199	0.0140	0.0020
	#2	0.3823	0.3057	0.0199	0.0140	0.0091
	#3	0.8601	0.6941	0.0199	0.0141	0.0020
	#4	0.5484	0.4338	0.0199	0.0141	0.0065
	#5	0.5004	0.3908	0.0199	0.0140	0.0065
	#6	0.7153	0.5668	0.0199	0.0141	0.0048
	#7	0.5836	0.4579	0.0199	0.0140	0.0035
	#8	0.4050	0.3298	0.0199	0.0140	0.0086
	#9	0.4982	0.3859	0.0199	0.0140	0.0065

- For the particular unicycle model, some explanations of resulting paths can be proposed. Based on shapes of paths generated with parameterizations #1, #2 and #7,#8 the differences in lengths of motion and inaccuracies in reaching goal points are easy to explain. In paths generated with parameterizations #1 and #7, the robot slightly reoriented itself at the very beginning and at the end of motion, but it has to move almost half of the path length before reaching a location where it changes the direction of motion (it reorients significantly). In paths generated with parameterizations #2 and #8, significant changes in direction of motion (and orientation changes) appears twice. The

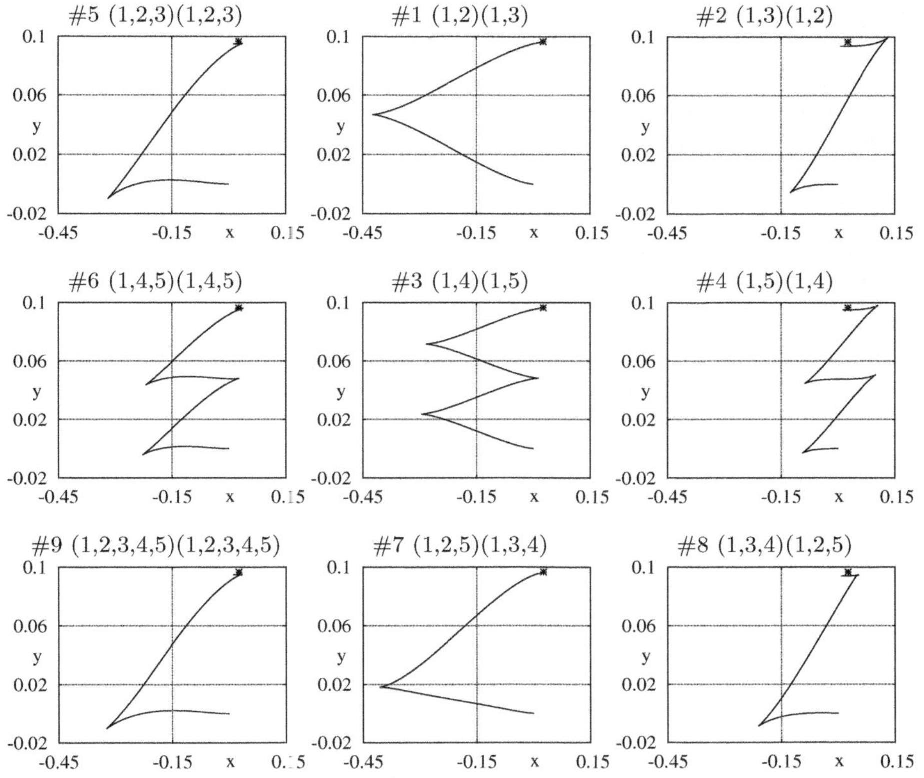

Fig. 2. The xy-plane projection of paths towards $q_f = (0.026, 0.097, 0)^T$ using the quality function (17) and various parameterizations. The asterix shows the goal position.

first fast one allows to shorten a path length, while at the second one, which appears at the end of motion, the approximation of motion (8) (calculated at the initial configuration $\boldsymbol{q}_0$) seems to introduce extra inaccuracies.

5 Conclusions

In this paper parameterizations of controls in the Lie algebraic method of motion planning were tested using two quality functions and compared to with no optimization case. Obtained results were examined based on average lengths of generated paths and values of quality functions. It appears that the safest strategy in selecting a particular parameterization is to take a redundant parameterization containing full harmonics up to a certain degree. Apparently, in this case a problem of processing large data sets and expensive calculations is likely to happen. In order to get desirable results, optimizing the energy of motion is also advised.

References

1. LaValle, S.: Planning Algorithms. Cambridge University Press, Cambridge (2006)
2. Jakubiak, J., Tchon, K.: Motion planning in velocity affine mechanical systems. Int. J. Control **83**(9), 1965–1974 (2010)
3. Ratajczak, A., Tchon, K.: Parametric and non-parametric jacobian motion planning for non-holonomic robotic systems. J. Intell. Rob. Syst. **77**(3), 445–456 (2015). https://doi.org/10.1007/s10846-013-9880-0
4. Duleba, I.: Algorithms of Motion Planning for Nonholonomic Robots. WUST Publ. House, Wroclaw (1998)
5. Duleba, I.: Kinematic models of doubly generalized n-trailer systems. J. Intell. Rob. Syst. **94**(1), 135–142 (2019)
6. Strichartz, R.: The Campbell-Baker-Hausdorff-Dynkin formula and solutions of differential equations. J. Funct. Anal. **72**, 320–345 (1987)
7. Spivak, M.: A Comprehensive Introduction to Differential Geometry, 3rd edn. Publ or Perish Inc., Houstron (1999)
8. Duleba, I., Khefifi, W.: Pre-control form of the gCBHD formula for affine nonholonomic systems. Syst. Control Lett. **55**(2), 146–157 (2006)
9. Chow, W.: Über Systeme von linearen partiellen Differentialgleichungen erster Ordnung. Math. Ann. **117**(1), 98–105 (1939)
10. Nakamura, Y.: Advanced Robotics: Redundancy and Optimization. Addison-Wesley Publ., Boston (1991)

Planar Features for Accurate Laser-Based 3-D SLAM in Urban Environments

Krzysztof Ćwian, Michał R. Nowicki, Tomasz Nowak, and Piotr Skrzypczyński(✉)

Institute of Robotics and Machine Intelligence, Poznań University of Technology, ul. Piotrowo 3A, 60-965 Poznań, Poland
piotr.skrzypczynski@put.poznan.pl

Abstract. Simultaneous Localization and Mapping (SLAM) systems using 3-D laser data typically represent the map as an unstructured point cloud, which is inefficient in data association and does not allow one to use the map for reasoning about the observed scene. In this paper we describe a laser-based SLAM system that represents the map as a collection of 3-D planar and line segments, which provide a natural way of representing man-made environments. We demonstrate that this representation improves the accuracy of trajectory estimation and makes it possible to represent major objects as geometric shapes.

Keywords: Autonomous driving · SLAM · LiDAR · 3D map · Planar features

1 Introduction

Limited availability of the Global Positioning System (GPS) signals in urban environments creates a need for localization solutions that provide accurate trajectory estimates in both outdoor and indoor scenarios, and use only the vehicle's on-board sensors. While visual SLAM is researched intensively [13], applications to autonomous vehicles that need to work day and night, and under different weather conditions, often rely on laser scanners that provide reliable range measurements and are relatively independent of the environmental conditions. In particular, 3-D laser scanners are considered the sensors of choice in autonomous driving [3]. An example application of laser-based SLAM in vehicle's localization is the Advanced Driver Assistance System (ADAS) of a city bus that is developed at Poznań University of Technology (PUT) together with Solaris Bus & Coach (SBC). The ADAS assists bus drivers while maneuvering or parking in narrow urban streets, where the GPS signal may be degraded, or in GPS-denied environments, such as a bus depot building. Our application requires accurate pose estimates to allow the ADAS to compute maneuvers in tight spaces, and will benefit from representing the surrounding obstacles as polygons, because this representation is natively used in motion planning [6]. Therefore, we developed a laser-based SLAM system that uses high-level features extracted

A. Bartoszewicz et al. (Eds.): PCC 2020, AISC 1196, pp. 941–953, 2020.
https://doi.org/10.1007/978-3-030-50936-1_79

from laser scans which group the measured points (Fig. 1). Owing to this map representation our system easily can distinguish measurements that belong to the major scene features from spurious readings increasing the accuracy of data association. The main contributions of our work are: robust data association methods for creation and updating of the features, efficient map management procedures that make it possible to build large, feature-based maps, and thorough evaluation of the proposed solution in representative scenarios.

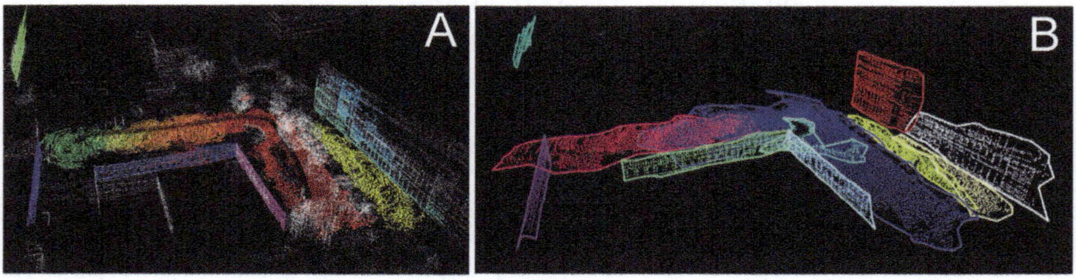

Fig. 1. Experiment at PUT campus with different map representations in laser-based SLAM: global unstructured point cloud (A) and planar features represented as 3-D polygons (B)

2 Related Work

Laser-based SLAM and laser-based odometry (i.e. incremental localization without creating a map) have been addressed by many authors in several application contexts. Early outdoor laser-based SLAM solutions used 2-D laser scanners assuming planar motion [2]. The availability of affordable 3-D laser scanners fostered research on laser-based SLAM that localizes the vehicles with respect to six degrees of freedom (6 d.o.f.) [1]. Those SLAM solutions either extract some geometric features from the scans [7] or employ raw point clouds. The most popular approaches are variants of the ICP (Iterative Closest Points) algorithm [10]. Unfortunately, ICP-like methods used for scan registration suffer from data association problems due to the need to re-establish the associations once the sensor pose is updated. An attempt to improve this situation is LOAM (Lidar Odometry And Mapping) [17], considered the state-of-the-art in laser-based localization, which semantically segments the acquired point clouds into planar objects and line segments (edges), but it does not build explicit geometric features. Then LOAM applies different variants of ICP to these classes (point-to-plane and point-to-line, respectively) in scan registration. The structure of the system that combines real-time scan-to-scan sensor pose tracking (i.e. odometry) and slower, but more accurate scan-to-map localization makes LOAM one of the most accurate real-time laser-based localization systems. Architecture similar to LOAM is also used by the more recent IMLS-SLAM [5] system, that improves the map representation using implicit moving least squares surfaces.

The LeGO-LOAM [12] proposes a number of practical improvements to LOAM, e.g. by modeling the ground plane, and distinguishing between points found on the ground and on other objects. LeGO-LOAM breaks down the pose optimization step by determining at first the sensor attitude (pitch and roll angles) with respect to the ground plane, and then computing the yaw angle and position. An early attempt to use planar features in laser-based SLAM was [15], while Salas-Moreno *et al.* [11] proposed to employ bounded planes and surfels for dense environment mapping. A similar map representation with bounded planar features is used by our PlaneLoc [16] which, however, utilizes RGB-D data and is a global localization system rather than full SLAM. An extended description of the experiments with the city bus is given in [4], but that work presents an earlier version of Plane-LOAM.

3 Proposed Solution

3.1 System Structure

We approach the problem of concurrent mapping and localization in an unknown environment in a way inspired by the LOAM architecture. This system uses a combination of scan-to-scan and scan-to-map data registration policies, that provides an effective trade-off between the real-time operation and high accuracy of trajectory estimation [17]. We have adopted this general scheme and the idea of using different ICP-like data association strategies in associating the scan points belonging to planar patches and edges (Fig. 2).

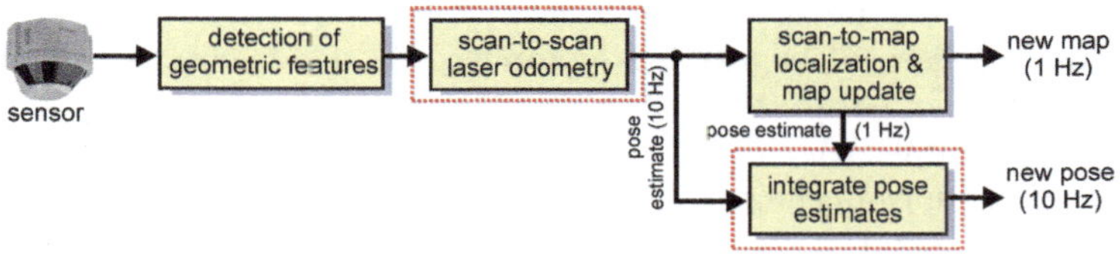

Fig. 2. Block scheme of the general Plane-LOAM architecture. Blocks in dot-line borders are adopted from LOAM without significant changes

In contrary to LOAM, which stores the map as points marked either as edges or planes, we build high-level geometric features in the form of 3-D planar patches and edges. As a result, our system, called Plane-LOAM, can represent the map of an environment with a limited number of features. This approach, especially applicable in environments with larger flat surfaces, allows us to conveniently manage all the features, to find correspondences between them, or even to detect some objects. In LOAM, the number of points creating a plane is fixed and equal to five, while in the proposed system their number varies depending on the size of the plane. A comparison of a wall representation in Plane-LOAM and in LOAM is presented in Fig. 3A. The map created by Plane-LOAM consists of both edge and planar features, as shown in Fig. 3B. The equation that represents a single planar feature:

$$Ax + By + Cz + D = 0, \tag{1}$$

where A, B, C, D are coefficients of the plane, is calculated using several nearby points collected from a single laser scan. Line segments are represented by six-dimensional Plücker coordinates $[\mathbf{l}_d, \mathbf{l}_m]$, where $\mathbf{l}_d$ is direction of the line, and $\mathbf{l}_m$ is moment of the line (both three-dimensional) with respect to global coordinates. Every time a new point is added to the given planar patch (or line segment), and the number of points forming the feature is smaller than the threshold (by default set to 30 points for patches and lines), the feature equation is updated.

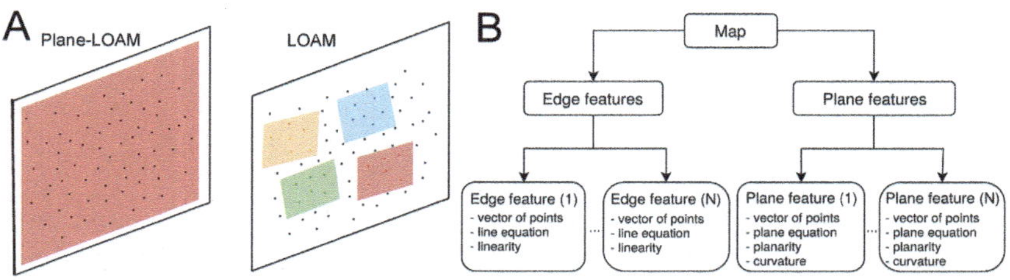

Fig. 3. High-level planar feature in Plane-LOAM compared to a point cloud with "plane" attributes in LOAM (A), and the structure of the environment map in Plane-LOAM (B)

Features have also additional properties: planarity and curvature in case of planar patches, and linearity in the case of line segments. The planarity (or linearity in the case of edges) p is computed as a percentage of points with the distance to the feature that is smaller than a certain threshold (0.2 m in the current implementation): $p = m/N \cdot 100\%$ where m is the number of points within a distance of 0.2 m from the plane (line), and N is the number of points belonging to the selected feature. Moreover, for planes the curvature c is calculated as proposed in [11]: $c = \lambda_{\min} \Big/ \sum_{i=1}^{3} \lambda_i$ using eigenvalues λ that are obtained by PCA (Principal Component Analysis) from the 3×3 covariance matrix that captures the dispersion of the set of all points belonging to the feature with respect to their centroid. Both these parameters are used for the management of features in the map (Sect. 3.3).

Management of the created planes and edges is accomplished in a few steps consisting of creating, updating, deleting and merging of features. All the mentioned steps are described in detail in the following sections on the example of planar features, as management of the edges is performed in an analogous way.

3.2 Creating Planar Features

The minimal number of points required to calculate the plane equation is three, but Plane-LOAM uses at least five points in initialization to reduce the possibility of error. For each planar feature the plane equation is calculated and all points added later must satisfy this equation. The process of creating new features and adding points to the existing ones consists of several steps. The first step is assigning new points to existing plane features. For every added scan point the algorithm finds three nearest features, with the point-to-plane distance below 0.3 m. The point-to-plane distance is calculated using the following equation:

$$d = \frac{Ax_0 + By_0 + Cz_0 + D}{\sqrt{A^2 + B^2 + C^2}}, \tag{2}$$

where: (x_0, y_0, z_0) are the point coordinates, and A, B, C, D are parameters of the supporting plane equation.

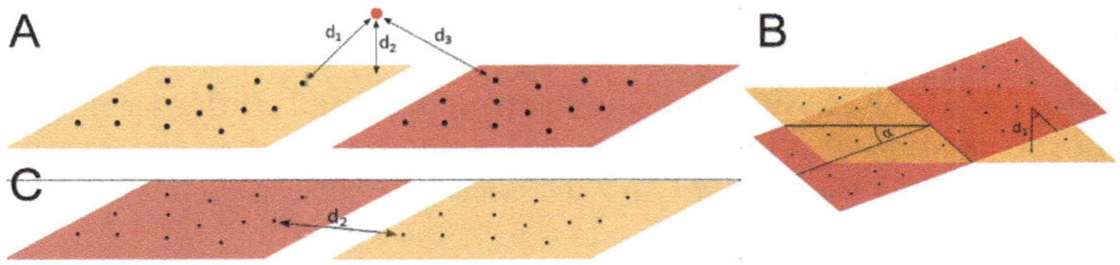

Fig. 4. Distances to the closest point d_1, to the closest plane d_2 and to the second closest plane point d_3 (A), angle between two planes α and exemplary point-to-plane distance d_1 (B), and distance d_2 between two points that belong to planes considered for matching (C)

Then the point-to-point distances to the nearest points belonging to each of the three considered features are computed to determine which planar patch is the closest one. In this case, the distance must be smaller than 1 m for the feature to be further considered. If there exists more than one planar feature complying with both of these requirements, the algorithm checks if one of them is considerably closer. This is achieved by comparing the distances d_{1min} and d_{3min} between the given scan point, and the closest points belonging to the first and second nearest features, respectively: $k_u = d_{1min}/d_{3min}$. If k_u is greater than 0.7, meaning that the distances are very similar, the given point should not be assigned to any feature at all, because the uniqueness of the match is insufficient, while incorrect correspondences may significantly reduce the system's accuracy. The conditions that must be met in order to add a new point to the existing feature are visualized in Fig. 4A. Each of them was designed for a particular purpose. The first one rejects points that do not comply with the plane equation. The second one forces the plane to be continuous and integral. The third one accepts only matchings with sufficient uniqueness. The threshold values in these parameters were obtained as a result of a series of simulations. By changing them, we can modify the size and the number of created planar features. The parameters used in the process of creating the linear features have the same purpose and similar geometric interpretation, which is shown in Fig. 5A.

If a new scan point can not be assigned to any existing feature, it is added to a group of points with less than five points. The only required condition is to be closer than 1 m to the nearest point in that group. Again, if that point cannot be assigned to any group of points, it creates a new one. The points not assigned to features are stored as initial accumulating structures (kernels), where other points from the processed laser scan are added, and eventually they can create new features.

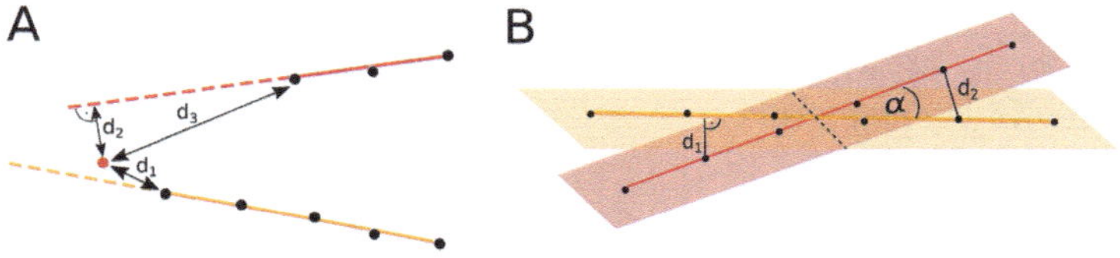

Fig. 5. Distances to the closest point d_1, to the line d_2 and to the second closest line point d_3 in the creation of a line segment feature (A), angle between two lines α, an exemplary point-to-line distance d_1 and distance between two points that belong to lines considered for matching d_2 (B)

3.3 Updating and Deleting Features

Some feature management procedures are implemented in Plane-LOAM in order to improve the efficiency of computations. Removing small features and merging co-planar ones decreases the number of processed features and increase system performance. Also applying a voxel grid to filter the points belonging to planar features results in reducing the computation time of map-related operations.

When all points from the processed laser scan are already assigned to features, the update step is performed. Firstly, all groups with less than five points are marked to be deleted at the end of this step. Such small features without the plane or line equation would not be useful in the next steps. Then for each planar feature, a voxel grid filter is applied to reduce the number of points. Next, the planarity (linearity) p and curvature c parameters are calculated. Features which do not meet the planarity (linearity) threshold set to 80% and curvature threshold set to 0.00015 are marked to be removed. The feature removal is performed once, at the end of the update step. This saves processing time due to implementation details.

3.4 Merging Features

The last stage of map maintenance is merging of the co-planar and overlapping planar features. To determine whether the features can be merged, we check if their supporting planes are parallel to each other and if the distance between the patches is small enough. At first, the angle α between two planes is calculated using the following equation:

$$\alpha = \arccos\left(\frac{\mathbf{n}_1 \cdot \mathbf{n}_2}{||\mathbf{n}_1|| \cdot ||\mathbf{n}_2||}\right) \tag{3}$$

where: $\mathbf{n_i}$ is the normal vector of the i-th plane. In the current Plane-LOAM version, the threshold for α is set to 10°. If this condition is met, we compute the mean residual error (using point-to-plane distance) between the points belonging to the first planar patch, and the supporting plane of the second considered feature. The same procedure is repeated for the points from the second feature, and the first supporting plane:

$$\varepsilon_1 = \frac{1}{N}\sum |d_i^1|,\ i = 1...N_1, \quad \varepsilon_2 = \frac{1}{N}\sum |d_i^2|,\ i = 1...N_2, \tag{4}$$

where: ε_1 and ε_2 are the first and the second residual errors, respectively, d_i^1 and d_i^2 are the distances between points belonging to the respective planar feature and the other's feature supporting plane, while N_1 and N_2 are numbers of points in the first and second feature, respectively. If both of these errors are smaller than 0.1 m, we continue the merging procedure. The considered angle α and exemplary distance d_1, required to calculate matching error are shown in Fig. 4B, while their counterparts for the very similar procedure of merging line segments are shown in Fig. 5B. The next condition ensures that the merged plane is consistent by checking the minimal distance between two points from two different planes (Fig. 4C). If the minimum distance d_2 is greater than the threshold 1 m, then the planar patches can not be merged.

Finally, the merged planar feature, despite passing all the previous tests, is checked for its planarity and curvature parameters. If those values exceed thresholds, the considered planar features cannot be merged, because in the next iteration it will be removed. The same procedure is executed for line segments using the linearity test and the same threshold values.

4 Matching Scans to the Map

In Plane-LOAM the algorithm that determines correspondences between the current laser scan points and the features already stored in the map has to find the plane or line equation to which a given point will be adjusted during the trajectory optimization step. The same conditions as in the process of creation of the features are used for finding the correspondences necessary for optimization. The only difference lies in the used values of parameters: the distance to an existing plane or line (d_1) must be smaller than 0.5 m, the distance to nearest point included in feature (d_2) must be smaller than 0.6 m, and the distance to the nearest point in the second closest feature (d_3) must be greater than 0.7 d_2.

During the optimization step, the laser scanner's pose is iteratively improved. Points belonging to the considered scan are registered to planar and linear features in the map using two variants of ICP based on the current pose estimate and are reevaluated at each iteration. As in LOAM, two different distance metrics are considered between the scanned points and their counterparts belonging to the features, which are assumed to be rotated and translated by the motion vector $[\mathbf{R}, \mathbf{t}]$. At the given timestamp k (scan index) the points $\mathbf{p}_k$ that are associated with planar patches are registered using point-to-plane metrics:

$$d_{\mathcal{P}} = f_{\mathcal{P}}\left(\mathbf{p}_{k,i}, [\mathbf{R}, \mathbf{t}]_k\right), i \in \mathcal{P}_k, \tag{5}$$

where $\mathcal{P}$ is the set of all planar features, while the points that are matched to edges are registered using point-to-line metrics:

$$d_{\mathcal{L}} = f_{\mathcal{L}}\left(\mathbf{p}_{k,i}, [\mathbf{R}, \mathbf{t}]_k\right), i \in \mathcal{L}_k, \tag{6}$$

where $\mathcal{L}$ represents all linear features in the map. Stacking together all the constraints defined by (5) and (6) we get the vector function $\mathbf{d} = \mathbf{f}\left(\mathbf{p}_k, [\mathbf{R}, \mathbf{t}]_k\right)$, which has a row corresponding to each point associated with a feature, while rows of $\mathbf{d}$ correspond to the residual distances. The motion estimate $[\mathbf{R}, \mathbf{t}]^*$ is computed using the Levenberg-Marquardt optimization algorithm:

$$[\mathbf{R}, \mathbf{t}]^* = \underset{\mathbf{R},\mathbf{t}}{\operatorname{argmin}}\left(\mathbf{f}\left(\mathbf{p}_k, [\mathbf{R}, \mathbf{t}]_k\right) - \mathbf{d}\right). \tag{7}$$

Optimization is repeated until either the translation and rotation increment computed in the following iterations is smaller than the given thresholds, or the maximum number of iterations (set to 10) is reached.

5 Experiments and Results

5.1 Sensors, Scenarios and Methodology

The Plane-LOAM system has been thoroughly tested in order to asses the accuracy of trajectory estimation and to determine the feasibility of our new map representation in the context of autonomous vehicle navigation. For that purpose, a number of laser scan sequences (datasets) were collected using different setups and different scenarios.

Fig. 6. Sensor setup used on vehicles (A), mounted on a car (B) and on a city bus (C), and the wearable setup with VLP-16 (D)

The 3-D laser sensor used in the majority of sequences is Sick MRS6124, which has been chosen for the city buses ADAS system in our previous tests [9]. The Sick scanner was mounted in a portable setup (Fig. 6A) together with a GPS receiver, camera, and inertial measurements unit (the two last sensors not used in the presented tests). These sensors were calibrated to a common reference frame and mounted either on a car (Fig. 6B) or a city bus provided by SBC (Fig. 6C). However, in the tests implemented using a mobile robot and a wearable sensory setup (Fig. 6D) the more compact Velodyne VLP-16 sensor was used. This allowed us to demonstrate that Plane-LOAM is feasible for 3-D sensors with different parameters and different field of view. The datasets used for the assessment of Plane-LOAM performance are shown in Table 1.

Table 1. Characteristics of the datasets used in Plane-LOAM evaluation

Name	Duration [s]	Length [m]	Sensor	Environment	Platform
LAAS	126	47	Velodyne	Indoor	Wearable
PUT-robot	210	163	Velodyne	Outdoor	Robot
PUT-car	167	324	Sick	Outdoor	Car
town-centre	100	748	Sick	Outdoor	Bus
town-service	228	595	Sick	Outdoor	Bus
town-suburbs	135	767	Sick	Outdoor	Bus

The indoor sequence was acquired while the last author visited the LAAS-CNRS lab in Toulouse, France. This experiment allowed us to collect data in a fully controlled environment with the ground truth trajectory obtained from a very accurate motion capture system. The first outdoor sequence was recorded at the PUT campus using a mobile robot and the VLP-16 sensor. As we focus on the Sick sensor for vehicle navigation, in this paper we show the only example map visualizations from this experiment (cf. Fig. 1). The PUT-car dataset was recorded also at PUT campus, but this time the Sick sensor was mounted on a car maneuvering between buildings. The next three sequences (town-center, town-service, town-suburbs) were recorded in a small town located nearby Poznań, with the sensors placed on the roof of a bus. The town-center dataset was collected in the city center including narrow streets with buildings on each side. The town-suburbs dataset covers a route on the outskirts of the same town. The town-service sequence was obtained at the SBC service facility.

The accuracy of the estimated trajectories is evaluated using the well-known error metrics introduced in [14]. The Absolute Trajectory Error (ATE) is the Euclidean distance between the corresponding points of the estimated and the ground truth trajectories. Prior to computing ATE these trajectories have to be aligned by finding a transformation that minimizes the distance between the two rigid sets of points representing the recorded sensor poses. This operation is performed by the valuation script provided by the authors of [14]. To asses, the accuracy of the whole trajectory, the Root Mean Squared Error (RMSE) of all the ATE values is computed. Moreover, for the LAAS dataset, we show the rotational RPE (Relative Pose Error), which is a relative rotational error between the successive sensor poses on the estimated trajectory. For other datasets, these metrics can not be used, as the GPS ground truth does not contain orientation information.

5.2 Accuracy in Controlled Environment

The accuracy of sensor trajectory estimation by Plane-LOAM was evaluated in a large indoor environment available at LAAS-CNRS, and compared with the ground-truth trajectory coming from the Qualisys motion capture system. The sequence covers a 30 m^2 rectangular area, and includes motion in all three axes, as the environment contained a few stairs. The ground truth data includes a full 6 d.o.f. poses acquired at the frequency of 180 Hz.

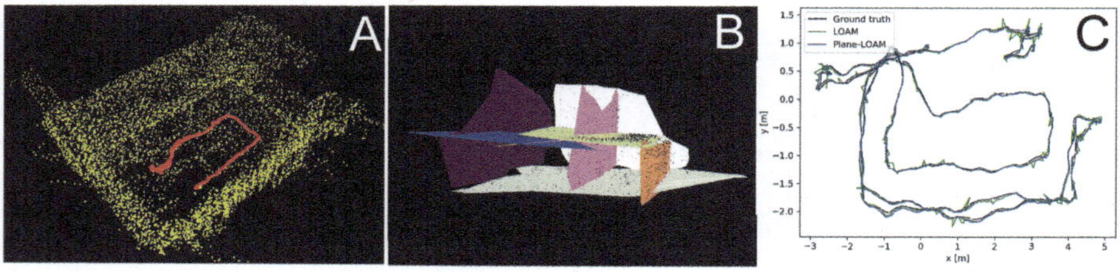

Fig. 7. LAAS-CNRS experiment: global map represented as point cloud (A) and as planar patches (B), and comparison of the localization errors relative to the motion capture system ground-truth shown in XY plane (C)

Figure 7A shows an unordered point cloud registered with the estimated sensor trajectory, while Fig. 7B demonstrates the Plane-LOAM map that gives a rough picture of the environment's layout. The ground-truth, LOAM and Plane-LOAM trajectories seen in the XY plane are shown in Fig. 7C. From these plots, it can be noticed that the Plane-LOAM trajectory is much smoother than the one estimated by LOAM, which translates to smaller localization error as shown in Table 2.

Table 2. Comparison of ATE errors for the LAAS and PUT-car datasets

SLAM system	LAAS				PUT-car		
	ATE_{RMSE}	ATE_{min}	ATE_{max}	RPE_{RMSE}	ATE_{RMSE}	ATE_{min}	ATE_{max}
LOAM	0.072 m	0.003 m	0.455 m	6.60°	7.01 m	3.09 m	14.52 m
Plane-LOAM	0.049 m	0.004 m	0.185 m	6.51°	3.70 m	1.65 m	7.26 m

5.3 Accuracy in Practical Applications

Initial outdoor experiments were performed at PUT campus car park using the sensors attached to a car's roof[1]. The sensor trajectory estimated by Plane-LOAM compared to the results of LOAM and the ground truth trajectory from GPS is shown in Fig. 8A.

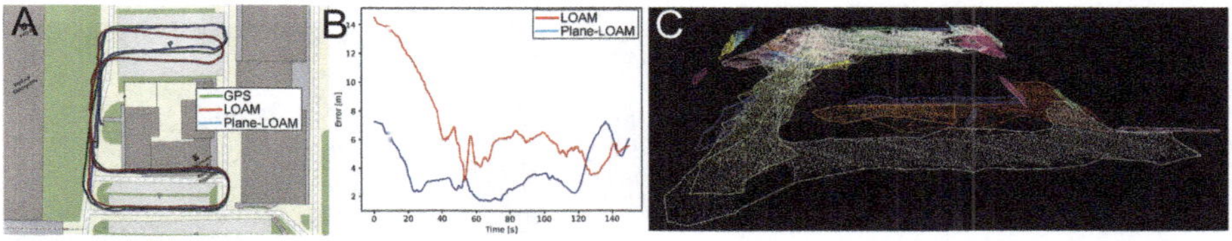

Fig. 8. PUT-car experiment: comparison of the estimated trajectories (A), position error relative to GPS data for LOAM and Plane-LOAM (B), and a view of the global map made of planar features (C)

The plot in Fig. 8B compares the positional errors with respect to the ground truth for Plane-LOAM and LOAM, while Fig. 8C depicts the planar patches stored in Plane-LOAM map. It can be observed that LOAM has significantly higher error for most of the trajectory. However, the biggest discrepancy at the very beginning of the trajectory, despite the fact that both trajectories start from the same position, is a result of trajectory alignment performed by the script computing ATE values. Statistics for the ATE measure are given in Table 2.

Table 3. Comparison of ATE errors for the datasets recorded at a small town

SLAM system	Town-centre		Town-service		Town-suburbs	
	ATE_{RMS}	ATE_{max}	ATE_{RMS}	ATE_{max}	ATE_{RMS}	ATE_{max}
LOAM	5.57 m	10.14 m	2.80 m	5.93 m	7.33 m	12.29 m
Plane-LOAM	3.33 m	4.92 m	1.94 m	4.46 m	3.40 m	6.30 m

The final practical experiments were conducted in a small town nearby Poznań using the sensory setup mounted on a city bus[2]. Basic statistics for the ATE values are provided for all three scenarios in Table 3. As can be seen, Plane-LOAM outperforms LOAM in all these cases with respect to both the RMS and maximum position error values. It was also observed that LOAM has a much bigger drift in elevation (z axis) for sequences where a flat, almost featureless ground plane was observed, e.g. an asphalt road. The smallest discrepancy

---
[1] A video is available at https://youtu.be/NGi4rnJSzlM.
[2] The town-center experiment is shown at https://youtu.be/Mj569vpXq9w.

between Plane-LOAM and LOAM was noticed for the town-service dataset. It might be caused by the fact, that during this test the bus was driving with a significantly lower velocity, around $6\frac{\mathrm{km}}{\mathrm{h}}$. Trajectories from this experiment are shown in Fig. 9A, and the ATE errors are plotted in Fig. 9B. Planar patches stored in the global map clearly show the main structures of the environment (Fig. 9C).

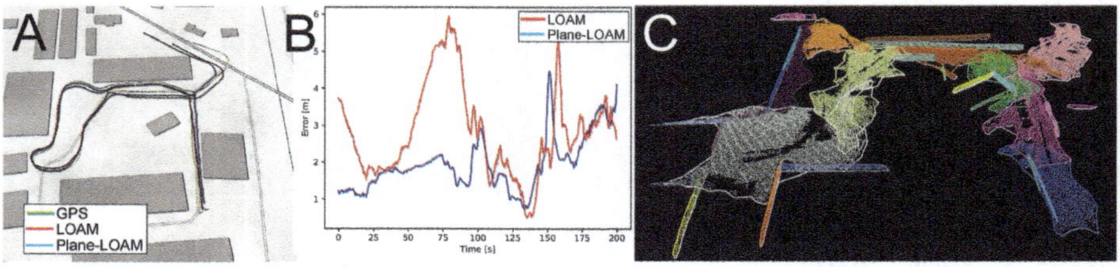

Fig. 9. Comparison of Plane-LOAM, LOAM and GPS trajectories recorded in the SBC service facility (A), position errors relative to the GPS trajectory in this experiment (B), and the global map of planar features showing the site layout (C)

6 Conclusions

This paper is focused on demonstrating the gains on trajectory estimation accuracy obtained due to the new map representation and structure in a localization and mapping system, which is in general based on the proven LOAM architecture. We have noticed improved accuracy in all the experiments, although the gain depends on the environment. Also, the speed of the vehicle makes a difference – at small speeds the ICP in LOAM has a better initial guess for data association, while at higher speeds the more sophisticated data association in Plane-LOAM demonstrates its advantage. Plane-LOAM estimates the full 6 d.o.f. sensor pose, and unlike LeGO-LOAM does not rely on heuristics that depend on the scenario.

Acknowledgements. This work was funded by the National Centre for Research and Development grant POIR.04.01.02-00-0081/17. The experiment in LAAS-CNRS was supported by the H2020 730994 grant TERRINet. The authors would like to thank Jan Wietrzykowski for his essential help in collecting the LAAS dataset.

References

1. Będkowski, J., Röhling, T., Hoeller, F., Shulz, D., Schneider, F.E.: Benchmark of 6D SLAM (6D Simultaneous Localization and Mapping) algorithms with robotic mobile mapping systems. Found. Comput. Decis. Sci. **42**(3), 275–295 (2017)
2. Bosse, M., Zlot, R.: Map matching and data association for large-scale two-dimensional laser scan-based slam. Int. J. Robot. Res. **27**(6), 667–691 (2008)

3. Bresson, G., Alsayed, Z., Yu, L., Glaser, S.: Simultaneous localization and mapping: a survey of current trends in autonomous driving. IEEE Trans. Intell. Veh. **2**(3), 194–220 (2017)
4. Ćwian, K.: Feature-Based Laser Simultaneous Localization and Mapping for Automotive Applications. Agencja Wydawnicza Impuls, Kraków (2019)
5. Deschaud, J.: IMLS-SLAM: scan-to-model matching based on 3D data. In: Proceedings of IEEE International Conference on Robotics and Automation, Brisbane, pp. 2480–2485 (2018)
6. Kicki P., Gawron T., Michałek M.: Machine learning approach to constrained path planning for intelligent articulated buses, PP-RAI 2019, Wroclaw (2019)
7. Kallasi, F.L., Rizzini, D., Caselli, S.: Fast keypoint features from laser scanner for robot localization and mapping. IEEE Robot. Autom. Lett. **1**(1), 176–183 (2016)
8. Ma L., Kerl C., Stückler J., Cremers D.: CPA-SLAM: consistent plane-model alignment for direct RGB-D SLAM. In: Proceedings of IEEE International Conference on Robotics and Automation,Stockholm, pp. 1285–1291 (2016)
9. Nowicki M., Nowak T., Skrzypczyński P.: Laser-based localization and terrain mapping for driver assistance in a city bus. In: Szewczyk, R., et al. (eds.) Automation 2019 Progress in Automation, Robotics and Measurement Techniques, AISC 920, pp. 502–512. Springer (2019)
10. Pomerleau, F., Colas, F., Siegwart, R.: A review of point cloud registration algorithms for mobile robotics. Found. Trends Robot. **4**(1), 1–104 (2015)
11. Salas-Moreno R.F., Glocken B., Kelly P.H.J., Davison A.J.: Dense planar SLAM. In: IEEE International Symposium on Mixed and Augmented Reality (ISMAR), Munich, pp. 157–164 (2014)
12. Shan, T., Englot, B.: LeGO-LOAM: lightweight and ground-optimized lidar odometry and mapping on variable terrain. In: Proceedings IEEE/RSJ International Conference on Intelligent Robots & Systems, Madrid, pp. 4758–4765 (2018)
13. Skrzypczyński, P.: Mobile robot localization: where we are and what are the challenges? In: Szewczyk, R., et al. (eds.) Automation 2017. Innovations in Automation, Robotics and Measurement Techniques, AISC 550, pp. 249–267. Springer (2017)
14. Sturm, J., Engelhard, N., Endres, F., Burgard, W., Cremers, D.: A benchmark for the evaluation of RGB-D SLAM systems. Proceedings of IEEE/RSJ International Conference on Intelligent Robots & Systems, Vilamoura, pp. 573–580 (2012)
15. Weingarten, J., Siegwart, R.: 3D SLAM using planar segments. In: IEEE/RSJ International Conference on Intelligent Robots and Systems, Beijing, pp. 3062–3067 (2006)
16. Wietrzykowski, J., Skrzypczyński, P.: PlaneLoc: probabilistic global localization in 3-D using local planar features. Robot. Auton. Syst. **113**(3), 160–173 (2019)
17. Zhang, J., Singh, S.: Low-drift and real-time lidar odometry and mapping. Auton. Robots **41**(2), 401–416 (2017)

Control of a Set of Unicycle-Like Robots Using an Approximate Linearisation

Dariusz Pazderski(✉)

Institute of Automatic Control and Robotics, Poznan University of Technology, Poznań, Poland
dariusz.pazderski@put.poznan.pl

Abstract. The paper deals with an alternative approach to control of a group of unicycle-like robots. The proposed algorithm is based on an approximate linearisation taking advantage of a dynamic feedback employing a transverse function. As a result, unicycle-like robots can be roughly perceived as unconstrained planar systems for which typical potential-like navigation methods can be used. In this paper both position and orientation stabilisation problem are discussed and a basic navigation algorithm for dynamic and static circular obstacles is proposed. The considered control/navigation method is verified numerically for the selected simulation scenario.

Keywords: Mobile robotics · Nonholonomic systems · Navigation · Multi-agent systems · Transverse functions approach · Potential functions

1 Introduction

Control and navigation of nonholonomic systems are essential issues in mobile robotics. The most popular type of mobile platforms is based on a differential drive due to its relatively high mobility and simplicity, [1]. Such a structure can be considered as the first-order three-dimensional system known as the unicycle.

A fundamental problem with a significant practical impact is the navigation in the presence of geometry constraints imposed on a workspace, [4]. So far many algorithms based on analytical, combinatorial and sampling approaches have been proposed to cover this task. Many analytical methods are usually derived from the concept of *potential functions* introduced in robotics by Khatib, [2]. A formal analysis of this method was carried out by Rimon and Koditschek, who introduced the so-called navigation function ensuring one global minimum for star obstacles, [13]. The potential function paradigm has proved to be an effective tool for control design also for phase-constrained systems, [3,16].

Despite that, it turns out that the application of potential-like navigation methods is challenging for nonholonomic systems. It is worth noting paper [15] where local minima problem for a class of nonholonomic systems is investigated.

A. Bartoszewicz et al. (Eds.): PCC 2020, AISC 1196, pp. 954–966, 2020.
https://doi.org/10.1007/978-3-030-50936-1_80

One of the most important control tool is based on linearisation of nonholonomic kinematics taking advantage of a static or dynamic feedback, [1,12]. Consequently, a common approach to investigate control of multi-agent systems is based on the assumption that agents are represented by simple linear dynamics. However, this assumption can be justified only in specific operation conditions due to various limitations of classic linearisation techniques.

In this paper we consider a new algorithm to control a set of unicycle-like robots with collision avoidance mechanism using the *approximate linearisation* based on a dynamic feedback. The essential idea is based on the application of the *transverse functions approach* proposed by Morin and Samson, [5–7]. Here we extend the results which were previously obtained for a single unicycle-like robot, [9,11]. Basically, it is assumed that a trajectory of the unicycle is coupled with a trajectory of a virtual unconstrained planar system. In such a case, it is possible to navigate a set of virtual robots based on a strategy developed for holonomic (linear) systems and to accomplish the task originally defined for unicycle robots. Moreover, one can also consider the stabilisation problem with respect to the full configuration which can be challenging when typical linearisation methods are used, [8].

The paper is organised as follows. In Sect. 2 basic notation is defined. In the next section a controller for one unicycle robot is discussed. In Sect. 4 a navigation algorithm for multiple robots is formulated while in Sect. 5 results of numerical simulations are presented. The last section concludes the paper.

2 Preliminaries

Consider two planar configurations $g = [g_x \; g_y \; g_\theta]^\top$ and $h = [h_x \; h_y \; h_\theta]^\top$, respectively, where subscripts x and y stand for position coordinates while subscript θ denotes angular coordinates. Additionally, the position coordinates will be also denoted using subscript xy, namely $g_{xy} = [g_x \; g_y]^\top$ and $h_{xy} = [h_x \; h_y]^\top \in \mathbb{R}^2$.

Assume that g and h are elements of Lie group $G \simeq SE(2)$ with the following group operation

$$\forall g,\, h \in G,\; gh := g + T(g_\theta)h, \tag{1}$$

where $T(\theta) := \begin{bmatrix} R(\theta) & 0 \\ 0 & 1 \end{bmatrix} \in \mathbb{R}^{3\times 3}$ with $R \in SO(2)$ being the rotation matrix. The inverse element of g, denoted by $g^{-1} \in G$, satisfies $gg^{-1} = g^{-1}g = e$, with $e := 0 \in \mathbb{R}^3$ being neutral (identity) element of group G. In the considered case one can easily verify that

$$g^{-1} = -T^\top(g_\theta)g. \tag{2}$$

The fundamental group diffeomorphisms include: left translation $l_g : G \to G, h \to gh$, right translation $r_g : G \to G, h \to hg$ and conjugation $\phi_g : G \to G, h \to l_g(r_{g^{-1}}(h)) = r_{g^{-1}}(l_g(h)) = ghg^{-1}$. Differentials of these maps can be computed as follows: $dl_g(h) := \frac{d}{dh}l_g(h) = T(g_\theta)$, $dr_g(h) := \frac{d}{dh}r_g(h) = I_{3\times 3} + \frac{dT(h_\theta)}{dh_\theta}g$ and $d\varphi_g(h) := \frac{d}{dh}ghg^{-1} = T(g_\theta) - \frac{dT(h_\theta)}{dh_\theta}g$, respectively.

Next, consider Lie algebra $\mathfrak{g}$ of Lie group G which consists of left-invariant vector fields X_i on G such that the following holds

$$\forall g, h \in G, dl_g(h)X_i(h) = X_i(gh). \tag{3}$$

We chose a set of the following left-invariant and linearly independent vector fields

$$X_1(g) := \begin{bmatrix} \cos g_\theta \\ \sin g_\theta \\ 0 \end{bmatrix}, \; X_2(g) := \begin{bmatrix} 0 \\ 0 \\ 1 \end{bmatrix}, \; X_3(g) := [X_2, X_1](g) = \begin{bmatrix} -\sin g_\theta \\ \cos g_\theta \\ 0 \end{bmatrix}, \tag{4}$$

where $[\cdot,\cdot]$ stands for the Lie bracket of vector fields, to define a basis of the control Lie algebra. Applying vector-matrix notation the basis can be represented by: $X(g) := [X_1\ X_2\ X_3]\,(g) \in \mathbb{R}^{3\times 3}$.

In order to describe the action of group G on its own algebra $\mathfrak{g}$ the following adjoint operator is defined $Ad : G \times \mathfrak{g} \to \mathfrak{g}$, while $Ad(g)X_i := d\varphi_g(e)X_i$ with $X_i \in \mathfrak{g}$. In addition, to simplify notation we use the conjugation of Ad which satisfies: $Ad^X(g) = X^{-1}(e)Ad(g)X(e)$.

3 Decoupling Controller for the Unicycle

3.1 Control Systems on $G \simeq SE(2)$

Taking advantage of the notation introduced above, we define the following unconstrained system

$$\dot{g}^* = X(g^*)\,u^*, \tag{5}$$

where $g^* \in G$ is the configuration and $u^* = [u_1\ u_2\ u_3]^\top \in \mathbb{R}^3$ is the input. From (5) and assuming that $u_3 \equiv 0$, one can easily derive the unicycle kinematics as

$$\dot{g} = X(g)\,Cu, \tag{6}$$

where $g \in G$, $u = [u_1\ u_2]^\top \in \mathbb{R}^2$, $C := [I_{2\times 2}\ 0]^\top \in \mathbb{R}^{3\times 2}$ and $u := [u_1\ u_2]^\top \in \mathbb{R}^2$ is the input.

3.2 Control Law

In this paper the main concept of the control approach applied for the unicycle is based on the so-called *transverse function* on the unit circle S^1 which can be defined as follows.

Definition 1 (Based on [6,9]). *Let $f : S^1 \to G$ be a smooth function which satisfies the following transversality condition*

$$\forall \alpha \in S^1,\ \|f(\alpha)\| \leq \delta \tag{7}$$

with $\delta > 0$, *and*

$$\forall \alpha \in S^1, \det \left[X_1(f(\alpha)) \; X_2(f(\alpha)) \; \frac{\partial f}{\partial \alpha}(\alpha) \right] \neq 0. \tag{8}$$

Then f *is a transverse function with respect to control vector fields* X_1 *and* X_2.

Since f can also depend on other variables apart from α, its time derivative can be written as: $\dot{f} = \frac{\partial f}{\partial \alpha}\dot{\alpha} + \frac{\partial f}{\partial t}$. To facilitate computations, we express the derivative $\frac{\partial f}{\partial \alpha}$ in basis X and define the following

$$A(\alpha) := X^{-1}(f)\tfrac{\partial f}{\partial \alpha}. \tag{9}$$

Thus, one can obtain

$$\dot{f} = X(f)A\dot{\alpha} + \tfrac{\partial f}{\partial t}. \tag{10}$$

Next, we impose a virtual geometry constraint to couple trajectories of the unicycle (6) and the omnidirectional kinematics (5). To be more precise, it is assumed that the distance between g and g^* is determined by the transverse function as

$$f := (g^*)^{-1}g. \tag{11}$$

Equivalently, one can write that

$$g^* = gf^{-1}. \tag{12}$$

Then, taking time derivative of g^* one can obtain the following open-loop dynamics

$$\dot{g}^* = X(g^*)Ad^X(f)\left(\bar{C}(\alpha)\bar{u} - \tfrac{\partial f}{\partial t}\right), \tag{13}$$

where $\bar{u} := \left[\bar{u}^\top \; \dot{\alpha}\right]^\top \in \mathbb{R}^3$ is the extended input and $\bar{C}(\alpha) := \left[C \; A(\alpha)\right]^\top \in \mathbb{R}^{3\times 3}$ is the square matrix.

Now consider the following proposition.

Proposition 1. *System* (6) *is approximately decoupled by applying the following dynamic feedback*

$$\bar{u} = \bar{C}^{-1}(\alpha)\left(X^{-1}(g^*)Ad^X(f^{-1})w + \tfrac{\partial f}{\partial \alpha}\right), \tag{14}$$

where $w = \left[w_x \; w_y \; w_\theta\right]^\top \in \mathbb{R}^3$ *is the auxiliary input.*

Proof. From the transversality condition (8) one can show that $\bar{C}$ is the invertible matrix. Consequently, dynamics (13) can be fully decoupled as

$$\dot{g}^* = w. \tag{15}$$

Recalling (11) one can state that $\forall t \geq 0$ trajectory g is in a vicinity of g^* determined by the transverse function f.

Remark 1. From (11) and recalling definitions (1), (2) one can prove that coordinates of trajectories $g(t)$ and $g^*(t)$ satisfy

$$g_{xy}(t) - g^*_{xy}(t) = R(g^*_\theta)f_{xy}(\alpha) \tag{16}$$

and

$$g_\theta(t) - g^*_\theta(t) = f_\theta(\alpha). \tag{17}$$

3.3 Selection of the Transverse Function

In order to complete the design of the controller (14) one has to use a particular form of the transverse function f. Here we take advantage of the formula investigated in [9] and define the following

$$\bar{f}(\alpha) := \begin{bmatrix} \varepsilon_1 \sin\alpha \\ \varepsilon_1\varepsilon_2\varepsilon_3 \sin 2\alpha \\ \varepsilon_2 \cos\alpha \end{bmatrix}, \tag{18}$$

where ε_i, $i = 1, 2, 3$, are parameters. Without the loss of generality, it can be assumed that all parameters are positive. Moreover, recalling the detailed analysis presented in [9] one can state that the transversality condition (8) is satisfied for $\varepsilon_1 \in \mathbb{R}_+$, $\varepsilon_2 \in (0, \pi)$ and $\varepsilon_3 \in \left(0, \frac{1}{2}\right)$.

However, the transverse function $f = \bar{f}$ can be further modified in order to improve the design flexibility. In this paper we consider a translation of (18) on Lie group G, namely we apply the following formula

$$f(\alpha, \alpha^*) := h(\alpha^*)\bar{f}(\alpha), \tag{19}$$

where

$$h(\alpha^*) := \left[0\ 0\ -\varepsilon_2^* \cos\alpha^*\right], \tag{20}$$

while $\alpha^* \in S^1$ is an auxiliary variable and $\varepsilon_2^* \geq 0$ is a parameter. It can be proved, cf. [7,10], that for $\bar{f}$ being a transverse function, f inherits the same characteristics. Similarly as in (10), time derivative of f can be represented by

$$\dot{f} = dl_h(\bar{f})\dot{\bar{f}} + dr_{\bar{f}}(h)\dot{h} = X(f)A\dot{\alpha} + \tfrac{\partial f}{\partial t}, \tag{21}$$

where $A(\alpha) = X^{-1}(\bar{f})\frac{\partial \bar{f}}{\partial \alpha}$ and

$$\tfrac{\partial f}{\partial t} := dr_{\bar{f}}(h)\dot{h} = \left[0\ 0\ \varepsilon_2^* \sin\alpha^*\right]^\top \dot{\alpha}^*. \tag{22}$$

The supremum of the Euclidean distance between current positions of the unicycle and the corresponding virtual robot is denoted by ρ and can be computed based on (16). Recalling (18) and (19) one can prove that

$$\rho := \sup_{\alpha} \left\| g_{xy}(t) - g_{xy}^*(t) \right\| = \sup_{\alpha} \varepsilon_1 \sin\alpha \sqrt{1 + 4\varepsilon_2^2\varepsilon_3^2 \cos^2\alpha}. \tag{23}$$

3.4 Position Control

Considering (17) one can conclude that for $g_\theta^* =$ const the orientation variable satisfies $g_\theta \in [-\varepsilon_2 + \zeta, \varepsilon_2 + \zeta]$, where $\zeta := g_\theta^* - \varepsilon_2^* \cos\alpha_2^*$. In particular, the interesting result is obtained when $\varepsilon_2 \geq \frac{\pi}{2}$. In such a case a feasible change of orientation g_θ exceeds π. Assuming that g_θ defines the inclination angle of a line on a plane, one can conclude that any line can be defined, cf. Fig. 1. When $\varepsilon_2 < \frac{\pi}{2}$ this property does not hold. Correspondingly, for $\varepsilon_2 > \frac{\pi}{2}$ it is possible to approximate arbitrarily trajectory on G without using oscillatory controls. In such a case one enables a *position control mode* in which the orientation of the unicycle is considered as an auxiliary variable (in spite of it, the distant between g_θ and g_θ^* is still bounded due to (17)).

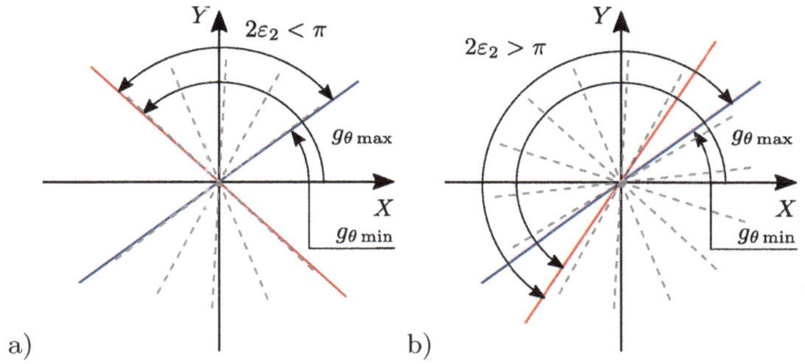

Fig. 1. Feasible ranges of linear translations for the constrained orientation: a) $\varepsilon_2 \in (0, \frac{\pi}{2})$, b) $\varepsilon_2 \in [\frac{\pi}{2}, \pi)$

3.5 Stabilisation of the Orientation

For point-to-point motion tasks the particular orientation of the vehicle at a goal may be required. Hence, it is important to address this issue with respect to the control strategy investigated in Sects. 3.2–3.4.

Assume that the desired orientation is denoted by θ_d. Then making $g_\theta^* = \theta_d$ and decreasing value of f_θ one can stabilise orientation of the unicycle with the required accuracy. However, as it is stated in Sect. 3.4, the selection of relatively high value of parameter ε_2 makes it possible to attenuate oscillatory inputs. Consequently, the improvement of the accuracy with respect to the orientation can be achieved not by simply decreasing ε_2 but by varying the parameter α^*, similarly as it is proposed in [10].

Proposition 2. *Using the following adaptation rule*

$$\dot{\alpha}^* = k_\alpha \tilde{\alpha}, \tag{24}$$

where $\tilde{\alpha} := \alpha - \alpha^*$ *and* $k_\alpha > 0$*, guarantees that for* $\varepsilon_2^* = \varepsilon_2$ *and* $w = 0$

$$\lim_{t\to\infty} \tilde{\alpha}(t) = 0 \text{ and } \lim_{t\to\infty} (g_\theta^* - g_\theta)(t) = 0. \tag{25}$$

Proof. Assuming that A from (9) can be decomposed as: $A = [A_1\ A_2]^\top$, where $A_1 \in \mathbb{R}^2$ and $A_2 \in \mathbb{R}$, the inverse of $\bar{C}$ satisfies

$$\bar{C}^{-1} = \begin{bmatrix} I_{2\times 2} & -\frac{A_1}{A_2} \\ 0 & -A_2 \end{bmatrix}. \tag{26}$$

Thus, the last component of $\bar{u}$ in (14) for $w = 0$ can be written as: $\dot{\alpha} = -A_2\,[0\ 0\ 1]\,\frac{\partial f}{\partial t}$. Recalling computations investigated in [9], the term $A_2(\alpha)$ can be represented by: $A_2(\alpha) = \varepsilon_1\varepsilon_2\gamma(\alpha)$, while $\forall \alpha \in S^1$, $\gamma(\alpha) > 0$ for positive parameters ε_i, $i = 1, 2, 3$, for which $\bar{f}$ satisfies the tranversality condition. Consequently, taken into account (22), the following auxiliary dynamics can be considered

$$\dot{\alpha} = -\varepsilon_1\varepsilon_2\varepsilon_2^*\gamma(\alpha)\sin\alpha^*\dot{\alpha}^*. \tag{27}$$

Taking time derivative of $\tilde{\alpha}$ one has $\dot{\tilde{\alpha}} = \dot{\alpha} - \dot{\alpha}^*$. Then, applying (27) and (24) one obtains:

$$\dot{\tilde{\alpha}} = -\left(1 + \varepsilon_1\varepsilon_2\varepsilon_2^*\gamma(\alpha)\sin\alpha^*\right)k_\alpha\tilde{\alpha}. \tag{28}$$

Clearly, when $\sup_\alpha \varepsilon_1\varepsilon_2\varepsilon_2^*\gamma(\alpha) < 1$ one can conclude that the dynamic system (28) is exponentially stable at $\tilde{\alpha} = 0$. As a result $f_\theta \to 0$ which implies: $g_\theta \to g_\theta^*$.

4 Extension to Multi-agent Case

Let $\mathcal{W}$ denote a planar workspace occupied by m separated (non-colliding) circular obstacles $\mathcal{O}_i$, $i = 1, \ldots, m$. It is assumed that $m_r \leq m$ obstacles correspond to unicycle-like robots while $m - m_r$ obstacles are fixed. Coordinates of the centre of ith, $i = 1, \ldots, m$, obstacle are denoted by $p_i \in \mathbb{R}^2$, while its radius is r_i. Additionally, we also consider $\bar{r}_i > r_i$ in order to introduce a safety margin.

The basic task is to find control inputs for m_r unicycle-like robots such that the robots converge to the desired goals and simultaneously avoid collisions (with themselves and the static obstacles). The position coordinates of the goal defined for j^{th} robot is denoted by $p_{d,i}$ while the desired orientation is $\theta_{d,i}$.

4.1 Holonomic Agents

Here we relax the preliminary assumption that the robots are subject to non-holonomic constraints. Conversely, we assume that the robots are treated as unconstrained disks on a plane and no orientation control is required.

In order to describe potentials of goals and obstacles we take advantage of a logarithmic measure. Consider i^{th} robot with respect to j^{th} obstacle, $i \neq j$. In such a case the repulsive potential can be defined as follows

$$V_{r,j}(p_i) = -\frac{\kappa_{r,j}}{2}\log\left(\|p_i - p_j\|^2 - (r_j + \bar{r}_i)^2\right), \tag{29}$$

where $\kappa_{r,i} > 0$. Similarly, the attractor for i^{th} robot takes the following form

$$V_{a,i}(p_i) = \frac{\kappa_{a,i}}{2}\log\left(\|p_i - p_{d,i}\|^2\right), \tag{30}$$

while $\kappa_{a,i} > 0$. The total potential for the ith robot satisfies

$$V_j(p_j) = V_{a,j}(p_j) - \sum_{1\leq i\leq m,\ i\neq j} V_{r,i}(p_j), \tag{31}$$

while its gradient can be computed as

$$\frac{\partial V_i}{\partial p_i} = \kappa_i \frac{(p_i - p_{d,i})^\top}{\|p_i - p_{d,i}\|^2} - \sum_{1\leq i\leq m,\ i\neq j} \kappa_{r,i}\frac{(p_i - p_j)^\top}{\|p_i - p_j\|^2 - (r_j + \bar{r}_i)^2}. \tag{32}$$

It can be proved that function V_i does not have local minima if $\kappa_{a,i}$ is made large enough (cf. also the concept of navigation function introduced by Rimon and Koditschek, [14]). In such a case $p_{d,i}$ becomes an attractor of trajectory $p_i(t)$ which is the solution of the following differential equation

$$\dot{p}_i = -\left(\frac{\partial V_i}{\partial p_i}\right)^{\top}, \tag{33}$$

where $p_i(0)$ is the feasible initial condition (namely no collisions occur at $t = 0$).

However, the side effect of the logarithmic attractor defined by (30) is the appearance of singularity at $p_i = p_{d,i}$. To overcome this issue, one can consider the following smooth approximation of the first term (attractive part of the gradient) in (32)

$$\frac{(p_i - p_{d,i})^{\top}}{\|p_i - p_{d,i}\|^2} \approx \frac{(p_i - p_{d,i})^{\top}}{\|p_i - p_{d,i}\|^2 + \epsilon}, \tag{34}$$

where $\epsilon > 0$ is a constant. Nonetheless, employing (34) affects the minimum of function V. As a result an equilibrium $p_{eq,i}$ of (33) lies in a vicinity of $p_{d,i}$ a such that: $\lim_{\epsilon \to 0^+} p_{eq,i} = p_{d,i}$.

4.2 Nonholonomic Agents

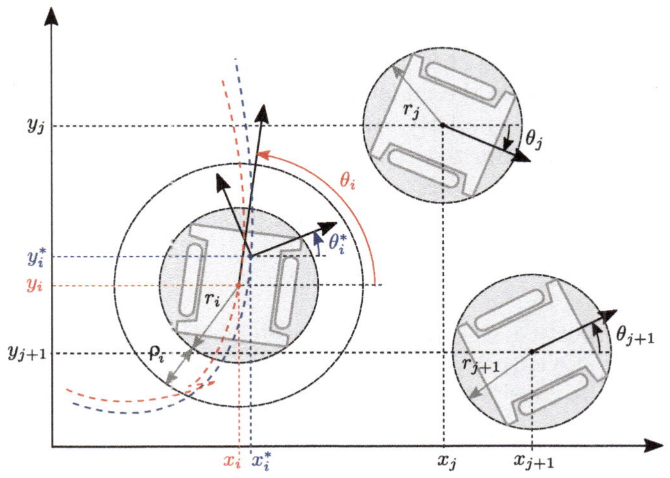

Fig. 2. Illustration of the geometry for nonholonomic robots: coordinates of the controlled unicycle robot and its corresponding virtual system are denoted by subscript i. Other vehicles are considered as moving obstacles.

The essential property of the decoupling controller (14) is the ability to employ a navigation technique designed for a holonomic system. Now we propose a modification of the method presented in Sect. 4.1.

Let us consider i^{th} unicycle-like vehicle with configuration $g_i \in G$ and compute configuration $g_i^* \in G$ of a companion virtual system using formula (12), cf. Fig. 2. Next, we apply the feedback (14) which, according to (15), brings the unconstrained linear control system: $\dot{g}_i^* = w_i$.

Using the notation from Sect. 4.1 we assume that position of the controlled robot is determined by

$$p_i = g_{xy,i}^* \in \mathbb{R}^2, \tag{35}$$

while the position of jth moving obstacle, $j = 1, 2, \ldots, m_r$, $j \neq i$, satisfies

$$p_j = g_{xy,i} \in \mathbb{R}^2. \tag{36}$$

Positions of other stationary obstacles are defined by $p_j = [p_{x,j}\ p_{y,j}]^\top \in \mathbb{R}^2$, $j = m_r + 1, \ldots, m$, while $p_{x,j}$ and $p_{y,j}$ are constants.

It is worth emphasising that although one can treat i^{th} controlled robot as the virtual system $\dot{p}_j = \tilde{w}_j = [w_{x_i}\ w_{y,i}]^\top$, the moving obstacles are represented by real vehicles (not the virtual ones). Hence, the collision detection is realised taking into account position of a virtual robot with respect to real obstacles. In order to do this one can increase the radius $\bar{r}_i$ to reflect the difference between trajectories $p_i(t)$ and $g_{xy}(t)$. Taking into account of (23) one obtains

$$\bar{r}_i = r_i + \rho_i. \tag{37}$$

In order to reduce the orientation error near the goal one can employ the concept introduced in Sect. 3.5. Since the stability of dynamics (28) is ensured for $w = 0$ one can disable the adaptation of α^* when a robot is in a significant distance from the goal. Similarly, $g_{\theta^*,j}$, which represents the orientation of the virtual system, can be adjusted gradually when the vehicle approaches the goal. For this purpose, taking into account i^{th} virtual robot and the selected goal $p_{d,i}$, we define the following smooth switching function

$$\kappa_\theta(p_i) = \tanh(\sigma \left\|p_i - p_{d,i}\right\|^{-2}), \tag{38}$$

where $\sigma > 0$ is a positive constant. It can be shown that $\lim_{\|p_i - p_{d,i}\| \to \infty} \kappa_\theta(p_i) = 0$ and $\lim_{\|p_i - p_{d,i}\| \to 0} \kappa_\theta(p_i) = 1$. Then the following proposition for the orientation control can be considered.

Proposition 3. *The orientation of ith robot can be stabilised at a neighbourhood of the desired point using the adaptation law (24) with $k_\alpha = k_{\alpha,i} := c_i \kappa_\theta(p_i)$, where $c_i > 0$, and using the following stabiliser: $w_{\theta,i} = -c_i\,(g_{\theta,i} - \theta_{d,i})$.*

5 Simulation Results

In order to verify properties of the proposed algorithm, numerical simulations in Matlab environment have been conducted. In simulations four nonhoholonomic vehicles are taken into account. Their initial configurations are: $g_1(0) = [0\ 0\ \frac{\pi}{2}]^\top$, $g_2(0) = [2\ 1\ -\frac{\pi}{2}]^\top$, $g_3(0) = [2\ 0\ \pi]^\top$ and $g_4(0) = [0\ 1\ -\frac{3}{4}\pi]^\top$.

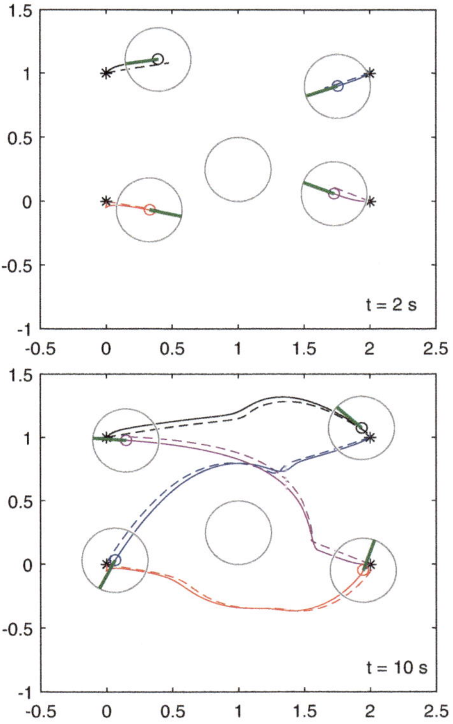

Fig. 3. Trajectories obtained in simulation S1. Positions of four unicycle robots and corresponding virtual robots are denoted by continuous and dashed curves, respectively. The following colours are used to distinguished the robots: red - 1, blue - 2, magenta - 3, black - 4. The orientations are denoted by green segments inside circles.

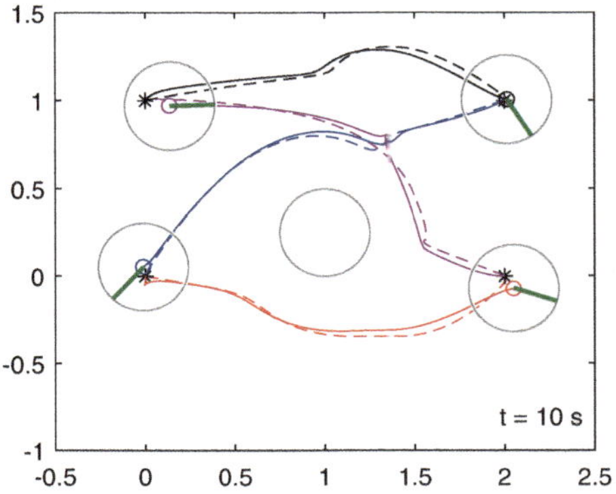

Fig. 4. Trajectories obtained in simulation S2. Positions of four unicycle robots and corresponding virtual robots are denoted by continuous and dashed curves, respectively. The following colours are used to distinguished the robots: red - 1, blue - 2, magenta - 3, black - 4. The orientations are denoted by green segments inside circles.

Table 1. Position and orientation errors in simulation S2 recorded at $t = 60$ s.

	$\|\|p_i - p_{d,i}\|\|$ [m]	$\|\|g_{xy,i} - p_{d,i}\|\|$ [m]	$\|g_{\theta,i} - \theta_{d,i}\|$ [rad]
Robot 1	0.0113	0.0216	$-9.96 \cdot 10^{-5}$
Robot 2	0.0117	0.0195	$-1.1 \cdot 10^{-6}$
Robot 3	0.0107	0.0524	$-2.63 \cdot 10^{-4}$
Robot 4	0.0109	0.0126	$-1.39 \cdot 10^{-5}$

Moreover, the desired positions are selected such that the following transitions are expected: $1 \to 3$, $2 \to 1$, $3 \to 4$ and $4 \to 2$ (the notation $i \to k$ means that i^{th} robot goes to the point where k^{th} robot is placed initially). Additionally, a stationary obstacle with the centre at $p_5 = [1\ 0.25]^\top$ is considered. The robots and the static obstacle are described by disks with the same radius $r_i = 0.2\,\text{m}$, $i = 1, \ldots, 5$. Parameters of the transverse function (18), (19) for each robot are selected as follows: $\varepsilon_{1,i} = 0.1$, $\varepsilon_{2,i} = \varepsilon_2^* = \frac{\pi}{2}$, $\varepsilon_{3,i} = 0.25$ and initial condition $\alpha_i(0) = 0$. It has been checked numerically that $\rho_i = \varepsilon_{1,i}$. The parameter of potential functions are chosen as: $\kappa_{a,i} = 0.25$, $\kappa_{r,j} = 0.01$ and $\epsilon = 0.1$.

In simulation S1 orientation of the robots is not stabilised, namely control input $w_{\theta,i} = 0$. Conversely, in simulation S2 a more complex control strategy based on Proposition 3 is used. Here, the following parameters are assumed: $\sigma = 0.2$ and $c_i = 0.25$ and $\theta_{d,i} = 0$.

The obtained results indicate that the motion task in both cases is accomplished properly. From Fig. 3 and 4 one can observe that trajectories of virtual robots converge to neighbourhoods of the desired points more smoothly. The non-zero residual errors obtained for these robots in the steady state result from the approximation (34). Basically, the trajectories at $t = 2$ s and $t = 5$ s are almost the same in both simulations, hence for the second scenario the corresponding figures are not presented. However, in simulation S2 orientations of the robots converge to the assumed values (cf. orientation of green segments at $t = 10$ s and $t = 50$ s). The steady state errors estimated based on the results obtained at $t = 60$ s are collected in Table 1.

6 Conclusions

The presented control strategy makes it possible to use different navigation methods originally developed for holonomic systems for unicycle-like vehicles. In addition, restrictions imposed by classic linearisation techniques such as the difficulty of stabilising the robot orientation and the requirement of non-zero linear velocity, [8], can be overcome.

However, the approach considered in the paper has also some disadvantages. The first drawback is the requirement of increasing the radius of obstacles due to the difference between trajectories of the real and virtual robots. The second is the complexity: in order to obtain satisfactory performance of the controller and to stabilise the position and orientation of vehicles, additional techniques, which are designed based on the stability of the augmented dynamics, are needed.

The proposed control methodology can be further developed and enhanced. In particular, it would be important to design a similar algorithm for more complex nonholonomic vehicles (e.g. car-like vehicles) where a path curvature is constrained. One can also consider an application of the method for moving targets. In near future, we also plan to conduct experimental verification of the algorithm.

References

1. Campion, G., Bastin, G., D'Andrea-Novel, B.: Structural properties and classification of kinematic and dynamic models of wheeled mobile robots. IEEE Trans. Robot. Autom. **12**(1), 47–62 (1996)
2. Khatib, O.: Real-time obstacle avoidance for manipulators and mobile robots. Int. J. Robot. Res. **5**(1), 90–98 (1986)
3. Kowalczyk, W., Michałek, M., Kozłowski, K.: Trajectory tracking control with obstacle avoidance capability for unicycle-like mobile robot. Bull. Pol. Acad. Sci.-Tech. Sci. **60**(3), 537–546 (2012)
4. LaValle, S.M.: Planning Algorithms. Cambridge University Press, Cambridge (2006)
5. Morin, P., Samson, C.: A characterization of the Lie algebra rank condition by transverse periodic functions. SIAM J. Control Optim. **40**, 1227–1249 (2001)
6. Morin, P., Samson, C.: Practical stabilization of driftless systems on Lie groups: the transverse function approach. IEEE Trans. Autom. Control **48**(9), 1496–1508 (2003)
7. Morin, P., Samson, C.: Control of nonholonomic mobile robots based on the transverse function approach. IEEE Trans. Robot. **25**(5), 1058–1073 (2009)
8. Oriolo, G., De Luca, A., Venditteli, M.: WMR control via dynamic feedback linearization: design, implementation and experimental validation. IEEE Trans. Control Syst. Technol. **10**(6), 835–852 (2002)
9. Pazderski, D.: Application of transverse functions to control differentially driven wheeled robots using velocity fields. Bull. Pol. Acad. Sci.-Tech. Sci. **64**(4), 831–851 (2016)
10. Pazderski, D.: Waypoint following for differentially driven wheeled robots with limited velocity perturbations. Asymptotic and practical stabilization using transverse function approach. J. Intell. Rob. Syst. **85**(3), 553–575 (2017)
11. Pazderski, D.: A robust smooth controller for a unicycle-like robot. Arch. Control Sci. **28**(1), 155–183 (2018)
12. Poonawala, H.A., Spong, M.W.: From nonholonomy to holonomy: time-optimal velocity control of differential drive robots. In: 2015 10th International Workshop on Robot Motion and Control (RoMoCo), pp. 97–102 (2015)
13. Rimon, E., Koditschek, D.: Exact robot navigation using artificial potential fields. IEEE Trans. Robot. Autom. **8**(5), 501–518 (1992)
14. Rimon, E., Koditschek, D.E.: The construction of analytic diffeomorphisms for exact robot navigation on star worlds. In: Proceedings of the IEEE International Conference on Robotics and Automation (ICRA), pp. 21–26 (1989)
15. Urakubo, T.: Stability analysis and control of nonholonomic systems with potential fields. J. Intell. Robot. Syst. **89**(1), 121–137 (2017)

16. Urakubo, T., Okuma, K., Tada, Y.: Feedback control of a two wheeled mobile robot with obstacle avoidance using potential functions. In: Proceedings of the IEEE/RSJ International Conference on Intelligent Robots and Systems (IROS), Sendai, Japan, pp. 2428–2433 (2004)

Formation Control of Non-holonomic Mobile Robots - Tuning the Algorithm

Wojciech Kowalczyk(✉) and Krzysztof Kozłowski

Poznań University of Technology, Piotrowo 3A, 60-965 Poznań, Poland
{wojciech.kowalczyk,krzysztof.kozlowski}@put.poznan.pl

Abstract. This paper presents the tuning methodology for the system of multiple two-wheeled mobile robots moving in formation. The procedure was applied to the trajectory tracking algorithm combined with collision avoidance based on the Artificial Potential Functions (APFs). Robots mimic motion of the virtual leader with a certain displacement avoiding collisions with each other and with circular shaped, static obstacles present in the environment. The results obtained during the computations are visualized to enable evaluation of the sensitivity of the closed-loop system to parameter selection. Then, the results of the simulation for the set of best parameters are discussed.

Keywords: Robot formation · Nonholonomic robot · Tuning algorithm · Path following · Artificial Potential Function

1 Introduction

First works concerning the problem of collision avoidance in multi-agent system were published by Leitmann and Skowronski in 1977 [8] and in 1980 [9]. In 1986 Khatib proposed the new control algorithm [2] in which he combined attracting (to the goal) and repelling (from the obstacles) interactions. The novelty was the use of *Artificial Potential Functions* (APFs), similar to models of intermolecular interactions. The control is based on the gradient of combined attracting APF and one or more repelling APFs. In this paper repelling APFs are used to avoid collisions in multi-robot system.

In recent years many articles concerned the problem of multiple mobile robot control, based on the kinematic model of the mobile platforms [12], [4], and the ones taking into account their dynamic behavior [1,6].

This paper presents control algorithms for a group of differentially driven mobile robots moving in formation. They execute trajectory tracking task. The algorithm is based on paper [10] and presented in detail, including Lyapunov stability proof in [5].

Section 2 presents formulation of the control problem. Section 3 describes control algorithm for the formation of non-holonomic mobile robots. Section 4 describes scaling procedure for the robots wheel controls. Section 5 presents details of the formation motion scenario used in the tuning method. Section 6

A. Bartoszewicz et al. (Eds.): PCC 2020, AISC 1196, pp. 967–978, 2020.
https://doi.org/10.1007/978-3-030-50936-1_81

shows tuning procedure and set of the best coefficients for the closed-loop system that was found. Simulation results for these settings are presented in Sect. 7. In the last section concluding remarks are given.

2 Problem Description

The task of the formation is to follow virtual leader that moves with desired linear and angular velocities $[v_l \quad \omega_l]^T$. A kinematic model of the i-th differentially-driven mobile robot ($i = 1 \ldots N$, N - number of robots) is given by the following equation:

$$\dot{q}_i = \begin{bmatrix} \cos\theta_i & 0 \\ \sin\theta_i & 0 \\ 0 & 1 \end{bmatrix} u_i, \tag{1}$$

where vector $q_i = [x_i \quad y_i \quad \theta_i]^\top$ denotes the pose and x_i, y_i, θ_i are position coordinates and orientation of the robot with respect to a global, fixed coordinate frame. Vector $u_i = \left[v_i \; \omega_i\right]^\top$ is the control vector consisting of linear v_i and angular ω_i velocity control signals of the platform.

The robots are expected to imitate the motion of the virtual leader. They should have the same velocities as the virtual leader. The position coordinates $[x_l \;\; y_l]^T$ of the virtual leader are used as a reference position for the individual robots but each of them have different spatial displacement with respect to the leader:

$$\begin{aligned} x_{id} &= x_l + d_{ix} \\ y_{id} &= y_l + d_{iy}, \end{aligned} \tag{2}$$

where vector $[d_{ix} \quad d_{iy}]^T$ is the desired displacement of the i-th robot. As the robots position converge to the desired values their orientations θ_i converge to the orientation of the virtual leader θ_l.

3 Control Algorithm

The goal of the control is to drive the formation along the desired trajectory avoiding collisions between agents and other obstacles. Achieving control goals is equivalent to bringing the following quantities to zero:

$$\begin{aligned} p_{ix} &= x_{id} - x_i \\ p_{iy} &= y_{id} - y_i \\ p_{i\theta} &= \theta_l - \theta_i. \end{aligned} \tag{3}$$

Collision avoidance behaviour is based on the APFs. This concept originally was proposed in [2]. All robots are surrounded by APFs that raise to infinity near objects border r_j (j - number of the robot/obstacle) and decreases to zero at some distance R_j, $R_j > r_j$.

One can introduce the following function [3]:

$$B_{aij}(l_{ij}) = \begin{cases} 0 & \text{for} \quad l_{ij} < r_j \\ e^{\frac{l_{ij}-r_j}{l_{ij}-R_j}} & \text{for } r_j \leq l_{ij} < R_j \\ 0 & \text{for} \quad l_{ij} \geq R_j \end{cases}, \tag{4}$$

that gives output $B_{aij}(l_{ij}) \in \langle 0, 1)$. Distance between the i-th and the j-th robot is defined as the Euclidean length $l_{ij} = \left\| [x_j \ y_j]^\top - [x_i \ y_i]^\top \right\|$.

Scaling the function given by Eq. (4) within the range $\langle 0, \infty)$ can be given as follows:

$$V_{aij}(l_{ij}) = \frac{B_{aij}(l_{ij})}{1 - B_{aij}(l_{ij})}, \tag{5}$$

that is used later to avoid collisions.

Further descriptive terms 'collision area' or 'collision region' are used for locations fulfilling conditions $l_{ij} < r_j$. The range $r_j < l_{ij} < R_j$ is called 'collision avoidance area' or 'collision avoidance region'.

Assumption 1. $\forall \{i, j\}, i \neq j, \ ||[x_{id} \ \ y_{id}]^T - [x_{jd} \ \ y_{jd}]^T|| > R_j$.

Assumption 2. *If robot i gets into the avoidance region of any other robot/obstacle j, $j \neq i$ its desired trajectory is temporarily frozen ($\dot{x}_{id} = 0$, $\dot{y}_{id} = 0$). If the robot leaves the avoidance area its desired coordinates are immediately updated. As long as the robot remains in the avoidance region its desired coordinates are periodically updated at certain discrete instants of time. The time period t_u of this update process is relatively large in comparison to the main control loop sample time.*

The system error expressed with respect to the coordinate frame fixed to the robot is described below:

$$\begin{bmatrix} e_{ix} \\ e_{iy} \\ e_{i\theta} \end{bmatrix} = \begin{bmatrix} \cos\theta_i & \sin\theta_i & 0 \\ -\sin\theta_i & \cos\theta_i & 0 \\ 0 & 0 & 1 \end{bmatrix} \begin{bmatrix} p_{ix} \\ p_{iy} \\ p_{i\theta} \end{bmatrix}. \tag{6}$$

Using the above equations and non-holonomic constraint $\dot{y}_i \cos(\theta_i) - \dot{x}_i \sin(\theta_i) = 0$ the error dynamics between the leader and the follower are as follows:

$$\begin{aligned} \dot{e}_{ix} &= e_{iy}\omega_i - v_i + v_l \cos e_{i\theta} \\ \dot{e}_{iy} &= -e_{ix}\omega_i + v_l \sin e_{i\theta} \\ \dot{e}_{i\theta} &= \omega_l - \omega_i. \end{aligned} \tag{7}$$

One can introduce the position correction variables that consist of position error and collision avoidance terms:

$$\begin{aligned} P_{ix} &= p_{ix} - \hat{w}_{ix} \\ P_{iy} &= p_{iy} - \hat{w}_{iy}, \end{aligned} \tag{8}$$

where

$$\hat{w}_{ix} = \sum_{j=1,j\neq i}^{N+M} \hat{w}_{ijx}, \quad \hat{w}_{iy} = \sum_{j=1,j\neq i}^{N+M} \hat{w}_{ijy}$$

are components of the consolidated collision avoidance vector $\hat{w}_i = [\hat{w}_{ix} \ \hat{w}_{iy}]^T$, and

$$\hat{w}_{ijx} = \frac{\partial V_{aij}}{\partial x_i}, \quad \hat{w}_{ijy} = \frac{\partial V_{aij}}{\partial y_i}$$

are components of the j-th obstacle APF's gradient $\hat{w}_{ij} = [\hat{w}_{ijx} \ \hat{w}_{ijy}]^T$ with respect to the global coordinate frame computed in the location of the i-th robot, M - number of static obstacles. V_{aij} depends on x_i and y_i according to equation (5). It is assumed that robots avoid collisions with each other and other obstacles present in the taskspace (only circle-shaped can occur). The correction variables are transformed to the local coordinate frame fixed in the mass centre of the robot:

$$\begin{bmatrix} E_{ix} \\ E_{iy} \\ e_{i\theta} \end{bmatrix} = \begin{bmatrix} \cos\theta_i & \sin\theta_i & 0 \\ -\sin\theta_i & \cos\theta_i & 0 \\ 0 & 0 & 1 \end{bmatrix} \begin{bmatrix} P_{ix} \\ P_{iy} \\ p_{i\theta} \end{bmatrix}. \tag{9}$$

Taking into account Eqs. (9), (8) and some transformations [5] gradient of the APF can be expressed with respect to the local coordinate frame fixed to the i-th robot:

$$\begin{bmatrix} \frac{\partial V_{aij}}{\partial e_{ix}} \\ \frac{\partial V_{aij}}{\partial e_{iy}} \end{bmatrix} = \begin{bmatrix} -\cos\theta_i & -\sin\theta_i \\ \sin\theta_i & -\cos\theta_i \end{bmatrix} \begin{bmatrix} \frac{\partial V_{aij}}{\partial x_i} \\ \frac{\partial V_{aij}}{\partial y_i} \end{bmatrix}. \tag{10}$$

Equation (9) using (8) and (10) can be rewritten as follows:

$$\begin{aligned} E_{ix} &= p_{ix}\cos(\theta_i) + p_{iy}\sin(\theta_i) + w_{ix} \\ E_{iy} &= -p_{ix}\sin(\theta_i) + p_{iy}\cos(\theta_i) + w_{iy}, \\ e_{i\theta} &= p_{i\theta} \end{aligned} \tag{11}$$

where

$$w_{ix} = \sum_{j=1,j\neq i}^{N+M} w_{ijx}, \quad w_{iy} = \sum_{j=1,j\neq i}^{N+M} w_{ijy} \tag{12}$$

are components of the consolidated collision avoidance vector $w_i = [w_{ix} \ w_{iy}]^T$, M - number of static obstecles, and

$$w_{ijx} = \frac{\partial V_{aij}}{\partial e_{ix}}, \quad w_{ijy} = \frac{\partial V_{aij}}{\partial e_{iy}} \tag{13}$$

are components of the j-th obstacle APF's gradient $w_{ij} = [w_{ijx} \ w_{ijy}]^T$ with respect to the local coordinate frame (fixed to the robot) computed in the location of the i-th robot.

Each derivative of the APF is transformed from the global coordinate frame to the local coordinate frame fixed to the robot. Finally, correction variables expressed with respect to the local coordinate frame are as follows:

$$\begin{aligned}E_{ix} &= e_{ix} + w_{ix} \\ E_{iy} &= e_{iy} + w_{iy}.\end{aligned} \tag{14}$$

Control algorithm from [10] for N robots extended by the collision avoidance is as follows:

$$\begin{aligned}v_i &= v_l + c_2 E_{ix} \\ \omega_i &= \omega_l + h(t, E_{iy}) + c_1 e_{i\theta},\end{aligned} \tag{15}$$

where $h(t, E_{iy})$ is bounded, depends linearly on E_{iy}, and continuously differentiable function. It must be properly chosen to ensure persistent excitation of the reference angular velocity [11]. Positive constants c_1 and c_2 are design parameters.

Assumption 3. *If the value of the linear control signal v_i is less than the considered threshold value v_t, i.e. $|v_i| < v_t$ (v_t - positive constant), it is replaced by a new value $\tilde{v}_i = S(v_i)v_t$ where*

$$S(v_i) = \begin{cases} -1 \ for \ v_i < 0 \\ 1 \ \ for \ v_i \geq 0 \end{cases}. \tag{16}$$

By partial substitution of (15) into (7) one can express error dynamics as follows:

$$\begin{aligned}\dot{e}_{ix} &= e_{iy}\omega_i - c_2 E_{ix} + v_l(\cos e_{i\theta} - 1) \\ \dot{e}_{iy} &= -e_{ix}\omega_i + v_l \sin e_{i\theta} \\ \dot{e}_{i\theta} &= -h_i(t, E_{iy}) - c_1 e_{i\theta}.\end{aligned} \tag{17}$$

The stability proof for the closed-loop system is presented in [5].

4 Wheel Velocity Scaling Procedure

Platform controls are transformed to the robots wheel controls using the following transformations:

$$\begin{aligned}\omega_{iR} &= \frac{d}{2\tilde{r}}\omega_i + \frac{1}{\tilde{r}}v_i \\ \omega_{iL} &= -\frac{d}{2\tilde{r}}\omega_i + \frac{1}{\tilde{r}}v_i,\end{aligned} \tag{18}$$

where $\tilde{r} = 0.0245$ m is radius of the wheels and $d = 0.148$ m is distance between wheels. The diameter of the robot is 0.17 m. To model the system the mechanical parameters of *MTracker* robot developed at Poznan University of Technology were used.

The desired wheel velocities are scaled down when at least one of the wheels exceeds the assumed limitation. The scaled control signal u_{iws} is calculated using equation $u_{iws} = s_i u_{iw}$ where

$$s_i = \begin{cases} \frac{\omega_{max}}{\omega_{io}} & \text{if} \quad \omega_{io} > \omega_{max} \\ 1 & \text{otherwise} \end{cases}, \tag{19}$$

and $\omega_{io} = \max\{|\omega_{iR}|, |\omega_{iL}|\}$ where ω_{iR} and ω_{iL} denote right and left wheel angular velocities respectively, ω_{max} is the predefined maximal angular velocity for each wheel.

In the analysis presented further ω_{max} was set to $12.15\frac{rd}{s}$. It is a half of the physically achievable velocity of the *MTracker* mobile platform wheels.

5 Formation Motion Scenario

The proposition of motion scenario that is composed of various stages present in the typical formation task has the following properties:

1. Robots initial positions are located far from the desired ones,
2. During the transient state robots have to avoid collision with static obstacles and other robots,
3. Desired trajectory for the formation consists of straight sections and arcs.

The number of robots is $N = 9$ and the number of static obstacles $M = 7$. Initial coordinates of mobile platforms were set using pseudo-random generator (they were generated only once; the same values were used in all numerical computations). The number of robots was a compromise between the requirement to obtain good settings for a complex case, and the available computing power/time. The goal for the robots was to build ring-shaped formation (diameter 10 m). Virtual leader is located in the middle of the ring. Initially most of the robots have to pass through a 'barrier' composed of seven static circle shaped obstacles with a diameter $r = 0.3$ m and the range of APF $R = 1.2$ m (the radii of robots and the ranges of their APFs are the same). The distances between static obstacles are 2.5 m, there are only 0.1 m wide slots between the neighbouring obstacle APFs. Persistent excitation was given by the equation: $h(t, E_{iy}) = \phi(t)\tanh(E_{iy})$, where the function $\phi(t)$ is non smooth pulse function of an amplitude 0.5, period of 4 s and the duty cycle w_{dc}. It introduces persistent excitation that is necessary to stabilize the system in the y direction (with respect to the local coordinate frame). Authors believe that in many applications selected scenario may be considered as representative.

6 Tuning the Algorithm

The tuned parameters were: c_1, c_2, and w_{dc}. There are still other parameters that could be tuned (i.e. the period and the amplitude of the function $\phi(t)$). Authors decided to leave them constant to limit parameter space size and reduce the computation cost.

The following criterion was used to tune the algorithm:

$$\Gamma_e = \min \tilde{\Gamma}_e = \min \int_0^T \sum_{i=1}^{N} \left(e_{ix}^2 + e_{iy}^2 + e_{i\theta}^2\right) dt. \tag{20}$$

The tuning process was conducted in three stages:

1. Coarse tuning - the system was simulated for all combinations of c_1, c_2 in the range $\langle 0.1, 2.0\rangle$ logarithmically equally spaced (values $\{0.1, 0.1648, 0.2714, 0.4472, 0.7368, 1.2139, 2.0\}$), and the duty cycle w_{dc} of $\phi(t)$ in the range from 10% to 90% with linear spacing 10%. The ranges of c_1, c_2, w_{dc} were chosen in accordance to the authors' experience. The resulting Γ_e was obtained for the following parameters: $c_1 = 0.4472$, $c_2 = 0.7368$, and $w_{dc} = 90\%$. The last value lies on the border of the tested range, which was unexpected.

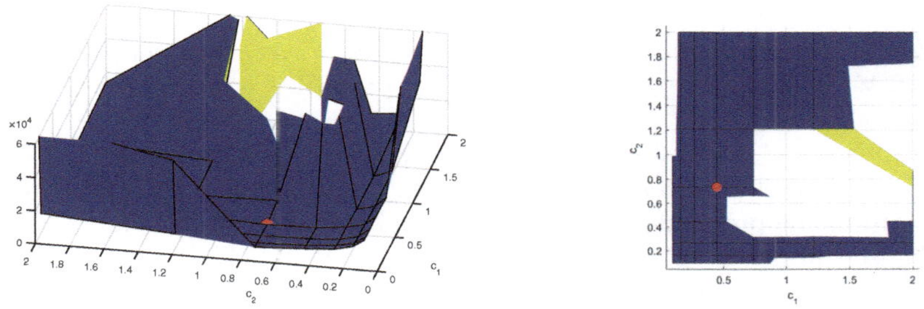

(a) Perspectiveview (b) Topview

Fig. 1. $\tilde{\Gamma}_e$ as a function of c_1 and c_2; $w_{dc} = 90\% = const.$

2. More accurate tuning - the system was simulated for all combinations of c_1 and c_2 in the range $\langle 0.2714, 1.2139\rangle$ logarithmically equally spaced ($\{0.2714, 0.3662, 0.4941, 0.6667, 0.8996, 1.2139\}$), and the duty cycle w_{dc} of $\phi(t)$ in the range from 80% to 96% with linear spacing 4%. The resulting Γ_e was obtained for the following parameters: $c_1 = 0.3662$, $c_2 = 0.6667$, and $w_{dc} = 96\%$.
3. Precise tuning - the system was simulated for all combinations of c_1 in the range $\langle 0.2714, 0.4941\rangle$ logarithmically equally spaced ($\{0.2714, 0.3060, 0.3449, 0.3888, 0.4383, 0.4941\}$), c_2 in the range $\langle 0.4941\ 0.8996\rangle$ logarithmically equally spaced ($\{0.4941, 0.5570, 0.6279, 0.7079, 0.7980, 0.8996\}$), and the duty cycle w_{dc} of $\phi(t)$ in the range from 96% to 99% with linear spacing 1%. The resulting Γ_e was obtained for the following parameters: $c_1 = 0.4941$, $c_2 = 0.7980$, and $w_{dc} = 98\%$.

The computations were executed using Matlab/Simulink running on the notebook equipped with Intel i5-8265U CPU 1.8 GHz, 24 GB RAM. In the first stage the number of iterations was 441 and its execution took over 48 h. In the second and third stages the number of iterations were 180 and 144, respectively. The execution times were proportionately smaller. The ranges selected in the first stage were made in accordance to the authors' experience. It turned out that they were accurate in case of c_1 and c_2 but in case of w_{dc} the best value was out

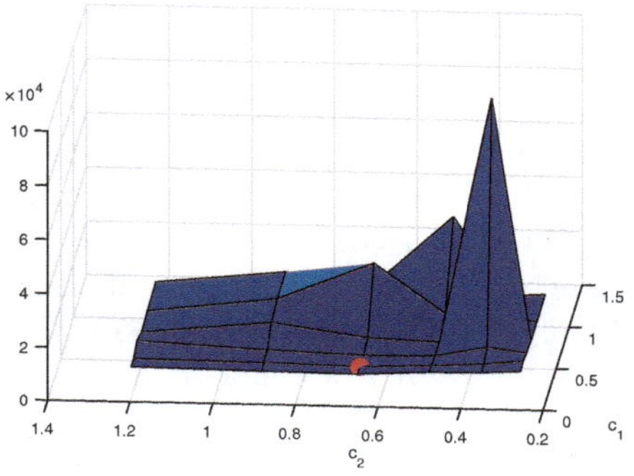

Fig. 2. $\tilde{\Gamma}_e$ as a function of c_1 and c_2, $w_{dc} = 96\% = const.$

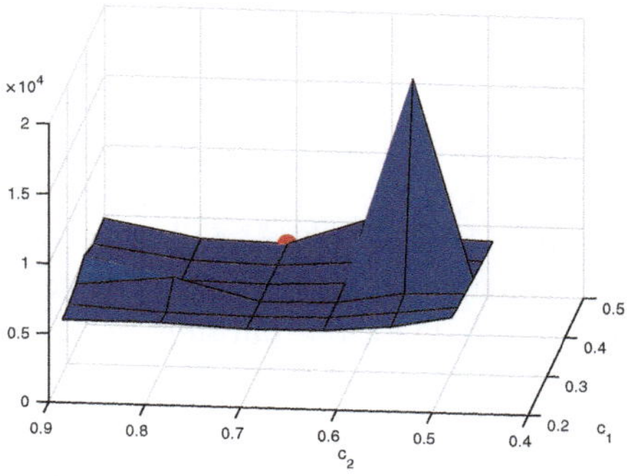

Fig. 3. $\tilde{\Gamma}_e$ as a function of c_1 and c_2, $w_{dc} = 98\% = const.$

of the initially selected range. In the second and third stages the most promising ranges from the previous stage were tested with higher resolution.

Figure 1 presents $\tilde{\Gamma}_e = f(c_1, c_2)$ for $w_{dc} = const. = 90\%$. Note that the whole parameter space is three dimensional and cannot be shown in 3*d* figure. Red marker represents the minimum found. Empty and yellow areas in the surface represent combinations of c_1 and c_2 for which $\tilde{\Gamma}_e$ was extremely high. This area was neglected in the further analysis. Figures 2 and 3 represent close surrounding of the coefficient space for which the best result was obtained in the previous step.

The authors compared the proposed tuning procedure with the one available in Matlab program. The function `fminsearch` runs nonlinear programming solver that can be applied to multivariable function. It uses the simplex search method [7]. The algorithm is not guaranteed to converge to a global minimum.

The results obtained were as follows: $c_1 = 0.2910$, $c_2 = 0.8521$ and $w_{dc} = 99.9\%$ and the tuning execution took almost 21 h. The value of the criterion Γ_e was worse by 1.71%. The advantages of the method proposed in this paper in comparison to `fminsearch` are as follows: it allows the analysis of the parameter space for the method, and evaluation of its sensitivity to the parameter selection. It can be easily done using presented graphs (Figures 1, 2 and 3). The second reason is the fact that there is no guarantee that the result obtained using `fminsearch` built-in optimizer is the global minimum. Our method does not provide such a guarantee, but reduces the risk as we search the entire parameter space.

Drawing further conclusions one can state that despite the difference of the absolute values of the settings obtained using both methods, the system behaves well which confirms its low sensitivity to design parameters.

7 Simulation Results

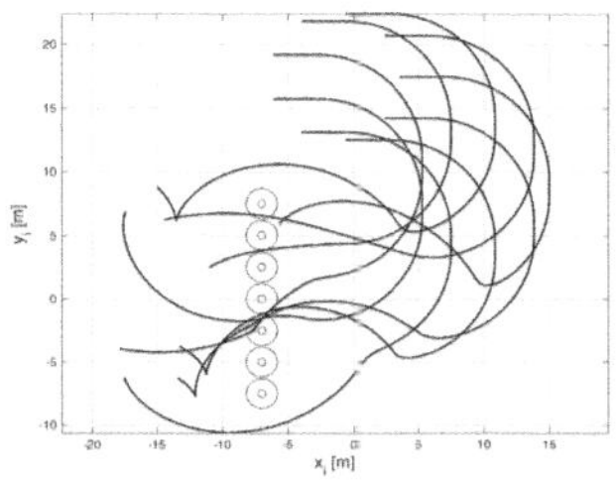

(a) Locations of robots in XY-plane

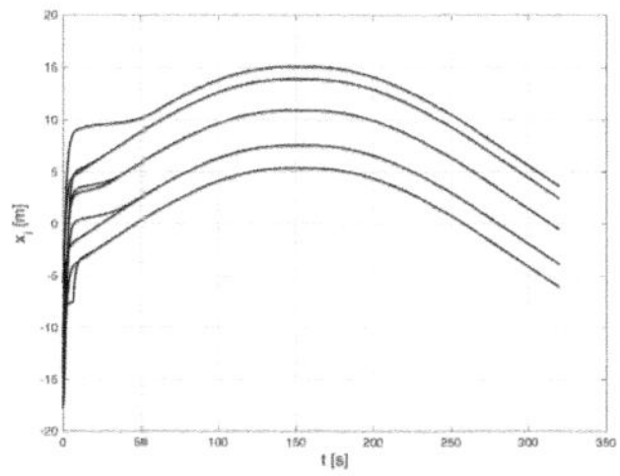

(b) x coordinates as a function of time

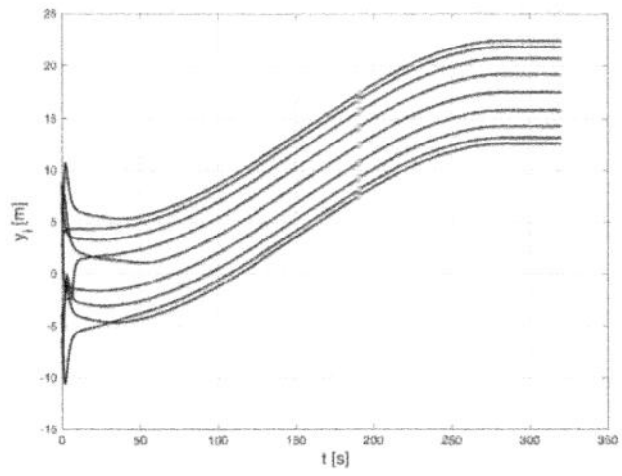

(c) y coordinates as a function of time

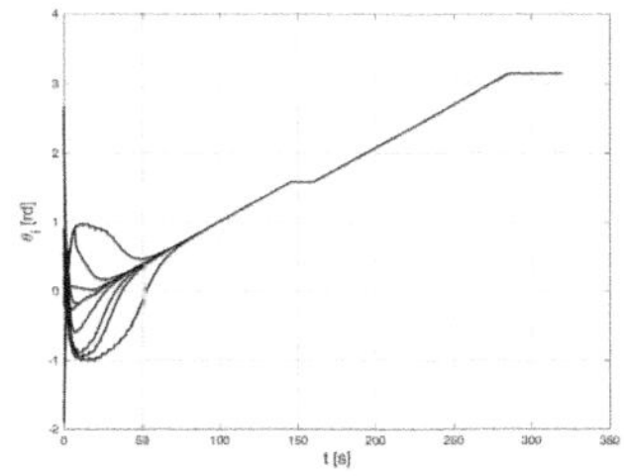

(d) Robot orientation as a function of time

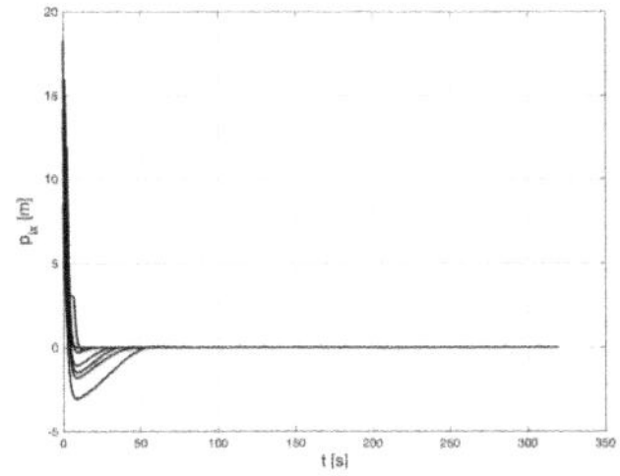

(e) Error in x coordinates as a function of time

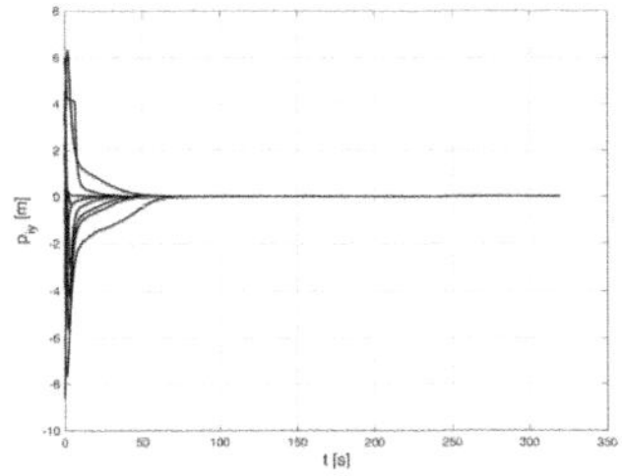

(f) Error in y coordinates as a function of time

This section presents simulation results for the test scenario and the algorithm coefficients set to the best values obtained in Sect. 6. Their values were as follows:

$$c_1 = 0.4941, \quad c_2 = 0.7980, \quad w_{dc} = 98\%. \tag{21}$$

Virtual leader starts the motion at the origin and initially moves along x axis with linear velocity $v_l = 0.1\frac{m}{s}$. After 20 s the angular velocity $\omega_l = 0.0125\frac{rd}{s}$ is applied still maintaining the previous linear speed. Then, after next 145.66 s its angular velocity is set to zero. Robot is moving straight along x axis (negative direction) up to 320 s.

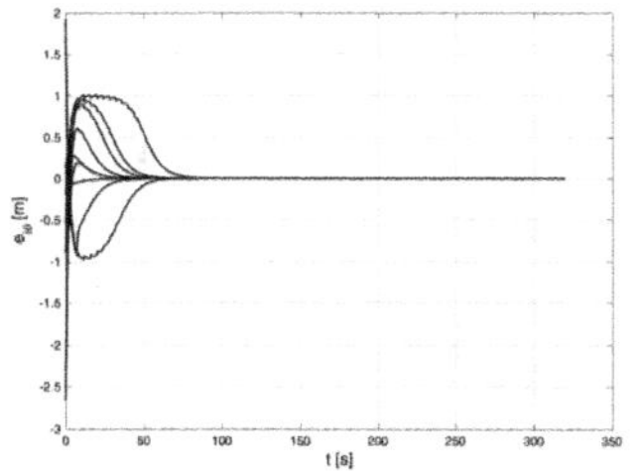

(g) Orientation error as a function of time

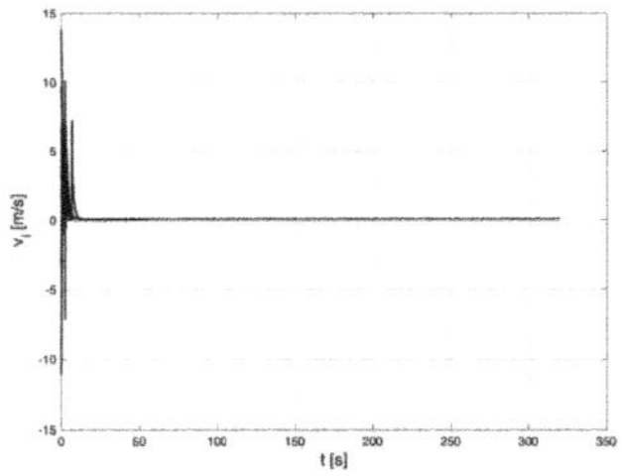

(h) Linear velocity control

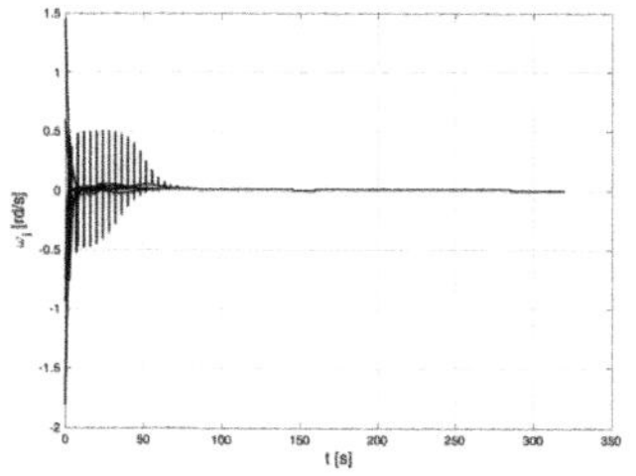

(i) Angular velocity control

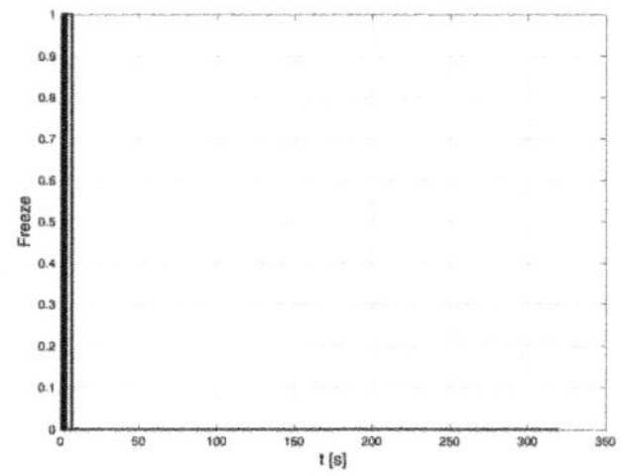

(j) 'Freeze' signal

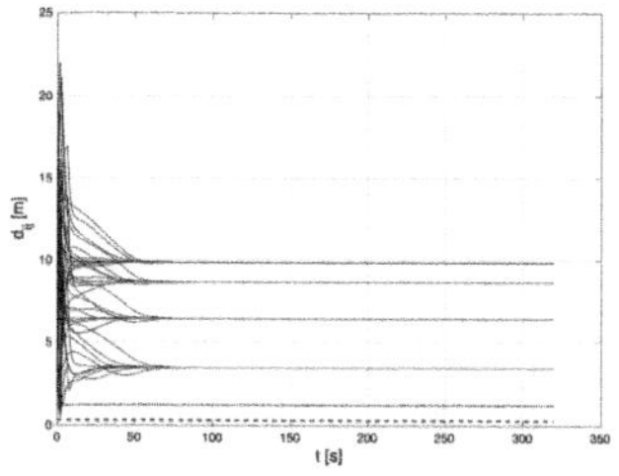

(k) Distances between robots

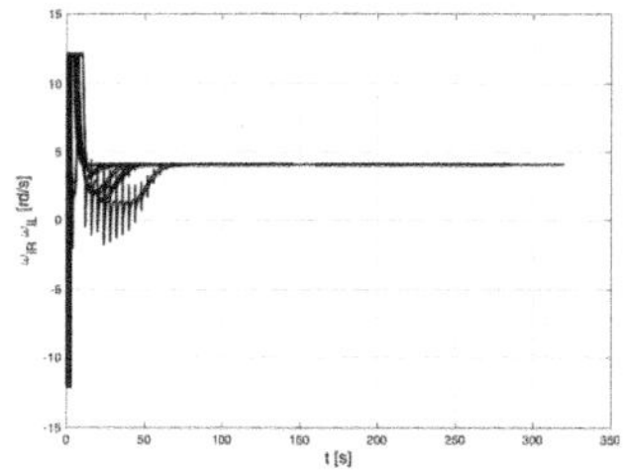

(l) Wheel velocities

Fig. 4. Numerical simulation: trajectory tracking for $N = 9$ robots

The parameter $t_u = 1\,\mathrm{s}$ was used. Figure 4a shows motion of robots in xy-plane. Figures 4b, 4c, 4d present time graphs of x_i, y_i and θ_i robots coordinates, respectively. One can see that robots reach their reference signals in about 80 s. Figures 4e and 4f show position errors expressed in the global coordinate frame. Figure 4g presents orientation errors as a function of time.

Figure 4h presents plots of linear velocity controls before scaling. There are several peaks in the transient state. These control values are not realizable in the real systems, and they are scaled down at the wheel control level. Figure 4i shows angular controls for the robots. One can clearly see the sequence of straight motion periods and movement along arcs periods. Figure 4j shows plot of the 'freeze' signals (the reference signals was frozen if this signal is set to 1 and unfrozen otherwise). The periods of 'freezing' reference signal occur for the first 10s of the transient state. Figure 4k shows relative distances between robots. This graph is not very readable in the initial stage because it contains $N(N-1) = 72$ signals, but it is clear that no pair of robots is getting distance close to $2r = 0.6$ m (dotted line). This minds that no collision occurred. Figure 4l presents plot of wheel velocities. Their values are limited to $\omega_{max} = 12.15\frac{rd}{s}$.

The coefficients (21) resulted in the convergence faster by 20% in comparison to the case presented in [5], even though the wheel velocities were not limited there.

8 Conclusion

This paper presents tuning procedure for the control of a group of differentially-driven mobile robots that tracks the desired trajectory. Robots avoid collision with each other and other obstacles existing in the taskspace. Wheel velocities are scaled down to values realizable by the physical actuators. Then the tuning procedure is applied. Simulation results for the formation of 9 robots tracking desired trajectory in the environment with 7 static obstacles show the effectiveness of the algorithm. It was done for the best settings obtained during the tuning procedure. Authors plan to verify obtained algorithm settings for different motion scenarios and conduct tests of the presented control method on a real two-wheeled mobile robots in the near future.

Acknowledgements. This work is supported by statutory grant 09/93/DSPB/0811.

References

1. Do, D.: Formation tracking control of unicycle-type mobile robots with limited sensing ranges. IEEE Trans. Control Sys. Technol. **16**(3), 527–538 (2008)
2. Khatib, O.: Real-time obstacle avoidance for manipulators and mobile robots. Int. J. Robot. Res. **5**(1), 90–98 (1986)
3. Kowalczyk, W., Michałek, M., Kozłowski, K.: Trajectory Tracking control with obstacle avoidance capability for unicycle-like mobile robot. Bull. Pol. Acad. Sci. Tech. Sci. **60**(3), 537–546 (2012)
4. Kowalczyk, W., Kozłowski, K.: Leader-Follower control and collision avoidance for the formation of differentially-driven mobile robots. In: 23rd International Conference on Methods and Models in Automation and Robotics (MMAR 2018), pp. 27-30, Miedzyzdroje, Poland, August 2018
5. Kowalczyk, W., Kozłowski, K.: Trajectory tracking and collision avoidance for the formation of two-wheeled mobile robots. Bull. Pol. Acad. Sci. Tech. Sci. **67**(5), 915–924 (2019)
6. Kowalczyk, W., Kozłowski, K., Tar, J.: Trajectory tracking for multiple unicycles in the environment with obstacles. In: 19th International Workshop on Robotics in Alpe-Adria-Danube Region (RAAD 2010), pp. 451–456, Budapest (2010). https://doi.org/10.1109/RAAD.2010.5524544

7. Lagarias, J.C., Reeds, J.A., Wright, M.H., Wright, P.E.: Convergence properties of the Nelder-Mead simplex method in low dimensions. SIAM J. Optim. **9**(1), 112–147 (1998)
8. Leitmann, G., Skowronski, J.: Avoidance control. J. Optim. Theory Appl. **23**(4), 581–591 (1977)
9. Leitmann, G.: Guaranteed avoidance strategies. J. Optim. Theory Appl. **32**(4), 569–576 (1980)
10. Loria, A., Dasdemir, J., Alvarez Jarquin, N.: Leader–follower formation and tracking control of mobile robots along straight paths. IEEE Trans. Control Syst. Technol. **24**(2), 727–732 (2016). https://doi.org/10.1109/TCST.2015.2437328
11. Loria, A., Panteley, E., Teel, A.: Relaxed persistency of excitation for uniform asymptotic stability. IEEE Trans. Autom. Control **46**(12), 1363–1368 (2001)
12. Mastellone, S., Stipanovic, D., Spong, M.: Formation Control and collision avoidance for multi-agent non-holonomic systems: theory and experiments. Int. J. Robot. Res. **27**, 107–126 (2008)

Local Path Planning for Autonomous Mobile Robot Based on APF-BUG Algorithm with Ground Quality Indicator

Kamil Wyrąbkiewicz(✉), Tomasz Tarczewski, and Łukasz Niewiara

Nicolaus Copernicus University, 87-100 Torun, Poland
kamillo@umk.pl

Abstract. In this paper, an enhanced Artificial Potential Field (AFP) algorithm applied to autonomous mobile robot is presented. The proposed solution is extended by an additional BUG algorithm and a ground quality indicator. The modification allows to avoid local minima in path planning caused by complex terrain obstacles. The developed algorithm takes into account the substrate quality, classifying a poor ground as an obstacle. It was implemented and tested in Matlab software, utilizing the track planning algorithm as a state machine. The simulation environment enables graphical presentation of the chosen path and the arrangement of moving area. The article presents and discusses a basic issue in local path planning of autonomous mobile platforms, i.e. the ground quality, which is ignored in classical algorithms. This is a non-trivial problem, which impacts the success rate of getting the final destination. The developed algorithm is extended by a BUG rule and ground quality indicator which allows to avoid immobilization of the platform.

Keywords: Artificial Potential Fields · Bug algorithm · Ground quality indicator · Autonomous mobile robots · Path planning

1 Introduction

1.1 Model of the Mobile Robot

Most of the mobile robots can be classified into remote controlled and autonomous. Since the remote control requires data exchange in real time, this type of steering is limited by the distance between the platform and beacon station. Therefore, both of the mentioned techniques may be used on the earth. Another situation occurs for vehicles applied for space exploration, i.e. mars rovers, which are a kind of mobile robots (Autonomous Mobile Robot – Mars Rover) and are designed for exploration of other planets [14]. In such a case, the distance and location of the research object makes it unable to employ real-time steering. This implicates a need of autonomous control employment.

Mobile platforms - Mars Rovers due to its application, imposes the use of autonomous control [3, 6]. The control method of an autonomous mobile robot – mars rover (AMR-MR) is strictly limited by the transmission rate of exchanged data, which is determined

A. Bartoszewicz et al. (Eds.): PCC 2020, AISC 1196, pp. 979–990, 2020.
https://doi.org/10.1007/978-3-030-50936-1_82

by the time delay in signal transmitting and receiving between the earth and mars. In addition, this time is varying and it is correlated with the distance between the planets. In general, it is determined by the position on their own orbits. AMR-MR must be designed for exploration task, which includes moving in the unknown environment. The construction of the drive system and suspension is determined through the working area of the mobile platform. Usually, these are also equipped in mechanical manipulators for manual tasks [14].

The realization of path planning of an autonomous mobile platform is non-trivial task, and it requires knowledge about the movement area. In general, two cases may occur: (i) the trajectory is realized in a known environment or, (ii) in a unknown one. The existing methods can be divided in two groups, corresponding to the knowledge about the movement area. The presented paper is focused on local path planning for an autonomous mobile platform in an unknown environment. The literature discuss some navigation and moving algorithms employed in such operating conditions [1–5, 7–9]. It should be noted, that to the best Authors knowledge, any of existing algorithms doesn't take into account ground quality indicator. Since the autonomous mobile robot may encounter different types of ground (i.e. gravel, rocks, water, ice) it is important to cope with such kind of the obstacles during local path planning.

In order to reach the goal localization, an autonomous operation requires employment of advanced control algorithms [10]. This task is more difficult without knowledge about the operating area. In such a case additional sensors and extended path planning should be employed.

The first chapter presents preliminary information about the robot model. In the second chapter information about APF and BUG algorithm is presented. The third chapter contains the proposed Ground Quality Indicator (GQI) algorithm. The fourth chapter contains examples and simulation results of the proposed algorithm. The last chapter contains conclusions.

2 Mobile Robot Control Algorithm

2.1 Artificial Potential Fields

The proposed solution is based on Artificial Potential Fields algorithm [10], and it is inspired by a classical interaction between charged particles. The chosen movement direction of the platform is based on the Coulomb force. At this stage it was assumed that the current position of the robot and destination should be known. It may be obtained using advanced positioning systems like: GPS, GLONASS, and Galileo, equipped with a ground signal transmitter [11]. Based on the Coulomb law, the value of the electrostatic interaction is given by [12]:

$$F_w = -\frac{kq_1q_2}{r^2} \tag{1}$$

where: k – interactions constant, q_1 – electric charge of the first particle, q_2 – electric charge of the second particle, r – distance between charged particles.

In respect to (1), two similar charges will repel, while two antagonistic particles will be attracted. In such a case, it should be assumed, that the given charge of the robot and obstacles is negative, while the target should be positive. This assumption allows to attract the platform by the goal, and repel by the obstacles. The considered interaction gives as a resultant a trajectory presented on Fig. 1.

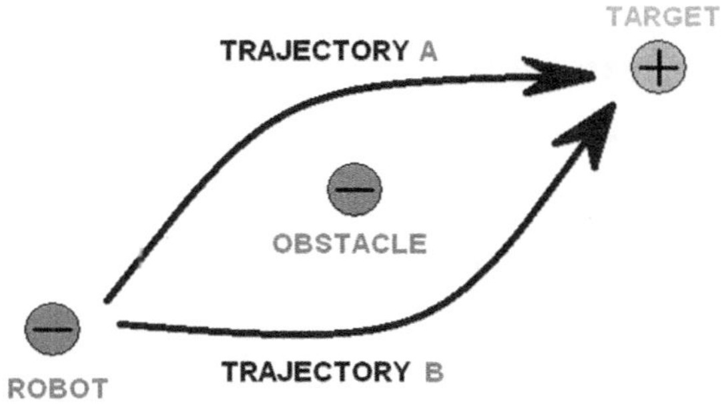

Fig. 1. The principle of artificial potential fields algorithm

In general, multiple obstacles can be present in a real environment, which makes it even more difficult to reach the final destination. To overcome this issue, the algorithm should taking into account all potential obstacles in the relation given in (1). Therefore the Coulomb-law may be extended to a following form:

$$\vec{F}_W = -\frac{k_{RT} q_R q_T}{r_{RT}^2} \frac{\vec{r}_{RT}}{r_{RT}} + \sum_{i=1}^{n} \frac{k_{RO} q_R q_O}{r_{ROi}^2} \frac{\vec{r}_{ROi}}{r_{ROi}} \tag{2}$$

where: R – mobile robot, T – target, O – obstacles, k_{RT} – interactions constant between mobile robots and target, k_{RO} – interactions constant between mobile robots and obstacles, q_R – mobile robot charge, q_O – obstacles charge, q_T – target charge, r_{RT} – distance between mobile robot and target, $\overrightarrow{r_{RT}}$ – directional vector from mobile robot to target, r_{ROi} – distance between mobile robot and obstacles, $\overrightarrow{r_{ROi}}$ – directional vector from mobile robot to obstacles.

In order to obtain simulation results, artificial potential field (APF) algorithm was implemented in Matlab environment [9]. As it was described in [6] an APF method suffers by the presence of local minima, which causes a risk of getting stuck in the actual position. In order to eliminate the described drawback, APF method can be enhanced by introducing a "Bug" method. This method involves bypassing of obstacles along its edges [7]. These will be described in the following subsection.

2.2 Extended Bug Algorithm

In order to overcome local minima on the track of the mobile robot, a "Bug" algorithm is employed [12, 13]. The extension introduces 4 additional states into the classical AFP algorithm (Table 1). In the initial step of the Bug method, the current position of the mobile robot is captured. This allows for determination of the distance between the platform and the target position – disRT (Fig. 2).

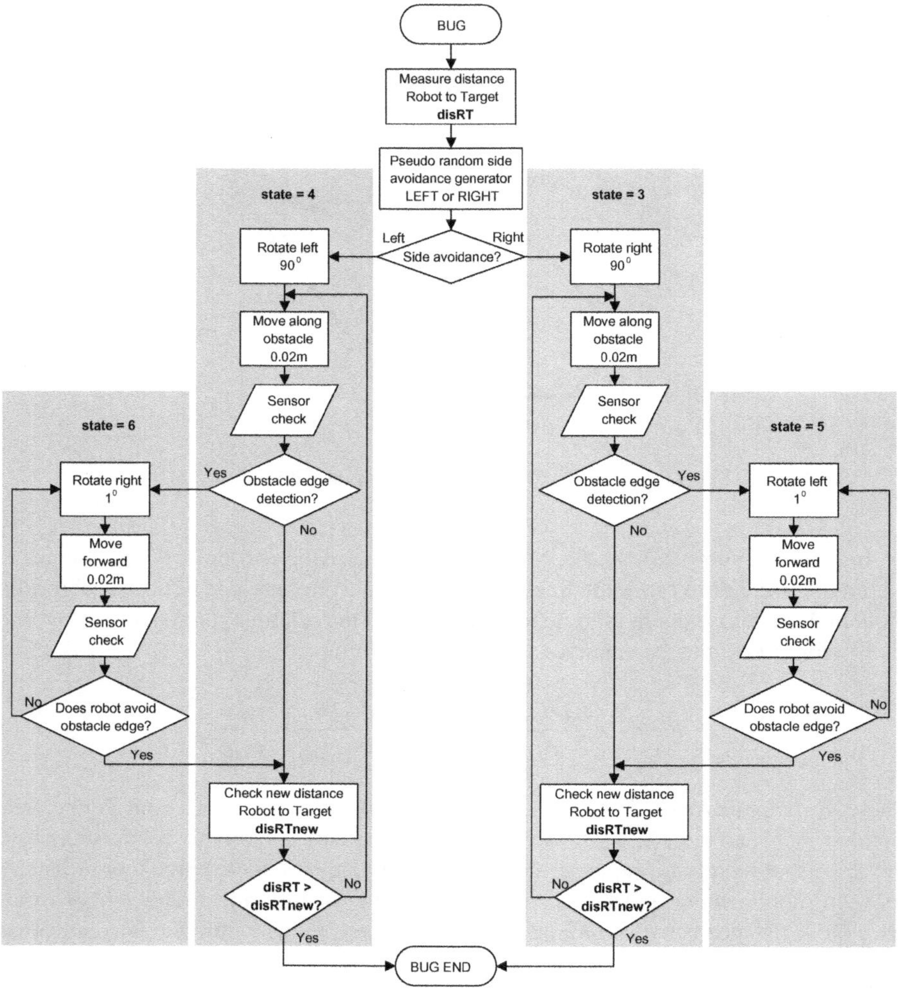

Fig. 2. The flow diagram of the Bug algorithm

The Bug algorithm (Fig. 2) is activated near obstacles, enabling to find a path if a local minimum occurs for APF method. In the initialization stage the distance between the robot and target is calculated. After that, a random function determines the movement direction, in order to get around obstacle. In such a case the state of the algorithm can take one of following values: 3 (right) or 4 (left). In one of this states, the robot moves parallel to the surface of the obstacle, maintaining a constant distance to it. The movement is realized until the platform reach a corner. In such a case the robot goes into state 5 or 6, which means a rotation counter- or clockwise, respectively. The rotation is realized in steps, 1 degree for each one until the current distance value to the target (disRTnew) is not smaller than the value at the initializing step (disRT). When this condition occurs,

Table 1. The states of the path planning algorithm.

No	State variable	Robot state	Comments
1.	0	Robot stop	
2.	1	Robot move forward	Move forward
3.	2	APF algorithm	New robot direction is calculated using artificial potential fields
4.	3	BUG algorithm right side	Move the robot longwise the obstacle in the right direction
5.	4	BUG algorithm left side	Move the robot longwise the obstacle on the left direction
6.	5	BUG algorithm right side rotate left	Move the robot along an arc in the right direction
7.	6	BUG algorithm left side rotate right	Move the robot along an arc in the left direction

the Bug algorithm is finished and the control algorithm goes back to APF (state 2). The proposed extension enables in a relatively simple way to bypass different types of obstacles.

2.3 Introduction of Ground Quality

Autonomous movement in an unknown area is a non-trivial task, since it requires advanced path planning and control algorithm. It should be mentioned, that in practice the quality of ground surface becomes an important issue. In extreme situations, it may be unable to reach the desired destination, if for example the robot will mire in the mud. In such a case the ground surface may be treated as two-dimensional obstacle, which is invisible for classical ultrasonic or optical sensors. In some cases, the employment of image analysis may also failure due to lack of knowledge about the friction coefficient. The initial recognition of ground quality may be realized by using information from wheels (e.g. angular velocity and torque). In such a case, the control algorithm can take into account the surface quality, which will allow to reach the target position. Therefore, the proposed APF-Bug algorithm has been enhanced by information related to the quality of the surface. Respective information was obtained by using ground identification method. In such a case, it was necessary to introduce additional states (Table 2) into the control system, corresponding to the above mentioned extension (Fig. 3).

3 Proposed Algorithm

The developed algorithm is realized using a state machine. The main states are presented in Tables 1 and 2. It is a combination of APF-Bug and ground quality identification (GQI). The flow chart of proposed improvement is shown in Fig. 4.

Table 2. The states of ground quality detect of the robot algorithms.

No	State variable	Robot state name	Comments
8.	7	Robot Ground Detect	Robot detect low quality surface
9.	8	Robot backward	Move backward
10.	9	Robot Ground Arc Left	Robot move arc style in left side
11.	10	Robot Ground Arc Right	Robot move arc style in right side

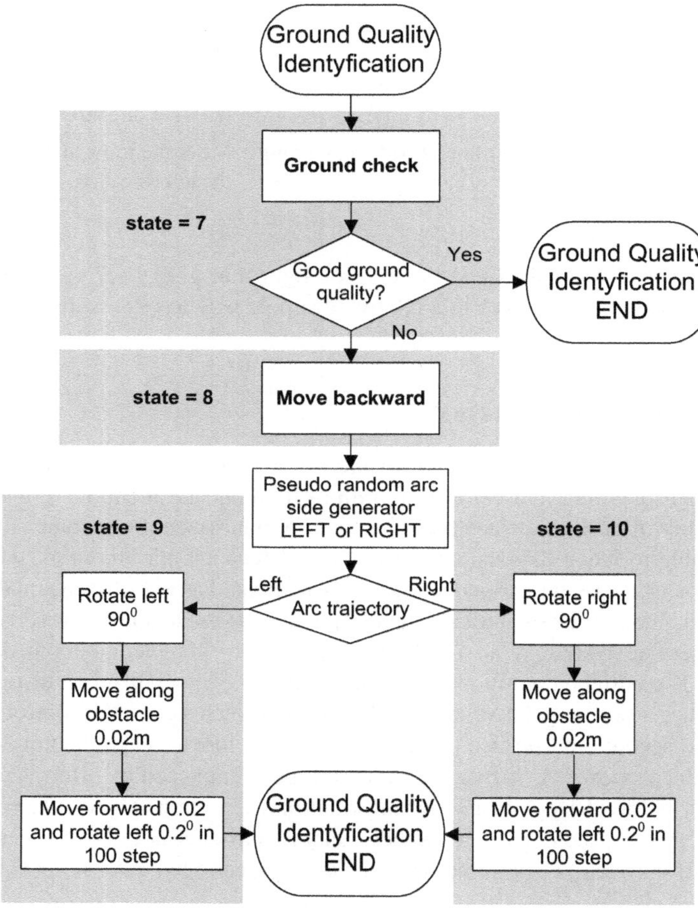

Fig. 3. The flow diagram of basics ground identification method

Mainly, the path is planned by using the APF algorithm, which is the core of the proposed solution. Therefore, the robot mainly operates in state no 2. Whereas, if a local minimum appears or the GQI detects a low quality of the ground, a state transition

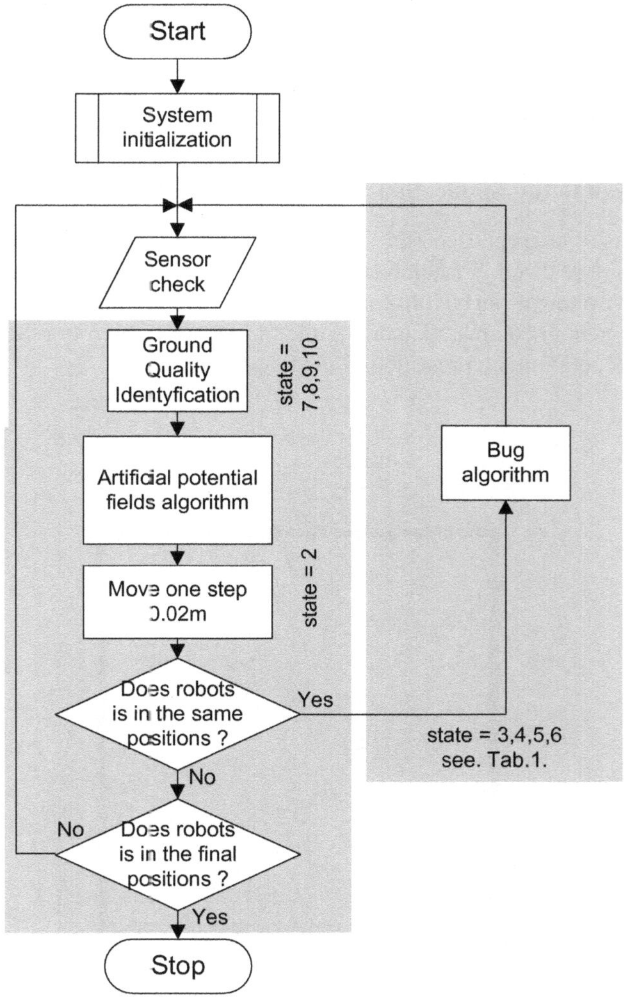

Fig. 4. The flow diagram of the proposed algorithm

occurs. In such a case, the algorithm goes into Bug or GQI stage, which corresponds to states no 3, 4 or 7, respectively. This means that the robot try to bypass an unknown obstacle to reach the target position. In the first case, the platform moves along the edge, so far as a corner will be detected. During operation in a Bug mode, the state is changed using additional states 5 and 6, depending on the movement direction. This occurs for vertical oriented obstacles, such as walls, trees, stones etc. In a case of low ground quality, the platform has to stop and move backwards, to leave the poor surface. In this operation mode, the robot has to bypass a horizontal oriented obstacle of unknown shape. In general, multiple periods of backwards movement may occur, as long as the algorithm moves around the area of low ground quality. In this operation mode, the state of the algorithm changes between states from 7 to 10.

4 Application and Simulation Results

4.1 Data Input to the Application

The initial tests of the proposed APF-BUG-GQI algorithm were realized using a developed application realized in Matlab software. Detailed information about the prepared simulation environment can be found in [12, 13]. It simulates the behaviour of the mobile platform in a predefined area, visualizing the trajectory of the robot. Several results of the algorithm operation are presented in Figs. 5, 6, 7 and 8.

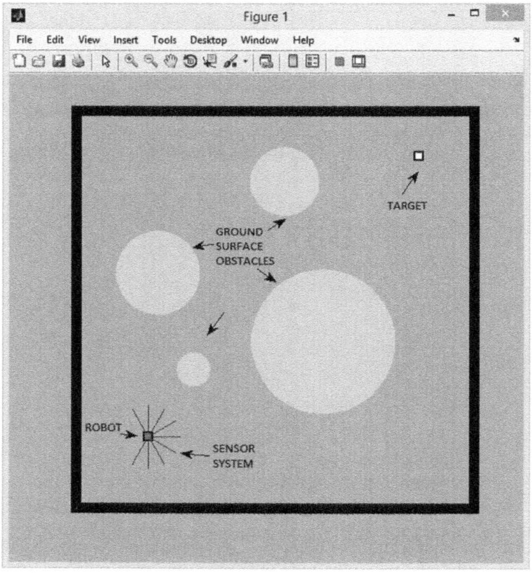

Fig. 5. The principle of artificial potential fields algorithm

4.2 Simulation with Interpretation

Environment with few obstacles is shown in Figs. 6, 7 and 8, respectively. Each of them contains arrows with local states of the mobile robot, according to Table 1 and Table 2.

In Fig. 6 the mobile robot began the movement at the START point (state 2). The platform moves forward to central obstacle. When the robot approaches it, the actual state changes to state no 3. In this position, a local minimum appears, because the platform is unable to follow the trajectory defined by the simple AFP algorithm. Therefore, the state no 3 initiates the BUG subroutine. In effect, the robot drives around the wall. When the platform reaches the end of the central obstacle, the subroutine goes into state no 5 and

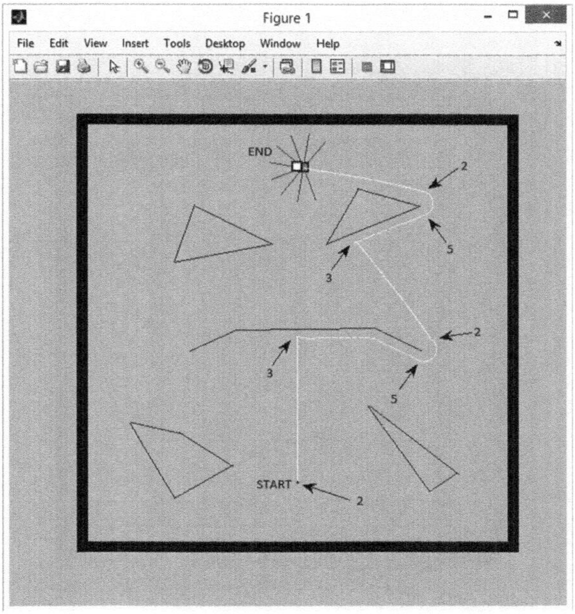

Fig. 6. Movement around 3D obstacles

the vehicle rotates to the left. Next, the subroutine ends and the movement is realized according to the APF algorithm until it encounters an another unavoidable obstacle, i.e. in the upper part of the map on the right side. At this point, the platform reaches the next wall and the BUG algorithm starts again – the state changes from no 2 to no 3. On the edge of the wall, the state changes again in order to rotate to the left (state no 5). After bypassing the last obstacle, the robot changes state to state no 3 and finally reaches the final point - END.

In Fig. 7 the considered case was extended by adding two additional areas with poor quality of the substrate. The mobile robot begins the movement at the START point (state no 2) and moves forward to the central obstacle. In the halfway, the platform enters an area with limited driving possibility. The quality indicator goes low and executes the GQI subroutine (state no 7). At first, the vehicle moves backward (state no 8). Later the mobile robot turns right and bypasses the difficult area by the right side. The state changes and the trajectory is generated using classical AFP algorithm. After reaching of the wall, the BUG subroutine is executed and the platform drives around it, like in the above case. In the next step the platform reaches a region with poor ground quality again, which executes the GQI subroutine for proper path generation. The states switch twice between no 7, 8 and 10, until the ground quality indicator goes again high. The state changes again to state no 2 and AFP method is realized. The vehicle reaches the final obstacle and the sensors detect it (state no 4), the wall is driven around by the left side. On the corner the robot turns right (state no 6) and after that it switches again to state no 2 – AFP subroutine is executed and finally the END point is reached.

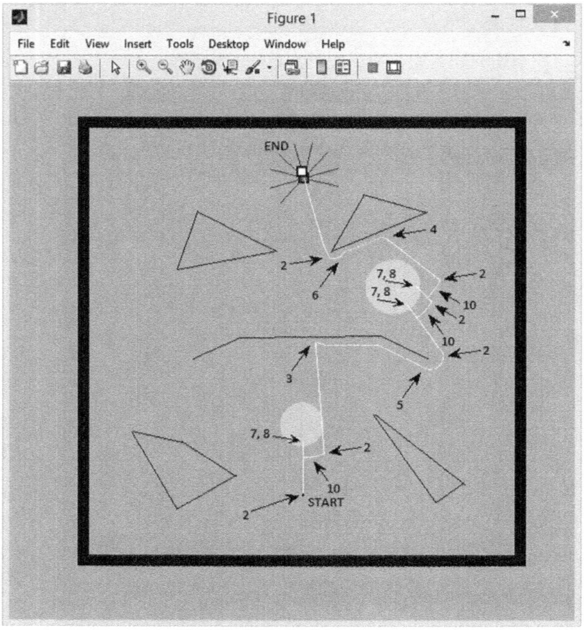

Fig. 7. Movement around ground obstacles and 3D obstacles

In Fig. 8 an another situation is investigated. In this case two regions of poor substrate quality and a single wall occurs. The mobile robot begins the movement at the START point (state no 2) and moves forward to the END point. The mobile platform drives into the poor ground which triggers the quality indicator low – the state changes to state no 7. Then the robot tries to move backwards (state no 8). After the reversing it drives along an arch to the right side (state no 10). The mobile platform drives forward to the END point using the APF algorithm (state no 2), a difficult ground area occurs again (state no 7). The mobile robots try to avoid the difficult ground area – the state switches between state no 8 and 10. Finally, the robot goes around this terrain (the ground quality indicator becomes high) and drives next to the END point using AFP algorithm (state no 2). Then the robot reaches the vertical wall and tries to bypass it using the BUG algorithm (state no 3). The platform drives around the obstacle on the right side. On the corner the state changes to state no 5 and the vehicle turns left to the direction of destination point. After that, the robot continues moving according to the APF algorithm (state no 2) but it finds another difficult ground area (state no 7). The robot bypasses the ground obstacle in one attempt and drives to the END point according to APF algorithm.

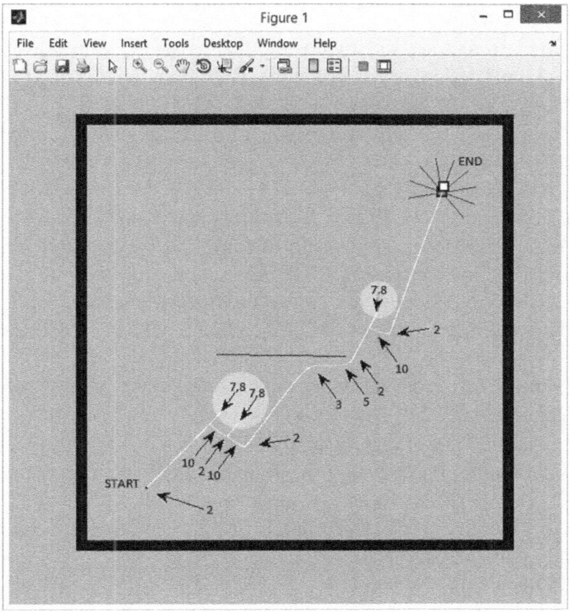

Fig. 8. Movement around ground obstacles and 3D obstacles

5 Summary

The first algorithmic approach to the identification of ground quality by an autonomous mobile robot is presented in this paper. The proposed solution has been implemented all of the three algorithms (APF-BUG-GQI) in the Matlab environment. Each of the presented examples shows the correctness and success of the proposed solution.

Information about the quality of the ground can be determined on wheel speed and load torque changes, also using advanced estimation techniques and artificial intelligence methods. It is planned to conduct experimental tests using an autonomous mobile robot and in a real environment.

References

1. Kozłowski, K., Kowalczyk, W.: Artificial potential based control for a large scale formation of mobile robots. In: Proceedings of the Fourth International Workshop on Robot Motion and Control (2004)
2. Skrzypczyński, P., Belter, D., Łabęcki, P.: Adaptive motion planning for autonomous rough terrain traversal with a walking robot. J. Field Rob. **33**(3), 337–370 (2016)
3. Weerakoon, T., Ishii, K., Nassiraei, A.A.F.: An artificial potential field based mobile robot navigation method to prevent from deadlock. J. Artif. Intell. Soft Comput. Res. **5**(3), 189–203 (2015)
4. Chołodowicz, E., Figurowski, D.: Mobile robot path planning with obstacle avoidance using particle swarm optimization. Pomiary Automatyka Robotyka **21**(3), 59–68 (2017)

5. Kowalczuk, Z., Duzinkiewicz, K.: Planowanie trajektorii ruchu zespołu robotów mobilnych z zastosowaniem metody warstwicowej. Fuzzy logic system in wheeled mobile robots formation path planning. Pomiary Automatyka Robotyka **54**(3), 140–144 (2008)
6. Lacroix, S., Mallet, A., Bonnafous, D., Bauzil, G., Fleury, S., Herrb, M., Chatila, R.: Autonomous rover navigation on unknown terrains: functions and integration. Int. J. Rob. Res. **21**(10), 917–942 (2002)
7. Kala, R., Warwick, K.: Planning autonomous vehicles in the absence of speed lanes using an elastic strip. IEEE Trans. Intell. Transp. Syst. **14**(4), 1743–1752 (2013)
8. Jiang, D.Z., Min, W.Z.: Mobile robot path tracking in unknown dynamic environment. In: Robotics, Automation and Mechatronics, IEEE Conference (2008)
9. Adeli, H., Tabrizi, M.H.N., Mazloomian, A., Hajipour, E., Jahed, M.: Path planning for mobile robots using iterative artificial potential field method. IJCSI Int. J. Comput. Sci. Issues **8**(4), 28–32 (2011)
10. Siegwart, R., Nourbakhsh, I.R.: Introductions to autonomous mobile robots. Massachusetts Institute of Technology, pp. 272–274 (2004)
11. Grabowska, E.: Application of terrestrial signal sources in supporting GPS system in tasks of engineering surveying. Prace Naukowe Politechniki Warszawskiej. Geodezja **43**, 37–53 (2008)
12. Wyrąbkiewicz, K., Tarczewski, T., Grzesiak, L.: Artificial potential fields algorithm for Mars rover path planning in an unknown environment. Poznan Univ. Technol. Acad. J. Electr. Eng. **80**, 183–189 (2014)
13. Wyrąbkiewicz, K., Tarczewski, T., Grzesiak, L.: Artifical potential fields with extended Bug algorithm for Mars rover path planning in an unknown environment. Comput. Appl. Elect. **12**, 422–433 (2014)
14. Washington, R., Golden, K., Bresina, J., Smith, D.E., Anderson, C., Smith, T.: Autonomous rovers for mars exploration. In: NASA Ames Research Center. 1999 IEEE Aerospace Conference. Proceedings (2002)

Computational Aspects and Applications of Advanced Control Algorithms

Tuning of Nonlinear MPC Algorithm for Vehicle Obstacle Avoidance

Robert Nebeluk(✉) and Maciej Ławryńczuk

Institute of Control and Computation Engineering, Warsaw University of Technology, ul. Nowowiejska 15/19, 00-665 Warsaw, Poland
R.Nebeluk@elka.pw.edu.pl, M.Lawrynczuk@ia.pw.edu.pl

Abstract. This work is concerned with tuning a nonlinear Model Predictive Control (MPC) algorithm. Typically, the weighting coefficients associated with the predicted control errors are constant for the consecutive sampling instants over the prediction horizon. This work discusses a tuning procedure whose objective is to find a set of coefficients which scale the influence of the consecutive predicted control errors. The coefficients are determined using an original simulation-based method. In order to demonstrate effectiveness of the method, an MPC algorithm for vehicle obstacle avoidance is developed. It is shown that the discussed tuning methods makes it possible to obtain much better control quality in comparison with the classical approach.

Keywords: Model Predictive Control · Tuning · Control system performance · Obstacle avoidance

1 Introduction

In Model Predictive Control (MPC) algorithms, at each of the consecutive discrete sampling instants, the best possible control policy is calculated on-line as a result of an optimisation process [16,17]. Typically, predicted control errors, i.e. the differences between the desired set-points and the predicted values of the controlled variables (process outputs) are minimised. Additionally, excessive increments of the manipulated variables (process inputs) may be penalised. Two great advantages of MPC algorithms are: the unique ability of taking into account constraints and very efficient control of Multiple Input Multiple Output (MIMO) processes. That is why MPC algorithms have been used for years in process control industrial applications, e.g. [3,9,13,20]. Currently, MPC algorithms are used in embedded systems, characterised by very short sampling times, of the order of milliseconds [1].

Actual performance of MPC heavily depends on tuning parameters. In general, two types of parameters must be selected. Firstly, it is necessary to find lengths of prediction and control horizons, but it may be accomplished using straightforward tuning methods, which depends on process dynamics [11,12]. Secondly, coefficients of the MPC cost-function which is minimised at every

A. Bartoszewicz et al. (Eds.): PCC 2020, AISC 1196, pp. 993–1005, 2020.
https://doi.org/10.1007/978-3-030-50936-1_83

sampling instant must be tuned. In particular, in the case of MIMO processes with strong cross-couplings this task is not easy as different set of parameters may result in completely different MPC action and process performance. In two early works [14,15] an analytical expression for calculation of penalties associated with increments of the manipulated variables has been formulated by approximating the process behaviour by a first-order-plus-dead-time block. A tuning method based on the Robust Performance Number (RPN) is described in [18]. An original approach which dynamically calculates set-points for MPC to accommodate user-defined output importance is described in [19]. The method may be regarded as more intuitive than selecting values for the MPC weighing coefficients. Nevertheless, the most frequent approach is to optimise off-line parameters of the MPC cost-function. Multi-objective performance optimisation by means of the goal attainment approach is considered in [2]. A thorough comparison of a few heuristic optimisation algorithms is reported in [10] (Particle Swarm Optimisation (PSO), firefly algorithm, grey wolf optimiser and Jaya algorithm have been used). An application of the PSO optimisation method to find parameters of MPC with model uncertainty is considered in [4]. An alternative tuning method, in which reference trajectory shaping is used to influence performance of the MPC-controlled process is described in [6].

Typically, the weighting coefficients associated with the predicted control errors are constant for the consecutive sampling instants over the prediction horizon. This work discusses a tuning procedure whose objective is to find a set of coefficients which scale the influence of the consecutive predicted control errors. Unlike the typical approach, these coefficients vary within the prediction horizon in order to increase control accuracy. The coefficients are determined using an original simulation-based method. In order to demonstrate effectiveness of the method, an MPC algorithm for vehicle obstacle avoidance is developed.

2 Multiple-Input Multiple-Output MPC Problem Formulation

It is assumed that the general MIMO processes has n_{u} inputs (manipulated variables), n_{x} state variables and n_{y} outputs (controlled variables). Hence, the process input vector is $u = [u_1 \ldots u_{n_{\mathrm{u}}}]^{\mathrm{T}}$, the state vector is $x = [x_1 \ldots x_{n_{\mathrm{x}}}]^{\mathrm{T}}$ and the output vector is $y = \left[y_1 \ldots y_{n_{\mathrm{y}}}\right]^{\mathrm{T}}$. For presentation, in this work two notation methods are used: scalars and vectors.

The vector of decision variables calculated in MPC at each sampling instant, $k = 0, 1, 2, \ldots$, has the length of $n_{\mathrm{u}}N_{\mathrm{u}}$ and consists of the current and future increments of the manipulated variables

$$\triangle \boldsymbol{u}(k) = \begin{bmatrix} \triangle u(k|k) \\ \vdots \\ \triangle u(k+N_{\mathrm{u}}-1|k) \end{bmatrix} \tag{1}$$

where the symbol $\triangle u(k+p|k)$ stands for increments of the manipulated variables for the future sampling instant $k+p$ calculated at the current instant k, N_{u} is the control horizon and $\triangle u(k+p|k) = 0$ for $p \geq N_{\rm u}$. The decision variables of MPC (1) are calculated as a result of minimisation of the MPC cost-function

$$J(k) = \sum_{p=1}^{N} \sum_{n=1}^{n_{\rm y}} \psi_{p,n} \left(y_n^{\rm sp}(k+p|k) - \hat{y}_n(k+p|k)\right)^2 + \sum_{p=0}^{N_{\rm u}-1} \sum_{n=1}^{n_{\rm u}} \lambda_{p,n} \left(\triangle u_n(k+p|k)\right)^2 \quad (2)$$

The symbols $y_n^{\rm sp}(k+p|k)$ and $\hat{y}_n(k+p|k)$ denote the set-point trajectory and the predicted trajectory for the sampling instant $k+p$ calculated at the instant k, which means that the first part of the MPC cost-function measures the future control errors for all process outputs over the prediction horizon N. These errors are weighted by means of the coefficients $\psi_{p,n} \geq$, $p = 1, \ldots, N$, $n = 1, \ldots, n_{\rm y}$. The second part of the MPC cost-function is a penalty term used to penalise excessive changes of the manipulated variables. These changes are weighted by means of the coefficients $\lambda_{p,n} > 0$, $p = 1, \ldots, N_{\rm u}$, $n = 1, \ldots, n_{\rm u}$. The MPC cost-function (2) may be conveniently expressed in the matrix-vector forms

$$J(k) = \sum_{p=1}^{N} \|y^{\rm sp}(k+p|k) - \hat{y}(k+p|k)\|^2_{\boldsymbol{\Psi}_p} + \sum_{p=0}^{N_{\rm u}-1} \|\triangle u(k+p|k)\|^2_{\boldsymbol{\Lambda}_p} \quad (3)$$

where the matrices $\boldsymbol{\Psi}_p = \mathrm{diag}(\psi_{p,1}, \ldots, \psi_{p,n_{\rm y}})$ and $\boldsymbol{\Lambda}_p = \mathrm{diag}(\lambda_{p,1}, \ldots, \lambda_{p,n_{\rm u}})$ are of dimensionality $n_{\rm y} \times n_{\rm y}$ and $n_{\rm u} \times n_{\rm u}$, respectively. Typically, constant coefficients $\psi_{1,n} = \ldots = \psi_{N,n}$ are used for the whole prediction horizon and constant coefficients $\lambda_{0,n} = \ldots = \lambda_{N_{\rm u}-1,n}$ for the whole control horizon. Equation (3) may be further transformed to

$$J(k) = \|\boldsymbol{y}^{\rm sp}(k) - \hat{\boldsymbol{y}}(k)\|^2_{\boldsymbol{\Psi}} + \|\triangle \boldsymbol{u}(k)\|^2_{\boldsymbol{\Lambda}} \quad (4)$$

where the set-point and predicted trajectories of the controlled variables are vectors of length $n_{\rm y}N$

$$\boldsymbol{y}^{\rm sp}(k) = \begin{bmatrix} y^{\rm sp}(k+1|k) \\ \vdots \\ y^{\rm sp}(k+N|k) \end{bmatrix}, \ \hat{\boldsymbol{y}}(k) = \begin{bmatrix} \hat{y}(k+1|k) \\ \vdots \\ \hat{y}(k+N|k) \end{bmatrix} \quad (5)$$

The weighting matrices $\boldsymbol{\Psi} = \mathrm{diag}(\boldsymbol{\Psi}_1, \ldots, \boldsymbol{\Psi}_N)$ and $\boldsymbol{\Lambda} = \mathrm{diag}(\boldsymbol{\Lambda}_0, \ldots, \boldsymbol{\Lambda}_{N_{\rm u}-1})$ are of dimensionality $n_{\rm y}N \times n_{\rm y}N$ and $n_{\rm u}N_{\rm u} \times n_{\rm u}N_{\rm u}$, respectively.

Although at each sampling instant as many as $n_{\rm u}N_{\rm u}$ future increments of the manipulated variables (1) are calculated, only the increments for the current sampling instant are applied to the process, i.e. $u(k) = \triangle u(k|k) + u(k-1)$. At the next sampling instant, $k+1$, the measurements of the process outputs are updated and the whole optimisation procedure is repeated.

Typically, the MPC cost-function is minimised subject to constraints. The rudimentary constrained MPC optimisation problem is

$$\begin{aligned} &\min_{\triangle \boldsymbol{u}(k)} \{J(k)\} \\ &\text{subject to} \\ &u^{\min} \leq u(k+p|k) \leq u^{\max},\ p=0,\ldots,N_{\mathrm{u}}-1 \\ &-\triangle u^{\max} \leq \triangle u(k+p|k) \leq \triangle u^{\max},\ p=0,\ldots,N_{\mathrm{u}}-1 \\ &y^{\min} \leq \hat{y}(k+p|k) \leq y^{\max},\ p=1,\ldots,N \end{aligned} \tag{6}$$

The symbols $u^{\min}$ and $u^{\max}$ denote minimal and maximal values of the manipulated variables, $\triangle u^{\max}$ stands for the maximal change rate of the manipulated variables, $y^{\min}$ and $y^{\max}$ denote minimal and maximal predicted values of the controlled variables. The matrix-vector form of the MPC optimisation problem (6) is

$$\begin{aligned} &\min_{\triangle \boldsymbol{u}(k)} \{J(k)\} \\ &\text{subject to} \\ &\boldsymbol{u}^{\min} \leq \boldsymbol{J}\triangle\boldsymbol{u}(k)+\boldsymbol{u}(k-1) \leq \boldsymbol{u}^{\max} \\ &-\triangle\boldsymbol{u}^{\max} \leq \triangle\boldsymbol{u}(k) \leq \triangle\boldsymbol{u}^{\max} \\ &\boldsymbol{y}^{\min} \leq \hat{\boldsymbol{y}}(k) \leq \boldsymbol{y}^{\max} \end{aligned} \tag{7}$$

where the matrix $\boldsymbol{J}$ of dimensionality $n_{\mathrm{u}}N_{\mathrm{u}} \times n_{\mathrm{u}}N_{\mathrm{u}}$ and the vector of length $n_{\mathrm{u}}N_{\mathrm{u}}$ are

$$\boldsymbol{J} = \begin{bmatrix} \boldsymbol{I}_{n_{\mathrm{u}}\times n_{\mathrm{u}}} & \boldsymbol{0}_{n_{\mathrm{u}}\times n_{\mathrm{u}}} & \cdots & \boldsymbol{0}_{n_{\mathrm{u}}\times n_{\mathrm{u}}} \\ \boldsymbol{I}_{n_{\mathrm{u}}\times n_{\mathrm{u}}} & \boldsymbol{I}_{n_{\mathrm{u}}\times n_{\mathrm{u}}} & \cdots & \boldsymbol{0}_{n_{\mathrm{u}}\times n_{\mathrm{u}}} \\ \vdots & \vdots & \ddots & \vdots \\ \boldsymbol{I}_{n_{\mathrm{u}}\times n_{\mathrm{u}}} & \boldsymbol{I}_{n_{\mathrm{u}}\times n_{\mathrm{u}}} & \cdots & \boldsymbol{I}_{n_{\mathrm{u}}\times n_{\mathrm{u}}} \end{bmatrix},\ \boldsymbol{u}(k-1) = \begin{bmatrix} u(k-1) \\ \vdots \\ u(k-1) \end{bmatrix} \tag{8}$$

the vectors of length $n_{\mathrm{u}}N_{\mathrm{u}}$ are

$$\boldsymbol{u}^{\min} = \begin{bmatrix} u^{\min} \\ \vdots \\ u^{\min} \end{bmatrix},\ \boldsymbol{u}^{\max} = \begin{bmatrix} u^{\max} \\ \vdots \\ u^{\max} \end{bmatrix},\ \triangle\boldsymbol{u}^{\max} = \begin{bmatrix} \triangle u^{\max} \\ \vdots \\ \triangle u^{\max} \end{bmatrix} \tag{9}$$

and the vectors of length $n_{\mathrm{y}}N$ are

$$\boldsymbol{y}^{\min} = \begin{bmatrix} y^{\min} \\ \vdots \\ y^{\min} \end{bmatrix},\ \boldsymbol{y}^{\max} = \begin{bmatrix} y^{\max} \\ \vdots \\ y^{\max} \end{bmatrix} \tag{10}$$

3 MPC Tuning Procedure

It is postulated that the sequences of the coefficients $\psi_{1,n}, \ldots, \psi_{N,n}$ (for the consecutive process outputs, $n = 1, \ldots, n_y$) form Gauss-like functions. It has been demonstrated in [7] that such functions improve most the control performance for nonlinear processes and other shapes such as bell–shaped, triangular etc. produce similar results to Gauss ones. At first, a very general approximation is used

$$\psi_{p,n} = \begin{cases} K_n & \text{if } p = m_n \\ 1 & \text{if } p \neq m_n \end{cases} \tag{11}$$

The function (11) is characterised by two parameters: m_n and K_n. The first one defines the chosen sampling instant within the prediction horizon for which the weighting parameter is tuned (and has the value of K_n), the rest of them have default values 1. When the parameters m_n and K_n are selected, the trajectory of the weighs is calculated from the Gauss function

$$\psi_{p,n} = K_n \exp\left(-\left(\frac{p - m_n}{a_n}\right)^2\right) \tag{12}$$

where a_n defines the spread. The parameters of each trajectory are chosen based on the value of total control error, for each controlled variable, described as

$$E_n = \sum_{k=1}^{k_{\max}} (y_n^{\text{sp}}(k) - y_n(k))^2 \tag{13}$$

where $k_{\max}$ is the last time instant of the simulation.

The detailed steps of the tuning procedure are:

1. The best trajectory (11) is found, i.e. its parameters m_n and K_n are determined.
   a) The best value of the parameter m_n can be found by performing a few experiments with an assumed constant value of the parameter K_n and then changing the value of the parameter m_n until the lowest value of the total control error is achieved. It is recommended to start the tests from $m_n = N/2$ and analyse the results obtained in some neighbourhood of this value first.
   b) The best value of the parameter K_n can be found by performing a few experiments until the lowest value of the total control error is achieved. During these experiments the chosen parameter m_n is used.
2. The best trajectory (12) is found, i.e. its parameter a_n is determined. During these experiments the chosen parameters m_n and K_n are used. It is recommended to start from a thin shape of a chosen trajectory of the weighting coefficients and slowly increase its width. The value of the parameter a_n is chosen for which the lowest value of the total control error is achieved.
3. The best trajectory (12) is used in the considered MPC algorithm.

For MIMO processes the above steps are repeated for the consecutive outputs, i.e. for $n = 1, \dots, n_{\mathrm{y}}$, in order to find each trajectory of weighting parameters. Each trajectory is found separately, the other ones must remain unchanged.

4 Simulation Results

4.1 Vehicle Model

The continuous-time model of the considered vehicle is defined by the following differential equations [8]

$$\dot{x}_{\mathrm{veh}}(t) = -v(t)\sin(\theta(t))\theta(t) + \cos(\theta(t))v(t) \tag{14}$$

$$\dot{y}_{\mathrm{veh}}(t) = v(t)\cos(\theta(t))\theta(t) + \sin(\theta(t))v(t) \tag{15}$$

$$\dot{\theta}(t) = \frac{\tan(\delta(t))}{C_{\mathrm{L}}}v(t) + \frac{v(t)(\tan^2(\delta(t))+1)}{C_{\mathrm{L}}}\delta(t) \tag{16}$$

$$\dot{v}(t) = 0.5T \tag{17}$$

The vehicle has a rectangular shape with a length of $C_{\mathrm{L}} = 5$ m and width of 2 m. The state variables are: x_{veh} – global X position of the vehicle centre, y_{veh} – global Y position of the vehicle centre, θ – heading angle of the vehicle (0 when facing east, counter-clockwise positive), v – speed of the vehicle (positive). y_{veh} and v are the controlled variables. The manipulated variables are: T – throttle (positive when accelerating, negative when decelerating), δ – steering angle (0 when aligned with car, counter-clockwise positive). All things considered, the state, manipulated and controlled vectors are: $x = [x_{\mathrm{veh}}\ y_{\mathrm{veh}}\ \theta\ v]^{\mathrm{T}}$, $u = [T\ \delta]^{\mathrm{T}}$ and $y = [y_{\mathrm{veh}}\ v]^{\mathrm{T}}$, respectively. The initial state is $[0,\ 0,\ 0,\ 20]$, the initial vector of the manipulated variables is $[0,\ 0]$, which means that the vehicle drives east at a constant speed of 20 metres per second. From the continuous-time model (14)–(17) a discrete-time one is obtained using the zero-order holder method and next used in MPC.

4.2 Vehicle Obstacle Avoidance MPC Problem Formulation

The following obstacle avoidance problem is considered:

a) The road is straight and has 3 lanes.
b) Each lane is 4 m wide.
c) The vehicle drives in the middle of the centre lane when not avoiding any obstacles,
d) The obstacle is a non-moving object in the middle of the centre lane with the same size as the vehicle..

e) The vehicle passes an obstacle only from the left (fast) lane
f) A virtual safe zone around the obstacle is created which makes it possible that the vehicle does not get too close to the obstacle when passing it. The safe zone is centred on the obstacle. Its length is 10 m, its width is 8 m.
g) The lidar can detect an obstacle 30 m in front of the vehicle.

In the minimised MPC cost-function the predicted control errors of two controlled variables, i.e. y_{veh} and v, are taken into account. Taking into account the lower and upper bound on the position y_{veh}, one obtains the constrains

$$x_2^{\min} - \leq \hat{x}_2(k+p|k) \leq x_2^{\max} \tag{18}$$

for $p = 1, \ldots, N$, where $x_2^{\min} = -6$ m, $x_2^{\max} = 6$ m. Moreover, the obstacle avoidance constraint is necessary [8]

$$c_{\text{x}}\hat{x}_1(k+p|k) - \hat{x}_2(k+p|k) + c_{\text{i}} \leq 0 \tag{19}$$

for $p = 1, \ldots, N$, where c_{x} and c_{i} are constraint slope and constraint intercept factors. To eliminate the infeasibility problems, all constraints are formulated as soft ones

$$x_2^{\min} - \varepsilon^{\min}(k) \leq \hat{x}_2(k+p|k) \leq x_2^{\max} + \varepsilon^{\max}(k) \tag{20}$$

and

$$c_{\text{x}}\hat{x}_1(k+p|k) - \hat{x}_2(k+p|k) + c_{\text{i}} - \varepsilon^{\text{x}}(k) \leq 0 \tag{21}$$

where $\varepsilon^{\min}(k) \geq 0$, $\varepsilon^{\max}(k) \geq 0$ and $\varepsilon^{\text{x}}(k) \geq 0$ are additional decision variables of the MPC optimisation problem. The resulting MPC optimisation problem is

$$\begin{aligned}
&\min_{\substack{\triangle \boldsymbol{u}(k) \\ \varepsilon^{\min}(k) \\ \varepsilon^{\max}(k) \\ \varepsilon^{\text{x}}(k)}} \left\{ J(k) = \left\|\boldsymbol{y}^{\text{sp}}(k) - \hat{\boldsymbol{y}}(k)\right\|_{\boldsymbol{\Psi}}^2 + \lambda \left\|\triangle \boldsymbol{u}(k)\right\|_{\boldsymbol{\Lambda}}^2 \right. \\
&\qquad\qquad \left. + \ \rho^{\min}(\varepsilon^{\min}(k))^2 + \rho^{\max}(\varepsilon^{\max}(k))^2 + \rho^{\text{x}}(\varepsilon^{\text{x}}(k))^2 \right\} \\
&\text{subject to} \\
&\boldsymbol{u}^{\min} \leq \boldsymbol{J}\triangle \boldsymbol{u}(k) + \boldsymbol{u}(k-1) \leq \boldsymbol{u}^{\max} \\
&\triangle \boldsymbol{u}^{\min} \leq \triangle \boldsymbol{u}(k) \leq \triangle \boldsymbol{u}^{\max} \\
&\boldsymbol{x}_2^{\min} - \boldsymbol{\varepsilon}^{\min}(k) \leq \hat{\boldsymbol{x}}_2(k) \leq \boldsymbol{x}_2^{\max} + \boldsymbol{\varepsilon}^{\max}(k) \\
&c_{\text{x}}\hat{\boldsymbol{x}}_1(k) - \hat{\boldsymbol{x}}_2(k) + c_{\text{i}}\mathbf{1}_{N\times 1} - \boldsymbol{\varepsilon}^{\text{x}}(k) \leq \mathbf{0}_{N\times 1} \\
&\varepsilon^{\min}(k) \geq 0, \ \varepsilon^{\max}(k) \geq 0, \ \varepsilon^{\text{x}}(k) \geq 0
\end{aligned} \tag{22}$$

where $\rho^{\min} > 0$, $\rho^{\max} > 0$, $\rho^{\text{x}} > 0$ are constants and the vectors of length N are: $\boldsymbol{\varepsilon}^{\min}(k) = \varepsilon^{\min}(k)\,[1\ldots 1]^{\text{T}}$, $\boldsymbol{\varepsilon}^{\max}(k) = \varepsilon^{\max}(k)\,[1\ldots 1]^{\text{T}}$, $\boldsymbol{\varepsilon}^{\text{x}}(k) = \varepsilon^{\text{x}}(k)\,[1\ldots 1]^{\text{T}}$. The nonlinear MPC optimisation problem (22) is solved by means of the Sequential Quadratic Programming method.

5 Tuning of Vehicle Obstacle Avoidance MPC

Constant parameters are: $N = 25$, $N_\mathrm{u} = 5$, $\lambda_1 = 1$, $\lambda_2 = 10000$. The parameters $\psi_{p,1}$ and $\psi_{p,2}$ are chosen by means of two methods. In the first one they have been arbitrarily set to $\psi_{p,1} = \psi_{p,2} = 1$, $p = 1, \ldots, N$ (no tuning), in the second one the tuning procedure described in Sect. 3 has been used.

During tuning the coefficients $\psi_{1,1}, \ldots, \psi_{N,1}$ and $\psi_{1,2}, \ldots, \psi_{N,2}$ are selected. The first step is to find the best values of the parameters m_1 and m_2 for an assumed constant values of the parameters K_1 and K_2. The parameters m_1 and m_2 are changed for each experiment and the obtained predicted control error is analysed. Let E_1 and E_2 denote the predicted errors for the first and the second process outputs, respectively. To reduce the number of tests it is recommended to start from $m_1 = m_2 = N/2$ and analyse at first the results obtained in some neighbourhood of these values. However, to show how the predicted errors depend on the different value of parameters m_1 and m_2 the results for all possible experiments are shown in Fig. 1. The trajectory $\psi_{p,1}$ uses the constant parameter $K_1 = 20$ while the trajectory $\psi_{p,2}$ during these tests remains constant. The error E_1 for the first controlled variable has bigger priority than the second one because the goal is to have efficient obstacle avoidance. The parameter $m_1 = 14$ is chosen for which the lowest value of E_1 is obtained. In the second step the parameter K_1 is found, the parameter $m_1 = 14$ is constant. Figure 2 shows the errors for different values of the parameter K_1. The parameter $K_1 = 25$ is chosen instead of $K_1 = 70$ for which the lowest value of the error E_1 is obtained. Larger values of K_n may result in large overshoot and always increases the amplitudes of the manipulated variables which may lead to infeasibility [7]. The error E_1 increases for $K_1 = 30$ and the parameter then has larger value, hence $K_1 = 25$ is chosen. Finally, the parameter a_1 of the trajectory $\psi_{p,1}$ is found, using the obtained parameters $m_1 = 14$ and $K_1 = 25$ which are constant. Figure 3 shows the influence of the parameter a_1 on the predicted errors. The parameter $a_1 = 10$ is chosen for which the lowest value of the error E_1 is obtained.

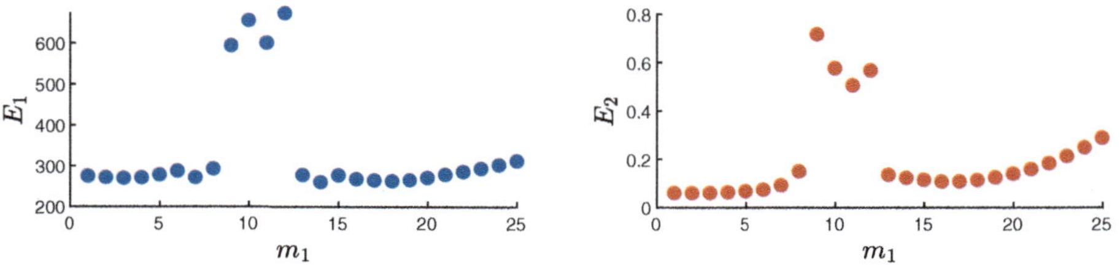

Fig. 1. Tuning the trajectory $\psi_{p,1}$: comparison of control system performance for different values of the parameter m_1; $K_1 = 20$

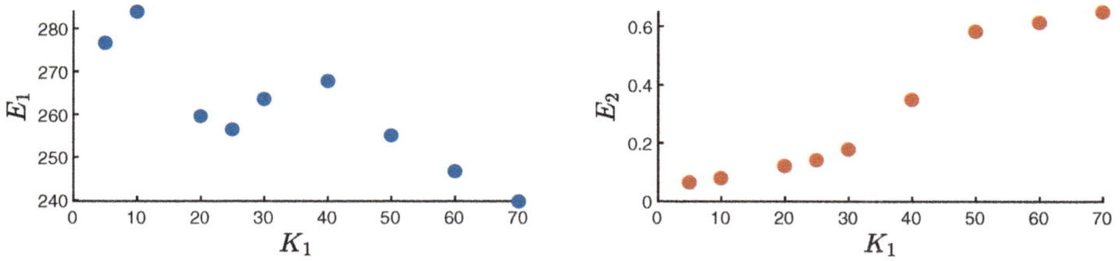

Fig. 2. Tuning the trajectory $\psi_{p,1}$: comparison of control system performance for different values of the parameter K_1; $m_1 = 14$

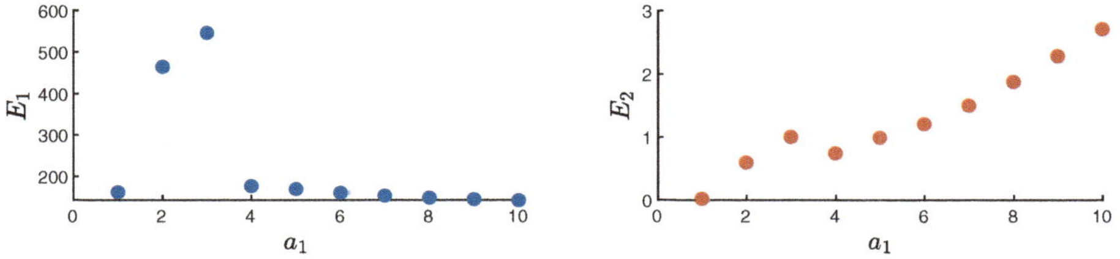

Fig. 3. Tuning the trajectory $\psi_{p,1}$: comparison of control system performance for different values of the parameter a_1; $K_1 = 25$, $m_1 = 14$

The trajectory $\psi_{p,2}$ is found similarly. Figure 4 shows the influence of the parameter m_2 on the predicted control errors, the assumed parameter $K_2 = 20$ is used. The parameter $m_2 = 1$ is chosen for which the lowest value of E_1 is obtained. In the second step the parameter K_2 is found, the parameter $m_2 = 1$ is constant. Figure 5 shows the errors for different values of the parameter K_2. The parameter $K_2 = 1$ is chosen for which the lowest value of the error E_1 is obtained. Finally, the parameter a_2 is found, the parameters $m_2 = 1$ and $K_2 = 1$ are constant. Figure 6 shows the influence of the parameter a_2 on the predicted errors. The value $a_2 = 1$ is chosen for which the lowest value of the error E_1 is obtained. The influence on total errors by changing each parameter of the trajectories is shown and this relationship is nonlinear as shown in Fig. 4.

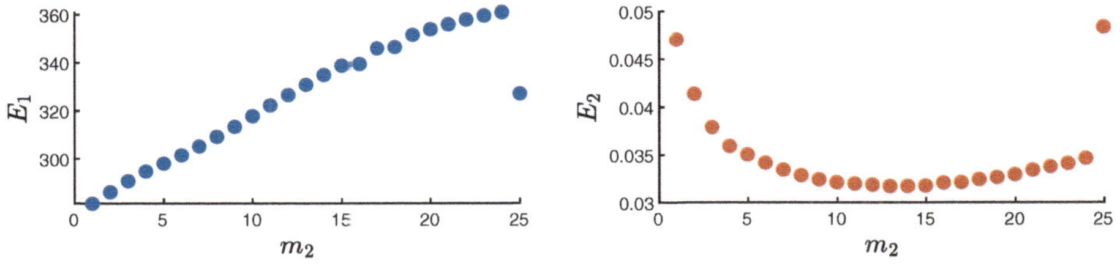

Fig. 4. Tuning the trajectory $\psi_{p,2}$: comparison of control system performance for different values of the parameter m_2; $K_2 = 20$

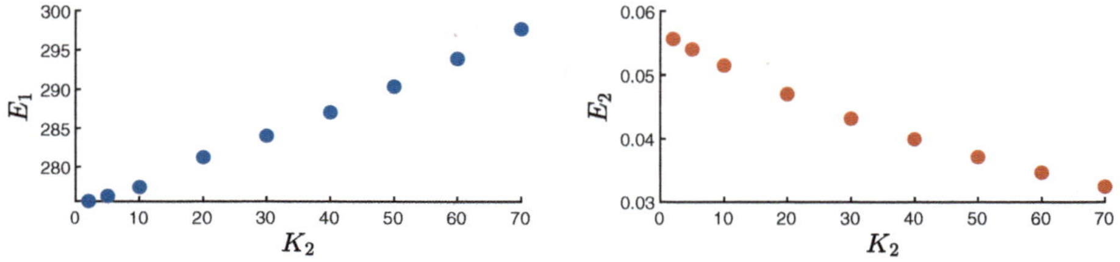

Fig. 5. Tuning the trajectory $\psi_{p,2}$: comparison of control system performance for different values of the parameter K_2; $m_2 = 1$

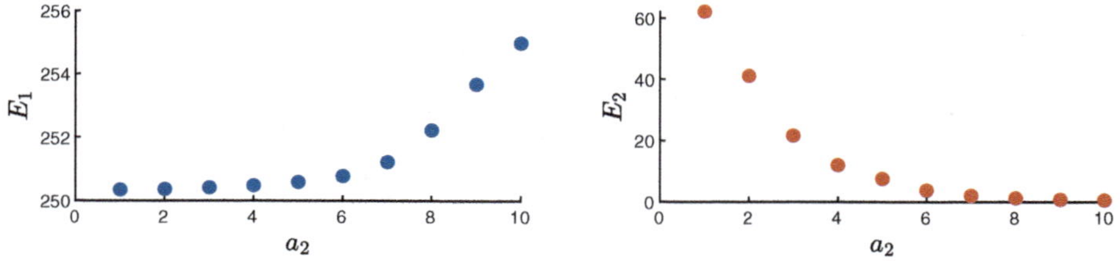

Fig. 6. Tuning the trajectory $\psi_{p,2}$: comparison of control system performance for different values of the parameter a_2; $K_2 = 1$, $m_2 = 1$

All things considered, Fig. 7 depicts shapes of the tuned trajectories $\psi_{p,1}$ and $\psi_{p,2}$. Simulation results before and after tuning are shown in Fig. 8. Two observations may be made. Firstly, the vehicle really bypasses the obstacle. Secondly, the tuned algorithm gives better control performance, i.e. its trajectories are faster and characterised by smaller overshoot. The improvement of control quality is guaranteed, regardless of different obstacle position or set-point trajectory of the first controlled variable since the parameters of trajectories $\psi_{p,1}$ and $\psi_{p,2}$ are chosen mostly for the lowest value of the control error E_1.

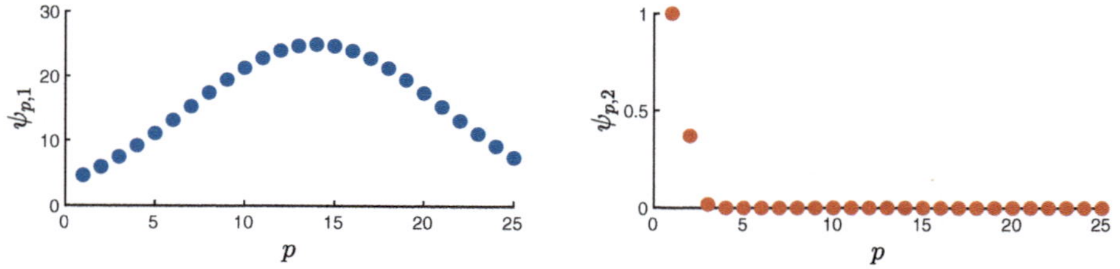

Fig. 7. The trajectories $\psi_{p,1}$ and $\psi_{p,2}$ which result in the best control performance

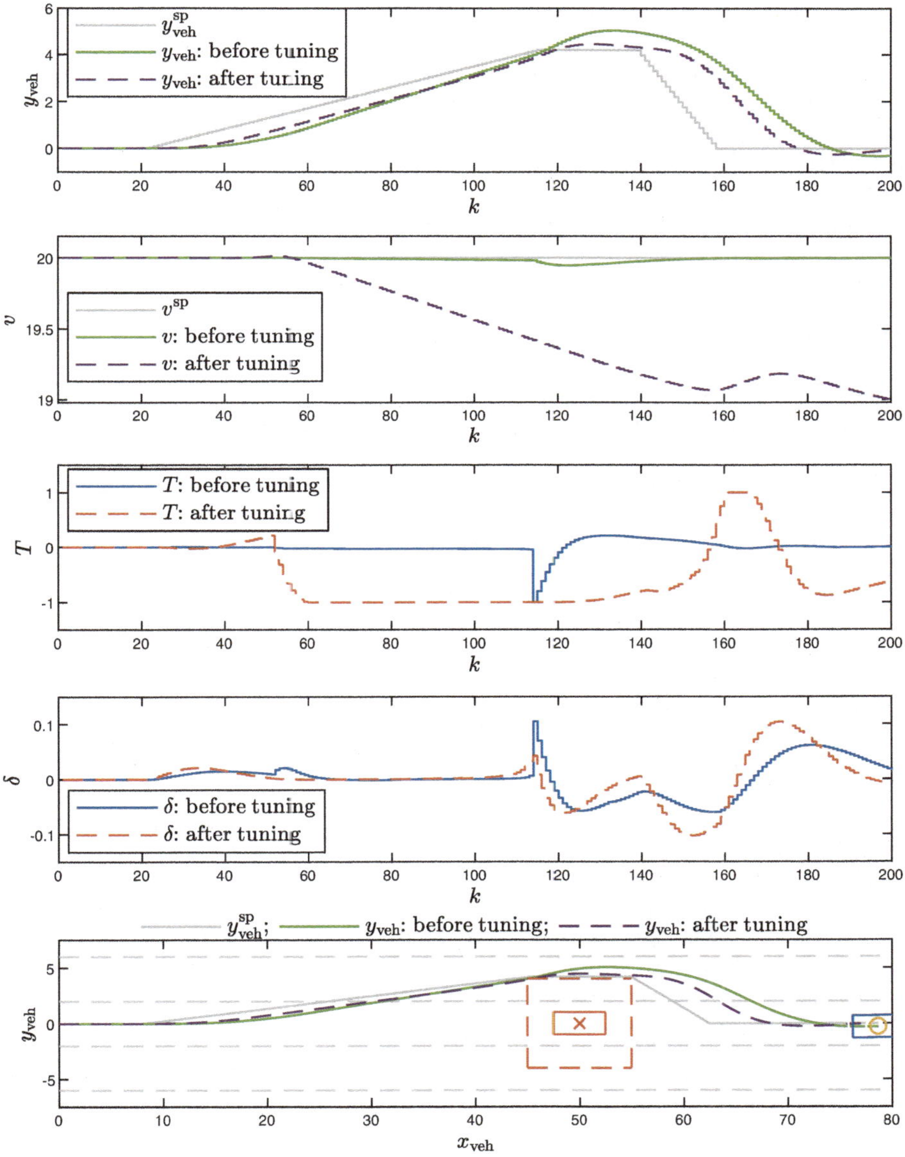

Fig. 8. Simulation results before and after tuning

6 Conclusions

This work presents a very efficient tuning approach of a nonlinear MPC algorithm. Next, its usefulness is demonstrated for vehicle obstacle avoidance. Although the discussed tuning method is quite simple, it results in much better control quality in comparison with the classical approach in which all predicted control errors over the prediction horizon are taken into account in the same way. It must be emphasised that during tuning numerical optimisation is not performed, the tuning method relies on simulations only. In future, it is planned to develop a computationally efficient MPC algorithm for the considered process in which simple quadratic optimisation is used on-line in place of nonlinear optimisation [5].

References

1. Chaber, P., Ławryńczuk, M.: Fast analytical model predictive controllers and their implementation for STM32 arm microcontroller. IEEE Trans. Industr. Inf. **15**, 4580–4590 (2019)
2. Exadaktylos, V., Taylor, C.J.: Multi-objective performance optimisation for model predictive control by goal attainment. Int. J. Control **83**, 1374–1386 (2010)
3. Grosso, J.M., Ocampo-Martinez, C., Puig, V.: Reliability-based economic model predictive control for generalised flow-based networks including actuators' health-aware capabilities. Int. J. Appl. Math. Comput. Sci. **26**, 361–654 (2016)
4. Júnior, G.A., Martins, M.A.F., Kalid, R.: A PSO-based optimal tuning strategy for constrained multivariable predictive controllers with model uncertainty. ISA Trans. **53**, 560–567 (2014)
5. Ławryńczuk, M.: Computationally Efficient Model Predictive Control Algorithms: A Neural Network Approach. Studies in Systems, Decision and Control, vol. 3. Springer, Cham (2014)
6. Nebeluk, R., Marusak, P.: Influencing predictive control system performance by reference trajectory shaping. Pomiary Automatyka Robotyka **23**, 21–30 (2019). (in Polish)
7. Nebeluk, R., Marusak, P.: Efficient MPC algorithms with variable trajectories of parameters weighting predicted control errors. Arch. Control Sci. (in review)
8. Obstacle Avoidance Using Adaptive Model Predictive Control: Model Predictive Control Toolbox for MATLAB. https://www.mathworks.com/help/mpc/ug/obstacle-avoidance-using-adaptive-model-predictive-control.html
9. Pour, F.K., Puig, V., Ocampo-Martinez, C.: Multi-layer health-aware economic predictive control of a pasteurization pilot plant. Int. J. Appl. Math. Comput. Sci. **28**, 97–110 (2018)
10. Sawulski J., Ławryńczuk M.: Optimisation-based tuning of dynamic matrix control algorithm for multiple-input multiple-output processes. In: Proceedings of the 23th IEEE International Conference on Methods and Models in Automation and Robotics MMAR 2018, pp. 160–165. Międzyzdroje, Poland (2018)
11. Scattolini, R., Bittanti, S.: On the choice of the horizon in long-range predictive control-some simple criteria. Automatica **26**, 915–917 (1990)
12. Seborg, D.E., Edgar, T.F., Mellichamp, D.A., Doyle III, F.J.: Process Dynamics and Control. Wiley, New York (2011)

13. Seybold, L., Witczak, M., Majdziek, P., Stetter, R.: Towards robust predictive fault-tolerant control for a battery assembly unit. Int. J. Appl. Math. Comput. Sci. **25**, 849–862 (2015)
14. Shridhar, R., Cooper, D.J.: A tuning strategy for unconstrained multivariable model predictive control. Ind. Eng. Chem. Res. **37**, 4003–4016 (1998)
15. Shridhar, R., Cooper, D.J.: A tuning strategy for unconstrained SISO model predictive control. Ind. Eng. Chem. Res. **36**, 729–746 (1997)
16. Tatjewski, P.: Offset-free nonlinear Model Predictive Control with state-space process models. Arch. Control Sci. **27**, 595–615 (2017)
17. Tatjewski, P.: Advanced Control of Industrial Processes. Structures and Algorithms. Springer, London (2007)
18. Trierweiler, J.O., Farina, L.A.: RPN tuning strategy for model predictive control. J. Process Control **13**, 591–598 (2003)
19. Yamashita, A.S., Alexandre, P.M., Zanin, A.C., Odloak, D.: Reference trajectory tuning of model predictive control. Control Eng. Pract. **50**, 1–11 (2016)
20. Zhou, F., Peng, H., Zhang, G., Zeng, X.: A robust controller design method based on parameter variation rate of RBF-ARX model. IEEE Access **7**, 160284–160294 (2019)

DMC Algorithm with Laguerre Functions

Piotr Tatjewski(✉)

Institute of Control and Computation Engineering, Faculty of Electronics and Information Technology, Warsaw University of Technology, Nowowiejska 15/19, 00-665 Warsaw, Poland
P.Tatjewski@ia.pw.edu.pl

Abstract. The paper is concerned with development and analysis of the Dynamic Matrix Control (DMC) discrete-time model predictive control algorithm with parametrisation of the control input trajectories by sets of Laguerre functions. First the appropriate formulation of the algorithm is developed. The main difference between it and the standard DMC formulation is that coefficients of the approximation by the Laguerre functions, instead of control input values, are the decision variables of the DMC optimization problem. Then the proposed DMCL (DMC with Laguerre functions) algorithm is applied to a multivariable benchmark problem to investigate its properties and to provide a concise comparison with the standard DMC algorithm.

Keywords: Process control · Model predictive control · DMC algorithm · Laguerre functions

1 Introduction

The MPC is now a well established advanced control technology, represented by a variety of successful control algorithms and software packages applied in practice, see, e.g., [1–4,6–10,10,11,15]. The DMC (Dynamic Matrix Control) algorithm was one of the very first MPC algorithms developed and applied in practice, it is still one of the most popular MPC solutions in the process industries. The DMC algorithm uses nonparametric process models in the form of discrete-time unit step responses. This is one of the reasons of its popularity, the step response models are relatively easy to identify during the standard on-line experiments. On the other hand, these models can have long horizons of dynamics, i.e., large numbers of sampling intervals before the step responses stabilize. In particular, this happens if the models have relatively short and different delay times, when compared to the dominant time constants. Then usually the sampling interval is chosen relatively small to capture accurate modeling of time delays and to avoid too large additional delay stemming from the sampling procedure. Therefore, long prediction horizons can then result in the DMC algorithm together with also relatively long control horizons. This means increased computational burden.

A. Bartoszewicz et al. (Eds.): PCC 2020, AISC 1196, pp. 1006–1017, 2020.
https://doi.org/10.1007/978-3-030-50936-1_84

The use of polynomial models is a way to simplify the representation of complex dynamical modeling, using Laguerre functions is here a popular solution, see, e.g., [13]. In MPC applications, the Laguerre functions were used first of all for simplifying process models, to speed up the use of these models in the prediction. The application of Laguerre functions for representation of control input trajectories can be found in [12, 14], mainly for state-space process models. Recently, this approach has been applied in nonlinear MPC for more efficient optimization with linearized models in predictive structures of GPC type [5]. In this paper the use of the Laguerre functions for the parametrization of predicted control input trajectories in the DMC algorithm will be proposed and analysed. The presented simulation results of an example benchmark MIMO process confirm the supposition that this may be computationally more efficient, first of all for processes with long step responses (long horizons of dynamics) and thus relatively long control horizons.

The structure of the paper is as follows. In Sect. 2 the standard DMC algorithm will be briefly recalled. In Sect. 3 the use of the Laguerre functions to parametrize the control input trajectories over the prediction horizon will be presented and the DMCL (DMC with Laguerre functions) algorithm will be formulated. In Sect. 4 the efficiency of the DMCL algorithm will be investigated on a MIMO benchmark example problem, including a concise comparison of the DMCL and DMC algorithms. Conclusions will be the last part of the paper.

2 Dynamic Matrix Control Algorithm

The DMC algorithm was one of the first MPC algorithms, it is still very popular in the process industries. The principle of the MPC is well known, different formulations of the MPC algorithms can be found in many papers and books, see, e.g., [6, 8–10, 15]. In this paper the DMC algorithm will be further developed, therefore its standard formulation will be first given.

The principle of the MPC is to evaluate current process control input signals by minimizing, at each sampling instant k, a performance function (cost function) over a future prediction horizon of N samples. The following performance function is one of the most widely used in process control applications:

$$J(k) = \sum_{p=1}^{N} \|[y^{sp}(k+p|k) - y(k+p|k)\|^2_{\mathbf{\Psi}} + \sum_{p=0}^{N_u-1} \|\triangle u(k+p|k)\|^2_{\mathbf{\Lambda}}\,, \tag{1}$$

where $\|x\|^2_{\mathbf{R}}=x^T\mathbf{R}x$, $\mathbf{\Psi} \geq \mathbf{0}$ and $\mathbf{\Lambda} > \mathbf{0}$ are square diagonal scaling matrices of dimensions corresponding to the dimensions n_y and n_u of the process controlled output and control input vectors, respectively (a simpler formulation of (1) is often used in theoretical considerations, with one scaling scalar λ only, i.e., $\mathbf{\Psi} = \mathbf{I}$ and $\mathbf{\Lambda} = \lambda\mathbf{I}$). In the formulation (1), $N_u \leq N$ denotes the length of the control horizon, $y^{sp}(k+p|k)$ and $y(k+p|k)$ are set-point (reference) and process output vectors predictions for a future sample $k+p$, but calculated at the current sample k, $p = 1, \ldots, N$.

In most standard DMC formulations control input increments on the control horizon are the decision variables,

$$\triangle u(k+p|k) = u(k+p|k) - u(k+p-1|k), \quad p = 0, \ldots, N_u - 1,$$

thus the vector of decision variables, denoted by $\triangle \mathbf{u}(k)$, is

$$\triangle \mathbf{u}(k) = [\triangle u(k|k)^T \ \triangle u(k+1|k)^T \cdots \triangle u(k+N_u-1|k)^T]^T. \tag{2}$$

We assume that the optimization of $J(k)$ is subject to simple constraints:

$$-\triangle u_{max} \leq \triangle u(k+p|k) \leq \triangle u_{max}, \quad p = 0, \ldots, N_u-1, \tag{3}$$

$$u_{min} \leq u(k+p|k) \leq u_{max}, \ p = 0, \ldots, N_u-1, \tag{4}$$

$$y_{min} \leq y(k+p|k) \leq y_{max}, \ p = 1, \ldots, N. \tag{5}$$

More general form of the constraints, including any linear functions of all variables used, is possible, but avoided here for simplicity.

Denoting composite vectors of set-points and predicted outputs on the prediction horizon by $\mathbf{y}^{sp}(k)$ and $\mathbf{y}^{pr}(k)$, respectively,

$$\mathbf{y}^{sp}(k) = [\ y^{sp}(k+1|k)^T \ \cdots \ y^{sp}(k+N|k)^T]^T, \tag{6}$$

$$\mathbf{y}^{pr}(k) = [\ y(k+1|k)^T \ \cdots \ y(k+N|k)^T]^T, \tag{7}$$

we can formulate, in a compact form, *the DMC optimization problem* which calculates the optimal control trajectory:

$$\begin{aligned} &\min_{\triangle \mathbf{u}(k)} \ \{ \ J(k) = \|\mathbf{y}^{sp}(k) - \mathbf{y}^{pr}(k)\|^2_{\underline{\boldsymbol{\Psi}}} + \|\triangle \mathbf{u}(k)\|^2_{\underline{\boldsymbol{\Lambda}}} \ \} \\ &\text{subject to}: \ (3), (4) \text{ and } (5), \end{aligned} \tag{8}$$

where

$$\underline{\boldsymbol{\Psi}} = \mathrm{diag}(\overbrace{\boldsymbol{\Psi}, \ \ldots, \ \boldsymbol{\Psi}}^{N \text{ times}}), \quad \underline{\boldsymbol{\Lambda}} = \mathrm{diag}(\overbrace{\boldsymbol{\Lambda}, \ \ldots, \ \boldsymbol{\Lambda}}^{N_u \text{ times}}) \tag{9}$$

and the predicted output trajectory $\mathbf{y}^{pr}(k)$ is calculated using the process model, in the form of unit step responses.

For MIMO processes, it is most convenient to formulate the overall process step response model in the matrix form, i.e., consisting of D matrices $\mathbf{S}_l$, each corresponding to one sampling instant l, covering the horizon of dynamics D of the process (i.e., number of sampling periods needed for the outputs to stabilize after the step input change),

$$\mathbf{S}_l = \begin{bmatrix} s_l^{11} & s_l^{12} & s_l^{13} & \cdots & s_l^{1n_u} \\ s_l^{21} & s_l^{22} & s_l^{23} & \cdots & s_l^{2n_u} \\ s_l^{31} & s_l^{32} & s_l^{33} & \cdots & s_l^{3n_u} \\ \vdots & \vdots & \vdots & \ddots & \vdots \\ s_l^{n_y1} & s_l^{n_y2} & s_l^{n_y3} & \cdots & s_l^{n_yn_u} \end{bmatrix}_{n_y \times n_u}, \quad l = 1, 2, ..., D,$$

where s_l^{ij} denotes l-th element of the response of the i-th process output on the unit step change of the j-th process input, $i = 1, \dots, n_y$, $j = 1, \dots, n_u$, see [10]. Using this model, the following prediction formulae can be obtained (assuming the case without measured disturbances) [10]:

$$\mathbf{y}^{pr}(k) = \mathbf{M}\,\triangle\mathbf{u}(k) + \mathbf{y}(k) + \mathbf{M}^P \triangle\mathbf{u}^P(k), \tag{10}$$

where

$$\mathbf{M} = \begin{bmatrix} \mathbf{S}_1 & 0 & 0 & \cdots & 0 \\ \mathbf{S}_2 & \mathbf{S}_1 & 0 & \cdots & 0 \\ \mathbf{S}_3 & \mathbf{S}_2 & \mathbf{S}_1 & \cdots & 0 \\ \vdots & \vdots & \vdots & \ddots & \vdots \\ \mathbf{S}_{N_u} & \mathbf{S}_{N_u-1} & \mathbf{S}_{N_u-2} & \cdots & \mathbf{S}_1 \\ \mathbf{S}_{N_u+1} & \mathbf{S}_{N_u} & \mathbf{S}_{N_u-1} & \cdots & \mathbf{S}_2 \\ \vdots & \vdots & \vdots & \vdots & \vdots \\ \mathbf{S}_N & \mathbf{S}_{N-1} & \mathbf{S}_{N-2} & \cdots & \mathbf{S}_{N-N_u+1} \end{bmatrix} \tag{11}$$

is the dynamic matrix,

$$\mathbf{M}^P = \begin{bmatrix} \mathbf{S}_2 - \mathbf{S}_1 & \mathbf{S}_3 - \mathbf{S}_2 & \mathbf{S}_4 - \mathbf{S}_3 & \cdots & \mathbf{S}_D - \mathbf{S}_{D-1} \\ \mathbf{S}_3 - \mathbf{S}_1 & \mathbf{S}_4 - \mathbf{S}_2 & \mathbf{S}_5 - \mathbf{S}_3 & \cdots & \mathbf{S}_{D+1} - \mathbf{S}_{D-1} \\ \mathbf{S}_4 - \mathbf{S}_1 & \mathbf{S}_5 - \mathbf{S}_2 & \mathbf{S}_6 - \mathbf{S}_3 & \cdots & \mathbf{S}_{D+2} - \mathbf{S}_{D-1} \\ \vdots & \vdots & \vdots & \ddots & \vdots \\ \mathbf{S}_{N+1} - \mathbf{S}_1 & \mathbf{S}_{N+2} - \mathbf{S}_2 & \mathbf{S}_{N+3} - \mathbf{S}_3 & \cdots & \mathbf{S}_{N+D-1} - \mathbf{S}_{D-1} \end{bmatrix}, \tag{12}$$

$$\mathbf{y}(k) = \begin{bmatrix} y(k) \\ y(k) \\ y(k) \\ \vdots \\ y(k) \end{bmatrix}, \quad \triangle\mathbf{u}^P(k) = \begin{bmatrix} \triangle u(k-1) \\ \triangle u(k-2) \\ \triangle u(k-3) \\ \vdots \\ \triangle u(k-(D-1)) \end{bmatrix}, \tag{13}$$

where the vector $\mathbf{y}(k)$ consists of N repetitions of the vector $y(k)$.

Once the optimization problem (8) with the predictions (10) has been solved, the first element $\triangle\hat{u}(k|k)$ of the optimal control trajectory is used only to calculate the current process input $u(k) = u(k-1) + \triangle\hat{u}(k|k)$. After the next measurement (at the next sampling instant) the whole DMC procedure is repeated (receding horizon strategy).

We shall assume $n_u = n_y$ in the paper, which in the linear case always yields a unique solution of the MPC optimization problem. However, this can be easily generalized to the case $n_u > n_y$, not unusual in MPC applications, by appropriate augmentation of the performance function [10].

3 The DMC Algorithm with Laguerre Functions

It is common to define the Laguerre functions by their transfer functions – the transfer function of the Laguerre function of order n is

$$\mathcal{L}_n(z) = \frac{\sqrt{1-a^2}}{z-a}\left(\frac{1-az}{z-a}\right)^{(n-1)}, \tag{14}$$

where a is a scaling factor, $0 \le a < 1$, see, e.g., [13,15]. Then the Laguerre function of order n is

$$l_n(k) = \mathcal{Z}^{-1}(\mathcal{L}_n(z)). \tag{15}$$

Let us define a set of n_L Laguerre functions of increasing order

$$l(p) = [l_1(p) \cdots l_{n_L}(p)]^T, \quad p = 0, ..., N-1. \tag{16}$$

Then, due to the structure of these functions, it can be easily seen that

$$l(p+1) = \mathbf{A}\, l(p), \tag{17}$$

where

$$\mathbf{A} = \begin{bmatrix} a & 0 & 0 & \cdots & 0 \\ \beta & a & 0 & \cdots & 0 \\ -\alpha\beta & \beta & a & \cdots & 0 \\ \alpha^2\beta & -\alpha\beta & \beta & \cdots & 0 \\ \vdots & \vdots & \vdots & \vdots & \vdots \\ (-\alpha)^{n_L-2}\beta & (-\alpha)^{n_L-3}\beta & \cdots & \beta & a \end{bmatrix}, \tag{18}$$

with the initial value

$$l(0) = \sqrt{1-a^2}\begin{bmatrix} 1 \\ -a \\ a^2 \\ -a^3 \\ \vdots \\ (-a)^{n_L-1} \end{bmatrix} \tag{19}$$

and where $\beta = 1 - a^2$, see, e.g., [15].

Let us assign to every component u_j of the process control input vector $u = [u_1 \cdots u_{n_u}]^T$ a set of n_L Laguerre functions $l_1^j, l_2^j, \ldots, l_{n_L}^j$, $j = 1, \ldots, n_u$ (the numbers n_L can be different for different components of u, but they are usually taken equal and thus we assume this numbers equal, for simplicity). Next, parameterize the components of the DMC decision control vector $\triangle\mathbf{u}(k)$ (2) in the following way

$$\triangle u_j(k+p|k) = \sum_{m=1}^{n_L} l_m^j(p)c_m^j(k) = l^j(p)^T c^j(k),$$

$$j = 1, ..., n_u, \quad p = 0, \ldots, N-1, \tag{20}$$

where

$$l^j(p) = [l_1^j(p) \ \cdots \ l_{n_L}^j(p)]^T \tag{21}$$

is defined according to (16) and $c^j(k)$ denotes the vector of coefficients of the Laguerre functions,

$$c^j(k) = [c_1^j(k) \ c_2^j(k) \ \cdots \ c_{n_L}^j(k)]^T, \ \ j = 1, \ldots, n_u. \tag{22}$$

Therefore, we have

$$\triangle u(k+p|k) = \begin{bmatrix} l^1(p)^T c^1(k) \\ l^2(p)^T c^2(k) \\ \vdots \\ l^{n_u}(p)^T c^{n_u}(k) \end{bmatrix}, \ \ p = 0, ..., N-1. \tag{23}$$

Further, if we define the full vector of coefficients $\mathbf{c}(k)$,

$$\mathbf{c}(k) = [c^1(k)^T \ c^2(k)^T \ \cdots \ c^{n_u}(k)^T]^T \tag{24}$$

and N matrices

$$\mathbf{L}(p) = \text{diag}\left(l^1(p)^T, l^2(p)^T, ..., l^{n_u}(p)^T\right), \ \ p = 0, ..., N-1, \tag{25}$$

then we can write (23) in the form

$$\triangle u(k+p|k) = \mathbf{L}(p)\, \mathbf{c}(k), \ \ p = 0, ..., N-1. \tag{26}$$

In this way we get finally the formula

$$\triangle \mathbf{u}(k) = \begin{bmatrix} \triangle u(k|k) \\ \triangle u(k+1|k) \\ \vdots \\ \triangle u(k+N-1|k) \end{bmatrix} = \begin{bmatrix} \mathbf{L}(0) \\ \mathbf{L}(1) \\ \vdots \\ \mathbf{L}(N-1) \end{bmatrix} \mathbf{c}(k) = \mathbf{L}\, \mathbf{c}(k). \tag{27}$$

It should be pointed out that the matrix $\mathbf{L}$ depends only on the value of the scaling factor a, it does not depend on time, thus it is evaluated off-line during the design phase. Certainly, different scaling factors a^j can be assumed for different components u_j of the control vector u, but this would only influence the off-line evaluation of the Laguerre functions l^j and thus matrix $\mathbf{L}$.

Substituting (27) into the prediction equation (10), into the second term in the performance function (8) and into the inequality constraints (3),(4),(5) converts the DMC optimization problem into the problem with optimization with respect to the vector of the Laguerre coefficients $\mathbf{c}(k)$. The DMC algorithm with this optimization problem, with the control input parametrization using Laguerre functions, will be further denoted by the acronym DMCL.

The dimension of the vector $\mathbf{c}(k)$ is $n_c = n_u \cdot n_L$ and is not dependent on the control horizon N_u, whereas the dimension of the decision vector $\triangle \mathbf{u}(k)$ in the standard DMC algorithm is $n_u \cdot N_u$. Therefore, if $n_L < N_u$, dimensionality of the DMC optimization problem (solved on-line at every sampling instant) is diminished. As n_L is not dependent on N_u, we assume $N_u = N$ in the DMCL algorithm. Moreover, if $N_u = N$ and the prediction horizon N is sufficiently long, the Laguerre functions are orthonormal (see, e.g., [15]) and this can be taken into account. Doing that when using (27), the second term in the performance function in (8) can be transformed to the following simple form:

$$\|\triangle \mathbf{u}(k)\|^2_{\underline{\mathbf{\Lambda}}} = \mathbf{c}(k)^T\mathbf{L}^T\underline{\mathbf{\Lambda}}\mathbf{L}\mathbf{c}(k) = \mathbf{c}(k)^T\underline{\mathbf{\Lambda}}_{\mathbf{L}}\mathbf{c}(k) = \|\mathbf{c}(k)\|^2_{\underline{\mathbf{\Lambda}}_L}, \tag{28}$$

where

$$\underline{\mathbf{\Lambda}}_L = \mathrm{diag}(\lambda_1\mathbf{I}_{n_L}\ \lambda_2\mathbf{I}_{n_L}\ \cdots\ \lambda_{n_u}\mathbf{I}_{n_L}), \tag{29}$$

with $\mathbf{I}_{n_L}$ denoting the identity matrix of dimension n_L and where the weights λ_i are elements of the weighting matrix $\mathbf{\Lambda}$ in the initial performance function (1), $\mathbf{\Lambda} = \mathrm{diag}(\lambda_1, \lambda_2, \ldots, \lambda_{n_u})$. Thus, the performance function of the DMCL algorithm takes the form

$$J(k) = \left\|\mathbf{y}^{sp}(k) - \left[\mathbf{M}\mathbf{L}\mathbf{c}(k) + \mathbf{y}(k) + \mathbf{M}^P\triangle\mathbf{u}^P(k)\right]\right\|^2_{\underline{\mathbf{\Psi}}} + \|\mathbf{c}(k)\|^2_{\underline{\mathbf{\Lambda}}_L}. \tag{30}$$

Certainly, when there is one common weighting factor λ for all components of the control input vector, i.e., $\mathbf{\Lambda} = \lambda\mathbf{I}_{n_u}$, then (28) reduces to $\lambda\|\mathbf{c}(k)\|^2$.

4 Efficiency of the DMCL Algorithm

As discussed in the previous section, the dimensionality n_c of the decision vector $\mathbf{c}(k)$ does not depend on the control horizon and thus we took both control and prediction horizons equal, $N_u = N$. On the other hand, $n_c = n_u \cdot n_L$ and thus depends on the number of Laguerre functions n_L taken to represent the trajectory of each component u_i of the process control input vector u, over the prediction horizon. It is known from the practice of model approximations by the Laguerre functions that taking rather small number of these functions leads usually to satisfactory results, it will be also confirmed in the example analysed further in this section. Therefore, the DMCL algorithm should be an effective solution in cases when longer control horizons are appropriate in the standard DMC algorithm. Therefore, we shall select a benchmark example corresponding to this case, to analyse the properties of the DMCL algorithm and to compare the results with those obtained using the standard formulation of the DMC algorithm.

Example

We shall apply and analyse the DMCL algorithm for the Wood-Berry (WB) methanol-water distillation column example, a well known benchmark for MIMO process control [16]. The 2×2 continuous time transfer function model for composition control in this column is as follows:

$$\begin{bmatrix} Y_1(s) \\ Y_2(s) \end{bmatrix} = \begin{bmatrix} \frac{12.8e^{-s}}{16.7s+1} & \frac{-18.9e^{-3s}}{21s+1} \\ \frac{6.6e^{-7s}}{10.9s+1} & \frac{-19.4e^{-3s}}{14.4s+1} \end{bmatrix} \begin{bmatrix} U_1(s) \\ U_2(s) \end{bmatrix} + \begin{bmatrix} \frac{3.8e^{-8s}}{14.9s+1} \\ \frac{4.9e^{-3s}}{13.2s+1} \end{bmatrix} F(s). \quad (31)$$

A discrete-time model is needed, therefore the original model (31) must be discretized. We have chosen the sampling period $T_p = 1$, to get the discrete-time model with accurate representation of the delay times. Using for discretization the Matlab function `c2d` we got the following result:

$$\begin{bmatrix} Y_1(z) \\ Y_2(z) \end{bmatrix} = \begin{bmatrix} \frac{0.744z^{-1}}{z-0.9419} & \frac{-0.8789z^{-3}}{z-0.9535} \\ \frac{0.5786z^{-7}}{z-0.9123} & \frac{-1.302z^{-3}}{z-0.9329} \end{bmatrix} \begin{bmatrix} U_1(z) \\ U_2(z) \end{bmatrix} + \begin{bmatrix} \frac{0.2467z^{-8}}{z-0.9351} \\ \frac{0.3575z^{-3}}{z-0.927} \end{bmatrix} F(z). \quad (32)$$

Step responses of this model are depicted in Fig. 1.

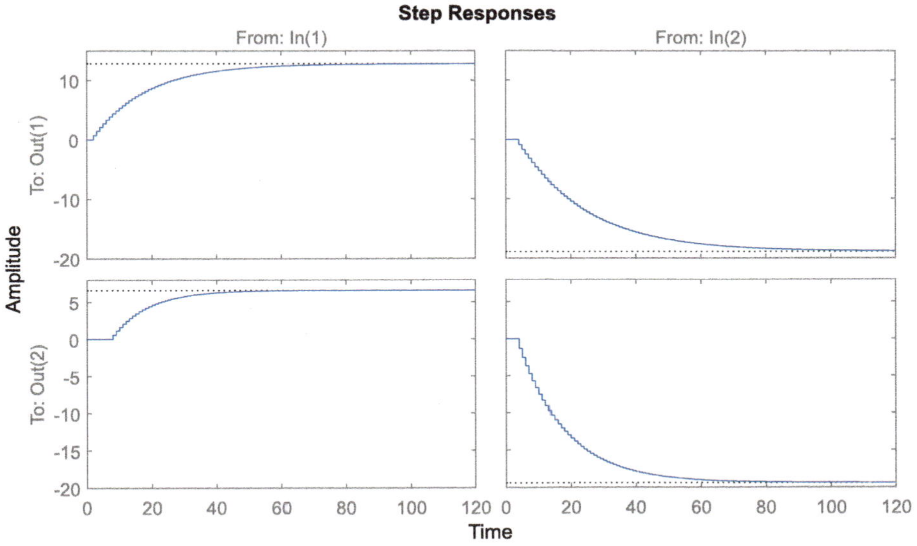

Fig. 1. Input-output step responses of the discrete-time Wood-Berry column model.

The key factor influencing efficiency of the DMCL algorithm is the number n_L of the Laguerre functions modeling each control input trajectory (recall we assumed it is the same for each component of the control vector). Therefore, the influence of n_L on the results of a single optimization of the DMCL performance function (30) was first investigated. The scaling factor $a = 0.55$ was taken, as values of a around this value were found to be appropriate. Sample results, for the demanding case with the unit step changes of both output reference values (set-points) at the current sample time, are presented in Fig. 2 and Fig. 3.

The shapes of the trajectories stabilize with the increase of n_L, the differences between the trajectories for $n_L = 4$ and $n_L = 5$ are small. This indicates that the choice $n_L = 5$ should be safely satisfactory, even $n_L = 4$ should suffice. Let us notice that $n_L = 5$ corresponds to the control horizon $N_u = 5$ in the standard DMC algorithm, resulting in the same dimensionality of both optimization problems.

Results of a sample simulation of the feedback control system with the DMCL algorithm are shown in Fig. 4, for the scenario with unit step changes of the reference values (at sampling instants $k = 2$, $k = 60$ and $k = 120$) and a step change of the disturbance F (see (32)) at sampling instant $k = 200$. The prediction horizon was assumed $N = 20$, which seems to be the shortest value assuring that the controlled outputs stabilize, following the inspection of Figs. 2 and 3. The horizon of dynamics was assumed $D = 80$, shorter values were decreasing the quality of the control. This coincides with the inspection of the step response model, see Fig. 1.

To compare efficiency of the DMCL and standard DMC algorithms, a series of simulations were performed, for different values of n_L and N_u. Recall that computational effort, measured by dimensionality of the optimization problem, is equal for both algorithms for $n_L = N_u$. Each simulation was performed for the reference and disturbance scenario as in Fig. 4. To measure the control quality, the associated value of the integrated squared control error (ISE) was calculated. The results, for the values of n_L and N_u from 3 to 10 are depicted in Fig. 5 (numbers of Laguerre functions n_L smaller than 3 are not reasonable). It can be easily seen from this comparison that the DMCL algorithm is a sound

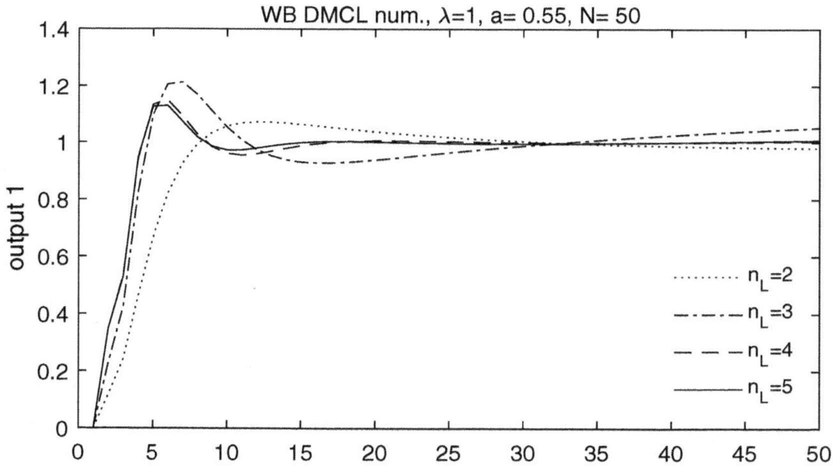

Fig. 2. Optimized trajectories of the first output, after unit step changes of both set-points, for different numbers of the Laguerre functions n_L.

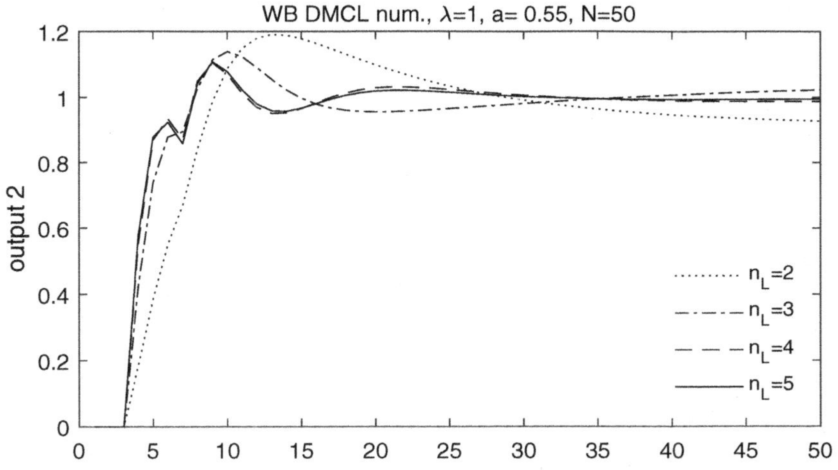

Fig. 3. Optimized trajectories of the second output, after unit step changes of both set-points, for different numbers of the Laguerre functions n_L.

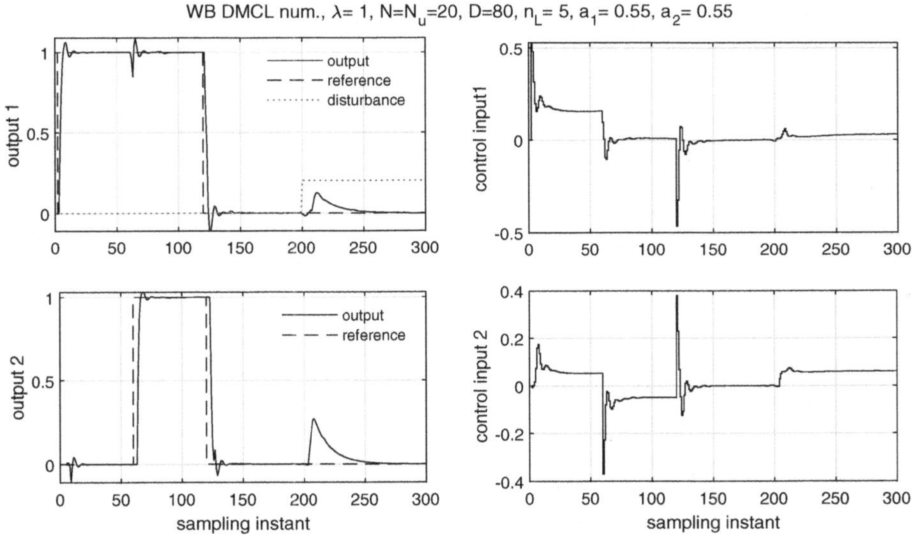

Fig. 4. Trajectories of the process outputs and inputs in a simulation of the DMCL feedback control system under step changes of the reference values and the disturbance

alternative to the classical DMC algorithm with the process control inputs as the decision variables. It achieves good control quality for $n_L = 4$ Laguerre functions and, certainly, keeps it when increasing n_L (for each control input), whereas the standard DMC algorithm achieves the same level of quality for the control horizon $N_u = 6$ and, certainly, also keeps it for increasing values. It should

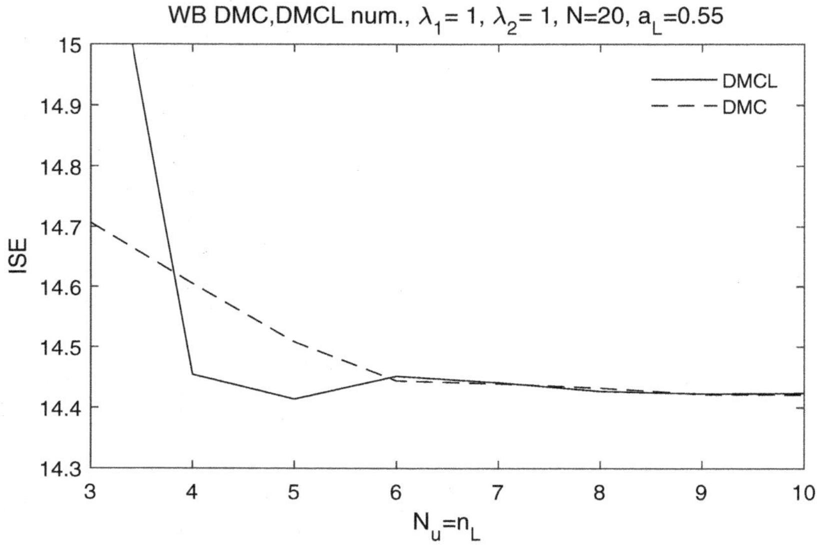

Fig. 5. Comparison of the control quality of the DMCL and DMC algorithms for the WB column example, measured by ISE values.

be noticed that the dynamics of the considered column is representative, with different delays and time constants leading to long horizons of dynamics. However, it is not most demanding - the transfer functions are without zeros in the numerators.

5 Conclusions

Development and analysis of the DMCL (DMC with Laguerre functions) algorithm, the DMC model predictive control algorithm with parametrisation of the control input trajectories by sets of the Laguerre functions, was the aim of the paper. The appropriate formulation of the algorithm was developed, where coefficients of the approximation by the Laguerre functions instead of process control input values are the decision variables of the MPC optimization problem. Then the developed DMCL algorithm was applied to a benchmark MIMO process, the Wood-Berry distillation column. Sample results of an extensive simulation study were presented showing that the DMCL algorithm is a sound alternative to the standard DMC algorithm. It offers the possibility to deliver equivalent results more effectively, with lower computational effort, especially for problems with long prediction and control horizons. More extensive investigation, including comparisons of both algorithms for other problems, including those with even more demanding dynamics, are topics of further research.

References

1. Blevins, T.L., McMillan, G.K., Wojsznis, W.K., Brown, M.W.: Advanced Control Unleashed. The ISA Society, Research Triangle Park (2003)
2. Blevins, T.L., Wojsznis, W.K., Nixon, M.: Advanced Control Foundation. The ISA Society, Research Triangle Park (2013)
3. Camacho, E., Bordons, C.: Model Predictive Control. Springer, London (1999)
4. Ławryńczuk, M.: Computationally Efficient Model Predictive Control Algorithms: A Neural Network Approach, Studies in Systems, Decision and Control, vol. 3. Springer, Heidelberg (2014)
5. Ławryńczuk, M.: Nonlinear model predictive control for processes with complex dynamics: parametrisation approach using Laguerre functions. Int. J. Appl. Math. Comput. Sci. **30**(1), 35–46 (2020)
6. Maciejowski, J.: Predictive Control. Prentice Hall, Harlow (2002)
7. Qin, S., Badgwell, T.: A survey of industrial model predictive control technology. Control Eng. Pract. **11**, 733–764 (2003)
8. Rawlings, J.B., Mayne, D.Q.: Model Predictive Control: Theory and Design. Nob Hill Publishing, Madison (2009)
9. Rossiter, J.: Model-Based Predictive Control. CRC Press, Boca Raton (2003)
10. Tatjewski, P.: Advanced Control of Industrial Processes. Springer, London (2007)
11. Tatjewski, P.: Advanced control and on-line process optimization in multilayer structures. Ann. Rev. Control **32**, 71–85 (2008)
12. Valencia-Palomo, G., Rossiter, J.: Using Laguerre functions to improve efficiency of multi-parametric predictive control. In: Proceedings of the 2010 American Control Conference, Baltimore (2010)
13. Wahlberg, B.: System identification using the Laguerre models. IEEE Trans. Autom. Control **36**(5), 551–562 (1991)
14. Wang, L.: Discrete model predictive controller design using Laguerre functions. J. Process Control **14**, 131–142 (2004)
15. Wang, L.: Model Predictive Control System Design and Implementation using MATLAB. Springer, London (2009)
16. Wood, R., Berry, M.: Terminal composition control of a binary distillation column. Chem. Eng. Sci. **28**(9), 1707–17 (1973)

Hardware-In-the-Loop Simulations of a GPC-Based Controller in Different Types of Buildings Using Node-RED

Dariusz Bismor(✉), Karol Jabłoński, Tomasz Grychowski, and Sławomir Nas

Silesian University of Technology, ul. Akademicka 2A, 44-100 Gliwice, Poland
Dariusz.Bismor@polsl.pl

Abstract. This paper presents the outcome of a project aiming at the development of a building simulator designed using Matlab® and Simulink® environments. The objective was to be able to test different control algorithms using the *hardware-in-the-loop* principle in a fast, flexible and effective way. The developed software allows one to simulate continuous dynamics of particular parameters in various time horizons, including a whole year simulation. The simulator has been used to test one of the model predictive control algorithms, implemented in the Node-RED environment and installed on the Beaglebone Black board.

Keywords: Hardware-in-the-loop · HVAC simulation · MPC · GPC control · Node-RED

1 Introduction

Development of energy-effective control algorithms for heating, ventilation and air-conditioning (HVAC) systems requires testing arrangemets which are close to real operating conditions [1]. Given the very slow dynamics, tests in physical buildings are very expensive and difficult to perform. To overcome this difficulty, a simulator of a building with HVAC system has been developed, using Matlab/Simulink® software. The simulator allows to simulate the most important values, such as temperature of the walls, room temperatures, and carbon dioxide concentration necessary do design and test control algorithms. To be as realistic as possible, all building partitions are simulated as multi-layered, according to their actual construction.

The simulator has a flexible structure, allowing to simulate any building, provided the building construction project is available. For testing purposes, three exemplary buildings were implemented: single-story single-family house (built in two different technologies), one floor of a multi-story office building, meeting modern thermal requirements, and an industrial (or commercial) hall, using modern technologies.

The simulator was used to implement and test advanced temperature control algorithms, e.g. Model Predictive Control (MPC) algorithms, or even neural

A. Bartoszewicz et al. (Eds.): PCC 2020, AISC 1196, pp. 1018–1029, 2020.
https://doi.org/10.1007/978-3-030-50936-1_85

network-based and fuzzy algorithms [2–4]. As MPC algorithms are adaptive, an identification subsystem was implemented, as well. The control algorithms are capable of adjusting the temperature using either underfloor heating system, convectors, or ventilation air.

2 Mathematical Models

2.1 Modeling of Building Elements

In the developed simulator, individual elements of a building are modeled using continuous differential equations with lumped parameters, which are frequently used to simulate buildings behavior [5].

Temperature of air in the i-th room and its changes are described by the equation resulting from the energy balance:

$$\frac{\mathrm{d}T_i}{\mathrm{d}t} = \frac{1}{C_p}\left[Q_p + Q_v + Q_e + Q_g + Q_d + Q_w\right]; \tag{1}$$

where C_p is the heat capacity of air in the room, Q_p is the energy supplied by heating systems (if present), Q_v is the energy supplied by the air from a ventilation system, Q_e is the energy emitted by persons staying in the room, Q_g is the energy transferred through windows, and Q_w is the energy radiated from the walls.

Temperature of the j-th layer of n-th wall in i-th room (floor and ceiling are also considered as walls) is modeled by the following equation:

$$\frac{\mathrm{d}T^{\mathrm{w}}_{inj}}{\mathrm{d}t} = \frac{A_{ni}}{C_{ni}}\left[U^{\mathrm{iw}}_{inj}\left(T^{\mathrm{w}}_{in(j+1)} - T^{\mathrm{w}}_{inj}\right) + U^{\mathrm{ow}}_{inj}\left(T^{\mathrm{w}}_{in(j-1)} - T^{\mathrm{w}}_{inj}\right) + \frac{Q}{A_{ni}}\right]; \tag{2}$$

where A_{ni} is the area of the wall, C_{ni} is its heat capacity, U^{iw}_{inj}, U^{ow}_{inj} are the inner and outer thermal conductivities of j-th layer of the wall, and Q is an additional heat delivered to the wall, for example from sun radiation or floor heating. Thus, the layers are numbered starting from outside of the room, and when temperatures of both the innermost and outermost layers are calculated, the inside and outside temperatures are considered, respectively (e.g., T^{w}_{in0} is the temperature outside the n-th wall in i-th room).

The changes in the concentration of the carbon dioxide in the i-th room are modeled using the following equation, based on the mass balance:

$$\frac{\mathrm{d}c_i}{\mathrm{d}t} = \frac{1}{V_i}((c_v - c_i)q_v + mq_e); \tag{3}$$

where c_v is the CO_2 concentration of the supply air, q_v is the mass flow of the supplied air, q_e is the amount of CO_2 produced by one person staying in the room per time unit, m is the number of people in the room, and V_i is the volume of the i-th room.

2.2 Solar Radiation Modeling

Influence of solar irradiation is one of the factors of greatest importance in simulation of thermal conditions in buildings [6]. Unfortunately, simulation of this influence is difficult due to its random character, dependent on weather and air condition (cloud overcast, clarity of the air, etc.).

The simulator calculates solar irradiance assuming the solar constant (the intensity of the solar radiation hitting one square meter of the Earth above the atmosphere) is $1367\,\mathrm{W/m^2}$. Further, the incidence angle of the sun radiation is calculated, considering the building geographical position. Moreover, to calculate the irradiance at the surface of the Earth, the influence of the atmosphere must be taken into account. The atmosphere is 100–500 km thick and consists of gas molecules, particles and dust of different concentration [7]. The next factor that needs to be included is the air clarity in subsequent months of a year. The influence of this factor is of the greatest importance in Poland, where the air pollution is the most severe in Europe. Therefore, average air clarity data were obtained from Institute of Meteorology and Water Management.

The last factor considered was cloud simulation. Unlike the air clarity, simulation of the average overcast would be a huge simplification for HVAC systems simulations. This is because the amount of solar energy in highly-windowed buildings may be so high, that cooling may be necessary during a cloudless day, while heating may be required during a cloudy day with the same outside temperature. Therefore, the simulator introduces several overcast strategies, which are selected on a random basis and in a way which results in the average overcast equal to the required (given by average historical data).

Finally, the influence of windows and shutters must be considered. This was accomplished by a simple attenuation factor, with values ranging from 0 to 1. However, such factor is flexible enough to allow even for simulation of control system adjusting the shutter position based on current light and thermal data.

3 Simulation Software

Due to the high complexity, the simulator code was divided into many source files prepared in various technologies and programming languages. Such structure allows for the efficient operation of critical simulation parts and for easy modification of particular modules. For the highest performance, mathematical models of individual elements were implemented as Matlab s-functions (*system functions*) in the C language. The main simulation component was designed as a Simulink library, and data preparation was carried out using Linux shell scripts. The whole simulation is lunched by a Matlab script, which assigns various simulation parameters, including numerous parameters defining the building thermal data. Both individual functions and data have been ordered in a hierarchical way.

In order to be able to simulate buildings of any geometry, it was necessary to prepare a data organization system that allows storing parameters of objects arranged a possibly intuitive manner, and then transferring them to the graphically programmed Simulink model. Most of the necessary data is stored in vectors of structures and in cell arrays. Such organization allows room subsystems to easily extract data to be used in simulation. All input data should also be placed in a format readable for Simulink.

Meteorological data available from the Institute of Meteorology and Water Management are used for the purposes of the simulation. The appropriate location and time are selected by a shell script from the complete archive, then the data is imported into the Matlab workspace. On the basis of time and geographical latitude information, data on the position of the sun in the sky and cloud cover are generated, as described in Sect. 2.2.

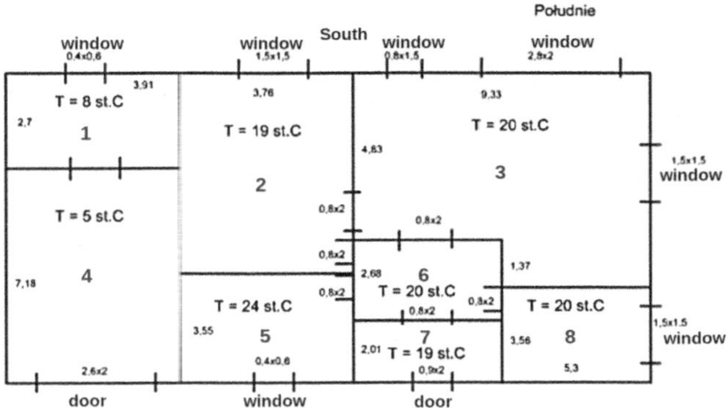

Fig. 1. Schematic diagram of the single-family house.

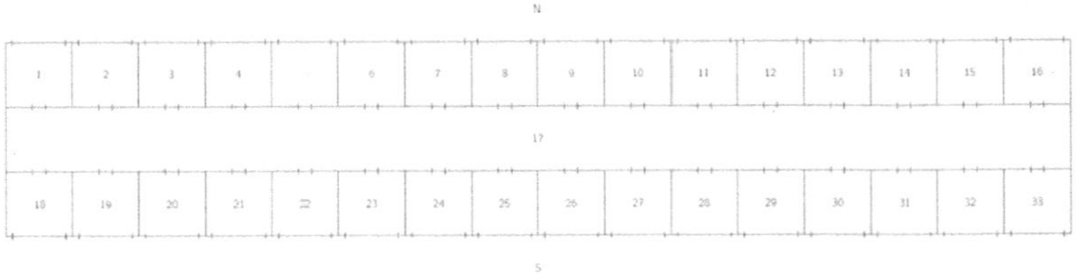

Fig. 2. Schematic diagram of the office building floor.

3.1 Single Family House

The first simulation building implemented was a single-story, single-family house; the schematic diagram of the house is presented in Fig. 1. The house is divided into eight separate rooms, two of which are unheated: room no. 1 (storage) and room no. 4 (garage). The remaining six rooms have a total area of 126.6 m^2. The outer walls are 38 cm thick, while the inner walls are 24 cm thick.

Thermal parameters of all walls (including floor and roof) were modeled twice, using two different technologies. In the first, "concrete" technology, outer walls were assumed to be erected using autoclaved cellular concrete (24 cm), and isolated using 12 cm of rock wool. Inner walls were similar, but without additional isolation. In the second, "wooden" technology, outer walls were assumed to be erected using the timber-framing method, with 20 cm of rock wood isolation and 2.5 cm of chipboard. In both the cases, all the parameters were taken from constraction projects of such houses.

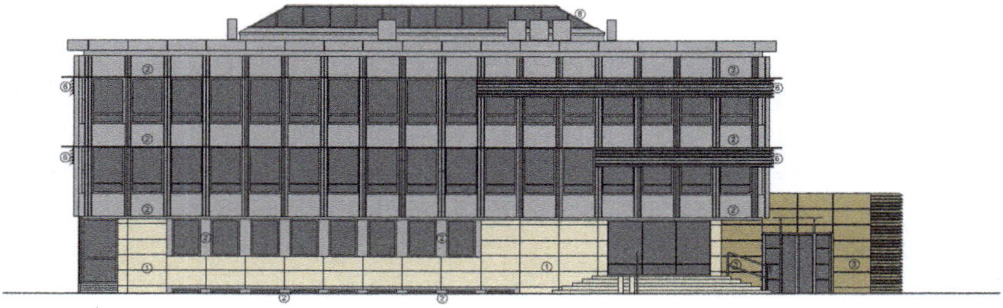

Fig. 3. Visualization of an industrial hall.

3.2 Office Building

A single floor of an office building was modeled as a second example (Fig. 2). The floor consists of 32 identical offices, and an unheated hall. The office building was assumed to be constructed using prefabricated large concrete slabs W-70, and all the parameters were taken from adequate ISO standards, as PN-EN ISO 12524:2003 and PN-EN ISO 6946:2008.

3.3 Industrial Hall

The last example modeled was an industrial (or commercial) hall, visualized in Fig. 3. The hall size is approx. 32 m by 50 m, and the height is approx. 7 m. The total area is therefore 1600 m^2. For control purposes, 6 distinct sectors of equal area (approx. 16×16 m) were implemented, with individual heating and temperature measurement. However, no internal walls or other partitions were modeled. The hall is assumed to be heated only with ventilation air.

4 MPC Controller

Although different control techniques were tested (relay control, PIDs, fuzzy control, etc.), in this publication we only describe the results of MPC control, which were the most satisfactory and were finally implemented in Node-RED.

From the variety of Model-Predictive Control algorithms, a Generalized Predictive Control (GPC) without contrainst [8] was selected based on authors experience. The literature on GPC is vast, and therefore the algorithm will not be detailed here. However, it should be emphasized that this type of control requires a parametric model of the controlled plant, in the ARIX (Auto-Regressive Incremental eXogenous) form. As the developed controller was aimed at being flexible enough to be applied in different types of buildings, with different thermal characteristics, the on-line identification of the model needed to be implemented as well. There are two major identification algorithms that are frequently used for on-line identification: Weighted Recursive Least Squares (WRLS) [9] and Least Mean Squares and its variants [10,11]. Although both the algorithms were tested, WRLS was selected for final tests due to its faster convergence.

The GPC algorithm was implemented using the C++ language, using the *boost uBLAS* matrix template class library. The developed code was flexible enough to be used in Simulink, and then in Node-RED, without any modifications. The first step—the simulation of the whole control system in Simulink only—was an obvious attitude towards fixing implementation bugs. The developed code was used to create an s-function, which is a source code representation of a Simulink block. The s-function accepted the following inputs: the set point temperature, the representative house temperature (control value), the outside temperature, the sun radiation power, and the vector of parameters of the identified building model. The output was the calculated heating system temperature.

4.1 Migration to Node-RED

Node-RED is a graphical programming environment for event-driven applications, build on Node.js [12]. Node.js is an event-driven JavaScript runtime, ideally suited for running on low-cost hardware, such as Raspberry Pi or Beaglebone Black.

Once the GPC controller was proved to work in Simulink, the C++ code was migrated to Node-RED. To achieve this without any modification of the code, a wrapper class has been created, using the N-API library [13]. N-API is a library for building Node.js native add-ons, which is designed to separate add-ons from changes in the underlying JavaScript engine. One of the major features of the N-API is that all JavaScript (and thus Node-RED) values are abstracted behind an opaque type named `napi_value`. Therefore, one of the main goals of the created wrapper class was to translate the C++ value into this abstract type. Of course, the required abstract interface needed to be filled, as well. Altogether, the wrapper class has a public constructor and an init function, six private functions and two private class members.

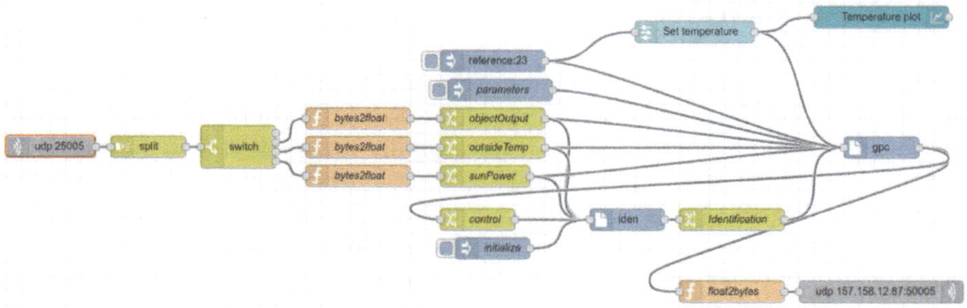

Fig. 4. Node-RED flow used in simulations.

To be able to use the GPC as a Node-RED block (referred to as a "node"), some additional JavaScript files needed to be created. All the software was uploaded to a Beaglebone development board, and then the new module was compiled, using the accompanying `npm` program. After this last step, a new node labeled "gpc" was available in the Node-RED environment. The same procedure applied to the identification class resulted in a node labeled "ident".

Finally, a way of communication between the Simulink installed on the PC and the Node-RED operating on the Beaglebone board had to be developed. After considering several options, the UDP protocol over an Internet connection was selected. To establish this type of communication, all data to be sent needed to be serialized on the sending side, and deserialized on the receiving end. In Simulink, a block called "Byte pack" was used to serialize the following important values: the CO_2 concentration, the building representative temperature, the outside temperature and the sun radiation power. The data thus serialized was send using the "UDP Send" block, configured to send it to the IP address of the Beaglebone board.

On the Beaglebone side, the data was received by "udp in" node, and then split into 8-byte long parts, using the "split" node (see Fig. 4). This results in an array of four eight-byte long fields, each representing a floating-point number. The array was divided using the "switch" node, and then each individual field was converted to a floating-point number using simple function denoted as the "bytes2float" node (the CO_2 concentration was not converted, as carbon dioxide concentration control was not included in this study). The data was then properly labeled using the "change" node, and sent to the "ident" and "gpc" nodes, as noted on the diagram (referred to as a "flow") in Fig. 4.

The value calculate by the "gpc" node is serialized using a simple "float2bytes" function, and then sent to the IP address of the PC with Simulink, using the "udp out" node. The data is received by "UDP Receive" Simulink block, and the value is retrieved using "Byte Unpack" block.

It should be emphasized that the sampling period selected for the GPC controller was equal to half an hour, or 1800s. With such long sampling period, the UDP communication with the Beaglebone board is fast enough to work

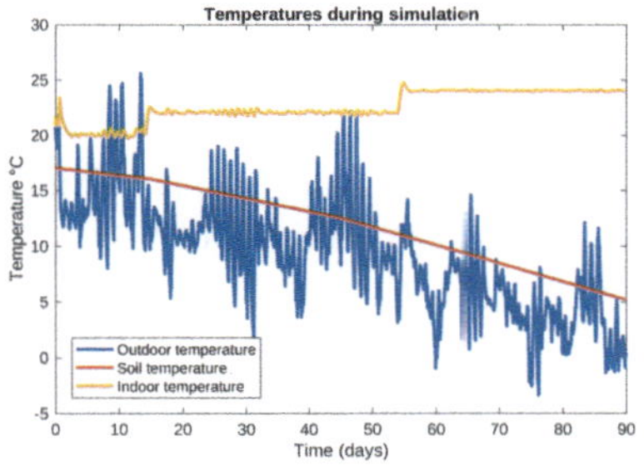

Fig. 5. Outdoor, soil, and indoor temperatures in simulations of a single-family house.

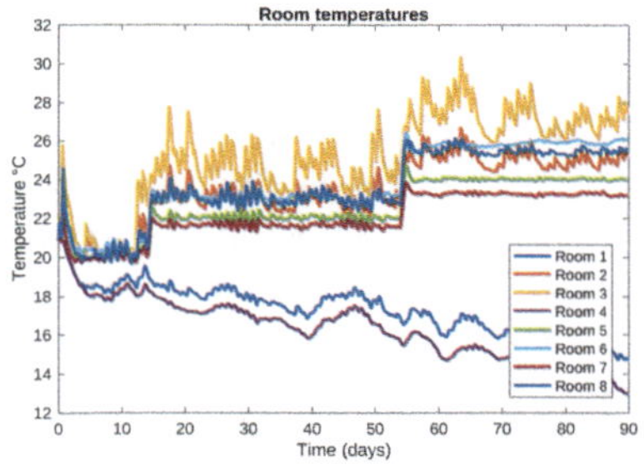

Fig. 6. Room temperatures for the single-family house.

efficiently in the majority of cases. Only in very simple simulations the Simulink must be slowed down in order not to loose any data. In this study, this was only necessary in the case of the industrial hall, and the Simulink was slowed down by selecting a sufficiently low value of the maximum step size.

5 Simulation Results

The simulation results presented below use outside air temperature values taken from meteorological data for Katowice, starting from 1 September 2017 (day 1 on time axis). The average values of supply air streams were taken from appropriate norms, and the supply air temperature was assumed to be 20 °C.

5.1 Single Family House

The results presented below are from simulations of the single family house designed using the "wooden" technology. The house was heated using an underfloor heating system.

Figure 5 shows the most important temperatures during the simulation period—90 days starting from September, 1st. The blue curve represents the outdoor temperature, which exhibits daily oscillations, but also shows a steady fall as autumn progresses. This fall is even better visible in the soil temperature, depicted in red curve. The yellow curve shows the selected indoor temperature, or process value. The indoor temperature needs to be selected as a house representative temperature. It could be a mean temperature from all heated rooms, but such temperature is influenced by the sun radiation. Therefore, in the case of this simulation, room 5 temperature was selected as representative. As room 5 window faces north, the influence of the sun is minimal.

The setpoint value at the beginning of the simulation was equal to 20 °C, and after a short period (2 days) when the identification was improving the model parameters, the indoor temperature achieves 20 °C. This value is maintained with an error not exceeding 0.6 °C. On the 14th day of the simulation, the setpoint temperature was increased to 22 °C (using the Node-RED dashboard). The indoor temperature reached this value after only around 12 h, and was maintained with the error not exceeding 0.4 °C. Finally, the setpoint temperature was increased to 24 °C on the 54th day of simulation. This value was achieved indoor with 0.7 °C overshoot, and maintained with the error not exceeding 0.1 °C—it was much easier to keep the error small as the season progressed and the outdoor temperature and the solar radiation did not influence the heating system very much.

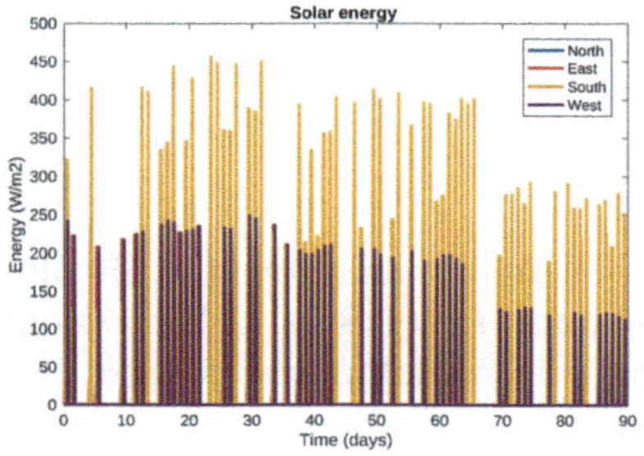

Fig. 7. Simulated solar radiation for the single family house.

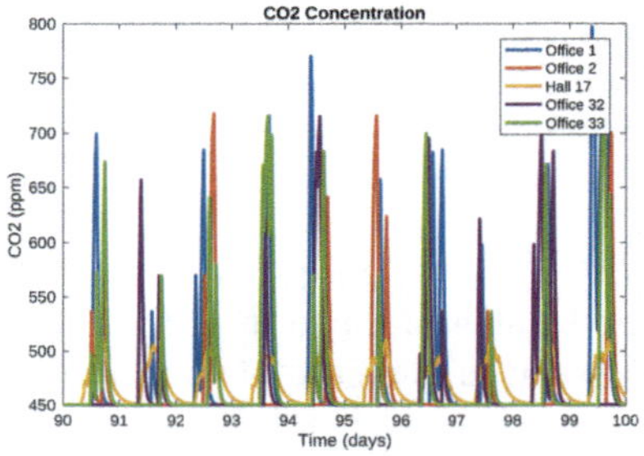

Fig. 8. Carbon dioxide concentration in selected rooms of the office building.

Figure 6 shows the temperatures in each separate room of the house. It needs to be emphasized that there was no additional control of separate room temperatures, and therefore the resulting mean room temperatures are different from the setpoint. The largest difference, and also the largest temperature variations are concerned with room 3, which is a room with large windows and garden doors. Therefore, this room is additionally heated with solar radiation to a substantial degree. Of course, under real conditions, this situation would be corrected e.g. by decreasing the floor heating system flow, or by using a separate room temperature control system.

The remaining rooms show a reasonable degree of convergence with the setpoint temperature, except for rooms 7 and 8, which are unheated, and in which temperature gradually decreases as winter approaches.

Figure 7 shows the simulated solar radiation on different elevations of the house. As the simulated period is during an autumn, no northern elevation sun radiation is observed. However, the sun energy on the southern elevation reaches 450 W/m^2—a value which influences temperature in highly-windowed room 3, as can be noticed in Fig. 6.

5.2 Office Building

In case of the office building it was assumed that the floor was divided into four separate zones, each containing eight offices. The GPC controller was used to control one zone only, consisting of rooms 26–33. The remaining rooms were using on-off control, with constant heating input temperature. The office was assumed to be heated using convection wall heaters.

Presence of office workers was simulated in office building simulation. Workers were assumed to arrive at the building around 8 am, visit different offices in a random manner, but always going throughout the hall, and leave the office around 6 pm. Each person was assumed to produce 80 W of heating energy, and to emit 20 l/h of CO_2. Therefore, the best way to confirm for presence of workers is to check the CO_2 concentration plot in Fig. 8. As can be noticed, the workers rise the carbon dioxide concentration from 450 ppm (a value assumed in the inlet air) to almost 800 ppm, when several persons remain in the same office.

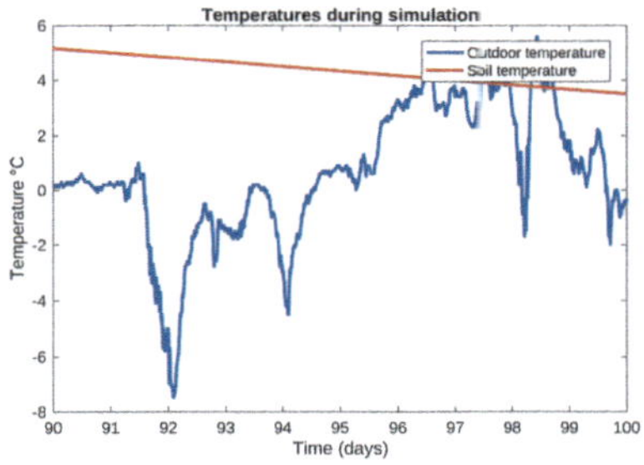

Fig. 9. Outdoor and soil temperatures in simulations of the office building.

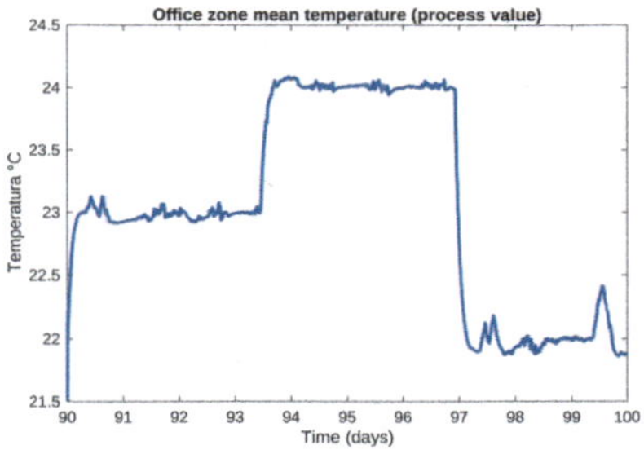

Fig. 10. Zone mean temperature during simulation of the office building

Figure 9 shows outdoor temperatures during the simulation, which starts on 30th of November. Thus, the outdoor temperature varies in range from −4 °C to 4 °C. The setpoint temperature was initially set to 23 °C, raised up to 24 °C on the fourth day of the simulation, and lowered down to 22 °C on the seventh day. The office mean temperature generally follows the setpoint temperature very well, as can be seen in Fig. 10. The only exception is on the last day of the simulation, where the mean temperature was higher than the setpoint by around 0.4 °C. This was caused by coincidence of two reasons: aggregation of a large number of workers in the zone offices (as can be noticed in Fig. 8), and high solar radiation (not shown).

5.3 Industrial Hall

The industrial hall was heated by warm air distribution only. As previously mentioned, the hall was divided into six zones, but no partitions between them were simulated. The hall representative temperature (process value) was assumed to be the mean temperature of all the zones. The simulated period started on November 30th; therefore, the outdoor temperatures were similar to those presented in Fig. 9. It was assumed that twenty workers are present in the hall 24 h/7 days in a week.

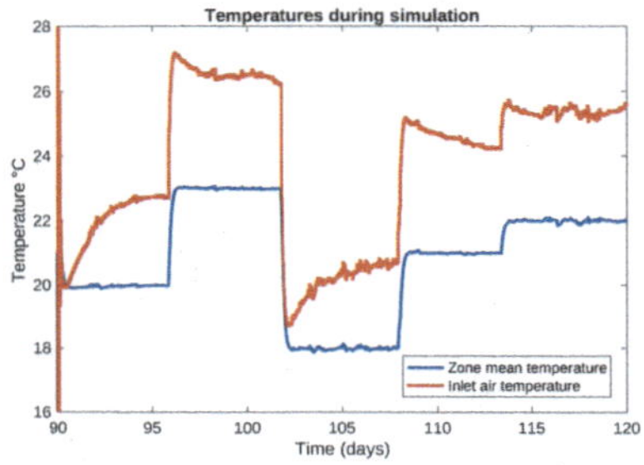

Fig. 11. Indoor temperatures during industrial hall simulation.

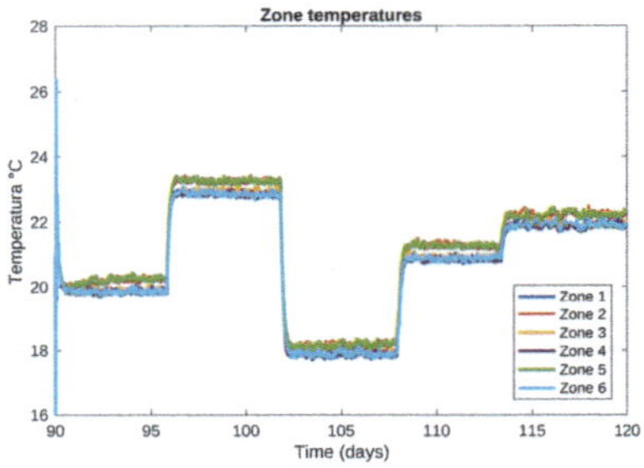

Fig. 12. Zone temperatures during industrial hall simulations.

Figure 11 presents the mean temperature and the ventilation air temperature during the simulation. After a short initial period, when the model parameters were adjusted, the mean temperature strictly follows the setpoint temperature (set in the Node-RED environment)—the error did not exceed 0.15 °C. Such small error was mainly due to the fact that the energy supplied by the workers and by sun radiation was small compared to the overall energy demand. Figure 12 presents temperatures in all the six zones. The temperatures are very similar, and the small difference comes from the fact that two zones are "internal", while four zones are "corner zones".

6 Conclusions

The aim of the work presented in this paper was to create the building simulation system which allows to test different control algorithms in the *hardware-in-the-loop* simulations. The created simulator main functions include: simulation of heat transfer through building compartments of various types, simulation of dynamics of room temperature changes based on energy balance, simulation of dynamics of changes in carbon dioxide concentration in rooms, simulation of supply air at selected temperature, simulation of selected heating systems (convector heaters and heating floor). The designed software has a modular construction, allowing easy creation of schemes of any buildings, even with high levels of complexity. It is also possible to easily extend the program with new systems.

Although different control algorithms were tested, this paper reports the results obtained using a Generalized Predictive Control algorithm. For high performance, the algorithm was implemented in C++ language, and initially tested in Simulink. After major bugs were corrected, the algorithm was migrated to the Node-RED environment, using the N-API library. The Node-RED was installed on Beaglebone Black, and both the building simulator and the GPC controller were connected using the UDP protocol. The simulations confirmed flexibility and good performance of the proposed solution.

Acknowledgment. The work described in this paper is within the project "Synergiczny system automatyki budynkowej zintegrowany z układami optymalizacji komfortu i klimatu w budynkach—SSAB", which is co-sponsored by WND-RPSL.01.02.00-24-0853/17-001, Regional Operating Program for Silesian Voivodship for years 2014–2020.

References

1. Andersen, K.K., Madsen, H., Hansen, L.H.: Modelling the heat dynamics of a building using stochastic differential equations. Energy Build. **31**(1), 13–24 (2000). https://doi.org/10.1016/S0378-7788(98)00069-3
2. Lawrynczuk, M.: Accuracy and computational efficiency of suboptimal nonlinear predictive control based on neural models. Appl. Soft Comput. **11**(2), 2202–2215 (2011). https://doi.org/10.1016/j.asoc.2010.07.021
3. Lawrynczuk, M.: Suboptimal nonlinear predictive control based on multivariable neural Hammerstein models. Appl. Intell. **32**(2, SI), 173–192 (2010). https://doi.org/10.1007/s10489-010-0211-x
4. Jabłoński, K., Grychowski, T.: Fuzzy inference system for the assessment of indoor environmental quality in a room. Indoor Built Environ. **27**(10), 1415–1430 (2018). https://doi.org/10.1177/1420326X17728097
5. Hudson, G., Underwood, C.: A simple building modelling procedure for MATLAB/SIMULINK. In: Proceedings of the International Building Performance and Simulation Conference, Kyoto Japan, vol. 2, pp. 777–783. Citeseer (1999)
6. Perera, D., Winkler, D., Skeie, N.O.: Multi-floor building heating models in MATLAB and Modelica environments. Appl. Energy **171**, 46–57 (2016). https://doi.org/10.1016/j.apenergy.2016.02.143
7. Freitas, S., Catita, C., Redweik, P., Brito, M.: Modelling solar potential in the urban environment: State-of-the-art review. Renew. Sustain. Energy Rev. **41**, 915–931 (2015). https://doi.org/10.1016/j.rser.2014.08.060
8. Lawrynczuk, M.: Nonlinear state-space predictive control with on-line linearisation and state estimation. Int. J. Appl. Math. Comput. Sci. **25**(4), 833–847 (2015). https://doi.org/10.1515/amcs-2015-0060
9. Söderström, T., Stoica, P.: System Identification. Prentice Hall International, Inc., New York (1989)
10. Bismor, D.: Extension of LMS stability condition over a wide set of signals. Int. J. Adapt. Control Signal Process. **29**(5), 653–670 (2015). https://doi.org/10.1002/acs.2500
11. Bismor, D., Pawelczyk, M.: Stability conditions for the leaky LMS algorithm based on control theory analysis. Arch. Acoust. **41**(4), 731–740 (2016). https://doi.org/10.1515/aoa-2016-0070
12. Node-red (2019). https://nodered.org/
13. N-API documentation (2019). https://nodejs.org/api/n-api.html

Evaluation of the Control Improvement Benefits for Campaign Profiles - Nitric Acid Production Example

Paweł D. Domański[1(✉)], Sebastian Golonka[2], Piotr Marusak[1], Bartosz Moszowski[2], and Ewa Wolff[2]

[1] Institute of Control and Computation Engineering, Warsaw University of Technology, ul. Nowowiejska 15/19, 00-665 Warsaw, Poland
{p.domanski,p.marusak}@ia.pw.edu.pl
[2] Grupa Azoty, Zakłady Azotowe Kędzierzyn S.A., Mostowa 30A, BOX 163, 47-220 Kędzierzyn-Koźle, Poland
{sebastian.golonka,bartosz.moszowski,ewa.wolff}@grupaazoty.com

Abstract. Control system performance significantly contributes to the overall process efficiency. Thereby, appropriate improvement measures are required to measure it. Many practical approaches use KPIs (Key Performance Indicators) in form of single numbers. These measures are further used to estimate possible improvement. However, there are situations where a single number cannot be used as the process efficiency measure. The performance of the nitric acid production installation is such an example. Instead of the single number, there is used a set of numbers in form of the campaign curve considered in the time domain. This problem is addressed and effectively solved. The proposed approach may be also used for other types of processes whose performance is non-stationary and is described by any type of the curve defined in time or any other variable domain.

Keywords: Control quality · Same limit · Performance · Nitric acid campaigns

1 Introduction

Industrial processes are complex and variable systems with many internal and external cross-correlations. These facts cause a lot of challenges for the control engineers. Control system performance issues are mostly addressed by the base control functionality almost always using single element or cascaded loops equipped with the PID algorithm. Properly selected and tuned base control philosophy enables plant operation and brings financial benefits [9]. Further improvement can be reached through the implementation of the Advanced Process Control (APC) and/or Process Optimization (PO) [14,19].

Any control rehabilitation project, similarly to other investment initiatives requires financial justification. Estimation methods to calculate the tangible

A. Bartoszewicz et al. (Eds.): PCC 2020, AISC 1196, pp. 1030–1042, 2020.
https://doi.org/10.1007/978-3-030-50936-1_86

results of the control system improvement have been investigated for years. The respective approaches have been proposed [3,21,27]. There are three well established approaches called: *same limit*, *same percentage* and *final percentage* rules [2,3]. The most popular one, called *the same method algorithm*, assumes that the mitigation of the process fluctuations will bring quantitative results [1], because the process may be safely shifted closer to the constraints and can operate with higher efficacy. The approach has been validated in several industrial projects in chemistry [9–11,13]. The effect of the minimized variance due to the better control has been further investigated towards economic performance indexes, like a quadratic function, a linear function with constraints and the so-called *cliff-tent* performance function [3]. The methodology has started with the Gaussian assumption and has been extended towards fat-tailed distributions [7,8,12].

Apart from above methodology industry have developed several other, simplified customized approaches. One of them is called the Data Reduction Method [6,20] and uses histogram. It is interesting to notice that the method uses median as the robust mean value estimator, and thus assumes non-Gaussian behavior in data. Another approach is strictly heuristic and is called the Best Operator Method [6]. It is based on the comparison of the historical operational data.

The same limit approach can by successfully applied to any process, which may be described by the single benefit variable. However, there are situations, when the process cannot be described by any single stationary variable/relation. There are many processes in chemical engineering, for which efficiency varies in time. Performance often depends on the catalyst fitness degrading with time. Such processes are often described by campaign curves, which show production efficiency change in time.

This paper solves that problem. The proposed methodology enables evaluation of the time-depended indexes in form of the campaign curves and gives the tools to asses their performance and predict tangible improvements due to the control system modernization. The approach is practically validated with the real process example of the nitric acid (HNO_3) production. The paper starts with the presentation of non-Gaussian statistics (Sect. 2). It is followed by the definition of the proposed method (Sect. 3). The methodology is verified on the industrial data (Sect. 4) and the paper concludes in Sect. 5 with final remarks and directions for further research.

2 Non-Gaussian Statistics

α-stable PDF belongs to the family of distributions with a characteristic function:

$$S_{\alpha,\beta,\delta,\gamma}(x) = \begin{cases} e^{-\gamma^{\alpha}|x|^{\alpha}\{1-i\beta sign(x)\tan(-\frac{\pi\alpha}{2})\}+i\delta x} & \text{for } \alpha \neq 1 \\ e^{-\gamma|x|\{1+i\frac{2}{\pi}\beta sign(x)\ln|x|\}+i\delta x} & \text{for } \alpha = 1 \end{cases} \tag{1}$$

$0 < \alpha \leq 2$ is called *stability* index, $|\beta| \leq 1$ is a *skewness*, $\delta \in \mathbb{R}$ is *location* and $\gamma > 0$ is *scale*. The family of α-stable distributions is a rich, including following functions:

– normal Gauss distributions $N(\mu, \sigma^2)$ given with $S_{2,\beta,\frac{\sigma}{\sqrt{2}},\mu}$,
– Cauchy PDF with scale γ and location δ given by $S_{1,0,\gamma,\delta}$.

Cauchy PDF is an example of the fat-tailed PDF (2). The shape for values further from mean does not decay so fast as it is with normal one. It is a symmetric function. Location factor $\delta \in \mathbb{R}$ informs about the position, while scaling $\gamma > 0$ reflects broadness.

$$PDF_{\delta,\gamma}(x) = \frac{1}{\pi\gamma}\left(\frac{\gamma^2}{(x-\delta)^2 + \gamma^2}\right) \tag{2}$$

Pearson distribution is a family of unimodal continuous PDFs that satisfy the following differential Eq. (3). Pearson described twelve families of distributions as solutions to the equation. These density functions include function skewness into histogram fitting, what extends classical estimation.

$$f'(x) = (x-d)\frac{f(x)}{ax^2 + bx + c} \tag{3}$$

The existence of outliers in data and the resulting fat tails in their distribution poses the main challenge during non-Gaussian data analysis. There are two approaches to that subject. In the first one the outliers are considered to sustain an important information. On the other hand, we may adopt the opposite assumption. The outliers are irrelevant and should be removed from the data [5]. Once the data are *clean* and free of the outliers we may use the classical approach with the normal Gaussian measures. This assumption has been applied in the proposal and evaluation of the *robust statistics*. They were introduced long ago [15], but works of Huber [16] gave them a new application feedback. They achieve good performance for data having various probability distributions, especially for normal ones. Robust methods have been developed to estimate location, scale, and regression parameters for time series affected by outliers and this feature is the most interesting in our case. The fact that the robust approach is not sufficiently used in control is probably connected with the fact that they have appeared and are mostly investigated in the research context of the chemical engineering. M-estimators with logistic psi-function for location and scale has been applied.

3 Proposed Methodology

First of all, one have to define the measure for the production performance. It is considered that this parameter depends on time. In our case the starting point is the moment of the catalyst exchange, called campaign startup. The installation starts and continues operation until the decision to terminate the campaign, i.e. to stop production and the exchange of the catalyst. Once the new catalyst is in place the new campaign starts. Due to various disturbances the efficiency curve might not be smooth. Some installation failures may happen, causing the index value for some moments deceived, just wrong. Such incidents, called the *artifacts*, have to be identified and removed from the data set.

The campaign shape is approximated by a preselected curve. There are many types of the trends and detrending approaches. In case of the nitric acid production the use of the second order polynomial seems to be a rational selection. Next, we remove the identified polynomial trend from data. We obtain residuum time series for the campaign installation performance index. Now it should have mean value close to zero. During the assessment of more campaigns than one, we average them. We calculate the average efficiency value for each day. Thus the mean campaign is evaluated. The histogram is calculated for the resulting time series of the average campaign. Its broadness reflects operational performance quality, in majority depended on the control quality.

Single histogram brings our method to the point, when we may directly apply any version of *the same limit* rule with any probability density function (PDF). Original algorithm uses normal distribution for a variable informing about economic performance. The result of this method gives improvement in form of the single number, which can be interpreted in the financial domain. The improvement zone can be easily calculated. The maximum (the worst) and minimum (the best) campaigns can be estimated similarly. We select the maximum or minimum values for each day. These datasets may be fitted with the polynomial indicating operating zone limitations. The validation of the methodology is addressed with the industrial example of the nitric acid production.

3.1 The Same Limit Algorithm

The algorithm is based on the evaluation of normal distribution for any variable. Improvement is evaluated as follows (see Fig. 1):

1. Evaluate histogram of the selected variable or the performance index.
2. Fit normal distribution to the obtained histogram which is described by two parameters: mean value and standard deviation σ.
3. It is assumed that mean value (M_{improv} for the improved system and M_{now} for the original one) is kept within the same distance from limitation. We shift the mean value towards the constraint. For the confidence level of 95% it is equal to $a = 1.65$. The mean value for the improved operation is estimated. Standard deviation σ_1 relates to the original system and σ_2 to the improved one.

$$M_{improv} = M_{now} \cdot a \cdot (\sigma_1 - \sigma_2) \tag{4}$$

4. Finally percentage improvement is calculated on basis of the following equation:

$$\Delta M = 100 \cdot \frac{M_{improv} - M_{now}}{M_{improv}} \tag{5}$$

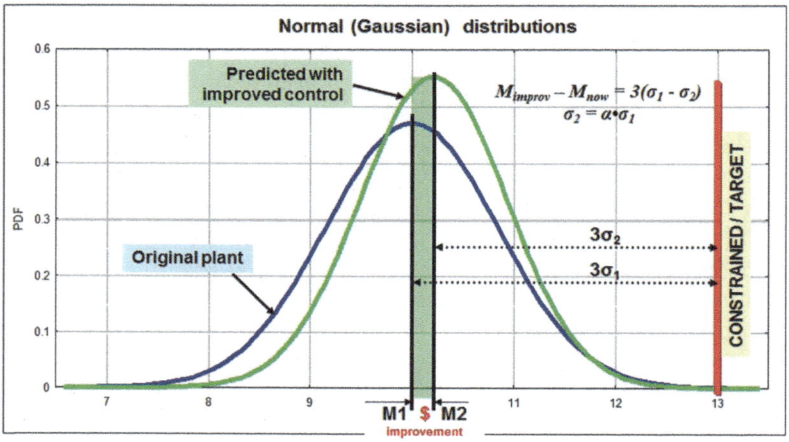

Fig. 1. Graphical representation of the Gauss-based *the same limit* method

4 Industrial Validation

Weak nitric acid is produced in separate installation using ammonia as the main input product [4,18,22,23] and consists of the following sub-processes:

1. air compression,
2. preparation of the air-ammonia mixture and ammonia oxidation,
3. preparation of the de-mineralized water and further of the high pressure steam,
4. absorption of the nitrogen oxides in the water to produce nitric acid,
5. nitric acid de-aeration and its storage.

Control aspect of the nitric acid production is reflected rather rarely in the literature, with the main impact put on the N2O emissions [17,24,25]. Production efficiency depends on the catalyst [26] and is calculated as a ratio between ammonia mass inflow and nitric acid mass outflow

$$\eta_{HNO_3} = \frac{\dot{m}_{NH_3}}{\dot{m}_{HNO_3}} \left[\frac{kg}{Mg}\right], \tag{6}$$

where: $\dot{m}_{NH_3}$ - ammonia mass inflow to the installation, $\dot{m}_{HNO_3}$ - nitric acid mass outflow. Data are rescaled and normalized to simplify the analysis and the comparison.

At first, raw production data are reviewed. They are presented in Fig. 2. There is a lot of wrong values in the time trends. Such *artifacts* frequently occur in industrial data. They are removed as they are erroneous and they may disturb evaluation. The preprocessed data time series is sketched in Fig. 3. Mean campaign is obtained through data averaging for each day (Fig. 4). Campaign operating zone is evaluated through minimal or average value data time series and further polynomial fitting. Second order polynomials are used. Next, we remove polynomial trend (Fig. 5) and calculate the histogram of the detrended data (Fig. 6). Improvement potential can be then evaluated.

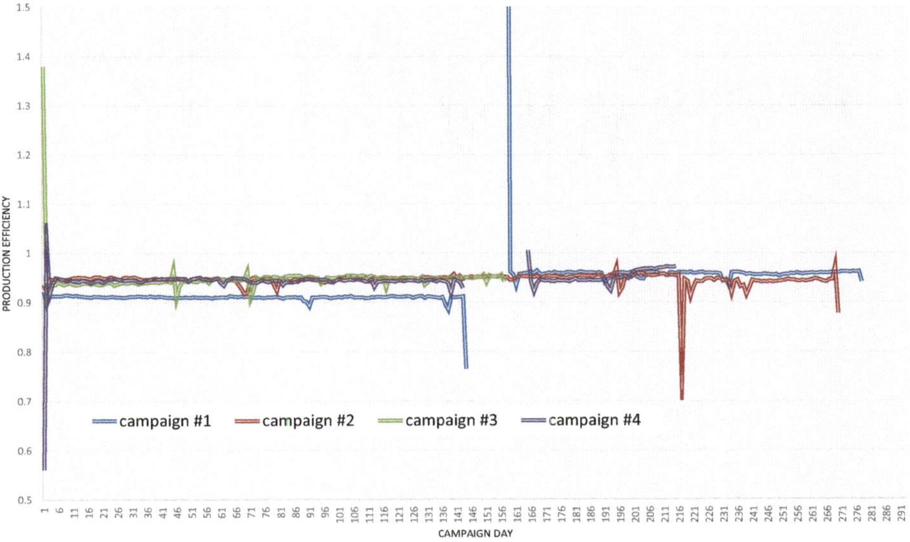

Fig. 2. Raw production data for four campaigns

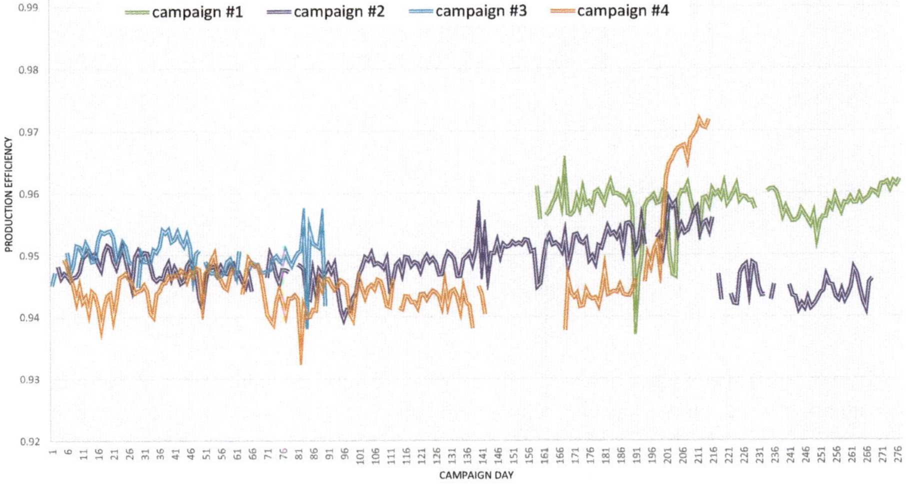

Fig. 3. Preprocessed production data for four campaigns

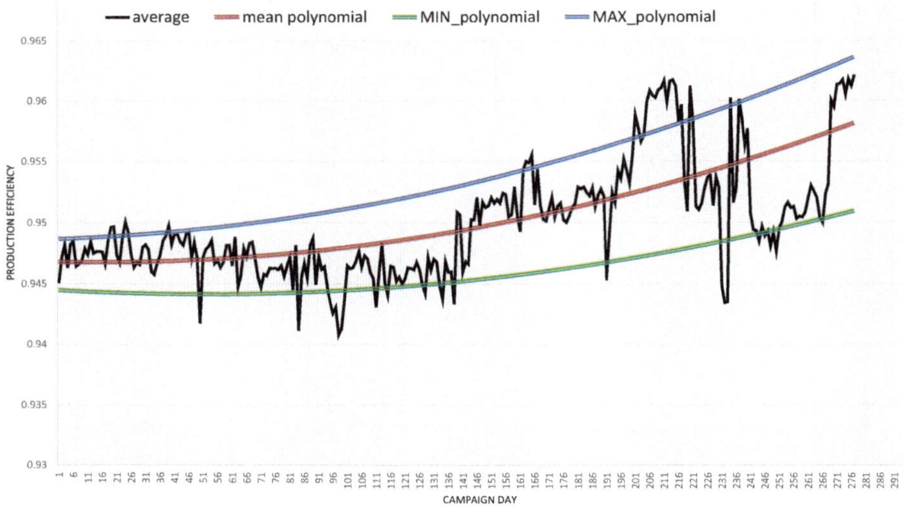

Fig. 4. Mean campaign and its operating zone

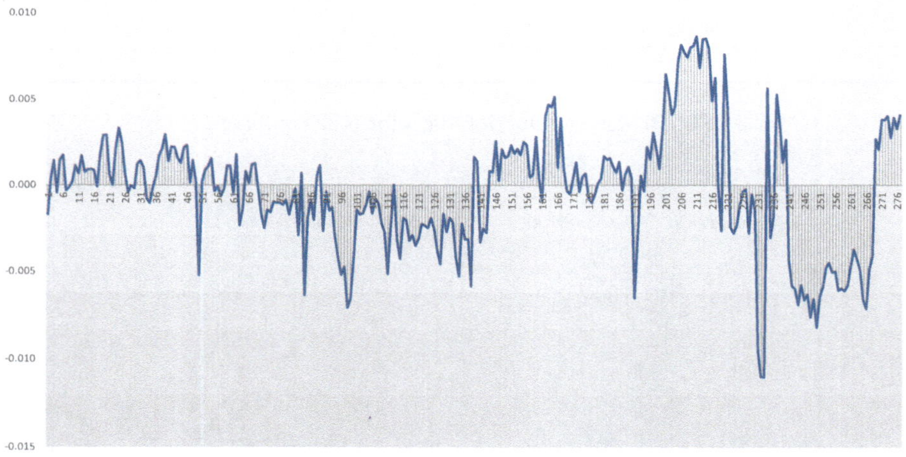

Fig. 5. Detrended mean campaign

In current research various distribution functions have been fitted to the histogram (black curves in the respective figures), namely Gauss PDF in Fig. 7, Cauchy in Fig. 8, Huber in Fig. 9, Pearson in Fig. 10 and Lévy α-stable in Fig. 11. The fitting is done using maximum likelihood estimation in all cases.

Next, the potential benefits associated with the control improvement have been evaluated for each PDF (the curve after improvement is drawn in red).

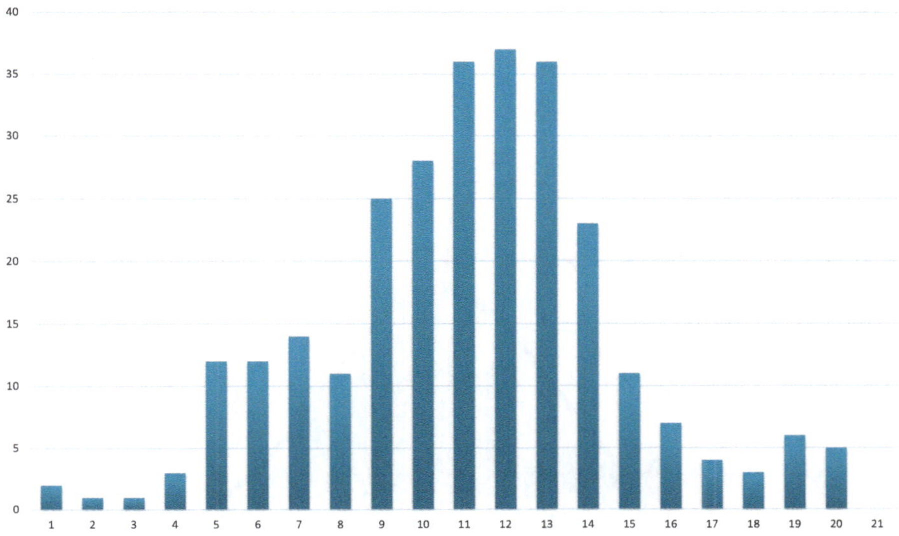

Fig. 6. Histogram for the detrended mean campaign

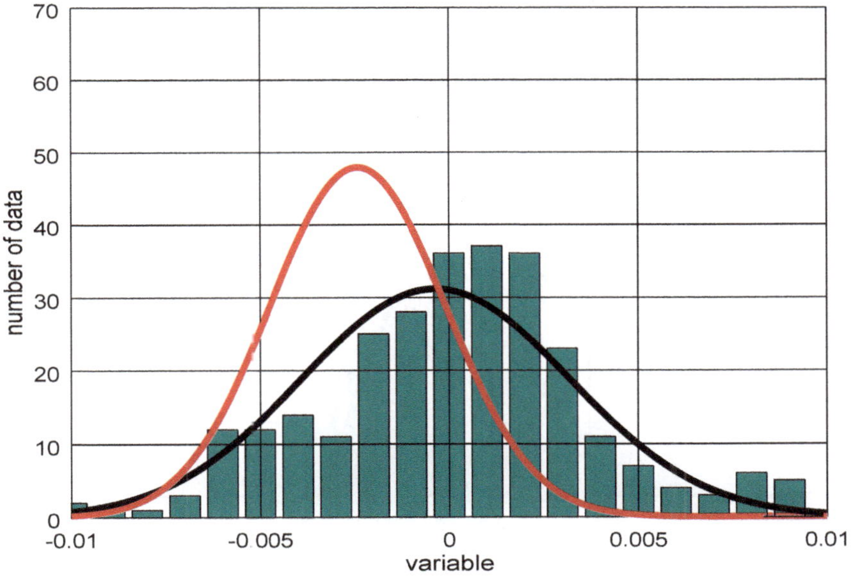

Fig. 7. Detrended campaign histogram (Gauss PDFs)

Additionally, for α-stable PDF, process control improvement after symmetrical tuning with zero skewness is added (Fig. 12).

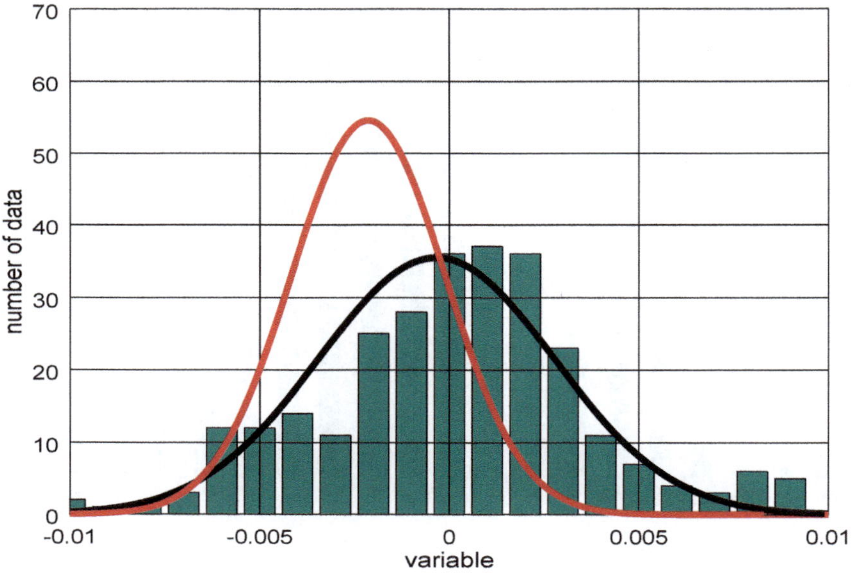

Fig. 8. Detrended campaign histogram (Huber PDFs)

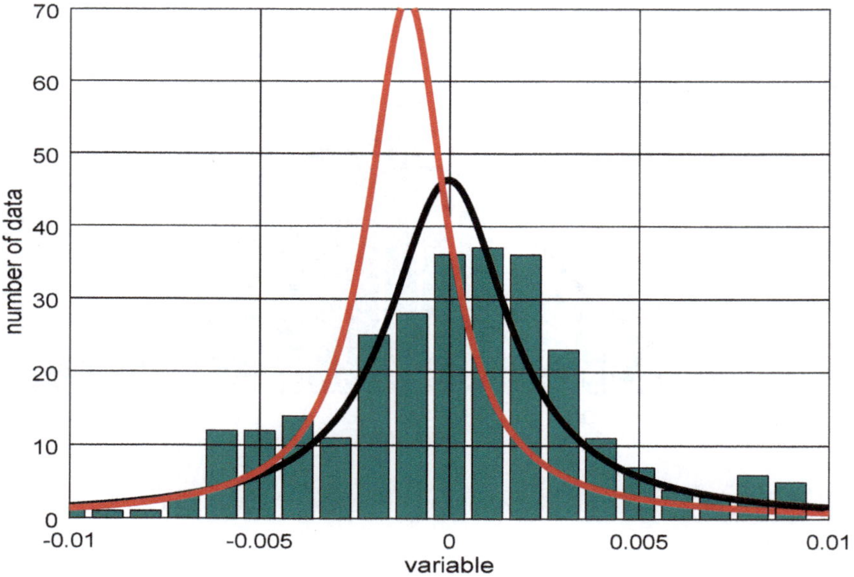

Fig. 9. Detrended campaign histogram (Cauchy PDFs)

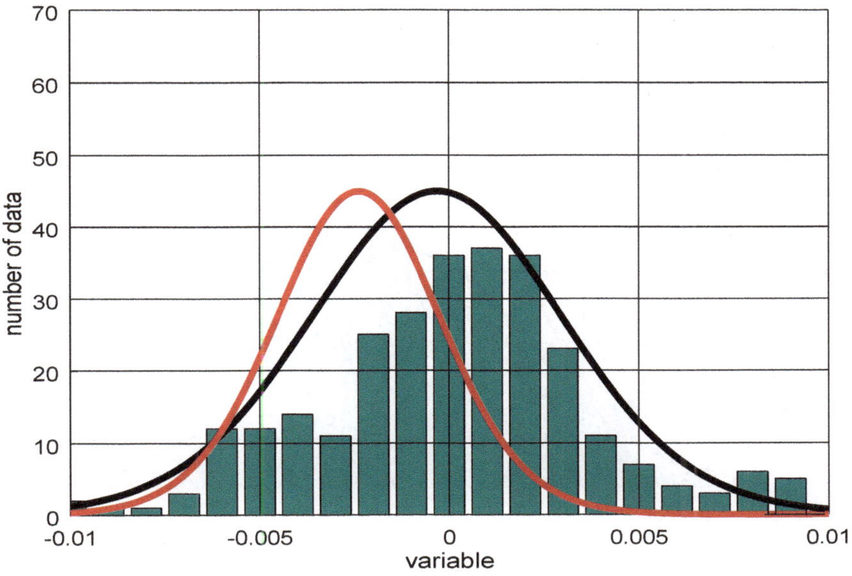

Fig. 10. Detrended campaign histogram (Pearson PDFs)

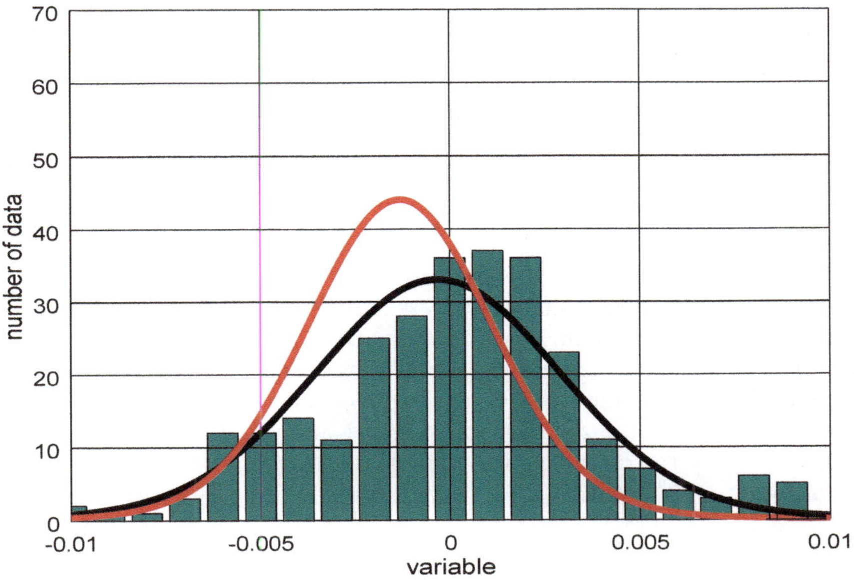

Fig. 11. Detrended campaign histogram (Lévy PDFs – L1)

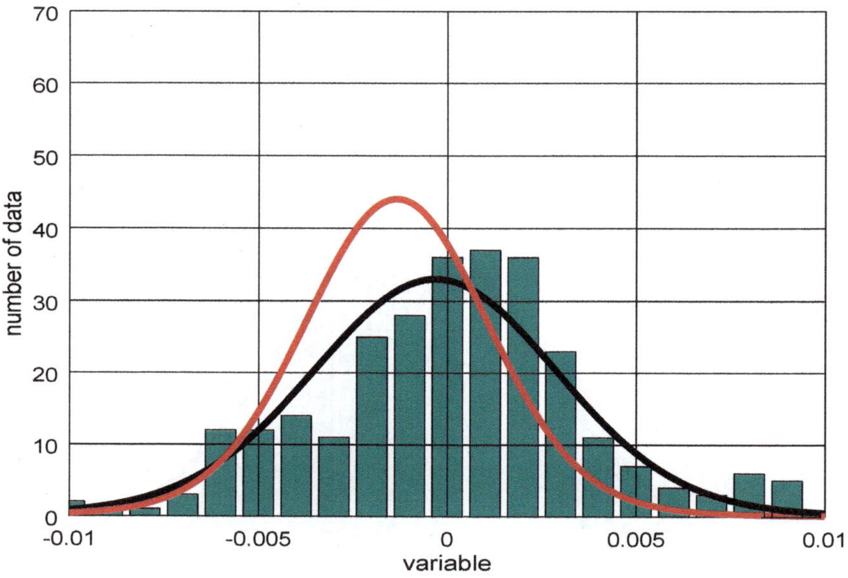

Fig. 12. Detrended campaign histogram (Lévy PDFs – L2)

Expected improvement benefit of each index (expressed in percentage and calculated in relation to the average value of the index before detrending) using all approaches is given in Table 1.

Table 1. Improvement of the indexes in percentage

Gauss	Huber	Cauchy	Pearson	L1	L2
−0.21677	−0.19047	−0.11614	−0.21677	−0.10361	−0.10245

One can easily observe that the Gauss and Pearson PDFs are the most optimistic ones (the largest numbers of the predicted benefit) and the α-stable PDF is the most pessimistic one, especially in the second case, when skewness was reduced to 0 after the improvement; though the difference between both cases is merely visible (compare red curves in Figs. 11 and 12). Cauchy approach gives similar results to the α-stable ones, while Huber is close to the Gauss, being less optimistic.

5 Conclusions and Further Research

The paper presents novel methodology for the estimation of the benefits that appear as the result of the control system rehabilitation. The proposed method applies to the situations, when the performance description is in the form of the time curve, called as the campaign in the chemical engineering. The approach is based on *the same limit* rule. The methodology has been used during the industrial project for the nitric acid production installation. The approach has enabled to evaluate the efficiency of the installation during campaigns and then to estimate improvement potential.

References

1. Ali, M.K.: Assessing economic benefits of advanced control. In: Process Control in the Chemical Industries, pp. 146–159. Chemical Engineering Department, King Saud University, Riyadh, Kingdom of Saudi Arabia (2002)
2. Bauer, M., Craig, I.K.: Economic assessment of advanced process control - a survey and framework. J. Process Control **18**(1), 2–18 (2008)
3. Bauer, M., Craig, I.K., Tolsma, E., de Beer, H.: A profit index for assessing the benefits of process control. Ind. Eng. Chem. Res. **46**(17), 5614–5623 (2007)
4. Clarke, S.I., Mazzafro, W.J., Updated by Staff: Nitric Acid. American Cancer Society (2005)
5. Daszykowski, M., Kaczmarek, K., Heyden, Y.V., Walczak, B.: Robust statistics in data analysis – a review: basic concepts. Chemometr. Intell. Lab. Syst. **85**(2), 203–219 (2007)
6. Dolenc, J.: Estimating benefits from process automation. 2007 Emerson Global Users Exchange, Grapevine, TX (2007). https://www.emersonautomationexperts.com/presentations/EstimatingBenefitsFromProcessAutomation_JohnDolenc.pdf
7. Domański, P.D.: Non-Gaussian assessment of the benefits from improved control. In: Preprints of the IFAC World Congress 2017, Toulouse, France, pp. 5092–5097 (2017)
8. Domański, P.D.: Control Performance Assessment: Theoretical Analyses and Industrial Practice. Springer, Cham (2020)
9. Domański, P.D., Golonka, S., Jankowski, R., Kalbarczyk, P., Moszowski, B.: Control rehabilitation impact on production efficiency of ammonia synthesis installation. Ind. Eng. Chem. Res. **55**(39), 10366–10376 (2016)
10. Domański, P.D., Golonka, S., Marusak, P.M., Moszowski, B.: Robust and asymmetric assessment of the benefits from improved control - industrial validation. IFAC-PapersOnLine **51**(18), 815–820 (2018). 10th IFAC Symposium on Advanced Control of Chemical Processes ADCHEM 2018
11. Domański, P.D., Ławryńczuk, M., Golonka, S., Moszowski, B., Matyja, P.: Multicriteria loop quality assessment: a large-scale industrial case study. In: Proceedings of IEEE International Conference on Methods and Models in Automation and Robotics MMAR, Miedzyzdroje, Poland, pp. 99–104 (2019)
12. Domański, P.D., Marusak, P.M.: Estimation of control improvement benefit with α-stable distribution. In: Kacprzyk, J., Mitkowski, W., Oprzedkiewicz, K., Skruch, P. (eds.) Trends in Advanced Intelligent Control, Optimization and Automation, vol. 577, pp. 128–137. Springer, Cham (2017)
13. Dziuba, K., Góra, R., Domański, P.D., Ławryńczuk, M.: Wielokryterialna ocena jakości regulacji procesu wytwarzania amoniaku. In: Zalewska, A. (ed.) PI Konferencja Naukowa Innowacje w Przemyśle Chemicznym, Warszawa, pp. 80–90 (2018)
14. Gabor, J., Pakulski, D., Domański, P.D., Świrski, K.: Closed loop NOx control and optimization using neural networks. In: IFAC Symposium on Power Plants and Power Systems Control, Belgium, Brussels, pp. 188–196 (2000)

15. Hawkins, D.M.: Identification of Outliers. Chapman and Hall, London, New York (1980)
16. Huber, P.J., Ronchetti, E.M.: Robust Statistics, 2nd edn. Wiley, Hoboken (2009)
17. Kayaert, A.: Nox control technology in nitric acid production plants. In: Schneider, T., Grant, L. (eds.) Air Pollution by Nitrogen Oxides, Studies in Environmental Science, vol. 21, pp. 687–698. Elsevier, Amsterdam (1982)
18. Kozioł, K., Sterkowicz, H., Biskupski, A., Żak, K.: New plant for neutralization of nitric acid with ammonia in grupa azoty zaklady azotowe kedzierzyn. operational experience. Przemysł Chemiczny **92**, 2211–2216 (2013)
19. Laing, D., Uduehi, D., Ordys, A.: Financial benefits of advanced control. Benchmarking and optimization of a crude oil production platform. In: Proceedings of American Control Conference, vol. 6, pp. 4330–4331 (2001)
20. Latour, P.R., Sharpe, J.H., Delaney, M.C.: Estimating benefits from advanced control. ISA Trans. **25**(4), 13–21 (1986)
21. Marlin, T.E., Perkins, J.D., Barton, G.W., Brisk, M.L.: Benefits from process control: results of a joint industry-university study. J. Process Control **1**(2), 68–83 (1991)
22. Moszowski, B., Wajman, T., Sobczak, K., Inger, M., Wilk, M.: The analysis of distribution of the reaction mixture in ammonia oxidation reactor. Pol. J. Chem. Technol. **21**(1), 9–12 (2019)
23. Moszowski, B., Wolff, E.: Experience gained during the mechanical and technological start-up of the nitric acid production plant. Przemysł Chemiczny **92**, 2207–2210 (2013)
24. Pérez-Ramirez, J., Kapteijn, F., Schőffel, K., Moulijn, J.A.: Formation and control of N2O in nitric acid production: where do we stand today? Appl. Catal. B **44**(2), 117–151 (2003)
25. Rigo, H.G., Mikucki, W.J., Davis, M.L.: Control of nitrogen oxide emissions for nitric acid plants. In: Barrekette, E.S. (ed.) Pollution. Environmental Science Research, vol. 2, pp. 278–287. Springer, Boston (1973)
26. Stefanova, M., Chuturkova, R.: Research of the efficiency of a secondary catalyst for nitrous oxide emission reduction at a nitric acid plant. Pol. J. Environ. Stud. **23**(5), 1875–1880 (2014)
27. Wei, D., Craig, I.: Development of performance functions for economic performance assessment of process control systems. In: AFRICON 2009, pp. 1–6 (2009)

A New Fuzzy Logic Decoupling Scheme for TITO Systems

Paweł Dworak[1(✉)] and Sandip Ghosh[2]

[1] West Pomeranian University of Technology in Szczecin,
ul. 26 Kwietnia 10, Szczecin, Poland
pawel.dworak@zut.edu.pl

[2] Indian Institute of Technology, Banaras Hindu University,
Varanasi 221005, UP, India

Abstract. In the paper fuzzy logic methods for dynamic decoupling of multi-input multi-output (MIMO) dynamical systems are analysed. A structure of the fuzzy precompensator, which may be used instead classical ideal and inverted decoupling control schemes, is presented. The proposal is illustrated by series of numerical simulations.

Keywords: Dynamic decoupling · MIMO systems · Fuzzy logic

1 Introduction

Despite many years of development control of multi-input multi-output (MIMO) dynamic systems enjoys a continuous attention. One of the most challenging requirement to satisfy in practice in MIMO control systems is a dynamic decoupling problem. Thus many researchers still analyze the problem and propose new ideas to solve it. [22,23] gives necessary and sufficient conditions for the existence of diagonal, block-diagonal, and triangular decoupling controllers for non-square plants and systems with non-unity feedback, with one or two degree-of-freedom controller configuration. Further, [24] has proposed a condition to check the existence of one-degree-of-freedom block decoupling controller. Parameterization of block decoupling controllers along with solving an optimal problem is proposed in [17]. [9] considers MIMO, proper, lumped and linear time invariant systems and gives analytical expressions of the Input/output (I/O) decoupling problem by the use of two-parameter stabilizing control.

Due to the plant model uncertainties and/or plant nonlinear behaviour the ideal dynamic decoupling is a very difficult task. Many of proposed control methods utilize advance techniques and controllers including optimisation methods, fuzzy logic and model predictive controllers. Apart of that many of them simply boils down to minimization of the coupling effects instead of full decoupling.

In [15], a robust decoupling controller for uncertain MIMO systems has been proposed, where uncertainty of model parameters and the desired performance is taken into account and the min-max non-convex optimization problem is used in

A. Bartoszewicz et al. (Eds.): PCC 2020, AISC 1196, pp. 1043–1054, 2020.
https://doi.org/10.1007/978-3-030-50936-1_87

the controller design. In [2,6,7] switching, fuzzy and neural decoupling controllers are constructed in order to control the nonlinear MIMO systems. [19] presents a survey on decoupling control based on multiple plant models.

Practically realizable dynamic decoupling control systems have been created for specific two-input two-output (TITO) plants. In [25] an automatic two step procedure for tuning of PID controller for a two-input two-output (TITO) process is presented. Tuning PID controllers procedures in a decentralised control systems are also presented in [10,11,28]. [1,8,21,27,31] solve the problem with the use of MPC algorithms, [4,9,12,13,15] use classical, simplified or inverted decoupling control schemes. Authors of [12,13] utilise fuzzy logic algorithms to solve the task.

In the paper a new fuzzy logic decoupling scheme for TITO systems is presented. The fuzzy algorithm here mimics the classic TITO control schemes. For the sake of simplicity of presentation we compare it with and use to decouple linear plants. However, its construction allows one to adapt and being used not only for linear, perfectly modeled plants but, first of all, uncertain and nonlinear ones and in case the ideal decoupling is not realizable. Here the scheme of the fuzzy block and stability of the system are presented and analysed. The obtained results are compared with the classical control schemes to show that it allows us to obtain good results when a dynamic decoupling comes into effect.

The remaining portion of the paper is organized as follows. In Sect. 2 we present a classical approach to control MIMO plants. Then, in Sect. 3 a scheme of the proposed fuzzy block is described. Stability analysis of the analysed control scheme is described in Sect. 4. Result of simulations of selected TITO plants are presented and discussed in Sect. 5. The paper ends with conclusion and some final remarks.

2 Classical Approach to Decoupling of TITO Plans

The classic decoupling systems used in practice are usually constructed for dynamic TITO objects described by the transmittance matrix

$$\mathbf{P}(s) = \begin{bmatrix} p_{11}(s) \; p_{12}(s) \\ p_{21}(s) \; p_{22}(s) \end{bmatrix} \tag{1}$$

with particular transfer functions $p_{ij}(s)$ of the form

$$p_{ij}(s) = \frac{K_{ij}}{T_{ij}s + 1} e^{-s\tau}, i, j = 1, 2 \tag{2}$$

A typical approach to control such TITO dynamic system is a decentralized control which means that the TITO system is treated like two single-input single-output (SISO) systems. Such separated loops are easier to analyze and tune. Interactions between them are usually neglected and treated as a disturbance. An example of such control scheme is presented in Fig. 1.

To minimize coupling effects, one has to appropriate pair of the plant inputs and outputs. To do that, one can use several interaction measuring methods. In

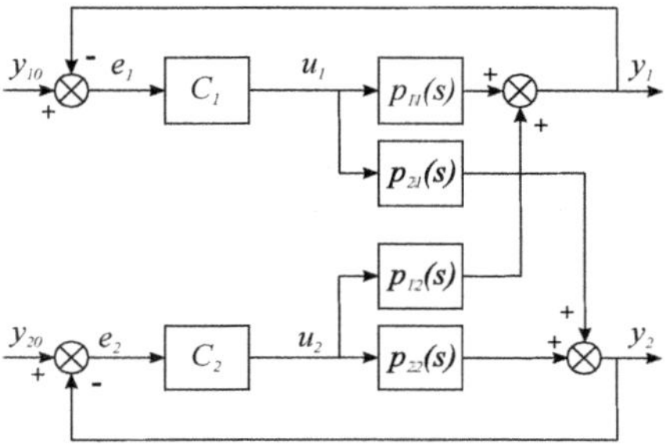

Fig. 1. Structure of the decentralised control system

general case, for MIMO systems, the most popular is the Relative Gain Array (RGA) [3] and its further modifications [16]. Apart from RGA there are also Hankel Interaction Index Array (HIIA) [29], Participation Matrix (PM) [26] and so on.

In case of TITO plants the perfect decoupling seems to be quite easy to solve by using some easy-to-calculate precompensators. Its general scheme is presented in Fig. 2. Calculating elements $d_{ij}(s)$ of the decoupler one may shape the transfer function between plants outputs $y_i, i = 1, 2$ and its "new inputs" $q_j, j = 1, 2$. Of all the possibilities, two (Fig. 3), called simplified and inverted decoupling, are utilized most frequently. Calculating elements

$$d_{12}(s) = \frac{-p_{12}(s)}{p_{11}(s)}, d_{21}(s) = \frac{-p_{21}(s)}{p_{22}(s)} \tag{3}$$

one obtain the following transfer functions of the decoupled system

$$\mathbf{K}(s) = \begin{bmatrix} p_{11}(s) - \frac{p_{12}(s)p_{21}(s)}{p_{22}(s)} & 0 \\ 0 & p_{22}(s) - \frac{p_{12}(s)p_{21}(s)}{p_{11}(s)} \end{bmatrix} \tag{4}$$

for the simplified and

$$\mathbf{K}(s) = \begin{bmatrix} p_{11}(s) & 0 \\ 0 & p_{22}(s) \end{bmatrix} \tag{5}$$

in the case of inverted decoupling.

Remark 1. In case of plants with more inputs and outputs finding a decoupling controller or schemes like in Fig. 2, is much more complicated and usually needs more advanced techniques. Additionally, we have to remember that the above methods are quite easy to implement for square plants i.e., plants with the same number of inputs and outputs. In case of the right or left invertible plants with more inputs or outputs respectively, we cannot divide the system into several separated SISO control loops. In such cases either a dedicated dynamic decoupling technique or MPC algorithms may be used.

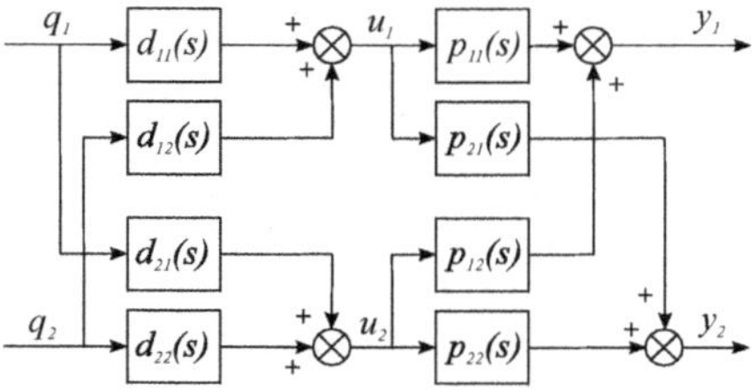

Fig. 2. Structure of the classic decoupler

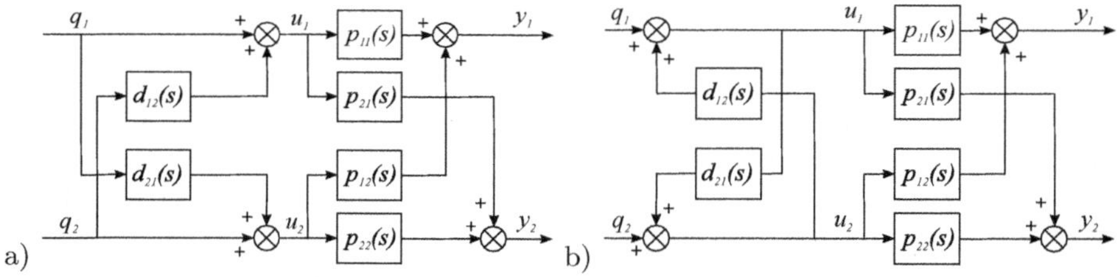

Fig. 3. Structure of the a) simplified and b) inverted decoupler

3 Fuzzy Lead-Lag Element

As we see the ideal decoupling of the TITO plant is theoretically possible with the use of a simple lead-lag element. However, in practice due to the model uncertainties and/or plant nonlinearity such task is not realizable. Instead of full decoupling we should try to find methods for considerable reductions of loops interactions. To realize that we need more flexible element than a classic lead-lag one, an element which, on one side, would be easy, on the other side, more universal and elastic in configuration. To realize such task we propose to use fuzzy logic algorithms, and to create a new fuzzy lead-lag element. Lead-lag elements are quite popular in control systems as they allow to shape the frequency characteristic. One may also easily find a fuzzy logic algorithms used to adapt parameters of a classic lead-lag compensators [18,20]. In a paper we propose a completely new idea of replacing an classic lead-lag element by a fuzzy one, as in Fig. 4.

Structure of the proposed fuzzy lead-lag element is presented in the Fig. 5. Its functionality is to mimic behaviour of the classical lead-lag element. So it has one input and one output. A fuzzy inferring rules utilise two signals q and q_p, first an input of the lead-lag element, second a feedback form the element output. The inferring output is an increment of the output signal. That is why an integral $1/s$ is used after the fuzzy block. The whole structure is complemented by two gains: K_{out} and K_{in} which together with the shape of the membership functions of fuzzy input and output sets are used to tune the element behaviour. In this work we assume that the parameters we use to tune the element are gains K_{out}

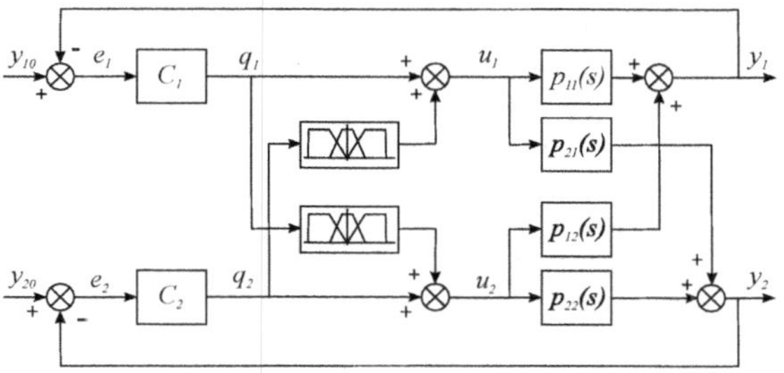

Fig. 4. Structure of the control system; simplified decoupling with a fuzzy lead-lag element

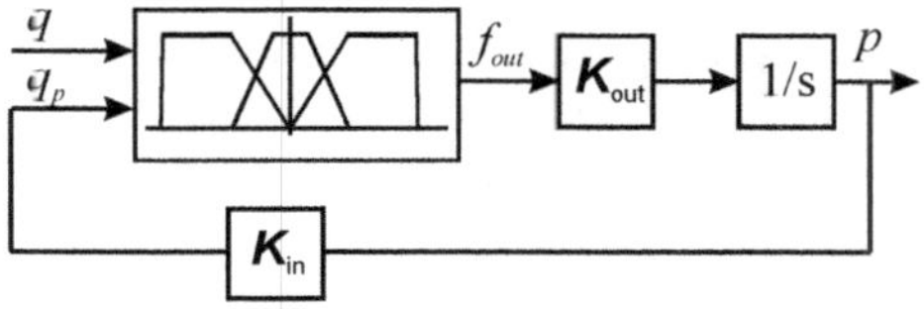

Fig. 5. Structure of the fuzzy lead-lag element

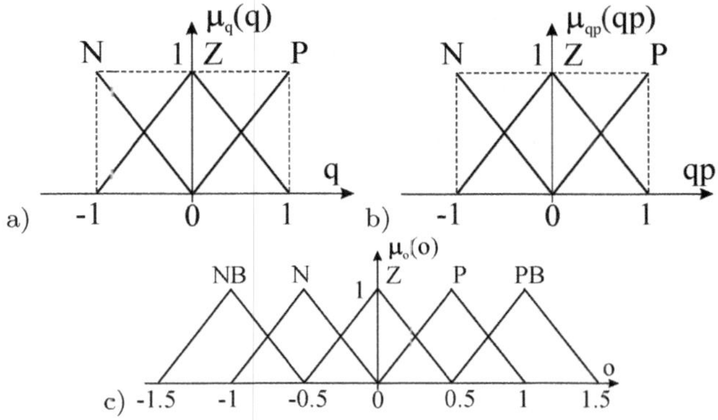

Fig. 6. Membership functions of a) input 'q', b) input 'qp' and c) output 'o'

and K_{in} only, whereas all other parameters are constant. The adapted fuzzy membership functions of the element inputs and output are presented in Fig. 6.

Three fuzzy sets for inputs fuzzification, five for output description together with a typical triangle membership functions are enough to construct a fully functional element. The inferring rules adopted here are presented in Table 1. According to these rules the positive q signal increases the element output, which then is decreased by a negative feedback q_p. After the transient, the output remains constant. Its value and also signal rise and fall rates depend on the gains K_{out} and K_{in}.

Table 1. Inferring rules of the fuzzy decoupler

qp/p	N	Z	P
N	Z	P	PB
Z	N	Z	P
P	NB	N	S

4 Stability Analysis

To analyse stability conditions of the synthesized control system we use a structural decomposition method proposed by [32], used also successfully to analyse the same problems by [10,11,28]. In this method the $n \times n$ multivariable system is decomposed into n SISO systems. Each SISO system, the process seen from the free input to the free output is analysed with assumption the other $n-1$ loops are closed with $n-1$ controllers. According to the scheme presented in Fig. 7 we obtain two transfer functions

$$p_1 = p_{11} - (p_{12}K_2(1 + p_{22}K_2)^{-1}p_{21} \tag{6}$$

for the first input and first output and

$$p_2 = p_{22} - (p_{21}K_1(1 + p_{11}K_1)^{-1}p_{12} \tag{7}$$

for the second input and output. K_1 and K_2 stand for the controllers that close the first and second loop respectively. The structural decomposition allows one to draw some important conclusions about decentralized control system, first of all about stability, which can be obtained from n SISO systems. According to the theorem given in [28] the decentralized system is stable if each individual decomposed SISO system is stable.

As it was shown in [14] a TITO system with the inverted decoupling scheme (Fig. 3) may be decomposed and presented in the following form for the first

$$p_1 = p_{11} - p_{12}(K_2 + d_{21})(1 + p_{22}(K_2 + d_{21})^{-1}p_{21} \tag{8}$$

and second SISO system

$$p_2 = p_{22} - p_{21}(K_1 + d_{12})(1 + p_{11}(K_1 + d_{12})^{-1}p_{12} \tag{9}$$

and the system stability can be checked by analyzing these transfer functions denominators, i.e. $1+p_{11}(K_1+d_{12})$ and $1+p_{22}(K_2+d_{21})$ respectively. Similarly we can construct such transfer functions for other decoupling schemes, ideal or simplified decoupling ones presented in Figs. 2 and 3. Such analysis may be even simpler if the integrator element in the structure of the fuzzy lead-lag element is saturated. Then one can analyse some specific cases with minimal and maximal values of these elements.

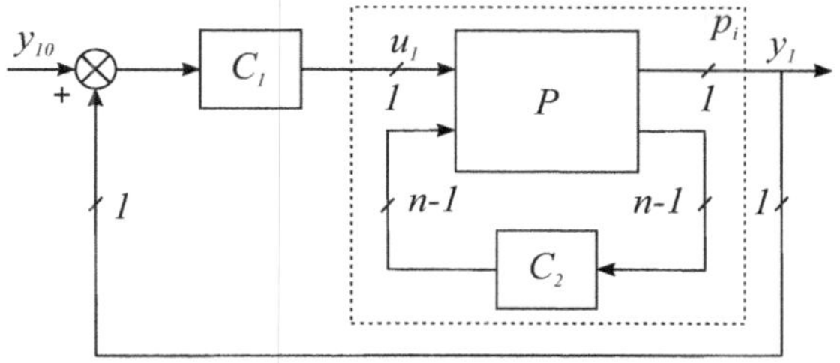

Fig. 7. Structural decomposition

5 Simulation Results

To present the efficacy of the proposed scheme we present two examples of TITO systems which are successfully decoupled with the use of proposed fuzzy lead-lag element. First one is controlled by two fuzzy PID like controllers tuned to obtain aperiodic transients, whereas the second one by two classical PI controllers.

Example 1. Let us assume a linear TITO plant described by a transfer functions matrix

$$\mathbf{P}(s) = \begin{bmatrix} \frac{1.2}{1+10s}e^{-5s} & \frac{0.6}{1+30s}e^{-10s} \\ \frac{0.5}{1+20s}e^{-10s} & \frac{1.5}{1+15s}e^{-4s} \end{bmatrix} \tag{10}$$

A classic decoupler for the simplified decoupling scheme presented in Fig. 3a, is calculated with the use of Eqs. 3, and takes the form

$$\mathbf{D}(s) = \begin{bmatrix} 1 & 0.5 \cdot \frac{1+10s}{1+30s}e^{-5s} \\ \frac{1}{3} \cdot \frac{1+15s}{1+20s}e^{-6s} & 1 \end{bmatrix} \tag{11}$$

Result of simulation carried out with such classic decoupler is presented in Fig. 8c. Replacing in this scheme both lead-lag elements by their fuzzy counterparts with the structure presented in Fig. 5, the membership functions presented in Fig. 6 and appropriately tuned K_{out} and K_{in} gains allows us to operate the system as in Fig. 8d with significantly reduced, almost zero, loops interactions. These results can be compared to the ones obtained with a distributed control system (Fig. 8a) and with static decouplers (Fig. 8b).

Example 2. As the second example, to show the effects of the discussed control methods, we have chosen, a well-known in the literature, distillation column model [30]

$$\mathbf{P}(s) = \begin{bmatrix} \frac{12.8e^{-s}}{16.7s+1} & \frac{-18.9e^{-3s}}{21s+1} \\ \frac{6.6e^{-7s}}{10.9s+1} & \frac{-19.4e^{-3s}}{14.4s+1} \end{bmatrix} \tag{12}$$

It is a frequently used model for which there are a lot of proposals of different control systems, thus may be used as a reference one. The presented in Fig. 9 simulations have been carried out with the PI controllers with gains $k_1 = 0.085$, $1/T_1 = 0.013$ for the first and $k_2 = -0.1$, $1/T_1 = -0.008$ for the controller in

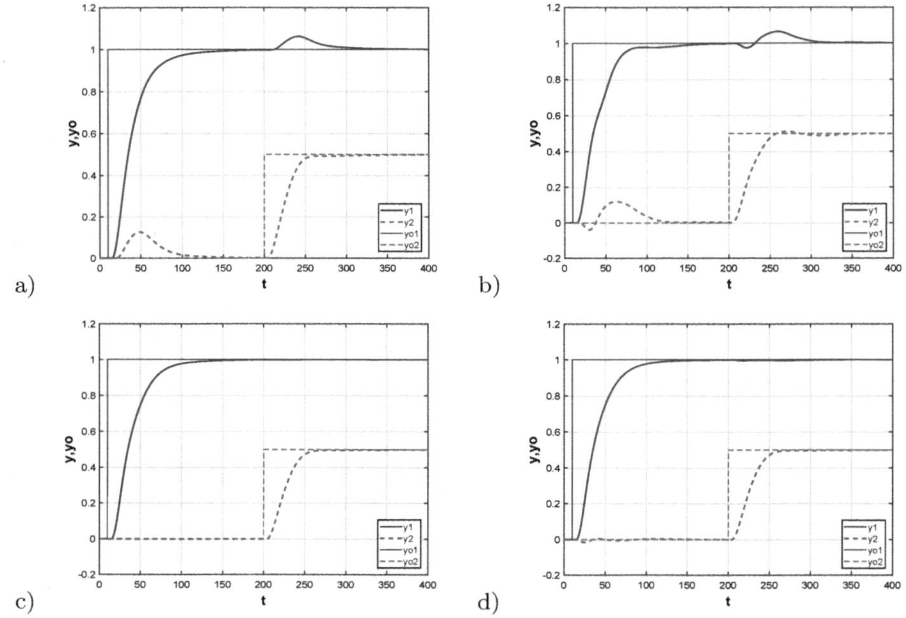

Fig. 8. Results of simulations for plant (10) in: a) closed loop with fuzzy controllers, b) closed loop with fuzzy controllers and static decoupler, c) closed loop with fuzzy controllers and dynamic decoupler, d) closed loop with fuzzy controllers and dynamic fuzzy decoupler

second loop, so the controllers are tuned similarly to most cases proposed in the literature. A distributed control system works as presented in Fig. 9a. The rest of results have been obtained with a inverted decoupling scheme (Fig. 3b). Elements of the dynamic decoupler have been calculated with the use of Eqs. 3 and takes the form

$$d_{12}(s) = \frac{18.9(16.7s+1)e^{-2s}}{12.8(21s+1)}, d_{21}(s) = \frac{6.6(14.4s+1)e^{-4s}}{19.4(10.9s+1)} \tag{13}$$

Results of simulations obtained with an inverted decoupler with a static and dynamic elements (13) are presented in Figs. 9b and 9c respectively. As expected, static decoupler elements do not reduce mutual loop interactions. A classic ideal decoupler gives very good results when a plant model is known perfectly. As we see in Fig. 9d a significant reduction of loop interactions is also possible with the use of a presented fuzzy lead-lag element.

Remark 2. A fuzzy lead-lag element may be tuned to mimic, in this paper, a decoupler calculated with the Eq. (3). However, depending on our knowledge on the plant, here its delays, one can tune the fuzzy elements for whole models obtained with (3) or without delays which can be implemented independently. In this example the fuzzy lead-lag compensators has been tuned to replace elements (13) without delays, which is realized with the gains $K_{out} = 25.88$, $K_{in} = 0.70$ and $K_{out} = 10.13$, $K_{in} = 2.87$ respectively. In case the elements d_{12} and d_{21} are modeled with delays we find the gains $K_{out} = 0.82$, $K_{in} = 0.69$ and $K_{out} = 0.22$, $K_{in} = 2.85$ respectively and the obtained results (Fig. 10) are slightly worse than those presented in Fig. 9.

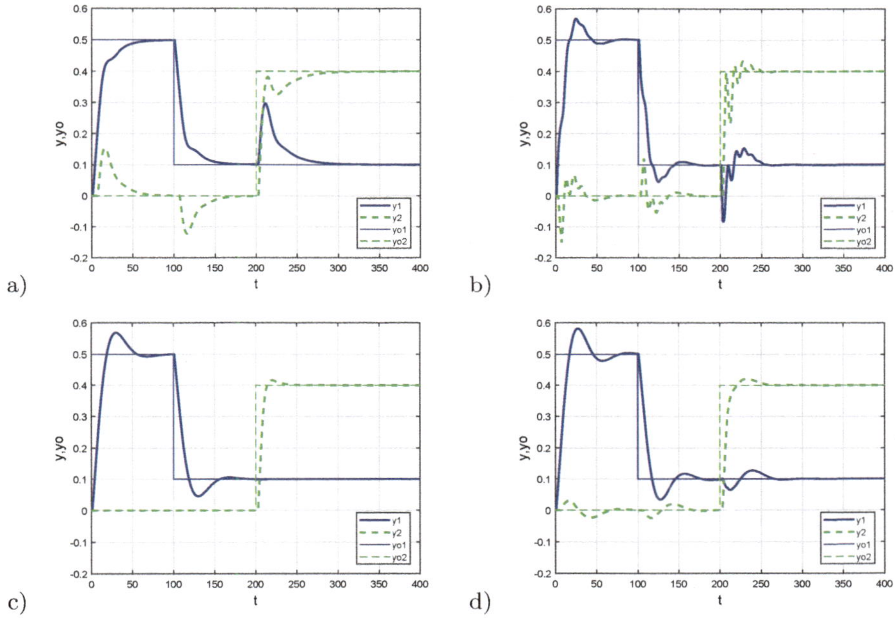

Fig. 9. Results of simulation for plant (12) in: a) closed loop with PI controllers, b) closed loop with PI controllers and static decoupler, c) closed loop with PI controllers and dynamic inverted decoupler, d) closed loop with PI controllers and dynamic inverted fuzzy decoupler, without delays

6 Conclusion

The presented in the paper fuzzy algorithm mimics the classic lead-lag element, thus it allows to realize decoupling objectives for TITO dynamic plants in the same way as in classical control schemes. However, its construction, due to features of fuzzy systems, allows one to adapt to work not only with linear, perfectly modeled plants but also uncertain and nonlinear ones. These properties may be very useful for dynamic decoupling of the nonlinear TITO (MIMO) plants which may still be treated like an open question. They also give a chance for construction of an adaptive dynamic decoupling compensator. As numerical simulations

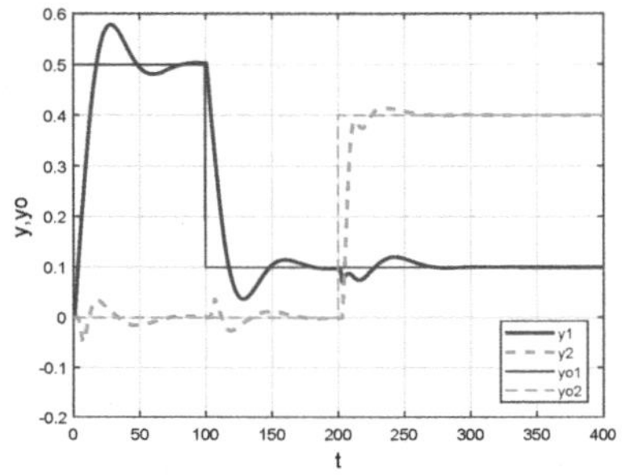

Fig. 10. Results of simulation for plant (12) in closed loop with PI controllers and dynamic inverted fuzzy decoupler, with delays

show, the proposed algorithm successfully realize decoupling objective and allows significant reduction of the input-output interactions. Apart of that it may be easily implemented and run in any industrial controller.

References

1. Arousi, F.: Predictive control algorithms for linear and nonlinear processes. Ph.D. thesis, Budapest, Hungary (2009)
2. Bańka, S., Dworak, P., Jaroszewski, K.: Design of a multivariable neural controller for control of a nonlinear MIMO plant. Int. J. Appl. Math. Comput. Sci. **24**(2), 357–369 (2014)
3. Bristol, E.H.: On a new measure of interaction for multivariable process control. IEEE Trans. Autom. Control AC **11**, 133–134 (1966). https://doi.org/10.1109/TAC.1966.1098266
4. Chiu, Ch.-S.: A dynamic decoupling approach to robust T-S fuzzy model-based control. IEEE Trans. Fuzzy Syst. **22**(5) 1088–1100 (2014). https://doi.org/10.1109/TFUZZ.2013.2280145
5. Dworak, P.: Squaring down plant model and I/O grouping strategies for a dynamic decoupling of left-invertible MIMO plants. Bull. Pol. Acad. Sci. **62**(3), 471–479 (2014). https://doi.org/10.2478/bpasts-2014-0050
6. Dworak, P.: A type of fuzzy T-S controller for a nonlinear MIMO dynamic plant. Elektronika ir Elektrotechnika **20**(5), 8–14 (2014). https://doi.org/10.5755/j01.eee.20.5.7091
7. Dworak, P., Brasel, M.: Improving quality of regulation of a nonlinear MIMO dynamic plant. Elektronika ir Elektrotechnika **19**(7), 3–6 (2013)
8. Dworak, P., Goyal, J.K., Aggarwal, S., Ghosh, S.: Effective use of MPC for dynamic decoupling of MIMO systems. Elektronika ir Elektrotechnika **25**(2), 3–8 (2019)
9. Galindo, R.: Input/output decoupling of square linear systems by dynamic two-parameter stabilizing control. Asian J. Control **18**(6), 2310–2316 (2016). https://doi.org/10.1002/asjc.1285
10. Garrido, J., Vazquez, F., Morilla, F.: An extended approach of inverted decoupling. J. Process Control **21**(1), 55–68 (2011)
11. Garrido, J., Vazquez, F., Morilla, F.: Centralized inverted decoupling control. Ind. Eng. Chem. Res. **52**(23), 7854–7866 (2013)

12. Hamdy, M., Ramadan, A.: Design of Smith predictor and fuzzy decoupling for MIMO chemical processes with time delays. Asian J. Control **19**(1), 57–66 (2017). https://doi.org/10.1002/asjc.1338
13. Hamdy, M., Ramadan, A., Abozalam, B.: Comparative study of different decoupling schemes for for TITO binary distillation column via PI controller. IEEE/CAA J. Autom. Sin. **5**(4), 869–877 (2018)
14. Hamdy, M., Ramadan, A., Abozalam, B.: A novel inverted fuzzy decoupling scheme for MIMO systems with disturbance: a case study of binary distillation column. J. Intell. Manuf. **29**, 1859–1871 (2018). https://doi.org/10.1007/s10845-016-1218-x
15. Hariz, M.B., Bouani, F.: Synthesis and implementation of a robust fixed low-order controller for uncertain systems. Arab. J. Sci. Eng. **41**(9), 3645–3654 (2016). https://doi.org/10.1007/s13369-016-2247-7
16. Khaki-Sedigh, A., Moaveni, B.: Springer. Control Configuration Selection for Multivariable Plants (2009). https://doi.org/10.1007/978-3-642-03193-9
17. Kucera, V.: Optimal decoupling controllers revisited. Control Cybern. **42**(1), 139–154 (2013)
18. Kumar, C.H., Parvatheedevi, P.: Fuzzy lead-lag controller used in control of flexible AC transmission system devices. Int. J. Adv. Res. Electr. Electron. Instrum. Eng. **4**(3), 1750–1758 (2015)
19. Liu, G., Wang, Z., Mei, C., Ding, Y.: A review of decoupling control based on multiple models. In: 24th Chinese Control and Decision Conference, pp. 1077–1081 (2012). https://doi.org/10.1109/CCDC.2012.6244171
20. Mota Sousa, F.M., Barbosa Amara, V.M., Fonseca, R.R.: Adaptive fuzzy feedforward-feedback controller applied to level control in an experimental prototype. IFAC-PapersOnLine **52**(1), 219–224 (2019)
21. Oblak, S., Skrjanc, I.: Multivariable fuzzy predictive functional control of a MIMO nonlinear system. In: IEEE International Symposium on Intelligent Control, Limassol, Cyprus, pp. 1029–1034 (2005). https://doi.org/10.1109/.2005.1467155
22. Park, K.H., Choi, G.H.: Necessary and sufficient conditions for the existence of decoupling controllers in the generalized plant model. J. Electr. Eng. Technol. **6**, 706–712 (2011). https://doi.org/10.5370/JEET.2011.6.5.706
23. Park, K.H.: Parameterization of decoupling controllers in the generalized plant model. IEEE Trans. Automat. Contr. **57**(4), 1067–1070 (2012). https://doi.org/10.1109/TAC.2011.2173410
24. Park, K.H.: A simple existence condition of one-degree-of-freedom block decoupling controllers. Automatica **51**, 14–17 (2015). https://doi.org/10.1016/j.automatica.2014.10.072
25. Pereira, R.D.O., Veronesi, M., Visioli, A., Normey-Rico, J.E.: Implementation and test of a new autotuning method for PID controllers of TITO processes. Control Eng. Pract. **58**, 171–185 (2017). https://doi.org/10.1016/j.conengprac.2016.10.010
26. Salgado, M.E., Conley, A.: MIMO interaction measure and controller structure selection. Int. J. Control **77**(4), 367–383 (2007). https://doi.org/10.1080/00207170420001976З1
27. Schmitz, U., Haber, R., Arousi, F., Bars, R.: Decoupling predictive control by error dependent tuning of the weighting factors. In: Process Control Conference, pp. 131–140 (2007)
28. Vazquez, F., Morilla, F.: Tuning decentralized PI controllers for MIMO systems with decouplings. In: 15th IFAC World Congress (2002)
29. Wittenmark, B., Salgado, M.E.: Hankel-norm based interaction measure for input-output pairing. In: Proceedings of the 15th Triennial World Congress, CD-ROM (2002). https://doi.org/10.3182/20020721-6-ES-1901.01625

30. Wood, R.K., Berry, M.W.: Terminal composition control of a binary distillation column. Chem. Eng. Sci. **28**(16), 1707–1710 (1973). https://doi.org/10.1016/0009-2509(73)80025-9
31. Zermani, M.A., Feki, E., Mami, A.: Self-tuning weighting factor to decoupling control for incubator system. Int. J. Inf. Technol. Control Autom. **2**(3), 67–83 (2013)
32. Zhu, Z.X.F.: Structural analysis and stability conditions of decentralized control systems. Ind. Eng. Chem. Res. **35**(3), 736–745 (1996)

Impact of the Lost Samples on Performance of the Discrete-Time Control System

Filip Russek and Paweł D. Domański(✉)

Institute of Control and Computation Engineering, Warsaw University of Technology, ul. Nowowiejska 15/19, 00-665 Warsaw, Poland
filip.russek@gmail.com , p.domanski@ia.pw.edu.pl

Abstract. Exceeding controller calculation time and consecutive loss of control samples impact discrete-time control system performance. It is obvious that lost samples deteriorate loop performance and it is expected that higher number of such losses should increase the deterioration gain. However, it is unclear how this impact occurs with different control structures and what is the possible scale of lost performance. Three discrete control systems: PD, PID and Internal Model Control (IMC) have been considered in the paper and then subjected to simulations with disturbances. The disruption model of the lost samples has been modeled on the basis of statistical anomalies estimated using fat-tailed distributions. Magnetic levitation process MagLev has been used as a simulation case study. This process is strongly non-linear and unstable, which makes it easier to observe the phenomenon of control disruptions. Obtained results show that simple predictive Internal Model Control strategy is the most robust against lost samples phenomenon, while PID and PD controllers are not able to track setpoint changes properly and lose stability much faster.

Keywords: Discrete-time control · PID · IMC · Lost samples · Magnetic levitation

1 Introduction

Performance of discrete-time (so called digital) control systems has attracted researchers for many years. It has been obvious from the beginning that incorporation of the sampling will cause deterioration of the control system performance as we are loosing some information. Furthermore, too long sampling period leads towards worse control performance [10]. Thus, selection of the sampling period T_p in digital control systems is crucial. It cannot be too large as we lose information and performance. It also cannot be too short as we need time to evaluate control signal or due to communication delays inside of the networked system. Sampling period selection issue for digital control may be found in many papers [11,12], together with model-based approaches [8].

A. Bartoszewicz et al. (Eds.): PCC 2020, AISC 1196, pp. 1055–1066, 2020.
https://doi.org/10.1007/978-3-030-50936-1_88

Embedded loop delay can lead to significant performance degradation [2]. The aspect of the induced delays in the networked systems has been evaluated by many researchers with two main focuses. The main emphasis of the theoretical research has been put on stability of systems with delays [1,3], especially when these delays start to exceed the sampling period. In general, if the control delay exceeds the sampling period, the system becomes multi-rate, i.e. its performance deteriorates as the plant updates do not correspond to the output samples [15]. Time delay exceeding sampling period in practice results in lost samples [16]. We will even fail to stabilize the system, if we have long periods with lost samples [13]. Missing or timeout measurements, sensor failures, communication traffic problems can cause lost samples as well [9]. The subject of lost samples can be generalized to the vacant sampling, i.e. situation when the measurement is lost. A vacant control signal can be interpreted as a random distortion to the control signal and strange disturbance in an overall procedure.

One may find different solutions to the lost sample problem. Some authors [2] propose to use simulation tools to properly design a control system to avoid or minimize the effect. In fact it means that the algorithm is sensitive during loop parametrization to avoid the effect. Another approach is to make sampling period slightly different from the control period [13], making the loop asynchronous. This approach require loop, often hardware-based, modifications. Oversampling is suggested in other cases [6].

Although the literature is rich with many researchers addressing the subject, there is no systematic analysis how the degree of sampling vacancy deteriorates the control and how different control algorithms are affected with lost samples. This paper uses the opportunity to perform such an analysis. Description starts in Sect. 2 with an introduction to the magnetic levitation. Sect. 3 presents considered controllers. Introduction is followed by simulations in Sect. 4. Section 5 concludes the results and identifies areas for further research.

2 Magnetic Levitation Process

Magnetic levitation process (MagLev) has been selected as nonlinear control benchmark [7]. The plant consists of a ferromagnetic ball kept in the air between two electromagnets EM1 and EM2 (see Fig. 1). The upper EM1 provides a vertical force F_{em1} overcoming the gravity, while the lower EM2 is used mainly for ball horizontal position stabilization F_{em2}. The distance from magnets is considered as Process Variable (PV). Considered process is very fast, nonlinear and unstable.

The analysis is performed for the simulation process. The following non linear discrete-time equations [4] have been obtained after Euler discretization with a sampling period $T_p = 0.001$ [sec].

$$\begin{aligned}
x_1(k+1) &= x_2(k) \cdot T_p + x_1(k) \\
x_2(k+1) &= -\frac{T_p}{m}\left\{F_{em1}(k) + F_{em2}(k)\right\} + T_p \cdot g + x_2(k) \\
x_3(k+1) &= \frac{T_p}{f_i\left(x_1(k)\right)}\left[k_i \cdot u_1(k) + c_i - x_3(k)\right] + T_p \cdot g + x_3(k) \\
x_4(k+1) &= \frac{1}{f_i\left(x_d - x_1(k)\right)}\left[k_i \cdot u_2(k) + c_i - x_4(k)\right] + x_4(k) \\
y(k) &= x_1(k),
\end{aligned} \tag{1}$$

where x_1 is a ball distance from EM1, x_2 – ball acceleration, x_3 – EM1 current, x_4 – EM2 current, y – the output and

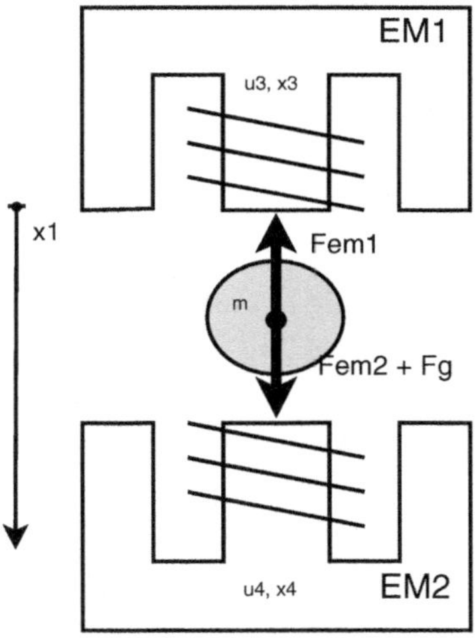

Fig. 1. MagLev system with two electromagnets EM1 and EM2 and a levitating ball [7]

$$\begin{aligned}
F_{em1}(k) &= x_3^2(k)\frac{F_{emP1}}{F_{emP2}}\exp\left(-\frac{x_1(k)}{F_{emP2}}\right), \\
F_{em2}(k) &= x_4^2(k)\frac{F_{emP1}}{F_{emP2}}\exp\left(-\frac{x_d - x_1(k)}{F_{emP2}}\right), \\
f_i\left(x_1(k)\right) &= \frac{f_{iP1}}{f_{iP2}}\exp\left(-\frac{x_1(k)}{F_{iP2}}\right).
\end{aligned}$$

Model parameters are included in Table 1.

Table 1. MagLev model parameters

Parameter	Value	Unit	Parameter	Value	Unit
m	0.0571	[kg]	g	9.81	$[m/s^2]$
F_{emP1}	$1.7521 \cdot 10^{-2}$	[H]	c_i	0.0243	[A]
F_{emP2}	$5.8231 \cdot 10^{-3}$	[m]	k_i	2.5165	[A]
f_{iP1}	$1.4142 \cdot 10^{-4}$	$[m \cdot s]$	x_{3MIN}	0.03884	[A]
f_{iP2}	$4.5626 \cdot 10^{-3}$	[m]	u_{MIN}	0.00498	–

3 Control Algorithms

Three different digital controllers are considered in the analysis: PD, PID and IMC. All the controllers share the same sampling period $T_p = 0.001$ [sec] and their parameters has been designed experimentally minimizing control error and settling time. PD controller uses the following control rule (2) with parameters: $k_p = 55$, $T_d = 4$ and $u_o = 0.3611$.

$$u(k) = k_p \cdot \varepsilon(k) + T_d \left(\frac{\varepsilon(k) - \varepsilon(k-1)}{T_p} \right) + u_o. \tag{2}$$

Parallel PID digital control rule (3) uses the following parameters: $k_p = 125$, $T_i = 85$, $K_i = 1.48$, $T_d = 6.5$, $K_d = 0.91$.

$$\begin{aligned} u(k) &= k_p \cdot \varepsilon(k) + K_d \left(\frac{T_d (\varepsilon(k) - \varepsilon(k-1))}{T_p} \right) + u_I(k), \\ u_I(k) &= u_I(k-1) + K_i \left(\frac{2T_2 (\varepsilon(k-1) - \varepsilon(k))}{T_i} \right). \end{aligned} \tag{3}$$

Internal Model Control is realized according to the general scheme [5] presented in Fig. 2. Feedforward and predictor modules are in form of a digital

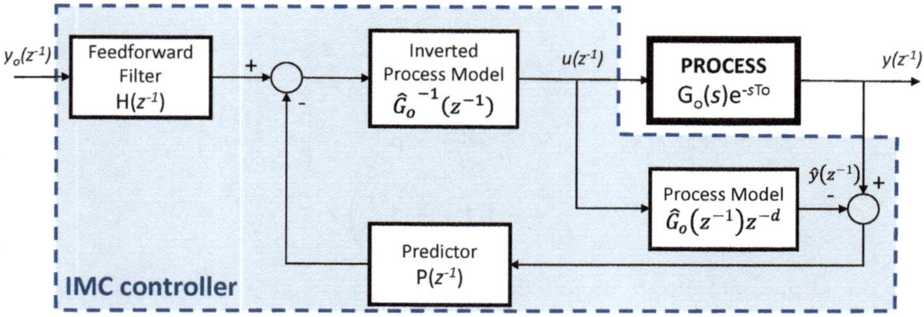

Fig. 2. IMC control rule diagram

second order transfer function described by a time constant T_q and damping ratio ζ. Predictor $P(z^{-1})$ uses $T_q = 3.0$ and $\zeta = 0.9$, while a feedforward filer $H(z^{-1})$ has $T_q = 2.0$ and $\zeta = 0.7$. Internal discrete time model is in form (4) [14].

$$G(z) = \frac{1.078^{-6} \cdot z + 1.077^{-6}}{z^2 - 1.996 \cdot z + 0.9965}, \tag{4}$$

All controllers give stable operation without an overshoot. PD and PID algorithms have steady state error. Comparison of the obtained steady state errors and settling times is presented in Table 2. Actually in the considered analysis, the controller performance is not crucial as such. The effect of lost samples is assessed in next paragraph.

Table 2. Comparison of the controllers

	IMC	PID	PD
Settling time	0.04	0.065	0.287
Steady state error	0.30%	3.97%	13.51%

A specific measure in form of mean square error between process output with lost samples y^{lost} and a perfect one y^{norm} (number of lost samples is equal to zero) is used to measure the sensitivity of a control loop against lost samples:

$$MSE^{lost} = \sum_{k=0}^{N} \left(y^{norm} - y^{lost}\right)^2. \tag{5}$$

4 Simulations

Process is modeled using discrete-time MagLev equations (1). The control goal is to stabilize ball position, i.e. process controlled variable (CV) using EM2 voltage, denoted as manilulated variable (MV). The analysis starts with the simulations of a loop with constant setpoint. Additive disturbance is added before the process in form of the α-stable distributed noise having stability factor $\alpha = 1.70$, which introduces outliers into the loop. Exemplary trends for three cases showing number of lost samples equal to *75*, *425* and *775* out of total number equal to *1500* are presented. PD control time trend is presented in Fig. 3, PID in Fig. 4 and IMC in Fig. 5.

Next plots compare considered controllers in a single plot. Figure 6 presents situation with *425* samples lost and Fig. 7 trends for *775* lost samples. It is perfectly clear that IMC control outperforms PD/PID in the robustness against lost samples.

Further comparison in Fig. 8 presents the diagram showing relationship between control performance measure and the number of lost samples. For better

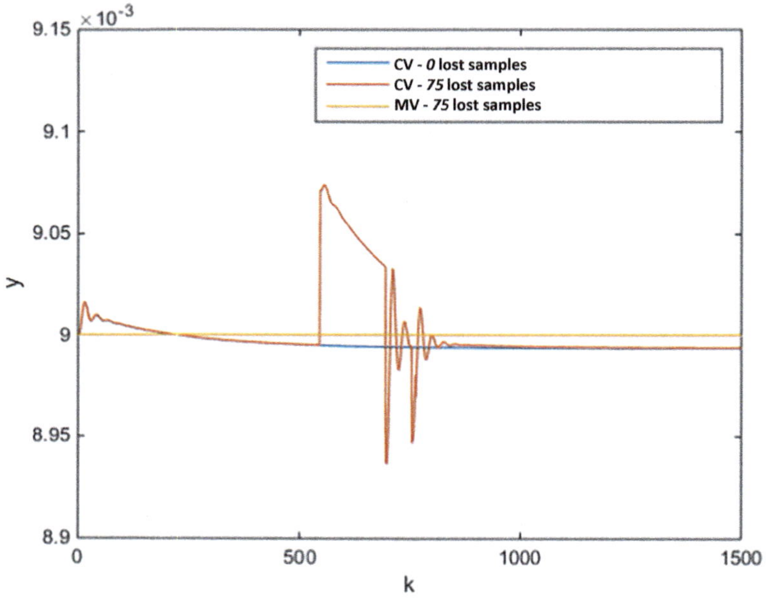

Fig. 3. Exemplary trends for PD control (number of lost samples = 75)

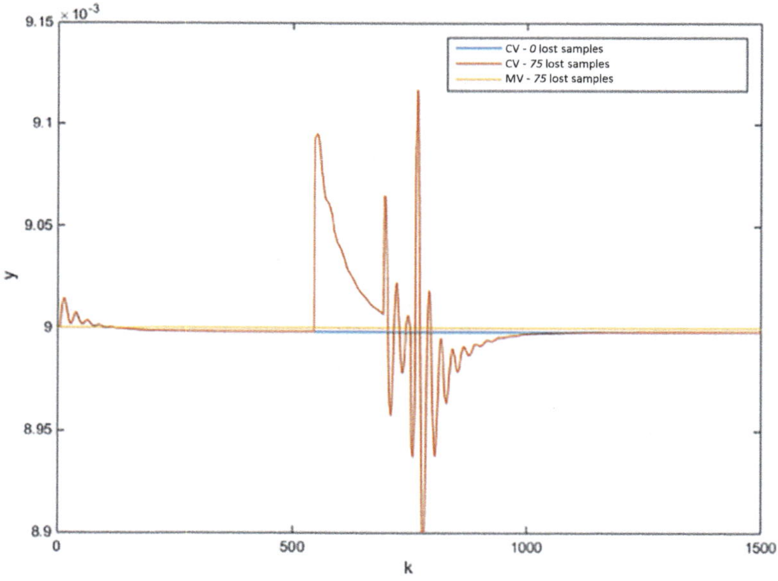

Fig. 4. Exemplary trends for PID control (number of lost samples = 75)

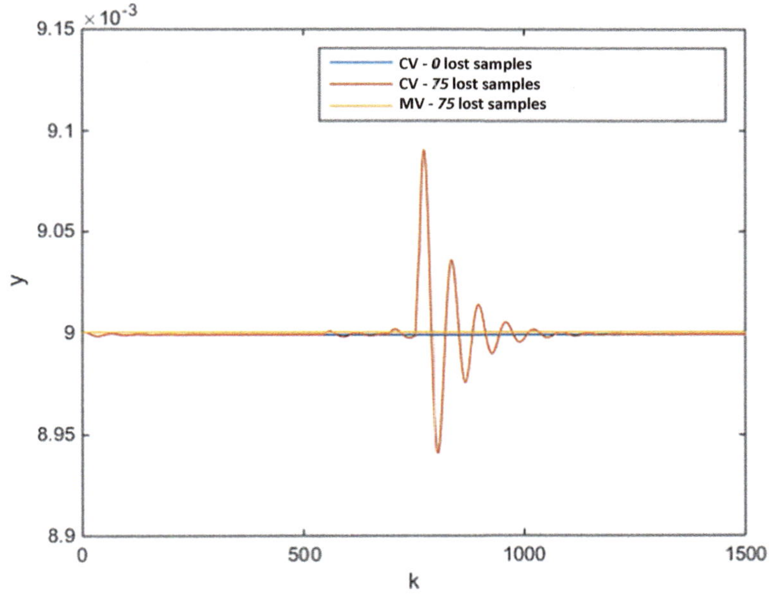

Fig. 5. Exemplary trends for IMC control (number of lost samples = 75)

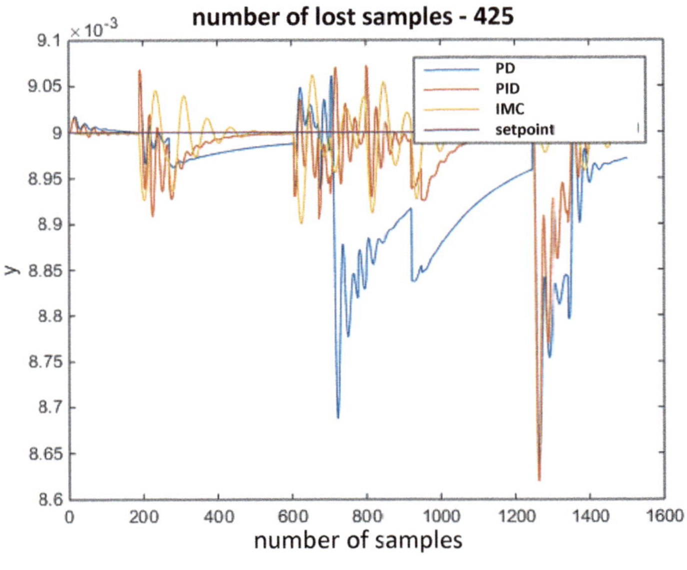

Fig. 6. Exemplary trends comparing PD, PID and IMC control (number of lost samples = 425)

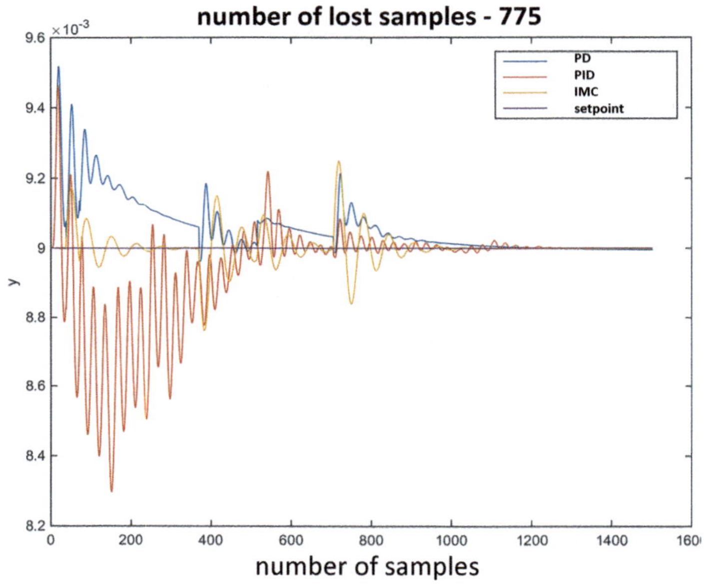

Fig. 7. Exemplary trends comparing PD, PID and IMC control (number of lost samples = 775)

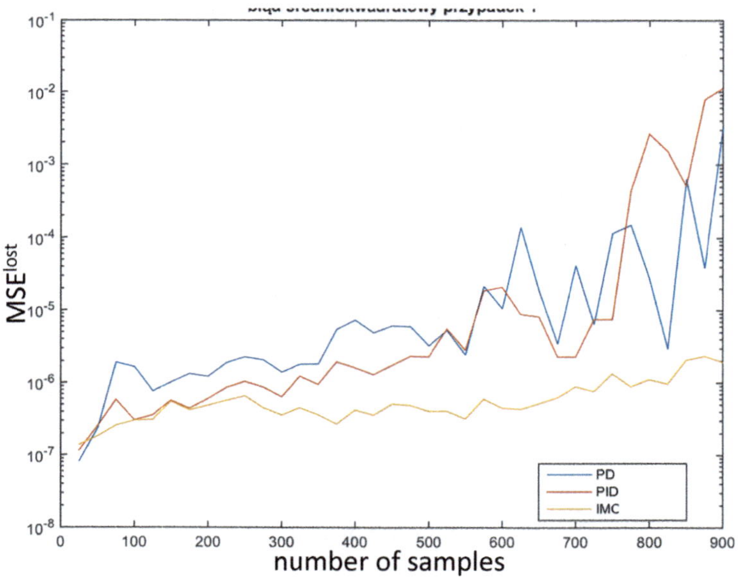

Fig. 8. General comparison of PD, PID and IMC control – case 1

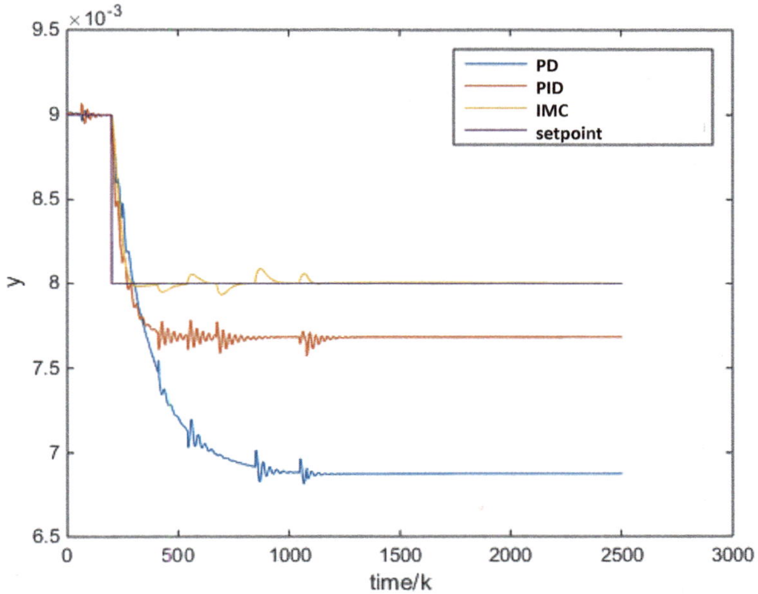

Fig. 9. Step response trends comparing PD, PID and IMC (number of lost samples = 125)

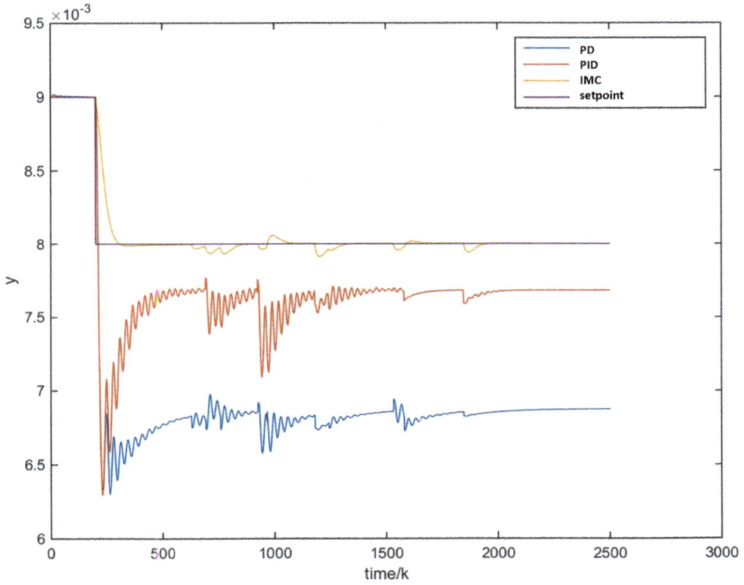

Fig. 10. Step response trends comparing PD, PID and IMC (number of lost samples = 600)

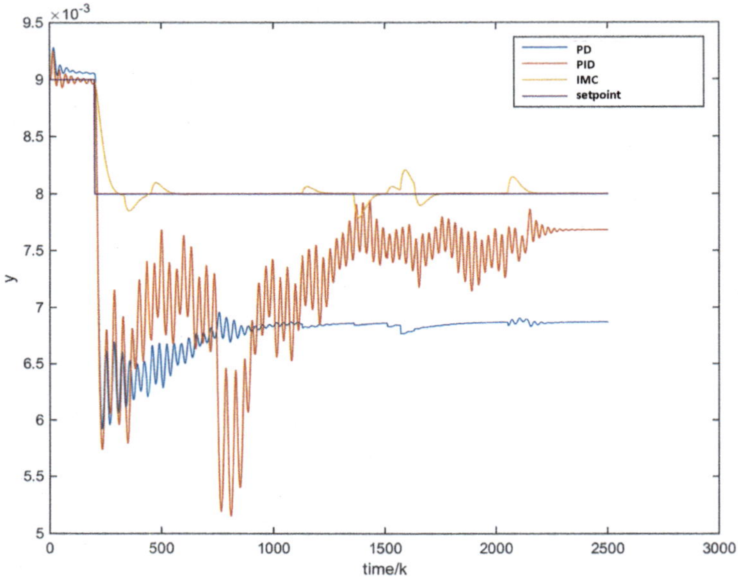

Fig. 11. Step response trends comparing PD, PID and IMC (number of lost samples = 1155)

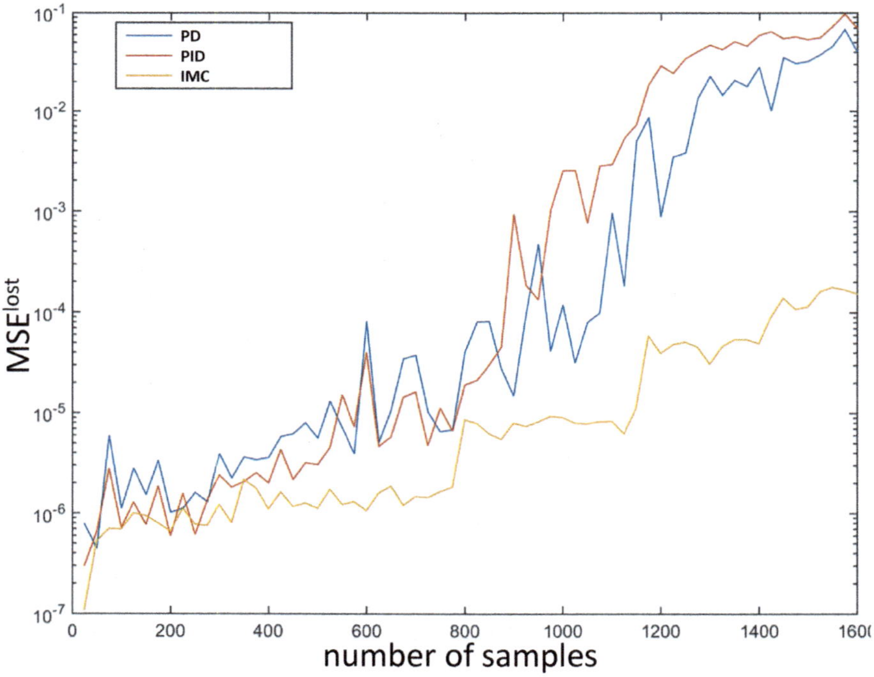

Fig. 12. General comparison of PD, PID and IMC control – case 1

visualization the index MSE^{lost} axis is logarithmic. The superiority of the IMC controller is evident, both in control performance and its robustness against lost samples.

The second simulation case addresses similar analysis, but in case of the system step response. Exemplary trends for three cases showing number of lost samples equal to *125*, *600* and *1155* out of total number equal to *2500* are presented. Comparison of the considered controllers is presented in Figs. 9, 10. Obtained results are quite similar to the previous case. Apart from zero steady state, IMC control is also dynamically significantly more robust as well.

It is also observable on the general comparison between the controllers showing relationship between the means square error and the number of lost samples. For sake of reality the steady state error has been subtracted for PD and POD control. Thereby, the comparison takes into consideration only the effect of lost samples and neglects the impact of a non-zero steady state control error.

5 Conclusions and Further Research

The paper presents the analysis of the sensitivity of selected control algorithms (PD, PID and IMC) against an effect of lost samples, which may appear in discrete-time control. The results are straightforward. Simple predictive control rule of the IMC outperforms PD/PID controllers. IMC controller is faster and has no steady state error. Simultaneously it is much more robust against an effect of lost samples. As it is shown the robustness is higher by at least one order of magnitude.

Though the results are promising, further research is required. It would be worth to validate the results with real-time control and with other processes, especially having significant transportation delay. In such case the application of predictive control strategy will be justified and various predictive controllers might be compared.

References

1. Branicky, M.S., Phillips, S.M., Zhang, W.: Stability of networked control systems: explicit analysis of delay. In: Proceedings of the 2000 American Control Conference, vol. 4, pp. 2352–2357 (2000)
2. Cervin, A., Henriksson, D., Lincoln, B., Eker, J., Arzen, K.E.: How does control timing affect performance? Analysis and simulation of timing using Jitterbug and TrueTime. IEEE Control Syst. Mag. **23**(3), 16–30 (2003)
3. Cloosterman, M., van de Wouw, N., Heemels, M., Nijmeijer, H.: Robust stability of networked control systems with time-varying network-induced delays. In: Proceedings of the 45th IEEE Conference on Decision and Control, pp. 4980–4985 (2006)
4. Czerwiński, K., Ławryńczuk, M.: Identification of discrete-time model of active magnetic levitation system. In: Mitkowski, W., Kacprzyk, J., Oprzędkiewicz, K., Skruch, P. (eds.) Trends in Advanced Intelligent Control, Optimization and Automation, pp. 599–608. Springer International Publishing, Cham (2017)

5. Domański, P.D.: Zastosowanie metod jakościowych do modelowania i projektowania układów regulacji. Ph.D. thesis, Instytut Automatyki i Informatyki Stosowanej, Politechnika Warszawska (1996)
6. Halevi, Y., Ray, A.: Integrated communication and control systems: part i - analysis, and part ii - design considerations. ASME J. Dyn. Syst. Meas. Control **110**(4), 367–381 (1988)
7. INTECO: Magnetic Levitation System User Guide, Kraków, Poland (2015)
8. Khosla, P.K.: Effect of sampling rates on the performance of model-based control schemes. In: Schweitzer, G., Mansour, M. (eds.) Dynamics of Controlled Mechanical Systems, pp. 271–284. Springer, Heidelberg (1989)
9. Koller, G., Sauter, T., Rauscher, T.: Effects of network delay quantization in distributed control systems. In: 5th IFAC International Conference on Fieldbus Systems and their applications 2003, IFAC Proceedings, Aveiro, Portugal, 7–9 July 2003, vol. 36, no. 13, pp. 291–298 (2003)
10. Laskawski, M., Wciślik, M.: Sampling rate impact on the tuning of PID controller parameters. Int. J. Electron. Telecommun. **62**(1), 43–48 (2016)
11. Levine, W.S.: The Control Handbook. Jaico Publishing House (1996)
12. MacGregor, J.F.: Optimal choice of the sampling interval for discrete process control. Technometrics **18**(2), 151–160 (1976)
13. Nilsson, J.: Real-time control systems with delays. Ph.D. thesis, Department of Automatic Control, Lund Institute of Technology, Sweden (1998)
14. Piłat, A.: Badania porównawcze dyskretnego regulatora pid dla aktywnego zawieszenia magnetycznego. Automatyka **14**(2), 181–196 (2010)
15. Samaranayake, L., Leksell, M., Alahakoon, S.: Relating sampling period and control delay in distributed control systems. In: EUROCON 2005 - The International Conference on "Computer as a Tool", vol. 1, pp. 274–277 (2005)
16. Torngren, M.: Fundamentals of implementing real-time control applications in distributed computer systems. Real Time Syst. **14**(3), 219–250 (1998)

Fast Nonlinear Model Predictive Control Algorithm with Neural Approximation for Embedded Systems: Preliminary Results

Patryk Chaber(✉)

Institute of Control and Computation Engineering, Warsaw University of Technology, ul. Nowowiejska 15/19, 00-665 Warsaw, Poland
P.Chaber@ia.pw.edu.pl

Abstract. This work presents preliminary results of research concerned with a fast nonlinear Model Predictive Control (MPC) algorithm implemented in an embedded system. In order to obtain a computationally efficient solution, a linear approximation of the predicted trajectory of the controlled variables is calculated for each sampling instant on-line which leads to a quadratic optimisation problem. Furthermore, the matrix of derivatives, which defines the linearised trajectory, is not determined analytically, but it is calculated (approximated) by a specially trained neural network. In order to show effectiveness of the discussed approach, a dynamic process with two inputs and two outputs is considered for which not only simulation results, but also results of real experiments performed in an embedded system based on a microcontroller are given.

Keywords: Embedded systems · Microcontrollers · Model Predictive Control · Neural networks · Nonlinear control

1 Introduction

In simple classical controllers, such as Proportional-Integral-Derivative one (PID) or linear Quadratic Regulator (LQR), the control law is fixed, while its parameters are calculated off-line. In advanced Model Predictive Control (MPC) algorithms, the values of the manipulated variables are determined at each of the consecutive discrete sampling instants from an optimisation procedure [10]. Typically, future (predicted) control errors are minimised. Such an optimisation-based formulation makes it possible to easily control numerous types of process which are difficult to handle by means of the classical algorithms, namely: Multiple-Input Multiple-Output, non-minimum phase (e.g. with delays), nonlinear. Furthermore, it is possible to easily take into account different types of constraints as they are simply a part of the optimisation program. Because of the mentioned advantages, the MPC algorithms have been applied for different technological processes, e.g. [3,8,13]. Typically, for industrial processes MPC algorithms are

A. Bartoszewicz et al. (Eds.): PCC 2020, AISC 1196, pp. 1067–1078, 2020.
https://doi.org/10.1007/978-3-030-50936-1_89

implemented using Programmable Logic Controllers (PLC) [12]. Currently, due to availability of very powerful hardware platforms, MPC algorithms may be implemented for fast processes with short sampling times (in embedded systems) using microcontrollers [1] or Field Programmable Gate Arrays (FPGA) [11]. The whole development process may be greatly shortened if software implementation is obtained using automatic code generation tools [2].

If a nonlinear model is used for prediction, a nonlinear optimisation problem at each sampling instant must be solved on-line. To reduce computational burden, MPC algorithms with on-line linearisations may be used [6]. In such approaches a linear approximation of the model or the predicted trajectory is used, which leads to easy to solve quadratic optimisation problems. In particular, the MPC algorithms with on-line linearisation of the predicted trajectory are reported to give better control quality than the simple algorithms with successive model linearisation [6]. Typically, the linearised trajectory is determined analytically at each sampling instant from model equations. In order to further speed up calculations the matrix of derivatives, which defines the linearised trajectory, may be approximated by a specially trained neural network [7].

This work is concerned with a fast nonlinear MPC algorithm with Nonlinear Prediction and Linearisation along the Trajectory with Neural Approximation (MPC-NPLT-NA). Implementation details of the algorithm are given, training of the neural approximator is discussed. Both simulation results and results of experiments in a microcontroller-based embedded system are reported.

2 MPC Problem Formulation

Let n_{u} and n_{y} denote the number of process inputs (manipulated variables) and outputs (controlled variables), respectively, i.e. $u = [u_1 \dots u_{n_{\mathrm{u}}}]^{\mathrm{T}}$, $y = [y_1 \dots y_{n_{\mathrm{y}}}]^{\mathrm{T}}$. In MPC the vector of decision variables calculated at each sampling instant, $k = 0, 1, 2, \dots$, has the length of $n_{\mathrm{u}} N_{\mathrm{u}}$ and consists of the current and future increments of the manipulated variables

$$\triangle \boldsymbol{u}(k) = \begin{bmatrix} \triangle u(k|k) \\ \vdots \\ \triangle u(k + N_{\mathrm{u}} - 1|k) \end{bmatrix} \tag{1}$$

where $\triangle u(k+p|k)$ denotes increments of the manipulated variables for the future sampling instant $k+p$ calculated at the current instant k, N_{u} is the control horizon. The rudimentary MPC optimisation problem is

$$\begin{aligned} &\min_{\triangle \boldsymbol{u}(k)} \left\{ \sum_{p=1}^{N} \| y^{\mathrm{sp}}(k+p|k) - \hat{y}(k+p|k) \|^2_{\boldsymbol{\Psi}_p} + \sum_{p=0}^{N_{\mathrm{u}}-1} \| \triangle u(k+p|k) \|^2_{\boldsymbol{\Lambda}_p} \right\} \\ &\text{subject to} \\ &u^{\min} \leq u(k+p|k) \leq u^{\max},\ p = 0, \dots, N_{\mathrm{u}} - 1 \\ &-\triangle u^{\max} \leq \triangle u(k+p|k) \leq \triangle u^{\max},\ p = 0, \dots, N_{\mathrm{u}} - 1 \end{aligned} \tag{2}$$

The symbols $y^{\text{sp}}(k+p|k)$ and $\hat{y}(k+p|k)$ denote the set-point trajectory and the predicted trajectory for the sampling instant $k+p$ calculated at the instant k, the prediction horizon is N. The weighting matrices $\boldsymbol{\Psi}_p = \text{diag}(\psi_{p,1}, \ldots, \psi_{p,n_\text{y}})$ and $\boldsymbol{\Lambda}_p = \text{diag}(\lambda_{p,1}, \ldots, \lambda_{p,n_\text{u}})$ are of dimensionality $n_\text{y} \times n_\text{y}$ and $n_\text{u} \times n_\text{u}$, respectively (notation $\|x\|_{\boldsymbol{\theta}}^2 = x^\text{T}\boldsymbol{\theta}x$ was used). The minimal and maximal values of the manipulated variables are denoted by $u^{\min}$ and $u^{\max}$, respectively, the maximal change rate of the manipulated variables is $\triangle u^{\max}$. At each sampling instant, having calculated $n_\text{u}N_\text{u}$ future increments of the manipulated variables over the control horizon (1), only the increments for the current sampling instant are applied to the process, i.e. $u(k) = \triangle u(k|k) + u(k-1)$. At the next sampling instant, $k+1$, the measurements of the process outputs are updated and the whole optimisation procedure is repeated.

3 MPC Algorithm with Nonlinear Prediction and Linearisation Along the Trajectory (MPC-NPLT)

When a nonlinear model of the process is used for prediction, i.e. to calculate the values of $\hat{y}(k+p|k)$, the rudimentary MPC optimisation problem (2) becomes a nonlinear task, which must be solved on-line. For fast system with short sampling periods of the millisecond order it is always a challenge. A computationally efficient alternative is to use the MPC-NPLT algorithm [6] with an advanced linearisation method. Let the trajectory of the current and future values of the manipulated variables corresponding to the calculated increments (Eq. (1)) be

$$\boldsymbol{u}(k) = \begin{bmatrix} u(k|k) \\ \vdots \\ u(k+N_\text{u}-1|k) \end{bmatrix} \tag{3}$$

Linearisation is performed along some assumed trajectory of the manipulated variables

$$\boldsymbol{u}^{\text{traj}}(k) = \begin{bmatrix} u^{\text{traj}}(k|k) \\ \vdots \\ u^{\text{traj}}(k+N_\text{u}-1|k) \end{bmatrix} \tag{4}$$

which has length $n_\text{u}N_\text{u}$. Using the nonlinear model it is possible to calculate for the assumed trajectory $\boldsymbol{u}^{\text{traj}}(k)$ the corresponding trajectory of the controlled variables

$$\hat{\boldsymbol{y}}^{\text{traj}}(k) = \begin{bmatrix} \hat{y}^{\text{traj}}(k+1|k) \\ \vdots \\ \hat{y}^{\text{traj}}(k+N|k) \end{bmatrix} \tag{5}$$

which is the vector of length $n_\text{y}N$. A linear approximation of the nonlinear predicted trajectory of the controlled variables $\hat{\boldsymbol{y}}(k)$ with respect to the trajectory of the future manipulated variables $\boldsymbol{u}(k)$ is

$$\hat{\boldsymbol{y}}(k) = \hat{\boldsymbol{y}}^{\text{traj}}(k) + \boldsymbol{H}(k)(\boldsymbol{u}(k) - \boldsymbol{u}^{\text{traj}}(k)) \tag{6}$$

where

$$\boldsymbol{H}(k)=\left.\frac{\mathrm{d}\hat{\boldsymbol{y}}(k)}{\mathrm{d}\boldsymbol{u}(k)}\right|_{\substack{\hat{\boldsymbol{y}}(k)=\hat{\boldsymbol{y}}^{\mathrm{traj}}(k)\\ \boldsymbol{u}(k)=\boldsymbol{u}^{\mathrm{traj}}(k)}}=\begin{bmatrix}\frac{\partial\hat{y}^{\mathrm{traj}}(k+1|k)}{\partial u^{\mathrm{traj}}(k|k)} & \cdots & \frac{\partial\hat{y}^{\mathrm{traj}}(k+1|k)}{\partial u^{\mathrm{traj}}(k+N_{\mathrm{u}}-1|k)}\\ \vdots & \ddots & \vdots\\ \frac{\partial\hat{y}^{\mathrm{traj}}(k+N|k)}{\partial u^{\mathrm{traj}}(k|k)} & \cdots & \frac{\partial\hat{y}^{\mathrm{traj}}(k+N|k)}{\partial u^{\mathrm{traj}}(k+N_{\mathrm{u}}-1|k)}\end{bmatrix} \tag{7}$$

is the matrix of the derivatives of the predicted trajectory of the controlled variables $\hat{\boldsymbol{y}}^{\mathrm{traj}}(k)$ with respect the assumed trajectory of the manipulated variables $\boldsymbol{u}^{\mathrm{traj}}(k)$. The matrix $\boldsymbol{H}(k)$ has dimensionality $n_{\mathrm{y}}N\times n_{\mathrm{u}}N_{\mathrm{u}}$. The linear approximation (6) of the nonlinear trajectory of the controlled variables may be expressed as a function of MPC decision variables (increments $\triangle\boldsymbol{u}(k)$)

$$\hat{\boldsymbol{y}}(k)=\boldsymbol{H}(k)\boldsymbol{J}\triangle\boldsymbol{u}(k)+\hat{\boldsymbol{y}}^{\mathrm{traj}}(k)+\boldsymbol{H}(k)(\boldsymbol{u}(k-1)-\boldsymbol{u}^{\mathrm{traj}}(k)) \tag{8}$$

where the matrix of dimensionality $n_{\mathrm{u}}N_{\mathrm{u}}\times n_{\mathrm{u}}N_{\mathrm{u}}$ and the vector of length $n_{\mathrm{u}}N_{\mathrm{u}}$ are

$$\boldsymbol{J}=\begin{bmatrix}\boldsymbol{I}_{n_{\mathrm{u}}\times n_{\mathrm{u}}} & \boldsymbol{0}_{n_{\mathrm{u}}\times n_{\mathrm{u}}} & \cdots & \boldsymbol{0}_{n_{\mathrm{u}}\times n_{\mathrm{u}}}\\ \boldsymbol{I}_{n_{\mathrm{u}}\times n_{\mathrm{u}}} & \boldsymbol{I}_{n_{\mathrm{u}}\times n_{\mathrm{u}}} & \cdots & \boldsymbol{0}_{n_{\mathrm{u}}\times n_{\mathrm{u}}}\\ \vdots & \vdots & \ddots & \vdots\\ \boldsymbol{I}_{n_{\mathrm{u}}\times n_{\mathrm{u}}} & \boldsymbol{I}_{n_{\mathrm{u}}\times n_{\mathrm{u}}} & \cdots & \boldsymbol{I}_{n_{\mathrm{u}}\times n_{\mathrm{u}}}\end{bmatrix},\quad \boldsymbol{u}(k-1)=\begin{bmatrix}u(k-1)\\ \vdots\\ u(k-1)\end{bmatrix} \tag{9}$$

When linearisation is carried out along the trajectory $\boldsymbol{u}^{\mathrm{traj}}(k)$ defined using the manipulated variables applied to the process at the previous sampling instant, i.e. $\boldsymbol{u}^{\mathrm{traj}}(k)=\boldsymbol{u}(k-1)$, Eq. (8) reduces to

$$\hat{\boldsymbol{y}}(k)=\boldsymbol{H}(k)\boldsymbol{J}\triangle\boldsymbol{u}(k)+\hat{\boldsymbol{y}}^{\mathrm{traj}}(k) \tag{10}$$

Using the prediction Eq. (10), the rudimentary MPC optimisation problem (2) becomes the following quadratic optimisation task

$$\begin{aligned}&\min_{\triangle\boldsymbol{u}(k)}\left\{J(k)=\left\|\boldsymbol{y}^{\mathrm{sp}}(k)-\boldsymbol{H}(k)\boldsymbol{J}\triangle\boldsymbol{u}(k)-\hat{\boldsymbol{y}}^{\mathrm{traj}}(k)\right\|_{\boldsymbol{\Psi}}^{2}+\left\|\triangle\boldsymbol{u}(k)\right\|_{\boldsymbol{\Lambda}}^{2}\right\}\\ &\text{subject to}\\ &\boldsymbol{u}^{\min}\leq\boldsymbol{J}\triangle\boldsymbol{u}(k)+\boldsymbol{u}(k-1)\leq\boldsymbol{u}^{\max}\\ &-\triangle\boldsymbol{u}^{\max}\leq\triangle\boldsymbol{u}(k)\leq\triangle\boldsymbol{u}^{\max}\end{aligned} \tag{11}$$

where the vectors of length $n_{\mathrm{u}}N_{\mathrm{u}}$ are

$$\boldsymbol{u}^{\min}=\begin{bmatrix}u^{\min}\\ \vdots\\ u^{\min}\end{bmatrix},\quad \boldsymbol{u}^{\max}=\begin{bmatrix}u^{\max}\\ \vdots\\ u^{\max}\end{bmatrix},\quad \triangle\boldsymbol{u}^{\max}=\begin{bmatrix}\triangle u^{\max}\\ \vdots\\ \triangle u^{\max}\end{bmatrix} \tag{12}$$

The weighting matrices $\boldsymbol{\Psi}=\mathrm{diag}(\boldsymbol{\Psi}_1,\ldots,\boldsymbol{\Psi}_N)$ and $\boldsymbol{\Lambda}=\mathrm{diag}(\boldsymbol{\Lambda}_0,\ldots,\boldsymbol{\Lambda}_{N_{\mathrm{u}}-1})$ are of dimensionality $n_{\mathrm{y}}N\times n_{\mathrm{y}}N$ and $n_{\mathrm{u}}N_{\mathrm{u}}\times n_{\mathrm{u}}N_{\mathrm{u}}$, respectively.

4 MPC-NPLT Algorithm with Neural Approximation (MPC-NPLT-NA)

In the classical MPC-NPLT algorithm [6] all entries of the matrix $\boldsymbol{H}(k)$, the structure of which is defined by Eq. (7), are calculated analytically on-line, using model equations. An alternative is to employ a neural approximator for this purpose [7]. In that approach, at the consecutive sampling instant, the matrix $\boldsymbol{H}(k)$ is approximated by a neural network. Its inputs define the current operating point of the process and are given by the most recent values of the manipulated variables, i.e. $u(k-1)$, $u(k-2), \ldots$, and the controlled ones, i.e. $y(k)$, $y(k-1), \ldots$. The outputs of the network approximate all partial derivatives $y^{\mathrm{traj}}(k+N|k)/\partial u^{\mathrm{traj}}(k+r|k)$ for $p=1,\ldots,N$, $r=0,\ldots,N_{\mathrm{u}}-1$. In order to collect data sets necessary to train and evaluate the neural network, the classical MPC-NPLT algorithm is first simulated for various operating conditions (different set-points) and all necessary variables are recorded.

5 Results of Simulations and Experiments

In simulations and experiments a nonlinear process with two inputs and two outputs ($n_{\mathrm{u}}=n_{\mathrm{y}}=2$) is considered. It has the Wiener structure, i.e. a linear dynamic block is followed by a nonlinear steady-state one [4,5]. The first part of the process is described by the following difference equations

$$\begin{aligned} v_1(k) &= b_1^{1,1}u_1(k-1) + b_1^{1,1}u_1(k-2) + b_3^{1,1}u_1(k-3) + b_4^{1,1}u_1(k-4) \\ &\quad + b_1^{1,2}u_2(k-1) + b_1^{1,2}u_2(k-2) + b_3^{1,2}u_1(k-3) + b_4^{1,2}u_1(k-4) \\ &\quad - a_1^1 v_1(k-1) - a_2^1 v_1(k-2) - a_3^1 v_1(k-3) - a_4^1 v_1(k-4) \qquad (13) \\ v_2(k) &= b_1^{2,1}u_1(k-1) + b_1^{2,1}u_1(k-2) + b_3^{2,1}u_1(k-3) + b_4^{2,1}u_1(k-4) \\ &\quad + b_1^{2,2}u_2(k-1) + b_1^{2,2}u_2(k-2) + b_3^{2,2}u_1(k-3) + b_4^{2,2}u_1(k-4) \\ &\quad - a_1^2 v_2(k-1) - a_2^2 v_2(k-2) - a_3^2 v_2(k-3) - a_4^2 v_2(k-4) \qquad (14) \end{aligned}$$

where v_1 and v_2 are the outputs of the first part of the process (and inputs of the second part). The values of the parameters are: $a_1^1=-2.4261$, $a_2^1=2.2073$, $a_3^1=-8.9252\times10^{-1}$, $a_4^1=1.3534\times10^{-1}$, $a_1^2=-2.6461$, $a_2^2=2.6197$, $a_3^2=-1.1500$, $a_4^2=1.8888\times10^{-1}$, $b_1^{1,1}=1.8041e-01$, $b_2^{1,1}=-8.9618\times10^{-2}$, $b_3^{1,1}=-9.0393\times10^{-2}$, $b_4^{1,1}=4.7540\times10^{-2}$, $b_1^{1,2}=3.6082\times10^{-2}$, $b_2^{1,2}=-1.7924\times10^{-2}$, $b_3^{1,2}=-1.8079\times10^{-2}$, $b_4^{1,2}=9.5081\times10^{-3}$, $b_1^{2,1}=2.5387\times10^{-2}$, $b_2^{2,1}=-1.4361\times10^{-2}$, $b_3^{2,1}=-1.4406\times10^{-2}$, $b_4^{2,1}=8.3563\times10^{-3}$, $b_1^{2,2}=3.1734\times10^{-1}$, $b_2^{2,2}=-1.7951\times10^{-1}$, $b_3^{2,2}=-1.8008\times10^{-1}$, $b_4^{2,2}=1.0445\times10^{-1}$. The second part of the process are described by the equations

$$y_1(k) = g_1(v_1(k)) = -\exp -v_1(k) + 1 \qquad (15)$$

$$y_2(k) = g_2(v_2(k)) = \frac{1}{30}v_2(k) + \frac{1}{150}v_2^3(k) \qquad (16)$$

The constraints are defined by: $u^{\min}=-1$, $u^{\max}=1$, $\triangle u^{\max}=0.1$.

5.1 Simulations

At first, the MPC-NPLT algorithm has been implemented in MATLAB to generate data sets to train the neural approximator. The horizons are $N = N_u = 20$, the weighting coefficients are all equal to 1, i.e. $\boldsymbol{\Psi} = \boldsymbol{I}_{n_y N \times n_y N}$ and $\boldsymbol{\Lambda} = \boldsymbol{I}_{n_u N_u \times n_u N_u}$.

Figure 1 presents a part of data (1000 samples in total) collected during simulation, i.e. the manipulated variables, the controlled variables vs. the required set-points. Additionally, time required to perform specific parts of the MPC-NPLT algorithm are given (in the consecutive sampling instants), i.e. the time necessary to calculate derivatives, the time necessary to solve the MPC-NPLT optimisation task and the overall time. There is no significant dependence between calculation time and the operating point of the process, thus no specific set-point trajectory has been designed. Maximal time values are not conclusive because the operating system randomly elongated some iterations of the MPC-NPLT algorithm, therefore only average calculation time values are given.

Figure 2 depicts derivatives determined using trained neural network based on validation data. Only a part (200 samples) of the data is presented. Both training and validation data contained a total of 1000 samples.

Neural networks, specifically of the Multi Layer Perceptron (MLP) type with one hidden layer, have been trained based on the matrices of derivatives calculated in an analytic way. An automatic differentiation could be used for this purpose in case of complex derivatives equations to minimise effort to obtain trained approximators. A variety of number of hidden neural networks have been tested. Simple networks with 3 hidden nodes served best, while still maintaining low enough number of weights of the network. Only the approximators which calculate the first two columns of the matrix of derivatives have been trained, because the other ones $\partial y(k+p|k)/\partial u(k+r|k) = \partial y(k+p+1|k)/\partial u(k+r+1|k)$, which greatly limits the number of weight of the neural network. For training the Broyden-Fletcher-Goldfarb-Shanno minimisation algorithm has been used, training has been repeated many times with randomly initialised weights. Figure 2 shows the values of first two columns. Each plot represents the derivative of one output signal over one input signal.

Tables 1, 2 and 3 shows the average time (in milliseconds) required to analytically calculate different parts of the MPC-NPLT algorithm (at one sampling instant). Similarly, Tables 4 and 5 reports the average time in the MPC-NPLT-NA algorithm with neural approximation.

Table 1 shows the average time required to analytically calculate the matrix of derivatives. Average time of calculation depends linearly on both control, and prediction horizons, although control horizon has visibly greater influence on its value. It is worth noting that this relation strongly depends on the structure of the used model, and therefore on the way the derivatives are defined.

Table 2 reports the average time required to perform minimisation in the MPC-NPLT algorithm. An interior-point algorithms is used to determine a solution of the MPC-NPLT minimisation problem (11). Nevertheless, the computational complexity of the minimisation task is strongly dependent on the number

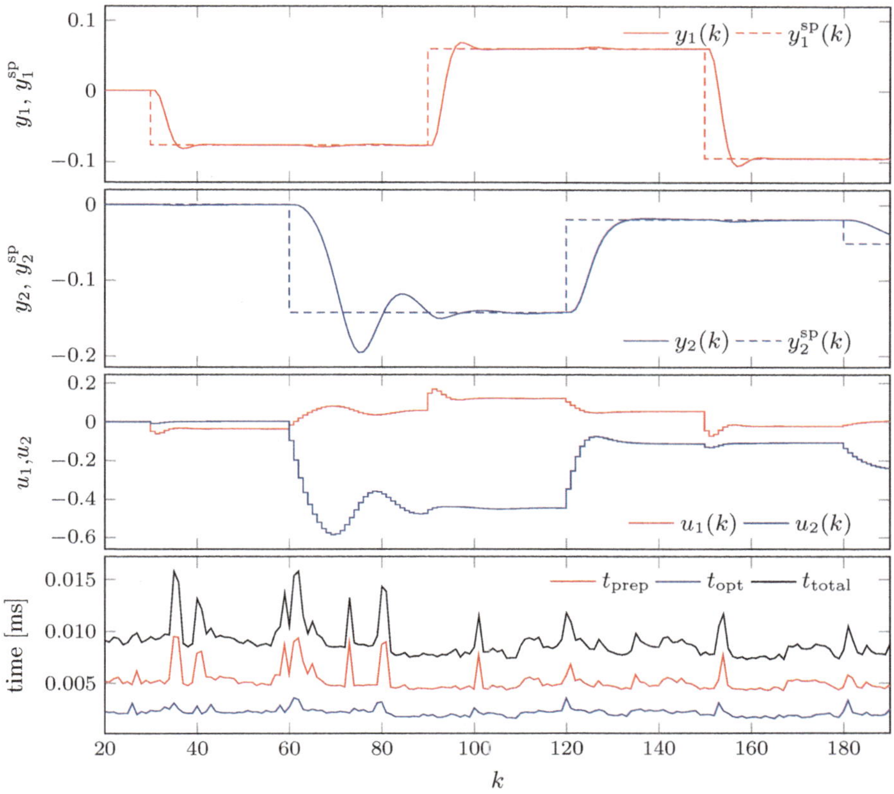

Fig. 1. Results of simulations performed in MATLAB: the trajectories obtained when the process is controlled by the MPC-NPLT algorithm ($N = N_u = 20$); the bottom panel shows calculation time (calculation of derivatives, solution of the optimisation task and the overall time)

of decision variables, namely $n_u N$. The average calculation time increases exponentially with the increase of the control horizon.

Table 3 shows the average time required to execute all calculations of the MPC-NPLT algorithm using analytical derivatives. In this measurements all calculations required to perform a single iteration of MPC-NPLT algorithm are included, e.g. determining matrices for quadratic programming task, calculating the predicted trajectory of the controlled variables, etc., and also already discussed derivatives calculation and minimisation task. It can be noticed that with the increase of prediction horizon, the ratio between time spent calculating derivatives and the time used to perform one execution of the algorithm increases exponentially. The significance of other operations than calculating derivatives and minimisation drops with increase of both control and prediction

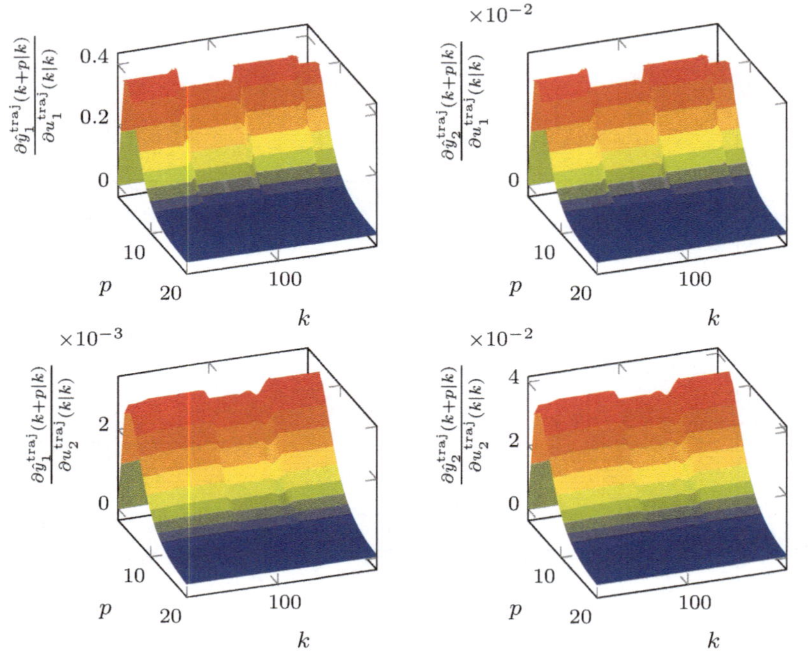

Fig. 2. The values of derivatives during a single experiment ($N = N_{\mathrm{u}} = 20$)

horizons. Therefore, for larger horizons the bottleneck of the control algorithm is the calculation of derivatives.

Table 4 details the average time required to calculate the matrix of derivatives in the MPC-NPLT-NA algorithm with neural approximation. The time of derivatives' approximation using neural network increases linearly with both control and prediction horizon. It is worth noting that the major influence on the average calculation time has control horizon. Although discussed relation between time required to approximate derivatives using neural network is similar to the according relation from MPC-NPLT, the rate at which time increases with the increase of horizons is greatly decreased by using approximation.

Table 5 shows the average time required to execute all calculations of the MPC-NPLT-NA algorithm with neural approximation. Compared to the MPC-NPLT algorithm, the relation between considered time measurements and horizons is similar, i.e. there is an exponential increase of time with an increase of control horizon due to the time required to perform minimisation (time of minimisation is the same for MPC-NPLT and MPC-NPLT-NA). In contrary to the MPC-NPLT algorithm the ratio of time required to approximate derivatives to a time required to execute a single iteration of MPC-NPLT-NA algorithm decreases with the increase of control horizons. It is clearly visible for large horizons that the minimisation time is almost as long as the total time of calculations performed in one sampling instant. It is worth noting that although measurements show that the time required to perform other matrix calculations is insignificant in context of total iteration time, this may be the result of extensive MATLAB optimisation of those operations. Thus, it is expected to have more impact on the average iteration time in embedded system.

Table 1. The influence of the control and prediction horizons on the average time (in milliseconds) required to analytically calculate the matrix of derivatives (at one sampling instant)

N	N_u				
	5	10	20	50	100
5	7.4405×10^{-1}	—	—	—	—
10	1.2197	1.6771	—	—	—
20	2.3001	3.0738	4.5843	—	—
50	5.3048	8.1634	1.4029×10^{1}	2.5964×10^{1}	—
100	9.9484	1.3944×10^{1}	2.6604×10^{1}	5.0245×10^{1}	1.0112×10^{2}

Table 2. The influence of the control and prediction horizons on the average time (in milliseconds) required to perform minimisation in the MPC-NPLT algorithm (at one sampling instant)

N	N_u				
	5	10	20	50	100
5	1.5928	—	—	—	—
10	1.4702	1.5908	—	—	—
20	1.5777	1.6360	2.2578	—	—
50	1.8492	2.1047	2.9988	2.0597×10^{1}	—
100	1.9468	1.9789	3.0188	2.1632×10^{1}	9.4917×10^{1}

5.2 Experiments in Embedded System

Finally, the designed MPC-NPLT-NA algorithm has been implemented in an embedded system. Both the controller and the emulated process have been implemented using STM32F746I microcontrollers included on WaveShare Open746-I Development boards. Communication between the controller and the process is performed with the use of Digital-Analogue and Analogue-Digital converters, thus making the signal prone to noise. The quadratic optimisation problem (11) is solved by means of the OSQP software [9]. Results of experiments performed in the embedded system, i.e. the trajectories obtained when the process is controlled by the MPC-NPLT-NA algorithm are depicted in Fig. 3. Due to a low

Table 3. The influence of the control and prediction horizons on the average time (in milliseconds) required to execute all calculations of the MPC-NPLT algorithm using analytical derivatives (at one sampling instant)

N	N_u				
	5	10	20	50	100
5	3.0459	—	—	—	—
10	3.5631	4.1490	—	—	—
20	5.4684	6.2767	8.2717	—	—
50	1.0612×10^1	1.4462×10^1	2.1987×10^1	5.0818×10^1	—
100	1.8281×10^1	2.2380×10^1	3.8653×10^1	7.9689×10^1	2.0444×10^2

Table 4. The influence of the control and prediction horizons on the average time (in milliseconds) required to calculate matrix of derivatives in the MPC-NPLT-NA algorithm with neural approximation (at one sampling instant)

N	N_u				
	5	10	20	50	100
5	1.5755×10^{-1}	—	—	—	—
10	1.3719×10^{-1}	1.4085e-01	—	—	—
20	1.0185×10^{-1}	1.3874×10^{-1}	2.0101×10^{-1}	—	—
50	1.2723×10^{-1}	2.0630×10^{-1}	2.6130×10^{-1}	5.9073×10^{-1}	—
100	1.7793×10^{-1}	2.0950×10^{-1}	3.1339×10^{-1}	6.3600×10^{-1}	1.0311

Table 5. The influence of the control and prediction horizons on the average time (in milliseconds) required to execute all calculations of the MPC-NPLT-NA algorithm with neural approximation (at one sampling instant)

N	N_u				
	5	10	20	50	100
5	3.1146	—	—	—	—
10	3.2448	2.6666	—	—	—
20	2.9150	3.1314	3.8303	—	—
50	4.6303	5.7097	5.9211	2.5763×10^1	—
100	9.9297	7.8670	9.6796	2.7790×10^1	9.9340×10^1

memory capacity, the horizons have been shortened ($N = 5$, $N_u = 3$) compared to the implementation in MATLAB ($N = N_u = 20$), which visibly affects the quality of control. Nevertheless, the controlled variables always reach their set-points, and the control signal is smooth, i.e. the measurement noise does not induce aggressive changes of the manipulated variables.

There is a vivid difference in calculation time. MATLAB executes entire iteration of MPC-NLPT algorithm ($N = N_{\mathrm{u}} = 20$) in microseconds, while with embedded implementation as many as tens of milliseconds are required to perform a single iteration of MPC-NPLT-NA ($N = 5$, $N_{\mathrm{u}} = 3$). The reason is twofold: the difference in computational power and the ability to optimise matrix operations. On the microcontroller, only simple acceleration of matrix operations with Digital Signal Processing can be used, whereas it is MATLAB's main focus to perform those operations in the least possible time. Nevertheless, the results obtained with embedded system, despite its much lower computational power, show that it is clearly possible to perform complex tasks including approximating derivatives, performing linearisation and finding the solution of a minimisation problem in tens of milliseconds.

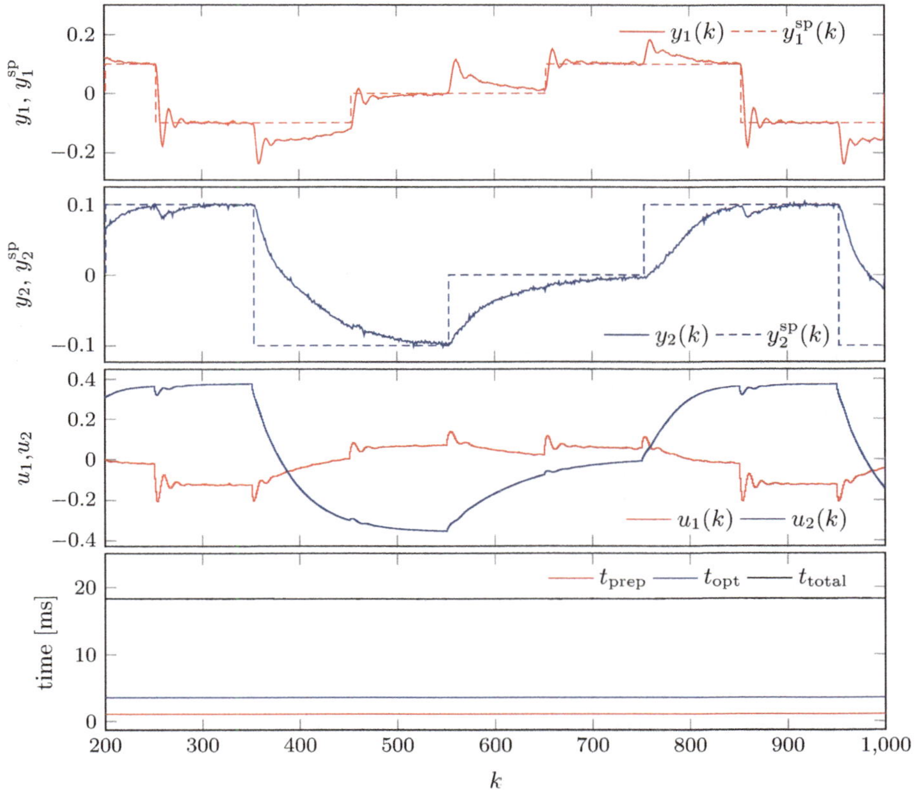

Fig. 3. Results of experiments performed in the embedded system: the trajectories obtained when the process is controlled by the MPC-NPLT-NA algorithm with neural approximation ($N = 5$, $N_{\mathrm{u}} = 3$)

6 Summary

The neural approximator is used in nonlinear MPC algorithm to shorten calculation time. The derivatives of the predicted controlled variable with respect to the future trajectory of the manipulated variable is not calculated analytically for each sampling instant on-line, but approximated by a classical neural network. The results of simulation and experiments performed in a microcontroller-based embedded system confirm usefulness of the discussed approach.

References

1. Chaber, P., Ławryńczuk, M.: Fast analytical model predictive controllers and their implementation for STM32 ARM microcontroller. IEEE Trans. Industr. Inf. **15**, 4580–4590 (2019)
2. Chaber, P., Ławryńczuk, M.: AutoMATiC: Code generation of model predictive control algorithms for microcontrollers. IEEE Trans. Industr. Inf. **16**(7), 4547–4556 (2020). https://doi.org/10.1109/TII.2019.2946842
3. Grosso, J.M., Ocampo-Martinez, C., Puig, V.: Reliability-based economic model predictive control for generalised flow-based networks including actuators' health-aware capabilities. Int. J. Appl. Math. Comput. Sci. **26**, 361–654 (2016)
4. Janczak, A., Korbicz, J.: Two-stage instrumental variables identification of polynomial Wiener Systems with invertible nonlinearities. Int. J. Appl. Math. Comput. Sci. **29**, 571–580 (2019)
5. Janczak, A.: Identification of Nonlinear Systems Using Neural Networks and Polynomial Models. A Block-Oriented Approach. Lecture Notes in Control and Information Sciences, vol. 310. Springer, Heidelberg (2004)
6. Ławryńczuk, M.: Computationally Efficient Model Predictive Control Algorithms: A Neural Network Approach. Studies in Systems, Decision and Control, vol. 3. Springer, Cham (2014)
7. Ławryńczuk, M.: Explicit nonlinear predictive control algorithms with neural approximation. Neurocomputing **129**, 570–584 (2014)
8. Pour, F.K., Puig, V., Ocampo-Martinez, C.: Multi-layer health-aware economic predictive control of a pasteurization pilot plant. Int. J. Appl. Math. Comput. Sci. **28**, 97–110 (2018)
9. Stellato, B., Banjac, G., Goulart, P., Bemporad, A., Boyd, S.: OSQP: an operator splitting solver for quadratic programs. arXiv e-prints https://arxiv.org/abs/1711.08013 (2017)
10. Tatjewski, P.: Advanced Control of Industrial Processes. Structures and Algorithms. Springer, London (2007)
11. Wojtulewicz, A., Ławryńczuk, M.: Implementation of multiple-input multiple-output dynamic matrix control algorithm for fast processes using field programmable gate array. IFAC-PapersOnLine **51**, 324–329 (2018)
12. Wojtulewicz, A., Ławryńczuk, M.: Computationally efficient implementation of dynamic matrix control algorithm for very fast processes using programmable logic controller. In: 2018 23rd International Conference on Methods & Models in Automation & Robotics (MMAR), pp. 579–584 (2018)
13. Zhou, F., Peng, H., Zhang, G., Zeng, X.: A robust controller design method based on parameter variation rate of RBF-ARX model. IEEE Access **7**, 160284–160294 (2019)

Hardware Accelerators for Fast Implementation of DMC and GPC Control Algorithms Using FPGA and Their Applications to a Servomotor

Andrzej Wojtulewicz(✉)

Institute of Control and Computation Engineering, Warsaw University of Technology, ul. Nowowiejska 15/19, 00-665 Warsaw, Poland
andrwoj@gmail.com

Abstract. This work presents hardware accelerators implemented in the Field Programmable Gate Array (FPGA) which perform fast calculations for Model Predictive Control (MPC) algorithms. Two MPC algorithms are considered: Dynamic Matrix Control (DMC) and Generalized Predictive Control (GPC). The hardware accelerator-based DMC and GPC algorithms are applied to a servomotor.

Keywords: Field programmable gate array · Model predictive control · Servomotor

1 Introduction

Model Predictive Control (MPC) algorithms [10] are implemented in industrial applications using Programmable Logic Controllers (PLC) [12]. Typically, process control industrial applications are characterised by relatively long sampling periods, of the order of tens of seconds or minutes. Currently, due to progress in electronics and availability of powerful microcontrollers, MPC algorithms may be also used in embedded systems which require much shorter sampling periods, of the order of single or tens of milliseconds [1,2].

In the most advanced MPC algorithms an optimisation problem is solved at each sampling instant to find the vector of decision variables. If a linear model is used for prediction, a quadratic optimisation problem is solved. Implementation of interior point and active set optimisation algorithms for MPC using the FPGA is described in [7]. Furthermore, in order to speed up calculations some parallel operations possible in FPGA may be used [8]. If a nonlinear model is used for prediction, a nonlinear optimisation task must be solved on-line. For this purpose the gradient-based Sequential Quadratic Programming (SQP) algorithm [6] or even the heuristic Particle Swarm Optimisation (PSO) method [13] may be used. In order to shorten the calculation time, analytical (explicit) MPC algorithms may be considered. In these approaches the constraints are not taken

A. Bartoszewicz et al. (Eds.): PCC 2020, AISC 1196, pp. 1079–1091, 2020.
https://doi.org/10.1007/978-3-030-50936-1_90

into account during calculation. The classical explicit MPC structure consists of a great number of linear control laws, the actually used one is selected taking into account the current process state. FPGA implementation of such an explicit MPC algorithm is described in [5].

This work describes implementation of analytical MPC algorithms using the FPGA. In the explicit formulation only the value of the manipulated variable for the current sampling instant is calculated [10]. In order to obtain short calculation time, calculations are performed by specialised hardware accelerators. Two MPC algorithms are considered: Dynamic Matrix Control (DMC) and Generalized Predictive Control (GPC). The hardware accelerator-based DMC and GPC algorithms are applied to a servomotor.

2 Model Predictive Control Formulations

At each discrete sampling instants, $k = 0, 1, 2, \ldots$, the DMC algorithm calculates on-line future increments of the manipulated variable (the process input)

$$\triangle \boldsymbol{u}(k) = [\triangle u(k|k) \ \ldots \ \triangle u(k + N_\mathrm{u} - 1|k)]^\mathrm{T} \tag{1}$$

where N_u is the control horizon. The decision variables (1) are found as a result of an optimisation process. The most typical minimised cost-function is

$$J(k) = \sum_{p=1}^{N} \|y^\mathrm{sp}(k+p|k) - \hat{y}(k+p|k)\|^2 + \lambda \sum_{p=0}^{N_\mathrm{u}-1} \|\triangle u(k+p|k)\|^2 \tag{2}$$

where $y^\mathrm{sp}(k+p|k)$ and $\hat{y}(k+p|k)$ denote the set-point trajectory and the predicted values of the controlled variable (the process output), $N \geq N_\mathrm{u}$ is the prediction horizon, $\lambda > 0$ is a penalty factor. Although N_u increments (1) are calculated, only the first one is actually applied to the process. At the next sampling instant the whole calculation procedure is repeated.

The predicted values of the process output are calculated using a dynamic model of the controlled process. In this work two model structures and two resulting MPC algorithms are considered: the step response model, which leads to Dynamic Matrix Control (DMC) algorithm, and the difference equation, which leads to Generalized Predictive Control (GPC) algorithm [10].

2.1 Dynamic Matrix Control Algorithm

If the process is stable with no integration, the step-response model is defined by the step-response coefficients $s_1, \ldots, s_D$, where D is named the horizon of dynamics. The MPC cost-function (2) in the vector-matrix notation is

$$J(k) = \|\boldsymbol{y}^\mathrm{sp}(k) - \hat{\boldsymbol{y}}(k)\|^2 + \|\triangle \boldsymbol{u}(k)\|^2_{\boldsymbol{\Lambda}} \tag{3}$$

where the vectors of length N are $\boldsymbol{y}^{\mathrm{sp}}(k) = [y(k+1|k) \ \ldots \ y(k+N|k)]^{\mathrm{T}}$ and $\hat{\boldsymbol{y}}(k) = [\hat{y}(k+1|k) \ \ldots \ \hat{y}(k+N|k)]^{\mathrm{T}}$, $\boldsymbol{\Lambda} = \mathrm{diag}(\lambda, \ldots, \lambda)$ is a matrix of dimensionality $N_{\mathrm{u}} \times N_{\mathrm{u}}$. Since the MPC cost-function (3) is quadratic in terms of the decision vector $\triangle \boldsymbol{u}(k)$, the optimal control increments are

$$\triangle \boldsymbol{u}(k) = \boldsymbol{K}(\boldsymbol{y}^{\mathrm{sp}}(k) - \boldsymbol{y}^{0}(k)) \tag{4}$$

When constraints are necessary, the calculated signals are projected onto the admissible set of constraints. The matrix of dimensionality $N_{\mathrm{u}} \times N$ is

$$\boldsymbol{K} = (\boldsymbol{M}^{\mathrm{T}}\boldsymbol{M} + \boldsymbol{\Lambda})^{-1}\boldsymbol{M}^{\mathrm{T}} = \begin{bmatrix} \overline{\boldsymbol{K}}_{1,1} \\ \vdots \\ \overline{\boldsymbol{K}}_{N_{\mathrm{u}},1} \end{bmatrix} = \begin{bmatrix} K_{1,1} & \ldots & K_{1,N} \\ \vdots & \ddots & \vdots \\ K_{N_{\mathrm{u}},1} & \ldots & K_{N_{\mathrm{u}},N} \end{bmatrix} \tag{5}$$

The step-response matrix

$$\boldsymbol{M} = \begin{bmatrix} s_1 & 0 & \ldots & 0 \\ s_2 & s_1 & \ldots & 0 \\ \vdots & \vdots & \ddots & \vdots \\ s_N & s_{N-1} & \ldots & s_{N-N_{\mathrm{u}}+1} \end{bmatrix} \tag{6}$$

is of dimensionality $N \times N_{\mathrm{u}}$. The free trajectory is

$$\boldsymbol{y}^{0}(k) = \boldsymbol{y}(k) + \boldsymbol{M}^{\mathrm{p}} \triangle \boldsymbol{u}^{\mathrm{p}}(k) \tag{7}$$

where the second step-response matrix of dimensionality $N \times (D-1)$ is

$$\boldsymbol{M}^{\mathrm{p}} = \begin{bmatrix} s_2 - s_1 & \ldots & s_D - s_{D-1} \\ s_3 - s_1 & \ldots & s_{D+1} - s_{\mathrm{D}-1} \\ \vdots & \ddots & \vdots \\ s_{N+1} - s_1 & \ldots & s_{N+D-1} - s_{D-1} \end{bmatrix} = \left[\boldsymbol{M}_1^{\mathrm{p}} \ \ldots \ \boldsymbol{M}_{D-1}^{\mathrm{p}}\right]^{\mathrm{T}} \tag{8}$$

the vectors $\boldsymbol{y}(k) = [y(k) \ \ldots \ y(k)]^{\mathrm{T}}$ and $\triangle \boldsymbol{u}^{\mathrm{p}}(k) = [\triangle u(k-1) \ldots \triangle u(k-(\mathrm{D}-1))]^{\mathrm{T}}$ are of length N and $D-1$, respectively.

In the classical DMC algorithm, in each sampling period as many as N_{u} future control increments $\triangle \boldsymbol{u}(k)$ are calculated from Eq. (4). In the computationally efficient compact version of the DMC algorithm only the first of them is found, which is actually applied to the process. If $y^{\mathrm{sp}}(k+1|k) = \ldots = y^{\mathrm{sp}}(k+N|k)$, from Eq. (4), the explicit DMC control law is obtained

$$\triangle u(k|k) = K^{\mathrm{e}}(y^{\mathrm{sp}}(k) - y(k)) - \sum_{i=1}^{D-1} K_i^{\mathrm{u}} \triangle u(k-i) \tag{9}$$

where

$$K^{\mathrm{e}} = \sum_{p=1}^{N} K_{1,p} \tag{10}$$

$$K_i^{\mathrm{u}} = \overline{\boldsymbol{K}}_1 \boldsymbol{M}_i^{\mathrm{p}} \tag{11}$$

2.2 Generalized Predictive Control Algorithm

The difference equation model is

$$y(k) = \sum_{i=1}^{n_{\mathrm{B}}} b_i u(k-i) - \sum_{i=1}^{n_{\mathrm{A}}} a_i y(k-i) \tag{12}$$

It may be proved that if the model (12) is used for prediction, the optimal control increments are given by Eq. (4), but the elements of the free trajectory are calculated recurrently from [10]

$$y^0(k+p|k) = \sum_{j=1}^{n_{\mathrm{B}}} e_j(p) u(k-j) + \sum_{j=0}^{n_{\mathrm{A}}} f_j(p) y(k-i) \tag{13}$$

for $p = 1, \ldots, N$. If the set-point variable is constant over the prediction horizon, the explicit GPC control law is obtained

$$\triangle u(k|k) = K^{\mathrm{e}} y^{\mathrm{sp}}(k) - \sum_{j=1}^{n_{\mathrm{B}}} K_j^{\mathrm{u}} \triangle u(k-j) - \sum_{j=0}^{n_{\mathrm{A}}} K_j^{\mathrm{y}} \triangle y(k-j) \tag{14}$$

where K^{e} is defined by Eq. (10) and

$$K_j^{\mathrm{u}} = \sum_{p=1}^{N} k_{1,p} e_j(p) \tag{15}$$

$$K_j^{\mathrm{y}} = \sum_{p=1}^{N} k_{1,p} f_j(p) \tag{16}$$

3 Hardware Accelerators for Implementation of MPC

In the past, the direct method of speeding up calculations was to increase the microprocessors' frequency. In the first years of the XXI century, this trend stopped, i.e. frequency is not increased (or the increase is very slow), but microprocessors offer multiple cores. Sequential programming is no more effective, programs must utilise parallel mechanisms and multiple cores. The same possibility is offered by the FPGA circuits. The FPGA structures are frequently used for signal processing [11]. In many cases, the algorithm can be accelerated more than 100 times by using process decomposition and parallelisation of calculations [3,4].

The developed system for predictive control based on the FPGA structure consists of two layers: software and hardware, mutually dependent. Let us discuss a consistent, universal approach to the implementation of MPC algorithms using the FPGA system. Typically, the NIOS II processor is used as a computational unit. In the described approach dedicated hardware accelerators implemented in programmable logic are used. Their objective is to accelerate calculations of analytical MPC algorithms using parallel operation of logic circuits. The overall structure of the control system with FPGA system is shown in Fig. 1. Its most important components are:

- Terasic Evaluation DE2i-150 board with Cyclone IV FPGA,
- dedicated interface boards for input output signals,
- the controlled process,
- PC with MATLAB script to record data and for visualisation.

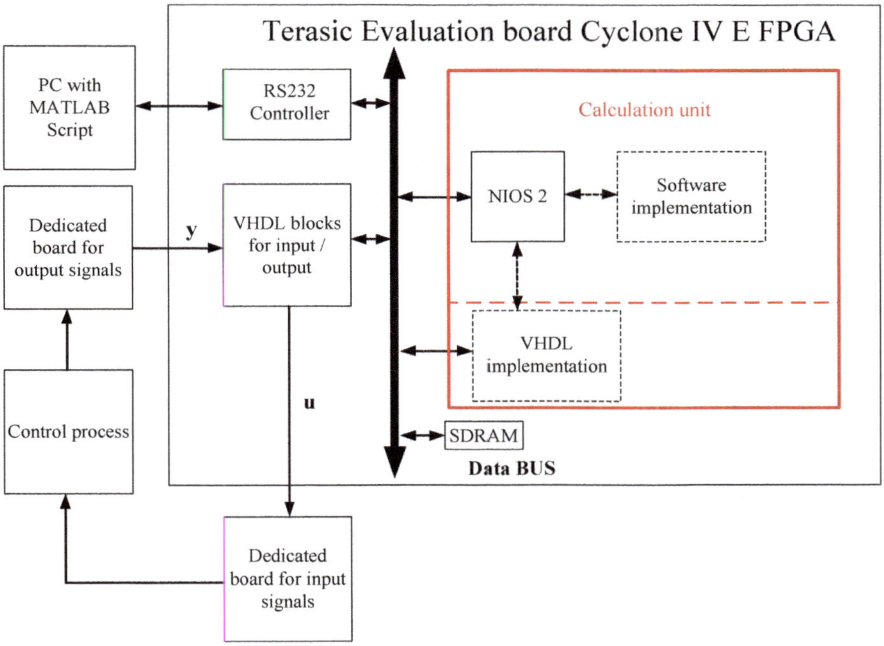

Fig. 1. General structure of control system using FPGA

The hardware part of the system consists of the evaluation board with FPGA, the dedicated interface board and the controlled process. The general idea of system control is based on NIOS II soft processor. The software is developed in C language to realise software algorithms, interconnect with VHDL implementation of accelerators, communication with PC to gather data from VHDL signal processing blocks.

3.1 DMC Algorithm

In this section the hardware implementation of the accelerator for the DMC algorithm is discussed. The current change of the manipulated variable is calculated from Eq. (9), where the constant coefficients (calculated off-line) are given by Eqs. (10) and (11). In implementation of such a formula using a microcontroller in C language, a *for loop* is used, in which multiplication is performed first, followed by summation to a temporary variable. In the case of the hardware

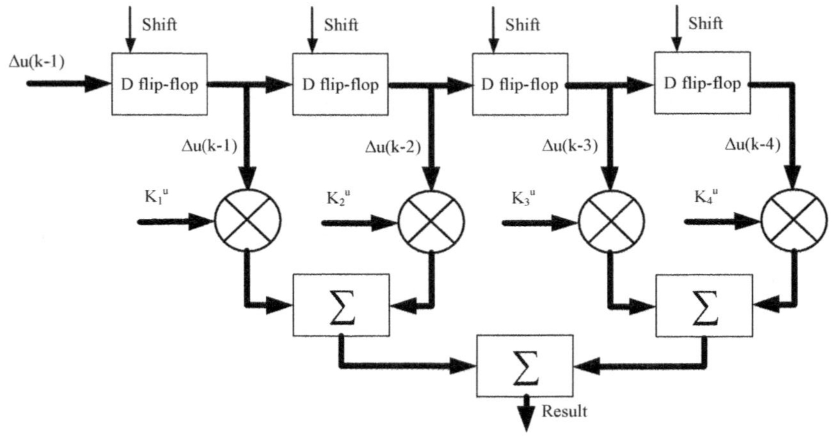

Fig. 2. The accelerator block for the DMC algorithm

FPGA-based solution, it is possible to perform all multiplications in parallel, and then add up the intermediate results. To implement this task in the hardware version, a dedicated Accelerator block has been prepared, which is shown in Fig. 2. The system input accepts the $\triangle u(k-i)$ register, which is provided by the NIOS II processor. The new value is downloaded to the buffer when the Shift signal changes, which starts the D-type flip-flops. The next values of the manipulated variable change are shifted to the next flip-flops. Next, these values go to systems multiplying them by the corresponding coefficients $\boldsymbol{K}_i^{\mathrm{u}}$. In the last stage, all values are added together and the result is returned, which goes back to the NIOS II processor. All process values are of float type. The maximum time for performing all calculations is $1 + M + DS$ clock cycles, where 1 cycle is necessary to load the new value $\triangle u(k-i)$, M cycles are necessary to perform multiplications, S cycle are used to carry out the single summation. It is worth noting that the structure of logic circuits is crucial for the efficiency of calculations. Adding systems continue to run sequentially. This can be modified and parallel calculations can be made for the respective results pairs from the multiplication systems. This creates a full binary tree that is more beneficial in terms of computational complexity.

3.2 GPC Algorithm

In this section the hardware implementation of the accelerator for the DMC algorithm is discussed. The current change of the manipulated variable is calculated from Eq. (14), where the constant coefficients (calculated off-line) are given by Eqs. (10), (15) and (16). The dedicated GPC Accelerator, shown in Fig. 3, is similar to the DMC one. It is possible to perform all multiplications in parallel, and then add up the intermediate results. The system input accepts registers $u(k-i)$ and $y(k)$, which are supplied from the NIOS II processor. New values are downloaded to buffers when the Shift signal changes, which starts the D-type flip-flops. Subsequent values of the control signal and process values are shifted to the next flip-flops. Then, these values go to the multiplication systems by the corresponding coefficients $\boldsymbol{K}_i^{\mathrm{u}}$ and $\boldsymbol{K}_i^{\mathrm{y}}$ respectively. In the last stage, all values are added together and the result is returned, which goes back to the NIOS II processor. The maximum duration of all calculations takes $1 + M + \max(\lceil \log_2(n_{\mathrm{B}}) \rceil, \lceil \log_2(n_{\mathrm{A}} + 1) \rceil)S + S$ clock cycles, where: loading new values of manipulated and control signals takes 1 cycle, it takes M cycles to perform multiplications and S cycles to carry out the single summation.

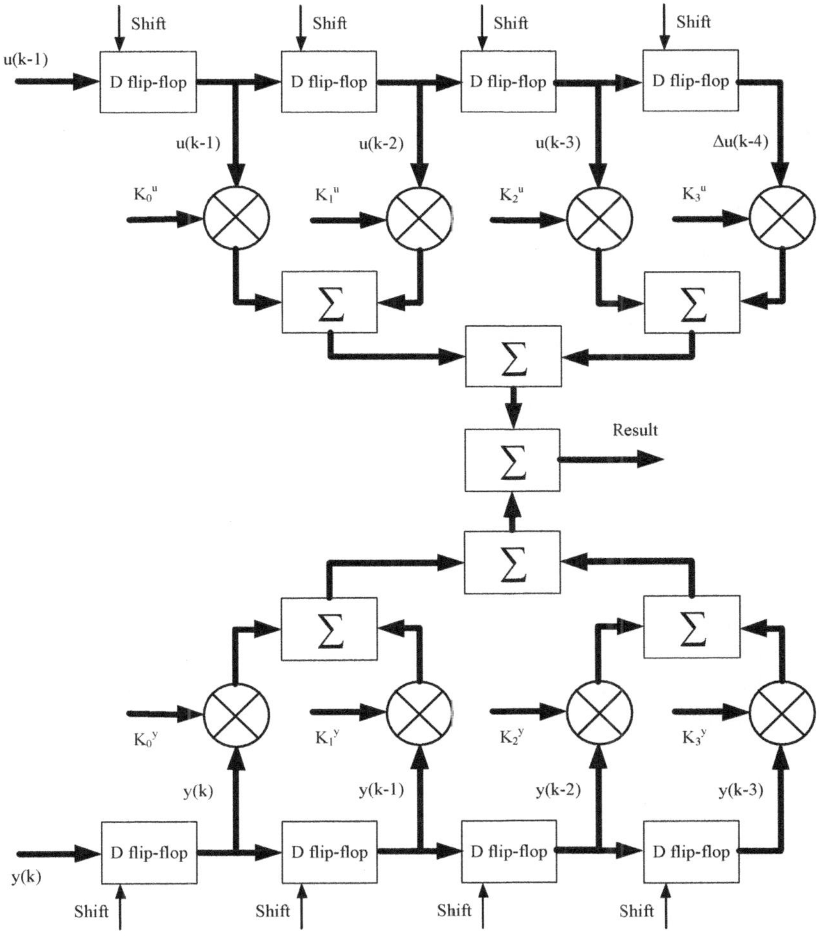

Fig. 3. The accelerator block for the GPC algorithm

Fig. 4. The servo process: the amplifier and the motor

4 Experiments

4.1 Process Description and Experimental Set-Up

The Intel Cyclone IV FPGA evaluation board is used. It offers as many as 150000 logic cells. As the controlled process an industrial drive set consisting of an amplifier and a low-power servomotor and low inertia (produced by Mitsubishi Electric) is considered. The amplifier is built of dedicated electronic circuits based on the semiconductor technologies. It controls a synchronous motor with permanent magnets with very high-speed dynamics and high precision position control, which is realised by means of an encoder with a resolution of 131072 pulses per one rotation of shaft. Figure 4 shows the servo amplifier and the motor.

The amplifier control system has a complicated structure based on the internal model and PID structure. Three basic methods of motor control are possible: a) the output torque control method, implemented by the output current control loops on the motor, b) speed control, in which the set analog signal directly translates into engine speed, c) position control, where the control consists of digital pulses fed to the amplifier's input and counted by a fast counter.

Additionally, a dedicated interface board is used with several possibilities of signal connections. Built-in galvanic isolation protects sensitive digital inputs and outputs of FPGA chip. Interfaces available on board are: 3V digital inputs and outputs, 24 V digital inputs and outputs, line driver inputs dedicated for encoders, analog inputs and analog outputs. Thanks to the board configuration it is possible to connect system to classical automation industry signal standards. Figure 5 shows the entire test stand.

The servo process is used in the SISO (Single Input Single Output) configuration. The manipulated variable is the motor rotation speed. The controlled variable is the actual rotation speed of motor based on high resolution encoder. The sampling time is 50 ms. The manipulated variable is limited to $u =< -1, 1 >$.

Fig. 5. The complete experimental set-up: 1 – the Terasic FPGA DE2i-150 development board, 2 – the servo motor, 3 – the dedicated hardware interface board, 4 – power supply, 5 – PC with Quartus Prime and MATLAB environment

4.2 Results of Experiments

The DMC and GPC algorithms are tuned in the following way:

1. The horizon of dynamics D is selected (only for the DMC algorithm).
2. The penalty factor λ is tuned.
3. Long prediction and control horizons are initially used (N and N_{u}).
4. The length of the horizons is gradually reduced, quality of control is monitored.
5. The prediction horizon is chosen and the lengths of the control horizon is gradually reduced, quality of control is monitored and the best value is chosen.
6. For chosen horizons some modifications of the λ parameter are performed, quality of control is monitored and the best value is chosen.

The following parameters have been found: $D = 30$, $N = 5$, $N_{\mathrm{u}} = 4$, $\lambda = 1$.

Figure 6 shows the normalised step response of the process. The horizon of dynamics D is approx. 1400 ms. Parameters of the model (12) are: $a_1 = -3.1737$, $a_2 = 4.0289$, $a_3 = -2.5573$, $a_4 = 0.8116$, $a_5 = -0.1030$, $b_1 = 0.0001$, $b_2 = 0.0021$, $b_3 = 0.0037$, $b_4 = 0.0010$, $b_5 = 0.00002$.

Figure 7 shows the process trajectories when the fast analytical DMC algorithm is used (calculations are carried out by the hardware accelerator). The time of calculation performed at each sampling instant is 0.2 ms. Figure 8 depicts the process trajectories when the fast analytical GPC algorithm is used. The obtained trajectories are similar, but due to stronger feedback in the GPC control law (Eq. (14)) than in the DMC one (Eq. (9)), the GPC algorithm is characterised by high variability of the manipulated variable. The time of calculation

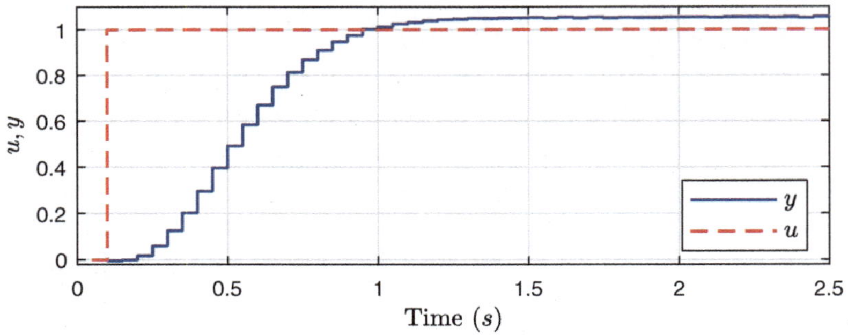

Fig. 6. The normalised step response

performed at each sampling instant of the GPC algorithm is 0.1 ms. It is because the GPC model is very compact when compared with the step-response one used in DMC.

Both DMC and GPC algorithms have been implemented in two additional versions: the classical analytical one, in which the whole vector of the decision variable is calculated at each sampling instant (Eq. (4)), and the numerical one, in which at each sampling instant a quadratic optimisation problem with constraints is solved. Table 1 compares calculation time. Based on performed experiments the following conclusions can be made:

1. The model used in the GPC algorithm typically has several times fewer coefficients than the step response model used in DMC algorithms. As a result, all versions of the GPC algorithms require shorter calculation time than the corresponding DMC ones.
2. The fast analytical algorithms allow to achieve the shortest calculation time, the classical analytical ones are a little slower, the most computationally demanding are the numerical ones.
3. The calculation time of analytical algorithms is constant, independent of the operating point, setpoints, disturbances, etc.

5 Summary

Hardware accelerators implemented in the FPGA may be efficiently used for fast implementation of MPC algorithms. For the chosen parameters and the considered servo, time of calculations necessary at one sampling instant is 0.2 and 0.1 ms for the analytical DMC and GPC algorithms. The described approach may be used for controlling other fast processes, with sampling time of the millisecond order. Two issues may be mentioned as future research directions: FPGA-based implementation of nonlinear MPC algorithms [9] and automatic code generation of the MPC accelerators [1,2].

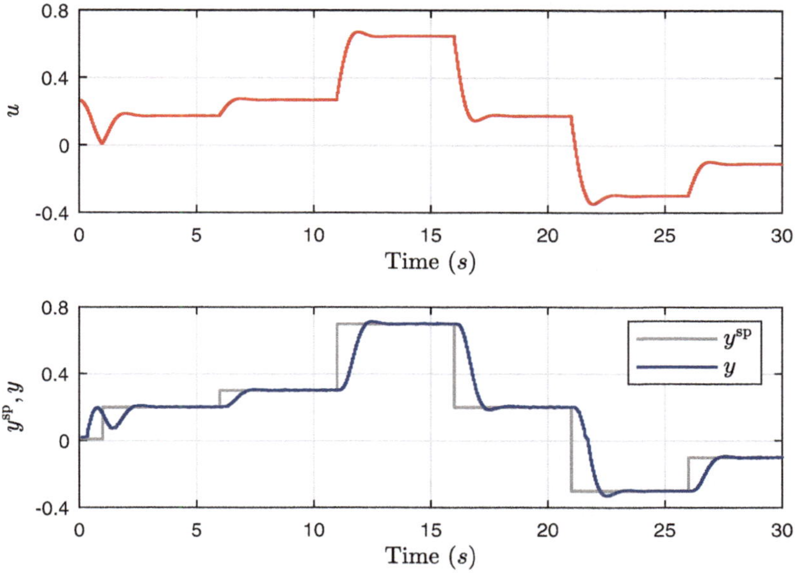

Fig. 7. Process trajectories controlled by the fast analytical DMC algorithm

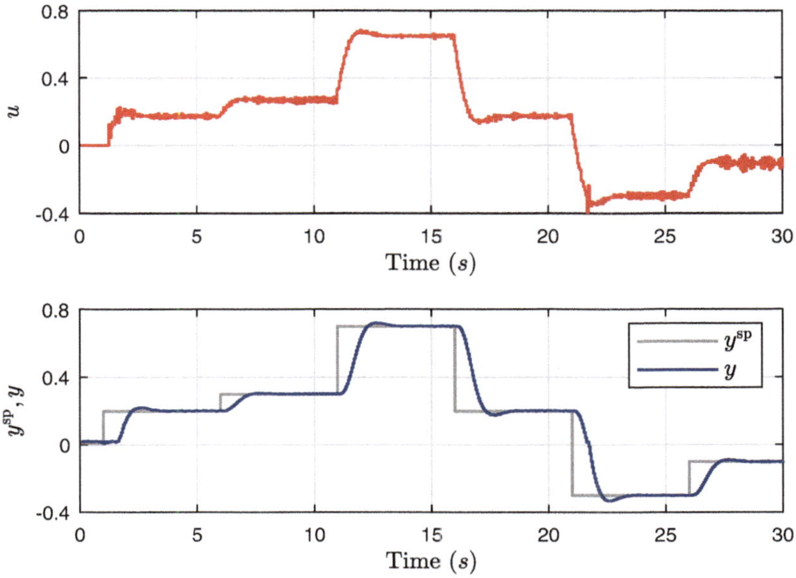

Fig. 8. Process trajectories controlled by the fast analytical GPC algorithm

Table 1. The comparison of calculation time for different versions of the DMC and GPC algorithms (at one sampling instant)

Algorithm	$T_{\text{calculation}}$ (ms)
DMC fast analytical	0.2
DMC analytical	0.8
DMC numerical	4.7
GPC fast analytical	0.1
GPC analytical	0.4
GPC numerical	4.4

References

1. Chaber, P., Ławryńczuk, M.: AutoMATiC: code generation of model predictive control algorithms for microcontrollers. IEEE Transactions on Industrial Informatics (2019). accepted for publication
2. Chaber, P., Ławryńczuk, M.: Fast analytical model predictive controllers and their implementation for STM32 arm microcontroller. IEEE Trans. Ind. Inform. **15**, 4580–4590 (2019)
3. Craven, S., Athanas, P.: Examining the Viability of FPGA Supercomputing. EURASIP J. Embedded Syst. 093652 (2006)
4. Edwards, M.D., Forrest, J., Whelan, A.E.: Acceleration of software algorithms using hard-ware/software co-design techniques. J. Syst. Architect. **42**, 697–707 (1996)
5. Johansen, T.A., Jackson, W., Schreiber, R., Tondel, P.: Hardware synthesis of explicit model predictive controllers. IEEE Trans. Control Syst. Technol. **15**, 191–197 (2007)
6. Knagge, G., Wills, A., Mills, A., Ninness, B.: ASIC and FPGA implementation strategies for model predictive control. In: 2009 European Control Conference (ECC), pp. 144–149, Budapest, Hungary (2009)
7. Lau, M. K. S., Yue, S. P., Ling, K. V., Maciejowski, J. M.: A comparison of interior point and active set methods for FPGA implementation of model predictive control. In: European Control Conference (ECC), pp. 156–161, Budapest, Hungary (2009)
8. Ling, K.V., Wu, B.F., Maciejowski, J.M.: Embedded model predictive control (MPC) using a FPGA. IFAC Proc. Volumes **41**, 15250–15255 (2008)
9. Ławryńczuk, M.: Computationally Efficient Model Predictive Control Algorithms: A Neural Network Approach. Studies in Systems, Decision and Control, vol. 3. Springer, Cham (2014)
10. Tatjewski, P.: Advanced Control of Industrial Processes. Structures and Algorithms. Springer, London (2007)
11. Wain, R., Bush, I., Guest, M., Deegan, M., Kozin, I., Kitchen, C.: An overview of FPGAs and FPGA programming. Initial experiences at Daresbury, Council for the Central Laboratory of the Research Councils (2006)
12. Wojtulewicz, A., Ławryńczuk, M.: Computationally efficient implementation of dynamic matrix control algorithm for very fast processes using programmable logic controller. 2018 23rd International Conference on Methods & Models in Automation & Robotics (MMAR), 579-584 (2018)

13. Xu, F., Chen, H., Jin, W., Xu, Y.: FPGA implementation of nonlinear model predictive control. In: 2014 26th Chinese Control and Decision Conference (CCDC), pp. 108–113, Changsha, China (2014)

Semi-automated Synthesis of Control System Software Through Graph Search

Tomasz Gawron(✉) and Krzysztof Kozłowski

Institute of Automation and Robotics, Faculty of Control, Robotics and Electrical Engineering, Poznan University of Technology, ul.Piotrowo 3a, 61-138 Poznań, Poland
{tomasz.gawron,krzysztof.kozlowski}@put.poznan.pl

Abstract. As the field of automation and robotics develops, control software driving the robotic systems becomes more and more complex. Currently, a predominant approach to modularization and structuring of software in the robotics community, which is embodied by ROS (Robot Operating System) and similar frameworks, is to decompose software into a set of functional modules (i.e. nodes, processes, classes). Particular modules comprising the system are chosen manually by the programmer and they exchange data using loosely coupled publish/subscribe communication mechanisms. While such an approach is helpful and currently highly developed, it seems to lack the means of verifying correctness of the whole control software system (i.e. is there a source of data for all input signals of a given module?, are its outputs actually used by other modules in the system?). Manual specification of system structure and verification of its correctness can be tedious and prone to errors. To help alleviate this, we propose a system composition algorithm utilizing graph search methods. It processes a set of modules implemented by the programmer and produces a graph of modules representing the system with input and output signals specified by the programmer. Such an algorithm can be used with nearly arbitrary control software frameworks.

Keywords: C++ · Software engineering · Control systems · Program synthesis

1 Introduction

It is well known that software development costs and time are the dominating factors in modern robotic projects. Therefore, it is beneficial to create methods for efficient development of correct control software. In this paper we develop a system composition algorithm, which outputs a graph of modules representing a system with programmer-specified input and output signals utilizing a defined set of modules defined by the programmer. Such an algorithm makes

This work was partially supported by Poznan University of Technology under the grant 33/32/SIGR/0003 and National Science Centre (NCN) under the grant No.2014/15/B/ST7/00429.

A. Bartoszewicz et al. (Eds.): PCC 2020, AISC 1196, pp. 1092–1103, 2020.
https://doi.org/10.1007/978-3-030-50936-1_91

it possible to endow an arbitrary control/robotics software framework with an ability to automatically generate parts of the system structure which can be implied from the other parts already defined by the programmer. This greatly reduces the probability of programmer making a mistake in definition of the system and shortens development time. It also facilitates code reuse, since the algorithm can precisely detect certain parts of the system, which have been already implemented in the readily available modules. Furthermore, our algorithm ensures that all modules used in the system will have their input and output signals correctly connected. This feature is usually not present in current software frameworks relying on publish/subscribe mechanisms. Its lack can be a source of many delays and debugging problems, which are especially severe in distributed systems, such as most modern robotic systems.

The paper is structured as follows. In Sect. 2 we state the problem and strictly define the input and output data for the proposed system composition algorithm. The introduced abstractions are independent of particular software implementation. In Sect. 3 we refer to the current state of the art in robotics/control software frameworks and outline a connection between our work and literature on program synthesis. In Sect. 4 we define and analyze the proposed algorithm. Finally, we conclude with Sect. 5, which contains results obtained with application of the proposed algorithm in a motion control system for a telescope (see Fig. 4 and [7] for a description) and future directions for development of our approach.

2 Problem Statement

Before formulating the main problem, let us introduce our definition of a software module. Note that in the sequel we also introduce some notation necessary for reasoning about types of signals processed by modules. For simplicity of exposition, we do not use a full type theory based notation utilized in various computer science literature. Similarly to programming languages, if $i \in \mathcal{I}$ we say that $\mathcal{I}$ is a *type* of i (neglecting some differences between sets and types, which are mostly inconsequential in our case). Furthermore, $o \in \mathcal{O}(s), s \in \mathcal{S}$ means that $\mathcal{O}$ is a parametric type, that is the set $\mathcal{O}$ is dependent on value of s. We say that $\mathcal{O}!$ is a set of all instances of $\mathcal{O}(s)$, that is a union of $\mathcal{O}(s)$ for all s.

We distinguish two types of modules: specific and generic. A specific module m is a Mealy automaton, which has an output o_x defined by mapping

$$o_x = m_x(i_x, s_x): \quad \mathcal{I}_x \times \mathcal{S}_x \mapsto \mathcal{O}_x, \tag{1}$$

and a transition function

$$s_x(t+1) = q_x(s_x(t), o_x(t)), \tag{2}$$

where $o_x \in \mathcal{O}_x$ is an output signal, $i_x \in \mathcal{I}_x$ is an input signal, and $s \in \mathcal{S}_x$ is an observable state of the module. Module m_x processes inputs i_x into outputs o_x. During the processing it can modify its state s_x using the transition function q_x. Initial value of the state is also defined in the module. Consult Fig. 1 for clarification of this data flow. A simple example of a specific module is an integrator block defined for scalar values.

Remark 1. The sets used in definitions of modules are defined according to the grammar of the programming language used to implement the system, hence our name for them: *types*. Thus, usually language compiler supports the user in definition of proper types and conversion rules between them (those rules can be actually expressed as subset equality relationships). However, the type definitions can be constructed arbitrarily and their particulars are transparent to our algorithm. It is only required the following basic operations are well defined: inclusion $\in$ and subset equality $\subseteq$.

A specific module m can be an *instance* of a generic module g with an output

$$o = g(i, s): \quad \mathcal{I}(k) \times \mathcal{S}(k) \times \mapsto \mathcal{O}(s, k), \tag{3}$$

and transition function

$$s(t+1) = q(s(t), o(t)), \tag{4}$$

for some parameter $k \in \mathcal{K}$. As an example, consider an integrator block, which is defined for vectors of an arbitrary size. Note that $\mathcal{K}$ can be a set of types itself (usually called a *kind*), which allows for definition of modules such as, for example, a sorting module capable of sorting sequences of arbitrary types.

Problem 1. System composition problem
Given

- a countable set of specific modules $\mathcal{M}$,
- a countable set of generic modules $\mathcal{G}$,
- a set of j indicated specific modules $\mathcal{F} = \{m_1, \dots m_j\}$,
- a set of n non-parametric system input signals $\mathcal{I} = \{\mathcal{I}_1, \dots \mathcal{I}_n\}$
- and a set of c non-parametric system output signals $\mathcal{O} = \{\mathcal{O}_1, \dots \mathcal{O}_c\}$,

construct a directed graph $(\mathcal{G}, \mathcal{V})$ representing a composed system with vertex set $\mathcal{V} \subseteq \mathcal{M} \cup \mathcal{G}! \cup \mathcal{F}$ and edge set $\mathcal{E}$ with the minimal possible number of vertices. It shall possess following properties:

P1. Every edge $e(m_1, m_2) \in \mathcal{E}$ connects the modules as follows: the output o_1 is connected the to input i_2 and $\mathcal{O}_1 \subseteq \mathcal{I}_2$.
P2. There is a set $\mathcal{M}_O \subseteq \mathcal{V}$ such that output types of modules from $\mathcal{M}_O$ comprise all c elements of $\mathcal{O}$.
P3. In $\mathcal{E}$ there is 1 or more outgoing edges from every module $m_x \in \mathcal{V}$ with output type $\mathcal{O}_x \notin \mathcal{O}$.
P4. There is a set $\mathcal{M}_I \subseteq \mathcal{V}$ such that input types of modules from $\mathcal{M}_I$ comprise all n elements of $\mathcal{I}$.
P5. In $\mathcal{E}$ there are 0 incoming edges to every module $m_x \in \mathcal{V}$ with input type $\mathcal{I}_x \in \mathcal{I}$ and exactly 1 incoming edge for every other module in $\mathcal{V}$.
P6. Indicated modules are contained in the graph, that is $\mathcal{F} \subseteq \mathcal{M}$.
P7. The vertex set $\mathcal{V}$ contains a minimal possible number of generic modules from the set $\mathcal{G}$. That is, one cannot find $\mathcal{V}$ with lower number of modules from the set $\mathcal{G}$ that still satisfies properties P1-P6.

Property P1 ensures that all graph edges are directed from outputs to input signals and that those signals are compatible, i.e. output of a module can be converted to the input of a module, which is connected to it. Property P2 guarantees that the composed system will generate specified output signals. In light of Property P1, one concludes that Property P3 ensures that no redundant output signals will be generated. If an output signal was not specified in the problem and it is not connected to any input, then it is redundant. Property P4 ensures that the composed system handles all specified input signals. While, one could argue that a system ignoring some inputs and generating all outputs could still be functional, we argue that in such a case a redundant input specification should be simply removed. Property P5 guarantees that all module inputs are connected to correct signals. Property P6 simply forces inclusion of all the specified modules, even if their inclusion does not influence the generation of outputs specified in the problem. Property P7 makes solutions of Problem 1 more intuitive and greatly reduces the set of those solutions in general. It ensures that all the possibilities of composing the system with specific modules from $\mathcal{M}$ are exhausted before any generic modules are used. We argue that a specific module, i.e. the one which not generic, hence not parameterized by type k, should always take priority over a generic module, since it is dedicated to process the input signal at hand.

It is worth noting that every subgraph of graph $(\mathcal{V}, \mathcal{E})$ can be defined as a specific module itself. It means that system structures resulting from solution of Problem 1 are composable. As a consequence, the definition of set $\mathcal{F}$ from definition of Problem 1, gives the user a great flexibility in influencing the resulting structure of the composed system. One can add entire subsystems composed by hand (or by execution of our algorithm) to set $\mathcal{F}$. Therefore, it can be easily decided by the user which subsystems of the whole system should be composed automatically. If the results of automatic composition are undesirable or not intuitive, the user can easily guide the composition algorithm by adding particular modules to the set $\mathcal{F}$. Observe, that solution of Problem 1 corresponds only to selection of modules and input/output matching of modules. The specific kind of processing done by the modules is not considered in Problem 1. The types of inputs and outputs must be specific enough to carry information not only about format of the data, but also semantic information about actual meaning of the data. Therefore, one can say, that the kind of processing done by the module, and a contract specifying its behavior is actually encoded in the types of its inputs and outputs. Problem 1 can have multiple equivalent correct solutions. It is not determined which one of those solutions will be chosen by our method.

An exemplary module and a graph representing system structure have been illustrated in Fig. 1. System shown in Fig. 1 consists of 5 modules. The system takes an input i_2 and computes output signals o_3, o_4 and o_5. Note that since certain module outputs are connected to inputs of other modules, the following type relations occur

$$\mathcal{O}_1 \subseteq \mathcal{I}_3,\ \mathcal{O}_2 \subseteq \mathcal{I}_4, \mathcal{O}_3 \subseteq \mathcal{I}_5, \mathcal{O}_3 \subseteq \mathcal{I}_1. \tag{5}$$

It is also worth pointing out, that our problem description allows for internal feedback connections in the resulting system and for routing a single output to inputs of multiple modules, though routing multiple outputs to a single input signal is forbidden.

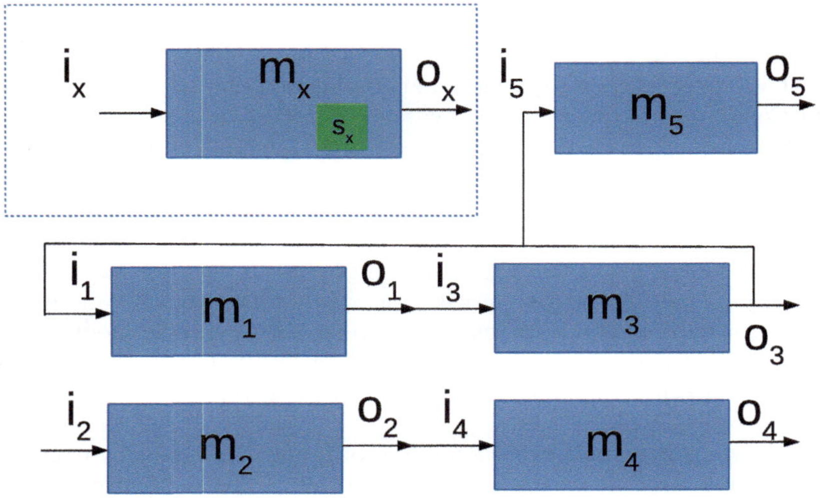

Fig. 1. An exemplary module m_x with its state s_x (top left) and an exemplary system composed from 5 modules.

Again, in Problem 1 we implicitly assume that signal type definitions are rich enough, to drive the correct composition of the system. That is, if two signal types are equal to each other, one can freely connect interchange modules generating those signals. Therefore, in practice, those types should be specific. For example, while a type defined by "is an integer value" is not specific enough in most cases, the type "scaled motor current value" is probably specific enough. One should observe a visualization of the graph resulting from the solution of Problem 1 and verify that type definitions are specific enough. This should be the case for any well-defined module, and one could treat such a situation as algorithm that solves Problem 1 uncovering the erroneous signal definitions in a specific module. Tools for such an interactive usage of our algorithm are currently in development.

Remark 2. Note that our module definition and problem statement is general enough to facilitate construction of systems with more control-oriented abstractions such as e.g. Embodied Agents [13], which might be beneficial in the future.

3 Related Work

In recent years ROS [5,8] has become a de facto standard in robotics research community. Its industrial version and rapid development of ROS2 shows that it is widely supported. However, it mainly provides process-level abstractions (nodes, topics, services) and thread-level abstractions (nodelets in ROS and components in ROS2). While its publish/subscribe model, standardized compilation tools and various run-time tools represent a huge leap forward, it does not ensure that nodes are correctly connected together and provided with required signals. OROCOS (Open Robot Control Software, see [2]) was designed from the ground up to work with mission-critical control system. OROCOS systems consist of modules called components. Each component is a C++ object with variables representing input and output ports. In OROCOS composition of a control system from components (modules) is done with separate API, which requires manual specification of which components should be present in the system and how they should be connected. While this can be tedious (and could be partially eliminated by our algorithm), it allows OROCOS to verify if all modules are connected and synchronized.

It is also worth noting that MIT (Massachusetts Institute of Technology) has developed the ETL [1] project, which (among other things) defines a publish/subscribe API for passing messages between embedded system modules with characteristics similar to ROS. There is also the Drake [12] C++ library, which uses a different, although a bit more limiting paradigm. It provides facilities to define and simulate dynamical systems and solve various optimization problems. It ensures correctness and simplifies system composition by utilization of a fixed system structure, rather than dynamical one considered in this work.

In [4] a stream-oriented Haskell interface to ROS called ROSHask has been proposed. It promotes a functional, highly modular approach to building ROS nodes. A node is generated automatically based on the specification of operations performed on input topics. The boilerplate code for defining a node executable, initializing ROS infrastructure and setting up publishers/subscribers is therefore no longer written by the developer. ROSHask ensures that system modules are correctly connected, since only such system structure specifications are possible in ROSHask. However, it still requires manual specification of system structure and it does not capture module interactions spanning more than one ROS node. In fact one can find similar solutions in other domains (see e.g. [3,10]).

There is also a notable attempt to build general program synthesis solutions, such as Haskell library for composition of embedded C programs, which is called CoPilot [11]. It utilizes monads and software-transactional memory to make implementation of provably correct C programs possible. Automatization of program synthesis at such a low level can be hard to control and hard to learn for developers. It is usually limited to relatively simple systems. One can find a review of program synthesis research in [6]. According to this review, one can argue that our approach is classified as enumerative search method on a module level.

4 System Composition Algorithm

To solve Problem 1 we propose Algorithm 1, which essentially searches a graph representing the space of all possible graphs representing potential system structures, to find a structure satisfying requirements of Problem 1. To explain further, as shown in Fig. 2, during the search we explore an implicit graph, in which *every vertex* is a graph, which is a candidate for solution of Problem 1, i.e. a $(\mathcal{V}, \mathcal{E})$. Each edge in this implicit graph connects two system structures, which differ only by inclusion of one additional module. Introducing the upper bound $\overline{V} > 0$ on the number modules in the system (which is a good engineering requirement to have anyway), we define an additive cost functional $J(\mathcal{V})$ minimized by Algorithm 1. It is dependent on the type of modules included in $\mathcal{V}$ and can be expressed as follows:

$$J(\mathcal{V}) = |\mathcal{V} \cap \mathcal{M}| + \overline{V}|\mathcal{V} \cap \mathcal{G}!|, \tag{6}$$

where operator $|\cdot|$ denotes cardinality of a set in this context. Cost (6) penalizes systems with a large number of modules and system structures including many generic modules.

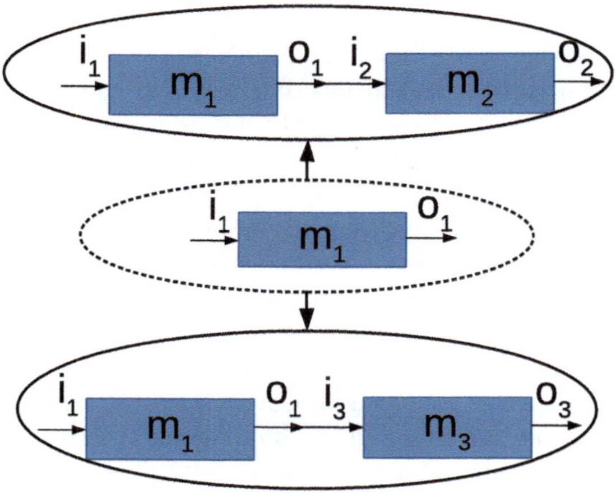

Fig. 2. A fragment of search space explored by Algorithm 1. Each candidate system graph corresponds to a graph node. Node denoted by dashed line (the middle node) is a parent of the other 2 nodes visible in the figure.

Algorithm 1. The system composition algorithm, which solves Problem 1.

```
$(\mathcal{V}, \mathcal{E}, J(\mathcal{V})) \leftarrow (\emptyset, \overline{V}^2, 0)$
$\mathcal{Q} \leftarrow \emptyset$
for all $\mathcal{I}_i \in \mathcal{I}$ do
    $\mathcal{A}_i \leftarrow$ all modules from $\mathcal{M}$ with input type $\mathcal{I}_i$
    if $\mathcal{A}_i = \emptyset$ then
        $\mathcal{A}_i \leftarrow$ all modules from $\mathcal{G}!$ with input type $\mathcal{I}_i$
        if $\mathcal{A}_i = \emptyset$ then
            return $(\mathcal{V}, \mathcal{E})$ solving Problem 1 does not exist
for all $\mathcal{O}_i \in \mathcal{O}$ do
    $\mathcal{B}_i \leftarrow$ all modules from $\mathcal{M}$ with output type $\mathcal{O}_i$
    if $\mathcal{B}_i = \emptyset$ then
        $\mathcal{B}_i \leftarrow$ all modules from $\mathcal{G}!$ with output type $\mathcal{O}_i$
        if $\mathcal{B}_i = \emptyset$ then
            return $(\mathcal{V}, \mathcal{E})$ solving Problem 1 does not exist
$\mathcal{A} = \mathcal{A}_1 \times \ldots \times \mathcal{A}_n, \quad \mathcal{B} = \mathcal{B}_1 \times \ldots \times \mathcal{B}_k, \quad \mathcal{V}_s = \mathcal{A} \times \mathcal{B}$
for all $\mathcal{V}_c \in \mathcal{V}_s$ do
    $\mathcal{V}_c \leftarrow \mathcal{V}_c \cup \mathcal{F}$
    remove element repetitions from $\mathcal{V}_c$
    $\mathcal{Q} \leftarrow$ push graph $(\mathcal{V}_c, \emptyset, J(\mathcal{V}_c))$ (see (3))
while $\mathcal{Q} \neq \emptyset$ do
    sort $\mathcal{Q}$ by cost $J$
    $(\mathcal{V}_c, \mathcal{E}_c, J(\mathcal{V}_c)) \leftarrow$ pop from $\mathcal{Q}$
    if $J(\mathcal{V}_c) \geq J(\mathcal{V}) \quad \vee \quad |V_c| > \overline{V}$ then
        continue
    feasible = true, $(\mathcal{V}_n, \mathcal{E}_n, J(\mathcal{V}_n)) \leftarrow (\mathcal{V}_c, \mathcal{E}_c, J(\mathcal{V}_c))$
    for all modules $m \in \mathcal{V}_c$ do
        if $m$ has an unconnected input $i \notin \mathcal{I}$ or output $o \notin \mathcal{O}$ then
            feasible = false
            find all compatible signal from modules in $\mathcal{V}_c$
            if no compatible outputs found then
                find unvisited module with matching signals in $\mathcal{M}$ and mark it visited
                add found module to $\mathcal{V}_n$, add connections to $\mathcal{E}_n$, break
            if no matching unvisited module found then
                find unvisited module with matching signals in $\mathcal{G}!$ and mark it visited
                add found module to $\mathcal{V}_n$, add connections to $\mathcal{E}_n$
    if feasible = true then
        if $J(\mathcal{V}_c) < J(\mathcal{V})$ then
            $(\mathcal{V}, \mathcal{E}, J(\mathcal{V})) \leftarrow (\mathcal{V}_c, \mathcal{E}_c, J(\mathcal{V}_c))$
    else
        if $(\mathcal{V}_c, \mathcal{E}_c) \neq (\mathcal{V}_n, \mathcal{E}_n)$ then
            $\mathcal{Q} \leftarrow$ push graph $(\mathcal{V}_c, \mathcal{E}_c, J(\mathcal{V}_c))$ (see (3))
            $\mathcal{Q} \leftarrow$ push graph $(\mathcal{V}_n, \mathcal{E}_n, J(\mathcal{V}_n))$ (see (3))
if $(\mathcal{V}, \mathcal{E}) = (\emptyset, \emptyset)$ then
    $(\mathcal{V}, \mathcal{E})$ solving Problem 1 does not exist
else
    return $(\mathcal{V}, \mathcal{E})$
```

Algorithm 1 proceeds by carrying out a Dijkstra-like search. However, there is a notable difference. The result we are after is a goal vertex in the problem space graph (i.e. the graph of system structure graphs), instead of the whole path between initial and final vertex. Furthermore, it is not trivial to construct the initial vertex in problem space. After initialization from lines 1–2, we construct such an initial state in lines 3–19. We collect sets $\mathcal{A}$ and $\mathcal{B}$ representing all the modules, which can be connected to inputs $\mathcal{I}$ and outputs $\mathcal{O}$ specified in Problem 1. Neglecting the set of prescribed modules $\mathcal{F}$ considered in line 17, The cartesian product of $\mathcal{A}$ and $\mathcal{B}$ is a set of all initial vertex sets $\mathcal{V}_c$ allowing for satisfaction of requirements stated in Problem 1. Note, that such a set might be empty, which is equivalent to a guarantee that solution to Problem 1 does not exist. Next, in lines 16–19 we make sure that there are no repeated modules in candidate sets, and push those sets to the priority queue $\mathcal{Q}$. At this moment, we enter line 20 where the search procedure is carried out. We prioritize queued nodes by cost J and terminate their expansion early if a maximal number of modules was exceed or the cost J is larger than the best currently known solution. After that, in lines 26–36 we assess the feasibility of the currently processed graph. At the same time, we try to find connections between the module that can be made, to make the graph feasible w.r.t. Problem 1. We prioritize connections to specific modules. Note that each time a graph is popped from queue $\mathcal{Q}$ we try to add as many edges as possible, but we stop after adding a single vertex. Additional vertices are added later, because we requeue the graph after expansion if the expansion was successful (see lines 41–43). We update the best solutions online, as shown in lines 37–39. Let us now perform a short analysis showing that Algorithm 1 indeed solves Problem 1.

Remark 3. Description of Algorithm 1 has been tailored to show the main concepts and logical flow of the algorithm. Its computationally efficient implementation should utilize various additional data structures to efficiently look for modules, which can be connected to the graph. Similarly, and additional map used to verify, which modules have been visited and a high performance priority queue shall be used in production implementation.

Theorem 1. *Algorithm 1 solves Problem 1.*

Proof. Let us begin by showing that Algorithm 1 terminates in finite time. Note that the set of possible vertices, and by implication edges, is finite (bounded by $\overline{V}$), therefore its exhaustive exploration takes finite time. Note that after the initial phase (iterating over a finite set of modules), the queue $\mathcal{Q}$ is expanded only in lines 41–42. This expansion stops after visiting a finite number of modules (hence the visitation marking in our algorithm). This shows that our algorithm terminates in finite time. From this execution flow, it is also evident that it explores the space of all possible modules and all possible connections, therefore it is complete. Since we explore all the solutions, and each feasible solution is compared to the current best one in line 37, our algorithm is optimal with respect to the cost J. This implies, by the construction of cost J, that our algorithms builds systems with minimal number of generic modules first (single generic module increases J by $\overline{V}$, which exceeds the maximum number of specific modules) and minimal number of overall modules after that, as required in Problem 1. Since the basic properties of Algorithm 1 have been established, we turn to properties P1-P6, since P7 is a consequence of optimality with respect to cost J. P1 is satisfied by construction - we only check such connections. P2 and P4 are satisfied by sets $\mathcal{A}$ and $\mathcal{B}$ computed in our algorithm. Inspection of lines 26–35 shows that properties P3 and P5 are satisfied by all solutions marked as feasible. In terms of P6, we force indicated modules into the graph at line 17. Since the graph can only expand in size, indicated modules $\mathcal{F}$ will be present in all the solutions.

5 Exemplary Results and Future Works

In Fig. 3 we show a subsystem-level fragment of a final system structure, which was composed based only on the specification of input signals $\mathcal{I}$, output signals $\mathcal{O}$ and modules $\mathcal{F}$ denoted by dashed lines. Other modules were automatically picked by Algorithm 1 from a set of 20 specific and generic modules corresponding to the initial set of modules provided by developers of the advanced telescope motion control system described in [7] and shown in Fig. 4. The modules were defined as C++ classes and templates annotated by macros, which automatically register metadata of modules in our system. In this particular case Algorithm 1 has been running for 0.03s and evaluated 93 potential solutions of Problem 1 before finding a correct solution. One can easily find that some generic modules were correctly used by the algorithm, and at the same time, due to the prioritization of specific modules over generic modules the result is intuitive. The control system structure was correctly adapted to the requirement stating than an observer module should be present. To satisfy this requirement a generic multiplexer module was used to adapt the intermediate signal structure and produce a correct input for the position controller module. It is also worth noting that the necessary presence of a current controller and two motor driver modules has been automatically inferred from system requirements, thanks to Algorithm 1. While this is only a moderately complex system, one can argue that it demonstrates basic applicability of the proposed algorithm. Similarly, only a subset of provided modules has been used, as should be the case in most practical systems, which reuse code libraries from other past projects. The particular modules from the set $\mathcal{M}$ supplied by the developers, were defined and implemented in C++ using a simple library. However, its description is outside of the scope of this paper.

In the future, we will focus on extending this approach with additional soft constraints referring to various traits of signals such as frequency of updates. Furthermore, one could endow our approach with a specification of mathematical invariants (possibly checked using and SMT solver like [9]), which should be respected by particular modules or the whole system. For example, Linear Temporal Logic specifications for system output signals would be most useful.

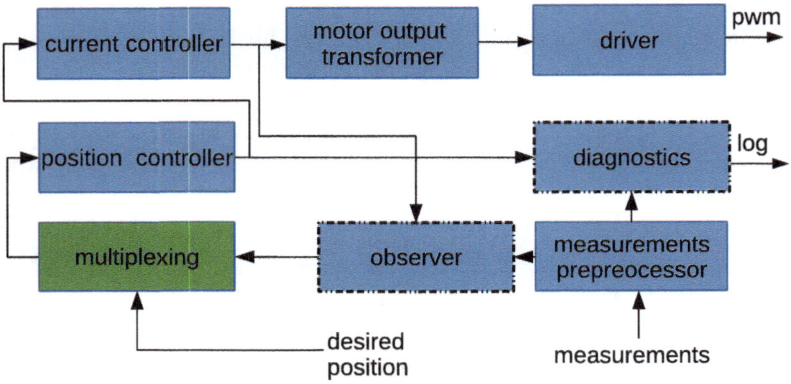

Fig. 3. A subsystem-level fragment of a telescope motion control system composed using Algorithm 1. The green multiplexing block is not a module. It represents a primitive operation of joining two signals into a combined signal, which was omitted from our considerations for simplicity of exposition.

Fig. 4. The telescope described in [7]. Its motion control system has been used as an example driving the development and verification of Algorithm 1.

References

1. Embedded template library. http://www.etlcpp.com/. Accessed 20 Jan 2020
2. Bruyninckx, H., Soetens, P., Koninckx, B.: The real-time motion control core of the Orocos project. In: IEEE International Conference on Robotics and Automation, pp. 2766–2771 (2003)
3. Chupin, G., Nilsson, H.: Functional reactive programming, restated. In: Proceedings of the 21st International Symposium on Principles and Practice of Programming Languages 2019, PPDP 2019. Association for Computing Machinery, New York (2019). https://doi.org/10.1145/3354166.3354172
4. Cowley, A., Taylor, C.J.: Stream-oriented robotics programming: the design of roshask. In: 2011 IEEE/RSJ International Conference on Intelligent Robots and Systems, pp. 1048–1054 (2011). https://doi.org/10.1109/IROS.2011.6095033
5. Garber, L.: Robot OS: a new day for robot design. Computer **46**(12), 16–20 (2013). https://doi.org/10.1109/MC.2013.434
6. Gulwani, S., Polozov, O., Singh, R.: Program synthesis. Foundations Trends® Program. Lang. **4**(1–2), 1–119 (2017). https://doi.org/10.1561/2500000010
7. Kozlowski, K., Pazderski, D., Krysiak, B., Jedwabny, T., Piasek, J., Kozlowski, S., Brock, S., Janiszewski, D., Nowopolski, K.: High precision automated astronomical mount. In: Szewczyk, R., Zieliński, C., Kaliczyńska, M. (eds.) Automation 2019, pp. 299–315. Springer, Cham (2020)
8. Maruyama, Y., Kato, S., Azumi, T.: Exploring the performance of ROS2. In: 2016 International Conference on Embedded Software (EMSOFT), pp. 1–10 (2016). https://doi.org/10.1145/2968478.2968502
9. de Moura, L., Bjørner, N.: Z3: an efficient smt solver. In: Ramakrishnan, C.R., Rehof, J. (eds.) Tools and Algorithms for the Construction and Analysis of Systems, pp. 337–340. Springer, Heidelberg (2008)
10. Perez, I., Bärenz, M., Nilsson, H.: Functional reactive programming, refactored. In: Proceedings of the 9th International Symposium on Haskell, Haskell 2016, pp. 33–44. Association for Computing Machinery, New York (2016). https://doi.org/10.1145/2976002.2976010
11. Pike, L., Wegmann, N., Niller, S., et al.: Copilot: monitoring embedded systems. Innovations Syst. Softw. Eng. **9**(1), 235–255 (2013). https://doi.org/10.1007/s11334-013-0223-x
12. Tedrake, R., The Drake Development Team: Drake: model-based design and verification for robotics (2019). https://drake.mit.edu
13. Zieliński, C., Winiarski, T.: General specification of multi-robot control system structures. Bull. Polish Acad. Sci. Tech. Sci. **58**(1), 15–28 (2010). https://doi.org/10.2478/v10175-010-0002-x

Model Predictive Control of a Dynamic System with Fast and Slow Dynamics: Implementation Using PLC

Sebastian Plamowski(✉)

Institute of Control and Computation Engineering, Warsaw University of Technology, ul. Nowowiejska 15/19, 00-665 Warsaw, Poland
S.Plamowski@ia.pw.edu.pl

Abstract. The article presents Model Predictive Control (MPC) of a multivariable laboratory process using Programmable Logic Controller (PLC). The Dynamic Matrix Control (DMC) MPC algorithm is used in which a step-response model of the process is used for prediction. Two main practical issues are discussed. Firstly, it is shown how to deal with DMC control of a dynamical systems in which some variables react quickly and some slowly. In order to deal with fast and slow dynamics, a decomposed control structure based on the DMC algorithm is used. Secondly, the decomposed DMC control structure is implemented using the PLC controller, taking into account the PLC resources. Results of real laboratory experiments are present to show effectiveness of the discussed method.

Keywords: Model Predictive Control (MPC) · Dynamic Matrix Control (DMC) · Programmable Logic Controller (PLC) · Multiple-Input Multiple-Output (MIMO) control · Decomposition

1 Introduction

Model Predictive Control (MPC) algorithms [1, 4, 9] are frequently used in industrial applications [3, 8], especially as a technique for efficient control of Multiple-Input Multiple-Output (MIMO) processes. Compared with the classical PID controller, MPC algorithms give very good control quality and tuning is much easier. Furthermore, in MPC all necessary constraints may be taken into account and also can be applied to nonlinear objects [5]. In large-scale industrial applications MPC algorithms are implemented in Distributed Control Systems (DCS) using Programmable Logic Controllers (PLC) [11] or specialized industrial controllers [6] whereas in embedded systems micro-controllers [2] or Field Programmable Gate Arrays (FPGA) [10] are used as hardware platforms.

This work discusses implementation of the Dynamic Matrix Control (DMC) MPC algorithm for a MIMO laboratory process. The DMC algorithm uses for prediction a step-response model of the controlled process. Unfortunately, some MIMO processes have two types of variables: some of them react quickly and some slowly. Hence, application of

A. Bartoszewicz et al. (Eds.): PCC 2020, AISC 1196, pp. 1104–1115, 2020.
https://doi.org/10.1007/978-3-030-50936-1_92

the classical DMC algorithm leads to difficulties since some input-output channels need a huge number of step response coefficients. In order to solve the problem, a decomposed control structure based on the DMC algorithm is presented. Secondly, it is shown how to implement the algorithm using the PLC controller with limited resources, which in the general MIMO cases is always an issue, but is even more important in the case of systems with fast and slow dynamics due to memory limits. Efficiency of the discussed approach is shown using a MIMO laboratory process, results of real experiments are reported.

2 DMC Problem Formulation

Let n_u and n_y denote the number of process inputs (manipulated variables) and outputs (controlled variables), respectively, i.e. $u = \left[u_1 \ldots u_{n_\mathrm{u}}\right]^T, y = \left[y_1 \ldots y_{n_\mathrm{y}}\right]^T$. In MPC the vector of decision variables calculated at each sampling instant, $k = 0, 1, 2, \ldots,$ has the length of $n_\mathrm{u}N_\mathrm{u}$ and consists of the current and future increments of the manipulated variables

$$\Delta \boldsymbol{u}(k) = \begin{bmatrix} \Delta u(k|k) \\ \vdots \\ \Delta u(k + N_u - 1|k) \end{bmatrix} \tag{1}$$

where $\Delta u(k+p|k)$ denotes increments of the manipulated variables for the future sampling instant $k+p$ calculated at the current instant k, N_u is the control horizon. The rudimentary MPC optimization problem is

$$\min_{\Delta \boldsymbol{u}(k)} \left\{ \sum\nolimits_{p=1}^{N} \left\| y^{\mathrm{sp}}(k+p|k) - \hat{y}(k+p|k) \right\|_{\boldsymbol{\Psi}_p}^2 + \sum\nolimits_{p=1}^{N_u-1} \left\| \Delta u(k+p|k) \right\|_{\boldsymbol{\Lambda}_p}^2 \right\} \tag{2}$$

subject to

$$\begin{gathered} u^{\min} \le u(k+p|k) \le u^{\max}, p = 0, \ldots, N_u - 1 \\ -\Delta u^{\max} \le u(k+p|k) \le \Delta u^{\max}, p = 0, \ldots, N_u - 1 \end{gathered}$$

The symbols $y^{\mathrm{sp}}(k+p|k)$ and $\hat{y}(k+p|k)$ denote the set-point trajectory and the predicted trajectory for the sampling instant $k+p$ calculated at the instant k, the prediction horizon is N. The weighting matrices $\boldsymbol{\Psi}_p = \mathrm{diag}\left(\psi_{p,1}, \ldots, \psi_{p,n_\mathrm{y}}\right)$ and $\boldsymbol{\Lambda}_p = \mathrm{diag}\left(\lambda_{p,1}, \ldots, \lambda_{p,n_\mathrm{u}}\right)$ are of dimensionality $n_\mathrm{y} \times n_\mathrm{y}$ and $n_\mathrm{u} \times n_\mathrm{u}$, respectively. The minimal and maximal values of the manipulated variables are denoted by $u^{\min}$ and $u^{\max}$, respectively, the maximal change rate of the manipulated variables is $\Delta u^{\max}$. At each sampling instant, having calculated $n_\mathrm{u}N_\mathrm{u}$ future increments of the manipulated variables over the control horizon (Eq. (1)), only the increments for the current sampling instant are applied to the process.

The basic optimization problem (2) is the same in all MPC algorithms, implementation details differ because for different types of dynamical models different prediction equations are derived [1, 4, 9]. Provided that a linear model is used for prediction, one obtains a Quadratic Programming (QP) optimization problem which may be solved by available solvers. Although such an approach is preferred, it strongly impacts memory resources and CPU capacity. In typical industrial applications MPC with on-line quadratic optimization is practically possible for rather moderate numbers of process inputs and outputs [6]. Furthermore, the prediction horizon and particularly the control horizon have significant impact on the resulting computational complexity and thus the time necessary to complete calculations at each sampling instant on-line. Additionally, in the DMC algorithm the impact of the horizon of process dynamics D, which defines the number of model step-response coefficients may be seriously taken into account [6].

In order to reduce the amount of CPU resources and speed up calculations, analytical version of MPC algorithms are used [9]. In that approach the constraints are not taken into account during optimization, but the obtained analytical result is projected onto the admissible set determined by the constraints. Taking into account the DMC algorithm, the increments of the manipulated variables for the current sampling instant are

$$\Delta U(k) = K^{e}\left[Y^{\mathrm{sp}}(k) - Y(k)\right] - \sum\nolimits_{j=1}^{D-1} K_{j}^{u} \Delta U(k-j) \tag{3}$$

Where K^{e} is a $n_{\mathrm{u}} \times n_{\mathrm{y}}$ matrix, K_{j}^{u} is a $n_{\mathrm{u}} \times n_{\mathrm{u}}$ matrix, $\left[Y^{\mathrm{sp}}(k) - Y(k)\right]$ is a vector of length n_{y} and $\Delta U(k-j)$ is a vector of length n_{u}.

The analytical version of the DMC algorithm and other MPC methods may be successfully implemented using hardware platforms with limited resources, e.g. microcontrollers [2], FPGAs [10] and PLCs [11].

3 The MIMO Laboratory Process

3.1 The Process Set-Up

The considered laboratory process is shown in Fig. 1. It simulates in laboratory environment the behavior of the fan heating system in a multi-storey building (or any aeration system, e.g. an installation supplying primary and secondary air to the combustion chamber of an energy boiler).

The process allows simultaneous control 7 Controlled Variables (CV): the temperatures T1, T2, T3 and Tout, where Tout is auxiliary temperature after the heater, and air flows F1, F2 and F3 on three levels. 8 Manipulated Variables (MV) are possible: heater power, fan power, three hot air dampers ADh1, ADh2, ADh3 and three cold air dampers ADc1, ADc2, ADc3. Table 1 lists all process variables.

3.2 Static Characteristics of the Process

To better understanding the properties of the considered laboratory process and interactions between variables the static characteristics between all MVs and CVs variables have been determined experimentally. The tests have been performed under the same test conditions: air dampers and fan have been operated at 50% and the heater at 20%. The operating range of the air dampers has been set at (30%–100%), below 30% the damper either has not moved or has been completely closed. The heater control range

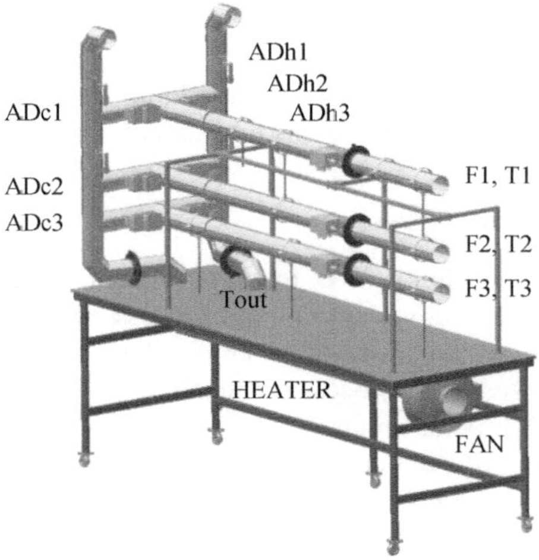

Fig. 1. The laboratory process

Table 1. The list of process variables

Symbol	Description	PV/MV
FAN	Fan power	MV
HEATER	Heater power	MV
ADh1	Air Damper hot collector level 1	MV
ADc1	Air Damper cold collector level 1	MV
ADh2	Air Damper hot collector level 2	MV
ADc2	Air Damper cold collector level 2	MV
ADh3	Air Damper hot collector level 3	MV
ADc3	Air Damper cold collector level 3	MV
Tout	Temperature after heater	CV
T1	Temperature on level 1	CV
F1	Flow air on level 1	CV
T2	Temperature on level 2	CV
F2	Flow air on level 2	CV
T3	Temperature on level 3	CV
F3	Flow air on level 3	CV

has been specified at 8-30% range. Below 8% the heater does not operate and above 30% the temperature is too high and thermal protection is activated.

The obtained characteristics show that the static properties of the process are almost linear. Moreover, strong interactions between the consecutive MVs and CVs are observed, which motivates the use of MPC algorithms, not simple PID ones. Additionally, some detailed observations are possible. There are strong dependences between variables on different levels caused by hot and cold air dampers. The heater does not impact the flow. The fan impacts mainly the flow and indirectly the temperature.

3.3 Step-Responses of the Process

In this work the DMC algorithm is used, in which a step-response model is necessary. Step responses of the consecutive input-output process channels have been obtained experimentally. The consecutive inputs of the process have been excited by a step signal and the resulting responses of all process outputs have been recorded. That procedure has been repeated for all process inputs.

The step-responses have been collected under the same test conditions, i.e. when the air dampers and fan have operated at 50% and the heater at 20%. During the test the values of MVs have been forced one by one separately about + 10%, except for the heater where due to the relatively small range of the control signal a + 5% step has been applied. Directions and amplitudes of the steps have been adjusted to the actual levels of process signals. There are two types of variables: fast and slow. The fast variables relate to flows and slow variables relate to temperatures. The flow values stabilize within a few seconds after changes of air damper position (Fig. 2, Fig. 4 and Fig. 3) and within a dozen of seconds after the fan load changes (Fig. 4 and Fig. 5).

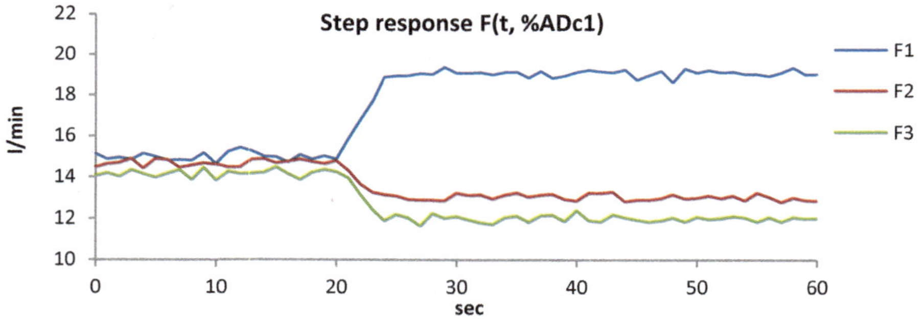

Fig. 2. The step response of flows after changes ADc1 position from 50% to 60%

The temperatures stabilize within 20 min after changes of air damper position (Fig. 3) and approximately within 23 min after the fan load changes (Fig. 5).

4 The Control Structure

4.1 Resource Considerations

The presented step-responses of the process presented in Figs. 2, 3, 4 and 5 confirm that the process has fast and slow dynamic properties. An attempt to apply one MIMO MPC algorithm to the considered process would require the use of a controller with 6 (or 7

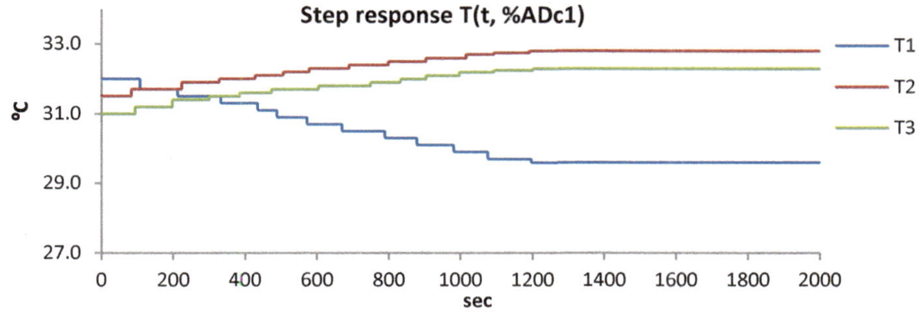

Fig. 3. The step response of temperatures after changes ADc1 position from 50% to 60%

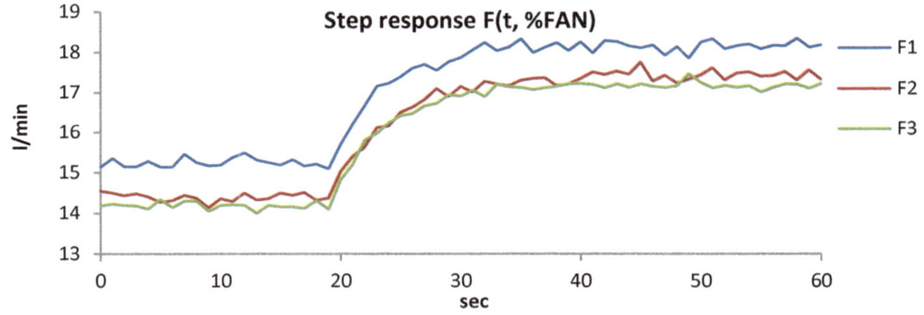

Fig. 4. The step response of flows after changes fan power from 50% to 60%

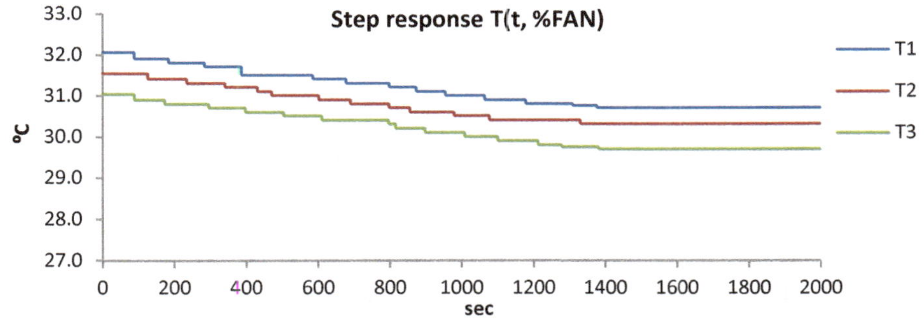

Fig. 5. The step response of temperatures after changes fan power from 50% to 60%

considering Tout) CV signals and 8 MV signals. Such a controller would have to work in a control loop corresponding to the dynamics of fast variables, i.e. with sampling time Tp equal to 1 s and a prediction and control horizon corresponding to the dynamics of slow variables. As a result, the control horizon would have to be set to 60 s and the prediction horizon to 1200 s (20 min). These requirements must be compared with available resources and technical limits of the PLC controller.

The control law of the analytical version of the DMC algorithm is given by Eq. (3). Its coefficients are calculated only once in off-line mode and are defined in the PLC program. These calculations do not impact PLC resources and performance. Implementation of the control law requires memory for the coefficients and memory for past values of MVs. The amount of the required memory depends on the values of tuning parameters of DMC algorithm. It depends mainly on the number MVs (n_u) and the length of the horizon of dynamics D, the impact of the number of CVs (n_y) is negligible. The coefficients can be stored in slower file registers, but the past values of MVs should be saved to fast registers. The number of coefficients for different values of tuning parameters is summarized in Table 2.

Table 2. The number of coefficients of DMC control law (Eq. (3))

$n_u \backslash D$	10	50	100	1000	1200
1	10	50	100	1000	1200
2	40	200	400	4000	4800
3	90	450	900	9000	10800
4	160	800	1600	16000	19200
5	250	1250	2500	25000	30000
6	360	1800	3600	36000	43200
7	490	2450	4900	49000	58800
8	640	3200	6400	64000	76800
9	810	4050	8100	81000	97200
10	1000	5000	10000	100000	120000

The number of coefficients determines the number of ADD and MUL instructions executed on CPU of the PLC. The number of past MVs stored in fast registers is defined by the number of MVs and D parameters and it is equal $n_u D$.

In the considered process the number of MVs is equal 8, number of CV is equal 7 and the length of the horizon of dynamic D is 1200. For these parameters the number of coefficients is 76800 and the number of past MVs is 9600. These values significantly exceed capabilities of the modern PLCs. Hence, an alternative control structure is necessary.

4.2 The Controller Decomposition

In this work the general MIMO control problem is decomposed into smaller subproblems. For this purpose, the author's original method based on the aggregated variables is be used. In this case, the key issue is to separate the control system into flow and temperature control. Independent treatment of fast and slow variables allows the appropriate selection of the calculation interval and the configuration parameters of the controllers. Therefore, the control structure is split into two parts based on independent multidimensional DMC

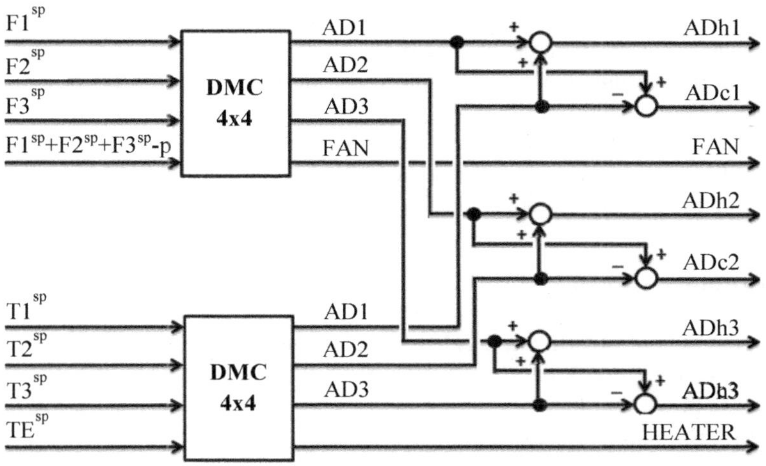

Fig. 6. The decomposed control structure

controllers operating with different sampling periods. The resulting decomposed control structure is depicted in Fig. 6.

The key issue at the decomposition stage is to correctly define the role of the individual variables. In the case of the considered process, the fan is responsible for the flows and the heater for the temperatures (for thermal energy). The air dampers should only be used for deviation temperatures and flows between the levels.

Changing the heater's control does not affect the flows, while the temperatures depend on both control variables: heater's and fan's power. Additionally, air dampers affect flow and temperature and should be as widely open as possible to minimize energy consumption. These considerations lead to the control structure shown in Fig. 6.

The fast controller is responsible for controlling the flow using the fan and air dampers. In order to reduce the number of variables it is assumed that the controller calculates one value for the hot and cold damper on a given level (presented in Fig. 6 as AD1, AD2 and AD3 variables). To minimize energy and keep the air dampers open as possible, the aggregate variable SUM($F1^{sp}$, $F2^{sp}$, $F3^{sp}$) is used in the control structure. The sum of flows depends only on the fan power because it can be assumed (with some approximation) that the air dampers change distribution between floors, and they do not change the total amount of the flow. Based on that assumption, the set-point for the variable SUM($F1^{sp}$, $F2^{sp}$, $F3^{sp}$) can be intentionally artificially lowered by the p parameter (set to small value $p = 0.01$). In this manner the structure optimizes the operating point, i.e. minimizing losses by reducing the fan rotation and keeping the air dampers maximally open. The final value of F1, F2 and F3 in steady state is set lightly (depends on p value) below the set point

The second controller is responsible for controlling the temperature with the heater and air dampers. In this case the aggregated variable TE (Thermal Energy) is defined

$$TE^{sp} = (T1^{sp} - T_{amb})F1^{sp} + (T2^{sp} - T_{amb})F2^{sp} + (T3^{sp} - T_{amb})F3^{sp} \quad (4)$$

where T_{amb} is ambient temperature. The new TE variable is used as a master variable for the heater. The air dampers change distribution the hot and cold air between floors, and they do not change the total amount of the thermal energy in the system. Thermal energy can be changed only by the heater. The values for dampers calculated by the second controller are added as corrections to the values calculated by the first one.

Since the static characteristics are practically linear, a linear DMC algorithm without the need to compensate for non-linearity can be used.

5 Implementation Using PLC

The discussed control structure has been implemented using Mitsubish FX5 PLC controller with the GX Works as a development software environment. Due to limited resources of the PLC both DMC algorithms are realized as the analytical DMC algorithm, the control law is given by Eq. (3).

The fast flow controller works with the sampling time 1 s, tuning parameters are set as follows: the horizon of dynamics $D = 15$, the prediction horizon $N = 12$, the control horizon $N_u = 4$. The fast controller utilizes the following resources:

- memory for coefficient: $n_u \times n_u \times D = 4 \times 4 \times 15 = 240$ (float elements),
- memory for past MVs: $n_u \times D = 4 \times 15 = 60$ (float elements),
- CPU for execution: 240 ADD and 240 MUL instructions.

The temperature control controller works with the sampling time 1 min, tuning parameters are set as follows: the horizon of dynamics $D = 23$, the prediction horizon $N = 20$, the control horizon $N_u = 6$. The slow controller utilizes the following resources:

- memory for coefficient: $n_u \times n_u \times D = 4 \times 4 \times 23 = 368$ (float elements),
- memory for past MVs: $n_u \times D = 4 \times 23 = 92$ (float elements),
- CPU for execution: 368 ADD and 368 MUL instructions.

The step-response model has been obtained for the operating point using the data from the step-responses (Figs. 2, 3, 4 and 5), the values have been smoothed and normalized to unit step response. For aggregated variables, the step-response relation has been recalculated corresponding to aggregation rules.

6 Results of Experiments

The decomposed control structure has been tested in the conditions of changing the set-points for fast (flow) and slow (temperature) variables. The obtained experimental results are presented in Fig. 7, Fig. 8, Fig. 9 and Fig. 10. The structure works correctly, the setpoints are achieved quickly without overshoot. The changes of the fan power caused by the changes of the flow set-point at individual levels do not impact the flow on the other levels. Also, the tests for temperatures, as shown in Fig. 9 and Fig. 10 are very good, the interactions between levels are decoupled. Due to space limitation of this works, presentation is limited only to the first and the second levels. The results obtained for the third floor are also very good, i.e. the discussed decomposed control structure results in fast transient responses, the required set-points are reached quickly, without significant overshoot.

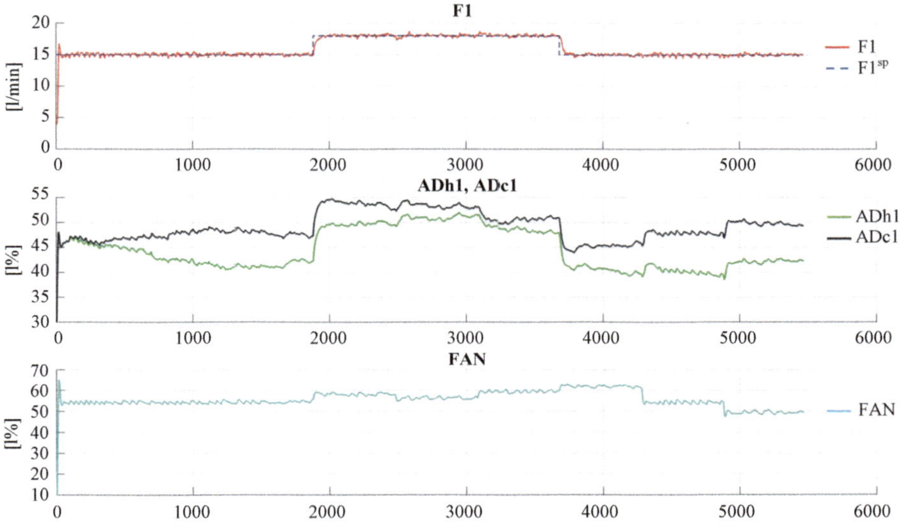

Fig. 7. Test for flow for the level 1

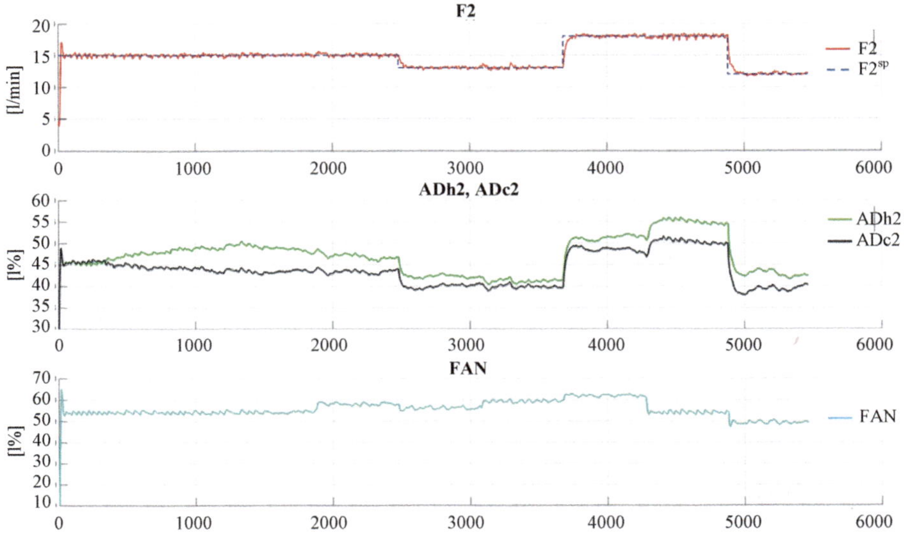

Fig. 8. Test for flow for the level 2

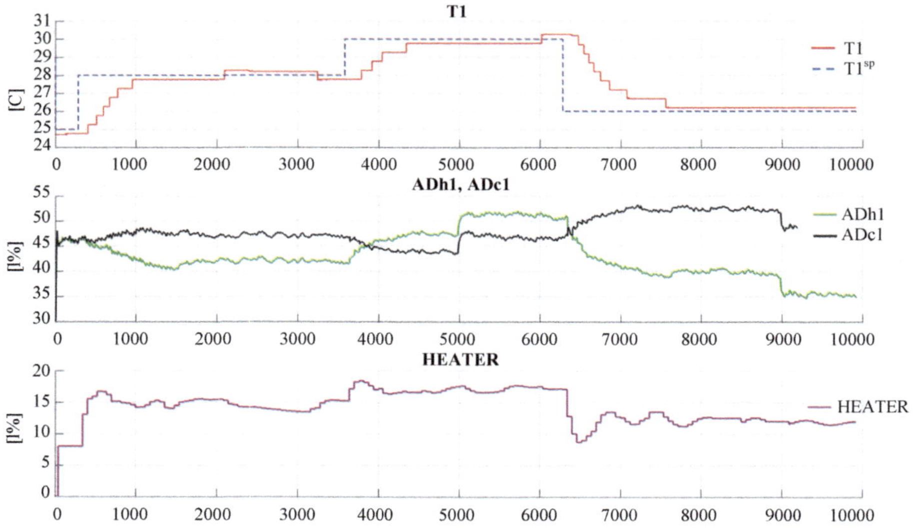

Fig. 9. Test for temperature for the level 1

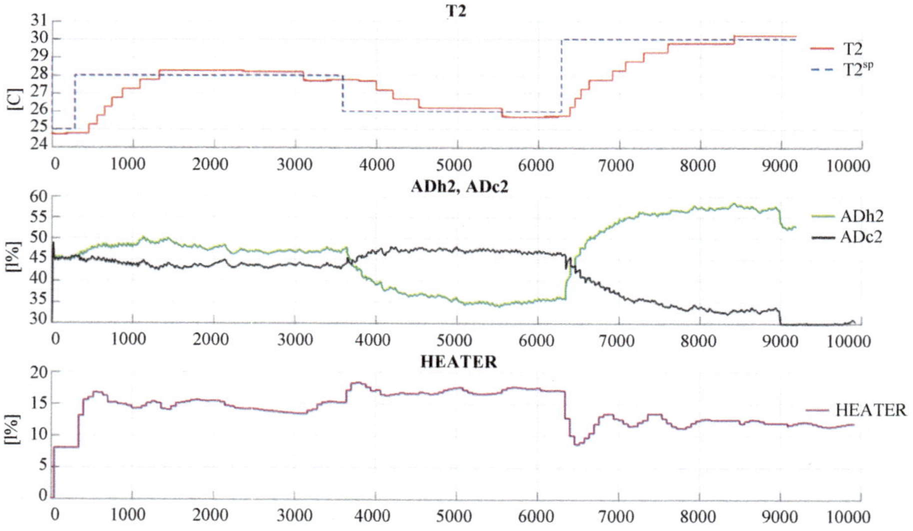

Fig. 10. Test for temperature for the level 2

7 Conclusions

The classical implementation of MIMO DMC algorithm may be in practice impossible when the considered process has fast and slow dynamics. One DMC algorithm should use the shortest sampling instant of the process to capture its fast dynamical properties.

Unfortunately, it means that for the slow dynamics a huge number of step-response coefficients must be used. Taking into account the currently available industrial hardware platforms (PLCs), implementation is not possible due to memory limitations.

The designed control structure based on the DMC algorithm solves the problem. Separate DMC sub-algorithms are used for the slow and the fast parts of the process, respectively. This work demonstrates effectiveness of the discussed control structure for the MIMO laboratory process. Additionally, some aggregated variables have been introduced for efficient implementation. The discussed decomposed structure effectively solves the problem of interactions between loops. It would be interesting to compare the author's method to known methods based on singular perturbation techniques and hierarchical control structure where controllers run in different control frequencies according to the dynamic of the variables [7, 12]. Such work will be carried out at the next stage of research.

References

1. Camacho, E.F., Bordons, C.: Model Predictive Control. Springer, London (1999). https://doi.org/10.1007/978-0-85729-398-5
2. Chaber, P., Ławryńczuk, M.: Fast analytical model predictive controllers and their implementation for STM32 ARM microcontroller. IEEE Trans. Industr. Inf. **15**, 4580–4590 (2019)
3. Forbes, M.G., Patwardhan, R., Hamadah, H., Gopaluni, R.: Model predictive control in industry: challenges and opportunities. IFAC-PapersOnLine **48**(8), 531–538 (2015)
4. Ławryńczuk, M.: Computationally Efficient Model Predictive Control Algorithms: a Neural Network Approach. Studies in Systems, Decision and Control, vol. 3. Springer, Cham (2014). https://doi.org/10.1007/978-3-319-04229-9
5. Marusak, P., Kuntanapreeda, S.: A neural network-based implementation of an MPC algorithm applied in the control systems of electromechanical plants. In: 8th TSME–International Conference on Mechanical Engineering, TSME–ICoME (2017)
6. Plamowski, S.: Implementation of DMC algorithm in embedded controller - resources, memory and numerical modifications. In: Polish Control Conference, pp. 335–343 (2017)
7. Roshany-Yamchi, S., Cychowski, M., Negenborn, R.R., De Schutter, B., Delaney, K., Connell, J.: Kalman filter-based distributed predictive control of large-scale multi-rate systems: application to power net-works. IEEE Trans. Control Syst. Technol. **21**(1), 27–39 (2013)
8. Qin, S.J., Badgwell, T.A.: A survey of industrial model predictive control technology. Control Eng. Practice **11**, 733–764 (2003)
9. Tatjewski, P.: Advanced Control of Industrial Processes. Springer, London (2007). https://doi.org/10.1007/978-1-84628-635-3
10. Wojtulewicz, A., Ławryńczuk, M.: Implementation of multiple-input multiple-output dynamic matrix control algorithm for fast processes using field programmable gate array. In: 15th IFAC Conference on Programmable Devices and Embedded Systems PDeS, Ostrava, Czech Republic, pp. 324–329 (2018)
11. Wojtulewicz, A., Ławryńczuk, M.: Computationally efficient implementation of dynamic matrix control algorithm for very fast processes using programmable logic controller. In: Proceedings of the 23th IEEE International Conference on Methods and Models in Automation and Robotics MMAR 2018, Międzyzdroje, Poland, pp. 579–584 (2018)
12. Zhou, Z., De Schutter, B., Lin, S., Xi, Y.: Two-level hierarchical model-based predictive control for large-scale urban traffic networks. IEEE Trans. Control Syst. Technol. **25**(2), 496–508 (2017)

Modeling, Identification and Control of Variable-Parameter Systems

Stabilizability of Linear Discrete Time-Varying Systems

Artur Babiarz(✉) and Adam Czornik

Department of Automatic Control and Robotics, Silesian University of Technology, Akademicka 2A, 44-100 Gliwice, Poland
{artur.babiarz,adam.czornik}@polsl.pl

Abstract. For linear discrete time-varying systems we discuss the relation between stabilizability, controllability and finiteness of quadratic cost functional. The role of the existence of global and bounded solutions of the discrete time-varying Riccati equation for stabilizability is also explained.

Keywords: Discrete time-varying system · Controllability · Stabilizability · Riccati equation

1 Introduction

One of the basic objectives of control systems is stability and the following two important questions are connected to this problem. When there is a control for a given system, preferably in the form of a feedback, that stabilizes the system and how to determine such a feedback. For linear systems with constant coefficients, these problems are well studied. The key role here is played by Kalman decomposition [15,16], which shows that the system is stabilizing if and only if its non-controllable part is stable. A number of methods of synthesis of a stabilizing feedback have also been developed for stationary systems [2,6, 9,10,20,23,24]. It should be noted that for controllable stationary systems, we can place the closed-loop system poles in any way by selecting the appropriate feedback. This is what the classic pole placement theorem says [8]. In addition, for stationary systems, relationships between stabilization and linear quadratic problem over an infinite time interval and the associated Riccati equation is well known [18].

For time-varying systems analogical questions are much less developed. At first there is now simple analogue of the Kalman decomposition theorem. An interesting discussion on this issue for discrete time systems can be found in [7] and [25]. The results about discrete time versions of the pole placement

The research presented here was done by authors as parts of the projects funded by the National Science Centre in Poland granted according to decision DEC-2017/25/B/ST7/02888 and Polish Ministry for Science and Higher Education funding for statutory activities 02/990/BK_19/0121.

A. Bartoszewicz et al. (Eds.): PCC 2020, AISC 1196, pp. 1119–1131, 2020.
https://doi.org/10.1007/978-3-030-50936-1_93

theorem for time-varying systems are discussed in the series of papers [3–5]. Some relations between discrete time-varying Riccati equation and stabilizability are presented in monographs [12,14,17].

In this paper we extend the existing results about the problems of asymptotic and uniform exponential stabilizability of linear discrete time-varying systems and its relations to finite cost conditions, algebraic time-varying discrete Riccati equation, complete and uniform complete controllability to zero, in several directions. The paper is organized as follows. In Sect. 2 we present the definitions and notations used by us, concepts of controllability and stabilizability together with they properties are presented in Sect. 3. Section 4 is devoted to the LQ problem and properties of discrete time-varying Riccati equation. Section 5 contains the main contribution of the paper and here we present theorems about asymptotic and uniform exponential stabilizability. Conclusions are formulated in Sect. 6.

2 Notation

Denote by $\mathbb{N} = \{1, 2, ...\}$ the set of natural numbers, symbol $a := b$ means that a is defined to be equal to b, $\mathbb{N}_0 := \mathbb{N} \cup \{0\}$, for $k \in \mathbb{N}$ we define $\mathbb{N}_k := \{k, k+1, k+2, ...\}$. By $\mathbb{R}$ we denote the set of real numbers, by $\mathbb{R}^n$ the n-dimensional Euclidean space with Euclidean norm $\|\cdot\|$ and by $\mathbb{R}^{n\times m}$ the set of matrices of size n by m with real entries. For a matrix $A \in \mathbb{R}^{n\times n}$ by $\|A\|$ we denote the spectral norm. $GL_n(\mathbb{R})$ is the subset of $\mathbb{R}^{n\times n}$ consisting of invertible matrices. The identity matrix of size n by n and zero matrix of size n by m are denoted by I_n and $0_{n\times m}$, respectively. A matrix $A \in \mathbb{R}^{n\times m}$ we identify with the linear operator from $\mathbb{R}^m$ to $\mathbb{R}^n$ and its kernel and image are denoted by $KerA$ and ImA, respectively. For a normed space X and $k_0, k_1 \in \mathbb{N}\cup\{\infty\}$ symbol $l^\infty([k_0, k_1), X)$ $(l^2([k_0, k_1), X))$ denote the set of all bounded (summable with squares) sequences from $[k_0, k_1)$ to X. Finally for two symmetric matrices Q, $R \in \mathbb{R}^{n\times n}$ notation $Q \prec R$ $(Q \preceq R)$ denotes that $R - Q$ is positive (non-negative) definite.

Consider a linear time-varying system described by the following equation

$$x(k+1) = A(k)x(k) + B(k)u(k), \tag{1}$$

where $A : \mathbb{N}_0 \to \mathbb{R}^{n\times n}$, $B : \mathbb{N}_0 \to \mathbb{R}^{n\times m}$ and $u : \mathbb{N}_0 \to \mathbb{R}^n$ is a control signal and $(k_0, x_0) \in \mathbb{N}_0 \times \mathbb{R}^n$ is an initial condition, $x(k_0) = x_0$. The solution of system (1) with the initial condition $x(k_0) = x_0$, $(k_0, x_0) \in \mathbb{N}_0 \times \mathbb{R}^n$ and a control $u : \mathbb{N}_0 \to \mathbb{R}^n$ will be denoted by $x(\cdot, k_0, x_0, u)$. We will also consider the homogeneous system

$$x(k+1) = A(k)x(k). \tag{2}$$

The solution of this system with the initial condition $x(k_0) = x_0$, $(k_0, x_0) \in \mathbb{N}_0 \times \mathbb{R}^n$ will be denoted by $x(\cdot, k_0, x_0)$. Sometimes we will assume that A is a Lyapunov sequence i.e. $A : \mathbb{N}_0 \to GL_n(\mathbb{R})$ and A, A^{-1} are bounded.

The transition matrix $\Phi_A(k, l)$, $k \geq l$, $k, l \in \mathbb{N}_0$ of system (2) is defined as follows

$$\Phi_A(k, l) = A(k-1)\cdots A(l),\ k > l \text{ and } \Phi_A(k, k) = I_n.$$

If all the matrices $A(k)$, $k \in \mathbb{N}_0$ are invertible then we define

$$\Phi_A(l,k) = \Phi_A^{-1}(k,l)$$

for $k,\, l \in \mathbb{N}_0$ and $k \geq l$.

Using the variation of constant formula (see [1]) we obtain

$$x(k;k_0,x_0,u) = \Phi_A(k,k_0)x_0 + \sum_{j=k_0}^{k-1} \Phi_A(k,j+1)B(j)u(j), \quad k > k_0. \tag{3}$$

In our further consideration we will use Cauchy-Schwarz inequality (see [22]).

Lemma 1. *For any two sequences a_i, b_i, $i = 1, ..., n$ of real numbers we have*

$$\left| \sum_{i=1}^{n} a_i b_i \right| \leq \sqrt{\sum_{i=1}^{n} a_i^2 \sum_{i=1}^{n} b_i^2}. \tag{4}$$

3 Controllability and Stabilizability

In this paper we will use the following concepts of stability and stabilizability (see [19]).

Definition 1. *System* (2) *is called:*

1. *asymptotically stable*

$$\Leftrightarrow \forall k_0 \in \mathbb{N}_0 \;:\; \lim_{k\to\infty} \Phi_A\left(k,k_0\right) = 0;$$

2. *exponentially stable*

$$\begin{aligned}\Leftrightarrow \forall k_0 \in \mathbb{N}_0\, \exists\, M \geq 1 \exists\, \beta > 0\, \forall\, k \in \mathbb{N}_{k_0} \;:\; &\left\|\Phi_A\left(k,k_0\right)\right\| \\ &\leq M \exp\left(-\beta\left(k-k_0\right)\right);\end{aligned}$$

3. *uniformly exponentially stable*

$$\begin{aligned}\Leftrightarrow \exists\, M \geq 1 \exists\, \beta > 0\, \forall k_0 \in \mathbb{N}_0\, \forall k \in \mathbb{N}_{k_0} \;:\; &\left\|\Phi_A\left(k,k_0\right)\right\| \\ &\leq M \exp\left(-\beta\left(k-k_0\right)\right).\end{aligned}$$

We now introduce the main concepts of stabilizability under investigation. They all correspond of the notion of stability.

Definition 2. *System* (1) *is called: asymptotically (exponentially, uniformly exponentially) stabilizable if there exists $F(\cdot) : \mathbb{N}_0 \to \mathbb{R}^{m\times n}$ such that the closed loop system*

$$x(k+1) = \left(A(k) + B(k)F\left(k\right)\right)x(k) \tag{5}$$

is asymptotically (exponentially, uniformly exponentially) stable. The appropriate feedback will be called asymptotically (exponentially, uniformly exponentially) stabilizing feedback. In case that in any of the above we may choose F as a bounded sequence, we say that system (1) *is (completely exponentially etc.) stabilizable by bounded feedback.*

In the literature there are many different concepts of controllability (see e.g. [11,17]) for time-varying systems. We will use the following one.

Definition 3. *System* (1) *is called* T*-uniformly completely controllable if*

$$\exists\, \eta > 0\, \forall\, k_0 \in \mathbb{N}_0$$

$$\forall\, x_0,\, x_1 \in \mathbb{R}^n\, \exists\, u \in l^2\left([k_0, k_0 + T), \mathbb{R}^m\right) :\, x(k_0 + T, k_0, x_0, u) = x_1$$

and

$$\|u(k)\| \leq \eta\left(\|x_1\| + \|x_0\|\right)$$

for all $k \in [k_0, k_0 + T)$*. If system* (1) *is* T*-uniformly completely controllable for certain* $T \in \mathbb{N}$ *then we will say that it is uniformly completely controllable. If we take in the above definition* $x_1 = 0$ *then we obtain definition of uniform complete controllability to zero.*

4 LQ Problem

Let us consider together with system (1) with initial condition $(k_0, x_0) \in \mathbb{N}_0 \times \mathbb{R}^n$ and control $u \in l^2([k_0, k_1); \mathbb{R}^m)$ the following cost functional on the interval $[k_0, k_1)$

$$J(k_1, k_0, x_0, u) := \sum_{j=k_0}^{k_1-1} \left(\|x(j, k_0, x_0, u)\|^2 + \|u(j)\|^2\right) \tag{6}$$

and on the infinite interval $[k_0, \infty)$

$$J(\infty, k_0, x_0, u) := \sum_{j=k_0}^{\infty} \left(\|x(j, k_0, x_0, u)\|^2 + \|u(j)\|^2\right). \tag{7}$$

Let us also introduce the following notation

$$V(k_1, k_0, x_0) := \inf_{u \in l^2([k_0,k_1);\mathbb{R}^m)} J(k_1, k_0, x_0, u), \tag{8}$$
$$V(\infty, k_0, x_0) := \inf_{u \in l^2([k_0,\infty);\mathbb{R}^m)} J(\infty, k_0, x_0, u).$$

The problems of finding controls which realize the above infima are known as linear quadratic (LQ) problems and their solutions are closely related to the following Riccati equation

$$\begin{gathered} P(k) = I_n + G(k)^T G(k) + \\ \left(A(k) - B(k)G(k)\right)^T P(k+1)\left(A(k) - B(k)G(k)\right) \end{gathered} \tag{9}$$

where

$$G(k) = \left(I_m + B(k)^T P(k+1) B(k)\right)^{-1} B(k)^T P(k+1) A(k). \tag{10}$$

The next theorem presents the solution of the LQ problem on finite interval [17, Theorem 6.28].

Theorem 1. *Consider the optimization problem* (1) *and* (6) *on the interval* $[k_0, k_1)$ *with* $0 \leq k_0 < k_1 < \infty$. *Then the Riccati Eq. (9) with terminal condition* $P(k_1) = 0$ *has a unique symmetric solution* $P(\cdot, k_1, k_0) : [k_0, k_1] \to \mathbb{R}^{n\times n}$. *This solution has the following properties:*

(i) the function

$$u : [k_0, k_1) \to \mathbb{R}^m,\ k \mapsto u(k) = -G(k)x(k)$$

is for all $x_0 \in \mathbb{R}^n$ *the unique solution of the optimization problem* (1) *and* (6). *Moreover,*

$$\forall k \in [k_0, k_1] \forall x_0 \in \mathbb{R}^n : V(k_1, k, x_0) = x_0^T P(k, k_1, k_0) x_0.$$

(ii)

$$\forall\, k \in [k_0, k_1) : 0_{n\times n} \prec P(k, k_1, k_0)$$

(iii)

$$\forall\, k_0 \leq k \leq k_1 < k_2 : P(k, k_1, k_0) \preceq P(k, k_2, k_0)$$

Let us consider the following condition for system (1):

$$(A1) \qquad \begin{array}{c} \forall k_0 \in \mathbb{N}_0 \, \exists C(k_0) \geq 0 \\ \forall x_0 \in \mathbb{R}^n \exists u \in l^2([k_0, \infty); \mathbb{R}^m) : J(\infty, k_0, x_0, u) \leq C(k_0)\|x_0\|^2. \end{array} \qquad (11)$$

The following theorem contains conditions equivalent to $(A1)$.

Theorem 2. *For system* (1) *the following conditions are equivalent to* $(A1)$*:*

(A1′)

$$\forall\, (k_0, x_0) \in \mathbb{N}_0 \times \mathbb{R}^n \exists u \in l^2\left([k_0, \infty); \mathbb{R}^m\right) : J(\infty, k_0, x_0, u) < \infty$$

(A1″)

$$\forall\, (k_0, x_0) \in \mathbb{N}_0 \times \mathbb{R}^n \exists u \in l^2\left([k_0, \infty); \mathbb{R}^m\right) : x(\cdot, k_0, x_0, u) \in l^2\left([k_0, \infty); \mathbb{R}^n\right).$$

Proof. Implications $(A1) \Rightarrow (A1')$ and $(A1'') \Rightarrow (A1')$ are trivial. We will show now the implication $(A1') \Rightarrow (A1)$. Let us fix $(k_0, x_0) \in \mathbb{N}_0 \times \mathbb{R}^n$ and denote $x_0 = \sum_{i=1}^n x_i^0 e_i$, where $e_1, \ldots, e_n \in \mathbb{R}^n$ is the standard Euclidean basis of $\mathbb{R}^n$. From the condition $(A1')$ we have

$$\forall i \in \{1, \ldots, n\} \, \exists u_i \in l^2\left([k_0, \infty); \mathbb{R}^m\right) : J\left(\infty, k_0, e_i, u_i\right) < \infty.$$

Define the control as follows $u(\cdot) := \sum_{i=1}^n x_i^0 u_i(\cdot)$. We have

$$\begin{aligned} x(k, k_0, x_0, u) &= \sum_{i=1}^n x_i^0 \left[\Phi_A(k, k_0)e_i + \sum_{j=k_0}^{k-1} \Phi_A(k, j+1)B(j)u_i(j) \right] \\ &= \sum_{i=1}^n x_i^0 x(k, k_0, e_i, u_i). \end{aligned}$$

From the Cauchy-Schwarz inequality (4) with $b_i = 1$, $i = 1, \ldots, n$ we obtain $\left(\sum_{i=1}^{n} a_i\right)^2 \leq n\left(\sum_{i=1}^{n} a_i^2\right)$ for any real numbers a_i, $i = 1, \ldots, n$. Using this inequality we get

$$J(\infty, k_0, x_0, u) = \sum_{j=k_0}^{\infty} \left(\|x(j, k_0, x_0, u)\|^2 + \|u(j)\|^2\right) =$$

$$\sum_{j=k_0}^{\infty} \left(\| \sum_{i=1}^{n} x_i^0 x(j, k_0, e_i, u_i)\|^2 + \| \sum_{i=1}^{n} x_i^0 u_i(j)\|^2 \right) \leq$$

$$n \max_{i \in \{1,\ldots,n\}} |x_i^0|^2 \sum_{i=1}^{n} \sum_{j=k_0}^{\infty} \left(\|x(j, k_0, e_i, u_i)\|^2 + \| \sum_{i=1}^{n} u_i(j)\|^2 \right) \leq$$

$$n \sum_{i=1}^{n} J(\infty, k_0, e_i, u_i) \|x_0\|^2$$

Defining $C(k_0) := n \sum_{i=1}^{n} J(\infty, k_0, e_i, u_i)$ we have

$$J(\infty, k_0, x_0, u) \leq C(k_0)\|x_0\|^2.$$

Now we will show that $(A1') \Rightarrow (A1'')$. For each $(k_0, x_0) \in \mathbb{N}_0 \times \mathbb{R}^n$ there exists $u \in l^2([k_0, \infty); \mathbb{R}^m)$ such that, $J(\infty, k_0, x_0, u) < \infty$. Therefore

$$\sum_{k=k_0}^{\infty} \|x(k, k_0, x_0, u)\|^2 \leq J(\infty, k_0, x_0, u) < \infty,$$

and $x(\cdot, k_0, x_0, u) \in l^2([k_0, \infty); \mathbb{R}^n)$. The proof is completed.

Now we will show a sufficient condition for $(A1)$ to hold.

Lemma 2. *If system* (1) *is completely controllable to zero then the system satisfies condition* $(A1)$.

Proof. Let us fix arbitrary $(k_0, x_0) \in \mathbb{N}_0 \times \mathbb{R}^n$. By the assumption about complete controllability to zero, there exists $k_1 \geq k_0$ and $v \in l^2([k_0, k_1); \mathbb{R}^m)$ such that $x(k_1, k_0, x_0, v) = 0$. If we define the control $u : [k_0, \infty) \to \mathbb{R}^m$ in the following way

$$u(k) := \begin{cases} v(k), & k \in [k_0, k_1) \\ 0, & k \in [k_1, \infty) \end{cases},$$

then $u \in l^2([k_0, \infty); \mathbb{R}^m)$ and

$$J(\infty, k_0, x_0, u) = \sum_{j=k_0}^{\infty} \left(\|x(j, k_0, x_0, u)\|^2 + \|u(j)\|^2\right)$$

$$= \sum_{j=k_0}^{k_1-1} \left(\|x(j, k_0, x_0, v)\|^2 + \|v(j)\|^2 \right) < \infty,$$

The last inequality means that condition $(A1')$ holds and therefore by Theorem 2 condition $(A1)$ is also satisfied. The proof is completed.

Next theorem presents a sufficient condition for existence of a global solution of the Riccati Eq. (9).

Theorem 3. *For system* (1) *we have the following chain of implications:*

$$(A1) \Rightarrow (12) \Rightarrow (13),$$

where for each $k_0 \in \mathbb{N}_0$ *the constant* $C(k_0)$ *is given by* $(A1)$ *and*

$$\forall\, 0 \le k_0 < k_1 < \infty\, \forall\, x_0 \in \mathbb{R}^n : \|x_0\|^2 \le V(k_1, k_0, x_0) \le C(k_0)\|x_0\|^2 \tag{12}$$

$$V(k_1, k_0, x_0) := \inf_{u \in l^2([k_0, k_1); \mathbb{R}^m)} J(k_1, k_0, x_0, u);$$

$$\left. \begin{array}{c} \textit{For each } k_0 \in \mathbb{N}_0 \textit{ there exists exactly one global solution} \\ \Pi(\cdot, k_0) : \mathbb{N}_{k_0} \to \mathbb{R}^{n \times n} \\ \textit{of equation (9) with the initial condition} \\ \Pi(k_0, k_0) := \lim_{k_1 \to \infty} P(k_0, k_1, k_0). \textit{ For each } k \ge k_0 \\ \textit{the solution is symmetric, positive definite and satisfies} \\ \Pi(k, k_0) = \lim_{k_1 \to \infty} P(k, k_1, k_0). \end{array} \right\} \tag{13}$$

Proof. $(A1) \Rightarrow (12)$ For any k_0, $k_1 \in \mathbb{N}$, $k_0 < k_1$ and $x_0 \in \mathbb{R}^n$ we have

$$\forall\, u \in l^2([k_0, k_1); \mathbb{R}^m) : J(k_1, k_0, x_0, u) =$$
$$\sum_{k=k_0}^{k_1-1} \left(\|x(k, k_0, x_0, u)\|^2 + \|u(k)\|^2 \right) \ge \|x_0\|^2$$

and in particular

$$V(k_1, k_0, x_0) = \inf_{u \in l^2([k_0, k_1); \mathbb{R}^m)} J(k_1, k_0, x_0, u) \ge \|x_0\|^2.$$

By the condition $(A1)$ there exists a control $u \in l^2([k_0, k_1); \mathbb{R}^m)$ such that,

$$V(k_1, k_0, x_0) \le J(k_1, k_0, x_0, u) \le J(\infty, k_0, x_0, u) \le C(k_0)\|x_0\|^2.$$

(12)$\Rightarrow$(13) For each $x_0 \in \mathbb{R}^n$ and k, $k_0 \in \mathbb{N}_0$, $k_0 \le k$ let us define sequences $a^k_\cdot : [k, \infty) \to \mathbb{R}$ by $a^k_j := (x_0)^T P(k, j, k_0) x_0$. By Theorem 1 these sequences are monotonically nondecreasing and by (12) bounded. Therefore there exists a limit

$$\Pi(k, k_0) := \lim_{j \to \infty} P(k, j, k_0).$$

and matrices $\Pi(k, k_0)$, $k_0 \le k$ are symmetric and positive definite solutions of (10). The proof is completed.

5 Stabilizability

In this paragraph we present the main results of this paper that describe relations between conditions $(A1)$, $(A2)$, existence of a global positive definite solution of the Riccati equation, uniform complete controllability to zero and exponential (uniform exponential stability). We will start with discussion of problems of asymptotic, and uniform exponential stabilizability of system (1).

Theorem 4. *If the Riccati Eq. (9) has a global positive definite solution, then system* (1) *is asymptotically stabilizable.*

Proof. Let us consider the feedback $F(\cdot) : \mathbb{N}_0 \to \mathbb{R}^{m\times n}$ defined by

$$F(k) = (I_m + B(k)^T \Pi(k+1,0)B(k))^{-1}B(k)^T \Pi(k+1,0)A(k), \tag{14}$$

where $\Pi(\cdot, 0)$ is the global solution of the Riccati Eq. (9). We will show that the closed loop system

$$x(k+1) = [A(k) - B(k)F(k)]\, x(k), \quad x(k_0) = x_0 \tag{15}$$

is asymptotically stable. By (10) and (13) we have

$$\begin{gathered} \Pi(k,0) = I_n + F(k)^T F(k) + \\ (A(k) - B(k)F(k))^T\, \Pi(k+1,0)\,(A(k) - B(k)F(k))\,. \end{gathered} \tag{16}$$

Let $x_F(\cdot, k_0, x_0)$ be the solution of (15) with initial condition $(k_0, x_0) \in \mathbb{N}_0 \times \mathbb{R}^n$. Let us write shortly $\xi(\cdot) := x_F(\cdot, k_0, x_0)$. By (16), with this notation, we get

$$\begin{gathered} \forall\, k \geq k_0 : \xi(k+1)^T \Pi(k+1,0)\xi(k+1) - \xi(k)^T \Pi(k,0)\xi(k) = \\ -\|\xi(k)\|^2 - \|F(k)\xi(k)\|^2 \leq -\|\xi(k)\|^2. \end{gathered}$$

Moreover, for $k_1 > k_0$ we have

$$\xi(k_1+1)^T \Pi(k_1+1,0)\xi(k_1+1) - x_0^T \Pi(k_0,0)x_0$$

$$= \sum_{k=k_0}^{k_1} \xi(k+1)^T \Pi(k+1,0)\xi(k+1) - \xi(k)^T \Pi(k,0)\xi(k)$$

$$\leq - \sum_{k=k_0}^{k_1} \|\xi(k)\|^2$$

and therefore

$$\sum_{k=k_0}^{k_1} \|\xi(k)\|^2 \leq x_0^T \Pi(k_0,0)x_0.$$

Since the right hand side does not depend on k_1, then $\xi(\cdot) \in l^2([k_0,\infty); \mathbb{R}^n)$ and therefore $\lim_{k\to\infty} \|\xi(k)\| = 0$. For the linear systems it means that it is asymptotically stable [5, p.120]. The proof is completed.

From the last theorem and Theorem 3 we get the following result.

Corollary 1. *If the condition* $(A1)$ *is satisfied, then system* (1) *is asymptotically stabilizable.*

It may happens that the constant $C(k_0)$ from condition $(A1)$ depends on k_0 and tends to infinity when $k_0 \to \infty$. Therefore looking for conditions for stabilizability by bounded feedback one may consider the following straightening of condition $(A1)$:

$$(A2) \qquad \begin{array}{c} \exists C \geq 0 \\ \forall (k_0, x_0) \in \mathbb{N}_0 \times \mathbb{R}^n \exists u \in l^2([k_0, \infty); \mathbb{R}^m) : J(\infty, k_0, x_0, u) \leq C\|x_0\|^2. \end{array}$$

Below we show an analogue of Lemma 2 for condition $(A2)$ and uniform complete controllability.

Lemma 3. *If system* (1) *is uniformly completely controllable to zero and has bounded coefficients, then it satisfies condition* $(A2)$.

Proof. By the assumption about uniform complete controllability to zero, there exist $T \in \mathbb{N}$, $\eta > 0$ such that for all $(k_0, x_0) \in \mathbb{N}_0 \times \mathbb{R}^n$ there exists $v \in l^2([k_0, k_0 + T); \mathbb{R}^m)$ such that $x(k_0 + T, k_0, x_0, v) = 0$ and $\|v(k)\| < \eta\|x_0\|$ for all $k \in [k_0, k_0 + T]$. Let us select such a control v for an arbitrary fixed $(k_0, x_0) \in \mathbb{N}_0 \times \mathbb{R}^n$. If we define the control $u : [k_0, \infty) \to \mathbb{R}^m$ in the following way

$$u(k) := \begin{cases} v(k), & k \in [k_0, k_0 + T) \\ 0, & k \in [k_0 + T, \infty) \end{cases},$$

then $u \in l^2([k_0, \infty); \mathbb{R}^m)$ and

$$J(\infty, k_0, x_0, u) = \sum_{j=k_0}^{\infty} \left(\|x(j, k_0, x_0, u)\|^2 + \|u(j)\|^2\right)$$

$$= \sum_{j=k_0}^{k_0+T-1} \left(\|x(j, k_0, x_0, v)\|^2 + \|v(j)\|^2\right) \leq$$

$$\sum_{j=k_0}^{k_0+T-1} \|x(j, k_0, x_0, v)\|^2 + T\eta^2\|x_0\|^2. \tag{17}$$

Denote

$$C = \max\left\{1, \sup_{k \in \mathbb{N}} \|A(k)\|, \sup_{k \in \mathbb{N}} \|B(k)\|\right\}.$$

By the variation of constant formulae (3) we have

$$\|x(j, k_0, x_0, u)\| = \left\|\Phi_A(j, k_0)x_0 + \sum_{i=k_0}^{j-1} \Phi_A(j, i+1)B(i)u(i)\right\| \leq$$

$$\left(C^{j-k_0}+\eta\sum_{i=k_0}^{j-1}C^{j-i}\right)\|x_0\|\leq\left(C^{T-1}+\eta\sum_{i=k_0}^{k_0+T-2}C^{k_0+T-1-i}\right)\|x_0\|\leq$$

$$\left(C^{T-1}+\eta\sum_{i=k_0}^{k_0+T-2}C^{T-1}\right)\|x_0\|=\left(C^{T-1}+\eta\left(T-1\right)C^{T-1}\right)\|x_0\| \tag{18}$$

From (17) and (18) we get the conclusion of the lemma.

The next result describes relations between condition $(A2)$ and exponential stabilizability.

Theorem 5. *System (1) satisfies condition $(A2)$ if and only if it is uniformly exponentially stabilizable by bounded feedback.*

Proof. Suppose that $F(\cdot)\in l^{\infty}\left([0,\infty);\mathbb{R}^{m\times n}\right)$ is a feedback such that the closed loop system (5) is uniformly exponentially stable. For any $(k_0,x_0)\in\mathbb{N}_0\times\mathbb{R}^n$ denote the solution of the closed loop system with initial condition $x(k_0)=x_0$ by $x_F(\cdot,k_0,x_0):\mathbb{N}_{k_0}\to\mathbb{R}^n$ and define a control $u(\cdot)=F(\cdot)x_F(\cdot,k_0,x_0)$. By the definition of uniform exponential stability we know that there exist $\gamma>0$ and $\lambda\in[0,1)$ such that

$$\forall\, k\geq k_0:\|x_F(k,k_0,x_0)\|\leq\gamma\lambda^{k-k_0}\|x_0\|,$$

and therefore

$$\|u(k)\|\leq\alpha\gamma\lambda^{k-k_0}\|x_0\|,$$

where $\alpha=\sup_{k\in\mathbb{N}_0}\|F\|$. Now the cost functional may be estimated as follows

$$\begin{aligned}J(\infty,k_0,x_0,u)&=\sum_{j=k_0}^{\infty}\left(\|x_F(j,k_0,x_0,u)\|^2+\|u(j)\|^2\right)\\&\leq\sum_{j=k_0}^{\infty}\left(\gamma^2\lambda^{2(k-k_0)}\|x_0\|^2+\alpha^2\gamma^2\lambda^{2(k-k_0)}\|x_0\|^2\right)\\&=\frac{\gamma^2\left(1+\alpha^2\right)}{1-\lambda^2}\|x_0\|^2.\end{aligned}$$

We see that condition $(A2)$ is satisfied with $C=\frac{\gamma^2\left(1+\alpha^2\right)}{1-\lambda^2}$.

Suppose now, that the condition $(A2)$ is satisfied. Let $\Pi(\cdot,0)$ be the global solution of Riccati Eq. (9) defined in Theorem 3 and define $F(\cdot)$ by (14). Notice that by the proof of Theorem 4 the feedback $-F(\cdot)$ is a asymptotically stabilizable feedback. Notice also that by point (12) of Theorem 3 with $C(k_0)=C$, we have

$$\forall\, 0\leq k_0<k_1\leq\infty\,\forall\, x_0\in\mathbb{R}^n:\|x_0\|^0\leq V(k_1,k_0,x_0)\leq C\|x_0\|^2 \tag{19}$$

and

$$\forall k \in \mathbb{N}_0 \forall x_0 \in \mathbb{R}^n : \|x_0\|^2 \leq (x_0)^T \Pi(k,0) x_0 \leq C\|x_0\|^2,$$

therefore

$$\forall k \in \mathbb{N}_0 : I_n \preceq \Pi(k,0) \preceq C I_n. \tag{20}$$

By (9) we get

$$\forall k \in \mathbb{N}_0 : (A(k) - B(k)F(k))^T \Pi(k+1,0)\,(A(k) - B(k)F(k)) - \Pi(k,0)$$

$$= -I_n - F(k)^T F(k) \preceq -I_n$$

and therefore by Theorem 23.3 from [21], closed loop system (5) is uniformly exponentially stable. It remains to show that the feedback $-F(\cdot)$ is bounded. We have

$$\forall k \in \mathbb{N}_0 : I_m \preceq I_m + B(k)^T \Pi(k+1,0) B(k).$$

Since for positive definite matrices $P, Q \in \mathbb{R}^{n\times n}$ the relation $P \preceq Q$ implies inequalities $\|P\| \leq \|Q\|$ and $\left\|Q^{-1}\right\| \leq \left\|P^{-1}\right\|$ (see [13, Corollary 7.7.4]), then by (20) we obtain

$$\forall k \in \mathbb{N}_0 : \left(\| \left(I_m + B(k)^T \Pi(k+1,0) B(k)\right)^{-1} \| \leq 1 \wedge \|\Pi(k,0)\| \leq C\right)$$

and therefore

$$\forall k \in \mathbb{N}_0 : \|F(k)\| = \left\|\left(I_m + B(k)^T \Pi(k+1,0) B(k)\right)^{-1} B(k)^T \Pi(k+1,0) A(k)\right\|$$

$$\leq \|B(k)\| \|\Pi(k+1,0)\| \|A(k)\| \leq \left(\sup_{k\in\mathbb{N}_0} \|B(k)\|\right) C \left(\sup_{k\in\mathbb{N}_0} \|A(k)\|\right).$$

The proof is completed.

6 Conclusions

In this paper we have investigated the problem of stabilizability of discrete time-varying linear systems and its relations to controllability, finiteness cost condition and discrete time-varying Riccati Equation. The main results say: if the Riccati equation has a global positive definite solution, then the system is stabilizable; the nonuniform finite cost condition $(A1)$ implies asymptotic stabilizability; uniform complete controllability to zero together with boundness of coefficients imply uniform finite cost condition $(A2)$; $(A2)$ is equivalent to uniform exponential stabilizability by bounded feedback.

References

1. Agarwal, R.P.: Difference Equations and Inequalities: Theory, Methods, and Applications. CRC Press, Boca Roton (2000)
2. Alexandridis, A., Galanos, G.: Optimal pole-placement for linear multi-input controllable systems. IEEE Trans. Circ. Syst. **34**(12), 1602–1604 (1987)
3. Babiarz, A., Banshchikova, I., Czornik, A., Makarov, E., Niezabitowski, M., Popova, S.: Proportional local assignability of Lyapunov spectrum of linear discrete time-varying systems. SIAM J. Control Optim. **57**(2), 1355–1377 (2019)
4. Babiarz, A., Banshchikova, I., Czornik, A., Makarov, E.K., Niezabitowski, M., Popova, S.: Necessary and sufficient conditions for assignability of the Lyapunov spectrum of discrete linear time-varying systems. IEEE Trans. Autom. Control **63**(11), 3825–3837 (2018)
5. Babiarz, A., Czornik, A., Makarov, E., Niezabitowski, M., Popova, S.: Pole placement theorem for discrete time-varying linear systems. SIAM J. Control Optim. **55**(2), 671–692 (2017)
6. Bhattacharyya, S., de Souza, E.: Pole assignment via Sylvester's equation. Syst. Control Lett. **1**(4), 261–263 (1982)
7. Bittanti, S., Bolzern, P.: On the structure theory of discrete-time linear systems. Int. J. Syst. Sci. **17**(1), 33–47 (1986)
8. Dickinson, B.: On the fundamental theorem of linear state variable feedback. IEEE Trans. Autom. Control **19**(5), 577–579 (1974)
9. Franklin, G.F., Powell, J.D., Emami-Naeini, A.: Feedback Control of Dynamic Systems. Prentice Hall Press, Upper Saddle River (2014)
10. Furuta, K., Kim, S.: Pole assignment in a specified disk. IEEE Trans. Autom. Control **32**(5), 423–427 (1987)
11. Gaishun, I.: Discrete-time systems. In: Natsionalnaya Akademiya Nauk Belarusi. Institut Matematiki Minsk (2001)
12. Halanay, A., Ionescu, V.: Time-Varying Discrete Linear Systems: Input-Output Operators. Riccati Equations. Disturbance Attenuation, vol. 68. Birkhäuser, Basel (2012)
13. Horn, R.A., Johnson, C.R.: Matrix Analysis. Cambridge University Press, Cambridge (2012)
14. Ichikawa, A., Katayama, H., et al.: Linear Time Varying Systems and Sampled-Data Systems, vol. 265. Springer, Heidelberg (2001)
15. Kalman, R.E.: On the general theory of control systems. In: Proceedings of the First International Congress on Automatic Control (1960)
16. Kalman, R.E.: Mathematical description of linear dynamical systems. J. Soc. Ind. Appl. Math. Ser. A Control **1**(2), 152–192 (1963)
17. Kwakernaak, H., Sivan, R.: Linear Optimal Control Systems, vol. 1. Wiley, Hoboken (1972)
18. Lancaster, P., Rodman, L.: Algebraic Riccati Equations. Oxford Science (1995)
19. Ludyk, G.: Stability of Time-Variant Discrete-Time Systems, vol. 5. Springer, Heidelberg (2013)
20. Reza Moheimani, S.O., Petersen, I.R.: Quadratic guaranteed cost control with robust pole placement in a disk. IEE Proc. Control Theory Appl. **143**(1), 37–43 (1996)
21. Rugh, W.J.: Linear System Theory, vol. 2. Prentice Hall, Upper Saddle River (1996)

22. Steele, J.: The Cauchy-Schwarz Master Class: An Introduction to the Art of Mathematical Inequalities. MAA problem books series. Cambridge University Press, Cambridge (2004)
23. Sugimoto, K.: Partial pole placement by LQ regulators: an inverse problem approach. IEEE Trans. Autom. Control **43**(5), 706–708 (1998)
24. Sugimoto, K., Yamamoto, Y.: On successive pole assignment by linear-quadratic optimal feedbacks. Linear Algebra Appl. **122–124**, 697–723 (1989)
25. Weiss, L.: Controllability, realization and stability of discrete-time systems. SIAM J. Control **10**(2), 230–251 (1972)

Controllability of Higher Order Linear Systems with Multiple Delays in Control

Jerzy Klamka(✉)

Silesian University of Technology, 44-100 Gliwice, Poland
Jerzy.klamka@polsl.pl

Abstract. In the present chapter finite-dimensional dynamical control systems described by linear higher-order ordinary differential state equations with multiple point delays in control are considered. Using algebraic methods, necessary and sufficient conditions for relative controllability in a given time interval for linear dynamical system with multiple point delays in control are formulated and proved. This condition is generalization to relative controllability case some previous results concerning controllability of linear dynamical systems without multiple point delays in the control. Proof of the main result is based on necessary and sufficient controllability condition for linear systems without delays in control. Simple numerical example, which illustrates theoretical result is also given. Finally, some remarks and comments on the existing results for controllability of dynamical systems with delays in control are also presented.

Keywords: Controllability · Linear systems · Higher order systems

1 Introduction

Controllability similarly as observability and stability is one of the fundamental concepts in modern mathematical control theory. This is a qualitative property of dynamical control systems and is of particular importance in control theory. Systematic study of controllability was started at the beginning of sixties in the last century, during IFAC Congress in 1960 in Moscow, when the general concept of controllability was proposed and discussed. This concept was based on the description in the form of state space equations for both time-invariant and time-varying linear continuous-time and discrete-time control systems. Without going into details roughly speaking, controllability generally means, that it is possible to steer dynamical control system from an arbitrary initial state to an arbitrary final state using the set of admissible controls.

It should be mentioned, that in the literature there are many different definitions and concepts of controllability, which strongly depend on the one hand on a class of dynamical control systems and on the other hand on the form of admissible controls. For example for delayed control systems it is generally necessary to distinguish between relative controllability and absolute controllability. Moreover, for linear control systems controllability is strongly connected with minimum energy control.

A. Bartoszewicz et al. (Eds.): PCC 2020, AISC 1196, pp. 1132–1140, 2020.
https://doi.org/10.1007/978-3-030-50936-1_94

Controllability problems for different types of dynamical systems require the application of numerous mathematical concepts and methods taken directly from differential geometry, functional analysis, topology, matrix analysis and theory of ordinary and partial differential equations and theory of difference equations. In the paper we will use mainly simple differential state-space models of dynamical systems, which provide a robust and universal method for studying controllability of various classes of systems.

In the present paper finite-dimensional dynamical control systems described by linear higher-order ordinary differential state equations with multiple point delays in control are considered. Using algebraic methods, necessary and sufficient condition for relative controllability in a given time interval for linear dynamical system with multiple point delays in control is formulated and proved. This condition is generalization to relative controllability case some previous results concerning controllability of linear dynamical systems without multiple point delays in the control (see for example [1–5] for more details). Proof of the main result is pure algebraic and based on the necessary and sufficient controllability condition presented in the paper [1] for linear systems without delays in control. Simple numerical example, which illustrates theoretical result is also given. Finally, some remarks and comments on the existing results for controllability of dynamical systems with delays in control are also presented.

2 System Description

In this chapter we study the mathematical model of the linear higher-order control system with multiple constant point delays in the control described by the following ordinary differential state equation

$$x^{(N)}(t) = \sum_{i=0}^{i=N-1} A_i x^{(i)}(t) + \sum_{j=0}^{j=M} B_j u(t-h_j) \quad \text{for} \quad t \in [0, T], \tag{1}$$

with zero initial conditions:

$$x(0) = 0 \quad u(t) = 0 \text{ for } \quad t \in [-h, 0) \tag{2}$$

where
the state $x(t) \in R^n = X$,
the control $u(t) \in R^m = U$,
A is $n \times n$ dimensional constant matrix,
$B_j, j = 0,1,2,\ldots,M$. are $n \times m$. dimensional constant matrices,

$$0 = h_0 < h_1 < \ldots < h_j < \ldots < h_M = h$$

are constant delays.

It is generally assumed, that he set of admissible controls for the dynamical control system (1) is unconstrained and is Banach space $U_{ad}= L_{\infty}([0, T], R^m)$.

Then, for a given admissible control $u(t)$ there exists a unique solution $x(t; u)$ for $t \in [0,T]$, of the state equation (1) with zero initial condition (2). Moreover, the dynamical system (1) is equivalent to the first-order system [2, 4]

$$z^{(1)}(t) = Cz(t) + \sum_{j=0}^{j=M(T)} D_j u(t - h_j) \tag{3}$$

where

$$M(t) = j, \quad for \quad h_j < t \leq h_{j+1}, \quad j = 0, 1, 2, \ldots, (M-1)$$
$$M(t) = M, \quad for \quad h_M < t$$

$$C = \begin{bmatrix} 0 & I & 0 & \ldots & 0 \\ 0 & 0 & I & \ldots & 0 \\ \vdots & \vdots & \vdots & \ddots & \vdots \\ 0 & 0 & 0 & \ldots & I \\ A_0 & A_1 & A_2 & \ldots & A_{N-1} \end{bmatrix}$$

$$z(t) = \begin{bmatrix} x(t) \\ x^{(1)}(t) \\ x^{(2)}(t) \\ \vdots \\ x^{(n-1)}(t) \end{bmatrix} \qquad \tilde{B}_j = \begin{bmatrix} 0 \\ 0 \\ \vdots \\ 0 \\ B_j \end{bmatrix}$$

Hence, the general solution of the differential state equation (1) is given by the following integral formula

$$x(t) = Z\exp(Ct)x(0) + Z\int_0^t \exp(C(t-\tau)) \sum_{j=0}^{j=M(t)} \tilde{B}_j u(t-h_j)d\tau$$

for $h_{M(T)} < t \leq h_{M(T)}$

where $Z = \left[I|0|0|\ldots|0 \right]$

or equivalently

$$x(t) = Z\exp(Ct)x(0) + Z\int_0^t \exp(C(t-\tau))D_j(t)v(\tau)d\tau$$

where

$$D_j(t) = \begin{bmatrix} 0 & 0 & \dots & 0 & 0 \\ 0 & 0 & \dots & 0 & 0 \\ \dots & \dots & \ddots & 0 & 0 \\ 0 & 0 & \dots & 0 & 0 \\ B_0 & B_1 & \dots & B_{M(t)-1} & B_{M(t)} \end{bmatrix} \quad v(t) = \begin{bmatrix} u(t) \\ u(t-h_1) \\ \vdots \\ u(t-h_{M(t)-1}) \\ u(t-h_{M(t)}) \end{bmatrix}$$

In the next part of this section, we shall introduce certain notations and present some important facts from the general controllability theory of higher-order linear differential equations.

As was mentioned in the introduction, for the linear dynamical control system with multiple point delays in the control (1), it is possible to define many different concepts of controllability. In the sequel we shall focus our attention on the relative controllability in the given time interval $[0, T]$. In order to do that, first of all let us introduce the notion of the attainable set at time $T > 0$ from zero initial conditions (2), denoted shortly by $K_T(U_c)$ and defined as follows [2, 3].

$$K_T(U_c) = \{x \in X : x = x(T, u),\ u(t) \in U_c \ for\ a.e.t \in [0, T]\} \tag{4}$$

where $x(t, u)$, $t > 0$ is the unique solution of the Eq. (1) with zero initial conditions (2) and a given admissible control $u \in L_\infty([0, T], R^m)$.

Now, using the concept of the attainable set given by the relation (4), let us recall the well known (see e.g. [2, 3] for details) definition of relative controllability in $[0, T]$ for dynamical system (1).

Definition 1. Dynamical control system (1) is said to be U_c-globally relative controllable in $[0, T]$ if $K_T(U_c) = R^n$.

Applying the Laplace transformation to the right-hand side of Eq. (1) yields the nxn matrix polynomial

$$L(s) = Is^N - \sum_{i=0}^{i=N-1} A_i s^i$$

where $s \in C$ is a complex variable.

Moreover, let $\sigma(L)$ denotes the spectrum of the polynomial matrix $L(s)$, namely,

$$\sigma(L) = \{s \in C : det\, L(s) = 0\}.$$

Now, for delayed system (1) let us consider polynomial matrix $L_T(s)$ depending on T and defined as follows

$$L_T(s) = \left[L(s) \middle| D(T)s^{nN-1} \middle| D(T)s^{nN-2} \middle| \dots \middle| D(T)s^2 \middle| D(T)s \middle| D(T) \right] \tag{5}$$

In the special case $n = 2$ for second order dynamical systems, i.e. for $N = 2$ we have

Let $p_1(s)$, $p_2(s)$,...,$p_k(s)$,...,$p_K(s)$, where

$$K = \binom{n + mM(T)N}{n}$$

are determinants of all possible *nxn*-dimensional submatrices of the matrix $L_T(s)$, which in fact are scalar polynomials of degree no more than *nN*. Therefore, for every $k = 1, 2,\ldots,K$ we have

$$p_k(s) = \left[\, 1 \middle| s \middle| s^2 \middle| \ldots \middle| s^{nN} \,\right] r_k$$

where r_k is the vector of the corresponding coefficients of $p_k(s)$.

Finally, following method given in the paper [1] let us define Plücker matrix for the system (1) as follows

$$P(L(s), D(T)) = [r_1, \; r_2 \ldots \; r_{K-1}, \; r_K]$$

3 Controllability Conditions

At the beginning of this section, for completeness of considerations, let us recall known in the literature necessary and sufficient conditions for relative controllability (see e.g. [1, 2], and [3] for more details).

Lemma 1. The following statements are equivalent.

(i) *System (1) is relatively controllable on [0, T].*
(ii) *rank* $[D(T), CD(T), C^2D(T),\ldots,C^{nN-1}D(T)] = nN$
(iii) *rank* $[L(s), D(T)] = n$ *for all* $s \in \sigma(L)$

Now, taking onto account the above considerations let us formulate the main result of this paper.

Theorem 1. The higher-order linear dynamical control system (1) is relatively controllable on time interval *[0, T]* if and only if the Plücker matrix *P(L(s), D(T))* has full row rank, i.e.,

$$rank\, P(L(s),\; D(T)) = nN + 1 \tag{6}$$

Proof. First of all, let us observe, that for dynamical systems with multiple delays in control (1) control we have piece-wise constant matrix *D(t)* instead constant matrix *B* as for dynamical systems without delays in control. Therefore, following [1], the $(nN + 1)xK$ dimensional matrix *P(L(s), D(T))* essentially depends on *T*, and for $h_{M(T)} < t \leq h_{M(T)}$ has the form (5).

Thus, using necessary and sufficient condition for controllability of systems without delays in control [1] we obtain equality (6).

For the very special case of second order dynamical control systems, i.e., for $N = 2$, from Theorem 1 follows next Corollary 1.

Corollary 1. The second-order dynamical system (1) is relatively controllable on *[0, T]* if and only if the Plücker matrix *P(L(s), D(T))* has full row rank, i.e.,

$$rank\, P(L(s),\ D(T)) = 2n + 1 \tag{7}$$

Remark 1. It is well known (see e.g. [2, 4] for more details), that for dynamical systems with delays absolute controllability and relative controllability strongly depend on the length of the time interval *[0, T]*. This fact is shown in the next simple illustrative example.

4 Example

In order to explain theoretical results let us consider the following rather simple illustrative example. Let the state equations of the second-order linear finite-dimensional dynamical control system defined on a given time interval *[0, T]*, have the following form

$$\begin{aligned} x_1^{(2)}(t) &= -x_2^{(1)}(t) - x_1(t) + x_2(t) + u(t) \\ x_2^{(2)}(t) &= -x_1^{(1)}(t) + x_1(t) + x_2(t) + u(t-h) \end{aligned} \tag{8}$$

Therefore, in this case

$$n = 2,\ m = 1,\ M = 1,\ 0 < h,\quad x(t) = (x_1(t), x_2(t))^{tr} \in R^2 \quad U = R,$$

and using the notations given in the previous sections matrices A_0, A_1 and B_0, B_1 have the following form

$$A_0 = \begin{bmatrix} -1 & 1 \\ +1 & +1 \end{bmatrix} \quad A_1 = \begin{bmatrix} 0 & -1 \\ -1 & 0 \end{bmatrix} \quad B_0 = \begin{bmatrix} 1 \\ 0 \end{bmatrix} \quad B_1 = \begin{bmatrix} 0 \\ 1 \end{bmatrix}$$

Therefore, matrix *L(s)* has the following form

$$L(s) = Is^2 - A_1 s - A_0 = \begin{bmatrix} s^2+1 & s-1 \\ s-1 & s^2-1 \end{bmatrix}$$

Hence,

$$\begin{aligned} L_T(s) &= \left[L(s) \middle| D(T)s \middle| D(T) \right] = \left[L(s) \middle| B_0 s \middle| B_0 \right] \\ &= \begin{bmatrix} s^2+1 & s-1 & s & 1 \\ s-1 & s^2-1 & 0 & 0 \end{bmatrix} \end{aligned}$$

for $0 < T < h$
In this case we have $n = 2$, $m = 1$, $M(T) = 1$, $N = 2$.
Hence $K = 6$.

Thus, we have *6* determinants of *2 * 2*- dimensional submatrices of the following simple form:

$$p_1(s) = s^4 - s^2 + 2s - 2 = \begin{bmatrix} 1 \; s \; s^2 \; s^3 \; s^4 \end{bmatrix} \begin{bmatrix} -2 \\ 2 \\ -1 \\ 0 \\ 1 \end{bmatrix} =$$

$$= \begin{bmatrix} 1 \; s \; s^2 \; s^3 \; s^4 \end{bmatrix} r_1$$

and hence $r_1 = [-2, 2, -1, 0, 1]^{tr}$

Similarly, without going into details it can be shown that the next determinants can be expressed as follows:

$$\begin{aligned} p_2(s) &= -s^2 + 1, \quad r_2 = [1, 0, -1, 0, 0]^{tr} \\ p_3(s) &= -s + 1, \quad r_3 = [1, -1, 0, 0, 0]^{tr} \\ p_4(s) &= -s^3 + s, \quad r_4 = [0, 1, 0, -1, 0]^{tr} \\ p_5(s) &= -s^2 + 1, \quad r_5 = [1, 0, -1, 0, 0]^{tr} \\ p_6(s) &= 0, \quad r_6 = [0, 0, 0, 0, 0]^{tr} \end{aligned}$$

Therefore,

$$rankP(L(S), B_0) = rank \begin{bmatrix} -2 & 1 & 1 & 0 & 1 & 0 \\ 2 & 0 & -1 & 1 & 0 & 0 \\ -1 & -1 & 0 & 0 & -1 & 0 \\ 0 & 0 & 0 & -1 & 0 & 0 \\ 1 & 0 & 0 & 0 & 0 & 0 \end{bmatrix} = 4 < 5 = nN + 1$$

Thus, the above control system is not relatively controllable on the time interval *[0, T]* for $T < h$.

The same result we have using eigenvalue approach. Indeed, in this case $s = 1$ is an eigenvalue of the polynomial matrix *L(s)*.

Moreover, we have

$$rank\begin{bmatrix} L(s) | B_0 \end{bmatrix} = rank \begin{bmatrix} s^2 + 1 & s - 1 & 1 \\ s - 1 & s^2 - 1 & 0 \end{bmatrix}$$

$$= rank\begin{bmatrix} L(1) | B_0 \end{bmatrix} = rank \begin{bmatrix} 2 & 0 & 1 \\ 0 & 0 & 0 \end{bmatrix} = 1 < 2$$

Hence, we have the same conclusion that dynamical system (1) is not relatively controllable on *[0, T]* for $T < h$.

However, for $T > h$ the situation is quite different. In this case for longer time interval *[0, T]* we have matrix $L_T(s)$ in the following form

$$L_T(s) = \begin{bmatrix} L(s) | D(T)s | D(T) \end{bmatrix} = \begin{bmatrix} L(s) | B_1 s | B_0 s | B_1 | B_0 \end{bmatrix}$$

$$= \begin{bmatrix} s^2+1 \; s-1 \; 0 \; s \; 0 \; 1 \\ s-1 \; s^2-1 \; s \; 0 \; 1 \; 0 \end{bmatrix}$$

In this case we have $n = 2$, $m = 1$, $M(T) = 2$, $N = 2$. Hence, $K = 15$. Thus, we have *15* determinants of $2 * 2$ dimensional submatrices.

Computing determinant $p_k(s)$ and coefficients vectors r_k we form 5×15-dimensional matrix $P(L(s), B_1, B_0)$, for which

$$rankP(L(S), B_1, B_0) = 5 = nN + 1$$

Hence, system is relatively controllable on *[0, T]* for $h < T$.

The same conclusion can be deduced using eigenvalue approach. Since $s = 1$ is an eigenvalue of the matrix *L(s,)* then we have

$$rank\left[L(s) \middle| B_1 \middle| B_0 \right] = rank\begin{bmatrix} s^2+1 \; s-1 \; 0 \; 1 \\ s-1 \; s^2-1 \; 1 \; 0 \end{bmatrix}$$

$$= rank\left[L(1) \; B_1 \; B_0 \right] = rank\begin{bmatrix} 2\,0\,0\,1 \\ 0\,0\,1\,0 \end{bmatrix} = 2$$

and system is relatively controllable on *[0, T]* for $h < T$.

This example illustrates that relative controllability depends on the length of time interval *[0, T]*.

5 Concluding Remarks

In this paper necessary and sufficient conditions for relative controllability on o given time interval for higher-order finite-dimensional dynamical control systems with multiple point delays in the control have been formulated and proved. In the proof of the main result previous has been used. The important feature of the new rank condition is that it does not require computation of eigenvalues of the considered dynamical system. These conditions extend to the case of relative controllability and dynamical control systems with delays in control the results published in [1] for linear systems without delays in control.

Acknowledgment. The work is supported by National Science Centre in Poland under grant: "Modelling, optimization and control for structural reduction of device noise", DEC-2017/25/B/ST7/02236.

References

1. Kalogeropoulos, G., Psarrakos, P.: A note on the controllability of higher-order linear systems. Appl. Math. Lett. **17**, 1375–1380 (2004)
2. Klamka, J.: Controllability of Dynamical Systems. Kluwer Academic Publishers, Dordrecht (1991)

3. Klamka, J.: Controllability of dynamical systems - a survey. Arch. Control Sci. **2**(3–4), 281–307 (1993)
4. Klamka, J.: Controllability of dynamical systems: a survey. Bull. Pol. Acad. Sci. Tech. Sci. **61**(2), 221–229 (2013)
5. Klamka, J.: Constrained controllability of second order dynamical systems with delay. Control Cybern. **42**(1), 111–121 (2013)

Minimum Fuel Resource Distribution in Multidimensional Logistic Networks Governed by Base-Stock Inventory Policy

Przemysław Ignaciuk(✉) and Łukasz Wieczorek

Lodz University of Technology, 215 Wólczańska St., 90-924 Łódź, Poland
przemyslaw.ignaciuk@p.lodz.pl, lukasz.wieczorek.1@edu.p.lodz.pl

Abstract. The paper addresses the problem of fuel-efficient resource redistribution in logistic networks in the phase preceding the engagement of market relations. In the considered class of systems, two types of entities – external sources and controlled nodes – form a complex interconnection structure. The flow of resources is governed using the classical base-stock inventory policy deployed in a distributed way. The optimization objective is to dynamically adjust the reference stock level at the controlled nodes so that excessive goods traffic is avoided while preparing the network for the customer demand in a latter, active market phase. The discrete-time finite-horizon optimization problem is solved analytically, which allows one to express the pattern of reference stock adaptation in a straightforward to implement, closed form. The derivations are validated by numerical tests.

Keywords: Optimal control · Logistic networks · Base-stock policy

1 Introduction

In recent years, the expansion of international cooperation has brought significant development of supply chains. The overseas logistic corridors have been extended and new ones have been created. In the literature, the majority of works addressing the logistic network operation assume already being in an active market situation, i.e., when a timely response to the customer demand is expected. The problem of establishing the initial state, before entering into direct contact with the environment, is often overlooked. However, in the current business reality, the preparatory stage is crucial to gain competitive advantage in the active phase.

Up to now, most of the research related to the supply chain management has considered specific models, subject to serious structural limitations, e.g.,

- single-stage systems – where the retailer is directly linked with the supplier [1, 2],
- chain systems, which may contain multiple intermediate nodes, arranged in a serial connection between the goods source and the retailer [3, 4],
- arborescent systems – comprising parallel, disjoint paths of goods flow [5–8].

A. Bartoszewicz et al. (Eds.): PCC 2020, AISC 1196, pp. 1141–1151, 2020.
https://doi.org/10.1007/978-3-030-50936-1_95

Meanwhile, the current logistic systems involve more complex, multi-echelon interconnection structures [9], which should not be treated with simplified methods due to undesirable side effects, e.g., excessive costs or loss in customer relations [10].

In this work, the problem of goods redistribution in multi-echelon logistic networks is addressed from a formal perspective. The considered class encompasses systems with arbitrary interconnection topology with goods reflow subject to non-negligible delay. The emphasis is placed on the preparatory phase, i.e., the way the goods are distributed among the nodes before the actual market interaction commences. The flow is governed by the popular, base-stock (BS) inventory policy [11]. The optimization objective is to determine the policy key parameter – the reference stock level – so that the costs of reflow are minimized within a given time horizon.

The optimization problem is solved analytically. The obtained closed-form expression for the stock level adjustment is easy to implement and does not involve substantial computational effort. The dynamic adaptation within the established time frame allows one to place the logistic company in a favorable situation with respect to the competition, at the same time, reducing the costs in the initial phase.

2 System Model

The purpose of the undertaken optimization problem is to adjust the reference stock level so that the logistic company may conveniently enter into active market environment, and face the external demand, under fuel-efficient resource redistribution.

2.1 Interconnection Structure

The considered class of systems involves interaction among two types of actors:

- external sources – that provide the goods for the network yet are not influenced by the customer demand directly,
- controlled nodes – that supply both their neighbors (as intermediate sources) and try to satisfy the external demand imposed by customers (as retailers).

The network topology encompasses N controlled nodes and M external sources. The nodes are connected with each other by unidirectional links. Each interconnection is characterized by a pair of attributes:

- the contribution factor that determines the part of the resources required by the ordering node to be acquired from a given supplier,
- the lead-time delay associated with a given transport channel (this delay is independent of the order size).

For practical reasons, the topologies with separated nodes, i.e., without connection to a supplier, as well as with self-suppling nodes, are excluded.

2.2 Node Interactions

Let $t = 0, 1, 2, \ldots, H$ quantify the duration of time, where H designates the planning horizon, i.e., the time of the preparatory phase assigned to the logistic company. The on-hand stock level at node i in period t is expressed as

$$x_i(t+1) = x_i(t) + \underbrace{\sum_{j=1}^{N+M} \varphi_{ji} q_i(t - \beta_{ji})}_{\text{incoming orders}} - \underbrace{\sum_{j=1}^{N} \varphi_{ij} q_j(t)}_{\text{outgoing orders}}, \tag{1}$$

where:

- φ_{ij} is the contribution factor of supplier i for controlled node j, $\sum_{j=1}^{N+M} \varphi_{ji} = 1$,
- β_{ij} is the lead-time delay of shipments passed from node i to j,
- $q_i(t)$ is the quantity of goods ordered by controlled node i in period t from its suppliers (both external sources and controlled nodes).

2.3 State-Space Description

In order to treat the considered control problem in an analytical way, it is convenient to express the system dynamics in a matrix-vector form. The variable dependencies may be grouped as

$$\mathbf{x}(t+1) = \mathbf{x}(t) + \sum_{k=1}^{L} \mathbf{\Phi}_k \mathbf{q}(t-k), \tag{2}$$

where:

- L is the maximum lead-time delay between any two interconnected nodes,
- $\mathbf{x}(t) = [x_1(t), x_2(t), \ldots, x_N(t)]^{\mathrm{T}}$ is the vector of stock level at the controlled nodes,
- $\mathbf{q}(t) = [q_1(t), q_2(t), \ldots, q_N(t)]^{\mathrm{T}}$ is the vector of resource quantity ordered by the controlled nodes from their suppliers,
- $\mathbf{\Phi}_k$ is a matrix containing the information about in-transit shipments,

$$\mathbf{\Phi}_k = \begin{bmatrix} \sum_{i:\alpha_{i1}=k} \varphi_{i1} & \alpha_{12} & \cdots & \alpha_{1N} \\ \alpha_{21} & \sum_{i:\alpha_{i2}=k} \varphi_{i2} & \cdots & \alpha_{2N} \\ \vdots & \vdots & \ddots & \vdots \\ \alpha_{N1} & \alpha_{N2} & \cdots & \sum_{i:\alpha_{iN}=k} \varphi_{iN} \end{bmatrix}, \tag{3}$$

where the diagonal entries represent the incoming shipments, realized with delay $k = 1, 2, \ldots, L$, and the off-diagonal ones

$$\alpha_{ij} = \begin{cases} -\varphi_{ij}, & \text{if } \beta_{ij} = k, \\ 0, & \text{otherwise.} \end{cases} \tag{4}$$

For the convenience of further analysis, two auxiliary matrices will be introduced:

$$\boldsymbol{\Phi} = \sum_{k=1}^{L} \boldsymbol{\Phi}_k, \tag{5}$$

and a hollow matrix corresponding to the internal shipments sent between the controlled nodes

$$\boldsymbol{\Phi}_+ = \begin{bmatrix} 0 & \varphi_{12} & \dots & \varphi_{1N} \\ \varphi_{21} & 0 & \dots & \varphi_{2N} \\ \vdots & \vdots & \ddots & \vdots \\ \varphi_{N1} & \varphi_{N2} & \dots & 0 \end{bmatrix}. \tag{6}$$

Under the assumption of complete order partitioning: $\sum_{j=1}^{N+M} \varphi_{ji} = 1$, it holds $\boldsymbol{\Phi}_+ = \mathbf{I} - \boldsymbol{\Phi}$, where $\mathbf{I}$ is an identity matrix of appropriate dimensions.

2.4 BS Inventory Control Policy

In order to control the flow of resources in the network, the classical BS inventory management policy is considered. It is implemented in a distributed way, i.e., separately at each node. The key parameter of this strategy is the reference stock level to be set at the controlled nodes. At the end of each period, a controlled node generates the ordering signal to its suppliers (both external sources and other controlled nodes) according to

$$\mathbf{q}(t) = \mathbf{x}_{\text{REF}}(t) - \mathbf{x}(t) - \sum_{j=1}^{L} \sum_{k=j}^{L} \boldsymbol{\Phi}_k \mathbf{q}(t-j), \tag{7}$$

where $\mathbf{x}_{\text{REF}}(t) = \left[x_{\text{REF},\,1}(t), x_{\text{REF},\,2}(t), \dots, x_{\text{REF},\,N}(t)\right]^{\text{T}}$ is the vector of reference stock levels of N controlled nodes.

3 Optimization Problem

3.1 Problem Statement

In the nominal case, i.e., when unobstructed flow of resources is observed, the optimization problem may be solved analytically. If the initially gathered stock does not permit such free goods redistribution, the BS policy operates correctly, yet with additional waiting time for the goods to reach intermediate nodes.

Assuming the initial stock at level $\mathbf{x}(0)$, and no *a priori* generated orders, i.e., $\mathbf{q}(t) = 0$ for $t < 0$, the task is to reach level $\mathbf{x}(H)$ within the preparatory phase of H periods so that imposed cost criteria are fulfilled. The objective considered in this work is to minimize the effort associated with goods management exercised by each controlled node. In the adopted approach, a controlled node is responsible for the resources since the moment of placing an order to effectuating the sales for the resources it requests from suppliers, shipments in-transit, and goods held in stock. Formally, the optimization problem may be described as

$$\min J(\mathbf{x}_{\mathrm{REF}}(t)) = \frac{1}{2}\sum_{t=0}^{H-1} \boldsymbol{\Psi}^{\mathrm{T}}(t)\mathbf{W}\boldsymbol{\Psi}(t), \tag{8}$$

where $\mathbf{W}$ is a positive definite matrix of weighting coefficients prioritizing the choice of supply channels and $\boldsymbol{\Psi}(t)$ is the quantity of resources that the nodes are responsible for,

$$\boldsymbol{\Psi}(t) = \underbrace{\mathbf{q}(t)}_{\text{ordered quantity}} + \underbrace{\sum_{j=1}^{L}\sum_{k=j}^{L} \boldsymbol{\Phi}_k \mathbf{q}(t-j)}_{\text{in-transit shipments}} + \underbrace{\mathbf{x}(t)}_{\text{on-hand stock}}. \tag{9}$$

The considered problem is difficult to solve analytically owing to the delays related to the goods transshipment. For that reason, an alternative, equivalent system description will be proposed. Let $\mathbf{y}(t)$ denote the inventory position of the controlled nodes (the sum of the on-hand stock level and in-transit orders),

$$\mathbf{y}(t) = \mathbf{x}(t) + \sum_{j=1}^{L}\sum_{k=j}^{L} \boldsymbol{\Phi}_k \mathbf{q}(t-j). \tag{10}$$

Lemma 1. The dynamics of $\mathbf{y}(t)$ can be described by

$$\mathbf{y}(t+1) = \mathbf{y}(t) + \boldsymbol{\Phi}\mathbf{q}(t). \tag{11}$$

Proof. Directly from the definition of $\mathbf{y}$, under zero initial input, $\mathbf{y}(0) = \mathbf{x}(0)$. Thus, applying (2) to (11), one has

$$\mathbf{y}(1) = \mathbf{x}(1) + \sum_{j=1}^{L}\sum_{k=j}^{L} \boldsymbol{\Phi}_k \mathbf{q}(1-j) = \mathbf{x}(0) + 0 + \sum_{k=1}^{L} \boldsymbol{\Phi}_k \mathbf{q}(0) = \mathbf{y}(0) + \boldsymbol{\Phi}\mathbf{q}(0). \tag{12}$$

Therefore, (11) is satisfied at $t = 0$. Afterwards, for any period $t > 0$, the following relation can be established:

$$\begin{aligned}
\mathbf{y}(t+1) &= \mathbf{x}(t+1) + \sum_{j=1}^{L}\sum_{k=j}^{L} \boldsymbol{\Phi}_k \mathbf{q}(t+1-j) \\
&= \mathbf{x}(t) + \sum_{k=1}^{L} \boldsymbol{\Phi}_k \mathbf{q}(t-k) + \sum_{j=1}^{L}\sum_{k=j}^{L} \boldsymbol{\Phi}_k \mathbf{q}(t+1-j) \\
&= \mathbf{x}(t) + \boldsymbol{\Phi}_1\mathbf{q}(t-1) + \boldsymbol{\Phi}_2\mathbf{q}(t-2) + \ldots + \boldsymbol{\Phi}_{L-1}\mathbf{q}(t-L+1) + \boldsymbol{\Phi}_L\mathbf{q}(t-L) \\
&\quad + [\boldsymbol{\Phi}_1 + \ldots + \boldsymbol{\Phi}_L]\mathbf{q}(t) + [\boldsymbol{\Phi}_2 + \ldots + \boldsymbol{\Phi}_L]\mathbf{q}(t-1) + \ldots + \boldsymbol{\Phi}_L\mathbf{q}(t-L+1) \\
&= \mathbf{x}(t) + [\boldsymbol{\Phi}_1 + \ldots + \boldsymbol{\Phi}_L]\mathbf{q}(t-1) + [\boldsymbol{\Phi}_2 + \ldots + \boldsymbol{\Phi}_L]\mathbf{q}(t-2) + \ldots \\
&\quad + \left[\boldsymbol{\Phi}_{L-1} + \boldsymbol{\Phi}_L\right]\mathbf{q}(t-L+1) + \boldsymbol{\Phi}_L\mathbf{q}(t-L) + [\boldsymbol{\Phi}_1 + \ldots + \boldsymbol{\Phi}_L]\mathbf{q}(t)
\end{aligned}$$

$$= \mathbf{x}(t) + \sum_{j=1}^{L}\sum_{k=j}^{L} \mathbf{\Phi}_k \mathbf{q}(t-j) + \Phi\mathbf{q}(t)$$

$$= \mathbf{y}(t) + \mathbf{\Phi}\mathbf{q}(t), \tag{13}$$

which ends the proof.

Note that by applying (10) into (7), the replenishment signals generated using the BS policy

$$\mathbf{q}(t) = \mathbf{x}_{\mathrm{REF}}(t) - \mathbf{y}(t). \tag{14}$$

In turn, substituting (10) and (14) into (9) yields

$$\mathbf{\Psi}(t) = \mathbf{x}_{\mathrm{REF}}(t). \tag{15}$$

Using variable **y**, optimization problem (8) can thus be reformulated as

$$\min J\left(\mathbf{x}_{\mathrm{REF}}(t)\right) = \frac{1}{2}\sum_{t=0}^{H-1} \mathbf{x}_{\mathrm{REF}}^{\mathrm{T}}(t)\mathbf{W}\mathbf{x}_{\mathrm{REF}}(t) \tag{16}$$

subject to the constraint

$$\mathbf{y}(t+1) = \mathbf{\Phi}_{+}\mathbf{y}(t) + \mathbf{\Phi}\mathbf{x}_{\mathrm{REF}}(t). \tag{17}$$

Theorem 1. Optimization problem (16) with constraint (17) is convex.

Proof. An optimization problem is convex when both the objective function and all the constraints are convex.

The power function $()^{\varepsilon}$ on R_{++} (positive real numbers) is convex if the exponent $\varepsilon \leq 0$ or $\varepsilon \geq 1$. All the linear functions are convex, as well. Hence, both the quadratic cost functional (16) and system constraint (17) are convex. This observation concludes the proof.

3.2 Problem Solution

For problem (16), the Hamiltonian can be defined as

$$\mathbf{H}(t) = \frac{1}{2}\mathbf{x}_{\mathrm{REF}}^{\mathrm{T}}(t)\mathbf{W}\mathbf{x}_{\mathrm{REF}}(t) + \boldsymbol{\lambda}^{\mathrm{T}}(t+1)\left[\mathbf{\Phi}_{+}\mathbf{y}(t) + \mathbf{\Phi}\mathbf{x}_{\mathrm{REF}}(t)\right], \tag{18}$$

where $\boldsymbol{\lambda}^{\mathrm{T}}(t+1)$ is the row vector of Lagrange multipliers.

The necessary conditions are as follows [12]:

- state equation

$$\mathbf{y}(t+1) = \frac{\partial \mathbf{H}(t)}{\partial \boldsymbol{\lambda}(t+1)} = \mathbf{\Phi}_{+}\mathbf{y}(t) + \mathbf{\Phi}\mathbf{x}_{\mathrm{REF}}(t), \tag{19}$$

- costate equation

$$\boldsymbol{\lambda}(t) = \frac{\partial \mathbf{H}(t)}{\partial \mathbf{y}(t)} = \boldsymbol{\Phi}_+^{\mathrm{T}} \boldsymbol{\lambda}(t+1), \tag{20}$$

- stationarity condition

$$0 = \frac{\partial \mathbf{H}(t)}{\partial \mathbf{x}_{\mathrm{REF}}(t)} = \frac{1}{2}\left(\mathbf{W} + \mathbf{W}^{\mathrm{T}}\right)\mathbf{x}_{\mathrm{REF}}(t) + \boldsymbol{\Phi}^{\mathrm{T}}\boldsymbol{\lambda}(t+1) = \mathbf{W}\mathbf{x}_{\mathrm{REF}}(t) + \boldsymbol{\Phi}^{\mathrm{T}}\boldsymbol{\lambda}(t+1). \tag{21}$$

First, solving (21) for $\mathbf{x}_{\mathrm{REF}}$, yields

$$\mathbf{x}_{\mathrm{REF}}(t) = -\mathbf{W}^{-1}\boldsymbol{\Phi}^{\mathrm{T}}\boldsymbol{\lambda}(t+1). \tag{22}$$

Note that since $\mathbf{W}$ is positive definite its inverse does exist.

Then, substituting (22) into (19), results in

$$\begin{aligned} \mathbf{y}(t+1) &= \boldsymbol{\Phi}_+\mathbf{y}(t) - \boldsymbol{\Phi}\mathbf{W}^{-1}\boldsymbol{\Phi}^{\mathrm{T}}\boldsymbol{\lambda}(t+1) \\ &= \boldsymbol{\Phi}_+\mathbf{y}(t) - \mathbf{A}\boldsymbol{\lambda}(t+1) \end{aligned} \tag{23}$$

where $\mathbf{A}$ is an auxiliary matrix $\mathbf{A} \triangleq \boldsymbol{\Phi}\mathbf{W}^{-1}\boldsymbol{\Phi}^{\mathrm{T}}$.

(20) is a homogeneous difference equation. Its solution with respect to the terminal condition is as follows

$$\boldsymbol{\lambda}(t) = \left(\boldsymbol{\Phi}_+^{\mathrm{T}}\right)^{H-t}\boldsymbol{\lambda}(H). \tag{24}$$

Substituting (24) into (23), gives

$$\mathbf{y}(t+1) = \boldsymbol{\Phi}_+\mathbf{y}(t) - \mathbf{A}\left(\boldsymbol{\Phi}_+^{\mathrm{T}}\right)^{H-t-1}\boldsymbol{\lambda}(H). \tag{25}$$

Solution of (25) in terms of the initial inventory position $\mathbf{y}(0)$ may be expressed as

$$\mathbf{y}(t) = \boldsymbol{\Phi}_+^t\mathbf{y}(0) - \sum_{k=0}^{t-1}\boldsymbol{\Phi}_+^{t-k-1}\mathrm{A}\left(\boldsymbol{\Phi}_+^{\mathrm{T}}\right)^{H-k-1}\boldsymbol{\lambda}(H). \tag{26}$$

The initial state $\mathbf{y}(0)$ and the final state $\mathbf{y}(H)$ are fixed, so their first derivatives are equal to zero. Hence, one can vary neither $\mathbf{y}(0)$, nor $\mathbf{y}(H)$, in determining the constrained minimum for the optimal control problem under consideration.

According to (26), the final inventory position may be calculated as

$$\mathbf{y}(H) = \boldsymbol{\Phi}_+^H\mathbf{y}(0) - \mathbf{B}\boldsymbol{\lambda}(H), \tag{27}$$

where

$$\mathbf{B} \triangleq \sum_{k=0}^{H-1}\boldsymbol{\Phi}_+^{H-k-1}\mathrm{A}\left(\boldsymbol{\Phi}_+^{\mathrm{T}}\right)^{H-k-1}$$

$$= \sum_{k=0}^{H-1} \mathbf{\Phi}_+^{H-k-1} \mathbf{\Phi} \mathbf{W}^{-1} \mathbf{\Phi}^{\mathrm{T}} \left(\mathbf{\Phi}_+^{\mathrm{T}}\right)^{H-k-1}. \tag{28}$$

Then, the Lagrange multiplier vector can be expressed using $\mathbf{y}(0)$ and $\mathbf{y}(H)$ as

$$\boldsymbol{\lambda}(H) = -\mathbf{B}^{-1}\left(\mathbf{y}(H) - \mathbf{\Phi}_+^H \mathbf{y}(0)\right). \tag{29}$$

By applying (29) into (24),

$$\boldsymbol{\lambda}(t) = -\left(\mathbf{\Phi}_+^{\mathrm{T}}\right)^{H-t} \mathbf{B}^{-1}\left[\mathbf{y}(H) - \mathbf{\Phi}_+^H \mathbf{y}(0)\right], \tag{30}$$

and finally, using (22), the optimal control

$$\mathbf{x}_{\mathrm{REF}}^*(t) = \mathbf{W}^{-1} \mathbf{\Phi}^{\mathrm{T}} \left(\mathbf{\Phi}_+^{\mathrm{T}}\right)^{H-t-1} \mathbf{B}^{-1}\left[\mathbf{y}(H) - \mathbf{\Phi}_+^H \mathbf{y}(0)\right]. \tag{31}$$

4 Numerical Study

In order to evaluate the closed-form optimal solution (31), the logistic system visualized in Fig. 1 is considered. In the examined topology, elements 1–7 represent controlled nodes and elements 8–10 denote the external sources. The parameters, i.e., the contribution factor and the lead-time delay, of node interconnections are listed in Table 1.

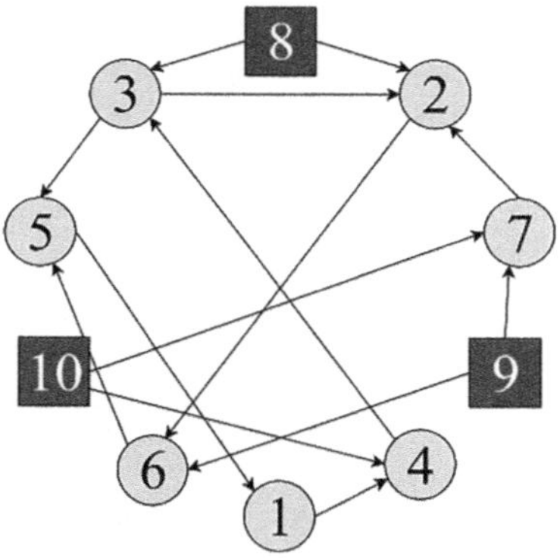

Fig. 1. Network topology (1–7 controlled nodes, 8–10 external sources).

In the analyzed scenario, one aims to accumulate enough resources at the network nodes, during the preparatory phase of $H = 30$ periods, so that it may operate properly in active market conditions. The initial stock is assumed empty.

According to [13], when in the active market relations the maximum expected demand $\mathbf{d}_{\max} = [d_1, d_2, \ldots, d_N]^{\mathrm{T}}$, the stock required to ensure full demand satisfaction may be calculated as

$$\mathbf{x}(H+1) = \left(\mathbf{I} + \sum_{k=1}^{L} k \mathbf{\Phi}_k\right) \mathbf{\Phi}^{-1} \mathbf{d}_{\max}. \tag{32}$$

Table 1. Interconnection parameters.

Link (from node i to j)	Parameters (φ_{ij}, β_{ij})
1 to 4	(0.6, 1)
2 to 6	(0.2, 4)
3 to 2	(0.3, 2)
3 to 5	(0.9, 1)
4 to 3	(0.3, 2)
5 to 1	(1.0, 2)
6 to 5	(0.1, 4)
7 to 3	(0.4, 4)
8 to 2	(0.3, 1)
8 to 3	(0.7, 3)
9 to 6	(0.8, 2)
9 to 7	(0.8, 3)
10 to 4	(0.4, 4)
10 to 7	(0.2, 2)

In addition, taking into account no initial input, the inventory position at the beginning of the active market phase at $t = H + 1$, satisfies $\mathbf{y}(H + 1) = \mathbf{x}(H + 1)$.

With the highest expected demand

$$\mathbf{d}_{\mathrm{max}} = [430, 320, 350, 375, 300, 325, 340]^{\mathrm{T}}, \tag{33}$$

the inventory position that should be reached at the end of the preparatory phase

$$\mathbf{y}(H) = [2824, 1434, 6996, 2726, 2855, 1886, 1915]^{\mathrm{T}} \text{ units.} \tag{34}$$

Figure 2 depicts the evolution of the decision variable $\mathbf{x}_{\mathrm{REF}}(t)$. The plot indicates that the reference stock levels do not change before $t = 10$ and thus the controlled nodes do not generate orders for their suppliers. Afterwards, the flow of goods proceeds to reach the desired state in period H. The preparatory phase yields the optimal control set as

$$\mathbf{x}_{\mathrm{REF}}(H) = [5127, 1862, 5286, 6086, 4210, 1710, 1754]^{\mathrm{T}} \tag{35}$$

with the performance index equal to $1.16 \cdot 10^8$. The desired objectives are fulfilled and the logistic company is ready to enter into active business relations with the market environment.

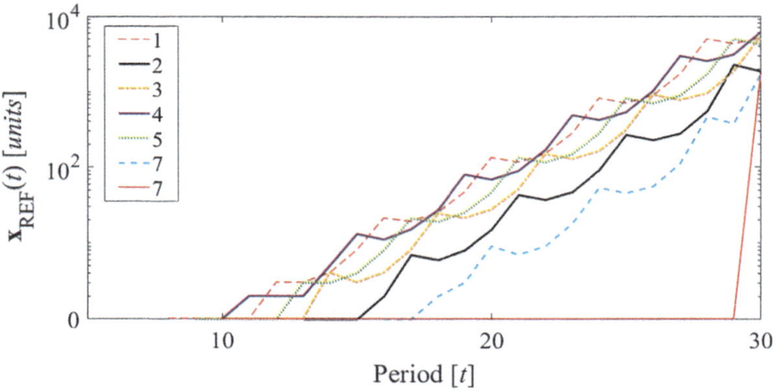

Fig. 2. Decision variable (reference stock level) at the controlled nodes.

5 Conclusions

The paper investigates the preparatory phase of the resource distribution process in networked structures that aims to place the logistic company into an active market situation in a competitive way. A multi-echelon topology with arbitrary interconnection pattern and delayed transportation channels is considered. A fuel-efficient optimization problem is formulated and solved analytically through an alternative state-space description that circumvents the intricacy of delay related events. The reference stock level of the base-stock policy is adjusted so that the operational costs of handling the goods transfer are minimized. The validity of formal content is confirmed via numerical tests. In the further study, non-linear inventory management strategies, as well as sensitivity aspects, will be considered.

References

1. Kiesmüller, G.P., de Kok, A.G., Dabia, S.: Single item inventory control under periodic review and a minimum order quantity. Int. J. Prod. Econ. **133**(1), 280–285 (2011)
2. Ignaciuk, P., Bartoszewicz, A.: Dead-beat and reaching-law-based sliding-mode control of perishable inventory systems. Bull. Pol. Acad. Sci. Tech. Sci. **59**(1), 39–49 (2011)
3. Ignaciuk, P.: Discrete inventory control in systems with perishable goods – a time-delay system perspective. IET Control Theory Appl. **8**(1), 11–21 (2014)
4. Lai, X., Chen, Z., Giri, B.C., Chiu, C.H.: Two-echelon inventory optimization for imperfect production system under quality competition environment. Math. Probl. Eng. **2015** (2015). Article ID 326919, 11 pages
5. Ignaciuk, P.: Dead-time compensation in continuous-review perishable inventory systems with multiple supply alternatives. J. Process Control **22**(5), 915–924 (2012)
6. Ignaciuk, P.: LQ optimal and robust control of perishable inventory systems with multiple supply options. IEEE Trans. Autom. Control **58**(8), 2108–2113 (2013)
7. Dominguez, R., Cannella, S., Framinan, J.M.: The impact of the supply chain structure on bullwhip effect. Appl. Math. Model. **39**(23–24), 7309–7325 (2015)
8. Bartoszewicz, A., Latosiński, P.: Sliding mode control of inventory management systems with bounded batch size. Appl. Math. Model. **66**, 296–304 (2019)

9. de Kok, T., Grob, C., Laumanns, M., Minner, S., Rambau, J., Schade, K.: A typology and literature review on stochastic multi-echelon inventory models. Eur. J. Oper. Res. **269**(3), 955–983 (2018)
10. Cattani, K.D., Jacobs, F.R., Schoenfelder, J.: Common inventory modeling assumptions that fall short: arborescent networks, Poisson demand, and single-echelon approximations. J. Oper. Manag. **29**(5), 488–499 (2011)
11. Cannella, S.: Order-Up-To policies in information exchange supply chains. Appl. Math. Model. **38**(23), 5553–5561 (2014)
12. Lewis, F.L., Vrabie, D.L., Syrmos, V.L.: Optimal Control, 3rd edn. Wiley, Hoboken (2012)
13. Ignaciuk, P., Wieczorek, L.: Networked base-stock inventory control in complex distribution systems. Math. Probl. Eng. **2019** (2019). Article ID 3754367, 14 pages

Modeling of Fractional-Order Systems

Accuracy Estimation of the Fractional, Discrete-Continuous Model of the One-Dimensional Heat Transfer Process

Krzysztof Oprzędkiewicz(✉) and Klaudia Dziedzic

Department of Automatic Control and Robotics, AGH University, al. A. Mickiewicza 30, 30-059 Krakow, Poland
{kop,kdz}@agh.edu.pl

Abstract. In the paper a new, state space, finite dimensional, non integer order model of a one-dimensional heat transfer process is considered. The proposed model uses a well known finite difference method and fractional Caputo operator to express the time derivative. The second order backward difference describes the derivative along the length. The analytical formula of the step response is given. Accuracy and convergence of the proposed model are numerically analyzed and compared to previously proposed state space model using semigroup approach. Results of simulations point that the good accuracy of the proposed model can be achieved for its relatively low order.

Keywords: Non-integer order systems · Heat transfer equation · Finite difference · Caputo operator

1 Introduction

The modeling of processes and phenomena hard to analyse with the use of other tools is one of main areas of application non integer order calculus. Non integer models for many physical phenomena were presented by various Authors, for example [3,4,6,7,23,27]. Analysis of anomalous diffusion problem with the use of fractional order approach and semigroup theory was presented for example by [24]. An observability problem for fractional order systems was presented for example by [10]. Minimum energy control for FO descriptor systems was analysed for example by [25].

Heat transfer processes can also be modeled using the non integer order approach. This problem has been investigated for example by [1,5,12,13]. The use of Caputo-Fabrizio operator in modeling of heat transfer processes was discussed by [26], the use operators with non singular kernel to modeling of thermal processes was deeply analysed in paper [2].

This paper is intended to propose and analyse a new, fractional order, finite dimensional, state-space model for heat transfer process in one dimensional plant. The proposed model uses well known finite difference approach, but the derivative with respect to time is described by fractional order difference.

A. Bartoszewicz et al. (Eds.): PCC 2020, AISC 1196, pp. 1155–1166, 2020.
https://doi.org/10.1007/978-3-030-50936-1_96

The paper is organized as follows: preliminaries recall elementary ideas from fractional calculus as well as the known model of the heat transfer in the form of Partial Differential Equation (PDE). Next the proposed fractional order, discrete-continuous model (FDC model) is presented. The analytical formula of the step response is also given and proved. Finally the accuracy and convergence of the model are numerically tested with the use of experimental results. Comparison to the fractional model using semigroup approach, previously proposed by authors is also presented.

2 Preliminaries

A presentation of elementary ideas is started with a definition of a non integer-order, integro-differential operator. It was given for example by [4,9,11,23]:

Definition 1 *(The elementary non integer order operator). The non integer-order integro-differential operator is defined as follows:*

$$ {}_aD_t^\alpha f(t) = \begin{cases} \frac{d^\alpha f(t)}{dt^\alpha} & \alpha > 0 \\ f(t) & \alpha = 0 \\ \int\limits_a^t f(\tau)(d\tau)^\alpha & \alpha < 0 \end{cases} . \tag{1} $$

where a and t denote time limits for operator calculation, $\alpha \in \mathbb{R}$ denotes the non integer order of the operation.

The integro-differential operator can be described by various definitions, given by Grünwald and Letnikov (GL), Riemann and Liouville (RL) and Caputo (C). In this paper the C definition is applied (see for example [4,9,11,23]):

Definition 2 *(The Caputo definition of the FO operator).*

$$ {}_0^CD_t^\alpha f(t) = \frac{1}{\Gamma(M-\alpha)} \int\limits_0^\infty \frac{f^{(M)}(\tau)}{(t-\tau)^{\alpha+1-M}} d\tau. \tag{2} $$

where $M-1 < \alpha < M$ denotes the non integer order of operation and $\Gamma(..)$ is the Gamma function.

For the Caputo operator the Laplace transform can be defined (see for example [8]):

Definition 3 *(The Laplace transform for Caputo operator).*

$$ \begin{aligned} &\mathcal{L}({}_0^CD_t^\alpha f(t)) = s^\alpha F(s), \ \ \alpha < 0 \\ &\mathcal{L}({}_0^CD_t^\alpha f(t)) = s^\alpha F(s) - \sum_{k=0}^{M-1} s^{\alpha-k-1} {}_0D_t^k f(0), \\ &\alpha > 0, \ \ M-1 < \alpha \leq M \in \mathbb{Z}. \end{aligned} \tag{3} $$

Consequently, the inverse Laplace transform for non integer order function is expressed as follows [11]:

$$\mathcal{L}^{-1}[s^{\alpha}F(s)] =_0 D_t^{\alpha} f(t) + \sum_{k=0}^{M-1} \frac{t^{k-1}}{\Gamma(k-\alpha+1)} f^{(k)}(0^+) \qquad (4)$$
$$M-1 < \alpha < M, \quad M \in \mathbb{Z}.$$

Finally the linear state equation needs to be recalled. It is as follows:

$$\begin{aligned} {}_0D_t^{\alpha} x(t) &= Ax(t) + Bu(t) \\ y(t) &= Cx(t) \end{aligned} . \qquad (5)$$

where $\alpha \in (0,1)$ denotes the fractional order of the state equation, $x(t) \in \mathbb{R}^N$, $u(t) \in \mathbb{R}^L$, $y(t) \in \mathbb{R}^P$ are the state, control and output vectors respectively, A, B, C are the state, control and output matrices, respectively.

3 The Experimental Plant and Its Fundamental Model

The simplified scheme of the considered heat plant is shown in Fig. 1. It has a form of a thin copper rod heated with an electric heater x_u long, located at one end of rod. An output temperature is measured using Pt-100 RTD sensors x_s long attached in points: 0.29, 0.50 and 0.73 of rod length. More details of the construction are given in the section "Experimental Results". The fundamental mathematical model describing the heat conduction in the plant is the partial differential equation of the parabolic type with the homogeneous Neumann boundary conditions at the ends, the homogeneous initial condition, the heat exchange along the length of rod and distributed control and observation. This equation with integer orders of both differentiations has been considered in papers [14–17]. The non integer order model with respect to time, employing Caputo operator was given in [18], its properties were analyzed also in [22].

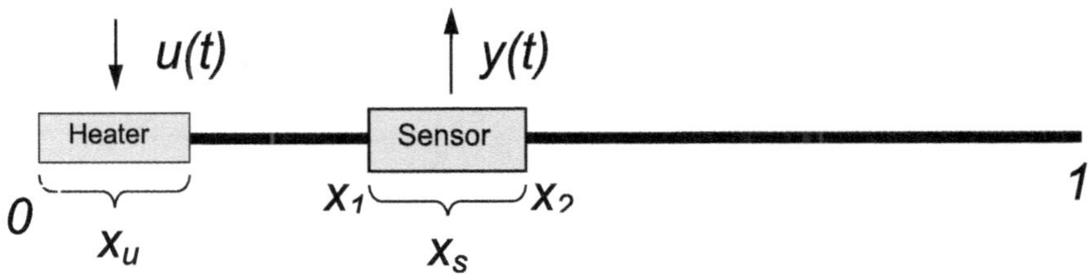

Fig. 1. The simplified scheme of the experimental system

It takes the following form:

$$\begin{cases} {}_0^C D_t^\alpha Q(x,t) = a_w \dfrac{\partial^2 Q(x,t)}{\partial x^2} - R_a Q(x,t) + b(x)u(t) \\ \dfrac{\partial Q(0,t)}{dx} = 0, t \geq 0 \\ \dfrac{\partial Q(1,t)}{dx} = 0, t \geq 0 \\ Q(x,0) = Q_0, 0 \leq x \leq 1 \\ y(t) = k_0 \int_0^1 Q(x,t)c(x)dx. \end{cases} \tag{6}$$

In (6) α denotes the non integer order of the system, a_w, R_a denote coefficients of heat conduction and heat exchange, k_0 is a steady-state gain of the model, $b(x)$ and $c(x)$ are heater and sensor functions in the following form:

$$b(x) = \begin{cases} 1, x \in [0, x_u] \\ 0, x \notin [0, x_u] \end{cases} . \tag{7}$$

$$c(x) = \begin{cases} 1, x \in [x_1, x_2] \\ 0, x \notin [x_1, x_2] \end{cases} . \tag{8}$$

The fundamental model of the heat transfer presented above is the base to propose the fractional, discrete-continuous (FDC) model.

4 The FDC Model of the Plant

Divide the length of the rod into N short sections $\Delta x = \frac{1}{N}$ long. Consequently the first and second derivative along length in Eq. (6) can be approximated by 1'st and 2'nd order differential quotients, analogically, as it was done in [14]:

$$\frac{\partial Q(x,t)}{\partial x} \approx \frac{Q(x+\Delta x,t) - Q(x,t)}{\Delta x}. \tag{9}$$

$$\frac{\partial^2 Q(x,t)}{\partial x^2} \approx \frac{Q(x+\Delta x,t) - 2Q(x,t) + Q(x-\Delta x,t)}{\Delta x}. \tag{10}$$

Next introduce the following notation:

$$Q_n(t) = Q(n\Delta x, t) \; n = 0, 1, ..., N. \tag{11}$$

Using the above convention the functions of heater and sensor (7) and (8) are expressed as:

$$b_n = \begin{cases} 1, n\Delta x \in [0, x_u] \\ 0, n\Delta x \notin [0, x_u] \end{cases} \quad n = 0, 1, ..., N. \tag{12}$$

$$c_n = \begin{cases} 1, x \in [x_1, x_2] \\ 0, n\Delta x \notin [x_1, x_2] \end{cases} \quad n = 0, 1, ..., N. \tag{13}$$

Using the approximations (9), (10) with notation (11), (12) and (13) to heat Eq. (6) we obtain:

$$\begin{cases} Q_0(t) = Q_1(t) \\ \ldots \\ {}_0^C D_t^{\alpha} Q_n(t) = a_w \frac{Q_{n+1}(t) - 2Q_n(t) + Q_{n-1}(t)}{\Delta x} - R_a Q_n(t) + b_n u(t), \\ n = 1, \ldots, N-1 \\ \ldots \\ Q_{N-1}(t) = Q_N(t) \\ Q_n(0) = 0, \ n = 0, 1, \ldots, N \\ y(t) = k_0 \Delta x \sum\limits_{n=0}^{N} Q_n(t) c_n. \end{cases} \tag{14}$$

The Eq. (14) can be expressed as the following, fractional order, finite dimensional state equation:

$$\begin{cases} {}_0^C D_t^{\alpha} Q(t) = AQ(t) + Bu(t) \\ y(t) = CQ_n(t) \end{cases} . \tag{15}$$

where $Q(t) = [Q_1(t) \ldots Q_n(t)]^T \in \mathbb{R}^N$ is the state vector, $u(t) \in \mathbb{R}$ is the control, $y(t) \in \mathbb{R}$ is the output, the state, control and output matrices A, B, and C are defined as follows:

$$A = d \begin{bmatrix} -1 - R & 1 & \ldots & 0 \\ 1 & -2 - R, & 1 \ldots \ldots & 0 \\ \ldots & \ldots & \ldots \ \ldots & \ldots \\ 0 & & \ldots \ 1 & -1 - R \end{bmatrix}_{N \times N} . \tag{16}$$

where:

$$\begin{aligned} R &= R_a \frac{\Delta x^2}{a_w} \\ d &= \frac{a_w}{\Delta x^2} \end{aligned} . \tag{17}$$

$$B = [b_0, b_1, \ldots, b_N]^T . \tag{18}$$

In (18) b_n are defined by (12).

$$C = [c_0, c_1, \ldots, c_N] . \tag{19}$$

In (19) c_n are defined by (13).

The eigenvalues of the state matrix (16) are as follows:

$$\lambda_n = d[-2(1 - \cos(\phi_n) - R], \ n = 1, \ldots, N. \tag{20}$$

where:

$$\phi_n = \frac{(n-1)\pi}{N}, \quad n = 1, \ldots, N. \tag{21}$$

The eigenvector p_n associated to n-th eigenvalue meets the following equation:

$$(A - \lambda I)p_n = 0, \ p_n \neq 0, \ n = 1, ..., N. \tag{22}$$

The spectrum of the state matrix A contains only single, purely real eigenvalues. This implies that each eignevalue is associated to single eigenvector with components described as follows:

$$\begin{cases} p_{2,n} = -(2c\cos\phi_{n-1})p_{1,n} = 0 \\ \cdots \\ p_{m-2,n} - 2\cos\phi_n p_{m-1,n} + p_{m,n} = 0 \\ \cdots \\ p_{N-1,n} - p_{N,n}(2\cos\phi_n - 1) = 0 \end{cases} \tag{23}$$

The solution of the Eq. (23) yields:

$$p_{2,n} = p_{1,n}(2\cos\phi_n - 1). \tag{24}$$

$$p_{m,n} = \begin{cases} p_{1,n}\left(\cos(m-1)\phi_n + \frac{\cos\phi_n - 1}{\sin\phi_n}\sin(m-1)\phi_n\right) & \phi_n \neq 0 \\ p_{1,n} \quad \phi_n = 0 \\ m = 3, ..., N \end{cases} . \tag{25}$$

The eigenvectors p_n can be normalized:

$$P_n = \frac{p_n}{||p_n||}, \ n = 1, .., N. \tag{26}$$

where:

$$||p_n|| = \sqrt{\sum_{m=1}^{N} p_{m,n}^2}, \ n = 1, .., N. \tag{27}$$

Consequently the matrix P transforming the system to the Jordan canonical form takes the following form:

$$P = [P_1 ... P_N]. \tag{28}$$

and the Jordan canonical form of the state Eq. (15) is as follows:

$$\begin{cases} {}_0^C D_t^\alpha Q^*(t) = A^* Q^*(t) + B^* u(t), \\ y(t) = C^* Q_n^*(t). \end{cases} \tag{29}$$

where:

$$\begin{aligned} A^* &= P^{-1}AP = diag\{\lambda_1, ..., \lambda_N\}, \\ B^* &= P^{-1}B = [b_1^*, ..., b_N^*]^T, \\ C^* &= CP. \end{aligned} \tag{30}$$

Assume the homogenous initial condition $Q_n(0) = 0, n = 1, ..., N$ and the control in the shape of the Heaviside function: $u(t) = 1(t)$. Then the $Q_n^*(t)$ component of the transformed state Eq. (29) is as follows:

$$Q_n^*(t) = \frac{E_\alpha(\lambda_n t) - 1}{\lambda_n} b_n^*, \ n = 1, .., N. \tag{31}$$

In (31) $E_\alpha(..)$ denotes the one parameter Mittag-Leffler function. Consequently the step response in the $n - th$ point of the rod takes the following form:

$$y(t) = C^* Q^*(t). \tag{32}$$

where C^* is expressed by (30), $Q^*(t) = [Q_1^*(t), ..., Q_n^*(t)]^T$, $Q_n^*(t)$ is described by (31).

5 Experimental Results

Experiments were executed with the use of the experimental system shown in Fig. 2. The length of rod is equal 260 [mm]. The control signal in the system is the standard current 0–20 [mA] given from analog output of the PLC. This signal is amplified to the range 0–1.5 [A] and it is the input for the heater. The temperature distribution along the rod is measured using standard Pt-100 RTD sensors. In the considered case the size and location of sensors are following:

$$\begin{cases} x = 0.29: & x_1 = 0.26, \ x_2 = 0.32 \\ x = 0.50: & x_1 = 0.47, \ x_2 = 0.53 \\ x = 0.73: & x_1 = 0.70, \ x_2 = 0.76 \end{cases}$$

Signals from the sensors are directly read by analog inputs of the PLC in Celsius degrees. Data from PLC are collected by SCADA application. The whole system is connected via PROFINET. The temperature distribution with respect to time and length is shown in Fig. 3. The step response of the model was tested in time range from 0 to $T_f = 300$ [s] with sample time 1 [s], parameters were calculated via minimization of the MSE (mean square error) cost function (33) using MATLAB *fminsearch* function.

To accuracy estimation the typical MSE cost function was applied:

$$MSE = \frac{1}{3K_s} \sum_{j=1}^{3} \sum_{k=1}^{K_s} \left(y_{e_j}^+(k) - y_j^+(k) \right)^2. \tag{33}$$

In (33) K_s denotes the number of collected samples for one sensor, $y_{e_j}^+(k)$ and $y_j^+(k)$ are step responses of plant and model in k-th time moments and at j-th output, $j = 1, 2, 3$.

The parameters a_w, R_a, α and β were estimated via minimization the cost function (33) using MATLAB function *fminsearch*. Results are given in the Table 1. The step response of the proposed model compared to experimental results is shown in the Fig. 4.

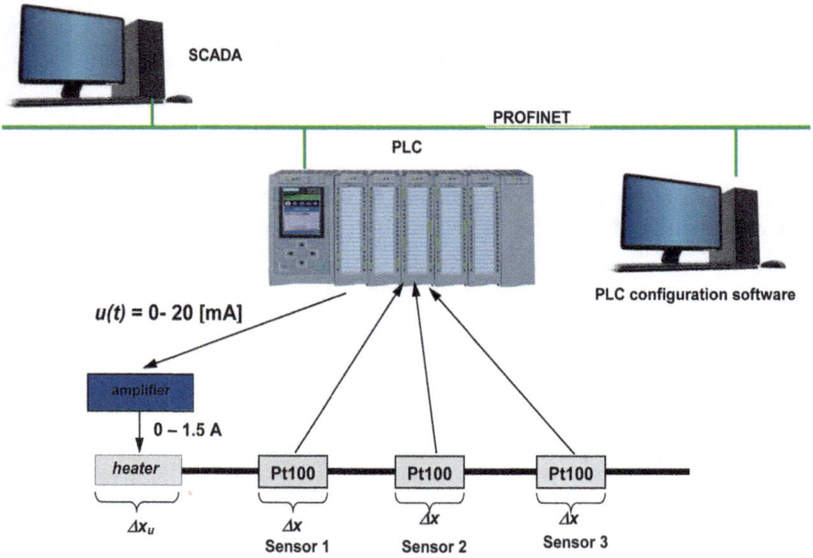

Fig. 2. The construction of the experimental system

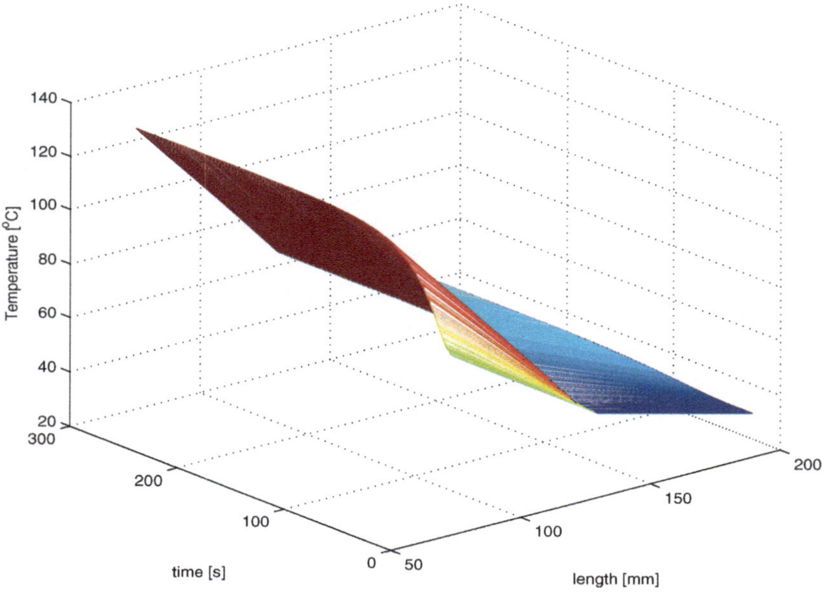

Fig. 3. The spatial-time temperature distribution in the plant

Table 1. Parameters of the model for different orders N

N	a_w	R_a	α	Cost function (33)
8	0.0003	0.0371	0.9614	0.1495
10	0.0003	0.1601	0.9289	0.0257
12	0.0003	0.0430	0.9260	0.0479
14	0.0002	0.0638	0.8791	0.0627
16	0.0001	0.1073	0.8525	0.0480
18	0.0003	0.0528	0.8999	0.0291
20	0.0002	0.0747	0.8600	0.0822
22	0.0003	0.0532	0.9033	0.0279
24	0.0003	0.0584	0.8877	0.0343
26	0.0003	0.0555	0.8927	0.0308
28	0.0003	0.0568	0.8870	0.0415
30	0.0003	0.0491	0.9111	0.0334

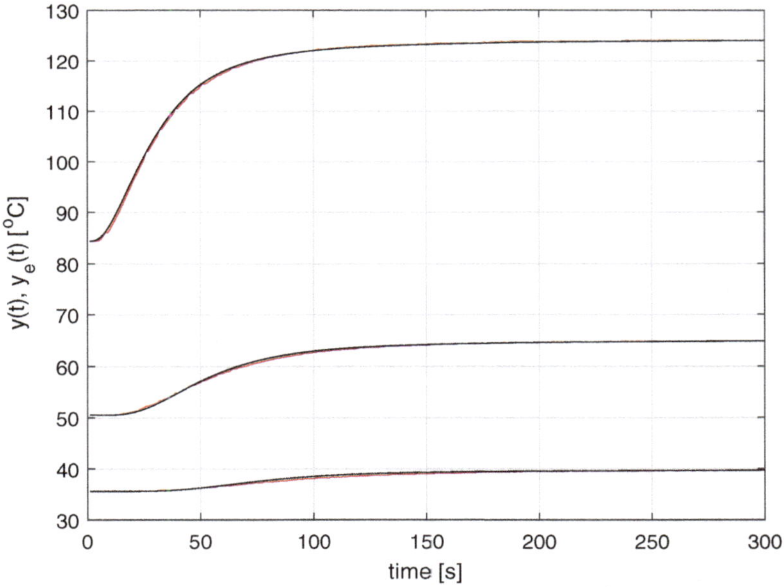

Fig. 4. Comparison model vs experiment for $N = 22$, red line - experiment, black line - FDC model.

Next the comparison of the proposed model to the analogical model using semigroup approach need to be shown. The fractional, "semigroup", infinite dimensional model (FID model) is given in papers: [18,19,22]. In the considered case only model using the integer order derivative with respect to length should

Table 2. Cost function MSE (33) for different N and compared models

N	FID model	FDC model
8	0.1434	0.1495
10	0.0801	0.0257
12	0.1315	0.0479
14	0.0743	0.0627
16	0.0646	0.0480
18	0.0748	0.0291
20	0.0504	0.0822
22	0.0242	0.0279
24	0.0347	0.0343
26	0.0382	0.0308
28	0.0264	0.0415
30	0.0283	0.0334

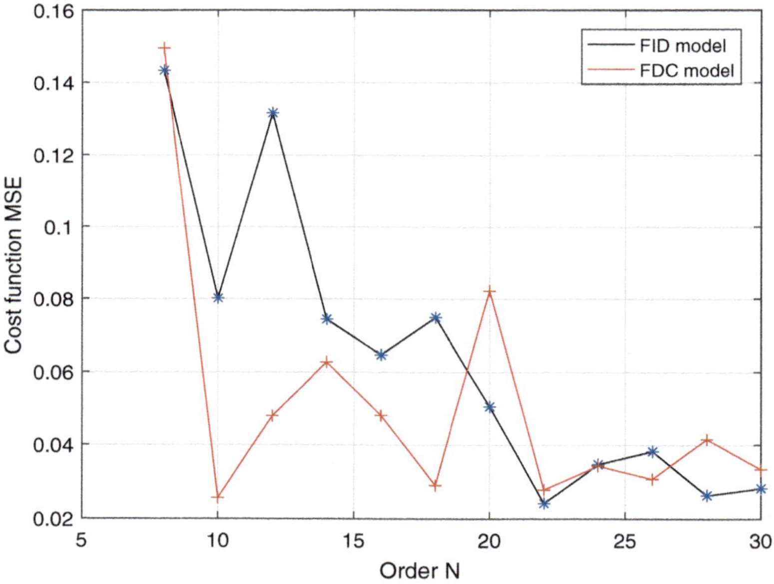

Fig. 5. Accuracy and convergence of infinite dimensional vs discrete-continuous model.

be considered. Results are shown in Table 2 and illustrated by Fig. 5. Numerical results employed here can be found in the paper [22], Table 1. From the Fig. 5 it can be concluded that the convergence of the proposed, FDC model is a little bit better than FID model. This implies that for lower orders of model $N = 10, ..., 18$

(size of state equation) the proposed model assures better accuracy in the sense of the cost function (33) than FID model. For higher orders, $N > 18$ the accuracy of the both models is practically the same, the infinite-dimensional model seems to be a little bit better.

6 Final Conclusions

The main final conclusion from the paper is that the proposed model is able to precisely describe the considered heat plant. The accuracy and convergence are satisfying. It assures good accuracy for relatively low order.

The spectrum of interesting problems associated to the proposed model is broad. Firstly deeper analysis of accuracy and convergence should be done with respect to approach given in [20,21]. The next issue is to employ the discrete fractional Riesz operator to describe the derivative along the length in the model. The positivity should be also analysed as well as the construction of fully discrete model with respect to the both time and space coordinates. An another interesting issue is comparing the Nelder-Mead simplex method to other optimization methods during parameter identification. It may lead to better results. Biologically inspired methods, for example PSO or GWO can be applied here.

Acknowledgment. This paper was sponsored by AGH project no 16.16.120.773.

References

1. Almeida, R., Torres, D.F.M.: Necessary and sufficient conditions for the fractional calculus of variations with Caputo derivatives. Commun. Nonlinear Sci. Numer. Simul. **16**(3), 1490–1500 (2011)
2. Atangana, A., Baleanu, D.: New fractional derivatives with non-local and non-singular kernel: theory and application to heat transfer. Therm. Sci. **20**(2), 763–769 (2016)
3. Caponetto, R., Dongola, G., Fortuna, L., Petras, I.: Fractional order systems: modeling and control applications. In: Chua, L.O. (ed.) World Scientific Series on Nonlinear Science, pp. 1–178. University of California, Berkeley (2010)
4. Das, S.: Functional Fractional Calculus for System Identyfication and Control. Springer, Heidelberg (2010)
5. Dlugosz, M., Skruch, P.: The application of fractional-order models for thermal process modelling inside buildings. J. Building Phys. **1**(1), 1–13 (2015)
6. Dzielinski, A., Sierociuk, D., Sarwas, G.: Some applications of fractional order calculus. Bull. Pol. Acad. Sci. Tech. Sci. **58**(4), 583–592 (2010)
7. Gal, C., Warma, M.: Elliptic and parabolic equations with fractional diffusion and dynamic boundary conditions. Evol. Equ. Control Theory **5**(1), 61–103 (2016)
8. Kaczorek, T.: Selected Problems of Fractional Systems Theory. Springer, Heidelberg (2011)
9. Kaczorek, T.: Singular fractional linear systems and electrical circuits. Int. J. Appl. Math. Comput. Sci. **21**(2), 379–384 (2011)

10. Kaczorek, T.: Reduced-order fractional descriptor observers for a class of fractional descriptor continuous-time nonlinear systems. Int. J. Appl. Math. Comput. Sci. **26**(2), 277–283 (2016)
11. Kaczorek, T., Rogowski, K.: Fractional Linear Systems and Electrical Circuits. Bialystok University of Technology, Bialystok (2014)
12. Kochubei, A.: Fractional-parabolic systems. Preprint arXiv:1009.4996 [math.ap] (2011)
13. Mitkowski, W.: Approximation of fractional diffusion-wave equation. Acta Mechanica et Automatica **5**(2), 65–68 (2011)
14. Oprzedkiewicz, K.: The discrete-continuous model of heat plant. Automatyka **2**(1), 35–45 (1998). (in Polish)
15. Oprzedkiewicz, K.: The interval parabolic system. Arch. Control Sci. **13**(4), 415–430 (2003)
16. Oprzedkiewicz, K.: A controllability problem for a class of uncertain parameters linear dynamic systems. Arch. Control Sci. **14**(1), 85–100 (2004)
17. Oprzedkiewicz, K.: An observability problem for a class of uncertain-parameter linear dynamic systems. Int. J. Appl. Math. Comput. Sci. **15**(3), 331–338 (2005)
18. Oprzedkiewicz, K., Gawin, E.: A non-integer order, state space model for one dimensional heat transfer process. Arch. Control Sci. **26**(2), 261–275 (2016)
19. Oprzedkiewicz, K., Gawin, E., Mitkowski, W.: Modeling heat distribution with the use of a non-integer order, state space model. Int. J. Appl. Math. Comput. Sci. **26**(4), 749–756 (2016)
20. Oprzedkiewicz, K., Mitkowski, W.: A memory-efficient noninteger-order discrete-time state-space model of a heat transfer process. Int. J. Appl. Math. Comput. Sci. (AMCS) **28**(4), 649–659 (2018)
21. Oprzedkiewicz, K., Mitkowski, W., Gawin, E., Dziedzic, K.: The Caputo vs. Caputo-Fabrizio operators in modeling of heat transfer process. Bull. Pol. Acad. Sci. Tech. Sci. **66**(4), 501–507 (2018)
22. Oprzędkiewicz, K., Gawin, E., Mitkowski, W.: Parameter identification for non integer order, state space models of heat plant. In: MMAR 2016: 21st International Conference on Methods and Models in Automation and Robotics, Międzyzdroje, Poland, 29 August–01 September 2016, pp. 184–188 (2016)
23. Podlubny, I.: Fractional Differential Equations. Academic Press, San Diego (1999)
24. Popescu, E.: On the fractional cauchy problem associated with a feller semigroup. Math. Rep. **12**(2), 181–188 (2010)
25. Sajewski, L.: Minimum energy control of descriptor fractional discrete-time linear systems with two different fractional orders. Int. J. Appl. Math. Comput. Sci. **27**(1), 33–41 (2017)
26. Salti, N.A., Karimov, E., Kerbal, S.: Boundary-value problems for fractional heat equation involving Caputo-Fabrizio derivative. New Trends Math. Sci. **4**(4), 79–89 (2016)
27. Sierociuk, D., Skovranek, T., Macias, M., Podlubny, I., Petras, I., Dzielinski, A., Ziubinski, P.: Diffusion process modeling by using fractional-order models. Appl. Math. Comput. **257**(1), 2–11 (2015)

Global Stability of Positive Discrete-Time Standard and Fractional Nonlinear Systems with Scalar Feedbacks

Tadeusz Kaczorek and Andrzej Ruszewski(✉)

Faculty of Electrical Engineering, Bialystok University of Technology, Wiejska 45D, 15-351 Białystok, Poland
a.ruszewski@pb.edu.pl

Abstract. The global stability of positive discrete-time standard and fractional orders nonlinear systems with scalar feedbacks is investigated. New sufficient conditions for the global stability of these classes of positive nonlinear systems are established. The effectiveness of these new stability conditions is demonstrated on simple examples of positive nonlinear discrete-time systems with scalar feedbacks.

Keywords: Global stability · Fractional order systems · Positive systems · Nonlinear systems · Discrete-time systems · Feedback

1 Introduction

In positive systems inputs, state variables and outputs take only nonnegative values for any nonnegative inputs and nonnegative initial conditions [1,5,6]. Examples of positive systems are industrial processes involving chemical reactors, heat exchangers and distillation columns, storage systems, compartmental systems, water and atmospheric pollutions models. A variety of models having positive behavior can be found in engineering, management science, economics, social sciences, biology and medicine, etc. Linear positive electrical circuits have been addressed in [2,13]. An overview of state of the art in positive systems theory is given in the monographs [1,5,6,9,18].

Mathematical fundamentals of the fractional calculus are given in the monographs [9,18,21,24]. The fractional positive linear systems have been investigated in [7,8,25]. The stability problem of fractional discrete-time linear systems have been considered in [3,26] in the case of positive systems and in [22,23] in the case of standard systems. The stability of fractional positive nonlinear systems have been analyzed in [11,12,14,17]. The global stability of nonlinear continuous-time standard and fractional positive systems have been analyzed in [10]. The global stability of nonlinear systems with negative feedbacks and positive not necessary asymptotically stable linear parts has been investigated in [15,16]. The absolute stability of a class of discrete Lur'e control system on an infinite-dimensional Hilbert space has been investigated in [4].

A. Bartoszewicz et al. (Eds.): PCC 2020, AISC 1196, pp. 1167–1175, 2020.
https://doi.org/10.1007/978-3-030-50936-1_97

In this paper the global stability of discrete-time nonlinear standard and fractional positive systems with scalar feedbacks will be addressed.

The paper is organized as follows. In Sect. 2 the basic definitions and theorems concerning the positive standard and fractional discrete-time linear systems are recalled. New sufficient conditions for the global positive discrete-time standard nonlinear systems with scalar feedbacks are established in Sect. 3. Similar sufficient conditions for fractional positive discrete-time nonlinear systems are given in Sect. 4. Concluding remarks are given in Sect. 5.

The following notation will be used: $\Re$ - the set of real numbers, $\Re^{n\times m}$ - the set of $n\times m$ real matrices, $\Re_+^{n\times m}$ - the set of $n\times m$ real matrices with nonnegative entries and $\Re_+^n = \Re_+^{n\times 1}$, I_n - the $n\times n$ identity matrix.

2 Preliminaries

2.1 Positive Discrete-Time Linear Systems

Let us consider the discrete-time linear system

$$x_{i+1} = Ax_i + Bu_i, \quad i = 0, 1, \dots, \tag{1}$$

$$y_i = Cx_i + Du_i, \tag{2}$$

where $x_i \in \Re^n$, $u_i \in \Re^m$, $y_i \in \Re^p$ are the state, input and output vectors and $A \in \Re^{n\times n}$, $B \in \Re^{n\times m}$, $C \in \Re^{p\times n}$, $D \in \Re^{p\times m}$.

Definition 1 [9]. *The system (1), (2) is called (internally) positive if $x_i \in \Re_+^n$ and $y_i \in \Re_+^p$ for any initial conditions $x_0 \in \Re_+^n$ and all inputs $u_i \in \Re_+^m$, $i \in Z_+$.*

Theorem 1 [9]. *The system (1), (2) is positive if and only if*

$$A \in \Re_+^{n\times n}, \quad B \in \Re_+^{n\times m}, \quad C \in \Re_+^{p\times n}, \quad D \in \Re_+^{p\times m}. \tag{3}$$

Definition 2 [9]. *The positive system (1), (2) with $u_i = 0$ is called asymptotically stable (the matrix $A \in \Re_+^{n\times n}$ is Schur) if*

$$\lim_{i\to\infty} x_i = 0. \tag{4}$$

Theorem 2 [9]. *The positive system (1), (2) is asymptotically stable if and only if one of the equivalent conditions is satisfied:*

1. *All coefficient of the characteristic polynomial*

$$p_A(z) = \det[I_n(z+1) - A] = z^n + a_{n-1}z^{n-1} + ... + a_1 z + a_0 \tag{5}$$

are positive, i.e. $a_k > 0$ for $k = 0, 1, ..., n-1$.

2. *All principal minors of the matrix*

$$\bar{A} = I_n - A = \begin{bmatrix} \bar{a}_{11} & \dots & \bar{a}_{1n} \\ \vdots & \dots & \vdots \\ \bar{a}_{n1} & \dots & \bar{a}_{nn} \end{bmatrix} \tag{6}$$

are positive, i.e.

$$|a_{11}| > 0, \quad \begin{vmatrix} \bar{a}_{11} & \bar{a}_{12} \\ \bar{a}_{21} & \bar{a}_{22} \end{vmatrix} > 0, \; \dots, \quad \det \bar{A} > 0. \tag{7}$$

3. *There exists strictly positive vector* $\lambda^T = [\lambda_1 \cdots \lambda_n]^T$, $\lambda_k > 0$, $k = 1, \dots, n$ *such that*

$$[A - I_n]\lambda < 0. \tag{8}$$

2.2 Positive Discrete-Time Fractional Linear Systems

In this paper the following definition of the fractional discrete-time derivative will be used [9]

$$\Delta^\alpha x_i = \sum_{k=0}^{\infty} (-1)^k \binom{\alpha}{k} x_{i-k}, \tag{9}$$

where $\alpha \in \Re$ is the order of the fractional difference and

$$\binom{\alpha}{k} = \begin{cases} 1 & \text{for } k = 0 \\ \frac{\alpha(\alpha-1)\dots(\alpha-k+1)}{k!} & \text{for } k = 1, 2, \dots \end{cases} \tag{10}$$

Consider the fractional discrete-time linear system, describe by the state-space equations

$$\Delta^\alpha x_{i+1} = Ax_i + Bu_i, \quad i \in Z_+ = 0, 1, \dots, \tag{11}$$

$$y_i = Cx_i + Du_i, \tag{12}$$

where $x_i \in \Re^n$, $u_i \in \Re^m$, $y_i \in \Re^p$ are the state, input and output vectors and $A \in \Re^{n \times n}$, $B \in \Re^{n \times n}$, $C \in \Re^{p \times n}$, $D \in \Re^{p \times m}$.

Using the definition (9) we may write the Eqs. (11), (12) in the form

$$x_{i+1} = A_\alpha x_i + \sum_{k=2}^{\infty} c_k x_{i-k+1} + Bu_i, \tag{13}$$

$$y_i = Cx_i + Du_i, \tag{14}$$

where $A_\alpha = A + I_n\alpha$ and

$$c_k = (-1)^{k+1} \binom{\alpha}{k}.$$

Definition 3 [9]. *The system (13), (14) is called (internally) positive fractional system if* $x_i \in \Re_+^n$ *and* $y_i \in \Re_+^p$ *for any initial conditions* $x_0 \in \Re_+^n$ *and all inputs* $u_i \in \Re_+^m$, $i \in Z_+$.

Theorem 3 [9]. *Let* $0 < \alpha < 1$. *Then the fractional system (13), (14) is positive if and only if*

$$A + I_n\alpha \in \Re_+^{n\times n}, \quad B \in \Re_+^{n\times m}, \quad C \in \Re_+^{p\times n}, \quad D \in \Re_+^{p\times m}. \tag{15}$$

3 Standard Positive Nonlinear Systems

Consider the nonlinear feedback system shown in Fig. 1 which consists of the positive linear part, the nonlinear element with characteristic $u = f(e)$ and scalar feedbacks. The linear part is described by the equations

$$x_{i+1} = Ax_i + Bu_i, \tag{16}$$

$$y_i = Cx_i, \tag{17}$$

where $x_i \in \Re^n$, $u_i \in \Re^m$, $y_i \in \Re^p$ are the state, input and output vectors and $A \in \Re^{n\times n}$, $B \in \Re^{n\times 1}$, $C \in \Re^{1\times n}$. The characteristic of the nonlinear element is shown in Fig. 2 and it satisfies the condition

$$0 \le \frac{f(e)}{e} \le k. \tag{18}$$

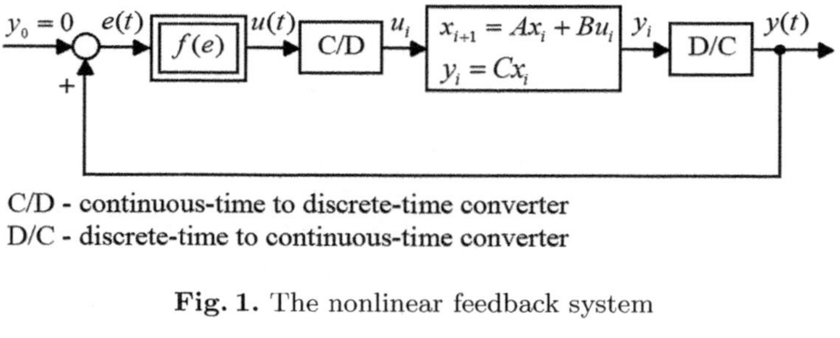

Fig. 1. The nonlinear feedback system

Theorem 4. *The standard nonlinear discrete-time system consisting of the positive linear part and the nonlinear element satisfying the condition (18) is globally stable if the matrix*

$$A_1 = A + kBC \in \Re_+^{n\times n} \tag{19}$$

is asymptotically stable.

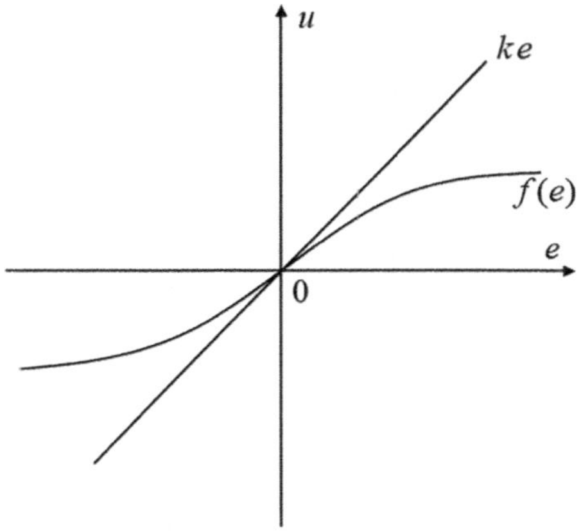

Fig. 2. The characteristic of nonlinear element

Proof. The proof will be accomplished by the use of the Lyapunov method [19, 20]. As the Lyapunov function $V(x_i)$ we choose

$$V(x_i) = \lambda^T x_i \geq 0 \quad \text{for} \quad x_i \in \Re_+^n, \quad i = 0, 1, \ldots, \tag{20}$$

where $\lambda \in \Re_+^n$ is strictly positive vector.

Using (20), (16) and (17) we obtain

$$\begin{aligned} \Delta V(x_i) &= V(x_{i+1}) - V(x_i) = \lambda^T (Ax_i + Bu_i) - \lambda^T x_i \\ &= \lambda^T (Ax_i + Bf(e_i)) - \lambda^T x_i \leq \lambda^T [(A - I_n) + kBC] x_i \end{aligned} \tag{21}$$

since $u_i = f(e_i) \leq ke_i = kCx_i$.

From (21) it follows that $\Delta V(x_i) < 0$ if the condition (19) is satisfied.

Example 1. Consider the nonlinear system with positive linear part described by (16), (17) for

$$A = \begin{bmatrix} 0.4 & 0.2 \\ 0.3 & 0.5 \end{bmatrix}, \quad B = \begin{bmatrix} 0.5 \\ 0.5 \end{bmatrix}, \quad C = \begin{bmatrix} 0.2 & 0.4 \end{bmatrix} \tag{22}$$

and the nonlinear element with characteristic shown in Fig. 2.

Using (22) and (19) we obtain

$$\begin{aligned} A_1 = A + kBC &= \begin{bmatrix} 0.4 & 0.2 \\ 0.3 & 0.5 \end{bmatrix} + k \begin{bmatrix} 0.5 \\ 0.5 \end{bmatrix} \begin{bmatrix} 0.2 & 0.4 \end{bmatrix} \\ &= \begin{bmatrix} 0.4 + 0.1k & 0.2 + 0.2k \\ 0.3 + 0.1k & 0.5 + 0.2k \end{bmatrix}. \end{aligned} \tag{23}$$

Using (5) for the matrix (23) we obtain the characteristic polynomial of the form

$$p_A(z) = \det[I_2(z+1) - A_1] = z^2 + (1.1 - 0.3k)z + 0.24 - 0.25k. \quad (24)$$

The matrix (23) satisfies the condition (19) and it is Schur matrix for $0 < k < 0.96$. Therefore, the nonlinear system is globally stable for $0 < k < 0.96$.

4 Fractional Positive Nonlinear Systems

Consider the nonlinear feedback system shown in Fig. 3 which consists of the positive linear part, the nonlinear element with characteristic $u = f(e)$ and the scalar gain feedback. The linear part is described by the equations

$$x_{i+1} = A_\alpha x_i + \sum_{k=2}^{\infty} c_k x_{i-k+1} + Bu_i, \quad (25)$$

$$y_i = Cx_i, \quad (26)$$

where $x_i \in \Re^n$, $u_i \in \Re^m$, $y_i \in \Re^p$ are the state, input and output vectors and $A \in \Re^{n\times n}$, $B \in \Re^{n\times 1}$, $C \in \Re^{1\times n}$, $A_\alpha = A + I_n\alpha$, $\alpha \in (0,\ 1)$. The characteristic of the nonlinear element is shown in Fig. 2 and it satisfies the condition (18).

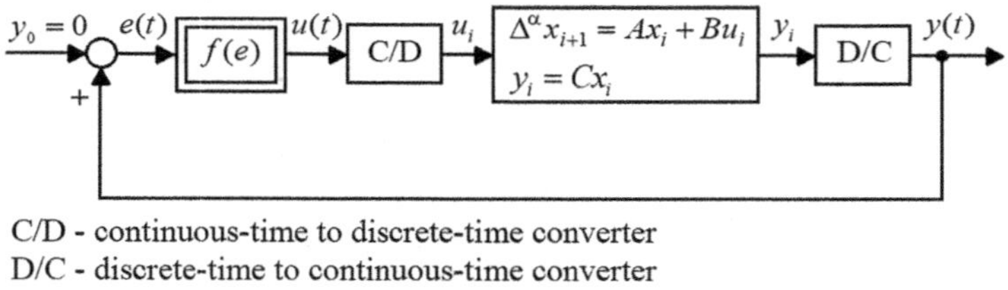

Fig. 3. The fractional nonlinear feedback system

Theorem 5. *The fractional nonlinear discrete-time system consisting of the positive linear part and the nonlinear element satisfying the condition (18) is globally stable if the matrix*

$$A_1 = A + kBC \in \Re_+^{n\times n} \quad (27)$$

is asymptotically stable.

Proof. As the Lyapunov function $V(x_i)$ we choose

$$V(x_i) = \lambda^T x_i \geq 0 \quad \text{for} \quad x_i \in \Re_+^n, \quad i = 0, 1, \ldots, \quad (28)$$

where $\lambda \in \Re_+^n$ is strictly positive vector.

Using (28), (25) and (26) we obtain

$$\begin{aligned}\Delta V(x_i) &= V(x_{i+1}) - V(x_i) = \lambda^T x_{i+1} - \lambda^T x_i \\ &= \lambda^T (A_\alpha x_i + \sum_{k=2}^{\infty} c_k x_{i-k+1} + Bu_i) - \lambda^T x_i \\ &= \lambda^T [(A_\alpha - I_n) x_i + \sum_{k=2}^{\infty} c_k x_{i-k+1} + Bf(e_i)] \\ &\leq \lambda^T \{[A_\alpha - I_n + kBC] x_i + \sum_{k=2}^{\infty} c_k x_{i-k+1}\}.\end{aligned} \tag{29}$$

Taking into account that

$$\sum_{k=2}^{\infty} c_k x_{i-k+1} \leq \sum_{k=2}^{\infty} c_k x_i = (1-\alpha) x_i$$

we obtain

$$\Delta V(x_i) = \lambda^T \{A + I_n \alpha - I_n + kBC + (1-\alpha) I_n\} x_i = \lambda^T [A + kBC] x_i < 0.$$

Example 2. Consider the nonlinear discrete-time system shown in Fig. 3 with the matrices

$$A = \begin{bmatrix} 0.5 & 0.1 \\ 0.2 & 0.4 \end{bmatrix}, \quad B = \begin{bmatrix} 0.1 \\ 0.2 \end{bmatrix}, \quad C = \begin{bmatrix} 0.1 & 0 \end{bmatrix} \tag{30}$$

and the characteristic $f(e)$ of the nonlinear element satisfying the condition (18) for $k = 1$.

Using (30) for $k = 1$ we obtain

$$A_1 = A + kBC = \begin{bmatrix} 0.5 & 0.1 \\ 0.2 & 0.4 \end{bmatrix} + \begin{bmatrix} 0.1 \\ 0.2 \end{bmatrix} [0.1 \; 0] = \begin{bmatrix} 0.51 & 0.1 \\ 0.22 & 0.4 \end{bmatrix} \in \Re_+^{n \times n}. \tag{31}$$

The matrix (31) is Schur matrix since the characteristic polynomial

$$\det(I_2 z - A_1) = \begin{vmatrix} z - 0.51 & -0.1 \\ -0.22 & z - 0.4 \end{vmatrix} = z^2 + 0.91z + 0.182 \tag{32}$$

has the zeros $z_1 = 0.613$, $z_2 = 0.297$. Therefore, the matrix (31) is asymptotically stable and the nonlinear system is globally stable for $k = 1$.

5 Concluding Remarks

The global stability of positive discrete-time standard and fractional orders nonlinear feedback systems has been investigated. New sufficient conditions for the global stability of the class of positive nonlinear systems are established. The effectiveness of these new stability conditions is demonstrated on simple examples of positive nonlinear discrete-time systems. The considerations can be extended to nonlinear positive systems with interval matrices of the linear parts.

Acknowledgment. This work was supported by National Science Centre in Poland under work No. 2017/27/B/ST7/02443.

References

1. Berman, A., Plemmons, R.J.: Nonnegative Matrices in the Mathematical Sciences. SIAM, Philadelphia (1994)
2. Borawski, K.: Modification of the stability and positivity of standard and descriptor linear electrical circuits by state feedbacks. Electr. Rev. **93**(11), 176–180 (2017). https://doi.org/10.15199/48.2017.11.36
3. Busłowicz, M., Kaczorek, T.: Simple conditions for practical stability of positive fractional discrete-time linear systems. Int. J. Appl. Math. Comput. Sci. **19**(2), 263–269 (2009). https://doi.org/10.2478/v10006-009-0022-6
4. Grabowski, P.: Absolute stability criteria for infinite-dimensional discrete Lur'e systems with application to loaded electric distortionless RLCG -transmission line. J. Differ. Equ. Appl. **19**, 304–331 (2013). https://doi.org/10.1080/10236198.2011.639366
5. Farina, L., Rinaldi, S.: Positive Linear Systems. Theory and Applications. Wiley, New York (2000)
6. Kaczorek, T.: Positive 1D and 2D Systems. Springer, London (2002)
7. Kaczorek, T.: Positive linear systems with different fractional orders. Bull. Pol. Acad. Sci. Techn. **58**(3), 453–458 (2010). https://doi.org/10.2478/v10175-010-0043-1
8. Kaczorek, T.: Positive linear systems consisting of n subsystems with different fractional orders. IEEE Trans. Circuits Syst. **58**(6), 1203–1210 (2011). https://doi.org/10.1109/TCSI.2010.2096111
9. Kaczorek, T.: Selected Problems of Fractional Systems Theory. Springer, Heidelberg (2011)
10. Kaczorek, T.: Analysis of positivity and stability of discrete-time and continuous-time nonlinear systems. Comput. Probl. Electr. Eng. **5**(1), 11–16 (2015)
11. Kaczorek, T.: Stability of fractional positive nonlinear systems. Arch. Control Sci. **25**(4), 491–496 (2015). https://doi.org/10.1515/acsc-2015-0031
12. Kaczorek, T.: Analysis of positivity and stability of fractional discrete-time nonlinear systems. Bull. Pol. Acad. Sci. Techn. **64**(3), 491–494 (2016). https://doi.org/10.1515/bpasts-2016-0054
13. Kaczorek, T.: Superstabilization of positive linear electrical circuit by state-feedbacks. Bull. Pol. Acad. Sci. Techn. **65**(5), 703–708 (2017). https://doi.org/10.1515/bpasts-2017-0075
14. Kaczorek, T.: Absolute stability of a class of fractional positive nonlinear systems. Int. J. Appl. Math. Comput. Sci. **29**(1), 93–98 (2019). https://doi.org/10.2478/amcs-2019-0007
15. Kaczorek, T.: Global stability of nonlinear feedback systems with positive linear parts. Int. J. Nonlinear Sci. Numer. Simul. **20**(5), 575–579 (2019). https://doi.org/10.1515/ijnsns-2018-0189
16. Kaczorek, T.: Global stability of positive standard and fractional nonlinear feedback systems. Bull. Pol. Acad. Sci. Techn. **68**(2), 285–288 (2020). https://doi.org/10.24425/bpasts.2020.133112
17. Kaczorek, T., Borawski, K.: Stability of positive nonlinear systems. In: 22nd International Conference Methods and Models in Automation and Robotics, Poland, pp. 564–569 (2017). https://doi.org/10.1109/MMAR.2017.8046890

18. Kaczorek, T., Rogowski, K.: Fractional Linear Systems and Electrical Circuits. Springer, Cham (2015)
19. Lyapunov, A.M.: General Problem of Stable Movement. Gostechizdat, Moscow (1963). (in Russian)
20. Leipholz, H.: Stability Theory. Academic Press, New York (1970)
21. Ostalczyk, P.: Discrete Fractional Calculus. World Scientific, River Edge (2016)
22. Ruszewski, A.: Stability of discrete-time fractional linear systems with delays. Arch. Control Sci. **29**(3), 549–567 (2019). https://doi.org/10.24425/acs.2019.130205
23. Ruszewski, A.: Practical and asymptotic stabilities for a class of delayed fractional discrete-time linear systems. Bull. Pol. Acad. Sci. Techn. **67**(3), 509–515 (2019). https://doi.org/10.24425/bpasts.2019.128426
24. Sabatier, J., Agrawal, O.P., Machado, J.A.T.: Advances in Fractional Calculus, Theoretical Developments and Applications in Physics and Engineering. Springer, London (2007)
25. Sajewski, Ł.: Decentralized stabilization of descriptor fractional positive continuous-time linear systems with delays. In: 22nd International Conference on Methods and Models in Automation and Robotics, Poland, pp. 482–487 (2017). https://doi.org/10.1109/MMAR.2017.8046875
26. Sajewski, Ł.: Stabilization of positive descriptor fractional discrete-time linear systems with two different fractional orders by decentralized controller. Bull. Pol. Acad. Sci. Techn. **65**(5), 709–714 (2017). https://doi.org/10.1515/bpasts-2017-0076

Discrete-Time Switched Models of Non-linear Fractional-Order Systems

Stefan Domek(✉)

Department of Control Engineering and Robotics, West Pomeranian University of Technology Szczecin, ul. Sikorskiego 37, 70-313 Szczecin, Poland
stefan.domek@zut.edu.pl

Abstract. In the paper methods for modeling complex, non-linear dynamical systems of non-integer order, using the so-called switched models that are based on the dynamic change of local linear models, depending on the value of an appropriately chosen switching function are discussed. Such a multimodel approach enables a relatively simple description of properties of many complex processes encountered in technology, especially in electrical engineering, automation and robotics, but also in nature, biology, medicine and, for example, in economics. Although the theory of switched systems has been developing intensively since over a dozen or so years, many issues and problems have not been solved yet. This is particularly true for systems of non-integer order. In the paper three type of discrete-time dynamical switched models of fractional systems: Fuzzy Takagi-Sugeno Model (FO FTS), Piecewise Linear Model (FO PWL) and a new one – Mixed Logical Dynamical Model (FO MLD) are proposed. Some examples of applications of such models in control are given.

Keywords: Switched models · Non-integer order calculus · Fractional order systems · MLD models

1 Introduction

The vast majority of the phenomena, processes and objects surrounding us show a non-linear nature. It is often so much strong that it is not enough to use linearized models to describe adequately their properties. In turn, determining and subsequent use of non-linear models is usually very difficult. This applies to phenomena and processes from various fields of life, science and technology, for example, biological, social, socio-cognitive, economic, transport, information and many technological processes and systems, not to mention control and monitoring systems, including solutions based on artificial intelligence.

One of the methods to overcome these difficulties consists in replacing a complex non-linear model by a set of local linear models valid for small areas around various operating points. The idea of this approach boils down to switching over active models in time so that the generalized modeling error does not exceed specified bounds (for example, in terms of a chosen norm) and to obtain a reduction in the computational

A. Bartoszewicz et al. (Eds.): PCC 2020, AISC 1196, pp. 1176–1188, 2020.
https://doi.org/10.1007/978-3-030-50936-1_98

complexity of the description at the same time. By this means a non-linear time-invariant process (NLTI) can be treated as a linear time-variant one (LTV). The instantaneous properties of the modeled non-linear process are then described by a quasilinear switched model consisting of a set of local linear submodels [1].

The switched systems have been the subject of intensive studies for decades [2]. It has been shown that they can model effectively various complex dynamical systems, including systems with perturbed parameters, chaos, multiple limit cycles and others. They also make it possible to better analyze systems present in modern technology, such as adaptive wide area networks, fault-tolerant systems, systems with multiple sampling periods, etc. It has also been shown that there is a large class of non-linear control processes that can be stabilized by switched local linear controllers, whereas this cannot be done by static state feedback [3].

On the other hand, it should be noted that also the theory of fractional order systems (FOS) has been intensively developed for the past decades [4, 5]. As follows from the research works conducted worldwide, the system description by means of non-integer differential equations is one of the more effective modeling methods and opens up new possibilities of modeling real properties of many complex phenomena and industrial processes. In automation and robotics, as in the case of integer-order models, the description by means of fractional-order models can be used indirectly for tuning or directly for synthesis of linear control algorithms [6, 7].

However, determining and subsequent use of non-linear fractional-order models is very difficult, if not more so than in the case of integer-order processes. Therefore, an idea was conceived to replace complex fractional-order non-linear time-invariant models (FO NLTI) by switched fractional-order models made up of a set comprising fractional-order linear time-variant models (FO LTV) [8].

The paper is structured as follows: in Sect. 2 the basics of fractional-order difference calculus and, based on them, discrete-time dynamical state space models of non-integer order are recalled. Then, the concept of switched models, their types and areas of use are reminded; in Sect. 3 selected non-integer order discrete-time switched state space models are defined, and examples of the use of discrete-time non-integer order switched models in control are given. Finally, the whole is summarized in Conclusions.

2 The Basic Concept

2.1 The Non-integer Order Difference Calculus

Let us consider the known integer-order $n \geq 0$ differential operator of the function $f(t),\ t \in \mathbb{R},\ t \geq 0,\ {}_{t_0}D_t^n f(t) = \frac{d^n f(t)}{dt^n}$, which is equivalent to the n-fold integral operator on the interval $[t_0, t]$, ${}_{t_0}I_t^n f(t) = {}_{t_0}D_t^{-n} f(t), n > 0$.

The generalized differential operator of $\alpha \in \mathbb{R}$ order of the function on the interval $[t_0, t]$ can be written as ${}_{t_0}D_t^{\alpha} f(t)$ with the remark that the generalized integral operator on the interval $[t_0, t]$ is written as ${}_{t_0}I_t^{\alpha} f(t)$, where ${}_{t_0}I_t^{\alpha} f(t) = {}_{t_0}D_t^{-\alpha} f(t)$.

There are known several definitions of the operator ${}_{t_0}D_t^{\alpha} f(t)$ proposed by various researchers, which differ in properties and/or the range of applicability. However, the most popular and most adopted are three of them: Riemann-Liouville's, Caputo's and

Grünwald-Letnikov's, ones [5, 9]. The Grünwald-Letnikov definition of the fractional-order derivative is particularly popular for reasons of application, especially to digital control systems, where it is natural to use discretized function values $f(t)$ taken with a sampling period for the purpose of computations h.

Definition 1 [4]. A derivative of fractional order $\alpha \in \mathbb{R}$ of function $f(t)$ is defined according to Grünwald and Letnikov as follows

$$ {}_{t_0}^{GL}D_t^{\alpha} f(t) = \lim_{h \to 0} h^{-\alpha} \sum_{j=0}^{\left\lfloor \frac{t-t_0}{h} \right\rfloor} c_j^{\alpha} f(t - jh) \tag{1} $$

where the symbol $\lfloor \cdot \rfloor$ denotes the integer part,

$$ c_j^{\alpha} = (-1)^j \binom{\alpha}{j}, \quad j = 0, 1, 2, \ldots \quad \text{or} \quad c_j^{\alpha} = \left(1 - \frac{1+\alpha}{j}\right) c_{j-1}^{\alpha}, \quad c_0^{\alpha} = 1 \tag{2} $$

and the so-called generalized Newton symbol is given by

$$ \binom{\alpha}{j} = \begin{cases} 1 & \text{for } j = 0 \\ \frac{\alpha(\alpha-1)\ldots.(\alpha-j+1)}{j!} & \text{for } j = 1, 2, 3, \ldots \end{cases} \tag{3} $$

In practice computer control systems are most commonly used, i.e. discrete control algorithms and discrete models of controlled plants are considered, and thereby discrete functions defined at discrete time instants $t \in \mathbb{Z}$. In such a case the fractional-order difference calculus represents an equivalent of the fractional-order differential calculus. Hence, based upon (1), the following definition may be introduced:

Definition 2 [10]. A discrete fractional-order difference of a discrete function is defined by

$$ {}_{t_0}\Delta_t^{\alpha} f(t) = \sum_{j=0}^{t-t_0} c_j^{\alpha} f(t - j), \quad \alpha \in \mathbb{R}, \quad t \in \mathbb{Z} \tag{4} $$

with the most commonly adopted simplified notation $t_0 = 0$ as

$$ \Delta^{\alpha} f(t) = \sum_{j=0}^{t} c_j^{\alpha} f(t - j) \tag{5} $$

If it is considered that the number of summands in the sum (5) is unavoidably finite in practice, then the following approximation is adopted most commonly

$$ \Delta^{\alpha} f(t) \approx \sum_{j=0}^{L} c_j^{\alpha} f(t - j) \tag{6} $$

where the number L of the samples $f(t)$ (equivalent to the length of memory where the samples are stored in practical realizations) should be chosen so that the truncation error does not exceed a given value. It is possible to satisfy, taking into account that the coefficients (2) decrease with the increase of j, that is the effect of samples being distant in time is becoming smaller and smaller.

2.2 Discrete-Time Dynamic State Space Models of Non-integer Order

Dynamic systems of non-integer order can be modeled in many ways, with transfer function and state space descriptions being most popular.

The nonlinear discrete-time model of fractional order in state variables can be introduced on the basis of the integer-order model [4]:

$$x(t+1) = \boldsymbol{f}(x(t), u(t)), \quad t \in \mathbb{Z} \tag{7}$$

Denoting

$$\boldsymbol{f}_d(x(t), u(t)) = \boldsymbol{f}(x(t), u(t)) - x(t) \tag{8}$$

we get

$$\Delta^1 x(t+1) = \boldsymbol{f}_d(x(t), u(t)) \tag{9}$$

Hence, by analogy, we can write:

Definition 3 [10]. A nonlinear discrete-time model of fractional order α in state variables is given by nonlinear state and output equations

$$\Delta^\alpha x(t+1) = \boldsymbol{f}_d(x(t), u(t)), \quad x(0) = x_0, \quad t \in \mathbb{Z} \tag{10}$$

$$y(t) = \boldsymbol{g}(x(t), u(t)) \tag{11}$$

where the individual vectors $x(t) \in X \subseteq \mathbb{R}^n$, $u(t) \in U \subseteq \mathbb{R}^m$, $y(t) \in Y \subseteq \mathbb{R}^p$ denote the model state, input and output, respectively, x_0 is the initial state and $t \in \mathbb{Z}$ denotes a discrete-time independent variable (consecutive sample instants).

In the linear case, by analogy with integer-order models, it may be introduced the definition of linear discrete-time models of fractional order:

Definition 4 [10]. A linear discrete-time model of fractional order α in state variables is given by the state and output equations

$$\Delta^\alpha x(t+1) = \boldsymbol{A_d} x(t) + \boldsymbol{B} u(t), \quad x(0) = x_0, \quad t \in \mathbb{Z} \tag{12}$$

$$y(t) = \boldsymbol{C} x(t) + \boldsymbol{D} u(t) \tag{13}$$

where $\boldsymbol{A} \in \mathbb{R}^{n \times n}$ is the state matrix, $\boldsymbol{B} \in \mathbb{R}^{n \times m}$, $\boldsymbol{C} \in \mathbb{R}^{p \times n}$ are the input and output matrices, matrix $\boldsymbol{D} \in \mathbb{R}^{p \times m}$ is equal to zero in the most common case,

$$\boldsymbol{A}_d = \boldsymbol{A} - \boldsymbol{I}_n \tag{14}$$

is the so-called complemented state matrix and $\boldsymbol{I}_n \in \mathbb{R}^{n \times n}$ denotes the identity matrix.

According to (6), we can write the approximated linear discrete-time model of fractional order α in state variables (12) as:

$$x(t+1) = \boldsymbol{A}_d(t)x(t) + \boldsymbol{B}(t)u(t) - \sum\nolimits_{i=1}^{L} c_i^\alpha(t)x(t+1-j) \tag{15}$$

2.3 Concept of Switched Models

The method of describing a non-linear process by a set of switched linear local models assumes that the entire operating area of the process is divided into subareas for which local linear models are to be determined. In switched models a special supervisory system determines an appropriately selected switching signal that defines the instantaneous degree of activity of each local model. That is why they are classified with the so-called hybrid systems that combine dynamic systems with logic ones [11, 12].

Elements of the lower layer that model local properties of the process are described by relationships typical of linear dynamic discrete-time systems, while elements of the higher layer, which governs the operation of the whole model, are described by laws of logic. In the basic approach these are classic binary logic laws, in the broader approach these are also fuzzy logic laws [10, 13]. The decision variable is given by the following definition:

Definition 5 [12]. Let $\Sigma = \{(t_0, j_0), \ldots, (t_k, j_k), \ldots\}, t_k \in \mathbb{Z}, j_k \in \mathbb{Z}_+$, denote a switching sequence of S local submodels, where $0 \le t_0 < t_1 < t_2 \ldots < t_k < \ldots$, with t_0 being the initial time instant, t_k being the k-th switching time instant, and j_k being the k-th set of active submodels.

The switching signal $\sigma(t)$ is of the form of a piecewise constant vector of weighting variables $\sigma_j(t)$

$$\sigma(t) = \begin{bmatrix} \sigma_1(t) \\ \sigma_2(t) \\ \vdots \\ \sigma_S(t) \end{bmatrix} \quad \text{for} \quad t \in [t_k, t_{k+1}) \qquad \sum_{j=1}^{S} \sigma_j(t) = 1 \tag{16}$$

with

$\sigma_j(t) \in \{0, 1\}$ for models switched according to classic binary logic laws,
$\sigma_j(t) \in [0 \ldots 1]$ for models switched according to fuzzy logic laws.

The switching signal $\sigma(t)$ can be in general a function of time, its past values, state variables, outputs and inputs of the modeled process, as well as a function of an external auxiliary signal. In practice, there can be distinguished various particular classes of switched models depending on the specific form of the switching law [12].

The need for an adequate description of switched systems in automation was recognized a few dozen years ago, when systems containing relays and other elements described by logic laws, important from an engineering point of view, were considered. They were commonly used, for example, in mechanical and power systems with on/off switches or valves, in transport systems with gears or speed selectors, etc., where discontinuous two-position and three-position controllers and continuous PID controllers with variable structure found wide application. It turned out then that there exists a large class of non-linear systems that can be stabilized by switching the control algorithm, but they cannot be stabilized through a continuous static state feedback. Moreover, it also turned out that switched systems have some specific features. For example, the system as a whole may not be stable for some switching strategies, even when all switched subsystems are stable. And vice versa, proper selection of the switching strategy can provide stability to the system in which the switched subsystems are unstable [14].

The above properties of switched systems are also the reason that formulating conditions for their stability, controllability or observability, being difficult in general, depends additionally on the scenario of changes and restrictions imposed on the switching signal. From this point of view, switched systems can be divided into two groups. Switched systems where no restrictions are imposed on time variability of the switching signal $\sigma(t)$ are called the systems with arbitrary switching. Otherwise, switched systems are called the systems with restricted switching, i.e. the systems in which restrictions are imposed on the switching signal $\sigma(t)$ relating to, for example, switching instants, minimum and maximum length of time between switching instants, sequences of changes in value, membership of a specific class of function.

2.4 Switched Discrete-Time Dynamic State Space Models of Non-integer Order

Taking into account (13), (15) and (16) we can write the instantaneous linear discrete-time local state space model of non-integer order in the form:

$$x(t+1) = \boldsymbol{A}_d(t)|_{\sigma(t)} x(t) + \boldsymbol{B}(t)|_{\sigma(t)} u(t) - \sum_{i=1}^{L} c_i^{\alpha|_{\sigma(t)}}(t) x(t+1-j) \tag{17}$$

$$y(t) = \boldsymbol{C}(t)|_{\sigma(t)} x(t), \qquad t \in \mathbb{Z} \tag{18}$$

where $\cdot|_{\sigma(t)}$ denotes the value of the matrix (or parameter) at the time instant t, dependent in a known way on the instantaneous switching signal $\sigma(t)$.

The main idea in the methods of switched modeling of the non-linear discrete-time fractional-order systems is the assumption that the linear model (17), (18) accurately describes the local, instantaneous dynamic properties of the non-linear system, as well as the non-linear model (10), (11).

However, it is important to define the method of the selection of active local models described by the matrices $\boldsymbol{A}_d(t)|_{\sigma(t)}$, $\boldsymbol{B}(t)|_{\sigma(t)}$, $\boldsymbol{C}(t)|_{\sigma(t)}$ and non-integer orders $\alpha|_{\sigma(t)}$.

3 Fractional-Order Discrete-Time Switched State Space Models

As was written above, modern technologies have caused a considerable interest in the study of dynamic non-linear fractional-order processes or of hybrid processes of a heterogeneous discrete-time (or continuous-time), non-integer order dynamics, and of logical nature at the same time. In such cases, the use of switched models allows us to approximate a complex system description with a set of simpler linear difference (differential) equations. According to an adopted method of approximation, various switched model can be obtained. Fuzzy Takagi-Sugeno models (FTS), Piecewise Linear models (PWL) or Piecewise Affine models and Mixed Logical Dynamical models (MLD) are the most popular for the integer order systems [10, 11]. In the next subsection such models for non-integer order processes are introduced.

3.1 Fractional-Order Switched Fuzzy Takagi-Sugeno Models (FO FTS)

In classic fuzzy integer-order Takagi-Sugeno models, the antecedents of rules are defined by the laws of fuzzy logic, while the consequents take the form of linear equations describing the components of local models. The structure of the equations is identical but their parameters are different. Depending on the adopted types of fuzzy logic operators AND and OR, a quasi-linear switching model is obtained, the structure of which is constant, and its parameters are continuously adjusted within the range determined by local models adopted in the consequents [13].

The proposed here discrete-time switched fractional-order fuzzy Takagi-Sugeno model has the form (17), (18), with

$$\begin{aligned} \boldsymbol{A}_d(t)|_{\sigma(t)} &= \sum\nolimits_{j=1}^{S} \tilde{\sigma}_j(t)\boldsymbol{A}_{d,j} \\ \boldsymbol{B}(t)|_{\sigma(t)} &= \sum\nolimits_{j=1}^{S} \tilde{\sigma}_j(t)\boldsymbol{B}_j \\ \boldsymbol{C}(t)|_{\sigma(t)} &= \sum\nolimits_{j=1}^{S} \tilde{\sigma}_j(t)\boldsymbol{C}_j \\ c_i^{\alpha|_{\sigma(t)}} &= \sum\nolimits_{j=1}^{S} \tilde{\sigma}_j(t)c_{i,j}^{\alpha} \end{aligned} \tag{19}$$

where the weight coefficients $\tilde{\sigma}_j(t)$ at the time-instant t is defined as

$$\tilde{\sigma}_j(t) = \frac{\sigma_j(t)}{\sum_{i=1}^{S}\sigma_i(t)}, \quad j = 1, 2, \ldots, S, \quad \sum\nolimits_{j=1}^{S} \tilde{\sigma}_j(t) = 1 \tag{20}$$

with $\sigma_j(t)$ determining the so-called degree of activation of individual models.

Rules of Local Models Selection

Switching of S local discrete-time fractional-order models is determined by defining of S fuzzy rules $R^j,\ j \in [1, S]$ with an antecedent in the form of a fuzzy logical product for each j-th fuzzy rule:

$$R^j\text{: IF } \{z_1(t) \subset M_{i,1} \text{ AND } z_2(t) \subset M_{i,2} \text{ AND } \cdots \text{ AND } z_q(t) \subset M_{i,q}\} \tag{21}$$

and a consequent determined by means of the fuzzy product operator:

$$\sigma_j(t) = \prod\nolimits_{i=1}^{m_k} \mu_{M_{i,k}}(z_k(t)) \tag{22}$$

where $z_k(t) \subset M_{i,k}$ denotes membership of the so-called premise variable $z_k(t)$ to the fuzzy set $M_{i,k}$ from among m_k collections with membership function $\mu_{M_{i,k}}(z_k(t))$:

$$\forall z_k \ \exists\left\{\mu_{M_{1,k}}, \mu_{M_{2,k}}, \ldots, \mu_{M_{m_k,k}}\right\} \sum\nolimits_{i=1}^{m_k} \mu_{M_{i,k}}(z_k(t)) = 1 \cap \mu_{M_{i,k}}(z_k(t)) \geq 0 \tag{23}$$

As elements of the premise variable vector:

$$z(t) = \left[z_1(t), z_2(t), \ldots, z_q(t)\right]^T \tag{24}$$

process states $x_k(t)$, inputs $u_k(t)$ and/or other measurable external variables which may constitute grounds for making decisions under the fuzzy rules system can be used.

Example

Let us consider the non-linear flexible inverted pendulum [7]

$$D_t^1 x_1(\mathrm{t}) = x_2(\mathrm{t}) \tag{25}$$

$$D_t^{1.2} x_2(t) = \frac{g\ \sin(x_1(t)) - aml(x_2(t))^2 \sin(2x_1(t))/2 - a\ \cos(x_1(t))u}{4l/3 - aml\ \cos^2 x_1(t)}$$

where $x_1(t)$ denotes the angle of the pendulum from the vertical and $x_2(t)$ is the angular velocity, g is the gravity constant, m is the mass of the pendulum, m_c is the mass of the cart, $2l$ is the length of the pendulum, and and u is the force applied to the cart; $a = 1/(m_c + m)$, (Fig. 1):

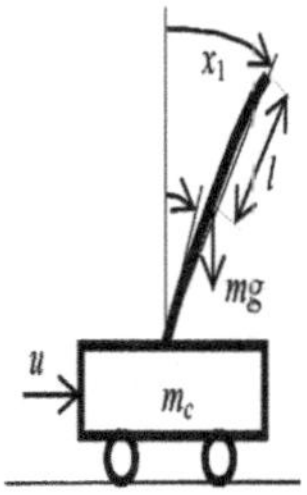

Fig. 1. Inverted pendulum with a flexible pole.

Such a non-linear flexible pendulum can be modelled by the proposed FO FTS model. Assuming two fuzzy sets and individual membership functions $M_{1,1}$, $M_{2,1}$ with process state x_1 as the premise variable (Fig. 2):

$$M_{1,1} = -\frac{1}{1 + \exp[-15(x_1 - \pi/8)]} + \frac{1}{1 + \exp[-15(x_1 + \pi/8)]}$$

$$M_{2,1} = 1 - M_{1,1} \tag{26}$$

we get two linear local discrete-time fractional-order models:

– around angle $x_1 = 0$

$$\boldsymbol{A}_{d1} = \begin{bmatrix} -1 & 1 \\ \frac{g}{4l/3 - aml} & -1 \end{bmatrix}, \quad \boldsymbol{B}_1 = \begin{bmatrix} 0 \\ \frac{-a}{(4l/3 - aml)} \end{bmatrix}$$

$$\boldsymbol{C}_1 = \begin{bmatrix} 1\,0 \\ 0\,1 \end{bmatrix}, \quad \left\{\alpha_{1,1},\ \alpha_{2,1}\right\} = \{1.0,\ 1.2\} \tag{27}$$

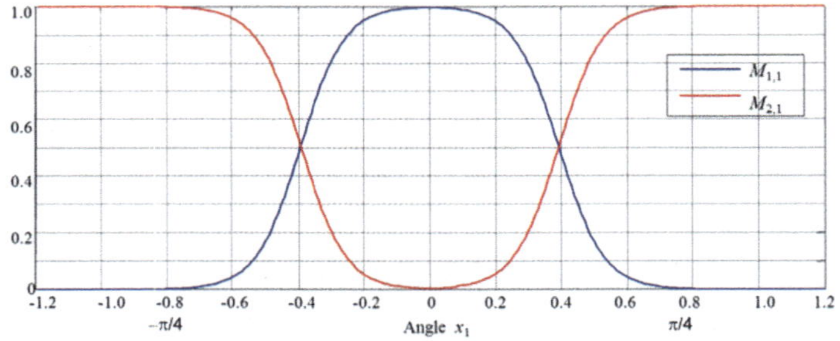

Fig. 2. The fuzzy sets and individual membership functions $M_{1,1}, M_{2,1}$.

– around angle $x_1 = \pm\pi/4$

$$A_{d2} = \begin{bmatrix} -1 & 1 \\ \frac{4g\sin(\pi/4)}{\pi(4l/3 - aml\cos^2(\pi/4))} & -1 \end{bmatrix}, \quad B_2 = \begin{bmatrix} 0 \\ \frac{-a\cos(\pi/4)}{(4l/3 - aml\cos^2(\pi/4))} \end{bmatrix}$$

$$C_2 = \begin{bmatrix} 1 & 0 \\ 0 & 1 \end{bmatrix}, \quad \{\alpha_{1,2}, \alpha_{2,2}\} = \{1.0, 1.2\} \tag{28}$$

It should be noted that the number of rules and therefore local models can be increased, e.g. to 4, for greater accuracy of the object (25) approximation.

3.2 Fractional-Order Switched Piecewise Linear Models (FO PWL)

The proposed discrete-time switched fractional-order piecewise linear model has the form (17), (18), with

$$A_d(t)|_{\sigma(t)} = A_{d,j}$$

$$B(t)|_{\sigma(t)} = B_j$$

$$C(t)|_{\sigma(t)} = C_j$$

$$c_i^{\alpha|_{\sigma(t)}} = c_{i,j}^{\alpha} \tag{29}$$

where $j \in \{1, 2, \ldots, S\}$ is the subscript of the switched function $\sigma_j(t)$ for which the selection of local models condition is true at the time-instant t.

Rules of Local Models Selection

For piecewise linear models the linear space $P = U \times X$ made up of the input space $u(t) \subset U \subseteq \mathbb{R}^m$ and the state space $x(t) \subset X \subseteq \mathbb{R}^n$ of the nonlinear process is divided into S convex polyhedrons $P_1, P_2, \ldots, P_S$:

$$P_j \subset U_j \times X_j \subseteq P, \quad P = \bigcup_{i=1}^{S} P_i, \quad \bigcap_{i=1}^{S} P_i = \emptyset \tag{30}$$

defined by matrices S_j^x, S_j^u, S_j^0 and linear matrix inequalities (LMI)

$$S_j^x x(t) + S_j^u u(t) \le S_j^0, \quad j = 1, 2, 3, \ldots, S \tag{31}$$

in such a way that the switching signal

$$\sigma_j(t) = 1, \quad \sum_{\substack{i=1 \\ i \neq j}}^{S} \sigma_i(t) = 0 \tag{32}$$

if and only if the vectors $u(t)$, $x(t)$ are inside the polyhedron P_j for which the inequality (31) holds.

Example
Let matrices that define polyhedrons in matrix inequalities (31) for individual polyhedrons are the following [10]:

$$S_1^x = S_2^x = S_3^x = S_4^x = \begin{bmatrix} -1 & 0 \\ 1 & 0 \\ 0 & 0 \\ 0 & 0 \end{bmatrix}$$

$$S_1^u = S_2^u = \begin{bmatrix} 0 \\ -0.3 \\ -1 \\ 1 \end{bmatrix}, \qquad S_3^u = S_4^u = \begin{bmatrix} 0.3 \\ 0 \\ -1 \\ 1 \end{bmatrix}$$

$$S_1^0 = \begin{bmatrix} 0 \\ 0.4 \\ 0 \\ 0.8 \end{bmatrix}, \quad S_2^0 = \begin{bmatrix} 0 \\ 0.4 \\ -0.8 \\ 1 \end{bmatrix}, \quad S_3^0 = \begin{bmatrix} -0.4 \\ 1 \\ 0 \\ 0.8 \end{bmatrix}, \quad S_4^0 = \begin{bmatrix} -0.4 \\ 1 \\ -0.8 \\ 1 \end{bmatrix}$$

Figure 3 shows an example of the instant membership of the vectors $u(t)$, $x(t)$ to polyhedrons P_1, P_2, P_3, P_4.

3.3 Fractional-Order Mixed Logical Dynamical Models (FO MLD)

The mixed logical dynamical integer-order models, introduced in the nineties of the last century, generalize a wide set of models, among which there are linear hybrid systems, finite state machines, some classes of discrete event systems, constrained linear systems, and nonlinear systems whose nonlinearities can be approximated by piecewise linear functions [11, 15].

According to techniques used in integer-order case, we propose fractional-order mixed logical dynamical models in the form, which is different from the model (17), (18). The switched matrices $A_d(t)|_{\sigma(t)}$, $B(t)|_{\sigma(t)}$, $C(t)|_{\sigma(t)}$, $D(t)|_{\sigma(t)}$ (if different from zero) and non-integer orders $\alpha|_{\sigma(t)}$ are replaced by real constant matrices A_d, B_1, C, D_1,

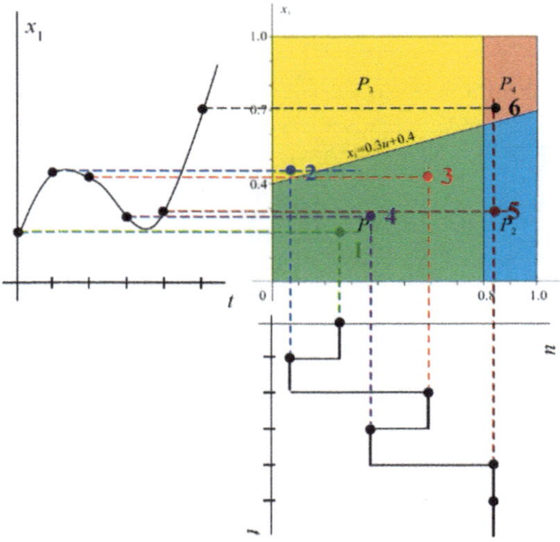

Fig. 3. Membership of the vectors $u(t), x(t)$ to polyhedrons P_j.

auxiliary real constant matrices $\boldsymbol{B}_2, \boldsymbol{B}_3, \boldsymbol{D}_2, \boldsymbol{D}_3$ and properly selected vectors of auxiliary variables $\delta(t), z(t)$:

$$\begin{gathered} x(t+1) = \boldsymbol{A}_d(t)x(t) + \boldsymbol{B}_1(t)u(t) - \sum_{i=1}^{L} c_i^{\alpha}(t)x(t+1-j) + \\ \boldsymbol{B}_2(t)\delta(t) + \boldsymbol{B}_3(t)z(t) \end{gathered} \tag{33}$$

$$y(t) = \boldsymbol{C}(t)x(t) + \boldsymbol{D}_1(t)u(t) + \boldsymbol{D}_2(t)\delta(t) + \boldsymbol{D}_3(t)z(t) \tag{34}$$

$$\boldsymbol{E}_2\delta(t) + \boldsymbol{E}_3 z(t) \le \boldsymbol{E}_1 u(t) + \boldsymbol{E}_4(t)x(t) + \boldsymbol{E}_5 \tag{35}$$

where $x(t) = \begin{bmatrix} x_c(t) \\ x_l(t) \end{bmatrix}$ is the extended state vector $x_c(t) \in \mathbb{R}^{n_c}$, $x_l(t) \in \{0, 1\}^{n_l}$, $y(t) = \begin{bmatrix} y_c(t) \\ y_l(t) \end{bmatrix} \in \mathbb{R}^{p_c} \times \{0, 1\}^{p_l}$ is the output vector, $u(t) = \begin{bmatrix} u_c(t) \\ u_l(t) \end{bmatrix} \in \mathbb{R}^{m_c} \times \{0, 1\}^{m_l}$ is the input vector, $z(t) \in \mathbb{R}^{r_c}$ and $\delta(t) \in \{0, 1\}^{r_l}$ are auxiliary variables, $\boldsymbol{E}_i$ denote real constant matrices, $\boldsymbol{E}_5$ is a real vector, $n_c > 0$, and $p_c, m_c, r_c, n_l, p_l, r_l \ge 0$. Inequalities (35) must be interpreted componentwise. Thus, the switching of local models in (17), (18), is replaced in the model (33), (34) by a set of suitable linear inequalities (35) with auxiliary continuous and logical variables.

It is known that the class of MLD models includes the following important classes of systems [11]:

- linear hybrid systems;
- sequential logical systems (Finite State Machines, Automata) (if $p_c, m_c, r_c = 0$);
- nonlinear dynamical systems, where the nonlinearity can be expressed through combinational logic (if $n_l = 0$);

- some classes of discrete event systems (if $n_c, p_c = 0$);
- constrained linear systems (if $n_l, m_l, p_l, r_l, r_c = 0$);
- linear systems (if $n_l, m_l, p_l, r_l, r_c = 0, \boldsymbol{E}_1, \boldsymbol{E}_4, \boldsymbol{E}_5 = 0$).

Analogically, the proposed FO MLD model (33)–(35) includes additionally:

- non-integer order linear hybrid systems;
- non-integer order nonlinear dynamical systems;
- non-integer order constrained linear systems;
- non-integer order linear systems.

4 Conclusions

In the paper methods for modeling complex, non-linear dynamical systems of non-integer order, using the so-called switched models are discussed. Three type of discrete-time dynamical switched models of fractional systems: Fuzzy Takagi-Sugeno Model (FO FTS), Piecewise Linear Model (FO PWL) and Mixed Logical Dynamical Model (FO MLD) are proposed and described.

The new proposed fractional-order mixed logical dynamical model includes a wide class of non-integer order systems and can be applied to modeling, higher-level control and optimization of complex systems. However, for a more complete opinion, a further analysis of the computational complexity of the proposed method and of the determination of parameters of the models vs. other switched fractional-order models is needed.

References

1. Mäkilä, P.M., Partington, J.R.: On linear models for nonlinear systems. Automatica **39**, 1–13 (2003)
2. Fang, L., Lin, H., Antsaklis, P.J.: Stabilization and performance analysis for a class of switched systems. In: Proceedings of the 43rd IEEE Conference on Decision Control, Atlantis, pp. 1179–1180 (2004)
3. Lin, H., Antsaklis, P.J.: Stability and stabilizability of switched linear systems: a survey of recent results. IEEE Trans. Autom. Control **54**(2), 308–322 (2009)
4. Kaczorek, T.: Selected Problems of Fractional Systems Theory. Springer, Heidelberg (2011)
5. Podlubny, I.: Fractional Differential Equations. Academic Press, San Diego (1999)
6. Chen, Y.Q., Petráš, I., Xue, D.: Fractional order control. In: American Control Conference, St. Louis, pp. 1397–1410 (2009)
7. Domek, S.: Switched state model predictive control of fractional-order nonlinear discrete-time systems. In: Pisano, A., Caponetto, R. (eds.) Advances in Fractional Order Control and Estimation, Asian J. Control, Special Issue **15**(3), 658–668 (2013)
8. Domek, S.: Piecewise affine representation of discrete in time, non-integer order systems. In: Mitkowski, W., Kacprzyk, J., Baranowski, J. (eds.) Advances in the Theory and Applications of Non-integer Order Systems. LN in Electrical Engineering, vol. 257, pp. 149–160. Springer, Heidelberg (2013)

9. Monje, C.A., Chen, Y.Q., Vinagre, B.M., Xue, D., Feliu, V.: Fractional Order Systems and Controls. Springer-Verlag, London (2010)
10. Domek, S.: Fractional-order calculus in model predictive control. West Pomeranian University of Technology Academic Press, Szczecin (2013). (in Polish)
11. Bemporad, A., Morari, M.: Control of systems integrating logic, dynamics, and constraints. Automatica **35**, 407–427 (1999)
12. Domek, S.: Switched models of non-integer order. PWN, Warszawa (2019). (in Polish)
13. Ying, H.: Fuzzy Control and Modeling: Analytical Foundations and Applications. Wiley-IEEE Press, Piscataway (2000)
14. Sun, Z., Ge, S.S.: Switched Linear Systems: Control and Design. Springer, London (2005)
15. Raman, R., Grossmann, I.E.: Relation between MILP modeling and logical inference for chemical process synthesis. Comput. Chem. Eng. **15**(2), 73–84 (1991)

SoftFRAC - Matlab Library for Realization of Fractional Order Dynamic Elements

Waldemar Bauer(✉), Jerzy Baranowski, Pawel Piątek, Katarzyna Grobler-Dębska, and Edyta Kucharska

Department of Automatic Control and Robotics, AGH University of Science and Technology, Al. Mickiewicza 30, 30-059 Kraków, Poland
{bauer,jb,ppi,grobler,edyta}@agh.edu.pl

Abstract. Nowadays, a realization of non-integer (fractional) order elements on a digital platform is a widely researched problem. The theory of such dynamic components is relatively well grounded. However, many problems of implementation on a digital platform are still open. Popular methods of implementation completely fail when used in real-time control applications. A need for efficient, numerically robust and stable implementation is obvious. These types of controllers and filters can be used in areas like telemedicine, biomedical engineering, signal processing, control, and many others. In this paper, the authors present the basic level of preliminary implementation of Matlab library for a realization of fractional order dynamic elements.

Keywords: Real time computation · Fractional systems · Approximation · Computation software · SoftFRAC

1 Introduction

The research of non-integer systems usage blooms. While it gravitates towards modeling and design aspects, we shouldnt neglect their implementation. For computer simulation, the popular FOMCON toolbox [20] covers most of the needs. The real-time control applications need more care especially in the aspects of numerical stability and errors accumulation. In this paper we present initial results on a new library - SoftFRAC - which will offer real-time ready solutions.

Popular methods for implementing fractional systems are Oustaloup filter method [14] and continuous fraction expansion (CFE, see eg. [15,17]). Discretizing them however causes instability, especially at high sampling frequencies or approximation orders [16]. For Oustaloup method, thanks to its structural properties, it is possible to remedy that issue. We discussed its sensitivity and stability problems during discretization in [4,5] and developed a new method - Time Domain Oustaloup approximation. Other avenues for warranting numerical stability require different methods. One approach focused on using Laguerre

A. Bartoszewicz et al. (Eds.): PCC 2020, AISC 1196, pp. 1189–1198, 2020.
https://doi.org/10.1007/978-3-030-50936-1_99

functions to create an approximation of impulse response [2,19], which was used among the others in [8–10,13,18]. Other structurally stable approaches include diffusive realizations, that use quadratures, such works (see [6,12]) result in a numerical stability, however require high orders for accuracy.

The main contribution of the paper is the introduction and announcement of the new library - SoftFRAC - which will implement those numerically stable methods allowing future use in multiple environments and applications and allow code generation. We describe briefly discussed algorithms and we present objective implementation in Matlab, inheriting from Control Toolbox standard.

We organize the rest of the paper as follows. First, we present our algorithms that constitute the core of the library. Then we present the implementation concepts, especially their object oriented aspects. Finally we present conclusions and future work.

2 Implementation of Non-integer System in Digital Environment

In this section we present three efficient methods for approximating fractional systems, that retain stability when discretized.

2.1 Time Domain Approximation

The non-integer transfer function s^γ have to be approximated with an integer order function. The Oustaloup continuous integer order approximation is given by Eq. (1) (see [14]):

$$s^\gamma \approx K \prod_{k=1}^{N} \frac{s+\omega_k'}{s+\omega_k}, \quad \gamma > 0, \tag{1}$$

where poles, zeros and gain can be evaluated as:

$$\begin{aligned} \omega_k' &= \omega_{min}\omega_u^{(2k-1-\gamma)/N} \\ \omega_k &= \omega_{min}\omega_u^{(2k-1+\gamma)/N} \\ K &= \omega_{max}^\gamma \\ \omega_u &= \sqrt{\frac{\omega_{max}}{\omega_{min}}} \end{aligned} \tag{2}$$

Approximation is designed for frequencies range $\omega \in [\omega_{min}, \omega_{max}]$ and N is the order of the approximation. As one can observe choosing a wide band of approximation results in large ω_u and high order N result in spacing of poles spacing from close to $-\omega_h$ to those very close to $-\omega_b$. This spacing is not linear (there is a grouping near $-\omega_b$) and causes problems in discretisation process.

This approach is to realize every block of the transfer function (1) in form of a state space system. Those first order systems will be then collected in a single triangular matrix resulting in full matrix realization. This continuous system of differential equations will be then discretized.

For zero initial condition we can transform Oustaloup approximation to state space system (see [7]):

$$\dot{\mathbf{x}} = \begin{bmatrix} A_1 & 0 & 0 & \dots & 0 \\ B_2 & A_2 & 0 & \dots & 0 \\ B_3 & B_3 & A_3 & \dots & 0 \\ \vdots & \vdots & \vdots & \ddots & \vdots \\ B_N & B_N & \dots & B_N & A_N \end{bmatrix} \mathbf{x} + \begin{bmatrix} KB_1 \\ KB_2 \\ KB_3 \\ \vdots \\ KB_N \end{bmatrix} u \tag{3}$$
$$y = \begin{bmatrix} 1 \, 1 \dots 1 \, 1 \end{bmatrix} \mathbf{x} + Ku$$

where

$$A_k = -\omega_k$$
$$B_k = \omega_k' - \omega_k$$
$$C_k = 1$$
$$D_k = 1$$

or in brief

$$\dot{\mathbf{x}} = \mathbf{A}\mathbf{x} + \mathbf{B}u$$
$$y = \mathbf{C}\mathbf{x} + Du \tag{4}$$

What can be immediately observed is that the matrix **A** is triangular. This is an extremely important in the case of this problem, as all its eigenvalues (poles of transfer function (1) are on its diagonal, so there is no need for eigenvalue products, which would lead to rounding errors. That is why discretization of (3) should have a structure preserving property.

2.2 LIRA – Laguerre Impulse Response Approximation Method

Due to the need to obtain approximation of non-integer order systems preserving numerical stability, a new method was developed, for the first time in [1].

Designed method can be applied for transfer functions in the form

$$\hat{g}(s) = \frac{q_m s^{\gamma_m} + q_{m-1} s^{\gamma_{m-1}} + \dots q_0}{s^{\sigma_n} + p_{n-1} s^{\sigma_{n-1}} + \dots p_0} \tag{5}$$

for q_k, γ_k, p_{l-1}, $\sigma_l \in \mathbb{R}$, $k = 1, 2, \dots, n$, $l = 1, 2, \dots, n$ and $m > n$.

Whose time domain response to a signal u is given by the convolution operator (for more details see [11])

$$y(t) = (u * g)(t) = \int_0^t u(t - \theta) g(\theta) \mathrm{d}\theta \tag{6}$$

where $g(t)$ is an inverse Laplace transform of $\hat{g}(s)$. it was shown that for bounded $g(t) \in \mathcal{L}_1(0,\infty) \cap \mathcal{L}_2(0,\infty)$ solution (6) can be approximated by a solution of a system of linear ordinary differential equation.

Approximating system has a form (for approximation order $n+1$, $\mathbf{x} = [\xi_0 \dots \xi_n]^\top$):

$$\begin{cases} \dot{\mathbf{x}}(t) = \mathbf{A}\mathbf{x}(t) + \mathbf{B}u(t) \\ y(t) = \mathbf{C}\mathbf{x}(t) \end{cases} \tag{7}$$

with matrices $\mathbf{A} = [a_{ij}]$, $a_{ij} = \begin{cases} -\mu, & i = j \\ -2\mu, & i > j \\ 0, & \text{otherwise} \end{cases}$, $\mathbf{B} = [b_i]$, $b_i = \sqrt{2\mu}$, $\mathbf{C} = [c_j]$, $c_j = \beta_j$ where β_k is given

$$\beta_k = \frac{\sqrt{2\mu}}{k!} \sum_{j=0}^{k} \binom{k}{j} c_j^k(\mu) \hat{g}^{(k-j)}(\mu) \tag{8}$$

where c_j^k are given recursively $c_j^k(\mu) = \frac{k-j+1}{2\mu} c_{j-1}^k$, $c_0^k(\mu) = (2\mu)^k$, $j = 0, 1, \dots, k$. The point of the method is to approximate an impulse response of system (5) with a finite series of orthonormal Laguerre functions. In [2] it was shown, that the approximation is convergent both in $\mathcal{L}_1$ and $\mathcal{L}_2$ function space norms.

It was also shown, that parameter μ should be chosen as

$$\mu = \arg\max \sum_{k=0}^{n} \beta_k^2(\mu) \tag{9}$$

2.3 Diffusive Realization of Non-integer Order Integrator

Diffusive realization of non-integer order integrator was researched it is based on the relationship, see [21]:

$$\frac{1}{s^\gamma} = \int_0^\infty \frac{\sin\gamma\pi}{\pi} \frac{1}{x^\gamma} \frac{1}{s+x} \mathrm{d}x \tag{10}$$

In this way it is possible to remove the non-integer power of a complex variable s and use the approximation

$$\int_0^\infty \frac{\sin\gamma\pi}{\pi} \frac{1}{x^\gamma} \frac{1}{s+x} \mathrm{d}x \approx \sum_{i=0}^{n} \frac{b_i}{s+x_i} \tag{11}$$

where the coefficients of the sum are calculated on the basis of numerical quadratures. In the works [3,6] the convergence of such approximation was analyzed using both quadrature for finite [6] and infinite [3] intervals.

Analysis of these results has concluded that since the approximation of (11) is the sum of the first order lowpass filters, the approximation quality will depend on their time constants. Because we are interested in good frequency representation for different orders of magnitude, the following transformation has been proposed

$$x = 10^{\theta}$$

For the interval $x \in [10^{\delta_1}, 10^{\delta_2}]$ (the unconstrained space can not be used because of the large scattering of values) integration is reduced to

$$\int_{\delta_1}^{\delta_2} \frac{\sin\gamma\pi}{\pi} \log(10) 10^{\theta(1-\gamma)} \frac{1}{s+10^{\theta}} d\theta. \tag{12}$$

This integral, although the integrand is more complex, allows the balance of all frequencies (high and low) to be balanced.

The Gauss-Legendre quadrature and Clenshaw-Curtis quadrature were analyzed in this context. In both cases the formulas for approximation coefficients have the same form, i.e.

$$\begin{aligned} x_j &= 10^{x_j^*} \\ b_j &= \frac{\sin\gamma\pi}{\pi} \log(10) 10^{x_j^*(1-\gamma)} w_j^* \end{aligned} \tag{13}$$

where for $*$ appropriately GL or CC has to be inserted. The diference is in determination of weights and nodes of quadratures. In case of Gaussa-Legendre quadrature they are

$$\begin{aligned} x_j^{GL} &= \lambda(T) \\ w_j^{GL} &= 2v_{1j}^2 \end{aligned} \tag{14}$$

where

$$T = \frac{1}{2} \begin{bmatrix} 0 & \frac{1}{\sqrt{1-2^{-2}}} & \cdots & 0 & 0 \\ \frac{1}{\sqrt{1-2^{-2}}} & 0 & \cdots & 0 & 0 \\ \vdots & \vdots & \vdots & \ddots & \vdots & \vdots \\ 0 & 0 & \cdots & 0 & \frac{1}{\sqrt{1-n^{-2}}} \\ 0 & 0 & \cdots & \frac{1}{\sqrt{1-n^{-2}}} & 0 \end{bmatrix} = V \mathrm{diag}(\lambda(T)) V^{-1}$$

while $V = [v_{ij}]_{n\times n}$. The disadvantage of this algorithm is the computational complexity of the complexity of solving the eigenproblem of matrix T.

In case of Clenshaw-Curtis, nodes are

$$x_j^{CC} = \cos j\frac{\pi}{n}, \quad j = 0,\ 1,\ \ldots,\ n \tag{15}$$

and are Chebyshev nodes of second kind. Weights $\mathbf{w}^{CC} = \left[w_0^{CC}\ w_1^{CC}\ \ldots\ w_n^{CC}\right]$ are computed using inverse fast fourier transform

$$\mathbf{w}^* = \mathrm{ifft}(\mathbf{g} + \mathbf{v}) \tag{16}$$

where $\mathbf{v} = [v_k]_{1\times n+1}$, $\mathbf{g} = [g_k]_{1\times n+1}$, and

$$\begin{cases} v_k = \dfrac{2}{1-4k^2}, \; k = 0,\; 1, \ldots, \left[\dfrac{n}{2}\right] - 1, \\ v_{[n/2]} = \dfrac{n-3}{2[n/2]-1} - 1, \\ v_{n-k} = v_k, \; k = 1,\; 2, \ldots, \left[\dfrac{n-1}{2}\right], \end{cases} \tag{17}$$

$$\begin{cases} g_k = -w_0^{CC}, \; k = 0,\; 1, \ldots, \left[\dfrac{n}{2}\right] - 1, \\ v_{[n/2]} = w_0^{CC}[(2 - \mathrm{mod}(n,2))n - 1], \\ v_{n-k} = g_k, \; k = 1,\; 2, \ldots, \left[\dfrac{n-1}{2}\right], \end{cases} \tag{18}$$

while $w_0^{CC} = (n^2 - 1 + \mathrm{mod}(n,2))^{-1}$. The advantage of this quadrature is that it significantly reduces the computational complexity of the coefficients, equal to the complexity of the fast Fourier transform.

3 Implementation of Matlab Library to Realization Fractional Order Dynamic Elements

Object-oriented programming (OOP) refers to a computer programming in which programmers define not only the data type of data structure but also the operations (functions) that can apply to the data structure. This approach improves the ability to manage software complexity, particularly important when developing and maintaining large applications and data structures. The concepts and rules used in object-oriented programming provide these important benefits:

- The concept of a data class makes it possible to define subclasses of data objects that share some or all of the main class characteristics is called inheritance. This property of OOP reduces development time and ensures more accurate coding.
- Since a class defines only the data it needs to be concerned with, when an instance of that class (an object) is run, the code will not be able to accidentally access other program data.
- The concept of data classes allows a programmer to create any new data type that is not already defined in the language itself.

OOP capabilities of the MATLAB language enable develop complex technical computing applications. In Matlab environment we can define classes and apply standard object-oriented design patterns, inheritance, encapsulation, and reference behavior without engaging in the low-level housekeeping tasks required by other languages.

3.1 SoftFRAC

Developed library SoftFRAC on this moment gives user functionality to create state-space fractional model and transfer function fractional model of the non-integer dynamic element s^{α} based on approximations described in second section. State-space fractional model class (`ssf`) inherit from standard state-space model class (`ss`) and have all functionality of this class. While transfer function fractional model class (`ssf`) inherit form transfer function model. In both cases to standard properties added variables described approximation:

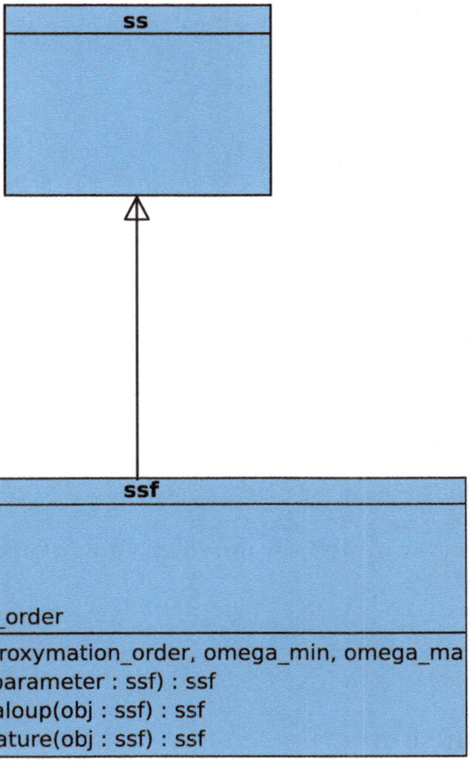

Fig. 1. Inherit and structure of class `ssf`.

- alpha - derivative order,
- omega_min - approximation frequency lower band,
- omega_max - approximation frequency upper band,
- approximation_order - approximation order

In classes we implemented the method to convert between the approximation methods described in Sect. 2. Because we used inheritance mechanism in implementation, the user can use all functionalities of parent classes ss and tf. In particular, user can use this realization to Simulink simulation, system behavior analysis and easy plotting of dynamic characteristics. Additional to we have added to the classes a construction method for easy conversion between those types, from `ssf` to `ssf` and vice versa. UML diagram of the `ssf` class implementation is presented in the Fig. 1.

Interface implementation of `ssf` class presents Matlab code:

```
classdef ssf < ss
   properties
       alpha
       omega_min
       omega_max
       approximation_order
   end

   methods
      function obj = ssf(alpha ,...
               approximation_order , omega_min ,...
               omega_max, Ts)
      end

      function obj = convert2Lira(obj)
      end

      function obj = convert2Oustaloup(obj)
      end

      function obj = convert2Qudrature(obj)
      end
    end
end
```

4 Conclusion

This paper present early stages of realization of the project SoftFRAC "Development of efficient computing software for simulation and application of non-integer order systems". The aim of the project is to develop a commercialization concept and carry out the development work to use the results of the already completed project "Designing and applying subsystems of a fraction of order in control systems". The essence of the planned commercialization is software that allows the implementation of non-integer order systems (regulators, filters) in real automation and signal processing systems. Competing (and available in open source form) software is characterized by either low accuracy or lack of numerical stability. SoftFRAC relies on algorithms developed as part of the completed project, which are characterized by high numerical robustness and scalable accuracy. The initial potential recipients were defined as academic units (supporting research on fractional differential equations and methods of designing control systems) and innovative industry (new methods of signal processing and robust control).

Presented here are only initial results on object-based implementation in Matlab. Our main goal however is to develop a fully multi-platform library, using additionally Python and C allowing rapid prototyping, code generation and real time control applications. Finished software will include graphical user interfaces, for model exploration, analysis and conversion. Our intention is, that the results of the project at least in some form will be freely available to the academic community.

Acknowledgment. Work partially realized in the project "Development of efficient computing software for simulation and application of non-integer order systems", financed by National Centre for Research and Development with TANGO programme.

References

1. Bania, P., Baranowski, J.: Laguerre polynomial approximation of fractional order linear systems. In: Mitkowski, W., Kacprzyk, J., Baranowski, J. (eds.) Advances in the Theory and Applications of Non-integer Order Systems: 5th Conference on Non-integer Order Calculus and Its Applications, Cracow, Poland, pp. 171–182. Springer (2013)
2. Bania, P., Baranowski, J., Zagórowska, M.: Convergence of Laguerre impulse response approximation for non-integer order systems. Math. Prob. Eng. **2016**, 13 (2016). https://doi.org/10.1155/2016/9258437. Article ID 9258437
3. Baranowski, J.: Quadrature based approximations of non-integer order integrator on finite integration interval. In: Babiarz, A., Czornik, A., Klamka, J., Niezabitowski, M. (eds.) Theory and Applications of Non-integer Order Systems, *Lecture Notes in Electrical Engineering*, vol. 407, pp. 11–20. Springer International Publishing (2017). https://doi.org/10.1007/978-3-319-45474-0_2
4. Baranowski, J., Bauer, W., Zagórowska, M.: Stability properties of discrete time-domain oustaloup approximation. In: Domek, S., Dworak, P. (eds.) Theoretical Developments and Applications of Non-Integer Order Systems, *Lecture Notes in Electrical Engineering*, vol. 357, pp. 93–103. Springer International Publishing (2016). https://doi.org/10.1007/978-3-319-23039-9_8
5. Baranowski, J., Bauer, W., Zagórowska, M., Dziwiński, T., Piątek, P.: Time-domain Oustaloup approximation. In: 2015 20th International Conference On Methods and Models in Automation and Robotics (MMAR), pp. 116–120. IEEE (2015)
6. Baranowski, J., Zagórowska, M.: Quadrature based approximations of non-integer order integrator on infinite integration interval. In: 2016 21st International Conference On Methods and Models in Automation and Robotics (MMAR) (2016)
7. Bauer, W., Baranowski, J., Dziwiński, T., Piątek, P., Zagórowska, M.: Stabilisation of magnetic levitation with a $PI^{\lambda}D^{\mu}$ controller. In: 2015 20th International Conference On Methods and Models in Automation and Robotics (MMAR), pp. 638–642. IEEE (2015)

8. De Keyser, R., Muresan, C., Ionescu, C.: An efficient algorithm for low-order direct discrete-time implementation of fractional order transfer functions. ISA Trans. **74**, 229–238 (2018)
9. Kapoulea, S., Psychalinos, C., Elwakil, A.: Single active element implementation of fractional-order differentiators and integrators. AEU - Int. J. Electron. Commun. **97**, 6–15 (2018)
10. Kawala-Janik, A., Bauer, W., Al-Bakri, A., Haddix, C., Yuvaraj, R., Cichon, K., Podraza, W.: Implementation of low-pass fractional filtering for the purpose of analysis of electroencephalographic signals. Lect. Notes Electr. Eng. **496**, 63–73 (2019)
11. Monje, C.A., Chen, Y., Vinagre, B.M., Xue, D., Feliu, V.: Fractional-order systems and controls: Fundamentals and applications. Advances in Industrial Control. Springer-Verlag, London (2010)
12. Monteghetti, F., Matignon, D., Piot, E.: Time-local discretization of fractional and related diffusive operators using gaussian quadrature with applications. Appl. Numer. Math. (2018)
13. Mozyrska, D., Wyrwas, M.: Stability of linear systems with Caputo fractional-, variable-order difference operator of convolution type (2018)
14. Oustaloup, A., Levron, F., Mathieu, B., Nanot, F.M.: Frequency-band complex noninteger differentiator: characterization and synthesis. IEEE Trans. Circuits Syst. I Fundam. Theory Appl. **47**(1), 25–39 (2000)
15. Petráš, I.: Fractional-Order Nonlinear Systems: Modeling, Analysis and Simulation. Nonlinear Physical Science. Springer (2011)
16. Piątek, P., Zagórowska, M., Baranowski, J., Bauer, W., Dziwiński, T.: Discretisation of different non-integer order system approximations. In: 2014 19th International Conference on Methods and Models in Automation and Robotics (MMAR), pp. 429–433. IEEE (2014)
17. Podlubny, I.: Fractional Differential Equations: An Introduction to Fractional Derivatives, Fractional Differential Equations, to Methods of Their Solution and Some of Their Applications. Mathematics in Science and Engineering. Elsevier Science (1998)
18. Rydel, M., Stanisławski, R., Latawiec, K., Gałek, M.: Model order reduction of commensurate linear discrete-time fractional-order systems. IFAC-PapersOnLine **51**(1), 536–541 (2018)
19. Stanisławski, R., Latawiec, K.J., Gałek, M., Łukaniszyn, M.: Modeling and identification of fractional-order discrete-time laguerre-based feedback-nonlinear systems. In: Latawiec, K.J., Łukaniszyn, M., Stanisławski, R. (eds.) Advances in Modelling and Control of Non-integer-Order Systems, Lecture Notes in Electrical Engineering, vol. 320, pp. 101–112. Springer International Publishing (2015)
20. Tepljakov, A., Petlenkov, E., Belikov, J.: FOMCON: a MATLAB toolbox for fractional-order system identification and control. Int. J. Microelectron. Comput. Sci. **2**, 51–62 (2011)
21. Trigeassou, J., Maamri, N., Sabatier, J., Oustaloup, A.: State variables andtransients of fractional order differential systems. Comput. Math. Appl. **64**(10), 3117–3140 (2012). https://doi.org/10.1016/j.camwa.2012.03.099. http://www.sciencedirect.com/science/article/pii/S0898122112003173. Advances in FDE, III

New Implementation of Discrete-Time Fractional-Order PI Controller by Use of Model Order Reduction Methods

Rafał Stanisławski(✉), Marek Rydel, and Krzysztof J. Latawiec

Department of Electrical, Control and Computer Engineering, Opole University of Technology, ul. Prószkowska 76, 45-758 Opole, Poland
{r.stanislawski,m.rydel,k.latawiec}@po.edu.pl

Abstract. The paper presents new results in implementation of a discrete-time fractional-order PI controller by use of computationally simple and accurate Model Order Reduction-based approximation of a fractional-order integrator. The main advantage of the introduced method is elimination of the steady-state control error, the feature outperforming other finite-length implementations of discrete-time fractional-order integrators. Simulation experiments confirm the effectiveness of the presented methodology, both in terms of high accuracy and computational effectiveness of the introduced approximation.

Keywords: Fractional-order PID controller · Model order reduction · Discrete time system

1 Introduction

Various PID and PID-like controllers have attracted considerable research interest due to their huge practical importance. Most of the closed control loops in industrial environments are based on this class of controllers. Although this concept of control is a well-known area of science and technology, current industrial controllers offer many extensions, e.g. gain scheduling, self-adaptation of parameters for time-varying processes, autotuning and many other features.

On the other hand, increasing research interest in fractional-order systems also affects the area of PID controllers. In this field, we can find plenty of papers, which show that specific properties of fractional-order derivatives and integrators can contribute to an improved PID-like control performance [5,13,26]. Moreover, in fractional-order PID controllers, we have two more parameters that can be tuned to improve the properties of the closed-loop system [3,4,19], including the robustness aspects [7,10]. Also, various papers implement fractional-order PID controllers in classical environments, like gain scheduling [1,24], internal model control [12,16], auto-tuning [8,14], realization of discrete-time control [6,11] and many others.

A. Bartoszewicz et al. (Eds.): PCC 2020, AISC 1196, pp. 1199–1209, 2020.
https://doi.org/10.1007/978-3-030-50936-1_100

The main problem in implementation of fractional-order PID controllers is the fact that the realization of fractional-order integrator and derivative/difference leads to the issue of infinite computational complexity. Therefore, in practical applications we have to use finite-length approximations of fractional-order integro-differentials. In this context, in the case of continuous-time systems, the Oustaloup approximators have often been used [2,25]. These approximators model fractional-order derivatives/differences as IIR filters and a number of works confirm the good performance of this approach. In discrete-time case, various methods have been used to approximate fractional-order differences/integrators, including PSE-based [21,22] and CFE-based approximations [17,18]. Also, Model Order Reduction (MOR) algorithms can be effectively used in this area [9,20,23]. A variety of the above mentioned approximations may affect properties of fractional-order PID-like controller (see e.g. [15]).

In this paper, we offer a new method for implementation of the discrete-time fractional-order PI controller. In particular, we propose a model of fractional-order integrator obtained by the use of MOR methods. The introduced approximation leads to elimination of steady-state control error, which can be observed when using practical (finite-length) implementations of fractional-order PI/PID controllers.

The paper is organized as follows. Having shortly stated the approximation problem in fractional-order PID-like controllers in Sect. 1, Sect. 2 presents a formulation of the problem and a simple motivating example. Section 3 introduces a new, MOR-based approximation method for the synthesis of the fractional-order integrator. A numerical example of Sect. 4 confirms the usefulness of the proposed method, and the conclusion of Sect. 5 summarizes the achievement of the paper.

2 Problem Formulation

It is well known that the fractional-order generalization of continuous-time PID controller can be defined as

$$u(t) = \left(K_p + K_i\mathcal{D}^{-\lambda} + K_d\mathcal{D}^{\mu}\right) e(t) \tag{1}$$

where K_p, K_i and K_d are the controller parameters, $e(t)$ and $u(t)$ are controller input and output, respectively, and $\mathcal{D}$ is the fractional-order derivation/integration operator. Note that $\mathcal{D}^{-\lambda}$ denotes integrator of fractional order $\lambda > 0$ and $\mathcal{D}^{\mu}$ denotes derivative of fractional-order $\mu > 0$. The operator $\mathcal{D}$ is usually described by the use of one of three definitions, that are Riemman-Liuville, Caputo or Grünwald-Letnikov ones. In a designing process of the fractional-order PID controller, we have to determine five controller parameters in terms of K_p, K_i, K_d, and fractional orders λ and μ.

Discrete-time version of fractional-order PID controller can be defined as follows

$$u_k = \left(K_p + K_i\Delta^{-\lambda} + K_d\Delta^{\mu}\right) e_k \tag{2}$$

with e_k and u_k are the system input and output, respectively, $k = 0, 1, \ldots$ is the discrete time, and $\Delta^{-\lambda}$ and Δ^{μ} denote discrete-time integrator and difference, respectively. The fractional-order difference can be defined by use of discrete-time Grünwald-Letnikov difference as

$$\Delta^{\mu} e_k = h^{-\mu} \sum_{j=0}^{k} (-1)^j \binom{\mu}{j} e_{k-j} \tag{3}$$

where h and $\binom{\mu}{j}$ denote the sampling time and the Newton binomial, respectively. Note that the Newton binomial coefficients can be also calculated in recursive way (see e.g. [21]). In general case, the $\mathcal{Z}$-transform of Eq. (3) is as follows

$$\Delta^{\mu}(z) = h^{-\mu} \sum_{j=0}^{\infty} (-1)^j \binom{\mu}{j} z^{-j} \tag{4}$$

The fractional-order integrator $\Delta^{-\lambda}$ can still be calculated by the use of Eqs. (3) and (4) where μ is substituted by $-\lambda$. Alternatively, the operator $\Delta^{-\lambda}$ can also be obtained from the IIR filter as

$$\Delta^{-\lambda}(z) = \frac{1}{\Delta^{\lambda}(z)} = \frac{1}{h^{-\lambda} \sum_{j=0}^{\infty} (-1)^j \binom{\lambda}{j} z^{-j}} \tag{5}$$

Note that, in a general form, both fractional-order difference (4) and integrator (5) cannot be implemented in practice, due to infinite sum in the equations. Therefore in practical applications, we have to use finite-length approximations of the fractional-order difference and integrator, respectively

$$\Delta_L^{\mu}(z) = h^{-\mu} \sum_{j=0}^{L} (-1)^j \binom{\mu}{j} z^{-j} \tag{6}$$

$$\Delta_L^{-\lambda}(z) = \frac{1}{h^{-\lambda} \sum_{j=0}^{L} (-1)^j \binom{\lambda}{j} z^{-j}} \tag{7}$$

The same results as in Eqs. (6) and (7) can be obtained by the use of the Power Series Expansion (PSE) to the Euler-based discretization scheme of fractional-order derivative and integrator

$$\Delta_L^{\pm r}(z) = \left(\frac{1}{h}\right)^{\pm r} PSE\left\{(1 - z^{-1})^{\pm r}\right\}_L \tag{8}$$

with $\pm r$ being the derivative/integrator order. Note that the discrete-time realization of fractional-order derivative/integrator can be realized through other discretization schemes, like Tustin- and Al-Alaoui-based ones. Also, we can apply other finite-length approximation approaches, e.g. based on Continuous Fraction Expansion [18,22].

It is important to note that implementation of finite-length approximations to fractional-order derivatives and integrators in the PID controllers affects the control process. One of the problems is presented in the following motivating example.

2.1 Motivating Example

Consider the discrete-time system $G(z) = \frac{0.01}{z-0.99}$. The system is under closed-loop control using the discrete-time fractional-order PI (PI^{λ}) controller, with $K_p = 1.3$, $K_i = 2.6$, $\lambda = 0.8$ and sampling period $h = 0.01$. The block diagram of the closed-loop system is presented in Fig. 1.

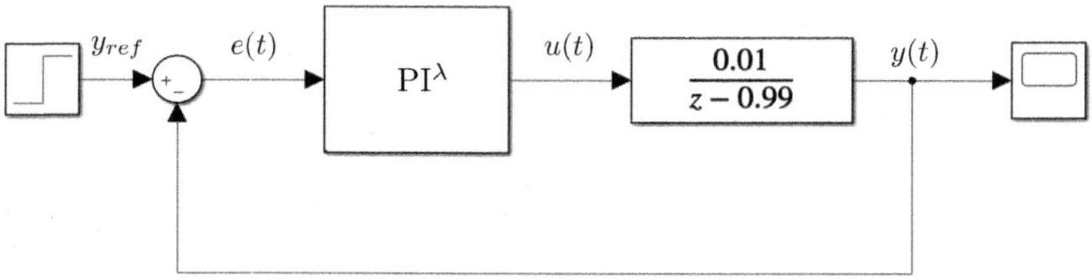

Fig. 1. Block diagram of the closed-loop system.

The fractional-order integrator $\Delta^{-\lambda}$ is realized both by use of infinite (as in Eq. (5)) and finite (as in Eq. (7) with $L = 1000$) implementation lengths, respectively. The time plots of the output signal y_k, $k = 0, ..., 10000$, from the closed-loop system for the input $y_{ref}(t) = \mathbf{1}(t)$ are presented in Fig. 2.

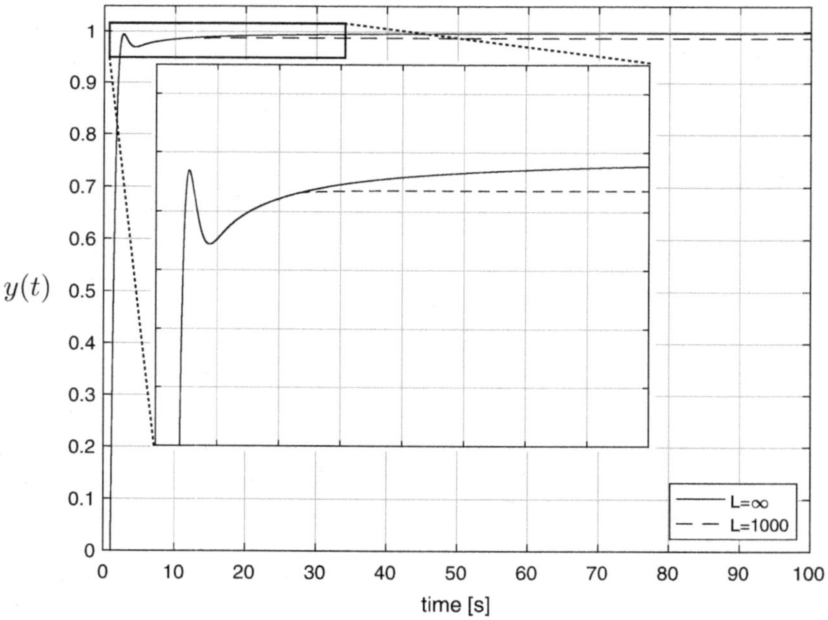

Fig. 2. Time plots of the output signal y_k.

It can be seen from Fig. 2 that finite-length implementation of the fractional-order integrator affects the steady-state output of the closed-loop system. In the considered example, the relative steady-state error is approximately equal to 1.3%. The error depends on several parameters, such as fractional order λ, K_i, but in general, an increasing implementation length L decreases the steady-state error. The presented results are also considered in Refs. [22]. Note that the example considers the PI$^\lambda$ controller, but the results can be immediately extended to the fractional-order PID controller case.

Problem Formulation: One of the most important problems in the implementation of discrete-time fractional-order PI/PID controllers in industrial environments is the necessity toe use finite-length approximation of fractional-order integrator/difference with the relatively low value of L (especially in case of use of low computing power hardware). This approximation leads to a) steady-state error of the closed-loop control process and b) low approximation accuracy of fractional-order integrator/difference used in PI/PID controller. In the paper, we propose a new concept to cope with the problem.

3 Main Result

In this section, we propose a new approximation of fractional-order integrator, which leads to the elimination of the steady-state error of the fractional-order PI/PID controller. On this basis, we use the Model Order Reduction methods to simplification of the calculation process of the fractional-order PI controller.

3.1 Approximation of Fractional-Order Integrator

It is well known that the fractional-order derivative/integrator operator satisfies the following condition

$$\mathcal{D}^{-\lambda_1}(\mathcal{D}^{\lambda_2}f(t)) = \mathcal{D}^{\lambda_2}(\mathcal{D}^{-\lambda_1}f(t)) = \mathcal{D}^{\lambda_2-\lambda_1}f(t) \tag{9}$$

The same properties are satisfied for the discrete-time counterpart of fractional-order difference/integration operator

$$\Delta^{-\lambda_1}(\Delta^{\lambda_2}f_k) = \Delta^{\lambda_2}(\Delta^{-\lambda_1}f_k) = \Delta^{\lambda_2-\lambda_1}f_k \tag{10}$$

On the basis of the above Eq. (10) we can present fractional-order integrator $\Delta^{-\lambda}$ as a combination of classical (integer-order) integrator and fractional-order derivative. Assuming that the fractional order $\lambda \in (0,1)$ we obtain

$$_I\Delta^{-\lambda}e_k = \Delta^{-1}(\Delta^{1-\lambda}e_k) = \Delta^{1-\lambda}(\Delta^{-1}e_k) \tag{11}$$

where $\Delta^{1-\lambda}$ is the fractional-order difference and Δ^{-1} is the discrete-time first-order integrator. Finally, taking into account that $\Delta^{-1}(z) = \frac{h}{1-z^{-1}}$, we arrive at

$$_I\Delta^{-\lambda}(z) = h^\lambda \sum_{j=0}^{\infty}(-1)^j\binom{1-\lambda}{j}\frac{z^{-j}}{1-z^{-1}} \tag{12}$$

Again, the fractional-order integrator ${}_I\Delta^{-\lambda}(z)$ cannot be practically implemented due to the infinite sum in Eq. (12), therefore we have to use of finite-length approximation

$$ {}_I\Delta_L^{-\lambda}(z) = h^{\lambda} \sum_{j=0}^{L} (-1)^j \binom{1-\lambda}{j} \frac{z^{-j}}{1-z^{-1}} \tag{13} $$

The impulse responses of the finite-length implementation of fractional-order integrators $\Delta_L^{-\lambda}(z)$ and ${}_I\Delta_L^{-\lambda}(z)$ for $\lambda = 0.8$, $h = 0.1$ and various L are presented in Fig. 6.

It can be seen from Fig. 3 that for $k \leq L$ impulse responses of both approximations ${}_I\Delta^{-\lambda}(z)$ and $\Delta^{-\lambda}(z)$ are equivalent to fractional-order integrator $\Delta^{-\lambda}(z)$. However, for $k > L$ the impulse response of ${}_I\Delta_L$ is constant (as for classical, first-order integrator), but responses of $\Delta_L^{-\lambda}$ tends to 0. This specific properties of the proposed integrator ${}_I\Delta_L$ can be also confirmed in frequency-domain characteristic presented in Fig. 4.

It can be seen from Fig. 4 that the magnitude characteristic changes $-20/\lambda$ Db/dec for $\omega > 0.05$ and -20 Db/dec for $\omega < 0.05$.

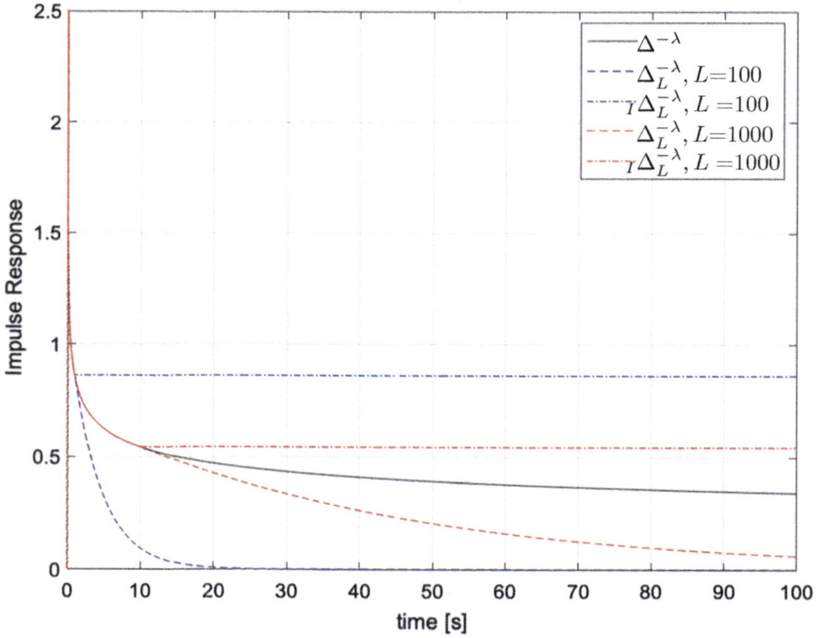

Fig. 3. Impulse response od fractional-order integrators.

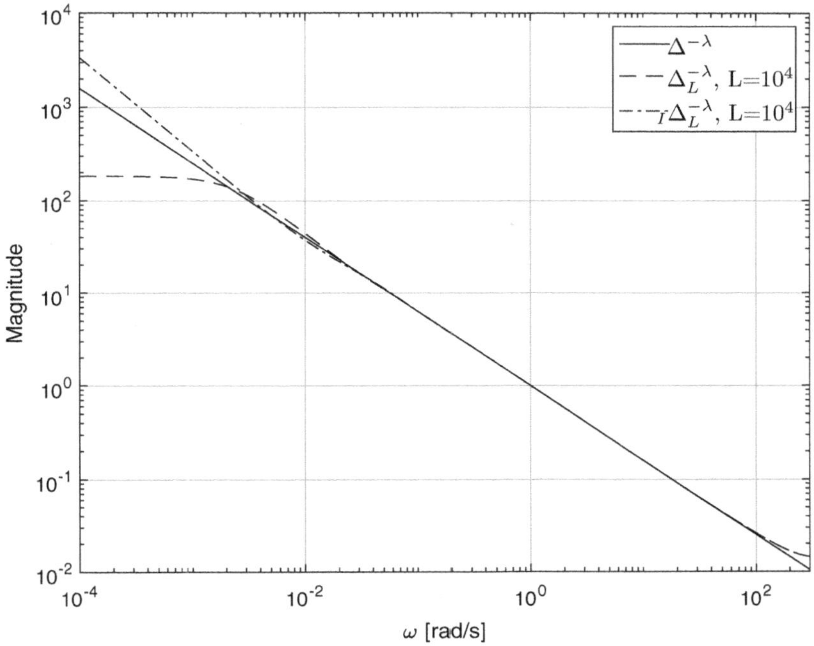

Fig. 4. Frequency-response fractional-order integrators.

3.2 Implementation of Model Order Reduction Method

Section 3.1 has proposed a new concept in approximation of fractional-order integrator. But still, in the implementation of the fractional-order PI controller, we have to use very high L in ${}_I\Delta_L(z)$ to obtain satisfactory approximation accuracy – usually, the value of L changes between 1000 to 10000. Therefore, the fractional-order regulator is still hardly implementable in the industrial environments due to high computational complexity. To cope with the problem, we can present Eq. (13) in the following form

$$ {}_I\Delta_L^{-\lambda}(z) = h^{\lambda}(1 - z^{-1})^{-1}\Delta_L^{1-\lambda}(z) \tag{14} $$

Now, we can use one of Model Order Reduction methods in terms of the so-called FIR-BT technique proposed in Ref. [23], to modeling of fractional-order difference $\Delta_L^{1-\lambda}(z)$. As a result, we obtain low order approximation of fractional-order difference $\Delta_L^{1-\lambda}(z)$ in the state space form of integer, arbitrary given order n

$$ \tilde{x}_{k+1} = \tilde{A}_r\tilde{x}_k + \tilde{B}_r u_k \tag{15} $$

$$ \tilde{y}_k = \tilde{C}_r\tilde{x}_k + D_r u_k \tag{16} $$

with $\tilde{A}_r \in \Re^{n\times n}$, $\tilde{B}_r \in \Re^{n\times 1}$, $\tilde{C}_r \in \Re^{1\times n}$ and $\tilde{D}_r \in \Re$. Finally, the fractional-order difference $\Delta_L^{1-\lambda}(z)$ can be modeling as

$$ \Delta_L^{1-\lambda}(z) \approx \Delta^{1-\lambda}(z) = \tilde{C}_r(Iz - \tilde{A}_r)\tilde{B}_r + \tilde{D}_r \tag{17} $$

A detailed procedure for the use of the so-called FIR-BT method has been presented in Ref. [23]. Exemplary parameters of FIR-BT-based model $\Delta_k^{1-\lambda}(z)$ of discrete-time fractional-order integrator ${}_I\Delta^{-\lambda}(z)$ for $\lambda = 0.8$, $n = 6$ and $L = 10^4$ are as follows

$$\left[\begin{array}{c|c} \tilde{A}_r & \tilde{B}_r \\ \hline \tilde{C}_r & \tilde{D}_r \end{array}\right] =$$

$$\left[\begin{array}{cccccc|c}
0.738 & 0.2023 & -0.4251 & -0.1158 & -2.188e\text{-}2 & 0.3683 & 0.2879 \\
2.348e\text{-}2 & 0.9333 & 0.2191 & 7.135e\text{-}2 & -4.57e\text{-}2 & -9.069e\text{-}2 & -0.2541 \\
-2.791e\text{-}2 & 7.783e\text{-}2 & 0.6295 & -6.905e\text{-}2 & 1.569e\text{-}2 & 0.1255 & 0.759 \\
8.665e\text{-}2 & -2.304e\text{-}2 & -6.4e\text{-}2 & 0.9804 & 2.424e\text{-}2 & -4.548e\text{-}2 & 0.1549 \\
-0.3208 & 0.1408 & -0.1934 & 4.652e\text{-}3 & 0.8429 & 0.3342 & 6.167e\text{-}2 \\
0.2478 & -0.2026 & 0.5083 & 0.1108 & 3.951e\text{-}2 & 0.6226 & -0.4852 \\ \hline
-5.517e\text{-}3 & 6.929e\text{-}2 & -0.5579 & -6.086e\text{-}2 & 1.344e\text{-}2 & 0.1053 & 2.512
\end{array}\right]$$

Note that, to obtain an approximation of the fractional-order difference $\Delta_k^{1-\lambda}(z)$, we can use other Model Order Reduction techniques. For instance, we can use the Frequency-Weighted method based on the impulse response of fractional-order difference by the so-called FIR-FW algorithm (see Ref. [20]).

4 Numerical Example

Consider a continuous-time integer-order system in the transfer function form [25]

$$G(s) = \frac{1}{(s+1)^3} \tag{18}$$

discretized with the sampling period $h = 0.01$. The system is under closed-loop control as in Fig. 1 using discrete-time fractional-order PI (PI^λ) controller with $K_p = 1$, $K_i = 0.5$ and $\lambda = 0.8$. To implement the PI^λ controller we use both the 'classical' approximation of fractional-order integrator $\Delta_L^\lambda(z)$ and ${}_I\Delta_k(z)$ introduced in this paper (see Subsect. 3.2). Figure 6 presents the impulse response of the actual PI controller vs. its both approximations.

It can be seen from Fig. 5 that impulse response of PI^λ with actual fractional-order integrator and its approximation Δ_L^λ tends to 0 for $t \to \infty$. In contrast, PI^λ controller with ${}_I\Delta_k^{-\lambda}$ does not tend to 0 as for classical, integer-order PI controller.

Figure 6 presents the outputs of the closed-loop system under the fractional-order PI controller containing actual integrator and its approximations.

It can be seen from Fig. 6 that implementation of ${}_I\Delta_L^\lambda$ and finally ${}_I\Delta_k^{-\lambda}$ into fractional-order PI controller leads to elimination of steady-state error. Moreover, using model order reduction techniques to the fractional-order difference (in fractional-order integrator ${}_I\Delta_k^{-\lambda}$) significantly simplifies the computational process without significant changes in dynamical properties of the controller.

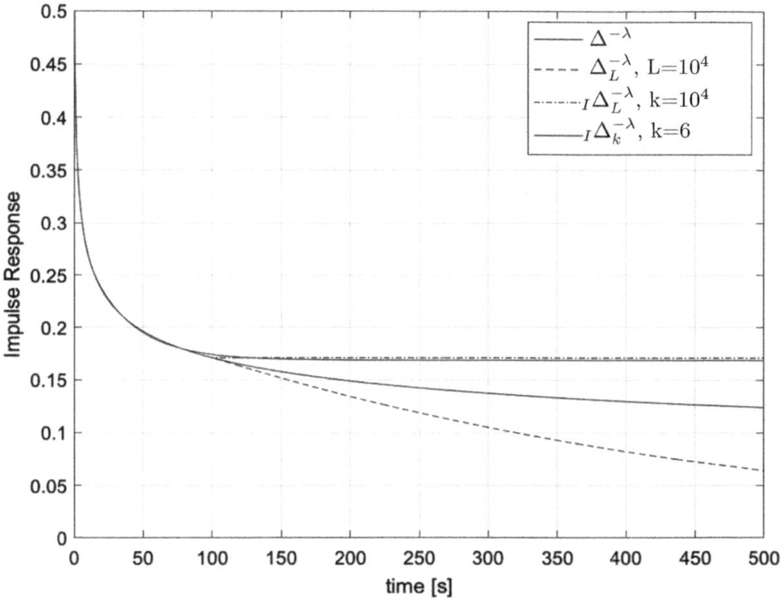

Fig. 5. Impulse responses of discrete-time fractional-order PI controller and its implementations.

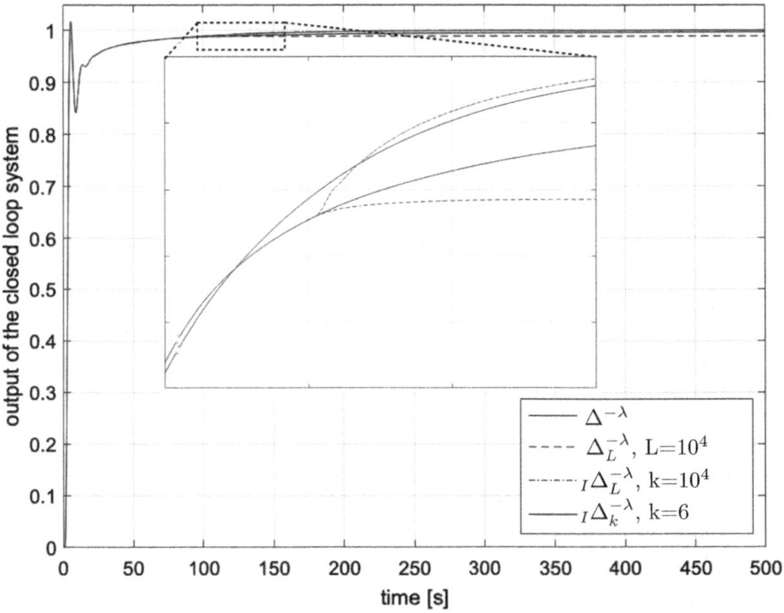

Fig. 6. Time plots of the closed-loop system outputs.

5 Conclusion

This paper has presented new results in implementation of discrete-time fractional-order PI controller by use of computationally simple and accurate approximation of fractional-order integrator. The main advantage of the proposed method is elimination of the steady-state control error, which usually occurs in other implementations of PI^{λ} controllers. Simulation experiments confirm the effectiveness of the introduced methodology, both in terms of high accuracy and computational simplicity.

The paper considers the PI^{λ} controller only, but the results can be immediately extended to the more general, fractional-order PID controller case.

References

1. Ahmed, M.F., Dorrah, H.T.: Design of gain schedule fractional PID control for nonlinear thrust vector control missile with uncertainty. Automatika **59**(3–4), 357–372 (2018). https://doi.org/10.1080/00051144.2018.1549696
2. Baranowski, J., Bauer, W., Zagorowska, M.: Stability properties of discrete time-domain Oustaloup approximation. In: Theoretical Developments and Applications of Non-integer Order Systems. Lecture Notes in Electrical Engineering. Springer (2016)
3. Barbosa, R., Tenreiro, J.A., Ferreira, I.M.: Tuning of PID controllers based on Bode's ideal transfer function. Nonlinear Dyn. **38**(1–4), 305–321 (2004)
4. Cervera, A., Baños, A., Monje, C.A.: Tuning of fractional PID controllers by using QFT. In: Proceedings of the 32nd Annual Conference of the IEEE Industrial Electronics Society (IECON 2006), Paris, France (2006)
5. da Costa, J.S.: An introduction to fractional control. In: Control, Robotics and Sensors, Institution of Engineering and Technology (2012). https://digital-library.theiet.org/content/books/ce/pbce091e
6. Hosseinnia, S.H., Tejado, I., Milanés, V., Villagrá, J., Vinagre, B.M.: Experimental application of hybrid fractional-order adaptive cruise control at low speed. IEEE Trans. Control Syst. Technol. **22**(6), 2329–2336 (2014)
7. HosseinNia, S.H., Tejado, I., Vinagre, B.M.: A method for the design of robust controllers ensuring the quadratic stability for switching systems. J. Vibr. Control **20**(7), 1085–1098 (2014). https://doi.org/10.1177/1077546312470480
8. Keyser, R.D., Muresan, C.I., Ionescu, C.M.: A novel auto-tuning method for fractional order PI/PD controllers. ISA Trans. **62**, 268–275 (2016). sI: Control of Renewable Energy Systems. http://www.sciencedirect.com/science/article/pii/S0019057816000392
9. Krajewski, W., Viaro, U.: A method for the integer-order approximation of fractional-order systems. J. Franklin Inst. **351**(1), 555–564 (2014)
10. Liu, L., Zhang, S.: Robust fractional-order PID controller tuning based on Bode's optimal loop shaping. Complexity **2018**(6570560) (2018)
11. Lopes, A.M., Machado, J.T.: Discrete-time generalized mean fractional order controllers. IFAC-PapersOnLine **51**(4), 43–47 (2018). 3rd IFAC Conference on Advances in Proportional-Integral-Derivative Control PID 2018. http://www.sciencedirect.com/science/article/pii/S240589631830315X

12. Maâmar, B., Rachid, M.: IMC-PID-fractional-order-filter controllers design for integer order systems. ISA Trans. **53**(5), 1620–1628 (2014). iCCA 2013. http://www.sciencedirect.com/science/article/pii/S0019057814000962
13. Monje, C., Chen, Y., Vinagre, B., Xue, D., Feliu, V.: Fractional-order Systems and Controls: Fundamentals and Applications. Series on Advances in Industrial Control. Springer, London (2010)
14. Monje, C.A., Vinagre, B.M., Feliu, V., Chen, Y.: Tuning and auto-tuning of fractional order controllers for industry applications. Control Eng. Pract. **16**(7), 798–812 (2008). http://www.sciencedirect.com/science/article/pii/S0967066107001566
15. Mozyrska, D., Ostalczyk, P.: Generalized fractional-order discrete-timeintegrator. Complexity **2017**(3452409) (2017)
16. Muresan, C.I., Dutta, A., Dulf, E.H., Pinar, Z., Maxim, A., Ionescu, C.M.: Tuning algorithms for fractional order internal model controllers for time delay processes. Int. J. Control **89**(3), 579–593 (2016)
17. Oprzędkiewcz, K., Mitkowski, W., Gawin, E.: The plc implementation of fractional-order operator using cfe approximation. In: Advances in Intelligent Systems and Computing, vol. 550. Springer (2017)
18. Oprzedkiewicz, K., Stanisławski, R., Gawin, E., Mitkowski, W.: A new algorithm for a CFE-approximated solution of a discrete-time non integer-order state equation. Bull. Pol. Acad. Sci. Tech. Sci. **65**(4), 429–437 (2017)
19. Padula, F., Visioli, A.: Tuning rules for optimal PID and fractional-order PID controllers. J. Process Control **21**(1), 69–81 (2011). http://www.sciencedirect.com/science/article/pii/S0959152410001927
20. Rydel, M., Stanisławski, R.: A new frequency weighted Fourier-based method for model order reduction. Automatica **88**, 107–112 (2018)
21. Stanisławski, R., Latawiec, K.J.: Normalized finite fractional differences - the computational and accuracy breakthroughs. Int. J. Appl. Math. Comput. Sci. **22**(4), 907–919 (2012)
22. Stanisławski, R., Latawiec, K.J., Łukaniszyn, M.: A comparative analysis of Laguerre-based approximators to the Grünwald-Letnikov fractional-order difference. Math. Problems Eng. **2015**, 1–10 (2015). Article ID: 512104
23. Stanisławski, R., Rydel, M., Latawiec, K.J.: Modeling of discrete-time fractional-order state space systems using the balanced truncation method. J. Franklin Inst. **354**(7), 3008–3020 (2017)
24. Tejado, I., HosseinNia, S.H., Vinagre, B.M., Chen, Y.: Efficient control of a smartwheel via internet with compensation of variable delays. Mechatronics **23**(7), 821–827 (2013). 1. Fractional Order Modeling and Control in Mechatronics 2. Design, control, and software implementation for distributed MEMS (dMEMS). http://www.sciencedirect.com/science/article/pii/S0957415813000767
25. Tejado, I., Vinagre, B.M., Traver, J.E., Prieto-Arranz, J., Nuevo-Gallardo, C.: Back to basics: meaning of the parameters of fractional order PID controllers. Mathematics **7**(6) (2019). https://www.mdpi.com/2227-7390/7/6/530
26. Vilanova, R., Visioli, A.: PID Control in the Third Millennium. Lessons Learned and New Approaches. Advances in Industrial Control. Springer, London (2012)

Autonomous Vehicles and Embedded Artificial Intelligence

Autonomous Delivery Robot AQUILO

Marek Długosz(✉), Paweł Węgrzyn, and Michał Roman

Department Automatic Control and Robotics,
AGH University of Science and Technology, Kraków, Poland
mdlugosz@agh.edu.pl, {pwegrzy,mroman}@student.agh.edu.pl

Abstract. The design and structure of the AQUILO Autonomous Delivery Robot are presented in this paper. The AQUILO robot's task is to transport small deliveries in such buildings as offices, schools, hospitals. The AQUILO robot is autonomous, that is, after starting a task it works without human supervision or control. Individual components used in AQUILO robot construction, for example drives, sensors as well as control algorithms are described in this paper. The proprietary system for stair detection used in the AQUILO robot to assure its safe operation is also described. The conclusions also contain information about further works on AQUILO robot development.

Keywords: Autonomous robot · Control · Office robot · Programming

1 Introduction

In recent years, a very rapid development of autonomous robots has been observed. More and more companies offer such robots both for industrial and civil applications. This the result of several factors, primarily from scientific advances in robotics and the fall in prices of the robot's basic subassemblies. It is predicted that in the future autonomous robots will take over more and more operations from humans, e.g. shopping, the transport of consignments, etc. [14,25]. Autonomous vacuum cleaners and lawnmowers are currently available.

One of the main tasks that must be undertaken by autonomous robots is the ability to move between a starting point and an end point. Obviously, this must be performed in a manner safe for humans, the robot and other objects. This seemingly simple and plain operation for humans is highly complex for a robot that has to perform a number of various tasks simultaneously, e.g. localization, obstacle detection, path planning and trajectory planning (schedule the movement along the planned path).

As early as in the 1980's autonomous mobile robots were built to be used in factories, warehouses, logistics centres, etc. [1]. One of the first robots of this type, named FIRST, was designed to deliver medicines, meals and other goods

The AQUILO robot was made with financial support of the AGH Center of Technology Transfer as part of programme Incubator of Innovation.

A. Bartoszewicz et al. (Eds.): PCC 2020, AISC 1196, pp. 1213–1224, 2020.
https://doi.org/10.1007/978-3-030-50936-1_101

up to 300 kg in hospitals [11]. The next robot designed for hospitals was HelpMate constructed by Transitions Research Corporation [9,10]. The HelpMate robot could navigate crowded corridors, avoid obstacles, people, and its task was to assist in delivering medicines and meals. Another experimental construction was a robot named Xsavier at NASA laboratories [21]. The Xavier robot's task was to deliver consignments in office rooms. The next autonomous robot was RHINO that, despite being designed to participate in autonomous robotics competitions, also performed the task of autonomous driving in rooms [8]. In recent years there has been a significant increase in the number of prototypes of autonomous delivery robots made by companies or startups, especially in USA. One of the better known constructions is the KIVA transport robot used, for example, in Amazon warehouses [13]. Another robot, named SAVIOKE was designed to deliver consignments in various locations (e.g. hotels, factories) [16]. Among similar autonomous robots used to deliver consignments Aethon and FeetchRobotic can be listed [6,26].

In this paper the design and completion of the autonomous delivery robot called AQUILO, which delivers small consignments, for example, in office rooms, is described. The aim of the design was to build a fully functional autonomous robot that could achieve its task safely. This paper presents the robot's basic components, such as the drive system, power supply, control system, the sensors used, safety systems and also the algorithms used. The paper is organised as follow, Sect. 2 presents the main assumptions which should be fulfilled by robot, Sect. 3 presents mechanical and drive construction of robot, Sect. 4 presents architecture of control system which detailed description of low level and high level control system. The next sections describe used sensors (Sect. 5), navigation system and software (Sect. 6), safety systems used in AQUILO robot (Sect. 7) and the last Sect. 8 presents conclusions.

2 Robot Design and Assumptions

The main task performed by the AQUILO autonomous robot is to deliver small consignments in such buildings as office blocks, hotels, hospitals, public administration buildings, etc. The AQUILO autonomous robot should be able to automatically establish maps of the rooms in which it is moving, plan the optimal delivery path and avoid any obstacles that it may encounter. The AQUILO robot must be safe, it cannot drive into anyone or other obstacles, its behaviour should not cause anxiety among users (it should not accelerate abruptly or move at excessive speed), and at any time it should be possible to take control of robot operations. In addition, the AQUILO robot should be equipped with an intuitive interface to communicate with people and to display various informative messages or robot operating status. The AQUILO robot must be equipped with a drive that allows manoeuvring in confined spaces, e.g. a lift, the robot's working time between consecutive loading should be a minimum of 10 h and the load which is carried should be approx. 30 kg. An important aspect which was also taken into account at the AQUILO robot designing stage was that servicing the robot should be as easy as possible (e.g. battery cell charging), and modifications or extensions at a later stage should be no obstacle. The final conceptual design of the AQUILO robot is shown in Fig. 1(a).

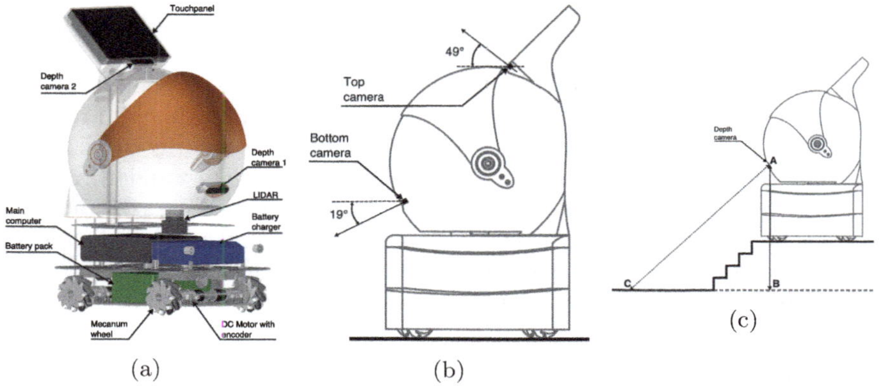

Fig. 1. The AQUILO robot, (a) the final concept, (b) location of the INTEL D415 cameras on the AQUILO robot, (c) the way in which the stairs in front of the robot are detected by the *Stairs Detection System*.

3 Drive and Mechanical Construction

While considering the requirements presented above, in particular the maximum manoeuvring in confined spaces, it was decided that mecanum wheels should be used [23]. The special feature of this drive is that it allows independent movements in both the x and y axis, as well as rotation around its own axis (3 degrees of freedom). Each of the robot's mecanum wheels must have its own motor allowing the orientation and rotational velocity of the wheel to be adjusted independently of the other wheels. The disadvantage of the mecanum wheels is the weight (for larger wheels) and the relatively small ground clearance. In the AQUILO robot 100 mm mecanum wheels and DC motors series 3257 from FAULHABER were used. Basic motor parameters: nominal voltage 24 V, maximum output speed 5900 RPM, encoder resolution 1024 pulses per revolution. DC motors are commonly used in robot constructions due to their high torque, compact size, affordable price and easy control [4]. The motors used are integrated with a gearbox and encoders to create an entire unit. Motors are controlled by using a PWM signal generated by a low-noise control system described in more detail in the Sect. 4.1.

4 Control System Architecture

The control system of the AQUILO robot is divided into three independent levels. Each level performs a specified group of its intended tasks. Individual levels can exchange data between themselves.

The AQUILO robot has two operating modes: manual and autonomous. In the manual mode, the robot's speed and the movement direction are controlled with the use of a joystick, pad or a different control device. This operating mode is used, for example, for collecting the data necessary to create maps of rooms. In the autonomous mode the AQUILO robot executes a selected algorithm of operation, e.g. the autonomous delivery of consignments, random wandering, etc. In this operating mode it is possible to have an impact on the robot's operations by using a robot interface (touch screen) that may display various types of information, questions related to the tasks being executed. When the robot is working in the autonomous mode, the operating mode can be switched to manual at any time.

4.1 Low Level Control

The lowest control level is executed by a specially designed unit based on the STM32 processor. This unit (hereinafter referred to as *ARE Robot Cape*) performs low-level tasks such as: generating PWM signals for motor controllers (4 outputs), reading data from encoders (4 inputs), speed and rotation direction control for each motor, adaptive lighting control, fan speed control and individual cell voltage measurement. In addition to this, *ARE Robot Cape* contains: 20 GPIO inputs/outputs, 2 relay outputs (30VDC or 277VAC), 2 relay outputs for lidar safety zones, SPI barometer support (e.g. BM280, BM160), 3 PWM 5 kHz universal outputs, supporting JSNH04T ultrasonic sensors and 4 optically isolated outputs. The *ARE Robot Cape* device works with Pololu motor controllers, thus enabling different types of motors to be connected. The main purpose of *ARE Robot Cape* is to correctly control the mecanum drive that requires the accurate determination of angular velocity for each wheel [23]. The direction and the absolute value of the velocity can be sent to *ARE Robot Cape* through a serial port to convert them into the rotational speed and orientations of all wheels. To control wheel rotational speed properly, the feedforward control is used with a stabilizing *PID* controller [2,3]. The parameters of *PID* controllers were selected numerically by minimizing a specified quality index, while the proper selection of parameters ensuring the stability of the control system was verified by coverage tests [22]. The use of fractional-order *PID* controllers is being considered for the future. They are more and more commonly used for controlling and modelling, and the results are very encouraging, see for example [20]. Another task performed by *ARE Robot Cape* is the rotational speed measurement of each wheel by using encoders. Properly converted and rescaled values of linear velocity and displacements by using wheel dimensions are then sent to a serial port. In addition, *ARE Robot Cape* reads, on an ongoing basis, the values of all the sensors which are connected, rescales and converts these readouts, and finally sends all data to a serial port to be read and used by other devices.

4.2 Mid-Level Control

The mid-level control of the AQUILO robot is performed by the *Robot Operation System* (*ROS*) framework that works on a PC with the Ubuntu operating system [12]. The *ROS* framework is a set of software libraries, modules and programs that allow advanced robotic control software to be developed. *ROS* was released under the BSD license, and as such is an open source software and free for both commercial and research use. The main advantages of the *ROS* framework include a multitude of ready-to-use modules, standardized communication between modules, multithreading, advanced test and simulation environment, the possibility of programming in Python or C++, a lot of ready-to-use applications and modules, and a huge community of developers and commercial companies supporting its development.

The main tasks of mid-level control of the AQUILO robot include: room map building, determining optimal paths, moving along the planned path, filtering and analysing data from lidar and depth cameras, avoiding encountered obstacles, communication and data exchange with the low-level control system and support for the communication interface with its operational logics. All of these tasks are carried out by appropriate *ROS* framework modules that were adequately configured and suited to the functional needs of the AQUILO robot.

4.3 High-Level Control–Management Control

The purpose of high level control is to provide an application programming interface (*API*) for operating a robot, managing a group of robots, gathering and analysing data sent by robots and global task planning for a group of robots. Until now this control level has not been implemented yet.

4.4 Power Supply

The AQUILO robot is powered by a Li-Pol cell package with a total capacity of 54 Ah and a voltage of 24 V. The selected cell package ensures the robot's continuous operation for 12–15 h. When working, the voltages on individual batteries are monitored continuously and if they become too low or unbalanced, the robot displays an appropriate message and stops working. To charge the cell package a microprocessor-controlled charger built in the AQUILO robot is used. Thus, the AQUILO robot does not require a special charging station. It is planned to add the ability of automatic wireless charging or an in docking station in the future.

5 Sensors

The AQUILO autonomous robot is provided with various sensors to be able to determine its position on a map, as well as to avoid obstacles encountered on its path. Figure 2 presents the layout and range of the AQUILO robot sensors.

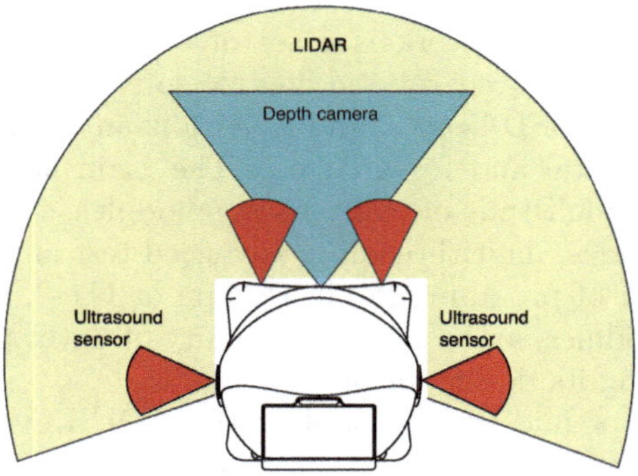

Fig. 2. The layout and range of the AQUILO robot sensors.

The basic sensor used by the AQUILO robot is HOKUYO UST-10LN lidar which has the following parameters: measurement steps 1081, angular resolution 0.25°, detection angle 270°, detection range 10 m, response 25 ms.

The lidar used is characterized by the fact that it is fully encapsulated together with its rotating laser head, thus significantly improving its durability and making it easy to use. The lidar is directly connected to the host computer. It is used to create maps of rooms, robot localization and obstacle detection.

Another sensors type used in the AQUILO robot are two INTEL®Real Sense^{TM} Depth D415 depth cameras. The special qualities of these cameras are that, beside normal images, they can give a depth image of a frame. Therefore, control algorithms have information about the real distances of different objects from the robot in the camera's field of view. The depth cameras are mounted at different angles, as shown in Fig. 1(b).

The lower camera is used by the *Stair Detection System* (abbr. *SDS*) designed for the detection of stairs (downwards) that the robot may encounter during work. Stairs, as map components, are not marked in maps created by the AQUILO robot (this has to be done manually and is an arduous task). The *SDS* system detects stairs on an ongoing basis and prevents the AQUILO robot from going down the stairs as that could surely lead to severe damage to the robot. The upper camera is used in the robot interface system. It can be used, for example, for recognizing faces of delivery receivers and senders to make authorization as simple as possible. Another functionality is the implementation of eye tracking of the user and displaying this movement on the AQUILO robot screen, thus representing robot–human interaction and cooperation.

The lidar, despite its advantages such as rapid response and accuracy, it is not 100% reliable. Therefore, in order to raise the robot's safety level the decision was made to install ultrasonic distance sensors, which enable the detection of obstacles at a distance of 20 cm–6 m away, on the robot itself. They are used in the robot's safety system to prevent it from colliding with other obstacles. The ultrasonic sensor arrangement is presented in Fig. 2.

Apart from the sensors mentioned above, the AQUILO robot is also equipped with a barometer used to estimate height, e.g. when the robot is travelling in a lift. To control the power supply system, the voltages on cells are continuously measured with an ADC converter.

6 Navigation

This is one of basic functionalities of the AQUILO robot. It allows the robot to move safely indoors. Navigation is based on room maps. These maps are built by a *ROS* node called *SLAM Gmapping.* A new map can be created in two ways. Firstly, the user operates the robot manually by using a pad and tries to steer it over the area which the robot will later work in. Secondly, the AQUILO robot, using an exploration algorithm *Frontier Exploration* implemented in *ROS*, tries to steer itself over the area of its planned movements [24]. In both cases, whilst on the move, the robot scans (with the lidar) the space around it and records data. These data are used to create a room map used by the AQUILO robot. The first method of map making is faster and more accurate - the user can precisely cover the whole area. However, it requires human manual control. The second method scans the space automatically, but it is slower, as the robot moves slowly over an unknown area. In addition, maps built using the second method may be less accurate since the robot cannot reach all the places in the scanned area.

To work properly the robot needs to know its position on the map at any time. To determine the current position on the map lidar measurements are used and data from the robot's wheel odometry, are combined together by using the Kalman filter algorithm [18]. The position is computed by using an adaptive Monte Carlo algorithm [12, 15]. There is also another method of robot localization which uses only depth cameras and the *Fast Sampling Plane Filtering* algorithm [5].

One of functionalities of the AQUILO robot is that it has the ability to use a lift. Thanks to this, the AQUILO robot is mobile throughout the entire building, on all floors. Map implementation in *ROS* allows only the use of 2D maps on one level, thus it has become necessary to develop an additional module named *Maps Repository Module* (abbr. *MRM*) to manage and switch the robot's maps depending on which level and part of the building the robot is in. For each building *MRM* creates a directed graph, whose nodes represent levels with its maps, while the edges indicate the possibility of switching from one map to another (that is, for example, changing floor in a building). Figure 3 presents an example of a graph with maps of individual building levels or zones.

Arrows at the edges indicate that the robot can go directly from one node to another. Should the AQUILO robot need to move from a point in map `MAP 1` to a

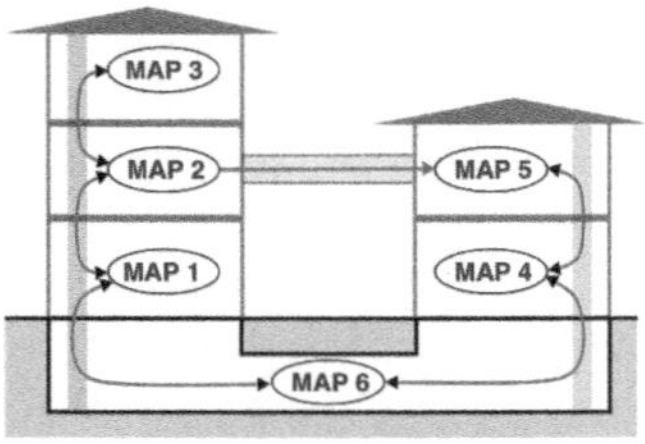

Fig. 3. The maps graph used by AQUILO robot.

point in `MAP 2`, it has to determine a path that also runs across the area of `MAP 6` because there is no direct connection between the points in `MAP 1` and in `MAP 2` (there is no oriented edge connecting the appropriate nodes of the map graph). The use of graph theory to manage area maps also enable the determination of various path variants with different cost identifying standards, e.g. the shortest distance, moving time, number of map changes [7].

7 Safety

The AQUILO autonomous robot is equipped with different systems assuring its correct and safe operation. The ideal safety system should be reliable, operable in any conditions and assure 100% efficiency. Unfortunately, such safety systems do not exist. However, to ensure an appropriate safety level, redundant safety systems are used, thus enabling the risk of an unwanted or dangerous situation to be significantly reduced.

7.1 Collision Avoidance

The main danger that may occur when the AQUILO robot is working is collision with other objects. The AQUILO robot contains three independent systems operating to prevent such events. The first anti-collision system consists of an algorithm detecting obstacles along the AQUILO robot's path and an algorithm modifying its path to avoid the detected obstacle. This system is implemented in the mid-level control layer and only works if the robot moves automatically. The system is very accurate, but to work properly it uses quite a lot of host computer resources. In the case of host computer failure or stopping the execution of the main control algorithm, this system also stops working, which is its weak point.

In manual control, the user is responsible for ensuring the collision-free movement of the robot. However, if for some reason the manual controls performed by the user could lead to a collision, this is avoided by the system active in the low-level control layer. The lidar used in the robot allows so called safety zones to be defined and if violation of such safety zone is detected, the lidar generates a Hi signal on one of its outputs. This signal is used in the AQUILO robot to disconnect motors from the power supply with a relay. The advantage of this system is its simplicity and operational reliability. Moreover, this system works fully independently on the mid-level control, thus its operation is not affected by either control computer or the execution of the main control algorithm. The disadvantage of this system is its two-state action, i.e. it can only stop the AQUILO robot.

Both the safety systems mentioned above use data from the lidar. However, dangerous situations can occur when the lidar does not detect an obstacle properly, e.g. glass doors. To avoid such events the AQUILO robot is provided with ultrasonic distance sensors. If an obstacle is detected at close distance by the ultrasonic sensors, then the AQUILO robot's drive is switched off. This system is implemented in the low-level control layer, and the distance is measured by the *ARE Robot Cape* device in which an appropriate algorithm is implemented. Using ultrasonic sensors to calculate distances is not very accurate and relatively slow (if several such sensors are used), however these sensors are capable of detecting such obstacles as glass doors.

7.2 Stair Avoidance

One of obstacles undetectable by the lidar and ultrasonic sensors are stairs. These are dangerous construction elements for a robot because it can fall down the stairs and cause injury to people at the bottom and damage to the robot itself. In order to ensure that the AQUILO robot does not fall down the stairs, the *Stairs Detection System* was designed and built. The system uses a depth camera oriented downwards analysing the space in front of the robot on an ongoing basis. Figure 1(c) presents the way in which the stairs in front of the robot are detected. The depth camera located at point A is directed downwards and measures the distance of the segment AC. Knowing the angle $\angle CAB$, distance AB can be calculated by using the Pythagorean theorem. If this value is greater than the set limit value, the *SDS* reports that stairs have been detected. The *SDS* is implemented in the mid-level control layer and when stairs are detected it stops the robot's movement towards this point. Stair detection systems are known in robotics with the difference that they are used in autonomous robot vacuum cleaners or stairs climbing robots [17]. Another method used for stair detection is marking them in a map as an obstacle (e.g. wall) or placing special markings which inform the robot about stairs. Such solutions have the following disadvantages: each map has to be edited by the user who must mark all the stairs, and if the localization of the robot in the map is inaccurate, it may not notice the stairs, and the necessity to install additional equipment [19]. The *SDS* designed and used in the AQUILO robot is free of these drawbacks.

7.3 Power Supply

When the AQUILO robot is working, voltages on individual battery cells are monitored continuously. If the measured voltage is smaller than the allowable value or the voltage difference between individual cells is too large, the AQUILO robot goes into failure mode and displays the appropriate message. In this case, it is impossible for the AQUILO robot to move. This protection prevents the battery pack used in the AQUILO robot from damage.

7.4 Drives

The drives of the AQUILO robot can be disconnected from the power supply with a mechanical switch at any time. This simple and reliable method can stop robot movement. Other robot systems such as computers and sensors work normally.

7.5 Other Protections

Apart from the safety systems described above, the AQUILO robot has had a number of other protection systems implemented. The AQUILO robot does not start to perform any tasks or interrupts a task if it finds that one of the sensors (lidar, depth camera, encoders) has stopped working or sent incorrect data.

The AQUILO robot also stops working when it is unable to follow the plotted path. Such a situation can occur, for example, if one of the drives ceases to work or when detected obstacles make it impossible to determine a trajectory or move along the planned path. In the latter case, the AQUILO robot will go into so called *recovery* mode after some time. In this mode the AQUILO robot tries to locate itself in the map once again and to determine a new feasible path.

In manual control mode, when the operator is responsible for the correct and safe operation of the AQUILO robot, its protection systems are also active. Operation with a wireless pad is possible only if one of the specified buttons is pressed down continuously. This protects against unauthorised use of the robot. When the AQUILO robot is in manual control mode all obstacle avoidance systems, *SDS* and power monitoring system are also active. Reversing the AQUILO robot is normally impossible (since the robot has no sensors detecting obstacles behind it), but is possible only if the AQUILO robot is in a special service mode when all the safety systems are switched off. Going into this service mode is possible only if a specified combination of pad buttons is selected, and this combination is sufficiently complex to avoid unauthorised use.

The AQUILO robot is also equipped with Adaptive Lighting System (abbr. *ALS*). This system additionally lights up the area in front of the AQUILO robot which improves the operation and accuracy of the *SDS* depth camera.

8 Conclusions

The design and structure of the AQUILO Autonomous Delivery Robot are presented in this paper. The robot has been designed and built by an AGH employee and student team. The AQUILO robot proved that it is possible to design and build various autonomous robots that will assist humans in performing different simple operations (e.g. transporting small deliveries). The advantage of the AQUILO robot is its modular structure thus making it possible to easily and cheaply adapt it to performing other tasks while maintaining all the advantages of the original construction such as manoeuvring, working time and protection level.

In the AQUILO robot mecanum wheels are used which, in combination with the *ARE Robot Cape* device, ensure very good manoeuvring even in confined spaces.

The multi-level control system used in the AQUILO robot works as assumed and expected. The control system is fully scalable, i.e. it is very easy to adapt it to large size autonomous robots. This was achieved partly due to the *ROS* framework that is becoming a standard in such constructions.

To perceive the area surrounding the robot, the lidar and depth camera were used. The use of data fusion algorithms along with data from the lidar and encoders ensure a very accurate and reliable localization of the AQUILO robot. A dozen of various protection systems were used in the AQUILO robot to ensure the safety of the robot itself and its users.

Further development plans assume the completion of the highest level control layer to ensure the full integration and cooperation of any number of AQUILO robots. Further development works on behavioural algorithms and artificial intelligence algorithms that will allow the AQUILO robot to perform more and more complex tasks are also planned. The team is also working on the commercialisation of the AQUILO robot construction. Its capabilities are presented during fairs and shows as well as to potential investors.

References

1. Ardiny, H., Witwicki, S., Mondada, F.: Construction automation with autonomous mobile robots: a review. In: 2015 3rd RSI International Conference on Robotics and Mechatronics (ICROM), pp. 418–424. IEEE (2015)
2. Aström, K.J., Murray, R.M.: Feedback Systems: An Introduction for Scientists and Engineers. Princeton University Press, Princeton (2010)
3. Åström, K.J., Wittenmark, B.: Computer-Controlled Systems: Theory and Design. Courier Corporation, New York (2013)
4. Baranowski, J., Długosz, M., Mitkowski, W.: Remarks about DC motor control. Arch. Control Sci. **18**(3), 289–322 (2008)
5. Biswas, J., Veloso, M.: Depth camera based indoor mobile robot localization and navigation. In: 2012 IEEE International Conference on Robotics and Automation, pp. 1697–1702. IEEE (2012)
6. Bloss, R.: Mobile hospital robots cure numerous logistic needs. Ind. Robot Int. J. **38**(6), 567–571 (2011)
7. Bondy, J.A., Murty, U.S.R., et al.: Graph theory with applications, vol. 290. Citeseer (1976)
8. Buhmann, J., Burgard, W., Cremers, A.B., Fox, D., Hofmann, T., Schneider, F.E., Strikos, J., Thrun, S.: The mobile robot Rhino. AI Mag. **16**(2), 31–31 (1995)
9. Evans, J.M.: Helpmate: An autonomous mobile robot courier for hospitals. In: Proceedings of IEEE/RSJ International Conference on Intelligent Robots and Systems (IROS 1994), vol. 3, pp. 1695–1700. IEEE (1994)

10. Evans, J.M., Krishnamurthy, B.: Helpmate®, the trackless robotic courier: a perspective on the development of a commercial autonomous mobile robot. In: Autonomous Robotic Systems, pp. 182–210. Springer (1998)
11. Katevas, N.: Mobile Robotics in Healthcare, vol. 7. IOS Press, Amsterdam (2001)
12. Koubâa, A.: Robot Operating System (ROS). Springer, Cham (2017)
13. Li, J.T., Liu, H.J.: Design optimization of amazon robotics. Autom. Control Intell. Syst. **4**(2), 48–52 (2016)
14. Manyika, J.: A future that works: AI, automation, employment, and productivity. Technical report, McKinsey Global Institute Research (2017)
15. Marder-Eppstein, E., Berger, E., Foote, T., Gerkey, B., Konolige, K.: The office marathon: robust navigation in an indoor office environment. In: International Conference on Robotics and Automation (2010)
16. Markoff, J.: Beep, Says the Bellhop. New York Times (2014)
17. Mihankhah, E., Kalantari, A., Aboosaeedan, E., Taghirad, H.D., Ali, S., Moosavian, A.: Autonomous staircase detection and stair climbing for a tracked mobile robot using fuzzy controller. In: 2008 IEEE International Conference on Robotics and Biomimetics, pp. 1980–1985. IEEE (2009)
18. Moore, T., Stouch, D.: A generalized extended Kalman filter implementation for the robot operating system. In: Proceedings of the 13th International Conference on Intelligent Autonomous Systems (IAS-13). Springer (2014)
19. Murai, R., Sakai, T., Kawano, H., Matsukawa, Y., Kitano, Y., Honda, Y., Campbell, K.C.: A novel visible light communication system for enhanced control of autonomous delivery robots in a hospital. In: 2012 IEEE/SICE International Symposium on System Integration (SII), pp. 510–516. IEEE (2012)
20. Oprzędkiewicz, K., Gawin, E., Gawin, T.: Real-time PLC implementations of fractional order operator. In: Conference on Automation, pp. 36–51. Springer (2018)
21. Simmons, R., Goodwin, R., Haigh, K.Z., Koenig, S., O'Sullivan, J.: A layered architecture for office delivery robots (1997)
22. Skruch, P., Długosz, M., Mitkowski, W.: Mathematical methods for verification of microprocessor-based PID controllers for improving their reliability. Eksploatacja i Niezawodność **17** (2015)
23. Taheri, H., Qiao, B., Ghaeminezhad, N.: Kinematic model of a four Mecanum wheeled mobile robot. Int. J. Comput. Appl. **113**(3), 6–9 (2015)
24. Trivun, D., Šalaka, E., Osmanković, D., Velagić, J., Osmić, N.: Active slam-based algorithm for autonomous exploration with mobile robot. In: 2015 IEEE International Conference on Industrial Technology (ICIT), pp. 74–79. IEEE (2015)
25. West, D.M.: The Future of Work: Robots, AI, and Automation. Brookings Institution Press, Washington (2018)
26. Wise, M., Ferguson, M., King, D., Diehr, E., Dymesich, D.: Fetch and freight: standard platforms for service robot applications. In: Workshop on Autonomous Mobile Service Robots (2016)

Customizable Inverse Sensor Model for Bayesian and Dempster-Shafer Occupancy Grid Frameworks

Jakub Porębski[1,2(✉)]

[1] Faculty of Electrical Engineering, Automatics, Computer Science and Biomedical Engineering, Department of Automatic Control and Robotics, AGH University of Science and Technology, Kraków, Poland
jakpor@agh.edu.pl
[2] Aptiv Services Poland S.A., Kraków, Poland

Abstract. Occupancy grid mapping is an important component in a road scene understanding for autonomous driving. It can encapsulate data from heterogeneous sensor sources like radars, LiDARs, cameras and ultrasonics. At the core of occupancy grid (OG) generation, there is usually an inverse sensor model (ISM), which infers the occupancy representation from the sensor readings. Traditional ISMs are characterized by a very rigid structure, suited only for one sensor type, and specific occupancy grid representation. This paper proposes a novel ISM framework, which offers a separation between free and occupied space, supporting both Bayesian and Dempster-Shafer OG representations. The framework is especially useful when dealing with multiple different sensors where custom or preselected probability distribution can be applied. The presented ISM architecture is modular and flexible, which is described in an illustrative example of application customized for different detection sources.

Keywords: Inverse sensor modeling · Occupancy grid · Environment mapping methods · Bayesian inference · Dempster-shafer evidence theory · Sensor models

1 Introduction

One of the most popular frameworks for an environment model definition in robotics is the occupancy grid map [4,11]. It is suited to perform sensor fusion and has been successfully utilized for a range of tasks including localization [10, 13] and path-planning [14].

The occupancy grid is a multidimensional spatial lattice, where each cell stand for an independent portion of a space [3]. The task of this environment model is to predict the probability that each grid cell is either occupied or free based on sensor observations. For that purpose forward or inverse sensors are implemented.

A. Bartoszewicz et al. (Eds.): PCC 2020, AISC 1196, pp. 1225–1236, 2020.
https://doi.org/10.1007/978-3-030-50936-1_102

The forward sensor model gives the probability of observing a reading based on map knowledge [15]. Forward models have the advantage that they can be determined experimentally and can characterize a sensor in a more straightforward manner. Forward model reflects the spatial structure and the spatial disposition of obstacles in the modeled physical world. However, the expectation-maximization algorithm utilized in this type of models requires accumulation of sensor data and the convergence could be too computationally complex for real-time constrained applications.

The other commonly used approach for probability estimation in occupancy grids is the inverse sensor model (ISM). This method infers the occupancy (cause) from the sensor readings (effect) [12]. In contrast to the forward modeling, the ISM may be given by the arbitrary distribution profiles which calculation is performed in real-time. The main limitation of this paradigm is the independence hypothesis of sensor measurements. It allows incremental data fusion limiting the required computation. However, the detection independence lead to occlusions and may be the cause of sensor conflicts [16].

In this work a design approach of the inverse sensors model is proposed, which separates the free and occupancy probabilities computation. Presented approach is fully customizable and allows easy adaptation for the multi-state occupancy grids (e.g. utilizing Dempster-Shafer evidence theory). Moreover, separation of free-space and occupancy through utilization of two separate maps lead to elimination of sensor conflicts as presented in [5].

The paper is structured as follows: the variety of features that are available for the inverse sensor modeling is described in Sect. 2; Sect. 3 presents a unifying framework of the inverse sensor model, which is customizable for the application needs; Sect. 4 shows experimental evaluation of the ISM compared to existing solutions. Summary and the development possibilities are given in Sect. 5.

2 Probabilistic Sensor Models

In the late-80s Elfes proposed a Bayesian inverse sensor model designed for the ultrasonic sensors [4]. The Bayesian approach computes the ISM utilizing the sensor model, therefore expressing measurement uncertainties given the physical location of the sensed obstacle. This approach requires the enumeration of all possible grid configurations which causes the exponential computation complexity.

In order to create the ISM with a linear complexity, the analytical approach has been proposed [1]. It approximates the sensor model using continuous function defined over the distance from the sensor. It is based on the Gaussian distribution [8] or on the power function [21].

The further simplification for the analytical ISM has been presented in [6,18,20], which approximates the occupancy depending on the distance from detection. This approach uses simple analytical functions like linear approximation or 3-valued spline [18,20].

Another way of acquiring an inverse sensor model is to learn the sensor-specific characteristics from the labeled data. The idea is to train a neural network ISM in a controlled environment where the occupancy state of cells and the measurements are known [19]. This method connects the linear complexity with sensor modeling, however the huge volume of training data is required to capture all relevant use-cases of the occupancy grid.

The main disadvantage of different analytical approaches is the tendency to improperly hide the occupancy depending on the order of incoming data [15]. This problem is called the sensor conflict and is caused by incremental fusion of opposing information like occupied and free-space. For that purpose Foroughi et al. in [5] introduced two independent grid maps: one for occupied and one for free-space accumulation. This approach requires separation of cell states, which could be performed already in the ISM.

3 Inverse Sensor Model Framework

Range sensors such as laser-based sensors, stereo-camera, radars and ultrasonic sensors are commonly used in robotics and also for autonomous vehicles. Each detection from these sensors is able to provide two not complementary types of information. Firstly range measurements sense the presence of object. Secondly the reflected ray carry the information about the possible traversal space to the target.

In common implementations of the inverse sensor models those two information are packed in one distribution [4,7,8,21]. Nevertheless, various grid applications require separation of free and occupancy data to solve sensor conflicts [5] or to estimate additional environment parameters [17].

In order to address this problem, the proposed framework unravels the ISM distribution by splitting the process into two separate paths, which internally accumulate two independent information. This implementation of the ISM could be paralleled since reflection and traversal space paths are independent, as presented in Fig. 1.

Moreover, the sensors utilized in the occupancy grids could be divided based on the angular uncertainty. The LiDAR sensor provides an accurate point cloud with huge amount of points. In that case the ISM computation has to be as fast as possible to preserve the computational constraints. For that purpose the different methods of 1D or hit point ISM has been proposed which neglect the angular uncertainties of the detections [6,18,20,21].

On the other hand, the radar or ultrasonic sensors have significant both radial and angular uncertainties, however they provide smaller number of detections in time period. In that case the ISM could be more accurate by application of 2D distribution as presented in [4,6,8].

The processing flow presented in Fig. 1 consists of cell selection (Sect. 3.2), probability calculation (Sect. 3.3) and mapping of internal ISM probabilities into occupancy grid map representation (Sect. 3.4).

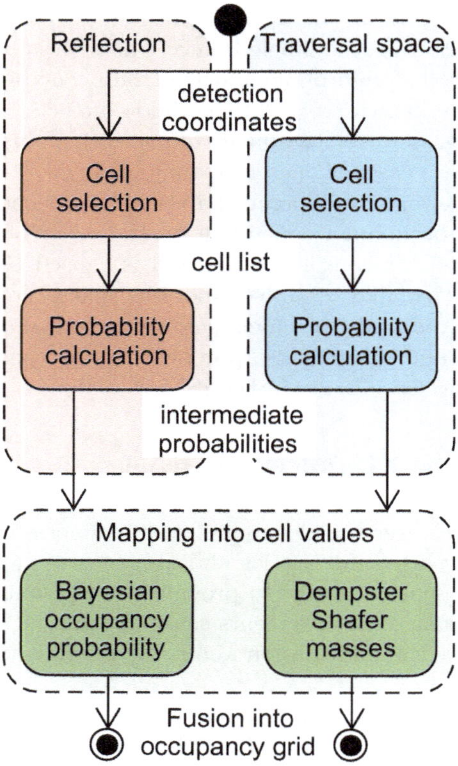

Fig. 1. Dual inverse sensor model framework information flow.

3.1 Representation of Information

Each processing path accumulates different probability type on separate intermediate grid. The intermediate grids **IG** are matrices, where each element describes the portion of a space called cell area c.

In the occupancy grid model cells are independent entities and most of the mathematical operations on the grid does nod depend on cell position. In all that cases, to simplify the notation cell indexes will be omitted meaning that the operation has to be applied for all cells in the grid.

Values of the reflection grid $\mathbf{IG}_\mathrm{R}$ express probability of a cell area c being an origin for any of processed sensor detection d. Respectively the traversal space data $\mathbf{IG}_\mathrm{T}$ describe probability of a cell area c being free from obstacles f.

$$\mathbf{IG}_\mathrm{R}(c) = p(c|d) \qquad \mathbf{IG}_\mathrm{T}(c) = p(c|f) \tag{1}$$

Intermediate grids store information only between each fusion iteration. In that period incoming detection data d^t are fused into intermediate grids

$\mathbf{IG}^t_\mathrm{R}$, $\mathbf{IG}^t_\mathrm{T}$. The fusion rule for each grid could be different and should depict the kind of information provided by sensors. Each consecutive reflection from some obstacle is a new evidence so it should increase the $\mathbf{IG}^t_\mathrm{R}$ probability. For that grid, the De Morgan's fusion method is used (2). On the other hand, the free-space information $\mathbf{IG}^t_\mathrm{T}$ depend mostly on different sensor capabilities instead on the number of traversed rays. To meet that assumption a maximum policy could be applied (3).

$$\mathbf{IG}^{t+1}_\mathrm{R}(c) = 1 - (1 - \mathbf{IG}^t_\mathrm{R}(c))(1 - p(c|d^t)) \tag{2}$$

$$\mathbf{IG}^{t+1}_\mathrm{T}(c) = \max(\mathbf{IG}^t_\mathrm{T}(c), p(c|f^t)) \tag{3}$$

3.2 Cell Selection

The computation of ISM for each detection on whole occupancy grid is excessive for real time application. To limit the number of cells, where the ISM has to be applied, cell selection policy is proposed.

The simplest approach is to apply the ISM only for the cell, where the detection is located. The cell hit by the detection receives some occupancy probability (usually maximum). Traversal space for the hit-point selection might be efficiently selected using Bresenham line algorithm [2].

For the 2D approach, cell selection should also reduce the area of ISM influence. In that case one of the reasonable solution is to select cells that lie in the 3σ area range. The procedure of selecting relevant cells is presented in algorithm 1. The example results of selection are depicted in Fig. 2.

Algorithm 1. Cell selector for 2D ISM

for all detections in scan **do**
    Find a cell hit by detection
    Define the rectangle area around detection to scan. The rectangle is an upper bound of selected cells. It can be a square with side length of $\max(3\sigma_r, 3\sigma_\phi)$, where σ_r and σ_ϕ are radial and axial detection standard deviations.
    for each cell in the rectangle around the hit cell **do**
        if the cell is within the 3σ ellipse range **then**
            Cell is selected and probability will be computed (see section 3.3)
        end if
    end for
end for

Common approach for the two-dimensional ray casting is to perform multiple iterations of the Bresenham line algorithm. However, this solution is susceptible to the Moire effect and could omit some cells at long ranges [1,5,6]. To eliminate this problem simple triangle rasterization algorithm such as Barycentric or Bresenham triangle filler could be used (Fig. 2).

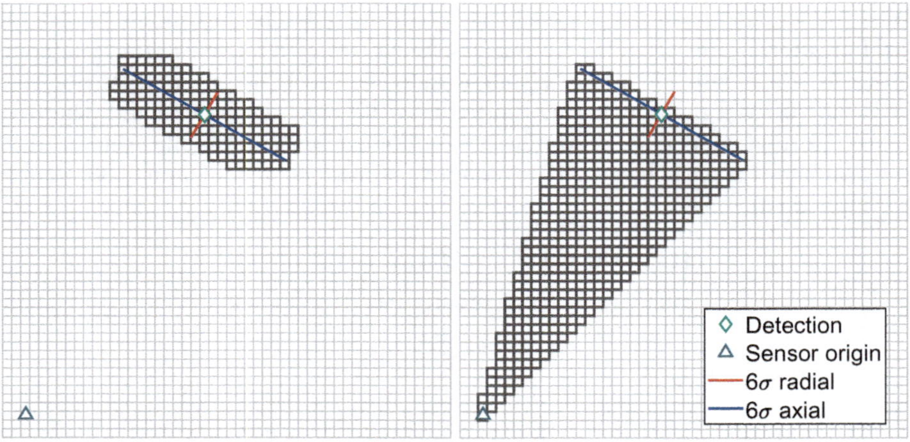

Fig. 2. Examples of cell selectors for detection reflection origin (left) and corresponding triangle free-space area (right). Dark border of cell means that the cell was selected for processing. Detection parameters: $r = 8m$, $\phi = 30°$, $\sigma_r = 0.2\,\text{m}$, $\sigma_\phi = 5°$. Grid resolution 0.2 m.

3.3 Probability Calculation

Having selected cells, the probabilities have to be computed. In this step information from the sensor model should be extracted to determine the distribution. The recent reports present both Gaussian and exponential models, which could be utilized to calculate sensor specific probability density function (pdf).

$$\text{Gaussian:} \quad \text{pdf}(x, \mu, \sigma) = \frac{1}{\sqrt{2\pi\sigma^2}} e^{\frac{-(x-\mu)^2}{2\sigma^2}} \tag{4}$$

$$\text{Exponential:} \quad \text{pdf}(x, \mu, \lambda) = \begin{cases} \lambda e^{-\lambda(x-\mu)} & x \geq \mu \\ 0 & x < \mu \end{cases} \tag{5}$$

The probability model is continuous which contradicts to the discrete cell representation. Hence, to measure probability from the pdf function, the integration over the cell area has to be performed. Calculation of an accurate integral is computationally infeasible, therefore the approximation is performed. The common approach is to use cell center as a reference point [9], however when the grid resolution is comparable to the detection uncertainty, the monotonic character of the distribution may be broken. In order to preserve the distribution shape, in the presented inverse sensor model, the pdf integral is approximated by maximum of the density function in the measured cell multiplied by cell area.

$$p(c)_{2D} = \iint_c \text{pdf}(x, \mu, \sigma) dx^2 \approx \max(\text{pdf}(c)) \times \text{cell area} \tag{6}$$

For the one dimension case, the continuous density function (*cdf*) could be calculated analytically, so the integration step may be defined as substitution:

$$p(c) = \int_c \text{pdf}(x, \mu, \sigma) = \text{cdf}(c_{end}) - \text{cdf}(c_{begin}) \tag{7}$$

In presented ISM framework, the probability of detection is calculated for every selected cell. The sanity check of the probability correctness is that sum of probabilities for all cells in the list should converge to one ($\sum_c p(c) \approx 1$) as presented in Fig. 3.

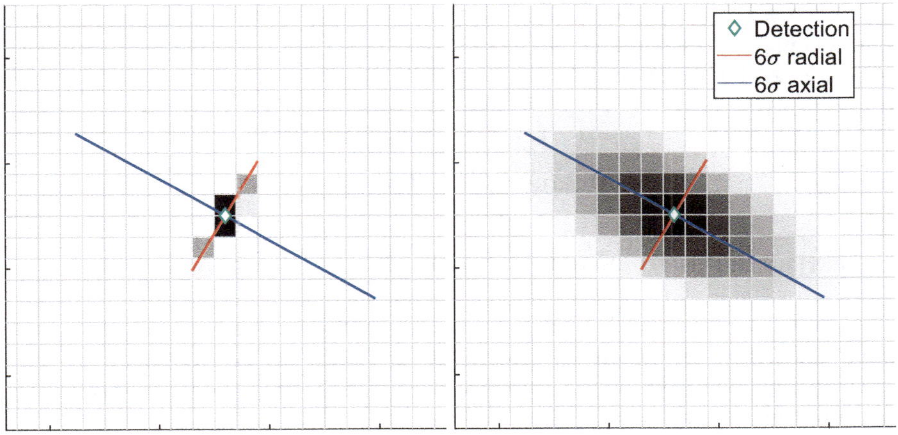

Fig. 3. Examples of probability calculation for detection scattering center using 1D (left) and 2D (right) Gaussian model with maximum integration method. Colormap not in scale.

The probabilities and integrals computed in this step are usually very small. In that case, beneficial for computation precision might be usage of logarithmic scale to store and further compute the occupancy beliefs.

The other approaches for the probability determination are the analytical ISMs described in Sect. 2. Those solutions bypass the probability distribution and integration [18,20], but they still could be integrated into presented framework with separated traversal and occupancy paths.

3.4 Mapping into Cell Values

Intermediate probabilities represent two non complementary evidences about occupied and free-space. That kind of representation is used in Dempster-Shafer based approaches. Using frame of discernment with states occupied and free ($\Omega = \{\mathrm{O}, \mathrm{F}\}$), the values determined on each intermediate grid could be fused independently into Dempter-Shafer grid using any combination rule. The evidence masses grids m are created directly from intermediate grids:

$$\mathbf{m}_{\mathrm{O}}(c) = \mathbf{IG}_{\mathrm{R}}(c) \qquad\qquad \mathbf{m}_{\mathrm{F}}(c) = \mathbf{IG}_{\mathrm{T}}(c) \tag{8}$$

The situation became more complex during conversion from intermediate grids into Bayesian framework. Traversal space intermediate grid $\mathbf{IG}_{\mathrm{T}}$ represents occupancy probability $\mathbf{p}_{\mathrm{occ}}$ below the neutral 0.5 value (9). Respectively reflection grid $\mathbf{IG}_{\mathrm{R}}$ should be converted into higher probabilities as presented in (10).

$$\mathbf{p}_{\mathrm{occ}}(c) = 0.5\ (\mathbf{IG}_{\mathrm{R}}(c) + 1) \tag{9}$$

$$\mathbf{p}_{\mathrm{occ}}(c) = 0.5\ (-\mathbf{IG}_{\mathrm{T}}(c) + 1) \tag{10}$$

4 Experimental Evaluation of the ISM Framework

The presented framework shows many customization and modification possibilities, which allow adaptation to different sensor characteristics. In order to demonstrate flexibility of the framework, the ISM was tuned to imitate model presented in [4].

One of the major benefits of the presented framework is the ability to minimize conflicts and occlusions which are ubiquitous in traditional iterative inverse sensor model architectures. The common, single-probability ISM was compared to disjointed ISM in a simple occlusion scenario.

4.1 Customization Possibilities

For the two-dimensional occupancy grid, the Gaussian distributions are widely used for both traversal and reflection probability calculation [4,8,21]. One of the examples of that distribution usage is the ultrasonic inverse sensor model presented by Elfes and Matthies in [4]. Based on their results, proposed ISM framework was able to mimic the distribution as presented in Fig. 4. The results were achieved using 2D Gaussian distribution in Polar coordinates for the traversal space and Cartesian 2D distribution for the reflection scattering center. Exact shape was tuned by modifying the standard deviations of those two separated distributions.

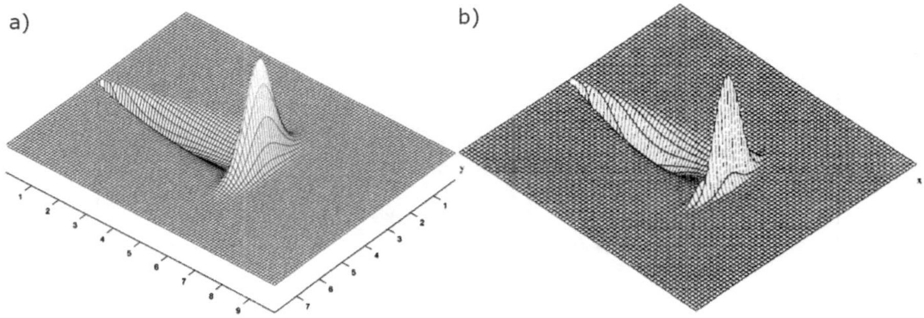

Fig. 4. Comparison of the ISM occupancy probability distributions: a) presented work, b) Elfes and Matthies [4].

4.2 Comparison with Traditional ISM Frameworks

The main disadvantage of the incremental inverse sensor model approaches is the tendency to hide the occupancy depending on the order of incoming data [15]. In common ISM framework each detection is a separate entity. Reflection and traversal space probabilities derived from that detection are immediately fused into the occupancy grid. In the proposed approach all detections from the same measurement time period could be processed together, enabling to solve the conflicting probabilities during intermediate fusion steps.

For the evaluation, a simple scenario with an obstacle in front of the wall was prepared. The difference in elevation or sensors' differing reflection characteristics could result in a detection set as presented on Fig. 5a, thus causing a sensor conflict. The incremental sensor model might detect closer obstacle, but the result depend on the order of detections. If the detections from the obstacle are processed before the wall's ones, the conflicting occupancy might completely disappear (Fig. 5c). On the other hand in the most favoring case the obstacle will be visible on the occupancy grid (Fig. 5e). In real case the detection order may depend on different factors or it could be random (Fig. 5d). Presented sensor conflict introduces unexpected behavior which cannot ensure required determinism of the occupancy grid.

The proposed architecture processes the occupancy and free-space probabilities in two, separate paths. Whole set of detections is processed on disjoint intermediate grids disconnecting the order of detections from the output grid. Moreover, by splitting non complementary occupancy and free-space information, the sensor conflict might be easily resolved as presented on Fig. 5b. The best result from single-probability ISM (Fig. 5e) is comparable to the proposed split-probabilities ISM (Fig. 5b).

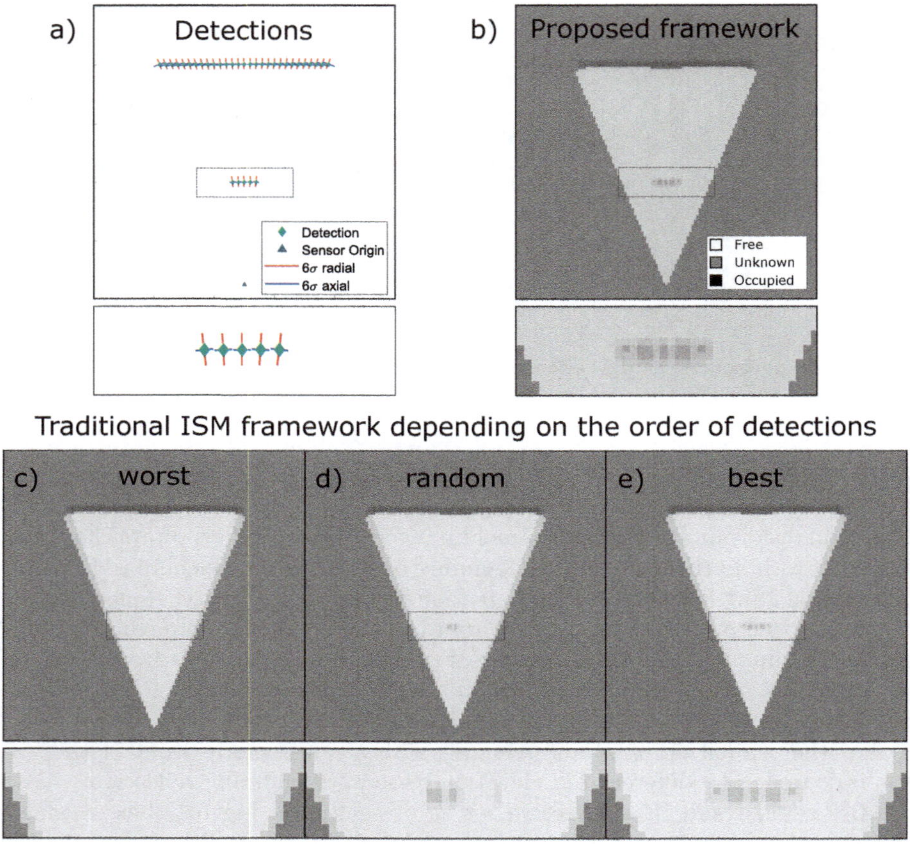

Fig. 5. Comparison of the proposed frameworks and traditional ISM for the obstacle in front of the wall (a). Below each image the enlarged obstacle area is presented. Proposed approach successfully solves sensor conflict while being independent on the detection order (b). In traditional approach result depends on the order of detections in scan (c-e). The worst case scenario accumulate closer detection first, while the best processes them from the furthest to the nearest (Sect. 4.2).

5 Summary and Further Work

This paper introduces a design approach of the inverse sensor model, which separates the free and occupancy probabilities. The proposed framework unifies various existing inverse sensor models in order to fit requirements for different types of detections. It is suited for both the Bayesian inference and Dempster-Shafer evidence theory which prove its customization abilities.

The presented inverse sensor model architecture is a comprehensive and extendable solution enabling optimization of operating parameters for each sensor in order to describe real measurement and increase occupancy grid accuracy.

The architecture was tested and compared to existing 2D inverse sensor models, however it can be also adapted into 3D or multidimensional occupancy grids.

Proposed ISM algorithm has similar computational complexity to traditional ISM frameworks. Nevertheless, described information flow expose a lot of parallelization opportunities, which are not possible using other common approaches. Those capability allows adaptation for the applications with high computational complexity constraints, preserving the high precision of modeling.

The proposed inverse model assumes that the sensor and vehicle position are known. In the further work localization uncertainties of the sensor origin should be taken into consideration in calculation of the occupancy distribution. The intermediate grid methodology could be extended to describe more detection states.

Acknowledgment. Research was funded by Polish Ministry of Science and Higher Education (MNiSW) Project No. 0014/DW/2018/02 and carried out in cooperation of Aptiv Services Poland S.A. – Technical Center Kraków and AGH University of Science and Technology – Faculty of Electrical Engineering, Automatics, Computer Science and Biomedical Engineering.

Special thanks to Aptiv coworkers: Krzysztof Kogut, Maciej Różewicz and Dariusz Cieślar for exceptional ideas and support.

References

1. Andriamahefa, T.R.: Integer occupancy grids: a probabilistic multi-sensor fusion framework for embedded perception. Ph.D. thesis, Université Grenoble Alpes (2017)
2. Bresenham, J.E.: Algorithm for computer control of a digital plotter. IBM Syst. J. **4**(1), 25–30 (1965)
3. Dietmayer, K.C., Reuter, S., Nuss, D.: Representation of fused environment data. In: Handbook of Driver Assistance Systems: Basic Information, Components and Systems for Active Safety and Comfort, pp. 1–30 (2014)
4. Elfes, A., Matthies, L.: Sensor integration for robot navigation: combining sonar and stereo range data in a grid-based representataion. In: 26th IEEE Conference on Decision and Control 1987, vol. 26, pp. 1802–1807. IEEE (1987)
5. Foroughi, M., Iurgel, U., Ioffe, A., Doerr, W.: Free space grid for automotive radar sensors. In: FAST-zero 2015: 3rd International Symposium on Future Active Safety Technology Toward zero traffic accidents (2015)
6. Homm, F., Kaempchen, N., Ota, J., Burschka, D.: Efficient occupancy grid computation on the GPU with lidar and radar for road boundary detection. In: 2010 IEEE Intelligent Vehicles Symposium (IV), pp. 1006–1013. IEEE (2010)
7. Joubert, D.: Adaptive occupancy grid mapping with measurement and pose uncertainty. Ph.D. thesis. Stellenbosch University, Stellenbosch (2012)
8. Konolige, K.: Improved occupancy grids for map building. Auton. Robots **4**(4), 351–367 (1997)
9. Markiewicz, P., Kogut, K., Różewicz, M., Skruch, P., Starosolski, R.: Occupancy grid fusion prototyping using automotive virtual validation environment. In: Proceedings of the 6th International Conference on Control, Mechatronics and Automation, pp. 81–85. ACM (2018)

10. Milstein, A.: Occupancy grid maps for localization and mapping. In: Motion Planning, IntechOpen (2008)
11. Moravec, H.P.: Sensor fusion in certainty grids for mobile robots. AI Mag. **9**(2), 61 (1988)
12. Pathak, K., Birk, A., Poppinga, J., Schwertfeger, S.: 3D forward sensor modeling and application to occupancy grid based sensor fusion. In: Intelligent Robots and Systems, 2007. IROS 2007. IEEE/RSJ International Conference on, pp. 2059–2064. IEEE (2007)
13. Stachniss, C.: Exploration and mapping with mobile robots. Ph. D. thesis, University of Freiburg (2006)
14. Stepan, P., Kulich, M., Preucil, L.: Robust data fusion with occupancy grid. IEEE Trans. Syst. Man Cybern. Part C (Appl. Rev.) **35**(1), 106–115 (2005)
15. Thrun, S.: Learning occupancy grid maps with forward sensor models. Auton. Robots **15**(2), 111–127 (2003)
16. Thrun, S., Burgard, W., Fox, D.: Probabilistic Robotics. MIT press, Cambridge (2005)
17. Valente, M., Joly, C., de La Fortelle, A.: Grid matching localization on evidential slam. In: 2018 15th International Conference on Control, Automation, Robotics and Vision (ICARCV), pp. 1477–1483. IEEE (2018)
18. Weiss, T., Schiele, B., Dietmayer, K.: Robust driving path detection in urban and highway scenarios using a laser scanner and online occupancy grids. In: 2007 IEEE Intelligent Vehicles Symposium, pp. 184–189. IEEE (2007)
19. Weston, R., Cen, S., Newman, P., Posner, I.: Probably unknown: deep inverse sensor modelling in radar (2018). arXiv preprint arXiv:1810.08151
20. Wurm, K.M., Hornung, A., Bennewitz, M., Stachniss, C., Burgard, W.: Octomap: A probabilistic, flexible, and compact 3D map representation for robotic systems. In: Proceedings of the ICRA 2010 Workshop on Best Practice in 3D Perception and Modeling for Mobile Manipulation, vol. 2 (2010)
21. Yguel, M., Aycard, O., Laugier, C.: Efficient GPU-based construction of occupancy grids using several laser range-finders. In: 2006 IEEE/RSJ International Conference on Intelligent Robots and Systems, pp. 105–110. IEEE (2006)

BlurNet: Keeping Collected Data Private with a Neural Network Based Pipeline

Daniel Dworak[1,2(✉)]

[1] AGH University of Science and Technology, Kraków, Poland
ddworak@agh.edu.pl
[2] APTIV Services Poland S.A., Kraków, Poland

Abstract. Data collection process is required to keep personal information of third parties private. This mandatory obligation has been recently enforced in a growing number of countries by appropriate laws, which forbid recording certain types of data. On the other hand, machine learning solutions based on neural networks, which are used in automotive industry, require a vast amount of data to learn from. The article addresses the problem of collecting camera frames in the form of a video file, where registered pedestrian faces and vehicle license plates are visible to a degree. This allows one to associate them with a particular person, hence they contain personal data. The pipeline for object detection and blurring algorithm is proposed. Processing images is being done by proposed BlurNet neural network based on YOLOv3 architecture. The article describes adaptations for a given data format, as well as modifications to improve the accuracy. Blurring algorithm is described as well. In order to maintain sufficient level of privacy, conducted experiments provide numerical answer regarding the performance of such solution.

Keywords: Object detection · Neural network · Faces · License plates · Blurring · Image processing · Data collection · Automotive · Privacy

1 Introduction

Modern cars have a variety of active safety systems and the development in a field of autonomous driving is blooming. This is in a high correlation to the recent progress in machine learning techniques, especially with image processing problems. Neural networks are state-of-the-art tools to be applied, particularly ones with convolutional layers. Neural nets learn from data and with sophisticated mechanisms, they tend to work the better, the more data is feed into them during a training process. When the performance increases proportionally to a number of examples in a training dataset, the data collection process becomes more important then ever. Cars equipped with dedicated sensors; cameras, RADARs and LiDARs, as well as professional logging systems, are driving among other daily road users to capture and record scenarios from real life situations. Such scenarios contain data in a form of images, point clouds, internal

A. Bartoszewicz et al. (Eds.): PCC 2020, AISC 1196, pp. 1237–1248, 2020.
https://doi.org/10.1007/978-3-030-50936-1_103

states and configuration parameters, and are used in the training process to deliver a functional system, later to be deployed in production cars.

One of a problems that come with a data acquisition process is to keep others personal information private. The problem is especially significant with camera recordings, which are easy to read by humans. During capturing of video images, the sensitive data like a human face or a vehicle license plate number is gathered and can be later used for the identification of such entities. One thing is to respect others privacy out of a decency, which is already a good reason to raise this issue. On the other hand, the digital privacy topic tends to become more and more popular in today's world and as a result, many public institutions and governments require certain regulations in this domain for a data collection process to be allowed to take place.

Obvious way of handling this problem is to ask, whether one agrees to be part of the process. This is, however, not a viable option when it comes to automotive industry. There are simply too many other road users, that the car is passing by every minute. It is not feasible to contact each of them and ask for a permission. The other way to avoid privacy issues in collected video recordings is to blur all essential data out of the image, making it unrecognizable, thus preventing from further use with regards to personal information. The article describes how such a pipeline, based on a neural network detector, could be assembled.

In the article, a solution to de-identification problem is presented. Based on a well known architecture, a neural network is used as an object detection algorithm. For this specific application however, the data format and objects constrains require additional modifications to the network in order for it to perform well. This modifications are the main subject of this article. Another subject is blurring method, which has been also discussed and tuned. Together, those parts create an usable de-identification pipeline for the use in the automotive industry.

The structure of this article is as follows. In the next section a literature overview with comparable methods is presented. Then, in Sect. 3 the data used for the training is described. Sections 4 and 5 contain details about the neural network architecture and the blurring algorithm respectively, which are then assembled in a complete pipeline. Following a theoretical part, in Sect. 6 undertaken experiments are defined and presented. The results are interpreted and discussed in Sect. 7.

2 Literature Review

Technological advance brings new challenges in keeping the privacy aspect. This global concern, which has arisen in recent years, results in international regulations regarding data collection, storage and redistribution [1]. Especially in a field of big data, researchers around the world emphasize the importance of the security manner [2].

Traditionally, hiding personal information is done by altering parts of the image where characteristic data occurs. Methods described in [3,4] varies from simple cropping and blacking out specific parts, through blurring and pixel-wise adjusting, to even more sophisticated methods like k-anonymity and similarity-based metrics.

More novel approach is described in [5,6], where authors proposed usage of Generative Adversarial Networks (GANs) to accomplish the task of hiding personal data. GANs are used to replace an original human face with one that is generated artificially. This way, all high level features of the face are preserved, but new image does not contain any actual real-life person.

In case of using such a solution in the automotive industry, another problems occur, taking into account the number of personal data present in the scene (pedestrians and cars), as well as high frames-per-second requirement for ADAS systems. To be able to process a vast amount of information, an efficient detection algorithm is needed. Among all available methods, neural networks are ones with the highest performance-to-speed ratio. There are few different approaches to a detection task in the image with neural networks.

One of the most proven two-stage detector is [7]. Region proposal is a mechanism to significantly reduce the number of a classification passes through the network, by disregarding obviously mismatched parts of the image. Further improvements to this method has been shown in [8,9]. Additionally, more advanced Focal loss function, proposed in [10], improves the performance and the speed even greater.

On the other hand, a different class of so-called single-shot detectors is described in [11] and [12]. Instead of utilizing classification to various parts of an image, those algorithms use regression mechanism to predict the confidence and the position of all objects at once. This approach is much faster, but tends to be less accurate. To improve the performance, some modifications are described in [13] and [14], which introduces the anchor boxes and the multi-scale training.

3 Dataset

Data input for the whole system are camera frames. In order to work for real-life solutions, the frames come from an actual automotive grade sensor mounted inside the car, which records video logs during test drives. Because later pipeline works on single frames, entire log is split into single shot images. The original format from a test car camera is encoded on 12-bits arrays, that means the pixel values range from 0 to 4095. Additionally the color space is very unique - RCCC, which stands for Red-Clear-Clear-Clear. Without going into much details, the hardware part responsible for capturing light beams has only red and gray filters, and so only those color saturation is captured.

Preprocessing done on the data consists of two parts. Firstly, the pseudo-colorization is being applied for better distinction over grayscale frames. With the usage of color exposure time and prepared look-up tables, the RCCC frames are converted to standard RGB ones. Second process is motivated purely by the neural network input shape. Because original camera images are high resolution, rescaling operation is needed. In contrary to just adjusting image size, more sophisticated method is applied. The aspect ratio of the image is preserved, as

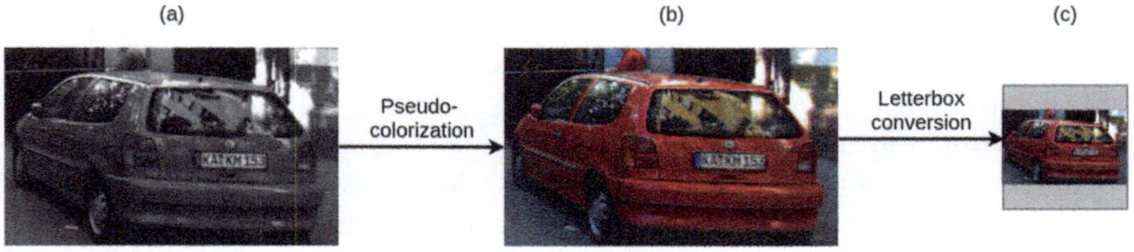

Fig. 1. Data preprocessing. (a) an original RCCC image, (b) a pseudo-colorized RGB image, (c) a letterbox input image. Image source [16]

higher dimension of the image is determine and the other is filled with neutral gray values, creating a letterbox squared image to be processed later on. Data prepocesing pipeline is shown in Fig. 1.

Dataset consists of 24000 labeled train images and 4000 labeled test images. The neural network responsible for faces and license plates detection is trained and validated on the former set, then all presented results and KPIs are with respect to the latter one. Examples from both sets has been manually labeled by human annotators, who were instructed to mark all sensitive cases of those classes that are recognizable by the them. That puts a bias on what the network should learn, as some distant unreadable license plates or people with their heads turned away from the camera are not a part of the labeled data.

4 Network Architecture

Detection of personal data, such as faces and license plates, in the proposed solution is being done by a neural network. As a base model, YOLOv3 network has been chosen, for numerous reasons. First of all, it is a proven solution for an object detection task. The implementation is widely known and available in many frameworks. But most of all, the deciding factor is that as a single-shot detector, YOLOv3 architecture has a significantly lower inference time, which is crucial in automotive system.

Some modifications were necessary, to improve the performance in this particular application. Original YOLOv3 network preprocessing changes dimensions of an input to a letterbox image of a size of 416×416 pixels. This is undesirable in case of a high resolution camera image from a car sensor, as resizing is essentially equal to losing information. Especially when labeled objects are only few pixels wide. Addressing this issue, additional layers has been added to process an image in a higher resolution. This modification allows to change an input shape to 832×832 pixels.

On top of that, the number of classes has been changed for only related ones and anchor boxes has been adjusted to match most common average dimensions of the labels in the dataset. U-shaped network architecture, which enables multi-scale training, outputs the detections at three different scales. At each one of them, 3 anchor shapes has been selected by the k-mean clustering algorithm (Fig. 2).

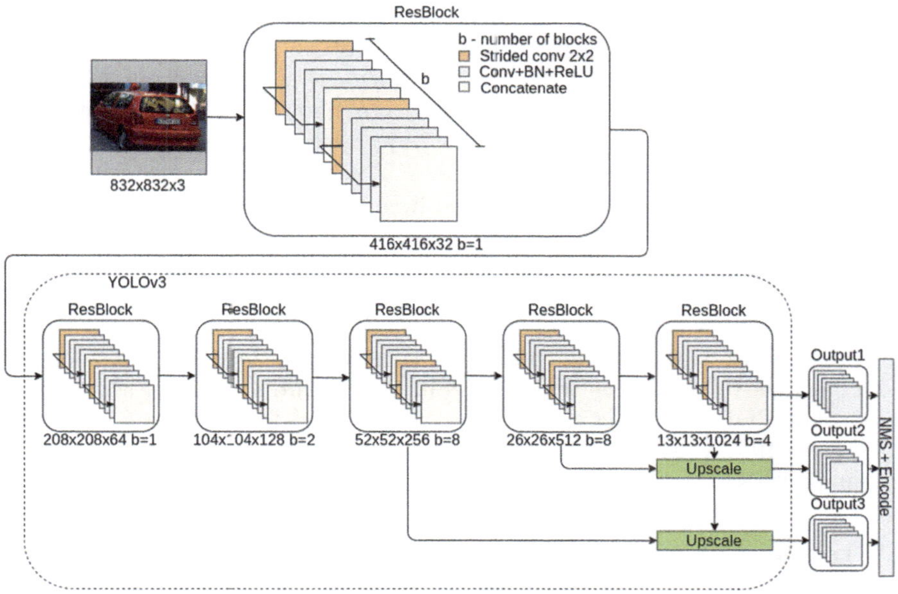

Fig. 2. BlurNet high-resolution architecture.

The regression loss function has been tuned as well. The authors of the original YOLOv3 paper presented weighted focal loss, which has been extended to distinct confidence of an detected object from empty grid cell better by tuning weights ratio between each loss part. During research, best weights has been chosen: $\lambda_{pos} = 1$, $\lambda_{size} = 1$, $\lambda_{obj} = 10$, $\lambda_{noobj} = 5$, $\lambda_{class} = 2$. Additional modification to loss function is normalization of object loss and no object loss by dividing corresponding parts by the number of grid cells with detected objects ($\sum \mathbb{M}^{obj}$) and without them ($\sum \mathbb{M}^{noobj}$) respectively. Formula for loss function is as follows:

$$
\begin{aligned}
Loss = {} & \frac{\lambda_{pos}}{\sum \mathbb{M}^{obj}} \sum_{i=0}^{S^2} \sum_{j=0}^{B} \mathbb{M}_{ij}^{obj} [(x_i - \hat{x}_i)^2 + (y_i - \hat{y}_i)^2] \\
& + \frac{\lambda_{size}}{\sum \mathbb{M}^{obj}} \sum_{i=0}^{S^2} \sum_{j=0}^{B} \mathbb{M}_{ij}^{obj} [(\sqrt{w_i} - \sqrt{\hat{w}_i})^2 + (\sqrt{h_i} - \sqrt{\hat{h}_i})^2] \\
& + \frac{\lambda_{obj}}{\sum \mathbb{M}^{obj}} \sum_{i=0}^{S^2} \sum_{j=0}^{B} \mathbb{M}_{ij}^{obj} (C_i - \hat{C}_i)^2 + \frac{\lambda_{noobj}}{\sum \mathbb{M}^{noobj}} \sum_{i=0}^{S^2} \sum_{j=0}^{B} \mathbb{M}_{ij}^{noobj} (C_i - \hat{C}_i)^2 \\
& + \frac{\lambda_{class}}{\sum \mathbb{M}^{obj}} \sum_{i=0}^{S^2} \sum_{j=0}^{B} \mathbb{M}_{ij}^{obj} \sum_{c \in classes} (p_i(c) - \hat{p}_i(c))^2
\end{aligned}
$$

where, $\mathbb{M}^{obj}$ is a mask with values 1 where object is labeled in target grid cell and $\mathbb{M}^{noobj} = 1 - \mathbb{M}^{obj}$. Prediction is encoded via tuples of $(b_x, b_y, b_w, b_h, conf)$ with respect to specific grid cell (c_x, c_y) and anchor (a_w, a_h), where $b_x = \sigma(x_i) + c_x$, $b_y = \sigma(y_i) + c_y$, $b_w = a_w \exp(w_i)$, $b_h = a_h \exp(h_i)$ and $conf = \sigma(C_i) * \sigma(p_i)$.

All those actions result in a single-shot object detection model called BlurNet, which has over 61.5M trainable parameters. The model is optimize with a given dataset and predicts pedestrian faces and license plates with higher performance on this particular data format.

The training process for both, original YOLOv3 and BlurNet, has been as similar as possible. Each network has been trained for 100 epochs, with early stopping mechanism. ADAM optimizer [15] was used. Learning rates have to be adjusted due to differences in loss functions values, but for comparison they have been set to the values that results in the networks converge around same epoch. For YOLO learning rate of $\alpha = 3 * 10^{-5}$ and for BlurNet learning rate of $\alpha = 3 * 10^{-4}$ was used. Learning rate decay was also incorporated into the training with same value of $decay = 1 * 10^{-6}$ for both cases.

5 Blurring Algorithm

The de-identification process on detected instances of private data is being done by blurring the parts of the image. This is not as sophisticated method as for example mentioned GANs approach, but it requires much less time to be performed. Additionally, the way the blurring is done, assures that the data can not be recreate.

The method is based on frequency domain analysis. Whole process can be divided into stages, as shown in Fig. 3.

First stage is an input RGB image. Further, the main transformation from RGB domain to frequency domain is done with usage of the two dimensional Fast Fourier Transform (FFT2D). Then, a blurring filter, given by formula:

$$(y - img\_size_y/2)^2 + (x - img\_size_x/2)^2 <= r^2$$

is created and applied to an image in frequency domain. Finally, masked results are transferred back to RGB domain by the inverted FFT operation.

Advantages of this approach allows to process each frame with multiple detections to be blurred relatively fast, when comparing to the network inference time. The method prevents reconstruction of a private data from blurred parts, as vast majority of frequencies, which are responsible for details, are being discarded.

Some additional features are embedded into this algorithm. Detection bounding boxes offset ensures, that minor errors in network predictions does not affect final results, which are properly blurred faces and license plates. To achieve smooth blurring, the surrounding area is taken into account when performing FFT and the resulting blurred image is then cropped to the size of the detection.

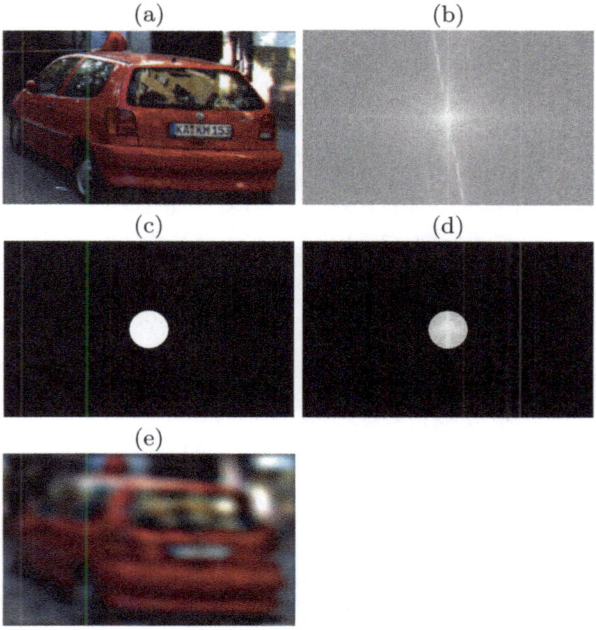

Fig. 3. Stages of blurring algorithm. (a) an original image, (b) FFT2D of the image, (c) a filter mask in the frequency domain, (d) a filtered image in the frequency domain, (e) a filtered image transformed back in the RGB domain. Image source [16].

6 Results

6.1 Network Modifications

One of the key elements of proposed solution is the object detection algorithm. The task is performed by a neural network. To evaluate the results, standard metrics, common in this field, were used [17]. Basic KPIs are *Precision* and *Recall*. For one score metrics, combinations of those two were calculated, namely $F1_{score}$ and $AveragePrecision$ (AP).

Firstly, a comparison between the BlurNet and the original YOLOv3 has been done. The BlurNet has been modified to better match an input data from a car sensor. The anchor boxes has been chosen based on labels in the dataset, to fit average sizes of them to the best degree. The results, with respect to mentioned metrics are shown in Table 1. While most of the metrics are very similar, there is a slight difference in *Precision* score between the two solutions. For license plates it changes from 0.76 to 0.82 and for pedestrian faces from 0.72 to 0.83. Other metrics are rather the same, and the differences could be attributed to somehow uneven initial weights.

Table 1. Network modifications results

Combined results				
Network	Precision	Recall	F1	mAP
YOLOv3	0.68	0.41	0.51	0.41
BlurNet	0.82	0.44	0.56	0.47
BlurNet HR	0.43	0.87	0.58	0.67
License plates class				
Network	Precision	Recall	F1	AP
YOLOv3	0.76	0.61	0.68	0.55
BlurNet	0.82	0.68	0.74	0.60
BlurNet HR	0.62	0.87	0.72	0.74
Pedestrian faces class				
Network	Precision	Recall	F1	AP
YOLOv3	0.72	0.35	0.47	0.31
BlurNet	0.83	0.37	0.52	0.33
BlurNet HR	0.40	0.88	0.55	0.60

Another aspect to compare, perhaps the most significant one, is the change of the resolution of an input image, which is being processed by the network. As faces and license plates are relatively small objects in the scene, preserving more details should improve performance by far. The actual results are shown in the table below. As expected, the performance has increased for a high resolution version of the BlurNet. This is especially true for *Recall* metric, which changed from 0.68 to 0.87 for license plates and more then double the value for a pedestrian faces class from 0.37 to 0.88. On the other hand, *Precision* of this solution is much lower, 0.62 down from 0.82 and 0.40 down from 0.83. That causes the combined metrics, $F1_{score}$ and AP, to be close, but still superior.

6.2 Blurring Algorithm

In terms of the blurring algorithm, presented results can be discussed with respect to theirs quality. The quality could be understood as a level of de-identification. The higher a blurring quality is, the more the image become unrecognizable to a human.

Filters shapes and sizes in frequency domain are some of the parameters to this algorithm. As shown in Fig. 4, one can consider a square based or a circular based filters, where the side length or the radius may differ from very narrow to very wide.

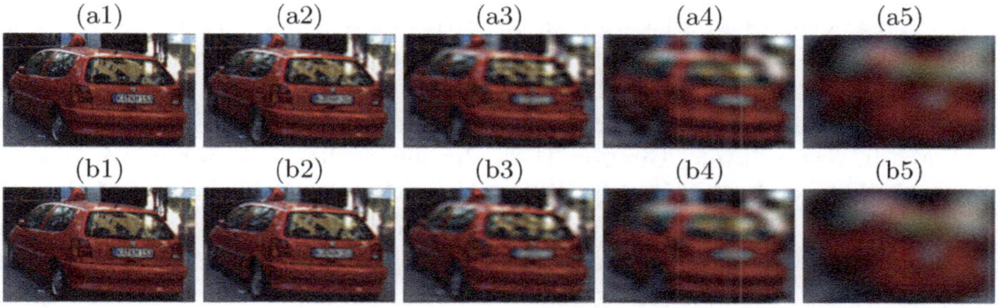

Fig. 4. Different filter kernels. (a1) and (b1) are original images, Top raw is a blurring done with a circle shaped kernels with a radius of 32, 16, 8 and 4 pixels respectively for (a2)–(a5). Bottom row is blurring done with square shaped kernels with a side of 64, 32, 16 and 8 pixels respectively for (b2)–(b5). Image source [16].

Another factor to the quality of blurring taken into account is the surrounding area of the detection. During a calculation of the Fourier Transformation of the image part with a detection, additional pixels around it are considered with some given offset. The value of this offset could change the output image quite significantly, even cropped back to the original size of the bounding box, as shown in Fig. 5.

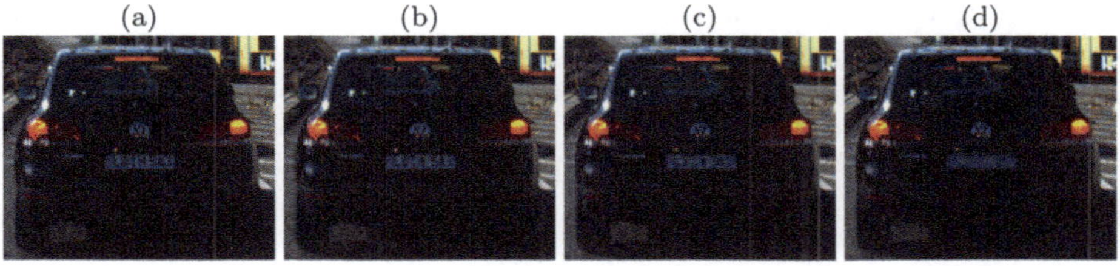

Fig. 5. Blurring effect of the additional area around the detection (a license plate). From (a) to (d) offset value is respectively 0, 10, 20 and 50 pixels. Image source [16].

7 Discussion

Proposed modifications to the network serves the purpose of increasing the performance of the predictions for pedestrian faces and license plates. By limiting classes the network can predict to only two mentioned, the number of the features per class at each layer of the network drastically increase, as the connections no longer need to distinguish an object out of many (50+ for YOLOv3) different classes. Better suited anchor boxes, which are calculated based on clustered labels, help to increase precision of the detections, which is shown when comparing original YOLOv3 and BlurNet. It is due to the fact, that this network architecture does not predict objects directly, but rather it predicts a deviation between the object and the most similar anchor. The more similar anchors are to the labels, the more precise detections will be.

The ability to process images with a higher resolution is beneficial even more. As shown in results table, BlurNet High Resolution has superior metrics values over a normal version. The only difference is that an input image of the HR version is twice the size and some additional layers has been added to the beginning of the network architecture. Because target objects are relatively small when compare to a whole image size, doubling an input size allows to grasp on more details, which in the end results in a better overall performance. The only drawback of this solution is decrease in *Precision*. Low *Precision* means that out of predicted detections only some are actual objects. On the other hand, high *Recall* means that most of the labels has been detected. Moreover, high AP score, which is the most sophisticated one, suggests that the solution is performing very well. This strange behavior, led to careful data examples examination. The results were that the network predicts objects which are actual faces and license plates, but were not part of labels due to a far distance and a low visibility, which does not reveal any private data. Nonetheless, the network learned to recognize those, even if it was not its primary task. Moreover, the simple blurring algorithm has been presented with various parameters configurations. The different filters shapes and sizes has been shown along with corresponding results they provide. The difference between square and circle filter shape are quite modest. Among all the options, circular shaped kernel with radius depending on detection size is the most suitable, as it provides high quality and smooth blurring. The effects of adding surrounding area to blurring FFT input has also been shown. The results present that the additional data from an original image transform the detection area with a smoother transition at the borders and a better overall blur. Judging by how well blurred part matches the nearby elements in the final image, an offset of 50 pixels has been chosen as the one (Fig. 6).

(a) (b) (c)

Fig. 6. Results of the pipeline. (a) an original image, (b) network detections, (c) blurred detections. Image source [16].

8 Concluding Remarks

Presented results show that the de-identification pipeline with the adaptation of YOLOv3 network architecture and a blurring algorithm performs within satisfaction boundaries on given data. The detection network has been adjusted to high resolution input in order to perform better on detecting desired objects for application specific situations and the sensor readings format. This change improves results significantly, which leads to the conclusion that applying processing of higher resolution images could be beneficial, whenever application allows to do so. The drawback is that inference time increases, which is undesirable in some cases. Altering anchor boxes size to given dataset is another small adjustment, which is quite straight forward to apply, but could increase performance even more, thus when fine tuning or reusing existing solution it is highly recommended to do so. Moreover, customizing loss function helps achieving converge sooner and with BlurNet parameterized loss, gives ability to test various configurations of the same architecture.

Blurring algorithm has been tested in various different parameters configurations and the configuration with best blurring quality has been chosen, base on visual effects. The algorithm, which uses FFT filtering proves to work very fast and efficient, which was its main purpose. Additional features, like the offset mechanism, results with better quality of blurring. Although this has not been tested in the scope of this work, it is highly possible that the smoother blurring may have minor impact on other perception algorithms in ADAS systems.

Acknowledgment. Research is partially funded by Polish Ministry of Science and Higher Education (MNiSW) Project No. 0014/DW/2018/02 and carried out in cooperation of Aptiv Services Poland S.A. – Technical Center Kraków and AGH University of Science and Technology – Faculty of Electrical Engineering, Automatics, Computer Science and Biomedical Engineering.

References

1. UE: General Data Protection Regulation. https://ec.europa.eu/commission/priorities/justice-and-fundamental-rights/data-protection/2018-reform-eu-data-protection-rules_en. Accessed 05 June 2019
2. Xu, L., Jiang, C., Wang, J., Yuan, J., Ren, Y.: Information security in big data: privacy and data mining. IEEE Access **2**, 1149–1176 (2014). https://doi.org/10.1109/ACCESS.2014.2362522
3. Newton, E., Sweeney, L., Malin, B.: Preserving privacy by de-identifying face images. IEEE Trans. Knowl. Data Eng. **17**, 232–243 (2005). https://doi.org/10.1109/TKDE.2005.32
4. Muraki, T., Oishi, S., Ichino, M., Echizen, I., Yoshiura, H.: Anonymizing face images by using similarity-based metric. In: 2013 International Conference on Availability, Reliability and Security, pp. 517–524 (2013). https://doi.org/10.1109/ARES.2013.68
5. Ren, Z., Lee, Y.J., Ryoo, M.S.: Learning to anonymize faces for privacy preserving action detection. CoRR abs/1803.11556 (2018). http://arxiv.org/abs/1803.11556

6. Wu, Y., Yang, F., Xu, Y., Ling, H.: Privacy-protective-GAN for privacy preserving face de-identification. J. Comput. Sci. Technol. **34**, 47–60 (2019). https://doi.org/10.1007/s11390-019-1898-8
7. Girshick, R.B., Donahue, J., Darrell, T., Malik, J.: Rich feature hierarchies for accurate object detection and semantic segmentation. CoRR abs/1311.2524 (2013). http://arxiv.org/abs/1311.2524
8. Girshick, R.B.: Fast R-CNN. CoRR abs/1504.08083 (2015). http://arxiv.org/abs/1504.08083
9. Ren, S., He, K., Girshick, R., Sun, J.: Faster R-CNN: towards real-time object detection with region proposal networks, pp. 91–99 (2015). http://papers.nips.cc/paper/5638-faster-r-cnn-towards-real-time-object-detection-with-region-proposal-networks.pdf
10. Lin, T., Goyal, P., Girshick, R.B., He, K., Dollár, P.: Focal loss for dense object detection. CoRR abs/1708.02002 (2017). http://arxiv.org/abs/1708.02002
11. Liu, W., Anguelov, D., Erhan, D., Szegedy, C., Reed, S.E., Fu, C., Berg, A.C.: SSD: single shot multibox detector. CoRR abs/1512.02325 (2015). http://arxiv.org/abs/1512.02325
12. Redmon, J., Divvala, S.K., Girshick, R.B., Farhadi, A.: You only look once: unified, real-time object detection. CoRR abs/1506.02640 (2015). http://arxiv.org/abs/1506.02640
13. Redmon, J., Farhadi, A.: Yolo9000: better, faster, stronger. arXiv preprint arXiv:1612.08242 (2016)
14. Redmon, J., Farhadi, A.: Yolov3: an incremental improvement. arXiv (2018)
15. Kingma, D., Ba, J.: Adam: a method for stochastic optimization. In: International Conference on Learning Representations (2014)
16. Geiger, A., Lenz, P., Stiller, C., Urtasun, R.: Vision meets robotics: the KITTI dataset. Int. J. Robot. Res. (IJRR) **32**, 1231–1237 (2013)
17. Powers, D.A.: Evaluation: from precision, recall and f-measure to ROC, informedness, markedness & correlation. J. Mach. Learn. Technol. **2**, 2229–3981 (2011). https://doi.org/10.9735/2229-3981

Generic Sensor Model for Object Detection Algorithms Validation

Kamil Lelowicz[1,2], Michał Jasiński[1,2(✉)], and Marcin Piątek[2]

[1] Department of Automatic Control and Robotics, AGH University of Science and Technology, al. A. Mickiewicza 30, 30-059 Kraków, Poland
{lelowicz,jasinski}@agh.edu.pl
[2] APTIV Services Poland S.A., ul. Podgorki Tynieckie 2, 30-399 Kraków, Poland
{kamil.lelowicz,michal.jasinski,marcin.piatek}@aptiv.com

Abstract. In the artificial intelligence era of perception algorithms being used in the state-of-the-art Advance Driver Assistance Systems, algorithm validation is not an easy task, mainly due to the amount of sensor data that must be processed at once. To provide the highest possible safety level of the solution, algorithm performance must be assessed in various difficult conditions. To address this problem, a novel Generic Sensor Model algorithm is presented, which can be directly incorporated in performance analysis of the perception algorithm in highly occluded driving scenarios. In this paper, the Generic Sensor Model algorithm is comprehensively described and its usefulness and robustness are proven with a set of experiments.

Keywords: Sensors · Mathematical modeling · Simulations · Automotive

1 Introduction

The 3D object detection algorithms are under heavy development, since they are a key component in the Advanced Driver-Assistance Systems. The state-of-the-art solutions are based on Artificial Intelligence algorithms [3,8,9,11] and they usually process data from sensors like cameras, radars and lidars, which makes them immensely complicated. To train an algorithm and to prove its robustness in various difficult conditions one should gather and label data from hundred millions of miles of test drives [7].

Usually 3D object detection algorithms predict a 3D Oriented Bounding Box for a given object and the quality of predictions is mostly evaluated using well-known metrics [6]. 3D Oriented Bounding Box is defined as a cuboid in three dimensional space given as:

- position of the cuboid center (x, y, z)
- orientation (yaw, pitch, roll)
- size (length, width, height).

A. Bartoszewicz et al. (Eds.): PCC 2020, AISC 1196, pp. 1249–1260, 2020.
https://doi.org/10.1007/978-3-030-50936-1_104

However, as shown in [5,12] poor algorithms performance in scenes with highly occluded objects is still an open issue and therefore to improve the assessment process of a given solution an additional information about the object visibility should be incorporated.

Checking how much an object is visible (not occluded) in the 3D space is a non-trivial task. Nevertheless, a 3D occlusions problem is well-known in the computer graphics literature in the image rendering process and for applications not requiring outstanding photorealistism it is usually solved using the ray-casting technique [10].

The contribution of this paper is an introduction to Generic Sensor Model algorithm, which calculates 3D occlusions directly from a high-level environment description consisting of 3D Oriented Bounding Boxes, thanks to which it can be directly used for labelled datasets analysis. Furthermore, the algorithm explicitly delivers the information about the projection size of an object (an area that a sensor is able to see) and the percentage value of the visibility of the object after including occlusions. Thanks to these properties it is possible to introduce a novel evaluation method for the perception algorithms, in which it is analyzed what should be the size and the visibility of an object from a sensor perspective. Finally, the presented algorithm is highly optimized and parametrized, hence one can focus either on very good accuracy or extremely low latency depending an application.

The article is organized as follows. In Sect. 2, the details of the presented algorithm are depicted. In Sect. 3, the results from set of experiments are presented showing how the algorithm could be parametrized in terms of the accuracy and the execution time. Finally, summary of the work is described and a further research plan is outlined.

2 Algorithm

The problem how to estimate the percentage value of the visible part of the Oriented Bounding Boxes was solved by the algorithm that defines a grid of rays (projection lines) coming from the sensor towards objects with the given density. Along this lines the algorithm calculates intersections with the objects and searches for the closest one. The ratio of the firsts intersections (the closest to the sensor) to all intersections for every object defines the visibility ratio that is the output of the algorithm. The algorithm consists of multiple steps that lead to the visibility ratio.

2.1 Calculate Vertices and Surfaces of Oriented Bounding Boxes

In this step, Cartesian and then radial coordinates are calculated for all vertices of each Oriented Bounding Box. The result of this step is the minimal and maximal radial coordinates for every Oriented Bounding Box i.e.:

- $\alpha_{min}^{(i)}$ minimal azimuth angle for i-th Oriented Bounding Box vertices
- $\alpha_{max}^{(i)}$ maximal azimuth angle for i-th Oriented Bounding Box vertices
- $\theta_{min}^{(i)}$ minimal elevation angle for i-th Oriented Bounding Box vertices
- $\theta_{max}^{(i)}$ maximal elevation angle for i-th Oriented Bounding Box vertices

These values define a projection area, which will be used in the upcoming steps during the intersections search.

In this step, also the parameters of each Oriented Bounding Box surface are calculated. These are:

- Cartesian coordinate of middle point of each surface (surface barycenter)
- Width and length of each surface

2.2 Grid Definition

During this step a grid consisting of rays, that are lines coming from the beginning of the coordinate system $(0, 0, 0)$ towards infinity, is defined. The grid is defined for a given area and density. The area of the grid is defined in radial coordinate system where border angles are defined as:

- $\min\{\alpha_{min}^{(i)}\}$ lowest azimuth angle from all Oriented Bounding Boxes projections
- $\max\{\alpha_{min}^{(i)}\}$ highest azimuth angle from all Oriented Bounding Boxes projections
- $\min\{\theta_{min}^{(i)}\}$ lowest elevation angle from all Oriented Bounding Boxes projections
- $\max\{\theta_{max}^{(i)}\}$ highest elevation angle from all Oriented Bounding Boxes projections

It means that the grid is defined only for the area where the Oriented Bounding Boxes exist, which is very important due to computational efficiency of the algorithm.

2.3 Intersections Calculations

For each Oriented Bounding Box two variables are defined:

- Intersection counter: stores the number of intersections proportional to the area of an Oriented Bounding Box projection
- Visible intersection counter: holds the number of intersections visible to the sensor proportional to the area of a visible projection

Then, for every ray in the grid with each surface of every Oriented Bounding Box intersections are calculated. Calculating intersection is time consuming. Thus, firstly it is checked whether a ray intersects the projection area of an Oriented Bounding Box defined by its azimuth and elevation angles. In case it does not it is assumed that there is no intersection between the ray and the Oriented Bounding Box. In the opposite case the following steps are repeated for every ray-surface pair:

1. Ray is transformed to the Cartesian coordinates system with the origin located in the middle of a given surface and the XY plane laying on the surface
2. Ray intersection with the XY plane for $Z = 0$ is calculated
3. It is checked whether the intersection is within the limits defined by the length and width of the surface.
4. If the above condition is met, the intersection coordinates are transformed back to the sensor coordinate system

If the ray intersects at least one surface of the Oriented Bounding Box, the intersection counter is incremented by 1. When all intersections for the given ray, and all possible Oriented Bounding Boxes, are calculated and checked and the ray is within the Field Of View of the sensor, the visible intersection counter of the Bounding Box, for which the intersection is closest to the origin of the sensor coordinate system, is incremented by 1.

2.4 Occlusion Calculation

When all intersections for all rays are checked the percentage value of the Bounding Box visible part is defined as:

$$v = \frac{a_v}{a} \cdot 100\% \tag{1}$$

where

- v percentage value of the visible part
- a intersection counter
- a_v visible intersection counter

Figure 1 present the results of the algorithm, where the blue color indicates that the objects, visible at least in 70%, and the intersections are seen by the sensor.

3 Experiments

The presented algorithm was implemented as a C++ library that can be either integrated with any given driving virtual environment for real-time simulations or used as an offline tool for huge datasets analysis. As the target platform Ubuntu 18.04 was selected mainly due to availability of simulators, like [4], and possibility of using system-specific libraries. All experiments were carried out on an Intel *i9-9900K* CPU with 64 GB RAM.

Commonly a real-time simulator is desire to obtain at least 20 frames per second similarly to real perceptions algorithms implemented in cars [2]. For offline analysis of thousands of scenes the requirements are getting worse and it is desirable to achieve as many frames per second as possible. Therefore, the main goal for the experiments was to test the time performance of the proposed solution.

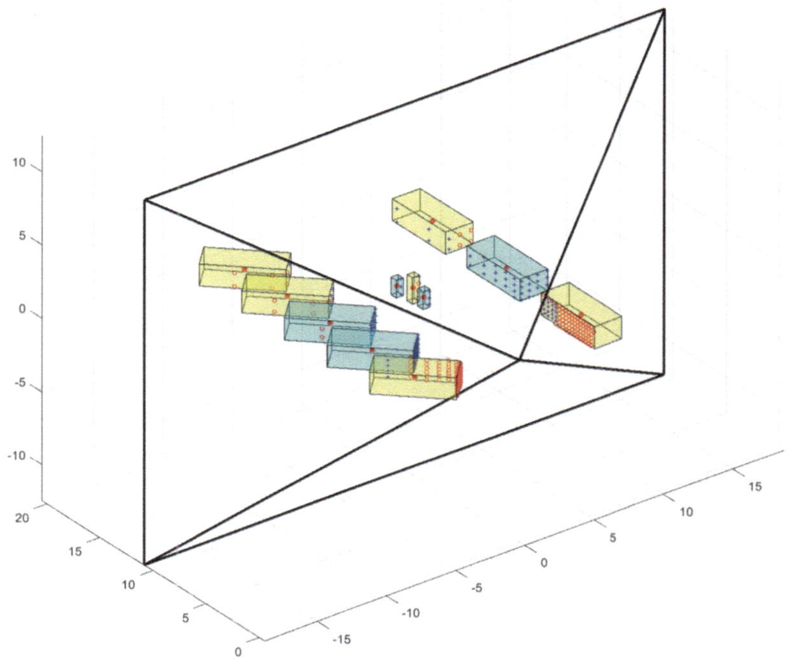

Fig. 1. The result of the algorithm with the sensor Field Of View (black lines) – Oriented Bounding Boxes (blue are visible to the sensor, yellow are not visible, visibility ratio is defined to 70%) and intersections (blue are visible to the sensor, red are not visible).

3.1 Predefined Scenes

To evaluate the performance of the algorithm two types of scenes were prepared:

- Highway: 7 lanes witch configurable number of cars on each lane.
- Parking: cars on both side of the scene with addition of three pedestrians located in the front of the sensor. On the right side of the scene the cars are parked in parallel, while on the left side they stand diagonally.

Figure 2 illustrate an example of both scenes. Objects marked as green are visible at least in 70%. The black borders of the pyramid indicate the field-of-view (FOV) of the sensor, which for every scene was set to 120° horizontally and 60° vertically.

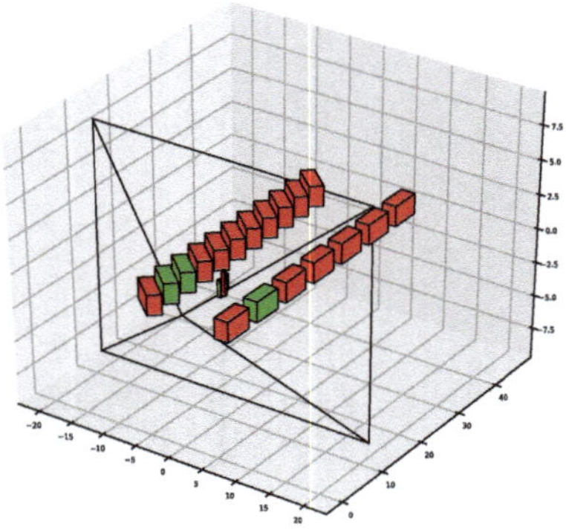

(a) Parking.

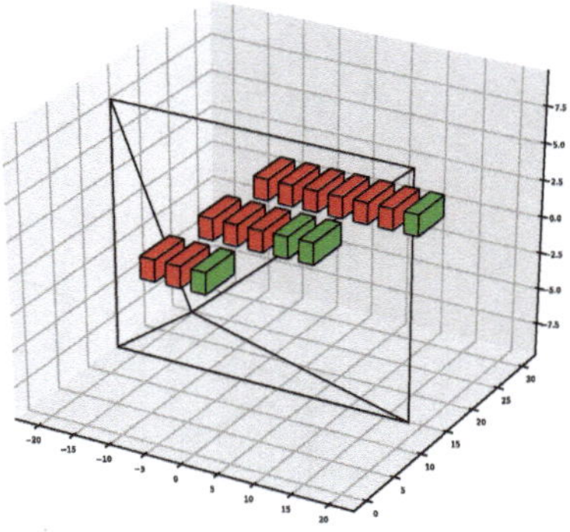

(b) Highway

Fig. 2. Examples of scenes. Green objects are visible at least in 70%. Borders of pyramids indicates sensor FOV.

3.2 Grid Density

A parameter having strong influence on the performance of the algorithm (in terms of both speed and accuracy) is the grid density. For instance, in case of a sparse grid computational complexity of the algorithm will significantly decrease leading to overall low latency, but with dropped precision in object occlusion calculation. On the other hand, selecting a too dense grid will lead to very good accuracy, however at the expense of the poor time performance. Therefore, the aim is to find a sustainable solution. Increasing the density should lead to improved information about an object occlusion. Theoretically, ignoring the fact that numbers have finite representation in a computer, for the infinite density one can get an ideal accuracy in the objects occlusions calculation. However, at some point further density increase will not give significant improvement in the accuracy. In the experiment the densest grid that was assumed to provide almost perfect information about objects occlusion was selected at 10^{-3} rad. Figure 3 illustrate an absolute percentage error of the occlusions calculated for the scene, which consisted of 43 objects viewed from different angles and with varying occlusions. The error is expressed with respect to the occlusion value that could be obtained using above mentioned density.

In the Fig. 3 the error for the density 10^{-2} reaches up to 20% and for $1.7 \cdot 10^{-4}$ the maximum error is only 0.2%. Hence, increasing the density over 10^{-3} does not bring any significant gain in the precision. Considering the fact that doubling the density value causes the fourfold increase in the total number of rays and hence the increase of the computational complexity of the algorithm, it was concluded that the density equal to 10^{-3} will provide a good compromise between speed and accuracy of the algorithm.

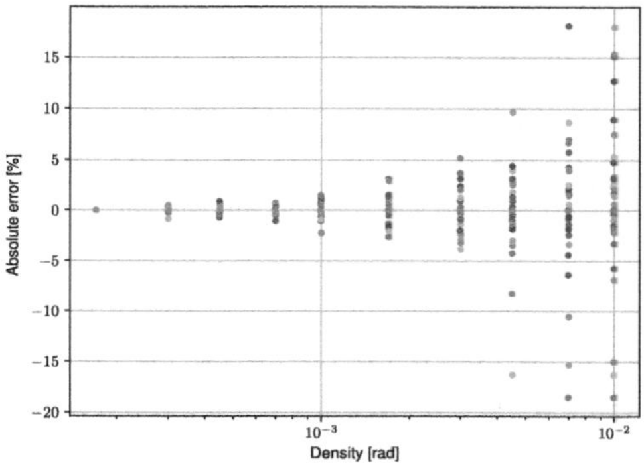

Fig. 3. Objects occlusions precision expressed as a function of density, where each object is represented using different dot.

3.3 Impact of the Number of Rays and Number of Objects

The size of the analyzed field of view is calculated dynamically. Depending on the position of objects in the scene, the smallest and the largest angle for objects seen by the sensor is calculated. A grid is created on this basis. The smallest and the largest angle for all objects is taken as the grid size. In order to determine the time dependence on the number of rays, the highway scene was used. In the scene, objects were placed in the front of the sensor and in that manner each rays would pass through exactly one object. The experiment was conducted for different grid densities and different number of objects in scene. The result are presented in the Fig. 4a. As expected, the time dependence of the algorithm execution is directly proportional to the total number of rays. Furthermore, the time needed to create and place objects in the scene by the algorithm, although they are created dynamically on the heap, is negligibly small. The perceivable slight deviations from the straight are likely to be caused by the computer memory access tasks. On the basis of the tests carried out, it can be concluded that the total processing time for a single ray intersecting only one object is approximately $3.23 \cdot 10^{-7}$ s.

In the case of a real scene, the distribution of cars on the stage is more complicated. To determine the algorithm behavior in such situations the parking scene was used. In addition, in the parking scene there are also pedestrians placed in the front of the sensor and moving perpendicular to the road. The Fig. 4b shows the time dependence of the algorithm execution depending on the number of objects in the parking scene. When there are only 5 objects in the scene, the cars are still on both sides of the road and very close to the sensor. Because of that, the grid size remains the same for all the test cases. At the beginning, the execution time increases very quickly, and then the rate of the growth stabilizes. This is due to the fact, that at first the scene is filled with objects. Then the next objects are added from the background of the scene. They are mainly occluded by objects already present in the scene. Furthermore, their projection

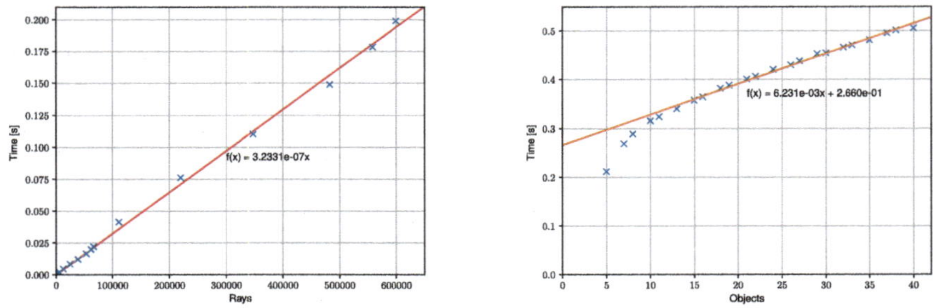

(a) Execution time with respect to number of rays

(b) Execution time with respect to number of objects

Fig. 4. Execution time with respect to scene size

size is getting smaller. It causes that the increase of the time calculation is not significant anymore and remains constant. Omitting scenes with small number of objects, one can calculate the dependence of the execution time on the number of objects, assuming that an additional object will be occluded. Thus, for a grid with the density equal to 0.001° additional occluded car size object increase calculation time by approximately 0.00623 s.

3.4 Impact of the Number of Threads

In order to improve the algorithm execution time, its multi-threaded version has been implemented using *Native POSIX Thread Library* (NPTL). The applied solution allows to use as many threads as the operating system is able to handle For instance, in the hardware used for the experiments it was possible to run even 32000 threads. The implementation was tested in terms of the performance depending on the size and type of the scene, grid density and the number of threads being used.

Figure 5 shows obtained results. The processor used for the tests has 8 physical cores and due to the hyperthreading the number of logical cores is equal to 16. Analyzing the obtained results one can notice that the calculations are scaled perfectly up to 8 threads, i.e. the actual number of cores. Further increasing the number of threads does not always bring benefits. This is particularly evident in the Figs. 5a and 5b. With large number of threads, the time needed to create and join all threads becomes significant and notably increases the total calculation time. If enough threads are used, this phenomenon is observed regardless of the grid density. However, increasing the number of threads above the number of actual cores in a dense grid still result in faster calculation times and moreover scaling is independent from the number of objects in the scene. The biggest gain from large number of threads is while the grid is very dense and there are a lot of objects in the scene. This can be mainly seen in the Fig. 5e.

The special case is when 12 threads were used. In most charts, particular in the Figs. 5e, 5f, 5e and 5f one can see an increase in the calculation time. This

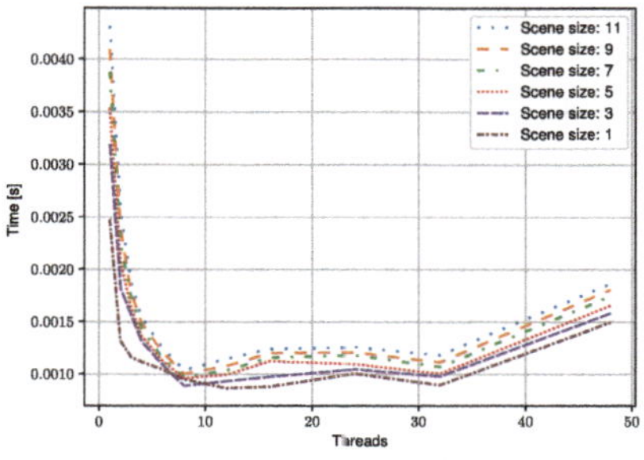

(a) Parking, density 0.01

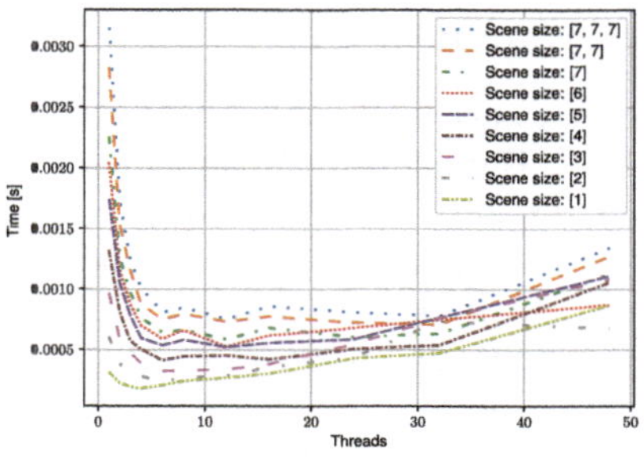

(b) Highway, density 0.01

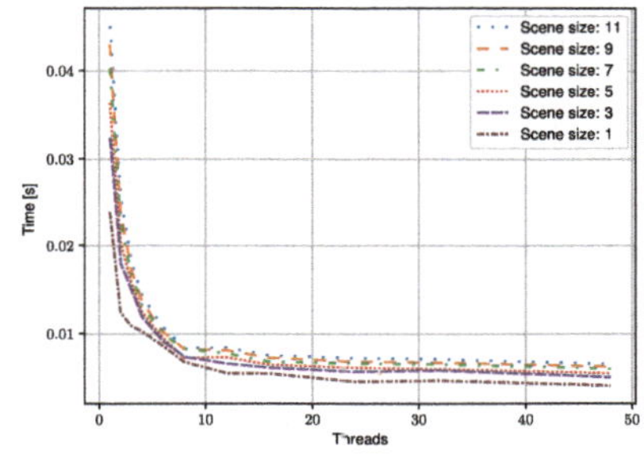

(c) Parking, density 0.003

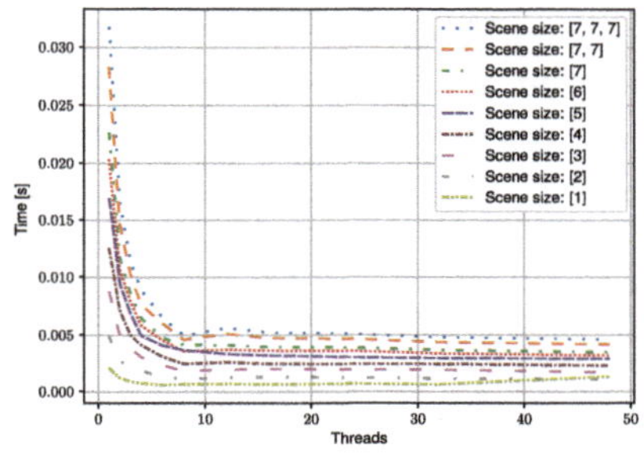

(d) Highway, density 0.003

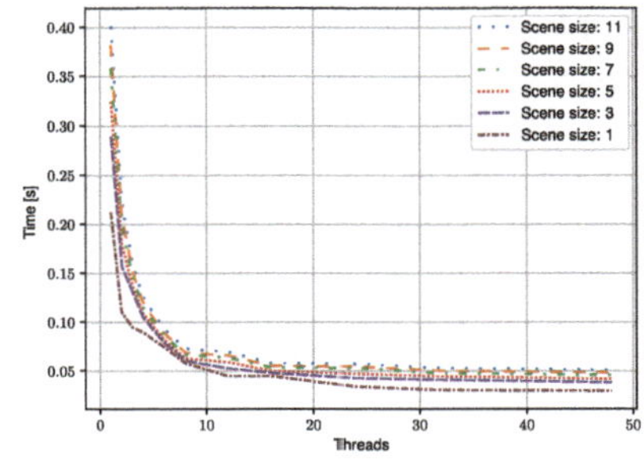

(e) Parking, density 0.001

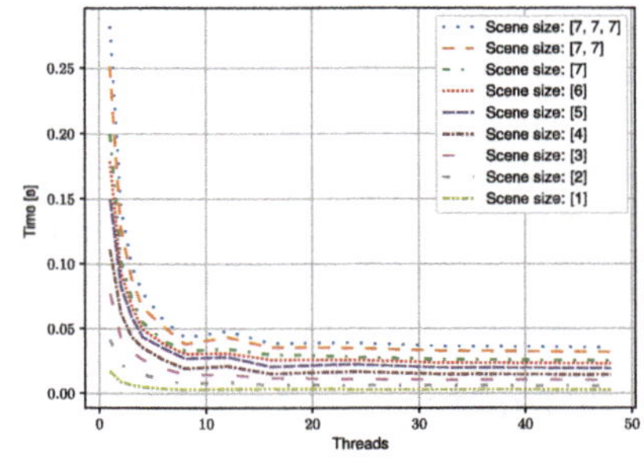

(f) Highway, density 0.001

Fig. 5. Execution time with respect to number of threads

is likely to be caused by to the fact, that scheduler of the Linux kernel is unable to ideally balance threads for a CPU with the hyperthreading for the certain number of threads. In our case, when the number of threads is grater than 8 and less than 15. Under these conditions, the kernel scheduler was using most of the time 6 or 7 out of 8 physically available cores at the time, which most likely caused an increase in the calculation time.

3.5 Amhdal's Law

One can notice the increase in the efficiency of calculations using more than 16 threads. Having only 16 logical cores one should not get accelerated by creating more threads. This is a result of the way the grid is created, as well as the way the calculations between the threads are distributed. Grid is always created as the smallest convex region that contains all objects in a field of view. This means that some part of rays does not point to any object. Thus, the calculation time for these rays is faster. The grid is always divided into evenly distributed vertical pieces and transmitted to the threads for the calculation. This results in an uneven distribution of calculations on the threads and hence some of them will finish faster. However, the system scheduler is able to ensure that all the cores will work with a high load causing a gain during the execution of the algorithm. According to the Amhdal's law [1] the speedup in the latency of the execution of a task at fixed workload that can be expected for a system whose resources are improved is given by the following formula:

$$S(s) = \frac{1}{(1-p) + \frac{p}{s}} \tag{2}$$

where:

- S theoretical improvement in the given task execution time.
- s the number of times that the execution time of the part of the task resulted from increased computational resources is improved.
- p part of the execution time that can be improved.

The Fig. 6 shows the comparison between the obtained results and Amhdal's law assuming that 90% of the part of the calculations were paralleled. Considering the influence of the scheduler it can be seen that the implementation of the algorithm behaves exactly as if 90% of its calculations were paralleled. In case of creating too many threads, the time of creating and joining threads is significant which affects the increase of the calculation time.

4 Summary

In this paper a novel algorithm was presented, which one can use as a robust tool for the validation of the perception algorithms in highly occluded driving scenarios. The main goal of this work was mainly the performance analysis of the proposed solution.

As it was described, the main advantage of the solution is an explicitly provided information about the visibility ratio of an object calculated with reference to a given sensor frame. Such information aims to be extremely useful during the evaluation of the real algorithm, because one can analyze what should be the size and visibility of an object so that it is being detected.

What is more, the results from the experiments proved the robustness and the scalability of the presented method in terms of both real-time performance and

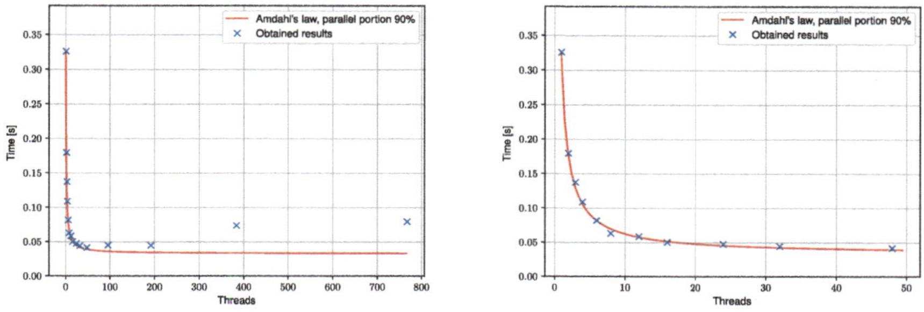

(a) Whole **(b)** Zoomed

Fig. 6. Amhdal's law

accuracy. As a result of its various adjustment possibilities, the algorithm can be comprehensively used either in the real-time simulations or in the offline analysis of the data, where the accuracy is much more important than the execution time.

Further research plan is to incorporate the presented method in the analysis of a given open source dataset to develop a set of metrics for perception algorithms validation purposes.

Acknowledgment 1. Research was funded by Polish Ministry of Science and Higher Education Project No. 0014/DW/2018/02 and carried out in cooperation of Aptiv Services Poland S.A. – Technical Center Kraków and AGH University of Science and Technology – Faculty of Electrical Engineering, Automatics, Computer Science and Biomedical Engineering.

Acknowledgment 2. The algorithm presented in this paper has been submitted in the European Patent Office on April 11th, 2019 as the following patent application "Method for estimating visibility of objects" with the following number 19168632.8.

References

1. Amdahl, G.M.: Validity of the single processor approach to achieving large scale computing capabilities. In: Proceedings of the April 18-20, 1967, Spring Joint Computer Conference, AFIPS '67 (Spring), pp. 483–485. Association for Computing Machinery, New York (1967). https://doi.org/10.1145/1465482.1465560
2. Caesar, H., Bankiti, V., Lang, A.H., Vora, S., Liong, V.E., Xu, Q., Krishnan, A., Pan, Y., Baldan, G., Beijbom, O.: nuscenes: A multimodal dataset for autonomous driving. arXiv preprint arXiv:1903.11027 (2019)
3. Chen, X., Ma, H., Wan, J., Li, B., Xia, T.: Multi-view 3d object detection network for autonomous driving (2016)
4. Dosovitskiy, A., Ros, G., Codevilla, F., Lopez, A., Koltun, V.: Carla: An open urban driving simulator (2017)

5. Gaidon, A., Wang, Q., Cabon, Y., Vig, E.: Virtual worlds as proxy for multi-object tracking analysis. In: The IEEE Conference on Computer Vision and Pattern Recognition (CVPR) (2016)
6. Hossin, M., Sulaiman, M.N.: A review on evaluation metrics for data classification evaluations. Int. J. Data Mining Knowl. Manage. Process **5**, 1–11 (2015). https://doi.org/10.5121/ijdkp.2015.5201
7. Kalra, N., Paddock, S.M.: Driving to safety: How many miles of driving would it take to demonstrate autonomous vehicle reliability? Transp. Res. Part A: Policy Practice **94**, 182–193 (2016). https://doi.org/10.1016/j.tra.2016.09.010. http://www.sciencedirect.com/science/article/pii/S0965856416302129
8. Ku, J., Mozifian, M., Lee, J., Harakeh, A., Waslander, S.: Joint 3d proposal generation and object detection from view aggregation. In: 2018 IEEE/RSJ International Conference on Intelligent Robots and Systems (IROS), pp. 1–8 (2018). https://doi.org/10.1109/IROS.2018.8594049
9. Lang, A.H., Vora, S., Caesar, H., Zhou, L., Yang, J., Beijbom, O.: Pointpillars: fast encoders for object detection from point clouds (2018)
10. Roth, S.D.: Ray casting for modeling solids. Comput. Graph. Image Process. **18**(2), 109–144 (1982). https://doi.org/10.1016/0146-664X(82)90169-1. http://www.sciencedirect.com/science/article/pii/0146664X82901691
11. Simon, M., Milz, S., Amende, K., Gross, H.M.: Complex-YOLO: An Euler-Region-Proposal for Real-Time 3D Object Detection on Point Clouds: Munich, Germany, 8–14 September 2018, Proceedings, Part I, pp. 197–209. Springer, Cham (2019). https://doi.org/10.1007/978-3-030-11009-3_11
12. Zeeshan Zia, M., Stark, M., Schindler, K.: Explicit occlusion modeling for 3d object class representations. In: The IEEE Conference on Computer Vision and Pattern Recognition (CVPR) (2013)

Safe and Goal-Based Highway Maneuver Planning with Reinforcement Learning

Mateusz Orłowski[1,2], Tomasz Wrona[1,2], Nikodem Pankiewicz[1,2(✉)], and Wojciech Turlej[1,2]

[1] AGH University of Science and Technology, Krakow, Poland
[2] Aptiv, Krakow, Poland

mateusz.orlowski@aptiv.com, tomasz.wrona@aptiv.com, nikodem.pankiewicz@aptiv.com, wojciech.turlej@aptiv.com

Abstract. As autonomous driving moves closer to a real-world application, more and more attention is being paid to the motion planning part of the system. To handle vastness of possible road scenarios, negotiate with other road users and generate an intelligent control strategy in a constantly changing environment, data-driven techniques and artificial intelligence methods seem to be the approach of choice. In this paper, we present reinforcement learning (RL) agent which is embedded in a deterministic, safety envelope. The agent is responsible for generating high-level maneuvers, such as a lane following or a lane change. The primary goal of the agent is to reach a given lane in a given distance, while traveling on a highway. The selected maneuver is then executed with use of deterministic methods utilizing concept of Responsible-Sensitive Safety (RSS) framework, which formalizes safety constrains in a form of mathematical model. The proposed solution has been evaluated in two environments: one in which the agent receives a predefined reward for getting to a correct lane and second, in which it is rewarded for doing this in a time-optimal manner. We have evaluated the proposed solution against an another RL-based agent, which is steering vehicle by low-level control signals, such as acceleration and steering angle.

Keywords: Autonomous driving · Behavior planning · Maneuver planning · Deep reinforcement learning

1 Introduction

Maneuver planning tends to be a feasible building block to solve autonomous driving problem in a highway environment. It provides a broad range of safety guarantees which is crucial when aiming for a production-grade technology. It has to work reliably on public roads and not only as a technology demonstrator in staged scenarios. However, tuning rules that drive these maneuvers requires tremendous amount of work for human engineers and can easily lead to sub-optimal solutions or even dangerous ones. Fortunately, this manual labor can be successfully replaced with data-driven approaches such as machine learning.

A. Bartoszewicz et al. (Eds.): PCC 2020, AISC 1196, pp. 1261–1274, 2020.
https://doi.org/10.1007/978-3-030-50936-1_105

Maneuver planning methods stay in contrast to end-to-end methods such as [2], where authors trained a deep convolutional neural network in supervised manner to directly produce a steering command given raw image data. While this approach can achieve acceptable performance in certain environments, it lacks safety mechanisms necessary to guarantee safety of generated trajectories.

One way to address this concern is to introduce a set of rules and constrains that self-driving car has to obey to be considered safe, and assure its observance by the autonomous driving system. This approach allows to separate algorithms responsible for safety from behavior and path planning components and contain them within a highly transparent safety framework module.

Usage of deterministic safety framework as a safety measure allows to define robust architecture in which behavior planning module based on reinforcement learning (RL) is used to choose high-level maneuvers that fulfill predefined goals in efficient manner. Output from behavior planner is used to generate trajectory that fulfill a short-term task defined by chosen maneuver in a safe way. If safe execution of the maneuver is not possible in current environment, maneuver choice performed by behaviour planner is suppressed and replaced by predefined minimum risk maneuver.

2 Related Work

2.1 Reinforcement Learning and Maneuver Planning

An application of reinforcement learning in maneuver planning domain is recently a broadly studied subject. In [26] authors summarized recent advances in an application of reinforcement learning algorithms to autonomous driving domain. One of the examples is lane merging from on-ramp presented in [27], where authors utilized Deep Q-Network (DQN) and Long Short-Term Memory (LSTM) network. An another interesting example is [17] which leverages DQN to train an agent to overtake other vehicles. These two works however, focus only on single road situations.

More general approach was presented in [13] where DQN was used to control a car on a three-lane highway. The agent was allowed to either change lane or accelerate and decelerate in his current lane, and it was promoted for avoiding collisions, keeping higher speeds and avoiding unnecessary maneuvers. To assure safety, the authors used their own set of safety rules. The motivation behind that is, that according to them, RSS assumes that each vehicle on the road is an autonomous vehicle. We disagree with this statement. Similar work was done in [16], where the authors also used DQN to train a highway pilot. Their safety framework again consists of simple rules, which overrides selected action with a fallback action if certain conditions are met. DQN was also used in [14] to teach an agent to drive as fast as possible on a two-lane highway. It was simply promoted for keeping high speed and punished for causing accidents, thus, no safety frameworks were introduced.

2.2 Safety Consideration

Rapid progress in a field of path planning for automotive applications, in a pursuit of provably safe advanced driver assistance systems, resulted in significant advances in area of trajectory generation safety. A number of safety frameworks were proposed to gather in-depth understanding of a dynamic, partially observed scene and provide data necessary to navigate in various driving environments in a safe manner. Such frameworks can support a decision-making process by evaluating a risk of certain maneuvers, setting boundaries for generated trajectories and predicting most plausible outcomes of near-future scenarios.

Two main research directions can be recognized in this area: a prediction-based risk evaluation and a worst-case-scenario safety envelope calculation.

Prediction-based approaches typically focus on providing continuous quality ratings, either applied to certain trajectories or states of the ego vehicle. Such ratings can be computed based on description of static and dynamic environment surrounding an ego vehicle, using distances to other road users [28] or a probabilistic estimation of collision risk [1]. A continuous function provides convenient input for a trajectory planning module, that can guide an ego vehicle within low risk areas using potential fields [21], or model-based methods, where risk function is used to formulate inequality constraints of optimization problem [20] or as a penalty term in a cost function [29].

The main drawback of continuous risk function approaches is typically a lack of safety guarantees. The continuous formulation provides valuable insight in the safety of an environment and possible states, but cannot be used to design a provably safe system without arbitrary assumptions regarding an allowed risk level. In order to formulate the problem in a more deterministic manner, Fail-Safe Motion Planning framework was proposed in [12]. In this work authors propose to plan a nominal trajectory based on prediction of the dynamic environment evolution, while also planning a fail-safe trajectory based on a deterministic worst case scenario prediction. If on a new planning step a fail-safe trajectory cannot be planned, the previous fail-safe trajectory is executed, enabling a safe behavior in all reasonable situations.

Similar concepts are also used in safety frameworks proposed by MobilEye [24] and NVIDIA [18]. Both frameworks attempt to formalize unwritten common-sense safe driving rules and local traffic laws in order to derive a definition of unsafe conditions. Such formulation allows to calculate a deterministic safety envelope, which constrains available ego trajectories to states that enable safe recovery from dangerous situations, under arbitrary assumptions regarding behavior of other road users.

While discussed approaches derive constraints for trajectory generation modules based on deterministic rules, additional considerations may be introduced to address complex aspects of behavior planning with conflicting objectives. Approach proposed by Aptiv in [4] focuses on including cultural, ethic and legal aspects of driving in a behavior planning problem by introducing hierarchical sets of prioritized rules and a formal way of their aggregation. This solution may be especially useful in a deterministic planning approach, where comfort considerations and progression towards a goal of a current task can be included within the described algorithm, forming a uniform decision-making system.

2.3 Our Contribution

In this work we present our safe and goal-based highway maneuver planner trained with Deep Recurrent Q-Learning (DRQN) algorithm and compare it with a naive baseline trained with Recurrent Soft Actor-Critic (RSAC) algorithm, which uses low-level control signals to steer a car. We demonstrate an advantage of using high-level decisions as an action space for our agent in term of safety, sanity and comfort.

This paper is divided as follows. In Sect. 3 we introduce a necessary notation for our maneuver planning algorithms, as well as a reinforcement learning framework we use. In Sect. 4 we present our proprietary environment for simulating realistic road scenarios and describe a setup used to train the maneuver planner in it. Finally, in Sect. 5 we show outcomes of our experiments.

3 Preliminaries

3.1 Path Planning System Architecture

Up till now a broad range of different approaches for path planning in autonomous driving was presented, along with a functional breakdown in a system architecture. Because of an inherent difficulty of path planning design, all architectures bring limitations, especially in corner-cases. An exhaustive survey on different approaches to path planning may be found in [19].

There exists an architecture scheme on which most of the approaches are based. This typical system architecture is presented in Fig. 1.

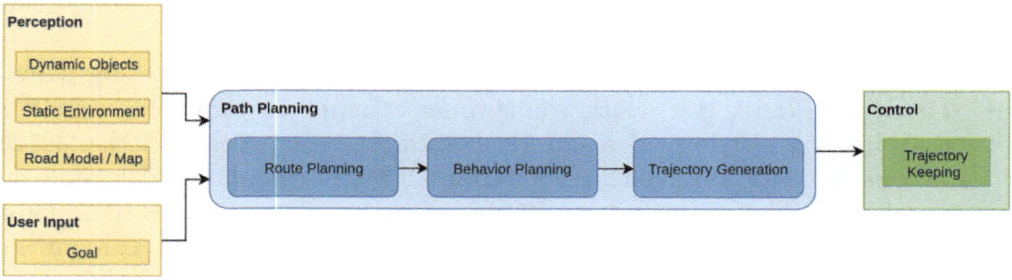

Fig. 1. The typical path planning system architecture.

An input to a path planning module consists of list of dynamic objects, a static environment representation and a road model definition. All of these information come from perception module, and user, i.e. passenger, is only responsible for setting a desired destination.

A path planning system is built from three major functional components: route planning, behavior planning and a trajectory generation module. Route planning works similar to car navigation and returns a sequence of lane-level goals, which when followed allows reaching a final destination. In literature, mission planning is another popular name for this module. Behavior planning chooses a high-level semantic maneuver to be performed (e.g. lane following, lane changing or stopping). In some approaches, this module returns also a set of additional signals such as speed, lane bias and lead-car distance [28]. The trajectory generation module plans a trajectory of a host vehicle to navigate in a partially known environment in a safe and efficient manner, returning a continuous reference trajectory. The module handles basic aspects of an ego locomotion planning such as speed keeping, adjusting velocity in proximity of other vehicles and precise positioning on an ego's lane.

The generated trajectory is then passed to a control block, which operates directly on a steering angle, throttle and braking signals with aim to keep the given reference trajectory.

Safety of the ego's behavior is ensured by a safety framework based on rules proposed in [24]. Safety framework is responsible for both suppression of unsafe maneuvers and execution of emergency responses if a dangerous situation is detected.

3.2 Reinforcement Learning Setting

We consider a standard reinforcement learning setting [25] in which an agent interacts with an environment $\mathcal{E}$ in a series of discrete time-steps. In each time-step t, based on en environment observation $o_t \in \mathcal{O}$ (which is strictly related to en environment state $s_t \in \mathcal{S}$), agent takes an action $a_t \in \mathcal{A}$. Subsequently, the environment transfers to a next state s_{t+1} according to an unknown state transition probability p, while providing the agent with a new observation o_{t+1} and a reward $r_t \in \mathcal{R}$.

A commonly used approach is to model the environment as Markov Decision Process (MDP), defined by a tuple $(\mathcal{S}, \mathcal{A}, p, r)$, in which a state transition function depends only on a current state and an action $p(s_{t+1}, r_t | s_t, a_t)$, and an agent has access to a complete environment state $(\forall_t, o_t = s_t)$. This definition represents the assumption that the future (s_{t+1}) is conditionally independent from the past $(s_{0:t-1}, a_{0:t-1})$ given present (s_t, a_t), or i.e. that the present contains all the information about the past.

While a Markov assumption is convenient, it is often unrealistic in real-world applications. In these settings, a framework called Partially Observable Markov Decision Processes (POMDP) might be used, where an agent does not have a full access to information about an environment. However, most of the current state-of-the-art approaches address this problem by utilizing recurrent neural networks instead of feed-forward ones, expecting that a time-tracking of series of observations will minimize the partial observability issue [8].

Deep Q-Learning. As the goal of an agent is to maximize its future rewards, Q-learning algorithm focuses on an estimation of a state-action value function $Q(s, a)$, which value, for a given state s and a given action a, represents an expected sum of discounted future rewards. If the estimation of state-action value function is correct, then an optimal policy is defined simply as acting greedily $a_t = \arg\max_a Q(s_t, a)$.

One of the first successful utilization of deep neural networks in a field of reinforcement learning was Deep Q-learning algorithm used to play a set of Atari games based on raw pixel data [15]. In this algorithm, a state-action value function $Q(s, a)$ is approximated by a deep neural network (Deep Q-Network, DQN) $Q(s, a|\theta)$, where θ are the network's parameters.

In DQN, an agent interacts with an environment either using its greedy policy $\mu(s) = \arg\max_a Q(s, a)$ or by picking a random action. At each time step, the agent decides on which policy to use, by taking the first one with probability $1 - \epsilon$ and the second with probability ϵ. After taking a step, a tuple describing a transition (s_t, a_t, r_t, s_{t+1}) is stored in a replay buffer $\mathcal{D}$. These transitions are sampled in a form of minibatches to drive an optimization, which is based on minimizing a temporal difference error δ_{err}:

$$\delta_{err} = Q(s_t, a_t) - (r_t + Q(s_{t+1}, \arg\max_a Q(s_{t+1}, a))) \quad (1)$$

Soft Actor-Critic. (SAC) algorithm [6,7] is state of the art off-policy algorithm. Similarly to DQN, it leverages replay buffer $\mathcal{D}$ for sample efficiency. To use stored samples, algorithm approximates Q-function $Q(s, a)$ by using Bellman equation. Policy is optimized by maximization of expected Q-function $Q(s, a)$ values and entropy objective.

Entropy maximization helps exploring action space at current state as well as reaching states not yet visited. In latest version of algorithm [7], automatic method for determining exploration coefficient α was introduced.

Objective for Q-function optimization (minimization) $J_Q(\theta)$ is defined in Eq. 2.

$$J_Q(\theta) = \mathbb{E}_{(s_t,a_t)\sim\mathcal{D}}[\frac{1}{2}(Q_\theta(s_t, a_t) - \hat{Q}(s_t, a_t))^2], \quad (2)$$

with

$$\hat{Q}(s_t, a_t) = r(s_t, a_t) + \gamma\mathbb{E}_{s_{t+1}\sim p}[V_{\bar{\psi}}(s_{t+1})])] \quad (3)$$

where the target value function $V_{\bar{\psi}}(s)$, with $\bar{\psi}$ as exponentially moving average of the value network weights, is implicitly parameterized through soft Q-function parameters in Eq. 4,

$$V(s_t) = \mathbb{E}_{a_t\sim\pi}[Q(s_t, a_t) - \alpha \log \pi(a_t|s_t)]. \quad (4)$$

Policy function $\pi_\phi(s)$ can be learned by directly minimizing objective in Eq. 5,

$$J_\pi(\phi) = \mathbb{E}_{s_t\sim\mathcal{D}}[\mathbb{E}_{a_t\sim\pi_\theta}[\alpha \log(\pi_\theta(a_t|s_t)) - Q_\theta(s_t, a_t)]]. \quad (5)$$

4 Highway Maneuver Planning with Reinforcement Learning

4.1 Highway Environment

While there are multiple publicly available open-source simulators, such as SUMO [11], CARLA [5] or AirSim [23], none of them fits perfectly to our needs, which are: a large complexity of road scenarios, low computational and memory footprints (it has to run efficiently in multiple independent instances on a single machine) and machine learning-friendliness (it has to export efficient Python API). Based on that research, we decided to create our own proprietary environment (Fig. 2), which fills this gap.

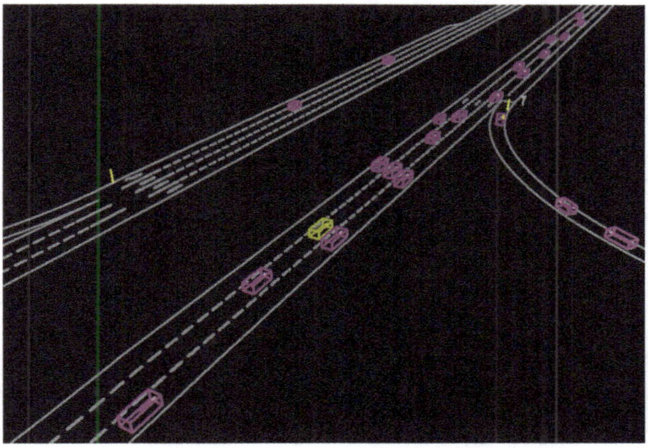

Fig. 2. Our proprietary simulator - on the picture: an ego vehicle (in yellow) is approaching an on-ramp intersection on a busy highway; other road users are marked in magenta.

With this simulator we are able to train agents on various road scenarios consisting of varying number of lanes, on-ramps and off-ramps. These scenarios are populated with intelligent road users, which realistically react to our and each other behaviors.

We wrapped the simulator with classical OpenAI Gym interface [3] and defined three environments: `Highway`, `HighwayTimeOptimal` and `HighwayController`. In these environments an agent has to solve two problems. First of them is to navigate on a multi-lane highway to reach a randomly designated goal. The goal is specified by a precisely defined point on a chosen lane in on average one kilometer away from a starting position of the agent. The environment is a typical goal-based environment and it was supposed to mimic ordinary driving with car navigation. The agent gets a sparse reward for reaching the appropriate place appointed by the goal. In case of a failure, it does not get any reward nor punishment, but the current episode is terminated.

The second problem is a time-optimal navigation. The task is to reach a goal as fast as possible, without violating traffic rules. The goal is defined in the same way as in `Highway` environment. The only difference is in a reward shaping mechanism, where the major part of the reward the agent receives for reaching the goal, but the final value is proportional to the mean speed achieved throughout the whole episode. The reward is carefully tuned, so the agent achieves a maximum reward, if it travels the whole episode with the speed limit. In addition, the agent is encouraged to increase its speed by receiving a small reward for changing its velocity. Finally, for collision, exiting the highway or taking an invalid action (e.g. changing lane to the left when being on the most left lane) it gets a penalty of `-1`.

During experiments we noticed that an agent flickers actions a lot. To overcome this undesirable behavior, we added another component that punishes the agent for keeping away from the center of his lane.

The last environment was introduced to create a baseline for RL agents that uses low-level control signal as their outputs. In this environment, the agent is allowed to directly control a steering angle and acceleration. There are no explicit safety checks, but a reward signal is shaped to promote sane behaviors (keeping safe distances, avoiding collisions) and discourage bad ones (speeding, leaving the road).

State Space. In order to achieve the next level of driving automation, vehicles are being equipped with multiple sensors which are able to provide various and necessary data about cars' surroundings. In this paper, we focus solely on a path planning problem and to abstract away from perception problems, we assume that a high-level representation of the surroundings (our reinforcement learning observation) is already available. Here we describe, how this representation looks in our setup.

Firstly, to define dynamic objects, we run target selection algorithms to pick interesting vehicles around an ego car.

A basic target selection consists of 7 vehicles and was used in all our experiments, due to its simplicity and realistic assumptions. An extended target selection was used only to check, whether extending visibility to the two additional lanes will result with better overtaking maneuvers. Our brief experiments proved this concept, but due to the unrealistic assumptions of this target selection algorithm, we focused solely on the basic version.

Each vehicle is represented by its relative lateral and longitudinal distance $(d_{lat}^{(i)}, d_{lng}^{(i)})$, velocity $(v_{lat}^{(i)}, v_{lng}^{(i)})$ and acceleration $(a_{lat}^{(i)}, a_{lng}^{(i)})$ with respect to an ego vehicle. The ego is represented by its lateral distance from the center of its lane d_{lat}, rotation ϕ and normalized velocity and acceleration in both lateral and longitudinal directions $(v_{lat}, v_{lng}, a_{lat}, a_{lng})$ respectively.

Secondly, we define a road model in a form of a binary vector, which represents an availability of all lanes on a highway in a place where an ego is located.

Action Space. In case of Highway and HighwayTimeOptimal environments, an action space consists of three high-level actions: follow lane, change lane to the left and change lane to the right. As discussed in previous sections, in order to ensure safety of the system, all actions are planned within the safety envelope defined by RSS-based safety framework. To achieve the safety goals and avoiding negative impact on a user experience and comfort, a deterministic path planning algorithm is used to generate piece-wise constant jerk longitudinal and lateral trajectories with acceleration saturation.

Lane following behavior is performed using adaptive cruise control (ACC) algorithm that abides safety framework rules, i.e. it keeps at least the safe distances to other road users.

To achieve safe lane change and prevent dangerous cut-ins, lane change action is divided into two sub-actions: prepare to lane change and actual lane change. Actual lane change is performed only after confirmation, that a large enough gap between road users on the target lane is present, and that an ego is positioned in respect to adjacent lane vehicles in such way, that distances to both front and rear vehicles are safe. If these conditions are not fulfilled, prepare to lane change sub-action is performed, which consists of positioning the ego vehicle longitudinally within the gap between adjacent lane vehicles and reaching predefined lateral lane bias necessary to negotiate the space for lane change.

In HighwayController environment, action space is composed of two continuous values: a steering angle and an acceleration.

4.2 Training

To solve Highway and HighwayTimeOptimal environments, we evaluated three methods: DQN, DQN with frame stacking and DRQN. DQN and DRQN shared the same state and action representation, whereas for DQN with frame stacking we stacked last 20 observations and used them directly as an input to a neural network. In case of DRQN, a single LSTM layer was inserted into its Q-function network.

Due to available computing resources training was done on a single machine with single optimizer and 64 actors collecting experience and storing them in Prioritized Replay Memory [22]. The number of actors is equal to the number of hyper-threaded cores available on our training machines. Each actor was sending its trajectories at the end of the episode and then it pulled latest parameters from the optimizer. Furthermore, one actor was used as an evaluator, which was computing a running mean score from the last 100 episodes/policies (one episode for one policy). At the end of the training, we took the most promising policy and evaluated it for another 200 episodes.

During the experiments, we observed that agents are likely to flicker with their actions. We addressed this problem in two ways. One was already presented in Sect. 4.1, and the other was to introduce sequential ϵ-greedy actions. The idea is that whenever agent has to pick a random action (due to its exploration strategy), it picks it as usual with ϵ probability, but than it repeats it by N steps, where $N \sim U(L_{min}, L_{max})$ is sampled at the beginning of each ϵ-greedy sequence, and L_{min}, L_{max} are hyper-parameters that have to be carefully tuned. For example, if the simulation sampling time is 0.5 s, and the lane change maneuver takes at least 4 s, the agent needs at least 8 time-steps to complete the maneuver.

Each training lasted for about 24 h, which translates to 9,000 episodes and 350,000 time-steps per each actor. We approximate that during the whole training our agents drove about 500,000 km. These numbers can be easily improved by introducing scalable, multi-node training architectures such as [9,10].

In addition, we ran an experiment with an RSAC agent outputting a low-level control signal for the ego vehicle. With this experiment, we wanted to show that training an agent to avoid collisions, keep safe distances and perform other interactions with road uses in a sane manner is hard. Even tough the agent learnt to successfully reach a goal as fast as possible, it still occasionally violated safety rules, caused an accident or simply exceeded a speed limit.

5 Results

Policies were evaluated as it was described in the previous section and the scores are shown in Table 1. Each method was able to solve both environments: `Highway` and `HighwayTimeOptimal`, but agent trained with DRQN algorithm performed best. Furthermore, DRQN and DQN-stack provide significant improvement over DQN, when solving `HighwayTimeOptimal` environment. This is mainly due to the overtaking maneuvers that DRQN and DQN-stack agents learned, when history of previous states was in some form available to them.

Table 1. Mean score statistics for DQN, DQN-stack and DRQN agents in `Highway` and `HighwayTimeOptimal` environments.

`Highway`	DQN	DQN-stack	DRQN
Moving mean score on 100 episodes	0.83	0.78	0.79
Mean score on 200 episodes	0.77	0.71	**0.81**
`HighwayTimeOptimal`			
Moving mean score on 100 episodes	0.90	0.90	0.95
Mean score on 200 episodes	0.73	0.79	**0.80**

Figure 3 presents training progress of DQN, DQN-stack and DRQN agents. In case of `Highway` environment, DQN obtained its top performance very quickly, after about 8 h of training. DQN-stack and DRQN needed about 15 h. This difference in training time becomes insignificant, when it comes to `HighwayTimeOptimal` environment, where each algorithm converges in similar time, i.e. 15 h.

In order to better understand the bias of our policies, we further evaluated them with additional performance indicators (KPIs). We checked in how many cases there was a collision with our agent, how many times an agent traveled a desired distance, but eventually landed on a wrong lane, and how many times it

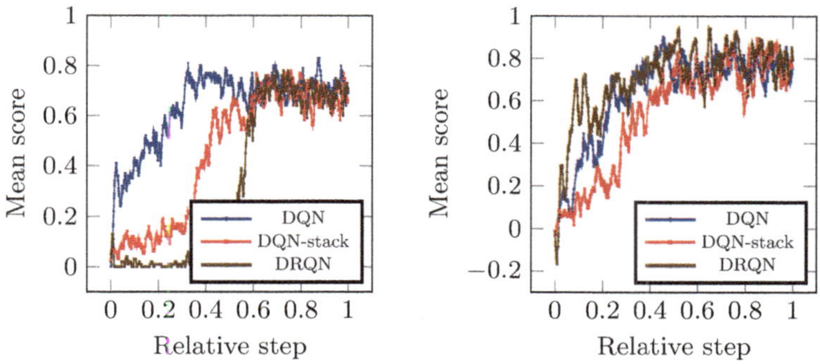

Fig. 3. Mean scores calculated as moving averages over 100 last policies for `Highway` (left) and `HighwayTimeOptimal` (right) environments. Different algorithms had different time steps, thus horizontal axis was normalized and presents relative time steps scaled to 24 h.

tried to change lane to a non-existing one. These values are shown in Table 2 as percentage values.

Table 2. Detailed performance indicators (KPIs) of DQN, DQN-stack, DRQN and RSAC agents in `Highway` and `HighwayTimeOptimal` environments.

`Highway`	DQN	DQN-stack	DRQN	RSAC
Collisions	2.19	3.33	3.13	
Goal reached	87.19	89.63	**94.38**	
Illegal lane change	9.69	5.56	**0.63**	
`HighwayTimeOptimal`				
Collisions	3.06	1.53	0.00	7.14
Goal reached	77.55	82.65	**83.67**	57.69
Illegal lane change	18.37	15.31	15.82	**7.69**

Collisions are caused by our imperfect target selection method, which fails in a situation, when our agent is changing to the lane, which is at the same time, the target lane for changing lane maneuver of the second agent. In this case, our agent misses the second agent in his targets list.

The goal reached metric shows the real potential of our agents. It presents in how many cases, an agent missed the goal by staying at a wrong lane. Our experiments show that this is usually caused by inadequate action space definition, which will be improved in our further research. To clarify this, let's consider an example, when an ego agent travels on the left lane of a crowdy two-lane highway. In front of him there is a car, and on the right lane there are multiple cars travelling in a close distance to each other. In this case, with current action space and our path planning algorithm, it is impossible for the ego agent to change lane. The only option he has is to keep trying by staying at his lane and periodically trying to perform the maneuver. Ultimately, it ends reaching the desired distance, but misses the right lane.

Illegal lane change happens, when an agent is on the most left or most right lane and tries to perform lane change maneuver - to the left or to the right, respectively. Our evaluation shows that history of states helps discouraging this behavior. This is also related to our path planning algorithm, which will continue with lane change maneuver, after first one is finished, unless agent explicitly order him to follow the lane. Our state representation lacks information about last performed action, thus we believe that memory, either in the form of stacked frames or hidden state of LSTM, helps agent to pick the right sequence of actions.

Table 2 presents also the score of RSAC agent in `HighwayTimeOptimal` environment. Its collision rate is significantly higher than for the other agents, and in this case, this is partially fault of the agent, as he is discouraged from performing dangerous maneuvers, but still allowed to do them. In case of a significant lower rate of illegal lane changes, this is due to the fact, that for maneuver planning, an episode is terminated once the agent *decides* to change lane to a non-existing one, but in case of an RSAC agent, an episode ends when the vehicle actually leaves the road. Finally, visual inspection of an RSAC agent behavior showed that the agent usually obey to defined safety rules, but from time to time, in order to maximize a time efficiency, especially at the end of an episode, it violates them.

6 Conclusion

We showed that applying reinforcement learning to maneuver planning, DQN in particular, together with safety envelope tends to be a promising building block for the development of self-driving vehicles. However, with limited action space agent was often ending up in situations where none of its available actions was yielding success in the end. While investigating behavior of an RSAC agent working on a low-level control interface, it was observed that agent had much better control on car behavior. By avoiding restriction of the action space, objectives may be satisfied in straight-forward way resulting in intelligent and assertive behaviors while negotiation with others at the same time. However, as the results show, the agent also suffers from higher collision rate, as safety aspect is only an another part of a general objective function. Based on that, we can conclude that simultaneously satisfying some form of safety guarantees and using reinforcement learning with low-level control signals might be a challenging task.

By analysing outcomes of both high-level and low-level control approaches, our further research will focus on extending our high-level action space to give agent more flexibility over vehicle behavior and transparency, while consequently keeping responsibility for safety outside reinforcement learning module.

References

1. Althoff, M., Stursberg, O., Buss, M.: Model-based probabilistic collision detection in autonomous driving. IEEE Trans. Intell. Transp. Syst. **10**(2), 299–310 (2009). https://doi.org/10.1109/TITS.2009.2018966. http://ieeexplore.ieee.org/document/4895669/
2. Bojarski, M., Del Testa, D., Dworakowski, D., Firner, B., Flepp, B., Goyal, P., Jackel, L.D., Monfort, M., Muller, U., Zhang, J., Zhang, X., Zhao, J., Zieba, K., et al.: End to End Learning for Self-Driving Cars (2016). http://arxiv.org/abs/1604.07316
3. Brockman, G., Cheung, V., Pettersson, L., Schneider, J., Schulman, J., Tang, J., Zaremba, W.: OpenAI Gym. CoRR **abs/1606.01540** (2016). http://arxiv.org/abs/1606.01540
4. Censi, A., Slutsky, K., Wongpiromsarn, T., Yershov, D., Pendleton, S., Fu, J., Frazzoli, E.: Liability, Ethics, and Culture-Aware Behavior Specification using Rulebooks, pp. 8536–8542. Institute of Electrical and Electronics Engineers (IEEE) (2019). https://doi.org/10.1109/icra.2019.8794364
5. Dosovitskiy, A., Ros, G., Codevilla, F., Lopez, A., Koltun, V.: CARLA: An Open Urban Driving Simulator (2017). URL http://arxiv.org/abs/1711.03938
6. Haarnoja, T., Zhou, A., Abbeel, P., Levine, S.: Soft Actor-Critic: Off-Policy Maximum Entropy Deep Reinforcement Learning with a Stochastic Actor (2018). http://arxiv.org/abs/1801.01290
7. Haarnoja, T., Zhou, A., Hartikainen, K., Tucker, G., Ha, S., Tan, J., Kumar, V., Zhu, H., Gupta, A., Abbeel, P., Levine, S.: Soft Actor-Critic Algorithms and Applications (2018). http://arxiv.org/abs/1812.05905
8. Hausknecht, M., Stone, P.: Deep recurrent q-learning for partially observable MDPs. In: AAAI Fall Symposium - Technical Report, vol. FS-15-06, pp. 29–37. AI Access Foundation (2015)
9. Horgan, D., Quan, J., Budden, D., Barth-Maron, G., Hessel, M., van Hasselt, H., Silver, D.: Distributed prioritized experience replay. In: International Conference on Learning Representations (2018). https://openreview.net/forum?id=H1Dy--0Z
10. Kapturowski, S., Ostrovski, G., Dabney, W., Quan, J., Munos, R.: Recurrent experience replay in distributed reinforcement learning. In: International Conference on Learning Representations (2019). https://openreview.net/forum?id=r1lyTjAqYX
11. Lopez, P.A., Behrisch, M., Bieker-Walz, L., Erdmann, J., Flotterod, Y.P., Hilbrich, R., Lucken, L., Rummel, J., Wagner, P., Wiebner, E.: Microscopic Traffic Simulation using SUMO. In: IEEE Conference on Intelligent Transportation Systems, Proceedings, ITSC, vol. 2018-November, pp. 2575–2582. Institute of Electrical and Electronics Engineers Inc. (2018). https://doi.org/10.1109/ITSC.2018.8569938
12. Magdici, S., Althoff, M.: Fail-safe motion planning of autonomous vehicles. In: 2016 IEEE 19th International Conference on Intelligent Transportation Systems (ITSC), pp. 452–458. IEEE, Rio de Janeiro, Brazil (2016). https://doi.org/10.1109/ITSC.2016.7795594. http://ieeexplore.ieee.org/document/7795594/
13. Makantasis, K., Kontorinaki, M., Nikolos, I.: A Deep Reinforcement-Learning-based Driving Policy for Autonomous Road Vehicles (2019). http://arxiv.org/abs/1907.05246
14. Mirchevska, B., Pek, C., Werling, M., Althoff, M., Boedecker, J.: High-level decision making for safe and reasonable autonomous lane changing using reinforcement learning. In: IEEE Conference on Intelligent Transportation Systems, Proceedings, ITSC, vol. 2018-November, pp. 2156–2162. Institute of Electrical and Electronics Engineers Inc. (2018). https://doi.org/10.1109/ITSC.2018.8569448

15. Mnih, V., Kavukcuoglu, K., Silver, D., Graves, A., Antonoglou, I., Wierstra, D., Riedmiller, M.: Playing Atari with Deep Reinforcement Learning (2013). http://arxiv.org/abs/1312.5602
16. Nageshrao, S., Tseng, E., Filev, D.: Autonomous Highway Driving using Deep Reinforcement Learning (2019). http://arxiv.org/abs/1904.00035
17. Ngai, D.C.K., Yung, N.H.C.: A multiple-goal reinforcement learning method for complex vehicle overtaking maneuvers. IEEE Trans. Intell. Transp. Syst. **12**, 509–522 (2011). https://doi.org/10.1109/TITS.2011.2106158
18. Nistér, D., Lee, H.L., Ng, J., Wang, Y.: The Safety Force Field. Technical report
19. Paden, B., Cap, M., Yong, S.Z., Yershov, D., Frazzoli, E.: A survey of motion planning and control techniques for self-driving urban vehicles. IEEE Trans. Intell. Veh. **1**(1), 33–55 (2016). https://doi.org/10.1109/tiv.2016.2578706
20. Qian, X., de La Fortelle, A., Moutarde, F.: A hierarchical Model Predictive Control framework for on-road formation control of autonomous vehicles. In: 2016 IEEE Intelligent Vehicles Symposium (IV), pp. 376–381. IEEE, Gotenburg, Sweden (2016). https://doi.org/10.1109/IVS.2016.7535413. http://ieeexplore.ieee.org/document/7535413/
21. Rasekhipour, Y., Khajepour, A., Chen, S.K., Litkouhi, B.: A potential field-based model predictive path-planning controller for autonomous road vehicles. IEEE Trans. Intell. Transp. Syst. **18**(5), 1255–1267 (2017). https://doi.org/10.1109/TITS.2016.2604240. http://ieeexplore.ieee.org/document/7576661/
22. Schaul, T., Quan, J., Antonoglou, I., Silver, D.: Prioritized experience replay. arXiv preprint arXiv:1511.05952 (2015)
23. Shah, S., Dey, D., Lovett, C., Kapoor, A.: AirSim: High-Fidelity Visual and Physical Simulation for Autonomous Vehicles (2017). http://arxiv.org/abs/1705.05065
24. Shalev-Shwartz, S., Shammah, S., Shashua, A.: On a Formal Model of Safe and Scalable Self-driving Cars (2017). http://arxiv.org/abs/1708.06374
25. Sutton, R.S., Barto, A.G.: Introduction to Reinforcement Learning, 1st edn. MIT Press, Cambridge (1998)
26. Talpaert, V., Sobh, I., Ravi Kiran, B., Mannion, P., Yogamani, S., El-Sallab, A., Perez, P.: Exploring applications of deep reinforcement learning for real-world autonomous driving systems. In: VISIGRAPP 2019 - Proceedings of the 14th International Joint Conference on Computer Vision, Imaging and Computer Graphics Theory and Applications, vol. 5, pp. 564–572. SciTePress (2019). https://doi.org/10.5220/0007520305640572
27. Wang, P., Chan, C.Y.: Formulation of deep reinforcement learning architecture toward autonomous driving for on-ramp merge. In: IEEE Conference on Intelligent Transportation Systems, Proceedings, ITSC, vol. 2018-March, pp. 1–6. Institute of Electrical and Electronics Engineers Inc. (2018). https://doi.org/10.1109/ITSC.2017.8317735
28. Wei, J., Snider, J.M., Gu, T., Dolan, J.M., Litkouhi, B.: A behavioral planning framework for autonomous driving. In: IEEE Intelligent Vehicles Symposium, Proceedings, pp. 458–464. Institute of Electrical and Electronics Engineers Inc. (2014). https://doi.org/10.1109/IVS.2014.6856582
29. Xu, W., Wei, J., Dolan, J.M., Zhao, H., Zha, H.: A real-time motion planner with trajectory optimization for autonomous vehicles. In: 2012 IEEE International Conference on Robotics and Automation, pp. 2061–2067. IEEE (2012)

Well Convergent and Computationally Efficient Quaternion Loss

Kamil Lelowicz[1,2](✉) and Jakub Derbisz[2]

[1] Department of Automatic Control and Robotics, AGH University of Science and Technology, al. A. Mickiewicza 30, 30-059 Kraków, Poland
lelowicz@agh.edu.pl

[2] APTIV Services Poland S.A., ul. Podgorki Tynieckie 2, 30-399 Kraków, Poland
{kamil.lelowicz,jakub.derbisz}@aptiv.com

Abstract. Rotation estimation, i.e. the ability to predict angles describing 3D positioned object, is an omnipresent problem in computer vision, computer graphics and 3D object detection task in automotive industry. Deep learning algorithms usually parameterise rotation using only the yaw angle. This paper presents detailed comparison of several 3D rotation distance functions using quaternions. We propose a computationally efficient quaternion loss function for neural network training. We conclude that function respects the topology of $SO(3)$ and is bi-invariant. We also show the geometrical representation of presented functions. Lastly, we evaluate the effectiveness of the proposed loss function and compare its performance with other methods using a public large-scale dataset.

Keywords: Optimization · Quaternions · Sensors · Neural networks · Automotive

1 Introduction

In 3D object detection task in automotive industry one of the key issues for a neural network to learn is the ability to predict angles describing 3D positioned object. Most of the works up to recent creation of nuScenes [4] dataset consisted of building machine learning algorithms with loss being described by single number which would be the yaw angle only. This is mostly because availability and popularity of Kitti dataset [7] in computer vision research community concentrated on automotive problems. However, complete rotation of a 3D object is actually described by yaw-pitch-roll angles. Euler angles or even rotation matrices have nevertheless some difficulties, as problems with composition, gimbal lock, compact representation or numerical stability and efficiency, which are resolved by quaternion representation, and as concluded in [9], quaternions are both spatially and computationally more efficient to use for 3D rotations.

For example 3×3 rotation matrix requires to store 9 floating points number, in case of quaternion we need to represent special orthogonal group which is 3 dimensional object in 4 dimensional space. Moreover the multiplication of

A. Bartoszewicz et al. (Eds.): PCC 2020, AISC 1196, pp. 1275–1286, 2020.
https://doi.org/10.1007/978-3-030-50936-1_106

rotation matrix can lead to non-orthogonal matrix, which means that they no longer describe the rotation. This is a challenging problem during optimization, because one has to maintain the orthogonality of the matrix. Six constrains have to be met. Each row has to be unit, and columns have to be mutually orthogonal. In case of quaternions one has to assure only one constrain, the unit magnitude of quatrernion. Renormalizing a quaternion is much more straightforward than reorthogonalizing of a matrix. In nuScenes [4] the ground truth for object's rotation is described with a quaternion. Now the problem is the creation of such loss function that would be able to predict yaw-pitch-roll angles and do this using quaternions only, without manipulating equivalent representations because operations on quaternions are very efficient, as stated. The problems are finding the appropriate loss function that would allow neural network to learn the rotation angles of 3D bounding boxes around the object, and determining what kind of function would allow the neural network to learn it efficiently and accurately. Moreover, loss function should prevent the overfitting and consequently allow for good generalization of obtained neural network. The function should be also well convergent to allow get result fast. These properties are often opposite, which causes difficulties in training neural networks.

The contributions of this paper are detailed comparison and evaluation in 3D object detection task three quaternion loss functions, showing their properties and that they can be taken as valid angle yaw-pitch-roll loss functions as opposed to the plain and direct approach.

The article is organized as follows. Firstly in Sect. 2 the brief overview of quaternions as a rotations are depicted. This is then followed by definition of distance function in Sect. 3. The properties of distance functions will be studied. In the next Sect. 4 non-quaternion distance functions will be describe. In Sect. 5 the numerical results from learning a machine learning model with respect to angle loss behaviour will be presented. Finally, summary of the work is described.

2 Quaternions as Rotations

Euler's definition of rotation [6] is as follows

Theorem 1. *Let Q, $Q^{'} \in \mathbb{R}^3$ then there exists an axis $d \in \mathbb{R}^3$ and an angle of rotation $\alpha \in (-\pi, \pi\rangle$ such that Q yields Q' when rotated by α about u.*

It is worth to mention that definition does not state uniqueness, but only states the existence. In literature [5] the combination of a rotation definition and a corresponding mathematical representation is called a rotational modality. This article discuss rotation defined by Euler's theorem represented by quaternions. Quaternions, denoted by $\mathbb{H}$, defined by Hamilton [8] are nonzero division ring. In general form a quaternion can be expressed as:

$$\mathbf{q} \equiv (a, b, c, d) \equiv (a, \mathbf{v}) \tag{1}$$

Special case are unit quaternions S^3, where $S^n = \{\mathbf{q} \in \mathbb{R}^{n+1}, ||\mathbf{q}||^2 = 1\}$ and $|| \cdot ||$ is 2-norm. A rotation by an angle α about axis defined by a unit vector

$\mathbf{u} = (u_x, u_y, u_z)$ can be expressed by unit quaternion as follows:

$$\mathbf{q} = \pm(cos\frac{\alpha}{2}, \mathbf{u}sin\frac{\alpha}{2}) \tag{2}$$

Unit quaternion can also written as a 2 × 2 unitary matrix [15]:

$$\begin{bmatrix} a + b\mathbf{i} & c + d\mathbf{i} \\ -c + d\mathbf{i} & a - b\mathbf{i} \end{bmatrix} \tag{3}$$

The Special Unitary group $SU(2)$ can be parmetrized by unit quatrenions giving isomorphism. Moreover, $SU(2)$ doubly covers the Special Orthogonal Group $SO(3)$. It means that for each rotation in $SO(3)$ exists a pair of members of $SU(2)$. The ambiguity is the result of choice of rotations axis in the logarithmic map. Since unit quaternions $\mathbf{q}$ and $-\mathbf{q}$ describe the same rotation it does not affect the result.

We will now give few definitions that will be used later. The exponential of a quatrnion $\mathbf{q}$ can be expressed using power series [1]:

$$e^{\mathbf{q}} = \sum_{n=0}^{\infty} \frac{\mathbf{q}^n}{n!} = e^a \left(cos||\mathbf{v}||, \frac{\mathbf{v}}{||\mathbf{v}||} sin||\mathbf{v}|| \right) \tag{4}$$

Then the rotation defined by quaterion can be expressed in a shorter form:

$$\mathbf{q} = \pm(cos\frac{\alpha}{2}, \mathbf{u}sin\frac{\alpha}{2}) = e^{\frac{\alpha}{2}(0,\mathbf{u})} \tag{5}$$

The logarithm is defined as inverse function of exponent [1]:

$$ln\, \mathbf{q} = \left(ln||\mathbf{q}||, \frac{\mathbf{v}}{||\mathbf{v}||} arccos\frac{a}{||\mathbf{q}||} \right) \quad for \quad ||\mathbf{q}|| \neq 0 \tag{6}$$

For unit quaternions one can simplify formula to:

$$ln\, \mathbf{q} = \left(0, \frac{\mathbf{v}}{||\mathbf{v}||} arccos\ a \right) \quad for \quad ||\mathbf{q}|| = 1 \tag{7}$$

Quaternion conjunction is defined as follows:

$$\mathbf{q}^* = (a, -b, -c, -d) \tag{8}$$

Moreover for unit quaternions conjunction is equal quatrnion inverse:

$$\mathbf{q}^{-1} = \frac{\mathbf{q}^*}{||\mathbf{q}||} \tag{9}$$

3 Distance Function

Metric in mathematics is a function that describes the distance between each pair in a set. Formally metric or distance function is a map:

$$d : X \times X \rightarrow \mathbb{R}^+ \cup \{0\} \tag{10}$$

where for all $x, y, z \in X$ following axioms are satisfied:

$$d(x,y) = 0 \iff x = y \tag{11a}$$
$$d(x,y) = d(y,x) \tag{11b}$$
$$d(x,y) \leq d(x,z) + d(z,y) \tag{11c}$$

However, it is worth noting that it is not a rare situation when our intuitive concept of measuring distance is not satisfied with given conditions. As mentioned above, each rotation can be described by two unit quaternions $\mathbf{q}$ and $-\mathbf{q}$. The distance between the two same rotations should be 0. In case of unit quaternions $d(\mathbf{q}, -\mathbf{q}) = 0 \nRightarrow \mathbf{q} = -\mathbf{q}$ Thus, axiom (11a) can not be meet. Nevertheless, the mapping from S^3 to SO(3) is 2 to 1. The "distance" on unit quaternion can become metric on 3D rotations.

The most obvious function used to construct a metric on Euclidean space is simple 2-norm. The function using $L2$ norm can be written as:

$$\begin{aligned} d_1 &: S^3 \times S^3 \rightarrow \mathbb{R}^+ \cup \{0\} \\ d_1(\mathbf{q}_1, \mathbf{q}_2) &= \frac{1}{2}||\mathbf{q}_1 - \mathbf{q}_2||^2 \\ &= \frac{1}{2}\left((a_1 - a_2)^2 + (b_1 - b_2)^2 + (c_1 - c_2)^2 + (d_1 - d_2)^2\right) \\ &= \frac{1}{2}\left(a_1^2 + a_2^2 + b_1^2 + b_2^2 + c_1^2 + c_2^2 + d_1^2 + d_2^2\right) \\ &\quad - a_1a_2 - b_1b_2 - c_1c_2 - d_1d_2 \\ &= 1 - \mathbf{q}_1 \cdot \mathbf{q}_2 \end{aligned} \tag{12}$$

where $\cdot$ is a dot product. Unfortunately $d_1(\mathbf{q}, -\mathbf{q}) = 2$. Also, the distance between two quaternions can be measured as the shortest geodesic path on the Riemannian manifold S^3 [2]:

$$\begin{aligned} d_2 &: S^3 \times S^3 \rightarrow \mathbb{R}^+ \cup \{0\} \\ d_2(\mathbf{q}_1, \mathbf{q}_2) &= ||ln\ \mathbf{q}_1^{-1}\mathbf{q}_2|| = ||(0, \tfrac{\mathbf{v}}{||\mathbf{v}||} arccos\ \mathbf{q}_1 \cdot \mathbf{q}_2)|| = arccos\ \mathbf{q}_1 \cdot \mathbf{q}_2 \end{aligned} \tag{13}$$

However also in this case $d_2(\mathbf{q}, -\mathbf{q}) = \pi \neq 0$. Thus, this function is not well defined on $SO(3)$.

To overcome this problem Ravni and Roth [11] proposed following function similar to (12):

$$\begin{aligned} d_3 &: S^3 \times S^3 \rightarrow \mathbb{R}^+ \cup \{0\} \\ d_3(\mathbf{q}_1, \mathbf{q}_2) &= \min\{||\mathbf{q}_1 - \mathbf{q}_2||, ||\mathbf{q}_1 + \mathbf{q}_2||\} \end{aligned} \tag{14}$$

It can be easily proved that (14) do not meet only axiom (11a). Thus, function d_2 is pseudometric [3] on unit quaternions, but in SO(3) it is a metric. In [16] was proposed modification of function (13):

$$\begin{aligned} d_4 &: S^3 \times S^3 \rightarrow \mathbb{R}^+ \cup \{0\} \\ d_4(\mathbf{q}_1, \mathbf{q}_2) &= \min\{arccos\,(\mathbf{q}_1 \cdot \mathbf{q}_2)\,, \pi - arcccos\,(\mathbf{q}_1 \cdot \mathbf{q}_2)\} \end{aligned} \tag{15}$$

Since both presented function used *min* term in their formulas they are not computationally efficient. Thus, using them as a target function during optimization can be cumbersome. To overcome this, the distance d_3 can be rewritten as follow:

$$\begin{aligned} d_4' &: S^3 \times S^3 \rightarrow \mathbb{R}^+ \cup \{0\} \\ d_4'(\mathbf{q}_1, \mathbf{q}_2) &= arccos||\mathbf{q}_1 \cdot \mathbf{q}_2|| \end{aligned} \tag{16}$$

One last step is to eliminate the inverse cosinus function. Arcus cosinus is strictly decreasing, so in order to minimize the function one has to maximize the argument of the function. In other words minimize the minus argument. After normalization such that the function will only take non-negative values one will get:

$$\begin{aligned} d_5 &: S^3 \times S^3 \rightarrow \mathbb{R}^+ \cup \{0\} \\ d_5 &: (\mathbf{q}_1, \mathbf{q}_2) = 1 - ||\mathbf{q}_1 \cdot \mathbf{q}_2|| \end{aligned} \tag{17}$$

It is also possible to modify function (14) to get the same distance:

$$\begin{aligned} d_5' &: S^3 \times S^3 \rightarrow \mathbb{R}^+ \cup \{0\} \\ d_5'(\mathbf{q}_1, \mathbf{q}_2) &= \min\{\frac{1}{2}||\mathbf{q}_1 - \mathbf{q}_2||^2, \frac{1}{2}||\mathbf{q}_1 + \mathbf{q}_2||^2\} \\ &= \min\{1 - \mathbf{q}_1 \cdot \mathbf{q}_2, 1 + \mathbf{q}_1 \cdot \mathbf{q}_2\} \\ &= 1 - ||\mathbf{q}_1 \cdot \mathbf{q}_2|| \end{aligned} \tag{18}$$

This function is also a psedometric in S^2, and metric in $SO(3)$ [9] and was used by Kuffner [10] for 3D Rigid Body Path Planning. Nevertheless, one will start with 2-norm on quaternion or the shortest great arc in S^3. Both of these approaches provide computationally efficient map $1 - |\mathbf{q}_1 \cdot \mathbf{q}_2|$. The only problem is lack of smoothness due to existence of absolute term.

To obtain rotation $\mathbf{q}_2$ from $\mathbf{q}_1$, the $\mathbf{q}_1$ should be multiplied by another rotation $\mathbf{p} \in S^3$, in such manner

$$\mathbf{q}_2 = \mathbf{q}_1 \mathbf{p} \tag{19}$$

which brings us to

$$\mathbf{p} = \mathbf{q}_1^{-1} \mathbf{q}_2 = \mathbf{q}_1^* \mathbf{q}_2 \tag{20}$$

Let $\mathbf{p} = (a_p, \mathbf{v}_p) = (cos\frac{\gamma}{2}, \mathbf{u}^T sin\frac{\gamma}{2})$ then using definition of quaternion multiplication one can get:

$$cos\frac{\gamma}{2} = a_p = a_1 a_2 + b_1 b_2 + c_1 c_2 + d_1 d_2 = \mathbf{q}_1 \cdot \mathbf{q}_2 \tag{21}$$

Therefore distance (13) is equal to:

$$||ln\ \mathbf{q}_1^{-1} \mathbf{q}_2|| = arccos\ \mathbf{q}_1 \cdot \mathbf{q}_2 = \frac{\gamma}{2} \tag{22}$$

where γ is simply a non-negative angle of rotation that yields orientation $\mathbf{q}_2$ from $\mathbf{q}_1$ rotated by γ about $\mathbf{u}$ or yields orientation $\mathbf{q}_1$ from $\mathbf{q}_2$ rotated by γ

about $-\mathbf{u}$. On the other hand the shortest distance in S^n between two points $\mathbf{w}_1, \mathbf{w}_2$ is simply the angle β between these vectors. Since

$$\mathbf{w}_1 \cdot \mathbf{w}_2 = ||\mathbf{w}_1||\,||\mathbf{w}_2|| cos\beta = cos\beta \tag{23}$$

The shortest arc in S^n is equal then to [5]:

$$\beta = arccos\ \mathbf{w}_1 \cdot \mathbf{w}_2 = \frac{\gamma}{2} \tag{24}$$

Hence, this results show the way how to interpolate two orientations. Interpolated orientations could lay on the shortest arc between two points in S^3. This type of interpolation is called Slerp - spherical linear interpolation [5]. Results can be used to define another distance function. Alternatively γ instead of β can be minimized. In general

$$cos\ \gamma = 2cos^2\frac{\gamma}{2} - 1 = 2\left(\mathbf{q}_1 \cdot \mathbf{q}_2\right)^2 - 1 \tag{25}$$

Thus, function will be in following form:

$$\begin{aligned} d_6 &: S^3 \times S^3 \to \mathbb{R}^+ \cup \{0\} \\ d_6(\mathbf{q}_1, \mathbf{q}_2) &= arccos\ (2\left(\mathbf{q}_1 \cdot \mathbf{q}_2\right)^2 - 1) \end{aligned} \tag{26}$$

As was mentioned above the arcus cosinus function is strictly decreasing function. Consequently, only term $2\left(\mathbf{q}_1 \cdot \mathbf{q}_2\right)^2 - 1$ have to be taken in to account. Furthermore, after normalization one will get following function:

$$\begin{aligned} d_6 &: S^3 \times S^3 \to \mathbb{R}^+ \cup \{0\} \\ d_6(\mathbf{q}_1, \mathbf{q}_2) &= 1 - \left(\mathbf{q}_1 \cdot \mathbf{q}_2\right)^2 \end{aligned} \tag{27}$$

In this case the distance is measured as nonlinear strictly increasing function of shortest arc β, hence (27) is also a distance function in $SO(3)$. The formulation of function (27) gives rise to:

Proposition 1. *The distance function defined by*

$$\begin{aligned} d_6 &: S^3 \times S^3 \to \mathbb{R}^+ \cup \{0\} \\ d_6(\mathbf{q}_1, \mathbf{q}_2) &= 1 - \left(\mathbf{q}_1 \cdot \mathbf{q}_2\right)^2 \end{aligned} \tag{28}$$

is bi-invariant.

Proof. Distance is left-invariant if:

$$d_6(\mathbf{q}_3\mathbf{q}_1, \mathbf{q}_3\mathbf{q}_2) = d_6(\mathbf{q}_1, \mathbf{q}_2) \tag{29}$$

Respectively right-invariant if :

$$d_6(\mathbf{q}_1\mathbf{q}_3, \mathbf{q}_2\mathbf{q}_3) = d_6(\mathbf{q}_1, \mathbf{q}_2) \tag{30}$$

For left-invariant there is:

$$\begin{aligned} d_6(\mathbf{q}_3\mathbf{q}_1, \mathbf{q}_3\mathbf{q}_2) &= 1 - (\mathbf{q}_3\mathbf{q}_1 \cdot \mathbf{q}_3\mathbf{q}_2)^2 \\ &= 1 - \left(\frac{(\mathbf{q}_3\mathbf{q}_1)^{-1}\mathbf{q}_3\mathbf{q}_2}{2} + \frac{\left((\mathbf{q}_3\mathbf{q}_1)^{-1}\mathbf{q}_3\mathbf{q}_2\right)^*}{2} \right)^2 \\ &= 1 - \left(\frac{\mathbf{q}_1^{-1}\mathbf{q}_3^{-1}\mathbf{q}_3\mathbf{q}_2}{2} + \frac{(\mathbf{q}_1^{-1}\mathbf{q}_3^{-1}\mathbf{q}_3\mathbf{q}_2)^*}{2} \right)^2 \\ &= 1 - \left(\frac{\mathbf{q}_1^{-1}\mathbf{q}_2}{2} + \frac{(\mathbf{q}_1^{-1}\mathbf{q}_2)^*}{2} \right)^2 \\ &= 1 - (\mathbf{q}_1 \cdot \mathbf{q}_2)^2 \\ &= d_6(\mathbf{q}_1, \mathbf{q}_2) \end{aligned} \tag{31}$$

Similar for right-invariant:

$$\begin{aligned} d_6(\mathbf{q}_1\mathbf{q}_3, \mathbf{q}_1\mathbf{q}_3) &= 1 - (\mathbf{q}_1\mathbf{q}_3 \cdot \mathbf{q}_1\mathbf{q}_3)^2 \\ &= 1 - \left(\frac{\mathbf{q}_1\mathbf{q}_3(\mathbf{q}_2\mathbf{q}_3)^{-1}}{2} + \frac{\left(\mathbf{q}_1\mathbf{q}_3(\mathbf{q}_2\mathbf{q}_3)^{-1}\right)^*}{2} \right)^2 \\ &= 1 - \left(\frac{\mathbf{q}_1\mathbf{q}_3\mathbf{q}_3^{-1}\mathbf{q}_2^{-1}}{2} + \frac{(\mathbf{q}_1\mathbf{q}_3\mathbf{q}_3^{-1}\mathbf{q}_2^{-1})^*}{2} \right)^2 \\ &= 1 - \left(\frac{\mathbf{q}_1\mathbf{q}_2^{-1}}{2} + \frac{(\mathbf{q}_1\mathbf{q}_2^{-1})^*}{2} \right)^2 \\ &= 1 - (\mathbf{q}_1 \cdot \mathbf{q}_2)^2 \\ &= d_6(\mathbf{q}_1, \mathbf{q}_2) \end{aligned} \tag{32}$$

Since distance function is both left- and right-invariant, hence is bi-invariant. □

Thus, our measure of the distance between two orientations is independent of the choice of coordinate system used. Moreover:

Proposition 2. *The distance function*

$$\begin{gathered} d_6 : S^3 \times S^3 \to \mathbb{R}^+ \cup \{0\} \\ d_6(\mathbf{q}_1, \mathbf{q}_2) = 1 - (\mathbf{q}_1 \cdot \mathbf{q}_2)^2 \end{gathered} \tag{33}$$

is equivalent to the following defined on $SO(3) \times SO(3)$*:*

$$\begin{gathered} d_7 : SO(3) \times SO(3) \to \mathbb{R}^+ \cup \{0\} \\ d_7(R_1, R_2) = \tfrac{1}{4}||I - R_1^T R_2||_F^2 \end{gathered} \tag{34}$$

where $||\cdot||_F$ *is Frobenious norm.*

Proof. The function (34) can be rewritten as:

$$\begin{aligned} d_7(R_1, R_2) &= \frac{1}{4}||I - R_1^T R_2||_F^2 \\ &= \frac{1}{4}\left(tr\left(I - R_1^T R_2\right)\right) \\ &= \frac{1}{4}\left(3 - tr\left(R_1^T R_2\right)\right) \end{aligned} \tag{35}$$

Let R has a rotation angle γ about axis $\mathbf{u}$, then for all $R \in SO(3)$ the trace is equal to $tr(R) = 1 + 2cos\ \gamma$ [9]. Admittedly, $R_1^T R_2 \in SO(3)$ and describe rotation form $R1$ to $R2$. In consequence:

$$\begin{aligned} d_7(R_1, R_2) &= \frac{1}{2}\left(1 - cos\ \gamma\right) \\ &= \frac{1}{2}\left(2 - 2cos^2\ \frac{\gamma}{2}\right) \\ &= \left(1 - cos^2\ \frac{\gamma}{2}\right) \end{aligned} \tag{36}$$

Taking into account (21), function (27) after mapping arguments to $SO(3)$ is equivalent to (36). □

4 Non-quaternion Losses

Let $(\alpha_1, \beta_1, \gamma_1)$ and $(\alpha_2, \beta_2, \gamma_2)$ be sets of Euler angles.

$$\begin{gathered} d_8 : E \times E \rightarrow \mathbb{R}^+ \cup \{0\} \\ d_8((\alpha_1, \beta_1, \gamma_1), (\alpha_2, \beta_2, \gamma_2)) = \sqrt{d(\alpha_1, \alpha_1)^2 + d(\beta_1, \beta_2)^2 + d(\gamma_1, \gamma_2)^2} \end{gathered} \tag{37}$$

where $d(x, y) = \min\{|x-y|, 2\pi-|x-y|\}$ and $E \in \mathbb{R}^3$. However, it is not a distance function on $SO(3)$ since under a particular order of rotation Euler angles $(0, 0, \pi)$ and $(\pi, \pi, 0)$ represent the same orientation, yet the distance between them is non-zero. Additional conditions to domain E have to be impose to overcome this problem: $\alpha, \gamma \in \langle -\pi, \pi), \beta \in \langle \frac{-\pi}{2}, \frac{\pi}{2})$. Nevertheless it should be noticed, that function (37) is not bi-invariant. Consequently, two nearby rotations may have a large distance value, while two distant rotations may have a smaller distance value. Thus, this metric does not have the property of quaternion metric shown in Proposition 1.

The distance between two orientation can be expressed also using rotation matrix as in function (34). Nonetheless, we have to assure that predicted matrices from neural network will be orthogonal. Thus, the six constraints must be maintained during the computations. Each row must be a unit vector and the columns must be mutually orthogonal. Moreover, for each predicted matrix we need to perform reorthogonalization which required to use single value decomposition algorithm. In comparison quaternion normalization only requires dividing by square root of sum of squares of its elements. Furthermore as was shown in proposition 2 the quaternion distance function is equivalent to the distance function directly defined on $SO(3)$. Thus, there is no need to use directly function (34) which causes computationally expensive calculation.

5 Experiments

We made experiments using three angle loss functions on quaternions. The experiments were done while training one of well performing neural network models for 3D object detection task in automotive industry. We concentrated on behaviour of part of the loss function responsible for learning rotation. The neural network was trained on nuScenes dataset. The inputs to the neural network were: lidar with 32 channels, front camera with resolution 1600×1200 and front long-range radar. Detailed information about dataset and sensors can be found in [4]. The neural network architecture was based on *Yolo v3* [12–14], however for 3D object detection task, its output related to 3D position of an object was of the form: $(object\ class, center_x, center_y, center_z, width, length, height, a, b, c, d)$, in camera's 3D coordinates, where $\mathbf{q} = (a, b, c, d)$ is a quaternion describing rotation. The quaternion output from the network is normalised to ensure that it is unit and determines the rotation. There was also returned 2D part in projected view (as in Yolo) of the form: $(left, right, top, bottom)$. All these values complemented introduced loss function. Training of the model was performed using roughly 22000 nuScenes samples, the test set consisted of about 2400 test samples that came from scenes different than those constituted training samples. Test samples contained about 4200 cars to be detected. The training of neural networks with different angle loss functions was performed until the KPI metrics presented in Table 1 were achieved (for different loss functions, the measures of detected objects given by these KPIs were comparable). Detailed KPI metrics description can be found in [4].

Table 1. KPI metrics for 3D bounding box object detection.

KPI	Result
Recall	0.80
Precision	0.70
F1 Score	0.75
Mean average precision	0.70

The loss function that was minimized while training the network consisted of weighted sum with three parts:

1. 2D part as in *Yolo v3* loss [14],
2. Squared L2 distances between 3D coordinates,
3. Quaternion angle loss for learning yaw-pitch-roll.

We mainly focused on evaluation of performance of quaternion loss:

$$\min_{\mathbf{q} \in S^3} d(\mathbf{p}, \mathbf{q}) \tag{38}$$

The experiments kept fixed underlying neural network and hyperparameters (as having fixed learning rate for all). Only the quaternion angle loss function was changed. The best performing loss functions gave neural networks that could be trained further using smaller learning rate, minimizing angle loss even more.

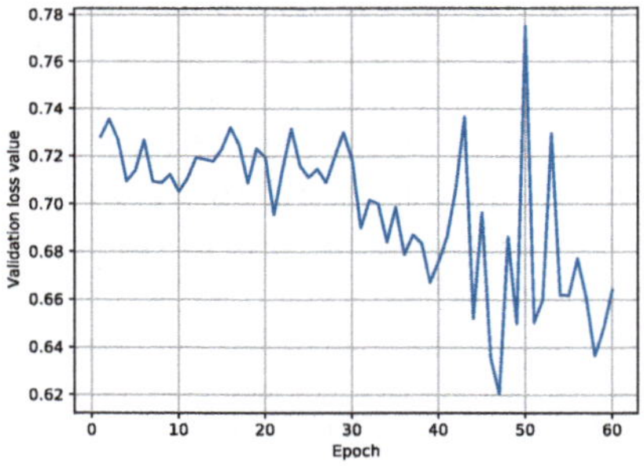

(a) Validation loss for function (12).

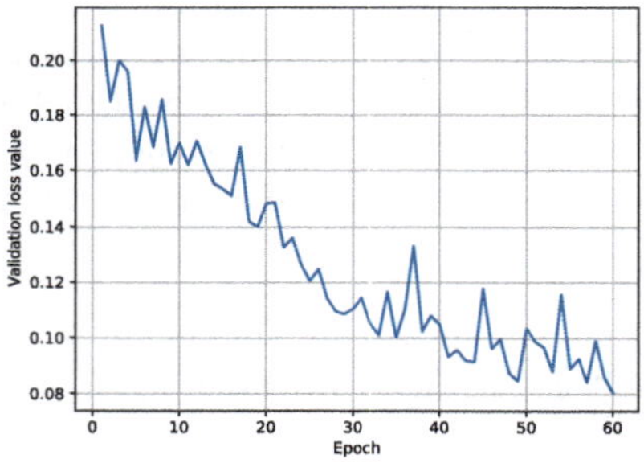

(b) Validation loss for function (17).

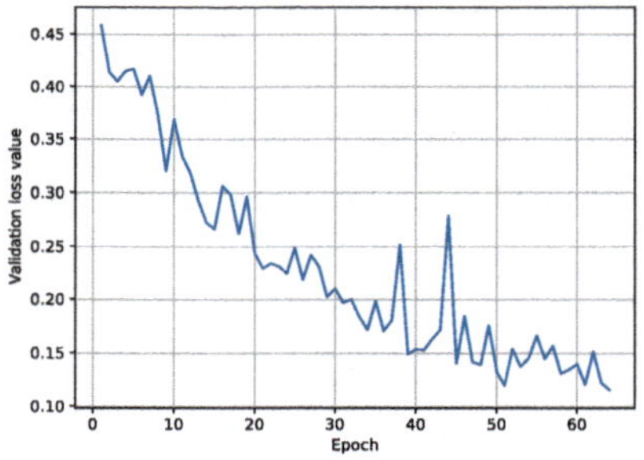

(c) Validation loss for function (27).

Fig. 1. Algorithm results.

Three experiments were carried out. Figure 1 shows graphs of quaternion validation loss for each function. The first angle loss function we tried was plainly and directly taking squared L2 norm (12), where for our loss, squared L2 norm was transformed into the form of $1 - \mathbf{p} \cdot \mathbf{q}$. In Fig. 1a one can observe that the results of learning with such loss function are bad and the oscillating loss is unstable and has not reached reasonable value. The reason for this is that $\mathbf{q}$ and $-\mathbf{q}$ are the same rotation that is not differentiated by this function. The second experiment was to take the same formula as above with added absolute value to the dot product. Results are presented in Fig. 1b. The last experiment was to take the function (27) as the quaternion loss function and results are shown in Fig. (1c). When comparing the results graphs for the second and third experiment, it can be seen that for function (27) we have a greater relative change in the value of the loss function. This is most likely due to differences in shapes between loss functions. Both functions converge, nevertheless the function (27) provides better convergence.

To be able to compare the obtained values of the function, it has been mapped to the same space using the function (24). Table 2 shows the performance of each distance function on validation data set. As expected, the first function tested

Table 2. Best results.

Distance function	Mean distance function value	β [rad]	β [^{o}]
$1 - \mathbf{p} \cdot \mathbf{q}$	0.631	1.193	68.34
$1 - \|\|\mathbf{p} \cdot \mathbf{q}\|\|$	0.067	0.368	21.09
$1 - (\mathbf{p} \cdot \mathbf{q})^2$	0.088	0.301	17.26

gives the worst results. The function (17) was found to be more than 20% worse than the function (27). When analyzing the results, attention should be paid to inaccuracies in the labelling of objects, which are result of the fact that it is done manually by people. Thus, they are significant. Achieving values very close to zero for loss function would mean overfitting of neural networks.

6 Summary

Several functions were reported for estimating distance function for given rotation and relationships between these functions were shown. The analysis showed that the proposed function respects the topology of $SO(3)$ and is bi-invariant. Also the geometrical interpretations of two functions were shown. Comparative experiments have been carried out in order to check which of the functions will allow to achieve the best results for the current network. The results are good, confirming that the proposition is a worthy candidate for simple and polynomial loss function working well with learning rotations described by quaternions. Analyzing the Table 2, it is possible to formulate an open question whether a function with a scalar product raised to an even greater power than 2 has certain interpretation allowing it also to be a distance function which shows better properties when training neural networks.

Acknowledgment. Research was funded by Polish Ministry of Science and Higher Education Project No. 0014/DW/2018/02 and carried out in cooperation of Aptiv Services Poland S.A. - Technical Center Kraków and AGH University of Science and Technology - Faculty of Electrical Engineering, Automatics, Computer Science and Biomedical Engineering.

References

1. Altmann, S.: Rotations, Quaternions, and Double Groups. Dover Publications (1986)

2. Angulo, J.: Riemannian l (p) averaging on lie group of nonzero quaternions. Adv. Appl. Clifford Algebras **24** (2014). https://doi.org/10.1007/s00006-013-0432-2
3. Arkhangel'skii, A.V., Pontryagin, L.S.: General Topology I. Springer, Heidelberg (1990)
4. Caesar, H., Bankiti, V., Lang, A.H., Vora, S., Liong, V.E., Xu, Q., Krishnan, A., Pan, Y., Baldan, G., Beijbom, O.: nuScenes: a multimodal dataset for autonomous driving (2019)
5. Dam, E.B., Koch, M., Lillholm, M.: Quaternions, interpolation and animation. Tech. rep. 1, Department of Computer Science University of Copenhagen, Denmark (1998)
6. Euler, L.: Decouverte d'un nouveau principe de mecanique. In: Opera Omnia 2, vol. 5, pp. 81–108. Orell Füssli (1753)
7. Geiger, A., Lenz, P., Stiller, C., Urtasun, R.: Vision meets robotics: the kitti dataset. I. J. Robotic Res. **32**(11), 1231–1237 (2013). http://dblp.uni-trier.de/db/journals/ijrr/ijrr32.html#GeigerLSU13
8. Hamilton, W.R.: Lectures on quaternions. Hodges and Smith, Dublin (1853)
9. Huynh, D.Q.: Metrics for 3D rotations: comparison and analysis. J. Math. Imaging Vis. **35**(2), 155–164 (2009)
10. Kuffner, J.: Effective sampling and distance metrics for 3D rigid body path planning. In: Proceedings - IEEE International Conference on Robotics and Automation, vol. 4, pp. 3993–3998 (2004). https://doi.org/10.1109/ROBOT.2004.1308895
11. Ravani, B., Roth, B.: Smooth invariant interpolation of rotations. J. Mech. Transm. Autom. Des. **105**(3), 460–467 (1983)
12. Redmon, J., Divvala, S., Girshick, R., Farhadi, A.: You only look once: unified, real-time object detection. In: Proceedings of the IEEE Conference on Computer Vision and Pattern Recognition, pp. 779–788 (2016)
13. Redmon, J., Farhadi, A.: Yolo9000: better, faster, stronger. In: Proceedings of the IEEE Conference on Computer Vision and Pattern Recognition, pp. 7263–7271 (2017)
14. Redmon, J., Farhadi, A.: Yolov3: an incremental improvement. arXiv preprint arXiv:1804.02767 (2018)
15. Rossmann, W.: Lie Groups: An Introduction Through Linear Groups. No. 9780199202515 in OUP Catalogue. Oxford University Press (2006). https://ideas.repec.org/b/oxp/obooks/9780199202515.html
16. Wunsch, P., Winkler, S., Hirzinger, G.: Real-time pose estimation of 3D objects from camera images using neural networks. In: Proceedings of International Conference on Robotics and Automation, vol. 4, pp. 3232–3237 (1997). https://doi.org/10.1109/ROBOT.1997.606781

VGG Based Unsupervised Anomaly Detection in Multivariate Time Series

Grzegorz Jabłoński[1,2(✉)]

[1] Aptiv, Kraków, Poland
grzegorz.jablonski@aptiv.com
[2] AGH University of Science and Technology, Kraków, Poland

Abstract. Anomaly detection in time series data is a known problem, but recent growth in the number of units that can produce data require models that work on unlabelled and diverse types of data. We propose to adapt the neural network introduced by Simonyan and Zisserman in 2015 called VGG16 and used to detect and classify objects in images. We show that the VGG16 architecture with 2-dimensional convolutions replaced with 1-dimensional version could be a building block of an autoencoder approach to detect anomalies. Additionally we show that the proposed model achieves results that are similar or better than of classical anomaly detection methods.

Keywords: Anomaly detection · Time series · Neural networks

1 Introduction

An increasing number of computing units available in nearly all aspects of life is a major challenge when it comes to monitoring their usage and health. Anomaly detection methods are proposed as a solution to indicate an unexpected change in behaviour of such units. Accurate anomaly detection can prompt response depending on the type of the monitored resources. One of the economic sectors that can benefit from the development of the anomaly detection models is automotive industry. Modern cars consists of many ECUs (Electronic Control Unit) that govern the behaviour of engine, entertainment systems, advanced driver assistance systems or autonomous driving in future. The product life-cycle requires monitoring of such systems and detection of software and hardware failures, which may result in anomalous behaviour. The list of possible failures is wide but examples include software bugs appearing after some time of the ECU usage or hardware failures caused by a wrong design.

1.1 Multivariate Time Series Data

The simplest data collected during monitoring of any computing unit usually include standard resource monitoring values such as CPU (central processing unit) usage, individual core usage, memory usage, network usage, sensors usages,

A. Bartoszewicz et al. (Eds.): PCC 2020, AISC 1196, pp. 1287–1296, 2020.
https://doi.org/10.1007/978-3-030-50936-1_107

temperature, voltage etc. We call such a collection of values time series. Specifically time series is a sequence of data points ordered in time. The dimension of time series is the cardinality of the data in each time point. If the dimension is one then we call the time series univariate, for the larger dimensions we call the time series multivariate.

1.2 Challenges

There are many challenges in developing an efficient anomaly detection method for time series:

- Unlabelled data. To accurately locate anomalies one has to process millions or billions of time series points. It is impossible economically and technically to manually analyse the data and label the suspicious fragments. Moreover, the data can be changing in time, so historical labelled data can be insufficient for training a system.
- Amount of data. As stated before the amount of data that can be processed is increasing. That applies to not only the number of units that can be monitored, but also the data that each unit is producing. There is a number of possible resources to monitor for each computing unit, such as memory usage, CPU usage, network usage, disk/external resources usage etc.
- Generalization. For each time series different types of behaviour is considered normal and anomalous. Thus the method should adapt to the data, without the unnecessary manual adjusting.

1.3 Our Method

To overcome listed challenges we propose to use an autoencoder model based on the VGG16 [9] neural network architecture. An autoencoder is a neural network that learns to reconstruct the input. Internally the autoencoder represents the input using a latent code that usually has smaller capacity that the input forcing the network to reconstruct the input approximately. The autoencoder consists of two main parts: encoder network that maps the input to the latent code and decoder that maps the latent code to the output. The VGG16 architecture is a convolutional deep neural network that achieves high accuracy in object detection in images. The VGG16 neural network consists of 16 trainable layers, and the representation depth could be used not only in image analysis task but also in the sequential data.

1.4 Paper Overview

We present research on anomaly detection using neural network in Sect. 2. In Sect. 3 we describe technical details of the problem and our proposed method. In Sect. 4 we present comparison of our method with classical anomaly detection methods on simulated data. In Sect. 5 we discuss the results and future possibilities to upgrade our network.

2 Related Work

Based on how the available data is used to train a model we can divide anomaly detection methods into three groups: i) statistical methods ii) supervised machine learning models iii) unsupervised machine learning models. The first and the longest researched group consists of large group of models that try to model the normal distribution of data and find discrepancies using different statistics such as mean value in the easiest case. The second group consists of all methods that belong to the classification group of ML models. This is a vast group, but heavily depends on the labelled data. The last group consists of models that are trained without the labels. We are mostly interested in this type of models because of the type of the data. We refer to the survey article on deep learning models in anomaly detection [2] for a broad overview of the area. The state of the art results include LSTM-AE [8], generative adversarial networks [4] and convolutional auto encoder for sound data [6].

3 Method Overview

Let $\mathcal{X} \subset \mathbb{R}^{n \times m}$ be a multivariate time series data of dimension m and n time points. We let $x_i^j \in \mathcal{X}$ to be ith value in jth time point. For simplicity we assume x^j to be a vector of all values in time point j, and x_i vector of all ith values in all time points. Additionally we use notation $a..b$ to select range of values for time and data dimension. We define time delay embedding of window length $L \in \mathbb{N}$ to be set of ranges of data points of length L for each time point that is $E_j = x^{j..(j+L)}$. Let $E = \{E_1, .., E_{n-L}\}$. Time delayed embedding is often called sliding windows methods, with the window of length L sliding in time to select range of data. A sample time delay embedding is shown in Fig. 1.

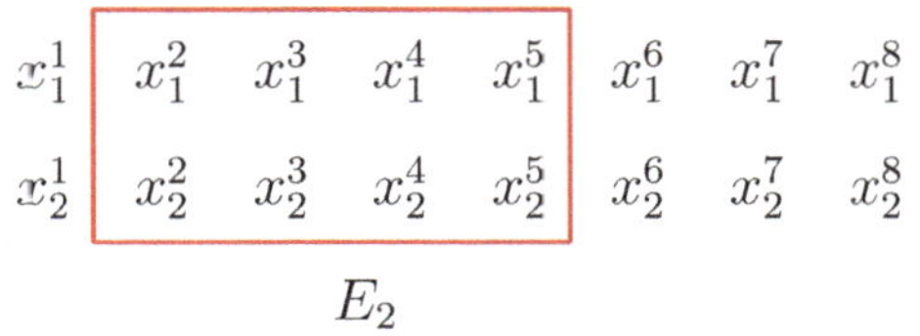

Fig. 1. Time delay embedding for 2 dimensional time series of length 8. Shown is window E_2 of size 4 at timepoint 2.

3.1 Convolutional Networks and VGG

Convolutional neural network is a class of neural network originally designed to process images. One of their characteristics is translation invariance and shared weights. These features made them very popular and successful in deep learning community. One of the most successful models based on CNN is a VGG16 [9] architecture which significantly outperformed previously proposed models at the publishing time. The model achieves 92.7% test accuracy in ImageNet, which is a dataset of over 14 milion images with 1000 classes. VGG16 was composed of blocks of 2 or 3 convolutional layers followed by pooling layer. The pooling layers in VGG16 select maximal values in a kernel of size 2 thus reducing the width and height of the input by half. For the detailed description of the convolutional and pooling layers we refer the reader [3].

3.2 Autoencoder Model

The autoencoder model consists of two parts represented by two functions: the encoder $\mathcal{E} : \mathcal{X} \to \mathcal{C}$, where $\mathcal{C} \subset \mathbb{R}^k$ is the latent code and the decoder $\mathcal{D} : \mathcal{C} \to \mathcal{X}$. The autoencoder part reconstructs the input with respect to some loss function $\mathcal{L}$. The graphical representation of autoencoder architecture is depicted in Fig. 2. If the dimension of the latent code $\mathcal{C}$ is smaller than of the input to encoder, then the encoder network is forced to find representation that minimizes reconstruction error. Usually this leads to approximating the input, and for the anomalous data the approximation should be different from the input.

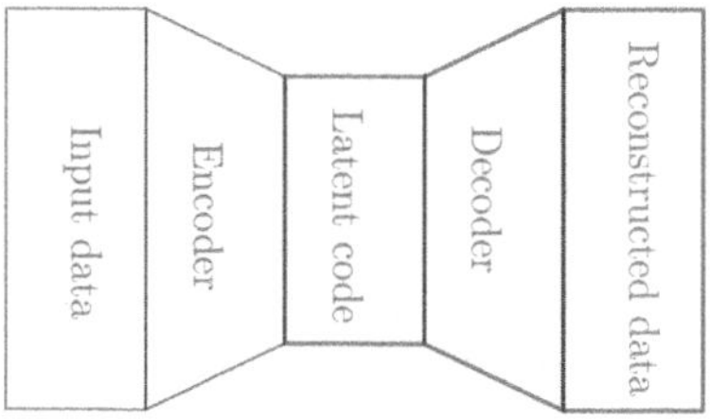

Fig. 2. Overview of the autoencoder model.

3.3 VGG-CAE Architecture

We decided to base our encoder architecture on the VGG16 model. We were motivated by the wide range of applications the VGG16 model was successfully adopted in the image analysis tasks.

The input of our network is made of the time delayed embedding of time series with window length $L = 64$ and the number of input channels m. The encoder has 4 blocks, each block consists of two convolutional layers and max pooling layer. The convolutional layers has 64, 128 or 256 channels, and the kernel of size 3. The max pooling layer has kernel size 2, and stride 2 which halves the size of the input. Finally the last layer is fully connected and has output of size C. For the details see Fig. 3.

The decoder has mirrored architecture of encoder. The first layer is fully connected, followed by three blocks consisting of convolutional and upsampling layer. Each convolutional layer has kernel of size 3 and the upsampling layer doubles the size of input by rewriting each value two times. The last layer is convolutional with m channels and size L. All layers use ReLU activation function except the last layer that uses linear activation. For the details see Fig. 4.

In our experiment the size of the latent code equals to the length L of the time delayed embedding.

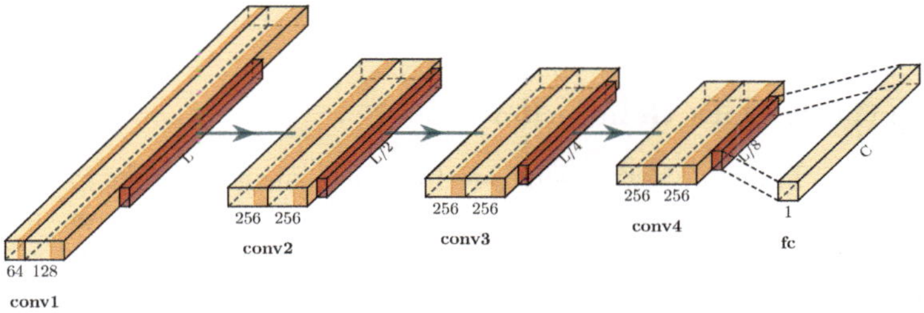

Fig. 3. Encoder architecture. Yellow boxes represent convolutional layers, red boxes are max pooling layers. The number below the convolution layer represent number of channels. On the right size of each layer block is the size of the last layer in the block with respect to the size of the input L (the length of the sliding window). Last layer is fully connected and has the output of size C representing the latent code size.

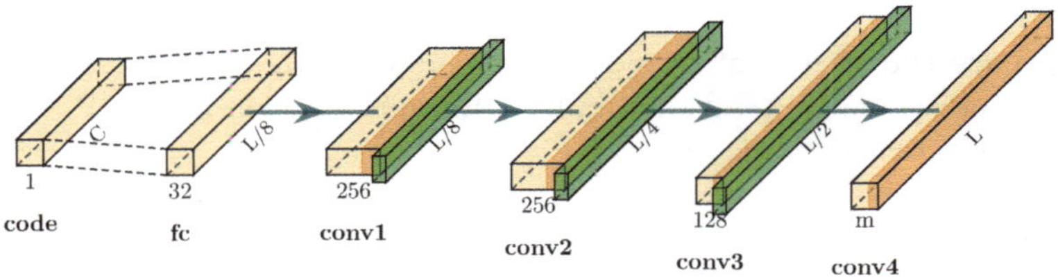

Fig. 4. Decoder architecture. Colours of layers and captions are same as in Fig. 3. Red boxes are upsampling layers.

Training. We implemented our model using TensorFlow framework. During training we used the Adam optimization method with learning rate 0.001, batch size was set to 8, and the model was trained for 5 epochs. We used the mean squared error as loss.

Thresholding. In order to define anomaly likelihood of a data point we first define the reconstruction error. The reconstruction error for a given time point i and window length L is: $e_i = \|E_i - \mathcal{E}(\mathcal{D}(E_i))\|$. We define anomaly likelihood as the difference between maximal reconstruction error obtained and the reconstruction error e_i scaled to $[0, 1]$ interval.

4 Experiments

We test our model with artificially generated datasets. Such an approach let us define and control all parameters of training and testing sets.

4.1 Synthetic Data Generation

Normal Data. The time series generated consists of $m = 5$ features, that is for each time point t we have a vector of five observed values. We assume observations are independent, and each value x_i is randomly drawn from the Gaussian distribution with variance $\sigma_i \in [0.2, 0.5, 1]$ and mean value $\mu_i \in [-1, 0, 1]$.

Anomalies. To simulate possible anomalies in the time series we inject values drawn from a different distribution. For a given time range $x^{s..t}$ where $t - s$ is length of the anomaly we replace values at randomly chosen dimensions with values drawn from uniform distribution from $[0, 1]$ interval. The contamination fraction c represent fraction of all time points that has been altered. In our experiment c ranges in $[0.05, 0.1, 0.15]$ (Fig. 5).

4.2 Classical Methods

We compare the results of proposed method with classical methods used for anomaly detection. We used the implementation provided by the PyOD library [10]. The selected methods are:

- kNN. For a datapoint it's distance to k th nearest neighbour is used as the outlying score. See [7] for details.
- IsolationForest. The IsolationForest builds a decision tree to 'isolate' observations. The path length from the root of the tree to the leaf that isolates the sample is used to calculate the outlying score. See [5] for details.
- Fully Connected Autoencoder. This model uses autoencoder principle to calculate reconstruction error. Instead of using convolutional layers in encoder and decoder as in our model it uses fully connected layers.
- PCA. Principal Component Analysis is a linear dimensionality reduction using Singular Value Decomposition. One can project the data to a lower dimensional space spanned by selected eigenvectors of covariance matrix of the data. Outliers will have higher distance to the eigenvectors with high eigenvalues due to the fact that they capture most of the variance in the data. See [1] for details.

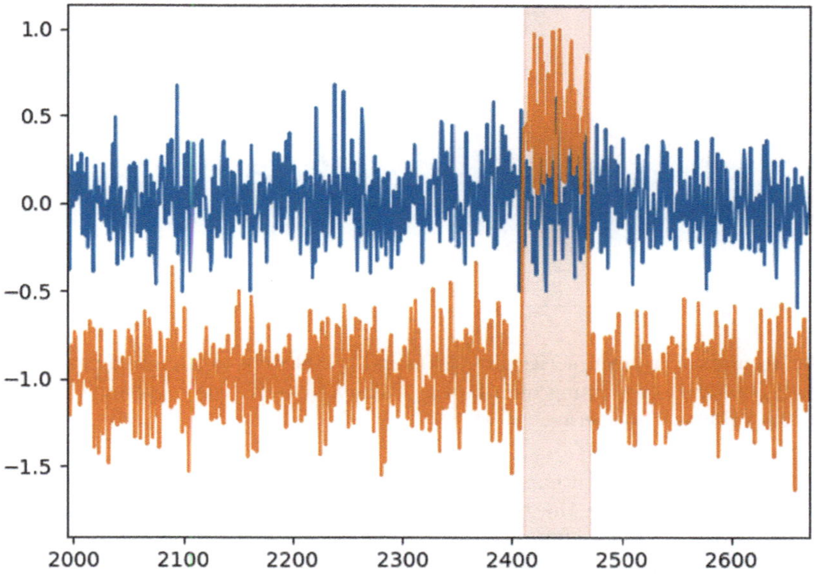

Fig. 5. Generated data. We show 2 dimensional time series with one visible anomalous region marked with dimmed background. The X axis represent arbitrarily chosen time range, the Y axis is the value of the time series.

4.3 Unsupervised Anomaly Detection

In the first experiment we generate data with various contamination levels $c \in [0.05, 0.1, 0.15]$. For each value c we generate training and testing set each of size $n = 10'000$ time points and dimension $m = 5$. We choose the window of length $L = 64$. For each tested model we first train it on the training dataset, with value c supplied to the model whenever necessary. Next we compute the statistics on the test dataset. We compute two statistics: Area Under Curve (AUC) and precision at rank k that is the precision among k anomalies with highest score, where k is true number of anomalies in the data. For the proposed model called Convolutional Autoencoder (CAE) we show the reconstructed data in the early stages of a training of the model in Fig. 6. The model first learns to reconstruct the average value of the data, and next the details of the time series (Table 1).

4.4 Semi-supervised Anomaly Detection

The second experiment we performed differs from the first one only by the test data. We generate the test data without anomalies, therefore during training the anomaly detection method does not see possible anomalies. We call this approach semi-supervised as it requires labelling or generating correct data for training phase. Similarly to the first experiment we use AUC as our main descriptor and

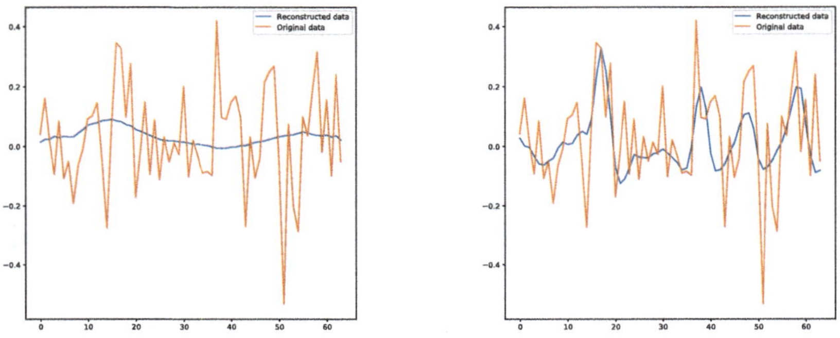

Fig. 6. Reconstructed time series after 5000 and 25000 steps of training neural network. One step consisted of training on a batch of size 8. The X axis represents time, the Y axis is the value of time series.

Table 1. Results of the first experiment. We train models on the data with anomalies, and calculate statistics on the test data. The reported precision values are precision at rank k, where k is the number of true anomalies in the test data. The FC-AE stands for fully connected autoencoder and CAE for convolutional autoencoder that is our method. Highlighted in green are best results for each contamination level and statistic.

	$c = 0.05$		$c = 0.1$		$c = 0.15$	
Method	AUC	Precision	AUC	Precision	AUC	Precision
kNN	0.964	0.508	0.935	0.472	0.928	0.499
IForest	0.944	0.475	0.909	0.495	0.893	0.505
PCA	0.963	0.466	0.932	0.534	0.926	0.489
FC-AE	0.963	0.483	0.939	0.533	0.926	0.489
CAE (our method)	0.964	0.460	0.932	0.556	0.932	0.526

show the results in Table 2. The precision at rank is highest for our method, and the AUC is close to other methods.

4.5 Running Time Comparison

We measure the time needed for training of each of described models for time series of length 10 and 20 thousands and 5 dimensions. The reported values were averaged over 7 runs and are presented in Table 3. The calculations were executed on laptop with Intel Core i7 2,60 GHz CPU with 12 cores and Quadro P1000 GPU. Our method is slower than IsolationForest and PCA method during training. For the detection our approach runs similarly to other except kNN which is significantly slower.

Table 2. Results of the second experiment. The models are trained on data without anomalies. We do not run experiments for different c values, because the training is not affected, therefore the results should not change.

	$c = 0.05$	
Method	AUC	Precision
kNN	0.962	0.523
IForest	0.916	0.525
PCA	0.966	0.527
FC-AE	0.966	0.525
CAE	0.960	0.531

Table 3. Running time comparison.

	Running time [s]			
Method	10k training	10k detection	20k training	20k detection
kNN	33.3	34	146	151
IsolationForest	1.49	0.47	3.19	1.02
PCA	0.40	0.20	1.04	0.37
FC-AE	36	0.19	84	0.702
VGG-CAE	32.6	0.36	68	0.61

5 Conclusions and Future Works

We devised an architecture of neural network autoencoder that outperforms results of classical methods. The work presented identified main challenges such as long training time. In future we would like to incorporate binary values and select optimal encoder-decoder architecture for different sizes of windows. Moreover, we would like to use a more sophisticated method of choosing thresholds, that do not require manual settings. Possible solution include analysing histograms of reconstruction error and setting the threshold at values that could be inferred from the histogram.

References

1. Aggarwal, C.C.: Outlier analysis. In: Data Mining. Springer, Cham (2015)
2. Chalapathy, R., Chawla, S.: Deep learning for anomaly detection: a survey. arXiv preprint arXiv:1901.03407 (2019)
3. Goodfellow, I., Bengio, Y.: Deep Learning. MIT Press, Cambridge (2016)
4. Li, D., Chen, D., Goh, J., Ng, S.K.: Anomaly detection with generative adversarial networks for multivariate time series. arXiv preprint arXiv:1809.04758 (2018)
5. Liu, F.T., Ting, K.M., Zhou, Z.H.: Isolation forest. In: 2008 Eighth IEEE International Conference on Data Mining, pp. 413–422. IEEE (2008)

6. Oh, D., Yun, I.: Residual error based anomaly detection using auto-encoder in SMD machine sound. Sensors **18**(5), 1308 (2018)
7. Ramaswamy, S., et al.: Efficient algorithms for mining outliers from large data sets. In: SIGMOD 2000: Proceedings of the 2000 ACM SIGMOD International Conference on Management of Data, Dallas, Texas, United States, pp. 427–438. ACM (2000)
8. Shipmon, D.T., Gurevitch, J.M., Piselli, P.M., Edwards, S.T.: Time series anomaly detection; detection of anomalous drops with limited features and sparse examples in noisy highly periodic data. arXiv preprint arXiv:1708.03665 (2017)
9. Simonyan, K., Zisserman, A.: Very deep convolutional networks for large-scale image recognition. arXiv preprint arXiv:1409.1556 (2014)
10. Zhao, Y., Nasrullah, Z., Li, Z.: PyOD: a Python toolbox for scalable outlier detection. J. Mach. Learn. Res. (JMLR) **20**(96), 1–7 (2019)

PC-Based Simulation Environment for the Engine Control Optimiser Hardware-in-the-Loop Testing

Jakub Sawulski(✉) and Maciej Ławryńczuk

Institute of Control and Computation Engineering, Warsaw University of Technology, Nowowiejska 15/19, 00-665 Warsaw, Poland
J.Sawulski@ia.pw.edu.pl

Abstract. This paper describes a motion simulator for testing a device that optimises fuel consumption along vehicle journey in real-time. The specific case of an original vehicle engine control system is considered. The PC application based on the Qt framework is written in Python. It allows to quickly run Hardware-in-the-Loop tests of optimiser instead of running them on the real test track. By using the CAN interface, tests with connection to simulator are interchangeable with those made using full-scale vehicle. This significantly shortens the entire process of developing an automatic vehicle engine control system. The open and modular architecture of the source code allows any modification and development of the application depending on the user-specific demands.

Keywords: Car simulator · Fuel consumption optimisation · Hardware-in-the-Loop · Optimiser-simulator architecture · Python · PyQt · SciPy

1 Introduction

Nowadays, more and more driving activities are being taken over by the automatic control systems called Advanced Driver Assistance Systems (ADAS). The result of this trend is a significant increase in the amount and complexity of the automotive embedded software. Tests carried out for any software development level are an important part of the process. One of them are software/hardware integration tests. In the case of hardware understood as a full-size vehicle, task can be at least logistically difficult and time consuming. For this reason, more and more validation tasks are carried out in virtual environments using various types of simulators. This approach is widely known as Hardware-in-the-Loop (HIL) simulation [5].

Various commercial simulators available with HIL capabilities have been developed (e.g., Mechanical Simulation *CarSim*, Siemens *Prescan*) [6]. Apart from their pure commercial licensing model, they use quite a lot of hardware resources due to the graphics rendering of the vehicle and environment.

A. Bartoszewicz et al. (Eds.): PCC 2020, AISC 1196, pp. 1297–1308, 2020.
https://doi.org/10.1007/978-3-030-50936-1_108

Fig. 1. The Orion car

This feature is not crucial if the considered task involves only the motion simulation. For HIL applications, it is usually necessary to depend on 3rd party tools like MATLAB/Simulink or to use standard automotive development modules supplied, among others, by dSPACE or National Instruments.

On the other hand, the open source solutions, like CARLA [2], are available. CARLA provides similar capabilities to commercial simulators, but does not provide support for HIL testing out of the box. Additionally, to make well-defined vehicle model in CARLA, a large set of parameters has to be provided, even if they are not directly related to fuel consumption. All listed environments are exhaustively used, e.g., for collision avoidance or sensor fusion algorithms development, which are tasks certainly distinct from the engine controller synthesis. Hence, it has been decided to develop a dedicated simulation environment.

This work discusses architecture and implementation of open environment for motion simulation of car used in parallel with engine control strategy optimiser executed on separated hardware platform [8]. The whole system is used to perform on-line experiments for an autonomous engine control system of an original fuel-efficient car which takes part in the international Shell Eco-marathon (SEM) competition [7].

2 Orion Vehicle

Apart from the open architecture of the simulator, the main goal is that it should be equivalent of a specific vehicle, The Orion car shown in Fig. 1 is considered in this work. It was built in 2018–2019 by Student Association for Vehicle Aerodynamics at the Warsaw University of Technology, aiming to take part in the SEM competition series. It fulfils all requirements to be allowed for start in Urban Concept vehicles category [10].

Shell company has organised dedicated subcompetiton for autonomous vehicles since 2018. It is supposed to be a response to the emerging trends in the global automotive market. For this reason, many teams put a lot of effort into the development of subsystems enabling autonomous driving.

In 2016, as a result of the Electronic Diagnostic System project, the Orion ancestor, the PAKS car was equipped with a fully integrated electronic vehicle diagnostics and control system. It allows logging of a wide range of parameters measured during the journey, as well as wireless transmission of these data using a telemetry channel. This system has been adopted into newly designed car.

Additionally, as many actuators as possible have been replaced or upgraded allowing for automatic control instead of manual one. Especially, the system has been enhanced by merging into it new devices such as high-level engine control unit and high current switch board. This fact is important in case of automatic control of vehicle engine. This expanded configuration allows for throttle manipulation and engine start programatically, without the driver intervention.

3 Optimiser-Simulator Architecture

Developing the fully autonomous engine controller is an essential component for making Orion car fully autonomous. The whole process consists of several steps, from the modelling vehicle at the beginning, through algorithms design and hardware implementation to the test of real vehicle equipped with a controller. The significantly time consuming activity during the whole process is to test proposed controller solution interoperability. Those evaluations can be made based on the real car test drives or simulations.

In order to examine an engine control algorithm, the most important feature to be given back is dynamics of a vehicle longitudinal motion. As far as the vehicle dynamics mathematical model is known, some testing of those features can be moved into simulation. It is necessary that the model is accurate, but some modelling errors must be taken into account. Using the simulator, it should be easy to test the control algorithm's robustness to statically inaccurate parameter values or time-dependent deviations.

The main idea of the Optimiser-Simulator pair is depicted in Fig. 2. In general, the main assumption is that the simulator should be 1:1 equivalent of the Orion car from the optimiser perspective.

3.1 Optimiser

Although the controller development process is of out this work scope, some of its properties, mainly connected with interfacing it, are essential to exist altogether with other vehicle system components. In general, the optimiser is a subsystem of the car, which continuously performs computation of the energy optimal control trajectory of the engine over the established horizon. In contrast to Model Predictive Control algorithms [12], it is assumed that it is not relative short and receding, but instead, it covers the entire desired route and is shortened in line with the distance travelled so far. Assumed binary engine control strategy causes that the complete steering trajectory is determined only by the time

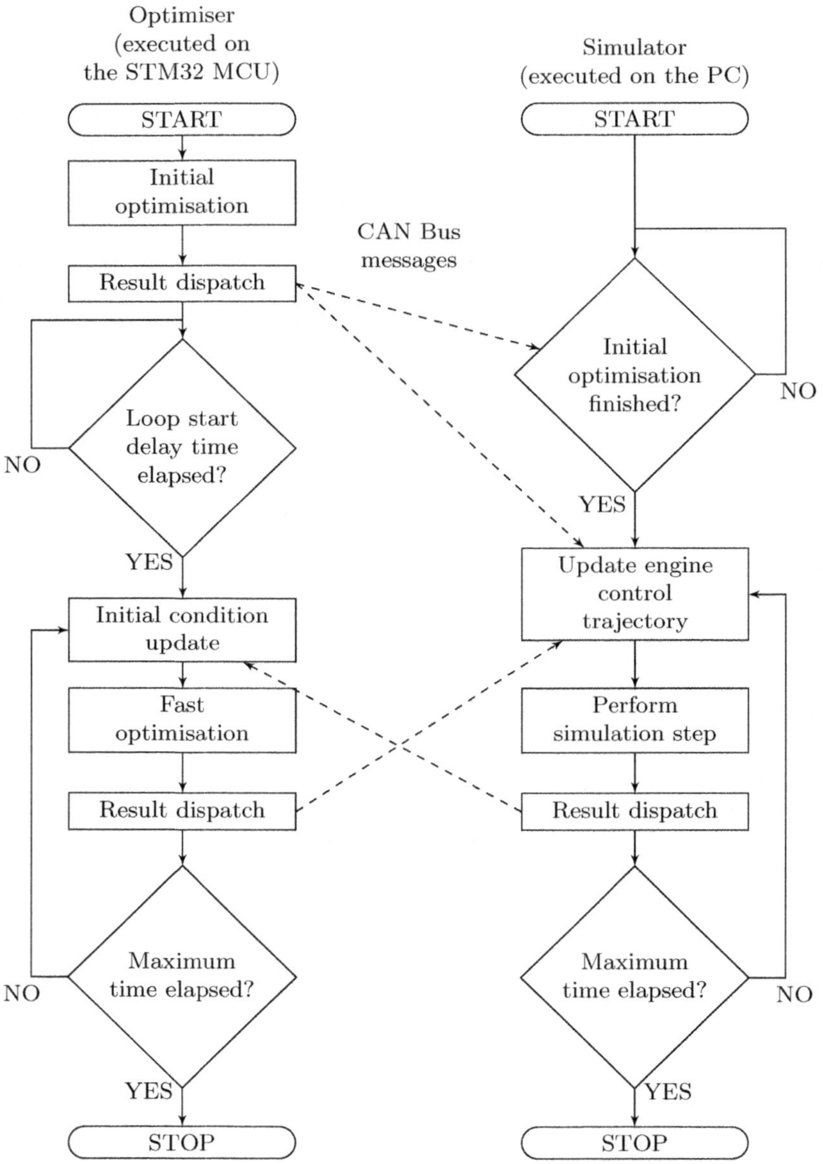

Fig. 2. Optimiser-Simulator data flow diagram

points at which the control value changes. This driving strategy is called Pulse and Glide (PnG) or Burn and Coast and leads to significant improvement of fuel efficiency [7].

Trajectory computation may be performed in many different ways: on the on-board processor, on the cloud architecture via reliable data-link or in any other way. Despite of technology used, the behaviour of optimiser is constrained by other elements integrated already in car control system. It should be stated that the process of obtaining optimal control trajectory for internal combustion engine is quite complex (it is a non-linear constrained optimisation problem).

The problem formulation in the context of the SEM competition requires to pass one lap of the track (approx. 1660 m) with maximum time allowed (approx. 240 s). It cannot be assured that it is possible to pass the distance exactly in the same way that is predicted by the optimiser due to modeling errors, disturbances etc. In order to improve system robustness the approximation of the optimal trajectory should be iteratively refreshed. This process is called Fast Optimisation as it should return result in limited time. In contrast, for the optimisation task performed before starting the route, called Initial Optimisation, there is no time limit imposed.

As a result, during the simulator design it is assumed that this device is able to read all data needed to obtain an actual vehicle state from the on-board communication bus in the way compliant with its specification. Those values are the initial point for the following optimisation process. The second task is to return a new solution (the point of the next throttle value change) every predefined real-time interval and put it into the bus. Those features are based on previously developed optimiser prototype based on the STM32 microcontroller described in [8].

3.2 Simulator

The second part of the designed system (Fig. 2.) is a simulator. As its primary objective is to test the optimiser for many configurations and its disturbance robustness, it has been assumed that this should be a GUI PC application. Although usually no user interface is created for the development toolkit applications, one can suspect that this will reduce an overall effort for complex system development. There are also the frameworks available with a large number of predefined widgets resulting in further programming effort reduction.

In addition, in order to be able to test the optimiser in the same configuration as in the real vehicle, it has to perceive the simulator in the same way as the set of all other communicating devices installed and acting in the car. Thus, the application must be equipped with an interface with the on-board data bus.

As a result the following software requirements have been established:

1. the application should contain a replaceable module: low level dynamics simulator,
2. the application should be based on a multi-platform widget framework,
3. the application should perform simulation step-by-step on the user demand and automatically through time-spans,
4. the application should support import and export capability for vehicle parameters used for simulation,

5. the application should be able to communicate with the optimiser via hardware USB-CAN converter.

4 Simulator Implementation

Due to the requirements of cross-platform and GUI, it has been decided that the application should be written in the Python programming language. Its recent popularity allows for using ready-made software collections, often distributed as the source code, on a free license basis. This allows one to modify chosen components in case of incorrect operation or if there is some specific functionality lack.

In addition, Python is an interpreted language which allows to quickly test the resulting application without the need for relatively time-consuming building process. It is also possible to build application for distribution as an executable file. Unlike MATLAB, Python runtime is relatively lightweight and it does not require commercial licensing. The rest of this chapter introduces ways to implement individual software modules.

4.1 Car Motion Simulator Module Based on SciPy

The main implementation task is to develop a vehicle dynamics simulator. Its implementation depends on the modelling method of the vehicle. In the considered case, a significantly simplified model in the form of the system of equations is used. It takes into account the dynamics of longitudinal movement and fuel consumption of the vehicle resulting in the following continuous-time differential equations

$$\begin{cases} \dot{x}_1 = x_2 & \text{(a)} \\ \dot{x}_2 = \left(\frac{p_2 u^2}{m} - \frac{\rho S c_\mathrm{x}}{2m}\right) {x_2}^2 + \left(\frac{p_1 u^2}{m} - g f_1\right) x_2 & \\ \qquad + \frac{p_0 u^2}{m} - g f_0 - g \sin(\alpha(x_1)) & \text{(b)} \\ \dot{x}_3 = (r_2 {x_2}^2 + r_1 x_2 + r_0) u & \text{(c)} \end{cases} \tag{1}$$

where the subsequent markings indicate:

- manipulated variable:
  - u – the normalised throttle opening ($0 \leq u \leq 1$),
- state variables:
  - x_1 – vehicle position (in m),
  - x_2 – vehicle speed (in m/s),
  - x_3 – instantaneous fuel consumption (in mL/s),
- parameters:
  - m – vehicle mass (in kg),
  - ρ – atmospheric air density (in kg/m^3),
  - S – vehicle cross sectional area (in m^2),
  - c_x – vehicle drag coefficient,

  - g – standard gravity (in m/s^2),
  - p_{n} – propulsion force curve coefficients (in Ns$^{(\mathrm{n})}$/m$^{(\mathrm{n+1})}$),
  - f_{n} – rolling resistance curve coefficients (in Ns$^{(\mathrm{n})}$/m$^{(\mathrm{n+1})}$),
  - r_{n} – fuel consumption curve coefficients (in mLs$^{(\mathrm{n-1})}$/m$^{(\mathrm{n})}$),
- parameter function:
  - $\alpha(x)$ – route slope as a function of position (in °).

The additional parameter r_{start} (in mL) characterises the amount of fuel used for engine start-up. A more detailed description of the model is provided in [7].

The model used is a classical system of ordinary differential equations. There are widely known solvers of this class problems, e.g. Runge-Kutta methods. There are ready made implementations of those usually distributed as a part of comprehensive numerical methods toolkits. One of them is SciPy: a Python package that allows calculations from a wide range engineering and scientific problems [4]. The SciPy library for the basic matrix manipulations depends on NumPy library. NumPy, in turn, wraps over BLAS and LAPACK algebra libraries ensuring high computation performance. In the considered case, the `solve_ivp` routine is used. By default, it uses the RK45 Dormand-Prince method [3].

Vehicle dynamics integration is designed as a specialised method of the simulator wrapper class. This architecture allows for introducing additional software layer for easy model substitution to any other vehicle. The class is equipped with an interface for getting any vehicle state values and setting desired throttle state: in the considered case, the only control variable for the vehicle.

4.2 Predefined Parameter Set Handling

As far as the vehicle model (1) is characterised by a meaningful set of the parameters it is helpful to enhance simulation environment with a tool for handling it. This feature will be extensively used in the case of optimiser testing for parameter uncertainty robustness.

The implementation of the parameter sets is based on JSON files. JavaScript Object Notation (JSON) is an open standard for human readable data format [11]. It allows for writing and reading arbitrary size data collections and modifying them with any text editor. By using standardised data storage format it is possible to access them with ready-to-use serialisation and deserialisation mechanisms.

4.3 GUI: Qt Widget Toolkit

Graphical User Interface (GUI) of the simulator application is based on a Python binding for C++ Qt library called PyQt. In addition to the interface development related parts, the Qt framework provides mechanisms for creating a multi-threaded applications using the event loop structure. A rich palette of expansion modules gives the opportunity to quickly develop an application [9]. One such module is a QtCharts that allows to use various types of charts to visualise data.

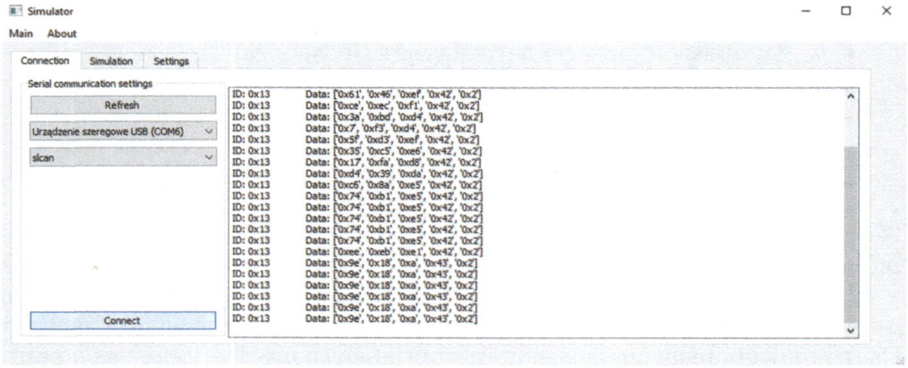

Fig. 3. Communication settings tab

The GUI is created using the QtDesigner tool. It is a WYSIWYG editor that allows one to compose the content of application windows using drag and drop technique. In the next step, the pyuic tool converts window design a Python source file which is responsible for building graphical interface at the runtime.

The main window of the app is organised into tabs. Two substantial ones are shown in Fig. 3 and Fig. 4. The first one is dedicated for configuration of the connection with the USB-CAN converter and displaying raw data received from this interface. After establishing connection, it is not allowed for changing the COM port used. The second one is the simulation control panel. It consist of the simulation controls as well as fields and graphs showing the current speed, location, fuel consumption and throttle of the vehicle. The simulation controls allow for make single simulation step, start and stop simulation performed in the continuous mode and reset the simulator which overrides all data with the initial values.

Information about the desired throttle position received from the bus is rewritten directly to the control field. It is also possible to manually override this value. This allows one to manually inject disturbance into the system during optimiser testing.

The parameter sets importing/exporting features are accessible under menu bar and they allow to specify file path for data import or export.

4.4 CAN Bus Protocol

The simulator should be fully complimentary with real car communication, so it has to be fully compliant with the Orion onboard communication bus specification. As stated before, it is based on the CAN. CAN bus protocol was developed in 1980s by Robert Bosch GmbH focusing on the automotive industry. Standard specify layers 1 (physical) and 2 (data link) of the seven-layer OSI model. CAN allows to connect up to 30 devices (nodes) with bitrate up to 1 Mib/s. All nodes communicate over twisted pair with a 120 Ω characteristic impedance. Instead of

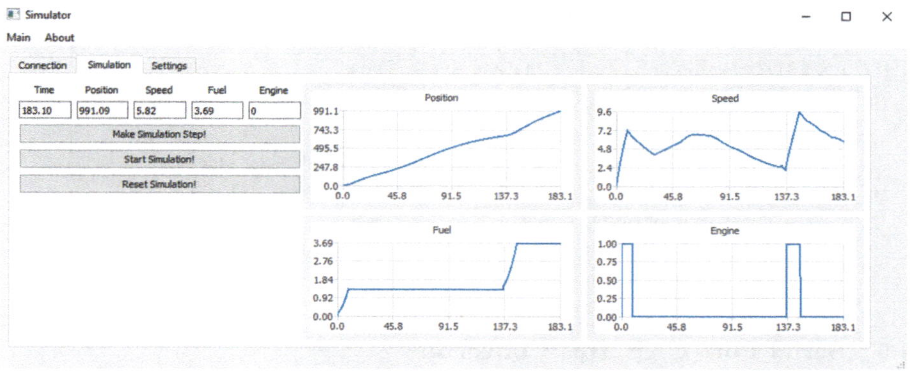

Fig. 4. Simulation tab

classically defined voltage logic levels, it uses "dominant" and "recessive" state which are determined by voltage difference between wires in pair.

From the viewpoint of software engineering, CAN is message-based protocol, what leads to obligation of using specific framing protocol. Each data-frame consist of (not only) numerical ID and from 0 to 8 payload bytes. The main drawback of bus multi-master architecture is that the potential medium access conflicts need to be resolved. CAN bus prioritises messages in order of descending IDs. Every device is obligated to listen on the bus during sending message. Thanks to "recessive" and "dominant" states, when collision occurs, node sending message with higher ID is able to detect that the bus is taken by node with lower ID message and is obligated to stop transmission until the bus is free. This concept may lead to starvation of messages with higher ID. Its influence on the bus real-time capabilities have to be considered but its out of the scope of this work. This phenomenon can be imitated by flooding the bus with random messages injected by dedicated module of the simulator.

Table 1 presents the Orion vehicle CAN bus specification subset. The data shown only apply to values that are exchanged by the optimiser during its operation time. The data transferred from the on-board installation to the optimiser are: speed and position of the vehicle, as well as the lap number, allowing to determine the moments of passing through the start/finish line in order to correct odometric measurements.

In the opposite direction, information about the next time point at which the control value changes is sent. As a result of using the PnG driving strategy, it is not necessary to send information about the desired value of the control amplitude. It is determined by the number of switches made between periods of the engine off state and the acceleration with the maximum power.

Table 1. CAN bus frames used definition

ID	EXT ID?	Frame name	DLC	Bytes	Data	Type	Factor	Unit
18	No	Front-left wheel measurements	4	0:1	Speed	uint16_t	10	m/h
				2:3	Total distance	uint16_t	1	m
19	No	Next engine switching point	4	0:3	Switch time	Float	1	s
32	No	Lap	1	0	Lap number	uint8_t	1	–

4.5 Serial Port-CAN Bus Converter

As far as the CAN bus connection of the simulator is vital, an appropriate interface has to be engaged. There are some commercially available professional solutions, but in this case of research, cheap, open-source converter has been chosen.

The UCCB CAN Converter [1] is a simple device based on STM32F042 series microcontroller. From USB interface side it registers as virtual COM port device with fixed 115200 baud rate and implements SL-CAN protocol. It allows to transmit and receive can frames with Standard and Extended IDs with adjustable CAN bus bitrate and bit timing.

Although there is a popular python-can library, communication with the converter could not be established, probably due to errors in the implementation of multi-threading mechanisms. For this reason, the interface's software part is based on the pyserial library and multithreading support provided by the Qt framework. After setting the serial port responsible for communication with the converter, a timer is started, which periodically polls the serial port for incoming data. When the complete CAN frame is received, the callbacks responsible for GUI refresh and update of the simulation state are called.

4.6 Time Synchronisation Between Real-Time and Non Real-Time Environment

HIL testing is usually performed with real-time constraints for the plant (simulator). The main problem noticed during testing the cooperation of the simulator and the optimiser based on the STM32F446 microcontroller is the time drift between them. This problem is caused by the inability to force the PC simulator application to be executed with real-time constraints because its execution is supervised by a general purpose operating system.

At a simulated lap travel time of about 240 s, the time drift could reach about 15 s (approx. 6%), while the CPU load throughout the simulation does not exceed 35%. For this reason, it has been decided to introduce the optimiser clock synchronisation mechanism relative to the simulator clock.

Alike the rest of the communication mechanisms, it is implemented using the CAN bus. Using the CAN frame with STD_ID=0 (maximum possible priority), a

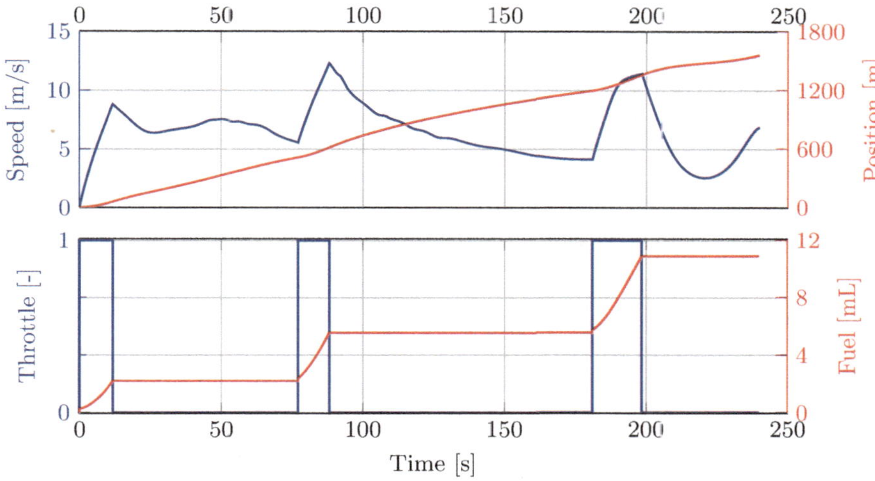

Fig. 5. Trajectories for engine state changes at [0.0; 11.8; 77.2; 88.2; 181.1; 198.5]

message containing the current time value calculated by the simulator is sent out every 0.1 s allowing to accurately determine the initial condition for optimisation.

The lack of the time synchronisation causes a mismatch between the position and speed values with the current time, which leads to the situation when the optimiser obtained mismatched solutions. The vehicle mean speed was artificially inflated to achieve solutions that met the lap time limit.

The clock synchronisation has been introduced to solve the problem. A similar solution will also be introduced in the Orion vehicle, which will receive an additional device ensuring a stable source of time for all on-board devices.

5 Conclusions

As a result of this work, the Orion car simulator has been developed. The simulation environment allows one to perform optimiser tests which determine the optimal way to use vehicle's engine. Figure 5 shows the trajectory obtained after performing the simulation for arbitrarily selected motor state switching points. The obtained waveform is consistent with the results obtained in [7] and [8]. It leads to conclusion that simulator has been implemented without major failures. Additionally, during the tests of connected simulator and optimiser, such potential problems caused by the synchronisation of the clocks of both devices are possible. They have been removed using a simple clock synchronisation mechanism, which will also be introduced to the target electronic system of the vehicle. It sufficiently imitates real-time operation without using a dedicated RT hardware platform and RTOS.

Compared with the commercially available simulators with extensive functionalities, the prepared application is lightweight and multiplatform. It implements a set of functions allowing for HIL testing of the developed optimiser using the CAN bus. Additionally, it can be easily expanded with additional modules.

In the future research, developed software will be exhaustively used for performing test scenarios for robustness assessment of engine control optimiser. All software developed for simulator is available on: https://github.com/kubasaw/carSim and the documentation is available on: https://carsim.readthedocs.io/.

References

1. CANDevices.com.pl. https://ucandevices.github.io/uccb.html. Accessed 14 Jan 2010
2. Dosovitskiy, A., Ros G., Codevilla F., Lopez A., Koltun V., CARLA: an Open Urban Driving Simulator. In: 1st Conference on Robot Learning (2017). arXiv: 1711.03938
3. Dormand, J.R., Prince, P.J.: A family of embedded Runge-Kutta formulae. J. Comput. Appl. Math. **6**, 19–26 (1980). https://doi.org/10.1016/0771-050x(80)90013-3
4. Johansson, R.: Numerical Python, Scientific Computing and Data Science Applications with Numpy, SciPy and Matplotlib. Apress, Berkeley, CA (2019). https://doi.org/10.1007/978-1-4842-4246-9
5. Raikwar, S., Jijyabhau Wani, L., Arun Kumar, S., Sreenivasulu Rao, M.: Hardware-in-the-Loop test automation of embedded systems for agricultural tractors. Measurement **133**, 271–280 (2019). https://doi.org/10.1016/j.measurement.2018.10.014
6. Rosique, F., Navarro, P.J., Fernández, C., Padilla, A.: A systematic review of perception system and simulators for autonomous vehicles research. Sensors **19**, 648 (2019). https://doi.org/10.3390/s19030648
7. Sawulski, J., Ławryńczuk, M.: Optimization of control strategy for a low fuel consumption vehicle engine. Inf. Sci. **493**, 192–216 (2019). https://doi.org/10.1016/j.ins.2019.04.047
8. Sawulski, J., Ławryńczuk, M.: Real-time optimisation of control strategy for a low fuel consumption vehicle engine using STM32 microcontroller: preliminary results. In: 2019 18th European Control Conference (ECC). IEEE (2019). https://doi.org/10.23919/ECC.2019.8796299
9. Scholl, M.V., Rocha, C.R.: Embedded SCADA for small applications. IFAC-PapersOnLine **49**, 246–253 (2016). https://doi.org/10.1016/j.ifacol.2016.10.559
10. Shell Eco-marathon. https://www.shell.com/make-the-future/shell-ecomarathon.html. Accessed 14 Jan 2010
11. Smith, B.: Beginning JSON. Apress, Berkeley, CA (2015). https://doi.org/10.1007/978-1-4842-0202-9
12. Tatjewski, P.: Advanced Control of Industrial Processes, Structures and Algorithms. Springer-Verlag London (2007). https://doi.org/10.1007/978-1-84628-635-3

The Concept of the Control System for the A-EVE Autonomous Electric Vehicle

Marek Długosz(✉), Michał Roman, and Paweł Węgrzyn

AGH University of Science and Technology, Al Mickiewicza 30,
30-059 Kraków, Poland
mdlugosz@agh.edu.pl

Abstract. The article describes the concept of the control system for the prototype EVE autonomous electric vehicle. The developed design of the control system for an autonomous vehicle consists of several subsystems whose correct operation, appropriate data exchange, and properly selected and configured low-level and high-level control algorithms allow its implementation in autonomous driving tasks. The article described individual subsystems of the control scheme, such as: the measurement system, low-level and high-level control system, and also the safety system which guarantees safe operation of the EVE autonomous vehicle.

Keywords: Autonomous · Autonomous vehicle · Electrical vehicle · Control systems

1 Introduction

Control system architectures in autonomous vehicles are highly complex systems. An autonomous vehicle, in addition to the implementation of typical control tasks such as the stabilization of the pre-set speed, must make several decisions related to the drive itself and the behaviour of the vehicle in road traffic (e.g. mode of behaviour at intersections, making the decision to change the lane, etc.) [2]. Analysing the existing architectures of control systems for prototype autonomous vehicles, one can notice that in most of them the control system is divided into two systems: high level control and low level control, [4,5,12,19,20]. This type of control architecture is used, among others, in autonomous vehicles participating in DARP [1] competitions such as Stanley Stanford [17], BOSS [18] or TALOS-Mit [3].

A multi-level control system architecture allows effective implementation of complex control systems as well as their easy modification, change or debugging and testing. Decomposition of low level controller to independent subsystem give an ability to verify used control algorithms [15] or testing new type of controllers, see [9,11]. The article describes the design and implementation of a two-level prototype control system for an autonomous vehicle called A-EVE. The assumption was to design and implement a fully functional vehicle control system which allows further development and testing of various aspects related to the

A. Bartoszewicz et al. (Eds.): PCC 2020, AISC 1196, pp. 1309–1320, 2020.
https://doi.org/10.1007/978-3-030-50936-1_109

steering and control of autonomous vehicles. An important part of the described control system is the low level controller, which was designed and implemented as part of the engineering project especially for autonomous vehicle EVE.

2 Description of the Prototype A-EVE Vehicle

The EVE vehicle (abbreviation ***E**lectrical* ***VE**hicle*) is a small electric vehicle constructed by the company Aptiv (formerly Delphi). Based on the cooperation agreement between AGH University of Science and Technology in Kraków and Aptiv, it was made available/transferred to the research team for the purpose of designing and building a prototype autonomous vehicle. To emphasize the fact that the prototype will be an autonomous vehicle, the abbreviation/acronym EVE was changed to A-EVE (***A**utonomous–**E**lectrical* ***VE**hicle*).

The A-EVE is a one-man electric vehicle powered by three batteries connected in a series. A three-phase electric motor (with 750 W power output and BI-36-2 motor controller), mounted on the rear axle of the vehicle, was used for propulsion. Front wheels turning is carried out via a *drive-by-wire* system, i.e. the readings from the steering wheel rotation change cause the motor with a linear gear to rotate, which turns the front wheels at an angle proportional to the angle of the steering wheel (Ackerman kinematics). Besides, there is an algorithm implemented in the steering wheel - *force feedback* - giving the impression of a real steering system and thanks to which the position of the steering wheel, and thus the front wheels, returns to the zero position. The A-EVE vehicle is equipped with two independent braking systems. The first one is an electrical system (electric motors clamp the brake shoes on the rear axle). The second one is a hydraulic system which brakes only the front wheels. Both systems are controlled by the brake pedal, with the difference being that the electric braking system can also be controlled by the computer (generating a control signal for brake actuators). The mass of the vehicle is approx. 350 kg, maximum speed 20 km/h (electronically limited due to the lack of shock damping in the suspension). The Fig. 1 presents a picture of the A-EVE vehicle.

3 Control System

The A-EVE vehicle uses a two-level control system which includes a high level controller and a low level controller. The structure of the control system, together with the input signals and sensors, is presented in the Fig. 4. Such a structure of the control system allows for optimal division of the tasks carried out by individual controllers (ECU1 high level controller and ECU2 low level controller), and also it provides a lot of flexibility if changes are necessary in the control system. The ECU1 high level controller communicates with the ECU2 low level controller via a USB bus. The ECU1 hight level controller is also connected to an LCD monitor with a touch screen, which is used for displaying the current status of the A-EVE vehicle and for facilitating management (Fig. 2).

Fig. 1. The A-EVE vehicle.

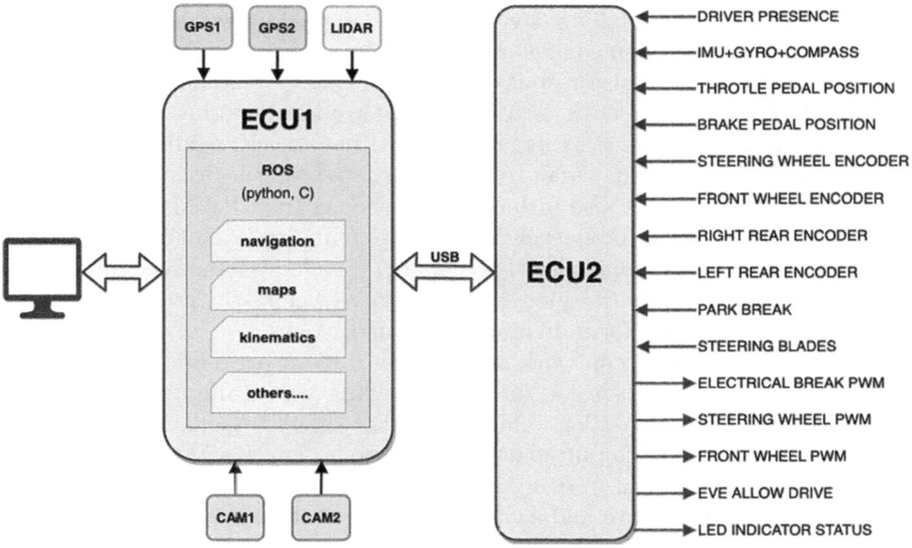

Fig. 2. The A-EVE control architecture diagram.

3.1 High Level Control

The high level control algorithms are implemented using a laptop computer with an Intel i5 processor, having 4 cores/8 threads, 16 GB RAM and 250 GB SSD (ECU1 controller). The A-EVE vehicle uses the ROS (***R****obot* ***O****peration* ***S****ystem*) framework in the Kinetic version [13,16]. The ECU1 controller performs higher-order control tasks such as determining global routes, determining local directions, analysing data from connected sensors such as depth cameras, LIDAR, GPSs, implementations of various types of safety algorithms for example emergency stopping, avoiding obstacles. Additionally in ECU1 are implemented different algorithms type of ADAS (***A****dvanced* ***D****river* ***A****ssistance* ***S****ystem*) [6,7] and data fusions algorithms [8]. The ECU1 controller also implements all kinds of "behavioural" actions of the A-EVE vehicle. All the control data generated by the ECU1 controller is sent via the USB-CDC protocol to the ECU2 controller, e.g. pre-set linear speed, pre-set steering wheel turning angle. The ECU2 controller sends the data necessary for the operation of higher-level control algorithms, e.g. linear speed, steering angle to the ECU1 controller.

3.2 Low Level Control

The low level control algorithms are implemented along with appropriate peripheral devices by a dedicated microcomputer which was designed and made as part of the engineering project [14]. The Fig. 4 shows a flow chart of the ECU2 low level controller. The ECU2 controller uses the STM32 NUCLEOF767 processor [10]. The ECU2 controller implements basic control algorithms such as speed controller, dynamics modelling, steering wheel angle adjustment together with the steering wheel position controller (drive-by-wire system and force-feedback). The ECU2 controller is also responsible for the correct reading and conversion of the data from devices such as: rear wheel encoders, steering angle encoder, encoder for measuring the steering angle of the front wheels, reading the position of acceleration and braking pedals, reading the data from the driver presence sensor. The ECU2 controller also includes generating correct PWM control signals (e.g. for the main engine or steering wheel motor) and relaying the data reading to the ECU1 high level controller. The paper [14] includes a detailed description of the ECU2 controller with exact specifications as well as the requirements and implementation method for individual functional blocks.

The ECU2 controller can work in one of four operating modes: manual control, automatic control using autonomous driving algorithms, control via the manipulator (pad) and A-EVE vehicle stop. For safety reasons a principle was adopted whereby if in the automatic control mode, any reaction of the driver, e.g. a steering wheel turn, pressing the brake pedal, etc. will cause the ECU2 controller to exit this state and switch to the state of manual control. In other words, manual control has the highest priority, thanks to which it is possible to immediately stop automatic control if the driver/supervisor finds that it is incorrect.

A-EVE vehicle security systems were also implemented in the ECU2 controller. For more detailed information on the A-EVE vehicle safety system, see Sect. 4 (Fig. 3).

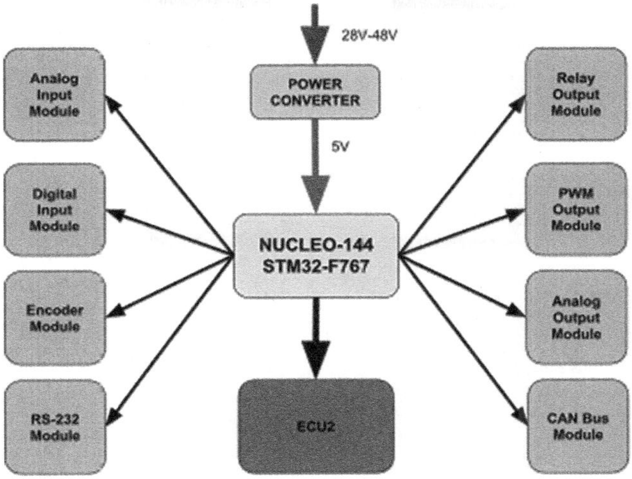

Fig. 3. Block diagram of ECU2 controller.

Programming the ECU2 Controller. The main microcontroller program was written in the C language using the HAL library provided by STMicroelectronics. The STM32CubeMX program was used to generate the hardware configuration, allowing the entry of microcontroller register configuration using an intuitive GUI. A decision was made to use the automatic code generation offered by the Matlab package using the Simulink Coder tool to quickly and easily implement various types of controllers. The Fig. 4 shows block diagrams of the ECU2 controller.

The code, generated by the Matlab packet, is cyclically executed in the main loop of the program with a frequency of 1 kHz. Besides, the ECU2 controller uses an interrupt mechanism to perform other tasks (e.g. counting ticks from encoders, reading the status of various switches connected to the inputs). The state machine logic switching the ECU2 controller into one of four states was implemented using the StateFlow toolbox of the Matlab software, Fig. 5.

Each of the states in which ECU2 include a sub-program responsible for controlling in the current state. The Fig. 6 presents an example of a control sub-program executed in the automatic mode. The use of Matlab graphical tools to create control software greatly simplifies and speeds up the process of designing and implementing various types of control algorithms performed in the ECU2 controller. In addition, we can be assured that the automatically generated code is formally correct and error-free.

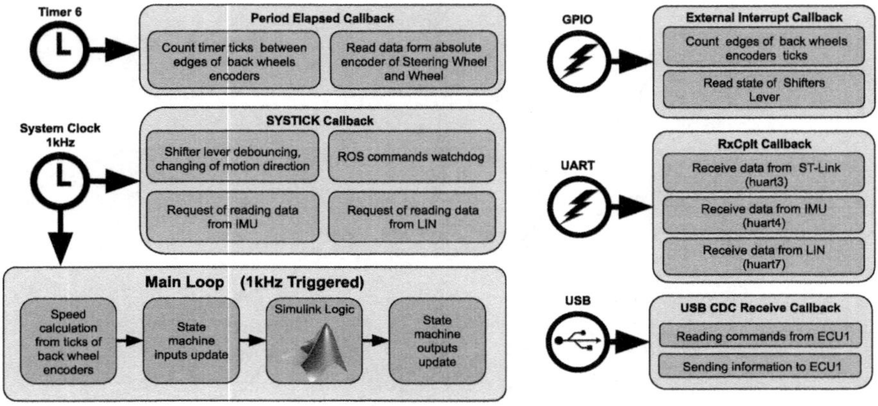

Fig. 4. Block diagram of software of the ECU2 controller.

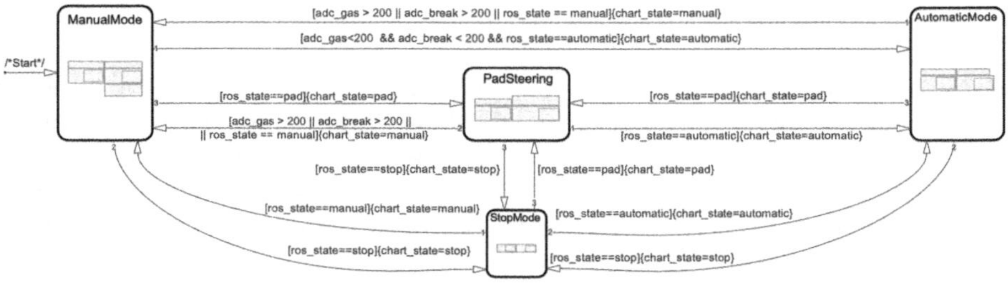

Fig. 5. The state machine of the ECU2 controller.

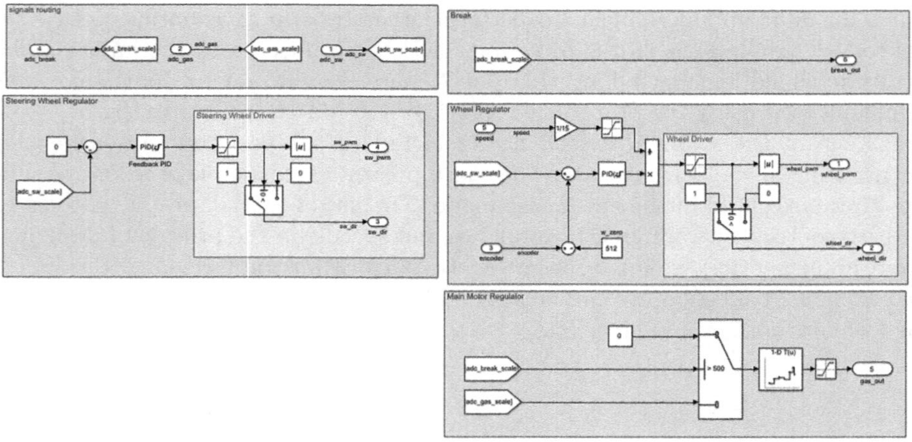

Fig. 6. Subprogram of automatic driving state of A-EVE vehicle.

The Impact of Compilation Flags on ECU2 Computation Time. One of the most important operating parameters of each discrete control system (and the ECU2 controller is a such one) is the time needed to calculate the appropriate control signals. In the ideal case, this time should be as short as possible and also unchanging. While compiling a program to an executable form on the processor, there exists a possibility to set various compiler options which can affect the response time of the discrete control system. In the Debug mode, the compiler generates additional information in the machine code which allows for tracking the progress of program execution using the software called Debugger. The amount of additionally generated code corresponds to the detail level of the information which can be obtained in the course of code Debugging. Unfortunately, as the debugging precision increases, the size of an executable file and its execution time increases.

The graph Fig. 7 shows the results of the execution time measurements for an ECU2 controller program depending on the setting of various configuration options of the program compilation.

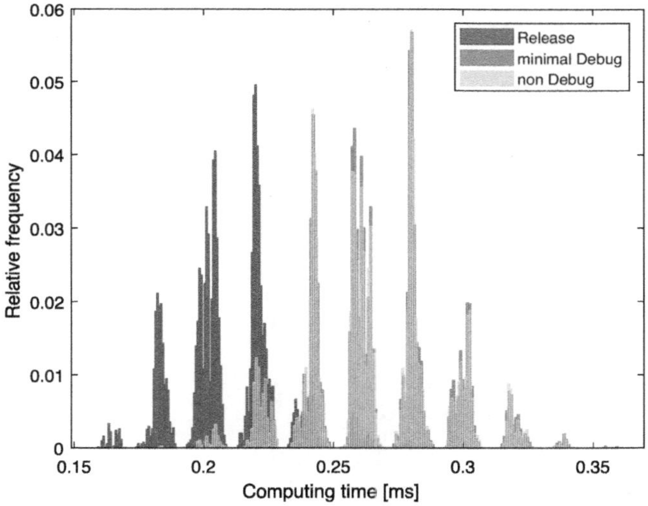

Fig. 7. The ECU2 software execution time depend on different compile flags.

The following program compilation configurations were tested:

1. Release mode – default settings of the SW4STM32 environment,
2. Debug mode – the `-g1` and `-Og` flags set up (minimum level of debug information and optimisation for the Debug mode)
3. Only the `-O0` flag set up (no debug information, optimisation disabled)

Computations in the Release mode last between 160 μs and 280 μs, while disabling code optimisation and enabling the minimal debug mode noticeably increase the computation time which is between 200 µs a 340 µs. This difference is caused in one case by the lack of optimisation and in the other by the excessive code which is made up by the additional information about program execution. The distribution of the program step execution time resembles the multi-modal distribution, which is most likely caused by the number of interrupts invoked during computations, as well as the time of computations for a specific set of data. It can also be seen that between the program compiled while using the `-g1` and `-Og` flags and the program compiled while using the `-O0` flag there are no noticeable differences in the computation time distribution.

Influence of Compilation Flags on the Stability of the Computation Step. Unlike in the previous case, this test measured the time between successive cycles of starting the main loop of the ECU2 controller program. The compilation flags settings are identical to the previous subsection because the same input data was used. The Fig. 8 presents histograms of measured computation step values for various configuration settings of the compiler. It can

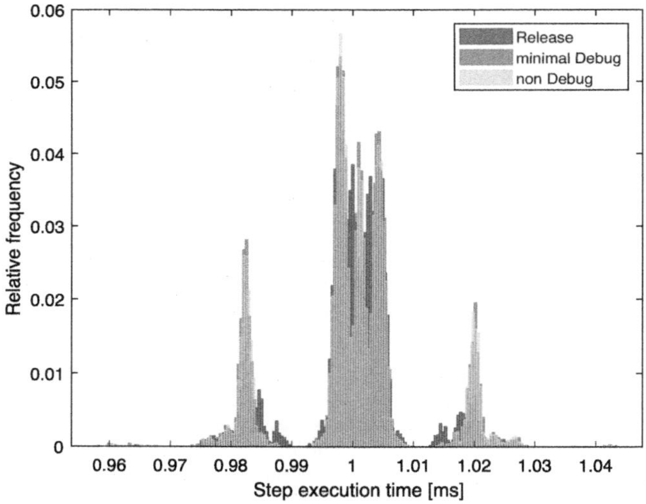

Fig. 8. The stability of computations step of the ECU2 controller in term of the different compile flags.

be seen that, unlike computation time, compilation flags do not significantly affect the stability of the computational step. This is because the computations are triggered by a hardware clock, found inside the processor, which causes the `HAL_SYSTICK_Callback()` interruption with a frequency of 1kHz. Due to the fact that it is operated by a separate and independent system timer, the method of generating program code does not affect the stability of this interruption.

However, the stability of invoking a step can be disturbed by another higher-priority interruption that will suspend the execution of the `HAL_SYSTICK_Callback()` interruption for the duration of its own code execution. That is why it is so important that the duration of interruption execution be as short as possible.

4 Safety Systems

Due to the size of the vehicle, there is a considerable risk of damage resulting from unintentional driving, caused by an error in the code of the tested algorithms. To increase the level of security, several safeguards connected in series were used, which in the event of specific situations disconnect the main motor relay at the lowest possible level (cutting off the relay control circuit). All of the assumed conditions must be met in order to drive the A-EVE vehicle. All of the presented safety features are located in the main relay coil circuit. The block diagram of the safety system is shown in the Fig. 9.

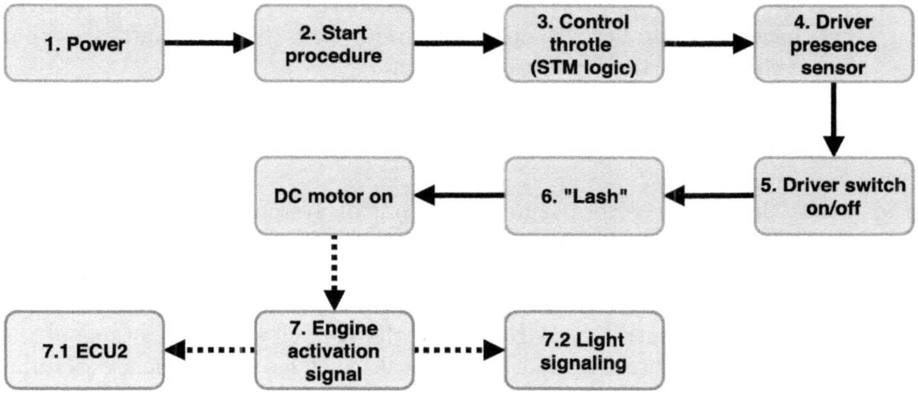

Fig. 9. Safety system diagram of the A-EVE vehicle.

1. Power supply – the EVE motor controller is powered directly by a set of 3 car batteries, the main relay motor coil is powered by a 24 V converter, thanks to which the A-EVE vehicle cannot drive if the converter is not powered.
2. The start procedure – is executed directly by the main engine controller, it requires the throttle pedal to not be pressed when enabling the power supply of the main engine controller. Otherwise, it will not be possible to start the engine, even if the power supply of the main engine controller is turned on. The car's start switch is another of its protective features. If there is no key or a wrong one is used, the main motor coil cannot be controlled.
3. Throttle control with ECU2 – the logic contained in ECU2 reads the position of the throttle pedal, the set speed, the signal of switching on the main relay,

and the pressure sensor in the driver's seat. Based on these signals, it controls the main engine controller. If all of the listed conditions are not met, no signal is sent to the engine controller.

4. Driver presence sensor – this is a pressure sensor located in the driver's seat. If a driver is not present, driving the A-EVE vehicle is not possible.
5. Driver switch – this safety feature uses the switch on the left side of the steering wheel, which, like other protections, is located in the control circuit of the main relay coil.
6. Lanyard type switch – which can be found in the back on the right side. This switch is made up from a jumper made of a thin flexible wire, which is stretched between two contacts. One of the permanent contacts is a magnet which holds the jumper in place while the vehicle on the move. This jumper is also located directly in the circuit of the main relay coil.
7. Signalling – uses the main relay coil signal:
   7.1 ECU2 – the control signal of the main relay coil is read by ECU2 and saved in the `EVE_INFO_ENGINE_ERR` bit and it is sent to the ECU1 hight level controller.
   7.2 Light – flashing warning area located on the left pillar of the vehicle, connected in parallel with the main relay coil. The area indicates that the vehicle is in the ready-to-drive state.

5 Conclusion

The article presents a description of the control system which was designed and used in the experimental autonomous vehicle called A-EVE. A decision was made to use a two-level control system which consists of the ECU1 high level controller and ECU2 low level controller. The task of the ECU1 high level controller is to execute control tasks related to the behaviour and reactions of the A-EVE vehicle. It performs, among others, tasks such as: localisation, determining the path of movement, obstacle detection, obstacle avoidance, supervision of the operation of all the components of the A-EVE vehicle. The task of the second ECU2 controller (low level controller) is to directly control individual components of the A-EVE vehicle (e.g. stabilisation of longitudinal speed or the turning angle of the front wheels), reading, scaling and calibrating the data from input sensors e.g. (encoders). A very important function also executed by ECU2 is the protection algorithm, thanks to which operating the A-EVE vehicle is safe. The proposed architecture of the control system allows for its easy adaptation or change. The use of automatic code generation tools and the selection of the ROS framework as well as of the popular programming languages C and Python allows a quick and effective implementation of various algorithms for controlling and functioning of the A-EVE vehicle.

In the next work stages, there are plans to add a separate controller that will perform the task of multi-sensor data fusion because the quality of operations of each control system significantly depends on the data (their quantity, accuracy). Another of the planned work stages consists in the design and implementation of the next version of the ECU2 controller with the possibility of using it not only in autonomous vehicles but also in autonomous robots.

References

1. Behringer, R., Sundareswaran, S., Gregory, B., Elsley, R., Addison, B., Guthmiller, W., Daily, R., Bevly, D.: The darpa grand challenge - development of an autonomous vehicle. In: IEEE Intelligent Vehicles Symposium 2004, pp. 226–231 (2004). https://doi.org/10.1109/IVS.2004.1336386
2. Cheng, H.: Autonomous intelligent vehicles: theory, algorithms, and implementation. In: Advances in Computer Vision and Pattern Recognition. Springer, London (2011). http://cds.cern.ch/record/2197195
3. Fletcher, L., Teller, S., Olson, E., Moore, D., Kuwata, Y., How, J., Leonard, J., Miller, I., Campbell, M., Huttenlocher, D., Nathan, A., Kline, F.R.: The MIT-Cornell collision and why it happened. J. Field Rob. **25**(10), 775–807 (2008). https://doi.org/10.1002/rob.20266. https://onlinelibrary.wiley.com/doi/abs/10.1002/rob.20266
4. Jo, K., Kim, J., Kim, D., Jang, C., Sunwoo, M.: Development of autonomous Car–Part 1: distributed system architecture and development process. IEEE Trans. Industr. Electron. **61**(12), 7131–7140 (2014). https://doi.org/10.1109/TIE.2014.2321342
5. Jo, K., Kim, J., Kim, D., Jang, C., Sunwoo, M.: Development of autonomous Car–Part 2: a case study on the implementation of an autonomous driving system based on distributed architecture. IEEE Trans. Industr. Electron. **62**(8), 5119–5132 (2015). https://doi.org/10.1109/TIE.2015.2410258
6. Komorkiewicz, M., Kryjak, T., Chuchacz-Kowalczyk, K., Skruch, P., Gorgoń, M.: FPGA based system for real-time structure from motion computation. In: 2015 Conference on Design and Architectures for Signal and Image Processing (DASIP), pp. 1–7 (2015). https://doi.org/10.1109/DASIP.2015.7367241
7. Komorkiewicz, M., Turek, K., Skruch, P., Kryjak, T., Gorgon, M.: FPGA-based hardware-in-the-loop environment using video injection concept for camera-based systems in automotive applications. In: 2016 Conference on Design and Architectures for Signal and Image Processing (DASIP), pp. 183–190 (2016). https://doi.org/10.1109/DASIP.2016.7853817
8. Markiewicz, P., Kogut, K., Różewicz, M., Skruch, P., Starosolski, R.: Occupancy grid fusion prototyping using utomotive virtual validation environment. In: Proceedings of the 6th International Conference on Control, Mechatronics and Automation - ICCMA 2018 (2018). https://doi.org/10.1145/3284516.3284540
9. Mitkowski, W., Zagórowska, M., Bauer, W.: Comparative analysis of dc motor control system. Appl. Mech. Mater. **817**, 111–121 (2016). https://doi.org/10.4028/www.scientific.net/amm.817.111
10. Noviello, C.: Masterig STM32. Leanpub (2018)
11. Oprzedkiewcz, K., Mitkowski, W., Gawin, E.: The PLC implementation of fractional-order operator using CFE approximation. Adv. Intell. Syst. Comput., 22–33 (2017). https://doi.org/10.1007/978-3-319-54042-9_2
12. Pérez, J., Milanés, V., Onieva, E.: Cascade architecture for lateral control in autonomous vehicles. IEEE Trans. Intell. Transp. Syst. **12**(1), 73–82 (2011). https://doi.org/10.1109/TITS.2010.2060722

13. Quigley, M., Gerkey, B., Smart, W.: Programming Robots with ROS: A Practical Introduction to the Robot Operating System. O'Reilly Media (2015). https://books.google.pl/books?id=G3v5CgAAQBAJ
14. Roman, M.: Design and implementation of autonomous vehicle controller compatible with ROS. Master's thesis, AGH University of Science and Technology (2019)
15. Skruch, P., Długosz, M., Mitkowski, W.: Mathematical methods for verification of microprocessor-based PID controllers for improving their reliability. Eksploatacja i Niezawodność **17**(3), 327–333 (2015)
16. Straszheim, T., Gerkey, B., Cousins, S.: The ros build system. IEEE Rob. Autom. Mag. **18**(2), 18–122 (2011). https://doi.org/10.1109/MRA.2011.941630
17. Thrun, S., Montemerlo, M., Dahlkamp, H., Stavens, D., Aron, A., Diebel, J., Fong, P., Gale, J., Halpenny, M., Hoffmann, G., Lau, K., Oakley, C., Palatucci, M., Pratt, V., Stang, P., Strohband, S., Dupont, C., Jendrossek, L.E., Koelen, C., Markey, C., Rummel, C., van Niekerk, J., Jensen, E., Alessandrini, P., Bradski, G., Davies, B., Ettinger, S., Kaehler, A., Nefian, A., Mahoney, P.: Stanley: the robot that won the DARPA grand challenge. J. Field Rob. **23**(9), 661–692 (2006). https://doi.org/10.1002/rob.20147. https://onlinelibrary.wiley.com/doi/abs/10.1002/rob.20147
18. Urmson, C., Baker, C., Dolan, J., Rybski, P., Salesky, B., Whittaker, W., Ferguson, D., Darms, M.: Autonomous driving in traffic: boss and the urban challenge. AI Mag. **30**(2), 17 (2009). https://doi.org/10.1609/aimag.v30i2.2238. https://www.aaai.org/ojs/index.php/aimagazine/article/view/2238
19. Veres, S.M., Molnar, L., Lincoln, N.K., Morice, C.P.: Autonomous vehicle control systems – a review of decision making. Proc. Inst. Mech. Eng. Part I J. Syst. Control Eng. **225**(2), 155–195 (2011). https://doi.org/10.1177/2041304110394727
20. Xu, L., Wang, Y., Sun, H., Xin, J., Zheng, N.: Design and implementation of driving control system for autonomous vehicle. In: 17th International IEEE Conference on Intelligent Transportation Systems (ITSC), pp. 22–28 (2014). https://doi.org/10.1109/ITSC.2014.6957660

Robust Control – Design, Analysis, Applications

ESO Architectures in the Trajectory Tracking ADR Controller for a Mechanical System: A Comparison

Krzysztof Łakomy(✉), Radosław Patelski, and Dariusz Pazderski

Poznań University of Technology, Piotrowo 3A, 60-965 Poznań, Poland
{krzysztof.pi.lakomy,radoslaw.z.patelski}@doctorate.put.poznan.pl, dariusz.pazderski@put.poznan.pl

Abstract. Proper operation of the Active Disturbance Rejection (ADR) controller requires a precise determination of the so-called total disturbance affecting the considered dynamical system, usually estimated by the Extended State Observer (ESO). The observation quality of total disturbance has a significant impact on the control error values, making room for a potential improvement of control system performance using different structures of ESO. In this article, we provide a quantitative comparison between the Luenberger and Astolfi/Marconi (AM) observers designed for three different extended state representations and utilized in the trajectory tracking ADR controller designed for a mechanical system. Included results were obtained in the simple simulation case, followed by the experimental validation on the main axis of a telescope mount.

Keywords: Active Disturbance Rejection Control (ADRC) · Extended State Observer (ESO) · Trajectory tracking · Mechanical system

1 Introduction

Research interest in the Active Disturbance Rejection (ADR) control method, initially introduced in [4], has been gradually growing in recent years resulting in its frequent use in the domains of industrial- [10,14,18] and mobile robotics [13]. Instead of relying on the precise model of the control object, the ADR method depends on a feedforward cancellation of the so-called total disturbance, usually estimated by the Extended State Observer (ESO) with a single additional state, see [3]. Since the control quality obtained with the ADR-based controller is highly dependent on the total disturbance estimation accuracy, we may improve control performance by choosing the structure of ESO that is more suitable to the expected type of disturbance. Instead of increasing the observer gains to estimate quickly varying disturbances, one can implement an ESO augmented

This work was partially supported by grants No. 33/32/SIGR/0003 and No. 2014/15/B/ST7/00429.

A. Bartoszewicz et al. (Eds.): PCC 2020, AISC 1196, pp. 1323–1335, 2020.
https://doi.org/10.1007/978-3-030-50936-1_110

by multiple states, see [11], or use a Resonant Extended State Observer (RESO) [10] in the presence of total disturbance including an oscillatory component. An ESO is most commonly implemented according to a high-gain Luenberger observer design method [7], what usually results in a strong amplification of the sensor measurement noise. To decrease the impact of measurement noise Astolfi and Marconi introduced a limited gain power observer for the systems in the canonical observability form [1], which was generalized to a wider class of objects in [17].

In this article, we would like to compare the results obtained for three aforementioned observer structures, corresponding to a conventional ESO augmented by a single and multiple states, and a specific extended state representation utilized in RESO. Chosen structures will be implemented using two different observer architectures, i.e., a high-gain Luenberger observer and the observer proposed by Astolfi/Marconi (AM). Since most of the observer architectures are presented in separate articles describing the expected control performance with various methods of analysis, a quantitative comparison between different observer types with a clearly defined comparison criterion is necessary. In our considerations, we focus on a trajectory tracking control task performed by a mechanical system under the presence of the external perturbations and measurement noise. Conducted simulations were made on a second-order linear dynamics, while the experimental studies were made on the main axis of a telescope mount, which in general can be modeled with nonlinear differential equations.

Notation: Throughout this paper, we assume that $\hat{x}$ correspond to the estimate of signal x, by marking $x \in \mathscr{C}^n$ we assume that signal x is at least n-times differentiable, $\boldsymbol{A} \succ 0$ means that matrix $\boldsymbol{A}$ is positive definite matrix in the sense that $\boldsymbol{x}^\top \boldsymbol{A}\boldsymbol{x} > 0$ for any vector $\boldsymbol{x}$, the norm of a matrix $\boldsymbol{A}$ is defined as $\|\boldsymbol{A}\| \triangleq \sup\{\|\boldsymbol{A}\boldsymbol{x}\| : \boldsymbol{x} \in \mathbb{R}^n \text{ and } \|\boldsymbol{x}\| = 1\}$, $\lambda_{\min}(\boldsymbol{A})$ and $\lambda_{\max}(\boldsymbol{A})$ are respectively the minimal and maximal eigenvalues of matrix $\boldsymbol{A}$, while $\boldsymbol{0}$ and $\boldsymbol{I}$ represent zero and identity matrices of the appropriate order.

2 Preliminaries

Let us consider a mechanical system having a single degree of freedom with the dynamics expressed as

$$J(q)\ddot{q} + h(q, \dot{q}) + h_m(q, \dot{q}) + \tau^* = \tau, \tag{1}$$

where $q \in \mathbb{R}$ corresponds to a system configuration, $J(q)$ is a system inertia, $h_m(q, \dot{q})$ corresponds to a known (or modeled) part of the system dynamics, $h(q, \dot{q})$ aggregates the unmodeled dynamical phenomena, $\tau^* \in \mathscr{C}^3$ is a bounded external disturbance with bounded derivatives, and τ is a control signal. To solve the trajectory tracking motion task, we first have to define the desired trajectory $q_d \in \mathscr{C}^5$ included in the control error definition $e \triangleq q_d - q$, which dynamics can be derived upon (1) and written down in a form

$$\ddot{e} = \ddot{q}_d - \ddot{q} = \underbrace{\ddot{q}_d - \frac{1}{J}(\tau - h_m - h - \tau^*) + \frac{1}{\hat{J}}(\tau - \hat{h}_m)}_{f(\ddot{q}_d, q, \dot{q}, \tau, \tau^*)} - \frac{1}{\hat{J}}(\tau - \hat{h}_m) \tag{2}$$

for $\hat{J} := \hat{J}(q)$ and $\hat{h}_m := \hat{h}_m(\hat{\dot{q}})$ being respectively the estimates of the system inertia and modeled dynamics, while $f(\cdot)$ represents the lumped total-disturbance. According to Eq. (2), we propose the trajectory tracking controller

$$\tau \triangleq \hat{h}_m + \hat{J}(\hat{f} + k_p e + k_d \hat{\dot{e}}), \tag{3}$$

where $k_p, k_d > 0$ are the gains of a PD controller. After a substitution of control law (3) into Eq. (2), we obtain a closed-loop dynamics

$$\ddot{e} = -k_p e - k_d \dot{e} - \tilde{f} + k_d \tilde{\dot{e}} \tag{4}$$

that may be treated as a linear system perturbed by signals $\tilde{f} \triangleq f - \hat{f}$ and $\tilde{\dot{e}} \triangleq \dot{e} - \hat{\dot{e}}$. According to (4), we may see that the control performance depends on the precise estimations of f and $\dot{e}$ that should be provided by the observer.

To design ESO, we firstly need to define an extended state vector $\boldsymbol{x} \triangleq [e\ \dot{e}\ f]^\top$, expressed here in the error domain (see [12]), which dynamics can be derived upon (2) and described with the state-space equations

$$\begin{cases} \dot{\boldsymbol{x}} = \boldsymbol{A}_3 \boldsymbol{x} - \frac{1}{\hat{J}} \boldsymbol{b}_3 (\tau - \hat{h}_m) + \boldsymbol{d}_3 \dot{f} \\ y = \boldsymbol{c}_3 \boldsymbol{x} + w \end{cases}, \tag{5}$$

where w is a bounded measurement noise, y is a system output, while $\boldsymbol{A}_n \triangleq \begin{bmatrix} \boldsymbol{0}^{n-1\times 1} & \boldsymbol{I}^{n-1} \\ 0 & \boldsymbol{0}^{1\times n-1} \end{bmatrix}$, $\boldsymbol{b}_n \triangleq \begin{bmatrix} 0 \\ 1 \\ \boldsymbol{0}^{n-2\times 1} \end{bmatrix}$, $\boldsymbol{c}_n \triangleq \begin{bmatrix} 1\ \boldsymbol{0}^{1\times n-1} \end{bmatrix}$, and $\boldsymbol{d}_n \triangleq \begin{bmatrix} \boldsymbol{0}^{n-1} \times 1 \\ 1 \end{bmatrix}$. A standard Luenberger ESO estimating the values of $\boldsymbol{x}$ and designed according to the Eqs. (5) is expressed as

$$\begin{cases} \dot{\hat{\boldsymbol{x}}} = \boldsymbol{A}_3 \hat{\boldsymbol{x}} - \frac{1}{\hat{J}} \boldsymbol{b}_3 (\tau - \hat{h}_m) + \boldsymbol{l}_3 (\hat{y} - y) \\ \hat{y} = \boldsymbol{c}_3 \hat{\boldsymbol{x}} \end{cases}, \tag{6}$$

where $\boldsymbol{l}_3 \triangleq [3\omega_o\ 3\omega_o^2\ \omega_o^3]^\top$ is the observer gain vector dependent on a single parameter ω_o. The quality of estimation can be determined by analyzing the dynamics of the observation error $\tilde{\boldsymbol{x}} \triangleq \boldsymbol{x} - \hat{\boldsymbol{x}}$ derived upon (5) and (6), i.e.,

$$\dot{\tilde{\boldsymbol{x}}} = \underbrace{(\boldsymbol{A}_3 - \boldsymbol{l}_3 \boldsymbol{c}_3)}_{\boldsymbol{H}_{\tilde{x}}} \tilde{\boldsymbol{x}} + \boldsymbol{d}_3 \dot{f} + \boldsymbol{l}_3 w. \tag{7}$$

Let us introduce a positive-definite function $V_{\tilde{x}} \triangleq \frac{1}{2} \tilde{\boldsymbol{x}}^\top \boldsymbol{P}_{\tilde{x}} \tilde{\boldsymbol{x}}$ bounded by $\frac{1}{2}\lambda_{\min}(\boldsymbol{P}_{\tilde{x}}) \|\tilde{\boldsymbol{x}}\|^2 \leq V_{\tilde{x}} \leq \frac{1}{2}\lambda_{\max}(\boldsymbol{P}_{\tilde{x}}) \|\tilde{\boldsymbol{x}}\|^2$, where a symmetric matrix $\boldsymbol{P}_{\tilde{x}} \succ 0$

is a solution of Lyapunov equation $\boldsymbol{H}_{\tilde{x}}\boldsymbol{P}_{\tilde{x}} + \boldsymbol{P}_{\tilde{x}}\boldsymbol{H}_{\tilde{x}}^\top + \omega_o \boldsymbol{I} = \boldsymbol{0}$. The time derivative $\dot{V}_{\tilde{x}}$ is bounded by

$$\begin{aligned} \dot{V}_{\tilde{x}} &= \frac{1}{2}\tilde{\boldsymbol{x}}^\top (\boldsymbol{H}_{\tilde{x}}^\top \boldsymbol{P}_{\tilde{x}} + \boldsymbol{P}_{\tilde{x}}\boldsymbol{H}_{\tilde{x}})\tilde{\boldsymbol{x}} + \tilde{\boldsymbol{x}}^\top \boldsymbol{P}_{\tilde{x}}\boldsymbol{d}_3 \dot{f} + \tilde{\boldsymbol{x}}^\top \boldsymbol{P}_{\tilde{x}}\boldsymbol{l}_3 w \\ &\leq -\frac{1}{2}(1-\nu_{\tilde{x}})\omega_o \|\tilde{\boldsymbol{x}}\|^2 + \|\tilde{\boldsymbol{x}}\| \left(\|\boldsymbol{P}_{\tilde{x}}\| \, |\dot{f}| + \omega_o^3 \|\boldsymbol{P}_{\tilde{x}}\| \, |w| - \frac{1}{2}\nu_{\tilde{x}}\omega_o \|\tilde{\boldsymbol{x}}\| \right) \end{aligned} \tag{8}$$

and fulfills the relation

$$\dot{V}_{\tilde{x}} \leq -\frac{1}{2}(1-\nu_{\tilde{x}})\omega_o \|\tilde{\boldsymbol{x}}\|^2 \text{ when } \|\tilde{\boldsymbol{x}}\| \geq \frac{2\|\boldsymbol{P}_{\tilde{x}}\|}{\nu_{\tilde{x}}\omega_o}|\dot{f}| + \frac{2\omega_o^2\|\boldsymbol{P}_{\tilde{x}}\|}{\nu_{\tilde{x}}}|w| \tag{9}$$

for some majorization constant $\nu_{\tilde{x}} \in (0,1)$. According to Th 5.1 from [6], multi-input ISS procedure utilized in [9] and [15], and to the relation (9), a time response of the dynamics (7) is bounded by

$$\forall_{t\geq 0} \|\tilde{\boldsymbol{x}}(t)\| \leq c_1 \|\tilde{\boldsymbol{x}}(0)\| e^{-\gamma_{\tilde{x}} t} + \frac{2\|\boldsymbol{P}_{\tilde{x}}\|}{\nu_{\tilde{x}}\omega_o} \sup_{t\geq 0}|\dot{f}(t)| + \frac{2\omega_o^2\|\boldsymbol{P}_{\tilde{x}}\|}{\nu_{\tilde{x}}} \sup_{t\geq 0}|w(t)|, \tag{10}$$

where $c_1 = \sqrt{\lambda_{\max}(\boldsymbol{P}_{\tilde{x}})/\lambda_{\min}(\boldsymbol{P}_{\tilde{x}})}$ and $\gamma_{\tilde{x}} = \omega_o(1-\nu_{\tilde{x}})/(2\lambda_{\max}(\boldsymbol{P}_{\tilde{x}}))$. In the result (10) we can see two components involving the perturbation signals, where the first one depends on the derivative of total disturbance while the second one depends on the measurement noise $w(t)$. It can be noticed, that as ω_o increases, the latter of two should outweigh the former, due to the presence of ω_o^2 factor. In other words, there should exist the value ω_o that minimizes the overall impact of perturbations on the observation error (10), as it was presented in the work [7].

Let us now consider a closed-loop dynamics of combined control error $\boldsymbol{\varepsilon} \triangleq [e \ \dot{e}]^\top$ expressed as

$$\dot{\boldsymbol{\varepsilon}} = \boldsymbol{H}_\varepsilon \boldsymbol{\varepsilon} + \boldsymbol{b}_2 \boldsymbol{k}\tilde{\boldsymbol{x}}, \tag{11}$$

where $\boldsymbol{H}_\varepsilon = \begin{bmatrix} 0 & 1 \\ -k_p & -k_d \end{bmatrix}$ and $\boldsymbol{k} = [0 \ k_d \ 1]$. Let us propose a positive-definite function $V_\varepsilon \triangleq \frac{1}{2}\boldsymbol{\varepsilon}^\top \boldsymbol{P}_\varepsilon \boldsymbol{\varepsilon}$ bounded by $\frac{1}{2}\lambda_{\min}(\boldsymbol{P}_\varepsilon)\|\boldsymbol{\varepsilon}\|^2 \leq V_\varepsilon \leq \frac{1}{2}\lambda_{\max}(\boldsymbol{P}_\varepsilon)\|\boldsymbol{\varepsilon}\|^2$, where a symmetric matrix $\boldsymbol{P}_\varepsilon \succ 0$ is a solution of Lyapunov equation $\boldsymbol{H}_\varepsilon \boldsymbol{P}_\varepsilon + \boldsymbol{P}_\varepsilon \boldsymbol{H}_\varepsilon^\top + \rho_\varepsilon \boldsymbol{I} = \boldsymbol{0}$ for some $\rho_\varepsilon > 0$. The time derivative $\dot{V}_\varepsilon$ is bounded by

$$\begin{aligned} \dot{V}_\varepsilon &= \frac{1}{2}\boldsymbol{\varepsilon}^\top (\boldsymbol{H}_\varepsilon^\top \boldsymbol{P}_\varepsilon + \boldsymbol{P}_\varepsilon \boldsymbol{H}_\varepsilon)\boldsymbol{\varepsilon} + \boldsymbol{\varepsilon}^\top \boldsymbol{P}_\varepsilon \boldsymbol{b}_2 \boldsymbol{k}\tilde{\boldsymbol{x}} \\ &\leq -\frac{1}{2}(1-\nu_\varepsilon)\rho_\varepsilon \|\boldsymbol{\varepsilon}\|^2 + \|\boldsymbol{\varepsilon}\| \left(\|\boldsymbol{P}_\varepsilon\| \, \|\boldsymbol{k}\| \, \|\tilde{\boldsymbol{x}}\| - \frac{1}{2}\nu_\varepsilon \rho_\varepsilon \|\boldsymbol{\varepsilon}\| \right) \end{aligned} \tag{12}$$

and fulfills the relation

$$\dot{V}_\varepsilon \leq -\frac{1}{2}(1-\nu_\varepsilon)\rho_\varepsilon \|\boldsymbol{\varepsilon}\|^2 \text{ when } \|\boldsymbol{\varepsilon}\| \geq \frac{2\|\boldsymbol{P}_\varepsilon\| \, \|\boldsymbol{k}\|}{\nu_\varepsilon \rho_\varepsilon} \|\tilde{\boldsymbol{x}}\| \tag{13}$$

for some majorization constant $\nu_\varepsilon \in (0,1)$. A time response of the dynamics (11), according to the relation (13), is bounded by

$$\forall_{t\geq 0} \|\boldsymbol{\varepsilon}(t)\| \leq \sqrt{\frac{\lambda_{\max}(\boldsymbol{P}_\varepsilon)}{\lambda_{\min}(\boldsymbol{P}_\varepsilon)}} \|\boldsymbol{\varepsilon}(0)\| e^{-\gamma_\varepsilon t} + \frac{2\|\boldsymbol{P}_\varepsilon\| \|\boldsymbol{k}\|}{\nu_\varepsilon \rho_\varepsilon} \sup_{t\geq 0} \|\tilde{\boldsymbol{x}}(t)\|, \tag{14}$$

where $\gamma_\varepsilon = \rho_\varepsilon(1-\nu_\varepsilon)/(2\lambda_{\max}(\boldsymbol{P}_\varepsilon))$. The final value of control errors depends on the magnitude of observation error $\tilde{\boldsymbol{x}}(t)$, that according to (10) is determined by the values of total disturbance derivative $\dot{f}(t)$, the magnitude of a measurement noise $w(t)$, and the observation gain parameter ω_o. In the latter part of this article, we will introduce the observer architectures alternative to (6) that could possibly improve the estimation quality and noise attenuation.

3 Alternative Observer Architectures

To introduce the first of alternative observer structures, we need to assume that the total disturbance has a sinusoidal component, and can be rewritten as a sum

$$f = f_o + f_r, \tag{15}$$

where f_o refers to the oscillatory part of the disturbance, while f_r contains the residual disturbances not included in f_o. We assume that the angular frequency of disturbance oscillations can be determined and is marked as ω_r, thus we may model the resonant component of the total disturbance as a harmonic oscillator

$$\ddot{f}_o + \omega_r^2 f_o = 0 \tag{16}$$

and consider it as a part of the new extended state dynamics

$$\begin{cases} \dot{\boldsymbol{\chi}}^* = \boldsymbol{A}_5^* \boldsymbol{\chi}^* - \frac{1}{J}\boldsymbol{b}_5(\tau - \hat{h}_m) + \boldsymbol{d}_5^* \dot{f}_r \\ y^* = \boldsymbol{c}_5 \boldsymbol{\chi}^* + w \end{cases}, \tag{17}$$

where $\dot{\boldsymbol{\chi}}^* \triangleq [e \ \dot{e} \ f \ \dot{f}_o \ \ddot{f}_o]^\top$ is an extended state, y^* is an output of the system, while $\boldsymbol{A}_5^* = \begin{bmatrix} \boldsymbol{0}^{4\times 1} & \boldsymbol{I}^4 \\ 0 & [0 \ 0 \ -\omega_r^2 \ 0] \end{bmatrix}$, and $\boldsymbol{d}_5^* = [0 \ 0 \ 1 \ 0 \ 0]^\top$. Based on the dynamics (17), we can define the equations of RESO in a form

$$\begin{cases} \dot{\hat{\boldsymbol{\chi}}}^* = \boldsymbol{A}_5^* \hat{\boldsymbol{\chi}}^* - \frac{1}{J}\boldsymbol{b}_5(\tau - \hat{h}_m) + \boldsymbol{l}_5(\hat{y}^* - y^*) \\ \hat{y}^* = \boldsymbol{c}_3 \hat{\boldsymbol{\chi}}^* \end{cases}, \tag{18}$$

where $\boldsymbol{l}_5 \triangleq [5\omega_o \ 10\omega_o^2 \ 10\omega_o^3 \ 5\omega_o^4 \ \omega_o^5]^\top$ is a vector of observer gains.

The rank of observer (18) is equal to $n = 5$, so to effectively compare it with an observer designed in the same manner as (6), we would like to introduce the new extended state vector $\dot{\boldsymbol{\chi}} \triangleq [e \ \dot{e} \ f \ \dot{f} \ \ddot{f}]^\top$, augmented by the additional two variables comparing to $\boldsymbol{x}$, satisfying rank$(\boldsymbol{\chi})$ = rank$(\boldsymbol{\chi}^*) = 5$. The dynamics of $\boldsymbol{\chi}$ is expressed as

$$\begin{cases} \dot{\boldsymbol{\chi}} = \boldsymbol{A}_5\boldsymbol{\chi} - \frac{1}{J}\boldsymbol{b}_5(\tau - \hat{h}_m) + \boldsymbol{d}_5\dddot{f} \\ \mathscr{y} = \boldsymbol{c}_5\boldsymbol{\chi} + w \end{cases}, \tag{19}$$

where $\mathscr{y}$ is the observer output. Comparing to $\boldsymbol{x}$ consisting of the combined vector $\boldsymbol{\varepsilon}$ and the total disturbance f, vector $\boldsymbol{\chi}$ takes also the first and second derivatives of total disturbance as the additional states, inducing $\dddot{f}$ to be a perturbation of (19). The Luenberger ESO designed according to (19) is described as

$$\begin{cases} \dot{\hat{\boldsymbol{\chi}}} = \boldsymbol{A}_5\hat{\boldsymbol{\chi}} - \frac{1}{J}\boldsymbol{b}_5(\tau - \hat{h}_m) + \boldsymbol{l}_5(\mathscr{y} - \hat{\mathscr{y}}) \\ \hat{\mathscr{y}} = \boldsymbol{c}_5\hat{\boldsymbol{\chi}} \end{cases}, \tag{20}$$

where $\boldsymbol{l}_5 \triangleq [5\omega_o \; 10\omega_o^2 \; 10\omega_o^3 \; 5\omega_o^4 \; \omega_o^5]^\top$ is a vector of observer gains.

All of the aforementioned observer structures are designed according to the Luenberger architecture. Besides changing the extended state definition between observers, we can also change the architecture of the designed ESOs using, for example, an Astolfi/Marconi observer design technique presented in [17]. According to [1], the use of this method should result in the smaller amplification of the high-frequency noise measurement, especially for the systems with a high relative degree. To be able to utilize AM observer design method, let us write down the generalized form of dynamics (5), (19), and (17) as

$$\begin{cases} \dot{z}_i = \phi_i(\boldsymbol{z}_i, z_{i+1}, \tau) + g_i(\zeta), \; 1 \le i < n \\ \dot{z}_n = \phi_n(\boldsymbol{z}_n) + g_n(\zeta) \end{cases}, \tag{21}$$

where $\boldsymbol{z}_i = [z_1, ..., z_i]^\top$ for $1 \le i \le n$ and ($\boldsymbol{z}_n \triangleq [z_1, ..., z_n]^\top = \boldsymbol{x}$, $\zeta = \dot{f}$) for the dynamics (5), ($\boldsymbol{z}_n = \boldsymbol{\chi}$, $\zeta = \dddot{f}$) for the dynamics (19) , and ($\boldsymbol{z}_n = \boldsymbol{\chi}^*$, $\zeta = \dot{f}_r$) for the dynamics (17). The Astolfi/Marconi observer, according to [1] and [17], can be designed as

$$\begin{cases} \dot{\boldsymbol{\xi}}_i &= \begin{bmatrix} \phi_i(\hat{\boldsymbol{z}}_i, \boldsymbol{b}_2^\top\boldsymbol{\xi}_i, \tau) + \omega_o\kappa_{i,1}\epsilon_i \\ \phi_{i+1}(\hat{\boldsymbol{z}}_{i+1}, \boldsymbol{b}_2^\top\boldsymbol{\xi}_{i+1}, \tau) + \omega_o^2\kappa_{i,2}\epsilon_i \end{bmatrix}, \; i \in \{1, ..., n-2\} \\ \dot{\boldsymbol{\xi}}_{n-1} &= \begin{bmatrix} \phi_{n-1}(\hat{\boldsymbol{z}}_{n-1}, \boldsymbol{b}_2^\top\boldsymbol{\xi}_{n-1}, \tau) + \omega_o\kappa_{n-1,1}\epsilon_{n-1} \\ \phi_n(\hat{\boldsymbol{z}}_n, \tau) + \omega_o^2\kappa_{n-1,2}\epsilon_{n-1} \end{bmatrix} \\ \hat{\boldsymbol{z}}_n = \boldsymbol{L}\boldsymbol{\xi} \end{cases}, \tag{22}$$

where $\boldsymbol{\xi} \triangleq [\boldsymbol{\xi}_1^\top \; ... \; \boldsymbol{\xi}_{n-1}^\top]^\top$ is the observer state, $\boldsymbol{\kappa}_i \triangleq [\kappa_{i,1}, \kappa_{i,2}]^\top$ contains the design parameters, while $\boldsymbol{L} = \text{blkdiag}(\boldsymbol{I}^2, \underbrace{\boldsymbol{b}_2, \; ..., \; \boldsymbol{b}_2}_{(n-2) \text{ times}})$ is a transformation matrix between $\boldsymbol{\xi}$ and $\boldsymbol{z}_n$. The estimation errors are defined as follows: $\epsilon_1 \triangleq e - \boldsymbol{c}_2\boldsymbol{\xi}_1$ and $\epsilon_i \triangleq \boldsymbol{b}_2^\top\boldsymbol{\xi}_{i-1} - \boldsymbol{c}_2\boldsymbol{\xi}_i$ for $i \in \{2, ..., n-1\}$.

Remark 1. The idea of AM observer structure (22) relies on the use of $(n-1)$-level cascade of two-dimensional observers, where the first element of the next cascade level is the estimate of the second element taken from the previous one. Such architecture should result in the additional filtering of high-frequency signals providing a better noise attenuation characteristics (at least for the systems with high values of rank n).

Remark 2. Observers (6) and (18) and their AM equivalents work properly when the signal $\dot{f}$ is bounded, while observer (20) and its AM equivalent need bounded $\dddot{f}$. According to the dynamics (2), controller (3), and assumption that the desired trajectory is at least $q_d \in \mathscr{C}^5$, we may claim that $\dot{f}$ and $\dddot{f}$ are bounded, as long as $\boldsymbol{\varepsilon}$ is in some compact set.

4 Simulation Results

The simulation study was conducted on the second order dynamical system described by transfer function $G(s) = 1/(s+1)^2$. Two different scenarios have been considered and both concerned following the trajectory designed as the step function with a step time $t = 7.5$ s and filtered with $G_f(s) = 1/(0.5s+1)^5$ under the external disturbance $\tau^* = 2.5\sin(15t)$ applied after $t = 5$ s. Filter G_f is introduced to meet the requirement of $q_d \in \mathscr{C}^5$ mentioned in Remark 2. The first scenario presents the results without the measurement noise w, while the second one a case including white measurement noise with the variance $\sigma_w = 10^{-5}$. According to the controller structure (3), we assumed that $\hat{h}_m \equiv 0$ and the parameters $\hat{J} = 1$, $k_p = 4$, $k_d = 4$. In the presented results, we have used following abbreviations considering particular observers: ESO $n = 3$ for (6), ESO $n = 5$ for (20), RESO for (18), and we have added an AM prefix to refer to their equivalents designed with (22). The values of $\boldsymbol{\kappa}_i$ from (22) were calculated with the tuning procedure presented in [1] resulting in ($\boldsymbol{\kappa}_1 = [0.8\ 0.48]^\top$, $\boldsymbol{\kappa}_2 = [0.8\ 0.16]^\top$) for AM ESO $n = 3$ and ($\boldsymbol{\kappa}_1 = [0.6\ 0.36]^\top$, $\boldsymbol{\kappa}_2 = [0.6\ 0.135]^\top$, $\boldsymbol{\kappa}_3 = [0.6\ 0.06]^\top$, $\boldsymbol{\kappa}_4 = [0.6\ 0.025]^\top$) for AM ESO $n = 5$ and RESO. A resonant angular frequency visible in (16) was set to $\omega_r = 15$ rad/s, while the values ω_o for all of the observers were chosen in a way to provide the same value of integral control cost criterion $J_e = \frac{1}{T_{\text{sim}}}\int_0^{T_{\text{sim}}} |e(t)|dt$ for the simulation time $T_{\text{sim}} = 20$ s. Tables 1 and 2 contain the values of ω_o and J_u for all of the considered cases, together with the control cost $J_u = \frac{1}{T_{\text{sim}}}\int_0^{T_{\text{sim}}} |\tau(t)|^2 dt$ and the average quality of total disturbance estimation $J_f = \frac{1}{T_{\text{sim}}}\int_0^{T_{\text{sim}}} |\tilde{f}(t)|dt$. The values of control error $e(t)$ and control signal $\tau(t)$ are also presented in Figs. 1 and 2. To avoid a peaking phenomenon appearance in the figures, the controller is inactive during the observer transient stage and turned on after time $t = 1$ s (see [5]).

In the case without the measurement noise, selected values of ω_o for Luenberger observers result in almost identical outcomes of the J_u and J_f values comparing to their AM equivalents (the only exception is the J_f for AM RESO, see Table 1, caused by a more dynamic transient stage that could be seen in the control error representation in Fig. 1). A specific structure of RESO allows

Table 1. Results obtained for the first scenario, without measurement noise

Observer type	ω_o	J_e	J_u	J_f
ESO $n = 3$	490.03	0.01	59.77	2.16
ESO $n = 5$	68.58	0.01	65.96	2.28
RESO	27.32	0.01	59.31	0.34
AM ESO $n = 3$	818.86	0.01	59.95	2.17
AM ESO $n = 5$	340.27	0.01	65.31	2.30
AM RESO	129.61	0.01	60.70	1.51

Table 2. Results obtained for the second scenario, with measurement noise

Observer type	ω_o	J_e	J_u	J_f
ESO $n = 3$	586.76	0.01	623.71	69.21
ESO $n = 5$	76.94	0.01	94.58	16.17
RESO	31.52	0.01	60.06	3.04
AM ESO $n = 3$	1057.46	0.01	844.89	92.89
AM ESO $n = 5$	352.48	0.01	91.02	18.09
AM RESO	140.34	0.01	60.41	2.77

the control system to perform much better than ESO $n = 5$ in the presence of sinusoidal disturbances, leading to the possibility of selecting smaller ω_o values to obtain the same control quality J_e. As a consequence of lower ω_o, the level of measurement noise amplification in RESO is smaller than in the case of ESO $n = 5$ (see J_u values in Tables 1 and 2). Control cost obtained for ESO $n = 5$ and RESO is approximately the same for the Astolfi/Marconi and Luenberger observers (see a zoomed subplot of control signals in Fig. 2). In the scenario with the measurement noise, the use of AM ESO $n = 3$ observer results in a greater control cost comparing to its Luenberger equivalent, what may be an indication that this observer architecture works well for the systems with higher rank (at least in the case of the selected trajectory and comparison criterion).

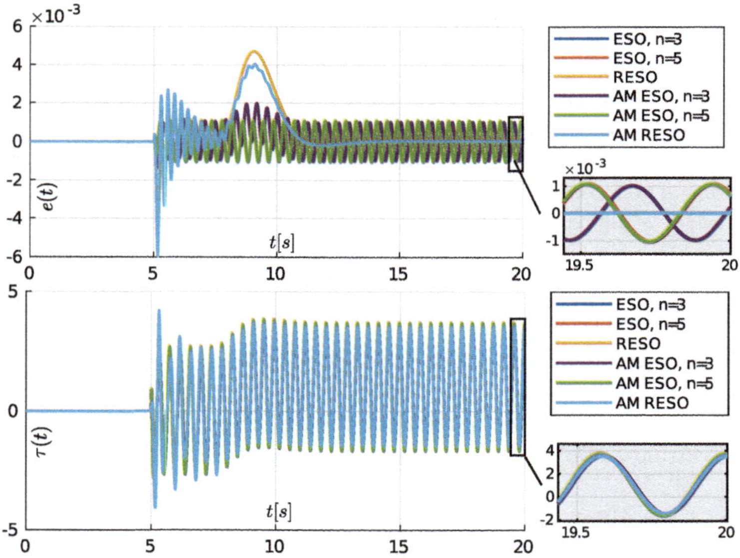

Fig. 1. Control error and control signal for the case without measurement noise

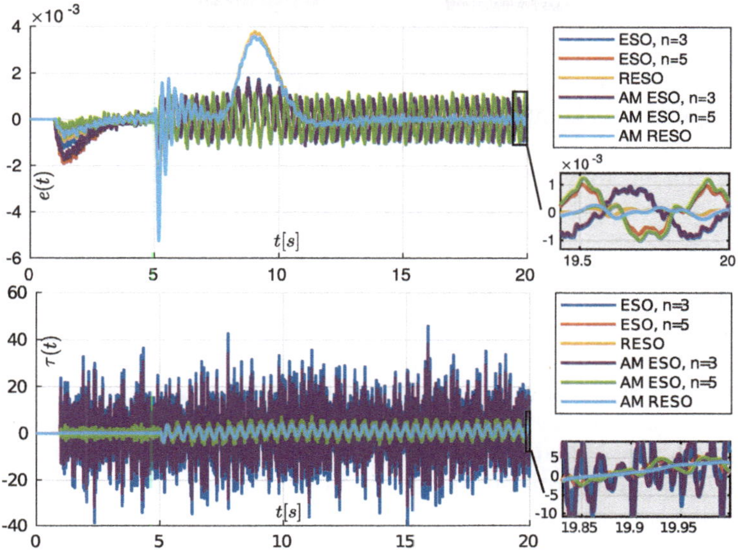

Fig. 2. Control error and control signal for the case with measurement noise

5 Experimental Validation

Experiments were conducted on the robotic mount of the astronomic telescope in a class 0.5-m that was presented in Fig. 3. A kinematic structure of the mount can be described as a 2 degree-of-freedom manipulator with perpendicular revolute joints. Each joint is gearlessly driven by permanent-magnet synchronous motors whose angular position is measured using a high-precision 32-bit rotary encoder, see [8]. Both motors are controlled by independent cascade controller with the position and current parts separated. The current loop of each axis contains a PI regulator and has much faster dynamics comparing to the loop associated with the mechanical part of the control system. In the experiments, we assume the influence of current control to be negligible and focus on the angular position control of the described system. To express the model of each telescope mount axis with the Eq. (1), we assign q to be the angular position of a joint,

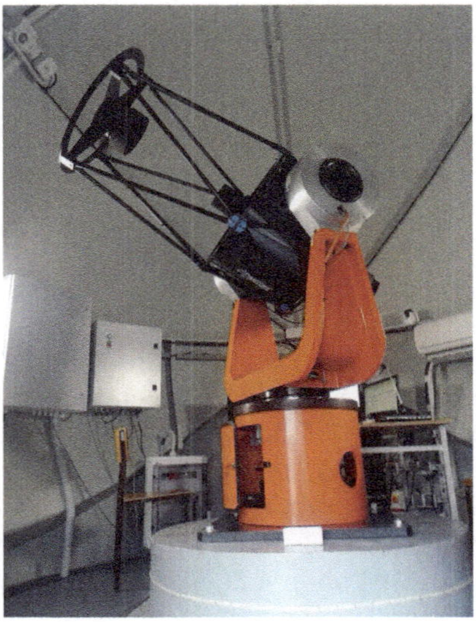

Fig. 3. Astronomic telescope setup

τ to the desired torque applied to the axis, while h_m represents friction forces estimated according to the LuGre model [2] with parameter values identified in [16]. In the experiments, we only considered the vertical (lower) axis of the mount, while the horizontal one was stabilized in a fixed upward position. The identified value of vertical axis inertia is equal to $\hat{J} = 30\ \mathrm{m} \cdot \mathrm{s}^2$. Preliminary results for the most common ADR controller using a 3rd order ESO (6) resulted in the observation of the oscillatory disturbances with the angular frequency of $\omega_r \approx 46.9$ rad that was set up as resonant frequency of RESO and AM RESO observers. We performed the experiments for a slowly-varying sinusoidal reference trajectory $q_d = 2.89 \cdot 10^{-4} \sin(0.4\pi t)$. Following such a slow trajectory is commonly required during astronomic observations (a maximal velocity of q_d corresponds to the fivefold velocity of stars observed on a night sky) and demands an extremely precise control performance to be able to follow a chosen object. The tuning of all parameters was performed according to the comparison criterion chosen in the simulations. Values ω_o for each observer were chosen to obtain comparable integral errors J_e for all attempts. Due to the dominant character of the oscillatory disturbance τ^*, it was not possible to obtain a similar control quality for RESO and ESO observers, so we have chosen the gains of RESO and RESO AM to be equal to ESO $n = 5$ and its AM equivalent. Values of $\boldsymbol{\kappa}_i$ were chosen as in the simulations. Obtained gains and values of J_e and J_u are presented in Table 3. The results of the experiments are presented in Fig. 4. The initial conditions $q(0)$ may strongly differ between particular experiments due to the very small range of angular positions generated by a trajectory generator (suitable for the astronomic observations), thus all plots are shown for $t \geq 5$ s, when all of the transient stages have already passed.

Table 3. Results obtained in the experiments

Observer type	ω_o	J_e $[10^{-6}]$	J_u
ESO $n = 3$	140	2.60	6.39
ESO $n = 5$	80	2.61	7.25
RESO	80	1.20	7.16
AM ESO $n = 3$	250	2.56	6.1
AM ESO $n = 5$	450	2.62	7.03
AM RESO	450	1.56	4.77

The use of an extremely precise encoder resulted in the consistency between the outcome of experiments and simulations without introduced measurement noise for the Luenberger ESOs. In the experiments, we obtained comparable control performance for 3rd and 5th order ESOs with a significantly lower gain value chosen for ESO $n = 5$ followed by a slightly increased control cost J_u. Likewise, resonant observer designed according to (18) proved superior to the conventional variants resulting in much better compensation of sinusoidal perturbation visible in the control error frequency spectrum presented in Fig. 4. Experiments performed with AM observers, although produced results similar to their Luenberger counterparts in the matter of chosen integral criteria, required different relative ω_o values comparing to the simulation results and had in a slightly smaller values of J_u. While the parameters of AM ESO for $n = 3$ were easily adjusted to provide desired performance, AM ESO $n = 5$ required significantly higher ω_o gain (despite the premises drew from simulations). Furthermore, AM RESO tended to cause unwanted vibrations around the resonant frequency of the mechanical structure (visible as a peak in the frequency spectrum of control error around $\omega \approx 75$ rad/s) leading to a destabilization of the control system for some gains ω_o. Whether such behavior is inherent to discussed observer architecture or was caused by the unusual character of the plant remains yet uncertain. Still, for each of the AM observers, it was possible to obtain a satisfactory tracking quality comparable with its Luenberger counterpart by correctly choosing observer gain.

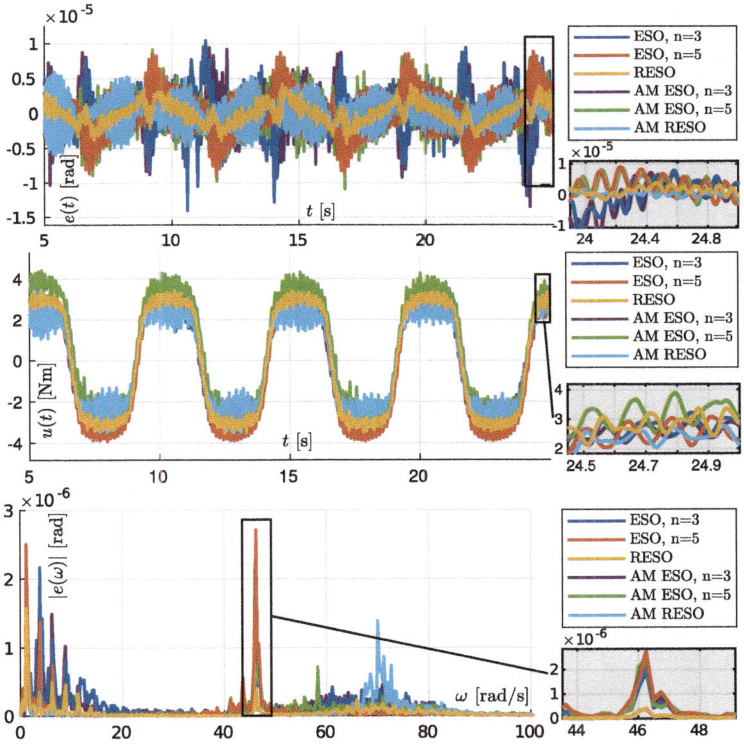

Fig. 4. Tracking error, control signal and error spectrum obtained in the experiment

6 Conclusions

Experimental and simulation results presented in this paper proved that it is possible to achieve the same control quality using different observer structures and architectures implemented as a part of the ADR controller designed for a mechanical system. Comparing to the 3rd and 5th order ESOs, the outcomes obtained with RESOs had smaller control cost values both for Luenberger and

Astolfi/Marconi architectures when the system was perturbed with a significant sinusoidal disturbance. In the case of simulation with high measurement noise values, we have presented that the Luenberger architecture has better characteristics than the AM one for 3rd order ESOs, while for 5th order ESOs their outcome is comparable.

References

1. Astolfi, D., Marconi, L.: A high-gain nonlinear observer with limited gain power. IEEE Trans. Autom. Control **60**(11), 3059–3064 (2015)
2. Canudas Wit, C., Olsson, H., Astrom, K.J., Lischinsky, P.: A new model for control of systems with friction. IEEE Trans. Autom. Control **40**(3), 419–425 (1995). https://doi.org/10.1109/9.376053
3. Gao, Z.: Active disturbance rejection control: a paradigm shift in feedback control system design. In: 2006 American Control Conference, p. 7 (2006)
4. Han, J.: From pid to active disturbance rejection control. IEEE Trans. Industr. Electron. **56**(3), 900–906 (2009)
5. Huang, Y., Xue, W.: Active disturbance rejection control: methodology and theoretical analysis. ISA Trans. **53**(4), 963–976 (2014). Disturbance Estimation and Mitigation
6. Khalil, H.K.: Nonlinear Systems, 3rd edn. Prentice Hall, Upper Saddle River (2002)
7. Khalil, H.K., Praly, L.: High-gain observers in nonlinear feedback control. Int. J. Robust Nonlinear Control **24**(6), 993–1015 (2014)
8. Kozlowski, K., Pazderski, D., Krysiak, B., Jedwabny, T., Piasek, J., Kozlowski, S., Brock, S., Janiszewski, D., Nowopolski, K.: High precision automated astronomical mount. In: Szewczyk, R., Zieliński, C., Kaliczyńska, M. (eds.) Automation 2019, pp. 299–315. Springer, Cham (2020)
9. Łakomy, K., Michałek, M.M.: Robust output-feedback vfo-adr control of underactuated spatial vehicles a task of following the non-parametrized path (2020). arXiv:2001.01963 [eess.SY]
10. Madonski, R., Ramirez-Neria, M., Stanković, M., Shao, S., Gao, Z., Yang, J., Li, S.: On vibration suppression and trajectory tracking in largely uncertain torsional system: an error-based ADRC approach. Mech. Syst. Signal Process. **134**, 106300 (2019)
11. Martinez-Vazquez, D.L., Rodriguez-Angeles, A., Sira-Ramirez, H.: Robust GPI observer under noisy measurements. In: 2009 6th International Conference on Electrical Engineering, Computing Science and Automatic Control (CCE), pp. 1–5 (2009)
12. Michałek, M.M.: Robust trajectory following without availability of the reference time-derivatives in the control scheme with active disturbance rejection. In: 2016 American Control Conference (ACC), pp. 1536–1541 (2016)
13. Michałek, M.M., Łakomy, K., Adamski, W.: Robust output-feedback cascaded tracking controller for spatial motion of anisotropically-actuated vehicles. Aerosp. Sci. Technol. **92**, 915–929 (2019)
14. Patelski, R., Pazderski, D.: Tracking control for a cascade perturbed control system using the active disturbance rejection paradigm. Arch. Control Sci. **29**(2), 387–408 (2019)
15. Peng, Z., Wang, J.: Output-feedback path-following control of autonomous underwater vehicles based on an extended state observer and projection neural networks. IEEE Trans. Syst. Man Cybern. B Cybern. Syst. **48**(4), 535–544 (2018)

16. Piasek, J., Patelski, R., Pazderski, D., Kozłowski, K.: Identification of a dynamic friction model and its application in a precise tracking control. Acta Polytechnica Hungarica **16**(10), 83–99 (2019)
17. Wang, L., Astolfi, D., Marconi, L., Su, H.: High-gain observers with limited gain power for systems with observability canonical form. Automatica **75**, 16–23 (2017)
18. Xue, W., Madonski, R., Lakomy, K., Gao, Z., Huang, Y.: Add-on module of active disturbance rejection for set-point tracking of motion control systems. IEEE Trans. Ind. Appl. **53**(4), 4028–4040 (2017)

Active Disturbance Rejection Control of High-Order Flat Underactuated Systems: Mass-Spring Benchmark Problem

Rafal Madonski[1(✉)], Mario Ramirez-Neria[2], and Wojciech Giernacki[3]

[1] Energy Electricity Research Center, International Energy College, Jinan University, Zhuhai 519070, Guangdong Province, People's Republic of China
rafal.madonski@jnu.edu.cn

[2] Mechatronics Department, Universidad Politecnica del Valle de Mexico, UPVM, Tultitlan de Mariano Escobedo, 54910 Fuentes del Valle, Mexico

[3] Institute of Robotics and Machine Intelligence, Faculty of Control, Robotics and Electrical Engineering, Poznan University of Technology, 60-965 Poznan, Poland
http://uav.put.poznan.pl

Abstract. A custom active disturbance rejection control (ADRC) is proposed for trajectory tracking in uncertain underactuated systems. By utilizing a particular property of the differentially flat plant model, a robust output-feedback controller with a novel cascade of extended state observers (ESO) is introduced. It effectively deals with the problem of governing high-order plants without over-amplification of the measurement noise, typically seen in conventional single high-gain observer-centered control approaches. The proposed solution is based on full utilization of the information already available about the governed system, without necessity for additional measurement devices. In order to be easily implementable, it assumes only limited knowledge of the system model and is expressed in an industry familiar error-based form with a straightforward tuning method. A representable two-mass three-spring benchmark problem is used throughout the paper to convey the proposed idea and evaluate it experimentally on a laboratory testbed.

Keywords: ADRC · Disturbance rejection · Underactuated system

1 Introduction

The methodology of active disturbance rejection [1] was already shown to be an effective way of formulating and solving the problem of unknown interactions with underactuated mass and unknown resonant modes in the class of mechanical systems [2]. It introduced a more general definition of disturbance (denoted as *total disturbance*), which aggregates both the external forces acting on the

A. Bartoszewicz et al. (Eds.): PCC 2020, AISC 1196, pp. 1336–1347, 2020.
https://doi.org/10.1007/978-3-030-50936-1_111

system and the internal, unknown dynamics. At the same time, trajectory planning can be realized taking advantage of the differential flatness property of the controlled system (if applicable), which allows to characterize the complete dynamics in terms of a flat output [3]. System flatness combined with the concept of active disturbance rejection control, ADRC [4], can be utilized to solve the trajectory tracking problem in underactuated mechanical systems. In such framework, the effect of underactuated dynamics (including possible arising resonance) on the governed output is treated as a part of the total disturbance to be estimated (by means of an extended state observer, or ESO) and simultaneously canceled. Some successful applications of such approach were recently reported in [5–7]. Such control scheme, even though practically appealing thanks to its relative simplicity, usually faces issues with fast and precise disturbance estimation, associated with high-frequency measurement noises and limited sensing capabilities, especially for high-order systems.

Motivated by this, a custom ADRC structure is proposed here. The main part is the utilization of a particular property of flat, mechanical, underactuated systems, which enables straightforward calculation of even order time-derivatives of the output signal just with position measurement [2]. This means that only the first-order time-derivatives of the output need to be estimated for control synthesis. It allows a virtual decomposition of the governed high-order plant and subsequent design of cascaded observers, which reduces the problem of over-amplification of measurement noise, which is a well-known issue with conventional single high-gain observer-based controllers (details in [8]). The proposed solution is an alternative to conventional approaches, which are mostly based on extra modeling, placing additional sensors allowing reduced-order observer design, time-varying observer gains, or filtration through internal states.

Additionally, a particular formulation of ADRC, dedicated to trajectory following tasks is adopted here [9]. It simplifies the original method and reduces its application restrictions (related to the availability of certain signals for control synthesis), while retaining robustness and adaptability of original ADRC.

The validity of the proposed control structure is evaluated here through multi-criteria experimental tests, including robustness verification and quantitative comparison with a standard, single observer-based approach. The control topology, although applicable to a larger class of systems, is introduced and exemplified throughout this paper using only a case of an underactuated two-mass-three-spring mechanical system [10], which is a well-known control benchmark.

2 Prerequisites

2.1 Plant Model

The considered system consists of two masses suspended on an anti-friction linear bearings and connected with a spring (with positive coefficient κ_2[N/m]). The first mass (m_1[kg]) is actuated by a linear DC motor, while the second mass (m_2[kg]), serves as an underactuated dynamics. Both masses are constrained with additional springs (with positive coefficients κ_1[N/m] and κ_3[N/m]), which are attached to the opposite sides of the testbed base. A diagram of the platform is shown in Fig. 1, in which terms x_1[m] and x_2[m] represent corresponding mass linear positions (system outputs) and u[N] represents the input force (control signal). The mechanism has thus fewer number of control inputs than its degrees of freedom, making it an underactuated mechanical system.

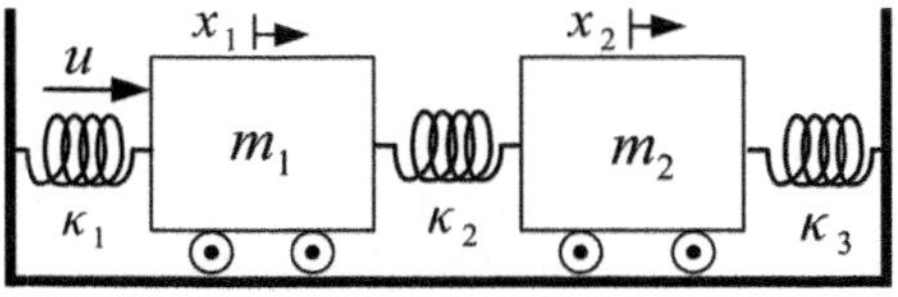

Fig. 1. Diagram of the considered two-mass three-spring rectilinear mechanism.

The mechanical part of the system can be formulated in simplified LTI form:

$$\begin{cases} m_1\ddot{x}_1 + (\kappa_1 + \kappa_2)x_1 - \kappa_2 x_2 = u, \\ m_2\ddot{x}_2 - \kappa_2 x_1 + (\kappa_2 + \kappa_3)x_2 = 0, \end{cases} \tag{1}$$

and with arbitrarily selected state variables $\boldsymbol{x} = [x_1\ \dot{x}_1\ x_2\ \dot{x}_2]^\top$, it can be alternatively expressed in state-space as:

$$\dot{\boldsymbol{x}} = \boldsymbol{A}\boldsymbol{x} + \boldsymbol{B}u,\ y = \boldsymbol{C}\boldsymbol{x}, \tag{2}$$

$$\boldsymbol{A} = \begin{bmatrix} 0 & 1 & 0 & 0 \\ -\frac{\kappa_1+\kappa_2}{m_1} & 0 & \frac{\kappa_2}{m_1} & 0 \\ 0 & 0 & 0 & 1 \\ \frac{\kappa_2}{m_2} & 0 & -\frac{\kappa_2+\kappa_3}{m_2} & 0 \end{bmatrix},\ \boldsymbol{B} = \begin{bmatrix} 0 \\ \frac{1}{m_1} \\ 0 \\ 0 \end{bmatrix},\ \boldsymbol{C} = \begin{bmatrix} 0 \\ 0 \\ 1 \\ 0 \end{bmatrix}^\top.$$

The above plant model is deliberately assumed to be a idealized linear version of the actual system. The simplified description is used here anyway. It is due to the fundamental design principle of ADRC, which states that as long as the modeling discrepancies are effectively mitigated in real time, an exact mathematical model of the plant is not needed [1].

2.2 Control Objective

The system model (2) requires a control law (u[N]) such that it forces output (x_2[m]) to track a harmonic trajectory (x_2^*[m]) starting from its initial position ($x_2(0) = 0$m). The designed controller has to be robust against the effects of numerous uncertainties, including unmodeled dynamics, parametric mismatches, and external disturbances. Following assumptions also apply.

- A1. The control design cannot use derivatives in the controller in order to avoid over-amplification of the measurement noise.
- A2. Both the analytical form of target signal x_2^* and its higher order derivatives are unavailable for controller synthesis.
- A3. Signals x_1, x_2 are measurable.
- A4. Structure of system model (2) is known, but its parameters are uncertain.

3 Proposed Control Design

3.1 Model Transformation Using Flatness

Due to limited space, we restrict ourselves only to key design steps, whereas details (including definitions) can be found in [3]. In order to verify the flatness system (2), its controllability matrix $\boldsymbol{C}_k = \left[\boldsymbol{B}\ \boldsymbol{AB}\ \boldsymbol{A}^2\boldsymbol{B}\ \boldsymbol{A}^3\boldsymbol{B}\right]$ is analyzed first. The considered linear system model is controllable with $\det(\boldsymbol{C}_k) = \kappa_2^2/\left(m_1^4 m_2^2\right)$, hence it is flat. Its standard flat output could be explicitly calculated as:

$$y_F = \left[0\,0\,0\,1\right]\boldsymbol{C}_k^{-1}\boldsymbol{x} = \frac{m_1 m_2}{\kappa_2}x_2. \tag{3}$$

However, to simplify later design stages, the flat output is chosen here as $x_\gamma = \rho y_F$, with $\rho = \kappa_2/(m_1 m_2)$. It is done deliberately to make it coincide exactly with the position x_2. Now, all the state variables can be obtained as differential functions of the flat output x_γ, which gives:

$$x_1 = \frac{\kappa_2+\kappa_3}{\kappa_2}x_\gamma + \frac{m_2}{\kappa_2}\ddot{x}_\gamma, \qquad x_2 = x_\gamma, \tag{4}$$

$$\dot{x}_1 = \frac{\kappa_2+\kappa_3}{\kappa_2}\dot{x}_\gamma + \frac{m_2}{\kappa_2}x_\gamma^{(3)}, \qquad \dot{x}_2 = \dot{x}_\gamma, \tag{5}$$

$$u = \frac{m_1 m_2}{\kappa_2}x_\gamma^{(4)} + \frac{1}{\kappa_2}(\kappa_1 m_1 + \kappa_2 m_1 + \kappa_2 m_2 + \kappa_3 m_1)\ddot{x}_\gamma + \frac{1}{\kappa_2}(\kappa_1\kappa_2 + \kappa_1\kappa_3 + \kappa_2\kappa_3)x_\gamma. \tag{6}$$

Next, the flat output and its consecutive time-derivatives can be parametrized by the system variables and their time-derivatives as:

$$x_\gamma = x_2, \qquad \dot{x}_\gamma = \dot{x}_2, \tag{7}$$

$$\ddot{x}_\gamma = \frac{\kappa_2}{m_2}x_1 - \frac{\kappa_2+\kappa_3}{m_2}x_2, \tag{8}$$

$$x_\gamma^{(3)} = \frac{\kappa_2}{m_2}\dot{x}_1 - \frac{\kappa_2+\kappa_3}{m_2}\dot{x}_2. \tag{9}$$

Now, the dynamics between system input and flat output can be written as:

$$x_\gamma^{(4)} = \frac{\kappa_2}{m_1 m_2}u(t) - \frac{1}{m_1 m_2}(\kappa_1\kappa_2 + \kappa_1\kappa_3 + \kappa_2\kappa_3)x_\gamma - \frac{1}{m_1 m_2}(\kappa_1 m_2 + \kappa_2 m_1 + \kappa_2 m_2 + \kappa_3 m_1)\ddot{x}_\gamma. \tag{10}$$

Although the flat output only uses x_2, we can see that x_1 appears in the high order dynamics in (8)–(9) due to underactuated nature of the system. It is also important to note that the second time-derivative of the flat output (8) can be described only using x_1 and x_2 (A3). This, together with (8) and (10), allows to virtually decompose the considered system model into a cascade of two subsystems. The input to the first subsystem is u and the output is $\ddot{x}_\gamma$. Signal $\ddot{x}_\gamma$ is also used as an auxiliary input to the second subsystem, which has the dynamics of pure integrators with output x_γ, being also the output of the entire system model. The introduced decomposition with the flatness-based cascade structure simplifies the controller design, as it eliminates the necessity of computation of higher-order time-derivatives of the flat output (which is the case in many conventional approaches), that would be inevitably affected in practice by high-frequency sensor noises in both measured positions x_1 and x_2.

System (10) can also be represented alternatively as:

$$x_\gamma^{(4)} = \frac{\kappa_2}{m_1 m_2} u(t) + d(t), \tag{11}$$

in which bounded term $d(t)$ aggregates the neglected internal disturbance and some unmodeled influence of external perturbation $\psi(t)$:

$$\begin{aligned} d(t) = &- \frac{1}{m_1 m_2}(\kappa_1\kappa_2 + \kappa_1\kappa_3 + \kappa_2\kappa_3) x_\gamma \\ &- \frac{1}{m_1 m_2}(\kappa_1 m_2 + \kappa_2 m_1 + \kappa_2 m_2 + \kappa_3 m_1)\ddot{x}_\gamma + \psi(t), \end{aligned} \tag{12}$$

which makes (11) being represented in a canonical form of disturbed integrators, which the ADRC approach is most conveniently implemented for.

3.2 Expression in Error Domain

Application of standard ADRC scheme (as seen in [4]) to (11) would lead to a robust feedback-linearization of the original system dynamics. An outer-loop state-feedback controller with additional feedforward term would be designed for the resultant integral chain. This step would normally assume full knowledge of the consecutive time-derivatives of the desired trajectory up to the order of the linearized dynamics. However, this cannot be directly used here as it would violate assumptions A1 and A2.

That is why, strategy from [9] is used which assumes reformulation of the considered control objective in error domain. It means that higher-order time-derivatives of the reference trajectory can be treated as part of the disturbance to be on-line estimated and mitigated. First, tracking error $e(t) \triangleq x_\gamma^*(t) - x_\gamma(t)$ is defined recalling that $x_\gamma^*(t) \equiv x_2^*(t)$ and $x_\gamma(t) \equiv x_2(t)$. Its dynamics writes:

$$e^{(4)} \triangleq x_\gamma^{*(4)} - x_\gamma^{(4)} \overset{(11)}{=} -bu(t) + f(t), \tag{13}$$

where $b = \kappa_2/(m_1 m_2)$ is the input gain and $f(t) = bu^*(t) + d^*(t) - d(t) + \varphi(t)$ is the disturbance lumping the effects of both external disturbances and unmodeled dynamics, such as viscous friction, nonlinear terms contained in $\varphi(t)$, internal disturbances $d(t)$, unknown desired control $u^*(t)$, and desired internal disturbance $d^*(t)$. Furthermore, the desired closed-loop error dynamics can become a part of the overall uncertainty. Hence, terms $k_1\dot{e}$, $k_2\ddot{e}$, and $k_3e^{(3)}$ are added to both sides of (13), resutling in:

$$e^{(4)} + k_3e^{(3)} + k_2\ddot{e} + k_1\dot{e} = k_3e^{(3)} + k_2\ddot{e} + k_1\dot{e} - bu(t) + f(t), \tag{14}$$

which ca be further represented as:

$$e^{(4)} + k_3e^{(3)} + k_2\ddot{e} + k_1\dot{e} = -bu(t) + F(t), \tag{15}$$

where $F(t) = k_3e^{(3)}+k_2\ddot{e}+k_1\dot{e}+f(t)$ is now the bounded *total disturbance*, which aggregates all the external and internal dynamics that affect the system, including the effects of feedback derivatives terms $k_1\dot{e}$, $k_2\ddot{e}$, $k_3e^{(3)}$, with $\{k_1, k_2, k_3\} > 0$ being the feedback controller coefficients (design parameters).

Finally, a following control action can be proposed:

$$u(t) = \left(u_0 + \hat{F}(t)\right)/b, \quad b > 0, \tag{16}$$

where $\hat{F}(t)$ is the estimation of $F(t)$ (to be discussed later). By making the prescribed closed-loop dynamics a part of $F(t)$, the outer-loop controller u_0 in (16) can be limited to just a proportional static output-feedback action $u_0 = k_0e$, for $k_0 > 0$. Such design provides simplicity, desired in industrial applications, and at the same time makes (16) satisfying assumptions A1–A3. The resultant closed-loop error dynamics is obtained by substituting (16) in (15), giving:

$$e^{(4)} + k_3e^{(3)} + k_2\ddot{e} + k_1\dot{e} + k_0e = F(t) - \hat{F}(t). \tag{17}$$

If we assume that $\hat{F} = F$, the closed-loop tracking error characteristic polynomial can be matched with a Hurwitz polynomial: $s^4+k_3s^3+k_2s^2+k_1s+k_0 := (s+\omega_c)^4$ and the control gains can be directly calculated using a pole-placement method as: $k_3 = 4\omega_c, k_2 = 6\omega_c^2, k_1 = 4\omega_c^3, k_0 = \omega_c^4$, with an arbitrarily chosen desired closed-loop bandwidth ω_c (for details see [11]).

3.3 Cascaded Observers

We first introduce a change of coordinates $e^{(i-1)} = z_i$, for $i = 1, 2, 3, 4$, making the system (15) represented as:

$$\Psi: \ \dot{z}_4 + k_3z_4 + k_2z_3 + k_1z_2 = -bu(t) + F(t), \tag{18}$$

or alternatively as:

$$\Psi: \begin{cases} \dot{z}_1 = z_2, \quad \dot{z}_2 = z_3, \quad \dot{z}_3 = z_4, \\ \dot{z}_4 = -\,bu(t) - k_1z_2 - k_2z_3 - k_3z_4 + F(t). \end{cases} \tag{19}$$

One of the disadvantages of standard ADRC for (19) is that the total disturbance $F(t)$ would have to be estimated using input and output signals assuming the considered system is accurately described with a canonical model of integrators solely perturbed with a matched, input-additive disturbance term. However, with the high relative order of the plant, the number of dynamic components that do not fit this idealized description increases, which adds to the overall modeling uncertainty. Noise is another factor, which given the high-gain form of ESO, negatively effects the total disturbance reconstruction precision.

To tackle this issue, a specialized cascade structure of ESOs (inspired by [6,7]) is proposed here that fully utilizes the information already available about the controlled system. First, the total disturbance is alternatively interpreted as:

$$F(t) = F_a(t) + F_r(t), \tag{20}$$

which consists of $F_a(t) = \alpha_0 + \alpha_1 t + \alpha_2 t^2 + \alpha_3 t^3 + \ldots + \alpha_m t^m \approx F(t)$, which is a polynomial approximation of total disturbance $F(t)$, and a residual term of this approximation denoted as $F_r(t)$. We can now use the mentioned flatness property of certain underactuated mechanical systems,. Here, it makes the considered plant model Ψ (19) virtually decomposable into two subsystems:

$$\Psi_1 : \begin{cases} \dot{z}_1 = z_2, \\ \dot{z}_2 = z_3, \end{cases} \tag{21}$$

$$\Psi_2 : \begin{cases} \dot{z}_3 = z_4, \\ \dot{z}_4 = -bu(t) - k_1 z_2 - k_2 z_3 - k_3 z_4 + F_a(t) + F_r(t), \end{cases} \tag{22}$$

where $z_1 \equiv e$, $z_3 \equiv \ddot{e}$ are both measurable according to A3. With two subsystems and two lumped disturbances, two ESOs have to be designed next.

ESO for (full) system $\boldsymbol{\Psi}$. The residual total disturbance $F_r(t)$ is assumed to be augmented state of the full system (21)–(22) and defined as $z_5 = F_r(t)$. To reconstruct z_5, it is temporarily assumed that $F_a(t)$ is available, hence (21)–(22) can be expressed in extended state form, with $\boldsymbol{z}_1 = [z_1\ z_2\ z_3\ z_4\ z_5]$, as:

$$\bar{\Psi} : \ \dot{\boldsymbol{z}}_1 = \boldsymbol{A}_1 \boldsymbol{z}_1 + \boldsymbol{B}_1 \boldsymbol{U}_1 + \boldsymbol{E}_1 \dot{F}_r(t), \ z_1 = \boldsymbol{C}_1 \boldsymbol{z}_1, \tag{23}$$

$$\boldsymbol{A}_1 = \begin{bmatrix} 0 & 1 & 0 & 0 & 0 \\ 0 & 0 & 1 & 0 & 0 \\ 0 & 0 & 0 & 1 & 0 \\ 0 & -k_1 & -k_2 & -k_3 & 1 \\ 0 & 0 & 0 & 0 & 0 \end{bmatrix}, \boldsymbol{B}_1 = \begin{bmatrix} 0 & 0 \\ 0 & 0 \\ 0 & 0 \\ -b & 1 \\ 0 & 0 \end{bmatrix}, \boldsymbol{E}_1 = \begin{bmatrix} 0 \\ 0 \\ 0 \\ 0 \\ 1 \end{bmatrix}, \boldsymbol{C}_1 = \begin{bmatrix} 1 \\ 0 \\ 0 \\ 0 \\ 0 \end{bmatrix}^\top,$$

$$\boldsymbol{U}_1 = \begin{bmatrix} u(t) \\ F_a(t) \end{bmatrix}$$

For the above full system, the error-based ESO can be designed as:

$$\hat{\bar{\Psi}} : \ \dot{\hat{\boldsymbol{z}}}_1 = \boldsymbol{A}_1 \hat{\boldsymbol{z}}_1 + \boldsymbol{B}_1 \hat{\boldsymbol{U}}_1 + \boldsymbol{L}_1 \epsilon_1, \ \hat{z}_1 = \boldsymbol{C}_1 \hat{\boldsymbol{z}}_1, \tag{24}$$

where $\hat{\boldsymbol{z}}_1 = [\hat{z}_1\ \hat{z}_2\ \hat{z}_3\ \hat{z}_4\ \hat{z}_5]^\top$ is the estimated state, $\boldsymbol{L}_1 = [l_1\ l_2\ l_3\ l_4\ l_5]^\top$ is the observer gains vector (design parameter), $\epsilon_1 \triangleq z_1 - \hat{z}_1$ is the observer estimation error. Term $\hat{\boldsymbol{U}}_1 = [u(t)\ \hat{F}_a(t)]^\top$ represents the input vector and it includes the estimate of $\hat{F}_a(t)$ which is constantly feeding information about the disturbance to the observer (thus reducing overall modeling uncertainty), allowing selection of small observer gains and consequently reducing the estimation noise. The residual term is estimated as $\hat{z}_5 = \hat{F}_r(t)$. The state matrix $\boldsymbol{A}_1$ contains coefficients k_1-k_3, which are embedded in the observer (as shown in Sect. 3.2). The higher-order target time-derivatives are thus not needed to be known (available) in advance as they are here reconstructed by the observer. This conveniently satisfies assumptions A1 and A2.

To choose the observer gains $\boldsymbol{L}_1$, a bandwidth parametrization-based methodology from [11] can be used here as well. We start by deriving:

$$\boldsymbol{H}_1 = \boldsymbol{A}_1 - \boldsymbol{L}_1\boldsymbol{C}_1 = \begin{bmatrix} -l_1 & 1 & 0 & 0 & 0 \\ -l_2 & 0 & 1 & 0 & 0 \\ -l_3 & 0 & 0 & 1 & 0 \\ -l_4 & -k_1 & -k_2 & -k_3 & 1 \\ -l_5 & 0 & 0 & 0 & 0 \end{bmatrix}.$$

By placing the roots of the characteristic polynomial in the left-hand side of the complex plane, the characteristic estimation error polynomial is determined as:

$$\begin{aligned} P_1(s) = \det(s\boldsymbol{I} - \boldsymbol{H}_1) &= s^5 + (k_3 + l_1)\, s^4 \\ &+ (k_2 + l_2 + k_3 l_1)\, s^3 + (k_1 + l_3 + k_2 l_1 + k_3 l_2)\, s^2 \\ &+ (l_4 + k_1 l_1 + k_2 l_2 + k_3 l_3)\, s + l_5. \end{aligned} \tag{25}$$

Next, the characteristic polynomial $P_1(s)$ is matched with a stable polynomial $(s + \omega_{o1})^5$, resulting in: $l_1 = 5\omega_{o1} - k_3$, $l_2 = 10\omega_{o1}^2 - k_2 - l_1 k_3$, $l_3 = 10\omega_{o1}^3 - k_1 - l_1 k_2 - l_2 k_3$, $l_4 = 5\omega_{o1}^4 - l_1 k_1 - l_2 k_2 - l_3 k_3$, and $l_5 = \omega_{o1}^5$, where ω_{o1} is the user-defined desired bandwidth of the observer $\hat{\bar{\Psi}}$ (24).

ESO for subsystem $\boldsymbol{\Psi}_2$. In this case, the approximated disturbance is assumed to be the augmented state of subsystem Ψ_2 (22) and defined as $z_5 = F_a(t)$. To estimate z_5, it is temporarily assumed that $F_r(t) = 0$, hence (22) can be expressed, with $\boldsymbol{z}_2 = [z_3\ z_4\ z_5]$, as:

$$\bar{\Psi}_2 :\ \dot{\boldsymbol{z}}_2 = \boldsymbol{A}_2\boldsymbol{z}_2 + \boldsymbol{B}_2\boldsymbol{U}_2 + \boldsymbol{E}_2\dot{F}_a(t),\ z_3 = \boldsymbol{C}_2\boldsymbol{z}_2, \tag{26}$$

$$\boldsymbol{E}_2 = \begin{bmatrix} 0 \\ 0 \\ 1 \end{bmatrix}, \boldsymbol{A}_2 = \begin{bmatrix} 0 & 1 & 0 \\ -k_2 & -k_3 & 1 \\ 0 & 0 & 0 \end{bmatrix}, \boldsymbol{B}_2 = \begin{bmatrix} 0 & 0 \\ -b & -k_1 \\ 0 & 0 \end{bmatrix}, \boldsymbol{U}_2 = \begin{bmatrix} u(t) \\ z_2 \end{bmatrix}, \boldsymbol{C}_2 = \begin{bmatrix} 1 \\ 0 \\ 0 \end{bmatrix}^\top,$$

Foe the above subsystem, a reduced-order ESO can be designed as:

$$\hat{\bar{\Psi}}_2 :\ \dot{\tilde{\boldsymbol{z}}}_2 = \boldsymbol{A}_2\tilde{\boldsymbol{z}}_2 + \boldsymbol{B}_2\tilde{\boldsymbol{U}}_2 + \boldsymbol{L}_2\epsilon_2,\ \tilde{z}_3 = \boldsymbol{C}_2\tilde{\boldsymbol{z}}_2, \tag{27}$$

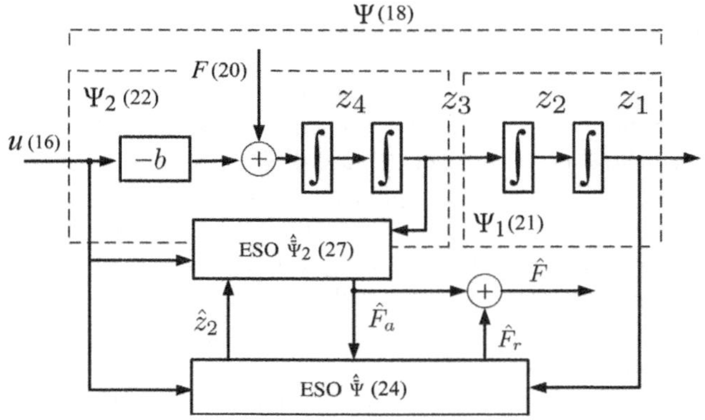

Fig. 2. The proposed ADRC scheme with cascade ESOs structure.

where $\tilde{\boldsymbol{z}}_2 = [\tilde{z}_3\ \tilde{z}_4\ \tilde{z}_5]^\top$ is the estimated state vector, $\boldsymbol{L}_2 = [\bar{l}_1\ \bar{l}_2\ \bar{l}_3]^\top$ is the observer gains vector (design parameter), $\epsilon_2 \triangleq z_3 - \tilde{z}_3$ is the observer estimation error, and $\tilde{\boldsymbol{U}}_2 = [u(t)\ \hat{z}_2]$ is the input vector containing $\hat{z}_2$, which is estimated by the observer $\hat{\hat{\Psi}}$ (24). One can notice that the approximated total disturbance is obtained through $\tilde{z}_5 = \hat{F}_a(t)$.

To select the observer gains, the same pole-placement approach is used and the estimation error state matrix is first calculated as:

$$\boldsymbol{H}_2 = \boldsymbol{A}_2 - \boldsymbol{L}_2\boldsymbol{C}_2 = \begin{bmatrix} -\bar{l}_1 & 1 & 0 \\ -\bar{l}_2 - k_2 & -k_3 & 1 \\ -\bar{l}_3 & 0 & 0 \end{bmatrix}, \tag{28}$$

with its characteristic polynomial being:

$$\begin{aligned} P_2(s) &= \det(s\boldsymbol{I} - \boldsymbol{H}_2) \\ &= s^3 + \left(k_3 + \bar{l}_1\right) s^2 + \left(k_2 + \bar{l}_2 + k_3\bar{l}_1\right) s + \bar{l}_3. \end{aligned}$$

The above is matched with a Hurwitz polynomial $(s + \omega_{o2})^3$, from which the observer gains are chosen as: $\bar{l}_1 = 3\omega_{o2} - k_3$, $\bar{l}_2 = 3\omega_{o2}(1 - k_3) - k_3^2 - k_2$, and $\bar{l}_3 = \omega_{o2}^3$, where ω_{o2} is the user-defined bandwidth of observer $\hat{\hat{\Psi}}_2$ (27).

The introduced cascaded observers $\hat{\hat{\Psi}}$ (24) and $\hat{\hat{\Psi}}_2$ (27) reconstruct the phase variables and the total disturbance at the same time, where the latter is estimated as a combination $\hat{F}(t) = \tilde{z}_5 + \hat{z}_5 = \hat{F}_a(t) + \hat{F}_r(t)$, with $\hat{z}_5 = \hat{F}_r(t)$ and $\tilde{z}_5 = \hat{F}_a(t)$. Such merged estimated total disturbance is being compensated for through (16). Block diagram of the entire control system is shown in Fig. 2.

4 Experimental Validation

Two hardware experiments (Table 1) are conducted to verify the effectiveness of the introduced control scheme with cascade error-based ESOs.

Table 1. Details of performed experimental tests.

Test	Implementation details
E1	A=0.01, T=5, m_1=2.38, m_2=1.53, f_d=0
E2	C1 (SOC without f_d), C2 (SOC with f_d), C3 (COC with f_d)

First experiment (E1) uses a target trajectory $x_2^* = A\sin(2t\pi/T)$ with zero initial conditions $x_1(0)=0$, $x_2(0)=x_\gamma(0)=0$, amplitude $A[m]$, and period $T[s]$. The design parameters are selected as $\omega_{o1}=60\,\text{rad/s}$, $\omega_{o2}=5\omega_{o1}$, and $\omega_c=30\,\text{rad/s}$, from which exact values of observers and controller tuning gains can be straightforwardly calculated using the pole-placement approach introduced in the previous section. For the considered experimental platform, these values can be considered as a *moderate* choice, resulting from a compromise between observer/controller convergence and its sensitivity to high-frequency sensor noise. In general, bandwidth ω_{o2} can be set significantly higher than ω_{o1}, since the former is designed for lower order subsystem, which for the used pole-placement tuning results in lower, however sufficient, observer gains.

Second experiment (E2) quantitatively compares the proposed cascade ESO-based ADRC (denoted for the purpose of the experiment as 'COC') with a standard single ESO-based ADRC (denoted as 'SOC') in three scenarios with (w/) or without (w/o) the influence of extra damper (f_d) connected with spring κ_3. No retuning takes place here after E2.

Figure 3 shows the results of experiment E1. The obtained control performance is shown through trajectory tracking accuracy (a), from which one can see that the flat output $x_\gamma(t) \equiv x_2(t)$ follows the target sinusoidal trajectory $x_\gamma^*(t) \equiv x_2^*(t)$. Consequently, the feedback error (b) is kept within acceptable (arbitrarily chosen) margin $\left(|e| \leq 6 \times 10^{-4} m\right)$ and the produced control signal (c) did not exceed hardware upper and lower limits $\left(|u| \leq 10\text{N}\right)$. The convergences of estimation errors ϵ_1 and ϵ_2 to close vicinity of zero are depicted in (d)-(e) and (f)-(g), respectively. Since the actual value of total disturbance $F(t)$, including its components $F_a(t)$ and $F_r(t)$, are not available in the real plant, their estimates are shown in subplots (f)-(k) in comparison with their 'real' counterparts designed in a dedicated numerical simulation. This, together with the observers error convergence shown above, validates the disturbance estimation part.

Figure 4 shows the outcomes of experiment E2. The nominal performance of SOC (C1) was *a priori* tuned to resemble that of COC from experiment E1 in terms amplitudes of control signal and feedback error. The SOC scheme was experimentally tested (C1-C2) and from the obtained results it can be seen that it did not perform satisfactory when extra damping was introduced (C2). It is especially seen in subfigures (a) and (b), which are the trajectory tracking and tracking error, respectively. Then the nominal version of COC approach (from experiment E1) was used and from the obtained trajectory tracking (c) and tracking error (d), it is clearly seen that it remained relatively immune to

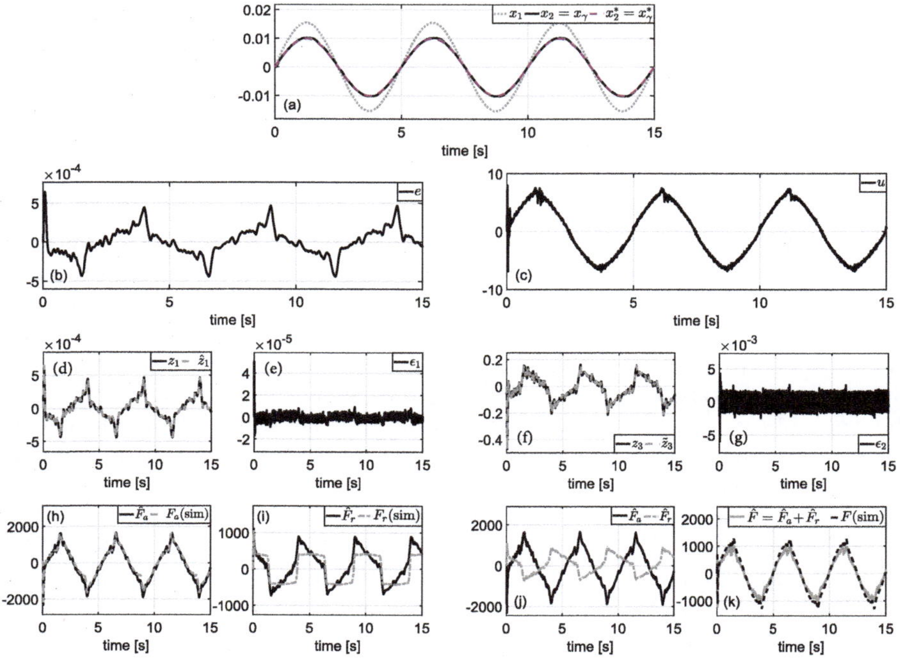

Fig. 3. [E1] The obtained results in terms of: desired trajectory tracking (a), tracking error (b), control signal (c), first observer estimation error (d)–(e), second observer estimation error (f)–(g), estimations of decomposed total disturbance terms and their 'real' values from simulation (h)–(i), estimated total disturbance and its 'real' value from simulation (j)–(k).

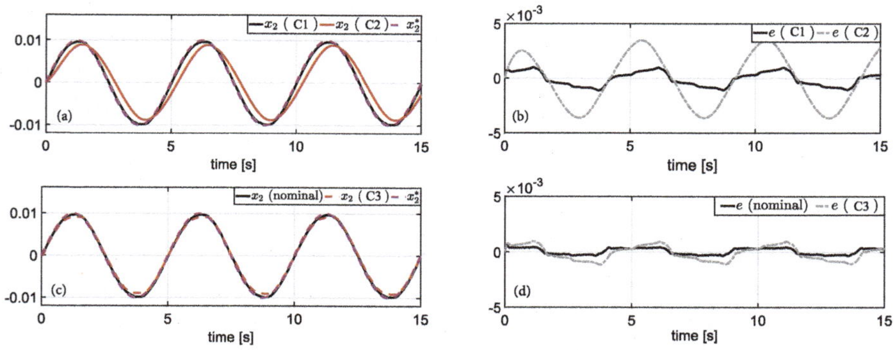

Fig. 4. [E2] Comparison study with resultant error performance indices: nominal (IAE = 20.01), C1 (IAE = 89.91), C2 (IAE = 308.39), C3 (IAE = 42.07).

the applied extra damping (C3). Through direct quantitative comparison of the two control structures (with the assumed tuning), the proposed COC method performed relatively better in terms of tracking precision, which is additionally confirmed by the calculated integral absolute error (IAE) indices.

5 Conclusions

A customized active disturbance rejection control design has been introduced for trajectory tracking in high-order underactuated mechanical systems. It is based on recognizing and fully utilizing the information which are available about the controlled system. By taking advantage of the differential flatness property of the plant model and applying a cascade combination of error-based extended state observers, the resultant data-driven framework was shown to be an alternative to the standard single-observer output-based ADRC structures. The proposed approach was shown to enhance the disturbance-rejection capabilities of the standard ADRC and reduce its sensitivity to measurement noise, thus increasing its practical appeal. Further work will focus on applying the proposed control algorithm to quadrotor UAVs [12].

References

1. Gao, Z.: On the centrality of disturbance rejection in automatic control. ISA Trans. **53**, 850–857 (2014)
2. Sira-Ramirez, H., et al.: Active Disturbance Rejection Control of Dynamic Systems: A Flatness Based Approach. Butterworth-Heinemann, Oxford (2018)
3. Sira-Ramirez, H., Agrawal, S.K.: Differentially Flat Systems. CRC Press, Boca Raton (2004)
4. Han, J.: From PID to active disturbance rejection control. IEEE Trans. Ind. Electron. **56**, 900–906 (2009)
5. Ramírez-Neria, et al.: Linear active disturbance rejection control of underactuated systems: the case of the Furuta pendulum. ISA Trans. **53**, 920–928 (2014)
6. Ramírez-Neria, et al.: On the linear control of underactuated nonlinear systems via tangent flatness and active disturbance rejection control: the case of the ball and beam system. J. Dyn. Syst. Meas. Control **138** (2016)
7. Ramírez-Neria, et al.: On the tracking of fast trajectories of a 3DOF torsional plant: a flatness based ADRC approach (early access). Asian J. Control (2020)
8. Madonski, R., Herman P.: On the usefulness of higher-order disturbance observers in real control scenarios based on perturbation estimation and mitigation. In: International Workshop on Robot Motion and Control, pp. 252–257 (2013)
9. Michalek, M.M.: Robust trajectory following without availability of the reference time-derivatives in the control scheme with active disturbance rejection. In: American Control Conference, pp. 1536–1541 (2016)
10. Zhang H., Shao, S., Gao, Z.: An active disturbance rejection control solution for the two-mass-spring benchmark problem. In: American Control Conference, pp. 1566–1571 (2016)
11. Gao, Z.: Scaling and bandwidth-parameterization based controller tuning. In: American Control Conference, pp. 4989–4996 (2003)
12. Giernacki, W.: Iterative learning method for in-flight auto-tuning of UAV controllers based on basic sensory information. Appl. Sci. **9**(4), 648 (2019)

Error-Based Active Disturbance Rejection Altitude/Attitude Control of a Quadrotor UAV

Momir Stankovic[1](✉), Rafal Madonski[2], Stojadin Manojlovic[1], Taki Eddine Lechekhab[1], and Davorin Mikluc[1]

[1] Military Academy, University of Defense in Belgrade, 11000 Belgrade, Serbia
Momir_Stankovic@yahoo.com

[2] Energy Electricity Research Center, International Energy College, Jinan University, Zhuhai 519070, People's Republic of China

Abstract. The paper focuses on controlling a quadrotor unmanned aerial vehicle (UAV). The highly nonlinear dynamics of the system, together with its under-actuated nature and strong cross-couplings, make the quadrotor control a challenging problem. To solve it, a robust strategy based on a concept of active disturbance rejection control (ADRC) is proposed. Its particular, error-based version is used to minimize sensing requirements and thus allows a more practical realization of the altitude/attitude trajectory following task without the availability of reference time derivatives. Three distinct variations of the error-based ADRC algorithms are derived and numerically tested. The findings of this research form a guide for end-users on how to select appropriate ADRC structure for quadrotor control depending on specific performance requirements and working conditions.

Keywords: ADRC · Quadrotor UAV · Trajectory-following · Altitude/attitude control

1 Introduction

From the control point of view, quadrotors are challenging systems to govern due to their nonlinear and cross-coupled dynamics, model uncertainty as well as under-actuated configuration. This issue has led to many control algorithms being proposed in the literature (see review in [1]). Recently, approaches based on traditional PID and LQR controllers for quadrotor UAV were presented in [2]. An iterative learning method for auto-tuning of UAV controllers was shown in [3]. In [4], a model predictive control (MPC) was designed for obstacle avoidance task. Also, different adaptive controllers for trajectory tracking tasks were introduced in [5, 6]. However, robustness against parametric uncertainties and disturbance rejection performance of traditional controllers are often limited. The various LQR and MPC algorithms are valuable solutions in many cases, but their control performances are mostly dependent on the precision of plant mathematical model. Furthermore, the robust adaptive control approaches, like [5, 6], provide high performance of different external and internal disturbances rejection, but their complex structure makes them difficult for practical hardware realization and implementation.

A. Bartoszewicz et al. (Eds.): PCC 2020, AISC 1196, pp. 1348–1358, 2020.
https://doi.org/10.1007/978-3-030-50936-1_112

Hence, in order to solve the challenging problem of quadrotor control, a robust strategy, based on a concept of active disturbance rejection control (ADRC), is utilized in this paper. The basic idea of the ADRC is to treat unmodeled dynamics, parameter perturbations and external disturbances as a single, aggregated total disturbance [7]. By including such lumped disturbance as an additional state variable, it can be estimated using an extended state observer (ESO) and rejected in real time. In this way, the system can be virtually transformed into an integral-chain form of r-th order, where r is the relative order of the system. In such framework, a relatively simple controller can be used even for a complex, disturbed system (like quadrotor), as long as the total disturbance is being simultaneously reconstructed and its effect on the controlled output is being effectively compensated for.

The high performance of a conventional ADRC algorithm in case of trajectory-tracking was shown in [8], but in this case, derivatives of reference signals had to be provided (preprogrammed) at the controller synthesis stage. However, in case of remote control of quadrotors, these signals are often not available in advance, which limits the application of conventional ADRC structure. Therefore, this research proposes the use of a recently introduced customized version of ADRC expressed in feedback error-based domain [9], which minimizes signal requirements and thus allows a potentially more practical solution to the quadrotor trajectory tracking task. This concept has been generalized and proved theoretically in [10] and applied to solve some challenging robotic control problems [11, 12]. In this paper, three distinct variations of the error-based ADRC algorithms are derived and tested in a simulation case study. The goal of the conducted study is to form guidelines for end-users for selecting specific ADRC design for quadrotor control, depending on the given control objective and working conditions.

The rest of the paper is organized as follows. The mathematical model of the quadrotor is presented in Sect. 2. In the next section, the error-based ADRC algorithm is derived. The multi-criteria simulation is described in Sect. 4. Finally, Sect. 5 concludes the work.

2 Quadrotor Mathematical Model

A typical quadrotor configuration is shown in Fig. 1 with two reference frames: the Earth-fixed frame $E : (O_E; x_E; y_E; z_E)$ and the body-fixed frame $B : (O_B; x_B; y_B; z_B)$. The Earth-fixed frame can be treated as an inertial reference, where position and attitude of the body-fixed frame, relative to the Earth-fixed frame, determine flight position and attitude of thequadrotor. Three translational $\boldsymbol{\xi} = [x \;\; y \;\; z\,]^{\mathrm{T}}$ and three angular coordinates (Euler angles) $\boldsymbol{\kappa} = [\varphi \;\; \theta \;\; \psi\,]^{\mathrm{T}}$ determine the absolute position of the quadrotor in space. Transition matrix from E to B frame is determined by roll angle ϕ, pitch angle θ and yaw angle ψ, and it writes:

$$\mathbf{T}_{BE} = \begin{bmatrix} c\phi c\psi & c\theta s\psi & -s\theta \\ s\phi s\theta c\psi - c\phi s\psi & s\phi s\theta s\psi + c\phi c\psi & s\phi c\theta \\ c\phi s\theta c\psi + s\phi s\psi & c\phi s\theta s\psi - s\phi c\psi & c\phi c\theta \end{bmatrix} \tag{1}$$

with cosine and sine functions abbreviated with ‘c’ and ‘s’, respectively.

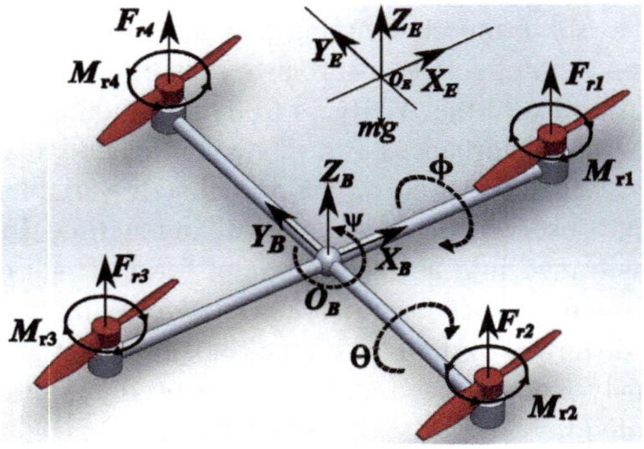

Fig. 1. Quadrotor configuration.

Due to a symmetrical configuration, the body inertia matrix of the quadrotor is in diagonal form $\mathbf{J} = \text{diag}\{ I_x \;\; I_y \;\; I_z \}$. Assuming constant mass m and constant rotational inertias $I_x,\ I_y,\ I_z$, the dynamics of the quadrotor can be represented with two Newton-Euler matrix equations:

$$m\dot{\boldsymbol{\nu}}_{\mathrm{E}} = \mathbf{F}_E \ , \quad \boldsymbol{\nu}_{\mathrm{E}} = \dot{\boldsymbol{\xi}} \tag{2}$$

$$\mathbf{J}\dot{\boldsymbol{\Omega}}_B = \mathbf{M}_B. \tag{3}$$

Equation (2) describes the quadrotor translational dynamics, with the resultant force acting on the quadrotor $\mathbf{F}_E$ given with respect to the E frame. Equation (3) describes the quadrotor rotational dynamics, with the quadrotor absolute angular velocity vector $\boldsymbol{\Omega}_{\mathrm{B}}$ and the resultant torque acting on the quadrotor $\mathbf{M}_B$ given with respect to the B frame. The resultant force $\mathbf{F}_E$ consists of the control force generated by four rotating propellers, the gravitational force and the aerodynamic drag force. Each of the four propellers generate force $F_{ri} = k_n\omega_i^2$, for $i = 1, 2, 3, 4$, which is proportional to the square of the angular velocity of the i-th rotor. As depicted in Fig. 1, the resultant control force $F_r = \sum_1^4 F_{ri}$ enables movement of the quadrotor along the z_B axis. Moving along the y_B axis is achieved with differential force $F_{r4} - F_{r2}$, which causes the change of roll angle ϕ. Analogously, moving along x_B axis is achieved with differential force $F_{r3} - F_{r1}$, which causes the change of pitch angle θ. In contrast to the control force, the gravitational and the aerodynamic drag forces are disturbing factors. The components of the gravitational force in E frame are $F_g = -m\begin{bmatrix} 0 & 0 & g \end{bmatrix}^{\mathrm{T}}$. Due to small speeds of the quadrotor, the aerodynamic drag force can be described with proportional relation to the translational speed of the quadrotor, and its components in B frame are given with $F_d = -\text{diag}\{ k_x \;\; k_y \;\; k_z \}\mathbf{T}_{BE}\boldsymbol{\nu}_E$, where $k_x,\ k_y,\ k_z$ are drag coefficients. In addition, with net torque $M_r = \sum_1^4 M_{ri}$ the quadrotor is balanced about z_B axis, as can be seen in Fig. 1. By changing the rotational speeds of the rotors, each propeller generates variable torque, which is proportional to the thrust force $M_{ri} = k_m F_{ri}$, for $i = 1, 2, 3, 4$, where k_m is the torque coefficient. The net torque causes the rotation of the quadrotor about the z_B axis, hence the yaw angle ψ changes.

From a mechanical point of view, the quadrotor is an underactuated system, with six degrees of freedom, and four control inputs:

$$\begin{bmatrix} U_1 \\ U_2 \\ U_3 \\ U_4 \end{bmatrix} = \begin{bmatrix} F_{r1} + F_{r2} + F_{r3} + F_{r4} \\ F_{r4} - F_{r2} \\ F_{r3} - F_{r1} \\ M_{r2} + M_{r4} - M_{r1} - M_{r3} \end{bmatrix}. \tag{4}$$

With the above modeling, it can be concluded that the altitude of the quadrotor is controlled with the first control input U_1, while the attitude of the quadrotor is controlled with the control inputs U_2, U_3, U_4, i.e. roll, pitch and yaw angles, respectively. Additional approximation can be introduced, if the perturbations from the hover condition are small. In that case, the absolute angular velocity vector of the quadrotor in B frame can be represented as:

$$\boldsymbol{\Omega}_B = \begin{bmatrix} p & q & r \end{bmatrix}^{\mathrm{T}} = \begin{bmatrix} \dot{\phi} & \dot{\theta} & \dot{\psi} \end{bmatrix}^{\mathrm{T}}. \tag{5}$$

Now, based on (2) and (3), the complete model of the quadrotor can be derived as:

$$\begin{aligned} \ddot{x} &= [\cos\phi\sin\psi + \cos\phi\sin\theta\cos\psi]U_1/m - k_x\dot{x}/m \\ \ddot{y} &= [-\sin\phi\cos\psi + \cos\phi\sin\theta\sin\psi]\, U_1/m - k_y\dot{y}/m \\ \ddot{z} &= [\cos\phi\cos\theta)]U_1/m - k_z\dot{z}/m - g \\ \ddot{\phi} &= [I_y - I_z]\dot{\theta}\dot{\psi}/I_x + U_2 d/I_x \\ \ddot{\theta} &= [I_z - I_x]\dot{\phi}\dot{\psi}/I_y + U_3 d/I_y \\ \ddot{\psi} &= [I_x - I_y]\dot{\phi}\dot{\theta}/I_z + U_4/I_z \end{aligned} \tag{6}$$

3 Error-Based ADRC Design and Tuning

3.1 Input-Output Representation

By including the external disturbances w_z, w_ϕ, w_θ, w_ψ in each control channel, the altitude/attitude dynamics of the quadrotor can be represented as:

$$\begin{aligned} \ddot{z} &= -K_z\dot{z}/m - g + w_z + b_z U_1 \\ \ddot{\phi} &= \dot{\theta}\dot{\psi}[I_y - I_z]/I_x + w_\phi + b_\phi U_2 \\ \ddot{\theta} &= \dot{\phi}\dot{\psi}[I_z - I_x]/I_y + w_\theta + b_\theta U_3 \\ \ddot{\psi} &= \dot{\phi}\dot{\theta}[I_x - I_y]/I_z + w_\psi + b_\psi U_4 \end{aligned} \tag{7}$$

where $b_z = [\cos(\phi)\cos(\theta)]/m$, $b_\phi = d/I_x$, $b_\theta = d/I_y$, $b_\psi = 1/I_z$ are the input gains in each channel. Based on the concept of ADRC, (7) can be reformulated as:

$$\begin{aligned} \ddot{z} &= f_z(\dot{z}, w_z, \Delta b_z U_1) + \hat{b}_z U_1 \\ \ddot{\varphi} &= f_\varphi(\dot{\theta}, \dot{\psi}, w_\varphi, \Delta b_\varphi U_2) + \hat{b}_\varphi U_2 \\ \ddot{\theta} &= f_\theta(\dot{\varphi}, \dot{\psi}, w_\theta, \Delta b_\theta U_3) + \hat{b}_\theta U_3 \\ \ddot{\psi} &= f_\psi(\dot{\varphi}, \dot{\theta}, w_\psi, \Delta b_\psi U_4) + \hat{b}_\psi U_4 \end{aligned} \tag{8}$$

with $f_z(\dot{z}, w_z, \Delta b_z U_1)$, $f_\varphi(\dot{\theta}, \dot{\psi}, w_\varphi, \Delta b_\varphi U_2)$, $f_\theta(\dot{\varphi}, \dot{\psi}, w_\theta, \Delta b_\theta U_3)$ and $f_\psi(\dot{\varphi}, \dot{\theta}, w_\psi, \Delta b_\psi U_4)$ representing the total disturbances in each channel. Next, the error-based ADRC design for the roll channel will be described, noting that similar procedure can be utilized for other channels as well.

3.2 Control Task Reformulation

In the conventional ADRC design, the total disturbance is rejected through the inner loop using its estimate provided by the ESO. The resultant dynamics is governed by a control law in the outer loop, which is responsible for achieving the desired trajectory tracking performance. This task can be effectively realized utilizing the reference time-derivatives up to the relative order of the system (for details see [13]). The reference time-derivatives can be calculated if the reference is known in advance, or by using a tracking differentiator [7]. However, if the reference is not known in advance, the effect of the unknown time-derivatives can be incorporated into the total disturbance and estimated by the ESO, by using a special, error-based form of the ADRC, derived next for the roll control channel.

If one defines the roll angle trajectory-following error as $e_\phi = \phi_r - \phi$, the roll channel dynamics, from (8) can be given in an alternative error-based form as:

$$\ddot{e}_\phi = \ddot{\phi}_r - \ddot{\phi} = \ddot{\phi}_r - f_\phi - \hat{b}_\phi U_2 = \tilde{f}_\phi - \hat{b}_\phi U_2 \tag{9}$$

where $\tilde{f}_\phi$ incorporates the unknown time derivative $\ddot{\phi}_r$. Extension of (9) with $k_1 \dot{e}_\phi$ $(k_1 > 0)$ on both sides leads to:

$$\ddot{e}_\phi + k_1 \dot{e}_\phi = k_1 \dot{e}_\phi + \tilde{f}_\phi - \hat{b}_\phi U_2 = F_\phi - \hat{b}_\phi U_2, \tag{10}$$

with F_ϕ defined as the total disturbance in the quadrotor roll channel. The main task of the ESO now is fast and accurate estimation of the total disturbance $\hat{F}_\phi$. In an idealistic case of "perfect estimation", i.e. $\hat{F}_\phi = F_\phi$, the control input:

$$U_2 = (\hat{F}_\phi + k_0 e_\phi)/\hat{b}_\phi, \quad k_0 > 0 \tag{11}$$

results in the trajectory-following error dynamics:

$$\ddot{e}_\phi + k_1 \dot{e}_\phi + k_0 e_\phi = 0. \tag{12}$$

The closed-loop system performance, in the sense of trajectory-following error regulation in the presence of total disturbance F_ϕ, can be shaped with design parameters k_0 and k_1, which can be obtained using pole placement method [14], by comparing:

$$(s + \omega_c)^2 = s^2 + k_1 s + k_0, \tag{13}$$

where ω_c represents a desired system closed-loop bandwidth (design parameter).

3.3 Observer Design and Its Variations

For the roll angle trajectory-following error dynamics (10), the state coordinates can be selected as $\mathbf{x}_{ESO} = \left[e_\varphi \ \dot{e}_\varphi \ F_\varphi \right]^{\mathrm{T}}$, with total disturbance F_ϕ being the additional state variable. In that case, the constant total disturbance F_ϕ can be directly estimated and then compensated. By analyzing the structure of the total disturbance, one can notice that in case of the considered quadrotor its dynamics is more complex. In order to increase the overall efficiency of the control structure, the so-called generalized ESOs (GESOs) are suggested for more complex total disturbance estimation and rejection. Extending the state vector with the total disturbance first ($\dot{F}_\phi$) and second ($\ddot{F}_\phi$) time-derivatives as extra state variables, the coordinates for the roll angle trajectory-following error dynamics (10) can be selected in an alternative fashion as: $\mathbf{x}_{GESO_1} = \left[e_\varphi \ \dot{e}_\varphi \ F_\varphi \ \dot{F}_\varphi \right]^{\mathrm{T}}$ or $\mathbf{x}_{GESO_2} = \left[e_\varphi \ \dot{e}_\varphi \ F_\varphi \ \dot{F}_\varphi \ \ddot{F}_\varphi \right]^{\mathrm{T}}$. Now, the roll angle trajectory-following error dynamics, for all three defined sets of coordinates, can be represented in the general state space form as:

$$\begin{aligned} \dot{\mathbf{x}} &= \mathbf{A}\mathbf{x} + \mathbf{b}U_2 + \mathbf{h}F_\phi^{(h)} \\ e_\phi &= \mathbf{c}^{\mathrm{T}}\mathbf{x} \end{aligned}, \tag{14}$$

with $h = 1, 2, 3$ denoting the h-th derivative of F_ϕ. The elements of the $(2+h)$-th order square matrix $\mathbf{A}$ are given as:

$$a_{i,j} = \begin{cases} -k_1\,, & \text{for } i = j = 2 \\ 1\,, & \text{for } i = j-1 \\ 0\,, & \text{otherwise} \end{cases} \tag{15}$$

while the other matrices are:

$$\mathbf{b} = \left[0 \ -\hat{b}_\phi \ 0_{1\times h} \right]^{\mathrm{T}}, \quad \mathbf{c} = \left[1 \ 0_{1\times(2+h-1)} \right]^{\mathrm{T}}, \quad \mathbf{h} = \left[0_{1\times(2+h-1)} \ 1 \right]^{\mathrm{T}}. \tag{16}$$

For the system (14), following error-based ESOs of $(2 + h)$-th order can be proposed:

$$\begin{aligned} \dot{\mathbf{z}} &= \mathbf{A}\mathbf{z} + \mathbf{B}U_2 + \mathbf{l}(e_\phi - \hat{e}_\phi) \\ \hat{e}_\phi &= \mathbf{c}^{\mathrm{T}}\mathbf{z} \end{aligned} \tag{17}$$

with different possible sets of state variables estimates:

$$\begin{aligned} ESO &: \ \mathbf{z} = \left[\hat{e}_\phi \ \hat{\dot{e}}_\phi \ \hat{F}_\phi \right]^{\mathrm{T}} \\ GESO_1 &: \ \mathbf{z} = \left[\hat{e}_\phi \ \hat{\dot{e}}_\phi \ \hat{F}_\phi \ \hat{\dot{F}}_\phi \right]^{\mathrm{T}} \\ GESO_2 &: \ \mathbf{z} = \left[\hat{e}_\phi \ \hat{\dot{e}}_\phi \ \hat{F}_\phi \ \hat{\dot{F}}_\phi \ \hat{\ddot{F}}_\phi \right]^{\mathrm{T}} \end{aligned}. \tag{18}$$

Each ESO has its potential advantages and disadvantages (to be analyzed later).

Based on the following estimation error state matrix:

$$\mathbf{H} = \mathbf{A} - \mathbf{lc} = \left\{ \underbrace{\begin{bmatrix} -l_1 & 1 & 0 \\ -l_2 & -k_1 & 1 \\ -l_3 & 0 & 0 \end{bmatrix}}_{ESO}, \underbrace{\begin{bmatrix} -l_1 & 1 & 0 & 0 \\ -l_2 & -k_1 & 1 & 0 \\ -l_3 & 0 & 0 & 1 \\ -l_4 & 0 & 0 & 0 \end{bmatrix}}_{GESO_1}, \underbrace{\begin{bmatrix} -l_1 & 1 & 0 & 0 & 0 \\ -l_2 & -k_1 & 1 & 0 & 0 \\ -l_3 & 0 & 0 & 1 & 0 \\ -l_4 & 0 & 0 & 0 & 1 \\ -l_5 & 0 & 0 & 0 & 0 \end{bmatrix}}_{GESO_2} \right. \tag{19}$$

observer gains can be calculated using the pole placement [14], by comparing:

$$(s + \omega_0)^{2+h} = \det(s\mathbf{I} - \mathbf{H}), \tag{20}$$

with $\omega_o = \omega_c/\varepsilon$, for $\varepsilon \in (0, 1)$ being the desired observer bandwidth (to be designed).

Following the above derivation of the roll angle error-based ADRC, the tracking controllers for the other channels can be designed analogously, noting that each channel requires its own set of design parameters, namely estimations of the input gains as well as the desired control and observer bandwidths.

4 Simulation Results

The comparison study is realized through Matlab/Simulink simulations, using quadrotor dynamic model (6), with values of parameters gathered in Table 1. The controllers coefficients are tuned as described in Sect. 3, choosing $\omega_c = 2$ rad/s for altitude controller and $\omega_c = 20$ rad/s for attitude controllers. The observer bandwidths in all structures are fixed at $\omega_0 = 10\omega_c$. The altitude and attitude tracking accuracy and disturbance rejection performance are the selected control quality criteria.

Table 1. Quadrotor model parameters.

Symbol	Definition	Value [unit]
k_n	Constant that relates angular velocity and thrust	5.7 10^{-8} [N/(rpm)^2]
k_m	Constant that relates thrust and torque	1.6 10^{-8} [m]
m	Quadrotor mass	0.5 [kg]
I_x	Rotational inertia of x-axis	0.0196 [kgm^2]
I_y	Rotational inertia of y-axis	0.0196 [kgm^2]
I_z	Rotational inertia of z-axis	0.0264 [kgm^2]
g	Gravitational acceleration	9.81 [m/s^2]
d	Quadrotor level arm	0.25 [m]
k_x, k_y, k_z	Aerodynamic drag coefficients	0.1 [N/(m/s)]

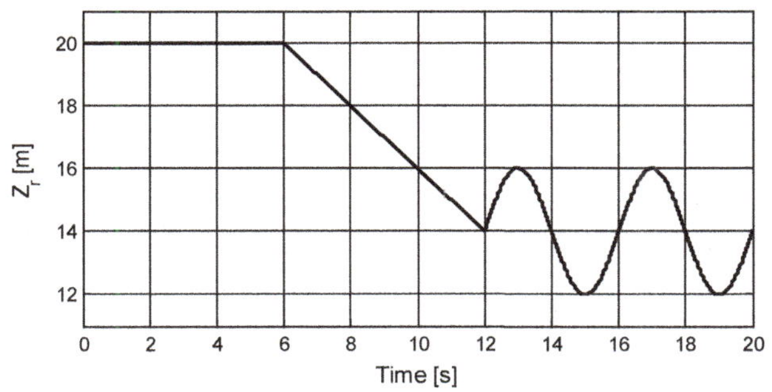

Fig. 2. Altitudechannel reference trajectory.

4.1 Altitude Control Performance

The control task is to follow a complex altitude reference trajectory, shown in Fig. 2.

In this case, full knowledge of the model parameter b_z is assumed in all control schemes with ESO, GESO1 and GESO2, i.e. $\hat{b}_z = b_z$. The obtained altitude tracking errors, for all control systems, are presented in Fig. 3.

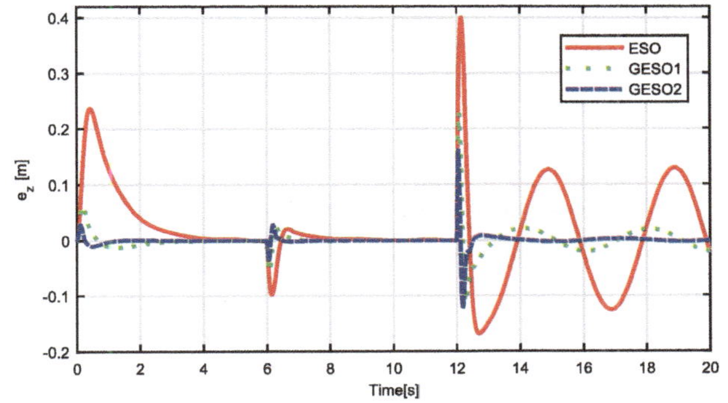

Fig. 3. Altitude tracking errors for different observer structures.

One can see that at $0 \leq t < 12$ s, that is during the constant-and ramp-type references, all controllers achieve minimal steady state error, and in the transients, both GESO-based controllers have better performance than the ESO-based controller. Furthermore, in the case of sinusoidal reference signal ($t \geq 12$ s), all the control structures generate non-zero tracking error, but it is evident that, as the order of the observer is increased, improvement in reference trajectory tracking is achieved as well. In order to analyze the control systems robustness, stability margins are calculated, based on Bode diagrams of the systems loop transfer functions [8]. The results, being positive gain margin (GM^+), negative gain margin (GM^-) and phase margin (PM), are gathered in Table 2. As expected, one can see that by increasing the observer order, the gain and phase margins decrease.

Table 2. Robustness indices for different observer-based control structures.

Observer structure	GM^- [dB]	GM^+ [dB]	PM [deg]
ESO	−22.2	18.3	52.8
GESO1	−16.1	13.3	35.1
GESO2	−8.5	11.1	26.1

4.2 Attitude Control Performance

This case deals with attitude tracking in the presence of external disturbances, internal uncertainties, and measurement noises, all introduced in the attitude channels. The reference signals are presented collectively in Fig. 4. The external disturbance is defined as a step function $w_\theta = 1$ and introduced in pitch channel starting at $t \geq 1$ s. The Gaussian zero mean white noise, with sampling time $T_s = 10^{-3}$ s and variance 10^{-7}, is used as a model of output measurement noise. Additionally, system parametric uncertainty is assumed by choosing $\hat{b}_\phi = 0.75 b_\phi$. The obtained pitch channel tracking results are depicted in Fig. 5, with corresponding control signals shown in Fig. 6.

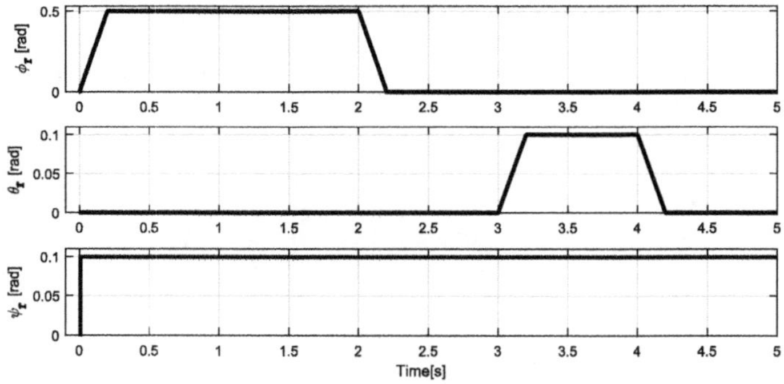

Fig. 4. The reference attitude trajectories.

It can be noticed, from Fig. 5, that all controllers provide minimal steady state tracking errors and rejections of the cross-coupled disturbances (see zoomed part in Fig. 5), latter which are caused by references changing in the other two control channels (roll and yaw). Nevertheless, the GESO2 enables better tracking during the transient stage, than GESO1 and ESO schemes. One can see, from Fig. 6, that the measurement noise affects the control signals visibly. The control algorithm with GESO2is most sensitive to noise in this case, which can potentially limit its practical deployment in highly noise-corrupted industrial environments. On the other hand, the ESO-based structure is the least sensitive to high-frequency measurement noise. This is due tothe fact that GESOs has larger gains than ESO for the same value of selected observer bandwidth.

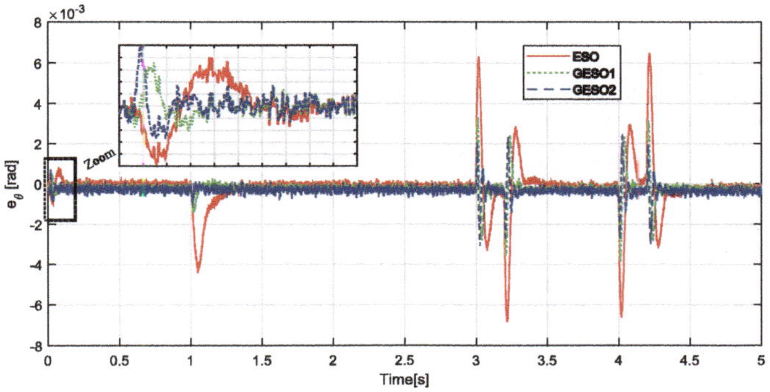

Fig. 5. Pitch channel tracking errors.

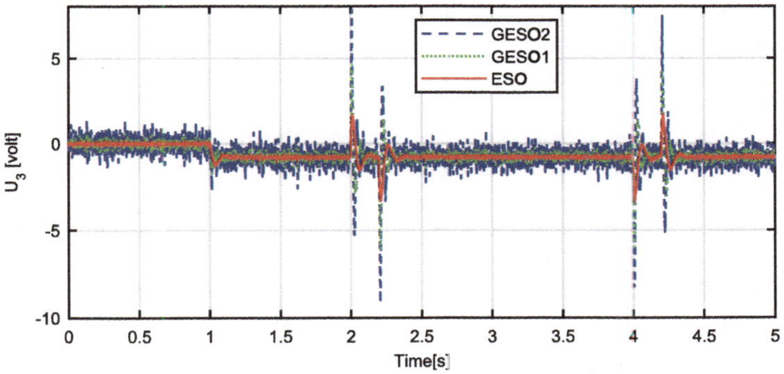

Fig. 6. Pitch channel control signals.

5 Conclusion

In this work, the quadrotor altitude/attitude trajectory-following task has been realized using an error-based form of ADRC, in which the assumed, aggregated disturbance, including effects of unknown reference time-derivatives, unmodeled dynamics, parameter perturbations, and external disturbances, has been simultaneously estimated by a dedicated observer and compensated on-the-fly. Three distinct observers have been considered in this work. Through numerical simulations, it has been shown that the choice of the observer order represents a trade-off between desired trajectory tracking and disturbance rejection performance, on one side, and robustness indices and measurement noise sensitivity, on the other. The observer type in ADRC should thus be a conscious decision taking into account unique characteristics of the given control problem, amount and type of uncertainties as well as environmental conditions, like the severity of high-frequency measurement noise.

References

1. Emran, J., Homayoun, N.: A review of quadrotor: an underactuated mechanical system. Ann. Rev. Control **46**, 165–180 (2018)
2. Chen, X., Zhang, G., Lu, C., Cheng, J.: Quadrotor aircraft attitude control algorithm based on improved UKF. In: IOP Conference Series: Earth and Environmental Science, vol. 233, no. 4, p. 042037 (2019)
3. Giernacki, W.: Iterative learning method for in-flight auto-tuning of UAV controllers based on basic sensory information. Appl. Sci. **9**(4), 648 (2019)
4. Greatwood, C., Richards, A.G.: Reinforcement learning and model predictive control for robust embedded quadrotor guidance and control. Auton. Robots **43**(7), 1681–1693 (2019). https://doi.org/10.1007/s10514-019-09829-4
5. Santos, M.C.P., Rosales, C.D., Sarapura, J.A., Sarcinelli-Filho, M., Carelli, R.: An adaptive dynamic controller for quadrotor to perform trajectory tracking tasks. J. Intell. Rob. Syst. **93**(1–2), 5–16 (2019)
6. Pérez-Alcocer, R., Moreno-Valenzuela, J.: Adaptive control for quadrotor trajectory tracking with accurate parametrization. IEEE Access **7**, 53236–53247 (2019)
7. Han, J.: From PID to active disturbance rejection control. IEEE Trans. Industr. Electron. **56**(3), 900–906 (2009)
8. Stanković, M.R., Manojlović, S.M., Simić, S.M., Mitrović, S.T., Naumović, M.B.: FPGA system-level based design of multi-axis ADRC controller. Mechatronics **40**, 146–155 (2016)
9. Michałek, M.: Robust trajectory following without availability of the reference time-derivatives in the control scheme with active disturbance rejection. In: American Control Conference, pp. 1536–1541 (2016)
10. Madonski, R., Shao, S., Zhang, H., Gao, Z., Yang, J., Li, S.: General error-based active disturbance rejection control for swift industrial implementations. Control Eng. Pract. **84**, 218–229 (2019)
11. Michałek, M., Łakomy, K., Adamski, W.: Robust output-feedback cascaded tracking controller for spatial motion of anisotropically-actuated vehicles. Aerosp. Sci. Technol. **92**, 915–929 (2019)
12. Madonski, R., et al.: On vibration suppression and trajectory tracking in largely uncertain torsional system: an error-based ADRC approach. Mech. Syst. Signal Process. **134**, 106300 (2019)
13. Xue, W., Huang, Y.: Performance analysis of active disturbance rejection tracking control for a class of uncertain LTI systems. ISA Trans. **58**, 133–154 (2015)
14. Gao, Z.: Scaling and bandwidth-parameterization based controller tuning. In: American Control Conference, vol. 6, pp. 4989–4996 (2006)

The Novel Approach to the Tuning of the Reduced-Order Active Disturbance Rejection Controller for Second-Order Processes

Paweł Nowak(✉), Patryk Grelewicz, and Jacek Czeczot

Silesian University of Technology, Akademicka 2A, 44-100 Gliwice, Poland
{pawel.nowak,patryk.grelewicz,jacek.czeczot}@polsl.pl

Abstract. In this paper, the novel approach to the tuning of the Reduced-order Active Disturbance Rejection Controller (RADRC) for the second-order processes is presented. In particular, this work shows that the independent use of five RADRC tuning parameters potentially provides the improvement in the control performance without any deterioration of the quality of the manipulated signal. The concept is based on the zero-pole cancellation technique assuming the second-order model of the process. Then, after theoretical analysis, the conclusions of the practical significance are drawn, and the tuning method is suggested. The theoretical considerations are confirmed by the validation based on the nonlinear laboratory pneumatic system. The validation stage consists of both simulation and experimental tests. The results show an improvement in robustness and a control quality compared to the conventional ADRC controller tuned using by the conventional method and to the conventional PID controller.

Keywords: Reduced-order Active Disturbance Rejection Controller (RADRC) · Second-order process · Tuning rules · Practical verification

1 Introduction

The Active Disturbance Rejection Control (ADRC) is a very promising advanced control algorithm, which seems to be an interesting alternative for the most popular PID controller [1, 2]. The concept of ADRC is unique, and it is based on the assumption that the internal dynamics and external disturbances are treated as an additional state variable called generalized disturbance. This additional state variable is estimated online by application of Extended State Observer (ESO). Then it is used to reduce the complex, nonlinear, time-varying and uncertain process to the simple cascade-integral form, which can be easily controlled by the conventional PD controller.

The ADRC algorithm has many different versions. From the process automation point of view, one of the most interesting is the so-called Reduced-order ADRC (RADRC), which is based on the Reduced-order ESO (RESO) [3]. The advantage of this controller is a simpler implementation and the reduction of the phase lag comparing to the conventional ADRC approach [4]. The bottleneck of this controller is a small number of reliable tuning methods.

A. Bartoszewicz et al. (Eds.): PCC 2020, AISC 1196, pp. 1359–1370, 2020.
https://doi.org/10.1007/978-3-030-50936-1_113

In general, the RADRC requires tuning of the RESO and of the PD^n controller where n denotes the order of RADRC. The number of the tuning parameters of the RADRC depends on its order n. For the example of the second-order RADRC (RADRC2) controller considered in this work, there are four independent parameters and additionally, the so-called scaling factor b_0. Such a large number of parameters significantly complicates the tuning of the RADRC2. To overcome these difficulties, Gao proposed the simplified but also very effective method of parameterization [5]. This method is very popular and used in many works. However, the important question arises whether the RADRC controller could be easily tuned without using the parameterization method proposed by Gao and whether this different tuning can provide any improvement in the control performance.

The results shown in this work partially answer this question. This article presents the theoretical analysis of the performance of the RADRC2 for the second-order processes. It starts with a general description of the RADRC controller. Then it is shown that one can locate the zeroes of this controller at the desired locations. This allows to reduce the number of the tuning parameters of the controller and consequently to simplify its tuning. Theoretical analysis is supported by simulation results and experimental tests.

2 Short Introduction to the Reduced Second-Order ADRC (RADRC2)

In this section, the RADRC2 algorithm derived based on the linear second-order model is shortly described. Let us consider the general form of the second-order stable linear process model as:

$$\ddot{y} = -\frac{b}{a}\dot{y} - \frac{1}{a}y + \frac{k}{a}u + \frac{k}{a}d, \tag{1}$$

where y is the controller output, u is the control signal, d is the load disturbance and $a > 0$, $b > 0$, $k > 0$ are the model parameters. Equation (1) can be rearranged to the form:

$$\ddot{y} = f + b_0 u, \tag{2}$$

where $f = -\frac{b}{a}\dot{y} - \frac{1}{a}y + \left(\frac{k}{a} - b_0\right)u$. In ADRC methodology, the function f is called the total disturbance, and according to the definition proposed in [1], it contains information about internal states and external disturbances as well. Parameter b_0 is a rough approximation of high-frequency gain of the plant. $\left(\frac{k}{a} - b_0\right)u$ is an error of approximation and therefore is added to the function f.

The value of function f can be estimated on-line based on the RESO. The application of RESO requires to represent the process model in the extended space state form, in which the function f is represented by the additional state variable. Consequently, the state vector has the form of $x = [x_1, x_2, x_3]^T = [y, \dot{y}, f]^T$.

It is assumed that the output of the process is measurable, and we only need to estimate x_2 and x_3. Let us define a vector $z = [z_2, z_3]^T$, which contains estimates of the

x_2 and x_3 variables respectively. Therefore, the following form of the state observer can be proposed:

$$\begin{aligned} \dot{\tilde{z}}_2 &= -l_1\tilde{z}_2 + \tilde{z}_3 + (l_2 - l_1^2)y + b_0 u \\ \dot{\tilde{z}}_3 &= -l_2\tilde{z}_2 - l_1 l_2 y \\ z_2 &= \tilde{z}_2 + l_1 y \\ z_3 &= \tilde{z}_3 + l_2 y \end{aligned}, \qquad (3)$$

where, $\tilde{z}_2$ and $\tilde{z}_3$ are the auxiliary variables used to avoid the numerical integration of y, l_1 and l_2 are the observer gains (tuning parameters). For the well-tuned observer, the state variable z_3 accurately tracks the function f, and it can be used to formulate the control law:

$$u = (u_0 - z_3)/b_0. \qquad (4)$$

Readers should note that under the assumption that $f \approx z_3$, the original process modelled by (1) is reduced to the simple double integrator form:

$$\ddot{y} = u_0, \qquad (5)$$

Furthermore, the control signal u_0 can be computed by the 2DOF PD controller:

$$u_0 = k_p(SP - y) - k_d z_2. \qquad (6)$$

Remark 1

The condition $f \approx z_3$ requires a convergence of the ESO, which is fulfilled under certain assumptions related to function f. In particular, it has to be continuously differentiable, and its derivative has to be bounded [6]. We assumed that signal d is constant, so these assumptions are fulfilled in our case.

Remark 2

In the majority of cases, the parameters l_1, l_2 and k_p, k_d are selected based on the so-called observer bandwidth ω_o and the controller bandwidth ω_c respectively, according to the tuning rules proposed in [5]:

$$l_1 = 2\omega_o;\ l_2 = \omega_o^2;\ k_p = \omega_c^2;\ k_d = 2\omega_c. \qquad (7)$$

In this paper, the formulas (7) are not used, and the original tuning is preserved. Thus, there are five independent tuning parameters: l_1, l_2, k_p, k_d, and b_0.

Assuming zero initial conditions and applying the Laplace transform to (1), (3), (4) and (6), the control system with the RADRC2 can be presented in the form of a two-degree of freedom block diagram shown in Fig. 1, where:

$$\begin{aligned} F_N &= \tfrac{1}{l_2}s^2 + \tfrac{l_1}{l_2}s + 1;\ F_D = C_N = \tfrac{k_d l_1 + l_2 + k_p}{k_p l_2}s^2 + \tfrac{k_d l_2 + k_p l_1}{k_p l_2}s + 1; \\ C_D &= \tfrac{b_0}{k_p l_2}s^2 + \tfrac{b_0(l_1 + k_d)}{k_p l_2}s;\ P_D = as^2 + bs + 1 \end{aligned} \qquad (8)$$

Readers should note that these transfer functions neither represent observer nor controller directly but result from mathematical calculations. Such form does not show the general concept of this controller but is useful for further derivations.

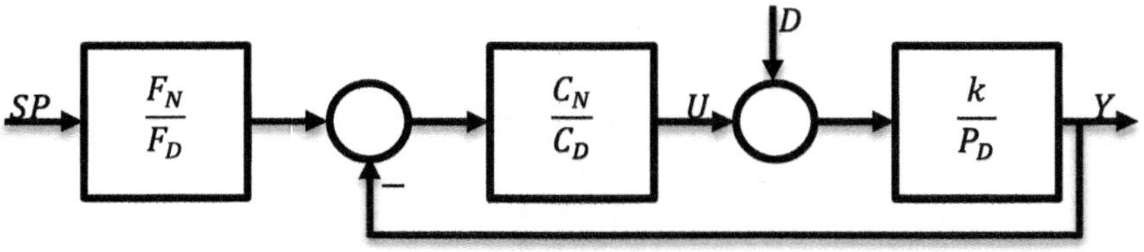

Fig. 1. Block diagram of the control system with RADRC2.

3 Main Results

The RADRC2 controller presented in the previous section has five independent tuning parameters. Thus, any tuning by the trial and error method will be difficult and time-consuming. This difficulty can be overcome by the application of the zero-pole cancellation technique, which results in the reduction in the number of tuning parameters. For this purpose, the characteristic equation can be written for the closed-loop system shown in Fig. 1:

$$\chi = kC_N + C_D P_D = 0. \tag{9}$$

The fourth-order of the characteristic Eq. (9) makes the analytical determination of its roots very complex. At the same time, it is worth noting that the C_N polynomial has the same order as the P_D polynomial, so the parameters of C_N can be adjusted to ensure that $C_N = P_D$. The coefficients of the polynomial C_N depend on four parameters, but only two of them can be adjusted independently. Arbitrarily l_1 and l_2 were chosen as the independent parameters, and it leads to the following relation:

$$l_1 = \frac{bk_p^2 - k_d k_p}{ak_p^2 + k_d^2 - k_p - k_d k_p b};\quad l_2 = \frac{k_p^2}{ak_p^2 + k_d^2 - k_p - k_d k_p b} \tag{10}$$

Remark 3
When RADRC2 tuning is based on (7), the relationship $C_N = P_D$ is not always possible because C_N always has a pair of complex roots. In other words, zero-pool cancellation is possible only in the cases of oscillatory processes. An additional degree of freedom to the form of (10) allows for higher flexibility in the tuning of RADRC2.

After assuming that the parameters $l_1 > 0$ and $l_2 > 0$ must be positive, the following limits can be derived:

$$k_d < bk_p, \tag{11}$$

$$k_d^2 - k_d k_p b + ak_p^2 - k_p > 0. \tag{12}$$

The limitation (12) has a more complex form than (11), but it can be shown that if (12) has a pair of real roots, both of them are positive. Additionally, the smaller one is an active constraint. In a case when (12) has a couple of complex roots, this limitation is always fulfilled, so the only active constraint is (11).

Using (10), Eq. (9) can be written in the form:

$$\chi = P_D(k + C_D) = \frac{b_0}{k_p l_2}\left(as^2 + bs + 1\right)\left(s - \frac{-(k_d + l_1) + \sqrt{(k_d + l_1)^2 - \frac{4l_2k_pk}{b_0}}}{2}\right)$$
$$\left(s - \frac{-(k_d + l_1) - \sqrt{(k_d + l_1)^2 - \frac{4l_2k_pk}{b_0}}}{2}\right). \quad (13)$$

Because, $k_d > 0$ and $l_l > 0$, all roots of the characteristic equation have negative real parts and the system is always stable. Additionally, one can use the parameter b_O to locate all roots of $C_D + 1$ binomial at a single point:

$$b_0 = \frac{4l_2k_pk}{(k_d + l_1)^2} \quad (14)$$

Remark 4
There are two interpretations of the meaning of b_O. In some works, e.g. [2] b_O is considered to be only model-based parameter in the ADRC design, and it has a physical interpretation. On the other hand, some works, e.g. [7] show the strict impact of parameter b_O on the quality of control. For this reason, it should rather be treated as an additional tuning parameter. This interpretation of parameter b_O is also used in this work.

Finally, the characteristic equation can be rewritten in the following form:

$$\chi = k\left(as^2 + bs + 1\right)\left(\frac{2}{k_d + l_1}s + 1\right)^2. \quad (15)$$

This equation shows that increasing the values of k_d and l_l results in the faster performance of the control system. However, on the other hand, the response is limited by the process poles (1).

The suggested location of the poles of the closed-loop system allows for the reduction of the initial five RADRC2 tuning parameters to only two of them, namely to k_p and k_d. Therefore, the level of tuning complexity should be similar to the tuning of the conventional PI controller. The guidance on how to select the k_p and k_d parameters are presented below.

Based on Fig. 1, the following transfer functions can be determined:

$$\frac{Y}{SP} = \frac{kF_N}{P_D(C_D + k)} = \frac{\frac{1}{l_2}s^2 + \frac{l_1}{l_2}s + 1}{(as^2 + bs + 1)\left(\frac{2}{k_d+l_1}s + 1\right)^2}, \quad (16)$$

$$\frac{Y}{SP} = \frac{kC_D}{P_D(C_D + k)} = \frac{\frac{4ks}{k_d+l_1}\left(\frac{1}{k_d+l_1}s + 1\right)}{(as^2 + bs + 1)\left(\frac{2}{k_d+l_1}s + 1\right)^2}. \quad (17)$$

In formula (17), one can see that the increment of the value of k_d results in the decrease of the gain of the transfer function. Thus, a smaller over-regulations can be expected in transients. Additionally, after assuming that the value of $k_p >> k_d$, based on (10), one can notice that:

$$l_1 \approx \frac{b}{a}; l_2 \approx \frac{1}{a} \tag{18}$$

By substituting (18) to (16), it can be seen that the zeros of this transfer function approach the poles introduced by the process dynamics. Thus, the tracking dynamics will be determined mainly by the second pair of poles depending on the value of the tuning parameter k_d. Summing up, the following conclusions can be drawn:

- For a small value of the k_p parameter – by changing the value of the k_d parameter, one can influence the value of the over-regulation and, to a small extent, the regulation time, for the load disturbance rejection.
- For a significant value of the parameter k_p – by changing the value of the k_d parameter, one can influence the speed of the tracking response. However, at the same time, the over-regulation in tracking may occur.

Finally, a simple *rule of thumb* tuning method can be proposed. One should start by setting the value of the parameter k_d to ensure satisfying disturbance rejection. In this case, the limitations (11) and (12) must be taken into account. Then, one should choose the parameter k_p to ensure satisfying tracking performance. The parameters l_1, l_2 and b_0 must be calculated by the formulas (10) and (14).

The presented formulas require knowledge of the process model in the form of the second-order (SO) system. However, in practice, the FOPDT (First-order plus Dead Time) modelling is used more. In [7], one can find an easy method on how to determine the required SO model based on the prior knowledge on the process dynamics in the form of the FOPDT model.

The practical verification of this method by simulations and experiment will be presented in the next section.

4 Verification

In this section, the description of the pneumatic setup used for final validation is presented. A more detailed description of this setup can be found in works [8, 9]. Then, the representative simulation results are shown based on the realistic model of the real setup. Finally, the experimental verification based on the real process is presented.

4.1 The Laboratory Pneumatic Setup

The laboratory pneumatic setup is presented in Fig. 2. It consists of two serially connected pneumatic tanks. Relative pressures at each tank are denoted as p_1, p_2. The system is manipulated by supplying air pressure p_0. The air flows between the tanks through the constant pneumatic resistance R_{pa}, R_{pb}, whose values are unknown. From the second

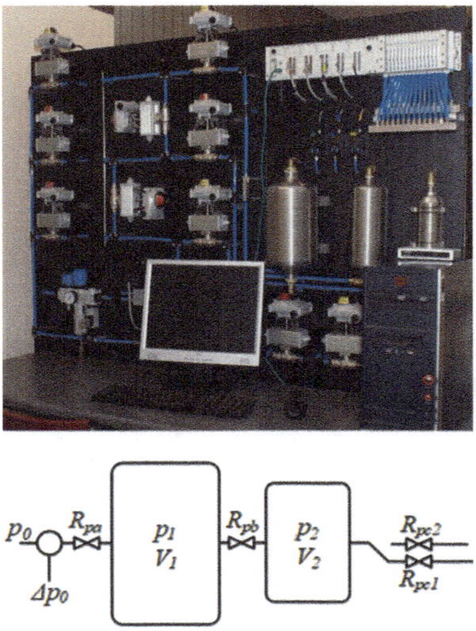

Fig. 2. Pneumatic laboratory setup and its simplified scheme.

tank, the air flows out through the one of two switched pneumatic resistances R_{pc1} or R_{pc2}.

The control goal for the setup is defined as stabilizing the relative pressure $y = p_2 = \mathrm{SP}$ by manipulating the supplying pressure $u = p_0$. Potentially, the setup can be disturbed in two ways. The first possibility is by the load disturbance adjusted by adding the additional pneumatic pressure $D = \Delta p_0$ to the inlet of the first tank. The second disturbance is switching between two pneumatic resistances R_{pc1} or R_{pc2} at the outlet from the second tank. Readers should note that, apart from disturbing the process, switching the pneumatic resistance between R_{pc1} and R_{pc2} has also a significant impact on the dynamical properties of the process.

As stated in Sect. 3, the proposed method requires knowledge of the SO model. In this case, the model describing the pneumatic setup is taken directly from [9] and has a form:

$$\frac{p_2}{p_0} = \frac{k}{(T_1 s + 1)(T_2 s + 1)} = \frac{0.7}{(10.7s + 1)(4.1s + 1)}. \tag{19}$$

4.2 Simulation Tests

Simulation tests are carried out first because they allow changing the tuning parameters in a wide range (without the influence of non-linearity and measuring noise). Also, system robustness can be easily verified.

The analysis presented in Sect. 3 shows that the parameter k_d has a very strong impact on the control performance of RADRC2. In this Section, the representative simulation has been carried out in order to verify this conclusion. Assuming the perfect knowledge about the real process (19), the influence of adjusting the values of the parameters k_p k_d is presented respectively in Figs. 3 and 4. The simulations were carried out assuming the perfect knowledge of the process SO dynamics. In both simulations, the scenario consists of the initial step change of the setpoint SP from 1 to 1.2 bar. Then, at 45 s, the load disturbance of $D = -0.8$ is applied to the process.

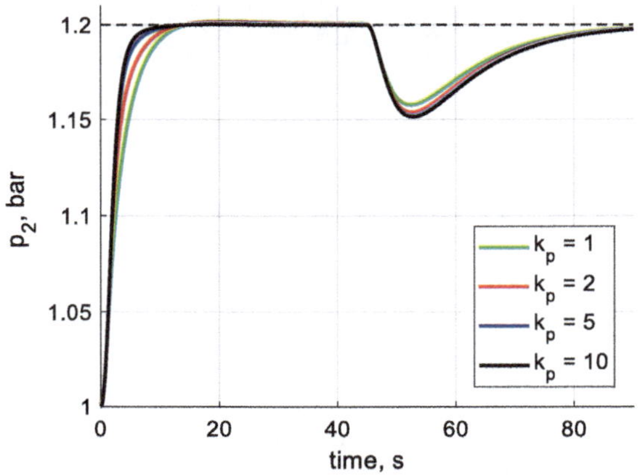

Fig. 3. Simulation results of the performance of the closed-loop control system for constant k_d = 2 and different denoted values of k_p

Figure 3 shows that from the practical viewpoint, the influence of the k_p parameter on the control performance is insignificant. As k_p increases, for tracking, the settling time decreases, and the over-regulation increases slightly. At the same time, Fig. 4 shows that the k_d parameter has a very significant impact on the over-regulation for disturbance rejection. It is worth noting that for $k_d = 10$, the disturbance rejection is very good. However, in practice, such a high value of the parameter k_d results in a very strong sensitivity on the measurement noise that significantly deteriorates the quality of the manipulated signal. For this reason, in practical cases, the value of k_d parameter should be limited. In Fig. 5 it is shown that for a sufficiently large k_p, the adjustment of k_d also has the impact on the tracking dynamics. These results confirm the conclusions drawn in Sect. 3.

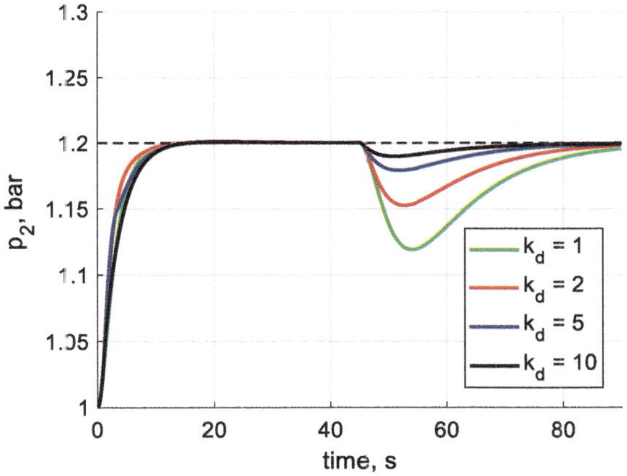

Fig. 4. Simulation results of the performance of the closed-loop control system for constant k_p = 3 and different denoted values of k_d

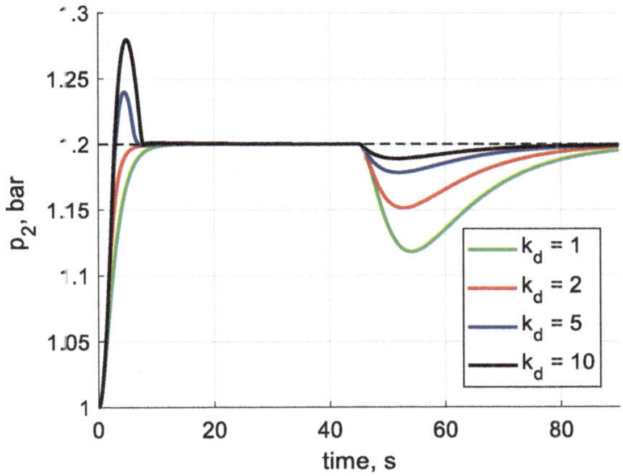

Fig. 5. Simulation results of the performance of the closed-loop control system for constant k_p = 30 and different denoted values of k_d

Finally, robustness of the control system to uncertainty on the model parameters was tested, and the results are shown in Fig. 6. The values of the more significant time constant and the process gain was changed by 50% as it is shown in the legend. The results show the high robustness of the control system. In practice, the differences in the control performance between the considered variations of the real processes are not very significant.

4.3 Experimental Results

The final validation was carried out experimentally on basis of the pneumatic setup described at beginning of this Section. The results are presented in Fig. 7. This time,

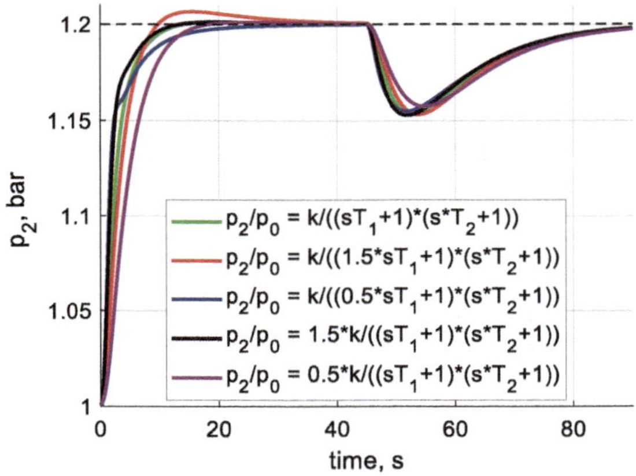

Fig. 6. Simulation results of the performance of the closed-loop control system for constant $k_p = 2$ and constant $k_d = 2$, in the presence of the different model uncertainty

the experiment scenario is more complex, and it consists of 6 stages. Each stage takes 45 s. First, SP is changed from 1 to 1.2, then the load disturbance $D = -0.8$ is applied. Stages 3 and 4 represent the switching between the pneumatic resistances R_{pc1} and R_{pc2} at the outlet from the second tank. Then, SP has changed again from 1.2 to 1, and the load disturbance is changed to $D = 0$. Finally, another switching between the pneumatic resistances R_{pc1} and R_{pc2} is applied to the process. It is worth noting that switching between the valves representing two different pneumatic resistances R_{pc1} and R_{pc2} results in the significant changes both in the process dynamics and in its statics. Thus, the results can experimentally confirm the robustness of the control system.

During the experiments, three different control systems were compared. The green line represents the considered RADRC2-based control system tuned by $k_p = 20$ and by $k_d = 1.5$. Based on the (10) and (14) the rest of tuning parameters were: $l_1 = 0.34$, $l_2 = 0.023$, $b_0 = 0.39$. For comparison, the results for the control system with the conventional ADRC tuned as it is shown in [9] are presented with the red line. In this case the tuning parameters are $\omega_c = 0.25$, $\omega_o = 1.25$, $b_0 = k/(T_1T_2) = 0.016$. Finally, the results of the conventional PID controller tuned by AMIGO method are presented by the black line ($k_r = 4.28$, $T_i = 6.79$, $T_d = 0.98$).

Figure 7 shows that RADRC controller has the same disturbances rejection properties as ADRC. Additionally, it provides better tracking properties. It should be mentioned, that in the case of switching the valves R_{PC1} and R_{PC2} there are some oscillations in the process output with the ADRC controller. This effect is not present in the control loop with RADRC, because of its better robustness. Similar conclusions can be derived based on the control signal. In this case, for better clarity, only ADRC and RADR control signal were shown.

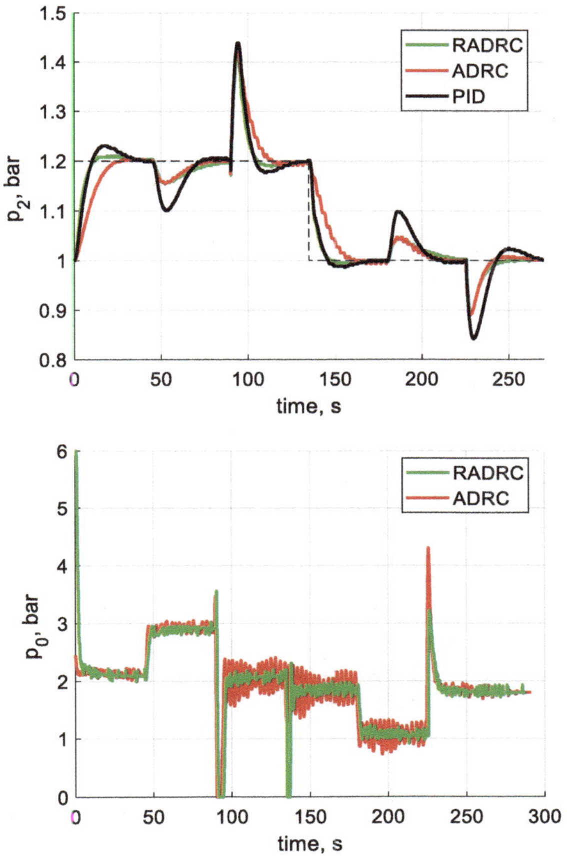

Fig. 7. Experimental validation of the performance of the closed-loop control systems. Comparison between RADRC2, ADRC and PID.

5 Conclusion

This paper describes the results of preliminary analysis of possibility of a different approach to the tuning of RADRC2. In comparison to the conventional tuning of ADRC, the major difference results from the fact that the zeros of the controller are freely located by adjusting two tuning parameters. The theoretical analysis shows that potentially, such an approach can improve the control performance, especially in terms of the disturbance rejection. This conclusion was confirmed by the corresponding simulation and practical validation.

The suggested concept is promising, but it requires further analysis. Especially, there is a lack of any reliable tuning rule. The one suggested in this paper is rather the rule of thumb, not the systematic approach that is required in practice. The future works will include deriving of such a tuning rule based on the simplified model of the process.

Acknowledgements. Paweł Nowak and Jacek Czeczot were financed by the grant from the Silesian University of Technology - subsidy for maintaining and developing the research potential in 2020. Patryk Grelewicz was financed by the European Union through the European Social Fund (grant POWR.03.02.00-00-I029).

References

1. Han, J.: From PID to active disturbance rejection control. IEEE Trans. Industr. Electron. **56**(3), 900–906 (2009). https://doi.org/10.1109/TIE.2008.2011621
2. Gao, Z.: Active disturbance rejection control: a paradigm shift in feedback control system design. In: American Control Conference, pp. 2399–2405 (2006). https://doi.org/10.1109/acc.2006.1656579
3. Huang, Y., Xue, W.: Active disturbance rejection control: methodology and theoretical analysis. ISA Trans. **53**(4), 963–976 (2014). https://doi.org/10.1016/j.isatra.2014.03.003
4. Xingling, S., Honglun, W.: Back-stepping active disturbance rejection control design for integrated missile guidance and control system via reduced-order ESO. ISA Trans. **57**, 10–22 (2015). https://doi.org/10.1016/j.isatra.2015.02.013
5. Gao, Z.: Scaling and bandwidth-parameterization based controller tuning. In: Proceedings of the 2003 American Control Conference, pp. 4989–4996 (2003). https://doi.org/10.1109/ACC.2003.1242516
6. Zheng, Q., Gao, L.Q., Gao, Z.: On estimation of plant dynamics and disturbance from input-output data in real time. In: IEEE International Conference on Control Applications, pp. 1167–1172 (2007). https://doi.org/10.1109/cca.2007.4389393
7. Chen, X., Li, D., Gao, Z., Wang, C.: Tuning method for second-order active disturbance rejection control. In: Proceedings of the 30th Chinese Control Conference, pp. 6322–6327 (2011)
8. Fratczak, M., Nowak, P., Klopot, T., Czeczot, J., Bysko, S., Opilski, B.: Virtual commissioning for the control of the continuous industrial processes; case study. In: 20th International Conference on Methods and Models in Automation and Robotics (MMAR), pp. 1032–1037 (2015). https://doi.org/10.1109/mmar.2015.7284021
9. Nowak, P., Czeczot, J.: Practical verification of active disturbance rejection controller for the pneumatic setup. In: 22nd International Conference on Methods and Models in Automation and Robotics (MMAR), pp. 19–24 (2017). https://doi.org/10.1109/mmar.2017.8046791

Comparison of Robustness of Selected Speed Control Systems Applied for Two Mass System with Backlash

Bartlomiej Wicher(✉) and Stefan Brock

Poznan University of Technology, ul. Piotrowo 3A, Poznan, Poland
{bartlomiej.wicher,stefan.brock}@put.poznan.pl

Abstract. The article compares robustness of five selected speed control systems to the changes in the controlled plant, which is a two mass system with backlash introduced between the motor and the linking shaft. Authors propose a test procedure which will take into account typical dynamical and steady states occurring in electrical drives. Moreover, modified quality indicators also are proposed in order to measure the influence of varying object parameter to the control process. During the research, authors changed the moment of inertia of the motor and the load as well as the backlash width. The results are summarized in tables in order to enable clear comparison between investigated methods.

Keywords: Two-mass system · Backlash · ADRC · 2DOF-PID · Torsional vibration · Damping · ISE · FFT · Spectrum

1 Introduction

Nowadays electric drives are commonly used in variety of applications, including industry, home appliances, automotive etc. The load is often connected with the propelling motor with shaft, gearbox and/or clutches. All these factors lead to the situation where resonance phenomena may occur in propulsion systems as well as, as a result of material wear or inaccuracy of individual components, backlash may be present in connections between individual components of the drive system [1–4]. Moreover, in some applications the parameters of the controlled plant can vary in wide range (e.g. mills). The variability of the control object is a challenge for the control system, because in the vast majority of applications, control systems do not implement the mechanism of adaptation, hence it is not possible to correct the current settings of the control system. Various implementation of PI and 2DOF controllers are investigated in order to achieve satisfying results of controlling speed of resonance system [5, 6]. Also more advanced methods are implemented and developed, such as Active Disturbance Rejection Control (ADRC) [7]. Authors of these multiple methods usually provide methods [4–6], [8]. The article compares the robustness of five selected speed control systems to the variability of the moment of inertia and the backlash introduced to the two-mass system.

A. Bartoszewicz et al. (Eds.): PCC 2020, AISC 1196, pp. 1371–1382, 2020.
https://doi.org/10.1007/978-3-030-50936-1_114

2 Model of the Object

The investigated plant model is presented in Fig. 1. The plant consists of two-mass system with backlash block situated between the motor and the shaft. Parameters of the object were kept constant during each single simulation. Mathematical equations describing the system without backlash are shown below (1) [1, 2]:

$$\begin{cases} J_1\dot{\omega}_1 = T_1 - T_T \\ J_2\dot{\omega}_2 = T_T - T_2 \\ T_T = k_W(\theta_1 - \theta_2) + D_W(\omega_1 - \omega_2) \end{cases} \quad (1)$$

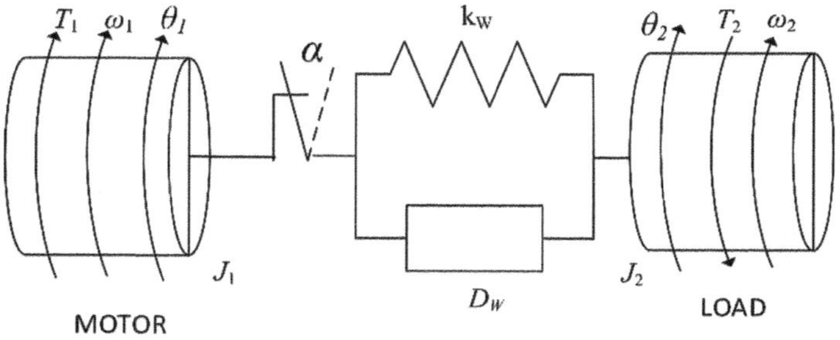

Fig. 1. Diagram of mechanical part of the system ω_1 – motor speed, ω_2 – load speed, θ_1 – motor position, θ_2–load position J_1 – motor moment of inertia, J_2 – load moment of inertia, k_W – stiffness coefficient, D_W – damping factor, α – backlash width. T_T is the torsional torque, T_1 is the motor electromagnetic torque, T_2 is the load torque.

The backlash model is described by the set of Eqs. (2) [6]:

$$\dot{\Theta}_b = \begin{cases} \max\left(0, \dot{\Theta}_d + \frac{k_W}{D_W}(\Theta_d - \Theta_b)\right) & if : \Theta_b = -\frac{1}{2}\alpha \ (T_T \le 0) \\ \dot{\Theta}_d + \frac{k_W}{D_W}(\Theta_d - \Theta_b) & if : |\Theta_b| < \frac{1}{2}\alpha \ (T_T = 0) \\ \min\left(0, \dot{\Theta}_d + \frac{k_W}{D_W}(\Theta_d - \Theta_b)\right) & if : \Theta_b = \frac{1}{2}\alpha \ (T_T \ge 0) \end{cases} \quad (2)$$

Where the Θ_d is the total dislocation of the shafts ends, the Θ_S is the torsional angle and the Θ_b is actual position within backlash angle (limited to $-0.5\alpha < \Theta_b < 0.5\alpha$).

3 Controllers Description

Overall control structure is presented in Fig. 2. The mechanical system was modified by changing the moments of inertia as well as the backlash width. The torque control loop remain the same during the test. Different controllers have been used in the "Controller" block. Depending on the structure of the investigated controller, different feedback signals were connected to this block.

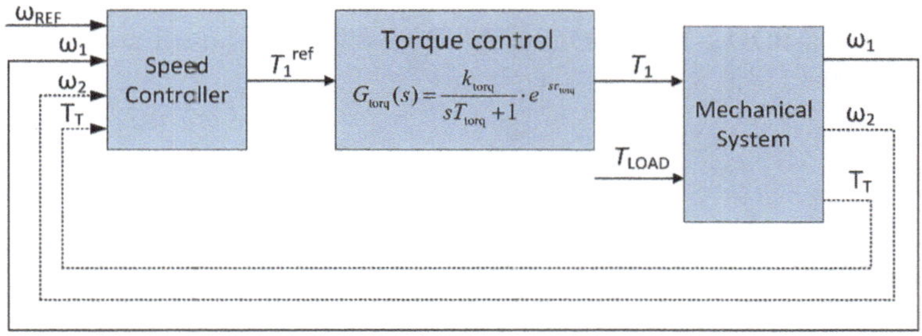

Fig. 2. General control structure of investigated system. The dashed lines represent optional feedback signals utilized when PI with additional feedback signals was used.

3.1 Active Disturbance Rejection Control – Feedback from Motor Side

The ADRC controller is presented in Fig. 3. The idea of this control system is to use Extended State Observer (ESO) in order to decouple all disturbances and unmodelled dynamics from the actual controller, which will "see" the object as a simple integrator of *n-th* order. Is this particular case the ESO together with the Rejector will simplify the system to *1-st* order integrator, so the Controller has only one parameter: proportional gain. The gains of the ESO were specified using the pole placement method [8].

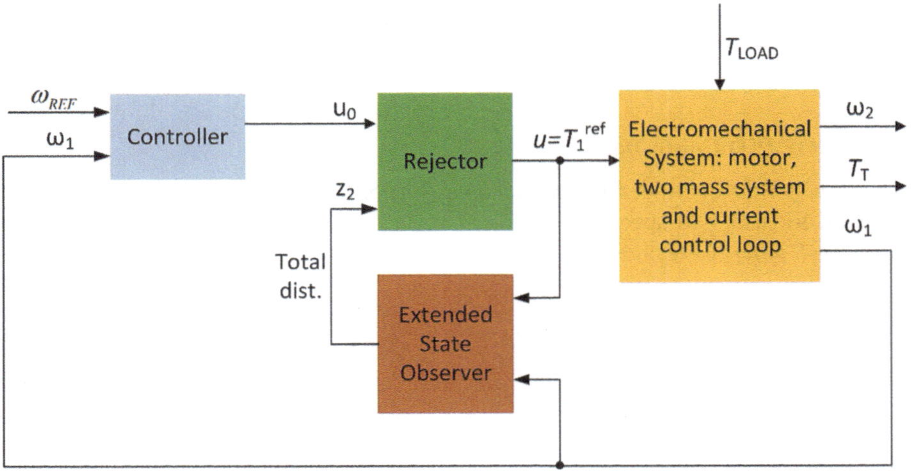

Fig. 3. The structure of ADRC controller with feedback from motor side.

3.2 Active Disturbance Rejection Control – Feedback from Load Side

The second speed control structure based on ADRC method [2, 3] is presented in Fig. 4. The difference between the method described above is that the feedback signal is taken from load side, as a consequence, the Controller "sees" *3-rd* order integrator, therefore its structure is a PDD^2. The ESO and the Controller were tuned according the pole placement method [8].

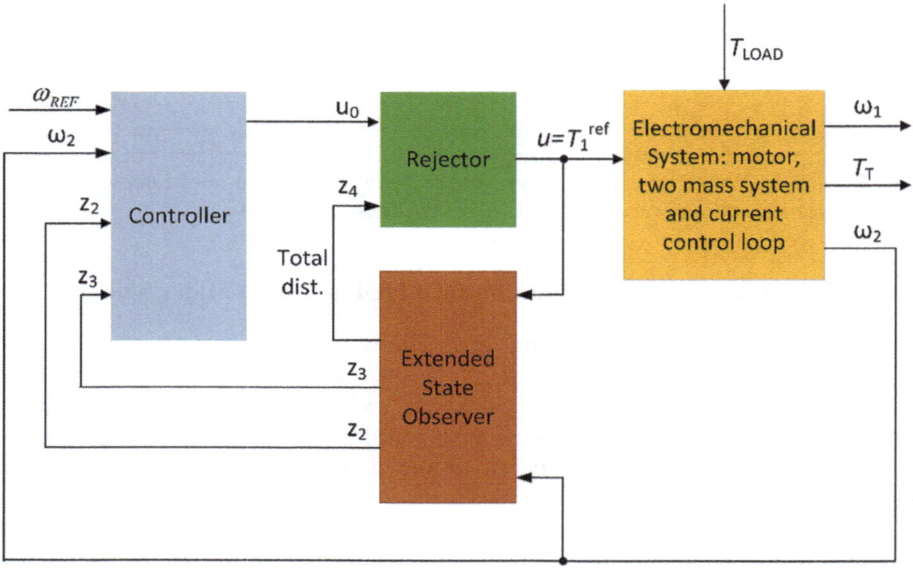

Fig. 4. The structure of ADRC controller – feedback from load side.

3.3 2-DOF PID Controller

The 2-DOF PID controller is presented in Fig. 5. It structure and parameters were designed for use as a speed controller of motor side for the two-mass system. The gains K_I, K_P, K_D were chosen according to the method described in [6].

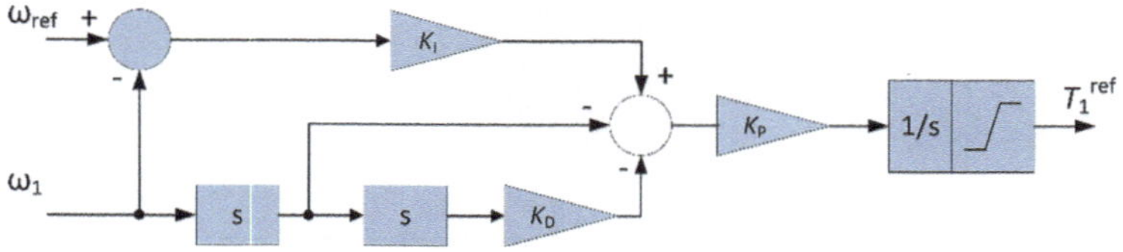

Fig. 5. 2-DOF PID controller

3.4 PI with Additional Feedback Signals – Derivative of the Difference of Motor and Load Speed

This controller utilizes two measurement signals – the speed of the motor and the speed of the load. The gains were calculated according to the equations described in [6] (Fig. 6).

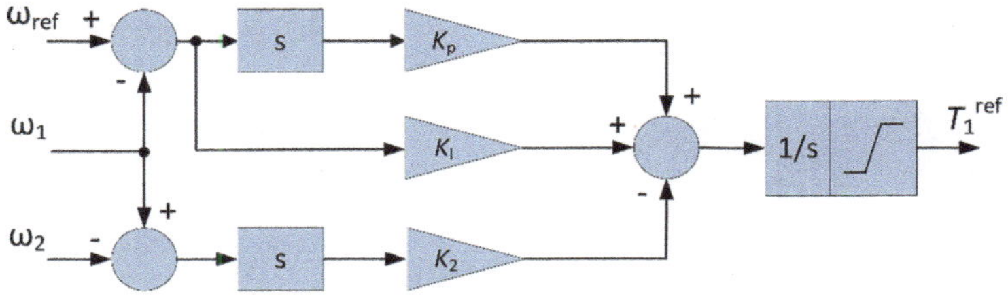

Fig. 6. PI controller with additional feedback from the derivative of the difference motor and load speed

3.5 PI with Additional Feedback Signals – Derivative of Torsional Torque

This type of controller utilizes additional feedback signal – the torsional torque (T_T). The gains of this controller were calculated according the method in [6] (Fig. 7).

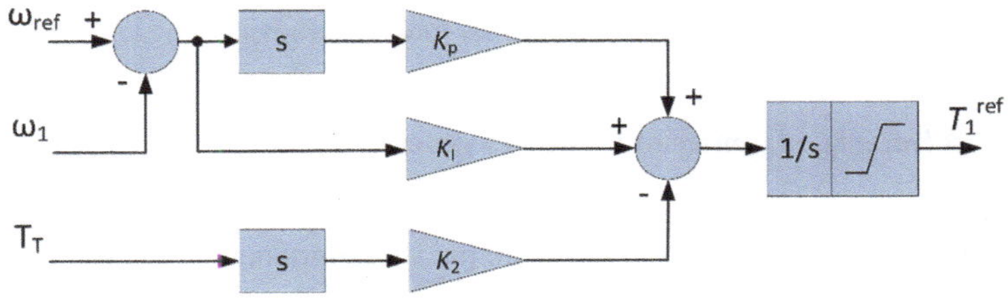

Fig. 7. PI with additional feedback from derivative of torsional torque

4 Test Scenarios, Quality Indicators

4.1 Test Procedure

The robustness test consisted of two parts. The first part concerns the study of the impact of the moment of inertia on the quality of the speed control process. The second part concerns the study of the effect of backlash on the quality of the speed control process. Moments of inertia of motor and load part change within the range of [0.02; 20], this range was the result of preliminary test, that were conducted in the range of [0.001; 100], but the system ran into very large oscillations for extremely low values of moments of inertia or the control process was dominated by saturation occurred in controllers, therefore the final range has been reduced. Both parts of tests are evaluated on the basis of two quality indicators, which are described in paragraph 4.2. The speed reference value and the load torque change according to the Fig. 8. The values of reference speed and load torque are chosen in this way so the system does not work into saturation region with the nominal moments of inertia.

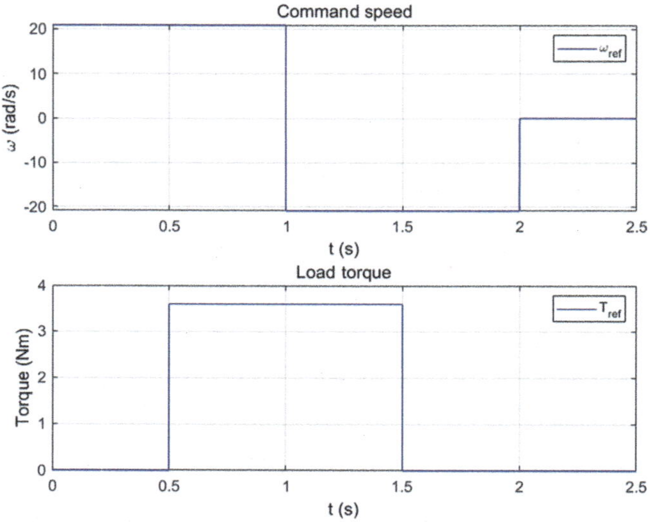

Fig. 8. Reference speed and load torque waveforms during tests.

4.2 Determining the Quality of Control Process

Authors decide to use two indicators for evaluation of the quality control process: modified ISE and the relative energy of high frequency components in control signal. The modified ISE is described by formula (3):

$$ISE_{LOG} = log_{10}\left(\sum\nolimits_{i=1}^{N} e^2(i)\right) \quad (3)$$

Where $e^2(i)$ is the control error in i-th sample for investigated controller, N is the number of samples during simulation.

The relative energy of high frequency components is calculated according to formula (4):

$$E_{HF\_REL} = \frac{\sum\nolimits_{n=n_B}^{n=\lfloor N/2 \rfloor} U^2(n)}{\sum\nolimits_{n=0}^{n_B-1} U^2(n)} \quad (4)$$

Where N is the number of FFT samples, n_B is number of boarder component of spectrum $n_B = \lfloor f_0/\Delta f \rfloor$ with boarder frequency f_0, $U(n)$ is the module of n-th spectrum component, $n = 0$ identifies the mean value component. Authors decided to use this additional quality indicator assuming that it will help to determine if the system is approaching to the instability region, because the control signal is not affected by the low – pass filtration of the object, so the high frequency oscillations will be first observed in control signal than the output signal. The higher value of the E_{HF_REL} the more high frequency oscillations in control signal is present, which can indicate, that the system is approaching to the boarder of the stability.

5 Results and Conclusions

Following tables show the results of simulations performed in Matlab/Simulink environment. The sampling time $T_S = 20\ \mu s$ is equal for all simulations. The initial parameters are corresponding to the laboratory stand which will be used for experimental verification: $J_1 = 0.00162\ kg \cdot m^2$, $J_2 = 0.006\ kg \cdot m^2$, $k_W = 29.42\ Nm/rad$, $D_W = 0.01\ Nm \cdot s/rad$. The actual values of moments of inertia are presented in each row and column in Tables 1, 2, 3, 4, 5, 6, 7, 8, 9 and 10 as a factor of nominal moment of inertia. The boarder frequency was set to *40* Hz as a compromise between the frequency of oscillations that can occur during normal operation of the control system and the oscillations that occur when the system approaches to the stability limit.

Table 1. Values of logarithm of ISE for selected J1 and J2 coefficients for PI with additional feedback signals – derivative of the difference of motor and load speed.

LOAD		J2 COEFFICIENT									
		0,02	0,05	0,1	0,2	0,5	1	2	5	10	20
J1 COEFFICIENT	0,02	6,18	6,15	6,09	5,98	5,69	5,34	4,87	4,13	3,54	3,04
	0,05	1,88	1,85	1,84	1,85	1,89	1,93	2,02	2,19	2,28	2,70
	0,1	1,86	1,80	1,76	1,74	1,77	1,85	1,99	2,28	2,43	2,70
	0,2	1,85	1,80	1,76	1,74	1,77	1,85	1,99	2,28	2,42	2,70
	0,5	1,84	1,79	1,76	1,74	1,77	1,86	2,00	2,28	2,41	2,69
	1	1,85	1,80	1,76	1,74	1,78	1,86	2,00	2,28	2,41	2,69
	2	1,89	1,83	1,78	1,76	1,79	1,88	2,01	2,28	2,40	2,68
	5	1,96	1,93	1,89	1,84	1,85	1,92	2,04	2,29	2,39	2,67
	10	2,03	2,02	2,00	1,96	1,94	1,98	2,09	2,29	2,38	2,64
	20	2,13	2,12	2,12	2,11	2,11	2,10	2,17	2,31	2,40	2,64
MOTOR		J2 COEFFICIENT									
		0,02	0,05	0,1	0,2	0,5	1	2	5	10	20
J1 COEFFICIENT	0,02	6,18	6,15	6,09	5,98	5,70	5,34	4,87	4,17	3,67	3,37
	0,05	2,38	2,63	2,65	2,66	2,67	2,67	2,67	2,68	2,66	2,51
	0,1	1,57	1,57	1,58	1,59	1,62	1,67	1,75	1,95	2,11	2,50
	0,2	1,57	1,57	1,58	1,59	1,62	1,67	1,76	1,95	2,11	2,50
	0,5	1,58	1,58	1,59	1,60	1,63	1,68	1,76	1,95	2,11	2,50
	1	1,59	1,60	1,60	1,62	1,65	1,69	1,77	1,96	2,11	2,50
	2	1,62	1,63	1,63	1,64	1,67	1,72	1,79	1,97	2,11	2,51
	5	1,70	1,70	1,71	1,72	1,74	1,78	1,84	2,00	2,12	2,52
	10	1,80	1,80	1,81	1,81	1,83	1,87	1,92	2,04	2,15	2,52
	20	1,95	1,96	1,96	1,96	1,98	2,00	2,04	2,12	2,19	2,53

Analyzing the data shown in Tables 1, 2, 3, 4 and 5 it can be seen that all control systems fail when motor moment of inertia drops below 5% of nominal value, it is worth to emphasize that it happens gradually. When it comes to decrease the load moment of inertia, most of the control systems remain stable, except for the ADRC load feedback control system. Increasing the motor and load moment of inertia leads to smooth increase the motor and load ISE without sudden change as it was observed during reduction. Once again the exception applies to the ADRC load feedback control system. From the ISE point of view authors would recommend the 2-DOF PID controller because it ensures robustness comparable to the PI with additional feedback signals, being simpler in application. The ADRC control systems are less robust but ensure better ISE when the moments of inertia are equal to those assumed during design process, it applies especially for the ADRC with feedback from the load side.

Table 2. Values of logarithm of ISE for selected J1 and J2 coefficients for PI with additional feedback signals – derivative of torsional torque.

LOAD		J2 COEFFICIENT									
		0,02	0,05	0,1	0,2	0,5	1	2	5	10	20
J1 COEFFICIENT	0,02	5,40	5,42	5,44	5,43	5,18	5,35	4,87	4,13	3,54	3,04
	0,05	1,90	1,87	1,83	1,80	1,81	1,88	2,03	2,23	2,35	2,41
	0,1	1,90	1,86	1,83	1,80	1,81	1,88	2,03	2,23	2,35	2,41
	0,2	1,89	1,86	1,83	1,80	1,81	1,88	2,03	2,23	2,35	2,41
	0,5	1,88	1,85	1,82	1,80	1,82	1,89	2,03	2,23	2,34	2,41
	1	1,89	1,86	1,83	1,81	1,82	1,90	2,04	2,23	2,34	2,42
	2	1,93	1,90	1,87	1,84	1,84	1,92	2,05	2,23	2,34	2,42
	5	2,00	1,99	1,97	1,95	1,93	1,97	2,09	2,24	2,34	2,42
	10	2,09	2,08	2,08	2,07	2,04	2,04	2,16	2,28	2,36	2,45
	20	2,22	2,22	2,22	2,21	2,19	2,19	2,25	2,35	2,42	2,53
MOTOR		J2 COEFFICIENT									
		0,02	0,05	0,1	0,2	0,5	1	2	5	10	20
J1 COEFFICIENT	0,02	5,41	5,43	5,44	5,43	5,19	5,35	4,89	4,20	3,75	3,52
	0,05	1,56	1,56	1,56	1,56	1,58	1,60	1,65	1,74	1,95	2,12
	0,1	1,56	1,56	1,56	1,57	1,58	1,60	1,65	1,74	1,96	2,13
	0,2	1,56	1,56	1,57	1,57	1,59	1,61	1,65	1,75	1,97	2,15
	0,5	1,58	1,58	1,58	1,59	1,60	1,62	1,66	1,76	1,99	2,17
	1	1,60	1,60	1,60	1,61	1,62	1,64	1,68	1,77	2,00	2,20
	2	1,64	1,65	1,65	1,65	1,66	1,68	1,72	1,81	2,03	2,24
	5	1,76	1,76	1,76	1,76	1,77	1,79	1,81	1,89	2,10	2,31
	10	1,89	1,89	1,90	1,90	1,90	1,92	1,94	1,99	2,16	2,36
	20	2,08	2,08	2,08	2,08	2,09	2,10	2,10	2,14	2,23	2,43

Table 3. Values of ISE for selected J1 and J2 coefficients for 2-DOF PID Controller

LOAD		J2 COEFFICIENT									
		0,02	0,05	0,1	0,2	0,5	1	2	5	10	20
J1 COEFFICIENT	0,02	6,41	6,37	6,30	6,16	5,81	5,41	4,91	4,15	3,55	3,05
	0,05	2,00	1,96	1,93	1,91	1,93	1,98	2,08	2,29	2,44	2,75
	0,1	2,00	1,96	1,93	1,91	1,93	1,98	2,08	2,29	2,44	2,75
	0,2	2,00	1,96	1,93	1,91	1,93	1,98	2,08	2,29	2,44	2,75
	0,5	1,99	1,95	1,93	1,92	1,93	1,99	2,08	2,29	2,44	2,75
	1	2,00	1,96	1,93	1,92	1,94	1,99	2,09	2,29	2,44	2,74
	2	2,04	1,99	1,95	1,93	1,95	2,00	2,10	2,30	2,44	2,74
	5	2,09	2,07	2,04	2,00	1,99	2,04	2,13	2,31	2,44	2,72
	10	2,14	2,14	2,13	2,10	2,07	2,10	2,17	2,34	2,45	2,71
	20	2,23	2,23	2,23	2,22	2,21	2,20	2,25	2,38	2,49	2,69
MOTOR		J2 COEFFICIENT									
		0,02	0,05	0,1	0,2	0,5	1	2	5	10	20
J1 COEFFICIENT	0,02	6,41	6,37	6,30	6,16	5,81	5,41	4,92	4,20	3,71	3,43
	0,05	1,82	1,83	1,83	1,84	1,86	1,89	1,95	2,09	2,20	2,56
	0,1	1,83	1,83	1,83	1,84	1,86	1,89	1,95	2,09	2,20	2,56
	0,2	1,83	1,83	1,83	1,84	1,86	1,89	1,95	2,09	2,20	2,56
	0,5	1,83	1,84	1,84	1,85	1,87	1,90	1,96	2,09	2,20	2,56
	1	1,84	1,85	1,85	1,86	1,88	1,91	1,96	2,10	2,21	2,57
	2	1,86	1,86	1,87	1,87	1,89	1,92	1,98	2,11	2,21	2,56
	5	1,91	1,91	1,92	1,92	1,94	1,97	2,02	2,13	2,23	2,56
	10	1,98	1,99	1,99	1,99	2,01	2,03	2,08	2,18	2,27	2,56
	20	2,10	2,10	2,11	2,11	2,12	2,14	2,17	2,24	2,34	2,56

When it comes to analyzing the relative energy of high frequency components in control signal (Tables 6, 7, 8, 9 and 10), the results confirm the results obtained from ISE and could be an alternative way of determining the stability of control loop. Additionally,

Table 4. Values of ISE for selected J1 and J2 coefficients for Active Disturbance Rejection Control – feedback from load side

LOAD		J2 COEFFICIENT									
		0,02	0,05	0,1	0,2	0,5	1	2	5	10	20
J1 COEFFICIENT	0,02	7,09	6,52	5,99	5,42	4,63	4,00	3,34	2,52	2,16	2,27
	0,05	7,30	6,90	6,44	5,91	5,15	4,53	3,87	2,93	2,37	2,27
	0,1	7,20	6,92	6,52	6,02	5,27	4,65	3,98	2,93	2,03	2,27
	0,2	6,90	6,71	6,33	5,81	5,01	4,33	3,54	1,83	2,02	2,27
	0,5	6,31	6,16	5,67	4,95	3,87	2,75	1,67	1,83	2,02	2,27
	1	5,78	5,60	4,99	4,09	2,49	1,62	1,68	1,82	2,02	2,28
	2	5,19	5,01	4,22	3,04	1,61	1,64	1,73	2,85	2,78	2,28
	5	4,38	4,10	2,95	1,70	1,68	2,85	3,97	4,20	3,95	3,43
	10	3,70	3,27	1,93	1,78	2,08	3,70	4,48	4,56	4,27	3,79
	20	2,96	2,19	2,00	1,91	2,62	3,82	4,46	4,54	4,31	3,88
MOTOR		J2 COEFFICIENT									
		0,02	0,05	0,1	0,2	0,5	1	2	5	10	20
J1 COEFFICIENT	0,02	7,15	6,82	6,71	6,69	6,70	6,71	6,72	6,72	6,65	4,02
	0,05	7,31	7,00	6,78	6,69	6,69	6,69	6,69	6,67	6,58	3,47
	0,1	7,20	6,96	6,69	6,50	6,44	6,43	6,42	6,33	3,35	3,13
	0,2	6,89	6,72	6,39	6,07	5,86	5,79	5,70	2,79	2,75	2,96
	0,5	6,30	6,15	5,69	5,09	4,59	4,30	2,02	2,24	2,81	3,07
	1	5,77	5,59	4,99	4,22	3,51	1,77	1,82	2,55	3,01	3,21
	2	5,18	4,99	4,22	3,26	1,68	1,67	2,41	4,88	5,32	3,56
	5	4,36	4,08	2,97	1,68	1,64	3,30	4,49	5,39	5,71	5,75
	10	3,68	3,24	1,82	1,71	2,02	3,35	4,36	5,13	5,44	5,53
	20	2,93	2,03	1,87	1,83	2,13	2,99	3,75	4,52	4,87	4,99

Table 5. Values of ISE for selected J1 and J2 coefficients for Active Disturbance Rejection Control – feedback from motor side

LOAD		J2 COEFFICIENT									
		0,02	0,05	0,1	0,2	0,5	1	2	5	10	20
J1 COEFFICIENT	0,02	5,04	5,09	5,09	5,07	4,96	4,80	4,52	3,96	3,44	3,00
	0,05	4,35	4,37	4,38	4,36	4,31	4,22	4,06	3,68	3,28	2,93
	0,1	3,29	3,29	3,30	3,30	3,29	3,25	3,17	2,99	2,81	2,73
	0,2	1,89	1,83	1,80	1,78	1,82	1,90	2,04	2,30	2,45	2,78
	0,5	1,85	1,82	1,79	1,78	1,82	1,91	2,05	2,30	2,45	2,77
	1	1,87	1,81	1,79	1,79	1,83	1,92	2,05	2,30	2,44	2,77
	2	1,94	1,87	1,82	1,81	1,85	1,93	2,07	2,31	2,44	2,76
	5	2,01	1,99	1,96	1,91	1,91	1,98	2,11	2,33	2,45	2,75
	10	2,09	2,09	2,08	2,05	2,02	2,06	2,16	2,36	2,45	2,72
	20	2,21	2,21	2,21	2,20	2,20	2,19	2,26	2,39	2,51	2,70
MOTOR		J2 COEFFICIENT									
		0,02	0,05	0,1	0,2	0,5	1	2	5	10	20
J1 COEFFICIENT	0,02	5,49	5,53	5,54	5,54	5,51	5,47	5,43	5,39	5,38	5,38
	0,05	4,80	4,83	4,83	4,83	4,82	4,79	4,76	4,71	4,68	4,67
	0,1	4,04	4,07	4,07	4,07	4,07	4,07	4,06	4,04	4,03	4,03
	0,2	1,65	1,65	1,66	1,67	1,71	1,77	1,86	2,05	2,18	2,58
	0,5	1,66	1,66	1,67	1,68	1,72	1,77	1,87	2,05	2,18	2,58
	1	1,67	1,68	1,69	1,70	1,73	1,79	1,88	2,06	2,18	2,58
	2	1,70	1,71	1,72	1,73	1,76	1,81	1,90	2,07	2,18	2,58
	5	1,79	1,80	1,80	1,81	1,84	1,88	1,96	2,11	2,21	2,59
	10	1,91	1,91	1,91	1,92	1,94	1,98	2,04	2,17	2,25	2,58
	20	2,07	2,08	2,08	2,08	2,10	2,12	2,16	2,23	2,34	2,57

Table 6. Relative energy of high frequency components for PI with additional feedback signals – derivative of the difference of motor and load speed.

		J2 COEFFICIENT									
		0,02	0,05	0,1	0,2	0,5	1	2	5	10	20
J1 COEFFICIENT	0,02	3,1E+01	3,3E+01	3,7E+01	4,6E+01	8,4E+01	1,9E+02	5,5E+02	2,9E+03	1,1E+04	3,8E+04
	0,05	7,2E+00	1,4E+01	1,4E+01	1,4E+01	1,5E+01	1,4E+01	1,2E+01	6,5E+00	2,5E+00	1,6E-02
	0,1	1,1E-02	1,2E-02	4,0E-03	3,0E-03	2,9E-03	2,8E-03	2,2E-03	8,4E-04	4,6E-04	6,4E-04
	0,2	1,1E-02	1,2E-02	3,5E-03	2,5E-03	2,4E-03	2,2E-03	1,8E-03	6,9E-04	3,9E-04	6,3E-04
	0,5	1,1E-02	1,2E-02	4,1E-03	3,0E-03	2,9E-03	2,8E-03	2,2E-03	8,5E-04	4,7E-04	7,0E-04
	1	1,1E-02	1,3E-02	5,2E-03	4,1E-03	4,0E-03	3,8E-03	3,0E-03	1,1E-03	6,1E-04	7,1E-04
	2	8,7E-03	1,1E-02	6,2E-03	5,4E-03	5,2E-03	4,8E-03	3,7E-03	1,4E-03	7,6E-04	7,9E-04
	5	6,3E-03	7,5E-03	5,8E-03	5,5E-03	5,2E-03	4,6E-03	3,4E-03	1,3E-03	7,9E-04	7,8E-04
	10	4,5E-03	4,8E-03	4,3E-03	4,2E-03	3,8E-03	3,3E-03	2,4E-03	1,1E-03	7,0E-04	7,1E-04
	20	2,2E-03	2,2E-03	2,1E-03	2,1E-03	1,9E-03	1,7E-03	1,3E-03	8,6E-04	5,1E-04	5,2E-04

Table 7. Relative energy of high frequency components PI with additional feedback signals – derivative of torsional torque.

		J2 COEFFICIENT									
		0,02	0,05	0,1	0,2	0,5	1	2	5	10	20
J1 COEFFICIENT	0,02	2,5E+01	2,6E+01	2,7E+01	2,3E+01	1,2E+01	1,8E+02	5,2E+02	2,6E+03	7,7E+03	1,5E+04
	0,05	1,1E-02	1,6E-02	2,6E-03	1,4E-03	1,2E-03	1,1E-03	7,5E-04	3,0E-04	2,2E-04	7,6E-05
	0,1	1,1E-02	1,6E-02	2,2E-03	9,8E-04	8,5E-04	7,6E-04	5,2E-04	2,1E-04	1,8E-04	5,5E-05
	0,2	1,1E-02	1,6E-02	2,3E-03	1,0E-03	8,9E-04	7,9E-04	5,4E-04	2,1E-04	1,9E-04	6,2E-05
	0,5	1,2E-02	1,6E-02	2,6E-03	1,4E-03	1,2E-03	1,1E-03	7,6E-04	2,9E-04	2,6E-04	9,3E-05
	1	1,0E-02	1,5E-02	3,1E-03	1,8E-03	1,7E-03	1,5E-03	1,0E-03	4,0E-04	3,4E-04	1,4E-04
	2	9,1E-03	1,4E-02	3,2E-03	2,1E-03	2,0E-03	1,7E-03	1,2E-03	4,8E-04	3,8E-04	1,8E-04
	5	7,9E-03	1,3E-02	3,0E-03	2,1E-03	1,9E-03	1,6E-03	1,1E-03	4,8E-04	4,4E-04	3,2E-04
	10	6,1E-03	1,0E-02	2,3E-03	1,6E-03	1,4E-03	1,2E-03	8,2E-04	4,4E-04	4,1E-04	3,8E-04
	20	3,1E-03	5,3E-03	1,2E-03	8,3E-04	7,4E-04	6,4E-04	5,3E-04	3,7E-04	2,4E-04	5,9E-04

Table 8. Relative energy of high frequency components for 2-DOF PID Controller

		J2 COEFFICIENT									
		0,02	0,05	0,1	0,2	0,5	1	2	5	10	20
J1 COEFFICIENT	0,02	3,2E+01	3,5E+01	4,0E+01	5,3E+01	1,1E+02	2,8E+02	8,8E+02	4,8E+03	1,8E+04	5,4E+04
	0,05	7,6E-03	8,2E-03	1,1E-03	1,9E-04	1,0E-04	9,4E-05	8,0E-05	3,8E-05	1,2E-05	3,5E-04
	0,1	7,5E-03	8,1E-03	1,0E-03	1,1E-04	1,6E-05	6,7E-06	4,2E-06	2,7E-06	1,1E-06	3,5E-04
	0,2	7,6E-03	8,2E-03	1,1E-03	1,1E-04	1,8E-05	9,4E-06	6,6E-06	3,2E-06	1,4E-06	3,4E-04
	0,5	7,3E-03	8,0E-03	1,2E-03	1,4E-04	4,3E-05	3,3E-05	2,7E-05	1,1E-05	5,7E-06	3,4E-04
	1	5,4E-03	7,3E-03	1,4E-03	1,9E-04	1,0E-04	9,2E-05	7,7E-05	3,2E-05	1,5E-05	3,2E-04
	2	2,5E-03	5,1E-03	1,2E-03	2,4E-04	1,8E-04	1,7E-04	1,4E-04	6,3E-05	2,6E-05	3,3E-04
	5	6,5E-04	1,6E-03	4,3E-04	2,2E-04	2,1E-04	1,9E-04	1,5E-04	7,5E-05	3,3E-05	3,3E-04
	10	2,8E-04	5,2E-04	2,2E-04	1,8E-04	1,6E-04	1,5E-04	1,1E-04	7,2E-05	4,4E-05	3,5E-04
	20	1,1E-04	1,5E-04	1,0E-04	9,5E-05	8,9E-05	7,9E-05	6,3E-05	5,2E-05	1,1E-04	2,7E-04

it shows that the increase in ISE while lowering moments of inertia comes from excessive high frequency oscillations in control signal, whereas the increase of ISE while increasing moments of inertia comes from the limits of the dynamics of control system.

Table 9. Relative energy of high frequency components for Active Disturbance Rejection Control – feedback from load side

		J2 COEFFICIENT									
		0,02	0,05	0,1	0,2	0,5	1	2	5	10	20
J1 COEFFICIENT	0,02	5,4E+01	5,6E+01	5,7E+01	5,7E+01	5,9E+01	6,2E+01	6,5E+01	6,1E+01	1,0E+01	1,1E-02
	0,05	6,1E+01	6,6E+01	6,8E+01	6,9E+01	7,2E+01	7,3E+01	7,6E+01	7,8E+01	3,5E+01	7,2E-03
	0,1	6,6E+01	7,3E+01	7,4E+01	7,6E+01	7,6E+01	8,1E+01	7,7E+01	7,2E+01	3,1E-02	6,3E-03
	0,2	7,1E+01	7,7E+01	7,4E+01	6,8E+01	6,5E+01	6,8E+01	6,3E+01	9,6E-02	1,6E-02	5,8E-03
	0,5	7,3E+01	7,3E+01	6,2E+01	5,5E+01	5,1E+01	4,9E+01	2,7E-01	3,9E-02	1,3E-02	5,5E-03
	1	7,5E+01	7,0E+01	5,9E+01	5,0E+01	4,6E+01	5,8E-01	1,2E-01	2,9E-02	1,2E-02	5,3E-03
	2	7,7E+01	7,1E+01	5,2E+01	4,5E+01	1,0E+00	2,0E-01	5,1E-02	3,0E-02	2,5E-02	6,0E-03
	5	8,1E+01	7,1E+01	4,6E+01	1,3E+00	1,7E-01	4,2E-02	6,8E-02	4,7E-02	2,6E-02	7,4E-03
	10	8,6E+01	6,5E+01	2,3E+00	2,5E-01	2,7E-02	1,1E-01	6,4E-02	5,2E-02	1,4E-02	5,5E-03
	20	7,6E+01	3,3E+00	7,7E-01	8,6E-02	3,2E-02	3,3E-02	5,0E-02	2,3E-02	1,2E-02	5,6E-03

Table 10. Relative energy of high frequency components for Active Disturbance Rejection Control – feedback from motor side

		J2 COEFFICIENT									
		0,02	0,05	0,1	0,2	0,5	1	2	5	10	20
J1 COEFFICIENT	0,02	4,5E+01	4,6E+01	4,8E+01	5,1E+01	6,2E+01	8,6E+01	1,6E+02	5,5E+02	1,8E+03	5,1E+03
	0,05	4,1E+01	4,2E+01	4,3E+01	4,4E+01	4,9E+01	5,8E+01	7,9E+01	1,8E+02	4,4E+02	1,1E+03
	0,1	3,2E+01	3,3E+01	3,3E+01	3,4E+01	3,6E+01	3,8E+01	4,3E+01	5,9E+01	8,2E+01	1,0E+02
	0,2	1,2E-02	9,9E-03	2,7E-03	1,9E-03	1,8E-03	1,7E-03	1,4E-03	5,9E-04	2,3E-04	4,6E-04
	0,5	1,1E-02	9,1E-03	2,1E-03	1,1E-03	1,0E-03	9,7E-04	7,9E-04	3,1E-04	1,3E-04	4,5E-04
	1	7,0E-03	8,3E-03	2,4E-03	1,2E-03	1,1E-03	1,0E-03	8,4E-04	3,2E-04	1,3E-04	4,8E-04
	2	2,9E-03	5,4E-03	2,2E-03	1,2E-03	1,1E-03	1,0E-03	8,2E-04	3,0E-04	1,3E-04	4,6E-04
	5	1,2E-03	1,9E-03	1,1E-03	9,4E-04	8,9E-04	7,9E-04	6,0E-04	2,2E-04	1,1E-04	4,3E-04
	10	7,6E-04	9,0E-04	7,2E-04	6,8E-04	6,3E-04	5,5E-04	4,1E-04	1,6E-04	1,2E-04	5,2E-04
	20	3,6E-04	3,7E-04	3,5E-04	3,4E-04	3,1E-04	2,6E-04	1,9E-04	1,1E-04	2,0E-04	4,2E-04

The influence of varying backlash is presented in Tables 11 and 12. For motor side – the backlash has significant influence (in terms of ISE) on the control process from the width of 10 degrees, the exception is the ADRC with feedback from load side, which is affected by the backlash starting from very small backlash width. For the load side – the ISE begins to increase for all controllers even for the smallest investigated amount of backlash. The relative high frequency components are not significantly affected by the backlash, except for the ADRC with feedback from load side.

All in all, it can be concluded that:

- the ADRC load side feedback controller ensures best values of quality indicators if the object parameters match the ones the system was designed for,
- the PI with additional feedback signals along with the 2-DOF PID ensure satisfying robustness of the control system,
- the relative energy of high frequency components in the control signal can be used as a marker of the control loop stability.

Table 11. Logarithm of ISE for selected backlash width for motor and load side.

		Backlash width (deg.)										
		0	0,1	0,2	0,5	1	2	5	10	20	30	50
LOAD	w1w2	1,86	2,03	2,15	2,19	2,53	2,80	2,66	2,62	3,19	3,53	3,93
	T_TORS	1,90	2,13	2,17	2,42	2,55	2,91	2,80	2,65	3,24	3,55	3,93
	2DOF-PID	1,99	2,09	2,14	2,23	2,55	2,75	2,62	2,64	3,20	3,55	3,94
	ADRC_MOT	1,92	2,06	2,11	2,20	2,53	2,76	2,62	2,62	3,17	3,52	3,92
	ADRC_LOAD	1,62	1,65	1,70	2,40	3,19	3,86	4,55	5,00	5,41	5,63	5,89
		Backlash width (deg.)										
		0	0,1	0,2	0,5	1	2	5	10	20	30	50
MOTOR	w1w2	1,69	1,70	1,70	1,69	1,75	1,90	1,75	1,69	2,57	3,00	3,41
	T_TORS	1,64	1,65	1,64	1,65	1,64	2,02	1,86	1,64	2,66	3,08	3,45
	2DOF-PID	1,91	1,91	1,91	1,92	1,96	1,99	1,94	1,91	2,64	3,06	3,46
	ADRC_MOT	1,79	1,79	1,78	1,80	1,86	1,91	1,84	1,78	2,50	2,91	3,35
	ADRC_LOAD	1,77	2,72	3,19	3,97	4,60	5,07	5,66	6,09	6,49	6,72	6,99

Table 12. Relative energy of high frequency components in control signal as a function of selected backlash width.

	alfa (deg.)									
	0	0,1	0,2	0,5	1	2	5	10	20	50
w1w2	0,0038	0,0039	0,0039	0,0038	0,0035	0,0052	0,0035	0,0037	0,0103	0,0165
T_TORS	0,0015	0,0019	0,0016	0,0018	0,0020	0,0053	0,0040	0,0017	0,0080	0,0143
2DOF-PID	0,0001	0,0003	0,0003	0,0004	0,0008	0,0016	0,0008	0,0006	0,0082	0,0143
ADRC_MOT	0,0010	0,0013	0,0013	0,0014	0,0019	0,0027	0,0017	0,0013	0,0118	0,0162
ADRC_LOAD	0,5790	1,7648	2,6698	3,8150	6,9905	9,2767	6,7542	0,3395	0,4096	0,5526

References

1. Wicher, B.: Model of ADRC speed control system for complex mechanical object with backlash. In: 22nd International Conference on Methods and Models in automation and Robotics, 28–31 August 2017, Miedzyzdroje, Poland (2017)
2. Wicher, B., Brock, S.: Active disturbance rejection control based load side speed controller for two mass system with backlash. In: 2018 IEEE 18th International Power Electronics and Motion Control Conference (PEMC), 26–30 August. Budapest, Hungary (2018)
3. Nordin, M., Gutman, P.-O.: Controlling mechanical systems with backlash—a survey. Automatica **38**(10), 1633–1649 (2002)
4. Gao, Z., Zhao, S.: An active disturbance rejection based approach to vibration suppression in two-inertia systems. In: American Control Conference, Baltimore, MD, USA (2010)
5. Muszynski, R., Deskur, J.: Damping of torsional vibrations in high-dynamic industrial drives. IEEE Trans. Industr. Electron. **57**(2), 544–552 (2010)
6. Szabat, K., Orlowska–Kowalska, T.: Vibration suppression in a two-mass drive system using pi speed controller and additional feedbacks—comparative study. IEEE Trans. Indust. Electron. **54**(2), 1193–1206 (2007)
7. Han, J.: From PID to active disturbance rejection control. IEEE Trans. Industr. Electron. **56**(3), 900–906 (2009)
8. Gao, Z.: Scaling and bandwidth patametrization based controller tuning. In: American Control Conference, Denver, Colorado (2003)

Discrete-Time Active Disturbance Rejection Control: A Delta Operator Approach

Mario Ramírez-Neria[1(✉)], Alberto Luviano-Juárez[2], Norma Lozada-Castillo[2], Gilberto Ochoa-Ortega[1], and Rafal Madonski[3]

[1] Mechatronics Department, Universidad Politecnica del Valle de Mexico, Av. Mexiquense S/N, C.P. 54910 Tultitlan Estado de México, Mexico
mramirezn.ctrl@gmail.com, gochoa79@gmail.com
[2] UPIITA-IPN, Av. IPN 2580 Col. Barrio La Laguna Ticomán, Mexico City, Mexico
aluvianoj@ipn.mx, gioconda@esfm.ipn.mx
[3] Energy Electricity Research Center, International Energy College, Jinan University, Zhuhai 519070, P. R. China
rafal.madonski@jnu.edu.cn

Abstract. The extended state observer (ESO) is a key component in any active disturbance rejection control (ADRC) scheme. Its discrete implementation is thus crucial in practical applications. In this paper, a discrete delta operator ESO is introduced and analyzed. As a starting point, a discrete shift operator ESO is used for transformation from continuous-time to delta operator discrete-time. A quantitative comparison between shift operator and delta operator ESOs (with different sampling time Δ[s]) is conducted here using a servomechanism system. The obtained numerical results show the proposed delta operator observer to work better than the shift operator observer in certain aspects.

Keywords: ADRC · Discrete extended state observer · Delta operator

1 Introduction

The extended state observer (ESO) is the key component in any active disturbance rejection control (ADRC) scheme [4]. Its discrete implementation is thus crucial in practical applications. When the controller/observer part is implemented for the purpose of real-time experiment, hardware with finite precision is often required. At the same time, certain aspects are important to designers during control implementation, such as minimum precision and sampling rate required to provide desired performance. High sampling rate is usually preferred in experiments since better approximation of continuous-time can be achieved [7].

This practically important problem of effective discretization of ADRC led to the development of various techniques in the literature. For instance, several conventional discrete-time variants of ADRC were examined and extended

A. Bartoszewicz et al. (Eds.): PCC 2020, AISC 1196, pp. 1383–1395, 2020.
https://doi.org/10.1007/978-3-030-50936-1_115

in [5]. Other example is [9], where the ESO was formulated and implemented as a current discrete estimator using typical discretization methods, such as a predictive discrete estimator with shift operator discretization. The goal was to maintain stable operation even at lower sampling rates. In the same work, several different methodologies and samples rates were implemented and compared. In general, some traditional discrete-time algorithms can be ill-conditioned when are applied to data taken at sampling rates that are high relative to the dynamics of the underlying continuous-time processes that is being sampled. This fact led to the development of so-called delta operator strategy [8], capable of unifying both continuous and discrete-time formulations. The delta operator creates a rapprochement between continuous and discrete dynamic system models and establishes a natural framework to investigate the behavior of discrete dynamic models in fast sampling limit [10].

Hence, a discrete delta operator ESO is proposed here. After a brief review of delta operator in the next section, Sect. 3 shows transformation from continuous-time to delta operator discrete-time based on the discrete shift operator ESO. Section 4 contains simulation results where the shift operator ESO is compared with the proposed delta operator ESO using a range of sampling rates. Finally, concluding remarks are given in Sect. 5.

2 Review of Delta Operator

In this section, some key concepts related to the delta operator are recalled first. Details on it can be found in [8]. Specific information on the delta operator in the context of state estimation and identification can be found in [1,6].

Definition 1. *Domain of possible non-negative "times" $\Omega^+(\Delta)$ is:*

$$\Omega^+(\Delta) = \begin{cases} \mathbb{R}^+ \cup \{0\} & : \Delta = 0, \\ \{0, \Delta, 2\Delta, 3\Delta, \ldots\} & : \Delta \neq 0 \end{cases}, \tag{1}$$

where Δ is the sampling period (discrete-time), or $\Delta = 0$ in continuous-time.

Definition 2. *Function $x(t)$, $t \in \Omega^+(\Delta)$, is given as $x(t) : \Omega^+ \to \mathbb{C}$.*

Definition 3. *The Δ operator is defined as*[1]*:*

$$\Delta x(t) \triangleq \frac{x(t+\Delta) - x(t)}{\Delta}, \quad \Delta \neq 0,$$

and satisfies a following condition:

$$\lim_{\Delta \to 0^+} \Delta(x(t)) = \frac{dx(t)}{dt}.$$

[1] Expressions 'delta operator' and 'Δ operator' will be used interchangeably throughout the work.

Definition 4. *The unified integration operation* $\mathbf{S}$ *is given as follows:*

$$\mathbf{S}_{t_1}^{t_2} x(\tau)d\tau = \begin{cases} \int_{t_1}^{t_2} x(\tau)d\tau & : \Delta = 0 \\ \Delta \sum\limits_{l=t_1/\Delta}^{l=t_2/\Delta-1} x(l\Delta) & : \Delta \neq 0 \end{cases}, \quad t_1, t_2 \in \Omega^+.$$

Definition 5. *The generalized exponential matrix (GEM) is defined as:*

$$E(A,t,\Delta) = \begin{cases} e^{At} & : \Delta = 0 \\ (I + A\Delta)^{t/\Delta} & : \Delta \neq 0 \end{cases},$$

where I *is an identity matrix and* $A \in \mathbb{C}^{n\times n}$. *The GEM is the fundamental matrix of* $\Delta x(t) = Ax(t)$. *Hence, the solution of system:*

$$\Delta x(t) = Ax(t);\; x(0) = x_0, \tag{2}$$

is $x(t) = E(A,t,\Delta)x_0$. *In a similar fashion, for system:*

$$\Delta x(t) = Ax(t) + Bu(t),\; x(0) = x_0, \tag{3}$$

the solution $x(t)$ *is given by:*

$$x(t) = E(A,t,\Delta)x_0 + \mathbf{S}_0^t E(A, t-\tau-\Delta, \Delta)Bu(\tau)d\tau.$$

One of the main features of the Δ operator is the stability region obtained in relation to sampling period. A following definition supports this characteristic:

Definition 6 (Stability boundary). *The solution of (2) is said to be asymptotically stable if, and only if, for all* x_0, $x(t) \to 0$ *as time elapses. The stability arises if and only if* $E(A,t,\Delta) \to 0$ *as* $t \to \infty$ *if and only if every eigenvalue of* A, *given by* λ_i, $i = 1, \ldots, n$ *satisfies:*

$$Re\{\lambda_i\} + \frac{\Delta}{2}|\lambda_i|^2 < 0. \tag{4}$$

The stability boundary is a circle with center $(-1/\Delta, 0)$ *and radius* $1/\Delta$ *- see Fig. 1. If all the eigenvalues* $\lambda_i, i = 1\ldots, n$ *of* A *satisfy condition (4), then the solution is asymptotically stable.*

3 From Continuous ESO to Discrete-Time Delta Operator

In this part, a step-by-step transformation from continuous to discrete time is shown. Let us first consider a following second order perturbed system:

$$\ddot{y}(t) = \varphi(t, y(t), \dot{y}(t)) + \psi(t) + bu(t) \tag{5}$$

where $y(t)$ is the output, $u(t)$ is the input, b is the system constant input gain, $\varphi(t, y(t), \dot{y}(t))$ is the endogenous dynamics, and $\psi(t)$ represents the exogenous disturbance. Following [12], both endogenous and exogenous dynamics can be lumped into a total disturbance term $f(t, y(t), \dot{y}(t), \psi(t))$. The system (5) can be now represented alternatively as a perturbed linear system:

$$\ddot{y}(t) = f(\cdot) + bu(t), \tag{6}$$

which is both observable and controllable.

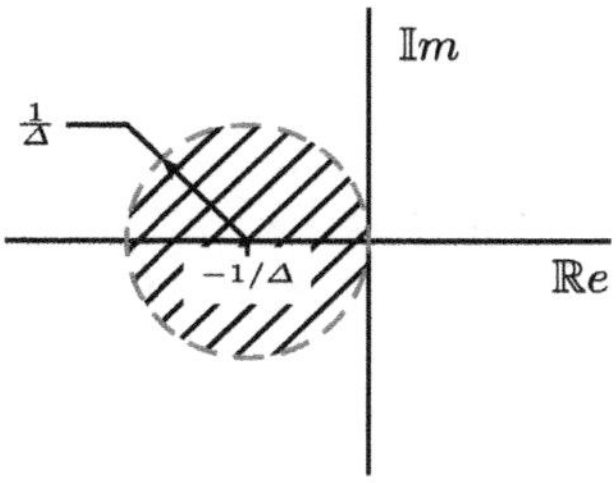

Fig. 1. Stability region for the unified operator.

3.1 Continuous ESO Design

By arbitrarily selecting state variables as $x_1(t) = y(t)$, $x_2(t) = \dot{y}(t)$, and $x_3(t) = f(\cdot)$, the system (6) in state space writes:

$$\dot{x}_1(t) = x_2(t), \quad \dot{x}_2(t) = x_3(t) + bu(t), \quad \dot{x}_3(t) = \dot{f}. \tag{7}$$

The extended state space model (7) represented in matrix form gives:

$$\begin{aligned} \dot{\mathbf{x}}(t) &= \mathbf{A}\mathbf{x}(t) + \mathbf{B}u(t) + \mathbf{E}\dot{f}, \\ y(t) &= \mathbf{C}\mathbf{x}(t) + \mathbf{D}u(t). \end{aligned} \tag{8}$$

where $\mathbf{A} = \begin{bmatrix} 0 & 1 & 0 \\ 0 & 0 & 1 \\ 0 & 0 & 0 \end{bmatrix}$, $\mathbf{B} = \begin{bmatrix} 0 \\ b \\ 0 \end{bmatrix}$, $\mathbf{E} = \begin{bmatrix} 0 \\ 0 \\ 1 \end{bmatrix}$ $\mathbf{C} = \begin{bmatrix} 1 & 0 & 0 \end{bmatrix}$, and $\mathbf{D} = 0$.

The state vector $\mathbf{x} = \begin{bmatrix} x_1 & x_2 & x_3 \end{bmatrix}^\top$ contains the extended state $x_3(t) = f(\cdot)$, which denotes solely the total disturbance term. Here, we assume that $f(\cdot)$ is smooth with bounded time derivatives, in this specific case $\dot{f}(\cdot) \approx 0$.

We can now construct a continuous time ESO for system (8) as:

$$\begin{aligned} \dot{\hat{\mathbf{x}}}(t) &= \mathbf{A}\hat{\mathbf{x}}(t) + \mathbf{B}u(t) + \mathbf{L}(y(t) - \hat{y}(t)), \\ \hat{y}(t) &= \mathbf{C}\hat{\mathbf{x}}(t), \end{aligned} \tag{9}$$

which can be rewritten using the output variable as:

$$\begin{aligned} \dot{\hat{\mathbf{x}}}(t) &= [\mathbf{A} - \mathbf{LC}]\, \hat{\mathbf{x}}(t) + \mathbf{B}u(t) + \mathbf{L}y(t), \\ \hat{y}(t) &= \mathbf{C}\hat{x}(t). \end{aligned} \tag{10}$$

The observer gains $\mathbf{L}$ can be straightforwardly computed by matching the poles of the characteristic equation with those of a Hurwitz polynomial as:

$$\lambda_o(s) = |s\mathbf{I} - (\mathbf{A} - \mathbf{LC})| := (s + \omega_0)^3, \tag{11}$$

which results in $\mathbf{L} = \left[3\omega_0,\ 3\omega_0^2,\ \omega_0^3\right]^T$, where $\omega_0 > 0$ is the desired observer bandwidth (design parameter).

3.2 Discretization of Shift Operator ESO

The continuous state space model in (8) is first discretized by applying a zero-order hold (ZOH) model [9], which results in:

$$\begin{aligned}\mathbf{x}((n+1)\Delta) &= \mathbf{\Phi}(\Delta)\mathbf{x}(n\Delta) + \mathbf{\Gamma}(\Delta)u(n\Delta),\\ y(n\Delta) &= \mathbf{H}\mathbf{x}(n\Delta),\end{aligned} \tag{12}$$

where Δ[s] is the sampling time, $\mathbf{\Phi}(\Delta)$ is the transition matrix, and $\mathbf{\Gamma}(\Delta)$ is the input distribution matrix. To compute $\mathbf{\Phi}(\Delta)$ and $\mathbf{\Gamma}(\Delta)$, let us first define:

$$\mathbf{R}(\Delta) = \int_0^{\Delta} \mathbf{\Phi}(\tau)d\tau = \Delta\sum_{k=0}^{\infty}\frac{(\mathbf{A}\Delta)^k}{(k+1)!}, \tag{13}$$

which is the integral of the nonsingular transition matrix, and its computation plays a key role in shifting invariant discretization. The transition matrix is:

$$\mathbf{\Phi}(\Delta) = e^{\mathbf{A}\Delta} = \sum_{k=0}^{\infty}\frac{(\mathbf{A}\Delta)^k}{(k)!}, \tag{14}$$

or according to (13), it can be simplified as:

$$\mathbf{\Phi}(\Delta) = \mathbf{I} + \mathbf{A}\mathbf{R}(\Delta) = \mathbf{I} + \mathbf{R}(\Delta)\mathbf{A} \tag{15}$$

On the other hand, the input distribution matrix can be computed as:

$$\mathbf{\Gamma}(\Delta) = \mathbf{R}(\Delta)\mathbf{B} = \Delta\sum_{k=0}^{\infty}\frac{(\mathbf{A}\Delta)^k}{(k+1)!}\mathbf{B}. \tag{16}$$

Now, a shift operator ESO discretization is proposed from the discrete state space model (12) and it gives:

$$\begin{aligned}\hat{\mathbf{x}}((n+1)\Delta) &= \mathbf{\Phi}(\Delta)\hat{\mathbf{x}}(n\Delta) + \mathbf{\Gamma}(\Delta)u(n\Delta) + \mathbf{L_p}\left(y(n\Delta) - \hat{y}(n\Delta)\right),\\ \hat{y}(n\Delta) &= \mathbf{H}\hat{\mathbf{x}}(n\Delta),\end{aligned} \tag{17}$$

The observer gain vector can be defined as $\mathbf{L_p} = \mathbf{\Phi}(\Delta)\mathbf{L_c}$ with $\mathbf{L_c} = \left[l_1\ l_2\ l_3\right]^T$, which reduces the discrete shift operator ESO (17) to:

$$\begin{aligned}\hat{\mathbf{x}}((n+1)\Delta) &= \left(\mathbf{\Phi}(\Delta) - \mathbf{\Phi}(\Delta)\mathbf{L_c}\mathbf{H}\right)\hat{\mathbf{x}}(n\Delta) + \mathbf{\Gamma}(\Delta)u(n\Delta) + \mathbf{\Phi}(\Delta)\mathbf{L_c}y(n\Delta),\\ \hat{y}(n\Delta) &= \mathbf{H}\hat{x}(n\Delta).\end{aligned} \tag{18}$$

The observer gain vector $\mathbf{L_c}$ can be determined by matching the poles of the discrete characteristic polynomial with those of a polynomial whose roots are located in the unit disk, that is:

$$P_o(z) = |z\mathbf{I} - (\mathbf{\Phi}(\Delta) - \mathbf{\Phi}(\Delta)\mathbf{L_c}\mathbf{H})| := (z - \beta)^3, \quad \beta < 1. \tag{19}$$

The relation between discrete and continuous ESO poles is defined by:

$$\beta = e^{-\omega_0 \Delta}, \tag{20}$$

and from (14) and (16), the matrices for (18) are $\mathbf{H} = \mathbf{C} = \begin{bmatrix} 1\,0\,0 \end{bmatrix}$ and:

$$\mathbf{\Phi}(\Delta) = \sum_{k=0}^{\infty} \frac{(\mathbf{A}\Delta)^k}{k!} = \begin{bmatrix} 1 & \Delta & \frac{\Delta^2}{2} \\ 0 & 1 & \Delta \\ 0 & 0 & 1 \end{bmatrix}, \quad \mathbf{\Gamma}(\Delta) = \Delta \sum_{k=0}^{\infty} \frac{(\mathbf{A}\Delta)^k}{(k+1)!}\mathbf{B} = \begin{bmatrix} b\frac{\Delta^2}{2} \\ b\Delta \\ 0 \end{bmatrix},$$

$$\mathbf{L}_c = \begin{bmatrix} 1 - \beta^3 \\ (1-\beta)^2(1+\beta)\frac{3}{2\Delta} \\ (1-\beta)^3\frac{1}{\Delta^2} \end{bmatrix}$$

3.3 Discretization of Delta Operator ESO

In order to discretize the continuous state space model (8) in terms of the delta operator, we follow [10] and hence we first subtract $\mathbf{x}(n\Delta)$ from both sides of the shift operator state (12) and then divide it by the sampling period Δ, so that the delta operator model can be expressed as:

$$\begin{aligned} \Delta\mathbf{x}(n\Delta) &= \frac{\mathbf{x}((n+1)\Delta) - \mathbf{x}(n\Delta)}{\Delta} = \frac{\mathbf{\Phi}(\Delta)\mathbf{x}(n\Delta) + \mathbf{\Gamma}(\Delta)u(n\Delta) - \mathbf{x}(n\Delta)}{\Delta} \\ y(n\Delta) &= \mathbf{H}\mathbf{x}(n\Delta), \end{aligned} \tag{21}$$

where $\Delta\mathbf{x}(n\Delta)$ is the difference quotient (see [2,8]). After some mathematical manipulations, the delta operator state model can be reduced to:

$$\begin{aligned} \Delta\mathbf{x}(n\Delta) &= \mathbf{\Phi}_\Delta(\Delta)\mathbf{x}(n\Delta) + \mathbf{\Gamma}_\Delta(\Delta)u(n\Delta), \\ \mathbf{x}((n+1)\Delta) &= \mathbf{x}(n\Delta) + \Delta\left[\Delta\mathbf{x}(n\Delta)\right], \\ y(n\Delta) &= \mathbf{H}\mathbf{x}(n\Delta), \end{aligned} \tag{22}$$

where $\mathbf{\Phi}_\Delta(\Delta) = (\mathbf{\Phi}(\Delta) - \mathbf{I})/\Delta$ is the delta transition matrix (see [3,11]) and $\mathbf{\Gamma}_\Delta(\Delta) = \mathbf{\Gamma}(\Delta)/\Delta$ is the delta input distribution. The average value of the transition matrix $\mathbf{R}_\Delta(\Delta)$ is:

$$\mathbf{R}_\Delta(\Delta) = \frac{\mathbf{R}(\Delta)}{\Delta} = \sum_{k=0}^{\infty} \frac{(\mathbf{A}\Delta)^k}{(k+1)!}. \tag{23}$$

Terms $\mathbf{\Gamma}_\Delta(\Delta)$ and $\mathbf{R}_\Delta(\Delta)$ can be related as:

$$\mathbf{\Phi}_\Delta(\Delta) = \mathbf{R}_\Delta(\Delta)\mathbf{A} = \mathbf{A}\mathbf{R}_\Delta(\Delta) \tag{24}$$

and the same goes for the delta input distribution $\mathbf{\Gamma}_{\Delta}(\Delta) = \mathbf{R}_{\Delta}(\Delta)\mathbf{B}$. By means of $\mathbf{R}_{\Delta}(\Delta)$, the delta operator state model (22) can be simplified:

$$\begin{aligned} \Delta\mathbf{x}(n\Delta) &= \mathbf{R}_{\Delta}(\Delta)\left[\mathbf{A}\mathbf{x}(n\Delta) + \mathbf{B}u(n\Delta)\right], \\ \mathbf{x}((n+1)\Delta) &= \mathbf{x}(n\Delta) + \Delta\left[\Delta\mathbf{x}(n\Delta)\right], \\ y(n\Delta) &= \mathbf{H}\mathbf{x}(n\Delta). \end{aligned} \tag{25}$$

Some properties of $\mathbf{R}_{\Delta}(\Delta)$, important in the context of this work, are:

- By applying the Leibnitz's rule, $\lim_{\Delta\to 0} \mathbf{R}_{\Delta}(\Delta) = \mathbf{I}$. It is important to remark that by design, $\lim_{\Delta\to 0} \mathbf{\Phi}_{\Delta}(\Delta) = \mathbf{A}$, $\lim_{\Delta\to 0} \mathbf{\Gamma}_{\Delta}(\Delta) = \mathbf{B}$, and $\lim_{\Delta\to 0} \Delta\mathbf{x}(n\Delta) = \frac{d\mathbf{x}}{dt}$. Consequently, for fast sampling rates ($\Delta \to 0$), the delta operator state model (25) tends to the continuous state space model in (8).
- The matrix $\mathbf{R}_{\Delta}(\Delta)$ is never singular.

Now, from (25), a delta operator ESO discretization is proposed as:

$$\begin{aligned} \Delta\hat{\mathbf{x}}(n\Delta) &= \mathbf{R}_{\Delta}(\Delta)\left[\mathbf{A}\hat{\mathbf{x}}(n\Delta) + \mathbf{B}u(n\Delta) + \mathbf{L_c}(y(n\Delta) - \hat{y}(n\Delta))\right], \\ \hat{\mathbf{x}}((n+1)\Delta) &= \hat{\mathbf{x}}(n\Delta) + \Delta\left[\Delta\hat{\mathbf{x}}(n\Delta)\right], \\ \hat{y}(n\Delta) &= \mathbf{H}\hat{\mathbf{x}}(n\Delta), \end{aligned} \tag{26}$$

and can be rewritten as:

$$\begin{aligned} \Delta\hat{\mathbf{x}}(n\Delta) &= \mathbf{R}_{\Delta}(\Delta)\left[\left(\mathbf{A} - \mathbf{L_c}\mathbf{H}\right)\hat{\mathbf{x}}(n\Delta) + \mathbf{B}u(n\Delta) + \mathbf{L_c}y(n\Delta)\right], \\ \hat{\mathbf{x}}((n+1)\Delta) &= \hat{\mathbf{x}}(n\Delta) + \Delta\left[\Delta\hat{\mathbf{x}}(n\Delta)\right], \\ \hat{y}(n\Delta) &= \mathbf{H}\hat{\mathbf{x}}(n\Delta). \end{aligned} \tag{27}$$

The gain observer vector $\mathbf{L_c}$ can be computed by placing the poles of the discrete delta operator characteristic polynomial in the γ domain:

$$P_o(\gamma) = \left|\gamma\mathbf{I} - \mathbf{R}_{\Delta}(\Delta)\left(\mathbf{A} - \mathbf{L_c}\mathbf{H}\right)\right| := (\gamma + \beta)^3. \tag{28}$$

The poles of the discrete delta operator ESO and those of the continuous ESO poles can be related as:

$$\beta = -\frac{1 - e^{\omega_0 \Delta}}{\Delta}. \tag{29}$$

Finally, the matrices for ESO (27) can be computed as $\mathbf{H} = \mathbf{C} = \begin{bmatrix} 1 \; 0 \; 0 \end{bmatrix}$ and:

$$\mathbf{R}_{\Delta}(\Delta) = \sum_{k=0}^{\infty} \frac{(\mathbf{A}\Delta)^k}{(k+1)!} = \begin{bmatrix} 1 & \frac{\Delta}{2} & \frac{\Delta^2}{6} \\ 0 & 1 & \frac{\Delta}{2} \\ 0 & 0 & 1 \end{bmatrix}, \quad \mathbf{L}_c = \begin{bmatrix} 3\beta - \frac{3\Delta\beta^2}{2} + \frac{\Delta^2\beta^3}{3} \\ 3\beta^2 - \Delta\beta^3 \\ \beta^3 \end{bmatrix}. \tag{30}$$

3.4 Control Rule Design

The control law in ADRC can be synthesized for both shift operator ESO and delta operator ESO using the estimated vector $\hat{\mathbf{x}}(n\Delta)$. The total disturbance cancellation is achieved using the estimated state variable $\hat{x}_3(n\Delta) = \hat{f}(n\Delta)$. The feedback control for the remaining system dynamics is realized with $u_0(n\Delta)$. Hence, the two-channel control rule has the form:

$$u(n\Delta) = \frac{u_0(n\Delta) - \hat{x}_3(n\Delta)}{b}, \quad (31)$$

with

$$u_0(n\Delta) = \omega_c^2(r(n\Delta) - \hat{x}_1(n\Delta)) - 2\omega_c\hat{x}_2(n\Delta). \quad (32)$$

and $r(n\Delta)$ being the reference motion profile.

4 Simulation Validation

4.1 Plant Model Description

For our tests, we consider a DC servomechanism expressed with a following second order mathematical model:

$$J\ddot{\theta}(t) + V_f\dot{\theta}(t) = \tau(t) = k_a V(t), \quad (33)$$

where $\theta(t)$ is the angular position, $\tau(t)$ is the input torque, $u(t)$ is the control input voltage, J is the motor and load inertia while, V_f is the viscous friction, and k_a is the amplifier gain. The above servomechanism model can be rewritten as:

$$\ddot{\theta}(t) + a\dot{\theta}(t) = bu(t), \quad (34)$$

with $a = V_f/J$ and $b = k_a/J$. The exact model parameters are selected as where $a = 0.45$ and $b = 31$, which emulates a real system identified in [13]. Now, let us assume that (34) is affected by an exogenous disturbances $\psi(t)$:

$$\ddot{\theta}(t) = -a\dot{\theta}(t) + \psi(t) + bu(t). \quad (35)$$

By defining the total disturbance function as $f(t) = -a\dot{\theta}(t) + \psi(t)$, the simplified linear perturbed system can be obtained as:

$$\ddot{\theta}(t) = f(t) + bu(t). \quad (36)$$

We arbitrarily select state variables as $x_1(t) = \theta(t)$, $x_2(t) = \dot{\theta}(t)$ and $x_3(t) = f(t)$. The extended state space model of (36) can be thus expressed as:

$$\begin{aligned} \dot{\mathbf{x}}(t) &= \mathbf{A}\mathbf{x}(t) + \mathbf{B}u(t) + \mathbf{E}\dot{f}(t), \\ y(t) &= \mathbf{C}\mathbf{x}(t) + \mathbf{D}u(t), \end{aligned} \quad (37)$$

where $\mathbf{x} = \begin{bmatrix} x_1 \\ x_2 \\ x_3 \end{bmatrix}$, $\mathbf{A} = \begin{bmatrix} 0\,1\,0 \\ 0\,0\,1 \\ 0\,0\,0 \end{bmatrix}$, $\mathbf{B} = \begin{bmatrix} 0 \\ b \\ 0 \end{bmatrix}$, $\mathbf{E} = \begin{bmatrix} 0 \\ 0 \\ 1 \end{bmatrix}$, and $\mathbf{C} = \begin{bmatrix} 1\,0\,0 \end{bmatrix}$.

Following the design steps from previous section, the servomechanism model (35) can be discretized by the ZOH. Hence, the discrete shift operator ESO and the discrete delta operator ESO can be both implemented using the extended state space (37), which is done next.

4.2 Shift Operator ESO

A discrete shift operator ESO for system (37) can be designed as in (18):

$$\begin{aligned}\hat{\mathbf{x}}((n+1)\Delta) &= (\mathbf{\Phi}(\Delta) - \mathbf{\Phi}(\Delta)\mathbf{L_c}\mathbf{H})\,\hat{\mathbf{x}}(n\Delta) + \mathbf{\Gamma}(\Delta)u(n\Delta) + \mathbf{\Phi}(\Delta)\mathbf{L_c}y(n\Delta)\\ \hat{y}(n\Delta) &= \mathbf{H}\hat{x}(n\Delta). \end{aligned} \tag{38}$$

Using (37), (14), (16) and (19), we can compute the matrices needed for the implementation of discrete observer:

$$\mathbf{\Phi}(\Delta) = \begin{bmatrix} 1 & \Delta & \frac{\Delta^2}{2} \\ 0 & 1 & \Delta \\ 0 & 0 & 1 \end{bmatrix}, \ \mathbf{\Gamma}(\Delta) = \begin{bmatrix} b\frac{\Delta^2}{2} \\ b\Delta \\ 0 \end{bmatrix},$$

$$\mathbf{H} = \begin{bmatrix} 1 \, 0 \, 0 \end{bmatrix}, \ \mathbf{L}_c = \begin{bmatrix} 1-\beta^3 \\ (1-\beta)^2(1+\beta)\frac{3}{2\Delta} \\ (1-\beta)^3\frac{1}{\Delta^2} \end{bmatrix}.$$

4.3 Delta Operator ESO

The proposed discrete delta operator ESO from (37) can be obtained using (27):

$$\begin{aligned}\Delta\hat{\mathbf{x}}(n\Delta) &= \mathbf{R}_\Delta(\Delta)\left[(\mathbf{A}-\mathbf{L_c}\mathbf{H})\,\hat{\mathbf{x}}(n\Delta) + \mathbf{B}u(n\Delta) + \mathbf{L_c}y(n\Delta)\right],\\ \hat{\mathbf{x}}((n+1)\Delta) &= \hat{\mathbf{x}}(n\Delta) + \Delta\left[\Delta\hat{\mathbf{x}}(n\Delta)\right],\\ \hat{y}(n\Delta) &= \mathbf{H}\hat{\mathbf{x}}(n\Delta).\end{aligned} \tag{39}$$

The matrices needed to implement the proposed discrete observer (39) can be directly computed using (37), (23), and (28):

$$\mathbf{R}_\Delta(\Delta) = \begin{bmatrix} 1 & \frac{\Delta}{2} & \frac{\Delta^2}{6} \\ 0 & 1 & \frac{\Delta}{2} \\ 0 & 0 & 1 \end{bmatrix}, \ \mathbf{H} = \begin{bmatrix} 1 \, 0 \, 0 \end{bmatrix}, \ \mathbf{L}_c = \begin{bmatrix} 3\beta - \frac{3\Delta\beta^2}{2} - \frac{\Delta^2\beta^3}{3} \\ 3\beta^2 - \Delta\beta^3 \\ \beta^3 \end{bmatrix}.$$

4.4 Comparison Study

In order to compare the performance of shift operator ESO with the performance of the proposed delta operator ESO, similar design parameters are set in both cases for the controller part ($\omega_c = 10$, $b = 31$) and the observer part ($\omega_0 = 90$).

Test #1 (for $\Delta = 0.0001$). In the first test, we set a relatively small sampling time $\Delta = 0.0001$ s. Figure 2a shows the reference step signal $r(n\Delta)$ and the resultant ADRC performance using the tested discrete observers. The results with shift operator ESO $\theta_S(n\Delta)$ are plotted in blue and with the proposed delta operator ESO $\theta_D(n\Delta)$ in red. We can notice that both ESOs have similar performance for the given sampling time. Figure 2b shows the respective control signals

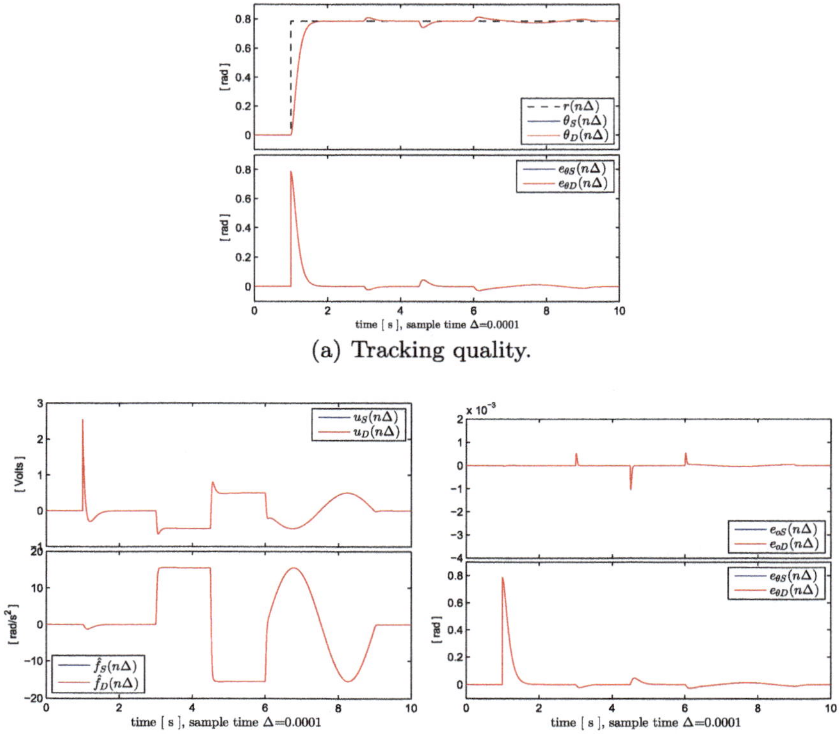

(a) Tracking quality.

(b) Controls and estimated disturbances.

(c) Estimation and tracking errors.

Fig. 2. Results of Test #1.

$u_S(n\Delta)$, $u_D(n\Delta)$ and total disturbance estimations $\hat{f}_S(n\Delta)$, $\hat{f}_D(n\Delta)$. The estimation errors $e_{oS}(n\Delta)$, $e_{oD}(n\Delta)$ and the tracking errors $e_{\theta S}(n\Delta)$, $e_{\theta D}(n\Delta)$ are shown in Fig. 3a. From the obtained results of the considered specific scenario, we can conclude that when $\Delta \to 0$, the performance of both shift and delta operator ESOs tend to the continuous ESO.

Test #2 (for $\Delta = 0.001$). In the second test, we increase the sampling time to $\Delta = 0.001$ s. Figure 3b shows that the shift operator ESO estimation error $e_{oS}(n\Delta)$ increases a little in comparison with the proposed delta operator ESO $e_{oD}(n\Delta)$. Consequently, the tracking error $e_{\theta S}(n\Delta)$ also increases in comparison with $e_{\theta D}(n\Delta)$.

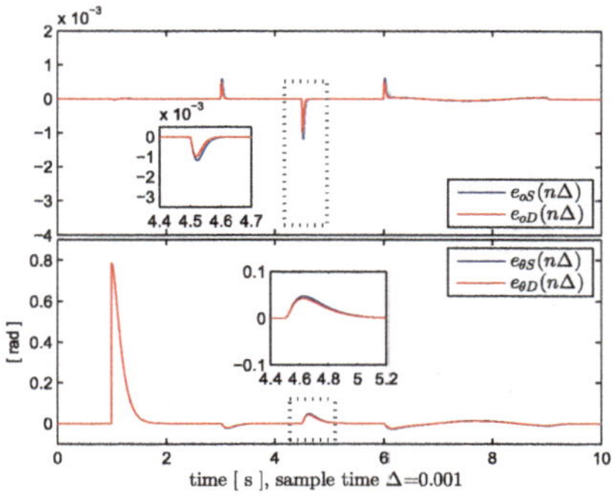

(a) Estimation and tracking errors

Fig. 3. Results of Test #2.

Test #3 (for $\Delta = 0.005$ and $\Delta = 0.006$). The sampling time was first set to $\Delta = 0.005$ s. Figure 4a shows the estimation error $e_{oS}(n\Delta)$ which increases in comparison with $e_{oD}(n\Delta)$ and with respect to Tests #1 and #2. The tracking error $e_{\theta S}(n\Delta)$ also increases in comparison with $e_{\theta D}(n\Delta)$. For $\Delta = 0.006$ s, the observation performance $e_{oS}(n\Delta)$ decreases, degrading the trajectory error $e_{\theta S}(n\Delta)$ as result (Fig. 4b). At the same time, the amplitude of $e_{oD}(n\Delta)$ has remained as comparable level to previous tests and, as a consequence, the tracking error $e_{\theta D}(n\Delta)$ also retains similar amplitude for all tests.

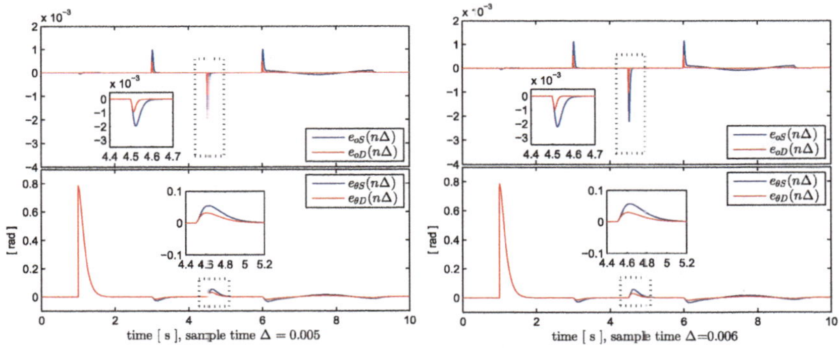

(a) Estimation and tracking errors. (b) Estimation and tracking errors.

Fig. 4. Results of Test #3.

Test #4 (for $\Delta = 0.01$). Finally, we set $\Delta = 0.01$ s and the observer error $e_{oS(n\Delta)}$ has higher amplitude affecting the performance of the tracking error $e_{\theta_S}(n\Delta)$, and although $e_{oD(n\Delta)}$ is smaller in amplitude, it presents undesired oscillations Fig. 5. Is important to remark that if we increase the sampling time to $\Delta > 0.01$ s, the delta operator ESO becomes unstable. Hence, we can conclude that, in this specific case study, the proposed delta operator ESO only works for relatively high sampling rates $\Delta \leq 0.01$ s.

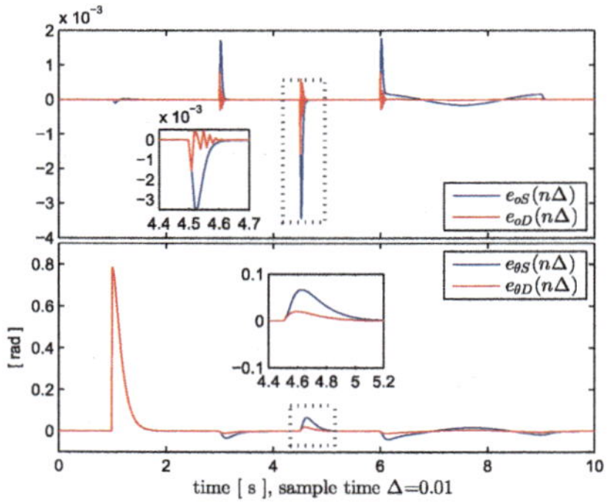

(a) Estimation and tracking errors.

Fig. 5. Results of Test #4.

5 Conclusion

A discrete ESO using delta operator was proposed in this work. It was compared in simulation environment with a shift operator ESO for different sampling times Δ[s]. The obtained numerical results showed that the introduced delta operator works better than shift operator ESO in cases with high sampling rate ($\Delta \leq 0.01$ s). At the same time, the proposed approach was found to be relatively easy to implement with only minor matrices needed to be designed.

References

1. Cortés-Romero, J., Luviano-Juárez, A., Sira-Ramírez, H.: A delta operator approach for the discrete-time active disturbance rejection control on induction motors. Math. Prob. Eng. **2013**, 9 (2013)
2. Feuer, A., Goodwin, G.: Sampling in Digital Signal Processing and Control. Springer Science & Business Media (2012)
3. Goodwin, G.C., Middleton, R.H., Poor, H.V.: High-speed digital signal processing and control. Proc. IEEE **80**(2), 240–259 (1992)
4. Han, J.: From PID to active disturbance rejection control. IEEE Trans. Industr. Electron. **56**(3), 900–906 (2009)
5. Herbst, G.: Practical active disturbance rejection control: bumpless transfer, rate limitation, and incremental algorithm. IEEE Trans. Industr. Electron. **63**(3), 1754–1762 (2016)
6. Luviano-Juárez, A., Cortés-Romero, J., Sira-Ramírez, H.: Algebraic identification and control of an uncertain DC motor using the delta operator approach. In: International Conference on Electrical Engineering Computing Science and Automatic Control, pp. 482–487 (2010)

7. Madonski, R., Herman, P.: On the usefulness of higher-order disturbance observers in real control scenarios based on perturbation estimation and mitigation. In: International Workshop on Robot Motion and Control, pp. 252–257 (2013)
8. Middleton, R.H., Goodwin, G.C.: Digital Control and Estimation: A Unified Approach. Prentice Hall, Englewood Cliffs (1990)
9. Miklosovic, R., Radke, A., Gao, Z.: Discrete implementation and generalization of the extended state observer. In: American Control Conference, pp. 1–6 (2006)
10. Neuman, C.P.: Properties of the delta operator model of dynamic physical systems. IEEE Trans. Syst. Man Cybern. **23**(1), 296–301 (1993)
11. Shor, M., Perkins, W.: Reliable control in the presence of sensor/actuator failures: a unified discrete/continuous approach. In: Conference on Decision and Control, pp. 1601–1606 (1991)
12. Sira-Ramirez, H., Luviano-Juárez, A., Ramírez-Neria, M., Zurita-Bustamante, E.W.: Active Disturbance Rejection Control of Dynamic Systems: A Flatness Based Approach. Butterworth-Heinemann (2018)
13. Villafuerte, R., Mondié, S., Garrido, R.: Tuning of proportional retarded controllers: theory and experiments. IEEE Trans. Control Syst. Technol. **21**(3), 983–990 (2012)

Fault-Tolerant Control and Design

Fault-Tolerant Design of a Balanced Two-Wheel Scooter

Ralf Stetter[1(✉)], Marcin Witczak[2], and Markus Till[1]

[1] Ravensburg-Weingarten University (RWU), Weingarten, Germany
ralf.stetter@rwu.de
[2] Institute of Control and Computational Engineering, University of Zielona Góra, Zielona Góra, Poland

Abstract. Fault-tolerant control is since several years a heavily researched scientific field that was successfully applied in numerous cases. In the last years this concept was accompanied by fault-tolerant design. It intends to enhance the controllability and diagnosability of technical systems through intelligent design as well as to increase the fault-tolerance of technical systems through inherently fault-tolerant design characteristics such as redundancy. The approaches, methods and tools of fault-tolerant design were applied to a balanced two-wheel scooter on different levels, ranging from a conscious requirements management to consciously chosen redundant elements on the most concrete level - the product geometry. On the functional level a virtual decision engine is presented, which allows the generation of correction factors for the control system.

Keywords: Fault-tolerant control · Fault-tolerant design

1 Introduction

The complexity of modern technical systems such as cars, aircrafts or production systems seems to be ever increasing. With each new product generation new functionalities as well as design and quality features arrive. However, with this rising complexity also the possibilities for faults increase. In this context, a fault is understood as an unpermitted deviation of one or more characteristic property or parameter of the system from the acceptable, regular condition [1], in contrast to failure which indicates a catastrophic event. The term "fault-tolerant control" summarizes algorithms and systems intended to accommodate the effects of faults and to prevent failure. Numerous research groups worldwide are aiming to expand the knowledge in this area and to develop innovative algorithms and systems. A concise review of their results can be found in the publications [1–4]. It is possible to enable and ease fault-tolerant control of technical systems by means of certain design characteristics such as an easy access to energy flows which have to be measured in order to detect faults. Additionally, certain design aspects can increase the fault-tolerance on their own (i.e. without employing

A. Bartoszewicz et al. (Eds.): PCC 2020, AISC 1196, pp. 1399–1410, 2020.
https://doi.org/10.1007/978-3-030-50936-1_116

fault-tolerant control), e.g. redundant elements. Both approaches can be summarized under the notion "fault-tolerant design"; this notion will be explained in detail in the next section. In later sections of this paper, fault-tolerant design will be discussed on the example of a balanced two-wheel scooter. This scooter is one of the use cases of large scale research project exploring the digital product lifecycle (DiP). Several variants were developed and serve for exploring all elements of a product-lifecycle including design, production and operation. Two realized variants are shown in Fig. 1 (compare also [5,6]).

Fig. 1. Balanced two-wheel scooters.

The main section of this paper is Sect. 3; it discusses several design aspects which constitute a fault-tolerant design of the balanced two-wheel scooter. The final section of the paper is a conclusion and summary.

2 Fault-Tolerant Design

Under the notion "fault-tolerant design" strategies, methods, algorithms, tools and insights are summarized which share the objective to support the synthesis of technical systems which are fault-tolerant because of their diagnosability and controllability but also because of their inherent fault-tolerant design qualities [4]. Up to now, only few research initiatives in this field can be identified and the orientation in this field is still challenging. Research initiatives in this area can be based on the existing rich body of research concerning systematic design and integrated product development [7–11]; however, fault-tolerant design was not yet a point of main emphasis in this field. A rather small body of research was so far directly concerned with fault-tolerant design. The general importance of this field is underlined by Rouissi and Hoblos [12]; they emphasize that the ability of a system to accommodate faults must be achieved by means of a conscious fault-tolerant design. Initial research works concern clearly demarcated fields such as wireless sensor networks [13], voting logic and redundant actuation devices [14], chip design [15], design of frequency converters [16] or artificial intelligence [17]. An initial approach to structure this field is made by Stetter [4]; it is further discussed in [18]. In this publications, it is proposed to base a structure for fault-tolerant design on the well-known models of product concretization in design science (e.g. Ponn and Lindemann [7]). The different levels for characteristics of fault-tolerant design are depicted in Fig. 2 (compare also [4]).

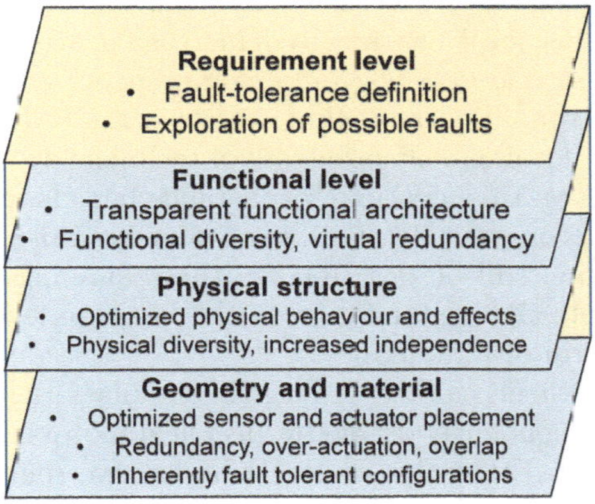

Fig. 2. Levels for characteristics of fault-tolerant design.

The design characteristics for fault-tolerant design for the balanced two-wheel scooter in the subsequent section are discussed in a sequence according to this model.

3 Characteristics of Fault-Tolerant Design

3.1 Fault-Tolerant Design on the Requirements Level

Requirements can be defined as the purpose, goals, constraints, and criteria associated with the development process of a technical system [19]. It is important to note that requirements are one of the most important factors in industrial system development (compare e.g. [20]). Four of ten top risks in projects are connected with requirements [21]. Analyses of industrial system development projects showed that only 52% of the originally allocated requirements appear in the final released version of the system [22]. The efficient and effective development of complex systems requires a conscious management of requirements which is usually called Requirements Management (RM) [23–25]. Current research integrates requirements into the digital product life-cycle; this integration is realised by using graph-based design languages based on UML [26]. A prerequisite for fault-tolerant control and design is a requirements management concerning monitoring; this field is researched as well [27]. The exploration of requirements is an essential step of fault-tolerant design and needs to include all fault possibilities, the expected and probable faults, the required level of fault-tolerance and the required form and amount of redundancy [4].

3.2 Design Characteristics on the Functional Level

It is of paramount importance to have a complete overview about the functionality of a product for all conscious design activities. The function level is the most abstract level which describes the technical solutions that will realize the requirements. This level has received increasing attention in the research community in the last years [28, 29] and a comprehensive framework was developed - the integrated function modelling framework (IFM) - which can also be integrated in digital development processes of technical systems (compare [29]). Fault-tolerant design on this level can for instance be achieved by means of analytical redundancies (compare [4]). For the scooter example, a virtual sensor and a fuzzy decision engine will be described in the remaining part of this section. It is important to note that redundancy on the function level is the highest and most independent form of redundancy.

A balanced two-wheel scooter is basically a platform equipped with two wheels which are independently driven by electrical motors. For the discussion in this paper, it is assumed that the scooter contains two sensors. The first sensor measures the inclination angle of the scooter and the person (α). The velocity of the wheels of the scooter is measured by a second sensor and from this the velocity of the scooter is calculated ($\dot{z}$). For this discussion it is assumed that both wheels have the same speed and consequently the scooter is moving in a straight line. Further assumptions are that both wheels of the scooter have continuously contact with the ground and do not experience slipping and that the driver is not expressing any unusual or unexpected behavior (and can therefore be modelled as rigid body). The two main parameters are shown in Fig. 3.

A dynamic analysis of the balanced two-wheel scooter is similar to the analysis of an inverted pendulum. This analysis results in a state space model that is based on the considerations of Grasser et al. 2002 [30] and the simplifications of Younis and Abdelati [31] (compare also [32]). From a continuous form a discrete from can be generated by means of the Euler methods. The discrete state space model including faults and disturbances can be formulated in the subsequent form:

$$\boldsymbol{x}_{k+1} = \boldsymbol{A}_k \boldsymbol{x}_k + \boldsymbol{B}_k \boldsymbol{u}_k + \boldsymbol{B}_k f_{a,k} + \boldsymbol{W}_1 \boldsymbol{w}_{1,k}, \tag{1}$$

$$\boldsymbol{y}_k = \boldsymbol{C}\boldsymbol{x}_k + \boldsymbol{C}_f f_{s,k} + \boldsymbol{W}_2 \boldsymbol{w}_{2,k}, \tag{2}$$

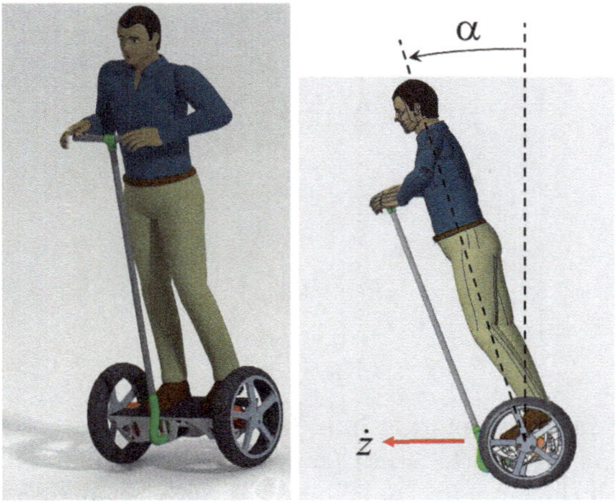

Fig. 3. Central parameters of a balanced two-wheel scooter.

with

$$\boldsymbol{A}_k = I + T_s \cdot \boldsymbol{A}, \quad \boldsymbol{B}_k = T_s \cdot \boldsymbol{B}, \tag{3}$$

where $\boldsymbol{x}$ is the state vector, $\boldsymbol{y}$ is the measurement vector and $\boldsymbol{u}$ is the input vector. The vector $f_{a,k}$ denotes an actuator fault whereas the vector $f_{s,k}$ denotes a sensor fault. $\boldsymbol{w}_{1,k}$ stands for an exogenous disturbance vector (which includes the discretization error) with a known distribution matrix $\boldsymbol{W}_1$. $\boldsymbol{w}_{2,k}$ denotes the measurement uncertainties with a known distribution matrix $\boldsymbol{W}_2$. The state includes the distance covered by the scooter ($\boldsymbol{z}$), the velocity of the scooter ($\dot{\boldsymbol{z}}$), the inclination angle of the scooter (α) and the angular velocity of this angle ($\dot{\alpha}$). The input $\boldsymbol{u}$ is the voltage applied to the electrical motors. The system matrices of this state space model are:

$$A = \begin{bmatrix} 0 & 1 & 0 & 0 \\ 0 & K_{x1} & K_{phi1} & 0 \\ 0 & 0 & 0 & 1 \\ 0 & K_{x2} & K_{phi2} & 0 \end{bmatrix} \tag{4}$$

$$B = \begin{bmatrix} 0 \\ K_{xu} \\ 0 \\ K_{phiu} \end{bmatrix} \tag{5}$$

where

$$K_{x1} = \frac{2K_m K_e (M_p l_r - Ip - M_p l^2)}{Rr^2 P_1}$$

$$K_{phi1} = \frac{M_p^2 g l^2}{P_1}$$

$$K_{x2} = \frac{2K_m K_e (r P_2 - M_p l)}{R r^2 P_1}$$

$$K_{phi2} = \frac{M_p g l P_2}{P_1}$$

$$P_1 = I_p P_2 + 2M_p l^2 (M_w \frac{I_w}{r^2})$$

$$P_2 = 2M_w + \frac{2I_w}{r^2} + M_p$$

By means of input estimation it is possible to realize an input estimation from the measurement of the state (compare e.g. [33]). It is then possible to compare this with the actually applied voltage and to generate a residual, which can be used for fault identification purposes:

$$z_{u,k} = \boldsymbol{u}_k - \hat{\boldsymbol{u}}_k. \tag{6}$$

In a fuzzy decision engine this residual information can be combined with some elements of the state, in this case the inclination angle. A fuzzy decision engine can combine information resulting from residuals with information about the state and experts knowledge in order to allow a well-founded decision whether a certain fault is present. This kind of decision engine can also be expanded into a fuzzy virtual actuator [34]. As stated above, the fuzzy decision engine is able to decide whether a certain fault is present. For the residual z and elements of the state (in this case the inclination angle α), membership functions μ_z and μ_α are derived that enable an initial evaluation of this input information. For this kind of objective, trapezoidal membership functions were found to be appropriate, because they represent the given input information in an efficient manner [35]. It is either possible to find the width of the membership functions by analysing the maximum and minimum values of the residuals or the state variables; i.e. the values of the parameters that define the width and inclination can be determined experimentally. For the accommodation of process noises, disturbances and mismatches between the plant and the analytical models, the core is an interval around zero; the size of this interval can be found through the analysis of experimental data [35]. In the development of the fuzzy decision engine, the discussion of design and control experts lead to the insight that the residuals are a very good fault indication but that a combination with other measured state variables in a fuzzy logic system can be very fruitful. For the scooter, the expert's knowledge concerning two given faults (one sensor fault; one actuator fault) was captured with three membership functions for the residual z_i concerning the input and the state variable α - μ_α - which serve as input of the fuzzy system. To give an example: the first input membership function μ_z was composed in the subsequent form:

$$\mu_z = \begin{cases} 1, & z < c \\ \frac{d-z}{d-c}, & c \leq z \leq d \\ 0, & z > d, \end{cases} \tag{7}$$

where c and d are parameters which are found based on the knowledge of experts or on experimental data. This membership function aims at characterising the situation that the respective residual indicates a fault and that the value of the residual is negative. The other membership functions were defined in analogous form. A similar structure was also chosen for the output variables. Figure 4 sh the structure of the resulting fuzzy interference system and membership functions for the evaluation of the residual and membership function for the output of one fault.

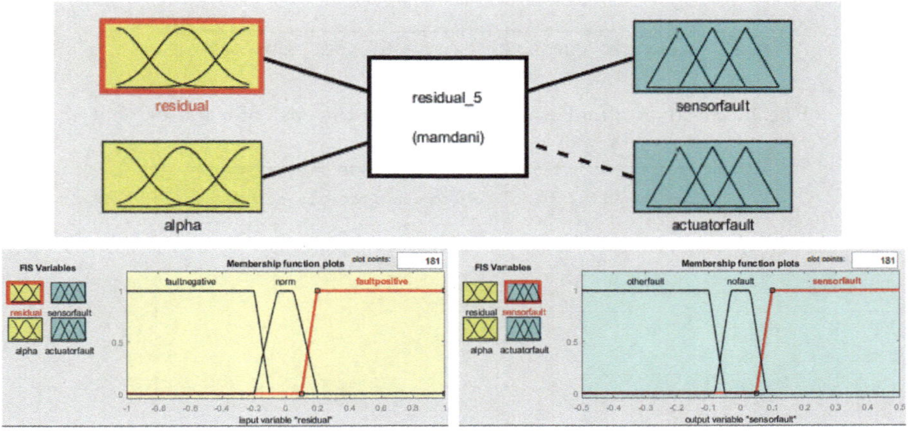

Fig. 4. Structure of the fuzzy inference system.

Two fault scenarios may be distinguished in the case of the scooter: the first one is a sensor fault giving a wrong voltage reading. The second one is an actuator fault of one or two motors which will directly result in an effect on the inclination angle. Both cases were simulated; for the sensor fault the residual $z_{u,k}$ and the state variable α are shown in Fig. 5. The result of the fuzzy decision engine are depicted in Fig. 6.

The blue line is the result of the output membership function which is intended to detect the respective sensor fault; a positive value indicates the information concerning the residual AND the state variable leads to the conclusion that the sensor fault is present (see Fig. 6). The red line is the result of the output membership function which is intended to detect the respective actuator fault; a negative value indicates that the information concerning the residual AND the state variable leads to the conclusion that the actuator fault is not

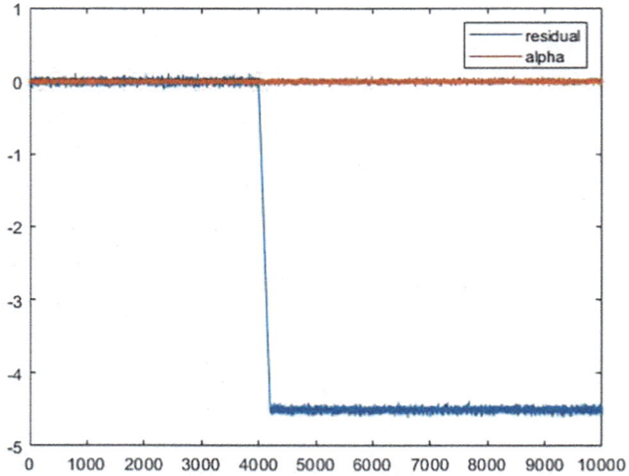

Fig. 5. Residual z and state variable α in the case of a sensor fault.

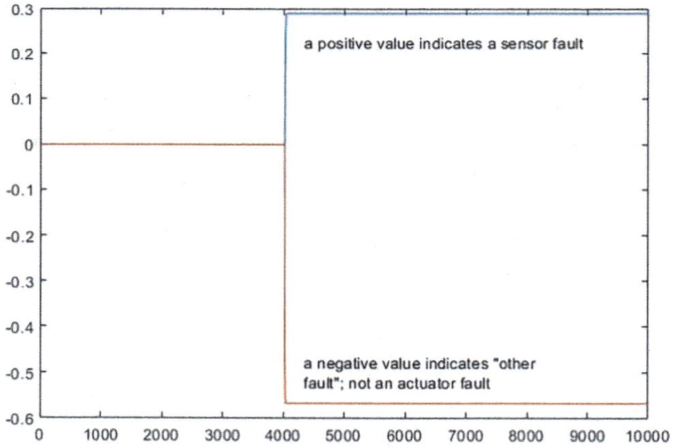

Fig. 6. Output of the fuzzy interference system for the sensor fault scenario.

present, but that another fault is present (see Fig. 6). It is clearly visible that for this fault the presence of the fault is immediately detected, thus underlining the effectiveness of the fuzzy decision engine. The introduction of this kind of analytical redundancy can be a high level measure for increasing the fault-tolerance of the scooter.

3.3 Design Characteristics on the Physical Level

Recent research indicates an increasing importance of abstract physic representations [36,37] This abstraction level describes how the desired functionality of a technical system is realized by means of physical (and chemical) phenomena. In spite of the merit of this level, it is omitted in this paper due to space restrictions.

3.4 Design Characteristics on the Level of Geometry and Material

The most concrete description level for a technical system concerns the geometry and material of the system and its components as well as the structure of these components. For the given scooter example, the discussion will concentrate again at the steering system. For a driver of a balanced two-wheel scooter, who expresses his steering wishes using a long lever, a reverse force in this lever is desirable. The realization of this force is a challenge due to unfavorable lever lengths (the lever length for the driver is much longer than the possible lever length for elastic elements which cause the reverse force). Due to this fact and the possibility of misuse forces that may lead to a fault, a fault-tolerant design is highly desirable. In this case it is possible to realize a certain degree of fault-tolerance by means of redundant elements, in this case the springs (compare [6] - Fig. 7).

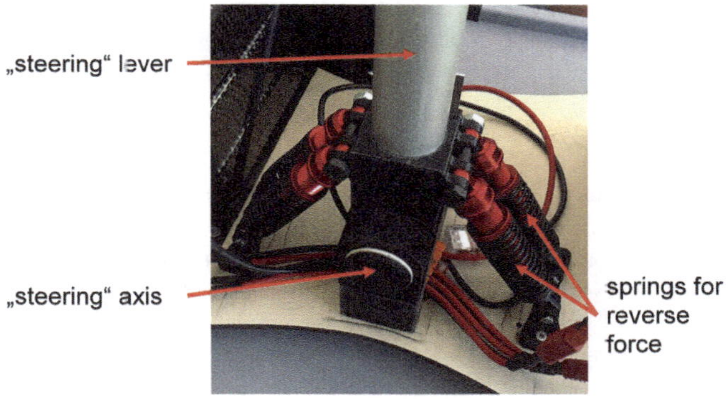

Fig. 7. Increased fault-tolerance by redundant springs in the steering system.

Another important design characteristic that increases the fault-tolerance is over-actuation. Frequently, the term “over-actuation” is used for describing the applications of more actuators than actually necessary for controlling motion systems [38,39]. It is important to note that over-actuation can also be understood as the use of stronger actuators than necessary [40]. In many cases, over-actuated technical systems dispose of a superior controllability; this design characteristics can furthermore increase the fault-tolerance, because of the possibility to use the over-actuation potential for the compensation of the effects of faults [40]. In the given scooter design, a certain amount of over-actuation is necessary in the drive motors in order to achieve a satisfactory controllability and to allow the compensation of faults such as a slippery surface or sudden moves of the driver. In such systems it is inevitable that some potential user mistakes are not forseen [41].

4 Conclusion and Outlook

The focus of this paper were design characteristics of a balanced two-wheel scooter, which can contribute to an enhanced fault-tolerance of this technical system. This description served for discussing the approaches of fault-tolerant design on different levels. It becomes apparent that a holistic approach towards fault-tolerance goes beyond rather simplistic considerations of redundancy. A key factor is a strong interconnection between design characteristics and diagnosis algorithms. It was found during this example that it is of paramount importance to connect the algorithm development with the system development. The combination of fault-tolerant control with consciously developed design characteristics with the identical aim can lead to technical systems with superior behavior in the case of a fault. Further research is planned, which will concentrate on an expansion of the fault-tolerant design methodology.

Acknowledgements. The project "digital product life-cycle" (ZaFH DiP) is supported by a grant from the European Regional Development Fund and the Ministry of Science, Research and the Arts of Baden-Württemberg, Germany (information under: https://efre-bw.de/). The work was additionally partially supported by the National Science Centre, Poland under Grant: UMO-2017/27/B/ST7/00620.

References

1. Blanke, M., Kinnaert, M., Lunze, J., Staroswiecki, M.: Diagnosis and Fault-Tolerant Control. Springer-Verlag, New York (2016)
2. Witczak, M.: Fault Diagnosis and Fault-Tolerant Control Strategies for Non-linear Systems. Lecture Notes in Electrical Engineering, vol. 266. Springer, Heidelberg (2014)
3. Ding, S.: Model-Based Fault Diagnosis Techniques: Design Schemes, Algorithms, and Tools. Springer-Verlag, Heidelberg (2008)
4. Stetter, R.: Fault-Tolerant Design and Control of Automated Vehicles and Processes: Insights for the Synthesis of Intelligent Systems. Springer-Verlag, Cham (2020)
5. Wünsch, F., Ramsaier, M., Breckle, T., Stetter, R., Till, M., Rudolph, S.: Executable cost-sensitive product development of a self-balancing two-wheel scooter with graph-based design languages. In: Marjanovic, D., et al. (eds.) Proceedings of the 15th International Design Conference DESIGN 2018, Dubrovnik (2018)
6. Schuster, J., Pahn, F.: Entwicklung und Bau zweier konzeptionell unterschiedlicher Segways. Bachelor-thesis, Ravensburg-Weingarten University (RWU) (2018)
7. Ponn, J., Lindemann, U.: Konzeptentwicklung und Gestaltung technischer Produkte. Springer, Heidelberg (2011)
8. Cross, N.: Engineering Design Methods: Strategies for Product Design. Wiley, Hoboken (2008)
9. Ehrlenspiel, K., Meerkamm, H.: Integrierte Produktentwicklung. Zusammenarbeit Denkabläufe, Methodeneinsatz. Carl Hanser Verlag, Munich (2013)
10. Lindemann, U.: Methodische Entwicklung technischer Produkte. Springer, Heidelberg (2009)

11. Pahl, G., Beitz, W., Feldhusen, J., Grote, K.H.: Engineering Design: A Systematic Approach. Springer, London (2007)
12. Rouissi, F., Hoblos, G.: Fault tolerant sensor network design with respect to diagnosability properties. In: Proceedings of the 8th IFAC Symposium on Fault Detection, Supervision and Safety of Technical Processes (SAFEPROCESS), pp. 1120–1124 (2012)
13. Shirazipourazad, S., Sen, A., Bandyopadhyay, S.: Fault-tolerant design of wireless sensor networks with directional antennas. Pervasive Mob. Comput. **13**, 258–271 (2014)
14. Oh, Y.G., Jeong, J.K., Lee, J.J., Lee, Y.H., Baek, S.M., Lee, S.J.: Fault-tolerant design for advanced diverse protection system. Nucl. Eng. Technol. **45**(6), 795–802 (2013)
15. Hsieh, T.-Y., Li, K.-H., Chung, C.-C.: A fault-analysis oriented re-design and cost-effectiveness evaluation methodology for error tolerant applications. Microelectron. J. **66**, 48–57 (2017)
16. Vedachalam, N., Umapathy, A., Ramadass, G.A.: Fault-tolerant design approach for reliable offshore multi-megawatt variable frequency converters. J. Ocean Eng. Sci. **1**, 226–237 (2016)
17. Porter, R., Ronen, A., Shoham, Y., Tennenholtz, M.: Fault tolerant mechanism design. Artif. Intell. **45**(6), 1783–1799 (2013)
18. Stetter, R., Göser, R., Gresser, S., Till, M., Witczak, M.: Fault-tolerant design of a shifting system for autonomous driving. Accepted for Presentation at the 16th International Design Conference DESIGN (2020)
19. Bühne, S., Herrmann, A.: Handbuch Requirements Management nach IREB Standard. Aus- und Weiterbildung zum IREB Certified Professional for Requirements Engineering Advanced Level "Requirements Management". IREB e.V. (2015)
20. Bernard, R., Irlinger, R.: About watches and cars: winning R&D strategies in two branches. Presentation at the International Symposium "Engineering Design - The Art of Building Networks", Garching, 4th April 2016 (2016)
21. Hruschka, P.: Business Analysis und Requirements Engineering: Produkte und Prozesse nachhaltig verbessern. Hanser, Cincinnati (2014)
22. Ebert, C., Jastram, M.: ReqIF: seamless requirements interchange format between business partners. IEEE Softw. **29**(5), 82–87 (2012)
23. Sharif Ullah, A.M.M., Sato, M., Watanabe, M., Mamunur Rashid, M.: Analysis of Kano-model-based customer needs for product development. Int. J. Autom. Technol. **10**(2), 132–143 (2016)
24. IBM: Rational DOORS. https://www.ibm.com/de-de/marketplace/requirements-management. Accessed 21 Mar 2020
25. Carrillo de Gea, J.M., Nicolas, J., Fernandez Aleman, J.L., Toval, A., Ebert, C., Vizca, A.: Requirements engineering tools: capabilities, survey and assessment. Inf. Softw. Technol. **54**(10), 1142–1157 (2012)
26. Holder, K., Zech, A., Ramsaier, M., Stetter, R., Niedermeier, H.-P., Rudolph, S., Till, M.: Model-based requirements management in gear systems design based on graph-based design languages. Appl. Sci. **7** (2017)
27. Stetter, R., Witczak, M.: Requirements management for monitoring and control. In: Proceedings of the 15th European Workshop an Advanced Control and Diagnosis (ACD), Bologna, Italy, 21st–22nd November 2019 (2019)
28. Eisenbart, B., Gericke, K., Blessing, L.T.M., McAloone, T.C.: A DSM-based framework for integrated function modelling: concept, application and evaluation. Res. Eng. Des. **28**(1), 25–41 (2016)

29. Ramsaier, M., Holder, K., Zech, A., Stetter, R., Rudolph, S., Till, M.: Digital representation of product functions in multicopter design. In: Maier, A., et al. (eds.) Proceedings of the 21st International Conference on Engineering Design (ICED 17) Vol 1: Resource Sensitive Design, Design Research Applications and Case Studies, Vancouver, Canada, 21–25 Aug 2017, pp. 369–378 (2017)
30. Grasser, F., D'Arrigo, A., Colombi, S., Rufer, A.: JOE: a mobile, inverted pendulum. IEEE Trans. Industr. Electron. **40**(1), 107–114 (2002)
31. Younis, W., Abdelati, M.: Design and implementation of an experimental segway model. In: Proceedings of the 2nd Mediterranean Conference on Intelligent Systems and Automation, Zarzis, Tunisia (2009)
32. van der Veen, J.: Stabilization and trajectory tracking of a segway. University of Groningen, Faculty of Science and Engineering (2018)
33. Stetter, R., Witczak, M., Pazera, M.: Virtual diagnostic sensors design for an automated guided vehicle. Appl. Sci. **8**(5), 702 (2018)
34. Stetter, R.: A virtual fuzzy actuator for the fault-tolerant control of a rescue vehicle. In: The IEEE International Conference on Fuzzy Systems (FUZZ-IEEE) (2020)
35. Mendonca, L.F., Sousa, J., da Costa, J.M.G.: Fault isolation using fuzzy model-based observers. IFAC Proc. Vol. **39**(13), 735–740 (2006)
36. Albers, A., Wintergerst, E.: The contact and channel approach (C&C2-A): relating a system's physical structure to its functionality. In: Chakrabarti, A., Blessing, L. (eds.) An Anthology of Theories and Models of Design: Philosophy, Approaches and Empirical Explorations, pp. 61–72. Springer, London (2014)
37. Ramsaier, M., Stetter, R., Till, M., Rudolph, S.: Abstract physics representation of a balanced two-wheel scooter in graph-based design languages. Accepted for Presentation at the 16th International Design Conference DESIGN (2020)
38. Ryll, M., Buelthoff, H.H., Giordano, P.R.: Overactuation in UAVs for enhanced aerial manipulation: a novel quadrotor concept with tilting propellers. In: Proceedings of the 6th International Workshop on Human-Friendly Robotics (2013)
39. Schneider, M.G.E., van de Molengraft, M.J.G., Steinbuch, M.: Benefits of over-actuation in motion systems. In: Proceeding of the 2004 American Control Conference (2004)
40. Stetter, R., Simundsson, A.: Design for control. In: Proceedings of ICED 2017, vol. 4, pp. 149–158 (2017)
41. Dubrova, E.: Fault-Tolerant Design. Springer, New York (2013)

A Fuzzy Logic Approach to Remaining Useful Life Estimation of Ball Bearings

Marcin Witczak[1], Bogdan Lipiec[1(✉)], Marcin Mrugalski[1], and Ralf Stetter[2,3]

[1] Institute of Control and Computation Engineering, University of Zielona Gora, 65-246 Zielona Gora, Poland
{m.witczak,b.lipiec,m.mrugalski}@issi.uz.zgora.pl
[2] Faculty of Mechanical Engineering, University of Applied Sciences Ravensburg-Weingarten, Weingarten, Germany
stetter@hs-weingarten.de
[3] Steinbeis Transfer Center Automotive Systems, Ravensburg, Germany
https://www.issi.uz.zgora.pl/?en

Abstract. The paper deals with the development of a modelling and prediction scheme capable of estimating a remaining useful life of ball bearings. In particular, a multiple model-based Takagi-Sugeno scheme is developed, which is able to follow the system degradation over the time and predict it in the future. Contrarily to the typical framework, multiple models designed with historical data are used to support diagnostic decisions. In particular, health status determination of the currently operating bearing is supported by the knowledge gathered from the preceding bearings, which went through the run-to-failure process. In both historical and actual bearing cases an efficient modelling scheme with low computational burden is proposed. It is also shown how to exploit it for predicting the bearings remaining useful life. Finally, the proposed approach is applied to data gathered from the PRONOSTIA Platform, designed for the purpose of IEEE Data Challenge pertaining remaining useful life estimation of ball bearings.

Keywords: Remaining useful life · Fuzzy logic · Degradation · Uncertainty intervals

1 Introduction

The development of industry and engineering requires implementation of more advanced diagnostic systems. More efficient diagnostic systems result in an increased production, which in turn affects the company's profits. For this reason, the development of monitoring systems for remaining life of various machinery is so important [7]. Condition monitoring techniques are used to investigate a wide range of machinery, such as wind turbines [14] and high-speed trains [4] as well cars, tractors, forklifts, etc. All of them contain bearings, which are critical for their performance. The ball bearing is one of the most common element

A. Bartoszewicz et al. (Eds.): PCC 2020, AISC 1196, pp. 1411–1423, 2020.
https://doi.org/10.1007/978-3-030-50936-1_117

used in almost every factory. Degradation of ball bearings is dangerous effect of machine operation. Vibration-based techniques have proven to be one of the most effective ways for bearing fault diagnosis among various techniques [13] and thus have become popular.

Remaining Useful Life (RUL) prediction is a broad research area, that includes the study about many components. Two most important categories are: RUL of batteries [8,10,12] and ball bearings. In both cases, systematic maintenance and replacements are required so that the factory can run smoothly and without unnecessary delays. RUL is a part of planning maintenance and replacements. Using this information, it is possible to correctly plan the times of shutdowns, which would not impair the production flow. The most common approach applied to predicting RUL is based on the exponential degradation model [3,6]. In spite of the incontestable appeal of such approaches, they inherit one common drawback pertaining the lack of usage of historical data for supporting decisions about the current health status of the equipment being diagnosed. Indeed, as can be observed in recent developments [6], the exponential model is designed exclusively based on the data of an exemplary system without taking into account the history associated with another system of the same structure. To tackle this drawback, this paper presents a new way to determining RUL of ball bearings. This method is based on multiple-model Takagi-Sugeno scheme [11,17,18]. In the proposed approach, determination of the health status of ball bearings is supported by historical data gathered during run-to-failure usage of other bearings of the same type. Thus, apart from modelling the degradation of the currently operating bearing the degradation determination is supported with the historical models. As a result, one historical model, which is operates closely to the current one is selected and used for predictions purposes.

The proposed approach is validated with IEEE PRONOSTIA Data Challenge [9] containing a set of run-to-failure data representing horizontal and vertical vibrations of bearings. The data set of contains vibration signals of six bearings from three different conditions generated with the test-stand portrayed in Fig. 1. Each test had different load and rotation speed. This allowed to test

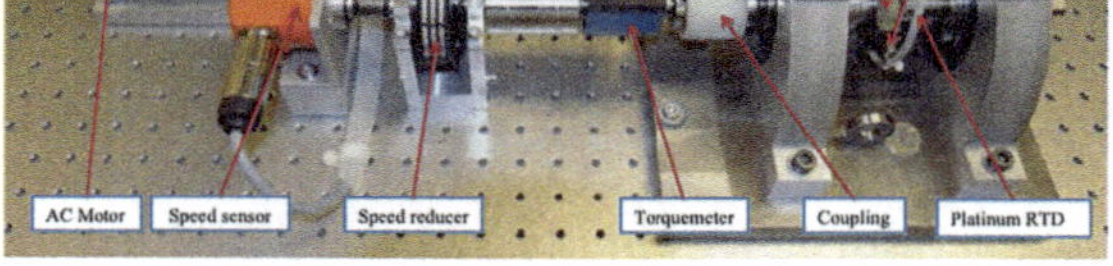

Fig. 1. PRONOSTIA Platform used during all tests.

bearings in various working conditions. Every test last until vibration signal (expressed with either horizontal or vertical acceleration exceeds 20 g [9].

The paper is organized as follows. Section 2.1 presents the proposed approach for designing Takagi-Sugeno models using historical run-to-failure data. Section 3 presents a way of designing a model of currently operating bearing along with determination of its RUL. Finally, the proposed approach is validated with PRONOSTIA benchmark.

2 A Novel Fuzzy Logic Approach

As mentioned in the introductory part of the paper, exponential models proven to be a powerful tool for mimicking the degradation process of various components [1,20], including ball bearings PRONOSTIA benchmark [15,16]. To facilitate further deliberations, the system being considered in this paper is a ball bearing defined within PRONOSTIA benchmark.

Thus, the idea behind a novel approach proposed in this section is to couple together a general exponential modelling approach with the available historical data gathered from different but the same type ball bearings. For that purpose a fuzzy logic framework is to be employed and carefully described in the remaining part of this section.

Let us start with reminding a general structure of exponential degradation model (see, e.g., [3,6] and the references therein):

$$x_k = \phi + \theta_k \exp\left(\beta_k t_k + \sigma_k \mathcal{B}(t_k) - \frac{\sigma_k^2}{2} t_k\right), \tag{1}$$

where k stands for the sample number, x_k is the degradation parameter being modelled (e.g, a vibration), ϕ is a known constant, which can be interpreted as minimum bound of the degradation parameter, while β, σ and θ are unknown parameters shaping the behaviour of the exponential model. Finally, t_k stands for the time at a given sample k, $\sigma\mathcal{B}(t_k)$ denotes a Brownian motion obeying a normal distribution $\mathcal{N}(0, \sigma^2 t_k)$. Subsequently, the model (1) undergoes a logarithmic transform of the classical form:

$$z_k = \ln(x_k - \phi) = \ln(\theta_k) + \left(\beta_k - \frac{\sigma_k^2}{2}\right) t_k + \sigma_k \mathcal{B}(t_k), \tag{2}$$

which can be written in a compact regressor-like form:

$$z_k = r_k^T p_k, \tag{3}$$

with

$$r_k = [1, t_k]^T, \quad p_k = \left[\ln(\theta), \beta - \frac{\sigma^2}{2}\right]^T. \tag{4}$$

where $r_k \in \mathbb{R}^2$ stands for the regressor vector, while $p_k \in \mathbb{R}^2$ denotes a time-varying parameter vector.

The seminal paper of Gebraeel et al. [3] proposes an elegant Bayesian framework for estimating p_k. However, the drawback of this initial framework was that no rules for starting such estimation were provided. Indeed, if the degradation signal x_k is below a certain level then this makes no sense to perform parameter p_k updates as the resulting model exhibits a constant linear behaviour. To tackle this problem, Li et al. [6] proposed an improved approach. In spite of the incontestable appeal of this approach, it inherits the drawbacks of the initial developments, i.e., it does not use the historical data gathered from different ball bearings of the same type.

To overcome the above difficulties, the following research questions has to be answered:

Q1: How to preserve the historical data and use it for supporting diagnostics and health assessment of the currently performing bearing?
Q2: How to incorporate the model (3) within the framework answering **Q2**?
Q3: How to use the developments concerning **Q1** and **Q2** for assessing RUL of a currently performing ball bearing?

The answers to the above question are provided in the subsequent part of this section.

2.1 Fuzzy Logic-Based Historical Data Framework

The objective of this section is to provide the answer to **Q1**. Let us start by simplifying the model (3) by setting $\phi = 0$. As indicated in [3,6], this transformation does not cause any lost of generality as ϕ is known and can be easily subtracted from x_k. Under such an assumption $z_k = x_k$.

Since z_k is a degradation signal, it is natural that it is bounded as follows:

$$\underline{z} \leq z_k \leq \bar{z}, \tag{5}$$

where $\underline{z} \geq 0$ stands from the minimum level from which z_k is perceived as the one expressing a degradation. Contrarily, it is simply indicating that the degradation is at the zero level and the bearing is in a fully healthy state. On the other hand, $\bar{z} > 0$ is treated as the maximum allowable degradation level. This means that it can be used to calculate the so-called Time-To-Failure (TTF), an expected time from current z_k to $\bar{z}$. Having the interval (5), it is possible to divide it into n classes each corresponding to a different health status of the bearing. For the convenience, these classes can be uniformly distributed over (5) with a spread:

$$s_i = \underline{z} + (i-1)\frac{\bar{z} - \underline{z}}{n-1}, \quad i = 1, \ldots, n. \tag{6}$$

These classes will form the fundamentals for embedding the so-called membership function [19] associated with each of the classes. Without a loss of generality, it is proposed to use triangular membership functions shaped as follows:

$$\begin{aligned} &a_i = b_{i-1}, \quad b_i = s_i, \quad c_i = b_{i+1}, \quad i = 2, \ldots, n-1, \\ &a_1 = b_1, \quad c_n = b_n, \end{aligned} \tag{7}$$

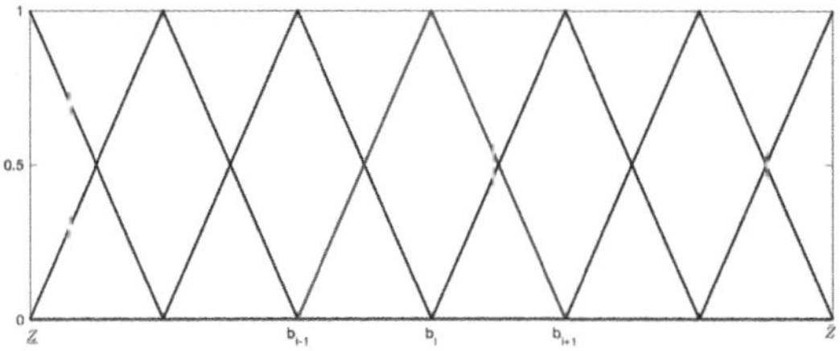

Fig. 2. Membership functions for degradation modelling

where a_i, b_i, c_i are the parameters defining ith membership function. The above process is realized as shown in Fig. 2. The choice of such membership functions is not accidental and the resulting profits will be exhibited in the sequel.

Having the membership function, the problem of degradation modelling boils down to form a Takagi-Sugeno [17] structure with the sub-systems given by:

$$\text{IF } z_k \in M_{z,i} \text{ THEN } z_k = r_k^T p^i + v_k, \quad i = 1, \ldots, n. \tag{8}$$

where $M_{z,i}$ stands for the fuzzy set associated with ith triangular membership function defined by (7) while v_k denotes modelling and measurement uncertainty. Thus, the system (8) can be written into a condensed form

$$z_k = \sum_{i=1}^{n} \mu_i(z_k) \left(r_k^T p^i + v_k \right), \tag{9}$$

$$\sum_{i=1}^{n} \mu_i(z_k) = 1, \quad \mu_i(z_k) \geq 0,$$

where $\mu_i(z_k)$ $(i = 1, \ldots, n)$ stands for the normalised ith rule firing strength calculated according to the triangular shape (7). Alternatively, the system (9) can be written as:

$$z_k = \bar{r}_k^T \bar{p} + v_k, \tag{10}$$

where

$$\bar{r} = [\mu_1(z_k) r_k^T, \mu_2(z_k) r_k^T, \ldots, \mu_n(z_k) r_k^T]^T, \quad \bar{p} = [(p^1)^T, \ldots, (p^n)^T]^T. \tag{11}$$

As it can observed, the behaviour of (10) depends on $2n$ parameters contained within $\bar{p} \in \mathbb{R}^{2n}$. Although for a reasonable number of degradation classes n this number is still relatively low, some of $\mu_l(z_k)$ can be inactive (equal zero) until the degradation enters a given class. This causes that some of regressor elements can be permanently equal to zero, which is a very unappealing phenomenon from the viewpoint of parameter estimation. To tackle this problem, let us exploit the properties of triangular membership functions portrayed in Fig. 2. Indeed, it can be observed that for any x_k at most 2 membership functions are active. This causes that (9) reduces to:

$$\begin{aligned} z_k &= \mu_i(z_k)r_k^T p^i + \mu_{i+1}(z_k)r_k^T p^{i+1} + v_k, \\ &\mu_i(z_k) + \mu_{i+1}(z_k) = 1, \quad \mu_i(z_k) \geq 0, \quad i = 1, \ldots, n-1. \end{aligned} \tag{12}$$

Having such a structure, it is possible to impose assumptions on v_k. In this paper, instead of deliberating about a statistical nature of v_k, which is modelled using Brownian motion in [3,6], it is proposed to assume that

$$|v_k| \leq \bar{v}. \tag{13}$$

where $\bar{v}$ is a known upper bound of v_k, which can be interpreted as a maximum discrepancy between the degradation process (12) and its model:

$$\begin{aligned} z_{m,k} &= \mu_i(z_k)r_k^T p^i + \mu_{i+1}(z_k)r_k^T p^{i+1}, \\ &\mu_i(z_k) + \mu_{i+1}(z_k) = 1, \quad \mu_i(z_k) \geq 0, \quad i = 1, \ldots, n-1. \end{aligned} \tag{14}$$

where $z_{m,k}$ stands for the degradation model output. Thus, the historical covering *run-to-failure* of $l = 1, \ldots, n_h$ can be stored using a set of models:

$$\begin{aligned} z_{m,k} &= \mu_i(z_k)r_k^T p_l^i + \mu_{i+1}(z_k)r_k^T p_l^{i+1}, \\ &\mu_i(z_k) + \mu_{i+1}(z_k) = 1, \quad \mu_i(z_k) \geq 0, \quad i = 1, \ldots, n-1. \end{aligned} \tag{15}$$

The advantage of such an approach is that instead of storing a possibly large data set, the degradation parameter vector $\bar{p}_l = [(p_l^1)^T, \ldots, (p_l^n)^T]^T$ is stored instead. The objective of the remaining part of this paper is to provide an algorithm for estimating $\bar{p}_l$, i.e., obtaining its estimate. To tackle this problem various recursive parameter estimation strategies can be employed, which assume the so-called bounded error [2,5] but Modified quasi-Outer Bounding Ellipsoid algorithm (MOBE) [2] can be perceived as an optimal candidate. Indeed, its simplicity is similar to the one of the celebrated Recursive Least-Square (RLS), which motivates its practical application within the algorithm that is proposed below. For the sake of notation simplicity, the index l (denoting lth bearing) is neglected as the same algorithm has to be performed for all $l = 1, \ldots, n_h$ bearings.

Step 0: Set $k = 1$, set $p_0^j = 0$, $j = 1, \ldots, n$ and $P_0^i = \rho\mathbf{I}$, $i = 1, \ldots, n-1$, where $\rho > 0$ stands for a sufficiently large positive constant.
Step 1: Determine a list of active sub-models

$$i = \{j : \mu_j(z_k) + \mu_{j+1}(z_k) = 1, j = 1, \ldots, n-1\}. \tag{16}$$

Step 2: Form temporary regressor and parameter vector:

$$r_k^i = [\mu_i(z_k)r_k^T, \mu_{i+1}(z_k)r_k^T]^T, \quad w_k^i = [(p_k^i)^T, (p_k^{i+1})^T]^T, \tag{17}$$

Step 3: Calculate

$$e_k^i = z_k - (r_k^i)^T w_{k-1}^i, \tag{18}$$

Step 4: If $|e_k^i| > \bar{v}$ then

$$g_k^i = (r_k^i)^T P_k^i r_k^i, \tag{19}$$

$$\lambda_k^i = \frac{g_k^i}{\frac{|e_k^i|}{\bar{v}} - 1}, \tag{20}$$

$$P_k^i = (\lambda_k^i)^{-1} \left(P_{k-1}^i - P_{k-1}^i r_k^i (r_k^i)^T P_{k-1}^i (g_k^i)^{-1} \right), \tag{21}$$

$$w_k^i = w_{k-1}^i + P_k^i r_k^i e_k^i, \tag{22}$$

else

$$P_k^i = P_{k-1}^i, \tag{23}$$

$$w_k^i = w_{k-1}^i. \tag{24}$$

Step 5: Set $k = k + 1$ and go to *Step 1.*

The crucial computational advantage of MOBE algorithm is that it does not update parameters when the absolute value of the current estimation error is lower than $\bar{v}$.

Finally, the overall scheme is implemented according to the scheme presented in Fig. 3. Let us start with describing a mechanism determining the so-called First Prediction Time (FPT) t_{FPT} from, which the degradation prediction is started. For that purpose two crucial parameters of the degradation signals are calculated, namely, Kurtosis and Root Mean Square (RMS). Note that in the IEEE PRONOSTIA benchmark, the vibration signal is used as a degradation one, which is illustrated in Fig. 3. As observed in several papers on degradation modelling (see, e.g., [3,6] and the references therein), Kurtosis is a measure sensitive to incipient faults [18] but its is not very useful in assessing their time varying nature, i.e., degradation development. Contrarily, RMS provides enough information to express the development of vibration energy. Thus, it is beneficial to couple this measures in a way proposed in [3]. Thus, the approach proposed in this paper employs the Kurtosis-based strategy for determining t_{FPT} while the degradation models and the associated parameter updates are realised using the above-proposed *5-STEP* algorithm.

The entire scheme is illustrated in Fig. 3 and can be characterized as follows:

1. a healthy historical data of a bearing (or all bearings of the same type) is used to determinate average Kurtosis μ_K and standard deviation σ_K;
2. A $3\sigma_K$ rule is used to form an allowable Kurtosis uncertainty interval

$$\mathbb{K} = [\mu_K - 3\sigma_K, \mu_K + 3\sigma_K]. \tag{25}$$

3. The actual measurement data is used to calculate Kurtosis m_f where f stands for the sample number corresponding to the actual time t_f.

4. The Kurtosis m_f is validated on a window $m_f, \ldots, m_{f+n_w}$, i.e., the first element m_{f+j}, $j = 0, \ldots, n_w$ within the window for which the remaining ones satisfy $m_{f+j} \notin \mathbb{K}$, is set as the one corresponding to FPT, i.e., $t_{FPT} = t_j$.
5. Once t_{FPT} is initiated, the process of designing a degradation model is started according to the proposed *5-STEP* algorithm.
6. It should be noted that Prediction of RUL is not used during the historical model construction phase and it will be exploited in the subsequent section.

Finally, it can be conclude that the above algorithm (cf. Fig. 3) results in a set of models corresponding to the run-to-failure of n_h bearings. As a result, a possibly large data corresponding to n_h run-to-failure bearings was translated into a set of $n_h \times 2 \times n$ degradation model parameters. Note that irrespective of the size of the data sets, the above number remains unchanged.

Having the above appealing results, the objective of the subsequent section is to incorporate the run-to-failure-based models for determining the remaining useful life of a currently operating bearing.

3 Remaining Useful Life Prediction

The objective of this section is to develop a framework exploiting a set of n_h run-to-failure models for supporting the decision about the health status of a currently operating one. The structure of the proposed algorithm is portrayed in Fig. 3. Thus, for the currently operating bearing undergoes the same procedure for model and parameter updating as described in the preceding section (cf. Fig. 3 and the 5-STEP algorithm).

The difference is that the 5-STEP algorithm is extended with an additional task concerning RUL prediction. The entire process can be summarized as follows:

Step 0: Set $k = 1$, set $p_0^j = 0$, $j = 1, \ldots, n$ and $P_0^i = \rho\mathbf{I}$, $i = 1, \ldots, n-1$, where $\rho > 0$ stands for a sufficiently large positive constant.
Step 1: Determine a list of active sub-models

$$i = \{j : \mu_j(z_k) + \mu_{j+1}(z_k) = 1, j = 1, \ldots, n-1\}. \tag{26}$$

Step 2: Form temporary regressor and parameter vector:

$$r_k^i = [\mu_i(z_k) r_k^T, \mu_{i+1}(z_k) r_k^T]^T, \quad w_k^i = [(p_k^i)^T, (p_k^{i+1})^T]^T, \tag{27}$$

Step 3: Calculate

$$e_k^i = z_k - (r_k^i)^T w_{k-1}^i, \tag{28}$$

Step 4: If $|e_k^i| > \bar{v}$ then

$$g_k^i = (r_k^i)^T P_k^i r_k^i, \tag{29}$$

$$\lambda_k^i = \frac{g_k^i}{\frac{|e_k^i|}{\bar{v}} - 1}, \tag{30}$$

$$P_k^i = (\lambda_k^i)^{-1} \left(P_{k-1}^i - P_{k-1}^i r_k^i (r_k^i)^T P_{k-1}^i (g_k^i)^{-1} \right), \tag{31}$$

$$w_k^i = w_{k-1}^i + P_k^i r_k^i e_k^i, \tag{32}$$

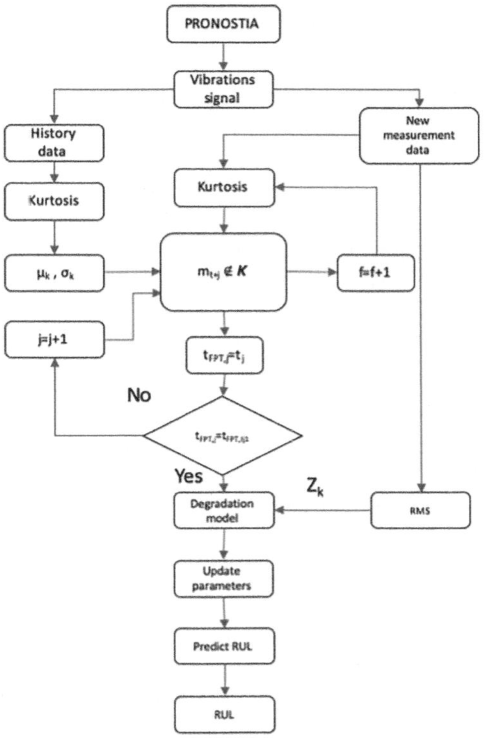

Fig. 3. Flowchart of the proposed framework

else

$$P_k^i = P_{k-1}^i, \tag{33}$$

$$w_k^i = w_{k-1}^i. \tag{34}$$

Step 5: Find the closest historical model

$$l = \arg \min_{l=1,\dots,n_h} Q_l^i. \tag{35}$$

Step 6: Predict time-to-failure t_f:

$$f = \frac{\bar{z} - p_{1,l}^n}{T_s p_{2,l}^n}, \quad t_f = T_s \times f. \tag{36}$$

Step 7: Calculate RUL

$$RUL_k = (f - k)T_s. \tag{37}$$

Step 8: Set $k = k + 1$ and go to *Step 1.*

Let us start with explaining *Step 5.* During this step two things are already available:

1. The current degradation class i is determined in *Step 1.*
2. A set of n_h historical models for class i

$$z_{m,k} = \mu_i(z_k) r_k^T p_l^i + \mu_{i+1}(z_k) r_k^T p_l^{i+1}, \qquad (38)$$
$$\mu_i(z_k) + \mu_{i+1}(z_k) = 1, \quad \mu_i(z_k) \geq 0, \quad l = 1, \ldots, n_h.$$

Thus, the squared differences between the actual model and the historical ones (38) can be easily determined as:

$$Q_l^i = \left(\mu_i(z_k) r_k^T (p_l^i - p^i) + \mu_{i+1}(z_k) r_k^T (p_l^{i+1} - p^{i+1})\right)^2, \; l = 1, \ldots, n_h. \qquad (39)$$

and hence, (35) provides lth historical model within ith degradation class, which is the closest to the currently operating one. Having lth historical model, let us proceed to *Step 6.* As it was already the maximum degradation signal bound (5) corresponds to the failure threshold. This simply means that at the failure moment, the lth model should satisfy:

$$\bar{z} = \mu_j(\bar{z}) r_f^T p_l^j + \mu_{j+1}(\bar{z}) r_f^T p_l^{j+1}, \qquad (40)$$
$$\mu_j(\bar{z}) + \mu_{j+1}(\bar{z}) = 1, \quad \mu_j(\bar{z}) \geq 0.$$

for some $i \leq j \leq n-1$ and $f > k$. From (6), it is is evident that $j = n-1$. Moreover, from Fig. 2, it is evident that $\mu_{n-1}(\bar{z}) = 0$ and $\mu_n(\bar{z}) = 1$, and hence, the remaining task is to determine $f > k$ for which (40) holds, i.e.,

$$\bar{z} = r_f^T p_l^n. \qquad (41)$$

To tackle this problem, the regressor form (4) has to be reminded, which yields $r_f = [1, t_f]^T$ with $t_f = T_s \times f$ where T_s stands for the known sampling time. Thus, (41) implies the failure time (36).

4 Experimental Validation

The objective of this section is to validate the proposed approach using PRONOSTIA benchmark. For that purpose the following set of parameters were used: $T_s = 10s$, $\underline{z} = 0.4$, $\bar{z} = 2$, $n = 10$, $\bar{v} = 0.1$, $\rho = 1e4$, $n_h = 5$. Having all these parameters and the historical data from n_h bearings the historical model determination was performed according to the approach proposed in Sect. 2.1. Figure 4 shows an exemplary degradation model resulting from the proposed approach. In this case, the first prediction time is $T_{FPT} = 13930$ s. as it can be observed, the proposed fuzzy-logic based approach possesses a satisfactory modelling quality. Subsequently, the RUL prediction algorithm proposed in Sect. 3 was examined on a bearing not included in a historical set of n_h bearings. Firstly, Fig. 5 shows the effect of choosing lth model for time-to-failure t_f prediction. In particular, the red one (lth) was selected while the green one shows another candidate solution contained in a set of n_h models. Based on this choice the time to failure was determined as $t_f = f \times T_s = 7500$ s. Finally, the predicted time is compare with the actual one (Fig. 6).

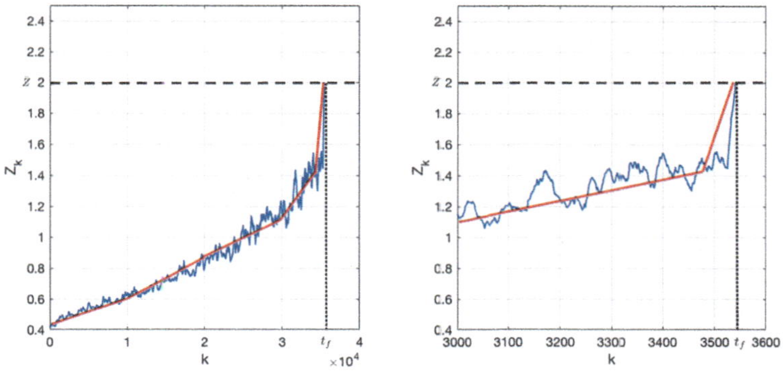

Fig. 4. Exemplary learning process

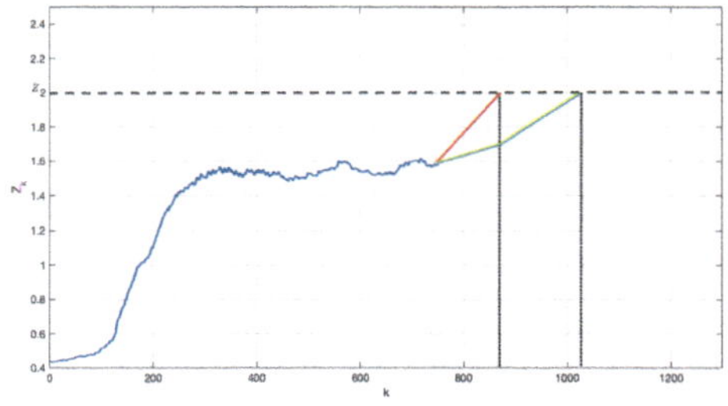

Fig. 5. RUL determination

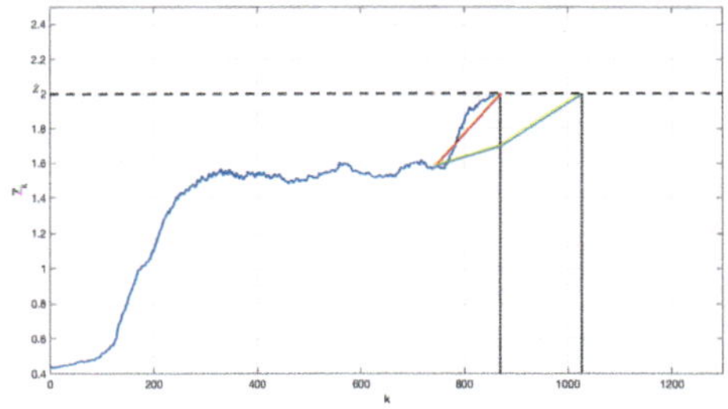

Fig. 6. RUL prediction v.s. real bearing failure

5 Concluding Remarks

The obtained results along with further comprehensive experiments (not listed in the paper due to the lack of space), clearly exhibit the benefits of the proposed approach. Another advantage of the proposed approach are low computational burden and a relatively easy implementation. This recommends its application to the so-called health-aware control systems, which constitutes the future research direction.

Acknowledgment. The work was supported by the National Science Centre of Poland under Grant: UMO-2017/27/B/ST7/00620.

References

1. Anis, M.D.: Towards remaining useful life prediction in rotating machine fault prognosis: an exponential degradation model. In: 2018 Condition Monitoring and Diagnosis (CMD), pp. 1–6 (2018). https://doi.org/10.1109/CMD.2018.8535765
2. Arablouei, R., Doğançay, K.: Modified quasi-OBE algorithm with improved numerical properties. Sig. Process. **93**(4), 797–803 (2013)
3. Gebraeel, N., Lawley, M., Li, R., Ryan, J.: Residual-life distributions from component degradation signals: a bayesian approach. IIE Trans. **37**(6), 543–557 (2005)
4. Hu, H., Tang, B., Gong, X., Wei, W., Wang, H.: Intelligent fault diagnosis of the high-speed train with big data based on deep neural networks. IEEE Trans. Ind. Inf. **13**(4), 2106–2116 (2017). https://doi.org/10.1109/TII.2017.2683528
5. Kraus, T., Mandour, G.I., Joachim, D.: Estimating the error bound in QOBE vowel classification. In: 2007 50th Midwest Symposium on Circuits and Systems, pp. 369–372 (2007). https://doi.org/10.1109/MWSCAS.2007.4488608
6. Li, N., Lei, Y., Lin, J., Ding, S.: An improved exponential model for predicting remaining useful life of rolling element bearings. IEEE Trans. Ind. Electron. **62**(12), 7762–7773 (2015)
7. Loutas, H., Roulias, D., Geogoulos, G.: Remaining useful life estimation in rolling bearings utilizing data-driven probabilistic E-support vectors regression. IEEE Trans. Reliab. **62**(4), 821–832 (2013). https://doi.org/10.1109/TR.2013.2285318
8. Miao, Q., Xie, L., Cui, H., Pecht, M.: Remaining useful life prediction of lithium-ion battery with unscented particle filter technique. Microelectron. Reliab. **53**(6), 805–810 (2012). https://doi.org/10.1016/j.microrel.2012.12.004
9. Nectoux, P.R.G., Medjaher, K., Ramasso, E., Morello, B., Zerhouni, N., Varnier., C.: Pronostia: an experimental platform for bearings accelerated life test. In: 2012 IEEE International Conference on Prognostics and Health Management, Denver, CO, USA (2012)
10. Pazera, M., Buciakowski, M., Witczak, M.: Robust multiple sensor fault-tolerant control for dynamic non-linear systems: application to the aerodynamical twin-rotor system. Int. J. Appl. Math. Comput. Sci. **28**(2), 297–308 (2018). https://doi.org/10.2478/amcs-2018-0021
11. Rutkowski, T., Łapa, K., Nielek, R.: On explainable fuzzy recommenders and their performance evaluation. Int. J. Appl. Math. Comput. Sci. **29**(3), 595–610 (2019). https://doi.org/10.2478/amcs-2019-0044
12. Saha, B., Goebel, K., Poll, S., Christophersen, J.: Prognostics methods for battery health monitoring using a Bayesian framework. IEEE Trans. Instrum. Meas. **58**(2), 291–296 (2009)
13. Si, X.S., Wang, W., Hu, C.H., Zhou, D.H.: Remaining useful life estimation - a review on the statistical data driven approaches. Eur. J. Oper. Res. **213**(1), 1–14 (2011). https://doi.org/10.1016/j.ejor.2010.11.018
14. Simani, S., Farsoni, S., Castaldi, P.: Data-driven techniques for the fault diagnosis of a wind turbine benchmark. Int. J. Appl. Math. Comput. Sci. **28**(2), 247–268 (2018). https://doi.org/10.2478/v10006-008-0046-3
15. Singleton, K.R., Strangas, E.G., Cui, H., Aviyente, S.: Extended Kalman filtering for remaining-useful-life estimation of bearings. IEEE Trans. Ind. Electron. **62**(3), 1781–1790 (2015). https://doi.org/10.1016/j.microrel.2012.12.004

16. Sutrisno, E., Oh, H., Vasan, A.S.S.: Estimation of remaining useful life of ball bearings using data driven methodologies. In: 2012 IEEE Conference on Prognostics and Health Management (PHM) (2012). https://doi.org/10.1109/ICPHM.2012.6299548
17. Tanaka, K., Sugeno, M.: Stability analysis and design of fuzzy control systems. Fuzzy Sets Syst. **45**(2), 135–156 (1992)
18. Witczak, M.: Fault Diagnosis and Fault-Tolerant Control Strategies for Non-Linear Systems. Lectures Notes in Electrical Engineering, vol. 266. Springer International Publisher. Heidelberg (2014)
19. Zadeh, L.A.: Knowledge representation in fuzzy logic. In: Zadeh, L.A. Fuzzy Sets, Fuzzy Logic, And Fuzzy Systems: Selected Papers, pp. 764–774. World Scientific (1996)
20. Zhang, L., Mu, Z., Sun, C.: Remaining useful life prediction for lithium-ion batteries based on exponential model and particle filter. IEEE Access **6**, 17729–17740 (2018)

Detection of State-Multiplicative Faults in Discrete-Time Linear Systems

Dušan Krokavec(✉) and Anna Filasová

Faculty of Electrical Engineering and Informatics, Department of Cybernetics and Artificial Intelligence, Technical University of Košice, Letná, 04200 Košice, Slovakia
{dusan.krokavec,anna.filasova}@tuke.sk

Abstract. The conditions for observer based residual filter design for linear discrete-time state-multiplicative systems are presented in the paper. With respect to the residual signal's time evolution, as well as to its robustness, the design problem is stated in terms of linear matrix inequalities (LMI). To expand the standard LMI formulation, norm bounds on disturbance and, in particular, a new characterisation of the norm boundaries of the multiplicative faults are projected into enhanced bounded real lemma structure of LMI. With given restrictions, the design steps are revealed in the example for projecting the state estimation error to fault residuals.

Keywords: Linear discrete-time systems · State-multiplicative faults · Luenberger observers · Residual filters · Linear matrix inequalities

1 Introduction

The problem of system state reconstruction and, especially, asymptotic state estimation in presence of disturbances, are essential in control (see, e.g. [2,14]) and model-based fault detection [8]. The opportunity of granting feasible solutions from LMI formulated observer design tasks allows one to solve the fault detection problems, where typical fault detection filters can find in [1,13]. The disturbance observation, fault estimation and related residual generation are applied for discrete-time systems in [4,5,9,11].

One of the main property, characterizing multiplicative faults is their affect on the system structure and their dependence on the system state variables [7]. The treatment of plant uncertainty so makes fault detection method different from the standard additive fault detection principle [10]. However, there are only few publications with the aim to address this problem in diagnosis, considering the direct connections of multiplicative faults with the system inputs and outputs [3, 6]. One approach to the treatment of multiplicative faults in control system which proposes the fault separation and associated limited state variable measurement is given in [12].

With interaction to the results presented above, the subjects of the paper are observer based fault residual structures for discrete-time linear systems with

A. Bartoszewicz et al. (Eds.): PCC 2020, AISC 1196, pp. 1424–1433, 2020.
https://doi.org/10.1007/978-3-030-50936-1_118

single multiplicative faults. Proposing one way of norm boundaries representation for the multiplicative faults, expanded Lyapunov function is used to admit tuning respecting faults and disturbances in the design step of the fault detection filter. Considering the system matrix element changes in nominal mode and in faulty regime after the occurrence of a single multiplicative fault, the enhanced bounded real lemma structure is preferred to decouple the Lyapunov matrix from the system matrix parameters. The goal is enough flexibility to guarantee dynamic properties of the observer structure, as well as satisfactory residual signal sensitivity and thresholds principle in fault detection.

The paper is organized as follows. Ensuing introduction in Sect. 1, and continuing task description and problem formulation in Sect. 2, one way of norm boundaries representation for the multiplicative faults is analyzed in Sect. 3. Enabling the internal properness and exploiting H_∞ norm approach, the design method is addressed in Sect. 4. In response, Sect. 5 shows the applicability of the method using an addressed simulation example and Sect. 6 gives some concluding remarks.

Throughout the paper, the following notations are used: $\boldsymbol{x}^{\mathrm{T}}$, $\boldsymbol{X}^{\mathrm{T}}$ denotes the transpose of the vector $\boldsymbol{x}$ and the matrix $\boldsymbol{X}$, respectively, for a square matrix $\boldsymbol{X} \prec 0$ means that $\boldsymbol{X}$ is symmetric negative definite matrix, the symbol $\boldsymbol{I}_n$ indicates the n-th order unit matrix, $\mathbb{R}$ denotes the set of real numbers, $\mathbb{R}^n$, $\mathbb{R}^{n \times r}$ refers to the set of all n-dimensional real vectors and $n \times r$ real matrices and $\mathcal{Z}_+$ is the set of positive integers.

2 Problem Formulation and Description

Using the time response approach, the considered discrete-time dynamical systems are represented using space-time description belonging to the following class of equations

$$\boldsymbol{q}(i+1) = \boldsymbol{F}\boldsymbol{q}(i) + \boldsymbol{F}^{\circ}_{\Delta}\boldsymbol{H}^{\circ}(i)\boldsymbol{q}(i) + \boldsymbol{G}\boldsymbol{u}(i)) + \boldsymbol{E}\boldsymbol{d}(i)\,, \tag{1}$$

$$\boldsymbol{z}(i) = \boldsymbol{C}\boldsymbol{q}(i) + \boldsymbol{o}_z(i)\,, \tag{2}$$

$$\boldsymbol{y}(i) = \boldsymbol{C}_y\boldsymbol{q}(i) + \boldsymbol{o}_y(i)\,, \tag{3}$$

where $\boldsymbol{q}(i) \in \mathbb{R}^n$, $\boldsymbol{u}(i) \in \mathbb{R}^r$, $\boldsymbol{y}(i), \boldsymbol{z}(i) \in \mathbb{R}^s$, stand for system state, system control input, controlled output and fault detection support measurement outputs, $\boldsymbol{d}(i) \in \mathbb{R}^p$ is unknown disturbance, $\boldsymbol{o}_z(i), \boldsymbol{o}_y(i) \in \mathbb{R}^m$ are measurement noise, the matrices $\boldsymbol{E} \in \mathbb{R}^{n \times p}$, $\boldsymbol{C} \in \mathbb{R}^{s \times n}$, $\boldsymbol{C}_y \in \mathbb{R}^{m \times n}$, $\boldsymbol{F}, \boldsymbol{F}^{\circ}_{\Delta} \in \mathbb{R}^{n \times n}$, $\boldsymbol{G} \in \mathbb{R}^{n \times r}$ are real finite valued, while the couple $(\boldsymbol{F}, \boldsymbol{C})$ is observable. The non-overlapping measurements can be represented by the condition

$$\boldsymbol{C}_y\boldsymbol{C}^{\mathrm{T}} = \boldsymbol{0} \tag{4}$$

if necessary. The instant $i \in \mathcal{Z}_+$ is used as a representation of the time-instant point $t_i = i\,T_s$, where T_s is sampling period.

To make multiplicative fault uncorrelated with state variables that are not exposed to multiplicative faults, the matrix $\boldsymbol{H}^{\circ}(i) \in \mathbb{R}^{n \times n}$ is diagonal, with elements representing unknown fault amplitudes. Adaptation to the global system state means that the number of columns of $\boldsymbol{H}^{\circ}(i)$ is $s \leq n$ and the column emplacement is reflected by the rank condition defined as $rank(\boldsymbol{F}^{\circ}_{\Delta}) \leq rank(\boldsymbol{H}^{\circ}(i))$. Moreover, it is assumed that the component fault terms of $\boldsymbol{H}^{\circ}(i)$ are bounded in L_2 norm sense and that all state variables, connected with nonzero diagonal components of $\boldsymbol{H}^{\circ}(i)$, are measurable.

In order to construct fault residuals, full order state observer, based on the parameters of the system (1), (2), is built as

$$\boldsymbol{q}_{ez}(i+1) = \boldsymbol{F}\boldsymbol{q}_{ez}(i) + \boldsymbol{G}\boldsymbol{u}(i) + \boldsymbol{J}(\boldsymbol{z}(i) - \boldsymbol{z}_e(i)), \tag{5}$$

$$\boldsymbol{z}_e(i) = \boldsymbol{C}\boldsymbol{q}_{ez}(i), \tag{6}$$

where $\boldsymbol{q}_{ez}(i) \in \mathbb{R}^{n}$ is the observer state vectors, $\boldsymbol{J} \in \mathbb{R}^{n \times s}$ is a matrix with entries in the prescribed real matrix space.

The objective is solvability of the parameters of (5) subject to the system (1), (2), with the goal to investigate design conditions such that the state variable estimation errors (and fault residuals)

$$\boldsymbol{e}_z(i) = \boldsymbol{q}(i) - \boldsymbol{q}_{ez}(i), \tag{7}$$

asymptotically converges towards zero vector in the fault-free regime when limiting $i \to \infty$ and are, generally, not equal to zero vector in fault occurrences.

3 State Multiplicative Faults

The generalization can be carried out in the time domain, where the relations to multiplicative fault detectability correspond to the sensor structure in the manner given by the following lemma.

Lemma 1. *Faults are manifested as multiplicative uncorrelated sensor faults if the fault model is defined as*

$$\boldsymbol{F}(i) = \boldsymbol{F}^{\circ}_{\Delta}\boldsymbol{H}^{\circ}(i)\boldsymbol{q}(i), \tag{8}$$

where

$$\boldsymbol{F}^{\circ}_{\Delta} = \begin{bmatrix} \boldsymbol{0} \; \boldsymbol{F}_{\Delta} \; \boldsymbol{0} \end{bmatrix}, \tag{9}$$

$$\boldsymbol{H}^{\circ}(i) = diag\begin{bmatrix} \boldsymbol{0} \; \boldsymbol{H}(i) \; \boldsymbol{0} \end{bmatrix}, \tag{10}$$

$$\boldsymbol{H}(i) = diag\begin{bmatrix} h_1(i) \; h_2(i) \cdots h_s(i) \end{bmatrix}, \tag{11}$$

$\boldsymbol{F}^{\circ}_{\Delta}, \boldsymbol{H}^{\circ}(i) \in \mathbb{R}^{n \times n}$, $\boldsymbol{F}_{\Delta} \in \mathbb{R}^{n \times s}$, $\boldsymbol{H}(i) \in \mathbb{R}^{s \times s}$ *and (2) is constructed in such a way that*

$$\boldsymbol{C} = \begin{bmatrix} \boldsymbol{0} \; \boldsymbol{I}_s \; \boldsymbol{0} \end{bmatrix}. \tag{12}$$

Proof. Considered model form (8) turns out directly that

$$\boldsymbol{F}_{\Delta}^{\circ}\boldsymbol{H}^{\circ}(i)\boldsymbol{q}(i) = \boldsymbol{F}_{\Delta}^{\circ}\boldsymbol{Q}_{d}(i)\boldsymbol{h}^{\circ}(i)\,, \tag{13}$$

where

$$\boldsymbol{Q}_{d}(i) = diag\left[q_1(i)\; q_2(i)\; \cdots\; q_n(i)\right], \tag{14}$$

$$\boldsymbol{h}^{\circ}(i) = \begin{bmatrix}\mathbf{0}\\ \boldsymbol{h}(i)\\ \mathbf{0}\end{bmatrix}, \quad \boldsymbol{h}^{\mathrm{T}}(i) = \left[h_1(i)\; h_2(i)\; \cdots\; h_s(i)\right] \tag{15}$$

and where the left side and right side forms of (13) are mutually "complementary". Defining

$$\boldsymbol{C}^{\circ} = diag\left[\mathbf{0}\; \boldsymbol{I}_s\; \mathbf{0}\right] = \begin{bmatrix}\mathbf{0}\\ \boldsymbol{C}\\ \mathbf{0}\end{bmatrix} = \boldsymbol{I}_{s/n}\,, \tag{16}$$

then, due to the intertwining properties, (16) implies

$$\boldsymbol{F}_{\Delta}^{\circ} = \boldsymbol{F}_{\Delta}^{\circ}\boldsymbol{I}_{s/n} = \boldsymbol{F}_{\Delta}^{\circ}\boldsymbol{C}^{\circ}, \tag{17}$$

which leads to the parametrization of (13) as

$$\boldsymbol{F}_{\Delta}^{\circ}\boldsymbol{I}_{s/n}\boldsymbol{Q}_{d}(i)\boldsymbol{h}^{\circ}(i) = \boldsymbol{F}_{\Delta}^{\circ}\boldsymbol{C}^{\circ}\boldsymbol{Q}_{d}(i)\boldsymbol{h}^{\circ}(i) = \boldsymbol{F}_{\Delta}^{\circ}\boldsymbol{Z}^{\circ}(i)\boldsymbol{h}^{\circ}(i)\,, \tag{18}$$

where

$$\boldsymbol{Z}^{\circ}(i) = diag\left[\mathbf{0}\; \boldsymbol{Z}_{d}(i)\; \mathbf{0}\right], \tag{19}$$

$$\boldsymbol{Z}_{d}(i) = diag\left[z_1(i)\; z_2(i)\; \cdots\; z_s(i)\right]. \tag{20}$$

The structure of the above used matrix product is simply of an particular open form given by

$$\boldsymbol{F}_{\Delta}^{\circ}\boldsymbol{H}^{\circ}(i)\boldsymbol{q}(i) = \left[\mathbf{0}\; \boldsymbol{F}_{\Delta}\; \mathbf{0}\right]\begin{bmatrix}\mathbf{0} & \mathbf{0} & \mathbf{0}\\ \mathbf{0} & \boldsymbol{Z}_{d}(i) & \mathbf{0}\\ \mathbf{0} & \mathbf{0} & \mathbf{0}\end{bmatrix}\begin{bmatrix}\mathbf{0}\\ \boldsymbol{h}(i)\\ \mathbf{0}\end{bmatrix}, \tag{21}$$

where $\boldsymbol{Z}_{d}(i)$ reflects all measured variable $\boldsymbol{z}(i)$ at the time instant i. This concludes the proof. ■

Remark 1. If the model for uncorrelated multiplicative faults is

$$\boldsymbol{F}(i) = \boldsymbol{F}_{\Delta}^{\circ}\boldsymbol{H}^{\circ}(i)\boldsymbol{q}(i) \tag{22}$$

where $\boldsymbol{F}_{\Delta}^{\circ}$ is of the block form (9) but with a zero column of matrix $\boldsymbol{F}_{\Delta}$, the problem is of the same formulation since there exist a permutation matrix $\boldsymbol{T} \in \mathbb{R}^{n\times n}$ such that with $\boldsymbol{F}_{\Delta}^{\bullet} \in \mathbb{R}^{s\times s}$, $\boldsymbol{H}^{\bullet}(i) \in \mathbb{R}^{s\times s}$,

$$\boldsymbol{T}^{\mathrm{T}}\boldsymbol{F}_{\Delta}^{\circ}\boldsymbol{T} = \left[\mathbf{0}\; \boldsymbol{F}_{\Delta}^{\bullet}\; \mathbf{0}\right], \tag{23}$$

$$\boldsymbol{T}^{\mathrm{T}}\boldsymbol{H}^{\circ}(i)\boldsymbol{T} = diag\left[\mathbf{0}\; \boldsymbol{H}^{\bullet}(i)\; \mathbf{0}\right], \tag{24}$$

$$\boldsymbol{C}^{\bullet} = \boldsymbol{C}\boldsymbol{T} = \left[\mathbf{0}\; \boldsymbol{I}_s\; \mathbf{0}\right]. \tag{25}$$

4 Residual Filter Design

In the following it is considered that on the observer (5)–(6) is build a fault residual filter, defined as

$$\boldsymbol{r}(i) = \boldsymbol{R}\boldsymbol{C}\boldsymbol{e}_z(i) \tag{26}$$

where $\boldsymbol{R}_z \in \mathbb{R}^{s \times s}$ is the filter gain matrix, optimized for H_∞ norms of $\boldsymbol{H}_d(z)$ and $\boldsymbol{H}_h(z)$, reflecting the mapping

$$\tilde{\boldsymbol{r}}_z(z) = \boldsymbol{H}_h(z)\tilde{\boldsymbol{h}}^\bullet(z)\,, \quad \tilde{\boldsymbol{r}}_z(z) = \boldsymbol{H}_d(z)\tilde{\boldsymbol{d}}(z)\,. \tag{27}$$

where $\boldsymbol{h}^\bullet(i) = \boldsymbol{Z}_d(i)\boldsymbol{h}(i)$.

Theorem 1. *The Luenberger observer (5), (6) is asymptotically stable if there exist positive definite symmetric matrices $\boldsymbol{P}, \boldsymbol{S} \in \mathbb{R}^{n \times n}$, matrices $\boldsymbol{R} \in \mathbb{R}^{s \times s}$, $\boldsymbol{Y} \in \mathbb{R}^{n \times s}$ and positive scalars $\gamma, \delta \in \mathbb{R}$ such that*

$$\boldsymbol{P} = \boldsymbol{P}^T > 0\,, \quad \boldsymbol{S} = \boldsymbol{S}^T > 0\,, \quad \gamma > 0\,, \quad \delta > 0\,, \tag{28}$$

$$\begin{bmatrix} -\boldsymbol{P} & * & * & * & * & * \\ \boldsymbol{S}\boldsymbol{F} - \boldsymbol{Y}\boldsymbol{C} & \boldsymbol{P} - 2\boldsymbol{S} & * & * & * & * \\ \boldsymbol{0} & \boldsymbol{F}_\Delta^{\circ\mathrm{T}}\boldsymbol{S} & -\delta\boldsymbol{I}_s & * & * & * \\ \boldsymbol{0} & \boldsymbol{E}^{\mathrm{T}}\boldsymbol{S} & \boldsymbol{0} & -\gamma\boldsymbol{I}_p & * & * \\ \boldsymbol{R}\boldsymbol{C} & \boldsymbol{0} & \boldsymbol{0} & \boldsymbol{0} & -\gamma\boldsymbol{I}_s & * \\ \boldsymbol{R}\boldsymbol{C} & \boldsymbol{0} & \boldsymbol{0} & \boldsymbol{0} & \boldsymbol{0} & -\delta\boldsymbol{I}_s \end{bmatrix} \prec 0\,, \tag{29}$$

When the above conditions are affirmative, it can be computed

$$\boldsymbol{J} = \boldsymbol{S}^{-1}\boldsymbol{Y} \tag{30}$$

and the residual generator gain matrix is defined directly by the matrix variable $\boldsymbol{R}$.

Hereafter, $*$ denotes the symmetric item in a symmetric matrix.

Proof. To fix optimisation with respect to residuals then, with a positive definite symmetric matrix $\boldsymbol{P} \in \mathbb{R}^{n \times n}$ and positive scalars $\gamma, \delta \in \mathbb{R}$ it can be nominated the following Lyapunov function

$$\begin{aligned} v(\boldsymbol{e}_z(i)) &= \boldsymbol{e}_z^{\mathrm{T}}(i)\boldsymbol{P}\boldsymbol{e}_z(i) + \delta^{-1}\sum_{j=0}^{i-1}(\boldsymbol{r}_z^{\mathrm{T}}(j)\boldsymbol{r}_z(j) - \delta^2\boldsymbol{h}^{\bullet\mathrm{T}}(j)\boldsymbol{h}^\bullet(j)) \\ &+ \gamma^{-1}\sum_{j=0}^{i-1}(\boldsymbol{r}_z^{\mathrm{T}}(j)\boldsymbol{r}_z(j) - \gamma^2\boldsymbol{d}^{\mathrm{T}}(j)\boldsymbol{d}(j)) \\ &> 0 \end{aligned} \tag{31}$$

and, evaluating the Lyapunov function difference, the resultant inequality must yield

$$\begin{aligned} \Delta v(\boldsymbol{e}_z(i)) &= \boldsymbol{e}_z^{\mathrm{T}}(i+1)\boldsymbol{P}\boldsymbol{e}_z(i+1) - \boldsymbol{e}_z^{\mathrm{T}}(i)\boldsymbol{P}\boldsymbol{e}_z(i) \\ &+ \gamma^{-1}\boldsymbol{r}_z^{\mathrm{T}}(i)\boldsymbol{r}_z(i) - \gamma\boldsymbol{d}^{\mathrm{T}}(i)\boldsymbol{d}(i) + \delta^{-1}\boldsymbol{r}_z^{\mathrm{T}}(i)\boldsymbol{r}_z(i) - \delta\boldsymbol{h}^{\bullet\mathrm{T}}(i)\boldsymbol{h}^\bullet(i) \\ &< 0\,. \end{aligned} \tag{32}$$

Since the observer (5), (6) state error relation is described by

$$\boldsymbol{e}_z(i+1) = \boldsymbol{F}_z\boldsymbol{e}(i) + \boldsymbol{F}^{\circ}_{\Delta}\boldsymbol{h}^{\bullet}(i) + \boldsymbol{E}\boldsymbol{d}(i)\,, \tag{33}$$

where

$$\boldsymbol{F}_z = \boldsymbol{F} - \boldsymbol{J}\boldsymbol{C}\,, \tag{34}$$

transforming (33) into a singular form such that

$$\boldsymbol{F}_z\boldsymbol{e}(i) + \boldsymbol{F}^{\circ}_{\Delta}\boldsymbol{h}^{\bullet}(i) + \boldsymbol{E}\boldsymbol{d}(i) - \boldsymbol{e}_z(i+1)) = \boldsymbol{0}\,, \tag{35}$$

then, with an arbitrary symmetric positive definite square matrix $\boldsymbol{S} \in \mathbb{R}^{n\times n}$ it yields

$$\boldsymbol{e}_z^{\mathrm{T}}(i+1)\boldsymbol{S}(\boldsymbol{F}_z\boldsymbol{e}(i) + \boldsymbol{F}^{\circ}_{\Delta}\boldsymbol{h}^{\bullet}(i) + \boldsymbol{E}\boldsymbol{d}(i) - \boldsymbol{e}_z(i+1)) = 0\,. \tag{36}$$

Adding (36) and its transposition to (32), this in the above relation leads to the condition

$$\begin{aligned}
&\boldsymbol{e}_z^{\mathrm{T}}(i+1)\boldsymbol{P}\boldsymbol{e}_z(i+1) + \boldsymbol{e}_z^{\mathrm{T}}(i)\boldsymbol{P}\boldsymbol{e}_z(i)\\
&+ \boldsymbol{e}_z^{\mathrm{T}}(i)(\gamma^{-1}\boldsymbol{C}^{\mathrm{T}}\boldsymbol{R}^{\mathrm{T}}\boldsymbol{R}\boldsymbol{C} + \delta^{-1}\boldsymbol{C}^{\mathrm{T}}\boldsymbol{R}^{\mathrm{T}}\boldsymbol{R}\boldsymbol{C})\boldsymbol{e}_z(i)\\
&- \gamma\boldsymbol{d}^{\mathrm{T}}(i)\boldsymbol{d}(i) - \delta\boldsymbol{h}^{\bullet\mathrm{T}}(i)\boldsymbol{h}^{\bullet}(i)\\
&+ \boldsymbol{e}_z^{\mathrm{T}}(i+1)\boldsymbol{S}(\boldsymbol{F}_z\boldsymbol{e}(i) + \boldsymbol{F}^{\circ}_{\Delta}\boldsymbol{h}^{\bullet}(i) + \boldsymbol{E}\boldsymbol{d}(i) - \boldsymbol{e}_z(i+1))\\
&+ (\boldsymbol{F}_z\boldsymbol{e}(i) + \boldsymbol{F}^{\circ}_{\Delta}\boldsymbol{h}^{\bullet}(i) + \boldsymbol{E}\boldsymbol{d}(i) - \boldsymbol{e}_z(i+1))^{\mathrm{T}}\boldsymbol{S}\boldsymbol{e}_z(i+1)\\
&< 0\,.
\end{aligned} \tag{37}$$

If new composed vector with the components $\boldsymbol{e}_z(i), \boldsymbol{h}^{\bullet}(i), \boldsymbol{d}(i), \boldsymbol{e}_z(i+1)$ is formed to be

$$\boldsymbol{e}_{\Xi}^{\mathrm{T}}(i) = \begin{bmatrix}\boldsymbol{e}_z^{\mathrm{T}}(i)\ \boldsymbol{e}_z^{\mathrm{T}}(i+1)\ \boldsymbol{h}^{\bullet\mathrm{T}}(i)\ \boldsymbol{d}^{\mathrm{T}}(i)\end{bmatrix}, \tag{38}$$

the inequality expressed in (37) can be rewritten as

$$\boldsymbol{e}_{\Xi}^{\mathrm{T}}(i)\boldsymbol{P}_{\Xi}\boldsymbol{e}_{\Xi}(i) < 0\,, \tag{39}$$

where negativity of (39) implies negative definiteness of the matrix

$$\boldsymbol{P}_{\Xi} = \begin{bmatrix}\gamma^{-1}\boldsymbol{C}^{\mathrm{T}}\boldsymbol{R}^{\mathrm{T}}\boldsymbol{R}\boldsymbol{C} + \delta^{-1}\boldsymbol{C}^{\mathrm{T}}\boldsymbol{R}^{\mathrm{T}}\boldsymbol{R}\boldsymbol{C} - \boldsymbol{P} & \boldsymbol{F}_z^{\mathrm{T}}\boldsymbol{S} & \boldsymbol{0} & \boldsymbol{0}\\ \boldsymbol{S}\boldsymbol{F}_z & \boldsymbol{P} - 2\boldsymbol{S} & \boldsymbol{S}\boldsymbol{F}^{\circ}_{\Delta} & \boldsymbol{S}\boldsymbol{E}\\ \boldsymbol{0} & \boldsymbol{F}^{\circ\mathrm{T}}_{\Delta}\boldsymbol{S} & -\delta\boldsymbol{I}_s & \boldsymbol{0}\\ \boldsymbol{0} & \boldsymbol{E}^{\mathrm{T}}\boldsymbol{S} & \boldsymbol{0} & -\gamma\boldsymbol{I}_p\end{bmatrix} \prec 0\,. \tag{40}$$

Constructing the strictly linear matrix inequality from the bilinear form (40) with respect to matrix variable $\boldsymbol{R}$ by using the Schur complement properties, the stability condition is performed as

$$\begin{bmatrix}-\boldsymbol{P} & \boldsymbol{F}_z^{\mathrm{T}}\boldsymbol{S} & \boldsymbol{0} & \boldsymbol{0} & \boldsymbol{C}^{\mathrm{T}}\boldsymbol{R}^{\mathrm{T}} & \boldsymbol{C}^{\mathrm{T}}\boldsymbol{R}^{\mathrm{T}}\\ \boldsymbol{S}\boldsymbol{F}_z & \boldsymbol{P} - 2\boldsymbol{S} & \boldsymbol{S}\boldsymbol{F}^{\circ}_{\Delta} & \boldsymbol{S}\boldsymbol{E} & \boldsymbol{0} & \boldsymbol{0}\\ \boldsymbol{0} & \boldsymbol{F}^{\circ\mathrm{T}}_{\Delta}\boldsymbol{S} & -\delta\boldsymbol{I}_s & \boldsymbol{0} & \boldsymbol{0} & \boldsymbol{0}\\ \boldsymbol{0} & \boldsymbol{E}^{\mathrm{T}}\boldsymbol{S} & \boldsymbol{0} & -\gamma\boldsymbol{I}_p & \boldsymbol{0} & \boldsymbol{0}\\ \boldsymbol{R}\boldsymbol{C} & \boldsymbol{0} & \boldsymbol{0} & \boldsymbol{0} & -\gamma\boldsymbol{I}_s & \boldsymbol{0}\\ \boldsymbol{R}\boldsymbol{C} & \boldsymbol{0} & \boldsymbol{0} & \boldsymbol{0} & \boldsymbol{0} & -\delta\boldsymbol{I}_s\end{bmatrix} \prec 0\,. \tag{41}$$

To reflect (34) it can write

$$SF_z = SF - SJC \tag{42}$$

and with the new matrix variable

$$Y = SJ\,, \tag{43}$$

(41) implies (29). This concludes the proof. ∎

Using this technique in conjunction with the norm-bound structure outlined in the preceding section is a first step towards accepting the opportunity associated with single multiplicative fault detection and estimation.

5 Illustrative Example

On this level, the system is described by (1), (2) with $s = 2$, $Ts = 0.02$ s, the disturbance and measurement noise from the normal distribution with zero mean and standard deviation 0.05 and 0.0001, respectively, the model parameters

$$F = \begin{bmatrix} 0.9324 & 0.0 & 0.1109 & 0.0990 \\ 0.0062 & 0.9197 & 0.0226 & 0.0002 \\ 0.0185 & 0.0 & 0.8744 & 0.0867 \\ 0.0001 & 0.0428 & 0.0001 & 0.9467 \end{bmatrix}, \quad F^{\circ}_{\Delta} = \begin{bmatrix} 0 & 0 & 1 & 0 \\ 0 & 0 & 0 & 0 \\ 0 & 1 & 0 & 0 \\ 0 & 0 & 0 & 0 \end{bmatrix}, \quad E = \begin{bmatrix} 0.0547 \\ 0.0621 \\ 0.0725 \\ 0.0267 \end{bmatrix},$$

$$G = \begin{bmatrix} 0.0081 & 0.0043 \\ 0.0110 & 0.0041 \\ 0.0028 & 0.0063 \\ 0.0025 & 0.0034 \end{bmatrix}, \quad h(i) = \begin{bmatrix} \mathbf{0} \\ h_1(i) \\ h_2(i) \\ \mathbf{0} \end{bmatrix}, \quad C = \begin{bmatrix} 0 & 1 & 0 & 0 \\ 0 & 0 & 1 & 0 \end{bmatrix}, \quad C_y = \begin{bmatrix} 1 & 0 & 0 & 0 \\ 0 & 0 & 0 & 1 \end{bmatrix}$$

and the matrix of multiplicative fault amplitudes

$$H^{\circ}(i) = diag\begin{bmatrix} 0 & h_1(i) & h_2(i) & 0 \end{bmatrix}, \quad \overline{h}_j = 0.001, \ \underline{h}_j = 0 \text{ for } j = 1, 2\,.$$

The system is stable and the system working point in simulation is set up by the control law

$$u(i) = N_w w\,, \text{ for all } i\,,$$

which assigns the system output to the desired steady state value $y_o = w$, where

$$N_w = (C_y(I_n - F)^{-1}G)^{-1} = \begin{bmatrix} -18.7567 & 77.7863 \\ 25.3041 & -101.1829 \end{bmatrix}, \quad w = \begin{bmatrix} 1 \\ 2 \end{bmatrix}.$$

Implementation (28), (29) in SeDuMi package toolbox [15] decomposes the feasible design problem into the set of LMI variables P, S, Y, R, γ, δ, where fixing $R = I_2$ serves receivable

$$\gamma = 5.2718\,, \qquad \delta = 9.6415\,,$$

$$\boldsymbol{P} = \begin{bmatrix} 0.9819 & -0.1712 & -0.1666 & -0.1081 \\ -0.1712 & 4.4900 & 0.0418 & -0.1213 \\ -0.1666 & 0.0418 & 3.6999 & -0.5402 \\ -0.1081 & -0.1213 & -0.5402 & 4.8638 \end{bmatrix} \succ 0\,,$$

$$\boldsymbol{S} = \begin{bmatrix} 1.0462 & -0.1626 & -0.1675 & -0.1314 \\ -0.1626 & 4.5076 & 0.0348 & -0.1170 \\ -0.1675 & 0.0348 & 4.1581 & -0.4544 \\ -0.1314 & -0.1170 & -0.4544 & 4.8897 \end{bmatrix} \succ 0\,,$$

$$\boldsymbol{Y} = \begin{bmatrix} 0.0147 & 0.1351 \\ 3.6430 & 0.0660 \\ -0.0051 & 3.3747 \\ 0.2095 & 0.0901 \end{bmatrix}$$

It can be shown simply that (30) gives

$$\boldsymbol{J} = \begin{bmatrix} 0.1500 & 0.2790 \\ 0.8153 & 0.0210 \\ 0.0053 & 0.8340 \\ 0.0669 & 0.1039 \end{bmatrix}$$

and using (34), the observer dynamics, expressing the summarized behaviour of all estimated variables, is defined by the system matrix and its eigenvalues

$$\boldsymbol{F}_z = \begin{bmatrix} 0.9324 & -0.1500 & -0.1681 & 0.0990 \\ 0.0062 & 0.1044 & 0.0016 & 0.0002 \\ 0.0185 & -0.0053 & 0.0404 & 0.0867 \\ 0.0001 & -0.0241 & -0.1038 & 0.9467 \end{bmatrix}, \quad \rho(\boldsymbol{F}_z) = \left\{ \begin{array}{c} 0.0537 \\ 0.1055 \\ 0.9323 \pm 0.0129\,\mathrm{i} \end{array} \right\}.$$

To carry out simulations, demonstrating fault residual filter properties, single multiplicative faults with formal description by step-like activation of $h_1(i)$ as $h_1(i) = 0.001$ while $h_2(i) = 0$, as well as when $h_2(i)$ is modified as $h_2(i) = 0.001$ while $h_1(i) = 0$, respectively, are considered in the own time scale from the time instant $t = 3$ s. To remove any additive dynamics in the residual filter responses, the system is stable and all initial states in simulation are setting to zero vectors.

The fault residuals time evaluations are depicted in Fig. 1a and Fig. 1b, which show that the residual filter design achieves very good time responses and stability performance. This clearly confirms the theoretic analysis and demonstrate resulting faster time responses if reflecting in design conditions also upper bound of H_∞ norm of the multiplicative fault transfer matrix function. Resulting δ and γ adjustment is approximately optimal because, in terms of the structure of the two last rows in (29), a delta increase leads to a gamma increase. By fixing the variable $\boldsymbol{R}$, the amplitude of the residual signal was set to values of the order of ten to minus three.

The system model parameters are defined to show cases where the parametric fault is acting on an element on the system matrix diagonal or on an element outside the diagonal of the matrix. The extraordinary conclusion is that parametric faults of values of the order of one thousandth can be detected in proposed residual stricture.

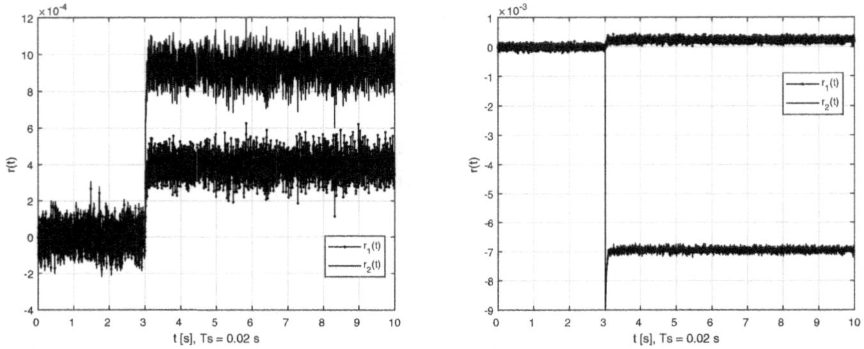

a) Residual signals evolution for h_1 b) Residual signals evolution for h_2

Fig. 1. System state and system state estimations

6 Concluding Remarks

The issue of the observer based fault residual structures for discrete-time linear systems with single multiplicative system faults, is the paper subject. Because the elements of the system matrix change after the occurrence of a multiplicative fault, the enhanced bounded real lemma structure is preferred, where the Lyapunov matrix $\boldsymbol{P}$ is decoupled from the matrix parameters of the system and the synthesis of the estimator gain matrix is formulated in relation to the slack matrix $\boldsymbol{S}$. These favorable synthesis starting points allow enough flexibility to guarantee stability and preferred dynamic properties of the observer structure, as well as satisfactory residual signal sensitivity. Design terms, feasible on standard numerical operations for linear matrix inequalities manipulation, are completely model based and smoothly convenient in use.

In the used configuration, the problem is resolved with respect to measurable state variables having input coincidence with multiplicative fault parameters structure. The number of residual signals may be equal to number of system control related outputs, defined for control system performance. In addition, due to partial elimination, estimates of the state and the output vector are substantially separated from disturbance disruptions.

This concept is also matched by the choice of system model in the illustrative example. In order to achieve the independence of the fault residual filter time responses from the control structure, a stable system is used for the simulations with the working mode setting by forced mode principle. Existence of potential set of residual filters, reflecting complementary segmentation of fault matrix structure, is related to provides the basis for the future directions of this research.

Acknowledgment. The work presented in this paper was supported by VEGA, the Grant Agency of the Ministry of Education and the Academy of Science of Slovak Republic under Grant No. 1/0608/17. This support is very gratefully acknowledged.

References

1. Chiang, L.H., Russell, E.L., Braatz, R.D.: Fault Detection and Diagnosis in Industrial Systems. Springer, London (2001). https://doi.org/10.1007/978-1-4471-0347-9
2. Ellis, G.: Observers in Control Systems. A Practical Guide. Academic Press, San Diego (2002)
3. Ferdowsi, H., Jagannathan, S.: Unified model-based fault diagnosis scheme for nonlinear discrete-time systems with additive and multiplicative faults. In: Proceedings 50th IEEE Conference on Decision and Control and European Control Conference CDC-ECC 2011, Orlando, USA, pp. 1570–1575 (2011). https://doi.org/10.1109/CDC.2011.6161292
4. Filasová, A., Krokavec, D., Liššinský, P.: Conditions with D-stability circle area in design of observer-based fault estimation. Appl. Math. Sci. **10**(35), 1705–1717 (2016). https://doi.org/10.12988/ams.2016.63112
5. Gao, Z.: Fault estimation and fault-tolerant control for discrete-time dynamic systems. IEEE Tran. Ind. Electron. **62**(4), 3874–3884 (2015). https://doi.org/10.1109/TIE.2015.2392720
6. Gao, C., Duan, G.: Robust adaptive fault estimation for a class of nonlinear systems subject to multiplicative faults. Circuits Syst. Sig. Proces. **31**(6), 2035–2046 (2012). https://doi.org/10.1007/s00034-012-9434-x
7. Gershon, E., Shaked, U., Yaesh, I.: H_∞ Control and Estimation of State-Multiplicative Linear Systems. Springer, London (2005). https://doi.org/10.1007/b103068
8. Gertler, J.: Fault Detection and Diagnosis in Engineering Systems. CRC Press, Boca Raton (1998)
9. Kim, K.S., Rew, K.H.: Reduced order disturbance observer for discrete-time linear systems. Automatica **49**(4), 968–975 (2013). https://doi.org/10.1016/j.automatica.2013.01.014
10. Korovin, S.K., Fomichev, V.V.: State Observers for Linear Systems with Uncertainty. Walter de Gruyter, Berlin (2009)
11. Krokavec, D., Filasová, A.: H_∞ norm principle in residual filter design for discrete-time linear positive systems. Eur. J. Control **45**, 17–29 (2019). https://doi.org/10.1016/j.ejcon.2018.10.001
12. Krokavec, D., Filasová, A.: On fault detection for discrete-time linear state-multiplicative systems with uncorrelated multiplicative faults. In: Proceedings 6th International Conference on Control, Decision and Information Technologies CoDIT 2019, Paris, France, pp. 894–899 (2019). https://doi.org/10.1109/CoDIT.2019.8820338
13. Lin, Y., Stadtherr, M.A.: Fault detection in continuous-time systems with uncertain parameters. In: Proceedings American Control Conference ACC 2007, New York, USA, pp. 3216–3221 (2007). https://doi.org/10.1109/ACC.2007.4282164
14. Luenberger, D.G.: Introduction to Dynamic Systems. Theory, Models and Applications. Wiley, New York (1979)
15. Peaucelle, D., Henrion, D., Labit, Y., Taitz, K.: User's Guide for SeDuMi Interface 1.04, LAAS-CNRS, Toulouse (2002)

Fault-Tolerant Tracking Control for Takagi–Sugeno Fuzzy Systems Under Actuator and Sensor Faults

Norbert Kukurowski[1](✉), Marcin Pazera[1], Marcin Witczak[1], Francisco-Ronay López-Estrada[2], and Teódulo Iván Bravo Cruz[2]

[1] Institute of Control and Computation Engineering, University of Zielona Góra, ul. Szafrana 2, 65-516 Zielona Góra, Poland
{n.kukurowski,m.pazera,m.witczak}@issi.uz.zgora.pl

[2] Tuxtla Gutiérrez Institute of Technology, National Technological Institute of Mexico, TURIX-Dynamics Diagnosis and Control Group, Carretera Panamericana Km 1080 SN, 29050 Tuxtla Gutierrez, Chiapas, Mexico
frlopez@ittg.edu.mx, teodulo_bravo@outlook.com

Abstract. The paper deals with the design of an active fault-tolerant control scheme, which is based on the actuator and sensor fault estimator. Thus, the paper starts with the development of such a fault estimation scheme capable of estimating these faults simultaneously. It is assumed that the estimator should have a desired H_∞ performance. Subsequently, a new fault-tolerant control law is proposed, which is based on a parallel digital twin of the system. Thus, the goal is to control the system in such a way as to follow the states of the references digital twin irrespective of the faults. Finally, the effectiveness of the proposed approach is verified with the laboratory three-tank system. In particular, the performance of the system is tested against a set of simultaneous actuator and sensor faults, respectively.

Keywords: Fault detection and diagnosis · Robust fault estimation · Takagi-Sugeno fuzzy system · Fault-tolerant tracking control

1 Introduction

Owing to the fact that industrial systems are increasingly complex, the presence of a damage is more probable. The most common ones are those that involve measurement tools (sensor faults) whose number is proliferating significantly due to the advent of IoT. Another dominant group of faults are those that affect system actuators (actuator faults). It is a well-known fact that an efficient way to achieve a good reliability is through the implementation of Fault Diagnosis and Isolation techniques (FDI) [3,5,21], which are responsible for monitoring the system and alerting where and when a fault occurs, as well the Fault-Tolerant Control (FTC) [4,14,22] whose purpose is to maintain the desired system performance in the presence of unappealing fault effects. In the literature one can find

A. Bartoszewicz et al. (Eds.): PCC 2020, AISC 1196, pp. 1434–1445, 2020.
https://doi.org/10.1007/978-3-030-50936-1_119

active FTC approaches [6,8], used for sensor's faults [7,13,18] and actuator's faults [2,9,17], being the application to both fault types the topic corresponding to this article. One of the methodologies to take into consideration, is the use of the output feedback and predictive compensation strategy for the tracking and optimization problems [20]. Other approaches work with a state estimation and sensor fault detection, isolation and fault estimation observers for descriptor-linear parameter varying systems. Such an approach uses the scheduling function depending on an unmeasurable state vector [10]. Another use a fault estimator for recovering the states and implement a novel FTC control [15]. Zhan, Yin et al. [27], proposed a integrated fault detection, diagnosis, and reconfigurable control scheme based on interacting multiple model (IMM) approach, where a IMM estimator is responsible for the fault detection and diagnosis. And [26] used a robust controller based on sliding-mode control scheme for a flexible spacecraft.

The objective of this paper is the development of a fault estimator which could estimate sensor and actuator faults at the same time and implement it within the FTC tracking control scheme. In particular, it tracks the state of a virtual twin of the system. Thus, the scheme aims at tracking the virtual system state irrespective of both actuator and sensor faults. Both estimator and FTC are developed under H_∞ performance [1,25]. Moreover, the nonlinear system model is represented in form of a set of linear local models with a Takagi-Sugeno technique [11,12].

The paper is organized as follows: Sect. 2 presents an novel estimator and FTC controller for both sensor and actuator faults. Subsequently, in Sect. 3, an illustrative example is presented to verify the correctness and performance of the proposed algorithm. Finally, Sect. 4 concludes the paper.

2 Fault-Tolerant Tracking Controller Design

Let us begin with describing a nonlinear system

$$\boldsymbol{x}_{f,k+1} = \boldsymbol{f}\left(\boldsymbol{x}_{f,k}, \boldsymbol{u}_{f,k}\right), \tag{1}$$

which can be properly rewritten by a Takagi-Sugeno (T–S) with faults and uncertainties:

$$\begin{aligned} \boldsymbol{x}_{f,k+1} &= \boldsymbol{A}(\alpha_k)\boldsymbol{x}_{f,k} + \boldsymbol{B}(\alpha_k)\boldsymbol{u}_{f,k} + \boldsymbol{B}(\alpha_k)\boldsymbol{f}_{a,k} + \boldsymbol{W}_1\boldsymbol{w}_{1,k} \qquad (2)\\ &= \sum_{i=1}^{M} h_i(\alpha_k)\left[\boldsymbol{A}^i\boldsymbol{x}_{f,k} + \boldsymbol{B}^i\boldsymbol{u}_{f,k} + \boldsymbol{B}^i\boldsymbol{f}_{a,k}\right] + \boldsymbol{W}_1\boldsymbol{w}_{1,k},\\ \boldsymbol{y}_{f,k} &= \boldsymbol{C}\boldsymbol{x}_{f,k} + \boldsymbol{C}_f\boldsymbol{f}_{s,k} + \boldsymbol{W}_2\boldsymbol{w}_{2,k}, \qquad (3) \end{aligned}$$

with

$$h_i(\alpha_k) \geq 0, \quad \forall i = 1, \ldots, M, \qquad \sum_{i=1}^{M} h_i(\alpha_k) = 1, \tag{4}$$

where $\boldsymbol{x}_{f,k} \in \mathbb{X} \subset \mathbb{R}^n$, $\boldsymbol{u}_{f,k} \in \mathbb{R}^r$, $\boldsymbol{y}_{f,k} \in \mathbb{R}^m$ indicate the state, input and output, respectively. Hence, $\boldsymbol{f}_{a,k} \in \mathbb{F}_a \subset \mathbb{R}^r$ denotes the actuator fault while

$\boldsymbol{f}_{s,k} \in \mathbb{F}_s \subset \mathbb{R}^{n_s}$ is the sensor fault. Furthermore, $\boldsymbol{C}_f$ signifies the sensor fault distribution matrix with $\text{rank}(\boldsymbol{C}_f) = n_s$ and satisfied relation $r + n_s \leq m$, which means that is impossible to estimate more faults than the measured outputs. Additionally, $\boldsymbol{W}_1$ and $\boldsymbol{W}_2$ denote the distribution matrices. Thus, $\boldsymbol{w}_{1,k}$ and $\boldsymbol{w}_{2,k}$ indicate exogenous disturbance vectors for the process and measurement uncertainties, respectively. The activation functions $h_i(\cdot)$ depend on the vector of premise variables $\alpha_k = \left[\alpha_k^1, \alpha_k^2, \ldots, \alpha_k^p\right]^T$, which is assumed to depend on measurable variables, e.g. system outputs and known inputs [19]. For the sake of brevity, the following notation is used: $\boldsymbol{A}(\alpha) = \sum_{i=1}^{M} h_i(\alpha_k) \boldsymbol{A}_i$, $\boldsymbol{B}(\alpha) = \sum_{i=1}^{M} h_i(\alpha_k) \boldsymbol{B}_i$.

The purpose is to develop a state and fault estimator as well as FTC tracking controller which will be able to minimize the error between a reference state $\boldsymbol{x}_k$ and the state of a possibly faulty system $\boldsymbol{x}_{f,k}$ irrespective of the simultaneous actuator and sensor faults with no need for changing an already existed controller. Thus, the strategy can be summarized with a schematic diagram, which is presented in Fig. 1.

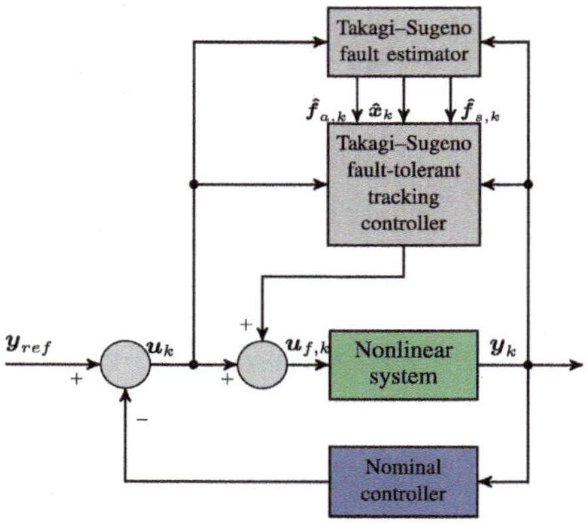

Fig. 1. Scheme of the tracking fault-tolerant control

Let us suppose that an initial controller is already employed in the system. It may be for example a simple state feedback controller of the form

$$\boldsymbol{u}_k = -\boldsymbol{K}\boldsymbol{x}_k. \tag{5}$$

Note that its design is beyond the scope of the paper and it can be realized with a large spectrum of available approaches. The objective is to develop an additional controller which will be able to minimize a tracking error. For that purpose let us consider the following reference model:

$$\boldsymbol{x}_{k+1} = \boldsymbol{A}(\alpha_k)\boldsymbol{x}_k + \boldsymbol{B}(\alpha_k)\boldsymbol{u}_k, \tag{6}$$

$$\boldsymbol{y}_k = \boldsymbol{C}\boldsymbol{x}_k. \tag{7}$$

where $\boldsymbol{x}_k \in \mathbb{R}^n$, $\boldsymbol{y}_k \in \mathbb{R}^m$ and $\boldsymbol{u}_k \in \mathbb{R}^r$ indicates the reference state, output and nominal control input, respectively. The essential idea is to define the control strategy for the possibly faulty and uncertain system (2)–(3) as follows

$$\boldsymbol{u}_{f,k} = -\hat{\boldsymbol{f}}_{a,k} + \boldsymbol{K}_1 \left(\boldsymbol{x}_k - \hat{\boldsymbol{x}}_{f,k}\right) + \boldsymbol{u}_k, \tag{8}$$

where $\hat{\boldsymbol{f}}_{a,k}$ signifies the actuator fault estimate. The important point to note here is that the state estimator $\hat{\boldsymbol{x}}_{f,k}$ can be used, whilst the $\boldsymbol{x}_{f,k}$ is not available. Let us define the state, actuator and sensor fault estimator:

$$\hat{\boldsymbol{x}}_{f,k+1} = \boldsymbol{A}(\alpha)\, \hat{\boldsymbol{x}}_k + \boldsymbol{B}(\alpha)\, \boldsymbol{u}_k + \boldsymbol{B}(\alpha)\, \hat{\boldsymbol{f}}_{a,k} + \boldsymbol{K}_x \left(\boldsymbol{y}_k - \boldsymbol{C}\hat{\boldsymbol{x}}_k - \boldsymbol{C}_f \hat{\boldsymbol{f}}_{s,k}\right), \tag{9}$$

$$\hat{\boldsymbol{f}}_{a,k+1} = \hat{\boldsymbol{f}}_{a,k} + \boldsymbol{K}_a \left(\boldsymbol{y}_k - \boldsymbol{C}\hat{\boldsymbol{x}}_k - \boldsymbol{C}_f \hat{\boldsymbol{f}}_{s,k}\right), \tag{10}$$

$$\hat{\boldsymbol{f}}_{s,k+1} = \hat{\boldsymbol{f}}_{s,k} + \boldsymbol{K}_s \left(\boldsymbol{y}_k - \boldsymbol{C}\hat{\boldsymbol{x}}_k - \boldsymbol{C}_f \hat{\boldsymbol{f}}_{s,k}\right). \tag{11}$$

Substituting (8) into (2) gives

$$\begin{aligned}\boldsymbol{x}_{f,k+1} = {} & \boldsymbol{A}(\alpha)\, \boldsymbol{x}_{f,k} - \boldsymbol{B}(\alpha)\, \hat{\boldsymbol{f}}_{a,k} + \boldsymbol{B}(\alpha)\, \boldsymbol{K}_1 \left(\boldsymbol{e}_k - \boldsymbol{e}_{f,k}\right) + \boldsymbol{B}(\alpha)\, \boldsymbol{u}_k \\ & + \boldsymbol{B}(\alpha)\, \boldsymbol{f}_{a,k} + \boldsymbol{W}_1 \boldsymbol{w}_{1,k},\end{aligned} \tag{12}$$

with:

$$\boldsymbol{e}_k = \boldsymbol{x}_k - \boldsymbol{x}_{f,k}, \qquad \boldsymbol{e}_{f,k} = \boldsymbol{x}_{f,k} - \hat{\boldsymbol{x}}_{f,k}, \tag{13}$$

where $\boldsymbol{e}_k$ and $\boldsymbol{e}_{f,k}$ indicate the tracking and estimation error, respectively. Thus

$$\begin{aligned}\boldsymbol{x}_{f,k+1} = {} & \boldsymbol{A}(\alpha)\, \boldsymbol{x}_{f,k} + \boldsymbol{B}(\alpha)\, \boldsymbol{e}_{a,k} + \boldsymbol{B}(\alpha)\, \boldsymbol{K}_1 \left(\boldsymbol{e}_k - \boldsymbol{e}_{f,k}\right) + \boldsymbol{B}(\alpha)\, \boldsymbol{u}_k \\ & + \boldsymbol{W}_1 \boldsymbol{w}_{1,k},\end{aligned} \tag{14}$$

where $\boldsymbol{e}_{a,k}$ denotes the actuator estimation error. Let us define the tracking and state estimation error:

$$\begin{aligned}\boldsymbol{e}_{k+1} = \boldsymbol{x}_{k+1} - \boldsymbol{x}_{f,k+1} = {} & \left(\boldsymbol{A}(\alpha) - \boldsymbol{B}(\alpha)\, \boldsymbol{K}_1\right) \boldsymbol{e}_k - \boldsymbol{B}(\alpha)\, \boldsymbol{e}_{a,k} \\ & - \boldsymbol{B}(\alpha)\, \boldsymbol{K}_1 \boldsymbol{e}_{f,k} - \boldsymbol{W}_1 \boldsymbol{w}_{1,k},\end{aligned} \tag{15}$$

$$\begin{aligned}\boldsymbol{e}_{f,k+1} = \boldsymbol{x}_{f,k+1} - \hat{\boldsymbol{x}}_{f,k+1} = {} & \left(\boldsymbol{A}(\alpha) - \boldsymbol{K}_x \boldsymbol{C}\right) \boldsymbol{e}_{f,k} + \boldsymbol{B}(\alpha)\, \boldsymbol{e}_{a,k} \\ & - \boldsymbol{K}_x \boldsymbol{C}_f \boldsymbol{e}_{s,k} + \boldsymbol{W}_1 \boldsymbol{w}_{1,k} - \boldsymbol{K}_x \boldsymbol{W}_2 \boldsymbol{w}_{2,k}.\end{aligned} \tag{16}$$

Subsequently, the actuator and sensor fault estimation error are described by:

$$\begin{aligned}\boldsymbol{e}_{a,k+1} = \boldsymbol{f}_{a,k+1} - \hat{\boldsymbol{f}}_{a,k+1} = {} & \boldsymbol{\varepsilon}_{a,k} + \boldsymbol{e}_{a,k} - \boldsymbol{K}_a \boldsymbol{C} \boldsymbol{e}_{f,k} - \boldsymbol{K}_a \boldsymbol{C}_f \boldsymbol{e}_{s,k} \\ & - \boldsymbol{K}_a \boldsymbol{W}_2 \boldsymbol{w}_{2,k},\end{aligned} \tag{17}$$

$$\begin{aligned}\boldsymbol{e}_{s,k+1} = \boldsymbol{f}_{s,k+1} - \hat{\boldsymbol{f}}_{s,k+1} = \boldsymbol{\varepsilon}_{s,k} + (\boldsymbol{I} - \boldsymbol{K}_s\boldsymbol{C}_f)\,\boldsymbol{e}_{s,k} - \boldsymbol{K}_s\boldsymbol{C}\boldsymbol{e}_{f,k}\\ - \boldsymbol{K}_s\boldsymbol{W}_2\boldsymbol{w}_{2,k},\end{aligned} \tag{18}$$

where

$$\boldsymbol{\varepsilon}_{a,k} = \boldsymbol{f}_{a,k+1} - \boldsymbol{f}_{a,k}, \qquad \boldsymbol{\varepsilon}_{s,k} = \boldsymbol{f}_{s,k+1} - \boldsymbol{f}_{s,k}.$$

Furthermore, let us combine (15)–(18) to achieve the following super–vectors:

$$\bar{\boldsymbol{e}}_k = \begin{bmatrix}\boldsymbol{e}_k\\ \boldsymbol{e}_{f,k}\\ \boldsymbol{e}_{a,k}\\ \boldsymbol{e}_{s,k}\end{bmatrix}, \qquad \bar{\boldsymbol{w}}_k = \begin{bmatrix}\boldsymbol{w}_{1,k}\\ \boldsymbol{w}_{2,k}\\ \boldsymbol{\varepsilon}_{a,k}\\ \boldsymbol{\varepsilon}_{s,k}\end{bmatrix}. \tag{19}$$

Hence

$$\begin{aligned}\bar{\boldsymbol{e}}_{k+1} = &\begin{bmatrix}\boldsymbol{A}(\alpha) - \boldsymbol{B}(\alpha)\boldsymbol{K}_1 & -\boldsymbol{B}(\alpha)\boldsymbol{K}_1 & -\boldsymbol{B}(\alpha) & \boldsymbol{0}\\ \boldsymbol{0} & \boldsymbol{A}(\alpha) - \boldsymbol{K}_x\boldsymbol{C} & \boldsymbol{B}(\alpha) & -\boldsymbol{K}_x\boldsymbol{C}_f\\ \boldsymbol{0} & -\boldsymbol{K}_a\boldsymbol{C} & \boldsymbol{I} & -\boldsymbol{K}_a\boldsymbol{C}_f\\ \boldsymbol{0} & -\boldsymbol{K}_s\boldsymbol{C} & \boldsymbol{0} & \boldsymbol{I} - \boldsymbol{K}_s\boldsymbol{C}_f\end{bmatrix}\bar{\boldsymbol{e}}_k\\ &+ \begin{bmatrix}-\boldsymbol{W}_1 & \boldsymbol{0} & \boldsymbol{0} & \boldsymbol{0}\\ \boldsymbol{W}_1 & -\boldsymbol{K}_x\boldsymbol{W}_2 & \boldsymbol{0} & \boldsymbol{0}\\ \boldsymbol{0} & -\boldsymbol{K}_a\boldsymbol{W}_2 & \boldsymbol{I} & \boldsymbol{0}\\ \boldsymbol{0} & -\boldsymbol{K}_s\boldsymbol{W}_2 & \boldsymbol{0} & \boldsymbol{I}\end{bmatrix}\bar{\boldsymbol{w}}_k,\end{aligned} \tag{20}$$

can be described in simpler form

$$\bar{\boldsymbol{e}}_{k+1} = \tilde{\boldsymbol{A}}_1(\alpha)\,\bar{\boldsymbol{e}}_k + \tilde{\boldsymbol{W}}_1\bar{\boldsymbol{w}}_k, \tag{21}$$

with:

$$\tilde{\boldsymbol{A}}_1(\alpha) = \begin{bmatrix}\bar{\boldsymbol{A}}_1(\alpha) & \bar{\boldsymbol{B}}(\alpha)\\ \boldsymbol{0} & \bar{\boldsymbol{A}}_2(\alpha)\end{bmatrix}, \quad \tilde{\boldsymbol{W}}_1 = \begin{bmatrix}-\boldsymbol{W}_1 & \boldsymbol{0}\\ \bar{\boldsymbol{W}}_1 & \bar{\boldsymbol{W}}_2\end{bmatrix}, \qquad \bar{\boldsymbol{W}}_2 = \bar{\boldsymbol{W}}_3 - \bar{\boldsymbol{K}}\tilde{\boldsymbol{W}}_2,$$

$$\bar{\boldsymbol{A}}_1(\alpha) = \boldsymbol{A}(\alpha) - \boldsymbol{B}(\alpha)\boldsymbol{K}_1, \quad \bar{\boldsymbol{B}}(\alpha) = \left[-\boldsymbol{B}(\alpha)\boldsymbol{K}_1 \; -\boldsymbol{B}(\alpha) \; \boldsymbol{0}\right],$$

$$\bar{\boldsymbol{W}}_1 = \left[\boldsymbol{W}_1^T \; \boldsymbol{0} \; \boldsymbol{0}\right]^T, \quad \bar{\boldsymbol{W}}_3 = \begin{bmatrix}\boldsymbol{0} & \boldsymbol{0} & \boldsymbol{0}\\ \boldsymbol{0} & \boldsymbol{I} & \boldsymbol{0}\\ \boldsymbol{0} & \boldsymbol{0} & \boldsymbol{I}\end{bmatrix}, \quad \tilde{\boldsymbol{W}}_2 = \left[\boldsymbol{W}_2 \; \boldsymbol{0} \; \boldsymbol{0}\right],$$

$$\bar{\boldsymbol{K}} = \left[\boldsymbol{K}_x^T \; \boldsymbol{K}_a^T \; \boldsymbol{K}_s^T\right]^T.$$

Finally, it can be observed that eigenvalues of $\tilde{\boldsymbol{A}}_1(\alpha)$ rely on the ones of $\bar{\boldsymbol{A}}_1(\alpha)$ and $\bar{\boldsymbol{A}}_2(\alpha)$ and for such a reason the FTC controller and estimator may be designed separately. Thus

$$\bar{\boldsymbol{A}}_2 = \begin{bmatrix}\boldsymbol{A}(\alpha) - \boldsymbol{K}_x\boldsymbol{C} & \boldsymbol{B}(\alpha) & -\boldsymbol{K}_x\boldsymbol{C}_f\\ -\boldsymbol{K}_a\boldsymbol{C} & \boldsymbol{I} & -\boldsymbol{K}_a\boldsymbol{C}_f\\ -\boldsymbol{K}_s\boldsymbol{C} & \boldsymbol{0} & \boldsymbol{I} - \boldsymbol{K}_s\boldsymbol{C}_f\end{bmatrix},$$

can be rewritten into following form

$$\boldsymbol{X}(\alpha) = \bar{\boldsymbol{A}}(\alpha) - \bar{\boldsymbol{K}}\bar{\boldsymbol{C}},$$

where:

$$\bar{\boldsymbol{A}}(\alpha) = \begin{bmatrix} \boldsymbol{A}(\alpha) & \boldsymbol{B}(\alpha) & \boldsymbol{0} \\ \boldsymbol{0} & \boldsymbol{I} & \boldsymbol{0} \\ \boldsymbol{0} & \boldsymbol{0} & \boldsymbol{I} \end{bmatrix}, \quad \bar{\boldsymbol{C}} = \begin{bmatrix} \boldsymbol{C} \, \boldsymbol{0} \, \boldsymbol{C}_f \end{bmatrix}.$$

Consequently

$$\tilde{\boldsymbol{e}}_{k+1} = \boldsymbol{X}(\alpha)\tilde{\boldsymbol{e}}_k + \bar{\boldsymbol{W}}_2\bar{\boldsymbol{w}}_k, \tag{22}$$

where $\tilde{\boldsymbol{e}}_k = \left[\boldsymbol{e}_{f,k}^T, \boldsymbol{e}_{a,k}^T, \boldsymbol{e}_{s,k}^T\right]^T$. Let us remind the Lyapunov candidate function

$$V_k = \tilde{\boldsymbol{e}}_k^T \boldsymbol{P}\tilde{\boldsymbol{e}}_k, \tag{23}$$

where $\boldsymbol{P} \succ 0$. Consequently stability condition can be described by

$$\Delta V_k + \tilde{\boldsymbol{e}}_k^T\tilde{\boldsymbol{e}}_k - \mu^2\bar{\boldsymbol{w}}_k^T\bar{\boldsymbol{w}}_k < 0, \tag{24}$$

with:

$$\Delta V_k = V_{k+1} - V_k, \qquad V_{k+1} = \tilde{\boldsymbol{e}}_{k+1}^T\boldsymbol{P}\tilde{\boldsymbol{e}}_{k+1}.$$

According to the mentioned considerations, the following theorem for the estimator design is proposed:

Theorem 1. *For a prescribed attenuation level μ of $\bar{\boldsymbol{w}}_k$, the $\mathcal{H}_\infty$ estimator design problem for the system (2)–(3) is solvable if there exist $\boldsymbol{N}$ and $\boldsymbol{P} \succ 0$, the following condition is satisfied:*

$$\begin{bmatrix} -\boldsymbol{P}+\boldsymbol{I} & \boldsymbol{0} & \bar{\boldsymbol{A}}^T(\alpha)\boldsymbol{P} - \bar{\boldsymbol{C}}^T\boldsymbol{N}^T \\ \boldsymbol{0} & -\mu^2\boldsymbol{I} & \bar{\boldsymbol{W}}_3^T\boldsymbol{P} - \tilde{\boldsymbol{W}}_2^T\boldsymbol{N}^T \\ \boldsymbol{P}\bar{\boldsymbol{A}}(\alpha) - \boldsymbol{N}\bar{\boldsymbol{C}} & \boldsymbol{P}\bar{\boldsymbol{W}}_3 - \boldsymbol{N}\tilde{\boldsymbol{W}}_2 & -\boldsymbol{P} \end{bmatrix} \prec 0. \tag{25}$$

Proof. Using (24) shows that

$$\tilde{\boldsymbol{e}}_{k+1}^T\boldsymbol{P}\tilde{\boldsymbol{e}}_{k+1} - \tilde{\boldsymbol{e}}_k^T\boldsymbol{P}\tilde{\boldsymbol{e}}_k + \tilde{\boldsymbol{e}}_k^T\tilde{\boldsymbol{e}}_k - \mu^2\bar{\boldsymbol{w}}_k^T\bar{\boldsymbol{w}}_k < 0. \tag{26}$$

From (22), the following inequality can be defined

$$\begin{aligned} &\tilde{\boldsymbol{e}}_k^T\left(\boldsymbol{X}^T(\alpha)\boldsymbol{P}\boldsymbol{X}(\alpha) - \boldsymbol{P} + \boldsymbol{I}\right)\tilde{\boldsymbol{e}}_k \\ &+ \tilde{\boldsymbol{e}}_k^T\left(\boldsymbol{X}^T(\alpha)\boldsymbol{P}\bar{\boldsymbol{W}}_2\right)\bar{\boldsymbol{w}}_k \\ &+ \bar{\boldsymbol{w}}_k^T\left(\bar{\boldsymbol{W}}_2^T\boldsymbol{P}\boldsymbol{X}(\alpha)\right)\tilde{\boldsymbol{e}}_k \\ &+ \bar{\boldsymbol{w}}_k^T\left(\bar{\boldsymbol{W}}_2^T\boldsymbol{P}\bar{\boldsymbol{W}}_2 - \mu^2\boldsymbol{I}\right)\bar{\boldsymbol{w}}_k < 0. \end{aligned} \tag{27}$$

Thus

$$\bar{\boldsymbol{v}}_k = \left[\tilde{\boldsymbol{e}}_k^T \ \bar{\boldsymbol{w}}_k^T\right]^T, \tag{28}$$

inequality (27) can be described as follows

$$\bar{\boldsymbol{v}}_k^T \begin{bmatrix} \boldsymbol{X}^T(\alpha)\boldsymbol{P}\boldsymbol{X}(\alpha) - \boldsymbol{P} + \boldsymbol{I} & \boldsymbol{X}^T(\alpha)\boldsymbol{P}\bar{\boldsymbol{W}}_2 \\ \bar{\boldsymbol{W}}_2^T\boldsymbol{P}\boldsymbol{X}(\alpha) & \bar{\boldsymbol{W}}_2^T\boldsymbol{P}\bar{\boldsymbol{W}}_2 - \mu^2\boldsymbol{I} \end{bmatrix} \bar{\boldsymbol{v}}_k \prec 0, \tag{29}$$

either in alternative form

$$\begin{bmatrix} \boldsymbol{X}^T(\alpha) \\ \bar{\boldsymbol{W}}_2^T \end{bmatrix} \boldsymbol{P} \left[\boldsymbol{X}(\alpha) \quad \bar{\boldsymbol{W}}_2\right] + \begin{bmatrix} -\boldsymbol{P} + \boldsymbol{I} & \boldsymbol{0} \\ \boldsymbol{0} & -\mu^2\boldsymbol{I} \end{bmatrix} \prec 0. \tag{30}$$

Accordingly, let us define the following inequality by using the Schur complement and multiplying it from left and right by $\mathrm{diag}(\boldsymbol{I}, \boldsymbol{I}, \boldsymbol{P})$

$$\begin{bmatrix} -\boldsymbol{P} + \boldsymbol{I} & \boldsymbol{0} & \boldsymbol{X}^T(\alpha)\boldsymbol{P} \\ \boldsymbol{0} & -\mu^2\boldsymbol{I} & \bar{\boldsymbol{W}}_2^T\boldsymbol{P} \\ \boldsymbol{P}\boldsymbol{X}(\alpha) & \boldsymbol{P}\bar{\boldsymbol{W}}_2 & -\boldsymbol{P} \end{bmatrix} \prec 0. \tag{31}$$

Consequently:

$$\boldsymbol{P}\boldsymbol{X}(\alpha) = \boldsymbol{P}\bar{\boldsymbol{A}}(\alpha) - \boldsymbol{P}\bar{\boldsymbol{K}}\bar{\boldsymbol{C}} = \boldsymbol{P}\bar{\boldsymbol{A}}(\alpha) - \boldsymbol{N}\bar{\boldsymbol{C}}, \tag{32}$$

$$\boldsymbol{P}\bar{\boldsymbol{W}}_2 = \boldsymbol{P}\bar{\boldsymbol{W}}_3 - \boldsymbol{P}\bar{\boldsymbol{K}}\tilde{\boldsymbol{W}}_2 = \boldsymbol{P}\bar{\boldsymbol{W}}_3 - \boldsymbol{N}\tilde{\boldsymbol{W}}_2, \tag{33}$$

proves the theorem. ∎

Finally, the design procedure is reduced to solve the LMIs, which are obtained by applying (25) to all vertices of $(\alpha) = \sum_{i=1}^{M} h_i(\alpha_k)\boldsymbol{A}_i$, $\boldsymbol{B}(\alpha) = \sum_{i=1}^{M} h_i(\alpha_k)\boldsymbol{B}_i$.

A as a result the following estimator gain matrices are defined

$$\bar{\boldsymbol{K}} = \begin{bmatrix} \boldsymbol{K}_x \\ \boldsymbol{K}_a \\ \boldsymbol{K}_s \end{bmatrix} = \boldsymbol{P}^{-1}\boldsymbol{N}. \tag{34}$$

Similarly, for the controller design, the following theorem is proposed:

Theorem 2. *For a prescribed attenuation level $\mu_c > 0$ of $\boldsymbol{e}_k$, the $\mathcal{H}_\infty$ controller design (8) is solvable if there exist matrices $\boldsymbol{N}$, $\boldsymbol{U}$, $\boldsymbol{P} \succ 0$ such that the following condition is satisfied:*

$$\begin{bmatrix} -\boldsymbol{P} & 0 & \boldsymbol{U}^T\boldsymbol{A}(\alpha)^T - \boldsymbol{N}^T\boldsymbol{B}(\alpha)^T & \boldsymbol{U}^T \\ 0 & -\mu_c^2\boldsymbol{I} & -\boldsymbol{W}_1^T & 0 \\ \boldsymbol{A}(\alpha)\boldsymbol{U} - \boldsymbol{B}(\alpha)\boldsymbol{N} & -\boldsymbol{W}_1 & \boldsymbol{P} - \boldsymbol{U} - \boldsymbol{U}^T & 0 \\ \boldsymbol{U} & 0 & 0 & -\boldsymbol{I} \end{bmatrix} \prec 0, \tag{35}$$

with $\boldsymbol{N} = \boldsymbol{K}_1\boldsymbol{U}$.

Proof. The proof can be constructed in a similar way as the one of *Theorem 1*. Indeed, it is straightforward to show that the existence condition reduces to (cf. (30) for the estimator design case):

$$\begin{bmatrix} \boldsymbol{A}^T(\alpha) \\ -\boldsymbol{W}_1^T \end{bmatrix} \boldsymbol{P} \begin{bmatrix} \boldsymbol{A}(\alpha) & -\boldsymbol{W}_1 \end{bmatrix} + \begin{bmatrix} -\boldsymbol{P}+\boldsymbol{I} & \boldsymbol{0} \\ \boldsymbol{0} & -\mu_c^2 \boldsymbol{I} \end{bmatrix} \prec 0, \tag{36}$$

which can be transformed into (35) using the approach of *Theorem 1* in [24], which completes the proof. ■

Finally, the design procedure is reduced to solve the LMIs, which are obtained by applying (35) to all vertices of $\boldsymbol{A}(\alpha) = \sum_{i=1}^{M} h_i(\alpha_k)\boldsymbol{A}_i$, $\boldsymbol{B}(\alpha) = \sum_{i=1}^{M} h_i(\alpha_k)\boldsymbol{B}_i$ and then determining $\boldsymbol{K}_1 = \boldsymbol{N}\boldsymbol{U}^{-1}$.

3 Illustrative Example

The objective of this section is to validate the proposed approach with the Multi-Tank (MT) system, which is illustrated in Fig. 2. A nonlinear model of the system is defined as follows [23]:

$$\begin{cases} H_{1,k+1} = H_{1,k} - c_1 H_{1,k}^{\alpha} + b u_k \\ H_{2,k+1} = H_{2,k} + (c_3 + c_4 H_{2,k})^{-1}\left(c_2 H_{1,k}^{\alpha} - c_5 H_{2,k}^{\alpha}\right) \\ H_{3,k+1} = H_{3,k} + \left(c_7 - (c_8 - H_{3,k})^2\right)^{-0.5}\left(c_6 H_{2,k}^{\alpha} - c_9 H_{3,k}^{\alpha}\right) \end{cases} \tag{37}$$

where H_1, H_2 and H_3 denote the water level in the first, second and third tank, respectively. Subsequently, the system is transformed into Takagi-Sugeno form which was developed by the authors in [16]. During experiments, a goal was to control the system in such a way that the water level should reach and keep the reference state $\boldsymbol{x}_k$ corresponding to the first tank level equal to 15 cm.

Let us define the following actuator and sensor fault scenario:

$$\boldsymbol{f}_{a,k} = \begin{cases} -0.27 \cdot \boldsymbol{u}_k & 70 \le t[s] \le 110 \\ 0 & \text{otherwise} \end{cases}, \tag{38}$$

$$\boldsymbol{f}_{s,k} = \begin{cases} \boldsymbol{y}_k - 0.1 & 60 \le t[s] \le 90 \\ 0 & \text{otherwise} \end{cases}, \tag{39}$$

as well as the sensor fault distribution matrix

$$\boldsymbol{C}_f = \begin{bmatrix} 0 \\ 1 \\ 0 \end{bmatrix}. \tag{40}$$

Figures 3a–3b show the performance of the fault estimator. It exhibits the actuator and sensor faults estimates along with the real faults. It can be noticed that

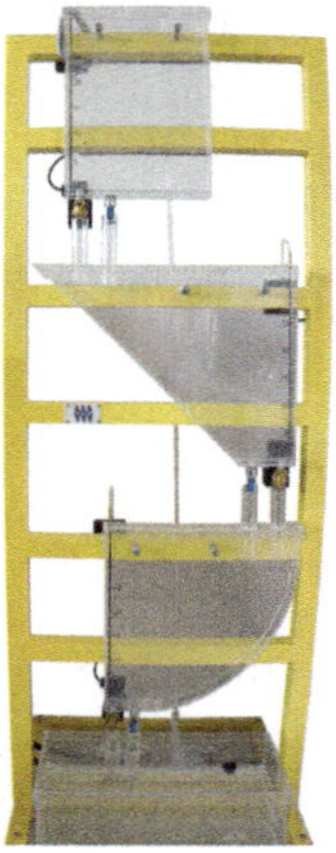

Fig. 2. The Multi-Tank system

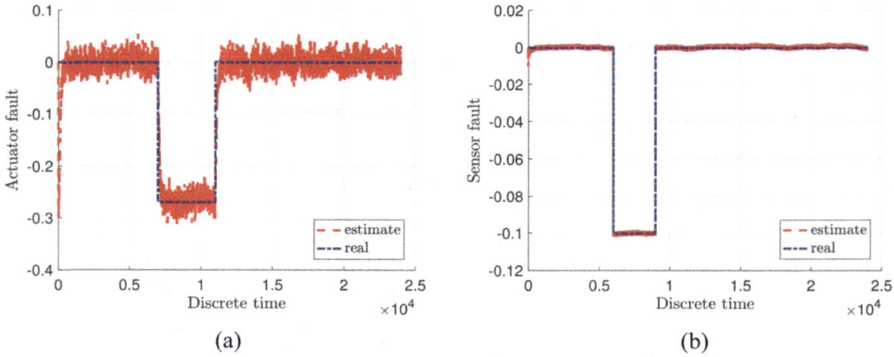

Fig. 3. Actuator fault with its estimate (a) and sensor fault with its estimate (b)

the quality of the fault estimation is very good in both cases, i.e., actuator and sensor ones. In the above scenario, the actuator fault is defined as an intermittent one, which can be perceived as a 27% loss of the actuator effectiveness. On the other hand, the sensor fault is defined as an temporary measurement bias for the water level in the second tank.

Figures 4a–4c present the response of the system with the application of the proposed FTC approach. Both in separate and simultaneous actuator and sensor faults, the proposed tracking FTC follows the reference state, i.e., x_f tracks x_k.

This is, of course, achieved with and additional control effort, which is portrayed in Fig. 5. It shows the comparison between the nominal control signal generated with the reference model and its fault-tolerant counterpart. It can be noticed that the control signal changes its shape accordingly to the fault estimate to compensate the loss of the actuator. Thus. it can be deduced that both actuator and sensor fault estimates contribute to the desired control behaviour.

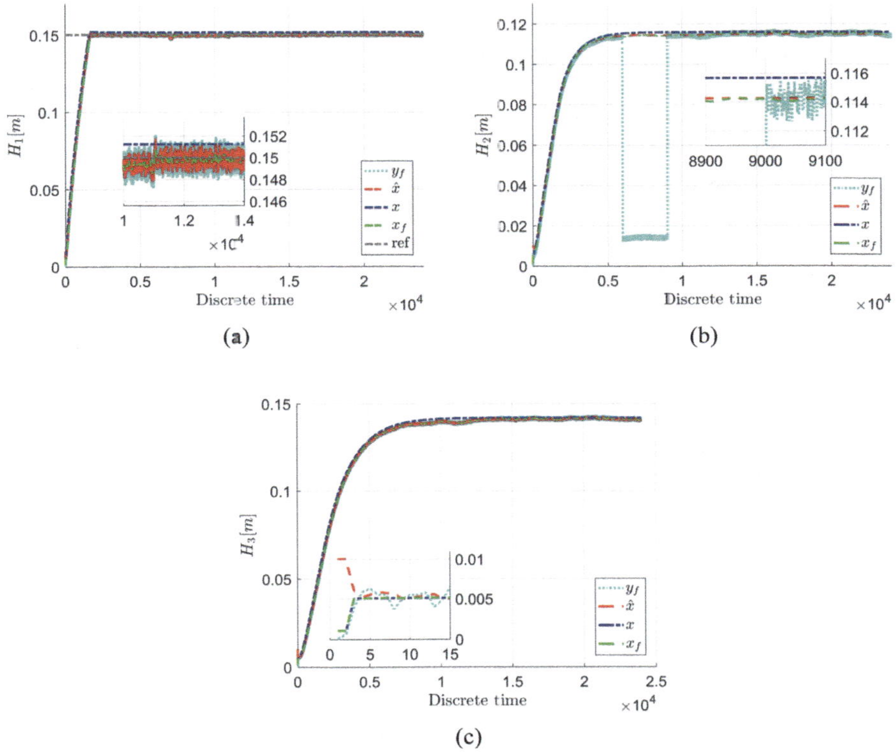

Fig. 4. Performance of the tracking FTC

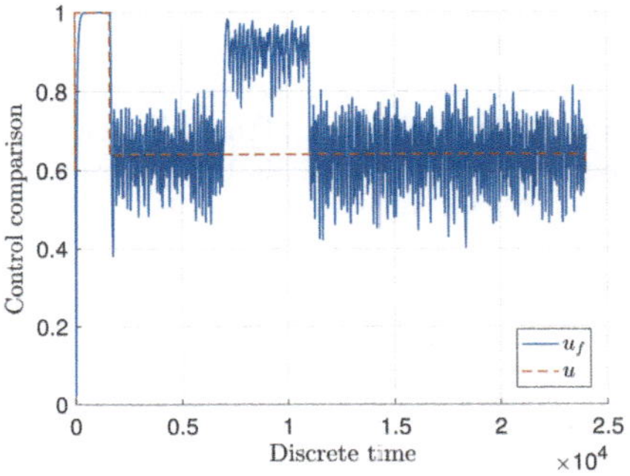

Fig. 5. Nominal and fault-tolerant control signals

4 Conclusions

The paper dealt with the problem of FTC for the Takagi-Sugeno systems in a presence of both actuator and sensor fault as well as process and measurement uncertainties. The positive feature of the proposed approach is that it can be applied to already existing control strategy, which means the former controller can be used as a nominal one and there is no need to change it. In order to do so, the fault estimator needs to be designed which allows obtaining the state estimate of the system as well as estimates of faults acting onto considered system. The important point is that the FTC controller operates independently and simultaneously to the nominal one. The paper provides the designing path for both estimator and FTC controller for the Takagi–Sugeno fuzzy systems. The gains are obtained after solving the linear matrix inequalities. The last part of the paper shows the illustrative example with the application of the proposed FTC to the nonlinear laboratory Multi-Tank system. The results of the state and fault estimation are very satisfactory, which implicates the quality of the new control law. The proposed FTC controller is able to control the system with minimization of the faults influence and it also minimizes the steady state error.

Acknowledgement. The work was supported by the National Science Centre, Poland under Grant: UMO-2017/27/B/ST7/00620.

References

1. Ahmad, M., Ali, A., Choudhry, M.A.: Fixed-structure H_∞ controller design for two-rotor aerodynamical system (TRAS). Arab. J. Sci. Eng. **41**(9), 3619–3630 (2016)
2. Aouaouda, S., Chadli, M., Righi, I.: Active FTC approach design for TS fuzzy systems under actuator saturation. In: 2019 6th International Conference on Control, Decision and Information Technologies (CoDIT), pp. 483–488. IEEE (2019)
3. Blanke, M., Kinnaert, M., Lunze, J., Staroswiecki, M.: Diagnosis and Fault-Tolerant Control. Springer, Heidelberg (2003)
4. Blanke, M., Schröder, J., Kinnaert, M., Lunze, J., Staroswiecki, M.: Diagnosis and Fault-Tolerant Control. Springer, Heidelberg (2006)
5. Ding, B.: Dynamic output feedback predictive control for nonlinear systems represented by a Takagi-Sugeno model. IEEE Trans. Fuzzy Syst. **19**(5), 831–843 (2011)
6. Edwards, C., Lombaerts, T., Smaili, H.: Fault Tolerant Flight Control: A Benchmark Challen. Lecture Notes in Control and Information Sciences. Springer, Heidelberg (2010)
7. Kommuri, S.K., Defoort, M., Karimi, H.R., Veluvolu, K.C.: A robust observer-based sensor fault-tolerant control for PMSM in electric vehicles. IEEE Trans. Industr. Electron. **63**(12), 7671–7681 (2016)
8. Lan, J., Patton, R.J.: A decoupling approach to integrated fault-tolerant control for linear systems with unmatched non-differentiable faults. Automatica **89**, 290–299 (2018)
9. Li, Y., Tong, S.: Adaptive neural networks decentralized FTC design for nonstrict-feedback nonlinear interconnected large-scale systems against actuator faults. IEEE Trans. Neural Netw. Learn. Syst. **28**(11), 2541–2554 (2016)

10. López-Estrada, F.R., Ponsart, J.C., Astorga-Zaragoza, C.M., Camas-Anzueto, J.L., Theilliol, D.: Robust sensor fault estimation for descriptor-LPV systems with unmeasurable gain scheduling functions: application to an anaerobic bioreactor. Int. J. Appl. Math. Comput. Sci. **25**(2), 233–244 (2015)
11. Maalej, S., Kruszewski, A., Belkoura, L.: Stabilization of takagi-sugeno models with non-measured premises: input-to-state stability approach. Fuzzy Sets Syst. **329**, 108–126 (2017)
12. Márquez, R., Guerra, T.M., Bernal, M., Kruszewski, A.: Asymptotically necessary and sufficient conditions for Takagi-Sugeno models using generalized non-quadratic parameter-dependent controller design. Fuzzy Sets Syst. **306**, 48–62 (2017)
13. de Oca, S.M., Rotondo, D., Nejjari, F., Puig, V.: Fault estimation and virtual sensor FTC approach for LPV systems. In: 2011 50th IEEE Conference on Decision and Control and European Control Conference, pp. 2251–2256. IEEE (2011)
14. Pazera, M., Buciakowski, M., Witczak, M.: Robust multiple sensor fault-tolerant control for dynamic non-linear systems: application to the aerodynamical twin-rotor system. Int. J. Appl. Math. Comput. Sci. **28**(2), 297–308 (2018)
15. Pazera, M., Witczak, M.: Towards robust simultaneous actuator and sensor fault estimation for a class of nonlinear systems: design and comparison. IEEE Access **7**, 97143–97158 (2019)
16. Rotondo, D., Witczak, M., Puig, V., Nejjari, F., Pazera, M.: Robust unknown input observer for state and fault estimation in discrete-time Takagi-Sugeno systems. Int. J. Syst. Sci. **47**(14), 1–16 (2016)
17. Sun, K., Sui, S., Tong, S.: Optimal adaptive fuzzy ftc design for strict-feedback nonlinear uncertain systems with actuator faults. Fuzzy Sets Syst. **316**, 20–34 (2017)
18. Tabbache, B., Rizoug, N., Benbouzid, M.E.H., Kheloui, A.: A control reconfiguration strategy for post-sensor FTC in induction motor-based EVs. IEEE Trans. Veh. Technol. **62**(3), 965–971 (2012)
19. Takagi, T., Sugeno, M.: Fuzzy identification of systems and its application to modeling and control. IEEE Trans. Syst. Man Cybern. B Cybern. **15**(1), 116–132 (1985)
20. Wang, T., Qiu, J., Gao, H.: Adaptive neural control of stochastic nonlinear time-delay systems with multiple constraints. IEEE Trans. Syst. Man Cybern. B Cybern. Syst. **47**(8), 1875–1883 (2016)
21. Witczak, M.: Modelling and Estimation Strategies for Fault Diagnosis of Non-Linear Systems. Springer, Heidelberg (2007)
22. Witczak, M.: Fault Diagnosis and Fault-Tolerant Control Strategies for Non-LInear Systems. Springer, Heidelberg (2014)
23. Witczak, M.: Fault Diagnosis and Fault-Tolerant Control Strategies for Non-Linear System. Lecture Notes in Electrical Engineering, vol. 266. Springer, Cham (2014)
24. Witczak, M., Buciakowski, M., Aubrun, C.: Predictive actuator fault-tolerant control under ellipsoidal bounding. Int. J. Adapt. Control Signal Process. **30**(2), 375–392 (2016)
25. Wu, H.N., Feng, S., Liu, Z.Y., Guo, L.: Disturbance observer based robust mixed H_2/H_∞ fuzzy tracking control for hypersonic vehicles. Fuzzy Sets Syst. **306**, 118–136 (2017)
26. Xiao, B., Hu, Q., Zhang, Y.: Adaptive sliding mode fault tolerant attitude tracking control for flexible spacecraft under actuator saturation. IEEE Trans. Control Syst. Technol. **20**(6), 1605–1612 (2011)
27. Zhang, Y., Jiang, J.: Bibliographical review on reconfigurable fault-tolerant control systems. Ann. Rev. Control **32**(2), 229–252 (2008)

Network-Based Approach to Increase Logical Reliability of a Vehicle E/E-Architecture

Mohamad Chamas[1], Jan Mehlstäubl[2(✉)], Steffen Eickhoff[3], and Kristin Paetzold[2]

[1] BMW Group, Munich, Germany
mohamad.chamas@bmw.de
[2] University of the Bundeswehr Munich, Neubiberg, Germany
{jan.mehlstaeubl,kristin.paetzold}@unibw.de
[3] University of Paderborn, Paderborn, Germany
steffen.e@mail.uni-paderborn.de

Abstract. The increasing number of functions in vehicles and their intensive interdependencies leads to a more complex e/e-architecture. This complexity increases the probability of error and influences system reliability significantly. Therefore, it is necessary to detect risks at early stages. In course of this, new methods are needed to carry out a reliability-oriented transfer of logical functions to the technical architecture. This paper presents an approach that derives a set of key figures for assessing logical reliability of functional concepts by using metrics from network theory. For this purpose, a control loop is introduced. After the functional design in a model-based environment, suitable metrics are applied. Based on the reference criterion Survival Probability, structural metrics are interpreted. With reference to the structural indicators, action strategies can be traced back to system model to increase logical reliability. First results showed that the logical reliability of functional concepts were improved by conceptual adjustments and partitioning recommendations.

Keywords: Logical reliability · Network theory · E/E-architecture · Fault Tree Analysis

1 Introduction

In times of faster development processes and more product functionalities due to trends such as autonomous driving, electro mobility or intelligent services, the interaction between functions is increasing significantly. The growing number of functions and their linkages result in systems with high complexity. In electric and electronic (e/e) development, this complexity is reflected in a growing number of on-board signals and interfaces between electronic control units [1].

The large number of interfaces results in a higher amount of possible causes of failure as well as error chains and affects system reliability significantly [2]. The level of detail in reliability analysis only reaches its peak in mature technical architecture phases due to insufficient information and high uncertainties. An indicator for reliability assessment

A. Bartoszewicz et al. (Eds.): PCC 2020, AISC 1196, pp. 1446–1457, 2020.
https://doi.org/10.1007/978-3-030-50936-1_120

on technical architecture is Mean-Time-To-Failure, which represents an expectation of the time until a system fails. With such indicators, a consideration of risk takes place only in technical architecture, but not in a logical one. One way to increase logical reliability of software architecture is refactoring [3]. Within the refactoring process, the internal software composition is restructured. Consequently, the partitioning process, which describes the allocation of functions to control units, represents an important lever to improve reliability in system architecture.

In industrial context, models are used to manage system complexity. Using model-based systems engineering (mbse), system models can be built in a formal environment through appropriate methods, tools and languages. This supports consistency, traceability, system understanding and enables the illustration of complex interrelations as well as the communication of essential content. However, mbse offers less methods for estimating and increasing logical reliability.

The aim of this work is to provide a procedure for logical reliability analysis with metrics of network theory based on a system model. This enables the generation of action strategies for increasing reliability of e/e-architecture as well as decision support of a suitable functional concept variant.

First, the paper provides a theoretical basis about system modelling, network theory, e/e-architecture, reliability, and risk analysis. Second, a literature review of related work is conducted. Section 4 introduces the developed approach for increasing logical reliability of an e/e-architecture. Subsequently, the approach is applied and evaluated on the functionality adaptive cruise control. Finally, the results of this paper are discussed critically and a conclusion is drawn.

2 Background

2.1 System Modelling with SysML

A graphical language that supports the mbse approach is the Systems Modeling Language (SysML). SysML provides the basis for specifying requirements, structure, behaviour, allocations, and constraints of complex mechatronic systems.

Chamas et al. [4] describe a method for modelling logical e/e-architecture of mechatronic systems in conceptual stage using SysML. In this, the activity diagram is used to describe logical interactions and the resulting system behaviour. The main development artefacts of activity diagrams are actions. They represent individual steps of functions. The connections between actions can be divided into control and object flows. Moreover, activity diagrams provide decision nodes and branches to map case distinctions and parallel processes. However, SysML does not provide any methods for the measurement of model complexity or reliability assessment.

2.2 Network Theory

In network theory, systems are described with nodes and their connections, the so-called edges. The meaning of the nodes and edges must be defined depending on the modelling context. For example, components of a system can be described with nodes and their mechanical connection with edges. Network theory offers various metrics to analyse characteristics of such networks. Maurer [5] uses several metrics to highlight the structural complexity of development processes. In the context of this work, the logical architecture of technical systems will be examined with a selection of these metrics. In the following relevant metrics are briefly presented:

- **Average Clustering Coefficient** measures local linkage, i.e. number of direct or indirect relations to neighbouring nodes.
- **Average Degree** summarises all input and output edges of a node.
- **Structural Robustness** indicates the impact of a change on an element.
- **Fitness Value** is an indicator for the integration possibility of a small graph into a larger graph.
- **Graph Energy** measures the structural complexity of the entire graph. This depends on the heterogeneity and the number of different elements as well as their connections.
- **Relative Centrality** indicates how often a node is on the shortest path between two other nodes.

2.3 E/E-Architecture

One focus of e/e-architecture deals with structuring and designing systems on communication level (see Fig. 1).

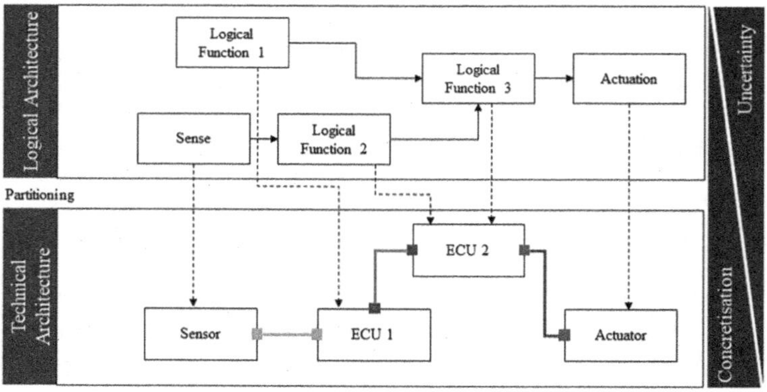

Fig. 1. Simplified overview on e/e-architecture

Two main levels are distinguished here [1]. Logical architecture (LA) takes place at early stages of product development. Logical sensing and actuator functions are described and put into context. With control flows, single building blocks of the LA are connected. However, compared to LA, technical architecture (TA) contains real components (e.g. ECU), ports and communication paths (e.g. CAN). In addition to the structural view, communication aspects like sender-receiver as well as collision and delay strategies are considered. Transferring the LA to TA is done through partitioning. Every abstract component in the first layer has to be realized by a component in the TA.

2.4 Reliability and Risk Analysis

Reliability of a system is defined as the "probability of survival against failures or malfunctions, which mask out one or more functions, or limit them in unspecified ways" [6]. As already explained, e/e-architecture has impacts on the reliability of a mechatronic system. Different levels of architecture (LA, TA) needs also a different view on reliability levels. For technical architecture with high concretisation, many hardware driven issues can be considered (Mean-Time-To-Failure, Failure in Time, Weibull-Distribution, etc.). Therefore, we will use the term technical reliability. Fowler et al. [3] describe the partitioning process through refactoring. Also parameters like distance d (number of communication paths), n (number of components in communication path, e.g. sensor), and b (width for bisection which describes redundancy) are necessary to take into account for partitioning LA on TA. Based on the logical architecture, the term logical reliability is introduced for determination of reliability in functional concepts.

Methods for risk analysis must be used continuously in the entire product development process to handle reliability. They include risk identification, root cause analysis, as well as quantification and assessment of risk [7]. Approaches that are suitable for risk analysis of logical architecture are the Failure Mode and Effect Analysis (FMEA) and the Fault Tree Analysis (FTA) [8, 9]. However, the FMEA and FTA require a high effort for creation, collection of data and organizational implementation. Moreover, the assessment of failure depends on the subjective view of the involved people.

3 Related Work

To reduce the effort for risk analysis and to objectify the results, traditional reliability methods can be linked with the methods of model-based system development. The following section presents previous approaches to analyse logical system reliability in connection with mbse.

An approach originates from Alt [10]. He describes the implementation of an FMEA with an existing system model. He generates a structure tree from the hierarchical relations of structure elements and their assigned functions. The existing dependencies between functions make it easier to identify malfunctions as well as causes and effects of errors. Biggs et al. [11] describe the SafeML modelling language, which can be used to summarise safety-related information with the associated specification in one system model. Potential risks are derived from classic reliability methods and are included in the system model. Cressent et al. [12] offer with their MeDISIS framework an extensive tool to combine SysML models with reliability methods. They fill an error database with possible malfunctions of the individual system elements and connect them to the system model. As a result, an FMEA and the associated malfunctions can be generated automatically from the system model. Kurtoglu et al. [13] focus with their Functional Failure Identification and Propagation (FFIP) method the propagation of failure. Multiple failures can be taken into account by combining individual errors as in an FTA. For each function, a so-called Function Failure Impact (FFI) metric is determined, which describes the severity of impact on the overall system and enables the iterative optimization of the system structure. A network-based approach originate from Chahin et al. [14]. They use metrics of network theory to evaluate the risk between two concept alternatives.

First, the functional system structure is described with nodes and subjectively weighted edges. In addition, a component graph is built in the same way. Subsequently, Chahin et al. [14] merge the functional and component graph and determine the risk potential of each node from the sum of weights of all outgoing edges.

From a high uncertainty in early phases to a high concretisation until start of production, the dilemma of late decisions is also mirrored in reliability assessment methods. All presented approaches deal with the coupling of system models with aspects of reliability. However, none of them meets the requirements of a fast and objective reliability analysis in early phases of product development. In addition, there is no approach to improve the management of complexity in logical architecture.

4 Findings - Approach to Increase Logical Reliability

The following section presents the developed approach to increase logical reliability in early stage of product development. Figure 2 gives an overview of the individual steps of the method. Starting with a system model, a suitable reference criterion must be defined. This reference criterion has to be a clearly interpretable and quantifiable indicator that enables a reliability assessment. In this approach, the Survival Probability R is used. After the interpretation of key figures, limit values must be defined. These limit values indicates whether an optimization of the system model is required or the partitioning process can be started. This enables the initiation of action strategies to optimize the system model. The individual steps are explained in the following subsections.

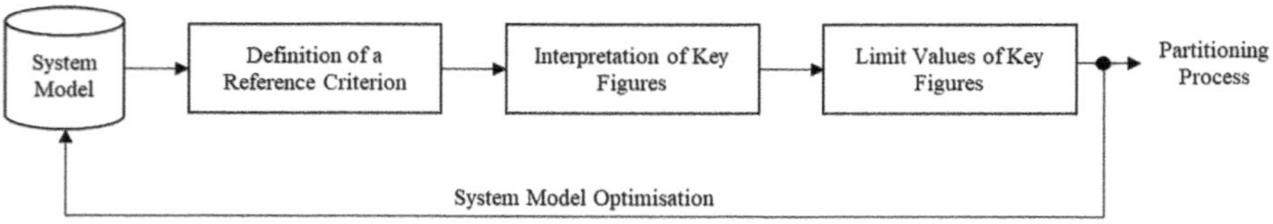

Fig. 2. Overview of the approach

4.1 Definition of a Reference Criterion

Survival Probability is originally used for reliability assessment in an FTA. The reliability of a network can be calculated by using failure probabilities of individual functions. If functions are in series, a malfunction can lead to a failure of an entire chain (OR-linkage). If functions are parallel, they have redundant paths and therefore, the reliability is higher (AND-linkage). In a functional concept modelled with SysML, there are usually both serial and parallel functional structures and every part of the activity diagram must be faultless. The Survival Probability for an overall system is calculated according to the Boolean operators as follows:
Serial:

$$R_s(t) = \prod_{i=1}^{n} R_i(t) \quad (1)$$

Parallel:

$$R_s(t) = 1 - \prod_{i=1}^{n} (1 - R_i(t)) \quad (2)$$

4.2 Interpretation of Key Figures

In order to identify dependencies between Survival Probability and metrics of network theory, 25 randomly generated structures were modelled with SysML. The Survival Probability was calculated for each of these structures. A constant value R was assumed for each action. Moreover, the metrics of network theory described in Sect. 2.2 were calculated for each randomly generated structure. For example, the Structural Robustness is calculated as follows:

$$\text{Structural Robustness} = 1 - \sum_{i=1}^{n} IEV(x_i) \quad (3)$$

$$IEV(x) = I'(x) * P(x) = \left(\frac{1}{N}\sum_{i=1}^{n} \frac{1}{d_{x,s}}\right)' * \left(\frac{1}{N}\sum_{i=1}^{n} \frac{1}{d_{s,x}}\right) \quad (4)$$

IEV(x): influence expected value; I(x): impact; P(x): impact probability; N: number of nodes; $d_{x,s}$: distance from node x to node s; $d_{s,x}$: distance from node s to node x

Figure 3 shows exemplary the dependence between the Survival Probability and the Structural Robustness.

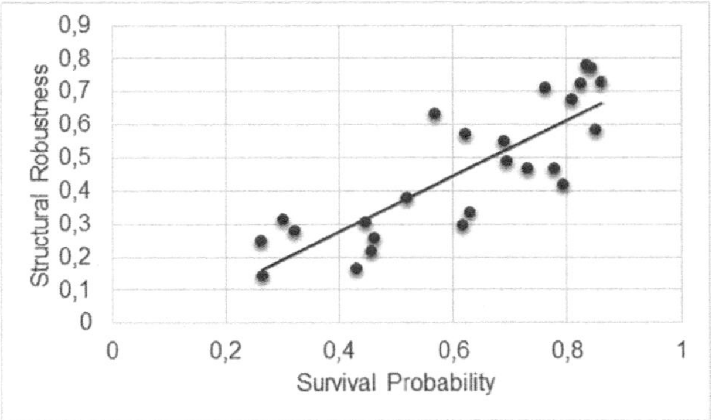

Fig. 3. Dependence between Structural Robustness and Survival Probability

A clear linear dependency between the two indicators can be seen. A high Survival Probability indicates a high Structural Robustness. The same procedure was followed for the other metrics. There it was also possible to identify linear dependencies. Structural Robustness is the only indicator that should be as high as possible. All other metrics should as low as possible to reach a high R.

4.3 Limit Values of Key Figures

In the previous section, a procedure to specify the dependencies between metrics of network theory and logical reliability was described. As a result, information about the logical reliability of a functional structure can be provided by simply calculating the metrics of network theory. A definition of limit values is required for the classification and the initiation of optimization measures. Therefore, 21 functional concepts with a high degree of maturity were used to determine these limit values. In several iteration loops, the concepts were enriched and continuously improved based on expert knowledge. Subsequently, the individual key figures were applied and calculated.

For the Structural Robustness and the Fitness Value, a correlation between their values and the size of the network was detected. In structures with less than 100 nodes, large fluctuations between the values occurred. If the number of nodes exceeds 100, the key figures converge to a constant limit. For this reason, logarithmic functions were used to describe their limit values. The constant limit values of the remaining size-independent key figures were calculated using the mean value and the standard deviation. The indicators, which must be as low as possible to achieve high reliability, are calculated using Eqs. (5) and (6). To calculate the indicators with a high target value, the signs of standard deviation must be inverted.

$$\text{Action Recommended} = \text{Mean Value} - \frac{1}{2}\,\text{Standard Deviation} \tag{5}$$

$$\text{Action Required} = \text{Mean Value} + \frac{1}{2}\,\text{Standard Deviation} \tag{6}$$

An overview of all key figures and their limit value is given in Table 1.

Table 1. Limit values of key figures (x: number of nodes)

Key figures	Action recommended	Action required
Average clustering	0,0117	0,0606
Average degree	2,3058	2,5875
Structural robustness	y = 0,7 + 0,05ln(x)	y = 0,62 + 0,05ln(x)
Fitness value	y = 2,2–0,298ln(x)	y = 2,4 – 0,298ln(x)
Graph energy	0,719	0,78
Relative centrality	0,348	0,8688

5 Case Study

Following, the vehicle function Adaptive Cruise Control (ACC) will be used to illustrate the application of the method. ACC controls the distance between two vehicles based on adjusting variables of positive and negative acceleration, realized by engine in case of positive or brakes and recuperation in case of negative acceleration. Figure 4 illustrates two different conceptual variants of ACC. The left one is the initial design, while the right one was redesigned by following the results of the key figures. The functional behaviour of both concepts is the same by analysing the output. First, actions determine and compare the distance of the ego-vehicle and front vehicle. Depending on whether the distance is too small or too large, braking torque or drive torque is controlled.

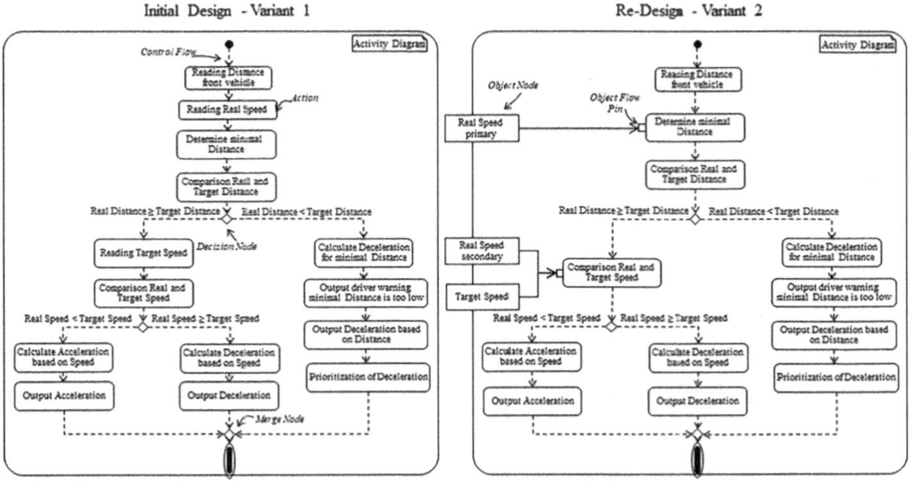

Fig. 4. Two different conceptual variants of ACC

The conceptual differences between the two variants can be seen by using signals with high integrity following ISO 26262 to determine actual speed and target speed in variant 2. The actions "Reading Target Speed" and "Reading Real Speed" were replaced by object nodes and object flows, which represents signals with higher integrity in the concept. Therefore, both actions are not part of the whole control flow anymore. The real speed is divided into two redundant signals (primary and secondary).

Table 2 shows the results of the key figures which were introduced before. Based on the determined limits of each indicator, it is obviously clear, that variant 1 shows greater need for actions than variant 2. Three of the indicators, Structural Robustness, Fitness Value and Relative Centrality exceed the limits for required action. The structural complexity of variant 1 is reduced by 10%. The impact on changes of elements is also improved by 5%. The most significant improvement has been reached for Relative Centrality, which has become more than 55% better.

With regard to the values of the key figures, variant 2 is more suitable than variant 1 in terms of logical reliability. This means, that taking safety mechanisms like redundancy have impact on structural behaviour of functional concepts, which also influences key figures. Accordingly, a fault tree analysis is done by comparing these two variants. In Fig. 5 fault trees of variant 1 and variant 2 are shown for the speed control.

The fault tree of distance control is here not shown for reasons of clarity. In fact, there are many points to be conducted. Following, two examples are figured out. First of all, it is obvious that the fault tree of the second variant has significantly less fault combinations

Table 2. Key figures for conceptual variant 1 and 2

Key figures	Action recommended	Action required	Variant 1 (x = 14)	Variant 2 (x = 12)
Average clustering	0,0117	0,0606	0	0
Average degree	2,3058	2,5875	1,875	1,875
Structural robustness	y(12) = 0,8354 y(14) = 0,832	y(12) = 0,744 y(14) = 0,752	0,7308	0,77
Fitness value	y(12) = 1,393 y(14) = 1,4136	y(12) = 1,593 y(14) = 1,6136	1,5233	1,433
Graph energy	0,719	0,78	0,8449	0,763
Relative centrality	0,348	0,8688	0,9231	0,588

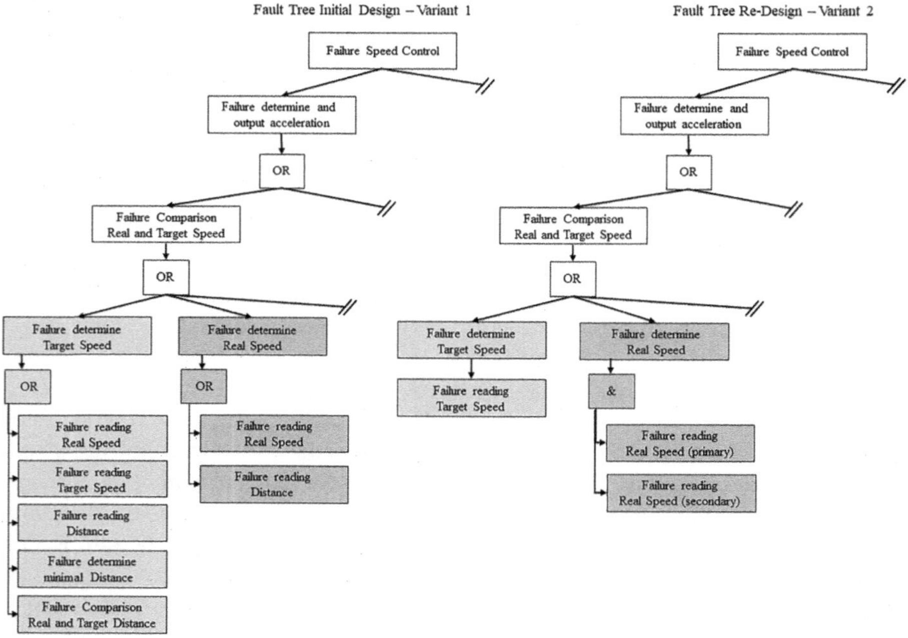

Fig. 5. Comparison of fault trees of both conceptual variants

than the first variant. Having a closer look on the failure for the determination of real vehicle speed, an improvement is done by replacing an AND-operator instead of an OR-operation in cases of the acceleration process as well as deceleration process. This was achieved by introducing two redundant real speed signals and the resolution of distance detection in the path. The distance-based function of the front car no longer affects the determination of real speed in the whole concept. Otherwise, in the first variant, reading real speed was the successor of the distance-based action of the front car.

A further positive issue is the reduction of fault possibilities for determination of target speed. Whereas in the first variant distance-based functions influences the determination target speed, in the second variant the failure possibilities are reduced only to failures of reading target speed. This can be explained by the fact that reading target speed in variant 1 was strictly connected to the chain before. In the second variant, the failure of determination target speed is limited to its source.

Both examples shows, that failures on real and target speed determination are no longer as critical as before, so that the logical reliability of the entire ACC function is not seriously affected anymore.

6 Discussion

6.1 Assumptions

Most of the methods for assessing reliability relate to technical architecture in late phase of product development. Approaches to estimate reliability of logical architecture are still rare and require a lot of effort for implementation. Moreover, it is difficult to imagine or even evaluate the reliability of logical architectures, even by quantitative approaches. Furthermore, the evaluation process depends on a subjective risk assessment of the people involved.

One of the main benefits of the introduced approach is the quick and cost-effective reliability analysis with metrics of network theory in case of an available system model. This enables an iterative assessment and optimization of the logical architecture. Another advantage is the classification of risk based on a set of key figures, which allows well-founded and objective conclusions about reliability. Furthermore, the key figures meets the requirements for identifying deficiencies in reliability of system structure through the transparent interpretation with the reference criterion Survival Probability. By introducing limit values, critical key figures can be identified and adjusted. In addition, concept variants can be compared and the most reliable alternative can be partitioned on the technical architecture. The increase in logical reliability was confirmed with the Adaptive Cruise Control function and the comparison of the fault trees of the two concept variants.

6.2 Limitations

The evaluation of functional concepts is limited to structural analysis. Therefore, it helps to get a quick view over architectural deficiencies. Nevertheless, there must be an evaluation for the individual functions in the form of occurrence, severity and frequency. This limitation is also linked to the equal treatment of the actions. To determine the relationship between a key figure and the reference criterion R (Survival Probability), every action has the same value for the calculation with Boolean operators. In our assumption, it was necessary to understand key figures behaviour without having more details about single actions. Further considerations needs an assessment of every action. For the calculation of the limit values, we considered 21 functional concepts with high maturity from the automotive domains of electrical power train with focus on charging, energy management and torque generation. As a complement for future work, we aim to scale this amount on two ways: Extending the amount of functional concepts from already accounted domain (here: electrical powertrain) and (2) extending the amount of functional concepts from further domains, e.g. infotainment and vehicle dynamics. This allows defining limit values more robustly based on a higher level of data. Additionally, we reduced our approach on a single reference criterion and in turn not an overall reliability assessment. For achieving this, further reference criteria from logical reliability has to be considered.

7 Conclusion

In order to handle the increasing number of functional interfaces and the system complexity, a method is essential to assess logical reliability in early stage of product development. The approach of this paper describes a network-based procedure to analyse and increase logical reliability of a system model. In this, the logical architecture is analysed with suitable metrics of network theory. The interpretation of these metrics with the reference criterion Survival Probability helps to draw direct conclusions to reliability. The systematic specification of limit values completes the set of key figures and enables monitored concept adjustments.

Future work will include the development and integration of options for assessing logical reliability at a functional level. This contains the distinction of the individual actions with regard to their reliability. In addition, the results of this work will be consolidated by using additional reference criteria to interpret the metrics of network theory.

References

1. Nörenberg, R.: Effizienter Regressionstest von E/E-Systemen nach ISO 26262. KIT Scientific Publishing, Karlsruhe (2012)
2. Fritzsche, R.: Erstellung von Parameter-Diagrammen in der Automobilindustrie. ATZ-Automobiltechnische Zeitschrift **108**, 492–497 (2006)
3. Fowler, M.: Refactoring: Improving the Design of Existing Code, 9th edn, p. 431. Addison-Wesley, Boston (2002)
4. Chamas, M.W., Paetzold, K.: Modeling of requirement-based effect chains of mechatronic systems in conceptual stage. IJEETC **7**, 127–134 (2018)
5. Maurer, M.: Structural awareness in complex product design. Zugl.: München, Techn. Univ., Diss, 2007, Verl. Dr. Hut, München (2007)
6. Verein Deutscher Ingenieure: Zuverlässigkeitskenngrößen: Verfügbarkeitskenngrößen (VDI4004). Beuth Verlag (1986)
7. Cottin, C., Döhler, S.: Auswahl und Überprüfung von Modellen. In: Risikoanalyse, pp. 309–368. Springer, Heidelberg (2013)
8. Ponn, J.: Absicherung der technischen Entwicklungsziele. In: Lindemann, U. (ed.) Handbuch Produktentwicklung, pp. 805–836. Hanser, München (2016)

9. Bertsche, B., Stohrer, M.: Zuverlässigkeit und Sicherheit. In: Lindemann, U. (ed.) Handbuch Produktentwicklung. Hanser, München (2016)
10. Alt, O.: Modellbasierte Systementwicklung mit SysML. Carl Hanser Fachbuchverlag, München (2012)
11. Biggs, G., Sakamoto, T., Kotoku, T.: A profile and tool for modelling safety information with design information in SysML. Softw. Syst. Model. **15**(1), 147–178 (2014). https://doi.org/10.1007/s10270-014-0400-x
12. Cressent, R., Idasiak, V., Kratz, F., David, P.: Mastering safety and reliability in a model based process. In: 2011 Proceedings-Annual Reliability and Maintainability Symposium. IEEE (2011)
13. Kurtoglu, T., Tumer, I.Y., Jensen, D.C.: A functional failure reasoning methodology for evaluation of conceptual system architectures. Res. Eng Design **21**(4), 209–234 (2010)
14. Chahin, A., Paetzold, K., Peterit, P., Grunow, C.: An approach for using the FMEA for the development of a new and revolutionary rotation piston engine. In: DS 85-2: Proceedings of NordDesign 2016, Trondheim, Norway, vol. 2, pp. 53–61 (2016)

Security-Oriented Fault-Tolerance in Systems Engineering: A Conceptual Threat Modelling Approach for Cyber-Physical Production Systems

Iris Gräßler[1(✉)], Eric Bodden[2], Jens Pottebaum[1], Johannes Geismann[2], and Daniel Roesmann[1]

[1] Heinz Nixdorf Institute – Product Creation, Paderborn University, Fürstenallee 11, 33102 Paderborn, Germany
{iris.graessler,jens.pottebaum,daniel.roesmann}@hni.upb.de
[2] Heinz Nixdorf Institute – Secure Software Engineering, Paderborn University, Fürstenallee 11, 33102 Paderborn, Germany
{eric.bodden,johannes.geismann}@hni.upb.de

Abstract. Faults in the realization and usage of cyber-physical systems can cause significant security issues. Attackers might exploit vulnerabilities in the physical configurations, control systems, or accessibility through internet connections. For CPS, two challenges are combined: Firstly, discipline-specific security measures should be applied. Secondly, new measures have to be created to cover interdisciplinary impacts. For instance, faulty software configurations in cyber-physical production systems (CPPS) might allow attackers to manipulate the correct control of production processes impacting the quality of end products. From liability and publicity perspective, a worst-case scenario is that such a corrupted product is delivered to a customer. In this context, security-oriented fault-tolerance in Systems Engineering (SE) requires measures to evaluate interdisciplinary system designs with regard to potential scenarios of attacks. The paper at hand contributes a conceptual threat modelling approach to cover potential attack scenarios. The approach can be used to derive both system-level and discipline-specific security solutions. As an application case, issues are focused on which attackers intend to exploit vulnerabilities in a CPPS. The goal is to support systems engineers in verification and validation tasks regarding security-oriented fault-tolerance.

Keywords: Systems engineering · Threat modelling · Scenario-based analysis

1 Motivation and Approach

Faults in the realization and usage of cyber-physical systems (CPS) can cause significant security issues. Fault tolerance is the ability of a system to tolerate faults without limitations in its performance of functions [1, 2]. The term 'fault' refers to unpermitted deviations of characteristics of a technical system. Thus, a fault represents a state of the system; fault tolerance requires fault detection, fault diagnosis and fault management by

A. Bartoszewicz et al. (Eds.): PCC 2020, AISC 1196, pp. 1458–1469, 2020.
https://doi.org/10.1007/978-3-030-50936-1_121

design [2]. Typical perspectives in design phases are focused on (potential) impacts of faults in terms of reliability, availability and safety [2].

At the same time, designers target security of a technical system. The system should withstand intentional attacks [3]. Assuming 'security' as a system characteristic, a fault-tolerant system needs to perform as usual even in the event of such an attack, without causing any failures or malfunctions. Attackers might exploit vulnerabilities in the physical configurations, control systems, or accessibility through internet connections. For CPS, two challenges arise: Firstly, discipline-specific security measures should be applied. Secondly, new measures have to be created to cover interdisciplinary impacts.

Cyber-Physical Production Systems (CPPS) [4] are treated as a specific field of research. Two levels have to be considered, the production process performed by the system and the product produced within that process. Faults might impact both levels. For instance, faulty software configurations in CPPS might allow attackers to manipulate the correct control of production processes, impacting the quality of end products. From liability and publicity perspective, a worst-case scenario is that such an attack is not detectable as a process deviation and, consequently, a corrupted product is delivered to a customer. Examples are already given in various branches [5].

In this context, security-oriented fault-tolerance in Systems Engineering (SE) requires measures to evaluate interdisciplinary system designs with regard to potential attack scenarios. Requirements of systemic threat modelling are documented in Sect. 3. An analysis of related work indicates that current methods and tools are specialized to single domains even when they are motivated by interdisciplinary fault analyses (Sect. 4). This paper contributes a conceptual threat modelling approach to cover potential attack scenarios from a holistic system's application perspective (Sect. 5). The approach can be used to derive both system-level and discipline-specific security solutions. As an application case, issues are focused on which attackers intend to exploit vulnerabilities in a CPPS. The goal is to support systems engineers in verification and validation tasks regarding security-oriented fault-tolerance.

2 State of the Art

This paper focuses on challenges in the design of Cyber-Physical Production Systems (CPPS) as a special type of integrated Cyber-Physical systems (CPS). The approach is driven by two main demands: enabling traceable fault-tolerance from system to component level and, at the same time, ensuring security-oriented fault-tolerance from component to system level. Domain and demands are concretized in the following sections.

2.1 CPS and CPPS

CPS are highly networked technical objects that contain embedded systems, can exchange digital information and can use other services. With appropriate sensors, they are linked with the environment, save available data, evaluate them with the help of services and influence the physical world with actuators (cf. Figure 1). They are connected by means of internet technologies [4].

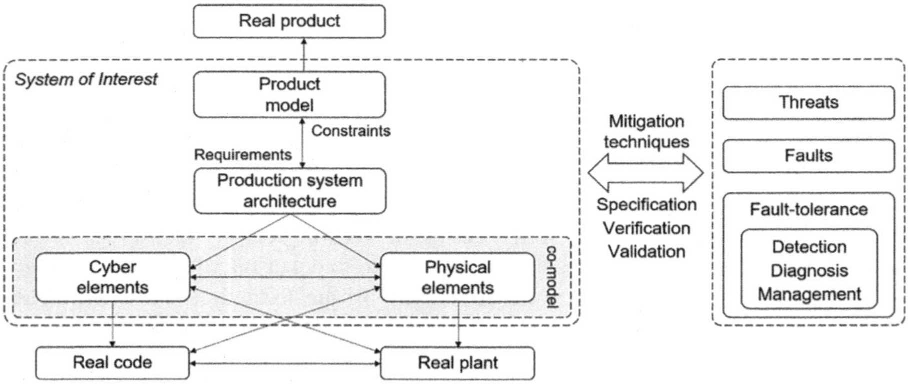

Fig. 1. Terminology and relationships of CPPS [4] and fault-tolerance [2]

Lee defines the term as follows: "Cyber-Physical Systems (CPS) is an integration of computation with physical processes. Embedded computers and networks monitor and control the physical processes, usually with feedback loops where physical processes affect computations and vice versa." [6] The term CPPS is derived from the definition of CPS. Accordingly, a CPPS is a production system that has the characteristics of a CPS. Production systems are socio-technical systems that transform input into output through value-adding and supporting processes. For this purpose, resources like machines, transport systems etc. are used. The objective of running a production system is the capability to produce the right products at the right time in the defined quality at reasonable costs. In contrast to conventional production systems, CPPS are communicate via internet. They are able to collect process information, which allows an integrated cognition and artificial intelligence [7]. The complete implementation of a CPPS may be entitled "Smart Factory". Materials, products and systems are equipped with sensors and actuators and ideally organized without human intervention [8].

Security in CPPS is mainly focused on issues related to information and communication technologies, especially smart grid. This is confirmed by Nguyen et al. who conducted a systematic study on approaches of Model Based Security Engineering [9]. Sadeghi et al. emphasize that security issues have to be addressed on all CPPS abstraction layers [5]. They elaborate on electronics, software and even humans. In addition to their approach, machines with their physical dimension should be included within CPPS system boundaries.

2.2 Security and Fault-Tolerance in Systems Engineering

Fault-tolerance implies detection, analysis and management of faults. These functionalities have to be integrated across all system-levels. Systemic fault-tolerance is mainly focused from the perspectives of safety, requiring system integrity. Typically, methods like Failure Modes and Effects (and Criticality) Analysis (FMEA/FMECA), Hazard Analysis as well as Event and Fault Tree Analysis (ETA/FTA) are combined for system safety and reliability in design phase [2]. In late 1990's, the Systems Security Engineering (SSE) project team identified capability levels and relevant process areas of SSE without specifying particular methods or processes [3]. As a result, SSE related practices were captured in ISO/IEC 21827. The model focuses specifically on information systems security. For the paper at hand, considerations of security risks, threats (including, for instance, threat agent's capabilities) and system security verification and validation (V&V) are relevant foundations.

Searching the library of the Design Society shows only very few publications related to fault-tolerance. Deyter et al. provide a case study of applying FTA on the principle solution using application scenarios in an early design phase of mechatronic systems [10]. Kolberg et al. propose a methodology to support design changes; they build up on six established "methods" from problem formulation to fault-tolerance [11]. Extending the literature review to google scholar, applications in specific branches can be added. Rostami et al. present brief insights into scenarios of threats propagating along globalized semiconductor supply chain [12]. Isaksson and Ritchey target system-level threats in their domain of nuclear facilities [13] where artefacts like the Design Basis Threat (DBT) are obligations.

2.3 Security-Oriented Fault-Tolerance

Security-orientation means coverage of various properties according to established security taxonomies (cf. [14]). The library of the Design Society does not contain any paper referring to both 'faults' or 'malfunctions' and 'security'.

In general, the reason for a fault is in the first place not important to the system designer. Hence, the term "fault" is used to cover all kind of faults. In contrast to safety, security-oriented faults describing faults caused by an (successful) attack to the system and are always based on a malicious intention. Since fault-tolerance describes the degree a system can handle faults without limitations to its behaviour, we argue that it is conceptual related to the term "threat modelling" [15] of the security domain. In threat modelling, potential threats to the system but also mitigations to these threats are defined. To determine which threats are relevant, security risk assessment is an essential part of a threat modelling approach. However, the only way to remove a threat completely from the system is to remove all affected system parts. Hence, when addressing potential threats, these threats are only mitigated and, therefore, a certain probability for these threats remain. Since it is not possible to prevent attacks, the goal of threat modelling is to find design decisions decreasing the probability of a successful attack and, therefore, increasing the systems security-oriented fault tolerance.

Moreover, following the concept of "defence-in-depth", a systematic threat modelling approach takes scenarios into account in which countermeasures are circumvented by an attacker or simply not correctly implemented. One essential part is to split the system into so-called "trust-zones" and to restrict privileges for each zone or system part to the minimum ("Principle of Least Privilege"). This technique helps to reduce the impact of an attack and, therefore, make the system more tolerant to certain attacks.

Hence, threat modelling can be used to systematically determine the degree a system is protected against threats or attacks respectively. In combination with a methodology ensuring continuous threat modelling along the whole engineering process, it is the root for security-oriented fault-tolerance.

In addition, when securing a system, it is always assumed that not all attacks can be known beforehand (for instance, if a used cryptographic library gets outdated). Thus, for some scenarios countermeasures are implemented that focus on detecting (and reporting) attacks instead of prevent them. Prominent examples are here Intrusion Detection Systems (IDS) or cryptographic signatures (e.g. to ensure the integrity of assets). Hence, threat modelling also covers threat scenarios that may occur in the future.

3 Requirements of Systemic Threat Modelling

When designing a method for security-oriented fault tolerance in the context of highly interdisciplinary development, there are several requirements such a method should take into account. They are derived from the different perspectives analyzed in Sect. 2. Firstly, there are requirements regarding the methodology itself focusing on the process as well as the context in which the method is applied. The method should

(R1) support all stages/steps of the engineering process, for instance, INCOSE Systems Engineering processes [16], VDI 2206V model of product and respective production system development (current state [17] and evolution [18]) up to operation and decommissioning stages,
(R2) cover informal and formal security requirements and threat descriptions, and
(R3) support interdisciplinary threat elicitation, risk analysis and mitigation.

Furthermore, it is essential that a method is not built from scratch but can be integrated into existing development frameworks. Therefore, the following requirements are important regarding the contextualization of the method. The method should

(R4) utilize standard vocabularies across disciplines, branches and system-levels,
(R5) be based on standard modelling methods/languages, and
(R6) take into account standard branch-specific scenarios documented in guidelines, directives or even norms.

In addition, a threat model can provide a backbone when collecting, persisting, and analyzing potential threats. The threat model should

(R7) cover both product security and CPPS security and their dependencies,
(R8) be extendable for new scenarios,
(R9) target all levels of system architectures bi-directionally (breaking down scenarios from system to component level, aggregating security assessments from component to system level),
(R10) allow for traceability of issues from faults to threats to misuse cases, and
(R11) be formalized to support derivation of Verification & Validation/test specifications.

4 Related Work

Since decades, scenario analysis is applied to inform decision making in a comprehensive and tangible way [19], including approaches supporting systems engineers [20]. Threat scenarios can be approached from three different perspectives: prognosis of potential future threats in general, modifications of an application scenario in terms of possible or even probable stories [21] as well as modifications of a system specification that describes a system state in case of an attack [22]. The library of Design Society does not contain any publication which covers both "threat" and "model*" as keywords. Three search results are only focused on threats regarding disruptive innovation, value-centric development and IP protection. Extending the review of related work on scenario based interdisciplinary approaches, google scholar indicates only partial overlaps. Kim and Cha apply a scenario based threat modelling approach on a broadband convergence Network [23] specifically focusing on information systems. Detecting insider threats is a special field of applications (cf., for instance, [24]).

SysML-Sec [25] is a model-driven engineering approach that "aimed at fostering the collaboration between system designers and security experts" [26] during all phases of the development of cyber-physical systems. It is based on SysML and provides customized SysML diagrams to describe security-related system parts. It also provides a methodology for a systematic development focusing on closing the gap between safety and security modelling. It covers steps for (security) requirements engineering, system design, design validation, and also (partially) code generation for the target system. In contrast to our approach, SysML-Sec focuses only on the development of CPS but does not take the special requirements of CPPS into account nor it takes the security requirements for the product into account.

5 Scenario-Based Security-Oriented Fault-Tolerance Validation

In this section, the proposed approach is presented based on the development steps shown in Fig. 2 and its application in a laboratory environment. The conceptual approach combines scenario-based and security-oriented development steps and iterative Model Based Systems Engineering methodologies. Threat modelling techniques are integrated into existing MBSE methods targeting, on the one hand, requirements **R4** and **R5** but also **R1** since the threat modelling is refined along the whole development cycle.

5.1 Conceptual Approach

Assuming a system specification is available after development iteration (S_i), its fault-tolerance shall be validated regarding identified threats [22]. S_i is specified by SysML referring to discipline specific co-models (like UML for software and STEP for mechanics) according to [4].

System-level security requirements are based on **security policies** and **global security requirements**. Global means system independent, often companywide specifications. Since these requirements are adopted specifically for product and production system in early planning steps, it is essential that they can be defined informally first and are getting refined at later development steps (cf. **R2**). These might differ in terms of branches, markets, countries and other aspects, for instance based on data privacy regulations like the GDPR. We assume that the security policies and security requirements are elicited using existing methods based on the combination of system assets and threat

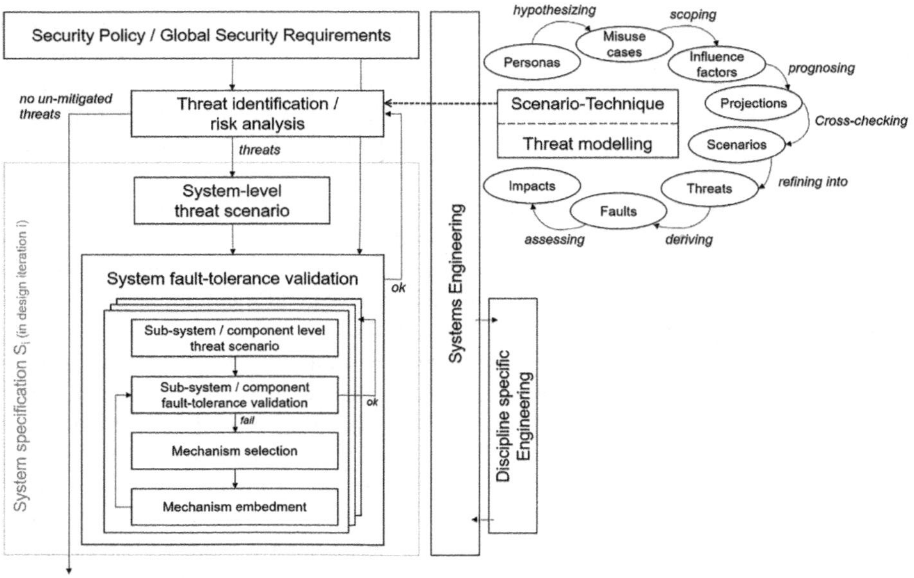

Fig. 2. Conceptual approach of fault-tolerance validation: Complementing Systems Engineering and discipline specific engineering by means of Scenario-Technique and threat modelling

categories like STRIDE [15] or common security properties like CIA [27] (cf. **R6**). On the one hand, they are complemented by discipline-specific security requirements. On the other hand, threat identification as an activity of risk analysis is performed (main path in Fig. 2). To ensure security on all levels, addressing a global security requirement has to be accomplished by different disciplines in combination (cf. **R3**).

The **Scenario Technique** is proposed as a method to estimate potential threats: Early phases of the scenario technique target artefacts like personas and abuse/misuse cases [28], future evolutions are estimated by influence factors and corresponding projections of possible futures which are aggregated into scenarios [29]. Influence factors are derived based on morphological analyses (cf. [13]) considering system-level security taxonomies (cf. [14]). **Scenarios** are used as descriptive models which are comprehensive for all stakeholders [19]. Scenarios are analyzed with regard to threats on system and sub-system/component levels. Resulting threats are formalized for uptake in engineering tools. Effects in terms of faults within the system/sub-system/component as well as impacts between systems (esp. production faults impacting products) can be derived.

A **system-level threat scenario** means a modification of S_i representing a threat causing a faulty state of the system. Once an initial set of threat scenarios are identified, these scenarios have to be refined to discipline-specific threats (describing which part of the scenario can be tackled by which discipline). Existing threat modelling techniques are integrated for each development stage (cf. [30]). Similar to [25], we suggest to use attack-defense graphs for documenting the refinement of threat scenarios into sub-scenarios and to define potential mitigation. This technique allows defining relations between the impacts on different system parts and different disciplines (cf. **R9**). Explicit modelling of relationships allows traceability along all analysis stages. This applies especially to traceability of threats and impacts between production system and materials/products handled within corresponding processes and transferred into customer use cases later. While a threat might refer to an element of the production system (e.g., a manufacturing cell in a company's factory), the impact might become active in product use at a customer's location (cf. Sect. 5.2). For this reason, we argue that it is essential to apply a holistic threat model covering both the production system and the product in one threat model. When eliciting or refining threats and mitigations, the security experts have to tag involved disciplines but also which part of the system or the product is affected by a threat or responsible for the correct implementation of a mitigation (cf. **R7**). Noteworthy is that security engineers of one discipline may define other disciplines as responsible for a mitigation during the **discipline-specific Engineering.**

Since it is unsuitable to review the whole threat model, it is important to enable discipline-specific views on the model showing only the relevant parts and impacts to the engineer [31]. Creating new threats or mitigations for another discipline will open up new threat model parts that have to be discussed at the beginning of the next iteration (S_{i+1}) by all involved disciplines before it becomes approved for the threat model. Correspondingly, at each iteration step, it can be determined if all desired mitigations are (designed to be) in place. If a specific threat is not mitigated sufficiently, it may be refined again or delegated to other disciplines as well in the next iteration. When all threat scenarios are sufficiently addressed (which is when the fault-tolerance is high enough), the threat is marked as mitigated and the iteration stops. If there are no un-mitigated threats, the threat modelling stops until changes to the threat model or the system models are made and a new iteration is necessary. Using iterations that are update whenever a model changes helps keeping the threat model up to date and to integrate new threats and mitigations when needed (cf. **R8**).

5.2 Application Case in Cyber-Physical Production Systems

An application case is simulated in a laboratory environment at Heinz Nixdorf Institute in Paderborn. Within a realistic factory environment, exemplary mechatronic products are produced like remote controlled cars (RC cars) and small drilling machines [7]. The lab is established as a complete production system with manufacturing cells, assembly station/line, material flow sub-system and software. All manufacturing cells include different manufacturing techniques (milling, turning and 3D printing) and industrial robots (portal robot, articulated robot and cobot). The software architecture spans from product configuration and Enterprise Resource Planning (ERP) to manufacturing execution and process control. For the analysis of security-oriented faults and corresponding fault-tolerance measures, a scenario is chosen which covers intentional security breaches in a production system producing RC cars, threats that are implied for products and impacts for end customers. Even though the example is far from real cars, correspondences between the lab case and real applications can be analyzed.

A sample scenario is defined as follows (cf. Fig. 3): Amongst others, a persona is defined regarding the intention to reduce trust in products of automotive industries by provoking technical issues up to accidents of private cars. A sample misuse case describes data manipulation causing the 3D printer to produce lower quality parts. Technically speaking, load-bearing capacity of parts is reduced by manipulating manufacturing parameters; standard non-destructive testing for quality assurance would not be capable of recognizing this threat before shipment to customers.

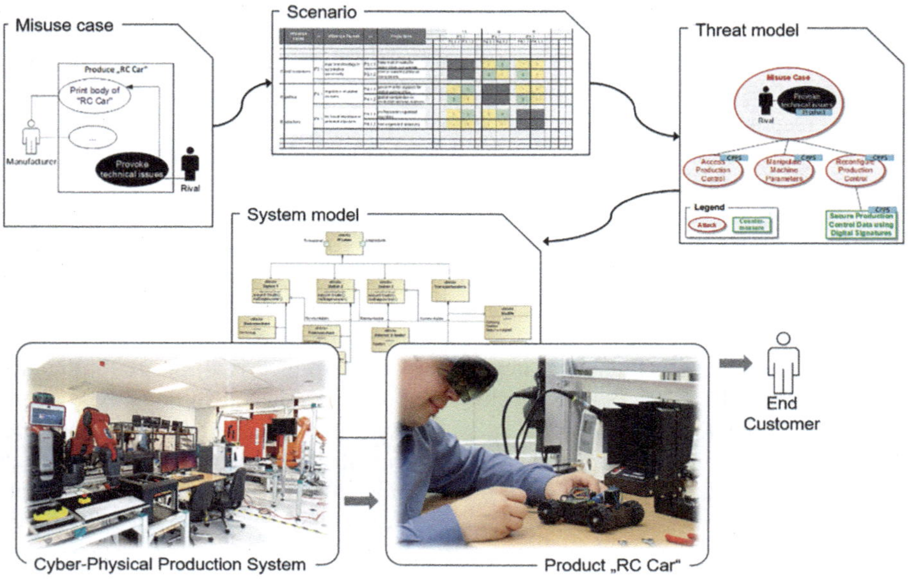

Fig. 3. Artefacts of scenario-based and security-oriented fault-tolerance validation

Scenario Technique is applied based on the assumption that relevancy, probability and business impact of misuse cases are influenced by various factors. Considering use cases as contextual frame for product usage, influence factors are collected in influence fields. Such fields subsume generic aspects like 'trust in technology in general' as well as specific one for automotive. In this sample scenario, 'interrelations with international politics' and 'technical expertise of potential attackers' were identified as relevant fields based on published studies in this domain. Cross-impact analysis results in clusters of active and/or passive factors. Projections are derived from statistics, simulations and expert workshops with emphasis on active factors. For instance, governments might tend to partnership or tend to global competition in protected national markets. End customers might keep trust in specific automotive companies or tend to switch to quickly to competitors benefitting from an attack. Hence, attackers can put higher pressures on companies. For realistic scenarios, consistency is checked for all combinations of projections. The aforementioned scenario results from assumptions that there is a significant risk of international attacks targeting impact on large groups of end customers. Based on these scenarios, threats are analysed upstream: Starting from targeted threats (car defects), the product creation process is investigated towards early production stages (delivery of raw material, printing of circuit boards, coding of software packages). Along this chain, potential structural and behavioural implications on system elements are annotated within the system model and its partial models.

Fault-tolerance is enabled by applying threat mitigation measures to potential threats. More specifically, defence-in-depth means to combine different measures to mitigate single or aggregated threats across all levels of the system architecture. With regard to the exemplary scenario, various measures could be combined in Systems Engineering: From a software engineering perspective, digital signatures for manufacturing data can avoid data breaches. From a mechanical engineering perspective, triggers can be applied to manufacturing execution data for fault detection and additional testing procedures can help to recognize affected products before shipment.

6 Summary and Outlook

The conceptual approach integrates approaches of misuse case modelling, scenario technique, threat modelling and Model Based Systems Engineering. The challenge is to derive inputs for fault-tolerance engineering from system-level to discipline specific levels of the system architecture. Vice versa, risk assessments and mitigation measures have to be aggregated from component to system level. To achieve both top-level requirements, the entire chain from conceptual modelling of scenarios to threats needs to be formalized in a way that systems engineers and developers are supported on their corresponding interaction levels. An application case is presented as a first validation in laboratory environment. The conceptual approach will be used to contribute to methodological frameworks, modelling languages and computational tools in future.

References

1. Potiron, K., El Fallah, S.A., Taillibert, P.: From Fault Classification to Fault Tolerance for Multi-agent Systems. Springer-Briefs in Computer Science. Springer, London (2013)
2. Isermann, R.: Fault-Diagnosis Systems: An Introduction from Fault Detection to Fault Tolerance. Springer, Heidelberg (2006)
3. Carnegie Mellon University: Systems Security Engineering Capability maturity Model: SSE-CMM Model Description Document (1999)
4. Fitzgerald, J., Larsen, P.G., Verhoef, M.: From embedded to cyber-physical systems: challenges and future directions. In: Fitzgerald, J., Larsen, P.G., Verhoef, M. (eds.) Collaborative Design for Embedded Systems: Co-modelling and Co-simulation, vol. 138, pp. 293–303. Springer, Berlin (2014)
5. Sadeghi, A.-R., Wachsmann, C., Waidner, M.: Security and privacy challenges in industrial internet of things. In: Proceedings of the 52nd Annual Design Automation Conference. ACM, New York, pp. 1–6 (2015)
6. Lee, E.A.: Cyber physical systems: design challenges. In: 11th IEEE International Symposium on Object and Component-Oriented Real-Time Distributed Computing, pp. 363–369. IEEE Computer Society, Los Alamitos (2008)
7. Gräßler, I., Pöhler, A., Pottebaum, J.: Creation of a learning factory for cyber physical production systems. Procedia CIRP **54**, 107–112 (2016). https://doi.org/10.1016/j.procir.2016.05.063

8. Frazzon, E.M., Hartmann, J., Makuschewitz, T., et al.: Towards socio-cyber-physical systems in production networks. In: Cunha, P.F. (ed.) 46th CIRP Conference on Manufacturing Systems 2013, vol. 7, pp. 49–54 (2013)
9. Nguyen, P.H., Ali, S., Yue, T.: Model-based security engineering for cyber-physical systems: a systematic mapping study. Inf. Softw. Technol. **83**, 116–135 (2017). https://doi.org/10.1016/j.infsof.2016.11.004
10. Deyter, S., Gausemeier, J., Kaiser, L., et al.: Modeling and analyzing fault-tolerant mechatronic systems. In: Norell Bergendahl, M., Grimheden, M., Leifer, L., et al. (eds.) Design Has Never Been This Cool: ICED 09, The 17th International Conference on Engineering Design, pp. 55–66. Design Society, Glasgow (2009)
11. Kolberg, E., Reich, Y., Levin, I.: Express engineering change management. In: Giess, M.P., Goh, Y.M., Lian Ding, L., et al. (eds.) ICED 07, The 16th International Conference on Engineering Design. Design Society (2007)
12. Rostami, M., Koushanfar, F., Rajendran, J., et al.: Hardware security: threat models and metrics. In: 2013 IEEE/ACM International Conference on Computer-Aided Design (ICCAD), pp. 819–823. IEEE Press, Piscataway (2013)
13. Isaksson, S., Ritchey, T.: Protection against sabotage of nuclear facilities: using morphological analysis in revising the design basis threat. Adaptation of the original paper 2010. In: 44th Annual Meeting of the Institute of Nuclear Materials Management (2003)
14. Schumacher, M., Fernandez-Buglioni, E., et al.: Security Patterns: Integrating Security and Systems Engineering. Wiley (2006)
15. Shostack, A.: Threat Modeling: Designing for Security. Wiley, Indianapolis (2014)
16. Walden, D.D., Roedler, G.J., Forsberg, K., et al. (eds.): Systems Engineering Handbook: A Guide for System Life Cycle Processes and Activities; INCOSE-TP-2003-002-04, 4th edn. Wiley, Hoboken (2015)
17. VDI: VDI 2206 Design methodology for mechatronic systems (VDI 2206) (2004)
18. Gräßler, I., Hentze, J., Bruckmann, T.: V-models for interdisciplinary systems engineering. In: Proceedings of the DESIGN 2018 15th International Design Conference. Faculty of Mechanical Engineering and Naval Architecture, University of Zagreb, Croatia, The Design Society, Glasgow, UK, pp. 747–756 (2018)
19. Börjeson, L., Höjer, M., Dreborg, K.-H., et al.: Scenario types and techniques: towards a user's guide. Futures **38**(7), 723–739 (2006). https://doi.org/10.1016/j.futures.2005.12.002
20. Gräßler, I., Hentze, J., Scholle, P.: Enhancing systems engineering by scenario-based anticipation of future developments. In: SoSE and Cyber Physical Systems (CPS), From Academia to Application and Back: 11th Systems of Systems Engineering Conference (SoSE). IEEE, Piscataway (2016)
21. Fitzgerald, J.: Developing & ranking threat scenarios. EDPACS **6**(3), 1–5 (1978). https://doi.org/10.1080/07366987809449432
22. Lotz, V.: Threat scenarios as a means to formally develop secure systems. J. Comput. Secur. **5**(1), 31–67 (1997)
23. Kim, Y.-G., Cha, S.: Threat scenario-based security risk analysis using use case modeling in information systems. Secur. Comm. Netw. **5**(3), 293–300 (2012). https://doi.org/10.1002/sec.321
24. Young, W.T., Memory, A., Goldberg, H.G., et al.: Detecting unknown insider threat scenarios. In: 2014 IEEE Security and Privacy Workshops: Proceedings, 17–18 May 2014, San Jose, California, USA. Conference Publishing Services, pp. 277–288. IEEE Computer Society, Los Alamitos (2014)
25. Roudier, Y., Apvrille, L.: SysML-Sec: a model driven approach for designing safe and secure systems. In: Hammoudi, S. (ed.) MODELSWARD 2015: Proceedings of the 3rd International Conference on Model-Driven Engineering and Software Development, pp. 655–664. IEEE, Piscataway (2015)

26. Apvrille, L., Roudier, Y.: Towards the model-driven engineering of secure yet safe embedded systems. Electron. Proc. Theor. Comput. Sci. **148**(4), 15–30 (2014). https://doi.org/10.4204/EPTCS.148.2
27. Andress, J.: The Basics of Information Security: Understanding the Fundamentals of InfoSec in Theory and Practice, 2nd edn. The Basics. Elsevier Science, Burlington (2014)
28. Røstad, L.: An extended misuse case notation: including vulnerabilities and the insider threat. In: 12th Working Conference on Requirements Engineering (REFSQ 2006): Foundation for Software Quality (2006)
29. Gräßler, I., Pottebaum, J., Scholle, P.: Integrated process and data model for agile strategic planning. In: Vajna, S. (ed.) 11th International Workshop on Integrated Design Engineering (2017)
30. Geismann, J., Gerking, C., Bodden, E.: Towards ensuring security by design in cyber-physical systems engineering processes. In: Kuhrmann, M., O'Connor, R.V., Houston, D. (eds.) Proceedings of the 2018 International Conference on Software and System Process - ICSSP 2018, pp. 123–127. ACM Press, New York (2018)
31. Völter, M., Stahl, T., Bettin, J., et al.: Model-Driven Software Development: Technology, Engineering, Management, 1 edn. Wiley Software Patterns Series. Wiley, s.l. (2013)

Robust Economic Model Predictive Control of Drinking Water Transport Networks Using Zonotopes

Khoury Boutrous(✉), Fatiha Nejjari, and Vicenç Puig

Advanced Control Systems Group, Technical University of Catalonia (UPC), Rambla Sant Nebridi 22, 08222 Terrassa, Spain
{boutrous.khoury,fatiha.nejjari,vicenc.puig}@upc.edu

Abstract. A robust economic Model Predictive Control (EMPC) approach is presented in this paper for the control of a Drinking Water Network (DWN) albeit the presence of uncertainties in the forecasted demands required for the predictive control design. The uncertain forecasted demand on the nominal MPC has the possibility of rendering the optimization process infeasible or degrade the controller performance. In this paper, the uncertainty on demand is considered unknown but bounded in a zonotopic set. Based on this uncertainty description, a robust MPC is formulated to ensure robust constraint satisfaction, performance and stability of the MPC for DWN to meet user requirements whilst ensuring lower operational cost for water utility operators.

Keywords: Model predictive control · Economic control · Robustness · Drinking water network · Zonotopes

1 Introduction

Model predictive control (MPC) stands out as the predominant modern controller used in industry, mainly due to its inherent ability to ensure closed-loop stability and constraint satisfaction whilst satisfying a cost function. Recently, the concept of incorporating directly an economic stage cost of industrial processes in a MPC design termed economic MPC has garnered interest with a plethora of applications as proposed in [11,12] showing promising results. This procedure involves an update of the generic cost function which normally involves tracking a set-point to one which explicitly involves economic terms such as energy and cost of production, enabling an improvement during transients and the ability to manipulate control variables to satisfy various economic requirements. The problem of stability in economic MPC has been extensively studied in [13]. As the name suggests, there exits economic entities (such as e.g. cost. price or demand) that serves as exogenous variables in the design of economic MPC. These variables are undoubtedly subject to stochastic variations which requires further control design considerations for a suitable operation. For example, in

A. Bartoszewicz et al. (Eds.): PCC 2020, AISC 1196, pp. 1470–1482, 2020.
https://doi.org/10.1007/978-3-030-50936-1_122

the design of an economic MPC with variable demand as done in [10], a forecast of demand to enable future predictions of states in the MPC optimization loop is required. But forecasted demand as a variable is subject to human behaviour which can be described as uncertain at best. Therefore, there is the need to ensure that controllers are built robust to design variations, which are inevitable in real life situations. Methods of stochastic MPC [5], the Min-Max Robust formulation [9], the tree-based method [8] and other proposed concepts have been successfully applied to problems of uncertainties in internal or exogenous parameters in MPC optimization formulations, but most of these methods yields computationally demanding problems. [6] provides a comprehensive overview of robust MPC, highlighting recent trends, limitations and proposed future research directions.

The prime objective of a DWN operation is to ensure that demand nodes in a network are served with quality water at desired pressures whilst minimising operational cost. EMPC in the area of water supply has been extensively studied. In [4], the large number of actuators (i.e. decision variables) in the DWN was parameterized to minimize the number of decision variables in the optimization loop ensuring a faster run-time. [10] designs a two level control scheme, where the upper level controls the complete system via a nonlinear MPC scheme and the lower level, a pump scheduling scheme through mixed integer programming, but these proposed schemes fails to account for uncertainties in the forecasted demand which may affect controller performance. Stochastic MPC for uncertainties at demand nodes of a DWN has been studied by [5] showing promising results of maintaining tractability and performance of the system subject to uncertainties but provides a somewhat computational expensive solution.

In this paper, a robust EMPC is designed for a DWN, specifically the case of the Barcelona Drinking Water Network, considering uncertainties in forecasted demand. The variations in demand are considered unknown but bounded, and are described to vary in a zonotopic set. Zonotopic sets show desirable characteristics of less complexity, flexibility and reliable computation of linear transformation and Minkowski sums compared to other geometric counterparts such as interval, ellipsoidal sets e.t.c [3]. The associated effect of the uncertainty in the demand on the states and inputs are described offline, taking advantage of the affine dependence of these variables on the demand as given in the model description of the DWN. Hence, a reachable set is constructed offline. The offline description of these sets provides a similar formulation of an optimization problem to the nominal controller, therefore offering similar optimization complexity. However, it must be noted that even though a robust MPC is achieved after this procedure, there is a certain degree of robustness for some magnitude of disturbance beyond which the optimization problem ceases to be feasible. The constraint formulations are then updated to ensure robust constraints satisfaction as well as a local LQR feedback control designed to mitigate the effect of the uncertainty on the predicted states as done in [2]. The rest of this paper is structured as follows: a mathematical preliminary to the concept of zonotopes will be given in Sect. 1. Then, a description of a linear DWN model to be used for the controller design is presented in Sect. 2. Section 3 presents the design of the robust MPC scheme and finally results and conclusions after application of the developed scheme is discussed in Sect. 4.

1.1 Mathematical Preliminaries

Definition 1. The Minkowski sum of two sets, $\mathcal{X}$ and $\mathcal{Y}$ is defined as $\mathcal{X} \oplus \mathcal{Y} = \{x + y : x \in \mathcal{X}, y \in \mathcal{Y}\}$.

Definition 2. A m-zonotope, $\mathcal{Z} \in \mathbb{R}^n$ is defined as the linear image of a m-dimensional hypercube, where m is the order of the zonotope and subsequently satisfies $(m \geq n)$. Given a vector $\mathbf{P} \in \mathbb{R}^n$ and a matrix $\mathbf{H} \in \mathbb{R}^{n \times m}$, a m-zonotopic set $\mathcal{Z}$, can be described as:

$$\mathcal{Z} = \mathbf{P} \oplus \mathbf{H}B^m = \{x \in \mathbb{R}^n | x = p + Hz, z \in B^m\}. \tag{1}$$

Which defines the Minkowski sum of the m-segments defined by m-columns of matrix $\mathbb{H} \in \mathbb{R}^n$. Given p, the center of the zonotope $\mathcal{Z}$ and $\mathbf{H} = \left[h_1, h_i, ...h_n\right]$ $\forall h_i \in \mathbb{R}^n$ as the generators of the zonotope. Properties governing computations of zonotopes are given as follows:

Property 1. The Minkowski sum of two zonotopes $\mathcal{Z}_1 = \mathbf{P}_1 \oplus \mathbf{H}_1 B^{n_1}$ and $\mathcal{Z}_2 = \mathbf{P}_2 \oplus \mathbf{H}_2 B^{n_2}$ is a zonotope defined by $\mathcal{Z} = (\mathbf{P}_1 + \mathbf{P}_2) \oplus [\mathbf{H}_1, \mathbf{H}_2] B^{n_1+n_2}$.

Property 2. The image of a zonotope $\mathcal{Z} = \mathbf{P} \oplus \mathbf{H}B^n$ by linear mapping matrix K can be calculated as $K\mathcal{Z} = K\mathbf{P} \oplus K\mathbf{H}B^n$.

2 DWN Control-Oriented Model

DWN comprises of a complex network of hydraulic elements that convey water from supply to demand [1]. These elements work interactively to satisfy demands at desired pressure and of good water quality during transportation. The network can thus be categorized into the supply, treatment, transport and distribution sub-levels. Active hydraulic elements (pumps and valves) control flow and/or pressure in specific paths in the network, interactively working with passive elements such as pipes and tanks which act as transport and storage elements respectively to satisfy the broader network requirements in terms of meeting demand as well as ensuring water supply by storing adequate water reserves in the network's tanks. There have been several attempts in literature to model DWNs which captures key dynamics at different levels of the drinking water network architecture. Through graph theory, some works consider flow directions of water at network nodes as well as interactions at the tanks [1] leading to a simple flow-based model description of the network, whilst other studies [5] consider both flow and pressure specifically taking into account the interactions when flow and hydraulic head equations are considered in the modelling process. Considering only the transport sub-level, the flow-based model offers an easier option to work with, largely due to its linearity but fails to captures key pressure dynamics which is important to present a complete mathematical behaviour of the network. Inclusion of pressure in the DWN dynamics introduces non-linearity from the pressure-flow affine equality equation into the optimization problem's constraints formulation resulting in a non-convex problem. Some works have been successful in designing nonlinear MPC [14] for the control of these nonlinear models, whilst works such as [15] considers nonlinear constraint relaxation to produce a set of linear inequality constraints for a linear Economic EMPC formulation. Despite its complexity, the nonlinear pressure-flow model offers a more realistic case to work with. The purpose of this paper is to only illustrate the ability of a set-based method (zonotope) of robust EMPC to ensure suitable controller performance under demand uncertainties, hence the comparatively less complex flow based model will be used. [7] presents a flow-based model on the Barcelona water network. Basic relationships between elements considering mass balance at tanks and equilibrium of flow directions at nodes, give rise to a discrete time dynamics as follows:

$$x_{k+1} = Ax_k + B_u u_k + B_d d_k, \tag{2}$$

$$0 = E_u u_k + E_d d_k. \tag{3}$$

$x_k \in \mathbb{R}_+{}^{n_x}$ is the vector of system states, denoting tank volumes at each time instant k. $u_k \in \mathbb{R}^{n_u}$ denotes the manipulated input from actuators affecting changes in states in combination with the non-negative model disturbance $d_k \in \mathbb{R}_+{}^{n_d}$, that represents the consumer demand. A, B_u, B_d, E_u and E_d are time-invariant matrices of suitable dimensions. From Eq. (3), it can be inferred that the control variable u does not take its value from $\mathbb{R}^{n_u}$ but in a linear variety. This inference enables an affine parameterisation of the control variables in terms of a minimum set of disturbance, mapping the control problem to a space with a smaller decision vector and with less computational burden due to the elimination of the equality constraint, (3) [10].

Assumption 1: *Considering that there are more variables than algebraic equations (i.e. $n_q < n_u$), the matrix E_u in (3) has a maximal rank. Therefore, it can be expressed in a reduced staggered form using the Gauss-Jordan elimination.*

From Assumption 1, the control variable is parameterized such that:

$$u_k = \tilde{P}\tilde{M}_1 \hat{u}_k + \tilde{P}\tilde{M}_2 d_k. \tag{4}$$

The model can be represented as a difference Eq. (5) by replacing (4) into Eq. (2), given as:

$$x_{k+1} = Ax_k + \hat{B}\hat{u}_k + \hat{B}_d d_k, \tag{5}$$

where $\hat{B} = B\tilde{P}\tilde{M}_1$ and $\hat{B}_d = B\tilde{P}\tilde{M}_2 + B_d$.
For an in-depth understanding on how the control variables are parameterized, reader is referred to [10].

3 Problem Formulation

The problem of designing a robust MPC controller must be such that the designed controller satisfies the tenets of robust stability, robust constraint satisfaction and robust performance for all realization of a system $\Sigma = f(u, x, d)$, albeit any variations of function variables. Assuming demand uncertainty in the system, the effects of these uncertainties on the exogenous demand variable d_k, $\Delta d_k \subseteq \delta d$ results in a subsequent variation in state $\Delta x_k \subseteq \delta X$ and input variables $\Delta u_k \subseteq \delta U$ as described in the affine relationships of Eqs. (2) and (3). Model variables, (d, x, u) can subsequently be decomposed into an uncertainty free and uncertain dependent component, with the latter involving a realization of variables at each time instant from bounded uncertainty sets, $(\delta d, \delta U, \delta X)$. Thus, the state and input uncertainty sets $(\delta X, \delta U)$ described as a zonotope are generated from the known zonotopic bounded set of the demand variation δd with the aid of the algebraic difference Eqs. (2)–(3).

Assumption 2: *The states x and demands d are considered observable at each time instant k and the pair $(A, \hat{B})$ is controllable.*

δd is bounded using a zonotope that can be formulated from a symmetric interval set considering bounded demand uncertainty such that $\Delta d_i \in [-\Delta d_i, \Delta d_i]$, where i denotes a particular demand node. The set δd can therefore be represented in a zonotopic form as:

$$\delta d_k = 0 \oplus H_d z_d. \tag{6}$$

Where 0 is a column vector of dimension n_d (n_d is the number of demand nodes), considered as the center the zonotope and H_d is a diagonal matrix of the generators represented as the bounds of variations at each demand node i: $z_d \in B^{n_d}$; $B = [-1\ 1]$. Considering that $\tilde{x}$, $\tilde{\hat{u}}$ and $\tilde{d}$ are the real dynamic states, inputs and demand respectively, which takes into account the uncertainty effects, appropriate decomposition of model variables is therefore given as: $\tilde{x} = x + \Delta x$, $\tilde{\hat{u}} = \hat{u} + \Delta\hat{u}$ and $\tilde{d} = d + \Delta d$.

The DWN model considering uncertainty in the demand variable is therefore given as from (2) and (3) as:

$$\tilde{x}_{k+1} = A\tilde{x}_k + \hat{B}\tilde{\hat{u}}_k + \hat{B}_d\tilde{d}_k, \tag{7}$$

$$0 = E_u\tilde{\hat{u}}_k + E_d\tilde{d}_k. \tag{8}$$

Nominal states $x \in \mathbb{R}_+{}^{n_x}$ and inputs $\hat{u} \in \mathbb{R}_+{}^{n_u}$ are considered bounded in a compact polyhedron $\mathbb{U}$ and $\mathbb{X}$ respectively, containing the origin in their interiors, with $\hat{u} \subseteq \mathbb{U}$ and $x \subseteq \mathbb{X}$. In the presence of uncertainty, it is desirable to generate a tube of trajectories, meaning a sequence of robust invariant reachable sets such that per every transition of states and inputs of the nominal system, the resultant states and inputs after effect of uncertainty remains in a closed and bounded set. A robust invariant tube, $\tilde{X}_k = \{\tilde{X}_0, \tilde{X}_1, ..., \tilde{X}_N\}$, $\forall\ \tilde{X}_k = x_k \oplus \delta X_k$ and subsequent control tube $\tilde{U}_k = \{\tilde{U}_0, \tilde{U}_1, ..., \tilde{U}_N\}$, $\forall\ \tilde{U}_k = \hat{u}_k \oplus \delta U_k$, is constructed either online or offline, with the offline procedure offering lesser computational burden. x_k and $\hat{u}_k$ are centers of the respective propagated state and control invariant tubes. The mismatch between nominal predicted states and real states resulting from uncertainties are mitigated by a local feedback controller K, in this case an LQR controller such that the selection of this feedback gain K satisfies a system with the assumption that $\tilde{d}_k = 0$

$$\Delta x_{k+1} = (A + \hat{B}K)\Delta x_k \tag{9}$$

with $\Delta x_k \subseteq \delta X_k$. The local controller ensures that the deviation of the system dynamics in the closed loop with $(A+\hat{B}K)$ is asymptotically stable. The primary aim is to have an optimal control problem, which keeps trajectories around the neighbourhood of the nominal optimal trajectory in the presence of uncertainties, satisfying the cost function under process constraints.

Assumption 3: *Considering that $(A + \hat{B}K)$ is strictly stable and $\tilde{x} = x + \Delta x$, such that the uncertain dynamic part $\Delta x_{k+i+1} = (A + \hat{B}K)\Delta x_{k+i} + \hat{B}_d\Delta d_{k+i}$. If $(A + \hat{B}K)\delta X \oplus \hat{B}_d\delta d \subseteq \delta X$, then it can be assumed that the transition of states from one time instant to another depends on the dynamics of the centers $x_{k+i+1} = Ax_{k+1} + \hat{B}\hat{u}_{k+1} + \hat{B}_d d_{k+1}$.*

3.1 Offline Computation of Zonotopic Reachable Sets

The feedback gain K is computed and kept constant at each time instant k throughout the prediction horizon of the MPC controller to minimize the deviation of the perturbed state as stated before. The following cost function is utilized for the design of an optimal local controller for state error minimization

$$J_\infty = \sum_{i=0}^{\infty}(\tilde{x}_k - x_k)^T Q(\tilde{x}_k - x_k) + \sum_{i=0}^{\infty}(\tilde{\hat{u}}_k)^T R(\tilde{\hat{u}}_k). \tag{10}$$

Where Q and R are positive definite matrices of appropriate dimensions, $\tilde{x}$ is the real state at time k from the real plant and $\tilde{\hat{u}}$, the actual inputs, with x_k as the nominal state prediction from the MPC at time instant k. From (7), the real state, $\tilde{x}_k$ (i.e. $\tilde{x}_k = x_k + \Delta x_k$) under conditions of uncertainty as stated before can be decomposed into:

Certain component: $\begin{cases} x_{k+1} = Ax_k + \hat{B}\hat{u}_k + \hat{B}_d d_k. \end{cases}$

Uncertain component: $\begin{cases} \Delta x_{k+1} = (A + \hat{B}K)\Delta x_k + \hat{B}_d\Delta d_k. \end{cases}$

where $\Delta\hat{u} = K\Delta x$, from the local LQR controller. From the uncertain component, a corresponding $N \in \mathbb{Z}_{>0}$ length of tube is computed, where N is the selected prediction horizon of the MPC controller. Therefore, realization of the deviation Δx from the set δX assuming that initial deviation, $\Delta x_0 = 0$ can be described as:

$$\delta X_{k+j} \subseteq \bigoplus_{i=1}^{j}(A + \hat{B}K)^{j-i}\hat{B}_d\delta d. \tag{11}$$

Given that $\delta d = 0 \oplus H_d B^{nd}$, and from properties 1 and 2 of the zonotope, it therefore follows that:

$$\delta X_{k+j} \subseteq 0 \oplus \Psi_{1,j} B^{n_d}, \tag{12}$$

$$\Psi_{1,j} = \bigoplus_{i=1}^{j} (A + \hat{B}K)^{j-i} \hat{B}_d H_d. \tag{13}$$

From (4) in the control parameterization as discussed in Sect. 2, the auxiliary control variable $\tilde{\hat{u}}$ at time instant k can be described as:

$$\tilde{\hat{u}}_k = \hat{u}_k + K\Delta x_k, \tag{14}$$

where $\hat{u}_k$ is the certain control variable at time k with associated uncertainty, $\Delta\hat{u}_k$. K is the local controller gain calculated from (10). From Eq. (4), and under decomposition into certain and uncertain parts, considering that the actual control variable $\tilde{\hat{u}}_k$ is a value from a set i.e. $\tilde{\hat{u}} \in \delta U$, the uncertain control set is given as:

$$\delta U_{k+j} \subseteq \tilde{P}\tilde{M}_1 K \delta X_{k+j} \oplus \tilde{P}\tilde{M}_2 \delta d, \tag{15}$$

Equation (15) is obtained considering the control parameterization (4). The sequence of cross-sections of the control tube can therefore be described in a zonotopic form as

$$\delta U_{k+j} \subseteq 0 \oplus [\tilde{P}\tilde{M}_1 K \Psi_{1,j}, \tilde{P}\tilde{M}_2 H_d] B^{2n_d}. \tag{16}$$

3.2 Terminal State Constraint Set

For robust stability and recursive feasibility, a terminal constraint set is formulated. A minimal terminal robust positive invariant set δX_f is constructed as an outer approximation of the exact equilibrium state set

$$\tilde{X}_* = \bigoplus_{j=0}^{\infty} (A + \hat{B}K)^j \hat{B}_d \delta d, \tag{17}$$

where $\tilde{X}_* \subseteq \delta X_f$. Given that $(A + \hat{B}K) = \hat{A}$ and $\hat{B}_d \delta d = \mathcal{W}$ and under the assumption that $\hat{A}$ is strictly stable, with $\mathcal{W} \subseteq \delta d$. An outer set approximation is assumed if there exist a certain $k \in \mathbb{Z}_{>0}$ such that, $(\hat{A})^k\, \mathcal{W} \subseteq \alpha\mathcal{W}, \forall\alpha = [0, 1)$. The infinite Minkowski sum of sets (17) under strict stability conditions ensures that convergence is guaranteed. Considering the infinite Minkowski sum,

$$\bigoplus_{j=0}^{\infty} (\hat{A})^j \mathcal{W} = \bigoplus_{j=0}^{k-1} (\hat{A})^j \mathcal{W} \oplus \bigoplus_{j=k}^{2k-1} (\hat{A})^j \mathcal{W} \oplus \bigoplus_{j=2k}^{3k-1} (\hat{A})^j \mathcal{W} \oplus ..., \tag{18}$$

Equation (18) can be simplified to achieve the condition $(\hat{A})^k\, \mathcal{W} \subseteq \alpha\mathcal{W}$ as follows:

$$\bigoplus_{j=0}^{\infty} (\hat{A})^j \mathcal{W} = \bigoplus_{j=0}^{k-1} (\hat{A})^j \mathcal{W} \oplus \bigoplus_{j=0}^{k-1} (\hat{A})^j (\hat{A})^k \mathcal{W} \oplus \bigoplus_{j=0}^{k-1} (\hat{A})^j (\hat{A})^{2k} \mathcal{W} \oplus ... \tag{19}$$

From $(\hat{A})^k\mathcal{W} \subseteq \alpha\mathcal{W}$, it can be stated that $(\hat{A})^{nk}\mathcal{W} \subseteq \alpha^n\mathcal{W}$. Taking $\bigoplus_{j=0}^{k-1}(\hat{A})^j\mathcal{W}$ to be equal to Π and considering Eq. (18), $\tilde{X}_*$ can be approximated from a truncation of (18) as:

$$\delta X_f = (1 + \alpha + \alpha^2 + ...)\Pi, \tag{20}$$

which results in an approximated set as given below:

$$\delta X_f \subseteq (1 - \alpha)^{-1}\Pi, \tag{21}$$

the set in a zonotope form, as described by the properties in Sect. 1.1 is given as:

$$\delta X_f \subseteq 0 \oplus (1 - \alpha)^{-1}\Psi_{2,j}B^{n_d}. \tag{22}$$

$$\Psi_{2,j} = \bigoplus_{j=0}^{k-1}(\hat{A})^j\hat{B}_d H_d.$$

Thus, the size of the set is dependent on the design parameter α. The terminal state of the real system, $\tilde{x}_n$ under uncertainty, belongs to an invariant set, $\tilde{X}_f \ \forall\tilde{x}_n \subseteq \tilde{X}_f$. The uncertain terminal state can henceforth be expressed in a decomposed form as:

$$x_n \oplus \delta X_f \subseteq \tilde{X}_f. \tag{23}$$

The constructed sequence of uncertain zonotopic sets will then be used in the sequel for the design of the robust EMPC by considering only alterations in the constraints.

3.3 Robust EMPC Formulation

From Assumption 2, the optimization problem for the economic cost minimization involves the minimization of the centers of the tube, the cost function therefore involves deterministic variables. According to [5], for the nominal MPC synthesis of a DWN, the objective function $\mathcal{L}(k, \hat{u}, x)$, involves three terms:

- To account for variable electricity at the pumps and production costs, the following term is included

$$\mathcal{J}_E(k) = (\alpha_1 + \alpha_2)^T\hat{u}(k), \tag{24}$$

  where α_1 is a fixed cost related to the water production and α_2 is the time varying electricity cost.
- A penalty equal to the sum of the squares of the deviation of the volume in each tank, i from a predefined safety threshold

$$\mathcal{J}_s(k) = \sum_{i=1}^{T} ||\varphi_i(k)||^2, \tag{25}$$

  where φ_i(k) denotes the deviation of the stored volume in tank, i of T tanks below the desired minimum volume.

– A penalty on the square of the flow variations in the actuators, i.e., the slew rate

$$\mathcal{J}_{\triangle U}(k) = ||\triangle \hat{u}(k)||^2, \tag{26}$$

where $\triangle \hat{u}(k) = \hat{u}(k) - \hat{u}(k-1)$.

The resultant objective function is given as follows:

$$\therefore \mathcal{L}(k, \hat{u}, x) = \lambda_1 \mathcal{J}_s(k) + \lambda_2 \mathcal{J}_{\triangle \hat{U}}(k) + \lambda_3 \mathcal{J}_E(k). \tag{27}$$

Where λ_1, λ_2 and λ_3 are design weights for each objective criterion.
The proposed robust approach only involves an update of the constraints considering $\mathbb{U} \oplus \delta U \subseteq \mathcal{U}$, $\mathbb{X} \oplus \delta X \subseteq \mathcal{X}$ and $X_f \oplus \delta X_f \subseteq \tilde{X}_f$

$$\min_{\hat{\mathbf{u}}(k), \mathbf{x}(k)} \sum_{i=0}^{N-1} \mathcal{L}(k, \hat{u}, x) \tag{28a}$$

$$s.t. \quad x_{k+i+1|k} = A x_{k+i|k} + \hat{B} \hat{u}_{k+i|k} + \hat{B}_d d_{K+i|k}, \tag{28b}$$

$$\hat{u}_{k+i|k} \subseteq \mathcal{U} \ominus \delta U, \tag{28c}$$

$$x_{k+i+1|k} \subseteq \mathcal{X} \ominus \delta X, \tag{28d}$$

$$x_{k+N|k} \subseteq X_f, \tag{28e}$$

$$\tilde{x_k} - x_k \subseteq \delta X. \tag{28f}$$

Considering that the optimization problem is feasible, i.e., there exists a non-empty solution given by the optimal sequence of control inputs $(\hat{u}^*(0), \hat{u}^*(1), ...\hat{u}^*(N-1))$, where N is the prediction horizon. From the principles of receding horizon, only the first control action $\hat{u}^*(0|k)$ of the sequence N values obtained from the solution of the MPC optimization problem is applied to the plant.

$$\hat{u}(k) = \hat{u}^*(0|k), \tag{29}$$

disregarding the rest of control actions. At the next time instant k, the optimization problem is solved again using the current measurements of states and disturbances, with the most recent forecast of the latter over the next future horizon. From the control parameterization (4), the control input to the plant at every time instant k is given by

$$u^*(0|k) = \tilde{P}\tilde{M_1}\hat{u}^*(0|k) + \tilde{P}\tilde{M_2}\tilde{d}(k) + K\Delta x(k). \tag{30}$$

Where $\tilde{d}(k)$ is the forecasted demand including uncertainty measured at time instant k.

4 Results

The proposed controller is applied to the aggregate model of the Barcelona drinking water network, which consists of 17 tanks, 61 control variables (26 pumps and 35 valves), 11 nodes and 25 demand points. As discussed in the modeling of the DWN, through control parametrization, the decision variables are reduced to 50 from 61 to ease computational burden and assist in formula ting the uncertainty sets. The weights λ, in the multi-objective problem, (27) are selected such that more priority is placed on the minimization of the economic cost against the other competing objective criteria of maintaining proper levels of safety volumes and control action smoothness. λ_1, λ_2 and λ_3 are chosen as 100, 10 and 0.1 respectively. Real demand data at different demand nodes are subsequently used as the forecasted model disturbance.

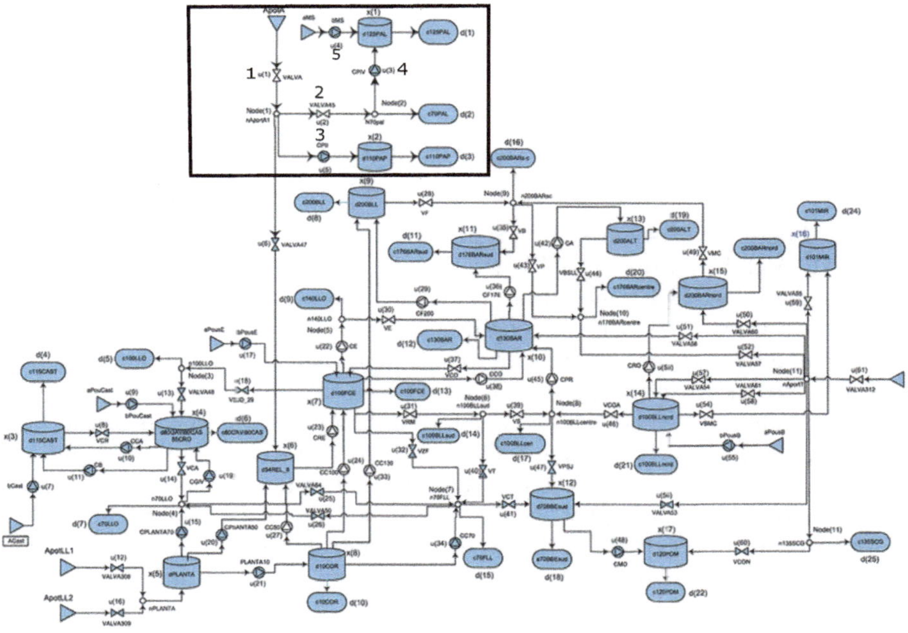

Fig. 1. Aggregate model of the Barcelona water network

4.1 Simulated Results

In the simulations, the additive demand uncertainty is taken as a variation between ±5% of maximum forecasted demand which is illustrated in Fig. 2 for one demand node C70PAL as shown in Fig. 1, from which the corresponding uncertain zonotopic set is formulated. The prediction horizon is 24 h, with a sampling time of 1 h. The MPC optimization problem (28) is solved with CPLEX® QP solver, with Yalmip and Matlab® R2017b (64 bits) using a PC with an Intel core i7 with 8 GB of RAM. To demonstrate the capabilities of the proposed controller, it is compared with a nominal EMPC controller of similar parameters but ignoring the demand uncertainties.

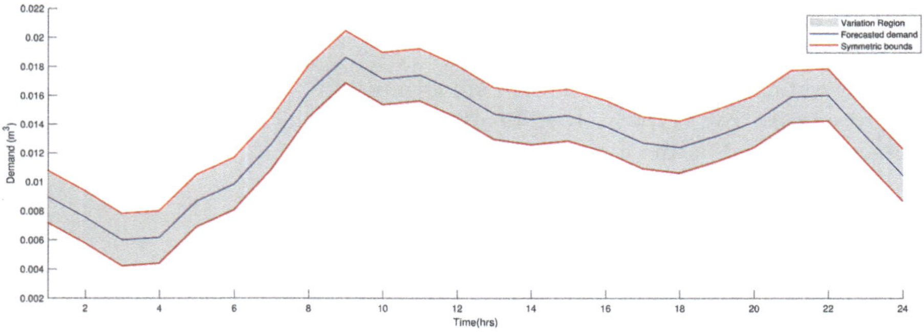

Fig. 2. Bounds on forecasted demand at C70PAL for 24 h

The dynamics of some selected elements, Tanks - (*d54REL*, *d110PAP* & *d125PAL*) and Actuators - (*VALMA*[1], *VALMA45*[2], *CPII*[3], *CPIV*[4] & *BMS*[5]) of the aggregate model with the actuators enumerated in order as listed in Fig. 1 are shown for purposes of comparison (Figs. 3 and 4) between the proposed robust and nominal EMPC controllers. The robust controller is augmented with a local controller to ensure that the deviation between nominal state and perturbed state is minimized, which can be tuned accordingly with Q and R of the LQR controller. From Fig. 3, the robust EMPC enables convergence to a neighbourhood of the nominal state in the presence of demand uncertainties, maintaining an almost similar behaviour of the network tank level as the nominal controller. The main control objective of the predictive controller is to reduce operational cost due to the actuator effort from the linear cost $\mathcal{J}_E(k)$. Summation of consumer demand occurs at some nodes, therefore there is an aggregation of additive uncertainties, Δd on actuators that effect flow when connected to nodes that has multiple demand points propagated to. From Fig. 4, the Robust EMPC shows similar control efforts as the nominal controller, especially with pumps not connected directly to demand points as the magnitude of uncertainty is comparatively less. For actuators connected directly to some nodes, for example, in the case of actuator *VALMA*[1], a marginally higher deviation is realised, this can be attributed to accumulated demand variations at node *NOP18* propagated to the actuator. Deviations are also due to the demand variations assumed measured for control inputs at each sampling time arising from the control parameterization (30). Also infeasible solutions were realised in the case of the nominal EMPC controller under the same uncertainty propagation.

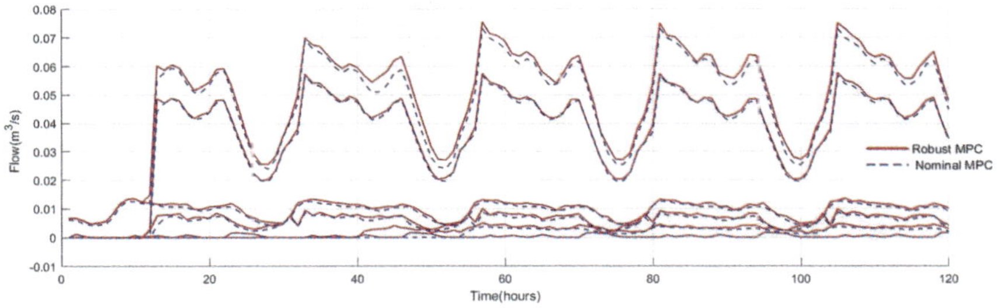

Fig. 3. Comparison of selected actuator dynamics for Robust and Nominal MPC

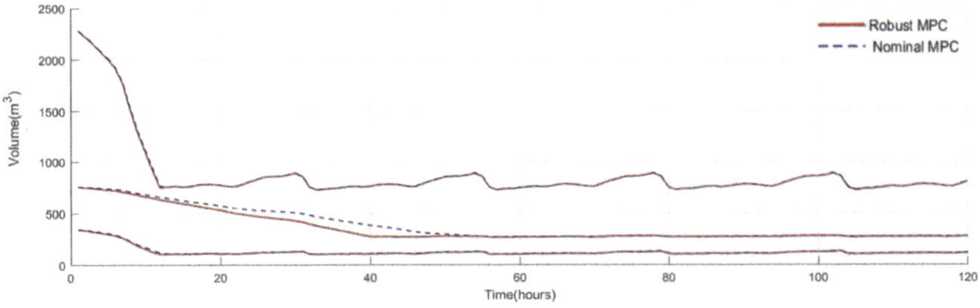

Fig. 4. Level dynamics of tanks for Robust and Nominal MPC

5 Conclusions

This paper seeks to address the problem of model uncertainty associated with the design of MPC, specifically in relation to demand uncertainties in a drinking water network. Uncertainties on demand are considered unknown but bounded in a zonotope. Then, a robust EMPC is designed based on this description, which shows similar behaviour when compared to a nominal EMPC without demand uncertainties satisfying intended control objectives and most importantly the robust EMPC preserves stability and feasibility. The optimization problem of the nominal EMPC is observed to be infeasible when subjected to the same magnitude of variation as the robust EMPC. As a future work, it is planned to compared the proposed approach with the chance-constraints approach introduced in [10] on DWNs.

References

1. Grosso, J., Ocampo-Martinez, C., Puig, V., Limon, D., Pereira, M.: Economic MPC for the management of drinking water networks. In: 2014 European Control Conference (ECC), pp. 2739–2744 (2014)
2. Pereira, M., Muñoz, D.P., Limon, D.: Robust economic model predictive control of a community micro-grid. In: 2016 IEEE 55th Conference on Decision and Control (CDC), pp. 2739–2744 (2017)

3. Le, V., Stoica, M., Alamo, T.: Zonotopes; From Guaranteed State Estimation to Control. Wiley, Hoboken (2013)
4. Cembrano, G., Quevedo, J., Puig, V., Pérez, R., Figueras, J., Escaler, I., Ramón, G., Barnet, G., Rodríguez, P., Casas, M.: A generic tool for real-time operational predictive optimal control of water networks. J. Int. Assoc. Water Pollut. Res. **64**(2), 448–459 (2011)
5. Wang, Y., Puig, V., Cembrano, G.: Non-linear economic model predictive control of water distribution networks. J. Process Control **55**(6), 23–34 (2017)
6. Bemporad, A., Morari, M.: Robust model predictive control: a survey. In: Garulli, A., Tesi, A. (eds.) Robustness in Identification and Control, vol. 245, pp. 207–226. Springer, London (1999)
7. Puig, V., Escobet, T., Sarrate, R., Quevedo, J.: Fault diagnosis and fault tolerant control in critical infrastructure systems. In: Kyriakides, E., Polycarpou, M. (eds.) Intelligent Monitoring, Control, and Security of Critical Infrastructure Systems. Studies in Computational Intelligence, vol. 565, pp. 263–299. Springer, Heidelberg (2010)
8. Velarde, P., Maestre, J.M., Ocampo-Martinez, C., Bordons, C.: Application of robust model predictive control to a renewable hydrogen-based microgrid. In: 2016 European Control Conference (ECC), pp. 1209–1214 (2016)
9. Löfberg, J.: Min-max approaches to robust model predictive control. Linköping Studies in Science and Technology. Dissertations (2003)
10. Grosso, J.M., Velarde, P., Ocampo-Martinez, C., Maestre, J.M., Puig, V.: Stochastic model predictive control approaches applied to drinking water networks. Optim. Control Appl. Methods **38**(4), 541–558 (2017)
11. Jain, T., Yamé, J.J.: IEEE Trans. Sustain. Energy **4**(4), 1696–1704 (2019)
12. Durand, H., Christofides, P.D.: Economic model predictive control for nonlinear processes incorporating actuator magnitude and rate of change constraints. In: 2016 American Control Conference (ACC), pp. 5068–5074 (2016)
13. Angeli, D., Amrit, R., Rawlings, J.B.: On average performance and stability of economic model predictive control. IEEE Trans. Autom. Control **57**(7), 1615–1626 (2012)
14. Wang, Y., Puig, V., Cembrano, G.: Non-linear economic model predictive control of water distribution networks. J. Process Control **56**(8), 23–34 (2017)
15. Wang, Y., Puig, V., Cembrano, G.: Economic model predictive control with nonlinear constraint relaxation for the operational management of water distribution networks. Energies **11**(4), 991–999 (2018)

Fault Detection and Isolation of Distributed Inverter-Based Microgrids

Horst Schulte(✉) and Alexander Pascal Cesarz

Department of Engineering I, Control Engineering Group,
University of Applied Sciences Berlin (HTW), Berlin, Germany
schulte@htw-berlin.de
http://home.htw-berlin.de/~schulte/

Abstract. This paper deals with a fault detection and isolation strategy for microgrids fed by voltage source inverters. To increase the reliability of microgrids, a model-based FDI concept for the detection of line and inverter faults is presented. First, a scalable overall state space model is presented to describe the power flow of a generic micogrid with an arbitrary number of droop controlled inverters. For this purpose, aggregated single inverter models are combined into a multi-inverter model, which is connected to a network of arbitrary topology. It is shown how this model serves as a basis to develop an observer-based FTC algorithm. Finally, simulation studies are carried out for a system with three droop controlled inverters, six network nodes and two load sinks. It will be shown how single network line faults and inverter faults can be detected and partially isolated with the centralized approach.

Keywords: FDI · FTC · Power systems · Micogrid

1 Introduction

To facilitate the integration of distributed generators (DGs) with renewable energy sources (RES), e.g. photovoltaics and wind turbines into the distribution grid, the concept of microgrid is proposed in [1]. A microgrid is a locally controllable subsystem of a larger electrical grid. It consists of several distributed generators (DG), loads and contains optionally storage devices. One main feature of a microgrid is that it can be operated either in (1) islanded mode, i.e. in a completely isolated manner from the main transmission system or (2) in grid-connected mode. Usually RES are connected via frequency inverters to the microgrid, that are operated either as current sources inverters or voltage source inverters (VSI). For current source operation it is assumed that the voltage curve is specified by the grid. The usual procedure of current control is done by setting the inverter voltage relative to the grid voltage [2–4]. This inverter operation is therefore also referred as grid following operation. However, with this method it is obviously not possible to form an AC network, since a sinusoidal grid voltage is assumed at the terminal of the converter. On the other hand, the operation

A. Bartoszewicz et al. (Eds.): PCC 2020, AISC 1196, pp. 1483–1495, 2020.
https://doi.org/10.1007/978-3-030-50936-1_123

VSI inverters does not require a given grid voltage. As a result, it is possible to form a three-phase microgrid that can be operated in islanded mode.

A well known decentralized method with which VSI are operated communication-less in microgrids is based on the droop-control concept. The fundamental principle is that the frequency of the output voltage is reduced proportionally to the active power (P-ω droop) and the voltage amplitude is reduced proportionally to the reactive power (Q-V droop). By drooping the voltage frequency and amplitude as described above, load sharing between DGs can be performed in an autonomic manner, which is similar to the power sharing between parallel synchronous generators (SGs) [5,6].

The investigation of the dynamics of microgrids with droop controlled inverters is mainly based on numerical simulations. Only a few papers deal with analytical modeling. A model with two droop controlled VSI is presented in [7] and [8]. In order to investigate the stability of practice-relevant microgrids with a high DG density, a graph-based method for modeling the passive grid structure and the modeling of inverters with droop control is proposed in [9]. Based on this model, it could be shown that an increasing number of inverters and short grid lines caused a poorly damped power flow dynamics. Further it was shown that changes of line parameters caused by short circuits or line breaks and change of inverter parameters, e.g. caused by sensor or actuator faults, can lead to instability of the entire microgrid [10].

In this paper, in order to be able to detect and isolate line and inverter faults systematically, a centralized fault detection and isolation (FDI) strategy based on the grid-multi-inverter model [10] is proposed. It is shown how this model is used for an observer-based FDI design that serves as a basis to develop a fault tolerant control (FTC) scheme. For conventional multimachine power systems with large synchronous generators (without inverter-interfaced DGs) or conventional grid following inverters FDI procedures have already been established: In [11] a distributed sensor FDI scheme is presented. Here it is assumed that each generator is interconnected with other generators through a transmission network. In [12] Wu, et al. propose an active fault detection and isolation scheme for islanding faults for distributed generation systems based on grid following current source inverters. It is a distributed scheme, in which each generating unit in the grid introduces the active detection signals and detects and isolates the faults in the grid, which is obtained using a Set-Membership filter. The scheme is evaluated with a simulation study. A more general approach without direct reference to power systems is pursued in [13]. In this paper, the problem of distributed FDI for leader–follower multi-agent systems with disturbances has been considered. For FDI purpose, an observer is designed which can estimate the overall follower's state of the leader–follower multi-model systems.

The paper is organized as follows: In Sect. 2 the considered micro grid structure and the state space model description of an arbitrary grid structure and a multi-inverter model consisting of droop controlled VSI is proposed. Based on these two models, a closed loop model is derived in Sect. 3. It contains a centralized, aggregated model for all controller and plant parameters such as droop coefficients, filter coefficients, and grid parameters. In Sect. 4 the observer design approach based on the centralized closed loop model is proposed. Finally, Sect. 5 shows the applicability of the method using an example consisting of a microgrid with three drooped controlled inverters, six network nodes and two load sinks.

2 Microgrid Structure and State Space Model Description

2.1 Grid Structure

An illustrating example of a microgrid structure is shown in Fig. 2. The network contains 7 lines, 6 nodes and three voltage source inverters. The lines are characterized by 7 admittance matrices $\mathbf{Y}_k$, $k = 1, \ldots, 7$ as a function of the line resistance R_k, line induction L_k and grid frequency ω

$$\mathbf{Y}_k = \mathbf{Y}_k(R_k, L_k, \omega) \ .$$

The VSI are directly connected without any losses to the nodes of the first column, where the output power flows are denote as $\mathbf{s}_{VSI,i}$, $i = 1, 2, 3$. The flows in the lines are denote as $\mathbf{s}_i, i = 1, \ldots, 7$ and the load flows are $\mathbf{s}_{L,i}, i = 1, 2$.

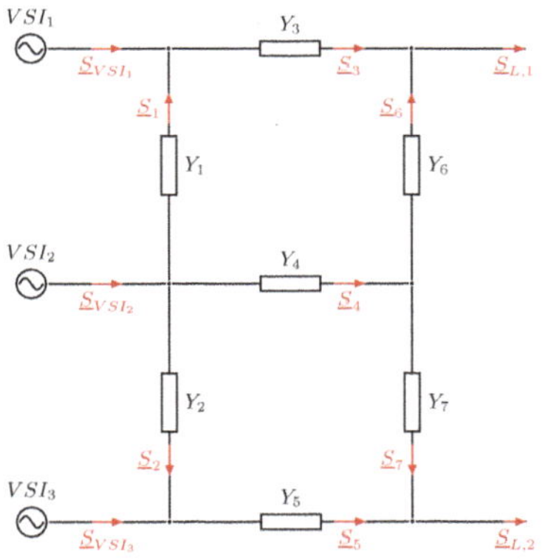

Fig. 1. Microgrid with distributed generators in the passive network

2.2 Single VSI Model

The state space model of a single droop-controlled voltage source inverter (VSI) [10] is given as follows

$$\begin{aligned} \dot{\mathbf{x}}_i &= \mathbf{A}_i \, \mathbf{x}_i + \mathbf{B}_{s,i} \, \mathbf{s}_{VSI,i} + \mathbf{B}_{f,i} \, \Delta\varphi_{POC,i} \ , \\ \mathbf{y}_i &= \begin{pmatrix} \mathbf{C}_{\mu,i} \\ \mathbf{C}_{f,i} \end{pmatrix} \mathbf{x}_i \ , \end{aligned} \tag{1}$$

where

$$\mathbf{A}_i = \begin{pmatrix} -1/\tau_{m,i} & 0 & 0 & 0 & 0 \\ 1/\tau_{VSI,i} & -1/\tau_{VSI,i} & 0 & 0 & 0 \\ 0 & 0 & -1/\tau_{m,i} & 0 & 0 \\ 0 & 0 & 1/\tau_{VSI,i} & -1/\tau_{VSI,i} & 0 \\ 0 & k_{P,i} & 0 & 0 & 0 \end{pmatrix},$$

$$\mathbf{B}_{s,i} = \begin{pmatrix} 1/\tau_{m,i} & 0 \\ 0 & 0 \\ 0 & 1/\tau_{m,i} \\ 0 & 0 \\ 0 & 0 \end{pmatrix}, \quad \mathbf{C}_{\mu,i} = \begin{pmatrix} 0 & -k_{D,i} & 0 & 0 & 1 \\ 0 & 0 & 0 & k_{Q,i} & 1 \end{pmatrix},$$

$$\mathbf{C}_{f,i} = \begin{pmatrix} 0 & k_{Q,i} & 0 & 0 & 0 \end{pmatrix}$$

with the state vector $\mathbf{x}_i = \left(P_{m,i}, P_{VSI,i}\,, Q_{m,i}\,, Q_{VSI,i}\,, \theta_i\right)^T$, the input vector $\mathbf{s}_{VSI,i} = \left(P_i\,, Q_i\right)^T$, and the third input as the difference between the grid frequency $\dot{\varphi}_{POC,i} = f_{POC,i}$ at the i'th point of connection (POC) and the nominal frequency $f_0 = \{50, 60\}$Hz with

$$\Delta\dot{\varphi}_{POC,i} = \dot{\varphi}_{POC,i} - f_0\,. \tag{2}$$

It should be noted that all active and reactive power variables P and Q in (1) are given as scaled per-unit variables (p.u.). The output vector is

$$\mathbf{y}_i = \big(\underbrace{\delta_{VSI,i}\,,\ \Delta v_{VSI,i}}_{=:\boldsymbol{\mu}^T_{VSI,i}}\,,\ \Delta\varphi_{VSI,i}\big)^T\,, \tag{3}$$

where δ_i denotes the voltage phase angle and Δv_i denotes the difference of voltage amplitude at the inverter terminal

$$\Delta v_{VSI,i} = v_{VSI,i} - v_0\,. \tag{4}$$

with the rated voltage v_0. For reasons of clarity $\delta_{VSI,i}$ and $\Delta v_{VSI,i}$ are combined in

$$\boldsymbol{\mu}_{VSI,i} = \left(\delta_{VSI,i}\,,\ \Delta v_{VSI,i}\right)^T, \tag{5}$$

compare (3). The third output $\Delta\varphi_{VSI,i}$ denotes the difference between the internal inverter frequency $\varphi_{VSI,i}$ and f_0 as nominal frequency

$$\Delta\dot{\varphi}_{VSI,i} = \dot{\varphi}_{VSI,i} - f_0\,. \tag{6}$$

Remarkable is that the frequency $\dot{\varphi}_{VSI,i}$ associated with the ith converter takes into account that in a purely inverter driven power grid each inverter adapts its injection frequency with high dynamics. As a consequence there is no common grid frequency during transient phases, rather in each point of connection, a different frequency $\dot{\varphi}_{POC,i}$ is possible. These will be represented by the multi-inverter model proposed in the next subsection. The state vector $\mathbf{x}_i$ contains the filtered active power measurement $P_{m,i}$, the reactive power measurement $Q_{m,i}$, and the drift angle θ caused by the integrated difference between the internal inverter frequency and grid frequency. A second first order filter with τ_{VSI} as a design parameter represents the dynamics of $P_{VSI,i}$ and $Q_{VSI,i}$, which are also included in $\mathbf{x}_i$. The input vector $\mathbf{s}_i$ contains the measured active P_i and reactive power Q_i at the inverter terminal. The VSI model coefficients in (1) are filter/design parameters: $\tau_{m,i}$ denotes the time constant of the first order measurement filter with P_i and Q_i as input signals, $\tau_{VSI,i}$ is a design parameter that represents an virtual converter inertia. The so-called droop coefficients are $k_{P,i}$ and $k_{Q,i}$, where $k_{P,i}$ specifies the active power-dependent frequency in percent related to the nominal frequency f_0 and $k_{Q,i}$ specifies the reactive power-dependent voltage related to the nominal voltage v_0. Additionally to the power dependent frequency caused by $k_{P,i} \cdot P_{VSI,i}$ a power dependent phase angle bias is added by $k_{D,i} \cdot P_{VSI,i}$ in the output equation of (1) by the first row of $\mathbf{C}_{\mu,i}$.

2.3 Multi VSI Model

We assume that the multigrid is formed from N_v single inverters that are numbered from $i = 1$ to N_v. In order to capture all active distributed generators in the grid all single VSI inverter models are block-diagonally combined into a large dynamic system with the augmented state vector

$$\tilde{\mathbf{x}} = \left(\mathbf{x}_1^T\,,\, \mathbf{x}_2^T\,,\, \cdots\,,\, \mathbf{x}_{N_v-1}^T\,,\, \mathbf{x}_{N_v}^T\right)^T\,, \tag{7}$$

the system matrix

$$\tilde{\mathbf{A}} = \operatorname{diag}\left(\mathbf{A}_1, \ldots, \mathbf{A}_{N_v}\right) \in \mathbb{R}^{5N_v \times 5N_v}\,, \tag{8}$$

the input matrices

$$\begin{aligned} \tilde{\mathbf{B}}_s &= \operatorname{diag}\left(\mathbf{B}_{s,1}, \ldots, \mathbf{B}_{s,N_v}\right) \in \mathbb{R}^{5N_v \times 2N_v}\,, \\ \tilde{\mathbf{B}}_f &= \operatorname{diag}\left(\mathbf{B}_{f,1}, \ldots, \mathbf{B}_{f,N_v}\right) \in \mathbb{R}^{5N_v \times N_v}\,, \end{aligned} \tag{9}$$

and output matrices

$$\begin{aligned} \tilde{\mathbf{C}}_\mu &= \operatorname{diag}\left(\mathbf{C}_{\mu,1}, \ldots, \mathbf{C}_{\mu,N_v}\right) \in \mathbb{R}^{2N_v \times 5N_v}\,, \\ \tilde{\mathbf{C}}_f &= \operatorname{diag}\left(\mathbf{C}_{f,1}, \ldots, \mathbf{C}_{f,N_v}\right) \in \mathbb{R}^{N_v \times 5N_v}\,. \end{aligned} \tag{10}$$

Using (7), (8), (9), and (10) a centralized state-space form of the multi VSI model is established with

$$\dot{\tilde{\mathbf{x}}} = \tilde{\mathbf{A}}\,\tilde{\mathbf{x}} + \tilde{\mathbf{B}}_s\,\tilde{\mathbf{s}}_{VSI} + \tilde{\mathbf{B}}_f\,\Delta\dot{\tilde{\varphi}}_{POC} \tag{11}$$

$$\tilde{\mathbf{y}} = \begin{pmatrix} \tilde{\mathbf{C}}_\mu \\ \tilde{\mathbf{C}}_f \end{pmatrix} \tilde{\mathbf{x}}\,, \tag{12}$$

where

$$\tilde{\mathbf{s}}_{VSI} = \begin{pmatrix} \mathbf{s}_{VSI,1} \\ \vdots \\ \mathbf{s}_{VSI,N_v} \end{pmatrix} \in \mathbb{R}^{2N_v} , \quad \Delta\dot{\tilde{\varphi}}_{POC} = \begin{pmatrix} \Delta\dot{\varphi}_{POC,1} \\ \vdots \\ \Delta\dot{\varphi}_{POC,N_v} \end{pmatrix} \in \mathbb{R}^{N_v} ,$$

$$\tilde{\boldsymbol{\mu}}_{VSI} = \begin{pmatrix} \mu_{VSI,1} \\ \vdots \\ \mu_{VSI,N_v} \end{pmatrix} \in \mathbb{R}^{2N_v} , \quad \Delta\dot{\tilde{\varphi}}_{VSI} = \begin{pmatrix} \Delta\dot{\varphi}_{VSI,1} \\ \vdots \\ \Delta\dot{\varphi}_{VSI,N_v} \end{pmatrix} \in \mathbb{R}^{N_v} .$$

For reasons of clarity and for the next step (12) is split into two output equations

$$\tilde{\boldsymbol{\mu}}_{VSI} = \tilde{\mathbf{C}}_{\mu}\, \tilde{\mathbf{x}} , \qquad \Delta\dot{\tilde{\varphi}}_{VSI} = \tilde{\mathbf{C}}_f\, \tilde{\mathbf{x}} . \tag{13}$$

The inverter phase change $\Delta\dot{\tilde{\varphi}}_{VSI}$ as the frequency at VSI terminals causes a deviation of the phases (frequency) at all POCs

$$\Delta\dot{\tilde{\varphi}}_{POC} = \mathbf{M}_f\, \Delta\dot{\tilde{\varphi}}_{VSI} , \tag{14}$$

where $\mathbf{M}_f \in \mathbb{R}^{N_v \times N_v}$ denotes the coupling matrix with the property $\mathbf{M}_f = \mathbf{M}_f^T$. In [9,10] two different concepts for the determination of $\mathbf{M}_f$ are discussed:

1. All frequencies at the POCs are represented by a single central grid frequency. Hence, $\mathbf{M}_{f,c}$ is uniform and denotes the central frequency coupling matrix.
2. The N_v frequencies at the POCs are different where the rate of change in the grid is different caused by varying line resistance between the inverters. Hence, $\mathbf{M}_{f,c}$ is not uniform.

By considering the coupling (14) we get with (13)

$$\Delta\dot{\tilde{\varphi}}_{POC} = \mathbf{M}_f\, \tilde{\mathbf{C}}_f\, \tilde{\mathbf{x}} \quad . \tag{15}$$

Finally, inserting (15) in (11) gives an aggregated multi VSI model in state space form

$$\begin{aligned} \dot{\tilde{\mathbf{x}}} &= (\tilde{\mathbf{A}} + \tilde{\mathbf{B}}_f\, \mathbf{M}_f\, \tilde{\mathbf{C}}_f)\tilde{\mathbf{x}} + \tilde{\mathbf{B}}_s\, \tilde{\mathbf{s}}_{VSI} , \\ \tilde{\boldsymbol{\mu}}_{VSI} &= \tilde{\mathbf{C}}_{\mu}\, \tilde{\mathbf{x}} \quad . \end{aligned} \tag{16}$$

2.4 Grid Model

In order to integrate the aggregated multi VSI model (16) into the complete grid structure, the so-called incidence matrix $\mathscr{K}$ is formulated. It contains the network information as representation of a directed network graph. The incidence matrix $\mathscr{K} \in \mathbb{R}^{2N_k \times 2N_l}$ defines which of the $n = 1, \ldots, N_k$ nodes are connected through the $k = 1, \ldots, N_l$ lines, whilst indicating the direction of positive power flow. Since the power flow is divided into reactive and active power the number of rows and columns of $\mathcal{K}$ double compared to the number of lines and nodes. For clarification, the $\mathcal{K}$ matrix is given for the example with $N_l = 7$ lines and $N_k = 6$ nodes shown in Fig. 2:

$$\mathcal{K} = \begin{pmatrix} \mathbf{I}_+ & \mathbf{0} & \mathbf{I}_- & \mathbf{0} & \mathbf{0} & \mathbf{0} & \mathbf{0} \\ \mathbf{I}_- & \mathbf{I}_- & \mathbf{0} & \mathbf{I}_- & \mathbf{0} & \mathbf{0} & \mathbf{0} \\ \mathbf{0} & \mathbf{I}_+ & \mathbf{0} & \mathbf{0} & \mathbf{I}_- & \mathbf{0} & \mathbf{0} \\ \mathbf{0} & \mathbf{0} & \mathbf{0} & \mathbf{I}_+ & \mathbf{0} & \mathbf{I}_- & \mathbf{I}_- \\ \mathbf{0} & \mathbf{0} & \mathbf{I}_+ & \mathbf{0} & \mathbf{0} & \mathbf{I}_+ & \mathbf{0} \\ \mathbf{0} & \mathbf{0} & \mathbf{0} & \mathbf{0} & \mathbf{I}+ & \mathbf{0} & \mathbf{I}_+ \end{pmatrix} \tag{17}$$

with

$$\mathbf{I}_+ = \begin{pmatrix} 1 & 0 \\ 0 & 1 \end{pmatrix}, \quad \mathbf{I}_- = \begin{pmatrix} -1 & 0 \\ 0 & -1 \end{pmatrix}, \quad \mathbf{0} = \begin{pmatrix} 0 & 0 \\ 0 & 0 \end{pmatrix}.$$

In addition to the network topology, the power flow in each kth line between node i and j is quantified by the linear approximation

$$\begin{pmatrix} P_k \\ Q_k \end{pmatrix} = \underbrace{\begin{pmatrix} -B_k & G_k \\ -G_k & -B_k \end{pmatrix}}_{\mathbf{Y}_k} \left[\underbrace{\begin{pmatrix} \delta_i \\ v_i \end{pmatrix}}_{\boldsymbol{\mu}_i} - \underbrace{\begin{pmatrix} \delta_j \\ v_j \end{pmatrix}}_{\boldsymbol{\mu}_j} \right], \tag{18}$$

where $\mathbf{Y}_k$ denotes the admittance matrix for the kth line with

$$G_k = \frac{R_k}{R^2 + (\omega L_k)^2} \qquad B_k = -\frac{\omega L_k}{R^2 + (\omega L_k)^2}, \tag{19}$$

where R_k denotes the line resistance and L_k the line induction. The power flow Eq. (18) is used under the three assumptions:

1. small angle difference holds whereby

$$\sin(\delta_i - \delta_j) \approx \delta_i - \delta_j, \quad \cos(\delta_i - \delta_j) \approx 1$$

2. small deviations of the line voltage amplitude from the nominal voltage.
3. frequency ω does not change in the grid. Thus G_k and B_k (19) are constant model parameters. The effect of significant frequency change on the FDI strategy will be investigated in a future work.

This allows to calculate the active and reactive power flow in all $k = 1, \ldots, N_l$ lines

$$\mathbf{s} = -\tilde{\mathbf{Y}}\, \mathcal{K}^T\, \tilde{\boldsymbol{\mu}} \tag{20}$$

with $\tilde{\boldsymbol{\mu}} = \left(\tilde{\boldsymbol{\mu}}_{VSI}^T\ \boldsymbol{\mu}_L^T\right)^T$, $\mathbf{s} = \left(\mathbf{s}_1^T, \ldots, \mathbf{s}_{N_l}^T\right)^T$, the power-voltage line equation of a single line $\mathbf{s}_k = \mathbf{Y}_k(\boldsymbol{\mu}_i - \boldsymbol{\mu}_j)$, the incidence matrix $\mathcal{K}$ and admittance matrix of the entire grid

$$\tilde{\mathbf{Y}} = \mathrm{diag}(\mathbf{Y}_1, \ldots, \mathbf{Y}_{N_k}) \in \mathbb{R}^{2N_l \times 2N_l}, \tag{21}$$

where $\boldsymbol{\mu}_L^T$ denotes the voltage phase and amplitude at all load nodes (cmp. Fig. 2). According to (20), the power flow in a microgrid of arbitrary structure which is represented by $\tilde{\mathbf{Y}}$ and $\mathscr{K}^T$ is caused by the voltages at all VSI and load nodes. The usage of Kirchhoff's law on the power flows results in

$$\mathbf{0}_{2N_k \times 1} = \mathscr{K}\mathbf{s} + \begin{pmatrix} \tilde{\mathbf{s}}_{VSI} \\ \mathbf{s}_L \end{pmatrix} , \tag{22}$$

where $\mathbf{s}_L$ denotes the power flow into the loads (cmp. Fig. 2). For further consideration, the incidence matrix $\mathscr{K}$ is divided into the sub-matrix[1] of the VSI side $\mathscr{K}_{VSI}$ and the load side $\mathscr{K}_L$

$$\mathscr{K} = \begin{pmatrix} \mathscr{K}_{VSI} \\ \mathscr{K}_L \end{pmatrix} , \quad \mathscr{K}^T = \begin{pmatrix} \mathscr{K}_{VSI}^T & \mathscr{K}_L^T \end{pmatrix} . \tag{23}$$

When $\mathbf{s}$ in (22) is replaced by (20), we get with (23) the compact form

$$\begin{pmatrix} \tilde{\mathbf{s}}_{VSI} \\ \mathbf{s}_L \end{pmatrix} = \underbrace{\begin{pmatrix} \boldsymbol{\Upsilon}_A & \boldsymbol{\Upsilon}_B \\ \boldsymbol{\Upsilon}_C & \boldsymbol{\Upsilon}_D \end{pmatrix}}_{\boldsymbol{\Upsilon}} \begin{pmatrix} \tilde{\boldsymbol{\mu}}_{VSI} \\ \boldsymbol{\mu}_L \end{pmatrix} \tag{24}$$

with

$$\Upsilon_A = \mathscr{K}_{VSI}\mathbf{Y}\mathscr{K}_{VSI}^T \quad , \qquad \Upsilon_B = \mathscr{K}_{VSI}\mathbf{Y}\mathscr{K}_L^T ,$$
$$\Upsilon_C = \mathscr{K}_L\mathbf{Y}\mathscr{K}_{VSI}^T \quad , \qquad \Upsilon_D = \mathscr{K}_L\mathbf{Y}\mathscr{K}_L^T .$$

From a system's perspective, (24) is not yet suitable as a network model. The right side should contain independent and the left side dependent variables. This is not the case here, because the load flow $\mathbf{s}_L$ is an external independent variable and is therefore on the wrong side in (24). On the other hand, the voltage $\boldsymbol{\mu}_L$ at the load nodes (magnitude and phase angle) are determined by the voltage of the VSIs $\tilde{\boldsymbol{\mu}}_{VSI}$ and the load flow $\mathbf{s}_L$. Hence, we need a network model of the form

$$\begin{pmatrix} \tilde{\mathbf{s}}_{VSI} \\ \boldsymbol{\mu}_L \end{pmatrix} = \begin{pmatrix} \mathbf{M}_A & \mathbf{M}_B \\ \mathbf{M}_C & \mathbf{M}_D \end{pmatrix} \begin{pmatrix} \tilde{\boldsymbol{\mu}}_{VSI} \\ \mathbf{s}_L \end{pmatrix} \tag{25}$$

where[2]

$$\begin{aligned}
\mathbf{M}_A &= \boldsymbol{\Upsilon}_A + \boldsymbol{\Upsilon}_B \left(\mathbf{I}_{N_l} - \mathbf{Z}_C\boldsymbol{\Upsilon}_B\right)^{-1} \mathbf{Z}_C\boldsymbol{\Upsilon}_A \\
\mathbf{M}_B &= \boldsymbol{\Upsilon}_B \left(\mathbf{I}_{N_l} - \mathbf{Z}_C\boldsymbol{\Upsilon}_B\right)^{-1} \mathbf{Z}_D \\
\mathbf{M}_C &= \left(\mathbf{I}_{N_l} - \mathbf{Z}_C\boldsymbol{\Upsilon}_B\right)^{-1} \mathbf{Z}_C\boldsymbol{\Upsilon}_A \\
\mathbf{M}_D &= \left(\mathbf{I}_{N_l} - \mathbf{Z}_C\boldsymbol{\Upsilon}_B\right)^{-1} \mathbf{Z}_D
\end{aligned} \tag{26}$$

[1] Note, this explicit mapping (VSI side/load side) is only possible if no intermediate nodes are allowed in the network.

[2] The inverse of $\boldsymbol{\Upsilon}$ in (24) can be calculated by the Schur complement, that results in $\mathbf{M}_{A,B,C,D}$.

with the identity matrix $\mathbf{I}_{N_l} \in \mathbb{R}^{N_l \times N_l}$ and

$$\mathbf{Z}_C = -\boldsymbol{\Upsilon}_D^{-1}\,\boldsymbol{\Upsilon}_C\,\mathbf{S}^{-1}$$
$$\mathbf{Z}_D = \boldsymbol{\Upsilon}_D^{-1} + \boldsymbol{\Upsilon}_D^{-1}\,\boldsymbol{\Upsilon}_C\,\mathbf{S}^{-1}\,\boldsymbol{\Upsilon}_B\boldsymbol{\Upsilon}_D^{-1}$$

the Schur complement

$$\mathbf{S} = \Upsilon_A - \Upsilon_B\Upsilon_D^{-1}\Upsilon_C.$$

With (25) and the corresponding matrices (26) the quasi stationary behavior of the network is completely described, which results in

$$\mathbf{s} = \left(\mathbf{Y}\mathscr{K}_{VSI}^T\;\mathbf{Y}\mathscr{K}_L^T\right)\begin{pmatrix}\tilde{\boldsymbol{\mu}}_{VSI}\\ \boldsymbol{\mu}_L\end{pmatrix}. \tag{27}$$

Note the grid model (25) is a static system without any dynamics. The dynamics of the centralized closed loop system presented in the following section are based entirely on the dynamics of the multi inverter model (16). Nevertheless, the dynamics can be significantly changed by the feedback in the network.

3 Centralized Nominal Closed Loop System

In this section, the centralized closed loop system consisting of the multi VSI and grid model is derived.

The network model (25) in decomposed form is given by

$$\tilde{\mathbf{s}}_{VSI} = \mathbf{M}_A\,\tilde{\boldsymbol{\mu}}_{VSI} + \mathbf{M}_B\,\mathbf{s}_L \tag{28}$$
$$\boldsymbol{\mu}_L = \mathbf{M}_C\,\tilde{\boldsymbol{\mu}}_{VSI} + \mathbf{M}_D\,\mathbf{s}_L \tag{29}$$

and the multi VSI model (16) in compact form consists of

$$\dot{\tilde{\mathbf{x}}} = \tilde{\mathbf{A}}'\tilde{\mathbf{x}} + \tilde{\mathbf{B}}_s\,\tilde{\mathbf{s}}_{VSI}\,, \tag{30}$$
$$\tilde{\boldsymbol{\mu}}_{VSI} = \tilde{\mathbf{C}}_\mu\,\tilde{\mathbf{x}} \tag{31}$$

with $\tilde{\mathbf{A}}' = \tilde{\mathbf{A}} + \tilde{\mathbf{B}}_f\,\mathbf{M}_f\,\tilde{\mathbf{C}}_f$. Inserting the network Eq. (28) into (30) yields

$$\dot{\tilde{\mathbf{x}}} = \tilde{\mathbf{A}}'\tilde{\mathbf{x}} + \tilde{\mathbf{B}}_s\left(\mathbf{M}_A\,\underbrace{\tilde{\boldsymbol{\mu}}_{VSI}}_{(31)} + \mathbf{M}_B\,\mathbf{s}_L\right), \tag{32}$$

and replacing $\tilde{\boldsymbol{\mu}}_{VSI}$ by (31) we get the differential equation of the centralized closed loop system

$$\dot{\tilde{\mathbf{x}}} = (\tilde{\mathbf{A}}' + \tilde{\mathbf{B}}_s\,\mathbf{M}_A\,\tilde{\mathbf{C}}_\mu)\tilde{\mathbf{x}} + \tilde{\mathbf{B}}_s\,\mathbf{M}_B\,\mathbf{s}_L\,. \tag{33}$$

The output equation follows from (29) and (31)

$$\boldsymbol{\mu}_L = \mathbf{M}_C\,\tilde{\mathbf{C}}_\mu\,\tilde{\mathbf{x}} + \mathbf{M}_D\,\mathbf{s}_L \tag{34}$$

Hence, (33) and (34) constitute the entire multi-grid system.

4 Observer-Based Residual Generation

For the detection and isolation of grid-line and VSI faults, the observer-based method of residual generation is used. The general approach will now be discussed.

4.1 Observer Structure

In this work, an observer bank with a following residual generation is used for fault isolation. To isolate N_f faults from each other, $j = 1, \ldots, N_f + 1$ observers must be designed. For each fault f_i, $i = 1, \ldots, N_f$ to be detected, an associated state space model is set up with which an observer is designed.

Structural investigations of the generic microgrid model (33), (34) has shown that the central state space model for describing the power dynamics is not fully observable. For this reason, the system is decomposed into an observable and non-observable subsystem. Using an equivalence transformation

$$(\,\mathbf{x}_{no}^T \quad \mathbf{x}_o^T\,)^T = \mathbf{T}\,\tilde{\mathbf{x}}\,, \qquad \mathbf{T} \in \mathbb{R}^{n \times n} \tag{35}$$

from the system (33), (34) we obtain

$$\begin{pmatrix} \dot{\mathbf{x}}_{no} \\ \dot{\mathbf{x}}_o \end{pmatrix} = \begin{pmatrix} \mathbf{A}_{no} & \mathbf{A}_{12} \\ \mathbf{0} & \mathbf{A}_o \end{pmatrix} \begin{pmatrix} \mathbf{x}_{no} \\ \mathbf{x}_o \end{pmatrix} + \begin{pmatrix} \mathbf{B}_{no} \\ \mathbf{B}_o \end{pmatrix} \mathbf{s}_L\,, \quad \boldsymbol{\mu}_L = (\mathbf{0}\ \mathbf{C}_0) \begin{pmatrix} \mathbf{x}_{no} \\ \mathbf{x}_o \end{pmatrix} + \mathbf{M}_D\,\mathbf{s}_L\,, \tag{36}$$

where $\mathbf{A}_{no}$, $\mathbf{B}_{no}$ denote the matrices of the unobservable subspace and $\mathbf{A}_o$, $\mathbf{B}_o$, $\mathbf{C}_o$ are related to the observable subspace. For the observable subspace an observer-bank can now be designed as follows

$$\begin{aligned} \dot{\hat{\mathbf{x}}}_{o,j} &= \mathbf{A}_{o,j}\,\hat{\mathbf{x}}_{o,j} + \mathbf{B}_{o,j}\,\mathbf{s}_L + \mathbf{L}\,(\boldsymbol{\mu}_L - \hat{\boldsymbol{\mu}}_{L,j}) \\ \hat{\boldsymbol{\mu}}_{L,j} &= \mathbf{C}_{o,j}\,\hat{\mathbf{x}}_{o,j} + \mathbf{M}_{D,j}\,\mathbf{s}_L \\ &\text{for} \quad j = 1, \ldots, N_f + 1 \qquad , \end{aligned} \tag{37}$$

where the indexes up to N_f represent the fault models and $N_F + 1$ is related to the nominal fault-free model.

4.2 Residual Generation Technique for Fault Isolation

Residuals are determined from the measured and estimated system outputs

$$\mathbf{r}_j(t) = \boldsymbol{\mu}_L(t) - \hat{\boldsymbol{\mu}}_{L,j}(t)\,, \qquad j = 1, \ldots, N_f + 1 \tag{38}$$

with $\mathbf{r}_{Nominal} := \mathbf{r}_{N_f+1}$ as the residual vector obtained from the nominal observer. For fault isolation, the L^2 norm of all residuals are compared with each other. In the faultless case it holds that

$$\lim_{t\to\infty} \|\mathbf{r}_{Nominal}(t)\| \leq \lim_{t\to\infty} \|\mathbf{r}_j(t)\|, \qquad j = 1, \ldots, N_f . \tag{39}$$

In case of a fault $k \in [1, N_f]$, however, the observer (37) which was designed based on that fault model k will provide the smallest residual (38).

$$\lim_{t\to\infty} \min_{j=1,\ldots,N_f} \mathbf{r}_j(t) = \lim_{t\to\infty} \mathbf{r}_k(t) \quad \rightarrow \quad \text{isolation of the } k\text{-th fault} \tag{40}$$

5 Illustrative Simulation Studies

5.1 Definition of the Faults

The objective is to detect and isolate four different grid line and two VSI faults for micogrids with a topology illustrated in Fig. 2. All considered faults are related to physical parameter changes that are distributed throughout the central state space model of the microgrid. Particularly grid line faults modelled by changes in R_k effects all matrices in (26). The parameter changes assigned to the faults in the microgrid are defined as follows:

1. line break occurs in Y_1: R_1 is changed from nominal value to $1 \cdot 10^6\ \Omega$.
2. line break occurs in Y_7: R_7 is changed from nominal value to $1 \cdot 10^6\ \Omega$.
3. closed circuit occurs in Y_1: R_1 is changed from nominal value to zero.
4. closed circuit occurs in Y_7: R_7 is changed from nominal value to zero.
5. fault occurs in VSI 1: $\tau_{VSI,1}$ is changed from nominal value to 2 sec
6. fault occurs in VSI 3: $\tau_{VSI,3}$ is changed from nominal value to 2 sec

5.2 Simulation Results

For representation the effectiveness of the observer bank (37) and the according residuals calculation (38) an occurring line break in Y_1 in the microgrid, see Fig. 2 is generated at $t = 15$ s. To smooth the steep transition in the residuals, a simple first order filter with $\tau = 1$ was used. Figure 2 clearly shows how the residuals calculated with the nominal observer r_N immediately increase strongly due to the fault in line Y_1. With the exception of r_1 calculated with observer $j = 1$, all residuals increase also.

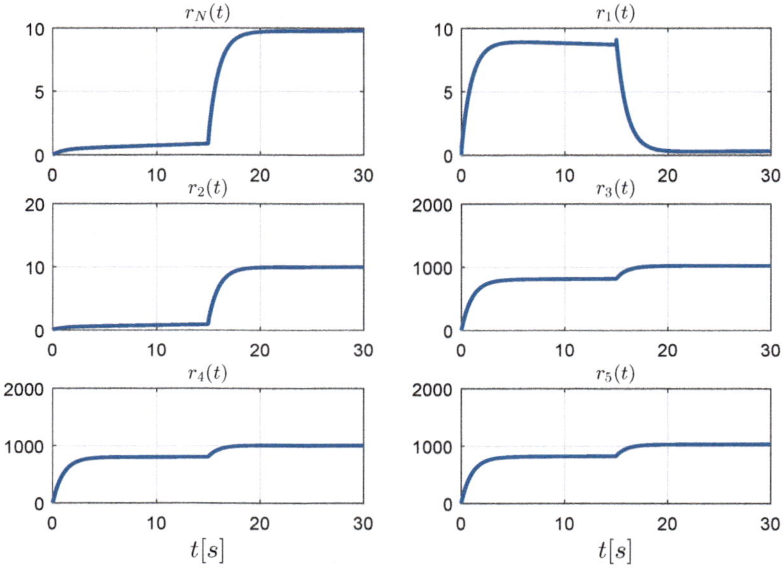

Fig. 2. Filtered residuals (38) for the occurring line break in Y_1 at $t = 15$ s

6 Conclusion

An FDI strategy of droop-controlled inverter-based microgrids was proposed. It was shown that line and converter faults can be modelled by parameter changes in a centralized state space model. The case study of the detection and isolation of a line fault was used to demonstrate the applicability by simulation. With this mechanism, which in some cases needs further investigation, the inverters can now be controlled in such a way that the faults effects in the network are reduced as much as possible by adaptation of the local virtual converter coefficient $\tau_{VSI,i}$, the droop coefficients $k_{P,i}$, $k_{Q,i}$ and by $k_{D,i}$.

References

1. Lasseter, R.H.: Microgrids. In: Proceedings of IEEE Power Engineering Society Winter Meeting, pp. 305–308 (2002)
2. Neacsu, D.O.: Power Switching Converters: Medium and High Power. CRC Press, Boca Raton (2006)
3. Yin, B., Oruganti, R., Panda, S., Bhat, A.: An output-power-control strategy for a three-phase PWM rectifier under unbalanced supply conditions. IEEE Trans. Industr. Electron. **55**(5), 2140–2151 (2008)
4. Song, H., Nam, K.: Dual current control scheme for PWM converter under unbalanced input voltage conditions. IEEE Trans. Ind. Electron. **46**(5), 953–959 (1999)
5. Chandorkar, M.C., Divan, D.M., Adapa, R.: Control of parallel connected inverters in standalone AC supply systems. IEEE Trans. Ind. Appl. **29**(1), 136–143 (1993)

6. Bevrani, H., Watanabe, M., Mitani, Y.: Power System Monitoring and Control. Wiley, Hoboken (2014)
7. Pietro, C.D., Vasca, F., Iannelli, L., Oliviero, F.: Decentralized synchronization of parallel inverters for train auxiliaries. In: Proceedings of Electrical Systems for Aircraft, Railway and Ship Propulsion, Bologna, Italy (2010)
8. Chandorkar, M., Divan, D., Adapa, R.: Control of parallel connected inverters in standalone ac supply systems. IEEE Trans. Ind. Appl. **29**(1), 136–143 (1993)
9. Jostock, M., Sachau, J., Tuttas, C.: LTI model of arbitrary voltage source inverter driven island grids. at–Automatisierungstechnik **61**(12), 818–830 (2013)
10. Jostock, M., Sachau, J., Hadji-Minaglou, J.-R.: Structured analysis of arbitrary island grid. In: IEEE PES Innovative Smart Grid Technologies Europe (ISGT Europe), Copenhagen, Denmark (2013)
11. Zhang, Q., Zhang, X., Polycarpou, M.M., Parisini, T.: Distributed sensor fault detection and isolation for multimachine power systems. Int. J. Robust Nonlinear Control **24** (8–9), 1403–1430 (2014). Special Issue: Fault Tolerant Control of Power Grids
12. Wu, Z., Yang, F., Han, Q.-L.: A novel islanding fault detection for distributed generation systems. Int. J. Robust Nonlinear Control **24**(8–9), 1431–1445 (2014). Special Issue: Fault Tolerant Control of Power Grids
13. Quan, Y., Chen, W., Wu, Z., Peng, L.: Distributed fault detection and isolation for leader-follower multi-agent systems with disturbances using observer techniques. Nonlinear Dyn. **93**(2), 863–871 (2018)
14. Gao, Z., Ding, S.X., Ma, Y.: Robust fault estimation approach and its application in vehicle lateral dynamic systems. Optimal Control Appl. Methods **28**, 143–156 (2007)

Autonomy of Surface and Underwater Marine Vessels

Ship Autopilot Software – A Case Study

Dariusz Rzońca(✉), Jan Sadolewski, Andrzej Stec, Zbigniew Świder, Bartosz Trybus, and Leszek Trybus

Department of Computer and Control Engineering, Rzeszów University of Technology, ul. W. Pola 2, 35-959 Rzeszów, Poland
{drzonca,js,astec,swiderzb,btrybus,ltrybus}@kia.prz.edu.pl

Abstract. Software of a ship autopilot, particularly its structure and size, developed in cooperation with a Dutch company is described. It is written in ST language, typical for PLCs and automation systems, in CPDev engineering environment. HMI displays are created by a graphic editor. The autopilot provides typical functionalities, such as heading control, track control, turn by radius and rate of turn, involving PID controllers. Wave-induced motions affecting the rudder are filtered out. Ship dynamics is identified by zig-zag or sinusoidal maneuvers. Control and HMI programs are executed by a runtime virtual machine.

Keywords: Ship autopilot · ST language · Control loops · HMI · Software measures

1 Introduction

It seems worthwhile to begin with history and motivation of this research. So 15 years ago, after encouragement by LUMEL Zielona Góra and support from a KBN grant, the authors began developing an engineering tool for programming PLCs and industrial controllers according to IEC 61131-3 standard [1]. The tool called CPDev (*Control Program Developer*) was implemented some time later in a small distributed system involving an 8-bit controller with remote I/Os [2]. Soon after that the authors were contacted by a Dutch company, Praxis Automation Technology B.V. [3] from Leiden, interested in application of the CPDev for programming controllers of its Mega-Guard ship automation and navigation system. At that time the controllers were programmed in a language not conformed with the IEC standard. The cooperation agreement was signed 10 years ago and since then the CPDev tool, growing considerably in the mean time [2, 4], after integration into Praxis design environment has been used for programming propulsion control, power management, fire control and other subsystems of the Mega-Guard. Typical subsystem consists of a controller, I/O unit (units) and operator panel.

However, there was no autopilot in the Mega-Guard at that time. So 5 years ago Praxis asked whether the authors could develop in CPDev an autopilot software with typical functionalities. After studying the well regarded books [5–7] and teaching texts, analyzing commercial autopilots from a few companies such as Kongsberg [8], Navico [9] and others, and counting on earlier experience with software for multifunction

A. Bartoszewicz et al. (Eds.): PCC 2020, AISC 1196, pp. 1499–1506, 2020.
https://doi.org/10.1007/978-3-030-50936-1_124

and adaptive controllers, the task of implementing typical functionalities seemed rather within reach. Progress in the subsequent research-and-development work has already reported [10, 11], including ship simulator [12, 13]. Currently the autopilot, called the Heading Control System (HCS) has been tested on "small sea" and announced on the Praxis web page [3].

Since algorithms of the developed autopilot are rather typical, the novelty of this paper lies in description of software structure and measures (size) of this particular solution. Such details are not disclosed in the manuals of the commercial autopilots but may be interesting from technical viewpoint. The software is written in ST language, as the other subsystems of the Mega-Guard. Next section briefly describes the HCS system. Section 3 reviews heading and track control loops, as well as filtering of wave-induced ship motions. Architecture and measures of control and HMI (Human Machine Interface) software are described in Sect. 4. Concluding remarks are given at the end.

2 Heading Control System

The HCS steers the ship heading, other words course, by controlling the rudder. The Praxis HCS [3] consists of the HC panel shown in Fig. 1 and Steering Control System (SCS) connected by redundant Ethernet. The panel runs the autopilot software providing rudder setpoint for the SCS.

Fig. 1. Heading control panel [3]

The following operating modes are available: HCS control, Track control, Turn by radius, Rate of Turn and Manual, if none of the three is selected. Functionalities and configuration parameters have been described in [11].

The SCS system consists of a controller, panels for steering/indication, and full/non follow up units for external manual control. Single or twin rudders are supported.

Sensors provide the HCS system with heading, position, speed, rudder angle, wind and draft data. The sensors are interfaced according to NMEA serial or Ethernet standards, originally to the Integrated Navigation System which distributes the data over the ship network.

3 Heading and Track Control Loops

While in the HCS mode the autopilot operates according to the diagram of Fig. 2. GYRO denotes heading measurement, OUT_HC the rudder angle (indication in Fig. 1) set by the SCS system according to setpoint provided by the PID controller.

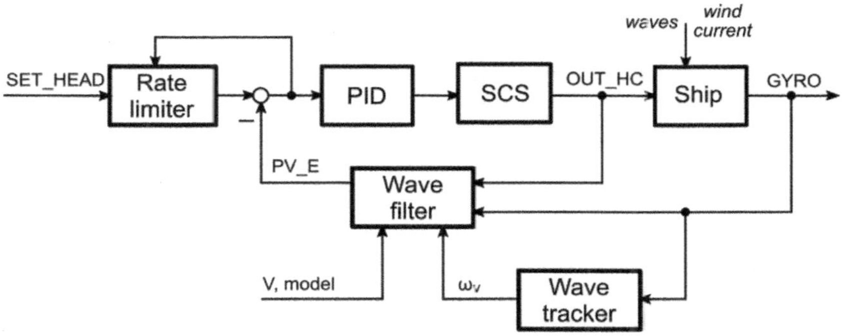

Fig. 2. Diagram of the HCS system

Rate limiter limits the rate of SET_HEAD step change to avoid over-steering (overshoot). While at rough sea the wave filter removes high frequency wave-induced component of GYRO leaving the low frequency PV_E (process variable) estimate of the heading. Therefore the output OUT_HC is not affected by waves. Wave tracker identifies wave frequency ω_v for the filter. The tracker switches itself off at calm sea according to a minimum wave parameter. Wind and draft measurements may be incorporated into the loop as feedforward.

PID controller settings are calculated using a simple Nomoto model of the ship [6], i.e.

$$\frac{\psi(s)}{\delta(s)} = \frac{k}{s(Ts+1)}, \tag{1}$$

where ψ denotes the heading (GYRO) and δ the rudder angle (OUT_HC). The parameters k, T depend on ship speed V, namely

$$k = k_0\frac{V}{V_0}, \quad T = T_0\frac{V_0}{V}, \tag{2}$$

with $\{k_0, T_0, V_0\}$ obtained by least-squares approximation from identification maneuvers (zig-zag, sinusoid) or set manually. In fact they are represented in Setup (Fig. 1) by a rudder ratio RR and counter rudder CR

$$RR = \frac{V_0}{Lk_0}, \quad CR = \frac{V_0}{L}T_0, \tag{3}$$

more familiar to marine community. L denotes the ship length.

Analytic formulae for PID settings derived in [10] by means of pole-zero cancellation and no overshoot requirement involve *k*, *T* and specification of controller sensitivity defined as the ratio $r = T/Tcl$ of the ship time constant *T* to closed-loop time constant *Tcl*. Default value $r = 2$ means that the autopilot operates twice as "fast" as the ship under manual control. Similar rules for PID settings can be found in [6] and [14].

Following [6, 15], the wave filter applies Luenberger observer algorithm defined by four-dimensional state-space model

$$A = \begin{bmatrix} 0 & 1 & 0 & 0 \\ 0 & -\frac{k}{T} & 0 & 0 \\ 0 & 0 & 0 & 0 \\ 0 & 0 & -\omega_v^2 & -2\xi\omega_v \end{bmatrix}, \quad B = \begin{bmatrix} 0 \\ 1 \\ 0 \\ 0 \end{bmatrix}, \quad C^T = \begin{bmatrix} 1 \\ 0 \\ 1 \\ 0 \end{bmatrix} \tag{4}$$

for some wave frequency ω_v, with $\xi << 1$. The observer splits the heading GYRO into low frequency PV_E and high frequency estimates. In the dynamic positioning area, observer-based estimation of ship velocities verified by experimental results has been described in [16].

Results of the wave filter simulated operation at rough sea are illustrated in Fig. 3 where the blue line denotes the rudder angle OUT_HC. When the filter is switched off the rudder moves in both directions trying to counteract motions of the ship. If it is switched on the rudder motions become negligible and may be completely removed by a dead zone. The other lines in Fig. 3 denote: red, GYRO; black, RATE OF TURN; green, SET HEADING (Trend display in the HC panel).

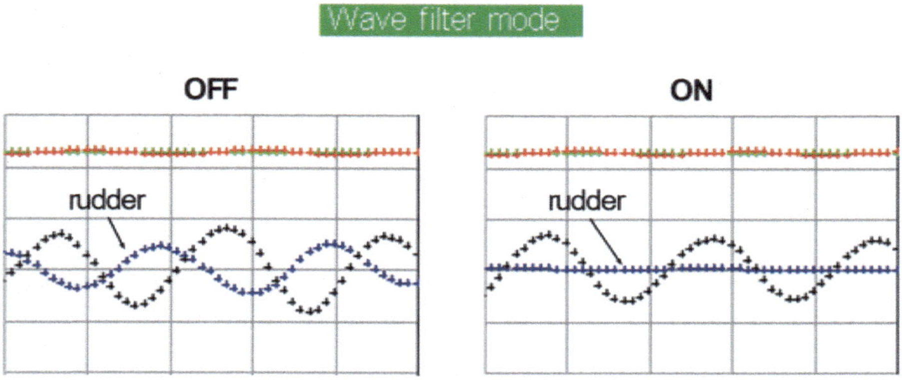

Fig. 3. Rudder for the wave filter off and on

Track control system shown in Fig. 4 is of cascade structure with the track controller as primary and heading controller (HCS) as secondary. The purpose is to keep the ship on a prescribed track (path) defined by waypoints on ECDIS chart. TRACK_REF denotes heading setpoint to the next waypoint. If by using the DGPS position the INS system detects a cross track error XTE from the track, the PID controller computes corresponding heading correction FD to the TRACK_REF, so as to bring the ship back on the track. The some ratio $r = T/Tcl$ determines sensitivity of the track controller.

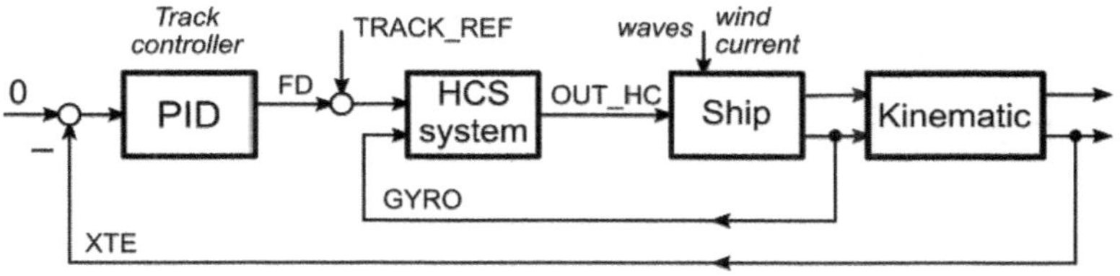

Fig. 4. Diagram of the track control system

4 Autopilot Software

The autopilot software has been developed in the CPDev environment [4]. The CPDev, a decade ago a tool providing ST programming only [2], has grown into almost complete control engineering environment supporting all IEC 61131-3 languages, HMI design, documentation generator, and research-oriented extensions [2]. It has a multiplatform capability what means that a program compiled into an intermediate code can be executed by a virtual machine running on different processors, from 8- up to 32/64-bit ones, and on FPGA [17]. HMI design tool called CPVis [18] enables development of controllers integrated with graphical panels, as the one in Fig. 1 and other from Praxis [3]. PLCs integrated with HMI panels are now offered by Beckhoff, Unitronix, Horner and others.

Naturally such devices execute both control and HMI software. The latter consists of two parts related to visualization and behavior. Visualization involves a set of displays designed by a graphical editor (here CPVis) from graphical objects imported from libraries. Behavior is determined by HMI programs that select displays and specify appearance of the objects. To do so, relevant variables are bound to attributes of the objects in a dialog window during display configuration. The control and HMI programs exchange data through common global variables. Extended CPDev runtime for the combined functionalities consists of the virtual machine for control and HMI runtime for visualization, both executed in parallel.

Software of the autopilot project consists of 9 programs shown in Fig. 5a where the AUTOPILOT executes control functions and the ones with VDP prefix handle visualization (number indicates display). The programs belong to a single task executed every 200 ms. Besides the programs, the project includes almost 20 function blocks and over 15 functions. Initial part of the project tree with several Program Organization Units (POUs) is in Fig. 5b. There are over 300 global variables, most of them grouped in subcatalogs according to logical relevance. Operation is governed by organizational variables and flags of which SYSTEM_MODE and CDISPLAY (current display) are the most important.

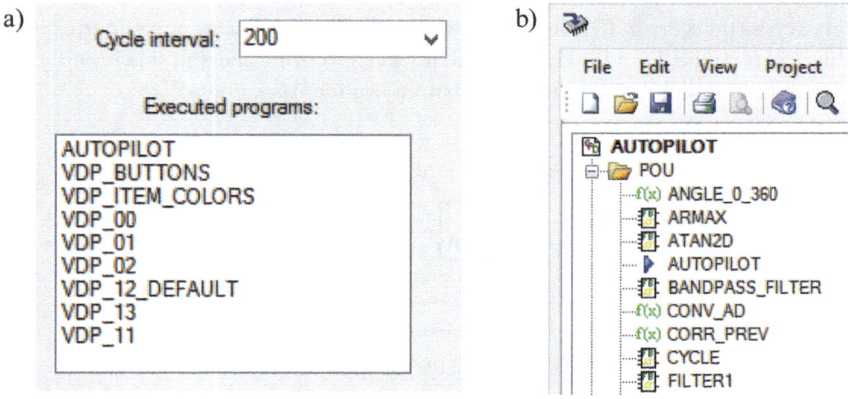

Fig. 5. a) Programs of the autopilot project; b) POUs in the project tree

Depending on the SYSTEM_MODE, relevant conditions and flags, the AUTOPILOT program calls appropriate function blocks according to the diagrams of Figs. 2, 4. Essential blocks are the following:

- WUTRCK tracks frequency and magnitude of wave-induced ship motions
- K_POLE calculates components of the wave filter gain
- KALMAN (observer) determines low frequency estimate PV_E of the heading
- SHIP_ID_ZZ/SS identifies ship dynamics by executing zig-zag or sinusoidal maneuver
- PID_HCSETT computes settings of the HC controller for given speed and sensitivity
- PID_TCSETT computes settings of the track controller
- PID_TC calculates course correction given the XTE cross track error
- PID_HC computes rudder angle setpoint for the SCS system.

Since all these blocks, except SHIP_ID_ZZ/SS, are normally called within the same cycle, execution of the autopilot software requires considerable processing power.

The VDP programs of Fig. 5a support the HMI involving 5 displays (MAIN, ADJUSTMENTS, STATUS/TUNINGS, TREND, XY CHART [11]). By monitoring the pushbuttons and knob (Fig. 1, lower part) they adjust values, change statuses or switch to another display. Each VDP program begins with verification whether the current display (CDISPLAY) is the same as the program number (num), so

IF CDISPLAY<>num THEN RETURN; END_IF

The MAIN display (Fig. 1, upper part) involves 10 animated objects and 5 texts (5" touchscreen). Overall there are over 150 objects and 50 texts in the current HMI. User-defined objects can be created by the CPVis tool.

For general orientation the autopilot software (Fig. 5a) involves about 6200 lines of ST code. Basic measures of software size are presented in Table 1 (from the PC simulator [11]). The runtime virtual machine applies 32-bit addressing.

Table 1. Software measures of the autopilot and HMI.

Autopilot	HMI
Project XML 210 kB	Displays 160 kB
Code memory 82 kB	Code memory 45 kB
Data memory 10 kB	Data memory 13 kB

5 Conclusions

An overview of functionalities and software of a ship autopilot developed in cooperation with the Praxis company from the Netherlands has been presented. The functionalities are typical, i.e. heading and track control loops involving PIDs, as well as wave filtering using an observer. Operation requires identification of ship dynamics in the form of the Nomoto model and specification of minimum wave to enable determination of wave frequency. The software is written in ST language of IEC 61131-3 standard appropriate for programming PLCs and automation systems. It is processed in CPDev environment suitable for controllers integrated with graphical panels. Since the other automation subsystems from Praxis have been designed in this environment, the uniformity is beneficial for practical reasons. Unlike in case of commercial autopilots, software measures of this particular autopilot are given.

Acknowledgments. This project is financed by the Minister of Science and Higher Education of the Republic of Poland within the "Regional Initiative of Excellence" program for years 2019 – 2022. Project number 027/RID/2018/19, amount granted 11 999 900 PLN.

References

1. IEC: IEC 61131-3 – Programmable controllers – Part 3: Programming languages (2003). 2013
2. Rzońca, D., Sadolewski, J., Stec, A.; Świder, Z., Trybus, B., Trybus, L.: Mini-DCS system programming in IEC 61131-3 Structured Text. J. Autom. Mob. Robot. Intell. Syst. **2**(3), 48–54 (2008). Developing a multiplatform control environment. JAMRIS **13**(4) (2019, in print)
3. Praxis Automation Technology B.V. http://www.praxis-automation.nl. Accessed 03 Jan 2020
4. CPDev Homepage. https://cpdev.kia.prz.edu.pl/. Accessed 03 Jan 2020
5. Lisowski, J.: Statek jako obiekt sterowania automatycznego. Wyd. Morskie, Gdańsk (1981). (Ship as Automatic Control Plant)
6. Fossen, T.I.: Guidance and Control of Ocean Vehicles, 4th edn. Wiley, Chichester (1999)

7. Śmierzchalski, R.: Automatyzacja i sterowanie statkiem. Wyd. PolitechnikiGdańskiej, Gdańsk (2013) (Ship Automation and Control)
8. Autopilot AP2000 Track Pilot. Kongsberg Maritime Ship Systems AS (2001)
9. Simrad AP70/AP80 Operator Manual. Navico Holding AS (2016)
10. Trybus, L., Świder, Z., Stec, A.: Tuning rules of conventional and advanced ship autopilot controllers. In: Szewczyk, R., Zieliński, C., Kaliczyńska, M. (eds.) Progress in Automation, Robotics and Measuring Techniques. AISC, vol. 350, pp. 303–311. Springer, Cham (2015)
11. Jamro, M., Rzońca, D., Sadolewski, J., Stec, A., Świder, Z., Trybus, B., Trybus, L.: Structure and functionalities of ship autopilot simulator. In: Szewczyk, R., Zieliński, C., Kaliczyńska, M. (eds.) Challenges in Automation, Robotics and Measurement Techniques. AISC, vol. 440, pp. 223–231. Springer, Cham (2016)
12. Stec, A.: Ship maneuvering model for autopilot simulator. In: Szewczyk, R., Zieliński, C., Kaliczyńska, M. (eds.) Progress in Automation, Robotics and Measuring Techniques. AISC, vol. 350, pp. 265–274. Springer, Cham (2015)
13. Marine Systems Simulator (2019). http://www.marinecontrol.org/Tutorial.html
14. Morawski, L., Pomirski, J.: Design of the robust PID course-keeping control system for ships. Pol. Marit. Res. **9**(1), 28–31 (2002)
15. Fossen, T.I., Perez, T.: Kalman filtering for positioning and heading control of ships and offshore rigs. IEEE Control Syst. Mag. **29**(6), 32–46 (2009)
16. Tomera, M.: Dynamic positioning system for a ship on harbour manoeuvring with different observers. Experimental results. Pol. Marit. Res. **21**(3), 13–24 (2014)
17. Hajduk, Z., Trybus, B., Sadolewski, J.: Architecture of FPGA embedded multiprocessor programmable controller. IEEE Trans. Industr. Electron. **62**(5), 2952–2961 (2015)
18. Jamro, M., Trybus, B.: IEC 61131-3 programmable human machine interfaces for control devices. In: 6th International Conference on Human System Interaction (HSI), Sopot, Poland, pp. 48–55 (2013)

Comparison of Different Course Controllers of Biomimetic Underwater Vehicle with Two Tail Fins

Michał Przybylski(✉), Piotr Szymak, Zygmunt Kitowski, and Paweł Piskur

Department of Naval Automatics, Polish Naval Academy, 81127 Gdynia, Poland
{p.szymak,m.przybylski,z.kitowski,p.piskur}@amw.gdynia.pl
http://wme.amw.gdynia.pl

Abstract. In recent times, we may notice some new designs of underwater vehicles which imitate living underwater organisms, e.g. a fish, a seal. These vehicles are called biomimetic. They are driven by an undulating propulsion, imitating a wavy motion of underwater creatures. The Biomimetic Underwater Vehicles (BUV) is a new control object in an underwater robotics. In the paper, problem of BUV's course control is taken into account. The classical Proportional–Integral–Derivative (PID) and Slide Mode (SM) controllers are compared. The settings of the controllers are tuned using three different optimization methods: the Genetic Algorithm (GA), the Particle Swarm Optimization (PSO) and the Pareto Search Algorithm (PSA). The research has been conducted using the mathematical model of mini CyberSeal driven by two side and two tail fins.

Keywords: Biomimetic Underwater Vehicle · Course control · Genetic Algorithm · Particle Swarm Optimization · Pareto Search Algorithm

1 Introduction

In recent years, a dynamical development of underwater robotics has been noticed. One of the latest innovative constructions in this field are autonomous biomimetic underwater vehicles (BUVs) [5,8,16]. They imitate underwater living organisms, e.g. fishes, marine mammals, etc. They can imitate both the construction and kinematics of motion [7]. BUVs are driven by undulating propulsion imitating real fins, e.g. of a fish – Fig. 1.

Such vehicles have the three following features: (1) their propulsion is more energy efficient and (2) more silent than classical propulsion based on rotary propellers, (3) they are very difficult to differentiate from the real inhabitants of underwater environment (larger secrecy and potential range of operation).

The BUV is driven using artificial fins similar to the ones used by the underwater creatures, especially fishes. The set of fins is called undulating or wavy propulsion. The fins can be located at the bow, stern or sides of the BUV's hull depending on its maneuverability and performance. Usually they are driven by the electric motors to make oscillating motion for generating proper thrust

A. Bartoszewicz et al. (Eds.): PCC 2020, AISC 1196, pp. 1507–1518, 2020.
https://doi.org/10.1007/978-3-030-50936-1_125

Fig. 1. The CyberFish version no. 3 in the swimming pool [7].

of the vehicle. The fin's motion is usually described by the following parameters: a frequency of oscillation, a maximal deflection called an amplitude and a neutral position being the zero position for oscillations. Usually the oscillations are described by a sinusoidal function, but also others functions can be used depending on a type and a purpose of the undulating propulsion. To control course of the BUV, it is usually equipped with one or more fins located at the stern moving in the horizontal surface.

In this paper, the control object which is the mini CyberSeal and then its mathematical model of motion is described based on classical Fossen model of motion of marine vehicle [2] with some improvements taking into account the new undulating propulsion system. This control object with its model of motion is described in details in the next section. Then, the controllers and their tuning methods are presented in the third section. In the next section, firstly research problem and then achieved results of research are described. In the last section, the conclusions and the plan for future research are formulated.

2 Control Object

2.1 Mini CyberSeal

The mini CyberSeal is a physical model of the larger BUV imitating a seal made in a scale. The main goal of the mini CyberSeal implementation before the larger counterpart building was to examine a new propulsion system consisting of two tail fins and two side fins. The second goal was to test different control algorithm for the mini CyberSeal, which can be useful for larger construction.

Construction of the mini CyberSeal is based on POM-C tube housing all the electronic components, sensors, li-ion 7.2 V, 10 Ah accumulator and servomotors – Fig. 2. CyberSeal is equipped with the following sensors: (1) a digital compass OS-5000 from Ocean Server company, (2) a depth sensor A-10 from Wika company with 0–1 bar range. Digital compass is used to obtain present value of a course. Fins are driven by the servomotors with the 1:1 angular gears. Servomotors are controlled by POLOLU-1353 miniMaestro servo controller via RS-232 interface. Communication is based on the wifi network supported by the TP-LINK TL-WR702N access point and server port WIZNET Wiz-145SR, providing four serial ports. Therefore, the microprocessor control system of mini CyberSeal can be located outside the BUV's hull. The mini CyberSeal was additionally equipped with internal artificial swim bladder responsible for a buoyancy control of the vehicle. Side and tail fins of the BUV were made from a polycarbonate and a rubber.

Fig. 2. The mini CyberSeal in the swimming pool [13].

The side fins can move in the range of ±45 [deg] with a speed up to 3 [Hz], while the tail fins can move in the range from approx. 80 [deg] outside to 12 [deg] inside according to the longitudinal axis of symmetry with the same speed up to 3 [Hz]. The tail fins are used to generate main thrust and additionally to change course, while the side fins can generate additional thrust and influence on change of depth by changing the BUV's trim. More details are include in [14].

2.2 Mathematical Model

Usually, an underwater vehicle is considered as a rigid body with the following features [3]: (1) it has three planes of symmetry, (2) it moves in six degrees of freedom, (3) it moves at a low speed in a viscous fluid. Motion of the vehicle is described by means of two reference systems: (1) the body fixed coordinate system associated with the vehicle $x_o y_o z_o$ and (2) the earth fixed coordinate system associated with the Earth xyz. The origin of the body fixed coordinate system O responds to the center of gravity of the vehicle, while its axes are defined as: (1) x_o is a longitudinal axis directed from the stern to the bow, (2) y_o is a transverse axis directed to the starboard, and (3) z_o is a perpendicular axis directed from the top to the bottom. Changes of the position of the body fixed coordinate system are described with respect to coordinate system associated with the Earth. Taking into account mentioned above assumptions, to simulate motion of BUV, a nonlinear model of underwater vehicle in 6 degrees of freedom was used. The motion of the vehicle is described by 6 differential equations, very often presented in the following compact matrix form [3]:

$$\mathbf{M}\dot{\nu} + \mathbf{D}(\nu)\nu + \mathrm{g}(\eta) = \tau \tag{1}$$

Here: ν is a vector of linear and angular velocities in the body fixed system, η is a vector of vehicle position coordinates and its Euler angles in the Earth fixed system, $\mathbf{M}$ is a matrix of inertia (the sum of the matrices of the rigid body mass and the accompanying mass of water), $\mathbf{D}(\nu)$ is a hydrodynamic damping matrix, $\mathrm{g}(\eta)$ is a vector of restoring forces and moments of forces of (gravity and buoyancy), τ is a vector of control signals (the sum of vector of forces and moments of force generated by propulsion system and by environmental disturbances).

The left side of the Eq. (1) includes forces and moments of force caused by an inertia of the body of the vehicle and an inertia of an accompanying masses of a viscous liquid, a hydrodynamic dumping of water environment, a balance of a gravity and a buoyancy. While the right side of the Eq. (1) represents the vector of forces and moments of force acting on the vehicle generated by a propulsion system and additional environmental disturbances, especially a sea current. The parameters of the matrices included in the left side of Eq. (1) were calculated based on the dependencies included in [3]. While the vector of forces and moments of force generated by the new type of propulsion system called undulating propulsion were calculated using equations included in [14]. The control object is equipped with two side fins and two tail fins. All the fins consist of one movable segment. Each fin generates a thrust with the value changing in time depending on the control parameters of the fin, especially an amplitude and a frequency of the fin oscillation. More details are included in [10]. Quite wide approach to the fish locomotion issues and control challenges is included in [1]. The mathematical model was verified with the real BUV in the laboratory conditions (in a swimming pool). Details of the verification tests are included in [11].

3 Methods

3.1 Controllers

Two types of classical controllers were used for the course control. The first controller is a proportional-integral-derivative action controller (PID) and the second is a slide mode controller (SM). The action of the PID controller is described by the following formula presented in the discrete form:

$$u(k) = k_p e(k) + k_i \sum_{k=1}^{k_{max}} e(k) + k_d \Delta e(k) \tag{2}$$

Where $u(k)$ is a control signal in k step of simulation. Variable $e(k)$ is an error signal in k step of simulation, while $\Delta e(k)$ is a change of error signals in k step of simulation, i.e. $e(k) - e(k-1)$. Constant quantities called gain factors are: k_p, k_i and k_d. While the sliding mode control (SM) is calculated using the following formulas also presented in the discrete form:

$$s(k) = \frac{e(k) + \lambda \Delta e(k)}{\varphi} \tag{3}$$

$$\text{if} \quad |s(k)| > 1, \quad \text{then} \quad s(k) = sign(s(k)) \tag{4}$$

$$u(k) = k_s s(k) \tag{5}$$

Where $u(k)$ is a control signal in k step of simulation and $s(k)$ is a switching function in k step of simulation. Variable $e(k)$ is an error signal in k step of simulation, while $\Delta e(k)$ is a change of error signal in k step of simulation, i.e. $e(k) - e(k-1)$. Constant settings of SM controller are: λ, φ, k_s, where λ is a gain factor of differentiate unit, φ is a thickness of a boundary layer and k_s is a gain factor of the switching function. Proper selection of the constants provide switching curve proper to dynamics of nonlinear controlled object.

The control signal u(k) in Eqs. (2)–(5) is moment of force relative to horizontal axis of symmetry.

3.2 Optimization Methods

To tune controllers' settings three following different optimization methods were used: GA, PSO, PSA. Some initial research on tuning course controllers' settings by means of PSO are included in [15].

The GA is a heuristic search that mimics the process of natural selection. It is based on an iterative evolutionary procedure involving selection of genotypes for reproduction based on their fitness, and then introducing genetically changed offspring into the next population. The changes are introduced into the offsprings by means of a mutation, a crossover and other genetic operators. The procedure is finished after achieving satisfactory genotypes (a set of features of an individual) which correspond to the phenotypes with high fitness function (the individual from a population) [4]. During the optimization process the initial population consisting of 40 individuals was generated using Matlab random generator. The individuals in the current generation are estimated using one of three fitness functions described in the next subsection. After calculation of the fitness function, reproduction algorithm creates children for the next generation. In the reproduction, the following operators were used: Rank fitness scaling, Stochastic uniform selection function, Crossover fraction equal to 0.8, Gaussian mutation function.

Fitness scaling converts the raw fitness scores that are returned by the fitness function to values in a range that is suitable for the selection function. The rank fitness scaling scales the raw scores based on the rank of each individual instead of its score. The rank of an individual is its position in the sorted scores. An individual with rank r has scaled score proportional to $\frac{1}{\sqrt{r}}$. Selection function specifies how the genetic algorithm chooses parents for the next generation. Stochastic uniform selection function lays out a line in which each parent corresponds to a section of the line of length proportional to its scaled value. The algorithm moves along the line in steps of equal size. At each step, the algorithm allocates a parent from the section it lands on. The first step is a uniform random number less than the step size. Crossover fraction specifies the fraction of the next generation, other than elite children, that are produced by crossover. Gaussian mutation function adds a random number taken from a Gaussian distribution with mean 0 to each entry of the parent vector. During the research, the GA was stopped when the maximum number of 100 generations was reached and/or when detection of no change in the best value of the fitness function for 50 maximum stall generations was achieved.

The inspiration for the PSO is flocks of birds or insects swarming. Each particle is attracted to some degree to the best location it has found so far, and also to the best location any member of the swarm has found. After some steps, the population can coalesce around one location, or can coalesce around a few locations, or can continue to move [12]. The PSO is a population-based algorithm and it is similar to the genetic algorithm. A collection of individuals called particles move in steps throughout a region. At each step, the algorithm evaluates the objective function at each particle. After this evaluation, the algorithm decides on the new velocity of each particle. The particles move, then the algorithm reevaluates [12].

Based on the literature [9,12] the following parameters of PSO were accepted: (1) MaxStallIterations (relative change in the best objective function value): 20, (2) MinNeighborsFraction (setting both the initial neighborhood size for each particle, and the minimum neighborhood size): 1, (3) SwarmSize: 100. Similar to GA, one of the key problem is to define properly the fitness function which is used for estimation of solutions encoded in particles. Three propositions of the function are included in the next subsection.

The next method PSA is usually used for multi-objective optimization, but it can also by used for single objective problem. The PSA uses pattern search on a set of points to search iteratively for nondominated points. It should satisfy all bounds and linear constraints at each iteration. Theoretically, the algorithm converges to points near the true Pareto front [2]. At the beginning PSA creates the initial set of 200 randomly selected points and then it checks if the points are feasible with respect to the bounds and linear constraints. If they are not feasible, the algorithm projects the initial points onto the linear subspace of linearly feasible points by solving a linear programming problem and it removes any duplicate points. Then, PSA divides points into two sets named 'archive' and 'iteratives'. Archive set contains nondominated points associated with a mesh size below 10^{-6} and satisfying all constraints to within 10^{-6}. PSA polls locations for each point in 'iterates'. The success is obtained if the polled points give at least one nondominated point. Then, PSA extends the poll in the successful directions repeatedly with doubling the mesh size to find a dominated point. If any nondominated point is received, the mesh size is halved. The algorithm is stopped when: (1) the mesh size exceeds the value (+Infinity), (2) the fitness function decreases to the value (−Infinity), (3) maximum number of iterations equal to 400 is received [6].

3.3 Fitness Functions

One of the key problem is to define proper fitness function, which will give the expected results. Therefore, in the paper three different functions were formulated. The first one is the simplest. It is a sum of absolute values of changes of error signals in all steps of simulation $\Delta e(k)$ (Eq. 6).

$$f_{fit1} = \sum_{k=1}^{k_{max}} |\Delta e(k)| \tag{6}$$

The second fitness function is an Integral of Squared Error (ISE) in discrete form (Eq. 7). The ISE is one of the basic integral indexes used to estimate control quality.

$$f_{fit2} = \sum_{k=1}^{k_{max}} \Delta e(k)^2 \tag{7}$$

The third proposed fitness function is based on combination of two direct control quality indexes:

$$f_{fit3} = \sum_{i=1}^{i_{max}} t_r(i) + k_M \sum_{i=1}^{i_{max}} M_p(i) \tag{8}$$

It takes into consideration rising times t_r in [s] and first overshoots M_p in [rad] for all i_{max} changes of desired course. Because of small value of M_p in [rad] comparing to t_r in [s] additional gain factor of the sum of first overshoots were introduced $k_M = 20$.

4 Research

4.1 Problem

The research problem was formulated in the following way. Having mathematical model of the BUV, two types of controllers (PID, SM), 3 optimization methods for the controllers' settings (GA, PSO, PSA) and three defined fitness function, the most effective combination of the controller and its tuning method with the fitness function should be find for the problem of controlling BUV's course. The controllers, the methods and the functions should be examined in both tuning and verifying processes. The tuning process was conducted for three different changes of desired course: (1) 90 [deg] - large course, (2) 30 [deg] - small course and (3) 90 [deg] and then 30 [deg] - two courses. The verifying process was carried out for 30 changes of course selected randomly from the range from 0 to 90 [deg]. The mean value of the fitness function was calculated for all the 30 verifying tests.

4.2 Results and Discussion

The initial results of research on tuning course controllers' settings by means of the PSO are included in [15]. In Tables 1, 2 and 3 detailed results of research was illustrated in the form of calculated fitness function respectively no. 1–3 for tuning (T) and verifying (V) BUV's course controllers for three changes of desired course. In the tables, the combination of type of controller and its tuning method was marked in the following way: type-method, e.g. PID-GA means that the PID controller was tuned by means of GA, SM-PSO means that the SM controller was tuned by means of PSO, etc.

The goal of the initial research was to determine lower and upper barriers for the controllers to avoid possible going into local minimas. The research was conducted by an expert using designed model of the mini CyberSeal, designed course controllers and the tuning methods. In the results of the initial research the following lower and upper barriers have been selected: (1) for PID settings $[k_p, k_d, k_i]$ - lower barrier equal to [20, 1000, −1] and upper barrier equal to [80, 3000, 1], (2) for SM settings $[\lambda, k_s, \varphi]$ - lower barrier equal to [−50, −20, −5] and upper barrier equal to [50, 20, 5]. The goal of the main part of research was to compare different controllers, the tuning methods with the fitness functions and additionally changes of desired course used in the tuning process. The only criteria of control quality was the value of fitness function which was minimized during optimization. Therefore, the terms 'best result' or 'the most effective' used in the further part of the paper relate to the situation where the smallest fitness function was obtained, i.e. the best control quality.

Analyzing the results received for tests using fitness function no. 1, it can be stated that: (1) PID-PSO and PID-PSA received the best results in the tuning process, while PID-GA in the verifying process taking into account all the changes of desired course, (2) SM-PSO received the best results in verifying process using large change of course in its tuning, (3) the smaller f_{fit1} was

Table 1. Values of fitness function no. 1 for tuning (T) and verifying (V) BUV's course controllers for 3 changes of desired course: small, large and two courses

Controller type	For large (T)	Course (V)	For small (T)	Course (V)	For two (T)	Courses (V)
PID-GA	82.4	116	49.3	104	333	101
SM-GA	200	100	79.8	643	519	152
PID-PSO	79.9	122	20.6	217	192	149
SM-PSO	193	97.1	63.7	531	481	150
PID-PSA	80.4	114	21.5	219	191	167
SM-PSA	112	411	73.3	743	488	137

obtained during tuning and verifying PID controller than SM controller, (4) the best verification of both controllers was received for controllers tuned using large change of desired course, (5) three optimization methods presents the similar efficiency.

Table 2. Values of fitness function no. 2 for tuning (T) and verifying (V) BUV's course controllers for 3 changes of desired course: small, large and two courses

Controller type	For large (T)	Course (V)	For small (T)	Course (V)	For two (T)	Courses (V)
PID-GA	81.9	66.6	10.1	93.4	298	35.9
SM-GA	157	53.3	20.3	2751	405	90.4
PID-PSO	80.3	101	6.56	299	156	288
SM-PSO	155	47.3	14.4	2532	385	88.4
PID-PSA	82.8	325	7.11	232	209	76.6
SM-PSA	81.9	2184	15.5	2467	385	93.3

Taking into consideration the results received for tests using fitness function no. 2, the following conclusions can be formulated: (1) PID-PSO and PID-PSA received the best results in the tuning process, while PID-GA in the verifying process taking into account all the changes of desired course, (2) similar as earlier the smaller f_{fit2} was obtained during tuning and verifying PID controller than SM controller, (3) the best verification of both controllers was received for controllers tuned using two changes of desired course, (4) three optimization methods presents the similar efficiency.

Considering the results received for tests with fitness function no. 3, it can be observed that: (1) PID-PSO and PID-PSA received the best results in the tuning process, while PID-GA in the verifying process taking into account all the changes of desired course, (2) PID-PSO received the best results in verifying process using two changes of course in its tuning, (3) in general the smaller f_{fit3} was obtained during tuning and verifying PID controller than SM controller, (4) the best verification of both controllers was received for controllers tuned using two changes of desired course, (5) three optimization methods present the similar efficiency.

Table 3. Values of fitness function no. 3 for tuning (T) and verifying (V) BUV's course controllers for 3 changes of desired course: small, large and two courses

Controller type	For large (T)	Course (V)	For small (T)	Course (V)	For two (T)	Courses (V)
PID-GA	5.80	17.8	5.96	14.2	19.4	8.81
SM-GA	25.2	28.1	16.9	23.8	57.7	22.1
PID-PSO	5.24	14.5	3.77	26.2	17.8	7.83
SM-PSO	21.2	27.1	18.7	27.4	55.9	41.9
PID-PSA	5.44	14.8	4.34	37.1	15.6	9.67
SM-PSA	21.4	26.7	17.8	23.1	31.3	23.1

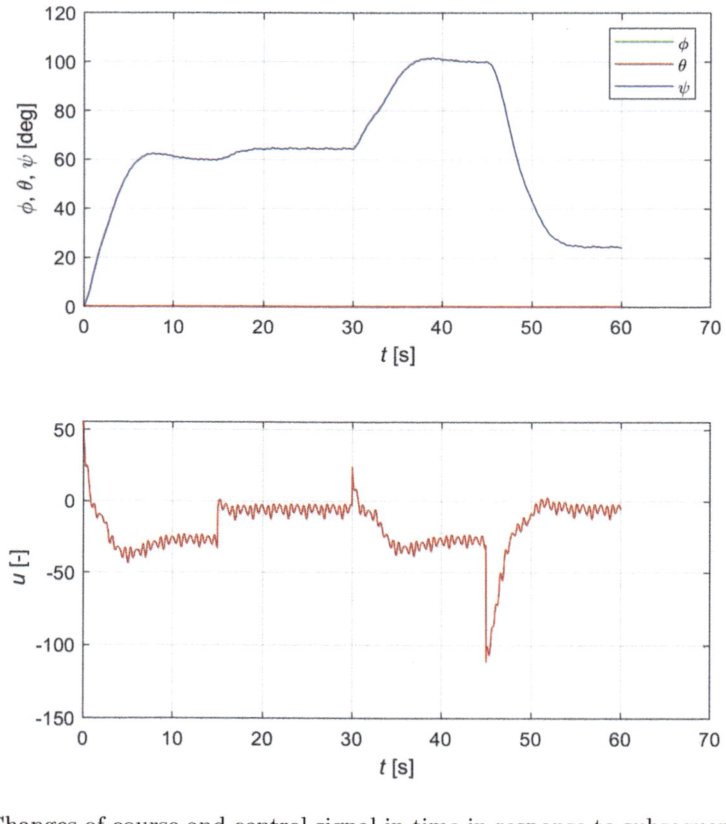

Fig. 3. Changes of course and control signal in time in response to subsequent desired courses: 60, 65, 100, 25 [deg] produced by PID-GA obtained using fitness function no. 3.

In Fig. 3 changes of course and control signal in time in response to subsequent desired courses: 60, 65, 100, 25 [deg] produced by PID-GA obtained using fitness function no. 3. As we can seen, the controller achieved desired courses quite quickly with slight overshoots.

5 Conclusions

In the paper, two classical controllers (PID, SM) with their settings tuned using three optimization methods (GA, PSO, PSA) used in problem of course control of BUV with two tail fins were compared. Moreover, three different fitness functions and changes of desired course used during tuning were tested.

Taking into account all the results of research, the following conclusions can be formulated. Better control quality was received for the PID controller than for the SM controller. Three optimization methods used to tune controllers' settings present the similar efficiency, i.e. comparable control quality of the tuned controllers were obtained. Although, PSO and PSA were better during tuning and verifying processes in general, GA allows us to receive the best controller in the verifying process for all three fitness functions. No the best solution was produced using small change of desired course during the tuning process. It means that two changes of desired course or at least large change of desired course should be used in tuning course controller of the BUV. It seems best to use a fitness function based on direct control quality indexes, e.g. f_{fit3} used in the research. Because there were no big differences in values for this function in contrast to the other two functions f_{fit1} and f_{fit2}.

As part of further research, testing other nonlinear controllers, e.g. fuzzy controller is considered. Moreover, an implementation of selected regulators in the BUV and their verification in real conditions are planned.

References

1. Colgate, J.E., Lynch, K.M.: Mechanics and control of swimming: a review. IEEE J. Oceanic Eng. **29**(3), 660–673 (2004)
2. Custódio, A.L., Madeira, J.F.A., Vaz, A.I.F., Vicente, L.N.: Direct multisearch for multiobjective optimization. SIAM J. Optim. **21**(3), 1109–1140 (2011)
3. Fossen, T.: Handbook of Marine Craft Hydrodynamics and Motion Control. Wiley, Hoboken (2011)
4. Goldberg, D.E.: Genetic Algorithms in Search, Optimization and Machine Learning. Addison Wesley, Massachusetts (1989)
5. Low, K.H.: Modelling and parametric study of modular undulating fin raysfor fish robots. Mech. Mach. Theory **44**, 615–632 (2009)
6. Mathworks: Matlab documentation. https://www.mathworks.com/help/matlab/
7. Malec, M., Morawski, M., Szymak, P., Trzmiel, A.: Analysis of parameters of traveling wave impact on the speed of biomimetic underwater vehicle. Solid State Phenomena **210**, 273–279 (2014)
8. Morawski, M., Słota, A., Zając, J., Malec, M., Krupa, K.: Hardware and low-level control of biomimetic underwater vehicle designed to perform ISR tasks. J. Mar. Eng. Technol. **16**(4), 227–237 (2017). https://doi.org/10.1080/20464177.2017.1387089
9. Pedersen, M.E.: Good parameters for particle swarm optimization. Hvass Laboratories, Luxembourg (2010)
10. Przybylski, M.: Mathematical model of biomimetic underwater vehicle. In: Proceedings of the 33rd International ECMS Conference on Modelling and Simulation, Italy, Caserta, vol. 33, no. 1, pp. 343–350 (2019)

11. Przybylski, M., Szymak, P.: Verification of the mathematical model of the biomimetic underwater vehicle. In: ModTech2020 International Conference on Modern Technologies in Industrial Engineering, Romania, Eforie Nord (2020, in press)
12. Shi, Y., Eberhart, R.C.: A modified particle swarm optimizer. In: Proceedings of IEEE International Conference on Evolutionary Computation, pp. 69–73 (1998)
13. Szymak, P., Praczyk, T., Pietrukaniec, L., Hożyń, S.: Laboratory stand for research on mini CyberSeal. Meas. Autom. Monit. **63**(7), 228–233 (2017)
14. Szymak, P.: Mathematical model of underwater vehicle with undulating propulsion. In: Third International Conference on Mathematics and Computers in Sciences and in Industry MCSI 2016, Greece, Chania, pp. 269–274 (2016)
15. Szymak, P., Piskur, P.: Using particle swarm optimization for tuning course controller of biomimetic underwater vehicle. In: 8th International Maritime Science Conference, pp. 255–262. University of Montenegro, Budva (2019)
16. Zhou, C., Low, K.H.: Design and locomotion control of a biomimetic underwater vehicle with fin propulsion. IEEE-ASME Trans. Mechatron. **17**, 25–35 (2019). https://doi.org/10.1109/TMECH.2011.2175004

Path Controller for Ships with Switching Approach

Mirosław Tomera(✉)

Department of Ship Automation, Gdynia Maritime University,
Morska 83 Str., 81-225 Gdynia, Poland
m.tomera@we.umg.edu.pl
http://atol.am.gdynia.pl/~tomera

Abstract. The work presents the algorithm for controlling the movement of a ship along a desired route. The planned desired route for a moving ship was defined as a set of way-points connected by straight lines. The ship's control is based on changes in the rudder angle, thus enabling the vessel to move along a given segment of the cruise route. The designed control algorithm is designed to minimize the heading error and cross-tracking error determined relative to the segment connecting two successive way-points. A different controller was used to minimize each of these errors. On the line segment, both controllers are switched on, while when there is a switch from one line segment to the next, only one controller is used, the one related to minimizing the course deviation. To obtain smooth control - after performing the return maneuver - make a skilful activation of the second controller, minimizing cross-tracking error. For this purpose, the control algorithm uses the appropriate switching logic with a scaled set signal from the controller minimizing the cross-tracking error. The performance quality of the developed algorithm for controlling the ship's motion was tested on the training ship Blue Lady at the Ship Handling Research and Training Centre located on Lake Silm at Kamionka near Iława, Poland.

Keywords: Ship steering · Underactuated control · Track-keeping · Experimental results

1 Introduction

In recent years, the number of autonomous vessels designed and built for civil, military and robotic research applications has increased [3]. This includes autonomous underwater vehicles, underwater gliders [8], and unmanned vessels floating on the water surface [2,12,23]. All these vessels have proved their usefulness in counteracting mines [4], autonomous underwater sampling [22], detection and monitoring of gas leaks [1], and tracking and inspection of offshore pipelines/cables [18,21]. They were also used for prevention of natural disasters [9] and environment protection [15].

An autonomous marine vessel moves along a predetermined desired route [6,16], which usually consists of a number of line segments connecting subsequent

A. Bartoszewicz et al. (Eds.): PCC 2020, AISC 1196, pp. 1519–1530, 2020.
https://doi.org/10.1007/978-3-030-50936-1_126

way-points. Hence, the control based on tracking subsequent way-points attracts attention of many researchers. The first way to control the vessel motion along such a desired route is to guarantee global asymptotic convergence of the heading error and cross-tracking error determined relative to the segment connecting two consecutive way-points [17]. Another method of control involves the line-of-sight (LOS) technique to control the ship along the desired route [5]. Subsequently, global κ-exponential way-point control of ship maneuvering was carried out in [7]. When determining subsequent way-points, many factors are taken into account, such as weather conditions and obstacle avoidance [10,11,14].

Conventional ships are usually equipped with one or two main propellers for forward speed control, and rudders for turning control. The minimum configuration for way-point tracking control comprises a single main propeller and a single rudder. This means that only two controls are available, and the ship should be considered underactuated for the task of 3 DOF tracking control [6].

The analyzed problem is not new and has already been considered in [19,20,24], among other publications. In [19], two algorithms were considered to control the ship motion along the desired route. They were the LQR state controller algorithm, and a control algorithm consisting of two separate channels with two component regulators: PD to minimize the heading error, and PI to minimize the cross-track error along a given segment. The main problem in these control systems was high oscillation of the ship course after the turning maneuver. Relatively good performance in this area was obtained by using PD and PI controllers connected in parallel. Another solution to this problem was proposed in [20], where a PI controller was activated after the turning maneuver. In this case the oscillation of the ship course was reduced, but this positive effect was accompanied with the appearance of a large first overshoot in the minimized cross-track error. This paper presents a solution based on scaled activation of the path controller associated with cross-track error minimization.

2 Formulating the Problem

The movement of a ship sailing on the water surface is described in three degrees of freedom. Two coordinate systems are used to describe it (Fig. 1). The first is the inertial frame (X_N, Y_N), associated with the water map, where the X_N axis points north and the Y_N axis points east. The second coordinate system (X_B, Y_B) is associated with the moving ship, and its origin is located on the waterline, at the point being the ship's center of gravity. The state variables x describing the ship movement are collected in two vectors [6]: $\eta = [x,\ y,\ \psi]^{\mathrm{T}}$ and $\nu = [u,\ v,\ r]^{\mathrm{T}}$. The components of vector η are defined in the inertial frame (X_N, Y_N), while those of vector ν are defined in the movable reference system (X_B, Y_B).

The ideal desired route will consist of a number (N) of segments. The heading ψ_k resulting from a given segment is a clockwise angle determined relative to the X_N axis

$$\psi_k = \operatorname{atan2}(y_{k+1} - y_k, x_{k+1} - x_k) \tag{1}$$

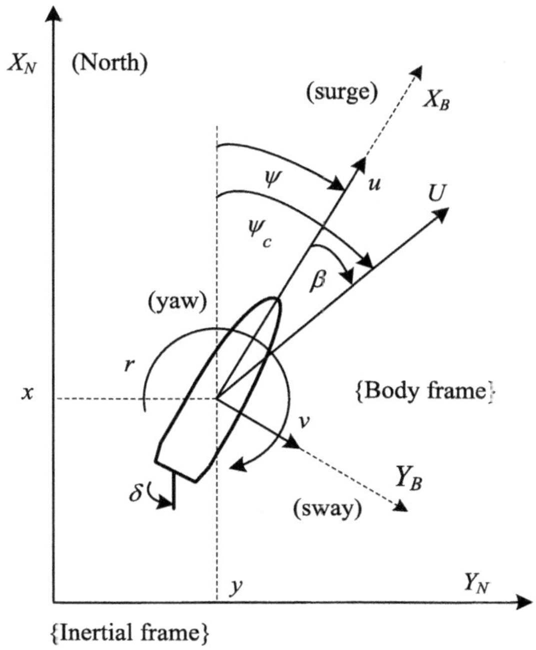

Fig. 1. Variables describing horizontal movement of the ship, (X_N, Y_N) – inertial frame, (X_B, Y_B) – moving coordinate system, (x, y) – position coordinates, ψ – ship heading, u – surge, v – sway, r – yaw, U – total velocity, δ – rudder angle, β – sideslip angle

A circle with radius R_k is created around each way-point (x_k, y_k). When at time t, the ship position (x, y) satisfies the condition

$$L_s = \sqrt{[x_{k+1} - x(t)]^2 + [y_{k+1} - y(t)]^2} \leq R_{k+1} \tag{2}$$

the desired route should be switched to the next segment.

In order to easily determine deviations of ship position relative to the realized segment of the desired path, a third reference coordinate system (X_R, Y_R) has been introduced. The origin of this reference system is located at the starting point (x_k, y_k) of the desired path segment, and the X_R axis coincides with the direction of this segment, determined by the other end located at point (x_{k+1}, y_{k+1}). The heading error ψ^r is defined as:

$$e_\psi(t) = \psi^r = \psi_k - \psi \tag{3}$$

while the ship's cross-track error from the desired path is calculated from the formula

$$e_y(t) = y^r = [x(t) - x_k]sin\psi_k - [y(t) - y_k]cos\psi_k \tag{4}$$

The control problem is to find an algorithm that will allow the ship to follow the ideal set route as precisely as possible (Fig. 2).

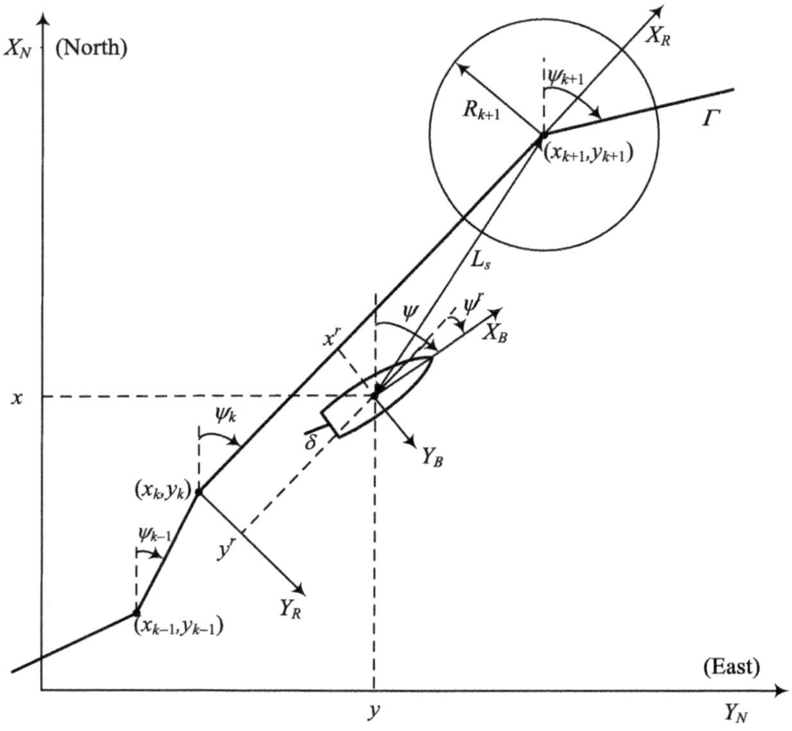

Fig. 2. The desired path of ship's motion and the defined reference coordinate systems.

3 Structure of the Control System

The implementation of the above control problem was carried out in the system shown in Fig. 3. The applied control system has a hierarchical structure, with one supervisory system and a block of switchable regulators. The input signal to this system is the desired route obtained from the system involved in determining the safe path of the ship. This route has a form of a broken line determined by coordinates of subsequent way-points (x_k, y_k).

3.1 Supervisory Switching System

Based on the measured ship position coordinates (x, y) and heading, in combination with the coordinates of two successive way-points (x_k, y_k) and (x_{k+1}, y_{k+1}) of the desired route, the decision maker in the supervisory system selects the operating mode for the ship motion control system. The switching logic used by the decision maker is shown in Fig. 4 in the form of a directed graph, where the nodes are the operating modes of the control system and the switching conditions are recorded on the branches.

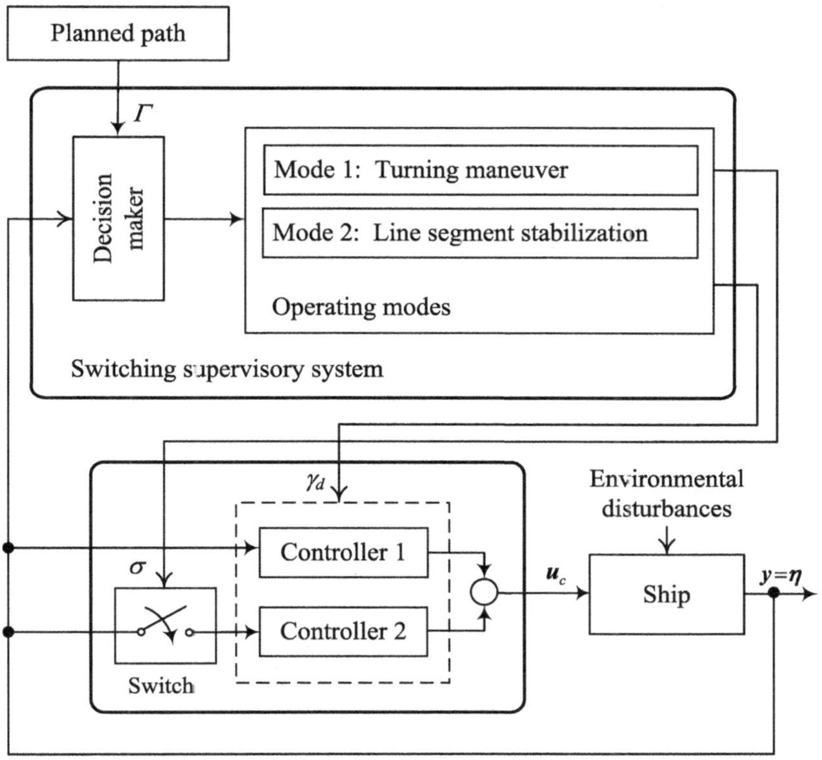

Fig. 3. Block diagram of ship motion control along a desired route

After activating the control system, the decision maker is in Mode 1, to which Regulator 1 is associated, and a turning maneuver is carried out to set the ship heading ψ close to the set course ψ_k resulting from the desired route segment. After fulfilling the condition given by the formula

$$|e_\psi| < 5 \qquad \& \qquad |\dot{e}_\psi| < 0.5 \tag{5}$$

the decision marker switches from Mode 1 to Mode 2, in which the ship motion along the linear segment of the desired route is stabilized using both controllers. Switching from Mode 2 to Mode 1 occurs when the ship is inside the circle R_{k+1} marked around the next way-point (x_{k+1}, y_{k+1}), as described by the relationship (2). The radius of the circle marked around the way-point is the function of the angle between two successive set headings at this way-point,

$$R_k = f\left(|\Delta\psi_k|\right) = f\left(|\psi_{k+1} - \psi_k|\right) \tag{6}$$

The relationship expressed by formula (6) was determined experimentally for the training ship Blue Lady at propeller revolutions $n_g = 440$ rpm. For this

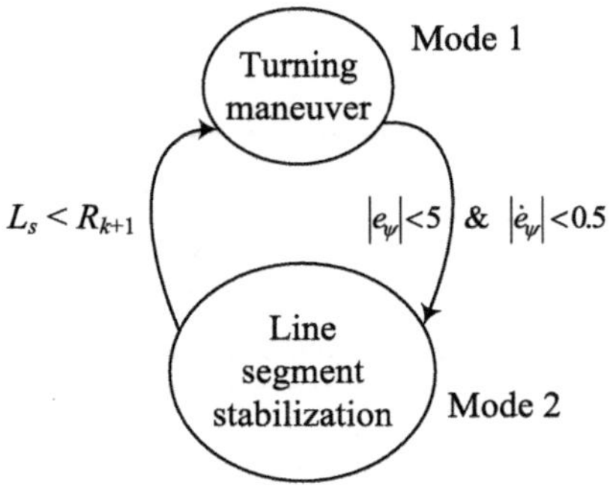

Fig. 4. Switching logic used in the decision maker

purpose, several turning maneuvers were carried out at different angles of course change on two consecutive linear segments of the desired route. The obtained test results are shown in Fig. 5.

In the diagram, the results of the experimental test are marked as asterisks. They were then approximated using the following function:

$$R_k = a_6|\Delta\psi_k|^6 + a_5|\Delta\psi_k|^5 + a_4|\Delta\psi_k|^4 + a_3|\Delta\psi_k|^3 + a_2|\Delta\psi_k|^2 + a_1|\Delta\psi_k| + a_0 \tag{7}$$

The values of the parameters of function (7), determined using the `polyfit` function in the Matlab toolbox [13], are given in Table 1.

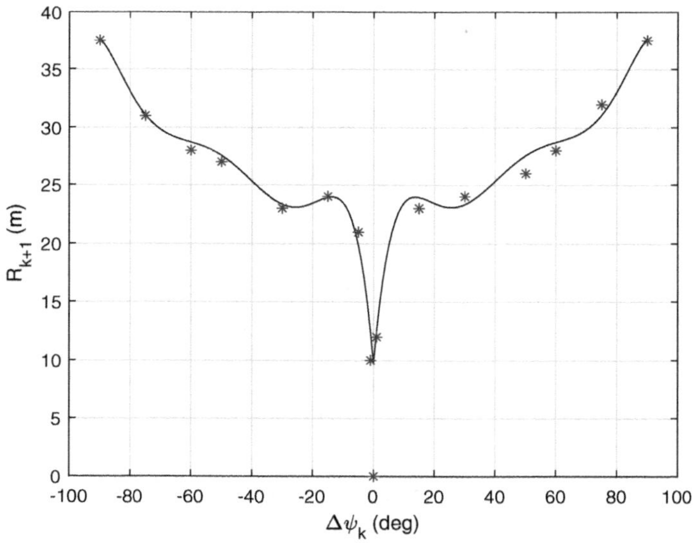

Fig. 5. Experimentally determined dependence for the radii of circle R_{k+1}

The setpoint vector passed from the switching supervisory system to the direct action controllers is as follows: $\gamma_d = [x_k, y_k, x_{k+1}, y_{k+1}]$.

Table 1. Values of parameters of function (7) describing radii of circle R_k

No	Variable	Value	No	Variable	Value
1	a_6	−4.036305e−09	5	a_2	0.2333698861
2	a_5	1.234118e−06	6	a_1	3.0135034920
3	a_4	−1.454676e−04	7	a_0	9.6875672868
4	a_3	0.0082975697			

3.2 Controller 1

In Mode 1, the turning maneuver was carried out using a linear PD controller described by the formula

$$\delta_c = u_{PD} = K_{P\psi} e_\psi + K_{D\psi} \dot{e}_\psi \tag{8}$$

where e_ψ is the heading error (3), and $\dot{e}_\psi = de_\psi/dt$ is its derivative.

3.3 Controller 2

In Mode 2, the linear segment of the desired path was stabilized by additionally activated Controller 2, whose control algorithm is given by formula (9). The task of Controller 2 is to minimize the cross-track error from the desired path

$$\delta_c = \mu \cdot u_{PI} = \mu \left(K_{Py} y^r + K_{Iy} y_I^r \right) \tag{9}$$

where μ is the scaling parameter, ranging between 0 and 1, $y^r = e_y$ is the cross-track error from the linear segment of the desired route (4), and y_I^r is the integral of the cross-track error, which can be saved in the form of an additional state in the controller

$$\dot{y}_I^r = y^r \tag{10}$$

To ensure a smooth transition of control when switching from Mode 1 to Mode 2, the setpoint signal determined in the second control path associated with minimizing the cross-track error from the desired route is scaled by the parameter μ, whose value changes over time, according to the graph shown in Fig. 6b. When at time t_o, the Mode in the control system changes from 1 to 2 (Fig. 6a), the value of this parameter changes linearly from 0 at time t_o to 1 at time $t_o + T$ (Fig. 6b).

The values of the parameters of the two component controllers (Controller 1 and Controller 2) determined in [19] are presented in Table 2.

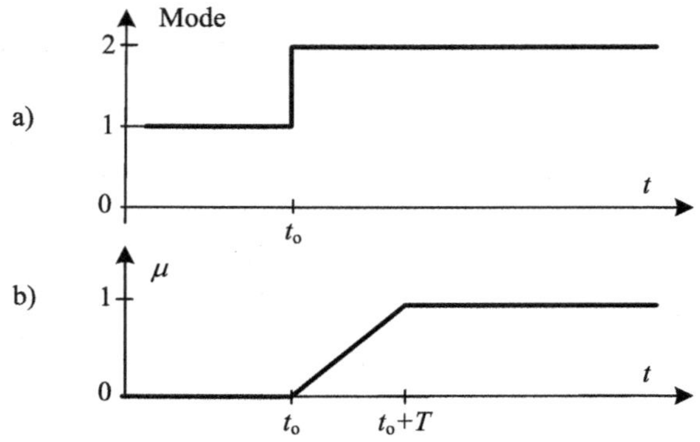

Fig. 6. Characteristic describing changes of parameter μ (9), for $T = 10$ s

Table 2. Parameters of component controllers

$K_{P\psi}$	$K_{D\psi}$	K_{Py}	K_{Iy}
1.60	19.92	2.125	0.05

4 Experimental Results

To check the correctness of the designed control system, both simulation and experimental tests were carried out. The experimental research was carried out on the training ship Blue Lady at the Ship Handling Research and Training Centre located on Lake Silm at Kamionka near Iława, Poland. The ship was in fully loaded condition. During the tests, the wind speed did not exceed 4 m/s. The rotational speed of the main propeller was 440 rpm. The results of tests of the designed control system with switchable structure and scaled output from the second control path, described in Sect. 3, were compared with those obtained for a controller with a similar structure, but without scaling of factor μ, which remained constant and equal to $\mu = 1$ during the tests [20]. The third control system included an algorithm in the form

$$\delta_c = u_{PD} + u_{PI} = K_{P\psi} e_\psi + K_{D\psi} \dot{e}_\psi + K_{Py} y^r + K_{Iy} y_I^r \tag{11}$$

The recorded values of the control errors: heading error (e_ψ), cross-track error from the desired route (e_y), and commanded rudder angle error (δ_c), are shown in Fig. 8 for the second segment of the desired route, between way-points 2 and 3.

The performance of control systems with the tested path controllers was assessed using performance indicators for the heading error (ψ_E), cross-track error (e_{y_E}), and commanded rudder angle (δ_E).

These indicators were considered in their discrete form as:

$$\psi_E = \frac{1}{N}\sum_{i=1}^{N}|e_{\psi i}| \qquad e_{y_E} = \frac{1}{N}\sum_{i=1}^{N}|e_{yi}| \qquad \delta_E = \frac{1}{N}\sum_{i=1}^{N}|\delta_{ci}| \tag{12}$$

The indicator δ_E was determined based on the commanded rudder blade deflection, as in the real-time system, measuring the actual angle of rudder blade was not possible during the tests (Fig. 8).

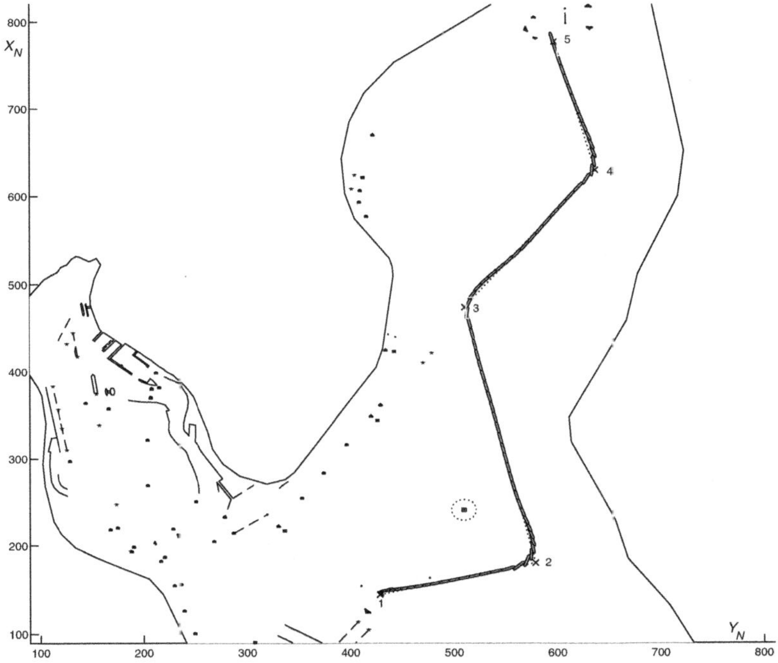

Fig. 7. Sample result of experimental test of ship motion control along a desired route, obtained for the control system with scaled switching

The values of performance indicators determined according to formula (11) for the desired route segment between way-points 2 and 3 are given in Table 3.

Table 3. Performance indicators for the desired route segment between way-points 2 and 3

	ψ_E	e_{yE}	δ_E
PDPI - scaled switching	10.1447	3.0409	7.1190
PDPI - unscaled switching	11.1444	3.8848	9.9954
PDPI - without switching	21.3736	6.1987	18.9114

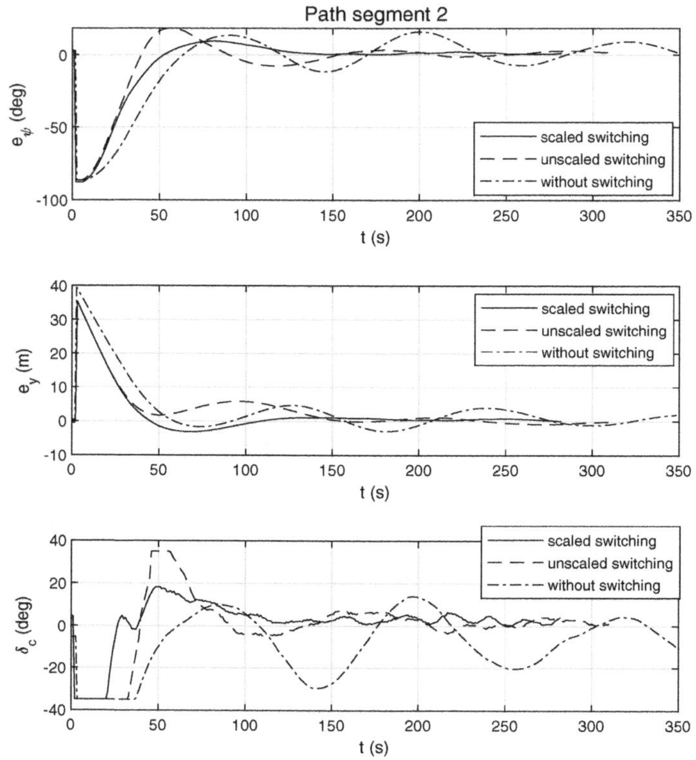

Fig. 8. Time-histories of control errors: heading (e_ψ), cross-track error from desired route (e_y), and commanded rudder angle (δ_c)

5 Conclusions

Experimental tests were carried out to assess the performance of the developed path controller with scaled switching. The obtained results were compared with those recorded using earlier versions of the controller. The first version did not contain any switching, and the two control paths associated with minimizing the heading error (e_ψ) and the cross-track error from a desired segment of the route (e_y) remained active during both turning maneuvers and stabilization along linear segments of the desired route. The more recent version had a switchable controller. During the turning, the controller minimizing the cross-track error from the desired route segment was switched off, and switched on again after the execution of the turning maneuver. In that case, the bump switching was a source of additional overshoots.

The obtained results of experimental tests allow to conclude that the best quality of ship motion control is obtained when using a switchable path controller with scaling of the output value to minimize the ship's cross-track error after switching that track on. This conclusion is confirmed by the time-histories of errors: heading error (e_ψ), cross-track error (e_y), and commanded rudder angle (δ_c), recorded for the selected desired route segment.

References

1. Blomberg, A.E.A., Sæbø, T.O., Hansen, R.E., Pedersen, R.B., Austeng, A.: Automatic detection of marine gas seeps using an interferometric sidescan sonar. IEEE J. Oceanic Eng. **42**(3), 590–602 (2017). https://doi.org/10.1109/JOE.2016.2592559
2. Caccia, M., Bibuli, M., Bono, R., Bruzzone, G., Bruzzone, G., Spirandelli, E.: Unmanned surface vehicle for coastal and protected waters applications: the Charlie project. Mar. Technol. Soc. J. **41**(2), 62–71 (2007). https://doi.org/10.4031/002533207787442259
3. Choyekh, M., Kato, N., Short, T., Ukita, M., Yamaguchi, Y., Senga, H., Yoshie, M., Tanaka, T., Kobayashi, E., Chiba, H.: Vertical water column survey in the Gulf of Mexico using autonomous underwater vehicle SOTAB-I. Mar. Technol. Soc. J. **49**(3), 88–101 (2015). https://doi.org/10.4031/MTSJ.49.3.8
4. Cornfield, S., Young, J.: Unmanned surface vehicles - game changing technology for naval operations. In: Roberts, G.N., Sutton, R. (eds.) Advances in Unmanned Marine Vehicles, pp. 311–328 (2006). https://doi.org/10.1049/PBCE069E_ch15. Chapter 15
5. Fossen, T.I., Breivik, M., Skjetne, R.: Line-of-sight path following of underactuated marine craft. In: Proceedings of the Sixth IFAC Conference Maneuvering and Control of Marine Crafts (MCMC), Girona, Spain, pp. 244–249. (2003). https://doi.org/10.1016/S1474-6670(17)37809-6
6. Fossen, T.I.: Handbook of Marine Craft Hydrodynamics and Motion Control. Wiley, New Jersey (2011)
7. Fredriksen, E., Pettersen, K.Y.: Global κ-exponential way-point maneuvering of ships: theory and experiments. Automatica **42**(4), 677–687 (2006). https://doi.org/10.1016/j.automatica.2005.12.020
8. Isa, K., Arshad, M., Ishak, S.: A hybrid-driven underwater glider model, hydrodynamics estimation, and an analysis of the motion control. Ocean Eng. **81**, 111–129 (2014). https://doi.org/10.1016/j.oceaneng.2014.02.002
9. Jorge, V.A.M., Granada, R., Maidana, R.G., Jurak, D.A., Heck, G., Negreiros, A.P.F., dos Santos, D.H., Gonçalves, L.M.G., Amory, A.M.: A Survey on unmanned surface vehicles for disaster robotics: main challenges and directions. Sensors **19**(3), 702 (2019). https://doi.org/10.3390/s19030702
10. Lazarowska, A.: Research on algorithms for autonomous navigation of ships. WMU J. Marit. Aff. **18**, 341–358 (2019)
11. Lisowski, J.: The sensitivity of state differential game vessel traffic model. Pol. Marit. Res. **23**(2), 14–18 (2016). https://doi.org/10.1515/pomr-2016-0015
12. Łebkowski, A.: Design of an autonomous transport system for coastal areas. TransNav: Int. J. Mar. Navig. Saf. Sea Transp. **12**(1), 117–124 (2018). https://doi.org/10.12716/1001.12.01.13
13. MathWorks. Technical computing software for engineers and scientists. The MathWorks, Inc. http://www.mathworks.com
14. Mohamed-Seghir, M.: Computational intelligence method for ship trajectory planning. In: 21st International Conference on Methods and Models in Automation and Robotics (MMAR), pp. 636–640 (2016). https://doi.org/10.1109/MMAR.2016.7575210

15. Norgren, P., Skjetne, R.: Using autonomous underwater vehicles as sensor platforms for ice-monitoring. Model. Ident. Control **35**(4), 263–277 (2014). https://doi.org/10.4173/mic.2014.4.4
16. Peng, Z., Wang, J., Wang, D.: Containment maneuvering of marine surface vehicles with multiple parameterized paths via spatial-temporal decoupling. IEEE/ASME Trans. Mechatron. **22**(2), 1026–1036 (2017). https://doi.org/10.1109/TMECH.2016.2632304
17. Pettersen, K.Y., Lefeber, E.: Way-point tracking control of ships. In: Proceedings of the 40th IEEE Conference Decision & Control, Orlando, Florida, USA, pp. 940–945 (2001). https://doi.org/10.1109/CDC.2001.980230
18. Szyrowski, T., Sharma, S.K., Sutton, R., Kennedy, G.A.: Subsea cable tracking in an un-certain environment using particle filters. J. Mar. Eng. Technol. **14**(1), 19–31 (2015). https://doi.org/10.1080/20464177.2015.1022381
19. Tomera, M.: Track-keeping of a physical model of the tanker along a specified route. Scientific J. Fac. Electr. Control Eng. Gdansk Uni. Technol. **51**(2016), 201–208 (2016). (in Polish)
20. Tomera, M., Alfuth, Ł.: Way-point path controller for ships, TransNav, Int. J. Mar. Navig. Saf. Sea Transp. Accepted for printing (2020)
21. Xiang, X., Jouvencel, B., Parodi, O.: Coordinated formation control of multiple autonomous underwater vehicles for pipeline inspection. Int. J. Adv. Robot. Syst. **7**(1), 75–84 (2010). https://doi.org/10.5772/7242
22. Xiang, X., Yu, C., Zheng, J., Xu, G.: Motion forecast of intelligent underwater sampling apparatus - part I: design and algorithm. Indian J. Geo-Mar. Sci. **44**(12), 1962–1970 (2015)
23. Zubowicz, T., Armiński, K., Witkowska, A., Śmierzchalski, R.: Marine autonomous surface ship - control system configuration. IFAC-PapersOnLine **52**(8), 409–415 (2019). https://doi.org/10.1016/j.ifacol.2019.08.100
24. Zwierzewicz, Z.: Robust and adaptive ship path-following control design with the full vessel model. In: 24th International Conference on Methods and Models in Automation and Robotics (MMAR), pp. 121–126 (2019). https://doi.org/10.1109/MMAR.2019.8864687

Autonomous Ship Utility Model Parameter Estimation Utilising Extended Kalman Filter

Anna Witkowska[1(✉)], Krzysztof Armiński[1], Tomasz Zubowicz[2], Filip Ossowski[1,2], and Roman Śmierzchalski[1]

[1] Department of Automatic Control, Gdańsk University of Technology, 80-233 Gdańsk, Poland
{anna.witkowska,roman.smierzchalski}@pg.edu.pl,
krzysztof.arminski@gmail.com
[2] Department of Electrical Engineering, Control Systems and Informatics, Gdańsk University of Technology, 80-233 Gdańsk, Poland
tomasz.zubowicz@pg.edu.pl
https://pg.edu.pl/anna.witkowska, https://pg.edu.pl/tomasz.zubowicz

Abstract. In this paper, a problem of autonomous ship utility model identification for control purposes is considered. In particular, the problem is formulated in terms of model parameter estimation (one-step-ahead prediction). This is a complex task due to lack of measurements of the parameter values, their time-variability and structural uncertainty introduced by the available models. In this work, authors consider and compare two utility models based on often utilised ship model structures with time-varying parameters identified recursively using the extended Kalman lter (EKF). The validation results have been obtained using simulation experiments in which the required information for the parameter estimation task had been generated using a cognitive model of B-481 ship. The results indicate the benefits and drawbacks, in terms of estimation accuracy and computational complexity, of using each of the investigated utility model structures.

Keywords: Extended Kalman filter · Ship parameter identification · Estimation

1 Introduction

Marine autonomous surface ships (MASS) are considered the future of maritime transport. Transport of products by sea is currently identified both as most efficient in terms of economic and ecological factors. Introducing autonomy to ship operation opens up new possibilities in the field of construction, design as well as means and manner of exploitation of the unit. Reduction or even lack of crew on board of a MASS has the potential to allow e.g. to improve storage capacity, reduce initial weight and operational costs while maintaining the same size of the vessel.

A. Bartoszewicz et al. (Eds.): PCC 2020, AISC 1196, pp. 1531–1542, 2020.
https://doi.org/10.1007/978-3-030-50936-1_127

Introducing autonomy to a given unit to obtain a MASS type behaviour requires significant amount of technical work related to i.e. advanced control system design. Notably, advanced control systems often rely on a knowledge of the unit dynamics (mathematical model) as well as rich measurement information feed. These two factors are considered critical to establish so-called situation awareness and acquire certain adaptation capabilities to enable operation in disturbed and uncertain operational—sea—conditions.

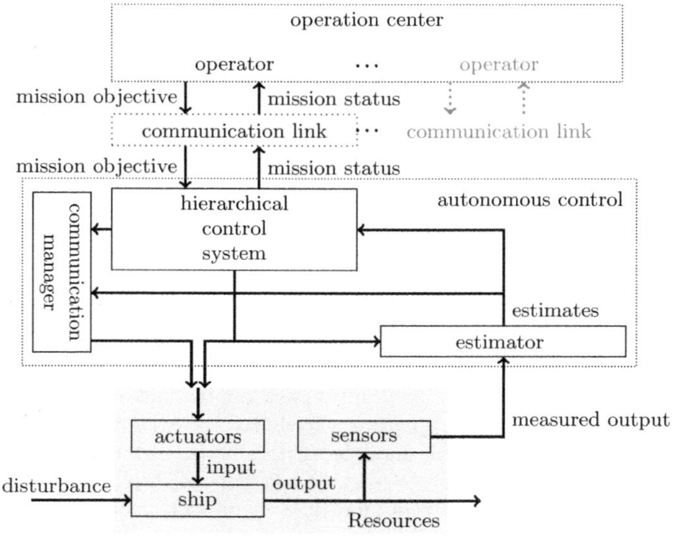

Fig. 1. Autonomous control structure

Currently, there is no comprehensive approach to the design of a MASS type control system. To that goal, in previous work, a general MASS control structure (Fig. 1) has been proposed in [16]. The proposed MASS control system enables autonomous operation understood in terms of enabling the unit to carry out mission objectives and adjust individual behavioural strategy concerning current operating state and conditions. To that goal a situation awareness is established by utilising advanced estimation and prediction techniques represented by an estimator block. The acquired information is redistributed in between the sub-systems by a dedicated communication manager. The tactics and strategy related to carrying out mission objectives is implemented by a hierarchical control system [16]. The reality of operating at sea forces one to focus not only on normal and disturbance operating conditions but, in principal predict, avoid or, in unfavourable situations, be able to mitigate the effects of emergency e.g. related to possible unit collision threat [10,13,14]. The proposed control system architecture has the potential to deliver theses functionalities. Typically, these tasks are carried out by, adjusting heading and speed of a unit, following specific control objectives, e.g.: stabilisation of ship's course and speed in straight-line motion; stabilisation of ship's angular speed during the course change over manoeuvre; changes of ship's speed during stabilisation on a given fixed course.

As it has been already mentioned, the construction of the autonomous control system requires certain advanced control structures and algorithms. These, on the other hand, are often based on an appropriate mathematical model, also including state and output constraints. It is also a fact, that due to high impact of hydro-dynamical factors depending on the steering objectives, distinct ship models or model parameter sets are required to describe MASS dynamics at distinct operational states and conditions. The fundamental division of mathematical ship models includes high-speed manoeuvring and low-speed manoeuvring, due to the significantly different dynamic properties of a ship moving at low and high speed. The higher speed is considered to be the speed above 2–3 knots—then it is controlled by a given trajectory with the course and/or speed [13]. Control of a vessel whose speed is lower than 2–3 knots refers to the task of dynamic positioning of the vessel—precise control of position and course [14]. In this work, authors focus on the course stabilisation task.

To achieve the goal, namely course stabilisation during the trajectory tracking, it is assumed that the resultant velocity of the ship is approximately constant, while the rudder angle deflection remains bound to a small range around its nominal position. This assumption makes it possible to reduce the order of the model describing the dynamics of the ship (e.g. neglecting the output of the drift angle and the angle of sway). Though, it results in a need to acquire multiple constant parameter value models (or single parameter varying model) to describe distinct operational modes of the unit (ship). In practice, the relationships of state variables, input and output signals are analysed, usually based on least squares (LS) e.g. [1] or weighted least squares (WLS), as well as recursive least squares (RLS) methods. Given the *a priori* set of information the model obtained using this approach typically achieves satisfactory prediction capabilities under the established operating conditions. Another branch of algorithms presented in literature focuses on the use of estimation methods [3], such as Kalman filter(KF) for linear models [8], extended Kalman filter (EKF) for linearised models [11]. Less frequently used are the unscented Kalman filter(UKF), which does not require model linearisation [9]. This also applies to an ensemble Kalman filter (ENKF) e.g. [2], which is based on the idea similar to Particle filter (PF) e.g. [7]. The advantage of the latter is that PF does not require Gaussian probability distributions, so throughout this assumption, it is considered less restrictive (or conservative) in contrast to e.g. ENKF. The relatively new approach is the EKF with many innovations (so-called multi-innovation EKF) and the forgetting factor [15]. This modification of KF increases identification accuracy and improves the convergence of the estimation error as indicated by the authors [15].

In this work, a utility model identification problem is addressed. The proposed approach is composed of two stages related to (mathematical) model structure selection and (subsequent) parameters identification. The obtained model is to sever control design purposes and in particular stabilising control of ship's course and speed during straight-line motion. The first stage, related to the structure selection is assumed to be performed based on available *a priori* knowledge of the process and requirements regarding control objectives. The second stage, related to parameter identification, is formulated using parametric optimisation framework originating from the field of operations research. Finally, under the presented circumstances, the formulated optimisation task is solved recursively, due to the non-stationary nature of the parameters, using EKF. The proposed approach is illustrated using two distinct utility model structures, which differ in complexity, number of state variables and estimated parameters. For comparative purposes, the experimental data was generated using a cognitive (mathematical) model of B-481 ship.

The contribution of this work is a comparative study of the state and parameter estimation and prediction performance obtained using two distinct utility model structures, both kept in the time-varying parameter model format.

The following parts of the manuscript are organised in the following manner. In Sect. 2 the model identification problem is formulated. Following, in Sect. 3 the EKF algorithm is introduced. Section 4 illustrates the proposed approach with numerical results. Section 5 concludes the paper.

2 Problem Formulation

Let $\mathbb{R}^n$ denote a n-dimensional vector space over a field of real numbers $\mathbb{R}$. Taking $\mathcal{F}_{\text{SS}}$ a vector field set up to represent the ship dynamics or its cognitive model and to serve as an input-output data source it follows from the discussion provided in the introduction, that the $\mathcal{F}_{\text{SS}} \stackrel{\text{def}}{=} \mathcal{F}_{\text{SS}}\left(\boldsymbol{x}(t), \boldsymbol{u}(t), \boldsymbol{d}(t), \boldsymbol{\theta}(t)\right)$, where at each time instant t: $(\boldsymbol{x}(t), \boldsymbol{u}(t), \boldsymbol{d}(t), \boldsymbol{\theta}(t)) \in (\mathbb{R}^{n_x}, \mathbb{R}^{n_u}, \mathbb{R}^{n_d}, \mathbb{R}^{n_\theta})$ yield: state, control input, disturbance input and parameter vectors, respectively, contained in corresponding spaces of appropriate dimensions. It follows, that including *a priori* knowledge on $\boldsymbol{\theta}(t)$, which indicates that at each t: $\boldsymbol{\theta}(t) \in \Omega_\theta$, where Ω_θ is a set of admissible parameters, and an assessment criterion $J(\boldsymbol{\theta}(t))$ the parameter identification task yields:

$$\boldsymbol{\theta}^*(t) = \arg \min_{\boldsymbol{\theta}(t) \in \Omega_{\text{FSS}}} J\left(\boldsymbol{\theta}(t)\right) \tag{1}$$

where: $\boldsymbol{\theta}^*(t)$ is the optimal time realisation of the parameters over considered time horizon and that at each t: $\Omega_{\text{FSS}} \equiv \Omega_\theta \times \mathcal{M}$. The latter construct is to indicate the fact that, apart from the *a priori* knowledge on $\boldsymbol{\theta}(t)$, the time realisation of the parameters is restrained by $\mathcal{F}_{\text{SS}}$ (ship dynamics) approximated by $\mathcal{M}$, which is essential.

A general $\mathcal{F}_{\text{SS}}$ (B-481 ship) configuration has been indicated in Fig. 2. It follows that, by dropping the time dependency to shorten the notation, the control input vector is considered as $\boldsymbol{u} \stackrel{\text{def}}{=} [\delta, n, H]^T$, where its components denote the rudder angle and propeller revolution and pitch, respectively. The $\boldsymbol{d} \stackrel{\text{def}}{=} [V_{\text{wsr}}, V_{\text{p}}, \gamma_{\text{w}}, \gamma_{\text{p}}, \gamma_{\text{f}}]^T$ and V_{wsr}, γ_{w} are the average wind speed and direction, respectively. γ_{f} denotes the wave direction, and V_{p}, γ_{p} denote the sea current speed and related direction, respectively. The vector $\boldsymbol{\eta} \stackrel{\mathrm{def}}{=} [x, y, z, \phi, \theta, \psi]^T \in \mathbb{R}^6$ consists of ship's position (x, y, z) and roll, pitch, yaw (heading) $\phi, \theta, \psi \in [0, 2\pi]$ in the Earth-fixed frame driven by $\boldsymbol{v}$. Finally, the measurement available quantities are collected into output vector $\boldsymbol{y} \in \mathbb{R}^{n_y}$. More details can be found in e.g. [16].

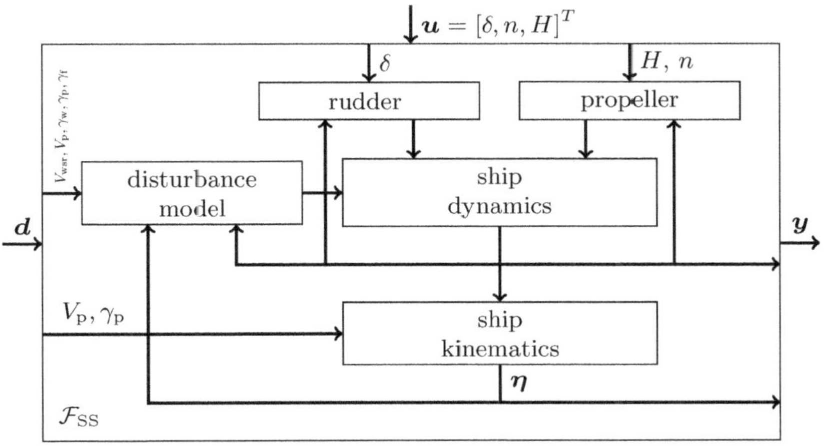

Fig. 2. A general $\mathcal{F}_{\mathrm{SS}}$ configuration

The presented research objectives have been met with respect to the following assumption.

Assumption 1. *The ship ($\mathcal{F}_{\mathrm{SS}}$) operates under constant speed and small rudder deviations from its nominal position.*

Assumption 2. *The ship model's ($\mathcal{M}$) structure is known and is given by C^1 class functions.*

Assumption 3. *The ship model's ($\mathcal{M}$) parameter vector ($\boldsymbol{\theta}$) is composed of unmeasurable, constant or low-frequency parameters.*

Under the Assumption 1 the pursued utility model is to serve stabilising heading and speed control system design. Assumption 2 enables one to conceive solution to (1) numerically. Finally, Assumption 3 implies a model for parameter variability.

A procedure proposed to provide a solution to the formulated problem using EKF has been presented in the following section.

3 Identification Procedure

Since a recursive—EKF—algorithm has been selected to solve (1), the problem at hand is discussed in a discrete time format. Hence, by taking a set of non-negative integers $\mathbb{Z}_+$ and T_s to denote a sample time with respect to Shannon Sampling Law, at each time instant $t = kT_\mathrm{s}$, $\forall k \in \mathbb{Z}_+$, it follows that:

$$\begin{aligned}(\boldsymbol{x}(k),\, \boldsymbol{u}(k),\, \boldsymbol{\theta}(k),\, \boldsymbol{y}(k)) &\stackrel{\text{def}}{=} (\boldsymbol{x}(kT_\mathrm{s}),\, \boldsymbol{u}(kT_\mathrm{s}),\, \boldsymbol{\theta}(kT_\mathrm{s}),\, \boldsymbol{y}(kT_\mathrm{s})) \\ &\in (\mathbb{R}^{n_x},\, \mathbb{R}^{n_u},\, \mathbb{R}^{n_\theta},\, \mathbb{R}^{n_y}) .\end{aligned} \tag{2}$$

3.1 General Model Structure

To mathematically describe the dynamics of the ship's unit ($\mathcal{F}_{\mathrm{SS}}$) a general state-space model in a discrete-time format is considered, as:

$$\mathcal{M}: \begin{cases} \boldsymbol{y}(k) &= \boldsymbol{H}\boldsymbol{x}(k) + \boldsymbol{v}(k) \\ \boldsymbol{x}(k+1) &= \boldsymbol{f}\left(\boldsymbol{x}(k), \boldsymbol{u}(k), \boldsymbol{\theta}(k)\right) + \boldsymbol{z}(k) \end{cases}, \tag{3}$$

with a stochastic system (process) disturbance vector $\boldsymbol{z}(k) \in \mathbb{R}^{n_z}$ and measurement noise vector $\boldsymbol{v}(k) \in \mathbb{R}^{n_v}$ given by:

$$\boldsymbol{z}(k) \sim \mathcal{N}(0, \boldsymbol{Z}(k)) \tag{4a}$$

$$\boldsymbol{v}(k) \sim \mathcal{N}(0, \boldsymbol{V}(k)) \tag{4b}$$

where: $\boldsymbol{Z}(k)$ and $\boldsymbol{V}(k)$ denote the corresponding covariance matrices of uncorrelated white noise Gaussian processes (4), $\boldsymbol{f}$ is a vector of nonlinear functions that, under Assumption 2, belong to C^1 and $\boldsymbol{H} \in \mathbb{R}^{n_y \times n_x}$ is an output matrix.

3.2 On-Line Parameter Estimation Using EKF

Taking (3) and considering Assumption 3 a one-step-ahead state prediction model is obtained:

$$\hat{\mathcal{M}}: \begin{cases} \hat{\boldsymbol{y}}(k) &= \boldsymbol{H}\hat{\boldsymbol{x}}(k) \\ \hat{\boldsymbol{x}}(k+1) &= \boldsymbol{f}\left(\hat{\boldsymbol{x}}(k), \boldsymbol{u}(k), \hat{\boldsymbol{\theta}}(k)\right) \\ \hat{\boldsymbol{\theta}}(k+1) &= \hat{\boldsymbol{\theta}}(k) \end{cases}, \tag{5}$$

were: $\hat{(\cdot)}$ is to denote an estimate of the underlying element $(\cdot)$. Consequently, by defining an extended state vector $\hat{\boldsymbol{\zeta}}(k) \stackrel{\text{def}}{=} [\hat{\boldsymbol{x}}^T(k),\, \hat{\boldsymbol{\theta}}^T(k)]^T$ and introducing a double time index notation, (5) is rewritten as:

$$\hat{\mathcal{M}}: \begin{cases} \hat{\boldsymbol{y}}(k|k-1) &= \boldsymbol{H}_{\mathrm{ex}}\hat{\boldsymbol{\zeta}}(k|k-1) \\ \hat{\boldsymbol{\zeta}}(k+1|k) &= \boldsymbol{f}_{\mathrm{ex}}\left(\hat{\boldsymbol{\zeta}}(k|k-1), \boldsymbol{u}(k)\right) \end{cases}, \tag{6}$$

where the structure of both $\boldsymbol{f}_{\text{ex}}$ and $\boldsymbol{H}_{\text{ex}}$ results directly form the above considerations.

Introducing a measurement feedback to (6), using the EKF estimation rules, in their classic form, enables one to calculate the estimate $\hat{\boldsymbol{\zeta}}(k)$ recursively. This is being done by invoking the following algorithm:

$$\hat{\boldsymbol{\zeta}}(k|k-1) = \boldsymbol{f}_{\text{ex}}\left(\hat{\boldsymbol{\zeta}}(k-1), \boldsymbol{u}(k-1)\right), \tag{7a}$$

$$\boldsymbol{P}(k|k-1) = \boldsymbol{A}(k-1)\boldsymbol{P}(k-1)\boldsymbol{A}^T(k-1) + \boldsymbol{Z}(k-1), \tag{7b}$$

$$\boldsymbol{K}(k) = \boldsymbol{P}(k|k-1)\boldsymbol{H}_{\text{ex}}^T(k)\left(\boldsymbol{H}_{\text{ex}}(k)\boldsymbol{P}(k|k-1)\boldsymbol{H}_{\text{ex}}^T(k) + \boldsymbol{V}(k)\right)^{-1}, \tag{7c}$$

$$\begin{aligned}\boldsymbol{P}(k) &= \left[I - \boldsymbol{K}(k)\boldsymbol{H}_{\text{ex}}(k)\right]\boldsymbol{P}(k|k-1)\left[I - \boldsymbol{K}(k)\boldsymbol{H}_{\text{ex}}(k)\right]^T, \\ &\quad + \boldsymbol{K}(k)\boldsymbol{V}(k)\boldsymbol{K}^T(k),\end{aligned} \tag{7d}$$

$$\hat{\boldsymbol{\zeta}}(k) = \hat{\boldsymbol{\zeta}}(k|k-1) + \boldsymbol{K}(k)\left(\boldsymbol{y}(k) - \boldsymbol{H}_{\text{ex}}\hat{\boldsymbol{\zeta}}(k|k-1)\right), \tag{7e}$$

$$\boldsymbol{A}(k) = \left.\frac{\partial \boldsymbol{f}_{\text{ex}}}{\partial \boldsymbol{\zeta}}\right|_{\hat{\boldsymbol{\zeta}}(k)}. \tag{7f}$$

The pair of Eqs. (7a) and (7b) constitutes a predictive part of the EKF, which on the basis of data from the previous time instant $k-1$ generates the prediction for the current time instant t of the state $\hat{\boldsymbol{\zeta}}(k|\,k-1)$ and the error covariance matrix $\boldsymbol{P}(k|\,k-1)$. In the next step, an update of the prediction is being made, on the basis of the current measurement feedback acquired at k. Then the current estimate of the state $\hat{\boldsymbol{\zeta}}(k)$ and the covariance matrix $\boldsymbol{P}(k)$ is determined. Both updates are performed by minimising the estimate covariance using the Kalman's gain matrix $\boldsymbol{K}(k)$. Additionally, in order to initialise the algorithm it is necessary to know the initial conditions $\hat{\boldsymbol{\zeta}}(0) = \boldsymbol{\zeta}_0$ and $\boldsymbol{P}(0) = \boldsymbol{P}_0$. These are identified from the *a priori* information acquired form the plant (ship).

Finally, it follows from the above description, that at each k the generated estimates $\hat{\boldsymbol{\zeta}}(k)$ are unbiased, i.e. $\mathrm{E}\left\{\hat{\boldsymbol{\zeta}}(k)\right\} = \mathrm{E}\left\{\boldsymbol{\zeta}(k)\right\}$ and optimal in the sense of minimal covariance $\boldsymbol{P}(k)$ of estimation error $\hat{\boldsymbol{e}}(k) \stackrel{\text{def}}{=} \boldsymbol{\zeta}(k) - \hat{\boldsymbol{\zeta}}(k)$.

4 Numerical Results

4.1 Experiment Setup, Methods and Tools

Depending on the simplification level, the two example model structures, namely $\mathcal{M}_1$ and $\mathcal{M}_2$, are taken into account for application of EKF filter. In particular, $\mathcal{M}_1(\boldsymbol{x}_{\mathcal{M}_1}, u, \boldsymbol{\theta}_{\mathcal{M}_1})$ SISO type model, where the control input is the rudder angle and the yaw angle as a state variable; $\mathcal{M}_2(\boldsymbol{x}_{\mathcal{M}_2}, u, \boldsymbol{\theta}_{\mathcal{M}_2})$ SIMO type model, where the control input is the rudder angle, the state vector created by the following variables: positions in x- and y- directions, yaw angle, surge, sway and yaw velocities. Moreover, under Assumption 1, a restriction in using a model $\mathcal{M}_1$ there are small changes in rudder angle, with a maximum amplitude of

$|\delta_{max}(t)| = 5°$. Model $\mathcal{M}_2$ can be used for both small and large amounts of excitation, with maximum amplitude $|\delta_{max}(t)| = 40°$. In both models the maximum speed has been assumed $|d_t\delta_{max}(t)| = 5°/\text{s}$, e.g. [5].

Mathematical Model $\mathcal{M}_1(\boldsymbol{x}_{\mathcal{M}_1}, \boldsymbol{u}_{\mathcal{M}_1}, \boldsymbol{\theta}_{\mathcal{M}_1})$. Let the structure of the ship mathematical model $\mathcal{M}_1(\boldsymbol{x}_{\mathcal{M}_1}, \boldsymbol{u}_{\mathcal{M}_1}, \boldsymbol{\theta}_{\mathcal{M}_1})$, implement the simplified Nomoto model of the vessel dynamics in 1 DOF [6,12], which deterministic part in continuous time domain is given by:

$$\mathcal{M}_1 : \begin{cases} \mathrm{d}_t\psi_{\mathcal{M}_1}(t) &= r_{\mathcal{M}_1}(t) \\ \mathrm{d}_t r_{\mathcal{M}_1}(t) &= a_{\mathcal{M}_1} r_{\mathcal{M}_1}(t) - b_{\mathcal{M}_1} r^3_{\mathcal{M}_1}(t) + d_{\mathcal{M}_1} + c_{\mathcal{M}_1}\delta_{\mathcal{M}_1} \end{cases}, \tag{8}$$

where: $\boldsymbol{x}_{\mathcal{M}_1}(t) = [\psi_{\mathcal{M}_1}(t), r_{\mathcal{M}_1}(t)]^T$—state vector: yaw angle, yaw velocity, $\boldsymbol{u}_{\mathcal{M}_1}(t) = \delta_{\mathcal{M}_1}(t)$—input vector, ruder angle, $\boldsymbol{\theta}_{\mathcal{M}_1}(t) = [a_{\mathcal{M}_1}, b_{\mathcal{M}_1}, d_{\mathcal{M}_1}, c_{\mathcal{M}_1}]^T$—unknown parameter vector. The model equations describe the angular acceleration $r'_{\mathcal{M}_1}(t)$ of the vessel in relation to the angular velocity $r_{\mathcal{M}_1}(t)$, the rudder angle $\delta_{\mathcal{M}_1}(t)$ and a vector of parameters $\boldsymbol{\theta}_{\mathcal{M}_1}(t)$, the value of which depends on the uncertainty resulting from states, environmental disturbances (eq. the directions and speeds of waves, currents and wind).

Finally, $\mathcal{M}_1(\boldsymbol{x}_{\mathcal{M}_1}, \boldsymbol{u}_{\mathcal{M}_1}, \boldsymbol{\theta}_{\mathcal{M}_1})$ is obtained by discretising (8) using forward finite difference method.

Mathematical Model $\mathcal{M}_2(\boldsymbol{x}_{\mathcal{M}_2}, \boldsymbol{u}_{\mathcal{M}_2}, \boldsymbol{\theta}_{\mathcal{M}_2})$. Consider $\mathcal{M}_2(\boldsymbol{x}_{\mathcal{M}_2}, \boldsymbol{u}_{\mathcal{M}_2}, \boldsymbol{\theta}_{\mathcal{M}_2})$ to represent the ship dynamics in 3DOF [4,5], which deterministic part in continuous time domain is given by:

$$\mathcal{M}_2 : \mathrm{d}_t\boldsymbol{x}_{\mathcal{M}_2}(t) = \boldsymbol{F}_{\mathcal{M}_2}(\boldsymbol{x}_{\mathcal{M}_2}, \boldsymbol{u}_{\mathcal{M}_2}, \boldsymbol{\theta}_{\mathcal{M}_2}), \tag{9}$$

where $\boldsymbol{x}_{\mathcal{M}_2}(t) \stackrel{\text{def}}{=} [x_{\mathcal{M}_2}(t), y_{\mathcal{M}_2}(t), \psi_{\mathcal{M}_2}(t), u_{\mathcal{M}_2}(t), v_{\mathcal{M}_2}(t), r_{\mathcal{M}_2}(t)]^T$ denotes the state vector (positions in x- and y- directions, yaw angle, surge, sway and yaw velocities), $\boldsymbol{u}_{\mathcal{M}_2}(t) \stackrel{\text{def}}{=} \delta_{\mathcal{M}_2}(t)$ is the input vector (ruder angle),

$$\boldsymbol{\theta}_{\mathcal{M}_2}(t) \stackrel{\text{def}}{=} [a_{\mathcal{M}_2 1}, a_{\mathcal{M}_2 2}, b_{\mathcal{M}_2 1}, \ldots, b_{\mathcal{M}_2 10}, c_{\mathcal{M}_2 1}, \ldots, c_{\mathcal{M}_2 15}, d_{\mathcal{M}_2 1}, \ldots, d_{\mathcal{M}_2 15}]^T \tag{10}$$

is the unknown parameter vector, $\boldsymbol{F}_{\mathcal{M}_2} \stackrel{\text{def}}{=} [f_{\mathcal{M}_2 1},\ f_{\mathcal{M}_2 2},\ \ldots,\ f_{\mathcal{M}_2 6}]^T$ and:

$$f_{\mathcal{M}_2 1} \stackrel{\text{def}}{=} a_{\mathcal{M}_2 1}\cos\psi_{\mathcal{M}_2}(t) + u_{\mathcal{M}_2}(t)\cos\psi_{\mathcal{M}_2}(t) - v_{\mathcal{M}_2}\sin\psi_{\mathcal{M}_2}(t), \tag{11a}$$

$$f_{\mathcal{M}_2 2} \stackrel{\text{def}}{=} a_{\mathcal{M}_2 2}\sin\psi_{\mathcal{M}_2}(t) + u_{\mathcal{M}_2}(t)\sin\psi_{\mathcal{M}_2}(t) + v_{\mathcal{M}_2}\cos\psi_{\mathcal{M}_2}(t), \tag{11b}$$

$$f_{\mathcal{M}_2 3} \stackrel{\text{def}}{=} r_{\mathcal{M}_2}(t), \tag{11c}$$

$$\begin{aligned} f_{\mathcal{M}_2 4} \stackrel{\text{def}}{=}\ & b_{\mathcal{M}_2 1}u_{\mathcal{M}_2}(t) + b_{\mathcal{M}_2 2}u^2_{\mathcal{M}_2}(t) + b_{\mathcal{M}_2 3}u^3_{\mathcal{M}_2}(t) + b_{\mathcal{M}_2 4}v^2_{\mathcal{M}_2}(t) \\ & + b_{\mathcal{M}_2 5}r^2_{\mathcal{M}_2}(t) + b_{\mathcal{M}_2 6}r_{\mathcal{M}_2}(t)v_{\mathcal{M}_2}(t) + b_{\mathcal{M}_2 7}\delta^2_{\mathcal{M}_2}(t) \\ & + b_{\mathcal{M}_2 8}u_{\mathcal{M}_2}(t)\delta^2_{\mathcal{M}_2}(t) + b_{\mathcal{M}_2 9}v_{\mathcal{M}_2}(t)\delta_{\mathcal{M}_2}(t) \\ & + b_{\mathcal{M}_2 10}u_{\mathcal{M}_2}(t)v_{\mathcal{M}_2}(t)\delta_{\mathcal{M}_2}(t) \end{aligned} \tag{11d}$$

$$
\begin{aligned}
f_{\mathcal{M}_2\,5} \stackrel{\text{def}}{=}\; & c_{\mathcal{M}_2\,1} v_{\mathcal{M}_2}(t) + c_{\mathcal{M}_2\,2} r_{\mathcal{M}_2}(t) + c_{\mathcal{M}_2\,3} v^3_{\mathcal{M}_2}(t) \\
& + c_{\mathcal{M}_2\,4} v^2_{\mathcal{M}_2}(t) r_{\mathcal{M}_2}(t) + c_{\mathcal{M}_2\,5} v_{\mathcal{M}_2}(t) u_{\mathcal{M}_2}(t) \\
& + c_{\mathcal{M}_2\,6} r_{\mathcal{M}_2}(t) u_{\mathcal{M}_2}(t) + c_{\mathcal{M}_2\,7} \delta_{\mathcal{M}_2}(t) \\
& + c_{\mathcal{M}_2\,8} \delta^3_{\mathcal{M}_2}(t) + c_{\mathcal{M}_2\,9} r_{\mathcal{M}_2}(t) \delta_{\mathcal{M}_2}(t) + c_{\mathcal{M}_2\,10} u^2_{\mathcal{M}_2}(t) \delta_{\mathcal{M}_2}(t) \\
& + c_{\mathcal{M}_2\,11} r_{\mathcal{M}_2}(t) \delta^2_{\mathcal{M}_2}(t) + c_{\mathcal{M}_2\,12} v^2_{\mathcal{M}_2}(t) \delta_{\mathcal{M}_2}(t) + c_{\mathcal{M}_2\,13} u^2_{\mathcal{M}_2}(t) \\
& + c_{\mathcal{M}_2\,14} u_{\mathcal{M}_2}(t) + c_{\mathcal{M}_2\,15},
\end{aligned} \tag{11e}
$$

$$
\begin{aligned}
f_{\mathcal{M}_2\,6} \stackrel{\text{def}}{=}\; & d_{\mathcal{M}_2\,1} v_{\mathcal{M}_2}(t) + d_{\mathcal{M}_2\,2} r_{\mathcal{M}_2}(t) + d_{\mathcal{M}_2\,3} v^3_{\mathcal{M}_2}(t) \\
& + d_{\mathcal{M}_2\,4} v^2_{\mathcal{M}_2}(t) r_{\mathcal{M}_2}(t) + d_{\mathcal{M}_2\,5} v_{\mathcal{M}_2}(t) u_{\mathcal{M}_2}(t) \\
& + d_{\mathcal{M}_2\,6} r_{\mathcal{M}_2}(t) u_{\mathcal{M}_2}(t) + d_{\mathcal{M}_2\,7} \delta_{\mathcal{M}_2}(t) + d_{\mathcal{M}_2\,8} \delta^3_{\mathcal{M}_2}(t) \\
& + d_{\mathcal{M}_2\,9} r_{\mathcal{M}_2}(t) \delta_{\mathcal{M}_2}(t) + d_{\mathcal{M}_2\,10} u^2_{\mathcal{M}_2}(t) \delta_{\mathcal{M}_2}(t) \\
& + d_{\mathcal{M}_2\,11} r_{\mathcal{M}_2}(t) \delta^2_{\mathcal{M}_2}(t) + d_{\mathcal{M}_2\,12} v^2_{\mathcal{M}_2}(t) \delta_{\mathcal{M}_2}(t) \\
& + d_{\mathcal{M}_2\,13} u^2_{\mathcal{M}_2}(t) + d_{\mathcal{M}_2\,14} u_{\mathcal{M}_2}(t) + d_{\mathcal{M}_2\,15}.
\end{aligned} \tag{11f}
$$

Finally, $M_2(x, u, \theta)$ is obtained by discretising (8) using forward finite difference method.

Experiment Description. Using $\mathcal{M}_1$ and $\mathcal{M}_2$ as a predictive part of EKF, two distinct estimators have been constructed. Both EKFs have been coupled with a cognitive model ($\mathcal{F}_{\text{SS}}$) of the B-481 ship to numerically asses their performance. The computed scenario assumed known rudder displacement profile (Fig. 3a) and negligible acting disturbances ($\boldsymbol{d}$).

4.2 Results and Discussion

In Fig. 3 the results regarding ship kinematics obtained during simulation experiments have been illustrated. In particular, in Fig. 3a the $\mathcal{F}_{\text{SS}}$ rudder displacement scenario has been depicted. In Fig. 3b the corresponding kinematic response of the unit ($\mathcal{F}_{\text{SS}}$), namely position change in XY coordinate frame, its measurement, estimate and prediction (using final estimate constant parameter set) using EKF equipped with $\mathcal{M}_2$ have been presented. In this case, the results from $\mathcal{M}_1$ are not included since $\mathcal{M}_1$ does not allow to track units position. Consequently, in Figs. 3c and 3d, the response related to ship's internal dynamics, namely angular velocity ($r_{\mathcal{M}_1}$, $r_{\mathcal{M}_2}$) and heading ($\psi_{\mathcal{M}_1}$, $\psi_{\mathcal{M}_2}$) have been shown, respectively.

The obtained results illustrate that both EKFs are capable of estimating both the angular velocity (Fig. 3c) and heading (Fig. 3d). Although, it can be found that the estimation error of the EKF equipped with, structurally more complex, $\mathcal{M}_2$ is lower than the one obtained using EKF based on $\mathcal{M}_1$. This fact results from the smaller modelling (structural) error (uncertainty) introduced by $\mathcal{M}_2$. The path estimation provided by EKF with $\mathcal{M}_2$ is satisfactory with negligible error Fig. 3b. In terms of the prediction capabilities, the constant parameter valued predictions made with $\mathcal{M}_1$ (in terms of angular velocity and heading) and $\mathcal{M}_2$ (in terms of angular velocity, heading and path) indicate a classical cone-like evolution of the prediction error due to structural uncertainty.

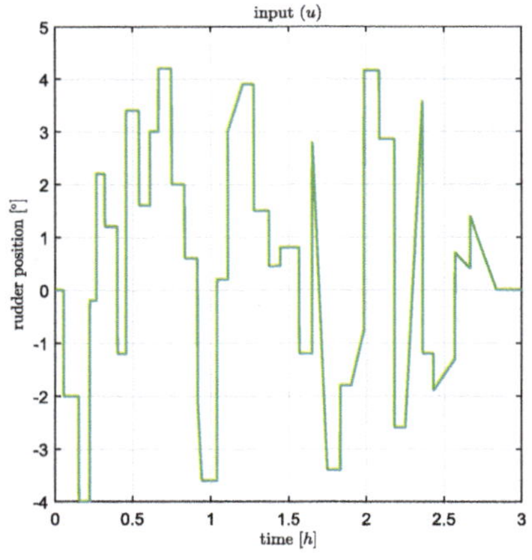

(a) Control input — rudder displacement

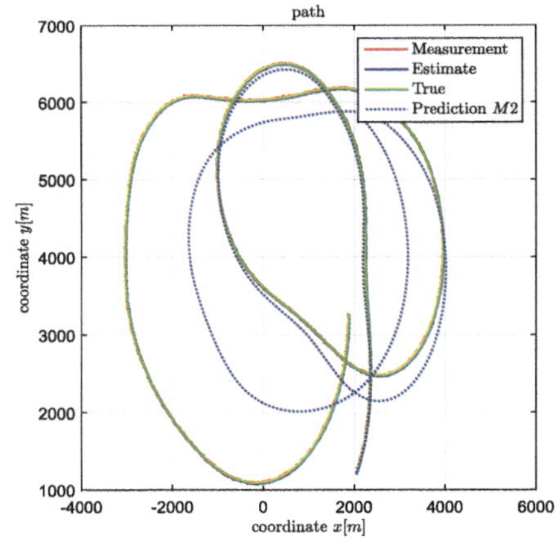

(b) Kinematic reaction — ship's path in XY coordinate frame

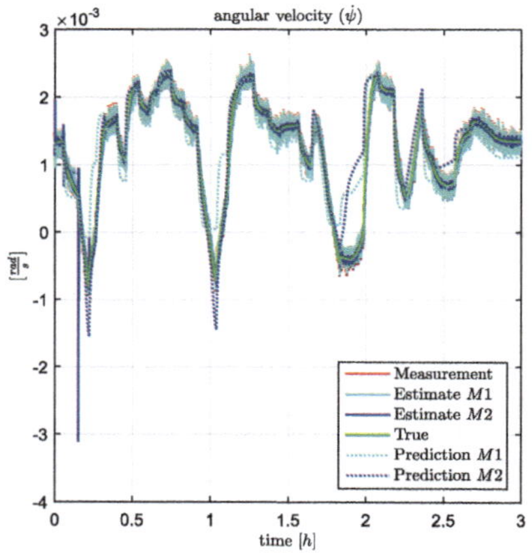

(c) Dynamic reaction — angular velocity

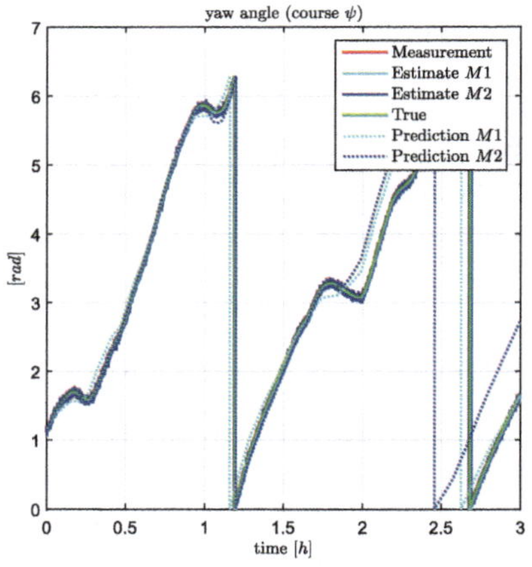

(d) Dynamic reaction — heading

Fig. 3. Ship kinematic and dynamic reaction to rudder displacement at constant speed and under negligible acting disturbances ($\boldsymbol{d}$)

As in the estimation case, the disproportion in the amount of uncertainty introduced by $\mathcal{M}_1$ and $\mathcal{M}_2$ can be observed. Again, the prediction made with $\mathcal{M}_2$ tend to be more accurate (Fig. 3c and Fig. 3d).

The estimated parameter time trajectories have been illustrated in Fig. 4. The model parameter time-variability is the result of both physical phenomenons introduced in $\mathcal{F}_{\mathrm{SS}}$ and model structural uncertainty. In case of EKF based on $\mathcal{M}_1$, the parameter time trajectories tend to converge in case of $d_{\mathcal{M}_1}$ and $c_{\mathcal{M}_1}$, while the parameters $a_{\mathcal{M}_1}$ and $b_{\mathcal{M}_1}$ remain variant over the whole experiment horizon (Fig. 4a). In the case of EKF based on $\mathcal{M}_2$, due to reduced impact of structural uncertainty, the parameters tend to converge as depicted in Fig. 4b. For clarity of presentation, due to a large number of parameters, the parameters trajectories have been normalised using their final values.

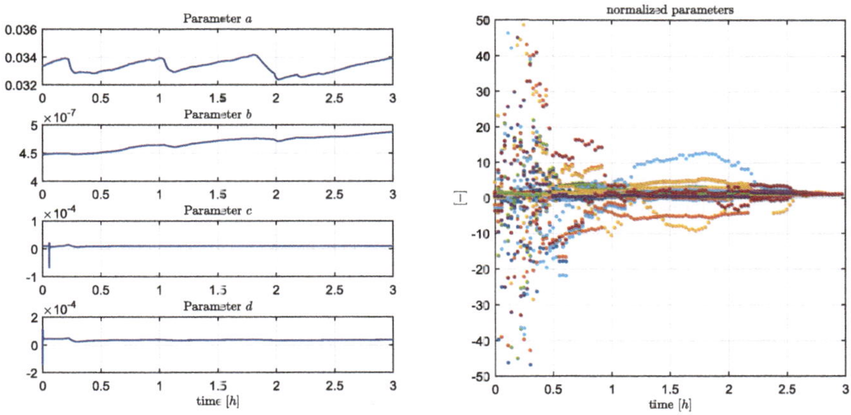

(a) Parameters contained in vector $\boldsymbol{\theta}_{\mathcal{M}_1}$ (b) Parameters contained in vector $\boldsymbol{\theta}_{\mathcal{M}_2}$

Fig. 4. Estimated parameter time trajectories

5 Conclusions

In this work, an approach to utility model parameter identification task has been presented by invoking an on-line recursive EKF algorithm for both state and parameter estimation. The utility model under investigation is assumed to serve control design purposes. In the experimental part two utility model structures, that differ substantially in the level of complexity, have been investigated. The obtained results have been compared in terms of estimation and prediction capabilities. It has been found that both models are capable to produce comparable angular velocity and heading estimates. The second, more complex structure, has been found to provide more accurate predictions. However, this comes at the price of a significantly larger number of parameters to be estimated in comparison to the first structure. (the second model parameter vector contains more than 10 times more elements).

References

1. Arminski, K., Zubowicz, T.: Robust identification of Quadrocopter model for control purposes. In: 22nd International Conference on Methods and Models in Automation and Robotics (MMAR), pp. 337–342. IEEE (2017)

2. Burgers, G., Jan van Leeuwen, P., Evensen, G.: Analysis scheme in the ensemble Kalman filter. Monthly Weather Rev. **126**(6), 1719–1724 (1998). https://doi.org/10.1175/1520-0493(1998)126⟨1719:ASITEK⟩2.0.CO;2
3. Daum, F.: Nonlinear filters: beyond the Kalman filter. IEEE A&E Aerosp. Electron. Syst. Mag. **20**, 57–69 (2005). https://doi.org/10.1109/MAES.2005.1499276
4. Fossen, T.I.: Handbook of Marine Craft Hydrodynamics and Motion Control. Wiley (2011)
5. Galbas, J.: Synteza układu sterowania precyzyjnego statkiem za pomocą sterów strumieniowych. Ph.D. thesis, Gdańsk University of Technology (1998). (in Polish)
6. Haro, M.: Stabilization of the ship's dynamic by the use of a dynamic controller. In: XXIII Automatic Journeys of the CEA-IFAC, pp. 157–168. Tenerife, Spain (2002)
7. Jaroś, K., Witkowska, A., Śmierzchalski, R.: Designing particle kalman filter for dynamic positioning. In: Kościelny, J.M., Syfert, M., Sztyber, A. (eds.) Advanced Solutions in Diagnostics and Fault Tolerant Control, pp. 157–168. Springer, Cham (2018)
8. Kalman, R.E.: A new approach to linear filtering and prediction problems. J. Basic Eng. **82**(1), 35–45 (1960). https://doi.org/10.1115/1.3662552
9. Khazraj, H., Faria da Silva, F., Bak, C.: A performance comparison between extended Kalman filter and unscented kalman filter in power system state estimation. In: Proceedings of the 2016 51st International Universities' Power Engineering Conference (2016). https://doi.org/10.1109/UPEC.2016.8114125
10. Lazarowska, A.: Ship's trajectory planning for collision avoidance at sea based on ant colony optimisation. J. Navig. **68**, 291–307 (2015). https://doi.org/10.1017/S0373463314000708
11. Shi, C., Zhao, D., Peng, J., Shen, C.: Identification of ship maneuvering model using extended Kalman filtering. In: Marine Navigation and Safety of Sea Transportation, vol. 3, pp. 329–334 (2009). https://doi.org/10.1201/9780203869345.ch59
12. Śmierzchalski, R.: Automatyzacja i sterowania statkiem. Wydawnictwo Politechniki Gdańskiej (2013)
13. Tomera, M.: Hybrid switching controller design for the maneuvering and transit of a training ship. Int. J. Appl. Math. Comput. Sci. **27**(1), 63–77 (2017)
14. Witkowska, A., Śmierzchalski, R.: Adaptive backstepping tracking control for an over-actuated DP marine vessel with inertia uncertainties. Int. J. Appl. Math. Comput. Sci. **28**(4), 679–693 (2018)
15. Xie, S., Chu, X., Liu, C., Liu, J., Mou, J.: Parameter identification of ship motion model based on multi-innovation methods. J. Marine Sci. Technol. (2019). https://doi.org/10.1007/s00773-019-00639-y
16. Zubowicz, T., Arminski, K., Witkowska, A., Śmierzchalski, R.: Marine autonomous surface ship - control system configuration. IFAC-PapersOnLine **52**(8), 409–415 (2019). https://doi.org/10.1016/j.ifacol.2019.08.100. 10th IFAC Symposium on Intelligent Autonomous Vehicles IAV 2019

Identification in a Laboratory Tunnel to Control Fluid Velocity

Pawel Piskur(✉), Piotr Szymak, and Joanna Sznajder

Polish Naval Academy, Institute of Electrical Engineering and Automatics, Smidowicza Street 69, 81-127 Gdynia, Poland
p.piskur@amw.gdynia.pl

Abstract. In this paper the method of fluid velocity control in the laboratory water tunnel is presented. The water tunnel is designed for measurements of undulating propulsion system for biomimetic underwater vehicle (BUV). The undulating propulsion system consists of the fin, servomotor, power supply and control unit. Due to the fluid structure interaction between the fin and water, special requirements arising from fluid mechanics theory must be met. It means that experimental measurements are reliable only if the generated propulsion force is measured for a dynamic state.

The aim of the fluid velocity control system is to compensate the undulating propulsion force by additional, external water pump. For constant fluid velocity an average value of the force generated by the undulating propulsion system should maintain at zero. For this reason, the external water pump is going to be controlled as a function of average value of the force generated by the undulating propulsion system. The fluid velocity as a function of the control signal for external pump should be identified due to the unknown characteristic caused mainly by resistance losses in the water tunnel and the nonlinear characteristic of the water pump.

This paper presents an identification problem to design the control system for maintaining the desired water velocity in the laboratory test stand.

Keywords: System identification · Nonlinear model · Undulating propulsion system · Biomimetic vehicle

1 Introduction

In recent years a rapid development of underwater vehicles has been noticed [1, 4, 17]. Among many types of underwater vehicles being developed, the biomimetic ones, with fish like movement, are becoming more popular due to their low hydroacoustic spectrum [7, 11]. This unique feature is desirable for harbour hidden inspection or observation among living underwater organisms [12, 16]. Another reason makes undulating propulsion more popular is a possibility to achieve a higher, than rotational propulsion system, energy efficiency [10, 14]. Here, the energy efficiency is understood as a ratio of a propulsion force to electric energy delivered to a driving system.

A. Bartoszewicz et al. (Eds.): PCC 2020, AISC 1196, pp. 1543–1552, 2020.
https://doi.org/10.1007/978-3-030-50936-1_128

The propulsion system in a biomimetic underwater vehicle (BUV) consists of a set of fins driven by servo motors and controlled by a microcontroller unit (MCU) [15]. The fish like movement can be reproduced with a fin made from a flexible material or as a connection of rigid body with degrees of freedom depends on a specific swim pattern [2]. Many links needed to accurately reproduce fish behavior makes vehicle model complex and its control techniques more complicated (Fig. 1a). Therefore, efforts are to be made to imitate the movement of fish with one piece of flexible fin (Fig. 1b). This provides to investigation of the fluid-structure interaction phenomena which depends on many construction factors [5, 8, 9].

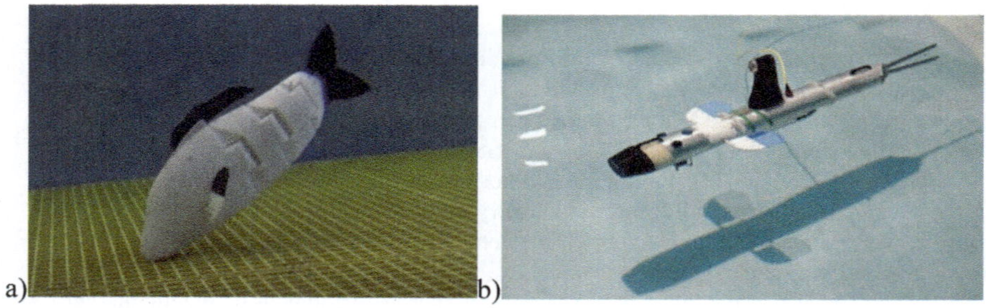

Fig. 1. Biomimetic underwater vehicle CyberFish, b) The torpedo-shape BUV with two side fins and one tail flexible fin [14]

Due to the nonlinearity of fluid structure interaction the experimental method is used to achieve characteristics of an undulating propulsion system. For that reason, the laboratory test stand was designed and equipped with specialized sensors. In Fig. 2 two water tunnels are depicted. On the left side the mini water tunnel designed in Polish Naval Academy laboratory is presented, while on the right side the bigger version of the tunnel is shown. In the test tunnel from the right side (Fig. 2b) there are more than 50 tons of water, that is why the preliminary tests and calibration process were provided in the laboratory mini tunnel, presented on the left side (Fig. 2a). The time of velocity control is needed not only for accuracy of the measurements but also for total time of tests provided. Both tunnels were equipped with specialized sensors for measuring the force of the fin's impact on the fluid. The fluid velocity was controlled using additional water pump and feedback signal from specialized high accuracy ultrasonic flowmeter.

For design and implementation of the control algorithms the identification must be provided. Here the identification process is realized for water pump with the fluid velocity losses in the water tunnel. Losses in the water tunnel depends mainly on its construction parameters: a length, a cross section area and corner angles. In addition the value of fluid (here water) velocity has impact on the pump capacity due to the nonlinear characteristic of the pump. The final result of identification process is the fluid velocity in the water tunnel as a function of control signal delivered to the water pump.

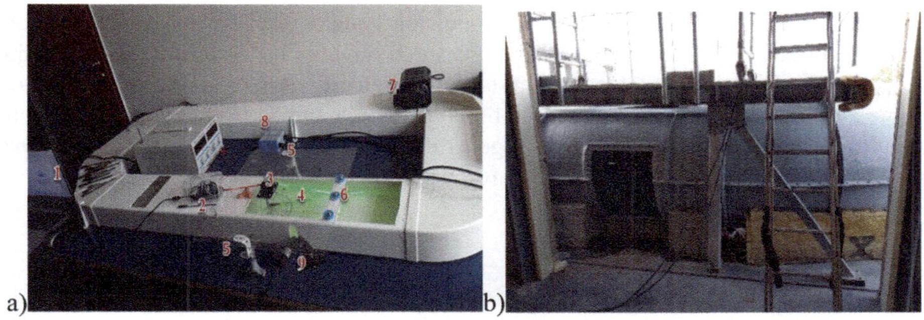

Fig. 2. Measurement stand: a) mini water tunnel (left side): 1 – computer, 2 – microcontroller, 3 – servomechanism, 4 – tested fin, 5 – strain gauges, 6 – ball bearings, 7 – water pump, 8 – flowmeter, 9 – laser; b) water tunnel for BUV measurements (right side)

Since the force generated by the undulating propulsion system has a sinusoidal characteristic, the average value should be taken into account [13]. The measurement conditions should ensure the average value of propulsion force generated by the undulating propulsion system close to zero while water flows around the fin. Then the controlled constant fluid velocity in water tunnel has the same value as for the fin movements in open water reservoir without water currents [3].

In the next section a method of analysis of undulating propulsion designed for biomimetic vehicles is presented. The aim of the research is to find the characteristics of undulating propulsions useful for the selection of construction and control system. Because the undulating propulsion is operated in the water environment, the analysis takes into account the fluid parameters and the interaction between the solid and the fluid as well. Considering the complexity of phenomena and non-linearity of the characteristics, a dimensional analysis was adopted to determine the mutual, selected relationships between the parameters of the undulating propulsion [18]. Fin kinematics depends on a combination of many factors. Therefore, the experimental method was selected to analyze the impact of construction parameters on the fin performance.

This paper is organized as follows. Section 2 describes the test stand and measurement algorithm for undulating propulsion system. Section 3 discusses the results of identification. Section 4 provides discussion on fluid velocity control. Finally, conclusions are presented in Sect. 5.

2 Methods and Materials

It was assumed that water is an incompressible fluid, with constant temperature, viscosity and density. The undulating propulsion system consists from a single fin driven by servomotor supplied from additional power supply. The control of a fin movement with different frequency and different maximal angle of attack is realized by MCU. The force F generated by undulating propulsion system is measured with specialized, precision strain gauges system. Information about the generated force is transmitted to the MCU. Due to the fluid mechanic requirements the velocity of fluid encircling the fin should compensate the force generated by the propulsion system. That is why, the propulsion force generated by the fin should not be measured directly, but as a function of a fluid velocity. The fluid velocity depends on the rotary velocity of additional, external water pump and losses in a water tunnel. The fin construction and algorithm control analysis should be made for constant velocity generated by the additional water pump. The schematic diagram of fluid velocity control is presented in Fig. 3.

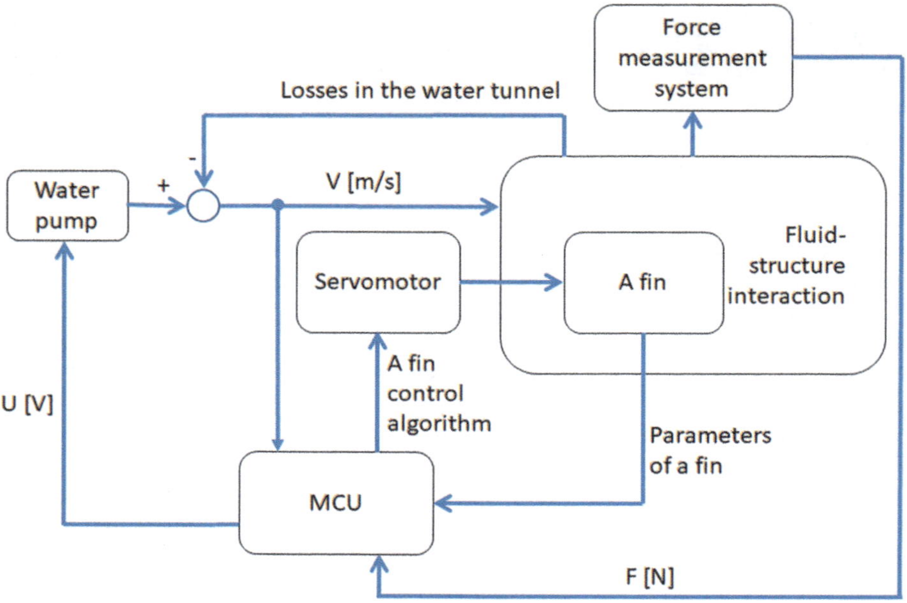

Fig. 3. Schematic diagram of the laboratory test stand

Due to the fluid mechanic theory a velocity of water should be proportional to the undulating propulsion system force generation. It means that the fluid velocity from external water pump should compensate the force generated by the analyzed fin.

A force generated from the fin has oscillatory characteristic [15], that is why the mean value of force in one cycle should be considered. Moreover external water pump has nonlinear characteristics and must be identified taking into account losses in whole water tunnel. For that reason a specialized ultrasonic flowmeter was applied in the water tunnel to measure the fluid velocity. The fluid velocity is a sum of water pump efficiency and losses in the water tunnel. The aim of the MCU is to control servomotor and external water pump depending on average force generated by the undulating propulsion system. The fluid velocity from the additional pump should be proportional to the force generated by undulating propulsion system.

For the linearized system the transmittance can be expressed by Eq. (1):

$$G(s) = \left. \frac{V(s)}{U(s)} \right|_{F_{avg}=0} \tag{1}$$

where:
V – is the fluid velocity;
U – is control signal form MCU sent to the water pump;
F_{avg} – is the mean value of force generated by undulating system.

For fluid structure interaction measurements the control system should be stable, it means that the controlled velocity should be constant in the specified period of time. That is why the model is linearized for constant water velocity and mean force values equal to zero ($F_{avg} = 0$).

2.1 Laboratory Test Stand

The laboratory water tunnel was made of polyethylene parts, with the following dimensions: 0.09 m height, 0.22 m width. The total length of the laboratory water tunnel is around 6 m. Different types of fins can be attached to the servomechanism (Dynamixel AX-12+), which is fixed to a transparent polycarbonate plate mounted on ball bearings. Measurements of the generated thrust are realized by two precision strain gauges installed differentially on both sides of the water tunnel.

Figure 4 presents a scheme of the system for measurement of the force in the longitudinal axis of symmetry X and the moment of force relative to the vertical axis of symmetry N generated by the undulating propulsion based on signals obtained from two strain gauges measuring forces F_1 and F_2.

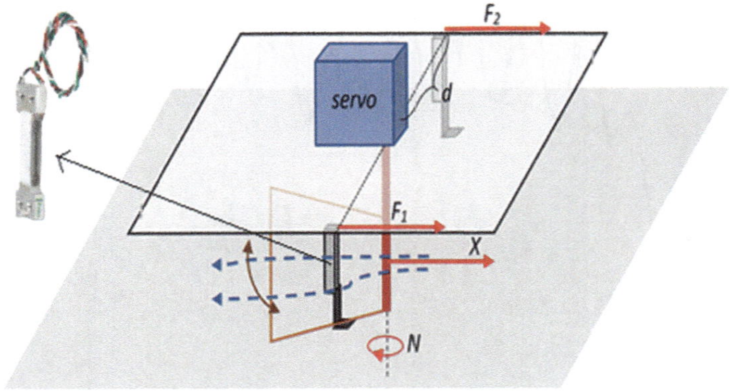

Fig. 4. Measurement of force X and moment of force N based on of forces F_1 and F_2 obtained from strain gauges

The signal obtained from the strain gauge is amplified with a Wheatstone bridge to obtain the appropriate signal that can be sent to the ADC converter in MCU.

The maximal fluid velocity obtained from external water pump was equal to 0.5 m/s. But the fluid velocity control range should be maintained with restriction to the laminar flow of fluid and the force generated by the undulating propulsion system. If the frequency of fin movement is changing then the fluid velocity should be changed proportionally to the average force generated by the fin (2).

$$F_{avg} = \frac{1}{n}\sum\nolimits_{i=1}^{n} X(i) \quad (2)$$

where:
n – is number of samples in the specified period of time.

In Fig. 5 forces acting between a fin measured in mini water tunnel for three different types of fin is presented. The frequency of control signal was the same for each fin and here was equal to 2.1 Hz. Differences in force results are due to differences of the fin stiffness. The designed mini water tunnel is to be used to test the impact of many factors on the generated propulsion force for many fins, while the larger tunnel is to be used to test BUV in real dimensions. For higher value of average propulsion force the higher velocity of water tunnel should be set to maintain the fluid-structure interaction force *F* close to zero. The undulating propulsion force depends on many construction and control parameters that's is why the fluid velocity should be controlled continuously.

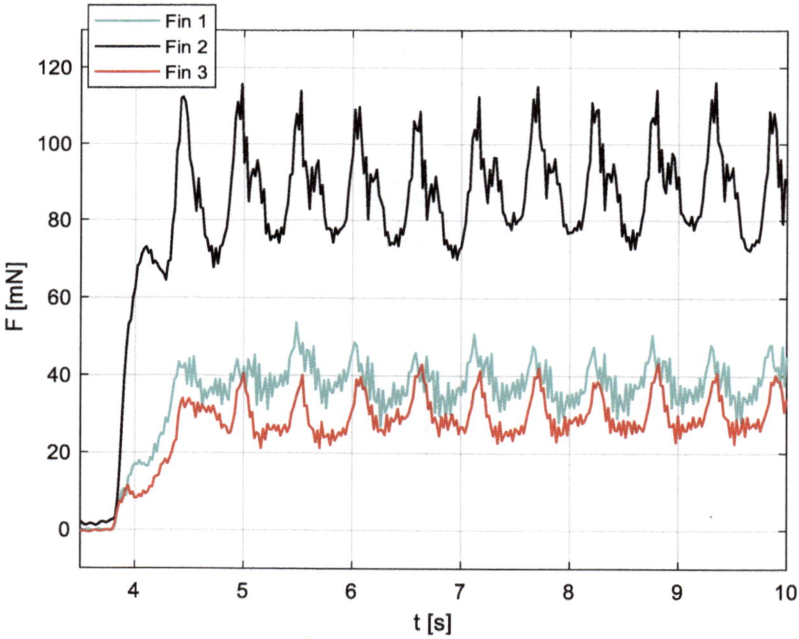

Fig. 5. Forces acting in longitudinal axis of symmetry measured for three different types of fin

3 Identification Results

Fluid velocity as a function of the control signal should be identified due to the unknown characteristic of the water pump and tunnel resistance. The water pump consist of an asynchronous motor with impeller. The asynchronous motor is control using voltage regulator. The range of the control signal is up to 5 V. In Fig. 6 the data used for verification process is depicted. The fluid velocity v was measured as a function of the signal control U. Data was recorded 5 times per second and the total measurement time was 240 s. It was assumed that operating point for linear model is to be close to the fluid velocity 0.2 m/s.

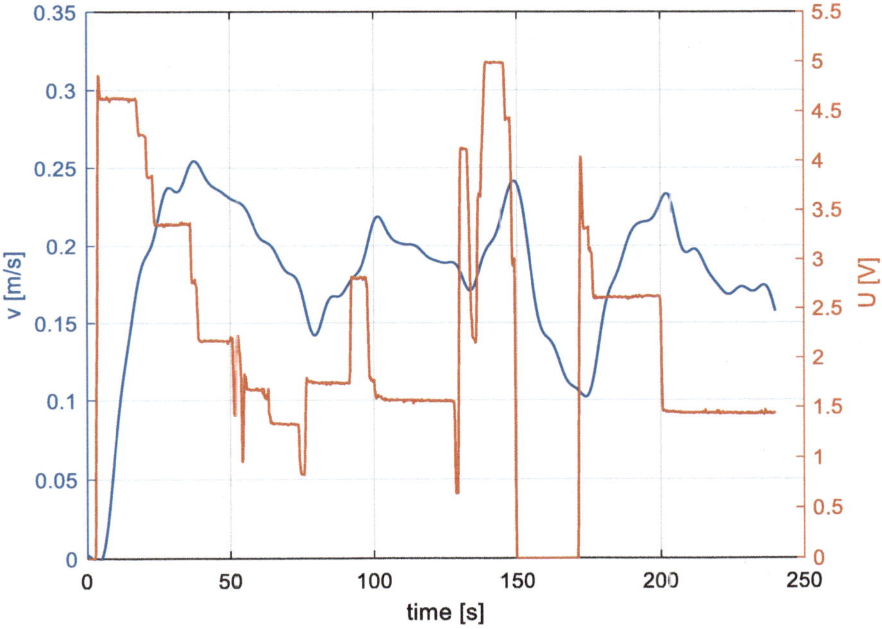

Fig. 6. Data used for identification process.

The identification process was carried out using Matlab - System Identification Toolbox [6]. In Fig. 7 the measured (black line) and the simulated model output (blue line) are depicted. After identification process the 87.91% accuracy was achieved. The best accuracy was achieved for 6 number of poles and 4 number of zeros. The continuous-time identified transfer function can be expressed with Eq. (3):

$$G(s) = \frac{0.002163\mathrm{s}^4 - 2.557\mathrm{e}^5 s^3 + 2.616e^{-5}\mathrm{s}^2 - 1.007e^{-7}\mathrm{s} + 5.999\mathrm{e}^{-8}}{\mathrm{s}^2 + 0.05781\mathrm{s} + 7.243\mathrm{e}^{-9}} \quad (3)$$

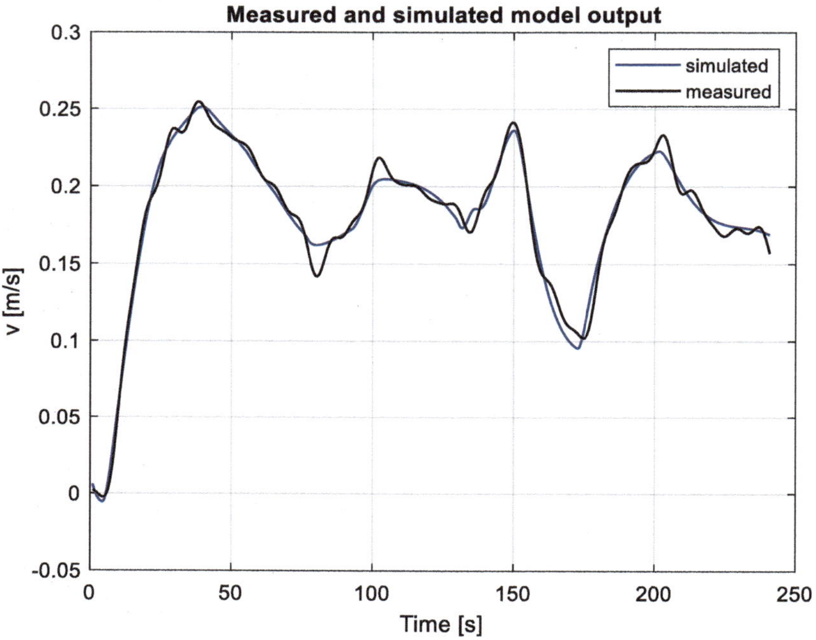

Fig. 7. The measured and simulated fluid velocity comparison

4 Future Research

According to the fluid structure interaction theory and measurements conditions for undulating propulsion system the fluid velocity should maintain constant. For this reason, the PID regulator is to be designed and implemented in the MCU after simulation tests. The scheme of the simulation model used for the simulation tests is presented in Fig. 8. The close loop control system consists of the identified system and PID controller.

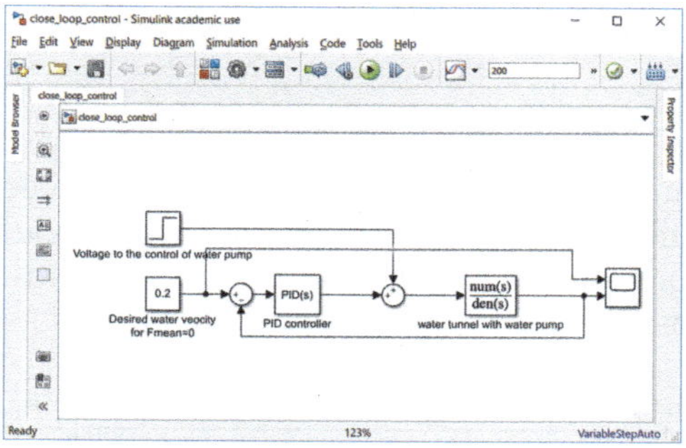

Fig. 8. The simulation model of fluid velocity control with PID controller

5 Conclusions

The identification results presented in this paper are one of the stages in analysis of the undulating propulsion system. A linear approximation of nonlinear system around the selected operating point is shown. The controller should be designed based on this approximation for the linear system around the selected operating point. The presented stand for laboratory tests shows that it is possible to determine the system parameters, therefore further testing of this control system should be provided.

The presented devices and the method of analysis of the underwater propulsion system will be further developed, and after successful verification in the laboratory water tunnel will be copied to a larger one.

Acknowledgment. The paper is supported by the Research Grant of the Polish Ministry of Defense entitled "Model studies of the characteristics of a undulating propulsion system".

References

1. Bazaz Behbahani, S.T.: Role of pectoral fin flexibility in robotic fish performance. J. Nonlinear Sci. 1155–1181 (2017). https://doi.org/10.1007/s00332-017-9373-6
2. Borazjani, I., Sotiropoulos, F.: Numerical investigation of the hydrodynamics of carangiform swimming in the transitional and inertial flow regimes. J. Exp. Biol. **211**, 1541–1558 (2008)
3. Tytella, E.D., Hsu, C.-Y.: The role of mechanical resonance in the neural control of swimming in fishes. Zoology **117**, 48–56 (2014). HHS Public Access
4. Zhang, F., Wang, J.: Gliding robotic fish for mobile sampling of aquatic environments. In: Proceedings of 11th IEEE International Conference on Networking, Sensing and Control, pp. 167–172 (2014)
5. Godoy-Diana, R.: Bio-inspired swimming and flying – Vortex dynamics and fluid/structure interaction. Mechanics of the fluids (2014)
6. https://www.mathworks.com/ (2019)
7. Koca, G.O., Bal, C., Korkmaz, D.: Three-dimensional modeling of a robotic fish based on real carp locomotion. Appl. Sci. (2018). https://doi.org/10.3390/app8020180
8. Krishnadas, A., Ravichandran, S.: Analysis of biomimetic cadual fin shapes for optimal propulsive efficiency. Ocean Eng. **153**, 132–142 (2018)
9. Malec, M., Morawski, M., Szymak, P., Trzmiel, A.: Analysis of parameters of traveling wave impact on the speed of biomimetic underwater vehicle. Solid State Phenom. **210**, 273–279 (2014)
10. Morawski, M., Malec, M.: Development of CyberFish – Polish Biomimetic Unmanned Underwater Vehicle BUUV, pp. 76–82. Trans Tech Publications, Zürich (2014). https://doi.org/10.4028/www.scientific.net/AMM.613.76
11. Morawski, M., Słota. A.: Hardware and low-level control of biomimetic underwater vehicle designed to perform ISR tasks. J. Mar. Eng. Technol. (2017). https://doi.org/10.1080/20464177.2017.1387089
12. Piskur, P., Szymak, P.: Algorithms for passive detection of moving vessels in marine environment. J. Mar. Eng. Technol. **16**, 377–385 (2017). https://doi.org/10.1080/20464177.2017.1398483
13. Przybylski, M.: Mathematical model of biomimetic underwater vehicle. In: Proceedings of the 33rd International ECMS Conference on Modelling and Simulation, Caserta, Italy, vol. 33, no. 1, pp. 343–350 (2019)

14. Szymak, P., Morawski, M., Malec, M.: Conception of research on bionic underwater vehicle with undulating propulsion. Solid State Phenom. **180**, 160–166 (2012). Mechatronics Systems, Mechanics and Materials, Book Series
15. Szymak, P., Przybylski, M.: Thrust measurement of biomimetic underwater vehicle with undulating propulsion. Sci. J. Pol. Naval Acad. (2018). https://doi.org/10.2478/sjpna-2018-0014
16. Piskur, P., Szymak, P., Jaskólski, K., Flis, L., Gasiorowski, M.: Hydroacoustic system in a biomimetic underwater vehicle to avoid collision with vessels with low-speed propellers in a controlled environment. Sensors **20**(4), 1–14 (2020). https://doi.org/10.3390/s20040968
17. Szymak, P., Praczyk, T.: Ground/Air Multisensor Interoperability, Integration, and Networking for Persistent ISR VII (2016). https://doi.org/10.1117/12.2225587
18. Szymak, P., Praczyk, T., Pietrukaniec, L., Hozyn, S.: Laboratory stand for research on mini CyberSeal. Measur. Autom. Monit. **63**(7), 228–233 (2017)

Vision-Based Modelling and Control of Small Underwater Vehicles

Stanisław Hożyń(✉)

Polish Naval Academy, Gdynia, Poland
s.hozyn@amw.gdynia.pl

Abstract. Modelling and control of underwater vehicles in most cases, demand their hydrodynamic parameters' identification, which is a timely and technically demanding task. Therefore, more convenient methods of utilising vision systems have been introduced. However, many solutions presented in the literature assume that a camera is mounted in the central part of a swimming pool. What is more, they are not applicable for trajectory determination, which constitutes an essential factor in devising a control system of autonomous vehicles. For that reason, a computer vision system has been designed and developed, which enables tracking a vehicle and determining its trajectory as well. The obtained results indicate that the developed system enables modelling and control of underwater vehicles under laboratory conditions.

Keywords: Mathematical model · Underwater vehicle · Vision system · Trajectory determination

1 Introduction

Small underwater vehicles have generated considerable recent research interest. They are utilised in a wide range of fields such as underwater inspection, seafloor mapping and surveying, studying marine creatures and divers' support [1, 2, 6, 9]. Additionally, they play an essential role in the development of larger vehicles since they serve as prototype models during preliminary tests [4, 7, 14, 20]. Using a scale model of a constructed vehicle enables testing various technical solutions under laboratory conditions. Therefore, as presented below, in Fig. 1a there is an autonomous underwater vehicle which was built in the Institute of Electrical Engineering and Automatics in Polish Naval Academy, as well as in Fig. 1b is its prototype. By using the prototype, the new type of propulsion imitating seal fins was constructed and tested [15, 16].

Modelling and control of underwater vehicles in most cases demand their hydrodynamic parameters' identification [3]. To conduct it, the experiments in the laboratory swimming pool are needed. They are time-consuming, expensive and technically challenging to perform. What is more, they are imprecise and should be repeated after every change in a vehicle's construction or equipment [22]. Another solution is based on the dynamic responses collected by inboard sensors [21]. Nevertheless, in the case of

A. Bartoszewicz et al. (Eds.): PCC 2020, AISC 1196, pp. 1553–1564, 2020.
https://doi.org/10.1007/978-3-030-50936-1_129

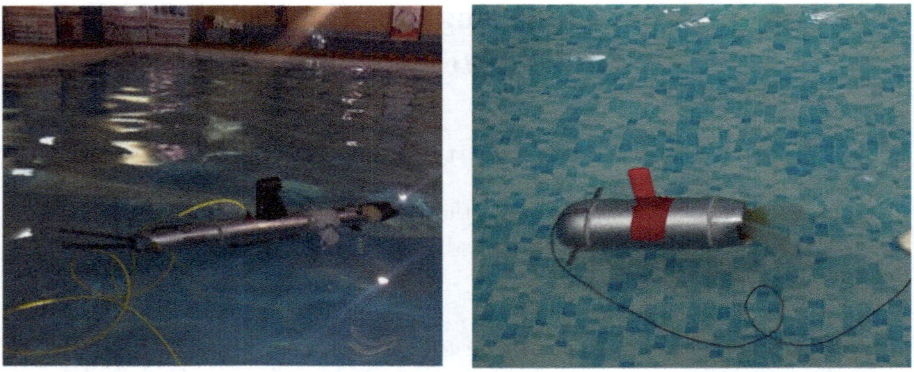

Fig. 1. a) Autonomous underwater vehicle, b) Prototype of an autonomous underwater vehicle.

small underwater vehicles, the required data may not be available due to the shortage of necessary sensors [13].

To overcome abovementioned difficulties, alternative approaches based on vision tracking methods have been developed [10, 11]. In these techniques, distinctive marks are put on a vehicle to calculate its displacement using a camera. Some researchers also utilise indicators on the bottom of a swimming pool to improve accuracy [12]. Unfortunately, these solutions assume that a camera is mounted in the central point above a swimming pool. It is challenging to comply with this restriction in smaller rooms equipped with large swimming pools where cameras can be mounted in the corners of the ceiling. In addition, they can be only utilised for parameters' identification, and they are not applicable to trajectory determination. Since trajectory determination is a vital factor in view of prototype's tests, we focused on devising a method, which is not only convenient for modelling but also control of small underwater vehicles in perspective of trajectory determination.

The purpose of the study is to describe and examine the vision-based method for modelling and control of small underwater vehicles. In the next section, the vision-based tracking system to measure the displacement of the vehicle is described. Then, modelling and control method are presented. In the end, results and conclusions are discussed.

2 Vision-Based Tracking System

The main task of the developed vision-based tracking system is to track an underwater vehicle in a swimming pool. Additionally, it should facilitate the installation of a camera in any space of a room. Therefore, to accomplish these ends, the following steps have to be included in the computer vision algorithm:

- camera calibration,
- perspective transformation,
- marker's segmentation, and
- displacement calculation.

Camera calibration plays a crucial role in view of accuracy [8]. Since each camera introduces distortion caused by its optical system, the calibration technique based on a chessboard pattern has been widely used in the literature [18]. In this attempt, there is assumed that the most commonly used correction is for the radial lens distortion that causes the actual image points to be displaced radially in the image plane. It can be approximated using the following expressions [8]:

$$x_v = \frac{x_u}{1 + k_1 r^2 + k_2 r^4 + k_5 r^6}, \quad y_v = \frac{y_u}{1 + k_1 r^2 + k_2 r^4 + k_5 r^6}, \quad r^2 = x_v^2 + y_v^2 \tag{1}$$

where: k_1, k_2, k_5 – intrinsic parameters of a camera defining radial distortion,
x_u, y_u – ideal pixel coordinates,
x_v, y_v – distorted coordinates

Another type of distortion is tangential distortion, caused by the not strictly collinear surface of the lens. It is usually written in the following form [8]:

$$x_t = 2k_3 x_u y_u + k_4(r^2 + 2x_u^2), \qquad y_t = k_3(r^2 + 2y_u) + 2k_4 x_u y_u \tag{2}$$

where: k_3, k_4 – tangential distortion,
x_t, y_t – distorted coordinates

After including lens distortion, the new point coordinate (x_d, y_d) is defined as follows:

$$\begin{aligned} x_d &= x_v + x_t \\ y_d &= y_v + y_t \end{aligned} \tag{3}$$

In order to calculate distortion coefficients, the chessboard pattern is placed in front of the camera. To increase accuracy, the pattern should be positioned in different directions in relation to the camera. Then, since the location of corner points in the chessboard pattern are known, the distortion coefficients can be calculated using methods presented in [19, 23].

Perspective transformation can also be determined using the same chessboard pattern. Based on the following homogenous image-plane pixel coordinates [5]

$$\begin{bmatrix} x' \\ y' \\ 1 \end{bmatrix} = \boldsymbol{H} \begin{bmatrix} x \\ y \\ 1 \end{bmatrix} \tag{4}$$

where x, y – coordinates of a point before the transformation, x', y' – new coordinates, $\boldsymbol{H}$ – transformation matrix defined as

$$\boldsymbol{H} = \begin{bmatrix} h_{11} & h_{12} & h_{13} \\ h_{21} & h_{22} & h_{23} \\ h_{31} & h_{32} & h_{33} \end{bmatrix} \tag{5}$$

non-homogenous image-plane pixel coordinates can be obtained from the following equations:

$$x' = \frac{h_{11}x + h_{12}y + h_{13}}{h_{31} + h_{32} + h_{33}}$$
$$y' = \frac{h_{21}x + h_{22}y + h_{23}}{h_{31} + h_{32} + h_{33}} \tag{6}$$

Assuming that $h_{33} = 1$, only four points are needed to calculate the transformation matrix. Due to the fact that the chessboard pattern contains more points (in this work 64 points) the RANSAC algorithm, which can remove gross errors during markers localisation, was utilised. Consequently, to calculate the transformation matrix, the chessboard pattern was placed on the water's surface in the central point of the swimming pool. This approach not only allowed to obtain the transformation matrix but also enabled the measurement of distances on the image. Since the geometrical characteristics of the chessboard pattern were known, the only simple calculation was needed to estimated dimensions which are represented by one pixel in the scene.

Segmentation algorithm, developed to detect a vehicle's position in a swimming pool, can be divided into the following parts:

- extraction of red pixels representing the vehicle's marker,
- thresholding,
- erosion,
- dilatation,
- elimination of the small blobs,
- localisation of the remaining blob in the image, and
- prediction of locations of the marker in the current image based on the previous image in case of not detecting the corresponding blob.

Because the laboratory can be prepared for the experiments, all red equipment should be removed from the view of the camera. Consequently, the segmentation task is easier because the only red object in the scene is the marker placed on the vehicle. However, since objects in other colours can contain a similar value of red pixels, the RGB colour space with double thresholding Min{R, G, B}, Max{R, G, B}, representing the lower and upper threshold, was utilised. Nevertheless, some random pixels, incorrectly classified as a marker area, still existed on the image. In view of this fact, morphological operations erosion and dilatation were utilised for removing them from the pictures. Additionally, if more than one blob remained on the picture, the perimeters of existed blobs were calculated, and the biggest one was taken into further consideration. If the marker had not been found in the current picture, its position would be determined using a linear estimator based on the previous one.

The displacement of the vehicle could be easily calculated since pixel dimensions were obtained in the previous steps. Therefore, the view of the scene was divided into columns and rows equivalent to the picture's resolution and dimensioned according to obtained results. Consequently, the change in the vehicle's position in consecutive frames was calculated in accordance with the Pythagorean theorem.

In order to determine the depth of a vehicle, the pressure sensor can be utilised. It is a very effective utilisation supporting other sensors each vehicle is equipped with. What is more, these sensors are relatively cheap and quite accurate.

The vision-based tracking system was also designed to determine the trajectory of a vehicle. To obtain this goal, communication between the tracking system and a vehicle had to be established. Consequently, the vehicle's heading and depth were sent to the system, while its velocity was calculated using the tracking module. The vehicle was equipped with heading and speed regulators, so the demanded setting as well the speed value were transmitted from the tracking system.

All abovementioned functionality demanded to develop a computer vision algorithm, which was designed and implemented into a window application in C++ programming language with the use of OpenCV and Qt libraries (Fig. 2).

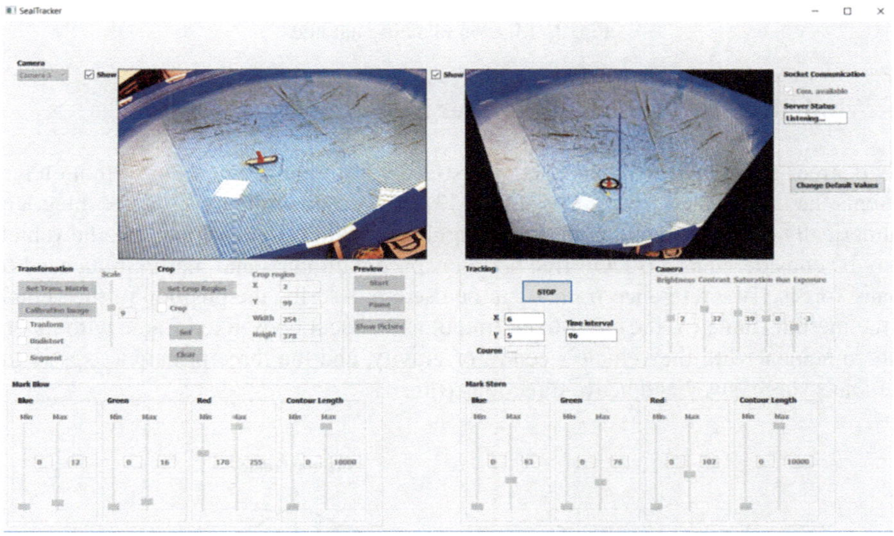

Fig. 2. The graphical user interface of the computer application.

Apart from facilitating convenient execution of the computer vision algorithm steps such as camera calibration and transformation matrix determination, it also enables to change thresholding values and the blob's marker length. Additionally, it was equipped with a mission planning module, in which the vehicle's trajectory between any numbers of points could be designed (Fig. 3). To obtain it, the communication module based on Ethernet connection and the TCP/IP protocol was devised.

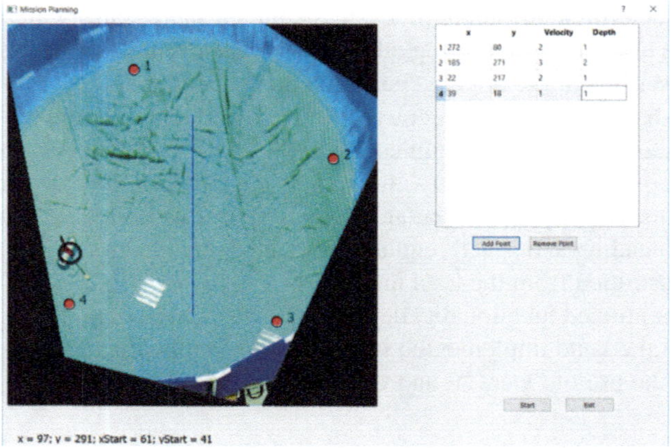

Fig. 3. Mission planning module.

3 Modelling of Small Underwater Vehicles

Six degrees of freedom nonlinear set of first-order differential equations of motion represents the underwater vehicle's dynamics [3]. As a result, equations may be integrated numerically to obtain vehicle linear and angular velocities. In this approach, the vehicle may be considered as a 6-DOF free body in space with mass and inertia influenced by many forces. Two reference frames can be used to describe the position of the vehicle – the inertial frame (or the earth-fixed frame) and a local body-fixed frame with its origin coincident with the vehicle's centre of gravity, and the three principal axes in the vehicle's surge, sway and heave directions (Fig. 4).

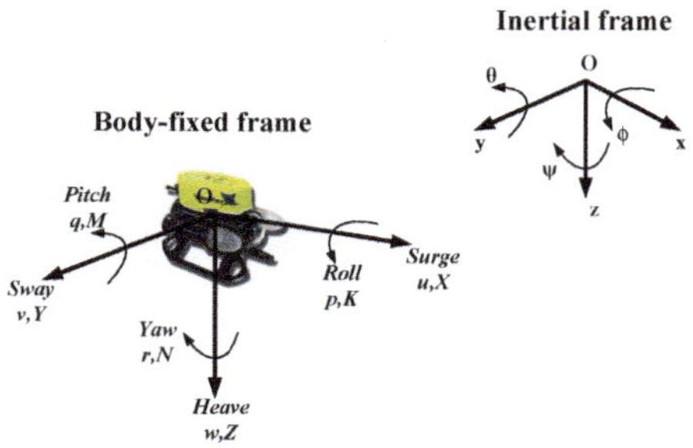

Fig. 4. Body-fixed and earth-fixed reference frames.

The mathematical model of underwater vehicles can be expressed with respect to the local body-fixed reference frame by nonlinear equations of motion (Fossen 2002)

$$\boldsymbol{M}\dot{\boldsymbol{v}} + \boldsymbol{C}(\boldsymbol{v})\boldsymbol{v} + \boldsymbol{D}(\boldsymbol{v})\boldsymbol{v} + \mathbf{g}(\boldsymbol{\eta}) = \boldsymbol{\tau} \tag{7}$$

where: $\boldsymbol{M}$ –inertia matrix (including added mass),
$\boldsymbol{C}(\boldsymbol{v})$ – matrix of Coriolis and centripetal terms,
$\boldsymbol{D}(\boldsymbol{v})$ – hydrodynamic damping and lift matrix,
$\boldsymbol{g}(\boldsymbol{\eta})$ – vector of gravitational forces and moments,
$\boldsymbol{\tau}$ – forces and moments acting on the vehicle in the body-fixed frame

The mathematical model of small underwater vehicles, due to their dimensions and velocities, can be significantly simplified. The main simplification concerns the following aspects [4, 11, 17]:

- vehicle moves with low speed; therefore, nonlinear Coriolis, centripetal, damping, restoring and buoyancy forces, as well as moments, can be linearized.
- The weight and buoyancy distribution forces the vehicle to return to the zero pitch and zero roll position.
- Vehicles are close to symmetric about horizontal and vertical planes. Moreover, it is operating at relatively low speeds so the system may be decoupled in two non-interactive subsystems: horizontal and vertical motions.
- The centre of buoyancy is equivalent to the centre of gravity in the horizontal plane.

Consequently, based on Eq. 7, the simplified mathematical model of this type of vehicle can be expressed in the following form:

$$\begin{aligned} \dot{u} &= a_1 u + a_2 r v + a_3 X \\ \dot{v} &= b_1 v + b_2 u r \\ \dot{r} &= c_1 r + c_2 u v + c_3 N \\ \dot{w} &= d_1 w + d_2 Z \end{aligned} \quad (8)$$

where $a_1, a_2, a_3, b_1, b_2, b_3, c_1, c_2$ are parameters to determine, u, v, w are liner velocities in the x, y, z directions and X, N, Z are: force in the x-direction, the moment around the z-axis and force in the z-direction, respectively (see Fig. 4).

4 Results

As mentioned previously, this work aimed to develop the vision-based system, which facilitates modelling and control of small underwater vehicles. To execute tests concerning the modelling part, a VideoRay Pro 4 underwater vehicle was utilised. The VideoRay Pro 4 is a small inspection-class ROV (Remotely Operated Vehicle) designed for underwater exploration at a maximum depth of 152 m. It carries the cameras, lights and sensors or accessories to the underwater places designated for a survey. Thrusters provide mobility, and the system is controlled from the surface using a control panel and hand controller. The control panel enables to send settings for thruster and receive data from the sensors as well. The maximum thrust provides a speed of the vehicle exceeding 2 m/s in the x-direction. The VideoRay Pro 4, utilized during experiments, is presented in Fig. 5.

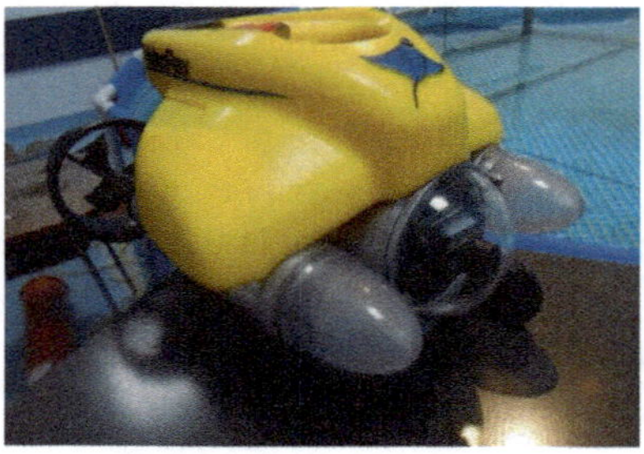

Fig. 5. Video Ray Pro 4.

For the sake of determining the mathematical model's parameters, an experiment in a laboratory swimming pool was executed. Firstly, the forces generated by the propellers were calculated. Then, the vehicle was remotely controlled by an operator, while its movement and the generated forces were saved to a file using the developed computer system. In the experiment, the generated forces strictly depended on the size and shape of the swimming pool. The operator had to design the movement of the vehicle in a way that allows keeping the vehicle in the swimming pool's area.

The obtained data was utilized to solve the Eq. (8) in order to determine the model's parameters using the Least Square Method. It is the commonly applied approach in regression analysis to approximate the solution of overdetermined systems. Consequently, the parameters listed in Table 1 were achieved.

Table 1. Parameters of the model.

Parameter	Value	Standard deviation[%]	Parameter	Value	Standard deviation [%]
a_1	−0,67	−5,7	a_2	0,17	6,3
b_1	−0,98	−12,3	b_2	0,58	11,5
c_1	−0,3	−6,3	c_2	17,54	4,2
d_1	−0,85	−4,3	d_2	0,12	3,9
a_3	−1,64	−7,6	c_3	1,9	8,5

Based on the determined parameters, a simulation model was developed using the Matlab/Simulink environment. Then, the experiment was repeated, and a new set of data was gathered. Finally, the comparison of the obtained results between the simulation model and the underwater vehicle for the new set of data was carried out.

The forces generated during the experiment are depicted in Figs. 6, 7 and 8. Figure 6 illustrates the force generated towards x-direction, Fig. 7 the force generated towards z-direction, and Fig. 8 the moment generated about the z-axis during the experiment.

The generated forces resulted in the vehicle's movement. The reached velocities in the x-direction, z-direction and around z-axis are presented in Fig. 9, 10 and 11.

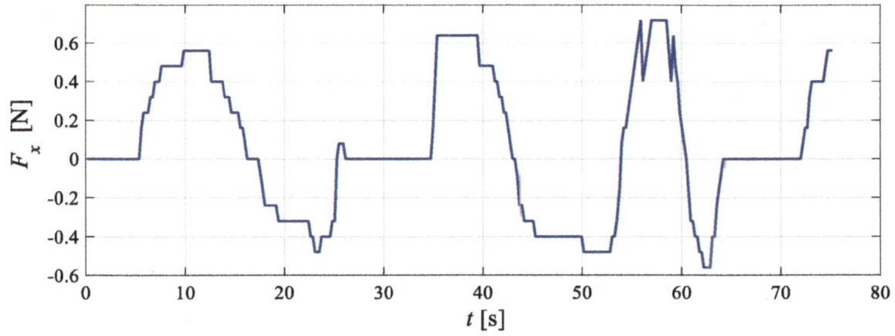

Fig. 6. Force generated towards *x*-direction.

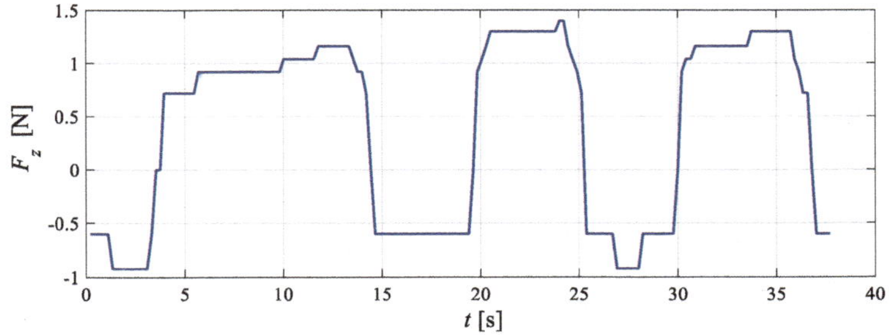

Fig. 7. Force generated towards *z*-direction moment generated about the z-axis.

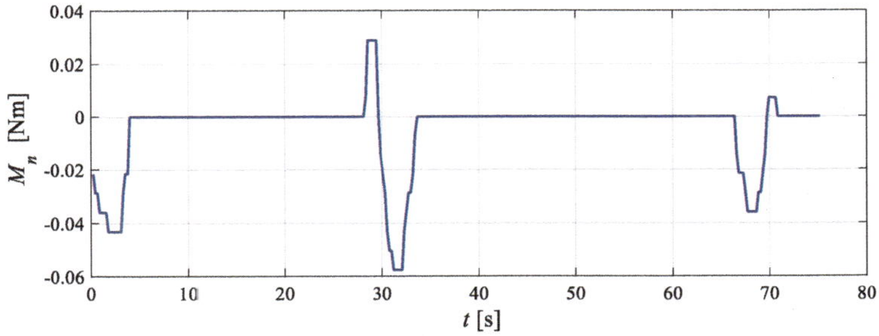

Fig. 8. Moment generated about the *z*-axis.

Comparing the obtained results, we see that the mathematical model tends to give predictions that are parallel to the data from the underwater vehicle. It should, however, be noted that at certain points of time, the prediction considerably differs from real data. This is due to the impact of the tether that impeded the vehicle's freedom of movement. Therefore, the achieved mean absolute errors equal to 0.03 m/s, 0.012 m/s and 0.1 rad/s for velocities in the x-direction, z-direction and around the z-axis, respectively, were considered as satisfactory.

The second part of the experiment was devoted to trajectory determination. For this purpose, an underwater vehicle with a new type of propulsion imitating seal fins was utilised. Because this kind of vehicle is a new solution, its manoeuvrability constituted a

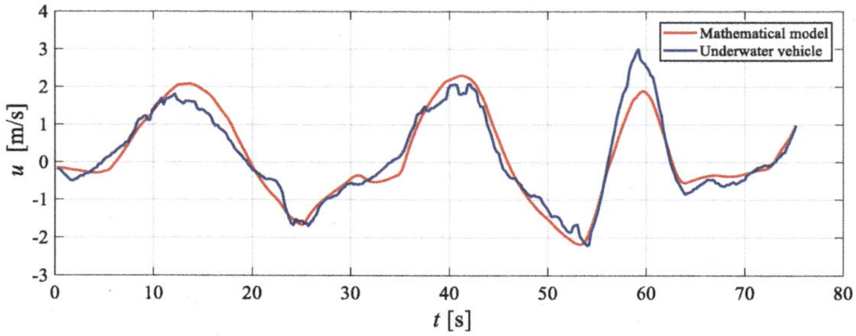

Fig. 9. Velocities in *u*-direction.

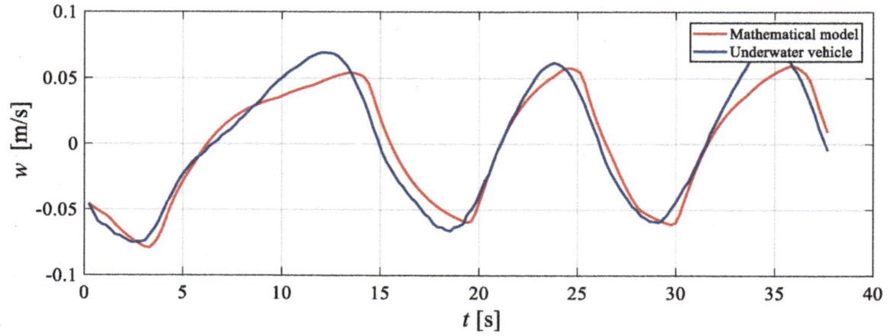

Fig. 10. Velocities in *z*-direction.

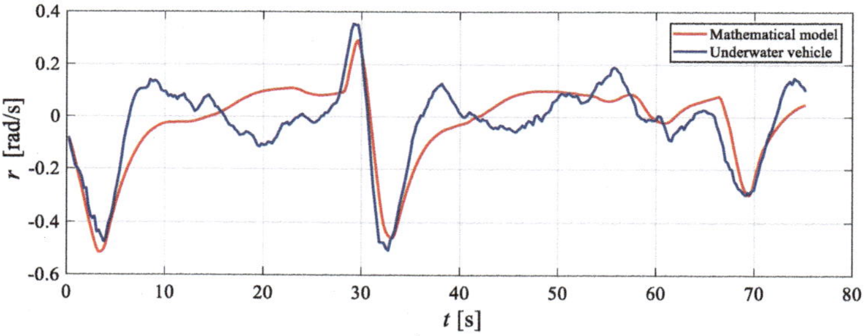

Fig. 11. Velocities around *z*-axis.

significant factor under construction. Therefore, the different trajectories were designed, and feasibility of the vehicle to follow them was tested. For example, Fig. 12 presents the trace of the vehicle travelling between designed points with determined velocity and depth.

It appears from the experiments that the developed vision-based tracking system facilitates designing trajectory of the vehicle as well as tracking it.

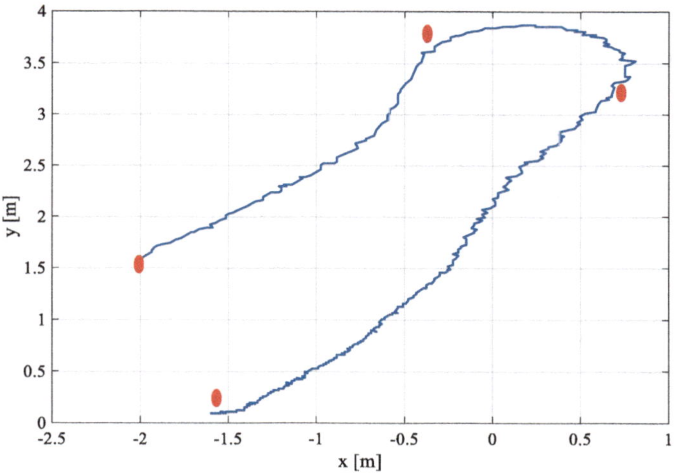

Fig. 12. Trace of the vehicle's trajectory.

5 Conclusions

Prior experimentation has documented the need for developing vision-based methods for modelling and control of small underwater vehicles. However, many solutions presented in the literature were based on the assumption that a camera is mounted in the central part of a swimming pool. What is more, they were not applicable for trajectory determination, which constitutes an essential factor in devising a control system of new vehicles. Therefore, the computer vision system has been designed and developed, which enables tracking a vehicle, and determines its trajectory as well.

Acknowledgement. The paper is supported by Project No. DOBR-BIO4/033/13015/2013, entitled "Autonomous underwater vehicles with silent undulating propulsion for underwater reconnaissance" financed by Polish National Centre of Research and Development.

References

1. Krupinski, S., Allibert, G., Hua, M.D., Hamel, T.: Pipeline tracking for fully-actuated autonomous underwater vehicle using visual servo control. In: 2012 American Control Conference (ACC), pp. 6196–6202 (2012)
2. Campos, E., Comby, F., Creuze, V., Torres, J., Antonio, E., Strauss, O.: AUV pipeline following by artificial vision. In: Congreso Internacional de Robotica Computation, pp. 251–256 (2015)
3. Chen, H.-H.H.: Vision-based tracking with projective mapping for parameter identification of remotely operated vehicles. Ocean Eng. **35**, 983–994 (2008). https://doi.org/10.1016/j.oceaneng.2008.03.001
4. Hozyn, S., Zak, B.: Local image features matching for real-time seabed tracking applications. J. Mar. Eng. Technol. (2017). https://doi.org/10.1080/20464177.2017.1386266

5. Praczyk, T., Hozyn, S., Bodnar, T., Pietrukaniec, L., Blaszczyk, M., Zablotny, M.: Concept and first results of optical navigational system. Trans. Marit. Sci. **8**, 46–53 (2019). https://doi.org/10.7225/toms.v08.n01.005
6. Hozyn, S., Zak, B.: A concept for application of a stereo vision method in control system of an underwater vehicle. Appl. Mech. Mater. **817**, 73–80 (2016). https://doi.org/10.4028/www.scientific.net/AMM.817.73
7. Szymak, P., Morawski, M., Malec, M.: Conception of research on bionic underwater vehicle with undulating propulsion. Solid State Phenom. **180**, 160–167 (2011). https://doi.org/10.4028/www.scientific.net/SSP.180.160
8. Govinda, L., Salgado-Jimenez, T., Bandala-Sanchez, M., Nava-Balanzar, L., Hernandez-Alvarado, R., Antonio, J.: Modelling, design and robust control of a remotely operated underwater vehicle. Int. J. Adv. Robot. Syst. **11**, 1 (2014). https://doi.org/10.5772/56810
9. Praczyk, T., Szymak, P., Naus, K., Pietrukaniec, L., Hozyn, S.: Report on research with biomimetic autonomous underwater vehicle — low level control. Sci. J. Pol. Nav. Acad. **212**, 105–123 (2018). https://doi.org/10.2478/sjpna-2018-0006
10. Praczyk, T., Szymak, P., Naus, K., Pietrukaniec, L., Hozyn, S.: Report on research with biomimetic autonomous underwater vehicle — navigation and autonomous operation. Sci. J. Pol. Nav. Acad. **213**, 53–67 (2019). https://doi.org/10.2478/sjpna-2018-0013
11. Fossen, T.I.: Marine Control Systems?: Guidance, Navigation and Control of Ships, Rigs and Underwater Vehicles. Marine Cybernetics, Trondheim (2002). ISBN 8292356002
12. Wang, W., Clark, C.M.: Modeling and simulation of the VideoRay Pro III underwater vehicle. In: Ocean. 2006 - Asia Pacific 2007 (2006). https://doi.org/10.1109/oceansap.2006.4393862
13. Viswanath, P., Mathew, M., Milgram, J., Von Alt, C., Prestero, T.: Verification of a Six-Degree of Freedom Simulation Model for the REMUS Autonomous Underwater Vehicle by in partial fulfillment of the requirements for the degrees of and at the Chairperson, Committee on Graduate Students Verification of a Six-Degree of F. Ocean Sci. pp. 1–9 (2001). https://doi.org/10.1575/1912/3040
14. Monroy-Anieva, J.A., Rouviere, C., Campos-Mercado, E., Salgado-Jimenez, T., Garcia-Valdovinos, L.G.: Modeling and control of a micro AUV: Objects follower approach. Sensors (Switzerland) **18**, 1–18 (2018). https://doi.org/10.3390/s18082574
15. Miskovic, N., Vukic, Z., Barisic, M.: Identification of coupled mathematical models for underwater vehicles. In: OCEANS 2007 – Europe, pp. 1–6. IEEE (2007)
16. Miskovic, N., Vukic, Z., Barisic, M.: Identification of unmanned underwater vehicles by self-oscillation method. Automatika **50**, 167–183 (2009)
17. Miskovic, N., Vukic, Z., Barisic, M.: Identification of underwater vehicles for the purpose of autopilot tuning. In: Underwater Vehicles, pp. 582–601. InTech, Vienna (2008). ISBN 9789537619497
18. Hozyn, S., Zak, B.: Distance measurement using a stereo vision system. Solid State Phenom. **196**, 189–197 (2013). https://doi.org/10.4028/www.scientific.net/SSP.196.189
19. Shortis, M.: Calibration techniques for accurate measurements by underwater camera systems. Sensors **15**, 30810–30826 (2015). https://doi.org/10.3390/s151229831
20. Zhang, Z.: A flexible new technique for camera calibration. IEEE Trans. Pattern Anal. Mach. Intell. **22**, 1330–1334 (2002). https://doi.org/10.1109/34.888718
21. Silvatti, A.P., Salve Dias, F.A., Cerveri, P., Barros, R.M.L.: Comparison of different camera calibration approaches for underwater applications. J. Biomech. **45**, 1112–1116 (2012). https://doi.org/10.1016/j.jbiomech.2012.01.004
22. Hartley, R., Zisserman, A.: Multiple View Geometry in Computer Vision, vol. 16. Australian National University, Canberra (2003)
23. Romanuke, V.V.: Interval uncertainty reduction via division-by-2 dichotomization based on expert estimations for short-termed observations. J. Uncertain Syst. **12**, 3–21 (2018)

Correction to: Intelligent Temperature and Vacuum Pressure Control System for a Thermionic Energy Converter

Bartosz Kania, Dariusz Kuś, Piotr Warda, and Jarosław Sikora

Correction to: Chapter "Intelligent Temperature and Vacuum Pressure Control System for a Thermionic Energy Converter" in: A. Bartoszewicz et al. (Eds.): *Advanced, Contemporary Control*, AISC 1196, https://doi.org/10.1007/978-3-030-50936-1_22

In the original version of the book, the abstract of Chapter 22 was published with an incorrect information, "The results of the cathode temperature tests in the range from 1050 K to 1207 K show that at any time the pressure value in a TEC vacuum chamber is kept within the specified range," which has now been corrected to "The results of the cathode temperature tests in the range from 340 K to 1273 K show that at any time the pressure value in a TEC vacuum chamber is kept within the specified range." The book and chapter have been updated with the changes.

The updated version of this chapter can be found at https://doi.org/10.1007/978-3-030-50936-1_22

A. Bartoszewicz et al. (Eds.): PCC 2020, AISC 1196, p. C1, 2020.
https://doi.org/10.1007/978-3-030-50936-1_130

Author Index

A. Bartoszewicz et al. (Eds.): PCC 2020, AISC 1196, pp. 1565–1568, 2020.
https://doi.org/10.1007/978-3-030-50936-1

GPSR Compliance

The European Union's (EU) General Product Safety Regulation (GPSR) is a set of rules that requires consumer products to be safe and our obligations to ensure this.

If you have any concerns about our products, you can contact us on ProductSafety@springernature.com

In case Publisher is established outside the EU, the EU authorized representative is:

Springer Nature Customer Service Center GmbH
Europaplatz 3
69115 Heidelberg, Germany

Batch number: 10761593

Printed by Printforce, the Netherlands